Fusillo

Nature Classification System

Herb & Vegetable Variety Checklist

1st Edition

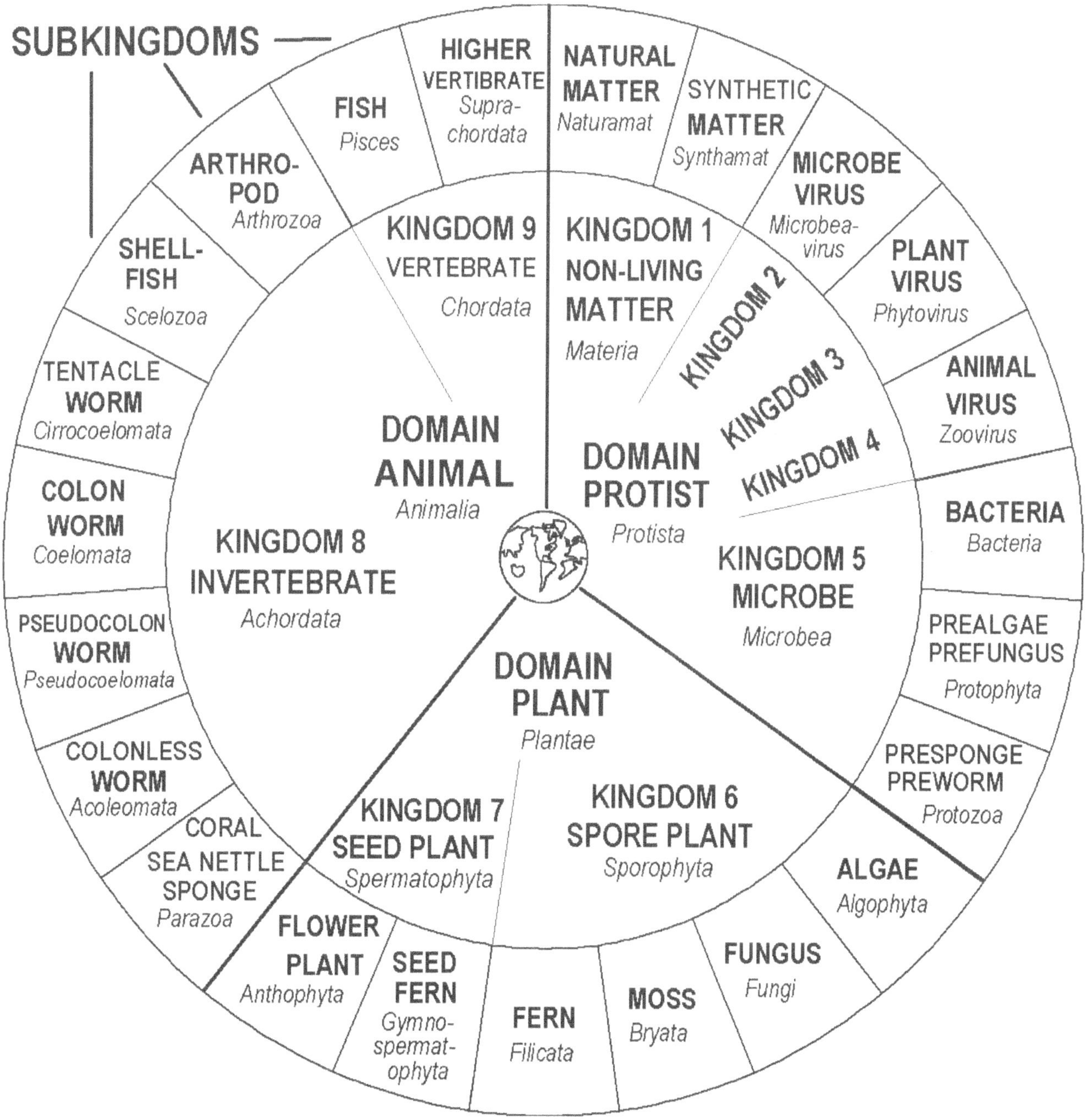

Lawrence J. Fusillo

Trademarks

Trademark names are used. Rather than put a trademark symbol after every occurrence of the trademarked name, the author used the names in an editorial fashion only, and to the benefit of the trademark owner, with no intention of infringement of the trademark. Where such designations appear in this book, they have been put in initial caps.

Warning and Disclaimer

The author of this work has added value to the underlying factual information herein by unique and original selection, by unique and original arrangement, and by unique and original classification of the information. Every effort has been made to make this book as complete and accurate as possible. However, no warranty of fitness is implied. The information provided is on an as-is basis.

Quantity Discounts

Check the publisher's web site, www.buybooksontheweb.com for sales on small quantities of books. The publisher offers discounts on this book when ordered in quantity for bulk purchases and special sales. For bulk sales within the United States please call Toll Free: 877-BUY-BOOK (877-289-2665), Local 610-941-9999, Fax 610-941-9959.

ISBN-10: 0-7414-9666-6
ISBN-13: 978-0-7414-9666-9

Published by:
INFINITY PUBLISHING
1094 New DeHaven Street, Suite 100
West Conshohocken, PA 19428-2713
Toll-free (877) BUY BOOK
Local Phone (610) 941-9999
Fax (610) 941-9959
Info@buybooksontheweb.com
www.buybooksontheweb.com

Printed in the United States of America

The paper used in this book meets the minimum requirements of the American National Standard for Information Sciences Permanence of Paper for Printed Library Materials ANSI Z39.48-1984

Published 1 August 2013

Bibliographic Data

LIBRARY OF CONGRESS BIBLIOGRAPHIC DATA	1. INTL. STD. BOOK NO. (ISBN) ISBN: 10:0-7414-9666-6	2. CIP Accession No.	3. NTIS Accession No.

4. TITLE AND SUBTITLE **Fusillo Nature Classification System** **Herb & Vegetable Variety Checklist 1st Edition**	5. Publication Date 1 August 2013 6. Performing Org. Code SCS
7. AUTHOR(S) Lawrence J. Fusillo	8. Performing Org. Rep. No. SCS-FNCS 09
9. PERFORMING ORGANIZATION NAME AND ADDRESS Lawrence J. Fusillo 280 Piccadilly Square, Apt E2 Athens, GA, 30605	10. Project/Task/Work No. WDI-WBS-09 11. Contract/Grant No. None
12. Sponsoring Organization Name and Complete Address (Street, City, State, ZIP) This work was performed without a contract, a sponsor, or a grant.	13. Period Covered Not Applicable 14. Sponsoring Agency Code Not Applicable

15. SUPPLEMENTARY NOTES

 This book *supercedes & replaces* "Systema Naturae, Ed. X. (Systema naturae per regna tria naturae, secundum classes, ordines, genera, species, cum characteribus, differentiis, synonymis, locis. Tomus I. Editio Decimal, reformata.)", by Carl Linnaeus, Holmiae publisher, 1758, SCS – Scientific Computer Software, Fusillo Nature Classification System Plant CD available from author for $40.

16. ABSTRACT (A 200 word or less summary of most significant information If document includes a significant bibliography or literature survey, mention it here)

All known herb and vegetable varieties have been assigned a Fusillo taxonomy category registration number, an official English name, a Latin Genus, a Latin species, and a Fusillo Latin variety name.

K = Kingdom, Sk = Subkingdom, Ph = Phylum, CL = Class

K1 *Materia* Nonliving Matter — K2 *Microbeavirus* Microbe Virus — K3 *Phytovirus* Plant Virus
K4 *Zoovirus* Animal Virus — K5 *Microbea* Microbe

K6 *Sporophyta*	Spore Plant				
Sk61 *Algophyta*	Algae				
Sk62 *Fungi*	Fungus				
Ph621 *Ascomycota*	Sac Fungus	Shelf Fungus, Lichen, Mold, Yeast			
Ph622 *Basidiomycota*	Club Fungus				
CL622.1 *Basidiomycetes*	Club Fungus	CL622.4	*Gastromycetes*	Cavity Fungus	
CL622.2 *Agaricomycetes*	Gill Mushroom	CL622.5	*Uredinomycetes*	Rust	
CL622.3 *Boletomycetes*	Nongill Mushroom	CL622.6	*Ustilaginomycetes*	Smut	
Sk63 *Bryophyta*	Moss				
Sk64 *Filicata*	Fern				
K7 *Spermatophyta*	Seed Plant				
Sk71 *Coniferophyta*	Conifer Plant				

Ph711 *Pinophyta* Pine — Ph712 *Cycadophyta* Cycad — Ph713 *Ginkgophyta* Ginkgo — Ph714 *Gnetophyta* Gnet

Sk72 *Anthophyta*	Flower Plant		
Ph721 *Monocotophyta*	Monocotyledon	Ph722 Monocotovegetophyta	Monocot Fruit, Herb, Vegetable
CL721.01 *Liliopsida*	Lily	CL722.01 *Liliopsida*	Banana, Breadfruit, Pinapple
CL721.02 *Commeliniopsida*	Grass	CL722.02 *Commelinopsida*	Corn, Barley, Oats, Rice, Rye, Wheat
CL721.03 *Pandanopsida*	Lilywood	CL722.03 *Pandoopsida*	Coconut, Date, Eggfruit
CL721.04 to CL721.12	Orchid	CL722.04 to CL723.12	Hyacinth Bulb, Vanilla Spice
Ph723 *Dicotophyta*	Dicotcotyledon	Ph724 Dicotovegetophyta	Dicot Fruit, Herb, Vegetable
CL723.01 *Asteropsida*	Aster	CL724.01 *Asteropsida*	Artichoke, Lettuce, Sunflower
CL723.02 *Caryophyllopsida*	Succulent	CL724.02 *Caryopyllopsida*	Beet, Rhubarb, Spinach
CL723.03 *Cornopsida*	Cornel	CL724.03 *Cornopsida*	Berry, Carrot, Celery, Parsnip, Turnip
CL723.04 *Gentiasolopsida*	Gentian, Sunplant	CL724.04 *Gentiasolopsida*	Eggplant, Pepper, Potato, Tomato
CL723.05 *Lamiopsida*	Mint	CL724.05 *Lamiopsida*	Olive, Mint
CL723.06 *Legumopsida*	Legume	CL724.06 *Legumopsida*	Green Bean, Kidney Bean, Lima Bean, Pea
CL723.07 *Magnoliopsida*	Magnolia	CL724.07 *Magnoliopsida*	Avocado, Cinnamon, Nutmeg, Sassafras
CL723.08 *Malvopsida*	Mallow	CL724.08 *Malvopsida*	Baobab, Cocoa, Fig, Grape, Okra
CL723.09 *Myrtopsida*	Myrtle	CL724.09 *Myrtoposida*	Guava, Macadamia Nut
CL723.10 *Nymphaeopsida*	Water Lily	CL724.10 *Nymphaeopsida*	Peppercorn
CL723.11 *Ranunculopsida*	Buttercup	CL724.11 *Ranunculopsida*	Barberry, Mayapple, Sausagefruit
CL723.12 *Rosopsida*	Rose	CL724.12 *Rosopsida*	Apple, Apricot, Cherry, Pear, Plum, Nut
CL723.13 *Rutopsida*	Rue	CL724.13 *Rutopsida*	Lemon, Lime, Orange, Tangerine
CL723.14 *Santalopsida*	Sandalwood	CL724.14 *Santalopsida*	Quandong Plum
CL723.15 *Theopsida*	Tea	CL724.15 *Theopsida*	Brazilnut, Kiwi, Mangosteen, Tea
CL723.16 *Violopsida*	Violet	CL724.16 *Violopsida*	Broccoli, Cabbage, Cauliflower
K8 *Achordata*	Invertebrate	K9 *Chordata*	Vertebrate Animal

17. KEY WORDS (six to twelve entries; alphabetical order; capitalize only the first letter of the first key word unless a proper name; separated by semicolons)

Botany; Classification; Horticulture, Dicotyledon; Fruit; Monocotyledon, Rose; Plant; Taxonomy

18. AVAILABILITY [X] Unlimited	19. Data CLASS	21. NO. OF PAGES
X For Commercial Distribution Do not release to NTIS X Order from Infinity Publishing, www.buybooksontheweb.com X Order from Barnes and Noble X Order from Borders X Order from Amazon.com	WORLD DATA INFRASTRUCTURE 20. Standards CLASS WORLD BASIC STANDARD	22. Price $39.95

Acknowledgements

Herb Species & Varieties

Herb species and varieties were compiled using the following books;

Bal1, 1997	**"Prescription for Nutritional Healing, 2nd Ed."**, J.A.Balch & P.A.Balch, ISBN:0-89529-727-2
Bal2, 2012	**"Prescription for Herbal Healing, 2nd Ed."**, P.A. Balch & S. Bell, ISBN:0-978-1-58333-452-2
Bliss, 1993	**"North American Dye Plants"**, by Anne Bliss, ISBN:0-934026-89-0
Bown, 1995	**"Herb Society of America Encyclopedia of Herbs"**, Deni Bown, ISBN:0-7894-0184-3
Cas1, 1993	**"Craft of the Dyer Colour from Plants and Lichens"**, K. Casselman, ISBN:0-486-27606-6
Cas2, 2001	**"Lichen Dyes, The New Source Book"**, by Karen Casselman, ISBN:0-486-41231-8
Clev, 1997	**"The Encyclopedia of Herbs & Spices"**, Clevey, Richmond, Morris, Mackley, ISBN:1-901289-06-0
Grae, 1974	**"Nature's Colors Dyes From Plants"**, by Ida Grae, ISBN:0-02-544950-8
Grue, 2xxx	**"Physician's Desk Reference (PDR) For Herbal Medicines 4th Ed."**, by J. Gruenwald et. al.
Kowa, 1998	**"Rodale's Illustrated Encyclopedia of Herbs"**, C.Kowalchik & W.H.Hylton, ISBN:0-87596-964-X
Krug, 1998	**"The Pocket Guide to Herbs"**, Anna Kruger, ISBN:1-85585-361-2
Murr, 2012	**"The Encyclopedia of Natural Medicine 3rd Ed."**, by M.T. Murray & J.E. Pizzorno, ISBN:978-1-4516-6300-6
Scal, 2002	**"Traditional Medicines from the Earth"**, by R. Scalzo & M. Cronin, ISBN:0-9707936-1-8
Wyk, 2004	**"Medicinal Plants of the World"**, Ben-Erik van Wyk & Michael Wink, ISBN:981-232-933-1

Vegetable Species - USA

Latin names for vegetable species were compiled using the following sources;

USDA-NPDC, 2009 The National Plant Data Center, National Resources Conservation Service (NRCS), US Dept. of Agriculture (USDA), Baton Rouge, LA 70874-4490 USA, **www.plants.usda.gov**

BONAP, 2009 Biota of North America Project (BONAP), Univ. of North Carolina, John Kartesz, custodian

RGB-Kew, 2010 Royal Botanical Gardens, Kew, England, www.kew.org

Royal Botanical Gardens Kew, Richmond, Surrey, UK is an organization dedicated to the preservation of selected plant *species*. Kew Gardens maintains a **W**orldwide **C**hecklist for **S**elected **P**lant (**WCSP**) *species* on **www.kew.org/wcsp**. Thanks go to R. Govaerts, the Kew Gardens dataset custodian, who listed worldwide vegetable *species*.

Vegetable Varieties - USA

English names for USA vegetable varieties were compiled using the following books

GSI-5, 1999 "Garden Seed Inventory Fifth Edition" An Inventory of Nursery Catalogs Listing All Vegetable Varieties available by Mail Order in the United States", complied by Seed Savers Exchange

GSI-6, 2004 "Garden Seed Inventory Sixth Edition" An Inventory of Nursery Catalogs Listing All Vegetable Varieties available by Mail Order in the United States, compiled by Seed Savers Exchange

Seed Savers Exchange is a non-profit organization whose members are dedicated to the preservation and distribution of heirloom seeds. Founded in 1975, the organization is located at Heritage Farm, near Decorah, Iowa. Seed Savers Exchange is the largest seed bank in the United States, with many thousands of varieties of heirloom seeds. The organization maintains an orchard that contains hundreds of 19th century apple varieties. Special thanks go to Joanne Thuente, of Seed Savers Exchange, who documented all known fruit varieties commercially available in the United States. Without her work, the USA vegetable variety checklist compiled in this book would not have been possible.

Vegetable Species & Varieties - Worldwide

For vegetable *species* and for vegetable *varieties* native to Canada, Mexico, C. & S. America, Africa, Europe, Asia and Australia, the following books were used:

Flora, 2004 "Flora" ISBN:0-74048-097-X, by Sean Hogan, et. al., revised 2003, reprinted 2004, 1584 pages

AHS, 2004 "The American Horticultural Society A-Z Encyclopeda of Garden Plants", ISBN:0-7566-0616-0, by Christopher Brickell & H. Marc Cathey, et. al., first published 1997, revised 2004, 1100 pages

Botanica, 2004 "Botanica The Illustrated A-Z of over 10,000 Garden Plants and how to cultivate them", ISBN:0-3-8331-1253-0, by Tony Rodd et. al., first published 1997, reprinted 2004, 1020 pages

RHBOV, 1994 "The Random House Book of Vegetables", ISBN:0-679-75024-X, by Roger Phillips & Martin Rix, 1994, 270 pages

Without the work of Govaerts, Hogan, Brickell, Cathey, Rodd, Wasson, Phillips & Rix the worldwide vegetable species and variety checklist compiled in this book would not have been possible.

Mushroom Varieties - USA

Kent H. McKnight and Vera B. McKnight provided edible mushroom varieties native the USA in the book entitled **"Peterson's Field Guide to Mushrooms North America"**, 1987, ISBN:0-395-91090-0.

Mushroom Varieties - Worldwide

P.M. Kirk, custodian of the Species Fungorum (**www.speciesfungorum.org**) dataset, provided Latin names for mushroom species and varieties worldwide. Giuseppe Pace provided English names mushroom species and varieties worldwide in his book entitled **"Mushrooms of the World"**, 1998, ISBN:1-55209-212-7.

Preface

(1) The Fusillo Nature Classification System is a bilingual "Latin/English System"
This is the first in the history of mankind, original publication of an **English Language Classification System**. This book is used by environmental protection organizations, teachers, & students as an English Language "Classification Decoding Key" to decoding Latin based Classification Systems from Latin into English. (All Catalogue of Life CoL species are in Latin.)

(2) The Fusillo Nature Classification System has no "Missing Categories"
The Fusillo Nature Classification is a Decanomial system for "variety, species, Genus, Family, Order, Class, Phylum, Subkingdom, Kingdom, & Domain". A complete system with each **SPECIES** in the Physical Universe properly assigned to all 8 higher taxonomic categories has never been published before. (CoL has missing categories.)

(3) The Fusillo Nature Classification System has no "Duplicate Families"
Single Genus Families are recognized when a Genus is truly distinct from all other Genera in the same Order. But when a Genus in a Single Genus Family is closely related to Genera in another Family in the same Order, that Genus has been moved to the closely related Family & the Single Genus Family has been deleted. (CoL has single Genus Families.)

(4) The Fusillo Nature Classification System is "Backward Compatible"
While scientists currently disagree on Family, Order, Class, Phylum, & Subkingdom assignments, there is universal agreement on the assignment of Latin GENUS & *species* names. The Fusillo Nature Classification System has been designed to be 100% backward compatible with Latin GENUS & *species* names. A few Genera have been split in two, retaining the old Latin GENUS name & adding a new Latin GENUS name. In these cases, the Latin *species* name provides 100% backward compatibility.

(5) The Fusillo Nature Classification System is an "Open Standard"
Fusillo Nature Classification System (FNCS) categories are on CD and are in universally readable MS Excel spreadsheet format (.xls). Purchase of a CD includes a personal use license for FNCS spreadsheet files for use on three sites; 1 home desktop, 1 mobile laptop, & 1 work PC. Import into database software for personal use is encouraged.

Latin Names for viruses, microbes, plants, & animals were obtained from the following books & CDs;

Title	Author/Publisher		Year
"*Systema Naturae per regna tria naturae, secundum classes, ordines, genera, species, cum characteribus,differentiis, synonymis, locis. 10th Edition*"	Carl Linnaeus/Holmiae		1758
"International Code of Nomenclature for Cultivated Plants"	Seattle Botanical Congress		1969
"A World Checklist of Mammalian Species 2nd Edition"	Corbet & Hill/Facts on File Publisher		1986
"Plant Families & Their Relationship" extracted from page 958 of "Botanica"	Dahlgren & Cronquist	ISBN:3-8331-1253-0	2004
"*Systema Naturae 2009 Edition*" on Catalog of Life (CoL) CD	UK Species 2000 Org.	(www.sp2000.org)	2010

English Names for viruses, microbes, plants, & animals were obtained from the following books;

Title	Author/Publisher		Year
"A World Checklist of Birds"	Monroe & Sibley	ISBN:0-300-07083-7	1993
"Latin Names Explained; Birds, Reptiles, Mammals	Gotch, A.F.	ISBN:0-8160-3377-3	1995
"Stearns Dictionary of Plant Names, Revised"	Stearn, W.	ISBN:0-88192-556-X	1996
"Latin Concise Dictionary" (Latin to English)	Harper Collins	ISBN:0-06-053690-X	1997
"The World of Primates"	Hook,P.	ISBN:0-517-16204-0	2000
"Encyclopedia of Aquatic Life"	Bannister & Campbell	ISBN:0-8160-1257-1	2002
"Stedmen's Medical Dictionary 28th Ed.	Stedman	ISBN:0-7817-6192-1	2005
"Virus Taxonomy 8th Report on Taxonomy of Viruses"	Fauquet,C./Elsevier	ISBN:0-12-249551-X	2005
"Encyclopedia of Virology 3rd Edition"	Mahy&Regenmortel	ISBN:0-12-373735-7	2008

The Fusillo Nature Classification System (FNCS) containing all living organisms was published in;

	Fusillo Nature Classification System (FNCS) Title	Author/Publisher			Year
1	"Anna Belle Nature Trail Guide 1st Edition"	Fusillo, L.J./Infinity	ISBN:0-7414-4201-9	Jul	2007
2	"Fusillo Nature Classification System Genus & Higher Taxa 1st Edition"	Fusillo, L.J./Infinity	ISBN:0-7414-4495-3	Jan	2008
3	"FNCS Nonliving Matter Species Checklist 1st Edition"	Fusillo, L.J./Infinity	ISBN:0-7414-4459-3	Mar	2008
4	"FNCS Virus Species Checklist 1st Edition"	Fusillo, L.J./Infinity	ISBN:0-7414-5214-6	Mar	2009
5	"FNCS Microbe Species Checklist 1st Edition"	Fusillo, L.J./Infinity	ISBN:0-7414-5475-0	Aug	2009
6	"FNCS Orchid Species Checklist 1st Edition"	Fusillo, L.J./Infinity	ISBN:0-7414-5668-0	Dec	2009
7	"FNCS Spore Plant Species Checklist 1st Edition"	Fusillo, L.J./Infinity	ISBN:0-7414-5878-0	May	2010
8	"FNCS Fruit Variety Checklist 1st Edition"	Fusillo, L.J./Infinity	ISBN:0-7414-6441-1	Mar	2011
9	"FNCS Herb & Vegetable Variety Checklist1st Edition"	Fusillo, L.J./Infinity	ISBN:0-7414-9666-6	Aug	2013
10	"FNCS Flower Variety Checklist 1st Edition"	Fusillo, L.J./Infinity		future	
11	"FNCS Wildflower Species Checklist 1st Edition"	Fusillo, L.J./Infinity		future	
12	"FNCS Invertebrate Coral, Nettle, Sponge, & Worm Species Checklist 1st Edition"	Fusillo, L.J./Infinity		future	
13	"FNCS Invertebrate Shellfish Species Checklist 1st Edition"	Fusillo, L.J./Infinity		future	
14	"FNCS Invertebrate Beetle Species Checklist 1st Edition"	Fusillo, L.J./Infinity		future	
15	"FNCS Invertebrate Spider & Insect Species Checklist 1st Edition"	Fusillo, L.J./Infinity		future	
16	"FNCS Vertebrate Fish Species Checklist"	Fusillo, L.J./Infinity		future	
17	"FNCS Vertebrate Amphibian, Reptile, Bird, & Mammal Species Checklist 1st Edition"	Fusillo, L.J./Infinity		future	

Lawrence J. Fusillo
280 Piccadilly Sq. Apt. E-2
Athens, GA 30605

TABLE OF CONTENTS

INTRODUCTION

There are three domains of non-living matter and living organisms in the physical universe. The Domain ***Protista*** **Protist** contains nonliving matter, and the microbes which are viruses, bacteria, protophytes (microscopic protoplants) and protozoans (microscopic animalcules). The Domain ***Plantae*** **Plant** contains the spore plants (plants that reproduce by spore) and the seed plants (plants that reproduce by seed). The Domain ***Animalia*** **Animal** contains the invertebrates (animals with no spinal cord & backbone) and the vertebrates (animals with a spinal cord & backbone)

There are nine kingdoms; 5 in the Protist Domain (1 *Materia* Nonliving Matter, 2 *Microbeavirus* Microbe Virus, 3 *Phytovirus* Plant Virus, 4 *Zoovirus* Animal Virus, and 5 Bacteria *Bacteria*), 2 in the Plant Domain (6 *Sporophyta* Spore Plant, and 7 *Spermatophyta* Seed Plant and 2 in the Animal Domain (8 *Achordata* Invertebrate, and 9 *Chordata* Vertebrate).

The Fusillo Nature Classification System makes a distinction between plants that are cultivated by Man and plants that only occur in the wild. Plants that are cultivated by Man include fruit plants, herb plants, vegetable plants, and flower plants. The Kingdom *Sporophyta* Spore Plant contains Subkingdom 1 *Algophyta* Algae, Subkingdom 2 *Fungi* Fungus, Subkingdom 3 *Bryata* Moss and Subkingdom 4 *Filicata* Fern. Edible Algae are included in Subkingdom 5 *Algovegetophyta* Cultivated Algae.

Genus species Nomeclature

Carl Linnaeus created the Genus species Latin Classification System for naming species and defining the human uses of nonliving matter, microbes plants and animals as a **COMMERCIAL SYSTEM** for enhancing commerce between the countries of Europe. In 1735 as he travelled through Europe he found that each province and town in a particular country had their own colloquial names or nicknames for minerals, microbes, herbs, fruits, vegetables, flowers, foliage landscape plants and animal meats. He sought to expedite commerce between provinces and countries by creating a universally recognized name for all non-living and living things. Since **CLASSIFICATION** was for **COMMERCE**, he included the **HUMAN USE** of the minerals as natural resources, microbes as antibiotics, plants as herbs, vegetables, fruits, flowers, or landscape foliage, and animals as meat or dairy. The **HUMAN USE** was part of the original Linnaeus Latin Classification System.

Carl Linnaeus [1707-1778], born in Sweden, was first trained as a physician and later became a professor of botany at Uppsala University in southeast Sweden. His book *Systema Naturae* (1st Edition 1735) began a series of encyclopedic works dealing with plants, animals, minerals, and medicinal herbs which classified and named the whole organic world, thereby providing the basis for the Latin classification system currently used by botanists. He was a prodigious writer and published works totaling more than 180 titles. His original system of classifying plants, based on the number of stamens and pistals, was expanded to include classification based on other plant characteristics.

His most permanent contribution to botany was his introduction in 1753 of his book *Species Planetarium* (Sp.Pl.) which presented a consistent binomial system of naming species by means of a single-word generic name (Genus) followed by a single-word specific epithet (species). His plant specimen collection is now at the Burlington House in London and belongs to the Linnaean Society of London. In the tenth edition of his book *Systema Naturae*, published in 1758, he added Latin two word binomial names for all known animals. By 1758, Carl Linnaeus had created Latin names for 7,700 species of plants and 4,400 species of animals.

In 2011, there were Latin names for 48,125 species of microbes, 346,392 species of plants and 1,091,761 species of animals. For microbes, there were 5,812 species of virus, 9,554 species of bacteria, 1,509 species of protophyte, and 31,250 species of protozoan, totaling 48,125. For plants, there were 33,500 species of algae, 53,600 species of fungus, 16,100 species of moss, 12,117 species of fern, 1,075 species of conifer, and 230,000 species of flower plant, totaling, 346,392. For animals there were 18,500 species of sponges, nettles and jellyfish, 130,300 species of worms, 128,650 species of shellfish, 760,000 species of insects, 32,000 species of fish, and 22,311 species of vertebrate animals (i.e. 10,075 species of amphibians, 7,034 species of reptiles, 9,702 species of birds, and 4,500 species of mammals), totaling 1,091,761.

Fusillo Nature Classification System (FNCS)

(Supercedes and Replaces the Linnaeus Classification System)

A	B	C	D	E	F	G	H	I	J
Regist. #	**GenusL**	**SpeciesL**	**VarietyL**	**Author, Year**	**Source**	**Location**	**VarietyE**	**SpeciesE**	**GenusE**

K	L	M	N	O	P	Q	R	S	T
Behavior	**Habit**	**Name Code**	**DomainL**	**KingdomL**	**SubkingdomL**	**PhylumL**	**ClassL**	**OrderL**	**FamilyL**

U	V	W	X	Y	Z	AA	AB	AC	AD
DomainE	**KingdomE**	**SubkingdomE**	**PhylumE**	**ClassE**	**OrderE**	**FamilyE**	**Status**	**Ecozone**	**Ecotype**

AE	AF	AG	AH	AI	AJ	AK	AL	AM	AN
Min Temp. Zone	**Max Temp. Zone**	**Height**	**Length**	**Silhouette**	**Feature 1**	**F 1 Color**	**Feature 2**	**F 2 Color**	**Feature 3**

AO	AP	AQ	AR	AS	AT	AU
F 3 Color	**Feature 4**	**F 4 Color**	**F 1 Photo**	**F 2 Photo**	**F 3 Photo**	**F 4 Photo**

FNCS Data Field Definitions

DATA FIELD	DEFINITION				Language
A Registration #	taxonomic number in format: **kspcofgsv**, **k** = kingdom #, **s** = subkingdom #, **p** = phylum #, **c** = class #, **o** = order #, **f** = family #, **gg** = genus #, **s.s** = species #, **vv** = variety #				
B GenusL	Latin Name for Genome (RNA/DNA genetic group)				**Latin**
C SpeciesL	Latin Name for a Specific Specimen				**Latin**
D VarietyL	Latin Name for Infraspecies, Subspecies or Variety				**Latin**
E **Author, Year**	scientist who first published species or variety, or book abbrev. and year of publication				**English**
F **Source**	Technical journal or book-page #-species/var. # or mail order company name abbrev.				**English**
G **Location**	indigenous (native) location by Subcontinent (Country) or by Country (City, State)				**English**
H **VarietyE**	**UNIQUE OFFICIAL VALID ENGLISH NAME** for Infraspecies, Subspecies or Variety				**English**
I **SpeciesE**	**UNIQUE OFFICIAL VALID ENGLISH NAME** for a Specific Specimen				**English**
J **GenusE**	**UNIQUE OFFICIAL VALID ENGLISH NAME** for a Genome (RNA/DNA genetic group)				**English**
K **Behavior**	**HUMAN USE: Microbe:** Antibiotic, Parasite, Pathogen, **plant:** herb, vegetable, fruit, flower, foliage, succulent, **animal:** Parasite, Pathogen, Predator, Symbiote				**English**
L **Habit**	**microbe:** Free Living, Symbiote, Parasite, Pathogen, **plant:** Annual, Biennial, Perennial, Trailing or Climbing Vine, Bush, Shrub, Tree, **animal:** Herbivore, Carnivore, Insectivore, Omnivore, Scavenger				**English**
M **Name Code**	Dataset Abbreviation-Sequence Number for Species or Infraspecies				**English**
N DomainL	Latin name for one of 3 Domains i.e. *Protista, Plantae, Animalia*				**Latin**
O Kingdom	Latin name for one of 9 Kingdoms				**Latin**
P SubkingdomL	Latin name for one of 38 Subkingdoms				**Latin**
Q PhylumL	Latin name for one of up to nine Phylums in a Subkindom				**Latin**
R ClassL	Latin name for one of up to 16 Classes in a Phylum				**Latin**
S OrderL	Latin name for one of up to 9 Orders in a Class				**Latin**
T FamilyL	Latin name for one of up to 9 Families in an Order				**Latin**
U DomainE	English name for one of 3 Domains i.e. Protist, Plant, Animal				**English**
V KingdomE	English name for one of 9 Kingdoms				**English**
W SubkingdomE	English name for one of 38 Subkingdoms				**English**
X English Phylum	English name for one of up to nine Phylums in a Subkindom				**English**
Y English Class	English name for one of up to 16 Classes in a Phylum				**English**
Z English Order	English name for one of up to 9 Orders in a Class				**English**
AA English Family	English name for one of up to 9 Families in an Order				**English**
AB **Status**	population: **A**bundant, **B**elow optimum, **C**onserved, **D**ecimated, **E**ndangered, **F**ailed, **G**enocide				**English**
AC **Ecozone**	ecozone # (by geographic area)				
AD **Ecotype**	ecotype # (by habitat type: desert, wetland, forest, inland, coastal plain)				**English**
AE **Min Temp. Zone**	minimum temperature (hardiness) zone #				
AF **Max Temp. Zone**	maximum temperature (heat) zone #				**English**
	Nonliving Matter	**Microbe**	**Plant**	**Animal**	
AG **Height**	Cleavage	colony width micrometers	min-max ft. or in.	min-max ft. or in.	**English**
AH **Length**	Fracture	colony length micrometers	min-max spread	min-max ft. or in.	**English**
AI **Silhouette**	crystal habit	colony shape	**plant** outline type	number of legs	**English**
AJ **Feature 1**	chemical formula	colony cell count	**leaf** shape & length	leg length	**English**
AK **F 1 Color**	appearance color	colony color	**leaf** color	fur/skin color	**English**
AL **Feature 2**	hardness mohs 1-10	cell shape & length	**flower** type, shape & length	ear/horn shape & length	**English**
AM **F 2 Color**	strike/streak color	colony color	**flower** color	ear/horn color	**English**
AN **Feature 3**	lattice structure	spore type & length	**fruit** type & length	foot or paw or hoof	**English**
AO **F 3 Color**	pigment color	spore color	**fruit** color	foot/paw/hoof color	**English**
AP **Feature 4**	specific gravity	cell wall thickness	**bark** texture	eye size & shape	**English**
AQ **F 4 Color**	Luster	cell wall color	**bark** color	eye color	**English**
AR **F 1 Photo**	aggregate	colony	**whole plant**	whole animal	**bitmap**
AS **F 2 Photo**	strike	cell	**leaf**	head	**bitmap**
AT **F 3 Photo**	pigment	spore	**flower**	foot or paw or hoof	**bitmap**
AU **F 4 Photo**	environment	cell wall	**fruit**	eye	**bitmap**

In order for a plant name to be accepted by the scientific community, the Latin name for the plant must be published in a technical journal or book. The naming of plants in Latin is governed by *The International Code of Botanical Nomenclature* which states that the earliest published Latin name for a species is the correct name and that species Latin name supercedes all later published synonym names for the same species.

The task of assigning English names to the Genus, species, and variety of herbs and vegetables was not as difficult as assigning English names to microbes, spore plants, non-flowering seed plants and flowering seed plants because English names for herbs and vegetables have already been standardized in the United States.

Higher Taxa Nomeclature

DOMAIN 1 Protista — Protist

KINGDOM 1 Materia
Latin: materia material

Nonliving Matter
SUBKINGDOMS

11 Naturamat	Natural Matter
12 Synthamat	Synthetic Matter

KINGDOM 2 Microbeavirus
Latin: microbea microbe
Latin: virus virulent organism

Microbe Virus
SUBKINGDOMS

21 Bacteriavirus	Bacteria Virus
22 Protophyteovirus	Protophyte Virus
23 Protozoafungivirus	Prefungus Protozoan Virus
24 Protozoaporiferavirus	Presponge Protozoan Virus
25 Protozoavermesvirus	Preworm Protozoan Virus

KINGDOM 3 Phytovirus
Latin: phyta phyton, plant
Latin: virus virulent organism

Plant Virus
SUBKINGDOMS

31 Algophytavirus	Algae Virus
32 Fungivirus	Fungus Virus
33 Bryatavirus	Moss Virus
34 Filicatavirus	Fern Virus
35 Gymnophytavirus	Seed Fern Virus
36 Anthophytavirus	Flower Plant Virus

KINGDOM 4 Zoavirus
Latin: zoa zoo, animal
Latin: virus virulent organism

Animal Virus
SUBKINGDOMS

41 Parazoavirus	Parazoan Virus
42 Acoelomatavirus	Colonless Worm Virus
43 Pseudocoelomatavirus	Pseudocolon Worm Virus
44 Coelomatavirus	Colon Worm Virus
45 Cirrocoelomatavirus	Tentacle Worm Virus
46 Scelozoavirus	Shellfish Virus
47 Arthropodavirus	Arthropod Virus
48 Piscesvirus	Fish Virus
49 Suprachordatavirus	Higher Vertebrate Virus

KINGDOM 5 Microbea
Latin: micro micron, small
Latin: biota biote, life form

Microbe
SUBKINGDOMS

51 Bacteria	Bacteria
52 Protophyta	Protophyte
53 Protozoaprefungi	Prefungus Protozoan
54 Protozoapreporifera	Presponge Protozoan
55 Protozoapreverma	Preworm Protozoan

DOMAIN 2 Plantae — Plant

KINGDOM 6 Sporophyta
Latin: spora spore
Latin: phyta phyton, plant

Spore Plant
SUBKINGDOMS

61 Algophyta	Algae
62 Fungi	Fungus
63 Bryata	Moss
64 Filicata	Fern (Spore Fern)

KINGDOM 7 Spermatophyta
Latin: sperma sperm, seed
Latin: phyta phyton, plant

Seed Plant
SUBKINGDOMS

71 Coniferophyta	Conifer (Seed Fern)
72 Anthophyta	Flower Plant

DOMAIN 3 Animalia — Animal

KINGDOM 8 Achordata
Latin: achordata chordless, spineless

Invertebrate Animal
SUBKINGDOMS

81 Parazoa	Parazoan (Sponge, Coral, Nettle)
82 Acoelomata	Colonless Worm
83 Pseudocoelomata	Pseudocolon Worm
84 Coelomata	Colon Worm
85 Cirracoelomata	Tentacle Worm
86 Scelozoa	Shellfish

	85 Arthropoda	Arthropod (Spider, Insect)
KINGDOM 9 Chordata	Vertebrate Animal	
Latin: chordata chord, spine	SUBKINGDOMS	
	91 Pisces	Fish
	PHYLUMS	
	911 Urochordata	Headless Fish
	912 Agnatha	Jawless Fish
	913 Chondrichthya	Cartilage Fish
	914 Osteichthya	Bone Fish
	92 Suprachordata	Higher Vertebrates
	PHYLUMS	
	921 Amphibia	Amphibian
	922 Reptilia	Reptile
	923 Aves	Bird
	924 Mammalia	Mammal

Herb Overview

Starting in 1980, there was widespread recognition that treating disease with manufactured synthetic drugs came with side effects that (a) could weaken the immune system making the body susceptible to additional opportunistic diseases and (b) poison the body because of high toxicity levels. A few herbs have side effects as well. The severity of the herb side effects can depend on dosage. In this book, when a herb has severe side effects, it is labeled as Poison ☠ as a warning, even if it is considered safe to take at low doses.

In 1998, the American public spent 2 trillion dollars on health care which was 17% of the gross domestic product (GDP).

By 1999, 2 percent of the American public had abandoned synthetic chemical drugs and had begun using Alternative Medicine herbs.

By 2010, 8 percent of the American public used herbs and herb extracts exclusively, 38 percent of the American public used herbs in conjunction with a physician's prescription for synthetic chemical drugs, and 54 percent still used synthetic chemical drugs exclusively.

Today, the Food and Drug Administration (FDA) of the federal government of the USA has standards for the preparation of all herbs and herb extracts. The FDA requires that no claims be made that a herb can "cure" or even "manage" a disease.

In this book, herbs are considered only to "suppress" an opportunistic disease caused by an invading foreign organism. No claims are made for disease "management" or disease "cure".

There is no original herb information in this book. All herb information is based on previous books published by qualified health care practitioners. Each herb is annotated with the source publication from which the information was taken. In addition the herb is labeled with the empirical or proven effectiveness stated in the source publication.

Empirical effectiveness is based on anecdotal observations. Proven effectivness is based on the amount and quality of scientific research that has been done. The categories of empirical effectiveness from lowest effectivness to highest effectivness are:

1 folk medicine, 2 traditional medicine, 3 ayurveda medicine,

The categories of scientific effectiveness from lowest effectivness to highest effectivness are:

4 statistical studies, 5 pharmalogical studies, 6 clinical studies and 7 clinical trials

Because there are over 5,000 herbs available in the United States choosing a herb for a particular illness or disease can be an impossible task. Therefore, included this book is a list of recommended herbs for each illness & disease. The herbs with the highest effectiveness were selected as recommendations in the sections entitled **"Medicinal Herbs Listed by Symptomatic Illness"** and **"Medicinal Herbs Listed by Disease"**. The source publication which recommended the herb is given with the herb selection for a symptomatic illness or disease.

According to "The Herb Society of America Encyclopedia of Herbs and their Uses", ISBN:0-7894-0184-3, page 233 under *Alisma plantago-aquatica* Ze Xie Water Plantain:

"Used in traditional Chinese medicine for kidney weakness, which manifests as deafness, tinnitius, and dizziness".

According to "The Herb Society of America Encyclopedia of Herbs and their Uses", ISBN:0-7894-0184-3, page 277 under *Eclipta prostrata* Prostrate Ecliptic Daisy:

"Internally for kidney and liver weakness (manifesting as tinnitus, premature graying of hair, poor teeth and eyesight, nervous disorders)".

In this book Folk Medicine, Traditional Medicine, Alternative Medicine, Auyrvedic Medicine, or Modern Medicine herb uses are considered valid and are combined and listed in use-abbreviation alphabetical order. The reference which cited the uses is given in the last column of the table below. To determine the area of Medicine, you must go back to the original references. Toxic herbs are listed in this book as economic (E) poison (PS☠).

WARNINGS

The Herb Society of America recommends that toxic herbs be administered by "Qualified Practioners Only". Kowalchik states for Folk Medicine herbs: "There appears to be no scientific basis for any of these practices.

Herb References

Manufacturers of herb extracts work with the American Botanical Council, the Herb Research Foundaton, the Herb Society of America, and the USA Pharmacopeia to develop the scientific techniques needed to verify the quality of herb extracts.

The German Commission E was established in 1978 to help physicians integrate herbs and herb extracts with conventional medicine. The work of the commission has been translated into English under the title:

Complete German Commision E Monographs: Theraputic Guide to Herbal Medicines

In the United States, comparable work was done to assist the American Medical Association and US physicians in integrating herbs and herb extracts into conventional medicine. This work is published in the book entitled:

Physician's Desk Reference for Herbal Medicines

The pharmacopeias of the United States are:

United States Pharmacopeia (USP)
The National Formulary
The American Herbal Pharmacopeia

The source publications for herbs used in this book are:

Individual Herb Uses

Bal2, 2012	**"Prescription for Herbal Healing, 2nd Ed."**, P.A. Balch & S. Bell, ISBN:0-978-1-58333-452-2
Bliss, 1993	**"North American Dye Plants"**, by Anne Bliss, ISBN:0-934026-89-0
Cas1, 1993	**"Craft of the Dyer Colour from Plants and Lichens"**, K. Casselman, ISBN:0-486-27606-6
Cas2, 2001	**"Lichen Dyes, The New Source Book"**, by Karen Casselman, ISBN:0-486-41231-8
Clev, 1997	**"The Encyclopedia of Herbs & Spices"**, Clevey, Richmond, Morris, Mackley, ISBN:1-901289-06-0
Grae, 1974	**"Nature's Colors Dyes From Plants"**, by Ida Grae, ISBN:0-02-544950-8
Bown, 1995	**"Herb Society of America Encyclopedia of Herbs"**, Deni Bown, ISBN:0-7894-0184-3
Krug, 1998	**"The Pocket Guide to Herbs"**, Anna Kruger, ISBN:1-85585-361-2
Kowa, 1998	**"Rodale's Illustrated Encyclopedia of Herbs"**, C.Kowalchik & W.H.Hylton, ISBN:0-87596-964-X
Wyk, 2004	**"Medicinal Plants of the World"**, Ben-Erik Van Wyk & Michael Wink, ISBN:981-232-933-1

Illness & Disease Herb Combinations

Bal1, 1997	**"Prescription for Nutritional Healing, 2nd Ed."**, J.A.Balch & P.A.Balch, ISBN:0-89529-727-2
Bal2, 2012	**"Prescription for Herbal Healing, 2nd Ed."**, P.A. Balch & S. Bell, ISBN:0-978-1-58333-452-2
Murr, 2012	**"The Encyclopedia of Natural Medicine 3rd Ed."**, by M.T. Murray & J.E. Pizzorno, ISBN:978-1-4516-6300-6
Scal, 2002	**"Traditional Medicines from the Earth"**, by R. Scalzo & M. Cronin, ISBN:0-9707936-1-8
Wyk, 2004	**"Medicinal Plants of the World"**, Ben-Erik Van Wyk & Michael Wink, ISBN:981-232-933-1

BOOKS USED BY Kowa, 1998

AHA	"The American Herb Association Quarterly", PO Box 353, Rescue CA, 95672
Barn, 1996	"Herbal Medicines – A Guide for Health Care Professionals 2nd Ed.", J. Barnes, L.A. Anderson, J.D. Phillipson, Rittenhouse Books, King of Prussia, PA
Griev, 1971	"A Modern Herbal" in 2 volumes, Mrs. M. Grieve, Dover Publications, New York.
HSA	"The Herbarist Annual", Herb Society of America, 2 Independence Ct., Concord, MA 01742
Mills, 1985	"The Dictionary of Modern Herbalism: A Comprehensive Guide to Practial Herbal Therapy"

BOOKS USED BY Wyk, 2004

Blum, 1998	"The Complete German Commission E Monographs", M. Blumenthal, ed., American Botanical Council, Austin, TX
Brow, 2002	"Dictionary of Medicinal Plants", R.G. Brown, Ivy, Publishing House, Raleigh
Chev, 2001	"Encyclopedia of Medicinal Plants", A. Chevallier, Dorling Kindersley, London, England
Duke, 2002	"CRC Handbook of Medicinal Herbs 2nd Ed.", James A. Duke, CRC Press, Boca Raton, FL.
Kapo, 2000	"CRC Handbook of Ayurvedic Medicinal Plants", L.D. Kapoor, CRC Press, Boca Raton, FL.
Will, 2002	"Major Herbs of Auyrveda", E.M. Williamson, editor, Churchill-Livingston, London, England

Herbal Physicians

By 2010, 8 percent of the American public used herbs and herb extracts exclusively, 38 percent of the American public used herbs in conjunction with a physician's prescription for synthetic chemical drugs, and 54 percent still used synthetic chemical drugs exclusively.

Medicinal healing without the use of synthetic chemical drugs is performed by Natural Medicine Pathological (Naturopathic) Physicians. Healing treatments include diet, nutritional supplements, **medicinal herbs**, acupuncture, biofeedback, meditation and homeopathy.

The Council on Naturopathic Medical Education (CNME) was created in 1978 to establish and administer educational programs and colleges of naturopathic medicine. There are three colleges and two universities in the United States that train naturopathic physicians with accreditation.

Baystr University 14500 Juanita Drive NE Kenmore, WA 98028-4966	 Phone: 425-823-1300 Web: www.bastyr.edu
Natural College of Natural Medicine 049 SW Porter Street Portland, OR 97201	 Phone: 503-552-1555 Web. www.ncnm.edu
Southwest College of Naturapathic Medicine 2140 E Broadway Road Tempe, AZ 85282	 Phone: 480-858-9100 Web: www.scnm.edu
National University of Health Sciences 200 E Roosevelt Lombard, IL 60148	 Phone: 630-629-2000 Web: www.nuhs.edu
University of Bridgeport College of Naturopathic Medicine Health Science Center 60 Lafayette Street Bridgeport, CT 06604	 Phone: 800-392-3582 Web: www.bridgeport.edu/naturopathy

The American Association of Naturopathic Physicians (AANP) is the national professional organization of licensed naturopathic physicians who are graduates of accredited institutions.

American Association of Naturopathic Physicians 4435 Wisconsin Ave, NW, Suite 403 Washington DC 20016	 Phone: 866-538-2267 Web:www.naturopathic.org

Naturopathic physicians are licensed 17 states and two territories.

Alaska	Idaho	North Dakota
Arizona	Kansas	Oregon
California	Maine	Utah
Connecticut	Minnesota	Vermont
District of Columbia	Montana	Washington
Hawaii	New Hampshire	Puerto Rico
		Virgin Islands

The American Association of Acupuncture and Oriental Medicine (AAAOM) is a national professional organization of over 10,000 licensed Chinese Medicine physicians who use traditional herb formulas.

American Association of Acupuncture and Oriental Medicine PO Box 96503 PMB 93504 Washington DC 20090-6503	 Phone: 866-455-7999 Web: www.aaamonline.org

The National Ayurvedic Medical Association is an association of Hindu Natural Medicine physicians.

National Ayurvedic Medical Association 620 Cabrillo Avenue Santa Cruz, CA 96505	 Phone: 800-669-8914 Web: www.ayurveda-nama.org

Herb Extract Mail-Order Companies

AkPharma Inc.	PO Box 111, Pleasantville NJ 08232	www.akpharma.com
Allovedic Remedies	140 Ethel Rd. West, Suite S-T, Piscataway NJ 08854	www.americanformulary.com
Ayush Herbs Inc.	2239 152nd Ave. NE, Redmond WA 98052	www.ayush.com
Backstore Vitality Web Inc.	13820 Stowe Drive, San Diego, CA 92064-8880	www.thebackstore.com
Bioherb Inc.	1584 Kensington Rd., San Marino, CA 91108	www.bioherbusa.com
Blue Light Inc.	530 West State St., Ithaca NY 14580	www.treasureofeast.com
Body Dynamics Inc.	2034 Statler Drive,Carrollton, TX 75007	www.bodydynamics.com
Botanical Laboratories Inc.	1441 West Smith Rd., Ferndale, WA 98248	www.botlab.com
Brion Herbs Corporation	9200 Jeronimo Rd., Irvine CA 92618	www.brionherbs.com
Cayenne Company Inc.	2235 East 38th St., Minneapolis, MN 55407-3083	www.cayennecompany.com
Country Life	180 Vanderbuilt Motor Pkwy, Hauppauge, NY 11788	www.country-life.com
Cyvex Nutrition	1851 Kaiser Avenue, Irvine, CA 92614	www.cyvex.com
East Earth Trade Winds	PO Box 493151, Redding, CA 96049-3151	www.eastearthtrade.com
Eleclectic Institute	36350 SE Industrial Way, Sandy, OR 97055	www.eclecticherb.com
EcoNugentics Inc.	2208 Northpoint Parkway, Santa Rosa, CA 95407	www.econugenics.com
Emerson Ecologies	7 Commerce Drive, Bedford, NH 03110	www.emersonecologies.com
Energique Inc.	201 Apple Boulevard, Woodbine, IA 51579	www.energiqueherbal.com
Food Science of Vermont	20 New England Dr., Essex Junction, VT 0545	www.foodscienceofvermont.com
Gaia Herbs Inc.	101 Gaia Herbs Dr., Brevard, NC 28712	www.gaiaherbs.com
Garden of Life	5500 Villiage Blvd., Suite 202, W. Palm Beach, FL 33407	www.gardenoflife.com
Ginco International	725 Cochran St., Unit C, Simi Valley, CA 93065	www.gincointernational.com
Golden Flower Chinese Herbs	2724 Vasser Place NE, Albequerque, NM 87107	www.gfherbs.com
Health From The Sun	1 Clock Tower Place, Suite 100, Maynard, MA 01754	www.healthfromthesun.com
Herb Pharm	PO Box 116, Williams, OR 97544	www.herb-pharm.com
The Herbal Apothecary	PO Box 598, Graton, CA 95444	www.herbalapothecaryonline.com
KPC Herbs Inc.	16 Goddard, Irvine, CA 92618	www.kpc.com
Lailab Inc.	1415 Hamlin Road, Durham, NC 277704	www.lalilab.com
Lane Labs USA Inc.	3 North Street, Waldwick, NJ 07463	www.lanelabs.com
Lily of the Desert Aloe Vera	1887 Geesling Road, Denton, TX 76208	www.lilyofthedesert.com
Mushroom Wisdom	1 Madison St., Suite F6, E. Rutherford, NJ 07073	www.maitake.com
Maypro Industries	2975 Westchester Ave., Purchase, NJ 10577	www.maypro.com
Mayway Trading Corp.	1338 Mandela Parkway, Oakland, CA 94607	www.mayway.com
Metagenics UltraBalance	100 Avenue La Pata, San Clemente, CA 92673	www.metagenics.com
Michael's Naturopathic	6003 Randolf Blvd., San Antonio, TX 78233	www.machaelshealth.com
NatraBio Neutraceutical	PO Box 5935, Bellingham, WA 98227-5935	www.narabio.com
Naturade Inc. NNC LLC	1 City West Blvd., Suite 1440, Orange, CA 92868	www.naturade.com
Nature's Answer	85 Commerce Drive, Hauppauge, NY 11788	www.naturesanswer.com
Nature's Best Inc.	195 Engineers Road, Hauppauge, NY 11788	www.naturesbest.com
Nature's Plus	548 Broadhollow Road, Melville, NY 11747	www.naturesplus.com
Nature's Way Products Inc.	3051 W. Maple Loop Dr., Suite 125, Lehi, UT 84043	www.naturesway.com
NA Herb & Spice (Oreganol)	PO Box 4885, Buffalo Grove, IL 60089	www.oreganol.com
Nuherbs	3820 Penniman Avenue, Oakland, CA 94619-1759	www.nuherbs.com
Neutraceutical Corporation	1400 Kearns Boulevard, Park City, UT 84060	www.neutraceutical.com
Optimum Nutrition Inc.	600 North Commerce Street, Aurora, IL 60504	www.optimumnutrition.com
Pines International Inc.	1992 East 1400 Road, Lawrence, KS 66044	www.wheatgrass.com
Planetary Herbals	PO Box 1760, Soquel, CA 95073	www.planetaryherbals.com
Prince of Peace Inc. (Ginseng)	3536 Arden Road, Haward, CA 94545	www.popus.com
Puritan's Pride*	1233 Montauk Highway, Oakdale, NY 11769-9001	www.puritan.com
Raintree Nutrition Inc.	3579 Hwy 50 E., Suite 222, Carson City, NV 89701	www.rain-tree.com
Rich Nature Labs Inc.	9700 Harbour Place, :Suite 128, Mukilteo, WA 98275	www.richnature.com
Solgar Inc.	500 Willow Tree Road, Leonia, NJ 07605	www.solgar.com
Source Naturals	23 Janis Way, Scotts Valley, CA 95066	www.sourcenaturals.com
Spring Wind Herbs Co.	2325 Fourth St., Suite 6, Berkley, CA 94710	www.springwind.com
Sundown Naturals	110 Orville Drive, Bohemia, NY 11716	www.sundownnaturals.com
Superior Trading Co. (Ginseng)	835 Washington St., San Francisco, CA 94108	www.superiortrading.com
Taiyo Intl. (Green Tea Extracts)	5960 Golden Hills Drive, Minneapolis, MN 55416	www.taiyointernational.com
Traditional Medicinals (Teas)	4515 Rose Road, Sebastapol, CA 95472	www.traditionalmedicinals.com
Twinlab Corporation	600 East Quality Drive, American Fork, UT 84003	www.twinlab.com
Vaxa International	600 N. Westshore Blvd. Suite 800, Tampa, FL 33609	www.vaxa.com
Wakunaga (Ginkgo, Garlic)	23501 Madero, Mission Viejo, CA 92691	www.kyolic.com
Zand Herbal Formulas	PO Box 5935, Bellingham, WA 98227-5935	www.zand.com

*Puritan's Pride is the company I use. They offer discount pricing and a wide range of herb extracts.

Herb Efficacy Verification

Most herbs sold in the United States are available only in "extract" form. The extraction process may result in a product which is "biologically inactive". A "biologically inactive" herb "extract" will not perform the medicinal function which the fresh herb plant material does. To be effective, the herb extract must be "biologically active". Usually, the extract is obtained not from the Whole Plant (WP) but from a specific herb plant part which contains the "biologically active" ingredients. The herb plant parts are:
Flower Bud (FL-BD), Leaf Bud (LF-BD), Bark(BK), Bulbroot (BR), Corm (CR), Flower (FL), Fruit (FR) Flower male Stamen or female Stigma (FS), Leaf (LF), Oil (OL), Pod (PD), Root Bark (RB), Rind or skin (RD), Rhizome (RZ), Resin (RN), Root (RT), Sap (SA), Seed (SD), Shoot (SH), Stalk (SL), Stem or branch (SM), Sprout (SR), Tuber Root (TU), and Wood (WD).

Usual herb extracts are liquid or powdered chemical compounds extracted from herb plant material using either water or a solvent such as alcohol or ethanol. As a result the extract contains only the water-soluble or solvent-soluble fraction of the plant material, usually about 20% of the total weight of the plant parts used. The non-soluble (fiberous) residue, about 80% of the total weight is discarded. Thus, the herb plant part material to soluble extract ratio is usually 2:10. Herb essential oil extracts have oils that are usually 1% of the total weight of the plant parts used. Thus, the herb plant part material to oil extract ratio is usually 1:100. Special herb extracts are liquids, powders, or essential oils extracted from herb plant material using a special extraction agent other than a solvent, such as CO2, so that desirable chemical compounds are extracted in high yield, while undersirable compounds are reduced to low concentrations or completely eliminated. The efficacy of herbs or herb extracts in suppressing aliment symptoms or opportunistic diseases is determined on one of five levels:

Level 1 Folk Medicine (FM)

The efficacy of fresh herb plant material in symptom or disease suppression has been determined empirically through many generations. Chinese emperor Shen-nung [3737-2697 BC] compiled the *Pen Tsao Pharmacopeia* which contained 314 herb preparations. Ayurveda is incorporated into most texts of the Vedas, on which the Hindu religion is based. Ayurveda evolved from 1000 BC to 1100 AD. Dioscorides wrote *De Materia Medica* in 512 AD with 500 herbs. John Gerard published *Herball or General Historie of Plantes* in 1597 AD which is still in print today. John Parkinson published *An Universall and Complete Herbal* in 1640 AD which contained 3,800 herbs.

Level 2 Traditional Medicine (TM)

Chemical studies have been performed on **Folk Medicine** herbs to determine which chemical compounds are "biologically active" compounds which are acting to suppress the symptom or disease. The fresh plant material is still used, but the "active ingredients" have been identified. The efficacy has been determined empirically.

Level 3 Ayurveda Traditional Hindu Medicine (HM)

Chemical studies have been performed on **Folk Medicine** herbs to determine which chemical compounds are "biologically active" compounds which are acting to suppress the symptom or disease. The fresh plant material is still used, but the "active ingredients" have been identified. The efficacy has been determined empirically. In the USA, the chemical studies are peformed at the Maharishi University in Fairfield Iowa.

Level 4 Herb Extract Statistical Studies (SS)

Statistical Studies (SS) are appropriate where many people have already been using the herb extract when it has been on the market for some time. By studying a large number of patients and a control group, it may be possible to show a statistically significant benefit. The results are published in a reputable technical journal. The fresh plant material is no longer used. Instead, herb plant **extracts** containing the "biologically active" compounds identified in **Traditional Medicine** are extracted using one of three extraction techniques. The result is either a liquid made commercial available in "bottle" form or a powder made commercially available in "pill" form. This is what is found in "Health Food" catalogues in the United States. In 2008, the USA Food & Drug Administration imposed Good Manufacturing Practices (GMP) for the preparation of herb **extracts**.
WARNING: THE EXTRACTION PROCESS may RENDER THE HERB "Biologically Inactive" which means that the extract does not perform the symptom suppression or disease suppression that the fresh herb plant material does. **Extract Integrity** – the extraction process should not break down the "biologically active" compounds in the fresh herb plant into derivative compounds. **Extract Purity** – extracts should be free from adulterations because they may dilute the product so that it becomes ineffective. Foreign organic matter is usually limited to 2%. Maximum lead levels are set at 10 mg/kg. Maximum cadmium levels are set at 0.3 mg/kg. Maximum Aldrin and Dieldrin pesticide content is set at .05 mg/kg. The extract should be free from detectable levels of strontium-90, iodine-131, caesium-134, caesium-137, and plutonium-239. The product should be free of the bacteria *Salmonella species* and *Escherichia coli*. Fungus content should not exceed .1mg/g. Aerobic bacteria content should not exceed 0.01ml/liter.

Level 5 Alternative Medicine Pharmalogical Studies (PS)

Alternative Medicine studies are Pharmalogical Studies. The results are published in a reputable technical journal. Studies are listed on pages 501 to 551 of the book entitled **"Prescription for Herbal Healing,2nd Ed."**,P.A. Balch & S. Bell,ISBN:0-978-1-58333-452-2. Herb extracts are applied: 1. in vitro – to a living cell outside the body, 2. in vivo – directly to living cells inside the body by injection and 3. ex vivo – extracts eaten and delivered to the blood stream via the digestive system and then individual cells are examined. In all three cases, cells are examined for disease suppression and detrimental side effects.

Level 6 Alternative Medicine Clinical Studies (CS)

Alternative Medicine studies are Clinical Studies. The results are published in a reputable technical journal. Studies are listed on pages 501 to 551 of the book entitled **"Prescription for Herbal Healing,2nd Ed."**,P.A. Balch & S. Bell,ISBN:0-978-1-58333-452-2. **Clinical Studies (CS)** are made by Alternative Medicine health care practitioners that support the efficacy of the herbal extract for a particular Health Food **Herb Extract** manufacturer. The more results that are reported by independent observers, the more plausible the claims become, especially if there is some physical evidence such as blood tests or pathology reports.

Level 7 Modern Medicine (MM) Clinical Trials (CT)

Modern Medicine **C**linical **T**rails (**CT**) in three phases are performed by PhD health care practitioners and MD medical doctors on **Alternative Medicine Herbs**, the results of the trails are published in a scientific technical journal. As of 2013, only a limited number of Herbs have had Clinical Trails. Herbs that have had Clinical Trials performed were identified on pages 28 to 350 of the book entitled **"Medicinal Plants of the World"**, Ben-Erik Van Wyk & Michael Wink, ISBN:981-232-933-1, 2004. In this book those herbs are identified in the section entitled "Medicinal Herbs listed by English Name".

In 2008, the Food and Drug Administration (FDA) imposed "Good Manufacturing Practices" standards on Herb extract manufacturers.

It is expected that in 2015, that the FDA will move to require that Herb extract bottles be labeled with medicinal uses for which there have been case studies with positive results published and that each medicinal use listed be labeled as "FDA Certified by Clinical Trials" or "Not FDA Certified".

Clinical Trials

A patient will be either assigned to a control group (placebo) or a treatment group (herb extract). Phase 1 Trails are done on a statistically significant number of patients, usually 60. If Phase 1 is successful (i.e. there were more successes in the treatment group than in the control group), Phase 2 Trails are done on a statistically significant number of patients, usually 600. If Phase 2 is successful (i.e. over 75% of the patients in the treatment group had favorable results), Phase 3 Trails are done on a relatively large number of patients, usually 6000. If 90% of the patients in the treatment group had favorable results, the herb extract is considered **C**er**t**ified (**CT**), but only for the particular manufacturer of the herb extract for which the Trials were performed. Just because one manufacturer "extract" results in a "biologically active" product does not necessarily mean that another manufacturer's "extract" product is "biologically active" Thus, the "biologically active" **C**er**t**ification (**CT**) is given only for one specific manufacturer.

USA Herb Pharmacopeias

Herb Pharmacopeia – a document which contains the necessary documentation for government approved medicinal herbs, medicinal herb extracts. It contains NO ingredients synthesized from chemicals.

American Herb Pharmacopeia PO Box 66809 Scotts Valley, CA 95607	 Phone: 831-461-6318 Web: www.herbal-ahp.org
United States Pharmacopeia 12601 Twinbrook Parkway Rockville, MD 20852-1790	 Phone: 800-227-8772 Web: www.usp.org

Phytocopeia (phyton is the Latin name for plant) – a dietary document which contains the necessary documentation for culinary, medicinal, and economic herbs, medicinal herb extracts, fruits, & vegetables. **Drug Pharmacopeia** – a list of synthetic drugs which are synthesized entirely from chemicals and which contain NO herb plant material or herb extracts. **Zoocopeia** – a list of invertebrate and vertebrate animals and their commercial uses. Carl Linneaus [1707-1778], a Swedish biologist, created the **Linnaeus Latin Classication System** to standardize COMMERCIAL USES of non-living mineral matter, microbes, plants and animals. He combined a Minerocopeia, Microbiocopeia, Phytocopeia, and Zoocopeia into the Linnaeus Latin Classification System. Lawrence J. Fusillo, the author of this book, created the **English** version of the Linnaeus Latin Classification System which he called the **Fusillo Nature Classification System.**

The **USA –Herb Pharmacopeias** contain the following **Fusillo Nature Classification System** categories;

(1) NAME The Latin Genus and species name of the herb, the English Genus name of the herb, and the English species name of the herb.

(2) ORIGIN The country and locality of the world that the herb is native to where the herb grows indigineously.

(3) PARTS USED The parts of the herb plant used: Flower Bud (FL-BD), Leaf Bud (LF-BD), Bark(BK), Bulbroot (BR), Corm (CR), Flower (FL), Fruit (FR) Flower male Stamen or female Stigma (FS), Leaf (LF), Oil (OL), Pod (PD), Root Bark (RB), Rind or skin (RD), Rhizome (RZ), Resin (RN), Root (RT), Sap (SA), Seed (SD), Shoot (SH), Stalk (SL), Stem or branch (SM), Sprout (SR), Tuber Root (TU), and Wood (WD).

(4) USES The symptomatic illnesses and opportunistic diseases which a herb has been reported to suppress. The culinary uses and the economic uses.

(5) PREPARATION & DOSAGE Instructions for the preparation and amount in grams (g) of fresh or dried medicinal herbs. Preparation can be: "cut-up part", "decoction", "juice", "maceration", "mouthwash", "oil", "ointment", "powder", "suppository", "syrup", "tincture", "tea", "tonic".

(6) ACTIVE INGREDIENTS The chemical formulas or chemical names for the biologically active constituents of a medicinal herb.

(7) EXTRACT PREPARATION Instructions for the extraction of the biologically active constituents of a medicinal herb.

(8) EFFICACY The proven effectivness of the herb.
Level 1 Folk Medicine – Author & Title of the folk medicine book in which the herb appears.
Level 2 Traditional Medicine (TM) – Author & Title of the traditional medicine book in which the herb appears.
Level 3 Ayurveda Medicine – Author & Title of the Hindu traditional medicine book in which the herb appears.
Level 4 Herb Extract Statistical Studies (SS) – Author. Article Title, and Journal Title of a reputable technical journal which reported the names of the biologically active constituents of the herb and the pharmacological effects of the biologically active constituents for use by **Herbal Health Food** companies.
Level 5 Alternative Medicine Pharmalogical Studies (PS) – Author, Article Title, and Journal Title of a reputable technical journal which reported pharmalogical results of the efficacy of the herb made by **Alternative Medicine** professionals usually in animal studies.
Level 6 Alternative Medicine Clinical Studies (CS) – Author, Article Title, and Journal Title of a reputable technical journal which reported clinical observations of the efficacy of the herb made by **Alternative Medicine** professionals.
Level 7 Modern Medicine (MM) Clinical Trials (CT) – Author, Article Title, and Journal Title of a reputable technical journal which reported the results of phase 1 (60 patients), phase 2 (600 patients), or phase 3 (6000 patients) clinical trials of herb efficacy tests containing a control (placebo) group and a test (active ingredient) group made by modern medicine professionals.

(9) TOXICITY A herb can be toxic if it a) induces a heart attack, b) is carcinogenic, or c) is a poison causing the nausea, vomiting or the shutdown of one or more vital organs.

(10) REGULATORY DOCUMENTS The documents which regulate the sale of herbs, herb extracts, and synthetics. The United States of America regulates the sale of ***herbs*** as ***dietary supplements***. All other countries of the world regulate the sale of ***herbs*** as ***herbal drugs***. The definition of a ***herbal drug*** is any herb or herb extract used in the treatment, suppression, or mitigation **but not cure** of a symptomatic illness or disease recognized in the official pharmacopeia of a nation. In the United States, a ***synthetic drug*** that contains no herbs or herb extracts and is used to **cure** a disease. The sale of ***synthetic drugs*** in the USA is regulated by the Food & Drug Administration (FDA). The sale of ***herbs*** in the USA is regulated by the Dietary Supplement, Health and Education Act of 1994.

A Drug Pharmacopeia contains the same categories as a Herb Pharmocopeia except that (7) becomes
(7) SYNTHETIC PREPARATION The Company name, Patent Title and Patent Number of the synthetic preparation, prepared entirely from chemicals with no herb parts or herb extracts.

Medicinal Herbs listed by English Name

for the suppression of **BUT NOT CURE** of

Symptomatic Illnesses – discomforts & pains in the body with no apparent cause.
Opportunistic Diseases – caused by invading foreign organisms.

References

All information in this section was extracted from the following references:

Bal2, 2012	**"Prescription for Herbal Healing, 2nd Ed."**, P.A. Balch & S. Bell, ISBN:0-978-1-58333-452-2
Wyk, 2004	**"Medicinal Plants of the World"**, Ben-Erik Van Wyk & Michael Wink, ISBN:981-232-933-1

The following is a list of **Herb Pharmacopeias** and **Herb Regulatory Documents** in use today:

Asia-Pharm.	Australia, China, and Japan, government registry lists known as national Pharmacopeas.
EEU-ESCOP	European Scientific Cooperative on Phytotherapy monograph series registry lists.
EEU-Comm.E	German Commision E monograph registry list.
UK-Pharm.	British Pharmacopoeia (published 1650)
UN-WHO	United Nations World Health Organization monograph series registry lists.
USA-DSHEA	Dietary Supplement, Health and Education Act of 1994, USA
USA-Pharm.	U.S. Pharmacopeia (published by the USA federal government 1820 to 1916)
CS	**C**linical **S**tudies have been published in a reputable technical journal.
CT	**C**ER**T**IFIED – Phase 1, 2 & 3 **C**linical **T**rials have been published in a technical journal.

Clinical Studies (**CS**) are listed in the References section pages 501-611 of **"Prescription for Herbal Healing, 2nd Ed.",** Clinical Trials (**CT**) are listed in the "Physician's Desk Reference (PDR) for Herbal Medicines".

Herb English Names

TM = Traditional Medicine, **SS** = Statistical Studies, **PS** = Pharmalogical Studies, **CS** = Clinical Studies, **CT** = Clinical Trials

Species	Genus	Latin Genus	Latin Species	Regulatory Documents	Reference
Abata	Cola Nut	*Cola*	*acuminata*	**PS** EEU-Comm.E	Wyk-109-01
Absinthe	Wormwood	*Artemisia*	*absinthum*	**PS** Asia-Pharm., EEU-Comm.E, EEU-ESCOP 4	Wyk-054-01
Acajou	Cashew	*Anacardium*	*occidentale*	**SS** (Bal2-508-04) USA-DSHEA	Wyk-398-07
Acerola	Malpighian Cherry	*Malpighia*	*glabra*	**PS** (Bal2-501-01) USA-DSHEA	Bal2-014-01
Adam's needle	Yucca	*Yucca*	*filamentosa*	**TM** USA-DSHEA	Wyk-432-03
Africa Cape	Aloe	*Aloe*	*ferox*	**PS** Asia-Pharm., EEU-Comm.E, EEU-ESCOP 5, UN-WHO 1	Wyk-040-01
African	Kigel Sausage Fruit	*Kigelia*	*africana*	**PS** USA-DSHEA	Wyk-185-01
African	Myrrh	*Commiphora*	*myrrha*	**PS** Asia-Pharm., EEU-Comm.E, EEU-ESCOP 6	Wyk-111-01
African	Sour Cherry	*Prunus-Cerasus*	*africana*	**CT** (Bal2-537-01) Asia-Pharm., UN-WHO 2	Wyk-256-01
African	Thyme	*Thymus*	*schimperi*	**TM** USA-DSHEA	Wyk-323-04
African	Wormwood	*Artemisia*	*afra*	**PS** USA-DSHEA	Wyk-055-01
African Potato	Star Grass	*Hypoxis*	*hemerocallidea*	**CS** USA-DSHEA	Wyk-176-01
Agrimony				See Thoroughwort Agrimony	
Alder	Thornless Buckthorn	*Frangula*	*alnus*	**PS** Asia-Pharm., EEU-Comm.E, EEU-ESCOP 5, UN-WHO 2	Wyk-268-01
Aleppo	Pine	*Pinus*	*halepensis*	**PS** (Bal2-533-05) USA-DSHEA	Bal2-105-01
Alexandrian	Senna	*Senna*	*alexandrina*	**PS** Asia-Pharm., EEU-Comm.E, EEU-ESCOP 5, UN-WHO 1	Wyk-298-01
Alfalfa	Medick	*Medicago*	*sativa*	**PS** USA-DSHEA	Wyk-201-01
All Heal	Mulberry Mistletoe	*Viscum*	*album*	**CS** Asia-Pharm., EEU-Comm.E	Wyk-342-01
Aloe				See Africa Cape Aloe	
Aloe				See Barbados Aloe	
Aloe				See Curaçao Aloe	
Aloe	Yucca	*Yucca*	*aloeifolia*	**TM** USA-DSHEA	Bal1-079-06
Alpine Ragwort	Groundsel	*Senecio*	*ovatus*	**PS** USA-DSHEA	Wyk-297-01
American	Avocado	*Persea*	*americana*	**PS** Asia-Pharm.	Wyk-232-01
American	Ginseng	*Panax*	*quinquefolius*	**PS** (Bal2-503-01)	Wyk-419-18
American	Licorice	*Glycyrrhiza*	*lepidota*	**PS** (Bal2-529-03) USA-DSHEA	Bal2-095-02
American	Prickly Ash	*Zanthoxylum*	*americanum*	**TM** (Bal2-535-02) USA-DSHEA	Bal2-115-01
American	Sweet Gum	*Liquidambar*	*styraciflua*	**TM** USA-DSHEA	Bal1-173-08
American	Walnut	*Juglans*	*nigra*	**TM** USA-DSHEA	Wyk-183-02
American	Witch Hazel	*Hamamelis*	*virginiana*	**CS** (Bal2-550-05) Asia-Pharm., EEU-Comm.E, EEU-ESCOP 5, UN-WHO 2	Wyk-164-01
American Mandrake	May Apple	*Podophyllum*	*peltatum*	**PS** Asia-Pharm., EEU-Comm.E	Wyk-248-01
Andiroba	Crabwood	*Carapa*	*guianensis*	**PS** (Bal2-503-02) USA-DSHEA	Bal2-022-01
Angelic				See Archangel Angelic	
Angelic				See Dong Quai Angelic	
Angelic				See Du Huo Angelic	
Angelic				See Tang Kuei Angelic	
Anise				See Chinese Anise	
Anise				See Japanese Anise	
Anise	Pimpernel	*Pimpinella*	*anisum*	**PS** Asia-Pharm., EEU-Comm.E, EEU-ESCOP 3	Wyk-240-01
Arabian	Coffee	*Coffea*	*arabica*	**CS** EEU-Comm.E	Wyk-108-01
Arawak	Guava	*Psidium*	*guajava*	**PS** USA-DSHEA	Wyk-258-01
Archangel	Angelic	*Angelica*	*archangelica*	**PS** Asia-Pharm., EEU-Comm.E(root)	Wyk-048-01
Arjune	Myrobalan Almond	*Terminalia*	*arjuna*	**PS** (Bal2-503-05) USA-DSHEA	Wyk-429-09
Artemis				See Absinthe Wormwood	
Artemis				See African Wormwood	
Artemis				See Chinese Wormwood	
Artemis				See Mugwort Wormwood	
Artemis				See Sweet Annie Wormwood	
Ashwagandha				See Winter Cherry Ashwagandha	
Asparagus				See Mediterranean Asparagus	
Atlantic White	Wood of Life Cedar	*Thuja*	*occidentalis*	**PS** Asia-Pharm.	Wyk-322-05
Atlas	Cedar	*Cedrus*	*atlantica*	**TM** USA-DSHEA	Bal1-067-04
Autumn Saffron	Meadow Crocus	*Colchicum*	*autumnale*	**PS** EEU-Comm.E	Wyk-110-01
Avocado				See American Avocado	
Babchi	Scurf Pea	*Psoralea*	*coryifolia*	**SS** (Bal2-536-01) USA-DSHEA	Wyk-423-21
Bachelor's Buttons	Tansy	*Tanacetum*	*vulgare*	**TM** Asia-Pharm. **induced abortion** ☠	Wyk-099-01
Balsam Pear	Bitter Melon	*Momordica*	*charantia*	**PS** (Bal2-506-05) USA-DSHEA	Bal2-033-01
Barbados	Aloe	*Aloe*	*barbadensis*	**PS** (Bal2-502-01) USA-DSHEA	Bal2-018-01
Barberry				See European Barberry	
Basket	Willow	*Salix*	*purpurea*	**TM** USA-DSHEA	Wyk-281-03
Bay	Willow	*Salix*	*pentandra*	**TM** USA-DSHEA	Bown-245-07
Bay Leaf	Laurel	*Laurus*	*noblis*	**PS** Asia-Pharm.	Wyk-188-01

Species	Genus	*Latin Genus*	*Latin Species*	Regulatory Documents	Reference
Bayberry				See Wax Myrtle Bayberry	
Bean				See Kidney Dry Bush Bean	
Bean				See Navy Dry Bush Bean	
Bearded	Skullcap	*Scutellaria*	*barbata*	**TM** (Bal2-541-02) USA-DSHEA	Bal2-129-02
Bedstraw				See Lady's Bedstraw	
Bedstraw				See Cleavers Bedstraw	
Beer	Hops	*Humulus*	*lupulus*	**PS** Asia-Pharm., EEU-Comm.E, EEU-ESCOP 4	Wyk-172-01
Bilberry	Whortleberry	*Vaccinium-vortabacca*	*myrtillus*	**PS** (Bal2-506-03) Asia-Pharm.,EEU-Comm.E	Wyk-333-01
Birch				See Black Birch	
Birch				See Silver Birch	
Birch				See Sweet Birch	
Birch				See White Birch	
Bird's Foot	Fenugreek	*Trigonella*	*foenum-graecum*	**PS** (Bal2-517-05) Asia-Pharm., EEU-Comm.E	Wyk-325-01
Birthwort	Snakeroot	*Aristolochia*	*clematitis*	**TM induced abortion** ☠	Wyk-051-01
Bitter	Candytuft	*Iberis*	*amara*	**PS** USA-DSHEA	Wyk-178-01
Bitter	Quasiwood	*Quassia*	*amara*	**PS** Asia-Pharm., EEU-Comm.E	Wyk-264-01
Bitter Melon				See Balsam Pear Bitter Melon	
Bitter Orange				See Seville Sour Orange	
Bittersweet Nightshade	Sunplant	*Solanum*	*dulcamara*	**CT** EEU-Comm.E	Wyk-304-01
Black	Birch	*Betula*	*nigra*	**TM** USA-DSHEA	Bal2-032-01
Black	Oak	*Quercus*	*velutina*	**TM** (Bal2-537-02) USA-DSHEA	Bal2-119-01
Black	Peppercorn	*Piper*	*nigrum*	**PS** USA-DSHEA	Wyk-244-01
Black	Willow	*Salix*	*nigra*	**TM** USA-DSHEA	Bown-245-06
Black Berried	Elderberry	*Sambucus*	*nigra*	**PS** Asia-Pharm., EEU-Comm.E,UN-WHO 2(flower)	Wyk-285-01
Black Cohosh	Bugbane	*Cimicifuga*	*racemosa*	**CS** Asia-Pharm., EEU-Comm.E, UN-WHO 2	Wyk-101-01
Black Haw	Tree Cranberry	*Viburnum-drysia*	*prunifolium*	**PS** Asia-Pharm.	Wyk-339-01
Black Horehound				See Cowbane Black Horehound	
Black Sampson	Coneflower	*Echinacea*	*angustifolia*	**TM** (Bal2-515-01) USA-DSHEA	Wyk-408-20
Black Seed	Mustard	*Brassica*	*nigra*	**PS** Asia-Pharm.	Wyk-070-01
Black Walnut				See American Walnut	
Blackberry				See Bramble Blackberry	
Blackseed	Love In A Mist	*Nigella*	*sativa*	**PS** USA-DSHEA	Wyk-216-01
Bladderwrack Kelp	Painted Dusky Ocher Algae	*Fucus*	*vesiculosis*	**TM** (Bal2-527-07) USA-DSHEA	Wyk-410-19
Blessed	Milk Thistle	*Silybum*	*marianum*	**CS** Asia-Pharm., EEU-Comm.E, UN-WHO 2	Wyk-300-01
Blessed Thistle	Knapweed	*Centaurea*	*benedicta*	**PS** Asia-Pharm., EEU-Comm.E	Wyk-087-01
Bloodroot	Cinquefoil	*Potentilla*	*erecta*	**PS** Asia-Pharm., EEU-Comm.E	Wyk-254-01
Blowball	Dandelion	*Taraxacum*	*officinale*	**PS** Asia-Pharm., EEU-Comm.E, EEU-ESCOP 2	Wyk-318-01
Blue	Cornflower	*Cyanus*	*segetum*	**TM** USA-DSHEA	Wyk-088-01
Blue Crown	Passion Flower	*Passiflora*	*caerulea*	**TM** USA-DSHEA	Bal2-110-02
Bog Bean	Buckbean	*Menyanthes*	*trifoliata*	**PS** Asia-Pharm., EEU-Comm.E	Wyk-208-01
Boldo	Peum	*Peumus*	*boldus*	**PS** Asia-Pharm., EEU-Comm.E, EEU-ESCOP 1	Wyk-236-01
Boneset	Eupatory	*Eupatorium*	*perfoliatum*	**TM** USA-DSHEA	Wyk-140-01
Borage				See Mediterranean Borage	
Bouncing Bet	Soapwort	*Saponaria*	*officinalis*	**PS** EEU-Comm.E (root, rhizome) poison ☠	Wyk-290-01
Brahmi	Water Hyssop	*Bacopa*	*monnieri*	**PS** (Bal2-507-04) USA-DSHEA	Bal2-036-01
Bramble	Blackberry	*Rubus-nigrabacca*	*fruticosus*	**TM** Asia-Pharm., EEU-Comm.E (leaf)	Wyk-277-01
Brazilian	Ipecac	*Carapichea*	*ipecacuanha*	**PS** Asia-Pharm.	Wyk-259-01
Brazilian Cocoa	Bread & Cheese	*Paullinia*	*cupana*	**PS** USA-DSHEA	Wyk-228-01
Brazilian Ginseng	Suma Amaranth	*Pfaffia*	*paniculata*	**TM** USA-DSHEA	Bal1-077-06
Brigham Tea	Joint Fir	*Ephedra*	*trifurca*	**TM** USA-DSHEA	Bown-278-06
Bromelain	Pineapple	*Ananas*	*comosus*	**PS** (Bal2-507-05) EEU-Comm.E	Wyk-046-01
Buchu				See Oval Leaf Buchu	
Buchu				See Round Leaf Buchu	
Buckeye	Horse Chestnut	*Aesculus*	*hippocastanum*	**CS** Asia-Pharm., EEU-Comm.E,EEU-ESCOP 4	Wyk-034-01
Buckthorn				See Purging Buckthorn	
Burdock				See Hare Burr Burdock	
Burnet				See Great Burnet	
Butcher's	Broom	*Ruscus*	*aculeatus*	**PS** (Bal2-508-03) EEU-Comm.E	Wyk-279-01
Butterbur				See Hybrid Butterbur	
Buttonweed	Mallow	*Malva*	*neglecta*	**TM** USA-DSHEA	Wyk-416-17
Cajueiro				See Acajou Cashew	
California	Eschscholtz Poppy	*Eschscholzia*	*californica*	**PS** (Bal2-509-02) USA-DSHEA	Wyk-138-01
Camphor Oil	Cinnamon	*Cinnamomum*	*camphora*	**PS** Pharm., EEU-Comm.E	Wyk-104-01
Cancer Bush	Balloon Pea	*Sutherlandia*	*frutescens*	**PS** USA-DSHEA	Wyk-313-01
Cancer Tree	Ribbed Leaf Tupelo	*Camptotheca*	*acuminata*	**CT** USA-DSHEA	Wyk-076-01
Candytuft				See Bitter Candytuft	
Cardamom				See India Cardamom	
Cardamom				See Tavoy Bastard Cardamom	
Carry Me Seed	Leaf Flower Spurge	*Phyllanthus*	*amarus*	**CS** (Bal2-511-01) USA-DSHEA	Wyk-420-17

Species	Genus	*Latin Genus*	*Latin Species*	Regulatory Documents	Reference
Cascara	Thornless Buckthorn	*Frangula*	*purshiana*	**PS** Bark is called **sagrada**, Asia-Pharm., EEU-Comm.E, EEU-ESCOP 5, UN-WHO 2	Wyk-269-01
Cascarilla	Croton	*Croton*	*eleuteria*	**TM** USA-DSHEA	Bown-269-03
Cashew				See Acajou Cashew	
Castor Oil	Castor Bean	*Ricinus*	*communis*	**PS** Asia-Pharm.	Wyk-274-01
Cat's Claw Guiana	Uncary	*Uncaria*	*guianensis*	**PS** (Bal2-509-05) USA-DSHEA	Bal2-043-02
Cat's Claw Paraguay	Uncary	*Uncaria*	*tomentosa*	**PS** (Bal2-509-05) USA-DSHEA	Wyk-330-01 Bal2-043-03
Catmint				See Jing Jie Split Catmint	
Catnip	Catmint	*Nepeta*	*cataria*	**PS** (Bal2-509-04) USA-DSHEA	Wyk-214-01
Catuaban	Tropical Redwood	*Erythroxylum*	*catuaba*	**PS** (Bal2-510-01) USA-DSHEA	Bal2-044-01
Cayenne Hot	Pepper	*Capsicum*	*annuum-longum-cayenneum*	**CS** (Bal2-510-02) USA-DSHEA	Bal2-044-02
Celery	Marshwort	*Apium*	*graveolens-dulce*	**TM** USA-DSHEA	Wyk-399-04
Cereal	Oats	*Avena*	*sativa*	**TM** EEU-Comm.E **(Oat Straw)**	Wyk-063-01
Chamomile				See German Chamomile	
Chanca Piedra				See Carry Me Seed Leaf Flower Spurge	
Chang Pu	Sweet Flag	*Acorus*	*gramineus*	**TM** USA-DSHEA	Bal2-015-01
Chaparral	Rush Pea	*Hoffmannseggia*	*divaricata*	**PS** (Bal2-511-02) USA-DSHEA	Wyk-414-16
Chasteberry				See Monk's Pepper Chasteberry	
Chebulic	Myrobalan Almond	*Terminalia*	*chebula*	**PS** USA-DSHEA	Wyk-320-01
Chen Pi				See East Asia Tangerine	
Chestnut				See Sweet Chestnut	
Chickweed				See Starweed Chickweed	
Chili Hot	Pepper	*Capsicum*	*annuum-frutescens-chilium*	**CT** Asia-Pharm., EEU-Comm.E	Wyk-079-01
China Kassod	Cinnamon	*Cinnamomum*	*aromaticum*	**PS** (Bal2-511-06)Pharm.,EEU-Comm.E,UN-WHO 1	Wyk-103-01
Chinese	Anise	*Illicium*	*verum*	**PS** Asia-Pharm., EEU-Comm.E	Wyk-180-01
Chinese	Balloon Bellflower	*Platycodon*	*grandiflorus*	**PS** UN-WHO 1	Wyk-247-04
Chinese	Boxthorn	*Lycium*	*chinense*	**PS** USA-DSHEA	Wyk-195-01
Chinese	Caterpillar Fungus	*Cordyceps*	*sinensis*	**TM** (Bal2-513-03) USA-DSHEA	Bal2-054-02
Chinese	Ginseng	*Panax*	*ginseng*	**CS** Asia-Pharm., EEU-Comm.E, UN-WHO 1	Wyk-224-01
Chinese	Goldthread	*Coptis*	*chinensis*	**PS** (Bal2-513-02) UN-WHO 1	Wyk-113-01
Chinese	Jojoba	*Simmondsia*	*chinensis*	**PS** USA-DSHEA	Wyk-301-01
Chinese	Licorice	*Glycyrrhiza*	*uralensis*	**PS** (Bal2-529-03) USA-DSHEA	Wyk-411-23
Chinese	Motherwort	*Leonurus*	*heterophyllus*	**TM** USA-DSHEA	Bal2-103-01
Chinese	Peony	*Paeonia*	*lactiflora*	**PS** UN-WHO 1	Wyk-223-01
Chinese	Rehmann Foxglove	*Rehmannia*	*glutinosa*	**TM** USA-DSHEA	Wyk-424-18
Chinese	Rhubarb	*Rheum*	*palmatum*	**PS** Asia-Pharm., EEU-Comm.E, EEU-ESCOP 6, UN-WHO 1	Wyk-270-01
Chinese	Senega Root			See Senega Root Milkwort	
Chinese	Star Magnolia	*Schisandra*	*chinensis*	**TM** (Bal2-540-03) USA-DSHEA	Wyk-426-19
Chinese	Wormwood	*Artemisia*	*capillaris*	**PS** (Bal2-504-02) USA-DSHEA	Bal2-026-01
Chinese Cucumber	Snake Gourd	*Trichosanthes*	*kirilowii*	**TM** USA-DSHEA	Bal1-102-09
Chinese Date	Jujube	*Ziziphus*	*jujuba*	**PS** USA-DSHEA	Wyk-350-01
Chinese	Thistle Daisy	*Atractylodes*	*macrocephala*	**TM** USA-DSHEA	Bown-246-02
Chinese Tuber	Larkspur Poppy	*Corydalis*	*yanhusuo*	**PS** (Bal2-513-05) USA-DSHEA	Wyk-406-19c
Chiretta	Green Gentian	*Frasera*	*chirata*	**TM** (Bal2-511-05) USA-DSHEA	Wyk-428-19
Cilantro	Coriander	*Coriandrum*	*sativum*	**PS** Asia-Pharm., EEU-Comm.E	Wyk-114-01
Cinnamon				See Camphor Oil Cinnamon	
Cinnamon				See China Kassod Cinnamon	
Cinnamon				See India Kassod Cinnamon	
Cinnamon				See Sri Lanka Cinnamon	
Citronella Grass	Spikelet Beard Grass	*Cymbopogon*	*nardus*	**PS** USA-DSHEA	Wyk-120-03
Clary	Sage	*Salvia*	*sclarea*	**PS** USA-DSHEA	Wyk-284-01
Cleavers	Bedstraw	*Galium*	*aparine*	**TM** USA-DSHEA	Krug-054-03
Cloves	Lilly Pilly	*Syzygium*	*aromaticum*	**PS** (Bal2-512-01) UN-WHO 2	Wyk-315-01
Cocaine Bolivia	Tropical Redwood	*Erythroxylum*	*coca*	**PS** USA-DSHEA	Wyk-137-01
Coffee				See Arabian Coffee	
Colforsin	Cole Mint	*Coleus*	*barbatus*	**TM** (Bal2-512-04) USA-DSHEA	Wyk-405-21
Colt's Foot				See Coughwort Colt's Foot	
Comfrey				See Knitbone Comfrey	
Coneflower				See Black Sampson Coneflower	
Coneflower				See Pale Purple Coneflower	
Coneflower				See Purple Coneflower	
Copiaban Balsam	Copal Gum	*Copaifera*	*langsdorffii*	**TM** (Bal2-513-01) USA-DSHEA	Wyk-406-15
Corn Mint	Mint	*Mentha*	*arvensis*	**PS** Asia-Pharm., EEU-Comm.E	Wyk-205-01
Cornfield	Hemp Nettle Mint	*Galeopsis*	*segetum*	**PS** Asia-Pharm., EEU-Comm.E	Wyk-151-01
Cornflower				See Blue Cornflower	

Species	Genus	Latin Genus	Latin Species	Regulatory Documents	Reference
Cornsilk				See Sweet Corn Zeatin Grass	
Couch Grass	Wild Rye	*Elymus*	*repens*	**TM** Asia-Pharm., EEU-Comm.E	Wyk-133-01
Coughroot				See Osha Alpine Lovage	
Coughwort	Colt's Foot	*Tussilago*	*farfara*	**PS** (Bal2-512-05) Asia-Pharm., EEU-Comm.E(leaf)	Wyk-328-01
Cow's Lip	Primrose	*Primula*	*veris*	**PS** Asia-Pharm., EEU-Comm.E, EEU-ESCOP 3	Wyk-255-01
Cowbane	Black Horehound	*Ballota*	*nigra*	**TM** Asia-Pharm.	Wyk-065-01
Crack	Willow	*Salix*	*fragilis*	**TM** USA-DSHEA	Wyk-281-02
Crampbark	Highbush Cranberry	*Viburnum-scrubia*	*opulus*	**TM** USA-DSHEA	Wyk-339-02
Cranberry				See Black Haw Tree Cranberry	
Cranberry				See Crampbark Highbush Cranberry	
Cranberry				See European Heath Cranberry	
Cranberry				See Moose Lowbush Cranberry	
Crown	Fritillary	*Fritillaria*	*thunbergii*	**TM** (Bal2-518-02) USA-DSHEA	Wyk-410-18b
Curaçao	Aloe	*Aloe*	*vera*	**PS** Asia-Pharm., EEU-Comm.E, UN-WHO 1	Wyk-041-01
Curly Cup Scaley	Gumweed	*Grindelia*	*squarrosa*	**PS** Asia-Pharm., EEU-Comm.E	Wyk-161-01
Curly Leaf	Parsley	*Petroselinum*	*crispum*	**PS** Asia-Pharm., EEU-Comm.E	Wyk-235-01
Cutch	Wattle	*Acacia*	*catechu*	**PS** Asia-Pharm.	Wyk-028-01
Damiana	Alder Hazel	*Turnera*	*diffusa*	**PS** Bal2-513-06) Asia-Pharm.	Wyk-327-01
Dan Shen				See Red Root Sage	
Dandelion				See Blowball Dandelion	
Dang Shen				See Poor Man's Ginseng Bonnet Bellflower	
Deadly	Nightshade	*Atropa*	*belladonna*	**PS** Asia-Pharm., EEU-Comm.E	Wyk-062-01
Devadaru	Cedar	*Cedrus*	*deodara*	**TM** USA-DSHEA	Wyk-403-13
Devil's Claw	Grapple Plant	*Harpagophytum*	*procumbens*	**CS** Asia-Pharm., EEU-Comm.E, EEU-ESCOP 2	Wyk-165-01
Devil's Dung	Giant Fennel	*Ferula*	*assa-foetida*	**PS** (Bal2-504-03) Asia-Pharm.	Wyk-143-01
Devil's Fuge	Mulberry Mistletoe	*Viscum*	*coloratum*	**TM** USA-DSHEA	Bal2-101-02
Dill				See Scented Dill	
Dog	Rosehip Rose Shrub	*Rosa-hipa-scrubia*	*canina*	**PS** Asia-Pharm., EEU-Comm.E (petal)	Wyk-275-01
Dong Quai	Angelic	*Angelica*	*sinensis*	**TM** (Bal2-514-04) UN-WHO 2	Wyk-048-02
Dragon's Blood	Croton	*Croton*	*lechleri*	**TM** (Bal2-539-03) USA-DSHEA	Bal2-124-01
Du Huo	Angelic	*Angelica*	*dahurica*	**PS** (Bal2-550-01) USA-DSHEA	Wyk-398-20b
Eagle Vine	Latex Milkvine	*Marsdenia*	*cundrango*	**PS** Asia-Pharm., EEU-Comm.E	Wyk-199-01
Earth Smoke	Fumitory	*Fumaria*	*officinalis*	**PS** Asia-Pharm., EEU-Comm.E	Wyk-148-01
East Asia	Gentian	*Gentiana*	*scabra*	**PS** (Bal2-519-01) USA-DSHEA	Bal2-072-02
East Asia	Lemon	*Citrus-limonus*	*limon*	**PS** Asia-Pharm.	Wyk-107-01
East Asia	Tangerine	*Citrus-tangiericus*	*reticulata*	**TM** (Bal2-510-03) USA-DSHEA	Wyk-405-09
Edible	Buckwheat	*Fagopyrum*	*esculentum*	**PS** USA-DSHEA	Wyk-142-01
Edible	Khat	*Catha*	*edulis*	**PS** USA-DSHEA	Wyk-084-01
Elderberry				See Black Berried Elderberry	
Elecampane				See Scabwort Elecampane	
Egyptian Privet	Henna	*Lawsonia*	*inermis*	**PS** USA-DSHEA	Wyk-190-01
English	Hawthorn Berry	*Crataegus*	*laevigata*	**PS** (Bal2-526-01) USA-DSHEA	Wyk-406-20b
English	Ivy	*Hedera*	*helix*	**CS** Asia-Pharm., EEU-Comm.E	Wyk-167-01
English	Lavender	*Lavandula*	*angustifolia*	**PS** (Bal2-528-03) Asia-Pharm., EEU-Comm.E	Wyk-189-01
English	Oak	*Quercus*	*rubur*	**PS** Asia-Pharm., EEU-Comm.E	Wyk-265-01
English	Walnut	*Juglans*	*regia*	**PS** Asia-Pharm., EEU-Comm.E	Wyk-183-01
English	Yew	*Taxus*	*baccata*	**CS** Asia-Pharm.	Wyk-319-01
English Mandrake	Bryony	*Bryonia*	*dioica*	**PS** USA-DSHEA	Wyk-072-01
English Serpentary	Knotweed	*Polygonum*	*bistorta*	**PS** USA-DSHEA	Wyk-233-01
Ephedra				See Brigham Tea Joint Fir	
Ephedra				See European Joint Fir	
Ephedra				See Horsetail Joint Fir	
Ephedra				See Ma Huang Joint Fir	
Ephedra				See Middle East Joint Fir	
Ephedra				See Mormon Tea Joint Fir	
Espinheira Santa	Mayten Holly	*Maytenus*	*chuchuhuasa*	**TM** (Bal2-517-01) USA-DSHEA	Bal2-065-01
Ethiopian	Siphonclaw Ginger	*Siphonochilus*	*aethiopicus*	**TM** USA-DSHEA	Wyk-302-01
Eucalpyt				See Tasmanian Blue Gum Tree	
Eurasian	Snowdrop Amaryllis	*Galanthus*	*nivalis*	**CS** USA-DSHEA	Wyk-149-01
European	Ash	*Fraxinus*	*excelsior*	**PS** Asia-Pharm., EEU-Comm.E	Wyk-147-01
European	Barberry	*Berberis*	*vulgaris*	**PS** (Bal2-506-02) Asia-Pharm.	Wyk-066-01
European	Gentian	*Gentiana*	*lutea*	**PS** Asia-Pharm., EEU-Comm.E, EEU-ESCOP 4	Wyk-155-01
European	Goldenrod	*Solidago*	*virga-aurea*	**PS** Asia-Pharm., EEU-Comm.E, EEU-ESCOP 2	Wyk-305-01
European	Heath Cranberry	*Vaccinium-cranabacca*	*oxycoccus*	**PS** USA-DSHEA	Wyk-334-01
European	Henbane	*Hyoscyamus*	*niger*	**CS** Asia-Pharm., EEU-Comm.E	Wyk-174-01
European	Hyssop	*Hyssopis*	*officinalis*	**PS** USA-DSHEA	Wyk-177-01
European	Joint Fir	*Ephedra*	*distachya*	**TM** USA-DSHEA	Wyk-134-02
European	Juniper	*Juniperus*	*communis*	**PS** Asia-Pharm., EEU-Comm.E, EEU-ESCOP 3	Wyk-184-01
European	Licorice	*Glycyrrhiza*	*glabra*	**PS** Asia-Pharm., EEU-Comm.E, UN-WHO 1	Wyk-160-01
European	Motherwort	*Leonurus*	*cardiaca*	**PS** Asia-Pharm., EEU-Comm.E	Wyk-191-01

Species	Genus	Latin Genus	Latin Species	Regulatory Documents	Reference
European	Olive	*Olea*	*europaea*	**PS** Asia-Pharm.	Wyk-219-01
European	Pasque Flower	*Pulsatilla*	*vulgaris*	**PS** USA-DSHEA poison ☠	Wyk-262-01
European	Peony	*Paeonia*	*officinalis*	**TM** Asia-Pharm.	Wyk-223-02
European	Pillatory	*Parietaria*	*officinalis*	**TM** USA-DSHEA	Wyk-226-01
European	Scopoli Sunplant	*Scopolia*	*carniolica*	**PS** EEU-Comm.E narcotic	Wyk-293-01
European	Skullcap	*Scutellaria*	*galericulata*	**TM** (Bal2-541-02) USA-DSHEA	Wyk-295-03
European Black	Currant	*Ribes*	*nigrum*	**PS** EEU-ESCOP 4	Wyk-273-01
European Tuber	Larkspur Poppy	*Corydalis*	*cava*	**PS** USA-DSHEA	Wyk-406-19a
Evening Star	Evening Primrose	*Oenothera*	*biennis*	**CS** Asia-Pharm., UN-WHO 2	Wyk-218-01
Eyebright				See Grove Eyebright	
False Unicorn Root	Devil's Bit	*Chamaelirium*	*luteum*	**TM** USA-DSHEA	Wyk-404-02
Fennel				See Sweet Fennel	
Fenugreek				See Bird's Foot Fenugreek	
Feverfew	Tansy	*Tanacetum*	*parthenium*	**CS** Asia-Pharm., EEU-ESCOP 2, UN-WHO 2	Wyk-098-01
Feverwort	Centaury	*Centaurium*	*erythreaea*	**PS** Asia-Pharm., EEU-Comm.E, EEU-ESCOP 6	Wyk-089-01
Fiber	Hemp	*Cannabis*	*sativa-sativa*	**-** Low Tetrahydrocannabinol hemp used for rope	Wyk-077-01
Field	Horsetail	*Equisetum*	*arvense*	**TM** Asia-Pharm., EEU-Comm.E	Wyk-136-01
Figwort				See Knotted Figwort	
Flavonoid				See Bilberry Whortleberry	
Flavonoid				See Blessed Milk Thistle	
Flavonoid				See English Hawthorn Berry	
Flavonoid				See Green Tea	
Flavonoid				See Maidenhair Ginkgo	
Flavonoid				See Quickthorn Hawthorn Berry	
Flavonoid				See Red Wine Grape	
Florist	Gold Mum	*Chrysanthemum*	*morifolium*	**CS** USA-DSHEA	Wyk-097-01
Frankincense				See India Oliban Frankincense	
Frankincense				See Sacred Oliban Frankincense	
French Lilac	Goat's Rue	*Galega*	*officinalis*	**PS** Asia-Pharm.	Wyk-150-01
French Maritime	Pine	*Pinus*	*pinaster*	**TM** USA-DSHEA	
Fritillary				See Crown Fritillary	
Fungus				See Chinese Caterpillar Fungus	
Garlic	Onion	*Allium*	*sativum*	**CT** (Bal2-518-03) Asia-Pharm., EEU-Comm.E, EEU-ESCOP 3, UN-WHO 1	Wyk-039-01
Gentian				See East Asia Gentian	
Gentian				See European Gentian	
German Chamomile	Mayweed Daisiette	*Matricaria*	*chamomilla*	**PS** (Bal2-510-03) Asia-Pharm., EEU-Comm.E, EEU-ESCOP 6, UN-WHO 1	Wyk-200-01
German Flag	Iris	*Iris*	*germanica*	**TM** Asia-Pharm.	Wyk-182-01
Ginger				See South Asia Ginger	
Ginkgo				See Maidenhair Ginkgo	
Ginseng				See American Ginseng	
Ginseng				See Chinese Ginseng	
Ginseng				See Poor Man's Ginseng Bonnet Bellflower	
Ginseng				See Siberian Eleuthero Ginseng	
Gipsywort	Bugleweed	*Lycopus*	*europaeus*	**CS** EEU-Comm.E	Wyk-196-01
Glaucous Collett	Yam	*Dioscorea*	*collettii hypoglauca*	**TM** USA-DSHEA	Bown-273-10
Globe Artichoke	Vegetable Thistle	*Cynara*	*scolymus*	**PS** Asia-Pharm., EEU-Comm.E	Wyk-121-01
Gold Bloom	Pot Marigold	*Calendula*	*officinalis*	**PS** (Bal2-509-01) Asia-Pharm., EEU-Comm.E, EEU-ESCOP 1, UN-WHO 2	Wyk-074-01
Goldenroot	Goldenseal	*Hydrastis*	*canadensis*	**PS** (Bal2-524-01) USA-DSHEA	Wyk-173-01
Goldenseal				See Goldenroot Goldenseal	
Goldthread				See Chinese Goldthread	
Gotu Kola	Hundredleaf	*Centella*	*asiatica*	**CS** Asia-Pharm., UN-WHO 1	Wyk-090-01
Grape Seed Oil				See Red Wine Grape	
Grape Skin				See Red Fox Grape	
Grape Skin				See Red Wine Grape	
Gravel Root	Eupatory	*Eupatorium*	*purpureum*	**TM** USA-DSHEA, Liver carcinogenic ☠	Wyk-140-02
Great	Burnet	*Sanguisorba*	*officinalis*	**PS** USA-DSHEA	Wyk-287-01
Great	Mullein	*Verbascum*	*thapsus*	**TM** USA-DSHEA	Wyk-336-03
Greater	Celandine	*Chelidonium*	*majus*	**PS** Pharm., EEU-Comm.E	Wyk-093-01
Greater	Pimpernel	*Pimpinella*	*major*	**PS** EEU-Comm.E (root)	Wyk-241-01
Greater	Plantain	*Plantago*	*major*	**TM** USA-DSHEA	Wyk-246-02
Grecian	Finger Foxglove	*Digitalis*	*lanata*	**PS** Asia-Pharm.	Wyk-125-01
Green	Tea	*Thea*	*sinensis-viridis*	**PS** (Bal2-524-03) Asia-Pharm.	Wyk-075-01
Grove	Eyebright	*Euphrasia*	*nemorosa*	**TM** (Bal2-517-03) Asia-Pharm.	Wyk-141-01
Guaiac Wood of Life				See West Indies Guaiac Wood of Life	
Guava				See Arawak Guava	
Guggul	Myrrh	*Commiphora*	*mukul*	**PS** (Bal2-525-01) USA-DSHEA	Wyk-406-06
Gum Ammoniac	Gift Parsley	*Dorema*	*ammoniaca*	**TM** Asia-Pharm.	Wyk-127-01

Species	Genus	*Latin Genus*	*Latin Species*	Regulatory Documents	Reference
Gum Arabic	Wattle	*Acacia*	*senegal*	**PS** Asia-Pharm.	Wyk-029-01
Hare Burr	Burdock	*Arctium*	*lappa*	**PS** Asia-Pharm.	Wyk-049-01
Hare's Ear	Thorow Wax	*Bupleurum*	*falcatum*	**PS** UN-WHO 1	Wyk-073-01
Hawthorn				See English Hawthorn Berry	
Hawthorn				See Quickthorn Hawthorn Berry	
Hawthorn				See Sharpthorn Hawthorn Berry	
Heal All	Selfheal	*Prunella*	*vulgaris*	**SS** (Bal2-535-03) USA-DSHEA	Wyk-423-07
Heart's Ease Pansy	Violet	*Viola*	*tricolor*	**PS** EEU-Comm.E, (external)	Wyk-341-01
Heath	Speedwell	*Veronica*	*officinalis*	**PS** USA-DSHEA	Wyk-338-01
Hemp				See Fiber Hemp	
Hemp				See Marijuana Hemp	
Henna				See Egyptian Privet Henna	
Herb Bennet	Avens	*Geum*	*urbanum*	**PS** Asia-Pharm.	Wyk-157-01
Herb Of Grace	Rue	*Ruta*	*graveolens*	**PS** USA-DSHEA	Wyk-280-01
Herb Of The Cross	Vervain	*Verbena*	*officinalis*	**PS** Asia-Pharm., EEU-Comm.E	Wyk-337-01
Herb Robert	Crane's Bill	*Geranium*	*robertianum*	**PS** USA-DSHEA	Wyk-156-01
High	Mallow	*Malva*	*sylvestris*	**PS** Asia-Pharm., EEU-Comm.E	Wyk-197-01
Ho She Wu	Knotweed	*Polygonum*	*multiflorum*	**TM** (Bal2-526-02) USA-DSHEA	Wyk-422-22
Hoelen	Mushroom			See Tuckahoe Wolf Foam Pore Fungus	
Holly Oregon Grape	Mahon Barberry	*Mahonia*	*nervosa*	**TM** USA-DSHEA	Wyk-416-13
Honey Bush	Cyclops Tea	*Cyclopia*	*intermedia*	**TM** USA-DSHEA	Wyk-119-01
Hops				See Beer Hops	
Horehound				See Cowbane Black Horehound	
Horehound				See White Horehound	
Horny Goatweed				See Inyo Kaku Horny Goatweed	
Horny Goatweed				See Yin Yang Yuo Horny Goatweed	
Horse Chestnut				See Buckeye Horse Chestnut	
Horseradish				See Red Cole Horseradish	
Horsetail				See Field Horsetail	
Horsetail	Joint Fir	*Ephedra*	*equisetina*	**TM** USA-DSHEA	Bown-278-02
Huang Qi	Milk Vetch	*Astragalus*	*membranaceus*	**CS** (Bal2-504-05) UN-WHO 1	Wyk-061-01
Hybrid	Butterbur	*Petasites*	*hybridus*	**CS** EEU-Comm.E (rhizome)	Wyk-234-01
Hyssop				See European Hyssop	
Iceland Moss	Heath Shield Lichen	*Cetraria*	*islandica*	**PS** Asia-Pharm.	Wyk-091-01
India	Cardamom	*Elettaria*	*cardamomum*	**PS** Asia-Pharm., EEU-Comm.E	Wyk-131-01
India	Neem	*Azadirachta*	*indica*	**CS** USA-DSHEA	Wyk-064-01
India Date	Tamarind	*Tamarindus*	*indica*	**PS** Asia-Pharm.	Wyk-317-01
India Kassod	Cinnamon	*Cinnamomum*	*tamala*	**TM** USA-DSHEA	Wyk-405-04
India Oliban	Frankincense	*Boswellia*	*serrata*	**PS** (Bal2-507-03) USA-DSHEA	Wyk-402-03
Indian Tobacco	Lobeline Bellflower	*Lobelia*	*inflata*	**PS** Asia-Pharm.	Wyk-194-01
Indochinese	Yam	*Dioscorea*	*oppositifolia*	**TM** USA-DSHEA	Wyk-126-02
Inkberry	Pokeberry	*Phytolacca*	*americana*	**PS** USA-DSHEA	Wyk-238-01
Inyo Kaku	Horny Goatweed	*Epimedium*	*brevicornum*	**PS** (Bal2-516-03) USA-DSHEA	Bal2-064-02
Ipecac				See Brazilian Ipecac	
Iporuru	Alchorne Spurge	*Alchornea*	*cordifolia*	**TM** (Bal2-527-04) USA-DSHEA	Wyk-397-09
Iranian	Saltwood Mustard	*Salvadora*	*persica*	**PS** USA-DSHEA	Wyk-282-01
Irish Moss	Cartilage Red Algae	*Chondrus*	*crispus*	**PS** USA-DSHEA	Wyk-096-01
Isidikawa	Yam	*Dioscorea*	*dregiana*	**TM** USA-DSHEA	Wyk-408-08
Ispaghula Psyllium Husk	Plantain	*Plantago*	*ovata*	**TM** Asia-Pharm., EEU-ESCOP 2	Wyk-245-04
Jambolan	Lilly Pilly	*Syzygium*	*cumini*	**PS** (Bal2-527-05) USA-DSHEA	Wyk-429-03
Jambul				See Jambolan Lilly Pilly	
Japanese	Anise	*Illicium*	*anisatum*	**PS** (Bal2-503-04) Anisatin convulsions ☠	Wyk-180-02
Japanese	Pagoda Tree	*Styphnolobium*	*japonicum*	**PS** Asia-Pharm. (rutin)	Wyk-312-01
Japanese	Yam	*Dioscorea*	*nipponica*	**TM** USA-DSHEA	Bown-273-12
Japanese Arrowroot	Kudzu	*Pueraria*	*lobata*	**PS** USA-DSHEA	Wyk-260-01
Java Seed	Kusam	*Brucea*	*javanica*	**PS** UN-WHO 1	Wyk-071-01
Jerusalem Cow's Lip	Lungwort	*Pulmonaria*	*officinalis*	**PS** Asia-Pharm.	Wyk-261-01
Jesuit's Bark	Quinine	*Cinchona*	*pubescens*	**CS** Asia-Pharm., EEU-Comm.E	Wyk-102-01
Jimsonweed	Thorn Apple	*Datura*	*stramonium*	**PS** Asia-Pharm.	Wyk-123-01
Jing Jie	Split Catmint	*Schizonepeta*	*tenuifolia*	**TM** (Bal2-541-01) USA-DSHEA	Wyk-426-20
Job's Tears	Reed Grain Grass	*Coix*	*lacryma-jobi*	**TM** (Bal2-511-03) USA-DSHEA	Bal2-051-01
Jojoba				See Chinese Jojoba	
Jujube				See Chinese Date Jujube	
Juniper				See European Juniper	
Kava Kava	Peppercorn	*Piper*	*methysticum*	**CT** Asia-Pharm., EEU-Comm.E, UN-WHO 2	Wyk-243-01
Kawaratake	Turkeytail Pore Fungus	*Coriolus*	*versicolor*	**SS** (Bal2-535-04) USA-DSHEA	Bal2-114-02
Khella	Bishop's Weed	*Ammi*	*visnaga*	**PS** (Bal2-528-01) Asia-Pharm.	Wyk-045-01
Kidney Dry Bush	Bean	*Phaseolus*	*vulgaris-bruscia-rena*	**TM** Asia-Pharm., EEU-Comm.E	Wyk-237-01

Species	Genus	Latin Genus	Latin Species	Regulatory Documents	Reference
King of Bitters	Water Willow	*Andrographis*	*paniculata*	**CS** (Bal2-503-03) USA-DSHEA	Wyk-398-15
Kinnikinnick	Bearberry	*Arctostaphylos*	*uva-ursi*	**PS** (Bal2-549-01) Asia-Pharm., EEU-Comm.E, EEU-ESCOP 5, UN-WHO 2	Wyk-050-01
Knitbone	Comfrey	*Symphytum*	*officinale*	**PS** (Bal2-512-06) Asia-Pharm., EEU-Comm.E(ext.)	Wyk-314-01
Knotted Figwort	Foxglove	*Scrophularia*	*nodosa*	**PS** Asia-Pharm.	Wyk-294-01
Kudzu				See Japanese Arrowroot Kudzu	
Lady's	Bedstraw	*Galium*	*verum*	**TM** Asia-Pharm.	Wyk-152-01
Lady's Mantle				See Lion's Foot Lady's Mantle	
Lake Baikal	Skullcap	*Scutellaria*	*baicalensis*	**PS** (Bal2-541-02) USA-DSHEA	Wyk-295-02
Large	Thyme	*Thymus*	*pulegoides*	**TM** USA-DSHEA	Wyk-323-03
Large	Trophycress	*Tropaeolum*	*majus*	**PS** EEU-Comm.E	Wyk-326-01
Large Gray	Willow	*Salix*	*cineria*	**TM** USA-DSHEA	Bown-245-04
Large Leaf Lime	Linden	*Tilia*	*platyphyllos*	**TM** USA-DSHEA	Wyk-324-02
Larkspur Poppy				See Chinese Tuber Larkspur Poppy	
Larkspur Poppy				See European Tuber Larkspur Poppy	
Larkspur Poppy				See M. East Tuber Larkspur Poppy	
Lemon	Beebrush	*Aloysia*	*triphylla*	**TM** USA-DSHEA	Wyk-042-01
Lemon				See East Asia Lemon	
Lemon	Thyme	*Thymus*	*citriodorus*	**TM** USA-DSHEA	Wyk-323-02
Lemon Balm	Melissa Balm	*Melissa*	*officinalis*	**PS** (Bal2-529-01) Asia-Pharm., EEU-Comm.E, EEU-ESCOP 2, UN-WHO 2	Wyk-204-01
Lemon Grass	Spikelet Beard Grass	*Cymbopogon*	*citratus*	**PS** Asia-Pharm.	Wyk-120-01
Leopard's Bane				See Mountain Leopard's Bane	
Lesser	Periwinkle	*Vinca*	*minor*	**PS** USA-DSHEA	Wyk-340-01
Lesser Galangal	Alpino Ginger	*Alpinia*	*officinarum*	**PS** Asia-Pharm., EEU-Comm.E	Wyk-043-01
Licorice				See American Licorice	
Licorice				See Chinese Licorice	
Licorice				See European Licorice	
Limaosinho	Mayten Holly	*Maytenus*	*ilicifolia*	**TM** (Bal2-517-01) USA-DSHEA	Bal2-065-02
Linden				See Large Leaf Lime Linden	
Linden				See Small Leaf Lime Linden	
Linseed Oil	Flax	*Linum*	*usitatissimum*	**CS** Asia-Pharm., EEU-Comm.E, EEU-ESCOP 1	Wyk-193-01
Lion's Foot	Lady's Mantle	*Alchemilla*	*vulgaris*	**PS** Asia-Pharm., EEU-Comm.E	Wyk-037-01
Long Needle Pitch	Pine	*Pinus*	*palustris*	**TM** USA-DSHEA	Bown-329-05
Long Stamen	Siphon Mint	*Orthosiphon*	*aristatus*	**PS** Asia-Pharm., EEU-Comm.E, EEU-ESCOP 1	Wyk-222-01
Lovage				See Love Parsley Lovage	
Love Parsley	Lovage	*Levisticum*	*officinale*	**PS** Asia-Pharm., EEU-Comm.E	Wyk-192-01
Lungwort				See Jerusalem Cow's Lip Lungwort	
M. East Tuber	Larkspur Poppy	*Corydalis*	*solida*	**PS** USA-DSHEA	Wyk-406-19b
Ma Huang	Joint Fir	*Ephedra*	*sinica*	**CS** Asia-Pharm., EEU-Comm.E, WHO 1	Wyk-134-01
Madagascar	Haronga	*Harungana*	*madagascarensis*	**PS** USA-DSHEA	Wyk-166-01
Maidenhair	Ginkgo	*Ginkgo*	*biloba*	**CT** EEU-Comm.E (selected comm. extracts only)	Wyk-158-01
Maitake	Grifol Meristic Downy Pore Fungus	*Grifola*	*frondosa*	**PS** (Bal2-531-01) USA-DSHEA	Bal2-097-01
Mallow				See Buttonweed Mallow	
Mallow				See High Mallow	
Mallow				See Musk Mallow	
Marijuana	Hemp	*Cannabis*	*sativa-indica*	**TM** (High Tetrahydrocannabinol) USA-DSHEA	Wyk-077-01
Marsh Mallow				See White Marsh Mallow	
Marsh Tea	Rosewood Heath	*Rhododendron*	*tomentosum*	**PS** USA-DSHEA	Wyk-272-01
Mat Grass	Knotweed	*Polygonum*	*aviculare*	**PS** EEU-Comm.E	Wyk-251-01
Maté	Ivy	*Ilex*	*paraguensis*	**PS** Asia-Pharm., EEU-Comm.E	Wyk-179-01
May	Lily of the Valley	*Convallaria*	*majalis*	**PS** Asia-Pharm., EEU-Comm.E	Wyk-112-01
Maypops	Passion Flower	*Passiflora*	*incarnata*	**PS** Asia-Pharm., EEU-Comm.E, EEU-ESCOP 4	Wyk-227-01
Mayten Holly				See Espinheira Santa Mayten Holly	
Mayten Holly				See Limaosinho Mayten Holly	
Mediterranean	Asparagus	*Asparagus*	*officinalis*	**PS** Asia-Pharm., EEU-Comm.E	Wyk-060-01
Mediterranean	Borage	*Borago*	*officinalis*	**PS** Asia-Pharm.	Wyk-068-01
Mediterranean	Myrtle	*Myrtus*	*communis*	**PS** Asia-Pharm.	Wyk-212-01
Mediterranean	Pomegranate	*Punica*	*granatum*	**PS** Asia-Pharm.	Wyk-263-01
Mediterranean	Rosemary	*Rosmarinus*	*officinalis*	**PS** Asia-Pharm., EEU-Comm.E, EEU-ESCOP 3	Wyk-276-01
Mediterranean	Thyme	*Thymus*	*vulgaris*	**PS** Asia-Pharm., EEU-Comm.E, EEU-ESCOP 1, UN-WHO 1	Wyk-323-01
Mediterranean Sweet	Basil	*Ocimum*	*basilicum-mediterranea*	**PS** UN-WHO 2	Wyk-217-01
Mengkudu	Tropical Mulberry	*Morinda*	*citrifolia*	**TM** (Bal2-533-01) USA-DSHEA	Bal2-102-01
Middle East	Joint Fir	*Ephedra*	*intermedia*	**TM** USA-DSHEA	Bown-278-04
Milfoil	Yarrow	*Achillea*	*millefolium*	**PS** (Bal2-550-06) Asia-Pharm., EEU-Comm.E	Wyk-030-01
Milk Vetch				See Huang Qi Milk Vetch	
Milk Thistle				See Blessed Milk Thistle	
Mistletoe				See All Heal Mulberry Mistletoe	

Species	Genus	Latin Genus	Latin Species	Regulatory Documents	Reference
Mistletoe				See Devil's Fuge Mistletoe	
Mistletoe				See Oak Mistletoe	
Mongolian	Rhubarb	*Rheum*	*rhabarbarum*	**TM** USA-DSHEA	Wyk-270-04
Monk's Pepper	Chasteberry	*Vitex*	*agnus-castus*	**CS** (Bal2-549-04) EEU-Comm.E	Wyk-343-01
Moose	Lowbush Cranberry	*Viburnum-bruscia*	*edule*	**TM** USA-DSHEA	Bal1-068-05
Mormon Tea	Joint Fir	*Ephedra*	*viridis*	**TM** USA-DSHEA	Kowa-184-01
Mother of	Thyme	*Thymus*	*serphyllum*	**TM** USA-DSHEA	Wyk-323-05
Motherwort				See Chinese Motherwort	
Motherwort				See European Motherwort	
Motherwort				See Siberian Motherwort	
Mountain	Leopard's Bane	*Arnica*	*montana*	**CT** (Bal2-504-01) Asia-Pharm., EEU-Comm.E, EEU-ESCOP 6	Wyk-053-01
Moutan	Peony	*Paeonia*	*suffruticosa*	**TM** USA-DSHEA	Bown-321-03
Mugo	Pine	*Pinus*	*mugo*	**TM** USA-DSHEA	Wyk-421-11
Mugwort	Wormwood	*Artemisia*	*vulgaris*	**PS** Asia-Pharm.	Wyk-057-01
Muira Puama Brazil	Potencywood	*Ptychopetalum*	*olacoides*	**TM** USA-DSHEA	Wyk-424-04a
Muira Puama Amazon	Potencywood	*Ptychopetalum*	*ovata*	**TM** (Bal2-533-03) USA-DSHEA	Bal2-104-01
Muira Puama Peru	Potencywood	*Ptychopetalum*	*uncinatum*	**TM** USA-DSHEA	Wyk-424-04b
Mullein				See Great Mullein	
Mullein				See Orange Mullein	
Musk	Mallow	*Malva*	*moschata*	**TM** USA-DSHEA	Bown-308-03
Myrrh				See African Myrrh	
Myrrh				See Guggul Myrrh	
Myrtle				See Mediterranean Myrtle	
Myrtle Flag	Sweet Flag	*Acorus*	*calamus*	**PS** Asia-Pharm.	Wyk-032-01
Narrow Leaf	Sage	*Salvia*	*officinalis*	**PS** Asia-Pharm., EEU-Comm.E, EEU-ESCOP 2	Wyk-283-01
Navy Dry Bush	Bean	*Phaseolus*	*vulgaris-bruscia-navalis*	**TM**	Wyk-237-01
New Jersey Tea	Redroot	*Ceanothus*	*americanus*	**TM** USA-DSHEA	Wyk-086-01
Noni	Tropical Mulberry	*Morinda*	*officinalis*	**TM** (Bal2-533-01) USA-DSHEA	Wyk-417-20
Northern	Thorow Wax	*Bupleurum*	*chinense*	**TM** USA-DSHEA	Bal2-038-01
Nut & Mace	Nutmeg	*Myristica*	*fragrans*	**PS** Asia-Pharm.	Wyk-210-01
Oak				See Black Oak	
Oak				See English Oak	
Oak				See White Oak	
Oak	Mistletoe	*Phoradendron*	*leucarpum*	**TM** USA-DSHEA	Kowa-389-02
Oats				See Cereal Oats	
Oil Gland Leaf	St. John's Wort	*Hypericum*	*perforatum*	**CS** (Bal2-545-02) Asia-Pharm., EEU-Comm.E, EEU-ESCOP 1, UN-WHO 2	Wyk-175-01
Old Man's Beard	Fringe Tree	*Chionanthus*	*virginicus*	**TM** USA-DSHEA	Wyk-095-01
Oleander				See Rose Bay Oleander	
Olive				See European Olive	
Onion				See Pamir-Alai Mountain Onion	
Opium	Poppy	*Papaver*	*somniferum*	**PS** Asia-Pharm.	Wyk-225-01
Orange				See Seville Sour Orange	
Orange	Mullein	*Verbascum*	*phlomoides*	**PS** Asia-Pharm., EEU-Comm.E (flower)	Wyk-336-01
Oregano				See Seasoning Oregano	
Oregon Grape				See Holly Oregon Grape	
Oregon Grape				See Pine Oregon Grape	
Oregon Grape				See Tall Oregon Grape	
Oriental	Sweet Gum	*Liquidambar*	*orientalis*	**TM** USA-DSHEA	Bal1-173-08
Osha	Alpine Lovage	*Ligusticum*	*porteri*	**TM** USA-DSHEA	Bal2-109-01
Ouabain	Strophanthin	*Strophanthus*	*gratus*	**PS** USA-DSHEA	Wyk-310-01
Oval Leaf	Buchu	*Agathosma*	*crenulata*	**TM** Asia-Pharm., EEU-Comm.E	Wyk-068-02
Pale Purple	Coneflower	*Echinacea*	*pallida*	**CT** Asia-Pharm., EEU-Comm.E, EEU-ESCOP 6, UN-WHO 1	Wyk-129-01
Pamir-Alai Mountian	Onion	*Allium*	*cepa*	**CT** Asia-Pharm., EEU-Comm.E, UN-WHO 1	Wyk-038-01
Papaya	Paw Paw	*Carica*	*papaya*	**TM** (Bal2-533-07) USA-DSHEA	Wyk-080-01
Pará	Fleckflower Cress	*Spilanthes*	*acmella*	**PS** USA-DSHEA	Wyk-307-01
Pasque Flower				See European Pasque Flower	
Passion Flower				See Blue Crown Passion Flower	
Passion Flower				See Maypops Passion Flower	
Patchouli	Beard Mint	*Pogostemon*	*cablin*	**PS** USA-DSHEA	Wyk-249-01
Pau D'Arco	Trumpet Tree	*Tabebuia*	*impetiginosa*	**PS** (Bal2-534-02) USA-DSHEA	Wyk-316-01
Pennyroyal	Mint	*Mentha*	*pulegium*	**TM** USA-DSHEA	Wyk-207-03
Peony				See Chinese Peony	
Peony				See European Peony	

Species	Genus	Latin Genus	Latin Species	Regulatory Documents	Reference
Peony				See Moutan Peony	
Peppermint	Mint	*Mentha*	*piperita*	**CS** Asia-Pharm., EEU-Comm.E, EEU-ESCOP 3, UN-WHO 2	Wyk-206-01
Pernambuco	Jaborandi	*Pilocarpus*	*jaborandi*	**CS** Asia-Pharm.	Wyk-239-01
Persian	Caraway	*Carum*	*carvi*	**PS** Asia-Pharm., EEU-Comm.E, EEU-ESCOP 6	Wyk-082-01
Peru Balsam	Myrrhwood	*Myroxylon*	*perierae*	**TM** USA-DSHEA	Wyk-211-02
Peruvian	Rhatany	*Krameria*	*lappacea*	**PS** Asia-Pharm., EEU-Comm.E	Wyk-186-01
Pillatory				See European Pillatory	
Pine				See Aleppo Pine	
Pine				See Long Needle Pitch Pine	
Pine				See Mugo Pine	
Pine				See Scotch Pine	
Pine Bark Extract	Pycnogenol			See French Maritime Pine	
Pine Oregon Grape	Mahon Barberry	*Mahonia*	*repens*	**TM** USA-DSHEA	Wyk-416-13
Pineapple				See Bromelain Pineapple	
Plantain				See Greater Plantain	
Plantain				See Ispaghula Psyllium Husk Plantain	
Plantain				See Psyllium Husk Plantain	
Plantain				See Ribwort Plantain	
Pleurisy Root	Milkweed	*Asclepias*	*tuberosa*	**PS** Asia-Pharm.	Wyk-058-01
Poison Nut	Strychnine	*Strychnos*	*nux-vomita*	**PS** Asia-Pharm. Poison ☠	Wyk-311-01
Pokeweed				See Inkberry Pokeberry	
Polysaccharide	Kureha			See Kawaratake Turkeytail Pore Fungus	
Pomegranate				See Mediterranean Pomegranate	
Poor Man's Ginseng	Bonnet Bellflower	*Codonopsis*	*pilosula*	**TM** (Bal2-511-02) USA-DSHEA	Wyk-405-17
Pot Marigold				See Gold Bloom Pot Marigold	
Potencywood				See Aphrodisiac Potencywood	
Potencywood				See Hooked Potencywood	
Potencywood				See Muira Puama Potencywood	
Prickly Ash				See American Prickly Ash	
Prickly Ash				See Szechuan Prickly Ash	
PSK				See Kawaratake Turkeytail Pore Fungus	
Psyllium Husk	Plantain	*Plantago*	*afra*	**PS** Asia-Pharm., EEU-ESCOP 5, UN-WHO 1	Wyk-245-01
Puccoon	Bloodroot	*Sanguinaria*	*canadensis*	**CS** Asia-Pharm.	Wyk-286-01
Pumpkin	Melon	*Cucurbita*	*maxima-pepo*	**PS** Asia-Pharm., EEU-Comm.E	Wyk-117-01
Purging	Buckthorn	*Rhamnus*	*catharticus*	**PS** Asia-Pharm., EEU-Comm.E	Wyk-267-01
Purging	Croton	*Croton*	*tiglium*	**TM** USA-DSHEA	Wyk-407-01
Purple	Coneflower	*Echinacea*	*purpurea*	**CS** Asia-Pharm., EEU-Comm.E, EEU-ESCOP 6, UN-WHO 1	Wyk-130-01
Pygeum				See African Sour Cherry	
Queen Anne's Lace	Carrot	*Daucus*	*carota*	**PS** Asia-Pharm.	Wyk-124-01
Queen of the Meadow	Meadowsweet	*Filipendula*	*ulmaria*	**PS** Asia-Pharm., EEU-Comm.E	Wyk-144-01
Quickthorn	Hawthorn Berry	*Crataegus*	*monogyna*	**CS** Asia-Pharm., EEU-Comm.E, EEU-ESCOP 6, UN-WHO 2	Wyk-115-01
Quinine				See Jesuit's Bark Quinine	
Raspberry				See Red Raspberry	
Red	Clover	*Trifolium*	*pratense*	**TM** USA-DSHEA	Wyk-430-04
Red	Raspberry	*Rubus-raspabacca*	*idaeus*	**TM** (Bal2-538-01) USA-DSHEA	Wyk-277-02
Red Cole	Horseradish	*Amoracia*	*rusticana*	**PS** Asia-Pharm., EEU-Comm.E	Wyk-052-01
Red Fox	Grape	*Vitis*	*labrusca-rubra*	**TM** (Bal2-538-02) USA-DSHEA	Bal2-121-01
Red Root	Sage	*Salvia*	*miltiorhiza*	**TM** (Bal2-514-01) USA-DSHEA	Wyk-426-04
Red Wine	Grape	*Vitis*	*vinifera-rubra*	**PS** (Bal2-538-02) USA-DSHEA	Wyk-344-01
Red Yeast Rice	Monad Sac Fungus	*Monascus*	*purpureus*	**TM** USA-DSHEA	Bal2-210-18
Rehmann Foxglove				See Chinese Rehmann Foxglove	
Reishi	Bracket Pore Fungus	*Ganoderma*	*lucidum*	**PS** (Bal2-538-03) USA-DSHEA	Bal2-122-01
Rhubarb				See Mongolian Rhubarb	
Ribwort	Plantain	*Plantago*	*lanceolata*	**PS** Asia-Pharm., EEU-Comm.E	Wyk-246-01
Rollbark	Warburg Pepperbark	*Warburgia*	*salutarius*	**PS** USA-DSHEA	Wyk-345-01
Roman	Chamomile	*Chamaemelum*	*nobile*	**PS** Asia-Pharm.	Wyk-092-01
Rooibos	Lathybush	*Aspalathus*	*linearis*	**PS** (Bal2-539-01) USA-DSHEA	Wyk-059-01
Rose Bay	Oleander	*Nerium*	*oleander*	**PS** Asia-Pharm.	Wyk-215-01
Roselle	Rose Mallow	*Hibiscus*	*sabdariffa*	**PS** Asia-Pharm., EEU-Comm.E	Wyk-170-01
Rosemary				See Mediterranean Rosemary	
Roseroot	Red Stonecrop	*Rhodiola*	*rosea*	**CS** Asia-Pharm.	Wyk-271-01
Rosy	Madagascar Periwinkle	*Catharanthus*	*roseus*	**CT** Asia-Pharm.	Wyk-085-01
Round Leaf	Buchu	*Agathosma*	*betulina*	**TM** Asia-Pharm., EEU-Comm.E	Wyk-068-01
Round Leaf	Sundew	*Drosera*	*rotundifolia*	**PS** Asia-Pharm., EEU-Comm.E	Wyk-128-01
Rowan	Mountain Ash	*Sorbus*	*aucuparia*	**PS** USA-DSHEA	Wyk-306-01
Sabal	Saw Palmetto	*Serenoa*	*repens*	**CS** Asia-Pharm., EEU-Comm.E, UN-WHO 2	Wyk-299-01
Sacred	Lotus Waterlily	*Nelumbo*	*nucifera*	**PS** USA-DSHEA	Wyk-213-01

Species	Genus	Latin Genus	Latin Species	Regulatory Documents	Reference
Sacred Oliban	Frankincense	*Boswellia*	*sacra*	**PS** USA-DSHEA	Wyk-069-01
Saffron	Crocus	*Crocus*	*sativus*	**PS** Asia-Pharm. **induced abortion** ☠	Wyk-116-01
Saffron	Distaff Thistle	*Carthamus*	*tinctorius*	**PS** Asia-Pharm.	Wyk-081-01
Sage				See Clary Sage	
Sage				See Narrow Leaf Sage	
Sage				See Red Root Sage	
Sandy	Everlasting Buttons	*Helichrysum*	*arenarium*	**PS** Asia-Pharm., EEU-Comm.E	Wyk-168-01
Sangre De Drago				See Dragon's Blood Croton	
Sarsaparilla Vera Cruz	Greenbrier	*Smilax*	*aristolochaefolia*	**PS** (Bal2-540-01) Asia-Pharm.	Wyk-303-01
Sarsaparilla Equador	Greenbrier	*Smilax*	*ferbifuga*	**TM** (Bal2-540-01) USA-DSHEA	Wyk-303-03
Sarsaparilla Australia	Greenbrier	*Smilax*	*glyciphylla*	**TM** USA-DSHEA	Bown-354-06
Sarsaparilla Jamaica	Greenbrier	*Smilax*	*ornata*	**TM** (Bal2-540-01) USA-DSHEA	Bal2-125-03
Sarsaparilla Honduras	Greenbrier	*Smilax*	*regelii*	**TM** (Bal2-540-01) USA-DSHEA	Wyk-303-04
Saw Palmetto				See Sabal Saw Palmetto	
Scabwort	Elecampane	*Inula*	*helenium*	**PS** (Bal2-516-02) Asia-Pharm.	Wyk-181-01
Scented	Dill	*Anethum*	*graveolens*	**PS** EEU-Comm.E (fruit)	Wyk-047-01
Scotch	Broom	*Cytisus*	*scoparius*	**PS** Asia-Pharm., EEU-Comm.E	Wyk-122-01
Scotch	Pine	*Pinus*	*sylvestris*	**PS** Asia-Pharm., EEU-Comm.E	Wyk-242-01
Seasoning	Oregano	*Origanum*	*vulgare*	**PS** Asia-Pharm., EEU-Comm.E	Wyk-221-01
Sea Squill	Urgin Squill	*Urginea*	*maritima*	**PS** Asia-Pharm., EEU-Comm.E	Wyk-331-01
Senega Root	Milkwort	*Polygala*	*senega*	**PS** (Bal2-511-04) Asia-Pharm., EEU-Comm.E, EEU-ESCOP 3, UN-WHO 2	Wyk-250-01
Seven Barks	Hydrange	*Hydrangea*	*arborescens*	**TM** USA-DSHEA	Wyk-413-07
Seville	Sour Orange	*Citrus-aurantius*	*aurantium*	**PS** (Bal2-507-01) Asia-Pharm.,EEU-Comm.E	Wyk-106-01
Sharpthorn	Hawthorn Berry	*Crataegus*	*oxyacantha*	**TM** USA-DSHEA	Bal2-083-03
Shepherd's Purse	Capsell Mustard	*Capsella*	*bursa-pastoris*	**PS** Asia-Pharm., EEU-Comm.E	Wyk-078-01
Shiitake	Lentinan Pinwheel Gill Mushroom	*Lentinula*	*edodes*	**TM** USA-DSHEA	Bal2-131-01
Siberian	Eleuthero Ginseng	*Eleutherococcus*	*senticosus*	**PS** Asia-Pharm., EEU-Comm.E, UN-WHO 2	Wyk-132-01
Siberian	Motherwort	*Leonurus*	*sibricus*	**TM** USA-DSHEA	Bown-302-11
Silver	Birch	*Betula*	*pendula*	**PS** (Bal2-506-04) EEU-Comm.E, EEU-ESCOP 1	Wyk-067-01
Silverweed	Cinquefoil	*Potentilla*	*anserina*	**PS** Asia-Pharm., EEU-Comm.E	Wyk-253-01
Silymarin				See Blessed Milk Thistle	
Skullcap				See Beaded Skullcap	
Skullcap				See European Skullcap	
Skullcap				See Lake Baikal Skullcap	
Skullcap				See Virginia Skullcap	
Slippery	Elm	*Ulmus*	*rubra*	**PS** (Bal2-543-03) Asia-Pharm.	Wyk-329-01
Sloe Blackthorn	Sour Cherry	*Prunus-Cerasus*	*spinosa*	**PS** Asia-Pharm., EEU-Comm.E	Wyk-257-01
Small Flower	Willow Evening Primrose	*Epilobium*	*parviflorum*	**TM** USA-DSHEA	Wyk-135-01
Small Leaf Lime	Linden	*Tilia*	*cordata*	**TM** (Bal2-547-02) Asia-Pharm., EEU-Comm.E	Wyk-324-01
Smooth	Rupturewort	*Herniaria*	*glabra*	**TM** Asia-Pharm., EEU-Comm.E	Wyk-169-01
Smooth	Sumac	*Rhus*	*glabra*	**TM** USA-DSHEA	Wyk-425-08
Snakeroot	Milkwood	*Rauvolfia*	*serpentina*	**PS** Asia-Pharm., EEU-Comm.E, UN-WHO 1	Wyk-266-01
Snow Fungus	Tremor Jelly	*Tremella*	*fuciformis*	**TM** (Bal2-543-04) USA-DSHEA	Bal2-135-01
Snowdrop Amaryllis				See Eurasian Snowdrop Amaryllis	
Soapwort				See Bouncing Bet Soapwort	
Soapwort	Baby's Breath	*Gypsophila*	*paniculata*	**PS** Asia-Pharm., EEU-Comm.E	Wyk-163-01
Sour Orange				See Seville Sour Orange	
South Africa Desert	Carrion Flower Cactus	*Hoodia*	*pilifera*	**CS** USA-DSHEA	Wyk-171-01
South Asia	Ginger	*Zingiber*	*officinale*	**CS** (Bal2-519-02) Asia-Pharm., EEU-Comm.E, EEU-ESCOP 1, UN-WHO 1	Wyk-349-01
Southern	Thorow Wax	*Bupleurum*	*scorzoneraefolium*	**TM** USA-DSHEA	Bal2-038-02
Soybean	Glycine Bean	*Glycine*	*max*	**PS** (Bal2-544-01) Asia-Pharm., EEU-Comm.E	Wyk-159-01
Spanish	Thyme	*Thymus*	*zygis*	**TM** USA-DSHEA	Wyk-323-06
Spearmint	Mint	*Mentha*	*spicata*	**SS** (Bal2-535-01) Asia-Pharm.	Wyk-207-01
Speedwell				See Heath Speedwell	
Spiny	Restharrow	*Ononis*	*spinosa*	**PS** Asia-Pharm., EEU-Comm.E, EEU-ESCOP 5	Wyk-220-01
Squaw Root	Blue Cohosh	*Caulophyllum*	*thalictroides*	**TM** USA-DSHEA	Wyk-403-13
Squaw Vine	Partidge Berry	*Mitchella*	*repens*	**TM** USA-DSHEA	Wyk-417-14
Sri Lanka	Cinnamon	*Cinnamomum*	*verum*	**PS** Pharm., EEU-Comm.E, UN-WHO 2	Wyk-105-01
St. John's Wort				See Oil Gland Leaf St. John's Wort	
Star Anise				See Chinese Anise	
Star Magnolia				See Chinese Star Magnolia	
Starweed	Chickweed	*Stellaria*	*media*	**PS** USA-DSHEA	Wyk-309-01
Stinging	Nettle	*Urtica*	*dioca*	**CS** Asia-Pharm., EEU-Comm.E, EEU-ESCOP 2 (leaf) & 4 (root), UN-WHO 2 (root)	Wyk-332-01
Succory	Chicory	*Cichorium*	*intybus*	**PS** USA-DSHEA	Wyk-100-01
Sweet	Birch	*Betula*	*lenta*	**TM** USA-DSHEA	Bal2-032-03
Sweet	Chestnut	*Castanea*	*sativa*	**TM** Asia-Pharm.	Wyk-083-01

Species	Genus	Latin Genus	Latin Species	Regulatory Documents	Reference
Sweet	Fennel	*Foeniculum*	*vulgare*	**PS** Asia-Pharm., EEU-Comm.E, EEU-ESCOP 1	Wyk-145-01
Sweet Annie	Wormwood	*Artemisia*	*annua*	**CS** USA-DSHEA	Wyk-056-01
Sweet Corn	Zeatin Grass	*Zea*	*mays-suavis*	**PS** (Bal2-513-04) Asia-Pharm,EEU-Comm.E(pollen)	Wyk-348-01
Sweet Flag				See Chang Pu Sweet Flag	
Sweet Flag				See Myrtle Flag Sweet Flag	
Syrian Rue	Pegan Yokeleaf	*Peganum*	*harmala*	**PS** USA-DSHEA, narcotic	Wyk-230-01
Szechuan	Prickly Ash	*Zanthoxylum*	*bungeanum*	**TM** (Bal2-535-02) USA-DSHEA	Bal2-115-02
Tabasco Hot	Pepper	*Capsicum*	*annuum-frutescens-tabascum*	**CT** Asia-Pharm., EEU-Comm.E	Wyk-079-02
Tall Oregon Grape	Mahon Barberry	*Mahonia*	*aquifolium*	**TM** USA-DSHEA	Wyk-416-13
Tansy				See FeverfewTansy	
Tansy				See Bachelor's Buttons Tansy	
Tasmanian Blue	Gum Tree	*Eucalyptus*	*globulus*	**PS** (Bal2-517-02) Asia-Pharm., EEU-Comm.E, EEU-ESCOP 6, UN-WHO 2	Wyk-139-01
Tavoy	Bastard Cardamom	*Amomum*	*villosum*	**TM** (Bal2-509-03) USA-DSHEA	Bal2-042-02
Tea				See Green Tea	
Tea Tree	Bottlebrush Myrtle	*Melaleuca*	*alternifolia*	**CS** Asia-Pharm., UN-WHO 2	Wyk-202-01
Thoroughwort	Agrimony	*Agrimonia*	*eupatoria*	**PS** (Bal2-501-02) EEU-Comm.E	Wyk-036-01
Three Leaf	Garlic Pear	*Crateva*	*nurvula*	**TM** (Bal2-549-03) USA-DSHEA	Wyk-406-21
Thyme				See African Thyme	
Thyme				See Large Thyme	
Thyme				See Lemon Thyme	
Thyme				See Mediterranean Thyme	
Thyme				See Mother of Thyme	
Thyme				See Spanish Thyme	
Tilden Flower				See Small Leaf Lime Linden	
Tolu Balsam	Myrrhwood	*Myroxylon*	*balsamum*	**PS** (Bal2-547-03) Asia-Pharm., EEU-Comm.E	Wyk-211-01
Trembling Aspen	Poplar	*Populus*	*tremuloides*	**PS** EEU-Comm.E (bud)	Wyk-252-01
Tuckahoe	Wolf Foam Pore Fungus	*Wolfiporia*	*extensa*	**PS** (Bal2-526-03) USA-DSHEA	Bal2-085-01
Turmeric	Curcumin	*Curcuma*	*longa*	**PS** Asia-Pharm., EEU-Comm.E, UN-WHO 1	Wyk-118-01
Twisted	Skeleton Ice Plant	*Sceletium*	*tortuosum*	**PS** USA-DSHEA	Wyk-292-01
Umckaloabo	Stork's Bill	*Pelargonium*	*sidoides*	**CS** USA-DSHEA	Wyk-231-01
Uva Ursi				See Kinnikinnick Bearberry	
Uzara	Milkbush	*Xysmalobium*	*undulatum*	**PS** EEU-Comm.E	Wyk-347-01
Valerian				See Vandal Root Valerian	
Vandal Root	Valerian	*Valeriana*	*officinalis*	**CS** (Bal2-549-02) Asia-Pharm., EEU-Comm.E, EEU-ESCOP 4, UN-WHO 1	Wyk-335-01
Varuna				See Three Leaf Garlic Pear	
Vervain				See Herb Of The Cross Vervain	
Violet Tree	Axe Wingpod	*Securidaca*	*longipedunculata*	**PS** USA-DSHEA	Wyk-296-01
Virginia	Skullcap	*Scutellaria*	*lateriflora*	**TM** (Bal2-541-02) Asia-Pharm.	Wyk-295-01
Wall	Germander	*Teucrium*	*chamaedrys*	**PS** USA-DSHEA	Wyk-321-01
Walnut				See American Walnut	
Walnut				See English Walnut	
Water Plantain				See Ze Xie Water Plantain	
Wax Myrtle	Bayberry	*Myrica*	*cerifera*	**PS** Asia-Pharm.	Wyk-209-01
West Indies	Guaiac Wood of Life	*Guajacum*	*officinale*	**PS** Asia-Pharm., EEU-Comm.E	Wyk-162-01
White	Birch	*Betula*	*alba*	**TM** USA-DSHEA	Bal2-032-04
White	Deadnettle	*Lamium*	*album*	**PS** Asia-Pharm., EEU-Comm.E (flower)	Wyk-187-01
White	Horehound	*Marrubium*	*vulgare*	**PS** Asia-Pharm., EEU-Comm.E	Wyk-198-01
White	Marsh Mallow	*Althaea*	*officinalis*	**PS** (Bal2-531-02) Asia-Pharm., EEU-Comm.E, EEU-ESCOP 1, UN-WHO 2	Wyk-044-01
White	Oak	*Quercus*	*alba*	**TM** USA-DSHEA	Bal1-078-04
White	Sandalwood	*Santalum*	*album*	**PS** EEU-Comm.E	Wyk-289-01
White	Willow	*Salix*	*alba*	**CS** Asia-Pharm., EEU-Comm.E, EEU-ESCOP 4	Wyk-281-01
White Walnut				See English Walnut	
Wild	Sweet Cherry	*Prunus-Avium*	*avium*	**TM** USA-DSHEA	Wyk-257-03
Wild	Yam	*Dioscorea*	*villosa*	**PS** (Bal2-550-02) Asia-Pharm.	Wyk-126-01
Willow				See Basket Willow	
Willow				See Bay Willow	
Willow				See Black Willow	
Willow				See Crack Willow	
Willow				See Large Gray Willow	
Willow				See White Willow	
Winter	Savory	*Satureja*	*montana*	**PS** USA-DSHEA	Wyk-291-01
Winter Cherry	Ashwagandha	*Withania*	*somnifera*	**PS** (Bal2-504-04) USA-DSHEA	Wyk-346-01
Wintergreen	Checkerberry	*Gaultheria*	*procumbens*	**PS** (Bal2-550-04) USA-DSHEA	Wyk-153-01
Witch Hazel				See American Witch Hazel	
Wolfsbane	Monkshood	*Aconitum*	*napellus*	**PS** USA-DSHEA	Wyk-031-01

Species	Genus	Latin Genus	Latin Species	Regulatory Documents	Reference
Wood	Sanicle	*Sanicula*	*europaea*	**PS** EEU-Comm.E	Wyk-288-01
Wood Betony	Hedge Nettle	*Stachys*	*officinalis*	**PS** USA-DSHEA	Wyk-308-01
Wood Sage	Germander	*Teucrium*	*scorodonia*	**TM** USA-DSHEA	Wyk-321-05
Woodland	Strawberry	*Fragaria*	*vesca*	**PS** USA-DSHEA	Wyk-146-01
Wormseed	Goosefoot	*Chenopodium*	*ambrosoides*	**PS** USA-DSHEA	Wyk-094-01
Yam				See Glaucous Collett Yam	
Yam				See Indochinese Yam	
Yam				See Isidikawa Yam	
Yam				See Japanese Yam	
Yam				See Wild Yam	
Yarrow				See Milfoil Yarrow	
Yellow	Dock	*Rumex*	*crispus*	**PS** Asia-Pharm.	Wyk-278-01
Yellow	Sweet Clover	*Melilotus*	*officinalis*	**PS** Asia-Pharm., EEU-Comm.E, EEU-ESCOP 4	Wyk-203-01
Yellow	Trumpet Jasmine	*Gelsemium*	*sempervirens*	**PS** USA-DSHEA Poison ☠	Wyk-154-01
Yellow Spring	Pheasant's Eye	*Adonis*	*vernalis*	**PS** Asia-Pharm., EEU-Comm.E	Wyk-033-01
Yerba Maté				See Maté	
Yin Yang Yuo	Horny Goatweed	*Epimedium*	*grandiflorum*	**TM** USA-DSHEA	Bal2-064-01
Yohimbe	Tubeflower Rubyroot	*Pausinystalia*	*johimbe*	**CS** (Bal2-551-01) Asia-Pharm.	Wyk-229-01
Ze Xie	Water Plantain	*Alisma*	*plantago-aquatica*	**TM** USA-DSHEA	Wyk-397-11

Herbs & Vegetables listed by Symptomatic Illness

All information in this section was extracted from the following references:

Bal1, 1997 **"Prescription for Nutritional Healing, 2nd Ed."**, J.A.Balch & P.A.Balch, ISBN:0-89529-727-2
Bal2, 2012 **"Prescription for Herbal Healing, 2nd Ed."**, P.A. Balch & S. Bell, ISBN:0-978-1-58333-452-2
Murr, 2012 **"The Encyclopedia of Natural Medicine 3rd Ed."**, by M.T. Murray & J.E. Pizzorno, ISBN:978-1-4516-6300-6
Scal, 2002 **"Traditional Medicines from the Earth"**, by R. Scalzo & M. Cronin, ISBN:0-9707936-1-8
Wyk, 2004 **"Medicinal Plants of the World"**, Ben-Erik Van Wyk & Michael Wink, ISBN:981-232-933-1

Cerebral Disorders

Anxiety HAN also known as **Stress**, all anxieties are based on a fear that you will fail to survive. In this book stress is a synonym for anxiety.
Herbs: *Angelica sinensis* (NSD, stress, Bal1), *Avena sativa* (DABA, Bal1), *Chamaemelum nobile* (NSD, Bal1), ***Eleutherococcus senticosus* (DCFS, Bal1, Bal2, Murr)**, *Foeniculum vulgare* (SDG, Bal1), *Ganoderma lucidum* (Bal2), *Ginkgo biloba* (DFRA, Bal1), *Glycyrrhiza glabra* (SDG, Bal1), ***Humulus lupulus* (NSD, Bal1, Wyk)**, ***Hypericum perforatum* (HAN, Bal1, Bal2, Scal, Wyk)**, *Leonurus cardiaca* (NSD, Bal1), *Matricaria chamomilla* (DALL, Bal2), *Melissa officinalis* (SDG, Bal1, Bal2), *Mentha piperita* (SDG, Bal1), *Nepeta cateria* (NSD, Bal1, Bal2), *Papaver californicum* (NSD, Bal2), ***Panax ginseng* (NSD, Bal2, Murr)**, *Panax quinquefolium* (Bal2), *Passiflora incarnata* (NSD, Bal1, Bal2), *Pinus pinaster* (DFRA, Murr), ***Piper methysticum* (NSD, Bal1, Bal2, Murr, Scal, Wyk)**, *Polygala senega* (NSD, stress, Bal1), *Rauvolfia serpentina* (NSD, Wyk), *Rehmannia* glutinosa (NSD, stress, Bal1), *Rhodiola rosea* (NSD, Murr), *Schisandra chinensis* (Bal2), *Scutellaria baicalensis* (HIN, NSD, Bal1, Bal2), *Scutellaria barata*(HIN, NSD, Bal1, Bal2), *Scutellaria lateriflora* (HIN, NSD, Bal1, Bal2), *Schizandra chinensis* (NSD, stress, Bal1), *Silybum marianum* (Bal1), *Tilia cordata* (NSD, Bal1, Bal2), *Uncaria guianensis* (Bal2), *Uncaria tomentosa* (Bal2), *Vaccinium-vortabacca myrtillis* (DFRA, Bal1, Bal2), *Valeriana officinalis* (HIN, NSD, Bal1, Bal2), *Viburnum-scrubia opulus* (NSD, Bal1), *Vitis vinifera* (DFRA, Murr), ***Withania somnifera* (HIN, Bal2, Murr)**, ***Zingiber officinale* (NSD, Bal1, Bal2)**, *Ziziphys jujuba* (NSD, stress, Bal1)
Herb Formulas: BUPLEURUM & OYSTER SHELL *Bupleurum chinense, Ostrea edulis* (shell), *Bupleurum scorzoneraefolium, Cinnamomum aromaticum, Panax ginseng, Pinellia turnata, Reum rhabarbarum, Scutellaria baicalensis, Scutellaria barata, Scutellaria lateriflora, Zingiber officinale, Zizyphus jubata, Wolfiporia extensa* (Bal2); COPTIS DECOCTION *Coptis chinensis, Gardenia angusta, Phellodendron amurense, Scutellaria baicalensis, Scutellaria barata, Scutellaria lateriflora* (Bal2); DONG QUAI & PEONY POWDER *Angelica sinensis, Paeonia lactiflora, Paeonia officinalis, Alisma plantago-aquatica, Atractylodes macrocephala, Ligusticum porteri, Wolfiporia extensa* (Bal2); FIVE INGREDIENT POWDER WITH PORIA *Alsima plantago-aquatica Atractylodes macrocephala Cinnamomum aromaticum, Polyporus brumalis, Polyporus squamosus, Polyporus tubaster, Polyporus umbellatus, Wolfiporia extensa,* (Bal2); FOUR SUBSTANCE DECOCTION *Angelica sinensis, Ligusticum porteri, Paeonia lactiflora, Paeonia officinalis, Rehmannia glutinosa* (Bal2); MIGRA-PROFEN *Piper methysticum* (70mg), *Piscidia erythrina* (bark, 20mg), *Rosmarinus officinalis* (leaf extract, 8mg), *Tanacetum parthenium* (leaf, 240mg), *Scutellaria baicalensis* (root, 110mg), *Valeriana officinalis* (root, 30mg), *Zingiber officinale* (rhizome extract, 8mg) (Scal); MINOR BUPLEURUM DECOCTION *Bupleurum chinense, Bupleurum scorzoneraefolium, Glycyrrhiza glabra, Panax ginseng, Pinellia turnata, Scutellaria baicalensis Scutellaria barata, Scutellaria lateriflora, Zingiber officinale, Ziziphus jujuba* (Bal2); PINELLIA & MAGNOLIA BARK DECOCTION *Pinellia turnata, Magnolia officinalis* (bark), *Perilla frutescens* (leaf), *Wolfiporia extensa, Zingiber officinale* (Bal2); SERENITY WITH KAVA KAVA *Artemisia vulgaris* (herb, 15mg), *Avena sativa* (seed, 27mg), *Humulus lupulus* (strobile, 24mg), *Matricaria chamomilla* (flower, 45mg), *Mentha piperita* (herb, 9mg), *Passiflora incarnata* (vine, 60mg), *Piper methysticum* (rhizome, 60mg), *Scutellaria lateriflora* (herb, 60mg) (Scal);
Vegetables (algae): Dulse (Bal1); **Vegetables (condiment):** Garlic (Bal1); **Vegetables (crucifer):** Broccoli (Bal1); **Vegetables (fruit):** Avocados (Bal1); **Vegetables (fungus):** Brewer's Yeast (Bal1); **Vegetables (grain):** Brown Rice, Buckwheat, Cereal, Corn, Millet, Oats, Quinoa, Rice Bran, Rye, Spelt, Tapioca, White Flour, White Rice, Wheat (Bal1); **Vegetables (leaf):** Chard, Kale, Lettuce, Spinach, Turnip Greens (Bal1, Murr); **Vegetables (legume):** Green Beans, Kidney Beans, Lentils, Lima Beans, Mung Beans, Peas, Pinto Beans, Soybeans, White Beans (Bal1, Murr); **Vegetables (sap):** Blackstrap Molasses (Bal1); **Vegetables (stalk):** Asparagus (Bal1);

(1) **Chronic Anxiety** a subjective feeling of uneasiness, nervousness.

(2) **Acute Anxiety** subjective feeling of dangerous threat, panic or terror.

(3) **Acute Chronic Anxiety** also known as **Depression** is characterized by despair, hopelessness, irritability, quickness to anger, loss of interenst in hobbies, feelings of worthlessness, feelings of inadequacy. Symptoms include chronic fatigue, either insomnia or excessive sleeping, headaches, back aches, and indigestion.
Herbs: *Crocus sativus* (Murr), ***Ginkgo biloba* (Bal2, Murr)**, ***Hypericum perforatum* (Bal1, Bal2, Murr)**, *Lavandula officinalis* (Murr), *Melissa officinalis, Morinda citrifolia, Piper methysticum* (Bal1, Bal2)
Herb Formulass: BIOTA SEED PILL *Thuja orientalis* (seed), *Acorus calamus, Angelica sinensis, Glycyrrhiza glabra, Lycium chinense, Ophiopogon japonicus, Rehmannia glutinosa, Scrophularia nodosa, Wolfiporia extensa* (Bal2); BUPLEURUM & OYSTER SHELL *Bupleurum chinense, Ostrea edulis* (shell), *Bupleurum scorzoneraefolium, Cinnamomum aromaticum, Panax ginseng, Pinellia turnata, Reum rhabarbarum, Scutellaria baicalensis, Scutellaria barata, Scutellaria lateriflora, Zingiber officinale, Zizyphus jubata, Wolfiporia extensa* (Bal2); COPTIS DECOCTION *Coptis chinensis, Gardenia angusta, Phellodendron amurense, Scutellaria baicalensis, Scutellaria barata, Scutellaria lateriflora* (Bal2); FRIGID EXTREMITIES DECOCTION *Bupleurum chinense, Citrus aurantium, Glycyrrhiza glabra, Paeonia officinalis* (Bal2); GINSENG DECOCTION *Panax ginseng, Angelica sinensis, Astragalus membranaceus, Atractylodes macrocephala, Bupleurum chinense, Cimicifuga racemosa,Citrus aurantium* (peel), *Glycyrrhiza glabra* (Bal2); LICORICE, WHEAT & JUJUBE DECOCTION *Glycyrrhiza glabra, Triticum aestivum, Ziziphys jubata* (Bal2); PINELLIA & MAGNOLIA BARK DECOCTION *Pinellia turnata, Magnolia officinalis* (bark), *Perilla frutescens* (leaf), *Wolfiporia extensa, Zingiber officinale* (Bal2); VACHA RASAYANUS *Acorus calamus* + sweeting herbs (Bal2); YOGARAJ GUGGULU *Commiphora mukul* + sweeting herbs, (Bal2);
Vegetables: "Avoid processed foods" (Bal2, page 278), **Vegetables:** Raw Nuts, Seeds (Murr), **Vegetables (legume):** Chick Peas, Green Beans, Kidney Beans, Lentils, Lima Beans, Mung Beans, Peas, Pinto Beans, Soybeans, Red Beans, White Beans (Murr); **Vegetables (grain):** Brown Rice, Buckwheat, Cereal, Corn, Millet, Oats, Quinoa, Rice Bran, Rye, Spelt, Tapioca, White Flour, White Rice, Wheat (Murr);

(4) **Manic Depressive Disorder** also known as **Bipolar Disorder**, is alternating periods of depression and mania. A person may go from feeling overwhelmed with misery and despair

(depression) to feeling invincible and elated (mania). Symptoms include loss of enthusiasm, attacks of rage, and loss of inhibition.
Herbs: *Hypericum perforatum* (St. John's Wort gives a feeling of well-being but does not address the underlying cause, Bal1)
Herb Formulas: PHYTO-PRŌZ SUPREME *Acorus calamus* (root, 3mg), *Avena sativa* (seed, 21mg), *Centella asiatica* (herb, 18mg), *Eleutherococcus senticosus* (root, 45mg), *Hypericum perforatum* (flower bud, 105mg), *Passiflora incarnata* (vine, 27mg), *Piper methysticum* (rhizome, 103mg), *Schisandra chinensis* (berry, 90mg), *Urtica dioica* (seed, 6mg), *Xanthoxylum clava-herculis* (bark, 9mg) (Scal);
Vegetables (legume): Chick Peas, Green Beans, Kidney Beans, Lentils, Lima Beans, Mung Beans, Peas, Pinto Beans, Soybeans, Red Beans, White Beans (Bal1); **Vegetables (grain, gluten-free):** Corn, Buckwheat, Millet (Bal1);

Fever HFV an elevation in body temperature above 98.6 degrees Fahrenheit. A part of the brain called the hypothalamus regulates body temperature.
Herbs: *Achillea millifolium* (Bal1), *Echinacea purpurea* (Bal1), *Glycyrrhiza glabra* (Bal1), *Hyssopis officinalis* (Bal1), *Lobelia inflata* (Bal1), *Nepeta cataria* (Bal1), *Phytolacca americana* (root, Bal1), *Prunus-Cerasus spinosa* (Bal1), *Tanacetum parthenium* (Bal1), *Taraxacum officinale* (Bal1), *Thymus vulgaris* (Bal1), *Trigonella foenum-graecum* (Bal1), *Zingiber officinale* (Bal1)

Headache HAC a narrowing of the blood vessels in the brain causing sharp stabbing pain, throbbing pain, or dull constant pain.

Types of Headache (adapted from Bal1, pages 300, 301)

Type of	Headache	Symptoms	Recommendation
Arthritis	Headache	Pain at the back of the neck and lower skull	Feverfew, Green Vegetables
Bilious	Headache	Dull pain in forehead. Throbbin pain in temples.	Raw Vegetables
Caffeine	Headache	Throbbing blood vessel pain in brain.	Diet of 75% Vegetables
Cluster	Headache	Throbbing pain on one side of head.	Ginkgo biloba extract
Exertion	Headache	Throbbing pain in brain after physical exercise, intercourse.	Diet of 75% Vegetables
Eyestrain	Headache	Throbbing pain behind the eyes.	Diet of 75% Vegetables
Fever	Headache	Throbbing blood vessel pain in brain.	Diet of 75% Vegetables
Hangover	Headache	Throbbing blood vessel pain in brain.	Diet of 75% Vegetables
Hunger	Headache	Throbbing blood vessel pain in brain.	Diet of 75% Vegetables
Hypertension	Headache	Dull pain in a localized area of the brain, aggravated by movement.	Diet of 75% Vegetables
Menstrual	Headache	Severe throbbing pain before, during, and after menstruation.	Diet of 75% Vegetables
Migraine	Headache	Severe throbbing pain often on one side of the head.	low carbohydrate, high protein
Sinus	Headache	Nagging pain in nasal and sinus area.	Vitamin A & C Vegetables
Temporal	Headache	Jabbing, burning, boring pain when chewing.	Steroid Vegetables
TMJ	Headache	Temporal Mandibular Joint pain above ear, or pain in face near jaw.	Diet of 75% Vegetables
Tension	Headache	Constant pain in one area or all over the head.	Diet of 75% Vegetables
Tic Douloureux	Headache	Short jabbing pains around mouth, jaw, or forehead	Diet of 75% Vegetables

National Headache Foundation
5252 North Western Avenue Phone: 800-843-2256
Chicago, IL 60625

American Council for Headache Education
875 Kings Highway Suite 200 Phone: 800-255-ACHE
Woodbury NJ 08096

Herbs: *Althaea officinalis* (Bal1), *Angelica sinensis* (migraine, Bal2), *Arctium lappa* (root, Bal1), *Artemisia vulgaris* (migraine, Bal1), *Capsicum annuum-longum-cayenneum* (regular & migraine, Bal1, Bal2), *Cimicifuga racemosa* (TM, Wyk), *Claviceps purpurea* (TM, Wyk), *Codonopsis pilosa* (BHP, Bal2), *Cyclamen europaeum* (TM, Wyk), *Digitalis purpurea* (TM, Wyk), *Ephedra trifurca* (Bal1), *Gelsemium sempervirens* (TM, Wyk), *Ginkgo biloba* (regular & migraine, Bal1, Bal2), *Glycrrhiza glabra* (regular & migraine, Bal1), *Hydrastis canadensis* (Bal1), *Iris versicolor* (TM, Wyk), *Lavandula angustifolia* (Bal1), *Lobelia inflata* (Bal1), *Matricaria chamomilla* (regular & migraine, Bal1, Bal2), *Mentha piperita* (regular & migraine, Bal1), *Mentha spicata* (Bal1), ***Petasites hybridus* (TM, migraine, Murr, Wyk)**, *Pinus pinaster* (bark, Murr), *Pueraria lobata* (neck pain, Bal2), *Quercus velutina* (migraine, quercetin, Bal2), *Rosmarinus officinalis* (regular & migraine, Bal1), *Salix alba* (regular & migraine, bark, Bal1, Bal2), *Salix cineria* (regular & migraine, bark, Bal1, Bal2), *Salix discolor* (regular & migraine, bark, Bal1, Bal2), *Salix fragilis* (regular & migraine, bark, Bal1, Bal2), *Salix nigra* (regular & migraine, bark, Bal1, Bal2), *Salix pentandra* (regular & migraine, bark, Bal1, Bal2), *Salix purpurea* (regular & migraine, bark, Bal1, Bal2), *Sanguinaria canadensis* (TM, Wyk), *Scutellaria baicalensis* (NSD, Bal1, Bal2), *Scutellaria barata* (NSD, Bal1, Bal2), *Scutellara lateriflora* (NSD, Bal1, Bal2), ***Tanacetum parthenium* (TM, migraine, Bal1, Bal2, Murr, Wyk)**, *Thymus vulgaris* (Bal1), *Tilia cordata* (BHP, Bal2), *Trigonella foenum-graecum* (Bal1), *Valeriana officinalis* (migraine, Bal1), *Vitis vinifera* (seed, Murr), ***Zingiber officinale* (migraine, Bal1, Murr)**
Herb Formulas: CINNAMON TWIG DECOCTION *Cinnamomum aromaticum, Paeonia officinalis, Paeonia suffruticosa, Prunus-Persica persica x communis* (seed) (Bal2); CINNAMON TWIG & KUDZU DECOCTION *Cinnamomum aromaticum, Paeonia officinalis, Paeonia suffruticosa, Prunus-Persica persica x communis* (seed), *Pueraria lobata* (Bal2); GASTRODIA & UNCARIA DECOCTION *Gastrodia-asiatica elata, Uncaria gambir, Uncaria guianensis, Uncaria tomentosa, Eucommia ulmoides, Gardenia angusta, Phoradendron leucarpum (Synonym Loranthus leucarpus), Polygonum aviculare, Scutellaria baicalensis, Scutellaria barata, Scutellaria lateriflora, Wolfia extensa*, Sichuan Ox Knee Root (Bal2) HEADACHE SALVE *Gaultheria procumbens* (oil), *Mentha piperita* (oil), *Zingiber officinale* (regular & migraine, Bal1); WHITE TIGER DECOCTION *Anemarrhena asphodeloides, Atractylodes macrocephala, Glycyrrhiza glabra*, Gypsum (CaSO4∗2H2O), Mirabilite (Na2SO4∗10H2O), *Oryza sativa, Panax ginseng, Rheum rhabarbarum* (Bal2);
Herb Formulas (migraine): BUPLEURUM & OYSTER SHELL *Bupleurum chinense, Ostrea edulis* (shell), *Bupleurum scorzoneraefolium, Cinnamomum aromaticum, Panax ginseng, Pinellia turnata, Reum rhabarbarum, Scutellaria baicalensis, Scutellaria barata, Scutellaria lateriflora, Zingiber officinale, Zizyphus jubata, Wolfiporia extensa* (Bal2); FOUR SUBSTANCE DECOCTION *Angelica sinensis, Ligusticum porteri, Paeonia lactiflora, Paeonia officinalis, Rehmannia glutinosa* (Bal2); GASTRODIA & UNCARIA DECOCTION *Gastrodia-asiatica elata, Uncaria gambir, Uncaria guianensis, Uncaria tomentosa, Eucommia ulmoides, Gardenia angusta, Phoradendron leucarpum (Synonym Loranthus leucarpus), Polygonum aviculare, Scutellaria baicalensis, Scutellaria barata, Scutellaria lateriflora, Wolfia extensa*, Sichuan Ox Knee Root (Bal2); MAJOR BUPLEURUM DECOCTION: *Bupleurum chinense, Citrus-aurantius*

aurantium, Paeonia officinalis, Pinellia turnata, Rheum rhabarabarum, Scutellaria baicalensis, Scutellaria barata, Scutellaria lateriflora, Zingiber officinale, Ziziphus jujuba (Bal2);
Vegetables: Almonds, Almond Milk, Watercress, Parsley, Fennel, Garlic (Bal1)

Insomnia HIN difficulty in falling asleep and difficulty in sleeping without interruption.
TM = traditional medicine, MM = modern medicine
Herbs: *Angelica sinensis* (TM, Wyk), *Aspalathus linearis* (Bal2), *Atropa belladonna* (TM, Wyk), *Avena sativa* (TM, Wyk), *Citrus aurantium* (TM, Wyk), *Corydalis cava* (TM, Wyk), *Crocus sativus* (TM, Wyk), *Eschscholtzia californica* (TM, Wyk), ***Humulus lupulus* (gamma-aminobutyric acid excitory Inhibitor, Bal1, Bal2, Wyk)**, ***Hypericum perforatum* (MM, Bal2, Wyk)**, *Lactuca sativa* (Bal2), *Lavandula angustifolia* (essential oils, Wyk), *Lycopus europaeus* (TM, Wyk), *Lycopus virginicus* (TM, Wyk), *Matricaria chamomilla* (Bal1, Bal2), ***Melissa officinalis* (essential oils, Bal2, Wyk)**, *Mentha piperita* (MM, Wyk), *Nardostachys grandiflora* (TM, Wyk), *Nepeta cateria* (Bal1), *Papaver californicum* (Bal1), ***Passiflora incarnata* (gamma-aminobutyric acid excitory Inhibitor, Bal1, Bal2, Wyk)**, ***Piper mythisticum* (MM, Bal1, Bal2, Wyk)**, *Piscidia piscipula* (TM, Wyk), *Rouvolfia serpentina* (MM, Wyk), *Scutellaria baicalensis* (Bal1) *Scutellaria barata* (Bal1), *Scutellaria lateriflora* (Bal1), *Shisandra chinensis* (Bal2), *Tilia cordata* (TM, Wyk)., ***Valeriana officinalis* (gamma-aminobutyric acid excitory Inhibitor, Bal1, Bal2, Murr, Scal, Wyk)**
Herb Formulas: BUPLEURUM DECOCTION: *Bupleurum chinense, Citrus-aurantius aurantium, Paeonia officinalis, Pinellia turnata, Rheum rhabarabarum, Scutellaria baicalensis, Scutellaria barata, Scutellaria lateriflora, Zingiber officinale, Ziziphus jujuba* (Bal2); CINNAMON TWIG & OYSTER SHELL *Cinnamomum aromaticum, Ostrea edulis* (shell), *Paeonia officinalis, Paeonia suffruticosa, Prunus-Persica persica x communis* (seed) (Bal2); COPTIS DECOCTION *Coptis chinensis, Gardenia angusta, Phellodendron amurense, Scutellaria baicalensis, Scutellaria barata, Scutellaria lateriflora* (Bal2); GASTRODIA & UNCARIA DECOCTION *Gastrodia-asiatica elata, Uncaria gambir, Uncaria guianensis, Uncaria tomentosa, Eucommia ulmoides, Gardenia angusta, Phoradendron leucarpum (Synonym Loranthus leucarpus), Polygonum aviculare, Scutellaria baicalensis, Scutellaria barata, Scutellaria lateriflora, Wolfia extensa,* Sichuan Ox Knee Root (Bal2); LICORICE, WHEAT & JUJUBE DECOCTION *Glycyrrhiza glabra, Triticum aestivum, Ziziphys jubata* (Bal2); PILL FOR DEAFNESS *Acorus calamus, Alisma plantago-aquatica, Cornus officinalis, Dioscorea oppositifolia, Dioscorea villosa,* Magnetite Fe_3O_4, *Paeonia suffruticosa, Rehmannia glutinosa, Schisandra chinensis, Wolfiporia extensa* (Bal2); SOUND SLEEP *Eschscholzia californica* (herb, 16mg), *Gamma Amino-butyric acid* (106mg), *Humulus lupulus* (strobile, 12mg), *L-glycine* (106mg), *Passiflora incarnata* (vine, 18mg), *Piper methysticum* (rhizome 42mg), *Scutellaria lateriflora* (herb, 12mg), *Valeriana officinalis* (root, 150mg) (Scal); WARM THE GALL BLADDER DECOCTION *Bambusa arundinacea, Citrus-aurantius aurantium, Glycyrrhiza glabra, Pinellia turnata, Wolfiporia extensa, Zingiber officinale* (Bal2);
Vegetables: Whole Grain Crackers (Bal1, Bal2)

Memory Loss HML the brain is surrounded by a membrane called the blood-brain barrier. If the blood is high in cholesterol and tryglycerides, the amount of nutrient rich blood that passes through the membrane decreases. Over time, this can result in the brain becoming mal-nourished causing memory loss. Free radicals that pass through the membrane can cause the neurons in the brain to short-circuit and miss-fire causing memory loss.
Herbs: *Allium sativum* (Bal2), *Bacopa monnieri* (Bal2), *Caulophyllum thalictroides* (Bal1), *Centella asiatica* (Bal2), *Crataegus laevigata* (Bal2), *Crataegus monogyna* (Bal2), ***Ginkgo biloba* (Bal1, Bal2, Scal)**, *Glycine max-soja* (lecithin, Bal2), *Illicium anisatum* (Bal1), *Illicium verum* (Bal1), *Panax ginseng* (Bal1), *Rosmarinus officinalis* (Bal1), *Withania somnifera* (Bal2)
Herb Formulas: BIOTA SEED PILL *Thuja orientalis* (seed), *Acorus calamus, Angelica sinensis, Glycyrrhiza glabra, Lycium chinense, Ophiopogon japonicus, Rehmannia glutinosa, Scrophularia nodosa, Wolfiporia extensa* (Bal2); BUPLEURUM & OYSTER SHELL *Bupleurum chinense, Ostrea edulis* (shell), *Bupleurum scorzoneraefolium, Cinnamomum aromaticum, Panax ginseng, Pinellia turnata, Reum rhabarbarum, Scutellaria baicalensis, Scutellaria barata, Scutellaria lateriflora, Zingiber officinale, Zizyphus jubata, Wolfiporia extensa* (Bal2); EIGHT INGREDIENT REHMANNIA PILL *Rehmannia glutinosa, Alisma plantago-aquatica, Astragalus membranaceus, Cornus officinalis, Dioscorea oppositifolia or villosa, Paeonia suffrutecosa, Schisandra chinensis, Wolfiporia extensa* (Bal2); GASTRODIA & UNCARIA DECOCTION *Gastrodia-asiatica elata, Uncaria gambir, Uncaria guianensis, Uncaria tomentosa, Eucommia ulmoides, Gardenia angusta, Phoradendron leucarpum (Synonym Loranthus leucarpus), Polygonum aviculare, Scutellaria baicalensis, Scutellaria barata, Scutellaria lateriflora, Wolfia extensa,* Sichuan Ox Knee Root (Bal2); MENTAL ALERTNESS *Avena sativa* (seed, 22mg), *Centella asiatica* (40mg), *Eleutherococcus senticosus* (root, 80mg), *Ginkgo biloba* (leaf, 40mg), *Mentha piperita* (20mg), *Polygonum multiflorum* (26mg), *Rosmarinus officinalis* (leaf extract, 8mg), *Voacanga* (seed, 10mg) (Scal);
Vegetables: Brewer's Yeast, Wheat Germ (Bal1); **Vegetables (grain):** Brown Rice, Buckwheat, Cereal, Corn, Millet, Oats, Quinoa, Rice Bran, Rye, Spelt, Tapioca, White Flour, White Rice, Wheat (Bal1); **Vegetables (legume):** Chick Peas, Green Beans, Kidney Beans, Lentils, Lima Beans, Mung Beans, Peas, Pinto Beans, Soybeans, Red Beans, White Beans (Bal1);

Circulatory System Disorders

Cardiovascular Disease also known as **Heart Disease**; disease of the heart, arteries and veins. It is the number 1 cause of death in the United States claiming more than 1 million lives every year.

Acidosis BAC a condition in which blood chemistry becomes overly acidic. Pathogenic viruses, bacteria, microbes, and fungus thrive in an overly acidic blood chemistry. Conversely, no disease causing organism can survive in an overly alkaline blood chemistry.
Herbs: *Humulus lupulus* (Bal1), *Salix alba* (Bal1), *Sambucus nigra* (bark, Bal1), *Zingiber officinale* (external, Bal1)
Vegetables: Avocado (*Persea americana*), Buckwheat (*Fagopyrum esculentum*), Corn (*Zea mays-suavis*), Lima Bean (*Phaseolus lunatus*), Maple Syrup (*Acer saccharum*), Millet (*Setaria italica*) Molasses (*Sorghum bicolor-saccharatum*), Soybean (*Glycine max*) (Bal1)

Anemia BAN a reduction in the number of red blood cells or the amount of hemoglobin in the blood.
Herbs: *Capsella bursa-pastoris* (Bal1), *Crataegus monogyna* (Bal1), *Hydrastis canadensis* (Bal1), *Mahonia aquifolium* (Bal1), *Medicago sativa* (Bal1), *Prunus-Avium avium* (Bal1), *Rubus-raspabacca idaeus* (Bal1), *Rumex crispus* (Bal1), *Tabebuia impetignosa* (Bal1), ***Taraxacum officinale* (Bal1, Bal2)**, ***Urtica dioca* (Bal1, Bal2)**, *Verbascum thapsis* (Bal1), ***Vitis vinifera* (grape seed extract, Bal1, Murr)**, *Vaccinium-vortabacca myrtillis* (Bal1)
Herb Formulas: FOUR SUBSTANCE DECOCTION *Angelica sinensis, Ligusticum porteri, Paeonia lactiflora, Paeonia officinalis, Rehmannia glutinosa* (Bal2)
Vegetables (condiments): Parsley (Bal1); **Vegetables (algae):** Kelp (Bal1); **Vegetables (crucifer):** Broccoli (Bal1); **Vegetables (fruit):** Squash (Bal1); **Vegetables (grain):** Brown Rice, Buckwheat, Cereal, Corn, Millet, Oats, Quinoa, Rice Bran, Rye, Spelt, Tapioca, White Flour, White Rice, Wheat (Bal1, Bal2); **Vegetables (leaf):** Chard, Kale,

Lettuce, Spinach, Turnip Greens (Bal1, Murr); **Vegetables (legume):** Peas, **Vegetables (stalk):** Asparagus, Okra, **Vegetables (root):** Yams (Bal1);

Angina BAA also known as **Angina Pectoris** and **Ischemia** irregular heart attacks, accompanied by intense pain and a feeling of constriction and anxiety caused by the bacteria *Bacteroides fusiformis vincentii, Borrelia vincentii, Streptococcus anginosus*.
TM = traditional medicine, MM = modern medicine
Herbs: *Alpinia officinarum* (TM, Wyk), ***Ammi visnaga* (MM, Bal2, Murr, Wyk)**, *Astragalus membranaceus* (Bal2), *Capsicum annuum-longum-cayenneum* (TM, Wyk), *Capsicum annuum-frutescens-chilli* (TM, Wyk), *Codonopsis pilosa* (Bal2), *Cordyceps sinensis* (Bal2), *Crataegus laevigata* (Bal2), ***Crataegus monogyna* (MM, Bal2, Wyk)**, ***Crataegus oxyacantha* (Murr)**, *Papaver somnifera* (MM, Wyk), *Terminalia arjuna* (Bal2), *Thea sinensis* (MM, Wyk), *Zingiber officinale* (TM, Wyk)
Herb Formulas: GASTRODIA & UNCARIA DECOCTION *Gastrodia-asiatica elata, Uncaria gambir, Uncaria guianensis, Uncaria tomentosa, Eucommia ulmoides, Gardenia angusta, Phoradendron leucarpum (Synonym Loranthus leucarpus), Polygonum aviculare, Scutellaria baicalensis, Scutellaria barata, Scutellaria lateriflora, Wolfia extensa*, Sichuan Ox Knee Root (Bal2)
Vegetables: Parsley, Spinach, Turnips (Bal2); **Vegetables (crucifer):** Broccoli, Brussels Sprouts, Cabbage, Cauliflower (Bal2, Murr); **Vegetables (fiber):** Flaxseed, Oat Bran, Pectin (Murr); **Vegetables (lily):** Onions, Garlic (Murr);

Arrhythmia BIH also known as **Cardiac Rhythm Disturbance** This includes irregular heartbeat (arrhythmia), rapid heartbeat (tachycardia; > 100 beats/min), and reduced heartbeat (brachycardia; < 60 beats/min).
Herbs: *Atropa belladonna* (Wyk), *Cinchona pubescens* (Wyk), ***Crataegus monogyna* (Murr, Wyk)**, *Cystisus scoparius* (Wyk), *Rauvolfia serpentina* (Wyk)

Arterial Sclerosis BAS also known as **Arterioscelrosis, Atherosclerosis, Intermittant Claudation** deposits composed of calcium buildup on the insides of artery walls. Peripheral veins and arteries become narrow or obstructed and circulation of the peripheral organs, such as legs, is reduced.
Herbs: *Aesculus hippocastanum* (BCR, Wyk), ***Allium sativum* (BHC, BHP, Bal2, Wyk)**, *Allium cepa* (BHC, Wyk), *Allium ursinum* (TM, Wyk), *Ammi visagna* (DFRA, Bal2), *Ananas comosus* (BHC, BHP, Bal2), *Andrographis paniculata* (BHC, Bal2), *Astragalus membranaceus* (BHC, Bal2), *Capsicum annuum-longum-cayenneum* (Bal1), *Coptis chinensis* (SGB, Bal2), *Coriolus versicolor* (DFRA, Bal2), *Crataegus laevigata* (DFRA, Bal1, Bal2), *Crataegus monogyna* (DFRA, Bal1, Bal2), *Curcuma longa* (DFRA, Bal2), *Cynara scolymus* (BHC, Wyk) *Eleutherococcus senticosus* (TM, Wyk), *Fagopyrum esculentum* (BCR, Wyk), ***Ginkgo biloba* (BCR, Bal1, Bal2, Wyk)**, *Glycine max* (BHC, Bal2), *Matricaria chamomilla* (DFRA, Bal2), *Medicago sativa* (BHC, Bal2), *Monascus purpureus* (BHC, Bal2), *Papaver somniferum* (BCR, Wyk), *Quercus velutina* (quercetin, DFRA, Bal2), *Rauvolfia serpentina* (BCR, Wyk), *Salvia miltiorhiza* (BHC, BHP, Bal2), *Scutellaria baicalensis* (SGB, Bal2), *Scutellaria barata* (SGB, Bal2), *Scutellaria lateriflora* (SGB, Bal2), *Silybum marianum* (DRA, Bal2), *Stellaria media* (Bal1), *Terminalia arjuna* (BHP, Bal2), ***Thea sinensis* (green, BCR, Bal2 Wyk)**, *Tremella fuciformis* (BHC, Bal2), *Vaccinium-vortabacca myrtillis* (BHC, Bal2), *Vinca minor* (BCR, Wyk), *Zanthoxylum americanum* (Bal2), *Zanthoxylum bungeanum* (Bal2), *Zingiber officinale* (BHC, Bal2)
Herb Formulas: DONG QUAI & PEONY POWDER *Angelica sinensis, Paeonia lactiflora, Paeonia officinalis, Alisma plantago-aquatica, Atractylodes macrocephala, Ligusticum porteri, Wolfiporia extensa* (Bal2);
Vegetables: Onions (Bal2)
Vegetables (grain): Brown Rice, Buckwheat, Cereal, Corn, Millet, Oats, Quinoa, Rice Bran, Rye, Spelt, Tapioca, White Flour, White Rice, Wheat (Bal1)

Cerebral Vascular Insufficiency a decreased blood supply to the brain. In the elderly, the carotid artery splits into two branches; one supplying the brain and one supplying the face and scalp. Sediment accumulates in the branches eventually reaching 90% blockage. Symptoms are short term memory loss, dizziness, headache, ringing in the ears, depression, blurred vision.
Herbs: *Ginkgo biloba* (Murr)
Vegetables: 3 servings per day (Murr)

Congestive Heart Failure the heart muscle is unable to pump blood fast enough to supply nutrients and oxygen to the body. About 550,000 new cases are diagonosed every year. Symptoms include fatigue and weakness. Causes are myocardial infarction (heart attack), diabetes, hyperthyroidism, and obesity.
Herbs: *Aconitum napellus* (Bal2), *Asparagus officinalis* (root, Bal2), *Astragalus membranaceus* (Bal2), *Commiphora mukul* (Bal2), ***Crataegus laevigata* (Bal2)**, ***Crataegus monogyna* (Bal2)**, ***Crataegus oxyacantha* (Murr)**, *Inula helenium* (Bal2), *Leonurus heterophyllus* (Bal2), *Pueraria lobata* (Bal2), ***Terminalia arjuna* (Bal2, Murr)**
Vegetables: "Reduce sodium intake by eliminating salted foods such as pickles, soy sauce, table salt", "In more severe cases, there may be a need to eliminate commercially prepared breads, canned vegetables and soups, cheese, and canned beets." (Bal2, page 266)

Congestive Heart Failure Risk (adapted from Murr, page 150)

Healthy Food & Exercise	Low Risk	Normal Risk	High Risk 1	High Risk 2	High Risk 3
Fruit Servings per Day	3	1-2	1	0-1	0
Vegetable Servings per Day	3	3	2	1	0
Total Servings per Day F & V	6	4-5	3	1-2	0
Exercise Hours per Week	> 4	3-4	2-3	1-2	0

Diabetes BDB glucose accumulates in the blood stream. The symptoms are fatigue, hunger, thirst and frequent urination. Virtually the entire vitamin & mineral content has been removed from white sugar, white breads, pastries, and many breakfast cereals. When these foods are eaten, blood sugar levels rise quickly putting a strain on the body's blood sugar control mechanism.
Herbs: *Agathosma betulina* (Bal1), *Agathosma crenulata* (Bal1), *Ananas comosus* (bromelain, Bal2), *Arctium lappa* (Bal2), *Arctostaphylos uva-ursi* (Bal1), *Atractylodes macrocephala* (Bal2), *Frasera chirata* (Bal2), *Gaylussacia baccata* (Bal1), *Ginkgo biloba* (Bal2), *Glycyrrhiza glabra* (Bal2), ***Gymnema silvestre* (Bal2, Murr)**, *Hydrastis canadensis* (Bal1), *Juniperus communis* (berry, Bal1, Bal2), ***Momordica charantia* (Bal2, Murr)**, *Panax ginseng* (Bal1, Bal2), *Phyllanthus amarus* (Bal2), *Scutellaria baicalensis* (Bal2),

Scutellaria barata (Bal2), *Scutellaria lateriflora* (Bal2), *Syzygium cuminii* (Bal2), *Tabebuia impetignosa* (Bal2), *Taraxacum officinale* (root, Bal1), *Thea sinensis-viridis* (Bal2), ***Trigonella foenum-graecum* (Bal2, Murr)**, *Vaccinium-vortabacca myrtillis* (Bal1)
COPTIS DECOCTION *Coptis chinensis, Gardenia angusta, Phellodendron amurense, Scutellaria baicalensis, Scutellaria barata, Scutellaria lateriflora* (Bal2); EIGHT INGREDIENT REHMANNIA PILL *Rehmannia glutinosa, Alisma plantago-aquatica, Astragalus membranaceus, Cornus officinalis, Dioscorea oppositifolia or villosa, Paeonia suffrutecosa, Schisandra chinensis, Wolfiporia extensa* (Bal2); MAJOR BUPLEURUM DECOCTION: *Bupleurum chinense, Citrus-aurantius aurantium, Paeonia officinalis, Pinellia turnata, Rheum rhabarabarum, Scutellaria baicalensis, Scutellaria barata, Scutellaria lateriflora, Zingiber officinale, Ziziphus jujuba* (Bal2); MINOR BUPLEURUM DECOCTION *Bupleurum chinense, Bupleurum scorzoneraefolium, Glycyrrhiza glabra, Panax ginseng, Pinellia turnata, Scutellaria baicalensis Scutellaria barata, Scutellaria lateriflora, Zingiber officinale, Ziziphus jujuba* (Bal2); SIX INGREDIENT REHMANNIA PILL *Rehmannia glutinosa, Alisma plantago-aquatica, Cornus officinalis, Dioscorea oppositifolia or villosa, Paeonia suffruticosa, Wolfiporia extensa* (Bal2); WHITE TIGER DECOCTION *Anemarrhena asphodeloides, Atractylodes macrocephala, Glycyrrhiza glabra,* Gypsum ($CaSO_4*2H_2O$), Mirabilite ($Na_2SO_4*10H_2O$), *Oryza sativa, Panax ginseng, Rheum rhabarbarum* (Bal2);
Vegetables: Raw, Brewer's Yeast, Oat Bran, Rice Bran (Bal1); **Vegetables (grain):** Brown Rice, Buckwheat, Cereal, Corn, Millet, Oats, Quinoa, Rice Bran, Rye, Spelt, Tapioca, White Flour, White Rice, Wheat (Bal1); **Vegetables (legume):** Chick Peas, Green Beans, Kidney Beans, Lentils, Lima Beans, Mung Beans, Peas, Pinto Beans, Soybeans, Red Beans, White Beans (Bal1); **Vegetables (lily):** Garlic, Onions (Murr), **Vegetables (root):** Beets, Onions, Parsnips, Potatoes, Radishes, Sweet Potatoes, Turnips (Bal1); **Vegetables (blood sugar stabilizers):** Garlic, Sauerkraut, Soybeans, Spirulina, Whole Grains, Legumes, Root Vegetables (Bal2)

(1) **Type 1 Diabetes** also known as **Juvenile Diabetes** usually starts in childhood. The body cannot produce enough insulin.
Herbs: *Pinus pinaster* (pine bark extract, Murr), *Gymnema sylvestre* (Murr), *Momordica charantia* (Murr), *Thea sinensis-viridis* (green tea extract, Murr), *Vitis vinifera* (grape seed extract, Murr)

(2) **Type 2 Diabetes** also known as **Adult Diabetes** starts in an adult. The cells in the body become insulin resistant. Cells in the skeletal muscles and liver do not respond to insulin and cannot absorb the needed sugars. This usually happens despite increased insulin production, so that the blood contains both too much insulin and too much glucose. Obesity or high blood pressure can restrict the flow of insulin to muscle cells. Too much glucose in the blood can cause numbness, sensitivity or pain.
Herbs: *Agrimonia eupatoria* (Bal2), ***Allium sativum* (Bal2, Murr)**, *Aloe vera* (Bal2), *Anacardium occidentale* (Bal2), *Angelica dahurica* (Bal2), *Astragalus membranaceus* (Bal2), *Morus alba* (extract, Murr), *Morus nigra* (extract, Murr), *Opuntia cochenillifera* (Bal2), *Rehmannia glutinosa* (Bal2), Touchi Extract (Murr)

(3) **Diabetic Ketoacidosis** a buildup of acidic keytone molecules in the blood as a result of burning of fat stores for energy.

(4) **Diabetic Nephropathy** the most severe form of kidney disease requiring dialysis.
Herbs: *Allium sativum* (Murr), *Crataegus laevigata* (extract, Murr), *Crataegus monogyna* (extract, Murr), *Hibiscus sabdariffa* (Murr), *Olea europaea* (leaf extract, Murr)

(5) **Diabetic Neuropathy** the loss of peripheral nerve functions characterized by tingling sensations, numbness, and burning pain. It starts in the feet and can spread elsewhere in the body.
Herbs: *Borago officinalis* (Murr), *Oenothera biennis* (Murr), *Ribes nigra* (Murr)

(6) **Diabetic Retinopathy** is an eye disorder that affect the retina. The retina is the "projection screen" at the back of the eye onto which the light that passes through the lens of the eye is thrown. The capillaries that nourish the retina break and leak blood plasma fluid or blood. The result is "bloodshot" eyes. Controlling blood sugar levels can combat this eye disease.
Herbs: *Crataegus laevigata* (extract, Bal2), *Crataegus monogyna* (extract, Bal2), *Ginkgo biloba* (Bal2), *Glycine max* (isoflavone concentrate, Bal2), *Phyllanthas amarus* (Bal2), *Pinus pinaster* (pine bark extract, Bal2), *Quercus velutina* (quercetin, Bal2), *Sizygium cuminii* Bal2), ***Vaccinium-vortabacca myrtillis* (bilberry extract, Bal2, Murr)**, ***Vitis vinifera* (grape seed extract, Bal2, Murr)**
Herb Formulas: EIGHT INGREDIENT REHMANNIA PILL *Rehmannia glutinosa, Alisma plantago-aquatica, Astragalus membranaceus, Cornus officinalis, Dioscorea oppositifolia or villosa, Paeonia suffrutecosa, Schisandra chinensis, Wolfiporia extensa* (Bal2);

Heart Attack BHA also known as **Myocardial Infarction**, the supply of blood to the heart by the coronary arteries is sharply reduced or cut off, portions of the heart muscle begin to die. The primary symptom is a deep, severe pain in the chest that can spread to the left arm, neck, jaw or area between the shoulder blades. Other symptoms include shortness of breath, sweating, nausea, and vomiting. Heart attack can be caused by severe Arterial Scelerosis where artieries become blocked, by severe Cardiac Arrhythmia, or by Aneurysm Rupture. Severe Arrhythmias result in 500,000 sudden deaths in the United States every year.
Herbs: *Allium sativum* (BCR, Bal2), *Andrographis paniculata* (BHC, Bal2), *Astragalus membranaceus* (BCR, Bal2), *Avena sativa* (NSD, straw, Bal1), *Borago officinalis* (minerals, seed, Bal1), *Capsicum annuum-longum-cayenneum* (NSD, & muscle strengthener, Bal1), *Centella asiatica* (BAA, Bal1), *Cimicifuga racemosa* (NSD, Bal1), *Citrus aurantea* (slows heartbeat, Bal2), *Cordyceps sinensis* (BHC, BHP, Bal2), *Crataegus laevigata* (muscle strengthener, Bal1, Bal2), *Crataegus monogyna* (muscle strengthener, Bal1, Bal2), *Curcuma longa* (NSD, Bal2), *Equisetum arvense* (minerals, Bal1), *Ginkgo biloba* (NSD & muscle strengthener, Bal1, Bal2), *Leonurus cardiaca* (muscle strengthener, Bal1), *Medicago sativa* (minerals, Bal1), *Melissa officinalis* (NSD, Bal2), *Panax ginseng* (BCR, Bal2), *Passiflora caerulea* (NSD, Bal1), *Passiflora incarnata* (NSD, Bal1), *Primula veris* (BAA, Bal1), *Pueraria lobata* (BCR, Bal2), *Rosmarinus officinalis* (BAA, Bal1), *Ruscus aculeatus* (muscle strengthener, Bal2), *Tabebuia impetignosa* (minerals, Bal1), *Salvia multiorhiza* (muscle strengthener, Bal1), *Scutellaria baicalensis* (NSD, Bal1, Bal2), *Scutellaria barata* (NSD, Bal1, Bal2), *Scutellaria lateriflora* (NSD, Bal1, Bal2), *Stachys officinalis* (NSD, Bal1), *Terminalia arjuna* (NSD, Bal2), *Thea sinensis-viridis* (DFRA, Bal1), *Urtica dioica* (minerals, Bal1), *Valeriana officinalis* (NSD, Bal1), *Zingiber officinale* (NSD & muscle strengthener, Bal1)
Herb Formulas: BUPLEURUM & OYSTER SHELL *Bupleurum chinense, Ostrea edulis* (shell), *Bupleurum scorzoneraefolium, Cinnamomum aromaticum, Panax ginseng, Pinellia turnata, Reum rhabarbarum, Scutellaria baicalensis, Scutellaria barata, Scutellaria lateriflora, Zingiber officinale, Zizyphus jubata, Wolfiporia extensa* (Bal2);
Vegetables: Almonds, Brewer's Yeast, Kelp, Sea Vegetables, Sesame Seeds (Bal1); **Vegetables (grain):** Brown Rice, Buckwheat, Cereal, Corn, Millet, Oats, Quinoa, Rice Bran, Rye, Spelt, Tapioca, White Flour, White Rice, Wheat,

Heart Insufficiency a lack of **Blood Circulation BCR**, results from an inability of the heart muscle to supply the body with adequate amounts of oxygen-rich blood.
TM = traditional medicine, MM = modern medicine
Herbs: *Achillea millifolium* (TM, Wyk), *Adonis vernalis* (glycosides, MM, Wyk), *Ammi visnaga* (MM, Wyk), *Apocynum cannabinum* (glycosides, MM, Wyk), *Arnica montana* (TM, Wyk), *Asclepias tuberosa* (Bal1), *Artemisia vulgaris* (Bal1), *Cinnamomum camphora*(MM, Wyk), *Convallaria majalis* (glycosides, MM, Wyk), ***Crateagus monogyna* (MM, flower & leaf, Scal, Wyk)**, *Crataegus oxycantha* (berry, Scal), *Cytisus scoparius* (TM, Wyk), *Digitalis purpurea* (glycosides, MM, Wyk), *Digitalis lanata* (glycosides, MM, Wyk), *Equisetum arvense* (Bal1), *Gentiana lutea* (Bal1), *Glycyrrhiza glabra* (TM, Wyk), *Helleborus niger* (glycosides, MM, Wyk), *Hydrastis canadensis* (Bal1), *Hypericum perforatum* (TM, Wyk), *Hyssopis officinalis* (Bal1), *Ilex paraguaensis* (TM, Wyk), *Lavandula angustifolia* (TM, Wyk), *Leonurus cardiaca* (MM, Wyk), *Nerium oleander* (glycosides, MM, Wyk), *Rosa-hipa-scrubia canina* (Bal1), *Rosmarinus officinalis* (TM, Wyk), *Stellaria media* (Bal1), *Strophanthus gratus* (glycosides, MM, Wyk), *Theobroma cacao* (TM, Wyk), *Thevetia peruviana* (glycosides, MM, Wyk), *Urginea maritima* (glycosides, MM, Wyk), *Valeriana officinalis* (TM, Wyk), *Viscum album* (TM, Wyk)
Vegetables: Brown Rice, Endive, Garlic, Lima Beans, Oat Bran, Onions, Peas, Psyllium Husk, Spinach (Bal1)

Heart Stimulation Deficiency BHS The coronary arteries supply blood to the heart. If the coronary arteries narrow, the amount of blood they supply to the heart may be insufficient to provide the oxygen the heart needs. Heart tissue may die causing dead patches in the heart. Hardening of the arteries (arterial sclerosis) leads to a narrowing of the coronary arteries.
Herbs: *Berberis vulgaris* (Bal1), *Capsicum annuum-longum-cayenneum* (Bal1), *Cimicifuga racemosa* (Bal1), *Crataegus monogyna* (Bal1), *Garcinia hanbury* (Bal1)*i*, *Crataegus monogyna* (Bal1), *Ginkgo biloba* (Bal1), *Panax ginseng* (Bal1), *Pfaffia paniculata* (Bal1), *Ruscus aculeatus* (Bal1), *Taraxacum officinale* (Bal1), *Valeriana officinalis* (Bal1)

Heat Exhaustion BHE also known as **Heatstroke**, the body's core temperature exceeds 100°F. Symptoms are dizziness, fainting, irregular heartbeat, nausea and vomiting.
Herbs: *Capsicum annuum-frutescens-chilium* (Bal2), *Capsicum annuum-longum-cayenneum* (Bal2), *Citrus-aurantius aurantium* (Bal2)

High Blood Pressure BHP the heart pumps blood in a way that causes the pressure against the walls of blood vessels to be abnormally high (above 140/90 mm Hg). When blood pressure is elevated, the heart has to work harder to pump blood to all of the tissues of the body. Excessive force that circulating blood exerts against the walls of arteries can cause the arteries to narrow can lead to arterial sclerosis, heart attack, kidney disease, memory loss, and stroke. **About 33.3% of American adults have high blood pressure.**

Stages of Blood Pressure
(adapted from Murr, page 150 & page 679)

Blood Pressure	Low	Normal	Stage 1 High	Stage 2 High	Stage 3 High
Systolic	< 125 mm Hg	125-134 mm Hg	135-149 mm Hg	150-164 mm Hg	> 164 mm Hg
Diastolic	< 90 mm Hg	90-94 mm Hg	95-104 mm Hg	105-114 mm Hg	> 114 mm Hg
Murr Recommendations	Vegetables (listed below)	Vegetables (listed below)	Vegetables & Herbs (listed below)	Vegetables & Herbs (listed below),CoQ10	Synthetic Drug Intervention

High Blood Pressure Risk
(adapted from Murr, page 150)

Healthy Food & Exercise	Low Risk	Normal Risk	High Risk 1	High Risk 2	High Risk 3
Fruit Servings per Day	3	1-2	1	0-1	0
Vegetable Servings per Day	3	3	2	1	0
Total Servings per Day F & V	6	4-5	3	1-2	0
Exercise Hours per Week	> 4	3-4	2-3	1-2	0

Herbs: *Adonis vernalis* (TM, Wyk), *Allium sativum* (TM, Wyk), *Ananas comosus* (bromelain, Bal1), *Angelica sinensis* (Bal2), *Capsicum annuum-longum-cayenneum* (Bal1), *Citrus-aurantius aurantium* (Bal2), *Coleus barbatus* (Bal2), *Convallaria majalis* (TM, Wyk), ***Crataegus monogyna* (TM, Bal2, Murr, Wyk)**, *Digitalis lanata* (TM, Wyk, page 353), *Digitalis purpurea* (TM, Wyk, page 353), *Epimedium brevicornum* (Bal2), *Epimedium grandiflorum* (Bal2), *Foeniculum vulgare* (Bal1), *Ganoderma lucidum* (Bal2), *Helleborus niger* (TM, Wyk), *Hibiscus sabdariffa* (Murr), *Humulus lupus* (NSD, Bal1), *Lentinula edodes* (Bal2), *Matricaria chamomilla* (Bal1), *Nerium oleander* (TM, Wyk), *Olea europaea* (Murr), *Panax ginseng* (Bal2), *Panax quinquefolium* (Bal2), *Petroselinum crispum* (Bal1), *Phyllanthus amarus* (Bal2), *Pinus pinaster* (pine bark extract, Bal2), *Rauvolfia serpentina* (TM, Wyk), *Rosmarinus officinalis* (Bal1), *Scutellaria baicalensis* (Bal2), *Scutellaria barata* (Bal2), *Scutellaria lateriflora* (Bal2),*Urginea maritima* (TM, Wyk), *Vaccinium-vortabacca myrtillus* (Bal2), *Valeriana officinalis* (NSD, Bal1), *Vitis vinifera* (grapeseed extract, Bal2), *Zea mays-saccharata* (cornsilk, Bal2)
Herb Formulas: <u>EIGHT INGREDIENT REHMANNIA PILL</u> *Rehmannia glutinosa, Alisma plantago-aquatica, Astragalus membranaceus, Cornus officinalis, Dioscorea oppositifolia or villosa, Paeonia suffrutecosa, Schisandra chinensis, Wolfiporia extensa* (Bal2)
Vegetables: Asparagus, Broccoli, Cabbage, Eggplant, Garlic, Kelp, Maitake Mushrooms, Shiitake Mushrooms, Reishi Mushrooms, Oat Bran, Peas, Squash, Sweet Potatoes (Bal1); **Vegetables (leaf):** Chard, Kale, Lettuce, Spinach, Turnip Greens (Bal1),
Vegetables (live juices): Beet Juice, Carrot Juice, Celery Juice, Parsley Juice, Spinach Juice (Bal1)
Vegetables (potassium, magnesium, fiber): Almonds, Artichokes, Beans, Broccoli, Carrots, Collard Greens, Filberts, Kale, Peanuts, Peas, Potatoes, Spinach, Squash, Sunflower Seeds, Sweet Potatoes, Tomatoes, Turnip Greens, Walnuts (Murr)
Vegetable Formulas: <u>KYO-GREEN</u> Barley & Wheat Grass Juice (Bal1)

High Cholesterol BHC Cholesterol is a fatty, waxy substance that is used for cell membrane creation, hormone production and nerve function. About 85% of the body's supply of cholesterol is made in the liver. High Cholesterol causes plaque in the arteries wheich impedes blood flow to the heart, brain, kidneys, genitals, and extremities. About 17% of the adult population of the United States has high cholesterol.

Stages of Cholesterol & Triglyceride
(adapted from Murr, page 150 & page 681)

Cholesterol & Tryglyceride	Low	Normal	Stage 1 High	Stage 2 High	Stage 3 High

High Density Lipoprotein (HDL)	> 75 mg/dl	74-64 mg/dl	64-55 mg/dl	54-35 mg/dl	< 35 mg/dl
Low Density Lipoprotein (LDL)	< 135 mg/dl	126-160 mg/dl	161-195 mg/dl	196-230 mg/dl	> 230 mg/dl
Total (HDL + LDL Cholesterol)	< 200 mg/dl	200-224 mg/dl	225-250 mg/dl	250-275 mg/dl	> 275 mg/dl
Tryglyceride	< 100 mg/dl	100-150 mg/dl	150-199 mg/dl	200-499 mg/dl	> 500 mg/dl

Herbs: *Allium sativum* (Bal2), *Capsicum-annuum-longum-cayenneum* (Bal1), *Commiphora mukul* (Bal2), *Cordyceps sinensis* (Bal2), *Crataegus laevigata* (Bal1, Bal2), *Crataegus monogyna* (Bal1, Bal2), *Cynara scolymus* (leaf, Bal2), *Dioscorea oppositifolia* (Bal2), *Dioscorea villosa* (Bal2), *Ferula assa-foetida* (Bal2), *Glycine max-soja* (lecithin, Bal2), *Hydrastis canadensis* (Bal1), *Lentinula edodes Scutellaria baicalensis* (Bal1), *Scutellaria barata* (Bal1), *Scutellaria lateriflora* (Bal1),, *Monascus purpureus* (Bal2), *Panax ginseng* (Bal2), *Scutellaria baicalensis* (Bal2), *Scutellaria barata* (Bal2), *Scutellaria lateriflora* (Bal2), *Silybum marianum* (Bal2), *Thea sinensis-viridis* (Bal2), *Tremella fuciformis* (Bal2), *Zingiber officinale* (Bal2)
Herb Formulas: BUPLEURUM & OYSTER SHELL *Bupleurum chinense, Ostrea edulis* (shell), *Bupleurum scorzoneraefolium, Cinnamomum aromaticum, Panax ginseng, Pinellia turnata, Reum rhabarbarum, Scutellaria baicalensis, Scutellaria barata, Scutellaria lateriflora, Zingiber officinale, Zizyphus jubata, Wolfiporia extensa* (Bal2); CHOLESTEROL VITAL MAINTENANCE *Chelidonium majus* top & root, 8mg), *Coleus forskholii* (root, 160mg), *Commiphora mukul* (resin, 150mg), *Cynara scolymus* (leaf, 200mg), *Dioscorea villosa* (rhizome, 22mg) (Scal); *Terminalia ajuna* (bark, 100mg), MINOR BUPLEURUM DECOCTION *Bupleurum chinense, Bupleurum scorzoneraefolium, Glycyrrhiza glabra, Panax ginseng, Pinellia turnata, Scutellaria baicalensis, Scutellaria barata, Scutellaria lateriflora, Zingiber officinale, Ziziphus jujuba* (Bal2);
Vegetables: Carrots, Garlic, Olive Oil (Bal1), Alfalfa Sprouts, Almonds, Canola Oil, Peanut Oil, Spirulina, Walnuts (Bal2);
Vegetables (fiber): Oat Bran, Oatmeal, Guar Gum, Pectin, Psyllium, Other Fiber Vegetables (Murr)
Vegetables (grain): Brown Rice, Buckwheat, Cereal, Corn, Millet, Oats, Quinoa, Rice Bran, Rye, Spelt, Tapioca, White Flour, White Rice, Wheat (Bal1); **Vegetables (live juices):** Beet Juice, Carrot Juice, Celery Juice (Bal1, Bal2);

Jaundice BJA also known as **Blood Bile**, a yellowing of the skin and eyes caused by a buildup of bilirubin in the blood. Bilirubin is a yellow brown substance released during the breakdown of red blood cells. The liver fails to filter out bilirubin as a result of **Liver Cirrhosis, Pernicious Anemia, Hepatitis,** and **Hemolysis**. Normally the bilirubin which the liver filters out is sent through bile ducts to the Gall Bladder. When the bile ducts are obstructed, the bilirubin passes from the liver back into the blood stream.
Herbs: *Arctium lappa* (root, Bal1), *Silybum marianum* (Bal1), *Taraxacum officinale* (Bal1), *Trifolium pratense* (Bal1)
Vegetables: "Eat only raw vegetables and fruits for one week. Then eat a diet consisting of 75 percent raw food for a month." (Bal1, page 357, recommendation #1); Vegetable (live juices): Beet Juice, Beet Greens Juice, Dandelion Extract Juice, Black Radish Extract Juice (Bal1)

Liver Cirrhosis is combated by a **Blood Liver Cleanse BLC**, the liver filters the impurities out of the blood and peforms lipid (fat) metabolism, protein biosynthesis, and glycogen storage. Liver cirrhosis is a destruction of liver cells by toxins, bacteria, fungi, or viruses.
TM = traditional medicine, MM = modern medicine
Herbs: *Agrimonia eupatoria* (TM, Wyk), *Aloe vera* (Bal1), *Arctium lappa* (root, Bal1, Bal2), *Berberis vulgaris* (Bal1), *Beta vulgaris* (TM, Wyk), *Chelidonium majus* (Bal1), *Chionanthus virginicus* (Bal1), *Chondrus crispus* (Bal1), *Cynara scolymus* (MM, Wyk), *Echinacea purpurea* (Bal1), *Eclipta prostrata* (TM, Wyk), *Equisetum arvense* (Bal1), *Foeniculum vulgare* (Bal1), *Glycine max* (MM, Wyk), *Helichrysum arenarium* (TM, Wyk), *Humulus lupulus* (Bal1), *Hydrastis canadensis* (Bal1), *Mahonia aquifolium* (Bal1), *Mahonia nervosa* (Bal1), *Mahonia repens* (Bal1), *Medicago sativa* (Bal1), *Phyllanthus amarus* (TM, Wyk), *Picrorhiza kurrooa* (TM, Wyk), *Pfaffia paniculata* (Bal1), *Rehmannia glutinosa* (TM, Wyk), *Rhaphanus sativus-niger* (Bal1), *Rosa-hipa canina* (Bal1), *Schisandra chinensis* (TM, Wyk), ***Silybum marianum* (MM, Bal1, Scal, Wyk)**, *Taraxacum officinale* (Bal1), *Thymus vulgaris* (Bal1), *Trifolium pratense* (Bal1, Bal2), *Wedelia calendulacea* (TM, Wyk)
Herb Formulas: CLEANSE MAINTENANCE *Alnus serrulata* (bark, 36mg), *Dicentra canadensis* (tuber, 40mg), *Podophyllum peltatum* (rhizome, 32mg), *Rumex crispus* (root, 40mg), *Scrophularia nodosa* (whole plant 32 mg) (Scal); HEP SUPPORT WITH ASTRAGALUS & MAITAKE *Astragalus membranaeus* (root, 200mg), *Bupleurum chinense* (root, 60mg), *Ganoderma lucidum* (30mg), *Glycyrrhiza glabra* (root, 15mg), *Glycyrrhiza uralensis* (root, 15mg), *Grifola frondosa* (20 mg), *Scutellaria baicalensis* (root, 100mg) (Scal); LIVER HEALTH *Curcuma longa* (root, 100mg), *Glycyrrhiza glabra* (root, 10mg), *Glycyrrhiza uralensis* (root, 10mg), *Schisandra chinensis* (berry, 100mg), *Scutellaria baicalensis* (56mg), *Silybum marianum* (seed, 100mg) (Scal);
Vegetables: Fresh Vegetable Juices, **Vegetables (fungus):** Brewer's Yeast, **Vegetables (grain):** Brown Rice, Buckwheat, Cereal, Corn, Millet, Oats, Quinoa, Rice Bran, Rye, Spelt, Tapioca, White Flour, White Rice, Wheat, **Vegetables (leaf):** Chard, Kale, Lettuce, Spinach, Turnip Greens, **Vegetables (legume):** Chick Peas, Green Beans, Kidney Beans, Lentils, Lima Beans, Mung Beans, Peas, Pinto Beans, Soybeans, Red Beans, White Beans, **Vegetables (sprout):** Alfalfa Sprouts (Bal1), **Vegetables (dark green):** Artichoke, Asparagus, Mild Bell Peppers, Broccoli, Brussels Sprouts, Chard, Collard Greens, Cornsalad Lettuce, Cucumbers, Green Beans, Kale, Leeks, Mustard Greens, Parsley, Peas, Seakale, Spinach, Zucchini Squash (Bal2), **Vegetables (yellow anti-oxidant):** Bell Peppers (yellow), Corn, Oca Tuber, Squash (Bal2), **Vegetables (orange):** Carrots, Pumpkin, Squash, Sweet Potato, Yam (Bal2)

Phlebitis also known as **Thrombophebitis**, an inflammation of a vein. Symptoms include swelling, pain, and bluish discoloration. This disorder occurs more frequently in women. It usually occurs over the age of 40.
Herbs: *Achillea millefolium* (Bal1), *Capsicum annuum-longum-cayenneum* (Bal1), *Crataegus laevigata* (Bal1), *Crataegus mongyna* (Bal1), *Ginkgo biloba* (Bal1), *Hamamelis virginiana* (Bal1), *Medicago sativa* (Bal1), *Plantago major* (Bal1), *Rosmarinus officinalis* (Bal1), *Rubus-raspabacca idaeus* (Bal1), *Ruscus aculeatus* (Bal1), *Scutellaria baicalensis* (Bal1), *Scutellaria barata* (Bal1), *Scutellaria lateriflora* (Bal1), *Tabebuia impetignosa* (Bal1), *Valeriana officinalis* (Bal1), *Zingiber officinale* (Bal1)
Vegetables: Soybeans, Nuts, Seeds (Bal1); **Vegetables (grain):** Brown Rice, Buckwheat, Cereal, Corn, Millet, Oats, Quinoa, Rice Bran, Rye, Spelt, Tapioca, White Flour, White Rice, Wheat (Bal1)

Stroke BST blood clots block arteries supplying oxygen to the brain. Symptoms include blurred vision, loss of bladder control, loss of bowel control, dizziness, paralysis, speech impediment, swallowing difficulties.
Herbs: *Allium sativum* (Murr), *Crataegus laevigata* (Bal2), *Crataegus monogyna* (Bal2), ***Ginkgo biloba* (Bal2, Murr)**, ***Pinus pinaster* (bark, Bal2, Murr)**, *Salvia miltiorhiza* (Bal2), ***Vitis vinifera* (seed, Bal2, Murr)**

Herb Formulas: RHUBARB & MOUTAN DECOCTION *Paeonia suffruticosa, Rheum rhabarbarum* (Bal2);

Varicose Veins BVV veins have tiny valves to prevent blood from flowing backwards. If the valves do not work properly blood accumulates in the veins stretching them. The veins become abnormally large, bulging and lumpy-looking, often bluish in color.
Herbs: ***Aesculus hippocastanum* (internal & external, Bal1, Bal2, Murr)**, ***Ananas comosus* (bromelain, Bal2, Murr)**, *Capsicum annuum-longum-cayenneum* (internal & external Bal2), ***Centella asiatica* (Bal1, Bal2, Murr)**, *Crataegus laevigata* (Bal1, Bal2), *Crataegus monogyna* (Bal1, Bal2), *Ginkgo biloba* (Bal1), *Hamamelis virginiana* (Bal2), ***Pinus pinaster* (bark, Bal2, Murr)**, *Quercus alba* (external, bark, Bal1), ***Ruscus aculeatus* (Bal1, Bal2, Murr)**, ***Vitis vinifera* (seed, Bal2, Murr)**
Vegetables: High Fiber Vegetables (Bal1)

Venous Insufficiency vein circulatory disorders, especially in the lower extremeties can lead to varicosis, throbophelbitis, and phlepbothrobmbosis. The endothelia of the veins become leaky and the valves of the veins lose their function.
Herbs: *Aesculus hippocastanum* (TM, Wyk), *Centella asiatica* (TM, Wyk), *Fagopyrum esculentum* (TM, Wyk), *Hamamelus virginiana* (TM, Wyk), *Melilotus officinalis* (TM, Wyk), *Ruscus aculeatus* (TM, Wyk), *Styphnolobium japonicum* (TM, Wyk)

Digestive System Disorders

Appetite SAP Anxiety, illness, and metal toxicity can cause poor apetite.
Herbs: *Carica papaya* (leaf, Bal1), *Centella asiatica* (Bal1), *Foeniculum vulgare* (seed, Bal1), *Mentha piperita* (Bal1), *Nepeta cataria* (Bal1), *Panax ginseng* (Bal1), *Serenoa repens* (berry, Bal1), *Zingiber officinale* (Bal1)

Crohn's Disease DCRO a chronic ulceration of the digestive tract from the mouth to the anus. This disorder is not contagious. Food allergies and free radical damage can aggrevate the symptoms.
Herbs: ***Aloe vera* (Bal1, Bal2, Murr)**, *Althea officinalis* (root, Bal1, Bal2), *Ananas comosus* (bromelain, Bal2), *Arctium lappa* (Bal1), *Boswellia serrata* (Murr), *Carica papaya* (Bal2), *Curcuma longa* (Murr), *Echinacea purpurea* (Bal1), *Foeniculum vulgare* (seeds, Bal2), *Glycyrrhiza glabra* (Bal1), *Hydrastis canadensis* (Bal1), *Hypericum perforatum* (Bal2), *Ilex paraguensis* (Bal1), *Mentha piperita* (Bal1, Bal2), *Pinus pinaster* (pine bark extract, Murr), *Plantago afra* (psyllium husk, Bal2), *Rosa-hipa-scrubia canina* (Bal1), *Silybum marianum* (Bal1, Bal2), *Tabebuia impetignosa* (Bal1), *Thea sinensis-viridis* (green tea extract, Murr), *Trifolium pratense* (Bal1), *Trigonella foenum-graecum* (Bal1), *Ulmus rubra* (bark extract, Bal1, Bal2), *Vitis vinifera* (grape seed extract, Murr)
Herb Formulas: GINSENG DECOCTION *Panax ginseng, Angelica sinensis, Astragalus membranaceus, Atractylodes macrocephala, Bupleurum chinense, Cimicifuga racemosa, Citrus aurantium* (peel), *Glycyrrhiza glabra* (Bal2); TRUE MAN'S DECOCTION *Angelica sinensis, Atractylodes macrocephala, Cinnamomum aromaticum, Glycyrrhiza glabra, Myristica fragrans, Papaver somniferum* (husk), *Paeonia officinalis, Panax ginseng, Saussurea lappa, Terminalia chebula* (Bal2); TRUE WARRIOR DECOCTION *Aconitum napellus, Atractylodes macrocephala, Paeonia officinalis, Zingiber officinale, Wolfiporia extensa* (Bal2);
Vegetables: "Be aware that raw fruits and vegetables may be irritating while Crohn's Disease is active." (Bal2, page 273);
Vegetables: "The diet should be high in dietary fiber and low in sugar and refined carbohydrates" (Murr, page 469, under Diet)
Vegetables: Organic Baby Foods, Steamed Vegetables, Well Cooked Brown Rice, Millet, and Oatmeal (Bal2), **Vegetables (fresh):** Carrots, Celery, Garlic, Parsnips, Turnips (Bal1); **Vegetables (fresh crucifer):** Broccoli, Brussels Sprouts, Cabbage, Cauliflower (Bal1); **Vegetables (fresh leaf):** Chard, Kale, Lettuce, Spinach, Turnip Greens (Bal1);

Constipation SCN stool is hard and stays in the lower colon making it difficult or impossible to go to the bathroom.
TM = traditional medicine
Herbs: *Ahnfeltia gigartinoides* (fiber, Wyk), *Aloe ferox* (laxative, Wyk), ***Aloe vera* (laxative, Bal1, Bal2, Wyk)**, *Astracantha gummifera* (fiber, Wyk), *Bryonia dioica* (TM, Wyk), *Citrullus colocynthis* (TM, Wyk), *Convolvulus scammonia* (TM, Wyk), *Ficus carica* (fiber, Wyk), *Foeniculum vulgare* (Bal1), *Frangula purshiana* (Bal1, Bal2), *Fucus vesiculosis* (Bal2), *Gelidium amansi* (fiber, Wyk), *Gelidium cartilagineum* (fiber, Wyk), ***Hibiscus sabdarrifa* (TM, Bal2, Wyk)**, ***Hydrastis canadensis* (laxative, Bal1, Wyk)**, *Ilex paraguensis* (Bal1, Bal2), *Ipomoea orizabensis* (TM, Wyk), *Ipomoea purga* (TM, Wyk), *Linum usitatissimum* (fiber, Wyk), *Medicago sativa* (Bal1), ***Plantago afra* (fiber, Bal2, Murr, Wyk)**, *Plantago arenaria* (fiber, Wyk), *Plantago ovata* (fiber, Wyk), *Prunus spinosa* (TM, Wyk), ***Rhamnus catharticus* (laxative, Bal2, Wyk)**, *Rhamnus frangula* (short term laxative, Murr, Wyk), *Rhamnus purshianus* (laxative, Wyk), ***Rheum rhabarbarum* (laxative, Bal1, Wyk)**, *Rheum officinal* (laxative, Wyk), *Ricinus communis* (laxative, Wyk), *Rosa-hipa canina* (TM, Wyk), ***Senna alexandrina* (short term laxative, leaf extract, Bal1, Bal2, Murr, Wyk)**, *Silybum marianum* (Bal1, Bal2), *Strychnos nux-vomica* (TM, Wyk), *Tamarindus indica* (fiber, Wyk), *Taraxacum officinale* (Bal2), *Triticum aestivum* (fiber, Wyk), *Zingiber officinale* (Bal2)
Herb Formulas: CINNAMON & PEONY DECOCTION *Cinnamomum aromaticum, Paeonia officinalis* (Bal2); TRIPHALA Amalaki, Bibhitaki, Haritaki (Bal2)
Vegetables: Asparagus (Bal1), Garlic (Bal1), Okra (Bal1), Peas (Bal1), Sweet Potatoes (Bal1); **Vegetables (crucifer):** Broccoli, Brussels Sprouts, Cabbage, Cauliflower (Bal1, Murr); **Vegetables (grain):** Brown Rice, Buckwheat, Cereal, Corn, Millet, Oats, Quinoa, Rice Bran, Rye, Spelt, Tapioca, White Flour, White Rice, Wheat (Bal1, Bal2); **Vegetables (leaf):** Chard, Kale, Lettuce, Spinach, Turnip Greens (Bal1, Murr); **Vegetables (legume):** Chick Peas, Green Beans, Kidney Beans, Lentils, Lima Beans, Mung Beans, Peas, Pinto Beans, Soybeans, Red Beans, White Beans (Bal2, Murr), **Vegetables (pectin fiber):** Beets (Bal2), Cabbage (Bal2), Carrots (Bal1, Bal2)

Diarrhea SDR an increase in the fluidity of bowel movements. It is usually accompanied by cramping and nausea. Microorganisms attack the digestive system causing a decrease in water absorption into the blood. Microorganisms include the bacteria *Escherichia coli, Faecalis aerogenes, Klebsiella pneumoniae, Salmonella paratyphi, Shigella dysenteriae*, the parasites *Entamoeba histolytica, Giardia lamblia, and Vibrio cholera.*
Herbs: *Agrimonia eupatoria* (Bal2), *Althea officinalis* (root, Bal2), *Ananas comosa* (Bal1), *Berberis aquifolium* (berberine, Murr), *Berberis vulgaris* (berberine, Murr), *Capsicum-anuum-longum-cayenneum* (Bal1), *Carica papaya* (Bal1), *Citrus-tangierica reticulata* (Bal2), *Coptis chinensis* (berberine, Murr), *Epimedium grandiflorum* (Bal2), *Hamamelis virginiana* (Bal2), *Hydrastis canadensis* (berberine, Murr), *Matricaria chamomilla* (Bal1), *Malus x-domestica* (Bal1), *Musa acuminata* (Bal1), *Plantago afra* (Bal2), *Plantago ovata* (Bal2), *Pinus pinaster* (pine bark extract, Murr) *Potentilla tormentilla* (root extract, Murr), *Rubus-nigrabacca fruticosus* (root bark, Bal1), *Rubus-raspabacca idaeus* (leaf, Bal1), *Tabebuia impetignosa* (Bal1), *Trigonella foenum-graecum* (Bal2), *Ulmus rubra* (Bal1), *Vaccinium-vortabacca myrtillus* (Bal2), *Vitis vinifera* (grape seed extract, Murr), *Zingiber officinale* (Bal1)
Herb Formulas: FIVE ACCUMULATION POWDER *Atractylodes macrocephala, Angelica dahurica, Citrus-aurantius aurantium, Cinnamomum aromaticum, Ephedra distachya, Glycyrrhiza glabra, Ligusticum porteri, Magnolia officinalis* (bark), *Paeonia officinalis,*

Pinellia turnata, Platycodon grandiflorus, Wolfiporia extensa, Zingiber officinale (Bal2); FIVE INGREDIENT POWDER WITH PORIA *Alsima plantago-aquatica Atractylodes macrocephala Cinnamomum aromaticum, Polyporus brumalis, Polyporus squamosus, Polyporus tubaster, Polyporus umbellatus, Wolfiporia extensa,* (Bal2); KUDZU DECOCTION *Pueraria lobata, Cinnamomum aromaticum, Ephedra sinica, Glycyrrhiza glabra, Paeonia lactiflora, Paeonia officinalis Zingiber officinalis, Ziziphus jujuba* (Bal2); LEDEBOURIELLA DECOCTION *Ledebouriella divaricata, Angelica sinensis, Atractylodes macrocephala, Ephedra distachya,* Epsom Salt ($MgSO_4 \cdot 7H_2O$), *Forsythia suspensa, Gardenia angusta* (fruit), *Glycyrrhiza glabra,* Gypsum ($CaSO_4 \cdot 2H_2O$), *Ligusticum porteri, Mentha piperita, Paeonia officinalis, Platycodon grandiflorus, Rheum rhabarbarum* (wine treated root), *Schizonepeta tenuifolia, Scutellaria baicallensis, Scutellaria barata, Scutellaria lateriflora,* Talc [$Mg_3(SiO_4)_{10}(OH)_2$]; TRUE MAN'S DECOCTION *Angelica sinensis, Atractylodes macrocephala, Cinnamomum aromaticum, Glycyrrhiza glabra, Myristica fragrans, Papaver somniferum* (husk), *Paeonia officinalis, Panax ginseng, Saussurea lappa, Terminalia chebula* (Bal2);
Vegetables: Brown Rice (Bal1), Oat Bran (Bal1), Potatoes (Bal2), White Rice (Bal1), Rice Bran (Bal1)

(1) **Allergic Diarrhea** includes Celiac Sprue Disease

(2) **Exudative Diarrhea** includes Crohn's Disease, Inflammatory Bowel Disease, Ulcerative Colitis

(3) **Infectious Diarrhea** is caused by viral infections (Viral Gastroenteritis) in 75% of infectious cases, by bacterial infection (Bacterial Gastroenteritis) in 15% of the cases, and by parasites (Parasitic Gastroenteritis) in 10% of the cases. Viruses are Homosapiensvirus corona-monoposi, *Homosapiensvirus pico-entero* (synonym *Enterovirus hominis*), and *Homosapiensvirus reo-rota* (Synonym *Rotavirus hominis*). Bacteria are *Campylobacter jejuni, Salmonella enterica, Shigella dysenteriae* and *Yersinia enterocoltica*. Parasites are *Cryptosporidum hominis, Entamoeba hystolytica, Giardia lamblia* and *Isospora hominis*

(4) **Osmotic Diarrhea** carbohydrate malabsorption due to lactose intolerance.

(5) **Secretory Diarrhea** caused by, lack of bile, toxin producing bacteria, and hormone producing tumors. Toxin producing bacteria are *Clostridium difficile, Escherichia coli, Staphylococcus aurea, Staphylococcus hominis, Vibrioi parahaemolytica,* and *Vibrio cholerae*.

Diverticultis SDV the mucus membrane lining of the colon becomes inflamed. Small, pouchlike growths called diverticula develop in the large intestine. Waste matter becomes trapped in them, causing infection, fever, chills and pain.
Herbs: *Achillea milifolium* (Bal1), *Aloe vera* (Bal1), *Capsicum annum-longum-cayenneum* (Bal1), *Carica papaya* (Bal1), *Glycyrrhiza glabra* (Bal1), *Hydrastis canadensis* (Bal1), *Matricaria chamomilla* (Bal1), *Meticago sativa* (Bal1), *Tabebuia impetignosa* (Bal1), *Trifolium pratense* (Bal1)

Flatulence SFL gas in the intestinal tract.
Herbs: *Achillea millefolium* (oil, Wyk), ***Angelica sinensis* (oil, Bal2, Wyk)**, *Artemisia vulgaris* (oil, Wyk), *Ballota nigra* (tea, Bal2), ***Carum carvi* (oil, Bal2, Wyk)**, *Cinnamomum aromaticum* (oil, Bal2), *Citrus-tangericus reticulata* (tea, Bal2), *Coriandrum sativum* (oil, Wyk), *Elettaria cardamomum* (oil, Wyk), ***Foeniculum vulgare* (oil, Bal2, Wyk)**, *Gentiana lutea* (bitters, Bal2), *Gentiana scabra* (bitters, Bal2), *Humulus lupulus* (Bal2), *Marrubium vulgare* (tea, Bal2), *Matricaria chamomilla* (oil, Wyk), *Mentha piperita* (oil, Wyk), *Pimpinella anisum* (oil, Wyk), *Rosmarinus officinalis* (oil, Wyk), *Taraxacum officinale* (tincture, Bal2)

Food Poisoning SFP also known as **Gastroenteritis** ingestion of contaminated food containing bacteria, viruses, parasites or mold. The result of eating food contaminated with pathogenic bacteria, viruses, or intestinal worms. microbe parasites, and mold which have toxins that disrupt the digestion process causing abdominal pain, diarrhea, cramping, fatigue, fever, and nausea.
Herbs: *Agrimonia eupatoria* (Bal2), *Foeniculum vulgare* (seed, Bal2), *Hydrastis canadensis* (Bal1), *Lobelia inflata* (Bal1), *Mentha piperita* (Bal2), *Silybum marianum* (Bal1, Bal2), *Syzygium aromaticum* (Bal2), *Trifolium pratense* (Bal1), *Ulmus rubra* (Bal2)
Herb Formulas: WHITE TIGER DECOCTION *Anemarrhena asphodeloides, Atractylodes macrocephala, Glycyrrhiza glabra,* Gypsum ($CaSO_4 \cdot 2H_2O$), Mirabilite ($Na_2SO_4 \cdot 10H_2O$), *Oryza sativa, Panax ginseng, Rheum rhabarbarum* (Bal2);
Vegetables (vomiting inducing): Ipecac Syrup (*Carapichea ipecacuana*) (Bal1)
Vegetables (liquid): Clear Broths, Gatorade, Ginger Ale (Bal2)

(1) **Colic** discomfort due to indigestion and gas caused by allergies, lactose intolerance, and infection usually in a child who is breast-feeding.
Herbs: *Foeniculum vulgare* (seed, Bal2), *Illicium anisatum* (Bal2), *Illicium verum* (Bal2), *Matricaria chamomilla* (Bal2), *Mentha piperita* (Bal2)
Vegetables: "nursing mothers avoid gas-forming foods, such as beans, broccoli, brussels sprouts, cauliflower, cucumbers, flaxseed and green, red, or yellow peppers" (Bal2, page 260)

(2) **Botulism** abdominal pain caused by the bacteria *Clostridium botulinus*.

(3) **Field Epsilon Proteanism** abdominal pain caused by the bacteria *Campylobacter hyointestinalis*

(4) **Gangrene DGNG** abdominal pain caused by the bacteria *Costridium perfringens*.
Herbs: *Capsicum annuum-longum-cayenneum* (Bal1), *Echinacea purpurea* (Bal1), *Ginkgo biloba* (Bal1), *Hydrastis canadensis* (Bal1), *Myrica cerifera* (Bal1), *Ruscus aculeatus* (circulation, Bal1)
Vegetables (germanium): Garlic, Onions, Shiitake Mushrooms (Bal1);
Vegetable Formulas: AGELESS COCKTAIL LIVE JUICE 4-5 fresh carrots, 3 sprigs fresh parsley 1 handful fresh spinach, 1 handful fresh kale, 1 fresh beet including tops, 1 clove garlic peeled, all put in juicer (Bal1, page 556); RAW POTATO LIVE JUICE 8 fresh potato skins with ½ inch of potato on the skin, 1 fresh carrot, 1 stalk fresh celery, 8 ounces bottled spring water, all put in juicer (Bal1, page 557 #1); CABBAGE LIVE JUICE 1/3 head of fresh cabbage cut in small pieces, 2 fresh carrots cut in small pieces, ¼ cup of bottled spring water, all put in juicer (Bal1, page 557 #2)

(5) **Gastritis** acid inflammation of the stomach lining. Symptoms include nausea and vomiting.
Herbs: *Berberis vulgaris* (Bal2), *Calendula officinalis* (Bal2), *Coptis chinensis* (Bal2), *Glycyrrhiza glabra* (Bal2), *Hydrastis canadensis* (Bal2), *Mahonia aquifolium* (Bal2), *Mahonia nervosa* (Bal2), *Mahonia repens* (Bal2), *Tabebuia impetignosa* (Bal2), *Ulex europaeus* (Bal2), *Uncaria guianensis* (Bal2), *Uncaria tomentosa* (Bal2)
Herb Formulas: AUGMENTED RAMBLING POWDER *Angelica sinensis, Atractylodes macrocephala, Bupleurum chinense, Gardenia angusta, Glycyrrhiza glabra, Paeonia lactiflora, Paeonia suffruticosa, Wolfiporia extensa* (Bal2); FIVE ACCUMULATION

POWDER *Atractylodes macrocephala, Angelica dahurica, Citrus-aurantius aurantium, Cinnamomum aromaticum, Ephedra distachya, Glycyrrhiza glabra, Ligusticum porteri, Magnolia officinalis* (bark), *Paeonia officinalis, Pinellia turnata, Platycodon grandiflorus, Wolfiporia extensa, Zingiber officinale* (Bal2); MAJOR CONSTRUCT MIDDLE DECOCTION *Panax ginseng, Hordeum vulgare* (maltose sugar), *Zanthoxylum americanum* or *bungeanum, Zingiber officinale* (Bal2); MINOR BUPLEURUM DECOCTION *Bupleurum chinense, Bupleurum scorzoneraefolium, Glycyrrhiza glabra, Panax ginseng, Pinellia turnata, Scutellaria baicalensis, Scutellaria barata, Scutellaria lateriflora, Zingiber officinale, Ziziphus jujuba* (Bal2);

(6) **Listeriosis** flu-like symptoms cause by the Meningitis Listeriosis Bacillid Firm Wall Bacteria *Listeria monocytogenes.*

(7) Norwalk Virus diarrhea and vomiting cause by the following virus species

Homosapiensvirus calici-noro 1 [M87661]	Human Calicivirus Norwalk
Homosapiensvirus calici-noro Alphatron [AF195847]	Human Calicivirus Norwalk Alphatron
Homosapiensvirus calici-noro DesertShield [U04469]	Human Calicivirus Norwalk Desert Shield
Homosapiensvirus calici-noro Hawaii [U07611]	Human Calicivirus Norwalk Hawaii
Homosapiensvirus calici-noro Lordsdale [X86557]	Human Calicivirus Norwalk Lordsdale
Homosapiensvirus calici-noro Mexico [U22498]	Human Calicivirus Norwalk Mexico
Homosapiensvirus calici-noro SnowMountain [L23831]	Human Calicivirus Norwalk Snow Mountain
Homosapiensvirus calici-noro Southampton [L07418]	Human Calicivirus Norwalk Southampton

(8) **Salmonellosis** abdominal pain cause by the Food Poisoning Salmon Enteric Gamma Protean Bacteria *Salmonella enterica typhimurium*

(9) **Scombriosis** abdominal pain caused by the Mackeral fish virus *Scombervirus.*

(10) **Staph Infection** diarrhea caused by the Round Bacillid Firm Wall Bacteria *Staphylococcus aureus and Staphylococcus hominis*

(11) **Trichinosis** fever, muscle pain and diarrhea caused by the Hairytail Aphasmid Roundworm parasite *Trichinella spiralis.*

Gall Bladder Inflammation **SGB** Bile is produced by the liver and transported to the intestine via bile ducts. Usually the stomach isn't digesting food, so unused bile is transported from the intestine to the gall bladder for storage and later use. Bacterial infection can inflame the gall bladder. Cholesterol, calcium carbonate, and bilirubin can inflame the gall bladder. The gall bladder makes stones out of these substances in an attempt to reduce inflammation. The gall bladder is "cleansed" by destroying the bacteria, and by neutralizing the inflammatory properties of cholesterol, calcium carbonate, and biliruben, thus stimulating the flow of bile.
TM = traditional medicine, MM = modern medicine
Herbs: *Achillea millefolium* (MM, Wyk), *Allium sativum* (TM, Wyk), *Angelica sinensis* (TM, Wyk), *Artemisia abrotanum* (TM, Wyk), *Artemisia absinthium* (MM, Wyk), *Atropa belladonna* (TM, Wyk), ***Berberis vugaris* (TM, Bal1, Wyk)**, *Calendula officinalis* (TM, Wyk), *Centaurea benedicta* (TM, Wyk), *Chelidonium majus* (TM, Wyk), *Cichorium intybus* (TM, Wyk), *Curcuma longa* (MM, Wyk), *Curcuma xanthorrhiza* (MM, Wyk), *Cynara scolymus* (MM, Wyk), *Dioscorea oppositifolia* (TM, Bal1), *Dioscorea villosa* (TM, Bal1), *Elettaria cardamomum* (TM, Wyk), *Equisetum arvense* (TM, Bal1), *Foeniculum vulgare* (TM, Bal1), *Fumaria officinalis* (MM, Wyk), *Gentiana lutea* (TM, Wyk), *Harpagophytum procumbens* (TM, Wyk), *Harungana madagascarensis* (TM, Wyk), *Helichrysum arenarium* (TM, Wyk), *Inula helenium* (TM, Wyk), *Marrubium vulgare* (TM, Wyk), *Medicago sativa* (Bal1), ***Mentha piperita* (MM, oil, Bal1, Wyk)**, *Nepeta cataria* (TM, Bal1), *Petroselinum crispum* (TM, Bal1), *Peumus boldus* (MM, Wyk), *Raphanus sativus* (TM, Wyk), *Scopolia carniolica* (TM, Wyk), ***Taraxacum officinale* (MM, Bal1, Wyk)**, *Viburnum-scrubia opulus* (TM, Bal1), *Veronicastrum virginicum* (TM, Wyk), ***Zingiber officinale* (TM, Bal1, Wyk)**
Herb Formulas: KOMBUCHA TEA Kombucha (a Lichen that is part algae, part fungus, part yeast, and part *Xylinum* bacteria), *Saccharum officinarum* (Sugar), *Thea sininensis* (Tea) "When placed in a mixture of ordinary sugar and tea for seven to ten days, the kombucha produces both a winey-tasting health restoring tea and a new 'baby'kombucha" (Bal1, page 557, under Making Kombucha Tea)
Vegetables: Beets (Bal1), "Eat a diet consisting of 75 percent raw foods." (Bal1, page 286, recommendation #3), "Avoid sugar and products containing sugar" (Bal1, page 286, recommendation #5)

Gall Stones SGS the bile ducts between the gall bladder and the liver become blocked by crystallized cholesterol. About 20 million Americans have had Gall Stones.
Herbs: *Chelidonium majus* (Bal2), ***Curcuma longa* (Bal2, Murr)**, *Cyamopsis tetragonoloba* (Guar Bean, gum, Murr), *Cynara scolymus* (Globe Artichoke, Murr), *Fagopyrum esculentum* (Buckwheat, Murr), *Gentiana lutea* (Bal2), *Gentiana scabra* (Bal2), ***Glycine max-soja* (lecithin, Bal2, Murr)**, *Medicago sativa* (Bal2), *Mentha piperita* (during attack, Bal2), ***Peumus boldus* (Boldo Peum, Bal2, Murr)**, *Pinus pinaster* (bark, Murr), *Plantago afra* (husk fiber, Murr), ***Silybum marianum* (Bal2, Murr)**, *Taraxacum officinale* (Murr), *Vitis vinifera* (seed, Murr), *Zanthoxylum americanum* (during attack, Bal2)
Herb Formulas: FRIGID EXTREMITIES DECOCTION *Bupleurum chinense, Citrus aurantium, Glycyrrhiza glabra, Paeonia officinalis* (Bal2); GENTIAN LONGDANCAO DECOCTION *Gentiana lutea, Gentiana scabra, Akebia quinata, Akebia trifoliate, Alisma plantago-aquatica, Angelica sinensis, Bupleurum chinense, Bupleurum scorzoneraefolium, Gardenia angusta, Plantago afra, Rehmannia glutinosa, Scutellaria baicalensis, Scutellaria barata, Sutellaria galericulata, Scutellaria lateriflora* (Bal2); MAJOR BUPLEURUM DECOCTION: *Bupleurum chinense, Citrus-aurantius aurantium, Paeonia officinalis, Pinellia turnata, Rheum rhabarabarum, Scutellaria baicalensis, Scutellaria barata, Scutellaria lateriflora, Zingiber officinale, Ziziphus jujuba* (Bal2);

Hemorrhoids (ointment) **OHM** dilations of the anal arteries which can be inflamed or bleeding.
Herbs: ***Aesculus hippocastanum* (TM, Bal2, Wyk)**, *Aloe vera* (external, Bal2), *Aesculus hippocastanum* (Murr), *Arnica montana* (TM, Wyk), *Atropa belladonna* (TM, Wyk), *Calendula officinalis* (external, Bal2), *Centella asiatica* (Murr), *Collinsonia canadensis* (root, Bal1), *Equisetum arvense* (Bal2), *Fagopyrum esculentum* (TM, Wyk), *Frangula alnus* (bark, Bal1), ***Hamamelus virginiana* (TM, external, Bal2, Wyk)**, ***Hypericum perforatum* (TM, external, Bal2, Wyk)**, *Matricaria chamomilla* (TM, Wyk), *Myroxylon balsamum* (TM, Wyk), *Petroselinum crispum* (Bal1), ***Pinus pinaster* (bark extract, Bal2, Murr)**, ***Plantago afra* (husk, Bal2, Murr)**, *Potentilla erecta* (TM, Wyk), *Quercus robur* (TM, Wyk), ***Ruscus aculeatus* (TM, Bal2, Murr, Wyk)**, *Sambucus nigra* (external, Bal1), ***Symphytum officinale* (TM, root, external paste, Bal1, Wyk)**, *Taraxacum officinale* (Bal2), *Verbascum thapsis* (external, Bal1), ***Vitis vinifera* (seed extract, Bal1, Bal2, Murr)**
Vegetables (fiber): Beets, Brazil Nuts, Carrots, Green Beans, Guar Gum, Oat Bran, Lima Beans, Peas, Psyllium Seed Husk, Wheat Bran (Bal1);**Vegetables (crucifer):** Broccoli, Brussels Sprouts, Cabbage, Cauliflower (Bal1); **Vegetables**

(grain): Brown Rice, Buckwheat, Cereal, Corn, Millet, Oats, Quinoa, Rice Bran, Rye, Spelt, Tapioca, White Flour, White Rice, Wheat (Bal1);

Heartburn SHB also known as **Acid Reflux**, a burning sensation in the stomach occurs when hydrochloric acid from the stomach backs up into the esophagus because the esophageal sphincter muscle fails to close. It can be caused by haital hernia, spicy foods, fried foods, alcohol, coffee, citrus fruits, chocolate, tomato sauce, ulcers, gall bladder infection, stress, or allergies.
Herbs: *Aloe vera* (juice, Bal1), *Althaea officinalis* (root, Bal1), *Carica papaya* (tea, Bal1), *Coptis chinensis* (tincture, Bal2), *Curcuma longa* (powder, Bal2), *Nepeta cataria* (tea, Bal1), *Foeniculum vulgare* (tea, Bal1), *Hydrastis canadensis* (tincture, Bal2), *Zingiber officinale* (tea, Bal1)
Vegetables: Raw Vegetables, Papaya, Pineapple, Raw Apple Cider Vinegar in Water, Raw Potato Juice (Bal1)

Hiatal Hernia SHH the stomach pushes upward thorough an opening in the diaphragm into the thorax as a result of blunt force trama or a congential defect. Symptoms include heartburn and belching. About 50% of Americans over the age of 40 have hiatal hernias. Ulcers often accompany haital hernias.
Herbs: *Aloe vera* (juice, Bal1), *Athaea officinalis* (root, Bal1), *Hydrastis canadensis* (Bal1), *Trifolium pratense* (Bal1), *Trigonella foenum-graecum* (Bal1), *Ulmus rubra* (Bal1)
Vegetables: "Include extra fiber in the diet." (Bal1, page 317, recommendation #4)

Hiccups sudden involuntary contractions of the diaphragm.
Herbs: *Alpinia galanga* (Bal2), *Zingiber officinale* (Bal2)

Indigestion SDG also known as **Dyspepsia**, a lack of bile secreted from the gall bladder and a lack of digestive enzymes in the digestive tract.
BP = bitter principles, DE = digestive enzymes, PP = pungent principles,
Herbs: *Acorus calamus* (PP, Wyk), *Allium cepa* (PP, Wyk), *Allium sativum* (PP, Wyk), *Allium ursinum* (PP, Wyk), *Aloe vera* (Bal1), *Alpinia officinarum* (PP, Wyk), *Ananas comusus* (DE, Wyk), *Artemisia absinthum* (BP, Wyk), *Brassica nigra* (PP, Wyk), ***Carica papaya* (DE, Bal1, Wyk)**, *Centaurea benedicta* (BP, Wyk), *Centaurium erythraea* (BP, Wyk), *Cichorium intybus* (BP, Wyk), *Cinchona pubescens* (BP, Wyk), *Citrus-aurantius aurantium* (BP, Wyk), *Citrus-reticulatus reticulata* (Bal2), *Ficus carica* (DE, Wyk), *Foeniculum vuglare* (Bal1, Bal2), *Gentiana lutea* (BP, Wyk), *Harungana madagascariensis* (BP, Wyk), *Hydrastis canadensis* (Bal1), *Iberis amara* (PP, Wyk), *Illicium anisatum* (Bal1), *Illicium verum* (Bal1), *Marrubium vulgare* (BP, Wyk), *Marsdenia cundurango* (BP, Wyk), *Matricaria chamomilla* (Bal1), *Medicago sativa* (Bal1), *Mentha piperita* (Bal1, Bal2), *Menyanthes trifoliata* (BP, Wyk), *Nepeta cataria* (Bal1), *Quassia amara* (BP, Wyk), *Sinapis alba* (PP, Wyk), *Trigonella foenum-graecum* (Bal1), *Tropaeolum majus* (PP, Wyk), *Ulmus rubra* (enema, Bal1), ***Zingiber officinale* (PP, Bal1, Bal2, Wyk)**
Herb Formulas: LIQUI-LIEVE *Coriandrum sativum* (leaf, 14mg), *Foeniculum vulgare* (seed, 16mg), *Fucus vesiculosus* (frond, 16mg), *Juniperis communis* (berry, 134mg), *Petroselinum crispum* (leaf, 108mg), *Rhamnus purshiana* (bark, 72mg), *Taraxacum officinale* (leaf & root, 200mg), *Trigonella foenum-graecum* (seed, 96mg) (Scal);
Vegetables: Parsley (Bal1), **Vegetables (grain):** Brown Rice, Buckwheat, Cereal, Corn, Millet, Oats, Quinoa, Rice Bran, Rye, Spelt, Tapioca, White Flour, White Rice, Wheat (Bal1);

Irritable Bowel SIB This is an inflammation of the lining of the colon. The mild form is **Irritable Bowel Syndrome**, the acute form is **Inflammatory Bowel Disease**. Irritable bowels, left untreated eventually get infected with bacteria and viruses that cause cancer.
TM = traditional medicine, MM = modern medicine
Herbs: *Achillea millefolium* (MM, Wyk), *Aloe vera* (Bal1), *Alpinia officinarum* (MM, Wyk), ***Althaea officinalis* (MM, Bal1, Wyk)**, *Angelica sinensis* (MM, Wyk), *Arctium lappa* (root, Bal1), *Atropa belladonna* (MM, Wyk), ***Calendula officinalis* (MM, Bal2, Wyk)**, *Carum carvi* (MM, Wyk), *Citrus aurantium* (TM, Wyk), *Coriandrum sativum* (MM, Wyk), *Crocus sativus* (TM, Wyk), *Ferula assa-foetida* (Bal2), *Foeniculum vulgare* (Bal2), ***Glycyrrhiza glabra* (MM, Bal1, Wyk)**, *Hydrastis canadensis* (Bal1), *Hypericum perforatum* (Bal2), *Juglans regia* (TM, Wyk), *Lavandula angustifolia* (TM, Wyk), *Levisticum officinale* (TM, Wyk), *Linum usitatissimum* (MM, Wyk), *Lobelia inflata* (Bal1), *Lythrum salicaria* (TM, Wyk), *Malva sylvestris* (MM, Wyk), ***Matricaria chamomilla* (MM, Bal1, Bal2, Wyk)**, *Medicago sativa* (Bal1, Bal2), ***Melissa officinalis* (MM, Bal2, Wyk)**, ***Mentha piperita* (MM, Bal1, Bal2, Murr, Wyk)**, *Mentzelia cordifolia* (MM, Wyk), *Plantago afra* (Bal2), *Rosa-hipa canina* (Bal1), *Rosmarinus officinalis* (Bal2), *Scutellaria baicalensis* (Bal1), *Scutellaria, barata* (Bal1), *Scutellaria, lateriflora* (Bal1), *Silybum marianum* (Bal1, Bal2), *Taraxacum officinale* (Bal2), *Tabebuia impetignosa, Thymus serpyllum* (TM, Wyk), *Trifolium pratense* (Bal1), *Trigonella foenum-graecum* (Bal1), *Ulmus rubra* (Bal1, Bal2), *Valeriana officinalis* (root, Bal1, Bal2), *Zingiber officinale* (Bal1, Bal2)
Herb Formulas: CALM THE STOMACH POWDER *Atractylodes macrocephala, Citrus-tangericus reticulata, Glycyrrhiza glabra, Magnolia officinalis* (bark) (Bal2); FOUR GENTLEMEN DECOCTION *Atractylodes macrocephala, Glycyrrhiza glabra, Panax ginseng, Wolfiporia extensa* (Bal2); MINOR CONSTRUCT MIDDLE DECOCTION *Hordeum vulgare* (malt), *Panax ginseng, Zingiber officinale* (Bal2); PINELLIA DECOCTION *Pinellia turnata, Perilla frutescens* (leaf), *Wolfiporia extensa, Zingiber officinale* (Bal2); ROBERT'S FORMULA *Geranium robertianum, Althaea officinalis* (root), *Baptisia tinctoria, Echinacea angustifolia, Hydrastis canadensis, Ulmus rubra* (Bal2);
Vegetables (fiber): Flaxseed, Oat Bran, Psyllium Husk (Bal1), **Vegetables (grain):** Brown Rice, Buckwheat, Cereal, Corn, Millet, Oats, Quinoa, Rice Bran, Rye, Spelt, Tapioca, White Flour, White Rice, Wheat (Bal1); **Vegetables (legume):** Chick Peas, Green Beans, Kidney Beans, Lentils, Lima Beans, Mung Beans, Peas, Pinto Beans, Soybeans, Red Beans, White Beans (Bal1);

Lactose Intolerance DLCI the inability to digest milk is caused by a lack of lactase, an enzyme manufactured in the small intestine that splits lactose into glucose and galactose. Symptoms include abdominal pain, diarrhea and gas.
Herbs: LactAid, a commercial formula that provides the enzyme lactase, is available in most pharmacies (Bal1, page 362, condsideration #5)
Vegetables (calcium): Blackstrap Molasses, Broccoli, Rhubarb, Soy Milk, Soy Cheese, Spinach (Bal1)

Morning Sickness nausea and vomiting during pregnancy which stimulates the production of estrogen. Estrogen thickens the lining of the uterus and stimulates the growth of blood vessels serving the placenta. About 60% of all pregnant women have morning sickness. The herbs listed below are intended to mask the symptoms but not inhibit the cause of morning sickness.
Herbs: *Matricaria chamomilla* (Bal2), *Rubus-raspabacca idaeus* (leaf, Bal1, Bal2), *Zingiber officinale* (Bal2)

Nausea a feeling of wanting to vomit.
Herbs: *Citrus-aurantius aurantium* (Bal2), *Frasera chirata* (Bal2), *Hypericum perforatum* (Bal2), *Zingiber officinale* (Bal2)

Peptic Ulcer SUL a spot in the lining of the stomach has eroded leaving an open wound inside the stomach (peptic ulcer). This can also occur in the gastrointestinal tract (gastric ulcer) and in the duodenum (duodenal ulcer) The bacteria *Heliobacter pylori* is associated with ulcer formation.
Herbs: ***Aloe vera* (Bal1, Bal2, Murr)**, *Althaea officinalis* (root, Bal1, Bal2), *Berberis vulgaris* (Bal2), *Cinnamomum aromaticum* (Bal2), *Citrus-aurantius aurantium* (Bal2), *Codonopsis pilosula* (Bal2), *Coptis chinensis* (Bal2), *Commiphora myrrha* (Bal1), ***Glycyrrhiza glabra* (Bal1, Bal2, Murr)**, *Humulus lupulus* (Bal1), *Hydrastis canadensis* (Bal1, Bal2), *Mahonia aquifolium* (root, Bal2), *Mahonia nervosa* (root, Bal2), *Mahonia repens* (root, Bal2), *Malva sylvestris* (Bal2), *Matricaria chamomilla* (Bal1, Bal2), *Medicago sativa* (Bal1), *Myrica cerifera* (Bal1), *Nepeta cateria* (Bal1), *Passiflora caerulea* (Bal1), *Passiflora incarnata* (Bal1), *Pinus pinaster* (bark extract, Murr), *Pistacia lentiscus* (Murr), ***Rheum rhabarbarum* (Bal2, Murr)**, *Salvia officinalis* (Bal1), *Syzygium aromaticum* (Bal2), *Ulmus rubra* (Bal1), *Uncaria guianensis* (Bal1, Bal2), *Uncaria tomentosa* (Bal1, Bal2), *Valeriana officinalis* (Bal1), *Vitis vinifera* (seed extract, Murr)
Herb Formulas: ASTRAGALUS DECOCTION *Astragalus membranaceus, Cinnamomum aromaticum, Glycyrrhiza glabra, Hordeum vulgare* (malt sugar), *Paeonia officinalis* (root), *Zingiber officinale, Ziziphus jujuba* (fruit) (Bal2); CALM THE STOMACH POWDER *Atractylodes macrocephala, Citrus-tangericus reticulata, Glycyrrhiza glabra, Magnolia officinalis* (bark) (Bal2); COPTIS DECOCTION *Coptis chinensis, Gardenia angusta, Phellodendron amurense, Scutellaria baicalensis, Scutellaria barata, Scutellaria lateriflora* (Bal2); FOUR GENTLEMEN DECOCTION *Atractylodes macrocephala, Glycyrrhiza glabra, Panax ginseng, Wolfiporia extensa* (Bal2); OPHIOPOGONIS DECOCTION *Ophiopogon japonicus* (bulb), *Glycyrrhiza glabra, Oryza sativa* (porridge), *Panax ginseng, Pinellia turnata, Ziziphus jujuba* (Bal2); WARM THE GALL BLADDER DECOCTION *Bambusa arundinacea, Citrus-aurantius aurantium, Glycyrrhiza glabra, Pinellia turnata, Wolfiporia extensa, Zingiber officinale* (Bal2);
Vegetables (soft): Avocados, Potatoes, Squash, Yams (Bal2); **Vegetables (dark green leaf):** Chard, Collard Greens, Cornsalad Lettuce, Kale, Mustard Greens, Parsley, Seakale, Spinach (Bal1); **Vegetables (grain):** Millet, White Rice (Bal2); **Vegetables (juice):** Alfalfa Juice, Barley Juice, Wheat Juice (Bal1);

Ulcerative Colitis SUL a chronic disorder in which mucus membranes lining the colon become inflamed and develop ulcers causing bloody diarrhea, pain, gas, bloating. Hard stools cause the mucus lining of the colon to bulge out into small pouch like projections called diverticuli.
Herbs: *Achillea millefolium* (Bal1), *Aloe vera* (Bal1), *Carica papaya* (Bal1), *Lobelia inflata* (enema, Bal1), *Matricaria chamomilla* (Bal1), *Medicago sativa* (Bal1), *Tanacetum parthenium* (Bal1), *Tabebuia impetignosa* (Bal1), *Taraxacum officinale* (Bal1), *Trifolium pratense* (Bal1), *Ulmus rubra* (Bal1)
Vegetables: Alfalfa Sprouts, Baby Food, Barley, Garlic, Lentils, Psyllium Husks (Bal1); **Vegetables (grain):** Brown Rice, Buckwheat, Cereal, Corn, Millet, Oats, Quinoa, Rice Bran, Rye, Spelt, Tapioca, White Flour, White Rice, Wheat (Bal1);

Vomiting the ejection of the stomach contents up through the throat and out of the mouth.
Herbs: *Codonopsis pilosula* (Bal2), *Foeniculum vulgare* (seed, Bal2), *Scutellaria baicalensis* (Bal2), *Scutellaria barata* (Bal2), *Sutellaria galericulata* (Bal2), *Scutellaria lateriflora* (Bal2), *Zingiber officinale* (Bal2)
Herb Formulas: FIVE INGREDIENT POWDER WITH PORIA *Alsima plantago-aquatica Atractylodes macrocephala Cinnamomum aromaticum, Polyporus brumalis, Polyporus squamosus, Polyporus tubaster, Polyporus umbellatus, Wolfiporia extensa,* (Bal2);

Worms DWRM Worms are intestinal parasites including Fluke, Hookworm, Pinworm, Roundworm, Tapeworm, Threadworm, and Whipworm. Most herbs that suppress intestinal parasites will also affect the host.
Herbs: *Artemisia annua* (Bal1), *Artemisia cina* (Bal1), *Artemisia vulgaris* (Bal1), *Chenopodium ambrosoides anthelminticum* (Bal1), *Chrysanthemum vulgare* (Bal1), *Cinchona succirubra* (Bal1), *Dryopteris filix-mas* (Bal1), *Carapichea ipicacuanha* (Bal1)

Ear Disorders

Ear Ache Infection DEAI also known as **Otitis Media**, bacteria infects the middle ear, three small bones behind the eardrum which causes ear pain, throbbing, chills, and fever when fluid builds up behind the eardum.
Herbs: *Allium sativum,* (external, oil, Bal1, Bal2), *Betula lenta* (xylitol extract, Murr), *Echinacea purpurea* (internal, extract, Bal1, Bal2), *Hydrastis canadensis* (internal, extract, Bal1, Bal2), *Hypericum perforatum* (external, oil, Bal2), *Olea europaea* (external, oil, Bal1, internal leaf extract, Bal2), *Pinellia turnata* (external, oil, Bal2), *Scutellaria biacalensis* (internal, extract, Bal2), *Scutellara barbata* (internal, extract, Bal2), *Scutellaria lateriflora* (internal, extract, Bal2), *Thea sinsensis-viridis* (internal, tea, Bal2), ***Verbascum thapsis* (external, oil, Bal2, Murr)**
Herb Formulas: GENTIAN LONGDANCAO DECOCTION *Gentiana lutea, Gentiana scabra, Akebia quinata, Akebia trifoliate, Alisma plantago-aquatica, Angelica sinensis, Bupleurum chinense, Bupleurum scorzoneraefolium, Gardenia angusta, Plantago afra, Rehmannia glutinosa, Scutellaria baicalensis, Scutellaria barata, Sutellaria galericulata, Scutellaria lateriflora* (Bal1); NEUROPATHIC EAR DROPS *Allium sativum* (oil, 0.05%), *Calendula officinals* (flowers, 28%), *Hypericum perforatum* (whole plant, 30%), *Lavandula officinalis* (oil, 5%), *Olea europaea* (oil, 10%), *Verbscum thapsis* (flowers, 25%), Vitamin E (1.95%) (Murr)
Vegetables: "Avoid the most common allergenic foods; wheat, dairy products, corn, oranges, peanut butter, and all simple carbohydrates, including sugar, fruits, and fruit juices" (Bal1, page 245, recommendation #1); "To help reduce and prevent the development of food allergies, do not repeat foods frequently." (Bal1, page 245, recommendation #2);
Vegetables: 'Because it is not usually possible to determine the exact allergen during acute otitis media, the most common allergenic foods should be eliminated from the diet: Wheat, Corn, Peanuts" (Murr, page 563, under Treatment Summary)

Hearing Loss DEAL occurs when the passage of sound waves into the ear and the conversion of sound waves to electrical energy is impaired reducing the amount of sound that the brain perceives. About 24 million American adults have hearing loss. About 30% of American adults over the age of 65 have hearing loss.
Herbs: *Achillea millefolium* (BCR, Bal1), *Allium sativum* (external oil ear drops, *Ananas comosus* (fresh pinapple, Bal1), *Arctium lappa* (BLC, root, Bal1), *Capsicum annuum-longum-cayenneum* (BCR, Bal1), *Commiphora myrrha* (BLC, gum, Bal1), *Curcuma longa* (BCR, Bal1), *Crataegus laevigata* (BLC, leaf & flower, Bal1), *Crataegus monogyna* (BLC, leaf & flower, Bal1), *Echinacea purpurea* (DABA, Bal1), *Ephedra distachya* (decongestant, Bal1), *Eucalyptus globulus* (decongestant, Bal1), *Ginkgo biloba* (BCR, Bal1), *Hydrastis canadensis* (BLC, Bal1), *Hyssopus officinalis* (decongestant, Bal1), *Matricaria chamomilla* (BCR, Bal1), *Myrica cerifera* (BLC, bark, Bal1), *Ruscus aculeatus* (BCR, Bal1), *Thymus vulgaris* (decongestant, Bal1), *Verbascum thapsis* (internal decongestant, external oil ear drops, Bal1), *Zingiber officinale* (BCR, Bal1)
Vegetables: Garlic, Kelp, Sea Vegetables (Bal1)

Meniere's DMNR also spelled **Ménière's,** impaired blood flow to the brain from clogged arteries resulting in dizziness, loss of balance, nausea and vomiting. In the semicircular canals of the inner ear, increased fluid retention causes pressure which disturbs hearing and balance.
Herbs: *Eleutherococcus senticosus* (Bal2), *Ginkgo biloba* (Bal1), *Glycyrrhiza glabra* (Bal2), *Hydrastis canadensis* (Bal2), *Ruscus aculeatus* (Bal1, Bal2), *Scutellaria baicalensis* (Bal2), *Scutellaria barata* (Bal2), *Scutellaria lateriflora* (Bal2)
Herb Formulas: BUPLEURUM & OYSTER SHELL *Bupleurum chinense, Ostrea edulis* (shell), *Bupleurum scorzoneraefolium, Cinnamomum aromaticum, Panax ginseng, Pinellia turnata, Reum rhabarbarum, Scutellaria baicalensis, Scutellaria barata, Scutellaria lateriflora, Zingiber officinale, Zizyphus jubata, Wolfiporia extensa* (Bal2); FIVE INGREDIENT POWDER WITH PORIA *Alsima plantago-aquatica Atractylodes macrocephala Cinnamomum aromaticum, Polyporus brumalis, Polyporus squamosus, Polyporus tubaster, Polyporus umbellatus, Wolfiporia extensa,* (Bal2); TRUE WARRIOR DECOCTION *Aconitum napellus, Atractylodes macrocephala, Paeonia officinalis, Zingiber officinale, Wolfiporia extensa* (Bal2);
Vegetables: Lecithin, Flavinoid Vegetables (Bal1)

Ringing in the Ears DEAT also known as **Tinnitis**, due to spontaneous wax buildup and eardrum edema, wax buildup and eardrum edema caused by the *tinnea* bacteria carried by the Tinnea moth, or the Tinnea virus or due to weakness in the liver and kidneys.
Herbs: *Achillea millifolium* (BCR, Bal1), *Arctium lappa* (BLC, root, Bal1), *Capsicum annuum-longum-cayenneum* (BCR, Bal1), *Commiphora myrrha* (BLC, gum, Bal1), *Cordyceps sinensis* (BCR, Bal2), *Crataegus laevigata* (BLC, leaf & flower, Bal1), *Crataegus monogyna* (BLC, leaf & flower, Bal1), *Curcuma longa* (BCR, Bal1), *Echinacea purpurea* (dizziness, Bal1), *Ephedra distachya* (decongestant, Bal1), *Eucalyptus globulus* (decongestant, Bal1), ***Ginkgo biloba* (BCR, Bal1, Bal2, Wyk)**, *Hydrastis canadensis* (BLC, Bal1), *Hyssopis officinalis* (decongestant, Bal1), *Matricaria chamomilla* (BCR, Bal1), *Myrica cerifera* (BLC, bark, Bal1), *Myrica gale* (BLC, bark, Bal1), *Ruscus aculeatus* (BCR, Bal1), *Thymus vulgaris*(decongestant, Bal1), *Verbascum phlomoides* (decongestant, Bal1), *Verbascum thapsis* (decongestant, Bal1), *Zingiber officinale* (BCR, Bal1)
Herbs: GENTIAN LONGDANCAO DECOCTION *Gentiana lutea, Gentiana scabra, Akebia quinata, Akebia trifoliate, Alisma plantago-aquatica, Angelica sinensis, Bupleurum chinense, Bupleurum scorzoneraefolium, Gardenia angusta, Plantago afra, Rehmannia glutinosa, Scutellaria baicalensis, Scutellaria barata, Sutellaria galericulata, Scutellaria lateriflora* (Bal1); SIX INGREDIENT REHMANNIA PILL *Rehmannia glutinosa, Alisma plantago-aquatica, Cornus officinalis, Dioscorea oppositifolia or villosa, Paeonia suffruticosa, Wolfiporia extensa* (Bal2);

Eye Disorders

Bags under the Eyes lack of sleep causes the skin under the eyes to loose elasticity.
Herbs: *Eleuthrococcus senticosus* (Bal2)
Vegetables: "Avoid salt, both table salt and that found in prepared foods", (Bal2, page 301, recommendation #1)

Bloodshot Eyes EYB the blood vessels on the surface of the eye become inflamed and congested with blood. They are caused by eye strain, fatigue, and improper diet.
Herbs: *Crataegus laevigata* (Bal2), *Crataegus monogyna* (Bal2), *Euphrasia nemorosa* (Bal2), *Rubus-raspabacca idaeus* (leaf, Bal1, Bal2), *Vaccinium-vortabacca myrtillus* (Bal2), *Verbascum thapsis* (Bal2)
Herb Formulas: AUGMENTED RAMBLING POWDER *Angelica sinensis, Atractylodes macrocephala, Bupleurum chinense, Gardenia angusta, Glycyrrhiza glabra, Paeonia lactiflora, Paeonia suffruticosa, Wolfiporia extensa* (Bal2); CIMICIFUGA & KUDZU DECOCTION *Cimicifuga racemosa, Pueraria lobata*

Blurred Vision EYV nearsightedness, farsightedness, and stigmatism result in fuzzy vision. A lack of the purple pigment rhodopsin can also cause fuzzy vision. A lack of adequate lighting can cause fuzzy vision.
Herbs: *Achillea millifolium* (Bal2), *Cimicifuga racemosa* (Bal2), *Rubus-raspabacca idaeus* (leaf, Bal1)
Herb Formulas: FOUR SUBSTANCE DECOCTION *Angelica sinensis, Ligusticum porteri, Paeonia lactiflora, Paeonia officinalis, Rehmannia glutinosa* (Bal2), GASTRODIA & UNCARIA DECOCTION *Gastrodia-asiatica elata, Uncaria gambir, Uncaria guianensis, Uncaria tomentosa, Eucommia ulmoides, Gardenia angusta, Phoradendron leucarpum (Synonym Loranthus leucarpus), Polygonum aviculare, Scutellaria baicalensis, Scutellaria barata, Scutellaria lateriflora, Wolfia extensa,* Sichuan Ox Knee Root (Bal2); VISION ENHANCEMENT *Astaxanthan* (2mg), *Lutein* (10mg), *Vaccinium-vortabacca myrtillis* (berry, 50mg) *Vitis vinifera* (seed, 20mg) (Scal)
Vegetables (potassium rich): Blackstrap Molasses, Brewer's Yeast, Brown Rice, Garlic, Kelp, Potatoes, Winter Squash, Wheat Bran, Yams (Bal2)

Cataracts DECT the lens of the eye thickens, becoming clouded or opaque. The most common form of cataracts is senile cataracts affecting people over the age of sixty five. This type of cataract is caused by free radical damage. With other types of cataracts free radicals are created by heavy metal poisoning, exposure to radiation, and the use of steroid drugs. Occasionally a cataract may swell causing secondary glaucoma.
Herbs: *Corydalis cava* (Bal2), *Corydalis yanhusuo* (Bal2), *Curcuma longa* (Bal2), *Gentiana lutea* (Bal2), *Gentiana scabra* (Bal2), *Ginkgo biloba* (Bal1), *Pinus pinaster* (pine bark extract, Murr), *Quercus velutina* (quercetin, Bal2), ***Vaccinium-vortabacca myrtillis* (Bal1, Bal2, Murr), *Vitis vinifera* (grape seed extract, Bal1, Bal2, Murr)**
Herb Formulas: EIGHT INGREDIENT REHMANNIA PILL *Rehmannia glutinosa, Alisma plantago-aquatica, Astragalus membranaceus, Cornus officinalis, Dioscorea oppositifolia or villosa, Paeonia suffrutecosa, Schisandra chinensis, Wolfiporia extensa* (Bal2)
Vegetables (legume amino acid): Chick Peas, Green Beans, Kidney Beans, Lentils, Lima Beans, Mung Beans, Peas, Pinto Beans, Soybeans, Red Beans, White Beans (Murr); **Vegetables (lutein & zeaxanthin):** Broccoli, Brussels Sprouts, Kale, Romaine Lettuce, Spinach, Turnip Greens (Bal2); **Vegetables (yellow anti-oxidant):** Bell Peppers (yellow), Corn, Oca Tuber, Squash (Murr)

Floaters are bits of debris floating within the vitrius humor of the eye, which is the jelly-like substance that fills the eyeball. The debris casts shadows over the retina which are seen as small black specks. Floaters tend to increase with age and are more noticeable when you are nearsighted.
Herbs: *Angelica sinensis* (Bal2), *Gingko biloba* (Bal2)

Glaucoma DEGC is characterized by increased inner eye pressure that can lead to reduced vision and eventually blindness.
TM = traditional medicine, MM = modern medicine
Herbs: *Aesculus hippocastanum* (TM, Wyk), *Allium cepa* (TM, Wyk), *Arnica montana* (TM, Wyk), *Atropa belladonna* (MM, Wyk), *Berberis vulgaris* (MM, Wyk), *Crataegus laevigata* (Bal2), *Crataegus monogyna* (Bal2), *Datura stramonium* (MM, Wyk), *Delphinium*

staphisagria (TM, Wyk), *Digitalis lanata* (TM, Wyk), ***Euphrasia nemorosa* (TM, Bal1, Bal2, Wyk)**, *Fagopyrum esculentum* (MM, rutin,Wyk), *Foeniculum vulgare* (TM, Wyk), ***Ginkgo biloba* (Bal1, Bal2, Murr)**, *Hamamelis virginiana* (TM, Wyk), ***Matricaria chamomilla* (TM, Bal1, Wyk)**, *Physostigma venenosum* (MM, Wyk), *Pilocarpus jaborandi* (MM, Wyk), *Pinus pinaster* (bark extract, Murr), *Prunus spinosa* (TM, Wyk), *Pueraria lobata* (Bal2), ***Rosa-hipa canina* (TM, Bal1, Wyk)**, *Rubus-raspabacca idaeus* (leaf, Bal1), *Ruta graveolens* (TM, Wyk), *Stellaria media* (Bal1), *Styphnolobium japonicum* (MM, Wyk), *Vaccinium-blauea angustifolium* (Bal1), *Vaccinium-blauea corymbosum* (Bal1), ***Vaccinium-vortabacca myrtillis* (Bal1, Bal2, Murr)**, *Vitis vinifera* (seed extract, Murr)
Herb Formulas: FIVE INGREDIENT POWDER WITH PORIA *Alsima plantago-aquatica Atractylodes macrocephala Cinnamomum aromaticum, Polyporus brumalis, Polyporus squamosus, Polyporus tubaster, Polyporus umbellatus, Wolfiporia extensa,* (Bal2);
Vegetables: "Avoid tobacco smoke, coffee, alcohol, nicotine, and all caffeine." (Bal1, page 290, recommendation #4)

Macular Degeneration DEMD the progressive destruction of the macula, the part of the retina responsible for fine vision.
Herbs: *Capsicum annuum-longum-cayenneum* (Bal1), *Euphrasia nemorosa* (Bal1), *Ginkgo biloba* (Bal2), *Myrica gale* (Bal1), *Pinus pinaster* (bark, Bal2), *Quercus velutina* (quercetin, Bal2), *Rubus-raspabacca idaeus* (Bal1), *Vaccinium-vortabacca myrtillis* (Bal1, Bal2), *Vitis vinifera* (seed, Bal2)
Vegetables (lycopene): Carrots, Green Peppers, Tomatoes (Murr); **Vegetables (zeaxanthin):** Corn, Paprika, Spinach (Murr); **Vegetables (lutein):** Carrots, Corn, Potatoes, Spinach, Tomatoes (Murr);

(1) **Wet Macular Degeneration** a degeneration of the macula, a part of the retina responsible for fine vision, due to hemorrhaging or leaking of fluid from a network of tiny blood vessels under the center of the retina. The hemorrhaging or leaking is caused by free radicals in the blood stream.

(2) **Dry Macular Degeneration** an atrophy of the rods and cones in the eye resulting in a loss in the sharpness of vision due to environmental toxins. A loss of curvature of the cornea resulting in a defocusing of the eye, due to aging.

Pink Eye OEY also known as **Conjunctivitis**, an inflammation of the conjunctiva, the membrane that lines the eyelid and most of the white of the eye. The eyes may appear swollen and bloodshot. Injury, bacteria, chlorine and other water borne chemicals can cause this eye disorder.
Herbs: *Berberis vulgaris* (Bal2), *Calendula officinalis* (Bal2), *Coptis chinensis* (Bal2), *Euphrasia nemorosa* (Bal1, Bal2), *Foeniculum vulgare* (Bal1, Bal2), *Hydrastis canadensis* (Bal1, Bal2), *Mahonia aquifolium* (root, Bal2), *Mahonia nervosa* (root, Bal2), *Mahonia repens* (root, Bal2), *Maricaria Recutita* (Bal1, Bal2)
Herb Formulas: COPTIS DECOCTION *Coptis chinensis, Gardenia angusta, Phellodendron amurense, Scutellaria baicalensis, Scutellaria barata, Scutellaria lateriflora* (Bal2); DONG QUAI & ALOE PILL *Angelica sinensis, Aloe vera* (Bal2); GENTIAN LONGDANCAO DECOCTION *Gentiana lutea, Gentiana scabra, Akebia quinata, Akebia trifoliate, Alisma plantago-aquatica, Angelica sinensis, Bupleurum chinense, Bupleurum scorzoneraefolium, Gardenia angusta, Plantago afra, Rehmannia glutinosa, Scutellaria baicalensis, Scutellaria barata, Sutellaria galericulata, Scutellaria lateriflora* (Bal1); MINOR BLUEGREEN DECOCTION *Asarum europaeum, Cinnamomum aromaticum* (bark), *Glycyrrhiza glabra, Paeonia lactiflora, Paeonia officinalis, Pinellia turnata, Schisandra chinensis, Zingiber officinale* (Bal2);

Gland Disorders

Adrenal Gland Impairment GAD The Adrenal glands are a pair of triangular –shaped organs that rest on top of the kidneys. Each gland is made up of two parts: the outer cortex and the inner medulla. The Adrenal gland outer cortex maintains the salt and water balance in the body, metabolizes carbohydrates, and regulates the blood sugar level. The Adrenal gland inner medulla produces adrenaline when the body is under stress.
IMSB = Immune System Booster
Herbs: *Astragalus membranaceus* (HAN, Bal1), *Echinacea purpurea* (IMSB, Bal1)*Eleutherococcus senticosus* (HAN, Bal1), *Silybum marianum* (BLC, Bal1)
Vegetables: Beans, Brown Rice, Garlic, Lettuce, Olive Oil, Onions, Peas, Pearl Barley (*Hordeum vulgare*), Shiitake Mushroom, Spinach (Bal1)

(1) **Addison's Disease DADD** is a condition where the Adrenal gland outer cortex is seriously underactive. It is a chronic disease that requires lifelong treatment. Poor nutritional habits, smoking, alcohol and drug abuse contribute to adrenal underactivity. Symptoms include fatigue, weakness, lethargy, dizziness, headaches, memory loss, and allergies.

(2) **Cushing's Syndrome DCSH** is a condition where the Adrenal gland outer cortex is seriously overactive. Long-term use of cortisone drugs causes the outer cortex to shrink and to be seriously overactive.

Dry Mouth GDM also known as **Xerostomia** is a condition in which the salivary glands fail to work properly. It is a side effect of more than 500 commonly used prescription and over-the-counter medications. Anti-depressants, anti-anxiety drugs, decongestants, high blood pressure medicines anti-diarrheals and muscle relaxants can cause dry mouth. Another cause of dry mouth is **Sjögren's Syndrome** which results from an attack on the salivary glands by an over active immune system.
Herbs: *Pilocarpus jaborandi* (pilocarpine, Bal2), *Polygonatum multiflorum* (Bal2), *Ulmus rubra* (Bal2), *Zanthoxylum americanum* (Bal2)
Herb Formulas: OPHIOPOGONIS DECOCTION *Ophiopogon japonicus* (bulb), *Glycyrrhiza glabra, Oryza sativa* (porridge), *Panax ginseng, Pinellia turnata, Ziziphus jujuba* (Bal2)
Vegetables: Wheat Germ (Bal2)
Fruits: Xylitol (the pentose sugar alcohol C5H12O5 occuring naturally in Plums, Raspberries and Strawberries, used as a sugar substitute, Bal2)

Hyperthyroidism DHET the thyroid gland produces too much thyroid hormone, resulting in an overactive metabolic state. Symptoms include being hot, insomnia, fatigue, frequent bowel movements, weakness, hair loss, hand tremors, rapid heartbeat and goiter. "Researchers in England studied 10 people who were being treated for Parkinson's Disease and found that all of them also had hyperthyroidism. Once the thyroid condition was treated the Parkinson's Disease improved dramatically." (Bal1, page 332, consideration #2)
Herbs: *Lycopus americanus* (Bal2), *Lycopus europaeus* (Bal2), *Leonurus cardiaca* (Bal2), *Leonurus heterophylla* (Bal2), *Leonurus sibricus* (Bal2)
Herb Formulas: BUPLEURUM & OYSTER SHELL *Bupleurum chinense, Ostrea edulis* (shell), *Bupleurum scorzoneraefolium, Cinnamomum aromaticum, Panax ginseng, Pinellia turnata, Reum rhabarbarum, Scutellaria baicalensis, Scutellaria barata, Scutellaria*

lateriflora, Zingiber officinale, Zizyphus jubata, Wolfiporia extensa (Bal2); EIGHT INGREDIENT REHMANNIA PILL *Rehmannia glutinosa, Alisma plantago-aquatica, Astragalus membranaceus, Cornus officinalis, Dioscorea oppositifolia or villosa, Paeonia suffrutecosa, Schisandra chinensis, Wolfiporia extensa* (Bal2); GENTIAN LONGDANCAO DECOCTION *Gentiana lutea, Gentiana scabra, Akebia quinata, Akebia trifoliate, Alisma plantago-aquatica, Angelica sinensis, Bupleurum chinense, Bupleurum scorzoneraefolium, Gardenia angusta, Plantago afra, Rehmannia glutinosa, Scutellaria baicalensis, Scutellaria barata, Sutellaria galericulata, Scutellaria lateriflora* (Bal2); HONEY FRIED LICORICE DECOCTION Vegetable Gelatin, *Cannabis sativa-indica, Cinnamomum aromaticum,Glycyrrhiza glabra, Ophiopogon japonicus* (bulb), *Panax ginseng, Rehmannia glutinosa, Zizyphus jujuba* (Bal2); MINOR BUPLEURUM DECOCTION *Bupleurum chinense, Bupleurum scorzoneraefolium, Glycyrrhiza glabra, Panax ginseng, Pinellia turnata, Scutellaria baicalensis, Scutellaria barata, Scutellaria lateriflora, Zingiber officinale, Ziziphus jujuba* (Bal2); OPHIOPOGONIS DECOCTION *Ophiopogon japonicus* (bulb), *Glycyrrhiza glabra, Oryza sativa* (porridge), *Panax ginseng, Pinellia turnata, Ziziphus jujuba* (Bal2)
Vegetables: Brewer's Yeast, Rutabaga, Soybeans, Turnips (Bal1); **Vegetables (crucifer):** Broccoli, Brussels Sprouts, Cabbage, Cauliflower (Bal1, Murr); **Vegetables (leaf):** Kale, Mustard Greens, Spinach (Bal1)

Hypoglycemia DHOG also known as **Low Blood Sugar**, an abnormally low level of glucose in the blood due to an oversecretion of insulin by the pancreas. Symptoms include fatigue, dizziness, headache, fainting spells, anxiety, cravings for sweets, night sweats, weakness in legs, swollen feet, tightness in chest and insomnia.
Herbs: *Dioscorea oppositifolia* (Bal1), *Dioscorea villosa* (Bal1), *Glycyrrhiza glabra* (Bal1), *Silybum marianum* (Bal1), *Taraxacum officinale* (root, Bal1), *Vaccinium-vortabacca myrtillus* (Bal1)
Vegetables (fiber): Beans, Broccoli, Brown Rice, Carrots, Jerusalem Arthichokes, Lentils, Potatoes, Raw Spinach, Squash, String Beans (Bal1); **Vegetables (legume):** Chick Peas, Green Beans, Kidney Beans, Lentils, Lima Beans, Mung Beans, Peas, Pinto Beans, Soybeans, Red Beans, White Beans (Murr);

Hypothyroidism DHOT the thyroid gland produces too little thyroid hormone, resulting in an underactive metabolic state. Symptoms include fatigue, loss of appetite, cold intolerance, slow heart rate, weight gain, infertility, muscle weakness, dry skin, constipation, anxiety, slow speech, goiter and druping swollen eyes.
Herbs: *Cimicifuga racemosa* (Bal1, Bal2), *Hydrastis canadensis* (Bal1), *Myrica cerifera* (Bal1), *Pinus pinaster* (bark extract, Murr), *Vitis vinifera* (seed extract, Murr)
Herb Formulas: FRIGID EXTREMITIES DECOCTION *Bupleurum chinense, Citrus aurantium, Glycyrrhiza glabra, Paeonia officinalis* (Bal2); KIDNEY CHI PILL *Aconitum napellus, Alisma plantago-aquatica, Cinnamomum aromaticum, Cornus officinalis, Dioscorea oppositifolia, Dioscorea villosa, Paeonia suffruticosa, Rehmannia glutinosa, Wolfiporia extensa* (Bal2); THYROID SUPPORT *Coleus forskolii* (root, 154mg), *Fucus vesiculosis* (frond, 56mg), *L-Tryosine* (300mg), *Nereocystis leuetkeana* (frond, 130mg), *Schisandra chinensis* (berry, 156mg), *Withania somnifera* (root, 120mg) (Scal); TRUE WARRIOR DECOCTION *Aconitum napellus, Atractylodes macrocephala, Paeonia officinalis, Zingiber officinale, Wolfiporia extensa* (Bal2);
Vegetables: Brussels Sprouts, Molasses, Parsley, Spinach, Turnips (Bal1); **Vegetables (crucifer):** Broccoli, Brussels Sprouts, Cabbage, Cauliflower, Kale, Mustard Greens (Bal1)

Lymphedema the lymphatic vessels are blocked, tissues retain fluid and are swollen.
Herbs: *Aesculus hippocastanum* (Bal2), *Ananas comosus* (bromelain, Bal2), *Pinus pinaster* (bark, Bal2), *Ruscus aculeatus* (Bal2), *Vitis vinifera* (seed, Bal2), *Zingiber officinale* (Bal2)

Obesity DOBE excess body fat caused by a lack of fat-burning metabolic hormones and proteins. About 40% of Americans are overweight. About 20% of Americans are obese. The herbs used for obesity can also be used for overweight.
Herbs: *Achillea millefolium* (Bal1), *Aloe vera* (juice, Bal1, Bal2), *Angelica dahurica* (Bal2), *Arctostaphylos uva-ursi* (Bal1), *Astragalus membranaceus* (Bal1, Bal2), *Avena sativa* (straw, Bal1), *Borago officinalis* (seed, Bal1), *Brassica nigra* (Bal1), *Capsicum-annuum-longum-cayenneum* (Bal1), *Cinnamomum aromaticum* (Bal1), *Cola acuminata* (Bal1), *Crataegus laevigata* (Bal1), *Crataegus monogyna* (Bal1), *Cyamopsis tetragonoloba* (Bal1), *Elettaria cardamomum* (Bal1), *Eleuthrococcus senticosus* (Bal1), *Ephedra distachya* (Bal1), *Equisetum arvense* (Bal1), *Eupatorium purpureum* (Bal1), *Foeniculum vulgare* (Bal1), *Fraxinus americana* (Bal1), *Fucus vesiculosis* (Bal1), *Garcinia cambogia* (Bal1, Bal2), *Glycyrrhiza glabra* (Bal1), *Hydrangea arborescens* (Bal1), *Hyssopis officinalis* (Bal1), *Ilex paraguensis* (Bal2), *Juniperus communis* (Bal1), *Medicago sativa* (Bal1), *Ruscus aculeatus* (Bal1), *Sinapis alba* (Bal1), *Smilax aristolochaefolia* (Bal1), *Smilax ferbifuga* (Bal1), *Smilax glyciphylla* (Bal1), *Smilax ornata* (Bal1), *Smilax regelii* (Bal1),*Taraxacum officinale* (Bal1, Bal2), *Thea sinensis-viridis* (Bal1), *Thymus vulgaris* (Bal1), *Trigonella foenum-graecum* (Bal1), *Zea mays-saccharata* (cornsilk, Bal1), *Zingiber officinale* (Bal1)
Herb Formulas: DIET-SLIM *Citrus-aurantius auranteum* (peel, 57mg), *Coleus forskohlii* (root, 138mg), *Corynanthe yohimbe* bark, 36mg), *Curcuma longa* (root, 90mg), *Fucus vesiculosus* (frond, 18mg), *Garcinia cambogia* (300mg), *Gymnema sylvestre* (leaf, 51mg), *Sambucus canadensis* (90mg), *Thea sinensis-viridis* (leaf, 150mg), *Zingiber officinale* (root extract, 12mg) (Scal); LEDEBOURIELLA DECOCTION *Ledebouriella divaricata, Angelica sinensis, Atractylodes macrocephala, Ephedra distachya,* Epsom Salt ($MgSO_4 \cdot 7H_2O$), *Forsythia suspensa, Gardenia angusta* (fruit), *Glycyrrhiza glabra,* Gypsum ($CaSO_4 \cdot 2H_2O$), *Ligusticum porteri, Mentha piperita, Paeonia officinalis, Platycodon grandiflorus, Rheum rhabarbarum* (wine treated root), *Schizonepeta tenuifolia, Scutellaria baicalensis, Scutellaria barata, Scutellaria lateriflora,* Talc [$Mg_3(SiO_4)_{10}(OH)_2$]; MAJOR BUPLEURUM DECOCTION: *Bupleurum chinense, Citrus-aurantius aurantium, Paeonia officinalis, Pinellia turnata, Rheum rhabarabarum, Scutellaria baicalensis, Scutellaria barata, Scutellaria lateriflora, Zingiber officinale, Ziziphus jujuba* (Bal2);
Vegetables: Baked Potatoes, Lentils, Sesame Seeds (Bal1); **Vegetables (grain):** Brown Rice, Buckwheat, Cereal, Corn, Millet, Oats, Quinoa, Rice Bran, Rye, Spelt, Tapioca, White Flour, White Rice, Wheat (Bal1); **Vegetables (legume):** Chick Peas, Green Beans, Kidney Beans, Lentils, Lima Beans, Mung Beans, Peas, Pinto Beans, Soybeans, Red Beans, White Beans (Bal1); **Vegetables (low-calorie):** Broccooli, Cabbage, Carrots, Cauliflower, Celery, Cucumbers, Kale, Lettuce, Onions, Radishes, Spinach, Turnips (Bal1)

Pancreatitis DPAN an obstruction of the pancreatic duct resulting from Gall Stones or a cancerous tumor. The pancreas is the gland which produces the hormones insulin and gulcagon which regulate blood sugar levels.
Herbs: *Cedrus atlantica* (Bal1), *Cedrus deodara* (Bal1), *Echinacea purpurea* (Bal1), *Gentiana lutea* (Bal1), *Gentiana scabra* (Bal1), *Glycyrrhiza glabra* (Bal1), *Hydrastis canadensis* (Bal1), *Taraxacum officinale* (Bal1)
Vegetables: "Eat a diet low in fat and sugar." (Bal1, page 419, recommendation #2)

Pellagra a gland malfunction that results in a Vitamin B deficiency.

Vegetables (vitamin B): Avocados, Brewer's Yeast, Broccoli, Collard Greens, Nuts, Peanut Butter, Potatoes, Sunflower Seeds, Tomatoes, Whole Grain Bread, Whole Grain Cereal (Bal1); **Vegetables (legume):** Chick Peas, Green Beans, Kidney Beans, Lentils, Lima Beans, Mung Beans, Peas, Pinto Beans, Soybeans, Red Beans, White Beans (Bal1);

Menstrual Disorders

General MNS Ben-Erik Van Wyk ISBN10:981-232-933-1, ISBN13:978-981-232-933-2, does not differentiate between the symptoms of abnormal menstruation. Therefore the herbs cited by Wyk are put into an all-inclusive **General** category. The purpose of the herb was put in parentheses when the purpose was available from other authors.
Herbs: *Achillia millefolium* (TM, Wyk), *Achemella vulgaris* (TM, Wyk), *Aloe vera* (TM, Wyk), *Aristolochia clematis* (TM, carcinogenic, Wyk), *Atropa belladonna* (TM, Wyk), *Brassica nigra* (TM, Wyk), *Chamaemelum nobile* (TM, Wyk), *Cheldonoium majus* (TM, Wyk), *Chrysanthemum vulgare* (TM, Wyk), *Cinchona pubescens* (TM, Wyk), *Claviceps purpurea* (TM, Wyk), *Cyclamen europaeum* (TM, Wyk), *Cytisus scoparius* (TM, Wyk), *Equisetum arvense* (TM, Wyk), *Erigeron canadensis* (TM, Wyk), *Hyoscyamus niger* (TM, Wyk), *Inula helenium* (TM, Wyk), *Juniperus communis* (TM, Wyk), *Lamium album* (TM, Wyk), *Matricaria recutitia* (TM, Wyk), *Melissa officinalis* (TM, Wyk), *Myristica fragrans* (TM, Wyk), *Petroselinum crispum* (TM, Wyk), *Polygonum aviculare* (TM, Wyk), *Polygonum hydropiper* (TM, Wyk), *Portentilla anserine* (TM, Wyk), *Pulsatilla vulgaris* (TM, Wyk), *Rhamnus frangula* (TM, Wyk), *Ricinus communis* (TM, Wyk), *Rosmarinus officinalis* (TM, Wyk), *Ruta graveolens* (TM, Wyk), *Senecio ovatus* (TM, carcinogenic, Wyk), *Senna alexandrina* (TM, Wyk), *Veratrum album* (TM, Wyk), *Viburnum prunifolium* (TM, Wyk), *Viscum album* (TM, Wyk)

(1) **Too Frequent Menstruation Cycle MCF** (Polymenorrhoea)
Herbs: *Morinda officinalis* (Noni Tropical Mulberry, Bal2)

(2) **Too Infrequent or Absent Menstruation Cycle MCI** (Amenorrhoea)
Herbs: *Salvia miltiorhiza* (Red Root Sage, Bal2)

(3) **Too Short Menstrual Period MPS** (Oligomenorrhoea)
Herbs: *Vitex agnus-castus* (TM, lengthens short periods, Bal2, Wyk)

(4) **Too Long Menstrual Period MPL**
Herbs: *Caulophyllum thalictroides* (Squaw Root Blue Cohosh, Bal2), *Cimicifuga racemosa* (TM, shortens long periods, Wyk)

(5) **Excessive Bleeding MEB** (Hypermenorrhoea, Menorrhagia)
Herbs: *Agrimonia eupatoria* (Bal2), *Capsella bursa-pastoris* (TM, stops excessive bleeding, Bal2, Wyk), *Cinnamomum verum* (TM, stops excessive bleeding, Bal2, Wyk), *Vitex agnus-castus* (Murr)

(6) **Insufficent Menstrual Bleeding MIB** (Hypomenorrhoea)

Climateric Menstrual Syn. CMS physical and emotional changes experienced when ovarian function diminishes in women over 45 prior to menopause. Physical changes include hot flushes, profuse perspiration, and insomnia. Emotional changes include depression.
HM = hormone mimetic, PS = psychosomatic, TM = traditional medicine
Herbs: *Cimicifuga racemosa* (HM, Wyk), *Eleutherococcus senticosus* (PS, Wyk), *Glycine max* (HM, Wyk), *Hypericum perforatum* (PS, Wyk), *Lycopus virginicus* (HM, Wyk), *Panax ginseng* (PS, Wyk), *Pulsatilla vulgaris* (TM, Wyk), *Rheum rhaponticum* (HM, Wyk), *Sanguinaria canadensis* (TM, Wyk), *Strychnos nux-vomica* (TM, Wyk), *Trifolium pratense* (HM, Wyk), *Viburnum prunifolium* (TM, Wyk) *Vitex agnus-castus* (HM, Wyk)

Estrogen Deficiency MES a symptom of menopause where the ovaries greatly reduced their estrogen production increasing the risk of osteoporosis and heart disease. Since the ovaries form estrogen from low-density lipoprotein (bad LDL cholesterol), the risk of high cholesterol also increases.
Herbs: *Lepedium mayenii* (Maca Peppergrass, Murr), *Trifolium pratense* (Red Clover, Bal2)
Herb Formulas: ESTRO-G: *Angelica gigas, Cynanchum wilfordii, Phlomis umbrosa* (Murr); PHYTO-ESTROGEN *Avena sativa* (seed, 21mg), *Cimicifuga racemosa* (rhizome, 27mg), *Hypericum perforatum* (flower bud, 12mg), *Medicago sativa* (leaf, 300mg), *Salvia officinalis* (12mg), *Taraxacum officinale* (leaf & root, 12mg), *Trifolium pratense* (blossom, 21mg), *Verbena hastata* (herb, 18mg), *Vitex agnus-castus* (berry, 90mg) (Scal)

Hot Flashes MHF a temporary sensation of heat experienced during menopause. Menopause usually starts at age 45 and is complete by age 55. However, hot flashes can continue to occur up to age 60.
Herbs: *Angelica sinensis* (Dong Quai Angelic, Bal1, Bal2), *Caulophyllum thalictroides* (Squaw Root Blue Cohosh, Bal2), *Centella asiatica* (Gotu Kola Hundredleaf, Bal1), *Cimicifuga racemosa* (Murr), *Hypericum perforatum* (Perforate St. Johns Wort, Murr), *Panax ginseng* (Chinese Ginseng, Bal2), *Trifolium pratense* (Red Clover, Murr), *Turnera diffusa* (Damiana Alder Hazel, Bal2), *Vitex agnus-castus* (Monk's Pepper Chasteberry, Bal2)
Herb Formulas: : CINNAMON TWIG & PORIA PILL *Cinnamomum aromaticum, Paeonia officinalis, Paeonia suffruticosa, Prunus-Persica persica x communis* (seed), *Wolfiporia extensa* (Bal2); DONG QUAI & PEONY POWDER *Angelica sinensis, Paeonia lactiflora, Paeonia officinalis, Alisma plantago-aquatica, Atractylodes macrocephala, Ligusticum porteri, Wolfiporia extensa* (Bal2);

Hysterectomy MHY surgical removal of the uterus. Conditions for which a hysterectomys is performed are: Fibroids 35%, Endometriosis 20%, Prolapse of the Uterus 20%, Other 25%. Symptoms include diminished sexual interest, anxiety, lack of estrogen. Over 600,000 hysterectomies are performed in the United States each year.
Herbs (estrogen): *Angelica sinensis* (Bal1), *Dioscorea oppositifolia* (Bal1), *Dioscorea villosa* (Bal1), *Foeniculum vulgare* (Bal1), *Glycyrrhiza glabra* (Bal1), *Illicium anisatum* (Bal1), *Illicium verum* (Bal1), *Panax ginseng* (Bal1), *Pfaffia paniculata* (Bal1), *Salvia officinalis* (Bal1), *Trifolium pratense* (Bal1), *Trigonella foenum-graecum* (Bal1)
Vegetables (fiber): Beans, Broccoli, Carrots, Jerusalem Arthichokes, Lentils, Potatoes, Raw Spinach, Squash, String Beans (Bal1); **Vegetables (grain):** Brown Rice, Buckwheat, Cereal, Corn, Millet, Oats, Quinoa, Rice Bran, Rye, Spelt, Tapioca, White Flour, White Rice, Wheat (Bal1);

Mastitis NMT also known as **Breast Pain, Mastodynia** caused by an infection of the breast contracted during breast feeding. **Mastodynia** occurs shortly before the onset of menstrual bleeding and is sometimes considered a symptom of Premenstrual Syndrome (PMS)
Herbs (nursing): *Aloe vera* (external, Bal2), *Althaea officinalis* (root, Bal2), *Bellis perennis* (TM, Wyk), *Calendula officinalis* (external, Bal2), *Coix lacryma-jobi* (Bal2), *Conium maculatum* (TM, Wyk), *Humulus lupulus* (TM, Wyk), *Hypericum perforatum* (external,

oil, Bal2), *Juglans regia* (TM, Wyk), *Phytolacca americana* (TM, Wyk), *Salvia officinalis* (TM, Wyk), *Schizonepeta tenuifolia* (external, Bal2), *Trigonella foenum-graecum* (Bal2), *Viola tricolor* (Bal2)
Herbs (not nursing): *Berberis vulgaris* (Bal2), *Coptis chinensis* (Bal2), *Hydrastis canadensis* (Bal2), *Mahonia aquifolium* (root, Bal2), *Mahonia nervosa* (root, Bal2), *Mahonia repens* (root, Bal2), *Matricaria chamomilla* (Krug), *Melaleuca alternifolia* (external, oil, Bal2)

Menopause **See Estrogen Deficiency, See Hot Flashes, See Vaginal Dryness.** The Kupperman Menopause Index (KMI) lists 11 symptoms for menopause.
Herb Formulas: DONG QUAI MENOPAUSE DECOCTION *Angelica sinensis, Alisma orientalis, Atractylodes chinensis, Ligusticum monnieri, Paeonia lactiflora, Sclerotium poriae* (Murr); ESTRO-G: *Angelica gigas, Cynanchum wilfordii, Phlomis umbrosa* (Murr)

Menstrual Pain NMP also known as **Dysmenorrhoea** Anti-dsymenorrhoics reduce menstrual pain and are often anti-spasmodic, analgesic, and astringent.
Herbs: *Amaranthus hypochondriacus* (Bal2), *Angelica sinensis* (Bal2)
Herb Formulas: AUGMENTED RAMBLING POWDER *Angelica sinensis, Atractylodes macrocephala, Bupleurum chinense, Gardenia angusta, Glycyrrhiza glabra, Paeonia lactiflora, Paeonia suffruticosa, Wolfiporia extensa* (Bal2); CINNAMON TWIG & PORIA PILL *Cinnamomum aromaticum, Paeonia officinalis, Paeonia suffruticosa, Prunus-Persica persica x communis* (seed), *Wolfiporia extensa* (Bal2);

Premenstrual Syndrome DPMS physical and emotional changes experienced several days before the onset of menstrual flow. Physical changes include abdominal pain, bloating, breast soreness, headache, insomnia, water retention,. Emotional changes include anxiety, depression, irritability, mood swings.hormonal imbalance, i.e. excessive levels of estrogen and inadequate levels of progesterone. Symptoms are bloating, acne, anxiety, backache, breast swelling, cramps, food cravings, fainting, fatigue, headache, insomnia and joint pain. Approximately 75% of women experience this disease.
Herbs: *Ananas comosus* (bromelain, Bal2), *Angelica sinensis* (Bal1, Bal2), *Angelica dahurica* (Bal1), *Asarum sieboldii* (Bal2), *Betula alba* (leaf, Bal2), *Betula nigra* (leaf, Bal2), *Betula lenta* (leaf, Bal2), *Betula pendula* (leaf, Bal2), *Centauria benedicta* (Bal1), *Chamaelirium luteum* (Bal1), *Cimicifuga racemosa* (Bal1, Bal2), *Crataegus laevigata* (Bal1), *Crataegus monogyna* (Bal1), *Crocus sativus* (Murr), *Dioscorea oppositifolia* (Bal1, Bal2), *Discoria villosa* (Bal1, Bal2), *Foeniculum vulgare* (seed, Bal1), *Fragaria vesca* (leaf, Bal1), *Ginkgo biloba* (Murr), *Hypericum perforatum* (Murr), *Leonurus heterophyllus* (Bal2), *Matricaria chamomilla* (Bal2), *Mentha piperita* (Bal1, Bal2), *Michella repens* (Bal1), *Piper methysticum* (Bal1), *Proboscidea louisianica* (unicorn root, Bal2), *Rosmarinus officinalis* (Bal1), *Rubus-raspabacca idaeus* (leaf, Bal1, Bal2), *Silybum marianum* (Bal1), *Smilax aristolochaefolia* (Bal1), *Smilax ferbifuga* (Bal1), *Smilax glyciphylla* (Bal1), *Smilax ornata* (Bal1), *Smilax regelii* (Bal1), *Tabebuia impetignosa* (Bal1), *Tanacetum parthenium* (Bal1), *Taraxacum officinale* (Bal1, Bal2), *Valeriana officinalis* (Bal1), *Viburnum-drysia prunifolium* (Bal1, Bal2), *Viburnum-scrubia opulus* (Bal1, Bal2), ***Vitex agnus-castus* (Bal2, Murr)**, *Zea mays-saccharata* (cornsilk, Bal1, Bal2)
Herb Formulas: AUGMENTED RAMBLING POWDER *Angelica sinensis, Atractylodes macrocephala, Bupleurum chinense, Gardenia angusta, Glycyrrhiza glabra, Paeonia lactiflora, Paeonia suffruticosa, Wolfiporia extensa* (Bal2); BUPLEURUM & CINNAMON TWIG DECOCTION *Bupleurum chinense, Cinnamomum aromaticum, Glycyrrhiza glabra, Paeonia officinalis, Panax ginseng, Pinellia turnata, Scutellaria baicalensis, Scutellaria barata, Scutellaria lateriflora, Zingiber officinale, Ziziphys jujuba* (Bal2); PMS DAY 1-14 *Angelica sinensis* (root, 164mg), *Cimicifuga racemosa* (root, 30mg), *Glycyrrhiza glabra* or *uralensis* (root, 20mg), *Mitchella repens* (berry, 16mg), *Serenoa repens* (berry, 20mg), *Taraxacum officinale* (leaf & root, 14mg), *Zingiber officinale* (rhizome extract, 8mg) (Scal); PMS DAY 14-28 *Angelica sinensis* (root, 54mg), *Cimicifuga racemosa* (rhizome, 30mg), *Dioscorea villosa* (tuber, 24mg), *Glycyrrhiza glabra* or *uralensis* (root, 30mg), *Hypericum perforatum* (flower bud, 24mg), *Leonurus cardiaca* (herb, 24mg), *Piper methysticum* (rhizome, 42mg), *Vitex agnus-castus* (berry, 90mg), *Zingiber officinale* (rhizome extract, 12mg) (Scal);
Vegetables: Breads, Cereals, Nuts, Seeds (Bal1); **Vegetables (legume):** Chick Peas, Green Beans, Kidney Beans, Lentils, Lima Beans, Mung Beans, Peas, Pinto Beans, Soybeans, Red Beans, White Beans (Bal1);

Uterine Prolapse the muscles that support the uterus no longer function. A portion of the uterus descends into the top of the vagina. In extreme cases, the uterus may protrude through the vaginal opening.
Herbs: *Viburnum-bruscia edule* (Bal1), *Viburnum-scrubia opulus* (Bal1)
Herb Formulas: HOCHU EKKI TO *Angelica sinensis, Astragalus membranacea, Atractylodes macrocephala, Bupleurum chinense, Cimicifuga racemosa, Citrus-aurantius aurantium, Glycyrrhiza glabra, Panax ginseng* (Bal2);
Vegetables (grain): Brown Rice, Buckwheat, Cereal, Corn, Millet, Oats, Quinoa, Rice Bran, Rye, Spelt, Tapioca, White Flour, White Rice, Wheat (Bal1);

Vaginal Dryness a result of menopause.
Herbs: *Angelica sinensis* (Bal1), *Aloe vera* (external, Bal1, Bal2), *Centella asiatica* (Bal1), *Cimicifuga racemosa* (Bal2), *Ulmus rubra* (external, Bal1)

Vaginitis an infection of the vaginal tract. Symptoms include increased volume of vaginal secretions, abnormal color of vaginal secretions, abnormal consistency of vaginal secretions, abnormal odor of vaginal secretions, vaginal itching, painful urination, and painful intercourse.

(1) **Bacterial Vaginitis DVGB** also known as **Non-Specific Vaginitis**, a shift in vaginal flora from a predominace of the bacteria *Lactobacillus acidopholis* to a type of bacteria, usually *Gardnerella vaginalis*, that degrades the mucins forming a natural barrier on the vaginal lining.
Herbs: *Bifidobacterium adolescentis* (internal probotic, Murr), *Bifidobacterium bifidum* (internal probotic, Murr), *Lactobacillus acidopholis* (internal probiotic, and external douche, Murr), *Pinus pinaster* (bark extract, Murr), *Vitis vinifera* (seed extract, Murr)

(2) **Chlamydia Vaginitis DSXC** Sexually transmitted *Chlamydia trachomatis*, a bacteria parasite that lives within human cells, is the most common sexually transmitted disease with 4 million new cases each year. Symptoms include genital inflammation, difficulty in urinating, painful intercourse, and itching. It affects 10% of sexually active females in the United States. In females 50,000 are rendered sterile each year.
Herbs: *Astragalus membranaceus* (Bal1, Bal2), *Echinacea purpurea* (Bal1), *Hydrastis canadensis* (Bal1, Bal2), *Tabebuia impetignosa* (Bal1), *Trifolium pratense* (Bal1)
Vegetables: Brown Rice, Fresh Vegetables, Raw Seeds (Bal1); **Vegetables (grain):** Brown Rice, Buckwheat, Cereal, Corn, Millet, Oats, Quinoa, Rice Bran, Rye, Spelt, Tapioca, White Flour, White Rice, Wheat (Bal1);

(3) **Fungal Vaginitis DVGF**, also known as **Vaginal Thrush, Yeast Infection**, a fungus *Candida albicans* overgrows in the vagina. Symptoms include a white cheesy discharge, intense itching, and burning.
Herbs: *Achillea millefolium* (Bal1), *Aloe vera* (juice, Bal1, Bal2), *Allium sativum* (allicin, Bal2), *Althaea officinalis* (root, Bal1), *Berberis vulgaris* (Bal1, Bal2), *Calendula officinalis* (Bal1), *Cinnamomum aromaticum* (internal, oil, Bal1, Bal2), *Coptis chinensis* (Ba2), *Echinacea purpurea* (Bal1, Bal2), *Hydrastis canadensis* (Bal1), *Mahonia aquifolium* (root, Bal2), *Mahonia nervosa* (root, Bal2), *Mahonia repens* (root, Bal2), ***Matricaria chamomilla* (Bal1, Krug)**, *Melaleuca alternifolia* (Bal1), *Tabebuia impetignosa* (Bal1), *Taraxacum officinale* (Bal1), *Usnea barbata* (Bal1)

(4) **Gonorrhea Vaginitis** DSXG a sexual disease caused by the bacteria *Neisseria gohorrhea*. Symptoms include painful discharge, bloody discharge, and pus filled discharge.
Herbs: *Astragalus membranaceus* (Bal2), *Berberis vulgaris* (Bal2), *Carapa guianensis* (Bal2), *Copiafera langsdorffii* (Bal2), *Coptis chinensis* (Bal2), *Epimedium brevicornum* (Bal2), *Epimedium grandiflorum* (Bal2), *Hydrastis canadensis* (Bal2), *Mahonia aquifolium* (root, Bal2), *Mahonia nervosa* (root, Bal2), *Mahonia repens* (root, Bal2), *Scutellaria baicalensis* (Bal2), *Scutellaria barata* (Bal2), *Scutellaria lateriflora* (Bal2), *Terminalia arjuna* (Bal2)
Herb Formulas: DONG QUAI & ALOE PILL *Angelica sinensis*, *Aloe vera* (Bal2); GENTIAN LONGDANCAO DECOCTION *Gentiana lutea, Gentiana scabra, Akebia quinata, Akebia trifoliate, Alisma plantago-aquatica, Angelica sinensis, Bupleurum chinense, Bupleurum scorzoneraefolium, Gardenia angusta, Plantago afra, Rehmannia glutinosa, Scutellaria baicalensis, Scutellaria barata, Sutellaria galericulata, Scutellaria lateriflora* (Bal2); EIGHT INGREDIENT REHMANNIA PILL *Rehmannia glutinosa, Alisma plantago-aquatica, Astragalus membranaceus, Cornus officinalis, Dioscorea oppositifolia or villosa, Paeonia suffruticosa, Schizandra chinensis, Wolfiporia extensa* (Bal2); PEONY & LICORICE DECOCTION *Paeonia officinalis, Glycyrrhiza glabra* (Bal2);

(5) **Herpes Vaginitis DSXH** also known as **Genital Herpes**. More than 30 million Americans have Herpes Simples Virus #2 (HSV-2). *Homosapiensvirus herpes-simplex-beta* is the most prevalent sexually transmitted disease in the United States. HSV-2 can cause inflammation of the liver with fever. Genital Herpes causes painful fluid filled blisters on the cervix or labia that can take up to three weeks to heal.
Herbs: *Aloe vera* (DAVR, external, Bal2), *Astragalus membranaceus* (DABI, Bal2), *Capsicum annuum-longum-cayenneum* (NSD, external, Bal1, Bal2), *Carica papaya* (NSD, external, Bal2), *Commipora myrrha* (Bal1), *Copiafera langsdorffii* (external, oil Bal2), *Echinacea purpurea* (Bal1), *Glycyrrhiza glabra* (DHSV, root, Bal1, Bal2), *Hydrastis canadensis* (internal tea, external extract, Bal1), *Hypericum perforatum* (external, Bal1, Bal2), *Juglans nigra* (external, Bal1), *Melaleuca alternifolia* (external, oil, Bal1), *Momordica charantia* (DAVR, Bal2), *Olea europaea* (DAVR, leaf, Bal2), *Syzygium aromaticm* (Bal2), *Thea sinensis* (external, Bal2), *Trifolium pratense* (Bal1), *Uncaria guianensis* (DAVR, Bal2), *Uncaria tomentosa* (DAVR, Bal2)
Vegetables: "Avoid alcohol, processed foods, colas, white flour products, sugar, refined carbohydrates, coffee and drugs to lessen the change of an outbreak." (Bal1, page 319, recommendation #1)

(6) **Leucorrhea Vaginitis DLKR** caused by the single-cell whip animalcule microbe *Trichomonas vaginalis*, *Monilia brunnea, carbonaria, laxa, polystroma, pruinosa*. The most frequent symptom is vaginal discharge with itching and burning. The discharge is frequently smelly, greenish yellow and frothy. This disease is common in women who are pregnant or who have diabetes.
Herbs: *Melaleuca alternifolia* (Bal1), *Tabebuia impetignosa* (Bal1)
Vegetables (grain, gluten free): Brown Rice, Millet (Bal1);

Nervous System Disorders

Arthritis NJA, OJA the inflammation of a joint characterized by pain, swelling, stiffness, deformity and diminished range of motion.

(1) **Ankylosing Spondylitis** has symptoms of spinal pain, stiffness, inflammation and changes in posture. It occurs most often in young men.

(2) **Gout NGT, OGT** (ointment) joint arthritis caused by too much uric acid in the blood, tissues, and urine. When it crystallizes uric acid takes the shape of a needle and jabs its way into the joints. The symptom is acute joint pain.
Herbs: *Alchornea cordifolia* (Bal2), *Ananas comosus* (bromelain, Bal2), ***Apium graveolens-dulce* (seed, Bal2, Murr)**, *Arctium lappa* (Bal1), *Betula pendula* (Bal1), *Capsicum annuum-longum-cayenneum* (internal & external, Bal1), ***Colchicum autumnale* (Bal1, Wyk)**, *Gaultheria procombens* (external, oil, Bal1), *Hyssopis officinalis* (Bal1), *Juniperus communis* (Bal1), *Medicago sativa* (Bal1), *Pinus pinaster* (bark, Murr), *Quercus velutinus* (quercetin, Bal2), *Smilax aristolochaefolia* (Bal2), *Smilax ferbifuga* (Bal2), *Smilax glyciphylla* (Bal2), *Smilax ornata* (Bal2), *Smilax regelii* (Bal2), *Uncaria guianensis* (Bal2), *Uncaria tomentosa* (Bal2), *Vaccinium-vortabacca myrtillis* (Bal2), *Vitis vinifera* (seed, Murr)
Vegetables: Celery Juice), Coffee, Nuts, Seeds, Raw Vegetables (Bal1); **Vegetables (grain):** Brown Rice, Buckwheat, Cereal, Corn, Millet, Oats, Quinoa, Rice Bran, Rye, Spelt, Tapioca, White Flour, White Rice, Wheat (Bal1);

(3) **Kawasaki Syndrome** has symptoms of fever, joint pain, rash on palms and soles, and heart complications.

(4) **Infectious Arthritis** caused by *Streptococcus, Staphylococcus, Gonococcus, Hemophilus,* and tubercule bacilli bacteria, several different viruses, and *Candida albicans* fungus. The infecting microbe travels through the bloodstream to the joint from the site of the original infection elsewhere in the body. Hence the term Infectious arthritis. Symptoms include redness, swelling, pain, tenderness, fever, chills, & body aches.

(5) **Osteoarthritis** the deterioration of the cartilage that covers the ends of the bones at joints. The sliding surfaces of the bone become pitted and irregular. The tendons and ligaments that hold the bone together become weakened & the joint itself becomes deformed. Bone spurs called osteophytes tend to develop. Osteoarthritis tends to stay localized in one joint of the body. Symptoms are stiffness and pain as a result of joint motion as well as a grating sensation and popping, clicking & banging noises.
Herbs: *Agathosma betulina* (Bal1), *Apium graveolens-dulce* (Bal1), *Arctium lappa* (Bal1), ***Boswellia serrata* (Bal2, Murr)**, *Capsicum annuum-longum-cayenneum* (Bal1, Bal2), *Copiafera langsdorfsi* (Bal2), *Crataegus laevigata* (Bal2), *Crataegus monogyna* (Bal2), ***Curcuma longa* (Bal2, Murr)**, *Ephedra trifurca* (Bal1), *Equisetum arvense* (Bal1), *Gaultheria procumbens* (Bal1), ***Harpagophytum procumbens* (Bal1, Bal2, Murr)**, *Hoffmanseggia divaricata* (Bal2), *Ligusticum porteri* (Bal2), *Medicago sativa* (Bal1, Bal2), *Petroselinum*

crispum (Bal1), *Pinus pinaster* (pine bark extract, Murr), *Salix alba* (bark, Bal2), *Salix discolor* (bark, Bal2), *Salix myricoides* (bark, Bal2), *Tanacetum parthenium* (Bal1, Bal2), *Uncaria guianensis* (Bal1, Bal2), *Uncaria tomentosa* (Bal1, Bal2), *Urtica dioica* (Bal1), *Vitis vinifera* (grape seed extract, Murr), *Withania somnifera* (Bal2), *Yucca aloeifolia* (Bal1, Bal2), *Zea mays-suavis* (Bal1), ***Zingiber officinale* (Bal1, Bal2, Murr)**
Herb Formulas: TRUE WARRIOR DECOCTION *Aconitum napellus, Atractylodes macrocephala, Paeonia officinalis, Zingiber officinale, Wolfiporia extensa* (Bal2);
Vegetables (condiment): Garlic, **Vegetables (grain):** Oat Bran, Rice, Rice Bran, Rye, Wheat, **Vegetables (leaf):** Chard, Kale, Lettuce, Spinach, Turnip Greens, **Vegetables (stalk):** Asparagus, **Vegetables (root):** Onions

(6) **Psoriatic Arthritis** has symptoms of spinal pain, stiffness, inflammation and changes in posture. It is accompanied by scaly patches of inflamed skin.

(7) **Reiter's Syndrome** has symptoms of spinal pain, stiffness, inflammation, and changes in posture. It occurs most often in middle aged men and is accompanied by urethritis and conjunctivitis.

(8) **Rheumatism NRH, ORH** Also known as **Rheumatoid Arthritis**. The immune system incorrectly identifies the synovial membranes of a joint that secret lubricating fluid as a hostile invading organism. The immune system destroys the synovial membranes and the cartilage and tissues around the joints become damaged or are destroyed. Scar tissue replaces the destroyed synovial membrane, cartilage & tissue & the spaces between the joints becomes narrower. The bones on either side of a joint eventual fuse together. Rheumatoid Arthritis occurs all over the body rather than in just one localized joint. Symptoms are stiffness, swelling, fatigue, anemia, weight loss, fever, and crippling pain as well as crinkling cellophane noise. Inflammation is reduced by inhibiting the enzymes of prostaglandin, and leucotriene formation. Another approach uses cortico-mimetic substances which induce cortisone formation and block inhibit phospholipase A2 production.
CM = Cortico Mimetic, ES = Enzyme Source, ISB = Immune System Booster, LEI = leucotrine enzyme inhibitor, PEI = prostaglandin enzyme inhibitor
Herbs: *Aconitum napellus* (Wyk), ***Ananas comosus* (ES, bromelain, Bal2, Murr, Wyk)**, *Arnica montana* (LEI, PEI, Wyk), ***Boswellia sacra* (PEI, Bal2, Wyk)**, *Brassica nigra* (LEI, Wyk), *Bryonia dioca* (LEI, Wyk), *Bupleurum chinense* (Bal2), *Bupleurum scorzoneraefolium* (Bal2), *Calendula officinalis* (CM, Wyk), ***Capsicum annuum-longum-cayenneum* (LEI, Bal2, Wyk)**, *Capsicum annuum-frutescens* (LEI, Wyk), *Cardiospermum halicacabum* (Wyk), *Carica papaya* (ES, Wyk), *Cinnamomum camphora* (LEI, Wyk), *Cinnamomum verum* (LEI, Wyk), *Commiphora mukul* (Wyk), *Croton tiglium* (LEI, Wyk), ***Curcuma longa* (ISB, Bal2, Murr)**, *Echinachea angustifolia* (ISB, Wyk), *Eucalyptus globulus* (LEI, Wyk), *Ficus carica* (ES, Wyk), *Filipendula ulmaria* (PEI, Wyk), *Gaultheria procumbens* (PEI, Wyk), *Gelsemium sempervirens* (Wyk), *Glycyrrhiza glabra* (CM, Wyk), *Gnaphalium polycephalum* (Wyk), *Harpagophytum procumbens* (PEI, Wyk), *Juniperus communis* (LEI, Wyk), *Lachnanthes tinctoria* (Wyk), *Matricaria chamomilla* (PEI, Wyk), *Phytolacca americana* (CM, Wyk), *Pinus mugo* (LEI, Wyk), *Pinus pinaster* (bark extract, Murr), *Pinus sylvestris* (LEI, Wyk), *Populus tremula* (PEI, Wyk), *Primula eliator* (PEI, Wyk), *Primula vera* (PEI, Wyk), *Rhododendron tomentosum* (LEI, Wyk), *Rosmarinus officinalis* (LEI, Wyk), *Ruta graveolens* (Wyk), *Salix alba* (PEI, Wyk), *Salix purpurea* (PEI, Wyk), *Similax regelii* (CM, Wyk), *Solanum dulcamara* (CM, Wyk), *Solidago virgaurea* (PEI, Wyk), *Spilanthes acmella* (LEI, Wyk), *Symphytum officinale* (Wyk), *Syzygium aromaticum* (LEI, Wyk), ***Tanacetum parthenium* (PEI, Bal2, Wyk)**, *Taraxaxum officinale* (LEI, Wyk), *Thymus vulgaris* (LEI, Wyk), *Uncaria guianensis* (NSD, Bal2), *Uncaria tomentosa* (NSD, Bal2), *Urtica dioica* (PEI, Wyk), *Verbena officinalis* (LEI, Wyk), *Viola tricolor* (PEI, Wyk), *Viscum album* (ISB, LEI, Wyk), *Vitis vinifera* (seed extract, Murr), *Withania somnifera* (CM, Wyk), ***Zingiber officinale* (Bal2, Murr)**
Herbs: FIVE INGREDIENT POWDER WITH PORIA *Alsima plantago-aquatica Atractylodes macrocephala Cinnamomum aromaticum, Polyporus brumalis, Polyporus squamosus, Polyporus tubaster, Polyporus umbellatus, Wolfiporia extensa,* (Bal2); FOUR EMPERORS EPHEDRA DECOCTION *Ephedra sinica, Cinnamomum aromaticum* (bark), *Glycyrrhiza glabra, Prunus armeniaca* (seed) (Bal2); KIDNEY CHI PILL *Aconitum napellus, Alisma plantago-aquatica, Cinnamomum aromaticum, Cornus officinalis, Dioscorea oppositifolia, Dioscorea villosa, Paeonia suffruticosa, Rehmannia glutinosa, Wolfiporia extensa* (Bal2); LEDEBOURIELLA DECOCTION *Ledebouriella divaricata, Angelica sinensis, Atractylodes macrocephala, Ephedra distachya,* Epsom Salt ($MgSO_4 * 7H_2O$), *Forsythia suspensa, Gardenia angusta* (fruit), *Glycyrrhiza glabra,* Gypsum ($CaSO_4 * 2H_2O$), *Ligusticum porteri, Mentha piperita, Paeonia officinalis, Platycodon grandiflorus, Rheum rhabarbarum* (wine treated root), *Schizonepeta tenuifolia, Scutellaria baicalensis, Scutellaria barata, Scutellaria lateriflora,* Talc [$Mg_3(SiO_4)_{10}(OH)_2$]; MINOR BUPLEURUM DECOCTION *Bupleurum chinense, Bupleurum scorzoneraefolium, Glycyrrhiza glabra, Panax ginseng, Pinellia turnata, Scutellaria baicalensis, Scutellaria barata, Scutellaria lateriflora, Zingiber officinale, Ziziphus jujuba* (Bal2); TRUE WARRIOR DECOCTION *Aconitum napellus, Atractylodes macrocephala, Paeonia officinalis, Zingiber officinale, Wolfiporia extensa* (Bal2);

Backache NBK Also known as **Lumbago**, the leading cause of back pain is simple muscle strain accompanied by a buildup of lactic acid and pyruvic acid. This results in a phenomenon called delayed onset muscle soreness (DOMS). Other causes of back pain include poor posture, improper footwear, calcium deficiency, slouching when sitting, and sleeping on a mattress that is too soft. A ruptured disc in the spinal cord can also cause severe intermittent or constant back pain.
Herbs: *Arctium lappa* (Bal1), *Avena sativa* (Bal1), *Equisetum arvense* (Bal1), *Medicago sativa* (Bal1), *Salix alba* (Bal1), *Ulmus rubra* (Bal1)
Herb Formulas: INFULA-PROFEN *Apium graveolens* (seed, 16mg), *Arctium lappa* (root & seed, 32mg), *Cucuma longa* (rhizome, 100mg), *Harpagophytum procumbens* (root, 300 mg), *Piscidia erythrina* (bark, 12mg), *Tanacetum parthenium* (top, 300mg), *Yucca filamentosa* (root, 46mg), *Urtica dioica* (leaf, 10mg), *Zingiber officinale* (rhizome extract, 8mg) (Scal);
Vegetables: Raw (Bal1, "eat no highly processed foods", "avoid all meats")

Bell's Palsy DBLP also known as idiopathic palsy, is a temporary weakening of the facial nerve. Symptoms include facial numbness or paryalysis.
Herbs: *Glycyrrhiza glabra* (Bal2), *Pueraria lobata* (Bal2), *Syzygium aromaticum* (Bal2)
Herb Formulas: OPHIOPOGONIS DECOCTION *Ophiopogon japonicus* (bulb), *Glycyrrhiza glabra, Oryza sativa* (porridge), *Panax ginseng, Pinellia turnata, Ziziphus jujuba* (Bal2)

Bone Fracture NBF a break or crack in a bone. The symptom is intense pain.
Herbs: *Ananas comosus* (bromelain, Bal1), *Arnica montana* (NSD, Bal2), *Boswellia serrata* (NSD, Bal2), *Carapa guianensis* (NSD, Bal2), *Curcuma longa* (external, Bal1), *Crataegus laevigata* (Bal2), *Crataegus monogyna* (Bal2), *Equisetum arvense* (Bal1), *Glycine max* (isoflavone concentrate, Bal2)

Vegetables: "Foods with preservatives should also be avoided due to their phosphorus content. Phosphorus can lead to bone loss." (Bal1, page 283, recommendation #2); **Vegetables (fresh, vitamin K):** Broccoli, Brussels Sprouts, Cabbage, Kale, Iceberg Lettuce, Romain Lettuce, Spinach (Bal2)

Bursitis NBU inflammation of the Bursa as a result of overuse, allergies to foods, airborne allergies, or calcium deposits. Bursa are small fluid filled sacs between tendons and bone used to cushion friction against bones and other tissues.
Herbs: ***Ananas comosus*** **(bromelain, Bal2, Murr)**, ***Curcuma longa*** **(Bal2, Murr)**, *Equisetum arvense* (Bal1, Bal2), *Zingiber officinale* (Bal2)
Vegetables: Raw (Bal1)

Carpal Tunnel Syndrome NCT When the median nerve in the wrist is twisted or damaged there can be faint tingling, mild numbness, or excruciating pain. The median nerve is responsible for sensation felt in the thumb, the palm, and the first three fingers of the hand.
Herbs: *Achillea millefolium* (Bal1), *Aloe vera* (external, Bal1), *Althaea officinalis* (internal, Bal1), *Ananas comosus* (internal, Bal2), *Arnica montana* (external, Bal2), *Capsicum annuum-longum-cayenneum* (external, Bal1, Bal2), *Curcuma longa* (external Bal2), *Eupatorium purpureum* (Bal1), *Gaultheria procumbens* (external, Bal1, Bal2), *Ginkgo biloba* (internal, Bal1), *Harpagophytum procumbens* (internal, Bal1, Bal2), *Hoffmanseggia divaricata* (external, Bal2), *Hypericum perforatum* (internal, Bal1, Bal2), *Ligusticum porteri* (external, Bal2), *Petroselinum crispum* (internal, Bal1), *Pinus pinaster* (internal, pine bark extract Murr), *Ruscus aculeatus* (internal, Bal1, Bal2), *Scutellaria baicalensis* (internal, Bal1, Bal2), *Scutellaria barata* (internal, Bal1, Bal2), *Scutellaria lateriflora* (internal, Bal1, Bal2), *Vitis vinifera* (internal, grape seed extract, Murr), *Withania somnifera* (internal, Bal2), *Yucca aloeifolia* (Bal1), *Zea mays-suavis* (internal, cornsilk, Bal1, Bal2)
Vegetables (crucifer): Broccoli, Brussels Sprouts, Cabbage, Cauliflower, **Vegetables (low oxalic acid):** Asparagus, Beets, Beet Greens, Chard, Parsley, Rhubarb, Sorrel, Spinach (Bal1)

Epilepsy DEPI recurring seizures caused by electrical disturbances in the nerve cells in a section of the brain. The condition is caused by genetic mutation. In (1) **Petite Mal Epilepsy** there is a blank stare lasting about 1 minute where the person appears to be daydreaming. In (2) **Atonic Epilepsy** a person loses consciousness for about ten seconds and falls to the ground. In (3) **Temporal Lobe Epilepsy** there is a blank stare, random motion, and chewing motion. There is no memory of this seizure. In (4) Grand Mall Epilepsy there are sudden cries, falls, rigidity and jerking of muscles, shallow breathing, and bluish skin lasting about 2 minutes. In (5) **Myoclonic Epilepsy** there are brief massive muscle jerks. In (6) **Jacksonian Epilepsy** jerking begins in the fingers and toes and progresses up through the body while the person remains conscious. In (7) **Sensory Epilepsy** the person sees, hears, or senses things that do not exist.
Herbs: *Acorus calamus* (Bal2), *Cimicifuga racemosa* (Bal1, Bal2), *Hyssopis officinalis* (Bal1), *Lobelia inflata* (Bal1), *Medicago sativa* (Bal1), *Polygala senega* (root, Bal2), *Silybum marianum* (Bal2), *Zingiber officinale* (Bal2)
Herb Formulas: BUPLEURUM & OYSTER SHELL *Bupleurum chinense, Ostrea edulis* (shell), *Bupleurum scorzoneraefolium, Cinnamomum aromaticum, Panax ginseng, Pinellia tumata, Reum rhabarbarum, Scutellaria baicalensis, Scutellaria barata, Scutellaria lateriflora, Zingiber officinale, Zizyphus jubata, Wolfiporia extensa* (Bal2); GASTRODIA & UNCARIA DECOCTION *Gastrodia-asiatica elata, Uncaria gambir, Uncaria guianensis, Uncaria tomentosa, Eucommia ulmoides, Gardenia angusta, Phoradendron leucarpum (Synonym Loranthus leucarpus), Polygonum aviculare, Scutellaria baicalensis, Scutellaria barata, Scutellaria lateriflora, Wolfia extensa,* Sichuan Ox Knee Root (Bal2);

Fibromyalgia DFMS a rheumatic **Syndrome** caused by improper brain chemistry, abnormalities in the immune system, chronic mercury poisoning from amalgam dental fillings, anemia, parasites, hypoglycemia, and hypothyroidism. Symptoms are burning, throbbing, shooting, and stabbing pains. Other symptoms include anxiety, palpitations, memory impairment, irritable bladder, skin sensitivity, dry eyes, dry mouth, dizziness, and impaired coordination. Muscles are tender to the touch at the lower vertebrate of the neck, at the second rib, around the upper part of the thigh, in the middle of the knee, at the base of the skull, in the upper back, in the mid back, on the side of the elbow, and on the buttocks. An associated sleep disorder often accompanies the disease in which deep sleep periods are interrupted by bouts of waking brain activity resulting in poor sleep. Chronic fatigue accompanies the disease and can range from mild to debilitating.
Herbs: *Allium sativum* (Bal1), *Arctium lappa* (Bal1), *Astragalus membranaceus* (Bal1), *Boswellia serrata* (Bal2), *Juglans nigra* (Bal1), *Capsicum annuum-longum-cayenneum* (Bal1), *Gaultheria procumbens* (Bal1), *Ginkgo biloba* (Bal1), *Glycyrrhiza glabra* (Bal1), *Hypericum perforatum* (Bal2), *Piper methysticum* (Bal2), ***Salix alba*** **(bark, Bal2)**, *Salix fragilis* (bark, Bal2), *Salix myricoides* (bark, Bal2), *Salix nigra* (bark, Bal2), *Salix pentandra* (bark, Bal2), *Salix purpurea* (bark, Bal2), *Scutellaria baicalensis* (Bal1), *Scutellaria barata* (Bal1), *Scutellaria lateriflora* (Bal1), *Silybum marianum* (Bal1), *Tabebuia impetignosa* (Bal1), *Taraxacum officinalis* (Bal1), *Trifolium pratense* (Bal1), *Uncaria guianensis* (Bal2), *Uncaria tomentosa* (Bal2), *Valeriana officinalis* (Bal1)
Vegetables: "Eat a well balanced diet of 50% raw foods and fresh "live" juices" (Bal2, page 276, recommendation #1), **Vegetables (grain):** Brown Rice, Millet (Bal2), **Vegetables:** Nuts, Seeds (Bal2)

Heel Spur NHS also known as **Bone Spur**, calcium deposits in unwanted areas of the body, usually the heel. Most people who have heel spurs are middle aged or overweight. Heel spurs can lead to the formation of tiny tumors at the heel nerve endings which can be very painful.
Herbs: *Arnica montana* (foot bath, Bal1), *Matricaria chamomilla* (foot bath, Bal1)
Vegetables: "Eliminate alcohol, coffee, and sugar from your diet. These inhibit the healing process and upset the body's mineral balance." (Bal1, page 310, recommendation #1)

Lupus DLUP This disease occurs in two forms; **Systemic Lupus Erythematosus** and **Discoid Lupus Erythematosus**. Both types of Lupus are auto-immune diseases where the body's immune system attacks its own tissues. The cause is an unidentified virus. The immune system attempts to create an antibody for the virus, but in the process mistakenly creates an antibody that destroys the body's own tissues. Symptoms include a butterfly shaped rash over the cheeks and nose that makes the face look like the face of a wolf (the Latin name for wolf is lupus).
Herbs: *Codonopsis pilosula* (Bal2), *Curcuma longa* (Murr), *Echinacea purpurea* (Bal1), *Glycyrrhiza glabra* (Bal2), *Hydrastis canadensis* (Bal1), *Matricaria chamomilla* (Bal2), *Medicago sativa* (Bal1), *Pinus pinaster* (bark extract, Murr), *Silybum marianum* (Bal1, Bal2), *Tabebuia impetignosa* (Bal1), *Tanacetum parthenium* (Bal1, Bal2), *Trifolium pratense* (Bal1), *Urtica dioica* (root, Bal2), *Vitis vinifera* (seed extract, Murr), *Wolfiporia extensa* (Bal2), *Yucca aloeifolia* (Bal1), *Zingiber officinale* (Murr)

Herb Formulas: FIVE INGREDIENT POWDER WITH PORIA *Alsima plantago-aquatica Atractylodes macrocephala Cinnamomum aromaticum, Polyporus brumalis, Polyporus squamosus, Polyporus tubaster, Polyporus umbellatus, Wolfiporia extensa,* (Bal2); GINSENG DECOCTION *Panax ginseng, Angelica sinensis, Astragalus membranaceus, Atractylodes macrocephala, Bupleurum chinense, Cimicifuga racemosa,Citrus aurantium* (peel), *Glycyrrhiza glabra* (Bal2); MINOR BUPLEURUM DECOCTION *Bupleurum chinense, Bupleurum scorzoneraefolium, Glycyrrhiza glabra, Panax ginseng, Pinellia turnata, Scutellaria baicalensis, Scutellaria barata, Scutellaria lateriflora, Zingiber officinale, Ziziphus jujuba* (Bal2);
Vegetables: Asparagus, Garlic, Oatmeal, Onions (Bal1); **Vegetables (grain):** Brown Rice, Buckwheat, Cereal, Corn, Millet, Oats, Quinoa, Rice Bran, Rye, Spelt, Tapioca, White Flour, White Rice, Wheat (Bal1); **Vegetables (leaf):** Chard, Kale, Lettuce, Spinach, Turnip Greens (Bal1);

Lyme Disease DLYM was first identified as a tick born illness in Lyme, Connecticut in 1975. Spirochete bacteria called *Borellia burgdorfei* are transmitted by the deer tick which occurs on white tailed deer, mice, birds, chipmunks, cows, horses, cats, dogs, lizards, and jackrabbits. The first symptom is a circular red ring symmetrically surrounding the tick bite location. Other symptoms include fatigue, flu-like fever, difficulty sleeping, muscle weakness, achiness, headache, stiff neck, back ache, nausea and vomiting. Small raised bumps occur on the skin and then fade. One month later facial paralysis can occur accompanied by enlargement of the spleen, lymph glands, & heart & irregular heartbeat & severe headache.
Herbs: *Althaea officinalis* (Bal1), *Crataegus monogyna* (Bal1), *Echinacea purpurea* (Bal1, Bal2), *Eleutherococcus senticosus* (Bal2), *Equisetum arvense* (Bal1), *Glycyrrhiza glabra* (Bal2), *Grifola frondosa* (Bal2), *Hydrastis canadensis* (Bal1, Bal2), *Lentinula edodes, Medicago sativa* (Bal1), *Panax ginseng* (Bal1, Bal2), *Pueraria lobata* (Bal2), *Scutellaria baicalensis* (Bal2), *Scutellaria barata* (Bal2), *Scutellaria lateriflora* (Bal2), *Silybum marianum* (Bal1), *Taraxacum officinalis* (Bal1), *Trifolium pratense* (Bal1, Bal2), *Uncaria gambir* (Bal2), *Uncaria guianensis* (Bal2), *Uncaria tomentosa* (Bal2)
Vegetables: Garlic (Bal1); **Vegetables (live juices):** Celery Juice, Parsley Juice, Spinach Juice (Bal1);
Vegetable Formulas: KYO-GREEN Barley & Wheat Grass Juice (Bal1)

Multiple Sclerosis DMSC a degenerative disease of the central nervous system. The myelin sheaths that cover nerves are destroyed and replaced with scar tissue resulting in the destruction of the nerves. Symptoms include dizziness, nausea and vomiting. In advanced stages symptoms are trouble walking, paralysis, and difficulty breathing.
Herbs: *Achillea millefolium* (Bal1), *Arctium lappa* (Bal1), *Echinacea purpurea* (Bal1), ***Ginkgo biloba* (Bal2, Murr)**, *Glycine max-soja* (lecithin, Bal2), *Hydrastis canadensis* (Bal1), *Hypericum perforatum* (Bal1), *Medicago sativa* (Bal1, Bal2), *Pinus pinaster* (bark extract, Murr), *Smilax aristolochaefolia* (Bal1), *Smilax ferbifuga* (Bal1), *Smilax glyciphylla* (Bal1), *Smilax ornata* (Bal1), *Smilax regelii* (Bal1), *Tabebuia impetignosa* (Bal1), *Taraxacum officinale* (Bal1), *Trifolium pratense* (Bal1), *Vitis vinifera* (seed extract, Murr)
Herb Formulas:); FIVE INGREDIENT POWDER WITH PORIA *Alsima plantago-aquatica Atractylodes macrocephala Cinnamomum aromaticum, Polyporus brumalis, Polyporus squamosus, Polyporus tubaster, Polyporus umbellatus, Wolfiporia extensa,* (Bal2); FOUR SUBSTANCE DECOCTION *Angelica sinensis, Ligusticum porteri, Paeonia lactiflora, Paeonia officinalis, Rehmannia glutinosa* (Bal2); GINSENG DECOCTION *Panax ginseng, Angelica sinensis, Astragalus membranaceus, Atractylodes macrocephala, Bupleurum chinense, Cimicifuga racemosa,Citrus aurantium* (peel), *Glycyrrhiza glabra* (Bal2); MINOR BUPLEURUM DECOCTION *Bupleurum chinense, Bupleurum scorzoneraefolium, Glycyrrhiza glabra, Panax ginseng, Pinellia turnata, Scutellaria baicalensis, Scutellaria barata, Scutellaria lateriflora, Zingiber officinale, Ziziphus jujuba* (Bal2);
Vegetables: Raw Nuts, Raw Sprouts, Seeds, Vegetable Oils (Bal1); **Vegetables (vitamin K, dark green leaf):** Chard, Collard Greens, Cornsalad Lettuce, Kale, Mustard Greens, Parsley, Seakale, Spinach (Bal1); **Vegetables (grain, gluten free):** Brown Rice, Millet (Bal1); **Vegetables (live juices):** Beet Juice, Carrot Juice, Celery Juice, Parsley Juice, Spinach Juice (Bal1);

Muscle Cramps NMC a tensing of the muscle usually in the calf or foot cause by an imbalance of calcium and magnesium in the body.
Herbs: *Angelica sinensis* (Bal1), *Crocus sativus* (Bal1), *Equisetum arvense* (Bal1), *Ginkgo biloba* (Bal1), *Lobelia inflata* (Bal1), *Medicago sativa* (Bal1), *Sambucus nigra* (Bal1), *Valeriana officinalis* (Bal1)
Vegetables: Brewer's Yeast, Cornmeal, Kelp (Bal1); **Vegetables (dark green anti-oxidant):** Artichoke, Asparagus, Mild Bell Peppers, Broccoli, Brussels Sprouts, Chard, Collard Greens, Cornsalad Lettuce, Cucumbers, Green Beans, Kale, Leeks, Mustard Greens, Parsley, Peas, Seakale, Spinach, Zucchini Squash (Bal1);

Muscle Soreness NSD pain from exercise or lifting excessive weight.
Herbs: *Arnica montana* (Bal2), *Capsicum-annuum-longum-cayenneum* (Bal2), *Desmodium affine* (Bal2), *Desmodium intortum* (Bal2), *Desmodium, tortuosum* (Bal2), *Desmodium triflorum* (Bal2), *Myroxylon balsamum* (Bal2)

Myasthenia Gravis muscle weakness caused by too few muscle nerve receptor sites. Symptoms are difficulty chewing, difficulty climbing stairs and difficulty lifting objects.
Herbs: *Astragalus membranaceus* (Bal2), *Glycine max-soja* (Bal2)
Herb Formulas: HOCHU EKKI TO *Angelica sinensis, Astragalus membranacea, Atractylodes macrocephala, Bupleurum chinense, Cimicifuga racemosa, Citrus-aurantius aurantium, Glycyrrhiza glabra, Panax ginseng* (Bal2);

Narcolepsy DNAR sleep attacks, cataplexy, sleep paralysis, and sleep hallucinations. A rare neurological disorder that affects 250,000 Americans.
Herbs: *Centella asiatica* (Bal1), *Ephedra distachya* (Bal1), *Ginkgo biloba* (Bal1), *Hypericum perforatum* (Bal1)
Vegetables: Brewer's Yeast, Kelp, Nuts, Sea Vegetables, Seeds (Bal1); **Vegetables (grain):** Brown Rice, Buckwheat, Cereal, Corn, Millet, Oats, Quinoa, Rice Bran, Rye, Spelt, Tapioca, White Flour, White Rice, Wheat (Bal1);**Vegetables (leaf):** Chard, Kale, Lettuce, Spinach, Turnip Greens (Bal1); **Vegetables (legume):** Chick Peas, Green Beans, Kidney Beans, Lentils, Lima Beans, Mung Beans, Peas, Pinto Beans, Soybeans, Red Beans, White Beans (Bal1);

Neuralgia NNR refers to pain along the nerves which occur in discrete attacks.
Herbs: *Aconitium napellus* (Wyk), *Mentha piperita* (Wyk), *Salix alba* (Wyk)

Neuritis NNR nerve degeneration resulting in muscle weakness, atrophy, loss of sensation, and loss of reflexes.
Herbs: *Achillea millefolium* (Bal1), *Althaea officinalis* (Bal1), *Calendula officinalis* (Bal1), *Humulus lupulus* (Bal1), *Hypericum perforatum* (Bal1), *Lactuca sativa* (Bal1), *Lobelia inflata* (Bal1), *Matricaria chamomilla* (Bal1), *Passiflora caerulea* (Bal1), *Passiflora incarnata* (Bal1), *Piper methysticum* (Bal1), *Rosmarinus officinalis* (Bal1), *Salix alba* (Bal1), *Scutellaria baicalensis* (Bal1), *Scutellaria*

barata (Bal1), *Scutellaria lateriflora* (Bal1), *Stachys officinalis* (Bal1), *Tanacetum parthenium* (Bal1), *Vaccinium-vortabacca myrtillus* (Bal1), *Valeriana officinalis* (Bal1), *Verbena hastata* (Bal1), *Yucca aloeifolia* (Bal1)
Vegetables: Nuts, Seeds (Bal1); **Vegetables (grain):** Brown Rice, Buckwheat, Cereal, Corn, Millet, Oats, Quinoa, Rice Bran, Rye, Spelt, Tapioca, White Flour, White Rice, Wheat (Bal1);

Restless Legs Syn. DRLS legs involuntarily flex at the knees during sleep, about once every ten minutes.
Herbs: *Aspalathus linearis* (Bal2), *Corydalis cava* (Bal2), *Corydalis yanhusuo* (Bal2), *Eschscholzia californica* (Bal2), *Passiflora caerulea* (Bal2), *Passiflora* (Bal2) *incarnata*, *Piper methysticum* (Bal2), *Schisandra chinensis* (Bal2), *Valeriana officinalis* (Bal2)
Herb Formulas: GASTRODIA & UNCARIA DECOCTION *Gastrodia-asiatica elata, Uncaria gambir, Uncaria guianensis, Uncaria tomentosa, Eucommia ulmoides, Gardenia angusta, Phoradendron leucarpum (Synonym Loranthus leucarpus), Polygonum aviculare, Scutellaria baicalensis, Scutellaria barata, Scutellaria lateriflora, Wolfia extensa,* Sichuan Ox Knee Root (Bal2)

Shin Splints pain in the shin muscles caused by strenuous exercise.
Herbs: *Arnica montana* (Bal2), *Carapa guianensis* (Bal2)

Sore Muscles NSM, OSM also known as **Sprain**, **Strain**, pain of muscle exertion during exercise due to lack of circulation and inflammatory hormones.
Herbs: *Aesculus hippocastanum* (external, Bal2), *Arnica montana* (external, Bal2), *Capsicum annuum-longum-cayenneum* (external, Bal2), *Desmodium affine* (Bal2), *Desmodium gangeticum* (Bal2), *Desmodium intortum* (Bal2), *Desmodium tortuosum* (Bal2), *Desmodium triflorum* (Bal2), *Mentha piperita* (external, Bal2), *Myroxylon balsamum* (external, Bal2)

Temp. Man. J. Syn. DTMJ also known as **Temporal Mandibular Joint Syndrome**, pain in the muscles and joints of the jaw, face, neck and shoulder. Symptoms are headaches, toothaches, dizziness, and ringing in the ears. About 10 million Americans have this disease.
Herbs: *Humulus lupulus* (Bal1), *Lobelia inflata* (Bal1), *Matricaria chamomilla* (Bal1), *Lactuca sativa* (Bal1), *Nepeta cateria* (Bal1), *Passiflora caerulea* (Bal1), *Passiflora incarnata* (Bal1), *Piper methysticum* (Bal1), *Rubus-raspavbacca idaeus* (Bal1), *Scutellaria baicalensis* (Bal1), *Scutellaria barata* (Bal1), *Scutellaria lateriflora* (Bal1), *Thymus vulgaris* (Bal1), *Valeriana officinalis* (Bal1), *Viola sororia* (Bal1)
Vegetables: Lightly Steamed (Bal1); **Vegetables (grain):** Brown Rice, Buckwheat, Cereal, Corn, Millet, Oats, Quinoa, Rice Bran, Rye, Spelt, Tapioca, White Flour, White Rice, Wheat (Bal1);

Tendinitis the connective tissue that connects the muscle to the bone becomes inflamed. The most frequently affected tendons are the achilles tendon at the back of the ankle, the front biceps tendons, the thumb tendons, the knee tendons, and the shoulder rotator cuff tendons.
Herbs: ***Ananas comosus* (bromelain, Bal2, Murr)**, *Capsicum annuum-longum-cayenneum* (Bal2), ***Curcuma longa* (Bal2, Murr)**, *Harpagophytum procumbens* (Bal2), *Pinus pinaster* (bark, Bal2), *Vitis vinifera* (seed, Bal2)

Toothache OTA severe pain from an inflammed root and an abscess at the bottom of the tooth caused by tooth decay penetrating the pulp chamber of the tooth.
Herbs: *Gaultheria procumbens* (Wintergreen Checkerberry, external, methyl salicylate, Bal2), *Piper methysticum* (Kava Kava Peppercorn, kavapyrone, Bal2), *Syzygium aromaticum* (Cloves Lilly Pilly, external, oil, Bal2), *Salix alba* (White Willow, bark, Bal2) , *Salix cineria* (Large Gray Willow, bark, Bal2), *Salix discolor* (Pussy Willow, bark, Bal2), *Salix fragilis* (Crack Willow, bark, Bal2), *Salix myricoides* (Bayberry Willow, bark, Bal2), *Salix nigra* (Black Willow, bark, Bal2), *Salix pendulina* (Weeping Willow, bark, Bal2), *Salix pentandra* (Laurel Willow, bark, Bal2), *Salix purpurea* (Basket Willow, bark, Bal2), *Thymus vulgaris* (external, oil, Bal2),

Psychosomatic Disorders

Alcoholism DALC alcohol is an addictive drug which causes excessive amounts of fat to accumulate in the liver which eventually leads to hepatitis which progresses to cirrhosis of the liver, and finally death. A chronic alcoholic's life span is usually shortened by 10 to 20 years depending on the severity of the addiction.
Herbs: *Arctium lappa* (Bal1), *Ganoderma lucidum* (Bal2), *Glycine max* (Bal2), *Medicago sativa* (Bal1), *Pinus pinaster* (pine bark extract, Murr), ***Pueraria lobata* (Bal2, Murr)**, ***Silybum marianum* (Bal1, Bal2, Murr)**, *Taraxacum officinalis* (Bal1), *Trifolium pratense*, (Bal1), *Valeriana officinalis* (Bal1), *Vitis vinifera* (grape seed extract, Murr)
Vegetables (raw leaf): Chard, Kale, Lettuce, Spinach, Turnip Greens (Bal1, Murr), **Vegetables (grain):** Brown Rice, Buckwheat, Cereal, Corn, Millet, Oats, Quinoa, Rice Bran, Rye, Spelt, Tapioca, White Flour, White Rice, Wheat (Bal1, Bal2, Murr),
Vegetables (legume): Chick Peas, Green Beans, Kidney Beans, Lentils, Lima Beans, Mung Beans, Peas, Pinto Beans, Soybeans, Red Beans, White Beans (Bal1, Murr), **Vegetables (vitamin A):** Dark Green, Yellow, Orange (Bal2)

Anorexia DANX also known as **Anorexia Nervosa**, an eating disorder characterized by a refusal to eat to the point of starvation. 95% of people who suffer from anorexia are women. Anorexia typically appears between age 12 and age 18. Two out of every one hundred females are anorexic. One out of every three females who suffer from anorexia have chronic anorexia which lasts their entire lives.
Herbs: *Carica papaya* (SAP, Bal1), *Centella asiatica* (SAP, Bal1), *Dioscorea oppositifolia* (BLC, Bal1), *Dioscorea villosa* (BLC, Bal1), *Foeniculum vulgare* (SAP, Bal1), *Mentha piperita* (SAP, Bal1), *Nepeta cateria* (SAP, Bal1), *Panax ginseng* (SAP, Bal1), *Serenoa repens* (SAP, Bal1), *Silybum marianum* (BLC, Bal1), *Taraxacum officinale* (BLC, Bal1), *Trifolium pratense* (BLC, Bal1)*Zingiber officinale* (SAP, Bal1)
Vegetables: Fresh & Raw (Bal1)

Bulimia DBUL "binge and purge" eating uncontrolled binge eating followed by "finger in the throat" vomiting or laxative induced extreme diarrhea. Substituting Psyllium Husk (Metamucil ®) during the "binge" phase for the high calorie foods that are usually eaten will provide a feeling of fullness without gaining weight. Thus, the primary reason for "purging" is taken away. It is almost impossible to vomit Psyllium Husk. Psyllium Husk reduces the diarrhea effects of extreme doses of laxative because it keeps the stool solid.
Herbs(Psyllium Husk): *Plantago afra* (Bal1)
Vegetables: Fiber

Drug Addiction DDAD cocaine, heroin, and morphined are addictive drugs. During the first one to ten days of withdrawal, addicts experience intense drug cravings, nausea, and vomiting.
Herbs: *Acorus gramineus* (Bal2), *Eleuthrococcus senticosus* (Bal1), *Valeriana officinalis* (Bal1)

Frigidity DFRI is the inability of a woman to experience pleasure from sexual intercourse because of low libido.
Herbs: *Centella asiatica* (Bal1), *Dioscorea oppositifolia* (dehydroepiandrosterone DHEA, Bal1), *Dioscorea villosa* (dehydroepiandrosterone DHEA, Bal1), *Eleuthrococcus senticosus* (Bal1), *Serenoa repens* (Bal1), *Smilax aristolochiaefolia* (Bal1), *Smilax febriguga* (Bal1), *Smilax ornata* (Bal1), *Smilax regelii* (Bal1), *Turnera diffusa* (Bal1)
Herb Formulas: WOMEN'S LIBIDO *Avena sativa* (seed, 16mg), *Chamaelirium luteum* (root, 16mg), *Epimedium grandiflorum* (herb, 26mg), *Juniperis basiliensis* (bark, 50mg), *Maytenus krukovit* (bark, 30mg), *Pfaffia paniculata* (root, 50mg), *Smilax officinalis* (root, 26mg), *Turnera diffusa* (leaf, 26mg), *Zingiber officinale* (root extract, 8mg) (Scal)
Vegetables: Alfalfa Sprouts, Avocados, Olive Oil, Pumpkin Seeds, Nuts, Soy Oil, Sesame Seeds, Sesame Oil, Sunflower Seeds, Wheat (Bal1);

Hangover headache following excessive consumption of alcohol.
Herbs: *Aloe vera* (prevention, Bal2), *Glycine max-soja* (lecithin, Bal2), *Panax ginseng* (tea, Bal2), *Polygala senega* (prevention, Bal2), *Scutellaria baicalensis* (extract, Bal2), *Scutellaria barata* (extract, Bal2), *Scutellaria lateriflora* (extract, Bal2)

Impotence DIPT also known as **Erectile Dysfunction**, the inability of a man to achieve or maintain an erection for sexual intercourse. An aphrodisiac is taken to counteract the effects of low libido and impotence.
Herbs: *Alchornia cordifolia* (Bal2), *Angelica archagelica* (Bal1), *Avena sativa* (Bal2), *Centella asiatica* (Bal1), *Dioscorea oppositifolia* (Bal1), *Dioscorea villosa* (Bal1), *Eleutherococcus senticosus* (Bal1, Bal2), *Epimedium grandiflorum* (Bal2), *Erythoxylon catuaba* (Bal2), *Eurycoma longifolia* (Murr), ***Ginkgo biloba* (Bal2, Murr)**, *Hydrangea arborescens* (root, Bal1), *Morinda citrifolia* (Bal2), *Morinda officinalis* (Bal2), ***Panax ginseng* (Bal2, Murr)**, ***Pausinystalia johimbe* (Bal2, Murr)**, *Pinus pinaster* (pollen extract Bal2, bark extract, Murr), *Prunus-Cerasus spinosa* (Bal1), *Ptychopetalum ovata* (Muira Puama Amazon Potencywood, Bal2), ***Ptychopetalum olacoides* (Muira Puama Brazil Potencywood, Bal2, Murr, Wyk)**, *Ptychopetalum uncinatum* (Muira Puama Peru Potencywood, Wyk), *Serenoa repens* (Bal1), *Smilax aristolochaefolia* (Bal1), *Smilax ferbifuga* (Bal1), *Smilax glyciphylla* (Bal1), *Smilax ornata* (Bal1), *Smilax regelii* (Bal1), *Tribulus terrestris* (Ground Puncturevine, Murr), *Trigonella foenum-graecum* (Murr), *Turnera diffusa* (Bal1, Bal2), *Urtica dioica* (Stinging Nettle, root, Bal2), *Vitis vinifera* (grape seed extract, Murr), *Withania somnifera* (Bal2)
Herb Formulas: MALE LIBIDO *Avena sativa* (seed, 15mg), *Corynanthe yohimbe* (bark, 30mg), *Epimedium grandiflorum* (54mg), *Lepedium meyenii* (tuber, 120mg), *Ptychopetalum olacoides* (root, 15mg), *Polygonum multiflorum* (root, 15mg), *Serenoa repens* (berry extract, 54mg), *Smilax officinalis* (root, 24mg), *Tribulus terrestris* (bark, 45mg) (Scal); SENSUALI TEA *Glycyrrhiza glabra, Smilax aristolochaefolia, Smilax ferbifuga, Smilax glyciphylla, Smilax ornata, Smilax regelii, Turnera diffusa* (Bal1); VIRILITY TWO *Centella asiatica,* Oak Grass, *Pausinystalia johimbe, Smilax aristolochaefolia, Smilax ferbifuga, Smilax glyciphylla, Smilax ornata, Smilax regelii, Turnera diffusa, Zingiber officinale*
Vegeables: Pumpkin Seeds (Bal1); **Vegetables:** Kyo-Green powdered drink, Sildinafil Citrate Extract (made synthetically as Viagra), Tadalafil Extract (made sythentically as Cialis), Vardenifil Extract (made synthetically as Levitra) (Bal2); **Vegetables (grain):** Brown Rice, Buckwheat, Cereal, Corn, Millet, Oats, Quinoa, Rice Bran, Rye, Spelt, Tapioca, White Flour, White Rice, Wheat (Murr); **Vegetables (legume):** Chick Peas, Green Beans, Kidney Beans, Lentils, Lima Beans, Mung Beans, Peas, Pinto Beans, Soybeans, Red Beans, White Beans (Murr);

Libido Deficiency is combated by **Aphrodisiac HAP** herbs for both men **and** women to remedy low sexual desire. (Estrogen herbs for **women** are listed under psychosomatic **Frigidity**. Testosterone herbs for **men** are listed under psychosomatic **Impotence**. Herbs that increase sex drive in general are listed below.
Herbs (aphrodesiac): *Illicium verum* (Chinese Anise, Bal2), ***Pausinystalia johimbe* (Yohimbe Tubeflower Rubyroot, Bal2, Wyk)**, *Panax ginseng* (Chinese Ginseng, Bal2), *Passiflora caerulea* (Blue Crown Passion Flower, Bal2), *Passiflora incarnata* (Maypops Passion Flower, Bal2), *Piper methysticum* (Kava Kava Peppercorn, Bal2), *Smilax aristolochaefolia* (Sarsaparilla, Vera Cruz Greenbrier, Bal2), *Smilax ferbifuga* (Sarasaparilla Equador Greenbrier, Bal2), *Smilax glyciphylla* (Sarsaparilla Australia Greenbrier, Bal2), *Smilax ornata* (Sarsaparilla Jamaica Greenbrier, Bal2), *Smilax regelii* (Sarsaparilla Honduras Greenbrier, Bal2), *Strychnos nux-vomitica* (Poison Nut Strychnine, Wyk), ***Turnera diffusa* (Damiana Alder Hazel, Bal2, Wyk)**, *Withania somnifera* (Winter Cherry Ashwagandha, Bal2)

Schizophrenia catatonia, delusions, hallucinations, violet outbursts.
Herbs: *Ginkgo biloba* (Bal1)
Vegetables: Brewer's Yeast, Peas, Sunflower Seeds (Bal1); **Vegetables (niacin):** Broccoli, Carrots, Corn, Potatoes, Tomatoes, Whole Wheat (Bal1);

Smoking Addiction DSMA 4000 chemicals have been identified as constituents of cigarette smoke. At least 43 of these are known to cause cancer in humans.
Herbs: *Capsicum annuum-longum-cayenneum* (Bal1), *Lobelia inflata* (NSD, Bal1), *Nepeta cateria* (NSD, Bal1), *Scutellaria biacalensis* (NSD, Bal1), *Scutellaria barata* (NSD, Bal1), *Scutellaria lateriflora* (NSD, Bal1), *Silybum marianum* (BLC, Bal1), *Taraxacum officinale* (BLC, root, Bal1), *Ulmus rubra* (RLC, Bal1), *Valeriana officinalis* (NSD, Bal1), *Zingiber officinale* (Bal1)
Herb Formulas: NICOTINE RELIEF *Acorus calamus* (root, 30mg), *Avena sativa* (seed, 38mg), *Glycyrrhiza glabra* (root, 158mg), *Glycyrrhiza uralensis* (root, 158mg), *Hypericum perforatum* (flower bud, 36mg), *Lobelia inflata* (herb & seed, 48mg), *Passiflora incarnata* (flowering vine, 26mg) (Scal);
Vegetables: Asparagus, Nuts, Seeds, Spinach, Sweet Potatoes, Turnips (Bal1); **Vegetables (crucifer):** Broccoli, Brussels Sprouts, Cabbage, Cauliflower (Bal1); **Vegetables (grain):** Brown Rice, Buckwheat, Cereal, Corn, Millet, Oats, Quinoa, Rice Bran, Rye, Spelt, Tapioca, White Flour, White Rice, Wheat (Bal1); **Vegetables (orange):** Bell Peppers, Carrots, Pumpkin, Squash, Sweet Potato, Yam (Bal1); **Vegetables (yellow):** Bell Peppers, Oca Tuber, Squash (Bal1);

Sweating Excessively also known as **Diaphoresis**, the sympathetic nervous system responds in an exaggerated manner to heat, exercise, strong emotions or fever.
Herbs: *Juglans nigra* (leaf, Bal2), *Juglans regia* (leaf, Bal2), *Melaleuca alternifolia* (Bal2), *Salvia officinalis* (Bal2), *Schisandra chinensis* (Bal2)
Herb Formulas: OYSTER SHELL POWDER *Astragalus membranacea, Ephedra distachya, Ostrea edulis, Triticum aestivum* (Bal2); SHIH QUA DA BU TANG *Angelica sinensis, Astragalus membranaceus, Atractylodes macrocephala, Cinnamomum aromaticum, Glycyrrhiza glabra, Ligusticum porteri, Paeonia officinalis, Panax ginseng, Rehmannia glutinosa, Wolfiporia extensa* (Bal2);

Respiratory System Disorders

General Lung Health

Herb Formulas: ALLER-LEAF *Achillea millefolium* (flower, 14mg), *Acorus calamus* (rhizome, 8mg), *Cucruma longa* (root & leaf, 132mg), *Euphrasia nemorosa* (132mg), *Hydrastis canadensis* (rhizome, 16mg), *Myrica cerifera* (bark, 100mg), *Scutellaria baicalensis* or *barata* or *lateriflora* (root, 36 mg), *Urtica dioica* (leaf, 106mg) (Sclazo)

Asthma RLA A lung disease characterized as an obstruction of bronchial airways. During an Asthma attack bronchial airways constrict impeding the outward passage of stale air. Symptoms are coughing wheezing, tightness in the chests and difficulty breathing.

TM = traditional medicine, MM = modern medicine

Herbs: *Allium cepa* (TM, Wyk), ***Ammi visnaga* (MM, Bal2, Wyk)**, *Aralia racemosa*(TM, Wyk), *Atropa belladonna* (MM, Wyk), *Carapichea ipecacuanha* (Bal2), ***Coleus barbatus, forskohlii* (Bal2, Murr)**, *Crataegus laevigata* (Bal2), *Crataegus monogyna* (Bal2), *Datura stramonium* (MM, Wyk), *Echinacea purpurea* (Bal1), *Ephedra distachya*(MM, Wyk), *Equisetum arvense* (Bal1), *Ginkgo biloba* (Bal1), ***Glycyrrhiza glabra* (Bal1, Bal2, Murr)**, *Grindelia robusta*(TM, Wyk), *Hedera helix* (Murr), *Hydrastis canadensis* (Bal1), *Hyoscyamus niger* (MM, Wyk), *Hyssopis officinalis* (Bal2), *Juniperus communis* (Bal1), ***Lobelia inflata* (TM, Bal2, Wyk)**, *Matricaria chamomilla* (Bal2), ***Pinus pinaster* (Bal2, Murr)**, *Quercus velutina* (quercetin, Bal2), *Sambucus nigra* (Bal2), *Scutellaria baicalensis* (Bal2), *Scutellaria barata* (Bal2), *Scutellaria lateriflora* (Bal2), *Solanum dulcamara*(TM, Wyk), *Tabebuia impetignosa* (Bal1), *Thea sinensis* (green, Bal1, Bal2, Murr), ***Tylophora indica, asthmatica* (Bal2, Murr)**, *Verbascum phlomoides* (Bal1), ***Vitis vinifera* (Bal2, Murr)**, *Ulmus rubra* (Bal1), *Zingiber officinale* (Bal2)

ARREST WHEEZING DECOCTION *Ephedra distachya, Ginkgo biloba* (nut), *Glycyrrhiza glabra, Prunus aremeniaca* (seed), *Tussilago farafara, Morus alba* (root), *Perilla frutescens* (seed), *Pinellia turnata, Scutellaria baicalensis, Scutellaria barata, Scutellaria lateriflora* (Bal2); FOUR EMPERORS EPHEDRA DECOCTION *Ephedra sinica, Cinnamomum aromaticum* (bark), *Glycyrrhiza glabra, Prunus armeniaca* (seed) (Bal2); GYPSUM EPHEDRA DECOCTION *Ephedra distachya*, Gypsum ($CaSO_4 * 2H_2O$), *Glycyrrhiza glabra, Prunus armeniaca* (seed) (Bal2); ASARUM & ACONITE EPHEDRA DECOCTION *Ephedra sinica, Aconitum napellus, Asarum europaeum* (Bal2); GINSENG DECOCTION *Panax ginseng, Angelica sinensis, Astragalus membranaceus, Atractylodes macrocephala, Bupleurum chinense, Cimicifuga racemosa,Citrus aurantium* (peel), *Glycyrrhiza glabra* (Bal2); MINOR BLUEGREEN DECOCTION *Asarum europaeum, Cinnamomum aromaticum* (bark), *Glycyrrhiza glabra, Paeonia lactiflora, Paeonia officinalis, Pinellia turnata, Schisandra chinensis, Zingiber officinale* (Bal2); MINOR BUPLEURUM DECOCTION *Bupleurum chinense, Bupleurum scorzoneraefolium, Glycyrrhiza glabra, Panax ginseng, Pinellia turnata, Scutellaria baicalensis, Scutellaria barata, Scutellaria lateriflora, Zingiber officinale, Ziziphus jujuba* (Bal2); OPHIOPOGONIS DECOCTION *Ophiopogon japonicus* (bulb), *Glycyrrhiza glabra, Oryza sativa* (porridge), *Panax ginseng, Pinellia turnata, Ziziphus jujuba* (Bal2)

Vegetables (beta-carotene): Carrots, Pumpkin, Squash, Sweet Potato, Yam (Bal2, Murr); **Vegetables (lily):** Garlic, Onion (Bal1, Murr); **Vegetables (grain):** Brown Rice, Buckwheat, Cereal, Corn, Millet, Oats, Quinoa, Rice Bran, Rye, Spelt, Tapioca, White Flour, White Rice, Wheat (Bal1);

Bronchitis RLB is caused by viral, bacterial, or fungal infection of the brochial tubes and is characterized by mucus production, cough, and laboured breathing. Saponins stimulate the vagus nerve. Iridoids have anti-inflammatory properties. Alkaloids have cough suppressant function (expectorant). Essential Oils stimulate the bronchial endothelia.

Herbs: *Aesculus hippocastanum* (saponins, Wyk), ***Ananas comosus* (bromelain extract, Bal2, Murr)**, *Asarum europaeum* (essential oils, Wyk), *Carapichea ipecacuanha* (alkaloids, Wyk), *Castanea sativa* (saponins, Wyk), *Cetraria islandica* (Bal1), *Cinnamomum camphora* (essential oils, Wyk), *Elutherococcus senticosus* (Bal1, Bal2), *Elymus repens* (Bal2), *Ephedra distachya* (alkaloids, Wyk), *Equisetum arvense* (Bal1), *Eucalyptus globulus* (essential oils, Wyk), *Fagus sylvatica* (Guaifenesin extract, Murr), *Foeniculum vulgare* (essential oils, Wyk), *Ganoderma lucidum* (Bal2), *Glechoma hederacea* (essential oils, Wyk), ***Glycyrrhiza glabra* (saponins, Murr, Wyk)**, *Gypsophila paniculata* (saponins, Wyk), ***Hedera helix* (saponins, Murr, Wyk)**, *Hydrastis canadensis* (alkaloids, Wyk), *Hyssopus officinalis* (essential oils, Wyk), ***Inula helenium* (essential oils, Bal2, Wyk)**, *Larix decidula* (essential oils, Wyk), *Ligusticum porteri* (Bal2), ***Lobelia inflata* (Bal2, Murr)**, *Mentha piperita* (essential oils, Wyk), *Myrtus communis* (essential oils, Wyk), *Panax quinquefolium* (Bal2), *Papaver somniferum* (alkaloids, Wyk), ***Pelargonium sidoides* (essential oils, Murr, Wyk)**, *Picea abies* (essential oils, Wyk), *Pimpinella anisum* (essential oils, Wyk), *Pinus mugo* (essential oils, Wyk), *Pinus pinaster* (pine bark extract, Murr), ***Plantago lanceolata* (saponins, Bal2, Wyk)**, *Polygala senega* (saponins, Wyk), *Primula eliator* (saponins, Wyk), *Primula veris* (saponins, Wyk), *Pulmonaria officinalis* (saponins, Wyk), *Rumex crispus* (saponins, Wyk), *Salvia officinalis* (essential oils, Wyk), *Sambucus nigra* (Bal2), *Saponaria officinalis* (saponins, Wyk), *Sinapis alba* (Bal2), *Thea sinensis* (alkaloids, Wyk), *Thymus vulgaris* (essential oils, Wyk), *Thymus serpyllum* (essential oils, Wyk), *Thymus zygis* (essential oils, Wyk), *Trigonella foenum-graecum* (Bal1, Bal2), *Tussilago farafara* (Bal2), *Verbena officinalis* (saponins, Wyk), *Veronica officinalis* (saponins, Wyk), *Viola tricolor* (saponins, Wyk), *Vitis vinifera* (grape seed extract, Murr)

Herb Formulas: FOUR EMPERORS EPHEDRA DECOCTION *Ephedra sinica, Cinnamomum aromaticum* (bark), *Glycyrrhiza glabra, Prunus armeniaca* (seed) (Bal2); KUDZU DECOCTION *Pueraria lobata, Cinnamomum aromaticum, Ephedra sinica, Glycyrrhiza glabra, Paeonia lactiflora, Paeonia officinalis Zingiber officinalis, Ziziphus jujuba* (Bal2); MINOR BLUEGREEN DECOCTION *Asarum europaeum, Cinnamomum aromaticum* (bark), *Glycyrrhiza glabra, Paeonia lactiflora, Paeonia officinalis, Pinellia turnata, Schisandra chinensis, Zingiber officinale* (Bal2); MINOR BUPLEURUM DECOCTION *Bupleurum chinense, Bupleurum scorzoneraefolium, Glycyrrhiza glabra, Panax ginseng, Pinellia turnata, Scutellaria baicalensis, Scutellaria barata, Scutellaria lateriflora, Zingiber officinale, Ziziphus jujuba* (Bal2); OPHIOPOGONIS DECOCTION *Ophiopogon japonicus* (bulb), *Glycyrrhiza glabra, Oryza sativa* (porridge), *Panax ginseng, Pinellia turnata, Ziziphus jujuba* (Bal2)

Vegetables: Garlic, Onions (Bal1, Bal2), **Vegetables (dark green anti-oxidant):** Artichoke, Asparagus, Mild Bell Peppers, Broccoli, Brussels Sprouts, Chard, Collard Greens, Cornsalad Lettuce, Cucumbers, Green Beans, Kale, Leeks, Mustard Greens, Parsley, Peas, Seakale, Spinach, Zucchini Squash (Murr); **Vegetables (legume anti-oxidant):** Chick Peas, Green Beans, Kidney Beans, Lentils, Lima Beans, Mung Beans, Peas, Pinto Beans, Soybeans, Red Beans, White Beans (Murr)

Cough (RLC) a cough is a reflex that helps to expel mucus or foreign bodies from the brochial tubes. Coughs can be silenced by deactivating the neuron "cough center" in the medulla oblongata.

Herbs: ***Althaea officinalis* (mucilage, Bal2, Wyk)**, *Drosera ramentacea* (expectorant, Wyk), ***Drosera rotundifolia* (expectorant, Bal2, Wyk)**, *Cetraria islandica* (mucilage, Wyk), *Echinacea purpurea* (DABA, DAVR, Bal2), *Ephedra distachya* (expectorant, Wyk), *Eryngium planum* (expectorant, Wyk), *Fritillaria thunbergii* (RST, Bal2), *Ganoderma lucidum* (expectorant, Bal2), *Lobelia inflata* (expectorant, Bal2), *Malva neglecta* (mucilage, Wyk), *Malva sylvestris* (mucilage, Wyk), *Mentha piperita* (expectorant, Bal2), *Papaver somniferum* (expectorant, Wyk), *Pinguicula vulgaris* (expectorant, Wyk), *Plantago lanceolata* (mucilage, Wyk), *Prunus avium* (expectorant, bark extract, Bal2), *Pulmonaria officinalis* (expectorant, SDR, Bal2), *Rhododendron tomentosum* (expectorant, Wyk), *Trigonella foenum-graecum* (mucilage, Wyk), ***Tussilago farafara* (mucilage, Bal2, Wyk)**, *Ulmus rubra* (RST, Bal2), ***Verbascum densiflorum* (mucilage, Bal2, Wyk)**

Herb Formulas: ECHINACEA SUPREME *Echinacea angustifolia* (root, 72mg), *Echinacea purpurea* (flowering top 120mg, root 72mg, seed 3mg) (Scal) ECHINACEA/GOLDENSEAL SUPREME *Berberis vulgaris* (root, 45mg), *Echinacea angustifolia* (root, 54mg), *Echinacea purpurea* (flowering top, 66mg root 66mg, seed 12mg), *Hydrastis canadensis* (root, 60mg), *Hypericum perforatum* (flower bud, 60mg), *Mahonia aquifolium* (root, 45mg), FOUR EMPERORS EPHEDRA DECOCTION *Ephedra sinica, Cinnamomum aromaticum* (bark), *Glycyrrhiza glabra, Prunus armeniaca* (seed) (Bal2); MINOR BLUEGREEN DECOCTION *Asarum europaeum, Cinnamomum aromaticum* (bark), *Glycyrrhiza glabra, Paeonia lactiflora, Paeonia officinalis, Pinellia turnata, Schisandra chinensis, Zingiber officinale* (Bal2);
Vegetables (leaf): Chard, Kale, Lettuce, Spinach, Turnip Greens, **Vegetables (orange beta-carotene):** Carrots, Pumpkin, Sweet Potato, Yam, **Vegetables (yellow anti-oxidant):** Bell Peppers (yellow), Corn, Oca Tuber, Squash (Bal2)

Nosebleed also known as **Epistaxis**, bleeding from inside the nose.
Herbs: *Agrimonia eupatoria* (Bal2), *Aloe vera* (Bal1), *Calendula officinalis* (Bal2), *Capsella bursa-pastoris* (Bal2), *Medicago sativa* (Bal2), *Pinus pinaster* (bark, Bal2), *Quercus alba* (bark, Bal1), *Tagates lucida* (Bal1), *Vitis vinifera* (seed, Bal2)
Herb Formulas: OPHIOPOGONIS DECOCTION *Ophiopogon japonicus* (bulb), *Glycyrrhiza glabra, Oryza sativa* (porridge), *Panax ginseng, Pinellia turnata, Ziziphus jujuba* (Bal2); WHITE TIGER DECOCTION *Anemarrhena asphodeloides, Atractylodes macrocephala, Glycyrrhiza glabra,* Gypsum ($CaSO_4 * 2H_2O$), Mirabilite ($Na_2SO_4 * 10H_2O$), *Oryza sativa, Panax ginseng, Rheum rhabarbarum* (Bal2);
Vegetables (vitamin K, crucifer): Broccoli, Brussels Sprouts, Cauliflower (Bal1);
Vegetables (vitamin K, dark green leaf): Chard, Collard Greens, Cornsalad Lettuce, Kale, Mustard Greens, Parsley, Seakale, Spinach (Bal1);

Croup RCR a narrowing of the vocal chords (larynx) or windpipe (trachea) due to swelling. A harsh wheezing noise is made when breathing in. It is usually occurs after bronchitis, a cold, or an allergy attack.
Herbs: *Echinachea purpurea* (Bal1), *Hydrastis canadensis* (Bal1), *Thymus vulgaris* (Bal1), *Trigonella foenum-graecum* (Bal1)

Emphysema RLE the lungs become inelastic after years of exposure to cigarette smoke and air pollution. The alveoli or air pouches stay permanently open making it difficult to exhale. Symptoms are coughing and breathlessness. Malnutrition is common in people with emphysema.
Herbs: *Allium sativum* (Bal1), *Amoracia rusticana* (Bal1), *Astragalus membranaceus* (Bal2), *Ephedra distachya* (Bal1), *Equisetum arvense* (Bal2), *Glycyrrhiza glabra* (Bal1), *Hedera helix* (RLC, Murr), *Lobelia inflata* (RLC, Murr), *Medicago sativa* (Bal1), *Pinus pinaster* (pine bark extract, Murr), *Quercus velutina* (quercetin, Bal2), *Rosmarinus officinalus* (Bal1), *Trigonella foenum-graecum* (Bal1), *Verbascum thapsis* (Bal1), *Vitis vinifera* (grape seed extract, Murr)
Vegetables: 50% Raw, 25% Vegetable Soup, Brown Rice, Millet, Whole Grain Cereals (Bal1)
Vegetables (protein drinks): Immunocal, Whey Milkshakes, Legume Smoothies (Bal2), "Any food you can get into your body if you are undernourished is encouraged" (Bal2, page 296, consideration #1)

Laryngitis RTL any acute infection of the throat can injure the mucosa lining the vocal cords. Sympoms include hoarseness, husky voice, and loss of voice.
Herbs: *Althaea officinalis* (root, Bal2), Bibitaki (Bal2), *Elymus repens* (Bal2), *Fritillaria thunbergii* (Bal2), *Hypericum perforatum* (Bal2), *Lobelia inflata* (Bal2), *Tussilago farafara* (Bal2)

Pneumonia DPNB, DPNV caused by viral, bacterial, or fungal infection of the brochial tubes and is characterized by mucus production, cough, and laboured breathing. Saponins stimulate the vagus nerve. Iridoids have anti-inflammatory properties. Alkaloids have cough suppressant function (expectorant). Essential Oils stimulate the bronchial endothelia.
Herbs: *Aesculus hippocastanum* (saponins, Wyk), *Ananas comosus* (Bal2), *Asarum europaeum* (essential oils, Wyk), *Carapichea ipecacuanha* (alkaloids, Wyk), *Castanea sativa* (saponins, Wyk), *Cetraria islandica* (Bal1), *Cinnamomum camphora* (essential oils, Wyk), *Echinacea purpurea* (Bal1), *Elutherococcus senticosus* (Bal1, Bal2), *Elymus repens* (Bal2), *Ephedra distachya* (alkaloids, Wyk), *Equisetum arvense* (Bal1), *Eucalyptus globulus* (essential oils, Wyk), *Foeniculum vulgare* (essential oils, Wyk), *Ganoderma lucidum* (Bal2), *Glechoma hederacea* (essential oils, Wyk), *Glycyrrhiza glabra* (saponins, Bal1, Wyk), *Gypsophila paniculata* (saponins, Wyk), *Hedera helix* (saponins, Wyk), *Hydrastis canadensis* (alkaloids, Bal1, Wyk), *Hyssopus officinalis* (essential oils, Wyk), ***Inula helenium* (essential oils, Bal2, Wyk)**, *Larix decidula* (essential oils, Wyk), *Ligusticum porteri* (Bal2), *Lobelia inflata* (Bal2), *Mentha piperita* (essential oils, Wyk), *Myrtus communis* (essential oils, Wyk), *Panax quinquefolium* (Bal2), *Papaver somniferum* (alkaloids, Wyk), *Pelargonium sidoides* (essential oils, Wyk), *Picea abies* (essential oils, Wyk), *Pimpinella anisum* (essential oils, Wyk), *Pimpinella anisum* (essential oils, Wyk), *Pinus mugo* (essential oils, Wyk), *Pinus pinaster* (pine bark extract, Murr), ***Plantago lanceolata* (saponins, Bal2, Wyk)**, *Polygala senega* (saponins, Wyk), *Primula eliator* (saponins, Wyk), *Primula veris* (saponins, Wyk), *Pulmonaria officinalis* (saponins, Wyk), *Rumex crispus* (saponins, Wyk), *Salvia officinalis* (essential oils, Wyk), *Sambucus nigra* (Bal2), *Saponaria officinalis* (saponins, Wyk), *Sinapis alba* (Bal2), *Thea sinensis* (alkaloids, Wyk), *Thymus vulgaris* (essential oils, Wyk), *Thymus serpyllum* (essential oils, Wyk), *Thymus zygis* (essential oils, Wyk), *Trigonella foenum-graecum* (Bal1, Bal2), *Tussilago farafara* (Bal2), *Verbena officinalis* (saponins, Wyk), *Veronica officinalis* (saponins, Wyk), *Viola tricolor* (saponins, Wyk), *Vitis vinifera* (grape seed extract, Murr), *Zingiber officinale* (Bal1)
Herb Formulas: FOUR EMPERORS EPHEDRA DECOCTION *Ephedra sinica, Cinnamomum aromaticum* (bark), *Glycyrrhiza glabra, Prunus armeniaca* (seed) (Bal2); KUDZU DECOCTION *Pueraria lobata, Cinnamomum aromaticum, Ephedra sinica, Glycyrrhiza glabra, Paeonia lactiflora, Paeonia officinalis Zingiber officinalis, Ziziphus jujuba* (Bal2); MINOR BLUEGREEN DECOCTION *Asarum europaeum, Cinnamomum aromaticum* (bark), *Glycyrrhiza glabra, Paeonia lactiflora, Paeonia officinalis, Pinellia turnata, Schisandra chinensis, Zingiber officinale* (Bal2); MINOR BUPLEURUM DECOCTION *Bupleurum chinense, Bupleurum scorzoneraefolium, Glycyrrhiza glabra, Panax ginseng, Pinellia turnata, Scutellaria baicalensis, Scutellaria barata, Scutellaria laterilfora, Zingiber officinale, Ziziphus jujuba* (Bal2); OPHIOPOGONIS DECOCTION *Ophiopogon japonicus* (bulb), *Glycyrrhiza glabra, Oryza sativa* (porridge), *Panax ginseng, Pinellia turnata, Ziziphus jujuba* (Bal2)
Vegetables (live juices): Celery Juice, Parsley Juice, Spinach Juice (Bal1);
Vegetable Formulas: SOLGAR Reishi Mushrooms, Maitake Mushrooms, Sea Algae (Bal1);

Sinusitis RSN there are 4 sets of open spaces within the bones of the skull that are the "plumbing system" of the head; 1) above the eyes – frontal sinuses, 2) between the nose and the cheekbones – maxillary sinuses, 3) behind the bridge of the nose – sphenoid sinuses, 4) in the upper nose – ethmoid sinuses. If the sinuses are too small, poorly positioned or inflamed they cannot handle the volume of mucus produced and become clogged.

Herbs: *Althaea officinalis* (Bal1), *Ananas comosus* (bromelain, Bal1, Bal2), *Ballota nigra* (Bal1, Bal2), *Berberis vulgaris* (Murr), *Citrus-aurantius aurantium* (Bal1), *Coptis chinensis* (Murr), ***Echinacea purpurea* (Bal1, Murr)**, *Ephedra distachya* (Bal1), ***Hydrastis canadensis* (Bal1, Murr)**, *Illicium anisatum* (Bal1, Bal2), *Ligusticum porteri* (Bal2), *Mahonia aquifolium* (Murr), *Marrubium vulgare* (Bal1, Bal2), *Pelargonium sidoides* (Murr), *Quercus velutina* (quercetin, Bal1), *Rosa-hipa canina* (Bal1), *Sambucus nigra* (Bal2), *Trifolium pratense* (Bal1), *Thymus vulgaris* (Bal1, Bal2), *Trigonella foenum-graecum* (Bal1), *Uncaria guianensis* (Bal1, Bal2), *Uncaria tomentosa* (Bal1, Bal2), *Verbascum phlomoides* (Bal1), *Zingiber officinale* (Bal1)
Herb Formulas: FENU-THYME *Trigonella foenum-graecum, Thymus vugaris* (Bal1);
Vegetables: Cayenne Pepper, Flaxseed Oil, Garlic, Raw Onion (Bal1)

(1) **Chronic Sinusitis DSNA** The sinuses are unable to handle the volume of mucus and become clogged due to growths in the nose, or exposure to irritant fumes, smells, and allergies.

(2) **Acute Sinusitis DSNB, DSNV** The sinuses are unable to handle the volume of mucus and become clogged due to bacterial, or viral *Homosapiensvirus pico-rhino*, infections.

Sore Throat RST also known as **Pharyngitis, Strep Throat** the mucosa of the throat are subject to viral, bacterial (*Streptococcus*), fungal, and viral infections that can lead to painful inflammation. Difficulty in swallowing, hoarseness and swollen tissues are symptoms of sore throat. Essential oils, glucosinolates, and tannins have antibacterial properties.
AI = anti-inflammatory, EO = essential oils, GL = glucosinates, TN = tannins
Herbs: *Achillea millefolium* (EO, Wyk), *Allium sativum* (Bal2), *Althaea officinalis* (AI, root, Bal1), *Amoracia rusticana* (GL, Wyk), *Berberis vulgaris* (Bal2), *Chamaemelum nobile* (EO, Wyk), *Commiphora erythraea* (EO, Wyk), ***Commiphora myrrha* (EO, Bal2, Wyk)**, *Coptis chinensis* (Bal2), ***Echinacea purpurea* (DABA, DAVR, Bal1, Bal2, Murr)**, *Eucalyptus golbulus* (EO, Wyk), *Foeniculum vulgare* (EO, Wyk), *Geum urbanum* (TN, Wyk), *Glycyrrhiza glabra* (AI, Bal1), *Hamamelis virginiana* (TN, Wyk), ***Hydrastis canadensis* (TN, DABA, DAVR, Bal1, Bal2, Murr, Wyk)**, *Krameria lappacea* (TN, Wyk), *Mahonia aquifolium* (root, Bal2), *Mahonia nervosa* (root, Bal2), *Mahonia repens* (root, Bal2), *Matricaria recuttia* (EO, Wyk), *Mentha piperita* (EO, Wyk), *Nepeta cateria* (HFV, enema, Bal1), *Pelargonium sidoides* (Murr), *Plantago lanceolata* (AI, Wyk), *Potentilla erecta* (TN, Wyk), *Pulmonaria officinalis* (AI, Bal1), *Quercus robur* (TN, Wyk), *Rheum palmatum* (TN, Wyk), *Rubus-raspabacca idaeus* (AI, leaf, Bal1), *Salvia officinalis* (EO, Wyk), *Salvia triloba* (EO, Wyk), *Santalum album* (EO, Wyk), *Solidago virgaurea* (AI, Wyk), *Syzygium aromaticum* (EO, Wyk), *Thymus vulgaris* (EO, Wyk), *Trigonella foenum-graecum* (AI, gargle, Bal1), *Tropaeolum majus* (GL, Wyk), *Ulmus rubra* (AI, Bal1), *Verbascum thapsis* (AI, Bal1), *Viola tricolor* (AI, Wyk), *Zingiber officinale* (Bal2)
Herb Formulas: LONICERA & FORSTYTHIA POWDER *Forsythia suspensia, Lonicera japonica* (Bal2);
Vegetables (live juices): Beet Juice, Carrot Juice, Celery Juice, Parsley Juice, Spinach Juice (Bal1);

Skin Disorders

Skin Inflammation DSIN also known as **Dermatitis**, an enhanced sebum production and inflammation due to bacterial infection from *Corynebacterium acnes* and *Staphylococcus epidermidis* bacteria. Skin inflammation can also be caused by the Candida fungus and the Herpes virus.
Herbs: *Commiphora myrrha* (internal, Bal1), *Hoffmanseggia divaricata* (external, Bal1), *Hydrastis canadensis* (internal, Bal1), *Rumex crispus* (external, Bal1), *Tabebuia impetignosa* (internal, Bal1), *Taraxacum officinale* (internal & external, Bal1), *Trifolium pratense* (internal, Bal1)

(1) **Acne DACN** a clogging of pores in the skin of the face with oil from sebaceous glands. The oil becomes trapped when bacteria *Corynebacterium acnes* and *Proprionibacterum acnes* multiply in the pores of the skin where the sebaceous glands are. Pores become filled with sebum below the surface of the skin and whiteheads appear.
Herbs (internal): *Allium sativum* (Bal1), *Arctium lappa* (root, Bal1), *Bellis perennis* (Wyk), *Berberis vulgaris* (Bal2), *Capsicum anuum-longum-cayenneum* (Bal1), *Chelidonium majus* (Wyk), *Coptis chinensis* (Bal2), *Echinacea angustifolia* (Bal2), *Echinacea purpurea* (Bal1, Bal2), *Equisetum arvense* (Wyk), *Filipendula ulmaria* (Wyk), *Fumaria officinalis* (Wyk), *Hydrastis canadensis* (Bal2), *Mahonia aquifolium* (root, Bal2, Wyk), *Medicago sativa* (Bal1), *Ononis repens* (Wyk), ***Pinus pinaster* (Murr)**, *Serenoa repens* (Bal2), *Silybum marianum* (Bal2), *Stellaria media* (Wyk), *Trifolium pratense* (Bal1), *Viola tricolor* (Wyk), *Vitex agnus-castus* (Bal2), ***Vitis vinifera-rubra* (Murr)**
Herbs (external): *Aloe ferox* (Wyk), *Calendula officinalis* (flower, Bal2), *Fragaria vesca* (leaf, Bal1) *Hamamelis virginiana* (bark, leaf, Bal2), *Hoffmanseggia divaricata* (Bal1), *Juglans nigra* (leaf, Bal2), ***Juglans regia* (leaf, Bal2, Wyk)**, *Lavandula angustifolia* (oil, Bal2), ***Melaleuca alternifolia* (oil, Bal1, Bal2, Murr)**, *Rumex crispus* (root, Bal1), *Taraxacum officinale* (root, Bal1)
Herb Formulas: CINNAMON TWIG & PORIA PILL *Cinnamomum aromaticum, Paeonia officinalis, Paeonia suffruticosa, Prunus-Persica persica x communis* (seed), *Wolfiporia extensa* (Bal2); TANG KUEI & PEONY POWDER *Archangelica sinensis, Paeonia officinalis* (Bal2); TRIPHALA Amalaki, Bibhitaki, Haritaki (Bal2)
Vegetables (high fiber): Gobishi Burdock Cereal (Bal2)
Vegetables (oxalic acid): Almonds, Beets, Chard, Spinach (Bal1, Bal2)
Vegetables (zinc): Soybeans, Sunflower Seeds (Bal1)

(2) **Allergic Dermatitis** scaling, flaking, thickening, and discoloration when the skin comes in contact with perfumes, cosmetics, rubber, creams, ointments, poison ivy, poison oak, and gold, silver, and nickel found in jewelry, and overexposure to sunlight.

(3) **Atopic Dermatitis** scaling, flaking, thickening, and discoloration of skin on the face, behind the knees, and in the bends of the elbows, that first appears spontaneously in early childhood.

(4) **Dandruff DDND** sebaceous glands become overactive causing the scalp to form scales which may itch and burn.
Herbs: *Arctium lappa* (external, Bal2), *Copiafera langsdorffii* (shampoo, Bal2), *Hoffmanseggia divaricata* (external, Bal1), *Hydrastis canadensis* (internal, Bal1), *Linum usitatisimum* (flaxeed oil, internal, Bal2), *Melaleuca alternifolia* (external, Bal2), *Taraxaxum officinale* (internal, Bal1), *Thymus vulgaris* (external, Bal1), *Trifolium pratense* (internal, Bal1)
Vegetables: Raw (Bal1)

(5) **Eczema** also known as **Atopic Dermatitis**, skin inflammation occurs in fluid filled blisters that seep, ooze, and crust over.
TM = traditional medicine, MM = modern medicine

Herbs: *Aloe vera* (external, Bal2), *Althaea officinalis* (external, root, Bal2), *Anacardium occidentale* (TM, Wyk), *Arctium lappa* (internal, gobashi cereal, Bal2), *Avena sativa* (oatmeal, external, Bal2, Murr), *Calendula officinalis* (external, Bal2), *Cardiospermum halicacabum* (TM, Wyk), *Chelidonium majus* (TM, Wyk), *Coleus barbatus* (internal, forskolin extract, Bal2), *Copiafera langsdorffii* (external, Bal2), *Croton lechleri* (external, Bal2), *Croton tiglium* (TM, Wyk), *Curcuma longa* (external, Bal2), *Daphne mezereum* (TM, Wyk), *Echinacea pallida* (MM, Wyk), *Euphrasia nemorosa* (TM, Wyk), *Fagopyrum esculentum* (TM, Wyk), *Galium verum* (TM, Wyk), ***Glycyrrhiza glabra* (external, Bal2, Murr)**, *Hamamelis virginiana* (MM, external, Bal2, Wyk), *Juglans regia* (TM, external, leaf, Bal2, Wyk), *Linum usitatissimum* (MM, Wyk), *Mahonia aquifolium* (internal, root Bal2), *Mahonia nervosa* (internal, root, Bal2), *Mahonia repens* (internal, root, Bal2), ***Matricaria chamomilla* (MM, external, Bal2, Murr, Wyk)**, *Nerium oleander* (TM, Wyk), *Oenothera biennis* (MM, Wyk), *Quercus robur* (MM, internal, quercetin extract, Bal2, Wyk), *Rosmarinus officinalis* (external, Bal2), *Smilax regeli* (MM, internal, Bal2, Wyk), *Solanum dulcamara* (MM, Wyk), *Stellaria media* (external, Bal2), *Urtica dioica* (TM, Wyk), *Viola tricolor* (MM, Wyk)
Herb Formulas: KUDZU DECOCTION *Pueraria lobata, Cinnamomum aromaticum, Ephedra sinica, Glycyrrhiza glabra, Paeonia lactiflora, Paeonia officinalis Zingiber officinalis, Ziziphus jujuba* (Bal2); LEDEBOURIELLA DECOCTION *Ledebouriella divaricata, Angelica sinensis, Atractylodes macrocephala, Ephedra distachya,* Epsom Salt ($MgSO_4 * 7H_2O$), *Forsythia suspensa, Gardenia angusta* (fruit), *Glycyrrhiza glabra,* Gypsum ($CaSO_4 * 2H_2O$), *Ligusticum porteri, Mentha piperita, Paeonia officinalis, Platycodon grandiflorus, Rheum rhabarbarum* (wine treated root), *Schizonepeta tenuifolia, Scutellaria baicalensis, Scutellaria barata, Scutellaria lateriflora,* Talc [$Mg_3(SiO_4)_{10}(OH)_2$];
Vegetables: Brown Rice, Millet (Bal2), **Vegetables:** "Food Allergy in susceptible Inividuals is the major cause of Eczema." (Murr, page 584, under Quick Review)

(6) **Herpetiform Dermatitis** severe itching due to intestinal and immune disorders triggered by the ingestion of dairy products and gluten.
Herbs: *Commiphora myrrha, Hoffmanseggia divaricata, Hydrastis canadensis, Rumex crispus, Tabebuia impetignosa, Taraxacum officinalis, Trifolium pratense*

(7) **Intertrigo** caused by the friction of skin rubbing against skin. It most often occurs in the groin, underarms and between the ribs and breasts.
Herbs: *Aloe vera* (Bal1), *Calendula officinalis* (Bal1), *Matricaria chamomilla* (Bal1), *Medicago sativa* (Bal1), *Melaleuca alternifolia* (Bal1), *Equisetum arvense* (Bal1), *Hordeum vulgare* (Bal1), *Petroselinum crispum* (root, Bal1), *Taraxacum officinale* (root, Bal1)
Vegetables: Garlic (Bal1)

(8) **Nummular Dermatitis** round lesions appear on the limbs caused by contact with nickel.

(9) **Seborrheic Dermatitis DSBR** also called **Seborrhea**, scaly patches of skin that result from a disorder of the sebaceous glands.
Herbs: *Aloe vera* (external, gel, Murr), *Hydrastis canadensis* (Bal1), ***Melaleuca alternifolia* (external, oil, Bal1, Murr)**, *Taraxicum officinale* (Bal1), *Trifolium pratense* (Bal1)
Vegetables:50% to 75% Raw Vegetables (Bal1);

Abscess (ointment) **OSA** pus accumulating under the skin breaks through the skin.
Herbs: *Arctium lappa* (internal, Bal1), *Astragalus membranaceus* (internal, Bal1), *Capsicum annuum-longum-cayenneum* (internal, Bal1), *Echinacea purpurea* (internal, Bal1), *Hydrastis canadensis* (internal, Bal1), *Lobelia inflata* (external, Bal1), *Melaleuca alternifolia* (external, Bal1), *Pffafia paniculata* (internal, Bal1), *Rumex crispus* (internal, Bal1), *Silybum marianum* (internal, Bal1), *Taraxacum officinale* (internal & external, Bal1), *Trifolium pratense* (internal, Bal1), *Ulmus rubra* (external, Bal1)
Vegetables (Algae): Kelp (Bal1)

Bruises (ointment) **OBR** occurs when tissues underlying the skin are injured. The skin is not broken but blood collects under the skin resulting in "black and blue" marks. The bruise may be painful when touched.
Herbs: *Achillea millifolium* (Bal2), *Aesculus hippocastanum* (Bal2), *Ananas comosus* (Bal2), *Arnica montana* (Bal2), *Equisetum arvense* (Bal1), *Juglans nigra* (Bal1), *Medicago Sativa* (Bal1, Bal2), *Oreganum vulgare* (Bal2), *Rosa-hipa-scrubia canina* (Bal1), *Rumex crispus* (Bal1), *Taraxacum officinale* (Bal1)
Vegetables (leaf): Chard, Kale, Lettuce, Spinach, Turnip Greens, **Vegetables (grain):** Buckwheat (Bal1)

Burns (ointment) **OBN** First degree burns affect only the outer layer of skin. Second degree burns extend just below the surface. Third degree burns extend to the entire thickness of the skin and possible to the underlying tissues. Burns are characterized by redness, blistering, and acute pain.
Herbs: *Aloe vera* (Bal1, Bal2), *Astragalus membranaceus* (internal, Bal2), *Berberis vulgaris* (Bal2), *Calendula officinalis* (Bal2), *Coptis chinensis* (Bal2), *Equisetum arvense* (Bal1), *Hamemelis virginiana* (Bal2), *Hydrastis canadensis* (Bal1, Bal2), *Hypericum perforatum* (Bal2), *Lavandula angustifolia, Mahonia aquifolium* (Bal2), *Mahonia nervosa* (Bal2), *Mahonia repens* (Bal2), *Myrica cerifera* (Bal1), *Quercus alba* (Bal1), *Rhus glabra* (Bal1), *Rubus-nigrabacca-bruscia fruticosis* (Bal1, Bal2), *Rubus-raspabacca idaeus* (Bal2), *Thea sinensis-nigra* (Bal1), *Thea sinensis-viridis,* (Bal1)*Ulmus rubra* (Bal1)
Vegetables: containing Vitamins A, B, C, D, E.

Corns & Warts

(1) **Corns DCLW, OCW** an excessive thickening of the skin (hyperkeratosis) resulting in calluses on the palm of the hand or the sole of the foot sometimes caused by *Staphylococcus* or *Streptococcus* bacteria.
Herbs: *Hydrastis canadensis* (Bal1), *Melaleuca alternifolia* Bal1)

(2) **Leprosy DCLW**

(3) **Warts DCLW, OCW,** small growths on the skin cause by *Homosapiensvirus papilloma*. Common warts can be flat or raised and are found on the hands, fingers, elbows, forarms, knees, face and around the fingernails. Plantar warts are bumpy white growths that resemble calluses and are found on the soles of the feet and the undersides of the toes. Genital Warts are soft moist growths and are found on the vagina, anus, penis, groin, and scrotum.
Herbs (external): *Aloe vera* (external, Bal1, Bal2), *Ananas comosus* (bromelain, Bal2), *Astragalus membranaceus* (Bal2), *Commiphora myrrha* (Bal1), *Gaultheria procumbens* (Bal1), *Hydrastis canadensis* (Bal1), *Juglans nigra* (Bal1, Bal2), *Melaleuca alternifolia* (Bal1), *Potentilla erecta* (Bal2), *Stellaria media* (Bal1), *Symphytum officinale* (external, Bal2), *Syzygium aromaticum* (Bal1), *Tabebuia impetignosa* (Bal1)
Vegetables: Asparagus, Garlic, Onions (Bal1);

Dry Skin (ointment) **ODS** a lack of natural oil secreted by the sebaceous glands in the skin. It most often affects women under the age of 35. Chapping an cracking are signs of dry skin.
Herbs: *Aloe vera* (external, Bal1), *Calendula officinalis* (external, Bal1), *Lavandula angustifolia* (external, Bal1), *Mentha piperita* (external, Bal1), *Symphytum officinale* (external, Bal1)
Vegetables (beta-carotene): Carrots, Pumpkin, Squash, Sweet Potato, Yam, **Vegetables (sulfur):** Garlic, Onions, Asparagus

Oily Skin the sebaceous glands produce more oil than is needed for proper lubrication of the skin.
Herbs: *Aloe vera* (external, Bal1), *Arctium lappa* (Bal1), *Avena sativa* (straw, Bal1), *Cymbopogon citratus* (Bal1), *Equisetum arvense* (Bal1), *Glycyrrhiza glabra* (Bal1), *Lavandula angustifolia* (external, Bal1), *Matricaria chamomilla* (Bal1), *Rosa-hipa canina* (Bal1), *Thymus vulgaris* (Bal1)
Vegetables: "Do not cook with oil, and do not eat any oils that have been subject to heat, where in processing or cooking." (Bal1, page 413, recommendation #2)

Perspiration (ointment) **OPS** also known as **Body Odor** (BO) and Hyperrhidrosis excessive sweating especially on the forehead and in the arm pits under the arms.
Herbs: *Atropa belladonna* (Wyk, TM), *Quercus robur* (Wyk, TM), *Salvia officinalis* (Wyk, TM), *Sanguinaria canadensis* (Wyk, TM), *Viola tricolor* (Wyk, TM)

Rash a redness of the skin due to cosmetics, molds, toxic chemicals, and wind chaffing.
Herbs: *Calendula officinalis* (external, Bal1), *Matricaria chamomilla* (external, Bal1), *Melaleuca alternifolia* (external, oil, Bal1), *Sambucus nigra* (external, flower, Bal1)
Herb Formulas: SKIN RASH POULTICE *Hoffmannseggia divaricata, Rumex crispus* (root), *Taraxacum officinale* (Bal1)

Rosacea DROS begins with red flushing of the nose and face and ends with blotchy red areas on the forehead, nose, cheekbones and chin. About 1 in 20 Americans have this disease. It is most common in women between the ages of 30 and 50.
Herbs: *Althaea, millefolium* (root, Bal1), *Arctium lappa* (root, Bal1), *Aloe vera* (external, Bal1), *Angelica sinensis* (Bal1), *Borago officinalis* (seed, Bal1), *Calendula officinalis* (Bal1), *Capsicum-annuum-longum-cayenneum* (Bal1), *Foeniculum vulgare* (seed, Bal1), *Medicago sativa* (Bal1), *Petroselinum crispum* (Bal1), *Pinus pinaster* (Murr), *Rosmarinus officinalis* (Bal1), *Rumex crispus* (root, Bal1), *Salvia officinalis* (Bal1), *Smilax aristolochaefolia* (Bal1), *Silybum marianum, Smilax ferbifuga* (Bal1), *Smilax glyciphylla* (Bal1), *Smilax ornata* (Bal1), *Smilax regelii* (Bal1), *Taraxaxum officinale* (root, Bal1), *Trifolium pratense* (Bal1), *Ulmus rubra* (Bal1), *Urtica dioica* (Bal1), *Vitis vinifera* (seed extract, Murr)
Vegetables: Raw (Bal1); **Vegetables (grain):** Brown Rice, Buckwheat, Cereal, Corn, Millet, Oats, Quinoa, Rice Bran, Rye, Spelt, Tapioca, White Flour, White Rice, Wheat (Bal1);

Scaly Skin OSS also known as **Psoriasis**, an inflammation of the skin resulting in scaly red patches of dry skin.
TM = traditional medicine, MM = modern medicine
Herbs: ***Aloe vera* (external, Bal2, Murr)**, *Althaea officinalis* (external, root, Bal2), *Ammi majus* (MM, Wyk), *Angelia sinensis* (Bal2), *Arctium lappa* (root, Bal1), *Andira araroba* (Synonym for *Vataireopsis araroba* Wyk), *Angelica sinensis* (MM, Wyk), *Berberis vulgaris* (Bal2), *Betula pendula* (TM, Wyk), ***Capsicum-annuum-longum-cayenneum* (external, Murr, Bal2)**, *Carapa guianensis* (external, Bal2), *Centella asiatica* (TM, Wyk), *Copiafera langsdorffii* (external, Bal2), *Coptis chinensis* (Bal2), *Curcuma longa* (Murr), *Fagus sylvatica* (TM, Wyk), *Glycine max-soja* (isoflavone, Bal2), *Glycyrrhiza glabra* (external, Bal2), ***Hydrastis canadensis* (Bal2, Murr)**, *Juniperus communis* (TM, Wyk), *Lavandula angustifolia* (external, Bal1, Bal2), *Levisticum officinale* (MM, Wyk), ***Mahonia aquifolium* (MM, root, Bal2, Wyk)**, *Matricaria chamomilla* (external, Bal2), *Psoralea coryifolia* (Bal2), *Rumex crispus* (Bal1), ***Silybum marianum* (Bal1, Bal2, Murr)**, ***Smilax aristolochiaefolia* (TM, Bal1, Bal2, Murr, Wyk)**, ***Smilax regelii* (TM, Bal1, Bal2, Murr, Wyk)**
Herb Formulas: COLON CLEANSE *Achillea millefolium, Allium sativum, Aloe vera, Althaea officinalis, Arctium lappa, Berberis vulgaris, Calendula officinalis, Echinacea purpurea, Equisetum arvense, Eupatorium perfoliatum, Foeniculum vulgare, Frangula purshiana, Glycyrrhiza glabra, Juglans cineria* (bark), *Inula helinium, Linum usitatissimum* (seed), *Lobelia inflata, Mentha piperita, Rheum rhabarbarum, Rubus-raspabacca idaea, Senna alexandrina, Tabebuia impetignosa, Trigonella foenum-graecum, Ulmus rubra* (Bal1); FOUR SUBSTANCE DECOCTION *Angelica sinensis, Ligusticum porteri, Paeonia lactiflora, Paeonia officinalis, Rehmannia glutinosa* (Bal2); PSORIASIS POULTICE *Hoffmannseggia divaricata, Rumex crispus, Taraxacum officinale* (Bal1);
Vegetables: Flaxseed Oil, Psyllium Husks, Soybean Oil (Bal1); **Vegetables (grain):** Brown Rice, Buckwheat, Cereal, Corn, Millet, Oats, Quinoa, Rice Bran, Rye, Spelt, Tapioca, White Flour, White Rice, Wheat (Bal1);

Sebaceous Cyst DSBC skin growths that contain a mixture of sebum oil and skin proteins.
Herbs: *Aloe vera* (external, Bal1), *Arctium lappa* (root, Bal1), *Hydrastis canadensis* (external, Bal1), *Melaleuca alternifolia* (external, oil, Bal1), *Silybum marianum* (Bal1), *Trifolium pratense* (Bal1)

Sunburn OSB excessive exposure to the sun's ultraviolet rays. The skin becomes red, and tender to the touch.
Herbs (external): *Aloe vera* (Bal1), *Calendula officinalis* (flower, Bal1), *Centella asiatica* (Bal1), *Equisetum arvense* (Bal1), *Hypericum perforatum* (Bal1), *Matricaria chamomilla* (Bal1), *Melaleuca alternifolia* (Bal1), *Symphytum officinale* (Bal1)
Vegetables (legume): Chick Peas, Green Beans, Kidney Beans, Lentils, Lima Beans, Mung Beans, Peas, Pinto Beans, Soybeans, Red Beans, White Beans (Bal1);

Ulcers (ointment) **OUL** open sores on the skin.

M. **Bedsores** are formed when pressure is exerted over bony areas of the body. Blood circulation is restricted causing the death of cells in the overlying tissue.

Herbs: *Calendula officinalis* (Bal1), *Commiphora myrrha* (Bal1), *Fagopyrum esculentum* (Bal1), *Hydrastis canadensis* (Bal1), *Pfaffia paniculata* (Bal1), *Symphytum officinale* (Bal1), *Tabebuia impetignosa* (Bal1), *Tilia cordata* (Bal1)
Vegetables: Raw, Oat Bran, Psyllium Husks (Bal1)

(2) Stress **Canker Sores** appear on the lips, gums, tongue, and the inside of cheeks. They start out as red inflammations with a yellow border. Fluid and white blood cells aggravate the inflammation causing the skin to burst into an open ulcer.

(3) Infectious **Canker Sores** appear on the lips, gums, tongue, and the inside of cheeks due to viruses. They start out as red inflammations with a yellow border. Bacteria invade the inflammation causing the skin to burst into an open ulcer.

Herbs: *Arctium lappa* (Bal1), *Cistus albiflorus* (Bal2), *Cistus crecticus* (Bal2), *Cistus ladaniferus* (Bal2), *Cistus maculatus* (Bal2), *Commiphora myrrha* (Bal2), ***Glycyrrhiza glabra* (deglycyrrhizinated licorice, Bal2, Murr),** *Hydrastis canadensis* (Bal1, Bal2), *Melaleuca alternifolia* (Bal1), *Pinus pinaster* (pine bark extract, Murr), *Quercus velutina* (quercetin, Bal2), *Rubus-raspabacca idaeus* (Bal1), *Tabebuia impetignosa* (Bal1), *Trifolium pratense* (Bal1), *Vitus vinifera* (grape seed extract, Murr)
Vegetables: Lettuce in salads (Bal1), Raw Onions (Bal1, Bal2)

(4) **Leg Ulcers** poor blood circulation in the legs causes dead patches of skin. The skin breaks producing an open sore.

Herbs: *Echinacea purpurea* (Bal1), *Hydrastis canadensis* (Bal1), *Medicago sativa* (Bal1), *Symphytum officinale* (external, Bal1), *Trifolium pratense* (Bal1)
Vegetables: Garlic (Bal1), **Vegetables (vitamin K, crucifer):** Broccoli, Brussels Sprouts, Cauliflower (Bal1); **Vegetables (vitamin K, dark green leaf):** Chard, Collard Greens, Cornsalad Lettuce, Kale, Mustard Greens, Parsley, Seakale, Spinach (Bal1);

Vitiligo DVTL also known as **Scaly Skin OSS**, an inflammation of the skin resulting in scaly white patches of dry skin.
TM = traditional medicine, MM = modern medicine
Herbs: ***Ammi majus* (MM, Bal2, Wyk)**, *Ammi visagna* (Bal2), *Andira araroba* (TM, Wyk), ***Angelica sinensis* (MM, Bal2, Wyk)**, *Betula pendula* (TM, Wyk), *Centella asiatica* (TM, Wyk), *Fagus sylvatica* (TM, Wyk), *Glycyrrhiza glabra* (Bal2), *Hypericum perforaium* (external, cream, Bal2), *Juniperus communis* (TM, Wyk), *Levisticum officinale* (MM, Wyk), *Mahonia aquifolium* (MM, Wyk), *Picrorhiza kurrooa* (Kutki Picroside Foxglove, Bal2), *Primula veris* (oil, Bal1), *Psoralea coryifolia* (Bal2), *Smilax aristolochiaefolia* (TM, Wyk), *Smilax regelii* (TM, Wyk)
Herbs: ASTRAGALUS DECOCTION *Astragalus membranaceus, Cinnamomum aromaticum, Glycyrrhiza glabra, Hordeum vulgare* (malt sugar), *Paeonia officinalis* (root), *Zingiber officinale, Ziziphus jujuba* (fruit) (Bal2);

Wounds OWN wounds are caused by cuts, stabs, abrasions and scratches.
Herbs: *Achillea millefolium* (Bal1), *Allum cepa* (Bal1), *Aloe vera* (external, Bal1, Bal2), *Ananas comosus* (internal, Bal1, Bal2), *Arnica montana* (Bal1), *Artemisia abrotanum* (Bal1), *Bellis perennis* (Bal1), *Berberis vulgaris* (external Bal2), *Calendula officinalis* (external, Bal1, Bal2), *Carapa guianensis* (external Bal2), *Carica papaya* (Bal1), *Centella asiatica* (internal, Bal1, Bal2), *Cinnamomum camphora* (Bal1), *Commiphora myrrha* (Bal1), *Coptis chinensis* (external, Bal2), *Croton lechleri* (external Bal2), *Echinacea pupurea* (Bal1), *Eucalyptus globulus* (Bal1), *Hamamelis virginiana* (Bal1), *Hydnocarpus kurzii* (Bal1), *Hydrastis canadensis* (external Bal2), *Hypericum perfoliatum* (Bal1), *Juglans regia* (Bal1), *Krameria lappacea* (Bal1), *Mahonia aquifolium* (external Bal2), *Mahonia nervosa* (external Bal2), *Mahonia repens* (external Bal2), *Marrubium vulgare* (Bal1), *Matricaria chamomilla* (external, Bal1, Bal2), *Melaleuca alternifolia* (Bal1), *Myroxylon balsamum* (Bal1), *Pinus pinaster* (internal, pine bark extract, Bal1, Bal2), *Quercus robur* (Bal1), *Rhododendron tomentosum* (Bal1), *Rosmarinus officinalis* (Bal1), *Ruta graveolens* (Bal1), *Symphytum officinale* (external. Bal1, Bal2), *Syzygium aromaticum* (Bal1), *Verbena officinalis* (Bal1), *Vitis vinifera* (internal, grape seed extract, Bal2)

Weak Hair, Nails DWHN nails are composed of the protein keratin. Health nail beds are pink. Weak nails are the result of nutritional deficiencies.
Herbs: *Arctium lappa* (Bal1), *Avena sativa* (Bal1), *Borago officinalis* (Bal1), *Centella asiatica* (Bal1), *Cimicifuga racemosa* (Bal1), *Cucurbita maxima-pepo* (seed, Bal1), *Curcuma longa* (Bal1), *Cymbopogon citratus* (Bal1), *Equisetum arvense* (Bal1), *Ginkgo biloba* (Bal1), *Linum usitatisimum* (Bal1), *Matricaria chamomilla* (Bal1), *Medicago sativa* (Bal1), *Petroselinum crispum* (Bal1), *Primula veris* (Bal1), *Rosmarinus officinalis* (Bal1), *Rumex crispus* (Bal1), *Ruscus aculeatus* (Bal1), *Salvia officinalis* (Bal1), *Smilax aristolochaefolia* (Bal1), *Smilax ferbifuga* (Bal1), *Smilax glyciphylla* (Bal1), *Smilax ornata* (Bal1), *Smilax regelii* (Bal1), *Taraxacum officinale* (Bal1)
Herb Formulas: SKIN & NAIL SUPPORT *Arctium lappa* (root, 17mg), *Centella asiatica* (leaf & root, 42mg), *Emblica officinalis* (triphala extract blend, 100mg), *Equisetum arvense* (aerial, 150mg, mineral ash 1mg), *Mahonia aquifolium* (root, 5mg), *Medicago sativa* (aerial, 60mg, mineral ash, 1mg), *Rumex crispus* (root, 26mg), *Scrophularia nodosa* (aerial, 6mg), *Terminalia bellerica* (triphala extract blend 50mg), *Terminalia chebula* (triphala extract blend, 50mg), *Urtica dioica* (leaf, 22mg, mineral ash 1mg) (Scal);
Vegetables: Oatmeal, Nuts, Seeds (Bal1); **Vegetables (grain):** Brown Rice, Buckwheat, Cereal, Corn, Millet, Oats, Quinoa, Rice Bran, Rye, Spelt, Tapioca, White Flour, White Rice, Wheat (Bal1); **Vegetables (legume):** Chick Peas, Green Beans, Kidney Beans, Lentils, Lima Beans, Mung Beans, Peas, Pinto Beans, Soybeans, Red Beans, White Beans (Bal1);

Wrinkles OSW a creasing of the skin due to a loss in elasticity. The cause of 90% of the cases of wrinkles is due to overexposure to the sun.
Herbs: *Aloe vera* (external, gel, Bal1, Bal2), *Arctium lappa* (root, Bal1), *Avena sativa* (straw, Bal1), *Calendula officinalis* (external, cream, Bal2), *Cucurbita maxima-pepo* (seed, Bal1), *Cymbopogon citratus* (Bal1), *Borago officinalis* (seed, Bal1), *Equisetum arvense* (Bal1), *Hamamelis virginiana* (external, cream, Bal2), *Linum usitatissimum* (seed Bal1), *Malpighia glabra* (Bal2), *Matricaria chamomilla* (Bal1), *Medicago sativa* (Bal1), *Petrosalinum crispum* (Bal1), *Symphytum officinale* (external, cream, allantoin, Bal2), *Thea sinensis* (external, cream, Bal2), *Thymus vulgaris* (Bal1), *Zingiber officinale* (Bal1)
Vegetables: Seeds, Nuts (Bal1); **Vegetables (grain):** Brown Rice, Buckwheat, Cereal, Corn, Millet, Oats, Quinoa, Rice Bran, Rye, Spelt, Tapioca, White Flour, White Rice, Wheat (Bal1); **Vegetables (legume):** Chick Peas, Green Beans, Kidney Beans, Lentils, Lima Beans, Mung Beans, Peas, Pinto Beans, Soybeans, Red Beans, White Beans (Bal1);

Urinary System Disorders

Urinary **Bladder Infection UBL** also known as **Cystitis**, infection is usually caused by some kind of bacteria. Nearly 85% of unrinary tract infections are caused by the bacterium *Escherichia coli*. *Chlamydia* may also cause bladder problems. Bladder infections are characterized by an urgent desire to empty the bladder.
Herbs: *Agathosma betulina* (Bal2), *Agathosma crenulata* (Bal1, Bal2), *Allium sativum* (Bal2), *Aloe vera* (Murr), *Althea officinalis* (Bal1, Bal2), *Arctium lappa* (Bal1), ***Arctostaphylos uva-ursi* (Bal1, Bal2, Murr)**, *Artemisia vulgaris* (Bal2), *Astragalus membranaceus* (Bal2), *Berberis vulgaris* (Bal2), *Betula alba* (Bal2), *Betula lenta* (Bal2), *Betula nigra* (Bal2), *Betula pendula* (Bal1, Bal2), *Centella asiatica* (Murr), *Coptis chinensis* (Bal2), ***Hydrastis canadensis* (Bal1, Bal2, Murr)**, *Juniperus communis* (Bal1), *Mahonia aquifolium* (Bal2), *Mahonia*

nervosa (Bal2), *Mahonia repens* (Bal2), *Pinus pinaster* (pine bark extract, Murr), *Piper methysticum* (Bal1), *Rosa-hipa-scrubia canina* (Bal1), *Ruscus aculeatus* (Bal2), *Taraxacum officinale* (Bal2), ***Vaccinium-cranabacca macrocarpon* (Bal1, Bal2, Murr)**, *Viburnum-drysia prunifolium* (Bal2), *Vitis vinifera* (grape seed extract, Murr), *Zea mays-suavis* (Bal1)
Vegetables: Celery, Parsley (Bal1)
Vegetables: "Avoid all simple sugars, refined carbohydrates, full strength fruit juice (diluted fruit juice is acceptable) and food allergens" (Murr, page 476,under Diet)

Endometriosis DEND inner uterine wall tissue is deposited outside the uterus, The tissues form cysts. Symptoms are bleeding between menstrual periods, unusual discharge, painful menstrual periods, heavy periods.
Herbs: *Allium sativum* (Bal1), *Angelica sinensis* (Bal1), *Arctium lappa* (root, Bal1), *Astragalus membranaceus* (Bal1), *Commiphora myrrha* (gum, Bal1), *Curcuma longa* (Bal2), *Dioscorea oppositifolia* (Bal2), *Dioscoria villosa* (Bal2), *Hydrastis canadensis* (Bal1), *Linum usitatissimum* (flaxseed oil, Murr), *Medicago sativa* (Bal1, Bal2), *Pinus pinaster* (pine bark extract, Murr), *Rubus-raspabacca idaeus* (leaf, Bal1), *Tabebuia impetignosa* (Bal1), *Thea sinensis-viridis* (Bal2), *Trifolium pratense* (Bal1), *Vitex agnus-castus* (Murr), *Vitis vinifera* (grape seed extract, Murr)
Herb Formulas: DONG QUAI & PEONY POWDER *Angelica sinensis, Paeonia lactiflora, Paeonia officinalis, Alisma plantago-aquatica, Atractylodes macrocephala, Ligusticum porteri, Wolfiporia extensa* (Bal2); FOUR SUBSTANCE DECOCTION *Angelica sinensis, Ligusticum porteri, Paeonia lactiflora, Paeonia officinalis, Rehmannia glutinosa* (Bal2)
Vegetables (crucifer): Broccoli, Brussels Sprouts, Cabbage, Cauliflower (Murr); **Vegetables (dark green anti-oxidant):** Artichoke, Asparagus, Mild Bell Peppers, Broccoli, Brussels Sprouts, Chard, Collard Greens, Cornsalad Lettuce, Cucumbers, Green Beans, Kale, Leeks, Mustard Greens, Parsley, Peas, Seakale, Spinach, Zucchini Squash (Bal2); **Vegetables (grain):** Brown Rice, Buckwheat, Cereal, Corn, Millet, Oats, Quinoa, Rice Bran, Rye, Spelt, Tapioca, White Flour, White Rice, Wheat (Murr); **Vegetables (yellow anti-oxidant):** Bell Peppers (yellow), Corn, Oca Tuber, Squash (Bal2);

Enlarged Prostate (UEP) Also known as **Benign Prostatic Hypertrophy** and **Prostatitis**, the prostate is enlarged by the growth of fibromuscular and epithelial tissues around the urethra. About 70% of men above age 75 have an enlarged prostate. The prostate gland encircles the urethra and forms seminal fluid which is the bulk of the ejaculate.
Herbs: *Agathosma crenulata* (UDR, Bal1), *Arctostaphylos uva-ursi* (UDR, Bal1, Bal2), *Capsicum annuum-longum-cayenneum* (Bal1), *Carica papaya* (NSD, Bal2), *Chamaelirium luteum* (Bal1), *Crateva nervula* (fat metabolizer, Bal2), *Cucurbita pepo* (TM, Wyk), *Eleuthrococcus senticosus* (tonic, Bal1), *Epilobium angustifolium* (TM, Wyk), *Epilobium parviflorum* (TM, Wyk), *Equisetum arvense* (blood in urine, Bal1), *Erythroxylum catuaba* (Bal2), *Humulus lupulus* (NSD, Wyk), *Hydrastis canadensis* (DASC, UDR, Bal1), *Hypericum perforatum* (NSD, Wyk), *Hypoxis hemerocallidea* (TM, Wyk), *Juniperus communis* (berries, UDR, Bal1), *Panax ginseng* (Bal1), *Petroselinum crispum* (UDR, Bal1), *Phleum pratense* (TM, Wyk), *Piper methysticum* (NSD, Wyk), *Populus tremula* (TM, Wyk), *Populus tremuloides* (TM, Wyk), ***Prunus-Cerasus africana* (TM, Bal1, Bal2, Murr, Wyk)**, *Pygeum africanum* (Synonym for *Prunus-Cerasus africana*), *Rhus aromatica* (TM, Wyk), *Scopolia carniolica* (TM, Wyk), *Scutellaria baicalensis* (Bal2), *Scutellaria barata* (Bal2), *Scutellaria lateriflora* (Bal2), ***Secale cereale* (TM, pollen extract, Murr, Wyk)**, ***Serenoa repens* (NSD, Bal1, Bal2, Murr, Scal, Wyk)**, *Ulmus rubra* (bark, UDR, Bal1), ***Urtica dioica* (root, growth hormone inhibitor, Bal2, Murr, Wyk)**, *Vaccinium-vortabacca myrtillus* (Bal2), *Valeriana officinalis* (NSD, Wyk), ***Zea mays-saccharata* (cornsilk, TM, Bal1, Wyk)**
Herb Formulas: EIGHT INGREDIENT REHMANNIA PILL *Rehmannia glutinosa, Alisma plantago-aquatica, Astragalus membranaceus, Cornus officinalis, Dioscorea oppositifolia or villosa, Paeonia suffrutecosa, Schisandra chinensis, Wolfiporia extensa* (Bal2); FIVE INGREDIENT POWDER WITH PORIA *Alsima plantago-aquatica Atractylodes macrocephala Cinnamomum aromaticum, Polyporus brumalis, Polyporus squamosus, Polyporus tubaster, Polyporus umbellatus, Wolfiporia extensa*, (Bal2); GRAVEL ROOT DECOCTION *Eupatorium purpureum* (root), *Althaea officinalis* (leaf), *Eryngium maritimum, Hydrangea arborescens* (root) Bal1; MINOR BUPLEURUM DECOCTION *Bupleurum chinense, Bupleurum scorzoneraefolium, Glycyrrhiza glabra, Panax ginseng, Pinellia turnata, Scutellaria baicalensis, Scutellaria barata, Scutellaria lateriflora, Zingiber officinale, Ziziphus jujuba* (Bal2);
Vegetables: Pumpkin Seeds (Bal1, Bal2)

Incontinence UIN also known as **Enurensis**, involuntary urination and bed-wetting can be caused by small urinary bladder size, urinary tract infection, or diabetes.
Herbs: *Agathosma crenulata* (Bal1, Bal2), *Avena sativa* (oat straw, Bal1, Bal2), *Equisetum arvense* (Bal2), *Petroselinum crispum* (Bal1, Bal2), *Plantago major* (Bal1), *Rubus-raspabacca idaeus* (Bal2), *Zea mays-saccharata* (cornsilk, Bal1, Bal2)
Herb Formulas: LICORICE, WHEAT & JUJUBE DECOCTION *Glycyrrhiza glabra, Triticum aestivum, Ziziphys jubata* (Bal2); MAJOR CONSTRUCT MIDDLE DECOCTION *Panax ginseng, Hordeum vulgare* (maltose sugar), *Zanthoxylum americanum* or *bungeanum, Zingiber officinale* (Bal2);
Vegetables: Brewer's Yeast, Spirulina, Sprouts (Bal1)

Kidney Infection also known as **Nephritis**, is combated by Urinary Kidney Cleanse **UKC** herbs. Impaired kidney function is caused by Diabetes, High Blood Pressure, Liver Cirrhosis, Lupus and toxins such as insect venom, poison mushrooms, snake venom and pesticides.
Herbs: Abuta (Bal2), *Agathosma betulina* (Bal1), *Agathosma crentulata* (Bal1), *Althaea officinalis* (root, Bal1, Bal2), *Arctostaphylos uva-ursi* (Bal1, Bal2), *Bupleurum falcatum* (Bal2), *Bupleurum, chinense* (Bal2), *Bupleurum scorzonaeraefolium* (Bal2), *Hydrangea arborescens* (Bal1), *Juniperus communis* (Bal1), *Solidago virga-aurea* (Bal1), *Taraxacum officinale* (root, Bal1, Bal2), *Trifolium pratense* (Bal1), *Urtica dioica* (Bal1), *Vaccinium-cranabacca oxycoccus* (Bal1), *Wolfiporia extensa* (Bal2), *Zea mays-saccharata* (cornsilk, Bal1)
Herb Formulas: FIVE INGREDIENT POWDER WITH PORIA *Alsima plantago-aquatica Atractylodes macrocephala Cinnamomum aromaticum, Polyporus brumalis, Polyporus squamosus, Polyporus tubaster, Polyporus umbellatus, Wolfiporia extensa*, (Bal2); MINOR BUPLEURUM DECOCTION *Bupleurum chinense, Bupleurum scorzoneraefolium, Glycyrrhiza glabra, Panax ginseng, Pinellia turnata, Scutellaria baicalensis, Scutellaria barata, Scutellaria lateriflora, Zingiber officinale, Ziziphus jujuba* (Bal2);
Vegetables: Asparagus, Celery, Cucumbers, Garlic, Papaya, Parsley Seeds, Potatoes, Watercress (Bal1); **Vegetables (legume):** Chick Peas, Green Beans, Kidney Beans, Lentils, Lima Beans, Mung Beans, Peas, Pinto Beans, Soybeans, Red Beans, White Beans (Bal1);

(1) **Bright's Disease** also known as **Nephrotic Syndrome**, kidney infection marked by blood protein in the urine, high blood pressure and edema caused by kidney capillary lesions.

(2) **Glomerulnephritis** an inflammation of glomeruli of the filtering units within the kidney caused by the bacteria *Streptococcus pyrogenes*

(3) **Hydronephrosis** an obstruction of urinary flow causing the kidney and renal pelvis to become filled with urine.

(4) **Pyelonephritis** also known as **Pyelonephrosis**, an inflammation of the kidneys and renal pelvis caused by bacterial infection.

(5) **Renal Tubular Acidosis** the kidneys fail to reabsorb bicarbonate normally causing impaired ammonia production and acid excretion.

Kidney Stones also known as **Renal Calculi**, calcium accumulation in the kidneys. Normally magnesium ammonium phosphate (NH4)MgPO4*6(H2O) calcium oxylate CaO4, cystine C6H12O4N2S2 and uric acid C5H4N4O3 remain suspended in solution. When the immune system is suppressed these substances may crystallize and clump together. The symptom is sharp pain in the lower abdomen and groin. Kidney stones are hereditary. Approximately 1 in 1000 Americans have kidney stones.
Herbs: ***Aloe Vera* (juice for calcium stones, Bal1, Bal2, Murr)**, *Althaea officinalis* (root, Bal1, Bal2), *Ammi visnaga* (Bal2), *Arctostaphylos uva-ursi* (Bal1), *Betula alba* (leaf, Bal2), *Betula lenta* (leaf, Bal2), *Betula nigra* (leaf, Bal2), *Betula pendula* (leaf, Bal2), *Crateva nervula* (Bal2), *Ginkgo biloba* (Bal1), *Hydrastis canadensis* (Bal1), *Phyllanthus amarus* (Bal2), *Senna alexandrina* (Murr)
Herb Formulas: POLYPORUS DECOCTION *Polyporus brumalis, Polyporus squamosus, Polyporus tubaster, Polyporus umbellatus, Alisma plantago-aquatica,* Kadinum Red Clay, Vegetable Gelatin, *Wolfiporia extensa* (Bal2);
Vegetables (vitamin A): Alfalfa Sprouts, Carrots, Pumpkin, Squash, Sweet Potatoes (Bal1)
Vegetables (beverage): Green Tea (*Thea sinensis-viridis*), Kinsenso Tea (*Lysimachia officinalis*), Lemon Juice (*Citrus-limonus limon*), Orange Juice (*Citrus-aurantius sinensis*) (Bal2)
Vegetables (low oxylate): Acorn Squash, Arugala, Coconut, Ketchup, Onions, Peas, Red Peppers, White Rice, Wild Rice, Zucchini Squash (Murr)

Urethritis UUR inflammation of the urinary tract which transports urine from the kidney to the urethra where it is expelled. Diuretic herbs have a rinsing and anti-iflammatory effect on the urinary tract.
Herbs: *Adonis vernalis* (TM, Wyk), *Agathosma betulina* (TM, Wyk), *Ammi visnaga* (analgesic, Wyk), *Angelica sinensis* (TM, Wyk), *Apium graveolens* (TM, Wyk), *Asparagus officinalis* (TM, Wyk), *Atropa belladonna* (analgesic, Wyk), *Berberis vulgaris* (analgesic, Wyk), *Betula pendula* (TM, Wyk), *Chelidonium majus* (analgesic, Wyk), *Convallaria majalis* (TM, Wyk), *Equisetum arvense* (TM, Wyk), *Fagopyrum esculentum* (TM, Wyk), *Genista tinctoria* (TM, Wyk), *Helichrysum arenarium* (TM, Wyk), *Helleborus niger* (TM, Wyk), *Herniaria glabra* (TM, Wyk), *Hibiscus sabdariffa* (TM, Wyk), *Juniperus communis* (TM, Wyk), *Lespedeza capitata* (TM, Wyk), *Levisticum officinale* (TM, Wyk), *Melilotus officinalis* (TM, Wyk), *Melissa officinalis* (analgesic, Wyk), *Ononis spinosa* (TM, Wyk), *Orthosiphon aristatus* (TM, Wyk), *Petasites hybridus* (analgesic, Wyk), *Petroselinum crispum* (TM, Wyk), *Phaseolus vulgaris* (TM, Wyk), *Pimpinella major* (TM, Wyk), *Polygonum hydropiper* (TM, Wyk), *Rhamnus frangula* (TM, Wyk), *Ribes nigrum* (TM, Wyk), *Rosa-hipa canina* (TM, Wyk), *Rubia tinctorium* (urulithiasis, Wyk), *Scopolia carnoiolica* (analgesic, Wyk), *Similax aristolochaefolia* (TM, Wyk), *Solidago canadensis* (TM, Wyk), *Solidago virgaurea* (TM, Wyk), *Taraxacum officinale* (TM, Wyk), *Urginea maratima* (TM, Wyk), *Urtica doica* (TM, Wyk), *Viola tricolor* (TM, Wyk)
Herb Formulas: URINARY SUPPORT *Arctostaphylos uva-ursi* (leaf, 370mg), *Chimaphila umbellata* (herb, 92mg), *Coptis chinensis* (rhizome, 50mg), *Echinacea purpurea* (flower head, root, seed, 30mg), *Usnea barbata* (herb, 14mg) (Scal);

Uterine Fibroids DUFB also known as **Uterine Myomas**, are tumors that form on the interior and exterior of the wall of the uterus. The cells of the tumor are abnormal muscle cells. About 25% of all women develop fibroid tumors. Fibroid tumors can cause heavy and frequent menstrual periods and can cause infertility. They are five times more common in African-Americans than in Caucasian-Americans. Over 30% of hysterectomies are done to remove fibroid tumors.
Herbs: *Cimicifuga racemosa, Cinnamomum aromaticum* (Bal2), *Ganoderma lucidum* (Bal2), *Pinus pinaster* (bark extract, Murr), *Salvia multiorhiza* (Bal2), *Thea sinensis-viridis* (Bal2), *Vitis vinifera* (seed extract, Murr)
AUGMENTED RAMBLING POWDER *Angelica sinensis, Atractylodes macrocephala, Bupleurum chinense, Gardenia angusta, Glycyrrhiza glabra, Paeonia lactiflora, Paeonia suffruticosa, Wolfiporia extensa* (Bal2); CINNAMON TWIG & PORIA PILL *Cinnamomum aromaticum, Wolfiporia extensa, Paeonia officinalis, Paeonia suffruticosa, Prunus-Persica persica x communis* (seed) (Bal2); DONG QUAI & PEONY POWDER *Angelica sinensis, Paeonia lactiflora, Paeonia officinalis, Alisma plantago-aquatica, Atractylodes macrocephala, Ligusticum porteri, Wolfiporia extensa* (Bal2); FOUR SUBSTANCE DECOCTION *Angelica sinensis, Ligusticum porteri, Paeonia lactiflora, Paeonia officinalis, Rehmannia glutinosa* (Bal2)
Vegetables: "Either use Wild Yam itself or use creams that are certified not to contain added progesterone" (Bal2, page 308, recommendation #3)

Herbs & Vegetables listed by Disease

All information in this section was extracted from the following references:

Bal1, 1997 **"Prescription for Nutritional Healing, 2nd Ed."**, J.A.Balch & P.A.Balch, ISBN:0-89529-727-2
Bal2, 2012 **"Prescription for Herbal Healing, 2nd Ed."**, P.A. Balch & S. Bell, ISBN:0-978-1-58333-452-2
Murr, 2012 **"The Encyclopedia of Natural Medicine 3rd Ed."**, by M.T. Murray & J.E. Pizzorno, ISBN:978-1-4516-6300-6
Scal, 2002 **"Traditional Medicines from the Earth"**, by R. Scalzo & M. Cronin, ISBN:0-9707936-1-8
Wyk, 2004 **"Medicinal Plants of the World"**, Ben-Erik Van Wyk & Michael Wink, ISBN:981-232-933-1

Bacterial Diseases

Bacterial Infection DABA is combated by Anti-Bacterial herbs. **Inflammation** is a natural reaction to bacterial infection. The affected tissues swell, redden, become warm and tender and may be painful.
Herbs: *Echinacea purpurea* (Bal1), *Equisetum arvense* (Bal1), *Hydrastis canadensis* (Bal1), *Medicago sativa* (Bal1), *Tabebuia impetignosa* (Bal1), *Trifolium pratense* (Bal1), *Vaccinium-vortabacca myrtillis* (Bal1), *Yucca filamentosa* (Bal1)
Vegetables: "Eat a diet coposed of 75 percent raw foods, and drink plenty of herbal teas and juices" (Bal1, page 346, recommendation #1)

Bad Breath DHAL also known as **Halitosis**, is combated by **Disease Anti-bacteria Mouthwash DAMW**, It is caused by tooth decay bacteria and gum disease bacteria.
Herbs: *Berberis vulgaris* (DAMW, Bal2), *Commiphora myrrha* (external, DAMW, Bal1), *Coptis chinensis* (DAMW, Bal2), *Crataegus laevigata* (DGUM, Bal2), *Crataegus monogyna* (DGUM, Bal2), *Curcuma longa* (DGUM, Bal2), *Foeniculum vulgare* (DHAL, seed, Bal2), *Hydrastis canadensis* (DAMW, Bal1, Bal2), *Mahonia aquifolium* (DAMW, Bal2), *Mahonia nervosa* (DAMW, Bal2), *Mahonia repens* (DAMW, Bal2), *Matricaria chamomilla* (DGUM, Bal2), *Medicago sativa* (chlorophyll, Bal1, Bal2), *Melaleuca alternifolia* (toothpaste, Bal2), *Mentha piperita* (DAMW, Bal1, Bal2), *Rosmarinus officinalis* (DAMW, Bal1), *Salvia officinalis* (DAMW, Bal1), *Uncaria guianensis* (DABA, Bal2), *Uncaria tomentosa* (DABA, Bal2)
Vegetables (chlorophyll): Parsley (Bal1, Bal2), "Eat at least 50 percent of your food raw every day." (Bal1, page 296, recommendation #1), "Do not use commercial mouthwashes. Most contain nothing more than flavoring, dye, and alcohol. While they may kill bacteria that cause bad breath, the bacteria soon return in greater force. (Bal1, page 296, recommendation #10)

Boil DBOI a round pus-filled nodule on the skin that results from infection by *Candia albicans, Proprionibacter acnes, Pseudomonas aeruginosa, Staphylococcus aureus, Streptococcus pyrogenes, Trichomonas vaginalis* and *Trichophyton mentagrophytes* bacteria (all of which are inhibited by *Melaleuca alternifolia* (Tea Tree Oil). Boils appear on the scalp, buttocks, face, and underarms. Infection begins in a hair follicle and the bacteria then bore deeper into the skin layers and the inflammation spreads.
Herbs: *Angelica sinensis* (internal, Bal2), *Arctium lappa* (internal, DABI, Bal1, Bal2), *Avena sativa* (DABA, Bal1), *Allium cepa* (DABA, Bal1), *Coptis chinensis* (internal & external, DABA, Bal2), *Crataegus laevigata* (internal, Murr), *Crataegus monogyna* (internal, Murr), *Echinacea purpurea* (external, DABA Bal1, Bal2), *Ginkgo biloba* (Murr), ***Hydrastis canadensis* (internal, Bal1, Bal2, Murr)**, *Mahonia aquifolium* (internal, DABA, Bal2), *Mahonia nervosa* (internal, DABA, Bal2), *Mahonia repens* (internal, DABA, Bal2), ***Melaleuca alternifolia* (external, DABA, Bal2, Murr)**, *Pffafia paniculata* (DABA, Bal1), *Pinus pinaster* (pine bark extract, Murr), *Schizonepeta tenuifolia* (internal, Bal2), *Tabebuia impetignosa* (external, DABI, Bal1), ***Silybum marianum* (BLC, Bal1, Murr)**, *Tabebuia impetignosa* (DABA, Bal2), *Taraxacum officinale* (BLC, Bal1), *Thea sinensis-viridis* (green tea, Murr), *Trifolium pratense* (internal, DABA, DABI, Bal1, Bal2), *Ulmus rubra* (external, DABA, Bal2), *Vaccinium-vortabacca myrtillus* (Murr), *Vitis vinifera* (grape seed extract, Murr)
Herb Formulas: <u>COPTIS DECOCTION</u> *Coptis chinensis, Gardenia angusta, Phellodendron amurense, Scutellaria baicalensis, Scutellaria barata, Scutellaria lateriflora* (Bal2)
Vegetables: Raw ("Use a cleansing fast to clear the system and rid the body of toxins that may cause boils.", Bal1, page 160, recommendation #1), **Vegetables (beta-carotene):** Carrots, Pumpkin, Squash, Sweet Potato, Yam (Bal2); **Vegetables (leaf):** Chard, Kale, Lettuce, Spinach (Bal2); **Vegetables:** "Illiminate from the diet any foods that may suppress immune system function (sugar, refined simple carbohydrates, and food allergins)." (Murr, page 354, Treatment Summary, paragraph 1)

Chlamydia DSXC Sexually transmitted *Chlamydia trachomatis* infection is the most common sexually transmitted disease with 4 million new cases each year. Symptoms include genital inflammation, difficulty in urinating, painful intercourse, and itching. In females 50,000 are rendered sterile each year. In males, an enlarged prostate, inflammation of the seminal vessels, and watery mucus urethral discharge may occur.
Herbs: *Astragalus membranaceus* (Bal1, Bal2), *Echinacea purpurea* (Bal1), *Hydrastis canadensis* (Bal1, Bal2), *Tabebuia impetignosa* (Bal1), *Trifolium pratense* (Bal1)
Vegetables: Brown Rice, Fresh Vegetables, Raw Seeds (Bal1); **Vegetables (grain):** Brown Rice, Buckwheat, Cereal, Corn, Millet, Oats, Quinoa, Rice Bran, Rye, Spelt, Tapioca, White Flour, White Rice, Wheat (Bal1);

Gonorrhea DSXG an infection of the penis and vagina caused by the bacteria *Neisseria gonorrhoeae*. Symptoms include painful urination, penis discharge, and vaginal discharge. There are 300,000 new cases of Gonorrhea reported each year in the United States.
Herbs: *Astragalus membranaceus* (Bal2), *Berberis vulgaris* (Bal2), *Carapa guianensis* (Bal2), *Copiafera langsdorffii* (Bal2), *Coptis chinensis* (Bal2), *Epimedium brevicornum* (Bal2), *Epimedium grandiflorum* (Bal2), *Hydrastis canadensis* (Bal2), *Mahonia aquifolium* (root, Bal2), *Mahonia nervosa* (root, Bal2), *Mahonia repens* (root, Bal2), *Scutellaria baicalensis* (Bal2), *Scutellaria barata* (Bal2), *Scutellaria lateriflora* (Bal2), *Terminalia arjuna* (Bal2)
Herb Formulas: <u>DONG QUAI & ALOE PILL</u> *Angelica sinensis, Aloe vera* (Bal2); <u>GENTIAN LONGDANCAO DECOCTION</u> *Gentiana lutea, Gentiana scabra, Akebia quinata, Akebia trifoliate, Alisma plantago-aquatica, Angelica sinensis, Bupleurum chinense, Bupleurum scorzoneraefolium, Gardenia angusta, Plantago afra, Rehmannia glutinosa, Scutellaria baicalensis, Scutellaria barata, Sutellaria galericulata, Scutellaria lateriflora* (Bal2); <u>EIGHT INGREDIENT REHMANNIA PILL</u> *Rehmannia glutinosa, Alisma plantago-aquatica, Astragalus membranaceus, Cornus officinalis, Dioscorea oppositifolia or villosa, Paeonia suffruticosa, Schizandra chinensis, Wolfiporia extensa* (Bal2); <u>PEONY & LICORICE DECOCTION</u> *Paeonia officinalis, Glycyrrhiza glabra* (Bal2);

Gum Disease DGUM also known as **Gingivitis, Periodontitis, Pyorrhea**, is caused by plaque from bacteria that adheres to the teeth. The gums become red, soft and shiny and bleed easily.
Herbs: *Aloe vera* (external, gel, Bal1, Bal2), *Centella asiatica* (Murr), *Commiphora myrrha* (Bal1), *Crataegus laevigata* (Bal1), *Crataegus monogyna* (Bal1), *Curcuma longa* (DAMW, Bal2), *Echinacea purpurea* (Bal1), *Hamamelis virginiana* (DAMW, Bal2), *Hydrastis*

canadensis (DAMW, Bal1, Bal2), *Melaleuca alternifolia* (external oil, Bal2), *Rosa-hipa canina* (Bal1), *Sanguinaria canadensis* (Murr), *Scutellaria baicalensis* (Bal2), *Scutellaria barata* (Bal2), *Sutellaria galericulata* (Bal2), *Scutellaria lateriflora* (Bal2), *Syzygium aromaticum* (external, oil, Bal1, Bal2), *Thea sinensis-viridis* (Bal2), *Thymus vulgaris* (external, oil, Bal2)
Herb Formulas: OPHIOPOGONIS DECOCTION *Ophiopogon japonicus* (bulb), *Glycyrrhiza glabra, Oryza sativa* (porridge), *Panax ginseng, Pinellia turnata, Ziziphus jujba* (Bal2);
Vegetables (grain): Brown Rice, Buckwheat, Cereal, Corn, Millet, Oats, Quinoa, Rice Bran, Rye, Spelt, Tapioca, White Flour, White Rice, Wheat (Bal1); **Vegetables (leaf):** Chard, Kale, Lettuce, Spinach (Bal1); **Vegetables (legume):** Chick Peas, Green Beans, Kidney Beans, Lentils, Lima Beans, Mung Beans, Peas, Pinto Beans, Soybeans, Red Beans, White Beans (Bal1);

Legionnaire's a lung an bronchial tube infection caused by the gamma protean bacteria *Legionella pneumophila*. The bacteria lives in water and is transmitted through water vapor in the air. The first symptoms are achiness, fatigue, headache and moderate fever. The advanced symptoms are chills, caughing, diarrhea, disorientation, nausea, vomiting, severe chest pain, shortness of breath, and high fever 105°F. People with compromised immune systems are at greater risk of contracting the disease.
Herbs: *Echinacea purpurea* (Bal1), *Eucalpytus globulus* (Bal1), *Hydrastis canadensis* (Bal1), *Nepeta cataria* (Bal1)
Vegetables: "Eat a diet consisting of 75 percent raw foods and very lightly steamed vegetables." (Bal1, page 376, recommendation #1)

Meningitis, Bacterial DMNB sore throat, red or purple skin rash, stiff neck, headache, high fever, chills, nausea, vomiting, and sensitivity to light cause by the bacteria *Elizabethkingia meningoseptica, Hemophilus influenzae, Listeria monocytogenes, Micrococcus tetragenus, Neisseria meningitides, and Streptococcus pneumoniae*.
Herbs: *Echinacea purpurea* (Bal1), *Hydrastis canadensis* (Bal1), *Hypericum perforatum* (Bal1), *Nepeta cateria* (enema, Bal1),
Vegetables: Nuts, Seeds (Bal1); **Vegetables (grain):** Brown Rice, Buckwheat, Cereal, Corn, Millet, Oats, Quinoa, Rice Bran, Rye, Spelt, Tapioca, White Flour, White Rice, Wheat (Bal1);

Pelvic Inflammation DPLV sexually transmitted infection of the female reproductive tract. It is caused by Chlamydia or Gonorrhea. More than 750,000 female Americans experience this disease every year. About 75,000 female Americans become infertile from this disease.
Herbs: *Berberis vulgaris* (Bal2), *Coptis chinensis* (Bal2), *Hydrastis canadensis* (Bal2), *Mahonia aquifolium* (root, Bal2), *Mahonia nervosa* (root, Bal2), *Mahonia repens* (root, Bal2), *Scutellaria baicalensis* (Bal2), *Scutellaria barata* (Bal2), *Sutellaria galericulata* (Bal2), *Scutellaria lateriflora* (Bal2)
Herb Formulas: DONG QUAI & PEONY POWDER *Angelica sinensis, Paeonia lactiflora, Paeonia officinalis, Alisma plantago-aquatica, Atractylodes macrocephala, Ligusticum porteri, Wolfiporia extensa* (Bal2); EIGHT INGREDIENT REHMANNIA PILL *Rehmannia glutinosa, Alisma plantago-aquatica, Astragalus membranaceus, Cornus officinalis, Dioscorea oppositifolia or villosa, Paeonia suffruticosa, Schizandra chinensis, Wolfiporia extensa* (Bal2); GENTIAN LONGDANCAO DECOCTION *Gentiana lutea, Gentiana scabra, Akebia quinata, Akebia trifoliate, Alisma plantago-aquatica, Angelica sinensis, Bupleurum chinense, Bupleurum scorzoneraefolium, Gardenia angusta, Plantago afra, Rehmannia glutinosa, Scutellaria baicalensis, Scutellaria barata, Sutellaria galericulata, Scutellaria lateriflora* (Bal2); PEONY & LICORICE DECOCTION *Paeonia officinalis, Glycyrrhiza glabra* (Bal2);

Rheumatic Fever DRMF affects children from ages 3 to 18. Develops following the bacteria *Streptococcus haemolytica* infection, strept throat, tonsillitis, scarlet fever or ear infection. Symptoms are pain, stiffness, and fever.
Herbs: *Arctium lappa* (Bal1), *Betula alba* (leaf, Bal1), *Betula lenta* (leaf, Bal1), *Betula nigra* (leaf, Bal1), *Betula pendula* (leaf, Bal1), *Commiphora myrrha* (gum, Bal1), *Crataegus laevigata* (leaf, Bal1), *Crataegus monogyna* (leaf, Bal1), *Echinacea purpurea* (Bal1), *Gaultheria procumbens* (external, oil, Bal1), *Hydrastis canadensis* (Bal1), *Lobelia inflata* (Bal1), *Myrica cerifera* (bark, Bal1), *Nepeta cateria* (Bal1), *Rumex crispus* (Bal1), *Salvia officinalis* (Bal1), *Silybum marianum* (Bal1), *Tabebuia impetignosa* (Bal1), *Urtica dioica* (Bal1), *Taraxacum officinale* (Bal1), *Trifolium pratense* (Bal1)
Vegetables: Flaxseed Oil, Garlic, Kelp (Bal1)

Syphilis DSXS the sexually transmitted bacteria *Treponema pallidum*. Stage 1 symptoms are sores on the penis, labia, cervix, anal area and mouth. Stage 2 symptoms are lesions on the palms, hands, and soles of the feet, rash, achy muscles and joints, warts on the genitals. Stage 3 symptoms include soft tumors which form anywhere on the body.
Herbs: *Aspalathus linearis* (Bal2), *Astragalus membranaceus* (Bal2), *Copiafera langsdorffii* (Bal2), *Ruscus aculeatus* (Bal2)

Tonsillitis DTNS also known as **Quinsy**, inflammation of the tonsils caused by the bacteria *Streptococcus haemolytica*. Symptoms include sore throat, difficulty swallowing, hoarseness, and coughing. From 1940 to 1980 every child's tonsils were surgically removed, even if the tonsils were healthy. Incidents of tonsillitis went to zero and today children do not have to have their tonsils removed as a safety precaution.
Herbs: *Echinacea purpurea* (Bal1), *Hydrastis canadensis* (Bal1), *Hypericum perforatum* (Bal1), *Matricaria chamomilla* (Bal1), *Nepeta cateria* (enema, Bal1), *Salvia officinalis* (Bal1), *Tabebuia impetignosa* (Bal1), *Thymus vulgaris* (Bal1), *Verbascum thapsis* (Bal1)
Herb Formulas: TONSILLITIS HOT INFUSION *Achillea millefolium, Mentha piperita, Sambucus nigra* (flower) (Bal1);
Vegetables: Vegetable Broths (Bal1)

Tooth Decay DTDC Also known as **Cavities**, bacteria combine with food to produce plaque that sticks to teeth. While in plaque, bacteria convert ingested sugars into acid. The acid leaches calcium and phosphate from the teeth. If unchecked the plaque can eat through the enamel, to the dentin (the body of the tooth) and finally can attact the pulp that contains the nerve of the tooth.
Herbs: *Achillea millefolium* (anti-inflammatory, Bal1), *Calendula officinalis* (anti-inflammatory, Bal1), *Gaultheria procumbens* (NSD, Bal1), *Hydrastis canadensis* (DABA, Bal1), *Hypericum perforatum* (NSD, Bal1), *Matricaria chamomilla* (Bal1), *Mentha piperita* (anti-inflammatory, Bal1), *Piper methysticum* (NSD, Bal1), *Salix alba* (NSD, Bal1), *Syzygium aromaticum* (anti-inflammatory, Bal1)
Vegetables: Raw (Bal1)

Tuberculosis DTBC inflammation of the bronchial tubes in the lungs caused by the bacteria *Mycobacterium tuberculosis*. Symptoms are coughing, loss of appetite, chest pain, and low grade fever.
Herbs: *Achillea millefolium* (Bal1), *Althaea officinalis* (Bal1), *Ballota nigra* (Bal1), *Bletilla striata* (Bal2), *Calendula officinalis* (Bal1), *Capsicum annuum-longum-cayenneum* (Bal1), *Commiphora myrrha* (Bal1), *Coptis chinesis* (Bal2), *Echinacea purpurea* (Bal1), *Elettaria cardamomum* (Bal2), *Eleutherococcus senticosus* (Bal2), *Ephedra distachys* (Bal1), *Foeniculum vulgare* (seed, Bal2), *Glycyrrhiza glabra*

(Bal1), *Hydrastis canadensis* (Bal1), *Inula helenium* (Bal1), *Lentinula edodes* (Bal2), *Lobelia inflata* (Bal1), *Marrubium vulgare* (Bal1), *Matricaria chamomilla* (Bal1), *Mentha piperita* (Bal1), *Ruscus aculeatus* (Bal1), *Tabebuia impetignosa* (Bal1), *Thymus vulgaris* (Bal1), *Verbascum thapsis* (Bal1)
Herb Formulas: MINOR BUPLEURUM DECOCTION *Bupleurum chinense, Bupleurum scorzoneraefolium, Glycyrrhiza glabra, Panax ginseng, Pinellia turnata, Scutellaria baicalensis, Scutellaria barata, Scutellaria lateriflora, Zingiber officinale, Ziziphus jujuba* (Bal2); OPHIOPOGONIS DECOCTION *Ophiopogon japonicus* (bulb), *Glycyrrhiza glabra, Oryza sativa* (porridge), *Panax ginseng, Pinellia turnata, Ziziphus jujba* (Bal2);
Vegetables: Alfalfa Sprouts, Asparagus Puree, Garlic, Nuts, Raw Seeds (Bal1); **Vegetables (grain):** Brown Rice, Buckwheat, Cereal, Corn, Millet, Oats, Quinoa, Rice Bran, Rye, Spelt, Tapioca, White Flour, White Rice, Wheat (Bal1); **Vegetables (live juices):** Beet Juice, Carrot Juice, Celery Juice, Parsley Juice, Potato Juice, Spinach Juice (Bal1);

Vaginitis DVGB inflammation of the female genitalia caused by uncontrolled growth of the bacteria *Gardnerella vaginitis*. Symptoms include burning, itching and abnormal vaginal discharge.
Herbs: *Achillea millefolium* (TM, Wyk), *Alchemilla vulgaris* (TM, Wyk), *Allium sativum* (internal & douche, Bal1, Bal2), *Aloe vera* (douch, Bal2), *Berberis vulgaris* (external, cream, Bal2), *Bryonia dioica* (TM, Wyk), *Calendula officinalis* (cream, douche, Bal1, Bal2), *Coptis chinensis* (external, cream, Bal2), *Croton lechleri* (douche, Bal2), ***Echinacea purpurea* (TM, douche, Bal1, Wyk)**, *Hydrastis canadensis* (douche, vaginal suppository, Bal1, Bal2), *Hypericum perforatum* (douche, Bal1), *Lamium album* (TM, Wyk), *Mahonia aquifolium* (external, root, Bal2), *Mahonia nervosa* (external, root, Bal2), *Mahonia repens* (external, root, Bal2), *Malva sylvestris* (TM, Wyk), *Matricaria chamomilla* (TM, Wyk), ***Melaleuca alternifolia* (vaginal suppository, Bal1, Bal2, Murr)**, *Plantago major* (douche, Bal1), *Polygonum aviculare* (TM, Wyk), *Quercus robur* (TM, Wyk), *Rosmarinus officinalis* (TM, Wyk), *Salvia officinalis* (TM, Wyk), *Symphytum officinale* (douche, Bal1), *Tabebuia impetignosa* (internal & douche, Bal2), *Thuja occidentalis* (TM, Wyk), *Urtica diolica* (TM, Wyk)
Vegetables: Brown Rice, Millet, Oat Bran (Bal1);

Cancer Diseases

Cancer (Tumor) **DCTM** an internal lump in the abdomen on one of the organs. Hoxey therapy employes internal and external herbal preparations, along with diet nutrition, vitamin and mineral supplements and psychological counseling to strengthen the body and fight the cancer. The Moerman regimen is a meatless high fiber diet that includes vitamin and mineral supplements. Dr. O Carl Simonton and Stephanie Matthews-Simonton developed a visualization technique to help patients increase the effectiveness of hteir immune systems. The following herbs are internal. For external herbs see under Skin Cancer. Cancer Rates by State from Highest Cancer Rate to Lowest Cancer Rate in 2010:
1. FL (highest cancer rate), 2. WV, 3.PA, 4. AR, 5. ME, 6. KY, 7. DC, 8. DE, 9. RI, 10. NJ. 11.MO, 12. OR, 13. TN, 14. MA, 15. NV, 16. ND, 17. OH, 18. AL, 19. IA, 20. MS, 21. MT, 22. OK, 23. TN, 24. WI, 25. KS, 26. IN, 27. IL, 28. SC, 29. AZ, 30. MD, 31. LA, 32. MI, 33. CT, 34. NY, 35. SD, 36. NE. 37. VT, 38. WA, 39. VA, 40. NH, 41. MN, 42. TX, **43. GA**, 44. ID, 45. WY, 46. CO, 47. CA, 48. NM, 49. HI, 50. UT, **51. AK** (lowest cancer rate)
TM = traditional medicine, MM = modern medicine
Herbs: *Allium sativum* (Murr), *Aloe vera* (cyclophosphamide, Bal2), *Angelica dahurica* (steroids, Bal2), *Astragalus membranaceus* (cyclophosphamide, methotrexate, vincristine, Bal1, Bal2), *Camptotheca acuminata* (MM, Wyk), *Catharanthus roseus* (MM, Wyk), *Cochicum autumnale* (MM, Wyk), *Coptis chinensis* (5-fluorouracil, Bal2), ***Coriolus versicolor* (cyclophosphamide, polysaccharides, TM Bal2, Wyk)**, *Curcuma longa* (cyclophosphamide, Bal2), *Dionaea muscipula* (TM, Wyk), ***Echinacea purpurea* (polysaccharides, TM Bal1, Wyk)**, *Fuscus vesiculosis* (DCTM, Bal2), *Ganoderma lucidum* (cyclophosphamide, Bal2), *Ginkgo biloba* (DCTM, Bal2), *Glycyrrhiza glabra* (steroids, Bal2), ***Lentinula edodes* (polysaccharides, 5-fluorouracil TM, Bal2, Wyk),** *Medicago sativa* (cyclophosphamide, Bal2), *Olea europeaea* (leaf extract, Bal2), *Panax ginseng* (methotrexate, mitomycin, vincristine, Bal2), *Pfaffia paniculata* (DABI, Bal1), *Pinus pinaster* (pine bark extract, Bal2), *Podophyllum peltatum* (MM, Wyk), *Podophyllum hexandrum* (MM, Wyk), *Quercus velutina* (Black Oak, quercetin, cisplatin, carboplatin, Bal2), *Schizandra chinensis* (Bal2), *Schizophyllum commune* (polysaccharides, TM Wyk), *Scutellaria baicalensis* (cyclophosphamide, Bal2), *Scutellaria barata* (cyclophosphamide, Bal2), *Scutellaria lateriflora* (cyclophosphamide, Bal2), *Silybum marianum* (Bal2), *Syzygium cumini* (Bal2), ***Tabebuia impetignosa* (DABI, TM, Bal1, Wyk)**, *Taraxacum officinale* (tissue repair, Bal1), *Taxus baccata* (MM, Wyk), *Taxus brevifolia* (MM, Wyk), ***Thea sinensis-viridis* (DFRA, Bal1, Bal2, Murr)**, *Trichosanthes kirilowii* (DCTM, Bal1), *Trifolium pratense* (DABI, Bal1, Bal2), *Uncaria guianensis* (Cat's Claw, Guiana, cisplatin, carboplatin, cyclophosphamide, mitomycin, thiotepa, Bal1, Bal2), *Uncaria tomentosa* (Cat's Claw Paraguay, cisplatin, carboplatin, cyclophosphamide, mytomycin, thiotepa, Bal1, Bal2), *Viscum album* (MM, Wyk), *Vitus vinifera* (grape seed extract, Bal2), *Withania somnifera* (cyclophosphamide, Bal2)
Herb Formulas: ESSIAC TEA *Arctium lappa* (root), *Rheum rhabarbarum* (root), *Rumex acetosella, Ulmus rubra* (bark) (Bal2); GINSENG DECOCTION *Panax ginseng, Angelica sinensis, Astragalus membranaceus, Atractylodes macrocephala, Cinnamomum aromaticum, Citrus-aurantius aurantium, Glycyrrhiza glabra, Paeonia officinalis, Polygala senega, Rehmannia gluinosa, Schisandra chinensis, Wolfiporia extensa* (Bal2); HOCHU EKKI TO *Angelica sinensis, Astragalus membranacea, Atractylodes macrocephala, Bupleurum chinense, Cimicifuga racemosa, Citrus-aurantius aurantium, Glycyrrhiza glabra, Panax ginseng* (Bal2); HOXLEY FORMULA *Arctium lappa, Berberis vulgaris, Frangula purshiana, Glycyrrhiza glabra, Phytolacca americana, Stillingia sylvatica, Trifolium pratense* (Bal2); PINELLIA DECOCTION *Pinellia turnata, Perilla frutescens* (leaf), *Wolfiporia extensa, Zingiber officinale* (Bal2); SHI QUA DA BU TANG *Angelica sinensis, Astragalus membranaceus, Atractylodes macrocephala, Cinnamomum aromaticum, Glycyrrhiza glabra, Ligusticum porteri, Paeonia officinalis, Panax ginseng, Rehmannia glutinosa, Wolfiporia extensa* (Bal2); SIX INGREDIENT REHMANNIA PILL *Rehmannia glutinosa, Alisma plantago-aquatica, Cornus officinalis, Dioscorea oppositifolia* or *villosa, Paeonia suffruticosa, Wolfiporia extensa* (Bal2);
Vegetables (condiment): Garlic (Bal2); **Vegetables (crucifer):** Broccoli, Brussels Sprouts, Cabbage, Cauliflower (Bal2, Murr); **Vegetables (fruit):** Squash (Bal2); **Vegetables (grain):** Brown Rice, Buckwheat, Cereal, Corn, Millet, Oats, Quinoa, Rice Bran, Rye, Spelt, Tapioca, White Flour, White Rice, Wheat (Bal1, Bal2); **Vegetables (legume):** Chick Peas, Green Beans, Kidney Beans, Lentils, Lima Beans, Mung Beans, Peas, Pinto Beans, Soybeans, Red Beans, White Beans (Bal2); **Vegetables (mushroom):** Maitake, Reishi, Shiitake (Bal1, Bal2); **Vegetables (orange):** Carrots, Pumpkin, Squash, Sweet Potato, Yam (Bal2); **Vegetables (root):** Onions (Bal2);

Cisplatin & Carboplatin – inhibits Testicular Cancer, **Bladder Cancer**, Cervix Cancer, Stomach Cancer, Prostate Cancer, Breast Cancer, Endometrial Cancer, and Lung Cancer.
Herbs: *Quercus nigra* (Black Oak, quercetin, cisplatin, carboplatin, Bal2), *Uncaria guianensis* (Cat's Claw Guiana, cisplatin, carboplatin, cyclophosphamide, mitomycin, thiotepa, Bal1, Bal2), *Uncaria tomentosa* (Cat's Claw Paraguay, cisplatin, carboplatin, cyclophosphamide, mytomycin, thiotepa, Bal1, Bal2)
Herbs Formulas: SHI QUA DA BU TANG *Angelica sinensis* (Don Quai Angelic), *Astragalus membranaceus* (Huang Qi Milk Vetch), *Atractylodes macrocephala* (Chinese Thistle Daisy), *Cinnamomum aromaticum* (China Kassod Cinnamon), *Glycyrrhiza glabra* (European Licorice), *Ligusticum porteri* (Osha Alpine Lovage), *Paeonia officinalis* (European Peony), *Panax ginseng* (Chinese Ginseng), *Rehmannia glutinosa* (Chinese Rehmann Foxglove), *Wolfiporia extensa* (Tuckahoe Wolf Foam Pore Fungus) (Bal2);

Cyclophosphamide – inhibits Leukemia, Hodgkin's Lymphoma, Multiple Myeloma, Scarcoma, Breast Cancer, Cervix Cancer, Lung Cancer, and Ovary cancer.
Herbs: *Aloe vera* (Curaçao Aloe, cyclophosphamide, Bal2), *Astragalus membranaceus* (Huang Qi Milk Vetch, cyclophosphamide, methotrexate, vincristine, Bal1, Bal2), ***Coriolus versicolor* (Kawaratake Turkeytail Pore Fungus, cyclophosphamide, polysaccharides, TM Bal2, Wyk)**, *Curcuma longa* (Turmeric Curcumin, cyclophosphamide, Bal2), *Ganoderma lucidum* (Reishi Bracket Pore Fungus, cyclophosphamide, Bal2), *Medicago sativa* (Alfalfa Medick, cyclophosphamide, Bal2), *Scutellaria baicalensis* (Lake Baikal Skullcap, cyclophosphamide, Bal2), *Scutellaria barata* (Bearded Skullcap, cyclophosphamide, Bal2), *Scutellaria lateriflora* (Virginia Skullcap, cyclophosphamide, Bal2), *Uncaria guianensis* (Cat's Claw Guiana, cisplatin, carboplatin, cyclophosphamide, mitomycin, thiotepa, Bal1, Bal2), *Uncaria tomentosa* (Cat's Claw Paraguay, cisplatin, carboplatin, cyclophosphamide, mytomycin, thiotepa, Bal1, Bal2) *Withania somnifera* (Winter Cherry Ashwagandha, cyclophosphamide, Bal2)
Herb Formulas: GINSENG DECOCTION *Panax ginseng* (Chinese Ginseng), *Angelica sinensis* (Don Quai Angelic), *Astragalus membranaceus* (Huang Qi Milk Vetch), *Atractylodes macrocephala* (Chinese Thistle Daisy), *Cinnamomum aromaticum*(China Kassod Cinnamon), *Citrus-aurantius aurantium* (Seville Sour Orange), *Glycyrrhiza glabra* (European Licorice), *Paeonia officinalis* (European Peony), *Polygala senega* (Seneca Root Milkwort), *Rehmannia gluinosa* (Chinese Rehmann Foxglove), *Schisandra chinensis* (Chinese Star Magnolia), *Wolfiporia extensa* (Tuckahoe Wolf Foam Pore Fungus) (Bal2); PINELLIA DECOCTION *Pinellia turnata* (Trefoil Crow Dipper), *Perilla frutescens* (Purple Beafsteak Mint, leaf), *Wolfiporia extensa* (Tuckahoe Wolf Foam Pore Fungus), *Zingiber officinale* (South Asia Ginger) (Bal2); SHI QUA DA BU TANG *Angelica sinensis* (Don Quai Angelic), *Astragalus membranaceus* (Huang Qi Milk

Vetch), *Atractylodes macrocephala* (Chinese Thistle Daisy), *Cinnamomum aromaticum* (China Kassod Cinnamon), *Glycyrrhiza glabra* (European Licorice), *Ligusticum porteri* (Osha Alpine Lovage), *Paeonia officinalis* (European Peony), *Panax ginseng* (Chinese Ginseng), *Rehmannia glutinosa* (Chinese Rehmann Foxglove), *Wolfiporia extensa* (Tuckahoe Wolf Foam Pore Fungus) (Bal2);

Doxorubicin – inhibits Leukemia, **Bladder Cancer**, Hodgkin's Lymphoma, Bone Cancer, Breast Cancer, Cervix Cancer, Endometrial Cancer, Head Cancer, Neck Cancer, Liver Cancer, Lung Cancer, Ovary Cancer, and Prostate Cancer.
Herbs: *Schisandra chinensis* (Chinese Star Magnolia, doxorubicin, Bal2), *Silybum marianum* (Blessed Milk Thistle, Bal2), *Syzygium cumini* (Jambolan Lilly Pilly, Bal2)
Herb Forumulas: ESSIAC TEA *Arctium lappa* (Hare Burr Burdock, root), *Rheum rhabarbarum* (Mongolian Rhubarb, root), *Rumex acetosella* (Sheep Sorrel Dock), *Ulmus rubra* (Slippery Elm, bark) (Bal2); HOCHU EKKI TO *Angelica sinensis* (Dong Quai Angelic), *Astragalus membranacea* (Huang Qi Milk Vetch), *Atractylodes macrocephala* (Chinese Thistle Daisy), *Bupleurum chinense* (Northern Thorow Wax), *Cimicifuga racemosa* (Black Cohosh Bugbane), *Citrus-aurantius aurantium* (Seville Sour Orange), *Glycyrrhiza glabra* (European Licorice), *Panax ginseng* (Chinese Ginseng) (Bal2); HOXLEY FORMULA *Arctium lappa* (Hare Burr Burdock), *Berberis vulgaris* (European Barberry), *Frangula purshiana* (Cascara Thornless Buckthorn, sagrada), *Glycyrrhiza glabra* (European Licorice), *Phytolacca americana* (Inkberry Pokeberry), *Stillingia sylvatica* (Queen's Delight Yaw Root), *Trifolium pratense* (Red Clover) (Bal2); SIX INGREDIENT REHMANNIA PILL *Rehmannia glutinosa* (Chinese Rehmann Foxglove), *Alisma plantago-aquatica* (Ze Xie Water Plaintain), *Cornus officinalis* (Cornelian Cherry Dogwood), *Dioscorea oppositifolia* (Indochinese Yam) *or villosa* (Wild Yam), *Paeonia suffruticosa* (Moutan Peony), *Wolfiporia extensa* (Tuckahoe Wolf Foam Pore Fungus) (Bal2);

5 –Fluorouracil – inhibits Breast Cancer, Colon Cancer, Pancreatic Cancer, Stomach Cancer, Head Cancer, Neck Cancer, and Ovary Cancer
Herbs: *Coptis chinensis* (Chinese Goldthread, 5-fluorouracil, Bal2), ***Lentinula edodes* (Shiitake Lentinan Pinwheel Gill Mushroom polysaccharides, 5-fluorouracil TM, Bal2, Wyk)**
Herb Formulas: SHI QUA DA BU TANG *Angelica sinensis* (Dong Quai Angelic), *Astragalus membranaceus* (Huang Qi Milk Vetch), *Atractylodes macrocephala* (Chinese Thistle Daisy), *Cinnamomum aromaticum* (China Kassod Cinnamon), *Glycyrrhiza glabra* (European Licorice), *Ligusticum porteri* (Osha Alpine Lovage), *Paeonia officinalis* (European Peony), *Panax ginseng* (Chinese Ginseng), *Rehmannia glutinosa* (Chinese Rehmann Foxglove), *Wolfiporia extensa* (Tuckahoe Wolf Foam Pore Fungus) (Bal2);

Methotrexate – inhibits Breast Cancer, Head, Cancer, Neck Cancer, Lung Cancer, Stomach Cancer, and Esophogeal Cancer
Herbs: *Astragalus membranaceus* (Huang Qi Milk Vetch, cyclophosphamide, methotrexate, vincristine, Bal1, Bal2), *Panax ginseng* (Chinese Ginseng, methotrexate, mytomycin, vincristine, Bal2)

Mytomycin – inhibits Stomach Cancer, Pancreatic Cancer, Anal Cancer, **Bladder Cancer**, Breast Cancer, Cervical Cancer, Colorectral Cancer, Head Cancer, Neck Cancer, and Lung Cancer
steroids – act as a catalyst for other cancer inhibitors
Herbs: *Panax ginseng* (Chinese Ginseng, methotrexate, mytomycin, vincristine, Bal2), *Uncaria guianensis* (Cat's Claw Guiana, cisplatin, carboplatin, cyclophosphamide, mitomycin, thiotepa, Bal1, Bal2), *Uncaria tomentosa* (Cat's Claw Paraguay, cisplatin, carboplatin, cyclophosphamide, mytomycin, thiotepa, Bal1, Bal2)
Herbs: SHI QUA DA BU TANG *Angelica sinensis* (Dong Quai Angelic), *Astragalus membranaceus* (Huang Qi Milk Vetch), *Atractylodes macrocephala*(Chinese Thistle Daisy), *Cinnamomum aromaticum* (China Kassod Cinnamon), *Glycyrrhiza glabra* (European Licorice), *Ligusticum porteri* (Osha Alpine Lovage), *Paeonia officinalis* (European Peony), *Panax ginseng* (Chinese Ginseng), *Rehmannia glutinosa* (Chinese Rehmann Foxglove), *Wolfiporia extensa* (Tuckahoe Wolf Foam Pore Fungus) (Bal2);

Steroids Cortisone, Dexamethoasone, Hydrocortisone, Methylprednisone, Prednisone, and Prednisolone, act as a catalyst to improve the effectiveness of cancer inhibiting herbs. Side effects include weight gain, swollen hands and feet, high blood pressure, cataracts, glaucoma, insomnia, and anxiety.
Herbs: *Angelica dahurica* (Du Huo Angelic, steroids, Bal2), *Glycyrrhiza glabra* (European Licorice, steroids, Bal2)

Bladder & Kidney Cancer DCBK blood in the urine; pain and burning with urination, increased frequency of urination.
Herbs (Bladder): *Allium sativum* (Garlic Onion, Bal2, page 219), *Astragalus membranaceus* (Huang Qi Milk Vetch, Bal2, page 219), *Eleuthrococcus senticosus* (Siberian Ginseng, Bal2, page 219), *Grifola frondosa* (Maitake Grifol Meristic Downy Pore Fungus, Bal2, page 219), *Quercus nigra* (Black Oak, quercetin extract, Bal2, page 219), *Thea sinensis-viridis* (Green Tea, Bal2, page 219), *Uncaria guianensis* (Cat's Claw Guiana, Bal2, page 219), *Uncaria tomentosa* (Cat's Claw Paraguay, Bal2, page 219), *Vitis vinifera-rubra* (Red Wine Grape, catechins, Bal2, page 219)
Herbs (Kidney): *Astragalus membranaceus* (Huang Qi Milk Vetch, Bal2, page 374), *Eleuthrococcus senticosus* (Siberian Ginseng, Bal2, page 374),*Ganoderma lucida* (Reishi Bracket Pore Fungus, Bal2, page 374), *Glycyrrhiza glabra* (European Licorice, Bal2, page 374), *Lentinula edodes* (Shiitake Lentinan Pinwheel Gill Mushroom, Bal2, page 374), *Uncaria guianensis* (Cat's Claw Guiana, Bal2, page 374), *Uncaria tomentosa* (Cat's Claw Paraguay, Bal2, page 374),
Herb Formulas (Bladder): COPTIS DECOCTION *Coptis chinensis* (Chinese Goldthread), *Gardenia angusta* (Cape Jasmine Alexander Rubyroot), *Phellodendron amurense* (Amur Corkbark), *Scutellaria baicalensis* (Lake Baikal Skullcap), *Scutellaria barata* (Bearded Skullcap), *Scutellaria lateriflora* (Virginia Skullcap) (Bal2), ESSIAC TEA *Arctium lappa* (Hare Burr Burdock, root), *Rheum rhabarbarum* (Mongolian Rhubarb, root), *Rumex acetosella* (Sheep Sorrel Dock), *Ulmus rubra* (Slippery Elm, bark) (Bal2);
Herb Formulas (Kidney): SHIH QUA DA BU TANG *Angelica sinensis, Astragalus membranaceus, Atractylodes macrocephala, Cinnamomum aromaticum, Glycyrrhiza glabra, Ligusticum porteri, Paeonia officinalis, Panax ginseng, Rehmannia glutinosa, Wolfiporia extensa* (Bal2);

Bone Cancer DCBN usually occurs in the knees and upper arm. It most often affects adolescents.
IMSB = immune system booster
Herbs: *Astragalus membranaceus* (IMSB, Bal2), *Bupleurum chinense* (interferon, Bal2), *Grifola frondosa* (IMSB, Bal2), *Lentinus elodes* (interferon, Bal2), *Eleutherococcus senticosus* (interferon, Bal2), *Psoralea corylifolia* (DCBN, Bal2), *Scutellaria baicalensis* (Lake Baikal Skullcap, IMSB, Bal2), *Scutellaria barata* (IMSB, Bal2), *Scutellaria lateriflora* (IMSB, Bal2)
Herb Formulas: OPHIOPOGONIS DECOCTION *Ophiopogon japonicus* (bulb), *Glycyrrhiza glabra, Oryza sativa* (porridge), *Panax ginseng, Pinellia turnata, Ziziphus jujba* (Bal2)
Vegetables (dark green): Kale, Spinach, **Vegetables (orange):** Carrots, Pumpkin, Sweet Potato, Yam, **Vegetables (yellow):** Squash (Bal2)

Breast Cancer DCBR Lumps and thickening in the breast, itching redness, and soreness of nipples.
IMSB = immune system booster
Herbs: *Allium sativum* (DCBR, Bal2), *Arctium lappa* (BLC, Bal1), *Astragalus membranaceus* (IMSB, Bal1, Bal2), *Curcuma longa* (Bal2), *Echinacea purpurea* (IMSB, Bal1), *Ginkgo biloba* (BCR, Bal1), *Glycine max* (isoflavone concentrate, Bal2), *Grifola frondosa* (IMSB, Bal2), *Hypericum perforatum* (DCBR, Bal2), *Lentinula edodes* (IMSB, Bal2), *Phoradendron leucocarpum* (Synonym for *Loranthus leucocarpus*, Bal2), *Pinus pinaster* (pine bark extract, Murr), *Pueraria lobata* (Bal2), *Quercus velutina* (quercetin, Bal2), *Silybum marianum* (BLC, DCBR, Bal1, Bal2), *Taraxacum officinale* (BLC, Bal1), ***Thea sinensis-viridis* (DFRA, Bal2, Murr)**, *Trifolium pratense* (BLC, Bal1), *Uncaria guianensis* (IMSB, Bal2), *Uncaria tomentosa* (IMSB, Bal2), *Viscum album, Viscum coloratum, Vitex agnus-castus* (Bal2), ***Vitis vinifera-rubra* (chatechins, Bal2, grape seed extract, Murr)**
CINNAMON TWIG & PORIA PILL *Cinnamomum aromaticum, Wolfiporia extensa, Paeonia officinalis, Paeonia suffruticosa, Prunus-Persica persica x communis* (seed) (Bal2); DONG QUAI & PEONY POWDER *Angelica sinensis, Paeonia lactiflora, Paeonia officinalis, Alisma plantago-aquatica, Atractylodes macrocephala, Ligusticum porteri, Wolfiporia extensa* (Bal2); ESSIAC TEA *Arctium lappa* (root), *Rheum rhabarbarum* (root), *Rumex acetosella, Ulmus rubra* (bark) (Bal2); HOXLEY FORMULA *Arctium lappa, Berberis vulgaris, Frangula purshiana, Glycyrrhiza glabra, Phytolacca americana, Stillingia sylvatica, Trifolium pratense* (Bal2);
Vegetables (beta-carotene): Carrots, Pumpkin, Squash, Sweet Potato, Yam (Bal1, Bal2); **Vegetables (crucifer):** Broccoli, Brussels Sprouts, Cabbage, Cauliflower (Bal1, Bal2, Murr); **Vegetables (grain): Brown Rice**, Buckwheat, Cereal, Corn, **Millet, Oats**, Quinoa, Rice Bran, Rye, Spelt, Tapioca, White Flour, White Rice, **Wheat**, (Bal1); **Vegetables (legume):** Green Beans, Kidney Beans, Lentils, Lima Beans, Mung Beans, Peas, Pinto Beans, Soybeans, White Beans (Bal1); **Vegetables (lily):** Garlic, Onions (Bal1);

Cervical & Uterine Cancer DCCU bleeding between menstrual periods, unusual discharge, painful menstral periods, heavy periods. The cervix is a small cylindrical female canal at the top of the uteris which passes sperm, menstrual blood, and a fetus at birth. **Cervical Dysplasia** is a condition where benign cancer cells exist in the cervix.
Herbs: *Aloe Vera* (Bal2), *Astragalus membranaceus* (Bal2), *Coriolus versicolor* (Bal2), *Curcuma longa* (Bal2), *Lentilula edodes* (Bal2), *Pinus pinaster* (cervical displasia, pine bark extract, Murr), ***Thea sinensis-viridis* (Bal2, Murr),** *Tremella fuciformis* (Bal2), *Vitis vinifera* (cervical displasia, grape seed extract, Murr)
Vegetables (crucifer): Broccoli, Brussels Sprouts, Cabbage, Cauliflower (Murr), **Vegetables (dark green anti-oxidant):** Artichoke, Asparagus, Mild Bell Peppers, Broccoli, Brussels Sprouts, Chard, Collard Greens, Cornsalad Lettuce, Cucumbers, Green Beans, Kale, Leeks, Mustard Greens, Parsley, Peas, Seakale, Spinach, Zucchini Squash (Bal1, Murr); **Vegetables (yellow anti-oxidant):** Bell Peppers (yellow), Corn, Oca Tuber, Squash (Bal2, Murr)

Colorectal Cancer DCCL rectal bleeding, blood in the stool, persistant diarrhea or persistant constipation.
Herbs: *Allium sativum* (Bal2), *Ananas comosus* (bromelain, Bal2), *Astragalus membranacea* (Bal2), *Berberis vulgaris* (Bal2), *Coptis chinensis* (Bal2), *Coriolus versicolor* (Bal2), *Curcuma longa* (Bal2), *Fucus vesiculosis* (Bal2), *Ganoderma lucidum* (Bal2), *Glycine max* (isoflavone concentrate, Bal2), *Grifola frondosa* (Bal2), *Hydrastis canadensis* (Bal2), *Lentinula edodes* (Bal2), *Mahonia aquifolium* (Bal2), *Mahonia nervosa* (Bal2), *Mahonia repens* (Bal2), *Quercus velutina* (quercetin, Bal2), *Thea sinensis-viridis* (Bal2)
MINOR BUPLEURUM DECOCTION *Bupleurum chinense, Bupleurum scorzoneraefolium, Glycyrrhiza glabra, Panax ginseng, Pinellia turnata, Scutellaria baicalensis, Scutellaria barata, Scutellaria lateriflora, Zingiber officinale, Ziziphus jujuba* (Bal2);
Vegetables (grain): Brown Rice, Buckwheat, Cereal, Corn, Millet, Oats, Quinoa, Rice Bran, Rye, Spelt, Tapioca, White Flour, White Rice, Wheat (Bal2); **Vegetables (legume):** Chick Peas, Green Beans, Kidney Beans, Lentils, Lima Beans, Mung Beans, Peas, Pinto Beans, Soybeans, Red Beans, White Beans (Bal2)

Endometrial Cancer DCEN inner uterine wall tissue is deposited outside the uterus, The tissues form cysts. Bleeding between menstrual periods, unusual discharge, painful menstrual periods, heavy periods.
Herbs: *Allium sativum* (Bal2), *Astragalus membranaceus* (Bal2), *Coriolus versicolor* (Bal2), *Glycine max* (isoflavone concentrate, Bal2), *Matricaria chamomilla* (Bal2), *Medicago sativa* (Bal2), *Thea sinensis-viridis* (Bal2), *Uncaria guianensis* (Bal2), *Uncaria tomentosa* (Bal2)
AUGMENTED RAMBLING POWDER *Angelica sinensis, Atractylodes macrocephala, Bupleurum chinense, Gardenia angusta, Glycyrrhiza glabra, Paeonia lactiflora, Paeonia suffruticosa, Wolfiporia extensa* (Bal2); CINNAMON TWIG & PORIA PILL *Cinnamomum aromaticum, Wolfiporia extensa, Paeonia officinalis, Paeonia suffruticosa, Prunus-Persica persica x communis* (seed) (Bal2); DONG QUAI & PEONY POWDER *Angelica sinensis, Paeonia lactiflora, Paeonia officinalis, Alisma plantago-aquatica, Atractylodes macrocephala, Ligusticum porteri, Wolfiporia extensa* (Bal2);
Vegetables (dark green anti-oxidant): Artichoke, Asparagus, Mild Bell Peppers, Broccoli, Brussels Sprouts, Chard, Collard Greens, Cornsalad Lettuce, Cucumbers, Green Beans, Kale, Leeks, Mustard Greens, Parsley, Peas, Seakale, Spinach, Zucchini Squash, **Vegetables (yellow anti-oxidant):** Bell Peppers (yellow), Corn, Oca Tuber, Squash (Bal2)

Hodgkin's Lymphoma DCLY also known as **Hodgkin's Disease**, enlarged rubbery lymph nodes, itching, night sweats, fever, weight loss.
Herbs: *Astragalus membranaceus* (Bal2), *Eleutherococcus senticosus* (Bal2)
Herb Formulas: SIX INGREDIENT REHMANNIA PILL *Rehmannia glutinosa, Alisma plantago-aquatica, Cornus officinalis, Dioscorea oppositifolia or villosa, Paeonia suffruticosa, Wolfiporia extensa* (Bal2)

Laryngeal Cancer DCLA persistent cough, hoarse throat.
Herbs: see under Tumor Cancer

Leukemia DCLK paleness fatigue, weight loss, repeated infections, easy bruising, bone and joint pain, nosebleeds.
Herbs: *Allium sativum* (Bal2), *Angelica sinensis* (Bal2), *Boswellia sacra* (Bal2), *Boswellia serrata* (Bal2), *Coriolus versicolor* (Bal2), *Crataegus laevigata* (Bal2), *Crataegus monogyna* (Bal2), *Curcuma longa* (Bal2), *Curcuma zedoaria* (root, Bal2), *Fucus vesiculosis* (Bal2), *Pueraria lobata* (Bal2), *Thea sinensis* (Bal2), *Vitis vinifera-rubra* (catechins, Bal2)
Herb Formulas: ESSIAC TEA *Arctium lappa* (root), *Rheum rhabarbarum* (root), *Rumex acetosella, Ulmus rubra* (bark) (Bal2); HOXLEY FORMULA *Arctium lappa, Berberis vulgaris, Frangula purshiana, Glycyrrhiza glabra, Phytolacca americana, Stillingia sylvatica, Trifolium pratense* (Bal2); MINOR BUPLEURUM DECOCTION *Bupleurum chinense, Bupleurum scorzoneraefolium, Glycyrrhiza glabra, Panax ginseng, Pinellia turnata, Scutellaria baicalensis, Scutellaria barata, Scutellaria lateriflora, Zingiber officinale, Ziziphus jujuba* (Bal2); POLYPORUS DECOCTION *Polyporus brumalis, Polyporus squamosus, Polyporus tubaster, Polyporus umbellatus, Alisma plantago-aquatica,* Kadinum Red Clay, Vegetable Gelatin, *Wolfiporia extensa* (Bal2);

Liver Cancer DCLV the liver may be enlarged and hard masses may be present. Symptoms include abdominal pain, loss of appetite, loss of weight.
Herbs: *Astragalus membranaceus* (Bal2), *Berberis vulgaris* (Bal2), *Cinnamomum aromaticum* (Bal2), *Coptis chinensis* (Bal2), *Ganoderma lucidum* (Bal2), *Hydrastis canadensis* (Bal2), *Glycine max-soja* (lecithin, Bal2), *Lentinula edodes* (Bal2), *Mahonia aquifolium* (root, Bal2), *Mahonia nervosa* (root, Bal2), *Mahonia repens* (root, Bal2), *Scutellaria baicalensis* (Bal2), *Scutellaria barata* (Bal2), *Scutellaria lateriflora* (Bal2)
Herb Formulas: ESSIAC TEA *Arctium lappa* (root), *Rheum rhabarbarum* (root), *Rumex acetosella, Ulmus rubra* (bark) (Bal2); MAJOR BUPLEURUM DECOCTION: *Bupleurum chinense, Citrus-aurantius aurantium, Paeonia officinalis, Pinellia turnata, Rheum rhabarabarum, Scutellaria baicalensis, Scutellaria barata, Scutellaria lateriflora, Zingiber officinale, Ziziphus jujuba* (Bal2); MINOR BUPLEURUM DECOCTION *Bupleurum chinense, Bupleurum scorzoneraefolium, Glycyrrhiza glabra, Panax ginseng, Pinellia turnata, Scutellaria baicalensis, Scutellaria barata, Scutellaria lateriflora, Zingiber officinale, Ziziphus jujube* (Bal2);

Lung Cancer DCLU persistent cough, sputum with blood, chest pain.
Herbs: *Aloe vera* (Bal2), *Astragalus membranaceus* (Bal2), *Capsicum annuum-longum-cayenneum* (Bal2), *Coriolus versicolor* (Bal2), *Curcuma longa* (Bal2), *Glycine max-soja* (isoflavone, Bal2), *Lentinula edodes* (Bal2), *Maytenus chuchuhuasa* (Bal2), *Panax ginseng* (Bal2), *Psoralea coryifolia* (Bal2), *Scutellaria baicalensis* (Bal2), *Scutellaria barata* (Bal2), *Scutellaria lateriflora* (Bal2), *Uncaria gambir* (Bal2), *Uncaria guianensis* (Bal2), *Uncaria tomentosa* (Bal2)
Herb Formulas: SIX INGREDIENT REHMANNIA PILL *Rehmannia glutinosa, Alisma plantago-aquatica, Cornus officinalis, Dioscorea oppositifolia or villosa, Paeonia suffruticosa, Wolfiporia extensa* (Bal2)

Mouth & Throat Cancer DCMT a chronic ulcer of the mouth, tongue, or throat that does not heal.
Herbs: see under Tumor Cancer

Ovarian Cancer DCOV tumors & pain in the ovaries.
Herbs: *Angelica sinensis* (ovarian cyst, Bal2), *Astragalus membranaceus* (Bal2), *Coriolus versicolor* (Bal2), *Curcuma longa* (Bal2), *Dioscorea oppositifolia* (ovarian cyst, Bal2), *Discorea villosa* (ovarian cyst, Bal2), *Glycine max-soja* (isoflavone, Bal2), *Maytenus chuchuhuasa* (Bal2), *Phoradendron leucarpum* (Bal2), *Quercus velutina* (quercetin, Bal2), *Silybum marianum* (ovarian cyst, Bal2), *Thea sinensis-viridis* (Bal2)
Herb Formulas (ovarian cancer): COPTIS DECOCTION *Coptis chinensis, Gardenia angusta, Phellodendron amurense, Scutellaria baicalensis, Scutellaria barata, Scutellaria lateriflora* (Bal2);
Herb Formulas (ovarian cancer): AUGMENTED RAMBLING POWDER *Angelica sinensis, Atractylodes macrocephala, Bupleurum chinense, Gardenia angusta, Glycyrrhiza glabra, Paeonia lactiflora, Paeonia suffruticosa, Wolfiporia extensa* (Bal2); CINNAMON TWIG & PORIA PILL *Cinnamomum aromaticum, Paeonia officinalis, Paeonia suffruticosa, Prunus-Persica persica x communis* (seed), *Wolfiporia extensa* (Bal2); DONG QUAI & PEONY POWDER *Angelica sinensis, Paeonia lactiflora, Paeonia officinalis, Alisma plantago-aquatica, Atractylodes macrocephala, Ligusticum porteri, Wolfiporia extensa* (Bal2); PEONY & LICORICE DECOCTION *Paeonia officinalis, Glycyrrhiza glabra* (Bal2);

Pancreatic Cancer DCPA tumors on the pancreas, usually fatal.
Herbs: see under Tumor Cancer

Polyps DCPL appearance of stalk-like growths in the lining of the stomach found during a colonoscopy.
Herbs: *Aloe vera* (Bal1), *Brassia nigra* (Bal1), *Capsicum-anuum-longum-cayenneum* (Bal1), *Cinnamomum aromaticum* (Bal1), *Elettaria cardamomum* (Bal1), *Garcinia cambogia* (Bal1), *Ruscus aculeatus* (Bal1), *Sinapis alba* (Bal1), *Thea sinensis-viridis* (Bal1), *Zingiber officinale* (Bal1)

Prostate Cancer DCPR weak or interrupted urine flow, continuous pain in lower back, pelvis, and upper thighs. The prostate is a gland at the base of the bladder the encircles the erethra, the tube through which urine flows to be expelled. About 80% of married men have prostate cancer to some degree. About 65% of unmarried men have prostate cancer to some degree.
Herbs: *Achillea millefolium* (Bal1), *Arctostaphylos uva-ursi* (Bal1), *Avena sativa* (straw, Bal1), *Coriolus versicolor* (Bal2), *Curcuma longa* (Bal2), *Dionaea muscipula* (Bal1), *Echinacea purpurea* (Bal1), *Eupatorium purpureum* (Bal1), *Glycine max-soja* (isoflavone, Bal2), *Glycyrrhiza glabra* (Bal1), *Hydrangea arborescens* (Bal1), *Hydrastis canadensis* (Bal1), *Petroselinum crispum* (root, Bal1), *Pfaffia paniculata* (Bal1), *Pinus pinaster* (bark extract, Murr), ***Prunus-Cerasus africana* (Bal1, Bal2, Murr)**, *Raphanus sativus-nigra* (Bal1), *Serenoa repens* (Bal1, Bal2), *Silybum marianum* (Bal2), *Tabebuia impetignosa* (Bal1), *Taraxacum officinale* (Bal1), *Thea sinensis-viridis* (Murr), *Trifolium pratense* (Bal1), *Turnera diffusa* (Bal1, Bal2), *Vitis vinifera* (seed extract, Murr), *Vitis vinifera-rubra* (catechins, Bal2)
Herb Formulas: GENTIAN LONGDANCAO DECOCTION *Gentiana lutea, Gentiana scabra, Akebia quinata, Akebia trifoliate, Alisma plantago-aquatica, Angelica sinensis, Bupleurum chinense, Bupleurum scorzoneraefolium, Gardenia angusta, Plantago afra, Rehmannia glutinosa, Scutellaria baicalensis, Scutellaria barata, Sutellaria galericulata, Scutellaria lateriflora* (Bal2); PEONY & LICORICE DECOCTION *Paeonia officinalis, Glycyrrhiza glabra* (Bal2);
Vegetables (crucifer): Broccoli, Brussels Sprouts, Cabbage, Cauliflower (Bal1); **Vegetables (grain):** Brown Rice, Buckwheat, Cereal, Corn, Millet, Oats, Quinoa, Rice Bran, Rye, Spelt, Tapioca, White Flour, White Rice, Wheat (Bal1); **Vegetables (orange):** Bell Peppers, Carrots, Pumpkin, Squash, Sweet Potato, Yam (Bal1, Bal2); **Vegetables (yellow):** Bell Peppers (yellow), Oca Tuber, Squash (Bal1, Bal2); **Vegetables (lycopene):** Tomatoes (Bal2); **Vegetables (leaf, vitamin A):** Chard, Kale, Lettuce, Spinach (Bal2)

Skin Cancer DCSK Also known as **Carcinoma**, **Melanoma**, a tumor or lump under the skin resembling a wart or ulcer that never heals, moles that change color, lesions that look like moles.
Herbs: *Aloe vera* (external, Bal2), *Althaea officinalis* (tissue repair, Bal1), *Arctium lappa* (tissue repair, Bal1), *Astragalus membranaceus* (DCTM, Bal1, Bal2), *Avena sativa* (tissue repair, Bal1), *Capsicum annuum-longum-cayenneum* (BLC, Bal1), *Chondrus crispus* (tissue repair, Bal1), *Coriolus versicolor* (Bal2), *Curcuma longa* (BLC, Bal1, Bal2), *Eleutherococcus senticosus* (Bal2), *Equisetum arvense* (tissue repair, Bal1), *Ganoderma lucidum* (Bal2), *Ginkgo biloba* (DFRA, Bal1), *Glycine max-soja* (isoflavone, Bal2), *Hydrastis canadensis* (BLC, Bal1), *Lentinula edodes* (Bal2), *Medicago sativa* (tissue repair, Bal1), *Melaleuca alternifolia* (external, Bal1), *Pinellia turnata* (Bal2), *Pueraria lobata* (Bal2), *Rosa-hipa-scrubia canina* (tissue repair, Bal1), *Rumex crispus* (tissue repair, Bal1), *Sanguinaria canadensis* (external, Bal2), *Schisandra chinensis* (Bal2), *Senecio ovatus* (external, Bal1), *Smilax aristolochaefolia* (BLC, Bal1), *Symphytum officinale* (external, Bal1), *Tabebuia impetignosa* (external, Bal1), *Taraxacum officinale* (tissue repair, Bal1), *Teucrium scorodonia* (external, Bal1), *Trichosanthes kirilowii* (DCTM, Bal1), *Uncaria guianensis* (Bal2), *Uncaria tomentosa* (Bal2), *Urtica dioica* (BLC, Bal1), *Vaccinium-vortabacca myrtillus* (BLC, Bal1), *Zingiber officinale* (BLC, Bal1)

Herb Formulas: AUGMENTED RAMBLING POWDER *Angelica sinensis, Atractylodes macrocephala, Bupleurum chinense, Gardenia angusta, Glycyrrhiza glabra, Paeonia lactiflora, Paeonia suffruticosa, Wolfiporia extensa* (Bal2); CINNAMON TWIG & PORIA PILL *Cinnamomum aromaticum, Paeonia officinalis, Paeonia suffruticosa, Prunus-Persica persica x communis* (seed), *Wolfiporia extensa* (Bal2);
Vegetables (beta-carotene): Carrots, Squash, Spinach, Sweet Potatoes; **Vegetables (crucifer):** Broccoli, Brussels Sprouts, Cabbage, Cauliflower, Kale, Turnips (Bal1)

Stomach Cancer DCST indigestion and pain after eating, weight loss.
Herbs: *Allium sativum* (Bal2), *Grifolia frondosa* (Bal2), *Lentinula edodes* (Bal2)
Herb Formulas: COPTIS DECOCTION *Coptis chinensis, Gardenia angusta, Phellodendron amurense, Scutellaria baicalensis, Scutellaria barata, Scutellaria lateriflora* (Bal2);

Testicular Cancer DCTS lumps, enlargement of testical, thickening of scrotum, scrotum pain, mild ache in lower abdomin or groin, enlargement or tenderness of breasts.
Herbs: see under Tumor Cancer

Free Radical Diseases

General DFRA free radicals are atoms that have an unpaired electron. Antioxidants provide the needed additional electron which neutralizes the harmful effects of free radicals. Natural antioxidants include Vitamin C, Vitamin E, Coenzyme Q-10 and Astaxanthin. Astaxanthin is a pink carotenoid that gives salmon, crabs, shrimp and flamingos their pink color.
Herbs: *Olea europaea* (leaf, Scal), *Oreganum vulgare* (oil, Scal)
Herb Formulas: ANTI-OXIDANT SUPREME *Crataegus spp.* (berry, 100mg), *Ginkgo biloba* (leaf, 20mg), *Rosmarinus officinalis* (leaf CO2 extract, 8mg), *Thea sinensis* (leaf, 100mg), *Vaccinium myrtillis* (berry, 50mg), *Xanthoxylum clava-herculis* (bark, 8mg) (Scal);

Age Spots DAGS also called liver spots, these are flat brown spots on the face, neck and hands. They are the result of lipofuscin accumulation caused by free radical damage in skin cells.
Herbs: *Arctium lappa* (Bal1), *Ginkgo biloba* (Bal1), *Glycyrrhiza glabra* (Bal1), *Panax ginseng* (Bal1), *Silybum marianum* (Bal1), *Trifolium pratense* (Bal1)

Alzheimer's DALZ Brain function declines to the point where mental disability results. Forgetfulness, fearfulness, depression, agitation, difficulty absorbing new information, non-recognition of friends and loved ones, and loss of normal emotional responses are typical symptoms.
Herbs: *Bacopa monneri* (Bal2), *Centella asiatica* (Bal2), *Crataegus laevigata* (Bal2), *Crataegus monogyna* (Bal2), *Curcuma longa* (Murr), ***Ginkgo biloba* (Bal1, Bal2, Murr)**, *Glycine max* (Bal2), *Huperzia serrata* (huperzine, Murr), *Rosa-hipa canina* (Bal2), *Ruscus aesculus* (Bal1, Bal2), *Withania somnifera* (Bal2)
Herb Formulas: BIOTA SEED PILL *Thuja orientalis* (seed), *Acorus calamus, Angelica sinensis, Glycyrrhiza glabra, Lycium chinense, Ophiopogon japonicus, Rehmannia glutinosa, Scrophularia nodosa, Wolfiporia extensa* (Bal2); BUPLEURUM & OYSTER SHELL *Bupleurum chinense, Ostrea edulis* (shell), *Bupleurum scorzoneraefolium, Cinnamomum aromaticum, Panax ginseng, Pinellia turnata, Reum rhabarbarum, Scutellaria baicalensis, Scutellaria barata, Scutellaria lateriflora, Zingiber officinale, Zizyphus jubata, Wolfiporia extensa* (Bal2); COPTIS DECOCTION *Coptis chinensis, Gardenia angusta, Phellodendron amurense, Scutellaria baicalensis, Scutellaria barata, Scutellaria lateriflora* (Bal2); EIGHT INGREDIENT REHMANNIA PILL *Rehmannia glutinosa, Alisma plantago-aquatica, Astragalus membranaceus, Cornus officinalis, Dioscorea oppositifolia or villosa, Paeonia suffruticosa, Schizandra chinensis, Wolfiporia extensa* (Bal2); GASTRODIA & UNCARIA DECOCTION *Gastrodia-asiatica elata, Uncaria gambir, Uncaria guianensis, Uncaria tomentosa, Eucommia ulmoides, Gardenia angusta, Phoradendron leucarpum (Synonym Loranthus leucarpus), Polygonum aviculare, Scutellaria baicalensis, Scutellaria barata, Scutellaria lateriflora, Wolfia extensa,* Sichuan Ox Knee Root (Bal2)
Vegetables (anti-oxidant spices): Cumin, Coriander, Fennel, Ginger, Turmeric (Bal2); **Vegetables (grain):** Oat Bran, Rice Bran (Bal1); **Vegetables (lecithin):** Black Eyed Peas, Brazil Nuts, Dandelion Flower, Fava Beans, Fenugreek Leaves, Lentils, Mung Beans, Sesame Seeds (Bal2), **Vegetables:** Cereals, Green Vegetables (Murr)

Dementia DDEM The pathology of dementia includes vein constriction, neurotransmitter difficiency, platelet aggregation, ischemia, and neuron degeneriation caused by free radicals in the brain.
Herbs: *Caulophyllum thalictroides* (Bal1),*Ginkgo biloba* (MM, Bal1, Wyk), *Galanthus nivalis* (MM, Wyk), *Illicium anisatum* (Bal1)
Vegetables: Bread, Brown Rice, Oat Bran, Psyllium Husks, Raw Nuts, Rice Bran, Whole Grain Cereal (Bal1)

Fungal Diseases

Fungal Infection combated by Anti-Fungal Disease **DAFU** herbs. Moist, itchy red patches which can occur anywhere on the body. Fungus is the evolutionary link or intelligent design link between the Plant Domain containing the Spore Plant Kingdom and the Seed Plant Kingdom, and the Animal Domain containing the Invertebrate Animal Kingdom and the Vertebrate Animal Kingdom. Fungus, while placed in the Spore Plant Kingdom, can be stationary (non-motile) and plant-like or animated (motile) and Animal-like. Most commonly the White Vaginal Saccharine Sugar Yeast Fungus *Candida albicans* and Tinea Fungi of the Ringworm Fungus Family *Arthrodermataceae* cause Fungal Skin infections.
Herbs: *Juglans nigra* (external, oil, Bal1), *Hydrastis canadensis* (external, Bal1), *Melaleuca alternifolia* (external, oil, Bal1), *Tabebuia impetignosa* (internal, tea, Bal1)
Vegetables: "Eat a diet of 60% to 70% raw foods" (Bal1, page 285, recommendation #1), "Do not eat any foods containing sugar or refined carbohydrates. Fungi thrive on sugar" (Bal1, page 285, recommendation #2), "Avoid cola drinks, grains, processed foods and fried greasy foods" (Bal1, page 285, recommendation #4)
Herb Formulas: NAIL SOAK DECOCTION *Hydrastis canadensis, Tabebuia impetignosa* (Bal1)

Athlete's Foot DALF Also known as *Tinea pedis*, it is the most common fungal disease of the skin. The fungi live off of the dead skin cells and calluses of the feet, especially between the toes. Symptoms include inflammation, burning, itching scaling, cracking, and blisters. It is highly contagious in warm, damp, places such as gyms and wimming pool locker rooms.
Herbs: ***Calendula officinalis* (Bal1, Bal2)**, *Corymbia citriodora* (Bal1), *Cymbopogon citratus* (Bal1), *Hydrastis canadensis* (Bal2), *Juglans nigra* (Bal2), ***Melaleuca alternifolia* (Bal1, Bal2)**, *Tabebuia impetignosa* (Bal1), *Thymus vulgaris* (Bal2)

Vegetables: Raw, **Vegetables (grain):** Brown Rice, Buckwheat, Cereal, Corn, Millet, Oats, Quinoa, Rice Bran, Rye, Spelt, Tapioca, White Flour, White Rice, Wheat (Bal1)

Candidaisis DCAN caused by the yeast-like parasitic fungus *Candida albicans*. The fungus can be found in the intestines, vagina, mouth, ears, nose and throat. Symptoms include constipation, diarrhea, irritable bowel, abdominal pain, headaches, bad breath, rectal itching, impotence, memory loss, mood swings, enlarged prostate, canker sores, heartburn, muscle and joint pain, sore throat, congestion, nagging cough, numbness in face, arms, legs, acne, night sweats, burning tongue, extreme fatigue, vaginitis, kidney and bladder infections, arthritis, depression, Hyperactivity, hypothyroidism, adrenal gland infection, and diabetes.
Herbs: *Ananas comosus* (bromelian, Bal1), ***Allium sativum*** **(DAFU, Bal1, Bal2, Murr)**, *Aloe vera* (Bal2), ***Berberis vulgaris*** **(berberine, Bal2, Murr)**, *Cinnamomum verum* (Bal1, Bal2), ***Coptis chinensis*** **(berberine, Bal2, Murr)**, *Corymbia citriodora* (Bal1), *Echinacea purpurea* (Bal2), *Ganoderma lucidum* (Bal2), *Grifola frondosa* (Bal1), *Hedera helix* (Bal1), ***Hydrastis canadensis*** **(berberine, Bal2, Murr)**, *Hymenaea oblongifolia* (Jatobá Copal Pea, Bal2), *Hypericum perforatum* (Bal1), *Lavandula angustifolia* (Bal2), ***Mahonia aquifolium*** **(berberine, Bal2, Murr)**, *Mahonia nervosa* (Bal2), *Mahonia repens* (Bal2), *Melaleuca alternifolia* (Murr), *Melia azederachta* (Bal1), *Mentha piperita* (Bal2), *Olea europaea* (leaf, Bal2), *Origanum vulgare* (Bal2), *Peltogyne confertiflora* (Jatobá Pelt Coral Pea, Bal2), *Quercus velutina* (quercetin, Bal1), *Rosmarinus officinalis* (Bal2), *Tabebuia impetignosa* (Bal1, Bal2), *Thymus vulgaris* (Bal1, Bal2)
Herb Formulas: GENTIAN LONGDANCAO DECOCTION *Gentiana lutea, Gentiana scabra, Akebia quinata, Akebia trifoliate, Alisma plantago-aquatica, Angelica sinensis, Bupleurum chinense, Bupleurum scorzoneraefolium, Gardenia angusta, Plantago afra, Rehmannia glutinosa, Scutellaria baicalensis, Scutellaria barata, Sutellaria galericulata, Scutellaria lateriflora* (Bal2);
Vegetables (grain, gluten free): Brown Rice, Millet (Bal1);

Jock Itch an inflammation of the groin area (where the thigh joins the abdomin).

Nail Fungus DNLF also known as **Paronychia**, the fingernails or toenails become raised above the surface of the nail bed and become lose as they separate from the skin.
Herbs: *Berberis vulgaris* (Bal2), *Coptis chinensis* (Bal2), *Echinacea angustifolia* (Bal2), *Hydrastis canadensis* (external Bal1), *Juglans nigra* (leaf, Bal2), *Mahonia aquifolium* (root, Bal2), *Mahonia nervosa* (root, Bal2), *Mahonia repens* (root, Bal2), *Melaleuca alternifolia* (external, oil, Bal2), *Sanguinaria canadensis* (Bal2), *Scutellaria baicalensis* (Bal2), *Scutellaria barata* (Bal2), *Sutellaria galericulata* (Bal2), *Scutellaria lateriflora* (Bal2), *Tabebuia impetignosa* (internal & external, Bal1, Bal2)

Oral Thrush DORT creamy looking white patches appear on the tongue and the mucous membranes of the mouth due to the yeast *Candida albicans*. This condition is most common in infants with under developed immune systems and in adults with compromised immune systems.

Ringworm DWRI moist, itchy red patches which can occur anywhere on the body caused by the Tinea Fungi of the Ringworm Fungus Family *Arthrodermataceae*.
Herbs: *Juglans nigra* (Bal2), *Juglans regia* (Bal2), *Melaleuca alternifolia* (Bal2), *Myroxylon balsamum* (Bal2), *Origanum vulgare* (Bal2), *Tabebuia impetignosa* (Bal2)

Genetic Diseases

Allergies DALL an inappropriate response by the immune system to a substance that is not normally harmful (i.e. animal hair, benzoic acid food additive, dust, insect venom, mold, nickel, lanolin, pollen, sulfite food additive). A non-toxic substance is wrongly identified by the immune system as an "invader" and white blood cells overreact causing damage to the body resulting in nasal congestion, coughing, wheezing, itching, hives, skin rashes, headache and fatigue.
Herbs: *Arctium lappa* (Bal1), *Echinacea purpurea* (Bal1), *Ephedra distachya* (Bal1), *Hydrastis canadensis* (Bal1), *Medicago sativa, Taraxacum officinalis* (Bal1), *Urtica dioica* (Scal)
Vegetables (condiments): Catsup, Garlic, Mustard, Parsley, Pepper, Pickles, **Vegetables (crucifer):** Broccoli, Brussels Sprouts, Cabbage, Cauliflower, **Vegetables (leaf):** Chard, Kale, Lettuce, Spinach, **Vegetables (legume):** Green Beans, Kidney Beans, Lentils, Lima Beans, Mung Beans, Peas, Pinto Beans, Soybeans, White Beans, **Vegetables (mushrooms):** Maitake Mushroom, Reishi Mushroom, Shiitake Mushroom, **Vegetables (orange):** Carrots, Pumpkin, Squash, Sweet Potato, Yam, **Vegetables (flowerhead):** Artichoke, **Vegetables (fruit):** Avocado, Cucumber, Eggplant, Olives, Peppers, Squash, Tomatoes, Zucchini, **Vegetables (grain):** Brown Rice, Buckwheat, Cereal, Corn, Millet, Oats, Quinoa, Rye, Spelt, Tapioca, White Flour, White Rice, Wheat, **Vegetables (root):** Beets, Onions, Parsnips, Potatoes, Radishes, Sweet Potatoes, Turnips, **Vegetables (sprout):** Alfalfa Sprouts, **Vegetables (stalk):** Asparagus, Celery, Okra, (Bal1)

(1) **Allergic Dermatitis** also known as Contact Dermatitis may be caused by contact with cosmetics, gold, nickel, perfumes, poison oak, poison ivy, rubber, silver or sunlight.
Herbs (skin rash external): *Hoffmannsegia divaricata* (Bal1), *Rumex crispus* (Bal1), *Taraxacum officinale* (Bal1)

(2) **Celiac Sprue DGCS** a gluten disorder caused by an intolerance of glutin, a component of wheat, barley, and oats. The lining of the intestine is damaged which impairs the body's ability to absorb vital nutrients. Because the disease impairs digestion, food allergies and diarrhea may also appear.

Gluten Intolerance Group
15110 10 Ave. SW, Suite A — Web: www.gluten.net
Seattle, WA 98166-1820

Celiac Sprue Association
PO Box 31700 — Web: www.csaceliacs.org
Omaha, NE 68131-0700

Celiac Disease Foundation
13251 Ventura Blvd., Suite 1 — Web: www.celiac.org
Studio City, CA 91604

Herbs: *Glycyrrhiza glabra* (Bal2), *Glycyrrhiza lepidota* (Bal2), *Glycyrrhiza uralensis* (Bal2), *Medicago sativa* (Bal1, Bal2), *Pinus pinaster* (pine bark extract, Murr), *Quercus velutina* (quercetin, Bal2), *Vitis vinifera* (grape seed extract, Murr)

Vegetables (gluten-free): Corn (Bal2), Buckwheat (Bal1, Bal2), Millet (Bal2), **Rice (Bal1, Bal2)**, Rice Bran (Bal1), **Vegetables (legume):** Chick Peas, Green Beans, Kidney Beans, Lentils, Lima Beans, Mung Beans, Peas, Pinto Beans, Soybeans, Red Beans, White Beans, **Vegetables (sap):** Blackstrap Molasses (Bal1)

(3) **Chemical Allergies** a reaction to foreign chemicals including carbolic acid, chlorine, coal, formaldehyde, gas, hair spray, oil, paint, or phenol. Symptoms include watery eyes, ringing in the ears, diahorrea, nausea, stomach upset,asthma, bronchitis and headache.
Herbs (skin rash external): *Calendula officinalis* (Bal1), *Matricaria chamomilla* (Bal1), *Melaleuca alternifolia* (Bal1), *Sambucus nigra* (Bal1)

(4) **Food Allergies** are caused by proteins in berries, chocolate, citrus fruits, corn, fish, nuts, or shellfish.
Herbs: *Agrimonia eupatoria* (Bal2), *Calendula officinalis* (Bal2), *Matricaria chamomilla* (Bal2), *Mentha piperita* (Bal2), *Pinus pinaster* (pine bark extract, Murr), *Quercus velutina* (quercetin, Murr), *Thymus vulgaris* (Bal2), *Zingiber officinalis* (Bal2), *Urtica dioca* (Bal2), *Vitis vinifera* (grape seed extract, Murr)
Herb Formulas: KUDZU DECOCTION *Pueraria lobata, Cinnamomum aromaticum, Ephedra sinica, Glycyrrhiza glabra, Paeonia lactiflora, Paeonia officinalis Zingiber officinalis, Ziziphus jujuba* (Bal2); MAJOR CONSTRUCT MIDDLE DECOCTION *Panax ginseng, Hordeum vulgare* (maltose sugar), *Zanthoxylum americanum* or *bungeanum, Zingiber officinale* (Bal2);
Vegetables (to combat seafood allergies): Perilla, Pickled Ginger (Bal2)

Potential Vegetable Allergies (adapted from Murr, pages 599, 600)

Family		Potential Vegetable Allergies
Asparagus	(Asparagaceae)	Asparagus
Celery	(Apiaceae)	Anise, Caraway, Carrot, Celery, Coriander, Cumin, Parsley
Crucifer	(Brassicaceae)	Broccoli, Brussels Sprouts, Cabbage, Cauliflower, Mustard, Radish, Turnip, Water Cress
Goosefoot	(Chenopodiaceae)	Beet, Chard, Spinach
Grain	(Graminaceae)	Barley, Corn, Oat, Rice, Rye, Wheat
Jointstem	(Polygonaceae)	Buckwheat, Rhubarb
Laurel	(Lauraceae)	Avocado, Camphor, Cinnamon
Legume	(Fabaceae)	Bean, Cocoa Bean, Lentil, Peanut, Pea, Soybean, Tamarind
Lettuce	(Lactucaceae)	Iceberg Lettuce, Romaine Lettuce
Onion	(Alliaceae)	Chive, Garlic, Leek, Onion
Sunflower	(Helianthaceae)	Sunflower Seed, Jerusalem Artichoke
Sunplant	(Solanaceae)	Bell Pepper, Chili Pepper, Eggplant, Potato, Tomato, Tobacco
Thistle	(Carduaceae)	Globe Artichoke

(5) **Hay Fever** also known as **Allergic Rhinitis** is an allergic response to pollen that affect the mucus membranes of the nose, eyes, and air passages. Symptoms include itchy eyes, runny nose, watery discharge from the eyes, and sneezing. About 22 million Americans suffer from seasonal Hay Fever.
Herbs: *Ballota nigra* (Bal1), *Ephedra distachya* (Bal1), *Ephedra equisetina* (Bal1), *Ephedra gerardiana* (Bal1), *Ephedra intermedia* (Bal1), *Ephedra sinica* (Bal1), *Ephedra trifurca* (Bal1), *Ephedra viridis* (Bal1), *Glycyrrhiza glabra* (Bal1), *Hydrastis canadensis* (Bal1), *Marrubium vulgare* (Bal1), *Medicago sativa* (Bal1), *Pinus pinaster* (bark, Murr), *Prunus-Avium serotina* (wild cherry bark, Bal1), *Thea sinensis-viridis* (extract, Murr), *Verbascum thapsis* (leaf, Bal1), *Urtica dioca* (Bal1), *Vitis vinifera* (seed, Murr)
Vegetables: Raw Nuts, Seeds (Bal1);**Vegetables (grain):** Brown Rice, Buckwheat, Cereal, Corn, Millet, Oats, Quinoa, Rice Bran, Rye, Spelt, Tapioca, White Flour, White Rice, Wheat (Bal1);

(6) **Hives DHVS**, also known as **Urticaria**, is characterized by sudden outbreaks of red, itchy welts on the skin. Allergic reaction causes histamine to be released into the skin which causes the swelling. Heptatitis B and Epstein-Barr viruses can also cause hives. The bacteria *Candida albicans* can trigger an outbreak of hives. The anti-biotic *Penicillin* can cause an outbreak of hives.

Types of Hives

(adapted from Murr, page 693)

Type of Hives	Scientific Term	Duration of Lesion	Symptoms
Scratching Hives	Dermatographic Urticaria	1 – 5 hours	Headache, Malaise
Overheating Hives	Cholinergic Urticaria	1/2 – 1 hours	Headache, Wheezing, Fainting
Cold Air & Bath Hives	Cold Urticaria	1 – 2 hours	Wheezing, Fainting
Sunlight Hives	Solar Urticaria	1/4 – 3 hours	Wheezing, Fainting
Pressure Hives	Pressure Urticaria	8 – 24 hours	Flu-like Symptoms
Heat Contact Hives	Heat Urticaria	1/2 – 1 hours	Digestive Upset, Wheezing, Fatigue
Water Contact Hives	Aquagenic Urticaria	1/2 – 1 hours	None
Vibration Hives	Angioedemic Urticaria	1 hours	Headache, Fainting
Exercise Hives	Muscular Urticaria	1/8 – 1/2 hours	Headache, Fainting, Flushing
Cold/Warm Air Hives	Familial Urticaria	48 hours	Headache, Tremor, Joint Pain

Herbs: *Aloe vera* (Bal2), *Echinacea purpurea* (Bal1), *Glycyrrhiza glabra* (Bal1), *Matricaria chamomilla* (Bal1, Bal2), *Medicago sativa* (Bal1, Bal2), *Panax ginseng* (Bal1), *Pinus pinaster* (bark extract, Murr), *Quercus velutina* (quercetin, Murr), *Rumex crispus* (Bal1), *Smilax aristolochaefolia* (Bal1), *Smilax ferbifuga* (Bal1), *Smilax glyciphylla* (Bal1), *Smilax ornata* (Bal1), *Smilax regelii* (Bal1), *Solanum nigrum* (Bal2), *Uncaria guianensis* (Bal1), *Uncaria tomentosa* (Bal1), *Urtica dioica* (Bal1, Bal2), *Vitis vinifera* (seed extract, Murr)
Herb Formulas: FOUR SUBSTANCE DECOCTION *Angelica sinensis, Ligusticum porteri, Paeonia lactiflora, Paeonia officinalis, Rehmannia glutinosa* (Bal2); LEDEBOURIELLA DECOCTION *Ledebouriella divaricata, Angelica sinensis, Atractylodes macrocephala, Ephedra distachya,* Epsom Salt ($MgSO_4 * 7H_2O$), *Forsythia suspensa, Gardenia angusta* (fruit), *Glycyrrhiza glabra,* Gypsum

($CaSO_4 * 2H_2O$), *Ligusticum porteri, Mentha piperita, Paeonia officinalis, Platycodon grandiflorus, Rheum rhabarbarum* (wine treated root), *Schizonepeta tenuifolia, Scutellaria baicallensis, Scutellaria barata, Scutellaria lateriflora,* Talc [$Mg_3(SiO_4)_{10}(OH)_2$];
Vegetables: "Especially avoid foods high in saturated fats, cholesterol, and sugar" (Bal1, page 329, recommendation #1)

(7) **Insect Allergies** ants, bumblebees, fleas, honeybees, hornets, mosquitoes, spiders, yellowjackets, wasps, can cause an allergy when the sting or bite. Insect Allergies affect 1 in every 200 Americans. Symptoms can be wheezing, tightness in the throat, nausea, diarrhea, hives, itching, pain swelling in the joints, vascular swelling, and respiratory distress.
Herbs (insect repellant): *Cedrus atlantica* (Bal1), *Cedrus deoara* (Bal1), *Cymbopogon narda* (Bal1), *Eucalpytus globulus* (Bal1), *Melaleuca alternifolia* (Bal1), *Mentha pulgeum* (Bal1), *Rosmarinus officinalis* (Bal1), *Ruta graveolens* (Bal1)

(8) **Poison Ivy, Oak Sumac DIOS** About 65% of Americans are sensitive to Poison Ivy, Poison Oak, and Poison Sumac plants. Symptoms are burning and itching, a red rash for poison ivy, blisters for poison oak.
Herbs: *Aloe Vera, Commiphora myrrha* (Bal1), *Echinacea purpurea* (Bal1), *Hydrastis canadensis* (Bal1), *Juglans nigra* (Bal1), *Quercus alba* (bark, Bal1), *Tagates lucida* (Bal1)
Vegetables: Vitamin A Vegetables, Vitamin E Vegetables (Bal1)

(9) **Respiratory Allergies** a supersensitive reaction in the lungs, throat and nasal passages from pollen, animal dander, feathers and airborne environmental elements.
Herbs: *Amoracia rusticana* (Bal2), *Aspalathus linearis* (Bal2), *Bupleurum chinense* (Bal2), *Bupleurum scorzoneraefolium* (Bal2), *Citrus recutita* (Bal2), *Eucalyptus globus* (Bal2), *Hypericum perforatum* (Bal2), *Matricaria chamomilla* (Bal2), *Pinus pinaster* (pine bark extract, Bal2), *Scutellaria baicalensis* (Bal2), *Scutellaria barata* (Bal2), *Scutellaria lateriflora* (Bal2), *Thymus vulgaris* (Bal2), *Zingiber officinalis* (Bal2), *Urtica dioica* (Bal2), *Vitis vinifera* (grape seed extract, Bal2)
Herb Formulas: FOUR EMPERORS EPHEDRA DECOCTION *Ephedra sinica, Cinnamomum aromaticum* (bark), *Glycyrrhiza glabra, Prunus armeniaca* (seed) (Bal2), ASARUM & ACONITE EPHEDRA DECOCTION *Ephedra sinica, Aconitum napellus, Asarum europaeum* (Bal2), KUDZU DECOCTION *Pueraria lobata, Cinnamomum aromaticum, Ephedra sinica, Glycyrrhiza glabra, Paeonia lactiflora, Paeonia officinalis Zingiber officinalis, Ziziphus jujuba* (Bal2)

Autism DAUT is characterized by a marked unresponsiveness to other people and to the surrounding environment. Symptoms include constant rocking, pounding feet while sitting, biting, pounding on their bodies and an inability to form attachments to others.
Herbs: ***Gingko biloba* (Bal1, Murr)**, *Pinus pinaster* (pine bark extract, Murr), *Vitis vinifera* (grape seed extract, Murr)
Vegetables: Raw (Bal1); **Vegetables (grain):** Brown Rice (Bal1); **Vegetables (legume):** Green Beans, Kidney Beans, Lentils, Lima Beans, Mung Beans, Peas, Pinto Beans, Soybeans, White Beans (Bal1, Murr); **Vegetables (root):** Potatoes (Bal1);

Breast Insufficiency DBRI is characterized by a small female breast. Several herbs are available for breast enlargement.
Herbs(Bal2-067-02): *Trigonella foenum-gracum*

Cystic Fibrosis DCYF is caused by a gene defect in a gene that encodes instructions for a protein that regulates the passage of saltin and out of the pancreas, digestive glands, respiratory glands and sweat glands, (exocrine glands). Symptoms are indigestion, difficulty breathing, and excessive loss of salt through the sweat glands. Other symptoms include clubbing of fingers, infertility foul smelling stools, and salty skin.
Herbs: *Achillea millefolium* (Bal1), *Echinacea purpurea* (Bal1), *Hydrastis canadensis* (Bal1), *Medicago sativa* (Bal1), *Zingiber officinale* (Bal1)

Down Syndrome DDWN a genetic defect that causes improper development of the brain in a fetus. It occurs in 1 of every 700 births. It is characterized by slow physical development, mental retardation, and flatted skull and facial features. Other features include slanted eyes, a depressed nose bridge, low-set ears, and a large furrowed tongue.
Herbs: *Allium sativum* (Bal1), *Fucus vesiculosus* (Bal1), *Glycine max* (lecithin extract, Bal1), *Zea mays-sacharata* (lecithin extract, Bal1)

Fibrocystic Breast DFBR if there is more fluid in breast tissues than the lymphatic system can transport out, small spaces in the breast fill with fluid. Fibrous tissue surrounds them and thickens to form a cyst. The cysts float freely in the breast and by pressing against a milk gland can stimulate the production of the pituitary hormone prolactin, which in turn stimulates milk production in the breast. Over 50% of adult females have this disease.
Herbs: *Dioscorea oppositifolia* (Bal2), *Dioscorea villosa* (Bal2), *Echinacea purpurea* (Bal1), *Fucus vesiculosus* (Bal2), *Glycine max* (Isoflavone concentrate, Bal2), *Hydrastis canadensis* (Bal1), *Mitchellia repens* (Bal1), *Oenothera biennis* (oil, Murr), *Phytolacca americana* (external, Bal1), *Pueraria lobata* (Bal2), *Salvia officinalis* (external, Bal1), *Tabebuia impetignosa* (Bal1), *Thea sinensis-viridis* (Bal2), *Trifolium pratense* (Bal1), *Verbascum thapsis* (Bal1), ***Vitex agnus-castus* (Bal2, Murr)**, *Zingiber officinale* (Bal2)
Herb Formulas: AUGMENTED RAMBLING POWDER *Angelica sinensis, Atractylodes macrocephala, Bupleurum chinense, Gardenia angusta, Glycyrrhiza glabra, Paeonia lactiflora, Paeonia suffruticosa, Wolfiporia extensa* (Bal2); DONG QUAI & PEONY POWDER *Angelica sinensis, Paeonia lactiflora, Paeonia officinalis, Alisma plantago-aquatica, Atractylodes macrocephala, Ligusticum porteri, Wolfiporia extensa* (Bal2);
Vegetables (germanium): Garlic, Onions, Shiitake Mushrooms (Bal1); **Vegetables (grain):** Brown Rice, Buckwheat, Cereal, Corn, Millet, Oats, Quinoa, Rice Bran, Rye, Spelt, Tapioca, White Flour, White Rice, Wheat (Bal1, Murr); **Vegetables (legume):** Chick Peas, Green Beans, Kidney Beans, Lentils, Lima Beans, Mung Beans, Peas, Pinto Beans, Soybeans, Red Beans, White Beans (Bal1), **Vegetables:** Whole Grain Cereals, Fresh High Fiber Vegetables (Bal2), **Vegetables (beta-carotene):** Carrots, Tomatoes, Winter Squash (Bal2)

Hair Loss OHL (ointment) also known as **Androgenetic Alopecia**, loss of scalp hair. The hair falls out in patches. It is caused by heredity, hormones and aging. Women experience hair thinning as they grow older. Men experience male pattern baldness.
Herbs: *Equisetum arvense* (Bal1), *Ginkgo biloba* (Bal2), *Glycyrrhiza glabra* (Bal1), *Malus x-domestica* (external, apple cider vinegar, Bal1), *Salvia officinalis* (external, tea, Bal1), *Thuja orientalis* (Bal2)

Herb Formulas: CINNAMON TWIG & LICORICE DECOCTION *Cinnamomum aromaticum, Glycyrrhiza galbra, Paeonia officinalis, Paeonia suffruticosa, Prunus-Persica persica x communis* (seed) (Bal2); ERECTILE STRENGTH & HAIR GROWTH *Avena sativa, Erythroxylum catuaba,* Gervâo, Mutamba, *Ptychopetalum olacoides, Ptychopetalum ovata, Ptychopetalum uncinatum, Smilax aristolochaefolia, Smilax ferbifuga, Smilax glyciphylla, Smilax ornata, Smilax regelii, Urtica dioica* (Raintree Nutrition); NERVOUS TENSION SHAMPOO *Carthamus tinctorius, Olea europaea* (Bal2); SEVEN TREASURE PILL *Amaranthus hypochondriacus, Angelica sinensis, Cuscuta epithymum, Lycium chinense, Polygonatum multiflorum, Psoralea coryifolia, Sesamum indicum* (seeds), *Wolfiporia extensa* (Bal2);
Vegetables: Soybeans (Bal1, Bal2); **Vegetables (biotin):** Brewer's Yeast, Brown Rice, Bulgur, Green Peas, Lentils, Oats, Sunflower Seeds, Walnuts (Bal1)

Hemophilia DHMP the blood does not clot normally because blood clotting proteins are not present in the blood. The National Hemophilia Foundation estimates that 450 male babies are born each year in the United States out of 4,500 live male births. According to their statistics 1 out of every 10 males has hemophilia. Women act as carriers for hemophilia but generally do not contract the disease. While females have two X chromosomes, males only have one. For a female to develop hemophilia both of her X chromosomes would have to carry the defective gene which is unlikely.
Herbs: *Medicago sativa* (vitamin K, Bal1)
Vegetables (vitamin K, crucifer): Broccoli, Brussels Sprouts, Cauliflower (Bal1); **Vegetables (vitamin K, dark green leaf):** Chard, Collard Greens, Cornsalad Lettuce, Kale, Mustard Greens, Parsley, Seakale, Spinach (Bal1);

Hyperactivity DHEA Also known as **Attention Deficit Hyperactivity Disorder (ADHD)** and **Attention Deficit Disorder (ADD)**; hyperactivity is the result of an over-active central nervous system. Symptoms are absentmindedness, clumsiness, difficulty solving problems, difficulty waiting, disorders of speech and hearing, easily frustrated, emotional instability, extreme distractability, forgetfulness, head knocking, inability to finish tasks, inability to sit still, lack of concentration, low tolerance for stress, self-destructive behavior, and temper tantrums.

Attention Deficit Disorder Association
PO Box 972 Phone: 800-487-2282
Mentor, OH 44061

Children With Attention Deficit Disorders
499 Northwest 70th Avenue, Suite 101 Phone: 305-587-3700
Plantation, FL 33317

Herbs: *Avena sativa* (Bal2), *Bacopa monnieri* (Bal2), *Chamaemelum nobile* (Bal1), *Crataegus laevigata* (Bal2), *Crataegus monogyna* (Bal2), ***Ginkgo biloba* (Bal2, Murr)**, *Humulus lupulus* (Bal1), *Lobelia inflata* (Bal1, Bal2), *Matricaria chamomilla* (Bal2), *Melissa officinalis* (Bal2), *Nepeta cateria* (Bal1), *Passiflora incarnata* (Bal1, Bal2), ***Pinus pinaster* (pine bark extract, Bal2, Murr)**, *Scutellaria baicalensis* (Bal1, Bal2), *Scutellaria barata* (Bal1, Bal2), *Scutellaria lateriflora* (Bal1, Bal2), *Stachys officinalis* (Bal1, Bal2), *Thymus vulgaris* (Bal1), *Valeriana officinalis* (Bal1, Bal2), ***Vitis vinifera* (grape seed extract, Bal2, Murr)**
Herb Formulas: SIX INGREDIENT REHMANNIA PILL *Rehmannia glutinosa, Alisma plantago-aquatica, Cornus officinalis, Dioscorea oppositifolia or villosa, Paeonia suffruticosa, Wolfiporia extensa* (Bal2)
Vegetables (grain): Brown Rice, Buckwheat, Cereal, Corn, Millet, Oats, Quinoa, Rice Bran, Rye, Spelt, Tapioca, White Flour, White Rice, Wheat (Bal2), **Vegetables (legume):** Chick Peas, Green Beans, Kidney Beans, Lentils, Lima Beans, Mung Beans, Peas, Pinto Beans, Soybeans, Red Beans, White Beans (Bal2), **Vegetables (anti-allergy):** Organic (Murr)

Motion Sickness DMSK the eyes, ears and muscle nerves send conflicting signals to the brain. Symptoms are dizziness, queasiness, cold sweats, nausea, and vomiting. It can occur while in any moving vehicle but most often occurs on ships when on choppy sea (also called sea sickness).
Herbs: *Ballota nigra* (Bal2), *Foeniculum vulgare* (seed, Bal2), *Mentha piperita* (Bal1, Bal2), *Zingiber officinale* (Bal1, Bal2)
Vegetables: Olives, Whole Grain Crackers (Bal1)

Wilson's the body is unable to metabolize copper as it should. Symptoms include bloody vomit, difficulty speaking, difficulty swallowing, difficulty walking, an enlarged spleen, jaundice, loss of appetite, fatigue, intellectual impairment, rigidity, muscle spasms, fluid accumulation in the abdomen and weight loss. It is a rare disease that affects 8500 Americans.
BLC = Blood Liver Cleanse, DCFS = Disease Chronic Fatigue Syndrome, DFRA = Disease Free Radical, HBF = Head Brain Function
Herbs: *Arctium lappa* (BLC, DCFS, Bal1), *Astragalus membranacea* (DCFS, Bal1), *Avena sativa* (HBF, straw, Bal1), *Capsicum annuum-longum-cayenneum* (BHP, DCFS, Bal1), *Centella asiatica* (HBF, Bal1), *Echinacea purpurea* (DCFS, Bal1), *Eleutherococcus senticosus* (DCFS, HBF, Bal1), *Ginkgo biloba* (HBF, Bal1), *Hydrastis canadensis* (DCFS, Bal1), *Hypericum perforatum* (DCFS, Bal1), *Lobelia inflata* (HBF, Bal1), *Medicago sativa* (HBF, Bal1), *Petroselinum crispum* (HBF, Bal1), *Pfaffia paniculata* (BLC, DCFS, Bal1), *Piper methysticum* (HBF, Bal1), *Rhapanus sativus-niger* (BLC, Bal1), *Scutellaria baicalensis* (HBF, Bal1), *Scutellaria barata* (HBF, Bal1), *Scutellaria lateriflora* (HBF, Bal1), *Silybum marianum* (BLC, DCFS, Bal1), *Tabebuia impetignosa* (DCFS, Bal1), *Taraxacum officinale* (BLC, DCFS, Bal1), *Trifolium pratense* (BLC, Bal1), *Uncaria guianensis* (DFRA, Bal1), *Uncaria tomentosa* (DFRA, Bal1), *Valeriana officinalis* (HBF, Bal1), *Vinca major* (HBF, Bal1), *Vinca minor* (HBF, Bal1)
Vegetables: Garlic, Onions (Bal1);

Malnutrition Diseases

Malabsorption DMLA the failure of the body to properly absorb vitamins, minerals, and other nutrients from food caused by a lack of digestive enzymes or damage to the intestinal walls from **Colitis, Crohn's Disease, Diverticulitis, Irritable Bowel Syndrome, Lactose Intolerance,** and **Parasites**.
Herbs: *Agathosma betulina* (Bal1), *Agathosma crenulata* (Bal1), *Aloe vera* (Bal1), *Chondrus crispus* (Bal1), *Foeniculum vulgare* (seed, Bal1), *Hydrastis canadensis, Medicago sativa* (Bal1), *Rheum rhabarbarum* (Bal1), *Rumex crispus* (Bal1), *Taraxacum officinale* (root, Bal1), *Urtica dioica* (Bal1), *Zingiber officinale* (Bal1)
Vegetables: Barley Malt, Brown Rice, Nut Milk, Oatmeal, Papaya, Soy Milk, Steamed Vegetables (Bal1)

Osteoporosis DOST deterioration of the bones causing them to become weak. Approximately 50% of women between the ages of 45 and 75 show signs of osteoporosis with 30%-40% bone loss. Vitamins C, D, E, and K, protein and calcium all play vital roles in rebuilding bone density and thickness. A lack of hormones can

inhibit the body from absorbing calcium properly. An estimated 200,000 out of 250,000 hip fractures occurring in the United States every year are due to osteoporosis.
Herbs: *Avena sativa* (straw, Bal1), *Cimicifuga racemosa* (Bal1), *Crataegus laevigata* (Bal2), *Crataegus monogyna* (Bal2), *Equisetum arvense* (Bal1), *Eupatorium perfoliatum* (Bal1), *Hordeum vulgare* (Bal1), *Glycine max-soja* (isoflavone, Bal2), *Medicago sativa* (Bal1, Bal2), *Petroselinum crispum* (Bal1), *Phytolacca americana* (Bal1), *Pinus pinaster* (bark extract, Murr), Por Huesos (Bal2), *Rosa-hipa canina* (Bal1), *Tanacetum parthenium* (Bal1, Bal2), *Taraxaxum officinale* (root, Bal1), *Thea sinensis-viridis* (Murr), *Urtica dioica* (Bal1), *Vitis vinifera* (seed extract, Murr), *Yucca aloeifolia* (Bal1)
Herb Formulas: EIGHT INGREDIENT REHMANNIA PILL *Rehmannia glutinosa, Alisma plantago-aquatica, Astragalus membranaceus, Cornus officinalis, Dioscorea oppositifolia or villosa, Paeonia suffrutecosa, Schisandra chinensis, Wolfiporia extensa* (Bal2);
Vegetables (calcium): Amaranth, Broccoli, Chestnuts, Chickweed, Dandelion Greens, Hazelnuts, Kelp, Molasses, Nettle, Oats, Sea Vegetables, Sesame Seeds, Soybeans, Turnip Greens, Watercress, Wheat Germ; (Bal1),
Vegetables (dark green leaf): Chard, Collard Greens, Cornsalad Lettuce, Kale, Mustard Greens, Parsley, Seakale, Spinach (Bal1, Murr);

Rickets DRCK also known as **Osteomalacia**, is a softening of the bones due to a Vitamin D deficiency. Symptoms are bowed legs, knock-knees, scoliosis, narrow rib cage, protruding breast bone, decayed teeth, and delayed walking.
Herbs: *Avena sativa* (straw, Bal1), *Equisetum arvense* (Bal1), *Taraxacum officinale* (root, Bal1), *Urtica dioica* (Bal1)
Vegetables: Calcium Vegetables, Nuts, Seeds, Vitamin D Vegetables (Bal1);

Microbial Diseases

Amoebaisis caused by amoeba parasite *Entamoeba histolytica*. Symptoms are frequent and painful bowel movements.
Herbs: *Carapichea ipecacuanha* (Bal2), *Psychotria ipecacuanha* (TM, Wyk)

Chagas DCHG a protozoan known as *Trypanosoma cruzi* attacks heart muscle tissue and joint muscle tissue and gastrointestinal tract muscle tissue.
Herbs: *Allium sativum* (Bal2), *Tabebuia impetiginosa* (TM, Bal2, Wyk)

Giardiasis abdominal pain caused by the protozoan parasite *Giardia lamblia.* The disease can persist for months or even years.
Herbs: *Berberis vulgaris* (Bal2), *Coptis chinensis* (Bal2), *Mahonia aquifolia* (Bal2), *Mahonia nervosa* (Bal2), *Mahonia repens* (Bal2)

Infertility DIFR failure to conceive. One in every five couples in the United States experience infertility. Causes are low sperm count (male) and failure to ovulate (female).
Herbs (male): *Angelica sinensis* (Bal2), *Astragalus membranaceus* (Bal1, Bal2), *Cimicifuga racemosa* (Bal2), *Epimedium brevicornum* (Bal2), *Epimedium grandiflorum* (Bal2), *Mucuna pruriens* (Murr), ***Panax ginseng* (Bal1, Bal2, Murr)**, *Pausinystalia johimbe* (Bal1), *Pygeum africanum* (Synonym for *Prunus-Cerasus africana* (Murr), *Serenoa repens* (Bal1), *Smilax aristolochaefolia* (Bal1), *Smilax ferbifuga* (Bal1), *Smilax glyciphylla* (Bal1), *Smilax ornata* (Bal1), *Smilax regelii* (Bal1), *Tribulus terrestris* (Murr), *Turnera diffusa* (Bal1), *Withania somnifera* (Murr)
Herb Formulas (male): CINNAMON TWIG & PORIA PILL *Cinnamomum aromaticum, Paeonia officinalis, Paeonia suffruticosa, Prunus-Persica persica x communis* (seed), *Wolfiporia extensa* (Bal2); HOCHU EKKI TO *Angelica sinensis, Astragalus membranacea, Atractylodes macrocephala, Bupleurum chinense, Cimicifuga racemosa, Citrus-aurantius aurantium, Glycyrrhiza glabra, Panax ginseng* (Bal2);
Herbs (female): *Angelica sinensis* (Bal1), *Centella asiatica* (Bal1), *Chamaelirium luteum* (Bal1), *Dioscorea oppositifolia* (Bal1), *Dioscoria villosa* (Bal1), *Glycyrrhiza glabra* (Bal1), *Panax ginseng* (Bal1), *Panax quinquefolius* (Bal2), *Turnera diffusa* (Bal1), ***Vitex agnus-castus* (Bal2, Murr)**
Herb Formulas (female):): CINNAMON TWIG & PORIA PILL *Cinnamomum aromaticum, Paeonia officinalis, Paeonia suffruticosa, Prunus-Persica persica x communis* (seed), *Wolfiporia extensa* (Bal2); FEM-RESTORE FORMULA *Aletris farinosa* (root, 20mg), *Ceanothus americanus* (root, 40mg), *Corydalis yanhusuo* (tuber, 16mg), *Fraxinus americana* (bark, 60mg), *Hydrastis canadensis* (rhizome, 20mg), *Lobelia inflata* (seed, 10mg), *Petasites hybridus* (root, 16mg), *Podophyllum peltatum* (rhizome, 30mg), *Viburnum prunifolium* (root, 20mg), *Zingiber officinale* (rhizome, 14mg) (Scal); PEONY & LICORICE DECOCTION *Paeonia officinalis, Glycyrrhiza glabra* (Bal2); WARM THE MENSES DECOCTION Animal Gelatin (donkey hide), *Angelica sinensis, Cinnamomum aromaticum, Evodia rutaecarpa, Glycyrrhiza glabra, Ligusticum porteri, Ophiopogon japonicus, Paeonia suffruticosa, Paeonia officinalis, Panax ginseng, Pinellia turnata* (Bal2)
Vegetables: "Do not eat Pumpkin Seeds" (Bal1, page 344, recommendation #5)

Malaria DMAL caused by a sporozoan protozoan parasite that infects mosquitoes. **Phylum *Myxozoa* Sporozoan; Class *Myxosporea*** species comprise the group traditionally known as sporozoans. They penetrate the skin or outer covering (integument) of any part of an animal body. They make their way to the intestine where release single nucleus (uninucleate) forms of amebas (called amebulinas). These amebulinas are carried by the blood to the different organs of the body where they penetrate the organ tissues. They also infest lungs, gills, muscles brain, bones and skin. They cause boils and tumors from one to two inches in diameter.Symptoms are flu-like. About 150,000 cases of Malaria are diagnosed in the United States every year.
Herbs: *Artemisia vulgaris* (Bal2), *Berberis vulgaris* (Bal2), *Cinchona pubescens* (quinine, Webster's Dictionary), *Coptis chinensis* (Bal2), *Lycium barbarum* (Bal2), *Lycium chinense* (Bal2), *Mahonia aquifolia* (Bal2), *Mahonia nervosa* (Bal2), *Mahonia repens* (Bal2)
Herb Formulas: MINOR BUPLEURUM DECOCTION *Bupleurum chinense, Bupleurum scorzoneraefolium, Glycyrrhiza glabra, Panax ginseng, Pinellia turnata, Scutellaria baicalensis, Scutellaria barata, Scutellaria lateriflora, Zingiber officinale, Ziziphus jujuba* (Bal2);

Scabies DSCB a persistant itchy rash caused by a tiny mite that burrows into the top layer of skin to lay its eggs. This results in groups of small red lumps.
Herbs (external): *Aloe vera* (Bal1), *Calendula officinalis* (Bal1), *Hydrastis canadensis* (Bal1), *Myroxylon perierae* (Bal1), *Symphytum officinale* (Bal1)
Vegetables: Cold Pressed Vegetable Oils (Bal1)

Trichomonaisis DSXT *Trichomonas vaginalis* a single cell protozoan with a whip tail. This sexually transmitted disease found in women between the ages of 16 and 35. Men experience urethral discharge. Women experience vaginal discharge.
Herbs: *Agrimonia eupatoria, Echinacea angustifolia*

Worms DWRM intestinal parasites.
ISB = immune system booster
Herbs: *Allium sativum* (ISB, Bal1), *Aloe vera* (juice, Bal1), *Artemisia vulgaris* (DWRM, Bal1), *Artemisia vulgaris* (DWRM, Bal1), *Calendula officinalis* (anal suppository, Bal1), *Capsicum annuum-longum-cayenneum* (ISB, Bal1), *Citrus-grapafructa paradisi* (DWRM, seed, Bal1), *Cucurbita maxima-pepo* (DWRM, Bal1), *Curcuma longa* (ISB, Bal1), *Ficus carica* (DWRM, Bal1), *Foeniculum vulgare* (SCC, seed, Bal1), *Franguls purshiana* (DWRM, Bal1), *Gentiana lutea* (DWRM, Bal1), *Gentiana scabra* (DWRM, Bal1), *Glycyrrhiza glabra* (SCC, root, Bal1), *Hamamelis virginiana* (anal suppository, Bal1), *Juglans cineria* (SCC, bark, Bal1), *Juglans nigra* (DWRM, Bal1, Bal2), *Linum usitatissimum* (SCC, seed, Bal1), *Rheum rhabarbarum* (DWRM, root, Bal1), *Senna alexandrina* (SCC, leaf, Bal1), *Tanacetum vulgare* (TM, Wyk), *Thymus vulgaris* (DWRM, Bal1), *Ulmus rubra* (DWRM, Bal1)
Vegetables: High Fiber Vegetables, Pumpkin Seeds, Sesame Seeds (Bal1); **Vegetables (grain):** Brown Rice, Buckwheat, Cereal, Corn, Millet, Oats, Quinoa, Rice Bran, Rye, Spelt, Tapioca, White Flour, White Rice, Wheat (Bal1);

(1) **Eelworm** *Anguillula aceti*, the minute vinegar eelworm.

(2) **Fluke DWFK genera** *Schistosoma, Schistosomatium, Austrobilharzia, Bilharziella, Dendrobilharzia, Gigantobilharzia, Heterobilharzia, Microbilharzia, Ornithobilharzia, Trichobilharzia* causes the disease Schistosomiasis. It burrows into the skin, matures as a larva, then migrates to the lungs and liver where it matures into the adult form. The adult form migrates to the bladder, intenstines, liver, lungs, rectum or spleen depending on the species.
Herbs: *Anacardium occidentale* (Bal2), *Dryopteris filix-mas* (TM, Wyk), *Juglans nigra* (DWRM, Bal2),

(3) **Hookworm DWHK** *Ancyclostoma duodenale, Necator americanus.* Symptoms include abdominal pain, nausea, anemia.
Herbs: *Chenopodium ambrosoides* (TM, Wyk)

(4) **Pinworm DWPN** *Enterobius vermicularis* about ¼" to ½" long. Symptoms include itching of the anal or vaginal area, insomnia, abdominal pain and nausea.
Herbs: *Ananas comosus* (bromelain, Bal2), *Artemisia vulgaris* (Bal2), *Carica papaya* (papain, Bal2), *Juglans nigra* (DWRM, Bal2),*Zanthoxylum americanum* (Bal2), *Zanthoxylum bungeanum* (Bal2)

(5) **Roundworm DWRO** *Ascaris lumbricoides, Trichinella spiralis.* **Round Worm** species, also known as nematodes (*Nematoda*), are the most diverse Phylum of life on earth. There are 80,000 species of Round Worms. They typically range in length from one hundeth of an inch to 9 feet. The longest is a 26 foot long whale parasite. Phasmids are sense organs found in the tail region of Class *Secernentea* round worms. Worms in the Class *Adenophorea* do not have phasmids. Many Phasmid Round Worms are parasites of vertebrates, insects, and plants. **Trichinosis** is the most familiar disease caused by Round Worms in the United States.
Herbs: *Artemisia cina* (TM, Wyk), *Chenopodium ambrosoides* (TM, Wyk), *Juglans nigra* (DWRM, Bal1),*Spigelia anthelmia* (Bal1),

(6) **Tapeworm DWTP**
Herbs: *Areca catechu* (Bal2), *Dryopteris filix-mas* (TM, Wyk), *Chenopodium ambrosoides* (TM, Wyk) *Juglans nigra* (DWRM, Bal2),

(7) **Threadworm**

(8) **Whipworm**

Poison Diseases

Aluminum Toxicity can lead to colic, rickets, gastroenteritis, osteoporosis, anemia, headach, liver cirrhosis, kidney failure, forgetfulness, speech disturbances, memory loss, and weak aching muscles.
Vegetables: "Eat a diet that is high in fiber." (Bal1, page 123, recommendation #1)

Arsenic Poisoning a metallic element found in pesticides, tobacco smoke, kelp, table salt, beer, seafood and drinking water. Symptoms are headache, fatigue, convulsions, vomiting diarrhea, muscle cramps, hair loss, dermatitis, and gastroenteritis. It affects the lungs, skin, kidneys and liver.
Herbs: *Allium sativum* (Bal1)
Vegetables: Onions, Garlic (Bal1); **Vegetables (legume):** Chick Peas, Green Beans, Kidney Beans, Lentils, Lima Beans, Mung Beans, Peas, Pinto Beans, Soybeans, Red Beans, White Beans (Bal1)

Cadmium Toxicity cadmium replaces zinc in the kidneys and liver. Symptoms are high blood pressure, dulled sense of smell, anemia, join soreness, hair loss, dry skin, scaley skin, and loss of appetite. It weakens the immune system by suppressing the production of T lymphocyte cells.
Herbs: *Allium sativum* (Bal1), *Medicago sativa* (Bal1)
Vegetables: High Fiber Vegetables (Bal1);

Chemical Poisoning disinfectants, hair sprays, paint, solvents, bedding, animal hair, household cleaning products, dust, plastics, wood furniture, cabinents, and petroleum products contain toxic chemicals and emit toxic vapors which can be breathed in, ingested, or absorbed through the skin. Symptoms are watery eyes, diarrhea, nausea, and ringing in the ears.
Herbs: *Vitis vinifera* (grape seed extract, Bal1)
Vegetables: High Fiber Vegetables, Almonds, Barley, Beets, Brazil Nuts, Brown Rice, Carrots, Garlic, Hazelnuts, Lentils, Onions, Spinach, Oatmeal (Bal1); **Vegetables (legume):** Chick Peas, Green Beans, Kidney Beans, Lentils, Lima Beans, Mung Beans, Peas, Pinto Beans, Soybeans, Red Beans, White Beans (Bal1)

Copper Toxicity the liver and brain contain the highest amounts of copper in the body. Too much copper can cause diarrhea, eczema, blood anemia, high blood pressure, kidney disease, nausea, liver cirrhosis, leukemia, sickle cell anemia, stomach pain, and damage to the central nervous system.

Herbs: Flavinoid Herbs, Rutin Herbs (Bal1)
Vegetables: Onions, Garlic (Bal1);

Food Toxicity fungicides, herbicides, insecticides, pesticides, preservatives, fertilizers, and ripening agents contain toxic chemicals. Herbs, Vegetables and the soil they are grown in are sprayed with these toxic chemicals.
Herbs: *Quercus velutina* (quercetin, Bal1), *Vitis vinifera* (grape seed extract, Bal1)
Vegetables: High Fiber Vegetables, Garlic, Oat Bran, Wheat Bran (Bal1);

Insect Bite DISB Biting insects include ants, fleas, gnats, mosquitos, and ticks. The bites of insects induce the release of mediators, such as histamine, bradykinin, or prostaglandin which interact with receptors in the skin causing itching from Pururitus. Mosquitos can also transmit Malaria, Yellow Fever and encephalitis viruses. Ticks can also transmit Babesiosis, Lyme Disease and Rocky Mountain Spotted Fever.
TM = traditional medicine, MM = modern medicine
Herbs: *Allium cepa* (TM, Wyk), *Aloe vera* (Bal2), *Calendula officinalis* (IR, Bal1, Bal2), *Carapa guianensis* (Bal2), *Cardiospermum halicacabum* (TM, Wyk), *Cinnamomum camphora* (MM, Wyk), *Copiafera langsdorffii* (Bal2), *Croton lechleri* (Bal2), *Cymbopogon nardus* (IR, Bal1), *Fagopyrum esculentum* (TM, Wyk), *Hydrastis canadensis* (IR, Bal1), *Lavandula angustifolia* (MM, Wyk), *Lobelia inflata* (Bal1), *Matricaria chamomilla* (MM, Wyk), *Melaleuca alternifolia* (IR, DASC, Bal1, Bal2), *Mentha piperita* (MM, Wyk), *Mentha pulegium* (IR, Bal1), *Mucuna puriens* (TM, Wyk), *Plantago lanceolata* (TM, Wyk)*Thymus vulgaris* (MM, Wyk), *Symphytum officinale* (Bal2), *Viola tricolor* (MM, Wyk)
Vegetables (mosquito repellant): Blackstrap Molasses, Brown Rice, Brewer's Yeast, Garlic, Wheat Germ (Bal1)

Insect **Sting** (ointment) **OST** Stinging insects include bees, hornets, yellow jackets, wasps, spiders, and ants. When an insect stings, it injects a poison venom through its stinger into the victim. Insect stings cause localized swelling and acute pain.
Herbs: *Calendula officinalis* (IR, Bal1), *Cymbopogon nardus* (IR, Bal1), *Hydrastis canadensis* (IR, Bal1), *Lobelia inflata* (OST, Bal1), *Melaleuca alternifolia* (IR, Bal1), *Mentha pulegium* (IR, Bal1), *Plantago major* (OST, Bal1), *Quercus alba* (OST, Bal1), *Symphytum officinale* (OST, Bal1), *Ulma rubra* (OST, Bal1)
Vegetables: Papaya (enzyme)

Iron Toxicity also known as **Hematochromatosis**, is a hereditary disease in which the body stores too much iron which degrades the cell membranes in the heart, liver and pancreas. Over 1 million people in the United States have this disease. Another 30 million Americans carry a gene that causes a milder form of the disease.
Herbs: *Silybum marianum* (Bal2)

Lead Poisoning lead is stored in the bones. It inhibits antioxidant enzymes, diminishing the immune system's ability to protect against free radical damage. It can damage the kidneys, liver , heart and nervous system. Symptoms include gastroenteritis, blue gums, diarrhea, anxiety loss of appetite, fatigue, tremors, seizures, gout, vertigo, insomnia, learning disabilities, and arthritis.
Herbs: *Aloe vera* (Bal1), *Medicago sativa* (Bal1)
Vegetables: Onions, Garlic, Kelp (Bal1); **Vegetables (legume):** Chick Peas, Green Beans, Kidney Beans, Lentils, Lima Beans, Mung Beans, Peas, Pinto Beans, Soybeans, Red Beans, White Beans (Bal1);

Mercury Toxicity mercury is found in the soil, water, fish, sewage sludge, fungicides, pesticides, cosmetics, dental fillings, fabric softeners, printer ink, paints, plastics, polishes, solvents, and wood preservatives. Symptoms include arthritis, dermatitis, dizziness, fatigue, gum disease, hair loss, insomnia, memory loss, muscle weakness, and excessive salivation.
Herbs: *Medicago sativa* (Bal1)
Vegetables: Garlic, Kelp, Onions (Bal1); **Vegetables (legume):** Chick Peas, Green Beans, Kidney Beans, Lentils, Lima Beans, Mung Beans, Peas, Pinto Beans, Soybeans, Red Beans, White Beans (Bal1);

Nickel Toxicity excess amounts of nickel can cause dermatitis and respiratory illness. It can contribute to heart attack.
Vegetables: Garlic, Kelp (Bal1);

Radioactive Iodine 131 DRSK is widespread in the environment due to atmospheric contamination of food and water. Nuclear reactors emit radioactive iodine 131 gas.
Herbs: *Vitis vinifera* (grape seed extract, Bal1)
Vegetables: Garlic, Kelp (Bal1); **Vegetables (crucifer):** Broccoli, Brussels Sprouts, Cabbage, Cauliflower (Bal1);

Radioactive Strontium 90 DRSK cattle feed is often made from crops irrigated with water contaminated with Sr-90. Symptoms include anemia, bone cancer, leukemia, and other forms of cancer.
Herbs: *Vitis vinifera* (grape seed extract, Bal1)
Vegetables: Garlic, Kelp, Lecithin, Seaweed, Sunflower Seeds (Bal1)

Snake Bite DSNB swelling and discoloration of the skin, racing pulse, shortness of breath, nausea, vomiting, shock, and convulsion, paralysis, unconsciousness and death. There are 23 poison snake species in the United States*.

Poisonous Snake Species (*USA)

Genus species	Discoverer	Native Location	Species	Genus
Agkistrodon bilineatus	Günther 1863	Mexico, C.America	Two Stripe Mexican	Moccasin
**Agkistrodon contortrix*	Linnaeus 1766	USA(AL,FL,**GA**,LA,MO,MS,NC,OK,SC,TN,TX)	Southern Copperhead	Moccasin
**Agkistrodon piscivorus*	Lacépède 1789	USA(AL, **GA**, NC, SC, VA)	Atlantic Cottonmouth	Moccasin
Agkistrodon taylori	Burger & Robertson 1951	Mexico (Tamaulipas, San Luis Potosi, Hidalgo)	Taylor Mexican	Moccasin
Crotalus aquilus	Klauber 1952	C. Mexico	Queretaran Dusky	Rattlesnake
**Crotalus atrox*	Baird & Girard 1853	USA(AR, AZ, CA, NM, NV, OK, TX), Mexico	Pacific Diamondback	Rattlesnake
Crotalus basiliscus	Cope 1864	Mexico(Sinaloa, Nayarit, Jalisco, Colima, Michoacan)	Basilisk	Rattlesnake
**Crotalus catalinensis*	Cliff 1954	E. Pac. Islands(Santa Catalina Island)	Catalina Island	Rattlesnake
**Crotalus cerastes*	Hallowell 1854	USA(AZ, CA, NV, UT), Mexico	Mohave Desert Sidewinder	Rattlesnake

Crotalus durissus	Linnaeus 1758	Mexico, C.America, S.America	Hard	Rattlesnake
Crotalus enyo	Cope 1861	Mexico(Baha California)	Lower California	Rattlesnake
**Crotalus horridus*	Linnaeus 1758	SC. & E. USA	Timber	Rattlesnake
Crotalus intermedius	Troschel 1865	Mexico(Hidalgo,Puebla,Tlaxcala,Veracruz,Oaxaca,Guerrero)	Mexican Smallhead	Rattlesnake
Crotalus lannomi	Tanner 1966	Mexico(Jalisco)	Autlan	Rattlesnake
**Crotalus lepidus*	Kennicott 1861	USA(AZ, NM, TX), Mexico	Mottled Rock	Rattlesnake
**Crotalus mitchellii*	Cope 1861	USA(CA, NV, UT), W.Mexico, E. Pac. Is.	Mitchell	Rattlesnake
**Crotalus molossus*	Baird & Girard 1853	USA(AZ, NM, TX), Mexico, E. Pac. Is.	Black Tailed	Rattlesnake
**Crotalus oreganus*	Holbrook 1840	Canada,USA(AZ,CA,OR,NM,MT,WA),Mexico	Northern Pacific	Rattlesnake
Crotalus polystictus	Cope 1865	C.Mexico	Mexican Lancehead	Rattlesnake
**Crotalus pricei*	Van Denburgh 1895	USA(AZ), Mexico	Pacific Twin Spotted	Rattlesnake
Crotalus pusillus	Klauber 1952	WC. Mexico (NE Michoacan, S Jalisco)	Tancitaran Dusky	Rattlesnake
**Crotalus rubur*	Cope 1892	USA(CA), Mexico(Baha CA), E. Pac. Is.	Red Diamondback	Rattlesnake
**Crotalus scutulatus*	Kennicott 1861	USA(AZ, CA, NM, NV, TX, UT), Mexico	Mohave	Rattlesnake
Crotalus stejnegeri	Dunn 1919	Mexico (Sierra Madre, W.Durango, S.Sinaloa)	Longtail	Rattlesnake
Crotalus tancitarensis	Alvarado-Diaz&Campbell 2004	SW Mexico (Cerro Tancítaro, Michoacán)	Tacitaro	Rattlesnake
**Crotalus tigris*	Kennicott 1859	USA(AZ), Mexico(Sonora)	Tiger	Rattlesnake
Crotalus tortugensis	Van & Denburgh 1921	Mexico (Isla Tortuga, Gulf of California)	Tortuga Island	Rattlesnake
Crotalus transversus	Taylor 1944	Mexico(Sierra Ajusco, NW. Morelos)	Cross Banded Mountain	Rattlesnake
Crotalus triseriatus	Wagler 1830	C. Mexico(WC Veracruz to SW Jalisco)	Central Plateau Dusky	Rattlesnake
Crotalus unicolor	Lidth De Jeude 1887	S.America(Aruba Is., Venezuela)	Aruba Island	Rattlesnake
Crotalus vegrandis	Klauber 1941	S.America(E Venezuela)	Uracoan	Rattlesnake
**Crotalus viridis*	Rafinesque 1818	Canada, USA(CA, OR, WA, MT), Mexico	Prairie	Rattlesnake
**Crotalus willardi*	Meek 1905	USA(AZ, NM) Mexico	Arizona Ridgenose	Rattlesnake
Leptodeira annulata	(Linnaeus,1758)	Mexico, C. & S. America	Banded	Cateye Snake
Leptodeira bakeri	Ruthven,1936	S. America (Aruba Is. & NE Venezuela)	Baker	Cateye Snake
Leptodeira frenata	(Cope,1886)	Mexico, C. America (Belize, Guatemala)	Rainforest	Cateye Snake
Leptodeira nigrofasciata	Günther,1868	Mexico, C. America (all countries)	Black Banded	Cateye Snake
Leptodeira punctata	Peters,1866	NW Mexico	Dotted	Cateye Snake
Leptodeira rubricata	Cope,1893	C. America (Costa Rica)	Red	Cateye Snake
**Leptodeira septentrionalis*	(Kennicott,1859)	S. USA, Mexico, C. America	Northern	Cateye Snake
Leptodeira splendida	(Günther,1895)	Mexico	Splendid	Cateye Snake
**Micruroides euryxanthus*	(Kennicott,1860)	S. USA (AZ, NM)	Arizona Yellow	Banded Coral Snake
Micrurus alleni	Schmidt 1936	C. America(all countries)	Allen	Coral Snake
Micrurus altirostris	Cope 1860	S. America(Argentina, Brazil, Uruguay)	Longbeak	Coral Snake
Micrurus ancoralis	Jan 1872	C. Amer. (Panama), S. Amer. (Columbia, Ecuador)	Regal	Coral Snake
Micrurus annellatus	Peters 1871	S. America (Boliva, Ecuador, Peru)	Ringed	Coral Snake
Micrurus averyi	Schmidt 1939	S. America (Amazonas,Pará,Brazil,S.Guyana)	Black Headed	Coral Snake
Micrurus bernadi	Cope,1887	Mexico (Hidalgo, Puebla, Veracruz)	Bernad	Coral Snake
Micrurus bocourti	Jan,1872	S. America (Colombia,Ecuador,Peru)	Bocourt	Coral Snake
Micrurus bogerti	Roze,1967	Mexico (Oaxaca)	Bogert	Coral Snake
Micrurus browni	Schmidt & Smith,1943	Mexico (Guerrero), C. Amer. (Guatemala)	Brown's	Coral Snake
Micrurus camilae	Renjifo & Lundberg,2003	S. America (Cordoba, Colombia)	Camilia	Coral Snake
Micrurus carinicauda	Schmidt,1936		Keel Tailed	Coral Snake
Micrurus catamayensis	Roze,1989	S. America (Catamayo, Equador)	Catamayo	Coral Snake
Micrurus clarki	Schmidt,1936	C. Amer. (Panama), S. Amer. (Colombia)	Clark	Coral Snake
Micrurus corallinus	Merrem,1820	S. America (Bahia, Misiones, Paraguay)	Painted	Coral Snake
Micrurus decoratus	Jan,1858		Decorated	Coral Snake
Micrurus diana	Roze,1983	S.America (Bolivia, Santiago)	Diana	Coral Snake
Micrurus diastema	Duméril, Bibron & Duméril,1854	Mexico, C. America (Guatemala)	Variable	Coral Snake
Micrurus dissoleucus	Cope,1860	Mexico, C. America (Guatemala)	Pigmy	Coral Snake
Micrurus distans	Kennicott,1860	W.Mexico	West Mexico	Coral Snake
Micrurus dumerili	Jan,1858	C.America(Panama),S.America(Colombia,Ecuador,Venezuela)	Dumeril	Coral Snake
Micrurus elegans	Jan,1858	Mexico(Chiapas,Oaxaca,Tabasco, Veracruz), C.Amer. (Guatemala)	Elegant	Coral Snake
Micrurus ephippifer	Cope,1886	Mexico (Oaxaca)	Oaxacan	CoralSnake
Micrurus filiformis	Günther,1859	S. America (Colombia, Peru)	Slender	Coral Snake
Micrurus frontalis	Duméril, Bibron & Duméril,1854	S.America(Bahia,Corrientes,Misiones,Uruguay)	Uruguay	Coral Snake
**Micrurus fulvius*	Linnaeus,1766	S. USA (AL, AR, FL,**GA**,LA, MS, TX)	Harlequin	Coral Snake
Micrurus hemprichii	Jan,1858	S.Amer. (Colombia,Ecuador,Guyana,Peru,Venezuela)	Hemprich	Coral Snake
Micrurus hippocrepis	Peters,1862	C. America (Belize, Guatemala)	Mayan	Coral Snake
Micrurus ibiboboca	Merrem,1820	S.America	Ibibo Boca	Coral Snake
Micrurus isozonus	Cope,1860	S. America (Colombia, Guyana, Roraima, Venezuela)	Venezuela Zoned	Coral Snake
Micrurus langsdorffi	(Wagler,1824)	S. America (Colombia, Ecuador, Peru, Venezuela)	Langsdorff	Coral Snake
Micrurus laticollaris	(Peters,1870)	Mexico (Colima, Guerrero, Jalisco, Michoacan, Morelos, Puebla)	Balsan	Coral Snake
Micrurus latifasciatus	Schmidt,1933	Mexico(Chiapas,Oaxaca), C.Amer.(Guatemala)	Broad Ringed	Coral Snake
Micrurus lemniscatus	Linnaeus,1758	S.Amer(Bolivia,Colombia,Ecuador,Guyana,Peru,Surinam,Venezuela)	South American	Coral Snake
Micrurus limbatus	Fraser,1964	Not Stated	Limba	Coral Snake
Micrurus margaritiferus	Roze,1967	S. America (Peru)	Speckled	Coral Snake
Micrurus medemi	Roze,1967	S. America (Colombia)	Federico Medem	Coral Snake
Micrurus mertensi	Schmidt,1936	S. America (Ecuador, Peru)	Robert Mertens	Coral Snake
Micrurus mipartitus	Duméril, Bibron & Duméril,1854	C. America (Panama), S. America (Colombia,Venezuela)	Red Tail	Coral Snake
Micrurus multifasciatus	Jan,1858	C. America (Nicaragua, Panama)	Many Banded	Coral Snake
Micrurus nebularis	Roze,1989	Mexico (Oaxaca)	Cloud Forest	Coral Snake
Micrurus nigrocinctus	Girard,1854	Mexico (Chiapas, Yucatan), C. America	Black Banded	Coral Snake
Micrurus pacaraimae	Morata De Carvalho,2002	S. America (Roraima)	Pacaraim	Coral Snake

Micrurus pachecogili	Campbell,2000	Mexico (Puebla,18°18mN, 97°31mW)	Pacheco-Gil	Coral Snake
Micrurus paraensis	Da Cunha & Nascimento,1973	S. America (Para, Surinam)	Para	Coral Snake
Micrurus peruvianus	Schmidt,1936	S. America (Amazonas, Ecuador, Peru)	Peru	Coral Snake
Micrurus petersi	Roze,1967	S. America (Ecuador)	Peters	Coral Snake
Micrurus proximans	Smith & Chrapliwy,1958	S. America (Nayarit)	Nayarit	Coral Snake
Micrurus psyches	Daudin,1803	S.Amer. (Colombia,Guyana,Para,Surinam,Venezuela)	Carib	Coral Snake
Micrurus putumayensis	Lancini,1962	S. America (Colombia, Peru)	Putumayo	Coral Snake
Micrurus pyrrhocryptus	Cope,1862	S. America (Bolivia, Mendoza, Paraguay)	Fire Cryptic	Coral Snake
Micrurus remotus	Roze,1987	S. America (Amazonas, Colombia, Venezuela)	Remote	Coral Snake
Micrurus renjifoi	Lamar,2003	S. America (Colombia)	Renjifo	Coral Snake
Micrurus ruatanus	Günther,1895	C. America (Bahia, Honduras)	Ruatan	Coral Snake
Micrurus sangilensis	Niceforo Maria,1942	S. America (Colombia)	Santander	Coral Snake
Micrurus scutiventris	Cope,1869	S. America (Colombia, Ecuador, Peru)	Scute Bellied	Coral Snake
Micrurus spixii	Wagler,1824	S. America (Amazonas, Bolivia, Colombia, Para, Peru, Tocantins, Venezuela)	Amazon	Coral Snake
Micrurus spurelli	Boulenger,1914	S. America (Colombia)	Spurell	Coral Snake
Micrurus steindachneri	Werner,1901	S. America (Ecuador)	Franz Steindachner	Coral Snake
Micrurus stewarti	Barbour & Amaral,1928	S. America (Panama)	Panamanian	Coral Snake
Micrurus stuarti	Roze,1967	C. America (Guatemala)	Stuart	Coral Snake
Micrurus surinamensis	Cuvier,1817	S. America (Bolivia, Colombia, Ecuador, Guyana, Peru, Surinam, Venezuela)	Aquatic	Coral Snake
Micrurus tamaulipensis	Lavin-Murcio & Dixon,2004	Mexico (Tamaulipas)	Tamaulipas	Coral Snake
**Micrurus tener*	Baird & Girard,1853	USA(AR, LA, TX), Mexico (Guanajuato, Queretaro, Amaulipas)	Texas	Coral Snake
Micrurus tricolor	Hoge,1956	S. America (Paraguay)	Tricolored	Coral Snake
Micrurus tschudii	Jan,1858	S. America (Bolivia, Ecuador, Peru)	Desert	Coral Snake
**Sistrurus catenatus*	Rafinesque 1818	Canada(Ontario), USA(IA,IL,IN,OH,MI,MN,MO,NY,PA,WI)	Massasauga	Sistern Rattlesnake
**Sistrurus miliarius*	Linnaeus 1766	USA(AL, **GA**, MS, NC, SC)	Pygmy	Sistern Rattlesnake
Sistrurus ravus	Cope 1865	Mexico (Puebla, Veracruz, Morelos,Tlaxcala)	Mexican Massasauga	Sistern Rattlesnake
**Trimorphodon biscutatus*	Duméril, Bibron & Duméril,1854	S. USA, Mexico, C. America	Costal	Lyre Snake
Trimorphodon tau	Cope,1870	Mexico	Mexican	Lyre Snake

Herbs: *Cimicifuga racemosa, Echinacea purpurea, Quercus alba* (external, Bal1), *Rumex crispus, Symphytum officinale* (external, Bal1), *Ulmus rubra* (external, Bal1)

Spider Bite DSPB produces numbness, redness, swelling, itching, cramping, sweating, headach, dizziness, nausea, and vomiting. The bite of the black widow spider causes severe abdominal pain.
Herbs: *Calendula officinalis* (Bal1), *Cinnamomum aromaticum* (Bal1), *Echinacea purpurea* (Bal1), *Ginkgo biloba* (Bal1), *Hydrastis canadensis* (Bal1), *Lavandula angustifolia* (Bal1), *Linum usitatissimum* (Bal1), *Lobelia inflata* (Bal1), *Melaleuca alternifolia* (Bal1), *Ocimum basilicum-mediterranea* (Bal1), *Rumex crispus* (Bal1), *Salvia officinalis* (Bal1), *Satureja montana* (Bal1), *Taraxacum officinale* (Bal1), *Thymus vulgaris* (Bal1), *Trigonella foenum-graecum* (Bal1)

X-Ray Radiation DRSK the National Academy of Science reported that exposure to 10 milliroentgens of radiation from X-rays increases the risk of sterility, tissue damage and cancer. Prolonged exposure to x-rays also destroys the body's immune system.
Herbs: *Hoffmannseggia divaricata*
Vegetables: Avocados, Buckwheat, Garlic, Kelp, Olive Oil, Safflower Oil (Bal1)

Viral Diseases

General Viral Disease is combated by **Anti-Viral Disease DAVR** herbs. In the past virologists named viruses by virus morphology. Viruses are host specific and today viruses are named according to the host that they affect. The 80 Human virus species are:

01 Human Adenovirus — *Homosapiensvirus adeno* (Adenoid Virus)
02 Human Anellovirus — *Homosapiensvirus anello* (Torque Teno)
03 Human Arenavirus — *Homosapiensvirus arena* (Rodent Excrement Crossover Virus)
04 Human Arterivirus — *Homosapiensvirus arterimonoposi* (Arteritis Virus)
05 Human Astrovirus — *Homosapiensvirus astro-mamo*
06 Human Birnavirus Aqua — *Homosapiensvirus birna-aqua* (Pancreatic Necrosis)
07 Human Birnavirus Avium — *Homosapiensvirus birna-avi* (Bursal Virus)
08 Human Bunyavirus Ortho — *Homosapiensvirus bunya-ortho* (Encephalitis)
09 Human Bunyavirus Hanta — *Homosapiensvirus bunya-hanta*
10 Human Bunyavirus Nairo — *Homosapiensvirus bunya-nairo* (Hemorrhagic Fever)
11 Human Bunyavirus Phlebo — *Homosapiensvirus bunya-phlebo*
12 Human Bunyavirus Mononega — *Homosapiensvirus bunya-mononega*
13 Human Calicivirus Noro — *Homosapiensvirus calici-noro* (Norwalk Virus)
14 Human Calicivirus Sapo — *Homosapiensvirus calici-sapo* (Sapporo Virus)
15 Human Coronavirus Monoposi — *Homosapiensvirus corona-monoposi* (Asthma, Bronchitis, Gastroenteritis)
16 Human Cornoavirus Toro — *Homosapiensvirus corona-toro*
17 Human Filovirus Marburg — *Homosapiensvirus filo-marburgi* (Marburg Virus)
18 Human Filovirus Ebola — *Homosapiensvirus filo-ebola* (Ebola Virus)
19 Human Flavivirus Monoposi — *Homosapiensvirus flavi-monoposi*
20 Human Flavivuris pesti — *Homosapiensvirus flavi-pesti*
21 Human Hepatitisvirus C — *Homosapiensvirus flavi-hepatitis-cappa* (Blood Transfusion Hepatitis)
22 Human Hepatitisvirus G — *Homosapiensvirus flavi-hepatitis-gamma*
23 Human Hepatitisvirus A — *Homosapiensvirus pico-hepatitis-alpha* (Infectious Hepatitis)
24 Human Hepatitisvirus D — *Homosapiensvirus pico-hepatitis-delta*
25 Human Hepatitisvirus B — *Homosapiensvirus hepatitis-beta* (Serum Hepatitis)
26 Human Hepatitisvirus E — *Homosapiensvirus hepatitis-epsilon*

27 Human Herpesvirus Simplex 1 *Homosapiensvirus herpes-simplex-alpha* (Cold Sores)
28 Human Herpesvirus Simplex 2 *Homosapiensvirus herpes-simplex-beta* (Genital Herpes)
29 Human Herpesvirus Varicello *Homosapiensvirus herpes-varicello* (Chicken Pox)
30 Human Herpesvirus Zoster *Homosapiensvirus herpes-varicello-zoster* (Shingles)
31 Human Herpesvirus Lympho *Homosapiensvirus herpes-lymphocrypto*
32 Human Herpesvirus Cyto *Homosapiensvirus herpes-cytomegalo*
33 Human Herpesvirus Roseolo *Homosapiensvirus herpes-roseolo*
34 Human Iridovirus *Homosapiensvirus irido* (Iridescent Virus)
35 Human Nodavirus *Homosapiensvirus noda-alpha*
36 Human Orthomyxovirus Flu *Homosapiensvirus orthomyxa-influenza*
37 Human Orthomyxovirus Thogo *Homosapiensvirus orthomyxa-thogoto*
38 Human Papillomavirus *Homosapiensvirus papilloma*
39 Human Paramyxovirus Rubula *Homosapiensvirus paramyxo-rubula* (Mumps, Rubul Parainfluenza)
40 Human Paramyxovirus Monega *Homosapiensvirus paramyxo-monega* (Menangle)
41 Human Paramyxovirus Avula *Homosapiensvirus paramyxo-avula* (Newcastle)
42 Human Paramyxovirus Respiro *Homosapiensvirus paramyxo-respiro* (Parainfluenza)
43 Human Paramyxovirus Henipa *Homosapiensvirus paramyxo-henipa*
44 Human Paramyxovirus Morbilli *Homosapiensvirus paramyxo-morbilli* (Measles)
45 Human Paramyxovirus Pneumo *Homosapiensvirus paramyxo-pnemo* (Pneumonia)
46 Human Paramyxovirus Monad *Homosapiensvirus paramyxo-mononega*
47 Human Parvovirus *Homosapiensvirus parvo* (Rheumatoid Arthritis)
48 Human Parvovirus Erythro *Homosapiensvirus parvo-erythro*
49 Human Parvovirus Adeno *Homosapiensvirus parvo-adeno*
50 Human Picovirus Entero *Homosapiensvirus pico-entero* (Coxsackie)
51 Human Picovirus Echo *Homosapiensvirus pico-entero-echo*
52 Human Picovirus Polio *Homosapiensvirus pico-polio* (Polio)
53 Human Picovirus Rhino *Homosapiensvirus pico-rhino* (Sinusitis Rhinovirus)
54 Human Picovirus Cardio *Homosapiensvirus pico-carido-encephalomylo* (Carditis Encephalomyelitis)
55 Human Picovirus Parecho *Homosapiensvirus pico-parecho*
56 Human Picovirus Kobu *Homosapiensvirus pico-kobu*
57 Human Picovirus Monoposi *Homosapiensvirus pico-monoposi*
58 Human Polyomavirus *Homosapiensvirus polyoma*
59 Human Poxvirus *Homosapiensvirus pox* (Auzduk, Cotia, Embu, Orf, Uasin Gishu, Yoka)
60 Human Poxvirus Molluscum *Homosapiensvirus pox-molluscum-contagiousum* (Soft Tumor Pox)
61 Human Poxvirus Variola *Homosapiensvirus pox-variola* (Smallpox)
62 Human Prion Virus *Homosapiensvirus prion-spoenc* (Spongiform Encephalitis, Creutzfeldt-Jakob)
63 Human Reovirus Ortho *Homosapiensvirus reo-ortho*
64 Human Reovirus Orbi *Homosapiensvirus reo-orbi* (Epizootic Hemorrhagic Virus)
65 Human Reovirus Rota *Homosapiensvirus reo-rota* (Rotavirus Diarrhea)
66 Human Reovirus Colti *Homosapiensvirus reo-colti* (Colt Salmon River Virus)
67 Human Reovirus Seado *Homosapiensvirus reo-seado*
68 Human Reovirus Phyto *Homosapiensvirus reo-phyto* (Wound Tumor Virus)
69 Human Retrovirus Gamma *Homosapiensvirus retro-gamma* (Leukemia, Reticuloendotheliosis, Sarcoma)
70 Human Retrovirus Delta *Homosapiensvirus retro-delta* (Lymphotropic Virus)
71 Human Retrovirus Lenti *Homosapiensvirus retro-lenti* (Human Immunodeficiency Virus HIV)
72 Human Retrovirus Spuma *Homosapiensvirus retro-spuma* (Foamy Virus)
73 Human Rhabdovirus Vesiculo *Homosapiensvirus rhabdo-vesiculo* (Vesicular Virus)
74 Human Rhabdovirus Lyssa *Homosapiensvirus rhabdo-lyssa* (Rabies)
75 Human Rhabdovirus Ephemero *Homosapiensvirus rhabdo-emephero* (Ephemeral Fever)
76 Human Rhabdovirus Novi *Homosapiensvirus rhabdo-novi* (Hemorrhagic Septicemia)
77 Human Rhabdovirus Mononega *Homosapiensvirus rhabdo-mononega*
78 Human Togavirus Alpha *Homosapiensvirus toga-alpha* (Sleeping Disease)
79 Human Togavirus Rubella *Homosapiensvirus toga-rubella* (German Measles)
80 Human Orthovirus *Homosapiensvirus ortho-monoposi*

Acq. Imm. Def. Syn DAID **Acquired Immune Deficiency Syndrome** (AIDS) occurs when the Human Immunodeficiency Virus (HIV) invades key immune cells called T-lymphocytes and multiplies it causes a breakdown in the body's immune system eventually leading to overwhelming infection, cancer, and death.
IMSB = immune system booster
EGF = endocrine gland function
Herbs: *Achyrocline satureoides* (TM, Wyk), ***Allium sativum* (DAVR, Bal2), *Aloe vera* (SDG, Bal1, Bal2),** *Ananas comosus* (protease inhibitor, Bal2), *Andrographis paniculata* (IMSB, Bal2), *Arctium lappa* (root, DASC, Bal1), *Arnica montana* (TM, Wyk), ***Astragalus membranaceus* (IMSB, Bal1, Bal2)**, ***Aspalathus linearis* (DHIV, Bal2),** *Baptisia tinctoria* (CS, Wyk), *Calendula officinalis* (TM, Wyk), *Capsicum annuum-longum-cayenneum* (IMSB, Bal1, Wyk), *Croton tiglium* (carcinogenic, NSD, Wyk), ***Curcuma longa* (DHIV, Bal2, Murr),** *Daphne mezereum* (IMSB, Wyk), *Dioscorea villosa* (rhizome, root, EGF, Bal1), *Echinacea angustifolia* (PS, Wyk), *Echinacea pallida* (PS, Wyk), *Echinacea purpurea* (leaf, root, DASC, Bal1, Wyk), ***Eleutherococcus senticosus* (root, BCR, Bal1, Wyk)**, *Epimedium brevicornum* (testosterone, Bal2), *Epimedium grandiflorum* (testosterone, Bal2), ***Erythroxylum catuaba* (DHIV, Bal2)**, *Eupatorium cannabinum* (carcinogenic, TM, Wyk), *Eupatorium perfoliatum* (TM, Wyk), *Euphorbia spp.* (carcinogenic, NSD, Wyk), *Ginkgo biloba* (leaf, BCR, Bal1), ***Glycyrrhiza glabra* (root, EGF, IMSB, Bal1, Bal2, Murr), *Grifola frondosa* (DHIV, Bal2),** *Hydrastis canadensis* (rhizome, root, DASC, Bal1), ***Hypericum perforatum* (flower, leaf, stem, DAVR, Bal1), *Lentinula edodes* (DHIV, Bal2)**, *Matricaria chamomilla* (TM, Wyk), *Pfaffia paniculata* (bark, berry, leaf, root, DASC, Bal1), ***Prunella vulgaris* (DHIV, Bal2),** *Rhapanus sativus-nigrus* (BLC, Bal1), *Rhus toxicodendron* (IMSB, Wyk), *Saccharomyces cerevisiae* (CS, Wyk), *Serenoa repens* (testosterone,

Bal2), ***Silybum marianum*** **(BLC, Bal1, Murr)**, *Sinapis nigra* (IMSB, Wyk), *Sutherlandia frutescens* (CS, Wyk), ***Tabebuia impetignosa*** **(inner bark, DASC, Bal1, Wyk)**, *Taraxacum officinale* (root, BLC, Bal1), *Thuja occidentalis* (CS, Wyk), *Trichosanthes kirilowii* (root, DCTM, Bal1) *Trifolium pratense* (flower, DASC, Bal1), ***Uncaria guianensis*** **(root, IMSB, Bal1, Bal2)**, ***Uncaria tomentosa*** **(root, IMSB, Bal1, Bal2)**, *Verbascum thapsis* (leaf, DASC, Bal1), *Viscum album* (CS, Wyk)
Herb Formulas: ASTRAGALUS SUPREME (IMSB, Scal): *Astragalus membranaceus* (root, 570mg), *Ligustrum lucidum* (fruit, 180mg), *Schizandra chinensis* (berry, 555 mg); CELL WELL (IMSB, Scal): *Bupleurum chinense* (root, 70mg), *Coleus barbatus* (root, 114mg), *Glycyrrhiza glabra* (root, 10mg), *Glycrrhiza uralensis* (root, 10mg), *Scutellaria baicalensis* or *barata* or *lateriflora* (88mg), *Tanacetum parthenium* (60mg), *Zingiber officinalis* (rhizome, CO2 extract), *Zizyphys jujube* (seed, 32mg) (Scal); CHINESE HIV CLINICAL TRIALS *Artemisia vulgaris, Astragalus membranaceus, Carthamus tinctorius, Glycyrrhiza glabra, Morus alba*, (DAVR, Bal2, page 349), MINOR BUPLEURUM DECOCTION *Bupleurum chinense, Bupleurum scorzoneraefolium, Glycyrrhiza glabra, Panax ginseng, Pinellia turnata, Scutellaria baicalensis, Scutellaria barata, Scutellaria lateriflora, Zingiber officinale, Ziziphus jujuba* (Bal2); SHIH QUA DA BU TANG *Angelica sinensis, Astragalus membranaceus, Atractylodes macrocephala, Cinnamomum aromaticum, Glycyrrhiza glabra, Ligusticum porteri, Paeonia officinalis, Panax ginseng, Rehmannia glutinosa, Wolfiporia extensa* (Bal2); TRUE MAN'S DECOCTION *Angelica sinensis, Atractylodes macrocephala, Cinnamomum aromaticum, Glycyrrhiza glabra, Myristica fragrans, Papaver somniferum* (husk), *Paeonia officinalis, Panax ginseng, Saussurea lappa, Terminalia chebula* (Bal2); TRUE WARRIOR DECOCTION *Aconitum napellus, Atractylodes macrocephala, Paeonia officinalis, Zingiber officinale, Wolfiporia extensa* (Bal2);
Vegetables (grain): Brown Rice, Millet, **Vegetables (legume):** Beans, Lentils, **Vegetables (crucifer):** Broccoli, Brussels Sprouts, Cabbage, Cauliflower, **Vegetables (orange colored):** Carrots, Pumpkin, Squash, Yam, **Vegetables (juice):** Beet Greens, Beet Root, Carrot, Garlic, Kale, Onion, Spinach, **Vegetables (mushrooms):** Maitake Mushroom, Reishi Mushroom, Shiitake Mushroom (Bal1)

Chickenpox DPXC the chickenpox virus, *Homosapiensvirus herpes-varicello*, causes headache and fever between 7 and 21 days after exposure. Twenty-four to Thirty-six hours later small round pimples appear on the face and body. Once the Herpes varicella virus has entered the body, it does not go away. It may lie dormant in the spinal chord and nerve ganglia for years. It is activated by a weakening ot the immune system and returns as Shingles.1
Herbs: *Arctium lappa* (BLC, Bal1), *Echinacea purpurea* (DAVR, Bal1), *Hydrastis canadensis* (DABI, Bal1), *Hypericum perforatum* (DAVR, Bal1), *Lithospermum erythrorhizon* (DPXC, Bal1), *Lonicera japonica* (DPXC, Bal1), *Nepeta cataria* (HFV, Bal1), *Tabebuia impetiginosa* (BLC, Bal1), *Zingiber officinale* (DABI, Bal1)
Vegetables: Avocado, Brewer's Yeast, Broth (Bal1)

Chronic Fatigue Syndrome DCFS can be linked with *Herpesvirus Epstein-Barr* that is also the cause of mononucleosis. Other causes include blood anemia, mercury poisoning from metal amalgam fillings, and *Candida albicans* fungus.The symptoms include vitality problems, a decrease of psychological and physical vigilance.
EGF = endocrine gland function, IMSB = immune system booster
Herbs: *Acacia catechu* (Bal1, page 270), *Acorus calamus* (pungent tonic, Wyk), *Alpinia officinarium* (pungent tonic, Wyk), *Arctium lappa* (BLC, Bal1), *Artemisia abrotanum* (bitter tonic, Wyk), *Astragalus membranaceus* (IMSB, Bal1), *Capsicum annuum-longum-cayenneum* (Bal1, page 270), *Centaurium erythraea* (bitter tonic, Wyk), *Centella asiatica* (Bal1, page 270), *Cinchhona pubescens* (bitter tonic, Wyk), *Citrus-aurauntius aurantium* (bitter tonic, Wyk), *Coffea arabica* (stimulant, Wyk), *Cola acuminata* (stimulant, Wyk), *Cyamopsis tetragonoloba* (Guarana was interpreted as Guar Bean, from Bal1, page 270), *Echinacea purpurea* (IMSB, Bal1, Bal2), ***Eleutherococcus senticosus*** **(adaptogen, extract, Bal1, Bal2, Murr, Scal, Wyk)**, *Gentiana lutea* (bitter tonic, Wyk), *Ginkgo biloba* (BCR, Bal1), *Glycyrrhiza glabra* (EGF, Bal1, Bal2), *Grifola frondosa* (IMSB, Bal2), *Harungana madagascarensis* (bitter tonic, Wyk), *Hydrastis canadensis* (DABI, Bal1), *Hypericum perforatum* (DAVR, Bal1, Bal2), *Ilex paraguarensis* (stimulant, Wyk), *Lentinula edodes* (HFV, Bal2), *Marsenia cundurango* (bitter tonic, Wyk), *Menyanthes trifoliata* (bitter tonic, Wyk), *Momordica charantia* (DHSV, Bal2), *Myristica fragrans* (pungent tonic, Wyk), ***Panax ginseng*** **(adaptogen, Bal2, Murr, Wyk)**, *Paullinia cupana* (stimulant, Wyk), ***Pinus pinaster*** **(IMSB, pine bark extract, Bal2, Murr)**, *Rhodiola rosea* (adaptogen, Murr), *Salvia multiorhiza* (NSD, Bal2), *Scutellaria baicalensis* (DABA, DAVR, HIN, Bal1, Bal2), *Scutellaria barata* (DABA, DAVR, HIN, Bal1, Bal2), *Scutellaria laterilfora* (DABA, DAVR, HIN, Bal1, Bal2), *Silybum marianum* (BLC, Bal1), *Sutherlandia frutescens* (adaptogen, Wyk), *Syzygium aromaticum* (Bal2), *Taraxacum officinale* (BLC, Bal1), *Theobroma cacao* (stimulant, Wyk), *Trifolium pratense* (BLC, Bal1), *Valeriana officinalis* (HIN, Bal1), ***Vitis vinifera*** **(IMSB, grape seed extract, Bal2, Murr)**, ***Withania somnifera*** **(adaptogen, Murr, Wyk)**, *Zingiber officinalis* (pungent tonic, Wyk)
Herb Formulas: ENERGY & VITALITY *Cola nitida* (nut, 42mg), *Eleutherococcus senticosus* (root, 90mg), *Ginkgo biloba* (leaf, 36mg), *Glycyrrhiza glabra* (root, 15mg), *Glycyrrhiza uralensis* (root, 15mg), *Panax ginseng* (root, 51mg), *Schisandra chinensis* (berry, 75mg), *Thea sinensis-viridis* (leaf, 90mg), *Urtica dioica* (seed, 6mg), *Xanthoxylum clava-herculis* (bark, 6mg) (Scal);
Vegetables: Raw (Bal1), **Vegetables:** Borage Oil, Evening Primrose Oil, Flaxseed Oil, Nuts (Bal2), **Vegetables:** Whole Organically Grown (Murr), **Vegetables (grain magnesium):** Brown Rice, Buckwheat, Cereal, Corn, Millet, Oats, Quinoa, Rice Bran, Rye, Spelt, Tapioca, White Flour, White Rice, Wheat (Bal1, Bal2), **Vegetables (leaf magnesium):** Chard, Kale, Lettuce, Spinach, Turnip Greens (Bal2)

Common **Cold DCLD** caused by a wide variety of viruses that infect the upper respiratory tract; the nasal passages, the sinuses, and the throat.
IMSB = immune system booser
Herbs: *Achillea millifolium* (essential oils, Bal1), *Allium cepa* (alkaloids, RLC, Bal1, Wyk), *Astragalus membranacea* (RRN, Bal2), *Ballota nigra* (RRN, Bal2), *Baptisia tinctoria* (DAVR, IMSB, Bal1, Wyk), *Cinchona pubescens* (alkaloids, RLC, Bal1, Wyk), *Cinnamomum camphora* (essential oils, Bal1, Wyk), *Commiphora myrrha* (essential oils, Bal1, Wyk), *Echinacea angustifolia* (IMSB, Wyk), *Echinacea pallida* (IMSB, Wyk), ***Echinacea purpurea*** **(extract, DAVR, IMSB, Bal1, Bal2, Murr, Wyk)**, *Ephedra distachya* (alkaloids, Bal1, Wyk), *Eucalyptus golbulus* (essential oils, Bal1, Wyk), *Euphrasia nemorosa* (alkaloids, RLC, Bal1, Wyk), *Foeniculum vulgare* (essential oils, Bal1, Wyk), *Galphimia glauca* (alkaloids, RLC, Bal1, Wyk), *Gentiana lutea* (alkaloids, RLC, Wyk), *Hydrastis canadensis* (DAVR, Bal1), *Hyssopis officinalis* (DAVR, RLC, Bal2), *Lentinula edodes* (IMSB, Bal2), *Ligusticum porteri* (ear infection, Bal2), *Lobelia inflata* (HFV, Bal1), *Luffa operculata* (alkaloids, RLC, Bal1, Wyk), *Marrubium vulgare* (RRN, Bal2), *Matricaria chamomilla* (essential oils, Bal1, Wyk), *Melaleuca alternifolia* (RST, Bal1, Bal2), *Mentha piperita* (essential oils, Bal1, Wyk), *Nepeta cateria* (HFV, Bal1), ***Pelargonium sidoides*** **(DAVR, IMSB, Bal1, Murr, Wyk)**, *Phytolacca americana* (alkaloids, RLC, Bal1, Wyk), *Picea albies* (essential oils, Bal1, Wyk), *Pimpinella anisum* (essential oils, Bal1, Wyk), *Pinus mugo* (essential oils, Bal1, Wyk), *Primula veris* (alkaloids, RLC, Bal1, Wyk), *Pueraria lobata* (NSD, Bal2), *Salvia officinalis* (essential oils, Bal1, Wyk), *Sambucus nigra* (DFRA, HFV, Bal2), *Tabebuia impetignosa* (essential oils, Bal1), *Thuja occidentalis* (DAVR, IMSB, Bal1, Wyk), *Thymus serpyllum* (essential oils, Bal1, Wyk), ***Thymus vulgaris*** **(essential oils, Bal1, Bal2, Wyk)**, *Tilia cordata* (RLC, RST, Bal2), *Trigonella foenum-graecum* (RRN, Bal2), *Ulmus rubra* (essential oils, RST, Bal1,

Bal2), *Uncaria guianensis* (IMSB, Bal2), *Uncaria tomentosa* (IMSB, Bal2), *Verbena officinals* (alkaloids, RLC, Bal1, Wyk), *Zingiber officinale* (essential oils, Bal1, Bal2)
Herb Formulas: LOMATIUM OSHA SUPREME *Allium sativum* (bulb, 40mg), *Berberis vulgaris* (root, 8mg), *Chondrus crispus* (20mg), *Echinacea angustifolia* (root, 3mg), *Echinacea purpurea* (root, 4mg, flowering top, 4mg, seed 1mg), *Grindelia robusta* (flower bud, 22mg), *Hydrastis canadensis* (rhizome, 8mg), *Hypericum perforatum* (flower bud, 6mg), *Hyssopus officinalis* (flower, 10mg), *Ligusticum porteri* (root, 34mg), *Lobelia inflata* (herb & seed, 6mg), *Lomatium dissectum* (root, 30mg), *Mahonia aquifolium* (root, 8mg) *Verbascum olympicum* (leaf, 8mg) (Scal);
Vegetables: Potato Peel Broth + Celery + Carrot (Bal1), Diluted Vegetable Juices (Murr), Diluted Clear Vegetable Soups (Murr)

(1) **Stress Common Cold** Symptoms include head congestion, sore throat, coughing, headache, fever, restlessness, sneezing, watery eyes, muscle aches and joint pains.

(2) **Infectious Common Cold** There are over 200 viruses that can cause the common cold which is an infection of the upper respiratory tract.

German Measles DMSG also known as **Rubella**, is caused a mild but contagious virus *Homosapiensvirus toga-rubella* (Synonym *Rubellavirus*) and affects the lymph glands in the neck and behind the ears. First day symptoms are coughing, fatigue, headache, mild fever, and stiffness in the neck. Second day symptoms include a rash on the face and neck spreading to the rest of the body. The virus runs it course in five to seven days.
Herbs: *Allium sativum* (enema, Bal1), *Hydrastis canadensis* (DAVR, extract placed under the tongue Bal1), *Lobelia inflata* (NSD, extract, Bal1), *Mentha piperita* (tea, Bal1), *Nepeta cataria* (enema, Bal1), *Syzygium aromaticum* (tea, Bal1)
Vegetables: Vegetable Broths (Bal1)

Hepatitis DHPT also known as **Steatoheptatitis**, an inflammation of the liver caused by viral infection. The liver becomes tender and enlarged and is unable to purify the blood properly. Toxins build up in the body and certain nutrients are not processed. Symptoms include fever, weakness, nausea, vomiting, headache, appetite loss, muscle aches, joint pains, drowsiness, dark urine, light colored stools, abdominal discomfort, and yellow skin jaundice. The forms are: Human Hepatitisvirus A *Homosapiensvirus pico-hepatitis-alpha* (Infectious Hepatitis), Human Hepatitisvirus B *Homosapiensvirus hepatitis-beta* (Serum Hepatitis), Human Hepatitisvirus C *Homosapiensvirus flavi-hepatitis-cappa* (Blood Transfusion Hepatitis), Human Hepatitisvirus D *Homosapiensvirus pico-hepatitis-delta,* Human Hepatitisvirus E *Homosapiensvirus heptatitis-epsilon,* and Human Hepatitisvirus G *Homosapiensvirus flavi-hepatitis-gamma.* All forms of Hepatitis are contagious.
Herbs: *Arctium lappa* (BLC, Bal1), ***Glycyrrhiza glabra* (DAVR, Bal1, Bal2, Murr)**, *Hydrastis canadensis* (Bal1), *Ligustrum lucidum* (Bal2), *Mentha piperita* (Bal2), *Pinus pinaster* (bark extract, Murr), *Raphanus sativus-niger* (Bal2), *Rumex crispus* (Bal1), *Schisandra chinensis* (Bal2), ***Silybum marianum* (silymarin, Bal1, Bal2, Murr)**, *Taraxacum officinale* (BLC, Bal1), *Thea sinensis-viridis* (Bal1) *Trifolium pratense* (Bal1), *Vitis vinifera* (seed extract, Murr), *Wolfiporia extensa* (Bal2)
Herbs Formulas: MINOR BUPLEURUM DECOCTION *Bupleurum chinense, Bupleurum scorzoneraefolium, Glycyrrhiza glabra, Panax ginseng, Pinellia turnata, Scutellaria baicalensis Scutellaria barata, Scutellaria lateriflora, Zingiber officinale, Ziziphus jujuba* (Bal2);
Vegetables: Raw for four weeks, Artichoke, Carrot Juice, Beet Juice (Bal1)
Vegetable Formulas: AGELESS COCKTAIL LIVE JUICE 4-5 fresh carrots, 3 sprigs fresh parsley 1 handful fresh spinach, 1 handful fresh kale, 1 fresh beet including tops, 1 clove garlic peeled, all put in juicer (Bal1, page 556); RAW POTATO LIVE JUICE 8 fresh potato skins with ½ inch of potato on the skin, 1 fresh carrot, 1 stalk fresh celery, 8 ounces bottled spring water, all put in juicer (Bal1, page 557 #1); CABBAGE LIVE JUICE 1/3 head of fresh cabbage cut in small pieces, 2 fresh carrots cut in small pieces, ¼ cup of bottled spring water, all put in juicer (Bal1, page 557 #2)

Herpes Cytovirus *Homosapiensvirus herpes-cytomegalo*

Herpes Lymphovirus *Homosapiensvirus herpes-lymphocrypto*

Herpes Roseolovirus *Homosapiensvirus herpes-roseolo*

Herpes Simplex Virus 1 DHSV causes mild skin outbreaks in the form of **Cold Sores** which are fever blisters of the mouth and lips. About 30% of the population in the United States has HSV-1. *Homosapiensvirus herpes-simplex-alpha* causes **Cold Sores**. Fever blisters first appear three to ten days after exposure. They last up to three weeks. The virus remains in the body and repeated outbreaks can be triggered by fever, cold or viral infection. Cold Sores are very contagous. There is always less discomfort with cold sores than with the initial outbreak. The virus can erupt later in life as Shingles.
Herbs: *Aloe vera* (DAVR, external, Bal2), *Callista nobilis* (cold sores, Bal2), *Carica papaya* (NSD, external, Bal2), *Copiafera langsdorffii* (external, oil Bal2), *Echinacea purpurea* (cold sores, Bal1), *Glycyrrhiza glabra* (DHSV, root, Bal1, Bal2), *Hydrastis canadensis* (cold, sores, Bal1), *Hypericum perforatum* (external, Bal1, Bal2), *Malpighia glabra* (cold sores, Bal2), *Melissa officinalis* (external, cold sores, Bal2), *Pinus pinaster* (bark extract, Murr), *Tabebuia impetignosa* (cold sores, Bal1), *Thea sinensis* (external, Bal2),*Trifolium pratense* (cold sores, Bal1), Vitis vinifera (seed extract, Murr)
Herb Formulas: GENTIAN LONGDANCAO DECOCTION *Gentiana lutea, Gentiana scabra, Akebia quinata, Akebia trifoliate, Alisma plantago-aquatica, Angelica sinensis, Bupleurum chinense, Bupleurum scorzoneraefolium, Gardenia angusta, Plantago afra, Rehmannia glutinosa, Scutellaria baicalensis, Scutellaria barata, Sutellaria galericulata, Scutellaria lateriflora* (Bal2); KUDZU DECOCTION *Pueraria lobata, Cinnamomum aromaticum, Ephedra sinica, Glycyrrhiza glabra, Paeonia lactiflora, Paeonia officinalis Zingiber officinalis, Ziziphus jujuba* (Bal2);
Vegetables: Raw (Bal1)

High Lysine Vegetables

(adapted from Murr, pages 662, 663)

Vegetable	Serving Size	Low Arginine	High Lysine
Green Beans	¾ cup	80 mg	80 mg
Lima Beans	3.5 oz	1,170 mg	1,470 mg
Mung Beans	3.5 oz	1,320 mg	1,930 mg
Red Beans	1/3 cup	340 mg	420 mg
Soy Beans	2/3 cup	620 mg	620 mg
Brewer's Yeast	3.5 oz	1,940 mg	3,510 mg

Herpes Simplex Virus 2 DSXH also known as **Genital Herpes**. More than 30 million Americans have HSV-2. *Homosapiensvirus herpes-simplex-beta* is the most prevalent sexually transmitted disease in the United States. HSV-2 can cause inflammation of the liver with fever. Genital Herpes causes painful fluid filled blisters that can take up to three weeks to heal.

Herbs: *Aloe vera* (DAVR, external, Bal2), *Astragalus membranaceus* (DABI, Bal2), *Capsicum annuum-longum-cayenneum* (NSD, external, Bal1, Bal2), *Carica papaya* (NSD, external, Bal2), *Commipora myrrha* (Bal1), *Copiafera langsdorffii* (external, oil Bal2), *Echinacea purpurea* (Bal1), *Glycyrrhiza glabra* (DHSV, root, Bal1, Bal2), *Hydrastis canadensis* (internal tea, external extract, Bal1), *Hypericum perforatum* (external, Bal1, Bal2), *Juglans nigra* (external, Bal1), *Melaleuca alternifolia* (external, oil, Bal1), *Momordica charantia* (DAVR, Bal2), *Olea europaea* (DAVR, leaf, Bal2), *Syzygium aromaticm* (Bal2), *Thea sinensis* (external, Bal2), *Trifolium pratense* (Bal1), *Uncaria guianensis* (DAVR, Bal2), *Uncaria tomentosa* (DAVR, Bal2)

Vegetables: "Avoid alcohol, processed foods, colas, white flour products, sugar, refined carbohydrates, coffee and drugs to lessen the change of an outbreak." (Bal1, page 319, recommendation #1)

Human Immune Deficiency DHIV also known as **Weakened Immune System**,
T-cells: white blood cells called Thymus Lymphocytes mature in the Thymus gland where they learn what cells are foreign to the body. The T-cell learns the size and shape of the foreign cells (antigens) and attempts to destroy them by secreting cytokine proteins. Interferon is one of the better known types of cytokine proteins.
B-cells: white blood cells called Bone Marrow Lymphocytes mature in the bone marrow and learn the size and shape of foreign cells and then learns to create proteins that have a shape and size that is the mirror image of the foreign cells (antigens). The antibody proteins latch-on the to the foreign cells and covers their reproductive structures so that they cannot replicate. If the foreign cell (or virus) has mutated the antibodies send the new shape and size information back to the B-cells so that they can reprogram their DNA to create new protein antibodies which are a "closer mirror image fit" to the foreign cells (or viruses). The virus *Homosapiensvirus retro-lenti* is a Human immunodeficiency Virus (HIV) which slows the production of T-cells and B-cells. A few herbs act as immune system boosters by stimulating increased production of T-cells and B-cells.
ISB = Immune System Booster

Herbs: *Allium sativum* (DHIV, Bal2), *Aloe vera* (SDG, Bal2), *Ananas comosus* (protease inhibitor, bromelain, Bal2), *Aspalathus linearus* (DHIV, Bal2), *Astragalus membranaceus* (ISB, Bal1, Bal2), *Ballota nigra* (ISB, Bal1), *Crataegus laevigata* (ISB, Bal1), *Crataegus monogyna* (ISB, Bal1), *Curcuma longa* (DHIV, Bal2), *Echinacea purpurea* (ISB, Bal1), *Erythoxylum catuaba* (DHIV, Bal2), *Gingko biloba* (BCR, DFRA, Bal1), *Glycyrrhiza glabra* (ISB, Bal1, Bal2), *Grifolia frondosa* (DHIV, Bal2), *Hydrastis canadensis* (ISB, Bal1), *Hypericum perforatum* (BLC, DAVR, Bal1), *Lentinuls edodes* (DHIV, Bal2), *Ligustrum lucidum* (ISB, Bal1), *Lycium chinense* (germanium, Bal1), *Marrubium vulgare* (ISB, Bal1), *Myrica cerifera* (ISB, Bal1), *Panax ginseng* (germanium, Bal1), *Pfaffia paniculata* (germanium, Bal1), *Prunella vulgaris* (DHIV, Bal2), *Rhaphanus sativus-niger* (BLC, Bal1), *Silybum marianum* (BLC, Bal1), *Taraxacum officinale* (BLC, Bal1), *Trifolium pratense* (ISB, Bal1), *Trigonella foenum-graecum* (ISB, Bal1), *Uncaria guianensis* (ISB, Bal2), *Uncaria tomentosa* (ISB, Bal2), *Wisteria frutescens* (germanium, Bal1)

Herbs: MINOR BUPLEURUM DECOCTION *Bupleurum chinense, Bupleurum scorzoneraefolium, Glycyrrhiza glabra, Panax ginseng, Pinellia turnata, Scutellaria baicalensis Scutellaria barata, Scutellaria lateriflora, Zingiber officinale, Ziziphus jujuba* (Bal2); SHIH QUA DA BU TANG *Angelica sinensis, Astragalus membranaceus, Atractylodes macrocephala, Cinnamomum aromaticum, Glycyrrhiza glabra, Ligusticum porteri, Paeonia officinalis, Panax ginseng, Rehmannia glutinosa, Wolfiporia extensa* (Bal2);

Influenza DFLU also known as **Flu** is caused by *Homosapiensvirus orthomyxa-influenza*. It is highly contagious. The virus mutates each year so there are many varieties of the Flu called strains. In the past the three most common strains were *Homosapiensvirus orthomyxo-influenza* A-HIND-PR/8/34, *Homosapiensvirus orthomyxo-influenza* B-LEE/40 and *Homosapiensvirus orthomyxo-influenza* C-California/78.

Herbs: *Achillea millefolium* (Bal1), *Allium sativum* (Bal2), *Capsicum-annuum-longum-cayenneum* (Bal1, Bal2), *Echinacea purpurea* (Bal1, Bal2), *Ephedra distachya* (Bal1), *Eucalyptus globulus* (external, Bal1), *Eupatorium perfoliatum* (Bal2), *Eupatorium purpureum* (Bal2), *Fritillaria thunbergii* (Bal2), *Glycyrrhiza glabra* (Bal2), *Hydrastis canadensis* (Bal1), *Illicium anisatum* (Bal2), *Illicium verum* (Bal2), *Ligusticum porteri*, *Lobelia inflata* (Bal1), *Mentha piperita* (Bal1), *Nepeta cataria* (Bal1, Bal2), *Pueraria lobata* (Bal2), *Sambucus nigra* (Bal2), *Scutellaria baicalensis* (Bal2), *Scutellaria barata* (Bal2), *Scutellaria lateriflora* (Bal2), *Tabebuia impetignosa* (Bal1), *Thea sinensis-viridis* (Bal2), *Tilia cordata* Lime Linden, *Tilia platyphyllos* Lime Linden, *Trigonella foenum-graecum* (Bal1), *Ulmus rubra* (Bal1), *Verbascum thapsis* (Bal2), *Zingiber officinale* (Bal1, Bal2)

Herb Formulas: FENU-THYME *Trigonella foenum-graecum, Thymus vugaris* (Bal1); KOMBUCHA TEA Kombucha (a Lichen that is part algae, part fungus, part yeast, and part *Xylinum* bacteria), *Saccharum officinarum* (Sugar), *Thea sininensis* (Tea) "When placed in a mixture of ordinary sugar and tea for seven to ten days, the kombucha produces both a winey-tasting health restoring tea and a new 'baby'kombucha" (Bal1, page 557, under Making Kombucha Tea)

Vegetables: "Consume plenty of fluids, especially fresh juices, herbal teas, soups, and quality water to prevent dehydration and help flush out the body." (Bal1, page 347, recommendation #1)

Measles DMSL also known as **Morbilla**, attacks the respiratory tract, lungs, and eyes and is caused by the virus *Homosapiensvirus paramyxo-morbilli*. Symptoms include fever, cough, sneezing, runny nose and red eyes, and a red rash on the forehead and ears. The rash can spread to other parts of the body.

Herbs: *Allium sativum* (enema, Bal1), *Glycyrrhiza glabra* (Bal2), *Lobelia inflata* (Bal1, Bal2), *Nepeta cateria* (enema, Bal1)

Herb Formulas: KUDZU DECOCTION *Pueraria lobata, Cinnamomum aromaticum, Ephedra sinica, Glycyrrhiza glabra, Paeonia lactiflora, Paeonia officinalis Zingiber officinalis, Ziziphus jujuba* (Bal2);

Vegetables: Vegetable Broths (Bal1)

Mononucleosis DMNN also known as **Epstein-Barr Virus** caused by *Homosapiensvirus herpes-mononucleosis*.

Herbs: *Astragalus membranaceus* (Bal1, Bal2), *Echinacea purpurea* (Bal1), *Gleditsia triacanthos* (Bal2), *Hydrastis canadensis* (Bal1), *Olea europaea* (leaf, Bal2), *Scutellaria baicalensis* (Bal2), *Scutellaria barata* (Bal2), *Scutellaria lateriflora* (Bal2), *Tabebuia impetignosa* (Bal1), *Taraxacum officinale* (Bal1)

Vegetables: Root Vegetables, Vegetable Soups (Bal1); **Vegetables (grain):** Brown Rice, Buckwheat, Cereal, Corn, Millet, Oats, Quinoa, Rice Bran, Rye, Spelt, Tapioca, White Flour, White Rice, Wheat (Bal1);

Meningitis, Viral DMNV sore throat, red or purple skin rash, stiff neck, headache, high fever, chills, nausea, vomiting, and sensitivity to light cause by the viruses *Homosapiensvirus arena lymphocytochoriomeningitis. Homosapiensvirus flavi-monoposi meningoencephalomyelitis.*

Herbs: *Echinacea purpurea* (Bal1), *Hydrastis canadensis* (Bal1), *Hypericum perforatum* (Bal1), *Nepeta cateria* (enema, Bal1),
Vegetables: Nuts, Seeds (Bal1); **Vegetables (grain):** Brown Rice, Buckwheat, Cereal, Corn, Millet, Oats, Quinoa, Rice Bran, Rye, Spelt, Tapioca, White Flour, White Rice, Wheat (Bal1);

Mumps DMUM infection of the parotid glands located at the jaw below the ear and caused by the virus *Homosapiensvirus paramyxo-rubula*. Symptoms include swelling of the glands, pain, headache and fever.
Herbs: *Achillea millefolium* (Bal1, Bal2), *Echinacea purpurea* (Bal1), *Eleutherococcus senticosus* (Bal2), *Mentha piperita* (Bal1), *Nepeta cateria* (Bal1), *Lobelia inflata* (Bal1), *Matricaria chamomilla* (Bal1), *Sambucus nigra* (flower, Bal1, Bal2), *Taraxacum officinale* (Bal1), *Ulmus rubra* (Bal1), *Verbascum thapsis* (Bal1)
Herb Formulas: KUDZU DECOCTION *Pueraria lobata, Cinnamomum aromaticum, Ephedra sinica, Glycyrrhiza glabra, Paeonia lactiflora, Paeonia officinalis Zingiber officinalis, Ziziphus jujuba* (Bal2); MINOR BUPLEURUM DECOCTION *Bupleurum chinense, Bupleurum scorzoneraefolium, Glycyrrhiza glabra, Panax ginseng, Pinellia turnata, Scutellaria baicalensis Scutellaria barata, Scutellaria lateriflora, Zingiber officinale, Ziziphus jujuba* (Bal2);
Vegetables: Raw (Bal1), "Eat mostly juiced or soft foods while your glands are swollen." (Bal2, page 411, recommendation #1)

Paget's Bone DPBN also known as **Osteitis Deformans**, a pattern of alternating remissions and flare-ups with the flare-ups becoming worse. Abnormal bone replaces normal bone in the pelvis, spine, thighs, skul, hips, shins, and upper arms. Deformities are bowed legs, barrel-shaped chest, bent spine, and enlarged forehead. Researchers suspect a virus *Homosapiensvirus ortho-monoposi osteosarcoma*.
Herbs: *Angelica sinensis* (Bal1), *Capsicum annuum-longum-cayenneum* (Bal1), *Cimicifuga racemosa* (Bal1), *Echinacea purpurea* (Bal1), *Equisetum arvense* (Bal1), *Eupatorium perfoliatum* (Bal1), *Glycyrrhiza glabra* (Bal1), *Humulus lupulus* (Bal1), *Hydrastis canadensis* (Bal1), *Hypericum perforatum* (Bal1), *Medicago sativa* (Bal1), *Passiflora caerulea* (Bal1), *Passiflora incarnata* (Bal1), *Petroselinum crispum* (Bal1), *Phytolacca americana* (Bal1), *Rosa-hipa canina* (Bal1), *Salix alba* (Bal1), *Scutellaria baicalensis* (Bal1), *Scutellaria barata* (Bal1), *Scutellaria lateriflora* (Bal1), *Tanacetum parthenium* (Bal1), *Taraxacum officinalis* (Bal1), *Urtica dioca* (Bal1), *Yucca aloeifolia* (Bal1), *Yucca filamentosa* (Bal1), *Valeriana officinalis* (Bal1)
Vegetables: Barley, Brewer's Yeast, Carob, Garlic, Kelp, Papaya, Whey (Bal1); **Vegetables (leaf):** Chard, Kale, Lettuce, Spinach, Turnip Greens (Bal1);

Parkinson's DPRK also known as **Paralysis Agitans**, a shaking disease accompanied by impaired speech and a fixed facial expression. It lasts 10 years and results in death.
Herbs: *Achillea millefolium* (Bal1), *Arctium lappa* (Bal1), *Capsicum annuum-longum-cayenneum* (Bal1), *Cimicifuga racemosa* (Bal1), *Crataegus laevigata* (Bal1), *Crataegus monogyna* (Bal1), *Eleutherococcus senticosus* (Bal1), *Faba mediapoda* (Synonym *Vicia faba*, Murr), ***Ginkgo biloba* (Bal1, Bal2, Murr)**, *Glycyrrhiza glabra* (Bal1), *Hydrastis canadensis* (Bal1), *Malpighia glabra* (Bal2), *Melissa officinalis* (Bal1), *Mucuna pruriens* (Murr), *Nepeta cateria* (Bal1), *Passiflora caerulea* (Bal1), *Passiflora incarnata* (Bal1), ***Pinus pinaster* (bark extract, Bal2, Murr)**, *Rumex crispus* (Bal1, Bal2), *Scutellaria baicalensis* (Bal1), *Scutellaria barata* (Bal1), *Scutellaria lateriflora* (Bal1), *Silybum marianum* (Bal1, Bal2), *Smilax aristolochaefolia* (Bal1), *Smilax ferbifuga* (Bal1), *Smilax glyciphylla* (Bal1), *Smilax ornata* (Bal1), *Smilax regelii* (Bal1), *Taraxacum officinale* (Bal1), *Thea sinensis-viridis* (Murr), *Trifolium pratense* (Bal1), *Valeriana officinalis* (Bal1), *Verbascum thapsis* (Bal1), ***Vitis vinifera* (seed extract, Bal2, Murr)**, *Zingiber officinale* (Bal1)
Vegetables: Almonds, Brazil Nuts, Chick Peas, Lentils, Lima Beans, Pecans, Pumpkin Seeds, Sesame Seeds (Bal1); **Vegetables (grain):** Brown Rice, Buckwheat, Cereal, Corn, Millet, Oats, Quinoa, Rice Bran, Rye, Spelt, Tapioca, White Flour, White Rice, Wheat (Bal1);

Raynaud's Phenomenon DRNP hypersensitivity to cold of the hands and feet. Small arteries in the fingers and toes contract depriving them of adequate amounts of oxygen. They become whitish or bluish in color. The disease is more common in women than in men.
Herbs: *Capsicum annuum-longum-cayenneum* (Bal1), *Ginkgo biloba* (Bal1), *Ruscus aculeatus* (Bal1), *Tabebuia impetignosa* (Bal1)
Vegetables: Flaxseed Nuts, Oil, Lecithin, Primrose Oil, Seeds (Bal1); **Vegetables (grain):** Brown Rice, Buckwheat, Cereal, Corn, Millet, Oats, Quinoa, Rice Bran, Rye, Spelt, Tapioca, White Flour, White Rice, Wheat (Bal1); **Vegetables (legume):** Green Beans, Kidney Beans, Lentils, Lima Beans, Mung Beans, Peas, Pinto Beans, Soybeans, White Beans (Bal1);

Reye's Syndrome DRYS primarily strikes children between the ages of 4 and 15. Symptoms are fever, severe vomiting, lethargy, memory lapses, paralysis in the arms and legs, double vision, speech impairment, hearing loss, convulsions, brain damage, coma and death as a result of cerebral edema or respiratory failure.
Herbs: *Aloe vera* (external, Bal1), *Arctostaphylos uva-ursi, Avena sativa* (straw, Bal1), *Calendula officinalis* (external, Bal1), *Crataegus laevigata, Crataegus monogyna, Dioscorea oppositifolia, Dioscorea villosa, Eleutherococcus senticosus, Eupatorium purpureum, Hydrangea arborescens, Hyssopis officinalis, Matricaria chamomilla* (internal & external, Bal1), *Medicago sativa, Mentha piperita, Nepeta cateria, Panax ginseng, Petroselinum crispum* (root, Bal1), *Silybum marianum, Tabebuia impetignosa, Zingiber officinale* (Bal1)
Vegetables: Flaxseed Oil, Garlic, Lecithin

SARS DSAR also known as **Severe Acute Respiratory Syndrome**, an impairment of the lungs due to viral infection.

Shingles (zoster) DSHZ crops of tiny extremely painful blisters appear on the abdomin between the bottom of the ribs and the navel caused by the virus *Homosapiensvirus herpes-varicello-zoster*. The blisters affect the nerve endings of the skin. An attack of Shingles is often preceded by three or four days of chills, fever and aches. Symptoms include numbness, tingling, shooting pains, fever, headache, chills and achiness. Shingles strikes 850,000 Americans each year. It is most common in people over age 50. Most cases last three weeks.
Herbs: *Astragalus membranaceus* (root, Bal1), *Capsicum-annuum-longum-cayenneum* (Bal1), *Echinacea purpurea* (Bal1), *Hydrastis canadensis* (Bal1), *Matricaria chamomilla* (Bal1), *Medicago sativa* (Bal1), *Rosa-hipa canina* (Bal1), *Silybum marianum* (Bal1), *Taraxacum officinale* (Bal1), *Valeriana officinalis* (Bal1)
Herb Formulas: STRESS & ITCHING TEASPOON TINCTURE *Avena sativa* (straw), *Hypericum perforatum, Scutellaria baicalensis, Scutellaria barata, Scutellaria lateriflora* (Bal1)
Vegetables: Brewer's Yeast, Garlic (Bal1); **Vegetables (grain):** Brown Rice, Buckwheat, Cereal, Corn, Millet, Oats, Quinoa, Rice Bran, Rye, Spelt, Tapioca, White Flour, White Rice, Wheat (Bal1);

Vertigo DVRT also known as dizziness, an impaired sense of balance such as sinking, falling, spinning when the ear and optical sensors are no longer synchronized due to psychological stress and viral infection.
Herbs: *Capsicum annuum-longum-cayennensis* (BCR, Bal1), *Ginkgo biloba* (BCR, Bal1, Wyk), *Ruscus aculeatus* (BCR, Bal1), *Taraxicum officinale* (BHP, Bal1), *Zingiber officinale* (BHP, Bal1)
Vegetables: No salt on Vegetables (Bal1)

Warts DCLW small growths on the skin caused by the human papilloma virus (HPV). Warts appear singly or inclusters. There are three types of warts: common warts, plantar warts and genital warts.
Herbs: *Aloe vera* (Bal1), *Chelidonium majus* (TM, Wyk), *Commiphora myrrha* (Bal1), *Gaultheria procumbens* (Bal1), *Hydrastis canadensis* (Bal1), *Juglans nigra* (Bal1), *Lycopodium annotinum* (TM, Wyk), *Melaleuca alternifolia* (Bal1), *Podophyllum peltatum* (MM, Wyk), *Stellaria media* (Bal1), *Syzygium aromaticum* (Bal1), *Tabebuia impetignosa* (Bal1), *Thuja occidentalis* (TM, Wyk)

Economic Herbs listed by Use

All information in this section was extracted from the following references:

Kowa, 1998 **"Rodale's Illustrated Encyclopedia of Herbs"**, C.Kowalchik & W.H.Hylton, ISBN:0-87596-964-X

Bath Herbs

Bath Herbs – Blood Circulation

Basil, Bay, Calendula Flower, Citronella, Fennel, Horseradish Root, Lavender Flower, Lemon Vervain, Lovage Root, Marjoram, Mint, Nettle, Pine Needle, Queen of the Meadow, Rosemary, Sage, Savory, Thyme, Vetiver Root

Bath Herbs – Footbath

Agrimony, Alder Bark, Burdock, Goat's Rue, Lavender Flower, Mustard Seed, Sage, Witch Hazel Bark & Leaf, Wormwood

Bath Herbs – Perfume

Angelica Root, Bay, Clove, Geranium, Jasmine Flower, Lavender Flower, Lemon Flower & Peel, Lovage Root, Mint, Myrtle Leaf & Flower, Orange Leaf, Flower, & Peel, Patchouli, Pennyroyal, Queen Of The Meadow, Rose Flower, Rosemary, Sandalwood, Southernwood

Bath Herbs – Soothing

Catnip, Chamomile Flower, Comfrey, Elderberry, Evening Primrose Flower, Hyssop, Jasmine Flower, Juniper Berry, Lemon Balm, Linden Flower, Marsh Mallow Root, Melilot, Mullein, Passionflower, Rose Flower, Slippery Elm Inner Bark, Tansy Flower, Valerian Root, Verbain, Violet

Bath Herbs – Tonic

Blackberry Leaf, Comfrey, Dandelion, Ginseng Root, Jasmine Flower, Nettle, Orange, Patchouli, Raspberry Leaf

Dye Herbs

Dye Herbs – Black Dye

Alder, Black Walnut, Yarrow

Dye Herbs – Blue Dye (multiple sources)

Blue Windflower, Blue Oriental Hyacinth, California Deer Brush Redroot, Tree Mallow, Dyer's Woad, Dyer's Indigo, Elderberry, Elecampane, Canterbury Bells Bellflower, German Flag Iris, Logwood Bloodwood, Oregon Grape, Polycarp Sunburst Schist Lichen, Purple Cabbage Mustard, Purple Dutch Crocus, Purple Stock Mustard, Red Flowering Maple Mallow, Red Hollyhock Marsh Mallow, Rose of Sharon Rose Mallow, Salal Purple Checkerberry, Stiff Leaf Cheesewood

Dye Herbs – Brown Dye

Burdock, Cascara Sagrada, Comfrey, Fennel, Geranium, Hops, Juniper, Madder, Onion, Pokeweed, Poplar

Dye Herbs – Gold Dye

Agrimony, Amaranth, Dock, Goldenrod, Lavender Cotton, Mullein, Onion, Plantain, Poplar, Ragwort, Safflower, Salsify, Yarrow

Dye Herbs – Gray Dye

Elderberry, Poplar, Raspberry, Sunflower, Yarrow

Dye Herbs – Green Dye (multiple sources)

Absinthe Wormwood, Alder Leaf Juneberry Serviceberry, Annual Sunflower, Big Sagebrush Wormwood, California Deer Brush Redroot, Canada Lettuce, Canterbury Bells Bellflower, Chinese Buckthorn, Crested Cup Lichen, Dark Red Snapdragon, Dahurian Buckthorn, Diffracted Beard Shield Lichen, Fern Leaf Dahl Mum, Field Horsetail, Fireweed Willow Evening Primrose, Fremont Horse Hair Shield Lichen, Green Freestone Plum, Hemp Dogbane, Indian Cotton Cudweed, Lady's Thumb Knotweed, Lamb's Quarters Goosefoot, Large Flower Sword Lily, Marsh Cotton Cudweed, Milfoil Yarrow, Orange African Daisy, Orange Eye Butterfly Bush, Oregon Grape Mahon Barberry, Peppermint Mint, Physodic Tube Shield Lichen, Portent Reindeer Cup Lichen, Povertyweed Ragweed, Prickly Lettuce, Red Hollyhock Marsh Mallow, Regal Stork's Bill, Ribwort Plantain, Running Juneberry Serviceberry, Russell Lupine, Spotted Joy Pye Weed Eupatory, Spreading Oak Shield Lichen, Spruce, Staves Acre Dolphin Larkspur, Succory Chicory, Sweet Everlasting Cotton Cudweed, Tasmanian Blue Gum Tree, Tumble Pigweed Amaranth, Virgin's Bower Leather Flower, White Leather Flower, White Sweet Clover, Wormseed Goosefoot

Dye Herbs – Orange Dye

Bloodroot, Pokeweed, Sorrel, Woad

Dye Herbs – Purple Dye

Blackberry, Geranium, Grape, Lady's Bedstraw

Dye Herbs – Red Dye (multiple sources)

Alder, Alkanet, Amaranth, Annatto, Beet, Boxwood, Dyer's Scatter Rug Ruffle Shield Lichen, Dyer's Tropical Mulberry, Dyer's Woolly Redroot Lily, Fragile Coral Ball Rim Lichen, Globe Coral Ball Rim Lichen, Gravel Root Eupatory, Iceland Moss Lichen, Lady's Bedstraw, Lamb's Quarters, Onion, Pernambuco Caesalpino, Perforated Scatter Rug Ruffle Shield Lichen, Pustule Toadskin Tripe Lichen, Red Hollyhock, Saffron Distaff Thistle, Sandalwood Barwood, Silique Cartilage Rim Lichen. Soridia Speckleback Shield Lichen, *Hypericum perforatum*, Tartarian Saucer Perforated Lichen

Dye Herbs – Tan Dye

Barberry, Onion, Oregon Grape, Raspberry, Sunflower, Sweet Woodruff, Uva Ursi

Dye Herbs – Yellow Dye
Agrimony, Barberry, Broom, Chamomile, Dandelion, Dock, Fennel, Fenugreek, Golden Marguerite, Goldenrod, Gridelia, Horseradish, Lady's Bedstraw, Onion, Safflower, Saffron, Sage, *Hypericum perforatum*,Salsify, Sunflower, Tansy, Yarrow

Hair Oil Herbs
Bay Rum Allspice, Chamomile, Coconut Feather Palm, Cornflower, Emblic, Eucalyptus, Henna, India Borage Spurflower Mint, Jaborandi, Jojoba, Kaffir Lime, Lemon Grass, Nettle, Olive Oil, Queensland Bean, Rose Mallow, Rosemary, Sage, Scotch Pine Oil, Sesame Seed Oil, Sticky Mallow, Tea Oil, Walnut, Ylang Ylang

Insect Repellant Herbs

Insect Repellent Herbs – Colorado Potato Beetle
Catnip, Coriander, Eucalpytus, Nasturtium, Onion, Tansy

Insect Repellent Herbs – Cucumber Beetle
Catnip, Marigold, Nasturtium, Radish, Rue

Insect Repellent Herbs – Flea Beetle
Catnip, Peppermint, Rue, Spearmint, Southernwood, Tansy, Tobacco, Wormwood

Insect Repellent Herbs – Japanese Beetle
Catnip, Chives, Garlic (Allium Sativum), Tansy

Insect Repellent Herbs – Mexican Bean Beetle
Margold, Rosemary, Savory

Insect Repellent Herbs – Ant
Peppermint, Spearmint,Tansy, Wormwood

Insect Repellent Herbs – Aphid
Catnip, Chives, Coriander Eucalyptus, Fennel, Garlic (*Allium sativum*), Larkspur, Marigold, Mustard, Nasturtium, Peppermint, Spearmint

Insect Repellent Herbs – Asparagus Beetle
Basil, Parsley, Tomato

Insect Repellent Herbs – Cabbage Butterfly
Southernwood

Insect Repellent Herbs – Cabbage Looper
Dill, Eucalyptus, Garlic (*Allium sativum*), Hyssop, Peppermint, Nasturtium, Onion, Pennyroyal, Sage, Southernwood, Spearmint, Thyme, Wormwood

Insect Repellent Herbs – Cabbage Maggot
Garlic (*Allium sativum*), Marigold, Radish, Sage, Wormwood

Insect Repellent Herbs – Cabbage Worm (imported)
Dill, Garlic (*Allium sativum*), Geranium, Hyssop, Peppermint, Nasturtium, Onion, Pennryoay, Sage, Southernwood, Tansy, Thyme

Insect Repellent Herbs – Carrot Fly
Leek, Lettuce, Onion, Rosemary, Sage, Tobacco, Wormwood

Insect Repellent Herbs – Codling Moth
Garlic (*Allium sativum*), Wormwood

Insect Repellent Herbs – Corn Earworm
Cosmos, Geranium, Marigold

Insect Repellent Herbs – Cutworm
Spiny Amaranth

Insect Repellent Herbs – Leafhopper
Geranium, Petunia

Insect Repellent Herbs – Peach Borer
Garlic (*Allium sativum*)

Insect Repellent Herbs – Spider Mite
Coriander

Insect Repellent Herbs – Squash Bug
Catnip, Nasturtium, Peppermint, Petunia, Radish, Spearmint, Tansy

Insect Repellent Herbs – Squash Vine Borer

Radish

Insect Repellent Herbs – Tomato Hornworm
Dill, Borage, Opal Basil

Insect Repellent Herbs – Whitefly
Nasturtium, Peppermint, Thyme, Wormwood

Perfume Herbs

Perfume Herbs – Citrus Scents
Beebalm, Lemon Balm, Lemon Grass, Lemon Thyme, Lemon Vervain, Orange Peel

Perfume Herbs – Earthy Scents
Bayberry, Chamomile, Clary, Lady's Bedstraw, Lavender Cotton, Marjoram, Oakmoss, Patchouli, Pine Needle, Rosemary, Sandalwood, Southernwood, Sweet Woodruff, Thyme, Vetiver, Wormwood

Perfume Herbs – Exotic Scents
Frankincense, Myrrh, Ylang Ylang

Perfume Herbs – Floral Scents
Acacia, Gardenia, Geranium, Hyacinth, Jasmine, Lavender, Lilac, Lily of the Valley, Orange Blossom, Rose, Sweet Pea, Violet

Perfume Herbs – Spicy Scents
Allspice, Angelica Root, Anise, Basil, Bay, Caraway, Cardamom, Carnation, Cinnamon, Clove, Coriander, Ginger, Hyssop, Mint, Nutmeg, Pennroyal, Vanilla

Rodent Repellant Herbs

Rodent Repellent Herbs – Mouse
Wormwood

Rodent Repellent Herbs – Mole
Castor Bean, Narcissus

Rodent Repellent Herbs – Rabbit
Garlic (*Allium sativum*), Marigold, Onion

Skin Lotion Herbs

Skin Lotion Herbs – Acne
Agrimony, Burdock, Clover, Fumitory, Horsetail, Iris, Lavender, Lemon Balm, Solomon's Seal, Southernwood, Tansy, White Birch Bark

Skin Lotion Herbs – Aftershave
Lavender, Sage

Skin Lotion Herbs – Antiseptic
Bay, Chamomile, Echinacea, Eucalyptus, Marjoram, Myrrh, Sorrel, Thyme

Skin Lotion Herbs – Blood Circulation
Beebalm, Calendula, Elderberry Flower, Horseradish, Lavender Lemon Vervain, Mint, Nettle, Plantain, Raspberry, Rosemary, Sage, Savory, Scented Geranium Flower

Skin Lotion Herbs – Cleansing
Chamomile, Fennel, Lady's Mantle, Linden, Lovage, Nettle, Parsley, Plantain

Skin Lotion Herbs – Moisturizing
Acacia Flower, Chamomile, Tussilago farafara (Coltsfoot), Comfrey, Licorice, Linden, Loosestrife, Marsh Mallow Root, Mint, Orange Flower, Slippery Elm

Skin Lotion Herbs – Skin Tightening
Agrimony, Bay, Bayberry Bark, Blackberry Leaf, Chamomile, Comfrey Leaf & Root, Dock, Elderberry Flower, Frankincense, Houseleek, Lady's Mantle, Lemon, Loosestrife, Mint, Myrrh, Nasturtium, Pansy, Primrose, Queen Of The Meadow, Raspberry, Rose, Rosemary, Sage, Violet, Wintergreen, Witch Hazel, Yarrow.

Herb Use Abbreviations

Culinary: flavoring (F), garnish (GN), seasoning (SN), spice (SP), vegetable (VG), **Medicinal:** head tonic (H), blood tonic (B), nervous system tonic (N), ointment (O), respiratory system tonic ®, stomach tonic (S), urinary system tonic (U), disease supressors **not cure** (D), **Economic:** cologne (CL), cosmetic (CM), detergent & soap (DS), dye (DYE), fiber (FI), gum resin (GR), hair oil (HO), insecticide (IC), incense (IN), insect repellent (IR), jewelery (JW), lipstick (LP), mucilage (MC), oleoresin (OR), perfume (PF), paint (PN), poison (PS), rubber (RU), solvent (SV), steroid drug (ST), wood feul (WF) wood lumber (WL), **see Key for complete abbreviations**

Key to Herb Plant Parts Used

WP Whole Plant	BD Bud (flower, leaf)	BK Bark	BR Bulbroot	CR Corm	FL Flower
FR Fruit	FS Stamen, Stigma	LF Leaf	NT Nut	PD Pod	RB Root Bark
RD Rind (or skin)	RZ Rhizome	RN Resin	RT Root	SA Sap	SD Seed
SH Shoot	SL Stalk	SM Stem (branch)	SR Sprout	TU Tuber Root	WD Wood

Key to Culinary Herb Uses

AF Alcoholic Bev. Flavoring	BF Beverage Flavoring	CF Candy, Confection Flav.	CO Cooking Oil
FC Food Coloring	FF Food Flavoring	FX Fixative & Thickener	**GF** Grain Flour
GN Garnish	JJ Jam, Jelly	JU Juice, OL Oil	PU Pulp, SY syrup
SN Seasoning	**SP** Spice	TE Tea	**VG** Vegetable

Key to Economic Herb Uses

CL Cologne	CM Cosmetic	DS Detergent, Soap	DYE Dye	FI Fiber	HO Hair Oil
GR Gum Resin	IC Insecticide	IN Incense	IR Insect Repellent	JW Jewelery	LP Lipstick
LS Landscaping	MC Mucilage	OR Oleoresin	PF Perfume	PN Paint	PS Poison
RU Rubber	SV Solvent	ST Steroid Drugs	WF Wood Fuel	WL Lumber	RP Rope

Key to Medicinal Herb Uses (NOT CURE)

		Medicinal Herb Uses			
BAA	Artery Angina	BAC	Blood Acidosis	BAK	Blood Alkalosis
BAN	Blood Anemia	BAS	Arterial Sclerosis	BCB	Chilblains (pernio)
BCR	Blood Circulation	BDB	Diabetes	BES	Enlarged Spleen
BHA	Heart Attack (myocar. inf.)	BHC	Blood High Cholesterol	BHE	Heart Edema (dropsy)
BHP	High Blood Pressure	BHS	Blood & Heart Stimulant	BIH	Irregular Heartbeat
BJA	Jaundice (Blood Bile)	BLC	Liver Cleanse (cirrhosis)	BLP	Low Blood Pressure
BPH	Vein Phlebitis	BSH	High Blood Sugar	BST	Blood Stroke
BTH	Blood Thrombosis	BVV	Varicose Veins	DABA	Antibacterial
DABI	Disease Antibiotic	DACN	Disease Acne	DADD	Disease Addison's
DAFU	Disease Antifungal	DAGS	Disease Age Spots	DAID	AIDS Ac. Im. Df. Syn.
DALC	Disease Alcoholism	DALF	Disease Athlete's Foot	DALL	Disease Allergies
DALZ	Disease Alzheimer's	DAMW	Mouthwash (stomatitis)	DAMY	Disease Aromatherapy
DANX	Disease Anorexia	DASC	Disease Antiseptic	DAUT	Disease Autism
DAVR	Disease Anti-Virus	DBOI	Disease Boil (carbuncle)	DBUL	Disease Bulimia
DCAN	Disease Candidaisis	DCBK	Cancer Bladder & Kidney	DCBR	Cancer Breast
DCCL	Cancer Colorectral	DCCU	Cancer Cervical & Uterine	DCEN	Cancer Endometriosis
DCFS	Chronic Fatigue Syndrome	DCHL	Disease Cholera	DCHR	Chorea (St. Vitus Dance)
DCLA	Cancer Laryngeal	DCLD	Disease Common Cold	DCLK	Cancer Leukemia
DCLU	Cancer Lung	DCLV	Cancer Liver	DCLW	Disease Corns CLW
DCLW	Disease Warts CLW	DCLY	Lymphoma (Hodgkin's)	DCMT	Cancer Mouth & Throat
DCOV	Cancer Ovarian	DCPA	Cancer Pancreatic	DCPL	Cancer Polyps
DCPR	Cancer Prostate	DCRO	Disease Crohn's	DCSH	Disease Cushing's
DCSK	Cancer Skin	DCST	Cancer Stomach	DCTM	Cancer Tumor
DCTS	Cancer Testicular	DCYF	Disease Cystic Fibrosis	DDEM	Disease Dementia (senility)
DDND	Disease Dandruff	DDPT	Disease Diphtheria	DDSN	Disease Down Syndrome
DEAI	Ear Ache Infection (otitis)	DEAL	Disease Ear Hearing Loss	DEAS	Disease Ear Ménière's
DEAS	Disease Ear Swimmer's	DEAT	Disease Ear Tinnitis	DECT	Disease Eye Cataracts
DEGC	Disease Eye Glaucoma	DEMD	Disease Eye Macular Deg.	DEPI	Disease Epilepsy
DERS	Disease Erysipelas	DFLU	Disease Influenza (Flu)	DFMS	Disease Fibromyalgia Syn.
DFRA	Disease Free Radical	DFRI	Disease Frigidity (female)	DGCS	Glutin Celiac Sprue
DGNG	Disease Gangrene	DGRD	Disease Dwarf Growth	DGRG	Disease Giant Growth
DGUM	Disease Gingivitis, Scurvy	DHAL	Halitosis (bad breath)	DHEA	Disease Hyperacidity
DHEA	Disease Hyperactivity	DHEG	Disease Hyperglycemia	DHET	Disease Hyperthyroidism
DHIV	Disease Immune Deficency	DHMP	Disease Hemophilia	DHMY	Disease Homeopathy
DHOG	Disease Hypoglycemia	DHOT	Disease Hypothyroidism	DHPT	Disease Hepatitis
DHSV	Disease Herpes Simplex	DHVS	Disease Hives	DHYF	Disease Hay Fever
DHYS	Disease Hysterectomy	DIFR	Disease Infertility	DIMP	Disease Impetigo
DIOS	Disease Ivy, Oak, Sumac	DIPT	Impotence (erectile disf.)	DISB	Insect Bite (pruritis)
DITG	Disease Intertrigo	DLCI	Lactose Intolerance	DLEG	Disease Legionnaire's
DLEI	Disease Leishmaniasis	DLKR	Disease Leukorrhea	DLUP	Disease Lupus
DLYM	Disease Lyme	DMAL	Disease Malaria (blackwater)	DMDF	Mediter. Fever (brucella)
DMHI	Menopause Hormone Imb.	DMNB	Meningitis Bacterial	DMNN	Disease Mononucleosis

Medicinal Herb Uses

DMNV	Meningitis Viral	DMSC	Disease Multiple Sclerosis	DMSG	German Measles (rubella)
DMSK	Disease Motion Sickness	DMSL	Disease Measles	DMUM	Disease Mumps
DNAR	Disease Narcolepsy	DOBE	Disease Obesity	DOLS	Disease Oily Skin
DOST	Disease Osteoporosis	DPAN	Disease Pancreatitis	DPBN	Disease Paget's Bone
DPLG	Disease Pellagra	DPNB	Pneumonia Bacterial	DPNV	Pneumonia Viral
DPRK	Disease Parkinson's	DPVI	Pelvic Inflammation	DPXC	Disease Chicken Pox
DPXH	Disease Oral Pox Herpes	DPXS	Disease Smallpox (variola)	DRAB	Disease Rabies
DRCK	Rickets (osteomalacia)	DRMF	Disease Rheumatic Fever	DRNP	Raynaud's Phenomenon
DROS	Disease Rosacea	DRSK	Radiation Sickness	DRYS	Reye's Syndrome
DSAD	Seasonal Affective Disord.	DSBC	Disease Sebaceous Cyst	DSBR	Disease Seborrhea
DSCB	Disease Scabies	DSCR	Disease Scrofula	DSEP	Disease Septicemia
DSFT	Disease Soft Teeth	DSHZ	Shingles (herpes zoster)	DSIN	Skin Infl.(dermat.,eczema)
DSNA	Disease Sinusitis Allergic	DSNB	Disease Sinusitis Bacterial	DSNB	Disease Snake Bite
DSNV	Disease Sinusitis Viral	DSPB	Disease Spider Bite	DSXC	Disease Sex Chlamydia
DSXG	Disease Gonorrhea (gleet)	DSXH	Disease Genital Herpes	DSXP	Disease Sex Priapitis
DSXS	Disease Sex Syphilis	DSXT	Disease Trichomonaisis	DSXW	Disease Sex Genital Warts
DSZP	Disease Schizophrenia	DTBC	Disease Tuberculosis	DTDC	Disease Tooth Decay
DTET	Disease Tetanus (lockjaw)	DTGR	Disease Tooth Grinding	DTMJ	Temp. Mand. Joint Syn.
DTND	Tendonitis (bursitis)	DTNS	Tonsillitis (quinsy)	DTYP	Disease Typhoid (fever)
DUFB	Disease Uterine Fibroids	DUWT	Disease Underweight	DVGB	Vaginitis Bacterial
DVGT	Vaginal Thrush	DVGV	Vaginitis Viral	DVTG	Disease Vertigo
DVTL	Disease Vitiligo	DWEL	Eelworm (nematode)	DWFK	Fluke (schistosomiasis)
DWHK	Disease Hookworm	DWHN	Disease Weak Hair, Nails	DWHP	Disease Whooping Cough
DWLD	Disease Wilson's	DWLS	Wilson's Syndrome	DWPN	Disease Pinworm
DWRG	Disease Ringworm	DWRM	Disease Worms (parasites)	DWRO	Disease Roundworm
DWTE	Thdworm Elephantiasis	DWTH	Disease Threadworm	DWTP	Disease Tapeworm
DWWP	Disease Whipworm	DYAW	Disease Yaws	HAC	Headache (clavus)
HAN	Anxiety (depression)	HAP	Head Aphrodisiac	HFN	Head Brain Function
HFV	Head Fever	HIN	Head Insomnia	HML	Head Memory Loss
HPS	Head Psychotropic	NBF	Nerve Bone Fracture	NBK	Nerve Backache (lumbago)
NCT	Nerve Carpal Tunnel Syn.	NDP	Nerve Debility, Paralysis	NGT	Nerve Gout
NHS	Nerve Heel, Bone Spur	NJA	Nerve Joint Arthritis	NMP	Nerve Menstrual Pain
NMS	Nerve Muscle Spasm (palsy)	NMT	Nerve Mastitis	NNC	Nerve Narcotic
NNR	Nerve Neuralgia (neuritis)	NOR	Nerve Orchitis	NRH	Nerve Rheumatism
NSC	Nerve Sciatica	NSD	Nerve Sedative (analgesic)	NSL	Nerve Scleroderma
NSP	Nerve Spondylosis	NTN	Nerve Tendinitis	NUC	Nerve Uterine Contraction
OBA	Ointment Blood Astringent	OBN	Ointment Burn	OBR	Ointment Bruise
OCB	Ointment Chilbains (pernio)	ODS	Ointment Dry Skin	OEY	Ointment Eye (conjunctivitis)
OFG	Ointment Fungus	OGN	Ointment Gingivitis (gum)	OGT	Ointment Gout
OHL	O. Hair Loss (alopecia)	OHM	O. Hemorrhoid (piles)	OJA	Ointment Joint Arthritis
OLC	O. Lung Congestion	OMC	Ointment Muscle Cramp	OMT	Ointment Mastitis
ONR	Ointment Neuralgia	OPC	Ointment Pericarditis	ORH	Ointment Rheumatism
OSA	Ointment Skin Abscess	OSB	Ointment Skin Sunburn	OSC	Ointment Sciatica
OSS	Scaly Skin (psoriasis)	OST	Ointment Insect Sting	OSW	Ointment Skin Wrinkles
OTA	Ointment Toothache	OTM	Ointment Tumor	OUL	Ointment Ulcer
OVV	Ointment Varicose Veins	OWN	Ointment Wound	RCR	Respiratory Croup
RLA	Respiratory Lung Asthma	RLB	Respirat. Lung Bronchitis	RLC	Respiratory Lung Cough
RLD	Res. Lung Dropsy (ascites)	RLE	Res. Lung Emphysema	RLP	Respiratory Lung Pleurisy
RRN	Respiratory Rhinitis (nose)	RSN	Respiratory Sinusitis	RST	Sore Throat (pharyngitis)
RTL	Res. Throat Laryngitis	SAD	Stomach Appendicitis	SAP	Stomach Apetite
SAR	Stomach Acid Reflux	SCC	Stomach Colon Cleanse	SCN	Constipation (laxative)
SCR	Stomach Cramps	SDG	Stomach Digestion	SDP	Stomach Dyspepsia
SDR	Stomach Diarrhea	SDV	Stomach Diverticulitis	SDY	Stomach Dysentery
SFL	Stomach Flatulence	SGB	S. Gall Bladder Cleanse	SGS	Stomach Gall Stones
SHH	Stomach Hiatal Hernia	SIB	Stomach Irritable Bowel	SPA	Stomach Pancreatis
SPN	Stomach Poison	SPR	Stomach Purgative	SUL	Stomach Ulcer (peptic)
UBL	U. Bladder Inf. (cystitis)	UDR	Urinary Diuretic	UEP	Urinary Enlarged Prostate
UIN	U. Incontinence (enuresis)	UKC	Kidney Cleanse (nephritis)	UKP	Urinary Kidney Pyelitis
UKS	Urinary Kidney Stone	UUR	Urinary Urethritis		

Vegetable Overview

The earliest archaeological evidence of vegetable use has been found in South America, Mexico, tropical Asia and the "Fertile Crescent" In Guitarrero Peru, there is archeological evidence for wild beans and squash in 10,000 BC. In Oaxaca Mexico, there is archaeological evidence for squash in 8500 BC. In Lake Titicaca and Chilica Canyon Peru, there is archaeological evidence for cultivated potatoes in 8000 BC. The "Fertile Crescent" stretches from Iraq and Syria to Egypt. In 8000 BC, the first cities in the "Fertile Crescent" grew wheat, barley, flax, peas, and lentils. Onions and carrots originated further east in Afghanistan and Western Asia. In China there is archeological evidence for yams and taro in 8000 BC. There is archaeological evidence for rice in Thailand in 5000 BC. Thre is archaeological evidence for Millet Foxtail Bristle Grass *Setaria italica* and Broomcorn Millet *Millium paniceum* in China in 5000 BC. In Mexico City Mexico there is archaeological evidence for cultivated maize in 5000 BC. In Tecahua Mexico there is archaeological evidence for cultivated beans in 3000 BC. The Israelites left behind cucumbers, melons, leeks, onions and garlic in Egypt after the Exodus in 1500 BC. The Romans grew asparagus, globe artichokes, cabbage, cauliflower, kale, and broccoli in 200 BC.

Edible spore plants include the lichens and the mushrooms. Classification is Domain 2 *Plantae* Plant, Kingdom 6 *Sporophyta* Spore Plant, Subkingdom 2 *Fungi* Fungus, Phylum 1 *Ascomycota* Sac Fungus, Class 3 *Lichinomycetes* Lichen (all lichens; edible, dye, and inedible), and Phylum 2 *Basidiomycota* Club Fungus, Class 2 *Agaricomycetes* Gill Mushroom(all gill mushrooms; edible, inedible, hallucinogenic and poison), and Class 3 *Boletomycetes* Nongill Mushroom (all nongill mushrooms; edible, inedible, hallucinogenic and poison).

Edible seed plants include the fruits, herbs, and vegetables. Classification is Domain 2 Plantae Plant, Kingdom 7 *Spermatophyta* Seed Plant, Subkingdom 1 *Coniferophyta* Conifer Plant (all cone bearing plants; includes edible pine nuts and a few herbs), Subkingdom 2 *Anthophyta* Flower Plant, Phylum 3 *Monocotovegetaphyta* Cultivated Monocot Flower Plants, and Phylum 4 *Dicotovegetophyta* Cultivated Dicot Flower Plants. The word "monocot" means that the plant seed initially sprouts one (mono) embryo leaf (cotyledon, abbreviated cot). The word "dicot" means that the plant seed initially sprouts two (di) embryo leaf (cotyledon, abbreviated cot).

Monocot vegetables include Banana, Breadfruit, Pinapple, Corn, Barley, Oats, Rice, Rye, Wheat, Coconut, Date, Eggfruit, Hyacinth Bulb, and Vanilla Spice.

Dicot cultivated plants comprise the majority of the cultivated plants and include the following;

Phylum 4 *Dicotovegetophyta* Cultivated Dicot Plants

Class 01 Cultivated Aster:	Artichoke, Lettuce, Sunflower
Class 02 Cultivated Succulent:	Beet, Rhubarb, Spinach, Pigeon Plum, Pitahaya, Prickly Pear
Class 03 Cultivated Cornel:	Berry, Carrot, Celery, Parsnip, Sapote, Turnip
	Berry: Blueberry, Cranberry, Huckleberry, Whortleberry
Class 04 Cultivated Gentian:	Natal Plum, Natal Orange, Numnum Plum, Goji Berry, Wolfberry
Class 04 Cultivated Sunplant:	Eggplant, Melon Pepper, Potato, Tomato
Class 05 Cultivated Mint:	Olive, Mint Leaf
Class 06 Cultivated Legume:	Green Bean, Kidney Bean, Lima Bean, Pea, Peanut
Class 07 Cultivated Magnolia:	Avocado, Cadberry, Cinnamon, Custard Apple, Nutmeg, Sassafras
Class 08 Cultivated Mallow:	Baobab, Breadfruit, Cocoa Bean, Fig, Grape, Mulberry, Okra
Class 09 Cultivated Myrtle:	Guava, Lillypilly Cherry, Pomegranate, Macadamia Nut
Class 10 Cultivated Lotus:	Peppercorn
Class 11 Cultivated Buttecup:	Akebe Bean, Barberry, Mayapple, Sausagefruit
Class 12 Cultivated Rose:	Apple, Apricot, Cherry, Pear, Plum, Quince, Nut, Berry
	Nut: Almond, Beechnut, Butternut, Hazelnut, Pecan, Hickorynut, Walnut
	Berry: Bayberry, Blackberry, Dewberry, Raspberry, Strawberry
Class 13 Cultivated Rue:	Acerola Cherry, Grapefruit, Kumquat, Lemon, Lime, Orange, Tangerine, Nut
	Calamondin, Citrange, Citron, Clementine, Kat, Pomelo, Tangelo
	Nut: Cashew, Mango, Pistachio, Mombin Plum
Class 14 Cultivated Sandalwood:	Quandong Plum
Class 15 Cultivated Tea:	Brazilnut, Kiwi, Mangosteen, Beverage Tea
Class 16 Cultivated Violet:	Broccoli, Brussels Sprouts, Cabbage, Cauliflower, Mustard Greens,
	Casaba, Cantaloupe, Honeydew Melon, Muskmelon, Watermelon

List of Vegetable Seed Suppliers

In the United States there are many commercial nurserys which will send vegetable seeds through the mail for planting. Two recent publications listing vegetable seed suppliers are;

"Garden Seed Inventory Sixth Edition", 2004

An Inventory of Nursery Catalogs Listing All Vegetable Varieties available by Mail Order in the United States, compiled by Seed Savers Exchange.

"Fusillo Knowledge Infrastructure Gardening By Mail 1st Edition", 2011

ISBN10:0—7414-6792-5, ISBN13:978-0-7414-6792-8

An Inventory of Nursery Vegetable Species available by Mail Order in the United States, compiled by Lawrence J. Fusillo, the author of this book.

HERBS

The author assumes no responsibility or liability for the typographical errors, transcription errors, or the reliability of information on herbs in this section. All information in this section was extracted from the references listed in the section Herb References in the Introduction.

HERBS are used for (1) culinary, (2) medicinal, and (3) economic purposes:

(1) **culinary** – alcoholic beverage flavoring, beverage flavoring, candy flavoring, cooking oil, food coloring, food flavoring, food fixative, grain for flour, garnish, jelly & jam, juice, pulp fruit, seasoning, **spice**, tea, **vegetable**.

(2) **medicinal** – the suppression of **BUT NOT CURE** of
- **Symptomatic Illnesses** – discomforts & pains in the body with no apparent cause.
- **Opportunistic Diseases** – caused by invading foreign organisms.

(3) **economic** – cologne, cosmetic, **DYE**, fiber, gum resin, hair oil, insecticide, incense, insect repellent, jewelery, lipstick, mucilage, oleoresin, perfume, paint, **POISON** (medicinal herbs which are **TOXIC** when taken in **EXCESS**), rubber, soap, solvent, steroid drugs, wood for fuel, wood for lumber building constriction.
There are three classes of HERB DYES; (1) Direct, (2) Mordant, and (3) Vat: (1) The Direct Dye class includes Turmeric, Safflower, some Lichens, and Annato. (2) In the Mordant Dye class, a chemical combines with the dye within the textile fiber to yield a lightfast dye. Mordants include alum, chrome, copper, iron, and tin. (3) In the Vat Dye class, insoluble dye is acted upon by either bacteria or chemicals to reduce the dye to solubility.

Medicinal Verification M(v): from lowest to highest
M(v): M(folk), M(traditional), M(ayurveda), M(SS) statistical studies, **M(PS)** pharmalogical studies, **M(CS)** clinical studies, **M(CT)** clinical trials

Latin Genus	L. Species	E. Species	English Genus	C:Culinary,M:Medicinal,E:Economic	Source
Abelmoschus	*moschatus*	Gumbo Okra	Musk Mallow	**C:**LF,SH:**VG**,**C:**SD:BF,**SP**,**M(traditional):**SD:DAMW,SDG,**M(traditional):**OL:HAP,OJA,OMC,**E:**BK:FI,**E:**RT:MC,**E:**OL:PF,**E:**WP:IC	Bown-226-01
Abelmoschus	*moschatus*	Gumbo Okra	Musk Mallow	**C:**SD,OL:AF,**M(traditional):**SD,OL:NMS,SDG,E:OL:PF	Wyk-395-01
Abies	*alba*	Silver	Fir	**M(traditional):**LF:OJA,OLG,RLC,SDR,SPR,UDR,OR:ONR,ORH,OWN,**E**:OL:SV,DYE(black)	Bown-226-02
Abies	*alba*	Silver	Fir	**M(traditional):**OL:NRH,RLB,RLC,UUR	Wyk-395-02
Abies	*balasmea*	Balm of Gilead	Fir	**C:**OL,FF,**M(traditional):**BK,LF:DASC,BCR,NJA,RLC,SDR,SPR,UDR,**M(traditional):**OL,OR:OBN,OLG,ORH,OMW,**E:**OL:CM,PF,DS	Bown-226-03
Abies	*balasmea*	Balm of Gilead	Fir	**E:**SM:DYE(alum-&-vinegar-gold,copper-sulfate-tan,chrome-warm-brown,tin-bright-gold,**E:**PD(cone):DYE(alum-&-chrome-brown,copper-sulfate-&-iron-gray,alum-&-vinegar-beige,baking-soda-tan)	Cas1-142-01
Abies	*siberica*	Siberian	Fir	**M(traditional):**OL:NRH,RLB,RLC,UUR	Wyk-395-02
Abroma	*augusta*	Devil's Cotton	Poison Divine	**M(traditional):**BK,RT:BDB	Wyk-395-03
Abronia	*umbellata*	Wild Four O'clock	Sand Vervain	**E:**LF,FL,SM:DYE(alum-golden-tan,chrome-dark-brown,copper-olive-brown,tin-yellow-brown,iron-dark-khaki)	Bliss-266-01
Abrus	*precatorius*	Licorice	Jequerty Pea	**C:**RT:FF,**M(traditional):**LF:OSC,RLC,RST,**M(traditional):**SD:NDP,DCLW,OSS,**E:**SD:JW,PS☠	Bown-226-04
Abrus	*precatorius*	Licorice	Jequerty Pea	**M(traditional):**SD:**contraceptive**, **induced abortion** ☠	Wyk-395-04
Abutilon	*hybridum*	Red Flowering	Maple Mallow	**E:**FL:DYE(alum-&-tin-navy-blue)	Grae-155-02
Abutilon	*indicum*	India	Maple Mallow	**M(traditional):**RT,SD:DSIN,OWN,SCN	Wyk-395-05
Acacia	*ancistrocarpa*	Barbed Fruit	Wattle	**M(traditional):**LF,SH:HAC	Bown-227-01
Acacia	*bivenosa*	Two Vein	Wattle	**M(traditional):**BK,RT:RLC,RTL	Bown-227-02
Acacia	*catechu*	Cutch	Wattle	**M(traditional):**LF,SH:DTDC,DAMW,DASC,OBA,OHM,OUL,RST,SDG,SDR,SDY,**E:**BK:DYE,**E:**WD:WL,WF	Bown-227-03
Acacia	*catechu*	Cutch	Wattle	**M(traditional):**BK,LF,SM,WD:DABI,DAMW,DBOI,DGUM,OAS,OUL,RST,SDR,SDY,SUL	Wyk-028-01, Wyk-395-06
Acacia	*farnesiana*	Cassie	Wattle	**C:**SD:CO,FF,**M(traditional):**BK:SDR,**M(traditional):**FL:OSS,**E:**BK,PD:DYE(black),**E:**FL:IC,PF	Bown-227-04
Acacia	*holosericea*	Silky Hair	Wattle	**M(traditional):**BK,RT:RLC,RTL,**E:**FL,SM:DYE(yellow)	Bown-227-05
Acacia	*monticola*	Mountain	Wattle	**M(traditional):**BK,RT:RLC,RTL,**E:**FL,SM:DYE(yellow)	Bown-227-06
Acacia	*pringlei californica*	California Pringle	Wattle	**E:**FL,SM:DYE(alum-yellow,chrome-gray-maize-yellow,chrome-light-gold-brown)	Grae-159-01
Acacia	*senegal*	Gum Arabic	Wattle	**C:**RN:FX,GR,**M(traditional):**RN:OBN,DCLW,OUL,RLC,RST,SDR,SDY	Bown-227-07
Acacia	*senegal*	Gum Arabic	Wattle	**M(traditional):**RN:DABA,DABI,DAFU,DSIN	Wyk-029-01, Wyk-395-07
Acacia	*tetragonophylla*	Tetragon Leaf	Wattle	**M(traditional):**BK,RT:RLC,RTL,**E:**FL,SM:DYE(y	Bown-227-08

Latin Genus	L. Species	E. Species	English Genus	C:Culinary,M:Medicinal,E:Economic	Source
				ellow)	
Acacia	*trachycarpa*	Thick Fruit	Wattle	**M(traditional):**LF,SH:BHA,**E:**FL,SM:DYE(yellow)	Bown-227-09
Acalypha	*indica*	India	Copperleaf Spurge	**M(traditional):**WP:DSIN,RLC,UDR	Wyk-395-08
Acanthosicyos	*horridus*			Synonym for *Sicyoshorrudus*	Wky-395-09
Acanthus	*mollis*	Soft	Bear's Breech	**M(traditional):**LF,RT:OWN,SIB	Wyk-395-10
Acer	*rubrum*	Red	Maple	**E:**LF:DYE(chrome-&-tin-tan),**E:**BK:DYE(chrome-brown,chrome-&-tin-dark-brown,alum-tan-rose,iron-gray-brown)	Cas1-175-01
Acer	*saccharum*	Sugar	Maple	**E:**LF:DYE(alum-yellow,alum-&-vinegar-beige, chrome-tan,iron-gray-green,copper-sulfate-soft-green)	Cas1-175-02
Acer	*spicatum*	Mountain	Maple	**E:**SM:DYE(alum-&-chrome-tan-red)	Cas1-175-03
Achillea	*decolorans*	English Mace	Yarrow	**C:**LF:**SN**	Clev-078-02
Achillea	*millefolium*	Milfoil	Yarrow	**M(traditional):**FR,LF:DAVR,HFV,OAS,SIB,UDR	Bal1-079-02
Achillea	*millefolium*	Milfoil	Yarrow	**M(folk):**LF,FL,SM:BLC,DCLD,DFLU,HIN,OHM,OWN,**M(traditional):**LF,FL,SM:DFRA,HFV,NMP,NMS,SDG,SDR,SFL,SIB,UDR,**M(ayurveda):**LF,FL,SM:BLC,**M(PS):**LF,FL,SM:BHP,BHS,DABA(*Staphylococcus aureus*),SAP,SDP,OBR	Bal2-156-01 Bal2-550-06
Achillea	*millefolium*	Milfoil	Yarrow	**C:**LF:**SN**,**M(traditional):**WP:BCR,BHP,BTH,DFLU,DMSL,HFV,NJA,NMP,NRH,OBA,OEY,OHM,OUL,OWN,SDG,SDP,SDR,SUL,UDR	Bown-227-10
Achillea	*millefolium*	Milfoil	Yarrow	**C:**LF:**SN**,**M(traditional):**WP(top):HFV,NRH,OAS,OTA	Clev-076-01
Achillea	*millefolium*	Milfoil	Yarrow	**M(traditional):**LF(top):NMS,NSD,OAS,OTA,OWN,SDR,SFL,**E:**FL:DYE(alum-yellow),**E:**WP:CM,DYE(iron-olive-green)	Kowa-175-13 Kowa-293-04 Kowa-516-01
Achillea	*millefolium*	Milfoil	Yarrow	**M(folk):**FL,SM:HFV,DCLD,DFLU,OAS,OBR,OWN	Krug-189-01
Achillea	*millefolium*	Milfoil	Yarrow	**M(PS):**WP:BHP,DABA,DCLD,HFV,NJA,NMS,UDR,SAP,SDP	Wyk-030-01, Wyk-395-11
Achillea	*millefolium*	Milfoil	Yarrow	**E:**FL:DYE(alum-chartreuse,alum-&-copper-tan-green,alum-&-iron-dark-green,alum-&-tin-yellow,chrome-&-tin-maize)	Grae-111-01
Achillea	*millefolium lanulosa*	Woolly Milfoil	Yarrow	**E:**LF,FL,SM:DYE(alum-mustard-yellow,chrome-copper,copper-green-gold,tin-bright-yellow,iron-gray-green)	Bliss-276-01
Achillea	*moschata*	Musk	Yarrow	**C:**LF:**SN**	Clev-078-03
Achyranthes	*aspera*	Devils' Horsewhip	Chaff Flower	**C:**LF:**VG**	Bown-227-11
Achyranthes	*bidentata*	Two Toothed	Chaff Flower	**C:**SD:GF,**M(traditional):**LF,RT,SM:BCR,BHP,BLC,NBK,NJA,NMP,SDG,UKC	Bown-227-12
Achyranthes	*bidentata*	Two Toothed	Chaff Flower	**M(traditional):**RT:DAMW,RLC,UDR	Wyk-395-11
Achyrocline	*satureoides*	Savory	Caffeic Aster	**M(traditional):**WP:DABA,NMS,NSD,SIB	Wyk-396-01
Acinos	*arvensis*	Field	Akinos Calamint	**C:**LF:**SN**,**M(traditional):**WP:SDG,UDG,**M(traditional):**OL:OBR,ONR,OSC,OTA	Bown-228-01
Acokanthera	*oppositifolia*	Bushman Poison	Wintersweet	**M(traditional):**SD:BHS,**E:**SD:PS☠	Wyk-396-02
Aconitum	*carmichaelii*	Carmichael	Monkshood	**M(traditional):**RT:BHS,NRH,NSD,OJA,ORH,UDR,UKC,**E:**RT:PS☠	Bown-228-02
Aconitum	*carmichaelii*	Carmichael	Monkshood	**M(traditional):**RT:NSD,ORH,**E:**RT:PS☠	Wyk-396-03
Aconitum	*ferox*	Deadly	Monkshood	**E:**RT:PS☠	Bown-228-03
Aconitum	*napellus*	Wolfsbane	Monkshood	**M(traditional):**RT:DHMY,BHS,HFV,NJA,NGT,NNR,NSD,OJA,OSC	Bown-228-04
Aconitum	*napellus*	Wolfsbane	Monkshood	**M(TM):**RT,TU:BHP,ONR,ORH,**E:**WP:PS☠	Kowa-001-01
Aconitum	*napellus*	Wolfsbane	Monkshood	**M(traditional):**RT,TU:NNR, NSC, **E:**WP:PS☠	Krug-008-01
Aconitum	*napellus*	Wolfsbane	Monkshood	**M(PS):**TU:HFV,NSD,ONR,ORH,RLC,**E:**TU:PS☠	Wyk-031-01, Wyk-396-04
Acorus	*calamus*	Myrtle Flag	Sweet Flag	**M(traditional):**RZ,OL:HBF,ONR,ORH,OUL,RLB,RSN,SDG,**E:**OL,RZ:PS☠	Bown-228-05
Acorus	*calamus*	Myrtle Flag	Sweet Flag	**C:**RZ:AF,CF(cinnamon),**E:**OL:PS☠(carcinogenic)	Kowa-476-01
Acorus	*calamus*	Myrtle Flag	Sweet Flag	**M(PS):**RZ:DWRM,NMS,RLA,SAP,SDG,SDY,SFL	Wyk-032-01
Acorus	*gramineus*	Chang Pu	Sweet Flag	**M(traditional):**RZ:DDAD, DEPI, DFRA, HAN, HIN, RLA, RLB, SDR, SDY	Bal2-015-01
Acorus	*gramineus*	Chang Pu	Sweet Flag	**M(traditional):**RZ:DABA,RLB,SDG,SIB	Bown-228-06
Acorus	*gramineus*	Chang Pu	Sweet Flag	**E:**RZ:PS☠	Wyk-396-05
Acronychia	*pedunculata*	Peduncle	Nailwort	**M(traditional):**BK,LF,RT:NSD,DSIN	Wyk-396-06
Actinogyra	*meuhlenbergii*			Synonym for *Umbilicaria meuhlenbergii*	Cas2-031-01
Adenophora	*liliifolia*	Lily Leaf	Ladybells	**C:**RT:**VG**	Bown-229-01
Adenophora	*stricta*	Fickle	Ladybells	**M(traditional):**RT:BHS,DTBC,RLB,RLC	Bown-229-02
Adenophora	*trachilioides*	Rough	Ladybells	**M(traditional):**RT:BHS,DTBC,RLB,RLC	Bown-229-03
Adiantum	*aethiopicum*	Australian	Maidenhair Fern	**M(traditional):**WP:RLB	Bown-229-04
Adiantum	*capillus-venerus*	Venus Hair	Maidenhair Fern	**M(traditional):**WP:DWRG,DDND,RLA,RLC,UD	Bown-229-05

Latin Genus	L. Species	E. Species	English Genus	C:Culinary,M:Medicinal,E:Economic R	Source
Adiantum	*capillus-venerus*	Venus Hair	Maidenhair Fern	**M(traditional):**WP:DHOG,NMP,RLC	Wyk-396-07
Adiantum	*pedatum*	Bird's Foot	Maidenhair Fern	**M(traditional):**WP:DWRG,DDND,RLA,RLC,UDR	Bown-229-06
Adlumia	*fungosa*	Allegheny	Foamleaf Fumitory Vine	**M(traditional):**WP:NSD,SDG	Wyk-396-08
Adonis	*vernalis*	Yellow Spring	Pheasant's Eye	**M(traditional):**WP:BHE,BHS,BIH,UDR	Bown-229-07
Adonis	*vernalis*	Yellow Spring	Pheasant's Eye	**M(PS):**WP(aerial):BHS,UBL,UKS	Wyk-033-01, Wyk-396-09
Aegle	*marmelos*	Stone Apple	Bael Quince	**M(traditional):**WP,FR:DEAI,OAS,SCN,SDP	Wyk-396-10
Aegopodium	*podagraria*	Bishopwort	Goutwort	**C:**LF:**SN**,**VG**,**M(traditional):**LF,RT:DHMY,NGT,NSC,OBN,OGT,OHM,OST,UDR	Bown-229-08
Aegopodium	*podagraria*	Bishopwort	Goutwort	**M(traditional):**WP,RT:NGT,NRH,OHM	Wyk-396-11
Aeollanthus	*gamwelliae*	Gamwell	Ninde	**C:**LF:**SN**,**E**:LF,FL,OL:PF,DS	Bown-229-09
Aeollanthus	*heliotropoides*	Heliotrope	Ninde	**E**:LF,FL,OL:PF,DS	Bown-229-10
Aeollanthus	*lamborayi*	Lamboray	Ninde	**E**:LF,FL,OL:PF,DS	Bown-229-11
Aeollanthus	*myrianthus*	Myriad Flower	Ninde	**E**:LF,FL,OL:PF,DS	Bown-229-12
Aeollanthus	*pubescens*	Downy Stem	Ninde	**E**:LF,FL,OL:PF,DS	Bown-229-13
Aerva	*javanica*	Java	Island Puzzle	**M(traditional):**LF,RT:DABI,DWRM,NNC,UDR	Wyk-396-12
Aesculus	*hippocastanum*	Buckeye	Horse Chestnut	**M(traditional):**BK,LF,SD:BVV,DSIN,HFV,NJA,NMP,NRH,OHM,RLC,**M(PS):**BK,LF,SD:BVV,OBR,OHM,OUL	Bal2-087-01 Bal2-527-03
Aesculus	*hippocastanum*	Buckeye	Horse Chestnut	**C:**SD:NT,**M(traditional):**BK,SD:BAS,BCB,BHA,BST,HFV,OBA,OHM,UDR,**E:**FR:CM	Bown-230-01
Aesculus	*hippocastanum*	Buckeye	Horse Chestnut	**C:**SD:NT,**M(folk):**SM-BK:HFV, **M(folk):**FR:NRH, **M(traditional):**FR:BCR,OHM,OVV	Krug-091-01
Aesculus	*hippocastanum*	Buckeye	Horse Chestnut	**M(CS):**SD:BVV,HFV,NRH,OHM	Wyk-034-01, Wyk-396-13
Aesculus	*hippocastanum*	Buckeye	Horse Chestnut	**E:**LF:DYE(alum-yellow,chrome-gold,iron-gray-tan),**E**:RD:DYE(tin-yellow-green,iron-gray, chrome-brown),**E:**FR:DYE(alum-&-vinegar-light-tan)	Cas1-119-01
Aethusa	*cynapium*	Dog Celery	Fool's Parsley	**M(traditional):**WP:SIB,**E:**WP:PS	Wyk-396-14
Aframomum	*angustifolium*	Madagascar	African Cardamom	**C:**FR,PU,**C:**SD:**SP**	Bown-230-02
Aframomum	*melegueta*	Grains of Paradise	African Cardamom	**C:**SD:**SP**,**M(traditional):**RT:HAP,NMS,SDG,SDY	Bown-230-03
Aframomum	*melegueta*	Grains of Paradise	African Cardamom	**M(traditional):**SD:SDP	Wyk-396-15
Agaricus	*campestris*	Meadow	Mushroom	**E:**WP:DYE(alum-soft-yellow)	Cas1-183-01
Agastache	*foeniculum*	Anise Fennel	Giant Hyssop	**C:**LF:**SN**,,**M(traditional):**LF,FL:RLB,RLC	Bown-230-04
Agastache	*foeniculum*	Anise Fennel	Giant Hyssop	**C:**LF:FF,TE	Clev-097-04
Agastache	*rugosa*	Wrinkled	Giant Hyssop	**C:**LF:FF,TE,**M(traditional):**LF,SM:DABA,HFV,NMS,SAP,SDG,SDP	Bown-230-05
Agastache	*rugosa*	Wrinkled	Giant Hyssop	**M(traditional):**WP:DAFU,SDG	Wyk-396-16
Agastache	*urticifolia*	Nettle Leaf	Giant Hyssop	**C:**LF:FF	Clev-097-05
Agathis	*australis*	Manila Copal	Kauri Pine	**E:**RN,OL:OR,PN	Wyk-396-17
Agathosma	*betulina*	Round leaf	Buchu	**C:**LF:AF,**M(traditional):**LF:DASC,NGT,NRH,RLC,SDG,UKC,UDR,**E:**LF:IR	Bown-230-07
Agathosma	*betulina*	Round leaf	Buchu	**C:**LF-OL:FF,TE,**M(traditional):**LF:NRH,OBR,OWN,SDG, UKC,UUR	Wyk-035-01, Wyk-396-18
Agathosma	*crenulata*	Oval Leaf	Buchu	**M(traditional):**LF:BDB,DVGB,DVGV,DGUM,RSN,SDG,SIB,UBL,UEP,UKC	Bal1-066-05
Agathosma	*crenulata*	Oval Leaf	Buchu	**C:**LF:AF,**M(traditional):**LF:DASC,NGT,NRH,RLC,SDG,UKC,UDR,**E:**LF:IR	Bown-230-08
Agave	*americana*	Century Plant	Agave	**C:**SL:**VG**,**M(traditional):**WP:BJA,HFV,OBN,SCN,SDG,SDY,SFL,UDR,**E:**SM:RP,**E:**LF:ST,**E:**RT:DS,**E:**WP:IC	Bown-231-01
Agave	*americana*	Century Plant	Agave	**C:**LF:AF,**M(traditional):**LF:SDG,**E:**LF:FI	Wyk-396-19
Agave	*sisalana*	Sisal	Agave	**E:**SM:RP,**E:**LF:ST	Bown-231-02
Ageratum	*conyzoides*	Billygoat	Flossflower	**M(traditional):**OL:DASC,OWN	Wyk-397-01
Aglaia	*argentea*	Silver	Grace Honeywood	**M(traditional):**LF,FL:HFV,DCLW	Bown-231-03
Aglaia	*edulis*	Fiji Lime	Grace Honeywood	**C:**FR:PU	Bown-231-04
Aglaia	*odorata*	Mock Lime	Grace Honeywood	**C:**LF:TE,**M(traditional):**LF,FL:HFV,**E:**FL:PF	Bown-231-05
Agrimonia	*eupatoria*	Thoroughwort	Agrimony	**M(traditional):**WP:BJA,BLC,DSIN,OAS,SDR,UBL,**M(PS):**WP:BDB	Bal2-016-01 Bal2-501-02
Agrimonia	*eupatoria*	Thoroughwort	Agrimony	**M(traditional):**WP:BLC,OBA,OHM,ORH,OSA,RST,SDP,SDR,SGS,SIB,UBL,UDR,UIN	Bown-231-06
Agrimonia	*eupatoria*	Thoroughwort	Agrimony	**M(TM):**WP:BJA,OAS,OWN,**E:**WP:DYE(yellow)	Clev-076-04
Agrimonia	*eupatoria*	Thoroughwort	Agrimony	**M(traditional):**WP:BLC,DAMW,DWTP,OAS,SGB,UKC,E:LF,SM:CM,DYE(alum-brass-yellow,chrome-gold)	Kowa-003-01 Kowa-173-01
Agrimonia	*eupatoria*	Thoroughwort	Agrimony	**M(folk):**WP:BLC,DAMW,DSNB,OWN,SGS	Krug-009-01
Agrimonia	*eupatoria*	Thoroughwort	Agrimony	**M(PS):**WP(aerial):OAS,OHM,OWN,RST,SDR,U	Wyk-036-01

Latin Genus	L. Species	E. Species	English Genus	C:Culinary,M:Medicinal,E:Economic	Source
				BL,UDR	
Agrimonia	*odorata*	Fragrant	Agrimony	**E:**WP:PF	Clev-076-05
Agrimonia	*pilosa*	Shaggy Speedwell	Agrimony	**M(traditional):**WP:OBA	Bown-231-07
Agrimonia	*procera*	Tall	Agrimony	**M(traditional):**WP:OAS,SDR,UDR	Wyk-397-02
Agrophora	*lyngei*			Synonym for *Umbilicaria lyngei*	Cas2-031-02
Agropyron	*repens*			See *Elymus repens*	Krug-058-02
Agrostemma	*githago*	Githago	Corn Cockle	**M(traditional):**SD:DAFU,**E:**SD:PS☠	Wyk-397-03
Ahnfeltia	*gigartinoides*	Limu Koeleele	Ahnfelt Gigartin Florid Red Algae	**M(traditional):**WP:SCN,**E:**WP:MC(agar)	Wyk-397-04
Ailanthus	*altissima*	Tree Of Heaven	Ailanto	**M(traditional):**BK:BIH,DMAL,DWTP,HFV,NMS,OBA,OHM,RLA,SDR,SDY	Bown-231-08
Ailanthus	*altissima*	Tree Of Heaven	Ailanto	**M(traditional):**BK:DASC,DCTM,DWRM,NMS,SDG,SDY	Wyk-397-05
Ajuga	*australis*	Australian	Bugle	**M(traditional):**WP:OWN	Bown-231-09
Ajuga	*bracteosa*	Bracted	Bugle	**M(traditional):**WP:NSD,OAS,OWN,SCN	Wyk-397-06
Ajuga	*chamaepitys*	Ground Pine	Bugle	**M(traditional):**LF:NGT,NRH,UDR	Bown-232-01
Ajuga	*chamaepitys*	Ground Pine	Bugle	**M(traditional):**LF:BHP,OAS	Clev-076-07
Ajuga	*decumbens*	Trailing	Bugle	**M(traditional):**WP:NSD,OAS,OWN,SCN	Wyk-397-06
Ajuga	*iva*	Mediterranean	Bugle	**M(traditional):**WP:DMAL	Bown-231-10
Ajuga	*remota*	African	Bugle	**M(traditional):**WP:BHP	Bown-231-11
Ajuga	*reptans*	Eurasian	Bugle	**M(traditional):**WP:OBA,OBR,DCLW,OWN	Bown-232-02
Ajuga	*reptans*	Eurasian	Bugle	**M(traditional):**LF:OAS,OWN	Clev-076-06
Akebia	*quinata*	Chocolate Vine	Sausage Melon	**C:**FR:PU,**M(traditional):**SM:DABA,DAFU,BCR,NJA,NRH,UDR	Bown-232-03
Akebia	*quinata*	Chocolate Vine	Sausage Melon	**M(traditional):**WP:SDG,UUR	Wyk-397-07
Akebia	*trifoliata*	Three Leaf	Sausage Melon	**C:**FR:PU,**M(traditional):**SM:DABA,DAFU,BCR,NJA,NRH,UDR	Bown-232-04
Albizia	*julibrissin*	Silky Siris	Albizzi Quiver Leaf	**M(traditional):**BK:BCR,DBOI,OBA,NSD,SAP,UDR,**M(traditional):**BK,FL:HIN,**M(traditional):**FL:HML,NSD,SDG	Bown-232-05
Albizia	*odoratissima*	Black Siris	Albizzi Quiver Leaf	**M(traditional):**BK,FL:DCLW,OUL	Bown-232-06
Alcea	*rosea*	Rose	Hollyhock	**M(traditional):**FL:RLC,RST,SIB,UBL,UDR	Bown-232-07
Alcea	*rosea*	Rose	Hollyhock	**M(traditional):**LF:OUL,UDR	Clev-079-01
Alcea	*rosea*	Rose	Hollyhock	**E:**FL:DYE(alum-&-tin-&-iron-brown,alum-&-tin-wine-red)	Grae-122-02
Alcea	*rosea*	Rose	Hollyhock	**E:**FL:DYE(alum-soft-beige-pink,iron-taupe,chrome-warm-tan,copper-sulfate-light-green,alum-&-tin-blue)	Cas1-152-01
Alchemilla	*alpina*	Alpine	Lady's Mantle	**M(traditional):**LF:DVGB,DVGV,NMP,OAS,SDR	Bown-232-08
Alchemilla	*arvensis*	Parsley Piert	Lady's Mantle	**M(traditional):**LF,FL,SH:UBL	Clev-077-03
Alchemilla	*vulgaris*	Lion's Foot	Lady's Mantle	**M(traditional):**LF,FL,SH:OAS,OEY	Clev-077-01
Alchemilla	*vulgaris*	Lion's Foot	Lady's Mantle	**M(folk):**LF,FL-SM:MEB,OAS,OWN,SDR	Krug-100-01
Alchemilla	*vulgaris*	Lion's Foot	Lady's Mantle	**M(PS):**WP(aerial):OWN,SDR,RST	Wyk-037-01, Wyk-397-08
Alchemilla	*xanthochlora*	Bear's Foot	Lady's Mantle	**M(traditional):**WP:DVGB,DVGV,NMP,SDR	Bown-232-09
Alchornea	*cordifolia*	Iporuru	Alchorne Spurge	**M(traditional):**BK:BDB,DCLD,DIFR,DIPT,HAP,NJA,NMS,NNC👽,NRH,RLC,SIB,UUR	Bal2-088-01 Bal2-527-04
Alchornea	*cordifolia*	Iporuru	Alchorne Spurge	**M(traditional):**LF,FR,RT:DABI(trypanosome),NMS,NNC👽	Wyk-397-09
Alectoria	*sarmentosa*	Runner	Witch Hair Shield Lichen	**E:**WP:DYE(Alectoronic-Acid-bright-orange, Alectoronic-Acid-&-Ammonia-taupe-violet)	Cas2-031-03
Aletris	*farinosa*	Unicorn Root	Mealy Rod Lily	**M(traditional):**RZ:DANX,NMP,NMS,SDP,SFL	Bown-233-01
Aletris	*farinosa*	Unicorn Root	Mealy Rod Lily	**M(traditional):**LF,RZ:NMP,SDP	Wyk-397-10
Alisma	*plantago-aquatica*	Ze Xie	Water Plantain	**M(traditional):**RT:BHC,BHP,DEAL,DEAT,DFRI,SDR,UKC(kidneyweaknessmanifestsasdeafness,tinnitusanddizziness)	Bown-233-02
Alisma	*plantago-aquatica*	Ze Xie	Water Plantain	**M(traditional):**RT:UDR,UUR	Wyk-397-11
Alkanna	*tinctoria*	Dyer's Bugloss	Alkanet	**C:**RT:FC,**M(traditional):**RT:DABA,OAS,OUL,**E:** RT:CM,DYE	Bown-233-03
Alkanna	*tuberculata*	Tubercle	Alkanet	**M(traditional):**RT:DABI,OAS,DSIN, **E:**RT:DYE(red)	Wyk-397-12
Alliaria	*petiolata*	Hedge	Garlic Mustard	**C:**LF:**SN**,**M(traditional):**LF,SM:OGT,ONR,ORH,DSIN,RLA,RLB	Bown-233-04
Alliaria	*petiolata*	Hedge	Garlic Mustard	**M(traditional):**WP:DCLD,DGUM(scurvy),OWN	Wyk-397-13
Allium	*ampeloprasum*	Elephant Garlic	Onion	**C:**BR:**SN**	Bown-234-01
Allium	*cepa*	Pamir-Alai Mountain	Onion	**C:**BR:**VG**,**M(traditional):**BR:BHP,RLB,SIB,UDR	Bown-234-02
Allium	*cepa*	Pamir-Alai Mountain	Onion	**C:**BR:**VG**,**M(traditional):**BR:BHC,BHP,DABA,DAFU,DASC,DWRM,HAC,NSD,OBN,OWN,RLA,RLC,SDG,UDR,UKC	Kowa-290-04 Kowa-399-01
Allium	*cepa*	Pamir-Alai	Onion	**M(CT):**BR:BAS,BDB,BHC,DABI,DCLD,DISB,OW	Wyk-038-01,

Latin Genus	L. Species	E. Species	English Genus	C:Culinary,M:Medicinal,E:Economic	Source
		Mountain		N,RLA,RLC,SAP,SDG,SDY	Wyk-397-14
Allium	*cepa-helva*	Yellow Skin	Onion	**E:**BR(skin):DYE(alum-yellow,tin-bright-orange, alum-&-tin-bright-yellow-orange,chrome-rust, chrome-&-tin-rich-brown,iron-ollve-green,copper-sulfate-kahki)	Cas1-186-02
Allium	*cepa-helva*	Yellow Skin	Onion	**E:**BR(skin):DYE(alum-&-iron-yellow-brown,alum-&-tin-tan-orange)	Grae-181-03
Allium	*cepa-helva*	Yellow Skin	Onion	**E:**BR(skin):DYE(alum-burnt-orange,alum-yellow, chrome-brass,tin-bright-orange,iron-brown)	Kowa-174-10
Allium	*cepa-rubra*	Red Skin	Onion	**E:**BR(skin):DYE(alum-henna-red,alum-maroon-red,alum-&-tin-khaki-green,alum-&-tin-olive-green)	Grae-182-02
Allium	*cepa-rubra*	Red Skin	Onion	E:BR(skin):DYE(alum-red-orange,chrome-gold,chrome-dark-tan,tin-tan-brown)	Kowa-174-10
Allium	*fistulosum*	Bunching	Onion	**C:**BR:**VG**,**M(traditional):**WP:DABI,BHC,HFV,RLC,SDG,UDR	Bown-234-03
Allium	*oleraceum*	Field Garlic	Onion	**C:**BR:**SN**(Garlicsubstitute)	Clev-077-05
Allium	*porrum*	Leek	Onion	**C:**SL:**VG**	Clev-152-01
Allium	*rosea*	Rosy Garlic		**C:**BR:**SN**(Garlicsubstitute)	Clev-077-08
Allium	*sativum*	Garlic	Onion	**M(traditional):**BR:BAS,BCR,BHP,BHS,BLC,DCLD,DFLU,HIN,NJA,RLA,RSN,SDG,SUL	Bal1-070-07
Allium	*sativum*	Garlic	Onion	**M(folk):**BR:DWHP,RLB,SFL,**M(traditional):**BR: BDB,DEAI,NJA,NMP,NSC,OCW,**M(ayurveda):**BR:HFV,NJA,RLB,SCN,**M(CS):**BR:BAS,BHA,BHC,DABA(*Helicobacter pylori*),DAFU,DCTM	Bal2-070-01 Bal2-518-03
Allium	*sativum*	Garlic	Onion	**C:**BR:**SN**,**M(traditional):**BR:DABA,DACN,BAS,BDB,BHA,BHC,BHP,DFLU,DWHP,HAP,HFV,OFG,RLB,SIB,SDY	Bown-234-04
Allium	*sativum*	Garlic	Onion	**C:**BR:**SN**,**M(traditional):**BR:BHP,DABI,RLC,SDG	Clev-077-04
Allium	*sativum*	Garlic	Onion	**C:**BR:**SN**,**M(traditional):**BR:BHP,DABA(staph, strep,typhus),DABI,DAFU,DASC,DCHL,DCST,DFLU,DTBC.DTYP,DWHP,DWRM,OUL,OWN,RST,SDY,SIB	Kowa-215-01 Kowa-287-03
Allium	*sativum*	Garlic	Onion	**C:**BR:**SN**,**M(folk):**BR:BHC,DABA,DASC,DCLD, OWN	Krug-079-01
Allium	*sativum*	Garlic	Onion	**M(CT):**BR:BHC,DASC,DAVR,DCLD,NMS,RLC	Wyk-039-01, Wyk-397-15
Allium	*schoenoprasum*	Chives	Onion	**C:**LF,SM:**SN**	Bown 234-05
Allium	*schoenoprasum*	Chives	Onion	**C:**LF,SM:**SN**	Clev-077-11
Allium	*schoenoprasum*	Chives	Onion	**C:**LF,SM:**SN**,**M(traditional):**BR-OL,SM-OL:BHP,DASC	Kowa-088-01
Allium	*schoenoprasum*	Chives	Onion	**C:**LF:**SN**	Krug-053-01
Allium	*schoenoprasum*	Chives	Onion	**E:**LF:DYE(alum-light-yellow,copper-sulfate-soft-yellow-green)	Cas1-121-01
Allium	*scorodoprasum*	Leek Garlic	Onion	**C:**LF,BR:**SN**	Bown-234-06
Allium	*tricoccum*	Ramps Leek	Onion	**C:**LF,SL,BR:**SN**,**M(traditional):**LF,SL,BR:DEAI, DWRM,RST	Bown-234-07
Allium	*triquetrum*	Three Cornered Garlic	Onion	**C:**BR:**SN**(Garlicsubstitute)	Clev-077-06
Allium	*tuberosum*	Garlic Chives	Onion	**C:**LF,FL:**SN**,**M(traditional):**LF,FL,SD,SL,BR:UBL,UIN,UKC	Bown-234-08
Allium	*ursinum*	Ramsons Chives	Onion	**C:**LF:**SN**,**M(traditional):**WP:BAS,BHP	Bown-234-09
Allium	*veneale*	Crow Garlic	Onion	**C:**BR:**SN**(Garlicsubstitute)	Clev-077-10
Alnus	*glutinosa*	Black	Alder	**M(traditional):**BK:DTDC,DAMW,NRH,ORH,RST,**M(traditional):**LF:ORH,**E:**BK:DYE(black,red)	Bown-235-01
Alnus	*glutinosa*	Black	Alder	**M(traditional):**BK,LF:DAMW,OAS,OWN	Wyk-397-16
Alnus	*incana tenuifolia*	Weak Hoary	Alder	**E:**LF,FR,WD:DYE(alum-tan-gold,chrome-chestnut-brown,copper-olive-gold,tin-gold,iron-dark-olive-green)	Bliss-032-01
Alnus	*rubra*	Red	Alder	**M(traditional):**BK:OAS	Bown-235-02
Alnus	*rugosa*	Speckled	Alder	**M(traditional):**BK:OAS	Bown-235-03
Alnus	*rugosa*	Speckled	Alder	**E:**LF:DYE(alum-yellow,alum-&-copper-sulfate-yellow-green,copper-sulfate-olive-green,iron-tan,tin-yellow-orange),**E:**BK,SM:DYE(chrome-&-baking-soda-tan-rose-brown,iron-dark-brown),**E**:RT:DYE(chrome-&-iron-gray-brown)	Cas1-090-01
Alnus	*serrulata*	Hazel	Alder	**M(traditional):**BK:OAS	Bown-235-04
Aloe	*aristata*	aristata	Aloe	**M(traditional):**LF-SA:DAFU,DWRM,OBN,OSB,DSIN,OWN, SAP,SCN,SDG,SPR,**E**:LF,SA:CM	Clev-078-04
Aloe	*barbadensis*	Barbados	Aloe	**M(PS):**LF-SA:DACN,DCTM,DHSV,DRSK,OBN, OHM,OSS,OWN,SCN,SUL	Bal2-018-01 Bal2-502-01
Aloe	*barbadensis*	Barbados	Aloe	**M(traditional):**LF-	Kowa-005-01

Latin Genus	L. Species	E. Species	English Genus	C:Culinary,M:Medicinal,E:Economic	Source
				SA:DABA,DASC,DCSK,DIOS,NSD,OBN, OSB,**E:**LF-SA:CM,DS	Kowa-279-01 Kowa-283-01
Aloe	*capensis*			Synonym for *Aloe ferox*	Bal2-018-02
Aloe	*ferox*	Africa Cape	Aloe	**M(PS):**LF-SA:DACN,DCTM,DHSV, DRSK,OBN, OHM,OSS,OWN,SCN,SUL	Bal2-018-02 Bal2-501-01
Aloe	*ferox*	Africa Cape	Aloe	**M(traditional):**LF-SA:DAFU,DWRM,OBN,OSB,DSIN,OWN, SAP,SCN,SDG,SPR,**E**:LF,SA:CM	Bown-235-06
Aloe	*ferox*	Africa Cape	Aloe	**M(PS):**LF-SA:NJA,OEY,DSIN,SCN,**E:**LF:CM	Wyk-040-01, Wyk-397-17
Aloe	*perryi*	Zanzibar	Aloe	**M(traditional):**LF-SA:DAFU,DWRM,OBN,OSB,DSIN,OWN, SAP,SCN,SDG,SPR,**E**:LF,SA:CM	Bown-235-07
Aloe	*succotrina*	Socotrine	Aloe	**M(traditional):**LF-SA:DAFU,DWRM,OBN,OSB,DSIN,OWN, SAP,SCN,SDG,SPR,**E**:LF,SA:CM	Bown-235-08
Aloe	*vera*	Curaçao	Aloe	**M(traditional):**LF-PU:DABA,DAFU,DAVR,OBN,DSIN,OWN, SCN,SDG	Bal1-065-01
Aloe	*vera*	Curaçao	Aloe	**M(PS):**LF-SA:DACN,DCTM,DHSV, DRSK,OBN, OHM,OSS,OWN,SCN,SUL	Bal2-018-03 Bal2-502-01
Aloe	*vera*	Curaçao	Aloe	**M(traditional):**LF-SA:DAFU,DWRM,OBN,OSB,DSIN,OWN, SAP,SCN,SDG,SPR,**E**:LF-SA:CM	Bown-235-09
Aloe	*vera*	Curaçao	Aloe	**M(traditional):**LF-SA:NMT,OBN,OWN	Clev-078-03
Aloe	*vera*	Curaçao	Aloe	**M(folk):**LF-JU:DIOS,DSIN,OBN,OWN	Krug-013-01
Aloe	*vera*	Curaçao	Aloe	**M(PS):**LF-SA:BDB,BHC,DCTM,OBN,DSIN,SCN	Wyk-041-01, Wyk-397-18
Aloysia	*triphylla*	Lemon	Beebrush	**C:**LF:**SN**,TE,**M(traditional):**LF-OL:DCLD,NMS,OAS,SDG	Bown-235-10
Aloysia	*triphylla*	Lemon	Beebrush	**C:**LF:**SN**,TE,**M(traditional):**LF-OL:HFV,NSD,SDG,**E:**OL:CL	Kowa-358-01
Aloysia	*triphylla*	Lemon	Beebrush	**M(traditional):**LF-OL:DCLD,HFV,NMS,NSD, RLA,SDG,SDR, SFL,SIB	Wyk-042-01, Wyk-397-19
Aloysia	*triphylla*	Lemon	Beebrush	**C:**LF:TE,**C:**LF,OL:BF,FF,**M(traditional):**LF:NSD, OTA,**E:**LF,OL:PF	Clev-101-03
Aloysia	*triphylla*	Lemon	Beebrush	**C:**LF,OL:BF,FF,**M(traditional):**LF-OL:NSD,SDG, **E:**LF,OL:DS,PF	Krug-176-01
Alpinia	*galanga*	Galangal	Alpino Ginger	**C:**OL:AF,BF,C:RT,RZ:**SP**,**M(traditional):**OL,RZ :SDG,SDR,UIN,UKC	Bown-235-11
Alpinia	*galanga*	Galangal	Alpino Ginger	**C:**RT,RZ:**SP**,**M(traditional):**RT,RZ:DABA,DHMY, DAMW,SFL	Clev-306-01
Alpinia	*officinarum*	Lesser Galangal	Alpino Ginger	**C:**RT,RZ:**SP**,**C:**OL:AF,BF,**M(traditional):**OL,RZ :DABA,DAFU,DGUM,DCSK,HFV,NRH,SDG,SIB, SUL	Bown-236-01
Alpinia	*officinarum*	Lesser Galangal	Alpino Ginger	**C:**RT,RZ:**SP**,**M(traditional):**RT,RZ:DABA,DHMY, DAMW,SFL	Clev-306-02
Alpinia	*officinarum*	Lesser Galangal	Alpino Ginger	**C:**RZ:**SP**,**M(PS):**RZ:SAP,SDG,SDP	Wyk-043-01, Wyk-397-20
Alpinia	*oxyphylla*	Black	Alpino Ginger	**M(traditional):**OL,RZ:SDG,SDR,UIN,UKC	Bown-235-12
Alstonia	*boone*	Twistflower Oleander	Devilwood	**M(traditional):**BK:SPN*(Strophanthus)*	Bown-236-02
Alstonia	*constricta*	Australian Quinine	Devilwood	**M(traditional):**BK:HFV,NMS,SCN	Wyk-398-01
Alstonia	*scholaris*	Dita Bark	Devilwood	**M(traditional):**BK:DMAL,DWRM,HFV,NMS,OAS ,SDR,SDY	Bown-236-03
Althaea	*officinalis*	White	Marsh Mallow	**M(traditional):**FL,LF,RT:HAC,RLC,SIB,RSN,RST,UB L,UDR,UKC	Bal1-074-02
Althaea	*officinalis*	White	Marsh Mallow	**M(traditional):**LF,RT:DAMW,OWN,RLC,RST,S DG,UUR, **M(PS):**LF,RT:RLC,RTL,SDG	Bal2-098-01 Bal2-531-02
Althaea	*officinalis*	White	Marsh Mallow	**C:**RT:FX,MC(marshmallow):**M(traditional):**LF:R LB,RLC,UBL,**M(traditional):**LF,RT:DBOI,DGNG, DGUM,DISB,OEY,OMT,OSA,**M(traditional):**RT: DABA,DWHP,RLA,RLB,RLC,SHH,SUL,UBL	Bown-236-04
Althaea	*officinalis*	White	Marsh Mallow	**C:**SH:**VG**, **M(traditional):**FL,LF:RLC ,SDG, **M(traditional):**RT:DSIN	Clev-078-05
Althaea	*officinalis*	White	Marsh Mallow	**C:**LF:FF,**C:**RT:CF,**M(traditional):**FL,LF,RT:RLC, RST,DSIN,UDR,UKS	Kowa-290-02 Kowa-379-01
Althaea	*officinalis*	White	Marsh Mallow	**C:**JU:CF,**M(folk):**LF,RT:RLB,RLC,RST,OST,OU L,OWN,SUL.UUR	Krug-114-01
Althaea	*officinalis*	White	Marsh Mallow	**M(PS):**LF:OBN,DSIN,OUL,RLC,**M(traditional):**R T:RLC,RST,SUL	Wyk-044-01, Wyk-398-02
Althaea	*rosea*			Synonym for *Alcea rosea*	
Amaranthus	*albus*	Tumble Pigweed	Amaranth	**E:**WP:DYE(alum-moss-green,alum-&-iron-forest-	Grae-102-03

Latin Genus	L. Species	E. Species	English Genus	C:Culinary,M:Medicinal,E:Economic	Source
				green,chrome-brass,chrome-&-iron-khaki-green)	
Amaranthus	*dubius*	Spleen	Amaranth	**M(traditional):**WP:OAS,SDR,UDR	Wyk-398-03
Amaranthus	*hypochondriacus*	Cereal Grain	Amaranth	**C:**LF:FC,**VG**,**C:**SD:**VG**,**M(traditional):**WP:OAS,OUL,OWN,SDR,**E:**WP:DYE(red)	Bown-236-05
Amaranthus	*hypochondriacus*	Cereal Grain	Amaranth	**C:**LF:**VG**, **M(folk):**LF:DAMW,OAS,OUL,SDR	Krug-014-01
Amaranthus	*polygamus*	Aphrodisiac	Amaranth	**M(traditional):**WP:OAS,OUL,OWN,SDR	Bown-236-06
Amaranthus	*retroflexus*	Redroot Pigweed	Amaranth	**C:**LF:**VG**,**C:**SD:NT,**M(traditional):**WP:OAS,OUL,OWN,SDR	Bown-236-07
Amaranthus	*retroflexus*	Redroot Pigweed	Amaranth	**E:**WP:DYE(alum-light-yellow,chrome-tan,copper-light-olive,tin-light-gold,iron-yellow-gray)	Bliss-206-01
Amaranthus	*retroflexus*	Redroot Pigweed	Amaranth	**E:**WP:DYE(alum-soft-yellow-green,tin-chartreuse,chrome-&-iron-brown)	Cas1-091-01
Amaranthus	*spinosus*	Spiny Pigweed	Amaranth	**M(traditional):**WP:OAS,OUL,OWN,SDR	Bown-236-08
Ambrosia	*ambrosoides*	Big Burr	Ragweed	**E:**LF,SL,SM:DYE(alum-gold,chrome-rust,copper–green-gold,tin-dull-gold,iron-dark-gray-green)	Bliss-202-01
Ambrosia	*artemisifolia*	Annual	Ragweed	**M(traditional):**LF:DALL	Wyk-398-04
Ambrosia	*tomentosa*	Povertyweed	Ragweed	**E:**WP:DYE(alum-light-yellow-green,chrome-gray-green,copper-ocher-green,tin-light-gold, iron-forest-green)	Bliss-192-01
Ambrosia	*trifida*	Giant	Ragweed	**E:**BD,FL,LF,SM:DYE(alum-yellow-green,chome-tan-gold,copper-green-gold,tin-tan-gold,iron-chartreuse-gray)	Bliss-204-01
Amelanchier	*alnifolia*	Alder Leaf Juneberry	Serviceberry	**E:**FL:DYE(alum-&-tin-warm-brown,alum-lemon-yellow,copper-sulfate-soft-green),**E:**LF:DYE (alum-yellow-green,copper-sulfate-green, chrome-brown,iron-gray,tin-brilliant-yellow-orange)	Cas1-160-02
Amelanchier	*stolonifera*	Running Juneberry	Serviceberry	**E:**FL:DYE(alum-&-tin-warm-brown,alum-lemon-yellow,copper-sulfate-soft-green),**E:**LF:DYE (alum-yellow-green,copper-sulfate-green,chrome–brown,iron-gray,tin-brillian-yellow-orange)	Cas1-161-01
Ammi	*majus*	Queen Anne's Lace	Bishop's Weed	**M(traditional):**SD:DVTL,OSS,UDR,**E:**SD:PS☠	Bown-236-09
Ammi	*visnaga*	Khella	Bishop's Weed	**M(traditional):**FR:BAA,BAS,BHS,DVTL,RLA, UKS	Bal2-091-01 Bal2-528-01
Ammi	*visnaga*	Khella	Bishop's Weed	**M(traditional):**SD:BAA,BAS,RLA,UKS,**E:**SD:PS ☠	Bown-236-10
Ammi	*visnaga*	Khella	Bishop's Weed	**M(PS):**FR:DAMW,NMS,RLA,RLB,SIB,UDR,UKS	Wyk-045-01, Wyk-398-05
Amomum	*aromaticum*	Bengal	Bastard Cardamom	**M(traditional):**FR:RLC,SDP	Wyk-398-06
Amomum	*villosum*	Tavoy	Bastard Cardamom	**M(traditional):**PD,OL:DABA,DAFU,DMAL,OUL, SAP,SDP, SDR,SFL	Bal2-042-02 Bal2-509-03
Amomum	*villosum xanthoides*	Grains of Paradise Tavoy	Bastard Cardamom	**C:**SD:AF,BF,**SP**,**M(traditional):**SD:SAP,SDG,SIB	Bown-237-01
Anacamptis	*morio*	Green Wing	Meadow Orchid	**M(traditional):**TU:SDR,**E:**TU:MC	Wyk-419-08
Anacardium	*occidentale*	Acajou	Cashew	**M(traditional):**BK,FR,NT,RN:DSNB,DVGB,DVGV,DWFK,**M(SS):**BK,FR,NT,RN:DLEI, **M(PS):**BK,FR,NT,RN:BDB, **E:**BK,FR,NT,RN:**contraceptive** ☠	Bal2-040-01 Bal2-508-04
Anacardium	*occidentale*	Acajou	Cashew	**C:**FR:BV,JJ,**C:**SD:NT,**M(traditional):**BK,FR,LF:SDR,**M(traditional):**BK:BSH,DHOG,**M(traditional):**BK,LF:DGUM,DMAL,OTA,**M(traditional):**LF:HFV,**M(traditional):**FR:DFLU,UDR,**M(traditional):**OL:DABA,DWRG,DCLW,**E:**OL:IR,RU	Bown-237-02
Anacardium	*occidentale*	Acajou	Cashew	**M(traditional):**BK,FR,LF:DASC,DHOG,DWRM, DSIN,OTA	Wyk-398-07
Anacyclus	*pyrethrum*	Pellitory	Mount Atlas Daisy	**M(traditional):**RT:BCR,NNR,OTA,RLC	Bown-237-03
Anacyclus	*pyrethrum*	Pellitory	Mount Atlas Daisy	**M(traditional):**RT:OTA	Wyk-398-08
Anadenanthera	*peregrina*	Peregrine	Cohoba	**M(traditional):**LF:NNC☠	Wyk-421-17
Anagallis	*arvensis*	Scarlet	Pimpernel	**M(traditional):**WP:BHE,BLC,DEPI,DTBC,DAVR, NRH,HFV,RLC,UDR,**E:**WP:PS☠	Bown-237-04
Anagallis	*arvensis*	Scarlet	Pimpernel	**M(traditional):**WP:DSIN,RLC,UDR,**E:**WP:PS☠	Wyk-398-09
Anamirta	*cocculus*	India Fish	Berry Moonseed	**M(traditional):**LF,FR:DWRG,DSIN, **E:**LF,FR:PS☠	Wyk-398-10
Ananas	*comosus*	Bromelain	Pineapple	**M(traditional):**FR-JU:BCR,BDB,BHS,DCBR, DCLK,DCLU,DCRO,DSNA,NMS,OBN,SDG,SUL, **M(PS):**FR-JU:DCTM,NJA,RSN	Bal2-037-01 Bal2-507-05
Ananas	*comosus*	Bromelain	Pineapple	**M(PS):**FR-JU,SM:SDG,UDR	Wyk-046-01, Wyk-398-11
Anaphalis	*margaritacea*	Margarite	Pearly Everlasting	**E:**FL,LF,SM:DYE(alum-yellow-gold,alum-&-copper-light-olive-green,chrome-orange-gold,chrome-&-copper-light-brown)	Grae-086-02
Anaphalis	*margaritacea*	Margarite	Pearly Everlasting	**E:**WP:DYE(alum-strong-yellow,alum-&-tin-chartreuse,chrome-gold,copper-sulfate-&-iron-	Cas1-191-02

Latin Genus	L. Species	E. Species	English Genus	C:Culinary,M:Medicinal,E:Economic	Source
				yellow-green)	
Anchusa	*officinalis*	European	Bugloss	**M(traditional):**WP:NSD,OWN,RLC,E:WP:PS☠	Wyk-398-12
Andira	*araroba*	Goa	Cabbage Bark	**M(traditional):**LF:DSIN	Wyk-398-13
Andira	*inermis*	Dog Almond	Cabbage Bark	**M(traditional):**BK:DWRM,NMS	Wyk-398-14
Andrographis	*paniculata*	King of Bitters	Water Willow	**M(traditional):**LF,SM:BDB,DCLD,DFLU,DHPT, DHSV,HFV,RST,SDG,SDR,SGB, **M(CS):**LF,SM:BAS,BHA,DCTM	Bal2-022-02 Bal2-503-03
Andrographis	*paniculata*	King of Bitters	Water Willow	**M(traditional):**WP:UKS	Wyk-398-15
Anemarrhena	*asphodeloides*	Zhi Mu	Sessile Anther Asphodel	**M(traditional):**RZ:DABA,DAFU,BSH,DTBC,HFV ,RLB,RLC,UDR,**E:**RZ:PS☠	Bown-238-01
Anemarrhena	*asphodeloides*	Zhi Mu	Sessile Anther Asphodel	**M(traditional):**RZ:HFV,DAMW	Wyk-398-16
Anemone	*blanda-blauea*	Blue Thimbleweed	Windflower	**E:**FL:DYE(alum-teal-blue,chrome-light-blue)	Grae-113-01
Anemone	*coronaria-blauea*	Blue Poppy	Windflower	**E:**FL:DYE(alum-teal-blue,chrome-light-blue)	Grae-113-01
Anemone	*cylindrica*	Cylindrical	Windflower	**M(traditional):**LF,RT:HAC,NRH,OTA,RLB	Wyk-398-17a
Anemone	*nemorosa*	Grove	Windflower	**M(traditional):**LF,RT:HAC,NRH,OTA,RLB	Wyk-398-17b
Anemone	*obtusiloba-blauea*	Blue Blunt Lobe	Windflower	**E:**FL:DYE(alum-teal-blue,chrome-light-blue)	Grae-113-01
Anemone	*pulsatilla*			Synonym for *Pulsatilla vulgaris*	Krug-133-01
Anemone	*ranunculoides*	Buttercup	Windflower	**M(traditional):**LF,RT:HAC,NRH,OTA,RLB	Wyk-398-17c
Anemone	*sylvestris*	Forest	Windflower	**M(traditional):**LF,RT:HAC,NRH,OTA,RLB	Wyk-398-17d
Anemone	*trullifolia-blauea*	Blue Himalayan	Windflower	**E:**FL:DYE(alum-teal-blue,chrome-light-blue)	Grae-113-01
Anethum	*graveolens*	Scented	Dill	**C:**LF:**SN**,**C:**OL:FF,**C:**SD:**SP**,**M(traditional):**LF, SD,OL:SDG,SFL,SHH,SIB,UDR,**E:**OL:DS	Bown-238-02
Anethum	*graveolens*	Scented	Dill	**C:**LF:**SN**,**C:**SD:**SP**,**M(traditional):**LF,SD,SM,O L:SDG,SFL, **E:**OL:PF	Clev-079-02, Clev-272-01
Anethum	*graveolens*	Scented	Dill	**C:**LF:**SN**,**C:**SD:**SP**,**M(traditional):**LF,SD:SAP, SDG,SFL,SIB	Kowa-160-01
Anethum	*graveolens*	Scented	Dill	**C:**LF:**SN**,**C:**SD:**SP**,**M(folk):**SD-OL:SIB	Krug-062-01
Anethum	*graveolens*	Scented	Dill	**C:**LF:**SN**,**C:**SD:**SP**,**M(traditional):**LF,SD:**M(PS):**FR:SDG,SDP,SFL,SIB,UDR	Wyk-047-01, Wyk-398-18
Anethum	*graveolens*	Scented	Dill	**E:**WP,FL:DYE(vinegar-beige-yellow,tin-strong-yellow-gold,chrome-dark-gold,copper-sulfate-&-iron-gray-green)	Cas1-135-02
Anethum	*sowa*	Asian	Dill	**C:**LF:**SN**	Clev-079-03
Angelica	*anomala*	Anomalous	Angelic	**M(traditional):**LF,RT,SD,SM:NMP	Bown-238-03
Angelica	*archangelica*	Archangel	Angelic	**M(traditional):**RT:BCR,BDB,DAVR,DCLD,DOB E,HAP,HFV,NBF,RLC,SAP,SFL,SIB,**M(PS):**RT: BAA,BHP, DABA(*Escherichia coli*),	Bal2-023-01 Bal2-152-01 Bal2-550-01
Angelica	*archangelica*	Archangel	Angelic	**C:**LF:**VG**,**C:**SL:**VG**,**C:**RT,SD:AF,**M(traditional) :**LF,RT,SD,SM:DANX,DFLU,HAC,HFV,NMP,NM S,ONR,ORH,RLB,RLC,RLP,SDG,SUL	Bown-238-04
Angelica	*archangelica*	Archangel	Angelic	**C:**LF,SH:AF,C:SL:**VG**,**M(traditional):**WP:SAP, SDG	Clev-079-04
Angelica	*archangelica*	Archangel	Angelic	**C:**SL:**VG**,**M(traditional):**LF,RT:HIN,**M(traditional) :**RT,SD:SDG,SFL,SIB,**M(traditional):**WP:HAC,HF V,NRH,DSIN,OTA,OWN**E:**RT-OL:CM	Kowa-011-01
Angelica	*archangelica*	Archangel	Angelic	**C:**SL:**VG**,**C:**SD:AF,**M(folk):**LF,RT:RLC,SDG	Krug-016-01
Angelica	*archangelica*	Archangel	Angelic	**C:**RT:AF,FF,**M(traditional):**RT:SAP,SDG,SFL	Wyk-048-01, Wyk-398-19
Angelica	*atropurpurea*	Dark Purple	Angelic	**C:**LF,SH:AF	Clev-079-05
Angelica	*atropurpurea*	Dark Purple	Angelic	**M(traditional):**LF,RT,SD,SM:NMP	Bown-238-05
Angelica	*dahurica*	Du Huo	Angelic	**M(PS):**RT:BDB,DABA,DAVR,HFV	Bal2-152-01 Bal2-550-01
Angelica	*dahurica*	Du Huo	Angelic	**M(traditional):**RT:BAN,NMP,SCN	Wyk-398-20a
Angelica	*gigas*	Giant	Angelic	**M(traditional):**LF,RT,SD,SM:NMP	Bown-238-06
Angelica	*glauca*	Glaucous	Angelic	**M(traditional):**LF,RT,SD,SM:NMP	Bown-238-07
Angelica	*keiskei*	Keiske	Angelic	**M(traditional):**LF,RT,SD,SM:NMP	Bown-238-08
Angelica	*polymorpha*	Dang Gui	Angelic	**M(traditional):**LF,RT,SD,SM:NMP	Bown-238-09
Angelica	*polymorpha*	Dang Gui	Angelic	**M(traditional):**RT:BAN,NMP,SCN	Wyk-398-20b
Angelica	*pubescens*	Downy	Angelic	**M(traditional):**LF,RT,SD,SM:NMP	Bown-238-10
Angelica	*sinensis*	Dong Quai	Angelic	**M(traditional):**RT:BAN,BAS,BHP,BHS,DALL,DB OI,DCLK,DINF,DUFB,NJA,NNR,NRH,NSD,OUL, SCN,UKC	Bal2-059-02 Bal2-514-04
Angelica	*sinensis*	Dong Quai	Angelic	**M(traditional):**RT:DABA,BAA,BLC,NJA,NMP,N NR,SCN	Bown-238-11
Aniba	*rosaeodora*	Rosewood	Cara Cara	**E:**OL:CF,PF,DS,**E:**WD:LM	Bown-239-01
Antennaria	*dioica*	Cat's Ear	Pussy Toes	**M(traditional):**WP:BLC,DHPT,DTNS,DVGV,DV GV,OAS,SDR,SGB,UDR	Bown-239-02
Antennaria	*dioica*	Cat's Ear	Pussy Toes	**M(traditional):**WP:BLC,RLC,SDR	Wyk-398-21
Anthemis	*cotula*	Chamomile	Dog Fennel Daisy	**M(traditional):**WP:DWRM,HFV,NMS	Wyk-398-22
Anthemis	*cotula*	Chamomile	Dog Fennel Daisy	**E:**WP:DYE(alum-deep-gold,chrome-tan-gold)	Grae-080-02
Anthemis	*cotula*	Chamomile	Dog Fennel Daisy	**E:**WP:DYE(alum-strong-yellow,tin-brilliant-	Cas1-116-03

Latin Genus	L. Species	E. Species	English Genus	C:Culinary,M:Medicinal,E:Economic	Source
				yellow,chrome-rich-gold,iron-yellow-green)	
Anthemis	*nobilis*			Synonym for *Chamaemelum nobilis*	Grae-176-02
Anthemis	*tinctoria*	Golden Maguerite	Dog Fennel Daisy	E:FL:DYE(alum-yellow-buff,chrome-gold-orange,tin-yellow)	Kowa-174-02
Anthoxanthum	*odoratum*	Sweet	Vernal Grass	**M(traditional):**FL:BCR,DHYF,HIN,NMS,OCB,OJA	Bown-239-03
Anthriscus	*cerefolium*	European	Chervil	**C:**LF:**SN**,**M(traditional):**LF:BLC,BJA,NRH,OEY,OHM,DSIN,RLC,UKC	Bown-239-04
Anthriscus	*cerefolium*	European	Chervil	**C:**LF:**SN**,**M(traditional):**LF:OWN,SDG	Clev-080-01
Anthriscus	*cerefolium*	European	Chervil	**C:**LF:**SN**,**M(traditional):**LF,FL:BHE,BHP,NGT,DSIN,RLC,RLP,SDG,UDR,UKS	Kowa-083-01
Anthriscus	*cerefolium*	European	Chervil	**C:**LF:**SN**,**M(folk):**LF:SAP,SDG,UDR	Krug-050-01
Anthriscus	*cerefolium*	European	Chervil	**C:**LF:**SN**,**M(traditional):**WP:DSIN,UDR	Wyk-399-01
Anthyllis	*vulneraria*	Vulnerable	Kidney Vetch	**M(traditional):**LF,FL:OWN,RLC	Wyk-399-02
Antirrhinum	*majus*	Dark Red	Snapdragon	**E:**FL:DYE(alum-&-cotton-pale-green,alum-&-jute-tan-gold,alum-&-wool-dark-green,chrome-&-wool-brown)	Grae-133-02
Aphanes	*arvensis*	Breakstone	Parsley Piert	**C:**LF:**SN**,**M(traditional):**LF:OAS,SGS,UDR,UKS	Bown-239-05
Aphanes	*arvensis*	Breakstone	Parsley Piert	**C:**LF:**SN**,**M(folk):**FL,LF:UDR,UKS,UUR	Krug-132-01
Aphanes	*arvensis*	Breakstone	Parsley Piert	**M(traditional):**WP:OAS,UDR,UUR	Wyk-399-03
Apium	*graveolens*	Wild Celery	Marshwort	**M(traditional):**WP:NGT,NJA,**M(traditional):**OL:OFG,OTM,RLA,RLB,SFL	Bown-240-01
Apium	*graveolens*	Wild Celery	Marshwort	**M(folk):**WP:NGT,NJA,NRH,**M(traditional):**SD:NRH,SDG	Krug-046-01
Apium	*graveolens-dulce*	Celery	Marshwort	**M(traditional):**SL-JU,RT,SD:BHP,DFRA,NJA,NMS,NSD,SAP,UKC	Bal1-067-05
Apium	*graveolens-dulce*	Celery	Marshwort	**C:**SD:**SP**,**C:**SL:**VG**,**M(traditional):**OL,SD:RLA,RLB,SFL	Clev-273-01
Apium	*graveolens-dulce*	Celery	Marshwort	**M(traditional):**LF,FR,RT:HAP,NGT,NSD,RLC,SDP,UDR	Wyk-399-04
Apocynum	*cannabinum*	Hemp	Dogbane	**M(traditional):**RT,RZ:BHA,DSXV,HFV,OHL,RLC,UDR,UKC	Bown-240-02
Apocynum	*cannabinum*	Hemp	Dogbane	**M(traditional):**RT:BHS,UDR	Wyk-399-05
Apocynum	*cannabinum*	Hemp	Dogbane	**E:**LF,FL,SM:DYE(alum-bright-gold,chrome-brass,copper-khaki-brass,iron-forest-green)	Bliss-108-01
Aquilaria	*agallocha*	Aloewood	Eaglewood	**M(traditional):**RN,WD:OAS,DSIN,SDP,UDR	Wyk-399-06
Aquilaria	*crassma*	Cambodia	Eaglewood	**C:**BK,WD:**SP**,**M(traditional):**BK,WD:HFV,NMS,NRH,OAS,OTM,RLB,SDG,**E:**RN:IN,PF	Bown-240-03
Aquilaria	*malaccensis*	Agarwood	Eaglewood	**C:**BK,WD:**SP**,**M(traditional):**BK,WD:HFV,NMS,NRH,OAS,OTM,RLB,SDG,**E:**RN:IN,PF	Bown-240-04
Aquilaria	*sinensis*	Chinese	Eaglewood	**C:**BK,WD:**SP**,**M(traditional):**BK,WD:HFV,NMS,NRH,OAS,OTM,RLB,SDG,**E:**RN:IN,PF	Bown-240-05
Aquilegia	*vulgaris*	European	Columbine	**M(traditional):**WP:DASC,**M(traditional):**RT,SUL,**E:**WP:PS☠	Clev-080-04
Aquilegia	*vulgaris*	European	Columbine	**M(traditional):**LF,SD:BJA,DGUM(scurvy),NSD,DSIN	Wyk-399-07
Arachis	*hypogaea*	Goober	Peanut	**M(traditional):**SD:DSIN,SCN	Wyk-399-08
Aralia	*chinensis*	Chinese	Aral	**C:**RT,RZ:NJA,NRH	Bown-240-06
Aralia	*hispida*	Hairy	Aral	**M(traditional):**RT,RZ:HFV	Bown-240-07
Aralia	*mandshurica*	Manchurian Spikenard	Aral	**M(traditional):**RT:DCLD,NSD,DSIN,RLC	Wyk-399-09
Aralia	*nudicaulis*	Wild Sarsaparilla	Aral	**C:**RT:TE,**M(traditional):**RT:DHMY,DBOI,HFV,OJA,RLC,UDR	Bown-240-08
Aralia	*racemosa*	American Spikenard	Aral	**M(traditional):**RT,RZ:HFV,NGT,NRH,RLB,RLC,UDR	Bown-240-09
Aralia	*spinosa*	Hercules Club	Aral	**M(traditional):**RT,RZ:HFV	Bown-240-10
Arbutus	*menziesii*	Pacific	Madrone	**E:**BK:DYE(alum-tan-rose-red,chrome-brass,copper-dark-camel,tin-tan,iron-medium-gray)	Bliss-155-01
Arbutus	*unedo*	Strawberry	Madrone	**M(traditional):**FL,FR:DASC,OAS,SDR,UUR	Wyk-399-10
Arceuthobium	*americanum*	Lodgepole Pine	Dwarf Mistletoe	**E:**WP:DYE(alum-beige-gold,chrome-green-gold,copper-dark-gold,tin-bright-gold,iron-gray-green)	Bliss-170-01
Arceuthobium	*californicum*	Sugar Pine	Dwarf Mistletoe	**E:**WP:DYE(alum-beige-gold,chrome-green-gold,copper-dark-gold,tin-bright-gold,iron-gray-green)	Bliss-170-01
Arctium	*lappa*	Hare Burr	Burdock	**M(traditional):**RT,SD:BLC,DBOI,NGT,SGB	Bal1-066-06
Arctium	*lappa*	Hare Burr	Burdock	**M(traditional):**RT,SD:,DAVR,DBOI,DCLD,DND,DMSL,DSBR,DTNS,ODS,OSS,OUL,RST,**M(PS):**RT,SD:BLC,DABA,DCBR,DFRA	Bal2-038-03 Bal2-508-02
Arctium	*lappa*	Hare Burr	Burdock	**C:**SL,RT:**VG**,**M(traditional):**RT,SM:DABA,DBOI,NGT,NRH,DSIN,OSS,**M(traditional):**SD:BSH,DCLD,DPNB,DPNV,RST	Bown-241-01

Latin Genus	L. Species	E. Species	English Genus	C:Culinary,M:Medicinal,E:Economic	Source
Arctium	*lappa*	Hare Burr	Burdock	**C**:SL:**VG**,**M(traditional)**:RT:ODS,DSIN,**M(traditional)**:SD:OCC,RST	Clev-080-06
Arctium	*lappa*	Hare Burr	Burdock	**C**:LF,RT:**VG**,**M(traditional)**:RT:BLC,DACN,ODS,DSIN,UDR	Kowa-056-01 Kowa-285-01
Arctium	*lappa*	Hare Burr	Burdock	**C**:RT,SL:**VG**,**M(folk)**:LF:DBOI,OBR,**M(folk)**:RT:DACN,OSS,**M(folk)**:SD:DWHN	Krug-040-01
Arctium	*lappa*	Hare Burr	Burdock	**M(PS)**:RT:DACN,DISB,ODS,DSIN,OWN,SDG,SIB	Wyk-049-01, Wyk-399-11
Arctium	*minus*	Beggar's Buttons	Burdock	**E**:LF,FR,SM:DYE(chome-light-green-gold,chrome-tan-green,copper-olive-green,tin-bright-gold,iron-dark-gray-green)	Bliss-070-01
Arctium	*minus*	Beggar's Buttons	Burdock	**E**:LF:DYE(alum-yellow,vinegar-tan,alum-&-iron-yellow-green,tin-strong-yellow)	Cas1-110-01
Arctoparmelia	*centrifuga*	Centrifuge	Arctic Ring Shield Lichen	**E**:WP:DYE(Usnic-Acid-&-Atranorin-&-Alectoronic-Acid-red-brown)	Cas2-036-19
Arctostaphylos	*columbiana*	Hairy British Columbia	Bearberry	**E**:LF,FL,FR:DYE(alum-tan-yellow,chrome-dark-gold,copper-green-brown,tin-bright-gold,iron-dark-charcoal-gray)	Bliss-158-01
Arctostaphylos	*manzanita*	Manzanite	Bearberry	**E**:LF:DYE(alum-buff-rosy-red)	Grae-139-02
Arctostaphylos	*uva-ursi*	Kinnikinnick	Bearberry	**M(traditional)**:LF:BDB,BHS,BLC,UBL,UEP,UKC,UDR	Bal1-078-02
Arctostaphylos	*uva-ursi*	Kinnikinnick	Bearberry	**M(traditional)**:LF:BLC,UDR,UKC, **M(PS)**:LF,FR:UBL,UUR	Bal2-148-01 Bal2-549-01
Arctostaphylos	*uva-ursi*	Kinnikinnick	Bearberry	**M(traditional)**:LF:DABA,OAS,DVGB,DVGV,UBL	Bown-241-02
Arctostaphylos	*uva-ursi*	Kinnikinnick	Bearberry	**M(traditional)**:LF:DASC,OAS,UBL,UDR,UKC,UKS,UUR,**E**:LF:DYE(direct-camel),E:WP:DYE(alum-&-iron-green)	Kowa-175-11 Kowa-293-01 Kowa-494-01
Arctostaphylos	*uva-ursi*	Kinnikinnick	Bearberry	**M(folk)**:LF:DASC,DIOS,OAS,UBL,UKS	Krug-027-01
Arctostaphylos	*uva-ursi*	Kinnikinnick	Bearberry	**M(PS)**:LF:UBL,UKC,UUR	Wyk-050-01, Wyk-399-12
Arctostaphylos	*uva-ursi*	Kinnikinnick	Bearberry	**E**:LF,FR,SM:DYE(alum-tan-gold,chrome-brown,copper-green-gold,tin-butter-yellow,iron-dark-olive-green)	Bliss-140-01
Arctotis	*acaulis-aurantia*	Orange Stemless	African Daisy	**E**:FL:DYE(alum-&-copper-green,chrome-&-copper-olive-green)	Grae-120-02
Arctotis	*hybrida-aurantia*	Orange Hybrid	African Daisy	**E**:FL:DYE(alum-&-copper-green,chrome-&copper-olive-green)	Grae-120-02
Areca	*catechu*	Betel Nut	Feather Palm	**M(traditional)**:RD:SCN,SFL,UDR,**M(traditional)**:SD:DMAL,DWRM,DWTP,OAS,SDY,**E**:RD:PS☠	Bown-241-03
Areca	*catechu*	Betel Nut	Feather Palm	**M(traditional)**:SD:DWRM	Wyk-399-13
Argania	*spinosa*	Spiny	Argan Sapote	**E**:SD,OL:CM,**E**:RN:GR	Wyk-399-14
Argemone	*mexicana*	Mexican	Prickly Poppy	**M(traditional)**:LF,SD:NSD,SPR	Wyk-399-15
Argemone	*polyanthemos*	Annual Blue Stem	Prickly Poppy	**E**:FL,LF,PD,SD,SM:DYE(alum-chartreuse,chrome-dark-olive-green,copper-dark-olive-green,tin-tan,iron-gray-green)	Bliss-196-01
Argyranthemum	*frutescens*	Yellow Paris Daisy	Marguerite	**E**:FL:DYE(chrome-gold,chrome-&-iron-mustard-green)	Grae-125-02
Arisaema	*amurense*	Amur	Hooded Spathe Ar Lily	**M(traditional)**:CR:DCCU,DEPI,DTET,NMS,RLC	Bown-241-04
Arisaema	*consanguineum*	Blood Root	Hooded Spathe Ar Lily	**M(traditional)**:CR:DCCU,DEPI,DTET,NMS,RLC	Bown-241-05
Arisaema	*consanguineum*	Blood Root	Hooded Spathe Ar Lily	**M(traditional)**:RZ:NSD,DSIN,**E**:RZ:PS☠	Wyk-399-16
Arisaema	*heterophyllum*	Variable Leaf	Hooded Spathe Ar Lily	**M(traditional)**:CR:DCCU,DEPI,DTET,NMS,RLC	Bown-241-06
Arisaema	*triphyllum*	Jack in the Pulpit	Hooded Spathe Ar Lily	**C**:SD:GF,**M(traditional)**:CR:DASC,DBOI,DSNB,DWHP,RLB,RLC,ORH	Bown-241-07
Aristolochia	*bracteata*	African	Snakeroot	**M(traditional)**:RT:DSNB	Bown-242-01
Aristolochia	*clematitis*	Birthwort	Snakeroot	**M(traditional)**:RT:DABA,DISB,DSNB,DSIN,OWN	Bown-242-02
Aristolochia	*clematitis*	Birthwort	Snakeroot	**M(traditional)**:RT:DSIN,OWN, **induced abortion** ☠	Wyk-051-01, Wyk-399-17
Aristolochia	*debilis*	Heartwort	Snakeroot	**M(traditional)**:FR:BHP,OHM,RLA,RLB,RLC,**M(traditional)**:RT:BHP,DISB,DSNB,NJA,SIB	Bown-242-03
Aristolochia	*fangui*	Chinese Birthwort	Snakeroot	**E**:RT:PS☠(kidney damage,carcinogenic)	Wyk-399-18
Aristolochia	*indica*	India Birthwort	Snakeroot	**M(traditional)**:RT:**induced abortion**☠	Bown-242-04
Aristolochia	*longa*	American	Snakeroot	**M(traditional)**:RT:DSNB	Bown-242-05
Aristolochia	*reticulata*	Red River	Snakeroot	**M(traditional)**:RT:DHSV,DMAL,DPNM,DTYP,NAR,NGT,NRH,OWN,RLP,UDR	Bown-242-06
Aristolochia	*rotunda*	European	Snakeroot	**M(traditional)**:RT:DSNB	Bown-242-07
Aristolochia	*serpentaria*	Serpentary	Snakeroot	**M(traditional)**:RT:DHSV,DMAL,DPNM,DTYP,NAR,NGT,NRH,OWN,RLP,UDR	Bown-242-08
Aristolochia	*serpentaria*	Serpentary	Snakeroot	**M(traditional)**:RT,RZ:DMAL,DSNB,SAP	Kowa-501-01
Aristolochia	*serpentaria*	Serpentary	Snakeroot	**M(folk)**:RT:DSNB,SPR	Krug-178-01
Armeria	*arenaria*	Jersey	Thrift	**M(traditional)**:FL:DASC,NSD,**E**:FL:PS☠(causes dermatitis)	Clev-081-02

Latin Genus	L. Species	E. Species	English Genus	C:Culinary,M:Medicinal,E:Economic	Source
Armeria	*maritima*	Sea Pink	Thrift	**M(traditional):**FL:DASC,NSD,**E:**FL:PS☠(causes dermatitis)	Clev-081-01
Amoracia	*rusticana*	Red Cole	Horseradish	**C:**LF:FF,**C:**RT:**SP**,**M(traditional):**LF,RT:HFV,NJA,NGT,NSC,OJA,OPC,OWN,RLP,**E:**LF,RT:PS☠	Bown-242-09
Amoracia	*rusticana*	Red Cole	Horseradish	**C:**RT:**SP**,**M(traditional):**LF,RT:BCR,DASC.NRH,SDG	Clev-081-03, Clev-274-01
Amoracia	*rusticana*	Red Cole	Horseradish	**C:**RT:**SP**,**M(traditional):**RT-OL:BCR,DABI, DCLD,NNR, RLC,RST,SAP,SDG,UDR	Kowa-289-02 Kowa-338-01
Amoracia	*rusticana*	Red Cole	Horseradish	**C:**RT:**SP**,**M(folk):**RT:BCB,BCR,OJA,ORH,SAP, SDG,UKC,UDR	Krug-092-01
Amoracia	*rusticana*	Red Cole	Horseradish	**C:**LF:**SP**,**M(PS):**RT:ORH,RLB,UUR	Wyk-052-01, Wyk-399-19
Arnica	*chamisonis*	Chamise	Leopard's Bane	**M(traditional):**FL,LF,RT:OWN,**E:**FL,LF,RT:PS☠	Wyk-399-20a
Arnica	*fulgens*	Shining	Leopard's Bane	**M(traditional):**FL,LF,RT:OWN,**E:**FL,LF,RT:PS☠	Wyk-399-20b
Arnica	*montana*	Mountain	Leopard's Bane	**M(traditional):**FL-OL,LF,RZ,RT:BPH,DCLD, DISB,HFV,NBF,NCT,NJA,NRH, OWN,RLB,RST, **M(PS):**FL-OL,LF,RZ,RT:DABI,DASC,NSD	Bal2-025-01 Bal2-504-01
Arnica	*montana*	Mountain	Leopard's Bane	**M(traditional):**FL:DABA,DAFU,BAS,BCB,BHA,BHS,OAS,OBR,OUL,RST,**E:**FL:PS☠	Bown-242-10
Arnica	*montana*	Mountain	Leopard's Bane	**M(traditional):**WP(top):OBR,RST,**E:**WP:PS☠	Clev-081-04
Arnica	*montana*	Mountain	Leopard's Bane	**M(traditional):**WP,FL(top):DHMY,NSD,OBR,**E:** WP:PS☠	Kowa-017-01 Kowa-283-02
Arnica	*montana*	Mountain	Leopard's Bane	**M(folk):**FL:OBR	Krug-018-01
Arnica	*montana*	Mountain	Leopard's Bane	**M(PS):**WP(top):DAMW,OBN,OBR,ORH,OWN, PS☠	Wyk-053-01, Wyk-400-01
Artemisia	*abrotanum*	Southernwood	Wormwood	**M(traditional):**LF:BLC,DWTH,HFV,NMS,OSC,OTM,SAP,SDG	Bown-243-01
Artemisia	*abrotanum*	Southernwood	Wormwood	**M(traditional):**LF:DASC,HAP,**E:**LF,SH:IR(mothballs)	Clev-082-01
Artemisia	*abrotanum*	Southernwood	Wormwood	**M(traditional):**LF:DACN,DASC,DCTM,DWRM,HAP,UDR,**E:**OL:IR(bee),PF,**E:**SM:DYE(tin-yellow)	Kowa-470-01
Artemisia	*abrotanum*	Southernwood	Wormwood	**M(folk):**WP:DIPT,**E:**LF-OL:IR(moth)	Krug-165-01
Artemisia	*abrotanum*	Southernwood	Wormwood	**C:**LF:**SP**,**M(traditional):**WP:DWRM	Wyk-400-02
Artemisia	*absinthium*	Absinthe	Wormwood	**M(traditional):**WP:BLC,DISB,DSNB,DWRO,OBR,SAP,SDG,UDR	Bown-243-02
Artemisia	*absinthium*	Absinthe	Wormwood	**M(traditional):**WP:BCR,BJA,DASC,DABA,DAFU,DCLD,DWRM,HFV,NMS,NRH,NSD,OBR,OJA, ONR,ORH,SDG,**E:**OL:IR,PS☠	Kowa-510-01
Artemisia	*absinthium*	Absinthe	Wormwood	**C:**WP:AF(vermouth),**M(folk):**WP(flowering):BCR,DWRM,ORH,SAP,SDG, facilitates childbirth, **E:**WP:IR	Krug-187-01
Artemisia	*absinthium*	Absinthe	Wormwood	**M(PS):**WP(top):DSIN,SAP,SDP,SGB,SIB	Wyk-054-01, Wyk-400-03
Artemisia	*absinthium*	Absinthe	Wormwood	**E:**WP:DYE(alum-yellow-green,chrome-gold,copper-sulfate-soft-green,iron-khaki)	Cas1-089-01
Artemisia	*afra*	African	Wormwood	**M(traditional):**WP:HFV,SDG	Bown-243-03
Artemisia	*afra*	African	Wormwood	**M(PS):**LF,SM:DCLD,DFLU,DWRM,HAC,NGT,NSD,RLA,RLC,RST,SCN,SDG,SFL,SIB,**E:**WP:PS☠	Wyk-055-01, Wyk-400-04
Artemisia	*annua*	Sweet Annie	Wormwood	**M(traditional):**WP:DABA,DMAL,HFV,OAS	Bown-243-04
Artemisia	*annua*	Sweet Annie	Wormwood	**M(CS):**WP:DMAL,HFV,SDP	Wyk-056-01, Wyk-400-05
Artemisia	*anomala*	Anomalous	Wormwood	**M(traditional):**WP:OBN,DSIN	Bown-243-05
Artemisia	*arborescens*	Bush	Wormwood	**E:**LF:PF	Bown-243-06
Artemisia	*californica*	California Mugwort	Wormwood	**E:**WP:DYE(alum-chartreuse,alum-&-iron-yellow-green,chrome-&-iron-gold)	Grae-098-01
Artemisia	*capillaris*	Chinese	Wormwood	**M(traditional):**WP(top):DWRM,SCN,SDR,UBL,UUR, **M(PS):**WP(top):DABI,BLC	Bal2-026-01 Bal2-504-02
Artemisia	*capillaris*	Chinese	Wormwood	**M(traditional):**LF,SH:BJA,BLC,DHPT,HFV,SGB, UDR	Bown-243-07
Artemisia	*caucasica*	Caucasian	Wormwood	**E:**LF:PF	Bown-243-08
Artemisia	*cina*	Levant	Wormwood	**M(traditional):**FL:DWRO	Bown-243-09
Artemisia	*cina*	Levant	Wormwood	**M(traditional):**FL:DWRM	Wyk-400-06
Artemisia	*dracunculoides*	Russian Tarragon	Wormwood	**C:**LF:**SP**,TE,**C:**OL:FF	Clev-082-03
Artemisia	*dracunculus*	French Tarragon	Wormwood	**C:**LF:**SP**,TE,**C:**OL:FF,**M(traditional):**LF,OL:DWRM,HFV,SDG,UDR,**E:**OL:PF,DS	Bown-243-10
Artemisia	*dracunculus*	French Tarragon	Wormwood	**C:**LF:AF,**SP**,**M(traditional):**LF:DSNB,HIN,OTA, SAP,SDG	Clev-082-02
Artemisia	*dracunculus*	French Tarragon	Wormwood	**C:**LF:**SP**,**M(traditional):**LF:DFRA,NRH,OTA,SAP,SFL,SIB,**E:**LF:DS,CM,PF	Kowa-482-01
Artemisia	*dracunculus*	French Tarragon	Wormwood	**C:**LF:**SP**,**M(folk):**RT:OTA	Krug-172-01

Latin Genus	L. Species	E. Species	English Genus	C:Culinary,M:Medicinal,E:Economic	Source
Artemisia	*frigida*	Prairie Sagebrush	Wormwood	**E:**LF,SL:DYE(alum-light-yellow-green,chrome-brass-gold,copper-olive,tin-yellow,iron-dark-moss-green)	Bliss-210-01
Artemisia	*lactiflora*	White Mugwort	Wormwood	**M(traditional):**LF,SM:BLC,**E:**LF,SM:PF	Bown-243-11
Artemisia	*ludoviciana*	White Sagebrush	Wormwood	**E:**LF,SM:PF	Bown-244-01
Artemisia	*maritima*	Sea	Wormwood	**M(traditional):**LF:DWRM	
Artemisia	*pontica*	Roman	Wormwood	**E:**LF:AF	Bown-244-03
Artemisia	*tridentata*	Big Sagebrush	Wormwood	**E:**FL-BD,LF,SM:DYE(alum-gold,-chrome-rust,copper-ocher,tin-bright-gold,iron-black-gray)	Bliss-212-01
Artemisia	*tridentata*	Big Sagebrush	Wormwood	**E:**FL-BD,LF,SM:DYE(alum-&-fleece-brilliant-yellow,alum-&-iron-sage-green,chrome-tan-gold,chrome-&-iron-dark-green	Grae-107-02
Artemisia	*verlotiorum*	Asian Mugwort	Wormwood	**M(traditional):**WP:NRH,SIB	Clev-082-08
Artemisia	*vulgaris*	Mugwort	Wormwood	**M(traditional):**LF:DWRM,HAC,HFV,NSD	Bal1-079-01
Artemisia	*vulgaris*	Mugwort	Wormwood	**C:**LF:**SN**,**M(traditional):**LF:DAFU,DWRO,DWTH,SAP,SDG,SDP,UDR	Bown-244-04
Artemisia	*vulgaris*	Mugwort	Wormwood	**C:**LF:AF,FF,**M(traditional):**LF:NRH,SDG,SFL,SIB,**E:**WP:IR	Clev-082-04
Artemisia	*vulgaris*	Mugwort	Wormwood	**E:**LF:CM,IR(moth)	Kowa-392-01
Artemisia	*vulgaris*	Mugwort	Wormwood	**C:**LF:AF,**M(folk):**LF,SM:MCF,MCI,OMC,ORH	Krug-120-01
Artemisia	*vulgaris*	Mugwort	Wormwood	**M(traditional):**LF:DWRO,DWTH,HAN,HIN,SAP,SDP	Wyk-057-01, Wyk-400-07
Arum	*maculatum*	Lords & Ladies	Ar Lily	**M(traditional):**RZ:OUL,SDG	Wyk-400-08
Arundo	*donax*	Giant	Reed	**M(traditional):**RZ:UDR	Wyk-400-09
Asarum	*canadense*	Canadian	Ginger Loch	**M(traditional):**RZ:DABI,NRH,RLA,RLC,SDG,UDR	Bown-244-05
Asarum	*europaeum*	European	Ginger Loch	**M(traditional):**RZ:DABA,NSD,UDR,RLC,SDG,**E:**RZ:PS☠	Wyk-400-10a
Asarum	*sieboldii*	Siebold	Ginger Loch	**M(traditional):**RZ:DABA,NSD,UDR,RLC,SDG,**E:**RZ:PS☠	Wyk-400-10b
Asclepias	*fruticosa*	Bushy	Milkweed	**M(traditional):**RT:BHS	Wyk-400-11
Asclepias	*speciosa*	Showy	Milkweed	**E:**FL,LF:DYE(alum-lemon-yellow,chrome-orange-brown,copper-olive-green,tin-bright-gold-yellow,iron-gray-green)	Bliss-166-01
Asclepias	*speciosa*	Showy	Milkweed	**E:**FL,LF:DYE(alum-&-copper-moss-green,chrome-&-copper-brass-green)	Grae-097-01
Asclepias	*tuberosa*	Pleurisy Root	Milkweed	**M(traditional):**RT:DPNB,DPNV,DRMF,HFV,NMS,OBR,ORH,OWN,OUL,RLA,RLB,RLC,RLP,SIB,**E:**RT:PS☠	Bown-244-06
Asclepias	*tuberosa*	Pleurisy Root	Milkweed	**M(folk):**RT:RLC	Krug-140-01
Asclepias	*tuberosa*	Pleurisy Root	Milkweed	**M(PS):**RT:DFLU,DPNB,RLB,RLP	Wyk-058-01, Wyk-400-12
Ascophyllum	*nodosum*	Knotted Wrack	Sac Painted Dusky Brown Ocher Algae	**C:**WP,TE,**M(traditional):**WP:iodine therapy	Wyk-400-13
Aspalathus	*linearis*	Rooibos	Lathybush	**C:**LF,SM:BF,**M(traditional):**LF,SM:DAVR,DSXS,HAN,HIN,**M(PS):**LF,SM:DAID,DALL,DFRA,NMS	Bal2-123-01 Bal2-539-01
Aspalathus	*linearis*	Rooibos	Lathybush	**C:**LF,SM:AF,BF,FF,**M(traditional):**LF,SM:DHYF,OSA,DSIN,RLA	Bown-244-07
Aspalathus	*linearis*	Rooibos	Lathybush	**M(PS):**LF,SM:DFRA,NMS,DSIN, **E:**LF,SM:CM	Wyk-059-01, Wyk-400-14
Asparagus	*cochinchinesis*	Cochin	Asparagus	**M(traditional):**TU:DABA,DDPT,DTBC,HFV,RLB,RLC,RRN,RST,UDR	Bown-245-01
Asparagus	*officinalis*	Vegetable	Asparagus	**C:**SL:**VG**,**M(traditional):**RZ,SL:BHE,BLC,DWFK,NGT,NRH,UBL,UKC,UKP	Bown-245-02
Asparagus	*officinalis*	Vegetable	Asparagus	**C:**SL:**VG**,**M(traditional):**RZ,SL:BHE,NRH,SCN,UKC	Clev-083-01
Asparagus	*officinalis*	Vegetable	Asparagus	**M(folk):**SM,RT:UDR,UKC	Krug-019-02
Asparagus	*officinalis*	Vegetable	Asparagus	**M(PS):**RZ:UDR,UKS,UUR	Wyk-060-01, Wyk-400-15
Asparagus	*officinalis*	Vegetable	Asparagus	**E:**WP:DYE(alum-light-golden-yellow,chrome-ocher,copper-yellow-green,tin-light-gold,iron-olive-green)	Bliss-038-01
Asparagus	*racemosus*	Shatavari	Asparagus	**M(traditional):**RZ:DFRI,DIFR,OJA,RLB,SDY,SUL	Bown-245-03
Asparagus	*racemosus*	Shatavari	Asparagus	**M(traditional):**RT:HAP,NMS,NSD	Wyk-400-16
Asperula	*odorata*			Synonym for *Galium odoratum*	Grae-189-02
Asperula	*tinctoria*	Dyer's	Woodruff	**E:**WP:DYE(yellow-orange)	Clev-093-06
Aspidosperma	*quebracho-blanco*	White	Quebracho	**M(traditional):**BK:BCR,HFV,NMS,RLA,RLB,RLE,**E:**BK:PS☠	Bown-245-04
Aspidosperma	*quebracho-blanco*	White	Quebracho	**M(traditional):**BK:HFV,RLA,RLB,RLC,OBN,OWN	Wyk-400-17
Asplenium	*scolopendrium*	Hart's Tongue	Spleenwort Fern	**M(traditional):**WP:OWN,RLC,SDG,UDR	Wyk-400-18

Latin Genus	L. Species	E. Species	English Genus	C:Culinary,M:Medicinal,E:Economic	Source
Aster	*amellus-purpurea*	Purple Italian	Daisy	**E:**LF,FL,SM:DYE(alum-green-yellow,chrome-dark-pumpkin-orange,copper-green-gold,tin-brightyellow,iron-dark-moss-green)	Bliss-042-01
Aster	*alpinus-purpurea*	Purple Alpine	Daisy	**E:**LF,FL,SM:DYE(alum-green-yellow,chrome-dark-pumpkin-orange,copper-green-gold,tin-brightyellow,iron-dark-moss-green)	Bliss-042-01
Aster	*frikartii-purpurea*	Purple Frickart	Daisy	**E:**LF,FL,SM:DYE(alum-green-yellow,chrome-dark-pumpkin-orange,copper-green-gold,tin-bright-yellow,iron-dark-moss-green)	Bliss-042-01
Aster	*novi-belgii-viola*	Violet Michelmas	Daisy	**E:**LF,SM:DYE(alum-yellow,copper-sulfate-&-iron-yellow-green,chrome-gray,tin-bright-yellow)	Cas1-095-01
Aster	*nemoralis*			Synonym for *Oclemenane moralis*	Cas1-095-02
Aster	*porteri*	White Porter	Daisy	**E:**LF,FL,SM:DYE(alum-gold,chrome-gold-brown,copper-green-gold,tin-bright-gold,iron-black-gold)	Bliss-044-01
Aster	*tartaricus*	Tartarian	Daisy	**M(traditional):**RT:DTBC,RLB,RLC	Bown-245-05
Astracantha	*gummifera*	Gum Tragacanth	Star Thorn	**C:**RN:MC,**E:**RN:GR	Wyk-400-19
Astragalus	*complanatus*	Flat	Milk Vetch	**M(traditional):**RT:BLC,UKC	Bown-245-06
Astragalus	*gummifer*	Gum	Milk Vetch	**C:**RN:GR	Bown-245-07
Astragalus	*gummifer*	Gum	Milk Vetch	**E:**RN:**contraceptive**☠	Kowa-283-06
Astragalus	*membranaceus*	Huang Qi	Milk Vetch	**M(traditional):**RT:DAID,DCFS,DCTM,DCLD,DFLU,SDG	Bal1-065-03
Astragalus	*membranaceus*	Huang Qi	Milk Vetch	**M(CS):**RT:BAA,BAS,BDB,BHA,BHS,BLC,DCLD,DCTM,DFRA,DIFR,SDG	Bal2-028-01 Bal2-504-05
Astragalus	*membranaceus*	Huang Qi	Milk Vetch	**M(traditional):**RT:BCR,BDB,BHP,BSH,OSA,UKC	Bown-246-01
Astragalus	*membranaceus*	Huang Qi	Milk Vetch	**M(CS):**RT:BDB,DCLD,OWN,UKS	Wyk-061-01
Astragalus	*mongolicus*	Mongolian	Milk Vetch	**M(traditional):**RT:BCR	Wyk-400-20
Atractylodes	*macrocephala*	Chinese	Thistle Daisy	**M(traditional):**RZ:DABA,BSH,DEGC,SDG,UDR	Bown-246-02
Atriplex	*halimus*	Saltbush	Orache	**M(traditional):**LF:BDB,RLC,SDP,UUR	Wyk-401-01
Atriplex	*hortensis*	Mountain Spinach	Orache	**C:**LF:**VG**,**M(traditional):**LF:DCFS	Bown-246-03
Atriplex	*hortensis*	Mountain Spinach	Orache	**C:**LF:**VG**,**M(traditional):**LF:DCFS	Clev-083-02
Atriplex	*hortensis*	Mountain Spinach	Orache	**M(traditional):**WP:BDB,DHOG,RLC,SDP,UUR	Wyk-401-01
Atriplex	*patula*	Spear	Orache	**C:**LF:**VG**	Clev-083-03
Atriplex	*prostrata*	Hastate	Orache	**C:**LF:**VG**	Clev-083-04
Atropa	*belladonna*	Deadly	Nightshade	**M(traditional):**WP:DHMY,BHA,DPRK,NMP,NMS,NNC💀,NSD,OMC,RLA,SGS,SUL,UKS,**E:**WP:PS☠	Bown-246-04
Atropa	*belladonna*	Deadly	Nightshade	**M(traditional):**WP:DEPI,DPRK,DWHP,HAN,NMS,NNC💀,SUL,UKS,**E:**WP:PS☠	Kowa-158-01 Kowa-280-01
Atropa	*belladonna*	Deadly	Nightshade	**M(folk):**FR,LF,RT:DEPI,DPRK,DWHP,NNC💀	Krug-028-01
Atropa	*belladonna*	Deadly	Nightshade	**M(PS):**LF,RT:HAP,HPS,NMS,NSD,RLA	Wyk-062-01, Wyk-401-02
Avena	*sativa*	Cereal	Oats	**M(traditional):**WP(top):HAN,HIN,NSD	Bal1-075-01
Avena	*sativa*	Cereal	Oats	**M(traditional):**WP(top):ADD,DCFS,DIPT,DSIN,HIN,NSD, **M(PS):**WP(top):HAP	Bal2-030-01 Bal2-506-01
Avena	*sativa*	Cereal	Oats	**C:**SD:**VG**,**M(traditional):**SD:DHSV,DMHI,DSHZ,OSS,DSIN	Bown-246-05
Avena	*sativa*	Cereal	Oats	**C:**SD:**VG**,**M(folk):**SD:DSIN,HAN,SDG	Krug-128-01
Avena	*sativa*	Cereal	Oats	**M(traditional):**RD(husk):BHC,**M(traditional):**SD(grain):CFS,**M(PS):**SM(straw):DSIN	Wyk-063-01, Wyk-401-03
Azadirachta	*indica*	India	Neem	**M(traditional):**BK,LF,OL,SD,RN:DAFU,BJA,DMAL,DTBC,DWRG,DWRM,HFV,NJA,NRH,NSD,OSA,DSIN,**E:**LF,OL:IC,**E:**RN:CM,DS	Bown-247-01
Azadirachta	*indica*	India	Neem	**M(PS):**BK,LF,SD,SM:DMAL,DWRM,OHM,DSIN,OWN,SDG,**E:**BK,LF,SD,SM:CM,DS	Wyk-064-01, Wyk-401-04
Bacopa	*monnieri*	Brahmi	Water Hyssop	**M(traditional):**WP(top):HML,**M(ayurveda):**WP(top):BLC,DSIN,DWHN,HFV,**M(PS):**WP(top):ADD,DALZ,DPRK,HAN,HML,NSD,SIB	Bal2-036-01, Bal2-507-04
Bacopa	*monnieri*	Brahmi	Water Hyssop	**M(traditional):**WP:NSD,RLB,SDP	Wyk-401-05
Baeomyces	*fungoides*			Synonym for *Dibaeis baeomyces*	Cas2-031-04
Ballota	*africana*	African	Black Horehound	**M(traditional):**WP:DCLD,HFV,HIN,RLC	Wyk-401-06
Ballota	*nigra*	Cowbane	Black Horehound	**M(traditional):**WP:NMS,RLB,SDP	Bown-247-02
Ballota	*nigra*	Cowbane	Black Horehound	**M(traditional):**WP,FL(top):HIN,NMS,NSD,RLC,SDP	Wyk-065-01, Wyk-401-07
Bambusa	*arundinacea*	Spiny	Bamboo	**M(traditional):**LF,RT:NMP,NMS,UDR	Wyk-401-08
Banisteriopsis	*caapi*	Ayahuasca	Banister Cherry	**M(traditional):**WD:NNC💀	Wyk-401-09
Baptisia	*alba*	White	False Indigo	**M(traditional):**RT:OWN	Bown-247-03
Baptisia	*australis*	Blue	False Indigo	**M(traditional):**RT:OWN	Bown-247-04
Baptisia	*bracteata*	Long Bract	False Indigo	**M(traditional):**RT:OWN	Bown-247-05
Baptisia	*tinctoria*	Dyer's	False Indigo	**M(traditional):**RT:DABA,DASC,HFV,OWN,SCN	Bown-247-06
Baptisia	*tinctoria*	Dyer's	False Indigo	**M(traditional):**RT:DABI,HFV,OWN,SCN	Wyk-401-10

Latin Genus	L. Species	E. Species	English Genus	C:Culinary,M:Medicinal,E:Economic	Source
Begonia	*rex-culturum*	Hybrid Rex	Begonia	**E:**FL:DYE(iron-gray-green,alum-&-tin-yellow-orange,chrome-gold)	Cas1-104-01
Belamcanda	*chinensis*	Chinese	Blackberry Lily	**M(traditional):**RZ:DABA,DAFU,DAVR,HFV,RLC,RST	Bown-247-07
Bellis	*perennis*	Perennial	English Daisy	**C:**FL-BD,FL,LF:**SN**,**M(traditional):**LF,FL:OAS,OBR,OWN,RLC	Bown-248-01
Bellis	*perennis*	Perennial	English Daisy	**M(traditional):**WP,FL:RLC,SDB,SGB	Wyk-401-11
Benincasa	*hispida*	Squash	Benincasa Melon	**C:**FR:**VG**,JC,**C:**SD:NT,**M(traditional):**FR:RLA,RLC,SUL,**M(traditional):**RD:BDB,HFV,UDR,**M(traditional):**SD:DWTP,DVGB,DVGV,RLC,**E:**RD,WX	Bown-248-02
Benincasa	*hispida*	Squash	Benincasa Melon	**M(traditional):**FR:DWRM,RLB,SCN,UDR	Wyk-401-12
Berberis	*aristata*	Wood Turmeric	Barberry	**M(traditional):**LF,FR,RT,SM:ABA,OEY,SDR	Bown-248-03
Berberis	*asiatica*	Daruharidra	Barberry	**M(traditional):**LF,FR,RT,SM:ABA,OEY,SDR	Bown-248-04
Berberis	*canadensis*	Canadian	Barberry	**E:**LF,SM:DYE(chrome-olive-green,chrome-&-tin-terra-cotta-brown)	Grae-147-01
Berberis	*vulgaris*	European	Barberry	**M(traditional):**LF,FR,RT:DSIN,RLB,SDG	Bal1-065-04
Berberis	*vulgaris*	European	Barberry	**M(traditional):**FR,RT-BK,SM-BK:BLC,DLEI,DMAL, DWHN,HFV,OWN,RSN,RST,SGS,UBL,UKS,UUR, **M(PS):**FR,RT-BK,SM-BK:DABA(*Pseudomonas, Salmonella,Shigella,Staphylococcus, Streptococcus, Vibrio*),SDR, SIB,SUL	Bal2-031-01 Bal2-506-02
Berberis	*vulgaris*	European	Barberry	**C:**FR:JJ,**M(traditional):**LF,FR,RT,SM:BHP,BLC,DHPT,DLEI,DMAL,HFV,NSD,SDY,SGS	Bown-248-05
Berberis	*vulgaris*	European	Barberry	**M(traditional):**FR:RST,**M(traditional):**RT,RZ:BLC,DABA,HFV,OAS,SCN,SDR,SPR,**M(traditional):**WP:NMS,NSD,**E:**BK,RT:DYE(tin-yellow),**E:**WP:DYE(iron-dark-green,alum-tan)	Kowa-020-01 Kowa-173-02 Kowa-283-01
Berberis	*vulgaris*	European	Barberry	**M(folk):**RB:DABA,SCN,**M(folk):**FR:DAMW,NSD	Krug-023-01
Berberis	*vulgaris*	European	Barberry	**M(PS):**FR,RB,RT:DHPT,BLC,OEY,SDG,SGB,SGS,UKC,UUR	Wyk-066-01, Wyk-401-13
Berberis	*vulgaris*	European	Barberry	**E:**LF,SM:DYE(chrome-olive-green,chrome-&-tin-terra-cotta-brown)	Grae-146-02
Bergenia	*crassifolia*	Elephant Ear	Paddle Saxifrage	**M(traditional):**LF,RT:DASC,OAS	Wyk-401-14
Beta	*vulgaris*	Mediterranean	Beet	**E:**BR:DYE(direct-deep-red,direct-magenta)	Grae-056-01
Beta	*vulgaris*	Mediterranean	Beet	**E:**LF:DYE(alum-&-tin-yellow,chrome-&-vinegar-tan)	Cas1-102-02
Beta	*vulgaris-conditiva*	Globe	Beet	**M(traditional):**BR:BLC,UKC	Wyk-401-15
Betula	*alba*	White	Birch	**M(traditional):**LF:DABA(*Staphylococcus aureus*),DBOI,DSCB,DSIN,NJA,NMS,NRH,ODS,OHL,OSS,UKC	Bal2-032-04
Betula	*alba*	White	Birch	**M(traditional):**LF:UDR	Wyk-401-16
Betula	*alleghaniensis*	Yellow	Birch	**E:**LF:DYE(alum-gold,tin-yellow,chrome-tan), **E:**BK (outer):DYE(copper-sulfate-&-chrome-tan-yellow, **E:**BK(inner):DYE(tin-orange,iron-purple-brown)	Cas1-104-02
Betula	*lenta*	Sweet	Birch	**M(traditional):**LF:DABA(*Staphylococcus aureus*),DBOI,DSCB,DSIN,NJA,NMS,NRH,ODS,OHL,OSS,UKC	Bal2-032-03
Betula	*lenta*	Sweet	Birch	**E:**OL:PF	Bown-248-06
Betula	*nigra*	Black	Birch	**M(traditional):**LF:DABA(*Staphylococcus aureus*),DBOI,DSCB,DSIN,NJA,NMS,NRH,ODS,OHL,OSS,UKC	Bal2-032-01
Betula	*nigra*	Black	Birch	**M(traditional):**OL:OSS,OWN	Bown-248-07
Betula	*pendula*	Silver	Birch	**M(traditional):**BK,LF,SA:DBOI,NJA,NSD,DSIN,UDR,UUR	Bal1-065-07
Betula	*pendula*	Silver	Birch	**M(traditional):**LF:DBOI,DSCB,NJA,NMS,NRH,ODS,OHL,OSS,UKC, **M(PS):**LF:DABA(*Staphylococcus aureus*),DSIN	Bal2-032-02 Bal2-506-04
Betula	*pendula*	Silver	Birch	**C:**SA:AF,**M(traditional):**BK,LF,OL:BAS,HFV,NGT,NJA,NRH,UBL,UDR,UKS,**M(traditional):**OL:OSS,OAS,OSA,DSIN,OSS	Bown-248-08
Betula	*pendula*	Silver	Birch	**C:**BK(inner),SM:BF,**M(traditional):**LF:NBK,NJA,NRH,DSIN,OWN	Kowa-044-01
Betula	*pendula*	Silver	Birch	**M(PS):**BK-OL:ODS,DSIN, **M(traditional):**LF:BHE,NGT,NRH, UBL,UUR,UDR	Wyk-067-01
Betula	*pubescens*	Downy	Birch	**M(traditional):**OL:OSS,OWN	Bown-248-09
Bidens	*pilosa*	Blackjack	Bur Marigold	**M(traditional):**WP:OAS,SDG,SDR,UDR	Wyk-401-17
Bidens	*tripartita*	Water Agrimony	Bur Marigold	**M(traditional):**WP:OAS,OHL,SIB,SUL,UDR	Bown-249-01
Biota	*orientalis*			Synonym for *Thuja orientalis*	Murr-161-01
Bixa	*orellana*	Lipstick Tree	Annatto	**C:**FR:PU,**C:**SD:**SP**,**M(traditional):**FR:DCMT,**M(traditional):**LF:DWRM,HFV,SIB,**M(traditional):**SD:OBN	Bown-249-02

Latin Genus	L. Species	E. Species	English Genus	C:Culinary,M:Medicinal,E:Economic	Source
Bixa	*orellana*	Lipstick Tree	Annatto	**C:**SD:**SP**,**M(traditional):**SD:HFV,SDY,**E:**SD-PU:DYE(oxblood-red),IR	Clev-275-01
Bixa	*orellana*	Lipstick Tree	Annatto	**M(traditional):**SD:OAS,OBN,OWN,E:CM,DYE	Wyk-401-18
Bletilla	*striata*	Striped	Little Blet Orchid	**M(traditional):**BR:ABA,DTBC,OAS,OBN,OSA,SUL	Bown-249-03
Boerhavia	*diffusa*	Red	Spiderling	**M(traditional):**LF,RT:DWRM	Wyk-401-19
Boletus	*edulis*	Edible	Bolete	**E:**WP:DYE(chrome-rust,chrome-&-tin-warm-brown)	Cas1-182-02
Borago	*officinalis*	Herb of Gladness	Borage	**M(traditional):**LF,SD:BCR,DMHI	Bal1-066-04
Borago	*officinalis*	Herb of Gladness	Borage	**C:**LF:AF,BF,**VG**,**C:**FL:**SN**,**M(traditional):**SD,OL:BLC,DAMW,DTBC,HFV,BHP,OSS,OEY,RLB,RLP,UDR,UKC	Bown-249-04
Borago	*officinalis*	Herb of Gladness	Borage	**C:**LF:**VG**,**C:**FL:**VG**,**C:**RT:AF,**M(traditional):**LF,SD:DCLD,HAN,RLC	Clev-083-05
Borago	*officinalis*	Herb of Gladness	Borage	**C:**LF:**VG**,**M(traditional):**FL:HFV,RLB,SDR,**M(traditional):**LF:DSIN	Kowa-052-01
Borago	*officinalis*	Herb of Gladness	Borage	**C:**LF:**VG**,**M(folk):**FL:HAN,CFS	Krug-038-01
Borago	*officinalis*	Herb of Gladness	Borage	**M(PS):**FL:HAN,UDR,NSD,RLC,**M(traditional):**SD:ODS,DSIN	Wyk-068-01, Wyk-401-20
Boswellia	*frereana*	Frere Oliban	Frankincense	**M(traditional):**RN:DASC,NSD,RLC	Wyk-401-21
Boswellia	*papyrifera*	Ethiopian	Frankincense	**M(traditional):**RN:DASC,NSD,RLC	Wyk-402-01
Boswellia	*sacra*	Sacred Oliban	Frankincense	**M(traditional):**RN:BCR,DAMW,DASC,DGUM,NMP,OBR,OSA,OSW,RLB,RLC,RRN,RST	Bown-250-01
Boswellia	*sacra*	Sacred Oliban	Frankincense	**M(PS):**BK-RN:BCR,DASC,DHMY,NSD,ORH,RLC	Wyk-069-01, Wyk-402-02
Boswellia	*serrata*	India Oliban	Frankincense	**M(traditional):**RN:DSNB,RLA,RLC,SIB, **M(PS):**RN:DCRO,NJA,SUL, **E:**OL,RN:CM,PF	Bal2-035-01 Bal2-507-03
Boswellia	*serrata*	India Oliban	Frankincense	**M(traditional):**RN:DASC,NJA,NSD,RLC	Wyk-402-03
Brassica	*alba*			Synonym for *Sinapis alba*	Clev-276-01
Brassica	*juncea-indica*	Light Brown	Mustard	**C:**LF:**VG**,**C**:SD:**SP**	Bown-250-02
Brassica	*juncea-indica*	Light Brown	Mustard	**C**:SD:**SP**	Clev-276-03
Brassica	*kaber*			Synonym for *Sinapis arvensis*	Cas1-183-01
Brassica	*nigra*	Dark Brown	Mustard	**M(traditional):**SD:OCC,OJA,SDG	Bal1-074-05
Brassica	*nigra*	Dark Brown	Mustard	**C:**FL,LF:**SN**,**C:**SD:**SP**,**M(traditional):**OL:BCB,ORH,RRN	Bown-250-03
Brassica	*nigra*	Dark Brown	Mustard	**C:**SD:**SP**	Clev-276-02
Brassica	*nigra*	Dark Brown	Mustard	**C:**SD:**SP**,**M(traditional):**SD:BCR,HAC,NRH,OTA,SDG,**M(traditional):**SD-OL:SCN	Kowa-393-01
Brassica	*nigra*	Dark Brown	Mustard	**C:**SD:**SP**,**M(folk):**SD:HAC,OJA,ORH,OTA,RLC	Krug-122-01
Brassica	*nigra*	Dark Brown	Mustard	**M(PS):**SD:BCR,SDG,**M(traditional):**SD-OL:ORH,DSIN, **E:**OL:PS☠	Wyk-070-01, Wyk-401-04
Brassica	*oleracea-capitata-purpurea*	Purple Cabbage	Mustard	**E:**LF:DYE(alum-blue,alum-&-tin-lavender)	Grae-175-03
Brickellia	*grandiflora*	Tassel Flower	Brickell Daisy	**E:**SM:DYE(alum-light-gold,chrome-gold-brown,copper-brass-gold,tin-bright-yellow-gold,iron-dark-yellow-gray)	Bliss-252-01
Brucea	*javanica*	Java Seed	Kusam	**M(traditional):**SD:DMAL,DVGT,DCLW,SDY	Bown-250-03
Brucea	*javanica*	Java Seed	Kusam	**M(PS):**FR,SD:DBOI,DMAL,DWRG,DWRO,DWTP,DWWP,SDY	Wyk-071-01, Wyk-401-05
Brunfelsia	*uniflora*	Vegetable Mercury	Brunfels	**M(traditional):**RT:DSXS,BJA,NRH	Bown-251-01
Bryocaulon	*divergens*	Divergent	Fox Hair Shield Lichen	**E:**WP:DYE(Olivetoric-Acid-orange)	Cas2-034-20
Bryonia	*alba*	White	Bryony	**M(traditional):**RT:BCR,NJA,NRH,NSD,OJA,RLA,RLB,RLP,SUL	Bown-251-02
Bryonia	*alba*	White	Bryony	**M(traditional):**RT:DHMY,ORH,**E:**RT:PS☠	Wyk-402-06
Bryonia	*dioica*	English Mandrake	Bryony	**M(traditional):**RT:BCR,NJA,NRH,NSD,OJA,RLA,RLB,RLP,SUL	Bown-251-03
Bryonia	*dioica*	English Mandrake	Bryony	**M(PS):**RT:DHMY,ORH,**E:**RT:PS☠	Wyk-072-01
Bryoria	*fremontii*	Fremont	Horse Hair Shield Lichen	**E:**WP:DYE(Vulpinic-Acid-tan-brown,Vulpinic-Acid-&-Ammonia-green)	Cas2-031-05
Bryoria	*fuscescens*	Fuscous	Horse Hair Shield Lichen	**E:**WP:DYE(Fumarprotocetraric-Acid-yellow)	Cas2-034-08
Buddleja	*davidii*	Orange Eye	Butterfly Bush	**E:**FL:DYE(alum-&-tin-orange-gold,alum-&-tin-gold-green,chrome-&-tin-gold-brown), **E:**FL,LF,SM:DYE(alum-&-iron-green,chrome-&-iron-black-green)	Grae-148-02
Bupleurum	*chinense*	Northern	Thorow Wax	**M(traditional):**RT:BCR,BLC,HFV,DALL,DCBN,DCLD,DMAL	Bal2-038-01 Bal2-508-01
Bupleurum	*falcatum*	Hare's Ear	Thorow Wax	**M(traditional):**RT:BCR,BLC,DAVR,DHSV,DMAL,OHM	Bown-251-04
Bupleurum	*falcatum*	Hare's Ear	Thorow Wax	**M(PS):**RT:BLC,DCLD,DFLU,DHPT,HFV	Wyk-073-01,

Latin Genus	L. Species	E. Species	English Genus	C:Culinary,M:Medicinal,E:Economic	Source
					Wyk-402-07
Bupleurum	*scorzoneraefolium*	Southern	Thorow Wax	**M(traditional):**RT:BCR,BLC,HFV,DALL,DCBN,DCLD,DMAL	Bal2-038-02
Bursera	*delpechiana*	Delpech	Torchwood	**M(traditional):**FR,OL,WD:ABI	Bown-251-05
Bursera	*glabrifolia*	Linalol	Torchwood	**C:**OL:FF,**M(traditional):**FR,OL,WD:DACN,DASC,NMS,DSIN,OWN,**E:**OL:PF	Bown-251-06
Bursera	*klugii*	Klug	Torchwood	**M(traditional):**FR,OL,WD:OTM	Bown-251-07
Bursera	*morelensis*	Morel	Torchwood	**M(traditional):**FR,OL,WD:OTM	Bown-251-08
Bursera	*pencillata*	India Lavender	Torchwood	**E:**OL:PF	Bown-251-09
Bursera	*simarda*	Ard Sima Rue	Torchwood	**E:**OL:PF	Bown-251-10
Bursera	*simarouba*	Sima Rue	Torchwood	**E:**OL:PF	Bown-251-11
Bursera	*spinosa*	Spiny	Torchwood	**E:**OL:PF	Bown-251-12
Butea	*monosperma*	Flame Of The Forest	Dhak	**M(traditional):**BK,LF,FL,SD:DAMW,DVGB,DVGV,DSIN,SDR	Wyk-402-08
Buxus	*macowanii*	Cape	Boxwood	**M(traditional):**LF:DHMY,**E:**LF:PS☠	Clev-084-02
Buxus	*sempervirens*	Evergreen	Boxwood	**M(traditional):**BK,LF,WD:AHM,DMAL,DWRM,HFV,NNC👽,NRH	Bown-251-13
Buxus	*sempervirens*	Evergreen	Boxwood	**M(traditional):**BK,LF,WD:DHMY,**E:**LF:DYE(red), PS☠	Clev-084-01
Caesalpinia	*echinata*	pernambuco	Caesalpino	**E:**WD:DYE(alum-blood-red,alum-&-copper-wine-red)	Grae-170-01
Calamintha	*grandiflora*	Savory	Calamint	**C:**LF:**SN**,TE	Bown-252-01
Calamintha	*nepeta*	Catnip	Calamint	**M(traditional):**WP:HIN,NMP,SDG	Bown-252-02
Calamintha	*nepeta*	Catnip	Calamint	**M(traditional):**WP:RLC,SFL	Wyk-402-09
Calamintha	*sylvatica*	Woodland	Calamint	**C:**LF:FF,**M(traditional):**WP:HIN,NMP,SDG	Bown-252-03
Calamintha	*sylvatica*	Woodland	Calamint	**M(traditional):**LF:RLC,**M(traditional):**WP:OBR	Clev-087-04
Calceolaria	*angustifolia*	Yellow	Ladies Purse	**E:**FL:DYE(alum-&-tin-maize-yellow,alum-&-tin-gold,chrome-&-tin-deep-orange)	Grae-130-03
Calendula	*officinalis*	Gold Bloom	Pot Marigold	**M(traditional):**FL:HFV,NNR,OSB,DSIN,OTA	Bal1-066-08
Calendula	*officinalis*	Gold Bloom	Pot Marigold	**M(traditional):**FL:DEAI,DWRM,OEY,OHM,OWN,SIB, **M(PS):**FL:DABA(*Helicobacter pylori, Staphylococcus aureus*),DAID,DCTM, **E:**FL:CM	Bal2-041-01 Bal2-509-01
Calendula	*officinalis*	Gold Bloom	Pot Marigold	**C:**FL:FC,GN,**M(traditional):**FL:BHS,BLC,DHMY,DALF,DASC,DHPT,DPXH,DPVI,DVGT,OEY,OGN,DSIN,SDG,SDV,SGB,SIB,SUL	Bown-252-04
Calendula	*officinalis*	Gold Bloom	Pot Marigold	**C:**FL:GN,**M(traditional):**FL:DASC,NMS,OWN,SIB,**E:**FL:DYE(saffron-orange,saffron-yellow)	Clev-084-03
Calendula	*officinalis*	Gold Bloom	Pot Marigold	**M(traditional):**FL:DABA,DFLU,DSXS,DTBC,HFV,NMS,OBN,,OBR,DSIN,OTA,OWN,SDG,E:FL:CM,DYE(alum-yellow)	Kowa-060-01 Kowa-284-01
Calendula	*officinalis*	Gold Bloom	Pot Marigold	**C:**FL:GN,**M(folk):**FL:EYB,EYV,HFV,OBN,OBR,OEY,OWN, **E:**FL:CM	Krug-111-01
Calendula	*officinalis*	Gold Bloom	Pot Marigold	**M(PS):**FL:NMS,OBN,ODS,OHM,DSIN,OWN,RST,SDG,SUL	Wyk-074-01, Wyk-402-10
Calendula	*officinalis*	Gold Bloom	Pot Marigold	**E:**FL:DYE(alum-yellow,vinegar-tan-yellow,salt-gold,tin-yellow-orange,iron-drab-brown)	Cas1-113-01
Callista	*nobilis*	Noble	Callisto Orchid	**M(traditional):**SM:HFV,RLC,SDG	Bown-273-01
Callistemon	*speciosus*	Albany	Bottlebrush	**E:**FL:DYE(alum-beige-green,alum-&-chrome-cinnamon-brown)	Grae-159-03
Calluna	*vulgaris*	Ling	Heather	**M(traditional):**WP:DHMY,OAS,NJA,NRH,RLC,SDR,UKC,UDR	Bown-252-05
Calluna	*vulgaris*	Ling	Heather	**M(traditional):**WP:SDR,UUR	Wyk-402-11
Caloplaca	*lactea*	Milky	Fire Dot Schist Lichen	**E:**WP:DYE(Lactic-Acid-&-Ammonia-pink)	Cas2-033-31
Caloplaca	*trachyphylla*	Rough Frond	Fire Dot Schist Lichen	**E:**WP:DYE(Anthraquinones-&-Ammonia-red-purple)	Cas2-031-06
Calotropus	*procera*	Apple of Sodom	Calotrope Milkwood	**M(traditional):**RT:cardinolide	Wyk-402-12
Caltha	*palustris*	Marsh	Fleshy Marigold	**M(traditional):**WP:DSIN	Wyk-402-13
Calvatia	*gigantea*	Australian	Giant Puffball	**E:**WP:DYE(alum-light-yellow,chrome-gold)	Cas1-182-03
Calycanthus	*occidentalis*	San Francisco	Spice Bush	**E:**FL:DYE(alum-&-copper-light-brown)	Grae-143-01
Calystegia	*sepium*	Hedge	False Bindweed	**E:**RT,SM:PF	Clev-088-07
Camellia	*japonica-rubra*	Red Japanese	Kamel Tea	**E:**FL:DYE(alum-&-iron-gray)	Grae-149-02
Camellia	*sinensis*			Synonym for *Thea sinensis*	Bown-252-05
Camellia	*sinensis*			Synonym for *Thea sinensis*	Wyk-075-01, Wyk-402-14
Campanula	*medium*	Canterbury Bells	Bellflower	**E:**FL:DYE(alum-pale-blue,chrome-medium-green)	Grae-113-02
Campanula	*rapunculoides*	Creeping Harebell	Bellflower	**E:**LF,FL,SL:DYE(alum-beige-rose,chrome-brown, copper-dark-green-gold,tin-tan-gold,iron-drab-gray-brown)	Bliss-096-01
Campanula	*rapunculus*	Rampion	Bellflower	**C:**LF:**SN**,**C:**RT:**VG**	Clev-084-04
Camptotheca	*acuminata*	Cancer Tree	Ribbed Leaf Tupelo	**M(CT):**LF:BLC,DCCL,DCOV,DCPA,ODS,SGB,SIB	Wyk-076-01, Wyk-402-15
Cananga	*odorata*	Fragrant	Ylang Ylang	**C:**OL:FF,**M(traditional):**FL:BHP,HAP,HFV,DHMY,DASC,DBOI,DMAL,OEY,OGT,**E:**OL:CM,HO,PF	Bown-253-01

Latin Genus	L. Species	E. Species	English Genus	C:Culinary,M:Medicinal,E:Economic	Source
Cananga	*odorata*	Fragrant	Ylang Ylang	**M(traditional):**OL:DASC,NSD,**E:**OL:PF	Wyk-402-16
Candelariella	*vitellina*	Egg Yellow	Goldspeck Candle Rim Lichen	**E:**WP:DYE(Calycin-yellow)	Cas2-034-09
Canella	*winterana*	White	Wild Cinnamon	**C:**BK:**SP**,**C:**LF:**SN**,**M(traditional):**BK,LF,OL:DABI,ORH,SCN,SDG,**E:**OL:IR,PF	Bown-253-02
Canella	*winterana*	White	Wild Cinnamon	**C:**BK:**SP**,**M(traditional):**BK:DASC,**E:**BK,IR,PS☠	Wyk-402-17
Cannabis	*sativa-sativa*	Fiber	Hemp	**E:**SM:RP	Bown-253-03
Cannabis	*sativa-indica*	Marajuana	Hemp	**M(traditional):**WP:DEGC,DUWT,NNC,NSD,SCN,DCLW	Bown-253-04
Cannabis	*sativa-indica*	Marajuana	Hemp	**M(PS):**FL,LF,RN,SD:DEGC,NMP,NRH,RLA,SAP,SCN	Wyk-077-01, Wyk-402-18
Cantharellus	*cibarius*	European	Chanterelle	**E:**WP:DYE(alum-&-chrome-yellow-orange, chrome-&-tin-rust)	Cas1-182-04
Capparis	*brevispina*	Short Spine	Caper	**C:**FL-BD:**SP**	Bown-253-05
Capparis	*decidua*	Deciduous	Caper	**C:**FL-BD:**SP**	Bown-253-06
Capparis	*rupestris*	Spineless	Caper	**C:**FL-BD:**SP**	Clev-280-01
Capparis	*spinosa*	Spiny	Caper	**C:**FL-BD:**SP**,**M(traditional):**RB:OAS,NGT,NRH,SDR,SIB, UDR,**M(traditional):**FL-BD:RLC,OEY	Bown-253-07
Capparis	*spinosa*	Spiny	Caper	**C:**FL-BD:**SP**,**M(traditional):**FL:OAS,DSIN,SCN,SDP	Wyk-402-19
Capparis	*zeylanica*	Sri Lanka	Caper	**C:**FL-BD:**SP**	Bown-253-08
Capsella	*bursa-pastoris*	Shepherd's Purse	Capsell Mustard	**M(traditional):**WP(aerial):OAS,SUL,UDR, **M(PS):**WP(aerial):OAS,OBN,OWN	Bal2-130-01 Bal2-542-01
Capsella	*bursa-pastoris*	Shepherd's Purse	Capsell Mustard	**C:**LF:**VG**,**M(traditional):**WP:BHP,DASC,OAS,OHM,OVV,OWN,UBL,UDR	Bown-254-01
Capsella	*bursa-pastoris*	Shepherd's Purse	Capsell Mustard	**C:**LF:**VG**,**M(folk):**WP:BVV,MEB,OAS,OWN,SDR,SUL	Krug-159-01
Capsella	*bursa-pastoris*	Shepherd's Purse	Capsell Mustard	**M(PS):**WP:OAS,OEY,SDR,SDY,UBL	Wyk-078-01, Wyk-402-20
Capsicum	*annuum-chinense-habanerum*	Habenero Hot	Pepper	**C:**FR:**SP**	Clev-284-05
Capsicum	*annum-conoides serranoi*	Serrano Cone Hot	Pepper	**C:**FR:**SP**	Clev-282-03
Capsicum	*annum-conoides mirasolii-cascabella*	Cascabel Mirasol Cone Hot	Pepper	**C:**FR:**SP**	Clev-283-05
Capsicum	*annuum-frutescens-chilium*	Chili Hot	Pepper	**C:**FR:**SP**,**M(traditional):**FR:BCB,BCR,BVV,DASC,HFV,NBK,ONR,RLA,RLP,SDG	Bown-254-02
Capsicum	*annuum-frutescens-chilium*	Chili Hot	Pepper	**M(CT):**FR,SD:OBK,OJA,OMS,ONR,ORH,RTL,SDP,SFL,SIB	Wyk-079-01a Wyk-402-21a
Capsicum	*annuum-frutescens-chilium anaheimensis*	Anaheim Chili Hot	Pepper	**C:**FR:**SP**	Clev-282-03
Capsicum	*annuum-frutescens-chilium poblano*	Poblano Chili Hot	Pepper	**C:**FR:**SP**	Clev-283-02
Capsicum	*annuum-frutescens-chilium guajilloi*	Guajillo Chili Hot	Pepper	**C:**FR:**SP**	Clev-283-04
Capsicum	*annuum-frutescens-chilium-aviculare chiltipinii*	Chiltipin Bird Chili Hot	Pepper	**C:**FR:**SP**	Clev-283-08
Capsicum	*annuum-frutescens-chilium-aviculare pequinii*	Pequin Bird Chil Hot	Pepper	**C:**FR:**SP**	Clev-283-06
Capsicum	*annuum-frutescens-tabascum*	Tabasco Hot	Pepper	**C:**FR:**SP**,**M(traditional):**FR:BCB,BCR,BVV,DASC,HFV,NBK,ONR,RLA,RLP,SDG	Bown-254-06
Capsicum	*annuum-frutescens-tabascum*	Tabasco Hot	Pepper	**C:**FR:**SP**,**M(folk):**FR:BCB,BCR,DASC,SAP,SDG,NJA	Krug-044-01
Capsicum	*annuum-frutescens-tabascum*	Tabasco Hot	Pepper	**M(CT):**FR,SD:OBK,OJA,OMS,ONR,ORH,RTL,SDP,SFL,SIB	Wyk-079-01b Wyk-402-21b
Capsicum	*annuum-longum-papricum*	Paprika Hot	Pepper	**C:**FR:**SP**	Clev-287-01
Capsicum	*annuum-longum-arienum pasilla*	Pasilla Banana Hot	Pepper	**C:**FR:**SP**	Clev-283-07
Capsicum	*annuum-longum-arienum pasilla-isleno-mulato*	Isleno Mulato Pasilla Banana Hot	Pepper	**C:**FR:**SP**	Clev-283-03
Capsicum	*annuum-longum-cayenneum*	Cayenne Hot	Pepper	**C:**FR:**SP**,**M(traditional):**FR:BCR,BHS,DCLD,NJA,NRH,RSN,RST,SDG,SUL	Bal1-067-03
Capsicum	*annuum-longum-cayenneum*	Cayenne Hot	Pepper	**M(folk):**FR:NMS,RST,BHS,BST,**M(ayurveda):**FR:DABI,DCHL,DFRI,BCR,BHE,HFV,NGT,NJA,NSC,RLC, **M(PS):**FR:DOBE	Bal2-044-01 Bal2-510-02
Capsicum	*annuum-longum-cayenneum*	Cayenne Hot	Pepper	**C:**FR:**SP**,**M(traditional):**FR:DCLD,OJA,ORH,OWN,RLA,RST,SDG,SDR,SFL,**M(traditional):**OL:OTA	Kowa-073-01 Kowa-284-04
Capsicum	*annuum-longum-jalapenum*	Jalapeño Hot	Pepper	**C:**FR:**SP**	Clev-285-03
Capsicum	*baccatum*	Finger	Pepper	**C:**FR:**VG**,**M(traditional):**FR:BCB,BCR,BVV,DA	Bown-254-04

Latin Genus	L. Species	E. Species	English Genus	C:Culinary,M:Medicinal,E:Economic	Source
				SC,HFV,NBK,ONR,RLA,RLP,SDG	
Capsicum	*cerasiforme-sucrum*	Sweet Cherry	Pepper	**C**:FR:**VG**	Clev-285-04
Capsicum	*chinense*	Torpedo	Pepper	**C**:FR:**VG**	Bown-254-05
Capsicum	*fasciculatum-sucrum*	Sweet Cone	Pepper	**C**:FR:**VG**	GSI6-322
Capsicum	*frutescens-sucrum*	Sweet Pencil	Pepper	**C**:FR:**VG**	GSI6-322
Capsicum	*glabrisculum*	Bird	Pepper	**C**:FR:**VG**	Bown-254-03
Capsicum	*grossum-sucrum*	Sweet Bell	Pepper	**C**:FR:**VG**	GSI6-316
Capsicum	*longum-pimentum-sucrum*	Sweet Pimento	Pepper	**C**:FR:**VG**	GSI6-322
Capsicum	*pubescens*	Hot Rocoto	Pepper	**C**:FR:**SP**	Bown-254-07
Carapa	*guianensis*	Andiroba	Crabwood	**M(traditional):**BK,LF:DISB,DWRM,HFV,OUL, **M(traditional):**SD-OL: NJA, M:FR-OL:RLC, **M(traditional):**NT-OL:DAGS,NJA,NRH,OSS, OWN, **M(PS):**BK:DABA, **M(PS):**FL:DCTM, **M(PS):**WD:DAFU	Bal2-022-01 Bal2-503-02
Carapichea	*ipecacuanha*	Brazilian	Ipecac	**M(traditional):**RT:DHMY,DWHP,HFV,RLB,RLC,SDY,**E:**RT:PS☠	Bown-258-02
Carapichea	*ipecacuanha*	Brazilian	Ipecac	**M(traditional):**RT:RLC,SPR	Kowa-281-06
Carapichea	*ipecacuanha*	Brazilian	Ipecac	**M(PS):**RT:RLB,RLC	Wyk-259-01, Wyk-423-22
Cardamine	*pratensis*	Meadow	Bitter Cress	**C:**LF:**VG**,**M(traditional):**LF:NMS,RLA	Bown-255-01
Carduus	*marianus*			Synonym for *Silybum marianum*	Bal2-099-02
Carduus	*natans*	Musk	Plumeless Thistle	**E:**FL(purple):DYE(alum-yellow-green,chrome-dark-brown,copper-dark-olive,tin-gold,iron-brown)	Bliss-258-01
Cardiospermum	*halicacabum*	Cacaba	Balloon Vine	**M(traditional):**LF,SD:DSIN,NSD	Wyk-402-22
Carex	*arenaria*	Sand	Sedge	**M(traditional):**RZ:UDR	Wyk-402-23
Carex	*californica*	California	Sedge	**E:**FL-SM:DYE(alum-tan-gold,chrome-dark-gold,copper-tan,tin-gold,iron-olive)	Bliss-220-01
Carica	*papaya*	Papaya	Paw Paw	**M(traditional):**BK,FR,SM:SAP,SDG,SIB	Bal1-075-03
				M(traditional):FR:SAP,SDG,SUL, **M(ayurveda):**FR:DWRM,OHM,RLA,RLB,UKC, **M(CS):**FR:DCTM,DHSV,DSHZ,UEP	Bal2-110-01 Bal2-533-07
Carica	*papaya*	Papaya	Paw Paw	**C:**FR:PU,**C:**SD:**SP**,**M(traditional):**FR,SA:SDG, **M(traditional):**LF,SD:DWRO,DWTH,**E:**FR:GR,IC	Bown-255-02
Carica	*papaya*	Papaya	Paw Paw	**C:**SD:**SP**,**M(traditional):**SD:OHM,SFL	Clev-287-01
Carica	*papaya*	Papaya	Paw Paw	**M(traditional):**FR:BHE,DABA,OWN,SDG,**E:**FR: shrinkproof	Wyk-080-01, Wyk-403-01
Carlina	*acaulis*	Stemless	Carline Thistle	**M(traditional):**RT:BLC,DACN,RLB,DSIN,SGB,UEP	Bown-255-03
Carlina	*acaulis*	Stemless	Carline Thistle	**M(traditional):**RT:DSIN,UDR	Wyk-403-02
Carthamus	*tinctorius*	Saffron	Distaff Thistle	**C:**FL:FC,**C:**OL:CO,**M(traditional):**FL,OL,SD:BAS,BCR,BHC,BHS,BJA,DMHI,DMSL,OBR,OJA,DSIN,OWN,**E:**FL:DYE(alum-yellow)	Bown-255-04
Carthamus	*tinctorius*	Saffron	Distaff Thistle	**C:**FL:FC,**E:**FL:DYE(alum-yellow,tin-rust,iron-brass)	Kowa-175-06 Kowa-436-01
Carthamus	*tinctorius*	Saffron	Distaff Thistle	**M(PS):**FL:BHS,HFV,OWN,**M(traditional):**SD-OL:BHC,DCRO	Wyk-081-01, Wyk-403-03
Carum	*ajowan*	Ajowan	Caraway	**C:**SD:**SP**,**M(traditional):**SD:RLA,SDG,**M(traditional):**OL,SD:SDR,SFL	Clev-288-01
Carum	*carvi*	Persian	Caraway	**C:**SD:**SP**,**C:**OL:FF,**C:**RT:**VG**,**M(traditional):**LF,OL,RT,SD:NMP,NMS,SDG,SDR,SFL,SHH,SIB,SUL,RLB,RTL,**E:**OL:PF	Bown-255-05
Carum	*carvi*	Persian	Caraway	**C:**SD:**SP**,**M(traditional):**SD:SDG,SFL,**E:**SD:CM,PF	Clev-289-01
Carum	*carvi*	Persian	Caraway	**C:**SD:**SP**,**C:**LF:GN,**C:**RT:**VG**,**M(traditional):**SD:SDG,SFL	Kowa-063-01
Carum	*carvi*	Persian	Caraway	**C:**SD:**SP**,M(folk):SD-OL:SFL,SDP,SIB	Krug-041-01
Carum	*carvi*	Persian	Caraway	**M(PS):**FR,OL:NMS,RLC,SAP,SCN,SDP,SFL	Wyk-082-01, Wyk-403-04
Cascara	*sagrada*			Synonym for *Frangula purshiana*	
Cassia	*fistula*	Golden Shower	Kassod Caesalpino	**M(traditional):**BK,RT:DSIN,**M(traditional):**FR:SPR	Wyk-403-05
Cassia	*senna*			Synonym for *Senna alexandrina*	Kowa-283-01
Castanea	*sativa*	Sweet	Chestnut	**C:**SD:NT,**M(traditional):**LF:DWHP,NRH,OAS,RLC,RST,SDR	Bown-256-01
Castanea	*sativa*	Sweet	Chestnut	**M(traditional):**LF:DAMW,DWHP,OAS,RLA,RLB,RLC,SDR	Wyk-083-01, Wyk-403-06
Castanospermum	*australe*	Black Bean	Moreton Bay Chestnut	**C:**SD:NT,**M(traditional):**SD:DHIV	Bown-256-02
Castanospermum	*australe*	Black Bean	Moreton Bay Chestnut	**M(traditional):**LF:DHIV	Wyk-403-07
Castilleja	*miniata*	Indian	Paintbrush Foxglove	**E:**FL(bracts):DYE(alum-bright-gold,chrome-ocher,copper-ocher,tin-bright-yellow,iron-dark-moss-green)	Bliss-136-01

Latin Genus	L. Species	E. Species	English Genus	C:Culinary,M:Medicinal,E:Economic	Source
Catalpa	*bignonioides*	Indian Bean	Catalpine	**M(traditional):**BK,FR:NSD,RLB,RLC	Wyk-403-08
Catha	*edulis*	Edible	Khat	**M(PS):**LF:DCFS,DMAL,DOBE,HFV,RLA,RLC	Wyk-084-01, Wyk-403-09
Catharanthus	*roseus*	Rosy	Madagascar Periwinkle	**M(traditional):**LF:BDB,BHP,BSH,DCLK,DCLY,OAS,SCN,SDG,RLA,**E:**LF:PS☠(Alkaloids)	Bown-256-03
Catharanthus	*roseus*	Rosy	Madagascar Periwinkle	**M(traditional):**LF:DASC,DCBR,DCLK,DCLY,DCTM(kidney),OAS,**E:**LF:PS☠(neurotoxin,immunotoxin)	Kowa-283-06 Kowa-492-01
Catharanthus	*roseus*	Rosy	Madagascar Periwinkle	**M(traditional):**LF:BDB,DCTM	Krug-137-02
Catharanthus	*roseus*	Rosy	Madagascar Periwinkle	**M(CT):**LF,RT:BDB,DCBR,DCCU,DCLU,DCLY,DCTM,NRH	Wyk-085-01, Wyk-403-10
Caulophyllum	*thalictroides*	Squaw Root	Blue Cohosh	**M(traditional):**RT:NMP,NMS,NSD	Bal1-066-02
Caulophyllum	*thalictroides*	Squaw Root	Blue Cohosh	**M(traditional):**RT,RZ:DCEN,DPVI,NGT,NJA,NRH,UDR	Bown-256-04
Caulophyllum	*thalictroides*	Squaw Root	Blue Cohosh	**M(traditional):**RT:DHMY,NMP,NMS,NRH,NSD,RLB,RST,**E:**RT:PS☠(damages heart muscle)	Kowa-050-01
Caulophyllum	*thalictroides*	Squaw Root	Blue Cohosh	**M(traditional):**RT:NMS,UDR	Wyk-403-11
Ceanothus	*americanus*	New Jersey Tea	Redroot	**C**:LF:TE,**M(traditional):**RT:DCLD,DDPT,DWHP,DTNS,NMS,OAS,OHM,RLC,RLB,RSN	Bown-256-05
Ceanothus	*americanus*	New Jersey Tea	Redroot	**C**:LF:TE,**M(traditional):**RT:BHP,OAS	Kowa-398-01
Ceanothus	*americanus*	New Jersey Tea	Redroot	**M(traditional):**RB,RT:HFV,NMS,OAS,OCC,DSIN,RST,**M(traditional):**LF:DAMW	Wyk-086-01, Wyk-403-12
Ceanothus	*integerrimus californicus*	California Deer Brush	Redroot	**E:**FL(blue):DYE(alum-&-tin-green)	Grae-138-03
Ceanothus	*sanguineus*	Red Stem	Redroot	**C**:LF:TE	Bown-256-06
Cedronella	*canariensis*	Canary Balm	Camphormint	**C:**LF:TE	Bown-256-07
Cedronella	*triphyllum*	Balm of Gilead	Camphormint	**C:**LF:TE	Bown-256-08
Cedrus	*atlantica*	Atlas	Cedar	**M(traditional):**LF:BCR,DAFU,DAVR,DCLW,RLC,UUR	Bal1-067-04
Cedrus	*atlantica*	Atlas	Cedar	**M(traditional):**WD,OL:BCR,DAFU,DTBC,OAS,DDND,OSA,RLB,**E:**OL:IR,PF,DS	Bown-257-01
Cedrus	*deodara*	Devadaru	Cedar	**M(traditional):**BK,OL,WD:DASC,NSD,RLB,RLC,UDR	Wyk-403-13a
Cedrus	*libani*	Lebanon	Cedar	**M(traditional):**BK,OL,WD:DASC,NSD,RLB,RLC,UDR	Wyk-403-13b
Ceiba	*pentandra*	Kapoc	Pochote Silk Cotton	**C:**OL:CO,**M(traditional):**RN:SDR,SDY,**M(traditional):**BK,LF:HAC,HFV,RLB,**M(traditional):**BK:OWN,**M(traditional):**LF:DERS,**E:**FR:FI	Bown-257-02
Centaurea	*benedicta*	Blessed Thistle	Knapweed	**M(traditional):**WP:BCR,BHS,BLC,SAP,lactation-aid	Bal1-066-01
Centaurea	*benedicta*	Blessed Thistle	Knapweed	**C:**RT:**VG**,**M(traditional):**WP:DABI,DANX,DASC,DABI,HFV,RLC,SAP,SDG,SDP,SFL,SDR,SIB,**E:**WP:PS☠	Bown-264-01
Centaurea	*benedicta*	Blessed Thistle	Knapweed	**C:**LF:**VG,C:**RT:**VG**,**M(traditional):**WP:BLC,SGB,SDG,**M(traditional):**LF:OCB,OBN,OWN	Clev-088-04
Centaurea	*benedicta*	Blessed Thistle	Knapweed	**C:**LF:**VG,M(folk):**WP(flowering):SAP,SDG,SPR	Krug-034-01
Centaurea	*benedicta*	Blessed Thistle	Knapweed	**M(PS):**WP:BDB,DCTM,NGT,NRH,OUL,OWN,SAP,SDP,SIB	Wyk-087-01, Wyk-403-14
Centaurea	*cyanus*			Synonym for *Cyanus segetum*	
Centaurea	*nigra*	Black	Knapweed	**M(traditional):**WP:DGUM,OWN,RST,UDR	Clev-085-04
Centaurea	*nigra*	Black	Knapweed	**E:**FL:DYE(alum-yellow,copper-sulfate-yellow-green,chrome-gold),**E:**WP:DYE(iron-green-gray,copper-sulfate-tan)	Cas1-161-01
Centaurea	*repens*			Synonym for *Rhaponticum repens*	Bliss-142-01
Centaurea	*scabiosa*	Greater	Knapweed	**M(traditional):**FL:OWN	Bown-257-05
Centaurium	*erythreaea*	Feverwort	Centaury	**M(traditional):**WP:BJA,BLC,DANX,DHPT,HFV,SAP,SDG,SDP,SGB	Bown-257-06
Centaurium	*erythreaea*	Feverwort	Centaury	**C:**WP:AF(Vermouth),**M(traditional):**WP:DHMY,DSIN,SDG	Clev-085-01
Centaurium	*erythreaea*	Feverwort	Centaury	**M(traditional):**WP(flowering):BCR,BLC,SAP,SGB	Krug-047-01
Centaurium	*erythreaea*	Feverwort	Centaury	**C:**WP:AF,BF,FF,**M(PS):**WP(aerial):BLC,HFV,OWN,SAP,SDG,SDP,SGB,UBL,UUR	Wyk-089-01, Wyk-403-16
Centaurium	*littorale*	Seaside	Centaury	**M(traditional):**WP:DHMY,SDG	Clev-084-02
Centaurium	*stellaris*	American	Centaury	**M(traditional):**WP:DHMY,SDG	Clev-084-03
Centella	*asiatica*	Gotu Kola	Hundredleaf	**M(traditional):**NT,RT,SD:BCR,BHS,BLC,DCFS,HAN,HAP,HIN,SAP,UKS	Bal1-071-05
Centella	*asiatica*	Gotu Kola	Hundredleaf	**M(traditional):**LF,SM:BJA,DSCB,SDR,SDY,UKS, **M(ayurveda):**LF,SM:DSIN,OCW,OUL **M(PS):**LF,SM:HFN	Bal2-080-01 Bal2-524-02
Centella	*asiatica*	Gotu Kola	Hundredleaf	**C:**LF:**VG**,**M(traditional):**WP:BVV,DMAL,DSXV,	Bown-258-01

Latin Genus	L. Species	E. Species	English Genus	C:Culinary,M:Medicinal,E:Economic	Source
				HFV,DCLW,OHM,ORH,OWN,**E:**WP:CM	
Centella	*asiatica*	Gotu Kola	Hundredleaf	**M(CS):**WP:OBN,OUL,OVV,OWN,SUL	Wyk-090-01, Wyk-403-17
Centranthus	*ruber*	Red	Soft Valerian	**M(traditional):**RT:NSD	Wyk-403-18
Cephaelis	*ipecacuanha*			Synonym for *Carapichea ipecacuanha*	Bown-258-02
Ceratonia	*sliqua*	Silique	Carob	**M(traditional):**FR:DGCS,DOBE,SDR	Wyk-403-19
Cercocarpus	*montanus*	Mountain	Mahogany Rosewood	**E:**LF,SM:DYE(alum-tan-gold,chrome-gold-brown,copper-tan-green,tin-soft-gold,iron-khaki)	Bliss-174-01
Cetraria	*californica*			Synonym for *Kaernefeltia californica*	Cas2-031-13
Cetraria	*canadensis*			Synonym for *Vulpicida canadensis*	Cas2-034-10
Cetraria	*cucullata*			Synonym for *Flavocetraria culcullata*	Cas2-031-07
Cetraria	*delisei*			Synonym for *Cetrariella delisei*	Cas2-031-08
Cetraria	*ericetorum*	Iceland	Heath Shield Lichen	**E:**WP:DYE(Lichesterinic-Acid-&-Ammonia-maize-tan)	Cas2-034-11
Cetraria	*islandica*	Iceland Moss	Heath Shield Lichen	**M(traditional):**WP:DABI,DBOI,DIMP,DTBC,RLB,RLC,RST,SIB	Bown-258-03
Cetraria	*islandica*	Iceland Moss	Heath Shield Lichen	**M(folk):**WP:DABI,DTBC	Krug-095-01
Cetraria	*islandica*	Iceland Moss	Heath Shield Lichen	**M(PS):**WP:DABA,OWN,RLC,RST,SAP	Wyk-091-01, Wyk-404-01
Cetraria	*islandica*	Iceland Moss	Heath Shield Lichen	**E:**WP:DYE(Fumarprotocetraric-Acid-yellow, Lichesterinic-Acid-brown,Lichesterinic-Acid-&-Ammonia-rusty-brown,Protolichesterinic-Acid-brown,Protolichesterinic-Acid-&-Ammonia-red)	Cas2-031-09
Cetraria	*nivalis*			Synonym for *Flavocetraria nivalis*	Cas2-031-10
Cetraria	*tilesii*			Synonym for *Vulpicida tilesii*	Cas2-034-14
Cetrariella	*delisei*	Delise	Alpine Heath Shield Lichen	**E:**WP:DYE(Gyrophoric-Acid-&-Ammonia-fuchsia)	Cas2-031-08
Chaenomeles	*speciosa*	Chinese	Flowering Quince	**C:**FR:JJ,**M(traditional):**FR:BCR,NJA,NMS,NRH, SDG	Bown-258-04
Chaenomeles	*speciosa*	Chinese	Flowering Quince	**M(traditional):**FR:NMS,NRH	Wyk-404-04
Chaenothecopsis	*rubescens*	Red	Gaping Theca Calico Sac Fungus	**E:**WP:DYE(Usnic-&-Vulpinic-Acid-gold-orange)	Cas2-034-17
Chamaelirium	*luteum*	False Unicorn Root	Devil's Bit	**M(traditional):**RT:DIFR,NMP,UEP	Bal1-070-02
Chamaelirium	*luteum*	False Unicorn Root	Devil's Bit	**M(traditional):**RT,RZ:DCEN,DMHI,DPVI,DWRM,OAS,SDG,UDR	Bown-258-05
Chamaelirium	*luteum*	False Unicorn Root	Devil's Bit	**M(traditional):**RT:OWN,OUL,UDR,UUR	Wyk-404-02
Chamaemelum	*arvensis*	Corn	Chamomile	**C:**FL,OL:TE,**M(traditional):**WP:BCR,SAP	Clev-086-01
Chamaemelum	*cotula*	Stinking	Chamomile	**C:**FL,OL:TE,**M(traditional):**WP:BCR,SAP	Clev-086-02
Chamaemelum	*nobile*	Roman	Chamomile	**M(traditional):**WP:HAC,HAN,HFV,HIN,NSD,SAP, SDG,SDV,SIB,UDR	Bal1-067-06
Chamaemelum	*nobile*	Roman	Chamomile	**C:**FL,OL:TE,**M(traditional):**FL:DHEA,HIN,NMP,SDG,SDP,SDV,**M(traditional):**OL:RLA,RLB,**E:**FL:DYE(gold-yellow),**E:**OL:HO	Bown-259-01
Chamaemelum	*nobile*	Roman	Chamomile	**M(traditional):**OL:BLC,OEY,SAP,NSD,**E:**OL:CM,PF,HO	Clev-085-05
Chamaemelum	*nobile*	Roman	Chamomile	**M(traditional):**FL-OL:DCLD,HAC,HAN,NMP,NMS,NSD, DSIN,SDG,**E:**FL-OL:CM,**E:**FL:DYE(alum-bright-yellow,chrome-yellow)	Kowa-079-01 Kowa-173-06 Kowa-285-02
Chamaemelum	*nobile*	Roman	Chamomile	**C:**FL-OL:AF,**M(folk):**FL(head):DASC,DSIN, DVGF,AC,HAN,HIN,NMP,NMT,OBN,OUL,SDG, **E:** FL-OL:CM	Krug-049-01
Chamaemelum	*nobile*	Roman	Chamomile	**C:**FL-OL:AF,**M(PS):**FL(head):NMS,NSD,SDP, SFL, **E:**FL-OL:CM,DS	Wyk-092-01, Wyk-404-03
Chelidonium	*majus*	Greater	Celandine	**M(traditional):**WP:BCR,BJA,DCSK,DCST,DHM, DHPT,DWRG,NGT,NJA,NMS,NRH,OBR,DCLW, OSS,OEY,OSA,OTM,RLB,RLC,SCN,SGB,SUL, **E:**WP:PS☠	Bown-259-02
Chelidonium	*majus*	Greater	Celandine	**M(traditional):**WP(top):NMS,**M(traditional):**SA:DCLW,**M(traditional):**RT:BLC,SGB	Clev-086-04
Chelidonium	*majus*	Greater	Celandine	**M(PS):**WP:BJA,DABI,NMS,NSD,SGB,SGS, **M(PS):**SA:DWRG,DCLW,OEY,DSIN,**E:**WP:PS☠	Wyk-093-01, Wyk-404-05
Chelone	*glabra*	Balmany	Turtlehead Foxglove	**M(traditional):**WP:BJA,BLC,DANX,SCN,SDG,SGS,SIB	Bown-259-03
Chelone	*glabra*	Balmany	Turtlehead Foxglove	**M(folk):**LF:DWRO,DWTH,BLC,OHM,OUL,SCN, SDG,SGB,SGS	Krug-022-01
Chelone	*glabra*	Balmany	Turtlehead Foxglove	**M(traditional):**WP:DWRM,SCN,SGB	Wyk-404-06
Chenopodium	*album*	Lamb's Quarters	Goosefoot	**C:**LF:SH:**VG**,**C:**SD:GF,**E:**LF,SH:DYE(red)	Clev-086-06
Chenopodium	*album*	Lamb's Quarters	Goosefoot	**E:**WP:DYE(vinegar-tan-yellow,alum-soft-yellow, copper-sulfate-soft-green,iron-gray,chrome-tan)	Cas1-162-01
Chenopodium	*album*	Lamb's Quarters	Goosefoot	**E:**LF,SM:DYE(alum-light-yellow,chrome-gray,copper-light-yellow-green,tin-creamy-yellow,iron-tan)	Bliss-126-01

Latin Genus	L. Species	E. Species	English Genus	C:Culinary,M:Medicinal,E:Economic	Source
Chenopodium	*ambrosoides*	Wormseed	Goosefoot	**C:**LF:**SN**,**M(traditional):**WP:DAFU,DWHK,DWRO,DWTP,NMS,OAS,RLA,RLC,SDY,**E:**OL:IC,**E:**LF,SH:PS☠	Bown-259-04
Chenopodium	*ambrosoides*	Wormseed	Goosefoot	**M(PS):**OL:DSNB,DWRM,DWHK	Wyk-094-01, Wyk-404-07
Chenopodium	*ambrosoides*	Wormseed	Goosefoot	**E:**WP:DYE(alum-&-copper-dark-green,chrome-&-copper-green-gold)	Grae-089-02
Chenopodium	*ambrosoides anthelminticum*	Mexican Tea Wormseed	Goosefoot	**M(traditional):**WP:DWHK,DWRO,DWTP	Bown-259-05
Chenopodium	*bonus-henricus*	Good King Henry	Goosefoot	**C:**LF(spinach),SH(asparagus):**VG**	Clev-086-05
Chenopodium	*botrys*	Ambrosia Quinoa	Goosefoot	**C:**SD:GF	Clev-086-09
Chenopodium	*californicum*	Green Leaf	Goosefoot	**E:**LF,SM:DYE(alum-yellow,chrome-brass,copper-light-olive,tin-light-yellow,iron-tan)	Bliss-124-01
Chenopodium	*leptophyllum*	Thin Leaf Quinoa	Goosefoot	**C:**SD:GF	Clev-086-10
Chenopodium	*nuttaliae*	Nuttal Quinoa	Goosefoot	**C:**SD:GF	Clev-086-11
Chenopodium	*quinoa*	Quinoa	Goosefoot	**C:**SD:GF,**M(traditional):**SD:RLC	Wyk-404-08
Chimaphila	*umbellata*	Ground Holly	Pipsissewa	**C:**LF:BV,**M(traditional):**WP:OAS,NJA,NRH,UDR,UEP,UKC,UKS,UUR	Bown-260-01
Chimaphila	*umbellata*	Ground Holly	Pipsissewa	**C:**LF:BV(rootbeer),**M(traditional):**LF:NRH,UKC	Kowa-415-01
Chimaphila	*umbellata*	Ground Holly	Pipsissewa	**M(traditional):**WP:DASC,OAS,UDR,UUR	Wyk-404-09
Chionanthus	*virginicus*	Old Man's Beard	Fringe Tree	**M(traditional):**RB:BES,BJA,BLC,DHPT,DMAL,HAC,OWN,SCN,SGB,SGS,SPA,UDR	Bown-260-02
Chionanthus	*virginicus*	Old Man's Beard	Fringe Tree	**M(traditional):**RB:BLC	Krug-078-01
Chionanthus	*virginicus*	Old Man's Beard	Fringe Tree	**M(traditional):**RB:BJA,BLC,DHMY,DHPT,HFV,OBR,OWN,SGB,SGS,UDR	Wyk-095-01, Wyk-404-10
Chondrodendron	*tomentosum*	Pareira Root	Cartilage Tree	**M(traditional):**RT,SM:HFV,OMC,SCN,UDR	Bown-260-03
Chondrodendron	*tomentosum*	Pareira Root	Cartilage Tree	**M(traditional):**RT,SM:NMS	Kowa-282-06
Chondrodendron	*tomentosum*	Pareira Root	Cartilage Tree	**M(traditional):**RT:NMS,UDR, **E:**RT:PS☠	Wyk-404-11
Chondrus	*crispus*	Irish Moss Seaweed	Cartilage Gigartin Florid Red Algae	**M(traditional):**WP:SCN,SIB,**E:**WP:CM(skinlotion)	Bal1-073-02
Chondrus	*crispus*	Irish Moss Seaweed	Cartilage Gigartin Florid Red Algae	**C:**MC:FX,**M(traditional):**WP:OSS,RLB,RLC,RST,SCN,SDP,SIB,UBL,**E:**MC:PN	Bown-260-04
Chondrus	*crispus*	Irish Moss Seaweed	Cartilage Gigartin Florid Red Algae	**M(folk):**WP:RLB,RLC	Krug-095-02
Chondrus	*crispus*	Irish Moss Seaweed	Cartilage Gigartin Florid Red Algae	**M(PS):**WP:RLB,RLC,SCN,**E:**WP:CM,DS	Wyk-096-01, Wyk-404-12
Chrysanthemum	*balsamita*			Synonym for *Tanacetum balsamita*	Wyk-404-13
Chrysanthemum	*cinerariifolium*			Synonym for *Pyrethrum cinerariifolium*	Wyk-404-14
Chrysanthemum	*frutescens*			Synonym for *Argyranthemum frutescens*	Grae-125-02
Chrysanthemum	*leucanthemum*			Synonym for *Leucanthemum vulgare*	Cas1-133-01
Chrysanthemum	*morifolium*	Florist	Gold Mum	**M(traditional):**FL:BAA,BCR,BHP,BLC,HFV	Bown-272-11
Chrysanthemum	*morifolium*	Florist	Gold Mum	**M(CS):**FL,LF:BHP,DABI,DACN,DBOI,HFV,OEY,DSIN	Wyk-097-01, Wyk-404-15
Chrysanthemum	*superbum*	Shasta Daisy		Synonym for *Leucanthemum superbum*	Cas1-206-01
Chrysanthemum	*parthenium*			Synonym for *Tanacetum parthenium*	Wyk-404-16
Chrysanthemum	*vulgare*			Synonym for *Tanacetum vulgare*	Wyk-404-17
Chrysanthemum	*zawadskii*			Synonym for *Dendranthema zawadskii*	Grae-114-02
Chrysopsis	*villosa*			Synonym for *Heterotheca villosa*	Bliss-120-01
Chrysothamnus	*pulchellus*	Pretty	Rabbit Brush	**E:**FL,LF,SM:DYE(alum-bright-gold,chrome-bright-rust,copper-green-gold,tin-bright-orange-gold-iron-very-dark-olive-green)	Bliss-200-01
Chrysothamnus	*viscidiflorus*	Sticky	Rabbit Brush	**E:**FL:DYE(alum-lemon-yellow,alum-&-tin-yellow-orange,chrome-gold-copper,chrome-&-tin-pumpkin-orange)	Grae-105-01
Cichorium	*endivia*	Endive	Chicory	**C:**LF:**VG**,**C:**RT,SD:BF(coffee)	Clev-087-03
Cichorium	*intybus*	Succory	Chicory	**C:**LF:**VG**,**C:**RT:BF(coffee),**M(traditional):**LF,RT:BLC,NGT,OHM,SCN,SGB,UDR	Bown-261-01
Cichorium	*intybus*	Succory	Chicory	**C:**LF:**VG**,**C:**RT,SD:BF(coffee),**M(traditional):**RT,SM:BAN,BLC,NGT,NRH,SGB	Clev-087-02
Cichorium	*intybus*	Succory	Chicory	**C:**LF:**VG**,**M(traditional):**RT:SCN,UDR	Kowa-085-01
Cichorium	*intybus*	Succory	Chicory	**C:**LF:**VG**,**M(folk):**RT:BLC,NMT,OEY,UDR	Krug-052-01
Cichorium	*intybus*	Succory	Chicory	**M(PS):**WP:BLC,NGT,NRH,SAP,SDP,SFL,SGB,SIB,UDR	Wyk-100-01, Wyk-404-18
Cichorium	*intybus*	Succory	Chicory	**E:**FL,LF,SM:DYE(alum-mustard-yellow,chrome-brass-gold,copper-yellow-green,tin-bright-yellow,iron-forest-green)	Bliss-080-01
Cichorium	*intybus*	Succory	Chicory	**E:**WP:DYE(copper-sulfate-yellow-green,alum-tan-yellow,iron-green,chrome-tan),**E:**RT:DYE(alum-light-brown,chrome-dark-brown,iron-deep-brown,copper-sulfate-&-tin-kahaki	Cas1-120-01
Cicuta	*virosa*	Cowbane	Water Hemlock	**M(traditional):**WP:DCLK,NMS,NRH,NSD	Wyk-404-19

Latin Genus	L. Species	E. Species	English Genus	C:Culinary,M:Medicinal,E:Economic	Source
Cimicifuga	*americana*	American	Bugbane	**M(traditional):**RZ:DASC,DCLD,DGUM,DMSL,HAC,HFV,RLC	Bown-261-02
Cimicifuga	*dahurica*	Dahurian	Bugbane	**M(traditional):**RZ:DASC,DCLD,DGUM,DMSL,HAC,HFV,RLC	Bown-261-03
Cimicifuga	*foetida*	Stinking	Bugbane	**M(traditional):**RZ:DASC,DCLD,DGUM,DMSL,HAC,HFV,RLC	Bown-261-04
Cimicifuga	*racemosa*	Black Cohosh	Bugbane	**M(traditional):**RT,RZ:BCR,BHC,BHP,DSNB,NBK,NJA,NMP,NSD	Bal1-065-08
Cimicifuga	*racemosa*	Black Cohosh	Bugbane	**M(traditional):**RT,RZ:DIFR,DSNB,HFV,NSD,RLB,RST,**M(CS):**RT,RZ:DIFR DUFB	Bal2-034-01 Bal2-507-02
Cimicifuga	*racemosa*	Black Cohosh	Bugbane	**M(traditional):**RZ:DHMY,DMHI,HFV,NJA,NRH,RLB,RLC,**E:**RZ:PS☠	Bown-261-05
Cimicifuga	*racemosa*	Black Cohosh	Bugbane	**M(CS):**RZ,RT:DEAT,NRH,NSD,RLC	Wyk-101-01, Wyk-404-20
Cinchona	*antioquiae*	Antioque Columbian	Quinine	**M(folk):**BK,RB:DMAL	Krug-138-01
Cinchona	*asperifolia*	La Paz Bolivian	Quinine	**M(folk):**BK,RB:DMAL	Krug-138-02
Cinchona	*barbacoensis*	Carachi Equadorian	Quinine	**M(folk):**BK,RB:DMAL	Krug-138-03
Cinchona	*boliviana*	Bolivian	Quinine	**M(folk):**BK,RB:DMAL	Krug-138-04
Cinchona	*calisaya*	Calisay Peruvian	Quinine	**M(traditional):**RB,SM:DMAL	Bown-261-06 Bown-261-07
Cinchona	*capuli*	Capul Equadorian	Quinine	**M(folk):**BK,RB:DMAL	Krug-138-06
Cinchona	*fruticosa*	Amazon Peruvian	Quinine	**M(folk):**BK,RB:DMAL	Krug-138-07
Cinchona	*glandulifera*	Gland Peruvian	Quinine	**M(folk):**BK,RB:DMAL	Krug-138-08
Cinchona	*hirsuta*	Hairy Peruvian	Quinine	**M(folk):**BK,RB:DMAL	Krug-138-09
Cinchona	*krauseana*	Krause Amazon Peruvian	Quinine	**M(folk):**BK,RB:DMAL	Krug-138-10
Cinchona	*lancifolia*	Venezuelan	Quinine	**M(folk):**BK,RB:DMAL	Krug-138-11
Cinchona	*ledgeriania*			Synonym for *Cinchona calisaya*	Bown-261-07
Cinchona	*lucumifolia*	Luce Equadorian	Quinine	**M(folk):**BK,RB:DMAL	Krug-138-12
Cinchona	*macrocalyx*	Large Calyx Peruvian	Quinine	**M(folk):**BK,RB:DMAL	Krug-138-13
Cinchona	*micrantha*	Tiny Flower Peruvian	Quinine	**M(folk):**BK,RB:DMAL	Krug-138-14
Cinchona	*mutisii*	Mutis Equadorian	Quinine	**M(folk):**BK,RB:DMAL	Krug-138-15
Cinchona	*nitida*	Shiny Huánuco Peruvian	Quinine	**M(folk):**BK,RB:DMAL	Krug-138-16
Cinchona	*officinalis*	Equadorian	Quinine	**M(traditional):**RB,SM:DMAL	Bown-261-08
Cinchona	*parabolica*	Parabolic Peruvian	Quinine	**M(folk):**BK,RB:DMAL	Krug-138-18
Cinchona	*pitayensis*	Pitaya Columbian	Quinine	**M(folk):**BK,RB:DMAL	Krug-138-19
Cinchona	*pubescens*	Jesuit's Bark	Quinine	**M(traditional):**RB,SM:DHMY,OAS,DMAL,HFV,NMS,NNR,OMC,RST	Bown-261-09
Cinchona	*pubescens*	Jesuit's Bark	Quinine	**M(traditional):**RB,RT:DMAL	Kowa-281-01
Cinchona	*pubescens*	Jesuit's Bark	Quinine	**M(folk):**BK,RB:DMAL	Krug-138-20
Cinchona	*pubescens*	Jesuit's Bark	Quinine	**M(CS):**BK:DMAL,SAP,SFL	Wyk-102-01, Wyk-405-01
Cinchona	*pyrifolia*	Pear Huénuco Peruvian	Quinine	**M(folk):**BK,RB:DMAL	Krug-138-21
Cinchona	*rugosa*	Wrinkled Equadorian	Quinine	**M(folk):**BK,RB:DMAL	Krug-138-22
Cinchona	*scrobiculata*	Cajamarca Peruvian	Quinine	**M(folk):**BK,RB:DMAL	Krug-138-23
Cinchona	*villosa*	Villose Peruvian	Quinine	**M(folk):**BK,RB:DMAL	Krug-138-24
Cinnamomum	*aromaticum*	China Kassod	Cinnamon	**M(folk):**BK,LF-OL:DCFS,SFL, **M(traditional):**BK,LF-OL:DUFB, SAP,SDG,SUL, **M(ayurveda):** BK,LF-OL:SDP, **M(PS):**BK,LF-OL:DABA,DAFU,DCLV	Bal2-050-02 Bal2-511-06
Cinnamomum	*aromaticum*	China Kassod	Cinnamon	**C:**BK:**SP**,**C:**FR:FF,**M(traditional):**BK:BAA,BCR,DCLD,DFLU,HFV,NJA,NMS,NRH,SAP,SDG,SDP,SDR,SFL,SIB,UKC,**E:**OL:CM	Bown-262-02
Cinnamomum	*aromaticum*	China Kassod	Cinnamon	**C:**BK:**SP**,**M(traditional):**OL:SDR,SFL	Clev-290-01
Cinnamomum	*aromaticum*	China Kassod	Cinnamon	**M(PS):**BK,FL:DCFS,DCLD,SAP,SDP,SDR,SFL	Wyk-103-01, Wyk-405-02
Cinnamomum	*burmanii*	E. Indies Kassod	Cinnamon	**C:**BK:**SP**,**E:**BK:IN	Bown-261-10
Cinnamomum	*camphora*	Camphor Oil	Cinnamon	**M(traditional):**LF,WD:DEPI,DWRM,BCR,HIN,N	Bown-262-01

Latin Genus	L. Species	E. Species	English Genus	C:Culinary,M:Medicinal,E:Economic	Source
				GT,NMP,NMS,NRH,OCB,OEY,OJA,OSA,OWN,RLA,RLB,RRN,RSN,SDG,**E**:OL:CM,**E**:LF,WD:PS☠	
Cinnamomum	*camphora*	Camphor Oil	Cinnamon	**M(traditional):**OL:DASC,DWRM	Kowa-280-04
Cinnamomum	*camphora*	Camphor Oil	Cinnamon	**M(PS):**WD-OL:BCR,DCLD,DFLU,DPNB,HFV,ORH,SDR	Wyk-104-01, Wyk-405-03
Cinnamomum	*cassia*			Synonym for *Cinnamomum aromaticum*	Bal2-050-02
Cinnamomum	*iners*	Wild	Cinnamon	**C**:BK:**SP**	Bown-261-11
Cinnamomum	*loureirii*	Saigon	Cinnamon	**M(traditional):**BK:BCR,BDB,DAFU,DOBE,OLC,SDG,SDR	Bal1-068-01
Cinnamomum	*loureirii*	Saigon	Cinnamon	**C**:BK:**SP**	Bown-262-03
Cinnamomum	*massoia*	Massoy Bark	Cinnamon	**C**:BK:**SP**,**E**:OL:PF	Bown-262-04
Cinnamomum	*oliveri*	Black Sassafras	Cinnamon	**C**:BK:**SP**	Bown-262-05
Cinnamomum	*tamala*	India Kassod	Cinnamon	**C**:BK:**SP**,**C**:LF:**SN**,**M(traditional):**OL,RT,WD:OUL	Wyk-405-04
Cinnamomum	*verum*	Sri Lanka	Cinnamon	**C**:BK:AF,BF,**SP**,**M(traditional):**BK:BCR,BHP,DCAN,DFLU,HFV,NJA,NRH,SDG,SDR,SIB,**E**:OL:PF	Bown-262-06
Cinnamomum	*verum*	Sri Lanka	Cinnamon	**C**:BK:**SP**,**M(traditional):**BK:OAS,SDR,SFL,**E**:LF-OL:PF	Clev-291-01
Cinnamomum	*verum*	Sri Lanka	Cinnamon	**C**:BK:**SP**,**M(traditional):**BK:DABA(Clostridium,Staphylococcus),DAFU(Aspergillus),OAS,SDG	Kowa-090-01
Cinnamomum	*verum*	Sri Lanka	Cinnamon	**M(PS):**BK(inner):DCLD,NRH,SAP,SDP,SDR,SFL	Wyk-105-01, Wyk-405-05
Cinnamomum	*zeylanicum*			Synonym for *Cinnamomum verum*	Bown-262-06
Cirsium	*arvense*	Canadian	Thistle	**E**:FL,LF,SM:DYE(alum-dull-chartreuse,chrome-dark-chocolate-brown,copper-dark-olive-green,tin-ocher,iron-gray-green)	Bliss-256-01
Cistus	*albiflorus*	White Flower	Rockrose	**E**:RN:OR,PF	Bown-262-07
Cistus	*creticus*	Crete	Rockrose	**E**:RN:OR,PF	Bown-262-08
Cistus	*ladaniferus*	Gum	Rockrose	**M(traditional):**LF,OL,RN:DABI,RLC,SDR,**E**:RN:OR,PF	Bown-262-09
Cistus	*maculatus*	Spotted	Rockrose	**E**:RN:OR,PF	Bown-262-10
Citrullus	*colocynthis*	Colocynth	Watermelon	**M(traditional):**FR:DCTM,NSD,OTM,SPR,**E**:FR:IC	Wyk-405-06
Citrus-aurantius	*aurantium*	Sour	Orange	**M(traditional):**FR,RD:BLC,DALL,DCLD,RLB,RLC,SAP,SCN,SIB,**M(PS):**FR,RD:DOBE,SDG,	Bal2-033-02 Bal2-507-01
Citrus-aurantius	*aurantium*	Sour	Orange	**C**:FR:JJ,**M(traditional):**LF,FL,FR,RD,SM:DABA,DAFU,BHP,RLC,SDG,SDR,SFL,SIB,UDR,**E**:OL:PF	Bown-263-01
Citrus-aurantius	*aurantium*	Sour	Orange	**M(PS):**RD:HIN,SAP,SDP,SFL	Wyk-106-01, Wyk-405-07
Citrus-aurantius	*aurantium bergamia*	Bergamot Sour	Orange	**C**:FL:FF,**C**:OL:TE,**M(traditional):**FL:SIB,**M(traditional):**OL:DVGN,HAP,NSD,**M(traditional):**RD:NMS,SDG	Bown-263-02
Citrus-citron	*hystrix kaffirensis*	Kaffir Lime	Citron	**C**:LF,RD:**SP**,**E**:FR-JU:DS,HO	Clev-292-01
Citrus-limon	*limon*	East Asia	Lemon	**C**:FR:BF,JJ,JU,**C**:OL:CF,**C**:RD:FF,**M(traditional):**FR:BCR,BVV,OSS,OHM,OST,RLB,RST,UDR,UKS,**E**:OL:PF,DS	Bown-263-03
Citrus-limon	*limon*	East Asia	Lemon	**M(folk):**FR-JU:DCLD,DGUM(scurvy),DMAL,OSB,RST,**E**: FR-JU:CM	Krug-102-01
Citrus-limon	*limon*	East Asia	Lemon	**M(PS):**FR-JU:DGUM(scurvy),**M(PS):**RD,OL:BCR,SDG	Wyk-107-01, Wyk-405-08
Citrus-limonius	*aurantiifolia*	South Asia	Lime	**C**:FR:BF,JJ,JU,**C**:RD:FF,**M(traditional):**FR:SDR,**M(traditional):**LF:HAC,OAS	Bown-262-11
Citrus-tangiericus	*reticulata*	East Asia	Tangerine	**M(traditional):**FR:BHS,DALL,SDG,SDR,SUL	Bal2-048-01 Bal2-510-03
Citrus-tangiericus	*reticulata*	East Asia	Tangerine	**C**:FR:PU,**M(traditional):**FR,RD,SD:BLC,DCBR,NBK,NMT,NOR,SDR,SFL,SGB,RLB	Bown-263-04
Citrus-tangiericus	*reticulata-clementinea*	Clementine	Tangerine	**C**:FR:PU,**M(traditional):**FR,RD,SD:BLC,DCBR,NBK,NMT,NOR,SDR,SFL,SGB,RLB	Bown-263-04
Citrus-tangiericus	*reticulata-clementinea*	Clementine	Tangerine	**M(traditional):**FR:SAP,SDG	Wyk-405-09
Citrus-tangiericus	*reticulata-mandarina*	Mandarin	Tangerine	**C**:FR:PU,**M(traditional):**FR,RD,SD:BLC,DCBR,NBK,NMT,NOR,SDR,SFL,SGB,RLB	Bown-263-04
Cladina				Synonym for *Cladonia*	
Cladonia	*alpestris*		Cup Lichen	Synonym for *Cladonia stellaris*	Cas2-034-15
Cladonia	*cristatella*	Crested	Cup Lichen	**E**:WP:DYE(Barbatic-Acid-&-Didymic-Acid-light-green)	Cas2-034-16
Cladonia	*implexa*		Cup Lichen	Synonym for *Cladonia portentosa-implexa*	Cas2-031-11
Cladonia	*portentosa*	Portent	Cup Lichen	**E**:WP:DYE(Perlatolic-Acid-leaf-green)	Cas2-034-17
Cladonia	*portentosa-implexa*	Impexed Portent	Cup Lichen	**E**:WP:DYE(Perlatolic-Acid-&-Ammonia-pink)	Cas2-031-11

Latin Genus	L. Species	E. Species	English Genus	C:Culinary,M:Medicinal,E:Economic	Source
Cladonia	*pyxidata*	Pyxie	Cup Lichen	**E:**WP:DYE(Fumarprotocetraric-Acid-gold)	Cas2-034-18
Cladonia	*rangiferina*	Reindeer	Cup Lichen	**E:**WP:DYE(Fumarprotocetraric-Acid-&-Atranorin-red-brown)	Cas2-034-19
Cladonia	*stellaris*	Star	Cup Lichen	**E:**WP:DYE(Perlatolic-Acid-&-Usnic-Acid-yellow)	Cas2-034-15
Clarkia	*pulchella*	Pretty	Farewell To Spring	**E:**FL:DYE(alum-&-iron-green-gray)	Grae-089-03
Claviceps	*purpurea*	Purple	Ergot	**M(traditional):**WP:HAC,NUC	Bown-263-05
Claviceps	*purpurea*	Purple	Ergot	**M(traditional):**WP:HAC,NUC	Wyk-405-10
Clematis	*armandii*	Armand	Leather Flower	**M(traditional):**WP:NJA,NRH,DSIN	Wyk-405-11a
Clematis	*chinensis*	Chinese	Leather Flower	**M(traditional):**RT:HFV,NJA,NMS,NRH,NSD,UDR	Bown-263-06
Clematis	*ligusticifolia*	White	Leather Flower	**E:**LF,FL,SM:DYE(alum-yellow,chrome-chestnut-brown,copper-light-tan-olive,tin-bright-gold,iron-dark-forest-green)	Bliss-086-01
Clematis	*recta*	Ground Virgin's Bower	Leather Flower	**M(traditional):**WP:NJA,NRH,DSIN	Wyk-405-11b
Clematis	*virginiana*	Virgin's Bower	Leather Flower	**E:**FL,SM:DYE(alum-yellow-green,iron-green,tin-chartreuse),**E:**LF,SM:DYE(iron-gray-green,chrome-brown,copper-sulfate-tan)	Cas1-123-01
Clematis	*vitalba*	Old Man's Beard	Leather Flower	**M(traditional):**WP:NJA,NRH,DSIN	Wyk-405-11c
Cleome	*serrulata*	Bee Plant	Spider Flower Caper	**E:**WP,FL:DYE(alum-ivory,chrome-light-gold,copper-light-ocher,tin-light-yellow,iron-tan)	Bliss-046-01
Clerodendron	*trichotomum*	Harlequin	Glory Bower	**M(traditional):**WP:NNC👽,NSD,SDG	Wyk-405-12
Clinopodium	*grandiflora*			Synonym for *Calamintha grandiflora*	Clev-088-02
Clinopodium	*nepeta*			Synonym for *Calamintha nepeta*	Clev-088-03
Clinopodium	*vulgare*	Wild Basil	Lesser Calamint	**M(traditional):**LF:OBR,RLC	Clev-088-01
Clutia	*tranvancorica*	Tranvancoria	Coralline	**E:**WD:DYE(alum-&-cotton-pale-apricot,alum-&-wool-brilliant-orange,alum-&-iron-dark-brown)	Grae-161-02
Cnicus	*benidictus*	St. Benidictine	Blessed Thistle	Synonym for *Centaurea benidicta*	Bown-264-01
Cnidium	*monnieri*	Monnier	Snow Parsley	**M(traditional):**SD,OL:DAFU,DVGB,DVGV,DSIN	Wyk-405-13
Coccotrema	*maritimum*			Synonym for *Pertusaria maritime*	Cas2-031-12
Cochlearia	*Amoracia*			See *Amoracia rusticana*	
Cochlearia	*officinalis*	European	Scurvy Grass	**M(traditional):**WP:DAMW,DGUM,BLC,SDG,UDR	Wyk-405-14
Cochlospermum	*gossypium*	Silk Cotton	Spiralseed Annato	**M(traditional):**RN-GR,RN-MC:SCN,UDR	Wyk-405-15
Cocos	*nucifera*	Asian	Coconut Feather Palm	**M(traditional):**OL:DSIN,**E:**OL:CM,HO	Wyk-405-16
Codonopsis	*pilosula*	Poor Man's Ginseng	Bonnet Bellflower	**M(traditional):**RT:HAC,HAN,SAP,SDR **M(PS):**RT:DLUP,SUL	Bal2-050-02 Bal2-511-02
Codonopsis	*pilosula*	Poor Man's Ginseng	Bonnet Bellflower	**M(traditional):**RT:RLC	Wyk-405-17
Codonopsis	*tangshen*	Dang Shen	Bonnet Bellflower	**C:**RT:**SN**,**M(traditional):**RT:BAN,BES,BHP,BSH,SAP,SDG	Bown-264-02
Coffea	*arabica*	Arabian	Coffee	**C:**SD:BF(coffee),**M(traditional):**SD:OBN,UDR	Bown-264-03
Coffea	*arabica*	Arabian	Coffee	**C:**SD:BF(coffee),**M(traditional):**SD:BJA,DCST,DFLU,DMAL,DVTG,HAC,HFV,RLA	Kowa-094-01
Coffea	*arabica*	Arabian	Coffee	**M(CS):**SD:DAMW,DCFS,DFLU,HFV,RST,SDR	Wyk-108-01, Wyk-405-18
Coffea	*arabica*	Arabian	Coffee	**E:**SD:DYE(vinegar-warm-tan,chrome-brown,iron-gray-tan)	Cas1-125-01
Coffea	*canephora*	African	Coffee	**C:**SD:BF,**M(traditional):**SD:OBN,UDR	Bown-264-04
Coffea	*liberica*	Malaysian	Coffee	**C:**SD:BF,**M(traditional):**SD:OBN,UDR	Bown-264-05
Coix	*lacryma-jobi*	Job's Tears	Reed Grain Grass	**C:**SD:**VG**,**M(traditional):**SD:DAVR,DWHN	Bal2-051-01
Coix	*lacryma-jobi*	Job's Tears	Reed Grain Grass	**C:**SD:**VG**,**M(traditional):**SD:BES,DABA,DAFU,HFV,NJA,NMS,NSD,SDR,**M(traditional):**SD:NRH	Bown-264-06
Cola	*acuminata*	Abata	Cola Nut	**C:**SD:BF,**E:**SD:DYE(Red)	Bown-264-07
Cola	*acuminata*	Abata	Cola Nut	**C:**SD:BF,**M(PS):**SD:DCFS,OAS,OWN,SDR	Wyk-109-01, Wyk-405-19a
Cola	*anomala*	Bamenda	Cola Nut	**C:**SD:BF,**E:**SD:DYE(Red)	Bown-264-08
Cola	*nitida*	Gbanja	Cola Nut	**C:**SD:BF,**M(traditional):**OAS,SAP,SDR,**E:**SD:DYE(Red)	Bown-264-09
Cola	*nitida*	Gbanja	Cola Nut	**C:**SD:BF,**M(traditional):**SD:DCFS,OAS,OWN,SDR	Wyk-405-19b
Cola	*pachycarpa*	Goora	Cola Nut	**C:**SD:BF,**E:**SD:DYE(Red)	Bown-264-10
Cola	*verticillata*	Owe	Cola Nut	**C:**SD:BF,**E:**SD:DYE(Red)	Bown-264-11
Colchicum	*autumnale*	Autumn Saffron	Meadow Crocus	**M(traditional):**CR,SD:BHE,DHMY,DMDF,NGT,NSL,RLA,**E:**CR,SD:PS☠	Bown-265-01
Colchicum	*autumnale*	Autumn Saffron	Meadow Crocus	**M(traditional):**CR,SD:NGT,NJA,**E:**WP:PS☠	Kowa-018-01 Kowa-279-03
Colchicum	*autumnale*	Autumn Saffron	Meadow Crocus	**M(PS):**CR,FL,SD:NGT	Wyk-110-01, Wyk-405-20
Coleus	*barbatus*	Colforsin	Cole Mint	**M(traditional):**LF,RT:BHP,BHS,DALL,DCTM,DEGC,DOBE,DSIN,HIN,OSS **M(CS):**LF,RT:RLA	Bal2-051-02 Bal2-512-04
Coleus	*barbatus*	Colforsin	Cole Mint	**M(traditional):**LF,RT:BHP,BHS,DEGC,RLA,RLB	Wyk-405-21

Latin Genus	L. Species	E. Species	English Genus	C:Culinary,M:Medicinal,E:Economic	Source
Coleus	*forskohlii*			Synonym for *Coleus barbatus*	Wyk-405-21
Collinsonia	*canadensis*	Stone Root	Horse Balm	**M(traditional):**RT,RZ:OAS,OHM,SDG,SDR,SIB,UBL,UDR,UKS	Bown-265-02
Collinsonia	*canadensis*	Stone Root	Horse Balm	**M(folk):**RT,RZ:UKS	Krug-166-02
Collinsonia	*canadensis*	Stone Root	Horse Balm	**M(traditional):**LF,RT:OWN,SDP,UDR,UKS	Wyk-406-01
Combretum	*caffrum*	Bushwillow	Combretwood	**M(traditional):**BK:DCTM	Wyk-406-02
Commiphora	*abyssinica*	Ethiopian	Myrrh	**M(traditional):**RN:DASC,**E:**RN:IN,OR,PF	Wyk-406-03
Commiphora	*foliacea*	Leafy	Myrrh	**E:**RN:IN,OR,PF	Bown-265-03
Commiphora	*gileadensis*	Balm of Gilead	Myrrh	**M(traditional):**RN:DASC,OAS,**E:**RN:IN,OR,PF	Bown-265-04
Commiphora	*gileadensis*	Balm of Gilead	Myrrh	**M(traditional):**RN:DAMY,DASC	Wyk-406-04
Commiphora	*guidottii*	Scented	Myrrh	**M(traditional):**RN:DAMY,DASC	Wyk-406-05
Commiphora	*habessinica*			Synonym for *Commiphora abbysinica*	Bown-265-05
Commiphora	*mukul*	Guggul	Myrrh	**M(traditional):**RN:BHC,DOBE,NJA **M(ayurveda):**RN:BCR,OHM, **M(PS):**RN:BHC,BHS,DACN	Bal2-082-01 Bal2-525-01
Commiphora	*mukul*	Guggul	Myrrh	**M(traditional):**RN:BHC,OWN,**E:**RN:IN,OR,PF	Bown-265-06
Commiphora	*mukul*	Guggul	Myrrh	**M(traditional):**RN:BHC,OWN	Wyk-406-06
Commiphora	*myrrha*	African	Myrrh	**M(traditional):**RN:DAMW,DASC,DGUM,DHAL,DSIN,OUL	Bal1-074-06
Commiphora	*myrrha*	African	Myrrh	**M(folk):**BK-RN:SDG,DASC **M(traditional):**BK-RN:DAMW,DGUM,NSD,OEY,OWN,RLC,RST, **M(ayurveda):**BK-RN:DAMW,DSIN,NMP,OUL,OWN, **M(PS):**BK-RN:BHC,DAMW,DASC,RST	Bal2-105-01 Bal2-533-04
Commiphora	*myrrha*	African	Myrrh	**M(traditional):**RN:BCR,DASC,DBOI,DEAI,DGUM,DTNS,HFV,NMS,OAS,OUL,OWN,RLB,RLC,RST,SDG,SDP,**E:**RN:IN,OR,PF	Bown-265-07
Commiphora	*myrrha*	African	Myrrh	**M(traditional):**RN:DASC,DGUM,OAS,RLC,RST,SDG,**E:**RN:IR(mosquitoe)	Kowa-290-03 Kowa-396-01
Commiphora	*myrrha*	African	Myrrh	**M(PS):**BK-RN:DAMW,DASC,DCLD,DGUM,DTNS,OAS,OUL,OWN,RLC,RST	Wyk-111-01, Wyk-406-07
Commiphora	*wrightii*	Gum Guggul	Myrrh	**M(traditional):**RN:NJA	Wyk-406-08
Comptonia	*peregrina*	Exotic	Sweet Fern	**M(traditional):**LF:DVGN,OAS,OST,SDR,SDY	Bown-265-08
Comptonia	*peregrina*	Exotic	Sweet Fern	**E:**LF:DYE(alum-bright-yellow,tin-brilliant-yellow,chrome-strong-gold,copper-sulfate-&-iron-rich-brown)	Cas1-212-01
Conium	*maculatum*	Poison Parsley	Hemlock	**M(traditional):**LF,FR,SD:DCBR,DCHR,DEPI,NNC👽,OHM,OMC,OMT,RLA,**E:**LF,FR,SD:PS☠	Bown-266-01
Conium	*maculatum*	Poison Parsley	Hemlock	**M(folk):**WP,SD:DCTM(external),DEPI,DHSV(external), NNC👽, **E:**WP:PS☠	Krug-087-01
Conium	*maculatum*	Poison Parsley	Hemlock	**M(traditional):**WP,SD:NMS,NSD, **E:**LF,FR,SD:PS☠	Wyk-406-09
Consolida	*regalis*	Regal	Field Larkspur	**M(traditional):**FL:UDR, **M(traditional):**SD:DWRM,NMS,NNC👽	Wyk-406-10
Convallaria	*majalis*	May	Lily of the Valley	**M(traditional):**WP:BAS,BHA,BHP,BHS,UDR,**E:**FL,LF,OL:PF	Bown-266-02
Convallaria	*majalis*	May	Lily of the Valley	**M(folk):**FL,LF:BHE,BHS,UDR	Krug-104-01
Convallaria	*majalis*	May	Lily of the Valley	**M(PS):**WP(aerial):BHS,**E:**WP(aerial):PS☠	Wyk-112-01, Wyk-406-11
Convolvulus	*arvensis*	Field	Bindweed	**C:**LF:**SN**,**M(traditional):**LF,FL:BLC,HFV,SCN	Clev-088-05
Convolvulus	*arvensis*	Field	Bindweed	**M(traditional):**WP:SCN	Wyk-406-12
Convolvulus	*arvensis*	Field	Bindweed	**E:**WP,FL:DYE(alum-gold,chrome-green-camel,copper-green-gold,tin-yellow,iron-dark-moss-green)	Bliss-048-01
Convolvulus	*arvensis*	Field	Bindweed	**E:**WP:DYE(alum-khaki-green,chrome-yellow,alum-&-copper-dull-green)	Grae-097-02
Convolvulus	*floridus*	Bois De Rose	Bindweed	**E:**RT,SM:PF	Clev-088-06
Convolvulus	*scammonia*	Scammony	Bindweed	**M(traditional):**RT:SCN	Wyk-406-13
Convolvulus	*sepium*			Synonym for *Calystegium sepium*	Clev-088-07
Conyza	*canadensis*	Canadian Fleabane	Horsetail Daisy	**M(traditional):**WP:DWRG,OAS,OSS,OHM,SDR,RLB,UDR,UKC,**E:**OL:IR	Bown-266-03
Conyza	*canadensis*	Canadian Fleabane	Horsetail Daisy	**M(traditional):**WP:DCLD,OAS,SDR,UDR, **E:**OL:IR	Wyk-406-14
Conyza	*canadensis*	Canadian Fleabane	Horsetail Daisy	**E:**FL,LF,SM:DYE(alum-bright-gold,chrome-brass,copper-green-gold,tin-vivid-gold,iron-dark-olive-green)	Bliss-134-01
Copaifera	*guyanensis*	Guyana	Copal Gum	**E:**RN:OR,PF,PS☠	Bown-266-04
Copaifera	*langsdorffii*	Copiaban Balsam	Copal Gum	**M(traditional):**RN:DABI,DDND,NJA,DSIN, **E:**RN:HO	Bal2-053-01 Bal2-513-01
Copaifera	*langsdorffii*	Copiaban Balsam	Copal Gum	**M(traditional):**RN:DABA,DSXG,DVGN,OAS,OCB,OSS,SDR,RLB,RLC,UBL,UDR,**E:**RN:OR,PF,PS☠	Bown-266-05
Copaifera	*langsdorffii*	Copiaban Balsam	Copal Gum	M**(traditional)**:LF,OL:DASC,DSIN,SDR,UDR	Wyk-406-15
Copaifera	*martii*	Marti	Copal Gum	**E:**RN:OR,PF,PS☠	Bown-266-06
Copaifera	*multijuga*	Coumarin	Copal Gum	**E:**RN:OR,PF,PS☠	Bown-266-07

Latin Genus	L. Species	E. Species	English Genus	C:Culinary,M:Medicinal,E:Economic	Source
Copaifera	*officinalis*	Jesuit's Balsam	Copal Gum	**E:**RN:OR,PF,PS☠	Bown-266-08
Copaifera	*reticulata*	Stinking	Copal Gum	**E:**RN:OR,PF,PS☠	Bown-266-09
Coprinus	*comatus*	Shaggy Mane	Inky Cap Gill Mushroom	**E:**WP:DYE(alum-gray,chrome-gray)	Cas1-182-05
Coptis	*chinensis*	Chinese	Goldthread	**M(PS):**RT:DABA(*Eschrichia coli*),BDB,UUR	Bal2-054-01 Bal2-513-02
Coptis	*chinensis*	Chinese	Goldthread	**M(traditional):**RT:BCR,DABA,DAVR,DEAI,HFV, NMS,OEY,OUL,SDY,SIB	Bown-267-01
Coptis	*chinensis*	Chinese	Goldthread	**M(PS):**RZ:OEY,SDR,SIB,SUL	Wyk-113-01, Wyk-406-16
Coptis	*groenlandica*	Greenland	Goldthread	**M(traditional):**RT:OUL	Bown-266-10
Coptis	*japonica*	Japanese	Goldthread	**M(traditional):**RT:OUL	Bown-266-11
Coptis	*teeta*	Teet	Goldthread	**M(traditional):**RT:OUL	Bown-266-12
Coptis	*trifolia*	India	Goldthread	**M(traditional):**RT:OUL	Bown-266-13
Cordyceps	*sinensis*	Chinese	Caterpillar Fungus	**M(traditional):**WP:BHS,DCLU,DCTM,DIPT,BHC ,UKC, **M(SS):**WP:DEAT	Bal2-054-02 Bal2-513-03
Cordyceps	*sinensis*	Chinese	Caterpillar Fungus	**M(traditional):**WP:DABA,DCMT,DTBC,RLC,UK C	Bown-267-02
Coreopsis	*auriculata*	Ear Leaf	Tickseed	**E:**FL,SM:DYE(alum-orange,chrome-orange-red)	Grae-115-03
Coreopsis	*calliopsidea*	Dyer's	Tickseed	**E:**FL,SM:DYE(alum-orange,chrome-red-orange)	Grae-081-01
Coreopsis	*gigantea*	Giant	Tickseed	**E:**FL,PD,SD,SM:DYE(alum-light-orange,chrome-deep-orange)	Grae-082-01
Coreopsis	*tinctoria*	Dyer's	Tickseed	**E:**SD:DYE(alum-gold-brown,chrome-olive-green)	Grae-116-01
Coreopsis	*tinctoria*	Dyer's	Tickseed	**E:**FL:DYE(alum-bright-yellow,tin-bright-orange-yellow,chrome-rusty-orange)	Cas1-128-01
Coriandrum	*sativum*	Cilantro	Coriander	**M(traditional):**SD:NJA,OHM,RLC,SAP,SDG,**E:** OL:PF	Bown-267-03
Coriandrum	*sativum*	Cilantro	Coriander	**C:**LF,SH:**SN**,**C:**RT:FF,**C:**SD:**SP**,**C:**SD-OL,AF, **M(traditional):**SD-OL:DABA,NNR,NRH,SIB,**E:**SD-OL,PF	Clev-293-01
Coriandrum	*sativum*	Cilantro	Coriander	**C:**LF:**SN**,**C:**RT:FF,**C:**SD:**SP**, **M(traditional):**SD:SDG,ORH	Kowa-125-01
Coriandrum	*sativum*	Cilantro	Coriander	**C:**LF:**SN**,**C:**SD:**SP**, **M(folk):**SD:SDG	Krug-057-01
Coriandrum	*sativum*	Cilantro	Coriander	**M(PS):**FR:OBN,OWN,SAP,SDP,**M(traditional):** OL:NJA,NRH	Wyk-114-01, Wyk-406-17
Coriolus	*versicolor*	Kawaratake	Turkeytail Pore Fungus	**M(SS):**WP:DCTM,DFRA,DHPT	Bal2-114-02 Bal2-535-04
Cornicularia	*californica*			Synonym for *Kaernefeltia californica*	Cas2-031-13
Cornicularia	*divergens*			Synonym for *Bryocaulon divergens*	Cas2-034-20
Cornus	*alternifolia*	Alternate Leaf	Dogwood	**E:**LF:DYE(alum-yellow,chrome-gold,vinegar-tan-beige,alum-&-vinegar-beige-pink,iron-&-chrome-brown,),**E:**RT:DYE(alum-yellow-gold,chrome-warm-tan)	Cas1-137-01
Cornus	*florida*	Flowering	Dogwood	**E:**LF:DYE(alum-yellow,chrome-gold,vinegar-tan-beige,alum-&-vinegar-beige-pink,iron-&-chrome-brown,),**E:**RT:DYE(alum-yellow-gold,chrome-warm-tan)	Cas1-137-02
Cornus	*officinalis*	Cornelian Cherry	Dogwood	**M(traditional):**FR:BLC,BHP,DABA,DAFU,OAS, UDR,UKC	Bown-267-04
Cornus	*officinalis*	Cornelian Cherry	Dogwood	**M(traditional):**LF:NMP	Wyk-406-18
Coronilla	*varia*	Variable	Crown Vetch	**E:**FL,LF,SM:DYE(alum-light-gold,chrome-brass-gold, copper-tan-green,tin-bright-gold,iron-mouse-brown)	Bliss-098-01
Corydalis	*cava*	European Tuber	Larkspur Poppy	**M(traditional):**TU:BHP,DECT,HAN,HIN,NNC👽, NSD, **M(folk):**TU:DPRK,DWRM,OUL,OWN	Bal2-055-02
Corydalis	*cava*	European Tuber	Larkspur Poppy	**M(traditional):**TU:BCR,DABA,HIN,NBK,NMS,N SD	Bown-267-05
Corydalis	*cava*	European Tuber	Larkspur Poppy	**M(traditional):**FL:NNC👽,NSD	Wyk-406-19a
Corydalis	*solida*	M. East Tuber	Larkspur Poppy	**M(traditional):**TU:BCR,DABA,HIN,NBK,NMS,N SD	Bown-268-01
Corydalis	*solida*	M. East Tuber	Larkspur Poppy	**M(traditional):**FL:NNC👽,NSD	Wyk-406-19b
Corydalis	*yanhusuo*	Chinese Tuber	Larkspur Poppy	**M(folk):**TU:DPRK,DWRM,OUL,OWN **M(traditional):**TU:BHP,DECT,HAN,HIN,NNC👽, NSD	Bal2-055-02 Bal2-513-05
Corydalis	*yanhusuo*	Chinese Tuber	Larkspur Poppy	**M(traditional):**FL:NNC👽,NSD	Wyk-406-19c
Corymbia	*citriodora*	Lemon Scented	Gum Myrtle	**M(traditional):**LF,OL,RN:DABA,DAFU,DBOI,DC AN,DHSV,DIMP,DSEP,OAS,DALF,DDND	Bown-280-03
Corymbia	*gummifera*	Kino	Gum Myrtle	**M(traditional):**LF,OL,RN:SDY,UBL	Bown-280-06
Cotoneaster	*apiculatus*	Cranberry	Quinceberry Wild Rose	**E:**FR:DYE(alum-tan-rose)	Grae-150-01
Cotula	*coronopifolia*	Crown Leaf	Brass Buttons	**E:**WP:DYE(chrome-deep-brass-gold)	Grae-080-01
Crataegus	*azarolus*	Azarole	Hawthorn Berry	**M(traditional):**European	Wyk-115-02
Crataegus	*cuneata*	Wedge Leaf	Hawthorn Berry	**M(traditional):**FR:SDR	Bown-268-02
Crataegus	*cuneata*	Wedge Leaf	Hawthorn Berry	**E:**FL:DYE(alum-yellow-green,chrome-gold-brown)	Grae-166-02

Latin Genus	L. Species	E. Species	English Genus	C:Culinary,M:Medicinal,E:Economic	Source
Crataegus	*laevigata*	English	Hawthorn Berry	M(folk):FL,FR,LF:NJA,M(traditional):FL,FR,LF:BCR,BHS,DALZ,HML,M(PS):FL,FR,LF:BAA,BAS,BHA,BHC,BHP,DADD,HAN	Bal2-083-01 Bal2-526-01
Crataegus	*laevigata*	English	Hawthorn Berry	C:LF:**SN**,M(traditional):FR:BCR,BHS	Bown-268-03
Crataegus	*laevigata*	English	Hawthorn Berry	M(traditional):FL,FR,LF:BCR,BHP,BHS,NMS	Kowa-275-01 Kowa-288-03
Crataegus	*macrosperma*	May Apple	Hawthorn Berry	E:FR:DYE(alum-soft-rich-yellow,chrome-strong-gold,alum-&-tin-bright-yellow-orange,iron-khaki)	Cas1-150-01
Crataegus	*monogyna*	Quickthorn	Hawthorn Berry	M(traditional):FL,FR,LF:BAN,BHS,BCR,BHC	Bal1-072-03
Crataegus	*monogyna*	Quickthorn	Hawthorn Berry	M(folk):FL,FR,LF:NJA,M(traditional):FL,FR,LF:BCR,BHS,DALZ,HML,M(PS):FL,FR,LF:BAA,BAS,BHA,BHC,BHP,DADD,HAN	Bal2-083-02 Bal2-526-01
Crataegus	*monogyna*	Quickthorn	Hawthorn Berry	M(traditional):FR:BCR,BHS	Bown-268-04
Crataegus	*monogyna*	Quickthorn	Hawthorn Berry	M(CS):FR:BHS,BHP	Krug-086-02
Crataegus	*monogyna*	Quickthorn	Hawthorn Berry	M(CS):LF,FL,FR:BHS	Wyk-115-01, Wyk-406-20
Crataegus	*oxyacantha*	Sharpthorn	Hawthorn Berry	M(folk):FL,FR,LF:NJA,M(traditional):FL,FR,LF:BCR,BHS,DALZ,HML,M(PS):FL,FR,LF:BAA,BAS,BHA,BHC,BHP,DADD,HAN	Bal2-083-03 Bal2-526-01
Crataegus	*oxyacantha*	Sharpthorn	Hawthorn Berry	M(CS):FR:BHS,BHP	Krug-086-01
Crataegus	*pentagyna*	Five Seed	Hawthorn Berry	M(traditional):European	Wyk-115-05
Crataegus	*pinnatifida*	Chinese	Hawthorn Berry	M(traditional):FR:BCR,BHP,DABA,SDG,SIB	Bown-268-05
Crataeva				Synonym for *Crateva*	
Crateva	*nurvula*	Three Leaf	Garlic Pear	M(traditional):BK,LF:UEP,UKC,UKS,UUR	Bal2-150-01 Bal2-549-03
Crateva	*nurvula*	Three Leaf	Garlic Pear	M(traditional):BK,LF:DSIN,RLC,UKS,UUR	Wyk-406-21
Crithmum	*maritimum*	Sea Fennel	Samphire	C:LF:**VG**,M(traditional):WP:DOBE,UKC	Bown-268-06
Crithmum	*maritimum*	Sea Fennel	Samphire	M(traditional):WP:DGUM(scurvy),SFL,UDR	Wyk-406-22
Crocus	*nudiflorus*	Nude Saffron	Crocus	E:FL:DYE(orange-yellow)	Bown-269-01
Crocus	*sativus*	Saffron	Crocus	C:FL:AF,FC,FF,**SP**,M(traditional):FL:BCR,BHP,BLC,SDG	Bown-269-02
Crocus	*sativus*	Saffron	Crocus	C:FL(stigma):**SP**,M(traditional):FL(stigma):NMS,NSD,SFL,E:FL(stigma):PF	Clev-294-01
Crocus	*sativus*	Saffron	Crocus	C:FL(stigma):**SP**,M(traditional):FL(stigma):DCLD,DCTM,DPXS,HAP,SAP	Kowa-437-01
Crocus	*sativus*	Saffron	Crocus	C:FL(stigma):**SP**,M(folk):FL(stigma):HAP,E:FL(stigma):DYE(orange)	Krug-151-01
Crocus	*sativus*	Saffron	Crocus	M(PS):FS:NMS,RLA,E:FS:PS☠	Wyk-116-01, Wyk-406-23
Crocus	*vernus purpurea*	Purple Dutch	Crocus	E:FL:DYE(alum-turquoise-blue,chrome-apple-green)	Grae-116-02
Croton	*eleuteria*	Cascarilla	Croton	M(traditional):BK:SDG	Bown-269-03
Croton	*lechleri*	Dragon's Blood	Croton	M(traditional):WD-RN:DABA,DAFU,DAVR,DCAN,DFLU,DHSV,DISB,OWN,RLC,SDR,SUL	Bal2-124-01 Bal2-539-03
Croton	*tiglium*	Purging	Croton	M(traditional):OL,SD:DBOI,DMAL,NMT,DCLW,OSS,OSA,SCN,SDY,SIB,E:OL,SD:PS☠	Bown-269-04
Croton	*tiglium*	Purging	Croton	M(traditional):OL,SD:SPR,E:OL,SD:PS☠(carcinogenic)	Wyk-407-01
Cryptantha	*virgata*	Miner's Candle	Cat's Eye	E:FL,LF,SM:DYE(alum-gray-tan,chrome-olive,copper-pea-soup-green,tin-tan-gold,iron-olive)	Bliss-168-01
Cryptotaenia	*canadensis*	Honewort	Wild Chervil	C:LF,SL:**VG**	Bown-269-05
Cucumis	*sativus*	Sweet	Cucumber	C:FR:**VG**,M(traditional):FR:OEY,OSB,UDR,M(traditional):SD:DWTP,E:FR:CM,DS,	Bown-269-06
Cucumis	*sativus*	Sweet	Cucumber	E:LF,SM:DYE(vinegar-beige-yellow,alum-pale-yellow,chrome-gold,tin-bright-yellow,iron-gray-tan)	Cas1-131-01
Cucurbita	*maxima*	Winter Squash	Melon	C:SD:NT,M(traditional):SD:DWFK,DWRO,DWTP,UDR,UEP	Bown-269-06
Cucurbita	*maxima*	Winter Neckless Squash	Melon	E:SM:DYE(vinegar-tan,iron-gray-green), E:RD:DYE(alum-yellow,chrome-yellow-orange, tin-bright-yellow,iron-yellow-gray,chrome-&-tin-rust)	Cas1-209-01
Cucurbita	*maxima-pepo*	Pumpkin	Melon	M(traditional):SD:UEP	Bal1-075-10
Cucurbita	*maxima-pepo*	Pumpkin	Melon	C:FR:pie,M(folk):SD:DACN,DWTP,UEP	Krug-144-01
Cucurbita	*maxima-pepo*	Pumpkin	Melon	M(PS):SD,OL:DWRO,DWTP,OWN,UBL,UDR,UEP	Wyk-117-01, Wyk-407-02
Cuminum	*cyminum*	Mediterranean	Cumin	C:SD:**SP**,M(traditional):OL:DABA,DABI,M(traditional):SD:BLC,HAP,OAS,SDG	Bown-270-01
Cuminum	*cyminum*	Mediterranean	Cumin	C:SD:**SP**,M(traditional):SD:SAP,SDR,SFL,SIB,E:OL:PF	Clev-295-01
Cuminum	*cyminium*	Mediterranean	Cumin	M(traditional):SD:DCLD,NSD,SDP,SFL	Wyk-407-03
Cupressus	*sempervirens*	Evergreen	Cypress	M(traditional):OL:NMS,DSIN,RLC	Wyk-407-04
Curcuma	*amada*	Mango	Curcumin	C:RZ,TU:**SP**	Bown-270-02
Curcuma	*aromatica*	Yu Jin Turmeric	Curcumin	M(traditional):RZ,TU:BCR,BJA,BLC,NMP,SDG,	Bown-270-03

Latin Genus	L. Species	E. Species	English Genus	C:Culinary,M:Medicinal,E:Economic	Source
				SGB	
Curcuma	*longa*	Turmeric	Curcumin	**M(traditional):**RZ:DABI,DCTM,DFRA,BHC,BLC	Bal1-078-01
Curcuma	*longa*	Turmeric	Curcumin	**M(traditional):**RZ:BLC,DECT,DGUM,DHAL,DWRG,OUL,OWN,SCN,SDG,SFL,SGB,SIB,**M(PS):** RZ:BAS,DAID,DCBR,DCCL,DCPA,NJA,SUL	Bal2-146-01 Bal2-547-04
Curcuma	*longa*	Turmeric	Curcumin	**C:**RZ,TU,FC,**SP**,**M(traditional):**RZ,TU:DWRG,BCR,BDB,BJA,BLC,NMP,OAS,SDG	Bown-270-04
Curcuma	*longa*	Turmeric	Curcumin	**C:**RZ:**SP**,**M(traditional):**RZ:DCLD,SDG,SUL,**E:** RZ:DYE	Clev-296-01
Curcuma	*longa*	Turmeric	Curcumin	**C:**RZ:**SP**,**M(PS):**RZ:SFL,SGB,SUL	Wyk-118-01, Wyk-407-05
Curcuma	*longa*	Turmeric	Curcumin	**E:**RZ,TU:DYE(vinegar-bright-gold-orange)	Grae-003-01
Curcuma	*xanthorhiza*	Java Turmeric	Curcumin	**M(traditional):**RZ:SDP,SFL	Wyk-118-03
Curcuma	*zedoaria*	Zedoary	Curcumin	**C:**RZ,TU,**SP**,**M(traditional):**RZ,TU:BCR,BLC,DCCU,NMS,RLC,SDG	Bown-270-05
Curcuma	*zedoaria*	Zedoary	Curcumin	**C:**RZ:**SP**,**M(traditional):**RZ:SDG,SFL,SIB,**E:**RZ:DYE	Clev-297-01
Curcuma	*zedoaria*	Zedoary	Curcumin	**M(traditional):**RZ,OL:SDG,SGB	Wyk-407-06
Cuscuta	*americana*	Devil's Hair	Dodder	**E:**WP:DYE(alum-soft-yellow,chrome-rust-gold,copper-olive-green,tin-bright-butter-gold,iron-green-brown)	Bliss-106-01
Cuscuta	*epithymum*	Thyme	Dodder	**M(traditional):**SD:BES,BLC,UKC	Bown-270-06
Cuscuta	*epithymum*	Thyme	Dodder	**M(traditional):**WP:SCN,SGB	Wyk-407-07
Cuscuta	*epithymum*	Thyme	Dodder	**E:**WP:DYE(alum-yellow,chrome-ocher)	Grae-084-03
Cuscuta	*japonica*	Japanese	Dodder	**M(traditional):**SD:BLC,DIPT,DVGN,SDR,UKC	Bown-270-07
Cyanus	*segetum*	Blue	Cornflower	**C:**FL:**VG**,**M(traditional):**FL:OAS,OEY,OUL,OWN,**E:**FL:HO	Bown-257-03
Cyanus	*segetum*	Blue	Cornflower	**M(traditional):**FL:BLC,DABA,DFRA,OEY,OWN,SDG,SGB,UKC,**M(traditional):**LF:NRH,**E:**FL:DYE	Wyk-088-01, Wyk-403-15
Cyanus	*segetum*	Blue	Cornflower	**E:**FL:DYE(alum-beige-yellow,tin-bright-yellow,chrome-tan-gold),**E:**WP:DYE(alum-&-tin-chartreuse,copper-sulfate-&-iron,gray-green)	Cas1-096-01
Cyclamen	*purpurascens*	European	Cyclamen	**M(traditional):**RZ:NGT,NMP,NRH,OTA	Wyk-407-08
Cyclopia	*falcata*	Honey Bush	Cyclops Tea	**M(traditional):**WP:DFRA	Wyk-407-09a
Cyclopia	*genistoides*	Broom Bush	Cyclops Tea	**M(traditional):**WP:DFRA	Wyk-407-09b
Cyclopia	*intermedia*	Honey Bush	Cyclops Tea	**C:**LF:TE,**M(traditional):**LF,SM:BDB,DFRA	Wyk-119-01, Wyk-407-09c
Cylcopia	*subternata*			Synonym for *Cyclopia falcata*	Wyk-407-09a
Cydonia	*oblonga*	Middle East	Quince	**M(traditional):**SD:RLB,SDR	Wyk-407-10
Cymbopogon	*citratus*	Lemon Grass	Spikelet Beard Grass	**M(traditional):**WP:DWHN,OAS,**E:**WP:PF	Bal1-073-06
Cymbopogon	*citratus*	Lemon Grass	Spikelet Beard Grass	**C:**LF:TE,**C:**SL:**SP**,**M(traditional):**LF,SM:DABA,DAFU,DALF,DSCB,DWRG,NMS,SDG,**E:**OL:CM,HO,PF,DS	Bown-270-08
Cymbopogon	*citratus*	Lemon Grass	Spikelet Beard Grass	**C:**LF-BD:**SN**,**C:**SL:**SP**,**M(traditional):**LF-BD,SL:BLC,DASC,**E:**OL:CM	Clev-089-02, Clev-298-01
Cymbopogon	*citratus*	Lemon Grass	Spikelet Beard Grass	**M(PS):**LF,OL:DCFS,DCLD,NSD,SDP	Wyk-120-01, Wyk-407-11
Cymbopogon	*flexuosus*	East Indies Lemon Grass	Spikelet Beard Grass	**C:**OL:FF,**M(traditional):**LF,SM:DABA	Clev-089-03
Cymbopogon	*martini*	Ginger Grass	Spikelet Beard Grass	**E:**OL:IR,PF,DS	Bown-270-09
Cymbopogon	*nardus*	Citronella Grass	Spikelet Beard Grass	**C:**LF,OL:TE,**E:**OL:PF	Clev-089-05
Cynanchum	*vincetoxicum*	White	Swallowwort	**M(traditional):**RT,RZ:UDR,SPR	Wyk-431-10
Cynara	*cardunculus*	Cardoon	Vegetable Thistle	**C:**SL:**VG**	Clev-089-06
Cynara	*scolymus*	Globe Artichoke	Vegetable Thistle	**C:**FL:**VG**,**M(traditional):**LF:BAS,BDB,BHC,BJA,BLC,BSH,DHPT,SGB	Bown-271-01
Cynara	*scolymus*	Globe Artichoke	Vegetable Thistle	**M(PS):**LF:BLC,BHC,SDP,SGB	Wyk-121-01, Wyk-407-12
Cynoglossum	*germanicum*	Green	Hound's Tongue	**M(traditional):**LF,RT:OBN,OBR,DSIN,RLB,RLC	Clev-090-02
Cynoglossum	*officinale*	Rats & Mice	Hound's Tongue	**M(traditional):**WP:DISB,OBN,OBR,OHM,DSIN,RLB,RLC,SDR,**E:**WP:PS☠(causes dermatitis)	Bown-271-02
Cynoglossum	*officinale*	Rats & Mice	Hound's Tongue	**M(traditional):**LF,RT:OBN,OBR,DSIN,RLB,RLC	Clev-090-01
Cynoglossum	*officinale*	Rats & Mice	Hound's Tongue	**M(traditional):**RT:NNR,OAS,DSIN,OWN	Wyk-407-13
Cyperus	*articulatus*	Ground Almond	Sedge	**M(traditional):**RZ,RT,TU:SDP	Bown-271-03
Cyperus	*esculentus*	Chufa	Sedge	**C:**TU:**VG**(ISBN:0-670-86752-7, page 209)	Flowerdew
Cyperus	*esculentus*	Chufa	Sedge	**M(traditional):**TU:SDP,SFL	Wyk-407-14
Cyperus	*longus*	Sweet Galingale	Sedge	**E:**RZ,RT,TU:PF	Bown-271-04
Cyperus	*rotundus*	Coco Root	Sedge	**M(traditional):**RZ,RT,TU:BLC,NMS,SDG	Bown-271-05
Cyperus	*rotundus*	Coco Root	Sedge	**M(traditional):**RZ,TU:HFV,lactation aid	Wyk-407-15
Cypripedium	*parviflorum*	Lesser Yellow	Lady's Slipper Orchid	**M(traditional):**RZ:HAC,HIN,NMS,NSD	Bown-271-06
Cypripedium	*pubescens*	Greater Yellow	Lady's Slipper Orchid	**M(traditional):**RZ:HAC,HIN,NMS,NSD	Bown-271-07
Cypripedium	*pubescens*	Greater Yellow	Lady's Slipper Orchid	**M(traditional):**RT:NSD	Wyk-407-16

Latin Genus	L. Species	E. Species	English Genus	C:Culinary,M:Medicinal,E:Economic	Source
Cytisus	*scoparius*	Scotch	Broom	**M(traditional):**WP:BHS,NNC👽,UDR,**E:**WP:PS☠	Bown-271-08
Cytisus	*scoparius*	Scotch	Broom	**E:**LF,FL:DYE(alum-green-yellow),**E:**FL:DYE (alum-bright-yellow,chrome-yellow),**E:**WP:PS☠(fatal)	Kowa-055-01 Kowa-173-05
Cytisus	*scoparius*	Scotch	Broom	**C:**LF-BD:**VG, M(folk):**SH:UDR, **E:**SM:Broom	Krug-039-01
Cytisus	*scoparius*	Scotch	Broom	**M(PS):**WP(top):BCR,BHS,DSNB,UDR	Wyk-122-01, Wyk-407-17
Dactylina	*arctica*	Arctic	Finger Shield Lichen	**E:**WP:DYE(Gyrophoric-Acid-&-Ammonia-fuchsia)	Cas2-031-14
Daemonorops	*draco*	Dragon's Blood	Demon Palm	**E:**RN:plaster	Wyk-407-18
Dahlia	*pinnata*	Fern Leaf	Dahl Mum	**E:**FL(orange-&-red):DYE(tin-bright-orange-yellow,chrome-&-tin-rust),**E:**FL(fuchsia):DYE (baking-soda-strong-yellow-green,alum-yellow, blue-virtriol-dark-gold,iron-green),**E:**FL (magenta -&-purple):DYE(iron-gray-green,copper-sulfate-&-tin-green,chrome-&-iron-brown)	Cas1-132-01
Dahlia	*pinnata*	Fern Leaf	Dahl Mum	**E:**FL(yellow):DYE(alum-&-tin-maize-yellow,alum-&-tin-&-iron-gold-brown),**E:**FL(yellow-red):DYE(chrome-burnt-orange),**E:**FL,SD:DYE(chrome-bright-orange)	Grae-118-01
Damiana				Synonym for *Turnera diffusa*	
Daphne	*genkwa*	Genkwa	Daphne Thymeline	**M(traditional):**FL-BD:BHE,DABA,DSCB,RLB,RLC,SCN, UDR,**E:**FL-BD:PS☠	Bown-272-01
Daphne	*gnidium*	Knidos	Daphne Thymeline	**E:**FL-BD:PS☠	Bown-272-02
Daphne	*laureola*	Spurge Laurel	Daphne Thymeline	**E:**FL-BD:PS☠	Bown-272-03
Daphne	*mezereum*	Mezereon	Daphne Thymeline	**M(traditional):**FL-BD:NRH,OUL,**E:**FL-BD:PS☠	Bown-272-04
Daphne	*mezereum*	Mezereon	Daphne Thymeline	**M(traditional):**BK:DSIN, **E:**BK:PS☠(carcinogenic)	Wyk-407-19
Datura	*inoxia metel*	Metel Prickly Bur	Thorn Apple	**E:**LF,FL,SD:PS☠	Bown-272-05
Datura	*inoxia meteloides*	Metel Prickly Burr	Thorn Apple	**E:**LF,FL,SD:PS☠	Bown-272-06
Datura	*stramonium*	Jimsonweed	Thorn Apple	**M(traditional):**LF,FL,SD:DPRK,NMS,NNC👽,OSA,RLA,**E:**LF,FL,SD:PS☠	Bown-272-07
Datura	*stramonium*	Jimsonweed	Thorn Apple	**M(traditional):**LF,FL,SD:BCR,DEPI,HAN,NMS, OHM,OJA,DSIN,OWN,RLA,**E:**LF,FL,SD:PS☠	Kowa-346-01
Datura	*stramonium*	Jimsonweed	Thorn Apple	**M(folk):**LF:DPRK,NNC👽,RLA	Krug-097-01
Datura	*stramonium*	Jimsonweed	Thorn Apple	**M(PS):**LF:DPRK,NMS,NSD,RLA,RLC	Wyk-123-01, Wyk-407-20
Daucus	*carota*	Queen Anne's Lace	Carrot	**M(traditional):**WP:NGT,SDG,UBL,UDR,UKS,**M(traditional):**SD:BHE,SFL,**E:**OL:CM,PF	Bown-272-08
Daucus	*carota*	Queen Anne's Lace	Carrot	**M(folk):**LF:UBL	Krug-182-01
Daucus	*carota*	Queen Anne's Lace	Carrot	**M(PS):**WP(aerial):DWRM,NGT,NMS,SDG,SFL, UBL,UDR,UKC	Wyk-124-01, Wyk-407-21
Daucus	*carota*	Queen Anne's Lace	Carrot	E:WP:DYE(alum-&-tin-yellow-green,copper-sulfate-beige,iron-yellow-gray,chrome-tan)	Cas1-114-01
Daucus	*sativa*	Mediterranean	Carrot	**C:**RT:**VG**,**M(traditional):**RT:Anticancer,Eyesight,Skin	Bown-272-09
Daucus	*sativa*	Mediterranean	Carrot	**E:**WP:DYE(alum-bright-yellow,copper-sulfate-dark-gold,chrome-orange-gold,iron-brown,tin-bright-chartreuse)	Cas1-114-02
Delphinium	*staphisagria*	Staves Acre	Dolphin Larkspur	**M(traditional):**FL:UDR),**M(traditional):**SD:DWRM,NMS,NNC👽,**E:**SD:IC,PS☠	Wyk-407-22
Delphinium	*staphisagria*	Staves Acre	Dolphin Larkspur	**E:**FL(purple):DYE(tin-bright-yellow,copper-sulfate-soft-green,iron-tan-pink)	Cas1-135-01
Dendranthema	*grandiflora*	Florist	Mum	**M(traditional):**FL:BAA,BCR,BHP,BLC,HFV	Bown-272-10
Dendranthema	*zawadskii*	Zawadsky	Mum	**E:**FL(purple):DYE(alum-gray-turquoise)	Grae-114-02
Dendrobium	*nobile*			Synonym for *Callista nobilis*	Bown-273-01
Dermatocarpon	*miniatum*	Lead	Stippleback Warty Lichen	**E:**WP:DYE(NoLichen-Acids-olive-brown)	Cas2-034-21
Derris	*elliptica*	Oval Leaf	Derr Peaflower	**E:**RT:IC	Wyk-408-01
Desmodium	*affine*	White Flower Amor Seco	Tick Trefoil	**M(traditional):**RT:NSD	Bal2-412-01
Desmodium	*gangeticum*	Ganges Amor Seco	Tick Trefoil	**M(traditional):**RT:BLC,DCLD,SDP	Wyk-408-02
Desmodium	*intortum*	Green Leaf Amor Seco	Tick Trefoil	**M(traditional):**RT:NSD	Bal2-412-01
Desmodium	*tortuosum*	Twisted Amor Seco	Tick Trefoil	**M(traditional):**RT:NSD	Bal2-412-01
Desmodium	*triflorum*	Three Flower Amor Seco	Tick Trefoil	**M(traditional):**RT:NSD	Bal2-412-01
Dianthus	*caryophyllus*	Carnation	Pink	**C:**FL:AF,BF,**VG**,**M(traditional):**HFV,**E:**OL:PF	Bown-273-02
Dianthus	*caryophyllus*	Carnation	Pink	**C:**FL:BF,FF	Clev-090-03
Dianthus	*chinensis*	Chinese	Pink	**M(traditional):**WP:BHP,DABA,HFV,SCN,SDG,UBL,UKS	Bown-273-03

Latin Genus	L. Species	E. Species	English Genus	C:Culinary,M:Medicinal,E:Economic	Source
Dianthus	*plumarius*	Cottage	Pink	**C:**FL:BF,FF	Clev-090-04
Dianthus	*superbus*	Fringed	Pink	**M(traditional):**WP:DABA,SDG	Bown-273-04
Dianthus	*superbus*	Fringed	Pink	**M(traditional):**WP:UDR,UUR	Wyk-408-03
Dibaeis	*baeomyces*	Baeomycesic	Earth Beret Cup Lichen	**E:**WP:DYE(Baeomycesic-Acid-&-Ammonia-pale-salmon)	Cas2-031-04
Dictamnus	*albus*	Burning Bush	Gas Plant	**M(traditional):**RB:BJA,DABA,DAFU,DMSG,DSCB,HFV,NJA,DSIN	Bown-273-05
Dictamnus	*albus*	Burning Bush	Gas Plant	**M(traditional):**RT:DWRM,NMS,UDR	Wyk-408-04
Digitalis	*grandiflora*	Yellow	Finger Foxglove	**M(traditional):**LF:BHS,**E:**LF:PS☠	Bown-273-06
Digitalis	*lanata*	Grecian	Finger Foxglove	**M(traditional):**LF:BHS,**E:**LF:PS☠	Bown-273-07
Digitalis	*lanata*	Grecian	Finger Foxglove	**M(traditional):**LF:BHS	Kowa-281-05
Digitalis	*lanata*	Grecian	Finger Foxglove	**M(PS):**LF:BHE,BHS,**E:**LF:PS☠	Wyk-125-01, Wyk-408-05
Digitalis	*lutea*	Straw	Finger Foxglove	**M(traditional):**LF:BHS,**E:**LF:PS☠	Bown-273-08
Digitalis	*purpurea*	Purple	Finger Foxglove	**M(traditional):**LF:BHS,**E:**LF:PS☠	Bown-273-09
Digitalis	*purpurea*	Purple	Finger Foxglove	**M(traditional)**:LF:BHS,BHP,**E:**FL:DYE(alum-chartreuse)	Kowa-197-01 Kowa-281-02
Digitalis	*purpurea*	Purple	Finger Foxglove	**M(folk)**:LF:BHS,BHE,BLC	Krug-077-01
Digitalis	*purpurea*	Purple	Finger Foxglove	**E:**FL:DYE(alum-chartreuse)	Grae-121-02
Dionaea	*muscipula*	Carolina	Venus Flytrap	**M(traditional):**WP:DFRA(known as Carnivora)	Wyk-408-06
Dioscorea	*batatas*			Synonym for *Dioscorea polystachya*	Wyk-408-07
Dioscorea	*collettii hypoglauca*	Glaucous Collett	Yam	**M(traditional):**RT,RZ,TU:NRH,SDG,UKC	Bown-273-10
Dioscorea	*composita*	Tepinapa	Yam	**E:contraceptive**☠	Kowa-282-04
Dioscorea	*dregiana*	Isidikawa	Yam	**M(traditional):**RZ:NSD	Wyk-408-08
Dioscorea	*hypoglauca*			Synonym for *Dioscorea collettii hypoglauca*	Bown-273-10
Dioscorea	*mexicana macrostachya*	Large Spike Mexican	Yam	**E:contraceptive**☠	Bown-273-11
Dioscorea	*macrostachya*			Synonym for *Dioscorea mexicana macrostachya*	Bown-273-11
Dioscorea	*nipponica*	Japanese	Yam	**M(traditional):**RT,RZ,TU:NRH,SDG,UKC	Bown-273-12
Dioscorea	*opposita*			Synonym for *Dioscorea oppositifolia*	Bown-274-01
Dioscorea	*oppositifolia*	Indochinese	Yam	**M(traditional):**RZ:BDB,NJA,NMP,NMS,NNR,NRH,NSD,RLC,SIB,UKS, **M(PS):**RZ:BHS	Bal2-152-02 Bal2-550-02
Dioscorea	*oppositifolia*	Indochinese	Yam	**M(traditional):**RT,RZ,TU:NRH,SDG,UKC	Bown-274-01
Dioscorea	*polystachya*	Cinnamon Vine	Yam	**M(traditional):**RZ:tonic	Wyk-408-07
Dioscorea	*quaternata*			Synonym for *Dioscorea villosa quaternata*	Bown-273-13
Dioscorea	*villosa*	Wild	Yam	**M(traditional):**RZ:BDB,NJA,NMP,NMS,NNR,NRH,NSD,RLC,SIB,UKS, **M(PS):**RZ:BHS	Bal2-152-03 Bal2-550-02
Dioscorea	*villosa*	Wild	Yam	**M(traditional):**RZ:DHOG,NMS,SGB,UKS	Bal1-078-05
Dioscorea	*villosa*	Wild	Yam	**M(traditional):**RT,RZ:DCRO,DWHP,NJA,NMP,NMS,RLA,RLB,SDV,SGB,SIB	Bown-274-02
Dioscorea	*villosa*	Wild	Yam	**M(folk):**RT:NMP,SIB	Krug-182-02
Dioscorea	*villosa*	Wild	Yam	**M(PS):**TU,RZ:NMS,NRH,RLC,SGB,SIB	Wyk-126-01, Wyk-408-09
Dioscorea	*villosa quaternata*	Four Leaf Wild	Yam	**M(traditional):**RT,RZ:DCRO,DWHP,NJA,NMP,NMS,RLA,RLB,SDV,SGB,SIB	Bown-273-13
Diospyros	*kaki*	Kaki	Persimmon	**C:**FR:JJ,PU,**M(traditional):**FR:BHP,OAS,OHM,RLB,RLC,SCN,SDR	Bown-274-03
Diospyros	*kaki*	Kaki	Persimmon	**M(traditional):**FR:HYE,HYO	Wyk-408-10
Diploschistes	*scruposus*	Scrup	Crater Nipple Tremor Oster Lichen	**E:**WP:DYE(Atranorin-&-Lecanoric-Acid-&-Ammonia-cherry)	Cas2-031-15
Dipsacus	*fullonum*	Venus Cup	Teasel	**M(traditional):**RT:OAS,SDR,UDR	Wyk-408-11
Dipsacus	*sylvestris*	Forest	Teasel	**E:**LF,FL,SM:DYE(alum-khaki,chrome-khaki,copper-khaki,tin-brown-gold,iron-dark-khaki)	Bliss-254-01
Dipteryx	*odorata*	Scented	Tonka Bean	**C:**SD:AF,BF,CF,**M(traditional):**SD:DWHP,**E:**SD:PF	Bown-274-04
Dipteryx	*odorata*	Scented	Tonka Bean	**E:**SD:PF	Wyk-408-12
Dodonaea	*viscosa*	Sticky	Hop Soapwood	**M(traditional):**LF:HFV,OAS,OTA,OWN,RST	Bown-274-05
Dorema	*ammoniaca*	Gum Ammoniac	Gift Parsley	**M(traditional):**RN:NMS,OJA,RLB,RLC,**E:**RN:GR,PF	Bown-275-01
Dorema	*ammoniaca*	Gum Ammoniac	Gift Parsley	**M(traditional):**RN:NJA,NMS,RLB,RLC,**E:**RN:PF	Wyk-127-01, Wyk-408-13
Dorstenia	*contrajerva*	Contrayerva	Dorsten Mulberry	**M(traditional):**RZ:HFV	Wyk-408-14
Drimia	*indica*	India	Bitter Squill	**M(traditional):**BR:BHE,**E:**BR:PS☠	Bown-275-02
Drimia	*maritima*	Seaside	Squill	**M(traditional):**BR:BHE,BHS,DSBR,DWHP,DDND,RLA,RLB,RLC,UDR,**E:**BR:PS☠	Bown-275-03
Drimia	*maritima*	Seaside	Squill	**M(traditional):**BR:BHS,RLC,UDR	Wyk-430-17
Drimys	*winteri*	Winter's Bark	Pepperwood	**C:**FR:PU(berry),**M(traditional):**BK:SDG,SIB	Bown-275-04
Drimys	*winteri*	Winter's Bark	Pepperwood	**M(traditional):**BK:SDG	Wyk-408-15
Drosera	*rotundifolia*	Round Leaf	Sundew	**M(traditional):**WP:DHMY,DFLU,DWHP,HAP,RLA,RLB,RLC,SIB,UDR	Bown-275-05
Drosera	*rotundifolia*	Round Leaf	Sundew	**M(PS):**WP:DWHP,NMS,DSIN,RLC,SDP	Wyk-128-01, Wyk-408-16
Dryobalanops	*aromatica*	Borneo	Camphor	**M(traditional):**OL,RN:DABA,DBOI,DCHL,DPNB,D	Bown-275-06

Latin Genus	L. Species	E. Species	English Genus	C:Culinary,M:Medicinal,E:Economic	Source
				PNV,DWRG,HFV,NMS,OEY,ORH,OSA,RST,**E:**OL:PF,**E:**WD:IC	
Dryopteris	*crassirhizoma*	Thick Rhizome	Buckler Fern	**M(traditional):**RZ:HFV,NNC👽,OAS	Bown-276-01
Dryopteris	*felix-mas*	Wood Fern	Buckler Fern	**M(traditional):**RZ:DABA,DAVR,DFLU,DMNB,DMNV,DMSL,DMUM,DPNB,DPNV,DWFK,DWTP,**E:**RZ:PS☠	Bown-276-02
Dryopteris	*felix-mas*	Wood Fern	Buckler Fern	**M(traditional):**RZ:DRCK,DWTP,NRH,OWN,**E:**RZ:PS☠	Clev-090-05
Dryopteris	*felix-mas*	Wood Fern	Buckler Fern	**M(traditional):**RZ:DWTP,DSIN, **E:**RZ:**induced abortion** ☠	Wyk-408-17
Dryopteris	*oreades*	Dwarf	Buckler Fern	**M(traditional):**RZ:NNC👽	Bown-276-03
Duboisia	*myoporoides*	Eye Plant	Dubois Corkwood	**M(traditional):**LF:NNC👽(atropine,hyocine,hyoscamine,scopolamine)	Bown-276-04
Duboisia	*myoporoides*	Eye Plant	Dubois Corkwood	**M(traditional):**LF:NNC👽,NMS,NSD	Wyk-408-18
Dulacia	*inopiflora*	Balsam	Potentwood	**M(traditional):**BK,RT,WD:HAP	Bown-276-05
Ecballium	*elaterium*	Mediterranean	Squirting Cucumber	**M(traditional):**FR:BHE,DSHZ,NRH,OJA,RSN,SPR,**E:**FR:PS☠	Bown-276-06
Ecballium	*elaterium*	Mediterranean	Squirting Cucumber	**M(traditional):**FR:SPR	Wyk-408-19
Echinacea	*angustifolia*	Black Sampson	Coneflower	**M(folk):**RT,RZ:DISB,DMEA,DSNB,DSXG,HAC,OBN,RLC,SDG,**M(traditional):**RT,RZ:DEAI,DFLU,OSA,OUL,OWN **M(PS):**RT,RZ:DCLD	Bal2-061-01 Bal2-515-01
Echinacea	*angustifolia*	Black Sampson	Coneflower	**M(traditional):**RT,RZ:OWN	Bown-276-07
Echinacea	*angustifolia*	Black Sampson	Coneflower	**M(folk):**RT,RZ:BLC,DABA(*Streptococcus*),DASC,DABI,DCTM,DSNB,DSXS,HAC,HFV,NRH,OHM,DSIN,OST,OWN,SDG,UBL	Kowa-176-01 Kowa-286-02
Echinacea	*angustifolia*	Black Sampson	Coneflower	**M(folk):**RT:DBOI,DISB,DSIN,DSNB,OBN,OSA	Krug-066-01
Echinacea	*angustifolia*	Black Sampson	Coneflower	**M(traditional):**RT:DABA,DAVR	Wyk-408-20
Echinacea	*pallida*	Pale Purple	Coneflower	**M(traditional):**RT,RZ:OWN	Bown-276-08
Echinacea	*pallida*	Pale Purple	Coneflower	**M(CT):**RT:DABA,DAVR,DCLD,DFLU	Wyk-129-01
Echinacea	*purpurea*	Purple	Coneflower	**M(traditional):**LF,RT:DAVR,DCLD,DFLU,DSNB,SIB	Bal1-069-03
Echinacea	*purpurea*	Purple	Coneflower	**M(folk):**RT,RZ:DISB,DMEA,DSNB,DSXG,HAC,OBN,RLC,SDG,**M(traditional):**RT,RZ:DCLD,DFLU,OSA,OUL,OWN,**M(CommE):**RT,RZ:DCAN,DCFS,DCLD,DFLUHFV,OBN,OWN,RLB,RLC,UUR	Bal2-061-02
Echinacea	*purpurea*	Purple	Coneflower	**M(traditional):**RT,RZ:DABA,DACN,DAFU,DAVR,DBOI,DGNG,DHSV,DSEP,OSA,OSS,OWN	Bown-276-09
Echinacea	*purpurea*	Purple	Coneflower	**M(CS):**FL-JU:DCLD,DSIN,OUL,OWN	Wyk-130-01, Wyk-408-21
Echinochloa	*crus-galli*	Barnyard	Cockspur Grass	**E:**LF,SL,SM:DYE(alum-tan-green,chrome-gray-green,copper-dark-mint-green,tin-gold,iron-gray-brown)	Bliss-264-01
Echium	*lycopsis*	Purple	Viper's Bugloss	**C:**FL:BF,CF,**C:**LF:**VG**,**M(traditional):**FL:DASC,**M(traditional):**WP:DSNB	Clev-091-02
Echium	*vulgare*	Blue Devil	Viper's Bugloss	**C:**FL:BF,CF,**C:**LF:**VG**,**M(traditional):**FL:DASC,**M(traditional):**WP:DSNB	Clev-091-01
Echium	*vulgare*	Blue Devil	Viper's Bugloss	**M(traditional):**RT:DEPI,OWN, **E:**RT:PS☠(carcinogenic)	Wyk-408-22
Eclipta	*alba*			Synonym for *Eclipta prostata*	Wyk-408-23
Eclipta	*prostrata*	Prostrate	Ecliptic Daisy	**M(traditional):**WP:OAS,BCR,BLC,DDPT,DHPT,SDG,UKC	Bown-277-01
Eclipta	*prostrata*	Prostrate	Ecliptic Daisy	**M(traditional):**WP:DAFU,OAS,DSIN,RLA	Wyk-408-23
Elaeagnus	*angustifolia*	Russian	Agnes Olive	**E:**LF:DYE(alum-yellow,tin-bright-gold,chrome-deep-gold-olive,iron-yellow-gray,vinegar-warm-beige)	Cas1-186-01
Elettaria	*cardamomum*	India	Cardamom	**C:**SD:**SP**,**M(traditional):**OL,SD:RLB,RLC,SDG,UKC**E:**OL:PF	Bown-277-02
Elettaria	*cardamomum*	India	Cardamom	**C:**PD,SD:**SP**,**M(traditional):**PD,SD:HAP,SDG,**E:**OL:PF	Clev-299-01
Elettaria	*cardamomum*	India	Cardamom	**C:**SD:**SP**,**M(traditional):**SD,OL:DAMW,SDG,SFL	Kowa-065-01
Elettaria	*cardamomum*	India	Cardamom	**C:**SD:**SP**,**M(PS):**FR,SD,OL:DABI,SGB, **M(PS)**:SD:DAMW,HAP,SDG,RLA,RLB,RLC,UUR	Wyk-131-01, Wyk-409-01
Eleutherococcus	*senticosus*	Siberian	Eleuthero Ginseng	**M(traditional):**RT:DCFS,DIPT,NJA,SAP,UKC, **M(PS):**RT:BHC,DAVR,DCTM,DMUM	Bal2-132-01 Bal2-543-01
Eleutherococcus	*senticosus*	Siberian	Eleuthero Ginseng	**M(traditional):**RB,RT:BCR,BSH,HIN	Bown-277-03
Eleutherococcus	*senticosus*	Siberian	Eleuthero Ginseng	**M(PS):**RT,RZ:DCFS,HAN	Wyk-132-01, Wyk-409-02
Elymus	*repens*	Couch Grass	Wild Rye	**M(traditional):**RZ,SD:RST,UKC,UKS	Bal2-056-01
Elymus	*repens*	Couch Grass	Wild Rye	**M(traditional):**RZ:BHC,NGT,NRH,UBL,UEP,UKC	Bown-277-04
Elymus	*repens*	Couch Grass	Wild Rye	**M(folk):**RZ:UBL,UEP	Krug-058-02
Elymus	*repens*	Couch Grass	Wild Rye	**M(traditional):**RT,RZ:DABI,NGT,NRH,UBL,UEP	Wyk-133-01,

Latin Genus	L. Species	E. Species	English Genus	C:Culinary,M:Medicinal,E:Economic	Source
				,UKC,UDR	Wyk-409-03
Embelia	*ribes*	False Pepper	Kilioe	**M(traditional):**LF:DWRM,SDP,UDR, E:LF:**contraceptive** ☠	Wyk-409-04
Entada	*phaseoloides*	Queensland Bean	Nicker Vine	**M(traditional):**SD:NSD,SDG,**E:**SD:HO	Wyk-409-05
Ephedra	*distachya*	European	Joint Fir	**M(traditional):**SM:DALL,DCLD,DOBE,HAN,OC C,RLA,RLB	Bal1-069-05
Ephedra	*distachya*	European	Joint Fir	**M(traditional):**SM:BHS,DALL,DFLU,RLA,UDR, UKC	Bown-278-01
Ephedra	*distachya*	European	Joint Fir	**M(traditional):**SM:BHS,RLA,RLB	Kowa-281-02
Ephedra	*distachya*	European	Joint Fir	**M(traditional):**WP,RT:RLB	Wyk-409-06
Ephedra	*equisetina*	Horsetail	Joint Fir	**M(traditional):**SM:BHE,NJA,RLA	Bown-278-02
Ephedra	*gerardiana*	Himalayan	Joint Fir	**M(traditional):**SM:BHE,NJA,RLA	Bown-278-03
Ephedra	*intermedia*	Middle East	Joint Fir	**M(traditional):**SM:BHE,NJA,RLA	Bown-278-04
Ephedra	*sinica*	Ma Huang	Joint Fir	**M(traditional):**SM:BHE,NJA,RLA	Bown-278-05
Ephedra	*sinica*	Ma Huang	Joint Fir	**M(CS):**SM:RLA,RLB,RRN,RSN	Wyk-134-01, Wyk-409-07
Ephedra	*trifurca*	Brigham Tea	Joint Fir	**M(traditional):**SM:UDR	Bown-278-06
Ephedra	*viridis*	Mormon Tea	Joint Fir	**C:**SM:TE,**M(traditional):**SM:DCLD,DSXG,DSXG ,HFV,UDR	Kowa-184-01
Ephedra	*viridis*	Mormon Tea	Joint Fir	**E:**SM:DYE(alum-light-tan,chrome-rosy-tan,copper-green-tan,tin-peach,iron-gray)	Bliss-172-01
Epigaea	*repens*	Trailing	Ground Heath	**M(traditional):**WP:OAS,UBL,UDR,UKC,UKS	Bown-278-07
Epilobium	*angustifolium*	Fireweed	Willow Evening Primrose	**M(traditional):**WP:OWN,RLC,SCN,SDG	Wyk-135-02
Epilobium	*angustifolium*	Fireweed	Willow Evening Primrose	**E:**FL,LF,SL:DYE(alum-yellow,chrome-rust,copper-green-gold,tin-bright-yellow,iron-dark-forest-green)	Bliss-116-01
Epilobium	*parviflorum*	Small Flower	Willow Evening Primrose	**M(traditional):**WP:UEP	Wyk-135-01, Wyk-409-08
Epimedium	*brevicornum*	Inyo Kaku	Horny Goatweed	**M(PS):**LF:DOST	Bal2-064-02
Epimedium	*grandiflorum*	Yin Yang Yuo	Horny Goatweed	**M(traditiona):**LF:DCFS,DCTM,DIFR,DIPT,HAP, HML,NBK,NJA	Bal2-064-01 Bal2-516-03
Epimedium	*sagittatum*	Barrenwort	Horny Goatweed	**M(traditional):**WP:BHP,BLC,DIPT,HAP,NBK,NJ A,RLA,RLB,RLC,**E:**WP:PS☠	Bown-278-08
Equisetum	*arvense*	Field	Horsetail	**M(traditional):**SM:BHE,BHS,DOST,DRCK,DSF T,DWHN,NJA,NBF,NMS,OBN,OWN,RLB,SGB,U DR	Bal1-072-06
Equisetum	*arvense*	Field	Horsetail	**C:**SM:**VG**,**M(folk):**SM:DTBC,DWHN,NGT,**M(tra ditional):**SM:NJA,OWN,UBL,UKC,RLB,RLE,SU L,**M(PS):**SM:DABA,OWN,UKS,UUR	Bal2-087-01
Equisetum	*arvense*	Field	Horsetail	**M(traditional):**SM:OAS,UBL,UEP,UUR	Bown-278-09
Equisetum	*arvense*	Field	Horsetail	**M(traditional):**SM:DTBC,SDR,UBL,**E:**SM:DYE(a lum-yellow-green,iron-green),PS☠	Kowa-341-01
Equisetum	*arvense*	Field	Horsetail	**M(folk):**SM:DWHN,RLC,UBL,UDR,UEP,UKS	Krug-093-01
Equisetum	*arvense*	Field	Horsetail	**M(traditional):**SM:BHE,OAS,DSIN,OWN,UDR,U KS,UUR	Wyk-136-01, Wyk-409-09
Equisetum	*arvense*	Field	Horsetail	**E:**SL:DYE(alum-bright-yellow,chrome-brass-gold,copper-green-gold,tin-bright-light-yellow,iron-moss-green)	Bliss-216-01
Equisetum	*arvense*	Field	Horsetail	**E:**LF:DYE(alum-&-copper-sulfate-soft-green,alum-light-green,copper-sulfate-yellow-green,chrome-dark-yellow,tin-brilliant-yellow,iron-gray-green)	Cas1-153-01
Equisetum	*hymale*	Scouring Rush	Horsetail	**M(traditional):**SM:OAS,OEY,UDR,UKC	Bown-278-09
Erigeron	*speciosus*	Showy	Fleabane Daisy	**E:**LF,FL,SM:DYE(alum-dark-yellow-green,chrome-camel,copper-dark-green-gold,tin-bright-gold,iron-green-black)	Bliss-222-01
Eriobotrya	*japonica*	Japanese	Loquat	**C:**FR,PU,**M(traditional):**LF,FR:DABA,DAVR,RL B,RLC	Bown-278-10
Eriodictyon	*californicum*	Mountain Balm	Woolly Tarwood	**M(traditional):**LF:DHYF,HFV,NMS,RLA,RLB,RS N,RTL	Bown-279-01
Eriodictyon	*californicum*	Mountain Balm	Woolly Tarwood	**M(traditional):**LF:RLC,UDR	Wyk-409-10
Eriogonum	*umbellatum nevadense*	Sulfur Flower Umbel	Woolly Buckwheat	**E:**FL,SL:DYE(alum-green-gold,chrome-rust,copper-dark-olive,tin-bright-gold,iron-green-black)	Bliss-238-01
Eriophyllum	*staechadifolium*	Seaside	Woolly Aster	**E:**FL:DYE(alum-bronze-gold,chrome-gold-brown)	Grae-057-01
Erodium	*cicutarium*	Redstem Filaree	Heron's Bill	**E:**WP:DYE(alum-beige,chrome-beige-green,copper-khaki,tin-beige-yellow,iron-charcoal-gray)	Bliss-236-01
Erodium	*texanum*	Texas Filaree	Heron's Bill	**E:**WP:DYE(alum-&-wool-brown-green,alum-&-cotton-green-gray,alum-&-jute-moss-green)	Grae-088-01
Eruca	*vesicaria*	Arugula	Rocket Cress	**C:**LF:**VG**,**C:**FL,OL,SD,PD:**SN**,**M(traditional):**L F:RLC	Bown-279-02
Eruca	*vesicaria sativa*	Sweet Arugula	Rocket Cress	**C:**LF:**SN**,**C:**PD:**VG**,**M(traditional):**LF:RLC,**M(t raditional):**SD:OBR	Clev-091-03

Latin Genus	L. Species	E. Species	English Genus	C:Culinary,M:Medicinal,E:Economic	Source
Eryngium	*aquaticum*	Button Snakeroot	Eryngo	**M(traditional):**RT:UKC	Bown-279-03
Eryngium	*campestre*	Field	Eryngo	**M(traditional):**RT:DWHP,DSIN,UUR	Bown-279-04
Eryngium	*campestre*	Field	Eryngo	**M(traditional):**RT:RLC,UDR,UUR	Wyk-409-11
Eryngium	*foetidum*	Culantro	Eryngo	**C:**LF,RT:**SN**,**M(traditional):**LF,RT:BHP,DEPI,HFV,NMS,SDG	Bown-279-05
Eryngium	*maritimum*	Sea Holly	Eryngo	**C:**RT:CF,JJ,**M(traditional):**RT:RLC,UBL,UDR,UEP,UKC,UUR	Bown-279-06
Eryngium	*planum*	Transylvania	Eryngo	**M(traditional):**RT:DWHP	Bown-279-07
Eryngium	*yuccifolium*	Rattlesnake	Eryngo	**M(traditional):**RT:UKC	Bown-279-08
Erysimum	*cheire*	Aegean	Wallflower	**M(traditional):**LF:RLC,RST,UDR	Wyk-409-12
Erysimum	*crepidifolium*	Hedge Mustard	Wallflower	**M(traditional):**WP:helveticoside	Wyk-409-13
Erythrina	*variegata*	India	Coral Tree	**M(traditional):**BK,LF:DSIN,SDP	Wyk-409-14
Erythrophleum	*suaveolens*	Sweet	Ordeal Tree	**M(traditional):**BK:BHS,**E:**BK:PS☠(arrow tips)	Wyk-409-15
Erythroxylum	*cataractacum*	Cocaine Waterfall	Tropical Redwood	**M(traditional):**LF:HPS,**E:**LF:PS☠	Bown-279-09
Erythroxylum	*catuaba*	Catuaban	Tropical Redwood	**M(traditional):**BK:DIPT,HAP,UEP,**M(PS):**BK:DABA(*Escherichia coli, Staphylococcus aureus*),DAID	Bal2-044-01 Bal2-510-01
Erythroxylum	*coca*	Cocaine Bolivia	Tropical Redwood	**C:**LF:BF(Coca Cola),**M(traditional):**LF:HPS,NNC💀	Bown-279-10
Erythroxylum	*coca*	Cocaine Bolivia	Tropical Redwood	**M(PS):**LF:DCFS	Wyk-137-01, Wyk-409-16
Erythroxylum	*novogranatense*	Cocaine Columbia	Tropical Redwood	**M(traditional):**LF:HPS,**E:**LF:PS☠	Bown-279-11
Eschscholzia	*californica*	California	Eschscholtz Poppy	**M(traditional):**WP(aerial):BLC,HIN,NNC💀,OTA,UBL, **M(PS):**WP(aerial):HAN,NSD	Bal2-042-01 Bal2-509-02
Eschscholzia	*californica*	California	Eschscholtz Poppy	**M(traditional):**WP:HIN,NMS,NNC💀,UDR	Bown-280-01
Eschscholzia	*californica*	California	Eschscholtz Poppy	**M(PS):**WP(aerial):BLC,HAN,HIN,NSD,OTA,SGB	Wyk-138-01, Wyk-409-17
Eucalyptus	*camaldulensis*	Red	Gum Tree	**M(traditional):**LF,OL,RN:HFV,OAS,OWN,RST,SDR	Bown-280-02
Eucalyptus	*citriodora*			Synonym for *Corymbia citriodora*	Bown-280-03
Eucalpytus	*cineria*	Ashy Stringy Bark	Gum Tree	**E:**LF:DYE(alum-maroon-red,direct-red-orange)	Grae-165-02
Eucalyptus	*coccifera*	Tasmanian Snow	Gum Tree	**E:**LF:DYE(alum-mustard-gold,alum-&-iron-dark-olive-green)	Grae-164-03
Eucalyptus	*dives*	Peppermint	Gum Tree	**M(traditional):**LF,OL:OAS,DASC,DCLD,DFLU,OJA,ONR,OSC,RLB,RST	Bown-280-04
Eucalyptus	*globulus*	Tasmanian Blue	Gum Tree	**M(traditional):**BK,LF,OL:BCR,DASC,DCLD,OCC,RLC	Bal1-069-06
Eucalyptus	*globulus*	Tasmanian Blue	Gum Tree	**M(folk):**LF-OL,SM-OL:BDB,BLC,DACN,DAMW,DFLU,DGUM,DSXG,DWHP,HFV,NNR,OUL,OWN,RLA,SAP,SGB,SIB,UBL,**M(traditional):**LF-OL,SM-OL:DAFU,DASC,DCLD,DALL,DISB,DTDC,HAC,OWN,RLB,RLC,**M(PS):**LF-OL,SM-OL:NRH,RLB,RLC	Bal2-065-01 Bal2-517-02
Eucalyptus	*globulus*	Tasmanian Blue	Gum Tree	**M(traditional):**LF,OL:DABA,DCLD,DFLU,HFV,NMS,OSA,OWN,RLB,RLC,RSN,**E:**OL:SV,**E:**WD:WL,**E:**LF,OL:PS☠	Bown-280-05
Eucalyptus	*globulus*	Tasmanian Blue	Gum Tree	**M(traditional):**BK-OL,LF-OL,RT-OL:DABA,DASC,DMAL,OAS,OSA,DSIN,RCR,RLA,RLB,RLC,**E:**LF-OL:CM	Kowa-185-01 Kowa-286-03
Eucalyptus	*globulus*	Tasmanian Blue	Gum Tree	**M(folk):**LF,OL:DASC,RLC,RST,**E:**WP:paper	Krug-069-01
Eucalyptus	*globulus*	Tasmanian Blue	Gum Tree	**M(PS):**LF,OL:DASC,DCLD,ORH,DSIN,RLB,RLC,RRN	Wyk-139-01, Wyk-409-18
Eucalyptus	*globulus*	Tasmanian Blue	Gum Tree	**E:**BK:DYE(alum-dark-khaki-green),**E:**LF:DYE(wool-camel-tan),**E:**SH:DYE(alum-&-iron-&-wool-green,alum-&-iron-&-cotton-gray)	Grae-163-02
Eucalyptus	*gummifera*			Synonym for *Corymbia gummifera*	Bown-280-06
Eucalyptus	*haemastoma*	Blood Kino	Gum Tree	**M(traditional):**LF,OL,RN:SDY,UBL	Bown-280-07
Eucalyptus	*leucoxylon*	White Iron Bark	Gum Tree	**E:**LF,PD:DYE(alum-burnt-orange,direct-terra-cotta)	Grae-165-01
Eucalyptus	*macarthurii*	Geranyl Acetate	Gum Tree	**E:**LF,OL:PF	Bown-280-08
Eucalyptus	*macrorhyncha*	Large Beak	Gum Tree	**M(traditional):**LF,OL,RN:BAS	Bown-280-09
Eucalyptus	*mannifera*	Manna	Gum Tree	**M(traditional):**SA:SCN	Bown-280-10
Eucalyptus	*polyanthemos*	Silver Dollar	Gum Tree	**E:**LF:DYE(alum-&-tin-brilliant-orange,tin-light-orange,chrome-&-tin-orange-red,chrome-&-tin-&-tartar-maroon-red)	Grae-163-01
Eucalyptus	*polybractea*	Many Bract	Gum Tree	**E:**LF,OL:CM,HO(eucalyptusoil)	Bown-280-11
Eucalyptus	*radiata australiana*	Australian Ray	Gum Tree	**E:**LF,OL:CM,HO(eucalyptusoil)	Bown-280-12
Eucalyptus	*smithii*	Smith	Gum Tree	**E:**LF,OL:CM,HO(eucalyptusoil)	Bown-280-13
Eucalyptus	*viminalis*	Sweet Manna	Gum Tree	**M(traditional):**SA:SCN	Bown-280-14
Eucommia	*ulmoides*	Duzhong	Gutta Percha	**M(traditional):**BK:BHP	Wyk-409-19
Eugenia	*caryophyllus*			Synonym for *Syzygium aromaticum*	Bal1-068-02
Eugenia	*jambolana*			Synonym for *Syzygium cumini*	
Euonymus	*atropurpureus*	Burning Bush	Spindlewood	**M(traditional):**BK,RB:BCR,SCN,SGB,UDR	Bown-280-15

Latin Genus	L. Species	E. Species	English Genus	C:Culinary,M:Medicinal,E:Economic	Source
Euonymus	*atropurpureus*	Burning Bush	Spindlewood	**M(traditional):**RB:BHE,SCN,**M(traditional):**FR:UDR,**M(traditional):**SD-OL:SPR,**E:**BK,LF,FR:PS☠	Kowa-502-01
Euonymus	*atropurpureus*	Burning Bush	Spindlewood	**M(traditional):**BK,RB:BLC,DSIN,SGB	Wyk-409-21
Euonymus	*atropurpureus*	Burning Bush	Spindlewood	**E:**BK:DYE(alum-&-tin-tan-orange,vinegar-tan-pink)	Cas1-110-02
Euonymus	*europaeus*	European	Spindlewood	**M(traditional):**BK,RB:BCB,BLC,DACN,OAS,OSA,OWN,SGB,UDR	Bown-281-01
Eupatorium	*ayapana*	Brazilian	Eupatory	**C:**LF:TE,**M(traditional):**WP:SDG	Clev-91-95
Eupatorium	*cannibium*	Hemp Agrimony	Eupatory	**M(traditional):**WP:BHE,DHMY,DFLU,HFV,NJA,NRH,UDR,PS☠	Bown-281-02
Eupatorium	*chinense*	Chinese	Eupatory	**M(traditional):**WP:DCLD,DDPT,NJA,**E:**WP:PS☠	Bown-281-03
Eupatorium	*fortunei*	Thoroughwort	Eupatory	**M(traditional):**WP:HFV,SDG,SDR,**E:**WP:PS☠	Bown-281-04
Eupatorium	*lindleyanum*	Lindley	Eupatory	**M(traditional):**WP:RLB,SDY,**E:**WP:PS☠	Bown-281-05
Eupatorium	*maculatum*	Spotted Joy Pye Weed	Eupatory	**E:**FL:DYE(alum-yellow,chrome-bright-gold,tin-sharp-yellow),**E:**WP:DYE(alum-yellow-green,copper-sulfate-&-iron-green,iron-gray),,**E:**WP:PS☠	Cas1-160-01
Eupatorium	*perfoliatum*	Boneset	Eupatory	**M(traditional):**LF,FL:HFV,OLC,RLB,SCN,PS☠	Bal1-066-03
Eupatorium	*perfoliatum*	Boneset	Eupatory	**M(traditional):**WP:DCLD,DFLU,HFV,OAS,RLB,SCN,PS☠	Bown-281-06
Eupatorium	*perfoliatum*	Boneset	Eupatory	**M(traditional):**WP,FL:DCLD,OMC,RLC	Clev-091-04
Eupatorium	*perfoliatum*	Boneset	Eupatory	**M(traditional):**WP:HFV,DCLD,DFLU,RLC(effectiveness questioned by pharmacologists)	Kowa-051-01 Kowa-283-05
Eupatorium	*perfoliatum*	Boneset	Eupatory	**M(folk):**WP(aerial):DFLU,HFV,SDG	Krug-037-01
Eupatorium	*perfoliatum*	Boneset	Eupatory	**M(traditional):**WP(aerial):DCLD,DFLU,HFV,NRH,RLB,RRN	Wyk-140-01, Wyk-409-22
Eupatorium	*purpureum*	Gravel Root	Eupatory	**M(traditional):**FL,RT:UDR,UEP,UUR	Bal1-072-01
Eupatorium	*purpureum*	Gravel Root	Eupatory	**M(traditional):**RT,RZ:UBL,UKC,UKS,UUR,**E:**WP:DYE(red),PS☠	Bown-281-07
Eupatorium	*purpureum*	Gravel Root	Eupatory	**M(folk):**RT:HAP,HFV,UDR,UKC **M(traditional):**RT:NGT,NRH,UBL,UEP	Krug-085-01
Eupatorium	*purpureum*	Gravel Root	Eupatory	**M(traditional):**WP:UBL,UDR,UEP,UKS,UUR	Wyk-410-01
Euphorbia-eurasia	*cyparissias*	Cypress	Eurasian Spurge	**M(traditional):**WP:SPR,UKC,UDR	Wyk-410-02
Euphorbia-eurasia	*esula*	Leafy	Eurasian Spurge	**E:**LF,FL,SM:DYE(alum-green-gold,chrome-tan-gold,copper-dark-green-gold,tin-bright-yellow,Iron-gold-black),**E:**WP:PSN	Bliss-150-01
Euphorbia-ameria	*hirta*	Asthmawort	American Spurge	**M(traditional):**WP:DASC,DHYF,NMS,RLA,RLB,RLC,RLE,SDY,**M(traditional):**JU:OBN,DCLW,**E:**WP:PS☠	Bown-281-08
Euphorbia-asiatica	*kansui*	Kansu	Asian Spurge	**E:**WP:PS☠	Bown-281-09
Euphorbia-afroeurasia	*lathyris*	Oil	Afroeurasian Spurge	**E:**OL:PS☠	Bown-281-10
Euphorbia-ameria	*marginata*	Snow On The Mountain	American Spurge	**E:**FL,LF,SM:DYE(alum-light-yellow,chrome-brass-gold,tin-bright-yellow,iron-tan-olive),PS☠	Bliss-224-01
Euphorbia-asiatica	*pekinensis*	Peking	Asian Spurge	**M(traditional):**WP:DABA,UDR,**E**:WP:PS☠	Bown-281-11
Euphorbia-asiatica	*pekinensis*	Peking	Asian Spurge	**M(traditional):**RT:SPR,UKC	Wyk-410-03
Euphorbia-ameria	*pulcherrima*	Poinsettia	American Spurge	**E:**LF:DYE(alum-green-brown)	Grae-156-02
Euphorbia-afra	*resinifera*	Resin	African Spurge	**E**:WP:PS☠	Bown-281-12
Euphorbia-afra	*resinifera*	Resin	African Spurge	**M(traditional):**RN:DSIN	Wyk-410-04
Euphrasia	*nemorosa*	Grove	Eyebright	**M(traditional):**WP:DALL,DHYF,OEY,RRN	Bal1-070-01
Euphrasia	*nemorosa*	Grove	Eyebright	**M(traditional):**WP(aerial):BLC,DHYF,OAS,OEY,RLC,RRN,RSN	Bal2-066-01 Bal2-517-03
Euphrasia	*nemorosa*	Grove	Eyebright	**M(traditional):**WP:DHSV,DHYF,OAS,OEY,DSIN,RRN,RSN	Bown-281-13
Euphrasia	*nemorosa*	Grove	Eyebright	**M(traditional):**WP:OEY	Kowa-187-01
Euphrasia	*nemorosa*	Grove	Eyebright	**M(folk):**WP:OEY	Krug-071-01
Euphrasia	*nemorosa*	Grove	Eyebright	**M(traditional):**WP(aerial):DHYF,OEY,DSIN,RLC,RSN,SDG	Wyk-141-01, Wyk-410-05
Euphrasia	*officinalis*	auct. Non L.		Synonym for *Euphrasia nemorosa*	Bown-281-13
Euphrasia	*officinalis*	L.		Synonym for *Euphrasia stricta*	Bown-281-13
Euphrasia	*stricta*	Drug	Eyebright	**M(traditional):**WP:NNC	Bown-281-13
Euryale	*ferox*	Prickly	Gorgon Waterlily	**C:**SD:NT,**M(traditional):**SD:DVGB,DVGV,SD:OAS,SDR	Bown-282-01
Eurycoma	*longifolia*	Longjack	Broadcomb	**M(traditional):**WP:DIPT	Murr-577-01
Evernia	*divaricata*	Spreading	Oak Shield Lichen	**E:**WP:DYE(Divaricatic-Acid-&-Ammonia-green)	Cas2-032-02
Evernia	*punastri*	Oak Moss	Oak Shield Lichen	**M(traditional):**WP:DABA,DTBC	Bown-282-02
Evernia	*punastri*	Oak Moss	Oak Shield Lichen	**E:**WP:DYE(Evernic-Acid-&-Atranorin-light-orange,Evernic-Acid-&-Atranorin-&-Ammonia-voilet-brown)	Cas2-032-01 Cas2-035-01
Evernia	*gyrophora*	Gyrophoric Acid	Oak Shield Lichen	**E:**WP:DYE(Gyrophoric-Acid-&-Ammonia-pink)	Cas2-031-16
Evodia	*rutaecarpa*	Rutecarpine	Evodiamine	**M(traditional):**LF:NSD,SDP	Wyk-409-20
Fabiana	*imbricata*	Pichi	Fabian Heath	**M(traditional):**WP:DSXG,UBL	Wyk-410-06
Fagopyrum	*esculentum*	Edible	Buckwheat	**C:**SD:GF,**M(traditional):**LF,FL,SD:BCB,BHP,OAS	Bown-282-03

Latin Genus	L. Species	E. Species	English Genus	C:Culinary,M:Medicinal,E:Economic	Source
Fagopyrum	*esculentum*	Edible	Buckwheat	**M(PS):**WP(aerial):BCR,BHE,BVV,DGUM,OHM	Wyk-142-01, Wyk-410-07
Fagus	*grandifolia*	American	Beech	**E:**LF:DYE(alum-yellow,alum-&-tin-rust,chrome-gold,copper-sulfate-tan,iron-gray-tan)	Cas1-102-01
Fagus	*sylvatica*	European	Beech	**C:**OL:CO,**M(traditional):**OL:DASC,RLB,RLC,**E:** WD:Creosote	Bown-282-04
Ferula	*assa-foetida*	Devil's Dung	Giant Fennel	**M(traditional):**RT-RN:BLC,DEPI,DWHP,DWRM,HAN, SCN,SFL,SIB,**E:**RT-RN:IR	Bal2-027-01 Bal2-504-03
Ferula	*assa-foetida*	Devil's Dung	Giant Fennel	**C:**RN:**SP**,**M(traditional):**RN:DWHP,DWRM,NMS,OJA,RLB,SCN,SDG,SDY,SFL,SIB,**E:**RN:GR	Bown-282-05
Ferula	*assa-foetida*	Devil's Dung	Giant Fennel	**C:**RN:**SP**,**M(traditional):**RN:DABA,HAN,RLB,SFL	Clev-092-07, Clev-301-01
Ferula	*assa-foetida*	Devil's Dung	Giant Fennel	**M(PS):**RT-RN:NSD,RLB,RLC,SDP,SFL	Wyk-143-01, Wyk-410-08
Ferula	*galbaniflua*			Synonym for *Ferula gummosa*	Clev-092-08
Ferula	*gummosa*	Galban Gum	Giant Fennel	**C:**RT-RN:FF,**M(traditional):**RN:DASC,DBOI,NMS,OSA, OWN,RLA,RLB,RLC,SDG,**E:**RN:GR,IN	Bown-283-01
Ferula	*gummosa*	Galban Gum	Giant Fennel	**E:**RN:IN	Clev-092-08
Ferula	*gummosa*	Galban Gum	Giant Fennel	**M(traditional):**RN:DABI,NMS,OWN,RLC,SDG,SFL,**E:**RN:PF	Wyk-143-02
Ficus	*benghalensis*	Banyan	Fig	**M(traditional):**BK,FR,LF,RT:OAS,DCLW,DSIN, SCN	Wyk-410-09a
Ficus	*carica*	Edible	Fig	**C:**FR:JJ,PU,**M(traditional):**FR:RLB,RLC,RST,SCN,**M(traditional):**LF:OEY,**M(traditional):**SA:DCLW	Bown-283-02
Ficus	*carica*	Edible	Fig	**M(traditional):**FR:SCN	Wyk-410-10
Ficus	*religiosa*	Peepul	Fig	**M(traditional):**BK,FR,LF,RT:OAS,DCLW,DSIN, SCN	Wyk-410-09b
Filipendula	*rubra*	Queen of the Prairie	Meadowsweet	**C:**RT:**VG**	Clev-092-03
Filipendula	*ulmaria*	Queen of the Meadow	Meadowsweet	**M(traditional):**WP:DDPT,DPNB,DFLU,DPNV,HFV,OAS,NJA,NRH,SIB,SDR,SDY,SUL,UBL, **E:**WP:DYE(green-yellow)	Bown-283-03
Filipendula	*ulmaria*	Queen of the Meadow	Meadowsweet	**C:**LF:AF,BF,**M(traditional):**FL:DFLU,HFV,NRH, SDG,**E:**OL:PF,**E:**FL:DYE(yellow),**E:**RT:DYE(black)	Clev-092-01
Filipendula	*ulmaria*	Queen of the Meadow	Meadowsweet	**C:**LF:AF,**M(folk):**FL:DCLD,DFLU,HFV,NJA,NRH,SDG	Krug-115-01
Filipendula	*ulmaria*	Queen of the Meadow	Meadowsweet	**M(PS):**FL:DASC,DCLD,HFV,NJA,NRH,OAS,UDR,SUL	Wyk-144-01, Wyk-410-11
Filipendula	*vulgaris*	Dropwort	Meadowsweet	**C:**RT:**VG**	Clev-092-02
Flavocetraria	*culcullata*	Hooded	Snow Heath Shield Lichen	**E:**WP:DYE(Protolichesterinic-Acid-&-Ammonia-red-purple)	Cas2-031-07
Flavocetraria	*nivalis*	Arctic	Snow Heath Shield Lichen	**E:**WP:DYE(No-Lichen-Acids-&-Ammonia-violet)	Cas2-031-10
Flavoparmelia	*caperata*	Caper	Green Shield Lichen	**E:**WP:DYE(Caperatic-Acid-yellow)	Cas2-034-12
Flavopunctelia	*soredica*	Soridia	Speckleback Shield Lichen	**E:**WP:DYE(Usnic-Acid-&-Lecanoric-Acid-&-Ammonia-red)	Cas2-032-22
Foeniculum	*azoricum*	Florence	Fennel	**C:**SL:**VG**	Clev-092-05
Foeniculum	*azoricum*	Florence	Fennel	**C:**SL:**VG**	Kowa-188-02
Foeniculum	*piperitum*	Cartucci	Fennel	**C:**SH:**VG**	Clev-092-06
Foeniculum	*vulgare*	Sweet	Fennel	**M(traditional):**FR,RT,SM:BLC,DOBE,OEY,SFL, SIB,UKC	Bal1-070-03
Foeniculum	*vulgare*	Sweet	Fennel	**C:**SD:**SP**,**M(folk):**SD:DSIN,DWTP,OEY**M(traditional):**SD:HFV,NMS,NSD,SAP,SDG,UDR, **M(PS):**SD:RLA,RLB,RLC,SDP	Bal2-067-01 Bal2-517-04
Foeniculum	*vulgare*	Sweet	Fennel	**M(traditional):**OL:DAMW,DGUM,RST,**M(traditional):**SD:SDG,SFL,SIB,**M(traditional):**RT:UDR, **E:**OL:PF,DS	Bown-283-04
Foeniculum	*vulgare*	Sweet	Fennel	**C:**SD:**SP**,TE,**C:**LF:**SN**,**C:**OL:FF,**C:**RT:**VG**,**C:** SH:GN,**SN**,**C:**SL,**SN**,**M(traditional):**OL,SD:DABA,DEAI,OTA,NRH,RLA,RLC,**E:**OL:DS,PF	Clev-092-04, Clev-302-01
Foeniculum	*vulgare*	Sweet	Fennel	**C:**SD:**SP**,**C:**LF:**SN**,**M(traditional):**LF,RT,SL,SD:DAMW,DCLD,NSD,RLC,RST,SDG,SFL,UDR, **E:**LF,FL:DYE(alum-mustard-yellow,chrome-gold-brown,iron-brown)	Kowa-173-11 Kowa-188-01 Kowa-286-04
Foeniculum	*vulgare*	Sweet	Fennel	**C:**SD:**SP**,**M(folk):**SD:DAMW,DOBE,SFL	Krug-072-01
Foeniculum	*vulgare*	Sweet	Fennel	**M(PS):**FR,OL:OEY,DSIN,RLC,RRN,SDP,SFL	Wyk-145-01, Wyk-410-12
Foeniculum	*vulgare*	Sweet	Fennel	**E:**FL,LF:DYE(alum-mustard-yellow,chrome-gold-brown)	Grae-087-02
Foeniculum	*vulgare dulce*			Synonym for *Foeniculum azoricum*	Clev-092-05

Latin Genus	L. Species	E. Species	English Genus	C:Culinary,M:Medicinal,E:Economic	Source
Foeniculum	*vulgare piperitum*			Synonym for *Foeniculum piperitum*	Clev-092-06
Forsythia	*suspensa*	Weeping	Golden Bell Privet	**M(traditional):**FR:BHS,DABA,DERS,DMUM,DTNS,HFV,OAS,OSA,SGB,UDR	Bown-284-01
Forsythia	*suspensa*	Weeping	Golden Bell Privet	**M(traditional):**LF:DABI	Wyk-410-13
Forsythia	*suspensa*	Weeping	Golden Bell Privet	**E:**FL:DYE(alum-yellow,tin-bright-yellow,chrome-&-tin-bright-gold,alum-&-daffodil-dull-yellow,tin-&-daffodil-deep-yellow,chrome-&-iron-&-daffodil-warm-tan)	Cas1-143-01
Fragaria	*moschata*	Musk	Strawberry	**M(traditional):**FR:NSD,SCN,**M(traditional):**LF:OAS,UDR,**E:**FR:PS☠	Clev-093-02
Fragaria	*vesca*	Woodland	Strawberry	**C:**FR:AF,BF,JJ,JU,PU,**C:**LF:TE,**M(traditional):**LF,RT:NGT,OAS,SCN,SDG,SDR,UDR,**M(traditional):**FR,JU:OSB,**E:**FR:PS☠	Bown-284-02
Fragaria	*vesca*	Woodland	Strawberry	**M(traditional):**LF:OAS,UDR,**M(traditional):**FR:NSD,SCN,**E:**FR:PS☠	Clev-093-01
Fragaria	*vesca*	Woodland	Strawberry	**M(PS):**LF,RZ:DAMW,OAS,NRH,SDR,UDR,UUR	Wyk-146-01, Wyk-410-14
Fragaria	*virginiana*	Large Fruit	Strawberry	**M(traditional):**FR:NSD,SCN,**M(traditional):**LF:OAS,UDR,**E:**FR:PS☠	Clev-093-03
Frangula	*alnus*	Alder	Thornless Buckthorn	**M(traditional):**BK(inner):BLC,DASC,OAS,SGB,SPR	Bown-340-04
Frangula	*alnus*	Alder	Thornless Buckthorn	**M(PS):**BK:SCN	Wyk-268-01, Wyk-424-20
Frangula	*purshiana*	Cascara	Thornless Buckthorn	**M(traditional):**BK(sagrada):DWRM,SCC,SCN	Bal1-066-09
Frangula	*purshiana*	Cascara	Thornless Buckthorn	**C:**BK(sagrada):AF,BF,CF,**C:**FL:honey,**C:**FR:PU,**M(traditional):**BK(sagrada):SCN,**E:**BK:PS☠(griping)	Kowa-066-01
Frangula	*purshiana*	Cascara	Thornless Buckthorn	**M(traditional):**BK(sagrada):BLC,OAS,SCN,SDG	Bown-340-06
Frangula	*purshiana*	Cascara	Thornless Buckthorn	**M(folk):**BK(sagrada):SCN,SDG	Krug-042-01
Frangula	*purshiana*	Cascara	Thornless Buckthorn	**M(PS):**BK(sagrada):SCN,SPR	Wyk-269-01, Wyk-425-01
Frasera	*carolinensis*	Carolina	Green Gentian	**M(traditional):**RT:OAS,SDG	Wyk-410-15
Frasera	*chirata*	Chiretta	Green Gentian	**M(traditional):**WP(aerial):BDB,SAP,SDG,SDP	Bal2-049-01 Bal2-511-05
Frasera	*chirata*	Chiretta	Green Gentian	**M(traditional):**WP:BLC,DMAL,HFV,SCN,SDG,SDP,SGB	Bown-357-06
Frasera	*chirata*	Chiretta	Green Gentian	**C:**WP:BF(bitters),**M(traditional):**WP:HFV,SCN,SDP	Wyk-428-19
Frasera	*japonica*	Japanese	Green Gentian	**C:**WP:FF(bitters)	Bown-357-07
Fraxinus	*americana*	White	Ash	**E:**LF:DYE(alum-soft-yellow,tin-bright-yellow,chrome-&-iron-gray,copper-sulfate-beige)	Cas1-094-01
Fraxinus	*bungeana*	Northern	Ash	**M(traditional):**SA:DABA,NSD,UDR	Bown-284-03
Fraxinus	*excelsior*	European	Ash	**M(traditional):**SA:NJA,NRH,SCN	Bown-284-04
Fraxinus	*excelsior*	European	Ash	**M(PS):**BK:HFV,NRH,**M(traditional):**LF:NJA,OUL,OWN,SCN,UBL	Wyk-147-01, Wyk-410-16
Fraxinus	*nigra*	Black	Ash	**E:**BK:DYE(alum-&-chrome-tan-rose,chrome-&-iron-brown)	Cas1-094-02
Fraxinus	*ornus*	Manna	Ash	**C:**SA,FF,**M(traditional):**BK-SA(manna):SCN	Bown-284-05
Fraxinus	*ornus*	Manna	Ash	**M(traditional):**BK-SA(manna):SCN	Wyk-147-02, Wyk-410-17
Fritillaria	*cirrhosa*	Tendril	Fritillary	**M(traditional):**BR:BHP,HFV,RLB,RLC	Bown-284-06
Fritillaria	*imperialis*	Imperial	Fritillary	**M(traditional):**BR:DCTM,HFV,NSD,OTM,RLC	Wyk-410-18a
Fritillaria	*pallidiflora*	Pale Flower	Fritillary	**M(traditional):**BR:BHP,HFV,RLB,RLC	Bown-284-07
Fritillaria	*thunbergii*	Crown	Fritillary	**M(traditional):**BR:DFLU,DPNB,RLB,RLC,RTL	Bal2-069-01
Fritillaria	*thunbergii*	Crown	Fritillary	**M(traditional):**BR:DCTM,HFV,NSD,OTM,RLC	Wyk-410-18b
Fritillaria	*verticillata*	Whorled	Fritillary	**M(traditional):**BR:BHP,DPNB,DPNV,HFV,OTM,RLB,RLC,**E:**BR:PS☠	Bown-284-08
Fucus	*serratus*	Hummertang	Rock Seaweed	**C:**WP:**VG**	
Fucus	*vesiculosis*	Bladderwrack	Rock Seaweed	**M(traditional):**WP:DCTM,DHOT,DOBE,DRSK,SCN	Bal2-090-01 Bal2-527-07
Fucus	*vesiculosis*	Bladderwrack	Rock Seaweed	**C:**WP:MC,**M(traditional):**WP:DABI,DHOT,DOBE,ORH,**E:**WP:MC(fertilizer)	Bown-285-01
Fucus	*vesiculosis*	Bladderwrack	Rock Seaweed	**M(folk):**WP:NRH	Krug-033-01
Fucus	*vesiculosis*	Bladderwrack	Rock Seaweed	**E:**WP:MC	Wyk-410-19
Fumaria	*officinalis*	Earth Smoke	Fumitory	**M(traditional):**WP,FL:BLC,DSIN,SCN,SDG,SGB,UDR	Bown-285-02
Fumaria	*officinalis*	Earth Smoke	Fumitory	**M(folk):**WP(flowering):BLC,DSIN	Krug-078-02
Fumaria	*officinalis*	Earth Smoke	Fumitory	**M(PS):**WP:NMS,ODS,DSIN,SCN,SDP,SGB,UDR	Wyk-148-01
Gaillardia	*aristata*	Paintspots	Blanket Sunflower	**E:**LF,FL,SL:DYE(alum-light-gold,chrome-green-gold,copper-green-gold,tin-bright-gold,iron-gray-brown)	Bliss-118-01

Latin Genus	L. Species	E. Species	English Genus	C:Culinary,M:Medicinal,E:Economic	Source
Galanthus	*nivalis*	Eurasian	Snowdrop Amaryllis	**M(CS):**BR:DALZ,NNR,**E:**BR:PS☠	Wyk-149-01
Galanthus	*warnowii*	Galanthamine	Snowdrop Amaryllis	**M(traditional):**BR:DALZ	Wyk-410-20
Galega	*officinalis*	French Lilac	Goat's Rue	**M(traditional):**WP:BDB,BSH,SCN,SDG,SPA,UDR	Bown-285-03
Galega	*officinalis*	French Lilac	Goat's Rue	**M(PS):**WP:BDB,DHOG,OUL,UDR, **E:**WP:cow lactation	Wyk-150-01, Wyk-410-21
Galeopsis	*segetum*	Cornfield	Hemp Nettle Mint	**M(PS):**WP:OAS,RLB,RLC,RRN,UDR	Wyk-151-01, Wyk-410-22
Galipea	*officinalis*	Cuspary Bark	Galipe Rue	**C:**BK:AF,BF,**M(traditional):**BK:BLC,HFV,NMS,SCN,SDR,SDY,SGB,**E:**BK:PS☠	Bown-285-04
Galipea	*officinalis*	Cuspary Bark	Galipe Rue	**M(traditional):**BK:HFV,NMS,SDG	Wyk-410-23
Galium	*aparine*	Cleavers	Bedstraw	**C:**LF,SM:**VG**,**C:**SD:TE,**M(traditional):**WP:BHP,DHPT,DTNS,HFV,OAS,OSS,DSIN,OTM,SCN,UBL,UDR	Bown-285-05
Galium	*aparine*	Cleavers	Bedstraw	**M(folk):**WP(flowering):DSIN,UDR	Krug-054-03
Galium	*aparine*	Cleavers	Bedstraw	**M(traditional):**WP:ODS,DSIN,OWN,UDR	Wyk-152-02, Wyk-411-01
Galium	*boreale*	Boreal Lady's	Bedstraw	**E:**FL,LF,SM:DYE(alum-gold,chrome-brass-gold,copper-light-olive,tin-bright-gold,iron-olive-green)	Bliss-146-01
Galium	*boreale*	Boreal Lady's	Bedstraw	**E:**FL,LF:DYE(yellow),**E:**RT:DYE(red)	Cas1-101-01
Galium	*odoratum*	Sweet Woodruff	Bedstraw	**C:**LF,SM:AF,**M(traditional):**WP:BJA,BPH,DHPT,HIN	Bown-286-01
Galium	*odoratum*	Sweet Woodruff	Bedstraw	**C:**LF,SM:AF,BF,**M(traditional):**WP:DSCB,OWN,E:WP:PS☠	Clev-093-04
Galium	*odoratum*	Sweet Woodruff	Bedstraw	**C:**FL,LF:AF,BF,**M(traditional):**WP:BHS,BJA,BLC,HAN,NMS,NSD,SDG,UDR,**M(traditional):**LF:OWN,**E:**LF,SM:DYE(alum-tan),**E:**RT:DYE(alum-red),IR	Kowa-175-09 Kowa-477-01
Galium	*odoratum*	Sweet Woodruff	Bedstraw	**C:**FL,LF:AF(brandy, white wine), **M(folk):**LF:OWN,SIB,SPR	Krug-186-01
Galium	*odoratum*	Sweet Woodruff	Bedstraw	**M(traditional):**WP:NSD,OWN,SIB	Wyk-411-02
Galium	*odoratum*	Sweet Woodruff	Bedstraw	**E:**LF,SM:DYE(alum-soft-tan,alum-&-iron-gray-green)	Grae-189-02
Galium	*verum*	Lady's	Bedstraw	**M(traditional):**WP:NMS,OAS,UBL,UDR,UKC,**E:**LF:DYE(yellow,),**E:**RT:DYE(red)	Bown-286-02
Galium	*verum*	Lady's	Bedstraw	**M(traditional):**WP:HIN,**E:**FL:DYE(alum-dull-yellow),alum-yellow,chrome-gold,tin-bright-yellow),**E:**RT:DYE(chrome-purple-red,iron-plum-purple)	Kowa-174-04 Kowa-349-01
Galium	*verum*	Lady's	Bedstraw	**M(folk):**LF,FL-SM:HAN,OAS,NSD	Krug-099-01
Galium	*verum*	Lady's	Bedstraw	**M(traditional):**WP:ODS,DSIN,OWN,UDR	Wyk-152-01, Wyk-411-03
Galphimia	*glauca*	Rain Of Gold	Gold Shower	**M(traditional):**LF:NSD,OWN,UUR	Wyk-411-04
Ganoderma	*lucidum*	Reishi	Bracket Pore Fungus	**M(traditional):**WP:BLC,DALL,HAN,RLB, **M(PS):**WP:BHC,BHP,DCLK,DCTM	Bal2-122-01 Bal2-538-03
Ganoderma	*lucidum*	Reishi	Bracket Pore Fungus	**M(traditional):**WP:BHC,BHP,BHS,BLC,BSH,DABA,DAVR,HIN,NJA,NSD,RLA,RLB,RLC	Bown-286-03
Garcinia	*cambogia*			Synonym for *Garcinia hanburyi*	
Garcinia	*hanburyi*	Hanbury	Garcin Gamboge	**M(traditional):**RN:SPR	Wyk-411-05
Gardenia	*angusta*	Cape Jasmine	Alexander Rubyroot	**C:**FL,FR:TE,**M(traditional):**FL,FR:BHP,BJA,DHPT,HFV,OAS,DSIN,OTA,OWN,**E:**FL,OL:PF	Bown-286-04
Gardenia	*angusta*	Cape Jasmine	Alexander Rubyroot	**M(traditional):**LF:DCLD,DFLU,OAS	Wyk-411-06
Garrya	*elliptica*	Wavy Leaf	Silk Tassel	**E:**FR,PD:DYE(alum-&-chrome-dark-gray)	Grae-142-02
Gastrodia-asiatica	*elata*	Tall	Asian Gastric Orchid	**M(traditional):**RZ:BHP,DEPI,DTET,DVTG,NJA,NMS,NSD	Bown-286-05
Gastrodia-asiatica	*elata*	Tall	Asian Gastric Orchid	**M(traditional):**RZ:BHP,HAC,NJA	Wyk-411-07
Gaultheria	*antipoda*	New Zealand Red	Checkerberry	**C:**FR:PU(BerryFruit)	Clev-093-08
Gaultheria	*procumbens*	Wintergreen	Checkerberry	**M(traditional):**LF,RT,SM:HAC,NJA,NRH,NSD,OTA	Bal1-078-07
Gaultheria	*procumbens*	Wintergreen	Checkerberry	**M(traditional):**LF-OL,FR:HAC,HFV,NRH,RST, **M(PS):**LF-OL,FR:NCT,OTA	Bal2-154-01 Bal2-550-04
Gaultheria	*procumbens*	Wintergreen	Checkerberry	**C:**LF:TE,**C:**FR:PU,**M(traditional):**OL:OJA,ONR,ORH,OSC,E:OL:PS☠	Bown-286-06
Gaultheria	*procumbens*	Wintergreen	Checkerberry	**M(traditional):**LF,OL:DAMW,OMC,RST	Clev-093-07
Gaultheria	*procumbens*	Wintergreen	Checkerberry	**C:**LF,OL:CF,**M(traditional):**FR,LF,OL:DSXG,HFV,NSD,OUL,RST,SDG,**E:**OL:CM	Kowa-506-01
Gaultheria	*procumbens*	Wintergreen	Checkerberry	**M(folk):**LF,OL:HFV,NJA,NSM,OJA,OSM,RST,SIB	Krug-184-01
Gaultheria	*procumbens*	Wintergreen	Checkerberry	**M(PS):**LF,OL:NJA,NMS,NRH,NSD,SDG, **E:**LF,OL:PS☠	Wyk-153-01, Wyk-411-08
Gaultheria	*shallon*	Salal Purple	Checkerberry	**E:**FR:DYE(alum-&-copper-dark-blue)	Grae-141-01
Gaylussacia	*baccata*	Black	Gay-Lussac Huckleberry	**E:**LF:DYE(alum-yellow,tin-brilliant-yellow,copper-sulfate-&-iron-yellow-green),**E:**FR:DYE(chrome-&-vinegar-tan-pink,iron-&-salt-pink-gray)	Cas1-108-02

Latin Genus	L. Species	E. Species	English Genus	C:Culinary,M:Medicinal,E:Economic	Source
Gelidium	*amansi*	Japanese Agar	Gel Florid Red Algae	**C:**WP:MC,**M(traditional):**WP:SCN	Bown-287-01
Gelidium	*amansi*	Japanese Agar	Gel Florid Red Algae	**M(traditional):**WP:SCN,**E:**WP:MC(agar)	Wyk-411-09
Gelidium	*cartilagineum*	American Agar	Gel Florid Red Algae	**E:**WP:MC(agar)	Bown-287-02
Gelsemium	*sempervirens*	Yellow	Trumpet Jasmine	**M(traditional):**RT,RZ:DMNB,DMNV,HFV,NMS,NNR,NSC,NSD,OTA	Bown-287-05
Gelsemium	*sempervirens*	Yellow	Trumpet Jasmine	**M(folk):**RT,RZ:DHMY(cold & flu)	Krug-190-02
Gelsemium	*sempervirens*	Yellow	Trumpet Jasmine	**M(PS):**RT,RZ:DWHP,NMS,NNR,NSD,OHM,RLA	Wyk-154-01, Wyk-411-10
Genista	*tinctoria*	Dyer's	Greenweed Broom	**M(traditional):**WP,FL:BHE,BHS,NGT,NRH,OSA,OSC,OTM,UDR	Bown-287-06
Genista	*tinctoria*	Dyer's	Greenweed Broom	**M(traditional):**FL,SH:BHE,SCN,DSIN,UDR,**E:**FL:DYE(yellow)	Clev-094-01
Genista	*tinctoria*	Dyer's	Greenweed Broom	**M(folk):**SD:NBF,SPR,**M(folk):**WP(flowering):BHE,OGT,UDR, **E:**FL:DYE(yellow)	Krug-065-01
Genista	*tinctoria*	Dyer's	Greenweed Broom	**M(traditional):**WP:SCN,UDR,**E:**FL:DYE(yellow)	Wyk-411-11
Gentiana	*lutea*	European	Gentian	**M(traditional):**RT:BLC,SGB,SGS **M(PS):**RT:SAP,SDP,SFL	Bal2-072-01 Bal2-519-01
Gentiana	*lutea*	European	Gentian	**C:**RT,RZ:AF,**M(traditional):**RT,RZ:BLC,DANX,HFV,SDG,SGB,SIB	Bown-287-07
Gentiana	*lutea*	European	Gentian	**M(traditional):**RT,RZ:BCR,SAP,SDG,**E:**RT:PS☠(hypertension,vomiting)	Kowa-220-01 Kowa-288-01
Gentiana	*lutea*	European	Gentian	**C:**RT:AF,**M(folk):**RT:BLC,SAP,SDG,SDP,SFL	Krug-080-01
Gentiana	*lutea*	European	Gentian	**M(PS):**RT,RZ:DANX,SAP,SDP,SFL,SGB	Wyk-155-01, Wyk-411-12
Gentiana	*macrophylla*	North Asia	Gentian	**M(traditional):**LF,RT:BCR,DMAL,DWRM,SAP,SDG	Bal1-070-08
Gentiana	*macrophylla*	North Asia	Gentian	**M(traditional):**RT:BJA,BLC,DALL,DHPT,HFV,NJA,NRH,NSD,SCN,SGB	Bown-287-06
Gentiana	*scabra*	East Asia	Gentian	**M(traditional):**RT:BLC,SGB,SGS **M(PS):**RT:SAP,SDP,SFL	Bal2-072-02 Bal2-519-01
Gentiana	*scabra*	East Asia	Gentian	**M(traditional):**RT:BLC,DEAT,OEY,SAP,SDG	Bown-288-01
Geranium	*maculatum*	American	Crane's Bill	**M(traditional):**WP,RT:DASC,DGUM,DVGB,DVGT,DVGV,OAS,OHM,OWN,RST,SDR,SDY,SIB,SUL	Bown-288-02
Geranium	*maculatum*	American	Crane's Bill	**M(traditional):**RT:OAS,SDR,SDY,SUL	Clev-094-02
Geranium	*maculatum*	American	Crane's Bill	**M(folk):**RT:DAMW,DGUM,OHM,OWN,SDR	Krug-015-01
Geranium	*maculatum*	American	Crane's Bill	**M(traditional):**RT:OWN,RST,SDP	Wyk-411-13
Geranium	*robertianum*	Herb Robert	Crane's Bill	**M(traditional):**WP:DGUM,DHSV,OAS,OSA,OWN,RST,SDR,SIB,SUL,UDR	Bown-288-03
Geranium	*robertianum*	Herb Robert	Crane's Bill	**M(traditional):**WP,RT:OAS,OWN	Clev-094-03
Geranium	*robertianum*	Herb Robert	Crane's Bill	**M(PS):**WP:OWN,DSIN,SDR,SIB,SUL,UUR	Wyk-156-01, Wyk-411-14
Geranium	*robertianum*	Herb Robert	Crane's Bill	**E:**WP:DYE(alum-gold-brown)	Grae-091-01
Geranium	*viscosissimum*	Sticky	Crane's Bill	**E:**FL,LF,SM:DYE(alum-gold,chrome-gold-brown, copper-dark-olive,tin-bright-yellow,iron-black-brown)	Grae-234-01
Geum	*rivale*	Chocolate	Avens	**C:**LF,RT:BF(chocolate flavoring)	Clev-095-02
Geum	*urbanum*	Herb Bennet	Avens	**M(traditional):**WP:DASC,DGUM,DVGB,DVGV,HFV,OAS,OHM,RST,SDG,SDR,SIB	Bown-288-04
Geum	*urbanum*	Herb Bennet	Avens	**C:**LF:FF,**C:**RT:**SP**(cloves),**M(traditional):**RT:BLC,DAMW,HFV,SAP,SDG	Clev-095-01
Geum	*urbanum*	Herb Bennet	Avens	**C:**RT:**SP**(cloves),**M(folk):**LF,RT:OHM,SDG,SDR	Krug-020-01
Geum	*urbanum*	Herb Bennet	Avens	**C:**RT,RZ:AF,**M(PS):**RT,RZ:BCB,DAMW,OHM,RST,SAP,SDP,SDR	Wyk-157-01, Wyk-411-15
Gillenia	*trifoliata*			Synonym for *Porteranthus trifoliatus*	Bown-288-05
Ginkgo	*biloba*	Maidenhair	Ginkgo	**M(traditional):**LF:BHS,DEAT,HAC,HAN,HFN,HML,DSIN,RLA,UKC	Bal1-071-02
Ginkgo	*biloba*	Maidenhair	Ginkgo	**M(traditional):**LF:BAA,DEAT,RLA,**M(PS):**LF:BHA,BST,DALZ,DCTM,DEAT,DEGC,DEMD,DIPT,DMSC,DPRK,DSAD,DSZP,HML, **M(CS):**LF:DVRT,HML,DEAT	Bal2-074-01 Bal2-520-01
Ginkgo	*biloba*	Maidenhair	Ginkgo	**C:**SD:NT,**M(traditional):**LF:BHS,BVV,DALL,DRNP,RLA,**M(traditional):**SD:RLA,RLC,UIN	Bown-288-06
Ginkgo	*biloba*	Maidenhair	Ginkgo	**M(CT):**LF:BCB,BCR,BHS,DALZ,HIN,**M(traditional):**SD:RLC	Wyk-158-01, Wyk-411-16
Gladiolus	*communis*	Whistling Jacks	Sword Lily	**M(traditional):**BR:DGUM(scurvy)	Wyk-411-17
Gladiolus	*grandiflorus*	Large Flower	Sword Lily	**E:**FL(lightshades):DYE(alum-&-vinegar-yellow,copper-sulfate-beige,chrome-soft-yellow-gold,tin-bright-yellow,rion-beige-gray),**E:**FL(dark-red,purple):DYE(iron-&-copper-sulfate-green,alum-&-tin-yellow-green)	
Glechoma	*hederacea*	Ground Ivy	Runaway Robin	**M(traditional):**WP:DGUM,OAS,OHM,RLB,RSN,	Bown-289-01

Latin Genus	L. Species	E. Species	English Genus	C:Culinary,M:Medicinal,E:Economic	Source
				RST,SIB,UBL	
Glechoma	*hederacea*	Ground Ivy	Runaway Robin	**M(traditional):**WP:DCLD,OWN,RLC,SDR,UDR	Wyk-411-18
Gleditsia	*triacanthos*	Honey	Locust	**M(traditional):**WP:DMNN	Bal2-406-02
Globularia	*alypum*	Globularin	Globe Daisy	**M(traditional):**LF:SPR,UDR	Wyk-411-19
Gloriosa	*superba*	Superb	Glory Lily	**M(traditional):**TU:DABA,DSCB,DCLW,**E:**TU:PS☠	Bown-289-02
Gloriosa	*superba*	Superb	Glory Lily	**M(traditional):**TU:colchicines	Wyk-411-20
Glycine	*max*	Soybean	Glycine Bean	**M(traditional):**SD:BAN,DALZ,HML,NJA,NSD,SGB,**M(PS):**SD:BAS,BHC,BLC,DCEN,DCLU,DCPR,DFRA,DOST,SGS	Bal2-135-02 Bal2-138-01 Bal2-544-01 Bal2-545-01
Glycine	*max*	Soybean	Glycine Bean	**M(PS):**SD:BCR,BHC,BLC,DCBR,DCPR,DHPT,DMHI,SAP	Wyk-159-01, Wyk-411-21
Glycyrrhiza	*glabra*	European	Licorice	**M(traditional):**RT:DALL,DCFS,DHSV,DHOG,HAN,HFV,NMS,RLA,RLB,RLE,SCC,SIB,SUL	Bal1-073-07
Glycyrrhiza	*glabra*	European	Licorice	**M(folk):**RT,RZ:DEPI,SAD,SCN,**M(traditional):**RT,RZ:DCLD,DSIN,BHS,SUL,**M(ayurveda):**RT,RZ:HAC,OEY,OWN,RLB,RST,SUL,**M(PS):**RT,RZ:BLC,DAID,DALL,DCFS,DCTM,DHOG,DHPT,DLUP,DSAR,DVRT,NJA,RLA,RLB,RLC,SUL	Bal2-095-01 Bal2-529-03
Glycyrrhiza	*glabra*	European	Licorice	**C:**RT:AF,BF,CF,**C:**RZ:**SP**,**M(traditional):**RT,RZ:DADD,DALL,DHSV,DSHZ,BLC,NJA,DSIN,RLA,RLB,RLC,SUL	Bown-289-03
Glycyrrhiza	*glabra*	European	Licorice	**C:**RT,RZ:AF,CF,**SP**,**M(traditional):**RT:HFV,RLB,RST,SCN,SUL,UBL,UKC,**E:**RT:PS☠(raisesbloodpressure)	Clev-095-03, Clev-303-01
Glycyrrhiza	*glabra*	European	Licorice	**C:**RT,RZ:AF,CF,**M(traditional):**RT,RZ:NJA,NMP,NRH,RLC,SCN,SUL,**E:**RT:CM,DS,PS☠	Kowa-290-01 Kowa-360-01
Glycyrrhiza	*glabra*	European	Licorice	**C:**RT:CF, **M(folk):**RT:MES,NMP,RLC,SDG,SUL	Krug-106-01
Glycyrrhiza	*glabra*	European	Licorice	**M(PS):**RT,RZ:DISB,OHM,OSB,RLC,SFL,SUL	Wyk-160-01, Wyk-411-22
Glycyrrhiza	*glabra-glandulifera*	Russian	Licorice	**M(traditional):**RT:RLC	Bown-289-07
Glycyrrhiza	*glabra-violacea*	Middle Eastern	Licorice	**M(traditional):**RT:RLC	Bown-289-06
Glycyrrhiza	*lepidota*	American	Licorice	**M(folk):**RT,RZ:DEPI,SAD,SCN,**M(traditional):**RT,RZ:DCLD,DSIN,BHS,SUL,**M(ayurveda):**RT,RZ:HAC,OEY,OWN,RLB,RST,SUL,**M(PS):**RT,RZ:BLC,DAID,DALL,DCFS,DCTM,DHOG,DHPT,DLUP,DSAR,DVRT,NJA,RLA,RLB,RLC,SUL	Bal2-095-02 Bal2-529-03
Glycyrrhiza	*lepidota*	American	Licorice	**M(traditional):**RT:RLC	Bown-289-04
Glycyrrhiza	*lepidota*	American	Licorice	**E:**LF,PD,SM:DYE(alum-yellow,chrome-brass,copper-camel,tin-bright-yellow,iron-moss-green)	Bliss-268-01
Glycyrrhiza	*uralensis*	Chinese	Licorice	**M(folk):**RT,RZ:DEPI,SAD,SCN,**M(traditional):**RT,RZ:DCLD,DSIN,BHS,SUL,**M(ayurveda):**RT,RZ:HAC,OEY,OWN,RLB,RST,SUL,**M(PS):**RT,RZ:BLC,DAID,DALL,DCFS,DCTM,DHOG,DHPT,DLUP,DSAR,DVRT,NJA,RLA,RLB,RLC,SUL	Bal2-095-03 Bal2-529-03
Glycyrrhiza	*uralensis*	Chinese	Licorice	**M(traditional):**RT:DACN,DADD,RLA,RLC,SUL	Bown-289-05
Glycyrrhiza	*uralensis*	Chinese	Licorice	**M(traditional):**RT,RZ:RLC	Wyk-411-23
Gnaphalium	*multiceps*			Synonym for *Pseudognaphalium affine*	Bown-289-09
Gnaphalium	*obtusifolium*	Sweet Everlasting	Cotton Cudweed	**M(traditional):**WP:DMUM,DTNS	Bown-289-10
Gnaphalium	*obtusifolium*	Sweet Everlasting	Cotton Cudweed	**E:**WP:DYE(alum-yellow,alum-&-iron-green)	Grae-082-02
Gnaphalium	*polycephalum*	Indian	Cotton Cudweed	**M(traditional):**WP:OUL	Bown-289-11
Gnaphalium	*polycephalum*	Indian	Cotton Cudweed	**E:**WP:DYE(alum-yellow,alum-&-iron-green)	Grae-082-02
Gnaphalium	*uliginosium*	Marsh	Cotton Cudweed	**M(traditional):**WP:DTNS,HAP,OAS,RTL,UDR	Bown-289-12
Gnaphalium	*uliginosium*	Marsh	Cotton Cudweed	**M(traditional):**WP:DASC,OAS,RST	Wyk-411-024
Gnaphalium	*uliginosium*	Marsh	Cotton Cudweed	**E:**WP:DYE(alum-yellow,alum-&-iron-green)	Grae-082-02
Gossypium	*arboreum*	Tree	Cotton	**E:**SD:FI	Bown-290-01
Gossypium	*barbadensis*	Sea Island	Cotton	**E:**SD:FI	Bown-290-02
Gossypium	*herbaceum*	Levant	Cotton	**C**:OL:CO,**M(traditional):**LF:DVGT,OBN,OBR,OSA,SIB,**M(traditional):**RB:NMP,**M(traditional):**SD:DHSV,DSCB,HFV,NOR,OWN,SDY,**E:**SD:FI,DS,**E:**FL:DYE(brown,yellow,HSA)	Bown-290-03
Gossypium	*hitsutum*	Upland	Cotton	**E:**SD:FI	Bown-290-04
Gossypium	*hitsutum*	Upland	Cotton	**M(traditional):**BK,SD:OAS,**E:**SD:FI,contraceptive	Wyk-412-01
Gracilaria	*verrucosa*	Australian Agar	Graceful Florid Red Algae	**E:**WP:MC	Bown-287-03
Gratiola	*officinalis*	European	Hedge Hyssop	**M(traditional):**WP:BES,BHE,BHS,BJA,BLC,DWRM,SPR,UDR,**E:**WP:PS☠	Bown-290-05
Gratiola	*officinalis*	European	Hedge Hyssop	**M(traditional):**WP:SPR,UDR	Wyk-412-02
Grevillea	*robusta*	Robust	Silk Oak	**E:**LF:DYE(alum-canary-yellow,alum-&-iron-olive-green)	Grae-167-03

Latin Genus	L. Species	E. Species	English Genus	C:Culinary,M:Medicinal,E:Economic	Source
Grifola	*frondosa*	Maitake	Grifol Meristic Downy Pore Fungus	**M(traditional):**WP:BDB,BLC,BHP,HAN, **M(PS):**WP:DAID,DCTM	Bal2-097-01 Bal2-531-01
Grindelia	*camporum*	Balsam	Gumweed	**M(traditional):**WP:DIOS,DWHP,NMS,NSD,OSA,DSIN,RLA,RLB,RLC,UBL	Bown-290-06
Grindelia	*camporum*	Balsam	Gumweed	**M((traditional):**WP:BAS,NMS,DSIN,OWN,RLA,RLB,RLC,RRN,UUR	Wyk-161-02
Grindelia	*camporum*	Balsam	Gumweed	**E:**FL,PD:DYE(alum-gold-yellow,alum-&-iron-olive-green),**E:**WP:PS☠(kidney poison)	Grae-090-01
Grindelia	*lanceolata*	Lance Leaf	Gumweed	**M(traditional):**WP:DIOS,RLB,**E:**WP:PS☠(kidney poison)	Bown-290-07
Grindelia	*lanceolata*	Lance Leaf	Gumweed	**E:**FL,PD:DYE(alum-gold-yellow,alum-&-iron-olive-green)	Grae-090-01
Grindelia	*robusta*	Robust	Gumweed	**M(traditional):**WP:NMS,RLC	Wyk-412-03
Grindelia	*squarrosa*	Curly Cup Scaley	Gumweed	**M(traditional):**WP:DIOS,NMS,RLB,RLC,**E:**WP:PS☠(kidney)	Bown-290-08
Grindelia	*squarrosa*	Curly Cup Scaley	Gumweed	**M(traditional):**WP:RLA,RLB,RLC	Kowa-239-01
Grindelia	*squarrosa*	Curly Cup Scaley	Gumweed	**M(PS):**WP:BAS,NMS,DSIN,OWN,RLA,RLB,RLC,RRN,UUR	Wyk-161-01
Grindelia	*squarrosa*	Recurved Leaf	Gumweed	**E:**FL,LF,SM:DYE(alum-yellow-green,chrome-green-gold,copper-light-chartreuse,tin-bright-yellow,iron-olive-green)	Bliss-128-01
Grindelia	*squarrosa*	Recurved Leaf	Gumweed	**E:**FL,PD:DYE(alum-gold-yellow,alum-&-iron-olive-green)	Grae-090-01
Guajacum	*officinale*	West Indies	Guaiac Wood of Life	**M(traditional):**RN,WD:BCR,DSXS,NGT,NJA,NRH,OGT,OJA,ORH,RLC,SCN,UDR,**E:**WD:WL	Bown-290-08
Guajacum	*officinale*	West Indies	Guaiac Wood of Life	**M(PS):**RN,WD:DFRA,NJA,NRH	Wyk-162-01, Wyk-412-04
Guarea	*rusbyi*	Cocillan	Guar Mahogany	**M(traditional):**BK:OAS,RLC,SPR	Wyk-412-05
Gutierrezia	*sarothrae*	Matchbrush	Snakeweed	**E:**FL,LF,SM:DYE(alum-green-gold,chrome-tan-rust,copper-olive-green,tin-bright-yellow-gold,iron-dark-olive)	Bliss-162-01
Gymnema	*sylvestre*	Gurmar	Miracle Milkweed	**M(traditional):**LF:DHOG	Wyk-412-06
Gymnocladus	*dioicus*	Kentucky	Coffee Bean	**C:**SD:BF(coffee substitute)	Wyk-412-07
Gynostemma	*pentaphyllum*	Sweet Tea Vine	Gospel Herb	**M(traditional):**WP:BCR,BDB,BLC,BSH,BHC,BHP,DCCL,NMS,NSD,RLA,RLB,SUL	Bown-291-01
Gypsophila	*paniculata*	Soapwort	Baby's Breath	**M(PS):**RT,RZ:DSIN,RLC,RRN	Wyk-163-01, Wyk-412-08
Gyrophora	*deusta*			Synonym for *Umbilicaria deusta*	
Haematomma	*lapponica*			Synonym for *Ophioparma lapponica*	Cas2-032-04
Haematomma	*ventosum*			Synonym for *Ophioparma ventosa*	Cas2-032-05
Haematoxylum	*campechianum*	Logwood	Bloodwood	**M(traditional):**WD:OAS,DYE	Wyk-412-09
Haematoxylum	*campechianum*	Logwood	Bloodwood	**E:**WD:DYE(alum-&-chalk-blue,alum-&-copper-blue,iron-sulfate-&-tartar-black-blue,tin-purple)	Grae-013-01 Grae-014-01
Hagenia	*abyssinica*	Ethiopian	Hagen Floret Rose	**M(traditional):**FL(female):DWTP	Wyk-412-10
Halopappus	*spinulosus*	Spiny	Goldenweed	**E:**FL,LF,SM:DYE(alum-gold,chrome-camel, copper-olive,tin-butter-gold,iron-dark-gray)	Bliss-230-01
Hamamelis	*virginiana*	American	Witch Hazel	**M(traditional):**BK,LF,SM:BPH,OAS,OHM,DSIN	Bal1-078-08
Hamamelis	*virginiana*	American	Witch Hazel	**M(traditional):**LF:BVV,DAMW,DGUM,OHM, **M(traditional):**BK,LF:SDR,**M(traditional):**BK,LF,SM:DISB,DSIN,OSB,OST,OWN	Bal2-155-01 Bal2-550-05
Hamamelis	*virginiana*	American	Witch Hazel	**M(traditional):**BK,LF,SM:BVV,DVGB,DVGV,OAS,OBN,OBR,OEY,OHM,DSIN,RST,SDR,SDY,SIB,SUL,**E:**BK,LF,SM:CM	Bown-291-02
Hamamelis	*virginiana*	American	Witch Hazel	**M(traditional):**BK,LF:OAS,OBR,OHM,OWN,OVV,**E:**BK,LF:PS☠(allergicreactions)	Clev-095-04
Hamamelis	*virginiana*	American	Witch Hazel	**M(traditional):**BK,LF:DISB,OAS,OBN,OBR,OHM,OSB,DSIN,**E:**OL:CM,HO	Kowa-293-03 Kowa-507-01
Hamamelis	*virginiana*	American	Witch Hazel	**M(folk):**BK,LF:BVV,DBOI,OAS,OBR,OHM,OUL,OWN,RST	Krug-185-01
Hamamelis	*virginiana*	American	Witch Hazel	**M(CS):**BK,LF,SM:BVV,DAMW,OAS,OBR,OHM,DSIN,OWN,RST,SDR	Wyk-164-01, Wyk-412-11
Haplopappus	*baylahuen*	Baylahuen	Paper Groundsel	**M(traditional):**WP:BLC,OAS,SGB	Wyk-412-12
Harpagophytum	*procumbens*	Devil's Claw	Grapple Plant	**M(folk):**TU:OWN,**M(traditional):**TU:BLC,DALL,NJA,SGB,UBL,UKC,**M(PS):**TU:NCT,NGT,NSD,NTN, **M(CS):**TU:SAP,SDG,NRH	Bal2-059-01 Bal2-514-03
Harpagophytum	*procumbens*	Devil's Claw	Grapple Plant	**M(traditional):**TU:OAS,NJA,NNR,NRH,NSD,NSP,OJA,ORH,SDG,SGB,SPA	Bown-291-03
Harpagophytum	*procumbens*	Devil's Claw	Grapple Plant	**M(CS):**TU:DBOI,NJA,NRH,OUL,SAP,SDG	Wyk-165-01, Wyk-412-13
Harungana	*madagascarensis*	Madagascar	Haronga	**M(PS):**BK,LF:SDP,SGB, **M(PS):**BK,SM-SA:DCLW,OWN	Wyk-166-01, Wyk-412-14

Latin Genus	L. Species	E. Species	English Genus	C:Culinary,M:Medicinal,E:Economic	Source
Hedeoma	floribunda	Oregano	American Pennyroyal	C:LF:FF,TE	Bown-291-04
Hedeoma	pulegoides	Squaw Mint	American Pennyroyal	M(traditional):WP:DCLD,DWHP,NMP,RLC,SDG,SFL,E:OL:IR,PS☠	Bown-291-05
Hedeoma	pulegoides	Squaw Mint	American Pennyroyal	M(folk):LF,OL:MCI,NMP E:OL:IR(flea),induced abortion☠	Krug-135-02
Hedeoma	pulegoides	Squaw Mint	American Pennyroyal	M(traditional):WP:SDG,E:OL:IC	Wyk-412-15
Hedera	helix	English	Ivy	M(traditional):LF:DABA,DIMP,DSCB,DWHP,HFV,NGT,NMS,NRH,OBN,DCLW,OJA,ONR,OSA,DSIN,OTA,RLB,RLC,	Bown-292-01
Hedera	helix	English	Ivy	M(traditional):FR:SPR,M(traditional):LF:HFV(external),OBR,OJA,E:RN:PN(varnish),E:WP:PS☠	Clev-096-01
Hedera	helix	English	Ivy	M(CS):LF:DCAN,DWRM,RLB,RLC, E:WD:DS,CM	Wyk-167-01, Wyk-412-16
Hedera	helix	English	Ivy	E:LF:DYE(alum-soft-yellow,tin-bright-yellow,chrome-tan)	Cas1-159-01
Helianthella	uniflora	Aspen	Dwarf Sunflower	E:FL,SM:DYE(alum-light-gold,chrome-red-brown,copper-tan-green,tin-light-yellow,iron-gray)	Bliss-242-01
Helianthus	annus	Annual	Sunflower	C:SD:GF,NT,C:SD-OL:CO,M(traditional):SD:RLB, M(traditional):WP:BHC,DMAL,DTBC,ORH	Bown-292-02
Helianthus	annus	Annual	Sunflower	C:SD-OL:CO,M(folk):SD,SD-OL:DCLD,RLB,RLC,UDR, E:FL,LF:IR(mosquito)	Krug-168-01
Helianthus	annus	Annual	Sunflower	E:FL,SM:DYE(alum-green-gold,chrome-brown,copper-green-gold,tin-gold,iron-green)	Bliss-244-01
Helianthus	annus	Annual	Sunflower	E:FL:DYE(alum-&-tin-yellow,chrome-gold,alum-&-baking-soda-yellow-green,alum-&-iron-chartreuse), E:LF,SL:DYE(alum-&-tin-yellow-green,iron-gray,copper-sulfate-bronze),E:SD:DYE(chrome-tan)	Cas1-211-01
Helianthus	tuberosus	Jerusalem Artichoke	Sunflower	M(traditional):TU:BDB	Wyk-412-17
Helichrysum	arenarium	Sandy	Everlasting Buttons	M(PS):FL:BLC,SDG,UBL,UDR	Wyk-168-01, Wyk-412-18
Helichrysum	cochinchinense	Cochin	Everlasting Buttons	C:LF:**SN**	Clev-096-03
Helichrysum	italicum	Curry Plant	Everlasting Buttons	C:SM:**SN**	Bown-292-03
Helichrysum	italicum	Curry Plant	Everlasting Buttons	C:LF,SM:BF,JJ,**SN**,E:OL:CM,IR	Clev-096-02
Helichrysum	italicum	Curry Plant	Everlasting Buttons	M(traditional):FL:DALL,DAVR,OCC	Wyk-168-03
Helichrysum	orientale	Immortelle	Everlasting Buttons	E:OL:PF	Bown-292-04
Helichrysum	petiolatum	Cudweed	Everlasting Buttons	E:FL,LF:DYE(alum-light-gold-yellow,alum-&-copper-khaki-green,chrome-brown,chrome-&-copper-light-brown)	Grae-085-01 Grae-086-01
Helichrysum	serphyllifolium	Hottentot Tea	Everlasting Buttons	C:LF:TE,E:LF:PS☠	Bown-292-05
Helichrysum	serphyllifolium	Hottentot Tea	Everlasting Buttons	C:LF:TE,E:LF:PS☠	Clev-096-04
Helichrysum	stoechas	Goldilocks	Everlasting Buttons	M(traditional):LF:DWRM	Clev-096-05
Helleborus	nigra	Black	Hellebore	M(traditional):WP:BJA,DCTM,NBK,NGT,NMS,DCLW,E:WP:PS☠(helleborin&helleborein)	Kowa-047-01
Helleborus	viridis	Green	Hellebore	M(traditional):RZ:BHS,DWRM,SCN	Wyk-412-19
Hemerocallis	fulva	Orange	Day Lily	E:FL:DYE(alum-yellow,alum-&-chrome-rust,chrome-&-tin-burnt-orange), E:RT:DYE(alum-&-tin-yellow-orange)	Cas1-173-01
Hemizonia	luzulaefolia	Luzule	Tarweed	E:WP:DYE(alum-&-wool-gold-yellow,alum-&-cotton-light-yellow,alum-&-iron-dark-brass-green)	Grae-109-02 Grae-110-01
Hepatica	noblis	American Liverwort	Liverleaf	M(traditional):WP:BLC,DABI,DWRG,OAS,OWN,RLB,SDG,SGB,UDR	Bown-292-06
Hepatica	noblis	American Liverwort	Liverleaf	M(traditional):WP:BLC,DSIN	Wyk-412-20
Heracleum	maximum	Cow Parsley	Cow Parsnip	M(traditional):WP:ORH,RLA,SDG	Bown-292-07
Heracleum	maximum-lanatum	Woolly Cow Parsley	Cow Parsnip	M(traditional):WP:DEPI	Bown-292-08
Heracleum	maximum-lanatum	Woolly Cow Parsley	Cow Parsnip	E:WP:DYE(alum-beige,chrome-mint-green,copper-olive-gold,tin-butter-yellow,iron-light-gray-brown)	Bliss-094-01
Heracleum	sphondylium	Keck	Cow Parsnip	C:SL,CF,C:SH:**VG**,M(traditional):WP:BHP,HAP,NSD,RLB,RLC,RTL,SDG	Bown-293-01
Heracleum	sphondylium	Keck	Cow Parsnip	M(traditional):WP:OUL,SDP	Wyk-412-21
Herniaria	glabra	Smooth	Rupturewort	M(traditional):WP(aerial):BLC,NJA,NRH,RLC,UBL,UKC,UDR,UUR	Wyk-169-01, Wyk-412-22
Herniaria	hirsuta	Hairy	Rupturewort	M(traditional):WP(aerial):BLC,NJA,NRH,RLC,UBL,UKC,UDR,UUR	Wyk-169-02
Herniaria	incana	Hoary	Rupturewort	M(traditional):WP(aerial):BLC,NJA,NRH,RLC,UBL,UKC,UDR,UUR	USDA-NPDC
Heteromeles	arbutifolia	Christmas Berry	Toyonberry	E:LF,SM:DYE(chrome-gold-brown)E:LF,FR:DYE(alum-&-iron-black-olive-green,alum-&-iron-dark-olive-green)	Grae-144-01

Latin Genus	L. Species	E. Species	English Genus	C:Culinary,M:Medicinal,E:Economic	Source
Heterotheca	*villosa*	Golden Woolly	False Golden Daisy	E:FL,LF,SM:DYE(alum-olive-gold,chrome-dark-copper,copper-dark-olive-gold,tin-deep-gold,iron-black-gold)	Bliss-120-01
Heuchera	*americana*	American Sanicle	Alumroot	M(traditional):RT:DVGB,DVGV,OAS,OWN,RST,SDR,SDY,SUL,E:RT:PS☠	Bown-293-02
Hibiscus	*kokio*	Red	Rose Mallow	E:FL:DYE(alum-dark-blue,chrome-dark-green)	Grae-152-02
Hibiscus	*rosa-sinensis*	Chinese	Rose Mallow	M(traditional):WP:FL:DSXS,OAS,NMP,NMS,RLC,UBL,E:FL,JU:CM,HO	Bown-293-03
Hibiscus	*sabdariffa*	Roselle	Rose Mallow	C:LF:TE,**VG**,C:FL:JJ,C:SD:NT,M(traditional):FL,LF,SD,SM:HFV,OAS,SDG,UKC,UDR	Bown-293-04
Hibiscus	*sabdariffa*	Roselle	Rose Mallow	C:FL(calyx):TE,M(PS):FL(calyx):BCR,DCLD,DSIN,RLC,SAP,SCN,UDR	Wyk-170-01, Wyk-412-23
Hibiscus	*sabdariffa altissima*	Hemp Roselle	Rose Mallow	E:SM:FI	Bown-293-05
Hibiscus	*syriacus*	Rose of Sharon	Rose Mallow	E:FL:DYE(alum-&-copper-&-tin-blue,chrome-&-tin-purple)	Grae-146-01
Hieracium	*pilosella*	Mouse Ear	Hawkweed	M(traditional):WP:DABI,DFLU,DMDF,NMS,OWN,RLA,RLB,RLC,SDR,SHH,UBL,UKC,UKS,UDR	Bown-293-07
Hieracium	*pilosella*	Mouse Ear	Hawkweed	M(traditional):WP:OAS,OWN,RLC	Wyk-412-24
Hierochloe	*odorata*	Vanilla Grass	Sweet Grass	C:LF:AF,C:OL:BF,CF,E:OL:PF	Bown-293-08
Hippophae	*rhamnoides*	Sallow Thorn	Sea Buckthorn	C:FR:BF,FF	Wyk-413-01
Hoffmannseggia	*divaricata*	Chaparral	Rush Pea	M(traditional):LF:DCTM,DFRA,OSB,DSIN,E:LF:PS☠	Bal1-067-07
Hoffmannseggia	*divaricata*	Chaparral	Rush Pea	M(traditional):LF,FL:DSNB,NJA,NMP,NSD,UDR, M(PS):LF,FL:DCTM,DFRA,NCT	Bal2-047-01 Bal2-511-02
Hoffmannseggia	*divaricata*	Chaparral	Rush Pea	M(traditional):WP:NRH,DSIN,SDP,UUR, E:WP:PS☠(liver)	Wyk-414-16
Hoodia	*pilifera*	South Africa Desert	Carrion Flower Column Cactus	M(CS):SM:DOBE	Wyk-171-01
Hoodia	*gordonii*	Ghaap	Carrion Flower Column Cactus	M(traditional):SM:DOBE	Wyk-413-02
Hordeum	*distichum*	Two Rowed	Barley	M(traditional):SD:DWRM,OBN,SCN	Bown-294-01
Hordeum	*polystichum*	Six Rowed	Barley	M(traditional):SD:DWRM,OBN,SCN	Bown-294-02
Hordeum	*vulgare*	Cereal	Barley	C:SD:BF,**SN**,C:SR:BF,**SN**,M(traditional):SD:SAP,SDG,M(traditional):SR:DHPT,RLC,SAP,SDG	Bown-294-03
Hordeum	*vulgare*	Cereal	Barley	M(traditional):SD:SIB	Wyk-413-03
Houttuynia	*cordata-chinensis*	Coriander Leaf	Houttuyn Lizard's Tail	C:LF:GN,**VG**,M(traditional):WP:DSNB,HFV,UDR	Bown-294-04
Houttuynia	*cordata-japnonica*	Orange Scented	Houttuyn Lizard's Tail	C:LF:GN,**VG**,M(traditional):WP:DSNB,HFV,UDR	Bown-294-05
Humulus	*lupulus*	Beer	Hops	M(traditional):FL,FR,LF:BCR,DHEA,HAN,HIN,NSD,OTA,OUL	Bal1-072-04
Humulus	*lupulus*	Beer	Hops	M(folk):FL:HAC,NSD,OUL,OWN,SAP, M(traditional):FL:HAN,HIN, M(PS):FL:SDG,HAN,HIN	Bal2-085-01 Bal2-527-01
Humulus	*lupulus*	Beer	Hops	C:FR:AF,C:OL:BF,FF,C:SH:**VG**,M(traditional):LF,FL,OL,SH:DABA,DHVS,DSXP,HIN,NMS,NSD,DSIN,OUL,SIB,UDR	Bown-294-06
Humulus	*lupulus*	Beer	Hops	C:FR(cone):AF,C:LF,SH:**VG**,M(traditional):FR:DBOI,HIN,NSD,E:FR:CM,E:SM:FI	Clev-096-06
Humulus	*lupulus*	Beer	Hops	C:FR(strobile):AF,M(traditional):FR(strobile):DASC,HIN,NSD,DSIN,SDG,E:SM(vine):FI	Kowa-288-04 Kowa-332-01
Humulus	*lupulus*	Beer	Hops	C:FR(strobile):AF,M(folk):FL,OL:HAN,SDG	Krug-090-01
Humulus	*lupulus*	Beer	Hops	M(PS):FR(cone):HAN,HIN,NSD,UDR	Wyk-172-01, Wyk-413-04
Huperzia	*selago*	Lycopodine	Fir Clubmoss	M(traditional):WP:DWRG,DWRM,SPR	Wyk-413-05
Huperzia	*serrata*	Hyperzine	Fir Clubmoss	M(traditional):WP:DALZ	Murr-288-01
Hyacinthus	*orientalis*	Oriental	Hyacinth	E:FL(blue):DYE(alum-&-tin-blue)	Grae-123-02
Hydnocarpus	*kurzii*	Chaulmoogra	Truffle Pod	M(traditional):OL,SD:DABI,DSCB,DSCR,DWRG,DWRM,HFV,NSD,DCLW,OSS,DSIN,E:OL,SD:PS☠	Bown-294-07
Hydnocarpus	*kurzii*	Chaulmoogra	Truffle Pod	M(traditional):OL:DSIN,DCLW	Wyk-413-06
Hydrangea	*arborescens*	Seven Barks	Hydrange	M(traditional):RT:DOBE,UDR,UEP,UKC,UKS	Bal1-072-07
Hydrangea	*arborescens*	Seven Barks	Hydrange	M(traditional):RT:BHE,DASC,NGT,NJA,UBL,UDR,UEP,UKS,UUR,E:RT:PS☠	Bown-294-08
Hydrangea	*arborescens*	Seven Barks	Hydrange	M(traditional):RT:UBL,UKS,UUR	Wyk-413-07
Hydrangea	*macrophylla*	Florist's	Hydrange	E:FL(white,pink):DYE(alum-beige,chrome-warm-brown,copper-sulfate-gray),E:FL(blue):alum-yellow,copper-sulfate-yellow-green)	Cas1-154-01
Hydrastis	*canadensis*	Goldenroot	Goldenseal	M(traditional):RT,RZ:BDB,BHP,DABA,DABI,DC	Bal1-071-04

Latin Genus	L. Species	E. Species	English Genus	C:Culinary,M:Medicinal,E:Economic	Source
				LD,DFLU,RST,SCN,SDG,SUL	
Hydrastis	*canadensis*	Goldenroot	Goldenseal	**M(traditional):**RT:DGUM,RST,SDG,SDR **M(PS):**RZ:BHS,DABA,DAFU,DCTM	Bal2-080-01 Bal2-524-01
Hydrastis	*canadensis*	Goldenroot	Goldenseal	**M(traditional):**RZ:DABA,DALL,DAFU,DGUM,DPVI,DVGB,DVGV,NMP,OAS,DSIN,OEA,OEY,RSN,SCN,SDG,SUL,**E:**RZ:PS☠	Bown-295-01
Hydrastis	*canadensis*	Goldenroot	Goldenseal	**M(traditional):**RT,RZ:BSH,DASC,HAN,NSD,OEY(staph),E:RT,RZ:DYE(chrome-orange,tin-pale-yellow)	Kowa-232-01
Hydrastis	*canadensis*	Goldenroot	Goldenseal	**M(folk):**RT,RZ:DAMW,DSIN,OEY,OUL,OWN,RST,SDG, **E:**LF-JU:DYE(yellow)	Krug-084-01
Hydrastis	*canadensis*	Goldenroot	Goldenseal	**M(PS):**RT,RZ:DABA,DAMW,OAS,SCN,SDG,SDR	Wyk-173-01, Wyk-413-08
Hygrophila	*spinosa*	Spiny	Swampweed	**M(traditional):**WP,RT:BLC,UDR,UUR	Wyk-413-09
Hymenaea	*oblongifolia*	Jatobá	Copal Pea	**M(traditional):**WP:DCAN	Bal2-486-09
Hyoscyamus	*albus*	Russian	Henbane	**M(traditional):**WP:NNC☻,E:WP:PS☠	Bown-295-02
Hyoscyamus	*albus*	Russian	Henbane	**M(traditional):**LF,RT,SD:NMS,NNC☻	Wyk-413-10
Hyoscyamus	*muticus*	Egyptian	Henbane	**M(traditional):**WP:NNC☻,E:WP:PS☠	Bown-295-03
Hyoscyamus	*niger*	European	Henbane	**M(traditional):**WP:DEAT,DWHP,NMS,NNC☻,NSD,ONR,ORH,OTA,RLA,**E:**WP:PS☠	Bown-295-04
Hyoscyamus	*niger*	European	Henbane	**M(folk):**LF:HIN,NSD, **M(folk):**SD:OTA,NRH, **E:**WP:PS☠	Krug-058-01
Hyoscyamus	*niger*	European	Henbane	**M(CS):**LF:NMS,NNC☻,NSD,OTA,RLA	Wyk-174-01
Hypericum	*androsaemum*	All Heal	St. John's Wort	**M(traditional):**WP:OWN	Clev-097-02
Hypericum	*calycinum*	Calyx	St. John's Wort	**E:**FL:DYE(alum-&-tin-yellow-orange,chrome-&-tin-pumpkin-orange)	Grae-135-01
Hypericum	*perforatum*	Oil Gland Leaf	St. John's Wort	**M(traditional):**FL,LF,SM:DAVR,DCAN,DHIV,DHRP,HAN,NSD	Bal1-076-07
Hypericum	*perforatum*	Oil Gland Leaf	St. John's Wort	**M(traditional):**WP:DABA,DAVR,DSIN,HAC,HANHIN,OBN,OWN,**M(PS):**WP:DAID,DCRO,DEAI,DHSV,NCT,OHM,SIB	Bal2-139-01 Bal2-545-02
Hypericum	*perforatum*	Oil Gland Leaf	St. John's Wort	**M(traditional):**WP:DASC,DSHZ,NSC,NSD,OAS,ONR,OSC,OBN,OBR,OWN,UIN,**E:**WP:CM	Bown-295-05
Hypericum	*perforatum*	Oil Gland Leaf	St. John's Wort	**C:**FL:FF,**M(traditional):**LF:BCR,DABA,NSD,OBN,OWN,SDG	Clev-097-01
Hypericum	*perforatum*	Oil Gland Leaf	St. John's Wort	**M(traditional):**LF,FL:DABA,NSD,OAS,OBR,OHM,SDG,SDR,SIB,SUL,**E:**FL:DYE(tin-orange-red),**E:**SM:DYE(alum-red-brown),**E:**WP(top):alum-yellow,chrome-bright-yellow)E:WP:PS☠(photosensitivity)	Kowa-175-08 Kowa-447-01
Hypericum	*perforatum*	Oil Gland Leaf	St. John's Wort	**M(folk):**WP(flowering)-JU:OBN,OUL,OWN	Krug-154-01
Hypericum	*perforatum*	Oil Gland Leaf	St. John's Wort	**M(CS):**FL,FL-SM:DWRM,HAN,SIB,OBN,OWN	Wyk-175-01, Wyk-413-11
Hypericum	*perforatum*	Oil Gland Leaf	St. John's Wort	**E:**WP:DYE(alum-yellow-green,copper-sulfate-bronze,chrome-gold,iron-&-tin-yellow-green)	Cas1-204-01
Hypericum	*perforatum*	Oil Gland Leaf	St. John's Wort	**E:**FL,LF:DYE(alum-mustard-gold,alum-&-copper-dark-yellow-green,chrome-raw-sienna,chrome-&-copper-light-brown)	Grae-092-01 Grae-093-01
Hypericum	*pyramidatum*	Great	St. John's Wort	**E:**WP:DYE(red)	Clev-097-01
Hypogymnia	*physodes*	Physodic	Tube Shield Lichen	**E:**WP:DYE(Physodalic-Acid-gold,Physodic-Acid-orange-brown)	Cas2-035-03
Hypoxis	*hemerocallidea*	African Potato	Star Grass	**M(CS):**TU-JU:OBN,**M(traditional):**LF,SM:UEP,UUR	Wyk-176-01, Wyk-413-12
Hyssopus	*officinalis*	European	Hyssop	**M(traditional):**WP:BCR,DEPI,DOBE,HFV,NGT,OCC,OWN,SFL	Bal1-073-01
Hyssopus	*officinalis*	European	Hyssop	**C:**LF:**SN**,**C:**OL:AF,**M(traditional):**WP,FL:HFV,OAS,OBR,OLC,RLB,RLC,SDG,SFL,SIB,**E:**OL:PF,PS☠	Bown-295-06
Hyssopus	*officinalis*	European	Hyssop	**C:**LF,FL:AF,FF,**M(traditional):**LF,FL:OBN,OST,RLC,RST,**E:**OL:PF	Clev-097-03
Hyssopus	*officinalis*	European	Hyssop	**C:**LF,FL:AF,FF,**M(traditional):**LF(top):BHE,BJA,HFV,NRH,OBR,OWN,RLB,RLC,RST,SPR,**E:**LF:CM	Kowa-289-03 Kowa-342-01
Hyssopus	*officinalis*	European	Hyssop	**M(folk):**LF,FL:OBR,OEY(black),RLB,RLC	Krug-094-01
Hyssopus	*officinalis*	European	Hyssop	**M(PS):**WP(aerial):DASC,RLB,RLC,RRN,**M(traditional):**OL:DAMW,BCR,OEY	Wyk-177-01, Wyk-413-13
Iberis	*amara*	Bitter	Candytuft	**M(PS):**SD:SDG,SGB,**M(traditional):**WP,FL:SDP,SIB	Wyk-178-01, Wyk-413-14
Ilex	*aquifolium*	English	Ivy	**M(traditional):**LF:DFLU,DMAL,HFV,NRH,OAS,RLB,UDR	Bown-296-01
Ilex	*aquifolium*	English	Ivy	**M(traditional):**LF:HFV,NMS	Wyk-413-15
Ilex	*opaca*	American	Ivy	**M(traditional):**LF:DWRM,HFV,NGT,NRH,RLB,SCN	Bown-296-02

Latin Genus	L. Species	E. Species	English Genus	C:Culinary,M:Medicinal,E:Economic	Source
Ilex	*paraguensis*	Maté	Ivy	**M(traditional):**WP:BLC,DALL,DOBE,SCN,SIB	Bal1-079-04
Ilex	*paraguensis*	Maté	Ivy	**M(traditional):**LF:DOBE,UDR **M(PS):**LF:DFRA	Bal2-099-01 Bal2-531-02
Ilex	*paraguensis*	Maté	Ivy	**C:**LF:TE,**M(traditional):**LF:HAC,NMS,NNR,NRH,NSD,UDR	Bown-296-03
Ilex	*paraguensis*	Maté	Ivy	**C:**LF:TE,**M(PS):**LF:DCFS,DOBE,UDR	Wyk-179-01, Wyk-413-16
Ilex	*verticillata*	Winterberry	Ivy	**M(traditional):**BK:BJA,DASC,DHPT,DHSV,HFV,OAS,DSIN,OUL,SCN	Bown-296-04
Ilex	*vomitoria*	Vomit	Ivy	**E:**LF:PS☠	Bown-295-05
Illicium	*anisatum*	Japanese	Anise	**M(traditional):**SD:RSN,SDG,lactation-aid,**E:**FR,PS☠	Bal1-065-02
Illicium	*anisatum*	Japanese	Anise	**E:**LF:PS☠	Bown-296-06
Illicium	*verum*	Chinese	Anise	**C:**RD(husk),SD:**SP**,**C:**OL:AF,BF,CF,FF,**M(traditional):**FR,OL:DABA,DAFU,NBK,RLC,SDG	Bown-296-07
Illicium	*verum*	Chinese	Anise	**C:**RD(husk),SD:**SP**,**M(traditional):**RD(husk),SD:DAMW,RLC,**E:**BK:IN	Clev-304-01
Illicium	*verum*	Chinese	Anise	**M(PS):**FR,OL:NMS,RLC,SDP	Wyk-180-01, Wyk-413-17
Impatiens	*pallida*	Jewelweed	Impatience	**M(traditional):**WP:SPR,UDR,**M(traditional):**JU:DIOS,DWRG,DCLW,OHM,**E:**WP:PS☠	Bown-296-08
Indigofera	*patens*	Spreading	Indigo	**M(traditional):**RT:OTA	Kowa-344-03
Indigofera	*suffruticosa*	Guatemalan	Indigo	**E:**LF,JU:DYE(blue)	Kowa-344-02
Indigofera	*tinctoria*	Dyer's	Indigo	**E:**LF,JU:DYE(blue),**M(traditional):**LF:BLC,HFV	Kowa-344-01
Indigofera	*tinctoria*	Dyer's	Indigo	**M(traditional):**LF:NNR,**E:**LF:DYE(indigo-blue)	Wyk-413-18
Indigofera	*tinctoria*	Dyer's	Indigo	**E:**LF(lye-&-hydrosulphite-blue)	Cas1-154-02
Indigofera	*tinctoria*	Dyer's	Indigo	**E:**LF:DYE(lye-&-urine-blue,henna-&-lime-blue,potash-&-vat-saxon-blue,date-&-urine-&-jar-indigo-blue,date-&-urine-&-vat-indigo-blue)	Grae-057-02 Grae-058-01 Grae-059-01
Inula	*brittanica-chinensis*	Chinese Starwort	Elecampane	**M(traditional):**FL:DABA,RLB,RLC,SDG,SFL	Bown-296-09
Inula	*cappa*	Caper	Elecampane	**M(traditional):**FL,OL:HAC,NRH,DSIN,RLB	Bown-296-10
Inula	*helenium*	Scabwort	Elecampane	**M(traditional):**RT:DPNB,DWRM,NMP,NSC,RLA,RLB,RLC,SDG,**M(PS):**RT:DABA DAFU	Bal2-063-02 Bal2-516-02
Inula	*helenium*	Scabwort	Elecampane	**C:**FL,OL:FF,**C:**RT:AF,CF,**M(traditional):**FL,OL,RT:DABA,DAFU,DHYF,DTBC,BDB,RLA,RLB,RLC,RLP,SDG,UDR	Bown-296-11
Inula	*helenium*	Scabwort	Elecampane	**C:**RT:CF,**M(traditional):**RT:NMS,RLA,RLB,RLC,SDG,UDR	Kowa-182-01
Inula	*helenium*	Scabwort	Elecampane	**C:**RT:AF(absinthe),CF,**M(folk):**RT:RLC	Krug-068-01
Inula	*helenium*	Scabwort	Elecampane	**M(PS):**RT,RZ:DWRM,DSIN,RLB,RLC,SDG,SFL,UUR	Wyk-181-01, Wyk-413-19
Ipoema	*arvensis*	Field	Morning Glory	**M(traditional):**RN,TU:DWRM,SCN,SIB,SPR	Bown-297-01
Ipoema	*digitata*	Finger	Morning Glory	**M(traditional):**RN,TU:HAP,UDR	Bown-297-02
Ipoema	*purga*	Jalap	Morning Glory	**M(traditional):**RN,TU:HFV,OAS	Bown-297-03
Ipoema	*purga*	Jalap	Morning Glory	**M(traditional):**RT,TU:SPR	Wyk-413-20
Ipoema	*turpeth*	Turpeth	Morning Glory	**M(traditional):**RN,TU:SPR	Bown-297-04
Ipoema	*violacea*	Violet	Morning Glory	**M(traditional):**SD:NNC☠	Wyk-413-21
Iris	*germanica*	German Flag	Iris	**C:**RZ:AF,FX,**M(traditional):**RZ:RLC	Wyk-182-01, Wyk-413-22
Iris	*germanica*	German Flag	Iris	**E:**FL:DYE(alum-&-tin-blue,chrome-&-tin-blue)	Grae-124-02
Iris	*germanica-florentina*	Florentine	Iris	**M(traditional):**RZ:OWN,RLC,SDR,SPR,UDR,**E:**RT:RZ:PF	Bown-297-05
Iris	*germanica-florentina*	Florentine	Iris	**M(traditional):**RZ:DACN,DISB,NMS,RLC,SDG,UDR	Kowa-406-01
Iris	*germanica-florentina*	Florentine	Iris	**M(folk):**RT:SPR, **E:**RT:CM,PF	Krug-096-01
Iris	*germanica-florentina*	Florentine	Iris	**C:**RZ:AF,FX,**M(traditional):**RZ:RLC	Wyk-182-02
Iris	*pallida*	Dalmation	Iris	**M(traditional):**RZ:SPR	Bown-297-06
Iris	*pallida*	Dalmation	Iris	**E:**FL:DYE(chrome-violet-blue)	Grae-124-01
Iris	*prismatica*	Slender Blue Flag	Iris	**E:**RZ:DYE(alum-tan-yellow,iron-soft-gray),**E:**FL:DYE(alum-yellow-copper-sulfate-yellow-green,iron-gray)	Cas1-158-01
Iris	*versicolor*	Harlequin Blue Flag	Iris	**M(traditional):**RZ:BLC,DACN,DASC,DHSV,DPVI,HAC,NJA,OSS,ORH,DSIN,OWN,SCN,SGB,SPR,UDR,**E:**RZ:PS☠	Bown-297-07
Iris	*versicolor*	Harlequin Blue Flag	Iris	**M(folk):**RZ:BLC,SDG	Krug-036-01
Iris	*versicolor*	Harlequin Blue Flag	Iris	**M(traditional):**RZ:RLC	Wyk-413-23

Latin Genus	L. Species	E. Species	English Genus	C:Culinary,M:Medicinal,E:Economic	Source
Isatis	*tinctoria*	Dyer's	Woad	**M(traditional):**LF,RT:DAVR,DERS,DFLU,DMNB,DMNV,DMUM,DVGT,HFV,OSA,RST	Bown-297-08
Isatis	*tinctoria*	Dyer's	Woad	**M(traditional):**LF:OAS,OWN	Clev-097-06
Isatis	*tinctoria*	Dyer's	Woad	**M(traditional):**LF:OAS,**E:**LF(young):DYE(direct-blue,alum-pink),**E:**LF:PS☠	Kowa-175-12 Kowa-508-01
Isatis	*tinctoria*	Dyer's	Woad	**E:**LF,SM:DYE(vat-indigo-blue)	Grae-008-01
Isatis	*tinctoria*	Dyer's	Woad	**E:**LF:DYE(vat-indigo-blue)	Cas1-220-01
Isidium	*westringii*			Synonym for *Pertusaria pseudocorallina*	Cas2-032-06
Iva	*xanthifolia*	Giant	Marsh Elder	**E:**LF,FL-BD,SM:DYE(alum-light-gold,chrome-tan,copper-olive,tin-yellow,iron-gray-green)	Bliss-160-01
Jacaranda	*procera*	Carob	Yacawood	**M(traditional):**LF:UDR	Wyk-413-24
Jasminum	*angustifolium*	Narrow Leaf	Jasmine	**M(traditional):**FL,OL:DWRG	Bown-298-01
Jasminum	*grandiflorum*	Royal	Jasmine	**M(traditional):**FL,OL,RT:DABA,DAVR,DCBR,DCLK,DCLY,DFRI,HAC,HFV,OAS,OEY,DSIN,UUR,**E:**FL,OL:PF	Bown-298-02
Jasminum	*grandiflorum*	Royal	Jasmine	**M(traditional):**FL,OL:HAN,NSD,**E:**FL,OL:PF	Wyk-414-01
Jasminum	*lanceolarium*	Lance Leaf	Jasmine	**M(traditional):**SM:DBOI,ORH,OSA,OWN	Bown-298-03
Jasminum	*odoratissimum*	Fragrant	Jasmine	**E:**FL,OL:PF	Bown-298-04
Jasminum	*officinale*	Poet's	Jasmine	**C:**OL:FF(maraschino),**M(traditional):**FL,OL:DAMY,E:FL,OL:PF	Bown-298-05
Jasminum	*sambac*	Arabian	Jasmine	**C:**FL,TE(GreenTea)	Bown-298-06
Jateorhiza	*miersii*	Miers	Colombo	**M(traditional):**RT:DAFU,SDG	Bown-298-07
Jateorhiza	*palmata*	Palm Leaf	Colombo	**M(traditional):**RT:BHP,DAFU,SDG,SDP,SDR,SDY	Bown-298-08
Jateorhiza	*palmata*	Palm Leaf	Colombo	**M(traditional):**RT:DWRM,OAS,SDG	Wyk-414-02
Jatropha	*curcas*	Physic Nut	Nettle Spurge	**M(traditional):**SD:SPR,**E:**SD:PS☠	Wyk-414-03
Juglans	*cinerea*	Butternut	Walnut	**M(traditional):**FR-NT:SCN,SDG	Bown-298-09
Juglans	*cinerea*	Butternut	Walnut	**E:**RD:DYE(alum-strong-yellow,chrome-&-tin-yellow-orange,tin-orange,iron-brown,copper-sulfate-gray-brown)	Cas1-112-01
Juglans	*nigra*	American	Walnut	**M(traditional):**BK,FR-NT,FR-RD,LF:DAFU,DHSV,DIOS,DWRM,OBR,DCLW,RST,SDG	Bal1-065-09
Juglans	*nigra*	American	Walnut	**M(traditional):**BK:DAFU,DHSV,DWRM,SCN,**M(traditional):**LF:DALF,DCTM,DHSV,DIPT,DSIN,DWRG,HAP,OSS,RLA,SCN,**M(traditional):**SD-OL:BLC,DWTP,SDY,SIB,**M(ayurveda):**LF:NRH,**M(PS):**LF:DSIN	Bal2-151-01
Juglans	*nigra*	American	Walnut	**E:**WD:WL	Bown-298-10
Juglans	*nigra*	American	Walnut	**M(folk):**FR:HAC,**E:**FR:DYE(brown),**E:**LF:IR(ants)	Krug-179-02
Juglans	*nigra*	American	Walnut	**M(traditional):**LF,FR:OAS,OWN,SDR	Wyk-414-04
Juglans	*nigra*	American	Walnut	**E:**RD(husk):alum-gold-brown,chrome-dark-brass,copper-coffee,tin-camel,iron-dark-charcoal-gray)	Bliss-054-01
Juglans	*nigra*	American	Walnut	**E:**LF:DYE(alum-yellow,tin-bright-yellow,chrome-gold,chrome-&-iron-brown), **E:**RD(husk):DYE(chrome-brown,chrome-&-iron-dark-brown,copper-sulfate-taupe,alum-warm-tan,alum-&-tin-bright-rusty-brown)	Cas1-218-01
Juglans	*nigra*	American	Walnut	**E:**LF,SM:DYE(wool-dark-gray,cotton-light-brown,jute-brown)	Grae-188-02
Juglans	*regia*	English	Walnut	**M(traditional):**BK:DAFU,DHSV,DWRM,SCN,**M(traditional):**LF:DALF,DCTM,DHSV,DIPT,DSIN,DWRG,HAP,OSS,RLA,SCN,**M(traditional):**SD-OL:BLC,DWTP,SDY,SIB,**M(ayurveda):**LF:NRH,**M(PS):**LF:DSIN	Bal2-151-01
Juglans	*regia*	English	Walnut	**C:**SD:NT,**M(traditional):**LF:RLA,RLC,SCN,UKS,**M(traditional):**FR:BAN,SDR,**M(traditional):**OL:OSS,OEY,OSA,DSIN,**E:**OL:CM,HO	Bown-298-11
Juglans	*regia*	English	Walnut	**M(folk):**FR:HAC,**E:**FR:DYE(brown),**E:**LF:IR(ants)	Krug-179-01
Juglans	*regia*	English	Walnut	**M(PS):**LF,RD(husk):DACN,DAFU,DAMW,OSB,DSIN,OUL,SDR,SIB,**E:**LF,RD(husk):IC(bedbug,lice)	Wyk-183-01, Wyk-414-05
Juglans	*regia*	English	Walnut	**E:**LF,FR(catkin):DYE(wool-light-gold-brown)	Grae-189-01
Juncus	*balticus*	Baltic Rush	Wiregrass	**E:**FL,SM:DYE(alum-bright-yellow,chrome-ocher,copper-olive,tin-yellow-gold,iron-moss-green)	Bliss-274-01
Julella	*lactea*			Synonym for *Pertusaria lactea*	
Juniperus	*communis*	European	Juniper	**M(traditional):**FR:DOBE,NGT,RLA,UBL,UDR,UEP,UKC	Bal1-073-03
Juniperus	*communis*	European	Juniper	**C:**FR,**SP**,**M(traditional):**FR,OL:DASC,NGT,NJA,NRH,ONR,ORH,SDG,SFL,SIB,UBL,UDR,UKC,UUR,**E:**FR:PF	Bown-299-01
Juniperus	*communis*	European	Juniper	**C:**FR(berry),**SP**,**M(traditional):**FR:BCR,DSNB,OGT,ORH,SIB,SFL,**E:**FR:IC,PF,PS☠	Clev-098-01, Clev-305-01

Latin Genus	L. Species	E. Species	English Genus	C:Culinary,M:Medicinal,E:Economic	Source
Juniperus	*communis*	European	Juniper	**C:**FR(berry),**SP**,**M(traditional):**FR:NJA,NRH,OBR,OUL,OWN,UDR,**E:**FR:PS☠	Kowa-347-01
Juniperus	*communis*	European	Juniper	**C:**FR(berry),**SP**,**M(traditional):**FR:DASC,NGT,OJA,ORH,UBL,UKC,UUR	Krug-098-01
Juniperus	*communis*	European	Juniper	**C:**FR(cone):AF,**M(PS):**FR(cone):BCR,NRH,OHM,DSIN,SDG,SFL,SDP,UDR,UUR	Wyk-184-01, Wyk-414-06
Juniperus	*oxycedrus*	Prickly	Juniper	**M(traditional):**RT:DWRM	Clev-098-03
Juniperus	*sabina*	Savin	Juniper	**E:**OL:PS☠(podophyllotoxin)	Bown-299-02
Juniperus	*sabina*	Savin	Juniper	**M(traditional):**WP:DCLW, **induced abortion** ☠	Wyk-414-07
Juniperus	*scopulorum*	Rocky Mountain	Junper	**M(traditional):**FR,OL:DCLD,DMUM,HAC,RLC,SDY	Bown-299-03
Juniperus	*virginiana*	Atlantic Red Cedar	Juniper	**M(traditional):**FR:DCLD,DMUM,HAC,RLC,SDY,E:OL:PS☠	Bown-299-04
Juniperus	*virginiana*	Atlantic Red Cedar	Juniper	E:OL:PF	Clev-098-04
Justicia	*adhatoda*	Malabar Nut	Justice Bear Breech	**M(traditional):**LF:DTBC,HFV,NMS,RLA,RLB,RLC	Bown-299-05
Justicia	*adhatoda*	Malabar Nut	Justice Bear Breech	**M(traditional):**FL:DABI,DWRM,RLB,RLC	Wyk-414-08
Kaempferia	*galanga*	Galangal	Kaempfer Ginger	**C:**RZ,SH:**SP**,**M(traditional):**RZ:DABA,HAC,DDND,OJA,OWN,RLB,RST,SDG,SDP,UDR,**E:**RZ:CM	Bown-299-06
Kaempferia	*pandurata*	Fiddle Leaf	Kaempfer Ginger	**C:**RZ,SH:**SP**	Bown-299-07
Kaempferia	*rotunda*	Round Leaf	Kaempfer Ginger	**C:**RZ,SH:**SP**	Bown-299-08
Kaernefeltia	*californica*	California	Thornbush Heath Shield Lichen	**E:**WP:DYE(Lichesterinic-&-Protolichesterinic-Acid-&-Ammonia-light-brown)	Cas2-031-13
Kalmia	*latifolia*	Mountain Laurel	Laurel Heath	**M(traditional):**LF:BAA,DEAT,DHSV,DSXS,HFV,OAS,DSIN,NNR,SDR,SUL	Bown-300-01
Kigelia	*africana*	African	Kigel Sausage Fruit	**M(PS):**FR:DSXS,ODS,ORH,OSB,DSIN,OUL,OWN	Wyk-185-01, Wyk-414-09
Kochia	*scoparia*	Firebush	Molly Goosefoot	**E:**FL,LF,SM:DYE(alum-beige,chrome-tan-olive,copper-tan-green,tin-light-tan-gold,iron-tan)	Bliss-144-01
Knautia	*arvensis*	Field	Buttons Pincushion	**M(traditional):**RT:OWN,**M(traditional):**WP:DHMY,DDND	Clev-118-01
Krameria	*lappacea*	Peruvian	Rhatany	**M(traditional):**RT:BCB,DVGB,DVGV,OAS,OHM,OWN,SDR,SUL	Bown-300-02
Krameria	*lappacea*	Peruvian	Rhatany	**M(PS):**RT:DAMW,OAS,OBN,OHM,OWN,SDR	Wyk-186-01, Wyk-414-10
Krameria	*triandra*			Synonym for *Krameria lappacea*	Wyk-186-02
Laburnum	*anagryoides*	Bean Clover	Golden Chain	**M(traditional):**LF,SD:SPR	Wyk-414-11
Lachnanthes	*tinctoria*	Dyer's	Woolly Redroot Lily	**M(traditional):**WP,TU:NNR,E:WP,TU:DYE(red)	Wyk-414-12
Lactarius	*deliciosus*	Delicious	Milkcap Brittle Gill Mushroom	**E:**WP:DYE(chrome-&-tin-orange-rust)	Cas1-182-06
Lactuca	*canadensis*	Canada	Lettuce	**M(traditional):**LF:NSD,RLC	Bown-300-03
Lactuca	*canadensis*	Canada	Lettuce	**E:**LF:DYE(alum-soft-yellow,chrome-gold,copper-sulfate-green,iron-avocado-green,tin-bright-yellow-green)	Cas1-163-01
Lactuca	*serriola*	Prickly	Lettuce	**M(traditional):**LF:DHEA,DWHP,HIN,NRH,NSD,RLB,RLC	Bown-300-04
Lactuca	*serriola*	Prickly	Lettuce	**E:**FL,LF,SM:DYE(alum-bright-yellow,chrome-bright-brass,copper-bright-yellow-green,tin-vivid-yellow,iron-dark-forest-green)	Bliss-194-01
Lactuca	*tartarica pulchella*	Blue Flowered Tartarian	Lettuce	**E:**FL,SM:DYE(alum-gold,chrome-chestnut-brown,copper-green-gold,tin-bright-gold,iron-very-dark-olive-green)	Bliss-064-01
Lactuca	*virosa*	Opium	Lettuce	**M(traditional):**LF:NSD,RLC	Bown-300-05
Lactuca	*virosa*	Opium	Lettuce	**M(folk):**LF,JU:HIN,DWHP,NNC👽,NSD,RLC	Krug-103-01
Lactuca	*virosa*	Opium	Lettuce	**M(traditional):**WP:NNC👽,NSD	Wyk-414-013
Lamium	*album*	White	Deadnettle	**C:**FL:TE,**M(traditional):**WP:DVGB,DVGV,OAS	Bown-300-06
Lamium	*album*	White	Deadnettle	**M(PS):**LF,FL,SM(top):BLC,BVV,OBR,DDND,DSIN,RLC,SDP,SFL,SIB,UDR,E:LF,FL,SM(top):CM,DS	Wyk-187-01, Wyk-414-14
Languas	*galanga*			Synonym for *Alpinia galanga*	Clev-306-01
Larix	*decidua*	European	Larch	**M(traditional):**BK:UUR,**M(traditional):**RN:DWTP,OAS,DSIN,OWN,RLB,SDR,UDR	Bown-301-01
Larix	*decidua*	European	Larch	**M(traditional):**BK,RN:DASC,OAS,DSIN,UDR,UUR	Wyk-414-15
Larix	*laricina*	American	Larch	**M(traditional):**RN:OAS,RLB,UDR	Bown-301-02
Larix	*laricina*	American	Larch	**E:**LF:DYE(chrome-warm-tan-yellow), **E:**SM:DYE(chrome-brown),**E:**PD(cone):DYE(vinegar-&-salt-tan),**E:**BK:DYE(iron-purple-gray)	Cas1-162-02
Larrea	*tridentata*			Synonym for *Hoffmannseggia divaricata*	Wyk-414-16
Lasallia	*papulosa*	Papp	Toadskin Tripe Lichen	**E:**WP:DYE(Anthraquinones-&-Ammonia-	Cas2-032-07

Latin Genus	L. Species	E. Species	English Genus	C:Culinary,M:Medicinal,E:Economic	Source
				magenta, Anthraquinones-&-Ammonia-pink)	
Lasallia	*pustulata*	Pustule	Toadskin Tripe Lichen	**E:**WP:DYE(Anthraquinones-&-Ammonia-red)	Cas2-032-08
Lathyrus	*japonicus*	Japanese	Beach Pea	E:WP:DYE(alum-yellow,tin-gold,copper-sulfate-yellow-green,chrome-&-alum-dark-gold,iron-yellow-gray,chrome-&-iron-taupe)	Cas1-190-02
Lathyrus	*odoratus*	Sweet Scented	Beach Pea	**E:**FL,SL:DYE(alum-ivory,chrome-light-tan, copper-light-gray-green,tin-butter-yellow,iron-tan)	Bliss-246-01
Laurus	*noblis*	Bay Leaf	Laurel	**C:**LF:**SP**,**M(traditional):**LF,OL:DSCB,OBR,DDND,ORH,OUL,SAP,SFL,SIB	Bown-301-03
Laurus	*noblis*	Bay Leaf	Laurel	**C:**LF:**SP**,**C:**FR,OL:AF(bay-rum),**M(traditional):**LF,DASC,OAS,OBN,OBR,SAP,**E:**FR,OL:PF	Clev-098-05, Clev-307-01
Laurus	*noblis*	Bay Leaf	Laurel	**C:**LF:**SP**,**M(traditional):**LF:NNC☠,OAS,SDG,SFL,UDR,**M(traditional):**OL:OBR,DSIN,UDR,**E:**OL:CM,IR	Kowa-036-01
Laurus	*noblis*	Bay Leaf	Laurel	**C:**LF:**SP**,**M(folk):**LF-OL:DSNB,NRH	Krug-025-01
Laurus	*noblis*	Bay Leaf	Laurel	**C:**LF:**SP**,**C:**OL:FF,**M(PS):**LF:SDG,SDP,SFL, **M(PS):**OL:AMY, **E:**OL:IC	Wyk-188-01, Wyk-414-17
Lavandula	*allardii*	Giant	Lavender	**E:**FL,OL:PF	Bown-301-04
Lavandula	*angustifolia*	English	Lavender	**M(traditional):**FL:HAC,HAN,OBN,ODS,DSIN	Bal1-073-05
Lavandula	*angustifolia*	English	Lavender	**M(traditional):**FL:DACN,DCFS,DSIN,HIN,NMS, NSD,OBR,OHL,OWN,SDG,SFL,UDR, **M(PS):**FL:DABA,DAFU,HAN	Bal2-093-01 Bal2-528-03
Lavandula	*angustifolia*	English	Lavender	**C:**FL:FF,JJ,**M(traditional):**FL,OL:BCR,DASC,DHAL,DISB,DSNB,DTBC,DVGB,DVGV,HFV,NMS,OBN,ONR,ORH,OSB,RLB,SDG	Bown-301-05
Lavandula	*angustifolia*	English	Lavender	**M(traditional):**FL-OL:DASC,HAC,RLC,NSD, **E:**LF,FL:IN	Clev-098-07
Lavandula	*angustifolia*	English	Lavender	**C:**LF:FF,JJ,**M(traditional):**FL-OL:DASC,DISB,NMS,NSD,OBR,SDG,**E:**LF:CM,IR	Kowa-350-01
Lavandula	*angustifolia*	English	Lavender	**M(traditional):**FL,OL:HAC,HAN,NSD,**M(traditional):**OL:DASC,OBN,OWN, **E:**OL:IR	Krug-101-01
Lavandula	*angustifolia*	English	Lavender	**M(PS):**FL:HIN,NMS,NSD,**M(traditional):**OL:AMY,DASC,OBN,OJA,OSB,OWN,**E:**OL:CM	Wyk-189-01, Wyk-414-18
Lavandula	*dentata*	French	Lavender	**E:**FL,OL:PF	Bown-301-06
Lavandula	*intermedia*	Lavandin	Lavender	**C:**FL:FF,JJ,**E:**FL,OL:PF	Bown-302-01
Lavandula	*lanata*	Woolly	Lavender	**E:**FL,OL:PF	Bown-302-02
Lavandula	*latifolia*	Broad Leaf	Lavender	**E:**FL,OL:PF	Bown-301-07
Lavandula	*multifida*	Branched	Lavender	**E:**FL,OL:PF	Bown-302-03
Lavandula	*officinalis*			Synonym for *Lavandula angustifolia*	Bal2-093-01
Lavandula	*pinnata*	Fern Leaf	Lavender	**E:**FL,OL:PF	Bown-302-04
Lavandula	*sawyersi*	Sawyers	Lavender	**E:**FL,OL:PF	Bown-302-05
Lavandula	*stoechas*	Spanish	Lavender	**M(traditional):**FL:DASC,**E:**FL,OL:IR,PF	Bown-302-06
Lavandula	*stoechas*	Spanish	Lavender	**E:**FL,OL:PF	Clev-099-03
Lavandula	*vera*			Synonym for *Lavandula angustifolia*	Bal2-093-01
Lavandula	*viridis*	Green	Lavender	**E:**FL,OL:PF	Bown-302-07
Lavatera	*arborea*	Mediterranean	Tree Mallow	**E:**FL:DYE(alum-blue,alum-&-tin-deep-blue)	Grae-095-01
Lawsonia	*inermis*	Egyptian Privet	Henna	**M(traditional):**LF,FL,OL:DCLW,SDY,**E:**OL:HO, E:LF,FL:DYE(orange-red)	Bown-302-08
Lawsonia	*inermis*	Egyptian Privet	Henna	**E:**LF:DYE(rich-red)	Krug-089-01
Lawsonia	*inermis*	Egyptian Privet	Henna	**M(PS):**LF:DASC,DEPI,DWTP,RST,OAS,DSIN,OWN,SDR,SDY,SUL,E:DS,DYE(henna-red-brown)	Wyk-190-01, Wyk-414-19
Lecanora	*muralis*	Wall Loving	Rim Lichen	**E:**WP:DYE(Usnic-Acid-&-Zeorin-tawny-olive)	Cas2-035-04
Lecanora	*tartarea*			Synonym for *Ochrolechia tartarea*	
Ledebouriella	*divaricata*	Spreading	Fang Feng	**M(traditional):**RT:DABI,NMS,NRH,NSD	Wyk-414-20
Ledum	*groenlandicum*	Labrador Tea	Delirium Heath	**M(traditional):**LF,SH:DSCB,OAS,DDND,DSIN,RLB,RLC,SDR,UDR,**E:**LF,SH:IC(mosquito)	Bown-302-09
Lentinula	*edodes*	Shiitake	Lentinan Pinwheel Gill Mushroom	**C:**WP:**VG**,**M(traditional):**WP:BHC,BHP,BHS,DAVR,DCFS,DCTM, **M(PS):**WP:DAID	Bal2-131-01 Bal2-529-02 Bal2-542-02
Lentinus	*edodes*			Synonym for *Lentinula edodes*	
Leontodon	*autumnalis*	Fall	Hawkbit Dandelion	**E:**FL:DYE(alum-yellow,tin-bright-gold,chrome-dark-gold,copper-sulfate-tan,)**E:**RT:DYE(vinegar-&-salt-tan,alum-gold,baking-soda-gold-brown,iron-gray,tin-bright-gold,chrome-brown)	Cas1-134-02
Leonurus	*cardiaca*	European	Motherwort	**M(folk):**WP(aerial):NMP,RLA,**M(traditional):**WP(aerial):BHS,**M(PS):**WP(aerial):BHS,DHET	Bal2-103-01 Bal2-533-02
Leonurus	*cardiaca*	European	Motherwort	**M(traditional):**WP:BCR,BHP,DABA,DAFU,NMS, NSD,UDR	Bown-302-10
Leonurus	*cardiaca*	European	Motherwort	**M(traditional):**WP:HAN(following childbirth)	Krug-119-01

Latin Genus	L. Species	E. Species	English Genus	C:Culinary,M:Medicinal,E:Economic	Source
Leonurus	*cardiaca*	European	Motherwort	**M(PS):**WP:BHS,DHET,HAN,NMS	Wyk-191-01, Wyk-414-21
Leonurus	*heterophylla*	Chinese	Motherwort	**M(folk):**WP(aerial):NMP,RLA,**M(traditional):**WP(aerial):BHS,**M(PS):**WP(aerial):BHS,DHET	Bal2-103-02 Bal2-533-02
Leonurus	*sibricus*	Siberian	Motherwort	**M(traditional):**WP:BCR,BHE,BHP,DABA,DAFU,OSA,DSIN,UKC,UKS,UDR	Bown-302-11
Lepidium	*meyenii*	Maca	Peppergrass	**M(traditional):**WP,RT:increases estrogen	Murr-791-01
Lepidium	*sativum*	Sweet	Peppergrass	**M(traditional):**WP,RT:DGUM,DWRM,UDR	Wyk-415-01a
Lepidium	*virginica*	Virginia	Peppergrass	**M(traditional):**WP,RT:DGUM,DWRM,UDR	Wyk-415-01b
Lepidium	*virginica*	Virginia	Peppergrass	**E:**FL,LF,SM:DYE(alum-beige-gold,chrome-mint-green,copper-ocher,tin-light-yellow,iron-olive)	Bliss-186-01
Leptadenia	*reticulata*	Jivanti	Thin Gland Milkweed	**M(traditional):**RT:DPXC,DTBC,HAP,RLC	Wyk-415-02
Leptospermum	*scoparium*	Red Flowered	New Zealand Teawood	**E:**FL,LF,SM:DYE(alum-yellow-green,chrome-yellow-green,alum-&-iron-black-green)	Grae-157-01 Grae-157-02
Lespedeza	*capitata*	Round Head	Bush Clover	**M(traditional):**WP:UDR	Wyk-415-03
Letharia	*columbiana*	Columbian	Staghorn Moss Shield Lichen	**E:**WP:DYE(Vulpinic-Acid-&-Atranorin-brilliant-yellow)	Cas2-035-05
Letharia	*vulpina*	Wolf	Staghorn Moss Shield Lichen	**E:**WP:DYE(Vulpinic-Acid-&-Atranorin-brilliant-lemon)	Cas2-035-06
Letharia	*vulpina*	Wolf	Staghorn Moss Shield Lichen	**E:**WP:DYE(direct-green-yellow)	Grae-004-01
Leucanthemum	*superbum*	Superb	Shasta Daisy	**E:**FL:DYE(alum-soft-yellow,chrome-yellow-gold,tin-bright-yellow)	Cas1-206-01
Leucanthemum	*vulgare*	Ox Eye	Shasta Daisy	**E:**WP:DYE(alum-soft-yellow-green,tin-bright-yellow-green,vinegar-beige,baking-soda-chartreuse,chrome-gold,iron-dull-pale-green)	Cas1-133-01
Levisticum	*chinensis*	Chinese	Lovage	**M(traditional):**LF,RT,SD:NMP	Clev-099-06
Levisticum	*officinale*	Love Parsley	Lovage	**C:**LF:FF,**C:**SH:**VG**,**C:**SL:CF,**C:**SD:FF,**M(traditional):**LF,RT,SM:NMS,NSD,OUL,RLC,RST,SAP,SDG,SFL,SIB,UBL,UKS,UDR	Bown-303-01
Levisticum	*officinale*	Love Parsley	Lovage	**C:**LF:FF,**C:**SL:CF,**C:**RT:**VG**,**C:**SD:FF,**M(traditional):**LF,RT,SM:SDG,SFL,**E:**LF,OL:PF	Clev-099-04
Levisticum	*officinale*	Love Parsley	Lovage	**C:**LF:FF,**C:**SL:CF,**C:**SD:FF,**M(traditional):**RT:BJA,DMAL,DOBE,NRH,RST,SDG,SIB,UDR,UKS	Kowa-369-01
Levisticum	*officinale*	Love Parsley	Lovage	**C:**LF,SM:FF,**C:**SD:**SN**,**M(folk):**RT,SD,SM:BJA,NRH,SDG,UDR,UKC,UKS	Krug-108-01
Levisticum	*officinale*	Love Parsley	Lovage	**C:**LF:AF,FF,**M(PS):**RZ,RT:BHE,RLC,SDG,SFL,UDR,UKS,UUR	Wyk-192-01, Wyk-415-04
Levisticum	*scoticum*	Scots	Lovage	**C:**LF:FF	Clev-099-05
Liatris	*chapmanii*	Liatrin	Blazing Star	**M(traditional):**LF,RT:DCST	Bown-303-02
Liatris	*punctata*	Snakeroot	Blazing Star	**M(traditional):**RT:DSIN	Bown-303-03
Liatris	*pycnostachya*	Kansas Purple Gayfeather	Blazing Star	**E:**SL:DYE(alum-creamy-yellow,chrome-light-camel,copper-light-olive,tin-lemon-yellow,iron-gray-green)	Bliss-058-01
Liatris	*scariosa*	Shrivelled	Blazing Star	**M(traditional):**WP:DSNB,UDR	Bown-303-04
Liatris	*spicata*	Magenta Gayfeather	Blazing Star	**M(traditional):**LF,RT:DABA,DSXG,OAS,RST,UKC,UDR,**E:**LF,RT:IR	Bown-303-05
Liatris	*squarrosa*	Scaly	Blazing Star	**M(traditional):**WP:DSNB,UDR	Bown-303-06
Ligusticum	*canbyi*	American	Alpine Lovage	**M(traditional):**WP:DCLD	Bown-303-07
Ligusticum	*porteri*	Coughroot	Alpine Lovage	**M(traditional):**RT:DABA,DCLD,DFLU,NJA,RLB,RSN,RST	Bal2-109-01 Bal2-533-06
Ligusticum	*porteri*	Coughroot	Alpine Lovage	**M(traditional):**SD,RT,OL:BCR,HFV,NMP,NMS,OTA,RLB,RLC,SDG,UKC	Bown-303-08
Ligusticum	*scoticum*	Scots Seaside	Alpine Lovage	**C:**LF,SL:**VG**,**M(traditional):**LF,SD,SM:BCR,NRH,SDG,UDR	Bown-303-09
Ligusticum	*sinense*	Chinese	Alpine Lovage	**M(traditional):**RT:DABA,BCR,BHP,BHS,HAC,NSD	Bown-303-10
Ligusticum	*sinense*	Chinese	Alpine Lovage	**M(traditional):**RT:HAC	Wyk-415-05
Ligustrum	*ibota*	Wax Stem	Privet	**E:**SM:WX	Clev-100-02
Ligustrum	*ibolium*	Wax Stem	Privet	**E:**LF:DYE(alum-bright-yellow,tin-chartreuse,chrome-gold,copper-sulfate-&-iron-bronze,iron-yellow-gray)	Cas1-198-01
Ligustrum	*japonicum*	Japanese	Privet	**C:**SD:BF(coffee)	Clev-100-03
Ligustrum	*lucidum*	Glossy	Privet	**M(traditional):**FR:BLC,DABA,DEAT,DECT,DHPT,DPRK,HIN,NBK,NRH,OTM,UDR,UKC	Bown-304-01
Ligustrum	*lucidum*	Glossy	Privet	**M(traditional):**FR:BLC,DABA	Clev-100-04
Ligustrum	*lucidum*	Glossy	Privet	**E:**FR,SM:DYE(chrome-&-aluminum-grass-green),**E:**FR:DYE(alum-&-iron-dark-blue-gray), **E:**LF:DYE(alum-yellow,alum-&-vinegar-dark-khaki-green	Grae-172-01 Grae-173-01 Grae-173-02
Ligustrum	*ovalifolium*	Oval Leaf	Privet	**E:**WP:LS	Clev-100-05
Ligustrum	*vulgare*	European	Privet	**E:**WD,WL,**E:**LF:DYE(yellow),**E:**WP:PS☠	Clev-100-01
Lilum	*candidum*	Madonna	Lily	**M(traditional):**BR,FL:BCB,OAS,OBN,OHL,OSA,DSIN,OUL,**E:**BR:MC	Bown-304-02

Latin Genus	L. Species	E. Species	English Genus	C:Culinary,M:Medicinal,E:Economic	Source
Lilum	*candidum*	Madonna	Lily	**M(traditional):**BR,FL:DEPI,RLC	Wyk-415-06
Lilum	*concolor*	Lungwort	Lily	**M(traditional):**BR,FL:RLB	Bown-304-03
Linaria	*vulgaris*	Butter & Eggs	Toadflax	**M(traditional):**WP:BHE,BLC,DHPT,OAS,OHM, DSIN,OUL,SCN,SGB,SIB,UDR	Bown-304-04
Linaria	*vulgaris*	Butter & Eggs	Toadflax	**M(traditional):**WP,FL:BJA,DHMY	Clev-100-06
Linaria	*vulgaris*	Butter & Eggs	Toadflax	**M(traditional):**WP:BLC,OHM,OWN,SCN,UDR	Wyk-415-07
Linaria	*vulgaris*	Butter & Eggs	Toadflax	**E:**FL,LF,SM:DYE(alum-yellow,chrome-ocher, copper-yellow-green,tin-bright-yellow,iron-tan-olive)	Bliss-072-01
Linaria	*vulgaris*	Butter & Eggs	Toadflax	**E:**WP:DYE(alum-&-copper-sulfate-yellow-green,tin-chartreuse,iron-green-gray)	Cas-111-01
Lindera	*benzoin*	Spice	Benjamin Laurel	**C:**BK,LF,SM:BF,FF,**C:**FR:**SP**,**M(traditional):**BK ,LF,FR,SM:BCR,DCLD,DWRM,SDY	Bown-304-05
Lindera	*glauca*	Joss Stick	Benjamin Laurel	**E:**BK,SM:IN	
Lindera	*strychnifolia*	Strychnine Leaf	Benjamin Laurel	**M(traditional):**BK,LF,SM:NMP	Bown-304-06
Linum	*catharticum*	Purging	Flax	**M(traditional):**WP:DHMY,SPR	Clev-101-02
Linum	*catharticum*	Purging	Flax	**M(traditional):**WP:SPR	Wyk-415-08
Linum	*usitatissimum*	Linseed Oil	Flax	**M(traditional):**SD,OL:DOST,DSFT,DWHN,SIB	Bal1-070-06
Linum	*usitatissimum*	Linseed Oil	Flax	**C:**SD,OL:CO,**M(traditional):**SD,OL:DBOI,NJA,OS A,OUL,RLB,RLC,RLP,RST,SCN,SIB,SDV,**E:**SM:FI	Bown-304-07
Linum	*usitatissimum*	Linseed Oil	Flax	**M(traditional):**SD,OL:DWRM,OBR,OWN,SCN,S DG,SIB,**E:**SD,OL:PN,**E:**SM:FI,**E:**WP:PS☠	Clev-101-01
Linum	*usitatissimum*	Linseed Oil	Flax	**C:**SD:FF(bread),**M(traditional):**SD-OL:DCLD,RLC,UUR	Kowa-194-01
Linum	*usitatissimum*	Linseed Oil	Flax	**M(folk):**SD:DBOI,DSIN,SCN,**M(folk):**SGS	Krug-076-01
Linum	*usitatissimum*	Linseed Oil	Flax	**M(CS):**SD:SCN,SIB,SDV	Wyk-193-01, Wyk-415-09
Lippia	*adoensis*	African	Lemon Vervain	**C:**LF:TE	Bown-305-01
Lippia	*alba*	White	Lemon Vervain	**C:**LF:**SN**,TE	Bown-305-02
Lippia	*citriodora*			Synonym for *Aloysia triphylla*	Clev-101-03
Lippia	*dulcis*	Cimmaron	Lemon Vervain	**C:**LF:FF,**M(traditional):LF**:RLC	Bown-305-03
Lippia	*graveolens*	Oregano	Lemon Vervain	**C:**LF:**SN**	Bown-305-04
Lippia	*javanica*	Java	Lemon Vervain	**M(traditional):**WP:HFV,**E:**OL:IR	Wyk-415-10
Lippia	*micromere*	West Indies Thyme	Lemon Vervain	**C:**LF:**SN**	Bown-305-05
Lippia	*palmeri*	Mexican Oregano	Lemon Vervain	**C:**LF:**SN**	Bown-305-06
Lippia	*pseudothea*	Brazilan Tea	Lemon Vervain	**C:**LF:TE	Bown-305-07
Liquidambar	*formosana*	Chinese	Sweet Gum	**M(traditional):**RN:OMC,ORH,OWN	Bown-305-08
Liquidambar	*orientalis*	Levant Storax	Sweet Gum	**M(traditional):**RN:BHS,BST,DASC,DSCB,DSIN, RLC	Bown-305-09
Liquidambar	*orientalis*	Levant Storax	Sweet Gum	**M(traditional):**RN:DSIN	Wyk-415-11
Liquidambar	*styraciflua*	American Storax	Sweet Gum	**C:**RN:FF,**M(traditional):**RN:DASC,DSCB,DVGB ,DVGV,DWRG,OHM,RLA,RLB,RLC,RST,UBL**E:** RN:PF	Bown-305-10
Liquidambar	*taiwaniana*	Taiwan	Sweet Gum	**M(traditional):**FR:BCR,DALL,NJA	Bown-305-11
Liriodendron	*tulipifera*	Tulip Poplar	Lyre Magnolia	**M(traditional):**BK,WD:DMAL,HFV	Wyk-415-12
Litchi	*chinensis*	Chinese	Lychee	**M(traditional):**SD:BDB,DCTM	Wyk-415-13
Lithospermum	*erythrorhizon*	Redroot	Gromwell	**M(traditional):**WP,RT:BCR,BHS,BLC,DBOI,DC SK,DHPT,DHSV,DMSL,DIOS,DPXC,DVGB,DVG V,HFV,OSA,DSIN,**E:**WP,RT:CM	Bown-305-12
Lithospermum	*officinale*	Stone Seed	Gromwell	**M:**SD:**antigonadotropic**☠, **antithyreotropic** ☠	Wyk-415-14
Lithospermum	*ruderale*	Puccoon	Gromwell	**M(traditional):**WP:**contraceptive** ☠	Bown-305-13
Lobaria	*pulmonaria*	Lungwort	Lobed Pelt Lichen	**M(traditional):**WP:OAS,DSIN,RLC	Wyk-415-15
Lobaria	*pulmonaria*	Lungwort	Lobed Pelt Lichen	**E:**WP:DYE(Stictic-Acid-&-Norstictic-Acid-apricot)	Cas2-035-07
Lobaria	*spathulata*	Spoon	Lobed Pelt Lichen	**E:**WP:DYE(Lobaric-Acid-&-Ammonia-light-red)	Cas2-032-11
Lobelia	*cardinalis*	Pukeweed	Lobeline Bellflower	**M(traditional):**WP:SPR	Bown-305-14
Lobelia	*chinensis*	Chinese	Lobeline Bellflower	**M(traditional):**WP:BHE,BJA,BLC,DAFU,DCST, DSNB,DWFK,HFV,DSIN,SDY,SIB,UKC,UDR,**E:** WP:PS☠	Bown-305-15
Lobelia	*chinensis*	Chinese	Lobeline Bellflower	**M(traditional):**WP:UDR	Wyk-415-16
Lobelia	*erinus*	Blue	Lobeline Bellflower	**E:**FL,SM:DYE(alum-&-copper-pastel-green,chrome-&-copper-pastel-green)	Grae-125-01
Lobelia	*inflata*	Indian Tobacco	Lobeline Bellflower	**M(traditional):**FL,LF,SD:BCR,DAVR,DCLD,DEP I,DFLU,HFV,NSD,RLA,RLB,RLC	Bal1-074-01
Lobelia	*inflata*	Indian Tobacco	Lobeline Bellflower	**M(traditional):**WP:DBOI,DWHP,ORH,OUL,RLA, RLB,RLC,RLP,**E:**WP:PS☠	Bown-306-01
Lobelia	*inflata*	Indian Tobacco	Lobeline Bellflower	**M(traditional):**WP:DISB,ORH,OBN,RLA,SDY,**E:** WP:PS☠	Kowa-364-01
Lobelia	*inflata*	Indian Tobacco	Lobeline Bellflower	**M(folk):**LF,SM:BHP,DDPT,DEPI,DIOS,DISB,DS MA,DWHP,RLA,SDY,SPR	Krug-107-01
Lobelia	*inflata*	Indian Tobacco	Lobeline Bellflower	**M(PS):**LF,SM:RLA,RLB,RLC	Wyk-194-01,

Latin Genus	L. Species	E. Species	English Genus	C:Culinary,M:Medicinal,E:Economic	Source
					Wyk-415-17
Lobelia	*siphilitica*	Syphiliswort	Lobeline Bellflower	**M(traditional):**WP:DSXS	Bown-305-16
Lobelia	*siphilitica*	Syphiliswort	Lobeline Bellflower	**M(traditional):**WP:DSXS	Wyk-415-18
Lobelia	*tupa*	Devil's Tobacco	Lobeline Bellflower	**M(traditional):**WP:NNC👽,OTA	Bown-305-17
Lobelia	*tupa*	Devil's Tobacco	Lobeline Bellflower	**M(traditional):**WP:NNC👽,OTA	Wyk-415-19
Lolium	*temulentum*	Narcotic	Darnel Rye	**M(traditional):**SD:NNC👽	Wyk-415-20
Lonicera	*caprifolium*	Woodbine	Honeysuckle	**M(traditional):**FL-BD,SM:RLC,SCN	Bown-306-02
Lonicera	*caprifolium*	Woodbine	Honeysuckle	**M(traditional):**FL:DASC,RLC	Clev-101-05
Lonicera	*caprifolium*	Woodbine	Honeysuckle	**M(traditional):**FL:DABI,RLB,UDR	Wyk-415-21
Lonicera	*involucata*	Twinberry	Honeysuckle	**E:**FR:DYE(tin-gray)	Grae-145-01
Lonicera	*japonica*	Japanese	Honeysuckle	**M(traditional):**SM:BHP,DABA,DHPT,DMUM,HFV,NJA,NMS,**M(traditional):**FL-BD:DBOI,DCBR,DMSL,DPXC,HFV,NMT,OEY,DSIN,OWN,RST,SIB,UUR,**M(traditional):**FL-BD,SM:DPNB,DPNV,SDY	Bown-306-03
Lonicera	*japonica*	Japanese	Honeysuckle	Chinesemedicine	Clev-101-06
Lonicera	*japonica*	Japanese	Honeysuckle	**M(traditional):**FL:DCTM,BHP,UEP	Wyk-415-22
Lonicera	*nigra*	Black	Honeysuckle	**M(traditional):**WP:DHMY	Clev-101-07
Lonicera	*peryclymenum*	Woodbine	Honeysuckle	**M(traditional):**FL-BD,SM:DASC,RLC,UDR	Bown-306-04
Lonicera	*peryclymenum*	Woodbine	Honeysuckle	**M(traditional):**BK:NJA,NRH,**M(traditional):**FL:BHS,RLA,**M(traditional):**LF:DSIN,**M(traditional):**RT:DWRM,**E:**FR:PS☠	Clev-101-04
Lophophora	*williamsii*	Peyote	Crested Rosette Cactus	**M(traditional):**WP:NNC👽	Wyk-416-01
Lophophytum	*leandri*	Leander	Crested Balan	**M(traditional):**TU:BLC,SDP	Wyk-416-02
Luffa	*cylindrica*	Sponge Gourd	Loofah	**C:**FR:**VG**,**C:**SD:CO,**M(traditional):**FR,OL,SD,SM:BCR,BLC,DBOI,DSHZ,NBK,NOR,NRH,OAS,OHM,**E:**FR:FI(sponge)	Bown-306-05
Luffa	*operculata*	Eyed	Loofah	**M(traditional):**FR:RLC,SCN,UDR	Wyk-416-03
Lupinus	*albus*	White	Lupine	**M(traditional):**SD:BDB,DSIN,UDR	Wyk-416-04a
Lupinus	*angustifolius*	Narrow Leaf	Lupine	**M(traditional):**SD:BDB,DSIN,UDR	Wyk-416-04b
Lupinus	*arboreus*	Yellow Bush	Lupine	**E:**FL:DYE(alum-&-tin-bright-yellow,Grae)	
Lupinus	*mutabilis*	Changeable	Lupine	**M(traditional):**SD:BDB,DSIN,UDR	Wyk-416-04c
Lupinus	*russellianus*	Russell	Lupine	**E:**WP:DYE(alum-yellow-green,copper-sulfate-&-iron-green)	Cas1-174-01
Lupinus	*shockleyi*	Purple Desert	Lupine	**E:**FL:DYE(chrome-bright-yellow-green, chrome-&-copper-dulll-green)	Grae-094-01
Lupinus	*succulentus*	Arroyo Purple	Lupine	**E:**FL:DYE(chrome-bright-yellow-green,chrome-&-copper-dull-green)	Grae-094-01
Lupinus	*texensis*	Texas Bluebonnet	Lupine	**E:**LF,SL:DYE(alum-light-yellow-green,chrome-brass-gold,copper-green-gold,tin-bright-yellow,iron-tan-gray)	Bliss-152-01
Lycium	*barbarum*	Matrimony Vine	Boxthorn	**M(traditional):**FR:BDB,DIPT,DVTG,NBK,**M(traditional):**RB:BHC,BHP,DABA,DTBC,HFV,DCLW,DSIN,RLA,RLC	Bown-306-06
Lycium	*barbarum*	Matrimony Vine	Boxthorn	**M(traditional):**FR:BCR,BLC,eyesight improvement	Wyk-416-05
Lycium	*chinense*	Chinese	Boxthorn	**M(PS):**FR:BCR,BLC,eyesight improvement	Wyk-195-01
Lycoperdon	*perlatum*	Pearl	Puffball	**C:**WP,**VG**,**M(traditional):**WP:OAS,OHM,OWN	Bown-306-07
Lycopersicon	*esculentum*	South American	Tomato	**C:**FR:**VG**, **M(traditional):**FR:DFRA	Wyk-416-06
Lycopersicon	*esculentum*	South American	Tomato	**E:**SM:DYE(vinegar-tan-yellow, alum-soft-yellow, tin-bright-yellow), E:SM,FR(green):alum-yellow-green,copper-yellow-green,iron-deep-green)	Cas1-215-01
Lycopodium	*clavatum*	Staghorn	Clubmoss	**M(traditional):**WP:DABA,DHMY,HFV,NRH,NSD,DSIN,SDG,SIB,UBL,UKC	Bown-307-01
Lycopodium	*clavatum*	Staghorn	Clubmoss	**M(traditional):**WP:NMS,NSD,UDR,UUR	Wyk-416-07
Lycopodium	*complanatum*	American Ground Pine	Clubmoss	**M(traditional):**WP:DABA,DHMY,HFV,NRH,NSD,DSIN,SDG,SIB,UBL,UKC	Bown-307-02
Lycopus	*americanus*	Horehound	Bugleweed	**M(traditional):**WP:DHET,DTBC,BHS,BSH,OAS,RLC	Bown-307-03
Lycopus	*europaeus*	Gipsywort	Bugleweed	**M(traditional):**WP:DHET,DTBC,BHS,BSH,OAS,RLC	Bown-307-04
Lycopus	*europaeus*	Gipsywort	Bugleweed	**M(CS):**WP:DHET,NMT,NSD	Wyk-196-01
Lycopus	*lucidus*	Glossy	Bugleweed	**M(traditional):**WP:NMP,OWN	Bown-307-05
Lycopus	*virginicus*	Virginia	Bugleweed	**M(traditional):**WP:DHET,DTBC,BHS,BSH,OAS,RLC	Bown-307-06
Lycopus	*virginicus*	Virginia	Bugleweed	**M(traditional):**WP:DHET,NMT,NSD	Wyk-416-08
Lysimachia	*mammilaria*	Moneywort	Primrose Loosestrife	**M(traditional):**WP:OAS,OWN,RLC,SDR	Wyk-416-09a
Lysimachia	*vulgaris*	Yellow	Primrose Loosestrife	**M(traditional):**WP:OAS,OWN,RLC,SDR	Wyk-416-09b
Lysimachia	*vulgaris*	Yellow	Primrose Loosestrife	**E:**FL:DYE(alum-beige-pink,chrome-camel,copper-green-gold,tin-bright-yellow-gold,iron-olive)	Bliss-278-01
Lythrum	*salicaria*	Purple	Loosestrife	**M(traditional):**WP:DABA,DCHL,DHPT,DIMP,DTYP,DVGB,DVGV,OAS,OEY,DSIN,OWN,SDR,S	Bown-307-07

Latin Genus	L. Species	E. Species	English Genus	C:Culinary,M:Medicinal,E:Economic	Source
				DY,SUL,UDR	
Lythrum	*salicaria*	Purple	Loosestrife	**M(traditional):**WP:DABA,DAMW,OEY	Clev-102-01
Lythrum	*salicaria*	Purple	Loosestrife	**M(traditional):**WP:DASC,OAS,DSIN,OWN,RLC	Wyk-416-10
Macleaya	*cordata*	Heart Leaf	Plume Poppy	**M(traditional):**WP:DABI,DAMW,RLC	Wyk-416-11
Maclura	*pomifera*	Osage Orange	Osage Mulberry	**E:**WD:DYE(alum-gold,chrome-sienna,copper-brass,tin-soft-gold,iron-dark-olive,),**E:**FR:DYE(alum –light-gold,chrome-green-gold,copper-tan-gold,tin-yellow,iron-tan),**E:**RB:DYE(alum-tan, chrome-drab-tan,copper-drab-olive,tin-tan-yellow,iron-drab-tan)	Bliss-180-01
Maclura	*pomifera*	Osage Orange	Osage Mulberry	**E:**WD:DYE(alum-intense-green-yellow,chrome-deep-burnt-orange,alum-&-iron-khaki-green,alum-&-chrome-intense-green-gold,alum-&-chrome-&-vinegar-yellow)	Grae-168-01 Grae-168-02 Grae-169-03
Magnolia	*acuminata*	Cucumber	Magnolia	**M(traditional):**BK:HFV,NRH	Bown-307-08
Magnolia	*grandiflora*	Loblolly	Magnolia	**M(traditional):**FL:NRH,**M(traditional):**BK:DABI, DMAL	Wyk-416-12a
Magnolia	*liliiflora*	Lily Flower	Magnolia	**M(traditional):**FL-BD,FL:BHP,DAFU,DCLD,RRN,RSN,NSD, **E:**FL,FL-BD:PS☠	Bown-307-09
Magnolia	*officinalis*	Chinese	Magnolia	**M(traditional):**BK,FL:DABA,DAFU,BHP,RLA,RLC,SDG,SDR	Bown-307-10
Magnolia	*officinalis*	Chinese	Magnolia	**M(traditional):**FL:NRH,**M(traditional):**BK:DABI, DMAL	Wyk-416-12b
Magnolia	*tripetala*	Umbrella	Magnolia	**M(traditional):**BK:HFV,NRH	Bown-307-11
Magnolia	*virginiana*	Sweet Bay	Magnolia	**M(traditional):**BK:DMAL,NGT,NRH,RLB	Bown-307-12
Mahonia	*aquifolium*	Tall Oregon Grape	Mahon Barberry	**M(traditional):**RT:BLC,DACN,ODS	Bal1-075-02
Mahonia	*aquifolium*	Tall Oregon Grape	Mahon Barberry	**M(traditional):**LF,RT-BK:DABA,BLC,OSS,SDR, SGB,SIB	Bal2-108-02
Mahonia	*aquifolium*	Tall Oregon Grape	Mahon Barberry	**C:**FR:JJ,**M(traditional):**FR,RB,RT:DHPT,OAS,DSIN,SDR,SIB,SGB	Bown-308-01
Mahonia	*aquifolium*	Tall Oregon Grape	Mahon Barberry	**C:**FR:JJ,**M(traditional):**FR,RT:BLC,DACN,DHMY,DHSV,ODS,DSIN,SGB,**E:**FR:DYE(alum-purple-blue),**E:**RT:DYE(alum-buff-brown,chrome-tan),**E:**WP:DYE(alum-khaki-yellow)	Kowa-175-01 Kowa-404-01
Mahonia	*aquifolium*	Tall Oregon Grape	Mahon Barberry	**M(folk):**RT:DSIN	Krug-119-02
Mahonia	*aquifolium*	Tall Oregon Grape	Mahon Barberry	**M(traditional):**BK:ODS,DSIN	Wyk-416-13
Mahonia	*aquifolium*	Tall Oregon Grape	Mahon Barberry	**E:**FR:DYE(alum-red-brown,chrome-olive,copper-brown,tin-khaki,iron-dark-brown-green)	Bliss-130-01
Mahonia	*aquifolium*	Tall Oregon Grape	Mahon Barberry	**E:**FR:DYE(alum-violet,alum-&-ammonia-dark-green,alum-&-tin-dark-blue-purple), **E**:LF:DYE(alum-&-copper-light-olive-green, chrome-&-copper-olive-green),**E:**RT:DYE(chrome-olive-green)	Grae-137-01 Grae-137-02 Grae-138-01 Grae-138-02
Mahonia	*nervosa*	Holly Oregon Grape	Mahon Barberry	**M(traditional):**LF,RT-BK:DABA,BLC,OSS,SDR, SGB,SIB	Bal2-108-03
Mahonia	*repens*	Pine Oregon Grape	Mahon Barberry	**M(traditional):**LF,RT-BK:DABA,BLC,OSS,SDR, SGB,SIB	Bal2-108-04
Mahonia	*repens*	Pine Oregon Grape	Mahon Barberry	**M(traditional):**FR,RT,RB:OAS,SGB	Bown-308-02
Mallotus	*philippinensis*	Philippine	Kamila	**M(traditional):**FR:DWRM,DSIN,SPR	Wyk-416-14
Malpighia	*glabra*	Acerola	Malpighian Cherry	**M(PS):**FR:BAA,DCLD,DCST,DEGC,DFLU,DPRK,RLB, **M(PS):**FR:DCMT,DHSV	Bal2-014-01 Bal2-501-01
Malpighia	*glabra*	Acerola	Malpighian Cherry	**M(traditional):**FR:25% vitamin C,highest vitamin C fruit	Wyk-416-15
Malus-domestica	*x domestica*	Orchard	Apple	**M(traditional):**FR:SDP,SDR	Wyk-416-16
Malus-pumila	*x pumila*	Orchard	Crab Apple	**E:**LF:DYE(alum-pale-yellow, chrome-gold, chrome-&-tin-rust, tin-strong-yellow-orange, iron-soft-gray), **E:**BK,RT:DYE(alum-tan-yellow, chrome-rosy-tan, chrome-&-iron-dark-brown, copper-sulfate-gray-brown), **E:**RD:DYE(alum-pale-yellow, chrome-soft-tan)	Cas1-092-01
Malva	*arborea*	Tree	Mallow	Synonym for *Lavatera arborea*	Grae-095-01 Grae-095-02
Malva	*moschata*	Musk	Mallow	**M(traditional):**FL,FR,LF:OAS,RLC	Bown-308-03
Malva	*neglecta*	Buttonweed	Mallow	**M(traditional):**FL,FR,LF:OAS,RLC	Bown-308-04
Malva	*neglecta*	Buttonweed	Mallow	**M(traditional):**FL:RLB	Clev-102-03
Malva	*neglecta*	Buttonweed	Mallow	**M(traditional):**FL,LF:RLC,**E:**FL:DYE	Wyk-416-17
Malva	*neglecta*	Buttonweed	Mallow	**E:**FL,LF,SM:DYE(alum-light-gray,chrome-tan-gold, copper-gray-green,tin-soft-yellow,iron-dark-gray-green)	Bliss-156-01
Malva	*parviflora*	Cheeseweed	Mallow	**C:**LF:**VG**,**C:**PD:**VG**	Clev-102-04
Malva	*parviflora*	Cheeseweed	Mallow	**E:**WP:DYE(alum-cream, alum-&-ammonia-bright-yellow, alum-&-iron-khaki-green)	Grae-095-03
Malva	*sylvestris*	High	Mallow	**C:**LF,SH:**VG**,**C:**FL:TE,**C:**PD:**SN**,**M(traditional):**FL,FR,LF:DBOI,DISB,OAS,OSA,RLA,RLB,RLC,RLE	Bown-308-05

Latin Genus	L. Species	E. Species	English Genus	C:Culinary,M:Medicinal,E:Economic	Source
				,RST,SIB	
Malva	*sylvestris*	High	Mallow	**C:**LF:**VG**,**C:**PD:**VG**,**M(traditional):**LF:OHM,OST,OUL,**M(traditional):**FL:DCLD,RLB,**E:**WP:PS☠	Clev-102-02
Malva	*sylvestris*	High	Mallow	**M(PS):**LF,FL:OWN,RLC,RST,SIB,UBL	Wyk-197-01
Malva	*verticillata*	Cluster	Mallow	**M(traditional):**FL,FR,LF:UDR,UUR	Bown-308-06
Malvaviscus	*arboreus*	Sleeping	Sticky Mallow	**E:**OL:HO	Bown-293-06
Mandragora	*autumnalis*	Autumn	Mandrake	**M(traditional):**RT:HAP,NMS,NNC👽,NSD	Wyk-416-18b
Mandragora	*officinarum*	Devil's Apple	Mandrake	**M(traditional):**RT:HAP,NSD,OUL,SPR	Bown-308-07
Mandragora	*officinarum*	Devil's Apple	Mandrake	**M(folk):**RT:DEPI,NNC👽,NRH,NSD, **E:**RT:PS☠	Krug-110-01
Mandragora	*officinarum*	Devil's Apple	Mandrake	**M(traditional):**RT:HAP,NMS,NNC👽,NSD	Wyk-416-18a
Manihot	*esculenta*	Tapioca	Manioc Potato	**M(traditional)**:RT:DSCB,**E:**RT:PS☠	Wyk-416-19
Manilkara	*zapota*	Chicle	Bulletwood Sapote	**C:**BK,RN:GR(chewing gum)	Wyk-416-20
Maranta	*arundinacea*	Arrowroot	Marant Ginger	**M(folk):**RZ:DISB,OWN	Krug-019-01
Maranta	*arundinacea*	Arrowroot	Marant Ginger	**M(traditional):**RZ:DSIN,SDP	Wyk-416-21
Marasmius	*oreades*	Fairy Ring	Pinwheel Gill Mushroom	**E:**WP:DYE(alum-yellow,chrome-tan)	Cas1-182-07
Marrubium	*vulgare*	White	Horehound	**M(traditional):**FL,LF:DHYF,RLB,RSN	Bal1-072-05
Marrubium	*vulgare*	White	Horehound	**M(folk):**WP(aerial):BJA,DAMW,DSIN,DTBC,DWHP,NMP,OUL,OWN,RLA,RLB,RST,SDR, **M(traditional):**WP(aerial):DCLD,RLC,RSN,SDG **M(PS):**WP(aerial):SAP,SFL	Bal2-086-01 Bal2-527-02
Marrubium	*vulgare*	White	Horehound	**M(traditional):**WP:BLC,DASC,DTYP,DWHP,NMS,OBR,OSA,RLA,RLB,RLC,SGB	Bown-308-08
Marrubium	*vulgare*	White	Horehound	**C:**LF:AF,CF,**M(traditional):**LF,OL:DASC,DCLD,HFV,RLC,RST,SPR,UDR	Kowa-289-01 Kowa-335-01
Marrubium	*vulgare*	White	Horehound	**M(folk):**WP(flowering):RLC, **M(folk):**LF:DCLD	Krug-181-01
Marrubium	*vulgare*	White	Horehound	**M(PS):**WP:RLB,RLC,SAP,SDG,SDP,SFL,SGB	Wyk-198-01, Wyk-416-22
Marrubium	*vulgare*	White	Horehound	**E:**LF,SM:DYE(alum-light-green-gold,chrome-tan,copper-olive,tin-gold,iron-charcoal-gray)	Bliss-090-01
Marsdenia	*cundrango*	Eagle Vine	Latex Milkvine	**M(traditional):**BK:DANX,SAP,SDG,SDP,**E:**SA:PS☠	Bown-309-01
Marsdenia	*cundrango*	Eagle Vine	Latex Milkvine	**M(PS):**BK:BLC,SAP,SDG,**E:**BK:PS☠	Wyk-199-01, Wyk-416-23
Marsdenia	*tinctoria*	Java Indigo	Latex Milkvine	**M(traditional):**BK:DIFR,**E:**WP:DYE(indigo),**E:**SA:PS☠	Bown-309-02
Matricaria	*chamomilla*	German Chamomile	Mayweed Daisiette	**M(traditional):**FL:HAN,NMP,OPS,SIB,SUL, **M(PS):**FL:ADD,DALL,DCEN,DFRA,OWN	Bal2-046-01 Bal2-510-03
Matricaria	*chamomilla*	German Chamomile	Mayweed Daisiette	**M(traditional):**FL:HIN,NMS,NSD,OBN,OHM,OMT,OSB,OUL,OWN,SDG,SIB	Bown-309-03
Matricaria	*chamomilla*	German Chamomile	Mayweed Daisiette	**C:**LF:TE,**M(traditional):**LF:NSD,SDG,**E:**FL:CM,E:OL:DS	Clev-103-01
Matricaria	*chamomilla*	German Chamomile	Mayweed Daisiette	**M(traditional):**FL:DCLD,HAC,HAN,NMS,NSD,SDG	Kowa-285-01
Matricaria	*chamomilla*	German Chamomile	Mayweed Daisiette	**M(folk):**FL:DASC,DSIN,DVGF,HAC,HAN,HIN,NMP,NMT,OBN,OUL,SDG	Krug-049-01
Matricaria	*chamomilla*	German Chamomile	Mayweed Daisiette	**M(PS):**FL,OL:DASC,HAN,NMS,DSIN,OWN,RRN,SDP,SDR,SFL,SIB	Wyk-200-01, Wyk-417-01
Matricaria	*chamomilla*	German Chamomile	Mayweed Daisiette	**E:**WP:DYE(alum-strong-yellow,tin-brilliant-yellow,chrome-rich-gold,iron-yellow-green)	Cas1-116-02
Matricaria	*maritima*			Synonym for *Tripleurospermum maritima*	Cas1-116-01
Matricaria	*recutita*			Synonym for *Matricaria chamomilla*	Cas1-116-02
Matteuccia	*struthiopteris*	Shuttlecock	Ostrich Fern	**E:**LF:DYE(alum-&-copper-sulfate-soft-lime-green,alum-&-tin-yellow-green,chrome-gold,alum-soft-gold,alum-&-tin-bright-gold,chrome-&-iron-brown)	Cas1-141-02
Matthiola	*incana*	Purple	Stock Mustard	**E:**FL:DYE(alum-blue,alum-&-tin-dark-blue,chrome-turquoise,chrome-&-tin-purple)	Grae-134-02 Grae-134-03
Maytenus	*chuchuhuasa*	Espinheira Santa	Mayten Holly	**M(traditional):**BK,LF:NBK,NJA,NRH,SIB,SUL	Bal2-065-01 Bal2-517-01
Maytenus	*ilicifolia*	Limaosinho	Mayten Holly	**M(traditional):**BK,LF:NBK,NJA,NRH,SIB,SUL	Bal2-065-02 Bal2-517-01
Medicago	*lupulina*	Lupine	Medick	**E:**LF,FL,PD,SM:DYE(alum-light-yellow-green,chrome-tan-gold,copper-gray-green,tin-lemon-yellow,iron-gray-green)	Bliss-052-01
Medicago	*sativa*	Alfalfa	Medick	**M(PS):**WP:DAFU,DCEN,DCTM,DGCS,DOST,SUL	Bal2-018-01 Bal2-501-03
Medicago	*sativa*	Alfalfa	Medick	**C:**LF:**VG**,**C:**SR:**SN**,**M(traditional):**WP:BAN,BCR,BHC,DUFB,OAS,SAP,SUL,UDR	Bown-309-04
Medicago	*sativa*	Alfalfa	Medick	**C:**LF:**VG**,**M(folk):**LF,SP:UDR	Krug-010-01
Medicago	*sativa*	Alfalfa	Medick	**M(PS):**LF,SD,SM,**SP**:BHC	Wyk-201-01,

Latin Genus	L. Species	E. Species	English Genus	C:Culinary,M:Medicinal,E:Economic	Source
					Wyk-417-02
Medicago	*sativa*	Alfalfa	Medick	**E:**WP:DYE(alum-light-yellow-green,chrome-mint-green,copper-light-sage-green,tin-lemon-yellow,iron-camel)	Bliss-034-01
Melaleuca	*alternifolia*	Tea Tree	Bottlebrush Myrtle	**M(traditional):**LF-OL,SM-OL:DAFU,DALF,DHSV,DISB,DSCB,DSPB,DVGB,DVGV,DACN,DCLW,DSIN,OWN,RST	Bal1-077-07
Melaleuca	*alternifolia*	Tea Tree	Bottlebrush Myrtle	**M(folk):**LF-OL,SM-OL:DGUM,DSIN,DTNS,OUL,RSN,RST,SIB,**M(traditional):**LF-OL,SM-OL:DBOI,DDND,DISB,NMT,RST,**M(PS):**LF-OL,SM-OL:DABA(*Eshcherichia coli, Saphylococcus aureus*),DACN,DALF,DCAN,DVGB,DWRG	Bal2-143-01 Bal2-546-03
Melaleuca	*alternifolia*	Tea Tree	Bottlebrush Myrtle	**M(traditional):**OL:DABA,DASC,DAFU,DALF,DAMW,DISB,DVGB,DVGT,DCLW,OWN,RLC,**E:**OL:DS	Bown-310-01
Melaleuca	*alternifolia*	Tea Tree	Bottlebrush Myrtle	**M(CS):**LF-OL:DACN,DAFU,DASC,DALF,DAMW,DISB,DTNS,DVGT,DSIN,OWN,RSN,RST	Wyk-202-01, Wyk-417-03
Melaleuca	*cajuputi*	Malaysian Whitewood	Bottlebrush Myrtle	**M(traditional):**OL:DASC,DWRO,NMS,RLC	Bown-309-05
Melaleuca	*leucadendra*	Australian Whitewood	Bottlebrush Myrtle	**C:**OL:CF,**M(traditional):**OL:BCB,DACN,DASC,DCLD,DTBC,DWRO,NMS,OGT,ONR,ORH,OTA,RLB,RLC,RRN,RSN,SIB,**E:**OL:IR,PF,DS	Bown-310-02
Melaleuca	*quinquenervia*	Five Nerve Niaouli Oil	Bottlebrush Myrtle	**M(traditional):**OL:DASC,**E:**OL:PF	Bown-309-06
Melaleuca	*viridiflora*	Green Flower Niaouli Oil	Bottlebrush Myrtle	**M(traditional):**OL:DASC,**E:**OL:PF	Bown-309-07
Melanelia	*fuliginosa*	Fulginous	Camouflage Shield Lichen	**E:**WP:DYE(Lecanoric-Acid-&-Ammonia-fuchsia)	Cas2-032-12
Melanelia	*glabratula*			Synonym for *Melanelia fuliginosa*	Cas2-032-12
Melanelia	*olivacea*	Olive	Camouflage Shield Lichen	**E:**WP:DYE(Fumarprotocetraric-Acid-tan)	Cas2-035-14
Melanelia	*stygia*	Stygid	Camouflage Shield Lichen	**E:**WP:DYE(Fumarprotocetraric-&-Caperatic-Acid-tan-brown	Cas2-035-08
Melia	*azederachta*	Chinaberry	Honey Mahogany	**M(traditional):**BK,LF,FR,RB:ADFU,DCAN,DVGB,DWRM,**E:**BK,LF,FR,RB:PS☠	Bown-310-03
Melianthus	*major*	Greater	Honey Nectar	**E:**FL:DYE(alum-&-tin-violet)	Grae-152-04
Melilotus	*alba*	White	Sweet Clover	**C:**LF:**VG**(fodder)	Clev-103-03
Melilotus	*alba*	White	Sweet Clover	**E:**FL,LF:DYE(alum-yellow,chrome-gold, copper-soft-green, tin-orange, iron-tan)	Cas1-124-01
Melilotus	*caerulea*	Blue	Sweet Clover	**C:**WP:FF	Bown-310-04
Melilotus	*officinalis*	Yellow	Sweet Clover	**C:**LF,SH:FF,TE,**M(traditional):**WP:BTH,BVV,DBOI,DERS,HAC,HIN,NMP.NMS,NNR,NSD,OBR,OEY,ORH,URD,**E:**WP:IR(moth)	Bown-310-05
Melilotus	*officinalis*	Yellow	Sweet Clover	**C:**FL,SD:FF,**M(traditional):**LF,SH:BTH,HIN,OWN,SIB,**E:**WP:PS☠	Clev-103-02
Melilotus	*officinalis*	Yellow	Sweet Clover	**M(PS):**LF,SM:BVV,DISB,NMS,OHM,SDG	Wyk-203-01, Wyk-417-04
Melilotus	*officinalis*	Yellow	Sweet Clover	**E:**FL,LF,SM:DYE(alum-light-yellow,chrome-nutmeg–brown,copper-green-gold,tin-yellow,iron-brown)	Bliss-280-01
Melissa	*officinalis*	Lemon Balm	Melissa Balm	**M(traditional):**WP-OL:BHP,DABA,DAVR,HAC,HAN,HFV,HIN,NMS,NRH,NSD,OST,OWN,SFL,SIB,**M(PS):**WP-OL:DHSV,HAN,HIN, **E:**WP-OL:IR	Bal2-094-01 Bal2-529-01
Melissa	*officinalis*	Lemon Balm	Melissa Balm	**C:**LF:AF,**SN**,TE,**M(traditional):**WP,OL:DABA,DAVR,DHSV,DHET,DISB,HAN,HFV,,NMS,NSD,OGT,SDG	Bown-310-06
Melissa	*officinalis*	Lemon Balm	Melissa Balm	**C:**LF:AF,FF,TE,**M(traditional):**LF,OL:BHP,DCLD,DFLU,HIN,SDG,**E:**OL:IR,PF	Clev-103-04
Melissa	*officinalis*	Lemon Balm	Melissa Balm	**C:**LF:FF,TE,**M(traditional):**LF-OL:DABA,DACN,DAVR,**E:**OL:CM,IR,WX(furniture polish)	Kowa-356-01
Melissa	*officinalis*	Lemon Balm	Melissa Balm	**C:**LF:FF, **M(folk):**LF,OL:HAN,NSD	Krug-021-01
Melissa	*officinalis*	Lemon Balm	Melissa Balm	**M(PS):**LF-OL:DABA,DAFU,DAVR,DHSV,HIN,SAP,SDG	Wyk-204-01, Wyk-417-05
Mentha	*aquatica*	Watermint	Mint	**M(traditional):**WP:DCLD,NMP,NMS,OAS,SDG,SDR,SIB,**E:**WP:PS☠	Bown-311-01
Mentha	*arvensis*	Corn Mint	Mint	**M(traditional):**WP:DABA,DCLD,DMSL,HAC,NMS,RST,SDG	Bown-311-02
Mentha	*arvensis*	Corn Mint	Mint	**E:**OL:CM	Kowa-282-01
Mentha	*arvensis*	Corn Mint	Mint	**M(PS):**LF-OL,SM-OL:DABA,ONR,SFL,SGB,SIB	Wyk-205-01, Wyk-417-06
Mentha	*arvensis pipirascens*	Japanese Corn Mint	Mint	**C:**OL:FF(menthol)	Bown-311-03 Clev-105-03

Latin Genus	L. Species	E. Species	English Genus	C:Culinary,M:Medicinal,E:Economic	Source
Mentha	*arvensis villosa*	American Corn Mint	Mint	C:LF:TE	Bown-311-04
Mentha	*gentilis*			Synonym for *Mentha gracilis*	Clev-105-04
Mentha	*gracilis*	Gingermint	Mint	C:LF,OL:FF	Bown-311-05
Mentha	*longifolia*	Horsemint	Mint	C:LF,OL:CF,**SN**	Bown-311-06
Mentha	*piperita*	Peppermint	Mint	M(traditional):WP,FL:HAC,NMS,NRH,SAP,SDG,SDR,SIB	Bal1-075-07
Mentha	*piperita*	Peppermint	Mint	M(traditional):WP(aerial),FL:HFV,DHAL,RLC,SDG,M(SS):LF:SDP,SIB,SFL,SGB,M(SS):WP(aerial)-OL:HAC,ONR,SGS,SIB	Bal2-173-01 Bal2-535-01
Mentha	*piperita*	Peppermint	Mint	M(traditional):LF,OL:DASC,DCLD,DFLU,DWRG,HIN,NMS,OBR,ONR,ORH,DSIN,RLA,RLC,RSN,SDG,SIB,SUL,E:OL:CM,IR,PS☠	Bown-311-07
Mentha	*piperita*	Peppermint	Mint	C:LF:BF,FF,TE,C:OL:AF,CF,M(traditional):LF:HIN,NSD,SDG,SIB,E:OL:CM	Clev-104-01
Mentha	*piperita*	Peppermint	Mint	C:LF:CF,M(traditional):LF:HAC,NMS,SDG,SFL,SIB	Kowa-291-03 Kowa-383-01
Mentha	*piperita*	Peppermint	Mint	M(folk):WP(flowering):DABA,NMP,NSD,OJA,OMC,SDG	Krug-116-01
Mentha	*piperita*	Peppermint	Mint	M(CS):LF:BLC,SFL,M(traditional):OL:HAC,DABA,DABI,DAVR,DAMW,NMS,NSD,SFL,SIB	Wyk-206-01, Wyk-417-07
Mentha	*piperita*	Peppermint	Mint	E:FL,LF,SM:DYE(alum-gold,chrome-camel,copper-brass-gold,tin-gold,iron-forest-green)	Bliss-184-01
Mentha	*piperita*	Peppermint	Mint	E:WP:DYE(alum-soft-yellow,copper-sulfate-tan-beige,chrome-gold,tin-yellow-orange,iron-tan-gray)	Cas1-177-01
Mentha	*piperita citrata*	Lemonmint Peppermint	Mint	C:OL:CM,PF(cologne)	Clev-105-05
Mentha	*pulegium*	European Pennyroyal	Mint	C:LF:FF,M(traditional):LF,OL:HFV,OAS,SDG,SIB,,E:OL:IR,DS	Bown-311-08
Mentha	*pulegium*	European Pennyroyal	Mint	C:LF:FF,M(traditional):LF,OL:DCLD,DISB,DSNB,HAC,SDG	Clev-104-02
Mentha	*pulegium*	European Pennyroyal	Mint	M(traditional):LF:SDG, E:OL:IR,**induced abortion**☠	Kowa-412-01
Mentha	*pulegium*	European Pennyroyal	Mint	M(folk):LF,OL:MCI,NMP E:OL:IR(flea),**induced abortion**☠	Krug-135-01
Mentha	*requenii*	Corsican	Mint	C:LF,OL:AF	Bown-311-09
Mentha	*requenii*	Corsican	Mint	C:LF,OL:AF(crèmedementhe)	Clev-105-06
Mentha	*rotundifolia*	Applemint	Mint	C:LF,OL:AF,BF,FF	Clev-104-03
Mentha	*satureioides*	Australian Pennyroyal	Mint	M(traditional):LF,OL:DCLD,NMP,NMS,SDG,SIB	Bown-311-10
Mentha	*smithiana*	Redmint	Mint	C:LF,OL:BF,JJ,FF,GN,TE	Bown-312-01
Mentha	*smithiana*	Redmint	Mint	C:LF,OL:BF,FF	Clev-105-07
Mentha	*spicata*	Spearmint	Mint	C:LF,OL:BF,JJ,FF,GN,TE,M(traditional):LF,OL:DAMW,NMS,SDG,SIB,SFL	Bown-312-02
Mentha	*spicata*	Spearmint	Mint	C:LF,OL:CF	Clev-105-01
Mentha	*spicata*	Spearmint	Mint	C:LF:AF,FF,M(folk):WP(flowering):DABA,NMP,NSD,OJA,OMC,SDG	Krug-117-01
Mentha	*spicata*	Spearmint	Mint	M(traditional):LF:SDG,SFL,M(traditional):OL:DABI,DAMW,RSN	Wyk-207-01, Wyk-417-08
Mentha	*suaveolens*	Pineapple	Mint	C:LF,OL:BF,JJ,CF,FF,GN,TE	Bown-312-03
Mentha	*suaveolens*	Pineapple	Mint	C:LF,OL:BF	Clev-105-08
Mentha	*villosa*	Bowles Spearmint	Mint	C:LF,OL:BF,CF	Bown-312-04
Mentzelia	*cordifolia*	Heart Leaf	Eveningstar	M(traditional):WP:NMS,SDG	Wyk-417-09
Mentzelia	*decapetala*	Sticktight	Eveningstar	E:FL,LF,SM:DYE(alum-tan,chrome-camel,copper-coffee,tin-rose-gold,iron-brown)	Bliss-112-01
Menyanthes	*trifoliata*	Bog Bean	Buckbean	M(traditional):LF:DANX,NJA,NRH,SCN,SDG,UDR,E:LF:PS☠	Bown-312-05
Menyanthes	*trifoliata*	Bog Bean	Buckbean	M(folk):LF:SAP,SDG	Krug-036-02
Menyanthes	*trifoliata*	Bog Bean	Buckbean	M(PS):LF:HFV,DCLD,NRH,DSIN,SAP,SDP,SGB	Wyk-208-01, Wyk-417-10
Mercurialis	*annua*	Annual	Mercury Spurge	M(traditional):WP:DWRM,RLC,SCN,UDR	Wyk-417-11a
Mercurialis	*perenna*	Dog	Mercury Spurge	M(folk):WP:DBOI,OCW,OEY,OWN,E:WP:PS☠	Krug-063-01
Mercurialis	*perenna*	Dog	Mercury Spurge	M(traditional):WP:DWRM,RLC,SCN,UDR	Wyk-417-11b
Mertensia	*ciliata*	Tall Chimingbells	Bluebells Borage	E:FL,SL:DYE(alum-gold,chrome-coffee-brown,copper-olive-green,tin-gold-yellow,iron-dark-charcoal-gray)	Bliss-060-01
Millettia	*reticulata*	Ji Xue Teng	Millett Ironwood	M(traditional):RT:NMP	Wyk-417-12
Mimusops	*globosa*	Globe	Mime Cherry	E:FR:dentistry	Wyk-417-13
Mitchella	*repens*	Squaw Vine	Partidge Berry	M(traditional):LF,SM:NMP,NSD	Bal1-077-05
Mitchella	*repens*	Squaw Vine	Partidge Berry	M(traditional):WP:NMP,OAS,UDR	Bown-312-06
Mitchella	*repens*	Squaw Vine	Partidge Berry	M(folk):WP(flowering):facilitating childbirth	Krug-166-01

Latin Genus	L. Species	E. Species	English Genus	C:Culinary,M:Medicinal,E:Economic	Source
Mitchella	*repens*	Squaw Vine	Partidge Berry	**M(traditional):**WP,FR:OAS,NMP	Wyk-416-14
Mangifera	*indica*	India	Mango	**C:**FR:**SP**	Clev-309-01
Momordica	*balsamina*	Balsam Apple	Bitter Melon	**M(traditional):**FR:NMS,SPR	Wyk-416-15
Momordica	*charantia*	Balsam Pear	Bitter Melon	**M(traditional):**FR:DCLD,DFLU,DWRM,OSS, **M(PS):**FR:BDB,DCTM	Bal2-033-01 Bal2-506-05
Momordica	*charantia*	Balsam Pear	Bitter Melon	**C:**FR:**VG**,**M(traditional):**FR,JU:SIB,SDY,**M(traditional):**LF:BJA,DWRM,HFV,**M(traditional):**FR:OBN,ODS,OHM	Bown-312-07
Momordica	*charantia*	Balsam Pear	Bitter Melon	**M(traditional):**FR,LF,SD:DHOG,DWRM,NMP,OWN	Wyk-417-16
Momordica	*cochinchinensis*	Balsam Cucumber	Bitter Melon	**M(traditional):**SD:NMT(external),OHM,DSIN,**E:**SD:PS☠	Bown-312-08
Momordica	*grosvenori*	Grosvenor	Bitter Melon	**M(traditional):**LF,FR:RLC	Bown-312-09
Monarda	*citriodora*	Lemon Beebalm	Bergamot	C:LF:**SN**,TE	Bown-313-01
Monarda	*citriodora-austromontana*	Oregano	Bergamot	C:LF:**SN**,TE	Bown-313-02
Monarda	*didyma*	Beebalm	Bergamot	**C:**FL:**SN**,**C:**LF:BF,TE,**M(traditional):**WP:HFV,RLC,SDG	Bown-313-03
Monarda	*didyma*	Beebalm	Bergamot	**M(traditional):**FL,LF:HAN,DACN,DASC,DCLD,OUL,OWN,SFL,**E:**FL,LF:PS☠(photosensitivity)	Clev-105-09
Monarda	*didyma*	Beebalm	Bergamot	**C:**FL:**SN**,**M(traditional):**WP:DABA,DAFU,DASC,DWHK,NMS,NMP,RLC,RST,SFL,**E:**OL:PF	Kowa-039-01
Monarda	*didyma*	Beebalm	Bergamot	**M(folk):**FL,LF:DCLD,HAC,HFV,NMP,RLC,RST,SFL	Krug-029-01
Monarda	*didyma*	Beebalm	Bergamot	**M(traditional):**WP:DCLD,SDG	Wyk-417-17a
Monarda	*fistulosa*	Horsemint	Bergamot	**C:**LF:FF,**C:**LF,SH:TE,**M(traditional):**WP:DCLD,HAC,HFV,RST,SDG	Bown-313-04
Monarda	*fistulosa*	Horsemint	Bergamot	**C**:LF:TE	Clev-105-10
Monarda	*fistulosa menthifolia*	Menthol Horsemint	Bergamot	**M(traditional):**LF:DASC	Bown-313-05
Monarda	*fistulosa menthifolia*	Menthol Horsemint	Bergamot	**E:**FL,LF,SM:DYE(alum-gold,chrome-brown,copper-olive,tin-gold,iron-dark-charcoal-gray)	Bliss-132-01
Monarda	*punctata*	Bee Balm	Bergamot	**M(traditional):**WP:DCLD,NBK,ONR,ORH,SDG,SDR,SIB,UDR	Bown-313-06
Monarda	*punctata*	Bee Balm	Bergamot	**M(traditional):**WP:DCLD,SDG	Wyk-417-17b
Monascus	*purpureus*	Red Yeast Rice	Monad Sac Fungus	**M(traditional):**WP:BHC	Bal2-210-18
Monsonia	*ovata*	Oval Leaf	Flamingo's Bill	**M(traditional):**WP:OAS,SDG	Wyk-417-18
Montia	*perfoliata*	Blinks	Minor's Lettuce	**C:**LF:**VG**(source of vitamin C)	Wyk-417-19
Morchella	*esculenta*	Edible	Morel	**E:**WP:DYE(chrome-warm-tan,chrome-&-iron-dark-brown)	Cas-182-08
Morella	*pensylvanica*	Pennsylvania	Morell Bayberry	**E:**LF:DYE(alum-strong-yellow,chrome-gold,tin-brilliant-yellow,iron-gray-green)	Cas-099-01
Morinda	*citrifolia*	Mengkudu	Tropical Mulberry	**M(traditional):**FR,LF,RT:BHP,HFV,RLC,SDG, **M(PS):**FR,LF,RT:DCTM	Bal2-102-01 Bal2-533-01
Morinda	*citrifolia*	Mengkudu	Tropical Mulberry	**C:**FR:**SP**,**M(traditional):**LF:DSNB,DTBC,OGT,OUL,**M(traditional):**FR:DGUM,SDG,SUL,**M(traditional):**RT:SCN	Bown-313-07
Morinda	*officinalis*	Noni	Tropical Mulberry	**M(traditional):**FR,LF,RT:BHP,HFV,RLC,SDG, **M(PS):**FR,LF,RT:DCTM	Bal2-102-02 Bal2-533-01
Morinda	*officinalis*	Noni	Tropical Mulberry	**M(traditional):**LF,FR,RT:UKC	Bown-313-08
Morinda	*officinalis*	Noni	Tropical Mulberry	**M(traditional):**FR,RT:HAP,UUR	Wyk-417-20
Morinda	*tinctoria*	Dyer's	Tropical Mulberry	**M(traditional):**LF,RT:OAS,**E:**LF,RT:DYE(red)	Bown-313-09
Moringa	*oleifera*	Ben Oil	Horseradish Tree	**C:**LF,FL,PD:**VG**,**C:**OL:FF,**C:**RT:**SP**,**C:**SD:NT,**M(traditional):**BK,RN:DSEP,DTBC,**M(traditional):**RT,JU:BES,BLC,RLA,NGT,NRH,UBL,UKS,**M(traditional):**BK,RN,RT:DBOI,DGUM,DSNB,OGT,DSIN,OUL,OWN,**E:**OL:DS,**E:**WD:rayon,cellophane	Bown-313-10
Moringa	*oleifera*	Ben Oil	Horseradish Tree	**M(traditional):**RT:DABI,SPR	Wyk-417-21
Moringa	*peregrina*	Edible Potato	Horseradish Tree	**C:**TU:**VG**	Bown-313-11
Moringa	*pterygosperma*	Wingseed	Horseradish Tree	**C:**LF,PD:**VG**,**C:**FL:**SN**	Bown-313-12
Morus	*alba*	White	Mulberry	**C:**FR:AF,JJ,PU,SY,**M(traditional):**FR:DEAT,SCN,**M(traditional):**LF:DCLD,DFLU,OEY,**M(traditional):**SM:BHP,NRH,**M(traditional):**RB:BDB,BHE,RLA,RLB,RLC,**E:**LF:silkwormfood,E:WD:WL	Bown-314-01
Morus	*alba*	White	Mulberry	**M(traditional):**BK,LF,FR:DCLD,RLC,**E:**LF:silkworm food	Wyk-417-22a
Morus	*mesozygia*	Canary Wood	Mulberry	**E:**WD:DYE(direct-maize-yellow,alum-turmeric-gold,alum-&-chrome-burnt-orange,chrome-terra-cotta,chome-&-tin-pumpkin-orange)	Grae-160-02 Grae-160-03 Grae-161-01
Morus	*nigra*	Black	Mulberry	**C:**FR:AF,JJ,PU,SY,**M(traditional):**FR:DEAT,SC	Bown-313-13

Latin Genus	L. Species	E. Species	English Genus	C:Culinary,M:Medicinal,E:Economic	Source
				N,**M(traditional):**LF:DCLD,DFLU,OEY,**M(traditional):**SM:BHP,NRH,**M(traditional):**RB:BDB,BHE,RLA,RLB,RLC,**E:**LF:silkwormfood,**E:**WD:WL	
Morus	*nigra*	Black	Mulberry	**M(traditional):**BK,LF,FR:DCLD,RLC, **E:**LF: silkworm food	Wyk-417-22b
Morus	*nigra*	Black	Mulberry	**E:**FR:DYE(alum-&-tin-red-violet,alum-&-tin-&-tartar-dark-purple,tin-purple),**E:**LF:DYE(alum-&-iron-yellow-green)	Grae-178-03 Grae-179-01
Mucuna	*pruriens*	Cow Itch	Velvet Bean	**C:SD:CF,M(traditional):**PD(hairs):DWRO,**M(traditional):**RT:BHE,DWTE,UDR,UKC,**M(traditional):**SD:HAP,**E:**PD(hairs):PS☠	Bown-314-02
Mucuna	*pruriens*	Cow Itch	Velvet Bean	**M(traditional):**FR:HAP,DWRM,DSIN	Wyk-417-23
Murraya	*koenigii*	Curry Leaf	Satinwood	**C:**LF:**SP**,**M(traditional):**LF:SAP,SCN,SDG,SDR,SIB	Bown-314-03
Murraya	*koenigii*	Curry Leaf	Satinwood	**C:**LF:**SP**	Clev-309-01
Murraya	*koenigii*	Curry Leaf	Satinwood	**C:**LF:**SP**,**M(traditional):**LF,FR:OBN,OWN,SDG	Wyk-418-01
Musa	*acuminata*	East Asian	Banana	**M(traditional):**FR:OAS,SDR,**M(traditional):**LF:DCLD,RLB	Wyk-418-02
Musa	*acuminata*	East Asian	Banana	**E:**RD:DYE(alum-beige,copper-sulfate-dark-brown,chrome-tan,tin-warm-gold,iron-gray-brown)	Cas1-097-01
Myrica	*californica*	Calfornia	Bayberry	**M(traditional):**WP:SDR,SIB	Bown-314-04
Myrica	*californica*	Calfornia	Bayberry	**E:**FR:DYE(chrome-gray-brown,chrome-&-tin-maroon-purple)	Grae-140-01
Myrica	*carolinensis*	Candleberry	Bayberry	**E:**FR:WX	Clev-106-02
Myrica	*cerifera*	Wax Myrtle	Bayberry	**C:**FR:**SN**,**M(traditional):**BK,RB:BCR,DABA,DCLD,DFLU,DVGB,DVGV,HFV,OAS,DDND,OHL,OSA,DSIN,OST,OUL,SDR,SIB	Bown-314-05
Myrica	*cerifera*	Wax Myrtle	Bayberry	**M(traditional):**RB:BCR,.SDR,**E:**FR:WX	Clev-106-03
Myrica	*cerifera*	Wax Myrtle	Bayberry	**M(traditional):**LF:DGUM,RST,NJA,**M(traditional):**RB:DISB,OBR,OUL,OWN,**E:**BK:CM,**E:**FR:WX	Kowa-038-01
Myrica	*cerifera*	Wax Myrtle	Bayberry	**M(folk):**RB:DSIN,OUL,RST,SDR, **E:**RB:CM	Krug-026-01
Myrica	*cerifera*	Wax Myrtle	Bayberry	**M(PS):**RB:DCLD,DFLU,DLKR,OAS,OUL,RST,SDR	Wyk-209-01, Wyk-418-03
Myrica	*gale*	Sweet Gale	Bayberry	**M(traditional):**RB:DHOT,BCR,HFV,OAS,SUL	Bal1-065-05
Myrica	*gale*	Sweet Gale	Bayberry	**C:**LF:**SN**,TE,**C:**WP:AF,**M(traditional):**WP:DASC,OAS,**E:**FR:WX	Bown-315-01
Myrica	*gale*	Sweet Gale	Bayberry	**C:**LF:FF,TE,**M(traditional):**LF:SDY,**E:**LF:IR(moth&flea)	Clev-106-01
Myrica	*gale*	Sweet Gale	Bayberry	**M(traditional):**LF:DWRG,DSIN	Wyk-418-04
Myrica	*gale*	Sweet Gale	Bayberry	**E:**LF:DYE(alum-yellow-green,iron-yellow-gray,chrome-brown),**E:**LF,SM:DYE(chrome-&-iron-strong-brown,copper-sulfate-&-iron-olive-green,alum-warm-tan)	Cas1-145-01
Myrica	*pensylvanica*			Synonym for *Morella pensylvanica*	Cas-099-01
Myrica	*rubra*	Oriental Red	Bayberry	**C:**SD:NT	Clev-106-04
Myristica	*fragrans*	Nut & Mace	Nutmeg	**C:**SD:**SP**,**M(traditional):**SD,OL:HIN,NMS,OAS,ORH,OTA,SDG,SDR,SDY,SIB,**E:**OL:PF,DS,WX,**E:**SD,OL:PS☠	Bown-315-02
Myristica	*fragrans*	Nut & Mace	Nutmeg	**C:**RD(aril),NT:**SP**,**M(traditional):**HAP,NNC☠,OAS,**E:**OL:PF	Clev-310-01
Myristica	*fragrans*	Nut & Mace	Nutmeg	**C:**RD(aril),SD:**SP**,**M(folk):**SD,Aril:HIN,NNC☠,SDG,SFL,**E:**SD,Aril: PS☠(overdose)	Krug-126-01
Myristica	*fragrans*	Nut & Mace	Nutmeg	**M(PS):**SD,OL:SDG,SDR	Wyk-210-01, Wyk-418-05
Myroxylon	*balsamum*	Tolu Balsam	Myrrhwood	**M(folk):**WD-RN:OWN,**M(traditional):**WD-RN:DABA,DAFU,DDND,DSCB,DWRM,OUL,OWN,RLB,RLC,RTL,SDR, **E:**WD-RN:CM,HO,PF	Bal2-145-02 Bal2-547-03
Myroxylon	*balsamum*	Tolu Balsam	Myrrhwood	**C:**OL,RN:BF,CF,FF,**M(traditional):**OL,RN,SD:DASC,RLC,**E:**OL:GR,PF	Bown-315-03
Myroxylon	*balsamum*	Tolu Balsam	Myrrhwood	**M(PS):**RN:DASC,OWN,RLC	Wyk-211-01, Wyk-418-06
Myroxylon	*perierae*	Peru Balsam	Myrrhwood	**M(folk):**WD-RN:OWN,**M(traditional):**WD-RN:DABA,DAFU,DDND,DSCB,DWRM,OUL,OWN,RLB,RLC,RTL,SDR, **E:**WD-RN:CM,HO,PF	Bal2-145-03
Myroxylon	*perierae*	Peru Balsam	Myrrhwood	**C:**RN,SD:FF,**M(traditional):**RN:SD:DASC,OBR,ORH,DSIN,OUL,RLA,RLB,RLC,SDR,**E:**RN,SD:CM,IN,PF,DS	Bown-315-04
Myroxylon	*perierae*	Peru Balsam	Myrrhwood	**M(traditional):**RN:DABA,DASC,DAFU,**E:**RN:CM	Kowa-279-04
Myroxylon	*perierae*	Peru Balsam	Myrrhwood	**M(traditional):**RN:OBN,OBR,OHM,OUL,OWN	Wyk-211-02
Myrrhis	*odorata*	Sweet Cicely	Myrrh	**C:**LF,RT:**VG**,**C:**SD:NT,**M(traditional):**LF,RT,SD:BAN,RLC,SDG	Bown-315-05
Myrrhis	*odorata*	Sweet Cicely	Myrrh	**C:**LF,RT:**VG**,**C:**SD:NT,**M(traditional):**WP:SCN,	Clev-106-05

Latin Genus	L. Species	E. Species	English Genus	C:Culinary,M:Medicinal,E:Economic	Source
				OWN	
Myrrhis	*odorata*	Sweet Cicely	Myrrh	**C:**LF,RT:**VG**,**M(traditional):**WP:SAP,SFL,**M(traditional):**RT:DASC	Kowa-474-01
Myrrhis	*odorata*	Sweet Cicely	Myrrh	**C:**LF:**VG**,**C:**SD:FF,**M(folk):**LF:DASC,DCTM, **M(folk):**SD:SDG	Krug-169-01
Myrrhis	*odorata*	Sweet Cicely	Myrrh	**M(traditional):**WP:RLC,UDR	Wyk-418-07
Myrtillocactus	*geometrizans*	Geometric	Myrtle Shrub Cactus	**C:**FR:PU,**M(traditional):**WP:BHS	Wyk-418-08
Myrtus	*communis*	Mediterranean	Myrtle	**C:**LF:**SN**,**C:**FR:**SP**,**M(traditional):**OL:DACN,DGUM,OHM,**M(traditional):**WP:DASC,DVGB,DVGV,OAS,RLB,RLC,RSN,UUR,**E:**OL:CM,PF,DS	Bown-316-01
Myrtus	*communis*	Mediterranean	Myrtle	**C:**LF:FF,**C:**FR:AF,**C:**FR-BD:FF,**M(traditional):**LF:DSIN,OWN,**M(traditional):**FR-JU:SDG,**E:**FL,LF:CM,DS,PF(eaud'ange)	Clev-106-06
Myrtus	*communis*	Mediterranean	Myrtle	**M(PS):**LF,OL:DACN,DAMY,DGUM,OAS,OHM,OWN,RLC,RSN,UUR	Wyk-212-01, Wyk-418-09
Myrtus	*ugni*			Synonym for *Ugni molinae*	Clev-106-07
Nabalus	*serpentarius*			Synonym for *Prenanthes serpentarius*	Wyk-418-10a
Narcissus	*pseudonarcissus*	Lesser Yellow	Daffodil	**E:**FL:DYE(alum-soft-yellow,alum-&-chrome-gold,tin-bright-yellow,chrome-&-iron-old-gold,iron-gold-gray)	Cas1-131-01
Narcissus	*pseudonarcissus*	Lesser Yellow	Daffodil	**E:**FL:DYE(alum-&-ammonia-bright-yellow,chrome-&-ammonia-deep-gold)	Grae-117-01
Nardostachys	*grandiflora*	Bible Oil	Spikenard	**M(traditional):**RT,OL:BHP,HIN,NMS,OAS,SDG, **E:**OL:CM(deodorant),PF	Bown-316-02
Nardostachys	*grandiflora*	Bible Oil	Spikenard	**M(traditional):**RT:NSD	Wyk-418-11
Nasturtium	*officinale*	Scurvy Grass	Watercress	**C:**LF:GN,**SN**,**M(traditional):**LF:BAN,BHE,NRH, OSA,SDG,SGB,RLB,RLC,UDR	Bown-316-03
Nasturtium	*officinale*	Scurvy Grass	Watercress	**C:**LF:**VG**,**M(traditional):**LF:OGT,ORH,RLC,**M(traditional):**SD:DWRM	Clev-107-01
Nasturtium	*officinale*	Scurvy Grass	Watercress	**C:**LF:**VG**,**M(folk):**LF,SM:DGUM(scurvy),DSIN, RLB,RLC,SDG,SPR	Krug-180-01
Nelumbo	*nucifera*	Sacred	Lotus Waterlily	**C:**RT:**VG**,**C:**SD:NT,**M(traditional):**RZ:SUL,**M(traditional):**SD:BHP,HIN,SDG,SDR,**M(traditional):**FL,SL:SUL	Bown-316-04
Nelumbo	*nucifera*	Sacred	Lotus Waterlily	**M(PS):**LF,FL:HFV,NSD,OHM,OAS,DSIN,SDR, **M(PS):**SD:HIN,NSD,SDG	Wyk-213-01, Wyk-418-12
Nepeta	*cataria*	Catnip	Catmint	**M(traditional):**LF:DCLD,DFLU,HFV,HAN,HIN,SAP,SDG	Bal1-067-01
Nepeta	*cataria*	Catnip	Catmint	**M(folk):**WP(top):DCLD,HFV,SIB,**M(traditional):**WP(top):DFLU,HAC,HIN,NNC☠,**M(PS):**WP(top):DABI	Bal2-043-01 Bal2-509-04
Nepeta	*cataria*	Catnip	Catmint	**C:**LF:**SN**,**M(traditional):**WP:HFV,HIN,NMS,NSD,OAS,SDG,SDR,SIB	Bown-317-01
Nepeta	*cataria*	Catnip	Catmint	**M(traditional):**LF:DCLD,HAP,NSD,OBR,OWN,SFL,SIB,**E:**LF,OL:PF	Clev-107-03
Nepeta	*cataria*	Catnip	Catmint	**C:**LF:FF,TE,**M(traditional):**LF(top):DCLD,DCTM,HAC,HFV,HIN,NSD,SDG,SFL	Kowa-071-01 Kowa-284-01
Nepeta	*cataria*	Catnip	Catmint	**M(folk):**LF:NNC☠,SDG,RLB	Krug-043-01
Nepeta	*cataria*	Catnip	Catmint	**M(PS):**LF,FL,OL:DALL,HAC,HAN,HIN,NJA,NRH,OEY,OHM,OTA,SDG,SDR,SIB	Wyk-214-01, Wyk-418-13
Nepeta	*cataria citriodora*	Lemon Catnip	Catmint	**C:**LF:TE	Bown-317-02
Nephroma	*expallidum*	Pale	Kidney Pelt Lichen	**E:**WP:DYE(Nephroarctin-brown)	Cas2-035-09
Nephroma	*helveticum*	Greek	Kidney Pelt Lichen	**E:**WP:DYE(Zeorin-orange-brown,Zeorin-&-Ammonia-dark-brown)	Cas2-032-13
Nerium	*oleander*	Rose Bay	Oleander	**M(PS):**LF:BHS,DSCB,DSIN	Wyk-215-01, Wyk-418-14
Nerium	*oleander*	Rose Bay	Oleander	**E:**FL:DYE(alum-light-gray-green,chrome-gray-green)	Grae-156-01
Nicotiana	*alata*	Jasmine	Tobacco	**E:**LF:smokingtobacco	Clev-107-04
Nicotiana	*attenuata*	Coyote	Tobacco	**E:**LF:smokingtobacco	Clev-107-06
Nicotiana	*benthamiana*	Bentham	Tobacco	**E:**LF:smokingtobacco	Clev-107-07
Nicotiana	*bigelovii*	Bigelow	Tobacco	**E:**LF:smokingtobacco	Clev-107-05
Nicotiana	*rustica*	Aztec	Tobacco	**E:**LF:smokingtobacco,IC	Clev-107-08
Nicotiana	*sanderae marona*	Maroon Sander	Tobacco	**E:**FL:DYE(alum-&-chrome-&-copper-&-tin-gray-green)	Grae-126-01
Nicotiana	*tabacum*	West Indies	Tobacco	**E:**LF:smokingtobacco	Clev-107-09
Nicotiana	*tabacum*	West Indies	Tobacco	**M(traditional):**WP:DWRM,SCN,**E:**WP:IC	Wyk-418-15
Nigella	*sativa*	Blackseed	Love In A Mist	**C:**SD:**SP**,**M(traditional):**SD,OL:NMP,NOR,OHM,OSA,RLB,SCN,SDG	Bown-317-03
Nigella	*sativa*	Blackseed	Love In A Mist	**C:**SD:**SP**,**M(traditional):**SD,OL:SDG,SFL,SIB	Clev-312-01
Nigella	*sativa*	Blackseed	Love In A Mist	**C:**SD:**SP**,**M(PS):**SD,OL:DASC,DWRM,HAC,NMS,RLA,SDG,SIB,UDR	Wyk-216-01, Wyk-418-16
Notopterygium	*Incisum*	Qiang Huo	Winged Angelic	**M(traditional):**RT:DCLD	Wyk-418-17

Latin Genus	L. Species	E. Species	English Genus	C:Culinary,M:Medicinal,E:Economic	Source
Nuphar	*luteum*	Yellow	Spatterdock Waterlily	**M(traditional)**:RT:HAP	Wyk-418-18
Nymphaea	*alba*	White	Waterlily	**M(traditional):**RZ:DASC,NSD,OAS	Bown-317-04
Nymphaea	*alba*	White	Waterlily	**M(traditional):**RT:HIN	Wyk-418-19
Nymphaea	*caerulea*	Blue Lotus	Waterlily	**M(traditional):**FL,FR,RT,SD:UBL,UKS	Bown-317-05
Nymphaea	*lotus*	Lotus	Waterlily	**C:**FR:**SN**,**C:**RT:GF,**VG**,**C:**SD:NT,**M(traditional):**RT:HAP,HFV,HIN,OAS,OHM,SDP,SDR,SIB,UBL,UDR,UKS,**M(traditional):**FR:DSNB	Bown-317-06
Nymphaea	*odorata*	Sweet Scented	Waterlily	**M(traditional):**RZ:DASC,DBOI,DCCU,DSXG,DTBC,DVGB,DVGV,OAS,OEY,RLB,RST,SDR,SDY,SIB,UBL,UEP	Bown-317-07
Nymphaea	*tuberosa*	Tuber	Waterlily	**M(traditional):**RZ:DASC,NSD,OAS	Bown-317-08
Ochrolechia	*parella*	Parelle	Saucer Perforated Lichen	**E:**WP:DYE(Orcinol-Depsides-&-Ammonia-red-purple)	Cas2-032-14
Ochrolechia	*tartarea*	Tartarian	Saucer Perforated Lichen	**E:**WP:DYE(Gyrophoric-Acid-copper,Gyrophoric-Acid-&-Ammonia-red)	Cas2-032-15
Ocimum	*americanum*			Synonym for *Ocimum canum*	Bown-318-06
Ocimum	*basilicum-afra*	African Camphor	Basil	**E:**OL:IR(camphor,mosquito)	Bown-318-01
Ocimum	*basilicum-anisum*	Anise	Basil	**C:**LF:**SN**	Kowa-025-02
Ocimum	*basilicum-asiatica*	Asian Eugenol	Basil	**C:**LF:**SN**	Bown-318-02
Ocimum	*basilicum-citriodora*	East Indies Lemon	Basil	**C:**SD:BF	Bown-318-03
Ocimum	*basilicum-corrugata groena*	Green Ruffles	Basil	**C:**LF:GN	Kowa-025-05
Ocimum	*basilicum-corrugata purpurea*	Purple Ruffles	Basil	**C:**LF:GN	Kowa-025-09
Ocimum	*basilicum-crispum*	Lettuce Leaf	Basil	**C:**LF:**SN**	Bown-318-04
Ocimum	*basilicum-crispum*	Lettuce Leaf	Basil	**C:**LF:**SN**	Kowa-025-04
Ocimum	*basilicum-europa*	European Cinnamon	Basil	**C:**LF:**SN**	Bown-318-05
Ocimum	*basilicum-europa*	European Cinnamon	Basil	**C:**LF:**SN**	Kowa-025-03
Ocimum	*basilicum-mediterranea*	Mediterranean Sweet	Basil	**C:**LF:**SN**,**M(traditional):**WP:DABA,DACN,DCLD,DFLU,DISB,DSNB,DWRM,HAC,HFV,HIN,NMS,DSIN,SDG,SIB**E:**OL:IR,PF	Bown-318-06
Ocimum	*basilicum-mediterranea*	Mediterranean Sweet	Basil	**C:**LF:AF,**SN**,**M(traditional):**WP:HAC,HAN,NSD,**E:**LF-OL:CM,DS,IR	Clev-108-01
Ocimum	*basilicum-mediterranea*	Mediterranean Sweet	Basil	**C:**LF:**SN**,**M(traditional):**WP:HAC,HAN,NSD,SDG,**M(traditional):**SD:DABA,**E:**OL:CM,DS,HO,PF	Kowa-022-01
Ocimum	*basilicum-mediterranea*	Mediterranean Sweet	Basil	**C:**LF:**SN**,**M(folk):**LF:DCLD,HAC,SDG	Krug-024-01
Ocimum	*basilicum-mediterranea*	Mediterranean Sweet	Basil	**M(PS):**WP:DABA,DWRM,OWN,SAP,SDG,SFL,UDR	Wyk-217-01, Wyk-418-20
Ocimum	*basilicum-minimum*			Synonym for *Ocimum minimum*	Bown-318-07
Ocimum	*basilicum-purpurescens darcaopala*	Dark Opal Purple	Basil	**C:**LF:GN	Bown-166-06 Kowa-025-10
Ocimum	*basilicum-reunuionensis*	Reunion Island Camphor	Basil	**E:**OL:IR(camphor,mosquito)	Bown-318-08
Ocimum	*basilicum-thyrsiflora*	Thyrse Flower	Basil	**C:**LF:**SN**(Thaicooking)	Kowa-025-11
Ocimum	*canum-camphorum*	Camphor Hoary	Basil	**E:**OL:IR(camphor,mosquito)	Bown-318-09
Ocimum	*canum-limonum*	Lemon Hoary	Basil	**C:**SD:BF	Bown-318-10
Ocimum	*canum-limonum*	Lemon Hoary	Basil	**C:**SD:BF	Kowa-025-01
Ocimum	*canum-spicatum*	Spicy Hoary	Basil	**C:**LF:**SN**	Bown-318-11
Ocimum	*gratissimum*	East Indies Spicy	Basil	**C:**LF:FF,TE,**M(traditional):**LF,OL:DIPT,DWRM,HAC,HFV,OBK,ORH,SDR,SDY,E:OL:IR(mosquito),PF	Bown-318-12
Ocimum	*gratissimum*	East Indies Spicy	Basil	**C:**LF:**SN**	Kowa-025-12
Ocimum	*kilimandshcaricum*	Mt. Kilimanjaro Camphor	Basil	**M(traditional):**WP:HFV,**E:**OL:IR(camphor,mosquito)	Bown-318-13
Ocimum	*kilimandshcaricum*	Mt. Kilimanjaro Camphor	Basil	**M(traditional):**WP:DCLD,RLC,**E:**OL:IR(camphor,mosquito)	Clev-108-02
Ocimum	*kilimandshcaricum*	Mt. Kilimanjaro Camphor	Basil	**M(traditional):**WP:DCLD,SDG	Kowa-025-13
Ocimum	*minimum*	Greek Bitter	Basil	**E:**WP:foliagepotoredgingherb	Bown-318-07
Ocimum	*minimum*	Greek Bitter	Basil	**M(traditional):**WP:HAC,HAN,NSD,**E:**LF-OL:CM,DS,IR	Clev-108-03
Ocimum	*minimum*	Greek Bitter	Basil	**E:**WP:foliagepotoredgingherb	Kowa-025-06
Ocimum	*sanctum*			Synonym for *Ocimum tenuiflorum*	
Ocimum	*tenuiflorum*	India Holy	Basil	**C:**LF:FF,**C:**SD:TE,:WP:DABA,DCLD,DFLU,HAC,HAP,HFV,NJA,NMS,NRH,OAS,DSIN,SDG,RSN,E:OL:IR,**E:**SM:rosary beads	Bown-318-14

Latin Genus	L. Species	E. Species	English Genus	C:Culinary,M:Medicinal,E:Economic	Source
Ocimum	*tenuiflorum*	India Holy	Basil	**M(traditional):**WP:DCLD,RLC,**E:**OL:IR(camphor,mosquito)	Clev-108-04
Ocimum	*tenuiflorum*	India Holy	Basil	**E:**WP:ornamental	Kowa-025-14
Ocimum	*tenuiflorum*	India Holy	Basil	**M(traditional):**LF,LF-JU:DSIN,OWN,RLC,SDG	Wyk-217-02, Wyk-418-21
Ocimum	*viride*	W. Africa Fever	Basil	**M(traditional):**LF:HFV	Clev-108-05
Oclemena	*nemoralis-pinca*	Pink Bog	Oclemen Aster	**E:**FL:DYE(alum-beige-yellow,chrome-tan,copper-sulfate-&-iron-gray-green)	Cas1-095-02
Oenanthe	*crocata*	Crocus	Water Dropwort	**M(traditional):**WP:NMS,**E:**WP:PS☠	Wyk-418-22
Oenothera	*biennis*	Evening Star	Evening Primrose	**M(traditional):**OL:BAS,BLC,DACN,DHEA,DMSC,NJA,ODS,DSIN,**E:**OL:CM	Bown-318-15
Oenothera	*biennis*	Evening Star	Evening Primrose	**C:**RT:**VG**,**C:**SD:**SP**,**M(traditional):**LF:OCC,RLC,**M(traditional):**RT:DBOI,OHM	Clev-108-06
Oenothera	*biennis*	Evening Star	Evening Primrose	**M(folk):**SD-OL:DDND,DPMS,DSIN,NMT,OSS, **M(PS):**SD-OL:DMSC	Krug-070-01
Oenothera	*biennis*	Evening Star	Evening Primrose	**M(CS):**SD,SD-OL:BCR,NRH,DSIN,SIB	Wyk-218-01, Wyk-419-01
Oenothera	*biennis*	Evening Star	Evening Primrose	**E:**FL:DYE(alum-yellow,tin-brilliant-yellow,chrome-bright-gold),**E:**WP:DYE(alum-&-copper-sulfate-yellow-green,iron-gray-green,alum-&-tin-chartreuse,chrome-tan)	Cas1-140-02
Oenothera	*erythrosepala*	Large Flower	Evening Primrose	**C:**RT,SH:**VG**	Clev-108-07
Oenothera	*villosa strigosa*	Bristle Hairy	Evening Primrose	**E:**LF,FL,SM:DYE(alum-light-yellow,chrome-camel,copper-olive-gray,tin-bright-yellow,iron-charcoal-gray)	Bliss-110-01
Okoubaka	*aubrevillei*	Aubreville	Okouba Sandalwood	**M(traditional):**BK:DSIN	Wyk-419-02
Oldenlandia	*diffusa*	Snake Needle Grass	Mille Grain	**M(traditional):**WP:BLC,BJA,DABA,DBOI,DCST,HFV,DSNB,OBR,OSA,RLA,RLC,UDR,UUR	Bown-319-01
Oldenlandia	*hedyotidea*	Sweet Grass	Mille Grain	**M(traditional):**LF,RT:RLC,**M(traditional):**RT:DSNB	Bown-319-02
Oldenlandia	*umbellata*	India Madder	Mille Grain	**E:**WP:DYE(forturbans)	Bown-319-03
Olea	*europaea*	European	Olive	**C:**FR:**VG**,**M(traditional):**LF:BHP,BHS, **M(PS):**LF:DABA,DAFU,DAID,DAVR,DCAN	Bal2-108-01
Olea	*europaea*	European	Olive	**C:**FR:**VG**,**M(traditional):**LF:BHP,**M(traditional)**:OL:DDND,ODS,SCN,SUL,**E:**OL:HO,**SP**	Bown-319-04
Olea	*europaea*	European	Olive	**C:**FR-OL:CO,**M(traditional):**FR-OL:OWN,SCN	Clev-109-01
Olea	*europaea*	European	Olive	**C:**FR-OL:CO,**M(folk):**FR-OL:DEAI,BHC,HFV,NSM,OWN,SCN	Krug-129-01
Olea	*europaea*	European	Olive	**M(PS):**LF:DHOT,BHP,NMS,UDR,**M(traditional):**OL:SCN, SGB	Wyk-219-01, Wyk-419-03
Olea	*europaea*	European	Olive	**E:**FR:DYE(alum-maroon,alum-&-aluminum-deep-purple),**E:**LF:DYE(alum-bright-yellow,alum-&-iron-olive-green),**E:**RD:DYE(alum-lavender,alum-&-copper-blue-gray,chrome-blue-green,chrome-&-copper-black)	Grae-179-02 Grae-180-01 Grae-180-02 Grae-181-01 Grae-181-02
Omphalodiscus	*virginis*			Synonym for *Umbilicaria virginis*	
Ononis	*spinosa*	Spiny	Restharrow	**M(PS):**RT:NGT,NRH,UDR,UKC,UUR	Wyk-220-01, Wyk-419-04
Operculina	*turpethum*	Turpeth Root	Lidpod Bindweed	**M(traditional):**RT:DWRM,SPR	Wyk-419-05
Ophioparma	*lapponica*	Lapland	Blood Spot Shield Lichen	**E:**WP:DYE(Hypothamnolic-Acid-&-Ammonia-purple-red)	Cas2-032-04
Ophioparma	*ventosa*	Vent	Blood Spot Shield Lichen	**E:**WP:DYE(Divaricatic-Acid-orange-brown,Divaricatic-Acid-&-Ammonia-deep-magenta)	Cas2-032-05
Ophiopogon	*japonicus*	Japanese Hyacinth	Mondo Grass	**M(traditional):**TU:HFV,HIN,NSD,RLB,RLC,SCN,SDG	Bown-319-05
Ophiopogon	*japonicus*	Japanese Hyacinth	Mondo Grass	**M(traditional):**TU:RLA,SCN	Wyk-419-06
Opuntia	*ficus-indica*	Tuna	Cholla Cactus	**M(traditional):**FL,FR:OAS,SDG	Wyk-419-07
Opuntia	*robusta*	Purple Prickly Pear	Cholla Cactus	**E:**FR:DYE(direct-magenta, poor lightfastness)	Grae-056-02
Orchis	*morio*			Synonym for *Anacamptis morio*	Wyk-419-08
Origanum	*dictamnus*	Hop Marjoram	Oregano	**C:**LF:**SN**,**C:**FL:TE,**C:**OL:AF(vermouth)	Bown-319-06
Origanum	*dictamnus*	Hop Marjoram	Oregano	**M(traditional):**WP:NSD,OWN	Clev-109-03
Origanum	*dictamnus*	Hop Marjoram	Oregano	**M(traditional):**WP:DCLD,DFLU,OWN,RLB,RLC,SDP,SIB	Wyk-419-09
Origanum	*heracleoticum*	Heracles	Oregano	**C:**LF:**SN**	Clev-109-05
Origanum	*majorana*	Sweet Marjoram	Oregano	**C:**LF,FL-SL:**SN**,**M(traditional):**WP:HAC,HIN,NMP,OJA,RLB,SDG	Bown-319-07
Origanum	*majorana*	Sweet Marjoram	Oregano	**C:**LF:AF,**SN**,TE,**M(traditional):**LF:DASC,OTA,RLC,RST,SDG,SIB,**E:**LF-OL:CM,PF	Clev-109-06
Origanum	*majorana*	Sweet Marjoram	Oregano	**C:**LF:**SN**,**M(traditional):**WP:DAFU,DFRA,SDG,	Kowa-174-08

Latin Genus	L. Species	E. Species	English Genus	C:Culinary,M:Medicinal,E:Economic	Source
				E:WP(top):CM,DYE(alum-green,chrome-olive-green)	Kowa-373-01
Origanum	*majorana*	Sweet Marjoram	Oregano	**C:**LF:**SN**,**M(folk):**LF:SDG	Krug-112-01
Origanum	*majorana*	Sweet Marjoram	Oregano	**M(traditional):**WP:DCLD,DFLU,RLB,RLC,SDP, SIB	Wyk-419-10
Origanum	*majoricum*	Hardy Marjoram	Oregano	**C:**LF,FL-SL:**SN**	Bown-320-01
Origanum	*prismaticum*	Prism	Oregano	**C:**LF:**SN**	Clev-109-04
Origanum	*onites*	Pot Marjoram	Oregano	**C:**LF,FL-SL:FF,**E:**LF,FL-SL:PF	
Origanum	*onites*	Pot Marjoram	Oregano	**C:**LF:AF,**SN**,TE,**M(traditional):**LF:DASC,OTA, RLC,RST,SDG,SIB,**E:**LF-OL:CM,PF	Clev-109-07
Origanum	*vulgare*	Seasoning	Oregano	**C:**LF,FL-SL:**SN**,**C:**OL:FF,TE,**M(traditional):**WP:DASC,DCLD,DFLU,NMS,NMP,SDG,SFL,**E:**OL:CL	Bown-320-02
Origanum	*vulgare*	Seasoning	Oregano	**C:**LF:AF,**SN**,TE,**M(traditional):**LF:DASC,OTA, RLC,RST,SDG,SIB,**E:**LF-OL:CM,PF	Clev-109-02
Origanum	*vulgare*	Seasoning	Oregano	**C:**LF:**SN**,**M(traditional):**LF:DISB(scorpion,spider),HAC,OTA,RLA,RLC,SDG,**E:**OL:CM	Kowa-401-01
Origanum	*vulgare*	Seasoning	Oregano	**C:**LF:**SN**,**M(folk):**LF:DAMW,DASC,DCLD,DFLU,NRH,RLC,SDG	Krug-113-01
Origanum	*vulgare*	Seasoning	Oregano	**M(PS):**WP:DAMY,DCLD,DFLU,DSIN,OWN,RLB, RLC,RRN,SDG,SDP,SIB	Wyk-221-01, Wyk-419-11
Orthocarpus	*barbatus*	Grand Coulee	Owl's Clover	**E:**WP:DYE(direct-lemon-yellow,alum-mustard, chrome-ocher,alum-&-iron-gray-yellow, chrome-&-iron-khaki-green)	Grae-101-02 Grae-102-01
Orthosiphon	*aristatus*	Long Stamen	Siphon Mint	**M(PS):**LF:UBL,UDR,UKC	Wyk-222-01, Wyk-419-12
Oryza	*sativa-brevisperma*		Short Grain Rice	**C:**SD:**VG**	Bown-320-03
Oryza	*sativa-mediasperma*		Medium Grain Rice	**C:**SD:**VG**,**M(traditional):**SD,SR:UDR,**M(traditional):**SR:SAP,SDG,**M(traditional):**RZ:HFV,DPNB,DPNV,DTBC	Bown-320-03
Oryza	*sativa-longasperma*		Long Grain Rice	**C:**SD:**VG**	Bown-320-03
Osmunda	*regalis*	Royal	Flowering Fern	**M(traditional):**RT:RLC,SDY,**E:**RT:orchidcompost	Clev-110-01
Oxalis	*acetosella*			Synonym for *Oxalis montana*	
Oxalis	*corniculata*	Creeping	Wood Sorrel	**E:**FL:DYE(alum-&-tin-maize-yellow,alum-&-tin-&-ammonia-neon-orange,chrome-&-tin-&-brown,chrome-&-tin-ammonia-deep-orange-red)	Grae-108-03 Grae-109-01
Oxalis	*montana*	Mountain	Wood Sorrel	**M(traditional):**WP:SDP	Wyk-419-13
Oxalis	*stricta*	Yellow	Wood Sorrel	**E:**LF,FL,SM:DYE(alum-light-beige,chrome-gray-tan,copper-beige,tin-yellow,iron-gray-green)	Bliss-182-01
Pachycarpus	*rigidus*	Stiff	Thick Pod Milkweed	**M(traditional):**RT:NMS,NSD	Wyk-419-14
Packera	*aurea*	Golden	Packer Ragwort	**M(traditional):**WP,RZ:DVGB,DVGV,OAS,UDR,**M(traditional):**WP,RZ,PS☠	Bown-352-01
Paeonia	*lactiflora*	Chinese	Peony	**M(traditional):**RT:DABA,BHP,BLC,HFV,NMS,NSD,OAS,DSIN	Bown-321-01
Paeonia	*lactiflora*	Chinese	Peony	**M(PS):**RT:DALL,DBOI,DEPI,HAC,HFV,BLC,NJA,NMS,NSD,RLC,OBN,DSIN,OWN,SDP	Wyk-223-01, Wyk-419-15
Paeonia	*lactiflora*	Chinese	Peony	**E:**FL:DYE(alum-yellow,tin-sharp-yellow-gold,chrome-strong-gold,copper-sulfate-bronze,iron-&-chrome-tan-orange)	Cas1-192-01
Paeonia	*officinalis*	European	Peony	**M(traditional):**RT:BVV,DEPI,DWHP,NMS,NSD, OHM,SGS,UDR,UKS	Bown-321-02
Paeonia	*officinalis*	European	Peony	**M(traditional):**RT,SD:DEPI	Wyk-419-16
Paeonia	*suffruticosa*	Moutan	Peony	**M(traditional):**RB:BHP,DABA,DALL,DBOI,HFV, NSD,DSIN,SIB,SUL	Bown-321-03
Paeonia	*veitchii*	Veitch	Peony	**M(traditional):**RT:DSIN	Bown-321-04
Palmaria	*palmata*	Dulse Purple	Hand Seaweed	**E:**WP:DYE(vinegar-warm-tan, baking-soda-taupe, alum-warm-beige, iron-gray-tan)	Cas1-138-01
Panax	*bipinnatifidus*	Russian	Ginseng	**M(traditional):**RT:DCFS,HAN	Wyk-224-02
Panax	*ginseng*	Chinese	Ginseng	**M(traditional):**RT:HAN,BDB,BCR,DCFS,DIFR,SAP,RLB	Bal1-071-03
Panax	*ginseng*	Chinese	Ginseng	**M(traditional):**RT:BLC,DIPT,HAP,HIN,NNR, OEY,SAP **M(PS):**RT:BDB,BHP,DCTM,DCFS, HAN, **M(CS):**RT:HAP,HFN	Bal2-078-01 Bal2-522-01
Panax	*ginseng*	Chinese	Ginseng	**M(traditional):**RT:BCR,BHC,BSH,HIN,SAP,**E:**RT:PS☠	Bown-321-05
Panax	*ginseng*	Chinese	Ginseng	**M(folk):**RT:HAN,HML	Krug-082-01
Panax	*ginseng*	Chinese	Ginseng	**M(CS):**RT:DCFS,HAN	Wyk-224-01, Wyk-419-17
Panax	*japonicus*	Japanese	Ginseng	**M(traditional):**RT:DCFS,HAN	Wyk-224-03
Panax	*notoginseng*	South China	Ginseng	**M(traditional):**RT:BCR	Bown-321-06

Latin Genus	L. Species	E. Species	English Genus	C:Culinary,M:Medicinal,E:Economic	Source
Panax	*pseudoginseng*	Nepal	Ginseng	**M(traditional):**RT:BAA,BCR,BHS,DABA,**M(traditional):**FL:DVTG	Bown-321-07
Panax	*quinquefolius*	American	Ginseng	**M(traditional):**RT:BHP,DCFS,DCLD,DFLU,DIFR,HAC,NRH,RLB,RLC,SCN,UBL, **M(PS):**RT:BDB,DCTM,DHEA	Bal2-020-01 Bal2-503-01
Panax	*quinquefolius*	American	Ginseng	**M(traditional):**RT:BCR	Bown-322-01
Panax	*quinquefolius*	American	Ginseng	**M(traditional):**RT:HAP,BAN,BDB,BHE,HAN,SUL	Kowa-224-01
Panax	*quinquefolius*	American	Ginseng	**M(traditional):**RT:HAP,HAN,HML	Krug-083-01
Panax	*quinquefolius*	American	Ginseng	**M(traditional):**RT:DCFS,HAN	Wyk-224-04, Wyk-419-18
Panax	*sokpayensis*	Sikkim	Ginseng		Kew-2010-07
Panax	*stipuleanatus*	Korean	Ginseng	**M(traditional):**RT:DRSK	Kowa-224-02
Panax	*stipuleanatus*	Korean	Ginseng	**M(traditional):**RT:DCFS,HAN	Wyk-224-05
Panax	*trifolius*	Three Leaf American	Ginseng		Kew-2010-09
Panax	*vietnamensis*	Vietnamese	Ginseng		Kew-2010-10
Panax	*wangianus*	Central China	Ginseng		Kew-2010-11
Panax	*zingiberensis*	Yunnan Ginger	Ginseng		Kew-2010-12
Pandanus	*odoratissimus*			Synonym for *Pandanus tectorius*	Bown-322-02
Pandanus	*tectorius*	Scented	Screw Pine	**C:**FR:PU,**C:**SD:NT,**M(traditional):**RT:UDR,**M(traditional):**LF-BD:DBOI,**E:**OL:CM	Bown-322-02
Papaver	*nudicaule*	Red-Orange Icelandic	Poppy	**E:**FL:DYE(alum-warm-beige,tin-yellow-gold, chrome-dark-gold,copper-sulfate-tan-pink,iron-warm-gray)	Cas1-197-01
Papaver	*nudicaule*	Red-Orange Icelandic	Poppy	**E:**FL:DYE(alum-&-tin-light-brick-red,alum-&-tin-&-tartar-beige),**E:**PD:DYE(alum-&-wool-mustard-yellow ,alum-&-cotton-light-brown,chrome-copper-brown)	Grae-129-03 Grae-130-01
Papaver	*orientale*	Great Scarlet Oriental	Poppy	**M(traditional):**FL,LF:NSD(codeine)	Kowa-281-04
Papaver	*rhoeas*	Corn	Poppy	**C:**SD:**SP**,**M(traditional):**FL:HIN,OAS,NSD,RLC,SDG	Bown-322-03
Papaver	*rhoeas*	Corn	Poppy	**M(traditional):**FL:RLC	Wyk-419-19
Papaver	*somniferum*	Opium	Poppy	**C:**SD:**SP**,**M(traditional):**FL:HIN,NNC☠,NSD,RLC,SDR,**E:**OL:DS,PN	Bown-322-04
Papaver	*somniferum*	Opium	Poppy	**C:**SD:**SP**,**M(traditional):**SD:DEAI,NSD,OTA,RLC	Clev-313-01
Papaver	*somniferum*	Opium	Poppy	**M(folk):**LF,PD:NMS,NNC☠,RLA	Kowa-282-05
Papaver	*somniferum*	Opium	Poppy	**C:**SD(ripe):**SP**,**M(folk):**SD(unripe):HAN,NMS,NNC☠,NSD,RLC	Krug-142-01
Papaver	*somniferum*	Opium	Poppy	**M(PS):**FR-SA:NNC☠,NSD,RLC	Wyk-225-01, Wyk-419-20
Parietaria	*judaica*	Spreading	Pillatory	**M(traditional):**WP:UBL,UDR,UKP,UKS	Bown-322-05
Parietaria	*judaica*	Spreading	Pillatory	**M(traditional):**WP:UBL,UDR,UKC,UKS,UUR	Wyk-226-02
Parietaria	*officinalis*	European	Pillatory	**M(traditional):**WP:UBL,UDR,UKC,UKS,UUR	Wyk-226-01, Wyk-419-21
Paris	*quadrifolia*	Four Leaf	Trinity Lily	**M(traditional):**RZ:DFLU,DSIN,OWN	Wyk-419-22
Parmelia	*arnoldii*			Synonym for *Parmotrema arnoldii*	Cas2-035-11
Parmelia	*caperata*			Synonym for *Flavoparmelia caperata*	Cas2-035-12
Parmelia	*clavulifera*	Club	Shield Lichen	**E:**WP:DYE(βOrcinolDepsidones-copper-rust)	Cas2-035-13
Parmelia	*fuliginosa*			Synonym for *Melanelia fuliginosa*	Cas2-032-17
Parmelia	*glabratula*			Synonym for *Melenelia fuliginosa*	Cas2-032-18
Parmelia	*olivacea*			Synonym for *Melanelia olivacea*	Cas2-035-14
Parmelia	*omphalodes*	Navelcap	Shield Lichen	**E:**WP:DYE(Salazinic-Acid-orange-red-brown)	Cas2-035-15
Parmelia	*perlata*			synonym for *Parmotrema perlatum*	Cas2-032-19
Parmelia	*pertusa*			Synonym for *Menegassia terebrata*	Cas2-032-20
Parmelia	*reticulata*			Synonym for *Rimelia reticulata*	Cas2-032-21
Parmelia	*rudecta*			Synonym for *Punctelia subflava-rudecta*	Cas2-035-16
Parmelia	*saxatilis*	Rock Loving	Shield Lichen	**E:**WP:DYE(Salazinic-Acid-bronze)	Cas2-035-17
Parmelia	*sulcata*	Furrowed Rock	Shield Lichen	**E:**WP:DYE(Salazinic-Acid-red-brown)	Cas2-035-18
Parmelia	*sulcata*	Furrowed Rock	Shield Lichen	**E:**WP:DYE(Salazinic-Acid-&-Ammonia-magenta)	Grae-079-02
Parmelia	*ulophyllodes*			Synonym for *Flavopunctelia soredica*	Cas2-032-22
Parmotrema	*arnoldii*	Arnold	Scatter Rug Ruffle Shield Lichen	**E:**WP:DYE(Anatorin-&-Alectoronic-Acid-gold)	Cas2-035-11
Parmotrema	*perforatum*	Perforated	Scatter Rug Ruffle Shield Lichen	**E:**WP:DYE(Alectoronic-Acid-&-Ammonia-red)	Cas2-032-23
Parmotrema	*perlatum*	Oak	Scatter Rug Ruffle Shield Lichen	**E:**WP:DYE(Stictic-Acid-russet-orange,Stictic-Acid-&-Ammonia-magenta-pink)	Cas2-032-19
Parmotrema	*perlatum*	Oak	Scatter Rug Ruffle Shield Lichen	**E:**WP:DYE(Stictic-Acid-deep-orange-russet, Stictic-Acid-&-Ammonia-bright-pink-magenta)	Grae-078-02 Grae-078-03
Parmotrema	*tinctorum*	Dyer's	Scatter Rug Ruffle Shield Lichen	**E:**WP:DYE(Lecanoric-Acid-gold,Lecanoric-Acid-&-Ammonia-red)	Cas2-032-25 Cas2-035-19
Passiflora	*caerulea*	Blue Crown	Passion Flower	**M(traditional):**WP,FL:BHP,HAN,NSD	Bal2-110-02

Latin Genus	L. Species	E. Species	English Genus	C:Culinary,M:Medicinal,E:Economic	Source
					Bal2-534-01
Passiflora	*incarnata*	Maypops	Passion Flower	**M(traditional):**WP,FL:DHEA,HAN,NNR,NSD	Bal1-075-05
Passiflora	*incarnata*	Maypops	Passion Flower	**M(traditional):**WP,FL:BHP,HAN,NSD	Bal2-110-03 Bal2-534-01
Passiflora	*incarnata*	Maypops	Passion Flower	**C:**FR:JJ,PU,**M(traditional):**WP:BHP,DSHZ,HAC,HIN,NMS,NSD,RLA,SIB	Bown-323-01
Passiflora	*incarnata*	Maypops	Passion Flower	**M(traditional):**FL(top):DCFS,HAC,HIN,NMS,NSD	Kowa-291-02 Kowa-410-01
Passiflora	*incarnata*	Maypops	Passion Flower	**C:**FR(granadilla):JU,**M(folk):**WP(flowering),FR:DEPI,HAN,HIN,NSD,**M(folk):**FL:HAP	Krug-134-01
Passiflora	*incarnata*	Maypops	Passion Flower	**M(PS):**WP:HIN,NSD,SIB	Wyk-227-01, Wyk-420-01
Pastinaca	*sativa*	European	Parsnip	**M(traditional):**FR,RT:NMS,SFL,UDR	Wyk-420-02
Pastinaca	*sativa*	European	Parsnip	**E:**LF:DYE(tin-clear-yellow,alum-yellow,chrome-gold,copper-sulfate-tan)	Cas1-189-02
Paullinia	*cupana*	Brazilian Cocoa	Bread & Cheese	**C:**SD:AF,BF,CF,TE,**M(traditional):**SD:DCFS,HAN,OAS	Bown-323-02
Paullinia	*cupana*	Brazilian Cocoa	Bread & Cheese	**C:**SD:BF,**M(PS):**SD:HAP,DCFS,SDR	Wyk-228-01, Wyk-420-03
Paullinia	*yoco*	Yoco	Bread & Cheese	**C:**SM:BF,TE	Bown-323-03
Pausinystalia	*lane-poolii*	Igbepo	Tubeflower Rubyroot	**M(traditional):**BK:DYAW,DSIN	Bown-323-04
Pausinystalia	*macroceras*	Big Horn	Tubeflower Rubyroot	**M(traditional):**BK:HAP	Bown-323-05
Pausinystalia	*tillesii*	Tilles	Tubeflower Rubyroot	**M(traditional):**BK:HAP	Bown-323-06
Pausinystalia	*johimbe*	Yohimbe	Tubeflower Rubyroot	**M(traditional):**BK:HAP	Bal1-079-05
Pausinystalia	*johimbe*	Yohimbe	Tubeflower Rubyroot	**M(folk):**BK:DCFS,DSIN **M(traditional):**BK:DIPT,HAP	Bal2-157-01 Bal2-551-01
Pausinystalia	*johimbe*	Yohimbe	Tubeflower Rubyroot	**M(traditional):**BK:BHS,DIPT,DFRI,HAP,**E:**BK:PS☠	Bown-323-07
Pausinystalia	*johimbe*	Yohimbe	Tubeflower Rubyroot	**M(traditional):**BK:BAA,BAS,BCR,BHP,DIPT,HAP,**E:**BK:PS☠	Kowa-520-01
Pausinystalia	*johimbe*	Yohimbe	Tubeflower Rubyroot	**M(CS):**BK:DFRI,DIPT,HAP,**E:**BK:PS☠	Wyk-229-01, Wyk-420-04
Peganum	*harmala*	Syrian Rue	Pegan Yokeleaf	**C:**SD:**SP**,**M(traditional):**FR,OL:DEPI,HAP,SDG,UDR	Bown-323-08
Peganum	*harmala*	Syrian Rue	Pegan Yokeleaf	**M(PS):**SD:DIPT,DPRK,NMS,NNC👽,NRH,OEY,DSIN,OWN	Wyk-230-01, Wyk-420-05
Peganum	*mexicana*	Garbancillo	Pegan Yokeleaf	**C:**SD:**SP**	USDA-NPDC
Pelargonium	*acetosum*	Sorrel Leaf	Stork's Bill	**C:**LF:**SN**	Bown-324-01
Pelargonium	*antidysentericum*	Dysentery	Stork's Bill	**M(traditional):**WP:SDY	Bown-324-02
Pelargonium	*betulinum*	Birch Leaf	Stork's Bill	**M(traditional):**WP:OLG,RLC,SDG	Bown-323-09
Pelargonium	*capitatum*	Wild Rose	Stork's Bill	**M(traditional):**WP:ODS,SDG,UBL,UKC,**E:**LF,OL:CM,PF	Bown-324-03
Pelargonium	*capitatum*	Wild Rose	Stork's Bill	**E:**LF:PF	Clev-110-03
Pelargonium	*citriodorum*	Lemon Citrus	Stork's Bill	**C:**LF:FF,TE	
Pelargonium	*citriodorum*	Lemon Citrus	Stork's Bill	**E:**LF:PF	Clev-110-06
Pelargonium	*citronellum*	Citron	Stork's Bill	**C:**LF:FF,TE	Bown-324-04
Pelargonium	*chlorinda*	Cedar Rose	Stork's Bill	**E:**LF:PF	Bown-324-05
Pelargonium	*crispum*	Lemon Scented	Stork's Bill	**C:**LF:FF,TE,**E:**LF:PF	Bown-324-06
Pelargonium	*crispum*	Lemon Scented	Stork's Bill	**E:**LF:PF	Clev-110-05
Pelargonium	*denticulatum*	Pine Scented	Stork's Bill	**E:**LF:PF	Kowa-457-02
Pelargonium	*domesticum*	Regal	Stork's Bill	**E:**FL:DYE(baking-soda-green,copper-sulfate-light-green,iron-tan,tin-&-alum-bright-green,chome-&-iron-gray),**E:**LF,FL:DYE(iron-olive-green,alum-&-iron-gray,chrome-khaki-brown,copper-sulate-green-gray)	Cas1-146-01
Pelargonium	*fragrans*	Nutmeg Scented	Stork's Bill	**C:**LF:BF,**M(traditional):**LF:OMC	Bown-324-11
Pelargonium	*fragrans*	Nutmeg Scented	Stork's Bill	**E:**LF:PF	Clev-110-07
Pelargonium	*graveolens*	Rose Scented	Stork's Bill	**C:**LF:FF,TE,**M(traditional):**WP:BCR,DACN,DASC,DTNS,DWRG,HAN,NMS,OBR,DSIN,**E:**OL:CM	Bown-324-14
Pelargonium	*graveolens*	Rose Scented	Stork's Bill	**C:**LF:BF,CF,TE,**E:**LF-OL:IR,PF	Clev-110-02
Pelargonium	*graveolens*	Rose Scented	Stork's Bill	**M(traditional):**OL(attar-of-roses):DAMY,**E:**OL:PF	Wyk-420-06
Pelargonium	*grossularioides*	Gooseberry	Stork's Bill	**E:**LF:PF	Kowa-457-04
Pelargonium	*hortorum*	Fish	Stork's Bill	**E:**FL:DYE(tin-&-tartar-dark-brown,alum-&-tin-&-tartar-green-brown),**E:**LF:DYE(alum-&-iron-purple)	Grae-121-03 Grae-122-01
Pelargonium	*mellissinum*	Lemon Balm	Stork's Bill	**E:**LF:PF	Kowa-457-05
Pelargonium	*nervosum*	Lime Scented	Stork's Bill	**E:**LF:PF	Clev-110-08
Pelargonium	*odoratissimum*	Apple Scented	Stork's Bill	**C:**LF:FF,TE,**M(traditional):**WP:DASC,OAS,OBR,ONR,SIB,SUL,RST,**E:**LF:IR,PF	Bown-324-17
Pelargonium	*odoratissimum*	Apple Scented	Stork's Bill	**E:**LF:PF	Clev-110-09
Pelargonium	*parviflorum*	Coconut Scented	Stork's Bill	**E:**LF:PF	Clev-110-10
Pelargonium	*quercifolium*	Almond Scented	Stork's Bill	**M(traditional):**LF:BHP,BHS,NRH,**E:**LF:IR,PF	Bown-324-20

Latin Genus	L. Species	E. Species	English Genus	C:Culinary,M:Medicinal,E:Economic	Source
Pelargonium	*radens*	Rasp Leaf	Stork's Bill	**M(traditional):**LF:OMC,**E:**LF:IR,PF	Bown-324-21
Pelargonium	*radens*	Rasp Leaf	Stork's Bill	**E:**LF:PF	Clev-110-04
Pelargonium	*radula*	Rose Lemon	Stork's Bill	**C:**LF:FF,**E:**OL:CM,IR(moth),PF	Bown-324-22
Pelargonium	*rapaceum*	Rape	Stork's Bill	**C:**TU:**VG**	Bown-324-24
Pelargonium	*sidoides*	Umckaloabo	Stork's Bill	**M(CS):**RT:DABI,DEAI,DTBC,RLB,RRN,**E:**OL:PF	Wyk-231-01, Wyk-420-07
Pelargonium	*tomentosum*	Peppermint Scented	Stork's Bill	**C:**LF:BF,JJ,TE	Bown-325-04
Pelargonium	*tomentosum*	Peppermint Scented	Stork's Bill	**E:**LF:PF	Clev-110-11
Pelargonium	*triste*	Sad	Stork's Bill	**C:**TU:**VG**	Bown-324-25
Peltigera	*aphthosa*	Aphthose	Pelt Lichen	**E:**WP:DYE(Gyrophoric-Acid-&-Zeorin-&-Ammonia-pink)	Cas2-032-26 Cas2-032-27
Peltigera	*canina*	Scented	Pelt Lichen	**E:**WP:DYE(No-Acid-gold, Ammonia-yellow)	Cas2-035-20
Peltigera	*venosa*	Veined	Pelt Lichen	**E:**WP:DYE(Zeorin-&-Tenuiorin-pink-gray,Zeorin-&-Tenuiorin-&-Ammonia-purple-pink)	Cas2-032-28
Peltogyne	*confertiflora*	Jatobá	Pelt Coral Pea	**M (traditiional):**WP:DCAN	Bal2-486-09
Penstemon	*rostriflorus*	Beaked Red	Beard Tongue	**E:**FL:DYE(alum-&-tartar-tan-brown,alum-&-copper-brown,chrome-&-copper-brown)	Grae-128-02 Grae-128-03
Penstemon	*superbus*	Superb Red	Beard Tongue	**E:**FL:DYE(alum-&-tartar-tan-brown,alum-&-copper-brown,chrome-&-copper-brown)	Grae-128-02 Grae-128-03
Pericallis	*cruenta*	Magenta	Pretty Ragwort	**E:**FL:DYE(alum-turquoise,chrome-green, alum-&-tin-dark-blue,chrome-&-tin-dark-blue, alum-&-tin-&-copper-wine-red,tin-&-copper-violet	Grae-114-03 Grae-115-01 Grae-115-02
Pericallis	*hybrida*	Florist	Pretty Ragwort	**E:**FL:DYE(alum-turquoise,chrome-green, alum-&-tin-dark-blue,chrome-&-tin-dark-blue, alum-&-tin-&-copper-wine-red,tin-&-copper-violet	Grae-114-03 Grae-115-01 Grae-115-02
Perilla	*frutescens*	Purple	Beefsteak Mint	**C:**LF,SD:FF,C:LF,OL:CF,**M(traditional):**SD:DABA,DALL,DCLD,NMS,RLA,RLB,RLC,SCN,**E:**SD,OL:PN	Bown-325-05
Perilla	*frutescens*	Purple	Beefsteak Mint	**C:**OL:CF	Wyk-420-08
Persea	*americana*	American	Avocado	**M(traditional):**LF: **contraception** ☠, **M(PS):**FR-OL:DAMY. **E:**FR-OL:CM,DS	Wyk-232-01, Wyk-420-09
Persea	*borbonia*	Red Bay	Avocado		USDA-NPDC
Persea	*humilis*	Silk Bay	Avocado		USDA-NPDC
Persea	*krugii*	Canela	Avocado		USDA-NPDC
Persea	*lingue*	Purple	Avocado	**E:**WD:WL	Flora-1012-3
Persea	*nubigena*	Guatemalan	Avocado		USDA-NPDC
Persea	*palustris*	Swamp	Avocado		USDA-NPDC
Persea	*schiedeana*	Coyo	Avocado		USDA-NPDC
Persea	*urbaniana*	Aquacatillo	Avocado		USDA-NPDC
Persicaria	*bistorta*			Synonym for *Polygonum bistorta*	Wyk-233-01, Wyk-420-10
Pertusaria	*corallina*	Coral	Warty Perforated Lichen	**E:**WP:DYE(Atranorin-&-Argopsin-&-Pannarin-&-Zeorin-&-Ammonia-red-purple)	Cas2-032-29
Pertusaria	*coriacea*			Synonym for *Umbilicaria coriacea*	Cas2-035-23
Pertusaria	*lactea*	Milky	Warty Perforated Lichen	**E:**WP:DYE(Atranorin-&-Argopsin-&-Pannarin-&-Zeorin-&-Ammonia-pink)	Cas2-033-31
Pertusaria	*maritima*	Seaside	Warty Perforated Lichen	**E:**WP:DYE(Stictid-Acid-&-Constictid-Acid-&-Ammonia-olive)	Cas2-031-12
Pertusaria	*pseudocorallina*	Lesser Coral	Warty Perforated Lichen	**E:**WP:DYE(Norstictic-Acid?-&-Ammonia-orange-pink)	Cas2-032-06
Petasites	*hybridus*	Hybrid	Butterbur	**M(CS):**LF:NSD,SIB,**M(traditional):**RZ:NMS	Wyk-234-01, Wyk-420-11
Petiveria	*alliacea*	Conga Root	Henweed	**M(traditional):**WP:DASC,DFLU,DPNB,DSXS,DWHP,HAC,HFV,NMS,NSD,RLA,RLB,SDR,UBL,UDR	Bown-325-06
Petroselinum	*crispum*	Curly Leaf	Parsley	**M(traditional):**FR,RT,SM:BHP,DCTM,DHAL,DOBE,DWRM,SDG,SFL,UEP,UKC	Bal1-075-04
Petroselinum	*crispum*	Curly Leaf	Parsley	**C:**LF:FF,GN,**M(traditional):**LF,RT:BAN,BHE,DANX,NJA,NMS,NRH,SDG,SIB,UBL,UDR,UEP,UKS,E:OL:PF	Bown-325-07
Petroselinum	*crispum*	Curly Leaf	Parsley	**C:**RT:**SN**,**M(traditional):**LF,RT:BJA,NRH,RLC,UKS,UUR,**E:**SD:PS☠	Clev-110-12
Petroselinum	*crispum*	Curly Leaf	Parsley	**C:**LF:FF,GN,**M(traditional):**LF,SD,SM:BHE,BJA,DAMW,RLA,RLC,SDG,SFL,UDR,UKC	Kowa-291-01 Kowa-407-01
Petroselinum	*crispum*	Curly Leaf	Parsley	**C:**LF:GN,**M(folk):**LF,RT,SD:BLC,NRH,SDG,SFL,UBL,UKC	Krug-131-01
Petroselinum	*crispum*	Curly Leaf	Parsley	**M(PS):**LF,RT:DSIN,SIB,UDR,UKS	Wyk-235-01, Wyk-420-12
Petroselinum	*crispum*	Curly Leaf	Parsley	**E:**LF,SM:DYE(alum-soft-yellow,chrome-bright-	Cas-189-01

Latin Genus	L. Species	E. Species	English Genus	C:Culinary,M:Medicinal,E:Economic	Source
				gold,tin-bright-yellow-green,iron-yellow-gray)	
Petunia	*atkinsiana*	Hybrid	Petunia	**E**:FL:DYE(alum-tan-yellow,copper-sulfate-bronze, chrome-gold,iron-yellow-green,tin-sharp-gold)	Cas1-192-02
Petunia	*atkinsiana-purpurea*	Purple Hybrid	Petunia	**E**:FL:DYE(chrome-&-walnut-leaf-light-khaki-green)	Grae-129-01
Petunia	*hybrida*			Synonym for *Petunia atkinsiana*	
Petunia	*rubra*	Red	Petunia	**E**:FL:DYE(chrome-&-marigold-flower-green-brown)	Grae-129-02
Peucedanum	*ostruthium*	Masterwort	Peuce Fennel	**M(traditional)**:RZ:HFV,NGT,NRH,SDG,UDR	Wyk-420-13
Peumus	*boldus*	Boldo	Peum	**M(traditional)**:BK,LF:DASC,BLC,DWRM,NRH,SGS,UGB,UDR,UUR	Bown-326-01
Peumus	*boldus*	Boldo	Peum	**M(PS)**:LF:DWRM,SDP,SGB,SIB	Wyk-236-01, Wyk-420-14
Pfaffia	*paniculata*	Brazilian Ginseng	Suma Amaranth	**M(traditional)**:BK,FR,LF,RT:BAN,BHP,BLC,DAID,DCFS,DCTM,HAN	Bal1-077-06
Pfaffia	*paniculata*	Brazilian Ginseng	Suma Amaranth	**M(traditional)**:RT:HAP,DCTM	Bown-321-09
Phaseolus	*vulgaris*	Kidney	Bean	**M(traditional)**:PD:NGT,DSIN,UBL,UKC	Wyk-420-15
Phaseolus	*vulgaris-bruscia-chili*	Chili Dry Bush	Bean	**M(traditional)**:PD:NGT,DSIN,UBL,UKC	Wyk-237-01
Phaseolus	*vulgaris-bruscia-navalis*	Navy Dry Bush	Bean	**M(traditional)**:PD:NGT,DSIN,UBL,UKC	Wyk-237-02
Phaseolus	*vulgaris-bruscia-pintoi*	Pinto Dry Bush	Bean	**M(traditional)**:PD:NGT,DSIN,UBL,UKC	Wyk-237-03
Phaseolus	*vulgaris-bruscia-rena*	Kidney Dry Bush	Bean	**M(traditional)**:PD:NGT,DSIN,UBL,UKC	Wyk-237-04
Phaseolus	*vulgaris-bruscia-rubra*	Red Dry Bush	Bean	**E**:SD:DYE(alum-beige-yellow,copper-sulfate-tan,chrome-brown-gray,tin-bright-gold,iron-gray-brown,copper-sulfate-&-iron-gray-green)	Cas1-100-01
Phaseolus	*vulgaris-bruscia-rubra*	Red Dry Bush	Bean	**E**:SD:DYE(alum-&-salt-terra-cotta-brown)	Grae-175-01
Phaseolus	*vulgaris-bruscia-snapa*	Round Snap Bush	Bean	**M(traditional)**:PD:NGT,DSIN,UBL,UKC	Wyk-237-05
Phaseolus	*vulgaris-bruscia-vaxa*	Wax Bush	Bean	**M(traditional)**:PD:NGT,DSIN,UBL,UKC	Wyk-237-06
Phaseolus	*vulgaris-vigna-snapa*	Round Snap Vine	Bean	**M(traditional)**:PD:NGT,DSIN,UBL,UKC	Wyk-237-07
Phellodendron	*amurense*	Amur	Corkbark	**M(traditional)**:BK:BHP,BJA,BLC,BSH,DABA,DBOI,HFV,OSA,SDR,SDY,SGB,SIB,UDR,UKC,UUR	Bown-326-02
Phellodendron	*amurense*	Amur	Corkbark	**M(traditional)**:BK:DABI,DMNB,DMNV,OEY,UUR	Wyk-420-17
Phlomis	*fruticosa*	Bushy	Jerusalem Sage	**C**:LF:TE	Clev-111-01
Phoradendron	*leucarpum*	Oak	Mistletoe	**M(traditional)**:WP:BHS,DCHR,DEPI,NSD,SDG,**E**:WP:PS☠	Kowa-389-02
Phoradendron	*serotinum*			Synonym for *Phoradendron leucarpum*	Kowa-389-02
Phormium	*tenax*	New Zealand	Flax Grass	**E**:FL:DYE(direct-brown),**E**:PD:DYE(alum-&-tin-bright-terra-cotta,alum-&-tin-&-ammonia-mauve)	Grae-150-02 Grae-151-01 Grae-152-01
Phragmites	*australis*	Carrizo	Hedge Reed	**M(traditional)**:RZ:DHAL,HFV,NSD,RLC,OTA,UDR,UUR	Bown-326-03
Phyllanthus	*amarus*	Carry Me Seed	Leaf Flower Spurge	**M(experimental)**:WP(aerial):DAID,DHPT-B,**M(SS)**:WP(aerial):BHP,UKS,**M(CS)**:WP(aerial):BDB	Bal2-047-01 Bal2-511-01
Phyllanthus	*amarus*	Carry Me Seed	Leaf Flower Spurge	**M(traditional)**:LF,FR:DHPT-B	Wyk-420-18
Phyllanthus	*emblica*	Emblic	Leaf Flower Spurge	**M(traditional)**:LF:BDB,SDP,**E**:OL:HO	Wyk-420-19
Phyllospora	*coralline*			Synonym for *Pertusaria corallina*	
Phyllostachys	*nigra*	Black	Timber Bamboo	**M(traditional)**:SA,SM:RLC,**M(traditional)**:LF,RT:HFV,**M(traditional)**:RT:DRAB	Bown-326-04
Phyllostachys	*pubescens*	Downy	Timber Bamboo	**C**:SH:**VG**	Bown-326-05
Physalis	*alkekengi*	Bladder	Ground Cherry	**C**:FR:PU,**M(traditional)**:FR,JU:HFV,NJA,NGT,NRH,UDR,UUR	Bown-326-06
Physalis	*alkekengi*	Bladder	Ground Cherry	**M(traditional)**:FR:UDR,UUR	Wyk-420-19
Physalis	*peruviana*	Gooseberry	Ground Cherry	**C**:FR:**VG**	Bown-326-07
Physcia	*americana*	American	Bladder Rim Lichen	**E**:WP:DYE(Atranorin-&-Triterpenes-yellow)	Cas2-036-01
Physostigma	*venosum*	Ordeal	Calabar Bean	**M(traditional)**:SD:DEGC,DEPI,DTET,NMS,NNC❢,NRH,**E**:SD:PS☠	Bown-327-01
Physostigma	*venosum*	Ordeal	Calabar Bean	**M(traditional)**:FR:DEGC,DTND,NJA	Kowa-280-03
Physostigma	*venosum*	Ordeal	Calabar Bean	**M(traditional)**:SD:DCHR,DEPI,DTET	Wyk-420-20
Phytolacca	*acinosa*	India	Pokeberry	**M(traditional)**:RT:DABA,DAFU,DBOI,RLC,UDR,UKC	Bown-327-02
Phytolacca	*americana*	Inkberry	Pokeberry	**M(traditional)**:FR,RT:DMUM,DTNS,NJA,NMT,OHM,OJA,DSIN,OUL,RLB,**E**:FR:PS☠	Bown-327-03
Phytolacca	*americana*	Inkberry	Pokeberry	**M(traditional)**:RT:NRH,**E**:FR:DYE(alum-pink-red,chrome-rust,tin-red),PS☠	Kowa-175-03 Kowa-417-01
Phytolacca	*americana*	Inkberry	Pokeberry	**M(folk)**:RT:NRH,SPR, **E**:FR:DYE(dark-blue)	Krug-140-02

Latin Genus	L. Species	E. Species	English Genus	C:Culinary,M:Medicinal,E:Economic	Source
Phytolacca	*americana*	Inkberry	Pokeberry	**M(PS):**FR,RT:DHMY,NRH,NSD,DSIN,SPR	Wyk-238-01, Wyk-420-21
Phytolacca	*dodecandra*	Soapberry	Pokeberry	**M(traditional):**FR:DWFK(bilharziasis,schistosomaisis)	Wyk-421-01
Picea	*abies*	Fir	Spruce	**M(traditional):**OL:DSIN,RLC	Wyk-421-02
Picea	*abies*	Fir	Spruce	**E:**LF:DYE(alum-yellow, tin-gold, chrome-bronze), **E:**SM:DYE(alum-bright-yellow,iron-green), **E:**PD (cone):DYE(alum-&-chrome-tan-red,iron-brown), **E:**BK:DYE(alum-tan-yellow, chrome-red-brown,iron-taupe)	Cas1-208-01
Picramnia	*antidesma*	Macary Bitter	Bitterbush	**M(traditional):**BK:DSXS,SDG	Wyk-421-03
Picrasma	*ailanthoides*	Japanese	Bitterwood	**C:**WD:AF,BF,**M(traditional):**WD:DCLK,DMAL,DWRO,DWTH,SAP,SDG,**E:**WD:IC(fly,spidermite,aphid),PS☠	Bown-327-04
Picrasma	*excelsa*	Jamaica Quassia	Bitterwood	**C:**WD:AF,BF,**M(traditional):**WD:DCLK,DMAL,DWRO,DWTH,SAP,SDG,**E:**WD:IC(fly,spidermite,aphid),PS☠	Bown-327-05
Picrasma	*excelsa*	Jamaica Quassia	Bitterwood	**M(folk):**BK:DWRM, **E:**WD:IC(aphids)	Krug-146-02
Picrasma	*excelsa*	Jamaica Quassia	Bitterwood	**M(traditional):**WD:DWRM,SDG,**E:**WD:IC	Wyk-421-04
Picrorhiza	*kurrooa*	Kutki	Picroside Foxglove	**M(traditional):**WP:BLC,SDG	Wyk-421-05
Pilocarpus	*jaborandi*	Pernambuco	Jaborandi	**M(traditional):**LF:BHS	Bown-327-06
Pilocarpus	*jaborandi*	Pernambuco	Jaborandi	**M(CS):**LF:DFLU,DTNS,HFV,ODS,RLB,**E:**LF:HO	Wyk-239-01
Pilocarpus	*microphyllus*	Maranham	Jaborandi	**M(traditional):**LF:DEGC,DSXS,BHS,OHL,OSS, **E:**LF:PS☠	Bown-328-01
Pilocarpus	*pinnatifolius*	Paraguay	Jaborandi	**M(traditional):**LF:BHS	Bown-327-07
Pilocarpus	*pinnatifolius*	Paraguay	Jaborandi	**M(traditional):**LF:DEGC	Wyk-421-06
Pilocarpus	*racemosus*	Acetillo	Jaborandi	**M(traditional):**LF:BHS	
Pilocarpus	*trachylophus*	Ceara	Jaborandi	**M(traditional):**LF:BHS	Bown-327-08
Pimenta	*dioica*	Pimento	Allspice	**C:**LF:TE,**C:**FR:**SP**,**M(traditional):**LF,FR,OL:DASC,NNC👽,NSD,SDG,SDR,SFL,**E:**FR,OL:CL,PF	Bown-328-02
Pimenta	*dioica*	Pimento	Allspice	**C:**FR(berry):**SP**,**M(traditional):**LF,OL:DASC,SFL,**E:**OL:PF	Clev-314-01
Pimenta	*dioica*	Pimento	Allspice	**M(folk):**OL:SDG,SFL	Krug-011-01
Pimenta	*dioica*	Pimento	Allspice	**C:**FR(berry):**SP**,**M(traditional):**FR:DASC,DCLD,SDG,**E:**OL:PF	Wyk-421-07
Pimenta	*racemosus*	Bay Rum	Allspice	**E:**LF:HO	Bown-328-03
Pimpinella	*anisum*	Aniseed	Pimpernel	**M(traditional):**SD-OL:DCLD,DFLU, DSNB, DSNV,HFV,RLB,RLC,RST,SAP,SIB,SDP, **M(PS):**SD-OL:DHAL	Bal2-023-02 Bal2-503-04
Pimpinella	*anisum*	Aniseed	Pimpernel	**C:**SD:**SP**,**C:**SD,OL:BF,CF,FF,**M(traditional):**LF,SD,OL:BCR,BLC,SDG	Bown-328-04
Pimpinella	*anisum*	Aniseed	Pimpernel	**C:**SD:**SP**,**M(traditional):**SD-OL:DASC,RLC	Clev-315-01
Pimpinella	*anisum*	Aniseed	Pimpernel	**C:**SD:**SP**,**M(traditional):**LF,SD,OL:RLC,SDG,SFL,**E:**OL:DS,PF	Kowa-015-01
Pimpinella	*anisum*	Aniseed	Pimpernel	**C:**SD:**SP**,**M(folk):**SD:DABI,DAMW,SDG,SFL	Krug-017-01
Pimpinella	*anisum*	Aniseed	Pimpernel	**M(PS):**FR,OL:DCLD,DFLU,RLB,RLC,SDG,SFL, SIB,**M(PS):**OL:DASC,NMS	Wyk-240-01, Wyk-421-08
Pimpinella	*major*	Greater	Pimpernel	**C:**LF:FF	Bown-328-05
Pimpinella	*major*	Greater	Pimpernel	**M(PS):**RT,RZ:DAMW,DASC,RLB,RLC,RST, **M(PS):**WP:BVV,OWN,SDG	Wyk-241-01, Wyk-421-09
Pimpinella	*saxifraga*	Saxifrage	Pimpernel	**C:**LF:FF,**C:**OL:AF,**M(traditional):**WP:DASC,DGUM,DMSL,NGT,NMS,OWN,RLC,RST,SDG,UBL, UDR,UKS	Bown-328-06
Pinellia	*ternata*	Trefoil	Crow Dipper	**M(traditional):**TU:RLC,SIB	Bown-329-01
Pinguicula	*vulgaris*	Bog Violet	Butterwort	**M(traditional):**LF:RLC,SPR	Wyk-421-10
Pinus	*caribaea*	Caribbean	Pine	**E:**RN:OR(turpentine)	Con-1141
Pinus	*halepensis*	Aleppo	Pine	**M(PS):**BK:BHS,DADD,DALL,DALZ,DCTM, DEMD,DFRA,OWN	Bal2-105-01 Bal2-533-05
Pinus	*halepensis*	Aleppo	Pine	**C:**RN:**SP**	Bown-329-02
Pinus	*maritima*			Synonym for *Pinus halipensis*	
Pinus	*massoniana*	Horsetail	Pine	**M(traditional):**OL:NSC,NRH,RLB	Bown-329-03
Pinus	*mugo*	Mugo	Pine	**M(traditional):**OL:RRN,RST,**E:**RN:OR(turpentine)	Wyk-421-11
Pinus	*mugo-montana*	Mountain Mugo	Pine	**E:**RN:OR(turpentine)	Con-3054
Pinus	*mugo-pumilio*	Dwarf Mugo	Pine	**M(traditional):**OL:DASC,ORH,RLA,RLB,RLC,RRN,**E:**OL:PF	Bown-329-04
Pinus	*mugo-rotundata*	Round Mugo	Pine	**E:**RN:OR(turpentine)	Con-5026
Pinus	*mugo-uncinata*	Hooked Mugo	Pine	**C:**SD:NT	Con-3158
Pinus	*palustris*	Long Needle Pitch	Pine	**M(traditional):**RN:DASC,DBOI,DWRG,RLB,OUL,**M(traditional):**OL,RN:ORH,**E:**RN:OR,SV(turpentine),WX(violinbowrosin)	Bown-329-05
Pinus	*pinaster*	French Martime	Pine	**M(traditional):**BK:DFRA(pycnogenol),**E:**RN:OR(turpentine)	Bown-329-06

Latin Genus	L. Species	E. Species	English Genus	C:Culinary,M:Medicinal,E:Economic	Source
Pinus	*pinaster-escarena*	Escarine	Pine	E:RN:OR(turpentine)	Con-5425
Pinus	*ponderosa*	Ponderosa	Pine	**E:**PD(cone):DYE(alum-warm-tan,chrome-ocher, copper-ocher-,tin-tan-peach,iron-dark-olive)	Bliss-190-01
Pinus	*ponderosa*	Ponderosa	Pine	**E:**LF(green):DYE(alum-beige-yellow), **E:**LF(brown):DYE(vinegar-tan, chrome-gold, chrome-&-iron-brown), **E:**PD(cone):DYE(alum-&-chrome-warm-tan), **E:**BK:DYE(alum-&-chrome-tan-red, alum-&-tin-pink-red)	Cas1-193-01
Pinus	*succinifera*	Amber Resin	Pine	**M(traditional):**RN:BHS,UKS,UUR	Bown-329-07
Pinus	*sylvestris*	Scotch	Pine	**M(traditional):**LF,SH,OL:DACN,DASC,DCFS,DPNB,OJA,ONR,ORH,OSC,RLA,RLB,RLC,SGB, UDR,UUR,**E:**OL:DS,HO	Bown-329-08
Pinus	*sylvestris*	Scotch	Pine	**M(PS):**LF-OL,SR:RLB,RLC,ONR,ORH	Wyk-242-01, Wyk-421-12
Pinus	*sylvestris*	Scotch	Pine	**E:**LF:DYE(alum-&-iron-olive-green)	Grae-171-01
Pinus	*tabuliformis*	Flat Needle	Pine	**M(traditional):**OL:NSC,NRH,RLB	Bown-329-09
Piper	*alba*	White	Peppercorn	**C:**FR:**SP**(dehuskedRed&OrangePeppercorn)	Clev-317-02
Piper	*angustifolia*	Matico Spinach	Peppercorn	**C:**LF:**SN**,**M(traditional):**LF:DABA	Bown-329-10
Piper	*auritum*	Makulan Spinach	Peppercorn	**C:**LF:**SN**,**M(traditional):**LF:DABA	Bown-329-11
Piper	*betle*	Betel	Peppercorn	**M(traditional):**OL:DABA,DDPT,DWRM,OSA	Bown-330-01
Piper	*betle*	Betel	Peppercorn	**M(traditional):**LF:RST,SDG	Wyk-421-13
Piper	*cubeba*	Tailed	Peppercorn	**C:**FR:**SP**,**M(traditional):**FR,OL:DASC,RLB,RLC,RSN,RST,SDG,SDY,UDR,**E:**OL:CL,PF	Bown-330-02
Piper	*cubeba*	Tailed	Peppercorn	**C:**FR:**SP**,**M(traditional):**FR:OTA,SAP	Clev-317-05
Piper	*groenum*	Green	Peppercorn	**C:**FR:**SP**,**M(traditional):**FR:OTA,SAP	Clev-317-03
Piper	*guinense*	Ashanti	Peppercorn	**M(traditional):**LF:DABA	Bown-329-12
Piper	*longum*	Long	Peppercorn	**C:**FR:**SP**,**M(traditional):**FR:HAC,DABI,DCLD,DEPI,NBK,NJA,NRH,NSC,NSD,RLA,RLB,RRN,SAR,SDG,SFL	Bown-330-03
Piper	*longum*	Long	Peppercorn	**C:**FR:**SP**,**M(traditional):**FR:OTA,SAP	Clev-317-04
Piper	*longum*	Long	Peppercorn	**M(traditional):**LF:RLA,SFL,UDR	Wyk-421-14
Piper	*methysticum*	Kava Kava	Peppercorn	**M(traditional):**RT:HAN,HIN,UDR,UUR	Bal1-073-04
Piper	*methysticum*	Kava Kava	Peppercorn	**M(folk):**RT,RZ:DOBE,DSXG,DSXS,HIN,RLA,NRH,SIB,**M(traditional):**RT,RZ:DASC,HAC,NSD,UBL,UEP,UUR,**M(PS):**RT,RZ:HAN,HIN	Bal2-089-02 Bal2-527-06
Piper	*methysticum*	Kava Kava	Peppercorn	**C:**RT:BF,**M(traditional):**RT,RZ:BCR,NJA,NRH, NMS,NSD,UGB,UDR,**E:**RT,RZ:PS☠	Bown-330-04
Piper	*methysticum*	Kava Kava	Peppercorn	**M(CT):**RT,RZ:HAN,NSD	Wyk-243-01, Wyk-421-15
Piper	*nigrum*	Black	Peppercorn	**C:**FR:**SP**,**M(traditional):**FR:DCHL,DEPI,DERS, HFV,RLC,RRN,RSN,SDG,SDR,SDY,SFL,SPN	Bown-330-05
Piper	*nigrum*	Black	Peppercorn	**C:**FR:**SP**,**M(traditional):**FR:OTA,SAP	Clev-317-01
Piper	*nigrum*	Black	Peppercorn	**C:**FR:**SP**,**M(folk):**FR:OSM,SDG,SFL	Krug-136-01
Piper	*nigrum*	Black	Peppercorn	**M(PS):**FR:SAP,SDG,SDP,**M(PS):**OL:OTA,ORH	Wyk-244-01, Wyk-421-16
Piper	*retrofractum*	Java Long	Peppercorn	**C:**FR:**SP**(HSA)	Bown-329-13
Piper	*rubra-aurantia*	Red & Orange	Peppercorn	**C:**FR:**SP**	Clev-317-05
Piptadenia	*peregrina*			Synonym for *Anadenanthera peregrina*	Wyk-421-17
Piscidia	*piscipula*	Jamaican Dogwood	Fishkill Stinkwood	**M(traditional):**BK:DWHP,HAC,HIN,NMP,NMS,NNR,NSD,OTA,RLA,RLC	Bown-330-06
Piscidia	*piscipula*	Jamaican Dogwood	Fishkill Stinkwood	**M(traditional):**BK,RT:HIN,NMS,NSD,RLA,RLC, **E:**BK,RT:PS☠	Wyk-421-18
Pistacia	*integerrima*	Crab's Claw	Pistachio	**M(traditional):**RT:HFV,RLC,SFL	Wyk-421-19
Pistacia	*lentiscus*	Mastic	Pistachio	**C:**RN,OL:BF,CF,**M(traditional):**RN,OL:DASC,DBOI,DWRG,ORH,OUL,RLC,UDR,**E:**RN,OL:OR,PF	Bown-331-01
Pistacia	*lentiscus*	Mastic	Pistachio	**M(traditional):**RN:DAMW,SDG, **E:**RN:dentistry	Wyk-421-20
Pistacia	*terebinthus*	Cyprus Turpentine	Pistachio	**C:**SD:NT,**E:**RN:OR,SV(turpentine)	Bown-331-02
Pisum	*sativum*	Sweet	Pea	**E:**PD:DYE(vinegar-beige,chrome-pale-tan,copper-sulfate-tan),**E:**SM:DYE(alum-yellow,tin-bright-yellow)	Cas1-190-01
Pittosporum	*crassifolium*	Stiff Leaf	Cheesewood	**E:**SD:DYE(alum-dark-blue)	Grae-171-02
Plantago	*afra*	Psyllium Husk	Plantain	**M(traditional):**SD(husk):OHM,SCN,SDR,SUL, UUR, **M(PS):**SD(husk):BDB,BHS,SCN,SIB,	Bal2-117-01 Bal2-536-02
Plantago	*afra*	Psyllium Husk	Plantain	**M(traditional):**SD(husk):SAR,SCC,SCN,SDR	Bown-331-07 Clev-111-07
Plantago	*afra*	Psyllium Husk	Plantain	**M(PS):**SD(husk):SAR,SCC,SCN,SDR	Wyk-245-01, Wyk-422-01
Plantago	*arenaria*	Dryland India Psyllium Husk	Plantain	**M(traditional):**SD(husk):SCN	Bown-331-03 Clev-111-08
Plantago	*asiatica*	Japanese	Plantain	**M(traditional):**WP:DBOI,DHPT,RLC,UDR,**M(tra**	Bown-331-04

Latin Genus	L. Species	E. Species	English Genus	C:Culinary,M:Medicinal,E:Economic	Source
		Psyllium Husk		**ditional):**SD:DVGT,OEY,RLC,SDR,UUR	
Plantago	*asiatica*	Japanese Psyllium Husk	Plantain	**M(traditional):**SD(husk):SCN	Wyk-422-02
Plantago	*coronopus*	Crow Foot	Plantain	**C:**LF:**VG**	Clev-111-03
Plantago	*indica*			Synonym for *Plantago arenaria*	Clev-111-08
Plantago	*lanceolata*	Ribwort	Plantain	**M(traditional):**LF:OWN,RLC	Clev-111-04
Plantago	*lanceolata*	Ribwort	Plantain	**M(PS):**LF:DAMW,DSIN,OWN,RLC,RST	Wyk-246-01, Wyk-422-03
Plantago	*lanceolata*	Ribwort	Plantain	**E:**LF,FL,SM:alum-yellow-green,chrome-chartreuse,copper-pea-soup-green,tin-bright-gold,iron-dark-forest-green)	Bliss-188-01
Plantago	*lanceolata*	Ribwort	Plantain	**E:**WP:DYE(alum-dull-gold,alum-vat-brown,chrome-dark-camel)	Grae-194-01 Grae-194-02
Plantago	*major*	Greater	Plantain	**M(traditional):**LF:DABI,DSNB,OST,OWN	Bal1-075-08
Plantago	*major*	Greater	Plantain	**M(traditional):**LF:DABA,DEAI,DHYF,DISB,DSHZ,OAS,OEY,OHM,OUL,OWN,RLA,RLB,RLC,RSN,SDR,SUL,UBL,UDR	Bown-331-05
Plantago	*major*	Greater	Plantain	**C:**LF,**VG**,**C:**SD:BF,**M(traditional):**LF:NSD,OWN,SCN,SDR,**M(traditional):**LF,RT:DAMW,OEY	Clev-111-02
Plantago	*major*	Greater	Plantain	**C:**LF,**VG**,**M(traditional):**LF:DISB,DIOS,OST,**M(traditional):**RT:OTA,**E:**WP:CM,DYE(alum-dull-gold,chrome-camel)	Kowa-175-02 Kowa-416-01
Plantago	*major*	Greater	Plantain	**M(folk):**LF:DIOS,DISB,DSIN,DSNB,HFV,OHM,OST,OUL,UKC,OWN	Krug-139-01
Plantago	*major*	Greater	Plantain	**M(traditional):**LF:DCLD,DFLU,OHM,UBL,UDR	Wyk-246-02, Wyk-422-04
Plantago	*major*	Greater	Plantain	**E:**LF:DYE(alum-soft-green,copper-sulfate-gold-bronze,chrome-yellow-green,iron-gray,tin-gray-green)	Cas1-194-01
Plantago	*media*	Hoary	Plantain	**E:**FL:PF	Clev-111-05
Plantago	*ovata*	Ispaghula Psyllium Husk	Plantain	**M(traditional):**SD(husk):SCN	Bown-331-06 Clev-111-06
Plantago	*ovata*	Ispaghula Psyllium Husk	Plantain	**M(traditional):**SD(husk):SCN	Wyk-422-05
Plantago	*psyllium*			Synonym for *Plantago afra*	Clev-111-07
Platismatia	*Glauca*	Glaucous	Rag Heath Shield Lichen	**E:**WP:DYE(Caperatic-Acid-&-Atranorin-apricot-yellow)	Cas2-036-02
Platycladus	*orientalis*			Synonym for *Thuja orientalis*	Bown-362-02
Platycodon	*grandiflorus*	Chinese	Balloon Bellflower	**C:**RT:**VG**,**M(traditional):**RT:DASC,DCLD,RLB,RLC,RLP,RST	Bown-331-08
Platycodon	*grandiflorus*	Chinese	Balloon Bellflower	**M(PS):**RT:BHP,DAVR,DTNS,RLA,RLB,RLC,RST,SUL	Wyk-247-01, Wyk-422-06
Plectranthus	*amboinicus*	India Borage	Spurflower Mint	**C:**LF:FF,TE,**M(traditional):**LF:DISB(scorpion),HAC,OBN,OBR,RLA,RLB,**E:**LF:HO	Bown-331-09
Plectranthus	*barbatus*	Bearded	Spurflower Mint	**M(traditional):**WP:BHS,DEGC,RLB,**E:**LF:CL	Bown-332-01
Plumbago	*europaea*	European	Leadwort	**M(traditional):**WP,RT:NSD,SDG	Wyk-422-07
Plumbago	*indica*	Whorled	Leadwort	**M(traditional): (traditional)**RT:OHM	Wyk-422-08
Podophyllum	*emodi*	Himalayan		Synonym for *Podophyllum hexandrum*	Bown-332-02
Podophyllum	*hexandrum*	Himalayan	May Apple	**M(traditional):**RN,RZ:SPR	Bown-332-02
Podophyllum	*hexandrum*	Himalayan	May Apple	**M(traditional):**RZ:DCTM,DCLW,SPR	Wyk-422-09
Podophyllum	*peltatum*	American Mandrake	May Apple	**M(traditional):**RN,RZ:DAVR,DCST,DCLW,SPR,**E:**IC,PS☠	Bown-332-03
Podophyllum	*peltatum*	American Mandrake	May Apple	**M(traditional):**RT:SCN,SPR,**E:**PS☠,**inducedabortion**☠	Kowa-282-02 Kowa-380-01
Podophyllum	*peltatum*	American Mandrake	May Apple	**M(PS):**RN,RZ:DCTM,DCLW,SPR	Wyk-248-01, Wyk-422-10
Pogostemon	*cablin*	Patchouli	Beard Mint	**M(traditional):**LF,OL:DAFU,DASC,DCLD,DHAL,DIMP,DSNB,HAC,HAP,HFV,NSD,OAS,DSIN,SDG,SDR,UDR	Bown-332-04
Pogostemon	*cablin*	Patchouli	Beard Mint	**M(PS):**LF,OL:DCLD,DFLU,HAC,HAP,HFV,DSIN,**M(traditional):**OL:HAN,NMP,**E:**OL:CM,DS,IR	Wyk-249-01, Wyk-422-11
Pogostemon	*comosus*	Tufted	Beard Mint	**M(traditional):**LF,OL:DASC,OAS,UDR	Bown-332-05
Pogostemon	*heyneanus*	Java Patchouli	Beard Mint	**M(traditional):**LF,OL:DASC,OAS,UDR	Bown-332-06
Polemonium	*caeruleum*	Charity	Jacob's Ladder	**M(traditional):**WP:DEPI,HAC,HFV,OAS	Bown-332-09
Polemonium	*caeruleum*	Charity	Jacob's Ladder	**M(traditional):**RT:RLC	Wyk-411-12
Polemonium	*reptans*	Creeping	Jacob's Ladder	**M(traditional):**RZ:DCLD,DTBC,OAS,RLB,RLC,RTL	Bown-332-10
Polygala	*amarella*	Dwarf	Milkwort	**M(traditional):**RT:RLC	Bown-332-11
Polygala	*senega*	Senega Root	Milkwort	**M(traditional):**RT:DALC(hangover),DEPI	Bal2-048-02 Bal2-511-04
Polygala	*senega*	Senega Root	Milkwort	**M(traditional):**RT:RCR,RLA,RLB,RLC	Bown-333-01

Latin Genus	L. Species	E. Species	English Genus	C:Culinary,M:Medicinal,E:Economic	Source
Polygala	*senega*	Senega Root	Milkwort	**M(PS):**RT:DSNB,RLA,RLB,RLC,RLE,RST	Wyk-250-01, Wyk-411-13
Polygala	*tenuifolia*	Siberian	Milkwort	**M(traditional):**RT:BHP,DABA,DBOI,HAN,HIN,RLB,RLC,NNC☠	Bown-333-02
Polygala	*vulgaris*	European	Milkwort	**M(traditional):**RT:RLC	Bown-332-12
Polygonatum	*multiflorum*	Dropberry	Solomon's Seal	**M(traditional):**RZ:BHS,DTBC,HAP,OAS,OBR,OHM,RLC,SIB	Bown-333-03
Polygonatum	*multiflorum*	Dropberry	Solomon's Seal	**M(folk):**RT:DSIN,OAS,OBR,**E:**RT:CM	Krug-163-01
Polygonatum	*multiflorum*	Dropberry	Solomon's Seal	**M(traditional):**RT:DABA,BLC,DIFR,OWN,RLC,UKC	Wyk-251-02, Wyk-422-14
Polygonatum	*odoratum*	Angular	Solomon's Seal	**M(traditional):**RZ:BHS,DTBC,HAP,OAS,OBR,OHM,RLC,SIB	Bown-333-04
Polygonum	*aviculare*	Mat Grass	Knotweed	**M(PS):**WP:DAMW,OAS,DSIN,RLB,RLC,RST	Wyk-251-01, Wyk-422-15
Polygonum	*aviculare*	Mat Grass	Knotweed	**E:**WP:DYE(alum-creamy-yellow,alum-&-ammonia-yellow,alum-&-copper-moss-green,chrome-brass-yellow,chrome-&-copper-brass-moss-green)	Grae-093-02 Grae-094-01
Polygonum	*aviculare*	Mat Grass	Knotweed	**E:**WP:DYE(alum-green-gold,chrome-tan,copper-olive,tin-apricot,iron-dark-olive-green)	Bliss-164-01
Polygonum	*bistorta*	English Serpentary	Knotweed	**M(traditional):**RZ:DCHL,DGUM,DVGB,DVGV,OAS,OHM,OUL,OWN,RST,SDR,SDY,SIB,UBL	Bown-333-05
Polygonum	*bistorta*	English Serpentary	Knotweed	**M(PS):**RZ:DAMW,OAS,DSIN,RST,SDR,SDY	Wyk-233-01, Wyk-420-10
Polygonum	*californica*	Lady's Thumb	Knotweed	**E:**LF,FL,SM:DYE(alum-green-gold,chrome-cinnamon,coppertan-green,tin-bright-gold,iron-very-dark-olive)	Bliss-148-01
Polygonum	*cuspidatum*	Japanese	Knotweed	**E:**LF,FL,SM:DYE(alum-yellow-camel,chrome-gold,copper-dark-camel,tin-bright-gold,iron-dark-gray-green)	Bliss-138-01
Polygonum	*hydropiper*	Red Knees	Knotweed	**M(traditional):**RZ:OAS	Bown-333-06
Polygonum	*hydropiper*	Red Knees	Knotweed	**M(traditional):**RZ:OAS,NRH	Wyk-422-16
Polygonum	*multiflorum*	He Shou Wu Fo T i	Knotweed	**M(traditiona):**RT:BHC,BLC,UKC	Bal2-084-01 Bal2-526-02
Polygonum	*multiflorum*	He Shou Wu Fo Ti	Knotweed	**M(traditional):**RT:BHC,BLC,BSH,DABA,OAS,OWN,SCN,UKC,**M(traditional):**SM:HIN,NSD,**M(traditional):**RT,SM:DWRG	Bown-333-07
Polygonum	*multiflorum*	He Shou Wu Fo Ti	Knotweed	**M(traditional):**RT:BLC,NSD,UKC	Wyk-422-17
Polygonum	*odoratum*	Vietnamese Coriander	Knotweed	**C:**LF:FF	Bown-333-08
Polypodium	*glycyrrhiza*	Licorice	Polypody Fern	**M(traditional):**RZ:DMSL,RLC	Bown-333-09
Polypodium	*virginianum*	Rock	Polypody Fern	**M(traditional):**RZ:DWTP,NJA,OLG,DSIN,OWN,RLB,RLC,SAP,SDG,UDR	Bown-333-10
Polypodium	*virginianum*	Rock	Polypody Fern	**C:**RZ:CF(sugar),**M(traditional):**RZ:DWRM,RLC,SCN,SGB	Wyk-422-18
Polypodium	*vulgare*			Synonym for *Polypodium virginianum*	Bown-333-10 Wyk-422-18
Populus	*alba*	White	Poplar	**M(traditional):**BK:BCB,BLC,DANX,HFV,OAS,OHM,OWN,NBK,NDP,NGT,NJA,NSD,SDG,UDR,UUR	Bown-334-01
Populus	*balsamifera*	Balm of Gilead	Poplar	**M(traditional):**LF-BD:RLC	Bown-334-02
Populus	*balsamifera*	Balm of Gilead	Poplar	**M(traditional):**BD:BHE,DSXG,HAC,HFV,NJA,NRH,NSD,OBN,OTA,RLC,RST,UKC	Kowa-418-01
Populus	*balsamifera*	Balm of Gilead	Poplar	**M(traditional):**BD:DFMS,OHM,OSB,OWN	Wyk-422-19
Populus	*deltoides*	Plains Cottonwood	Poplar	**E:**LF,SM:DYE(alum-bright-yellow,chrome-dark-brown,copper-ocher,tin-bright-orange-gold,iron-dark-olive)	Bliss-092-01
Populus	*candicans*	Ontario	Poplar	**M(traditional):**LF-BD:BCR,HFV,OAS,OJA,ORH,RLB,RLC,RSN	Bown-334-03
Populus	*grandidentata*	Big Tooth Aspen	Poplar	**E:**LF:DYE(alum-clear-yellow,tion-bright-yellow-orange,copper-sulfate-beige,chrome-gold,),**E:**SM:DYE(copper-sulfate-&-iron-gray,)	Cas1-094-04
Populus	*nigra*	Black	Poplar	**M(traditional):**BK:BCB,BLC,DANX,HFV,OAS,OHM,OWN,NBK,NDP,NGT,NJA,NSD,SDG,UDR,UUR	Bown-334-04
Populus	*nigra-italica*	Italian Black	Poplar	**E:**LF:DYE(alum-yellow,tin-bright-yellow-orange,chrome-bright-gold,tin-&-chrome-rust,chrome-&-iron-brown,copper-sulfate-brown)	Cas1-196-01
Populus	*tremula*	Quaking Aspen	Poplar	**M(traditional):**LF,BK:DCLD,NRH,NSD,SDR,UBL	Wyk-423-01
Populus	*tremuloides*	Trembling Aspen	Poplar	**M(traditional):**BK:BCB,BLC,DANX,HFV,OAS,OHM,OWN,NBK,NDP,NGT,NJA,NSD,SDG,UDR,UUR	Bown-334-05
Populus	*tremuloides*	Trembling Aspen	Poplar	**M(PS):**BK:DCLD,NJA,NRH,SDR,UBL,**M(traditional):**BK,LF:UEP,**M(PS):**LF-	Wyk-252-01, Wyk-423-02

Latin Genus	L. Species	E. Species	English Genus	C:Culinary,M:Medicinal,E:Economic	Source
				BD:NRH,OHM,OSB,OWN,RLB	
Populus	*tremuloides*	Trembling Aspen	Poplar	**E:**LF,SM:DYE(alum-gold-yellow,chrome-chestnut-brown,copper-olive-green,tin-vivid-gold,iron-dark-olive-green)	Bliss-040-01
Populus	*tremuloides*	Trembling Aspen	Poplar	**E:**LF:DYE(alum-clear-yellow,tion-bright-yellow-orange,copper-sulfate-beige,chrome-gold,),**E:**SM:DYE(copper-sulfate-&-iron-gray,)	Cas1-094-03
Poria	*cocos*			Synonym for *Wolfiporia extensa*	
Porteranthus	*trifoliatus*	Bowman's	Physic Root	**M(traditional):**RB:RLA,RLB,RLC,SCN,SDR,SPR	Bown-288-05
Portulaca	*grandiflora*	Moss Rose	Purslane	**M(traditional):**JU:DHPT,DISB,DSNB,OBN,DSIN	Bown-334-06
Portulaca	*oleracea*	Little Hogweed	Purslane	**C:**LF:**VG**,**M(traditional):**WP:DABA,DBOI,DSNB,HFV,NMT,OHM,DSIN,OST,SAD,SDY,SIB,UDR	Bown-334-07
Portulaca	*oleracea*	Little Hogweed	Purslane	**C:**LF:**VG**,**M(traditional):**LF:HAC,HFV,OCC,DSIN	Clev-112-01
Portulaca	*oleracea*	Little Hogweed	Purslane	**C:**LF:**VG**,**M(folk):**LF:HFV,NGT,UDR	Krug-145-01
Portulaca	*oleracea*	Little Hogweed	Purslane	**M(traditional):**WP:DABI,DWRM,SDG,UDR	Wyk-423-03
Portulaca	*oleracea*	Little Hogweed	Purslane	**E:**LF,FL,SM:DYE(alum-gray,chrome-tan,copper-light-gray-tan,tin-dull-gold,iron-mouse-gray)	Bliss-198-01
Portulaca	*oleracea*	Little Hogweed	Purslane	**E:**WP:DYE(alum-yellow,vinegar-beige,copper-sulfate-tan,alum-&-tin-yellow-green,chrome-gold-brown)	Cas1-198-02
Potentilla	*anserina*	Silverweed	Cinquefoil	**M(traditional):**LF:OAS,**M(traditional):**LF:CM	Kowa-420-02
Potentilla	*anserina*	Silverweed	Cinquefoil	**M(PS):**LF:DAMW,NMS,OAS,OHM,DSIN,RST,SDR,SIB	Wyk-253-01,Wyk-423-04
Potentilla	*aurea*	Golden	Cinquefoil	**E:**FL:DYE(alum-soft-yellow)	Cas1-122-01
Potentilla	*canadensis*	Dwarf	Cinquefoil	**M(folk):**LF,RT:HAP,OAS	Krug-054-01
Potentilla	*erecta*	Bloodroot	Cinquefoil	**M(traditional):**RT:DCRO,DSHZ,DVGB,DVGV,OAS,OBN,OHM,OSB,OUL,RST,SDR,SDV,SIB,SUL	Bown-334-08
Potentilla	*erecta*	Bloodroot	Cinquefoil	**M(traditional):**RT,RZ:DSXG,RST,**M(traditional):**LF:OAS,E:RT:DYE(chrome-red-brown,iron-purple-red)	Kowa-420-01
Potentilla	*erecta*	Bloodroot	Cinquefoil	**M(PS):**RZ:DAMW,OAS,DSIN,OWN,RST,SDR,SDY,SIB	Wyk-254-01,Wyk-423-05
Potentilla	*erecta*	Bloodroot	Cinquefoil	**E:**LF,FL,SM:DYE(alum-light-yellow,chrome-camel-green,copper-olive-green,tin-bright-yellow,iron-black-green)	Bliss-084-01
Potentilla	*neumannii*	Spring	Cinquefoil	**E:**RT:DYE(chrome-red-brown,iron-purple-red)	Kowa-175-04
Potentilla	*reptans*	Creeping	Cinquefoil	**M(folk):**LF,RT:HAP,OAS	Krug-054-02
Potentilla	*verna*			Synonym for *Potentilla neumannia*	Kowa-175-04
Poterium	*sanguisorba*			Synonym for *Sanguisorba minor*	Clev-112-02
Prenanthes	*alba*	White	Rattlesnake Root	**M(traditional):**WP:DSNB	Wyk-418-10b
Prenanthes	*serpentaria*	Cankerweed	Rattlesnake Root	**M(traditional):**WP:DSNB	Wyk-418-10a
Primula	*atrorubra*	Dark Red	Primrose	**E:**FL:DYE(alum-green-yellow,alum-&-iron-bright-avacado-green)	Grae-130-02
Primula	*eliator*	Ox Lip	Primrose	**M(traditional):**RT,RZ:RLB,RLC,**M(traditional):**FL:HAC,NSD	Wyk-423-06
Primula	*veris*	Cow's Lip	Primrose	**C:**FL:AF,CF,FF,TE,**M(traditional):**FL,RT:DWHP,HAC,HIN,NJA,NMS,NSD,OJA,ONR,OSB,RLB,RLC	Bown-335-01
Primula	*veris*	Cow's Lip	Primrose	**C:**FL:AF,**M(folk):**FL,RT,RZ:HAC,HIN,RLB,RLC	Krug-060-01
Primula	*veris*	Cow's Lip	Primrose	**M(PS):**FL,RT,RZ:DCLD,RLB,RLC,RRN,RST	Wyk-255-01
Primula	*vulgaris*	English	Primrose	**M(traditional):**SD,OL:BHP,DMSC,DOBE,NJA,NMS,DSIN	Bal1-075-09
Primula	*vulgaris*	English	Primrose	**C:**LF,FL:FF,**M(traditional):**WP:HAN,HIN,NGT,NMS,NRH,NSD,OJA,OWN,RLB,RLC,	Bown-335-02
Primula	*vulgaris*	English	Primrose	**E:**FL:DYE(alum-pale-yellow,copper-sulfate-bronze,chrome-warm-tan,tin-bright-yellow)	Cas1-197-02
Proboscidea	*louisianica*	Louisiana	Unicorn	**M(traditional):**RT:DPMS	Bal2-440-14
Prostanthera	*cineolifera*	Cineole	Mintbush	**M(traditional):**LF:DCLD,HAC	Bown-335-03
Prostanthera	*rotundifolia*	Round Leaf	Mintbush	**M(traditional):**LF:DABA,DAFU,DCLD,HAC	Bown-335-04
Prunella	*vulgaris*	Heal All	Selfheal	**M(traditional):**WP(aerial),FL:BLC,DAVR,OAS,OBR,OEY,OWN,SDR,**M(SS):**WP(aerial),FL:DAID,HSV	Bal2-116-01 Bal2-535-03
Prunella	*vulgaris*	Heal All	Selfheal	**C:**WP:FF,**M(traditional):**FL:BHP,BLC,DABA,DHEA,DMUM,HFV,NMT,OAS,OEY,SGB,UDR,**M(traditional):**WP:OBN,OBR,OHM,RST	Bown-335-05
Prunella	*vulgaris*	Heal All	Selfheal	**M(traditional):**WP:DAMW,OWN,RST,SUL,**M(traditional):**SD:NSD	Clev-112-04
Prunella	*vulgaris*	Heal All	Selfheal	**M(folk):**WP(flowering):DAMW,DASC,OAS,OWN,RST	Krug-157-01
Prunella	*vulgaris*	Heal All	Selfheal	**M(traditional):**WP:BCR,OAS,OWN	Wyk-423-07

Latin Genus	L. Species	E. Species	English Genus	C:Culinary,M:Medicinal,E:Economic	Source
Prunella	*vulgaris*	Heal All	Selfheal	**E:**FL,SM:DYE(alum-bright-olive-green)	Grae-108-02
Prunella	*vulgaris*	Heal All	Selfheal	**E:**WP:DYE(alum-soft-yellow,copper-sulfate-brown,chrome-orange-gold,iron-gray,tin-brilliant-gold)	Cas1-150-01
Prunus-Cerasus	*africana*	African	Sour Cherry	**M(folk):**BK:DIPT,OHL,SDG,UKC,**M(traditional):** BK:DMAL,HFV,**M(CS):**BK:UBL,UEP	Bal2-118-01 Bal2-537-01
Prunus-Cerasus	*africana*	African	Sour Cherry	**M(traditional):**SM-BK:UEP	Bown-335-07
Prunus-Cerasus	*africana*	African	Sour Cherry	**M(CT):**SM-BK:UEP	Wyk-256-01, Wyk-423-08
Prunus-Armeniaca	*armeniaca*	Wild	Apricot	**C:**FR:AF,JJ,JU,**M(traditional):**FR,OL,SD:DCST(laetrile),RLA,RLB,RLC,RLE,SCN,**E:**OL:CM,**E:**OL,SD:PS☠	Bown-336-01
Prunus-Armeniaca	*armeniaca*	Wild	Apricot	**M(traditional):**SD:RLC	Wyk-423-09
Prunus-Avium	*avium*	Wild	Sweet Cherry	**M(traditional):**FR:BAN	Bal1-129-03
Prunus-Avium	*avium*	Wild	Sweet Cherry	**M(traditional):**FR:OAS,SDR,UBL,UDR	Bown-335-06
Prunus-Avium	*avium*	Wild	Sweet Cherry	**M(traditional):**FR-SL:OAS,UDR	Wyk-423-10
Prunus-Avium	*avium-nigra*	Black	Sweet Cherry	**C:**FR:CF	Bown-335-06
Prunus-Cerasus	*x cerasus-maurus*	Morello	Sour Cherry	**M(traditional):**LF:OAS	Wyk-423-11
Prunus	*domestica*	Wild European	Plum	**C:**FR:CF,FF,**M(traditional):**FR(dried,prune):SCN	Bown-336-02
Prunus	*domestica*	Wild European	Plum	**E:**BK:DYE(alum-&-chrome-red-brown,alum-&-iron-taupe)	Cas1-195-01
Prunus	*domestica nana*	Dwarf Wild European	Plum	**E:**LF:DYE(alum-yellow,tin-bright-yellow-orange,,alum-&-chrome-rust,chrome-gold)	Cas1-195-02
Prunus	*domestica-frea-groena*	Green Freestone	Plum	**E:**LF:DYE(alum-&-copper-light-green,chrome-&-copper-yellow-green,chrome-&-iron-yellow-green)	Grae-182-03 Grae-183-01
Prunus	*domestica-frea-rubra*	Red Freestone	Plum	**E:**FR:DYE(chrome-&-ammonia-dark-olive-green), **E:**FR:DYE(alum-&-iron-green-gray) **E:**FR:DYE(chrome-&-tin-violet) **E:**LF:DYE(alum-&-tartar-olive-green, alum-&-tin-violet-purple, alum-&-tin-lavender)	Grae-183-02 Grae-183-03 Grae-184-01 Grae-184-02 Grae-184-03
Prunus-Amygdalus	*dulcis*	Sweet	Almond	**C:**SD:NT,**C:**OL:CF,**M(traditional):**OL,SD:SCN,SGS,UKS	Bown-336-02
Prunus-Amygdalus	*dulcis*	Sweet	Almond	**M(folk):**NT,OL:DSIN	Krug-012-01
Prunus-Amygdalus	*dulcis amara*	Bitter Sweet	Almond	**M(traditional):**SD:DCTM(laetrile),E:SD:PS☠	Wyk-423-12
Prunus-Cerasus	*japonica*	Oriental Bush	Sour Cherry	**M(traditional):**SD:BHP,HIN,SCN,UDR	Bown-336-04
Prunus-Cerasus	*laurocerasus*	Laurel	Sour Cherry	**C:**OL:FF,**M(traditional):**LF,OL:NMS,NSD,OEY,RLC,SDG,**E:**LF,OL:☠	Bown-336-05
Prunus-Cerasus	*laurocerasus*	Laurel	Sour Cherry	**M(traditional):**LF:RLC(cherry laurel water)	Wyk-423-13
Prunus-Mahaleb	*mahaleb*	Middle East	Mahaleb Cherry	**C:**SD:**SP**	Clev-322-01
Prunus-Armeniaca	*mume*	Mume	Apricot	**C:**FR:AF,**M(traditional):**FR(unripe):DAFU,DWRO,OAS,DCLW,RLC,SDR	Bown-336-06
Prunus-Armeniaca	*mume*	Mume	Apricot	**M(traditional):**FR:OAS,RLC,SDR	Wyk-423-14
Prunus-Persica	*persica*	Wild	Peach	**C:**FR:CF,**M(traditional):**BK,LF:DWHP,RLB,RLC,SIB,**M(traditional):**LF:DBOI,DMAL,OHM,DSIN,**M(traditional):**FL:BHE,SCN,**M(traditional):**SD:RLA,RLC,SCN,E:OL:CM	Bown-336-07
Prunus-Avium	*serotina*	Rum Black	Sweet Cherry	**M(traditional):**BK:DABA,DAVR,DWHP,NSD,OAS,NDP,NSD,RLB,RLC,SDG,SDP,SDR,SIB	Bown-336-08
Prunus-Avium	*serotina*	Rum Black	Sweet Cherry	**M(traditional):**BK:RLC(wild cherry cough drops)	Wyk-423-15
Prunus-Cerasus	*spinosa*	Sloe Blackthorn	Sour Cherry	**C:**FR:AF(gin),**M(traditional):**FL,FR:OAS	Bown-335-08
Prunus-Cerasus	*spinosa*	Sloe Blackthorn	Sour Cherry	**M(PS):**FL,FR:DAMW,DGUM,OAS,RST	Wyk-257-01, Wyk-423-16
Prunus-Chocacerasus	*virginiana melanocarpa*	Black Virginia	Chokecherry	**E:**LF,SM:DYE(alum-gold,chrome-chestnut-brown, copper-dark-gold,tin-pumpkin,iron-olive-green), **E:**FR,JU:alum-tan-pink,chrome-olive-green, copper-tan,tin-light-olive,iron-gray-green)	Bliss-082-01
Prunus-Chocacerasus	*virginiana melanocarpa*	Black Virginia	Chokecherry	**E:**LF:DYE(alum-yellow-green,copper-sulfate-beige,chrome-gold,iron-gray,tin-light-gold), **E:**FR:DYE(vinegar-&-salt-gray-tan,alum-&-tin-taupe-pink),**E:**BK:DYE(iron-gray,alum-purple-gray)	Cas1-118-01
Pseudevernia	*consocians*	Social	Least Antler Shield Lichen	**E:**WP:DYE(Lecanoric-Acid-&-Atranorin-orange, Lecanoric-Acid-&-Atranorin-&-Ammonia-pink)	Cas2-032-31 Cas2-036-03
Pseudevernia	*furfuracea*	Mealy	Least Antler Shield Lichen	**E:**WP:DYE(Lecanoric-Acid-&-Atranorin-vivid-orange)	Cas2-036-04
Pseudocyphellaria	*anomala*	Anomalous	Speckle Belly Lobed Pelt Lichen	**E:**WP:DYE(Stictic-Acid-&-Calycin-orange-brown)	Cas2-036-05
Pseudognaphalium	*affine*	Affinity	Mock Cotton Cudweed	**M(traditional):**WP:RLC	Bown-289-09
Pseudognaphalium	*obtusifolium*	Rabbit Tobacco	Mock Cotton Cudweed	**M(traditional):**WP:UDR	Wyk-423-17

Latin Genus	L. Species	E. Species	English Genus	C:Culinary,M:Medicinal,E:Economic	Source
Psidium	*guajava*	Arawak	Guava	**M(PS):**LF:BDB,DBOI,DMAL,HFV,OUL,OWN,RLC,SDR	Wyk-258-01, Wyk-423-18
Psilocybe	*mexicana*	Mexican	Smoothcap Ringstalk Gill Mushroom	**M(traditional):**WP:NNC	Wyk-423-19
Psophocarpus	*tetragonolobus*	Four Angled	Winged Bean	**M(traditional):**SD:BDB,SCN,UDR	Wyk-423-20
Psoralea	*coryifolia*	Babchi	Scurf Pea	**M(traditional):**SD:DSIN,OHL,OSS,**M(SS):**SD:DABA(*Staphylococcus aureus*),DVTL	Bal2-116-02 Bal2-536-01
Psoralea	*coryifolia*	Babchi	Scurf Pea	**M(traditional):**SD:DABA,DIPT,OAS,OHL,RLA,SDR,UDR,UKC,UUR	Bown-337-01
Psoralea	*coryifolia*	Babchi	Scurf Pea	**M(traditional):**SD:DVTG,ODS,DSIN	Wyk-423-21
Psoralea	*esculenta*	Scurf Potato	Scurf Pea	**C:**TU:**VG**	Bown-336-09
Psoralea	*tenuifolia*	Scrufy	Scurf Pea	**E:**LF,FL,SM:DYE(alum-gold,chrome-rust-gold,copper-green-gold,tin-light-yellow,iron-moss-green)	Bliss-218-01
Psychotria	*ipecacuanha*			Synonym for *Carapichea ipecacuanha*	Wyk-423-22
Ptelea	*trifoliata*	Wafer Ash	Hop Ruewood	**M(traditional):**RB:DABA,DWPN,DWRO,HFV,OWN,SDG	Bown-337-02
Ptelea	*trifoliata*	Wafer Ash	Hop Ruewood	**C:**FR:AF,**M(traditional):**LF:DWRM	Wyk-424-01
Pteridium	*aquilinum*	Aquline	Bracken Fern	**E:**LF:DYE(alum-beige-yellow,chrome-brass,copper-green-brass,tin-yellow,iron-gray)	Bliss-066-01
Pterocarpus	*dalbergoides*	Andaman Padauk	Barwood	**E:**WD:DYE(alum-salmon-pink, alum-&-copper-deep-terra-cotta-red)	Grae-169-02 Grae-169-03
Pterocarpus	*marsupium*	Malabar Kino	Barwood	**M(traditional):**SA:DVGB,DVGV,BDB,OAS,RST,SDR,SDY	Bown-337-03
Pterocarpus	*marsupium*	Malabar Kino	Barwood	**M(traditional):**BK:OAS,SDG	Wyk-424-02
Pterocarpus	*santalinus*	Sandalwood	Barwood	**M(traditional):**SA,WD:BDB,**E:**WD:DYE(red)	Bown-337-04
Pterocarpus	*santalinus*	Sandalwood	Barwood	**M(traditional):**LF:SDG, **E:**WD:DYE(red)	Wyk-289-02, Wyk-424-03
Pterocladia	*pinnata*	New Zealand Agar	Wing Branched Florid Red Algae	**E:**WP:MC(agar)	
Ptychopetalum	*olacoides*	Muira Puama Amazon	Potencywood	**M(traditional):**LF:HAP	Wyk-424-04a
Ptychopetalum	*ovata*	Muira Puama Brazil	Potencywood	**M(traditional):**LF:DCFS,DIPT,HAN,HAP	Bal2-104-01
Ptychopetalum	*uncinatum*	Muira Puama Peru	Potencywood	**M(traditional):**LF:HAP	Wyk-424-04b
Pueraria	*lobata*	Japanese Arrowroot	Kudzu	**M(traditional):**RT:BAA,BDB,BHP,DALC,DCLD,DOST,HAC,OSS,RRN,SDR,SIB, **M(PS):**RT:BHS,DALC,DCTM	Bal2-092-01 Bal2-528-02
Pueraria	*lobata*	Japanese Arrowroot	Kudzu	**C:**RT:FX,**M(traditional):**RT:DCLD,DFLU,DSNB, **M(traditional):**FL:SIB,SPN(alcohol)	Bown-337-05
Pueraria	*lobata*	Japanese Arrowroot	Kudzu	**M(PS):**RT:BHP,DCLD,DFLU,HFV,NMS,SDY,SIB	Wyk-260-01, Wyk-424-05
Pulmonaria	*officinalis*	Jerusalem Cow's Lip	Lungwort	**C:**LF:FF,**M(traditional):**WP,FL:OAS,OEY,OHM,OWN,RLB,RLC,SDR	Bown-337-06
Pulmonaria	*officinalis*	Jerusalem Cow's Lip	Lungwort	**C:**LF:FF,**M(traditional):**LF:DHMY,RLB,SIB,**M(traditional):**RT:OWN	Clev-113-01
Pulmonaria	*officinalis*	Jerusalem Cow's Lip	Lungwort	**M(folk):**WP(flowering):OAS,OWN,RLB,RLC,SDR	Krug-109-01
Pulmonaria	*officinalis*	Jerusalem Cow's Lip	Lungwort	**M(PS):**LF:OHM,OWN,RLC,SDG	Wyk-261-01, Wyk-424-06
Pulmonaria	*saccharata*	Bethlehem Sage	Lungwort	**C:**LF:FF,**M(traditional):**LF:DHMY,RLB,SIB,**M(traditional):**RT:OWN	Clev-113-02
Pulsatilla	*chinensis*	Chinese Anemone	Pasque Flower	**M(traditional):**WP,FL:DABA	Bown-338-01
Pulsatilla	*chinensis*	Chinese Anemone	Pasque Flower	**M(traditional):**RT:SDY	Wyk-424-07
Pulsatilla	*patens*	Bleeding Nose	Pasque Flower	**M(traditional):**WP,FL:RLC,UDR	Bown-338-02
Pulsatilla	*pratensis*	Meadow	Pasque Flower	**M(traditional):**WP,FL:DHMY	Bown-338-03
Pulsatilla	*vulgaris*	European	Pasque Flower	**M(traditional):**WP,FL:DHEA,DSEP,DWHP,HAC,HIN,NMS,NNR,NSD,DSIN,RLA,RLB,**E:**WP,FL:PS☠	Bown-338-04
Pulsatilla	*vulgaris*	European	Pasque Flower	**M(folk):**FL,LF:DPLV,OEY,NMP	Krug-133-01
Pulsatilla	*vulgaris*	European	Pasque Flower	**M(PS):**WP:DABI,DHMY,HAC,HIN,NMP,NMS,NNR,DSIN,RLC,SDG	Wyk-262-01, Wyk-424-08
Punctelia	*subflava–rudecta*	Rough Yellow	Speckled Shield Lichen	**E:**WP:DYE(Lecanoric-Acid-gold,Lecanoric-Acid-&-Ammonia-pink)	Cas2-035-16
Punica	*granatum*	Mediterranean	Pomegranate	**C:**FR:AF,CF,JU(grenadine),**C:**SD:**SP**,GN,**M(traditional):**BK,FR-JU,RD,RT,SD:DAVR,DVGB,DVGV,DWRM,OAS,RST,SDR,SDY	Bown-338-05
Punica	*granatum*	Mediterranean	Pomegranate	**C:**SD:**SP**,**M(traditional):**SD:HFV,RST,SDR	Clev-321-01

Latin Genus	L. Species	E. Species	English Genus	C:Culinary,M:Medicinal,E:Economic	Source
Punica	*granatum*	Mediterranean	Pomegranate	**C:**FR-JU:AF(grenadine),FF,**M(folk):**BK:DWTP,**M(folk):**RD:OAS,SDY,**M(folk):**LF:DABA,DSIN	Krug-141-01
Punica	*granatum*	Mediterranean	Pomegranate	**C:**FR:JU,**M(PS):**RB:DWTP,**M(PS):**PU,RD:SDG,SDR	Wyk-263-01, Wyk-424-09
Punica	*granatum*	Mediterranean	Pomegranate	**E:**FL:DYE(direct-orange-brown, alum-tan), **E:**RD:DYE(iron-dark-brown,direct-gray, direct-ooze-black **E:**RD,SD:DYE(direct-beige-apricot) **E:**RD:DYE(iron-&-lime-dark-brown)	Grae-185-01 Grae-185-02 Grae-185-03 Grae-185-04 Grae-186-01
Purshia	*tridentata*	Cliff Rose	Bitterbrush	**E:**LF,SM:DYE(light-tan-yellow,chrome-cocoa,copper-tan-green,tin-butter-yellow,iron-gray-green)	Bliss-050-01
Pycnanthemum	*flexuosum*	Appalachian	Mountain Mint	**M(traditional):**WP:DCLD,HAC,HFV,RSN,SDG	Bown-338-06
Pycnanthemum	*incanum*	Hoary	Mountain Mint	**M(traditional):**WP:DCLD,HAC,HFV,RSN,SDG	Bown-338-07
Pycnanthemum	*muticum*	Clustered	Mountain Mint	**C:**LF:FF	Bown-338-08
Pycnanthemum	*pilosum*	Downy	Mountain Mint	**C:**LF:FF	Bown-338-09
Pycnanthemum	*virginianum*	Virginia	Mountain Mint	**C:**LF,FL,FL-BD:FF,**M(traditional):**WP:HFV,NMS,SDG,SIB	Bown-338-10
Pyrethrum	*cinerariifolium*	Ashy Leaf	Pyrethrin Mum	**E:**FL:IC	Krug-146-01
Pyrethrum	*cinerariifolium*	Ashy Leaf	Pyrethrin Mum	**E:**FL:IC	Wyk-404-14
Pyrus-edulis	*communis*	European	Pear	**E:**LF:DYE(alum-soft-lemon-yellow,chrome-gold,alum-&-tin-bright-yellow-orange,copper-sulfate-beige,copper-sulfate-&-iron-tan)	Cas1-191-01
Quassia	*amara*	Bitter	Quasiwood	**M(PS):**WD:DABI,DWRM,SAP,SDG,**E:**WD:IC(fly)	Wyk-264-01, Wyk-424-10
Quercus	*alba*	White	Oak	**M(traditional):**BK:BVV,DASC,DIOS,OBN,OST,OWN,SDR	Bal1-078-04
Quercus	*alba*	White	Oak	**M(traditional):**BK:DASC,OAS(U.S.Pharmacopeia1916)	Bown-338-11
Quercus	*alba*	White	Oak	**E:**FR(acorn):DYE(chrome-tan-brown,iron-dark-brown,chrome-&-tin-gold-brown)	Cas1-089-03
Quercus	*borealis*	Red	Oak	**E:**BK:DYE(alum-gold,chrome-bright-gold,tin-strong-yellow,iron-&-copper-sulfate-gray,vinegar-&-tin-red-brown)	Cas1-089-02
Quercus	*robur*	English	Oak	**M(traditional):**BK:BVV,DASC,DGUM,DVGB,DVGV,DWRG,OAS,OBR,OHM,DSIN,OUL,RST,SDR,SDY,SUL	Bown-338-12
Quercus	*robur*	English	Oak	**C**:acorn:coffee,**M(folk):**LF:OWN,**M(traditional):**BK:DASC,OAS,RST,SDR	Krug-127-01
Quercus	*robur*	English	Oak	**M(PS):**BK:DASC,OAS,DSIN,OWN,SDR	Wyk-265-01, Wyk-424-11
Quercus	*robur*	English	Oak	**E:***Cynips* insect Oak Gall:DYE(alum-tan-yellow,chrome-green-brown,copper-green-brown,tin-light-yellow,iron-dark-gray)	Bliss-178-01
Quercus	*robur*	English	Oak	**E:**LF(summer):DYE(alum-yellow,copper-sulfate-tan,chrome-gold),E:LF(fall):DYE(alum-tan,alum-&-chrome-brown,tin-tan-rust)	Cas1-185-01
Quercus	*velutina*	Black	Oak	**M(traditional):**BK:DALL,DHVS,DFRA,RLE,**M(SS):**BK:BAS,DCTM,**M(PS):**BK:DECT,DEMD,DSIN,UEP	Bal2-119-01 Bal2-537-02
Quillaja	*saponaria*	Panama Bark	Soapbark Rosewood	**M(traditional):**BK:OAS,DDND,OUL,RLB,RLC,**E:**BK:PS☠	Bown-339-01
Quillaja	*saponaria*	Panama Bark	Soapbark Rosewood	**M(traditional):**BK:RLC,**E:**BK:DS	Wyk-424-12
Ramalina	*farinacea*	Mealy	Cartilage Rim Lichen	**E:**WP:DYE(Protocetraric-&-Salazinic-Acid-yellow,Protocetraric-&-Salazinic-Acid-&-Ammonia-pink-brown)	Cas2-033-04
Ramalina	*scopulorum*		Cartilage Rim Lichen	Synonym for *Ramalina subfraxina*	Cas2-033-05
Ramalina	*subfraxina*	Ash	Cartilage Rim Lichen	**E:**WP:DYE(Orsinols-yellow,Orsinols-&-Ammonia-red)	Cas2-033-05
Ranunculus	*acris*	Showy	Buttercup	**M(traditional):**WP:NGT,NRH,DCLW	Wyk-424-13
Ranunculus	*acris*	Showy	Buttercup	**E:**FL,LF,SM:DYE(alum-light-brass, chrome-gold, copper-khaki, tin-yellow, iron-gray-green)	Bliss-074-01
Ranunculus	*ficaria*	Pilewort	Buttercup	**M(traditional):**WP:OAS,OHM	Bown-339-02
Ranunculus	*ficaria*	Pilewort	Buttercup	**M(folk):**WP:OCW,OHM	Krug-045-01
Ranunculus	*ficaria*	Pilewort	Buttercup	**M(traditional):**SD:DGUM(scurvy),OHM	Wyk-424-14
Raphanus	*raphanistrum*	Wild	Radish	**E:**WP:DYE(alum-&-tin-soft-yellow, chrome-&-tin-bright-gold-yellow,copper-beige-tan, iron-gray)	Cas1-199-01
Raphanus	*sativus*	Sweet	Radish	**C:**LF:FF,**C:**RT:**VG**,**M(traditional):**SD:SAR,SDG,SDR,SFL	Bown-339-03
Raphanus	*sativus*	Sweet	Radish	**M(traditional):**BR:NMS,RLC,SDG,SGB	Wyk-424-15
Raphanus	*sativus*	Sweet	Radish	**E:**WP(top):DYE(alum-&-tin-soft-yellow, chrome-&-tin-bright-gold-yellow,copper-beige-tan, iron-green)	Cas1-199-02
Raphanus	*sativus-niger*	Black	Radish	**M(traditional):**BR:BLC	Bal1-205-05

Latin Genus	L. Species	E. Species	English Genus	C:Culinary,M:Medicinal,E:Economic	Source
Raphiolepis	*indica*	India	Needle Bracteole Hawthorn	**E:**FR:DYE(alum-dark-blue, chrome-turquoise alum-&-tin-purple)	Grae-153-01 Grae-154-01
Rauvolfia	*serpentina*	Snakeroot	Milkwood	**M(traditional):**RT:HAN,BHP,NSD	Bown-339-04
Rauvolfia	*serpentina*	Snakeroot	Milkwood	**M(traditional):**RT:DHEG,HAN,HIN,BHP	Kowa-283-02 Kowa-468-01
Rauvolfia	*serpentina*	Snakeroot	Milkwood	**M(PS):**RT:BLC,DISB,DSNB,HAN,HIN,BHP,NRH ,NSD,SCN	Wyk-266-01, Wyk-424-16
Rauvolfia	*tetraphylla*	American Snakeroot	Milkwood	**M(traditional):**RT:BPH,NSD	Wyk-424-17b
Rauvolfia	*vomitoria*	African Snakeroot	Milkwood	**M(traditional):**RT:BPH,NSD	Wyk-424-17a
Rehmannia	*glutinosa*	Chinese	Rehmann Foxglove	**M(traditional):**RT:BAN,BDB,DABA,BHS,BSH,HFV,OAS,UDR,UKC	Bown-340-01
Rehmannia	*glutinosa*	Chinese	Rehmann Foxglove	**M(traditional):**RT:BHP,BLC,HFV,OAS	Wyk-424-18
Reseda	*lutea*	Yellow	Mignonette	**E:**FL,LF,SM:DYE(brilliant_yellow,red_yellow)	Clev-113-04
Reseda	*luteola*	Weld	Mignonette	**E:**FL-OL:PF	Clev-113-03
Reseda	*luteola*	Weld	Mignonette	**E:**WP:DYE(alum-&-chalk-&-tartar-yellow)	Grae-014-03
Reseda	*odorata*	Scented	Mignonette	**M(traditional):**WP:DHYF,NSD,OWN,RLA,**E:**FL-OL:PF,	Clev-113-05
Rhamnus	*catharticus*	Purging	Buckthorn	**M(traditional):**BK,FR:DWRM,DSIN,SGS,SPR	Bown-340-02
Rhamnus	*catharticus*	Purging	Buckthorn	**M(PS):**FR:SCN	Wyk-267-01, Wyk-424-19
Rhamnus	*davurica*	Dahurian	Buckthorn	**E:**FR:DYE(green)	Bown-340-03
Rhamnus	*frangula*			Synonym for *Frangula purshiana*	Bown-340-04 Wyk-424-20
Rhamnus	*infectoria*	Avignon Berry	Buckthorn	**E:**FR:DYE(yellow)	Bown-340-05
Rhamnus	*purshiana*			Synonym for *Frangula purshiana*	Bown-340-06 Krug-042-01 Wyk-425-01
Rhamnus	*utilis*	Chinese	Buckthorn	**E:**FR:DYE(green)	Bown-340-07
Rhaponticum	*repens*	Creeping	Needle Knapweed	E:FL,LF,SM:DYE(alum-light-gold,chrome-brass-gold,copper-green-gold,tin-bright-yellow,iron-dark-khaki)	Bliss-142-01
Rheum	*australe*	Tibetan	Rhubarb	**M(traditional):**RZ:SCN	Bown-341-01
Rheum	*coreanum*	Korean	Rhubarb	**M(traditional):**RZ:SCN	Bown-341-04
Rheum	*emodi*	Himalayan	Rhubarb	**M(traditional):**RZ:SCN	Wyk-270-02
Rheum	*officinale*	Chinese	Rhubarb	**M(traditional):**RZ:SCN	Bown-341-02
Rheum	*officinale*	Chinese	Rhubarb	**M(traditional):**RZ:SCN	Wyk-425-02
Rheum	*palmatum*	Chinese	Rhubarb	**M(traditional):**RT,SL:BLC,DABI,DWRM,SCN,SGB,SIB	Bal1-076-03
Rheum	*palmatum*	Chinese	Rhubarb	**M(traditional):**RZ:BLC,DHMY,OAS,OBN,OHM, DSIN,SCN,SDG,SDR,SGB	Bown-341-03
Rheum	*palmatum*	Chinese	Rhubarb	**M(PS):**RZ:BLC,DABI,SCN,SDR	Wyk-270-01, Wyk-425-03
Rheum	*palmatum-x-coreanum*	Japanese	Rhubarb	**M(traditional):**RZ:SCN	Bown-341-04
Rheum	*rhabarbarum*	Mongolian	Rhubarb	**C:**SL:**VG**	Bown-341-05
Rheum	*rhaponticum*	False	Rhubarb	**M(traditional):**RZ:SCN	Wyk-270-05, Wyk-425-04
Rheum	*webbianum*	Webb	Rhubarb	**M(traditional):**RZ:SCN	Wyk-270-06
Rhodiola	*rosea*	Roseroot	Red Stonecrop	**M(CS):**RZ:DCFS,DINF	Wyk-271-01, Wyk-425-05
Rhododendron	*aureum*	Golden	Rosewood Heath	**M(traditional):**LF:HIN,NNR,NRH,UDR	Wyk-425-06a
Rhododendron	*campylocarpum*	Curved Pod	Rosewood Heath	**M(traditional):**LF:HIN,NNR,NRH,UDR	Wyk-425-06b
Rhododendron	*canadensis*	Canada	Rosewood Heath	**E:**LF:DYE(vinegar-beige, alum-bright-yellow, copper-gold, chrome-bright-rust, tin-bright-orange, iron-khaki-green	Cas1-202-01
Rhododendron	*canadensis*	Canada	Rosewood Heath	**E:**LF:DYE(alum-&-iron-gray-green)	Grae-156-03
Rhododendron	*ferrugineum*	Ferruginous	Rosewood Heath	**M(traditional):**LF:HIN,NNR,NRH,UDR	Wyk-425-06c
Rhododendron	*ponticum*	Asia Minor	Rosewood Heath	**M(traditional):**LF:HIN,NNR,NRH,UDR	Wyk-425-06d
Rhododendron	*tomentosum*	Marsh Tea	Rosewood Heath	**M(PS):**LF,SM:DHMY,DWHP,NNC👽,NRH,RLC	Wyk-272-01, Wyk-425-07
Rhodymenia	*palmata*			Synonym for *Palmaria palmata*	Cas1-138-01
Rhus	*aromatica*	Fragrant	Sumac	**M(traditional):**BK,FR,RT:OAS,DSIN,RST,SDG, UBL	Wyk-425-08a
Rhus	*chinensis*	Chinese	Sumac	**M(traditional):**RB,FR:OHM,OUL,RLC,SDR	Bown-341-06
Rhus	*coriaria*	Sicilian	Sumac	**C:**FR:BF,**SP**	Bown-341-07
Rhus	*coriaria*	Sicilian	Sumac	**C:**FR(berry):**SP**,**M(traditional):**FR:HFV,SDG,SIB	Clev-322-01
Rhus	*glabra*	Smooth	Sumac	**M(traditional):**RB:DASC,OAS,OHM,DSIN,OUL, SDR,SDY,**M(traditional):**FR:HFV,UUR	Bown-341-08
Rhus	*glabra*	Smooth	Sumac	**M(traditional):**BK,FR,RT:OAS,DSIN,RST,SDG,	Wyk-425-08b

Latin Genus	L. Species	E. Species	English Genus	C:Culinary,M:Medicinal,E:Economic	Source
				UBL	
Rhus	*hirta*	Staghorn	Sumac	**C**:FR:BF,**E**:RT:DYE(yellow)	Bown-341-10
Rhus	*hirta*	Staghorn	Sumac	**M(traditional)**:BK,FR,RT:OAS,DSIN,RST,SDG, UBL	Wyk-425-08c
Rhus	*hirta*	Staghorn	Sumac	**E**:FR,LF:DYE(alum-green-gold,chrome-chestnut –brown, copper-deep-brass-gold, tin-bright-yellow, iron-dark-charcoal-gray)	Bliss-240-01
Rhus	*hirta*	Staghorn	Sumac	**E**:LF:DYE(alum-yellow, , chrome-bright-gold, copper-bronze, tin-brilliant-yellow, iron-khaki)	Cas1-209-02
Rhus	*radicans*			Synonym for *Toxicodendron radicans*	Bown-341-09
Rhus	*toxicodendron*			Synonym for *Toxicodendron pubescens*	Wyk-425-09
Rhus	*trilobata*	Squawbush	Sumac	E:FR,LF,SM:DYE(alum-yellow,chrome-brown,copper –olive-tan, tin-bright-yellow, iron-dark-gray-green)	Bliss-232-01
Rhus	*typhina*			Synonym for *Rhus hirta*	Bown-341-10
Ribes	*nigrum*	European Black	Currant	**C**:FR:AF,BF,JJ,**M(traditional)**:LF,FR:DABA,DA MW,DCLD,OAS,RST	Bown-341-11
Ribes	*nigrum*	European Black	Currant	**C**:FR:JJ,**M(PS)**:LF:JA,RLC,SDR,**M(PS)**:LF,BD:N RH,UUR,**M(PS)**:FR:DCLD,DFLU	Wyk-273-01, Wyk-425-10
Ribes	*rubrum*	Nordic Red	Currant	**C**:FR:JJ	Wyk-273-02
Ribes	*rubrum*	Nordic Red	Currant	**E**:FR,LF,SM:DYE(alum-butter-yellow, chrome-coffee-brown, copper-dark-olive, tin-bright-gold, iron-dark-charcoal-gray)	Bliss-102-01
Ricinus	*communis*	Caster Oil	Castor Bean	**M(traditional)**:SD-OL:OEY,DSIN,SCN,SPR,**E**:SD-OL:CM,DS,FI,IR(cockroach),MC(fertilizer),PN(varnish),WX	Bown-342-01
Ricinus	*communis*	Caster Oil	Castor Bean	**M(traditional)**:SD-OL:SCN,OEY,E:SD-OL:CM,**contraceptive**☠(foam)	Kowa-069-01 Kowa-280-05
Ricinus	*communis*	Caster Oil	Castor Bean	**M(PS)**:SD-OL:DBOI,OWN,SCN,SPR	Wyk-274-01, Wyk-425-11
Rimelia	*reticulata*	Netted	Cracked Ruffle Shield Lichen	**E**:WP:DYE(Salazinic-Acid-&-Ammonia-pink)	Cas2-032-21
Robinia	*pseudacacia*	False Acacia	Locustwood	**M(traditional)**:BK,LF:HAC(migraine)	Wyk-425-12
Roccella	*babingtonii*	Babington	Orchil Comma Lichen	**E**:WP:DYE(Lecanoric-Acid-&-Rocellic-Acid-&-Ammonia-deep-purple)	Cas2-033-06
Roccella	*fuciformis*	Fucous	Orchil Comma Lichen	**E**:WP:DYE(Lecanoric-Acid-&-Rocellic-Acid-&-Ammonia-crimson)	Cas2-033-07
Roccella	*montagnei*	Montagne	Orchil Comma Lichen	**E**:WP:DYE(Lecanoric-Acid-&-Rocellic-Acid-&-Ammonia-purple)	Cas2-033-08
Roccella	*phycopsis*	Algae	Orchil Comma Lichen	**E**:WP:DYE(Lecanoric-Acid-&-Rocellic-Acid-&-Ammonia-crimson)	Cas2-033-09
Roccella	*tinctoria*	Dyer's	Orchil Comma Lichen	**E**:WP:DYE(Lecanoric-Acid-&-Rocellic-Acid-&-Ammonia-purple)	Cas2-033-10
Rorippa	*nasturtium-aquaticum*	Water Cress	Yellow Cress	**M(traditional)**:WP:SGB,UDR	Wyk-425-13
Rorippa	*palustris*	Marsh	Yellow Cress	**C**:LF:**VG**	Clev-107-02
Rosa-herita-bruscia	*agrestis*	Field	Bush Heritage Rose	**E**:LF,SH:DYE(alum-yellow-green, chrome-green-bronze, copper-gold, tin-bright-orange-yellow, iron-brown)	Cas1-203-02
Rosa-hipa-brusca	*alba*	White	Rosehip Rose Bush	**C**:FL-OL:FF,**E**:FL-OL:CM,DS,PF(roseoil)	Bown-342-02
Rosa-hipa-scrubia	*canina*	Brier	Rosehip Rose Shrub	**M(traditional)**:FR(hip):DASC,SDR,UBL	Bal1-076-04
Rosa-hipa-scrubia	*canina*	Brier	Rosehip Rose Shrub	**C**:FR(hip):AF,JJ,**M(traditional)**:FR(hip):DCLD,DFLU,DGUM(scurvy),OAS,SDR,SIB	Bown-342-03
Rosa-hipa-scrubia	*canina*	Brier	Rosehip Rose Shrub	**C**:FR(hip):AF,JJ,**M(traditional)**:FR(hip):DCLD,DFLU,OAS,SCN,SDG,UDR,**M(traditional)**:FL:DAMW,OAS,RST,**E**:OL:PF	Kowa-292-01 Kowa-422-01
Rosa-hipa-scrubia	*canina*	Brier	Rosehip Rose Shrub	**C**:FR(hip):JJ, **M(folk)**:FR(hip):UDR,UKS	Krug-064-01
Rosa-hipa-scrubia	*canina*	Brier	Rosehip Rose Shrub	**M(PS)**:FR(hip):SIB,**M(PS)**:SD:UUR,UDR	Wyk-275-01, Wyk-425-14
Rosa-elda-bruscia	*centifolia*	Provence	Old Garden Rose Bush	**C**:FL-OL:FF,**M(traditional)**:FL-OL:CM,DS,PF(roseoil)	Bown-342-04
Rosa-elda-bruscia	*centifolia*	Provence	Old Garden Rose Bush	**M(traditional)**:FL-OL(attar-of-roses):DAMW,DAMY,OEY	Wyk-425-15
Rosa-elda-bruscia	*damascena*	Damask	Old Garden Rose Bush	**C**:FL-OL:FF,**M(traditional)**:FL-OL:CM,DS,PF(roseoil)	Bown-342-05
Rosa-rubiginosae	*eglanteria*	Eglantine	Sweet Brier Rosehip Rose Shrub	**M(traditional)**:SD-OL:OBN,**E**:SD-OL:CM	Bown-342-06
Rosa-gallicae	*gallica-officinalis*	Apothecary	French Old Garden Rose Bush	**C**:FL:FF,**M(traditional)**:FL:DAMY,DCFS,DCLD,HAN,OEY,DSIN,RLB,RST,SDR,SIB	Bown-342-07
Rosa-neo-bruscia	*x thea*	Hybrid Tea	Bush Modern Rose	**E**:FL(fermented):DYE(alum-yellow, chrome-brown, copper-gold, tin-bright-yellow-orange, iron-gray)	Cas-203-01

Latin Genus	L. Species	E. Species	English Genus	C:Culinary,M:Medicinal,E:Economic	Source
Rosa-laevigatae	*laevigata*	Asian	Cherokee Climbing Old Garden Rose	**M(traditional):**FR(hip):DABA,DAVR,DINF,OAS, SDG,SDR,UUR	Bown-342-08
Rosa-chinenses	*odorata*	Tea	China Shrub Old Garden Rose	**E:**FL,LF,SM:DYE(alum-bright-yellow-green, chrome-gold-brown, copper-moss-green, tin-bright-yellow, iron-dark-forest-green)	Bliss-271-01
Rosa-cinnamomeae-hipa-scrubia	*rugosa*	Hedgehog	Cinnamon Rosehip Rose Shrub	**M(traditional):**FL:BCR,SAP,SDG	Bown-342-09
Rosmarinus	*officinalis*	Mediterranean	Rosemary	**M(traditional):**LF:BCR,BLC,BHP,DABA,DCTM,OAS,OCC,SDG	Bal1-076-05
Rosmarinus	*officinalis*	Mediterranean	Rosemary	**C:**LF:AF(Benedictine,Danziger,Goldwasser), **M(folk):**LF,OL:DCFS,DSIN,HAC,HML,NNR,NSD,SDG,**M(traditional):**LF,OL:DALZ,DCTM,DDND,OHL, **M(PS):**LF,OL:ORH,SDG,SDP,SIB	Bal2-123-02 Bal2-539-02
Rosmarinus	*officinalis*	Mediterranean	Rosemary	**C:**LF:**SN**,**M(traditional):**FL,LF,OL:DABI,BCR,HAN,NMS,SDG	Bown-243-01
Rosmarinus	*officinalis*	Mediterranean	Rosemary	**C:**LF:**SN**,**M(traditional):**LF:BLC,DAMW,HAC,HAN,ONR,ORH,DSIN,OWN,SDG,**E:**LF:HO,PS☠	Clev-113-06
Rosmarinus	*officinalis*	Mediterranean	Rosemary	**C:**LF:**SN**,**M(traditional):**FL,LF:DASC,HAC,HAN,HIN,NMS,NSD,SDG,**M(traditional):**OL:OBN, OBR,ORH,DSIN,**E:**LF,FL:DYE(alum-yellow-green)	Kowa-175-05 Kowa-292-02 Kowa-428-01
Rosmarinus	*officinalis*	Mediterranean	Rosemary	**C:**LF:**SN**,**M(folk):**LF:HAC,**M(folk):**OL:OGT,OSM,**E:**OL:HO	Krug-148-01
Rosmarinus	*officinalis*	Mediterranean	Rosemary	**M(PS):**LF,OL:DABA,DAFU,DWRM,BCR,HAC,NSD,SAP,SFL	Wyk-276-01, Wyk-425-16
Rosmarinus	*officinalis*	Mediterranean	Rosemary	**E:**FL,LF:DYE(alum-yellow-green)	Grae-186-02
Rubia	*cordifolia*	India Madder	Redroot Vine	**M(traditional):**RT,RZ:BCR,BJA,DABA,DHPT,DHSV,NRH,OAS,DSIN,RLB,RLC,SGS,SDY,UDR, UKS	Bown-243-02
Rubia	*peregrina*	Wild Madder	Redroot Vine	**E:**RT:DYE(pale-red)	Clev-114-02
Rubia	*tinctoria*	Dyer's Madder	Redroot Vine	**M(traditional):**LF,SM:BLC,DHMY,SCN,UBL,**M(traditional):**RT:OWN	Clev-114-01
Rubia	*tinctoria*	Dyer's Madder	Redroot Vine	**E:**RT:DYE(alum-lacquer-red,chrome-garnet-red,tin-orange,iron-brown)	Kowa-173-06 Kowa-371-01
Rubia	*tinctoria*	Dyer's Madder	Redroot Vine	**M(traditional):**RT:UBL,UDR,UKS,UUR	Wyk-425-17
Rubia	*tinctoria*	Dyer's Madder	Redroot Vine	**E:**WP:DYE(alum-&-chalk-red, chrome-&-chalk-red)	Grae-014-02
Rubus-nigrabacca-scrubia	*allegheniensis*	Allegheny	Blackberry Shrub	**E:**FR:DYE(sugar-&-flour-pink-tan, vinegar-orchid –purple), **E:**LF(fresh):DYE(alum-yellow, chrome-bright-gold, tin-soft-orange, **E:**LF(purple),SM: DYE(chrome-brown, iron-gray),**E:**SH:DYE(alum-&-copper-yellow-green, chrome-warm-gold-brown, tin-bright-yellow-green, iron-gray-green)	Cas1-105-01
Rubus-nigrabacca-scrubia	*allegheniensis*	Allegheny	Blackberry Shrub	**E:**FR:DYE(alum-&-tin-&-sugar-&-flour-violet)	Grae-175-02
Rubus-agrestis-bruscia	*chingii*	Ching	Brambleberry Bush	**M(traditional):**FR:BLC,DINF,DIPT,OAS,UKC	Bown-244-01
Rubus-raspabacca-bruscia	*coreanus*	Korean	Raspberry Bush	**M(traditional):**FR:BLC,DINF,DIPT,OAS,UKC	Bown-244-02
Rubus-nigrabacca-bruscia	*fruticosus*	Bramble	Blackberry Bush	**C:**FR:AF,JJ,**M(traditional):**LF,RB,RT:DAMW,DGUM,OAS,OHM,RST,SDR,SDY,UBL	Bown-244-03
Rubus-nigrabacca-bruscia	*fruticosus*	Bramble	Blackberry Bush	**C:**FR:AF,JJ, **M(folk):**LF:DGUM,OHM,RST,SDR	Krug-032-01
Rubus-nigrabacca-bruscia	*fruticosus*	Bramble	Blackberry Bush	**M(traditional):**LF:DAMW,OAS,RST,SDR	Wyk-277-01, Wyk-425-18
Rubus-rorabacca	*hispidus*	Swamp	Dewberry Bush	**M(traditional):**LF:SDR,SDY	Bown-244-04
Rubus-raspabacca-bruscia	*idaeus*	European Red	Raspberry Bush	**M(traditional):**BK,LF,RT:DOST,DSFT,DWHN,NMS	Bal1-076-02
Rubus-raspabacca-bruscia	*idaeus*	European Red	Raspberry Bush	**M(folk):**LF:DAMW,BCR,DSIN,OAS,RLC,RST, **M(traditional):**LF:BDB,BLC,DFLU,HFV,NMS,OBN,UDR,SDR	Bal2-120-01 Bal2-538-01
Rubus-raspabacca-bruscia	*idaeus*	European Red	Raspberry Bush	**M(traditional):**LF:DAMW,DTNS,OBN,OEY,OUL, OWN,SDR	Bown-244-05
Rubus-raspabacca-bruscia	*idaeus*	European Red	Raspberry Bush	**C:**FR:AF(framboise),JJ,**M(folk):**FR(vinegar):HFV,RST,**M(folk):**LF:DAMW,DGUM,OUL,SDR	Krug-147-01
Rubus-raspabacca-bruscia	*idaeus*	European Red	Raspberry Bush	**M(traditional):**LF:DAMW,OAS,RST,SDR	Wyk-277-02, Wyk-425-19
Rubus-raspabacca-bruscia	*idaeus*	European Red	Raspberry Bush	**E:**FR:DYE(sugar-&-flour-rose-tan, vinegar-rose-gray), **E:**LF(fresh):DYE(alum-yellow, chrome-bright-gold, tin-soft-orange, **E:**LF(purple),SM: DYE(chrome-brown, iron-gray),**E:**SH:DYE(alum-&-copper-yellow-green, chrome-warm-gold-	Cas1-201-01

Latin Genus	L. Species	E. Species	English Genus	C:Culinary,M:Medicinal,E:Economic	Source
				brown, tin-bright-yellow-green, iron-gray-green)	
Rubus-raspabacca-scrubia	*occidentalis*	American Black	Raspberry Shrub	**M(traditional):**LF:SDR,SDY	Bown-244-06
Rubus-raspabacca-scrubia	*odoratus*	Purple Flower	Raspberry Shrub	**M(traditional):**LF:SDR,SDY	Bown-244-07
Rubus-vinabacca-bruscia	*parviflorus*	Thimbleberry	Wineberry Shrub	**M(traditional):**LF:SDR,SDY	Bown-244-08
Rubus-nigrabacca-bruscia	*procumbens*	Creeping	Blackberry Bush	**M(traditional):**LF:SDR,SDY	Bown-244-09
Rubus-raspabacca-bruscia	*strigosus*	American Red	Raspberry Bush	**C:**FR:AF(framboise),JJ,**M(folk):**FR(vinegar):HFV,RST,**M(folk):**LF:DAMW,DGUM,OUL,SDR	Krug-147-02
Rubus-raspabacca-bruscia	*strigosus*	American Red	Raspberry Bush	**E:**FR:DYE(sugar-&-flour-rose-tan, vinegar-rose-gray), **E:**LF(fresh):DYE(alum-yellow, chrome-bright-gold, tin-soft-orange, **E:**LF(purple),SM:DYE(chrome-brown, iron-gray),**E:**SH:DYE(alum-&-copper-yellow-green, chrome-warm-gold-brown, tin-bright-yellow-green, iron-gray-green)	Cas1-201-02
Rudbeckia	*hirta*	Black Eyed Susan	Eyed Coneflower	**E:**FL:DYE(alum-light-gray-green, chrome-tan, copper-olive-green, tin-gold, iron-dark-olive-green)	Bliss-056-01
Rudbeckia	*hirta*	Black Eyed Susan	Eyed Coneflower	**E:**FL,LF:DYE(alum-&-copper-light-green-yellow, copper-avocado-green, iron-deep-olive-green)	Cas1-106-01
Rudbeckia	*hirta*	Black Eyed Susan	Eyed Coneflower	**E:**FL:DYE(chrome-olive-green)	Grae-121-01
Rumex	*acetosa*	Sorrel	Dock	**C:**LF:FF,**M(traditional):**LF:OAS,UDR,**E:**LF-JU:SV	Bown-244-10
Rumex	*acetosa*	Sorrel	Dock	**C:**LF:FF,**M(folk):**LF:DGUM(scurvy),DSIN,HFV	Krug-164-01
Rumex	*acetosa*	Sorrel	Dock	**M(traditional):**WP:DWRM,DSIN,SCN,UDR	Wyk-425-20a
Rumex	*acetosella*	Sheep Sorrel	Dock	**M(traditional):**RT:UDR,UUR	Bown-244-11
Rumex	*acetosella*	Sheep Sorrel	Dock	**C:**LF:FF	Clev-114-04
Rumex	*acetosella*	Sheep Sorrel	Dock	**M(traditional):**WP:DWRM,DSIN,SCN,UDR	Wyk-425-20b
Rumex	*acetosella*	Sheep Sorrel	Dock	**E:**WP:DYE(alum-&-copper-soft-yellow-green,chrome-tan,copper-&-iron-dull-green,tin-chartreuse)	Cas1-206-02
Rumex	*alpinus*	Monk's Rhubarb	Dock	**M(traditional):**RZ:SCN	Clev-114-05
Rumex	*alpinus*	Monk's Rhubarb	Dock	**E:**RT:DYE(direct-gold)	Grae-083-01
Rumex	*aquatica*	Pacific Water	Dock	**M(traditional):**RT:BAN,BJA,BLC,OAS,DSIN,SCN,SGB,**E:**RT:PS☠(nausea,dermatitis)	Bown-244-12
Rumex	*crispus*	Yellow	Dock	**M(traditional):**LF,RT:BAN,BLC,DHVS,ODS,DSIN,SCC	Bal1-079-03
Rumex	*crispus*	Yellow	Dock	**M(traditional):**RT:BAN,BJA,BLC,OAS,DSIN,SCN,SGB,**E:**RT:PS☠(nausea,dermatitis)	Bown-244-13
Rumex	*crispus*	Yellow	Dock	**C:**SM:**VG**,**M(traditional):**LF:DSIN,**M(traditional):**RT:DAMW,SCN	Clev-114-06
Rumex	*crispus*	Yellow	Dock	**M(traditional):**RT:DHVS,DWRG,SCN,**E:**RT:DYE(alum-deep-yellow),**E:**LF:DYE(alum-yellow,chrome-red,iron-dark-green)	Kowa-162-01 Kowa-173-09
Rumex	*crispus*	Yellow	Dock	**M(folk):**RT:DSIN,SCN	Krug-190-01
Rumex	*crispus*	Yellow	Dock	**M(PS):**RT:DBOI,DWRO,DWTP,BLC,NJA,DSIN,OUL,SCN	Wyk-278-01, Wyk-425-21
Rumex	*crispus*	Yellow	Dock	**E:**LF,SM:DYE(alum-beige,chrome-camel-brown,copper-light-gray-green,tin-gold,iron-gray-green)	Bliss-100-01
Rumex	*crispus*	Yellow	Dock	**E:**LF,SM:DYE(alum-yellow, chrome-gold, copper-&-tin-bright-yellow-green, copper-&-iron-gray-green), **E:**RT:DYE(chrome-gold-yellow, vinegar-tan, copper-&-iron-gray-green, **E:**WP:DYE(alum-warm-beige, chrome-gold, copper-brown, tin-bright-gold-yellow, iron-dark-brown,	Cas1-136-01
Rumex	*crispus*	Yellow	Dock	**E:**RT:DYE(alum-old-gold, iron-green-brown) **E:**FL:DYE(alum-beige-rose)	Grae-083-02 Grae-084-01 Grae-084-02
Rumex	*obtusifolius*	Bitter	Dock	**M(traditional):**RT:DSIN,SCN	Bown-244-14
Rumex	*patientia*	Herb Patience	Dock	**C:**LF:**VG**	Clev-114-07
Rumex	*rugosus*	Wrinkled Sorrel	Dock	**C:**LF:FF,**M(traditional):**LF:BLC,DACN,DSIN,**M(traditional):**RT:SDR,**E:**LF:PS☠	Clev-114-03
Rumex	*scutatus*	French Sorrel	Dock	**C:**LF:FF	Bown-244-15
Rumex	*scutatus*	French Sorrel	Dock	**C:**LF:FF	Clev-144-08
Rumex	*vesicarius*	Ruby	Dock	**C:**SH:FF	Clev-144-09
Ruscus	*aculeatus*	Butcher's	Broom	**M(traditional):**FL,FL-SM,SD:BCR,BHE,BPH,BVV,DEAT,DOBE,DRNP,DVGT,NCT,UBL,UKC	Bal1-066-07
Ruscus	*aculeatus*	Butcher's	Broom	**M(PS):**WP:BCR,BVV,OHM,UBL	Bal2-039-01 Bal2-508-03
Ruscus	*aculeatus*	Butcher's	Broom	**C:**SH:**VG**,**M(traditional):**WP:BCR,NGT,OHM,SCN,UBL,UDR,UKS,**E:**WP:PS☠(causeshighblood pressure)	Bown-245-01

Latin Genus	L. Species	E. Species	English Genus	C:Culinary,M:Medicinal,E:Economic	Source
Ruscus	*aculeatus*	Butcher's	Broom	**M(PS):**RT,RZ:BVV,OHM,SCN	Wyk-279-01, Wyk-425-22
Ruta	*graveolens*	Herb Of Grace	Rue	**C:**LF:AF,**M(traditional):**LF:DEAI,DEPI,DHMY,NMS,NRH,OEY,ONR,ORH,DSIN,SDG,SIB	Bown-245-02
Ruta	*graveolens*	Herb Of Grace	Rue	**C:**OL:AF,FF,**M(traditional):**LF:DHMY,DEPI,NSD,**E**:OL:CM,PF	Clev-115-01
Ruta	*graveolens*	Herb Of Grace	Rue	**M(traditional):**WP:SAP,SDG,SFL,SIB,**E:**WP:PS ☠,	Kowa-433-01
Ruta	*graveolens*	Herb Of Grace	Rue	**M(PS):**LF:DSNB,EYV,OEY,ORH,OST	Krug-150-01
Ruta	*graveolens*	Herb Of Grace	Rue	**M(PS):**WP:BCR,BHP,HFV,NJA,NMS,DSIN,OTA,SAP,SDP,**E:**WP:PS ☠,**induced abortion** ☠	Wyk-280-01, Wyk-426-01
Sagittaria	*latifolia*	Broadleaf Duct Potato	Arrowhead	**E:**LF:DYE(alum-soft-yellow,chrome-gold,tin-orange-rust,**E:**TU:DYE(chrome-old-gold,tin-orange-rust)	Cas1-093-01
Salicornia	*rubra*	Red Samphire	Pickleweed	**E:**WP:DYE(alum-&-iron-gray-green)	Grae-102-02
Salicornia	*maritima*	Slender Glasswort	Pickleweed	**E:**WP:DYE(alum-&-iron-gray-green)	Grae-102-02
Salicornia	*virginica*	Virginia Glasswort	Pickleweed	**E:**WP:DYE(alum-&-iron-gray-green)	Grae-102-02
Salix	*alba*	White	Willow	**M(traditional):**BK:HAC,NBK,NSD,OTA,OWN	Bal1-078-06
Salix	*alba*	White	Willow	**M(traditional):**BK:DCLD,NJA, **M(SS):**BK:DOST, NBK, **M(PS):**BK:HAC,HFV,NRH	Bal2-153-01 Bal2-550-03
Salix	*alba*	White	Willow	**M(traditional):**BK:HAC,HFV,NGT,NJA,NNR,NRH,SDR,SDY,**M(traditional):**LF:HFV,OAS,SIB	Bown-245-03
Salix	*alba*	White	Willow	**M(traditional):**BK:DASC,DHYF,HAC,NJA,NNR, **E:**BK:CM	Kowa-504-01
Salix	*alba*	White	Willow	**M(folk):**BK:HFV,NRH,NSD,**E:**BK:Aspirin	Krug-183-01
Salix	*alba*	White	Willow	**M(CS):**BK:DFLU,HAC,HFV,NRH,NSD,**E:**WD:WL	Wyk-281-01, Wyk-426-02
Salix	*alba*	White	Willow	**E:**LF:DYE(alum-yellow-green,chrome-gold,copper-green-yellow,tin-bright-yellow,iron-light-gray)	Cas1-219-01
Salix	*amygdaloides*	Peach Leaf	Willow	**E:**LF,SM:DYE(alum-light-bright-yellow, chrome-dark-copper, copper-green-gold, tin-bright-yellow, iron-olive-green)	Bliss-272-01
Salix	*cineria*	Large Gray	Willow	**M(traditional):**BK:HFV,NSD	Bown-245-04
Salix	*discolor*	Pussy	Willow	**M(traditional):**BK:DCLD,NJA, **M(SS):**BK:DOST, NBK, **M(PS):**BK:HAC,HFV,NRH	Bal2-153-05
Salix	*exigua*	Narrow Leaf	Willow	**E:**LF:DYE(alum-pale-yellow, chrome-gold, alum-&-iron-avocado, chrome-&-iron-deep-mustard-green)	Grae-142-01
Salix	*fragilis*	Crack	Willow	**M(traditional):**BK:HFV,NSD	Bown-245-05
Salix	*fragilis*	Crack	Willow	**M(traditional):**BK:DFLU,HAC,HFV,NRH,NSD	Wyk-281-02
Salix	*hindsiana*			Synonym for *Salix exigua*	Grae-142-01
Salix	*myricoides*	Bayberry	Willow	**M(traditional):**BK:DCLD,NJA, **M(SS):**BK:DOST, NBK, **M(PS):**BK:HAC,HFV,NRH	Bal2-153-02
Salix	*nigra*	Black	Willow	**M(traditional):**BK:HFV,NSD	Bown-245-06
Salix	*nigra*	Black	Willow	**M(traditional):**BK:DCLD,NJA, **M(SS):**BK:DOST, NBK, **M(PS):**BK:HAC,HFV,NRH	Bal2-153-03
Salix	*pendulina*	Weeping	Willow	**M(traditional):**BK:DCLD,NJA, **M(SS):**BK:DOST, NBK, **M(PS):**BK:HAC,HFV,NRH	Bal2-153-04
Salix	*pentandra*	Laurel	Willow	**M(traditional):**BK:HFV,NSD	Bown-245-07
Salix	*purpurea*	Basket	Willow	**M(traditional):**BK:HFV,NSD	Bown-245-08
Salix	*purpurea*	Basket	Willow	**M(traditional):**BK:DFLU,HAC,HFV,NRH,NSD	Wyk-281-03
Salsola	*kali*	Russian	Salt Thistle	**E:**FL,LF,SM:DYE(alum-light-gray-green, chrome-gray, copper-gray-green, tin-dull-gold, iron-gray-brown)	Bliss-208-01
Salvadora	*persica*	Iranian	Saltwood Mustard	**M(traditional):**LF:DAMW,NJA,NRH,OBR,OSA, **M(traditional):**LF,RB:DCLD,DWRM,OAS,RLC,UDR,UUR,**M(traditional):**WD:OWN,**M(traditional):**FR:SAP,SCN,**M(traditional):**FR,SD:BCR,**E:**LF:salt,**E:**SD:WX	Bown-246-01
Salvadora	*persica*	Iranian	Saltwood Mustard	**M(PS):**RT:DABA,DTDC(tooth decay)	Wyk-282-01, Wyk-426-03
Salvia	*clevelandii*	Blue Jim	Sage	**E:**LF:PF	Bown-246-02
Salvia	*coccinea*	Texas	Sage	**E:**LF:PF	Bown-246-03
Salvia	*dorisiana*	Fruit Scented	Sage	**C:**LF:**SN**,**E:**LF:PF	Bown-246-04
Salvia	*elegans*	Pineapple	Sage	**C:**LF:**SN**	Bown-246-05
Salvia	*fruticosa*	Greek	Sage	**C:**LF,OL:**SN**,TE,**M(traditional):**LF,OL:DFLU,NRH,RLC	Bown-246-06
Salvia	*hispanica*	Mexican	Sage	**C:**SD:BF(chia)	Bown-246-07
Salvia	*fruticosa*	Greek	Sage	**C:**LF:TE	Clev-116-04
Salvia	*horminoides*	Wild Clary	Sage	**C:**LF:FF,**M(traditional):**LF:DAMW,DSIN,**E:**LF:PF(muscatel)	Clev-116-03

Latin Genus	L. Species	E. Species	English Genus	C:Culinary,M:Medicinal,E:Economic	Source
Salvia	*lavandulifolia*	Spanish	Sage	**C**:OL:FF,**M(traditional)**:LF,OL:DASC,DINF,HAN,HFV,OAS,RLC,SDG,**E**:OL:CM,DS	Bown-246-08
Salvia	*microphylla*	Red Bush	Sage	**M(traditional)**:FL,LF:HFV	Bown-246-09
Salvia	*miltiorhiza*	Red Root	Sage	**M(traditional)**:RT:DHPT,DUFB,NSD, **M(PS)**:RT:BHS	Bal2-057-01 Bal2-514-01
Salvia	*miltiorhiza*	Red Root	Sage	**M(traditional)**:RT:DABI,DBOI,BCR,BHC,HIN,NMT,NMP,NSD,OAS,OUL	Bown-246-10
Salvia	*miltiorhiza*	Red Root	Sage	**M(traditional)**:RT:BCR,NSD	Wyk-426-04
Salvia	*officinalis*	Narrow Leaf	Sage	**M(traditional)**:LF:DTNS,DWHN,SDG,RST	Bal1-076-06
Salvia	*officinalis*	Narrow Leaf	Sage	**C**:LF,**SN**,TE,**M(traditional)**:FL,LF,OL:BLC,DAMW,DASC,DGUM,DISB,DFRI,DPRK(salivation),DTBC(perspiration),DVGB,DVGV,HAN,NMS,OAS,DSIN,RST,SDG,SFL,**E**:OL:CM,PF,**E**:FL,LF:PS☠	Bown-246-11
Salvia	*officinalis*	Narrow Leaf	Sage	**C**:LF:**SN**,**M(traditional)**:LF:BLC,DHMY,DASC,DTNS,HAN,RTL	Clev-115-02
Salvia	*officinalis*	Narrow Leaf	Sage	**C**:LF:**SN**,**M(traditional)**:LF,OL:BDB,BSH,DAMW,DCLD,NMS,OAS,OBR,OWN,RST,SDG,**E**:top:CM,DYE(alum-buff-yellow,chrome-deep-yellow,iron-green-gray)	Kowa-175-07 Kowa-292-03 Kowa-439-01
Salvia	*officinalis*	Narrow Leaf	Sage	**C**:LF:**SN**,**M(folk)**:LF,OL:DAMW,DABA,DASC,HAN,HIN,RST,SDG,**E**:OL:HO	Krug-152-01
Salvia	*officinalis*	Narrow Leaf	Sage	**C**:LF:**SN**,**M(PS)**:LF:BDB,DAMW,DASC,DGUM,NMS,RST,SDR,SFL	Wyk-283-01, Wyk-426-05
Salvia	*pomifera*	Apple	Sage	**C**:LF:FF,**C**:insectgall:CF	Bown-247-01
Salvia	*purpurea*	Purple Red	Sage	**M**:LF,OL:DGUM,RST,OUL	Krug-153-01
Salvia	*rutilans*	Pineapple	Sage	**C**:LF:BF,CF	Clev-116-05
Salvia	*sclarea*	Clary	Sage	**C**:LF:AF,FF,**C**:FL,FF,TE,**M(traditional)**:FL,LF,SD,OL:HAP,NMS,NSD,OAS,OEY,OUL,SAP,SDG,**E**:OL:CM,DS,PF(muscatel)	Bown-247-02
Salvia	*sclarea*	Clary	Sage	**C**:LF:FF,**M(traditional)**:LF:DAMW,DSIN,**E**:LF:PF(muscatel)	Clev-116-01
Salvia	*sclarea*	Clary	Sage	**C**:LF:AF,FF,TE,**M(traditional)**:LF:NMS,OAS,OEY,SDG,UKC,**E**:LF:CL,CM,DS,PF	Kowa-091-01
Salvia	*sclarea*	Clary	Sage	**C**:LF:AF(muscatel),**M(folk)**:SD:EYV,OEY,**E**:OL:PF	Krug-153-02
Salvia	*sclarea*	Clary	Sage	**C**:LF:BF,**M(PS)**:FL:DASC,DCLD,DHYF,OAS,SFL,**M(PS)**:FR:NSD,SCN,**E**:FR:DYE,**E**:LF:CM,PF	Wyk-284-01, Wyk-426-06
Salvia	*viridis*	Annual Clary	Sage	**C**:LF,SD:FF,**C**:OL:AF,**M(traditional)**:FL,LF,OL,SD:DGUM	Bown-247-03
Salvia	*viridis*	Annual Clary	Sage	**C**:LF:FF,**M(traditional)**:LF:DAMW,DSIN,**E**:LF:PF(muscatel)	Clev-116-02
Sambucus	*canadensis*	American	Elderberry	**C**:FL:BF,**C**:FR:CF,JJ	Clev-116-07
Sambucus	*canadensis*	American	Elderberry	**M(traditional)**:FR:SCN,UDR	Wyk-426-07
Sambucus	*canadensis*	American	Elderberry	**E**:LF:DYE(alum-tan,chrome-gold,tin-yellow,iron-brown)	Cas1-139-01
Sambucus	*ebulus*	Dwarf	Elderberry	**M(traditional)**:RZ:DHMY,**M(traditional)**:BK,FL:SPR,**E**:FR:PS☠	Clev-116-08
Sambucus	*nigra*	Black Berried	Elderberry	**M(traditional)**:BK,LF,RT:BCR,DFRA,HFV,SCN,**M(traditional)**:FL:DSIN	Bal1-069-04
Sambucus	*nigra*	Black Berried	Elderberry	**M(folk)**:FL:DCLD,HFV,RLC,**M(traditional)**:FL:DFLU,DHYF,RLA,RLB,RSN,**M(PS)**:FL:DCLD,HFV,RLB,RLC	Bal2-063-01 Bal2-516-01
Sambucus	*nigra*	Black Berried	Elderberry	**C**:FL:AF,FF,JJ,**C**:FR:AF,JJ,ketchup,**M(traditional)**:BK:NJA,SCN,**M(traditional)**:BK,LF:BCB,OBN,**M(traditional)**:FL:DAMW,OEY,DSIN,OWN,**M(traditional)**:FL,FR:DCLD,DFLU,HFV,RSN,**M(traditional)**:FR:NRH,**E**:FL:CM,**E**:LF:IC	Bown-247-04
Sambucus	*nigra*	Black Berried	Elderberry	**C**:FL:BF,**C**:FR:CF,JJ,**M(traditional)**:BK:DEPI,**M(traditional)**:BK,LF:SCN,**M(traditional)**:FL:OEY,**M(traditional)**:FR:DCLD,RLC,RSN,**M(traditional)**:LF:DSIN,**M(traditional)**:RT:UKC,**E**:FL:CM,E:LF:IR	Clev-116-06
Sambucus	*nigra*	Black Berried	Elderberry	**C**:FR:AF,JJ,**M(traditional)**:FL:SDG,**E**:FL:CM,**E**:FR:DYE(alum-violet,tin-blue-gray-chrome-blue),**E**:LF:DYE(alum-soft-yellow,chrome-deep-yellow),**E**:BK:DYE(iron-gray)	Kowa-173-10 Kowa-179-01
Sambucus	*nigra*	Black Berried	Elderberry	**M(folk)**:FR:DCLD,**M(folk)**:FL:DSIN,**E**:LF:IR	Krug-067-01
Sambucus	*nigra*	Black Berried	Elderberry	**M(PS)**:FL:DCLD,DHYF,RLC,**M(PS)**:FR:NSD,SCN,UDR	Wyk-285-01
Sambucus	*nigra*	Black Berried	Elderberry	**E**:FR:DYE(alum-&-tartar-mauve,alum-&-tartar-&-ammonia-gray-blue	Grae-136-01 Grae-136-02
Sambucus	*racemosa*	Redberry	Elderberry	**M(traditional)**:SD-OL:SPR	Clev-116-07
Sanguinaria	*canadensis*	Puccoon	Bloodroot	**M(traditional)**:RZ:BCB,BCR,DABI,DTDC(tooth-	Bown-247-05

Latin Genus	L. Species	E. Species	English Genus	C:Culinary,M:Medicinal,E:Economic	Source
				decay),HFV,NMS,DCLW,DSIN,RLB,RLC,RST,UDR,E:RT:DYE(orange-red),E:RZ:PS☠	
Sanguinaria	*canadensis*	Puccoon	Bloodroot	**M(traditional):**RZ:DCBR,DCLD,DCSK,DPNB,DPNV,DWHP,RLB,RLC	Kowa-048-01 Kowa-173-04
Sanguinaria	*canadensis*	Puccoon	Bloodroot	**E:**RT:DYE(orange, alum-rust, tin-pink)	Krug-035-01
Sanguinaria	*canadensis*	Puccoon	Bloodroot	**M(CS):**RZ:DABI,DAMW,DCTM,DDBA,DGUM,NMS,DCLW,OUL,RCR,RLA,RLB,RLC,RST,RTL	Wyk-286-01, Wyk-426-08
Sanguinaria	*canadensis*	Puccoon	Bloodroot	**E:**RT:DYE(alum-tan-orange,vinegar-tan-pink,tin-barn-red)	Cas1-108-01
Sanguisorba	*minor*	Salad	Burnet	**C:**LF:**SN**,**M(traditional):**LF:SDR,SUL,**M(traditional):**RT:OBN,OWN	Clev-112-02
Sanguisorba	*minor*	Salad	Burnet	**C:**LF:**SN**,**M(traditional):**LF:OAS,OWN,SDR	Kowa-449-01
Sanguisorba	*officinalis*	Great	Burnet	**C:**LF:**SN**,**M(traditional):**LF:SDR,SUL,**M(traditional):**RT:OBN,OWN	Clev-112-03
Sanguisorba	*officinalis*	Great	Burnet	**M(traditional):**LF,RT:DABI,DGUM,OAS,OBN,OHM,DSIN,SDR,SDY,SIB,SUL	Bown-248-01
Sanguisorba	*officinalis*	Great	Burnet	**M(PS):**RT:DAMW,OAS,OBN,OHM,DSIN,OUL,OWN,SDR,SUL	Wyk-287-01, Wyk-426-09
Sanicula	*europaea*	Wood	Sanicle	**M(PS):**WP:DAMW,DGUM,OAS,RLC,SDR,SIB,SFL,SUL,**M(traditional):**LF:OBN,OBR,OHM,DSIN,OWN	Wyk-288-01, Wyk-426-10
Santalum	*album*	White	Sandalwood	**M(traditional):**OL,RT,WD:DASC,HAN,HFV,NMS,NSD,OAS,DSIN,SDG,UDR,**E:**OL:DS,IN,PF,**E:**WD:WL	Bown-248-02
Santalum	*album*	White	Sandalwood	**M(PS):**WD:DAMY,DSXG,DSIN,UBL,UUR,SDG	Wyk-289-01, Wyk-426-11
Santalum	*fernadezianum*	Fernandez Island	Sandalwood	**M(traditional):**OL:PF,reportedextinctin1916	Bown-248-03
Santalum	*lanceolatum*	Plum Bush	Sandalwood	**M(traditional):**OL,RT,WD:DBOI,DSXG,DSIN	Bown-248-04
Santalum	*spicatum*	Australian	Sandalwood	**M(traditional):**OL:PF	Bown-248-05
Santolina	*chamaecyparis*	False Cypress	Lavender Cotton	**M(traditional):**LF,FL-SM:BJA,BLC,DISB,DWRM,DSIN,OST,OWN,SDG,SIB	Bown-248-06
Santolina	*chamaecyparis*	False Cypress	Lavender Cotton	**M(traditional):**LF:DWRM,NJA,NRH,**E:**LF:IR(moth)	Clev-117-01
Santolina	*chamaecyparis*	False Cypress	Lavender Cotton	**M(traditional):**LF:DWRM,OAS,**E:**FL,LF:DYE(alum-yellow,chrome-gold)	Kowa-174-05 Kowa-354-01
Santolina	*chamaecyparis*	False Cypress	Lavender Cotton	**E:**FL,LF:DYE(alum-yellow, chrome-sienna-gold, **E:**LF:DYE(alum-&-tin-pea-green,chrome-&-tin-green-brown	Grae-187-01 Grae-187-02
Santolina	*neopolitana*	Italian	Lavender Cotton	**M(traditional):**LF:DWRM,NJA,NRH,**E:**LF:IR(moth)	Clev-117-02
Saponaria	*officinalis*	Bouncing Bet	Soapwort	**M(traditional):**RZ,SM:BJA,BLC,NGT,DSIN,RLB,RLC,SCN,UDR,**E:**RZ:PS☠(destroys red blood cells, paralysis)	Bown-249-01
Saponaria	*officinalis*	Bouncing Bet	Soapwort	**M(traditional):**RT,RZ:RLC,SCN,SPR,UDR,**M(traditional):**LF,RT:DACN,DSIN,**E:**LF,RT:DS	Kowa-469-01
Saponaria	*officinalis*	Bouncing Bet	Soapwort	**M(folk):**RT,SM:DSIN,SPR,**E:**RT,SM:DS	Krug-162-01
Saponaria	*officinalis*	Bouncing Bet	Soapwort	**M(PS):**RT,RZ:DSIN,RLB,RLC, **E:**WP:DS	Wyk-290-01, Wyk-426-12
Saponaria	*officinalis*	Bouncing Bet	Soapwort	**E:**FL:DYE(alum-pale-yellow, chrome-soft-gray-green, copper-soft-gray-green, tin-light-yellow, iron-light-camel	Bliss-226-01
Saraca	*ascoa*	Askoa	Sarak	**M(traditional):**BK:NMP	Wyk-426-13
Sarothamnus	*scoparius*			See *Cytisus scoparius*	Krug-039-01
Sargassum	*fusiforme*	Gulfweed	Sargasso Seaweed	**M(traditional):**WP:BHE,BHP,DOBE,OAS,RLB,RLC,UDR	Bown-249-02
Sassafras	*albidum*	Fennel Scented	Sassafras	**C:**RT:BF(root-beer),FF,TE,**M(traditional):**RB:DACN,DASC,DSXG,DSXS,NJA,NMP,NRH,OEY,DSIN,SDG,SIB,UDR,**M(traditional):**OL:DISB,**E:**RT:PF,**E:**OL:PS☠	Bown-249-03
Sassafras	*albidum*	Fennel Scented	Sassafras	**C:**RT:BF(root-beer), **M(traditonal):**RB:DASC,DIOS,DSXS,HFV,NMS,OEY,OWN,**E:**OL:PS☠	Kowa-451-01
Sassafras	*albidum*	Fennel Scented	Sassafras	**M(folk):**BK,RB:HFV,DIOS,DSIN	Krug-155-02
Sassafras	*albidum*	Fennel Scented	Sassafras	**E:**BK:DYE(alum-&-ammonia-light-terra-cotta)	Grae-188-01
Sassafras	*albidum-molle*	Soft	Sassafras	**M(traditional):**BK,OL,RT:NGT,NRH,UDR, **E:**WP:PS☠(safrol)	Wyk-426-14
Satureja	*hortensis*	Summer	Savory	**C:**LF:AF,**SN**,**M(traditional):**FL,LF:DASC,HAP,OAS,OST,RLB,RLC,RST,SDG,SDR,SIB,**E:**LF:salt	Bown-249-04
Satureja	*hortensis*	Summer	Savory	**C:**LF:AF,**SN**,**M(traditional):**LF:SAP,SDG,SIB,**E**:SH:IR(moth)	Clev-117-03
Satureja	*hortensis*	Summer	Savory	**C:**LF:**SN**,**M(traditional):**LF:DASC,DWRM,OAS,	Kowa-453-01

Latin Genus	L. Species	E. Species	English Genus	C:Culinary,M:Medicinal,E:Economic	Source
				RLC,RST,SAP,SDG,SDR	
Satureja	*hortensis*	Summer	Savory	**C:**LF:**SN**,**M(folk):**LF:HAP,SFL	Krug-167-01
Satureja	*hortensis*	Summer	Savory	**M(traditional):**FL:RLA,RLC,**E:**OL:DS,CM	Wyk-291-02
Satureja	*montana*	Winter	Savory	**C:**LF:AF,**SN**,**M(traditional):**LF,SH:DASC,HAP,OAS,OST,RLB,RLC,RST,SDG,SDR,SIB,**E:**LF:salt	Bown-249-05
Satureja	*montana*	Winter	Savory	**C:**LF:AF,FF,**M(traditional):**LF:SAP,SDG,SIB,**E:**SH:IR(moth)	Clev-117-04
Satureja	*montana*	Winter	Savory	**M(PS):**FL:DASC,DAFU,SDG,SFL,SIB,**M(traditional):**OL:OBN,DSIN,OWN	Wyk-291-01, Wyk-426-15
Satureja	*spicigera*	Autumn	Savory	**C:**LF:AF,**SN**,**M(traditional):**LF:DASC,HAP,OAS,OST,RLB,RLC,RST,SDG,SDR,SIB,**E:**LF:salt	Bown-249-06
Satureja	*thymbra*	Spring	Savory	**C:**LF:**SN**,**M(traditional):**LF:DABA,RLB,RLC,SDG	Bown-250-01
Saussurea	*lappa*	Bur	Sawwort	**M(traditional):**OL,RT:BJA,DABA,DCHL,BHP,NMS,NRH,DSIN,RLA,RLC,SCN,SGB,**E:**OL:PF	Bown-250-02
Saussurea	*lappa*	Bur	Sawwort	**M(traditional):**OL,RT:RLA,RLB,RLC	Wyk-426-16
Scabiosa	*atropurpurea*	Mourning Bride	Pincushion Flower	**E:**FL:DYE(alum-bright-green)	Grae-132-01
Scabiosa	*arvensis*			Synonym for *Knautia arvensis*	Clev-118-01
Sceletium	*tortuosum*	Twisted	Skeleton Ice Plant	**M(PS):**WP:HAN,NNC☠,NSD,SIB	Wyk-292-01, Wyk-426-17
Schisandra	*chinensis*	Northern China	Star Magnolia	**M(traditional):**FR:BDB,DHPT,DHEA,HIN,HML,OAS,DSIN,RLA,RLC,SDR,UUR	Bown-250-03
Schisandra	*sphenanthera*	Southern China	Star Magnolia	**M(traditional):**FR:BDB,DHPT,DHEA,HIN,HML,OAS,DSIN,RLA,RLC,SDR,UUR	Bown-250-04
Schinus	*molle*	Peruvian	Mastic Peppercorn	**C:**FR:**SP**	Wyk-426-18a
Schinus	*terebinthifolius*	Pink	Mastic Peppercorn	**C:**FR:**SP**,**E:**FR:PS☠(>15peppercorns)	Clev-319-01
Schinus	*terebinthifolius*	Pink	Mastic Peppercorn	**C:**FR:**SP**	Wyk-426-18b
Schisandra	*chinensis*	Chinese	Star Magnolia	**M(traditional):**FR:BLC,DHPT,RLC,SDG,SDR,**M(PS):**FR:BHC,BHP,DCOV,DCSK,HAN,HIN	Bal2-127-01 Bal2-540-03
Schisandra	*chinensis*	Chinese	Star Magnolia	**M(traditional):**FR:BLC,NSD,DSIN	Wyk-426-19
Schizonepeta	*tenuifolia*	Jing Jie	Split Catmint	**M(traditional):**WP:DCLD,DSIN,HAC,RST,**M(PS):**WP:DBOI,NMT,OMT	Bal2-128-01 Bal2-541-01
Schizonepeta	*tenuifolia*	Jing Jie	Split Catmint	**M(traditional):**WP:DCLD	Wyk-426-20
Schoenocaulon	*officinale*	Tropical America	Feather Shank Lily	**M(traditional):**SD:DWRG,**E:**SD:IC	Wyk-426-21
Schoenoplectus	*acutus*	Hard Stem	Plaited Bulrush	**E:**FL,SL:DYE(alum-tan, chrome-gray-tan, copper-gray-green, tin-tan-gold, iron-gray-green	Bliss-068-01
Scopolia	*carniolica*	European	Scopoli Sunplant	**M(traditional):**RZ:HAN,NMS,NNC☠,SDR,SDY	Bown-250-05
Scopolia	*carniolica*	European	Scopoli Sunplant	**M(PS):**RZ:NMS,NNC☠,ORH	Wyk-293-01, Wyk-427-01
Scopolia	*lurida*	Lurid	Scopoli Sunplant	**M(traditional):**RZ:NNC☠	Bown-250-06
Scopolia	*pysaloides*	Bladder	Scopoli Sunplant	**M(traditional):**RZ:NNC☠	Bown-250-07
Scopolia	*tangutica*	Tangut	Scopoli Sunplant	**M(traditional):**RZ:NNC☠	Bown-250-08
Scrophularia	*auriculata*	Shoreline Figwort	Foxglove	**M(traditional):**WP:OUL,OWN	Bown-250-09
Scrophularia	*ningpoensis*	Ningpo Figwort	Foxglove	**M(traditional):**RT:BHP,BHS,DABA,DBOI,HFV,HIN,OSA,RLC,RST	Bown-250-10
Scrophularia	*nodosa*	Knotted Figwort	Foxglove	**M(traditional):**WP:BCR,DAFU,ODS,DSIN,OBN,OUL,OWN,NMT,SCN,UDR	Bown-251-01
Scrophularia	*nodosa*	Knotted Figwort	Foxglove	**M(folk):**WP(flowering):DBOI,DSIN,OWN	Krug-075-01
Scrophularia	*nodosa*	Knotted Figwort	Foxglove	**M(PS):**WP:NRH,ODS,DSIN,OBN,OSB,OUL,OWN,UDR	Wyk-294-01, Wyk-427-02
Scutellaria	*baicalensis*	Lake Baikal	Skullcap	**M(traditional):**WP(aerial):BCR,BHS,DCFS,DHEA,HAC,HAN,HIN,NMS,NRH,NSD	Bal1-077-03
Scutellaria	*baicalensis*	Lake Baikal	Skullcap	**M(traditional):**RT:BAS,BHP,DABA,DALL,DAVR,DCLD,DDPT,DHPT,DSHZ,HAN,HFV,RLA,UDR,**M(PS):**RT:DABA(*Corynebacterium diptheriae, Saphylococcus aureus, Streptococcus pneumoniae, Pseudomonas aeruginosa*), BDB,BHS,DADD,DCTM,DHYF,DFLU,DGUM,DSXG	Bal2-129-01 Bal2-541-02
Scutellaria	*baicalensis*	Lake Baikal	Skullcap	**M(traditional):**RT:BHC,BHP,BJA,DABA,DHPT,HFV,NMS,NSD,SDG,SDR,SDY,SIB,SUL,UDR	Bown-251-02
Scutellaria	*baicalensis*	Lake Baikal	Skullcap	**M(PS):**WP(aerial):BAS,BCR,DALL,DSIN	Wyk-295-02, Wyk-427-03
Scutellaria	*barbata*	Bearded	Skullcap	**M(traditional):**RT:BAS,BHP,DABA,DALL,DAVR,DCLD,DDPT,DHPT,DSHZ,HAN,HFV,RLA,UDR,**M(PS):**RT:DABA(*Corynebacterium diptheriae, Saphylococcus aureus, Streptococcus pneumoniae, Pseudomonas aeruginosa*), BDB,BHS,DADD,DCTM,DHYF,DFLU,DGUM,DSXG	Bal2-129-02 Bal2-541-02
Scutellaria	*barbata*	Bearded	Skullcap	**M(traditional):**WP:DCTM,DSNB,RST	Bown-251-03
Scutellaria	*galericulata*	European	Skullcap	**M(traditional):**WP:BCR,DABA	Bown-251-04

Latin Genus	L. Species	E. Species	English Genus	C:Culinary,M:Medicinal,E:Economic	Source
Scutellaria	*galericulata*	European	Skullcap	**M(traditional):**WP(aerial):HAC,HAN,HIN,SDG	Clev-118-03
Scutellaria	*galericulata*	European	Skullcap	**M(traditional):**WP(aerial):NMS,NSD	Wyk-295-03
Scutellaria	*lateriflora*	Virginia	Skullcap	**M(traditional):**RT:BAS,BHP,DABA,DALL,DAVR,DCLD,DDPT,DHPT,DSHZ,HAN,HFV,RLA,UDR,**M(PS):**RT:DABA(*Corynebacterium diptheriae, Saphylococcus aureus, Streptococcus pneumoniae, Pseudomonas aeruginosa*),BDB,BHS,DADD,DCTM,DHYF,DFLU,DGUM,DSXG	Bal2-129-03 Bal2-541-02
Scutellaria	*lateriflora*	Virginia	Skullcap	**M(traditional):**WP:HFV,HIN,NMS,NNR,NSD,UKC,E:WP:PS☠	Bown-251-05
Scutellaria	*lateriflora*	Virginia	Skullcap	**M(folk):**WP:HAC,Rabies Infected Dog Bite	Krug-160-01
Scutellaria	*lateriflora*	Virginia	Skullcap	**M(traditional):**WP(aerial):DEPI,HAN,HIN,NMS,NSD	Wyk-295-01, Wyk-427-04
Secale	*cereale*	Cereal	Rye	**M** Skullcap:FL(pollen):UEP	Wyk-427-05
Securidaca	*longipedunculata*	Violet Tree	Axe Wingpod	**M(PS):**RT:HAC,NRH,OTA,ORH,OWN,RLC,SCN,**contraceptive**☠,**E:**BK:PS☠(arrow tip poison)	Wyk-296-01, Wyk-427-06
Sedum	*acre*	Gold Moss	Stonecrop	**M(traditional):**LF:DHMY,OBN,OBR,OWN,**E:**LF:PS☠	Clev-118-04
Sedum	*acre*	Gold Moss	Stonecrop	**M(traditional):**WP:SPR,E:WP:**induced abortion** ☠	Wyk-427-07
Selenicereus	*grandiflorus*	Night Blooming	Moon Wax Cactus	**M(traditional):**FL,SM:BAA,BHE,BHS,HAC,NRH,NSD,UDR,UKC,**E:**FL,SM:PS☠(hallucination)	Bown-251-06
Selenicereus	*grandiflorus*	Night Blooming	Moon Wax Cactus	**M(traditional):**WP:BHS	Wyk-427-08
Semecarpus	*anacardium*	Cashew	Marking Nut Tree	**M(traditional):**FR:HAC,DCLW	Wyk-427-09
Sempervivum	*tectorum*	Hen & Chickens	Houseleek	**M(traditional):**LF:DHSZ,DISB,OAS,OBN,DCLW,OHM,OSB,DSIN,OST,UDR,**E:**LF:PS☠	Bown-251-07
Sempervivum	*tectorum*	Hen & Chickens	Houseleek	**C:**LF,SH:**VG**,**M(traditional):**LF:OBN,DCLW,DSIN,OST	Clev-119-01
Sempervivum	*tectorum*	Hen & Chickens	Houseleek	**M(traditional):**WP:DSIN	Wyk-427-10
Senecio	*aureus*			Synonym for *Packera aurea*	Bown-352-01
Senecio	*cruentus*			Synonym for *Pericallis cruenta*	Grae-114-03
Senecio	*hybridus*			Synonym for *Pericallis hybrida*	Grae-114-03
Senecio	*jacobaea*	Sticking Willie	Groundsel	**M(traditional):**WP,RZ:NJA,NMP,NRH,NSC	Bown-352-02
Senecio	*jacobaea*	Sticking Willie	Groundsel	**M(traditional):**WP:DSIN	Wyk-297-03
Senecio	*jacobaea*	Sticking Willie	Groundsel	**E:**FL,LF,SM:DYE(alum-bright-yellow-gold, chrome-brass, copper-green-brass, tin-bright-yellow, iron-dark-gray-green)	Bliss-250-01
Senecio	*jacobaea*	Sticking Willie	Groundsel	**E:**FL,LF,SM:DYE(alum-yellow,chrome-bright-gold,tin-copper-tan-yellow,bright-yellow,iron-gray-beige)	Cas1-200-02
Senecio	*jacobaea*	Sticking Willie	Groundsel	**E:**FL,LF,SM:DYE(alum-&-ammonia-bright-yellow, chrome-&-ammonia-brass-gold,alum-&-copper-chartreuse,chrome-&-copper-mustard-brown,alum-&-tin-orange,chrome-&-tin-light-burnt-orange	Grae-106-01 Grae-106-02 Grae-107-01
Senecio	*ovatus*	Alpine	Groundsel	**M(PS):**WP:OAS	Wyk-297-01, Wyk-427-11
Senecio	*robbinsii*			Synonym for *Packera schweinitziana*	Cas1-200-03
Senecio	*triangularis*	Arrowleaf	Groundsel	**E:**WP:DYE(alum-yellow-green, chrome-dark-yellow-green, copper-olive-green, tin-gold, iron-dark-olive-green)	Bliss-036-01
Senecio	*vulgaris*	Old Man in the Spring	Groundsel	**M(traditional):**WP,RZ:DVGB,DVGV,OAS,UDR,**M(traditional):**WP,RZ,PS☠	Bown-352-03
Senecio	*vulgaris*	Old Man in the Spring	Groundsel	**M(traditional):**WP:BCR	Wyk-297-02
Senna	*alata*	Emperor's Candlesticks	Senna	**M(traditional):**LF,PD:DWRG	Bown-352-04
Senna	*alexandrina*	Alexandrian	Senna	**M(traditional):**LF,PD:DABA,SCN,**E:**LF,PD:PS☠	Bown-352-05
Senna	*alexandrina*	Alexandrian	Senna	**M(traditional):**LF,PD:SCN	Kowa-283-01
Senna	*alexandrina*	Alexandrian	Senna	**M(traditional):**PD:SCN,SPR	Krug-158-01
Senna	*alexandrina*	Alexandrian	Senna	**M(PS):**LF,PD:SCN	Wyk-298-01, Wyk-427-12
Senna	*fistula*	Pudding Pipe	Senna	**M(traditional):**SM-PU:SCN	Bown-352-06
Senna	*laevigata*	Smooth	Senna	**C:**SD:BF(coffee)	Bown-352-07
Senna	*marilandica*	Maryland	Senna	**M(traditional):**LF,PD:SCN	Bown-352-08
Senna	*obtusifolia*	Java Bean	Senna	**M(traditional):**LF,PD:DWRG	Bown-352-09
Senna	*occidentalis*	Septic Weed	Senna	**C:**SD:BF(coffee)	Bown-352-10
Senna	*reticulata*	Netted	Senna	**M(traditional):**LF,PD:DABA,DABI,DAFU	Bown-352-11
Senna	*sericea*	Silky	Senna	**C:**SD:BF(coffee)	Bown-352-12
Senna	*sieberiana*	Sieber	Senna	**M(traditional):**LF,PD:DSXS	Bown-352-13
Senna	*sophira*	Algar Robill	Senna	**M(traditional):**LF,PD:DWRG	Bown-352-14
Senna	*surrattensis*	Glossy Shower	Senna	**M(traditional):**LF,PD:DSXS	Bown-352-15

Latin Genus	L. Species	E. Species	English Genus	C:Culinary,M:Medicinal,E:Economic	Source
Sequoia	sempervirens	California	Redwood	**E**:BK:DYE(alum-tan,chrome-light-brown)	Grae-173-03
Serenoa	repens	Sabal	Saw Palmetto	**M(traditional):**FR,SD:HAP,SAP,UDR,UEP,UUR	Bal1-077-02
Serenoa	repens	Sabal	Saw Palmetto	**M(folk):**FR:NMT,UBL,UUR,**M(traditional):**FR:DSIN,HAP,RLC,**M(PS):**FR:DCPR	Bal2-126-01 Bal2-540-02
Serenoa	repens	Sabal	Saw Palmetto	**M(traditional):**FR:DASC,DIMP,HAP,NSD,RLB,RLC,UBL,UDR,UEP	Bown-352-16
Serenoa	repens	Sabal	Saw Palmetto	**M(folk):**FR:DIPT,SDG,RLB,RLC,UEP	Krug-156-01
Serenoa	repens	Sabal	Saw Palmetto	**M(CS):**FR:UEP	Wyk-299-01, Wyk-427-13
Serenoa	serrulata			Synonym for *Serenoa repens*	Krug-156-01
Sesamum	indicum	India	Sesame	**C:**OL:CO,FF(salad-oil),**C:**SD:**SP**,**M(traditional):**SD:BLC,DTDC(tooth-decay),DOST,DSFT,NJA,NMS,OHL,RLC,SCN,UKC,**M(traditional):**LF:SDR,SDY,SIB,UBL,**M(traditional):**OL:SCN,**E:**OL:DS,HO,margarine	Bown-352-17
Sesamum	indicum	India	Sesame	**C:**SD:**SP**,**M(traditional):**SD:SCN,**E:**OL:DS,CM	Clev-323-01
Sesamum	indicum	India	Sesame	**C:**SD:**SP**,**M(traditional):**SD:DHOG	Wyk-427-14
Sicyos	horridus	Horrible	Burr Cucumber	**C:**SD:FF,**E:**SD:CM	Wky-395-09
Sida	cordifolia	Heart Leaf	Fan Petals Mallow	**M(traditional):**RT:NSD,**E:**RT:MC	Wyk-427-15a
Sida	retusa	Notched Leaf	Fan Petals Mallow	**M(traditional):**RT:NSD,**E:**RT:MC	Wyk-427-15b
Sida	rhombifolia	Cuban Jute	Fan Petals Mallow	**M(traditional):**RT:NSD,**E:**RT:MC	Wyk-427-15c
Silybum	marianum	Blessed	Milk Thistle	**M(traditional):**FR,LF,SD:BJA,BLC,DFRA,DHPT,ODS,SIB	Bal1-074-03
Silybum	marianum	Blessed	Milk Thistle	**C:**FL:**VG**,**M(traditional):**SD:BJA,BLC,DMAL,OSS,SGB,**M(PS):**SD:BDB,BLC,DHPT,SDP,SGB	Bal2-100-01 Bal2-531-04
Silybum	marianum	Blessed	Milk Thistle	**M(traditional):**WP:BJA,BLC,DHPT,NMS,SGB	Bown-353-01
Silybum	marianum	Blessed	Milk Thistle	**C:**LF,SH,SR:FF,**C:**FL(head),RT,SM:**VG**	Clev-119-02
Silybum	marianum	Blessed	Milk Thistle	**M(CS):**FR:BLC,SDP,SGB	Wyk-300-01, Wyk-427-16
Simarouba	cedron	Cedrin	Simaba	**M(traditional):**SD:DMAL,DSNB,HFV,NMS	Bown-353-02
Simarouba	cedron	Cedrin	Simaba	**M(traditional):**SD:DSNB,HFV	Wyk-427-17
Simmondsia	chinensis	Chinese	Jojoba	**M(traditional):**SD-OL:DACN,ODS,OSB,E:SD-OL:DS,lubricant	Bown-353-03
Simmondsia	chinensis	Chinese	Jojoba	**C:**SD-OL:CM,HO	Wyk-301-01, Wyk-427-18
Sinapis	alba	Yellow	Roman Mustard	**C:**SD:**SP**,**M(traditional):**SD,OL:BCB,BCR,DABI,DBOI,OBR,OJA,ONR,DSIN,RLB,RLC,SDG	Bown-353-04
Sinapis	alba	Yellow	Roman Mustard	**C:**SD:**SP**	Clev-276-01
Sinapis	alba	Yellow	Roman Mustard	**C:**SD:**SP**,**M(folk):**SD:DCLD,DFLU,HAC,OJA,ORH,OTA,RLC	Krug-123-01
Sinapis	alba	Yellow	Roman Mustard	**M(traditional):**SD:DCLD,ORH,SDP	Wyk-427-19
Sinapis	arvensis	Charlock	Roman Mustard	**E:**WP:DYE(alum-pale-yellow,copper-sulfate-pale-yellow-green,chrome-tan,iron-soft-gray,tin-yellow,alum-&-tin-gold)	Cas1-183-01
Siphonochilus	aethiopicus	Ethiopian	Siphonclaw Ginger	**M(traditional):**RZ:DCLD,DMAL,DVGT,HAC,HFV,RLA,RLC	Wyk-302-01, Wyk-427-20
Sisymbrium	altissimum	Tumble	Hedge Mustard	**E:**FL,LF,SM:DYE(alum-light-yellow, chrome-light-gray-green, copper-green-gold, tin-light-gold-beige, iron-camel)	Bliss-262-01
Sisymbrium	officinale	Mediterranean	Hedge Mustard	**C:**LF:FF,**M(traditional):**WP:RLB,RLC,RST,RTL,SCN,SDG,UDR,**E:**WP:PS☠(excessaffectstheheart)	Bown-354-01
Smilax	aristolochaefolia	Sarsaparilla Vera Cruz	Greenbrier	**M(traditional):**RT:BLC,DCFS,DFRI,DHVS,DIFR,DIPT,DRSK,NSD,UDR	Bal1-077-01
Smilax	aristolochaefolia	Sarsaparilla Vera Cruz	Greenbrier	**M(folk):**RT:NGT,ORH,**M(traditional):**RT:BCR,DSIN,DSXS,OSS,OWN,UDR,UUR	Bal2-125-01 Bal2-540-01
Smilax	aristolochaefolia	Sarsaparilla Vera Cruz	Greenbrier	**C:**RZ,TU:BF,**M(traditional):**RZ,TU:DINF,ODS	Bown-354-02
Smilax	aristolochaefolia	Sarsaparilla Vera Cruz	Greenbrier	**C:**RT:BF,**M(PS):**RT:HAP,DCLW,ODS,OJA,ORH	Wyk-303-01, Wyk-427-21
Smilax	china	China Root	Greenbrier	**M(traditional):**RZ,TU:BJA,DABI,DBOI,DCTM,DSXS,NGT,NRH,OAS,ODS,OUL,SIB,UDR,UUR	Bown-354-03
Smilax	china	China Root	Greenbrier	**C:**RZ,TU:BF	Wyk-303-02
Smilax	ferbifuga	Sarsaparilla Equador	Greenbrier	**M(folk):**RT:NGT,ORH,**M(traditional):**RT:BCR,DSIN,DSXS,OSS,OWN,UDR,UUR	Bal2-125-02 Bal2-540-01
Smilax	ferbifuga	Sarsaparilla Equador	Greenbrier	**C:**RZ,TU:BF,**M(traditional):**RZ,TU:DINF,ODS	Bown-354-04
Smilax	ferbifuga	Sarsaparilla Equador	Greenbrier	**C:**RZ,TU:BF	Wyk-303-03
Smilax	glabra	Cat	Greenbrier	**M(traditional):**RZ,TU:BJA,DBOI,DSXS,NJA,NR	Bown-354-05

Latin Genus	L. Species	E. Species	English Genus	C:Culinary,M:Medicinal,E:Economic	Source
				H,OUL,UUR	
Smilax	*glyciphylla*	Sarsaparilla Australia	Greenbrier	**C:**RZ,TU:BF	Bown-354-06
Smilax	*officinalis*	Sarsaparilla Amazon	Greenbrier	**M(folk):**RT:DSIN,DSXS,NRH	Krug-155-01
Smilax	*ornata*	Sarsaparilla Jamaica	Greenbrier	**M(folk):**RT:NGT,ORH,**M(traditional):**RT:BCR,DSIN,DSXS,OSS,OWN,UDR,UUR	Bal2-125-03 Bal2-540-01
Smilax	*regelii*	Sarsaparilla Honduras	Greenbrier	**M(folk):**RT:NGT,ORH,**M(traditional):**RT:BCR,DSIN,DSXS,OSS,OWN,UDR,UUR	Bal2-125-04 Bal2-540-01
Smilax	*regelii*	Sarsaparilla Honduras	Greenbrier	**C:**RZ,TU:BF,**M(traditional):**RZ,TU:BLC,DASC,DHSV,DSXS,DVGB,DVGV,DSIN,SDG	Bown-354-07
Smilax	*regelii*	Sarsaparilla Honduras	Greenbrier	**C:**RZ,TU:BF	Wyk-303-04
Smyrnium	*olusatrum*	Horse Parsley	Alexanders Lovage	**C:**LF,SH,SM,RT:**VG**,**C:**FL-BD:FF,**C:**SD:**SN**	Bown-354-08
Smyrnium	*olusatrum*	Horse Parsley	Alexanders Lovage	**C:**LF,FL-BD,SH,SM:FF,**C:**SD:**SP**, **M(traditional):**RT:UDR,**M(traditional):**LF:OWN	Clev-119-03
Solanum	*aculeastrum*	Soda Apple	Sunplant	**M(traditional):**FR:OTA	Wyk-304-02
Solanum	*carolinense*	Horse Nettle	Sunplant	**M(traditional):**RB,SM:DABA,NMS,NSD	Bown-354-09
Solanum	*dulcamara*	Bittersweet Nightshade	Sunplant	**M(traditional):**RB,SM:BJA,HFV,RLB,RLC,NRH, NSD,OAS,ORH,DSIN,OUL,SUL,UDR,**E:**RB,SM: PS☠(paralysis)	Bown-354-10
Solanum	*dulcamara*	Bittersweet Nightshade	Sunplant	**M(CT):**FL,SM:DISB,ORH,DSIN,RLA,RLB,RLC	Wyk-304-01, Wyk-428-01
Solanum	*lycopersicum*			Synonym for *Lycopersicon esculentum*	Cas1-215-01
Solanum	*melongena*	Eggplant	Sunplant	**M(traditional):**FR:DSIN	Wyk-304-03
Solanum	*nigrum*	Black Nightshade	Sunplant	**M(traditional):**WP:DHVS	Bal2-352-03
Solanum	*nigrum*	Black Nightshade	Sunplant	**M(traditional):**WP:OEY,OHM,DSIN,OUL	Wyk-428-02
Solanum	*nigrum*	Black Nightshade	Sunplant	**E:**WP,FR:DYE(alum-bright-yellow, chrome-dull-gold,alum-&-iron-khaki-green,chrome-&-iron-khaki-green,alum-&-copper-chartreuse,chrome-&-copper-deep-chartreuse	Grae-100-02 Grae-101-01
Solanum	*tuberosum*	Potato	Sunplant	**M(traditional):**TU:SUL	Wyk-304-04
Solanum	*xanthocarpum*	Yellow Pod	Sunplant	**M(traditional):**SD:RLC	Wyk-304-05
Solidago	*canadensis*	Canada	Goldenrod	**M(traditional):**LF,FL,SM:RST	Bown-355-01
Solidago	*canadensis*	Canada	Goldenrod	**M(traditional):**WP:DASC,UDR,UUR	Wyk-428-03
Solidago	*canadensis*	Canada	Goldenrod	**E:**WP:DYE(alum-mustard,chrome-tan-orange,chrome-&-iron-olive-brown)	Grae-089-01
Solidago	*juncea*	Early	Goldenrod	**E:**FL:DYE(alum-tan,chrome-gold,tin-yellow), **E:**WP:DYE(copper-yellow-green,iron-green), **E:**WP(except FL):DYE(copper-avocado-green, iron-olive-green),**E:**FL,LF:DYE(chrome-brown, copper-brown)	Cas1-148-01
Solidago	*nemoralis*	Gray	Goldenrod	**M(traditional):**LF:OAS,SFL,UDR,**M(traditional):** FL:SCN	Kowa-230-02
Solidago	*nemoralis*	Gray	Goldenrod	**E:**FL,SL:DYE(alum-bright-gold-yellow, chrome-rust, copper-bright-green-gold, tin-vivid-gold, iron-dark-olive-green)	Bliss-122-01
Solidago	*odora*	Sweet	Goldenrod	**M(traditional):**LF:OAS,UDR	Kowa-230-03
Solidago	*rigida*	Stiff	Goldenrod	**M(traditional):**LF,FL,SM:OST(beesting)	Bown-355-02
Solidago	*virga-aurea*	European	Goldenrod	**M(traditional):**LF,FL,SM:BLC,DFLU,DISB,DWHP,OAS,DSIN,OUL,OWN,RLC,RST,SDG,SFL,UKC,UKS,UUR	Bown-355-03
Solidago	*virga-aurea*	European	Goldenrod	**M(traditional):**WP:DHMY,HFV,OWN,SDG,UBL, UKC,UUR	Clev-120-01
Solidago	*virga-aurea*	European	Goldenrod	**M(traditional):**LF:DFLU,DGUM,DISB,DMAL,DMSL,HAC,NJA,DSIN,OWN,RST,UKC,**E:**FL:DYE(alum-yellow,chrome-gold,tin-bright-yellow),**E:**WP:DYE(iron-yellow-green)	Kowa-230-01 Kowa-174-03
Solidago	*virga-aurea*	European	Goldenrod	**M(folk):**WP(flowering):OWN	Krug-083-02
Solidago	*virga-aurea*	European	Goldenrod	**M(PS):**WP:DAMW,DASC,OWN,UBL,UDR,UKC, UKS	Wyk-305-01
Sonchus	*oleraceus*	Mediterranean	Sow Thistle	**E:**FL,LF,SM:DYE(alum-light-gold, chrome-tan, copper-green-gold, tin-gold, iron-moss-green)	Bliss-228-01
Sophora	*flavescens*	Yellow Pagoda	Necklace Pod	**M(traditional):**RT:BJA,DABA,DAFU,DCTM,DISB,DSCB,DSXS,DVGB,DWRG,DCLW,DSIN,SDR,SDY,UDR,UUR	Bown-355-04
Sophora	*japonica*	Japanese Pagoda	Necklace Pod	**M(traditional):**FL:BHC,BHP,DABA,NMS,OAS,**M(traditional):**FR:BLC,DWRM,**E:**FR:DYE(yellow)	Bown-355-05
Sophora	*pachycarpa*	Thick Pod	Necklace Pod	**M(traditional):**LF,SD:BCR,NMP	Wyk-428-04
Sophora	*secundiflora*	Mescal Bean	Necklace Pod	**M(traditional):**RT:DEAI	Bown-355-06

Latin Genus	L. Species	E. Species	English Genus	C:Culinary,M:Medicinal,E:Economic	Source
Sophora	*subprostrata*	Pigeon Pea	Necklace Pod	**M(traditional):**RT:DAMW,RST	Bown-355-07
Sorbus	*aucuparia*	Rowan	Mountain Ash	**C:**FR:AF,JJ,**M(folk):**FR-JU:OAS,OHM,RST,RTL	Krug-149-01
Sorbus	*aucuparia*	Rowan	Mountain Ash	**C:**FR:FX,**M(PS):**FR:BDB,DGUM(scurvy),NJA,OHM,SCN,SDR,UKC	Wyk-306-01, Wyk-428-05
Sorbus	*aucuparia*	Rowan	Mountain Ash	**E:**LF:DYE(alum-tan-gold,chrome-dark-bright-gold,copper-warm-tan,tin-bright-yellow-gold, iron-soft-gray-brown), **E:**FR:DYE(alum-salmon-pink, chrome-tan-rose)	Cas1-178-02
Sphaerophorus	*fragilis*	Fragile	Coral Ball Rim Lichen	**E:**WP:DYE(Hypothamnolic-Acid-&-Ammonia-red)	Cas2-033-11
Sphaerophorus	*globosus*	Globe	Coral Ball Rim Lichen	**E:**WP:DYE(Hypothamnolic-Acid-&-Ammonia-red)	Cas2-033-11
Sphagnum	*cymbilifolium*	Bog	Peat Moss	**M(traditional):**WP:DASC,OAS,OWN,**E:**WP:orchidsubstrate	Bown-355-08
Sphagnum	*japonica*	Japanese	Peat Moss	**M(traditional):**WP:DABI,DISB,DSIN,SDY	Bown-355-09
Spigelia	*anthelmia*	West Indies	Pinkroot	**M(traditional):**WP,RT:DWRM,SPR	Wyk-428-06
Spilanthes	*acmella*	Pará	Fleckflower Cress	**C:**LF:FF,**M(PS):**WP:DABI,DAMW,DCLD,DEAI,DFLU,DGUM,DHSV,DWRM,HAC,NSD,OTA,SDG	Wyk-307-01, Wyk-428-07
Spinachia	*oleracea*	South Asia	Spinach	**E:**LF:DYE(alum-yellow,chrome-gold,copper-yellow-green, tin-bright-yellow-green, iron-gray)	Cas1-207-01
Stachys	*officinalis*	Wood Betony	Hedge Nettle	**M(traditional):**LF:BHS,DHEA,NMS,NNR	Bal1-078-09
Stachys	*officinalis*	Wood Betony	Hedge Nettle	**M(traditional):**WP:BHP,DGUM,HAC,HAN,OAS, NNR,NSD,OBR,OUL,OWN,RSN,RST,SDG,SIB, **E:**WP:PS☠	Bown-355-10
Stachys	*officinalis*	Wood Betony	Hedge Nettle	**C:**LF:TE,**M(traditional):**LF:BHP,DWRM,HAC,HAN,OAS,RLA,RLB,RLC,RST,SDR,UKC,**E:**WP:DYE(alum-chartreuse)	Kowa-042-01 Kowa-173-03 Kowa-273-04
Stachys	*officinalis*	Wood Betony	Hedge Nettle	**M(folk):**WP(aerial):HAC	Krug-031-01
Stachys	*officinalis*	Wood Betony	Hedge Nettle	**M(PS):**WP:DAMW,DGUM,DHMY,HAC,HAN,NNR,NSD,SDG,SDR	Wyk-308-01, Wyk-428-08
Stachys	*officinalis*	Wood Betony	Hedge Nettle	**E:**WP:DYE(alum-&-salt-&-copper-chartruese-green)	Grae-090-02
Stachys	*palustris*	Marsh Woundwort	Hedge Nettle	**M(traditional):**WP:DASC,DVTG,NGT,NMS,OAS	Bown-355-11
Stachys	*palustris*	Marsh Woundwort	Hedge Nettle	**M(folk):**WP(flowering):DASC,NJA,NGT,NMS,OAS,OWN,SDR	Krug-188-01
Stellaria	*media*	Starweed	Chickweed	**M(traditional):**WP:BCR,DCLD,DCLW,OLC,DSIN,RLB,RLC	Bal1-067-08
Stellaria	*media*	Starweed	Chickweed	**C:**SM:**VG**,**M(traditional):**WP:DBOI,DVGB,DVGV,NRH,ODS,DSIN,OUL	Bown-356-01
Stellaria	*media*	Starweed	Chickweed	**M(traditional):**LF,SM:DHMY,NJA,DSIN,OWN	Clev-120-02
Stellaria	*media*	Starweed	Chickweed	**C:**SM:**VG**,**M(folk):**WP(aerial):DBOI,OHM,OSA	Krug-051-01
Stellaria	*media*	Starweed	Chickweed	**M(PS):**WP:BVV,DHMY,NRH,OHM,OBN,ODS,DSIN	Wyk-309-01, Wyk-428-09
Stemona	*japonica*	Japanese	Stemonvine	**M(traditional):**TU:DABA,DAFU,DTBC,DWHP,DWPN,RLB,RLC,SDY,**E:**TU:IC	Bown-356-02
Stemona	*sessilifolia*	Sessile Leaf	Stemonvine	**M(traditional):**TU:DABA,DAFU,DTBC,DWHP,DWPN,RLB,RLC,SDY,**E:**TU:IC	Bown-356-03
Stemona	*tuberosa*	Tuber	Stemonvine	**M(traditional):**TU:DABA,DAFU,DTBC,DWHP,DWPN,RLB,RLC,SDY,**E:**TU:IC	Bown-356-04
Stephania	*triandra*	Three Stamen	Crown Moonseed	**M(traditional):**RT:BCR,hypotensive,HFV	Wyk-428-10
Sterculia	*urens*	Stinging	Bottle Boatnut	**M(traditional):**RN:SCN	Wyk-428-11
Stereocaulon	*alpinum paschale*		Snow Foam Rim Lichen	**E:**WP:DYE(Lobaric-Acid-rust)	Cas2-036-09
Stereocaulon	*tomentosum*		Snow Foam Rim Lichen	**E:**WP:DYE(Stictic-Acid-orange)	Cas2-036-09
Stevia	*rebaudiana*	Sweet	Candyleaf	**C:**LF:CF(sugar,300 times sweeter than sucrose)	Wyk-428-12
Stillingia	*sylvatica*	Queen's Delight	Yaw Root	**M(traditional):**RT:BLC,DSXS,RLB,RLC,SCN,UDR	Bown-356-05
Stillingia	*sylvatica*	Queen's Delight	Yaw Root	**M(traditional):**RT:DSIN,SCN,SPR, **E:**RT:PS☠(carcinogenic)	Wyk-428-13
Strophanthus	*gratus*	Ouabain	Twistflower	**M(traditional):**SD:BAA,BHE,BHP,BHS,**E:**SD:PS☠(heartattack)	Bown-356-06
Strophanthus	*gratus*	Ouabain	Twistflower	**M(traditional):**SD:BHS	Kowa-283-03
Strophanthus	*gratus*	Ouabain	Twistflower	**M(PS):**SD:BHS, **E:**SD:PS☠(fish and arrow tip poison)	Wyk-310-01, Wyk-428-14
Strophanthus	*hispidus*	Hairy	Twistflower	**E:**SD:PS☠(fish and arrow tip poison)	Bown-356-07
Strophanthus	*intermedius*	Intermediate	Twistflower	**E:**SD:PS☠(fish and arrow tip poison)	Bown-356-08
Strophanthus	*kombe*	Kombe	Twistflower	**E:**SD:PS☠(fish and arrow tip poison)	Bown-356-09
Strophanthus	*kombe*	Kombe	Twistflower	**M(traditional):**SD:BHS, **E:**SD:PS☠(fish and arrow tip poison)	Wyk-310-02
Strophanthus	*strophanthus*	Eurasian	Twistflower	**E:**SD:PS☠(fishandarrowtippoison)	Bown-356-10
Strychnos	*nux-vomita*	Poison Nut	Strychnine	**M(traditional):**SD:DCFS,SAP,**E:**SD:PS☠(suffocation,death)	Bown-357-01
Strychnos	*nux-vomita*	Poison Nut	Strychnine	**M(PS):**SD:DHMY,**E:**SD:PS☠(rodent poison)	Wyk-311-01,

Latin Genus	L. Species	E. Species	English Genus	C:Culinary,M:Medicinal,E:Economic	Source
					Wyk-428-15
Styphnolobium	*japonicum*	Japanese	Pagoda Tree	**M(PS):**FL-BD:BCR,BVV,OHM,**E:**FL-BD:DYE(yellow)	Wyk-312-01, Wyk-428-16
Styrax	*benzoin*	Gum Benzoin	Snowbell	**C:**RN,FF,**M(traditional):**RN:DAMY,DASC,DCLD,OAS,RLB,RLC,RST,**E:**RN:CM,GR,PF	Bown-357-02
Styrax	*benzoin*	Gum Benzoin	Snowbell	**E:**RN:CM,DS,PF	Kowa-280-02
Styrax	*benzoin*	Gum Benzoin	Snowbell	**M(traditional):**RN:DASC,OAS,DSIN,RLC	Wyk-428-17
Styrax	*hypoglauca*	Gray Benzoin	Snowbell	**E:**RN:CM,GR,PF	Bown-357-03
Styrax	*officinalis*	Storax	Snowbell	**E:**RN:CM,GR,PF	Bown-357-04
Styrax	*tonkinensis*	Siam Benzoin	Snowbell	**E:**RN:CM,GR,PF	Bown-357-05
Succisa	*pratensis*	Daisy Fleabane	Devil's Bit	**C:**LF:**VG**,**M(traditional):**RT:OWN,SDR	Clev-118-02
Supraoxidea	*dismutasea*	Dismutase	Superoxide	**M(traditional):**??:DFRA	Bal1-045-03
Sutherlandia	*frutescens*	Cancer Bush	Balloon Pea	**M(PS):**LF,SM:BDB,BLC,DAID,DCLD,DCTM,DFLU,DTBC,HAN,HFV,NRH,OBN,DSIN,OWN,RLA,RLB,RLC,SAP,SDG,SDY,SIB,SUL,UUR	Wyk-313-01, Wyk-428-18
Swertia	*chirata*			Synonym for *Frasera chirata*	Bown-357-06 Wyk-428-19
Swertia	*japonica*			Synonym for *Frasera japonica*	Bown-357-07
Symphytum	*officinale*	Knitbone	Comfrey	**M(traditional):**LF,RT:DSCB, DSIN,OBN,OBR,OHM,OSB, OUL,OWN	Bal1-068-03
Symphytum	*officinale*	Knitbone	Comfrey	**M(folk):**RT:NRH,RLP,SDR, **M(traditional):**WP(aerial):DSIN,NBF,OBR, **M(PS):**WP(aerial):DCTM	Bal2-052-02 Bal2-512-06
Symphytum	*officinale*	Knitbone	Comfrey	**C:**LF:**VG**,**M(traditional):**LF,RT:BVV,OAS,NMT,NRH,ODS,OHM,OJA,DSIN,OUL,RLB,RLC,SIB,SUL,**E:**LF:PS☠	Bown-357-08
Symphytum	*officinale*	Knitbone	Comfrey	**C:**LF,SH(cooked):**VG**,**M(traditional):**LF:OBR,DSIN,OWN,RLB,**M(traditional):**RT:DHMY,SDG,SIB,E:LF,SH(raw),RT:PS☠	Clev-120-03
Symphytum	*officinale*	Knitbone	Comfrey	**C:**LF:**VG**,**M(traditional):**LF,RT:DABA,DASC,DBOI,DISB,OBN,OBR,OWN,ODS,DSIN,OWN,**E:**MC:CM,**E:**LF:DYE(iron-brown)	Kowa-103-01 Kowa-173-07 Kowa-285-01
Symphytum	*officinale*	Knitbone	Comfrey	**M(PS):**LF,RT,RZ:OBR,OWN	Krug-056-01
Symphytum	*officinale*	Knitbone	Comfrey	**M(PS):**RT,RZ:DBOI,OBR,OJA,OWN, **E:**RT,RZ:PS☠	Wyk-314-01, Wyk-428-20
Symphytum	*uplandicum*	Russian	Comfrey	**C:**LF:**VG**,**M(traditional):**LF,RT:BVV,OAS,NMT,NRH,ODS,OHM,OJA,DSIN,OUL,RLB,RLC,SIB,SUL,**E:**LF:PS☠	Bown-358-01
Symphytum	*uplandicum*	Russian	Comfrey	**C:**LF(cooked):**VG**	Clev-121-01
Symplocarpus	*foetidus*	Polecat	Skunk Cabbage	**M(traditional):**RT,RZ:DHYF,DWHP,NMS,NSD,RLA,RLB,RLC,UDR	Bown-358-02
Symplocarpus	*foetidus*	Polecat	Skunk Cabbage	**M(folk):**RT:NSD,RLC	Krug-160-02
Symplocarpus	*foetidus*	Polecat	Skunk Cabbage	**M(traditional):**WP:NMS,RLC	Wyk-428-21
Symplocos	*racemosa*	Lodh Bark	Sweetleaf	**M(traditional):**BR,RT:BDB,HFV,NMP,OEY,SDR	Wyk-428-22
Synsepalum	*dulciferum*	Sweet	Miraculous Berry	**C:**SD:CF(sugar)	Wyk-428-23
Syringa	*vulgaris*	European	Lilac	**M(traditional):**WP,FL:HFV,SDG	Wyk-429-01
Syringa	*vulgaris charles-jolyi*	Charles Joly Purple	Lilac	**E:**FL:DYE(alum-light-green,alum-ooze-blue-green) **E:**LF:DYE(alum-&-iron-dark-green,alum-&-iron-&-ammonia-dark-brown), **E:**SM:DYE(chrome-orange-yellow)	Grae-154-02 Grae-154-03 Grae-155-01
Syringa	*vulgaris masséna*	Masséna Purple	Lilac	**E:**FL:DYE(alum-light-green,alum-ooze-blue-green) **E:**LF:DYE(alum-&-iron-dark-green,alum-&-iron-&-ammonia-dark-brown), **E:**SM:DYE(chrome-orange-yellow)	Grae-154-02 Grae-154-03 Grae-155-01
Syringa	*vulgaris nightii*	Night Purple	Lilac	**E:**FL:DYE(alum-light-green,alum-ooze-blue-green) **E:**LF:DYE(alum-&-iron-dark-green,alum-&-iron-&-ammonia-dark-brown), **E:**SM:DYE(chrome-orange-yellow)	Grae-154-02 Grae-154-03 Grae-155-01
Syringa	*vulgaris sensation*	Sensation Purple	Lilac	**E:**FL:DYE(alum-light-green,alum-ooze-blue-green) **E:**LF:DYE(alum-&-iron-dark-green,alum-&-iron-&-ammonia-dark-brown), **E:**SM:DYE(chrome-orange-yellow)	Grae-154-02 Grae-154-03 Grae-155-01
Syringa	*vulgaris yankee-doodlei*	Yankee Doodle Purple	Lilac	**E:**FL:DYE(alum-light-green,alum-ooze-blue-green) **E:**LF:DYE(alum-&-iron-dark-green,alum-&-iron-&-ammonia-dark-brown), **E:**SM:DYE(chrome-orange-yellow)	Grae-154-02 Grae-154-03 Grae-155-01
Syzygium	*aromaticum*	Cloves	Lilly Pilly	**M(traditional):**FL-BD,OL:DASC,DWRM,SDG	Bal1-068-02
Syzygium	*aromaticum*	Cloves	Lilly Pilly	**M(traditional):**FL-BD,OL:DGUM, DHSV,HAC, SUL, **M(ayurveda):**FL-BD,OL: DANX,DHAL, OEY,OTA,SFL,SIB,**M(PS):**FL-BD,OL:DABA (*Pseudomonas aerugiosa, Shigella, Staphylococcus arureus, Streptococcus*	Bal2-050-01 Bal2-512-01

Latin Genus	L. Species	E. Species	English Genus	C:Culinary,M:Medicinal,E:Economic	Source
				pneumoniae),DABI,DAFU,DAVR	
Syzygium	*aromaticum*	Cloves	Lilly Pilly	**C:**FL-BD:**SP**,**M(traditional):**OL:DIMP,DISB,DWRM,OAS, OTA,SDG,SIB,**E:**OL:PF	Bown-358-03
Syzygium	*aromaticum*	Cloves	Lilly Pilly	**C:**FL-BD:**SP**,**M(traditional):**OL:DAMW,DASC,OTA,SDG,SFL,SIB,**E:**OL:toothpaste	Clev-300-01
Syzygium	*aromaticum*	Cloves	Lilly Pilly	**C:**FL-BD:**SP**,**M(traditional):**OL:NSD,**E:**OL:dentistry,IR(moth)	Kowa-093-01
Syzygium	*aromaticum*	Cloves	Lilly Pilly	**C:**FL-BD:**SP**,**M(PS):**OL:DAMW,NSD,ORH, OTA,RST,SFL, **E:**OL:IR	Wyk-315-01, Wyk-429-02
Syzygium	*cumini*	Jambolan	Lilly Pilly	**M(ayurveda):**FR,SD:BDB,**M(PS):**BK,LF,SD:BHP,**M(PS):**FR,SD:DCTM	Bal2-089-01 Bal2-527-05
Syzygium	*cumini*	Jambolan	Lilly Pilly	**M(traditional):**FL-BD:BDB	Bown-358-04
Syzygium	*cumini*	Jambolan	Lilly Pilly	**M(traditional):**BK:DAMW,DSIN,RST	Wyk-429-03
Tabebuia	*heptaphylla*	Seven Leaf	Trumpet Tree	**M(traditional):**BK,WD:DSXS,**E:**WD:WL	Bown-358-05
Tabebuia	*impetiginosa*	Pau D'Arco	Trumpet Tree	**M(traditional):**BK(inner):BCR,BLC,DABA,DAID, DALF,DALL,DCAN,DCTM,NRH,DCLW,RLC,SIB, SUL	Bal1-075-06
Tabebuia	*impetiginosa*	Pau D'Arco	Trumpet Tree	**M(traditional):**BK(inner):DABA,DAFU,DAVR,DCAN,DCHG,DCTM,**M(PS):**BK(inner):DMAL	Bal2-111-01 Bal2-534-02
Tabebuia	*impetiginosa*	Pau D'Arco	Trumpet Tree	**M(traditional):**BK(inner):DABI,DAFU,DCAN,DCTM,DHSV,DSCB,DSXS,HFV,NRH,DSIN,SUL, **E:**WD:WL	Bown-358-06
Tabebuia	*impetiginosa*	Pau D'Arco	Trumpet Tree	**M(PS):**BK(inner):DABA,DABI,DAFU,DCAN,DCTM,DSXS,HFV,NRH,DSIN,RLB,SIB	Wyk-316-01, Wyk-429-04
Tabebuia	*incana*	Hoary	Trumpet Tree	**M(traditional):**BK,WD:DCTM	Bown-358-07
Tabebuia	*insignis*	Insignia	Trumpet Tree	**M(traditional):**BK,WD:SUL	Bown-358-08
Tabebuia	*neochrysantha*	Gold Mum	Trumpet Tree	**M(traditional):**BK,WD:DCTM,SUL	Bown-358-09
Tabebuia	*rosea*	Rose	Trumpet Tree	**M(traditional):**BK,WD:DCTM	Bown-358-10
Tabebuia	*serratifolia*	Yellow Poui	Trumpet Tree	**M(traditional):**BK,WD:DCTM	Bown-358-11
Tabernaemontana	*divaricata*	Pinwheel Flower	Milkwood	**M(traditional):**LF,RT,WD:DWRM,DSIN,OTA, SDR	Wyk-429-05
Tagetes	*erecta*	Aztec	Marigold	**M(traditional):**WP:SCN,SIB	Bown-358-12
Tagetes	*lucida*	Sweet Mace	Marigold	**C:**LF,FL:**SN**,**M(traditional):**WP:BHP,DISB(scorpion),DMAL,HFV,NNC☠,SDG,SDR,SIB,UDR,**E:**WP:IN,IR(tick)	Bown-359-01
Tagetes	*lucida*	Sweet Mace	Marigold	**E:**FL:DYE(alum-yellow-tan,chrome-gold)	Kowa-174-07
Tagetes	*lucida*	Sweet Mace	Marigold	**E:**FL:DYE(alum-yellow,alum-&-vinegar-rich-gold, chrome-&-tin-burnt-orange, chrome-&-iron-brown, tin-orange)	Cas1-176-01
Tagetes	*lucida*	Sweet Mace	Marigold	**E:**FL:DYE(alum-&-tartar-yellow-&-tin-orange, alum-gold, alum-&-iron-dull-green)	Grae-125-03
Tagetes	*minuta*	Muster John Henry	Marigold	**C:**LF:**SN**,**M(traditional):**WP:DABI,NMS,OHM,, DSIN,SDG,SIB,SPR,UDR.**E:**OL:DWEL,DWRM(slug),IC,PF	Bown-359-02
Tagetes	*patula*	French	Marigold	**C:**LF:**SN**,**M(traditional):**WP:OEY,ORH,SDG,UDR,**E:**OL:PF	Bown-359-03
Tamarindus	*indica*	India Date	Tamarind	**C:**PD-PU:**SP**,**M(traditional):**PD-PU:DASC,OEY,OUL,SDG,SCN	Clev-324-01
Tamarindus	*indica*	India Date	Tamarind	**C:**FR:BF,FF,**M(traditional):**FR:BJA,DASC,HFV, OAS,OEY,ORH,OUL,RLA,SCN,SDG,SDY	Bown-359-04
Tamarindus	*indica*	India Date	Tamarind	**C:**FR-PU:BF,FF,**M(PS):**FR-PU:SAP,SCN,SDG, SGB	Wyk-317-01, Wyk-429-06
Tanacetum	*balsamita*	Alecost	Tansy	**C:**LF:FF,**M(traditional):**LF:BLC,OAS,OST,SDG, SCN,SDB	Bown-359-05
Tanacetum	*balsamita*	Alecost	Tansy	**C:**LF,FL:FF,**M(traditional):**LF,FL:DCLD,SDG	Clev-087-01
Tanacetum	*balsamita*	Alecost	Tansy	**C:**LF:BF,FF,**M(traditional):**LF:BLC,DASC,HFV, OAS,OUL,SCN,SDY,UDR,**E:**LF:CM	Kowa-128-01
Tanacetum	*balsamita*	Alecost	Tansy	**C:**LF:FF,**M(folk):**FL,LF:SDG	Krug-059-01
Tanacetum	*balsamita*	Alecost	Tansy	**C:**LF:AF(beer),**M(traditional):**LF:DWRM	Wyk-404-13
Tanacetum	*cinerariifolium*	Dalmation	Tansy	**E:**FL:IC	Bown-359-06
Tanacetum	*coccineum*	Red	Tansy	**E:**FL:IC	Bown-359-07
Tanacetum	*parthenium*	Feverfew	Tansy	**M(traditional):**BK,FL,LF:HAC,HFV,NJA,NSD,SAP,SIB	Bal1-070-05
Tanacetum	*parthenium*	Feverfew	Tansy	**M(folk):**WP(aerial):DAMW,DASC,DWRM,HAC,NSD,OWN,SDG,**M(traditional):**WP(aerial):DALL, DEAT,DISB,DVRT,HFV,NJA,NMP,OTA,RLA, **M(CS):**WP(aerial):HAC,NJA	Bal2-068-01 Bal2-518-01
Tanacetum	*parthenium*	Feverfew	Tansy	**M(traditional):**WP:DISB,HAC,HFV,NJA,NMS,N	Bown-359-08

Latin Genus	L. Species	E. Species	English Genus	C:Culinary,M:Medicinal,E:Economic	Source
				RH,OBR,SCN,SDG,E:LF:PS☠(dermatitis,mouth ulcer)	
Tanacetum	*parthenium*	Feverfew	Tansy	M(traditional):LF:OJA,M(traditional):WP:HAC, SDG,E:SM:IR(moth),E:LF:PS☠	Clev-121-02
Tanacetum	*parthenium*	Feverfew	Tansy	M(traditional):LF:HAC(migraine),NJA,SDG,E:LF,SM:DYE(chrome-green-yellow)	Kowa-173-12 Kowa-192-01 Kowa-287-02
Tanacetum	*parthenium*	Feverfew	Tansy	M(folk):LF:HAC,HIN,SDG, M(CT):LF:HAC(migraine)	Krug-074-01
Tanacetum	*parthenium*	Feverfew	Tansy	M(CS):LF,FL:HAC,HFV,NRH,DSIN	Wyk-098-01
Tanacetum	*vulgare*	Bachelor's Buttons	Tansy	C:LF:FF,M(traditional):WP:DSCB,DWRM,SDG, E:WP:PS☠	Bown-360-01
Tanacetum	*vulgare*	Bachelor's Buttons	Tansy	C:LF:FF,M(traditional):LF:OBR,OVV,M(traditional):WP:DWRM,E:WP:IR	Clev-121-03
Tanacetum	*vulgare*	Bachelor's Buttons	Tansy	C:LF:FF,M(traditional):LF,FL:DACN,OGT,OSB, E:FL:DYE(alum-green-yellow),E:LF:DYE(alum-yellow-green),E:WP(top):DYE(iron-dark-green),E:LF,FL:CM,IR(fly),PS☠	Kowa-175-10 Kowa-479-01
Tanacetum	*vulgare*	Bachelor's Buttons	Tansy	C:LF:FF,M(folk):LF,FL:DINF,DWRM,E:SM:IR(fly)	Krug-171-01
Tanacetum	*vulgare*	Bachelor's Buttons	Tansy	M(traditional):LF,FL:BLC,DWRO,DWTH,DWTP, HAC,NMS,ONR,ORH,DSIN,SAP,SDG,SFL,E:LF,FL:PS☠	Wyk-099-01
Tanacetum	*vulgare*	Bachelor's Buttons	Tansy	E:LF,FL,SM:DYE(alum-yellow-green, chrome-gold, copper-tan-green, tin-bright-yellow, iron-dark-moss-green)	Bliss-248-01
Tanacetum	*vulgare*	Bachelor's Buttons	Tansy	E:FL:DYE(alum-warm-tan-yellow, chrome-gold, tin-yellow), E:LF:DYE(chrome-yellow-green, iron-green), E:WP:DYE(alum-yellow-green, chrome-bronze, iron-olive-green)	Cas1-214-01
Taraxacum	*kok-sagyz*	Russian	Dandelion	E:SA:RU	Clev-121-05
Taraxacum	*officinale*	Blowball	Dandelion	C:RT:BF(coffee),M(traditional):WP:BAN,BHC,BJA,BLC,DBOI,DCBR,DHPT,NRH,OSA,SDG,UDR, UKC	Bal1-069-01
Taraxacum	*officinale*	Blowball	Dandelion	M(folk):LF,RT:BLC,DSIN,NGT,OHM,SDP,SGB, UBL,UKC,UDR,UURI,M(traditional):LF,RT:BLC, DOBE,NMT,OHM,SCN,SDG,SGS,UBL,UUR, M(ayurvedic):LF,RT:BLC,DTBC,NGT,SFL,SIB, SUL,UKC,M(PS):LF,RT:SAP,SDG,UUR	Bal2-058-01 Bal2-514-02
Taraxacum	*officinale*	Blowball	Dandelion	C:LF:FF,M(traditional):WP:DSCB,DWRM,SDG, E:OL:PS☠	Bown-360-02
Taraxacum	*officinale*	Blowball	Dandelion	C:LF,FL:AF,FF,C:RT:BF(coffee),M(traditional): LF:UDR,M(traditional):FL:OST,OWN,M(traditional):RT:BLC,SGB	Clev-121-04
Taraxacum	*officinale*	Blowball	Dandelion	C:LF,FL:AF(wine),C:RT:BF(coffee),M(traditional):RT-JU:BCR,BDB,BLC,DASC,NJA,OWN,SAP, SCN,SDG,E:FL:DYE(alum-soft-yellow,tin-yellow),E:WP:DYE(direct-magenta)	Kowa-141-01 Kowa-173-08 Kowa-286-01
Taraxacum	*officinale*	Blowball	Dandelion	C:LF:FF, C:LF,FL,RT:AF,M(folk):LF:UDR, M(folk):RT:BLC,SGB	Krug-061-01
Taraxacum	*officinale*	Blowball	Dandelion	C:LF:FF,C:RT:BF(coffee),M(PS):LF,RT:BDB,BLC,NRH,DSIN,SAP,SGB,SIB,UDR	Wyk-318-01, Wyk-429-07
Taraxacum	*officinale*	Blowball	Dandelion	E:LF,FL,RT:DYE(alum-beige, chrome-olive, copper-light-olive, tin-soft-gold, iron-gray-green)	Bliss-104-01
Taraxacum	*officinale*	Blowball	Dandelion	E:FL:DYE(alum-yellow,tin-bright-gold,chrome-dark-gold,copper-sulfate-tan)E:RT:DYE(vinegar-&-salt-tan,alum-gold,baking-soda-gold-brown,iron-gray,tin-bright-gold,chrome-brown)	Cas1-134-01
Taraxacum	*scapigera*	Australian	Dandelion	C:TU:**VG**	Clev-121-06
Taxus	*baccata*	English	Yew	E:WD:WL	Bown-360-03
Taxus	*baccata*	English	Yew	M(CS):BK,LF:DCBR,DCOV,DCTM	Wyk-319-01, Wyk-429-08
Taxus	*brevifolia*	Pacific	Yew	M(traditional):BK,LF:DCBR,DCLU,DCOV,E:BK, LF:PS☠	Bown-360-04
Taxus	*brevifolia*	Pacific	Yew	M(traditional):BK,LF:DCTM	Wyk-319-02
Taxus	*canadensis*	Canadian	Yew	M(traditional):LF:DFLU,E:LF:PS☠	Bown-360-05
Taxus	*media*	Intermediate	Yew	M(traditional):BK,LF:DCTM	Wyk-319-03
Taxus	*wallichiana*	Himalayan	Yew	M(traditional):BK,LF:DCTM	Wyk-319-04
Teloschistes	*flavicans*	Yellow	Orange Bushy Schist Lichen	E:WP:DYE(Anthraquinones-yellow-orange)	Cas2-036-10
Teloschistes	*velifera*	Velvet	Orange Bushy Schist Lichen	E:WP:DYE(Anthraquinones-&-Ammonia-purple-pink)	Cas2-033-12

Latin Genus	L. Species	E. Species	English Genus	C:Culinary,M:Medicinal,E:Economic	Source
Terminalia	*arjuna*	Arjune	Myrobalan Almond	**M(traditional):**BK,FR:BAA,BCR,NBF,OUL,SCN,SDR,SDY,SUL, **M(PS):**BK,FR:BHA,DACN	Bal2-024-01 Bal2-503-05
Terminalia	*arjuna*	Arjune	Myrobalan Almond	**M(traditional):**FR:SCN	Bown-360-06
Terminalia	*arjuna*	Arjune	Myrobalan Almond	**M(traditional):**BK:BHS,DASC,OAS,SCN,SDG,UUR	Wyk-429-09
Terminalia	*bellirica*	Belliric	Myrobalan Almond	**M(traditional):**FR:SCN	Bown-360-07
Terminalia	*bellirica*	Belliric	Myrobalan Almond	**M(traditional):**FR:DASC,OAS,OWN,SCN,SDG	Wyk-429-10
Terminalia	*chebula*	Chebulic	Myrobalan Almond	**M(traditional):**FR:DAMW,DGUM,DVGB,DVGV,DWRM,OAS,OHM,OUL,OWN,RLA,RLC,SCN,SDG,SDR,SDY	Bown-361-01
Terminalia	*chebula*	Chebulic	Myrobalan Almond	**M(PS):**FR:DASC,OAS,SCN,SDG	Wyk-320-01, Wyk-429-11
Tetraclinus	*articulata*	Jointed	Fourleaf Cypress	**E:**RN:dentistry	Wyk-429-12
Teucrium	*africanum*	African	Germander	**M(traditional):**WP:OHM,SDY	Wyk-321-02
Teucrium	*canadense*	American	Germander	**M(traditional):**WP:RLA,RLC	Bown-361-02
Teucrium	*chamaedrys*	Wall	Germander	**LF**:AF(liqueur,vermouth,wine),**M(traditional):**WP:DASC,DGUM,DSNB,HFV,OAS,DSIN,OWN,NJA,NRH,RLA,RLB,RLC,SAP,SDG,SDR,SGB	Bown-361-03
Teucrium	*chamaedrys*	Wall	Germander	**C:**LF:AF,TE,**M(traditional):**LF:SDG,SGB	Clev-122-01
Teucrium	*chamaedrys*	Wall	Germander	**C:**LF:AF(beer),**M(traditional):**LF:BHE,BJA,HAN,HFV,DASC,DTNS,NMS,OAS,OWN,RLA,RLB,RST,SDG,UDR	Kowa-222-01
Teucrium	*chamaedrys*	Wall	Germander	**M(PS):**WP:DAMW,DOBE,OAS,SDR,UDR, **E:**WP:PS☠	Wyk-321-01, Wyk-429-13
Teucrium	*divaricatum*	Spreading	Germander	**M(traditional):**WP:RLA,RLC	Bown-361-04
Teucrium	*fruticans*	Tall	Germander	**C:**LF:AF,TE	Clev-122-02
Teucrium	*lucidrys*	Lucid Wall	Germander	**M(traditional):**WP:RLA,RLC	Bown-361-05
Teucrium	*marum*	Cat Thyme	Germander	E:OL,RN:	Bown-361-06
Teucrium	*marum*	Cat Thyme	Germander	**M(traditional)**	Wyk-321-03
Teucrium	*montanum*	Mountain	Germander	**M(traditional)**	Wyk-321-04
Teucrium	*polium*	Poley	Germander	**M(traditional)**	Wyk-321-05
Teucrium	*scordium*	Water	Germander	**M(traditional)**	Wyk-321-06
Teucrium	*scorodonia*	Woodland Sage	Germander	**M(traditional):**LF:AF(beer)	Bown-361-07
Teucrium	*scorodonia*	Woodland Sage	Germander	**M(traditional):**LF:OWN,RSN,RST	Clev-122-03
Teucrium	*scorodonia*	Woodland Sage	Germander	**M(traditional)**	Wyk-321-07
Teucrium	*trifidum*	Three Cut	Germander	**M(traditional):**WP:OHM,SDY	Wyk-321-08
Thalictrum	*polycarpum*	Many Pod	Meadow Rue	**E:**WP:DYE(alum-&-ammonia-bright-yellow)	Grae-096-01
Thaumatococcus	*daniellii*	Danielle	Miracle Fruit Ginger	**C:**SD(aril):CF(sugar substitute)	Wyk-429-14
Thea	*sinensis-bohea*	Black	Tea	**C:**LF:TE,**C:**OL:FF,**M(traditional):**SH,OL:ABA,DHPT,DISB,OAS,OEY,SDR,SDY,SIB,UDR,**E:**OL:HO,PF	Bown-252-06
Thea	*sinensis-bohea*	Black	Tea	**M(PS):**LF:DSIN,SDR	Wyk-075-01
Thea	*sinensis-bohea*	Black	Tea	**E:**LF(Orange Pekoe):DYE(alum-beige,vinegar-light-tan,chrome-tan-brown,copper-taupe, iron-beige-gray)	Cas1-214-02
Thea	*sinensis-bohea*	Black	Tea	**E:**LF:DYE(vinegar-tan-rose, vinegar-&-iron-gray	Grae-004-02
Thea	*sinensis-viridis*	Green	Tea	**M(traditional):**LF:DCCL,DCFS,DCMT,DCSK,DCST,DFRA	Bal1-072-02
Thea	*sinensis-viridis*	Green	Tea	**M(traditional):**LF:DMAL,**M(ayurveda):**LF:DCFS,HAC,HFV,SAP,SDR,**M(PS):**LF:BAS,BDB,BHC,BHS,BLC,DCBR,DCCL,DCEN,DFRA,DGUM,DOBE,NGT,OCW,RLA	Bal2-082-01 Bal2-524-03
Thea	*sinensis-viridis*	Green	Tea	**M(PS):**LF:DFRA,UDR	Wyk-075-02
Theobroma	*cacao*	Chocolate	Cocoa Nut	**C:**FR:AF,CF,**M(traditional):**FR,SD:BAA,BHP,OBN,ODS,UDR,**E:**FR(butter):CM	Bown-361-08
Theobroma	*cacao*	Chocolate	Cocoa Nut	**C:**SD:CF(chocolate)	Wyk-429-15
Thermopsis	*montana*	Mountain	Gold Banner	**E:**FL,LF,SM:DYE(alum-light-green-gold, chrome-rust-brown, copper-green-gold, tin-bright-yellow, iron-green-brown	Bliss-114-01
Thevetia	*peruviana*	Peruvoside	Lucky Nut	**E:**SD:IC	Wyk-429-16
Thlaspi	*alliaceum*	Garlic	Pennycress	**C:**LF:FF,C:SD:**SP**	Clev-122-05
Thlaspi	*arvense*	Field	Pennycress	**C:**LF:FF,C:SD:**SP**	Clev-122-04
Thomasella	*lactea*			Synonym for *Pertusaria lactea*	
Thuja	*occidentalis*	Atlantic White	Wood of Life Cedar	**M(traditional):**BK,LF:DAFU,DAVR,DCTM,DVGV,BHS,OAS,DCLW,ODS,ORH,DSIN,RLB,UBL	Bown-362-01
Thuja	*occidentalis*	Atlantic White	Wood of Life Cedar	**M(folk):**BK(inner):MIB, **M(folk):**SM:OCW,RLB	Krug-173-01
Thuja	*occidentalis*	Atlantic White	Wood of Life Cedar	**M(PS):**LF,OL:DCLD,HFV,NRH,RLB,RLC,UBL,UDR,**M(traditional):**SM:DHMY, **E:**OL:**induced abortion** ☠	Wyk-322-01, Wyk-429-17
Thuja	*occidentalis*	Atlantic White	Wood of Life Cedar	**E:**LF(tip):DYE(alum-bright-yellow,chrome-gold, tin-yellow-orange, iron-tan-brown)	Cas1-115-01

Latin Genus	L. Species	E. Species	English Genus	C:Culinary,M:Medicinal,E:Economic	Source
Thuja	*orientalis*	Chinese	Wood of Life Cedar	**M(traditional):**LF:DABA,DMUM,NJA,OAS,DSIN,OHL,RLA,RLB,RLC,SDY,SUL,**M(traditional):**SD:HIN,NSD,SCN	Bown-362-02
Thymbra	*spicata*	Spike	Zatar	**C:**LF:FF	Kowa-521-01
Thymus	*caespititius*	Azores	Thyme	**C:**LF:**SN**	Bown-362-03
Thymus	*capitatus*	Conehead	Thyme	**C:**LF:**SN**,**E:**LF-OL:CL,DS	Bown-362-04
Thymus	*capitatus*	Conehead	Thyme	**E:**LF-OL:CL,PF	Clev-123-01
Thymus	*cilicius*	Cicilian	Thyme	**C:**LF:**SN**	Bown-362-05
Thymus	*citriodorus*	Lemon	Thyme	**C:**LF:**SN**,**M(traditional):**LF-OL:DAMY	Bown-362-06
Thymus	*citriodorus*	Lemon	Thyme	**E:**LF-OL:CL,PF	Clev-123-02
Thymus	*citriodorus*	Lemon	Thyme	**M(traditional):**WP:NMS,RLC,SDG	Wyk-323-02
Thymus	*herba-barona*	Caraway	Thyme	**C:**LF:**SN**	Bown-362-07
Thymus	*herba-barona*	Caraway	Thyme	**C:**LF:**SN**	Clev-123-03
Thymus	*mastichinus*	Marinade	Thyme	**C:**LF,OL:**SN**	Bown-362-08
Thymus	*mastichinus*	Marinade	Thyme	**C:**LF:**SN**	Clev-123-04
Thymus	*praecox*	Creeping	Thyme	**C:**LF:**SN**	Bown-363-01
Thymus	*pseudolanuginosus*	Woolly	Thyme	**C:**LF:**SN**	Bown-363-02
Thymus	*pulegoides*	Large	Thyme	**C:**LF:**SN**	Bown-363-03
Thymus	*pulegoides*	Large	Thyme	**M(traditional):**WP:NMS,RLC,SDG	Wyk-323-03
Thymus	*schimperi*	African	Thyme	**M(traditional):**WP:NMS,RLC,SDG	Wyk-323-04
Thymus	*serphyllum*	Mother of	Thyme	**C:**LF:**SN**,**M(traditional):**WP:DAMW,DASC,DGUM,DWHP,NMS,NSD,OMT,ORH,OSC,OWN,RLB,RLC,RST,RTL,SDG,SFL,SIB	Bown-363-04
Thymus	*serphyllum*	Mother of	Thyme	**E:**WP:LS(lawns)	Clev-123-05
Thymus	*serphyllum*	Mother of	Thyme	**M(traditional):**WP:NMS,RLC,SDG	Wyk-323-05
Thymus	*vulgaris*	Mediterranean	Thyme	**M(traditional):**FL,FR,LF:BHC,BLC,DASC,DCAN,HAC,HFV,RCR,SFL	Bal1-077-08
Thymus	*vulgaris*	Mediterranean	Thyme	**M(traditional):**LF:DABA,DAFU,DAMW,DALL,DAVR,DFRA,OBR,ORH,OWN,RLA,RLC,SDG, **M(PS):**LF:RLB,RLC	Bal2-144-01 Bal2-547-01
Thymus	*vulgaris*	Mediterranean	Thyme	**C:**LF:**SN**,**M(traditional):**WP:DASC,DAFU,DGUM,DTNS,DWHP,OAS,OJA,ORH,NMS,RLA,RLB,RLC,RTL,SDG,SDR,SIB,**M(traditional):**OL:DAMY	Bown-363-05
Thymus	*vulgaris*	Mediterranean	Thyme	**C:**LF:AF,**SN**,**M(traditional):**LF:SDG,**M(traditional):**LF-OL:DASC,DWRM,**E:**LF-OL:CL,PF	Clev-122-06
Thymus	*vulgaris*	Mediterranean	Thyme	**C:**LF:AF,**SN**,**M(traditional):**WP(top):DABA(salmonella,staph,strep),DAMW,DASC,DWHK,HAC,HAN,HFV,NMS,NSD,OMT,DSIN,SDG,SFL,SIB,**E:**LF:CM	Kowa-283-04 Kowa-292-04 Kowa-487-01
Thymus	*vulgaris*	Mediterranean	Thyme	**C:**LF:**SN**,**M(folk):**WP(flowering):DABA,DASC,RLC,SDG	Krug-174-01
Thymus	*vulgaris*	Mediterranean	Thyme	**M(PS):**LF,FL,OL:DABI,DAMW,DCLD,NMS,OWN,RLB,RLC,RST,SIB	Wyk-323-01, Wyk-429-18
Thymus	*zygis*	Spanish	Thyme	**M(traditional):**LF,FL,OL:DABI,DAMW,DCLD,NMS,OWN,RLB,RLC,RST,SIB	Wyk-323-06
Tilia	*cordata*	Small Leaf Lime	Linden	**C:**FL:CF(honey),**M(traditional):**FL:DCLD,DFLU,HAC,HFV,RLC,RST,SDG	Bal2-145-01 Bal2-547-02
Tilia	*cordata*	Small Leaf Lime	Linden	**M(traditional):**FL:BAS,BHP,DCLD,DFLU,HAC,NMS,RLC,SDG,UDR,UUR	Bown-363-06
Tilia	*cordata*	Small Leaf Lime	Linden	**M(traditional):**FL:DCLD,NMS,NSD,RLC,RST,SDG,UDR	Wyk-324-01, Wyk-429-20
Tilia	*cordata*	Small Leaf Lime	Linden	**E:**LF:DYE(alum-strong-yellow,chrome-gold,tin-brilliant-yellow,iron-taupe)	Cas1-098-01
Tilia	*europa*	European Lime	Linden	**M(folk):**FL:DCLD,DFLU,HAN,HIN,NSD,**E:**FL:CM	Krug-105-01
Tilia	*platyphyllos*	Large Leaf Lime	Linden	**M(traditional):**FL:NSD	Wyk-324-02
Tinospora	*cordifolia*	Gilo	Amruth Moonseed	**M(traditional):**SM:BLC,HFV,NRH,DSIN	Wyk-429-21
Toxicodendron	*pubescens*	Atlantic Poison Oak	Toxicwood Vine	**E:**WP:PS☠	Wyk-425-09
Toxicodendron	*radicans*	Atlantic Poison Ivy	Toxicwood Vine	**E:**WP:PS☠	Bown-341-09
Trachyspermum	*ammi*			Synonym for *Trachyspermum copticum*	Wyk-430-01
Trachyspermum	*copticum*	Thyme	Ajowan Caraway	**C:**SD:**SP**,**M(traditional):**SD,OL:BHE,DASC,DCHL,DCLD,DFLU,DVGB,NJA,NMS,NRH,ORH,RLA,RLC,SDG,SDR,SFL,SIB,UDR,**E:**SD:epoxy	Bown-364-01
Trachyspermum	*copticum*	Thyme	Ajowan Caraway	**C:**SD:**SP**,**M(traditional):**FR:OL:SFL	Wyk-430-01
Trachyspermum	*roxburghii*	Roxburgh	Ajowan Caraway	**M(traditional):**LF:DABI,BHS,RLA,SDG,SFL	Wyk-430-02
Tragopogon	*pratensis*	Meadow	Goat's Beard Salsify	**M(traditional):**RT:BDB,BLC,SGB	Wyk-430-03
Tragopogon	*pratensis*	Meadow	Goat's Beard Salsify	**E:**LF,FL,SD,SM:DYE(alum-beige-gold, chrome-camel, copper-olive, tin-gold, iron-gray)	Bliss-214-01
Tremella	*fuciformis*	Snow Fungus	Tremor Jelly	**M(traditional):**WP:BAS,BHC,BLC,DCTM	Bal2-135-01

Latin Genus	L. Species	E. Species	English Genus	C:Culinary,M:Medicinal,E:Economic	Source
					Bal2-543-04
Tribulus	*terrestris*	Ground	Puncturevine	**M(traditional):**WP:DIPT,DINF	Murr-577-04 Murr-743-03
Trichosanthes	*cucumeroides*	Cucumber	Snake Gourd	**E:**FR:DS	Bown-364-02
Trichosanthes	*foetidissima*	Buffalo	Snake Gourd	**E:**FR:DS	Bown-364-03
Trichosanthes	*japonica*	Japanese	Snake Gourd	**M(traditional):**FR:DABA,DAFU,BCR,RLC,SCN	Bown-364-04
Trichosanthes	*kirilowii*	Chinese Cucumber	Snake Gourd	**M(traditional):**PD,SD:DCTM	Bal1-102-09
Trichosanthes	*kirilowii*	Chinese Cucumber	Snake Gourd	**M(traditional):**FR:DABA,DAFU,DCTM,BCR,RLB,RLC,SCN,**M(traditional):**TU:BDB,RLC,OSA	Bown-364-05
Trichosanthes	*rosthornii*	Rosthorn	Snake Gourd	**M(traditional):**FR:DABA,DAFU,BCR,RLC,SCN	Bown-364-06
Trifolium	*alexandrinum*	Berseem	Clover	**C:**LF:**VG**(fodder)	Clev-123-07
Trifolium	*amabile*	Aztec	Clover	**C:**LF:**VG**	Clev-123-08
Trifolium	*hybridum*	Alsike	Clover	**C:**LF:**VG**(fodder)	Clev-123-09
Trifolium	*pratense*	Red	Clover	**M(traditional):**FL:BLC,DABA,DABI,DAID,DHIV,DSIN,SIB	Bal1-076-01
Trifolium	*pratense*	Red	Clover	**M(traditional):**FL:DCBR,DCLY,DCOV,DWHP,NGT,NMS,ODS,DSIN,RLC,UDR	Bown-364-07
Trifolium	*pratense*	Red	Clover	**C:**FL:AF,**M(traditional):**FL-OL:HIN,OBN,ODS,DSIN,RLC	Clev-123-06
Trifolium	*pratense*	Red	Clover	**M(traditional):**FL:DSIN	Wyk-430-04
Trifolium	*pratense*	Red	Clover	**E:**LF,FL,SM:DYE(alum-tan-gold,chrome-camel,copper-gray-brown,tin-peach-gold,iron-gray-brown)	Bliss-260-01
Trifolium	*pratense*	Red	Clover	**E:**FL:DYE(alum-strong-yellow-green,copper-soft-green,tin-brilliant-chartreuse)	Cas1-124-02
Trifolium	*repens*	White	Clover	**E:**FL:PF	Clev-123-10
Trigonella	*foenum-graecum*	Bird's Foot	Fenugreek	**M(traditional):**SD:HFV,OEY,RLA,RSN,SCN	Bal1-070-04
Trigonella	*foenum-graecum*	Bird's Foot	Fenugreek	**C:**SD:**SP**,**M(traditional):**SD:BHC,BHS,DSIN,OUL,SDG,DBRI,**M(ayurveda):**SD:DANX,HFV,RLB,RLC,SIB,**M(PS):**SD:BDB,DFRA,DSIN,SAP	Bal2-067-02 Bal2-517-05
Trigonella	*foenum-graecum*	Bird's Foot	Fenugreek	**C:**LF:**SN**,**C:**SD:**SP**,**C:**SR:**VG**,**M(traditional):**LF,SD:BDB,BSH,DALL,DCTM,DTBC,DWRM,HAP,HFV,NJA,NBK,NGT,NMP,DSIN,RLC,SDG,SCN,SIB,UDR	Bown-364-08
Trigonella	*foenum-graecum*	Bird's Foot	Fenugreek	**C:**LF:**SN**,**C:**SD:**SP**,**C:**SR:**VG**,**M(traditional):**LF,SD:BAN,BDB,BHP,SDG,**E:**LF,SD:DYE	Clev-325-01
Trigonella	*foenum-graecum*	Bird's Foot	Fenugreek	**C:**SD:**SP**,**M(traditional):**SD:DBOI,DCLD,HFV,RLA,RLC,SCN,SDG,SUL,**E:**SD:DYE(alum-yellow)	Kowa-191-01 Kowa-287-01
Trigonella	*foenum-graecum*	Bird's Foot	Fenugreek	**C:**SD:**SP**,**M(folk):**SD:DIPT,DWHN,HAP	Krug-073-01
Trigonella	*foenum-graecum*	Bird's Foot	Fenugreek	**M(PS):**SD:DANX,BDB,BHC,DBOI,DIPT,DOBE,DSIN,OUL,RLC,RST,SAP,SDG,SUL	Wyk-325-01, Wyk-430-05
Trillium	*erectum*	Birthroot	Wake Robin	**M(traditional):**RZ:DISB,OAS,DSIN,OST,OUL,RLC	Bown-365-01
Trillium	*erectum*	Birthroot	Wake Robin	**M(folk):**RT,RZ:HAP,MEB,SDR,SDY	Krug-030-01
Trillium	*erectum*	Birthroot	Wake Robin	**M(traditional):**RZ:NMP,NSD,childbirth	Wyk-430-06
Trillium	*kamtshaticum*	Kamschatka	Wake Robin	**E:**RZ:ST	Bown-364-09
Trillium	*tschonowskii*	Tschonowsky	Wake Robin	**E:**RZ:ST	Bown-364-10
Tripleurospermum	*maritima*	Chamomile	False Mayweed	**E:**WP:DYE(alum-strong-yellow,tin-brilliant-yellow,chrome-rich-gold,iron-yellow-green)	Cas1-116-01
Tropaeolum	*majus*	Large	Trophycress	**C:**LF,FL,FL-BD:FF,**M(traditional):**WP: DABA,DAFU,DGUM,DWHN,OHL,DSIN,RLC,UDR	Bown-365-02
Tropaeolum	*majus*	Large	Trophycress	**C:**LF:FF,**C:**FL:GN,**C:**FL-BD:**SN**,**M(traditional):**FL,LF,SD:DASC,SDG,**M(traditional):**SD:DBOI,DWRM	Clev-123-11
Tropaeolum	*majus*	Large	Trophycress	**C:**LF:FF	Kowa-397-01
Tropaeolum	*majus*	Large	Trophycress	**C:**LF:FF,**C:**FL-BD:**SN**,**M(folk):**LF:DGUM(scurvy) **M(folk):**LF,FL:DABA,RLB,UUR	Krug-124-01
Tropaeolum	*majus*	Large	Trophycress	**M(PS):**LF:DABI,DAFU,DCAN,UDR,UUR	Wyk-326-01, Wyk-430-07
Tropaeolum	*majus*	Large	Trophycress	**E:**FL:DYE(alum-beige-yellow,chrome-tan-gold,copper-yellow-green,tin-bright-gold,iron-gray)	Cas1-184-01
Tsuga	*canadensis*	Atlantic	Hemlock	**M(traditional):**BK,OL:RLC	Wyk-430-08
Tsuga	*canadensis*	Atlantic	Hemlock	**E:**SM(tip):DYE(alum-bright-yellow,chrome-bright-gold,chrome-&-iron-warm-brown,tin-brilliant-yellow)	Cas1-151-01
Tulipa	*praecox-monoflora*	Single Early Cup	Tulip	**E:**FL(faded yellow & white):DYE(chrome-gold,tin-yellow,iron-yellow-gray), **E:**FL(faded red & maroon):DYE(alum-&-vinegar-tan-red,chrome-&-copper-warm-brown,iron-gray)	Cas1-216-01
Turnera	*diffusa*	Damiana	Alder Hazel	**M(traditional):**LF:DFRI,HAP,SDG	Bal1-068-06
Turnera	*diffusa*	Damiana	Alder Hazel	**M(traditional):**LF:DFRI,DIPT,HAN,HAP,SAP	Bal2-057-01 Bal2-513-06

Latin Genus	L. Species	E. Species	English Genus	C:Culinary,M:Medicinal,E:Economic	Source
Turnera	*diffusa*	Damiana	Alder Hazel	**M(traditional):**WP:DCFS,DFRI,DIPT,DVGB,NMP,HAN,NSD,SAP,SCN,SDG,UEP,UUR	Bown-365-03
Turnera	*diffusa*	Damiana	Alder Hazel	**M(PS):**LF:DCFS,DIPT,HAN,HAP,UEP,**M(PS):**OL:DAMY	Wyk-327-01, Wyk-430-09
Turnera	*ulmifolia*	Sage Rose	Alder Hazel	**M(traditional):**WP:DCFS,DFRI,DIPT,DVGB,NMP,HAN,NSD,SAP,SCN,SDG,UEP,UUR	Bown-365-04
Tussilago	*farfara*	Coughwort	Colt's Foot	**M(traditional):**FL-BD:DPNB,RLB,RLC,RTL, **M(PS):**FL-BD:DABA	Bal2-052-01 Bal2-512-05
Tussilago	*farfara*	Coughwort	Colt's Foot	**C:**LF:**VG**,**C:**FL:AF(wine),**M(traditional):**FL,LF:DISB,DWHP,OAS,OUL,DSIN,NMS,RLA,RLB,RLC,RTL	Bown-365-05
Tussilago	*farfara*	Coughwort	Colt's Foot	**C:**FL:AF,**C:**RT:CF,**M(traditional):**LF:RLB,RTL, **M(traditional):**FL,LF:OUL,OWN	Clev-124-01
Tussilago	*farfara*	Coughwort	Colt's Foot	**C:**FL:AF(cleats-beer,clayt-wine),**C:**LF:**VG**, **M(traditional):**LF-MC:RST	Kowa-097-01
Tussilago	*farfara*	Coughwort	Colt's Foot	**M(folk):**FL,LF:RLA,RLB,RLC	Krug-055-01
Tussilago	*farfara*	Coughwort	Colt's Foot	**M(PS):**LF:DAMW,RLA,RLB,RLC,RST	Wyk-328-01, Wyk-430-10
Tussilago	*farfara*	Coughwort	Colt's Foot	**E:**FL:DYE(alum-yellow,chrome-dark-gold, copper-tan,tin-bright-gold,iron-beige-gray), **E:**LF:DYE(alum-yellow,chrome-tan,copper-gray, tin-yellow-green,iron-taupe	Cas1-126-01
Tylophora	*asthmatica*	Asthma	Tylophorine	**M(traditional):**LF:RLC,**E:**LF:PS☠	Wyk-430-11
Typha	*angustifolia*	Bulrush	Cattail	**M(traditional):**LF,FL(pollen),FR(hairs):BCR,DCLY,DWTP,NMP,OAS,OSA,OWN,SDR,UDR,**E:**LF,FR(hairs):FI	Bown-366-01
Typha	*angustifolia*	Bulrush	Cattail	**M(traditional):**FL(pollen):DTNS,OAS,DSIN	Wyk-430-12
Typha	*bungeana*			Synonym for *Typhalax mannii*	
Typha	*davidiana*	David	Cattail	**M(traditional):**LF,FL(pollen),FR(hairs):BCR,DCLY,DWTP,NMP,OAS,OSA,OWN,SDR,UDR,**E:**LF,FR(hairs):FI	Bown-366-03
Typha	*latifolia*	Reedmace	Cattail	**M(traditional):**LF,FL(pollen),FR(hairs):BCR,DCLY,DWTP,NMP,OAS,OSA,OWN,SDR,UDR,**E:**LF,FR(hairs):FI	Bown-366-04
Typha	*latifolia*	Reedmace	Cattail	**C:**LF,RT,SM:**VG, E:**LF:DYE(alum-beige-gold, chrome-gold, copper-olive-green, tin-creamy-gold, iron-olive-green)	Bliss-078-01
Typha	*latifolia*	Reedmace	Cattail	**E:**SH:DYE(alum-yellow,alum-&-copper-yellow-green,chrome-yellow-gold,tin-bright-yellow), **E:**RT:DYE(vinegar-tan,alum-&-salt-yellow, **E:**FL:DYE(alum-beige,chrome-gold,iron-brown)	Cas1-114-01
Typha	*laxmannii*	Laxmann	Cattail	**M(traditional):**LF,FL(pollen),FR(hairs):BCR,DCLY,DWTP,NMP,OAS,OSA,OWN,SDR,UDR,**E:**LF,FR(hairs):FI	Bown-366-02
Typha	*minima*	Lesser	Cattail	**M(traditional):**LF,FL(pollen),FR(hairs):BCR,DCLY,DWTP,NMP,OAS,OSA,OWN,SDR,UDR,**E:**LF,FR(hairs):FI	Bown-366-05
Typha	*orientalis*	Asian	Cattail	**M(traditional):**LF,FL(pollen),FR(hairs):BCR,DCLY,DWTP,NMP,OAS,OSA,OWN,SDR,UDR,**E:**LF,FR(hairs):FI	Bown-366-06
Typhonium	*giganteum*	Giant	Typhon Ar Lily	**M(traditional):**RZ:HAC,NSD	Wyk-430-13
Ugni	*molinae*	Chilean Murtillo	Cranberry Guava	**C:**FR:JJ	Clev-106-07
Ulex	*europaeus*	European	Gorse	**E:**FL:DYE(chrome-orange)	Grae-136-03
Ulmus	*americana*	American	Elm	**E:**LF:DYE(alum-yellow,vinegar-tan,baking-soda-yellow-green,chrome-gold), **E:**BK:DYE(alum-&-salt-tan,vinegar-tan-beige, chrome-gray-brown,iron-dark-gray-brown	Cas1-140-01
Ulmus	*fulva*			Synonym for *Ulmus rubra*	Krug-161-01
Ulmus	*rubra*	Slippery	Elm	**M(traditional):**BK(inner):DCLD,DFLU,RST,SDR, SUL	Bal1-077-04
Ulmus	*rubra*	Slippery	Elm	**M(traditional):**BK(inner):NGT,NRH,OBN,OWN, RST,SDG,SIB,SUL,**(PS):**BK(inner):DCRO,DCTM	Bal2-133-01 Bal2-543-03
Ulmus	*rubra*	Slippery	Elm	**M(traditional):**BK(inner):BCB,DBOI,OBN,OSA, OWN,RLC,RST,SCN,SIB,SUL	Bown-366-07
Ulmus	*rubra*	Slippery	Elm	**M(folk):**BK(inner):SDG,SDR	Krug-161-01
Ulmus	*rubra*	Slippery	Elm	**M(PS):**BK(inner):DAMW,DBOI,OBN,OSA,OUL, OWN,RST,SDR,SIB	Wyk-329-01, Wyk-430-14
Umbellularia	*californica*	California	Bay Laurel	**C:**LF:**SN**,**M(traditional):**LF:HAC,NNR,NSD,SIB	Bown-366-08
Umbellularia	*californica*	California	Bay Laurel	**E:**LF:DYE(alum-light-tan, chrome-beige, copper-light-avocado, tin-light-yellow, iron-charcoal-gray)	Bliss-076-01
Umbellularia	*californica*	California	Bay Laurel	**E:**FR:DYE(alum-&-tartar-beige-rose,chrome-&-tartar-khaki-green),**E:**FL:DYE(alum-beige-green)	Grae-165-03 Grae-166-01

Latin Genus	L. Species	E. Species	English Genus	C:Culinary,M:Medicinal,E:Economic	Source
Umbilicaria	*americana*	American	Tripe Lichen	**E:**WP:DYE(Gyrophoric-Acid-&-Ammonia-purple-magenta)	Cas2-033-13
Umbilicaria	*coriacea*	Leather	Tripe Lichen	**E:**WP:DYE(Depsides-&-Depsidones-&-Xanthones-chocolate-brown)	Cas2-036-11
Umbilicaria	*deusta*	Burned	Tripe Lichen	**E:**WP:DYE(Gyrophoric-Acid-&-Umbilicaric-Acid-&-Ammonia-purple-magenta)	Cas2-033-14
Umbilicaria	*deusta*	Burned	Tripe Lichen	**E:**WP:DYE(Gyrophoric-Acid-&-Umbilicaric-Acid-Ammonia-magenta-violet)	Grae-078-01
Umbilicaria	*hyperborea*	Hyperboreal	Tripe Lichen	**E:**WP:DYE(Gyrophoric-Acid-&-Umbilicaric-Acid-&-Ammonia-purple-magenta)	Cas2-033-15
Umbilicaria	*hyperborea*	Hyperboreal	Tripe Lichen	**E:**WP:DYE(Gyrophoric-Acid-&-Umbilicaric-Acid-&-Ammonia-magenta-violet)	Grae-078-02
Umbilicaria	*lyngei*	Lynge	Tripe Lichen	**E:**WP:DYE(Gyrophoric-Acid-brown,Gyrophoric-Acid-&-Ammonia-violet-brown)	Cas2-031-02
Umbilicaria	*meuhlenbergii*	Meuhlenberg	Tripe Lichen	**E:**WP:DYE(Gyrophoric-Acid-brown,Gyrophoric-Acid-&-Ammonia-red-pink)	Cas2-033-17
Umbilicaria	*papulosa*			Synonym for *Lasallia papulosa*	Cas2-033-18
Umbilicaria	*pensylvanica*			Synonym for *Lasallia pensylvanica*	Cas2-033-19
Umbilicaria	*phaea*	Dusky	Tripe Lichen	**E:**WP:DYE(Gyrophoric-Acid-&-Umbilicaric-Acid-&-Ammonia-purple-magenta)	Cas2-033-20
Umbilicaria	*phaea*	Dusky	Tripe Lichen	**E:**WP:DYE(Gyrophoric-Acid-&-Umbilicaric-Acid-&-Ammonia-magenta-violet)	Grae-078-03
Umbilicaria	*polyphylla*	Many Frond	Tripe Lichen	**E:**WP:DYE(Gyrophoric-Acid-&-Umbilicaric-Acid-&-Ammonia-purple-magenta)	Cas2-033-21
Umbilicaria	*polyphylla*	Many Frond	Tripe Lichen	**E:**WP:DYE(Gyrophoric-Acid-&-Umbilicaric-Acid-&-Ammonia-magenta-violet)	Grae-078-04
Umbilicaria	*polyrrhiza*	Many Rhizome	Tripe Lichen	**E:**WP:DYE(Gyrophoric-Acid-&-Umbilicaric-Acid-&-Ammonia-purple-magenta)	Cas2-033-22
Umbilicaria	*proboscidea*	Snout	Tripe Lichen	**E:**WP:DYE(Gyrophoric-Acid-&-Umbilicaric-Acid-&-Ammonia-purple-magenta)	Cas2-033-23
Umbilicaria	*proboscidea*	Snout	Tripe Lichen	**E:**WP:DYE(Gyrophoric-Acid-&-Umbilicaric-Acid-&-Ammonia-magenta-violet)	Grae-078-05
Umbilicaria	*pustulata*			Synonym for *Lasallia pustulata*	Cas2-033-24
Umbilicaria	*torreyfacta*	Torreyfact	Tripe Lichen	**E:**WP:DYE(Gyrophyoric-Acid-&-Stictic-Acid-brown,Gyrophyoric-Acid-&-Stictic-Acid-&-Ammonia-purple-magenta)	Cas2-033-25
Umbilicaria	*vellea*	Velvet	Tripe Lichen	**E:**WP:DYE(Gyrophoric-Acid-&-Umbilicaric-Acid-&-Ammonia-purple-magenta)	Cas2-033-26
Umbilicaria	*vellea*	Velvet	Tripe Lichen	**E:**WP:DYE(Gyrophoric-Acid-&-Umbilicaric-Acid-&-Ammonia-magenta-violet)	Grae-078-06
Umbilicaria	*virginis*	Virgin	Tripe Lichen	**E:**WP:DYE(Gyrophoric-Acid-&-Norstictic-Acid-&-Ammonia-purple-magenta)	Cas2-032-16
Umbellularia	*californica*	California	Umbel Laurel	**M(traditional):**LF,FR:DASC,OBR,OBN,SAP	Clev-098-06
Uncaria	*gambir*	Gambir	Uncary	**M(traditional):**WP:OAS,SDR,SDY	Wyk-430-15
Uncaria	*guianensis*	Cat's Claw Guiana	Uncary	**M(folk):**BK:SDR,SIB,**M(traditional):**BK:BCR,DCLD, **M(PS):**BK:DAID,DAVR,DCTM,DFRA,NJA, **E:**BK:**contraceptive** ☠,	Bal2-043-02 Bal2-509-05
Uncaria	*tomentosa*	Cat's Claw Paraguay	Uncary	**M(traditional):**BK,RT:DAID,DAVR,DFRA,NJA,SCC,SUL	Bal1-067-02
Uncaria	*tomentosa*	Cat's Claw Paraguay	Uncary	**M(folk):**BK:SDR,SIB,**M(traditional):**BK:BCR,DCLD, **M(PS):**BK:DAID,DAVR,DCTM,DFRA,NJA, **E:**BK:**contraceptive** ☠,	Bal2-043-03 Bal2-509-05
Uncaria	*tomentosa*	Cat's Claw Paraguay	Uncary	**M(PS):**SM-BK,RT:BCR,BDB,DAID,DAVR,DCFS,DCTM,DFRA,DHSV,NJA,RLA	Wyk-330-01, Wyk-430-16
Urceolaria	*scruposa*			Synonym for *Diploschistes scruposus*	Cas2-031-15
Urginea	*maritima*	Sea Squill	Urgin Squill	**M(PS):**BR:BHS,OWN,RLB,RLA,RLC,UDR,**E:**BR:PS☠	Wyk-331-01, Wyk-430-17
Urtica	*breweri*	American	Nettle	**C:**LF,SH:**VG**	Clev-124-03
Urtica	*dioca*	Stinging	Nettle	**M(traditional):**FL,LF,RT:BAN,DALL,DHYF,NJA,NSD,RLC,UDR,UKC	Bal1-074-07
Urtica	*dioca*	Stinging	Nettle	**C:**LF:**VG**,**M(folk):**LF,RT:BHE,DSBR,NGT,NRH,UEP,**M(traditional):**LF,RT:DALL,DHVS,DHYF,DTBC,NJA,RLC,UDR,**M(PS):**LF,RT:NRH,UKS,UUR	Bal2-142-01 Bal2-546-01
Urtica	*dioca*	Stinging	Nettle	**C:**LF:**VG**,**M(traditional):**WP:DISB,BAN,BHP,BSH,NJA,NGT,NRH,OAS,OBN,OHM,OGT,OHL,OJA,ONR,OSC,DSIN,UDR,**E:**LF:DYE(chlorophyll-green)	Bown-366-09
Urtica	*dioca*	Stinging	Nettle	**C:**LF,SH:AF, **VG**,**M(traditional):**LF:DHMY,OAS,RLC,SCN,**E:**LF:HO,**E:**SM:FI	Clev-124-02
Urtica	*dioca*	Stinging	Nettle	**C:**LF:**VG**,**M(traditional):**WP:RLA,**E:**OL:HO,**E:**WP:CM,DYE(alum-green-yellow,chrome-tan)	Kowa-174-09 Kowa-471-01
Urtica	*dioca*	Stinging	Nettle	**M(folk):**LF:BAN,DWHN,NJA,NGT,NRH	Krug-125-01

Latin Genus	L. Species	E. Species	English Genus	C:Culinary,M:Medicinal,E:Economic	Source
Urtica	*dioca*	Stinging	Nettle	**M(CS):**LF,RT:NRH,UKS,UUR,**M(traditional):**RT:UEP	Wyk-332-01
Urtica	*dioca*	Stinging	Nettle	**E:**WP:DYE(alum-yellow-green,chrome-gold,copper-strong-yellow-green,tin-chartreuse, iron-gray-green)	Cas1-184-02
Urtica	*pilulifera*	Roman	Nettle	**M(traditional):**WP:BPH,BSH,OAS,UDR	Bown-366-10
Urtica	*thunbergii*	Japanese	Nettle	**C:**LF,SH:**VG**	Clev-124-04
Urtica	*urens*	Burning	Nettle	**M(traditional):**WP:BPH,BSH,OAS,UDR	Bown-366-11
Urtica	*urens*	Burning	Nettle	**C:**LF,SH:AF,**VG**,**M(traditional):**LF:DHMY,OAS,RLC,SCN,**E:**LF:HO,**E:**SM:FI	Clev-124-05
Urtica	*urens*	Burning	Nettle	**M(folk):**LF:BAN,DWHN,NJA,NGT,NRH	Krug-125-02
Urtica	*urens*	Burning	Nettle	**M(traditional):**LF:NRH,**M(traditional):**RT:UEP	Wyk-430-18
Usnea	*articulata*		Beard Shield Lichen	**E:**WP:DYE(Usnic-Acid-ocher-gold)	Cas2-036-12
Usnea	*barbata*	Old Man's Beard	Beard Shield Lichen	**M(traditional):**WP:RLC,RST	Wyk-430-19
Usnea	*barbata*	Old Man's Beard	Beard Shield Lichen	**E:**WP:DYE(Barbatolic-Acid-orange-red-brown, Barbatolic-Acid-&-Ammonia-blue-purple)	Cas2-033-28 Cas2-036-13
Usnea	*barbata*	Old Man's Beard	Beard Shield Lichen	**E:**WP:DYE(Barbatolic-Acid-deep-orange-russet)	Grae-079-01
Usnea	*dasypoga*	Hairy Hanging	Beard Shield Lichen	**E:**WP:DYE(Salazinic-Acid-orange-brown)	Cas2-036-14
Usnea	*dasypoga*	Hairy Hanging	Beard Shield Lichen	**E:**WP:DYE(Salazinic-Acid-red-brown)	Cas2-036-15
Usnea	*diffracta*	Diffracted	Beard Shield Lichen	**E:**WP:DYE(Usnic-Acid-&-Atronin-green)	Cas2-036-16
Usnea	*diffracta*	Diffracted	Beard Shield Lichen	**E:**WP:DYE(Usnic-Acid-&-Ammonia-green)	Cas2-036-16
Usnea	*glabrata*	Smooth	Beard Shield Lichen	**E:**WP:DYE(Usnic-Acid-&-Ammonia-gray-green-yellow)	Cas2-033-29
Usnea	*hirta*	Hairy	Beard Shield Lichen	**E:**WP:DYE(Usnic-Acid-&-Ammonia-purple)	Cas2-033-30
Usnea	*rubescens*	Red	Beard Shield Lichen	Synonym for *Chaenothecopsis rubescens*	Cas2-036-17
Usnea	*subfloridana*	Florida	Beard Shield Lichen	**E:**WP:DYE(Salzinic-Acid-gray-olive)	Cas2-036-18
Vaccinium-blauea	*angustifolium*	Low Bush	Blueberry	**E:**LF:DYE(alum-tan-gold, chrome-rust, copper-dark-tan, tin-orange-gold, iron-dark-gray-green)	Bliss-062-01
Vaccinium-blauea	*corymbosum*	High Bush	Blueberry	**E:**LF:DYE(alum-yellow, tin-brilliant-yellow, copper-sulfate-&-iron-yellow-green), **E:**FR:DYE(chrome-&-vinegar-tan-pink, iron-&-salt-pink-gray)	Cas1-108-01
Vaccinium-cranabacca	*macrocarpon*	American Large	Heath Cranberry	**M(traditional):**FR-JU:UBL,UUR	Bal1-068-05
Vaccinium-cranabacca	*macrocarpon*	American Large	Heath Cranberry	**E:**FR-JU+sugar:DYE(vinegar-beige,alum-&-tin-tan-pink,chrome-gray,iron-taupe)	Cas1-130-01
Vaccinium-vortabacca	*myrtillus*	Bilberry	Whortleberry	**M(traditional):**LF:DECT,DEMD,DFRA,DHOG,HAN,UDR,UUR	Bal1-065-06
Vaccinium-vortabacca	*myrtillus*	Bilberry	Whortleberry	**M(traditional):**LF:BAA,BCR,DFRA,OUL,OWN,RST,SDR,**M(PS):**LF:BDB,BHS,BVV,DABA(*Escherichia coli, Staphylococcus aureus*),DECT,NMP,SUL	Bal2-032-01 Bal2-506-03
Vaccinium-vortabacca	*myrtillus*	Bilberry	Whortleberry	**C:**FR:AF,JJ,**M(traditional):**LF:BDB,**M(traditional):**FR:DGUM,BAN,BSH,OAS,OBN,OHM,DSIN,SDR,SDY,UDR,UUR	Bown-367-01
Vaccinium-vortabacca	*myrtillus*	Bilberry	Whortleberry	**M(PS):**FR:DAMW,RST,SDR,**M(PS):**LF:BCR,BDB,NGT,NJA,OHM,DSIN,SDG,UKC,UUR	Wyk-333-01, Wyk-430-20
Vaccinium-cranabacca	*oxycoccus*	European	Heath Cranberry	**M(traditional):**FR-JU:UBL,UUR	Bal1-068-05
Vaccinium-cranabacca	*oxycoccus*	European	Heath Cranberry	**E:**FR-JU+sugar:DYE(vinegar-beige,alum-&-tin-tan-pink,chrome-gray,iron-taupe)	Cas1-130-02
Vaccinium-cranabacca	*oxycoccus*	European	Heath Cranberry	**M(PS):**FR:DGUM(scurvy),UBL,UKC,UUR	Wyk-334-01, Wyk-430-21
Vaccinium	*vitis-idaea*	Lignonberry	Hyacinth Heath	**M(traditional):**FR:BDB,UBL,SDR,UUR	Bown-367-02
Valeriana	*celtica*	Celtic Spikenard	Valerian	**E:**WP:LS(rockgarden)	Clev-124-07
Valeriana	*edulis*	Edible	Valerian	**C:**RT:**VG**	Clev-124-08
Valeriana	*jatamansi*	Jatamans	Valerian	**M(traditional):**RT,RZ,OL:HAN	Bown-367-03
Valeriana	*mexicana*	Mexican	Valerian	**M(traditional):**RT,RZ,OL:HAN	Bown-367-04
Valeriana	*officinalis*	Vandal Root	Valerian	**M(traditional):**RT,RZ:BCR,BHP,DCFS,DCLD,HAN,HIN,NMS,NSD,SIB,SUL	Bal1-078-03
Valeriana	*officinalis*	Vandal Root	Valerian	**M(traditional):**RT:HAC,HAN,HIN,NNR,NSD,SDG,SIB, **M(PS):**RT:HAN,HIN	Bal2-148-01 Bal2-549-02
Valeriana	*officinalis*	Vandal Root	Valerian	**C:**RT,RZ,OL:AF,BF,CF,**M(traditional):**RT,RZ,OL:BHP,HAC,HAN,HIN,NMS,NSD,DSIN,OUL,SDG,**E:**WP:PS☠	Bown-367-05
Valeriana	*officinalis*	Vandal Root	Valerian	**M(traditional):**RT:HAC,HAN,HIN,NMS,NSD,OEY,SIB	Clev-124-06
Valeriana	*officinalis*	Vandal Root	Valerian	**M(traditional):**RT,RZ:NMS,NSD,,**E:**WP:PS☠(depression,dizziness,stupor,vomiting)	Kowa-293-02 Kowa-495-01
Valeriana	*officinalis*	Vandal Root	Valerian	**M(folk):**RT:HAN,HIN,DEPI,NSD	Krug-175-01
Valeriana	*officinalis*	Vandal Root	Valerian	**M(CS):**RT,RZ:HIN,NSD,SIB	Wyk-335-01, Wyk-430-22

Latin Genus	L. Species	E. Species	English Genus	C:Culinary,M:Medicinal,E:Economic	Source
Valeriana	*phu*	Cretan Spikenard	Valerian	**M(traditional):**RT:HAC,HAN,HIN,NMS,NSD,OEY,SIB	Clev-124-09
Vanilla-tropicameria	*fragrans*			Synonym for *Vanilla planifolia*	Clev-326-01
Vanilla-tropicameria	*planifolia*	Central America	Tropical American Vanilla Orchid	**C:**FR(bean):**SP**	Bown-368-01
Vanilla-tropicameria	*planifolia*	Central America	Tropical American Vanilla Orchid	**C:**FR(bean):**SP**,**M(traditional):**FR(bean):HAP,HFV,**E:**FR(bean):PF	Clev-326-01
Vanilla-tropicameria	*planifolia*	Central America	Tropical American Vanilla Orchid	**C:**FR(bean):**SP**	Wyk-430-23
Vanilla-tropicameria	*pompona*	West Indies	Tropical American Vanilla Orchid	**C:**FR(bean):**SP**	Bown-368-02
Vanilla-australasia	*tahitensis*	Tahiti	South Asian Vanilla Orchid	**C:**FR(bean):**SP**	Bown-368-03
Veratrum	*album*	White	False Hellebore	**M(traditional):**RT:BHP,DWRM,**E:**RT:IC,PS☠	Clev-125-02
Veratrum	*album*	White	False Hellebore	**M(traditional):**RZ:NRH,NSD,SPR, **E:**RZ:IC,PS☠	Wyk-430-24
Veratrum	*viride*	American Green	False Hellebore	**M(traditional):**RT:BHP,DWRM,**E:**RT:PS☠	Clev-125-01
Veratrum	*viride*	American Green	False Hellebore	**M(traditional):**RT:DHMY,BHP,BLC,NRH,NSC,UKC,**E:**RT:PS☠	Kowa-008-01 Kowa-279-02
Verbascum	*densiflorum*	Large Flowered	Mullein	**M(traditional):**WP:DASC,NSD,RLC,UDR	Bown-368-04
Verbascum	*densiflorum*	Large Flowered	Mullein	**M(traditional):**FL:DFLU,DSIN,OWN,RLB,RLC,UDR	Wyk-336-02
Verbascum	*nigrum*	Dark	Mullein	**M(traditional):**WP:DASC,NSD,RLC,UDR	Bown-368-05
Verbascum	*phlomoides*	Orange	Mullein	**M(traditional):**LF:DEAI,DHYF,HIN,NSD,DCLW,RLA,RLB,SCN	Bal1-074-04
Verbascum	*phlomoides*	Orange	Mullein	**M(traditional):**WP:DASC,NSD,RLC,UDR	Bown-368-06
Verbascum	*phlomoides*	Orange	Mullein	**M(PS):**FL:DFLU,DSIN,OWN,RLB,RLC,UDR	Wyk-336-01, Wyk-431-01
Verbascum	*thapsus*	Aaron's Rod	Mullein	**M(traditional):**WP:BAN	Bal1-129-08
Verbascum	*thapsus*	Aaron's Rod	Mullein	**M(traditional):**FL:DFLU,DSIN,OWN,RLB,RLC,UDR	Wyk-336-03
Verbascum	*thapsus*	Aaron's Rod	Mullein	**M(traditional):**WP:BCB,DASC,DBOI,DEAI,DFLU,DTBC,DTNS,DWHP,HAN,HIN,NSD,OHM,ORH,OWN,RLA,RLB,RLC,RTL,UDR,UUR	Bown-368-07
Verbascum	*thapsus*	Aaron's Rod	Mullein	**M(traditional):**FL,LF:RLA,RLC,**E:**LF,SM:candle wick	Krug-121-01
Verbascum	*thapsus*	Aaron's Rod	Mullein	**E:**LF,SL:DYE(alum-gray-yellow, chrome-bright-ocher, copper-olive, tin-light-yellow, iron-gray)	Bliss-176-01
Verbascum	*thapsus*	Aaron's Rod	Mullein	**E:**WP(top),FL:DYE(alum-soft-yellow, chrome-gold-bronze,copper-bronze,tin-bright-yellow,iron-gray)	Cas1-179-01
Verbascum	*thapsus*	Aaron's Rod	Mullein	**E:**LF:DYE(alum-&-ammonia-bright-yellow, chrome-&-ammonia-chartreuse), **E:**FL,LF:DYE(alum-&-copper-moss-green, chrome-&-copper-dark-yellow-green)	Grae-099-01 Grae-100-01
Verbena	*hastata*	Blue	Vervain	**M(traditional):**WP:BLC,RLA	Bown-368-08
Verbena	*hastata*	Blue	Vervain	**M(traditional):**WP:DCLD,RLC	Wyk-431-02
Verbena	*officinalis*	Herb of the Cross	Vervain	**M(traditional):**WP:BJA,BLC,DCFS,DGUM,DFLU,DMAL,DWFK,HAC,HAN,NSD,OAS,ONR,OSA,DSIN,RLA,SGB,UDR	Bown-368-09
Verbena	*officinalis*	Herb of the Cross	Vervain	**M(traditional):**WP:BLC,DHMY,HFV,NRH,NSD,OEY,DSIN,SDG,SIB	Clev-125-03
Verbena	*officinalis*	Herb of the Cross	Vervain	**M(traditional):**WP:HFV,NMS,OAS,**E:**WP:CM	Kowa-497-01
Verbena	*officinalis*	Herb of the Cross	Vervain	**M(folk):**WP:DSNB,HAC,OAS,OWN	Krug-177-01
Verbena	*officinalis*	Herb of the Cross	Vervain	**M(PS):**WP:BLC,DHMY,HFV,NRH,DSIN,OWN,RLA,RLC,RST,UBL,UDR,UKC	Wyk-337-01, Wyk-431-03
Veronica	*beccabunga*	European	Speedwell	**M(traditional):**WP:SPR	Clev-125-05
Veronica	*officinalis*	Heath	Speedwell	**M(traditional):**WP:OAS,NJA,NRH,DSIN,RLB,RLC,SDG,UDR	Bown-368-10
Veronica	*officinalis*	Heath	Speedwell	**M(traditional):**WP:BLC,DSIN,SDG	Clev-125-04
Veronica	*officinalis*	Heath	Speedwell	**M(PS):**WP:DAMW,DHMY,OWN,NJA,RLC,SAP,SDP,SGB	Wyk-338-01, Wyk-431-04
Veronicastrum	*virginicum*	Culver's Root	Tall Speedwell	**M(traditional):**RT,RZ:BLC,NMS,SCN,SDG,SGB	Bown-369-01
Veronicastrum	*virginicum*	Culver's Root	Tall Speedwell	**M(traditional):**RT,RZ:SDG,SPR	Clev-125-06
Veronicastrum	*virginicum*	Culver's Root	Tall Speedwell	**M(folk):**RT,RZ:SPR	Krug-033-01
Veronicastrum	*virginicum*	Culver's Root	Tall Speedwell	**M(traditional):**RT:BLC,SCN,UBL	Wyk-431-05
Vetiveria	*zizanoides*	Cuscus	Vetiver	**M(traditional):**RT:BCR,DASC,NSD,**E:**OL:CM,DS,PF,**E:**RT:FI,IR	Bown-369-02
Vetiveria	*zizanoides*	Cuscus	Vetiver	**E:**RT,OL:IC,PF	Wyk-431-06
Viburnum-bruscia	*cassinoides*	Swamp Haw	Lowbush Cranberry	**M(traditional):**FR-JU:UBL,UUR	Bal1-068-05

Latin Genus	L. Species	E. Species	English Genus	C:Culinary,M:Medicinal,E:Economic	Source
Viburnum-bruscia	*edule*	Moose	Lowbush Cranberry	**M(traditional):**FR-JU:UBL,UUR	Bal1-068-05
Viburnum-scrubia	*opulus*	Crampbark	Highbush Cranberry	**M(traditional):**FR-JU:NMS,NSD	Bal1-133-06
Viburnum-scrubia	*opulus*	Crampbark	Highbush Cranberry	**M(traditional):**BK:BHP,NMP,NMS,NSD,OAS,SCN	Bown-369-03
Viburnum-scrubia	*opulus*	Crampbark	Highbush Cranberry	**M(traditional):**BK:BHP,NMS,SIB	Wyk-339-02, Wyk-431-07
Viburnum-scrubia	*trilobum*	American	Highbush Cranberry	**M(traditional):**FR-JU:UBL,UUR	Bal1-068-05
Viburnum-drysia	*prunifolium*	Black Haw	Tree Cranberry	**M(traditional):**FR-JU:UBL,UUR	Bal1-068-05
Viburnum-drysia	*prunifolium*	Black Haw	Tree Cranberry	**M(traditional):**SM-BK,RB:BHP,NMP,NMS,NSD,OAS,RLA	Bown-369-04
Viburnum-drysia	*prunifolium*	Black Haw	Tree Cranberry	**M(PS):**SM-BK,RB:BHP,NMS,SIB	Wyk-339-01, Wyk-431-08
Vicia	*benghalensis*	Winter Hairy	Vetch	**E:**FL:DYE(alum-&-tin-deep-turquoise-green, chrome-&-tin-deep-turquoise-green, alum-&-copper-bright-yellow-green, chrome-&-copper-bright-yellow-green)	Grae-110-01 Grae-110-02
Vicia	*cracca*	Bird	Vetch	**E:**WP:DYE(alum-soft-yellow,chrome-tan,copper-gold, tin-bright-yellow, iron-green)	Cas1-217-01
Vinca	*minor*	Lesser	Periwinkle	**M(traditional):**WP:SDG,UDR	Bown-369-05
Vinca	*minor*	Lesser	Periwinkle	**M(traditional):**WP:BHP,OAS,DSIN,OUL,OWN,RST	Clev-126-01
Vinca	*minor*	Lesser	Periwinkle	**M(PS):**WP:DDEM,DVGT	Wyk-340-01, Wyk-431-09a
Vinca	*major*	Greater	Periwinkle	**M(traditional):**WP:BAS,DAMW,DVGB,DVGV,OAS,RST	Bown-369-06
Vinca	*major*	Greater	Periwinkle	**M(traditional):**WP:BHP,OAS,DSIN,OUL,OWN,RST	Clev-126-02
Vinca	*major*	Greater	Periwinkle	**M(folk):**WP(flowering):DGUM,MEB,OAS,OWN	Krug-137-01
Vinca	*major*	Greater	Periwinkle	**M(traditional):**WP:DDEM,DVGT	Wyk-340-02, Wyk-431-09b
Vincetoxicum	*hirundinaria*			Synonym for *Cynanchum vincetoxicum*	Wyk-431-10
Viola	*arvensis*	European Field Pansy	Violet	**E:**FL(blue,maroon,purple):DYE(vinegar-tan,alum-beige-pink,chrome-gold,iron-gray), **E:**FL(mixed): DYE(alum-soft-yellow,chrome-gold,copper-warm-tan, tin-brilliant-yellow,iron-soft-gray	Cas1-188-01
Viola	*bicolor*	American Field Pansy	Violet	**E:**FL(blue,maroon,purple):DYE(vinegar-tan,alum-beige-pink,chrome-gold,iron-gray), **E:**FL(mixed): DYE(alum-soft-yellow,chrome-gold,copper-warm-tan, tin-brilliant-yellow,iron-soft-gray	Cas1-188-02
Viola	*diffusa*	Diffuse	Violet	Chinese Medicine	Bown-370-01
Viola	*inconspicua*	Inconspicuous	Violet	Chinese Medicine	Bown-370-02
Viola	*odorata*	Sweet	Violet	**C:**FL:CF,**M(traditional):**FL,LF:DAMW,DAMY,DASC,DCBR,DCLU,DCST,RLA,RLB,RLC,RST,**E:**OL:PF	Bown-370-03
Viola	*odorata*	Sweet	Violet	**M(traditional):**FL,LF:RLC,RSN,**E:**FL:PF	Clev-126-04
Viola	*odorata*	Sweet	Violet	**C:**FL,LF:GN,**M(traditional):**FL,LF:DASC,RLC	Kowa-498-01
Viola	*odorata*	Sweet	Violet	**C:**FL:AF(parfait amour), **M(folk):**LF:HAC,HIN, **M(folk):**FL,LF:RLB,RLC,RST	Krug-170-01
Viola	*odorata*	Sweet	Violet	**M(traditional):**FL:RLC,**E:**FL:CM	Wyk-431-11
Viola	*odorata*	Sweet	Violet	**E:**FL,LF:DYE(alum-yellow,chrome-tan,copper-gold,tin-bright-yellow-green,iron-gray-green)	Cas1-217-01
Viola	*patrinii*	China	Violet	Chinesemedicine	Bown-370-04
Viola	*sororia*	Blue	Violet	**M(traditional):**WP:DCLD,HAC	Clev-126-05
Viola	*striata*	Striped	Violet	**M(traditional):**WP:DCTM	Bown-370-05
Viola	*tricolor*	Heart's Ease Pansy	Violet	**M(traditional):**WP:DWHP,HFV,NRH,DSIN,OUL,RLB,RLC,SCN,UDR,UUR	Bown-370-06
Viola	*tricolor*	Heart's Ease Pansy	Violet	**M(traditional):**WP:BHE,BHP,BHS,BLC,DACN,DCLD,NRH,DSIN,SDG	Clev-126-03
Viola	*tricolor*	Heart's Ease Pansy	Violet	**M(folk):**WP(flowering):DACN,DSIN,NJA,NGT,OSS,RLC,UDR,UUR	Krug-130-01
Viola	*tricolor*	Heart's Ease Pansy	Violet	**M(PS):**WP:DACN,DIMP,DISB,NRH,DSIN,RLC,RST,UDR	Wyk-341-01, Wyk-431-12
Viola	*tricolor*	Heart's Ease Pansy	Violet	**E:**FL:DYE(alum-&-copper-deep-yellow-green, alum-blue-green	Grae-127-01 Grae-128-01
Viola	*yezoensis*	Yezo China	Violet	**M(traditional):**WP:DABA,DBOI,DEAI,DERS,DMUM,DSNB,OEY,RST	Bown-370-07
Virola	*calophylloidea*	Calaba	Virol Nutmeg	**M(traditional):**BK,SD:NNC👽,OWN,SDP	Wyk-431-13a
Virola	*cebifera*	Ceba	Virol Nutmeg	**M(traditional):**BK,SD:NNC👽,OWN,SDP	Wyk-431-13b
Viscum	*album*	All Heal	Mulberry Mistletoe	**M(traditional):**FR:BHS,DEPI,NGT,NMS **M(traditional):**SM:DCFS,HAN, **M(traditional):**LF:BHP,BHS,DWHP,RLA,SDR,	Bal2-101-01 Bal2-532-01

Latin Genus	L. Species	E. Species	English Genus	C:Culinary,M:Medicinal,E:Economic	Source
				M(CS):LF,SM:DCTM,NJA	
Viscum	*album*	All Heal	Mulberry Mistletoe	**M(traditional):**LF,SM:BCB,BAS,BHP,BVV,DCH R,DCLU,DCOV,NMS,NSD,OJA,ORH,OUL,UDR	Bown-370-08
Viscum	*album*	All Heal	Mulberry Mistletoe	**M(traditional):**WP:BHS,DCHR,DEPI,NSD,SDG, **E:**FR:PS☠	Kowa-389-01
Viscum	*album*	All Heal	Mulberry Mistletoe	**M(folk):**LF,SM:BHP,DEPI,NSD, **E:**FR:PS☠	Krug-118-01
Viscum	*album*	All Heal	Mulberry Mistletoe	**M(CS):**WP:BCR,BHP,BHS,DCTM,NSD	Wyk-342-01, Wyk-431-14
Viscum	*capense*	Africa Cape	Mulberry Mistletoe	**C:**WP:TE	Wyk-431-15
Viscum	*coloratum*	Devil's Fuge	Mulberry Mistletoe	**M(traditional):**FR:BHS,DEPI,NGT,NMS **M(traditional):**SM:DCFS,HAN, **M(traditional):**LF:BHP,BHS,DWHP,RLA,SDR, **M(CS):**LF,SM:DCTM,NJA	Bal2-101-02 Bal2-532-01
Viscum	*coloratum*	Devil's Fuge	Mulberry Mistletoe	**M(traditional):**LF,SM:BCR	Bown-370-09
Vitellaria	*paradoxa*	Paradox	Shea Butterwood	**C:**SD-OL:margarine,**E:**SD-OL:cacaobutter	Wyk-431-16
Vitex	*agnus-castus*	Monk's Pepper	Chasteberry	**M(traditional):**FR:DACN,DCEN,HIN,SFL,UEP, **M(PS):**FR:DCPR,NMP,NMT	Bal2-150-02 Bal2-549-04
Vitex	*agnus-castus*	Monk's Pepper	Chasteberry	**M(traditional):**FR:NMP,NMS,OAS	Bown-370-10
Vitex	*agnus-castus*	Monk's Pepper	Chasteberry	**C:**SD:**SP**(pepper),**E:**SM:basketwicker	Clev-126-06
Vitex	*agnus-castus*	Monk's Pepper	Chasteberry	**M(CS):**FR:DHMY,NMP,lowers libido	Wyk-343-01, Wyk-431-17
Vitex	*canescens*	Hoary	Chasteberry	**M(traditional):**FR:DABA,NSD	Bown-370-11
Vitex	*negundo*	Chinese	Chasteberry	**M(traditional):**LF:DABA,DCBR,DMAL,DWRG,D SNB,NJA,**M(traditional):**SM:OJA,ORH,OTA,RS T,**M(traditional):**WP:DCLD,RLC,**M(traditional):** LF,FR,RT:RLA,SDG,**M(traditional):**FR,OL,RT:R LB,**M(traditional):**FR:HAC,OEY,OHM,**M(traditio nal):**LF:HAC,DSXG,**M(traditional):**RT:HFV,RLB ,**M(traditional):**LF- JU:OUL,**M(traditional):**OL:DSIN	Bown-371-01
Vitex	*negundo*	Chinese	Chasteberry	**C:**SD:**SP**(pepper),**E:**SM:basketwicker	Clev-126-07
Vitex	*negundo*	Chinese	Chasteberry	**M(traditional):**LF,RT:HML,NBK,NRH	Wyk-431-18
Vitex	*quinata*	Five Leaf	Chasteberry	**M(traditional):**FR:DABA,NSD	Bown-370-12
Vitex	*rotundifolia*	Round Leaf	Chasteberry	Chinesemedicine	Bown-370-13
Vitex	*trifolia*	Three Leaf	Chasteberry	Chinesemedicine	Bown-370-14
Vitis	*labrusca-concordensis*	Concord Fox	Grape	**E:**RD(skin):DYE(chrome-&-ammonia-dark- green-blue, chrome-&-copper-dark-blue, chrome-&-iron-dark-blue)	Grae-177-02 Grae-178-01 Grae-178-02
Vitis	*labrusca-rubra*	Red Fox	Grape	**M(traditional):**RD(skin):BHS,DCBK,DCBR,DCL K,DCMT,DCPR	Bal2-121-01 Bal2-538-02
Vitis	*vinifera-rubra*	Red Wine	Grape	**M(traditional):**SD:DFRA,**M(traditional):**RD(skin):BAN(Bal1-129-06)	Bal1-045-01
Vitis	*vinifera-rubra*	Red Wine	Grape	**M(traditional):**RD(skin):BHS,DCBK,DCBR,DCL K,DCMT,DCPR	Bal2-121-02 Bal2-538-02
Vitis	*vinifera-rubra*	Red Wine	Grape	**C:**LF:FF,**C:**FR:AF(wine),JJ,C:SD- OL:CO,mayonnaise,tartar,**M(traditional):**LF,SM: BCR,BHC,BHP,BVV,DAMW,DGUM,OAS,OEY,D SIN,RST,UDR,**M(traditional):**FR:BLC	Bown-371-02
Vitis	*vinifera-rubra*	Red Wine	Grape	**M(PS):**SD:DFRA	Wyk-344-01, Wyk-431-19
Vitis	*vinifera-rubra*	Red Wine	Grape	**E:**LF:DYE(alum-&-ammonia-&-tartar-intense- yellow)	Grae-177-02
Vitis	*vinifera-rubra*	Red Wine	Grape	**M(PS):**LF:BCR,OAS,SDR	Wyk-344-02
Variolaria	*lactea*			Synonym for *Pertusaria lactea*	Cas2-033-31
Variolaria	*orcina*	Orcinol	Variable Perforated Lichen	**E:**WP:DYE(Orcinol-&-Ammonia-purple)	Cas2-033-32
Voacanga	*africana*	African	Tabersonine	**M(traditional):**SD:semisynthesis of vincamine	Wyk-431-20a
Voacanga	*thouarsii*	Thouars	Tabersonine	**M(traditional):**SD:semisynthesis of vincamine	Wyk-431-20b
Vulpicida	*canadensis*			**E:**WP:DYE(Vulpinic-Acid-green-yellow)	Cas2-034-10
Vulpicida	*tilesii*			**E:**WP:DYE(Vulpinic-Acid-bright-yellow)	Cas2-034-14
Warburgia	*salutarius*	Rollbark	Warburg Pepperbark	**M(PS):**LF,BK,RB:DABI,DAMW,DCLD,DFLU,DM AL,DSXS,HAC,NRH,OTA,RLB,RLC,SUL	Wyk-345-01, Wyk-431-21
Warburgia	*ugandensis*	Ugandan	Warburg Pepperbark	**M(traditional):**BK:SPR	Wyk-431-22
Wasabia	*japonica*	Japanese	Wasabi	**C:**RT:FF:**M(traditional):**RT:SDG,fishpoisoninga ntedote	Bown-371-03
Wisteria	*frutescens*	American	Wisteria	**M(traditional):**LF:DHIV	Bal1-526-16
Wedelia	*calendulacea*	Margold	Wedel Oxeye Daisy	**M(traditional):**WP:BLC	Wyk-431-23
Withania	*somnifera*	Winter Cherry	Ashwagandha	**M(traditional):**WP:DALZ,DCFS,DCTM,HAP,HIN ,HML, **M(PS):**WP:BDB,DLUP,NCT,NJA	Bal2-028-01 Bal2-504-04
Withania	*somnifera*	Winter Cherry	Ashwagandha	**M(traditional):**RT:DCFS,DINF,DIPT,DMSC,HAP ,HIN,OAS,NJA,NSD	Bown-371-04

Latin Genus	L. Species	E. Species	English Genus	C:Culinary,M:Medicinal,E:Economic	Source
Withania	*somnifera*	Winter Cherry	Ashwagandha	**M(PS):**RT:DSXS,NSD,**M(PS):**LF:OHM,ORH,OSA,OWN	Wyk-346-01, Wyk-431-24
Wolfiporia	*cocos*			Synonym for *Wolfiporia extensa cocos*	
Wolfiporia	*extensa*	Tuckahoe	Wolf Foam Pore Fungus	**M(traditional):**WP:UKC,UDR, **M(PS):**WP:DABA,UKC	Bal2-085-01 Bal2-526-03
Wolfiporia	*extensa*	Tuckahoe	Wolf Foam Pore Fungus	**M(traditional):**WP:HAN,HIN,NSD,SDR,UDR	Bown-372-01
Woodfordia	*fruticosa*	Dhathaki	Woodford Loosestrife	**E:**FL,RT:fermenting agent	Wyk-432-01
Wyethia	*angustifolia*	California	Mule Ears	**E:**FL,LF,SM:DYE(alum-brass-gold)	Grae-098-02
Xanthium	*italicum*	Italian	Cocklebur	**E:**LF,SM,Burr:DYE(alum-gold, chrome-tan-olive, copper-brass-gold,tin-gold,iron-dark-gray-green)	Bliss-088-01
Xanthium	*strumarium*	Rough	Cocklebur	**M(traditional):**FR:DABA,DAFU,DISB,NBK,NJA, NMS,NRH,DCLW,RRN,RSN,RST	Bown-372-02
Xanthoparmelia	*centrifuga*			Synonym for *Arctoparmelia centrifuga*	
Xanthoparmelia	*conspersa*	Scattered	Yellow Shield Lichen	**E:**WP:DYE(Stictic-Acid-red-brown,Stictic-Acid-&-Ammonia-drab-apricot)	Cas2-036-20 Cas2-033-33
Xanthoparmelia	*lineola*	Lineole	Yellow Shield Lichen	**E:**WP:DYE(Salzinic-Acid-russet)	Cas2-036-21
Xanthoparmelia	*tartarica*	Tartarian	Yellow Shield Lichen	**E:**WP:DYE(Gyrophoric-Acid-copper)	Cas2-036-22
Xanthoria	*ectanea*	Ectane	Sunburst Schist Lichen	**E:**WP:DYE(Anthraquinones-&-Ammonia-slate-blue)	Cas2-034-01
Xanthoria	*ectanoides*	Lesser Ectane	Sunburst Schist Lichen	**E:**WP:DYEAnthraquinones-&-Ammonia-blue)	Cas2-034-02
Xanthoria	*elegans*	Elegant	Sunburst Schist Lichen	**E:**WP:DYE(Anthraquinones-&-Ammonia-mauve)	Cas2-034-03
Xanthoria	*parietina*	Parietin	Sunburst Schist Lichen	**E:**WP:DYE(Parietin-corn-yellow,Parietin-&-Ammonia-blue-violet)	Cas2-034-04 Cas2-036-23
Xanthoria	*polycarpa*	Polycarp	Sunburst Schist Lichen	**E:**WP:DYE(Anthraquinones-&-Ammonia-blue)	Cas2-034-05
Xysmalobium	*undulatum*	Uzara	Milkbush	**M(PS):**RT:HAC,NMS,OWN,SDR,SIB	Wyk-347-01, Wyk-431-02
Yucca	*aloeifolia*	Aloe	Yucca	**M(traditional):**RT:BLC,DOST,NJA	Bal1-079-06
Yucca	*filamentosa*	Adam's needle	Yucca	**M(traditional):**WP:SCN,**E:**WP:FI,ST	Wyk-432-03
Yucca	*glauca*	Soapweed	Yucca	**E:**FL:DYE(alum-yellow-green, chrome-green-gold, copper-pea-soup-green, tin-light-yellow, iron-tan	Bliss-282-01
Xanthoxylum				Synonym for *Zanthoxylum*	
Zanthoxylum	*acanthopodium*	Sikkim	Prickly Ash	**C:**FR:**SP**	Bown-372-03
Zanthoxylum	*americanum*	American	Prickly Ash	**M(traditional):**FR:BCR,DABI,DSCB,DWFK,DWRO, SGS	Bal2-115-01 Bal2-535-02
Zanthoxylum	*americanum*	American	Prickly Ash	**M(traditional):**BK,FR:BCR,HFV,NBK,NJA,NRH, OJA,ORH,OTA,SDG,SDR	Bown-372-04
Zanthoxylum	*americanum*	American		**M(folk):**BK:NRH,OTA, **M(folk):**BK,FR:BCB,BCR,NMS,SDG,SIB	Krug-143-01
Zanthoxylum	*armatum*	Indochina	Prickly Ash	**C:**FR:**SP**,**M(traditional):**BK,FR:OTA	Bown-372-05
Zanthoxylum	*bungeanum*	Szechuan	Prickly Ash	**M(traditional):**FR:BCR,DABI,DSCB,DWFK,DWRO,SGS	Bal2-115-02 Bal2-535-02
Zanthoxylum	*capense*	Africa Cape	Prickly Ash	Africanmedicine	Bown-372-06
Zanthoxylum	*clavaherculis*	Southern USA	Prickly Ash	**M(traditional):**BK,FR:BCR,HFV,NBK,NJA,NRH, OJA,ORH,OTA,SDG,SDR	Bown-372-07
Zanthoxylum	*piperitum*	Szechuan	Prickly Ash	**C:**LF:FF,**C:**FR:**SP**,**M(traditional):**BK:SDG	Bown-372-08
Zanthoxylum	*piperitum*	Szechuan	Prickly Ash	C:RD(husk):**SP**,**M(traditional):**BK:OTA,**M(traditional):**BK,FR:BLC,NRH,SDG	Clev-319-01
Zanthoxylum	*piperitum*	Szechuan	Prickly Ash	**C:**FR:**SP**,**M(traditional):**FR:DWRM,SDG	Wyk-431-04
Zanthoxylum	*planispinum*	China Winged	Prickly Ash	Chinesemedicine	Bown-372-09
Zanthoxylum	*schinifolium*	Pepper Leaf	Prickly Ash	**C:**LF:FF,**C:**FR:**SP**,**M(traditional):**BK:SDG	Bown-372-10
Zanthoxylum	*simulans*	Similar	Prickly Ash	**C:**LF:FF,**C:**FR:**SP**,**M(traditional):**BK:SDG	Bown-372-11
Zea	*mays-saccharata*	Sweet Corn	Zeatin Grass	**M(folk):**FS(cornsilk):BHE,DOBE,UDR,UEP, **M(traditional):**FS(cornsilk):UBL,UKC, **M(PS):**FS(cornsilk):NCT	Bal1-068-04
Zea	*mays-saccharata*	Sweet Corn	Zeatin Grass	**M(traditional):**FS(cornsilk):BHS,BLC,UBL,UKC, UDR, **M(PS):**FS(cornsilk):NCT	Bal2-055-01 Bal2-513-04
Zea	*mays-saccharata*	Sweet Corn	Zeatin Grass	**C:**FR:**VG**,**C:**OL:CO,**M(traditional):**FS(cornsilk): BSH,UBL,UDR,UEP,UKS	Bown-372-13
Zea	*mays-saccharata*	Sweet Corn	Zeatin Grass	**M(traditional):**FS(cornsilk):UDR,UUR	Krug-058-01
Zea	*mays-saccharata*	Sweet Corn	Zeatin Grass	**M(PS):**FS(cornsilk):NJA,NRH,UBL,UDR,**M(PS):** FL(pollen):SAP,UEP	Wyk-348-01, Wyk-431-05
Zea	*mays-saccharata*	Sweet Corn	Zeatin Grass	**E:**LF,husk:DYE(alum-soft-yellow,chrome-tan,copper-&-tin-soft-yellow-green,iron-gray-tan), **E:**cob:DYE(vinegar-&-chrome-pale-beige)	Cas1-129-01
Zea	*mays-saccharata-purpurea*	Purple Sweet Corn	Zeatin Grass	**E:**FR:DYE(alum-dark-purple-magenta)	Grae-177-01
Zingiber	*cassumunar*			Synonym for *Zingiber montanum*	Bown-373-01
Zingiber	*mioga*	Japanese	Ginger	**C:**RT,RZ:**SP**	Bown-373-02
Zingiber	*montanum*	Arunachal Pradesh	Ginger	**C:**RT,RZ:**SP**	Bown-373-01
Zingiber	*officinale*	South Asia	Ginger	**M(traditional):**RT,RZ:BCR,DFRA,HFV,NMS,OW	Bal1-071-01

Latin Genus	L. Species	E. Species	English Genus	C:Culinary,M:Medicinal,E:Economic	Source
				N,SCC,SDG	
Zingiber	*officinale*	South Asia	Ginger	**M(traditional):**RZ:DCLD,SAP,NJA,OAS,RLC, **M(ayurveda):**RZ:DANX,RST,SDP, **M(PS):**RZ:DWFK,NJA,**M(CS):**RZ:SAP,SDP	Bal2-072-01 Bal2-519-02
Zingiber	*officinale*	South Asia	Ginger	**C:**RT,RZ:**SP**,**M(traditional):**RT,RZ:BCR,BLC,DCLD,DFLU,DVRT,OBK,ORH,RLB,RLC,SDG,SDR,SIB,**E:**OL:PF	Bown-373-03
Zingiber	*officinale*	South Asia	Ginger	**C:**RT,RZ:**SP**,**M(traditional):**RT,RZ:BCR,DFLU,RLC,SDG,SFL,**E:**OL:PF	Clev-328-01
Zingiber	*officinale*	South Asia	Ginger	**C:**RT,RZ:**SP**,**M(traditional):**RT,RZ:BCR,RLB,RSN,SDG,SFL	Kowa-223-01 Kowa-288-02
Zingiber	*officinale*	South Asia	Ginger	**M(folk):**RZ:BCR,DCLD,DFLU,SDG, **M(CS):**RZ:DMSK	Krug-081-01
Zingiber	*officinale*	South Asia	Ginger	**M(CS):**RZ:HFV,NMS,RLC,SDP,SDR,SFL	Wyk-349-01, Wyk-431-06
Zingiber	*zerumbet*	Asian Bitter	Ginger	**C:**RT,RZ:**SP**	Bown-373-04
Zinnia	*elegans*	Florist	Zinn Daisy	**E:**FL:DYE(alum-yellow,chrome-bronze,copper-khaki,tin-bright-gold,iron-gray-green)	Cas1-222-01
Ziziphus	*jujuba*	Chinese Date	Jujube	**C:**FR:PU,**M(traditional):**SD:BAN,HAN,HIN,DCFS,SAP,SDR	Bown-373-05
Ziziphus	*jujuba*	Chinese Date	Jujube	**M(PS):**FR:DALL,RLC,**M(traditional):**SD:HIN,NSD	Wyk-350-01, Wyk-431-07
Ziziphus	*mucronata*	Buffalo Thorn	Jujube	**M(traditional):**FR:DALL,RLC,**M(traditional):**SD:NSD	Wyk-431-08
Ziziphus	*spinosa*	Suan Zau Ren	Jujube	**M(traditional):**SD:HIN,NSD	Wyk-431-09
Ziziphus	*vulgaris*	European	Jujube	**C:**FR:PU,**M(traditional):**SD:BAN,HAN,HIN,DCFS,SAP,SDR	Bown-373-06

VEGETABLES

A diet rich in vegetables is the best defense against chronic disease.

Vegetables

Color	Vegetables
Light Green	Avocado, Bok Choy, Broad Bean, Cabbage, Cardoon Celery, Celtuce, Chayote Gourd, Chinese Mustard Greens, Chives, Celery, Fennel, Iceplant Spinach, Kohlrabi, Lettuce, Lima Beans, Loofah Melon, Marsh Samphire, New Zealand Spinach, Perilla Leaf Garnish, Purslane Spinach, Tomatillo, Watercress, Water Spinach, Wintercress
Dark Green	Artichoke, Asparagus, Mild Bell Peppers, Broccoli, Brussels Sprouts, Chard, Collard Greens, Cornsalad Lettuce, Cucumbers, Green Beans, Kale, Leeks, Mustard Greens, Parsley, Peas, Seakale, Spinach, Zucchini Squash
Blue Green	Blue Cabbage, Blue Seakale, Chinese Chives
Brown Seeds	Buckwheat, Fennel
Orange	Bell Peppers, Butternut Squash, Carrots, Papaya, Pumpkin, Sweet Potato, Yam
Purple	Beets, Cabbage, Eggplant, Onions
Red	Kidney Beans, Bell Peppers, Onions, Radishes, Rhubarb, Runner Beans, Tomatoes
White	Adzuki Bean Sprouts, Anu Tuber, Arrowhead Corm, Bamboo Shoots, Burdock Root, Cassava Tuber (Tapioca), Chinese Artichoke Tuber, Fava Bean Sprouts, Navy Beans, Cauliflower, Evening Primrose Root, Florence Fennel Corm, Horseradish, Jerusalem Artichoke Tuber, Leek, Livingstone Potato, Mung Bean Sprouts, Onions, Parsnip, Potatoes, Sacred Lotus Rakkyo Bulb, Rhizome, Salsify Root, Scorzonera Root, Skirret Root, Taro Root, Turnips, Udo Shoots, Water Chestnut Tuber, White Radish, Yam Bean Tuber
Yellow	Bell Peppers (yellow), Corn, Oca Tuber, Squash

Anti-Cancer Vegetables

Phytochemical	Action	Vegetables
Carotene	Antioxidant	Carrots, Kale, Squash, Spinach, Sweet Potatoes, Tomatoes, Yams
Coumarin	Antioxidant catalyst	Beets, Carrots, Celery, Fennel
Dithiolthione Glucosinolate Thiocyanate	Detoxification	Crucifer (Brassicaceae) vegetables: Broccoli, Brussels Sprouts, Cabbage, Cauliflower, Kale, Spinach
Flavonoid	Anti-oxidant	Greens, Peppers, Tomatoes, Herbs: Bilberry Whortleberry, Blessed Milk Thistle, English Hawthorn Berry, Green Tea, Maidenhair Ginkgo, Quickthorn Hawthorn Berry, Red Wine Grape
Lignan	Anti-oxidant	Flaxseed, Flaxseed Oil
Limonoid	Detoxification	Celery
Polyphenal	Anti-oxidant	Chocolate, Green Tea, Red Wine
Sterol	Anti-carcinogen	Nuts, Seeds, Soybeans

DOMAIN 1 *Protista* Protist

Latin: *Planta* plant
Latin: *Phyta* phyton, plant

KINGDOMS

1 *Materia* Nonliving Matter
2 *Microbeavirus* Microbe Virus
3 *Phytovirus* Plant Virus
4 *Zoovirus* Animal Virus
5 *Microbea* Microbe

DOMAIN 2 *Plantae* Plant

Latin: *Planta* plant
Latin: *Phyta* phyton, plant

KINGDOMS

6 *Sporophyta* Spore Plant
7 *Spermatophyta* Seed Plant

KINGDOM 6 *Sporophyta* Spore Plant

Latin: *Sperma* sperm, seed
Latin: *Phyta* phyton, plant

SUBKINGDOMS

61 *Algophyta*	Algae
62 *Fungi*	Fungus
63 *Bryata*	Moss
64 *Filicata*	Fern

SUBKINGDOM 61 *Algophyta* Algae

Latin: *Algo* algae
Latin: *vegeta* vegetable
Latin: *Phyta* phyton, plant

PHYLUMS

611 *Bacillariophyta*	Diatom Brown Algae
612 *Chlorophyta*	Olive Green Algae
613 *Chrysophyta*	Golden Ocher Algae
614 *Cyanidophyta*	Blue Green Algae
615 *Ochrophyta*	Ocher Algae
616 *Prasinophyta*	Unicell Green Algae
617 *Rhodophyta*	Red Algae

Phylum 615 *Ochrophyta* Ocher Algae

Latin: *Ochro* ocher
Latin: *Phyta* phyton, plant

CLASSES

615.1 *Bodonophyceae*	Bodo Ocher Algae
615.2 *Bolidophyceae*	Bold Ocher Algae
615.3 *Craspedophyceae*	Fringed Ocher Algae
615.4 *Fragilarophyceae*	Fragile Ocher Algae
615.5 *Pelagophyceae*	Pelagic Ocher Algae
615.6 *Pinguiophyceae*	Greasy Ocher Algae
615.7 *Phaeophyceae*	Dusky Ocher Algae
615.8 *Raphidophyceae*	Needle Ocher Algae

Class 615.7 *Phaeophyceae* Dusky Ocher Algae

Latin: *Phaeo* faded, dusky
Latin: *Phycea* phyton, plant

ORDERS

615.71 *Phaeophytales*	Brown Dusky Ocher Algae
615.72 *Ectocarpales*	Ectocarp Dusky Ocher Algae
615.73 *Fucales*	**Painted Dusky Ocher Algae**
615.74 *Laminariales*	Flat Dusky Ocher Algae
615.75 *Scytosiphonales*	Scyte Siphon Ocher Algae
615.76 *Spacelariales*	Spacelar Dusky Ocher Algae

Order 615.73 *Fucales* Painted Dusky Ocher Algae

Latin: *Fuca* fuscous

FAMILIES

615.731 *Fucaceae*	**Painted Dusky Ocher Algae**
615.732 *Sargassaceae*	**Sargasso Seaweed**
615.733 *Sierococcaceae*	Round Painted Ocher Algae

Family 615.731 *Fucaceae* Painted Dusky Ocher Algae

Genus	Species	*Variety*	Native Location	Variety	Specimen	Genome	Habit
Fucus	*serratus*	.	Worldwide (Oceans, Lakes, Rivers)	.		Brown Kelp Rock Seaweed	Edible
Fucus	*serratus*	*abbreviatus*	Worldwide (Oceans, Lakes, Rivers)	Abbreviated		Brown Kelp Rock Seaweed	Edible
Fucus	*serratus*	*angustifrons*	Worldwide (Oceans, Lakes, Rivers)	Narrow Frond		Brown Kelp Rock Seaweed	Edible
Fucus	*serratus*	*angustus*	Worldwide (Oceans, Lakes, Rivers)	Narrow Leaf		Brown Kelp Rock Seaweed	Edible
Fucus	*serratus*	*arcticus*	Worldwide (Oceans, Lakes, Rivers)	Arctic		Brown Kelp Rock Seaweed	Edible
Fucus	*serratus*	*elongatus*	Worldwide (Oceans, Lakes, Rivers)	Elongated		Brown Kelp Rock Seaweed	Edible
Fucus	*serratus*	*grandifrons*	Worldwide (Oceans, Lakes, Rivers)	Grand Frond		Brown Kelp Rock Seaweed	Edible
Fucus	*serratus*	*integerrimus*	Worldwide (Oceans, Lakes, Rivers)	Entire		Brown Kelp Rock Seaweed	Edible

Genus	Species	Variety	Native Location	Variety	Specimen	Genome	Habit
Fucus	*serratus*	*laciniatus*	Worldwide (Oceans, Lakes, Rivers)	Lace Frond	Brown	Kelp Rock Seaweed	Edible
Fucus	*serratus*	*latifolius*	Worldwide (Oceans, Lakes, Rivers)	Wide Frond	Brown	Kelp Rock Seaweed	Edible
Fucus	*vesiculosus*	*vesiculosus*	N.Atl.Oc.,N.Sea,Eng.Channel,Med.Sea	Original	Bladderwrack	Kelp Rock Seaweed	Edible
Fucus	*vesiculosus*	*abbreviatus*	N.Atl.Oc.,N.Sea,Eng.Channel,Med.Sea	Abbreviated	Bladderwrack	Kelp Rock Seaweed	Edible
Fucus	*vesiculosus*	*acutus*	N.Atl.Oc.,N.Sea,Eng.Channel,Med.Sea	Acute	Bladderwrack	Kelp Rock Seaweed	Edible
Fucus	*vesiculosus*	*aestuarii*	N.Atl.Oc.,N.Sea,Eng.Channel,Med.Sea	Aestuar	Bladderwrack	Kelp Rock Seaweed	Edible
Fucus	*vesiculosus*	*alternans*	N.Atl.Oc.,N.Sea,Eng.Channel,Med.Sea	Alternate	Bladderwrack	Kelp Rock Seaweed	Edible
Fucus	*vesiculosus*	*angustifolius*	N.Atl.Oc.,N.Sea,Eng.Channel,Med.Sea	Narrow Leaf	Bladderwrack	Kelp Rock Seaweed	Edible
Fucus	*vesiculosus*	*angustifrons*	N.Atl.Oc.,N.Sea,Eng.Channel,Med.Sea	Narrow Frond	Bladderwrack	Kelp Rock Seaweed	Edible
Fucus	*vesiculosus*	*balticus*	N.Atl.Oc.,N.Sea,Eng.Channel,Med.Sea	Baltic Sea	Bladderwrack	Kelp Rock Seaweed	Edible
Fucus	*vesiculosus*	*ceratiformis*	N.Atl.Oc.,N.Sea,Eng.Channel,Med.Sea	Horned	Bladderwrack	Kelp Rock Seaweed	Edible
Fucus	*vesiculosus*	*chondriformis*	N.Atl.Oc.,N.Sea,Eng.Channel,Med.Sea	Cartilage	Bladderwrack	Kelp Rock Seaweed	Edible
Fucus	*vesiculosus*	*compressus*	N.Atl.Oc.,N.Sea,Eng.Channel,Med.Sea	Compressed	Bladderwrack	Kelp Rock Seaweed	Edible
Fucus	*vesiculosus*	*crispus*	N.Atl.Oc.,N.Sea,Eng.Channel,Med.Sea	Curly	Bladderwrack	Kelp Rock Seaweed	Edible
Fucus	*vesiculosus*	*divaricatus*	N.Atl.Oc.,N.Sea,Eng.Channel,Med.Sea	Spreading	Bladderwrack	Kelp Rock Seaweed	Edible
Fucus	*vesiculosus*	*elongatus*	N.Atl.Oc.,N.Sea,Eng.Channel,Med.Sea	Elongated	Bladderwrack	Kelp Rock Seaweed	Edible
Fucus	*vesiculosus*	*filiformis*	N.Atl.Oc.,N.Sea,Eng.Channel,Med.Sea	Thread Frond	Bladderwrack	Kelp Rock Seaweed	Edible
Fucus	*vesiculosus*	*flabellatus*	N.Atl.Oc.,N.Sea,Eng.Channel,Med.Sea	Fan Frond	Bladderwrack	Kelp Rock Seaweed	Edible
Fucus	*vesiculosus*	*fluviatilis*	N.Atl.Oc.,N.Sea,Eng.Channel,Med.Sea	River	Bladderwrack	Kelp Rock Seaweed	Edible
Fucus	*vesiculosus*	*foliaceus*	N.Atl.Oc.,N.Sea,Eng.Channel,Med.Sea	Leafy	Bladderwrack	Kelp Rock Seaweed	Edible
Fucus	*vesiculosus*	*gracillimus*	N.Atl.Oc.,N.Sea,Eng.Channel,Med.Sea	Slender	Bladderwrack	Kelp Rock Seaweed	Edible
Fucus	*vesiculosus*	*grandifrons*	N.Atl.Oc.,N.Sea,Eng.Channel,Med.Sea	Grand Frond	Bladderwrack	Kelp Rock Seaweed	Edible
Fucus	*vesiculosus*	*lanceolatus*	N.Atl.Oc.,N.Sea,Eng.Channel,Med.Sea	Lance Frond	Bladderwrack	Kelp Rock Seaweed	Edible
Fucus	*vesiculosus*	*laterifructus*	N.Atl.Oc.,N.Sea,Eng.Channel,Med.Sea	Wide Fruit	Bladderwrack	Kelp Rock Seaweed	Edible
Fucus	*vesiculosus*	*latus*	N.Atl.Oc.,N.Sea,Eng.Channel,Med.Sea	Wide	Bladderwrack	Kelp Rock Seaweed	Edible
Fucus	*vesiculosus*	*limicola*	N.Atl.Oc.,N.Sea,Eng.Channel,Med.Sea	Lime Dwelling	Bladderwrack	Kelp Rock Seaweed	Edible
Fucus	*vesiculosus*	*linearis*	N.Atl.Oc.,N.Sea,Eng.Channel,Med.Sea	Strap Frond	Bladderwrack	Kelp Rock Seaweed	Edible
Fucus	*vesiculosus*	*mytili*	N.Atl.Oc.,N.Sea,Eng.Channel,Med.Sea	Mytil	Bladderwrack	Kelp Rock Seaweed	Edible
Fucus	*vesiculosus*	*nanus*	N.Atl.Oc.,N.Sea,Eng.Channel,Med.Sea	Dwarf	Bladderwrack	Kelp Rock Seaweed	Edible
Fucus	*vesiculosus*	*plicatus*	N.Atl.Oc.,N.Sea,Eng.Channel,Med.Sea	Pleated	Bladderwrack	Kelp Rock Seaweed	Edible
Fucus	*vesiculosus*	*pseudoceranoides*	N.Atl.Oc.,N.Sea,Eng.Channel,Med.Sea	Lesser Horned	Bladderwrack	Kelp Rock Seaweed	Edible
Fucus	*vesiculosus*	*racemosus*	N.Atl.Oc.,N.Sea,Eng.Channel,Med.Sea	Raceme	Bladderwrack	Kelp Rock Seaweed	Edible
Fucus	*vesiculosus*	*rigidus*	N.Atl.Oc.,N.Sea,Eng.Channel,Med.Sea	Stiff	Bladderwrack	Kelp Rock Seaweed	Edible
Fucus	*vesiculosus*	*robustus*	N.Atl.Oc.,N.Sea,Eng.Channel,Med.Sea	Robust	Bladderwrack	Kelp Rock Seaweed	Edible
Fucus	*vesiculosus*	*rotundatus*	N.Atl.Oc.,N.Sea,Eng.Channel,Med.Sea	Round	Bladderwrack	Kelp Rock Seaweed	Edible
Fucus	*vesiculosus*	*sphaerocarpus*	N.Atl.Oc.,N.Sea,Eng.Channel,Med.Sea	Spherical Pod	Bladderwrack	Kelp Rock Seaweed	Edible
Fucus	*vesiculosus*	*spiralis*	N.Atl.Oc.,N.Sea,Eng.Channel,Med.Sea	Spiral	Bladderwrack	Kelp Rock Seaweed	Edible
Fucus	*vesiculosus*	*subecostatus*	N.Atl.Oc.,N.Sea,Eng.Channel,Med.Sea	Lesser Ribbed	Bladderwrack	Kelp Rock Seaweed	Edible
Fucus	*vesiculosus*	*subfusiformis*	N.Atl.Oc.,N.Sea,Eng.Channel,Med.Sea	Spindle	Bladderwrack	Kelp Rock Seaweed	Edible
Fucus	*vesiculosus*	*subglobosus*	N.Atl.Oc.,N.Sea,Eng.Channel,Med.Sea	Lesser Globe	Bladderwrack	Kelp Rock Seaweed	Edible
Fucus	*vesiculosus*	*tenuis*	N.Atl.Oc.,N.Sea,Eng.Channel,Med.Sea	Tenuous	Bladderwrack	Kelp Rock Seaweed	Edible
Fucus	*vesiculosus*	*terminalis*	N.Atl.Oc.,N.Sea,Eng.Channel,Med.Sea	Terminal	Bladderwrack	Kelp Rock Seaweed	Edible
Fucus	*vesiculosus*	*turgidus*	N.Atl.Oc.,N.Sea,Eng.Channel,Med.Sea	Turgid	Bladderwrack	Kelp Rock Seaweed	Edible
Fucus	*vesiculosus*	*vadorum*	N.Atl.Oc.,N.Sea,Eng.Channel,Med.Sea	Vador	Bladderwrack	Kelp Rock Seaweed	Edible
Fucus	*vesiculosus*	*volubilis*	N.Atl.Oc.,N.Sea,Eng.Channel,Med.Sea	Twining	Bladderwrack	Kelp Rock Seaweed	Edible

Family 615.732 *Sargassaceae* Sargasso Seaweed

Genus	Species	Variety	Native Location	Variety	Specimen	Genome	Habit
Sargassum	*fusiforme*	.	Atlantic Ocean	.	Spindle	Sargasso Seaweed	Edible

Order 615.74 *Laminariales* Flat Dusky Ocher Algae

Latin: *Laminaria* laminar

FAMILIES

615.741 *Laminariaceae* Flat Dusky Ocher Algae
615.742 *Alariaceae* Alar Flat Dusky Ocher Algae
615.743 *Chordaceae* Chord Flat Dusky Ocher Algae
615.744 *Lessoniaceae* Lesson Flat Dusky Ocher Algae
615.745 *Phyllariaceae* Phyllar Flat Dusky Ocher Algae

Family 615.741 *Laminariaceae* Flat Dusky Ocher Algae

Genus	Species	Variety	Native Location	Variety	Specimen	Genome	Habit
Laminaria	*saccharina*	*saccharina*	Worldwide (Oceans, Lakes, Rivers)	Original	Sweet	Flat Brown Seaweed	Edible
Laminaria	*saccharina*	*agardhi*	Worldwide (Oceans, Lakes, Rivers)	Agardh	Sweet	Flat Brown Seaweed	Edible
Laminaria	*saccharina*	*borealis*	Worldwide (Oceans, Lakes, Rivers)	Boreal	Sweet	Flat Brown Seaweed	Edible
Laminaria	*saccharina*	*bullata*	Worldwide (Oceans, Lakes, Rivers)	Bullet	Sweet	Flat Brown Seaweed	Edible
Laminaria	*saccharina*	*caperata*	Worldwide (Oceans, Lakes, Rivers)	Caper	Sweet	Flat Brown Seaweed	Edible
Laminaria	*saccharina*	*cuneifolia*	Worldwide (Oceans, Lakes, Rivers)	Wedge Frond	Sweet	Flat Brown Seaweed	Edible
Laminaria	*saccharina*	*eliptica*	Worldwide (Oceans, Lakes, Rivers)	Eliptic	Sweet	Flat Brown Seaweed	Edible
Laminaria	*saccharina*	*faeroensis*	Worldwide (Oceans, Lakes, Rivers)	Faro Island	Sweet	Flat Brown Seaweed	Edible

Genus	Species	*Variety*	Native Location	Variety	Specimen	Genome	Habit
Laminaria	*saccharina*	*grandis*	Worldwide (Oceans, Lakes, Rivers)	Grand	Sweet	Flat Brown Seaweed	Edible
Laminaria	*saccharina*	*groenlandica*	Worldwide (Oceans, Lakes, Rivers)	Greenland	Sweet	Flat Brown Seaweed	Edible
Laminaria	*saccharina*	*hieroglyphica*	Worldwide (Oceans, Lakes, Rivers)	Hieroglyphic	Sweet	Flat Brown Seaweed	Edible
Laminaria	*saccharina*	*lanceolata*	Worldwide (Oceans, Lakes, Rivers)	Lance Frond	Sweet	Flat Brown Seaweed	Edible
Laminaria	*saccharina*	*latifolia*	Worldwide (Oceans, Lakes, Rivers)	Broad Frond	Sweet	Flat Brown Seaweed	Edible
Laminaria	*saccharina*	*latissima*	Worldwide (Oceans, Lakes, Rivers)	Wide Frond	Sweet	Flat Brown Seaweed	Edible
Laminaria	*saccharina*	*linearis*	Worldwide (Oceans, Lakes, Rivers)	Belt Frond	Sweet	Flat Brown Seaweed	Edible
Laminaria	*saccharina*	*longissima*	Worldwide (Oceans, Lakes, Rivers)	Long	Sweet	Flat Brown Seaweed	Edible
Laminaria	*saccharina*	*maxima*	Worldwide (Oceans, Lakes, Rivers)	Great	Sweet	Flat Brown Seaweed	Edible
Laminaria	*saccharina*	*membranacea*	Worldwide (Oceans, Lakes, Rivers)	Membrane	Sweet	Flat Brown Seaweed	Edible
Laminaria	*saccharina*	*oblonga*	Worldwide (Oceans, Lakes, Rivers)	Oblong	Sweet	Flat Brown Seaweed	Edible
Laminaria	*saccharina*	*obovata*	Worldwide (Oceans, Lakes, Rivers)	Fiddle Frond	Sweet	Flat Brown Seaweed	Edible
Laminaria	*saccharina*	*phyllitis*	Worldwide (Oceans, Lakes, Rivers)	Phyllite	Sweet	Flat Brown Seaweed	Edible
Laminaria	*saccharina*	*phyllopus*	Worldwide (Oceans, Lakes, Rivers)	Phyllope	Sweet	Flat Brown Seaweed	Edible
Laminaria	*saccharina*	*prima*	Worldwide (Oceans, Lakes, Rivers)	Prime	Sweet	Flat Brown Seaweed	Edible
Laminaria	*saccharina*	*stackhousei*	Worldwide (Oceans, Lakes, Rivers)	Stackhouse	Sweet	Flat Brown Seaweed	Edible
Laminaria	*saccharina*	*sublaevis*	Worldwide (Oceans, Lakes, Rivers)	Smooth	Sweet	Flat Brown Seaweed	Edible
Laminaria	*saccharina*	*viridissima*	Worldwide (Oceans, Lakes, Rivers)	Green	Sweet	Flat Brown Seaweed	Edible
Laminaria	*saccharina*	*vittata*	Worldwide (Oceans, Lakes, Rivers)	Banded	Sweet	Flat Brown Seaweed	Edible

Family 615.744 *Lessoniaceae* Lesson Flat Dusky Ocher Algae

Genus	Species	*Variety*	Native Location	Variety	Specimen	Genome	Habit
Macrocystis	*pyrifera*	*pyrifera*	Worldwide (Oceans, Lakes, Rivers)	Original	Birnentang	Cyst Dusky Seaweed	Edible
Macrocystis	*pyrifera*	*humboldtii*	Worldwide (Oceans, Lakes, Rivers)	Humboldt	Birnentang	Cyst Dusky Seaweed	Edible
Macrocystis	*pyrifera*	*latifolia*	Worldwide (Oceans, Lakes, Rivers)	Broad Frond	Birnentang	Cyst Dusky Seaweed	Edible
Macrocystis	*pyrifera*	*luxurians*	Worldwide (Oceans, Lakes, Rivers)	Luxuriant	Birnentang	Cyst Dusky Seaweed	Edible
Macrocystis	*pyrifera*	*orbigniana*	Worldwide (Oceans, Lakes, Rivers)	Orbigni	Birnentang	Cyst Dusky Seaweed	Edible
Macrocystis	*pyrifera*	*pelagica*	Worldwide (Oceans, Lakes, Rivers)	Pelagic	Birnentang	Cyst Dusky Seaweed	Edible
Macrocystis	*pyrifera*	*planicaulis*	Worldwide (Oceans, Lakes, Rivers)	Flat Stem	Birnentang	Cyst Dusky Seaweed	Edible
Macrocystis	*pyrifera*	*tenuifolia*	Worldwide (Oceans, Lakes, Rivers)	Tenuous Frond	Birnentang	Cyst Dusky Seaweed	Edible
Nereocystis		*luetkeana*	Worldwide (Oceans, Lakes, Rivers)	.	Kelp	Nere Dusky Seaweed	Edible

Phylum 617 *Rhodophyta* Red Algae

Latin: *rhodo* red
Latin: *Phyta* phyton, plant

CLASSES

617.1 *Bangiophyceae* Bangion Red Algae
617.2 *Compsopogophyceae* Beard Red Algae
617.3 *Florideophyceae* Florid Red Algae
617.4 *Rhodellophyceae* Little Red Algae

Class 617.3 *Florideophyceae* Florid Red Algae

Latin: *Flora* floral, flowering
Latin: *Phycea* phyton, plant

ORDERS

617.31 *Forideales* Florid Red Algae
617.32 *Batrachospermales* Tadpole Florid Red Algae
617.33 *Bonnemaisoniales* Bonne Maison Red Algae
617.34 *Ceramiales* Ceram Florid Red Algae
617.35 *Coralliniales* Coral Florid Red Algae
617.36 *Gelidiales* Arctic Florid Red Algae
617.37 *Gigartinales* Gigartin Florid Red Algae
617.38 *Gracilariales* Graceful Florid Red Algae
617.39 *Nemaliales* Thread Florid Red Algae

Order 617.36 *Gelidiales* Arctic Florid Red Algae

Family 617.361 *Gelidiaceae* Arctic Florid Red Algae

Genus	Species	*Variety*	Native Location	Variety	Specimen	Genome	Habit
Gelidium	*Amansii*	.	Arctic Ocean	.	Amans	Arctic Florid Red Algae	Edible
Gelidium	*cartilagineum*	.	Arctic Ocean	.	Agar	Arctic Florid Red Algae	Edible
Pterocladia	*pinnata*	.	Arctic Ocean	.	Pinnate	Wing Branched Arctic Florid Red Algae	Edible

Order 617.37 *Gigartinales* Gigartin Florid Red Algae

Family 617.371 *Gigartinaceae* Gigartin Florid Red Algae

Genus	Species	*Variety*	Native Location	Variety	Specimen	Genome	Habit
Chondrus	*Crispus*	.	Atlantic Ocean	.	Irish Moss	Cartilage Gigartin Florid Red Algae	Edible
Chondrus	*Crispus*	*crispus*	Atlantic Ocean	Original	Irish Moss	Cartilage Gigartin Florid Red Algae	Edible
Chondrus	*Crispus*	*abbreviatus*	Atlantic Ocean	Abbreviated	Irish Moss	Cartilage Gigartin Florid Red Algae	Edible
Chondrus	*Crispus*	*aegagropilus*	Atlantic Ocean	Downy	Irish Moss	Cartilage Gigartin Florid Red Algae	Edible
Chondrus	*Crispus*	*aequalis*	Atlantic Ocean	Unequal	Irish Moss	Cartilage Gigartin Florid Red Algae	Edible
Chondrus	*Crispus*	*densus*	Atlantic Ocean	Dense	Irish Moss	Cartilage Gigartin Florid Red Algae	Edible
Chondrus	*Crispus*	*filiformis*	Atlantic Ocean	Filament	Irish Moss	Cartilage Gigartin Florid Red Algae	Edible
Chondrus	*Crispus*	*incurvatus*	Atlantic Ocean	Incurved	Irish Moss	Cartilage Gigartin Florid Red Algae	Edible
Chondrus	*Crispus*	*lacerus*	Atlantic Ocean	Lacerated	Irish Moss	Cartilage Gigartin Florid Red Algae	Edible
Chondrus	*Crispus*	*lonchophorus*	Atlantic Ocean	Lance Frond	Irish Moss	Cartilage Gigartin Florid Red Algae	Edible
Chondrus	*Crispus*	*lutescens*	Atlantic Ocean	Yellowish	Irish Moss	Cartilage Gigartin Florid Red Algae	Edible
Chondrus	*crispus*	*membranaceus*	Atlantic Ocean	Membrane	Irish Moss	Cartilage Gigartin Florid Red Algae	Edible
Chondrus	*crispus*	*patens*	Atlantic Ocean	Spreading	Irish Moss	Cartilage Gigartin Florid Red Algae	Edible
Chondrus	*crispus*	*planus*	Atlantic Ocean	Flat	Irish Moss	Cartilage Gigartin Florid Red Algae	Edible
Chondrus	*crispus*	*polychotomus*	Atlantic Ocean	Polychotome	Irish Moss	Cartilage Gigartin Florid Red Algae	Edible
Chondrus	*crispus*	*sarniensis*	Atlantic Ocean	Sarnian	Irish Moss	Cartilage Gigartin Florid Red Algae	Edible
Chondrus	*crispus*	*stellatus*	Atlantic Ocean	Star	Irish Moss	Cartilage Gigartin Florid Red Algae	Edible
Chondrus	*crispus*	*uncinatus*	Atlantic Ocean	Hooked	Irish Moss	Cartilage Gigartin Florid Red Algae	Edible
Chondrus	*crispus*	*virens*	Atlantic Ocean	Greenish	Irish Moss	Cartilage Gigartin Florid Red Algae	Edible

Order 617.38 *Gracilariales* Graceful Florid Red Algae

Family 617.381 *Graciliaceae* Graceful Florid Red Algae

Genus	Species	*Variety*	Native Location	Variety	Specimen	Genome	Habit
Gracilaria	*Gracilis*	.		.	Agar	Graceful Florid Red Algae	Edible
Gracilaria	*Gracilis*	*gracilis*		Original	Agar	Graceful Florid Red Algae	Edible
Gracilaria	*Gracilis*	*ecorticata*		Corticle	Agar	Graceful Florid Red Algae	Edible
Gracilaria	*Gracilis*	*procerrima*		Tall	Agar	Graceful Florid Red Algae	Edible
Gracilaria	*Gracilis*	*ramulosa*		Branching	Agar	Graceful Florid Red Algae	Edible
Gracilaria	*Gracilis*	*verrucosa*		Verrucose	Agar	Graceful Florid Red Algae	Edible

SUBKINGDOM 62 *Fungi* Fungus

Latin: *Fungi* fungus

PHYLUMS

621 *Ascomycota*	Sac Fungus
622 *Basidiomycota*	Club Fungus
623 *Zygomycota*	Multinucleate Fungus

When considering mushrooms and other fungus, the following two questions are the most important.
(1) Where do they grow; USA, Europe, Asia, or Africa?
(2) Are they; Edible-Excellent, Edible-Good, Edible-Fair, Edible-Poor, Edible-Toxic, Upalatable (of no alimentary interest), Inedible, Toxic, Poison, or Edibility Unknown?.

The following sources were used to answer these questions.

GSI-6 "Garden Seed Inventory 6th Ed." (ISBN:1-882424-59-X) lists edible mushrooms found in the USA.

MOTW "Mushrooms of the World" – (ISBN:1-55209-212-7) lists inedibility or edibility of fungus and mushrooms found not only in the USA but also in Europe.

PFGM "Petersons Field Guide to Mushrooms North America" (ISBN:0-395-91090-0) lists the inedibility or edibility of mushrooms found in the USA.

Edible-Toxic When authors say that a given mushroom is **"edible"** to some people and causes **"gastroenteritis"** for others, they are given this designation.

The edible mushrooms below are assumed edible only after being well cooked. The author of this book assumes no responsibility for the edibility of the mushrooms listed in this book. The reader uses the following list of fungus and mushrooms at his or her own risk.

Phylum 621 *Ascomycota* Sac Fungus

Latin: *Asco* ascus, spore sac
Latin: *Mycota* mycote, fungus

CLASSES

621.1 *Ascomycetes*	Sac Fungus
621.2 *Lichinomycetes*	Lichen
621.3 *Mucoromycetes*	Mold
621.4 *Hemiascomycetes*	Yeast

Class 621.1 *Ascomycetes* Sac Fungus

Latin: *Asco* ascus, spore sac
Latin: *Mycetes* mycete, fungus

ORDERS

621.101 *Ascomycetales*	Sac Fungus
621.102 *Anamorphales*	Anamorph Sac Fungus
621.103 *Chaetothyriales*	Bristle Sac Fungus
621.104 *Coronophorales*	Crown Sac Fungus
621.105 *Diaporthales*	Dual Porthe Sac Fungus
621.106 *Dothidales*	Doth Sac Fungus
621.107 *Helotales*	**Helot Sac Fungus**
621.108 *Hysteriales*	Hysteria Sac Fungus
621.109 *Microthyriales*	Thyre Sac Fungus
621.110 *Mycocaliciales*	Calico Sac Fungus
621.111 *Pezizales*	Peziz Sac Fungus
621.112 *Pezizavegetales*	**Peziz Sac Mushroom**
621.113 *Phyllachorales*	Choral Sac Fungus
621.114 *Pleosporales*	Pleospore Sac Fungus
621.115 *Rhytismatales*	Wrinkled Sac Fungus
621.116 *Sordariales*	Sordid Sac Fungus
621.117 *Trichosphaerales*	Hairy Sphere Sac Fungus
621.118 *Xylariales*	Wood Sac Fungus

Order 621.107 *Helotiales* Helot Sac Fungus

Latin: *Helotia* helot

FAMILIES

661.107.01 *Helotiaceae* **Helot Mushroom**
661.107.02 *Bulgariaceae* Bulgaria Helot Fungus
661.107.03 *Cudoniaceae* Cudon Helot Sac Fungus
661.107.04 *Dermataceae* Dermal Helot Sac Fungus
661.107.05 *Geoglossaceae* Earth Tongue Sac Fungus
661.107.06 *Hemiphacidiaceae* Half Lens Helot Sac Fungus
661.107.07 *Hyaloscyphaceae* Hyaline Scyph Sac Fungus
661.107.08 *Leotiaceae* **Slippery Cap Mushroom**
661.107.09 *Phacidiaceae* Lens Helot Sac Fungus
661.107.10 *Rutstroemiaceae* Rue Stroma Sac Fungus
661.107.11 *Sclerotiniaceae* Shell Helot Sac Fungus
661.107.12 *Vibrisseaceae* Vibrisse Helot Sac Fungus

Family 621.107.01 *Helotiaceae* Helot Sac Mushroom

Genus	Species	Variety	Native Location	Variety	Specimen	Genome	Habit
Pezizella	*Vulgaris*	.	MOTW, page 235	.	Ordinary	Peziz Helot Sac Fungus	Edible-Fair

Family 621.107.08 *Leiotiaceae* Slippery Cap Sac Mushroom

Genus	Species	Native Location	Specimen	Genome	Habit
Leotia	*Lubrica*	USA (open woods), PFGM, p. 33, MOTW, p. 239	Helot	Slippery Cap Sac Mushroom	Edible-Good
Leotia	*viscosa*	USA (in woods), PFGM, page 33	Winter	Slippery Cap Sac Mushroom	Unknown

Order 621.111 *Pezizales* Peziz Sac Fungus

Latin: *Peziza* peziz

FAMILIES

621.111.01 *Pezizaceae* Cup Peziz Fungus
621.111.02 *Ascobolaceae* Bowl Peziz Fungus
621.111.03 *Ascodesmidaceae* Fiber Cup Peziz Fungus
621.111.04 *Discinaceae* Disc Peziz Fungus
621.111.05 *Sarcosomataceae* Stalk Cup Peziz Fungus
621.111.06 *Terfeziaceae* Terfez Cup Peziz Fungus

Order 621.112 *Pezizavegetales* Peziz Sac Mushroom

Latin: *Peziza* peziz
Latin: *Muscheron* mushroom

FAMILIES

621.112.01 *Pezizavegetaceae* Peziz Sac Mushroom
621.112.02 *Discinaceae* False Morel
621.112.03 *Helvellaceae* Lorchel
621.112.04 *Morchellaceae* Morel
621.112.05 *Pyronemataceae* Fire Cup Sac Mushroom
621.112.06 *Tuberaceae* Truffle

Family 621.112.01 *Pezizavegetaceae* Peziz Sac Mushroom

Genus	Species	Variety	Native Location	Variety	Specimen	Genome	Habit
Peziza	*ammophia*	.	USA, PFGM, page 58		Sand Loving	Peziz Sac Mushroom	Unknown
Peziza	*apiculata*	.	USA, PFGM, page 47		Apex	Peziz Sac Mushroom	Unknown
Peziza	*badia*	.	USA, MOTW, page 238	.	Red Brown	Peziz Sac Mushroom	Edible-cooked
Peziza	*echinospora*	.	Europe (soil after fire)	.	Prickly Spore	Peziz Sac Mushroom	Edible-cooked
Peziza	*echinospora*	*assimilata*	Europe (soil after fire)	Original	Prickly Spore	Peziz Sac Mushroom	Edible-cooked
Peziza	*echinospora*	*umbrina*	Europe, MOTW, p. 238	Dark	Prickly Spore	Peziz Sac Mushroom	Edible-cooked
Peziza	*macrospora*		PGFM, page 47		Large Spore	Peziz Sac Mushroom	Unknown
Peziza	*nigritella*	.	MOTW, page 238	.	Black	Peziz Sac Mushroom	Unpalatable
Peziza	*palustris*		PFGM, page 46		Marsh	Peziz Sac Mushroom	Unknown
Peziza	*phyllogena*	.	USA, PFGM, page 56	.	Pig Ear	Peziz Sac Mushroom	Unknown
Peziza	*repanda*	.	USA, MOTW, page 238	.	Folded	Peziz Sac Mushroom	Unpalatable
Peziza	*saniosa*	.	Europe, MOTW, page 238	.	Pus	Peziz Sac Mushroom	Unpalatable
Peziza	*succosa*	.	Europe MOTW, p. 238	.	Juicy	Peziz Sac Mushroom	Edible-cooked

Genus	Species	Variety	Native Location	Variety	Specimen	Genome	Habit
Peziza	*trachycarpa*	.	USA, PFGM, page 57	.	.	Peziz Sac Mushroom	Unknown
Peziza	*vesiculosa*	.	USA, PFGM, page 58	.	Barnyard	Peziz Sac Mushroom	Edible-cooked
Disciotis	*maturescens*	.	.	.	.	Disc Peziz Sac Mushroom	Unknown
Disciotis	*venosa*	.	USA, MOTW, p. 235, PFGM, p. 37	.	Veiny	Disc Peziz Sac Mushroom	Edible-Fair
Disciotis	*venosa*	*venosa*	USA (forest)	Original	Veiny	Disc Peziz Sac Mushroom	Edible-Fair
Disciotis	*venosa*	*reticulata*	USA (forest)	Netted	Veiny	Disc Peziz Sac Mushroom	Edible-Fair
Plicaria	*leiocarpa*		USA, PFGM, page 57		Lion Spore	Plicar Peziz Sac Mushroom	Unknown
Sarcosphaera	*coronaria*	.	Europe (Pine - Spruce forest)	.	Crown	Fleshy Peziz Sac Mushroom	Poison
Sarcosphaera	*coronaria*	*coronaria*	Europe (Pine - Spruce forest)	Original	Crown	Fleshy Peziz Sac Mushroom	Poison
Sarcosphaera	*coronaria*	*crassa*	PFGM, page 58	Violet Star	Crown	Fleshy Peziz Sac Mushroom	Poison
Sarcosphaera	*coronaria*	*eximia*	Europe (Pine - Spruce forest)	Extrodinary	Crown	Fleshy Peziz Sac Mushroom	Poison

Family 621.112.02 *Discinaceae* False Morel

Genus	Species	Variety	Native Location	Variety	Specimen	Genome	Habit
Discina	*ancilis*	.	USA (conifer & hardwood forest)	.	Thick Cup	False Morel	Edible-Poor
Discina	*ancilis*	*ancilis*	USA (conifer & hardwood forest)	Original	Thick Cup	False Morel	Edible-Poor
Discina	*ancilis*	*perlata*	USA (conifer & hardwood forest), MOTW, p. 238, PFGM, p. 47	Pearl	Thick Cup	False Morel	Edible-Poor
Discina	*olympiana*		USA, PFGM, page 47		Olympia	False Morel	Unknown
Discina	*warnei*		USA, PFGM, page 47		Warne	False Morel	Unknown
Gyromitra	*ambigua*	.	USA, PFGM, pages 51, 54	.	Ambiguous	Round Mitre False Morel	Unknown
Gyromitra	*brunnea*	.	S. Canada, USA (decaying wood in hardwood forest), PFGM, p. 48	.	Brown	Round Mitre False Morel	Poison
Gyromitra	*caroliniana*	.	C. & E. USA (hardwood forest), PFGM, page 50	.	Carolina	Round Mitre False Morel	Edible
Gyromitra	*esculenta*	.	USA (conifer & hardwood forest), MOTW, page 234, PFGM, page 51	.	Deathsac	Round Mitre False Morel	Edible-Toxic
Gyromitra	*gigas*	.	W. USA (conifer mountain forest), PFGM, page 52	.	Snow Giant	Round Mitre False Morel	Edible-Good
Gyromitra	*infula*	.	USA (decaying wood, conifer & hardwood forest), MOTW, p. 234, PFGM, p. 53	.	Hooded Deathsac	Round Mitre False Morel	Edible-Toxic
Gyromitra	*leucoxantha*	.	USA, PFGM, page 47	.	Light Yellow	Round Mitre False Morel	Unknown
Gyromitra	*sphaerospora*	.	USA, PFGM, page 49	.	Round Spore	Round Mitre False Morel	Unknown
Pseudorhizina	*californica*	.	W. Canada, W. USA (conifer forest), MOTW, p. 234, PFGM, p. 49	.	California	Rhizid False Morel	Inedible

Family 621.112.03 *Helvellaceae* Lorchel

Genus	Species	Variety	Native Location	Variety	Specimen	Genome	Habit
Helvella	*acetabulum*	.	USA (open woods), MOTW, page 235, PFGM, page 43	.	Vinegar Cup	Lorchel	Edible
Helvella	*albella*	.	USA, PFGM, page 45	.	White	Lorchel	Unknown
Helvella	*atra*	.	USA, PFGM, page 45	.	.	Lorchel	Unknown
Helvella	*crispa*	.	USA (conifer & hardwood forest), MOTW, page 234, PFGM, page 44	.	Wavy Saddle Back	Lorchel	Edible
Helvella	*elastica*	.	USA (conifer & hardwood forest), MOTW, page 234, PFGM, page 45	.	Flexible	Lorchel	Edible
Helvella	*griseoalba*	.	USA, PFGM, page 44	.	Gray White	Lorchel	Unknown
Helvella	*lacunosa*	.	USA (conifer & hardwood forest), MOTW, page 234, PFGM, page 45	.	Elfin Saddle	Lorchel	Edible
Helvella	*lacunosa*	*lacunosa*	USA (conifer & hardwood forest)	Original	Elfin Saddle	Lorchel	Edible
Helvella	*lacunosa*	*sulcata*	USA (conifer & hardwood forest), PFGM, page 46	Furrowed	Elfin Saddle	Lorchel	Edible
Helvella	*latispora*		USA		Wide Spore	Lorchel	Unknown
Helvella	*latispora*	*latispora*	USA	Original	Wide Spore	Lorchel	Unknown
Helvella	*latispora*	*stevensii*	USA, PFGM, page 45	Stevens	Wide Spore	Lorchel	Unknown
Helvella	*macropus*	.	USA, MOTW, page 238	.	Large Foot	Lorchel	Unpalatable
Helvella	*monachella*	.	Europe, MOTW, page 234	.	Small Nun	Lorchel	Edible
Helvella	*queletii*	.	USA, PFGM, page 44	.	Quelet	Lorchel	Unknown
Helvella	*silvicola*	.	Europe	.	Forest	Lorchel	Unknown
Helvella	*silvicola*	*silvicola*	Europe	Original	Forest	Lorchel	Unknown
Helvella	*silvicola*	*auricula*	Europe, MOTW, page 235	Eared	Forest	Lorchel	Unknown
Helvella	*underwoodii*	.	Europe, MOTW, page 234	.	Underwood	Lorchel	Poison

Family 621.112.04 *Morchellaceae* Morel

Genus	Species	Variety	Native Location	Variety	Specimen	Genome	Habit
Morchella	*angusticeps*	.	USA (forests with Douglas Fir), MOTW, p. 237, PFGM, p. 37	.	Narrow Head	Morel	Edible-Good
Morchella	*atrotomentosa*	.	USA (after forest fires), PFGM, p. 40	.	Burnsite	Morel	Edible
Morchella	*canaliculata*	.	USA, PFGM, page 38	.	Channeled	Morel	Unknown

Genus	Species	Variety	Native Location	Variety	Specimen	Genome	Habit
Morchella	*costata*	.	USA (conifer forest), MOTW, p. 237	.	Ribbed	Morel	Edible-Excellent
Morchella	*crassipes*	.	C. & N. USA (open woods), MOTW, p. 237, PFGM, p. 39	.	Thick Footed	Morel	Edible-Excellent
Morchella	*crassistipa*		USA, PFGM, page 39		Thick Stippled	Morel	Unknown
Morchella	*deliciosa*	.	USA (hardwood & Douglas Fir forests), PFGM, page 40	.	Delicious	Morel	Edible-Excellent
Morchella	*deliciosa*	*deliciosa*	USA (hardwood & Douglas Fir forests)	Original	Delicious	Morel	Edible-Excellent
Morchella	*deliciosa*	*conica*	USA (hardwood & Douglas Fir forests), MOTW, page 237, PFGM, page 38	Black	Delicious	Morel	Edible-Excellent
Morchella	*elata*	.	USA (conifer forest), MOTW, p. 237, PFGM, p. 39	.	Tall	Morel	Edible-Excellent
Morchella	*elata*	*elata*	USA (conifer forest)	Original	Tall	Morel	Edible-Excellent
Morchella	*elata*	*purpurascens*	USA (conifer forest)	Purple	Tall	Morel	Edible-Excellent
Morchella	*esculenta*	.	USA (forest, grassland, orchard), PFGM, page 39	.	Edible	Morel	Edible-Excellent
Morchella	*esculenta*	*esculenta*	USA (forest, grassland, orchard)	Original	Edible	Morel	Edible-Excellent
Morchella	*esculenta*	*rotunda*	USA (meadows & broadleaf forest)	Round	Edible	Morel	Edible-Excellent
Morchella	*esculenta*	*umbrina*	USA (forest, grassland, orchard), MOTW, page 237	Dark	Edible	Morel	Edible-Excellent
Morchella	*hortensis*	.	Europe, MOTW, p. 237	.	Horticultural	Morel	Edible-Excellent
Morchella	*spongiola*	.	Europe, MOTW, p. 237	.	Spongy	Morel	Edible-Excellent
Morchella	*vulgaris*	.	Europe, MOTW, p. 237	.	Ordinary	Morel	Edible-Excellent
Disciotis	*venosa*	.	Europe, MOTW, p. 235	.	Cup	Disc Morel	Edible
Disciotis	*venosa*	*venosa*	Europe	Original	Cup	Disc Morel	Edible
Disciotis	*venosa*	*reticulata*	Europe	Netted	Cup	Disc Morel	Edible
Mitrophora	*semilibera*	.	USA, PFGM, page 40	.	Half Free	Mitre Morel	Edible-Good
Mitrophora	*semilibera*	*semilibera*	USA	Original	Half Free	Mitre Morel	Edible-Good
Mitrophora	*semilibera*	*hybrida*	USA	Hybrid	Half Free	Mitre Morel	Edible-Good
Mitrophora	*semilibera*	*rimosipes*	USA	Cow's Head	Half Free	Mitre Morel	Edible-Good
Ptychoverpa	*bohemica*	.	USA (rich soil in lowlands), PFGM, p. 42, MOTW, p.237	.	Early	Folded Bell Morel	Edible
Verpa	*conica*	.	USA (open woods), PFGM, p. 43	.	Cone	Bell Morel	Edible
Verpa	*conica*	*conica*	USA (open woods)	Original	Cone	Bell Morel	Edible
Verpa	*conica*	*relhanii*	USA (open woods)	Relhan	Cone	Bell Morel	Edible
Verpa	*digitaliformis*	.	USA (S. California), MOTW, page 237, PFGM, pages 42, 43	.	Finger	Bell Morel	Edible

Family 621.112.05 *Pyronemataceae* Fire Cup Sac Mushroom

Genus	Species	Variety	Native Location	Variety	Specimen	Genome	Habit
Aleuria	*Aurantia*	.	MOTW, p. 238, PFGM, p. 59	.	Orange	Fairy Fire Cup Sac Mushroom	Edible-Excellent
Aleuria	*Aurantia*	*aurantia*	C. & W. USA (logging roads, skid trails)	Original	Orange	Fairy Fire Cup Sac Mushroom	Edible-Excellent
Aleuria	*Aurantia*	*coccinea*	C. & W. USA (logging roads, skid trails)	Scarlet	Orange	Fairy Fire Cup Sac Mushroom	Edible-Excellent
Otidea	*Alutacea*	.	Europe, MOTW, page 235	.	Thin Leather	Ear Fire Cup Sac Mushroom	Unknown
Otidea	*Leporine*	.	USA (open woods soil & moss), MOTW, p. 235, PFGM, page 61	.	Yellow	Ear Fire Cup Sac Mushroom	Edible
Otidea	*onotica*	.	USA (broadleaf forest), MOTW, p. 235	.	Donkey	Ear Fire Cup Sac Mushroom	Edible
Tarzetta	*catinus*		USA, MOTW, page 235			Tarzette Cup Sac Mushroom	Unpalatable
Tarzetta	*catinus*	*catinus*	USA	Original	Cat	Tarzette Cup Sac Mushroom	Unpalatable
Tarzetta	*catinus*	*pustulata*	USA, PFGM, page 57	Pustule	Cat	Tarzette Cup Sac Mushroom	Unpalatable

Family 621.112.06 *Tuberaceae* Truffle

Genus	Species	Variety	Native Location	Variety	Specimen	Genome	Habit
Tuber	*aestivum*	.	Europe (broadleaf forest), MOTW, page 240	.	Summer	Truffle	Edible-Good
Tuber	*brumale*	.	Europe (Oak broadleaf forest), MOTW, page 240	.	Winter	Truffle	Edible
Tuber	*magnatum*	.	Europe (Oak broadleaf forest), MOTW, page 240	.	White	Truffle	Edible-Excellent
Tuber	*melanosporum*	.	Europe (Oak broadleaf forest), MOTW, page 240	.	Black Spore	Truffle	Edible-Excellent

Class 621.2 *Lichnomycetes* Lichen

Latin: *Lichina* lichen
Latin: *Mycetes* mycete, fungus

ORDERS

621.201 *Lichinales* **Lichen**
621.202 *Arthoniales* Comma Lichen
621.203 *Caliciales* Calico Lichen
621.204 *Capnodiales* Cap Lichen
621.205 *Cladoniales* **Cup Lichen**

621.206	***Lecanorales***	**Rim Lichen**
621.207	*Ostropales*	Oster Lichen
621.208	***Parmeliales***	**Shield Lichen**
621.209	***Peltigerales***	**Pelt Lichen**
621.210	*Pyrenulales*	Rash Lichen
621.211	*Teloschistales*	Schist Lichen

Order 621.201 *Lichinales* Lichen

Latin: *Lichina* lichen

FAMILIES

621.201.01	*Lichinaceae*	Lichen
621.201.02	*Agyriaceae*	Agrarian Lichen
621.201.03	*Gloeheppiaceae*	Glow Pelt Lichen
621.201.04	*Lepariaceae*	Leper Lichen
621.201.05	*Lichinotheliaceae*	Nipple Lichen
621.201.06	*Patellariaceae*	Patella Lichen
621.201.07	*Pertusariaceae*	Perforated Lichen
621.201.08	*Strigulaceae*	Bristle Lichen
621.201.09	*Thelenellaceae*	Thelen Lichen
621.201.10	*Trichotheliaceae*	Hairy Nipple Lichen
621.201.11	***Umbilicariaceae***	**Tripe Lichen**

Family 621.201.11 *Umbilicariaceae* Tripe

Genus	Species	Variety	Native Location	Variety	Specimen	Genome	Habit
Umbilicaria	*esculenta*	.		.	Edible	Tripe Lichen	Edible
Umbilicaria	*muhlenbergii*				Muhlenberg	Tripe Lichen	Edible

Order 621.205 *Cladoniales* Cup Lichen

Latin: *Cladon* clad, cup

FAMILIES

621.205.01	***Cladoniaceae***	**Cup Lichen**
621.205.02	*Baeomycetaceae*	Beret Cup Lichen
621.205.03	*Icmadophilaceae*	Candy Cup Lichen

Family 621.205.01 *Cladoniaceae* Cup Lichen

Genus	Species	Variety	Native Location	Variety	Specimen	Genome	Habit
Cladonia	*rangiferina*	.		.	Raindeer	Cup Lichen	Edible

Order 622.206 *Lecanorales* Rim Lichen

Latin: *Lecana* lichen

Latin: *Ora* rim

FAMILIES

621.206.01	*Lecanoraceae*	Rim Lichen
621.206.02	*Acarosporaceae*	Cracked Rim Lichen
621.206.03	*Arctotomiaceae*	Tumor Rim Lichen
621.206.04	*Bacidiaceae*	Dot Rim Lichen
621.206.05	*Biatoraceae*	Crust Dot Rim Lichen
621.206.06	*Briagantaeaceae*	Brick Fire Dot Rim Lichen
621.206.07	*Candelariaceae*	Candle Rim Lichen
621.206.08	*Catillariaceae*	Washed Rim Lichen
621.206.09	*Coccocarpiaceae*	Shell Rim Lichen
621.206.10	*Dactylosporaceae*	Finger Rim Lichen
621.206.11	*Ectolechiaceae*	Ecto Rim Lichen
621.206.12	*Heppiaceae*	Ruby Rim Lichen
621.206.13	***Hymeneliaceae***	**Membrane Rim Lichen**
621.206.14	*Lecidiaceae*	Tile Rim Lichen
621.206.15	*Loxosporeaceae*	Ragged Rim Lichen

621.206.16 *Magalariaceae* Forest Dot Rim Lichen
621.206.17 *Megalosporeaceae* Megalith Dot Rim Lichen
621.206.18 *Micaraceae* Mica Dot Rim Lichen
621.206.19 *Pannaraceae* Pan Rim Lichen
621.206.20 *Phylictidaceae* Whitewash Rim Lichen
621.206.21 *Physiaceae* Bladder Rim Lichen
621.206.22 *Pilocarpaceae* Fuzzy Rim Lichen
621.206.23 *Porpidiaceae* Boulder Rim Lichen
621.206.24 *Psoraceae* Scabby Rim Lichen
621.206.25 *Ramalinaceae* Cartilage Rim Lichen
621.206.26 *Rhizocarpaceae* Map Rim Lichen
621.206.27 *Sphaerophoraceae* Ball Rim Lichen
621.206.28 *Stereocaulaceae* Foam Rim Lichen
621.206.29 *Thelocarpaceae* Nipple Rim Lichen

Family 621.206.13 *Hymeneliaceae* Membrane Rim Lichen

Genus	Species	Variety	Native Location	Variety	Specimen	Genome	Habit
Aspicilia	*esculenta*	.		.	Edible	Membrane Rim Lichen	Edible

Order 621.208 *Parmeliales* Shield Lichen

Latin: *Parmelia* shield

FAMILIES

661.208.01 *Parmeliaceae* Shield Lichen
661.208.02 *Alectoriaceae* Hair Shield Lichen
661.208.03 *Cetrariaceae* Heath Shield Lichen
661.208.04 *Parmelariaceae* Lesser Shield Lichen

Family 621.208.02 *Alectoriaceae* Hair Shield Lichen

Genus	Species	Variety	Native Location	Variety	Specimen	Genome	Habit
Alectoria	*ochroleuca*	.		.	Light Ocher	Witch Hair Shield Lichen	Edible
Alectoria	*ochroleuca*	*ochroleuca*		Original	Light Ocher	Witch Hair Shield Lichen	Edible
Alectoria	*ochroleuca*	*citrina*		Citrine	Light Ocher	Witch Hair Shield Lichen	Edible
Alectoria	*ochroleuca*	*tenuior*		Tenuous	Light Ocher	Witch Hair Shield Lichen	Edible
Bryoria	*fremontii*				Fremont	Horse Hair Shield Lichen	Edible

Family 621.208.03 *Cetrariaceae* Heath Shield Lichen

Genus	Species	Variety	Native Location	Variety	Specimen	Genome	Habit
Cetraria	*Islandica*	.	Atl. Oc. Is. (Iceland)	.	Iceland	Heath Shield Lichen	Edible
Cetraria	*Islandica*	*Islandica*	Atl. Oc. Is. (Iceland)	Original	Iceland	Heath Shield Lichen	Edible
Cetraria	*Islandica*	*crispiformis*	Atl. Oc. Is. (Iceland)	Curly	Iceland	Heath Shield Lichen	Edible
Cetraria	*islandica*	*platyna*	Atl. Oc. Is. (Iceland)	Flat	Iceland	Heath Shield Lichen	Edible

Order 621.209 *Peltigerales* Pelt Lichen

Latin: *Peltigera* pelt bearing

FAMILIES

621.209.01 *Peltigeraceae* Pelt Lichen
621.209.02 *Collemataceae* Jelly Pelt Lichen
621.209.03 *Lobariaceae* Lobed Pelt Lichen
621.209.04 *Placynthiaceae* Cynthia Pelt Lichen

Family 621.209.03 *Lobariaceae* Lobed Pelt Lichen

Genus	Species	Variety	Native Location	Variety	Specimen	Genome	Habit
Lobaria	*pulmonaria*	.		.	Lungwort	Lobed Pelt Lichen	Edible
Lobaria	*pulmonaria*	*pulmonaria*		Original	Lungwort	Lobed Pelt Lichen	Edible
Lobaria	*pulmonaria*	*hypomela*		Dark Underside	Lungwort	Lobed Pelt Lichen	Edible
Lobaria	*pulmonaria*	*linita*		Linite	Lungwort	Lobed Pelt Lichen	Edible

Class 621.3 *Mucoromycetes* Mold

Latin: *Mucor* mucus, mold
Latin: *Mycetes* mycete, fungus

ORDERS

621.301	*Mucoromycetales*	Mold
621.302	*Eurotiales*	Downy Mold
621.303	***Hypocreales***	**Hypoocreal Mold**
621.304	*Laboulbeniales*	Insect Parasite Mold
621.305	*Microascales*	Sac Mold
621.306	*Mycosphaerales*	Sphere Mold
621.307	*Myriangiales*	Myriad Angle Mold
621.308	*Onygenales*	Onyx Mold
621.309	*Saccharomycetales*	Yeast Mold
621.310	*Taphrinales*	Taphrin Mold

Order 621.303 *Hypocreales* Hypocreal Mold

Latin: *Hypo* under
Latin: *Crea* cre

FAMILIES

621.303.01	***Hypocrealaceae***	**Hypocreal Mold**
621.303.02	*Bionectriaceae*	Biotic Cornicle Mold
621.303.03	*Ceratostomataceae*	Horn Hypocreal Mold
621.303.04	***Clavicipitaceae***	**Club Hypocreal Mold**
621.303.05	*Nectriaceae*	Cornicle Hypocreal Mold
621.303.06	*Nessliaceae*	Nessle Hypocreal Mold

Family 621.303.01 *Hypocrealaceae* Hypocreal Mold

Genus	Species	Variety	Native Location	Variety	Specimen	Genome	Habit
Hypomyces	*lactifluorum*	.	USA, PFGM, p. 29	.	Flowing Milk	**Mushroom Pimple**	Edible
Hypomyces	*tulasneanus*	.	USA	.	Tulasne	**Mushroom Pimple**	Edible
Hypomyces	*tulasneanus*	*tulasneanus*	USA	Original	Tulasne	**Mushroom Pimple**	Edible
Hypomyces	*tulasneanus*	*luteovirens*	USA, PFGM, p. 30	Yellow Green	Tulasne	**Mushroom Pimple**	Edible

Family 621.303.04 *Clavicipitaceae* Club Hypocreal Mold

Genus	Species	Variety	Native Location	Variety	Specimen	Genome	Habit
Claviceps	*purpurea*	.	Europe, HSA, page 109	.	Violet Black Sclerota	**Ergot Mold**	Edible
Claviceps	*purpurea*	*purpurea*	Europe	Original	Violet Black Sclerota	**Ergot Mold**	Edible
Claviceps	*purpurea*	*secalis*	Europe	Rye	Violet Black Sclerota	**Ergot Mold**	Edible
Claviceps	*purpurea*	*spartinae*	Europe	Cordgrass	Violet Black Sclerota	**Ergot Mold**	Edible
Cordyceps	*sinensis*		S. Asia (S. China, Himalaya)	.	Chinese	**Heart Ergot Mold**	Edible

Phylum 622 *Basidiomycota* Club Fungus

Latin: *Basidia* basidia
Latin: *Mycota* mycote, fungus

CLASSES

622.1 *Basidiomycetes*	**Club Fungus**
622.2 *Agaricomycetes*	**Gill Mushroom**
622.3 *Boletomycetes*	**Nongill Mushroom**
622.4 *Gastromycetes*	Cavity Fungus
622.5 *Lichinbasidiomycetes*	Club Lichen
622.6 *Uredinomycetes*	Rust Fungus
622.7 *Ustilaginomycetes*	Smut Fungus

Class 622.1 *Basidiomycetes* Club Fungus

Latin: *Basidia* basidia
Latin: *Mycetes* mycete, fungus

ORDERS

622.101 *Basidiomycetales*	Club Fungus
622.102 *Clavariales*	**Coral Fungus**
622.103 *Hymenochaetales*	**Membrane Bristle Fungus**
622.104 *Polyporales*	Pore Fungus
622.105 *Theloporales*	**Crust**
622.106 *Temellales*	**Jelly Fungus**

Order 622.102 *Clavariales* Coral Fungus

Latin: *Clavaria* clave, club

FAMILIES

622.121 *Clavariaceae*	**Coral Fungus**
622.122 *Acrospermaceae*	Acrospore Coral Fungus
622.123 *Ramariaceae*	**Branch Coral Fungus**
622.124 *Sparassidaceae*	**Lace Coral Fungus**

Family 622.102.01 *Claviariaceae* Coral Fungus

Genus	Species	Variety	Native Location	Variety	Specimen	Genome	Habit
Clavaria	*Argillacea*	.	Europe (sandy & peaty grounds in moss), MOTW, p. 222	.	Clay Club	Coral Fungus	Edible
Clavaria	*argillacea*	*argillacea*	Europe (sandy & peaty grounds in moss)	Original	Clay Club	Coral Fungus	Edible
Clavaria	*argillacea*	*pusilla*	Europe (sandy & peaty grounds in moss)	Puny	Clay Club	Coral Fungus	Edible
Clavaria	*atkinsoniana*	.	USA, PFGM, page 71	.	Atkinson	Coral Fungus	Unknown
Clavaria	*fragilis*	.	USA (on moss)	.	Fragile	Coral Fungus	.
Clavaria	*fragilis*	*fragilis*	USA (on moss)	Original	Fragile	Coral Fungus	Unpalatable
Clavaria	*fragilis*	*gracilis*	USA (on moss)	Slender	Fragile	Coral Fungus	Unpalatable
Clavaria	*fragilis*	*sphaerospora*	USA (on moss)	Round Spore	Fragile	Coral Fungus	Unpalatable
Clavaria	*fragilis*	*vermicularis*	USA (on moss), MOTW, page 222	Worm	Fragile	Coral Fungus	Unpalatable
Clavaria	*fumosa*	.	Europe (in woods), MOTW, p. 222, PFGM, p. 71	.	Smoky	Coral Fungus	Edible
Clavaria	*purpurea*	.	Canada, USA (conifer forest), PFGM, page 71	.	Purple	Coral Fungus	Edible
Clavaria	*rubicundula*		USA, PFGM, page 71		Red	Coral Fungus	Unknown
Clavaria	*vermicularis*	.	USA (moist soil), PFGM, page 71	.	White Worm	Coral Fungus	Edible
Clavaria	*zollingeri*	.	NC. & NE. USA (mossy soil), PFGM, page 71	.	Magenta	Coral Fungus	Unknown
Clavaria	*zollingeri*	*zollingeri*	NC. & NE. USA (mossy soil)	Original	Magenta	Coral Fungus	Unknown
Clavaria	*zollingeri*	*amethystina*	NC. & NE. USA (mossy soil), PFGM, page 72	Amethyst	Magenta	Coral Fungus	Unknown
Clavariadelphus	*cokeri*	.	NE. USA (New England), PFGM, page 73	.	Coker	Adelphi Coral Fungus	Edible
Clavariadelphus	*ligula*	.	Europe (conifer forest), MOTW, page 222, PFGM, page 73	.	Tongue	Adelphi Coral Fungus	Edible
Clavariadelphus	*lovejoyae*	.	W. USA (Rocky Mtns.), PFGM, p. 72	.	Lovejoy	Adelphi Coral Fungus	Edible
Clavariadelphus	*pistillaris*	.	USA, MOTW, page 221, PFGM, page 73	.	Large Club	Adelphi Coral Fungus	Edible-Poor
Clavariadelphus	*sachalinensis*	.	USA, PFGM, page 73	.	Sachalin	Adelphi Coral Fungus	Edible
Clavariadelphus	*truncatus*	.	USA (conifer forest), MOTW, p. 221, PFGM, p. 72	.	Flat Top	Adelphi Coral Fungus	Edible-Good

Clavariadelphus	truncatus	truncatus	USA (conifer forest)	Original	Flat Top	Adelphi Coral Fungus	Edible-Good
Clavariadelphus	truncatus	borealis	USA (conifer forest)	Boreal	Flat Top	Adelphi Coral Fungus	Edible-Good
Clavariadelphus	unicolor	.	USA, PFGM, page 72	.	Lilac Pink	Adelphi Coral Fungus	Edible
Clavicorona	piperata	.	USA	.	Pepper	Crown Coral Fungus	Unknown
Clavicorona	piperata	piperata	USA	Original	Pepper	Crown Coral Fungus	Unknown
Clavicorona	piperata	avellanea	USA, PFGM, page 73	Avella	Pepper	Crown Coral Fungus	Unknown
Clavicorona	pyxidata	.	USA (decaying stumps, Aspen, Willow, Cottonwood), MOTW, p. 219, PFGM, p. 73	.	Pyxie	Crown Coral Fungus	Edible-Laxative
Clavulinopsis	fusiformis	.	USA (in grass & along trails), MOTW, page 221	.	Spindle	Lesser Coral Fungus	Edible
Clavulinopsis	helvola	.	Europe (in moss), MOTW, page 222	.	Yellow	Lesser Coral Fungus	.
Macrotyphula	fistulosa	.	Europe (wood debris broadleaf forest), MOTW, p. 222	.	Hollow	Cat Tail Coral Fungus	Edible
Multiclavula	mucida		USA, PFGM, page 71			Multibranch Coral Fungus	Unknown

Family 622.102.03 *Ramariaceae* Branch Coral Fungus

Genus	Species	Variety	Native Location	Variety	Specimen	Genome	Habit
Ramaria	*abietina*	.	Europe (Fir conifer forest), MOTW, page 221	.	Fir	Branch Coral Fungus	Unpalatable
Ramaria	*abietina*	*abietina*	Europe (Fir conifer forest)	Original	Fir	Branch Coral Fungus	Unpalatable
Ramaria	*abietina*	*ochraceovirens*	PFGM, page 74	Ocher Green	Fir	Branch Coral Fungus	Unpalatable
Ramaria	*apiculata*	.	USA, PFGM, page 74	.	Apex	Branch Coral Fungus	Unknown
Ramaria	*apiculata*	*apiculata*	USA	Original	Apex	Branch Coral Fungus	Unknown
Ramaria	*apiculata*	*brunnea*	USA	Brown	Apex	Branch Coral Fungus	Unknown
Ramaria	*aurea*	.	USA (conifer forest), MOTW, p. 218, PFGM, p. 76	.	Golden	Branch Coral Fungus	Edible-Laxative
Ramaria	*botrytis*	.	USA (Beech open woods), MOTW, p 218, PFGM, p. 75	.	Red Tipped	Branch Coral Fungus	Edible
Ramaria	*botrytis*	*botrytis*	USA (Beech open woods)	Original	Red Tipped	Branch Coral Fungus	Edible
Ramaria	*botrytis*	*compactospora*	USA (Beech open woods)	Compact Spore	Red Tipped	Branch Coral Fungus	Edible
Ramaria	*botrytis*	*parvula*	USA (Beech open woods)	Small	Red Tipped	Branch Coral Fungus	Edible
Ramaria	*botrytis*	*rufescens*	USA (Beech open woods)	Blonde	Red Tipped	Branch Coral Fungus	Edible
Ramaria	*botrytoides*	.	USA, PFGM, page 75	.	Berry	Branch Coral Fungus	Unknown
Ramaria	*cauliflloriformis*	.	USA, PFGM, page 75	.	Cauliflower	Branch Coral Fungus	Unknown
Ramaria	*elegantissima*	.	Europe, MOTW, p. 218	.	Elegant	Branch Coral Fungus	Inedible
Ramaria	*flaccida*	.	USA, MOTW, page 221	.	Flaccid	Branch Coral Fungus	Inedible
Ramaria	*flava*	.	USA (mtn. Fir conifer & Beech broadleaf forests), MOTW, p. 218	.	Pale Yellow	Branch Coral Fungus	Edible-Laxative
Ramaria	*formosa*	.	USA (broadleaf forest), MOTW, p. 218, PFGM, pp. 75,76	.	Beautiful	Branch Coral Fungus	Edible-Laxative
Ramaria	*gelatinosa*	.	USA (conifer & broadleaf forests), MOTW, page 218	.	Gelatin	Branch Coral Fungus	Inedible
Ramaria	*gelatiniaurantia*	.	USA, PFGM, page 76	.	Orange Gelatin	Branch Coral Fungus	Unknown
Ramaria	*largentii*	.	USA, PFGM, page 75	.	Largent	Branch Coral Fungus	Unknown
Ramaria	*longispora*	.	USA, PFGM, page 76	.	.	Branch Coral Fungus	Unknown
Ramaria	*pallida*	.	Europe (broadleaf forest), MOTW, page 219	.	Pale	Branch Coral Fungus	Edible-Laxative
Ramaria	*rubrievanescens*	.	USA, PFGM, page 75	.	Red Vane	Branch Coral Fungus	Unknown
Ramaria	*rubripermanens*	.	USA, PFGM, page 75	.	Red Permanent	Branch Coral Fungus	Unknown
Ramaria	*strasseri*	.	USA, PFGM, page 75	.	Strasser	Branch Coral Fungus	Unknown
Ramaria	*stricta*	.	USA (on wood debris), MOTW, p. 219, PFGM, p. 75	.	Narrow	Branch Coral Fungus	Inedible
Ramaria	*stricta*	*stricta*	USA (on wood debris)	Original	Narrow	Branch Coral Fungus	Inedible
Ramaria	*stricta*	*concolor*	USA (on wood debris)	One Color	Narrow	Branch Coral Fungus	Inedible
Ramaria	*tsugina*	.	USA, PFGM, page 74	.	Hemlock	Branch Coral Fungus	Unkown

Order 622.103 *Hymenochaetales* Membrane Bristle Fungus

Latin: *Hymena* hymen, sheath membrane
Latin: *Chaeta* bristle

FAMILIES

622.103.01 *Hymenochaetaceae* Membrane Bristle Fungus
622.103.02 *Schizoporaceae* Split Pore Bristle Fungus

Family 622.103.01 *Hymenochaetaceae* Membrane Bristle Fungus

Genus	Species	Variety	Native Location	Variety	Specimen	Genome	Habit
Inonotus	*hispidus*	.	MOTW, page 212	.	Bristly	Fiber Ear Bristle Fungus	Edible
Inonotus	*hispidus*	Hispidus	Europe (broadleaf forest)	Original	Bristly	Fiber Ear Bristle Fungus	Edible
Inonotus	*hispidus*	Spongiosus	Europe (broadleaf forest)	Spongy	Bristly	Fiber Ear Bristle Fungus	Edible
Inonotus	*hispidus*	Velutinus	Europe (broadleaf forest)	Velvet	Bristly	Fiber Ear Bristle Fungus	Edible
Inonotus	*hispidus*	Villosus	Europe (broadleaf forest)	Villose	Bristly	Fiber Ear Bristle Fungus	Edible

Order 622.104 *Polyporales* Pore Fungus

Latin: *Poly* many
Latin: *pora* pore

FAMILIES

622.104.01 *Polyporaceae* **Pore Fungus**
622.104.02 *Albetrellaceae* **White Trellis Pore Fungus**
622.104.03 *Boreostereaceae* Bore Stereo Pore Fungus
622.104.04 *Cyphellaceae* Cypher Pore Fungus
622.104.05 *Cystostereaceae* Cyst Stereo Pore Fungus
622.104.06 *Epitheliaceae* Epiphyte Nipple Pore Fungus
622.104.07 *Fistulinaceae* **Ox Tongue Pore Fungus**
622.104.08 *Fomitopsidaceae* Foam Pore Fungus
622.104.09 *Ganodermataceae* **Shiny Skin Pore Fungus**
622.104.10 *Grammotheleaceae* Grass Nipple Pore Fungus
622.104.11 *Haplopilaceae* Downy Pore Fungus
622.104.12 *Hyphodermataceae* Hypha Skin Pore Fungus
622.104.13 *Meripilaceae* **Meristic Downy Fungus**
622.104.14 *Phanerochaetaceae* Bristle Pore Fungus
622.104.15 *Podoscyphaceae* Pod Scyph Pore Fungus
622.104.16 *Sistotremataceae* Sistine Tremor Pore Fungus
622.104.17 *Sparassidaceae* **Lace Coral Pore Fungus**
622.104.18 *Tubulicrinaceae* Tubule Lily Pore Fungus
622.104.19 *Xenasmataceae* Strange Pore Fungus

Family 622.104.01 *Polyporaceae* Pore Fungus

Genus	Species	Variety	Native Location	Variety	Specimen	Genome	Habit
Polyporus	*arcularius*	.	Europe, MOTW, page 201	.	Small Chest	Pore Fungus	Inedible
Polyporus	*arcularius*	*arcularius*	Europe	Original	Small Chest	Pore Fungus	Inedible
Polyporus	*arcularius*	*alveolarius*	Europe	Alveole	Small Chest	Pore Fungus	Inedible
Polyporus	*arcularius*	*exasperatus*	Europe	Exasperating	Small Chest	Pore Fungus	Inedible
Polyporus	*badius*	.	USA (on broadleaf trees)	.	Red Brown	Pore Fungus	Inedible
Polyporus	*badius*	*durus*	USA (on broadleaf trees)	Original	Red Brown	Pore Fungus	Inedible
Polyporus	*badius*	*perennis*	USA (on broadleaf trees)	Perennial	Red Brown	Pore Fungus	Inedible
Polyporus	*badius*	*picipes*	USA (on broadleaf trees), MOTW, page 202	Black Foot	Red Brown	Pore Fungus	Inedible
Polyporus	*brumalis*	.	Europe (on broadleaf trees), MOTW, page 201	.	Winter Solstice	Pore Fungus	Edible
Polyporus	*brumalis*	*brumalis*	Europe (on broadleaf trees)	Original	Winter Solstice	Pore Fungus	Edible
Polyporus	*brumalis*	*ciliatus*	Europe (on broadleaf trees)	Hairy	Winter Solstice	Pore Fungus	Edible
Polyporus	*brumalis*	*fasciculatus*	Europe (on broadleaf trees)	Bundle	Winter Solstice	Pore Fungus	Edible
Polyporus	*brumalis*	*fuscidulus*	Europe (on broadleaf trees)	Dark	Winter Solstice	Pore Fungus	Edible
Polyporus	*brumalis*	*hypocrateriformis*	Europe (on broadleaf trees)	Crater	Winter Solstice	Pore Fungus	Edible
Polyporus	*brumalis*	*umbilicatus*	Europe (on broadleaf trees)	Umbilicle	Winter Solstice	Pore Fungus	Edible
Polyporus	*ciliatus*	.	Europe, MOTW, page 201	.	Fringed	Pore Fungus	Unknown
Polyporus	*ciliatus*	*ciliatus*	Europe	Original	Fringed	Pore Fungus	Unknown
Polyporus	*ciliatus*	*nummulariformis*	Europe	Coin	Fringed	Pore Fungus	Unknown
Polyporus	*circinatus*		USA, MOTW, page 309	.	.	Pore Fungus	Unknown
Polyporus	*leptocephalus*	.	Europe	.	Light Head	Pore Fungus	Inedible
Polyporus	*leptocephalus*	*leptocephalus*	Europe	Original	Light Head	Pore Fungus	Inedible
Polyporus	*leptocephalus*	*elegans*	Europe, MOTW, page 202	Elegant	Light Head	Pore Fungus	Inedible
Polyporus	*leptocephalus*	*nigripes*	Europe	Black Foot	Light Head	Pore Fungus	Inedible
Polyporus	*melanopus*	.	Europe, MOTW, page 202	.	Black Foot	Pore Fungus	Inedible
Polyporus	*melanopus*	*melanopus*	Europe	Original	Black Foot	Pore Fungus	Inedible
Polyporus	*melanopus*	*cyathoides*	Europe	Cup	Black Foot	Pore Fungus	Inedible
Polyporus	*squamosus*	.	USA (on living or dead wood in Poplar, Willow & Elm deciduous forest), MOTW, page 202, PFGM, page 129	.	Dryad Saddle	Pore Fungus	Edible
Polyporus	*squamosus*	*squamosus*	USA (Poplar, Willow & Elm forest)	Original	Dryad Saddle	Pore Fungus	Edible
Polyporus	*squamosus*	*cellulosus*	USA (Poplar, Willow & Elm forest)	Cellulose	Dryad Saddle	Pore Fungus	Edible
Polyporus	*squamosus*	*fagicola*	USA (Poplar, Willow & Elm forest)	Beech	Dryad Saddle	Pore Fungus	Edible
Polyporus	*squamosus*	*juglandis*	USA (Poplar, Willow & Elm forest)	Walnut	Dryad Saddle	Pore Fungus	Edible
Polyporus	*squamosus*	*maculatus*	USA (Poplar, Willow & Elm forest)	Spotted	Dryad Saddle	Pore Fungus	Edible
Polyporus	*squamosus*	*maximus*	USA (Poplar, Willow & Elm forest)	Greater	Dryad Saddle	Pore Fungus	Edible
Polyporus	*squamosus*	*michelii*	USA (Poplar, Willow & Elm forest)	Michel	Dryad Saddle	Pore Fungus	Edible
Polyporus	*squamosus*	*polymorphus*	USA (Poplar, Willow & Elm forest)	Polymorphic	Dryad Saddle	Pore Fungus	Edible
Polyporus	*squamosus*	*rangiferinus*	USA (Poplar, Willow & Elm forest)	Raindeer	Dryad Saddle	Pore Fungus	Edible

Polyporus	*squamosus*	*rostkovii*	USA (Poplar, Willow & Elm forest)	Rostkov	Dryad Saddle	Pore Fungus	Edible
Polyporus	*tuberaster*	.	Europe	.	Tuber Star	Pore Fungus	Edible
Polyporus	*tuberaster*	*tuberaster*	Europe	Original	Tuber Star	Pore Fungus	Edible
Polyporus	*tuberaster*	*forquignonii*	Europe, MOTW, page 202	Forquignon	Tuber Star	Pore Fungus	Edible
Polyporus	*udus*	.	USA	.	.	Pore Fungus	Unknown
Polyporus	*udus*	*udus*	USA	.	.	Pore Fungus	Unknown
Polyporus	*udus*	*maculatus*	USA	.	.	Pore Fungus	Unknown
Polyporus	*umbellatus*	.	USA (on stumps & roots of Beech & Oak broadleaf forest), MOTW, p. 206, PFGM, p. 129	.	Umbrella	Pore Fungus	Edible-Good
Polyporus	*umbellatus*	*umbellatus*	USA (on stumps & roots of Beech & Oak broadleaf forest)	Original	Umbrella	Pore Fungus	Edible-Good
Polyporus	*umbellatus*	*ramosissimus*	USA (on stumps & roots of Beech & Oak broadleaf forest)	Very Branched	Umbrella	Pore Fungus	Edible-Good
Polyporus	*umbellatus*	*ramosus*	USA (on stumps & roots of Beech & Oak broadleaf forest)	Branched	Umbrella	Pore Fungus	Edible-Good
Polyporus	*varius*	.	Europe, MOTW, page 202	.	Varied	Pore Fungus	Inedible
Polyporus	*varius*	*varius*	Europe	Original	Varied	Pore Fungus	Inedible
Polyporus	*varius*	*lateralis*	Europe	Wide	Varied	Pore Fungus	Inedible
Polyporus	*varius*	*nummularius*	Europe	Coin	Varied	Pore Fungus	Inedible
Coriolus	*versicolor*		E. China & Japan (woods) ISBN:978-1-58333-452-2		Kawaratake	Turkeytail Pore Fungus	Edible
Faerberia	*carbonaria*	.	Europe (burned soil), MOTW, page 223	.	Ash Gray	Ferber Pore Fungus	Edible
Laetiporus	*sulphureus*	.	USA (on living & deadwood in conifer & hardwood forest), MOTW, p. 205, PFGM, p. 127	.	Sulphur Shelf	Bright Pore Fungus	Edible-Toxic
Laetiporus	*sulphureus*	*sulphureus*	USA (conifer & hardwood forest)	Original	Sulphur Shelf	Bright Pore Fungus	Edible-Toxic
Laetiporus	*sulphureus*	*citrinus*	USA (conifer & hardwood forest)	Citrine	Sulphur Shelf	Bright Pore Fungus	Edible-Toxic
Laetiporus	*sulphureus*	*coriaceus*	USA (conifer & hardwood forest)	Leather	Sulphur Shelf	Bright Pore Fungus	Edible-Toxic
Laetiporus	*sulphureus*	*imbricatus*	USA (conifer & hardwood forest)	Tiled	Sulphur Shelf	Bright Pore Fungus	Edible-Toxic
Laetiporus	*sulphureus*	*lingua-cervina*	USA (conifer & hardwood forest)	Deer Tongue	Sulphur Shelf	Bright Pore Fungus	Edible-Toxic
Laetiporus	*sulphureus*	*ramosus*	USA (conifer & hardwood forest)	Branched	Sulphur Shelf	Bright Pore Fungus	Edible-Toxic
Laetiporus	*sulphureus*	*tenax*	USA (conifer & hardwood forest)	Tough	Sulphur Shelf	Bright Pore Fungus	Edible-Toxic
Lentinus	*adherens*		See *Neolentinus adherens*				
Lentinus	*degener*		See *Neolentinus schaefferi*				
Lentinus	*edodes*	.	See *Lentinula edodes*	.	.		
Lentinus	*lepedius*	.	See *Neolentinus lepidius*	.	.		
Lentinus	*omphalodes*		See *Lentinellus michneri*				
Lentinus	*ponderosus*	.	See *Neolentinus ponderosus*	.	.		
Lentinus	*sajor-caju*	.	.	.	Sajor Caju	Sawgill Pore Fungus	Edible
Lentinus	*sajor-caju*	*sajor-caju*	.	Original	Sajor Caju	Sawgill Pore Fungus	Edible
Lentinus	*sajor-caju*	*americana*	USA	American Oyster	Sajor Caju	Sawgill Pore Fungus	Edible
Lentinus	*sajor-caju*	*densifolius*	.	Dense	Sajor Caju	Sawgill Pore Fungus	Edible
Lentinus	*sajor-caju*	*gray-oysteri*	.	Gray Oyster	Sajor Caju	Sawgill Pore Fungus	Edible
Lentinus	*sajor-caju*	*hollandica*	Europe (Netherlands)	Dutch Oyster	Sajor Caju	Sawgill Pore Fungus	Edible
Lentinus	*sajor-caju*	*indica*	S. Asia (India)	India Oyster	Sajor Caju	Sawgill Pore Fungus	Edible
Lentinus	*sajor-caju*	*phoenix*	S. Asia (India)	Phoenix Oyster	Sajor Caju	Sawgill Pore Fungus	Edible
Lentinus	*spretus*	.	See *Neolentinus suffrutescens*	.	.		
Lentinus	*squamosus*	.	USA, PFGM, page 161	.	Scaly Tiger	Sawgill Pore Fungus	Unknown
Lentinus	*sulcatus*	.	See *Heliocybe sulcata*	.	.		
Lentinus	*tigrinus*	.	USA (on Willow, Poplar & Beech logs & wood in deciduous forest), MOTW, page 126, PFGM, page 160	.	Tiger	Sawgill Pore Fungus	Edible
Lentinus	*vuplinus*		See *Lentinellus vulpinus*				
Neolentinus	*adhaerens*	.	Europe (Fir & Pine conifer forest), MOTW, page 126	.	Adherent	Modern Sawgill Pore Fungus	Indedible
Neolentinus	*adhaerens*	*adhaerens*	Europe (Fir & Pine conifer forest)	Original	Adherent	Modern Sawgill Pore Fungus	Indedible
Neolentinus	*adhaerens*	*pulverulentus*	Europe (Fir & Pine conifer forest)	Powder	Adherent	Modern Sawgill Pore Fungus	Indedible
Neolentinus	*lepideus*	.	USA (on stumps & logs in conifer forest), MOTW, p. 128, PFGM, p. 159	.	Scaly	Modern Sawgill Pore Fungus	Edible-Poor
Neolentinus	*ponderosus*	.	USA, PFGM, page 160	.	Giant	Modern Sawgill Pore Fungus	Edible
Neolentinus	*schaefferi*	.	Europe	.	Schaeffer	Modern Sawgill Pore Fungus	Indedible
Neolentinus	*schaefferi*	*schaefferi*	Europe	Original	Schaeffer	Modern Sawgill Pore Fungus	Indedible
Neolentinus	*schaefferi*	*degener*	Europe, MOTW, page 126	Degenerate	Schaeffer	Modern Sawgill Pore Fungus	Indedible
Neolentinus	*suffrutescens*	.	USA	.	Shrubby	Modern Sawgill Pore Fungus	Edible
Neolentinus	*suffrutescens*	*suffrutescens*	USA	Original	Shrubby	Modern Sawgill Pore Fungus	Edible
Neolentinus	*suffrutescens*	*spretus*	USA, PFGM, page 160	Scorned	Shrubby	Modern Sawgill Pore Fungus	Edible

Family 622.104.02 *Albatrellaceae* White Trellis Pore Fungus

Genus	Species	Variety	Native Location	Variety	Specimen	Genome	Habit
Albatrellus	*confluens*	.	USA (conifer forest), MOTW, page 205	.	Confluent	White Trellis Pore Fungus	Edible
Albatrellus	*confluens*	*confluens*	USA (conifer forest)	Original	Confluent	White Trellis Pore Fungus	Edible

Albatrellus	*confluens*	*artemidorus*	USA (conifer forest)	Wormwood	Confluent	White Trellis Pore Fungus	Edible
Albatrellus	*confluens*	*aurantius*	USA (conifer forest)	Orange	Confluent	White Trellis Pore Fungus	Edible
Albatrellus	*cristatus*	.	USA (on humus & rotten wood in Oak broadleaf forest), MOTW, page 204	.	Crested	White Trellis Pore Fungus	Inedible
Albatrellus	*ovinus*	.	USA (mountain Pine conifer forest), MOTW, p. 204	.	Sheep	White Trellis Pore Fungus	Edible-Good
Albatrellus	*ovinus*	*ovinus*	USA (mountain Pine conifer forest)	Original	Sheep	White Trellis Pore Fungus	Edible-Good
Albatrellus	*ovinus*	*carinthiacus*	USA (mountain Pine conifer forest)	Carinthian	Sheep	White Trellis Pore Fungus	Edible-Good
Albatrellus	*ovinus*	*fragilis*	USA (mountain Pine conifer forest)	Fragile	Sheep	White Trellis Pore Fungus	Edible-Good
Scutiger	*oregonensis*	.	USA (Fir & broadleaf forests)	.	Oregon	Bowl White Trellis Pore Fungus	Edible-Good
Scutiger	*oregonensis*	*oregonensis*	USA (Fir & broadleaf forests)	Original	Oregon	Bowl White Trellis Pore Fungus	Edible-Good
Scutiger	*oregonensis*	*pes-caprae*	USA (Fir & broadleaf forests), MOTW, page 204	Goat's Foot	Oregon	Bowl White Trellis Pore Fungus	Edible-Good

Family 622.104.07 *Fistulinaceae* Ox Tongue Pore Fungus

Genus	Species	Variety	Native Location	Variety	Specimen	Genome	Habit
Fistulina	*hepatica*	.	USA (on living or dead wood in hardwood forest, especially Oak), MOTW, page 213, PFGM, p. 129	.	Beafsteak	Ox Tongue Pore Fungus	Edible-Good
Fistulina	*hepatica*	*hepatica*	USA (hardwood forest, especially Oak)	Original	Beafsteak	Ox Tongue Pore Fungus	Edible-Good
Fistulina	*hepatica*	*buglossum*	USA (hardwood forest, especially Oak)	Beef Tongue	Beafsteak	Ox Tongue Pore Fungus	Edible-Good
Fistulina	*hepatica*	*bulliardii*	USA (hardwood forest, especially Oak)	Bulliard	Beafsteak	Ox Tongue Pore Fungus	Edible-Good

Family 622.104.09 *Fomitopsidaceae* Foam Pore Fungus

Genus	Species	Variety	Native Location	Variety	Specimen	Genome	Habit
Wolfiporia	*extensa*	*cocos*	Asia & USA on the roots of pines. ISBN 978-1-58333-452-2, p. 85	Hoelen	Tuckahoe	Wolf Foam Pore Fungus	Edible-Good

Family 622.104.09 *Ganodermataceae* Shiny Skin Pore Fungus

Genus	Species	Variety	Native Location	Variety	Specimen	Genome	Habit
Ganoderma	*applanatum*	.	USA, MOTW, page 213, PFGM, page 127	.	Artist's	Shiny Skin Pore Fungus	Indedible
Ganoderma	*applanatum*	*applanatum*	USA	Original	Artist's	Shiny Skin Pore Fungus	Indedible
Ganoderma	*applanatum*	*lipsiensis*	USA	Lipsian	Artist's	Shiny Skin Pore Fungus	Indedible
Ganoderma	*lucidum*	.	USA, MOTW, page 213	.	Reishi	Shiny Skin Pore Fungus	Edible
Ganoderma	*lucidum*	*lucidum*	USA	Original	Reishi	Shiny Skin Pore Fungus	Edible
Ganoderma	*lucidum*	*annulatum*	USA	Ringed	Reishi	Shiny Skin Pore Fungus	Edible
Ganoderma	*lucidum*	*castaneus*	USA	Chestnut	Reishi	Shiny Skin Pore Fungus	Edible
Ganoderma	*lucidum*	*crustatus*	USA	Encrusted	Reishi	Shiny Skin Pore Fungus	Edible
Ganoderma	*lucidum*	*dimidiatus*	USA	Divided	Reishi	Shiny Skin Pore Fungus	Edible
Ganoderma	*lucidum*	*flabelliformis*	USA	Fancap	Reishi	Shiny Skin Pore Fungus	Edible
Ganoderma	*lucidum*	*hemisphaericum*	USA	Round	Reishi	Shiny Skin Pore Fungus	Edible
Ganoderma	*lucidum*	*laccatus*	USA	Milky	Reishi	Shiny Skin Pore Fungus	Edible
Ganoderma	*lucidum*	*rubellum*	USA	Red Dragon	Reishi	Shiny Skin Pore Fungus	Edible
Ganoderma	*lucidum*	*rugosus*	USA	Wrinkled	Reishi	Shiny Skin Pore Fungus	Edible
Ganoderma	*lucidum*	*verniceus*	USA	Varnish	Reishi	Shiny Skin Pore Fungus	Edible
Ganoderma	*lucidum*	*vernicosus*	USA	Varnished	Reishi	Shiny Skin Pore Fungus	Edible
Ganoderma	*tsugae*	.	E. Asia (Japan), MOTW, p. 213	.	Tsuga	Shiny Skin Pore Fungus	Inedible

Family 622.104.13 *Meripilaceae* Meristic Downy Pore Fungus

Genus	Species	Variety	Native Location	Variety	Specimen	Genome	Habit
Meripilus	*giganteus*	.	USA (on stumps of Beech & Oak broadleaf trees), PFGM, page 129	.	Giant	Meristic Downy Fungus	Edible
Meripilus	*giganteus*	*giganteus*	USA (on stumps of Beech & Oak)	Original	Giant	Meristic Downy Fungus	Edible
Meripilus	*giganteus*	*acanthoides*	USA (on stumps of Beech & Oak)	Prickly	Giant	Meristic Downy Fungus	Edible
Meripilus	*giganteus*	*frondosus*	USA (on stumps of Beech & Oak)	Frond	Giant	Meristic Downy Fungus	Edible
Grifola	*frondosa*	.	S. Canada, USA (on living & dead wood in conifer & Oak hardwood forest), MOTW, page 208, PFGM, page 128	.	Hen Of The Woods	Braided Meristic Downy Fungus	Edible -Good
Grifola	*frondosa*	*frondosa*	S. Canada, USA (on living & dead wood in conifer & Oak hardwood forest)	Original	Hen Of The Woods	Braided Meristic Downy Fungus	Edible -Good
Grifola	*frondosa*	*cristatus*	S. Canada, USA (on living & dead wood in conifer & Oak hardwood forest)	Crested	Hen Of The Woods	Braided Meristic Downy Fungus	Edible -Good
Grifola	*frondosa*	*intybaceus*	S. Canada, USA (on living & dead wood	Chicory	Hen Of The	Braided Meristic Downy	Edible

			in conifer & Oak hardwood forest)			Woods Fungus	-Good
Grifola	*frondosa*	*maitakei*	S. Canada, USA (on living & dead wood in conifer & Oak hardwood forest)	Maitake		Hen Of The Braided Meristic Downy Woods Fungus	Edible -Good

Family 622.104.17 *Sparassidaceae* Lace Coral Pore Fungus

Genus	Species	Variety	Native Location	Variety	Specimen	Genome	Habit
Sparassis	*crispa*	.	USA (sandy soil under Virginia Pine), MOTW, p. 216, PFGM, p. 76	.	Ruffles	Lace Coral Pore Fungus	Edible-Excellent
Sparassis	*crispa*	*crispa*	USA (sandy soil under Virginia Pine)	Original	Ruffles	Lace Coral Pore Fungus	Edible-Excellent
Sparassis	*crispa*	*radicata*	USA (sandy soil under Virginia Pine)	Rooted	Ruffles	Lace Coral Pore Fungus	Edible-Excellent
Sparassis	*laminosa*	.	Europe (roadsides, ditches), MOTW, page 216	.	Plate	Lace Coral Pore Fungus	Edible-Excellent
Sparassis	*laminosa*	*laminosa*	Europe (roadsides, ditches)	Original	Plate	Lace Coral Pore Fungus	Edible-Excellent
Sparassis	*laminosa*	*herbstii*	Europe (roadsides, ditches)	Herbst	Plate	Lace Coral Pore Fungus	Edible-Excellent

Order 622.105 *Theloporales* Crust

Latin: *Thelo* crust
Latin: *Pora* pore

FAMILIES

622.105.01 *Theloporaceae* Groundwort Crust
622.105.02 *Corticiaceae* Corticle Wood Crust
622.105.03 *Meruliaceae* Wood Crust

Family 622.105.01 *Thelophoraceae* Groundwort Crust

Genus	Species	Native Location	Specimen	Genome	Habit
Polyozellus	*multiplex*	S. Canada, N. USA (conifer forest), PFGM, page 77	Blue	Chanterelle Groundwort Crust	Edible

Order 622.106 *Tremellales* Jelly Fungus

Latin: *Trema* tremor, jelly

FAMILIES

622.106.01 *Tremellaceae* Tremor Jelly
622.106.02 *Aporpiaceae* Perplexed Jelly
622.106.03 *Auriculariaceae* Ear Jelly
622.106.04 *Dacrymycetaceae* Tear Jelly
622.106.05 *Exidiaceae* Warty Jelly
622.106.06 *Sirobasidiaceae* Pitted Jelly

Family 622.106.01 *Tremellaceae* Tremor Jelly

Genus	Species	Variety	Native Location	Variety	Specimen	Genome	Habit
Tremella	*foliacea*	.	USA (on deadwood & tree stumps), PFGM, p. 67	.	Leaf	Tremor Jelly	Edible
Tremella	*foliacea*	*foliacea*	USA (on deadwood & tree stumps), PFGM, p. 67	Original	Leaf	Tremor Jelly	Edible
Tremella	*foliacea*	*frondosa*	USA (on deadwood & tree stumps), PFGM, p. 68	Frond	Leaf	Tremor Jelly	Edible
Tremella	*mesenterica*	.	USA (on wood), MOTW, page 14, PFGM, p. 67	.	Witch's Butter	Tremor Jelly	Edible
Tremella	*mesenterica*	*mesenterica*	USA (on wood)	Original	Witch's Butter	Tremor Jelly	Edible
Tremella	*mesenterica*	*lutescens*	USA (on wood), PFGM, page 68	Yellow	Witch's Butter	Tremor Jelly	Edible

Family 622.106.03 *Auriculariaceae* Ear Jelly

Genus	Species	Variety	Native Location	Variety	Specimen	Genome	Habit
Auricularia	*auricular*	.	USA (decaying wood in Elder, Walnut, Beech, Yellow Locust & Willow broadleaf forest)	.	Brown	Ear Jelly	Edible
Auricularia	*auricula*	*auricula*	USA (decaying wood in Elder, Walnut, Beech, Yellow Locust & Willow broadleaf forest)	Original	Brown	Ear Jelly	Edible
Auricularia	*auricula*	*judae*	USA (decaying wood in Elder, Walnut, Beech, Yellow Locust & Willow broadleaf forest), MOTW, page 231	Juda	Brown	Ear Jelly	Edible
Auricularia	*auricula*	*judae-lactea*	USA (decaying wood in Elder, Walnut, Beech, Yellow Locust & Willow broadleaf forest), PFGM, page 64	Milky Juda	Brown	Ear Jelly	Edible
Auricularia	*mesenterica*	.	Europe (on stumps & dead trunks in broadleaf forest), MOTW, page 231	.	Membrane	Ear Jelly	Inedible
Auricularia	*mesenterica*	*mesenterica*	Europe (on stumps & dead trunks in broadleaf forest)	Original	Membrane	Ear Jelly	Inedible
Auricularia	*mesenterica*	*corrugata*	Europe (on stumps & dead trunks in broadleaf forest)	Ruffled	Membrane	Ear Jelly	Inedible
Auricularia	*mesenterica*	*lobata*	Europe (on stumps & dead trunks in broadleaf forest)	Lobed	Membrane	Ear Jelly	Inedible
Auricularia	*mesenterica*	*tremelloides*	Europe (on stumps & dead trunks in	Tremor	Membrane	Ear Jelly	Inedible

			broadleaf forest)				
Auricularia	*polytricha*	.	USA, PFGM, page 64		.	Many Hair Ear Jelly	Unknown

Family 622.106.05 *Exidiaceae* Warty Jelly

Genus	Species	Variety	Native Location	Variety	Specimen	Genome	Habit
Guepinia	*helvelloides*	.	Europe (on moss, grass in confer forest), MOTW, page 233, PFGM, page 66	.	Apricot	Guepin Warty Jelly	Edible-Good
Guepinia	*helvelloides*	*helvelloides*	Europe (on moss, grass in confer forest)	Original	Apricot	Guepin Warty Jelly	Edible-Good
Guepinia	*helvelloides*	*rufus*	Europe (on moss, grass in confer forest)	Blonde	Apricot	Guepin Warty Jelly	Edible-Good
Phlogiotis	*helvelloides*	.	Synonym for *Guepinia helvelloides*	.	.	.	.
Pseudohydnum	*gelatinosum*	.	USA (decaying wood in Pine confer forest), MOTW, page 233, PFGM, page 66	.	Toothjelly	Truffle Warty Jelly	Unpalatable

Class 622.2 *Agaricomycetes* Gill Mushroom

Latin: *Asco* ascus, spore sac
Latin: *Mycetes* mycete, fungus

ORDERS

622.201	*Agaricales*	Gill Mushroom
622.202	*Russulales*	Brittle Gill Mushroom

Order 622.201 *Agaricales* Gill Mushroom

Latin: *Agarica* agaric, mushroom

FAMILIES

622.201.01	*Agaricaceae*	Gill Mushroom
622.201.02	*Bolbitaceae*	Bonnet Gill Mushroom
622.201.03	*Coprinaceae*	Cap Gill Mushroom
622.201.04	*Cortinariaceae*	Web Cap Gill Mushroom
622.201.05	*Entolomataceae*	Pinkgill Gill Mushroom
622.201.06	*Hydnangiaceae*	Tallowgill Gill Mushroom
622.201.07	*Hygrophoraceae*	Waxycap Gill Mushroom
622.201.08	*Lepiotaceae*	Parasol Gill Mushroom
622.201.09	*Marasmiaceae*	Pinwheel Gill Mushroom
622.201.10	*Mesophelliaceae*	Truffle Gill Mushroom
622.201.11	*Omphalinaceae*	Navalcap Gill Mushroom
622.201.12	*Pleurotaceae*	Oyster Gill Mushroom
622.201.13	*Pluteaceae*	Roofcap Gill Mushroom
622.201.14	*Psathyrellaceae*	Crumblecap Gill Mushroom
622.201.15	*Pterulaceae*	Wingcap Gill Mushroom
622.201.16	*Schizophyllaceae*	Splitgill Gill Mushroom
622.201.17	*Strophariaceae*	Ringstalk Gill Mushroom
622.201.18	*Tricolomataceae*	Cavalier Gill Mushroom
622.201.19	*Typhulaceae*	Typha Gill Mushroom

Family 622.201.01 *Agaricaceae* Gill Mushroom

Genus Species	Variety	Native Location	Variety	Specimen	Genome	Habit
Agaricus arvensis	.	USA (open woods & grassland)	.	Horse	Gill Mushroom	Edible-Excellent
Agaricus arvensis	*arvensis*	USA (open woods & grassland)	Original	Horse	Gill Mushroom	Edible-Excellent
Agaricus arvensis	*exquisitus*	USA (open woods & grassland)	Exquisite	Horse	Gill Mushroom	Edible-Excellent
Agaricus arvensis	*leucotrichus*	USA (open woods & grassland)	Light Hair	Horse	Gill Mushroom	Edible-Excellent
Agaricus augustus	.	USA (grasslands & conifer forest)	.	Prince	Gill Mushroom	Edible-Good
Agaricus augustus	*augustus*	USA (grasslands & conifer forest)	Original	Prince	Gill Mushroom	Edible-Good
Agaricus augustus	*perrarus*	USA (grasslands & conifer forest)	Perrar	Prince	Gill Mushroom	Edible-Good
Agaricus augustus	*subrufescens*	USA (grasslands & conifer forest)	Almond	Prince	Gill Mushroom	Edible-Good
Agaricus bernardii	.	USA (coastal marshes)	.	Parkway	Gill Mushroom	Inedible
Agaricus bernardii	*bernardii*	USA (Pensacola, FL)	Original	Parkway	Gill Mushroom	Edible
Agaricus bernardii	*ingratus*	USA	Ingrate	Parkway	Gill Mushroom	Edible
Agaricus bernardii	*maleolens*	USA	Soft	Parkway	Gill Mushroom	Edible
Agaricus bisporus	.	USA (manured soil, roadsides)	.	Campignon	Gill Mushroom	Edible-Good
Agaricus bisporus	*bisporus*	USA (manured soil, roadsides)	Original	Campignon	Gill Mushroom	Edible-Good
Agaricus bisporus	*albidus*	USA (manured soil, roadsides)	White	Campignon	Gill Mushroom	Edible-Good
Agaricus bisporus	*avellaneus*	USA (manured soil, roadsides)	Avella Vecchia	Campignon	Gill Mushroom	Edible-Good
Agaricus bisporus	*brunnescens*	USA (manured soil, roadsides)	Brown Button	Campignon	Gill Mushroom	Edible-Good
Agaricus bisporus	*cookeanus*	USA (manured soil, roadsides)	Cooke	Campignon	Gill Mushroom	Edible-Good
Agaricus bisporus	*hortensis*	USA (manured soil, roadsides)	Garden	Campignon	Gill Mushroom	Edible-Good
Agaricus bisporus	*portobelloi*	USA (manured soil, roadsides)	Portobello	Campignon	Gill Mushroom	Edible-Good
Agaricus bitorquis	.	USA (manured soil, roadsides)	.	Two Ring	Gill Mushroom	Edible
Agaricus bitorquis	*bitorquis*	USA (manured soil, roadsides)	Original	Two Ring	Gill Mushroom	Edible
Agaricus bitorquis	*edulis*	USA (manured soil, roadsides)	Edible	Two Ring	Gill Mushroom	Edible
Agaricus bitorquis	*rodmanii*	USA (Pensacola, FL)	Sheathed Stalk	Two Ring	Gill Mushroom	Edible
Agaricus bitorquis	*validus*	USA (manured soil, roadsides)	Valid	Two Ring	Gill Mushroom	Edible
Agaricus campestris	.	USA (lawns, pastures, roadsides)	.	Meadow	Gill Mushroom	Edible-Excellent
Agaricus campestris	*campestris*	USA (lawns, pastures, roadsides)	Original	Meadow	Gill Mushroom	Edible-Excellent
Agaricus campestris	*equestris*	USA (lawns, pastures, roadsides)	Horse	Meadow	Gill Mushroom	Edible-Excellent
Agaricus campestris	*fuscopilosellus*	USA (lawns, pastures, roadsides)	Fuscous Downy	Meadow	Gill Mushroom	Edible-Excellent
Agaricus campestris	*isabellinus*	USA (lawns, pastures, roadsides)	Isabelle	Meadow	Gill Mushroom	Edible-Excellent
Agaricus campestris	*radicatus*	USA (lawns, pastures, roadsides)	Radial	Meadow	Gill Mushroom	Edible-Excellent
Agaricus campestris	*squamulosus*	USA (lawns, pastures, roadsides)	Scaley	Meadow	Gill Mushroom	Edible-Excellent

Agaricus campestris	*villaticus*	USA (lawns, pastures, roadsides)	Rustic	Meadow	Gill Mushroom	Edible-Excellent
Agaricus comtulus	.	E. USA (on soil in various habitats)	.	Yellow Cap	Gill Mushroom	Edible
Agaricus comtulus	*comtulus*	E. USA (on soil in various habitats)	Original	Yellow Cap	Gill Mushroom	Edible
Agaricus comtulus	*niveolutescens*	E. USA (on soil in various habitats)	Snowy	Yellow Cap	Gill Mushroom	Edible
Agaricus dulcidulus	.	Europe (woodlands)	.	Sweet	Gill Mushroom	Edible
Agaricus dulcidulus	*dulcidulus*	Europe (woodlands)	Original	Sweet	Gill Mushroom	Edible
Agaricus dulcidulus	*amethystina*	Europe (woodlands)	Amethyst	Sweet	Gill Mushroom	Edible
Agaricus dulcidulus	*purpurellus*	Europe (woodlands)	Purple	Sweet	Gill Mushroom	Edible
Agaricus dulcidulus	*rubellus*	Europe (woodlands)	Red	Sweet	Gill Mushroom	Edible
Agaricus dulcidulus	*semotus*	Europe (woodlands)	Silky	Sweet	Gill Mushroom	Edible
Agaricus hondensis	.	USA	.	Honda	Gill Mushroom	Poison
Agaricus nivescens	.	USA (grass, open conifer woods)	.	Snowy Cap	Gill Mushroom	Edible-Excellent
Agaricus placomyces	.	USA (on forest soil)	.	Flat Top	Gill Mushroom	Poison
Agaricus placomyces	*placomyces*	USA (on forest soil)	Original	Flat Top	Gill Mushroom	Poison
Agaricus placomyces	*meleagris*	USA (on forest soil)	Grayscale	Flat Top	Gill Mushroom	Poison
Agaricus radicatus	.	Europe (meadows, gardens, dirt roads)	.	Rooted	Gill Mushroom	Poison
Agaricus rodmanii	.	USA (lawns, roadsides, trails)	.	Sheathed Stalk	Gill Mushroom	Edible-Good
Agaricus silvaticus	.	Canada, USA (on soil)	.	Forest	Gill Mushroom	Edible-Toxic
Agaricus silvaticus	*silvaticus*	Canada, USA (on soil)	Original	Forest	Gill Mushroom	Edible-Toxic
Agaricus silvaticus	*fagetorum*	Canada, USA (on soil)	Faget	Forest	Gill Mushroom	Edible-Toxic
Agaricus silvaticus	*haemorrhoidarius*	Canada, USA (on soil)	Bloody Agaric	Forest	Gill Mushroom	Edible-Toxic
Agaricus silvaticus	*pallidus*	Canada, USA (on soil)	Pale	Forest	Gill Mushroom	Edible-Toxic
Agaricus silvaticus	*sanguinarius*	Canada, USA (on soil)	Red	Forest	Gill Mushroom	Edible-Toxic
Agaricus silvaticus	*vinosobrunneus*	Canada, USA (on soil)	Wine Red Brown	Forest	Gill Mushroom	Edible-Toxic
Agaricus silvicola	.	USA (conifer & hardwood forest)	.	Woodland	Gill Mushroom	Edible-Good
Agaricus silvicola	*silvicola*	USA (conifer & hardwood forest)	Original	Woodland	Gill Mushroom	Edible-Good
Agaricus silvicola	*abruptibulbus*	USA (conifer & hardwood forest)	Flat Bulb	Woodland	Gill Mushroom	Edible-Good
Agaricus silvicola	*flavescens*	USA (conifer & hardwood forest)	Yellow	Woodland	Gill Mushroom	Edible-Good
Agaricus subrufescens	.	C. & E. USA	.	Small Spore Prince	Gill Mushroom	Edible-Good
Agaricus subrutilescens	.	NW. USA (on forest soil)	.	Woolly Stalk	Gill Mushroom	Edible-Toxic
Agaricus urinascens	.	W. USA (meadows, pastures)	.	Urine	Gill Mushroom	Edible-Good
Agaricus urinascens	*urinascens*	W. USA (meadows, pastures)	Original	Urine	Gill Mushroom	Edible-Good
Agaricus urinascens	*albertii*	W. USA (meadows, pastures)	Albert	Urine	Gill Mushroom	Edible-Good
Agaricus urinascens	*crocodilinus*	W. USA (meadows, pastures)	Crocodile	Urine	Gill Mushroom	Edible-Good
Agaricus urinascens	*macrosporus*	W. USA (meadows, pastures)	Large Spore	Urine	Gill Mushroom	Edible-Good
Agaricus urinascens	*schaefferianus*	W. USA (meadows, pastures)	Schaeffer	Urine	Gill Mushroom	Edible-Good
Agaricus urinascens	*stramineosquamulosus*	W. USA (meadows, pastures)	Straw Scaled	Urine	Gill Mushroom	Edible-Good
Agaricus urinascens	*stramineus*	W. USA (meadows, pastures)	Straw	Urine	Gill Mushroom	Edible-Good
Agaricus urinascens	*villaticus*	W. USA (meadows, pastures)	Rustic	Urine	Gill Mushroom	Edible-Good
Agaricus vaporarius	.	Europe (manured pastures, Fir conifer forest)	.	Heat Loving	Gill Mushroom	Edible
Agaricus vaporarius	*vaporarius*	Europe (manured pastures, Fir conifer forest)	Original	Heat Loving	Gill Mushroom	Edible
Agaricus vaporarius	*pseudovillaticus*	Europe (manured pastures, Fir conifer forest)	Villose	Heat Loving	Gill Mushroom	Edible
Agaricus xanthodermus	.	USA (meadows, gardens, broadleaf forests)	.	Yellow Stainer	Gill Mushroom	Inedible
Agaricus xanthodermus	*xanthodermus*	USA (meadows, gardens, broadleaf forests)	Original	Yellow Stainer	Gill Mushroom	Inedible
Agaricus xanthodermus	*griseus*	USA (meadows, gardens, broadleaf forests)	Gray Scale	Yellow Stainer	Gill Mushroom	Inedible
Agaricus xanthodermus	*lepiotoides*	USA (meadows, gardens, broadleaf forests)	Scaly Ear	Yellow Stainer	Gill Mushroom	Inedible
Agaricus xanthodermus	*pseudocretaceus*	USA (meadows, gardens, broadleaf forests)	Chalky	Yellow Stainer	Gill Mushroom	Inedible

Family 622.201.02 *Bolbitaceae* Bonnet Gill Mushroom

Genus	Species	Variety	Native Location	Variety	Specimen	Genome	Habit
Agrocybe	*acericola*	.	USA (decaying hardwood, especially Maple)	.	Maple	Earth Scale Bonnet Gill Mushroom	Edible-Toxic
Agrocybe	*aegerita*	.	Europe (on trunks of Poplar & other Broadleaf trees)	.	Southern	Earth Scale Bonnet Gill Mushroom	Edible-Excellent
Agrocybe	*aegerita*	*Aegerita*	Europe (on trunks of Poplar & other Broadleaf trees)	Original	Southern	Earth Scale Bonnet Gill Mushroom	Edible-Excellent
Agrocybe	*aegerita*	*populus*	Europe (on trunks of Poplar & other Broadleaf trees)	Poplar	Southern	Earth Scale Bonnet Gill Mushroom	Edible-Excellent
Agrocybe	*dura*	.	USA (lawns, pastures, orchards)	.	Cracked	Earth Scale Bonnet Gill Mushroom	Edible-Poor
Agrocybe	*erebia*	.	Canada, N. USA (conifer & hardwood forest, especially Poplar)	.	Leather	Earth Scale Bonnet Gill Mushroom	Edible
Agrocybe	*paludosa*	.	Europe (marshland)	.	Marsh	Earth Scale Bonnet Gill Mushroom	Unpalatable
Agrocybe	*pediades*	.	USA (pasture, grassland)	.	Round Top	Earth Scale Bonnet Gill Mushroom	Edible
Agrocybe	*pediades*	*pediades*	USA (pasture, grassland)	Original	Round Top	Earth Scale Bonnet Gill Mushroom	Edible
Agrocybe	*pediades*	*arenicola*	USA (pasture, grassland)	Sand Dwelling	Round Top	Earth Scale Bonnet Gill Mushroom	Edible
Agrocybe	*pediades*	*semiorbicularis*	USA (pasture, grassland)	Orb	Round Top	Earth Scale Bonnet Gill Mushroom	Edible
Agrocybe	*pediades*	*subpediades*	USA (pasture, grassland)	Lesser	Round Top	Earth Scale Bonnet Gill Mushroom	Edible

Agrocybe	*pediades*	*temulenta*	USA (pasture, grassland)	Temulent	Round Top	Earth Scale Bonnet Gill Mushroom	Edible
Agrocybe	*praecox*	.	Europe (grasslands near broadleaf trees)	.	Early	Earth Scale Bonnet Gill Mushroom	Edible
Agrocybe	*praecox*	*praecox*	Europe (grasslands near broadleaf trees)	Original	Early	Earth Scale Bonnet Gill Mushroom	Edible
Agrocybe	*praecox*	*gibberosa*	Europe (grasslands near broadleaf trees)	Humped	Early	Earth Scale Bonnet Gill Mushroom	Edible
Conocybe	*albipes*	.	USA	.	White	Ring Stalk Bonnet Gill Mushroom	Edible
Conocybe	*albipes*	*albipes*	USA	Original	White	Ring Stalk Bonnet Gill Mushroom	Edible
Conocybe	*albipes*	*lactea*	USA (Pensacola, FL), PFGM, page 308	Milk	White	Ring Stalk Bonnet Gill Mushroom	Edible
Conocybe	*albipes*	*pseudocrispa*	USA	Curled	White	Ring Stalk Bonnet Gill Mushroom	Edible
Conocybe	*arrhenii*	.	USA, PFGM, page 307	.	Wrinkle	Ring Stalk Bonnet Gill Mushroom	Poison
Conocybe	*filaris*	.	USA, PFGM, page 308	.	Filament	Ring Stalk Bonnet Gill Mushroom	Unknown
Conocybe	*fimicola*	.	USA, PFGM, page 308	.	Animal Dung	Ring Stalk Bonnet Gill Mushroom	Unknown
Conocybe	*intermedia*	.	USA (decayed logs & woody debris in hardwood forest), PFGM, p. 308	.	Intermediate	Ring Stalk Bonnet Gill Mushroom	Unknown
Conocybe	*pinguis*	.	USA (decayed logs & woody debris in hardwood forest), PFGM, p. 308	.	Fat	Ring Stalk Bonnet Gill Mushroom	Unknown
Conocybe	*rugosa*	.	USA, PFGM, page 308	.	.	Ring Stalk Bonnet Gill Mushroom	Unknown
Conocybe	*stercoraria*	.	USA (decayed logs & woody debris in hardwood forest), PFGM, p. 308	.	Filthy	Ring Stalk Bonnet Gill Mushroom	Unknown
Conocybe	*tenera*	.	USA, PFGM, page 308	.	Rusty Hood	Ring Stalk Bonnet Gill Mushroom	Unknown
Conocybe	*togularis*	.	USA, PFGM, page 308	.	Tog	Ring Stalk Bonnet Gill Mushroom	Unknown
Hebeloma	*alpinum*	.	Europe, MOTW, page 65	.	Alpine	Blunt Margin Bonnet Gill Mushroom	Inedible
Hebeloma	*birrum*	.	Europe (under conifers), MOTW, page 65	.	Birr	Blunt Margin Bonnet Gill Mushroom	Inedible
Hebeloma	*crustuliniforme*	.	USA (conifer & hardwood forest), PFGM, page 300	.	Poison Pie	Blunt Margin Bonnet Gill Mushroom	Poison
Hebeloma	*crustuliniforme*	*crustuliniforme*	USA (conifer & hardwood forest)	Original	Poison Pie	Blunt Margin Bonnet Gill Mushroom	Poison
Hebeloma	*crustuliniforme*	*microspermum*	USA (conifer & hardwood forest)	Small Spore	Poison Pie	Blunt Margin Bonnet Gill Mushroom	Poison
Hebeloma	*fragile*	.	Europe (in grass under Poplars, MOTW, page 65	.	Fragile	Blunt Margin Bonnet Gill Mushroom	Inedible
Hebeloma	*gregarium*	.	USA (in woods), PFGM, page 301	.	Gregarious	Blunt Margin Bonnet Gill Mushroom	Poison
Hebeloma	*helatum*	.	Europe (under conifers), MOTW, page 66	.	Helate	Blunt Margin Bonnet Gill Mushroom	Inedible
Hebeloma	*hiemale*	.	Europe, MOTW, page 66	.	Autumn	Blunt Margin Bonnet Gill Mushroom	Inedible
Hebeloma	*mesophaeum*	.	Canada, N. USA (conifer forest), PFGM, page 300	.	Dusky Poison	Blunt Margin Bonnet Gill Mushroom	Poison
Hebeloma	*mesophaeum*	*mesophaeum*	Canada, N. USA (conifer forest)	Original	Dusky Poison	Blunt Margin Bonnet Gill Mushroom	Poison
Hebeloma	*mesophaeum*	*crassipes*	Canada, N. USA (conifer forest)	Crass	Dusky Poison	Blunt Margin Bonnet Gill Mushroom	Poison
Hebeloma	*radicosum*	.	Europe (meadows & old stumps in broadleaf forest), MOTW, page 66	.	Rooted	Blunt Margin Bonnet Gill Mushroom	Edible-Poor
Hebeloma	*sacchariolens*	.	Europe, MOTW, page 66	.	Burnt Sugar	Blunt Margin Bonnet Gill Mushroom	Inedible
Hebeloma	*sinapizans*	.	C. USA (Spruce forest), PFGM, page 300	.	Mustard Flavor Poison	Blunt Margin Bonnet Gill Mushroom	Poison
Hebeloma	*sporadicum*	.	N. USA (Spruce forest), PFGM, 300	.	Sporadic	Blunt Margin Bonnet Gill Mushroom	Poison
Panaeolina		*foenisecii*	Canada, USA (lawns, pastures, parks), PFGM, page 282		Foenisec	Pan Bonnet Gill Mushroom	Edible

Family 622.201.03 *Coprinaceae* Cap Gill Mushroom

Genus	Species	Variety	Native Location	Variety	Specimen	Genome	Habit
Coprinus	*comatus*	.	USA (in loose or sandy soil rich in fertilizer), MOTW, p. 50, PFGM, p. 277	.	Shaggy Mane	Inky Cap Gill Mushroom	Edible-Excellent

Coprinus	*comatus*	*comatus*	USA (in loose or sandy soil rich in fertilizer)	Original	Shaggy Mane	Inky Cap Gill Mushroom	Edible-Excellent
Coprinus	*comatus*	*ovatus*	USA (in loose or sandy soil rich in fertilizer), MOTW, page 50	Oval	Shaggy Mane	Inky Cap Gill Mushroom	Edible-Excellent
Coprinus	*disseminatus*	.	Europe (on trunks of broadleaf trees), MOTW, p. 52	.	Scattered	Inky Cap Gill Mushroom	Unpalatable
Coprinus	*quadrifidus*	.	C. & E. USA (on logs & debris in hardwood forest), PFGM, page 280	.	Felt Scale	Inky Cap Gill Mushroom	Edible-Toxic
Coprinus	*silvaticus*	.	USA (in woods), MOTW, page 50	.	Forest	Inky Cap Gill Mushroom	Unpalatable
Coprinus	*sterquilinus*	.	USA, PFGM, page 277	.	Sterquil	Inky Cap Gill Mushroom	Unknown
Coprinellus	*domesticus*	.	USA (woodlands), MOTW, page 50	.	Domestic	Least Inky Cap Gill Mushroom	Unpalatable
Coprinellus	*micaceus*	.	USA (parks, lawns, fields & hardwood forest), MOTW, page 52, PFGM, page 278	.	Glistening	Least Inky Cap Gill Mushroom	Edible-Good
Coprinellus	*radians*	.	USA (Pensacola, FL), MOTW, page 52	.	Radiant	Least Inky Cap Gill Mushroom	Edible
Coprinopsis	*atramentaria*	.	USA (dead stumps & living tree trunks), MOTW, page 50, PFGM, page 276	.	Black Ink	Lesser Inky Cap Gill Mushroom	Edible-Good
Coprinopsis	*cinerea*	.	USA (on dung heaps & manured soil), PFGM, page 277	.	Gray Shag	Lesser Inky Cap Gill Mushroom	Edible-Excellent
Coprinopsis	*fuscescens*	.	Europe, MOTW, page 50	.	Fuscous	Lesser Inky Cap Gill Mushroom	Edible
Coprinopsis	*insignis*	.	Europe (manured ground, parks, gardens), MOTW, p. 50	.	Insignia	Lesser Inky Cap Gill Mushroom	Edible
Coprinopsis	*nivea*	.	USA (on dung of large animals), MOTW, page 50, PFGM, page 278	.	Snowy	Lesser Inky Cap Gill Mushroom	Edible-Good
Coprinopsis	*picacea*	.	Europe (broadleaf forest), MOTW, page 52	.	Magpie	Lesser Inky Cap Gill Mushroom	Edible

Family 622.201.04 *Cortinariaceae* Web Cap Gill Mushroom

Genus	Species	Variety	Native Location	Variety	Specimen	Genome	Habit
Cortinarius	*ahsii*	.	W. USA (on melted snowbanks in Rocky Mtns. conifer forest), PFGM, page 288	.	Smith's	Web Cap Gill Mushroom	Unknown
Cortinarius	*californicus*	.	USA (California), PFGM, p. 294	.	California	Web Cap Gill Mushroom	Unknown
Cortinarius	*cylindripes*	.	USA, PFGM, page 289	.	Cylindric	Web Cap Gill Mushroom	Unknown
Cortinarius	*gentilis*	.	USA (on moist mossy soil in conifer forest), PFGM, p. 290	.	Goldband	Web Cap Gill Mushroom	Poison
Cortinarius	*heliotropicus*	.	C. & E. USA (hardwood forest), PFGM, page 291	.	Heliotrope	Web Cap Gill Mushroom	Unknown
Cortinarius	*hercynicus*	.	USA, PFGM, page 295	.	Thistle	Web Cap Gill Mushroom	Unknown
Cortinarius	*iodiodes*	.	USA, PFGM, page 292	.	Iodine	Web Cap Gill Mushroom	Unknown
Cortinarius	*laniger*	.	Canada, N. USA (conifer forest), PFGM, page 292	.	Hoary	Web Cap Gill Mushroom	Unknown
Cortinarius	*orichalceus*	.	W. USA (conifer forest), PFGM, page 292	.	Copper Red	Web Cap Gill Mushroom	Unknown
Cortinarius	*renidens*	.	USA, PFGM, page 299	.	Shiny	Web Cap Gill Mushroom	Unknown
Cortinarius	*splendidus*	.	N. USA (Great Lakes to New England), PFGM, page 294	.	Splendid	Web Cap Gill Mushroom	Unknown
Dermocybe	*bolaris*	.	Europe, MOTW, page 75	.	Bolar	Skin Web Cap Gill Mushroom	Unknown
Dermocybe	*cinnabarina*	.	Europe (Beech woods), MOTW, page 75	.	Cinnibar	Skin Web Cap Gill Mushroom	Unknown
Dermocybe	*cinnamomea*	.	Europe (humid woods), MOTW, page 75	.	Cinnamon	Skin Web Cap Gill Mushroom	Suspect
Dermocybe	*cinnamomea*	*lutea*	Europe (humid woods)	.	.	Skin Web Cap Gill Mushroom	Suspect
Dermocybe	*cinnamomea*	*violacea*	Europe (humid woods)	.	.	Skin Web Cap Gill Mushroom	Suspect
Dermocybe	*malicoria*	.	Europe (Conifer woods), MOTW, page 75	.	Malicor	Skin Web Cap Gill Mushroom	Unknown
Dermocybe	*malicoria*	*malicoria*	Europe (Conifer woods)	Original	Malicor	Skin Web Cap Gill Mushroom	Unknown
Dermocybe	*malicoria*	*croceifolia*	Europe (Conifer woods)	Crocus	Malicor	Skin Web Cap Gill Mushroom	Unknown
Dermocybe	*orellana*	.	USA, MOTW, page 75, PFGM, page 292	.	Sorrel Deathcap	Skin Web Cap Gill Mushroom	Poison
Dermocybe	*orellana*	*orellana*	.	Original	Sorrel Deathcap	Skin Web Cap Gill Mushroom	Poison
Dermocybe	*orellana*	*rutilans*	.	Red	Sorrel Deathcap	Skin Web Cap Gill Mushroom	Poison
Dermocybe	*purpurea*	.	.	.	Purple	Skin Web Cap Gill Mushroom	Unknown
Dermocybe	*purpurea*	*purpurea*	.	Original	Purple	Skin Web Cap Gill Mushroom	Unknown
Dermocybe	*purpurea*	*phoenicea*	USA, MOTW, page 75,	Redcap	Purple	Skin Web Cap Gill Mushroom	Unknown

			PFGM, page 294				
Dermocybe	*rubella*	.	USA	.	Red	Skin Web Cap Gill Mushroom	Unknown
Dermocybe	*rubella*	*rubella*	USA	Original	Red	Skin Web Cap Gill Mushroom	Unknown
Dermocybe	*rubella*	*orellanoides*	USA	Orellan	Red	Skin Web Cap Gill Mushroom	Unknown
Dermocybe	*rubella*	*speciosissima*	USA, PFGM, page 293	Special	Red	Skin Web Cap Gill Mushroom	Unknown
Dermocybe	*sanguinea*	.	USA, MOTW, page 75, PFGM, page 294	.	Redgill	Skin Web Cap Gill Mushroom	Unknown
Dermocybe	*sanguinea*	*sanguinea*	USA	Original	Redgill	Skin Web Cap Gill Mushroom	Unknown
Dermocybe	*sanguinea*	*punicea*	USA	Punic	Redgill	Skin Web Cap Gill Mushroom	Unknown
Dermocybe	*semisanguinea*	.	USA, MOTW, page 75, PFGM, page 293	.	Lesser Redgill	Skin Web Cap Gill Mushroom	Unknown
Dermocybe	*veneta*	.	Europe, MOTW, page 75	.	Venetian	Skin Web Cap Gill Mushroom	Unknown
Dermocybe	*veneta*	*veneta*	Europe	Original	Venetian	Skin Web Cap Gill Mushroom	Unknown
Dermocybe	*veneta*	*montana*	Europe, MOTW, page 75	Mountain	Venetian	Skin Web Cap Gill Mushroom	Unknown
Dermocybe	*zinzberatus*	.	Europe, MOTW, page 75	.	Tuft	Skin Web Cap Gill Mushroom	Unknown
Inoloma	*alboviolaceum*	.	USA (Oak & Beech broadleaf forest), MOTW, page 74	.	White Violet	Fiber Margin Web Cap Gill Mushroom	Edible
Inoloma	*anomalum*	.	Europe (Pine conifer & Birch broadleaf forest), MOTW, page 74	.	Irregular	Fiber Margin Web Cap Gill Mushroom	Edible
Inoloma	*corrugatum*	.	USA (broadleaf forest), MOTW, page 77	.	Wrinkled	Fiber Margin Web Cap Gill Mushroom	Edible
Inoloma	*pholideum*	.	Europe, MOTW, page 68	.	Scaly	Fiber Margin Web Cap Gill Mushroom	Unpalatable
Inoloma	*traganum*	.	Europe (Fir conifer forest), MOTW, p. 72, PFGM, pp. 290, 295	.	He Goat	Fiber Margin Web Cap Gill Mushroom	Seasoning
Inoloma	*violaceum*	.	S. Canada, USA (hardwood forest), MOTW, p. 77, PFGM, p. 294	.	Violet	Fiber Margin Web Cap Gill Mushroom	Edible
Inoloma	*violaceum*	*violaceum*	S. Canada, USA (hardwood forest)	Original	Violet	Fiber Margin Web Cap Gill Mushroom	Edible
Inoloma	*violaceum*	*velatum*	S. Canada, USA (hardwood forest)	Fleecy	Violet	Fiber Margin Web Cap Gill Mushroom	Edible
Leucocortinarius	*bulbiger*	.	Europe (on sandy soil in conifer forest), MOTW, page 79	.	Bulb Bearing	Creamcap Web Cap Gill Mushroom	Edible
Leucocortinarius	*odorifer*	.	Europe (on sandy soil in conifer forest), MOTW, page 79	.	Perfumed	Creamcap Web Cap Gill Mushroom	Inedible
Myxacium	*collinitum*	.	Canada, N. USA (on duff in conifer & hardwood forest), MOTW, page 68, PFGM, page 289	.	Greased	Mucilage Web Cap Gill Mushroom	Edible
Myxacium	*collinitum*	*collinitum*	Canada, N. USA (on duff in conifer & hardwood forest)	Original	Greased	Mucilage Web Cap Gill Mushroom	Edible
Myxacium	*collinitum*	*muscigenum*	Canada, N. USA (on duff in conifer & hardwood forest)	Mushroom	Greased	Mucilage Web Cap Gill Mushroom	Edible
Myxacium	*livido-ochraceum*	.	Europe (under Beech & other broadleaf trees)	.	Livid Ocher	Mucilage Web Cap Gill Mushroom	Edible
Myxacium	*livido-ochraceum*	*livido-ochraceum*	Europe (under Beech & other broadleaf trees)	Original	Livid Ocher	Mucilage Web Cap Gill Mushroom	Edible
Myxacium	*livido-ochraceum*	*elatior*	Europe (under Beech & broadleaf trees), MOTW, p. 66, PFGM, p. 289	Tall	Livid Ocher	Mucilage Web Cap Gill Mushroom	Edible
Myxacium	*mucifluum*	.	Europe (broadleaf forest), MOTW, page 68	.	Mucus Secreting	Mucilage Web Cap Gill Mushroom	Edible
Myxacium	*mucifluum*	*mucifluum*	Europe (broadleaf forest)	Original	Mucus Secreting	Mucilage Web Cap Gill Mushroom	Edible
Myxacium	*mucifluum*	*coeruleipes*	Europe (broadleaf forest)	Blue	Mucus Secreting	Mucilage Web Cap Gill Mushroom	Edible
Myxacium	*mucifluum*	*pinicola*	Europe (broadleaf forest)	Pine	Mucus Secreting	Mucilage Web Cap Gill Mushroom	Edible
Myxacium	*mucosum*	.	USA, MOTW, page 68, PFGM, page 289	.	Mucus	Mucilage Web Cap Gill Mushroom	Edible
Myxacium	*salor*	.	Europe (Beech forest), MOTW, page 68	.	Sallow	Mucilage Web Cap Gill Mushroom	Edible
Myxacium	*triviale*	.	Europe (Willow forest), MOTW, page 68, PFGM, page 289	.	Vulgar	Mucilage Web Cap Gill Mushroom	Edible-Poor
Myxacium	*triviale*	*triviale*	Europe (Willow forest)	Original	Vulgar	Mucilage Web Cap Gill Mushroom	Edible-Poor
Myxacium	*triviale*	*repandum*	Europe (Willow forest)	Folded	Vulgar	Mucilage Web Cap Gill Mushroom	Edible-Poor
Phlegmacium	*caesiocanescens*	.	Europe	.	Blue Gray	Mucus Web Cap Gill Mushroom	Unknown
Phlegmacium	*caerulescens*	.	USA (conifer & broadleaf forests), MOTW, page 71	.	Violet Blue	Mucus Web Cap Gill Mushroom	Edible
Phlegmacium	*caerulescens*	*caerulescens*	USA (conifer & broadleaf forests)	Original	Violet Blue	Mucus Web Cap Gill Mushroom	Edible
Phlegmacium	*caerulescens*	*caesiocyaneum*	USA (conifer & broadleaf forests), MOTW, page 71	Pale	Violet Blue	Mucus Web Cap Gill Mushroom	Edible
Phlegmacium	*calochroum*	.	Europe (under Conifer & Beech Trees), MOTW, page 70	.	Calico Ocher	Mucus Web Cap Gill Mushroom	Unpalatable

Phlegmacium	calochroum	calochroum	Europe (under Beech Trees)	Original	Calico Ocher	Mucus Web Cap Gill Mushroom	Unpalatable
Phlegmacium	calochroum	coniferarum	Europe (under Conifer Trees)	Conifer	Calico Ocher	Mucus Web Cap Gill Mushroom	Unpalatable
Phlegmacium	calochroum	parvum	Europe (under Beech Trees)	Small	Calico Ocher	Mucus Web Cap Gill Mushroom	Unpalatable
Phlegmacium	cumatile	.	Europe, MOTW, page 69	.	Glaucous	Mucus Web Cap Gill Mushroom	Edible-Good
Phlegmacium	cyanopus	.	Europe (broadleaf forest), MOTW, page 72	.	Blue Foot	Mucus Web Cap Gill Mushroom	Unpalatable
Phlegmacium	delibutus	.	USA (Beech, Birch, Oak & Fir forest), MOTW, page 70, PFGM, page 289	.	Bluegill	Mucus Web Cap Gill Mushroom	Edible
Phlegmacium	dibaphus	.	Europe, MOTW, page 71	.	Two Color	Mucus Web Cap Gill Mushroom	Unknown
Phlegmacium	dibaphus	dibaphus	Europe	Original	Two Color	Mucus Web Cap Gill Mushroom	Unknown
Phlegmacium	dibaphus	nemorosus	Europe	Wood	Two Color	Mucus Web Cap Gill Mushroom	Unknown
Phlegmacium	elegantior	.	Europe, MOTW, page 70	.	Elegant	Mucus Web Cap Gill Mushroom	Edible
Phlegmacium	elegantior	quercus-ilicis	Europe	Original	Elegant	Mucus Web Cap Gill Mushroom	Edible
Phlegmacium	elegantior	fulmineum	Europe, MOTW, page 70	Exploding	Elegant	Mucus Web Cap Gill Mushroom	Edible
Phlegmacium	glaucopus	.	Canada, N. USA (conifer forest), MOTW, page 72, PFGM, page 291	.	Blue Foot	Mucus Web Cap Gill Mushroom	Edible
Phlegmacium	largus	.	Europe (broadleaf forest), MOTW, page 70	.	Large	Mucus Web Cap Gill Mushroom	Edible
Phlegmacium	largus	largus	Europe (broadleaf forest)	Original	Large	Mucus Web Cap Gill Mushroom	Edible
Phlegmacium	largus	lividoviolaceus	Europe (broadleaf forest)	Livid Violet	Large	Mucus Web Cap Gill Mushroom	Edible
Phlegmacium	muricinus	.	Europe (conifer forest), MOTW, page 69	.	Burnt Fingernail	Mucus Web Cap Gill Mushroom	Inedible
Phlegmacium	polymorphus	.	Europe (conifer forest)	.	Polymorph	Mucus Web Cap Gill Mushroom	Edible
Phlegmacium	polymorphus	polymorphus	Europe (conifer forest)	Original	Polymorph	Mucus Web Cap Gill Mushroom	Edible
Phlegmacium	polymorphus	multiformis	Europe (conifer forest), MOTW, page 70	Multiform	Polymorph	Mucus Web Cap Gill Mushroom	Edible
Phlegmacium	polymorphus	napus	Europe (conifer forest), MOTW, page 70	Nape	Polymorph	Mucus Web Cap Gill Mushroom	Edible
Phlegmacium	praestans	.	Europe (Birch & Oak forest), MOTW, page 69	.	Imposing	Mucus Web Cap Gill Mushroom	Edible-Good
Phlegmacium	praestans	praestans	Europe (Birch & Oak forest)	Original	Imposing	Mucus Web Cap Gill Mushroom	Edible-Good
Phlegmacium	praestans	berkeleyi	Europe (Birch & Oak forest)	Berkeley	Imposing	Mucus Web Cap Gill Mushroom	Edible-Good
Phlegmacium	purpurascens	.	USA (conifer & broadleaf forests), MOTW, page 69	.	Purple	Mucus Web Cap Gill Mushroom	Edible
Phlegmacium	purpurascens	purpurascens	USA (conifer & broadleaf forests)	Original	Purple	Mucus Web Cap Gill Mushroom	Edible
Phlegmacium	purpurascens	largusoides	USA (conifer & broadleaf forests)	Large	Purple	Mucus Web Cap Gill Mushroom	Edible
Phlegmacium	rufo-olivaceum	.	USA, PFGM, page 293	.	Rufous Olive	Mucus Web Cap Gill Mushroom	Unknown
Phlegmacium	rufo-olivaceum	rufo-olivaceum	USA	Original	Rufous Olive	Mucus Web Cap Gill Mushroom	Unknown
Phlegmacium	rufo-olivaceum	testaceum	USA	Test	Rufous Olive	Mucus Web Cap Gill Mushroom	Unknown
Phlegmacium	rufo-olivaceum	vinosum	USA	Wine	Rufous Olive	Mucus Web Cap Gill Mushroom	Unknown
Phlegmacium	sodagnitum	.	Europe, MOTW, page 71	.	Violet	Mucus Web Cap Gill Mushroom	Unknown
Phlegmacium	variicolor	.	Europe (mountain conifer forest), MOTW, page 70	.	Variable Color	Mucus Web Cap Gill Mushroom	Edible-Fair
Phlegmacium	variicolor	marginatum	Europe (mountain conifer forest)	Original	Variable Color	Mucus Web Cap Gill Mushroom	Edible-Fair
Phlegmacium	variicolor	nemorense	Europe (mountain conifer forest)	Wood	Variable Color	Mucus Web Cap Gill Mushroom	Edible-Fair
Phlegmacium	varium	.	USA (conifer forest), MOTW, page 70	.	Variegated	Mucus Web Cap Gill Mushroom	Edible-Good
Phlegmacium	varium	varium	USA (conifer forest)	Original	Variegated	Mucus Web Cap Gill Mushroom	Edible-Good
Phlegmacium	varium	decolorans	USA (conifer forest)	Discolore	Variegated	Mucus Web Cap Gill	Edible-Good

				d		Mushroom	
Rozites	*caperatus*	.	USA (conifer & hardwood forest), MOTW , page 62	.	Wrinkled	Nitecap Web Cap Gill Mushroom	Edible-Good
Telamonia	*armillata*	.	USA (Birch mixed conifer-hardwood forest), MOTW, page 77, PFGM, page 288	.	Cinnabar Bracelet	Belted Web Cap Gill Mushroom	Edible
Telamonia	*brunnea*	.	Europe (conifer forest), MOTW, page 77	.	Brown	Belted Web Cap Gill Mushroom	Unpalatable
Telamonia	*brunnea*	*brunnea*	Europe (conifer forest)	Original	Brown	Belted Web Cap Gill Mushroom	Unpalatable
Telamonia	*brunnea*	*glandicolor*	Europe (conifer forest)	Gland	Brown	Belted Web Cap Gill Mushroom	Unpalatable
Telamonia	*hinnulea*	.	Europe (woods & meadows), MOTW, page 77	.	Fawn	Belted Web Cap Gill Mushroom	Inedible
Telamonia	*hinnulea*	*hinnulea*	Europe (woods & meadows)	Original	Fawn	Belted Web Cap Gill Mushroom	Inedible
Telamonia	*hinnulea*	*radicata*	Europe (woods & meadows)	Rooted	Fawn	Belted Web Cap Gill Mushroom	Inedible
Telamonia	*ionophylla*	.	Europe	.	Violet Gill	Belted Web Cap Gill Mushroom	Edible-Poor
Telamonia	*ionophylla*	*ionophylla*	Europe	Original	Violet Gill	Belted Web Cap Gill Mushroom	Edible-Poor
Telamonia	*ionophylla*	*scutulata*	Europe, MOTW, page 78	Checkered	Violet Gill	Belted Web Cap Gill Mushroom	Edible-Poor
Telamonia	*torva*	.	Europe (Beech broadleaf forest), MOTW, page 78	.	Grim	Belted Web Cap Gill Mushroom	Edible-Poor

Family 622.201.05 *Entolomataceae* Pinkgill Gill Mushroom

Genus	Species	Variety	Native Location	Variety	Specimen	Genome	Habit
Clitopilus	*abortivus*	.	USA	.	Abortive	Depressed Pinkgill Gill Mushroom	Edible
Clitopilus	*prunulus*	.	USA (Oak open woodlands), MOTW, page 122	.	Plum	Depressed Pinkgill Gill Mushroom	Edible-Excellent
Clitopilus	*prunulus*	*prunulus*	USA (Oak open woodlands)	Original	Plum	Depressed Pinkgill Gill Mushroom	Edible-Excellent
Clitopilus	*prunulus*	*orcella*	USA (Oak open woodlands), MOTW, page 122	Orcelle	Plum	Depressed Pinkgill Gill Mushroom	Edible-Excellent
Rhodocybe	*truncata*	.	Europe (Fir conifer forest), MOTW, page 120	.	Blunt	Rosecap Pinkgill Gill Mushroom	Edible
Rhodophyllus	*aprilis*	.	Europe (Elm & Hornbeam forest), MOTW, page 83	.	April	Rosy Pinkgill Gill Mushroom	Inedible
Rhodophyllus	*clypeatus*	.	Europe (grasslands & shady places), MOTW, page 83	.	Round Shield	Rosy Pinkgill Gill Mushroom	Edible
Rhodophyllus	*clypeatus*	*clypeatus*	Europe (grasslands & shady places)	Original	Round Shield	Rosy Pinkgill Gill Mushroom	Edible
Rhodophyllus	*clypeatus*	*pallidogriseum*	Europe (grasslands & shady places)	Pale Gray	Round Shield	Rosy Pinkgill Gill Mushroom	Edible
Rhodophyllus	*mammosus*	.	Europe (alpine pastures), MOTW, page 83	.	Nippled Deathcap	Rosy Pinkgill Gill Mushroom	Poison
Rhodophyllus	*medius*	.	Europe (meadows & woods), MOTW, page 82	.	Medium	Rosy Pinkgill Gill Mushroom	Inedible
Rhodophyllus	*nitidus*	.	Europe (under Conifers), MOTW, page 82	.	Shiny Blue	Rosy Pinkgill Gill Mushroom	Unknown
Rhodophyllus	*pascuus*	.	Europe (alpine pastures), MOTW, page 83	.	Pasture Deathcap	Rosy Pinkgill Gill Mushroom	Poison
Rhodophyllus	*prunuloides*	.	Europe, MOTW, page 83	.	Plum	Rosy Pinkgill Gill Mushroom	Edible
Rhodophyllus	*prunuloides*	*prunuloides*	Europe	Original	Plum	Rosy Pinkgill Gill Mushroom	Edible
Rhodophyllus	*prunuloides*	*inopiliforme*	Europe	Inopil	Plum	Rosy Pinkgill Gill Mushroom	Edible
Rhodophyllus	*prunuloides*	*obscurum*	Europe	Obscure	Plum	Rosy Pinkgill Gill Mushroom	Edible
Rhodophyllus	*rhodopolius*	.	Europe (mountain Beech broadleaf forest), MOTW, page 82	.	Slender Deathcap	Rosy Pinkgill Gill Mushroom	Poison
Rhodophyllus	*rhodopolius*	*rhodopolius*	Europe (mountain Beech broadleaf forest)	Original	Slender Deathcap	Rosy Pinkgill Gill Mushroom	Poison
Rhodophyllus	*rhodopolius*	*nidorosus*	Europe (humid broadleaf forest), MOTW, page 82	Burnt Odor	Slender Deathcap	Rosy Pinkgill Gill Mushroom	Poison
Rhodophyllus	*scabiosus*	.	Europe (in heath), MOTW, page 82	.	Blue Scabby	Rosy Pinkgill Gill Mushroom	Unknown
Rhodophyllus	*sepium*	.	Europe (along hedges of thornbushes), MOTW, page 83	.	Hedge	Rosy Pinkgill Gill Mushroom	Edible
Rhodophyllus	*sinuatus*	.	Europe	.	Sinuous Deathcap	Rosy Pinkgill Gill Mushroom	Poison

Rhodophyllus sinuatus	*sinuatus*	Europe	Original	Sinuous Deathcap	Rosy Pinkgill Gill Mushroom	Poison
Rhodophyllus sinuatus	*eulividus*	Europe	Ordinary	Sinuous Deathcap	Rosy Pinkgill Gill Mushroom	Poison
Rhodophyllus sinuatus	*fertile*	Europe	Fertile	Sinuous Deathcap	Rosy Pinkgill Gill Mushroom	Poison
Rhodophyllus sinuatus	*lividoalbus*	Europe (meadows & woods), MOTW, page 82	Livid White	Sinuous Deathcap	Rosy Pinkgill Gill Mushroom	Poison
Rhodophyllus sinuatus	*lividocyanulus*	Europe	Livid Cyan	Sinuous Deathcap	Rosy Pinkgill Gill Mushroom	Poison
Rhodophyllus sinuatus	*lividus*	Europe (Beech, Oak & Yellow Locust broadleaf forest), MOTW, page 82	Livid	Sinuous Deathcap	Rosy Pinkgill Gill Mushroom	Poison
Rhodophyllus sinuatus	*roseus*	Europe	Rosy	Sinuous Deathcap	Rosy Pinkgill Gill Mushroom	Poison

Family 622.201.06 *Hydnangiaceae* Tallowgill Gill Mushroom

Genus	Species	Variety	Native Location	Variety	Specimen	Genome	Habit
Laccaria	*amethystina*	.	C. & E. USA (moist places in hardwood forest), MOTW, page 119, PFGM, page 155	.	Amethyst	Milky Tallowgill Gill Mushroom	Edible-Fair
Laccaria	*amethystina*	*amethystina*	C. & E. USA (moist places in hardwood forest)	Original	Amethyst	Milky Tallowgill Gill Mushroom	Edible-Fair
Laccaria	*amethystina*	*hudsonii*	C. & E. USA (moist places in hardwood forest)	Hudson	Amethyst	Milky Tallowgill Gill Mushroom	Edible-Fair
Laccaria	*laccata*	.	USA (grass, moist soil near rivers & lakes, Poplar forest), MOTW, page 119, PFGM, page 156	.	Deceiver	Milky Tallowgill Gill Mushroom	Edible-Fair
Laccaria	*laccata*	*laccata*	USA (grass, moist soil near rivers & lakes, Poplar forest)	Original	Deceiver	Milky Tallowgill Gill Mushroom	Edible-Fair
Laccaria	*laccata*	*affinis*	USA (grass, moist soil near rivers & lakes, Poplar forest)	Affine	Deceiver	Milky Tallowgill Gill Mushroom	Edible-Fair
Laccaria	*laccata*	*amethystea*	USA (grass, moist soil near rivers & lakes, Poplar forest)	Amethyst	Deceiver	Milky Tallowgill Gill Mushroom	Edible-Fair
Laccaria	*laccata*	*anglica*	USA (grass, moist soil near rivers & lakes, Poplar forest)	Angel	Deceiver	Milky Tallowgill Gill Mushroom	Edible-Fair
Laccaria	*laccata*	*farinacea*	USA (grass, moist soil near rivers & lakes, Poplar forest)	Farina	Deceiver	Milky Tallowgill Gill Mushroom	Edible-Fair
Laccaria	*laccata*	*moelleri*	USA (grass, moist soil near rivers & lakes, Poplar forest)	Moeller	Deceiver	Milky Tallowgill Gill Mushroom	Edible-Fair
Laccaria	*laccata*	*pallidifolia*	USA (grass, moist soil near rivers & lakes, Poplar forest)	Pale	Deceiver	Milky Tallowgill Gill Mushroom	Edible-Fair
Laccaria	*laccata*	*rosella*	USA (grass, moist soil near rivers & lakes, Poplar forest)	Rosy	Deceiver	Milky Tallowgill Gill Mushroom	Edible-Fair
Laccaria	*laccata*	*sandicina*	USA (grass, moist soil near rivers & lakes, Poplar forest), MOTW, page 119	Flesh	Deceiver	Milky Tallowgill Gill Mushroom	Edible-Fair
Laccaria	*laccata*	*scotica*	USA (grass, moist soil near rivers & lakes, Poplar forest)	Scotch	Deceiver	Milky Tallowgill Gill Mushroom	Edible-Fair
Laccaria	*laccata*	*subalpina*	USA (grass, moist soil near rivers & lakes, Poplar forest)	Alpine	Deceiver	Milky Tallowgill Gill Mushroom	Edible-Fair
Laccaria	*maritima*	.	USA (on sandy soil on lake, river & ocean beaches)	.	Seaside	Milky Tallowgill Gill Mushroom	Edible-Poor
Laccaria	*maritima*	*maritima*	USA (on sandy soil on lake, river & ocean beaches)	Original	Seaside	Milky Tallowgill Gill Mushroom	Edible-Poor
Laccaria	*maritima*	*trullisata*	USA (on sandy soil on lake, river & ocean beaches), MOTW, page 119, PFGM, page 157	Sandy	Seaside	Milky Tallowgill Gill Mushroom	Inedible
Laccaria	*ochropurpurea*	.	USA (lawns, parks, hardwood forest), PFGM, page 157	.	Ocher Purple	Milky Tallowgill Gill Mushroom	Edible
Laccaria	*proxima*	.	USA (grass, moist soil near rivers & lakes, Poplar forest), MOTW, p. 119	.	Near	Milky Tallowgill Gill Mushroom	Edible
Laccaria	*pumila*	.	USA	.	Low	Milky Tallowgill Gill Mushroom	Unknown
Laccaria	*pumila*	*altaica*	USA	Original	Low	Milky Tallowgill Gill Mushroom	Unknown
Laccaria	*pumila*	*striatula*	USA	Striped	Low	Milky Tallowgill Gill Mushroom	Unknown
Laccaria	*tortilis*	.	USA (humid & muddy soils), MOTW, page 119	.	Twisted	Milky Tallowgill Gill Mushroom	Unpalatable

Family 622.201.07 *Hygrophoraceae* Waxycap Gill Mushroom

Genus	Species	Variety	Native Location	Variety	Specimen	Genome	Habit
Hygrophorus	*agathosmus*	.	Europe (mountain Fir conifer forest), MOTW, page 168, PFGM, page 202	.	Almond	Ivorygill Waxycap Gill Mushroom	Edible
Hygrophorus	*agathosmus*	*agathosmus*	Europe (mountain Fir conifer forest)	Original	Almond	Ivorygill Waxycap Gill Mushroom	Edible
Hygrophorus	*agathosmus*	*aureofloccosus*	Europe (mountain Fir conifer forest), PFGM, page 213	Golden Fleece	Almond	Ivorygill Waxycap Gill Mushroom	Edible
Hygrophorus	*agathosmus*	*cerasinus*	Europe (mountain Fir conifer forest)	Red	Almond	Ivorygill Waxycap Gill Mushroom	Edible
Hygrophorus	*amarus*	.	W. USA (in woods), PFGM, page 206	.	Bitter	Ivorygill Waxycap Gill Mushroom	Unknown
Hygrophorus	*arbustivus*	.	Europe (Oak & Hazelnut broadleaf forest), MOTW, page 167	.	Tree Dwelling	Ivorygill Waxycap Gill Mushroom	Edible
Hygrophorus	*bakerensis*	.	USA (in woods), PFGM, pages 203, 212	.	Baker	Ivorygill Waxycap Gill Mushroom	Unknown
Hygrophorus	*calophyllus*	.	USA (in woods), PFGM, pages 204, 208	.	Pretty Gill	Ivorygill Waxycap Gill Mushroom	Unknown
Hygrophorus	*camarophyllus*	.	Canada, N. USA (Pine & Spruce conifer forest), MOTW, page 168, PFGM, page 203	.	Curved Gill	Ivorygill Waxycap Gill Mushroom	Edible-Good
Hygrophorus	*camarophyllus*	*camarophyllus*	Canada, N. USA (Pine & Spruce conifer forest)	Original	Curved Gill	Ivorygill Waxycap Gill Mushroom	Edible-Good
Hygrophorus	*camarophyllus*	*caprinus*	Canada, N. USA (Pine & Spruce conifer forest)	Goat	Curved Gill	Ivorygill Waxycap Gill Mushroom	Edible-Good
Hygrophorus	*capreolarius*	.	USA (mountain conifer forest), PFGM, page 206	.	Roebuck	Ivorygill Waxycap Gill Mushroom	Edible
Hygrophorus	*chrysodon*	.	USA (mountain conifer & mixed forest), PFGM, page 204	.	Golden Tooth	Ivorygill Waxycap Gill Mushroom	Edible-Fair
Hygrophorus	*chrysodon*	*chrysodon*	USA (mountain conifer & mixed forest)	Original	Golden Tooth	Ivorygill Waxycap Gill Mushroom	Edible-Fair
Hygrophorus	*chrysodon*	*leucodon*	USA (mountain conifer & mixed forest)	Light	Golden Tooth	Ivorygill Waxycap Gill Mushroom	Edible-Fair
Hygrophorus	*discoxanthus*	.	S. Canada, N. USA (conifer forest)	.	Yellow Disc	Ivorygill Waxycap Gill Mushroom	Edible
Hygrophorus	*discoxanthus*	*discoxanthus*	S. Canada, N. USA (conifer forest)	Original	Yellow Disc	Ivorygill Waxycap Gill Mushroom	Edible
Hygrophorus	*discoxanthus*	*chrysaspis*	S. Canada, N. USA (conifer forest), MOTW, page 166, PFGM, page 205	Golden Rough	Yellow Disc	Ivorygill Waxycap Gill Mushroom	Edible
Hygrophorus	*discoxanthu s*	*discoideus*	S. Canada, N. USA (conifer forest), PFGM, page 204	Clay	Yellow Disc	Ivorygill Waxycap Gill Mushroom	Edible
Hygrophorus	*eburneus*	.	Canada, USA (on soil in various habitats), MOTW, page 166, PFGM, page 205	.	Ivory White	Ivorygill Waxycap Gill Mushroom	Edible-Poor
Hygrophorus	*eburneus*	*eburneus*	Canada, USA (on soil in various habitats)	Original	Ivory White	Ivorygill Waxycap Gill Mushroom	Edible-Poor
Hygrophorus	*eburneus*	*carneipes*	Canada, USA (on soil in various habitats), MOTW, page 166	Flesh Foot	Ivory White	Ivorygill Waxycap Gill Mushroom	Edible-Poor
Hygrophorus	*eburneus*	*cossus*	Canada, USA (on soil in various habitats), PFGM, page 205	Woodworm	Ivory White	Ivorygill Waxycap Gill Mushroom	Edible-Poor
Hygrophorus	*eburneus*	*cossus-quercetorum*	Canada, USA (on soil in various habitats)	Oak Woodworm	Ivory White	Ivorygill Waxycap Gill Mushroom	Edible-Poor
Hygrophorus	*erubescens*	.	S. Canada, N. & C. USA (Pine & Spruce conifer forest), PFGM, page 206	.	Pinkcap	Ivorygill Waxycap Gill Mushroom	Inedible
Hygrophorus	*flavodiscus*	.	USA, PFGM, page 207	.	Yellow Disc	Ivorygill Waxycap Gill Mushroom	Unknown
Hygrophorus	*fulgeneus*	.	USA, PFGM, pages 203, 207	.	Smoky	Ivorygill Waxycap Gill Mushroom	Unknown
Hygrophorus	*gliocyclus*	.	S. Canada, N. & C. USA (Pine & Spruce conifer forest), PFGM, page 206	.	Slimy	Ivorygill Waxycap Gill Mushroom	Unknown
Hygrophorus	*glutinosus*	.	USA, PFGM, page 205	.	Glutinous	Ivorygill Waxycap Gill Mushroom	Unknown
Hygrophorus	*hedrychii*	.	Europe (broadleaf forest)	.	Hedrych	Ivorygill Waxycap Gill Mushroom	Edible-Poor
Hygrophorus	*hedrychii*	*hedrychii*	Europe (broadleaf forest)	Original	Hedrych	Ivorygill Waxycap Gill Mushroom	Edible-Poor
Hygrophorus	*hedrychii*	*melizeus*	Europe (broadleaf forest), MOTW, page 167	Honey	Hedrych	Ivorygill Waxycap Gill Mushroom	Edible-Poor
Hygrophorus	*hyacinthus*	.	Europe (in woods), MOTW, page 168	.	Hyacinth	Ivorygill Waxycap Gill Mushroom	Edible
Hygrophorus	*hypothejus*	.	USA (Pine & Fir conifer & mixed conifer-broadleaf forests), MOTW, page 168, PFGM, page 207	.	Winter Herald	Ivorygill Waxycap Gill Mushroom	Edible
Hygrophorus	*hypothejus*	*hypothejus*	USA (Pine & Fir conifer & mixed conifer-broadleaf forests)	Original	Winter Herald	Ivorygill Waxycap Gill Mushroom	Edible

Hygrophorus	*hypothejus*	*aureus*	USA (Pine & Fir conifer & mixed conifer-broadleaf forests), MOTW, page 167	Golden	Winter Herald	Ivorygill Waxycap Gill Mushroom	Edible
Hygrophorus	*hypothejus*	*expallens*	USA (Pine & Fir conifer & mixed conifer-broadleaf forests), MOTW, page 168	Pale	Winter Herald	Ivorygill Waxycap Gill Mushroom	Edible
Hygrophorus	*inocybiformis*	.	USA, PFGM, page 204	.	Inocybe	Ivorygill Waxycap Gill Mushroom	Unknown
Hygrophorus	*karstenii*	.	Europe (Fir conifer forest)	.	Karsten	Ivorygill Waxycap Gill Mushroom	Edible
Hygrophorus	*laurae*	.	USA	.	Laura	Ivorygill Waxycap Gill Mushroom	Unknown
Hygrophorus	*lindtneri*	.	USA, Europe (broadleaf trees)	.	Lindtner	Ivorygill Waxycap Gill Mushroom	Edible
Hygrophorus	*lindtneri*	*lindtneri*	Europe (broadleaf trees)	Original	Lindtner	Ivorygill Waxycap Gill Mushroom	Edible
Hygrophorus	*lindtneri*	*carpini*	Europe (broadleaf trees)	Carpin	Lindtner	Ivorygill Waxycap Gill Mushroom	Edible
Hygrophorus	*lindtneri*	*leucophaeus*	USA, Europe (broadleaf trees), PFGM, page 205	Light Dusky	Lindtner	Ivorygill Waxycap Gill Mushroom	Edible
Hygrophorus	*lucorum*	.	Europe (Larch broadleaf forest), MOTW, page 170	.	Sacred Wood	Ivorygill Waxycap Gill Mushroom	Edible
Hygrophorus	*marzuolus*	.	USA (conifer & Oak, Chestnut, and Beech broadleaf forests), Europe, MOTW, page 171, PFGM, page 207	.	March	Ivorygill Waxycap Gill Mushroom	Edible-Excellent
Hygrophorus	*nemoreus*	.	Europe (in woods), MOTW, page 170	.	Wood Loving	Ivorygill Waxycap Gill Mushroom	Edible-Poor
Hygrophorus	*nemoreus*	*nemoreus*	Europe (in woods)	Original	Wood Loving	Ivorygill Waxycap Gill Mushroom	Edible-Poor
Hygrophorus	*nemoreus*	*leporinus*	Europe (plains & mountain conifer woods)	Pretty	Wood Loving	Ivorygill Waxycap Gill Mushroom	Edible-Poor
Hygrophorus	*olivaceoalbus*	.	W. & N. USA (conifer forest), MOTW, page 168, PFGM, page 208	.	Olive Gray	Ivorygill Waxycap Gill Mushroom	Edible-Good
Hygrophorus	*persoonii*	.	Europe (Hornbeam & Beech broadleaf forest)	.	Persoon	Ivorygill Waxycap Gill Mushroom	Edible-Good
Hygrophorus	*persoonii*	*persoonii*	Europe (Hornbeam & Beech broadleaf forest)	Original	Persoon	Ivorygill Waxycap Gill Mushroom	Edible-Good
Hygrophorus	*persoonii*	*limacinus*	Europe (Hornbeam & Beech broadleaf forest)	Snail	Persoon	Ivorygill Waxycap Gill Mushroom	Edible-Good
Hygrophorus	*piceae*	.	Europe (mountain conifer forest), MOTW, page 166	.	Pine	Ivorygill Waxycap Gill Mushroom	Edible
Hygrophorus	*poetarum*	.	USA (broadleaf forest), MOTW, page 167, PFGM, page 209	.	Poet's	Ivorygill Waxycap Gill Mushroom	Edible-Good
Hygrophorus	*ponderatus*	.	USA, PFGM, pages 210, 211	.	Heavy	Ivorygill Waxycap Gill Mushroom	Unknown
Hygrophorus	*pudorinus*	.	W. & N. USA (mountain Fir & Red Pine conifer forest), MOTW, page 167, PFGM, page 209	.	Blushing	Ivorygill Waxycap Gill Mushroom	Edible
Hygrophorus	*purpurascens*	.	USA (mountain conifer forest), PFGM, page 206	.	Purple	Ivorygill Waxycap Gill Mushroom	Edible
Hygrophorus	*pustulatus*	.	USA (bogs & Pine & Fir conifer forest), MOTW, page 171, PFGM, page 213	.	Pustule	Ivorygill Waxycap Gill Mushroom	Edible
Hygrophorus	*pustulatus*	*pustulatus*	USA (bogs & Pine & Fir conifer forest)	Original	Pustule	Ivorygill Waxycap Gill Mushroom	Edible
Hygrophorus	*pustulatus*	*tenebratus*	USA (bogs & Pine & Fir conifer forest), MOTW, page 171	Pierced	Pustule	Ivorygill Waxycap Gill Mushroom	Edible
Hygrophorus	*pustulatus*	*tephroleucus*	USA (bogs & Pine & Fir conifer forest), PFGM, page 212	Light Ashen	Pustule	Ivorygill Waxycap Gill Mushroom	Edible
Hygrophorus	*russula*	.	N. & C. USA (plains Oak & Beech broadleaf forest), PFGM, page 209	.	False Brittlegill	Ivorygill Waxycap Gill Mushroom	Edible
Hygrophorus	*sordidus*	.	S. Canada, USA (open woods), PFGM, page 210	.	Hardwood	Ivorygill Waxycap Gill Mushroom	Edible
Hygrophorus	*speciosus*	.	S. Canada, USA (moist places under Larch conifer forest), PFGM, page 210	.	Larch	Ivorygill Waxycap Gill Mushroom	Unknown
Hygrophorus	*subalpinus*	.	W. USA (on melted snowbanks from Rocky Mtns. to pacific coast), PFGM, page 211	.	Subalpine	Ivorygill Waxycap Gill Mushroom	Edible-Poor
Hygrophorus	*tennesseensis*	.	W. USA (CA), E. USA (conifer forest), PFGM, page 212	.	Tennessee	Ivorygill Waxycap Gill Mushroom	Unknown
Camarophyllus	*borealis*	.	USA, PFGM, page 197	.	Boreal	Archgill Waxycap Gill Mushroom	Unknown
Camarophyllus	*cantharellus*	.	USA, PFGM, pages 198, 201	.	Chanterelle	Archgill Waxycap Gill Mushroom	Unknown
Camarophyllus	*cantharellus*	*cantharellus*	USA	Original	Chanterelle	Archgill Waxycap Gill Mushroom	Unknown

Camarophyllus	*cantharellus*	*lepida*	USA	Thin	Chanterelle	Archgill Waxycap Gill Mushroom	Unknown
Camarophyllus	*pratensis*	.	USA (meadows), MOTW, page 170, PFGM, page 197	.	Meadowcap	Archgill Waxycap Gill Mushroom	Edible-Excellent
Camarophyllus	*pratensis*	*berkeleyanus*	USA (grassland, open woods)	Berkeley	Meadowcap	Archgill Waxycap Gill Mushroom	Edible-Excellent
Camarophyllus	*pratensis*	*ortonii*	USA (grassland, open woods)	Orton Butter	Meadowcap	Archgill Waxycap Gill Mushroom	Edible-Excellent
Camarophyllus	*pratensis*	*pallidus*	USA (meadows)	Pale	Meadowcap	Archgill Waxycap Gill Mushroom	Edible-Excellent
Camarophyllus	*virgineus*	.	USA (meadows, conifer & deciduous forest), PFGM, page 197, MOTW, page 166	.	Virgin	Archgill Waxycap Gill Mushroom	Edible-Fair
Camarophyllus	*virgineus*	*virgineus*	USA (meadows, conifer & deciduous forest)	Original	Virgin	Archgill Waxycap Gill Mushroom	Edible-Fair
Camarophyllus	*virgineus*	*fuscescens*	USA (meadows)	Fuscous	Virgin	Archgill Waxycap Gill Mushroom	Edible-Fair
Camarophyllus	*virgineus*	*niveus*	USA (meadows), MOTW, page 166, PFGM, page 197	Snowy	Virgin	Archgill Waxycap Gill Mushroom	Edible-Fair
Camarophyllus	*virgineus*	*ochraceopa llidus*	USA (conifer & deciduous forest)	Pale Ocher	Virgin	Archgill Waxycap Gill Mushroom	Edible-Fair
Camarophyllus	*virgineus*	*roseipes*	USA (meadows)	Rosy	Virgin	Archgill Waxycap Gill Mushroom	Edible-Fair
Camarophyllus	*virgineus*	*subradiata*	USA (meadows)	Radiating	Virgin	Archgill Waxycap Gill Mushroom	Edible-Fair
Hygrocybe	*bresadolae*	.	Europe (meadows), MOTW, page 170	.	Bresadola	Slender Waxycap Gill Mushroom	Edible
Hygrocybe	*ceracea*	.	Europe, MOTW, page 173	.	Wax	Slender Waxycap Gill Mushroom	Edible
Hygrocybe	*ceracea*	*ceracea*	Europe	Original	Wax	Slender Waxycap Gill Mushroom	Edible
Hygrocybe	*ceracea*	*vitellinoides*	Europe	Yellow	Wax	Slender Waxycap Gill Mushroom	Edible
Hygrocybe	*chlorophana*	.	USA (conifer & deciduous forest), MOTW, page 173, PFGM, page 200	.	Greenish	Slender Waxycap Gill Mushroom	Edible
Hygrocybe	*chlorophana*	*chlorophan a*	USA (conifer & deciduous forest)	Original	Greenish	Slender Waxycap Gill Mushroom	Edible
Hygrocybe	*chlorophana*	*aurantiaca*	USA (conifer & deciduous forest)	Orange	Greenish	Slender Waxycap Gill Mushroom	Edible
Hygrocybe	*chlorophana*	*euroflavesc ens*	USA (conifer & deciduous forest)	Yellowish	Greenish	Slender Waxycap Gill Mushroom	Edible
Hygrocybe	*chlorophana*	*flavescens*	USA (conifer & deciduous forest), PFGM, page 199	Yellow	Greenish	Slender Waxycap Gill Mushroom	Edible
Hygrocybe	*coccinea*	.	USA (conifer & hardwood forest), MOTW, page 173, PFGM, page 198	.	Scarlet Hood	Slender Waxycap Gill Mushroom	Edible
Hygrocybe	*coccinea*	*coccinea*	USA (conifer & hardwood forest)	Original	Scarlet Hood	Slender Waxycap Gill Mushroom	Edible
Hygrocybe	*coccinea*	*crenata*	USA (conifer & hardwood forest)	Scalloped	Scarlet Hood	Slender Waxycap Gill Mushroom	Edible
Hygrocybe	*coccinea*	*umbonata*	USA (conifer & hardwood forest)	Penis	Scarlet Hood	Slender Waxycap Gill Mushroom	Edible
Hygrocybe	*conica*	.	USA (grasslands), MOTW, page 173, PFGM, page 198	.	Cone	Slender Waxycap Gill Mushroom	Edible-Toxic
Hygrocybe	*conica*	*conica*	USA (grasslands)	Original	Cone	Slender Waxycap Gill Mushroom	Edible-Toxic
Hygrocybe	*conica*	*chloroides*	USA (grasslands)	Green	Cone	Slender Waxycap Gill Mushroom	Edible-Toxic
Hygrocybe	*conica*	*cinereifolia*	USA (grasslands)	Cindery Gill	Cone	Slender Waxycap Gill Mushroom	Edible-Toxic
Hygrocybe	*conica*	*conicopalus tris*	USA (grasslands)	Marsh	Cone	Slender Waxycap Gill Mushroom	Edible-Toxic
Hygrocybe	*conica*	*nigrescens*	USA (grasslands), MOTW, page 173, PFGM, page 199	Black	Cone	Slender Waxycap Gill Mushroom	Edible-Toxic
Hygrocybe	*conica*	*olivaceonigr a*	USA (grasslands)	Olive & Black	Cone	Slender Waxycap Gill Mushroom	Edible-Toxic
Hygrocybe	*conica*	*pseudoconi ca*	USA (grasslands)	Lesser	Cone	Slender Waxycap Gill Mushroom	Edible-Toxic
Hygrocybe	*conica*	*tristis*	USA (grasslands)	Sad	Cone	Slender Waxycap Gill Mushroom	Edible-Toxic
Hygrocybe	*davisii*	.	USA, PFGM, page 200	.	Davis	Slender Waxycap Gill Mushroom	Unknown
Hygrocybe	*erinaceus*	.	USA, PFGM, page 199	.	Hedgehog	Slender Waxycap Gill Mushroom	Unknown
Hygrocybe	*glutinipes*	.	Europe	.	Glutin	Slender Waxycap Gill Mushroom	Edible

Hygrocybe	*glutinipes*	*glutinipes*	Europe	Original	Glutin	Slender Waxycap Gill Mushroom	Edible
Hygrocybe	*glutinipes*	*aurantioviscida*	Europe	Orange Sticky	Glutin	Slender Waxycap Gill Mushroom	Edible
Hygrocybe	*glutinipes*	*citrina*	Europe, MOTW, page 173	Lemon Yellow	Glutin	Slender Waxycap Gill Mushroom	Edible
Hygrocybe	*glutinipes*	*rubra*	Europe	Red	Glutin	Slender Waxycap Gill Mushroom	Edible
Hygrocybe	*hondurensis*	.	USA, PFGM, page 200	.	Honduras	Slender Waxycap Gill Mushroom	Unknown
Hygrocybe	*intermedia*	.	Europe, MOTW, page 170	.	Intermediate	Slender Waxycap Gill Mushroom	Edible
Hygrocybe	*laeta*	.	USA (damp soil, mossy places), MOTW, page 173, PFGM, page 200	.	Pinkgill	Slender Waxycap Gill Mushroom	Edible
Hygrocybe	*laeta*	*laeta*	USA (damp soil, mossy places)	Original	Pinkgill	Slender Waxycap Gill Mushroom	Edible
Hygrocybe	*laeta*	*davisii*	USA (damp soil, mossy places)	Davis	Pinkgill	Slender Waxycap Gill Mushroom	Edible
Hygrocybe	*laeta*	*flava*	USA (damp soil, mossy places)	Yellow	Pinkgill	Slender Waxycap Gill Mushroom	Edible
Hygrocybe	*laeta*	*houghtonii*	USA (damp soil, mossy places), PFGM, page 200	Houghton	Pinkgill	Slender Waxycap Gill Mushroom	Edible
Hygrocybe	*laeta*	*peckii*	USA (damp soil, mossy places), PFGM, page 200	Peck	Pinkgill	Slender Waxycap Gill Mushroom	Edible
Hygrocybe	*laeta*	*roseiceps*	USA (damp soil, mossy places), PFGM, page 200	Rosecap	Pinkgill	Slender Waxycap Gill Mushroom	Edible
Hygrocybe	*marchii*	.	USA, PFGM, page 200	.	March	Slender Waxycap Gill Mushroom	Unknown
Hygrocybe	*marginata*	.	USA (mixed conifer-deciduous forest), PFGM, page 200	.	Orangegill	Slender Waxycap Gill Mushroom	Edible
Hygrocybe	*miniata*	.	USA (conifer & deciduous forest), MOTW, page 173, PFGM, page 201	.	Vermilion	Slender Waxycap Gill Mushroom	Edible
Hygrocybe	*miniata*	*miniata*	USA (conifer & deciduous forest)	Original	Vermilion	Slender Waxycap Gill Mushroom	Edible
Hygrocybe	*miniata*	*strangulata*	USA (conifer & deciduous forest)	Strangled	Vermilion	Slender Waxycap Gill Mushroom	Edible
Hygrocybe	*nitrata*	.	USA, PFGM, page 199	.	Nitrate	Slender Waxycap Gill Mushroom	Unknown
Hygrocybe	*nitrata*	*nitrata*	USA	.	Nitrate	Slender Waxycap Gill Mushroom	Unknown
Hygrocybe	*nitrata*	*murinacea*	USA	.	Nitrate	Slender Waxycap Gill Mushroom	Unknown
Hygrocybe	*persistens*	.	USA	.	Persistant	Slender Waxycap Gill Mushroom	Unknown
Hygrocybe	*persistens*	*persistens*	USA	Original	Persistant	Slender Waxycap Gill Mushroom	Unknown
Hygrocybe	*persistens*	*acutoconica*	USA, PFGM, page 199	Sharp Cone	Persistant	Slender Waxycap Gill Mushroom	Unknown
Hygrocybe	*persistens*	*aurantiolute scens*	USA	Orange Yellow	Persistant	Slender Waxycap Gill Mushroom	Unknown
Hygrocybe	*persistens*	*constans*	USA	Constant	Persistant	Slender Waxycap Gill Mushroom	Unknown
Hygrocybe	*persistens*	*cuspidata*	USA, PFGM, page 199	Cusped	Persistant	Slender Waxycap Gill Mushroom	Unknown
Hygrocybe	*persistens*	*konradii*	USA	Konrad	Persistant	Slender Waxycap Gill Mushroom	Unknown
Hygrocybe	*persistens*	*konradii-silvatica*	USA	Forest Konrad	Persistant	Slender Waxycap Gill Mushroom	Unknown
Hygrocybe	*persistens*	*konradii-subglobispora*	USA	Globe Konrad	Persistant	Slender Waxycap Gill Mushroom	Unknown
Hygrocybe	*persistens*	*langei*	USA	Lange	Persistant	Slender Waxycap Gill Mushroom	Unknown
Hygrocybe	*persistens*	*parapersist ens*	USA	Lesser Persistant	Persistant	Slender Waxycap Gill Mushroom	Unknown
Hygrocybe	*persistens*	*pseudoconi ca*	USA	Lesser Cone	Persistant	Slender Waxycap Gill Mushroom	Unknown
Hygrocybe	*persistens*	*subconica*	USA	Cone	Persistant	Slender Waxycap Gill Mushroom	Unknown
Hygrocybe	*psittacina*	.	USA (on soil in various habitats), MOTW, page 173, PFGM, page 202	.	Parrot	Slender Waxycap Gill Mushroom	Edible
Hygrocybe	*psittacina*	*psittacina*	USA (on soil in various habitats)	Original	Parrot	Slender Waxycap Gill Mushroom	Edible
Hygrocybe	*psittacina*	*perplexa*	USA (on soil in various habitats)	Perplexed	Parrot	Slender Waxycap Gill Mushroom	Edible
Hygrocybe	*psittacina*	*perplexa-*	USA (on soil in various	Shadow	Parrot	Slender Waxycap	Edible

		sciophana	habitats)	Perplexed		Gill Mushroom	
Hygrocybe	*punicea*	.	USA (meadows & mountains), MOTW, page 170, PFGM, page 198	.	Scarlet	Slender Waxycap Gill Mushroom	Edible-Excellent
Hygrocybe	*punicea*	*punicea*	USA (meadows & mountains)	Original	Scarlet	Slender Waxycap Gill Mushroom	Edible-Excellent
Hygrocybe	*punicea*	*acutopunicea*	USA (meadows & mountains)	Sharp	Scarlet	Slender Waxycap Gill Mushroom	Edible-Excellent
Hygrocybe	*tahquamenonensis*	.	USA, PFGM, page 199	.	Ahquamenona	Slender Waxycap Gill Mushroom	Unknown
Hygrocybe	*turunda*	.	Europe, MOTW, page 173	.	Incense	Slender Waxycap Gill Mushroom	Edible
Limacella	*delicata*	.	USA	.	Delicate	Slimy Waxycap Gill Mushroom	Edible
Limacella	*delicata*	*delicata*	USA	Original	Delicate	Slimy Waxycap Gill Mushroom	Edible
Limacella	*delicata*	*glioderma*	USA, MOTW, page 32, PFGM, page 214	Glueskin	Delicate	Slimy Waxycap Gill Mushroom	Edible
Limacella	*delicata*	*vinosorubescens*	USA	Wine Red	Delicate	Slimy Waxycap Gill Mushroom	Edible
Limacella	*delicata*	*vinosorubescens-roseofloccosa*	USA	Fleecy Wine Red	Delicate	Slimy Waxycap Gill Mushroom	Edible
Limacella	*glischra*	.	USA, PFGM, page 213	.	Glistening	Slimy Waxycap Gill Mushroom	Unknown
Limacella	*guttata*	.	Europe (humid Beech & Fir woods), MOTW, page 32, PFGM, page 214	.	Droplet	Slimy Waxycap Gill Mushroom	Edible
Limacella	*guttata*	*guttata*	Europe (humid Beech & Fir woods)	Original	Droplet	Slimy Waxycap Gill Mushroom	Edible
Limacella	*guttata*	*fischeri*	USA (humid Beech & Fir woods), MOTW, page 32	American	Droplet	Slimy Waxycap Gill Mushroom	Edible
Limacella	*guttata*	*lenticularis*	USA (humid Beech & Fir woods)	Dewy	Droplet	Slimy Waxycap Gill Mushroom	Edible
Limacella	*guttata*	*megalodactylus*	Europe (humid Beech & Fir woods)	Large Finger	Droplet	Slimy Waxycap Gill Mushroom	Edible
Limacella	*illinita*	.	Europe (conifer forest), MOTW, page 32, PFGM, page 214	.	Greasy	Slimy Waxycap Gill Mushroom	Edible
Limacella	*illinita*	*illinita*	Europe (conifer forest)	Original	Greasy	Slimy Waxycap Gill Mushroom	Edible
Limacella	*illinita*	*argillacea*	Europe (conifer forest), PFGM, page 215	Argylle	Greasy	Slimy Waxycap Gill Mushroom	Edible
Limacella	*illinita*	*rubescens*	Europe (conifer forest), PFGM, page 215	Red	Greasy	Slimy Waxycap Gill Mushroom	Edible
Limacella	*irrorata*	.	Europe (sunny areas), MOTW, page 32	.	Dewy	Slimy Waxycap Gill Mushroom	Unknown

Family 622.201.08 *Lepiotaceae* Parasol Gill Mushroom

Genus	Species	Variety	Native Location	Variety	Specimen	Genome	Habit
Lepiota	*alluvina*	.	USA, MOTW, page 309	.	Alluvial	Parasol Gill Mushroom	Unknown
Lepiota	*aspera*	.	USA (grassland, conifer & hardwood forest), PFGM, page 242	.	Rough	Parasol Gill Mushroom	Edible
Lepiota	*aspera*	*aspera*	USA (grassland, conifer & hardwood forest)	Original	Rough	Parasol Gill Mushroom	Edible
Lepiota	*aspera*	*acutesquamosa*	USA (grassland, conifer & hardwood forest), MOTW, page 37, PFGM, page 241	Sharp Scaled	Rough	Parasol Gill Mushroom	Unpalatable
Lepiota	*aspera*	*friesii*	USA (grassland, conifer & hardwood forest)	Fries	Rough	Parasol Gill Mushroom	Edible
Lepiota	*aspera*	*furcata*	USA (grassland, conifer & hardwood forest)	Forked	Rough	Parasol Gill Mushroom	Edible
Lepiota	*brunneoincamata*	.	Europe (parks, grasslands), MOTW, page 38	.	Brown Deathcap	Parasol Gill Mushroom	Poison
Lepiota	*brunneoincamata*	*brunneoincamata*	Europe (parks, grasslands)	Original	Brown Deathcap	Parasol Gill Mushroom	Poison
Lepiota	*brunneoincamata*	*barlae*	Europe (parks, grasslands)	Barla	Brown Deathcap	Parasol Gill Mushroom	Poison
Lepiota	*brunneoincamata*	*patouillardii*	Europe (parks, grasslands)	Patouillard	Brown Deathcap	Parasol Gill Mushroom	Poison
Lepiota	*calcicola*	.	USA	.	Limestone	Parasol Gill Mushroom	Unknown
Lepiota	*calcicola*	*calcicola*	USA	Original	Limestone	Parasol Gill Mushroom	Unknown
Lepiota	*calcicola*	*hispida*	USA, MOTW, page 309	Hairy	Limestone	Parasol Gill Mushroom	Unknown
Lepiota	*clypeolaria*	.	USA (on soil in forest),	.	Shield	Parasol Gill	Edible

			MOTW, page 38, PFGM, page 242			Mushroom	
Lepiota	*clypeolaria*	*clypeolaria*	USA (on soil in forest)	Original	Shield	Parasol Gill Mushroom	Edible
Lepiota	*clypeolaria*	*colubrina*	USA (on soil in forest)	Snake	Shield	Parasol Gill Mushroom	Edible
Lepiota	*clypeolaria*	*ochraceosulfurescens*	USA (on soil in forest)	Ocher Sulfur	Shield	Parasol Gill Mushroom	Edible
Lepiota	*cristata*	.	USA, MOTW, page 38	.	Crested	Parasol Gill Mushroom	Inedible
Lepiota	*cristata*	*cristata*	USA	Original	Crested	Parasol Gill Mushroom	Inedible
Lepiota	*cristata*	*felinoides*	USA, MOTW, page 38	Feline	Crested	Parasol Gill Mushroom	Inedible
Lepiota	*cristata*	*pallidior*	USA	Pale	Crested	Parasol Gill Mushroom	Inedible
Lepiota	*cristata*	*subfelinoides*	USA	Feline	Crested	Parasol Gill Mushroom	Inedible
Lepiota	*erminea*	.	Europe	.	Mink	Parasol Gill Mushroom	Inedible
Lepiota	*erminea*	*erminea*	Europe	Original	Mink	Parasol Gill Mushroom	Inedible
Lepiota	*erminea*	*alba*	Europe, MOTW, page 38	White	Mink	Parasol Gill Mushroom	Inedible
Lepiota	*farinosa*	.	USA, MOTW, page 309	.	Mealy	Parasol Gill Mushroom	Unknown
Lepiota	*felina*	.	USA, MOTW, page 38	.	Feline	Parasol Gill Mushroom	Inedible
Lepiota	*fuscovinacea*	.	Europe, MOTW, page 38	.	Dark Wine Red	Parasol Gill Mushroom	Inedible
Lepiota	*helveola*	.	Europe (parks, grasslands), MOTW, page 38	.	Blonde Deathcap	Parasol Gill Mushroom	Poison
Lepiota	*hystrix*	.	Europe (rich humus soil), MOTW, page 37	.	Bristle	Parasol Gill Mushroom	Inedible
Lepiota	*jacobi*	.	USA	.	Jacob	Parasol Gill Mushroom	Unknown
Lepiota	*jacobi*	*jacobi*	USA	Original	Jacob	Parasol Gill Mushroom	Unknown
Lepiota	*jacobi*	*eriophora*	USA, PFGM, page 242	Woolly	Jacob	Parasol Gill Mushroom	Unknown
Lepiota	*jacobi*	*langei*	USA	Lange	Jacob	Parasol Gill Mushroom	Unknown
Lepiota	*magnispora*	.	USA (on soil)	.	Big Spore	Parasol Gill Mushroom	Inedible
Lepiota	*magnispora*	*magnispora*	USA (on soil)	Original	Big Spore	Parasol Gill Mushroom	Inedible
Lepiota	*magnispora*	*metulispora*	USA (on soil), MOTW, page 38	Cone Spore	Big Spore	Parasol Gill Mushroom	Inedible
Lepiota	*magnispora*	*ventriosospora*	USA (on soil)	Ventral Spore	Big Spore	Parasol Gill Mushroom	Inedible
Lepiota	*subgracilis*	.	Europe	.	Lesser Slender	Parasol Gill Mushroom	Inedible
Lepiota	*subgracilis*	*subgracilis*	Europe	Original	Lesser Slender	Parasol Gill Mushroom	Inedible
Lepiota	*subgracilis*	*kuehneriana*	Europe	Kuehner	Lesser Slender	Parasol Gill Mushroom	Inedible
Lepiota	*subgracilis*	*latispora*	Europe, MOTW, page 38	Wide Spore	Lesser Slender	Parasol Gill Mushroom	Inedible
Lepiota	*subgracilis*	*minor*	Europe	Minor	Lesser Slender	Parasol Gill Mushroom	Inedible
Lepiota	*subgracilis*	*ochraceosulfurescens*	Europe	Ocher Sulfur	Lesser Slender	Parasol Gill Mushroom	Inedible
Lepiota	*subgracilis*	*wasseri*	Europe	Wasser	Lesser Slender	Parasol Gill Mushroom	Inedible
Lepiota	*subincarnata*	.	Europe (parks, grasslands), MOTW, page 38	.	Flesh Deathcap	Parasol Gill Mushroom	Poison
Lepiota	*subincarnata*	*subincarnata*	Europe (parks, grasslands)	Original	Flesh Deathcap	Parasol Gill Mushroom	Poison
Lepiota	*subincarnata*	*josserandii*	Europe (parks, grasslands)	Josserand	Flesh Deathcap	Parasol Gill Mushroom	Poison
Lepiota	*umbonata*	.	Europe (broadleaf trees), MOTW, page 34	.	Penis	Parasol Gill Mushroom	Edible-Good
Chlorophyllum	*molybdites*	.	USA (meadows, lawns, pastures), MOTW, page 36, PFGM, page 240	.	Lead Gray	Greengill Parasol Gill Mushroom	Edible-Toxic
Chlorophyllum	*morganii*	.	USA (meadows, lawns, pastures), MOTW, page	.	Morgan Deathcap	Greengill Parasol Gill Mushroom	Poison

			36				
Chlorophyllum	*rhacodes*	.	USA (grasslands, fields, open woods), E. Europe, MOTW, page 36, PFGM, page 246	.	Shaggy	Greengill Parasol Gill Mushroom	Edible-Good
Chlorophyllum	*rhacodes*	*rhacodes*	USA (grasslands, fields, open woods), E. Europe	Original	Shaggy	Greengill Parasol Gill Mushroom	Edible-Good
Chlorophyllum	*rhacodes*	*bohemica*	USA (grasslands, fields, open woods), E. Europe	Bohemian	Shaggy	Greengill Parasol Gill Mushroom	Edible-Good
Chlorophyllum	*rhacodes*	*brunnea*	USA (grasslands, fields, open woods), E. Europe	Brown	Shaggy	Greengill Parasol Gill Mushroom	Edible-Good
Chlorophyllum	*rhacodes*	*hortensis*	USA (grasslands, fields, open woods), E. Europe, MOTW, page 37	Garden	Shaggy	Greengill Parasol Gill Mushroom	Edible-Good
Chlorophyllum	*rhacodes*	*olivieri*	USA (grasslands, fields, open woods), E. Europe	Olivier	Shaggy	Greengill Parasol Gill Mushroom	Edible-Good
Leucoagaricus	*americanus*	.	USA (grasslands, deciduous woods), MOTW, page 36, PFGM, page 244	.	American	Smoothcap Parasol Gill Mushroom	Edible
Leucoagaricus	*americanus*	*americanus*	USA (grasslands, deciduous woods)	Original	American	Smoothcap Parasol Gill Mushroom	Edible
Leucoagaricus	*americanus*	*bresadolae*	USA (grasslands, deciduous woods)	Bresadole	American	Smoothcap Parasol Gill Mushroom	Edible
Leucoagaricus	*brunneus*	.	USA, PFGM, page 246	.	Browning	Smoothcap Parasol Gill Mushroom	Unknown
Leucoagaricus	*carneifolius*	.	Europe, MOTW, page 38	.	Fleshgill	Smoothcap Parasol Gill Mushroom	Edible
Leucoagaricus	*carneifolius*	*carneifolius*	Europe	Original	Fleshgill	Smoothcap Parasol Gill Mushroom	Edible
Leucoagaricus	*carneifolius*	*densifolius*	Europe, MOTW, page 38	Dense	Fleshgill	Smoothcap Parasol Gill Mushroom	Edible
Leucoagaricus	*cinerascens*	.	Europe (on soil), MOTW, page 38	.	Cinder	Smoothcap Parasol Gill Mushroom	Edible
Leucoagaricus	*leucothites*	.	Europe	.	Leucothite	Smoothcap Parasol Gill Mushroom	Edible
Leucoagaricus	*leucothites*	*leucothites*	Europe	Original	Leucothite	Smoothcap Parasol Gill Mushroom	Edible
Leucoagaricus	*leucothites*	*holosericeus*	Europe, MOTW, page 38	Silky	Leucothite	Smoothcap Parasol Gill Mushroom	Edible
Leucoagaricus	*meleagris*	.	Europe, MOTW, page 49	.	Guinea Hen	Smoothcap Parasol Gill Mushroom	Inedible
Leucoagaricus	*naucinus*	.	USA (lawns, parks, pastures), MOTW, page 38, PFGM, page 243	.	Nut Shell	Smoothcap Parasol Gill Mushroom	Edible-Good
Leucoagaricus	*nympharum*	.	Europe (meadows, conifer forest)	.	Nymph	Smoothcap Parasol Gill Mushroom	Edible-Good
Leucoagaricus	*nympharum*	*nympharum*	Europe (meadows, conifer forest)	Original	Nymph	Smoothcap Parasol Gill Mushroom	Edible-Good
Leucoagaricus	*nympharum*	*puellaris*	Europe (meadows, conifer forest), MOTW, page 34	Girl	Nymph	Smoothcap Parasol Gill Mushroom	Edible-Good
Leucocoprinus	*badhamii*	.	Europe (woody debris), MOTW, page 36	.	Badham	Lightcap Parasol Gill Mushroom	Edible
Leucocoprinus	*birnbaumii*	.	USA (on rich soil in greenhouses & pots), PFGM, page 245	.	Pleated Deathcap	Lightcap Parasol Gill Mushroom	Poison
Leucocoprinus	*birnbaumii*	*birnbaumii*	USA (on rich soil in greenhouses & pots)	Original	Pleated Deathcap	Lightcap Parasol Gill Mushroom	Poison
Leucocoprinus	*birnbaumii*	*luteus*	USA (on rich soil in greenhouses & pots)	Yellow	Pleated Deathcap	Lightcap Parasol Gill Mushroom	Poison
Leucocoprinus	*cepistipes*	.	USA (on rich soil in greenhouses & pots), PFGM, page 245	.	Onion Stem	Lightcap Parasol Gill Mushroom	Edible
Leucocoprinus	*cepistipes*	*cepistipes*	USA (on rich soil in greenhouses & pots)	Original	Onion Stem	Lightcap Parasol Gill Mushroom	Edible
Leucocoprinus	*cepistipes*	*cheimonoceps*	USA (on rich soil in greenhouses & pots)	Cheimon	Onion Stem	Lightcap Parasol Gill Mushroom	Edible
Leucocoprinus	*cepistipes*	*rorulentus*	USA (on rich soil in greenhouses & pots)	Rorulent	Onion Stem	Lightcap Parasol Gill Mushroom	Edible
Leucocoprinus	*procera*	.	USA (lawns, pastures, grasslands, open woods), PFGM, page 245	.	Tall	Lightcap Parasol Gill Mushroom	Edible-Excellent
Macrolepiota	*excoriata*	.	USA (fields, pastures, paths), MOTW, page 37	.	Flayed	Large Parasol Gill Mushroom	Edible-Excellent
Macrolepiota	*fuliginosa*	.	USA (meadows, grainfields, mixed open woods), MOTW, page 34	.	Shiny	Large Parasol Gill Mushroom	Edible-Excellent
Macrolepiota	*fuliginosa*	*fuliginosa*	USA (meadows, grainfields, mixed open woods)	Original	Shiny	Large Parasol Gill Mushroom	Edible-Excellent
Macrolepiota	*fuliginosa*	*permixta*	USA (meadows, grainfields, mixed open woods), MOTW, page 34	Mixed	Shiny	Large Parasol Gill Mushroom	Edible-Excellent
Macrolepiota	*gracilenta*	.	USA, MOTW, page 34	.	Slender	Large Parasol	Unknown

						Gill Mushroom	
Macrolepiota	*mastoidea*	.	USA, MOTW, page 34	.	Breast	Large Parasol Gill Mushroom	Edible
Macrolepiota	*procera*	.	USA (meadows, grainfields, mixed open woods), MOTW, page 34	.	Tall	Large Parasol Gill Mushroom	Edible-Excellent
Macrolepiota	*procera*	*procera*	USA (meadows, grainfields, mixed open woods)	Original	Tall	Large Parasol Gill Mushroom	Edible-Excellent
Macrolepiota	*procera*	*pseudo-olivascens*	USA (meadows, grainfields, mixed open woods)	Olive	Tall	Large Parasol Gill Mushroom	Edible-Excellent
Melanophyllum	*echinatum*	.	Europe (humis soil), MOTW, page 37	.	Urchin	Dark Gill Mushroom	Inedible
Parasola	*auricoma*	.	Europe (woodlands)	.	Ear	Greater Parasol Gill Mushroom	Unpalatable
Parasola	*auricoma*	*auricoma*	Europe (woodlands)	Original	Ear	Greater Parasol Gill Mushroom	Unpalatable
Parasola	*auricoma*	*hansenii*	Europe (woodlands), MOTW, page 50	Hansen	Ear	Greater Parasol Gill Mushroom	Unpalatable
Parasola	*plicatilis*	.	USA (lawns & grasslands), PFGM, page 279	.	Umbrella Inky Cap	Greater Parasol Gill Mushroom	Edible
Phaeolepiota	*aurea*	.	Europe (Fir & broadleaf forest), MOTW, page 62	.	Golden	Dusky Parasol Gill Mushroom	Edible-Fair
Phaeolepiota	*aurea*	*aurea*	Europe (Fir & broadleaf forest)	Original	Golden	Dusky Parasol Gill Mushroom	Edible-Fair
Phaeolepiota	*aurea*	*spectabilis*	Europe (Fir & broadleaf forest)	Spectacular	Golden	Dusky Parasol Gill Mushroom	Edible-Fair

Family 622.201.09 *Marasmiaceae* Pinwheel Gill Mushroom

Genus	Species	Variety	Native Location	Variety	Specimen	Genome	Habit
Marasmius	*albogriseus*	.	USA, PFGM, page 166	.	Gray	Pinwheel Gill Mushroom	Unknown
Marasmius	*Alliaceus*	.	Europe, MOTW, page 92	.	Onion	Pinwheel Gill Mushroom	Unpalatable
Marasmius	*androsaceus*	.	USA (conifer forest), MOTW, page 92, PFGM, page 165	.	Horsehair	Pinwheel Gill Mushroom	Unpalatable
Marasmius	*Bellipes*	.	USA, PFGM, page 168	.	Bell Foot	Pinwheel Gill Mushroom	Unknown
Marasmius	*Capillaries*	.	USA, PFGM, page 168	.	Capillary	Pinwheel Gill Mushroom	Unknown
Marasmius	*Collinus*	.	Europe, MOTW, page 91	.	Hillside	Pinwheel Gill Mushroom	Inedible
Marasmius	*fulvoferrugineus*	.	USA, PFGM, page 168	.	Orange Brown	Pinwheel Gill Mushroom	.
Marasmius	*Oreades*	.	USA (lawns, parks, meadows, shrublands), MOTW, page 91, PFGM, page 166	.	Fairy Ring	Pinwheel Gill Mushroom	Edible-Excellent
Marasmius	*pallidocephalus*	.	USA, PFGM, page 165	.	Pale Cap	Pinwheel Gill Mushroom	Unknown
Marasmius	*personatus*	.	Europe (plant debris), MOTW, page 92	.	Pepper	Pinwheel Gill Mushroom	Seasoning
Marasmius	*plicatulus*	.	NW. USA (Sitka Spruce & Live Oak forest), PFGM, page 166	.	Magenta Velvet	Pinwheel Gill Mushroom	Edible
Marasmius	*prasiosmus*	.	Europe, MOTW, page 92	.	Leek	Pinwheel Gill Mushroom	Unpalatable
Marasmius	*pulcherripes*	.	USA, PFGM, page 168	.	Pretty Foot	Pinwheel Gill Mushroom	Unknown
Marasmius	*rotula*	.	USA (on decaying wood in conifer & deciduous forest), MOTW, page 92, PFGM, page 167	.	Little Wheel	Pinwheel Gill Mushroom	Unpalatable
Marasmius	*scorodonius*	.	USA (on plant debris), MOTW, page 92	.	Garlic Herb	Pinwheel Gill Mushroom	Edible
Marasmius	*siccus*	.	S. Canada, USA (on leaf mold & litter in confer & hardwood forest), MOTW, page 309, PFGM, page 168	.	Dry	Pinwheel Gill Mushroom	Unknown
Marasmius	*thujinus*	.	USA, PFGM, page 166	.	Thujin	Pinwheel Gill Mushroom	Unknown
Marasmius	*wynnei*	.	Europe (mountain woods)	.	Wynne	Pinwheel Gill Mushroom	Edible
Marasmius	*wynnei*	*wynnei*	Europe (mountain woods)	Original	Wynne	Pinwheel Gill Mushroom	Edible
Marasmius	*wynnei*	*carpathicus*	Europe (mountain woods)	Carpathian Mountain	Wynne	Pinwheel Gill Mushroom	Edible
Marasmius	*wynnei*	*globularis*	Europe (mountain woods), MOTW, page 91	Globular	Wynne	Pinwheel Gill Mushroom	Edible

Marasmius	*wynnei*	*suaveolens*	Europe (mountain woods)	Sweet	Wynne	Pinwheel Gill Mushroom	Edible
Armillaria	*albolonaripes*	.	NW. USA (conifer forest), PFGM, page 133	.	Scaley	Bracelet Pinwheel Gill Mushroom	Edible
Armillaria	*borealis*	.	USA, PFGM, page 168	.	Boreal	Bracelet Pinwheel Gill Mushroom	.
Armillaria	*mellea*	.	USA (on living or dead wood in conifer & hardwood forest), MOTW, page 110, PFGM, page 136	.	Honey	Bracelet Pinwheel Gill Mushroom	Edible-Good
Armillaria	*mellea*	*mellea*	USA (on living or dead wood in conifer & hardwood forest)	Original	Honey	Bracelet Pinwheel Gill Mushroom	Edible-Good
Armillaria	*mellea*	*glabra*	USA (on living or dead wood in conifer & hardwood forest)	Bald	Honey	Bracelet Pinwheel Gill Mushroom	Edible-Good
Armillaria	*mellea*	*maxima*	USA (on living or dead wood in conifer & hardwood forest)	Greater	Honey	Bracelet Pinwheel Gill Mushroom	Edible-Good
Armillaria	*mellea*	*minor*	USA (on living or dead wood in conifer & hardwood forest)	Lesser	Honey	Bracelet Pinwheel Gill Mushroom	Edible-Good
Armillaria	*mellea*	*sulphurea*	USA (on living or dead wood in conifer & hardwood forest)	Sulfur	Honey	Bracelet Pinwheel Gill Mushroom	Edible-Good
Armillaria	*tabescens*	.	USA (on living or dead wood in conifer & hardwood forest), MOTW, page 116, PFGM, page 137	.	Friendship	Bracelet Pinwheel Gill Mushroom	Edible
Armillaria	*tabescens*	*tabescens*	USA (on living or dead wood in conifer & hardwood forest)	Original	Friendship	Bracelet Pinwheel Gill Mushroom	Edible
Armillaria	*tabescens*	*monadelpha*	USA (on living or dead wood in conifer & hardwood forest)	United	Friendship	Bracelet Pinwheel Gill Mushroom	Edible
Armillaria	*tabescens*	*socialis*	USA (on living or dead wood in conifer & hardwood forest)	Social	Friendship	Bracelet Pinwheel Gill Mushroom	Edible
Armillaria	*zelleri*	.	Europe, MOTW, page 109	.	Zeller	Bracelet Pinwheel Gill Mushroom	Inedible
Flammulina	*velutipes*	.	USA (on living wood in Aspen, Willow & Elm hardwood forest), MOTW, page 91, PFGM, page 152	.	Velvet Shank	Flamegill Pinwheel Gill Mushroom	Edible-Good
Flammulina	*velutipes*	*velutipes*	USA (on living wood in Aspen, Willow & Elm hardwood forest)	Original	Velvet Shank	Flamegill Pinwheel Gill Mushroom	Edible-Good
Flammulina	*velutipes*	*enokii*	E. Asia, USA (on living wood in Aspen, Willow & Elm hardwood forest)	Enoki	Velvet Shank	Flamegill Pinwheel Gill Mushroom	Edible-Good
Flammulina	*velutipes*	*enotakei*	E. Asia, USA (on living wood in Aspen, Willow & Elm hardwood forest)	Enotake	Velvet Shank	Flamegill Pinwheel Gill Mushroom	Edible-Good
Flammulina	*velutipes*	*lactea*	USA (on living wood in Aspen, Willow & Elm hardwood forest)	Milky	Velvet Shank	Flamegill Pinwheel Gill Mushroom	Edible-Good
Flammulina	*velutipes*	*lactea-alba*	USA (on living wood in Aspen, Willow & Elm hardwood forest)	Milky White	Velvet Shank	Flamegill Pinwheel Gill Mushroom	Edible-Good
Lentinula	*edodes*	.	USA (mountain Chestnut, Holly, Oak broadleaf forest), MOTW, page 126	.	Shiitake	Lentinan Pinwheel Gill Mushroom	Edible-Good
Lentinula	*edodes*	*edodes*	USA (mountain Chestnut, Holly, Oak broadleaf forest)	Original	Shiitake	Lentinan Pinwheel Gill Mushroom	Edible-Good
Lentinula	*edodes*	*black-forestii*	USA (mountain Chestnut, Holly, Oak broadleaf forest)	Black Forest	Shiitake	Lentinan Pinwheel Gill Mushroom	Edible-Good
Lentinula	*edodes*	*bolshoi-breezei*	USA (mountain Chestnut, Holly, Oak broadleaf forest)	Bolshoi Breeze	Shiitake	Lentinan Pinwheel Gill Mushroom	Edible-Good
Lentinula	*edodes*	*CA-77A*	USA (mountain Chestnut, Holly, Oak broadleaf forest)	#77A Chinese	Shiitake	Lentinan Pinwheel Gill Mushroom	Edible-Good
Lentinula	*edodes*	*CA-77B*	USA (mountain Chestnut, Holly, Oak broadleaf forest)	#77B Chinese	Shiitake	Lentinan Pinwheel Gill Mushroom	Edible-Good
Lentinula	*edodes*	*CS-011*	USA (mountain Chestnut, Holly, Oak broadleaf forest)	#011 Chinese	Shiitake	Lentinan Pinwheel Gill Mushroom	Edible-Good
Lentinula	*edodes*	*CS-015*	USA (mountain Chestnut, Holly, Oak broadleaf forest)	#015 Chinese	Shiitake	Lentinan Pinwheel Gill Mushroom	Edible-Good
Lentinula	*edodes*	*CS-016*	USA (mountain Chestnut, Holly, Oak broadleaf forest)	#016 Chinese	Shiitake	Lentinan Pinwheel Gill Mushroom	Edible-Good
Lentinula	*edodes*	*CS-024*	USA (mountain Chestnut, Holly, Oak broadleaf forest)	#024 Chinese	Shiitake	Lentinan Pinwheel Gill Mushroom	Edible-Good
Lentinula	*edodes*	*CS-041*	USA (mountain Chestnut, Holly, Oak broadleaf forest)	#041 Chinese	Shiitake	Lentinan Pinwheel Gill Mushroom	Edible-Good
Lentinula	*edodes*	*CS-118*	USA (mountain Chestnut, Holly, Oak broadleaf forest)	#118 Chinese	Shiitake	Lentinan Pinwheel Gill Mushroom	Edible-Good
Lentinula	*edodes*	*CS-125*	USA (mountain Chestnut, Holly, Oak broadleaf forest)	#125 Chinese	Shiitake	Lentinan Pinwheel Gill Mushroom	Edible-Good
Lentinula	*edodes*	*CW-012*	USA (mountain Chestnut, Holly, Oak broadleaf forest)	#012 Chinese	Shiitake	Lentinan Pinwheel Gill Mushroom	Edible-Good
Lentinula	*edodes*	*CW-025*	USA (mountain Chestnut, Holly, Oak broadleaf forest)	#025 Chinese	Shiitake	Lentinan Pinwheel Gill Mushroom	Edible-Good
Lentinula	*edodes*	*CW-026*	USA (mountain Chestnut, Holly, Oak broadleaf forest)	#026 Chinese	Shiitake	Lentinan Pinwheel Gill Mushroom	Edible-Good

Lentinula	*edodes*	*CW-730*	USA (mountain Chestnut, Holly, Oak broadleaf forest)	#730 Chinese	Shiitake	Lentinan Pinwheel Gill Mushroom	Edible-Good
Lentinula	*edodes*	*donkui*	USA (mountain Chestnut, Holly, Oak broadleaf forest)	Donku	Shiitake	Lentinan Pinwheel Gill Mushroom	Edible-Good
Lentinula	*edodes*	*east-windii*	USA (mountain Chestnut, Holly, Oak broadleaf forest)	East Wind	Shiitake	Lentinan Pinwheel Gill Mushroom	Edible-Good
Lentinula	*edodes*	*felt-hatii*	USA (mountain Chestnut, Holly, Oak broadleaf forest)	Felt Hat	Shiitake	Lentinan Pinwheel Gill Mushroom	Edible-Good
Lentinula	*edodes*	*JC-stari*	USA (mountain Chestnut, Holly, Oak broadleaf forest)	Jupiter China Star	Shiitake	Lentinan Pinwheel Gill Mushroom	Edible-Good
Lentinula	*edodes*	*jupiteri*	USA (mountain Chestnut, Holly, Oak broadleaf forest)	Jupiter	Shiitake	Lentinan Pinwheel Gill Mushroom	Edible-Good
Lentinula	*edodes*	*night-velvetii*	USA (mountain Chestnut, Holly, Oak broadleaf forest)	Night Velvet	Shiitake	Lentinan Pinwheel Gill Mushroom	Edible-Good
Lentinula	*edodes*	*marsi*	USA (mountain Chestnut, Holly, Oak broadleaf forest)	Mars	Shiitake	Lentinan Pinwheel Gill Mushroom	Edible-Good
Lentinula	*edodes*	*mercuryi*	USA (mountain Chestnut, Holly, Oak broadleaf forest)	Mercury	Shiitake	Lentinan Pinwheel Gill Mushroom	Edible-Good
Lentinula	*edodes*	*neptunei*	USA (mountain Chestnut, Holly, Oak broadleaf forest)	Neptune	Shiitake	Lentinan Pinwheel Gill Mushroom	Edible-Good
Lentinula	*edodes*	*plutoi*	USA (mountain Chestnut, Holly, Oak broadleaf forest)	Pluto	Shiitake	Lentinan Pinwheel Gill Mushroom	Edible-Good
Lentinula	*edodes*	*QR*	USA (mountain Chestnut, Holly, Oak broadleaf forest)	Quality Run	Shiitake	Lentinan Pinwheel Gill Mushroom	Edible-Good
Lentinula	*edodes*	*saturnii*	USA (mountain Chestnut, Holly, Oak broadleaf forest)	Saturn	Shiitake	Lentinan Pinwheel Gill Mushroom	Edible-Good
Lentinula	*edodes*	*shiangkui*	USA (mountain Chestnut, Holly, Oak broadleaf forest)	Shiangku	Shiitake	Lentinan Pinwheel Gill Mushroom	Edible-Good
Lentinula	*edodes*	*snowcapii*	USA (mountain Chestnut, Holly, Oak broadleaf forest)	Snowcap	Shiitake	Lentinan Pinwheel Gill Mushroom	Edible-Good
Lentinula	*edodes*	*south-windii*	USA (mountain Chestnut, Holly, Oak broadleaf forest)	South Wind	Shiitake	Lentinan Pinwheel Gill Mushroom	Edible-Good
Lentinula	*edodes*	*venus*	USA (mountain Chestnut, Holly, Oak broadleaf forest)	Venus	Shiitake	Lentinan Pinwheel Gill Mushroom	Edible-Good
Lentinula	*edodes*	*west-windii*	USA (mountain Chestnut, Holly, Oak broadleaf forest)	West Wind	Shiitake	Lentinan Pinwheel Gill Mushroom	Edible-Good
Lentinula	*edodes*	*WR-046*	USA (mountain Chestnut, Holly, Oak broadleaf forest)	#046 West Run	Shiitake	Lentinan Pinwheel Gill Mushroom	Edible-Good
Lentinula	*edodes*	*WW-044*	USA (mountain Chestnut, Holly, Oak broadleaf forest)	#044 West Wind	Shiitake	Lentinan Pinwheel Gill Mushroom	Edible-Good
Micromphale	*foetidum*	.	USA, PFGM, page 170	.	Fetid	False Pinwheel Gill Mushroom	Unknown
Micromphale	*foetidum*	*rufocarneum*	USA	Original	Fetid	False Pinwheel Gill Mushroom	Unknown
Micromphale	*foetidum*	*venosum*	USA	Veined	Fetid	False Pinwheel Gill Mushroom	Unknown
Micromphale	*perforans*	.	Europe, MOTW, page 92	.	Perforated	False Pinwheel Gill Mushroom	Unpalatable
Oudemansiella	*mucida*	.	USA (Beech deciduous forest), MOTW, page 84	.	Slimy Mucus	Deep Root Pinwheel Gill Mushroom	Edible
Rhodocollybia	*butyracea*	.	S. Canada, N. USA (conifer forest), MOTW, page 88	.	Butter	Pink Pinwheel Gill Mushroom	Edible
Rhodocollybia	*butyracea*	*butyracea*	S. Canada, N. USA (conifer forest)	Original	Butter	Pink Pinwheel Gill Mushroom	Edible
Rhodocollybia	*butyracea*	*asema*	S. Canada, N. USA (conifer & mixed conifer-hardwood forest)	Aseme	Butter	Pink Pinwheel Gill Mushroom	Edible
Rhodocollybia	*maculata*	.	S. Canada, N. USA (on humus & decaying wood in conifer forest), MOTW, page 86	.	Spotted	Pink Pinwheel Gill Mushroom	Edible-Poor
Rhodocollybia	*maculata*	*maculata*	S. Canada, N. USA (on humus & decaying wood in conifer forest)	Original	Spotted	Pink Pinwheel Gill Mushroom	Edible-Poor
Rhodocollybia	*maculata*	*scorzonerea*	S. Canada, N. USA (on humus & decaying wood in conifer forest)	Scorzonera	Spotted	Pink Pinwheel Gill Mushroom	Edible-Poor
Rhodocollybia	*prolixa*	.	Europe (woods)	.	Spreading	Pink Pinwheel Gill Mushroom	Edible
Rhodocollybia	*prolixa*	*prolixa*	Europe (woods)	Original	Spreading	Pink Pinwheel Gill Mushroom	Edible
Rhodocollybia	*prolixa*	*distorta*	Europe (woods)	Distorted	Spreading	Pink Pinwheel Gill Mushroom	Edible
Strobilurus	*albipilata*	.	USA, PFGM, page 185	.	White Downy	Pinecone Pinwheel Gill Mushroom	Unknown
Strobilurus	*conigenoides*	.	USA, PFGM, page 184	.	Magnolia Cone	Pinecone Pinwheel Gill Mushroom	Unknown
Strobilurus	*esculentus*	.	USA, PFGM, page 91	.	Edible	Pinecone Pinwheel Gill Mushroom	Edible-Bitter

Strobilurus	*lignitilis*	.	USA, PFGM, page 185	.	Wood	Pinecone Pinwheel Gill Mushroom	Unknown
Strobilurus	*occidentalis*	.	USA, PFGM, page 184	.	Occidental	Pinecone Pinwheel Gill Mushroom	Unknown
Strobilurus	*stephanocystis*	.	Europe (on fallen pinecones)	.	Stephan Cyst	Pinecone Pinwheel Gill Mushroom	Unpalatable
Strobilurus	*stephanocystis*	*stephanocystis*	Europe (on fallen pinecones)	Original	Stephan Cyst	Pinecone Pinwheel Gill Mushroom	Unpalatable
Strobilurus	*stephanocystis*	*conigenus*	Europe (on fallen pinecones), MOTW, page 92	Pinecone	Stephan Cyst	Pinecone Pinwheel Gill Mushroom	Unpalatable
Strobilurus	*tenacellus*	.	Europe (on Pine & Spruce cones), MOTW, page 91	.	Tenacious	Pinecone Pinwheel Gill Mushroom	Inedible
Strobilurus	*trullisatus*	.	USA, PFGM, page 184	.	Plaster	Pinecone Pinwheel Gill Mushroom	Unknown
Strobilurus	*wyomingensis*	.	USA, PFGM, page 185	.	Wyoming	Pinecone Pinwheel Gill Mushroom	Unknown
Xerula	*longipes*	.	USA (on rotting wood in Oak broadleaf forest), MOTW, page 84, PFGM, page 179	.	Long Foot	Dry Pinwheel Gill Mushroom	Edible-Good
Xerula	*radicata*	.	USA (on rotting wood in Beech deciduous forest), MOTW, page 84, PFGM, page 179	.	Deep Root	Dry Pinwheel Gill Mushroom	Edible
Xerula	*radicata*	*radicata*	USA (on rotting wood in Beech deciduous forest)	Original	Deep Root	Dry Pinwheel Gill Mushroom	Edible
Xerula	*radicata*	*marginata*	USA (on rotting wood in Beech deciduous forest)	Margined	Deep Root	Dry Pinwheel Gill Mushroom	Edible
Xerula	*umbonata*	.	USA (N. Californa woods), MOTW, page 84	.	Penis	Dry Pinwheel Gill Mushroom	Unknown

Family 622.201.10 *Mesophelliaceae* Truffle Gill Mushroom

Genus	Species	Variety	Native Location	Variety	Specimen	Genome	Habit
Malajczukia	*amicorum*	.	S. Asia, Pac. Oc.Is., Australia	.	Amicable	Truffle Gill Mushroom	Unknown

Family 622.201.11 *Omphalinaceae* Navalcap Gill Mushroom

Genus	Species	Variety	Native Location	Variety	Specimen	Genome	Habit
Omphalina	*chrysophylla*	.	USA (decaying conifer wood) PFGM, page 177	.	Goldgill	Navalcap Gill Mushroom	Unknown
Omphalina	*postii*		N. USA (wet soil) PFGM, page 178		Tiny	Navalcap Gill Mushroom	Unknown

Family 622.201.12 *Pleurotaceae* Oyster Gill Mushroom

Genus	Species	Variety	Native Location	Variety	Specimen	Genome	Habit
Pleurotus	*columbinus*	.	USA (conifer & broadleaf forests), MOTW, page 125	.	Blue Capped	Oyster Gill Mushroom	Edible-Toxic
Pleurotus	*cornucopiae*	.	USA (on Oak, Elm, Poplar & Beech trunks), MOTW, page 123	.	Trumpet	Oyster Gill Mushroom	Edible-Good
Pleurotus	*cornucopiae*	*cornucopiae*	USA (on Oak, Elm, Poplar & Beech trunks)	Original	Trumpet	Oyster Gill Mushroom	Edible-Good
Pleurotus	*cornucopiae*	*aureus*	USA (on Oak, Elm, Poplar & Beech trunks)	Golden	Trumpet	Oyster Gill Mushroom	Edible-Good
Pleurotus	*cornucopiae*	*helvus*	USA (on Oak, Elm, Poplar & Beech trunks)	Yellow	Trumpet	Oyster Gill Mushroom	Edible-Good
Pleurotus	*cystidiosus*	.	USA, PFGM, page 182	.	Pinewood	Oyster Gill Mushroom	Edible
Pleurotus	*cystidiosus*	*cystidiosus*	USA	Original	Pinewood	Oyster Gill Mushroom	Edible
Pleurotus	*cystidiosus*	*formosensis*	USA	Beautiful	Pinewood	Oyster Gill Mushroom	Edible
Pleurotus	*djamor*	.	USA	.	D'Jamor	Oyster Gill Mushroom	Edible
Pleurotus	*djamor*	*djamor*	USA	Original	D'Jamor	Oyster Gill Mushroom	Edible
Pleurotus	*djamor*	*leptogrammus*	USA	Thin	D'Jamor	Oyster Gill Mushroom	Edible
Pleurotus	*djamor*	*moselei*	USA	Mosele	D'Jamor	Oyster Gill Mushroom	Edible
Pleurotus	*djamor*	*ninguidus*	USA	Ningu	D'Jamor	Oyster Gill Mushroom	Edible
Pleurotus	*djamor*	*pacificus*	USA	Pacific	D'Jamor	Oyster Gill Mushroom	Edible
Pleurotus	*djamor*	*pinca*	USA, GSI-6, page 258	Pink	D'Jamor	Oyster Gill	Edible

						Mushroom	
Pleurotus	*djamor*	*placentodes*	USA	Placenta	D'Jamor	Oyster Gill Mushroom	Edible
Pleurotus	*djamor*	*prometheus*	USA	Prometheus	D'Jamor	Oyster Gill Mushroom	Edible
Pleurotus	*djamor*	*salmoneostramineus*	USA	Salmon	D'Jamor	Oyster Gill Mushroom	Edible
Pleurotus	*djamor*	*scabellus*	USA	Rough	D'Jamor	Oyster Gill Mushroom	Edible
Pleurotus	*djamor*	*scabriusculus*	.	Scabre	D'Jamor	Oyster Gill Mushroom	Edible
Pleurotus	*dryinus*	.	USA (Oak broadleaf forest), MOTW, page 125	.	Oakwood	Oyster Gill Mushroom	Edible-Fair
Pleurotus	*dryinus*	*dryinus*	USA (Oak broadleaf forest)	Original	Oakwood	Oyster Gill Mushroom	Edible-Fair
Pleurotus	*dryinus*	*acerinus*	USA (Oak broadleaf forest)	Maple	Oakwood	Oyster Gill Mushroom	Edible-Fair
Pleurotus	*dryinus*	*albertinii*	USA (Oak broadleaf forest)	Albertini	Oakwood	Oyster Gill Mushroom	Edible-Fair
Pleurotus	*dryinus*	*corticatus*	USA (Oak broadleaf forest)	Bark	Oakwood	Oyster Gill Mushroom	Edible-Fair
Pleurotus	*dryinus*	*pometi*	USA (Oak broadleaf forest)	Pomet	Oakwood	Oyster Gill Mushroom	Edible-Fair
Pleurotus	*dryinus*	*spongiosus*	USA (Oak broadleaf forest)	Spongy	Oakwood	Oyster Gill Mushroom	Edible-Fair
Pleurotus	*dryinus*	*tephrotrichus*	USA (Oak broadleaf forest)	Ashy Braids	Oakwood	Oyster Gill Mushroom	Edible-Fair
Pleurotus	*dryinus*	*underwoodii*	USA (Oak broadleaf forest)	Underwood	Oakwood	Oyster Gill Mushroom	Edible-Fair
Pleurotus	*eryngii*	.	N. Africa, Europe (on sandy soil & Sea Holly Eyrngo remains), MOTW, page 123	.	King	Oyster Gill Mushroom	Edible-Excellent
Pleurotus	*eryngii*	*eryngii*	N. Africa, Europe (on sandy soil & Sea Holly Eyrngo remains)	Original	King	Oyster Gill Mushroom	Edible-Excellent
Pleurotus	*eryngii*	*elaeoselini*	N. Africa, Europe (sandy soil)	Olive Selin	King	Oyster Gill Mushroom	Edible-Excellent
Pleurotus	*eryngii*	*ferulae*	N. Africa, Europe (sandy soil on Giant Fennel remains), MOTW, page 123	Fennel	King	Oyster Gill Mushroom	Edible-Excellent
Pleurotus	*eryngii*	*fuscus*	N. Africa, Europe (sandy soil)	Fuscous	King	Oyster Gill Mushroom	Edible-Excellent
Pleurotus	*eryngii*	*nebrodensis*	N. Africa, Europe (mountain sandy soil, Sicily), MOTW, page 123	Nebrodi Mountain	King	Oyster Gill Mushroom	Edible-Excellent
Pleurotus	*flabellatus*	.	USA, GSI-6, page 258	.	Strawberry	Oyster Gill Mushroom	Edible
Pleurotus	*ostreatus*	.	USA (on logs & stumps in Cottonwood & Tulip Poplar deciduous forest), PFGM, page 181	.	Cottonwood & Tulip Poplar Forest	Oyster Gill Mushroom	Edible-Excellent
Pleurotus	*ostreatus*	*ostreatus*	USA (on logs & stumps in Cottonwood & Tulip Poplar deciduous forest)	Original	Cottonwood & Tulip Poplar Forest	Oyster Gill Mushroom	Edible-Excellent
Pleurotus	*ostreatus*	*albidus*	USA (on logs & stumps in Cottonwood & Tulip Poplar deciduous forest)	White	Cottonwood & Tulip Poplar Forest	Oyster Gill Mushroom	Edible-Excellent
Pleurotus	*ostreatus*	*blue-dolphinii*	USA (on logs & stumps in Cottonwood & Tulip Poplar deciduous forest)	Blue Dolphin	Cottonwood & Tulip Poplar Forest	Oyster Gill Mushroom	Edible-Excellent
Pleurotus	*ostreatus*	*fruitingii-cool-weatheri*	USA (on logs & stumps in Cottonwood & Tulip Poplar deciduous forest)	Cool Weather Fruiting	Cottonwood & Tulip Poplar Forest	Oyster Gill Mushroom	Edible-Excellent
Pleurotus	*ostreatus*	*fruitingii-warm-weatheri*	USA (on logs & stumps in Cottonwood & Tulip Poplar deciduous forest)	Warm Weather Fruiting	Cottonwood & Tulip Poplar Forest	Oyster Gill Mushroom	Edible-Excellent
Pleurotus	*ostreatus*	*glandulosus*	USA (on logs & stumps in Cottonwood & Tulip Poplar deciduous forest), MOTW, page 125	Gland	Cottonwood & Tulip Poplar Forest	Oyster Gill Mushroom	Edible-Excellent
Pleurotus	*ostreatus*	*gray-brownii*	USA (on logs & stumps in Cottonwood & Tulip Poplar deciduous forest)	Gray Brown	Cottonwood & Tulip Poplar Forest	Oyster Gill Mushroom	Edible-Excellent
Pleurotus	*ostreatus*	*gray-dovei*	USA (on logs & stumps in Cottonwood & Tulip Poplar deciduous forest)	Gray Dove	Cottonwood & Tulip Poplar Forest	Oyster Gill Mushroom	Edible-Excellent
Pleurotus	*ostreatus*	*gray-purplei*	USA (on logs & stumps in Cottonwood & Tulip Poplar deciduous forest)	Gray Purple	Cottonwood & Tulip Poplar Forest	Oyster Gill Mushroom	Edible-Excellent
Pleurotus	*ostreatus*	*opuntiae*	USA (on logs & stumps	Prickly	Cottonwood	Oyster Gill	Edible-

			in Cottonwood & Tulip Poplar deciduous forest)	Pear Cactus	& Tulip Poplar Forest	Mushroom	Excellent
Pleurotus	*ostreatus*	*salignus*	USA (on logs & stumps in Cottonwood & Tulip Poplar deciduous forest)	Willow	Cottonwood & Tulip Poplar Forest	Oyster Gill Mushroom	Edible-Excellent
Pleurotus	*ostreatus*	*strainii-21*	USA (on logs & stumps in Cottonwood & Tulip Poplar deciduous forest)	Strain #21	Cottonwood & Tulip Poplar Forest	Oyster Gill Mushroom	Edible-Excellent
Pleurotus	*ostreatus*	*strainii-23*	USA (on logs & stumps in Cottonwood & Tulip Poplar deciduous forest)	Strain #23	Cottonwood & Tulip Poplar Forest	Oyster Gill Mushroom	Edible-Excellent
Pleurotus	*ostreatus*	*revolutus*	USA (on logs & stumps in Cottonwood & Tulip Poplar deciduous forest)	Revolute	Cottonwood & Tulip Poplar Forest	Oyster Gill Mushroom	Edible-Excellent
Pleurotus	*ostreatus*	*roseus*	USA (on logs & stumps in Cottonwood & Tulip Poplar deciduous forest)	Rosy	Cottonwood & Tulip Poplar Forest	Oyster Gill Mushroom	Edible-Excellent
Pleurotus	*ostreatus*	*treei*	USA (on logs & stumps in Cottonwood & Tulip Poplar deciduous forest)	Tree	Cottonwood & Tulip Poplar Forest	Oyster Gill Mushroom	Edible-Excellent
Pleurotus	*pulmonarius*	.	Europe, MOTW, page 123	.	Brown	Oyster Gill Mushroom	Edible
Pleurotus	*pulmonarius*	*pulmonarius*	USA	Original	Brown	Oyster Gill Mushroom	Edible
Pleurotus	*pulmonarius*	*italiana*	Europe (Italy)	Italian	Brown	Oyster Gill Mushroom	Edible
Pleurotus	*sapidus*	.	USA, PFGM, page 182	.	Lilac Gray	Oyster Gill Mushroom	Edible
Pleurotus	*sapidus*	*sapidus*	USA	Original	Lilac Gray	Oyster Gill Mushroom	Edible
Pleurotus	*sapidus*	*floridana*	SE. USA (Florida)	Florida	Lilac Gray	Oyster Gill Mushroom	Edible
Hohenbuehelia	*angustata*	.	USA, PFGM, page 153	.	Narrow	Hohenbuehel Oyster Gill Mushroom	Unknown
Hohenbuehelia	*mastrucata*	.	USA, PFGM, page 153	.	Flat Breast	Hohenbuehel Oyster Gill Mushroom	Unknown
Hohenbuehelia	*petalodes*	.	USA (on decaying wood in conifer & deciduous forest), PFGM, page 153	.	Leaf	Hohenbuehel Oyster Gill Mushroom	Edible
Hohenbuehelia	*petalodes*	*petalodes*	USA (on decaying wood in conifer & deciduous forest)	Original	Leaf	Hohenbuehel Oyster Gill Mushroom	Edible
Hohenbuehelia	*petalodes*	*carbonaria*	USA (on decaying wood in conifer & deciduous forest)	Charcoal	Leaf	Hohenbuehel Oyster Gill Mushroom	Edible
Hohenbuehelia	*petalodes*	*geogenia*	USA (on decaying wood in conifer & deciduous forest)	Earth	Leaf	Hohenbuehel Oyster Gill Mushroom	Edible
Hohenbuehelia	*petalodes*	*queletii*	USA (on decaying wood in conifer & deciduous forest)	Quelet Earth	Leaf	Hohenbuehel Oyster Gill Mushroom	Edible
Hypsizygus	*ulmarius*	.	USA (Elm broadleaf forest), MOTW, page 125	.	Elm	Hyponic Viscose Oyster Gill Mushroom	Edible
Panellus	*mitis*	.	Europe (on deadwood in conifer forest), MOTW, page 128	.	Mild	Fancap Oyster Gill Mushroom	Inedible
Panellus	*serotinus*	.	USA (on wood in conifer & deciduous forest), MOTW, page 125, PFGM, page 180	.	Late	Fancap Oyster Gill Mushroom	Edible
Panellus	*serotinus*	*serotinus*	USA (on wood in conifer & deciduous forest)	.	Late	Fancap Oyster Gill Mushroom	Edible
Panellus	*serotinus*	*almeni*	USA (on wood in conifer & deciduous forest)	.	Late	Fancap Oyster Gill Mushroom	Edible
Panellus	*serotinus*	*flaccidus*	USA (on wood in conifer & deciduous forest)	.	Late	Fancap Oyster Gill Mushroom	Edible
Panellus	*stipticus*	.	USA (on logs & stumps in deciduous forest), MOTW, page 128, PFGM, page 180	.	Bitter	Fancap Oyster Gill Mushroom	Inedible
Panellus	*stipticus*	*stipticus*	USA (on logs & stumps in deciduous forest)	Original	Bitter	Fancap Oyster Gill Mushroom	Inedible
Panellus	*stipticus*	*albidotomentosus*	USA (on logs & stumps in deciduous forest)	White Woolly	Bitter	Fancap Oyster Gill Mushroom	Inedible
Panellus	*stipticus*	*farinaceus*	USA (on logs & stumps in deciduous forest)	Farina	Bitter	Fancap Oyster Gill Mushroom	Inedible
Panellus	*stipticus*	*lateralis*	USA (on logs & stumps in deciduous forest)	Wide	Bitter	Fancap Oyster Gill Mushroom	Inedible
Pleurocybella	*porrigens*	.	N. USA (on logs and stumps in Hemlock & conifer forest), PFGM, page 182	.	Angel's Wings	Manycap Oyster Gill Mushroom	Edible

Family 622.201.13 *Pluteaceae* Roofcap Gill Mushroom

Genus	Species	Variety	Native Location	Variety	Specimen	Genome	Habit
Pluteus	*admirabilis*	.	USA (conifer & mixed conifer-hardwood forest), PFGM, page 248	.	Yellow	Roofcap Gill Mushroom	Edible
Pluteus	*atromarginatus*	.	Europe (on live conifer wood)	.	Dark Margined	Roofcap Gill Mushroom	Edible
Pluteus	*atromarginatus*	*atromarginatus*	Europe (on live conifer wood)	Original	Dark Margined	Roofcap Gill Mushroom	Edible
Pluteus	*atromarginatus*	*nigrofloccosus*	Europe (on live conifer wood), MOTW, page 41	Black Fleece	Dark Margined	Roofcap Gill Mushroom	Edible
Pluteus	*atromarginatus*	*tricuspidatus*	Europe (on live conifer wood)	Three Cusp	Dark Margined	Roofcap Gill Mushroom	Edible
Pluteus	*aurantiorugosus*	.	Europe (on live broadleaf trees)	.	Orange Wrinkled	Roofcap Gill Mushroom	Unpalatable
Pluteus	*aurantiorugosus*	*aurantiorugosus*	Europe (on live broadleaf trees)	Original	Orange Wrinkled	Roofcap Gill Mushroom	Unpalatable
Pluteus	*aurantiorugosus*	*caloceps*	Europe (on live broadleaf trees)	Beautiful	Orange Wrinkled	Roofcap Gill Mushroom	Unpalatable
Pluteus	*aurantiorugosus*	*coccineus*	Europe (on live broadleaf trees), MOTW, page 41	Scarlet	Orange Wrinkled	Roofcap Gill Mushroom	Unpalatable
Pluteus	*cervinus*	.	Canada, USA (on decaying wood, sawdust), MOTW, page 41, PFGM, page 248	.	Deer	Roofcap Gill Mushroom	Edible
Pluteus	*cervinus*	*cervinus*	Canada, USA (on decaying wood, sawdust)	Original	Deer	Roofcap Gill Mushroom	Edible
Pluteus	*cervinus*	*atricapillus*	Canada, USA (on decaying wood, sawdust)	Dark Capillary	Deer	Roofcap Gill Mushroom	Edible
Pluteus	*cervinus*	*curtisii*	Canada, USA (on decaying wood, sawdust)	Curtis	Deer	Roofcap Gill Mushroom	Edible
Pluteus	*cervinus*	*eximius*	Canada, USA (on decaying wood, sawdust)	Exima	Deer	Roofcap Gill Mushroom	Edible
Pluteus	*chrysophaeus*	.	C. & E. USA (decaying wood in deciduous forest), PFGM, page 250	.	Sienna	Roofcap Gill Mushroom	Unknown
Pluteus	*chrysophaeus*	*chrysophaeus*	C. & E. USA (on decaying wood in deciduous forest)	Original	Sienna	Roofcap Gill Mushroom	Unknown
Pluteus	*chrysophaeus*	*galeroides*	C. & E. USA (on decaying wood in deciduous forest)	Galer	Sienna	Roofcap Gill Mushroom	Unknown
Pluteus	*chrysophaeus*	*luteovirens*	C. & E. USA (on decaying wood in deciduous forest)	Yellow Green	Sienna	Roofcap Gill Mushroom	Unknown
Pluteus	*chrysophaeus*	*xanthophaeus*	C. & E. USA (on decaying wood in deciduous forest)	Dusky Yellow	Sienna	Roofcap Gill Mushroom	Unknown
Pluteus	*cinereofuscus*	.	Europe (on humus under broadleaf trees), MOTW, page 41	.	Dark Gray	Roofcap Gill Mushroom	Unpalatable
Pluteus	*cinereofuscus*	*cinereofuscus*	Europe (on humus under broadleaf trees)	Original	Dark Gray	Roofcap Gill Mushroom	Unpalatable
Pluteus	*cinereofuscus*	*major*	Europe (on humus under broadleaf trees)	Greater	Dark Gray	Roofcap Gill Mushroom	Unpalatable
Pluteus	*cinereofuscus*	*olivaceus*	Europe (on humus under broadleaf trees)	Olive	Dark Gray	Roofcap Gill Mushroom	Unpalatable
Pluteus	*ephebeus*	.	Europe	.	Ephebe	Roofcap Gill Mushroom	Edible
Pluteus	*ephebeus*	*ephebeus*	Europe	Original	Ephebe	Roofcap Gill Mushroom	Edible
Pluteus	*ephebeus*	*lepiotoides*	Europe	Parasol	Ephebe	Roofcap Gill Mushroom	Edible
Pluteus	*ephebeus*	*murinus*	Europe, MOTW, page 41	Mouse Gray	Ephebe	Roofcap Gill Mushroom	Edible
Pluteus	*ephebeus*	*pearsonii*	Europe	Pearson	Ephebe	Roofcap Gill Mushroom	Edible
Pluteus	*flavofulgineus*	.	N. & C. USA (on decayed wood in deciduous forest), PFGM, page 250	.	Brownish Yellow	Roofcap Gill Mushroom	Unknown
Pluteus	*magnus*	.	USA, PFGM, page 249	.	Big Deer	Roofcap Gill Mushroom	Unpalatable
Pluteus	*nanus*	.	USA (on decaying wood in deciduous forest), MOTW, page 41, PFGM, page 252	.	Dwarf	Roofcap Gill Mushroom	Unpalatable
Pluteus	*nanus*	*nanus*	USA (on decaying wood in deciduous forest)	Original	Dwarf	Roofcap Gill Mushroom	Unpalatable
Pluteus	*nanus*	*griseopus*	USA (on decaying wood in deciduous forest)	Gray	Dwarf	Roofcap Gill Mushroom	Unpalatable
Pluteus	*petasatus*	.	USA (sawdust piles), PFGM, page 252	.	Sawdust	Roofcap Gill Mushroom	Edible
Pluteus	*petasatus*	*petasatus*	USA (sawdust piles)	Original	Sawdust	Roofcap Gill	Edible

						Mushroom	
Pluteus	petasatus	patricius	USA (sawdust piles), MOTW, page 41	Patrician	Sawdust	Roofcap Gill Mushroom	Edible
Pluteus	plautus	.	USA (on humus under broadleaf trees)	.	Flat Wide	Roofcap Gill Mushroom	Unpalatable
Pluteus	plautus	plautus	USA (on humus under broadleaf trees)	Original	Flat Wide	Roofcap Gill Mushroom	Unpalatable
Pluteus	plautus	alborugosus	USA (on humus under broadleaf trees)	White Wrinkled	Flat Wide	Roofcap Gill Mushroom	Unpalatable
Pluteus	plautus	boudieri	USA (on humus under broadleaf trees)	Boudier	Flat Wide	Roofcap Gill Mushroom	Unpalatable
Pluteus	plautus	depauperatus	USA (on humus under broadleaf trees)	Pauper	Flat Wide	Roofcap Gill Mushroom	Unpalatable
Pluteus	plautus	dryophiloides	USA (on humus under broadleaf trees)	Wood Loving	Flat Wide	Roofcap Gill Mushroom	Unpalatable
Pluteus	plautus	gracilis	USA (on humus under broadleaf trees)	Slender	Flat Wide	Roofcap Gill Mushroom	Unpalatable
Pluteus	plautus	granulatus	USA (on humus under broadleaf trees)	Grainy	Flat Wide	Roofcap Gill Mushroom	Unpalatable
Pluteus	plautus	punctipes	USA (on humus under broadleaf trees)	Dotted	Flat Wide	Roofcap Gill Mushroom	Unpalatable
Pluteus	podospileus	seticeps	USA, PFGM, page 250	.	.	Roofcap Gill Mushroom	Unknown
Pluteus	romellii	.	USA (on decaying wood in deciduous forest)	.	Romell	Roofcap Gill Mushroom	Unknown
Pluteus	romellii	romellii	USA (on decaying wood in deciduous forest)	Original	Romell	Roofcap Gill Mushroom	Unknown
Pluteus	romellii	lutescens	USA (on decaying wood in deciduous forest), PFGM, page 251	Brown	Romell	Roofcap Gill Mushroom	Unknown
Pluteus	romellii	splendidus	USA (on decaying wood in deciduous forest)	Splendid	Romell	Roofcap Gill Mushroom	Unknown
Pluteus	rugosidiscus	.	USA, PFGM, page 248	.	Wrinkled Disc	Roofcap Gill Mushroom	Unknown
Pluteus	salicinus	.	Europe (broadleaf trees), MOTW, page 41	.	Willow	Roofcap Gill Mushroom	Unpalatable
Pluteus	salicinus	salicinus	Europe (broadleaf trees)	Original	Willow	Roofcap Gill Mushroom	Unpalatable
Pluteus	salicinus	beryllus	Europe (broadleaf trees)	Beryl	Willow	Roofcap Gill Mushroom	Unpalatable
Pluteus	salicinus	floccosus	Europe (broadleaf trees)	Fleecy	Willow	Roofcap Gill Mushroom	Unpalatable
Pluteus	tomentosulus	.	USA, PFGM, page 249	.	Fluffy	Roofcap Gill Mushroom	Edible
Pluteus	umbrosus	.	USA, MOTW, page 309	.	Shady	Roofcap Gill Mushroom	Unknown
Pluteus	villosus	.	.	.	.	Roofcap Gill Mushroom	.
Pluteus	washingtonensis	.	USA, PFGM, page 249	.	Washington Deer	Roofcap Gill Mushroom	Edible
Amanita	amici	.	Europe	.	Amici	Amanite Roofcap Gill Mushroom	Suspect
Amanita	atkinsoniana	.	E. USA, PFGM, page 220	.	Atkinson	Amanite Roofcap Gill Mushroom	Unknown
Amanita	bisporigera	.	USA (Oak woods), MOTW, page 28, PFGM, page 239	.	Two Spore Death Angel	Amanite Roofcap Gill Mushroom	Poison
Amanita	brunnescens	.	C. & E. USA (conifer & deciduous forest), MOTW, page 18, PFGM, page 216	.	Cleft Foot Deathcap	Amanite Roofcap Gill Mushroom	Poison
Amanita	brunnescens	brunnescens	C. & E. USA (conifer & deciduous forest)	Original	Cleft Foot Deathcap	Amanite Roofcap Gill Mushroom	Poison
Amanita	brunnescens	pallida	C. & E. USA (conifer & deciduous forest), PFGM, page 217	Pale	Cleft Foot Deathcap	Amanite Roofcap Gill Mushroom	Poison
Amanita	caesarea	.	USA (hardwood forest), C. America, E. Asia (Japan), PFGM, page 217	.	Caesar's	Amanite Roofcap Gill Mushroom	Edible-Excellent
Amanita	caesarea	caesarea	USA (hardwood forest), C. America, E. Asia (Japan)	Original	Caesar's	Amanite Roofcap Gill Mushroom	Edible-Excellent
Amanita	caesarea	aurantia	USA (hardwood forest), C. America, E. Asia (Japan)	Orange	Caesar's	Amanite Roofcap Gill Mushroom	Edible-Excellent
Amanita	calyptrata	.	USA (conifer & mixed conifer-hardwood forest), PFGM, page 218	.	Hooded Grisette	Amanite Roofcap Gill Mushroom	Edible
Amanita	calyptroderma	.	USA (under Oak & Pine), PFGM, page 219	.	Capskin	Amanite Roofcap Gill Mushroom	Edible
Amanita	chlorinosma	.	E. USA (conifer & hardwood forest), PFGM, page 219	.	Chlorine Deathcap	Amanite Roofcap Gill Mushroom	Poison
Amanita	cinereoconia	.	SE. USA (deciduous forest), PFGM, page 220	.	Gray Dust	Amanite Roofcap Gill Mushroom	Edible

Amanita	cinereopannosa	.	USA, PFGM, page 221	.	Sticky Cap	Amanite Roofcap Gill Mushroom	Unknown
Amanita	citrina	.	C. & E. USA (conifer & deciduous forest), MOTW, page 26, PFGM, page 221	.	False Deathcap	Amanite Roofcap Gill Mushroom	Edible
Amanita	citrina	citrina	C. & E. USA (conifer & deciduous forest)	Original	False Deathcap	Amanite Roofcap Gill Mushroom	Edible
Amanita	citrina	alba	C. & E. USA (conifer & deciduous forest)	White	False Deathcap	Amanite Roofcap Gill Mushroom	Edible
Amanita	citrina	bulbosa	C. & E. USA (conifer & deciduous forest)	Bulbous	False Deathcap	Amanite Roofcap Gill Mushroom	Edible
Amanita	citrina	glabra	C. & E. USA (conifer & deciduous forest), MOTW, page 26	Bald	False Deathcap	Amanite Roofcap Gill Mushroom	Edible
Amanita	citrina	mappa	C. & E. USA (conifer & deciduous forest), MOTW, page 26	Map	False Deathcap	Amanite Roofcap Gill Mushroom	Edible
Amanita	cokeri	.	E. USA (woodlands & forest), PFGM, page 222	.	Solitary Deathcap	Amanite Roofcap Gill Mushroom	Poison
Amanita	cothurnata	.	USA, PFGM, page 229	.	Cothurn	Amanite Roofcap Gill Mushroom	Unknown
Amanita	crenulata	.	USA, MOTW, page 309	.	Scalloped	Amanite Roofcap Gill Mushroom	Unknown
Amanita	daucipes	.	C. & E. USA (conifer & mixed conifer-hardwood forest), PFGM, page 223	.	Turnip Blub	Amanite Roofcap Gill Mushroom	Unknown
Amanita	eliae	.	USA, MOTW, page 25	.	Elia	Amanite Roofcap Gill Mushroom	Edible
Amanita	farinosa	.	USA (sandy soil, trails, roads, open woods), PFGM, page 223	.	Mealycap	Amanite Roofcap Gill Mushroom	Unknown
Amanita	flavoconia	.	C. & E. USA (conifer & deciduous forest), MOTW, page 18, PFGM, page 224	.	Yellow Wart Deathcap	Amanite Roofcap Gill Mushroom	Poison
Amanita	flavorubescens	.	C. & E. USA (grassland & forest), PFGM, page 225	.	Yellow Blusher Deathcap	Amanite Roofcap Gill Mushroom	Poison
Amanita	franchetii	.	N. & C. USA (conifer forest), PFGM, page 216	.	Warty Deathcap	Amanite Roofcap Gill Mushroom	Poison
Amanita	frostiana	.	SE. USA, PFGM, page 225	.	Frost's	Amanite Roofcap Gill Mushroom	Edible
Amanita	gemmata	.	USA (conifer & hardwood forest), PFGM, page 225	.	Jeweled Deathcap	Amanite Roofcap Gill Mushroom	Poison
Amanita	gemmata	gemmata	USA (conifer & hardwood forest)	Original	Jeweled Deathcap	Amanite Roofcap Gill Mushroom	Poison
Amanita	gemmata	adnata	USA (conifer & hardwood forest)	Joined	Jeweled Deathcap	Amanite Roofcap Gill Mushroom	Poison
Amanita	gemmata	amici	USA (conifer & hardwood forest), MOTW, page 25	Amici	Jeweled Deathcap	Amanite Roofcap Gill Mushroom	Poison
Amanita	gemmata	exannulata	USA (conifer & hardwood forest)	Ringed	Jeweled Deathcap	Amanite Roofcap Gill Mushroom	Poison
Amanita	gemmata	junquillea	USA (conifer & hardwood forest), MOTW, page 25	Junquille	Jeweled Deathcap	Amanite Roofcap Gill Mushroom	Poison
Amanita	lepiotoides	.	Europe (in woods), MOTW, page 22	.	Scaly Ear	Amanite Roofcap Gill Mushroom	Unknown
Amanita	longipes	.	USA, PFGM, page 220	.	Long Foot	Amanite Roofcap Gill Mushroom	Unknown
Amanita	malleata	.	Europe (grass, meadows, on the edge of Oak woods)	.	Supple	Amanite Roofcap Gill Mushroom	Edible
Amanita	microlepis	.	USA, PFGM, pages 220, 233	.	Small Scale	Amanite Roofcap Gill Mushroom	Unknown
Amanita	muscaria	.	USA (under Birch, Fir, Larch), MOTW, page 19, PFGM, page 227	.	Fly Killer	Amanite Roofcap Gill Mushroom	Poison
Amanita	muscaria	muscaria	USA (under Birch, Fir, Larch)	Original	Fly Killer	Amanite Roofcap Gill Mushroom	Poison
Amanita	muscaria	aureola	USA (under Birch, Fir, Larch), MOTW, page 18	Golden	Fly Killer	Amanite Roofcap Gill Mushroom	Poison
Amanita	muscaria	aureola-puella	USA (under Birch, Fir, Larch)	Girl Golden	Fly Killer	Amanite Roofcap Gill Mushroom	Poison
Amanita	muscaria	circinnata	USA (under Birch, Fir, Larch)	Circular	Fly Killer	Amanite Roofcap Gill Mushroom	Poison
Amanita	muscaria	formosa	USA (under Birch, Fir, Larch), MOTW, page 18	Taiwan	Fly Killer	Amanite Roofcap Gill Mushroom	Poison
Amanita	muscaria	minor	USA (under Birch, Fir, Larch)	Lesser	Fly Killer	Amanite Roofcap Gill Mushroom	Poison
Amanita	muscaria	regalis	USA (under Birch, Fir, Larch), MOTW, page 18	Regal	Fly Killer	Amanite Roofcap Gill Mushroom	Poison

Amanita	*muscaria*	*regalis-emilii*	USA (under Birch, Fir, Larch)	Emil Regal	Fly Killer	Amanite Roofcap Gill Mushroom	Poison
Amanita	*muscaria*	*regalis-umbrina*	USA (under Birch, Fir, Larch)	Dark Regal	Fly Killer	Amanite Roofcap Gill Mushroom	Poison
Amanita	*nitida*	.	E. USA, PFGM, page 228	.	Shiny	Amanite Roofcap Gill Mushroom	Unknown
Amanita	*nivalis*	.	Europe (Alps Mtns.), MOTW, page 30	.	Snowy	Amanite Roofcap Gill Mushroom	Unpalatable
Amanita	*onusta*	.	E. USA (deciduous forest), PFGM, page 228	.	Loaded Deathcap	Amanite Roofcap Gill Mushroom	Poison
Amanita	*ovoidea*	.	S. Europe (calcareous soil), MOTW, page 22	.	Ovalcap	Amanite Roofcap Gill Mushroom	Edible
Amanita	*pachycolea*	.	W. USA, PFGM, page 238	.	Lumpy Deathcap	Amanite Roofcap Gill Mushroom	Poison
Amanita	*pantherina*	.	Canada, N. USA (conifer & mixed forest), MOTW, page 18, PFGM, page 229	.	Panther Deathcap	Amanite Roofcap Gill Mushroom	Poison
Amanita	*pantherina*	*pantherina*	Canada, N. USA (conifer & mixed forest)	Original	Panther Deathcap	Amanite Roofcap Gill Mushroom	Poison
Amanita	*pantherina*	*abietum*	Canada, N. USA (conifer & mixed forest), MOTW, page 18	Mountain Fir	Panther Deathcap	Amanite Roofcap Gill Mushroom	Poison
Amanita	*pantherina*	*robusta*	Canada, N. USA (conifer & mixed forest)	Robust	Panther Deathcap	Amanite Roofcap Gill Mushroom	Poison
Amanita	*parcivolvata*	.	E. USA (deciduous forest), PFGM, page 228	.	Flimsy Veil Deathcap	Amanite Roofcap Gill Mushroom	Poison
Amanita	*peckiana*	.	E. Canada, E. USA (hardwood forest), PFGM, page 230	.	Gluecap Deathcap	Amanite Roofcap Gill Mushroom	Poison
Amanita	*pelioma*	.	S. & E. USA (hardwood forest), PFGM, page 231	.	Olive Deathcap	Amanite Roofcap Gill Mushroom	Poison
Amanita	*phalloides*	.	USA (on soil in forest), MOTW, page 27, PFGM, page 231	.	Deathcap	Amanite Roofcap Gill Mushroom	Poison
Amanita	*phalloides*	*phalloides*	USA (on soil in forest)	Original	Deathcap	Amanite Roofcap Gill Mushroom	Poison
Amanita	*phalloides*	*alba*	USA (on soil in forest), MOTW, page 26	White	Deathcap	Amanite Roofcap Gill Mushroom	Poison
Amanita	*phalloides*	*citrina*	USA (on soil in forest), MOTW, page 26	White	Deathcap	Amanite Roofcap Gill Mushroom	Poison
Amanita	*phalloides*	*umbrina*	USA (on soil in forest), MOTW, page 26	Dark	Deathcap	Amanite Roofcap Gill Mushroom	Poison
Amanita	*phalloides*	*viridis*	USA (on soil in forest)	Green	Deathcap	Amanite Roofcap Gill Mushroom	Poison
Amanita	*polypyramis*	.	E. USA (conifer & hardwood forest), PFGM, page 232	.	Many Warts Deathcap	Amanite Roofcap Gill Mushroom	Poison
Amanita	*ponderosa*	.	Europe, MOTW, page 22	.	Heavy	Amanite Roofcap Gill Mushroom	Unknown
Amanita	*porphyria*	.	Canada, N. USA (conifer & mixed forest), MOTW, page 25, PFGM, page 233	.	Porphyry Deathcap	Amanite Roofcap Gill Mushroom	Poison
Amanita	*porphyria*	*porphyria*	Canada, N. USA (conifer & mixed forest)	Original	Porphyry Deathcap	Amanite Roofcap Gill Mushroom	Poison
Amanita	*porphyria*	*psuedoporphyria*	Europe (conifer & broadleaf forest), MOTW, page 25	Lesser	Porphyry Deathcap	Amanite Roofcap Gill Mushroom	Poison
Amanita	*porphyria*	*recutita*	Canada, N. USA (conifer & mixed forest)	Flayed	Porphyry Deathcap	Amanite Roofcap Gill Mushroom	Poison
Amanita	*ravenelii*	.	SE. USA (deciduous forest), PFGM, page 234	.	Pinecone	Amanite Roofcap Gill Mushroom	Edible
Amanita	*rhopalopus*	.	E. USA (on soil in woodlands), PFGM, page 234	.	Club Foot	Amanite Roofcap Gill Mushroom	Edible
Amanita	*rubescens*	.	USA (hardwood forest), MOTW, page 20, PFGM, page 235	.	Pink Blusher	Amanite Roofcap Gill Mushroom	Edible-Excellent
Amanita	*rubescens*	*rubescens*	USA (hardwood forest)	Original	Pink Blusher	Amanite Roofcap Gill Mushroom	Edible-Excellent
Amanita	*rubescens*	*alba*	USA (hardwood forest)	Whitish	Pink Blusher	Amanite Roofcap Gill Mushroom	Edible-Excellent
Amanita	*rubescens*	*annulosulphurea*	USA (hardwood forest), MOTW, page 20	Sulfur Ring	Pink Blusher	Amanite Roofcap Gill Mushroom	Edible-Excellent
Amanita	*rubescens*	*annulosulphurea-magnifica*	USA (hardwood forest)	Magnificent Sulfur Ring	Pink Blusher	Amanite Roofcap Gill Mushroom	Edible-Excellent
Amanita	*silvicola*	.	W. Canada, NW. USA (conifer forest), PFGM, page 236	.	Woodland	Amanite Roofcap Gill Mushroom	Unknown
Amanita	*smithiana*	.	W. USA (coastal), PFGM, pages 220, 235	.	Smith's	Amanite Roofcap Gill Mushroom	Unknown
Amanita	*spissa*	.	USA (under Oak & Pine), MOTW, page 20	.	Thick	Amanite Roofcap Gill Mushroom	Edible
Amanita	*spissa*	*spissa*	USA (under Oak & Pine)	Original	Thick	Amanite Roofcap Gill Mushroom	Edible

Amanita	*spissa*	*ampla*	USA (under Oak & Pine)	Ample	Thick	Amanite Roofcap Gill Mushroom	Edible
Amanita	*spissa*	*cariosa*	USA (under Oak & Pine)	Crumble	Thick	Amanite Roofcap Gill Mushroom	Edible
Amanita	*spissa*	*excelsa*	USA (under Oak & Pine), MOTW, page 20	Excelsior	Thick	Amanite Roofcap Gill Mushroom	Edible-Poor
Amanita	*spissa*	*valida*	USA (under Oak & Pine), MOTW, page 20	Valid	Thick	Amanite Roofcap Gill Mushroom	Suspect
Amanita	*spreta*	.	USA, PFGM, page 234	.	Spret	Amanite Roofcap Gill Mushroom	Unknown
Amanita	*strobiliformis*	.	USA (meadows, deciduous forest), MOTW, paqge 23	.	Egg Shaped	Amanite Roofcap Gill Mushroom	Edible
Amanita	*strobiliformis*	*strobiliformis*	USA (meadows, deciduous forest)	Original	Egg Shaped	Amanite Roofcap Gill Mushroom	Edible
Amanita	*strobiliformis*	*pellita*	USA (meadows, broadleaf Tree woods), MOTW, page 23	Furry	Egg Shaped	Amanite Roofcap Gill Mushroom	Edible
Amanita	*strobiliformis*	*solitaria*	USA (thin woods of broadleaf Trees, Pine & Oak)	Solitary	Egg Shaped	Amanite Roofcap Gill Mushroom	Edible
Amanita	*tenuifolia*	.	USA, PFGM, page 239	.	Slender Death Angle	Amanite Roofcap Gill Mushroom	Poison
Amanita	*tephrea*	.	E. USA (hardwood forest), PFGM, page 237	.	Ash Gray	Amanite Roofcap Gill Mushroom	Poison
Amanita	*vaginata*	.	Canada, USA (lawns & forests), PFGM, 237	.	Grisette Sheathed	Amanite Roofcap Gill Mushroom	Edible-Good
Amanita	*vaginata*	*vaginata*	Canada, USA (lawns & forests)	Original	Grisette Sheathed	Amanite Roofcap Gill Mushroom	Edible-Good
Amanita	*vaginata*	*alba*	Canada, USA (lawns & forests)	White	Grisette Sheathed	Amanite Roofcap Gill Mushroom	Edible-Good
Amanita	*vaginata*	*alba-fungites*	Canada, USA (lawns & forests)	White Fungus	Grisette Sheathed	Amanite Roofcap Gill Mushroom	Edible-Good
Amanita	*vaginata*	*albida*	Canada, USA (lawns & forests)	Whitish	Grisette Sheathed	Amanite Roofcap Gill Mushroom	Edible-Good
Amanita	*vaginata*	*grisea*	Canada, USA (lawns & forests)	Gray	Grisette Sheathed	Amanite Roofcap Gill Mushroom	Edible-Good
Amanita	*vaginata*	*livida*	Canada, USA (lawns & forests)	Liver Gray	Grisette Sheathed	Amanite Roofcap Gill Mushroom	Edible-Good
Amanita	*vaginata*	*plumbea*	Canada, USA (lawns & forests)	Lead Gray	Grisette Sheathed	Amanite Roofcap Gill Mushroom	Edible-Good
Amanita	*vaginata*	*violacea*	Canada, USA (lawns & forests)	Violet	Grisette Sheathed	Amanite Roofcap Gill Mushroom	Edible-Good
Amanita	*verna*	.	USA (conifer & hardwood forest), PFGM, page 238	.	Fool's Deathcap	Amanite Roofcap Gill Mushroom	Poison
Amanita	*verna*	*verna*	USA (conifer & hardwood forest)	Original	Fool's Deathcap	Amanite Roofcap Gill Mushroom	Poison
Amanita	*verna*	*virosa*	USA (conifer & hardwood forest)	Fetid	Fool's Deathcap	Amanite Roofcap Gill Mushroom	Poison
Amanita	*vernalis*		USA (conifer & hardwood forest), MOTW, page 25	.	Springtime	Amanite Roofcap Gill Mushroom	Edible
Amanita	*virosa*	.	USA (conifer & deciduous forest), MOTW, page 28, PFGM, pages 222, 231, 239	.	Destroying Angel	Amanite Roofcap Gill Mushroom	Poison
Amanitopsis	*battarrae*	.	Europe (meadows, Fir conifer forest)	.	Battarra	Lesser Amanite Roofcap Gill Mushroom	Edible-Good
Amanitopsis	*battarrae*	*battarrae*	Europe (meadows, Fir conifer forest)	Original	Battarra	Lesser Amanite Roofcap Gill Mushroom	Edible-Good
Amanitopsis	*battarrae*	*umbrinolutea*	Europe (meadows, Fir conifer forest), MOTW, page 32, PFGM, page 227	Dark Yellow	Battarra	Lesser Amanite Roofcap Gill Mushroom	Edible-Good
Amanitopsis	*ceciliae*	.	USA (grassland & forest)	.	Cecilia	Lesser Amanite Roofcap Gill Mushroom	Edible-Fair
Amanitopsis	*ceciliae*	*ceciliae*	USA (grassland & forest)	Original	Cecilia	Lesser Amanite Roofcap Gill Mushroom	Edible-Fair
Amanitopsis	*ceciliae*	*inaurata*	USA (grassland & forest), MOTW, page 30, PFGM, page 226	Gilded	Cecilia	Lesser Amanite Roofcap Gill Mushroom	Edible-Fair
Amanitopsis	*ceciliae*	*strangulata*	USA (grassland & forest), MOTW, page 30	Strangulat ed	Cecilia	Lesser Amanite Roofcap Gill Mushroom	Edible-Fair
Amanitopsis	*fulva*	.	USA, PFGM, page 238	.	.	Lesser Amanite Roofcap Gill Mushroom	Unknown
Amanitopsis	*lividopallescens*	.	USA (edge of woods, along forest paths), MOTW, page 30	.	Pale Livid	Lesser Amanite Roofcap Gill Mushroom	Edible-Excellent
Amanitopsis	*nivalis*	.	UK (Scotland), MOTW, page 30	.	Snowy	Lesser Amanite Roofcap Gill Mushroom	Edible
Amanitopsis	*oriena*	.	Europe (Alps Mtns.), MOTW, page 30	.	Highlander	Lesser Amanite Roofcap Gill Mushroom	Unpalatable
Amanitopsis	*volvata*	.	USA, MOTW, page 309	.	Volvate	Lesser Amanite Roofcap Gill Mushroom	Unknown

Aspidella	*echinocephala*	.	Europe (dry calcareous soil, edge of woods), MOTW, page 22, PFGM, page 223	.	Prickly Cap	Rough Roofcap Gill Mushroom	Edible-Good
Aspidella	*echinocephala*	*echinocephala*	Europe (dry calcareous soil, edge of woods)	Original	Prickly Cap	Rough Roofcap Gill Mushroom	Edible-Good
Aspidella	*echinocephala*	*aculeata*	Europe (dry calcareous soil, edge of woods)	Prickly	Prickly Cap	Rough Roofcap Gill Mushroom	Edible-Good
Aspidella	*echinocephala*	*solitaria*	Europe (dry calcareous soil, edge of woods), PFGM, page 223	Solitary	Prickly Cap	Rough Roofcap Gill Mushroom	Edible-Good
Aspidella	*echinocephala*	*vittadinii*	S. Europe (dry calcareous soil, edge of woods), MOTW, page 22	Vittadini	Prickly Cap	Rough Roofcap Gill Mushroom	Edible-Good
Volvariella	*bombycina*	.	USA (on dead logs & living tree trunks of deciduous trees), MOTW, page 41, PFGM, page 252	.	Silky	Sheath Roofcap Gill Mushroom	Edible-Excellent
Volvariella	*gloiocephala*	.	USA (gardens, fields, open woods), MOTW, page 40	.	Glowhead	Sheath Roofcap Gill Mushroom	Edible-Poor
Volvariella	*gloiocephala*	*gloiocephala*	USA (gardens, fields, open woods)	Original	Glowhead	Sheath Roofcap Gill Mushroom	Edible-Poor
Volvariella	*gloiocephala*	*speciosa*	USA (gardens, fields, open woods), PFGM, page 253	Big Sheath	Glowhead	Sheath Roofcap Gill Mushroom	Edible-Poor
Volvariella	*hypopithys*	.	Europe	.	Ape	Sheath Roofcap Gill Mushroom	Unpalatable
Volvariella	*hypopithys*	*hypopithys*	Europe	Original	Ape	Sheath Roofcap Gill Mushroom	Unpalatable
Volvariella	*hypopithys*	*biloba*	Europe	Two Lobe	Ape	Sheath Roofcap Gill Mushroom	Unpalatable
Volvariella	*hypopithys*	*plumulosa*	Europe, MOTW, page 41	Plume	Ape	Sheath Roofcap Gill Mushroom	Unpalatable
Volvariella	*hypopithys*	*pubescentipes*	Europe	Downy	Ape	Sheath Roofcap Gill Mushroom	Unpalatable
Volvariella	*media*	.	Europe, MOTW, page 41	.	Medium	Sheath Roofcap Gill Mushroom	Edible
Volvariella	*murinella*	.	Europe (rotten wood, moss, conifer forest), MOTW, page 41	.	Mouse Gray	Sheath Roofcap Gill Mushroom	Unpalatable
Volvariella	*pubipes*	.	Europe, MOTW, page 41	.	Downy	Sheath Roofcap Gill Mushroom	Unpalatable
Volvariella	*pusilla*	.	Europe (gardens, paths), MOTW, page 41	.	Puny	Sheath Roofcap Gill Mushroom	Unpalatable
Volvariella	*pusilla*	*pusilla*	Europe (gardens, paths)	Original	Puny	Sheath Roofcap Gill Mushroom	Unpalatable
Volvariella	*pusilla*	*parvula*	Europe (gardens, paths)	Small	Puny	Sheath Roofcap Gill Mushroom	Unpalatable
Volvariella	*surrecta*	.	Europe (on Clubfoot Funnelcap & Cloudy Funnelcap Cavallier mushrooms, on decomposing Cavallier mushrooms)	.	Surrect	Sheath Roofcap Gill Mushroom	Unpalatable
Volvariella	*surrecta*	*surrecta*	Europe (on Clubfoot Funnelcap & Cloudy Funnelcap Cavallier mushrooms, on decomposing Cavallier mushrooms)	Original	Surrect	Sheath Roofcap Gill Mushroom	Unpalatable
Volvariella	*surrecta*	*loveiana*	Europe (on Clubfoot Funnelcap & Cloudy Funnelcap Cavallier mushrooms, on decomposing Cavallier mushrooms)	Love's	Surrect	Sheath Roofcap Gill Mushroom	Unpalatable
Volvariella	*volvacea*	.	USA (on rotting wood & in greenhouses), PFGM, page 253	.	Paddy Straw	Sheath Roofcap Gill Mushroom	Edible
Volvariella	*volvacea*	*volvacea*	USA (on rotting wood & in greenhouses)	Original	Paddy Straw	Sheath Roofcap Gill Mushroom	Edible
Volvariella	*volvacea*	*fukurotakei*	E. Asia (Japan), USA (on rotting wood & in greenhouses)	Fukurotake	Paddy Straw	Sheath Roofcap Gill Mushroom	Edible
Volvariella	*volvacea*	*heimii*	USA (on rotting wood & in greenhouses)	Heim	Paddy Straw	Sheath Roofcap Gill Mushroom	Edible
Volvariella	*volvacea*	*laosensis*	SE. Asia (Laos), USA (on rotting wood & in greenhouses)	Laotian	Paddy Straw	Sheath Roofcap Gill Mushroom	Edible
Volvariella	*volvacea*	*philippinensis*	Pac. Is. (Philippines), USA (on rotting wood & in greenhouses)	Philippine	Paddy Straw	Sheath Roofcap Gill Mushroom	Edible

Family 622.201.14 *Psathyrellaceae* Crumblecap Gill Mushroom

Genus	Species	Variety	Native Location	Variety	Specimen	Genome	Habit
Psathyrella	*candolleana*	.	Canada, USA (on stumps & deadwood in hardwood forest), MOTW, page 52, PFGM, page 281	.	Fringed	Crumblecap Gill Mushroom	Edible-Good
Psathyrella	*candolleana*	*candolleana*	Canada, USA (on stumps &	Original	Fringed	Crumblecap Gill	Edible-Good

			deadwood in hardwood forest)			Mushroom	
Psathyrella	*candolleana*	*appendiculata*	Canada, USA (on stumps & deadwood in hardwood forest)	Appendage	Fringed	Crumblecap Gill Mushroom	Edible-Good
Psathyrella	*candolleana*	*egenula*	Canada, USA (on stumps & deadwood in hardwood forest)	Egen	Fringed	Crumblecap Gill Mushroom	Edible-Good
Psathyrella	*candolleana*	*microlepidota*	Canada, USA (on stumps & deadwood in hardwood forest)	Small Scale	Fringed	Crumblecap Gill Mushroom	Edible-Good
Psathyrella	*castaneifolia*	.	USA, PFGM, page 283	.	Chestnut	Crumblecap Gill Mushroom	Unknown
Psathyrella	*dilineata*	.	USA (on deadwood in hardwood forest), PFGM, page 282	.	Artist	Crumblecap Gill Mushroom	Unknown
Psathyrella	*gracilis*	.	S. Canada, N. USA (on moist soil in bogs & swamps), PFGM, page 283	.	Thimblecap	Crumblecap Gill Mushroom	Unknown
Psathyrella	*piluliformis*	.	USA (on decaying wood in hardwood forest)	.	Bracket	Crumblecap Gill Mushroom	Edible
Psathyrella	*piluliformis*	*piluliformis*	USA (on decaying wood in hardwood forest)	Original	Bracket	Crumblecap Gill Mushroom	Edible
Psathyrella	*piluliformis*	*hydrophila*	USA (on decaying wood in hardwood forest), MOTW, page 52, PFGM, page 283	Cluster	Bracket	Crumblecap Gill Mushroom	Edible
Psathyrella	*piluliformis*	*subpapillata*	USA (on decaying wood in hardwood forest)	Papilla	Bracket	Crumblecap Gill Mushroom	Edible
Psathyrella	*prona*		USA	.	Pron	Crumblecap Gill Mushroom	Unknown
Psathyrella	*prona*	*prona*	USA	Original	Pron	Crumblecap Gill Mushroom	Unknown
Psathyrella	*prona*	*albidula*	USA	White	Pron	Crumblecap Gill Mushroom	Unknown
Psathyrella	*prona*	*cana*	USA	Hoary	Pron	Crumblecap Gill Mushroom	Unknown
Psathyrella	*prona*	*cana-atomata*	USA, PFGM, page 283	Atomic	Pron	Crumblecap Gill Mushroom	Unknown
Psathyrella	*prona*	*orbitarum*	USA	Orb	Pron	Crumblecap Gill Mushroom	Unknown
Psathyrella	*prona*	*subatomata*	USA	Lesser Atomic	Pron	Crumblecap Gill Mushroom	Unknown
Psathyrella	*rigidipes*	.	USA, PFGM, page 285	.	Stiff	Crumblecap Gill Mushroom	Unknown
Psathyrella	*rugocephala*	.	USA, PFGM, page 285	.	Wrinked Head	Crumblecap Gill Mushroom	Unknown
Psathyrella	*sarcocephala*	.	Europe (on tree trunks), MOTW, page 52	.	Flesh Head	Crumblecap Gill Mushroom	Edible
Psathyrella	*spadicea*	.	Europe (on tree trunks), MOTW, page 52	.	Spade	Crumblecap Gill Mushroom	Unpalatable
Psathyrella	*ulignicola*	.	W. USA (on moist soil in Rocky Mtns.), PFGM, page 284	.	Aspen	Crumblecap Gill Mushroom	Unknown
Lacrymaria	*lacrymabunda*	.	USA (on gasslands, compost heaps, sawdust piles), MOTW, page 53	.	Weeping	Teardrop Crumblecap Gill Mushroom	Edible-Fair
Lacrymaria	*lacrymabunda*	*lacrymabunda*	USA (on gasslands, compost heaps, sawdust piles)	Original	Weeping	Teardrop Crumblecap Gill Mushroom	Edible-Fair
Lacrymaria	*lacrymabunda*	*gracillima*	USA (on gasslands, compost heaps, sawdust piles)	Slender	Weeping	Teardrop Crumblecap Gill Mushroom	Edible-Fair
Lacrymaria	*lacrymabunda*	*velutina*	USA (on gasslands, compost heaps, sawdust piles), MOTW, Page 53, PFGM, page 285	Weeping Widow	Weeping	Teardrop Crumblecap Gill Mushroom	Edible-Fair

Family 622.201.15 *Pterulaceae* Wingcap Gill Mushroom

Genus	Species	Variety	Native Location	Variety	Specimen	Genome	Habit
Pterula	*bambusae*	.		.	Bamboo	Wingcap Gill Mushroom	Unknown

Family 622.201.16 *Schizophyllaceae* Splitgill Gill Mushroom

Genus	Species	Variety	Native Location	Variety	Specimen	Genome	Habit
Schizophyllum	*commune*	.	USA (on decaying hardwood & conifer) PFGM, page 80	.	Hardwood & Conifer Forest	Splitgill Gill Mushroom	Inedible

Family 622.201.17 *Strophariaceae* Ringstalk Gill Mushroom

Genus	Species	Variety	Native Location	Variety	Specimen	Genome	Habit
Stropharia	*aeruginosa*	.	N. USA (on forest litter), MOTW, page 55, PFGM, page 262	.	Verdigris	Ringstalk Gill Mushroom	Edible-Poor
Stropharia	*aeruginosa*	*aeruginosa*	N. USA (on forest litter)	Original	Verdigris	Ringstalk Gill	Edible-Poor

						Mushroom	
Stropharia	aeruginosa	albocyanea	N. USA (on forest litter), MOTW, page 55	White Cyan	Verdigris	Ringstalk Gill Mushroom	Edible-Poor
Stropharia	aeruginosa	albocyanea-worthingtonii	N. USA (on forest litter)	Worthington White Cyan	Verdigris	Ringstalk Gill Mushroom	Edible-Poor
Stropharia	ambigua	.	W. & N. USA (mixed conifer-hardwood forest), MOTW, page 55, PFGM, page 263	.	Fringed	Ringstalk Gill Mushroom	Edible-Poor
Stropharia	coronilla	.	USA (pastures, lawns & grasslands), MOTW, page 54, PFGM, page 263	.	Crown Toadstool	Ringstalk Gill Mushroom	Poison
Stropharia	coronilla	coronilla	USA (pastures, lawns & grasslands)	Original	Crown Toadstool	Ringstalk Gill Mushroom	Poison
Stropharia	coronilla	obdurata	USA (pastures, lawns & grasslands)	Obdurate	Crown Toadstool	Ringstalk Gill Mushroom	Poison
Stropharia	hardii	.	S. & E. USA, PFGM, pages 264, 305	.	Lesser Crown Toadstool	Ringstalk Gill Mushroom	.
Stropharia	hornemannii	.	Canada, USA (on decaying logs in conifer forest), MOTW, page 55, PFGM, page 265	.	Luxuriant	Ringstalk Gill Mushroom	Inedible
Stropharia	hornemannii	hornemannii	Canada, USA (on decaying logs in conifer forest)	Original	Luxuriant	Ringstalk Gill Mushroom	Inedible
Stropharia	hornemannii	depilata	Canada, USA (on decaying logs in conifer forest)	Depilitate	Luxuriant	Ringstalk Gill Mushroom	Inedible
Stropharia	melanosperma	.	Europe (manured ground), MOTW, page 54	.	Dark Spore	Ringstalk Gill Mushroom	Inedible
Stropharia	rugosoannulata	.	USA (cultivated soil, open woodlands), PFGM, page 265	.	Rough Ring	Ringstalk Gill Mushroom	Edible-Good
Stropharia	rugosoannulata	rugosoannulata	USA (cultivated soil, open woodlands)	Original	Rough Ring	Ringstalk Gill Mushroom	Edible-Good
Stropharia	rugosoannulata	ferrii	USA (cultivated soil, open woodlands), MOTW, page 55	Ferri	Rough Ring	Ringstalk Gill Mushroom	Edible-Good
Stropharia	rugosoannulata	garden-giantii	USA (cultivated soil, open woodlands)	Garden Giant	Rough Ring	Ringstalk Gill Mushroom	Edible-Good
Stropharia	rugosoannulata	wine-redii	USA (cultivated soil, open woodlands)	Wine Red	Rough Ring	Ringstalk Gill Mushroom	Edible-Good
Stropharia	semiglobata	.	USA (on domestic animal dung), MOTW, page 54, PFGM, page 266	.	Manure Roundhead	Ringstalk Gill Mushroom	Edible-Poor
Stropharia	semiglobata	semiglobata	USA (on domestic animal dung)	Original	Manure Roundhead	Ringstalk Gill Mushroom	Edible-Poor
Stropharia	semiglobata	stercoraria	USA (on domestic animal dung)	Dung	Manure Roundhead	Ringstalk Gill Mushroom	Edible-Poor
Stropharia	squamosa	.	Canada, N. USA (on soil in various habitats), MOTW, page 55, PFGM, page 268	.	Scaly	Ringstalk Gill Mushroom	Edible-Poor
Stropharia	thrausta	.	USA, PFGM, page 268	.	Pumpkin	Ringstalk Gill Mushroom	Edible-Poor
Stropharia	umbonatescens	.	Europe (mountains), MOTW, page 54	.	Penis	Ringstalk Gill Mushroom	Inedible
Hypholoma	capnoides	.	W. Canada, W. USA (on living conifer wood), MOTW, page 56	.	Orange Stump	Fringed Margin Ringstalk Gill Mushroom	Edible
Hypholoma	epixanthum	.	Europe (mountain Pine & Beech stumps), MOTW, page 56	.	Blonde	Fringed Margin Ringstalk Gill Mushroom	Inedible
Hypholoma	fasciculare	.	USA (on logs & stumps in conifer & hardwood forest), MOTW, page 55	.	Sulfur Tuft Deathcap	Fringed Margin Ringstalk Gill Mushroom	Poison
Hypholoma	fasciculare	fasciculare	USA (on logs & stumps in conifer & hardwood forest)	Original	Sulfur Tuft Deathcap	Fringed Margin Ringstalk Gill Mushroom	Poison
Hypholoma	fasciculare	pusillum	USA (on logs & stumps in conifer & hardwood forest)	Puny	Sulfur Tuft Deathcap	Fringed Margin Ringstalk Gill Mushroom	Poison
Hypholoma	lateritium	.	USA (on logs & stumps in hardwood forest), MOTW, page 56	.	Brickcap	Fringed Margin Ringstalk Gill Mushroom	Edible-Poor
Hypholoma	lateritium	lateritium	USA (on logs & stumps in hardwood forest), MOTW, page 56	Original	Brickcap	Fringed Margin Ringstalk Gill Mushroom	Edible-Poor
Hypholoma	lateritium	permagnum	USA (on logs & stumps in hardwood forest)	Magnum	Brickcap	Fringed Margin Ringstalk Gill Mushroom	Edible-Poor
Hypholoma	lateritium	pomposum	USA (on logs & stumps in hardwood forest)	Pompous	Brickcap	Fringed Margin Ringstalk Gill Mushroom	Edible-Poor
Hypholoma	lateritium	schaefferi	USA (on logs & stumps in hardwood forest)	Schaeffer	Brickcap	Fringed Margin Ringstalk Gill Mushroom	Edible-Poor
Hypholoma	lateritium	squamosum	USA (on logs & stumps in hardwood forest)	Scaley	Brickcap	Fringed Margin Ringstalk Gill Mushroom	Edible-Poor
Hypholoma	lateritium	vulgaris	USA (on logs & stumps in hardwood forest)	Ordinary	Brickcap	Fringed Margin Ringstalk Gill Mushroom	Edible-Poor
Hypholoma	radicosum	.	Europe (on Pine stumps),	.	Many Roots	Fringed Margin Ringstalk	Inedible

			MOTW, page 56			Gill Mushroom	
Kuehneromyces	*mutabilis*	.	NW. USA (on decaying logs & stumps), MOTW, page 61, PFGM, page 272	.	Two Toned	Kuehner Ringstalk Gill Mushroom	Edible-Good
Pholiota	*abietis*	.	USA (on Fir Logs), PFGM, page 271	.	Fir Log	Scalecap Ringstalk Gill Mushroom	Unknown
Pholiota	*adiposa*	.	Europe (on Beech trees), MOTW, page 58, PFGM, page 271	.	Fat	Scalecap Ringstalk Gill Mushroom	Edible-Fair
Pholiota	*alnicola*	.	Europe (on Alder & Birch trees), MOTW, page 59	.	Alder	Scalecap Ringstalk Gill Mushroom	Unpalatable
Pholiota	*alnicola*	*alnicola*	Europe (on Alder & Birch trees)	Original	Alder	Scalecap Ringstalk Gill Mushroom	Unpalatable
Pholiota	*alnicola*	*apicrea*	Europe (on Alder & Birch trees)	Mild	Alder	Scalecap Ringstalk Gill Mushroom	Unpalatable
Pholiota	*alnicola*	*aromatica*	Europe (on Alder & Birch trees)	Aromatic	Alder	Scalecap Ringstalk Gill Mushroom	Unpalatable
Pholiota	*alnicola*	*salicicola*	Europe (on Alder & Birch trees)	Willow	Alder	Scalecap Ringstalk Gill Mushroom	Unpalatable
Pholiota	*alnicola*	*salicicola-amara*	Europe (on Alder & Birch trees)	Bitter Willow	Alder	Scalecap Ringstalk Gill Mushroom	Unpalatable
Pholiota	*astragalina*	.	Europe (mountain conifer forest), MOTW, page 59	.	Goldfinch	Scalecap Ringstalk Gill Mushroom	Unpalatable
Pholiota	*aurivella*	.	USA (on dead logs & living tree trunks of hardwood trees), MOTW, page 58, PFGM, page 270	.	Goldskin	Scalecap Ringstalk Gill Mushroom	Edible-Poor
Pholiota	*aurivella*	*aurivella*	USA (on dead logs & living tree trunks of hardwood trees)	Original	Goldskin	Scalecap Ringstalk Gill Mushroom	Edible-Poor
Pholiota	*aurivella*	*lilacifolia*	USA (on dead logs & living tree trunks of hardwood trees)	Lilac	Goldskin	Scalecap Ringstalk Gill Mushroom	Edible-Poor
Pholiota	*flammans*	.	S. Canada, N. USA (on decayed wood in conifer forest), MOTW, page 59, PFGM, page 272	.	Flame	Scalecap Ringstalk Gill Mushroom	Unpalatable
Pholiota	*fulvozonata*	.	USA, PFGM, page 268	.	Dark Zoned	Scalecap Ringstalk Gill Mushroom	Unknown
Pholiota	*hiemalis*	.	USA (on logs of Fir trees), PFGM, page 271	.	Winter Deathcap	Scalecap Ringstalk Gill Mushroom	Poison
Pholiota	*highlandensis*	.	W. USA (on burned wood & charred soil), PFGM, page 272	.	Highland	Scalecap Ringstalk Gill Mushroom	Unknown
Pholiota	*highlandensis*	*highlandensis*	W. USA (on burned wood & charred soil)	Original	Highland	Scalecap Ringstalk Gill Mushroom	Unknown
Pholiota	*highlandensis*	*carbonaria*	W. USA (on burned wood & charred soil), MOTW, page 59, PFGM, page 271	Charcoal	Highland	Scalecap Ringstalk Gill Mushroom	Unknown
Pholiota	*highlandensis*	*gigantea*	W. USA (on burned wood & charred soil)	Giant	Highland	Scalecap Ringstalk Gill Mushroom	Unknown
Pholiota	*highlandensis*	*persicina*	W. USA (on burned wood & charred soil)	Persian	Highland	Scalecap Ringstalk Gill Mushroom	Unknown
Pholiota	*lucifera*	.	Europe (on woody debris), MOTW, page 59	.	Glossy	Scalecap Ringstalk Gill Mushroom	Unpalatable
Pholiota	*marginella*	.	USA, PFGM, page 272	.	Margined	Scalecap Ringstalk Gill Mushroom	Unknown
Pholiota	*nameko*	.	USA, GSI-6, page 258	.	Nameko	Scalecap Ringstalk Gill Mushroom	Edible
Pholiota	*populnea*	.	Europe (on fallen Poplar & fallen broadleaf trees)	.	Poplar	Scalecap Ringstalk Gill Mushroom	Inedible
Pholiota	*populnea*	*comosa*	Europe (on fallen Poplar & fallen broadleaf trees)	Original	Poplar	Scalecap Ringstalk Gill Mushroom	Inedible
Pholiota	*populnea*	*destruens*	Europe (on fallen Poplar & fallen broadleaf trees), MOTW, page 61	Destroyer	Poplar	Scalecap Ringstalk Gill Mushroom	Inedible
Pholiota	*squarrosa*	.	USA (on decaying wood), MOTW, page 58, PFGM, page 272	.	Shaggy	Scalecap Ringstalk Gill Mushroom	Edible-Toxic
Pholiota	*squarrosa*	*squarrosa*	USA (on decaying wood)	Original	Shaggy	Scalecap Ringstalk Gill Mushroom	Edible-Toxic
Pholiota	*squarrosa*	*verruculosa*	USA (on decaying wood)	Verrucous	Shaggy	Scalecap Ringstalk Gill Mushroom	Edible-Toxic
Pholiota	*squarrosoides*	.	S. Canada, N. USA (on wood in hardwood forest), MOTW, page 58, PFGM, page 273	.	Sharp Scales	Scalecap Ringstalk Gill Mushroom	Edible
Pholiota	*versi*	.	C. & E. USA (on logs & debris in hardwood forest), PFGM, page 273	.	Straw	Scalecap Ringstalk Gill Mushroom	Unknown
Pholiota	*vernalis*	.	S. Canada, N. USA (on decaying wood), PFGM, page 273	.	Spring	Scalecap Ringstalk Gill Mushroom	Edible

Family 622.201.18 *Tricholomataceae* Cavalier Gill Mushroom

Genus	Species	Variety	Native Location	Variety	Specimen	Genome	Habit
Tricholoma	*acerbum*	.	Europe (Beech & Oak broadleaf forest), MOTW, page 97	.	Bitter	Cavalier Gill Mushroom	Edible-Bitter
Tricholoma	*albobrunneum*	.	Europe (mountain conifer forest), MOTW, page 94, PFGM, page 185	.	White & Brown	Cavalier Gill Mushroom	Edible-Toxic
Tricholoma	*album*	.	USA (Beech & Birch woods), MOTW, page 102	.	White	Cavalier Gill Mushroom	Inedible
Tricholoma	*album*	*album*	USA (Beech & Birch woods)	.	White	Cavalier Gill Mushroom	Inedible
Tricholoma	*album*	*raphanicum*	USA (Beech & Birch woods)	.	White	Cavalier Gill Mushroom	Inedible
Tricholoma	*argyraceum*	.	Europe, MOTW, page 98	.	Silver	Cavalier Gill Mushroom	Edible
Tricholoma	*argyraceum*	*argyraceum*	Europe	Original	Silver	Cavalier Gill Mushroom	Edible
Tricholoma	*argyraceum*	*inocybeoides*	Europe	Fiber Head	Silver	Cavalier Gill Mushroom	Edible
Tricholoma	*atrosquamosum*	.	Europe (humid woods), MOTW, page 100	.	Dark Squamule	Cavalier Gill Mushroom	Edible
Tricholoma	*atrosquamosum*	*atrosquamosum*	Europe (humid woods)	Original	Dark Squamule	Cavalier Gill Mushroom	Edible
Tricholoma	*atrosquamosum*	*nigromarginatum*	Europe (humid woods)	Black Margined	Dark Squamule	Cavalier Gill Mushroom	Edible
Tricholoma	*atroviolaceum*	.	USA, PFGM, pages 189, 193	.	Dark Violet	Cavalier Gill Mushroom	Unknown
Tricholoma	*aurantium*	.	S. Canada, N. USA (conifer & mixed forest), MOTW, page 94, PFGM, page 185	.	Golden	Cavalier Gill Mushroom	Unpalatable
Tricholoma	*batschii*	.	USA (conifer forest)	.	Batsch	Cavalier Gill Mushroom	Edible-Bitter
Tricholoma	*batschii*	*batschii*	USA (conifer forest)	Original	Batsch	Cavalier Gill Mushroom	Edible-Bitter
Tricholoma	*batschii*	*subannulatum*	USA (conifer forest), MOTW, page 93	Ringed	Batsch	Cavalier Gill Mushroom	Edible-Bitter
Tricholoma	*bufonium*	.	Europe, MOTW, page 104	.	Bufon	Cavalier Gill Mushroom	Inedible
Tricholoma	*caligatum*	.	USA (conifer, hardwood & mixed forest), PFGM, page 134	.	Misty Matsutake	Cavalier Gill Mushroom	Edible
Tricholoma	*cingulatum*	.	E. Asia (Willow humid woods, Japan), MOTW, page 100	.	Cing Matsutake	Cavalier Gill Mushroom	Edible-Good
Tricholoma	*colossus*	.	Europe (conifer forest), MOTW, page 104	.	Colossal	Cavalier Gill Mushroom	Edible
Tricholoma	*columbetta*	.	USA (in woods), MOTW, page 102	.	Small Dove	Cavalier Gill Mushroom	Edible-Good
Tricholoma	*equestre*	.	USA (conifer & hardwood forest), MOTW, page 100	.	Man On Horseback	Cavalier Gill Mushroom	Edible-Good
Tricholoma	*equestre*	*equestre*	USA (Pine conifer & hardwood forest)	Original	Man On Horseback	Cavalier Gill Mushroom	Edible-Good
Tricholoma	*equestre*	*auratum*	USA (Pine conifer & hardwood forest)	Eared	Man On Horseback	Cavalier Gill Mushroom	Edible-Good
Tricholoma	*equestre*	*flavovirens*	USA (Pine conifer & hardwood forest), PFGM, page 186	Yellow Green	Man On Horseback	Cavalier Gill Mushroom	Edible-Good
Tricholoma	*equestre*	*populinum*	USA (Pine conifer & hardwood forest)	Poplar	Man On Horseback	Cavalier Gill Mushroom	Edible-Good
Tricholoma	*fulvum*	.	S. Canada, N. USA (conifer forest)	.	Brownstain	Cavalier Gill Mushroom	Edible-Toxic
Tricholoma	*fulvum*	*fulvum*	S. Canada, N. USA (conifer forest)	Original	Brownstain	Cavalier Gill Mushroom	Edible-Toxic
Tricholoma	*fulvum*	*flavobrunneum*	S. Canada, N. USA (under Oak & Birch), MOTW, page 94, PFGM, page 185	Yellow Brown	Brownstain	Cavalier Gill Mushroom	Edible-Toxic
Tricholoma	*fulvum*	*nictitans*	S. Canada, N. USA (conifer forest)	Winking	Brownstain	Cavalier Gill Mushroom	Edible-Toxic
Tricholoma	*goniospermum*	.	Europe (mountain meadows & mountain woods), MOTW, page 105	.	Angled Spore	Cavalier Gill Mushroom	Edible
Tricholoma	*groanense*	.	Europe (heaths & moors of Lombardy), MOTW, page 98	.	Moor Deathcap	Cavalier Gill Mushroom	Poison
Tricholoma	*hordum*	.	Europe, MOTW, page 98	.	Swollen	Cavalier Gill Mushroom	Inedible
Tricholoma	*imbricatum*	.	N. USA (Pine & Larch conifer forest), MOTW, page 100, PFGM, page 187	.	Shingle Head	Cavalier Gill Mushroom	Edible

Tricholoma	*inamoenum*	.	W. USA (Rocky Mtns. to pacific coast in conifer forest), MOTW, page 104, PFGM, page 188	.	Irksome Deathcap	Cavalier Gill Mushroom	Poison
Tricholoma	*inamoenum*	*inamoenum*	W. USA (Rocky Mtns. to pacific coast in conifer forest)	Original	Irksome Deathcap	Cavalier Gill Mushroom	Poison
Tricholoma	*inamoenum*	*insigne*	W. USA (Rocky Mtns. to pacific coast in conifer forest)	Insignia	Irksome Deathcap	Cavalier Gill Mushroom	Poison
Tricholoma	*intermedia*	.	USA, PFGM, page 187	.	Intermediate	Cavalier Gill Mushroom	Unknown
Tricholoma	*lascivum*	.	USA (Beech & Birch woods), MOTW, page 102	.	Wanton	Cavalier Gill Mushroom	Edible
Tricholoma	*lascivum*	*lascivum*	USA (Beech & Birch woods)	Original	Wanton	Cavalier Gill Mushroom	Edible
Tricholoma	*lascivum*	*robustum*	USA (Beech & Birch woods)	Robust	Wanton	Cavalier Gill Mushroom	Edible
Tricholoma	*lavidanum*	.	Europe (conifer forest), MOTW, page 102	.	Soapy	Cavalier Gill Mushroom	Inedible
Tricholoma	*leucophyllum*	.	USA, PFGM, pages 187, 188	.	Light	Cavalier Gill Mushroom	Unknown
Tricholoma	*magnivelare*	.	Canada, USA (mountain conifer forest)	.	Matsutake	Cavalier Gill Mushroom	Edible
Tricholoma	*magnivelare*	*magnivelare*	Canada, USA (mountain conifer forest)	Original	Matsutake	Cavalier Gill Mushroom	Edible
Tricholoma	*magnivelare*	*arenicola*	Canada, USA (mountain conifer forest), PFGM, page 211	Sand Dwelling	Matsutake	Cavalier Gill Mushroom	Edible
Tricholoma	*magnivelare*	*ponderosum*	Canada, USA (mountain conifer forest), MOTW, page 93, PFGM, page 135	White	Matsutake	Cavalier Gill Mushroom	Edible
Tricholoma	*magnivelare*	*murrillianum*	Canada, USA (mountain conifer forest)	Murrill	Matsutake	Cavalier Gill Mushroom	Edible
Tricholoma	*matsutake*	.	E. Asia (Red Pine forest in Japan)	.	Oriental Matsutake	Cavalier Gill Mushroom	Edible
Tricholoma	*montanum*	.	USA (conifer forest), MOTW, page 97	.	Mountain Deathcap	Cavalier Gill Mushroom	Poison
Tricholoma	*myomyces*	.	USA (conifer forest), PFGM, page 189	.	Waxygill	Cavalier Gill Mushroom	Edible-Good
Tricholoma	*myomyces*	*myomyces*	USA (conifer forest)	Original	Waxygill	Cavalier Gill Mushroom	Edible-Good
Tricholoma	*myomyces*	*bisporigerum*	USA (conifer forest)	Double Spore	Waxygill	Cavalier Gill Mushroom	Edible-Good
Tricholoma	*myomyces*	*terreum*	USA (conifer forest), MOTW, pages 98, 100, PFGM, page 193	Ashen	Waxygill	Cavalier Gill Mushroom	Edible-Good
Tricholoma	*orirubens*	.	Europe (mountain humid woods), MOTW, page 100	.	Reddening	Cavalier Gill Mushroom	Edible
Tricholoma	*orirubens*	*orirubens*	Europe (mountain humid woods)	Original	Reddening	Cavalier Gill Mushroom	Edible
Tricholoma	*orirubens*	*horribile*	Europe (mountain humid woods)	Horrible	Reddening	Cavalier Gill Mushroom	Edible
Tricholoma	*pardinum*	.	Canada, N. USA (Fir conifer & Beech mixed forest), MOTW, page 98, PFGM, page 188	.	Tigertop Deathcap	Cavalier Gill Mushroom	Poison
Tricholoma	*pessundatum*	.	USA (conifer forest), MOTW, page 97	.	Ruined	Cavalier Gill Mushroom	Inedible
Tricholoma	*platyphyllum*	.	USA, PFGM, page 188	.	Flat Gill Deathcap	Cavalier Gill Mushroom	Poison
Tricholoma	*populinum*	.	USA, Europe (on sandy soil in Poplar forest), MOTW, page 97, PFGM, page 185	.	Poplar	Cavalier Gill Mushroom	Edible-Excellent
Tricholoma	*portentosum*	.	Canada, N. USA (conifer & mixed forest), MOTW, page 101, PFGM, page 189	.	Sooty Head	Cavalier Gill Mushroom	Edible-Good
Tricholoma	*psammopus*	.	Europe (Pine & Larch conifer forest), MOTW, page 100	.	Sandy Foot	Cavalier Gill Mushroom	Inedible
Tricholoma	*radotinense*	.	Europe (Fir conifer forest, Radotin, Czechoslovakia), MOTW, page 97	.	Radotin	Cavalier Gill Mushroom	Inedible
Tricholoma	*robustum*	.	Europe (conifer forest), MOTW, page 104	.	Robust	Cavalier Gill Mushroom	Edible
Tricholoma	*robustum*	*robustum*	Europe (conifer forest)	Original	Robust	Cavalier Gill Mushroom	Edible
Tricholoma	*robustum*	*focale*	Europe (conifer forest), MOTW, page 93	Scarf	Robust	Cavalier Gill Mushroom	Edible
Tricholoma	*salero*	.	Europe, MOTW, page	.	Salt	Cavalier Gill	Inedible

			94			Mushroom	
Tricholoma	saponaceum	.	Canada, N. USA (conifer & deciduous forest), MOTW, page 102, PFGM, page 190	.	Soapy	Cavalier Gill Mushroom	Edible-Poor
Tricholoma	saponaceum	saponaceum	Canada, N. USA (conifer & deciduous forest)	Original	Soapy	Cavalier Gill Mushroom	Edible-Poor
Tricholoma	saponaceum	atrovirens	Canada, N. USA (conifer & deciduous forest)	Dark Green	Soapy	Cavalier Gill Mushroom	Edible-Poor
Tricholoma	saponaceum	lavedanum	Canada, N. USA (conifer & deciduous forest)	Lather	Soapy	Cavalier Gill Mushroom	Edible-Poor
Tricholoma	saponaceum	lavedanum-boudieri	Canada, N. USA (conifer & deciduous forest)	Boudier Lather	Soapy	Cavalier Gill Mushroom	Edible-Poor
Tricholoma	saponaceum	napipes	Canada, N. USA (conifer & deciduous forest)	Thick Foot	Soapy	Cavalier Gill Mushroom	Edible-Poor
Tricholoma	saponaceum	squamosum	Canada, N. USA (conifer & deciduous forest)	Scaly	Soapy	Cavalier Gill Mushroom	Edible-Poor
Tricholoma	saponaceum	squamosum-ardosiacum	Canada, N. USA (conifer & deciduous forest)	Bright Scaly	Soapy	Cavalier Gill Mushroom	Edible-Poor
Tricholoma	saponaceum	sulphurinum	Canada, N. USA (conifer & deciduous forest)	Sulfur	Soapy	Cavalier Gill Mushroom	Edible-Poor
Tricholoma	scalpturatum	.	Europe, MOTW, page 98	.	Sculptured	Cavalier Gill Mushroom	Edible
Tricholoma	scalpturatum	scalpturatum	Europe	Original	Sculptured	Cavalier Gill Mushroom	Edible
Tricholoma	scalpturatum	chrysites	Europe	Golden	Sculptured	Cavalier Gill Mushroom	Edible
Tricholoma	sciodes	.	Europe, MOTW, page 96	.	Shadow	Cavalier Gill Mushroom	Inedible
Tricholoma	sejunctum	.	USA (in woods), MOTW, 101, PFGM, page 187	.	Disjointed	Cavalier Gill Mushroom	Inedible
Tricholoma	sejunctum	sejunctum	USA (in woods)	Original	Disjointed	Cavalier Gill Mushroom	Inedible
Tricholoma	sejunctum	squamuliferum	USA (in woods), MOTW, 101	Squamule	Disjointed	Cavalier Gill Mushroom	Inedible
Tricholoma	squarrulosum	.	Europe (humid woods), MOTW, page 100	.	Pustule	Cavalier Gill Mushroom	Edible
Tricholoma	sudum	.	Europe, MOTW, page 102	.	Dry Deathcap	Cavalier Gill Mushroom	Poison
Tricholoma	sulphurescens	.	USA (hardwood & mixed conifer-hardwood forest), MOTW, page 104	.	Sulfur	Cavalier Gill Mushroom	Edible-Bitter
Tricholoma	sulphurescens	sulphurescens	USA (hardwood & mixed conifer-hardwood forest)	Original	Sulfur	Cavalier Gill Mushroom	Edible-Bitter
Tricholoma	sulphurescens	resplendens	USA (hardwood & mixed conifer-hardwood forest), PFGM, page 189	Shiny	Sulfur	Cavalier Gill Mushroom	Edible-Bitter
Tricholoma	sulphureum	.	USA (deciduous forest), MOTW, page 104, PFGM, page 191	.	Sulfur Deathcap	Cavalier Gill Mushroom	Poison
Tricholoma	sulphureum	sulphureum	USA (deciduous forest)	Original	Sulfur Deathcap	Cavalier Gill Mushroom	Poison
Tricholoma	sulphureum	bufonum	USA (deciduous forest)	Toadstool	Sulfur Deathcap	Cavalier Gill Mushroom	Poison
Tricholoma	sulphureum	hemisulphureum	USA (deciduous forest)	Half	Sulfur Deathcap	Cavalier Gill Mushroom	Poison
Tricholoma	sulphureum	pallidum	USA (deciduous forest)	Pale	Sulfur Deathcap	Cavalier Gill Mushroom	Poison
Tricholoma	ustale	.	USA (on humid ravines in broadleaf forest), MOTW, page 97	.	Golden Red	Cavalier Gill Mushroom	Inedible
Tricholoma	ustale	ustale	USA (on humid ravines in broadleaf forest)	Original	Golden Red	Cavalier Gill Mushroom	Inedible
Tricholoma	ustale	fulvellum	USA (on humid ravines in broadleaf forest)	Blond	Golden Red	Cavalier Gill Mushroom	Inedible
Tricholoma	ustaloides	.	Europe, MOTW, page 97	.	Lesser Golden Red	Cavalier Gill Mushroom	Inedible
Tricholoma	vaccinum	.	Canada, N. USA (mountain Pine & Spruce conifer forest), MOTW, page 100, PFGM, page 191	.	Fuzztop	Cavalier Gill Mushroom	Edible-Toxic
Tricholoma	venenatum	.	USA (copses), MOTW, page 98, PFGM, page 190	.	Venum Deathcap	Cavalier Gill Mushroom	Poison
Tricholoma	virgatum	.	USA (Fir conifer & Beech mixed forests), MOTW, page 98, PFGM, page 192	.	Silver Streaks	Cavalier Gill Mushroom	Edible-Poor
Tricholoma	zelleri	.	Canada, N. USA (pine forest), PFGM, page 193	.	Zeller	Cavalier Gill Mushroom	Edible
Calocybe	fallax	.	USA, PFGM, page 194	.	Deceptive	Callisto Cap Cavalier	Unknown

						Gill Mushroom	
Calocybe	*gambosa*	.	Europe (meadows, bushes, open woods)	.	Hock	Callisto Cap Cavalier Gill Mushroom	Edible-Excellent
Calocybe	*gambosa*	*gambosa*	Europe (meadows, bushes, open woods), MOTW, page 105	Original	Hock	Callisto Cap Cavalier Gill Mushroom	Edible-Excellent
Calocybe	*gambosa*	*albella*	Europe (meadows, bushes, open woods), MOTW, page 105	White	Hock	Callisto Cap Cavalier Gill Mushroom	Edible-Excellent
Calocybe	*gambosa*	*georgii*	Europe (meadows, bushes, open woods), MOTW, page 105	St. George	Hock	Callisto Cap Cavalier Gill Mushroom	Edible-Excellent
Calocybe	*gambosa*	*graveolens*	Europe (meadows, bushes, open woods), MOTW, page 105	Scented	Hock	Callisto Cap Cavalier Gill Mushroom	Edible-Excellent
Calocybe	*gambosa*	*palumbina*	Europe (meadows, bushes, open woods), MOTW, page 105	Ringdove	Hock	Callisto Cap Cavalier Gill Mushroom	Edible-Excellent
Catathelasma	*imperiale*	.	W. USA (Rocky Mtns. to pacific coast in conifer forest), MOTW, page 109, PFGM, page 137	.	Commander	Cat Nipple Cavalier Gill Mushroom	Edible
Catathelasma	*ventricosum*	.	USA	.	Pot Belly	Cat Nipple Cavalier Gill Mushroom	Unpalatable
Clitocybe	*aeruginosa*	.	USA, PFGM, page 144	.	Copper	Funnelcap Cavalier Gill Mushroom	Unknown
Clitocybe	*albirhiza*	.	W. USA (Rocky Mtns. to pacific coast in conifer forest), PFGM, page 138	.	White Strings	Funnelcap Cavalier Gill Mushroom	Unknown
Clitocybe	*augeana*	.	USA, PFGM, page 140	.	Mealy Deathcap	Funnelcap Cavalier Gill Mushroom	Poison
Clitocybe	*bartelliae*	.	USA, PFGM, page 143	.	Bartelli	Funnelcap Cavalier Gill Mushroom	Unknown
Clitocybe	*candida*	.	USA (mountains, open woods, pastures), MOTW, page 112	.	White Giant	Funnelcap Cavalier Gill Mushroom	Edible-Fair
Clitocybe	*catinus*	.	USA, MOTW, page 308	.	Cat	Funnelcap Cavalier Gill Mushroom	Unknown
Clitocybe	*clavipes*	.	Canada, USA (conifer & hardwood forest), Mexico, MOTW, page 114, PFGM, page 139	.	Clubfoot	Funnelcap Cavalier Gill Mushroom	Edible-Toxic
Clitocybe	*clavipes*	*clavipes*	Canada, USA (conifer & hardwood forest), Mexico	Original	Clubfoot	Funnelcap Cavalier Gill Mushroom	Edible-Toxic
Clitocybe	*clavipes*	*comitialis*	Canada, USA (conifer & hardwood forest), Mexico	Assembly	Clubfoot	Funnelcap Cavalier Gill Mushroom	Edible-Toxic
Clitocybe	*costata*	.	Europe (conifer forest)	.	Ribbed	Funnelcap Cavalier Gill Mushroom	Unknown
Clitocybe	*costata*	*costata*	Europe (conifer forest)	Original	Ribbed	Funnelcap Cavalier Gill Mushroom	Unknown
Clitocybe	*costata*	*incilis*	Europe (conifer forest), MOTW, page 111	Brook	Ribbed	Funnelcap Cavalier Gill Mushroom	Unknown
Clitocybe	*dealbata*	.	USA (lawns, parks, fields), MOTW, page 111, PFGM, page 140	.	White Powder Deathcap	Funnelcap Cavalier Gill Mushroom	Poison
Clitocybe	*dealbata*	*dealbata*	USA (lawns, parks, fields)	Original	White Powder Deathcap	Funnelcap Cavalier Gill Mushroom	Poison
Clitocybe	*dealbata*	*sudorifica*	USA (lawns, parks, fields)	Sweat	White Powder Deathcap	Funnelcap Cavalier Gill Mushroom	Poison
Clitocybe	*dicolor*	.	Europe (woods, plains, mountains), MOTW, page 114	.	Two Color	Funnelcap Cavalier Gill Mushroom	Edible
Clitocybe	*fragrans*	.	USA (in woods), MOTW, page 111	.	Anise Scented	Funnelcap Cavalier Gill Mushroom	Edible
Clitocybe	*fragrans*	*fragrans*	USA (in woods)	Original	Anise Scented	Funnelcap Cavalier Gill Mushroom	Edible
Clitocybe	*fragrans*	*deceptiva*	USA (in woods)	Deceptive	Anise Scented	Funnelcap Cavalier Gill Mushroom	Edible
Clitocybe	*fragrans*	*depauperat a*	USA (in woods)	Imperfect	Anise Scented	Funnelcap Cavalier Gill Mushroom	Edible
Clitocybe	*fragrans*	*obsoleta*	USA (under Pine), MOTW, page 115	Imperfect	Anise Scented	Funnelcap Cavalier Gill Mushroom	Unpalatable
Clitocybe	*geotropa*	.	USA (grass, blackberry bushes, open woods), MOTW, page 112	.	Downward	Funnelcap Cavalier Gill Mushroom	Edible-Good
Clitocybe	*geotropa*	*geotropa*	USA (grass, blackberry bushes, open woods)	Original	Downward	Funnelcap Cavalier Gill Mushroom	Edible-Good
Clitocybe	*geotropa*	*maxima*	USA (grass, blackberry bushes, open woods), MOTW, page 112, PFGM, 141	Greater	Downward	Funnelcap Cavalier Gill Mushroom	Edible-Poor
Clitocybe	*gibba*	.	USA (conifer & deciduous forest), PFGM, page 141	.	Forest	Funnelcap Cavalier Gill Mushroom	Edible-Good
Clitocybe	*gibba*	*gibba*	USA (conifer & deciduous forest)	Original	Forest	Funnelcap Cavalier Gill Mushroom	Edible-Good
Clitocybe	*gibba*	*infundibuli formis*	USA (conifer & deciduous forest), MOTW, page 111	Funnel	Forest	Funnelcap Cavalier Gill Mushroom	Edible-Good
Clitocybe	*gibba*	*membranac*	USA (conifer &	Membrane	Forest	Funnelcap Cavalier	Edible-Good

		ea	deciduous forest)			Gill Mushroom	
Clitocybe	glaucoalba	.	USA, PFGM, page 144	.	Light Gray	Funnelcap Cavalier Gill Mushroom	Unknown
Clitocybe	graveolens	.	USA, PFGM, page 143	.	Scented	Funnelcap Cavalier Gill Mushroom	Unknown
Clitocybe	inornata	.	Europe (conifer forest), MOTW, page 112	.	Unadorned	Funnelcap Cavalier Gill Mushroom	Edible-Poor
Clitocybe	inornata	inornata	Europe (conifer forest)	Original	Unadorned	Funnelcap Cavalier Gill Mushroom	Edible-Poor
Clitocybe	inornata	zygophylla	Europe (conifer forest)	Yoke Gill	Unadorned	Funnelcap Cavalier Gill Mushroom	Edible-Poor
Clitocybe	morbifera	.	USA, PFGM, page 140	.	Morbid	Funnelcap Cavalier Gill Mushroom	Unknown
Clitocybe	nebularis	.	Canada, USA (conifer & mixed forest), Mexico, MOTW, page 114, PFGM, page 142	.	Cloudy	Funnelcap Cavalier Gill Mushroom	Edible-Toxic
Clitocybe	nebularis	nebularis	Canada, USA (conifer & mixed forest), Mexico	Original	Cloudy	Funnelcap Cavalier Gill Mushroom	Edible-Toxic
Clitocybe	nebularis	alba	Canada, USA (conifer & mixed forest), Mexico	White	Cloudy	Funnelcap Cavalier Gill Mushroom	Edible-Toxic
Clitocybe	nuda	.	USA (meadow, lawn, orchard, conifer & hardwood forest), PFGM, page 143	.	Blewit	Funnelcap Cavalier Gill Mushroom	Edible-Good
Clitocybe	odora	.	USA (conifer & hardwood forest), MOTW, page 115, PFGM, page 144	.	Anise	Funnelcap Cavalier Gill Mushroom	Edible-Fair
Clitocybe	odora	odora	USA (conifer & hardwood forest)	Original	Anise	Funnelcap Cavalier Gill Mushroom	Edible-Fair
Clitocybe	odora	alba	USA (conifer & hardwood forest), MOTW, page 115	White	Anise	Funnelcap Cavalier Gill Mushroom	Edible-Fair
Clitocybe	odora	pacifica	USA (conifer & hardwood forest), PFGM, page 144	Pacific	Anise	Funnelcap Cavalier Gill Mushroom	Edible-Fair
Clitocybe	odora	trogii	USA (conifer & hardwood forest)	Trog	Anise	Funnelcap Cavalier Gill Mushroom	Edible-Fair
Clitocybe	odora	virens	USA (conifer & hardwood forest)	Greenish	Anise	Funnelcap Cavalier Gill Mushroom	Edible-Fair
Clitocybe	odora	viridis	USA (Pine conifer & hardwood forest), MOTW, page 115	Green	Anise	Funnelcap Cavalier Gill Mushroom	Edible-Fair
Clitocybe	phyllophila	.	USA (on plant debris in woods)	.	Whitewash Deathcap	Funnelcap Cavalier Gill Mushroom	Poison
Clitocybe	phyllophila	phyllophila	USA (on plant debris in woods)	Original	Whitewash Deathcap	Funnelcap Cavalier Gill Mushroom	Poison
Clitocybe	phyllophila	cerussata	USA (on plant debris in woods), MOTW, page 111	White Lead	Whitewash Deathcap	Funnelcap Cavalier Gill Mushroom	Poison
Clitocybe	phyllophila	difformis	USA (manured soil)	Deformed	Whitewash Deathcap	Funnelcap Cavalier Gill Mushroom	Poison
Clitocybe	phyllophila	pithyophila	USA (on plant debris in woods)	Monkey Head	Whitewash Deathcap	Funnelcap Cavalier Gill Mushroom	Poison
Clitocybe	pseudoirina	.	USA, PFGM, page 142	.	Lesser Irina	Funnelcap Cavalier Gill Mushroom	Unknown
Clitocybe	rivulosa	.	USA (in woods), MOTW, page 111	.	Brook Deathcap	Funnelcap Cavalier Gill Mushroom	Poison
Clitocybe	rivulosa	rivulosa	USA (in woods)	Original	Brook Deathcap	Funnelcap Cavalier Gill Mushroom	Poison
Clitocybe	rivulosa	minor	USA (in woods)	Lesser	Brook Deathcap	Funnelcap Cavalier Gill Mushroom	Poison
Clitocybe	rivulosa	neptunea	USA (in woods)	Neptune	Brook Deathcap	Funnelcap Cavalier Gill Mushroom	Poison
Clitocybe	sinopica	.	Europe (Sinope, Greece)	.	Sinope	Funnelcap Cavalier Gill Mushroom	Unknown
Clitocybe	sinopica	sinopica	Europe (Sinope, Greece)	Original	Sinope	Funnelcap Cavalier Gill Mushroom	Unknown
Clitocybe	sinopica	subsinopica	Europe (Sinope, Greece)	Lesser	Sinope	Funnelcap Cavalier Gill Mushroom	Unknown
Clitocybe	squamulosa	.	USA, Europe (conifer forest), MOTW, page 111, PFGM, page 141	.	Squamule	Funnelcap Cavalier Gill Mushroom	Edible
Clitocybe	squamulosa	squamulosa	Europe (conifer forest)	Original	Squamule	Funnelcap Cavalier Gill Mushroom	Edible
Clitocybe	squamulosa	sinopicoides	Europe (conifer forest)	Sinope	Squamule	Funnelcap Cavalier Gill Mushroom	Edible
Clitocybe	subclavipes	.	USA, PFGM, page 139	.	Lesser Club Foot	Funnelcap Cavalier Gill Mushroom	Unknown
Clitocybe	subconnexa	.	S. Canada, N. & C. USA (forest soil), page 144	.	Clustered	Funnelcap Cavalier Gill Mushroom	Edible
Clitocybe	sudorifica	.	USA, PFGM, page 140	.	Dry	Funnelcap Cavalier Gill Mushroom	Unknown
Clitocybe	tarda	.	USA (lawn, greenhouse, open	.	Dingy	Funnelcap Cavalier	Unknown

			woods), PFGM, page 145		Bowlcap	Gill Mushroom	
Clitocybe	*tarda*	*tarda*	USA (lawn, greenhouse, open woods)	Original	Dingy Bowlcap	Funnelcap Cavalier Gill Mushroom	Unknown
Clitocybe	*tarda*	*alcalina*	USA (lawn, greenhouse, open woods), PFGM, page 145	Alkaline	Dingy Bowlcap	Funnelcap Cavalier Gill Mushroom	Unknown
Clitocybe	*vermicularis*	.	Europe (conifer forest), MOTW, page 111	.	Worm	Funnelcap Cavalier Gill Mushroom	Edible
Clitocybula	*abundans*	.	USA (on logs & decaying wood in hardwood & mixed conifer-hardwood forest), PFGM, page 146	.	Abundant	Bowlcap Cavalier Gill Mushroom	Unknown
Clitocybula	*familia*	.	USA (on decaying stumps in conifer forest), MOTW, page 88	.	Family	Bowlcap Cavalier Gill Mushroom	Edible
Collybia	*albiflavida*	.	USA, MOTW, page 308	.	Yellowish White	Coincap Cavalier Gill Mushroom	Unknown
Collybia	*cirrhata*	.	USA, PFGM, page 151	.	Cirrha	Coincap Cavalier Gill Mushroom	Unknown
Collybia	*cirrhata*	*cirrhata*	USA	Original	Cirrha	Coincap Cavalier Gill Mushroom	Unknown
Collybia	*cirrhata*	*amanitae*	USA	Amanite	Cirrha	Coincap Cavalier Gill Mushroom	Unknown
Collybia	*cookei*	.	USA, PFGM, page 151	.	Cooke	Coincap Cavalier Gill Mushroom	Unknown
Collybia	*iocephala*	.	USA (on forest decaying litter), PFGM, page 149	.	Purple	Coincap Cavalier Gill Mushroom	Unknown
Collybia	*leucocephala*	.	Europe, MOTW, page 86	.	Lightcap	Coincap Cavalier Gill Mushroom	Unknown
Collybia	*polyphylla*	.	USA, PFGM, page 148	.	Many Gill	Coincap Cavalier Gill Mushroom	Unknown
Collybia	*subnuda*	.	USA, PFGM, page 149	.	Lesser Nude	Coincap Cavalier Gill Mushroom	Unknown
Collybia	*tuberosa*	.	USA (on forest mushroom remains), PFGM, page 151	.	Appleseed	Coincap Cavalier Gill Mushroom	Unknown
Collybia	*tuberosa*	*tuberosa*	USA (on forest mushroom remains)	Original	Appleseed	Coincap Cavalier Gill Mushroom	Unknown
Collybia	*tuberosa*	*sclerotipes*	USA (on forest mushroom remains)	Hard	Appleseed	Coincap Cavalier Gill Mushroom	Unknown
Cystoderma	*ambrosii*	.	Europe, MOTW, page 40	.	Ambrose	Grainy Cap Cavalier Gill Mushroom	Unpalatable
Cystoderma	*amianthinum*	.	USA, Europe (in woods, on moors), MOTW, page 40, PFGM, page 152	.	Incorruptable	Grainycap Cavalier Gill Mushroom	Edible
Cystoderma	*amianthinum*	*amianthinum*	USA, Europe (in woods, on moors)	Original	Incorruptable	Grainy Cap Cavalier Gill Mushroom	Edible
Cystoderma	*amianthinum*	*album*	USA, Europe (in woods, on moors)	White	Incorruptable	Grainycap Cavalier Gill Mushroom	Edible
Cystoderma	*amianthinum*	*longisporum*	USA, Europe (in woods, on moors)	Long Spore	Incorruptable	Grainycap Cavalier Gill Mushroom	Edible
Cystoderma	*amianthinum*	*ruguloso reticulatum*	USA, Europe (in woods, on moors)	Wrinkled Net	Incorruptable	Grainycap Cavalier Gill Mushroom	Edible
Cystoderma	*carcharias*	.	Europe (conifer forest), MOTW, page 40	.	Rough	Grainycap Cavalier Gill Mushroom	Edible
Cystoderma	*cinnabarinum*	.	USA, MOTW, page 40, PFGM, page 152	.	Orange Red	Grainycap Cavalier Gill Mushroom	Unpalatable
Cystoderma	*cinnabarinum*	*cinnabarinum*	USA	Original	Orange Red	Grainycap Cavalier Gill Mushroom	Unpalatable
Cystoderma	*cinnabarinum*	*terrei*	USA	Earth	Orange Red	Grainycap Cavalier Gill Mushroom	Unpalatable
Cystoderma	*fallax*	.	USA, MOTW, page 40, PFGM, page 152	.	Ocher	Grainycap Cavalier Gill Mushroom	Unpalatable
Cystoderma	*granulosum*	.	USA, MOTW, page 40, PFGM, page 152	.	Grainy	Grainycap Cavalier Gill Mushroom	Unpalatable
Cystoderma	*haematites*	.	Europe, MOTW, page 40	.	Violet Red	Grainycap Cavalier Gill Mushroom	Unpalatable
Floccularia	*fusca*	.	USA, PFGM, page 134	.	Fuscous	Woolly Cavalier Gill Mushroom	Unknown
Floccularia	*luteovirens*	.	W. USA (Rocky Mtns. Aspen forest)	.	Yellow & Green	Woolly Cavalier Gill Mushroom	Edible
Floccularia	*luteovirens*	*luteovirens*	W. USA (Rocky Mtns. Aspen forest)	Original	Yellow & Green	Woolly Cavalier Gill Mushroom	Edible
Floccularia	*luteovirens*	*alba*	W. USA (Rocky Mtns. Aspen forest)	Whitish	Yellow & Green	Woolly Cavalier Gill Mushroom	Edible
Floccularia	*luteovirens*	*straminea*	W. USA (Rocky Mtns. Aspen forest)	Straw	Yellow & Green	Woolly Cavalier Gill Mushroom	Edible
Floccularia	*luteovirens*	*straminea-americana*	W. USA (Rocky Mtns. Aspen forest), PFGM, page 135	Yellow Bracelet	Yellow & Green	Woolly Cavalier Gill Mushroom	Edible
Floccularia	*pitkinensis*	.	USA, PFGM, page 134	.	Pitkin	Woolly Cavalier	Unknown

						Gill Mushroom	
Gymnopus	*acervatus*	.	USA (on stumps, logs, branches in conifer forest), MOTW, page 88, PFGM, page 147	.	Cluster	Nude Coincap Cavalier Gill Mushroom	Edible-Toxic
Gymnopus	*confluens*	.	S. Canada, USA (hardwood & mixed conifer-hardwood forest), MOTW, page 91	.	Tufted	Nude Coincap Cavalier Gill Mushroom	Edible
Gymnopus	*dryophilus*	.	USA (forest soil & decaying wood), MOTW, page 89, PFGM, page 149	.	Forest Friend	Nude Coincap Cavalier Gill Mushroom	Edible-Toxic
Gymnopus	*dryophilus*	*dryophilus*	USA (forest soil & decaying wood)	.	Forest Friend	Nude Coincap Cavalier Gill Mushroom	Edible-Toxic
Gymnopus	*dryophilus*	*alvearis*	USA (forest soil & decaying wood)	.	Forest Friend	Nude Coincap Cavalier Gill Mushroom	Edible-Toxic
Gymnopus	*dryophilus*	*aurata*	USA (forest soil & decaying wood)	.	Forest Friend	Nude Coincap Cavalier Gill Mushroom	Edible-Toxic
Gymnopus	*erythropus*	.	Europe (on decaying plants in woods)	.	Red Foot	Nude Coincap Cavalier Gill Mushroom	Edible
Gymnopus	*erythropus*	*erythropus*	Europe (on decaying plants in woods)	Original	Red Foot	Nude Coincap Cavalier Gill Mushroom	Edible
Gymnopus	*erythropus*	*bresadolae*	Europe (on decaying plants in woods), MOTW, page 88	Bresadola	Red Foot	Nude Coincap Cavalier Gill Mushroom	Edible
Gymnopus	*fuscopurpureus*	.	Europe, MOTW, page 92	.	Fuscous Purple	Nude Coincap Cavalier Gill Mushroom	Unpalatable
Gymnopus	*fusipes*	.	USA (on trunks & stumps of broadleaf trees), MOTW, page 86	.	Spindle	Nude Coincap Cavalier Gill Mushroom	Edible
Gymnopus	*fusipes*	*fusipes*	USA (on trunks & stumps of broadleaf trees)	Original	Spindle	Nude Coincap Cavalier Gill Mushroom	Edible
Gymnopus	*fusipes*	*contorta*	USA (on trunks & stumps of broadleaf trees)	Contorted	Spindle	Nude Coincap Cavalier Gill Mushroom	Edible
Gymnopus	*fusipes*	*oedematopus*	USA (on trunks & stumps of broadleaf trees)	Edema Foot	Spindle	Nude Coincap Cavalier Gill Mushroom	Edible
Gymnopus	*ocior*	.	Europe	.	Oci	Nude Coincap Cavalier Gill Mushroom	Edible
Gymnopus	*ocior*	*ocior*	Europe	Original	Oci	Nude Coincap Cavalier Gill Mushroom	Edible
Gymnopus	*ocior*	*exsculptus*	Europe	Dug Out	Oci	Nude Coincap Cavalier Gill Mushroom	Edible
Gymnopus	*ocior*	*funicularis*	Europe (woods, meadows), MOTW, page 89	Cord	Oci	Nude Coincap Cavalier Gill Mushroom	Edible
Gymnopus	*ocior*	*xanthopus*	Europe (woods, meadows), MOTW, page 89	Yellow Stalk	Oci	Nude Coincap Cavalier Gill Mushroom	Edible
Gymnopus	*peronatus*	.	USA, PFGM, page 149	.	Peron	Nude Coincap Cavalier Gill Mushroom	Unknown
Lepista	*caespitosa*	.	Europe (mountain pastures), MOTW, page 122	.	Tuft	Wine Pitcher Cavalier Gill Mushroom	Edible
Lepista	*caespitosa*	*caespitosa*	Europe (mountain pastures)	Original	Tuft	Wine Pitcher Cavalier Gill Mushroom	Edible
Lepista	*caespitosa*	*fasciculata*	Europe (mountain pastures), PFGM, page 145	Bundled	Tuft	Wine Pitcher Cavalier Gill Mushroom	Edible
Lepista	*densifolia*	.	USA, PFGM, page 145	.	Dense	Wine Pitcher Cavalier Gill Mushroom	Unknown
Lepista	*flaccida*	.	USA (conifer & broadleaf forests)	.	Flaccid	Wine Pitcher Cavalier Gill Mushroom	Edible
Lepista	*flaccida*	*flaccida*	USA (conifer &	Original	Flaccid	Wine Pitcher	Edible

			broadleaf forests)			Cavalier Gill Mushroom	
Lepista	*flaccida*	*gilva*	USA (Fir mountain conifer & broadleaf forest), MOTW, page 114	Whitish Yellow	Flaccid	Wine Pitcher Cavalier Gill Mushroom	Edible
Lepista	*flaccida*	*splendens*	USA (conifer & broadleaf forests)	Splendid	Flaccid	Wine Pitcher Cavalier Gill Mushroom	Edible
Lepista	*glaucocana*	.	Europe (Beech broadleaf forest), MOTW, pages 120, 122	.	Glaucous Hoary	Wine Pitcher Cavalier Gill Mushroom	Unpalatable
Lepista	*inversa*	.	USA (conifer & broadleaf forests), PFGM, page 141	.	Bowlcap	Wine Pitcher Cavalier Gill Mushroom	Edible-Poor
Lepista	*irina*	.	USA (on Beech & Pine forest soil & woody debris), MOTW, page 121, PFGM, page 141	.	Blushing Bowlcap	Wine Pitcher Cavalier Gill Mushroom	Edible-Toxic
Lepista	*nuda*	.	USA (in woods), MOTW, page 120	.	Nude	Wine Pitcher Cavalier Gill Mushroom	Edible-Excellent
Lepista	*nuda*	*nuda*	USA (in woods)	Original	Nude	Wine Pitcher Cavalier Gill Mushroom	Edible-Excellent
Lepista	*nuda*	*gracilis*	USA (in woods)	Slender	Nude	Wine Pitcher Cavalier Gill Mushroom	Edible-Excellent
Lepista	*nuda*	*lilacea*	USA (in conifer woods)	Lilac	Nude	Wine Pitcher Cavalier Gill Mushroom	Edible-Excellent
Lepista	*nuda*	*pruinosa*	USA (in woods)	Hoarfrost	Nude	Wine Pitcher Cavalier Gill Mushroom	Edible-Excellent
Lepista	*panaeolus*	.	USA (hill & mountain meadows), MOTW, page 122	.	Dazzling	Wine Pitcher Cavalier Gill Mushroom	Edible-Good
Lepista	*personata*	.	USA	.	Person	Wine Pitcher Cavalier Gill Mushroom	Unknown
Lepista	*sadica*	.	Europe (manured ground), MOTW, page 120	.	Sad	Wine Pitcher Cavalier Gill Mushroom	Edible
Lepista	*saeva*	.	Europe (humid meadows & pastures), MOTW, page 120, PFGM, page 143	.	Horrible	Wine Pitcher Cavalier Gill Mushroom	Edible-Good
Lepista	*sordida*	.	Europe (meadows, parks, gardens), MOTW, page 120	.	Sordid	Wine Pitcher Cavalier Gill Mushroom	Edible
Lyophyllum	*connatum*	.	Europe (humid places on decayed wood), MOTW, page 106	.	Conjoined	Loosecap Cavalier Gill Mushroom	Edible
Lyophyllum	*decastes*	.	USA (grassland, roadsides, sawdust), PFGM, page 163	.	Shimeiji	Loosecap Cavalier Gill Mushroom	Edible-Good
Lyophyllum	*decastes*	*decastes*	USA (grassland, roadsides, sawdust)	Original	Shimeiji	Loosecap Cavalier Gill Mushroom	Edible-Good
Lyophyllum	*decastes*	*aggregatum*	USA (grassland, roadsides, sawdust), MOTW, page 106	Fried Chicken	Shimeiji	Loosecap Cavalier Gill Mushroom	Edible-Good
Lyophyllum	*fumosum*	.	Europe (clearings, meadows, edges of woods)	.	Smoky	Loosecap Cavalier Gill Mushroom	Edible
Lyophyllum	*fumosum*	*fumosum*	Europe (clearings, meadows, edges of woods)	Original	Smoky	Loosecap Cavalier Gill Mushroom	Edible
Lyophyllum	*fumosum*	*cinerascens*	Europe (clearings, meadows, edges of woods)	Cinder	Smoky	Loosecap Cavalier Gill Mushroom	Edible
Lyophyllum	*fumosum*	*conglobatum*	Europe (clearings, meadows, edges of woods), MOTW, page 106	Globe	Smoky	Loosecap Cavalier Gill Mushroom	Edible
Lyophyllum	*loricatum*	.	Europe, MOTW, page 106	.	Armored	Loosecap Cavalier Gill Mushroom	Edible
Lyophyllum	*montanum*	.	W. USA (on melted snowbanks in Rocky Mtns.), PFGM, page 164	.	Snow Funnel	Loosecap Cavalier Gill Mushroom	Unknown
Megacollybia	*platyphylla*	.	USA (stumps, logs & debris in hardwood forest), PFGM, page 194, MOTW, page 87	.	Broadgill	Large Coincap Cavalier Gill Mushroom	Edible-Bitter
Melanoleuca	*alboflavida*	.	USA (roadsides, fields, open woods), PFGM, page 169	.	Yellowish	Dark & Light Cavalier Gill Mushroom	Edible
Melanoleuca	*brevipes*	.	USA (in woods), MOTW, page 109	.	Short Foot	Dark & Light Cavalier Gill Mushroom	Edible
Melanoleuca	*cognata*	.	Europe (grassy spots under conifer & Poplar), MOTW, page 106	.	Kindrid	Dark & Light Cavalier Gill Mushroom	Edible

Melanoleuca	*cognata*	*cognata*	Europe (grassy spots under conifer & Poplar)	Original	Kindrid	Dark & Light Cavalier Gill Mushroom	Edible
Melanoleuca	*cognata*	*nauseosa*	Europe (grassy spots under conifer & Poplar)	Nauseous	Kindrid	Dark & Light Cavalier Gill Mushroom	Edible
Melanoleuca	*cognata*	*robusta*	Europe (grassy spots under conifer & Poplar)	Robust	Kindrid	Dark & Light Cavalier Gill Mushroom	Edible
Melanoleuca	*exscissa*	.	Europe, MOTW, page 109	.	Torn	Dark & Light Cavalier Gill Mushroom	Edible
Melanoleuca	*exscissa*	*exscissa*	Europe	Original	Torn	Dark & Light Cavalier Gill Mushroom	Edible
Melanoleuca	*exscissa*	*cinerascens*	Europe	Cinder	Torn	Dark & Light Cavalier Gill Mushroom	Edible
Melanoleuca	*exscissa*	*kuehneri*	Europe	Kuehner	Torn	Dark & Light Cavalier Gill Mushroom	Edible
Melanoleuca	*grammopodia*	.	USA (mountains, pastures, meadows, gardens)	.	Lined Foot	Dark & Light Cavalier Gill Mushroom	Edible
Melanoleuca	*grammopodia*	*grammopodia*	USA (mountains, pastures, meadows, gardens)	Original	Lined Foot	Dark & Light Cavalier Gill Mushroom	Edible
Melanoleuca	*grammopodia*	*macrocarpa*	USA (mountains, pastures, meadows, gardens)	Large Fruit	Lined Foot	Dark & Light Cavalier Gill Mushroom	Edible
Melanoleuca	*grammopodia*	*obscura*	USA (mountains, pastures, meadows, gardens)	Obsure	Lined Foot	Dark & Light Cavalier Gill Mushroom	Edible
Melanoleuca	*melaleuca*	.	USA (in pastures, meadows & open woods)	.	Light Honey	Dark & Light Cavalier Gill Mushroom	Edible
Melanoleuca	*polioleuca*	.	Europe (clearings, pastures)	.	Polished	Dark & Light Cavalier Gill Mushroom	Edible
Melanoleuca	*polioleuca*	*polioleuca*	Europe (clearings, pastures)	Original	Polished	Dark & Light Cavalier Gill Mushroom	Edible
Melanoleuca	*polioleuca*	*vulgaris*	Europe (clearings, pastures), MOTW, page 106	European	Polished	Dark & Light Cavalier Gill Mushroom	Edible
Melanoleuca	*strictipes*	.	Europe (mountain pastures)	.	Thin Foot	Dark & Light Cavalier Gill Mushroom	Edible
Melanoleuca	*strictipes*	*strictipes*	Europe (mountain pastures)	Original	Thin Foot	Dark & Light Cavalier Gill Mushroom	Edible
Melanoleuca	*strictipes*	*cinereifolia*	Europe (mountain pastures)	Original	Thin Foot	Dark & Light Cavalier Gill Mushroom	Edible
Melanoleuca	*strictipes*	*cnista*	Europe (dry broadleaf forest), MOTW, page 106	Scraped	Thin Foot	Dark & Light Cavalier Gill Mushroom	Edible
Melanoleuca	*strictipes*	*evenosa*	Europe (mountain pastures), MOTW, page 106	Veinless	Thin Foot	Dark & Light Cavalier Gill Mushroom	Edible
Mycena	*epipterygia*	.	Europe (stumps & plant debris), MOTW, page 93, PFGM, page 172	.	Membrane	Fairy Helmet Cavalier Gill Mushroom	Unpalatable
Mycena	*epipterygia*	*epipterygia*	USA, Europe (stumps & plant debris)	Original	Membrane	Fairy Helmet Cavalier Gill Mushroom	Unpalatable
Mycena	*epipterygia*	*badiceps*	USA, Europe (stumps & plant debris)	Red Brown	Membrane	Fairy Helmet Cavalier Gill Mushroom	Unpalatable
Mycena	*epipterygia*	*candida*	USA, Europe (stumps & plant debris)	Virgin White	Membrane	Fairy Helmet Cavalier Gill Mushroom	Unpalatable
Mycena	*epipterygia*	*citrinella*	USA, Europe (stumps & plant debris)	Lemon Yellow	Membrane	Fairy Helmet Cavalier Gill Mushroom	Unpalatable
Mycena	*epipterygia*	*lignicola*	USA, Europe (stumps & plant debris)	Wood Dwelling	Membrane	Fairy Helmet Cavalier Gill Mushroom	Unpalatable
Mycena	*epipterygia*	*nutans*	USA, Europe (stumps & plant debris)	Nodding	Membrane	Fairy Helmet Cavalier Gill Mushroom	Unpalatable
Mycena	*epipterygia*	*plicatocrenata*	USA, Europe (stumps & plant debris)	Folded Scallop	Membrane	Fairy Helmet Cavalier Gill Mushroom	Unpalatable
Mycena	*epipterygia*	*rubescens*	USA, Europe (stumps & plant debris)	Red	Membrane	Fairy Helmet Cavalier Gill Mushroom	Unpalatable
Mycena	*epipterygia*	*splendidipes*	USA, Europe (stumps & plant debris)	Splendid	Membrane	Fairy Helmet Cavalier Gill Mushroom	Unpalatable
Mycena	*epipterygia*	*viscosa*	USA, Europe (stumps & plant debris)	Sticky	Membrane	Fairy Helmet Cavalier Gill Mushroom	Unpalatable
Mycena	*epipterygioides*	.	USA, PFGM, page 172	.	Lesser Membrane	Fairy Helmet Cavalier Gill Mushroom	Unknown
Mycena	*galericulata*	.	USA (stumps & plant debris), MOTW, page 93, PFGM, page 172	.	Fur Cap	Fairy Helmet Cavalier Gill Mushroom	Unpalatable
Mycena	*galericulata*	*galericulata*	USA (stumps & plant debris)	Original	Fur Cap	Fairy Helmet Cavalier Gill Mushroom	Unpalatable
Mycena	*galericulata*	*albida*	USA (stumps & plant debris)	White	Fur Cap	Fairy Helmet Cavalier Gill Mushroom	Unpalatable
Mycena	*galericulata*	*berkeleyi*	USA (stumps & plant debris)	Berkeley	Fur Cap	Fairy Helmet Cavalier Gill Mushroom	Unpalatable
Mycena	*galericulata*	*radicatella*	USA (stumps & plant debris)	Rooting	Fur Cap	Fairy Helmet Cavalier Gill Mushroom	Unpalatable
Mycena	*galericulata*	*rugosa*	USA (stumps & plant debris)	Wrinkled	Fur Cap	Fairy Helmet Cavalier Gill Mushroom	Unpalatable

Mycena	galericulata	rugulosiceps	USA (stumps & plant debris)	Ruffled	Fur Cap	Fairy Helmet Cavalier Gill Mushroom	Unpalatable
Mycena	galopus	.	Europe (stumps & plant debris), MOTW, page 93	.	Latex Foot	Fairy Helmet Cavalier Gill Mushroom	Unpalatable
Mycena	galopus	galopus	Europe (stumps & plant debris)	Original	Latex Foot	Fairy Helmet Cavalier Gill Mushroom	Unpalatable
Mycena	galopus	candida	Europe (stumps & plant debris)	Virgin	Latex Foot	Fairy Helmet Cavalier Gill Mushroom	Unpalatable
Mycena	galopus	candida-alba	Europe (stumps & plant debris)	Virgin White	Latex Foot	Fairy Helmet Cavalier Gill Mushroom	Unpalatable
Mycena	galopus	nigra	Europe (stumps & plant debris)	Black	Latex Foot	Fairy Helmet Cavalier Gill Mushroom	Unpalatable
Mycena	galopus	nigra-fusconigra	Europe (stumps & plant debris)	Dark Black	Latex Foot	Fairy Helmet Cavalier Gill Mushroom	Unpalatable
Mycena	galopus	nigra-leucogala	Europe (stumps & plant debris)	White Latex Black	Latex Foot	Fairy Helmet Cavalier Gill Mushroom	Unpalatable
Mycena	griseoviridis	.	USA, PFGM, page 172	.	Gray & Green	Fairy Helmet Cavalier Gill Mushroom	Unknown
Mycena	haematopus	.	USA (stumps & plant debris), MOTW, page 93, PFGM, page 173	.	Red Latex	Fairy Helmet Cavalier Gill Mushroom	Unpalatable
Mycena	haematopus	haematopus	USA (stumps & plant debris)	Original	Red Latex	Fairy Helmet Cavalier Gill Mushroom	Unpalatable
Mycena	haematopus	marginata	USA (stumps & plant debris)	Margined	Red Latex	Fairy Helmet Cavalier Gill Mushroom	Unpalatable
Mycena	inclinata	.	USA, Europe (stumps & plant debris), MOTW, page 93, PFGM, page 173	.	Slanted	Fairy Helmet Cavalier Gill Mushroom	Unpalatable
Mycena	inclinata	inclinata	USA, Europe (stumps & plant debris)	Original	Slanted	Fairy Helmet Cavalier Gill Mushroom	Unpalatable
Mycena	inclinata	calopus	USA, Europe (stumps & plant debris)	Beautiful	Slanted	Fairy Helmet Cavalier Gill Mushroom	Unpalatable
Mycena	latifolia	.	USA, MOTW, page 309	.	.	Fairy Helmet Cavalier Gill Mushroom	Unknown
Mycena	leaiana	.	USA (on wood of trees in deciduous forest), PFGM, page 174	.	Golden	Fairy Helmet Cavalier Gill Mushroom	Unknown
Mycena	leptocephala	.	USA, PFGM, page 171	.	Thincap	Fairy Helmet Cavalier Gill Mushroom	Unknown
Mycena	leptocephala	leptocephala	USA	Original	Thincap	Fairy Helmet Cavalier Gill Mushroom	Unknown
Mycena	leptocephala	alcalina	USA, PFGM, page 171	Alkaline	Thincap	Fairy Helmet Cavalier Gill Mushroom	Unknown
Mycena	lilacifolia	.	USA (on decaying stumps and logs in conifer forest), PFGM, page 174	.	Lilac Gill	Fairy Helmet Cavalier Gill Mushroom	Unknown
Mycena	maculata	.	USA, PFGM, page 173	.	Spotted	Fairy Helmet Cavalier Gill Mushroom	Unknown
Mycena	metata	.	USA	.	.	Fairy Helmet Cavalier Gill Mushroom	Unknown
Mycena	metata	metata	USA	.	.	Fairy Helmet Cavalier Gill Mushroom	Unknown
Mycena	metata	collariata	USA, MOTW, page 309	.	.	Fairy Helmet Cavalier Gill Mushroom	Unknown
Mycena	monticola	.	USA, PFGM, page 176	.	Mountain	Fairy Helmet Cavalier Gill Mushroom	Unknown
Mycena	overholtsii	.	W. USA (on Rocky Mtns. conifer wood), MOTW, page 92, PFGM, page 176	.	Overholts	Fairy Helmet Cavalier Gill Mushroom	Inedible
Mycena	parabolica	.	USA, PFGM, page 173	.	Parabolic	Fairy Helmet Cavalier Gill Mushroom	Unknown
Mycena	pelianthina	.	Europe, MOTW, page 92	.	Radish Odor	Fairy Helmet Cavalier Gill Mushroom	Inedible
Mycena	polygramma	.	Europe (stumps & plant debris), MOTW, page 93	.	Lined	Fairy Helmet Cavalier Gill Mushroom	Unpalatable
Mycena	polygramma	polygramma	Europe (stumps & plant debris)	Original	Lined	Fairy Helmet Cavalier Gill Mushroom	Unpalatable
Mycena	polygramma	candida	Europe (stumps & plant debris)	Virgin White	Lined	Fairy Helmet Cavalier Gill Mushroom	Unpalatable
Mycena	prolifera	.	USA, MOTW, page 309	.	.	Fairy Helmet Cavalier Gill Mushroom	Unknown
Mycena	pura	.	USA, Europe (woods, moss, on dead leaves), MOTW, page 92, PFGM, page 175	.	Pure	Fairy Helmet Cavalier Gill Mushroom	Edible
Mycena	pura	pura	USA, Europe (woods, moss, on dead leaves)	Original	Pure	Fairy Helmet Cavalier Gill Mushroom	Edible
Mycena	pura	alba	USA, Europe (woods,	White	Pure	Fairy Helmet Cavalier	Edible

			moss, on dead leaves)			Gill Mushroom	
Mycena	*pura*	*carnea*	USA, Europe (woods, moss, on dead leaves)	Flesh Color	Pure	Fairy Helmet Cavalier Gill Mushroom	Edible
Mycena	*pura*	*ianthina*	USA, Europe (woods, moss, on dead leaves)	Blue Violet	Pure	Fairy Helmet Cavalier Gill Mushroom	Edible
Mycena	*pura*	*lutea*	USA, Europe (woods, moss, on dead leaves)	Yellow	Pure	Fairy Helmet Cavalier Gill Mushroom	Edible
Mycena	*pura*	*multicolor*	USA, Europe (woods, moss, on dead leaves)	Multicolor	Pure	Fairy Helmet Cavalier Gill Mushroom	Edible
Mycena	*pura*	*pseudopura*	USA, Europe (woods, moss, on dead leaves)	Lesser	Pure	Fairy Helmet Cavalier Gill Mushroom	Edible
Mycena	*pura*	*purpurea*	USA, Europe (woods, moss, on dead leaves)	Purple	Pure	Fairy Helmet Cavalier Gill Mushroom	Edible
Mycena	*pura*	*roseoviolacea*	USA, Europe (woods, moss, on dead leaves)	Rose Violet	Pure	Fairy Helmet Cavalier Gill Mushroom	Edible
Mycena	*pura*	*violacea*	USA, Europe (woods, moss, on dead leaves)	Violet	Pure	Fairy Helmet Cavalier Gill Mushroom	Edible
Mycena	*rosella*	.	USA, PFGM, page 176	.	Roselle	Fairy Helmet Cavalier Gill Mushroom	Unknown
Mycena	*rosella*	*rosella*	USA	Original	Roselle	Fairy Helmet Cavalier Gill Mushroom	Unknown
Mycena	*rosella*	*rosea*	USA	Rosy	Roselle	Fairy Helmet Cavalier Gill Mushroom	Unknown
Mycena	*rosella*	*strobilina*	USA	Strobe	Roselle	Fairy Helmet Cavalier Gill Mushroom	Unknown
Mycena	*sanguinolenta*	.	Europe (stumps & plant debris), MOTW, page 93	.	Bloody	Fairy Helmet Cavalier Gill Mushroom	Unpalatable
Mycena	*sanguinolenta*	*sanguinolenta*	Europe (stumps & plant debris)	Original	Bloody	Fairy Helmet Cavalier Gill Mushroom	Unpalatable
Mycena	*sanguinolenta*	*cruenta*	Europe (stumps & plant debris)	Crucifix	Bloody	Fairy Helmet Cavalier Gill Mushroom	Unpalatable
Mycena	*texensis*	.	USA (Texas), PFGM, page 174	.	Texas	Fairy Helmet Cavalier Gill Mushroom	Unknown
Mycena	*tintinnabulum*	.	Europe (stumps & plant debris), MOTW, page 93	.	Little Bell	Fairy Helmet Cavalier Gill Mushroom	Unpalatable
Pseudoclitocybe	*brumalis*	.	USA (humid conifer forest), MOTW, page 118	.	Winter Solstice	Lesser Funnelcap Cavalier Gill Mushroom	Edible
Pseudoclitocybe	*cyathiformis*	.	USA (on soil & woody debris in humid woods, meadows fields), MOTW, page 118	.	Cup Shaped	Lesser Funnelcap Cavalier Gill Mushroom	Edible
Pseudoclitocybe	*expallens*	.	Europe (meadows), MOTW, page 118	.	Pale	Lesser Funnelcap Cavalier Gill Mushroom	Edible
Pseudoclitocybe	*pruinosa*	.	Europe (Fir conifer forest), MOTW, page 118	.	Hoarfrost	Lesser Funnelcap Cavalier Gill Mushroom	Edible
Pseudoclitocybe	*suaveolens*	.	Europe (in woods), MOTW, page 115	.	Sweet Anise	Lesser Funnelcap Cavalier Gill Mushroom	Edible
Pseudoclitocybe	*vibecina*	.	Europe, MOTW, page 118	.	Bruise	Lesser Funnelcap Cavalier Gill Mushroom	Edible
Pseudoclitocybe	*vibecina*	*vibecina*	Europe	Original	Bruise	Lesser Funnelcap Cavalier Gill Mushroom	Edible
Pseudoclitocybe	*vibecina*	*favrei*	Europe	Favre	Bruise	Lesser Funnelcap Cavalier Gill Mushroom	Edible
Pseudoclitocybe	*vibecina*	*langei*	Europe	Lange	Bruise	Lesser Funnelcap Cavalier Gill Mushroom	Edible
Pseudoclitocybe	*vibecina*	*orientalis*	Europe	Oriental	Bruise	Lesser Funnelcap Cavalier Gill Mushroom	Edible
Rhodopaxillus	.	.	Synonym for *Lepista*	.	.	.	.
Tephrocybe	*atrata*	.	Europe (grasslands, open woods, burned grounds), MOTW, page 86	.	Mourning	Ashcap Cavalier Gill Mushroom	Unpalatable
Tephrocybe	*coracina*	.	Europe (conifer forest), MOTW, page 86	.	Coracine	Ashcap Cavalier Gill Mushroom	Edible
Tephrocybe	*Inolens*	.	Europe (conifer forest), MOTW, page 86	.	Odorless	Ashcap Cavalier Gill Mushroom	Edible
Tephrocybe	*rancida*	.	Europe (near stumps of conifer & broadleaf trees), MOTW, page 86	.	Rancid	Ashcap Cavalier Gill Mushroom	Unknown
Tricholomopsis	*Decora*	.	Canada, N. USA (on logs in mountain Fir conifer forest), MOTW, page 96, PFGM, page 195	.	Yellow Rider	Broad Cavalier Gill Mushroom	Edible
Tricholomopsis	*Rutilans*	.	USA (on stumps, logs, branches in conifer forest), PFGM, page 194, MOTW, page 96	.	Red Rider	Broad Cavalier Gill Mushroom	Edible-Poor
Tricholomopsis	*Rutilans*	*rutilans*	USA (on stumps, logs, branches in conifer forest)	Original	Red Rider	Broad Cavalier Gill Mushroom	Edible-Poor
Tricholomopsis	*Rutilans*	*variegata*	USA (on stumps, logs, branches in conifer	Variegated	Red Rider	Broad Cavalier Gill Mushroom	Edible-Poor

			forest), MOTW, page 96				
Tricholomopsis	*sulphureoides*	.	USA, Europe (on stumps, logs, branches in conifer forest), PFGM, page 195	.	Sulfur	Broad Cavalier Gill Mushroom	Edible
Tricholomopsis	*sulphureoides*	*sulphureoides*	USA, Europe (on stumps, logs, branches in conifer forest)	Original	Sulfur	Broad Cavalier Gill Mushroom	Edible
Tricholomopsis	*sulphureoides*	*ornata*	USA, Europe (on stumps, logs, branches in conifer forest), MOTW, page 96	Ornate	Sulfur	Broad Cavalier Gill Mushroom	Edible

Order 622.202 *Russulales* Brittlegill Mushroom

Latin: *Russula brittle* **FAMILIES**

622.202.01 *Russulaceae* Brittlegill Mushroom
622.202.02 *Bondarzewiaceae* Bondarzev Brittlegill Mushroom
622.202.03 *Echinodontiaceae* Urchin Brittlegill Mushroom
622.202.04 *Lachnocladiaceae* Downy Brittlegill Mushroom
622.202.05 *Peniophoraceae* Pen Brittlegill Mushroom
622.202.06 *Stereaceae* Stereo Brittlegill Mushroom

Family 622.202.01 *Russulaceae* Brittlegill Mushroom

Genus	Species	Variety	Native Location	Variety	Specimen	Genome	Habit
Russula	*adusta*	.	USA (in woods), MOTW, page 146, PFGM, page 319	.	Burned	Brittlegill Mushroom	Edible-Poor
Russula	*aeruginea*	.	USA (Pine & Spruce conifer & Birch broadleaf forests), MOTW, page 152, PFGM, page 324	.	Copper Rust	Brittlegill Mushroom	Edible-Fair
Russula	*albella*	.	USA, PFGM, page 317	.	Small White	Brittlegill Mushroom	Unknown
Russula	*albida*	.	USA, PFGM, page 317	.	White	Brittlegill Mushroom	Unknown
Russula	*albidula*	.	E. USA (conifer & hardwood forest), PFGM, page 317	.	Whitish	Brittlegill Mushroom	Unknown
Russula	*albonigra*	.	USA, Europe (Beech broadleaf woods), MOTW, page 146, PFGM, page 319	.	Dark Gray	Brittlegill Mushroom	Edible-Poor
Russula	*alutacea*	.	USA (conifer & broadleaf forests), MOTW, page 155	.	Leather	Brittlegill Mushroom	Edible-Good
Russula	*amethystina*	.	Europe (conifer forest), MOTW, page 153	.	Amethyst	Brittlegill Mushroom	Edible
Russula	*amoena*	.	Europe (conifer forest), MOTW, page 151	.	Pleasant	Brittlegill Mushroom	Edible
Russula	*amoenolens*	.	USA, PFGM, page 323	.	.	Brittlegill Mushroom	Unknown
Russula	*aquosa*	.	Europe (conifer forest), MOTW, page 157	.	Watery	Brittlegill Mushroom	Inedible
Russula	*aquosa*	*aquosa*	Europe (conifer forest)	Original	Watery	Brittlegill Mushroom	Inedible
Russula	*aquosa*	*carminea*	Europe (conifer forest)	Carmine	Watery	Brittlegill Mushroom	Inedible
Russula	*atropurpurea*	.	USA (conifer & Oak broadleaf forests), MOTW, page 157	.	Dark Purple	Brittlegill Mushroom	Inedible
Russula	*atropurpurea*	*atropurpurea*	USA (conifer & Oak broadleaf forests)	Original	Dark Purple	Brittlegill Mushroom	Inedible
Russula	*atropurpurea*	*depallens*	USA (conifer & Oak broadleaf forests)	Becoming Pale	Dark Purple	Brittlegill Mushroom	Inedible
Russula	*atropurpurea*	*krombholzii*	USA (conifer & Oak broadleaf forests)	Krombholz	Dark Purple	Brittlegill Mushroom	Inedible
Russula	*atropurpurea*	*rubra*	USA (conifer & Oak broadleaf forests)	Red Center	Dark Purple	Brittlegill Mushroom	Inedible
Russula	*atropurpurea*	*undulata*	USA (conifer & Oak broadleaf forests)	Wavy	Dark Purple	Brittlegill Mushroom	Inedible
Russula	*atrorubens*	.	Europe (sandy woods), MOTW, page 157	.	Dark Red	Brittlegill Mushroom	Inedible
Russula	*aurea*	.	USA (humid broadleaf woods)	.	Golden	Brittlegill Mushroom	Edible-Fair
Russula	*aurea*	*aurea*	USA (humid broadleaf woods)	Original	Golden	Brittlegill Mushroom	Edible-Fair
Russula	*aurea*	*aurata*	USA (humid broadleaf woods), MOTW, page 156	Gilded	Golden	Brittlegill Mushroom	Edible-Fair
Russula	*aurora*	.	Europe, MOTW, page 146	.	Aurora	Brittlegill Mushroom	Edible
Russula	*aurora*	*aurora*	Europe	.	Aurora	Brittlegill Mushroom	Edible
Russula	*aurora*	*incarnata*	Europe	.	.	Brittlegill Mushroom	Edible
Russula	*aurora*	*rosea*	Europe	.	.	Brittlegill Mushroom	Edible
Russula	*aurora*	*velutipes*	Europe	.	.	Brittlegill Mushroom	Edible
Russula	*azurea*	.	Europe (conifer forest), MOTW, page 151	.	Violet Blue	Brittlegill Mushroom	Edible

Russula	*badia*	.	Europe (mountain Fir & Pine conifer forest), MOTW, page 160	.	Red Brown	Brittlegill Mushroom	Inedible
Russula	*betularum*	.	USA (Birch broadleaf forest), MOTW, page 160	.	Birch	Brittlegill Mushroom	Inedible
Russula	*brevipes*	.	USA (conifer & mixed conifer-hardwood forest), PFGM, page 323	.	Stubby	Brittlegill Mushroom	Edible
Russula	*brunneoviolacea*	.	Europe (Oak & Beech broadleaf forest), MOTW, page 148)	.	Violet Brown	Brittlegill Mushroom	Edible-Excellent
Russula	*brunneoviolacea*	*brunneoviolacea*	Europe (Oak & Beech broadleaf forest)	Original	Violet Brown	Brittlegill Mushroom	Edible-Excellent
Russula	*brunneoviolacea*	*pseudoviolacea*	Europe (Oak & Beech broadleaf forest)	Lesser	Violet Brown	Brittlegill Mushroom	Edible-Excellent
Russula	*caerulea*	.	Europe (Pine open woods & conifer forest), MOTW, page 153	.	Blue Violet	Brittlegill Mushroom	Inedible
Russula	*cerolens*	.	USA, PFGM, page 323	.	Cerolent	Brittlegill Mushroom	Unknown
Russula	*chloroides*	.	Europe (conifer & broadleaf forests), MOTW, page 144	.	Greenish	Brittlegill Mushroom	Edible
Russula	*cicatricata*	.	Europe	.	Scarred	Brittlegill Mushroom	Edible
Russula	*cicatricata*	*cicatricata*	Europe	Original	Scarred	Brittlegill Mushroom	Edible
Russula	*cicatricata*	*barlae*	Europe, MOTW, page 152	Barla	Scarred	Brittlegill Mushroom	Edible
Russula	*cicatricata*	*fusca*	Europe	Dark	Scarred	Brittlegill Mushroom	Edible
Russula	*claroflava*	.	Europe (Birch broadleaf forest), MOTW, page 162	.	Chrome Yellow	Brittlegill Mushroom	Edible
Russula	*claroflava*	*claroflava*	Europe (Birch broadleaf forest)	Original	Chrome Yellow	Brittlegill Mushroom	Edible
Russula	*claroflava*	*constans*	Europe (Birch broadleaf forest)	Uniform	Chrome Yellow	Brittlegill Mushroom	Edible
Russula	*claroflava*	*flava*	Europe (Birch broadleaf forest)	Light	Chrome Yellow	Brittlegill Mushroom	Edible
Russula	*compacta*	.	C. & E. USA (conifer & mixed conifer-hardwood forest), PFGM, page 318	.	Compact	Brittlegill Mushroom	Edible
Russula	*crassotunicata*	.	USA, PFGM, page 319	.	Crass Blunt	Brittlegill Mushroom	Unknown
Russula	*crustosa*	.	USA, PFGM, page 325	.	Encrusted	Brittlegill Mushroom	Edible
Russula	*cuprea*	.	Europe (conifer & broadleaf forests)	.	Copper	Brittlegill Mushroom	Inedible
Russula	*cuprea*	*cuprea*	Europe (conifer & broadleaf forests)	Original	Copper	Brittlegill Mushroom	Inedible
Russula	*cuprea*	*firmula*	Europe (Fir conifer & broadleaf forests), MOTW, pages, 151, 153	Firm	Copper	Brittlegill Mushroom	Inedible
Russula	*cuprea*	*urens*	Europe (conifer & broadleaf forests)	Burning	Copper	Brittlegill Mushroom	Inedible
Russula	*cyanoxantha*	.	USA (in woods), MOTW, page 148	.	Blue & Yellow	Brittlegill Mushroom	Edible-Excellent
Russula	*cyanoxantha*	*cyanoxantha*	USA (in woods)	Original	Blue & Yellow	Brittlegill Mushroom	Edible-Excellent
Russula	*cyanoxantha*	*cutefracta*	USA (in woods), MOTW, page 149	Fractured	Blue & Yellow	Brittlegill Mushroom	Edible-Excellent
Russula	*cyanoxantha*	*pallida*	USA (in woods)	Pale	Blue & Yellow	Brittlegill Mushroom	Edible-Excellent
Russula	*cyanoxantha*	*peltereaui*	USA (in woods)	Peltereau	Blue & Yellow	Brittlegill Mushroom	Edible-Excellent
Russula	*decolorans*	.	USA (Fir conifer forest), MOTW, page 162	.	Discolored	Brittlegill Mushroom	Edible
Russula	*delica*	.	USA (conifer & broadleaf forests), MOTW, page 144, PFGM, page 318	.	Weaned	Brittlegill Mushroom	Edible-Poor
Russula	*delica*	*delica*	USA (conifer & broadleaf forests)	Original	Weaned	Brittlegill Mushroom	Edible-Poor
Russula	*delica*	*flavispora*	USA (conifer & broadleaf forests)	Yellow Spore	Weaned	Brittlegill Mushroom	Edible-Poor
Russula	*densifolia*	.	USA (conifer & hardwood forest), MOTW, page 145, PFGM, page 319	.	Dense Gill	Brittlegill Mushroom	Edible-Toxic
Russula	*densifolia*	*densifolia*	USA (conifer & hardwood forest)	Original	Dense Gill	Brittlegill Mushroom	Edible-Toxic
Russula	*densifolia*	*densissima*	USA (conifer & hardwood forest)	Whitecap	Dense Gill	Brittlegill Mushroom	Edible-Toxic
Russula	*dissimilans*	.	USA, PFGM, page 319	.	Dissimilar	Brittlegill Mushroom	Unknown
Russula	*emetica*	.	USA (decayed wood, moss, bogs), MOTW, page 160, PFGM, page 320	.	Sickener Deathcap	Brittlegill Mushroom	Poison
Russula	*emetica*	*emetica*	USA (decayed wood, moss, bogs)	Original	Sickener Deathcap	Brittlegill Mushroom	Poison
Russula	*emetica*	*clusii*	USA (decayed wood, moss, bogs)	Clus	Sickener Deathcap	Brittlegill Mushroom	Poison
Russula	*emetica*	*gregaria*	USA (decayed wood, moss, bogs)	Gregariou s	Sickener Deathcap	Brittlegill Mushroom	Poison

Russula	exalbicans	.	Europe, MOTW, page 158	.	Excaliber	Brittlegill Mushroom	Inedible
Russula	farinipes	.	Europe (in woods), MOTW, page 162	.	Flour Foot	Brittlegill Mushroom	Inedible
Russula	fellea	.	Europe (Beech broadleaf woods), MOTW, page 162	.	Bitter	Brittlegill Mushroom	Inedible
Russula	flavipes	.	E. USA (grasslands & deciduous forest), PFGM, page 320	.	Golden	Brittlegill Mushroom	Unknown
Russula	foetens	.	USA (in humid woods), MOTW, page 164, PFGM, page 322	.	Fetid	Brittlegill Mushroom	Inedible
Russula	fragilis	.	USA (on soil & rotten wood in conifer, mixed & hardwood forest), MOTW, page 160, PFGM, page 321	.	Fragile Deathcap	Brittlegill Mushroom	Poison
Russula	fragilis	fragilis	USA (on soil & rotten wood in conifer, mixed & hardwood forest)	Original	Fragile Deathcap	Brittlegill Mushroom	Poison
Russula	fragilis	bataillei	USA (on soil & rotten wood in conifer, mixed & hardwood forest)	Bataille	Fragile Deathcap	Brittlegill Mushroom	Poison
Russula	fragilis	fallax	USA (on soil & rotten wood in conifer, mixed & hardwood forest)	Deceptive	Fragile Deathcap	Brittlegill Mushroom	Poison
Russula	fragilis	knauthii	USA (on soil & rotten wood in conifer, mixed & hardwood forest)	Knauth	Fragile Deathcap	Brittlegill Mushroom	Poison
Russula	fragilis	violascens	USA (on soil & rotten wood in conifer, mixed & hardwood forest)	Violet	Fragile Deathcap	Brittlegill Mushroom	Poison
Russula	fragrantissima	.	USA (conifer & hardwood forest), PFGM, page 321	.	Stinking	Brittlegill Mushroom	Edible
Russula	grata	.	Europe (broadleaf forest)	.	Pleasing	Brittlegill Mushroom	Inedible
Russula	grata	grata	Europe (broadleaf forest)	Original	Pleasing	Brittlegill Mushroom	Inedible
Russula	grata	laurocerasi	Europe (broadleaf forest), MOTW, page 164, PFGM, page 322	Cherry Laurel	Pleasing	Brittlegill Mushroom	Inedible
Russula	grisea	.	Europe (Broadleaf forest), MOTW, page 148	.	Violet Gray	Brittlegill Mushroom	Edible-Excellent
Russula	grisea	grisea	Europe (Broadleaf forest)	Original	Violet Gray	Brittlegill Mushroom	Edible-Excellent
Russula	grisea	palumbina	Europe (Broadleaf forest)	Ringdove	Violet Gray	Brittlegill Mushroom	Edible-Excellent
Russula	grisea	pictipes	Europe (Broadleaf forest)	Painted Foot	Violet Gray	Brittlegill Mushroom	Edible-Excellent
Russula	heterophylla	.	USA (conifer & broadleaf forests), MOTW, page 148	.	Variable Lamella	Brittlegill Mushroom	Edible-Good
Russula	heterophylla	heterophylla	USA (conifer & broadleaf forests)	Original	Variable Lamella	Brittlegill Mushroom	Edible-Good
Russula	heterophylla	adusta	USA (conifer & broadleaf forests)	Dusty	Variable Lamella	Brittlegill Mushroom	Edible-Good
Russula	heterophylla	livida	USA (conifer & broadleaf forests), MOTW, page 148	Yellow Spot	Variable Lamella	Brittlegill Mushroom	Edible-Good
Russula	heterophylla	virginea	USA (conifer & broadleaf forests), MOTW, page 148	Virgin White	Variable Lamella	Brittlegill Mushroom	Edible-Good
Russula	illota	.	Europe (in woods), MOTW, page 164	.	Dirty	Brittlegill Mushroom	Inedible
Russula	innocua	.	USA (Pine & Spruce conifer & Birch broadleaf forests)	.	Harmless	Brittlegill Mushroom	Edible-Poor
Russula	innocua	innocua	USA (Pine & Spruce conifer & Birch broadleaf forests)	Original	Harmless	Brittlegill Mushroom	Edible-Poor
Russula	innocua	smaragdina	USA (Pine & Spruce conifer & Birch broadleaf forests), MOTW, page 152	Emerald	Harmless	Brittlegill Mushroom	Edible-Poor
Russula	insignis	.	USA (grassy areas, open woods)	.	Insignia	Brittlegill Mushroom	Inedible
Russula	insignis	insignis	USA (grassy areas, open woods)	Original	Insignia	Brittlegill Mushroom	Inedible
Russula	insignis	depauperata	USA (grassy areas, open woods)	Imperfect	Insignia	Brittlegill Mushroom	Inedible
Russula	insignis	pectinata	USA (grassy areas, open woods)	Combed	Insignia	Brittlegill Mushroom	Inedible
Russula	integra	.	USA (plains & mountain conifer forest), MOTW, page 154	.	Entire	Brittlegill Mushroom	Edible-Good
Russula	integra	integra	USA (plains & mountain conifer forest)	Original	Entire	Brittlegill Mushroom	Edible-Good
Russula	integra	campestris	USA (plains & mountain conifer forest)	Field	Entire	Brittlegill Mushroom	Edible-Good
Russula	integra	purpurella	USA (plains & mountain conifer forest)	Purple	Entire	Brittlegill Mushroom	Edible-Good
Russula	ionochlora	.	USA (broadleaf forest), MOTW, page 148	.	Violet & Green	Brittlegill Mushroom	Edible-Excellent
Russula	lilacea	.	Europe (Oak, Beech & Hazelnut broadleaf forest), MOTW, page 151	.	Lilac	Brittlegill Mushroom	Edible
Russula	lilacea	lilacea	Europe (Oak, Beech & Hazelnut broadleaf forest)	Original	Lilac	Brittlegill Mushroom	Edible

Russula	*lilacea*	*carnicolor*	Europe (Oak, Beech & Hazelnut broadleaf forest)	Flesh Color	Lilac	Brittlegill Mushroom	Edible
Russula	*lilacepes*	.	USA, PFGM, page 324	.	Lilac Foot	Brittlegill Mushroom	Unknown
Russula	*linnaei*	.	USA (conifer & broadleaf forests), MOTW, page 146	.	Linnaeus	Brittlegill Mushroom	Inedible
Russula	*maculata*	.	USA (Beech broadleaf forest), MOTW, page 157	.	Spotted	Brittlegill Mushroom	Inedible
Russula	*maculata*	*maculata*	USA (Beech broadleaf forest)	Original	Spotted	Brittlegill Mushroom	Inedible
Russula	*maculata*	*elegans*	USA (Beech broadleaf forest)	Elegant	Spotted	Brittlegill Mushroom	Inedible
Russula	*mariae*	.	USA (in woods)	.	Maria	Brittlegill Mushroom	Edible-Seasoning
Russula	*melliolens*	.	Europe (broadleaf forest), MOTW, page 156	.	Honey Scented	Brittlegill Mushroom	Edible-Good
Russula	*melliolens*	*Melliolens*	Europe (broadleaf forest)	Original	Honey Scented	Brittlegill Mushroom	Edible-Good
Russula	*melliolens*	*Sapida*	Europe (broadleaf forest)	Tasty	Honey Scented	Brittlegill Mushroom	Edible-Good
Russula	*melzeri*	.	Europe (conifer forest)	.	Melzer	Brittlegill Mushroom	Edible
Russula	*melzeri*	*melzeri*	Europe (conifer forest)	Original	Melzer	Brittlegill Mushroom	Edible
Russula	*melzeri*	*laricina*	Europe (Larch conifer forest), MOTW, page 153	Larch	Melzer	Brittlegill Mushroom	Edible
Russula	*mustelina*	.	Europe (mountain Fir & Pine conifer forest), MOTW, page 164	.	Marten	Brittlegill Mushroom	Edible
Russula	*nauseosa*	.	USA (mountain conifer forest)	.	Nauseating	Brittlegill Mushroom	Edible
Russula	*nauseosa*	*nauseosa*	USA (mountain conifer forest), MOTW, page 156	Original	Nauseating	Brittlegill Mushroom	Edible
Russula	*nauseosa*	*fusca*	USA (mountain conifer forest)	Dark	Nauseating	Brittlegill Mushroom	Edible
Russula	*nigricans*	.	USA (in woods), MOTW, page 145, PFGM, page 319	.	Blackish Brown	Brittlegill Mushroom	Edible-Poor
Russula	*nigricans*	*nigricans*	USA (in woods)	Original	Blackish Brown	Brittlegill Mushroom	Edible-Poor
Russula	*nigricans*	*elephantina*	USA (in woods)	Snout	Blackish Brown	Brittlegill Mushroom	Edible-Poor
Russula	*nitida*	.	Europe (mountain conifer forest)	.	Shiny	Brittlegill Mushroom	Edible
Russula	*nitida*	*nitida*	Europe (mountain conifer forest)	Original	Shiny	Brittlegill Mushroom	Edible
Russula	*nitida*	*venosa*	Europe (mountain conifer forest), MOTW, page 160	Veined	Shiny	Brittlegill Mushroom	Edible
Russula	*nobilis*	.	Europe	.	Noble	Brittlegill Mushroom	Inedible
Russula	*nobilis*	*nobilis*	Europe	Original	Noble	Brittlegill Mushroom	Inedible
Russula	*nobilis*	*fageticola*	Europe (under Beech)	Beech	Noble	Brittlegill Mushroom	Inedible
Russula	*nobilis*	*mairei*	Europe, MOTW, page 160	Maire	Noble	Brittlegill Mushroom	Inedible
Russula	*ochroleuca*	.	USA (conifer forest), MOTW, page 162	.	Yellowish White	Brittlegill Mushroom	Inedible
Russula	*ochroleuca*	*ochroleuca*	USA (conifer forest)	Original	Yellowish White	Brittlegill Mushroom	Inedible
Russula	*ochroleuca*	*citrina*	USA (conifer forest)	Citrine	Yellowish White	Brittlegill Mushroom	Inedible
Russula	*ochroleuca*	*granulosa*	USA (conifer forest)	Grainy	Yellowish White	Brittlegill Mushroom	Inedible
Russula	*olivacea*	.	USA (conifer & broadleaf forests), MOTW, page 153	.	Olive	Brittlegill Mushroom	Edible-Good
Russula	*paludosa*	.	Europe (marshland, swampy ground, humid conifer forest), MOTW, pages 160, 162	.	Swamp	Brittlegill Mushroom	Edible-Good
Russula	*parazurea*	.	Europe (broadleaf forest), MOTW, page 152	.	Dull Cyan	Brittlegill Mushroom	Edible
Russula	*pectinata*		C. & E. USA (deciduous & mixed conifer-deciduous forest), MOTW, page 164, PFGM, page 323	.	Scalloped	Brittlegill Mushroom	Edible-Poor
Russula	*pectinatoides*	.	C. & E. USA (deciduous & mixed conifer-deciduous forest), PFGM, page 322	.	Lesser Scalloped	Brittlegill Mushroom	Edible-Poor
Russula	*pseudoaeruginea*	.	USA (Birch broadleaf forest)	.	Lesser Copper	Brittlegill Mushroom	Edible-Good
Russula	*pseudoaeruginea*	*pseudoaeruginea*	USA (Birch broadleaf forest)	Original	Lesser Copper	Brittlegill Mushroom	Edible-Good
Russula	*pseudoaeruginea*	*galochroa*	USA (Birch broadleaf forest), MOTW, page 128	Milky	Lesser Copper	Brittlegill Mushroom	Edible-Good
Russula	*pseudodelica*	.	Europe (conifer & broadleaf forests), MOTW, page 144	.	Herring	Brittlegill Mushroom	Edible
Russula	*pseudointegra*	.	Europe (Oak broadleaf forest), MOTW, page 154	.	Lesser Entire	Brittlegill Mushroom	Edible
Russula	*puellaris*	.	USA (Pine conifer & Birch & Alder broadleaf forests), page 161	.	Girl	Brittlegill Mushroom	Edible
Russula	*puellaris*	*puellaris*	USA (Pine conifer & Birch & Alder broadleaf forests)	Original	Girl	Brittlegill Mushroom	Edible
Russula	*puellaris*	*leprosa*	USA (Pine conifer & Birch & Alder broadleaf forests)	Elegant	Girl	Brittlegill Mushroom	Edible
Russula	*queletii*	.	Europe (mountain Spruce conifer forest), MOTW, page 158	.	Quelet	Brittlegill Mushroom	Inedible
Russula	*queletii*	*queletii*	Europe (mountain Spruce conifer forest)	Original	Quelet	Brittlegill Mushroom	Inedible
Russula	*queletii*	*flavovirens*	Europe (mountain	Yellow	Quelet	Brittlegill Mushroom	Inedible

			Spruce conifer forest)	Green			
Russula	*risigallina*	.	USA (in woods)	.	Risigall	Brittlegill Mushroom	Edible-Excellent
Russula	*risigallina*	*armeniaca*	USA (in woods)	Original	Risigall	Brittlegill Mushroom	Edible-Excellent
Russula	*risigallina*	*chamaeleontina*	USA (conifer forest), MOTW, page 156	Chameleon	Risigall	Brittlegill Mushroom	Edible-Excellent
Russula	*risigallina*	*lutea*	USA (in woods), MOTW, page 156	Lute Yellow	Risigall	Brittlegill Mushroom	Edible-Excellent
Russula	*risigallina*	*luteorosella*	USA (in woods)	Yellow Rose	Risigall	Brittlegill Mushroom	Edible-Excellent
Russula	*risigallina*	*ochracea*	USA (in woods)	Ocher	Risigall	Brittlegill Mushroom	Edible-Excellent
Russula	*risigallina*	*vitellina*	USA (in woods)	Egg Yellow	Risigall	Brittlegill Mushroom	Edible-Excellent
Russula	*romellii*	.	Europe (broadleaf forest), MOTW, page 154	.	Romeli	Brittlegill Mushroom	Edible
Russula	*rosea*	.	USA (conifer & broadleaf forests)	.	Rose	Brittlegill Mushroom	Edible-Poor
Russula	*rosea*	*rosea*	USA (conifer & broadleaf forests)	Original	Rose	Brittlegill Mushroom	Edible-Poor
Russula	*rosea*	*alba*	USA (conifer & broadleaf forests)	Whitish	Rose	Brittlegill Mushroom	Edible-Poor
Russula	*rosea*	*lactea*	USA (conifer & broadleaf forests)	Milky	Rose	Brittlegill Mushroom	Edible-Poor
Russula	*rosea*	*lepida*	USA (conifer & broadleaf forests), MOTW, page 146	Scaly	Rose	Brittlegill Mushroom	Edible-Poor
Russula	*sanguinaria*	.	USA (Pine conifer forest), MOTW, page 158	.	Blood Red	Brittlegill Mushroom	Inedible
Russula	*sanguinaria*	*sanguinaria*	USA (Pine conifer forest)	Original	Blood Red	Brittlegill Mushroom	Inedible
Russula	*sanguinaria*	*pseudorosacea*	USA (Pine conifer forest), MOTW, page 158	Lesser Rose	Blood Red	Brittlegill Mushroom	Inedible
Russula	*sanguinaria*	*rosacea*	USA (Pine conifer forest)	Rose	Blood Red	Brittlegill Mushroom	Inedible
Russula	*sanguinaria*	*sanguinea*	USA (Pine conifer forest)	Ordinary	Blood Red	Brittlegill Mushroom	Inedible
Russula	*sardonia*	.	USA (plains Pine forest), MOTW, page 158	.	Bitter	Brittlegill Mushroom	Inedible
Russula	*sardonia*	*sardonia*	USA (plains Pine forest)	Original	Bitter	Brittlegill Mushroom	Inedible
Russula	*sardonia*	*chrysodacryon*	USA (plains Pine forest)	Golden Tear	Bitter	Brittlegill Mushroom	Inedible
Russula	*sardonia*	*citrina*	USA (plains Pine forest)	Citrine	Bitter	Brittlegill Mushroom	Inedible
Russula	*sardonia*	*drimeia*	USA (plains Pine forest)	Acid	Bitter	Brittlegill Mushroom	Inedible
Russula	*sardonia*	*emeticiformis*	USA (plains Pine forest)	Vomit	Bitter	Brittlegill Mushroom	Inedible
Russula	*sardonia*	*flavovirens*	USA (plains Pine forest)	Yellow Green	Bitter	Brittlegill Mushroom	Inedible
Russula	*sardonia*	*mellina*	USA (plains Pine forest)	Honey	Bitter	Brittlegill Mushroom	Inedible
Russula	*sardonia*	*viridis*	USA (plains Pine forest)	Green	Bitter	Brittlegill Mushroom	Inedible
Russula	*silvestris*	.	USA (in woods), MOTW, page 160	.	Forest	Brittlegill Mushroom	Inedible
Russula	*silvestris*	*silvestris*	USA (in woods)	Original	Forest	Brittlegill Mushroom	Inedible
Russula	*silvestris*	*emeticella*	USA (in woods)	Vomit	Forest	Brittlegill Mushroom	Inedible
Russula	*solaris*	.	Europe (Beech broadleaf woods), MOTW, page 162	.	Sun	Brittlegill Mushroom	Inedible
Russula	*sororia*	.	Europe (conifer & broadleaf forest), MOTW, page 164	.	Fetid Sister	Brittlegill Mushroom	Inedible
Russula	*sororia*	*sororia*	Europe (conifer & broadleaf forest)	Original	Fetid Sister	Brittlegill Mushroom	Inedible
Russula	*sororia*	*consobrina*	Europe (conifer forest), MOTW, page 162	Cousin	Fetid Sister	Brittlegill Mushroom	Inedible
Russula	*sororia*	*intermedia*	Europe (broadleaf forest)	Intermediate	Fetid Sister	Brittlegill Mushroom	Inedible
Russula	*subfoetens*	.	USA, PFGM, page 322	.	Lesser Fetid	Brittlegill Mushroom	Unknown
Russula	*torulosa*	.	Europe (plains conifer forest), MOTW, page 158	.	Bulky	Brittlegill Mushroom	Edible
Russula	*torulosa*	*torulosa*	Europe (plains conifer forest)	Original	Bulky	Brittlegill Mushroom	Edible
Russula	*torulosa*	*fuscorubra*	Europe (plains conifer forest)	Fuscous Red	Bulky	Brittlegill Mushroom	Edible
Russula	*turci*	.	Europe (Fir & Pine conifer forest), MOTW, page 153	.	Baroness Turco-Lazzari	Brittlegill Mushroom	Edible
Russula	*variata*	.	C. & E. USA (hardwood & mixed conifer-hardwood forest), PFGM, page 323	.	Variable	Brittlegill Mushroom	Edible
Russula	*velenovskyi*	.	Europe (Beech & Birch broadleaf forest), MOTW, page 154	.	Velenovsky	Brittlegill Mushroom	Edible
Russula	*ventricosipes*	.	C. & E. USA (sandy soil near Pine), PFGM, page 324	.	Sand	Brittlegill Mushroom	Unknown
Russula	*vesca*	.	USA (plains, conifer & broadleaf forests), MOTW, page 151	.	Emaciated	Brittlegill Mushroom	Edible-Good
Russula	*vesca*	*vesca*	USA (plains, conifer & broadleaf forests)	Original	Emaciated	Brittlegill Mushroom	Edible-Good
Russula	*vesca*	*mitis*	USA (plains, conifer & broadleaf forests)	Mild	Emaciated	Brittlegill Mushroom	Edible-Good
Russula	*vinosa*	.	Europe (humid conifer forest)	.	Wine Color	Brittlegill Mushroom	Edible
Russula	*vinosa*	*vinosa*	Europe (humid conifer forest)	Original	Wine Color	Brittlegill Mushroom	Edible
Russula	*vinosa*	*brunnea*	Europe (humid conifer forest)	Brown	Wine Color	Brittlegill Mushroom	Edible
Russula	*vinosa*	*obscura*	Europe (humid conifer forest), MOTW, page 156	Obscure	Wine Color	Brittlegill Mushroom	Edible
Russula	*vinosa*	*purpurea*	Europe (humid conifer forest)	Purple	Wine Color	Brittlegill Mushroom	Edible
Russula	*violacea*	.	Europe (open Pine woods),	.	Violet	Brittlegill Mushroom	Inedible

			MOTW, page 161				
Russula	violacea	violacea	Europe (open Pine woods)	Original	Violet	Brittlegill Mushroom	Inedible
Russula	violacea	carneolilacina	Europe (open Pine woods)	Flesh Lilac	Violet	Brittlegill Mushroom	Inedible
Russula	violacea	viridis	Europe (open Pine woods)	Green	Violet	Brittlegill Mushroom	Inedible
Russula	violeipes	.	USA (conifer forest)	.	Violet Foot	Brittlegill Mushroom	Edible
Russula	violeipes	violeipes	USA (conifer forest)	Original	Violet Foot	Brittlegill Mushroom	Edible
Russula	violeipes	citrina	USA (conifer forest)	Citrine	Violet Foot	Brittlegill Mushroom	Edible
Russula	violeipes	chlora	USA (conifer forest)	Green	Violet Foot	Brittlegill Mushroom	Edible
Russula	violeipes	olivascens	USA (conifer forest)	Olive	Violet Foot	Brittlegill Mushroom	Edible
Russula	violeipes	punctata	USA (conifer forest), MOTW, page 162	Dotted	Violet Foot	Brittlegill Mushroom	Edible
Russula	virescens	.	C. & E. USA (deciduous & mixed conifer-deciduous forest), MOTW, page 149, PFGM, page 325	.	Green	Brittlegill Mushroom	Edible-Excellent
Russula	viscida	.	Europe (mountain Fir conifer forest), MOTW, page 160	.	Sticky	Brittlegill Mushroom	Edible-Poor
Russula	viscida	viscida	Europe (mountain Fir conifer forest)	Original	Sticky	Brittlegill Mushroom	Edible-Poor
Russula	viscida	artesiana	Europe (mountain Fir conifer forest)	Artes	Sticky	Brittlegill Mushroom	Edible-Poor
Russula	viscida	chrismantiae	Europe (mountain Fir conifer forest)	Christmanti	Sticky	Brittlegill Mushroom	Edible-Poor
Russula	viscida	occidentalis	Europe (mountain Fir conifer forest)	Occidental	Sticky	Brittlegill Mushroom	Edible-Poor
Russula	xerampelina	.	USA (conifer & mixed conifer-deciduous forest), MOTW, page 152, PFGM, page 325	.	Shellfish	Brittlegill Mushroom	Edible-Good
Russula	xerampelina	xerampelina	USA (conifer & mixed conifer-deciduous forest)	Original	Shellfish	Brittlegill Mushroom	Edible-Good
Russula	xerampelina	erythropus	USA (conifer & mixed conifer-deciduous forest)	Earth Red	Shellfish	Brittlegill Mushroom	Edible-Good
Russula	xerampelina	ochraceus	USA (conifer & mixed conifer-deciduous forest)	Ocher	Shellfish	Brittlegill Mushroom	Edible-Good
Lactarius	acerrimus	.	Europe (broadleaf forest), MOTW, page 140	.	Very Bitter	Milkcap Brittlegill Mushroom	Inedible
Lactarius	acris	.	Europe (broadleaf forest), MOTW, page 139	.	Bitter	Milkcap Brittlegill Mushroom	Inedible
Lactarius	aerumbonatus	.	Europe (conifer forest), MOTW, page 140	.	Knobless	Milkcap Brittlegill Mushroom	Inedible
Lactarius	affinis	.	S. Canada, USA (pastures, mixed conifer-hardwood forest), PFGM, page 326	.	Kindred	Milkcap Brittlegill Mushroom	Unknown
Lactarius	allardii	.	USA, PFGM, page 333	.	Allard	Milkcap Brittlegill Mushroom	Unknown
Lactarius	alnicola	.	USA, PFGM, page 335	.	Alder	Milkcap Brittlegill Mushroom	Unknown
Lactarius	alpinus	.	USA, PFGM, page 335	.	Alpine	Milkcap Brittlegill Mushroom	Unknown
Lactarius	angustifolius	.	USA (Texas), PFGM, page 334	.	Narrow Gill	Milkcap Brittlegill Mushroom	Unknown
Lactarius	argyllaceifolius	.	USA (hardwood forests), PFGM, page 338	.	Argyle Gill	Milkcap Brittlegill Mushroom	Unknown
Lactarius	aspideus	.	Europe, MOTW, page 136	.	Shield	Milkcap Brittlegill Mushroom	Inedible
Lactarius	atroviridis	.	C. & E. USA (conifer & hardwood forest), PFGM, page 327	.	Dark Spotted	Milkcap Brittlegill Mushroom	Unknown
Lactarius	aurantiacus	.	USA (conifer & broadleaf forests)	.	Orange	Milkcap Brittlegill Mushroom	Edible-Poor
Lactarius	aurantiacus	aurantiacus	USA (conifer & broadleaf forests)	Original	Orange	Milkcap Brittlegill Mushroom	Edible-Poor
Lactarius	aurantiacus	aurantiofulvus	USA (conifer & broadleaf forests)	Fulvous	Orange	Milkcap Brittlegill Mushroom	Edible-Poor
Lactarius	aurantiacus	mitissimus	USA (conifer & broadleaf forests), MOTW, page 142	Mild	Orange	Milkcap Brittlegill Mushroom	Edible-Poor
Lactarius	badiosanguineus	.	Europe (mountains), MOTW, page 140	.	Blood Red Brown	Milkcap Brittlegill Mushroom	Inedible
Lactarius	barrowsii	.	USA, PFGM, page 331	.	Barrows	Milkcap Brittlegill Mushroom	Unknown
Lactarius	bertillonii	.	S. Canada, C. & E. USA (humid broadleaf forest), MOTW, page 129	.	Bertillon	Milkcap Brittlegill Mushroom	Edible-Poor
Lactarius	bertillonii	bertillonii	S. Canada, C. & E. USA (humid broadleaf forest)	Original	Bertillon	Milkcap Brittlegill Mushroom	Edible-Poor
Lactarius	bertillonii	queletii	S. Canada, C. & E. USA (humid broadleaf forest)	Quelet	Bertillon	Milkcap Brittlegill Mushroom	Edible-Poor
Lactarius	blennius	.	USA (Beech & Oak broadleaf forest), MOTW, page 136	.	Mucilage	Milkcap Brittlegill Mushroom	Inedible
Lactarius	blennius	blennius	USA (Beech & Oak	Original	Mucilage	Milkcap Brittlegill	Inedible

			broadleaf forest)			Mushroom	
Lactarius	*blennius*	*virescens*	USA (Beech & Oak broadleaf forest)	Verdant	Mucilage	Milkcap Brittlegill Mushroom	Inedible
Lactarius	*blennius*	*viridis*	USA (Beech & Oak broadleaf forest)	Green	Mucilage	Milkcap Brittlegill Mushroom	Inedible
Lactarius	*caeruleitinctus*	.	SE. USA (Florida), PFGM, page 330	.	Blue Tint	Milkcap Brittlegill Mushroom	Unknown
Lactarius	*caespitosus*	.	W. USA (Spruce-Fir forest from Rocky Mtns. to pacific coast), PFGM, page 328	.	Gray	Milkcap Brittlegill Mushroom	Unknown
Lactarius	*camphoratus*	.	USA (Pine conifer & broadleaf forests), MOTW, page 144	.	Camphor	Milkcap Brittlegill Mushroom	Edible-Seasoning
Lactarius	*camphoratus*	*camphoratus*	USA (Pine conifer & broadleaf forests)	Original	Camphor	Milkcap Brittlegill Mushroom	Edible-Seasoning
Lactarius	*camphoratus*	*cimicarius*	USA (Pine conifer & broadleaf forests)	Cimicar	Camphor	Milkcap Brittlegill Mushroom	Edible-Seasoning
Lactarius	*camphoratus*	*terrei*	USA (Pine conifer & broadleaf forests)	Earth	Camphor	Milkcap Brittlegill Mushroom	Edible-Seasoning
Lactarius	*chelidonium*	.	USA, PFGM, page 331	.	Chelidon	Milkcap Brittlegill Mushroom	Unknown
Lactarius	*chrysorrheus*	.	S. Canada, USA (hardwood & mixed conifer-hardwood forest), Mexico, MOTW, page 130, PFGM, page 328	.	Gold Drop	Milkcap Brittlegill Mushroom	Inedible
Lactarius	*chrysorrheus*	*chrysorrheus*	S. Canada, USA (hardwood & mixed conifer-hardwood forest), Mexico	Original	Gold Drop	Milkcap Brittlegill Mushroom	Inedible
Lactarius	*chrysorrheus*	*theiogalus*	S. Canada, USA (hardwood & mixed conifer-hardwood forest), Mexico, MOTW, page 130	Yellow Latex	Gold Drop	Milkcap Brittlegill Mushroom	Inedible
Lactarius	*circellatus*	.	Europe (Hornbeam broadleaf forest), MOTW, page 139	.	Ringed	Milkcap Brittlegill Mushroom	Inedible
Lactarius	*citriolens*	.	Europe	.	Lemon	Milkcap Brittlegill Mushroom	Inedible
Lactarius	*citriolens*	*citriolens*	Europe	Original	Lemon	Milkcap Brittlegill Mushroom	Inedible
Lactarius	*citriolens*	*cilicioides*	Europe, MOTW, page 133a	Goat Hair	Lemon	Milkcap Brittlegill Mushroom	Inedible
Lactarius	*colorascens*	.	USA, PFGM, page 329	.	Colorful	Milkcap Brittlegill Mushroom	Unknown
Lactarius	*controversus*	.	USA (in grass of Poplar open woods), MOTW, page 131	.	Creamy Pink	Milkcap Brittlegill Mushroom	Edible-Bitter
Lactarius	*corrugis*	.	C. & E. USA (hardwood & mixed conifer-hardwood forest), MOTW, page 135, PFGM, page 329	.	Wrinkled	Milkcap Brittlegill Mushroom	Edible-Good
Lactarius	*deceptivus*	.	S. Canada, USA (Oak, Hemlock & mixed conifer-hardwood forest), PFGM, page 329	.	Cotton Roll	Milkcap Brittlegill Mushroom	Edible-Poor
Lactarius	*deliciosus*	.	USA (Pine & Juniper conifer forest), MOTW, page 134, PFGM, page 330	.	Saffron	Milkcap Brittlegill Mushroom	Edible-Good
Lactarius	*deliciosus*	*deliciosus*	USA (Pine & Juniper conifer forest)	Original	Saffron	Milkcap Brittlegill Mushroom	Edible-Good
Lactarius	*deliciosus*	*japonicus*	E. Asia (Japan Pine & Juniper conifer forest), MOTW, page 134	Pale	Saffron	Milkcap Brittlegill Mushroom	Edible-Good
Lactarius	*deterrimus*	.	USA, PFGM, page 331	.	Diterring	Milkcap Brittlegill Mushroom	Edible-Good
Lactarius	*fallax*	.	USA, PFGM, page 332	.	Deceptive	Milkcap Brittlegill Mushroom	Unknown
Lactarius	*fuliginosus*	.	USA (Oak broadleaf forest), MOTW, page 139	.	Smoky Sooty	Milkcap Brittlegill Mushroom	Inedible
Lactarius	*fulvissimus*	.	Europe, MOTW, page 142	.	Blonde Red	Milkcap Brittlegill Mushroom	Edible-Poor
Lactarius	*fulvissimus*	*fulvissimus*	Europe	.	Blonde Red	Milkcap Brittlegill Mushroom	Edible-Poor
Lactarius	*fulvissimus*	*britannicus*	Europe	Original	Blonde Red	Milkcap Brittlegill Mushroom	Edible-Poor
Lactarius	*fulvissimus*	*ichoratus*	Europe	Putrified	Blonde Red	Milkcap Brittlegill Mushroom	Edible-Poor
Lactarius	*fulvissimus*	*mitissimus*	Europe, MOTW, page 142	Mild	Blonde Red	Milkcap Brittlegill Mushroom	Edible-Poor
Lactarius	*fulvissimus*	*subsericatus*	Europe	Silky	Blonde Red	Milkcap Brittlegill Mushroom	Edible-Poor
Lactarius	*fumosus*	.	C. & E. USA (on forest	.	Smoky	Milkcap Brittlegill	Unknown

			soil), PFGM, page 331			Mushroom	
Lactarius	*gerardii*	.	USA, PFGM, page 332	.	Gerard	Milkcap Brittlegill Mushroom	Unknown
Lactarius	*glaucescens*	.	USA, PFGM, page 334	.	Glaucous	Milkcap Brittlegill Mushroom	Unknown
Lactarius	*glyciosmus*	.	USA (Birch broadleaf forest)	.	Sweet Odor	Milkcap Brittlegill Mushroom	Inedible
Lactarius	*glyciosmus*	*glyciosmus*	USA (Birch broadleaf forest)	Original	Sweet Odor	Milkcap Brittlegill Mushroom	Inedible
Lactarius	*glyciosmus*	*impolitus*	USA (Birch broadleaf forest), MOTW, page 136	Unclean	Sweet Odor	Milkcap Brittlegill Mushroom	Inedible
Lactarius	*gossypinus*	.	USA, PFGM, page 336	.	Cottony	Milkcap Brittlegill Mushroom	Unknown
Lactarius	*hatsudake*	.	E. Asia (Japan), MOTW, page 134	.	Hatsudake	Milkcap Brittlegill Mushroom	Edible-Poor
Lactarius	*helvus*	.	C. & E. USA (bogs, moist forest soil), MOTW, page 140, PFGM, page 327	.	Blonde Yellow	Milkcap Brittlegill Mushroom	Inedible-Toxic
Lactarius	*helvus*	*helvus*	C. & E. USA (bogs, moist forest soil)	Original	Blonde Yellow	Milkcap Brittlegill Mushroom	Inedible-Toxic
Lactarius	*helvus*	*aquifluus*	C. & E. USA (bogs, moist forest soil), MOTW, page 327	Waterdrop	Blonde Yellow	Milkcap Brittlegill Mushroom	Inedible-Toxic
Lactarius	*helvus*	*tomentosus*	C. & E. USA (bogs, moist forest soil)	Woolly	Blonde Yellow	Milkcap Brittlegill Mushroom	Inedible-Toxic
Lactarius	*helvus*	*tomentosus-marginatus*	USA, PFGM, page 330	Margined Woolly	Blonde Yellow	Milkcap Brittlegill Mushroom	Inedible-Toxic
Lactarius	*hepaticus*	.	Europe (plains), MOTW, page 140	.	Liver	Milkcap Brittlegill Mushroom	Inedible
Lactarius	*hygrophoroides*	.	C. & E. USA (hardwood forest), PFGM, page 331	.	Orange	Milkcap Brittlegill Mushroom	Edible-Good
Lactarius	*indigo*	.	Canada, USA (on forest soil), Mexico, MOTW, page 134, PFGM, page 331	.	Indigo	Milkcap Brittlegill Mushroom	Edible-Good
Lactarius	*involutus*	.	USA, MOTW, page 309	.	Involute	Milkcap Brittlegill Mushroom	Unknown
Lactarius	*kauffmanii*	.	USA, PFGM, page 328	.	Kauffman's	Milkcap Brittlegill Mushroom	Unknown
Lactarius	*lacunarum*	.	Europe	.	Sunken	Milkcap Brittlegill Mushroom	Edible-Poor?
Lactarius	*lacunarum*	*lacunarum*	Europe	Original	Sunken	Milkcap Brittlegill Mushroom	Edible-Poor?
Lactarius	*lacunarum*	*decipiens*	Europe, MOTW, page 142	Deceptive	Sunken	Milkcap Brittlegill Mushroom	Edible-Poor?
Lactarius	*lacunarum*	*rubescens*	Europe	Red	Sunken	Milkcap Brittlegill Mushroom	Edible-Poor?
Lactarius	*lignyotus*	.	USA (wet soil in bogs & mountain Fir conifer forest), MOTW, page 139, PFGM, page 332	.	Sooty	Milkcap Brittlegill Mushroom	Edible-Poor
Lactarius	*lilacinus*	.	Europe, MOTW, page 140	.	Lilac	Milkcap Brittlegill Mushroom	Inedible
Lactarius	*lilacinus*	*lilacinus*	Europe	Original	Lilac	Milkcap Brittlegill Mushroom	Inedible
Lactarius	*lilacinus*	*lateritioroseus*	Europe	Brick Red	Lilac	Milkcap Brittlegill Mushroom	Inedible
Lactarius	*maculatus*	.	USA, PFGM, page 339	.	Spotted	Milkcap Brittlegill Mushroom	Unknown
Lactarius	*mammosus*	.	Europe (mountain conifer forest)	.	Nippled	Milkcap Brittlegill Mushroom	Inedible
Lactarius	*mammosus*	*mammosus*	Europe (mountain conifer forest)	Original	Nippled	Milkcap Brittlegill Mushroom	Inedible
Lactarius	*mammosus*	*confusus*	Europe (mountain conifer forest)	Confused	Nippled	Milkcap Brittlegill Mushroom	Inedible
Lactarius	*mammosus*	*fuscus*	Europe (mountain conifer forest), MOTW, page 140	Cinnamon	Nippled	Milkcap Brittlegill Mushroom	Inedible
Lactarius	*mucidus*	.	USA (moist bogs & conifer forest), PFGM, page 332	.	Slimy	Milkcap Brittlegill Mushroom	Unpalatable
Lactarius	*neuhoffii*	.	C. & E. USA, PFGM, page 334	.	Neuhoff	Milkcap Brittlegill Mushroom	Unknown
Lactarius	*obscuratus*	.	Europe (humid Alder broadleaf forest)	.	Dark	Milkcap Brittlegill Mushroom	Inedible
Lactarius	*obscuratus*	*obscuratus*	Europe (humid Alder broadleaf forest)	Original	Dark	Milkcap Brittlegill Mushroom	Inedible
Lactarius	*obscuratus*	*cyathula*	Europe (humid Alder broadleaf forest), MOTW, page 140	Small Cup	Dark	Milkcap Brittlegill Mushroom	Inedible
Lactarius	*obscuratus*	*obnubilus*	Europe (humid Alder broadleaf forest)	Nubile	Dark	Milkcap Brittlegill Mushroom	Inedible

Lactarius	*obscuratus*	*radiatus*	Europe (humid Alder broadleaf forest)	Shiny	Dark	Milkcap Brittlegill Mushroom	Inedible
Lactarius	*olivaceoumbrinus*	.	USA, PFGM, page 328	.	Olive Pine Forest	Milkcap Brittlegill Mushroom	Unknown
Lactarius	*pallidus*	.	USA (Birch & Beech broadleaf forest), MOTW, page 139, PFGM, page 326	.	Pale	Milkcap Brittlegill Mushroom	Edible-Poor
Lactarius	*paradoxus*	.	USA, PFGM, page 337	.	Paradox	Milkcap Brittlegill Mushroom	Edible
Lactarius	*payettensis*	.	USA, PFGM, page 335	.	Payette	Milkcap Brittlegill Mushroom	Unknown
Lactarius	*peckii*	.	E. & S. USA (open deciduous woodlands), PFGM, page 333	.	Peck	Milkcap Brittlegill Mushroom	Unpalatable
Lactarius	*picinus*	.	Europe (mountain Fir conifer forest), MOTW, page 139	.	Mountain Fir	Milkcap Brittlegill Mushroom	Inedible
Lactarius	*piperatus*	.	E. USA (deciduous forest), MOTW, page 130, PFGM, page 333	.	Pepper	Milkcap Brittlegill Mushroom	Edible-Toxic
Lactarius	*piperatus*	*piperatus*	E. USA (deciduous forest)	Original	Pepper	Milkcap Brittlegill Mushroom	Edible-Toxic
Lactarius	*piperatus*	*pergamenus*	E. USA (deciduous forest), MOTW, page 130	Pergamen	Pepper	Milkcap Brittlegill Mushroom	Inedible
Lactarius	*populinus*	.	Europe (in grass in open Poplar woods), MOTW, page 131	.	Poplar	Milkcap Brittlegill Mushroom	Inedible
Lactarius	*porninsis*	.	Europe (Pornin, France, Larch conifer forest), MOTW, page 142	.	Pornin	Milkcap Brittlegill Mushroom	Edible
Lactarius	*pseudoaffinis*	.	USA, PFGM, page 327	.	Lesser Kindred	Milkcap Brittlegill Mushroom	Unknown
Lactarius	*pseudodeliciosus*	.	USA, PFGM, page 331	.	Lesser Delicious	Milkcap Brittlegill Mushroom	Unknown
Lactarius	*pseudomucidus*	.	USA, PFGM, page 333	.	Lesser Slimy	Milkcap Brittlegill Mushroom	Unknown
Lactarius	*pterosporus*	.	USA (under Hornbeam broadleaf forest), MOTW, page 139	.	Winged Spore	Milkcap Brittlegill Mushroom	Inedible
Lactarius	*pubescens*	.	USA, MOTW, page 133, PFGM, 338	.	Downy Deathcap	Milkcap Brittlegill Mushroom	Poison
Lactarius	*pubescens*	*pubescens*	USA	Original	Downy Deathcap	Milkcap Brittlegill Mushroom	Poison
Lactarius	*pubescens*	*cilicioides*	USA, MOTW, page 133b	Goat Hair	Downy Deathcap	Milkcap Brittlegill Mushroom	Poison
Lactarius	*pyrogalus*	.	Europe (Hazelnut broadleaf forest), MOTW, page 139	.	Burning Latex	Milkcap Brittlegill Mushroom	Inedible
Lactarius	*quieticolor*	.	Europe	.	Quiet Color	Milkcap Brittlegill Mushroom	Edible-Good
Lactarius	*quieticolor*	*quieticolor*	Europe	Original	Quiet Color	Milkcap Brittlegill Mushroom	Edible-Good
Lactarius	*quieticolor*	*hemicyaneus*	Europe, MOTW, page 134	Blue	Quiet Color	Milkcap Brittlegill Mushroom	Edible-Good
Lactarius	*quietus*	.	USA (Oak broadleaf forest), MOTW, page 142	.	Solitary	Milkcap Brittlegill Mushroom	Edible
Lactarius	*repraesentaneus*	.	W. Canada, W. USA (mountain Spruce forest). MOTW, page 133, PFGM, page 334	.	Shaggy Bear	Milkcap Brittlegill Mushroom	Inedible-Toxic
Lactarius	*resinus*	.	Europe (conifer & Birch broadleaf forest), MOTW, page 133	.	Conifer Resin	Milkcap Brittlegill Mushroom	Inedible-Toxic
Lactarius	*romagnesii*	.	USA	.	Romagnes	Milkcap Brittlegill Mushroom	Unknown
Lactarius	*romagnesii*	*romagnesii*	USA	Original	Romagnes	Milkcap Brittlegill Mushroom	Unknown
Lactarius	*romagnesii*	*speciosus*	USA, PFGM, page 334	Special	Romagnes	Milkcap Brittlegill Mushroom	Unknown
Lactarius	*rubrilacteus*	.	W. USA (Pine forest from Rocky Mtns. to pacific coast), PFGM, page 334	.	Red Juice	Milkcap Brittlegill Mushroom	Edible
Lactarius	*rufus*	.	N. USA (on soil and mossy places in conifer forest), MOTW, page 140, PFGM, page 334	.	Red Hot	Milkcap Brittlegill Mushroom	Edible-Seasoning
Lactarius	*rufus*	*rufus*	N. USA (on soil and mossy places in conifer forest)	Original	Red Hot	Milkcap Brittlegill Mushroom	Edible-Seasoning
Lactarius	*rufus*	*exumbonatus*	N. USA (on soil and mossy places in conifer forest)	Penis	Red Hot	Milkcap Brittlegill Mushroom	Edible-Seasoning
Lactarius	*rufus*	*mollis*	N. USA (on soil and mossy places in conifer forest)	Soft	Red Hot	Milkcap Brittlegill Mushroom	Edible-Seasoning
Lactarius	*rugatus*	.	Europe, MOTW, page 135	.	Corrugated	Milkcap Brittlegill Mushroom	Edible-Good

Lactarius	*salmonicolor*	.	USA (Pine & Juniper conifer forest), PFGM, page 331	.	Salmon	Milkcap Brittlegill Mushroom	Edible-Good
Lactarius	*salmonicolor*	*salmonicolor*	USA (Pine & Juniper conifer forest)	Original	Salmon	Milkcap Brittlegill Mushroom	Edible-Good
Lactarius	*salmonicolor*	*salmoneus*	USA (Pine & Juniper conifer forest)	Greater	Salmon	Milkcap Brittlegill Mushroom	Edible-Good
Lactarius	*salmonicolor*	*subsalmoneus*	USA (Pine & Juniper conifer forest)	Lesser	Salmon	Milkcap Brittlegill Mushroom	Edible-Good
Lactarius	*sanguifluus*	.	Europe (conifer forest), MOTW, page 134	.	Flowing Blood	Milkcap Brittlegill Mushroom	Edible-Good
Lactarius	*scrobiculatus*	.	Canada, N. USA (conifer forest), MOTW, page 133, PFGM, page 335	.	Spotstalk	Milkcap Brittlegill Mushroom	Toxic
Lactarius	*semisanguifluus*	.	Europe (in Pine woods), MOTW, page 134	.	Faded Ocher	Milkcap Brittlegill Mushroom	Edible-Good
Lactarius	*serifluus*	.	Europe (conifer & broadleaf forests), MOTW, page 144	.	Flowing Serum	Milkcap Brittlegill Mushroom	Edible-Poor
Lactarius	*sordidus*	.	USA, PFGM, page 328	.	Sordid	Milkcap Brittlegill Mushroom	Unknown
Lactarius	*spinosulus*	.	Europe (humid Birch & Elm broadleaf forest), MOTW, page 140	.	Spiny	Milkcap Brittlegill Mushroom	Inedible
Lactarius	*spinosulus*	*spinosulus*	Europe (humid Birch & Elm broadleaf forest)	Original	Spiny	Milkcap Brittlegill Mushroom	Inedible
Lactarius	*spinosulus*	*violaceus*	Europe (humid Birch & Elm broadleaf forest)	Violet	Spiny	Milkcap Brittlegill Mushroom	Inedible-Toxic
Lactarius	*subdulcis*	.	USA (Beech & Hornbeam broadleaf forest), MOTW, page 142	.	Mildly Sweet	Milkcap Brittlegill Mushroom	Unpalatable
Lactarius	*subpalustris*	.	USA, PFGM, page 339	.	Lesser Marsh	Milkcap Brittlegill Mushroom	Unknown
Lactarius	*subplinthogalus*	.	E. USA (hardwood & mixed conifer-hardwood forest), PFGM, page 336	.	Maryland	Milkcap Brittlegill Mushroom	Unknown
Lactarius	*subpurpureus*	.	E. USA (hardwood & mixed conifer-hardwood forest), PFGM, page 336	.	Purplish Red Wine	Milkcap Brittlegill Mushroom	Edible
Lactarius	*subvellereus*	.	USA, PFGM, page 340	.	Lesser Velvet	Milkcap Brittlegill Mushroom	Unknown
Lactarius	*tabidus*	.	Europe (Pine conifer & Birch broadleaf forest), MOTW, page 130	.	Corrupted	Milkcap Brittlegill Mushroom	Edible-Poor
Lactarius	*thyinos*	.	N. USA (Cedar woods, swamps & bogs), PFGM, page 337	.	Orange Ring	Milkcap Brittlegill Mushroom	Edible
Lactarius	*torminosus*	.	USA (on forest soil, usually near Birch), MOTW, page 133, PFGM, page 337	.	Powderpuff Deathcap	Milkcap Brittlegill Mushroom	Poison
Lactarius	*torminosus*	*torminosus*	USA (on forest soil, usually near Birch)	Original	Powderpuff Deathcap	Milkcap Brittlegill Mushroom	Poison
Lactarius	*torminosus*	*sublateritius*	USA (on forest soil, usually near Birch)	Smooth	Powderpuff Deathcap	Milkcap Brittlegill Mushroom	Poison
Lactarius	*trivialis*	.	W. USA (humid conifer forest), MOTW, page 139, PFGM, page 338	.	Ordinary	Milkcap Brittlegill Mushroom	Inedible
Lactarius	*turpis*	.	USA (in humid ravines of conifer & Birch broadleaf forest)	.	Ugly	Milkcap Brittlegill Mushroom	Inedible
Lactarius	*turpis*	*Turpis*	USA (in humid ravines of conifer & Birch broadleaf forest)	Original	Ugly	Milkcap Brittlegill Mushroom	Inedible
Lactarius	*turpis*	*plumbeus*	USA (in humid ravines of conifer & Birch broadleaf forest), MOTW, page 138	Lead Gray	Ugly	Milkcap Brittlegill Mushroom	Inedible
Lactarius	*uvidus*	.	USA (Birch, Aspen, Pine forest), MOTW, page 136, PFGM, page 339	.	Damp	Milkcap Brittlegill Mushroom	Inedible-Toxic
Lactarius	*vellereus*	.	S. Canada, C. & E. USA (conifer & mixed conifer-hardwood forest), MOTW, page 129, PFGM, page 339	.	Fleecy	Milkcap Brittlegill Mushroom	Edible-Poor
Lactarius	*vellereus*	*vellereus*	S. Canada, C. & E. USA (conifer & mixed conifer-hardwood forest)	Original	Fleecy	Milkcap Brittlegill Mushroom	Edible-Poor
Lactarius	*vellereus*	*albivellus*	S. Canada, C. & E. USA (conifer & mixed conifer-hardwood forest)	White	Fleecy	Milkcap Brittlegill Mushroom	Edible-Poor
Lactarius	*vellereus*	*deceptivus*	S. Canada, C. & E. USA (conifer & mixed conifer-hardwood forest), MOTW, page 129	Deceptive	Fleecy	Milkcap Brittlegill Mushroom	Edible-Poor

Lactarius	*vellereus*	*hometii*	S. Canada, C. & E. USA (conifer & mixed conifer-hardwood forest)	Homet	Fleecy	Milkcap Brittlegill Mushroom	Edible-Poor
Lactarius	*vellereus*	*velutinus*	S. Canada, C. & E. USA (conifer & mixed conifer-hardwood forest), MOTW, page 129	Velvet	Fleecy	Milkcap Brittlegill Mushroom	Edible-Poor
Lactarius	*vietus*	.	Europe (Pine conifer & Birch broadleaf forest), MOTW, page 136	.	Wilted	Milkcap Brittlegill Mushroom	Inedible
Lactarius	*villosus*	.	USA, PFGM, page 338	.	Villose	Milkcap Brittlegill Mushroom	Unknown
Lactarius	*vinaceorufescens*	.	USA, PFGM, page 329	.	Brownish Red	Milkcap Brittlegill Mushroom	Unknown
Lactarius	*violascens*	.	Europe (conifer forest), MOTW, page 136	.	Violet	Milkcap Brittlegill Mushroom	Edible
Lactarius	*volemus*	.	USA (in woods), MOTW, page 135, PFGM, page 340	.	Tawny	Milkcap Brittlegill Mushroom	Edible
Lactarius	*volemus*	*volemus*	USA (in woods)	Original	Tawny	Milkcap Brittlegill Mushroom	Edible
Lactarius	*volemus*	*ichoratus*	USA (in woods)	Ichorate	Tawny	Milkcap Brittlegill Mushroom	Edible
Lactarius	*volemus*	*lactifluus*	USA (in woods)	White Waterdrop	Tawny	Milkcap Brittlegill Mushroom	Edible
Lactarius	*volemus*	*oedematopus*	USA (in woods)	Demate	Tawny	Milkcap Brittlegill Mushroom	Edible
Lactarius	*volemus*	*subrugatus*	USA (in woods)	Winkled	Tawny	Milkcap Brittlegill Mushroom	Edible
Lactarius	*zonarioides*	.	Europe (mountain Spruce conifer forest), MOTW, pages 133, 140	.	Mountain Spruce	Milkcap Brittlegill Mushroom	Inedible-Toxic
Lactarius	*zonarius*	.	USA (plains & broadleaf forest), MOTW, page 133	.	Zoned	Milkcap Brittlegill Mushroom	Inedible-Toxic
Lactarius	*zonarius*	*zonarius*	USA (plains & broadleaf forest)	Original	Zoned	Milkcap Brittlegill Mushroom	Inedible-Toxic
Lactarius	*zonarius*	*insulsus*	USA (plains & broadleaf forest), MOTW, page 131	Tasteless	Zoned	Milkcap Brittlegill Mushroom	Inedible-Toxic

Family 622.202.02 *Bondarzewiaceae* Bondarzev Brittlegill Mushroom

Genus	Species	Variety	Native Location	Variety	Specimen	Genome	Habit
Bondarzewia	*montana*	.	Europe (on trunks in conifer forest), MOTW, page 209	.	Mountain	Bondarzev Brittlegill Mushroom	Edible
Bondarzewia	*montana*	*montana*	Europe (on trunks in conifer forest)	Original	Mountain	Bondarzev Brittlegill Mushroom	Edible
Bondarzewia	*montana*	*mesentericus*	Europe (on trunks in conifer forest)	Mesentery	Mountain	Bondarzev Brittlegill Mushroom	Edible

Class 622.3 *Boletomycetes* Nongill Mushroom

Latin: *boleto* bowl
Latin: *Mycetes* mycete, fungus

ORDERS

622.301 *Boletales* Bolete
622.302 *Cantharellales* Chanterelle
622.303 *Hydnales* Spinecap Nongill Mushroom

Order 622.301 *Boletales* Bolete

Latin: *boleto* bowl

FAMILIES

622.301.01 *Boletalaceae* Bolete
622.301.02 *Gomphidiaceae* Slimecap Bolete
622.301.03 *Gyroporaceae* Hollowstalk Bolete
622.301.04 *Paxillaceae* Brimcap Bolete
622.301.05 *Rhizopogonaceae* False Truffle Bolete
622.301.06 *Suillaceae* Slipperycap Bolete

Family 622.301.01 *Boletaceae* Bolete

Genus	Species	Variety	Native Location	Variety	Specimen	Genome	Habit
Boletus	*aereus*	.	USA, S. Europe (Oak & Chestnut woods), MOTW, page 180, PFGM, page 104	.	Bronze	Bolete Nongill Mushroom	Edible-Excellent
Boletus	*appendiculatus*	.	Europe (Birch & Linden broadleaf forest), MOTW, page 181	.	Appendage	Bolete Nongill Mushroom	Edible-Excellent
Boletus	*armeniacus*	.	Europe (broadleaf forest), MOTW, page 196	.	Apricot	Bolete Nongill Mushroom	Edible
Boletus	*badius*	.	E. USA (conifer & mixed conifer forest), MOTW, page 198, PFGM, page 103	.	Red Brown	Bolete Nongill Mushroom	Edible-Good
Boletus	*badius*	*badius*	E. USA (conifer & mixed conifer forest)	Original	Red Brown	Bolete Nongill Mushroom	Edible-Good
Boletus	*badius*	*castaneus*	E. USA (conifer & mixed conifer forest)	Chestnut	Red Brown	Bolete Nongill Mushroom	Edible-Good
Boletus	*betulicola*	.	.	.	.	Bolete Nongill Mushroom	.
Boletus	*bicolor*	.	USA	.	Two Color	Bolete Nongill Mushroom	Edible
Boletus	*byssinus*	.	.	.	.	Bolete Nongill Mushroom	.
Boletus	*calopus*	.	USA (mountain conifer open woods), MOTW, page 182, PFGM, pages 102, 105	.	Beautiful Foot Deathcap	Bolete Nongill Mushroom	Poison
Boletus	*calopus*	*calopus*	USA (mountain conifer open woods)	Original	Beautiful Foot Deathcap	Bolete Nongill Mushroom	Poison
Boletus	*calopus*	*olivaceus*	USA (mountain conifer open woods)	Olive	Beautiful Foot Deathcap	Bolete Nongill Mushroom	Poison
Boletus	*campestris*	.	USA, PFGM, page 105	.	Field	Bolete Nongill Mushroom	Unknown
Boletus	*chrysenteron*	.	USA (grass, moss, woods), MOTW, page 196, PFGM, page 105	.	Red Cracked	Bolete Nongill Mushroom	Edible
Boletus	*chrysenteron*	*chrysenteron*	USA (grass, moss, woods)	Original	Red Cracked	Bolete Nongill Mushroom	Edible
Boletus	*chrysenteron*	*communis*	USA (grass, moss, woods)	American	Red Cracked	Bolete Nongill Mushroom	Edible
Boletus	*clintonianus*	.	USA, MOTW, page 307	.	Clinton's	Bolete Nongill Mushroom	Unknown
Boletus	*dupainii*	.	Europe (broadleaf forest), MOTW, page 186	.	Dupain	Bolete Nongill Mushroom	Edible-Toxic
Boletus	*eastwoodiae*	.	USA (broadleaf forest), MOTW, page 187	.	Eastwood Deathcap	Bolete Nongill Mushroom	Poison
Boletus	*edulis*	.	USA (forest soil), MOTW, page 177, PFGM, page 104	.	King	Bolete Nongill Mushroom	Edible-Excellent
Boletus	*edulis*	*edulis*	USA (Fir conifer forest & Oak, Linden Beech & Hornbeam broadleaf forest)	Original	King	Bolete Nongill Mushroom	Edible-Excellent
Boletus	*edulis*	*albus*	USA (Fir conifer forest & Oak, Linden Beech & Hornbeam broadleaf forest), MOTW, page 177	White	King	Bolete Nongill Mushroom	Edible-Excellent
Boletus	*edulis*	*arcticus*	USA (Fir conifer forest & Oak, Linden Beech & Hornbeam broadleaf forest)	Arctic	King	Bolete Nongill Mushroom	Edible-Excellent
Boletus	*edulis*	*citrinus*	USA (Fir conifer forest & Oak, Linden Beech & Hornbeam broadleaf forest), MOTW, page 177	Citrine	King	Bolete Nongill Mushroom	Edible-Excellent

Boletus	*edulis*	*fuscoruber*	USA (Spruce conifer & Beech broadleaf forests), MOTW, page 180	Brown	King	Bolete Nongill Mushroom	Edible-Excellent
Boletus	*edulis*	*laevipes*	USA (Fir conifer forest & Oak, Linden Beech & Hornbeam broadleaf forest)	Smooth	King	Bolete Nongill Mushroom	Edible-Excellent
Boletus	*edulis*	*solidus*	USA (Fir conifer forest & Oak, Linden Beech & Hornbeam broadleaf forest)	Solid	King	Bolete Nongill Mushroom	Edible-Excellent
Boletus	*edulis*	*trisporus*	USA (Fir conifer forest & Oak, Linden Beech & Hornbeam broadleaf forest)	Triple Spore	King	Bolete Nongill Mushroom	Edible-Excellent
Boletus	*fechtneri*	.	Europe (humid Oak, Linden & Birch broadleaf forest)	.	Fechtner	Bolete Nongill Mushroom	Edible
Boletus	*fechtneri*	*fechtneri*	Europe (humid Oak, Linden & Birch broadleaf forest)	Original	Fechtner	Bolete Nongill Mushroom	Edible
Boletus	*fechtneri*	*pallescens*	Europe (humid Oak, Linden & Birch broadleaf forest), MOTW, page 183	Pale	Fechtner	Bolete Nongill Mushroom	Edible
Boletus	*ferrugineus*	.	USA (conifer forest)	.	Rusty	Bolete Nongill Mushroom	Edible-Good
Boletus	*ferrugineus*	*ferrugineus*	USA (conifer forest)	Original	Rusty	Bolete Nongill Mushroom	Edible-Good
Boletus	*ferrugineus*	*citrinovirens*	USA (conifer forest)	Yellow Green	Rusty	Bolete Nongill Mushroom	Edible-Good
Boletus	*ferrugineus*	*spadiceus*	USA (conifer forest), MOTW, page 196	Date Brown	Rusty	Bolete Nongill Mushroom	Edible-Good
Boletus	*flavus*	.	USA (Larch woods), MOTW, page 191	.	Yellow	Bolete Nongill Mushroom	Edible-Toxic
Boletus	*floridanus*	.	SE. USA (Florida Oak forest), PFGM, page 106	.	Florida	Bolete Nongill Mushroom	Unknown
Boletus	*fragrans*	.	USA (mountain & hill broadleaf woods), MOTW, page 181	.	Fragrant	Bolete Nongill Mushroom	Edible
Boletus	*fraternus*	.	E. USA (lawns, roadsides, open woods), PFGM, page 105	.	Summer Redcap Deathcap	Bolete Nongill Mushroom	Poison
Boletus	*frostii*	.	E. USA (Oak forest), MOTW, page184, PFGM, page 106	.	Frost's	Bolete Nongill Mushroom	Edible-Toxic
Boletus	*impolitus*	.	USA (Oak broadleaf woods), MOTW, page 181	.	Unclean	Bolete Nongill Mushroom	Edible
Boletus	*impolitus*	*impolitus*	USA (Oak broadleaf woods)	Original	Unclean	Bolete Nongill Mushroom	Edible
Boletus	*impolitus*	*suspectus*	USA (Oak broadleaf woods)	Suspect	Unclean	Bolete Nongill Mushroom	Edible
Boletus	*longicurvipes*	.	USA, PFGM, pages 109, 121	.	Long Curved	Bolete Nongill Mushroom	Unknown
Boletus	*lupinus*	.	Europe, MOTW, page 187	.	Wolf Deathcap	Bolete Nongill Mushroom	Poison
Boletus	*luridiformis*	.	Europe (Beech broadleaf forest)	.	Filthy	Bolete Nongill Mushroom	Edible-Fair
Boletus	*luridiformis*	*luridiformis*	Europe (Beech broadleaf forest)	Original	Filthy	Bolete Nongill Mushroom	Edible-Fair
Boletus	*luridiformis*	*discolor*	Europe (Beech broadleaf forest)	Discolored	Filthy	Bolete Nongill Mushroom	Edible-Fair
Boletus	*luridiformis*	*erythropus*	USA, Europe (Beech broadleaf forest), MOTW, page 184, PFGM, page 102	Red Foot	Filthy	Bolete Nongill Mushroom	Edible-Fair
Boletus	*luridus*	.	USA (in woods), MOTW, page 184, PFGM, pages 102, 105	.	Dirty	Bolete Nongill Mushroom	Edible-Toxic
Boletus	*luridus*	*luridus*	USA (in woods)	Original	Dirty	Bolete Nongill Mushroom	Edible-Toxic
Boletus	*luridus*	*rubeolarius*	USA (in woods)	Reddish	Dirty	Bolete Nongill Mushroom	Edible-Toxic
Boletus	*luridus*	*rubriceps*	USA (in woods)	Red	Dirty	Bolete Nongill Mushroom	Edible-Toxic
Boletus	*ornatipes*	.	C. & E. USA (trails, roads & clearings in hardwood forest), PFGM, page 107	.	Gold Stalk	Bolete Nongill Mushroom	Edible
Boletus	*parvulus*	.	USA, PFGM, page 106	.	Small	Bolete Nongill Mushroom	Unknown
Boletus	*pinicola*	.	Europe (Pine & Fir conifer & Chestnut & Beech broadleaf forests), MOTW, page 181	.	Pine Dwelling	Bolete Nongill Mushroom	Edible-Excellent
Boletus	*pinophilus*	.	USA, GSI	.	Pine Loving	Bolete Nongill Mushroom	Edible
Boletus	*projectellus*	.	C. & E. USA, PFGM, page 107	.	Projecting	Bolete Nongill Mushroom	Unknown
Boletus	*pseudosulphureus*	.	Europe (Beech broadleaf forest)	.	.	Bolete Nongill Mushroom	Edible-Toxic
Boletus	*pseudosulphureus*	*pseudosulphureus*	Europe (Beech broadleaf forest)	.	Original	Bolete Nongill Mushroom	Edible-Toxic
Boletus	*pseudosulphureus*	*junquilleus*	Europe (Beech broadleaf forest), MOTW, page 184	.	Jonquil	Bolete Nongill Mushroom	Edible-Toxic
Boletus	*pulverulentus*	.	Europe (Fir conifer & Chestnut & Beech broadleaf forests), MOTW, page 198	.	Powdery	Bolete Nongill Mushroom	Edible-Good
Boletus	*queletii*	.	Europe (broadleaf woods), MOTW, page 186	.	Quelet	Bolete Nongill Mushroom	Edible
Boletus	*queletii*	*queletii*	Europe (broadleaf woods)	Original	Quelet	Bolete Nongill Mushroom	Edible
Boletus	*queletii*	*lateritius*	Europe (broadleaf woods), MOTW, page 186	Brick Red	Quelet	Bolete Nongill Mushroom	Edible

Boletus	queletii	rubicundus	Europe (broadleaf woods)	Reddish	Quelet	Bolete Nongill Mushroom	Edible
Boletus	radicans	.	USA (Oak & Beech broadleaf forest)	.	Shiny	Bolete Nongill Mushroom	Inedible
Boletus	radicans	radicans	USA (Oak & Beech broadleaf forest)	Original	Shiny	Bolete Nongill Mushroom	Inedible
Boletus	radicans	albidus	USA (Oak & Beech broadleaf forest), MOTW, page 182	Whitish	Shiny	Bolete Nongill Mushroom	Inedible
Boletus	radicans	albus	USA (Oak & Beech broadleaf forest)	White	Shiny	Bolete Nongill Mushroom	Inedible
Boletus	radicans	amarus	USA (Oak & Beech broadleaf forest)	Bitter	Shiny	Bolete Nongill Mushroom	Inedible
Boletus	radicans	pachypus	USA (Oak & Beech broadleaf forest), MOTW, page 182	Fat Foot	Shiny	Bolete Nongill Mushroom	Inedible
Boletus	regius	.	Europe (humid Oak, Linden & Birch broadleaf forest), MOTW, page 183	.	Regal	Bolete Nongill Mushroom	Edible-Excellent
Boletus	reticulatus	.	Europe (Oak & Beech broadleaf forest), MOTW, page 178	.	Netted	Bolete Nongill Mushroom	Edible-Excellent
Boletus	reticulatus	reticulatus	Europe (Oak & Beech broadleaf forest)	Original	Netted	Bolete Nongill Mushroom	Edible-Excellent
Boletus	reticulatus	aestivalis	Europe (Oak & Beech broadleaf forest)	Summer	Netted	Bolete Nongill Mushroom	Edible-Excellent
Boletus	retipes	.	USA, PFGM, page 107	.	Net Foot	Bolete Nongill Mushroom	Unknown
Boletus	rhodopurpureus	.	Europe (Oak & Beech broadleaf woods), MOTW,page 186	.	.	Bolete Nongill Mushroom	Edible-Toxic
Boletus	rhodopurpureus	rhodopurpureus	Europe (Oak & Beech broadleaf woods)	.	.	Bolete Nongill Mushroom	Edible-Toxic
Boletus	rhodopurpureus	xanthopurpureus	Europe (Oak & Beech broadleaf woods)	.	.	Bolete Nongill Mushroom	Edible-Toxic
Boletus	rubellus	.	Europe (in woods)	.	Red	Bolete Nongill Mushroom	Edible
Boletus	rubellus	rubellus	Europe (in woods)	Original	Red	Bolete Nongill Mushroom	Edible
Boletus	rubellus	sanguineus	Europe (in woods)	Blood	Red	Bolete Nongill Mushroom	Edible
Boletus	rubellus	versicolor	Europe (in woods), MOTW, page 196	Multicolored	Red	Bolete Nongill Mushroom	Edible
Boletus	rubiginosus	.	Europe (Oak & Beech broadleaf forest), MOTW, page 178	.	Velvet Red Brown	Bolete Nongill Mushroom	Edible
Boletus	rubropunctus	.	USA (deciduous, conifer, mixed forest), PFGM, page 108	.	Red Dot	Bolete Nongill Mushroom	Unknown
Boletus	satanas	.	Europe (open broadleaf woods), MOTW, page 187	.	Satan Deathcap	Bolete Nongill Mushroom	Poison
Boletus	subfraternus	.	USA, PFGM, page 106	.	Lesser Fraternal	Bolete Nongill Mushroom	Unknown
Boletus	subtomentosus	.	Europe (grassy places in woods), MOTW, page 196	.	Downy	Bolete Nongill Mushroom	Edible-Poor
Boletus	subtomentosus	subtomentosus	Europe (grassy places in woods)	Original	Downy	Bolete Nongill Mushroom	Edible-Poor
Boletus	subtomentosus	marginalis	Europe (grassy places in woods)	Margined	Downy	Bolete Nongill Mushroom	Edible-Poor
Boletus	subtomentosus	lanatus	Europe (grassy places in woods)	Woolly	Downy	Bolete Nongill Mushroom	Edible-Poor
Boletus	subtomentosus	leguei	Europe (grassy places in woods)	Legue	Downy	Bolete Nongill Mushroom	Edible-Poor
Boletus	subtomentosus	striipes	Europe (grassy places in woods)	Striped Foot	Downy	Bolete Nongill Mushroom	Edible-Poor
Boletus	subtomentosus	xanthus	Europe (grassy places in woods)	Yellow	Downy	Bolete Nongill Mushroom	Edible-Poor
Boletus	torosus	.	Europe (Beech broadleaf forest), MOTW, page 184	.	Muscular	Bolete Nongill Mushroom	Edible-Toxic
Boletus	variipes	.	USA, PFGM, page 104	.	Variable Foot	Bolete Nongill Mushroom	Unknown
Boletus	xanthocyaneus	.	Europe (Oak & Beech broadleaf open woods)	.	Yellow & Blue	Bolete Nongill Mushroom	Edible-Toxic
Boletus	xanthocyaneus	xanthocyaneus	Europe (Oak & Beech broadleaf open woods)	Original	Yellow & Blue	Bolete Nongill Mushroom	Edible-Toxic
Boletus	xanthocyaneus	purpureus	Europe (Oak & Beech broadleaf open woods)	Purple	Yellow & Blue	Bolete Nongill Mushroom	Edible-Toxic
Boletus	zelleri	.	USA, GSI-5, page 410	.	Zeller	Bolete Nongill Mushroom	Edible
Aureoboletus	auriporus	.	USA, MOTW, page 307	.	.	Golden Bolete Nongill Mushroom	Unknown
Aureoboletus	gentilis	.	Europe (on grass in broadleaf forest)	.	Pretty	Golden Bolete Nongill Mushroom	Edible
Aureoboletus	gentilis	gentilis	Europe (on grass in broadleaf forest)	Original	Pretty	Golden Bolete Nongill Mushroom	Edible
Aureoboletus	gentilis	cramesinus	Europe (on grass in broadleaf forest), MOTW, page 195	Bright Red	Pretty	Golden Bolete Nongill Mushroom	Edible
Aureoboletus	gentilis	tenuipes	Europe (on grass in broadleaf forest)	Thin Foot	Pretty	Golden Bolete Nongill Mushroom	Edible
Austroboletus	betula	.	USA (Birch woods), PFGM, page 100	.	Birch	Southern Bolete Nongill Mushroom	Unknown
Boletellus	mirabilis	.	W. Canada, NC. & NW. USA (conifer forest), MOTW, page	.	Bragger	Shagnet Bolete Nongill Mushroom	Edible-Excellent

			181, PFGM, page 106					
Boletellus	*russellii*	.	C. & E. USA (hardwood forest), MOTW, page 196, PFGM, page 101	.	Russell	Shagnet Bolete Nongill Mushroom	Edible
Boletopsis	*leucomelaena*	.	USA (mountain conifer forest), MOTW, page 206	.	Black & White	False Bolete Nongill Mushroom	Edible-Poor
Buchwald oboletus	*lignicola*	.	Europe (conifer forest), MOTW, page 195	.	Wood Dwelling	Buchwald Bolete Nongill Mushroom	Edible
Buchwald oboletus	*sphaerocephalus*	.	Europe (near Pine trees)	.	Roundcap	Buchwald Bolete Nongill Mushroom	Edible
Buchwald oboletus	*sphaerocephalus*	*sphaerocephalus*	Europe (near Pine trees)	Original	Roundcap	Buchwald Bolete Nongill Mushroom	Edible
Buchwald oboletus	*sphaerocephalus*	*sulphureus*	Europe (near Pine trees), MOTW, page 194	Sulfur	Roundcap	Buchwald Bolete Nongill Mushroom	Edible
Chalciporus	*piperatus*	.	Canada, USA (conifer & Birch broadleaf forest), MOTW, page 195, PFGM, page 108	.	Pepper	Chalk Pore Bolete Nongill Mushroom	Edible-Toxic
Leccinum	*atrostipitatum*	.	USA, PFGM, page 111	.	Dark Stippled	Scaberstalk Bolete Nongill Mushroom	Unknown
Leccinum	*aurantiacum*	.	Canada, N. USA (Pine conifer & Aspen, Poplar & Birch broadleaf forests), MOTW, page 189	.	Orange	Scaberstalk Bolete Nongill Mushroom	Edible-Excellent
Leccinum	*aurantiacum*	*aurantiacum*	Canada, N. USA (Pine conifer & Aspen, Poplar & Birch broadleaf forests)	Original	Orange	Scaberstalk Bolete Nongill Mushroom	Edible-Excellent
Leccinum	*aurantiacum*	*populinum*	Canada, N. USA (Pine conifer & Aspen, Poplar & Birch broadleaf forests)	Poplar	Orange	Scaberstalk Bolete Nongill Mushroom	Edible-Excellent
Leccinum	*aurantiacum*	*rufum*	Canada, N. USA (Pine conifer & Aspen, Poplar & Birch broadleaf forests)	Aspen	Orange	Scaberstalk Bolete Nongill Mushroom	Edible-Excellent
Leccinum	*crocipodium*	.	Europe (Oak & Beech broadleaf forest), MOTW, page 190	.	Saffron Foot	Scaberstalk Bolete Nongill Mushroom	Edible-Poor
Leccinum	*crocipodium*	*crocipodium*	Europe (Oak & Beech broadleaf forest)	Original	Saffron Foot	Scaberstalk Bolete Nongill Mushroom	Edible-Poor
Leccinum	*crocipodium*	*nigrescens*	Europe (Oak & Beech broadleaf forest)	Black	Saffron Foot	Scaberstalk Bolete Nongill Mushroom	Edible-Poor
Leccinum	*crocipodium*	*tessellatus*	Europe (Oak & Beech broadleaf forest)	Mosaic	Saffron Foot	Scaberstalk Bolete Nongill Mushroom	Edible-Poor
Leccinum	*duriusculum*	.	USA (Poplar broadleaf forest), MOTW, page 189	.	Hard	Scaberstalk Bolete Nongill Mushroom	Edible-Good
Leccinum	*holopus*	.	USA, PFGM, page 112	.	Whole Foot	Scaberstalk Bolete Nongill Mushroom	Unknown
Leccinum	*holopus*	*holopus*	USA	Original	Whole Foot	Scaberstalk Bolete Nongill Mushroom	Unknown
Leccinum	*holopus*	*americanum*	USA	American	Whole Foot	Scaberstalk Bolete Nongill Mushroom	Unknown
Leccinum	*holopus*	*nucatum*	USA	Nucate	Whole Foot	Scaberstalk Bolete Nongill Mushroom	Unknown
Leccinum	*insigne*	.	C. & W. USA (plains & Rocky Mtns.), PFGM, page 111	.	Quaking Aspen	Scaberstalk Bolete Nongill Mushroom	Edible-Excellent
Leccinum	*lepidum*	.	Europe (coastal W. France), MOTW, page 190	.	Agreeable	Scaberstalk Bolete Nongill Mushroom	Edible
Leccinum	*leucophaeus*	.	Europe (Birch broadleaf forest), MOTW, page 189	.	Light & Dark	Scaberstalk Bolete Nongill Mushroom	Edible-Fair
Leccinum	*pseudoscabrum*	.	Europe (Hornbeam, Oak & Hazelnut broadleaf forest)	.	Scabby	Scaberstalk Bolete Nongill Mushroom	Edible-Fair
Leccinum	*pseudoscabrum*	*pseudoscabrum*	Europe (Hornbeam, Oak & Hazelnut broadleaf forest)	Original	Scabby	Scaberstalk Bolete Nongill Mushroom	Edible-Fair
Leccinum	*pseudoscabrum*	*carpini*	Europe (Hornbeam, Oak & Hazelnut broadleaf forest), MOTW, page 189	Hornbeam	Scabby	Scaberstalk Bolete Nongill Mushroom	Edible-Fair
Leccinum	*pseudoscabrum*	*oxydabilis*	Europe (Birch, Hornbeam, Oak & Hazelnut broadleaf forest), MOTW, page 189	Birch	Scabby	Scaberstalk Bolete Nongill Mushroom	Edible-Fair
Leccinum	*scabrum*	.	USA (Birch forest), PFGM, page 111	.	Birch	Scaberstalk Bolete Nongill Mushroom	Edible-Good
Leccinum	*scabrum*	*scabrum*	USA (Birch forest)	Original	Birch	Scaberstalk Bolete Nongill Mushroom	Edible-Good
Leccinum	*scabrum*	*avellaneum*	USA (Birch forest)	Almond	Birch	Scaberstalk Bolete Nongill Mushroom	Edible-Good
Leccinum	*scabrum*	*melaneum*	USA (Birch forest)	Dark	Birch	Scaberstalk Bolete Nongill Mushroom	Edible-Good
Leccinum	*scabrum*	*molle*	USA (Birch forest)	Soft	Birch	Scaberstalk Bolete Nongill Mushroom	Edible-Good
Leccinum	*scabrum*	*murinaceum*	USA (Birch forest)	Murine	Birch	Scaberstalk Bolete Nongill Mushroom	Edible-Good
Leccinum	*scabrum*	*olivaceosum*	USA (Birch forest)	Olive	Birch	Scaberstalk Bolete Nongill Mushroom	Edible-Good
Leccinum	*scabrum*	*subcinnamomeum*	USA (Birch forest)	Cinnamon	Birch	Scaberstalk Bolete Nongill Mushroom	Edible-Good
Leccinum	*versipelle*	.	Europe (Birch broadleaf forest)	.	Changing	Scaberstalk Bolete Nongill Mushroom	Edible
Leccinum	*versipelle*	*versipelle*	Europe (Birch broadleaf forest)	Original	Changing	Scaberstalk Bolete Nongill Mushroom	Edible
Leccinum	*versipelle*	*floccopus*	Europe (Birch broadleaf forest)	Fleece	Changing	Scaberstalk Bolete Nongill Mushroom	Edible
Leccinum	*versipelle*	*glandaceum*	Europe (Birch broadleaf forest)	Gland	Changing	Scaberstalk Bolete Nongill Mushroom	Edible

Leccinum	*versipelle*	*rufescens*	Europe (Birch broadleaf forest), MOTW, page 189	Blonde	Changing	Scaberstalk Bolete Nongill Mushroom	Edible
Leccinum	*versipelle*	*testaceoscabrum*	Europe (Birch broadleaf forest)	Scabre	Changing	Scaberstalk Bolete Nongill Mushroom	Edible
Leccinum	*vulpinum*	.	Europe (Spruce & Pine forest), MOTW, page 190	.	Wolf	Scaberstalk Bolete Nongill Mushroom	Edible
Porphyrellus	*porphyrosporus*	.	Europe (plains, mountains in Fir conifer woods), MOTW, page 200	.	Purple Spore	Chocolate Bolete Nongill Mushroom	Edible-Poor
Porphyrellus	*porphyrosporus*	*porphyrosporus*	Europe (plains, mountains in Fir conifer woods)	Original	Purple Spore	Chocolate Bolete Nongill Mushroom	Edible-Poor
Porphyrellus	*porphyrosporus*	*fuligineus*	Europe (plains, mountains in Fir conifer woods), MOTW, page 200	Sooty	Purple Spore	Chocolate Bolete Nongill Mushroom	Edible-Poor
Pseudoboletus	*parasiticus*	.	E. Canada, E. USA (on Earthball Puffballs), E. Asia (Japan), PFGM, page 107	.	Earthball Puffball Parasite	Least Bolete Nongill Mushroom	Edible
Strobilomyces	*confusus*	.	USA (hills, mountains in broadleaf & mixed conifer-broadleaf forests), PFGM, page 113	.	Confused Woods	Pinecone Bolete Nongill Mushroom	Unknown
Strobilomyces	*strobilaceus*	.	USA (hills, mountains in broadleaf & mixed conifer-broadleaf forests), MOTW, page 200	.	Old Man Of The Woods	Pinecone Bolete Nongill Mushroom	Edible-Poor
Strobilomyces	*strobilaceus*	*strobilaceus*	USA (hills, mountains in broadleaf & mixed conifer-broadleaf forests)	.	Old Man Of The Woods	Pinecone Bolete Nongill Mushroom	Edible-Poor
Strobilomyces	*strobilaceus*	*cinereus*	USA (hills, mountains in broadleaf & mixed conifer-broadleaf forests)	Original	Old Man Of The Woods	Pinecone Bolete Nongill Mushroom	Edible-Poor
Strobilomyces	*strobilaceus*	*floccopus*	USA (hills, mountains in broadleaf & mixed conifer-broadleaf forests), PFGM, page 112, MOTW, page 200	Fleecy Foot	Old Man Of The Woods	Pinecone Bolete Nongill Mushroom	Edible-Poor
Tylopilus	*chromapes*	.	SE. Canada, E. USA (Pine, Hemlock & Aspen forest), PFGM, page 121	.	Yellow Foot	Knob Bolete Nongill Mushroom	Edible
Tylopilus	*eximius*	.	USA, PFGM, page 124	.	Select	Knob Bolete Nongill Mushroom	Edible-Excellent
Tylopilus	*felleus*	.	S. Canada, C. & E. USA (sandy soil in conifer & hardwood forest), MOTW, page 179, PFGM, page 122	.	Bitter	Knob Bolete Nongill Mushroom	Edible-Bitter
Tylopilus	*felleus*	*felleus*	S. Canada, C. & E. USA (sandy soil in conifer & hardwood forest)	Original	Bitter	Knob Bolete Nongill Mushroom	Edible-Bitter
Tylopilus	*felleus*	*alutarius*	S. Canada, C. & E. USA (sandy soil in conifer & hardwood forest), MOTW, page 179	Thin Leather	Bitter	Knob Bolete Nongill Mushroom	Edible
Tylopilus	*indecisis*	.	C. & E. USA (hardwood forest), PFGM, page 123	.	Question	Knob Bolete Nongill Mushroom	Unpalatable
Tylopilus	*plumbeoviolaceus*	.	C. & E. USA (deciduous forest), PFGM, page 124	.	Gray Violet	Knob Bolete Nongill Mushroom	Edible-Bitter
Tylopilus	*rubrobrunneus*	.	USA, PFGM, pages 123, 124	.	Red Brown	Knob Bolete Nongill Mushroom	Unpalatable
Xanthoconium	*affine*	.	C. & E. USA (hardwood forest)	.	Related	Yellow Cone Bolete Nongill Mushroom	Edible
Xanthoconium	*affine*	*affine*	C. & E. USA (hardwood forest)	Original	Related	Yellow Cone Bolete Nongill Mushroom	Edible
Xanthoconium	*affine*	*maculosus*	C. & E. USA (hardwood forest), PFGM, page 103	Spotted	Related	Yellow Cone Bolete Nongill Mushroom	Edible
Xanthoconium	*separans*	.	USA, PFGM, page 104	.	Separated	Yellow Cone Bolete Nongill Mushroom	Unknown
Xerocomus	*amarellus*	.	Europe (conifer forest), MOTW, page 195	.	Amarelle	Dry Hair Bolete Nongill Mushroom	Inedible
Xerocomus	*badiorufus*	.	Europe (France), MOTW, page 198	.	.	Dry Hair Bolete Nongill Mushroom	Unknown

Family 622.301.02 *Gomphidiaceae* Slimecap Bolete

Genus	Species	Variety	Native Location	Variety	Specimen	Genome	Habit
Gomphidius	*glutinosus*	.	USA (conifer forest, especially Spruce) MOTW, page 173, PFGM, page 316	.	Sticky	Slimecap Bolete Nongill Mushroom	Edible-Good
Gomphidius	*gracilis*	.	Europe (mountain Larch), MOTW, page 173	.	Slender	Slimecap Bolete Nongill Mushroom	Edible
Gomphidius	*largus*	.	USA, PFGM, page 316	.	Large	Slimecap Bolete Nongill Mushroom	Unknown
Gomphidius	*maculatus*	.	USA (Larch mountain forest), MOTW, page 173	.	Spotted	Slimecap Bolete Nongill Mushroom	Edible

Gomphidius	*maculatus*	*maculatus*	USA (Larch mountain forest)	Original	Spotted	Slimecap Bolete Nongill Mushroom	Edible
Gomphidius	*maculatus*	*cookei*	USA (Larch mountain forest)	Cooke	Spotted	Slimecap Bolete Nongill Mushroom	Edible
Gomphidius	*nigricans*	.	Europe, MOTW, page 173	.	Black	Slimecap Bolete Nongill Mushroom	Edible
Gomphidius	*oregonensis*	.	NW. USA (Oregon), MOTW, page 308	.	Oregon	Slimecap Bolete Nongill Mushroom	Unknown
Gomphidius	*roseus*	.	Europe (Pine conifer forest), MOTW, page 174	.	Rosy	Slimecap Bolete Nongill Mushroom	Edible
Gomphidius	*subroseus*	.	USA, PFGM, page 316	.	Pink	Slimecap Bolete Nongill Mushroom	Edible
Gomphidius	*viscidus*	.	USA (Pine conifer forest), MOTW, page 174	.	Clammy	Slimecap Bolete Nongill Mushroom	Edible-Fair
Gomphidius	*viscidus*	*viscidus*	USA (Pine conifer forest)	Original	Clammy	Slimecap Bolete Nongill Mushroom	Edible-Fair
Gomphidius	*viscidus*	*giganteus*	USA (Pine conifer forest)	Giant	Clammy	Slimecap Bolete Nongill Mushroom	Edible-Fair
Gomphidius	*viscidus*	*testaceus*	USA (Pine conifer forest)	Brick Red	Clammy	Slimecap Bolete Nongill Mushroom	Edible-Fair
Chroogomphus	*helveticus*	.	Europe (Spruce & Pine forest), MOTW, page 174	.	Switzerland	Woolly Slimecap Bolete Nongill Mushroom	Edible
Chroogomphus	*leptocystis*	.	USA, PFGM, page 316	.	Thin Cyst	Woolly Slimecap Bolete Nongill Mushroom	Edible
Chroogomphus	*ochraceus*	.	USA, PFGM, page 315	.	Ocher	Woolly Slimecap Bolete Nongill Mushroom	Edible
Chroogomphus	*rutilus*	.	Canada, N. USA (conifer forest), MOTW, page 174	.	Brown	Woolly Slimecap Bolete Nongill Mushroom	Edible
Chroogomphus	*rutilus*	*rutilus*	Canada, N. USA (conifer forest)	Original	Brown	Woolly Slimecap Bolete Nongill Mushroom	Edible
Chroogomphus	*rutilus*	*britannicus*	Canada, N. USA (conifer forest)	Brittanicus	Brown	Woolly Slimecap Bolete Nongill Mushroom	Edible
Chroogomphus	*rutilus*	*corallinus*	Canada, N. USA (conifer forest)	Coral	Brown	Woolly Slimecap Bolete Nongill Mushroom	Edible
Chroogomphus	*tomentosus*	.	W. USA (conifer forest from Rocky Mtns. to pacific coast), PFGM, page 315	.	Orange Woollycap	Woolly Slimecap Bolete Nongill Mushroom	Edible
Chroogomphus	*vinicolor*	.	USA (Pine conifer forest), MOTW, page 174, PFGM, page 315	.	Wine Color	Woolly Slimecap Bolete Nongill Mushroom	Edible

Family 622.301.03 *Gyroporaceae* Hollow Stalk Bolete

Genus	Species	Variety	Native Location	Variety	Specimen	Genome	Habit
Gyroporus	*castaneus*	.	C. & E. USA (hardwood forest), MOTW, page 177, PFGM, page 110	.	Chestnut	Hollow Stalk Bolete Nongill Mushroom	Edible-Excellent
Gyroporus	*castaneus*	*castaneus*	C. & E. USA (hardwood forest)	Original	Chestnut	Hollow Stalk Bolete Nongill Mushroom	Edible-Excellent
Gyroporus	*castaneus*	*fulvidus*	C. & E. USA (hardwood forest)	Fulvous	Chestnut	Hollow Stalk Bolete Nongill Mushroom	Edible-Excellent
Gyroporus	*cyanescens*	.	E. USA (roads & trails in Chestnut & Oak hardwood forest), MOTW, page 174, PFGM, page 110	.	Cornflower	Hollow Stalk Bolete Nongill Mushroom	Edible-Excellent
Gyroporus	*cyanescens*	*cyanescens*	E. USA (roads & trails in hardwood forest)	Original	Cornflower	Hollow Stalk Bolete Nongill Mushroom	Edible-Excellent
Gyroporus	*cyanescens*	*constrictus*	E. USA (roads & trails in hardwood forest)	Constricted	Cornflower	Hollow Stalk Bolete Nongill Mushroom	Edible-Excellent
Gyroporus	*cyanescens*	*lacteus*	E. USA (roads & trails in hardwood forest)	Milky	Cornflower	Hollow Stalk Bolete Nongill Mushroom	Edible-Excellent
Gyroporus	*purpurinus*	.	USA, PFGM, page 110	.	Grayish Red	Hollow Stalk Bolete Nongill Mushroom	Unknown
Rubinoboletus	*ballouii*	.	E. USA (Oak & Beech hardwood forest), PFGM, page 121	.	Bittersweet	Ruby Hollow Stalk Bolete Nongill Mushroom	Edible-Bitter
Rubinoboletus	*rubinus*	.	Europe (broadleaf forest), MOTW, page 195	.	.	Ruby Hollow Stalk Bolete Nongill Mushroom	Inedible

Family 622.301.04 *Paxillaceae* Brimcap Bolete

Genus	Species	Variety	Native Location	Variety	Specimen	Genome	Habit
Paxillus	*corrugatus*	.	Europe (on decaying shrubs), MOTW, page 176	.	Wrinkled	Brimcap Bolete Nongill Mushroom	Unpalatable
Paxillus	*involutus*	.	USA (humid broadleaf forest), MOTW, page 175, PFGM, page 313	.	Naked	Brimcap Bolete Nongill Mushroom	Edible-Good
Paxillus	*involutus*	*involutus*	USA (humid broadleaf forest)	Original	Naked	Brimcap Bolete Nongill Mushroom	Edible-Good

Paxillus	*involutus*	*excentricus*	USA (humid broadleaf forest)	Excentric	Naked	Brimcap Bolete Nongill Mushroom	Edible-Good
Paxillus	*leptopus*	.	Europe (conifer, Alder broadleaf & mixed woods), MOTW, page 175	.	Thin Foot	Brimcap Bolete Nongill Mushroom	Edible
Gyrodon	*lividus*	.	USA (Alder, Birch & Hazelnut broadleaf forest), MOTW, page 201	.	Liver Gray	Roundtooth Brimcap Bolete Nongill Mushroom	Edible-Good
Gyrodon	*lividus*	*lividus*	USA (Alder, Birch & Hazelnut broadleaf forest)	Original	Liver Gray	Roundtooth Brimcap Bolete Nongill Mushroom	Edible-Good
Gyrodon	*lividus*	*brachyporus*	USA (Alder, Birch & Hazelnut broadleaf forest)	Short Pore	Liver Gray	Roundtooth Brimcap Bolete Nongill Mushroom	Edible-Good
Gyrodon	*lividus*	*sistotrema*	USA (Alder, Birch & Hazelnut broadleaf forest)	Firm Pore	Liver Gray	Roundtooth Brimcap Bolete Nongill Mushroom	Edible-Good
Gyrodon	*lividus*	*sistotremoides*	USA (Alder, Birch & Hazelnut broadleaf forest)	Fixed Pore	Liver Gray	Roundtooth Brimcap Bolete Nongill Mushroom	Edible-Good
Gyrodon	*merulioides*	.	E. USA (hardwood forest, Ash trees), PFGM, page 101	.	Shallow Pore	Roundtooth Brimcap Bolete Nongill Mushroom	Edible-Poor
Leucopaxillus	*gentianeus*	.	USA (on humus in conifer forest), PFGM, page 162	.	Bitter	Funnel Brimcap Bolete Nongill Mushroom	Unknown
Leucopaxillus	*giganteus*	.	USA (mountains, open woods, pastures), MOTW, page 112	.	Giant	Funnel Brimcap Bolete Nongill Mushroom	Edible-Fair
Leucopaxillus	*laterarius*	.	C. & E. USA (hardwood forest), PFGM, page 163	.	Early	Funnel Brimcap Bolete Nongill Mushroom	Edible-Bitter
Leucopaxillus	*paradoxus*	.	USA (on humus & decayed wood in conifer forest)	.	Paradox	Funnel Brimcap Bolete Nongill Mushroom	Unknown
Leucopaxillus	*paradoxus*	*paradoxus*	USA (on humus & decayed wood in conifer forest)	Original	Paradox	Funnel Brimcap Bolete Nongill Mushroom	Unknown
Leucopaxillus	*paradoxus*	*albissimus*	USA (on humus & decayed wood in conifer forest), PFGM, page 161	White	Paradox	Funnel Brimcap Bolete Nongill Mushroom	Unknown
Leucopaxillus	*tricolor*	.	Europe (Beech & Oak broadleaf forest), MOTW, page 97	.	Three Color	Funnel Brimcap Bolete Nongill Mushroom	Edible
Phylloporus	*rhodoxanthus*	.	USA (conifer & hardwood forest), MOTW, page 174, PFGM, page 314	.	Goldgills	Velvet Brimcap Bolete Nongill Mushroom	Edible
Tapinella	*atrotomentosa*	.	USA (logs & stumps in Pine & Spruce forest), MOTW, page 176, PFGM, page 313	.	Velvet Foot	Tapin Brimcap Bolete Nongill Mushroom	Unpalatable
Tapinella	*panuoides*	.	USA (logs & stumps in Pine forest), MOTW, page 176, PFGM, page 313	.	Tumor	Tapin Brimcap Bolete Nongill Mushroom	Edible
Tapinella	*panuoides*	*panuoides*	USA (logs & stumps in Pine forest)	Original	Tumor	Tapin Brimcap Bolete Nongill Mushroom	Edible
Tapinella	*panuoides*	*fagi*	USA (logs & stumps in Pine forest)	Fag	Tumor	Tapin Brimcap Bolete Nongill Mushroom	Edible
Tapinella	*panuoides*	*rubrosquamulosus*	USA (logs & stumps in Pine forest)	Red Scaled	Tumor	Tapin Brimcap Bolete Nongill Mushroom	Edible

Family 622.301.05 *Rhizopogonaceae* FalseTruffle Bolete

Genus	Species	Variety	Native Location	Variety	Specimen	Genome	Habit
Rhizopogon	*luteolus*	.	Europe (Pine conifer forest), MOTW, page 233	.	Ocher	False Truffle Bolete Nongill Mushroom	Edible-Poor
Rhizopogon	*luteolus*	*luteolus*	Europe (Pine conifer forest)	Original	Ocher	False Truffle Bolete Nongill Mushroom	Edible-Poor
Rhizopogon	*luteolus*	*induratus*	Europe (Pine conifer forest)	Hard	Ocher	False Truffle Bolete Nongill Mushroom	Edible-Poor
Truncocolumella	*citrina*	.	USA (Douglas Fir forest), PFGM, page 349	.	Citrine	Puffball FalseTruffle Bolete Nongill Mushroom	Unknown

Family 622.301.06 *Suillaceae* Slipperycap Bolete

Genus	Species	Variety	Native Location	Variety	Specimen	Genome	Habit
Suillus	*acidus*	.	USA, PFGM, page 117	.	Acidic	Slipperycap Bolete Nongill Mushroom	Unknown
Suillus	*albidipes*	.	Europe, MOTW, page 192	.	White Foot	Slipperycap Bolete Nongill Mushroom	Edible
Suillus	*americanus*	.	E. Canada, E. USA (White Pine forest), MOTW, page 192, PFGM, page 113	.	American	Slipperycap Bolete Nongill Mushroom	Edible
Suillus	*borealis*	.	N. USA, PFGM, page 114	.	Boreal	Slipperycap Bolete Nongill Mushroom	Edible
Suillus	*bovinus*	.	USA (mountain Pine woods), MOTW, page 194	.	Bovine	Slipperycap Bolete Nongill Mushroom	Edible-Poor
Suillus	*brevipes*	.	USA (Spruce & Pine forest), 114	.	Stubby Stalk	Slipperycap Bolete Nongill Mushroom	Edible

Suillus	*cavipes*	.	USA (mountain Larch forest), MOTW, page 199, PFGM, page 114	.	Hollow Stalk	Slipperycap Bolete Nongill Mushroom	Edible-Poor
Suillus	*cavipes*	*cavipes*	USA (mountain Larch forest)	Original	Hollow Stalk	Slipperycap Bolete Nongill Mushroom	Edible-Poor
Suillus	*cavipes*	*porosus*	USA (mountain Larch forest)	Porous	Hollow Stalk	Slipperycap Bolete Nongill Mushroom	Edible-Poor
Suillus	*collinitus*	.	USA (mountains), MOTW, page 192	.	Sticky	Slipperycap Bolete Nongill Mushroom	Edible
Suillus	*collinitus*	*collinitus*	USA (mountains)	Original	Sticky	Slipperycap Bolete Nongill Mushroom	Edible
Suillus	*collinitus*	*fluryi*	USA (mountains)	Flury	Sticky	Slipperycap Bolete Nongill Mushroom	Edible
Suillus	*cothurnatus*	.	USA, PFGM, page 117	.	Cothurn	Slipperycap Bolete Nongill Mushroom	Unknown
Suillus	*flavidus*	.	USA (Pine conifer forest), MOTW, page 191	.	Yellow	Slipperycap Bolete Nongill Mushroom	Edible
Suillus	*flavidus*	*flavidus*	USA (Pine conifer forest)	Original	Yellow	Slipperycap Bolete Nongill Mushroom	Edible
Suillus	*flavidus*	*pulchellus*	USA (Pine conifer forest)	Pretty	Yellow	Slipperycap Bolete Nongill Mushroom	Edible
Suillus	*granulatus*	.	USA (Pine conifer forest), MOTW, page 192, PFGM, page 115	.	Granulated	Slipperycap Bolete Nongill Mushroom	Edible-Fair
Suillus	*granulatus*	*granulatus*	USA (Pine conifer forest)	Original	Granulated	Slipperycap Bolete Nongill Mushroom	Edible-Fair
Suillus	*granulatus*	*lactifluus*	USA (Pine conifer forest)	Flowing Milk	Granulated	Slipperycap Bolete Nongill Mushroom	Edible-Fair
Suillus	*grevillei*	.	USA (Larch open woods), PFGM, page 117	.	Greville	Slipperycap Bolete Nongill Mushroom	Edible
Suillus	*grevillei*	*grevillei*	USA (Larch open woods)	Original	Greville	Slipperycap Bolete Nongill Mushroom	Edible
Suillus	*grevillei*	*annularius*	USA (Larch open woods)	Ringed	Greville	Slipperycap Bolete Nongill Mushroom	Edible
Suillus	*grevillei*	*badius*	USA (Larch open woods)	Red Brown	Greville	Slipperycap Bolete Nongill Mushroom	Edible
Suillus	*grevillei*	*elegans*	USA (Larch open woods), MOTW, page 191	Elegant	Greville	Slipperycap Bolete Nongill Mushroom	Edible
Suillus	*grevillei*	*proximus*	USA (Larch open woods), PFGM, page 117	Elegant	Greville	Slipperycap Bolete Nongill Mushroom	Edible
Suillus	*hirtellus*	.	USA, PFGM, page 119	.	Hairy	Slipperycap Bolete Nongill Mushroom	Unknown
Suillus	*lakei*	.	W. USA (Douglas Fir forest), PFGM, page 116	.	Lake's	Slipperycap Bolete Nongill Mushroom	Edible
Suillus	*lakei*	*lakei*	W. USA (Douglas Fir forest)	Original	Lake's	Slipperycap Bolete Nongill Mushroom	Edible
Suillus	*lakei*	*landkamme ri*	W. USA (Douglas Fir forest)	Landkam mer	Lake's	Slipperycap Bolete Nongill Mushroom	Edible
Suillus	*leptopus*	.	Europe (coastal Pine conifer forest), MOTW, page 192	.	Thin Foot	Slipperycap Bolete Nongill Mushroom	Edible
Suillus	*luteus*	.	S. Canada, USA (Pine conifer forest), MOTW, page 192, PFGM, page 116	.	Slippery Jack	Slipperycap Bolete Nongill Mushroom	Edible-Toxic
Suillus	*mitis*	.	USA, MOTW, page 192	.	Mild	Slipperycap Bolete Nongill Mushroom	Edible
Suillus	*neoalbidipes*	.	USA (Pine & White Pine forest), PFGM, page 117	.	Pale	Slipperycap Bolete Nongill Mushroom	Edible-Toxic
Suillus	*pictus*	.	E. USA (Eastern White Pine forest), E. Asia (Japan), MOTW, page 199, PFGM, page 118	.	Painted	Slipperycap Bolete Nongill Mushroom	Edible-Good
Suillus	*placidus*	.	Europe (Pine conifer forest), MOTW, page 182	.	Pleasant	Slipperycap Bolete Nongill Mushroom	Edible
Suillus	*plorans*	.	Europe (mountain conifer forest)	.	Weeping	Slipperycap Bolete Nongill Mushroom	Edible
Suillus	*ponderosus*	.	USA, PFGM, page 116	.	Heavy	Slipperycap Bolete Nongill Mushroom	Unknown
Suillus	*pseudobrevip es*	.	W. USA (Rocky Mtns. Lodgepole Pine forest), PFGM, page 119	.	Pine	Slipperycap Bolete Nongill Mushroom	Edible
Suillus	*pseudogranul atus*	.	SE. USA (Florida), PFGM, page 117	.	Lesser Grainy	Slipperycap Bolete Nongill Mushroom	Unknown
Suillus	*sibiricus*	.	E. Asia (Siberia), USA (Alaska), PFGM, pages 113, 120	.	Siberian	Slipperycap Bolete Nongill Mushroom	Edible
Suillus	*sibiricus*	*sibiricus*	E. Asia (Siberia), USA (Alaska)	Original	Siberian	Slipperycap Bolete Nongill Mushroom	Edible
Suillus	*sibiricus*	*helveticus*	E. Asia (Siberia), USA (Alaska), MOTW, page 191	Swiss	Siberian	Slipperycap Bolete Nongill Mushroom	Edible

Suillus	*subaureus*	.	USA, PFGM, page 119	.	Lesser Golden	Slipperycap Bolete Nongill Mushroom	Unknown
Suillus	*subluteus*	.	USA, PFGM, page 117	.	Lesser Yellow	Slipperycap Bolete Nongill Mushroom	Unknown
Suillus	*tomentosus*	.	Canada (NS), USA (two needle Pine forest)	.	Woollycap	Slipperycap Bolete Nongill Mushroom	Edible-Acidic
Suillus	*tridentinus*	.	Europe (Larch conifer forest), MOTW, page 191	.	Trentino	Slipperycap Bolete Nongill Mushroom	Edible
Suillus	*tridentinus*	*tridentinus*	Europe (Larch conifer forest)	Original	Trentino	Slipperycap Bolete Nongill Mushroom	Edible
Suillus	*tridentinus*	*aurantiporus*	Europe (Larch conifer forest)	Orange Pore	Trentino	Slipperycap Bolete Nongill Mushroom	Edible
Suillus	*umbonatus*	.	NW. USA (Pine forest), PFGM, page 120	.	Mountain	Slipperycap Bolete Nongill Mushroom	Edible
Suillus	*variegatus*	.	Europe (Pine forest), MOTW, page 194	.	Variegated	Slipperycap Bolete Nongill Mushroom	Edible
Suillus	*viscidus*	.	USA (Larch conifer forest), MOTW, page 191	.	Clammy	Slipperycap Bolete Nongill Mushroom	Edible
Suillus	*viscidus*	*viscidus*	USA (Larch conifer forest)	Original	Clammy	Slipperycap Bolete Nongill Mushroom	Edible
Suillus	*viscidus*	*aeruginascens*	USA (Larch conifer forest)	Copper Rust	Clammy	Slipperycap Bolete Nongill Mushroom	Edible
Suillus	*viscidus*	*bresadolae*	USA (Larch conifer forest), MOTW, page 191	Bresadola	Clammy	Slipperycap Bolete Nongill Mushroom	Edible
Suillus	*viscidus*	*flavogriseus*	USA (Larch conifer forest)	Yellow Gray	Clammy	Slipperycap Bolete Nongill Mushroom	Edible
Suillus	*viscidus*	*flavogriseus-nueschii*	USA (Larch conifer forest)	Nuesch Yellow Gray	Clammy	Slipperycap Bolete Nongill Mushroom	Edible
Suillus	*viscidus*	*laricinus*	USA (Larch conifer forest)	Larch	Clammy	Slipperycap Bolete Nongill Mushroom	Edible
Fuscoboletinus	*ochraceoroseus*	.	W. Canada, NW. USA (Larch forest), PFGM, page 109	.	Rosy	Fuscous Slipperycap Bolete Nongill Mushroom	Edible-Bitter

Order 622.32 *Cantharellales* Chanterelle

Latin: *canthar* chanter
Latin: *ellales* little

FAMILIES

622.321 *Cantharellaceae* Chanterelle
622.322 *Aphelariaceae* Simple Chanterelle
622.323 *Botryobasidiaceae* Ball Chanterelle
622.324 *Clavulinaceae* Club Chanterelle
622.325 *Gomphaceae* Bolt Chanterelle

Family 622.321 *Cantharellaceae* Chanterelle

Genus	Species	Variety	Native Location	Variety	Specimen	Genome	Habit
Cantharellus	*cibarius*	.	USA (conifer & hardwood forest), MOTW, page 224, PFGM, page 81	.	Edible	Chanterelle Nongill Mushroom	Edible-Good
Cantharellus	*cibarius*	*cibarius*	USA (conifer & hardwood forest)	Original	Edible	Chanterelle Nongill Mushroom	Edible-Good
Cantharellus	*cibarius*	*albidus*	USA (conifer & hardwood forest)	White	Edible	Chanterelle Nongill Mushroom	Edible-Good
Cantharellus	*cibarius*	*alborufescens*	USA (conifer & hardwood forest)	White & Rufous	Edible	Chanterelle Nongill Mushroom	Edible-Good
Cantharellus	*cibarius*	*bicolor*	USA (conifer & hardwood forest)	Two Color	Edible	Chanterelle Nongill Mushroom	Edible-Good
Cantharellus	*cibarius*	*edulis*	USA (conifer & hardwood forest)	Ordinary	Edible	Chanterelle Nongill Mushroom	Edible-Good
Cantharellus	*cibarius*	*ferruginascens*	USA (conifer & hardwood forest)	Ferruginous	Edible	Chanterelle Nongill Mushroom	Edible-Good
Cantharellus	*cibarius*	*flavipes*	USA (conifer & hardwood forest)	Yellow	Edible	Chanterelle Nongill Mushroom	Edible-Good
Cantharellus	*cibarius*	*nanus*	USA (conifer & hardwood forest)	Dwarf	Edible	Chanterelle Nongill Mushroom	Edible-Good
Cantharellus	*cibarius*	*neglectus*	USA (conifer & hardwood forest)	Neglected	Edible	Chanterelle Nongill Mushroom	Edible-Good
Cantharellus	*cibarius*	*pallens*	USA (conifer & hardwood forest)	Pale	Edible	Chanterelle Nongill Mushroom	Edible-Good
Cantharellus	*cibarius*	*pallidus*	USA (conifer & hardwood forest)	Pallid	Edible	Chanterelle Nongill Mushroom	Edible-Good
Cantharellus	*cibarius*	*rufipes*	USA (conifer & hardwood forest)	Blonde	Edible	Chanterelle Nongill Mushroom	Edible-Good

Cantharellus	*cibarius*	*salmoneus*	USA (conifer & hardwood forest)	Salmon	Edible	Chanterelle Nongill Mushroom	Edible-Good
Cantharellus	*cibarius*	*umbrinus*	USA (conifer & hardwood forest)	Umber	Edible	Chanterelle Nongill Mushroom	Edible-Good
Cantharellus	*cibarius*	*vulgaris*	USA (conifer & hardwood forest)	Ordinary	Edible	Chanterelle Nongill Mushroom	Edible-Good
Cantharellus	*cinereus*	.	Europe (Beech broadleaf forest), MOTW, page 223	.	Cinder Gray	Chanterelle Nongill Mushroom	Edible-Good
Cantharellus	*cinereus*	*cinereus*	Europe (Beech broadleaf forest)	Original	Cinder Gray	Chanterelle Nongill Mushroom	Edible-Good
Cantharellus	*cinereus*	*hydrolyps*	Europe (Beech broadleaf forest)	Waterdrop	Cinder Gray	Chanterelle Nongill Mushroom	Edible-Good
Cantharellus	*cinnabarinus*	.	SE. Canada, E. USA (hardwood forest), PFGM, page 82	.	Red	Chanterelle Nongill Mushroom	Edible
Cantharellus	*friesii*	.	Europe, MOTW, page 224	.	Fries	Chanterelle Nongill Mushroom	Edible
Cantharellus	*ignicolor*	.	USA, PFGM, page 84	.	Fire	Chanterelle Nongill Mushroom	Unknown
Cantharellus	*kauffmanii*	.	USA, PFGM, page 86	.	Kauffman	Chanterelle Nongill Mushroom	Unknown
Cantharellus	*lateritius*	.	USA (hardwood forests), PFGM, page 83	.	Smooth	Chanterelle Nongill Mushroom	Edible-Good
Cantharellus	*lutescens*	.	E. Canada, C. & E. USA (moss near bogs & springs, mountain conifer woods), MOTW, page 223, PFGM, page 83	.	Yellowish	Chanterelle Nongill Mushroom	Edible-Good
Cantharellus	*minor*	.	C. & E. USA (hardwood forest), PFGM, page 84	.	Small	Chanterelle Nongill Mushroom	Edible
Cantharellus	*muscigenus*		USA (on wood chips & moss), MOTW, page 223	.	Moss	Chanterelle Nongill Mushroom	Edible-Seasoning
Cantharellus	*odoratus*	.	USA, PFGM, pages 83, 84	.	Scented	Chanterelle Nongill Mushroom	Edible
Cantharellus	*pallidus*	.	Europe (Japan), MOTW, page 224	.	Pale	Chanterelle Nongill Mushroom	Inedible
Cantharellus	*subalbidus*	.	USA (Pensacola, FL), GSI-6, page 257	.	Lesser White	Chanterelle Nongill Mushroom	Edible
Cantharellus	*tubaeformis*	.	C. & E. USA (moss near bogs & springs, Pine conifer forest), MOTW, page 223, PFGM, page 84	.	Funnel	Chanterelle Nongill Mushroom	Edible-Good
Cantharellus	*tubaeformis*	*tubaeformis*	C. & E. USA (moss near bogs & springs, Pine conifer forest)	Original	Funnel	Chanterelle Nongill Mushroom	Edible-Good
Cantharellus	*tubaeformis*	*infundibuliformis*	C. & E. USA (moss near bogs & springs, Pine conifer forest)	Trumpet	Funnel	Chanterelle Nongill Mushroom	Edible-Good
Cantharellus	*tubaeformis*	*luteolus*	C. & E. USA (moss near bogs & springs, Pine conifer forest)	Yellow	Funnel	Chanterelle Nongill Mushroom	Edible-Good
Cantharellus	*tubaeformis*	*subramosus*	C. & E. USA (moss near bogs & springs, Pine conifer forest)	Branching	Funnel	Chanterelle Nongill Mushroom	Edible-Good
Craterellus	*caeruleofuscus*	.	USA (conifer & deciduous forest), PFGM, page 77	.	Dark Blue	Crater Chanterelle Nongill Mushroom	Unknown
Craterellus	*cornucopioides*	.	USA (conifer & hardwood forest), MOTW, page 223, PFGM, page 85	.	Horn Of Plenty	Crater Chanterelle Nongill Mushroom	Edible-Good
Craterellus	*cornucopioides*	*cornucopioides*	USA (conifer & hardwood forest)	Original	Horn Of Plenty	Crater Chanterelle Nongill Mushroom	Edible-Good
Craterellus	*cornucopioides*	*ochrosporus*	USA (conifer & hardwood forest)	Ocher Spore	Horn Of Plenty	Crater Chanterelle Nongill Mushroom	Edible-Good
Craterellus	*fallax*	.	USA (conifer & hardwood forest), PFGM, page 85	.	Deceptive	Crater Chanterelle Nongill Mushroom	Edible
Craterellus	*foetidus*	.	USA, PFGM, page 85	.	Fetid	Crater Chanterelle Nongill Mushroom	Unknown
Hygrophoropsis	*aurantiaca*	.	Europe (conifer forest), MOTW, page 118, PFGM, page 154	.	Orange	False Chanterelle Nongill Mushroom	Edible-Poor
Hygrophoropsis	*aurantiaca*	*aurantiaca*	Europe (conifer forest)	Original	Orange	False Chanterelle Nongill Mushroom	Edible-Poor
Hygrophoropsis	*aurantiaca*	*lactea*	Europe (conifer forest)	Milky	Orange	False Chanterelle Nongill Mushroom	Edible-Poor
Hygrophoropsis	*aurantiaca*	*nigripes*	Europe (on decaying stumps in conifer forest)	Black	Orange	False Chanterelle Nongill Mushroom	Edible-Poor
Hygrophoropsis	*aurantiaca*	*pallida*	Europe (conifer forest)	Pale	Orange	False Chanterelle Nongill Mushroom	Edible-Poor
Hygrophoropsis	*aurantiaca*	*robusta*	Europe (conifer forest)	Robust	Orange	False Chanterelle Nongill Mushroom	Edible-Poor
Hygrophoropsis	*aurantiaca*	*rufa*	Europe (conifer forest)	Blonde	Orange	False Chanterelle Nongill Mushroom	Edible-Poor
Hygrophoropsis	*olida*	.	Europe (Spruce conifer forest), MOTW, page 224	.	Scented	False Chanterelle Nongill Mushroom	Edible
Pseudocraterellus	*sinuosus*	.	Europe (broadleaf forest),	.	Curved	Lesser Crater	Edible

			MOTW, page 224			Chanterelle Nongill Mushroom	

Family 622.302.04 *Clavulinaceae* Club Chanterelle

Genus	Species	Variety	Native Location	Variety	Specimen	Genome	Habit
Clavulina	*cinerea*	.	USA (in woods), MOTW, page 219	.	Ashen	Club Chanterelle Nongill Mushroom	Edible-Laxative
Clavulina	*cinerea*	*cinerea*	USA (in woods)	Original	Ashen	Club Chanterelle Nongill Mushroom	Edible-Laxative
Clavulina	*cinerea*	*gracilis*	USA (in woods)	Slender	Ashen	Club Chanterelle Nongill Mushroom	Edible-Laxative
Clavulina	*coralloides*	.	USA (in woods, on footpaths)	.	Coral	Club Chanterelle Nongill Mushroom	Unpalatable
Clavulina	*coralloides*	*coralloides*	USA (in woods, on footpaths)	Original	Coral	Club Chanterelle Nongill Mushroom	Unpalatable
Clavulina	*coralloides*	*cristata*	USA (in woods, on footpaths), MOTW, page 221	Crested	Coral	Club Chanterelle Nongill Mushroom	Unpalatable
Clavulina	*coralloides*	*incarnata*	USA (in woods, on footpaths)	Flesh Color	Coral	Club Chanterelle Nongill Mushroom	Unpalatable
Clavulina	*coralloides*	*lappa*	USA (in woods, on footpaths)	Bur	Coral	Club Chanterelle Nongill Mushroom	Unpalatable
Clavulina	*coralloides*	*subrugosa*	USA (in woods, on footpaths)	Wrinkled	Coral	Club Chanterelle Nongill Mushroom	Unpalatable
Clavulina	*rugosa*	.	Europe (along forest paths), MOTW, page 221	.	Wrinkled	Club Chanterelle Nongill Mushroom	Edible
Clavulina	*rugosa*	*rugosa*	Europe (along forest paths)	Original	Wrinkled	Club Chanterelle Nongill Mushroom	Edible
Clavulina	*rugosa*	*alcyonaria*	Europe (along forest paths)	Elk	Wrinkled	Club Chanterelle Nongill Mushroom	Edible
Clavulina	*rugosa*	*canaliculata*	Europe (along forest paths)	Grooved	Wrinkled	Club Chanterelle Nongill Mushroom	Edible
Clavulina	*rugosa*	*fuliginea*	Europe (along forest paths)	Sooty	Wrinkled	Club Chanterelle Nongill Mushroom	Edible
Clavulina	*rugosa*	*herveyi*	Europe (along forest paths)	Hervey	Wrinkled	Club Chanterelle Nongill Mushroom	Edible

Family 622.302.05 *Gomphaceae* Bolt Chanterelle

Genus	Species	Variety	Native Location	Variety	Specimen	Genome	Habit
Gomphus	*clavatus*	.	S. Canada, N. USA (conifer forest), MOTW, page 221, PFGM, page 86	.	Pig's Ears	Bolt Chanterelle Nongill Mushroom	Edible-Good
Gomphus	*clavatus*	*clavatus*	S. Canada, N. USA (conifer forest)	Original	Pig's Ears	Bolt Chanterelle Nongill Mushroom	Edible-Good
Gomphus	*clavatus*	*truncatus*	S. Canada, N. USA (conifer forest)	Truncated	Pig's Ears	Bolt Chanterelle Nongill Mushroom	Edible-Good
Gomphus	*floccosus*	.	USA (conifer & hardwood forest), E. Asia, MOTW, page 224, PFGM, page 86	.	Scaly	Bolt Chanterelle Nongill Mushroom	Edible
Gomphus	*floccosus*	*floccosus*	USA (conifer & hardwood forest), E. Asia	Original	Scaly	Bolt Chanterelle Nongill Mushroom	Edible
Gomphus	*floccosus*	*bonarii*	USA (conifer & hardwood forest), E. Asia, PFGM, page 86	Bonar	Scaly	Bolt Chanterelle Nongill Mushroom	Edible
Gomphus	*floccosus*	*canadensis*	USA (conifer & hardwood forest)	Canada	Scaly	Bolt Chanterelle Nongill Mushroom	Edible
Gomphus	*floccosus*	*excavatus*	USA (conifer & hardwood forest), E. Asia	Excavated	Scaly	Bolt Chanterelle Nongill Mushroom	Edible
Gomphus	*floccosus*	*wilsonii*	USA (conifer & hardwood forest), E. Asia	Wilson	Scaly	Bolt Chanterelle Nongill Mushroom	Edible
Gomphus	*pseudoclavatus*	.	USA (hardwood forests), PFGM, page 86	.	Lesser Pig's Ears	Bolt Chanterelle Nongill Mushroom	Unknown

Order 622.303 *Hydnales* Spinecap Nongill Mushroom

Latin: *Hydna* truffle **FAMILIES**

622.303.01 *Hydnaceae* Truffle Spinecap Nongill Mushroom
622.303.02 *Auriscalpiaceae* Pinecone Spinecap Nongill Mushroom
622.303.03 *Bankeraceae* Banker Spinecap Nongill Mushroom
622.303.04 *Hericiaceae* Hedgehog Nongill Spinecap Mushroom
622.303.05 *Steccherinaceae* Steccher Spinecap Nongill Mushroom

Family 622.331 *Hydnaceae* Spinecap Nongill Mushroom

Genus	Species	Variety	Native Location	Variety	Specimen	Genome	Habit
Hydnum	*abietum*	.	USA (conifer forest), MOTW, page 216	.	Fir	Truffle Spinecap Nongill Mushroom	Inedible
Hydnum	*albidum*	.	E. Canada, E. USA (conifer & hardwood forest), PFGM, page 418	.	White	Truffle Spinecap Nongill Mushroom	Edible
Hydnum	*albomagnum*	.	USA, PFGM, page 89	.	Giant White	Truffle Spinecap Nongill Mushroom	Edible
Hydnum	*levigatum*	.	Europe (conifer forest), MOTW, page 214	.	Smooth	Truffle Spinecap Nongill Mushroom	Inedible
Hydnum	*repandum*	.	USA (in woods), MOTW, page 214, PFGM, page 89	.	Spreading	Truffle Spinecap Nongill Mushroom	Edible-Good
Hydnum	*repandum*	*repandum*	USA (in woods)	Original	Spreading	Truffle Spinecap Nongill Mushroom	Edible-Good
Hydnum	*repandum*	*abietinum*	USA (in woods)	Fir	Spreading	Truffle Spinecap Nongill Mushroom	Edible-Good
Hydnum	*repandum*	*album*	USA (in woods)	White	Spreading	Truffle Spinecap Nongill Mushroom	Edible-Good
Hydnum	*repandum*	*aurantium*	USA (in woods)	Orange	Spreading	Truffle Spinecap Nongill Mushroom	Edible-Good
Hydnum	*repandum*	*bicolor*	USA (in woods)	Two Color	Spreading	Truffle Spinecap Nongill Mushroom	Edible-Good
Hydnum	*repandum*	*bulbosum*	USA (in woods)	Bulb	Spreading	Truffle Spinecap Nongill Mushroom	Edible-Good
Hydnum	*repandum*	*clandestinum*	USA (in woods)	Clandestine	Spreading	Truffle Spinecap Nongill Mushroom	Edible-Good
Hydnum	*repandum*	*diffractum*	USA (in woods)	Diffracted	Spreading	Truffle Spinecap Nongill Mushroom	Edible-Good
Hydnum	*repandum*	*flavidum*	USA (in woods)	Yellow	Spreading	Truffle Spinecap Nongill Mushroom	Edible-Good
Hydnum	*repandum*	*medium*	USA (in woods)	Medium	Spreading	Truffle Spinecap Nongill Mushroom	Edible-Good
Hydnum	*repandum*	*pallidum*	USA (in woods)	Pale	Spreading	Truffle Spinecap Nongill Mushroom	Edible-Good
Hydnum	*repandum*	*washingtonianum*	USA (in woods)	Washington	Spreading	Truffle Spinecap Nongill Mushroom	Edible-Good
Hydnum	*rufescens*	.	USA (in woods), MOTW, page 214	.	Rufous	Truffle Spinecap Nongill Mushroom	Edible-Good
Hydnum	*rufescens*	*rufescens*	USA (in woods)	Original	Rufous	Truffle Spinecap Nongill Mushroom	Edible-Good
Hydnum	*rufescens*	*sulcatipes*	USA (in woods)	Furrowed	Rufous	Truffle Spinecap Nongill Mushroom	Edible-Good
Hydnum	*umbilicatum*	.	C. & E. USA (boggy soil in conifer forest), MOTW, page 214, PFGM, page 90	.	Depressed	Truffle Spinecap Nongill Mushroom	Edible

Family 622.332 *Auriscalpiaceae* Pinecone Spinecap Nongill Mushroom

Genus	Species	Variety	Native Location	Variety	Specimen	Genome	Habit
Auriscalpium	*vulgare*		Europe (pinecones), MOTW, page 217		Ordinary	Pinecone Spinecap Nongill Mushroom	Inedible
Auriscalpium	*vulgare*	*vulgare*	Europe (pinecones)	Original	Ordinary	Pinecone Spinecap Nongill Mushroom	Inedible
Auriscalpium	*vulgare*	*fechtneri*	Europe (pinecones)	Fechtner	Ordinary	Pinecone Spinecap Nongill Mushroom	Inedible
Lentinellus	*cochleatus*	.	USA (on old stumps in Beech broadleaf forest), MOTW, page 128, PFGM, page 158	.	Spiral	Pliable Pinecone Spinecap Nongill Mushroom	Edible-Poor
Lentinellus	*cochleatus*	*cochleatus*	USA (on old stumps in Beech broadleaf forest)	Original	Spiral	Pliable Pinecone Spinecap Nongill Mushroom	Edible-Poor
Lentinellus	*cochleatus*	*inolens*	USA (on old stumps in Beech broadleaf forest)	Odorless	Spiral	Pliable Pinecone Spinecap Nongill Mushroom	Edible-Poor
Lentinellus	*cochleatus*	*marcelianus*	USA (on old stumps in Beech broadleaf forest)	Marcel	Spiral	Pliable Pinecone Spinecap Nongill Mushroom	Edible-Poor
Lentinellus	*micheneri*	.	Europe (on buried wood debris)	.	Michener	Pliable Pinecone Spinecap Nongill Mushroom	Inedible
Lentinellus	*micheneri*	*micheneri*	Europe (on buried wood debris)	Original	Michener	Pliable Pinecone Spinecap Nongill Mushroom	Inedible
Lentinellus	*micheneri*	*americanus*	USA (on buried wood debris)	American	Michener	Pliable Pinecone Spinecap Nongill Mushroom	Inedible
Lentinellus	*micheneri*	*omphalodes*	USA, Europe (on buried wood debris), MOTW, page 126, PFGM, page 158	Depressed Cap	Michener	Pliable Pinecone Spinecap Nongill Mushroom	Inedible

Lentinellus	*micheneri*	*umbilicatus*	Europe (on buried wood debris)	Sunken Cap	Michener	Pliable Pinecone Spinecap Nongill Mushroom	Inedible
Lentinellus	*montanus*	.	USA, PFGM, page 159	.	Mountain	Pliable Pinecone Spinecap Nongill Mushroom	Edible
Lentinellus	*ursinus*	.	USA, PFGM, page 159	.	Bear	Pliable Pinecone Spinecap Nongill Mushroom	Unknown
Lentinellus	*ursinus*	*ursinus*	USA	Original	Bear	Pliable Pinecone Spinecap Nongill Mushroom	Unknown
Lentinellus	*ursinus*	*hispidus*	USA	Hairy	Bear	Pliable Pinecone Spinecap Nongill Mushroom	Unknown
Lentinellus	*ursinus*	*luxurians*	USA	Luxuriant	Bear	Pliable Pinecone Spinecap Nongill Mushroom	Unknown
Lentinellus	*ursinus*	*pelliculosa*	USA	Pellicule	Bear	Pliable Pinecone Spinecap Nongill Mushroom	Unknown
Lentinellus	*vulpinus*	.	USA, Europe, MOTW, page 126, PFGM, pages 158, 159	.	Fox Fur Colored	Pliable Pinecone Spinecap Nongill Mushroom	Inedible

Family 622.303.03 *Bankeraceae* Banker Spinecap Nongill Mushroom

Genus	Species	Variety	Native Location	Variety	Specimen	Genome	Habit
Bankera	*fuligineoalba*	.	Europe, MOTW, page 214, PFGM, page 88	.	Brown Margin	Banker Spinecap Nongill Mushroom	Unpalatable
Bankera	*fuligineoalba*	*fuligineoalba*	Europe	Original	Brown Margin	Banker Spinecap Nongill Mushroom	Unpalatable
Bankera	*fuligineoalba*	*macrodon*	Europe	Large Tooth	Brown Margin	Banker Spinecap Nongill Mushroom	Unpalatable
Bankera	*fuligineoalba*	*reticulatus*	Europe	Netted	Brown Margin	Banker Spinecap Nongill Mushroom	Unpalatable
Bankera	*violascens*	.	Canada, USA	.	Violet	Banker Spinecap Nongill Mushroom	Unpalatable
Bankera	*violascens*	*violascens*	Canada, USA	Original	Violet	Banker Spinecap Nongill Mushroom	Unpalatable
Bankera	*violascens*	*carnosa*	Canada, USA, PFGM, page 88	Fleshy	Violet	Banker Spinecap Nongill Mushroom	Unpalatable
Hydnellum	*aurantiacum*	.	USA (conifer forest), MOTW, page 217, PFGM, page 92	.	Orange Spine	Toothed Banker Spinecap Nongill Mushroom	Inedible
Hydnellum	*aurantiacum*	*aurantiacum*	USA (conifer forest)	Original	Orange Spine	Toothed Banker Spinecap Nongill Mushroom	Inedible
Hydnellum	*aurantiacum*	*complectipes*	USA (conifer forest)	Plaited	Orange Spine	Toothed Banker Spinecap Nongill Mushroom	Inedible
Hydnellum	*aurantiacum*	*floriforme*	USA (conifer forest)	Floral	Orange Spine	Toothed Banker Spinecap Nongill Mushroom	Inedible
Hydnellum	*aurantiacum*	*stohlii*	USA (conifer forest)	Stohl	Orange Spine	Toothed Banker Spinecap Nongill Mushroom	Inedible
Hydnellum	*caeruleum*	.	USA (conifer forest), MOTW, page 217, PFGM, page 93	.	Blue Spine	Toothed Banker Spinecap Nongill Mushroom	Inedible
Hydnellum	*caeruleum*	*caeruleum*	USA (conifer forest)	Original	Blue Spine	Toothed Banker Spinecap Nongill Mushroom	Inedible
Hydnellum	*caeruleum*	*alachuanum*	USA (conifer forest)	Alachuan	Blue Spine	Toothed Banker Spinecap Nongill Mushroom	Inedible
Hydnellum	*caeruleum*	*cyaneotinctum*	USA (conifer forest)	Cyan Tinted	Blue Spine	Toothed Banker Spinecap Nongill Mushroom	Inedible
Hydnellum	*caeruleum*	*ferrugipes*	USA (conifer forest), PFGM, page 93	Rusty Spine	Blue Spine	Toothed Banker Spinecap Nongill Mushroom	Inedible
Hydnellum	*caeruleum*	*inquinatum*	USA (conifer forest)	Inquinate	Blue Spine	Toothed Banker Spinecap Nongill Mushroom	Inedible
Hydnellum	*compactum*	.	Europe	.	Compact	Toothed Banker Spinecap Nongill Mushroom	Inedible
Hydnellum	*compactum*	*compactum*	Europe	Original	Compact	Toothed Banker Spinecap Nongill Mushroom	Inedible
Hydnellum	*compactum*	*acre*	Europe, MOTW, page 216	Acre	Compact	Toothed Banker Spinecap Nongill Mushroom	Inedible
Hydnellum	*compactum*	*mirabile*	Europe	Miracle	Compact	Toothed Banker Spinecap Nongill Mushroom	Inedible
Hydnellum	*compactum*	*tuberculosum*	Europe	Tubercule	Compact	Toothed Banker Spinecap Nongill Mushroom	Inedible
Hydnellum	*conigenum*	.	USA (conifer forest), PFGM, page 93	.	Funnel	Toothed Banker Spinecap Nongill Mushroom	Inedible
Hydnellum	*cruentum*	.	USA, PFGM, page 93	.	Blood	Toothed Banker Spinecap Nongill Mushroom	Unknown
Hydnellum	*cyanellum*	.	USA (N. California), PFGM, page 97	.	Bluish	Toothed Banker Spinecap Nongill Mushroom	Unknown
Hydnellum	*cyanopodium*	.	USA, PFGM, page 94	.	Blue Foot	Toothed Banker Spinecap Nongill Mushroom	Unknown
Hydnellum	*earlianum*	.	USA, PFGM, page 93	.	Earl	Toothed Banker Spinecap	Unknown

						Nongill Mushroom	
Hydnellum	*scleropodium*	.	USA, PFGM, page 93	.	Tough Stalk	Toothed Banker Spinecap Nongill Mushroom	Unknown
Hydnellum	*septentrionale*	.	USA, PFGM, page 93	.	Northern Spine	Toothed Banker Spinecap Nongill Mushroom	Unknown
Hydnellum	*suaveolens*	.	USA (conifer forest), MOTW, page 217, PFGM, page 94	.	Sweet Spine	Toothed Banker Spinecap Nongill Mushroom	Inedible
Hydnellum	*suaveolens*	*suaveolens*	USA (conifer forest)	Original	Sweet Spine	Toothed Banker Spinecap Nongill Mushroom	Inedible
Hydnellum	*suaveolens*	*boreale*	USA (conifer forest)	Boreal	Sweet Spine	Toothed Banker Spinecap Nongill Mushroom	Inedible
Hydnellum	*suaveolens*	*gravis*	USA (conifer forest)	Grave	Sweet Spine	Toothed Banker Spinecap Nongill Mushroom	Inedible
Hydnellum	*suaveolens*	*pullum*	USA (conifer forest)	Pull	Sweet Spine	Toothed Banker Spinecap Nongill Mushroom	Inedible
Sarcodon	*amarescens*	.	Europe, MOTW, page 216	.	Bitter	Fleshy Banker Spinecap Nongill Mushroom	Inedible
Sarcodon	*fennicus*	.	USA, Europe, MOTW, page 216, PFGM, page 96	.	Finland	Fleshy Banker Spinecap Nongill Mushroom	Inedible
Sarcodon	*fuligineoviolaceus*	.	USA, Europe, MOTW, page 216, PFGM, page 97	.	Smoky Violet	Fleshy Banker Spinecap Nongill Mushroom	Inedible
Sarcodon	*fuligineoviolaceus*	*fuligineoviolaceus*	Europe	Original	Smoky Violet	Fleshy Banker Spinecap Nongill Mushroom	Inedible
Sarcodon	*fuligineoviolaceus*	*talpa*	Europe	Mole	Smoky Violet	Fleshy Banker Spinecap Nongill Mushroom	Inedible
Sarcodon	*imbricatus*	.	USA (conifer & hardwood forest), MOTW, page 216, PFGM, page 96	.	Scaly	Fleshy Banker Spinecap Nongill Mushroom	Edible
Sarcodon	*imbricatus*	*imbricatus*	USA (conifer & hardwood forest)	Original	Scaly	Fleshy Banker Spinecap Nongill Mushroom	Edible
Sarcodon	*imbricatus*	*aspratus*	USA (conifer & hardwood forest)	Rough	Scaly	Fleshy Banker Spinecap Nongill Mushroom	Edible
Sarcodon	*leucopus*	.	Europe (conifer forest)	.	White Foot	Fleshy Banker Spinecap Nongill Mushroom	Inedible
Sarcodon	*leucopus*	*leucopus*	Europe (conifer forest)	Original	White Foot	Fleshy Banker Spinecap Nongill Mushroom	Inedible
Sarcodon	*leucopus*	*laevigatus*	Europe (conifer forest)	Smooth	White Foot	Fleshy Banker Spinecap Nongill Mushroom	Inedible
Sarcodon	*rimosus*	.	USA (conifer & mixed conifer forest), PFGM, page 96	.	Cracked	Fleshy Banker Spinecap Nongill Mushroom	Unknown
Sarcodon	*scabrosus*	.	USA (deciduous & hardwood forest), MOTW, page 216, PFGM, page 95	.	Scabby	Fleshy Banker Spinecap Nongill Mushroom	Inedible
Sarcodon	*squamosus*	.	Europe, MOTW, page 216	.	Scaly	Fleshy Banker Spinecap Nongill Mushroom	Edible-Poor

Family 622.303.04 *Hericiaceae* Hedgehog Spinecap Nongill Mushroom

Genus	Species	Variety	Native Location	Variety	Specimen	Genome	Habit
Hericium	*abietis*	.	USA (Olympia, WA), PFGM, page 91	.	Conifer Coral	Hedgehog Spinecap Nongill Mushroom	Edible
Hericium	*coralloides*	.	USA (on dead or living Beech, Elm, Oak & Ash hardwood trees), MOTW, page 214, PFGM, page 91	.	Coral	Hedgehog Spinecap Nongill Mushroom	Edible
Hericium	*coralloides*	*coralloides*	USA (on dead or living Beech, Elm, Oak & Ash hardwood trees)	Original	Coral	Hedgehog Spinecap Nongill Mushroom	Edible
Hericium	*coralloides*	*abietinum*	USA (Fir conifer forest)	Fir	Coral	Hedgehog Spinecap Nongill Mushroom	Inedible
Hericium	*coralloides*	*caput-ursi*	USA (on dead or living Beech, Elm, Oak & Ash hardwood trees), MOTW, page 214	Bear's Head	Coral	Hedgehog Spinecap Nongill Mushroom	Edible
Hericium	*coralloides*	*laciniatum*	USA (on dead or living Beech, Elm, Oak & Ash hardwood trees)	Lacy	Coral	Hedgehog Spinecap Nongill Mushroom	Edible
Hericium	*coralloides*	*ramosum*	USA (on dead or living Beech, Elm, Oak & Ash hardwood trees), PFGM, page 92	Comb	Coral	Hedgehog Spinecap Nongill Mushroom	Edible
Hericium	*coralloides*	*reichii*	USA (on dead or living Beech, Elm, Oak & Ash hardwood trees)	Reich	Coral	Hedgehog Spinecap Nongill Mushroom	Edible
Hericium	*erinaceus*	.	USA (cracks & knotholes of	.	Bearded	Hedgehog Spinecap	Edible-Fair

			deciduous trees), MOTW, page 214, PFGM, page 91			Nongill Mushroom	
Hericium	*erinaceus*	*erinaceus*	USA (cracks & knotholes of deciduous trees)	Original	Bearded	Hedgehog Spinecap Nongill Mushroom	Edible-Fair
Hericium	*erinaceus*	*caput-medusae*	USA (cracks & knotholes of deciduous trees), MOTW, page 214	Head Of Medusa	Bearded	Hedgehog Spinecap Nongill Mushroom	Edible-Fair
Hericium	*erinaceus*	*echinus*	USA (cracks & knotholes of deciduous trees)	Urchin	Bearded	Hedgehog Spinecap Nongill Mushroom	Edible-Fair
Hericium	*erinaceus*	*grande*	USA (cracks & knotholes of deciduous trees)	Grand	Bearded	Hedgehog Spinecap Nongill Mushroom	Edible-Fair
Hericium	*erinaceus*	*hystrix*	USA (cracks & knotholes of deciduous trees)	Porcupine	Bearded	Hedgehog Spinecap Nongill Mushroom	Edible-Fair
Hericium	*erinaceus*	*lion's-manei*	USA (cracks & knotholes of deciduous trees)	Lion's Mane	Bearded	Hedgehog Spinecap Nongill Mushroom	Edible-Fair
Hericium	*erinaceus*	*monkey-headii*	USA (cracks & knotholes of deciduous trees)	Monkey Head	Bearded	Hedgehog Spinecap Nongill Mushroom	Edible-Fair
Hericium	*erinaceus*	*pom-pomii*	USA (cracks & knotholes of deciduous trees)	Pom Pom	Bearded	Hedgehog Spinecap Nongill Mushroom	Edible-Fair
Hericium	*erinaceus*	*quercinum*	USA (cracks & knotholes of deciduous trees)	Oak	Bearded	Hedgehog Spinecap Nongill Mushroom	Edible-Fair
Hericium	*erinaceus*	*unguiculatum*	USA (cracks & knotholes of deciduous trees)	Claw	Bearded	Hedgehog Spinecap Nongill Mushroom	Edible-Fair

Class 622.4 *Gastromycetes* Cavity Fungus

Latin: *Gastro* gastric
Latin: *Mycetes* mycete, fungus

ORDERS

664.1 *Gastromycetales* Cavity Fungus
664.2 *Lycoperdales* Puffball
664.3 *Phallales* Horn Fungus

Order 622.42 *Lycoperdales* Puffball

Latin: *Lyco* wolf
Latin: Perdales flatulence

FAMILIES

622.402.01 *Lycoperdaceae* Puffball
622.402.02 *Geoastraceae* Earth Star Puffball
622.402.03 *Sclerodermataceae* Earthball Puffball
622.402.04 *Tulostomataceae* Stalked Puffball

Family 622.402.01 *Lycoperdaceae* Puffball

Genus	Species	Variety	Native Location	Variety	Specimen	Genome	Habit
Lycoperdon	*caudatum*	.	USA	.	Tailed	Puffball	Unknown
Lycoperdon	*caudatum*	*caudatum*	USA	Original	Tailed	Puffball	Unknown
Lycoperdon	*caudatum*	*pedicellatum*	USA, MOTW, page 309	Pedicel	Tailed	Puffball	Unknown
Lycoperdon	*curtisii*	.	USA, PFGM, page 355	.	Curtis	Puffball	Unknown
Lycoperdon	*echinatum*	.	E. Canada, E. USA (hardwood forest), MOTW, page 231, PFGM, page 354	.	Prickly	Puffball	Edible-Poor
Lycoperdon	*echinatum*	*echinatum*	E. Canada, E. USA (hardwood forest)	Original	Prickly	Puffball	Edible-Poor
Lycoperdon	*echinatum*	*hoylei*	E. Canada, E. USA (hardwood forest)	Hoyle	Prickly	Puffball	Edible-Poor
Lycoperdon	*echinatum*	*gemmatum*	E. Canada, E. USA (hardwood forest), MOTW, page 230	Gem Studded	Prickly	Puffball	Edible-Poor
Lycoperdon	*flavotinctum*	.	USA, PFGM, page 355	.	Yellow Tinted	Puffball	Unknown
Lycoperdon	*mammiforme*	.	USA	.	Nipple	Puffball	Unknown
Lycoperdon	*mammiforme*	*mammiforme*	USA	Original	Nipple	Puffball	Unknown
Lycoperdon	*mammiforme*	*velatum*	USA, MOTW, page 309	Velvet	Nipple	Puffball	Unknown
Lycoperdon	*marginatum*	.	USA (sandy soil in various habitats), PFGM, page 355	.	Naked	Puffball	Edible
Lycoperdon	*molle*	.	USA (conifer & hardwood forest), PFGM, page 355	.	Smooth	Puffball	Edible
Lycoperdon	*muscorum*	.	USA, PFGM, page 355	.	Musky	Puffball	Unknown
Lycoperdon	*nigrescens*	.	USA	.	Black	Puffball	Unknown
Lycoperdon	*nigrescens*	*nigrescens*	USA	Original	Black	Puffball	Unknown
Lycoperdon	*nigrescens*	*foetidum*	USA, PFGM, page 355	Fetid	Black	Puffball	Unknown
Lycoperdon	*nigrescens*	*perlatum*	USA	Perlate	Black	Puffball	Unknown
Lycoperdon	*oblongisporum*	.	USA, PFGM, page 356	.	Oblong Spore	Puffball	Unknown

Lycoperdon	*perlatum*	.	USA (roadsides, trails, open woods, forest), PFGM, page 355	.	Gem	Puffball	Edible-Good
Lycoperdon	*perlatum*	*perlatum*	USA (roadsides, trails, open woods, forest)	Original	Gem	Puffball	Edible-Good
Lycoperdon	*perlatum*	*gemmatum*	USA (roadsides, trails, open woods, forest)	Studded	Gem	Puffball	Edible-Good
Lycoperdon	*pulcherrimum*	.	USA (sandy soil in open hardwood woods), PFGM, page 356	.	Beautiful	Puffball	Edible
Lycoperdon	*pyriforme*	.	USA (on decaying logs, stumps, sawdust), MOTW, page 231, PFGM, page 356	.	Pear	Puffball	Edible-Good
Lycoperdon	*pyriforme*	*pyriforme*	USA (on decaying logs, stumps, sawdust)	Original	Pear	Puffball	Edible-Good
Lycoperdon	*pyriforme*	*tessellatum*	USA (on decaying logs, stumps, sawdust)	Mosaic	Pear	Puffball	Edible-Good
Lycoperdon	*umbrinum*	.	USA, Europe, MOTW, page 231, PFGM, page 355	.	Dark	Puffball	Inedible
Bovista	*aestivalis*	.	Europe	.	Summer	Globe Puffball	Unpalatable
Bovista	*aestivalis*	*aestivalis*	Europe	Original	Summer	Globe Puffball	Unpalatable
Bovista	*aestivalis*	*cepiformis*	Europe	Onion	Summer	Globe Puffball	Unpalatable
Bovista	*aestivalis*	*furfuraceum*	Europe, MOTW, page 230	Scaly	Summer	Globe Puffball	Unpalatable
Bovista	*minor*	.	C. & S. USA, PFGM, page 351	.	Lesser	Globe Puffball	Unknown
Bovista	*nigrescens*	.	Europe (mountain pastures), MOTW, page 231	.	Black	Globe Puffball	Edible
Bovista	*nigrescens*	*nigrescens*	Europe (mountain pastures)	Original	Black	Globe Puffball	Edible
Bovista	*nigrescens*	*globosum*	Europe (mountain pastures)	Globe	Black	Globe Puffball	Edible
Bovista	*pila*	.	USA (pastures, roadsides, open woodlands), PFGM, page 351	.	Pillar	Globe Puffball	Edible
Bovista	*plumbea*	.	Europe (mountain pastures), MOTW, page 231, PFGM, page 351	.	Lead Gray	Globe Puffball	Edible
Bovista	*plumbea*	*plumbea*	Europe (mountain pastures)	Original	Lead Gray	Globe Puffball	Edible
Bovista	*plumbea*	*ovalispora*	Europe (mountain pastures)	Oval Spore	Lead Gray	Globe Puffball	Edible
Bovista	*pusilla*	.	USA (pastures, roadsides)	.	Mini	Globe Puffball	Edible
Calbovista	*subsculpta*	.	W. USA (open woodland from Rocky Mtns. to pacific coast), PFGM, page 357	.	.	Beautiful Globe Puffball	Edible-Excellent
Calvatia	*booniana*	.	W. USA, PFGM, page 353	.	Pacific	Giant Puffball	Unknown
Calvatia	*coelatum*	.	Europe (mountain grass), MOTW, page 230	.	Chisled	Giant Puffball	Edible-Excellent
Calvatia	*cretacea*	.	USA, PFGM, page 353	.	Arctic Alpine	Giant Puffball	Unknown
Calvatia	*cyathiformis*	.	USA (pastures, lawns, parks, golf courses, grassy fields), PFGM, page 352	.	Vase	Giant Puffball	Edible
Calvatia	*fragilis*	.	USA, PFGM, page 352	.	Fragile	Giant Puffball	Unknown
Calvatia	*fragilis*	*fragilis*	USA	Original	Fragile	Giant Puffball	Unknown
Calvatia	*fragilis*	*pilula*	USA	Pile	Fragile	Giant Puffball	Unknown
Calvatia	*fumosa*	.	W. USA (Spruce-Fir forest from Rocky Mtns. to pacific coast), PFGM, page 352	.	Smoky	Giant Puffball	Unknown
Calvatia	*gigantea*	.	E. USA (meadows, fields, open woods), PFGM, page 353	.	Australian	Giant Puffball	Edible-Good
Calvatia	*hesperia*	.	USA, PFGM, page 353	.	Hesperian	Giant Puffball	Unknown
Calvatia	*leiospora*	.	USA, PFGM, page 352	.	Lion Spore	Giant Puffball	Unknown
Calvatia	*lepidophora*	.	USA, PFGM, page 353	.	Thin Prairie	Giant Puffball	Unknown
Calvatia	*maximum*	.	Europe (mountains, pastures, trails, open woods), MOTW, page 228	.	Greater	Giant Puffball	Edible-Excellent
Calvatia	*pachyderma*	.	USA, PFGM, page 353	.	Thick Skin	Giant Puffball	Unknown
Calvatia	*polygonia*	.	USA (prairie regions), PFGM, page 353	.	Warty Spore Prairie	Giant Puffball	Unknown
Calvatia	*sculpta*	.	W. USA (on duff in Rocky Mtns. conifer forest), PFGM, page 353	.	Sculptured	Giant Puffball	Unknown
Calvatia	*sigillata*	.	USA, PFGM, page 352	.	Sigillate	Giant Puffball	Unknown
Calvatia	*subcretacea*	.	USA, PFGM, pages 352, 353	.	Cretacean	Giant Puffball	Unknown
Handkea	*excipuliformis*	.	USA	.	Container	Handke Puffball	Edible
Handkea	*excipuliformis*	*excipuliformis*	USA	Original	Container	Handke Puffball	Edible
Handkea	*excipuliformis*	*elatum*	USA	Tall	Container	Handke Puffball	Edible
Handkea	*excipuliformis*	*saccatum*	USA, MOTW, page 230	Sack	Container	Handke Puffball	Edible
Handkea	*utriformis*	.	Europe	.	Wineskin	Handke Puffball	Edible-Good
Handkea	*utriformis*	*utriformis*	Europe	Original	Wineskin	Handke Puffball	Edible-Good
Handkea	*utriformis*	*bovista*	Europe	Puffy	Wineskin	Handke Puffball	Edible-Good

Handkea	*utriformis*	*caelatum*	Europe, MOTW, page 230	Chisled	Wineskin	Handke Puffball	Edible-Good
Handkea	*utriformis*	*sinclairii*	Europe	Sinclair	Wineskin	Handke Puffball	Edible-Good
Mycenastrum	*corium*	.	USA (pastures, barnyards, feedlots), PFGM, page 358	.	Leather	Giant Pasture Puffball	Edible
Vascellum	*pratense*	.	USA (pastures, lawns, parks, golf courses, grassy fields)	.	Pasture	Vase Puffball	Edible
Vascellum	*pratense*	*pratense*	USA (pastures, lawns, parks, golf courses, grassy fields)	Original	Pasture	Vase Puffball	Edible
Vascellum	*pratense*	*depressum*	USA (pastures, lawns, parks, golf courses, grassy fields), PFGM, page 357	Depressed	Pasture	Vase Puffball	Edible
Vascellum	*pratense*	*hiemale*	USA (pastures, lawns, parks, golf courses, grassy fields)	Winter	Pasture	Vase Puffball	Edible

Family 622.402.04 *Tulostomataceae* Stalked Puffball

Genus	Species	Variety	Native Location	Variety	Specimen	Genome	Habit
Pisolithus	*tinctorius*	.	USA, Europe (conifer forest), MOTW, page 233, PFGM, page 366	.	Tincture	Pea Rock Stalked Puffball	Edible
Podaxis	*argentinus*	.	USA, PFGM, page 365	.	Silver	Pod Stalked Puffball	Unknown
Podaxis	*farlowii*	.	USA, PFGM, page 365	.	Farlow	Pod Stalked Puffball	Unknown
Podaxis	*longii*	.	W. USA (on sandy soil along roadsides & under cactus), PFGM, page 364	.	Giant	Pod Stalked Puffball	Unknown
Podaxis	*microporus*	.	USA, PFGM, page 365	.	Tiny Pore	Pod Stalked Puffball	Unknown
Podaxis	*pistillaris*	.	W. USA (desert soil from S. Idaho to California), S. Asia (N. India), PFGM, page 365	.	Khumbi	Pod Stalked Puffball	Edible

Order 622.403 *Phallales* Horn Fungus

Latin: *Phallus* penis

FAMILIES

622.403.01 *Phallaceae* Stinkhorn Horn Fungus
622.403.02 *Clathraceae* Lattice Horn Fungus
622.403.03 *Hysterangiaceae* Uterus Horn Fungus

Family 622.403.02 *Clathraceae* Lattice Horn Fungus

Genus	Species	Variety	Native Location	Variety	Specimen	Genome	Habit
Clathrus	*columnatus*	.	USA, PFGM, page 345	.	Column	Stinkhorn Lattice Horn Fungus	Unknown
Clathrus	*ruber*	.	Europe, Asia, PFGM, page 345	.	Red	Stinkhorn Lattice Horn Fungus	Inedible
Clathrus	*ruber*	*ruber*	Europe, Asia	.	Red	Stinkhorn Lattice Horn Fungus	Inedible
Clathrus	*ruber*	*cancellatus*	Europe (broadleaf forest), MOTW, page 228	Lattice	Red	Stinkhorn Lattice Horn Fungus	Inedible
Clathrus	*ruber*	*kusanoi*	Asia (Japan, in woods), MOTW, page 224	Kusano	Red	Stinkhorn Lattice Horn Fungus	Inedible
Anthurus	*muellerianus*	.	Europe	.	Egg	Sponge Lattice Horn Fungus	Inedible
Anthurus	*muellerianus*	*muellerianus*	Europe	Original	Egg	Sponge Lattice Horn Fungus	Inedible
Anthurus	*muellerianus*	*aseroëformis*	Europe (humid woods), MOTW, page 224	Unpleasant	Egg	Sponge Lattice Horn Fungus	Inedible
Anthurus	*gardneri*	.	USA, PFGM, page 344	.	Gardner	Sponge Lattice Horn Fungus	Unknown
Pseudocolus	*fusiformis*	.	USA, PFGM, page 345	.	Stinky Squid	Dead Log Lattice Horn Fungus	Unknown
Pseudocolus	*schellenbergiae*	.	USA (woods), W. Indies (Cuba), Asia (Japan), MOTW, page 224	.	Schellenberg	Dead Log Lattice Horn Fungus	Inedible

SUBKINGDOM 63 *Bryata* Moss

There were no known Moss herbs or vegetables at the time of publication.

SUBKINGDOM 64 *Filicata* Fern

Latin: *filicata* filament, fern leaf

PHYLUMS

641 *Lycopodiophyta* Moss Fern
642 *Pteridophyta* Fern

Fusillo Fern Classification

Domain *Plantae*	Plant
Kingdom 6 *Sporophyta*	Spore Plant
Subkingdom 4 *Filicata*	Fern
Phylum 1 *Lycopodiophyta*	**Moss Fern**
Class 1 *Lycopodiopsida*	Moss Fern
Order 1 *Lycopodiales*	Club Moss
Family 1 *Lycopodiaceae*	Club Moss
Order 2 *Huperziales*	Fir Moss
Family 1 *Huperziaceae*	Fir Moss
Class 2 *Isosellopsida*	Quillwort & Spike Moss
Order 1 *Isoetales*	Quillwort
Family 1 *Isoetaceae*	Quillwort
Order 2 *Sellaginellales*	Spike Moss
Family 1 *Sellaginellaceae*	Spike Moss
Phylum 2 *Pteridophyta*	**Leaf Fern**
Class 1 *Protofilicopsida*	Primitive Fern
Order 1 *Equistetales*	Horsetail
Family 1 *Equisetaceae*	Horsetail
Order 2 *Psilotales*	Whisk Fern
Family 1 *Psilotaceae*	Whisk Fern
Class 2 *Parafilicopsida*	Wetland, Tropical, & Dryland Fern
Order 1 *Hydropterales*	Wetland Fern
Family 1 *Azollaceae*	Mosquito Pond Fern
Family 2 *Parkeriaceae*	Water Sprite Fern
Family 3 *Marsileaceae*	Water Clover Fern
Family 4 *Salviniaceae*	Floating Water Fern
Order 2 *Marattales*	Tropical Fern
Family 1 *Marattaceae*	Tropical Fern
Order 3 *Ophioglossales*	Dryland Fern
Family 1 *Ophioglossaceae*	Succulent Fern
Class 4 *Filicopsida*	Subtropical Fern
Order 1 *Filicales*	Fern
Family 1 *Pteridaceae*	Fern
Family 2 *Hymenophyllaceae*	Filmy Fern
Family 3 *Osmundaceae*	Flowering Fern
Family 4 *Schizeaceae*	Curly Grass Fern
Order 2 *Polypodiales*	Polypody Fern
Family 01 *Polypodiaceae*	Polypody Fern
Family 02 *Aspleniaceae*	Spleenwort Fern
Family 03 *Blechnaceae*	Misorus Fern
Family 04 *Cyathaceae*	Tree Fern
Family 05 *Dennstaedtiaceae*	Hay Scented Fern
Family 06 *Dicksoniaceae*	Dickson Fern
Family 07 *Drypopteridaceae*	Wood Fern
Family 08 *Gleichenaceae*	Forked, Scrambling, & Umbrella Fern
Family 09 *Grammatidaceae*	Marked Fern
Family 10 *Thelypteridaceae*	Delicate Fern
Family 11 *Vittariaceae*	Shoestring Fern

PHYLUM 642 *Pteridophyta* Fern

Latin: *filicata* filament, fern leaf

CLASSES

642.1 *Protofilicopsida* Primitive Fern
642.2 *Parafilicopsida* Wetland, Tropical, Dryland Fern
642.3 *Parafilicopsida* True Fern

Class 642.1 *Protofilicopsida* Primitive Fern

Latin: *filicata* filament, fern leaf

ORDERS

642.11 *Equisetales*	Horsetail
642.12 *Psilotales*	Whisk Fern

Order 642.11 *Equisetales* Fern

Latin: *filicata* filament, fern leaf

FAMILIES

642.111 *Equisetaceae*	Fern

Family 642.111 *Equisetaceae* Horsetail

Horsetail (*Equisetum*)

Genus	species	infraspecies	cultivar	specimen	Genome	Vegetable
Equisetum	*arvense*	.	.	Field	Horsetail	Stem

Class 642.3 *Filicopsida* True Fern

Latin: *filicata* filament, fern leaf

ORDERS

642.31 *Filicales*	Fern
642.32 *Polyopdales*	Polypody Fern

Order 682.31 *Filicales* Fern

Latin: *filicata* filament, fern leaf

FAMILIES

642.311 *Pteridaceae*	Fern
642.312 *Hymenophyllaceae*	Filmy Fern
642.313 *Osmumdaceae*	Flowering Fern
642.314 *Schizaeaceae*	Curly Grass Fern

Family 682.311 *Pteridaceae* Fern

Maidenhair Fern (*Adiantum*)

Genus	species	infraspecies	cultivar	specimen	Genome	Vegetable
Adiantum	*aethiopicum*	.	.	Ethiopian	Maidenhair Fern	Frond
Adiantum	*aleuticum*	.	.	Five Finger	Maidenhair Fern	Frond
Adiantum	*aleuticum*	*aleuticum*	Original	Five Finger	Maidenhair Fern	Frond
Adiantum	*aleuticum*	*calderi*	Calder	Five Finger	Maidenhair Fern	Frond
Adiantum	*aleuticum*	*subpumilum*	Dwarf	Five Finger	Maidenhair Fern	Frond
Adiantum	*capillus-veneris*	.	.	Venus Hair	Maidenhair Fern	Frond
Adiantum	*capillus-veneris*	*capillus-veneris*	Original	Venus Hair	Maidenhair Fern	Frond
Adiantum	*capillus-veneris*	*modestum*	Modest	Venus Hair	Maidenhair Fern	Frond
Adiantum	*capillus-veneris*	*protrusum*	Protuding	Venus Hair	Maidenhair Fern	Frond
Adiantum	*pedatum*	.	.	Bird's Foot	Maidenhair Fern	Frond
Adiantum	*pedatum*	*pedatum*	Original	Bird's Foot	Maidenhair Fern	Frond
Adiantum	*pedatum*	*asiaticum*	Asian	Bird's Foot	Maidenhair Fern	Frond
Adiantum	*pedatum*	*imbricatum*	Crowded	Bird's Foot	Maidenhair Fern	Frond
Adiantum	*pedatum*	*japonicum*	Japanese	Bird's Foot	Maidenhair Fern	Frond
Adiantum	*pedatum*	*laceratum*	Lace Leaf	Bird's Foot	Maidenhair Fern	Frond
Adiantum	*Pedatum*	*miss-sharplesi*	Miss Sharples	Bird's Foot	Maidenhair Fern	Frond
Adiantum	*Pedatum*	*montanum*	Mountain	Bird's Foot	Maidenhair Fern	Frond

KINGDOM 7 *Spermatophyta* Seed Plant

Latin: *Sperma* sperm, seed
Latin: *Phyta* phyton, plant

SUBKINGDOMS

71 *Gymnospermatophyta*	Seed Fern
72 *Anthophyta*	Flower Plant

SUBKINGDOM 71 *Coniferophyta* Conifer (Seed Fern)

Latin: *Antho* anther, flower
Latin: *Phyta* phyton, plant

PHYLUMS

711 *Cycadophyta*	Cycad
712 *Ginkgophyta*	Ginkgo
713 *Gnetophyta*	Gnet
714 *Pinophyta*	Pine

Class 711.01 *Cycadopsida*	**Cycad**
Class 711.02 *Cycadovegetpsida*	**Cultivated Cycad**
Order 1 *Cycadales*	Cycad
Family 1 *Cycadaceae*	Cycad
Family 2 *Stangeriaceae*	Fern Cycad
Family 3 *Zamiaceae*	Palm Cycad
Class 712.01 *Ginkgopsida*	**Ginkgo**
Class 712.02 *Ginkgovegetpsida*	**Cultivated Ginkgo**
Order 1 *Ginkgoales*	Ginkgo
Family 1 *Ginkgoaceae*	Ginkgo
Class 713.01 *Gnetopsida*	**Gnet**
Class 713.02 *Gnetovegetopsida*	**Cultivated Gnet**
Order 1 *Gnetales*	Gnet
Family 1 *Gnetaceae*	Gnet
Class 714.01 *Pinopsida*	**Pine**
Class 714.02 *Pinovegetpsida*	**Pinyon Pine**
Order 1 *Pinales*	Pine
Family 1 *Pinaceae*	Pine
Family 2 *Araucariaceae*	Monkey Puzzle Pine
Family 3 *Podocarpaceae*	Plum Pine
Family 4 *Sequoiaceae*	Sequoia Pine
Order 2 *Cedrales*	Cedar
Family 1 *Cedraceae*	Cedar
Family 2 *Cupressaceae*	Cypress
Order 3 *Ephedrales*	Ephedra
Family 1 *Ephedraceae*	Ephedra
Order 4 *Gnetales*	Gnet
Family 1 *Gnetaceae*	Gnet
Order 5 *Taxales*	Yew
Family 1 *Taxaceae*	Yew

Class 714.02 *Pinovegetopsida* Pinyon Pine

Latin: *Pino* pin frond, pine

ORDERS

714.021 *Pinales*	Pine
714.022 *Cedrales*	Cedar
714.023 *Ephedrales*	Ephedra
714.024 *Gnetales*	Gnet
714.025 *Taxales*	Yew

Order 714.021 *Pinales* Pine

Latin: *Pino* pin frond, pine

FAMILIES

714.021.1 *Pinaceae*	Pine
714.021.2 *Araucariaceae*	Monkey Puzzle Pine
714.021.3 *Podocarpaceae*	Plum Pine
714.021.4 *Sequoiaceae*	Sequoia Pine

FAMILY 714.021.1 *Pinaceae* Pine

Genus	Species	Specimen	Genome	C:Culinary, M:Medicinal, E:Economic
Pinus	*caribaea*	West Indies	Pine	E:OL,RN:OR,Turpentine
Pinus	*halepensis*	Aleppo	Pine	E:OL,RN:OR,Turpentine

Pinus	*Halepensis x pinaster*	Maritime Aleppo	Pine	**E**:OL,RN:OR,Turpentine
Pinus	*massoniana*	Masson Horsetail	Pine	**E**:OL,RN:OR,Turpentine
Pinus	*mugo*	Mugo Mountain	Pine	**E**:OL,RN:OR,Turpentine
Pinus	*palustris*	Long Needle Pitch	Pine	**E**:OL,RN:OR,Turpentine
Pinus	*pinaster*		Maritime	Pine **E**:OL,RN:OR,Turpentine
Pinus	*ponderosa*	Ponderosa	Pine	**E**:OL,RN:OR,Turpentine
Pinus	*resinosa*	Resin	Pine	**E**:OL,RN:OR,Turpentine
Pinus	*rotundata*	Dwarf Mugo Mtn.	Pine	**E**:OL,RN:OR,Turpentine
Pinus	*sylvestris*	Scotch	Pine	**E**:OL,RN:OR,Turpentine
Pinus	*tabuliformis*	Chinese Red	Pine	**E**:OL,RN:OR,Turpentine
Pinus	*Taeda*	Loblolly	Pine	**E**:OL,RN:OR,Turpentine

Genus	**Species**	**Specimen**	**Genome**	**C:Culinary, M:Medicinal, E:Economic**
Abies	*alba*	European Silver	Fir	**E**:OL,RN:OR,Turpentine

Order 714.022 *Cedrales* Cedar

Latin: *Cedra* cedar

FAMILIES

714.022.1 *Cedraceae* Cedar
714.022.2 *Cupressaceae* Cypress

FAMILY 714.022.1 *Cedraceae* Cedar

Genus	**Species**	**Specimen**	**Genome**	**C:Culinary, M:Medicinal, E:Economic**
Cedrus	*atlantica*	Atlantic	Cedar	**E**:OL:DS,PF, Cedar Oil

FAMILY 714.022.2 *Cupressaceae* Cypress

Genus	**Species**	**Specimen**	**Genome**	**C:Culinary, M:Medicinal, E:Economic**
Juniperus	*communis*	European	Juniper	**C**:FR:FF, **E**:OL:PF, Juniper Oil
Juniperus	*Sabina*	Savin	Juniper	**E**:OL:PF, Savin Juniper Oil
Juniperus	*scopulorum*	Rocky Mountain	Juniper	**E**:OL:PF, Rocky Mountain Juniper Oil
Juniperus	*virginiana*	Atlantic Red Cedar	Juniper	**E**:OL:PF, Red Cedar Oil

SUBKINGDOM 72 *Anthophyta* Flower Plant

Latin: *Antho* anther, flower	**PHYLUMS**		**SPECIES**
Latin: *Phyta* phyton, plant	721 *Monocotophyta*	Wild Monocot (except wild orchid)	25,000
		Wild Orchid	25,000
	722 *Dicotophyta*	Wild Dicot	60,000
	723 *Monocotovegetophyta*	Cultivated Monocot	1,000
	724 *Dicotovegetophyta*	Cultivated Dicot	25,000

PHYLUM 723 *Monocotovegetophyta* Cultivated Monocot

Latin: *mono* single	**CLASSES**	
Latin: *cota* cotyledon	722.01 *Liliopsida*	Lily
Latin: *vegeta* vegetation	722.02 *Commelinopsida*	Grass
Latin: *phyta* phyton, plant	722.03 *Pandanopsida*	Lilywood
	722.04 *Orchidafroeurasiopsida*	Afroeurasian Orchid
	722.05 *Orchidafropsida*	African Orchid
	722.06 *Orchidamalagasiopsida*	Madagascar Orchid
	722.07 *Orchidaeuropaopsida*	European Orchid
	722.08 *Orchidanordasia*	North Asian Orchid
	722.09 *Orchidaustalasiopsida*	South Asian Orchid
	722.10 *Orchidanordameriopsida*	North American Orchid
	722.11 *Orchidatropicameriopsida*	Tropical American Orchid
	722.12 *Orchidasudameriopsida*	Southern South American Orchid

Class 722.01 *Liliopsida*	Lily
Order 1 *Liliales*	Lily
Family 1 *Liliaceae*	Lily
Family 2 *Campynemaceae*	Field Lily
Family 3 *Colchicaceae*	Meadow Crocus
Family 4 *Corsiaceae*	Corsian Lily
Family 5 *Funkiaceae*	Plantain Lily
Family 6 *Luzuriagaceae*	Luzurian Lily
Family 7 *Melanthiaceae*	Honey Lily
Family 8 *Smilacaceae*	Greenbrier
Order 2 *Alismatales*	Water Plantain
Family 1 *Alismataceae*	Water Plantain
Family 2 *Aponogetonaceae*	Aponi Pond Grass
Family 3 *Butomaceae*	Water Gladiola
Family 4 *Cymodoceaceae*	Surf Grass
Family 5 *Hydrocharitaceae*	Water Charity
Family 6 *Limnocharitaceae*	Water Velvetleaf
Family 7 *Potamogetonaceae*	Pond Grass
Family 8 *Zosteraceae*	Eel Grass
Order 3 *Amaryllidales*	Amaryllis
Family 1 *Amaryllidaceae*	Amaryllis
Family 2 *Agavaceae*	Agave
Family 3 *Alliaceae*	Allyl Lily
Family 4 *Alstroemeriaceae*	South American Lily
Family 5 *Anthericaceae*	Spider Agave Lily
Family 6 *Hypoxidaceae*	Starflower Grass
Order 4 *Arales*	Aroid Lily
Family 1 *Araceae*	Ar Lily
Family 2 *Acoraceae*	Flag Lily
Family 3 *Anthuriaceae*	Flamingo Lily
Family 4 *Juncaginaceae*	Arrow Grass
Family 5 *Lemnaceae*	Duckweed
Family 6 *Philodendraceae*	Terrestrial Ar Lily
Family 7 *Ruppiaceae*	Widgeon Grass
Family 8 *Tofieldiaceae*	Tofield Ar Lily
Order 5 *Asparagales*	Asparagus Lily
Family 1 *Asparagaceae*	Asparagus
Family 2 *Blandfordiaceae*	Blandford Lily
Family 3 *Boryaceae*	Bory Lily
Family 4 *Iridaceae*	Iris
Family 5 *Petrosaviaceae*	Petrosav Lily
Family 6 *Tecophilaeaceae*	Tecophile Lily
Order 6 *Asteliales*	Sword Grass
Family 1 *Asteliaceae*	Sword Grass
Family 2 *Asphodelaceae*	Asphodel
Family 3 *Dasypogonaceae*	Beard Grass Lily
Family 4 *Hyacinthaceae*	Hyacinth
Family 5 *Philesiaceae*	Philesian Lily
Family 6 *Xanthorrhoeaceae*	Skirt Grass
Family 7 *Xeronemataceae*	Flax Grass
Order 7 *Bromeliales*	Bromelaid
Family 1 *Bromeliaceae*	Bromelaid
Family 2 *Haemodoraceae*	Redroot
Family 3 *Pontederiaceae*	Pickerelweed
Order 8 *Dioscoreales*	Yam Lily
Family 1 *Dioscoreaceae*	Yam Lily
Family 2 *Burmanniaceae*	BlueThread Lily
Family 3 *Convallariaceae*	Lily Of The Valley
Family 4 *Dracaenaceae*	Dragon Lily
Family 5 *Gloriosaceae*	Glory Vine
Family 6 *Nartheciaceae*	Bellflower Lily
Family 7 *Ruscaceae*	Broom
Family 8 *Uvulariaceae*	Merry Bells Lily
Order 9 *Zingiberales*	Ginger
Family 1 *Zingiberaceae*	Ginger
Family 2 *Cannaceae*	Cane Ginger
Family 3 *Costaceae*	Cost Ginger
Family 4 *Heliconiaceae*	Bird Of Paradise
Family 5 *Lowiaceae*	Lowian Ginger
Family 6 *Marantaceae*	Marant Ginger
Family 7 *Musaceae*	Banana
Family 8 *Strelitziaceae*	Strelitz Ginger

Class 722.02 *Commelinopsida*	Grass
Order 1 *Commelinales*	Teret Grass
Family 1 *Commelinaceae*	Teret Grass
Family 2 *Cyperaceae*	Sedge Grass
Family 3 *Eriocaulaceae*	Woolly Stem Grass
Family 4 *Hanguanaceae*	Hanguan Grass
Family 5 *Juncaceae*	Rush Grass
Family 6 *Mayacaceae*	Bog Moss Grass
Family 7 *Philydraceae*	Philydran Grass
Family 8 *Xyridaceae*	Yellow Eyed Grass
Order 2 *Graminales*	Grain
Family 1 *Avenaceae*	Oats
Family 2 *Hordeaceae*	Barley
Family 3 *Miliaceae*	Millet
Family 4 *Oryzaceae*	Rice
Family 5 *Secaleaceae*	Rye
Family 6 *Tripsacaceae*	Gammagrass (Corn)
Family 7 *Triticaceae*	Wheat
Order 3 *Parapoales*	Field Grass
Family 1 *Parapoaceae*	Field Grass
Family 2 *Anarthriaceae*	Jointless Grass
Family 3 *Centrolepidaceae*	Center Thin Grass
Family 4 *Ecdeiocoleaceae*	Ecdeion Grass
Family 5 *Rapateaceae*	Rapate Grass
Family 6 *Thurniaceae*	Thurn Grass
Order 4 *Poales*	Meadow Grass
Family 1 *Poaceae*	Meadow Grass
Family 2 *Agrostidaceae*	Bent Grass
Family 3 *Andropogonaceae*	Beard Grass
Family 4 *Chloridaceae*	Windmill Grass
Family 5 *Festucaceae*	Fescue Grass
Family 6 *Panicaceae*	Panic Grass
Family 7 *Phalaridaceae*	Canary Grass
Family 8 *Zoysiaceae*	Zoysia Grass

Class 722.03 *Pandanopsida*	Wood Lily
Order 1 *Pandanales*	Screw Pine
Family 1 *Pandanaceae*	Screw Pine
Family 2 *Cyclanthaceae*	Hat Plant
Family 3 *Stemonaceae*	Stemon Screw Pine
Family 4 *Triuridaceae*	Three Tail Screw Pine
Family 5 *Velloziaceae*	Velloze Screw Pine
Order 2 *Arecales*	Palm
Family 1 *Arecaceae*	Feather Palm
Family 2 *Arecavannaceae*	Fan Palm
Family 3 *Pararecaceae*	Frond Palm
Order 3 *Arundoales*	Reed
Family 1 *Arundoaceae*	Reed
Family 2 *Bambusaceae*	Bamboo
Family 3 *Joinvilleaceae*	Joinville Reed
Family 4 *Restionaceae*	Reston Reed
Family 5 *Sparganiaceae*	Bur Reed
Family 6 *Typhaceae*	Cat Tail Reed

Class 722.04 *Orchidafroeurasiopsida*	Afroeurasian Orchid
Order 1 *Orchidafroeurasiales*	Afroeurasian Orchid
Family 1 *Orchidafroeurasiaceae*	Afroeurasian Orchid
Family 2 *Habenariafroeurasiaceae*	Afroeurasian Strap Bog Orchid
Order 2 *Cypripediafroeurasiales*	Afroeurasian Ladyslipper
Family 1 *Cypripediafroeurasiaceae*	Afroeurasian Ladyslipper
Order 3 *Epidendrafroeurasiales*	Afroeurasian Reedstem
Family 1 *Epipogiaceae*	Epiphyte Bearded Orchid
Family 2 *Neottiafroeurasiaceae*	Afroeurasian Modern Twayblade
Family 3 *Nerviliafroeurasiaceae*	Afroeurasian Nervil Orchid
Order 4 *Spiranthafroeurasiales*	Afroeurasian Ladies Tresses
Family 1 *Spiranthafroeurasiaceae*	Afroeurasian Ladies Tresses

Class 722.05 *Orchidafropsida*	African Orchid
Order 1 *Orchidafrales*	African Orchid
Family 1 *Orchidafraceae*	African Orchid
Family 2 *Habenariafraceae*	African Strap Bog Orchid
Family 3 *Holothrixaceae*	Wholethread Orchid
Family 4 *Huttoneaceae*	Hutton Orchid
Order 2 *Disafrales*	African Hooded Orchid
Family 1 *Disafraceae*	African Hooded Orchid
Family 2 *Coriciafraceae*	African Coryce Orchid
Order 3 *Epiarethusafrales*	African Dragon's Mouth
Family 1 *Ancistrochilafraceae*	African Hooklip Orchid
Family 2 *Calanthafraceae*	African Callisto Orchid
Order 4 *Epidendafrales*	African Reedstem Orchid
Family 1 *Gastrodiafraceae*	African Cavity Orchid
Family 2 *Malaxafraceae*	African Adder's Mouth
Family 3 *Neottiafraceae*	African Modern Twayblade
Family 4 *Nerviliafraceae*	African Nervil Orchid
Order 5 *Podochilafrales*	African Pod Lip Orchid
Family 1 *Bulbophyllafraceae*	African Mobile Lip Orchid
Order 6 *Spiranthafrales*	African Ladies Tresses
Family 1 *Spiranthafraceae*	African Ladies Tresses
Family 2 *Goodiyerafraceae*	African Rattlesnake Orchid
Family 3 *Tropidiafraceae*	African Palm Orchid
Order 7 *Vandafrales*	African Vanda Orchid
Family 1 *Aerangafraceae*	African Nectar Spur
Family 2 *Angarecafraceae*	African Comet Orchid
Family 3 *Polystachyafraceae*	African Drooping Spike
Family 4 *Trichoglottafraceae*	African Hairytongue
Order 8 *Vandacymbafrales*	African Boat Lip Orchid
Family 1 *Cymbidiafraceae*	African Boat Lip Orchid
Family 2 *Cyrtopodafraceae*	African Archpod Orchid
Order 9 *Vanillafrales*	African Vanilla Orchid
Family 1 *Vanillafraceae*	African Vanilla Orchid

Class 722.06 *Orchidamalagasiopsida*	Madagascar Orchid
Order 1 *Orchidmalagasiales*	Madagascar Orchid
Family 1 *Orchidmalagasiaceae*	Madagascar Orchid
Family 2 *Habaenarimalagasiaceae*	Madagascar Strap Bog Orchid
Order 2 *Disamalagasiales*	Madagascar Hooded Orchid
Family 1 *Disamalagasiaceae*	Madagascar Hooded Orchid
Family 2 *Corycimalagasiaceae*	Madagascar Coryce Orchid
Order 3 *Epiarhethusimalagasiales*	Madagascar Dragon's Mouth
Family 1 *Calanthemalagasiaceae*	Madagascar CallistoOrchid
Order 4 *Epidendramalagasiales*	Madagascar Reedstem Orchid
Family 1 *Gastrodimalagasiaceae*	Madagascar Cavity Orchid
Family 2 *Malaxamalagasiaceae*	Madagascar Adder's Mouth
Family 3 *Nervilimalagasiaceae*	Madagascar Nervil Orchid
Order 5 *Podochilamalagasiales*	Madagascar Pod Lip Orchid
Family 1 *Bulbophyllamalagasiaceae*	Madagascar Mobile Lip Orchid
Order 6 *Spiranthamalagasiales*	Madagascar Ladies Tresses
Family 1 *Goodyeramalagasiaceae*	Madagascar Rattlesnake Orchid
Family 2 *Tropidimalagasiaceae*	Madagascar Palm Orchid
Order 7 *Vandamalagasiales*	Madagascar Vanda Orchid
Family 1 *Aergagamalagasiaceae*	Madagascar Nectar Spur Orchid
Family 2 *Angraecamalagasiaceae*	Madagascar Comet Orchid
Family 3 *Polystachymalagasiaceae*	Madagascar Drooping Spike
Family 4 *Trichoglottamalagasiaceae*	Madagascar Hairytongue Orchid
Order 8 *Vandacymbimalagasiales*	Madagascar Boat Lip Orchid
Family 1 *Cymbidimalagasiaceae*	Madagascar Boat Lip Orchid
Family 2 *Cyrtopodimalagasiaceae*	Madagascar Arch Pod Orchid
Order 9 *Vanillamalagasiales*	Madagascar Vanilla Orchid
Family 1 *Vanillamalagasiaceae*	Madagascar Vanilla Orchid

Class 722.07 *Orchidaeuropaopsida*	European Orchid
Order 1 *Orchideuropales*	European Orchid
Family 1 *Orchideuropaceae*	European Orchid
Family 2 *Habenarieuropaceae*	European Strap Bog Orchid
Order 2 *Epidendraeuropales*	European Reed Stem Orchid
Family 1 *Neottieuropaceae*	Eur. Modern Tway Blade Orchid

Class 722.08 *Orchidanordasiopsida*	North Asian Orchid
Order 1 *Orchidonordasiales*	North Asian Orchid
Family 1 *Orchidonordasiaceae*	North Asian Orchid
Order 2 *Cypripedinordasiales*	N. Asian Lady Slipper Orchid
Family 1 *Cypripedinordasiaceae*	N. Asian Lady Slipper Orchid
Order 3 *Epidendrnordasiales*	N. Asian Reed Stem Orchid
Family 1 *Calypsonordasiaceae*	N. Asian Calypso Orchid
Family 2 *Neottinordasiaceae*	N. Asian Modern Twayblade Orchid
Order 4 *Vanillanordasiales*	N. Asian Vanilla Orchid
Family 1 *Vanillanordasiaceae*	N. Asian Vanilla Orchid

Class 722.09 *Orchidaustalasiopsida*	South Asian Orchid
Order 1 *Orchidaustroasiales*	South Asian Orchid
Family 1 *Orchiaustralasiaceae*	South Asian Orchid
Family 2 *Habenaraustralasiaceae*	S. Asian Strap Bog Orchid
Family 3 *Saccolabiaustralasiaceae*	S. Asian Sac Lip Orchid
Family 4 *Wullschlaegelaustralasiaceae*	S. Asian Wullschlaege Orchid
Order 2 *Apostiaustralasirales*	South Asian Post Orchid
Family 1 *Apostiaceae*	South Asian Post Orchid
Order 3 *Cypripediaustralasirales*	S. Asian Lady Slipper Orchid
Family 1 *Cypripediaustralasiiaceae*	S. Asian Lady Slipper Orchid
Family 2 *Paphiopedilaceae*	S. Asian Pouch Lady Slipper Orchid
Order 4 *Disaustralasirales*	S. Asian Queen Disa Orchid
Family 1 *Disaaustralasiaceae*	S. Asian Queen Disa Orchid
Family 2 *Coryciaustralasiaceae*	S. Asian Coryce Disa Orchid
Order 5 *Diurales*	S. Asian Double Tail Orchid
Family 1 *Diuraceae*	S. Asian Double Tail Orchid
Family 2 *Acianthaceae*	South Asian Mosquito Orchid
Family 3 *Caledeniaceae*	S. Asian Spider Orchid
Family 4 *Cryptostylaceae*	S. Asian HiddenOrchid
Family 5 *Drakeaceae*	S. Asian Hammer Orchid
Family 6 *Megastylaceae*	S. Asian Mega Orchid
Family 7 *Prasophyllaceae*	S. Asian Leek Orchid
Family 8 *Pterostylaceae*	S. Asian Green Hood Orchid
Family 9 *Thelymitraceae*	S. Asian Sun Orchid
Order 6 *Epiarethusaustralasiales*	S. Asian Dragon's Mouth Orchid
Family 1 *Arethusaustralasiaceae*	S. Asian Dragon's Mouth Orchid
Family 2 *Ancistrochilaustralasiaceae*	S. Asian Ancestral Lip Orchid
Family 3 *Bletiaustralasiaceae*	South Asian Urn Orchid
Family 4 *Calantheaustralasiaceae*	S. Asian Callisto Orchid
Family 5 *Plocoglottceae*	S. Asian Ploce Lip Orchid
Family 6 *Tainiaceae*	South Asian Tain Orchid
Family 7 *Thuniaceae*	S. Asian Bamboo Reed Orchid
Order 7 *Epidendraustraasiales*	S. Asian Reed Stem Orchid
Family 1 *Agrostophyllaceae*	S. Asian Bent Grass Orchid
Family 2 *Calypsoaustralasiaceae*	S. Asian Calypso Orchid
Family 3 *Coelogyneaceae*	South Asian Angel Orchid
Family 4 *Gastrodiaustralasiaceae*	South Asian Cavity Orchid
Family 5 *Glomeraceae*	S. Asian Glomerule Orchid
Family 6 *Malaxaustralasiaceae*	S. Asian Adder's Mouth Orchid
Family 7 *Neottiaustralasiaceae*	S. Asian Modern Twayblade Orchid
Family 8 *Nerviliaustralasiaceae*	S. Asian Nervil Orchid
Family 9 *Rhizanthellaceae*	S. Asian Rhizome Orchid
Order 8 *Podochilales*	S. Asian Pod Lip Orchid
Family 1 *Podochilaceae*	S. Asian Pod Lip Orchid
Family 2 *Bulbophyllaustralasiaceae*	S. Asian Mobile Lip Orchid
Family 3 *Dentrobiaceae*	S. Asian Reed Orchid
Family 4 *Eriaceae*	S. Asian Woolly Orchid
Family 5 *Thelasaceae*	S. Asian Nipple Orchid
Order 9 *Spiranthaustralasiales*	S. Asian Ladies Tresses Orchid
Family 1 *Spiranthaustralasiaceae*	S. Asian Ladies Tresses Orchid
Family 2 *Crytpostyleaceae*	S. Asian Hidden Orchid
Family 3 *Goodyeraaustralasiaceae*	S. Asian Rattlesnake Orchid
Family 4 *Tropidiaustralasiaceae*	S. Asian Tropical Palm Orchid
Order A *Vandaustralasiales*	S. Asian Vanda Orchid
Family 1 *Vandaceae*	S. Asian Vanda Orchid
Family 2 *Vandahybridaceae*	S. Asian Hybrid Vanda Orchid
Family 3 *Aeridaceae*	S. Asian Foxtail Airplant Orchid
Family 4 *Phalaenopsaceae*	S. Asian Moth Orchid
Family 5 *Polystachyaustralasiaceae*	S. Asian Drooping Spike Orchid
Family 6 *Trichoglottaustralasiaceae*	S. Asian Hairy Tongue Orchid
Order B *Vandacymbaustralasiales*	S. Asian Boat Lip Orchid
Family 1 *Cymbidiaustralasiaceae*	S. Asian Boat Lip Orchid
Family 2 *Cyrtopodiaustralasiaceae*	S. Asian Boat Pod Orchid
Order C *Vanillaustralasiales*	S. Asian Vanilla Orchid
Family 1 *Vanillaustralasiaceae*	S. Asian Vanilla Orchid

Class 722.10 ***Orchidanordameriopsida***	**North American Orchid**
Order 1 *Orchidanordameriales*	North American Orchid
Family 1 *Orchidanordameriaceae*	North American Orchid
Family 2 *Galeraceae*	N. American Gale Orchid
Family 3 *Habenarinordameriaceae*	N. American Strap Bog Orchid
Order 2 *Cypripedinordameriales*	N. American Lady Slipper Orchid
Family 1 *Cypripedinordameriaceae*	N. American Lady Slipper Orchid
Order 3 *Epiarithusanordameriales*	N. American Dragon's Mouth Orchid
Family 1 *Arethusanordameriaceae*	N. American Dragon's Mouth Orchid
Family 2 *Calpogonaceae*	N. American Snap Lip Orchid
Order 4 *Epidendrinordameriales*	N. American Reed Stem Orchid
Family 1 *Calypsonordameriaceae*	N. American Calypso Orchid
Family 2 *Malaxinordameriaceae*	N. American Adder's Mouth Orchid
Family 3 *Neottinordameriaceae*	N. American Modern Twayblade Orchid
Family 4 *Triphoranordameriaceae*	N. American Nodding Caps Orchid
Order 5 *Spiranthenordameriales*	N. American Ladies Tresses Orchid
Family 1 *Spiranthenordameriaceae*	N. American Ladies Tresses Orchid
Family 2 *Goodyeranordameriaceae*	N. American Rattlesnake Orchid
Order 6 *Vanillanordameriales*	North American Vanilla Orchid
Family 1 *Vanillanordameriaceae*	North American Vanilla Orchid
Order 7 *Vandamaxillanordameriales*	North American Tiger Orchid
Family 1 *Maxillarinordameriaceae*	North American Tiger Orchid
Class 722.11 ***Orchidatropicameriopsida***	**Tropical American Orchid**
Order 1 *Orchidatropicameriales*	Tropical American Orchid
Family 1 *Orchidatropicameriaceae*	Tropical American Orchid
Family 2 *Chloratropicamereaceae*	Tropical American Green Orchid
Family 3 *Cryptocentraceae*	Tropical American Hidden Center Orchid
Family 4 *Habenaritropicameriaceae*	Tropical American Strap Bog Orchid
Family 5 *Rauhiellaceae*	Tropical American Rauhian Orchid
Family 6 *Wullschlaegelitropicameriaceae*	Tropical American Wullschlaege Orchid
Order 2 *Cypripeditropicameriales*	Tropical American Lady Slipper Orchid
Family 1 *Cypripeditropicameriaceae*	Tropical American Lady Slipper Orchid
Family 2 *Phragmipediaceae*	Tropical American Slipper Orchid
Family 3 *Selenipediaceae*	Tropical American Selene Slipper Orchid
Family 2 *Phragmipediaceae*	Tropical American Slipper Orchid
Order 3 *Epiarethusatropicameriales*	Tropical American Dragon's Mouth Orchid
Family 1 *Bletitropicameriaceae*	Tropical American Blet Orchid
Family 2 *Calanthetropicameriaceae*	Tropical American Callisto Orchid
Family 3 *Chysaceae*	Tropical American Pendent Raceme Orchid
Family 4 *Coeliaceae*	Tropical American Hollow Orchid
Family 5 *Sobraliaceae*	Tropical American Sobral Orchid
Order 4 *Epidendratropicameriales*	Tropical American Reed Stem Orchid
Family 1 *Epidendraceae*	Tropical American Reed Stem Orchid
Family 2 *Cryptarrhenaceae*	Tropical American Hidden Arrhen Orchid
Family 3 *Malaxitropicameriaceae*	Tropical American Adder's Mouth Orchid
Family 4 *Meiracylliaceae*	Tropical American Meiracyllian Orchid
Family 5 *Pleurothallitropicameriaceae*	Tropical American Bonnet Orchid
Family 6 *Triphoratropicameriaceae*	Tropical American Nodding Caps Orchid
Family 7 *Xerorchaceae*	Tropical American Xeric Orchid
Order 5 *Laeliales*	Tropical American Laely Orchid
Order 6 *Podochilotropicameriales*	Tropical American Pod Lip Orchid
Order 7 *Spiranthatropicameriales*	Tropical American Ladies Tresses Orchid
Order 8 *Vandatropicameriales*	Tropical American Vanda Orchid
Order 9 *Vandacymbotropicameriales*	Tropical American Cymb Vanda Orchid
Order A *Vandamaxillotropicameriales*	Tropical American Tiger Vanda Orchid
Order B *Vanillatropicameriales*	Tropical American Vanilla Orchid
Class 722.11 ***Orchidosudameriopsida***	**Southern South American Orchid**
Order 1 *Orchidosudameriales*	Southern South American Orchid
Family 1 *Chlorasudamereaceae*	Southern South American Green Orchid
Family 2 *Codonorchaceae*	Southern South American Codon Orchid
Family 3 *Habenarasudameriaeae*	Southern South American Strap Bog Orchid
Order 2 *Epidendrasudameriales*	Southern South American Reed Stem Orchid
Family 1 *Pluerothallasudamereaceae*	Southern South American Bonnet Orchid
Order 3 *Spiranthasudameriales*	Southern South American Ladies Tresses Orchid
Family 1 *Spiranthasudamereaceae*	Southern South American Ladies Tresses Orchid
Family 1 *Presscottisudamereaceae*	Southern South American Prescott Orchid

LAWN GRASSES

Lawn grasses are cultivated grasses and are classified as landscape herbs.

Poaceae: Kentucky Bluegrass (Poa pratensis), Rough Bluegrass (Poa trivialis), Blue Lawn Grass (Garnotia stricta)

Agrostidaceae: Redtop Bent Grass (Agrostis alba), Creeping Bent Grass (Agrostis palustris), Colonial Bent Grass (Agrostis tenuis), Three Awn Grass (Aristida), Bluejoint Reed Bent Grass (Calamagrostis canadensis), Pinegrass Reed Bent Grass (Calamagrostis rubescens), Ikali Sacaton Grass (Sporobolus airoides), Big Sacaton Grass (Sporobolus wrightii), Needle Grass (Stipa)

Andropogonaceae: Big Bluestem Beard Grass (Andropogon gerardi), Centipede Grass (Eremochloa ophiurioides), Chinese Silver Grass (Miscanthus sinensis), Little Bluestem Split Beard Grass (Schizachyrium scoparium)

Chloridaceae: Side Oats Grama Grass (Bouteloua curtipendula), Blue Grama Grass (Bouteloua gracilis), Hairy Grama Grass (Bouteloua hirsuta), Finger Buffalo Grass (Buchloe dactyloides), Bermuda Grass (Cynodon dactylon), Midway Bermuda Grass (Cynodon dactylon midwayii), Patriot Bermuda Grass (Cynodon dactylon patriotica), Tifway 419 Bermuda Grass (Cynodon dactylon x Cynodon transvaalensis germplasma), T-10 Bermuda Grass (Cynodon dactylon T-10), TifGreen #328 Bermuda Grass (Cynodon dactylon tifgreenii-328), TifSport Bermuda Grass (Cynodon dactylon tifsportii), TifSport #2 Bermuda Grass (Cynodon dactylon tifsportii-2)

Festucaceae: Sheep Fescue (Festuca ovina), Hard Sheep Fescue (Festica ovina duriscula), Red Fescue (Festuca rubra), Shade Red Fescue (Festuca rubra heterophylla), California Brome Grass (Bromus carinatus), Rescue Brome Grass (Bromus catharticus), Smooth Brome Grass (Bromus inermis), Silver Pampas Grass (Cortaderia selloana), Cock's Foot Orchard Grass (Dactylis glomerata)

Panicaceae: Guinea Panic Grass (Panicum maximum), Para Panic Grass (Panicum purpurescens), Switch Panic Grass (Panicum virgatum), Broad Leaf Carpet Grass (Axonopus compressus), Narrow Leaf Carpet Grass (Axonopus fissifolius), Bristle Basket Grass (Oplismenus hirtellus), Dallis Millet Panic Grass (Paspalum dilitatum), Crimson Fountain Grass (Pennisetum setaceum), St. Augustine Grass (Stenotaphrum secundatum)

Phalaridaceae: Reed Canary Grass (Phalaris arundinacea)

Zoysiaceae: Manila Zoysia Grass (Zoysia matrella)

Class 722.01 *Liliopsida* Lily

Latin: *Lilia* lily

ORDERS

722.011	*Liliales*	Lily
722.012	*Alismatales*	Water Plantain
722.013	*Amaryllidales*	Amaryllis
722.014	*Arales*	Ar Lily
722.015	*Asparagales*	Asparagus
722.016	*Bromeliales*	Bromelaid
722.017	*Dioscoreales*	Yam Vine
722.018	*Zingiberales*	Ginger

Order 722.011 *Liliales* Lily

Latin: *Lilia* lily

FAMILIES

722.011.1	*Liliaceae*	Lily
722.011.2	*Campynemaceae*	Field Lily
722.011.3	*Colchicaceae*	Meadow Crocus
722.011.4	*Corsiaceae*	Corsian Lily
722.011.5	*Funkiaceae*	Plantain Lily
722.011.6	*Luzuriagaceae*	Luzurian Lily
722.011.7	*Melanthiaceae*	Bunchflower Lily
722.011.8	*Smilacaceae*	Greenbrier

Family 722.011.1 *Liliaceae* Lily

Genus	Species	Specimen	Genome	C:Culinary, M:Medicinal, E:Economic
Fritillaria	*affinis*	Rice Grain	Fritillary Lily	C:SD(grain):**VG**
Fritillaria	*camschatcensis*	Kamchatka	Fritillary Lily	C:SD(grain):**VG**
Fritillaria	*latifolia*	Broad Leaf	Fritillary Lily	C:SD(grain):**VG**

Varieties

Rice Grain Fritillary (*Fritillaria affinis*)

Genus	species	infraspecies	Cultivar	Specimen	Genome
Fritillaria	*affinis*	*affinis*	Original	Rice Grain	Fritillary Lily
Fritillaria	*affinis*	*esculenta*	Edible	Rice Grain	Fritillary Lily
Fritillaria	*affinis*	*eximia*	Distinguished	Rice Grain	Fritillary Lily
Fritillaria	*affinis*	*floribunda*	Abundant Flower	Rice Grain	Fritillary Lily
Fritillaria	*affinis*	*gracilis*	Slender	Rice Grain	Fritillary Lily
Fritillaria	*affinis*	*lanceolata*	Lance Leaf	Rice Grain	Fritillary Lily
Fritillaria	*affinis*	*lunellii*	Lunell	Rice Grain	Fritillary Lily
Fritillaria	*affinis*	*micrantha*	Small Flower	Rice Grain	Fritillary Lily
Fritillaria	*affinis*	*multiflora*	Multiflower	Rice Grain	Fritillary Lily
Fritillaria	*affinis*	*mutica*	Blunt	Rice Grain	Fritillary Lily
Fritillaria	*affinis*	*parviflora*	Little Flower	Rice Grain	Fritillary Lily
Fritillaria	*affinis*	*phaeanthera*	Dusky Flower	Rice Grain	Fritillary Lily
Fritillaria	*affinis*	*tristulis*	Sad	Rice Grain	Fritillary Lily
Fritillaria	*camschatcensis*	*camschatcensis*	Original	Kamchatka	Fritillary Lily
Fritillaria	*camschatcensis*	*edulis*	Edible	Kamchatka	Fritillary Lily
Fritillaria	*camschatcensis*	*flavescens*	Yellow	Kamchatka	Fritillary Lily
Fritillaria	*camschatcensis*	*nigra*	Black	Kamchatka	Fritillary Lily
Fritillaria	*camschatcensis*	*quadrifoliata*	Four Leaf	Kamchatka	Fritillary Lily
Fritillaria	*camschatcensis*	*saranna*	Black Sarana	Kamchatka	Fritillary Lily
Fritillaria	*latifolia*	*latifolia*	Original	Broad Leaf	Fritillary Lily
Fritillaria	*latifolia*	*esculenta*	Edible	Broad Leaf	Fritillary Lily

Fritillaria	*latifolia*	*nobilis*	Noble	Broad Leaf	Fritillary Lily

Family 722.011.5 *Funkiaceae* Plantain Lily

Day Lily (*Hemerocallis esculenta*)

Genus	Species	Specimen	Genome	C:Culinary, M:Medicinal, E:Economic
Hemerocallis	*esculenta*	Edible	Day Lily	C:BR:**VG**

Family 722.011.8 *Smilaceae* Greenbrier

Genus	Species	Specimen	Genome	C:Culinary, M:Medicinal
Smilax	*aristolochaefolia*	Mexican Sarsaparilla	Greenbrier	**C**:RT:BF(root beer)
Smilax	*china*	China Root Sarsaparilla	Greenbrier	**C**:RT:BF(root beer)
Smilax	*glabra*	Assam Sarsaparilla	Greenbrier	**C**:RT:BF(root beer)
Smilax	*glycophylla*	Australian Sarsaparilla	Greenbrier	**C**:RT:BF(root beer)
Smilax	*purhampuy*	Tropical American Sarsaparilla	Greenbrier	**C**:RT:BF(root beer)
Smilax	*regelii*	Central American Sarsaparilla	Greenbrier	**C**:RT:BF(root beer)
Maianthemum	*oleraceum*	Vegetable	May Lily	**C**:BR:**VG**

Varieties

Sarsaparilla (*Smilax spp.*)

May Lily (*Maianthemum oleraceum*)

Genus	species	infraspecies	Cultivar	Specimen	Genome
Smilax	*china*	*china*	Original	China Root Sarsaparilla	Greenbrier
Smilax	*china*	*japonica*	Japanese	China Root Sarsaparilla	Greenbrier
Smilax	*china*	*microphylla*	Small Leaf	China Root Sarsaparilla	Greenbrier
Smilax	*china*	*pteropus*	Wing Foot	China Root Sarsaparilla	Greenbrier
Smilax	*china*	*recticaulis*	Square Stem	China Root Sarsaparilla	Greenbrier
Smilax	*china*	*taiheiensis*	Taihei	China Root Sarsaparilla	Greenbrier
Smilax	*china*	*taquetii*	Taquet	China Root Sarsaparilla	Greenbrier
Smilax	*china*	*thomsoniana*	Thomson	China Root Sarsaparilla	Greenbrier
Smilax	*china*	*variegata*	Variegated	China Root Sarsaparilla	Greenbrier
Smilax	*china*	*xanthocarpa*	Yellow Pod	China Root Sarsaparilla	Greenbrier
Smilax	*china*	*yanagitai*	Yanagita	China Root Sarsaparilla	Greenbrier
Smilax	*china*	*yanagitai-boninensis*	Bonin Yanagita	China Root Sarsaparilla	Greenbrier
Smilax	*china*	*yanagitai-kuru*	Kuru Yanagita	China Root Sarsaparilla	Greenbrier
Smilax	*glabra*	*glabra*	Original	Assam Sarsaparilla	Greenbrier
Smilax	*glabra*	*blinii*	Blin	Assam Sarsaparilla	Greenbrier
Smilax	*glabra*	*concolor*	One Color	Assam Sarsaparilla	Greenbrier
Smilax	*glabra*	*dunniana*	Dunn	Assam Sarsaparilla	Greenbrier
Smilax	*glabra*	*hookeri*	Hooker	Assam Sarsaparilla	Greenbrier
Smilax	*glabra*	*lanceolata*	Lance Leaf	Assam Sarsaparilla	Greenbrier
Smilax	*glabra*	*maculata*	Spotted	Assam Sarsaparilla	Greenbrier
Smilax	*glabra*	*mengmaensis*	Mengma	Assam Sarsaparilla	Greenbrier
Smilax	*glabra*	*trigona*	Three Angled	Assam Sarsaparilla	Greenbrier
Smilax	*purhampuy*	*purhampuy*	Original	Tropical American Sarsaparilla	Greenbrier
Smilax	*purhampuy*	*aequatoris*	Equatorial	Tropical American Sarsaparilla	Greenbrier
Smilax	*purhampuy*	*febrifuga*	Equadorian	Tropical American Sarsaparilla	Greenbrier
Smilax	*regelii*	*regelii*	Original	Central American Sarsaparilla	Greenbrier
Smilax	*regelii*	*ornata*	Honduran	Central American Sarsaparilla	Greenbrier
Smilax	*regelii*	*utilis*	Useful	Central American Sarsaparilla	Greenbrier
Maianthemum	*oleraceum*	*oleraceum*	Original	Vegetable	May Lily .
Maianthemum	*oleraceum*	*acuminatum*	Pointed	Vegetable	May Lily .
Maianthemum	*oleraceum*	*crassifolium*	Crass Leaf	Vegetable	May Lily .
Maianthemum	*oleraceum*	*mientienensis*	Mien Tien	Vegetable	May Lily .

Order 722.012 *Alismatales* Water Plantain

Latin: *Alisma* alism

FAMILIES

722.012.1 *Alismataceae*	**Water Plantain**
722.012.2 *Aponogetonaceae*	Hawthorn Pond Grass
722.012.3 *Butomaceae*	Water Oxcutter
722.012.4 *Cymodoceaceae*	Surf Grass
722.012.5 *Hydrocharitaceae*	Waterweed
722.012.6 *Limnocharitaceae*	Water Velvetleaf
722.012.7 *Potamogetonaceae*	Pond Grass
722.012.8 *Zosteraceae*	Eel Grass

Family 722.012.1 *Alismataceae* Water Plantain

Genus	Species	Specimen	Genome	C:Culinary, M:Medicinal
Sagittaria	*latifolia*	Broadleaf Duck Potato	Arrowhead	C:TU:**VG**
Sagittaria	*sagittifolia*	Arrowleaf Duck Potato	Arrowhead	C:TU:**VG**
Sagittaria	*trifolia*	Trefoil Duck Potato	Arrowhead	C:TU:**VG**

Varieties

Duck Potato (*Sagittaria spp.*)

Genus	species	infraspecies	Cultivar	Specimen	Genome
Sagittaria	*latifolia*	*latifolia*	Original	Broadleaf Duck Potato	Arrowhead
Sagittaria	*latifolia*	*angustifolia*	Narrow Leaf	Broadleaf Duck Potato	Arrowhead
Sagittaria	*latifolia*	*diversifolia*	Diverse	Broadleaf Duck Potato	Arrowhead
Sagittaria	*latifolia*	*esculenta*	Edible	Broadleaf Duck Potato	Arrowhead
Sagittaria	*latifolia*	*gigantea*	Giant	Broadleaf Duck Potato	Arrowhead
Sagittaria	*latifolia*	*glabra*	Smooth	Broadleaf Duck Potato	Arrowhead
Sagittaria	*latifolia*	*gracilis*	Slender	Broadleaf Duck Potato	Arrowhead
Sagittaria	*latifolia*	*hastata*	Notched	Broadleaf Duck Potato	Arrowhead
Sagittaria	*latifolia*	*longirostra*	Long Beak	Broadleaf Duck Potato	Arrowhead
Sagittaria	*latifolia*	*macrophylla*	Large Leaf	Broadleaf Duck Potato	Arrowhead
Sagittaria	*latifolia*	*major*	Greater	Broadleaf Duck Potato	Arrowhead
Sagittaria	*latifolia*	*obtusa*	Blunt Leaf	Broadleaf Duck Potato	Arrowhead
Sagittaria	*latifolia*	*ornithorhyncha*	Bird Beak	Broadleaf Duck Potato	Arrowhead
Sagittaria	*latifolia*	*planipes*	Flat	Broadleaf Duck Potato	Arrowhead
Sagittaria	*latifolia*	*pubescens*	Downy	Broadleaf Duck Potato	Arrowhead
Sagittaria	*latifolia*	*simplex*	Simple	Broadleaf Duck Potato	Arrowhead
Sagittaria	*latifolia*	*variabilis*	Variable	Broadleaf Duck Potato	Arrowhead
Sagittaria	*latifolia*	*viscosa*	Viscose	Broadleaf Duck Potato	Arrowhead
Sagittaria	*sagittifolia*	*sagittifolia*	Original	Duck Potato	Arrowhead
Sagittaria	*sagittifolia*	*acuminata*	Sharp Tip	Duck Potato	Arrowhead
Sagittaria	*sagittifolia*	*aequiloba*	Unequal Lobe	Duck Potato	Arrowhead
Sagittaria	*sagittifolia*	*angustata*	Narrow	Duck Potato	Arrowhead
Sagittaria	*sagittifolia*	*aquatica*	Aquatic	Duck Potato	Arrowhead
Sagittaria	*sagittifolia*	*bollei*	Bolle	Duck Potato	Arrowhead
Sagittaria	*sagittifolia*	*bulbosa*	Bulbous	Duck Potato	Arrowhead
Sagittaria	*sagittifolia*	*butomoides*	Cut Leaf	Duck Potato	Arrowhead
Sagittaria	*sagittifolia*	*divaricata*	Spreading	Duck Potato	Arrowhead
Sagittaria	*sagittifolia*	*gigantea*	Giant	Duck Potato	Arrowhead
Sagittaria	*sagittifolia*	*heterophylla*	Variable Leaf	Duck Potato	Arrowhead
Sagittaria	*sagittifolia*	*major*	Greater	Duck Potato	Arrowhead
Sagittaria	*sagittifolia*	*minor*	Lesser	Duck Potato	Arrowhead
Sagittaria	*sagittifolia*	*monoeca*	Monoecious	Duck Potato	Arrowhead
Sagittaria	*sagittifolia*	*pumila*	Low	Duck Potato	Arrowhead
Sagittaria	*sagittifolia*	*stratiotes*	Striped	Duck Potato	Arrowhead
Sagittaria	*sagittifolia*	*tenuior*	Tenuous	Duck Potato	Arrowhead
Sagittaria	*sagittifolia*	*vallisneriifolia*	Eelgrass Leaf	Duck Potato	Arrowhead
Sagittaria	*trifolia*	*trifolia*	Original	Trefoil Duck Potato	Arrowhead
Sagittaria	*trifolia*	*albida*	White	Trefoil Duck Potato	Arrowhead
Sagittaria	*trifolia*	*alismifolia*	Alism Leaf	Trefoil Duck Potato	Arrowhead
Sagittaria	*trifolia*	*angustifolia*	Narrow Leaf	Trefoil Duck Potato	Arrowhead
Sagittaria	*trifolia*	*caerulea*	Blue	Trefoil Duck Potato	Arrowhead
Sagittaria	*trifolia*	*diversifolia*	Diverse Leaf	Trefoil Duck Potato	Arrowhead
Sagittaria	*trifolia*	*doniana*	Don	Trefoil Duck Potato	Arrowhead
Sagittaria	*trifolia*	*edulis*	Edible	Trefoil Duck Potato	Arrowhead
Sagittaria	*trifolia*	*hastata*	Notched	Trefoil Duck Potato	Arrowhead
Sagittaria	*trifolia*	*heterophylla*	Variable Leaf	Trefoil Duck Potato	Arrowhead
Sagittaria	*trifolia*	*hirundinacea*	Snake	Trefoil Duck Potato	Arrowhead
Sagittaria	*trifolia*	*japonica*	Japanese	Trefoil Duck Potato	Arrowhead
Sagittaria	*trifolia*	*leucopetala*	Light Petal	Trefoil Duck Potato	Arrowhead
Sagittaria	*trifolia*	*longiloba*	Long Lobed	Trefoil Duck Potato	Arrowhead
Sagittaria	*trifolia*	*macrophylla*	Large Leaf	Trefoil Duck Potato	Arrowhead
Sagittaria	*trifolia*	*obtusa*	Blunt Leaf	Trefoil Duck Potato	Arrowhead
Sagittaria	*trifolia*	*retusa*	Notched	Trefoil Duck Potato	Arrowhead
Sagittaria	*trifolia*	*sinensis*	Chinese	Trefoil Duck Potato	Arrowhead
Sagittaria	*trifolia*	*subaequilonga*	Unequal Long	Trefoil Duck Potato	Arrowhead
Sagittaria	*trifolia*	*subhastata*	Spear	Trefoil Duck Potato	Arrowhead
Sagittaria	*trifolia*	*suitensis*	Suite	Trefoil Duck Potato	Arrowhead

Order 722.013 *Amaryllidales* Amaryllis

Latin: *Amaryllis* amaryllis

FAMILIES

722.013.1 *Amaryllidaceae* Amaryllis
722.013.2 *Agavaceae* Agave
722.013.3 *Alliaceae* Onion

722.013.4 *Alstroemeriaceae* South American Amaryllis
722.013.5 *Anthericaceae* Spider Agave Lily
722.013.6 *Hypoxidaceae* Starflower Grass

Family 722.013.2 *Agavaceae* Agave

Genus	Species	Specimen	Genome	C:Culinary, M:Medicinal, E:Economic
Agave	*americana*	Century Plant	Agave	C:SL:**VG** (stalk vegetable), E:RT:DS,FI
Agave	*sisalana*	Sisal	Agave	E:RT:FI,SD (hecogenin steroid drug)

Varieties

Century Plant Agave (*Agave americana*)

Genus	species	infraspecies	Cultivar	Specimen	Genome
Agave	*americana*	*americana*	Original	Century Plant	Agave
Agave	*americana*	*abrupta*	Abrupt	Century Plant	Agave
Agave	*americana*	*altissima*	Tall	Century Plant	Agave
Agave	*americana*	*complicata*	Complicated	Century Plant	Agave
Agave	*americana*	*cornelius*	Cornelius	Century Plant	Agave
Agave	*americana*	*expansa*	Expansive	Century Plant	Agave
Agave	*americana*	*felina*	Feline	Century Plant	Agave
Agave	*americana*	*fuerstenbergii*	Fuerstenberg	Century Plant	Agave
Agave	*americana*	*gracilispina*	Slender Spine	Century Plant	Agave
Agave	*americana*	*ingens*	Enormous	Century Plant	Agave
Agave	*americana*	*marginata*	Margined	Century Plant	Agave
Agave	*americana*	*mediopicta*	Striped	Century Plant	Agave
Agave	*americana*	*mediopicta-alba*	White Striped	Century Plant	Agave
Agave	*americana*	*melliflua*	Honey Sap	Century Plant	Agave
Agave	*americana*	*milleri*	Miller	Century Plant	Agave
Agave	*americana*	*oaxacensis*	Oaxaca	Century Plant	Agave
Agave	*americana*	*ornata*	Decorated	Century Plant	Agave
Agave	*americana*	*picta*	Painted	Century Plant	Agave
Agave	*americana*	*protamericana*	Protoamerica	Century Plant	Agave
Agave	*americana*	*ramosa*	Branching	Century Plant	Agave
Agave	*americana*	*rasconensis*	Rascon	Century Plant	Agave
Agave	*americana*	*spectabilis*	Spectacular	Century Plant	Agave
Agave	*americana*	*striata*	Striated	Century Plant	Agave
Agave	*americana*	*subtilis*	Subtle	Century Plant	Agave
Agave	*americana*	*subzonata*	Lesser Zoned	Century Plant	Agave
Agave	*americana*	*theometel*	Theometel	Century Plant	Agave
Agave	*americana*	*variegata*	Variegated	Century Plant	Agave
Agave	*americana*	*virginica*	Virginia	Century Plant	Agave
Agave	*americana*	*zonata*	Zoned	Century Plant	Agave
Agave	*sisalana*	*sisalana*	Original	Sisal	Agave
Agave	*sisalana*	*armata*	Spiny	Sisal	Agave
Agave	*sisalana*	*segurae*	Segura	Sisal	Agave

Family 722.013.3 *Alliaceae* Onion

Genus	Species	Specimen	Genome	C:Culinary, M:Medicinal, E:Economic
Allium	*altaicum*	Altai Mtn. Bunching	Onion	C:BR:**VG**
Allium	*ampeloprasum*	Elephant Leek	Onion	C:BR:**VG**
Allium	*ascalonicum*	Scallion Shallot	Onion	C:BR:**VG**
Allium	*canadense*	Meadow Garlic	Onion	C:BR:**VG**
Allium	*carinatum*	Keeled Garlic	Onion	C:BR:**VG**
Allium	*cepa*	Spring	Onion	C:BR:**VG**
Allium	*cepa-aggregatum*	Bunching Shallot	Onion	C:BR:**VG**
Allium	*cepa-alba*	White	Onion	C:BR:**VG**
Allium	*cepa-helva*	Yellow	Onion	C:BR:**VG**
Allium	*cepa-proliferum*	Topset	Onion	C:BR:**VG**
Allium	*cepa-purpurea-rubra*	Purple Red	Onion	C:BR:**VG**
Allium	*cepa-rubra*	Red	Onion	C:BR:**VG**
Allium	*chinense*	Rakkayo Shallot	Onion	C:BR:**VG**
Allium	*fistulosum*	Bunching	Onion	C:BR:**VG**
Allium	*fistulosum-ascalõnia*	Scallion Bunching	Onion	C:BR:**VG**
Allium	*galanthum*	Snowdrop	Onion	C:BR:**VG**
Allium	*mobilense*	Mobile Field Garlic	Onion	C:BR:**VG**
Allium	*odorum*	Leekflower Chives	Onion	C:BR:**VG**
Allium	*oleraceum*	Field Garlic	Onion	C:BR:**VG**
Allium	*oreoprasoides*	Lesser Mtn. Leek	Onion	C:BR:**VG**
Allium	*oreoprasum*	Mountain Leek	Onion	C:BR:**VG**
Allium	*oreoscordum*	Mountain Garlic	Onion	C:BR:**VG**
Allium	*oschaninii*	Oschanin Shallot	Onion	C:BR:**VG**
Allium	*porrum*	Leek	Onion	C:BR:**VG**
Allium	*potosiense*	San Luis Potosí	Onion	C:BR:**VG**

Allium	*praecox*	Early	Onion	C:BR:**VG**
Allium	*praemixtum*	Early Mixed Shallot	Onion	C:BR:**VG**
Allium	*pseudoampeloprasum*	Lesser Leek	Onion	C:BR:**VG**
Allium	*pskemense*	Pskem Shallot	Onion	C:BR:**VG**
Allium	*pulchellum*	Pretty Keeled Garlic	Onion	C:BR:**VG**
Allium	*ramosum*	Siberian Chives	Onion	C:BR:**VG**
Allium	*roseum*	Rose Garlic	Onion	C:BR:**VG**
Allium	*sativum*	Sweet Garlic	Onion	C:BR:**VG**
Allium	*sativum-dura-alba*	**White Hardneck Garlic**	Onion	C:BR:**VG**
Allium	*sativum-dura-brunnea*	**Brown Hardneck Garlic**	Onion	C:BR:**VG**
Allium	*sativum-dura-longicuspis*	**Long Cusp Hardneck Garlic**	Onion	C:BR:**VG**
Allium	*sativum-dura-marmorata*	**Marbled Hardneck Garlic**	Onion	C:BR:**VG**
Allium	*sativum-dura-ophion*	**Snake Hardneck Garlic**	Onion	C:BR:**VG**
Allium	*sativum-dura-ophioscorodon*	**Snaked Hardneck Garlic**	Onion	C:BR:**VG**
Allium	*sativum-dura-purpurea*	**Purple Hardneck Garlic**	Onion	C:BR:**VG**
Allium	*sativum-dura-rocambolea*	**Rocambole Hardneck Garlic**	Onion	C:BR:**VG**
Allium	*sativum-dura-rubra*	**Red Hardneck Garlic**	Onion	C:BR:**VG**
Allium	*sativum-molla-alba*	**White Softneck Garlic**	Onion	C:BR:**VG**
Allium	*sativum-molla-brunnea*	**Brown Softneck Garlic**	Onion	C:BR:**VG**
Allium	*sativum-molla-marmorata*	**Marbled Softneck Garlic**	Onion	C:BR:**VG**
Allium	*sativum-molla-purpurea*	**Purple Softneck Garlic**	Onion	C:BR:**VG**
Allium	*sativum-molla-rubra*	**Red Softneck Garlic**	Onion	C:BR:**VG**
Allium	*sativum-molla-scolymus*	**Artichoke Softneck Garlic**	Onion	C:BR:**VG**
Allium	*schoenoprasoides*	Wild Chives	Onion	C:BR:**VG**
Allium	*schoenoprasum*	Chives	Onion	C:BR:**VG**
Allium	*scorodoprasum*	Leek Garlic	Onion	C:BR:**VG**
Allium	*senescens*	Flat Leaf Chives	Onion	C:BR:**VG**
Allium	*sphaerocephalon*	Roundhead Garlic	Onion	C:BR:**VG**
Allium	*suaveolens*	Sweet	Onion	C:BR:**VG**
Allium	*tuberosum*	Garlic Chives	Onion	C:BR:**VG**
Allium	*ursinum*	Shaggy Garlic	Onion	C:BR:**VG**
Allium	*vavilovii*	Vavilov Shallot	Onion	C:BR:**VG**

Varieties

Shallot, Bunching (*Allium cepa-aggregatum*)

Genus	species	infraspecies	Cultivated Variety Cultivar	Specimen	Genome
Allium	*cepa-aggregatum*	*ambitionii*	Ambition	Bunching Shallot	Onion
Allium	*cepa-aggregatum*	*asian-purplei*	Asian Purple	Bunching Shallot	Onion
Allium	*cepa-aggregatum*	*atlantica*	Atlantic	Bunching Shallot	Onion
Allium	*cepa-aggregatum*	*atlasi*	Atlas	Bunching Shallot	Onion
Allium	*cepa-aggregatum*	*bonillai*	Bonilla Globe	Bunching Shallot	Onion
Allium	*cepa-aggregatum*	*breton-giant*	BretonGiant	Bunching Shallot	Onion
Allium	*cepa-aggregatum*	*british-columbiana*	British Columbia	Bunching Shallot	Onion
Allium	*cepa-aggregatum*	*brittony-princei*	Prince De Bretagne	Bunching Shallot	Onion
Allium	*cepa-aggregatum*	*brittony-redii*	Brittany Red	Bunching Shallot	Onion
Allium	*cepa-aggregatum*	*canada-yellowii*	CanadaYellow	Bunching Shallot	Onion
Allium	*cepa-aggregatum*	*chinensis*	Chinese	Bunching Shallot	Onion
Allium	*cepa-aggregatum*	*drittler-white-nestii*	Drittler White Nest	Bunching Shallot	Onion
Allium	*cepa-aggregatum*	*dutchii*	Dutch	Bunching Shallot	Onion
Allium	*cepa-aggregatum*	*dutchii-redii*	Dutch Red	Bunching Shallot	Onion
Allium	*cepa-aggregatum*	*ed's-redii*	Ed's Red	Bunching Shallot	Onion
Allium	*cepa-aggregatum*	*eldalba*	Old White	Bunching Shallot	Onion
Allium	*cepa-aggregatum*	*european-goldenii*	European Golden	Bunching Shallot	Onion
Allium	*cepa-aggregatum*	*evereadyi*	Eveready	Bunching Shallot	Onion
Allium	*cepa-aggregatum*	*forman-yellowii*	Forman Yellow	Bunching Shallot	Onion
Allium	*cepa-aggregatum*	*francoi*	French	Bunching Shallot	Onion
Allium	*cepa-aggregatum*	*frenchii*	French	Bunching Shallot	Onion
Allium	*cepa-aggregatum*	*frenchii-epicureana*	Epicurian French	Bunching Shallot	Onion
Allium	*cepa-aggregatum*	*frenchii-redii*	Red French	Bunching Shallot	Onion
Allium	*cepa-aggregatum*	*fresh-greenii*	Fresh Green	Bunching Shallot	Onion
Allium	*cepa-aggregatum*	*frogs-legsi*	FrogLegs	Bunching Shallot	Onion
Allium	*cepa-aggregatum*	*german-giantii*	German Giant	Bunching Shallot	Onion
Allium	*cepa-aggregatum*	*goldenii*	Golden	Bunching Shallot	Onion
Allium	*cepa-aggregatum*	*golden-snow-shoei*	Golden Snow Shoe	Bunching Shallot	Onion
Allium	*cepa-aggregatum*	*grandma-featherston'si*	Grandma Featherston's	Bunching Shallot	Onion
Allium	*cepa-aggregatum*	*grayi*	Gray	Bunching Shallot	Onion
Allium	*cepa-aggregatum*	*greeley-bunchingii*	Greeley	Bunching Shallot	Onion
Allium	*cepa-aggregatum*	*grisellei*	Griselle	Bunching Shallot	Onion
Allium	*cepa-aggregatum*	*hative-de-niortii*	N. Native White	Bunching Shallot	Onion
Allium	*cepa-aggregatum*	*hekoi*	Heko Green	Bunching Shallot	Onion
Allium	*cepa-aggregatum*	*holland-redii*	Holland Red	Bunching Shallot	Onion
Allium	*cepa-aggregatum*	*joe'si*	Joe's	Bunching Shallot	Onion
Allium	*cepa-aggregatum*	*jumbo-brittanyi*	Jumbo Brittany	Bunching Shallot	Onion
Allium	*cepa-aggregatum*	*ken'si*	Ken's	Bunching Shallot	Onion

Allium	*cepa-aggregatum*	*kirbyi*	Kirby	Bunching Shallot	Onion
Allium	*cepa-aggregatum*	*le-jardinii*	Le Jardin	Bunching Shallot	Onion
Allium	*cepa-aggregatum*	*longbow-redii*	Longbow Red	Bunching Shallot	Onion
Allium	*cepa-aggregatum*	*loretta-yellowii*	Loretta Yellow	Bunching Shallot	Onion
Allium	*cepa-aggregatum*	*louisiana-cajunii*	Louisiana Cajun	Bunching Shallot	Onion
Allium	*cepa-aggregatum*	*louisiana-joei*	Louisiana Joe	Bunching Shallot	Onion
Allium	*cepa-aggregatum*	*minnesotensis-vinteri*	Minnesota Winter	Bunching Shallot	Onion
Allium	*cepa-aggregatum*	*multiplier- yellowii*	Multiplier Yellow	Bunching Shallot	Onion
Allium	*cepa-aggregatum*	*odetta's-whitei*	Odetta's White	Bunching Shallot	Onion
Allium	*cepa-aggregatum*	*old-germanii*	Old German	Bunching Shallot	Onion
Allium	*cepa-aggregatum*	*peari*	Pear	Bunching Shallot	Onion
Allium	*cepa-aggregatum*	*pikant-redii*	Pikant Red	Bunching Shallot	Onion
Allium	*cepa-aggregatum*	*pink-seeds-blumii*	Pink Seeds Blum	Bunching Shallot	Onion
Allium	*cepa-aggregatum*	*potato-onionii*	Potato	Bunching Shallot	Onion
Allium	*cepa-aggregatum*	*potato-onionii-brownii*	Brown Potato	Bunching Shallot	Onion
Allium	*cepa-aggregatum*	*potato-onionii-gary-millwoodii*	Gary Millwood Potato	Bunching Shallot	Onion
Allium	*cepa-aggregatum*	*potato-onionii-grandpa-archori*	Grandpa Achor Potato	Bunching Shallot	Onion
Allium	*cepa-aggregatum*	*potato-onionii-redii*	Red Potato	Bunching Shallot	Onion
Allium	*cepa-aggregatum*	*potato-onionii-yellowii*	Yellow Potato	Bunching Shallot	Onion
Allium	*cepa-aggregatum*	*red-sunii*	Red Sun	Bunching Shallot	Onion
Allium	*cepa-aggregatum*	*ron-frerichii*	Ron Frerich	Bunching Shallot	Onion
Allium	*cepa-aggregatum*	*salem-yellowii*	Salem Yellow	Bunching Shallot	Onion
Allium	*cepa-aggregatum*	*santei*	Sante	Bunching Shallot	Onion
Allium	*cepa-aggregatum*	*sharonii*	Sharon	Bunching Shallot	Onion
Allium	*cepa-aggregatum*	*sweet-besanconensis*	Sweet Besancon	Bunching Shallot	Onion
Allium	*cepa-aggregatum*	*thai-purplei*	Thai Purple	Bunching Shallot	Onion
Allium	*cepa-aggregatum*	*tohono-o'odham-l'itoi'si*	O'odham Ito Tohono	Bunching Shallot	Onion
Allium	*cepa-aggregatum*	*tohono-papago-itoi*	Papago Ito Tohono	Bunching Shallot	Onion
Allium	*cepa-aggregatum*	*tohono-pima-itoi*	Pima Ito Tohono	Bunching Shallot	Onion
Allium	*cepa-aggregatum*	*walesensis*	Welsh	Bunching Shallot	Onion
Allium	*cepa-aggregatum*	*yellow-pear-bottlei*	Bottle Yellow Pear	Bunching Shallot	Onion
Allium	*cepa-aggregatum*	*yellow-pear-seedii*	Seed Yellow Pear	Bunching Shallot	Onion

Onion, White (*Allium cepa-alba*)

Genus	species	infraspecies	Cultivated Variety	Cultivar Specimen	Genome
Allium	*cepa-alba*	*albionii*	Albion Globe	White	Onion
Allium	*cepa-alba*	*aragonii*	Aragon Yellowskin	White	Onion
Allium	*cepa-alba*	*awahia*	Awahia Globe	White	Onion
Allium	*cepa-alba*	*barlettai*	Barletta White Pearl	White	Onion
Allium	*cepa-alba*	*bedfordshire-championii*	Bedfordshire Champion	White	Onion
Allium	*cepa-alba*	*benny-southportii*	Benny Southport	White	Onion
Allium	*cepa-alba*	*ben-shemenii*	Ben Shemen	White	Onion
Allium	*cepa-alba*	*bermuda-yellowii*	Bermuda	White	Onion
Allium	*cepa-alba*	*bianca-di-maggioi*	Bianca Di Maggio	White	Onion
Allium	*cepa-alba*	*blanc-hatif-de-parisi*	Blanc Hatif De Paris	White	Onion
Allium	*cepa-alba*	*blanco-duroi*	Blanco Duro	White	Onion
Allium	*cepa-alba*	*borettana*	Borettano Button	White	Onion
Allium	*cepa-alba*	*borettano-blanca*	Borettano Blanca	White	Onion
Allium	*cepa-alba*	*brunswick-rodoi*	Brunswick Rodo	White	Onion
Allium	*cepa-alba*	*burgundyi*	Burgundy	White	Onion
Allium	*cepa-alba*	*california-early-redii*	California Early Redskin	White	Onion
Allium	*cepa-alba*	*crystal-waxii*	Crystal Wax Bermuda	White	Onion
Allium	*cepa-alba*	*danvers-yellow-globei*	Danvers Yellowskin	White	Onion
Allium	*cepa-alba*	*downing-yellow-globei*	Downing Yellowskin	White	Onion
Allium	*cepa-alba*	*early-avivii*	Early Aviv	White	Onion
Allium	*cepa-alba*	*early-red-burgeri*	Early Redskin Burger	White	Onion
Allium	*cepa-alba*	*eary-yellow-globei*	Early Yellowskin	White	Onion
Allium	*cepa-alba*	*ebenezeri*	Ebenezer	White	Onion
Allium	*cepa-alba*	*eclipsei-1303*	Eclipse Bermuda	White	Onion
Allium	*cepa-alba*	*el-toroi*	El Toro Flat	White	Onion
Allium	*cepa-alba*	*everestii*	Everest	White	Onion
Allium	*cepa-alba*	*fresno-alba*	Frenso Flat	White	Onion
Allium	*cepa-alba*	*gladstonei*	Gladstone	White	Onion
Allium	*cepa-alba*	*granex-033*	Granex 033	White	Onion
Allium	*cepa-alba*	*granex-429*	Granex 429	White	Onion
Allium	*cepa-alba*	*grano*	Grano Cone	White	Onion
Allium	*cepa-alba*	*grano-hawkii*	Hawk Grano Cone	White	Onion
Allium	*cepa-alba*	*grano-neomexicana*	New Mexico Grano Cone	White	Onion
Allium	*cepa-alba*	*grano-prr*	Pink Root Resistant Grano Cone	White	Onion
Allium	*cepa-alba*	*grano-ringeri*	Ringer Grano Cone	White	Onion
Allium	*cepa-alba*	*grano-texana*	Texas Grano Cone	White	Onion
Allium	*cepa-alba*	*grano-texana-0438*	#0438 Texas Grano Cone	White	Onion
Allium	*cepa-alba*	*grano-texana-0502*	#0502 Texas Grano Cone	White	Onion
Allium	*cepa-alba*	*grano-texana-1015y*	#1015 Texas Grano Cone	White	Onion
Allium	*cepa-alba*	*grano-texana-yellowi*	Texas Yellowskin Grano Cone	White	Onion
Allium	*cepa-alba*	*imai-extra-early-yellowii*	Imai Extra Early	White	Onion
Allium	*cepa-alba*	*kelsae-sweet-giantii*	Kelsa Sweet Giant Cone	White	Onion
Allium	*cepa-alba*	*lancastriana*	Lancaster Football	White	Onion

Allium	*cepa-alba*	*lisbonii*	Lisbon Scallion	White	Onion
Allium	*cepa-alba*	*lisbonii-bunchingii*	Lisbon Bunching	White	Onion
Allium	*cepa-alba*	*maureenii*	Maureen	White	Onion
Allium	*cepa-alba*	*newburgii*	Newburg	White	Onion
Allium	*cepa-alba*	*numex-luna*	Numex Luna	White	Onion
Allium	*cepa-alba*	*owai*	Owa Globe	White	Onion
Allium	*cepa-alba*	*paris-silverskinii*	Paris Silverskin	White	Onion
Allium	*cepa-alba*	*pompeii*	Pompei Flat	White	Onion
Allium	*cepa-alba*	*pompeii-wonderi*	Wonder Of Pompei	White	Onion
Allium	*cepa-alba*	*portugalensis*	Portugal	White	Onion
Allium	*cepa-alba*	*pukekohe-longkeeperi*	Pukekohe Longkeeper	White	Onion
Allium	*cepa-alba*	*queenii*	White Queen	White	Onion
Allium	*cepa-alba*	*queen-silvatoi*	Queen Silvato	White	Onion
Allium	*cepa-alba*	*red-hamburgeri*	Redskin Hamburger	White	Onion
Allium	*cepa-alba*	*ringmasteri*	Ringmaster	White	Onion
Allium	*cepa-alba*	*rio-blanco-grande*	Rio Blanco Grande	White	Onion
Allium	*cepa-alba*	*robusta*	Robust	White	Onion
Allium	*cepa-alba*	*southern-white-globei*	Southern	White	Onion
Allium	*cepa-alba*	*southport-white-bunchingii*	Southport Bunching	White	Onion
Allium	*cepa-alba*	*southport-white-globei*	Southport Globe	White	Onion
Allium	*cepa-alba*	*southport-white-globei-404*	Southport Globe #404	White	Onion
Allium	*cepa-alba*	*southport-yellow-globei*	Southport Yellowskin	White	Onion
Allium	*cepa-alba*	*spanish-bunching*	Spanish Bunching	White	Onion
Allium	*cepa-alba*	*staroi*	Star Globe	White	Onion
Allium	*cepa-alba*	*stockton-early-redii*	Stockton Early Redskin	White	Onion
Allium	*cepa-alba*	*stockton-early-yellowii*	Stockton Early Yellowskin	White	Onion
Allium	*cepa-alba*	*stockton-redii*	Stockton Redskin	White	Onion
Allium	*cepa-alba*	*stuttgarter-diskosi*	Stutgartter Discos	White	Onion
Allium	*cepa-alba*	*sweet-spanishii*	Sweet Spanish	White	Onion
Allium	*cepa-alba*	*sweet-spanishii-ben-shemenii*	Ben Shemen Sweet Spanish	White	Onion
Allium	*cepa-alba*	*sweet-spanishii-bunchingii*	Bunching Sweet Spanish	White	Onion
Allium	*cepa-alba*	*sweet-spanishii-exhibitionii*	Exhibition Sweet Spanish	White	Onion
Allium	*cepa-alba*	*sweet-spanishii-jumboi*	Jumbo Sweet Spanish	White	Onion
Allium	*cepa-alba*	*sweet-spanishii-utahensis*	Utah Sweet Spanish	White	Onion
Allium	*cepa-alba*	*sweet-spanishii-valenciana*	Valencia Sweet Spanish	White	Onion
Allium	*cepa-alba*	*wallawallaensis*	Walla Walla	White	Onion
Allium	*cepa-alba*	*yellow-bermudaensis*	Yellowskin Bermuda	White	Onion
Allium	*cepa-alba*	*yellow-bermudaensis-excelii-986*	Exel #986 Yellowskin Bermuda	White	Onion
Allium	*cepa-alba*	*yellow-sweet-spanishii*	Yellowskin Sweet Spanish	White	Onion
Allium	*cepa-alba*	*yellow-sweet-spanishii-coloradoi-6*	Colorado #6 Sweet Spanish	White	Onion
Allium	*cepa-alba*	*yellow-sweet-spanishii-riversidensis*	Riverside Sweet Spanish	White	Onion
Allium	*cepa-alba*	*yellow-sweet-spanishii-utahensis*	Utah Yellowskin Sweet Spanish	White	Onion
Allium	*cepa-alba*	*yellow-sweet-spanishii-utahensis-dryi*	Dry Utah Yellowskin Sweet Spanish	White	Onion
Allium	*cepa-alba*	*yellow-sweet-spanishii-valencia*	Valencia Yellowskin Sweet Spanish	White	Onion
Allium	*cepa-alba*	*yellow-sweet-spanishii-valencia-burell'si*	Burell's Valencia Sweet Spanish	White	Onion
Allium	*cepa-alba*	*zittau-giantii*	Zittau Giant	White	Onion
Allium	*cepa-alba*	*zittau-giantii-zirus*	Zirius Zittau Giant	White	Onion

Onion, Yellow (*Allium cepa-helva*)

Genus	species	infraspecies	Cultivated Variety	Cultivar Specimen	Genome
Allium	*cepa-helva*	*alias-craigii*	Alisa Craig G	Yellow	Onion
Allium	*cepa-helva*	*australia-brownii*	Australian Brownskin	Yellow	Onion
Allium	*cepa-helva*	*borrettana*	Bronzeskin Borrettano	Yellow	Onion
Allium	*cepa-helva*	*borrettano-red-stripedii*	Red Striped Borrettano	Yellow	Onion
Allium	*cepa-helva*	*borrettano-yellowii*	Yellowskin Borrettano	Yellow	Onion
Allium	*cepa-helva*	*bountyi*	Bounty Cone	Yellow	Onion
Allium	*cepa-helva*	*breakthrui*	Breakthru	Yellow	Onion
Allium	*cepa-helva*	*buffaloi*	Buffalo Flat	Yellow	Onion
Allium	*cepa-helva*	*cipolle-di-rovatoi*	Cipolle Di Rovato	Yellow	Onion
Allium	*cepa-helva*	*clear-dawnii*	Clear Dawn	Yellow	Onion
Allium	*cepa-helva*	*colossalii*	Colossal	Yellow	Onion
Allium	*cepa-helva*	*contessai*	Contessa	Yellow	Onion
Allium	*cepa-helva*	*crossbowii*	Crossbow	Yellow	Onion
Allium	*cepa-helva*	*dorata-di-parmai*	Dorata Di Parma	Yellow	Onion
Allium	*cepa-helva*	*express-yellowii*	Express Flat	Yellow	Onion
Allium	*cepa-helva*	*first-earlyi*	First Early Flat	Yellow	Onion
Allium	*cepa-helva*	*giallo-di-milanoi*	Giallo Di Milano	Yellow	Onion
Allium	*cepa-helva*	*gold-coinii*	Gold Coin Flat	Yellow	Onion
Allium	*cepa-helva*	*golden-globei*	Golden Globe	Yellow	Onion
Allium	*cepa-helva*	*hygroi*	Hygro Globe	Yellow	Onion
Allium	*cepa-helva*	*marcelii*	Marcel Globe	Yellow	Onion
Allium	*cepa-helva*	*marionii*	Marion Globe	Yellow	Onion
Allium	*cepa-helva*	*marshall's-autumn-goldii*	Marshall's Autumn Gold	Yellow	Onion
Allium	*cepa-helva*	*marshall's-giant-fen-globei*	Marshall's Giant Fen Globe	Yellow	Onion
Allium	*cepa-helva*	*mercatoi*	Mercato	Yellow	Onion
Allium	*cepa-helva*	*new-york-earlyi*	New York Early	Yellow	Onion
Allium	*cepa-helva*	*numex-starlightii*	Numex Starlite	Yellow	Onion
Allium	*cepa-helva*	*pearls-of-heavenii*	Pearls Of Heaven Scallion	Yellow	Onion

Allium	*cepa-helva*	*piatta-of-bergamoi*	Piatta Of Bergamo	Yellow	Onion
Allium	*cepa-helva*	*reliancei*	Reliance Flat	Yellow	Onion
Allium	*cepa-helva*	*rijnsbergeri*	Rijnsburger	Yellow	Onion
Allium	*cepa-helva*	*senshyu-semi-globe-yellowii*	Senshyu	Yellow	Onion
Allium	*cepa-helva*	*sherkston-sweet-giantii*	Sherkston Sweet Giant	Yellow	Onion
Allium	*cepa-helva*	*sturonii*	Sturon Flat	Yellow	Onion
Allium	*cepa-helva*	*stuttgarteri*	Stuttgart Flat	Yellow	Onion
Allium	*cepa-helva*	*turboi*	TurboFlat	Yellow	Onion
Allium	*cepa-helva*	*williamette-sweetii*	Williamette Sweet Flat	Yellow	Onion
Allium	*cepa-helva*	*yellow-dutchii*	Yellow Dutch	Yellow	Onion
Allium	*cepa-helva*	*yellow-ebenezeri*	Yellow Ebenezer Flat	Yellow	Onion
Allium	*cepa-helva*	*yellow-globei*	Yellow Globe	Yellow	Onion
Allium	*cepa-helva*	*yellow-keeperi*	Yellow Keeper	Yellow	Onion
Allium	*cepa-helva*	*yellow-of-parmaensis*	Yellow Of Parma Torpedo	Yellow	Onion
Allium	*cepa-helva*	*yellow-rockii*	Yellow Rock	Yellow	Onion

Onion, Topset (*Allium cepa-proliferum*)

Genus	species	infraspecies	Cultivated Variety	Cultivar Specimen	Genome
Allium	*cepa-proliferum*	*amish-spreadingii*	Amish Spreading	Topset	Onion
Allium	*cepa-proliferum*	*catawissa-onionii*	Catawissa	Topset	Onion
Allium	*cepa-proliferum*	*catawissa-redii*	Catawissa Red	Topset	Onion
Allium	*cepa-proliferum*	*dawes-hillii*	Dawes Hill	Topset	Onion
Allium	*cepa-proliferum*	*egyptian-onionii*	Egyptian	Topset	Onion
Allium	*cepa-proliferum*	*everlastingii*	Everlasting	Topset	Onion
Allium	*cepa-proliferum*	*fleeneri*	Fleener	Topset	Onion
Allium	*cepa-proliferum*	*grandma-vadasii*	Grandma Vadas	Topset	Onion
Allium	*cepa-proliferum*	*heritage-sweet-whitei*	Heritage Sweet White	Topset	Onion
Allium	*cepa-proliferum*	*mccollum-whitei*	McCollum White	Topset	Onion
Allium	*cepa-proliferum*	*mccullar's-whitei*	McCullar's White	Topset	Onion
Allium	*cepa-proliferum*	*missouri-valley-redii*	Missouri Valley Red	Topset	Onion
Allium	*cepa-proliferum*	*moritz-egyptiana*	Moritz Egyptian	Topset	Onion
Allium	*cepa-proliferum*	*mrs.-cox'si*	Mrs. Cox's	Topset	Onion
Allium	*cepa-proliferum*	*nancy-coonii*	Nancy Coon	Topset	Onion
Allium	*cepa-proliferum*	*purple-topsetii*	Purple	Topset	Onion
Allium	*cepa-proliferum*	*volga-german-luftzwiebelii*	Volga German Luftzwiebel	Topset	Onion
Allium	*cepa-proliferum*	*walkingii*	Walking	Topset	Onion
Allium	*cepa-proliferum*	*white-topsetii*	White	Topset	Onion

Onion, Purple Red (*Allium cepa-purpurea-rubra*)

Genus	species	infraspecies	Cultivated Variety	Cultivar Specimen	Genome
Allium	*cepa-purpurea-rubra*	*cipollini-striped-redii*	Cipollini Striped	Purple Red	Onion
Allium	*cepa-purpurea-rubra*	*purpurettei*	Purplette	Purple Red	Onion
Allium	*cepa-purpurea-rubra*	*red-simianei*	Red Simiane Torpedo	Purple Red	Onion
Allium	*cepa-purpurea-rubra*	*red-torpedoi*	Red Torpedo	Purple Red	Onion
Allium	*cepa-purpurea-rubra*	*red-wethersfieldii*	Wethersfieldt	Purple Red	Onion
Allium	*cepa-purpurea-rubra*	*rossa-toscano-di-fulminei*	Rossa Toscano Di Fulmine	Purple Red	Onion
Allium	*cepa-purpurea-rubra*	*simionei-red-bottlei*	Simione Torpedo	Purple Red	Onion
Allium	*cepa-purpurea-rubra*	*southport-red-globei*	Southport Red Globe	Purple Red	Onion

Onion, Red (*Allium cepa-rubra*)

Genus	species	infraspecies	Cultivated Variety	Cultivar Specimen	Genome
Allium	*cepa-rubra*	*ampostai*	Amposta	Red	Onion
Allium	*cepa-rubra*	*benny's-redii*	Benny's	Red	Onion
Allium	*cepa-rubra*	*carmenii*	Carmen	Red	Onion
Allium	*cepa-rubra*	*cherokee-redii*	Cherokee	Red	Onion
Allium	*cepa-rubra*	*creolei-c5*	Creole #C-5	Red	Onion
Allium	*cepa-rubra*	*di-bassano-o-di-genova-morai*	Mora Di Bassano O Di Genova	Red	Onion
Allium	*cepa-rubra*	*di-bassano-o-di-genova-rossai*	Rossa Di Bassano O Di Genova	Red	Onion
Allium	*cepa-rubra*	*eglisau-redii*	Eglisau	Red	Onion
Allium	*cepa-rubra*	*flat-of-italyii*	Flat Of Italy Cipollo	Red	Onion
Allium	*cepa-rubra*	*genovesei*	Genovese Cipollo	Red	Onion
Allium	*cepa-rubra*	*gizai-6*	Giza #6	Red	Onion
Allium	*cepa-rubra*	*italian-blood-red-bottlei*	Italian Blood Red Torpedo	Red	Onion
Allium	*cepa-rubra*	*italian-red-earlyi*	Italian Early	Red	Onion
Allium	*cepa-rubra*	*kurena-redii*	Kurenai Flat	Red	Onion
Allium	*cepa-rubra*	*long-red-florencei*	Long Red Florence Torpedo	Red	Onion
Allium	*cepa-rubra*	*mammoth-redii*	Mammoth Red	Red	Onion
Allium	*cepa-rubra*	*north-holland-blood-redii*	North Holland	Red	Onion
Allium	*cepa-rubra*	*red-baronii*	Red Baron	Red	Onion
Allium	*cepa-rubra*	*red-bermudaensis*	Bermuda	Red	Onion
Allium	*cepa-rubra*	*red-creolei*	Red Creole	Red	Onion
Allium	*cepa-rubra*	*red-creolei-c5*	Red Creole #C5	Red	Onion
Allium	*cepa-rubra*	*red-eyesi*	Red Eyes	Red	Onion
Allium	*cepa-rubra*	*red-granoi*	Red Grano	Red	Onion
Allium	*cepa-rubra*	*red-long-of-tropea*	Red Long of Tropea Torpedo	Red	Onion

Allium	*cepa-rubra*	*redmanii*	Redman	Red	Onion
Allium	*cepa-rubra*	*redmatei*	Redmate Torpedo	Red	Onion
Allium	*cepa-rubra*	*rocardoi*	Rocardo Flat	Red	Onion
Allium	*cepa-rubra*	*rossa-di-firenzei*	Rossa Di Firenze	Red	Onion
Allium	*cepa-rubra*	*rossa-di-milanoi*	Rossa Di Milano	Red	Onion
Allium	*cepa-rubra*	*rossa-lunga-di-firenzei*	Rossa Lunga Di Firenze	Red	Onion
Allium	*cepa-rubra*	*rouge-de-florencei*	Rouge De Florence	Red	Onion
Allium	*cepa-rubra*	*rouge-pale-de-niortii*	Rouge Pale De Niort	Red	Onion
Allium	*cepa-rubra*	*rubyi*	Ruby Globe	Red	Onion
Allium	*cepa-rubra*	*savonai*	Savona Flat	Red	Onion
Allium	*cepa-rubra*	*shonan-redii*	Shonan	Red	Onion
Allium	*cepa-rubra*	*sweet-flat-redii*	Sweet Flat	Red	Onion

Onion, Red Purple (*Allium cepa-rubra-purpurea*)

Genus	species	infraspecies	Cultivated Variety	Cultivar Specimen	Genome
Allium	*cepa-rubra-purpurea*	.	.	Red Purple	Onion
Allium	*cepa-rubra-purpurea*	*red-globei*	Red Globe	Red Purple	Onion
Allium	*cepa-rubra-purpurea*	*red-valleyi*	Red Valley	Red Purple	Onion

Onion, Bunching (*Allium fistulosum*)

Genus	species	Infraspecies	Cultivated Variety	Cultivar Specimen	Genome
Allium	*fistulosum*	*asagi-bunchingii*	Asagi	Bunching	Onion
Allium	*fistulosum*	*beltsville-bunchingii*	Beltsville Nebuka	Bunching	Onion
Allium	*fistulosum*	*crimson-forestii*	Crimson Forest	Bunching	Onion
Allium	*fistulosum*	*deep-purplei*	Deep Purple	Bunching	Onion
Allium	*fistulosum*	*guardsmanii*	Guardsman	Bunching	Onion
Allium	*fistulosum*	*pacifica-22*	Pacific #22	Bunching	Onion
Allium	*fistulosum*	*pacifica-pearlii*	Pacific Pearl	Bunching	Onion
Allium	*fistulosum*	*paradei*	Parade	Bunching	Onion
Allium	*fistulosum*	*red-goddessi*	Red Goddess	Bunching	Onion
Allium	*fistulosum*	*red-stalkii*	Red Stalk	Bunching	Onion
Allium	*fistulosum*	*snowstormii*	Snowstorm	Bunching	Onion
Allium	*fistulosum*	*southport-bunchingii-k99*	K99 Southport	Bunching	Onion
Allium	*fistulosum*	*southport-bunchingii-k99-selectii*	Select K99 Southport	Bunching	Onion
Allium	*fistulosum*	*toga-red-bunchingii*	Red Toga	Bunching	Onion
Allium	*fistulosum*	*toga-sperling'si*	Sperling's Toga	Bunching	Onion
Allium	*fistulosum*	*white-knightii*	White Knight	Bunching	Onion
Allium	*fistulosum*	*yakko-summeri*	Yakko Summer	Bunching	Onion
Allium	*fistulosum*	*yoshimai*	Yoshima	Bunching	Onion

Scallion, Bunching (*Allium fistulosum-ascalonia*)

Genus	Species	infraspecies	Cultivated Variety	Cultivar Specimen	Genome
Allium	*fistulosum-ascalõnia*	*annual-green-bunchingii*	AnnualGreen	Scallion Bunching	Onion
Allium	*fistulosum-ascalõnia*	*carlsbad-bunchingii*	Carlsbad	Scallion Bunching	Onion
Allium	*fistulosum-ascalõnia*	*emerald-islei*	Emerald Isle	Scallion Bunching	Onion
Allium	*fistulosum-ascalõnia*	*edo-green-bunchingii*	Edo Green	Scallion Bunching	Onion
Allium	*fistulosum-ascalõnia*	*fukagawai*	Fukagawa	Scallion Bunching	Onion
Allium	*fistulosum-ascalõnia*	*he-shi-ko-bunchingii*	He Shi Ko	Scallion Bunching	Onion
Allium	*fistulosum-ascalõnia*	*he-shi-ko-hardy-whitei*	Hardy White He Shi Ko	Scallion Bunching	Onion
Allium	*fistulosum-ascalõnia*	*hikari-bunchingii*	Hikari	Scallion Bunching	Onion
Allium	*fistulosum-ascalõnia*	*ishikurai*	Ishikura	Scallion Bunching	Onion
Allium	*fistulosum-ascalõnia*	*ishikurai-improvedii*	Improved Ishikura	Scallion Bunching	Onion
Allium	*fistulosum-ascalõnia*	*ishikurai-santa-clausei*	Santa Clause Ishikura	Scallion Bunching	Onion
Allium	*fistulosum-ascalõnia*	*hikari-bunchingii*	Hikari	Scallion Bunching	Onion
Allium	*fistulosum-ascalõnia*	*iwatsuki-multistalkii-9*	Multistalk #9 Iwatsuki	Scallion Bunching	Onion
Allium	*fistulosum-ascalõnia*	*javelotii*	Javelot	Scallion Bunching	Onion
Allium	*fistulosum-ascalõnia*	*kinchoi*	Kincho	Scallion Bunching	Onion
Allium	*fistulosum-ascalõnia*	*koba-green-bunchingii*	Koba Green	Scallion Bunching	Onion
Allium	*fistulosum-ascalõnia*	*kujo-multistalkii*	Kujo Multistalk	Scallion Bunching	Onion
Allium	*fistulosum-ascalõnia*	*kuronobori*	Kuronobori	Scallion Bunching	Onion
Allium	*fistulosum-ascalõnia*	*nebuka-evergreen-white-bunchingii*	Evergreen White Nebuka	Scallion Bunching	Onion
Allium	*fistulosum-ascalõnia*	*nebuka-long-white-bunching*	Long White Nebuka	Scallion Bunching	Onion
Allium	*fistulosum-ascalõnia*	*nebuka-summeri*	Summer Nebuka	Scallion Bunching	Onion
Allium	*fistulosum-ascalõnia*	*non-bulb-bunchingii*	Non Bulb	Scallion Bunching	Onion
Allium	*fistulosum-ascalõnia*	*perennial-bunchingii*	Perennial	Scallion Bunching	Onion
Allium	*fistulosum-ascalõnia*	*red-beardii*	Red Beard	Scallion Bunching	Onion
Allium	*fistulosum-ascalõnia*	*sakata-long-whitei*	Long White Sakata	Scallion Bunching	Onion
Allium	*fistulosum-ascalõnia*	*savelii*	Savel White	Scallion Bunching	Onion
Allium	*fistulosum-ascalõnia*	*summer-islei*	Summer Isle	Scallion Bunching	Onion
Allium	*fistulosum-ascalõnia*	*tokyo-bunchingii*	Tokyo	Scallion Bunching	Onion
Allium	*fistulosum-ascalõnia*	*tokyo-feastii*	Feast Tokyo	Scallion Bunching	Onion
Allium	*fistulosum-ascalõnia*	*tokyo-long-whitei*	Long White Tokyo	Scallion Bunching	Onion
Allium	*fistulosum-ascalõnia*	*tokyo-natsuguroi*	Natsuguro Tokyo	Scallion Bunching	Onion
Allium	*fistulosum-ascalõnia*	*victoria-bunchingii*	Victoria	Scallion Bunching	Onion
Allium	*fistulosum-ascalõnia*	*welsh-onionii*	Welsh	Scallion Bunching	Onion

Allium	*fistulosum-ascalõnia*	*welsh-redii*	Red Welsh	Scallion Bunching	Onion
Allium	*fistulosum-ascalõnia*	*welsh-get-set-redii*	Get Set Red Welsh	Scallion Bunching	Onion
Allium	*fistulosum-ascalõnia*	*white-bunchingii*	White Lisbon	Scallion Bunching	Onion
Allium	*fistulosum-ascalõnia*	*white-flashii*	WhiteFlash	Scallion Bunching	Onion
Allium	*fistulosum-ascalõnia*	*white-saladii*	White Salad	Scallion Bunching	Onion
Allium	*fistulosum-ascalõnia*	*white-spear-bunchingii*	White Spear	Scallion Bunching	Onion
Allium	*fistulosum-ascalõnia*	*white-winter-bunchingii*	White Winter	Scallion Bunching	Onion

Chives, Leekflower (*Allium odorum*)

Genus	Species	infraspecies	Cultivated Variety Cultivar	Specimen	Genome
Allium	*odorum*	*beckerianum*	Becker	Leekflower Chives	Onion
Allium	*odorum*	*diaphanum*	Diaphane	Leekflower Chives	Onion
Allium	*odorum*	*hiro-habai*	Hiro Haba	Leekflower Chives	Onion
Allium	*odorum*	*lancipetalum*	Lance Petal	Leekflower Chives	Onion
Allium	*odorum*	*potaninii*	Potanin	Leekflower Chives	Onion
Allium	*odorum*	*tataricum*	Tartarian	Leekflower Chives	Onion
Allium	*odorum*	*tenderpolei*	Tartarian	Leekflower Chives	Onion
Allium	*odorum*	*umbellatum*	Umbel	Leekflower Chives	Onion
Allium	*odorum*	*weichanicum*	Weichan	Leekflower Chives	Onion

Leek (*Allium porrum*)

Genus	Species	infraspecies	Cultivated Variety Cultivar	Specimen	Genome
Allium	*porrum*	*alaskana*	Alaska 10."	Leek	Onion
Allium	*porrum*	*albana*	Albana ?.?"	Leek	Onion
Allium	*porrum*	*albescens*	White	Leek	Onion
Allium	*porrum*	*albinstari*	Albinstar 8.0"	Leek	Onion
Allium	*porrum*	*alita*	Alita 8.7"	Leek	Onion
Allium	*porrum*	*american-flag-largei*	Large American Flag 8.5"	Leek	Onion
Allium	*porrum*	*american-flag-londonensis*	London American Flag 8.5"	Leek	Onion
Allium	*porrum*	*american-flag-londonensis-broadii*	Broad London American Flag 8.5"	Leek	Onion
Allium	*porrum*	*american-flag-musselburghii-giantii*	Giant Musselburgh American Flag 12."	Leek	Onion
Allium	*porrum*	*american-flag-musselburghii-improvedii*	Giant Musselburgh American Flag 10."	Leek	Onion
Allium	*porrum*	*american-flag-musselburghii-mammoth-potii*	Mammoth Pot Musselburgh American Flag 8.5"	Leek	Onion
Allium	*porrum*	*american-flag-musselburghii*	Musselburgh American Flag 8.5"	Leek	Onion
Allium	*porrum*	*american-flag-pandora*	Pandora American Flag 8.5"	Leek	Onion
Allium	*porrum*	*arcadia*	Arcadia ?.?"	Leek	Onion
Allium	*porrum*	*ardea*	Ardea ?.?"	Leek	Onion
Allium	*porrum*	*arena*	Arena 7.0"	Leek	Onion
Allium	*porrum*	*argenta*	Argenta ?.?"	Leek	Onion
Allium	*porrum*	*arkansasensis*	Arkansas ?.?"	Leek	Onion
Allium	*porrum*	*ascendens*	Ascending	Leek	Onion
Allium	*porrum*	*autumn-giant-goliathii*	Goliath Autumn Giant 8.0"	Leek	Onion
Allium	*porrum*	*autumn-giant-mammothii*	Mammoth Autumn Giant 8.0"	Leek	Onion
Allium	*porrum*	*autumn-giant-provitei*	Porvite Autumn Giant 8.0"	Leek	Onion
Allium	*porrum*	*autumn-giant-september*	September Autumn Giant 8.0"	Leek	Onion
Allium	*porrum*	*autumn-giant-triumphatori*	Triumphator Autumn Giant 8.0"	Leek	Onion
Allium	*porrum*	*babbingtonii*	Babbington ?.?"	Leek	Onion
Allium	*porrum*	*banditii*	Bandit 6.0"	Leek	Onion
Allium	*porrum*	*bertolonii*	Bertolon ?.?"	Leek	Onion
Allium	*porrum*	*blizzardii*	Blizzard ?.?"	Leek	Onion
Allium	*porrum*	*bluastari*	Blaustar 6.7"	Leek	Onion
Allium	*porrum*	*blue-solaizei*	Blue Solaize 6.0"	Leek	Onion
Allium	*porrum*	*blue-solaizei-st.-victori*	St. Victor Blue Solaize 6.0"	Leek	Onion
Allium	*porrum*	*blue-solaizei-vernori*	Vernor Blue Solaize 6.0"	Leek	Onion
Allium	*porrum*	*broad-londonensis*	Broad London 9.0"	Leek	Onion
Allium	*porrum*	*bulbiferum*	Bulb Bearing	Leek	Onion
Allium	*porrum*	*bulgarian-giantii*	Bulgarian Giant 8.0"	Leek	Onion
Allium	*porrum*	*bulgarian-giantii-lincolnii*	Lincoln Bulgarian Giant 8.0"	Leek	Onion
Allium	*porrum*	*bulgarian-triumphii*	Bulgarian Triumph 8.0"	Leek	Onion
Allium	*porrum*	*bulgaricum*	Bulgarian	Leek	Onion
Allium	*porrum*	*byzantinum*	Byzantine	Leek	Onion
Allium	*porrum*	*carentani-giantii*	Giant Carentan 8.5"	Leek	Onion
Allium	*porrum*	*carentan-winteri*	Winter Carentan 8.5"	Leek	Onion
Allium	*porrum*	*carina*	Carina ?.?"	Leek	Onion
Allium	*porrum*	*castelstari*	Castelstar ?.?"	Leek	Onion
Allium	*porrum*	*caudatum*	Stalk	Leek	Onion
Allium	*porrum*	*chinese-flatii*	Chinese Flat ?.?"	Leek	Onion
Allium	*porrum*	*ciscoi*	Cisco 8.5"	Leek	Onion
Allium	*porrum*	*columbusensis*	Columbus 8.0"	Leek	Onion
Allium	*porrum*	*conora*	Conora ?.?"	Leek	Onion
Allium	*porrum*	*copenhagen-market-kontai*	Copenhagen Market Konta ?.?"	Leek	Onion
Allium	*porrum*	*dachenfeldtii-90-0255*	Dachenfeldt #90-0255 ?.?"	Leek	Onion
Allium	*porrum*	*danora*	Danora ?.?"	Leek	Onion
Allium	*porrum*	*dawn-giantii*	Dawn Giant 15."	Leek	Onion
Allium	*porrum*	*durabelii*	Durabel ?.?"	Leek	Onion
Allium	*porrum*	*duriaeanum*	Durian	Leek	Onion

Allium	*porrum*	*electra*	Electra 8.0"	Leek	Onion
Allium	*porrum*	*elephant-garlica*	Elephant Garlic 8.5"	Leek	Onion
Allium	*porrum*	*elephant-leekii*	Elephant 8.5"	Leek	Onion
Allium	*porrum*	*elina*	Elina 8.0"	Leek	Onion
Allium	*porrum*	*eskimo*	Eskimo ?.?"	Leek	Onion
Allium	*porrum*	*ester-cookii*	Ester Cook 8.5"	Leek	Onion
Allium	*porrum*	*falltimei*	Falltime 36."	Leek	Onion
Allium	*porrum*	*firena*	Firena ?.?"	Leek	Onion
Allium	*porrum*	*frostii*	Frost 6.0"	Leek	Onion
Allium	*porrum*	*gabiliana*	Gabilan 8.0"	Leek	Onion
Allium	*porrum*	*gasparrinii*	Gasparrin	Leek	Onion
Allium	*porrum*	*gennevilliersi*	Gennevilliers 8.0"	Leek	Onion
Allium	*porrum*	*gennevilliersi-furori*	Furor Gennevilliers 8.0"	Leek	Onion
Allium	*porrum*	*gennevilliersi-splendida*	Splendid Gennevilliers 8.0"	Leek	Onion
Allium	*porrum*	*giant-flagii*	Giant Flag 8.0"	Leek	Onion
Allium	*porrum*	*gracile*	Slender	Leek	Onion
Allium	*porrum*	*great-headedii*	Great Headed ?.?"	Leek	Onion
Allium	*porrum*	*halleri*	Haller	Leek	Onion
Allium	*porrum*	*hilari*	Hilari ?.?"	Leek	Onion
Allium	*porrum*	*hivori*	Hivor ?.?"	Leek	Onion
Allium	*porrum*	*holmense*	Holm	Leek	Onion
Allium	*porrum*	*imperialis*	Imperial ?.?"	Leek	Onion
Allium	*porrum*	*inegolii*	Inegol 12."	Leek	Onion
Allium	*porrum*	*invernoensis*	Inverno 10."	Leek	Onion
Allium	*porrum*	*jerseyi*	Jersey 8.0"	Leek	Onion
Allium	*porrum*	*jolantii*	Jolant 13."	Leek	Onion
Allium	*porrum*	*juane-du-poitoui*	Juane Du Poitou ?.?"	Leek	Onion
Allium	*porrum*	*kajakii*	Kajak ?.?"	Leek	Onion
Allium	*porrum*	*kilima*	Kilima 14."	Leek	Onion
Allium	*porrum*	*king-richardii*	King Richard 12."	Leek	Onion
Allium	*porrum*	*kurrat*	Kurrat	Leek	Onion
Allium	*porrum*	*laetum*	Bright	Leek	Onion
Allium	*porrum*	*lancelotii*	Lancelot 13."	Leek	Onion
Allium	*porrum*	*laurai*	Laura 10."	Leek	Onion
Allium	*porrum*	*lavi*	Lavi 9.0"	Leek	Onion
Allium	*porrum*	*leefallii*	Leefall 9.0"	Leek	Onion
Allium	*porrum*	*leeglooi*	Leegloo 7.0"	Leek	Onion
Allium	*porrum*	*leekoolii*	Leekool 8.5"	Leek	Onion
Allium	*porrum*	*leekwikii*	Leekwik 10."	Leek	Onion
Allium	*porrum*	*lineare*	Strap Leaf	Leek	Onion
Allium	*porrum*	*long-bulgarese-reuzenii*	Long Bulgarese Reuzen ?.?"	Leek	Onion
Allium	*porrum*	*long-de-mezieresi*	Long De Mezieres	Leek	Onion
Allium	*porrum*	*long-d'hiver-de-parisi*	Long D'Hiver De Paris ?.?"	Leek	Onion
Allium	*porrum*	*long-falli*	Long Fall 36."	Leek	Onion
Allium	*porrum*	*longina*	Longina 8.0"	Leek	Onion
Allium	*porrum*	*lyon-prizetakeri*	Lyon Prizetaker 8.0"	Leek	Onion
Allium	*porrum*	*molosensis*	Molos ?.?"	Leek	Onion
Allium	*porrum*	*oratoi*	Orato ?.?"	Leek	Onion
Allium	*porrum*	*otina*	Otina 8.0"	Leek	Onion
Allium	*porrum*	*otina-mammothii*	Mammoth Otina 8.0"	Leek	Onion
Allium	*porrum*	*panchoi*	Pancho ?.?"	Leek	Onion
Allium	*porrum*	*pinola*	Pinola ?.?"	Leek	Onion
Allium	*porrum*	*portuguese-garlic-clovei*	Portuguese Garlic Clove ?.?"	Leek	Onion
Allium	*porrum*	*prenora*	Prenora 8.0"	Leek	Onion
Allium	*porrum*	*primora*	Primor 8.0"	Leek	Onion
Allium	*porrum*	*purplei*	Purple ?.?"	Leek	Onion
Allium	*porrum*	*pylium*	Pyle ?.?"	Leek	Onion
Allium	*porrum*	*ramona*	Ramona 8.0"	Leek	Onion
Allium	*porrum*	*resei*	Rese 12."	Leek	Onion
Allium	*porrum*	*rikori*	Rikor 8.0"	Leek	Onion
Allium	*porrum*	*rivalis*	Rival 35."	Leek	Onion
Allium	*porrum*	*rustica*	Rustic 4.0"	Leek	Onion
Allium	*porrum*	*scopulicola*	Scoped	Leek	Onion
Allium	*porrum*	*scotlandensis*	Scotland 6.0"	Leek	Onion
Allium	*porrum*	*sherwoodii*	Sherwood 12."	Leek	Onion
Allium	*porrum*	*siegfriedii*	Siegfried ?.?"	Leek	Onion
Allium	*porrum*	*snowstari*	Snowstar ?.?"	Leek	Onion
Allium	*porrum*	*spectabile*	Spectacular ?.?"	Leek	Onion
Allium	*porrum*	*st.-catherine-deleigei*	St. Catherine De Liege ?.?"	Leek	Onion
Allium	*porrum*	*stora*	Stora ?.?"	Leek	Onion
Allium	*porrum*	*swiss-giant-coloma*	Swiss Giant Coloma ?.?"	Leek	Onion
Allium	*porrum*	*syriacum*	Syrian ?.?"	Leek	Onion
Allium	*porrum*	*tadorna*	Tadorna 7.5"	Leek	Onion
Allium	*porrum*	*tenori*	Tenor 8.0"	Leek	Onion
Allium	*porrum*	*thessalum*	Thessalonian ?.?"	Leek	Onion
Allium	*porrum*	*thori*	Thor ?.?"	Leek	Onion
Allium	*porrum*	*titanii*	Titan 8.0"	Leek	Onion
Allium	*porrum*	*tivii*	Tivi ?.?"	Leek	Onion
Allium	*porrum*	*toledoi*	Toledo 8.0"	Leek	Onion

Allium	*porrum*	*uniquei*	Unique 7.5"	Leek	Onion
Allium	*porrum*	*varna*	Varna 24."	Leek	Onion
Allium	*porrum*	*verina*	Verina 8.0"	Leek	Onion
Allium	*porrum*	*whitei*	White ?.?"	Leek	Onion
Allium	*porrum*	*wiedemannii*	Wiedemann	Leek	Onion
Allium	*porrum*	*winora*	Winora 6.0"	Leek	Onion
Allium	*porrum*	*winteri*	Winter 7.0"	Leek	Onion
Allium	*porrum*	*winteri-giantii*	Winter Giant 7.0"	Leek	Onion
Allium	*porrum*	*winteri-giantii-neomexicana*	New Mexico Winter Giant 7.0"	Leek	Onion
Allium	*porrum*	*winteri-giantii-vincentii*	Vincent Winter Giant 6.4"	Leek	Onion

Garlic, Hardneck (*Allium sativum-dura*)

Genus	Species	Infraspecies	Cultivated Variety Cultivar	Specimen	Genome
Allium	*sativum-dura-alba*	*baba-franchuk'si*	Baba Franchuk's 07 Clove	White Hardneck Garlic	Onion
Allium	*sativum-dura-alba*	*colorado-purplei*	Colorado Purple 09 Clove	White Hardneck Garlic	Onion
Allium	*sativum-dura-alba*	*fish-lakei-01(f1)*	Fish Lake #01 ?? Clove	White Hardneck Garlic	Onion
Allium	*sativum-dura-alba*	*fish-lakei-02(f2)*	Fish Lake #02 ?? Clove	White Hardneck Garlic	Onion
Allium	*sativum-dura-alba*	*fish-lakei-05(f5)*	Fish Lake #05 ?? Clove	White Hardneck Garlic	Onion
Allium	*sativum-dura-alba*	*fish-lakei-06(f6)*	Fish Lake #06 ?? Clove	White Hardneck Garlic	Onion
Allium	*sativum-dura-alba*	*fish-lakei-07(f7)*	Fish Lake #07 ?? Clove	White Hardneck Garlic	Onion
Allium	*sativum-dura-alba*	*fish-lakei-08(f8)*	Fish Lake #08 ?? Clove	White Hardneck Garlic	Onion
Allium	*sativum-dura-alba*	*fish-lakei-09(f9)*	Fish Lake #09 ?? Clove	White Hardneck Garlic	Onion
Allium	*sativum-dura-alba*	*fish-lakei-11(f11)*	Fish Lake #11 ?? Clove	White Hardneck Garlic	Onion
Allium	*sativum-dura-alba*	*fish-lakei-12(f12)*	Fish Lake #12 ?? Clove	White Hardneck Garlic	Onion
Allium	*sativum-dura-alba*	*fish-lakei-20(f20)*	Fish Lake #20 ?? Clove	White Hardneck Garlic	Onion
Allium	*sativum-dura-alba*	*fish-lakei-22(f22)*	Fish Lake #22 ?? Clove	White Hardneck Garlic	Onion
Allium	*sativum-dura-alba*	*fish-lakei-cubensis(fc)*	Fish Lake Cuba 19 Clove	White Hardneck Garlic	Onion
Allium	*sativum-dura-alba*	*georgian-crystal-gaterslebenii-6819*	Georgian Crystal 05 Clove	White Hardneck Garlic	Onion
Allium	*sativum-dura-alba*	*georgian-fire-gaterslebenii-6822*	Georgian Fire 07 Clove	White Hardneck Garlic	Onion
Allium	*sativum-dura-alba*	*japoi*	Japo ?? Clove	White Hardneck Garlic	Onion
Allium	*sativum-dura-alba*	*jovakii*	Jovak ?? Clove	White Hardneck Garlic	Onion
Allium	*sativum-dura-alba*	*kaskaskia-redii*	Kaskasia Red 06 Clove	White Hardneck Garlic	Onion
Allium	*sativum-dura-alba*	*kasahkistanensis*	Kazahkistan ?? Clove	White Hardneck Garlic	Onion
Allium	*sativum-dura-alba*	*polish-hardneckii*	Polish Porcelain 05 Clove	White Hardneck Garlic	Onion
Allium	*sativum-dura-alba*	*polish-jennii*	Polish Jenn 07 Clove	White Hardneck Garlic	Onion
Allium	*sativum-dura-alba*	*premium-whitei*	Premium 06 Clove	White Hardneck Garlic	Onion
Allium	*sativum-dura-alba*	*primii*	Prim 08 Clove	White Hardneck Garlic	Onion
Allium	*sativum-dura-alba*	*russian-hardneckii*	Russian ?? Clove	White Hardneck Garlic	Onion
Allium	*sativum-dura-alba*	*tibetensis*	Tibet ?? Clove	White Hardneck Garlic	Onion
Allium	*sativum-dura-alba*	*uzbek-turbanii*	Uzbek Turban ?? Clove	White Hardneck Garlic	Onion
Allium	*sativum-dura-alba*	*vostanii*	Vostani Porcelain ?? Clove	White Hardneck Garlic	Onion
Allium	*sativum-dura-brunnea*	*alison'si*	Alison's 09 Clove	Brown Hardneck Garlic	Onion
Allium	*sativum-dura-brunnea*	*atkin's-russianii*	Atkin's Russian 09 Clove	Brown Hardneck Garlic	Onion
Allium	*sativum-dura-brunnea*	*bogatyr-gaterslebenii-7204*	Bogatyr 07 Clove	Brown Hardneck Garlic	Onion
Allium	*sativum-dura-brunnea*	*burgundia*	Burgundy 10 Clove	Brown Hardneck Garlic	Onion
Allium	*sativum-dura-brunnea*	*chrysalis-rosei*	Chrysalis Rose 08 Clove	Brown Hardneck Garlic	Onion
Allium	*sativum-dura-brunnea*	*dushanbei*	Dushanbe ?? Clove	Brown Hardneck Garlic	Onion
Allium	*sativum-dura-brunnea*	*fisher-knofel-redii*	Fisher Knofel Red ?? Clove	Brown Hardneck Garlic	Onion
Allium	*sativum-dura-brunnea*	*fish-lakei-23(f23)*	Fish Lake #23 10 Clove	Brown Hardneck Garlic	Onion
Allium	*sativum-dura-brunnea*	*gigantus*	Gigantus 10 Clove	Brown Hardneck Garlic	Onion
Allium	*sativum-dura-brunnea*	*keeperi*	Keeper ?? Clove	Brown Hardneck Garlic	Onion
Allium	*sativum-dura-brunnea*	*krasnodar-whiteskinii*	Whiteskin Krasnodar 06 Clove	Brown Hardneck Garlic	Onion
Allium	*sativum-dura-brunnea*	*krasnodar-redskinii*	Redskin Krasnodar 07 Clove	Brown Hardneck Garlic	Onion
Allium	*sativum-dura-brunnea*	*new-mexico-topsetii*	New Mexico Topset ?? Clove	Brown Hardneck Garlic	Onion
Allium	*sativum-dura-brunnea*	*red-czar-gaterslebenii-6017*	Red Czar ?? Clove	Brown Hardneck Garlic	Onion
Allium	*sativum-dura-brunnea*	*thaii*	Thai 09 Clove	Brown Hardneck Garlic	Onion
Allium	*sativum-dura-brunnea*	*vici*	Vic ?? Clove	Brown Hardneck Garlic	Onion
Allium	*sativum-dura-brunnea*	*yampolskijii*	Yampolskij 05 Clove	Brown Hardneck Garlic	Onion
Allium	*sativum-dura-longicuspis*	*blazer-pristinksi-250662*	Blazer 04 Clove	Long Cusp Hardneck Garlic	Onion
Allium	*sativum-dura-longicuspis*	*brown-roseii*	Brown Rose 06 Clove	Long Cusp Hardneck Garlic	Onion
Allium	*sativum-dura-longicuspis*	*brown-tempest-pristinski-493098*	Brown Tempest 06 Clove	Long Cusp Hardneck Garlic	Onion
Allium	*sativum-dura-longicuspis*	*chatkalensis*	Chatkal ?? Clove	Long Cusp Hardneck Garlic	Onion
Allium	*sativum-dura-longicuspis*	*kitabensis*	Kitab 04 Clove	Long Cusp Hardneck Garlic	Onion
Allium	*sativum-dura-longicuspis*	*northei-3*	Northe #3 10 Clove	Long Cusp Hardneck Garlic	Onion
Allium	*sativum-dura-longicuspis*	*persian-stari*	Persian Star 07 Clove	Long Cusp Hardneck Garlic	Onion
Allium	*sativum-dura-longicuspis*	*pskemii*	Pskem 03 Clove	Long Cusp Hardneck Garlic	Onion
Allium	*sativum-dura-longicuspis*	*purple-tipii-ral-46*	Purple Tip 08 Clove	Long Cusp Hardneck Garlic	Onion
Allium	*sativum-dura-longicuspis*	*rosewood-pristiniskii-493099*	Rosewood Porcelain 04 Clove	Long Cusp Hardneck Garlic	Onion
Allium	*sativum-dura-longicuspis*	*wild-buffii*	Wild Buff 04Clove	Long Cusp Hardneck Garlic	Onion
Allium	*sativum-dura-marmorata*	*andidzhanskijii*	Andidzhansky ?? Clove	Marbled Hardneck Garlic	Onion
Allium	*sativum-dura-marmorata*	*armeniana*	Armenian ?? Clove	Marbled Hardneck Garlic	Onion
Allium	*sativum-dura-marmorata*	*bartlettii*	Bartlett ?? Clove	Marbled Hardneck Garlic	Onion
Allium	*sativum-dura-marmorata*	*beaulaensis*	Beaula ?? Clove	Marbled Hardneck Garlic	Onion
Allium	*sativum-dura-marmorata*	*blanakii*	Blanak ?? Clove	Marbled Hardneck Garlic	Onion
Allium	*sativum-dura-marmorata*	*brown-vesperi*	Brown Vesper ?? Clove	Marbled Hardneck Garlic	Onion
Allium	*sativum-dura-marmorata*	*italian-gourmetii*	Italian Gourmet ?? Clove	Marbled Hardneck Garlic	Onion
Allium	*sativum-dura-marmorata*	*choparskyi*	Choparsky 04 Clove	Marbled Hardneck Garlic	Onion
Allium	*sativum-dura-marmorata*	*dugansky-gaterslebenii-130*	Dugansky 06 Clove	Marbled Hardneck Garlic	Onion

Allium	sativum-dura-marmorata	fairlandensis	Fairland ?? Clove	Marbled Hardneck Garlic	Onion
Allium	sativum-dura-marmorata	flintii	Flint ?? Clove	Marbled Hardneck Garlic	Onion
Allium	sativum-dura-marmorata	floha-gaterslebenii-264	Floha 07 Clove	Marbled Hardneck Garlic	Onion
Allium	sativum-dura-marmorata	gemii (ral-27)	Gem ?? Clove	Marbled Hardneck Garlic	Onion
Allium	sativum-dura-marmorata	georgia-homesteadii	Georgia Homestead ?? Clove	Marbled Hardneck Garlic	Onion
Allium	sativum-dura-marmorata	germedorsi	Germedors ?? Clove	Marbled Hardneck Garlic	Onion
Allium	sativum-dura-marmorata	grandma-walkeri	Grandma Walker ?? Clove	Marbled Hardneck Garlic	Onion
Allium	sativum-dura-marmorata	gravesii	Graves ?? Clove	Marbled Hardneck Garlic	Onion
Allium	sativum-dura-marmorata	hoodii	Hood ?? Clove	Marbled Hardneck Garlic	Onion
Allium	sativum-dura-marmorata	iowensis	Iowa ?? Clove	Marbled Hardneck Garlic	Onion
Allium	sativum-dura-marmorata	kansasensis-roadsidei	Kansas Roadside ?? Clove	Marbled Hardneck Garlic	Onion
Allium	sativum-dura-marmorata	kelsoensis	Kelso ?? Clove	Marbled Hardneck Garlic	Onion
Allium	sativum-dura-marmorata	khabari	Khabar ?? Clove	Purple Hardneck Garlic	Onion
Allium	sativum-dura-marmorata	kyjevii	Kyjev Porcelain ?? Clove	Marbled Hardneck Garlic	Onion
Allium	sativum-dura-marmorata	lapanantkariensis	Lapanantkari Porcelain 05 Clove	Marbled Hardneck Garlic	Onion
Allium	sativum-dura-marmorata	legacyi	Legacy 04 Clove	Marbled Hardneck Garlic	Onion
Allium	sativum-dura-marmorata	marie's-specialis	Marie's Special ?? Clove	Marbled Hardneck Garlic	Onion
Allium	sativum-dura-marmorata	mexican-redii	Mexican Red ?? Clove	Marbled Hardneck Garlic	Onion
Allium	sativum-dura-marmorata	mexican-redii-silveri	Silverskin Mexican Red ?? Clove	Marbled Hardneck Garlic	Onion
Allium	sativum-dura-marmorata	mexicanoi-b	Mexicano B ?? Clove	Marbled Hardneck Garlic	Onion
Allium	sativum-dura-marmorata	monshanskij-gaterslebenii-131	Monshanskij 06 Clove	Marbled Hardneck Garlic	Onion
Allium	sativum-dura-marmorata	nations-roadii	Nations Road ?? Clove	Marbled Hardneck Garlic	Onion
Allium	sativum-dura-marmorata	ohio-wildii	Ohio Wild ?? Clove	Marbled Hardneck Garlic	Onion
Allium	sativum-dura-marmorata	paceri	Pacer ?? Clove	Marbled Hardneck Garlic	Onion
Allium	sativum-dura-marmorata	penasco-blue-leafii	Penasco Blue Leaf ?? Clove	Marbled Hardneck Garlic	Onion
Allium	sativum-dura-marmorata	pennsylvania-dutchii	Pennsylvania Dutch ?? Clove	Marbled Hardneck Garlic	Onion
Allium	sativum-dura-marmorata	red-janicei	Red Janice ?? Clove	Marbled Hardneck Garlic	Onion
Allium	sativum-dura-marmorata	rockroadii	Rockroad ?? Clove	Marbled Hardneck Garlic	Onion
Allium	sativum-dura-marmorata	rogersi	Rogers ?? Clove	Marbled Hardneck Garlic	Onion
Allium	sativum-dura-marmorata	russian-rosei	Russian Rose ?? Clove	Marbled Hardneck Garlic	Onion
Allium	sativum-dura-marmorata	shvelisiensis	Shvelisi 10 Clove	Marbled Hardneck Garlic	Onion
Allium	sativum-dura-marmorata	tangsiuinai	Tangsiuina ?? Clove	Marbled Hardneck Garlic	Onion
Allium	sativum-dura-marmorata	vekakii	Vekak 11 Clove	Marbled Hardneck Garlic	Onion
Allium	sativum-dura-marmorata	vladivostokensis-2	Vladivostok II 07 Clove	Marbled Hardneck Garlic	Onion
Allium	sativum-dura-marmorata	walkeri	Walker ?? Clove	Marbled Hardneck Garlic	Onion
Allium	sativum-dura-marmorata	wonha-gaterslebenii-6801	Wonha ?? Clove	Marbled Hardneck Garlic	Onion
Allium	sativum-dura-ophion	bai-pi-suanii	Bai Pi Suan 04 Clove	Snake Hardneck Garlic	Onion
Allium	sativum-dura-ophion	broadleaf-czech-gaterslebenii-146	Broadleaf Czech 05 Clove	Snake Hardneck Garlic	Onion
Allium	sativum-dura-ophion	chengduensis	Chengdu ?? Clove	Snake Hardneck Garlic	Onion
Allium	sativum-dura-ophion	chilean-purplei	Chilean Purple ?? Clove	Snake Hardneck Garlic	Onion
Allium	sativum-dura-ophion	chinese-purplei	Chinese Purple ?? Clove	Snake Hardneck Garlic	Onion
Allium	sativum-dura-ophion	chinese-purplei-brownii	Chinese Purple Brown ?? Clove	Snake Hardneck Garlic	Onion
Allium	sativum-dura-ophion	chinese-red-&-whitei	Chinese Red & White ?? Clove	Snake Hardneck Garlic	Onion
Allium	sativum-dura-ophion	chinese-stripei	Chinese Stripe ?? Clove	Snake Hardneck Garlic	Onion
Allium	sativum-dura-ophion	davidoi	Davido ?? Clove	Snake Hardneck Garlic	Onion
Allium	sativum-dura-ophion	dubnaensis-1	Dubna #1 ?? Clove	Snake Hardneck Garlic	Onion
Allium	sativum-dura-ophion	estonian-redii	Estonian Red 06 Clove	Snake Hardneck Garlic	Onion
Allium	sativum-dura-ophion	german-northern-whitei	N. White Porcelain German 04 Clove	Snake Hardneck Garlic	Onion
Allium	sativum-dura-ophion	german-porcelainii	Porcelain German 05 Clove	Snake Hardneck Garlic	Onion
Allium	sativum-dura-ophion	german-whitei	White Porcelain German 04 Clove	Snake Hardneck Garlic	Onion
Allium	sativum-dura-ophion	morado-gigantei	Gigante Morado 07 Clove	Snake Hardneck Garlic	Onion
Allium	sativum-dura-ophion	morado-redii	Red Morado ?? Clove	Snake Hardneck Garlic	Onion
Allium	sativum-dura-ophion	morado-spanishii	Spanish Morado 04 Clove	Snake Hardneck Garlic	Onion
Allium	sativum-dura-ophion	glacier-porcelainii	Glacier Porcelain ?? Clove	Snake Hardneck Garlic	Onion
Allium	sativum-dura-ophion	gypsy-redii	Gypsy Red Porcelain 05 Clove	Snake Hardneck Garlic	Onion
Allium	sativum-dura-ophion	hikkalda-zai-rai	Hikkalda Zai Rai ?? Clove	Snake Hardneck Garlic	Onion
Allium	sativum-dura-ophion	hokkaido-zai-taii	Hokkaido Zai Tai 08 Clove	Snake Hardneck Garlic	Onion
Allium	sativum-dura-ophion	idaho-specialis	Idaho Special ?? Clove	Snake Hardneck Garlic	Onion
Allium	sativum-dura-ophion	jain-shang-dongii	Jain Shang Dong ?? Clove	Snake Hardneck Garlic	Onion
Allium	sativum-dura-ophion	japanese-sakura-yellowii	Japanese Sakura Yellow 07 Clove	Snake Hardneck Garlic	Onion
Allium	sativum-dura-ophion	jiangsui	Jiangsu ?? Clove	Snake Hardneck Garlic	Onion
Allium	sativum-dura-ophion	kyjevii	Jiangsu ?? Clove	Snake Hardneck Garlic	Onion
Allium	sativum-dura-ophion	leningrad-gaterslebenii-684	Leningrad Porcelain 05 Clove	Snake Hardneck Garlic	Onion
Allium	sativum-dura-ophion	mahoganyi-pristinskii-493097	Pristinsky Mahogany ?? Clove	Snake Hardneck Garlic	Onion
Allium	sativum-dura-ophion	mahoganyi-fal-11	FAL #11 Mahogany ?? Clove	Snake Hardneck Garlic	Onion
Allium	sativum-dura-ophion	matterhornensis	Matterhorn 06 Clove	Snake Hardneck Garlic	Onion
Allium	sativum-dura-ophion	metechi-gaterslebenii-7081	Metechi 05 Clove	Snake Hardneck Garlic	Onion
Allium	sativum-dura-ophion	music-hotii	Hot Music 05 Clove	Snake Hardneck Garlic	Onion
Allium	sativum-dura-ophion	music-pink-stripedii	Pink Striped Music 04 Clove	Snake Hardneck Garlic	Onion
Allium	sativum-dura-ophion	music-porcelainii	Porcelain Music 08 Clove	Snake Hardneck Garlic	Onion
Allium	sativum-dura-ophion	music-porcelainii-filareei	Filaree Porcelain Music ?? Clove	Snake Hardneck Garlic	Onion
Allium	sativum-dura-ophion	music-pungens	Pungent Music ?? Clove	Snake Hardneck Garlic	Onion
Allium	sativum-dura-ophion	music-medium-hotii	Medium Hot Music 05 Clove	Snake Hardneck Garlic	Onion
Allium	sativum-dura-ophion	music-rich-flavori	Rich Flavor Music ?? Clove	Snake Hardneck Garlic	Onion
Allium	sativum-dura-ophion	music-sharp-flavori	Sharp Flavor Music ?? Clove	Snake Hardneck Garlic	Onion
Allium	sativum-dura-ophion	music-yuccaensis	Yucca Ridge Music 07 Clove	Snake Hardneck Garlic	Onion
Allium	sativum-dura-ophion	red-romanii	Red Roman ?? Clove	Snake Hardneck Garlic	Onion
Allium	sativum-dura-ophion	shandongii	Shandong ?? Clove	Snake Hardneck Garlic	Onion
Allium	sativum-dura-ophion	shandongii-purplei	Purple Shandong ?? Clove	Snake Hardneck Garlic	Onion

Allium	*sativum-dura-ophion*	*shandongii-tzanii*	Tzan Shandong ?? Clove	Snake Hardneck Garlic	Onion
Allium	*sativum-dura-ophion*	*shantung-purplei*	Shantung 07Clove	Snake Hardneck Garlic	Onion
Allium	*sativum-dura-ophion*	*siberiana*	Siberian 07 Clove	Snake Hardneck Garlic	Onion
Allium	*sativum-dura-ophion*	*volpei*	Volpe ?? Clove	Snake Hardneck Garlic	Onion
Allium	*sativum-dura-ophion*	*wildfirei*	Wildfire Porcelain 06 Clove	Snake Hardneck Garlic	Onion
Allium	*sativum-dura-ophion*	*zemo-gatersleben-6307*	Zemo Porcelain 06 Clove	Snake Hardneck Garlic	Onion
Allium	*sativum-dura-ophion*	*zouhii*	Zouh ?? Clove	Snake Hardneck Garlic	Onion
Allium	*sativum-dura-ophioscorodon*	*bill-angellii*	Bill Angell ?? Clove	Snaked Hardneck Garlic	Onion
Allium	*sativum-dura-ophioscorodon*	*early-redii*	Early Red ?? Clove	Snaked Hardneck Garlic	Onion
Allium	*sativum-dura-ophioscorodon*	*germanii(redii)*	Dull German Red 12 Clove	Snaked Hardneck Garlic	Onion
Allium	*sativum-dura-ophioscorodon*	*germanii-redii*	Bright German Red 10 Clove	Snaked Hardneck Garlic	Onion
Allium	*sativum-dura-ophioscorodon*	*korean-tanii*	Korean Tan 08 Clove	Snaked Hardneck Garlic	Onion
Allium	*sativum-dura-ophioscorodon*	*lanzerottii*	Lanzerotti ?? Clove	Snaked Hardneck Garlic	Onion
Allium	*sativum-dura-ophioscorodon*	*lavallai*	Lavalla ?? Clove	Snaked Hardneck Garlic	Onion
Allium	*sativum-dura-ophioscorodon*	*montana-rojai*	Montana Roja 11 Clove	Snaked Hardneck Garlic	Onion
Allium	*sativum-dura-ophioscorodon*	*ozarkana-old-homesteadii*	Old Homestead Ozark ?? Clove	Snaked Hardneck Garlic	Onion
Allium	*sativum-dura-ophioscorodon*	*ozarkana-tanii*	Tan Ozark ?? Clove	Snaked Hardneck Garlic	Onion
Allium	*sativum-dura-ophioscorodon*	*ozarkana-whitei*	White Ozark ?? Clove	Snaked Hardneck Garlic	Onion
Allium	*sativum-dura-ophioscorodon*	*russian-brownii*	Russian Brown10 Clove	Snaked Hardneck Garlic	Onion
Allium	*sativum-dura-ophioscorodon*	*turkish-hardneckii*	Turkish ?? Clove	Snaked Hardneck Garlic	Onion
Allium	*sativum-dura-ophioscorodon*	*wild-tanii*	Wild Tan ?? Clove	Snaked Hardneck Garlic	Onion
Allium	*sativum-dura-ophioscorodon*	*yugoslaviana*	Yugoslavian 12 Clove	Snaked Hardneck Garlic	Onion
Allium	*sativum-dura-purpurea*	*basque-turbanii-11*	Basque Turban 11 Clove	Purple Hardneck Garlic	Onion
Allium	*sativum-dura-purpurea*	*bavariensis*	Bavarian 08 Clove	Purple Hardneck Garlic	Onion
Allium	*sativum-dura-purpurea*	*belarusi*	Belarus ?? Clove	Purple Hardneck Garlic	Onion
Allium	*sativum-dura-purpurea*	*californica*	California ?? Clove	Purple Hardneck Garlic	Onion
Allium	*sativum-dura-purpurea*	*celaya-purplei*	Celaya ?? Clove	Purple Hardneck Garlic	Onion
Allium	*sativum-dura-purpurea*	*chanzhuensis*	Chanzhu 11 Clove	Purple Hardneck Garlic	Onion
Allium	*sativum-dura-purpurea*	*chrysalis-purplei*	Chrysalis 08 Clove	Purple Hardneck Garlic	Onion
Allium	*sativum-dura-purpurea*	*cichisdzhvari*	Cichisdzhvari ?? Clove	Purple Hardneck Garlic	Onion
Allium	*sativum-dura-purpurea*	*cichisdzhvari-redgrain-gaterslebenii-6820*	Redgrain Cichisdzhvari ?? Clove	Purple Hardneck Garlic	Onion
Allium	*sativum-dura-purpurea*	*de-vivoi*	De Vivo 05 Clove	Purple Hardneck Garlic	Onion
Allium	*sativum-dura-purpurea*	*dixon-strainii*	Dixon Strain ?? Clove	Purple Hardneck Garlic	Onion
Allium	*sativum-dura-purpurea*	*ferganskijii*	Fergansky 10 Clove	Purple Hardneck Garlic	Onion
Allium	*sativum-dura-purpurea*	*fish-lakei-03(f3)*	Fish Lake #03 06 Clove	Purple Hardneck Garlic	Onion
Allium	*sativum-dura-purpurea*	*fish-lakei-31(f31)*	Fish Lake #31 06 Clove	Purple Hardneck Garlic	Onion
Allium	*sativum-dura-purpurea*	*fish-lakei-32(f32)*	Fish Lake #32 06 Clove	Purple Hardneck Garlic	Onion
Allium	*sativum-dura-purpurea*	*fish-lakei-34(f34)*	Fish Lake #34 06 Clove	Purple Hardneck Garlic	Onion
Allium	*sativum-dura-purpurea*	*french-germinadori*	French Germinador 10 Clove	Purple Hardneck Garlic	Onion
Allium	*sativum-dura-purpurea*	*hornbyi*	Hornby 09 Clove	Purple Hardneck Garlic	Onion
Allium	*sativum-dura-purpurea*	*northern-quebecensis*	N. Quebec 07 Clove	Purple Hardneck Garlic	Onion
Allium	*sativum-dura-purpurea*	*purple-gazer-gaterslebenii-7092*	Glazer 07 Clove	Purple Hardneck Garlic	Onion
Allium	*sativum-dura-purpurea*	*purple-italian-easy-peelii*	Italian Easy Peel 08 Clove	Purple Hardneck Garlic	Onion
Allium	*sativum-dura-purpurea*	*purple-maxii*	Purple Max 09 Clove	Purple Hardneck Garlic	Onion
Allium	*sativum-dura-purpurea*	*purple-understreakii*	Understreak Red ?? Clove	Purple Hardneck Garlic	Onion
Allium	*sativum-dura-purpurea*	*puslinchii*	Puslinch 12 Clove	Purple Hardneck Garlic	Onion
Allium	*sativum-dura-purpurea*	*raceyi*	Racey 11 Clove	Purple Hardneck Garlic	Onion
Allium	*sativum-dura-purpurea*	*romanian-redii*	Romanian Red Porcelain 04 Clove	Purple Hardneck Garlic	Onion
Allium	*sativum-dura-purpurea*	*salt-springii*	Salt Spring 08 Clove	Purple Hardneck Garlic	Onion
Allium	*sativum-dura-purpurea*	*salt-springii-selectii*	Select Salt Spring 11 Clove	Purple Hardneck Garlic	Onion
Allium	*sativum-dura-purpurea*	*stein-mountainii*	Stein Mountain 08 Clove	Purple Hardneck Garlic	Onion
Allium	*sativum-dura-rocambolea*	*alexandriensis*	Alexandria 10 Clove	Rocambole Hardneck Garlic	Onion
Allium	*sativum-dura-rocambolea*	*brown-saxon-gaterslebenii-7116*	Brown Saxon 02 Clove	Rocambole Hardneck Garlic	Onion
Allium	*sativum-dura-rocambolea*	*carpathiana*	Carpathian 08 Clove	Rocambole Hardneck Garlic	Onion
Allium	*sativum-dura-rocambolea*	*colorado-blackii*	Colorado Black 06 Clove	Rocambole Hardneck Garlic	Onion
Allium	*sativum-dura-rocambolea*	*colorado-garlic-seed-foundation*	Colorado Garlic Seed Foundation 06 Clove	Rocambole Hardneck Garlic	Onion
Allium	*sativum-dura-rocambolea*	*colorado-leahii-99*	Colorado Leah #99 Porcelain ?? Clove	Rocambole Hardneck Garlic	Onion
Allium	*sativum-dura-rocambolea*	*creole-nativei*	Native Creole ?? Clove	Rocambole Hardneck Garlic	Onion
Allium	*sativum-dura-rocambolea*	*creole-pescadero-redii*	Pescadero Red Creole ?? Clove	Rocambole Hardneck Garlic	Onion
Allium	*sativum-dura-rocambolea*	*creole-redii*	Red Creole ?? Clove	Rocambole Hardneck Garlic	Onion
Allium	*sativum-dura-rocambolea*	*croatiana*	Croatian ?? Clove	Rocambole Hardneck Garlic	Onion
Allium	*sativum-dura-rocambolea*	*czechland-racei*	Czechland Race ?? Clove	Rocambole Hardneck Garlic	Onion
Allium	*sativum-dura-rocambolea*	*enid's-vertalia*	Enid's Vertalia ?? Clove	Rocambole Hardneck Garlic	Onion
Allium	*sativum-dura-rocambolea*	*french-redii*	Red French 08 Clove	Rocambole Hardneck Garlic	Onion
Allium	*sativum-dura-rocambolea*	*french-redii-asiana*	Red Asian French 08 Clove	Rocambole Hardneck Garlic	Onion
Allium	*sativum-dura-rocambolea*	*french-rocambolei*	Sweet French 08 Clove	Rocambole Hardneck Garlic	Onion
Allium	*sativum-dura-rocambolea*	*german-brownii*	German Brown ?? Clove	Rocambole Hardneck Garlic	Onion
Allium	*sativum-dura-rocambolea*	*germanyi*	Germany ?? Clove	Rocambole Hardneck Garlic	Onion
Allium	*sativum-dura-rocambolea*	*israelii*	Israeli 06 Clove	Rocambole Hardneck Garlic	Onion
Allium	*sativum-dura-rocambolea*	*kievensis*	Kiev ?? Clove	Rocambole Hardneck Garlic	Onion
Allium	*sativum-dura-rocambolea*	*killarneyi*	Killarney 09 Clove	Rocambole Hardneck Garlic	Onion
Allium	*sativum-dura-rocambolea*	*korean-rocambolea*	Korean 08 Clove	Rocambole Hardneck Garlic	Onion
Allium	*sativum-dura-rocambolea*	*l.-k.-mannii-c751*	L. K. Mann 08 Clove	Rocambole Hardneck Garlic	Onion
Allium	*sativum-dura-rocambolea*	*marinoi*	Marino 07 Clove	Rocambole Hardneck Garlic	Onion
Allium	*sativum-dura-rocambolea*	*montana-giantii*	Montana Giant 09 Clove	Rocambole Hardneck Garlic	Onion
Allium	*sativum-dura-rocambolea*	*mountain-rocambolea*	Mountain ?? Clove	Rocambole Hardneck Garlic	Onion
Allium	*sativum-dura-rocambolea*	*philipsi*	Phillips ?? Clove	Rocambole Hardneck Garlic	Onion
Allium	*sativum-dura-rocambolea*	*pitarelli*	Pitarelli ?? Clove	Rocambole Hardneck Garlic	Onion
Allium	*sativum-dura-rocambolea*	*purple-mirandai*	Purple Miranda 10 Clove	Rocambole Hardneck Garlic	Onion

Allium	*sativum-dura-rocambolea*	*purple-rocambolei*	Purple 08 Clove	Rocambole Hardneck Garlic	Onion
Allium	*sativum-dura-rocambolea*	*red-rezan-gaterslebenii-7205*	Red Rezan ?? Clove	Rocambole Hardneck Garlic	Onion
Allium	*sativum-dura-rocambolea*	*rogerroi*	Rogerro ?? Clove	Rocambole Hardneck Garlic	Onion
Allium	*sativum-dura-rocambolea*	*russian-giantii*	Russian Giant 09 Clove	Rocambole Hardneck Garlic	Onion
Allium	*sativum-dura-rocambolea*	*russian-redii*	Russian Red 11 Clove	Rocambole Hardneck Garlic	Onion
Allium	*sativum-dura-rocambolea*	*russian-redstreakii*	Russian Redstreak 08 Clove	Rocambole Hardneck Garlic	Onion
Allium	*sativum-dura-rocambolea*	*siberian-rocambolei*	Siberian 07 Clove	Rocambole Hardneck Garlic	Onion
Allium	*sativum-dura-rocambolea*	*slovakii*	Slovak ?? Clove	Rocambole Hardneck Garlic	Onion
Allium	*sativum-dura-rocambolea*	*spanish-rojai*	Spanish Roja 10 Clove	Rocambole Hardneck Garlic	Onion
Allium	*sativum-dura-rocambolea*	*special-idahoensis*	Special Idaho ?? Clove	Rocambole Hardneck Garlic	Onion
Allium	*sativum-dura-rocambolea*	*summit-rojai*	Summit Roja 10 Clove	Rocambole Hardneck Garlic	Onion
Allium	*sativum-dura-rocambolea*	*temptressi-gsf-65*	Temptress 06 Clove	Rocambole Hardneck Garlic	Onion
Allium	*sativum-dura-rocambolea*	*ukrainian-hotii*	Ukranian Hot Roja ?? Clove	Rocambole Hardneck Garlic	Onion
Allium	*sativum-dura-rocambolea*	*ukrainian-topsetii*	Ukranian Topset ?? Clove	Rocambole Hardneck Garlic	Onion
Allium	*sativum-dura-rocambolea*	*webber-greekii*	Webber Greek 06 Clove	Rocambole Hardneck Garlic	Onion
Allium	*sativum-dura-rocambolea*	*wild-rocambolei-ral-159*	Wild ?? Clove	Rocambole Hardneck Garlic	Onion
Allium	*sativum-dura-rocambolea*	*waddellii-rocambolei*	Waddell ?? Clove	Rocambole Hardneck Garlic	Onion
Allium	*sativum-dura-rocambolea*	*wisconsin-german-redii*	Wisconsin German Red ?? Clove	Rocambole Hardneck Garlic	Onion
Allium	*sativum-dura-rocambolea*	*wisconsin-whitei*	Wisconsin White ?? Clove	Rocambole Hardneck Garlic	Onion
Allium	*sativum-dura-rocambolea*	*zahadarai*	Zaharada Porcelain 05 Clove	Rocambole Hardneck Garlic	Onion
Allium	*sativum-dura-rubra*	*ail-rose-de-latreci*	Ail Rose Lautrec 08 Clove	Red Hardneck Garlic	Onion
Allium	*sativum-dura-rubra*	*belgiana*	Belgian ?? Clove	Red Hardneck Garlic	Onion
Allium	*sativum-dura-rubra*	*chesnok-red-gaterslebenii-6811*	Chesnok 10 Clove	Red Hardneck Garlic	Onion
Allium	*sativum-dura-rubra*	*czech-redii*	Czech ?? Clove	Red Hardneck Garlic	Onion
Allium	*sativum-dura-rubra*	*german-extra-hardyi*	German Extra Hardy 05 Clove	Red Hardneck Garlic	Onion
Allium	*sativum-dura-rubra*	*hungarian-redii*	Hungarian ?? Clove	Red Hardneck Garlic	Onion
Allium	*sativum-dura-rubra*	*mcdougall'si*	McDougall's 08 Clove	Red Hardneck Garlic	Onion
Allium	*sativum-dura-rubra*	*pearly-redii*	Pearly	Red Hardneck Garlic	Onion
Allium	*sativum-dura-rubra*	*shatili-gaterslebenii-7108*	Shatili 08 Clove	Red Hardneck Garlic	Onion
Allium	*sativum-dura-rubra*	*skuri-gaterslebenii-6038*	Skuri 10 Clove	Red Hardneck Garlic	Onion
Allium	*sativum-dura-rubra*	*susan-delafield'si*	Susan Delafield's 08 Clove	Red Hardneck Garlic	Onion
Allium	*sativum-dura-rubra*	*mavnivii*	Mavniv Ukrainian 07 Clove	Red Hardneck Garlic	Onion
Allium	*sativum-dura-rubra*	*western-rojoi*	Western Rojo ?? Clove	Red Hardneck Garlic	Onion

Garlic, Softneck (*Allium sativum-molla*)

Genus	Species	Infraspecies	Cultivated Variety Cultivar	Specimen	Genome
Allium	*sativum-molla-alba*	*achaldaba*	Achaldaba ?? Clove	White Softneck Garlic	Onion
Allium	*sativum-molla-alba*	*achatami*	Achatami ?? Clove	White Softneck Garlic	Onion
Allium	*sativum-molla-alba*	*argunii*	Arguni 15 Clove	White Softneck Garlic	Onion
Allium	*sativum-molla-alba*	*athensensis*	Athens 02 Clove	White Softneck Garlic	Onion
Allium	*sativum-molla-alba*	*baldur's-bestii*	Baldur's Best ?? Clove	White Softneck Garlic	Onion
Allium	*sativum-molla-alba*	*blossomii*	Blossom 07 Clove	White Softneck Garlic	Onion
Allium	*sativum-molla-alba*	*boliviana*	Bolivian 13 Clove	White Softneck Garlic	Onion
Allium	*sativum-molla-alba*	*california-early-whitei*	California Early 04 Clove	White Softneck Garlic	Onion
Allium	*sativum-molla-alba*	*california-giantii*	California Giant 04 Clove	White Softneck Garlic	Onion
Allium	*sativum-molla-alba*	*california-late-whitei*	California Late 04 Clove	White Softneck Garlic	Onion
Allium	*sativum-molla-alba*	*chet's-italiana*	Chet's Italian 15 Clove	White Softneck Garlic	Onion
Allium	*sativum-molla-alba*	*chilean-silveri*	Chilean Silver 16 Clove	White Softneck Garlic	Onion
Allium	*sativum-molla-alba*	*dominic'si*	Dominic's ?? Clove	White Softneck Garlic	Onion
Allium	*sativum-molla-alba*	*early-red-italianii*	Early Red Italian 16 Clove	White Softneck Garlic	Onion
Allium	*sativum-molla-alba*	*fish-lakei-04(f4)*	Fish Lake #04 ?? Clove	White Softneck Garlic	Onion
Allium	*sativum-molla-alba*	*fish-lakei-40(f40)*	Fish Lake #40 ?? Clove	White Softneck Garlic	Onion
Allium	*sativum-molla-alba*	*fish-lakei-41(f41)*	Fish Lake #41 ?? Clove	White Softneck Garlic	Onion
Allium	*sativum-molla-alba*	*fish-lakei-44(f44)*	Fish Lake #44 08 Clove	White Softneck Garlic	Onion
Allium	*sativum-molla-alba*	*fish-lakei-45(f45)*	Fish Lake #45 ?? Clove	White Softneck Garlic	Onion
Allium	*sativum-molla-alba*	*gilroyi*	Gilroy ?? Clove	White Softneck Garlic	Onion
Allium	*sativum-molla-alba*	*grizzlyi*	Grizzly ?? Clove	White Softneck Garlic	Onion
Allium	*sativum-molla-alba*	*hungarian-purple-stripei*	Hungarian Purple Stripe 10 Clove	White Softneck Garlic	Onion
Allium	*sativum-molla-alba*	*irbyi*	Irby 20 Clove	White Softneck Garlic	Onion
Allium	*sativum-molla-alba*	*italian-earlyi*	Italian Early ?? Clove	White Softneck Garlic	Onion
Allium	*sativum-molla-alba*	*italian-pungentii*	Italian Pungent ?? Clove	White Softneck Garlic	Onion
Allium	*sativum-molla-alba*	*italian-latei*	Italian Late 20 Clove	White Softneck Garlic	Onion
Allium	*sativum-molla-alba*	*jumboi*	Jumbo ?? Clove	White Softneck Garlic	Onion
Allium	*sativum-molla-alba*	*locatii*	Locati 11 Clove	White Softneck Garlic	Onion
Allium	*sativum-molla-alba*	*lotus*	Lotus 08 Clove	White Softneck Garlic	Onion
Allium	*sativum-molla-alba*	*maxatawnyi*	Maxatawny 12 Clove	White Softneck Garlic	Onion
Allium	*sativum-molla-alba*	*mother-of-pearlii*	Mother Of Pearl 20 Clove	White Softneck Garlic	Onion
Allium	*sativum-molla-alba*	*nichols-silverskinii*	Nichols Silverskin ?? Clove	White Softneck Garlic	Onion
Allium	*sativum-molla-alba*	*polish-softneckii*	Polish 12 Clove	White Softneck Garlic	Onion
Allium	*sativum-molla-alba*	*randl-colorado-whitei*	Randl Colorado 11 Clove	White Softneck Garlic	Onion
Allium	*sativum-molla-alba*	*s-&-h-silveri*	S & H Silver 18 Clove	White Softneck Garlic	Onion
Allium	*sativum-molla-alba*	*russian-infernoi*	Russian Inferno ?? Clove	White Softneck Garlic	Onion
Allium	*sativum-molla-alba*	*seely-hillensis*	Seely Hill ?? Clove	White Softneck Garlic	Onion
Allium	*sativum-molla-alba*	*sicilian-whitei*	Sicilian ?? Clove	White Softneck Garlic	Onion
Allium	*sativum-molla-alba*	*siciliano-creamyi*	Siciliano Creamy ?? Clove	White Softneck Garlic	Onion
Allium	*sativum-molla-alba*	*silverskin-prinstinskii-485592*	Pristinsky Silverskin ?? Clove	White Softneck Garlic	Onion
Allium	*sativum-molla-alba*	*silverskinii*	Silverskin 17 Clove	White Softneck Garlic	Onion
Allium	*sativum-molla-alba*	*silverwhitei*	Silverwhite 17 Clove	White Softneck Garlic	Onion

Allium	*sativum-molla-alba*	*singlei*	Single ?? Clove	White Softneck Garlic	Onion
Allium	*sativum-molla-alba*	*spanish-beniteei*	Spanish Benitee ?? Clove	White Softneck Garlic	Onion
Allium	*sativum-molla-alba*	*sweet-havenii*	Sweet Haven ?? Clove	White Softneck Garlic	Onion
Allium	*sativum-molla-alba*	*wedhamii*	Wedham ?? Clove	White Softneck Garlic	Onion
Allium	*sativum-molla-brunnea*	.	.	Brown Softneck Garlic	Onion
Allium	*sativum-molla-brunnea*	*molla-brunnea*	Original	Brown Softneck Garlic	Onion
Allium	*sativum-molla-brunnea*	*beijingensis*	Beijing 06 Clove	Brown Softneck Garlic	Onion
Allium	*sativum-molla-brunnea*	*guatemalana*	Guatemalan 07 Clove	Brown Softneck Garlic	Onion
Allium	*sativum-molla-brunnea*	*japanese-sakura-tanii*	Japanese Sakura Tan 07 Clove	Brown Softneck Garlic	Onion
Allium	*sativum-molla-brunnea*	*limburgii*	Limburg 10 Clove	Brown Softneck Garlic	Onion
Allium	*sativum-molla-brunnea*	*red-revel-pristinskii-250663*	Red Revel ?? Clove	Brown Softneck Garlic	Onion
Allium	*sativum-molla-brunnea*	*sicilian-goldii*	Gold Sicilian ?? Clove	Brown Softneck Garlic	Onion
Allium	*sativum-molla-brunnea*	*sicilian-silveri*	Silver Sicilian ?? Clove	Brown Softneck Garlic	Onion
Allium	*sativum-molla-brunnea*	*silverskin-tanii*	Tan Silverskin 12 Clove	Brown Softneck Garlic	Onion
Allium	*sativum-molla-brunnea*	*silverskin-pinkii*	Pink Silverskin 17 Clove	Brown Softneck Garlic	Onion
Allium	*sativum-molla-brunnea*	*singing-fallsii*	Singing Falls 11 Clove	Brown Softneck Garlic	Onion
Allium	*sativum-molla-brunnea*	*ukranian-danieloi*	Danielo Ukrainian 11 Clove	Brown Softneck Garlic	Onion
Allium	*sativum-molla-marmorata*	*breathlessi*	Breathless ?? Clove	Marbled Softneck Garlic	Onion
Allium	*sativum-molla-marmorata*	*frenchii*	French 12 Clove	Marbled Softneck Garlic	Onion
Allium	*sativum-molla-marmorata*	*korean-mad-dogii*	Korean Mad Dog 14 Clove	Marbled Softneck Garlic	Onion
Allium	*sativum-molla-marmorata*	*laconai-1*	Lacona #1 ?? Clove	Marbled Softneck Garlic	Onion
Allium	*sativum-molla-marmorata*	*lewandoski-virus-freei-14*	Lewandoski Virus Free 14 Clove	Marbled Softneck Garlic	Onion
Allium	*sativum-molla-marmorata*	*lillium*	Lillium ?? Clove	Marbled Softneck Garlic	Onion
Allium	*sativum-molla-marmorata*	*loiaconoi*	Loiacono ?? Clove	Marbled Softneck Garlic	Onion
Allium	*sativum-molla-marmorata*	*maitakei*	Maitake ?? Clove	Marbled Softneck Garlic	Onion
Allium	*sativum-molla-marmorata*	*mild-frenchii*	Mild French 14 Clove	Marbled Softneck Garlic	Onion
Allium	*sativum-molla-marmorata*	*nootka-rosea*	Nootka Rose 20 Clove	Marbled Softneck Garlic	Onion
Allium	*sativum-molla-marmorata*	*olomukensis*	Olomuk ?? Clove	Marbled Softneck Garlic	Onion
Allium	*sativum-molla-marmorata*	*sonorana*	Sonoran ?? Clove	Marbled Softneck Garlic	Onion
Allium	*sativum-molla-marmorata*	*susan-mariei*	Susan Marie ?? Clove	Marbled Softneck Garlic	Onion
Allium	*sativum-molla-marmorata*	*susanvillensis*	Susanville 13 Clove	Marbled Softneck Garlic	Onion
Allium	*sativum-molla-marmorata*	*sweet-mississippiensis*	Sweet Mississippi ?? Clove	Marbled Softneck Garlic	Onion
Allium	*sativum-molla-marmorata*	*tai-cangii*	Tai Cang 05 Clove	Marbled Softneck Garlic	Onion
Allium	*sativum-molla-marmorata*	*tajikistanensis*	Tajikisatan 07 Clove	Marbled Softneck Garlic	Onion
Allium	*sativum-molla-marmorata*	*vilnius*	Vilnius ?? Clove	Marbled Softneck Garlic	Onion
Allium	*sativum-molla-purpurea*	*aunt-louisei*	Aunt Louise ?? Clove	Purple Softneck Garlic	Onion
Allium	*sativum-molla-purpurea*	*beautyi*	Beauty ?? Clove	Purple Softneck Garlic	Onion
Allium	*sativum-molla-purpurea*	*cuban-purplei*	Cuban 07 Clove	Purple Softneck Garlic	Onion
Allium	*sativum-molla-purpurea*	*denman-islandica*	Denman Island 10 Clove	Purple Softneck Garlic	Onion
Allium	*sativum-molla-purpurea*	*early-italian-purplei*	Early Italian ?? Clove	Purple Softneck Garlic	Onion
Allium	*sativum-molla-purpurea*	*kulbaki-gatersleben-5865*	Kulbaki ?? Clove	Purple Softneck Garlic	Onion
Allium	*sativum-molla-purpurea*	*labera-purplei*	Labera Purple ?? Clove	Purple Softneck Garlic	Onion
Allium	*sativum-molla-purpurea*	*ontarioensis*	Ontario 06 Clove	Purple Softneck Garlic	Onion
Allium	*sativum-molla-purpurea*	*purple-queenii*	Queen ?? Clove	Purple Softneck Garlic	Onion
Allium	*sativum-molla-purpurea*	*spanishii*	Spanish Pink 12 Clove	Purple Softneck Garlic	Onion
Allium	*sativum-molla-purpurea*	*xianii*	Xian 10 Clove	Purple Softneck Garlic	Onion
Allium	*sativum-molla-rubra*	*ajo-rojoi*	Ajo Rojo ?? Clove	Red Softneck Garlic	Onion
Allium	*sativum-molla-rubra*	*bohemiensis*	Bohemian ?? Clove	Red Softneck Garlic	Onion
Allium	*sativum-molla-rubra*	*china-rosei*	China Rose ?? Clove	Red Softneck Garlic	Onion
Allium	*sativum-molla-rubra*	*chinese-pinkii*	Chinese Pink 12 Clove	Red Softneck Garlic	Onion
Allium	*sativum-molla-rubra*	*greekii*	Greek 16 Clove	Red Softneck Garlic	Onion
Allium	*sativum-molla-rubra*	*idaho-silveri*	Idaho Silver ?? Clove	Red Softneck Garlic	Onion
Allium	*sativum-molla-rubra*	*idita-redii*	Idita ?? Clove	Red Softneck Garlic	Onion
Allium	*sativum-molla-rubra*	*mexican-red-purple-stripei*	Mexican Purple Stripe ?? Clove	Red Softneck Garlic	Onion
Allium	*sativum-molla-rubra*	*mexican-red-silverskinii*	Mexican Silverskin ?? Clove	Red Softneck Garlic	Onion
Allium	*sativum-molla-rubra*	*rallyi*	Rally Turban 11 Clove	Red Softneck Garlic	Onion
Allium	*sativum-molla-rubra*	*rose-du-vari*	Rose Du Var 13 Clove	Red Softneck Garlic	Onion
Allium	*sativum-molla-rubra*	*st.-helensensis*	St. Helens 16 Clove	Red Softneck Garlic	Onion
Allium	*sativum-molla-rubra*	*silver-rosei*	Silver Rose 13 Clove	Red Softneck Garlic	Onion
Allium	*sativum-molla-rubra*	*tashkentii*	Tashkent 11 Clove	Red Softneck Garlic	Onion
Allium	*sativum-molla-rubra*	*tipitillai*	Tipatilla ?? Clove	Red Softneck Garlic	Onion
Allium	*sativum-molla-rubra*	*walla-walla-earlyi*	Early Walla Walla 16 Clove	Red Softneck Garlic	Onion
Allium	*sativum-molla-rubra*	*western-rosei*	Western Rose 15 Clove	Red Softneck Garlic	Onion
Allium	*sativum-molla-rubra*	*white-rosei*	White Rose ?? Clove	Red Softneck Garlic	Onion
Allium	*sativum-molla-scolymus*	*acropolisi*	Acropolis ?? Clove	Artichoke Softneck Garlic	Onion
Allium	*sativum-molla-scolymus*	*aftonii*	Afton ?? Clove	Artichoke Softneck Garlic	Onion
Allium	*sativum-molla-scolymus*	*ail-de-pays-gersi*	Ail De Pays Gers 10 Clove	Artichoke Softneck Garlic	Onion
Allium	*sativum-molla-scolymus*	*alanii*	Alan ?? Clove	Artichoke Softneck Garlic	Onion
Allium	*sativum-molla-scolymus*	*asian-tempestii*	Asian Tempest 06 Clove	Artichoke Softneck Garlic	Onion
Allium	*sativum-molla-scolymus*	*california-virus-freei-25*	California Virus Free #25 ?? Clove	Artichoke Softneck Garlic	Onion
Allium	*sativum-molla-scolymus*	*chamiskuri-gaterslebenii-6305*	Chamiskuri ?? Clove	Artichoke Softneck Garlic	Onion
Allium	*sativum-molla-scolymus*	*chinese-sativum*	Chinese 08 Clove	Artichoke Softneck Garlic	Onion
Allium	*sativum-molla-scolymus*	*french-purplei*	French Purple ?? Clove	Artichoke Softneck Garlic	Onion
Allium	*sativum-molla-scolymus*	*germidouri*	Germidour 07 Clove	Artichoke Softneck Garlic	Onion
Allium	*sativum-molla-scolymus*	*gomecari*	Gomecari ?? Clove	Artichoke Softneck Garlic	Onion
Allium	*sativum-molla-scolymus*	*gergory's-china-rosei*	Gregory's China Rose ?? Clove	Artichoke Softneck Garlic	Onion
Allium	*sativum-molla-scolymus*	*himalayan-redii*	Himalayan Red 02 Clove	Artichoke Softneck Garlic	Onion
Allium	*sativum-molla-scolymus*	*inchelium-redii*	Inchelium Red 15 Clove	Artichoke Softneck Garlic	Onion
Allium	*sativum-molla-scolymus*	*italian-purplei*	Italian Purple 09 Clove	Artichoke Softneck Garlic	Onion

Allium	*sativum-molla-scolymus*	*jameyi*	Jamey ?? Clove	Artichoke Softneck Garlic	Onion
Allium	*sativum-molla-scolymus*	*kellyi*	Kelly ?? Clove	Artichoke Softneck Garlic	Onion
Allium	*sativum-molla-scolymus*	*kettle-river-giantii*	Kettle River Giant 11 Clove	Artichoke Softneck Garlic	Onion
Allium	*sativum-molla-scolymus*	*lorz-italianii*	Lorz Italian 16 Clove	Artichoke Softneck Garlic	Onion
Allium	*sativum-molla-scolymus*	*lukakii*	Lukak ?? Clove	Artichoke Softneck Garlic	Onion
Allium	*sativum-molla-scolymus*	*machashi-gaterslebenii-5873*	Machashi ?? Clove	Artichoke Softneck Garlic	Onion
Allium	*sativum-molla-scolymus*	*maiskijii*	Maiskij 07 Clove	Artichoke Softneck Garlic	Onion
Allium	*sativum-molla-scolymus*	*mikulovskii*	Mikulovsky ?? Clove	Artichoke Softneck Garlic	Onion
Allium	*sativum-molla-scolymus*	*mucdi-gatersleben-5876*	Mucdi 12 Clove	Artichoke Softneck Garlic	Onion
Allium	*sativum-molla-scolymus*	*oklahoma-artichokei*	Oklahoma ?? Clove	Artichoke Softneck Garlic	Onion
Allium	*sativum-molla-scolymus*	*okrentii*	Okrent ?? Clove	Artichoke Softneck Garlic	Onion
Allium	*sativum-molla-scolymus*	*old-yellowi*	Okrent ?? Clove	Artichoke Softneck Garlic	Onion
Allium	*sativum-molla-scolymus*	*oregon-bluei*	Oregon Blue 15 Clove	Artichoke Softneck Garlic	Onion
Allium	*sativum-molla-scolymus*	*pioneeri*	Pioneer 15 Clove	Artichoke Softneck Garlic	Onion
Allium	*sativum-molla-scolymus*	*polish-whitei*	Polish White 11 Clove	Artichoke Softneck Garlic	Onion
Allium	*sativum-molla-scolymus*	*purple-cauldonii*	Purple Cauldron ?? Clove	Artichoke Softneck Garlic	Onion
Allium	*sativum-molla-scolymus*	*pyongyang-gaterslebenii-7041*	Pyongyang 09 Clove	Artichoke Softneck Garlic	Onion
Allium	*sativum-molla-scolymus*	*red-tochii*	Red Toch 14 Clove	Artichoke Softneck Garlic	Onion
Allium	*sativum-molla-scolymus*	*sam-loiaconoi*	Sam Loiacono 04 Clove	Artichoke Softneck Garlic	Onion
Allium	*sativum-molla-scolymus*	*sicilian-artichokei*	Siclian ?? Clove	Artichoke Softneck Garlic	Onion
Allium	*sativum-molla-scolymus*	*simoneti-gatersleben-5878*	Simoneti 12 Clove	Artichoke Softneck Garlic	Onion
Allium	*sativum-molla-scolymus*	*taia-gatersleben-6040*	Taia ?? Clove	Artichoke Softneck Garlic	Onion
Allium	*sativum-molla-scolymus*	*thermadronei*	Thermadrone ?? Clove	Artichoke Softneck Garlic	Onion
Allium	*sativum-molla-scolymus*	*transylvaniana*	Transylvanian 13 Clove	Artichoke Softneck Garlic	Onion
Allium	*sativum-molla-scolymus*	*trueheartii*	Trueheart ?? Clove	Artichoke Softneck Garlic	Onion
Allium	*sativum-molla-scolymus*	*udabmoi*	Udabmo 19 Clove	Artichoke Softneck Garlic	Onion
Allium	*sativum-molla-scolymus*	*turkish-softneckii*	Turkish ?? Clove	Artichoke Softneck Garlic	Onion
Allium	*sativum-molla-scolymus*	*white-seedlessi*	White Seedless ?? Clove	Artichoke Softneck Garlic	Onion
Allium	*sativum-molla-scolymus*	*wild-plum-whitei*	Wild Plum White ?? Clove	Artichoke Softneck Garlic	Onion
Allium	*sativum-molla-scolymus*	*zahorskyi*	Zahorsky ?? Clove	Artichoke Softneck Garlic	Onion

Chives (*Allium schoenoprasum*)

Genus	Species	Infraspecies	Cultivated Variety	Cultivar Specimen	Genome
Allium	*schoenoprasum*	*acutum*	Sharp Leaf	Chives	Onion
Allium	*schoenoprasum*	*albiflorum*	White Flower	Chives	Onion
Allium	*schoenoprasum*	*alpinum*	Alpine	Chives	Onion
Allium	*schoenoprasum*	*alvarense*	Alvar	Chives	Onion
Allium	*schoenoprasum*	*broteroi*	Brotero	Chives	Onion
Allium	*schoenoprasum*	*buhseanum*	Buhse	Chives	Onion
Allium	*schoenoprasum*	*caespitans*	Clump	Chives	Onion
Allium	*schoenoprasum*	*carneum*	Flesh	Chives	Onion
Allium	*schoenoprasum*	*ciboulea*	Large	Chives	Onion
Allium	*schoenoprasum*	*coloratum*	Colored	Chives	Onion
Allium	*schoenoprasum*	*duriminium*	Hard Lead	Chives	Onion
Allium	*schoenoprasum*	*foliosum*	Leafy	Chives	Onion
Allium	*schoenoprasum*	*forscatei*	Forscate	Chives	Onion
Allium	*schoenoprasum*	*glaucum*	Cyan	Chives	Onion
Allium	*schoenoprasum*	*grandea*	Grande	Chives	Onion
Allium	*schoenoprasum*	*gredense*	Greda	Chives	Onion
Allium	*schoenoprasum*	*grolaui*	Grolau Windowsill	Chives	Onion
Allium	*schoenoprasum*	*idzuense*	Idzu	Chives	Onion
Allium	*schoenoprasum*	*improvedii*	Improved	Chives	Onion
Allium	*schoenoprasum*	*latiorifolium*	Wide Leaf	Chives	Onion
Allium	*schoenoprasum*	*laurentianum*	Laurent	Chives	Onion
Allium	*schoenoprasum*	*lusitanicum*	Portugal	Chives	Onion
Allium	*schoenoprasum*	*montanum*	Mountain	Chives	Onion
Allium	*schoenoprasum*	*orosiae*	Oros	Chives	Onion
Allium	*schoenoprasum*	*palustre*	Marsh	Chives	Onion
Allium	*schoenoprasum*	*punctulatum*	Dotted	Chives	Onion
Allium	*schoenoprasum*	*purpurascens*	Purple	Chives	Onion
Allium	*schoenoprasum*	*raddeanum*	Radde	Chives	Onion
Allium	*schoenoprasum*	*reflexum*	Reflexed	Chives	Onion
Allium	*schoenoprasum*	*riparium*	Riverside	Chives	Onion
Allium	*schoenoprasum*	*roseum*	Rose	Chives	Onion
Allium	*schoenoprasum*	*sibiricum*	Siberian	Chives	Onion
Allium	*schoenoprasum*	*staroi*	Staro	Chives	Onion
Allium	*schoenoprasum*	*tenuifolium*	Fine Leaf	Chives	Onion
Allium	*schoenoprasum*	*udinicum*	Udin	Chives	Onion

Chives, Flat Leaf (*Allium senescens*)

Genus	Species	Infraspecies	Cultivated Variety	Cultivar Specimen	Genome
Allium	*senescens*	*albiflorum*	White Flower	Flat Leaf Chives	Onion
Allium	*senescens*	*andersonii*	Anderson	Flat Leaf Chives	Onion
Allium	*senescens*	*baicalense*	Baikal Lake	Flat Leaf Chives	Onion
Allium	*senescens*	*brevipedicellatum*	Short Pedicel	Flat Leaf Chives	Onion
Allium	*senescens*	*crispum*	Curly	Flat Leaf Chives	Onion

Allium	*senescens*	*danubiale*	Danube	Flat Leaf Chives	Onion
Allium	*senescens*	*flavescens*	Yellow	Flat Leaf Chives	Onion
Allium	*senescens*	*glaucum*	Cyan	Flat Leaf Chives	Onion
Allium	*senescens*	*illyricum*	Illyrian	Flat Leaf Chives	Onion
Allium	*senescens*	*kungii*	Kung	Flat Leaf Chives	Onion
Allium	*senescens*	*latifolium*	Broad	Flat Leaf Chives	Onion
Allium	*senescens*	*minor*	Lesser	Flat Leaf Chives	Onion
Allium	*senescens*	*narcissifolium*	Daffodil Leaf	Flat Leaf Chives	Onion
Allium	*senescens*	*purpureum*	Purple	Flat Leaf Chives	Onion
Allium	*senescens*	*serotinum*	Late Flowering	Flat Leaf Chives	Onion
Allium	*senescens*	*sessiliflorum*	Sessile Flower	Flat Leaf Chives	Onion

Chives, Garlic (*Allium tuberosum*)

Genus	Species	Infraspecies	Cultivated Variety Cultivar	Specimen	Genome
Allium	*tuberosum*	*americana*	American	Garlic Chives	Onion
Allium	*tuberosum*	*angulosum*	Angled	Garlic Chives	Onion
Allium	*tuberosum*	*annamensis*	Vietnamese	Garlic Chives	Onion
Allium	*tuberosum*	*argyi*	Argy	Garlic Chives	Onion
Allium	*tuberosum*	*chinense*	Chinese	Garlic Chives	Onion
Allium	*tuberosum*	*clarkei*	Clarke	Garlic Chives	Onion
Allium	*tuberosum*	*gow-choii*	Gow Choy	Garlic Chives	Onion
Allium	*tuberosum*	*mauvea*	Mauve	Garlic Chives	Onion
Allium	*tuberosum*	*nirai*	Nira	Garlic Chives	Onion
Allium	*tuberosum*	*roxburghii*	Roxburgh	Garlic Chives	Onion
Allium	*tuberosum*	*sulvia*	Sulvian	Garlic Chives	Onion
Allium	*tuberosum*	*uliginosum*	Marsh	Garlic Chives	Onion
Allium	*tuberosum*	*yezoense*	Yezo	Garlic Chives	Onion

Family 722.013.6 *Hypoxidaceae* Starflower Grass

Genus	Species	Specimen	Genome	C:Culinary, M:Medicinal, E:Economic
Hypoxis	*goetzei*	Goetze	Starflower Grass	C:LF:**VG**

Varieties

Goetze Starflower Grass (*Hypoxis goetzei*)

Genus	Species	infraspecies	Cultivar	Specimen	Genome
Hypoxis	*goetzei*	*goetzei*	Original	Goetze	Starflower Grass
Hypoxis	*goetzei*	*esculenta*	Edible	Goetze	Starflower Grass
Hypoxis	*goetzei*	*rubiginosa*	Ruby Red	Goetze	Starflower Grass
Hypoxis	*goetzei*	*turbinata*	Turbin	Goetze	Starflower Grass

Order 722.014 *Arales* Aroid Lily

Latin: *Ara* aroid, ar

FAMILIES

722.014.1 *Araceae* **Aroid Lily**
722.014.2 *Acoraceae* **Flag Aroid Lily**
722.014.3 *Anthuriaceae* Flamingo Lily
722.014.4 *Juncaginaceae* Arrow Grass
722.014.5 *Lemnaceae* Duckweed
722.014.6 *Philodendraceae* Terrestrial Ar Lily
722.014.7 *Ruppiaceae* Widgeon Grass
722.014.8 *Tofieldiaceae* Tofield Ar Lily

Family 722.014.1 *Araceae* Aroid Lily

Genus	Species	Specimen	Genome	C:Culinary, M:Medicinal, E:Economic
Arisaema	*amurense*	Amur River Basin	Cobra Aroid Lily	C:CR:**VG**, **E**:CR:PS☠
Arisaema	*consanguineum*	Blood Root	Cobra Aroid Lily	C:CR:**VG**, **E**:CR:PS☠
Arisaema	*dracontium*	Dragon Root	Cobra Aroid Lily	C:CR:**VG**, **E**:CR:PS☠
Arisaema	*heterophyllum*	Variable Leaf	Cobra Aroid Lily	C:CR:**VG**, **E**:CR:PS☠
Arisaema	*triphyllum*	Jack in the Pulpit	Cobra Aroid Lily	C:CR:**VG**, **E**:CR:PS☠
Colocasia	*esculenta*	Coco	Taro Yam	C:TU:**VG**

Varieties

Cobra Aroid Lily (*Arisaema spp.*)

Coco Taro Yam (*Colocasia esculenta*)

Genus	Species	infraspecies	Cultivar	Specimen	Genome

Arisaema	*amurense*	*amurense*	Original	Amur River Basin	Cobra Aroid Lily
Arisaema	*amurense*	*abense*	Aba	Amur River Basin	Cobra Aroid Lily
Arisaema	*amurense*	*atropurpureum*	Dark Purple	Amur River Basin	Cobra Aroid Lily
Arisaema	*amurense*	*denticulatum*	Toothed	Amur River Basin	Cobra Aroid Lily
Arisaema	*amurense*	*furusei*	Furuse	Amur River Basin	Cobra Aroid Lily
Arisaema	*amurense*	*komarovii*	Komarov	Amur River Basin	Cobra Aroid Lily
Arisaema	*amurense*	*magnidens*	Large Tooth	Amur River Basin	Cobra Aroid Lily
Arisaema	*amurense*	*purpureum*	Purple	Amur River Basin	Cobra Aroid Lily
Arisaema	*amurense*	*robustum*	Robust	Amur River Basin	Cobra Aroid Lily
Arisaema	*amurense*	*serratum*	Sawtooth	Amur River Basin	Cobra Aroid Lily
Arisaema	*amurense*	*variegatum*	Variegated	Amur River Basin	Cobra Aroid Lily
Arisaema	*amurense*	*violaceum*	Violet	Amur River Basin	Cobra Aroid Lily
Arisaema	*consanguineum*	.	.	Blood Root	Cobra Aroid Lily
Arisaema	*consanguineum*	*consanguineum*	Original	Blood Root	Cobra Aroid Lily
Arisaema	*consanguineum*	*alpestre*	Alpine	Blood Root	Cobra Aroid Lily
Arisaema	*consanguineum*	*brevipes*	Short	Blood Root	Cobra Aroid Lily
Arisaema	*consanguineum*	*davidianum*	Davidian	Blood Root	Cobra Aroid Lily
Arisaema	*consanguineum*	*divaricatum*	Spreading	Blood Root	Cobra Aroid Lily
Arisaema	*consanguineum*	*giraldii*	Girald	Blood Root	Cobra Aroid Lily
Arisaema	*consanguineum*	*giganteum*	Giant	Blood Root	Cobra Aroid Lily
Arisaema	*consanguineum*	*kelung-insulare*	Kelung Island	Blood Root	Cobra Aroid Lily
Arisaema	*consanguineum*	*kerrii*	Kerr	Blood Root	Cobra Aroid Lily
Arisaema	*consanguineum*	*latisectum*	Wide Section	Blood Root	Cobra Aroid Lily
Arisaema	*consanguineum*	*tatarinowii*	Tatarinow	Blood Root	Cobra Aroid Lily
Arisaema	*consanguineum*	*undulatum*	Wavy	Blood Root	Cobra Aroid Lily
Arisaema	*consanguineum*	*vituperatum*	Vituperative	Blood Root	Cobra Aroid Lily
Arisaema	*heterophyllum*	*heterophyllum*	Original	Variable Leaf	Cobra Aroid Lily
Arisaema	*heterophyllum*	*ambiguum*	Ambiguous	Variable Leaf	Cobra Aroid Lily
Arisaema	*heterophyllum*	*brachyspathum*	Short Spathe	Variable Leaf	Cobra Aroid Lily
Arisaema	*heterophyllum*	*koreanum*	Korean	Variable Leaf	Cobra Aroid Lily
Arisaema	*heterophyllum*	*kwangtungense*	Kwangtung	Variable Leaf	Cobra Aroid Lily
Arisaema	*heterophyllum*	*limprichtii*	Limpricht	Variable Leaf	Cobra Aroid Lily
Arisaema	*heterophyllum*	*manshuricum*	Manchurian	Variable Leaf	Cobra Aroid Lily
Arisaema	*heterophyllum*	*multisectum*	Multisection	Variable Leaf	Cobra Aroid Lily
Arisaema	*heterophyllum*	*nigropunctatum*	Black Dotted	Variable Leaf	Cobra Aroid Lily
Arisaema	*heterophyllum*	*stenospathum*	Thin Spathe	Variable Leaf	Cobra Aroid Lily
Arisaema	*heterophyllum*	*tacquetii*	Tacquet	Variable Leaf	Cobra Aroid Lily
Arisaema	*heterophyllum*	*takeoi*	Takeo	Variable Leaf	Cobra Aroid Lily
Arisaema	*dracontium*	*dracontium*	Original	Dragon Root	Cobra Aroid Lily
Arisaema	*dracontium*	*boscii*	Bosc	Dragon Root	Cobra Aroid Lily
Arisaema	*dracontium*	*plukenetii*	Plukenet	Dragon Root	Cobra Aroid Lily
Arisaema	*triphyllum*	*triphyllum*	Original	Jack in the Pulpit	Cobra Aroid Lily
Arisaema	*triphyllum*	*acuminatum*	Sharp Leaf	Jack in the Pulpit	Cobra Aroid Lily
Arisaema	*triphyllum*	*atrorubens*	Dark Red	Jack in the Pulpit	Cobra Aroid Lily
Arisaema	*triphyllum*	*bispadiceum*	Two Spathe	Jack in the Pulpit	Cobra Aroid Lily
Arisaema	*triphyllum*	*brasilianum*	Brasil	Jack in the Pulpit	Cobra Aroid Lily
Arisaema	*triphyllum*	*deflexum*	Deflexed	Jack in the Pulpit	Cobra Aroid Lily
Arisaema	*triphyllum*	*hastatum*	Arrow Leaf	Jack in the Pulpit	Cobra Aroid Lily
Arisaema	*triphyllum*	*montanum*	Mountain	Jack in the Pulpit	Cobra Aroid Lily
Arisaema	*triphyllum*	*obtusoquinatum*	Blunt	Jack in the Pulpit	Cobra Aroid Lily
Arisaema	*triphyllum*	*pallascens*	Pale	Jack in the Pulpit	Cobra Aroid Lily
Arisaema	*triphyllum*	*pusillum*	Puny	Jack in the Pulpit	Cobra Aroid Lily
Arisaema	*triphyllum*	*stewardsonii*	Stewardson	Jack in the Pulpit	Cobra Aroid Lily
Arisaema	*triphyllum*	*trispadiceum*	Three Spathe	Jack in the Pulpit	Cobra Aroid Lily
Arisaema	*triphyllum*	*viride*	Green	Jack in the Pulpit	Cobra Aroid Lily
Arisaema	*triphyllum*	*zebrinum*	Zebra	Jack in the Pulpit	Cobra Aroid Lily
Colocasia	*esculenta*	*esculenta*	Original	Coco	Taro Yam
Colocasia	*esculenta*	*acris*	Acrid	Coco	Taro Yam
Colocasia	*esculenta*	*aegyptiaca*	Egyptian	Coco	Taro Yam
Colocasia	*esculenta*	*antiquorum*	Antiquity	Coco	Taro Yam
Colocasia	*esculenta*	*aquatilis*	Aquatic	Coco	Taro Yam
Colocasia	*esculenta*	*ebiimo*	Ebimo	Coco	Taro Yam
Colocasia	*esculenta*	*euchlora*	Green	Coco	Taro Yam
Colocasia	*esculenta*	*fontanesii*	Fontanes	Coco	Taro Yam
Colocasia	*esculenta*	*gracilis*	Slender	Coco	Taro Yam
Colocasia	*esculenta*	*himalensis*	Himalaya	Coco	Taro Yam
Colocasia	*esculenta*	*multifolia*	Many Leaf	Coco	Taro Yam
Colocasia	*esculenta*	*neocaledonica*	New Caledonia	Coco	Taro Yam
Colocasia	*esculenta*	*nymphaeifolia*	Nymph	Coco	Taro Yam
Colocasia	*esculenta*	*oyasetage*	Oyasetag	Coco	Taro Yam
Colocasia	*esculenta*	*patens*	Spreading	Coco	Taro Yam
Colocasia	*esculenta*	*peltata*	Pelt	Coco	Taro Yam
Colocasia	*esculenta*	*purpurea*	Purple	Coco	Taro Yam
Colocasia	*esculenta*	*rosea*	Rose	Coco	Taro Yam
Colocasia	*esculenta*	*rotundifolia*	Round Leaf	Coco	Taro Yam
Colocasia	*esculenta*	*rupicola*	Rock Dwelling	Coco	Taro Yam
Colocasia	*esculenta*	*stolonifera*	Stolon	Coco	Taro Yam
Colocasia	*esculenta*	*tonoimo*	Tono Imo	Coco	Taro Yam
Colocasia	*esculenta*	*vera*	Verity	Coco	Taro Yam

Colocasia	*esculenta*	*violacea*	Violet	Coco	Taro Yam
Colocasia	*esculenta*	*virosa*	Viral	Coco	Taro Yam
Colocasia	*esculenta*	*yamamotoi*	Yamamoto	Coco	Taro Yam

Family 722.014.2 *Acoraceae* Flag Aroid Lily

Genus	Species	Specimen	Genome	C:Culinary, M:Medicinal, E:Economic
Acorus	*calamus*	Sweet	Flag Aroid Lily	C:RZ:CF,**VG**
Acorus	*calamus-angustatus*	Myrtle	Flag Aroid Lily	C:RZ:CF,**VG**
Acorus	*gramineus*	Grass Leaf	Flag Aroid Lily	C:RZ:CF,**VG**

Varieties

Sweet Flag (*Acorus calamus*)

Genus	Species	infraspecies	Cultivar	Specimen	Genome
Acorus	*calamus*	*calamus*	Original	Sweet Flag	Flag Aroid Lily
Acorus	*calamus*	*americanus*	American	Sweet Flag	Flag Aroid Lily
Acorus	*calamus*	*aromaticus*	Aromatic	Sweet Flag	Flag Aroid Lily
Acorus	*calamus*	*belangeri*	Belanger	Sweet Flag	Flag Aroid Lily
Acorus	*calamus*	*casia*	Casie	Sweet Flag	Flag Aroid Lily
Acorus	*calamus*	*commersonii*	Commerson	Sweet Flag	Flag Aroid Lily
Acorus	*calamus*	*commutatus*	Changing	Sweet Flag	Flag Aroid Lily
Acorus	*calamus*	*elatus*	Tall	Sweet Flag	Flag Aroid Lily
Acorus	*calamus*	*europaeus*	European	Sweet Flag	Flag Aroid Lily
Acorus	*calamus*	*flexuosus*	Flexible	Sweet Flag	Flag Aroid Lily
Acorus	*calamus*	*floridanus*	Florida	Sweet Flag	Flag Aroid Lily
Acorus	*calamus*	*griffithii*	Griffith	Sweet Flag	Flag Aroid Lily
Acorus	*calamus*	*latifolius*	Broad Leaf	Sweet Flag	Flag Aroid Lily
Acorus	*calamus*	*nilghirensis*	Nilghir	Sweet Flag	Flag Aroid Lily
Acorus	*calamus*	*odoratus*	Scented	Sweet Flag	Flag Aroid Lily
Acorus	*calamus*	*terrestris*	Terrestrial	Sweet Flag	Flag Aroid Lily
Acorus	*calamus*	*undulatus*	Wavy	Sweet Flag	Flag Aroid Lily
Acorus	*calamus*	*variegatus*	Variegated	Sweet Flag	Flag Aroid Lily
Acorus	*calamus*	*verus*	Verified	Sweet Flag	Flag Aroid Lily
Acorus	*calamus-angustatus*	.	.	Myrtle Sweet Flag	Flag Aroid Lily
Acorus	*calamus-angustatus*	*angustatus*	Original	Myrtle Sweet Flag	Flag Aroid Lily
Acorus	*calamus-angustatus*	*asiaticus*	Asian	Myrtle Sweet Flag	Flag Aroid Lily
Acorus	*calamus-angustatus*	*cochinchinensis*	Cochin	Myrtle Sweet Flag	Flag Aroid Lily
Acorus	*calamus-angustatus*	*crassispadix*	Crass Spadix	Myrtle Sweet Flag	Flag Aroid Lily
Acorus	*calamus-angustatus*	*rumphianus*	Rumph	Myrtle Sweet Flag	Flag Aroid Lily
Acorus	*calamus-angustatus*	*spurius*	Spurios	Myrtle Sweet Flag	Flag Aroid Lily
Acorus	*calamus-angustatus*	*tatarinowii*	Tatarinow	Myrtle Sweet Flag	Flag Aroid Lily
Acorus	*calamus-angustatus*	*terrestris*	Terrestrial	Myrtle Sweet Flag	Flag Aroid Lily
Acorus	*calamus-angustatus*	*triqueter*	Triquet	Myrtle Sweet Flag	Flag Aroid Lily
Acorus	*gramineus*	.	.	Grass Leaf	Flag Aroid Lily
Acorus	*gramineus*	*gramineus*	Original	Grass Leaf	Flag Aroid Lily
Acorus	*gramineus*	*humilis*	Low	Grass Leaf	Flag Aroid Lily
Acorus	*gramineus*	*japonica*	Japanese	Grass Leaf	Flag Aroid Lily
Acorus	*gramineus*	*macrospadiceus*	Large Spadix	Grass Leaf	Flag Aroid Lily
Acorus	*gramineus*	*ogonii*	Japanese Rush	Grass Leaf	Flag Aroid Lily
Acorus	*gramineus*	*pusillus*	Dwarf Japanese Rush	Grass Leaf	Flag Aroid Lily
Acorus	*gramineus*	*variegatus*	Variegated Japanes Rush	Grass Leaf	Flag Aroid Lily
Acorus	*gramineus*	*xiangyeus*	Xiangye	Grass Leaf	Flag Aroid Lily

Order 722.015 *Asparagales* Asparagus Lily

Latin: *Asparagus* asparagus
Latin: *Plia* pliant, softstem

FAMILIES

722.015.1 *Asparagaceae*	**Asparagus**
722.015.2 *Blandfordiaceae*	Blandford Asparagus Lily
722.015.3 *Boryaceae*	Bory Lily
722.015.4 *Iridaceae*	**Iris**
722.015.5 *Petrosaviaceae*	Petrosav Lily
722.015.6 *Tecophilaeaceae*	Tecophile Lily

Family 722.151 *Asparagaceae* Asparagus

Genus	Species	Specimen	Genome	C:Culinary, M:Medicinal, E:Economic
Asparagus	*acutifolius*	Sharp Leaf	Asparagus Lily	C:SH:**VG**
Asparagus	*cochinchinensis*	Chinese	Asparagus Lily	C:TU:**VG**
Asparagus	*officinalis*	**Mediterranean**	**Asparagus Lily**	C:SH:**VG**
Asparagus	*racemosus*	Shatavari	Asparagus Lily	C:RZ:**VG**

Varieties

Asparagus (*Asparagus officinalis*)

Genus	species	infraspecies	Cultivated Variety Cultivar	Specimen	Genome
Asparagus	*officinalis*	*altilis*	Tall	Mediterranean	Asparagus Lily
Asparagus	*officinalis*	*argenteuilensis-praecox*	Argenteuil Early	Mediterranean	Asparagus Lily
Asparagus	*officinalis*	*barr's-mammothii*	Barr's Mammoth	Mediterranean	Asparagus Lily
Asparagus	*officinalis*	*californica-500*	California #500	Mediterranean	Asparagus Lily
Asparagus	*officinalis*	*campestris*	Field	Mediterranean	Asparagus Lily
Asparagus	*officinalis*	*collinus*	Collin	Mediterranean	Asparagus Lily
Asparagus	*officinalis*	*conover's-colossalii*	Conover's Colossal	Mediterranean	Asparagus Lily
Asparagus	*officinalis*	*embuduensis*	Embudu	Mediterranean	Asparagus Lily
Asparagus	*officinalis*	*franklinii*	Franklin Male	Mediterranean	Asparagus Lily
Asparagus	*officinalis*	*giant-mammothii*	Giant Mammoth	Mediterranean	Asparagus Lily
Asparagus	*officinalis*	*laraci*	Larac	Mediterranean	Asparagus Lily
Asparagus	*officinalis*	*lucullus*	Lucullus Male	Mediterranean	Asparagus Lily
Asparagus	*officinalis*	*martha-washingtonii*	Martha Washington	Mediterranean	Asparagus Lily
Asparagus	*officinalis*	*mary-washingtonii*	Mary Washington	Mediterranean	Asparagus Lily
Asparagus	*officinalis*	*mary-washingtonii-2*	Mary Washington Improved	Mediterranean	Asparagus Lily
Asparagus	*officinalis*	*mary-washingtonii-500*	Mary Washington #500	Mediterranean	Asparagus Lily
Asparagus	*officinalis*	*mary-washingtonii-roberts-improvedii*	Mary Washington Roberts Improved	Mediterranean	Asparagus Lily
Asparagus	*officinalis*	*mary-washingtonii-roberts-strainii*	Mary Washington Roberts Strain	Mediterranean	Asparagus Lily
Asparagus	*officinalis*	*mary-washingtonii-rust-resistantii*	Mary Washington Rust Resistant	Mediterranean	Asparagus Lily
Asparagus	*officinalis*	*mary-washingtonii-UC-072*	Mary Washington Univ. Calif. #072	Mediterranean	Asparagus Lily
Asparagus	*officinalis*	*mary-washingtonii-UC-157*	Mary Washington Univ. Calif. #157	Mediterranean	Asparagus Lily
Asparagus	*officinalis*	*mary-washingtonii-UC-711*	Mary Washington Univ. Calif. #711	Mediterranean	Asparagus Lily
Asparagus	*officinalis*	*mary-washingtonii-35-vikingii*	Mary Washington Viking Vineland #35	Mediterranean	Asparagus Lily
Asparagus	*officinalis*	*mary-washingtonii-KB3-vikingii*	Mary Washington Viking #KB3	Mediterranean	Asparagus Lily
Asparagus	*officinalis*	*oxycarpus*	Sharp	Mediterranean	Asparagus Lily
Asparagus	*officinalis*	*polyphyllus*	Many Leaf	Mediterranean	Asparagus Lily
Asparagus	*officinalis*	*prostratus*	Prostrate	Mediterranean	Asparagus Lily
Asparagus	*officinalis*	*purpurea-bracta*	Purple Bract	Mediterranean	Asparagus Lily
Asparagus	*officinalis*	*purpurea-passion*	Purple Passion	Mediterranean	Asparagus Lily
Asparagus	*officinalis*	*purpurea-suavis*	Sweet Purple	Mediterranean	Asparagus Lily
Asparagus	*officinalis*	*saxonii*	Saxon Male	Mediterranean	Asparagus Lily
Asparagus	*officinalis*	*strictus*	Erect	Mediterranean	Asparagus Lily
Asparagus	*officinalis*	*suttonii-gigantea*	Sutton Giant	Mediterranean	Asparagus Lily

Order 6 Asteliales Sword Grass
- Family 1 *Asteliaceae* Sword Grass
- Family 2 *Asphodelaceae* Asphodel
- Family 3 *Dasypogonaceae* Beard Grass Lily
- Family 4 *Hyacinthaceae* Hyacinth
- Family 5 *Philesiaceae* Philesian Lily
- Family 6 *Xanthorrhoeaceae* Skirt Grass
- Family 7 *Xeronemataceae* Flax Grass

Order 7 *Bromeliales* Bromelaid
- Family 1 *Bromeliaceae* Bromelaid
- Family 2 *Haemodoraceae* Redroot
- Family 3 *Pontederiaceae* Pickerelweed

Order 722.018 *Dioscoreales* Yam

Latin: *Dioscorea* Dioscorides; a 1st century Greek physician

FAMILIES

- **722.018.1 *Dioscoreaceae*** **Yam**
- 722.018.2 *Burmanniaceae* Blue Thread Lily
- 722.018.3 *Convallariaceae* Lily Of The Valley
- 722.018.4 *Dracaenaceae* Dragon Lily
- 722.018.5 *Gloriosaceae* Glory Vine
- 722.018.6 *Nartheciaceae* Rod Lily
- 722.018.7 *Ruscaceae* Broom
- 722.018.8 *Uvulariaceae* Merry Bells Lily

Family 722.018.1 *Dioscoreaceae* Yam

Genus	Species	Specimen	Genome	C:Culinary, M:Medicinal, E:Economic
Dioscorea	*alata*	Greater	Yam	**C**:TU:**VG**
Dioscorea	*balcanica*	Balkan	Yam	**C**:TU:**VG**
Dioscorea	*batatas*			Synonym for *Dioscorea polystachya*
Dioscorea	*bulbifera*	Aerial	Yam	**C**:TU:**VG**
Dioscorea	*cayennensis*	Cayenne	Yam	**C**:TU:**VG**
Dioscorea	*communis*	Black Bryony	Yam	**C**:TU:**VG**
Dioscorea	*composita*	Tepinapa	Yam	**C**:TU:**VG**(Wyk)

Dioscorea	*deltoidea*	Nepal	Yam	C:TU:**VG**(Wyk)
Dioscorea	*elephantipes*	Elephant's Foot	Yam	C:TU:**VG**
Dioscorea	*esculenta*	Lesser	Yam	C:TU:**VG**
Dioscorea	*floribunda*	Abundant Flower	Yam	C:TU:**VG**(Wyk)
Dioscorea	*japonica*	Asian Mountain	Yam	C:TU:**VG**
Dioscorea	*mexicana*	Mexican	Yam	C:TU:**VG**(Wyk)
Dioscorea	*oppositifolia*	Indochinese	Yam	C:TU:**VG**
Dioscorea	*polystachya*	Cinnamon Vine	Yam	C:TU:**VG**
Dioscorea	*rotundata*			Synonym for *Discorea cayennensis*
Dioscorea	*trifida*	Cush Cush	Yam	C:TU:**VG**
Dioscorea	*villosa*	Wild	Yam	C:TU:**VG**(Wyk)

Varieties

Yam (*Dioscorea spp.*)

Genus	species	infraspecies	Cultivated Variety Cultivar	Specimen	Genome
Dioscorea	*alata*	*alata*	Original	Greater	Yam
Dioscorea	*alata*	*atropurpurea*	Dark Purple	Greater	Yam
Dioscorea	*alata*	*colocasiifolia*	Taro Root	Greater	Yam
Dioscorea	*alata*	*eburnea*	Ivory	Greater	Yam
Dioscorea	*alata*	*globosa*	Globe	Greater	Yam
Dioscorea	*alata*	*javanica*	Java	Greater	Yam
Dioscorea	*alata*	*purpurea*	Purple	Greater	Yam
Dioscorea	*alata*	*rubella*	Red	Greater	Yam
Dioscorea	*alata*	*sapinii*	Sapin	Greater	Yam
Dioscorea	*alata*	*tarri*	Tarr	Greater	Yam
Dioscorea	*alata*	*vera*	Vera	Greater	Yam
Dioscorea	*bulbifera*	*bulbifera*	Original	Aerial	Yam
Dioscorea	*bulbifera*	*anthropophagorum*	Ape	Aerial	Yam
Dioscorea	*bulbifera*	*crispata*	Curly	Aerial	Yam
Dioscorea	*bulbifera*	*elongata*	Elongated	Aerial	Yam
Dioscorea	*bulbifera*	*heterophylla*	Variable Leaf	Aerial	Yam
Dioscorea	*bulbifera*	*hoffa*	Hoff	Aerial	Yam
Dioscorea	*bulbifera*	*korrorensis*	Korror	Aerial	Yam
Dioscorea	*bulbifera*	*latifolia*	Broad Leaf	Aerial	Yam
Dioscorea	*bulbifera*	*longipetiolata*	Long Petiole	Aerial	Yam
Dioscorea	*bulbifera*	*perrieri*	Perrier	Aerial	Yam
Dioscorea	*bulbifera*	*pulchella*	Pretty	Aerial	Yam
Dioscorea	*bulbifera*	*rogersii*	Rogers	Aerial	Yam
Dioscorea	*bulbifera*	*rotunda*	Round Leaf	Aerial	Yam
Dioscorea	*bulbifera*	*sativa*	Sweet	Aerial	Yam
Dioscorea	*bulbifera*	*suavia*	Sugary	Aerial	Yam
Dioscorea	*bulbifera*	*sylvestris*	Forest	Aerial	Yam
Dioscorea	*bulbifera*	*tamifolia*	Bryony Leaf	Aerial	Yam
Dioscorea	*bulbifera*	*tenuiflora*	Tenuous Leaf	Aerial	Yam
Dioscorea	*bulbifera*	*vera*	Vera	Aerial	Yam
Dioscorea	*bulbifera*	*violacea*	Violet	Aerial	Yam
Dioscorea	*cayennensis*	*cayennensis*	Original	Cayenne	Yam
Dioscorea	*cayennensis*	*aculeata*	Prickly	Cayenne	Yam
Dioscorea	*cayennensis*	*berteroana*	Bertero	Cayenne	Yam
Dioscorea	*cayennensis*	*camerunensis*	Cameroon	Cayenne	Yam
Dioscorea	*cayennensis*	*moma*	Mum	Cayenne	Yam
Dioscorea	*cayennensis*	*occidentalis*	Occidental	Cayenne	Yam
Dioscorea	*cayennensis*	*pruinosa*	Plum Leaf	Cayenne	Yam
Dioscorea	*cayennensis*	*rotundata*	Round	Cayenne	Yam
Dioscorea	*communis*	*communis*	Original	Black Bryony	Yam
Dioscorea	*communis*	*baccifera*	Berry	Black Bryony	Yam
Dioscorea	*communis*	*canariensis*	Canary Island	Black Bryony	Yam
Dioscorea	*communis*	*cirrhosa*	Tendril	Black Bryony	Yam
Dioscorea	*communis*	*cordifolia*	Heart Leaf	Black Bryony	Yam
Dioscorea	*communis*	*cretica*	Crete	Black Bryony	Yam
Dioscorea	*communis*	*edulis*	Edible	Black Bryony	Yam
Dioscorea	*communis*	*norsa*	Norse	Black Bryony	Yam
Dioscorea	*communis*	*parviflora*	Small Flower	Black Bryony	Yam
Dioscorea	*communis*	*racemosa*	Raceme	Black Bryony	Yam
Dioscorea	*communis*	*subtriloba*	Lesser Three Lobe	Black Bryony	Yam
Dioscorea	*communis*	*triloba*	Three Lobe	Black Bryony	Yam
Dioscorea	*elephantipes*	*elephantipes*	Original	Elephant's Foot	Yam
Dioscorea	*elephantipes*	*montana*	Mountain	Elephant's Foot	Yam
Dioscorea	*elephantipes*	*testudinaria*	Tortoise Shell	Elephant's Foot	Yam
Dioscorea	*esculenta*	*esculenta*	Original	Lesser	Yam
Dioscorea	*esculenta*	*fasciculata*	Fascicle	Lesser	Yam
Dioscorea	*esculenta*	*fulvidotomentosa*	Fulvous Woolly	Lesser	Yam
Dioscorea	*esculenta*	*papillaris*	Papilla	Lesser	Yam
Dioscorea	*esculenta*	*papuana*	Papua	Lesser	Yam
Dioscorea	*esculenta*	*spinosa*	Spiny	Lesser	Yam
Dioscorea	*esculenta*	*tiliifolia*	Linden Leaf	Lesser	Yam
Dioscorea	*esculenta*	*tugui*	Tugu	Lesser	Yam

Dioscorea	*japonica*	*japonica*	Original	Asian Mountain Yam
Dioscorea	*japonica*	*belophylloides*	Bell Leaf	Asian Mountain Yam
Dioscorea	*japonica*	*fauriei*	Faurie	Asian Mountain Yam
Dioscorea	*japonica*	*goeringiana*	Goering	Asian Mountain Yam
Dioscorea	*japonica*	*kelungensis*	Kelung	Asian Mountain Yam
Dioscorea	*japonica*	*kiangsiensis*	Kiangsi	Asian Mountain Yam
Dioscorea	*japonica*	*nagarum*	Nagar	Asian Mountain Yam
Dioscorea	*japonica*	*neglecta*	Neglected	Asian Mountain Yam
Dioscorea	*japonica*	*oldhamii*	Oldham	Asian Mountain Yam
Dioscorea	*japonica*	*pilifera*	Downy	Asian Mountain Yam
Dioscorea	*japonica*	*pseudojaponica*	Lesser	Asian Mountain Yam
Dioscorea	*japonica*	*vera*	Vera	Asian Mountain Yam
Dioscorea	*mexicana*	*mexicana*	Original	Mexican Yam
Dioscorea	*mexicana*	*anconensis*	Ancona	Mexican Yam
Dioscorea	*mexicana*	*astrostigma*	Star Stigma	Mexican Yam
Dioscorea	*mexicana*	*bilbergiana*	Bilberg	Mexican Yam
Dioscorea	*mexicana*	*cocolmeca*	Cocolmec	Mexican Yam
Dioscorea	*mexicana*	*deamii*	Deam	Mexican Yam
Dioscorea	*mexicana*	*deppei*	Deppe	Mexican Yam
Dioscorea	*mexicana*	*leiboldiana*	Leibold	Mexican Yam
Dioscorea	*mexicana*	*macrophylla*	Large Leaf	Mexican Yam
Dioscorea	*mexicana*	*macrostachya*	Large Stack	Mexican Yam
Dioscorea	*mexicana*	*propinqua*	Related	Mexican Yam
Dioscorea	*mexicana*	*sessiliflora*	Leaf Flower	Mexican Yam
Dioscorea	*mexicana*	*tuerckheimii*	Tuerckheim	Mexican Yam
Dioscorea	*oppositifolia*	*oppositifolia*	Original	Indochinese Yam
Dioscorea	*oppositifolia*	*dukhunensis*	Dukhun	Indochinese Yam
Dioscorea	*oppositifolia*	*linnaei*	Linnaeus	Indochinese Yam
Dioscorea	*oppositifolia*	*thwaitesii*	Thwaites	Indochinese Yam
Dioscorea	*polystachya*	*polystachya*	Original	Cinnamon Vine Yam
Dioscorea	*polystachya*	*clavata*	Collar	Cinnamon Vine Yam
Dioscorea	*polystachya*	*daikok*	Daikok	Cinnamon Vine Yam
Dioscorea	*polystachya*	*decaisneana*	Decaisne	Cinnamon Vine Yam
Dioscorea	*polystachya*	*doryphora*	Dory	Cinnamon Vine Yam
Dioscorea	*polystachya*	*flabellata*	Fan Leaf	Cinnamon Vine Yam
Dioscorea	*polystachya*	*potaninii*	Potanin	Cinnamon Vine Yam
Dioscorea	*polystachya*	*rakuda*	Rakuda	Cinnamon Vine Yam
Dioscorea	*polystachya*	*rosthornii*	Rosthorn	Cinnamon Vine Yam
Dioscorea	*polystachya*	*swinhoei*	Swinhoe	Cinnamon Vine Yam
Dioscorea	*polystachya*	*tsukune*	Tsukune	Cinnamon Vine Yam
Dioscorea	*trifida*	*trifida*	Original	Cush Cush Yam
Dioscorea	*trifida*	*affinis*	Affine	Cush Cush Yam
Dioscorea	*trifida*	*angustifolia*	Narrow Leaf	Cush Cush Yam
Dioscorea	*trifida*	*articulata*	Articulated	Cush Cush Yam
Dioscorea	*trifida*	*brasiliensis*	Brazil	Cush Cush Yam
Dioscorea	*trifida*	*goyazensis*	Goyaz	Cush Cush Yam
Dioscorea	*trifida*	*palmata*	Palm Leaf	Cush Cush Yam
Dioscorea	*trifida*	*quinquelobata*	Five Lobe	Cush Cush Yam
Dioscorea	*trifida*	*ruiziana*	Ruiz	Cush Cush Yam
Dioscorea	*villosa*	*villosa*	Original	Wild Yam
Dioscorea	*villosa*	*cliffortiana*	Cliffort	Wild Yam
Dioscorea	*villosa*	*glabra*	Smooth	Wild Yam
Dioscorea	*villosa*	*glabrifolia*	Smooth Leaf	Wild Yam
Dioscorea	*villosa*	*glauca*	Yam	
Dioscorea	*villosa*	*hexaphylla*	Six Leaf	Wild Yam
Dioscorea	*villosa*	*hirticaulis*	Hairy Stalk	Wild Yam
Dioscorea	*villosa*	*laeviuscula*	Very Smooth	Wild Yam
Dioscorea	*villosa*	*lloydiana*	Lloyd	Wild Yam
Dioscorea	*villosa*	*longifolia*	Long Leaf	Wild Yam
Dioscorea	*villosa*	*megaptera*	Large Wing	Wild Yam
Dioscorea	*villosa*	*paniculata*	Panicle	Wild Yam
Dioscorea	*villosa*	*pruinosa*	Plum Leaf	Wild Yam
Dioscorea	*villosa*	*quaternata*	Four Leaf	Wild Yam
Dioscorea	*villosa*	*quinata*	Five Leaf	Wild Yam
Dioscorea	*villosa*	*repanda*	Wavy Margin	Wild Yam
Dioscorea	*villosa*	*sativa*	Sweet	Wild Yam
Dioscorea	*villosa*	*waltheri*	Walther	Wild Yam

Order 722.019 *Zingiberales* Ginger

Latin: *Zingiber* Ginger;

FAMILIES

722.019.1 *Zingiberaceae* Ginger
722.019.2 *Cannaceae* Cane Ginger
722.019.3 *Costaceae* Cost Ginger
722.019.4 *Heliconiaceae* Bird Of Paradise
722.019.5 *Lowiaceae* Low Ginger

722.019.6 *Marantaceae* Marant Ginger
722.019.7 *Musaceae* Banana
722.019.8 *Strelitziaceae* Strelitz Ginger

Family 722.019.1 *Zingiberaceae* Ginger

Genus	Species	Specimen	Genome	C:Culinary, M:Medicinal, E:Economic
Zingiber	*cassumunar*			Synonym for *Zingiber montanum*
Zingiber	*mioga*	Mioga	Ginger	C:TU:**VG**
Zingiber	*montanum*	Mountain	Ginger	C:TU:**VG**
Zingiber	*officinale*	South Asia	Ginger	C:TU:**VG**
Zingiber	*zerumbet*	Zerumbet	Ginger	C:TU:**VG**
Aframomum	*angustifolium*	Madagascar	African Cardamom	C:SD(grain):**VG**
Aframomum	*melegueta*	Guinea Grains	African Cardamom	C:SD(grain):**VG**
Alpinia	*galanga*	Galangal	Alpino Cardamom	C:RZ:**VG**
Alpinia	*officinarum*	Lesser Galangal	Alpino Cardamom	C:RZ:**VG**
Alpinia	*oxyphylla*	Black	Alpino Cardamom	C:RZ:**VG**
Amomum	*villosum*	Tavoy	Bastard Cardamom	C:SD(grain):**VG**
Elettaria	*brachycalyx*	Globe	Cardamom	C:SD(grain):**VG**
Elettaria	*cardamomum*	India	Cardamom	C:SD(grain):**VG**
Elettaria	*ensal*	Ensal	Cardamom	C:SD(grain):**VG**
Elettaria	*kapitensis*	Kapita	Cardamom	C:SD(grain):**VG**
Elettaria	*inearicrista*	Tassel	Cardamom	C:SD(grain):**VG**
Elettaria	*longipilosa*	Long Pilose	Cardamom	C:SD(grain):**VG**
Elettaria	*longituba*	Long Tube	Cardamom	C:SD(grain):**VG**
Elettaria	*multiflora*	Multiflower	Cardamom	C:SD(grain):**VG**
Elettaria	*rubida*	Ruby	Cardamom	C:SD(grain):**VG**
Elettaria	*stoloniflora*	Stolon Flower	Cardamom	C:SD(grain):**VG**
Elettaria	*surculosa*	Sucker Shoot	Cardamom	C:SD(grain):**VG**

Varieties

Ginger (*Zingiber officinale*)

Genus	species	infraspecies	Cultivated Variety	Specimen	Genome
Zingiber	*mioga*	*mioga*	Original	Mioga	Ginger
Zingiber	*mioga*	*echuanense*	Echuan	Mioga	Ginger
Zingiber	*mioga*	*mijooka*	Mijooka	Mioga	Ginger
Zingiber	*mioga*	*variegatum*	Variegated	Mioga	Ginger
Zingiber	*montanum*	*montanum*	Original	Mountain	Ginger
Zingiber	*montanum*	*anthorrhiza*	Flower Rhizome	Mountain	Ginger
Zingiber	*montanum*	*cassumunar*	Cassum	Mountain	Ginger
Zingiber	*montanum*	*luridum*	Lurid	Mountain	Ginger
Zingiber	*montanum*	*palamauense*	Palamau	Mountain	Ginger
Zingiber	*montanum*	*purpureum*	Purple	Mountain	Ginger
Zingiber	*montanum*	*subglabrum*	Smooth	Mountain	Ginger
Zingiber	*officinale*	*officinale*	Original	South Asia	Ginger
Zingiber	*officinale*	*aromaticum*	Aromatic	South Asia	Ginger
Zingiber	*officinale*	*cholmondeleyi*	Cholmondeley	South Asia	Ginger
Zingiber	*officinale*	*macrorhizonum*	Large Rhizome	South Asia	Ginger
Zingiber	*officinale*	*missionis*	Mission	South Asia	Ginger
Zingiber	*officinale*	*rubens*	Reddish	South Asia	Ginger
Zingiber	*officinale*	*rubrum*	Red	South Asia	Ginger
Zingiber	*officinale*	*sichuanense*	Sichuan	South Asia	Ginger
Zingiber	*zerumbet*	*zerumbet*	Original	Zerumbet	Ginger
Zingiber	*zerumbet*	*amaricans*	Amar	Zerumbet	Ginger
Zingiber	*zerumbet*	*aromaticum*	Aromatic	Zerumbet	Ginger
Zingiber	*zerumbet*	*blancoi*	Blanco	Zerumbet	Ginger
Zingiber	*zerumbet*	*darceyi*	Darcey	Zerumbet	Ginger
Zingiber	*zerumbet*	*littoralis*	Seaside	Zerumbet	Ginger
Zingiber	*zerumbet*	*magnum*	Magnum	Zerumbet	Ginger
Zingiber	*zerumbet*	*sylvestre*	Forest	Zerumbet	Ginger
Zingiber	*zerumbet*	*truncatum*	Truncated	Zerumbet	Ginger
Zingiber	*zerumbet*	*valenzuelae*	Valenzuel	Zerumbet	Ginger

African Cardamom (*Aframomum angustifolium, melegueta*)

Aframomum	*angustifolium*	*angustifolium*	.	East Africa Spice	African Cardamom
Aframomum	*angustifolium*	*baumannii*	Baumann	East Africa Spice	African Cardamom
Aframomum	*angustifolium*	*clusii*	Clus	East Africa Spice	African Cardamom
Aframomum	*angustifolium*	*erythrocarpum*	Red Pod	East Africa Spice	African Cardamom
Aframomum	*angustifolium*	*hanburyi*	Hanbury	East Africa Spice	African Cardamom
Aframomum	*angustifolium*	*madagascariense*	Madagascar	East Africa Spice	African Cardamom
Aframomum	*angustifolium*	*melegueta*	Guinea Grains	East Africa Spice	African Cardamom
Aframomum	*angustifolium*	*nemorosum*	Coin Cap	East Africa Spice	African Cardamom
Aframomum	*angustifolium*	*paludosa*	Marsh	East Africa Spice	African Cardamom
Aframomum	*angustifolium*	*sanguineum*	Blood Red	East Africa Spice	African Cardamom

Aframomum	*angustifolium*	*sansibaricum*	Zanzibar	East Africa Spice	African Cardamom
Aframomum	*melegueta*	*melegueta*	Original	West Africa Spice	African Cardamom
Aframomum	*melegueta*	*grandiflorum*	Grand Flower	West Africa Spice	African Cardamom
Aframomum	*melegueta*	*grana-paradisi*	Guinea Grains	West Africa Spice	African Cardamom
Aframomum	*melegueta*	*violacea*	Violet	West Africa Spice	African Cardamom
Alpinia	*galanga*	*galanga*	Original	Galangal	Alpino Ginger
Alpinia	*galanga*	*alba*	White	Galangal	Alpino Ginger
Alpinia	*galanga*	*bifida*	Split	Galangal	Alpino Ginger
Alpinia	*galanga*	*carnea*	Fleshy	Galangal	Alpino Ginger
Alpinia	*galanga*	*pyramidata*	Pyramid	Galangal	Alpino Ginger
Alpinia	*galanga*	*rheedei*	Rheed	Galangal	Alpino Ginger
Alpinia	*galanga*	*viridiflora*	Green Flower	Galangal	Alpino Ginger

Tavoy Bastard Cardamom (*Amomum villosum*)

Amomum	*villosum*	*villosum*	Original	Tavoy	Bastard Cardamom
Amomum	*villosum*	*echinosphaera*	Urchin Seed	Tavoy	Bastard Cardamom
Amomum	*villosum*	*nanum*	Dwarf	Tavoy	Bastard Cardamom
Amomum	*villosum*	*xanthioides*	Grains of Paradise	Tavoy	Bastard Cardamom

India Cardamom (*Elettaria cardamomum*)

Elettaria	*cardamomum*	*cardamomum*	Original	India	Cardamom
Elettaria	*cardamomum*	*ensal*	Ensal	India	Cardamom
Elettaria	*cardamomum*	*malabaricum*	Malabar	India	Cardamom
Elettaria	*cardamomum*	*minus*	Lesser	India	Cardamom
Elettaria	*cardamomum*	*racemosum*	Raceme	India	Cardamom
Elettaria	*cardamomum*	*minuscula*	Miniscule	India	Cardamom
Elettaria	*cardamomum*	*repens*	Creeping	India	Cardamom
Elettaria	*cardamomum*	*uncinatum*	Hooked	India	Cardamom
Elettaria	*ensal*	*ensal*	Original	Ensal	Cardamom
Elettaria	*ensal*	*major*	Greater	Ensal	Cardamom

Class 722.02 *Commelinopsida* Grass

Order 722.021 *Commelinales* Teret Grass

Latin: *Comma* concerning
Latin: *Linales* line grass

FAMILIES

722.021.1	*Commelinaceae*	Teret Grass
722.021.2	*Cyperaceae*	Sedge Grass
722.021.3	*Eriocaulaceae*	Woolly Stem Grass
722.021.4	*Hanguanaceae*	Hanguan Grass
722.021.5	*Juncaceae*	Rush Grass
722.021.6	*Mayacaceae*	Bog Moss Grass
722.021.7	*Philydraceae*	Philydran Grass
722.021.8	*Xyridaceae*	Yellow Eyed Grass

Family 722.021.2 *Cyperaceae* Sedge Grass

Genus	Species	Specimen	Genome C:Culinary, M:Medicinal
Cyperus	*articulatus*	Ground Almond	Sedge Grass **C**:RT,RZ,TU:**VG**
Cyperus	*esculentus*	Chufa	Sedge Grass **C**:RT,RZ,TU:**VG**
Cyperus	*longus*	Sweet Galingale	Sedge Grass **C**:RT,RZ,TU:**VG**
Cyperus	*rotundus*	Coco Root	Sedge Grass **C**:RT,RZ,TU:**VG**

Varieties

Ground Almond (*Cyperus articulatus*)

Genus	species	infraspecies	Cultivated Variety	Specimen	Genome
Cyperus	*articulatus*	*articulatus*	Original	Ground Almond	Sedge Grass
Cyperus	*articulatus*	*aphyllus*	Leafless	Ground Almond	Sedge Grass
Cyperus	*articulatus*	*autumnalis*	Autumn	Ground Almond	Sedge Grass
Cyperus	*articulatus*	*borbonicus*	Borbon	Ground Almond	Sedge Grass
Cyperus	*articulatus*	*conglomeratus*	Clustered	Ground Almond	Sedge Grass
Cyperus	*articulatus*	*cordobensis*	Cordoba	Ground Almond	Sedge Grass
Cyperus	*articulatus*	*erythrostachys*	Red Spike	Ground Almond	Sedge Grass
Cyperus	*articulatus*	*fistulosus*	Hollow Stem	Ground Almond	Sedge Grass
Cyperus	*articulatus*	*gymnos*	Naked	Ground Almond	Sedge Grass
Cyperus	*articulatus*	*interceptus*	Intercepting	Ground Almond	Sedge Grass
Cyperus	*articulatus*	*longispiculosus*	Long Spicule	Ground Almond	Sedge Grass
Cyperus	*articulatus*	*multiflorus*	Many Flower	Ground Almond	Sedge Grass
Cyperus	*articulatus*	*niloticus*	Nile	Ground Almond	Sedge Grass
Cyperus	*articulatus*	*nodosus*	Node	Ground Almond	Sedge Grass
Cyperus	*articulatus*	*pertenuis*	Pertenu	Ground Almond	Sedge Grass

Chufa (*Cyperus esculentus*)

Genus	species	infraspecies	Cultivated Variety	Specimen	Genome
Cyperus	*esculentus*	*esculentus*	Original	Chufa	Sedge Grass
Cyperus	*esculentus*	*angustispicatus*	Narrow Spike	Chufa	Sedge Grass
Cyperus	*esculentus*	*aureus*	Golden	Chufa	Sedge Grass
Cyperus	*esculentus*	*bahiensis*	Bahia	Chufa	Sedge Grass
Cyperus	*esculentus*	*buchananii*	Buchanan	Chufa	Sedge Grass
Cyperus	*esculentus*	*callistus*	Callisto	Chufa	Sedge Grass
Cyperus	*esculentus*	*chrysostachys*	Golden Spike	Chufa	Sedge Grass
Cyperus	*esculentus*	*cubensis*	Cuban	Chufa	Sedge Grass
Cyperus	*esculentus*	*cyclolepis*	Cyclic Thin	Chufa	Sedge Grass
Cyperus	*esculentus*	*damiettensis*	Damiette	Chufa	Sedge Grass
Cyperus	*esculentus*	*evolutus*	Untwined	Chufa	Sedge Grass
Cyperus	*esculentus*	*fresenii*	Fresen	Chufa	Sedge Grass
Cyperus	*esculentus*	*fulvescens*	Fulvous	Chufa	Sedge Grass
Cyperus	*esculentus*	*gracilescens*	Graceful	Chufa	Sedge Grass
Cyperus	*esculentus*	*gracilis*	Slender	Chufa	Sedge Grass
Cyperus	*esculentus*	*heermannii*	Heermann	Chufa	Sedge Grass
Cyperus	*esculentus*	*helodes*	Bog	Chufa	Sedge Grass
Cyperus	*esculentus*	*hydra*	Hydroponic	Chufa	Sedge Grass
Cyperus	*esculentus*	*leptostachyus*	Thin Spike	Chufa	Sedge Grass
Cyperus	*esculentus*	*lutescens*	Yellow	Chufa	Sedge Grass
Cyperus	*esculentus*	*macrostachyus*	Large Spike	Chufa	Sedge Grass
Cyperus	*esculentus*	*melanorhizus*	Dark Rhizome	Chufa	Sedge Grass
Cyperus	*esculentus*	*nervosus*	Nerve	Chufa	Sedge Grass
Cyperus	*esculentus*	*pallidus*	Pale	Chufa	Sedge Grass
Cyperus	*esculentus*	*phymatodes*	Tumor	Chufa	Sedge Grass
Cyperus	*esculentus*	*princeps*	Prince	Chufa	Sedge Grass
Cyperus	*esculentus*	*repens*	Creeping	Chufa	Sedge Grass

Cyperus	*esculentus*	*ruficomus*	Rufous	Chufa	Sedge Grass
Cyperus	*esculentus*	*sativus*	Sweet	Chufa	Sedge Grass
Cyperus	*esculentus*	*scirpoides*	Club Rush	Chufa	Sedge Grass
Cyperus	*esculentus*	*sieberianus*	Sieber	Chufa	Sedge Grass
Cyperus	*esculentus*	*sprucei*	Spruce	Chufa	Sedge Grass
Cyperus	*esculentus*	*tenorianus*	Tenor	Chufa	Sedge Grass
Cyperus	*esculentus*	*tuberosus*	Tuber	Chufa	Sedge Grass
Cyperus	*esculentus*	*variabilis*	Variable	Chufa	Sedge Grass

Sweet Galingale (*Cyperus longus*)

Genus	species	infraspecies	Cultivated Variety	Specimen	Genome
Cyperus	*longus*	*longus*	Original	Sweet Galingale	Sedge Grass
Cyperus	*longus*	*auratus*	Eared	Sweet Galingale	Sedge Grass
Cyperus	*longus*	*badius*	Bay	Sweet Galingale	Sedge Grass
Cyperus	*longus*	*difformis*	Difformed	Sweet Galingale	Sedge Grass
Cyperus	*longus*	*elongatus*	Elongated	Sweet Galingale	Sedge Grass
Cyperus	*longus*	*fenzelianus*	Fenzel	Sweet Galingale	Sedge Grass
Cyperus	*longus*	*guthnickii*	Guthnick	Sweet Galingale	Sedge Grass
Cyperus	*longus*	*heldreichianus*	Heldreich	Sweet Galingale	Sedge Grass
Cyperus	*longus*	*intermedius*	Intermediate	Sweet Galingale	Sedge Grass
Cyperus	*longus*	*longispiculatus*	Long Spicule	Sweet Galingale	Sedge Grass
Cyperus	*longus*	*lucanus*	Lucan	Sweet Galingale	Sedge Grass
Cyperus	*longus*	*myriostachyus*	Myriad Spike	Sweet Galingale	Sedge Grass
Cyperus	*longus*	*pallidus*	Pale	Sweet Galingale	Sedge Grass
Cyperus	*longus*	*parlatoris*	Parlator	Sweet Galingale	Sedge Grass
Cyperus	*longus*	*speciosus*	Showy	Sweet Galingale	Sedge Grass
Cyperus	*longus*	*speciosus*	Showy	Sweet Galingale	Sedge Grass

Coco Root (*Cyperus rotundus*)

Genus	species	infraspecies	Cultivated Variety	Specimen	Genome
Cyperus	*rotundus*	*rotundus*	Original	Coco Root	Sedge Grass
Cyperus	*rotundus*	*acutus*	Sharp	Coco Root	Sedge Grass
Cyperus	*rotundus*	*amaliae*	Amalie	Coco Root	Sedge Grass
Cyperus	*rotundus*	*bifax*	Cleft	Coco Root	Sedge Grass
Cyperus	*rotundus*	*brevibracteatus*	Short Bract	Coco Root	Sedge Grass
Cyperus	*rotundus*	*carinalis*	Keel	Coco Root	Sedge Grass
Cyperus	*rotundus*	*carinatus*	Keeled	Coco Root	Sedge Grass
Cyperus	*rotundus*	*centiflorus*	Many Flower	Coco Root	Sedge Grass
Cyperus	*rotundus*	*comosus*	Tufted	Coco Root	Sedge Grass
Cyperus	*rotundus*	*contractus*	Contracted	Coco Root	Sedge Grass
Cyperus	*rotundus*	*depallescens*	Pallid	Coco Root	Sedge Grass
Cyperus	*rotundus*	*disruptus*	Disrupted	Coco Root	Sedge Grass
Cyperus	*rotundus*	*divaricatus*	Spreading	Coco Root	Sedge Grass
Cyperus	*rotundus*	*elongatus*	Elongated	Coco Root	Sedge Grass
Cyperus	*rotundus*	*hydra*	Hydroponic	Coco Root	Sedge Grass
Cyperus	*rotundus*	*inconspicuus*	Inconspicuous	Coco Root	Sedge Grass
Cyperus	*rotundus*	*latifolius*	Broad Leaf	Coco Root	Sedge Grass
Cyperus	*rotundus*	*latimarginatus*	Wide Margined	Coco Root	Sedge Grass
Cyperus	*rotundus*	*macrostachyus*	Large Spike	Coco Root	Sedge Grass
Cyperus	*rotundus*	*major*	Greater	Coco Root	Sedge Grass
Cyperus	*rotundus*	*merkeri*	Merker	Coco Root	Sedge Grass
Cyperus	*rotundus*	*nubicus*	Nubian	Coco Root	Sedge Grass
Cyperus	*rotundus*	*pallidus*	Pale	Coco Root	Sedge Grass
Cyperus	*rotundus*	*platystachys*	Flat Spike	Coco Root	Sedge Grass
Cyperus	*rotundus*	*procerula*	Tall	Coco Root	Sedge Grass
Cyperus	*rotundus*	*retzii*	Retz	Coco Root	Sedge Grass
Cyperus	*rotundus*	*salsolus*	Thistle	Coco Root	Sedge Grass
Cyperus	*rotundus*	*spadiceus*	Chestnut Brown	Coco Root	Sedge Grass
Cyperus	*rotundus*	*subcapitatus*	Capped	Coco Root	Sedge Grass
Cyperus	*rotundus*	*taylorii*	Taylor	Coco Root	Sedge Grass
Cyperus	*rotundus*	*weinlandii*	Weinland	Coco Root	Sedge Grass

Order 722.022 *Graminales* Grain Grass

Latin: *Grama* grain

FAMILIES

722.022.1	*Avenaceae*	Oats
722.022.2	*Hordeaceae*	Barley
722.022.3	*Miliaceae*	Millet
722.022.4	*Oryzaceae*	Rice
722.022.5	*Secaleaceae*	Rye
722.022.6	*Tripsacaceae*	Gammagrass (Corn)
722.022.7	*Triticaceae*	Wheat

Family 722.022.1 *Avenaceae* Oats

Genus	Species	Specimen	Genome	C:Culinary, M:Medicinal, E:Economic
Avena	*abyssinica*	Ethiopian	Oats	C:SD(grain):**VG**
Avena	*aemulans*	Emulated	Oats	C:SD(grain):**VG**
Avena	*barbata*	Slender	Oats	C:SD(grain):**VG**
Avena	*brevis*	Short	Oats	C:SD(grain):**VG**
Avena	*byzantina*	Byzantine	Oats	C:SD(grain):**VG**
Avena	*chinensis*	Chinese	Oats	C:SD(grain):**VG**
Avena	*eriantha*	Woolly	Oats	C:SD(grain):**VG**
Avena	*fatua*	Flax Grass	Oats	C:SD(grain):**VG**
Avena	*longiglumis*	Long Glume	Oats	C:SD(grain):**VG**
Avena	*maroccana*	Moroccan	Oats	C:SD(grain):**VG**
Avena	*murphyi*	Murphy	Oats	C:SD(grain):**VG**
Avena	*nuda*	Naked	Oats	C:SD(grain):**VG**
Avena	*prostrata*	Prostrate	Oats	C:SD(grain):**VG**
Avena	*sativa*	**Cereal**	**Oats**	C:SD(grain):**VG**
Avena	*saxatilis*	Rock Dwelling	Oats	C:SD(grain):**VG**
Avena	*sterilis*	Animated	Oats	C:SD(grain):**VG**
Avena	*strigosa*	Lopsided	Oats	C:SD(grain):**VG**

Varieties

Oats (*Avena sativa*)

Genus	species	infraspecies	Cultivated Variety	Specimen	Genome
Avena	*sativa*	*byzantina*	Byzantine	Cereal	Oats
Avena	*sativa*	*hybrida*	Hybrid	Cereal	Oats
Avena	*sativa*	*nigra*	Black	Cereal	Oats
Avena	*sativa*	*orientalis*	Oriental	Cereal	Oats
Avena	*sativa*	*secunda*	Second	Cereal	Oats

Family 722.022.2 *Hordeaceae* Barley

Genus	Species	Specimen	Genome	C:Culinary, M:Medicinal, E:Economic
Hordeum	*arizonicum*	Arizona	Barley	C:SD(grain):**VG**
Hordeum	*brachyantherum*	Meadow	Barley	C:SD(grain):**VG**
Hordeum	*bulbosum*	Bulbous	Barley	C:SD(grain):**VG**
Hordeum	*depressum*	Dwarf	Barley	C:SD(grain):**VG**
Hordeum	*intercedens*	Bobtail	Barley	C:SD(grain):**VG**
Hordeum	*jubatum*	Foxtail	Barley	C:SD(grain):**VG**
Hordeum	*jubatum*	Foxtail	Barley	C:SD(grain):**VG**
Hordeum	*marinum*	Seaside	Barley	C:SD(grain):**VG**
Hordeum	*murinum*	Mouse	Barley	C:SD(grain):**VG**
Hordeum	*pusillum*	Puny	Barley	C:SD(grain):**VG**
Hordeum	*secalinum*	Lesser Cereal	Barley	C:SD(grain):**VG**
Hordeum	*spontaneum*	Spontaneous	Barley	C:SD(grain):**VG**
Hordeum	*vulgare*	**Cereal**	**Barley**	C:SD(grain):**VG**

Varieties

Barley (*Hordeum vulgare*)

Genus	species	infraspecies	cultivar	specimen	Genome	Vegetable
Hordeum	*vulgare*	*aegiceras*	Shieldhorn	Cereal	Barley	Grain
Hordeum	*vulgare*	*distichon*	Two Rank	Cereal	Barley	Grain
Hordeum	*vulgare*	*hexastichon*	Six Rank	Cereal	Barley	Grain
Hordeum	*vulgare*	*irregulare*	Irregular	Cereal	Barley	Grain
Hordeum	*vulgare*	*polystichum*	Irregular	Cereal	Barley	Grain
Hordeum	*vulgare*	*sativum*	Sweet	Cereal	Barley	Grain
Hordeum	*vulgare*	*trifurcatum*	Trident	Cereal	Barley	Grain

Family 722.022.3 *Miliaceae* Millet

Genus	Species	Specimen	Genome	C:Culinary, M:Medicinal, E:Economic
Milium	*amphicarpon*	Maidencane	Millet	C:SD(grain):**VG**
Milium	*effusum*	Woodland	Millet	C:SD(grain):**VG**
Milium	*glaucum*	Pearl	Millet	C:SD(grain):**VG**
Milium	*paniceum*	Broomcorn	Millet	C:SD(grain):**VG**
Milium	*pedicellare*	Pedicel	Millet	C:SD(grain):**VG**
Milium	*schmidtianum*	Schmidt	Millet	C:SD(grain):**VG**
Milium	*scrobiculatum*	Pitted Crowngrass	Millet	C:SD(grain):**VG**
Milium	*tenella*	Mediterranean	Millet	C:SD(grain):**VG**
Milium	*transcaucasicum*	Caucasian	Millet	C:SD(grain):**VG**
Milium	*vernale*	Spring	Millet	C:SD(grain):**VG**
Dactyloctenium	*aegyptium*	**African Millet**	Crow Foot Grass	C:SD(grain):**VG**
Echinochloa	*colona*	Jungle Ricegrass	Cock's Spur Grass	C:SD(grain):**VG**
Echinochloa	*crus-galli*	Barnyard Cock's Spur	Cock's Spur Grass	C:SD(grain):**VG**

Echinochloa	*crus-pavonis*	Barnyard Cock's Spur	Cock's Spur Grass	**C**:SD(grain):**VG**
Echinochloa	*esculenta*	**Japanese Millet**	Cock's Spur Grass	**C**:SD(grain):**VG**
Echinochloa	*frumentacea*	**Barnyard Millet**	Cock's Spur Grass	**C**:SD(grain):**VG**
Echinochloa	*muricata*	Rough Barnyard Millet	Cock's Spur Grass	**C**:SD(grain):**VG**
Echinochloa	*oryzoides*	Barnyard Ricegrass	Cock's Spur Grass	**C**:SD(grain):**VG**
Echinochloa	*turneriana*	Channel Millet	Cock's Spur Grass	**C**:SD(grain):**VG**
Echinochloa	*walteri*	Coastal Barnyard Millet	Cock's Spur Grass	**C**:SD(grain):**VG**
Eleusine	*africana*	African Millet	Goosegrass	**C**:SD(grain):**VG**
Eleusine	*coracana*	Ragi Finger Millet	Goosegrass	**C**:SD(grain):**VG**
Eleusine	*indica*	India	Goosegrass	**C**:SD(grain):**VG**
Eleusine	*major*	Great	Goosegrass	**C**:SD(grain):**VG**
Eleusine	*tristachya*	Three Spike	Goosegrass	**C**:SD(grain):**VG**
Eragrostis	*airoides*	Darnel	Lovegrass	**C**:SD(grain):**VG**
Eragrostis	*atropioides*	Hard Stem	Lovegrass	**C**:SD(grain):**VG**
Eragrostis	*atrovirens*	Thalian	Lovegrass	**C**:SD(grain):**VG**
Eragrostis	*bahiensis*	Bahia	Lovegrass	**C**:SD(grain):**VG**
Eragrostis	*barrelieri*	Barrelier	Lovegrass	**C**:SD(grain):**VG**
Eragrostis	*capillaris*	Lace	Lovegrass	**C**:SD(grain):**VG**
Eragrostis	*cilianensis*	Stink Grass	Lovegrass	**C**:SD(grain):**VG**
Eragrostis	*ciliaris*	Gopher Tail	Lovegrass	**C**:SD(grain):**VG**
Eragrostis	*cumingii*	Cuming	Lovegrass	**C**:SD(grain):**VG**
Eragrostis	*curtipedicellata*	Short Pedicel	Lovegrass	**C**:SD(grain):**VG**
Eragrostis	*curvula*	Weeping	Lovegrass	**C**:SD(grain):**VG**
Eragrostis	*cylindriflora*	Cylinder Flower	Lovegrass	**C**:SD(grain):**VG**
Eragrostis	*cyperoides*	Bristly	Lovegrass	**C**:SD(grain):**VG**
Eragrostis	*deflexa*	Pacific	Lovegrass	**C**:SD(grain):**VG**
Eragrostis	*echinochloidea*	African Urchin	Lovegrass	**C**:SD(grain):**VG**
Eragrostis	*elliottii*	Field	Lovegrass	**C**:SD(grain):**VG**
Eragrostis	*elongata*	Elongated	Lovegrass	**C**:SD(grain):**VG**
Eragrostis	*erosa*	Chihuahuan	Lovegrass	**C**:SD(grain):**VG**
Eragrostis	*fosbergii*	Fosberg	Lovegrass	**C**:SD(grain):**VG**
Eragrostis	*frankii*	Sandbar	Lovegrass	**C**:SD(grain):**VG**
Eragrostis	*gangetica*	Slim Flower	Lovegrass	**C**:SD(grain):**VG**
Eragrostis	*glomerata*	Cluster Pond	Lovegrass	**C**:SD(grain):**VG**
Eragrostis	*glutinosa*	Sticky	Lovegrass	**C**:SD(grain):**VG**
Eragrostis	*grandis*	Large Hawaiian	Lovegrass	**C**:SD(grain):**VG**
Eragrostis	*hirsuta*	Big Top	Lovegrass	**C**:SD(grain):**VG**
Eragrostis	*hypnoides*	Teal Creeping	Lovegrass	**C**:SD(grain):**VG**
Eragrostis	*intermedia*	Plains Millet	Lovegrass	**C**:SD(grain):**VG**
Eragrostis	*japonica*	Old World Pond	Lovegrass	**C**:SD(grain):**VG**
Eragrostis	*lehmanniana*	Lehmann	Lovegrass	**C**:SD(grain):**VG**
Eragrostis	*leptophylla*	Mountain	Lovegrass	**C**:SD(grain):**VG**
Eragrostis	*leptostachya*	Australian	Lovegrass	**C**:SD(grain):**VG**
Eragrostis	*lugens*	Mourning	Lovegrass	**C**:SD(grain):**VG**
Eragrostis	*lutescens*	Six Weeks	Lovegrass	**C**:SD(grain):**VG**
Eragrostis	*mauiensis*	Maui	Lovegrass	**C**:SD(grain):**VG**
Eragrostis	*mexicana*	Mexican	Lovegrass	**C**:SD(grain):**VG**
Eragrostis	*minor*	Little	Lovegrass	**C**:SD(grain):**VG**
Eragrostis	*monticola*	Kalamalo	Lovegrass	**C**:SD(grain):**VG**
Eragrostis	*obtusa*	Blunt	Lovegrass	**C**:SD(grain):**VG**
Eragrostis	*obtusiflora*	Alkali	Lovegrass	**C**:SD(grain):**VG**
Eragrostis	*palmeri*	Rio Grande	Lovegrass	**C**:SD(grain):**VG**
Eragrostis	*parviflora*	Small Flower	Lovegrass	**C**:SD(grain):**VG**
Eragrostis	*paupera*	Oahu	Lovegrass	**C**:SD(grain):**VG**
Eragrostis	*pectinacea*	Tufted	Lovegrass	**C**:SD(grain):**VG**
Eragrostis	*perplexa*	Perplexed	Lovegrass	**C**:SD(grain):**VG**
Eragrostis	*pilosa*	Downy	Lovegrass	**C**:SD(grain):**VG**
Eragrostis	*plana*	South African	Lovegrass	**C**:SD(grain):**VG**
Eragrostis	*prolifera*	Dominican	Lovegrass	**C**:SD(grain):**VG**
Eragrostis	*refracta*	Coastal Plain	Lovegrass	**C**:SD(grain):**VG**
Eragrostis	*reptans*	Creeping	Lovegrass	**C**:SD(grain):**VG**
Eragrostis	*scaligera*	Scaley	Lovegrass	**C**:SD(grain):**VG**
Eragrostis	*secundiflora*	Red	Lovegrass	**C**:SD(grain):**VG**
Eragrostis	*sessilispica*	Tumble	Lovegrass	**C**:SD(grain):**VG**
Eragrostis	*setifolia*	Bristle Leaf	Lovegrass	**C**:SD(grain):**VG**
Eragrostis	*silveana*	Silveus	Lovegrass	**C**:SD(grain):**VG**
Eragrostis	*spectabilis*	Purple	Lovegrass	**C**:SD(grain):**VG**
Eragrostis	*spicata*	Spike	Lovegrass	**C**:SD(grain):**VG**
Eragrostis	*superba*	Superb	Lovegrass	**C**:SD(grain):**VG**
Eragrostis	*swallenii*	Swallen	Lovegrass	**C**:SD(grain):**VG**
Eragrostis	*tef*	Teff	Lovegrass	**C**:SD(grain):**VG**
Eragrostis	*tracyi*	Sanibel Island	Lovegrass	**C**:SD(grain):**VG**
Eragrostis	*trichocolea*	Hairy Sheath	Lovegrass	**C**:SD(grain):**VG**
Eragrostis	*trichodes*	Sand	Lovegrass	**C**:SD(grain):**VG**
Eragrostis	*unioloides*	Chinese	Lovegrass	**C**:SD(grain):**VG**

Eragrostis	*urbaniana*	Urban	Lovegrass	C:SD(grain):**VG**
Eragrostis	*variabilis*	Kawelu	Lovegrass	C:SD(grain):**VG**
Eragrostis	*viscosa*	Sticky	Lovegrass	C:SD(grain):**VG**
Setaria	*faberi*	Chinese Millet	Foxtail Bristle Grass	C:SD(grain):**VG**
Setaria	*italica*	**Millet**	Foxtail Bristle Grass	C:SD(grain):**VG**
Sorghum	*almum*	Columbus Grass	Sorgograss	C:SD(grain):**VG**
Sorghum	*arundinaceum*	Shatter Cane	Sorgograss	C:SD(grain):**VG**
Sorghum	*bicolor*	**Millet**	Sorgograss	C:SD(grain):**VG**
Sorghum	*bicolor-caffrorum*	**Kaffir Millet**	Sorgograss	C:SD(grain):**VG**
Sorghum	*bicolor-durra*	**Durra Millet**	Sorgograss	C:SD(grain):**VG**
Sorghum	*bicolor-miliaceum*	**Great Millet**	Sorgograss	C:SD(grain):**VG**
Sorghum	*bicolor-saccharatum*	**Molasses**	Sorgograss	C:SD(grain):**VG**
Sorghum	*bicolor-technicum*	**Broomcorn**	Sorgograss	C:SD(grain):**VG**
Sorghum	*drummondii*	Chicken Corn	Sorgograss	C:SD(grain):**VG**
Sorghum	*halepense*	Johnson Grass	Sorgograss	C:SD(grain):**VG**
Sorghum	*propinquum*	Related Millet	Sorgograss	C:SD(grain):**VG**

Varieties

Millet (*Millium spp.*)

Genus	species	infraspecies	cultivar	specimen	Genome
Milium	*amphicarpon*	*amphicarpon*	Original	Maidencane	Millet
Milium	*amphicarpon*	*ciliatum*	Hairy	Maidencane	Millet
Milium	*amphicarpon*	*floridanum*	Florida	Maidencane	Millet
Milium	*amphicarpon*	*muehlenbergianum*	Goober Grass	Maidencane	Millet
Milium	*amphicarpon*	*purshii*	Blue	Maidencane	Millet
Milium	*effusum*	*effusum*	Original	Woodland	Millet
Milium	*effusum*	*adscendens*	Ascending	Woodland	Millet
Milium	*effusum*	*alpicola*	Alpine	Woodland	Millet
Milium	*effusum*	*aureum*	Bowle's Golden	Woodland	Millet
Milium	*effusum*	*candlestickii*	Candlestick	Woodland	Millet
Milium	*effusum*	*cisatlanticum*	Atlantic	Woodland	Millet
Milium	*effusum*	*coloratum*	Painted	Woodland	Millet
Milium	*effusum*	*confertum*	Crowded	Woodland	Millet
Milium	*effusum*	*dubium*	Dubious	Woodland	Millet
Milium	*effusum*	*elatius*	Tall	Woodland	Millet
Milium	*effusum*	*latifrons*	Broad Front	Woodland	Millet
Milium	*effusum*	*lerchenfeldianum*	Lerchenfeld	Woodland	Millet
Milium	*effusum*	*nepalense*	Nepal	Woodland	Millet
Milium	*effusum*	*subacaule*	Stemless	Woodland	Millet
Milium	*effusum*	*transsilvanicum*	Transylvania	Woodland	Millet
Milium	*effusum*	*variegatum*	Variegated	Woodland	Millet
Milium	*effusum*	*willdenowii*	Willdenow	Woodland	Millet
Milium	*glaucum*	*glaucum*	Original	Pearl	Millet
Milium	*glaucum*	*albicauda*	White Stem	Pearl	Millet
Milium	*glaucum*	*americanum*	American	Pearl	Millet
Milium	*glaucum*	*ancylochaete*	Bristle	Pearl	Millet
Milium	*glaucum*	*aureum*	Golden	Pearl	Millet
Milium	*glaucum*	*cereale*	Cereal	Pearl	Millet
Milium	*glaucum*	*cinereum*	Cinder	Pearl	Millet
Milium	*glaucum*	*deflexum*	Deflexed	Pearl	Millet
Milium	*glaucum*	*echinurus*	Urchin	Pearl	Millet
Milium	*glaucum*	*gambiense*	Gambian	Pearl	Millet
Milium	*glaucum*	*gibbosum*	Swollen	Pearl	Millet
Milium	*glaucum*	*giganteum*	Giant	Pearl	Millet
Milium	*glaucum*	*indicum*	India	Pearl	Millet
Milium	*glaucum*	*leonis*	Lion Tooth	Pearl	Millet
Milium	*glaucum*	*linnaei*	Linnaeus	Pearl	Millet
Milium	*glaucum*	*longipedunculata*	Long Peduncle	Pearl	Millet
Milium	*glaucum*	*macrostachyum*	Large Stack	Pearl	Millet
Milium	*glaucum*	*maiwa*	Maiwa	Pearl	Millet
Milium	*glaucum*	*malacochaete*	Soft Bristle	Pearl	Millet
Milium	*glaucum*	*megastachyum*	Giant Stack	Pearl	Millet
Milium	*glaucum*	*nigritarum*	Black	Pearl	Millet
Milium	*glaucum*	*plukenetii*	Plukenet	Pearl	Millet
Milium	*glaucum*	*pycnostachyum*	Dense Spike	Pearl	Millet
Milium	*glaucum*	*solitarium*	Solitary	Pearl	Millet
Milium	*glaucum*	*spicatum*	Spike	Pearl	Millet
Milium	*glaucum*	*typhoideum*	Cat Tail	Pearl	Millet
Milium	*glaucum*	*willdenowii*	Willdenow	Pearl	Millet
Milium	*paniceum*	.	.	Broomcorn	Millet
Milium	*paniceum*	*paniceum*	Original	Broomcorn	Millet
Milium	*paniceum*	*aerugineum*	Rusty	Broomcorn	Millet
Milium	*paniceum*	*agricola*	Farmland	Broomcorn	Millet
Milium	*paniceum*	*anthracinum*	Anthracine	Broomcorn	Millet
Milium	*paniceum*	*asperrimum*	Rough	Broomcorn	Millet
Milium	*paniceum*	*densepilosum*	Dense Downy	Broomcorn	Millet

Milium	*paniceum*	*digitaria*	Crab Grass	Broomcorn	Millet
Milium	*paniceum*	*esculentum*	Edible	Broomcorn	Millet
Milium	*paniceum*	*aquilum*	Eagle	Broomcorn	Millet
Milium	*paniceum*	*atrobrunneum*	Dark Brown	Broomcorn	Millet
Milium	*paniceum*	*coffeatum*	Coffe	Broomcorn	Millet
Milium	*paniceum*	*corsinum*	Corsine	Broomcorn	Millet
Milium	*paniceum*	*densobrunneum*	Dense Brown	Broomcorn	Millet
Milium	*paniceum*	*fasciculatum*	Browntop	Broomcorn	Millet
Milium	*paniceum*	*fuscum*	Fuscous	Broomcorn	Millet
Milium	*paniceum*	*glaucum*	Glaucous	Broomcorn	Millet
Milium	*paniceum*	*nicotianum*	Tobacco	Broomcorn	Millet
Milium	*paniceum*	*ruderale*	Ruderale	Broomcorn	Millet
Milium	*paniceum*	*spontaneum*	Spontaneous	Broomcorn	Millet
Milium	*paniceum*	*subvitellinotephrum*	Egg Yellow	Broomcorn	Millet
Milium	*paniceum*	*virescens*	Green	Broomcorn	Millet
Milium	*paniceum*	*repens*	Creeping	Broomcorn	Millet
Milium	*paniceum*	*ruderale*	Black Seeded	Broomcorn	Millet
Milium	*paniceum*	*sumatrense*	Sumatra	Broomcorn	Millet
Milium	*paniceum*	*texanum*	Texas	Broomcorn	Millet
Milium	*scrobiculatum*	*scrobiculatum*	Original	Pitted Crown Grass	Millet
Milium	*scrobiculatum*	*adelogaeum*	Adel	Pitted Crown Grass	Millet
Milium	*scrobiculatum*	*akoense*	Ako	Pitted Crown Grass	Millet
Milium	*scrobiculatum*	*amazonicum*	Amazon	Pitted Crown Grass	Millet
Milium	*scrobiculatum*	*auriculatum*	Ear	Pitted Crown Grass	Millet
Milium	*scrobiculatum*	*barbatum*	Beard	Pitted Crown Grass	Millet
Milium	*scrobiculatum*	*biglumaceum*	Two Glume	Pitted Crown Grass	Millet
Milium	*scrobiculatum*	*bispiculatum*	Two Spikelet	Pitted Crown Grass	Millet
Milium	*scrobiculatum*	*borbonicum*	Borbon	Pitted Crown Grass	Millet
Milium	*scrobiculatum*	*boscianum*	Bosc	Pitted Crown Grass	Millet
Milium	*scrobiculatum*	*cartilagineum*	Cartilage	Pitted Crown Grass	Millet
Milium	*scrobiculatum*	*coloratum*	Painted	Pitted Crown Grass	Millet
Milium	*scrobiculatum*	*commersoni*	Commerson	Pitted Crown Grass	Millet
Milium	*scrobiculatum*	*commutatum*	Changeable	Pitted Crown Grass	Millet
Milium	*scrobiculatum*	*confertum*	Crowded	Pitted Crown Grass	Millet
Milium	*scrobiculatum*	*coromandelinum*	Coromandel	Pitted Crown Grass	Millet
Milium	*scrobiculatum*	*deightonii*	Deighton	Pitted Crown Grass	Millet
Milium	*scrobiculatum*	*firmum*	Firm	Pitted Crown Grass	Millet
Milium	*scrobiculatum*	*frumentaceum*	Grain Bearing	Pitted Crown Grass	Millet
Milium	*scrobiculatum*	*gracillimum*	Slender	Pitted Crown Grass	Millet
Milium	*scrobiculatum*	*grande*	Grand	Pitted Crown Grass	Millet
Milium	*scrobiculatum*	*hirsutum*	Hairy	Pitted Crown Grass	Millet
Milium	*scrobiculatum*	*horneri*	Horner	Pitted Crown Grass	Millet
Milium	*scrobiculatum*	*jardinii*	Jardin	Pitted Crown Grass	Millet
Milium	*scrobiculatum*	*kora*	Kora	Pitted Crown Grass	Millet
Milium	*scrobiculatum*	*latifolium*	Broad Leaf	Pitted Crown Grass	Millet
Milium	*scrobiculatum*	*ledermannii*	Ledermann	Pitted Crown Grass	Millet
Milium	*scrobiculatum*	*mauritanicum*	Morroco	Pitted Crown Grass	Millet
Milium	*scrobiculatum*	*metabolon*	Metabolon	Pitted Crown Grass	Millet
Milium	*scrobiculatum*	*metzii*	Metz	Pitted Crown Grass	Millet
Milium	*scrobiculatum*	*minor*	Lesser	Pitted Crown Grass	Millet
Milium	*scrobiculatum*	*moratii*	Morati	Pitted Crown Grass	Millet
Milium	*scrobiculatum*	*orbiculare*	Kodo	Pitted Crown Grass	Millet
Milium	*scrobiculatum*	*palustre*	Marsh	Pitted Crown Grass	Millet
Milium	*scrobiculatum*	*polo*	Polo	Pitted Crown Grass	Millet
Milium	*scrobiculatum*	*polystachyum*	Many Stack	Pitted Crown Grass	Millet
Milium	*scrobiculatum*	*pseudo-orbiculare*	Orb	Pitted Crown Grass	Millet
Milium	*scrobiculatum*	*pubescens*	Downy	Pitted Crown Grass	Millet
Milium	*scrobiculatum*	*purpurascens*	Purple	Pitted Crown Grass	Millet
Milium	*scrobiculatum*	*serpens*	Serpent	Pitted Crown Grass	Millet
Milium	*scrobiculatum*	*turgidum*	Turgid	Pitted Crown Grass	Millet
Milium	*scrobiculatum*	*velutinum*	Velvet	Pitted Crown Grass	Millet
Milium	*scrobiculatum*	*virgatum*	Twiggy	Pitted Crown Grass	Millet
Milium	*scrobiculatum*	*zollingeri*	Zollinger	Pitted Crown Grass	Millet

African Millet (*Dactylotenium aegypticum*)

Genus	species	infraspecies	cultivar	specimen	Genome
Dactyloctenium	*aegyptium*	*aegyptium*	Original	African Millet	Crow Foot Grass
Dactyloctenium	*aegyptium*	*carolinianum*	Carolina	African Millet	Crow Foot Grass
Dactyloctenium	*aegyptium*	*cavarum*	Cavar	African Millet	Crow Foot Grass
Dactyloctenium	*aegyptium*	*ciliare*	Ciliated	African Millet	Crow Foot Grass
Dactyloctenium	*aegyptium*	*cruciatum*	Cross	African Millet	Crow Foot Grass
Dactyloctenium	*aegyptium*	*distachyum*	Two Stack	African Millet	Crow Foot Grass
Dactyloctenium	*aegyptium*	*erectum*	Erect	African Millet	Crow Foot Grass
Dactyloctenium	*aegyptium*	*figarei*	Figare	African Millet	Crow Foot Grass
Dactyloctenium	*aegyptium*	*guineensis*	Guinea	African Millet	Crow Foot Grass
Dactyloctenium	*aegyptium*	*meridionale*	Meridion	African Millet	Crow Foot Grass
Dactyloctenium	*aegyptium*	*mpuetense*	Mpuetu	African Millet	Crow Foot Grass
Dactyloctenium	*aegyptium*	*mucronatum*	Scalloped	African Millet	Crow Foot Grass
Dactyloctenium	*aegyptium*	*nukaviense*	Nukavi	African Millet	Crow Foot Grass

Dactyloctenium	*aegyptium*	*pectinatum*	Pectin	African Millet	Crow Foot Grass
Dactyloctenium	*aegyptium*	*prostratum*	Prostrate	African Millet	Crow Foot Grass
Dactyloctenium	*aegyptium*	*saccharinum*	Sweet	African Millet	Crow Foot Grass
Dactyloctenium	*aegyptium*	*sparsum*	Sparse	African Millet	Crow Foot Grass
Dactyloctenium	*aegyptium*	*viviparum*	Viviparous	African Millet	Crow Foot Grass

Hedgehog Millet (*Echinochloa spp.*)

Genus	species	infraspecies	cultivar	specimen	Genome
Echinochloa	*colona*	*colona*	Original	Jungle Rice Grass	Cock's Spur Grass
Echinochloa	*colona*	*angustatum*	Narrow	Jungle Rice Grass	Cock's Spur Grass
Echinochloa	*colona*	*aquaticum*	Water	Jungle Rice Grass	Cock's Spur Grass
Echinochloa	*colona*	*atroviolaceum*	Dark Violet	Jungle Rice Grass	Cock's Spur Grass
Echinochloa	*colona*	*brachiariaeforme*	Brachiar	Jungle Rice Grass	Cock's Spur Grass
Echinochloa	*colona*	*brizoides*	Quaking	Jungle Rice Grass	Cock's Spur Grass
Echinochloa	*colona*	*caesium*	Blue Green	Jungle Rice Grass	Cock's Spur Grass
Echinochloa	*colona*	*cumingianum*	Cuming	Jungle Rice Grass	Cock's Spur Grass
Echinochloa	*colona*	*cuspidata*	Cusp	Jungle Rice Grass	Cock's Spur Grass
Echinochloa	*colona*	*daltonii*	Dalton	Jungle Rice Grass	Cock's Spur Grass
Echinochloa	*colona*	*dichotomus*	Forked Pair	Jungle Rice Grass	Cock's Spur Grass
Echinochloa	*colona*	*divaricata*	Spreading	Jungle Rice Grass	Cock's Spur Grass
Echinochloa	*colona*	*echinochloa*	Uchin	Jungle Rice Grass	Cock's Spur Grass
Echinochloa	*colona*	*equitans*	Equator	Jungle Rice Grass	Cock's Spur Grass
Echinochloa	*colona*	*flaccida*	Flaccid	Jungle Rice Grass	Cock's Spur Grass
Echinochloa	*colona*	*geniculata*	Bent	Jungle Rice Grass	Cock's Spur Grass
Echinochloa	*colona*	*glauca*	Glaucous	Jungle Rice Grass	Cock's Spur Grass
Echinochloa	*colona*	*haematodes*	Blood Red	Jungle Rice Grass	Cock's Spur Grass
Echinochloa	*colona*	*hookeri*	Hooker	Jungle Rice Grass	Cock's Spur Grass
Echinochloa	*colona*	*humile*	Low	Jungle Rice Grass	Cock's Spur Grass
Echinochloa	*colona*	*incerta*	Dubious	Jungle Rice Grass	Cock's Spur Grass
Echinochloa	*colona*	*leiantha*	Lion Flower	Jungle Rice Grass	Cock's Spur Grass
Echinochloa	*colona*	*longifolia*	Long Leaf	Jungle Rice Grass	Cock's Spur Grass
Echinochloa	*colona*	*maculatum*	Spotted	Jungle Rice Grass	Cock's Spur Grass
Echinochloa	*colona*	*margaritaceum*	Magarite	Jungle Rice Grass	Cock's Spur Grass
Echinochloa	*colona*	*minor*	Lesser	Jungle Rice Grass	Cock's Spur Grass
Echinochloa	*colona*	*musei*	Muse	Jungle Rice Grass	Cock's Spur Grass
Echinochloa	*colona*	*muticus*	Blunt	Jungle Rice Grass	Cock's Spur Grass
Echinochloa	*colona*	*numidianum*	Numidia	Jungle Rice Grass	Cock's Spur Grass
Echinochloa	*colona*	*petiveri*	Petiver	Jungle Rice Grass	Cock's Spur Grass
Echinochloa	*colona*	*prorepens*	Lesser Creeping	Jungle Rice Grass	Cock's Spur Grass
Echinochloa	*colona*	*repens*	Creeping	Jungle Rice Grass	Cock's Spur Grass
Echinochloa	*colona*	*subverticillata*	Lesser Whorled	Jungle Rice Grass	Cock's Spur Grass
Echinochloa	*colona*	*terrestris*	Terrestrial	Jungle Rice Grass	Cock's Spur Grass
Echinochloa	*colona*	*tetrastichum*	Five Stick	Jungle Rice Grass	Cock's Spur Grass
Echinochloa	*colona*	*vivipara*	Viviparous	Jungle Rice Grass	Cock's Spur Grass
Echinochloa	*colona*	*zonalis*	Banded	Jungle Rice Grass	Cock's Spur Grass
Echinochloa	*crus-galli*	*crus-galli*	Original	Barnyard Millet	Cock's Spur Grass
Echinochloa	*crus-galli*	*aristata*	Grain	Barnyard Millet	Cock's Spur Grass
Echinochloa	*crus-galli*	*frumentacea*	Awn	Barnyard Millet	Cock's Spur Grass
Echinochloa	*crus-galli*	*mitis*	Mild	Barnyard Millet	Cock's Spur Grass
Echinochloa	*crus-galli*	*muricata*	Hardshell	Barnyard Millet	Cock's Spur Grass
Echinochloa	*crus-galli*	*vittata*	Striped	Barnyard Millet	Cock's Spur Grass
Echinochloa	*esculenta*	*esculenta*	Original	Japanese Millet	Cock's Spur Grass
Echinochloa	*esculenta*	*utilis*	Useful	Japanese Millet	Cock's Spur Grass
Echinochloa	*frumentacea*	*frumentacea*	Original	India Millet	Cock's Spur Grass
Echinochloa	*frumentacea*	*chiwapa*	Chipawa	India Millet	Cock's Spur Grass
Echinochloa	*frumentacea*	*edulis*	Original	India Millet	Cock's Spur Grass
Echinochloa	*muricata*	*muricata*	Original	Rough Barnyard Millet	Cock's Spur Grass
Echinochloa	*muricata*	*coarctata*	Arctic	Rough Barnyard Millet	Cock's Spur Grass
Echinochloa	*muricata*	*ludoviciana*	Louisiana	Rough Barnyard Millet	Cock's Spur Grass
Echinochloa	*muricata*	*michauxii*	Michaux	Rough Barnyard Millet	Cock's Spur Grass
Echinochloa	*muricata*	*microstachya*	Tiny Spike	Rough Barnyard Millet	Cock's Spur Grass
Echinochloa	*muricata*	*multiflora*	Many Flower	Rough Barnyard Millet	Cock's Spur Grass
Echinochloa	*muricata*	*pungens*	Pungent	Rough Barnyard Millet	Cock's Spur Grass
Echinochloa	*muricata*	*wiegandii*	Wiegand	Rough Barnyard Millet	Cock's Spur Grass
Echinochloa	*oryzoides*	*oryzoides*	Original	Barnyard Ricegrass	Cock's Spur Grass
Echinochloa	*oryzoides*	*coarctata*	Arctic	Barnyard Ricegrass	Cock's Spur Grass
Echinochloa	*oryzoides*	*erecta*	Upright	Barnyard Ricegrass	Cock's Spur Grass
Echinochloa	*oryzoides*	*grandiflora*	Grand Flower	Barnyard Ricegrass	Cock's Spur Grass
Echinochloa	*oryzoides*	*hispidula*	Hairy	Barnyard Ricegrass	Cock's Spur Grass
Echinochloa	*oryzoides*	*hostii*	Host	Barnyard Ricegrass	Cock's Spur Grass
Echinochloa	*oryzoides*	*macrocarpa*	Large Grain	Barnyard Ricegrass	Cock's Spur Grass
Echinochloa	*oryzoides*	*mirabile*	Wonderful	Barnyard Ricegrass	Cock's Spur Grass
Echinochloa	*oryzoides*	*phyllopogon*	Bearded	Barnyard Ricegrass	Cock's Spur Grass
Echinochloa	*oryzoides*	*retzii*	Retz	Barnyard Ricegrass	Cock's Spur Grass
Echinochloa	*oryzoides*	*riukiuensis*	Riukiu	Barnyard Ricegrass	Cock's Spur Grass
Echinochloa	*oryzoides*	*rubra*	Red	Barnyard Ricegrass	Cock's Spur Grass
Echinochloa	*walteri*	*walteri*	Original	Coastal Barnyard Millet	Cock's Spur Grass
Echinochloa	*walteri*	*breviseta*	Short Bristle	Coastal Barnyard Millet	Cock's Spur Grass
Echinochloa	*walteri*	*hispida*	Hairy	Coastal Barnyard Millet	Cock's Spur Grass

Echinochloa	*walteri*	*laevigata*	Smooth	Coastal Barnyard Millet	Cock's Spur Grass
Echinochloa	*walteri*	*longiaristata*	Long Awn	Coastal Barnyard Millet	Cock's Spur Grass
Echinochloa	*walteri*	*longiseta*	Long Bristle	Coastal Barnyard Millet	Cock's Spur Grass

Goosegrass Millet (*Eleusine spp.*)

Genus	species	infraspecies	cultivar	specimen	Genome
Eleusine	*africana*	*africana*	Original	African Millet	Goose Grass
Eleusine	*africana*	*aegyptiaca*	Egyptian	African Millet	Goose Grass
Eleusine	*africana*	*pectinata*	Pectin	African Millet	Goose Grass
Eleusine	*coracana*	*coracana*	Original	Ragi Finger Millet	Goose Grass
Eleusine	*coracana*	*dragon's-clawi*	Dragon's Claw	Ragi Finger Millet	Goose Grass
Eleusine	*indica*	*indica*	Original	India	Goose Grass
Eleusine	*indica*	*ara*	Ar	India	Goose Grass
Eleusine	*indica*	*dissecta*	Dissected	India	Goose Grass
Eleusine	*indica*	*distachya*	Two Stack	India	Goose Grass
Eleusine	*indica*	*distans*	Remote	India	Goose Grass
Eleusine	*indica*	*domingensis*	Domingo	India	Goose Grass
Eleusine	*indica*	*geminata*	Gem	India	Goose Grass
Eleusine	*indica*	*glabra*	Smooth	India	Goose Grass
Eleusine	*indica*	*gonantha*	Gonad Flower	India	Goose Grass
Eleusine	*indica*	*gouinii*	Gouin	India	Goose Grass
Eleusine	*indica*	*gracilis*	Slender	India	Goose Grass
Eleusine	*indica*	*inaequalis*	Unequal	India	Goose Grass
Eleusine	*indica*	*japonica*	Japanese	India	Goose Grass
Eleusine	*indica*	*macrosperma*	Large Seed	India	Goose Grass
Eleusine	*indica*	*major*	Greater	India	Goose Grass
Eleusine	*indica*	*marginata*	Margined	India	Goose Grass
Eleusine	*indica*	*monostachya*	Single Stack	India	Goose Grass
Eleusine	*indica*	*oligostachya*	Small Stack	India	Goose Grass
Eleusine	*indica*	*pectinata*	Pectin	India	Goose Grass
Eleusine	*indica*	*polydactyla*	Many Finger	India	Goose Grass
Eleusine	*indica*	*repens*	Creeping	India	Goose Grass
Eleusine	*indica*	*rigidifolia*	Stiff Leaf	India	Goose Grass
Eleusine	*indica*	*sandaensis*	Sanda	India	Goose Grass
Eleusine	*indica*	*scabra*	Scabre	India	Goose Grass
Eleusine	*indica*	*spicata*	Spike	India	Goose Grass
Eleusine	*indica*	*textilis*	Textile	India	Goose Grass

Lovegrass Millet (*Eragrostis spp.*)

Genus	species	infraspecies	cultivar	specimen	Genome
Eragrostis	*cilianensis*	*cilianensis*	Original	Stink Grass	Love Grass
Eragrostis	*cilianensis*	*argentina*	Argentina	Stink Grass	Love Grass
Eragrostis	*cilianensis*	*articulata*	Articulated	Stink Grass	Love Grass
Eragrostis	*cilianensis*	*cachectica*	Cachectian	Stink Grass	Love Grass
Eragrostis	*cilianensis*	*compacta*	Compact	Stink Grass	Love Grass
Eragrostis	*cilianensis*	*costata*	Ribbed	Stink Grass	Love Grass
Eragrostis	*cilianensis*	*flava*	Yellow	Stink Grass	Love Grass
Eragrostis	*cilianensis*	*flexuosa*	Flexible	Stink Grass	Love Grass
Eragrostis	*cilianensis*	*glandulifera*	Gland	Stink Grass	Love Grass
Eragrostis	*cilianensis*	*insularis*	Island	Stink Grass	Love Grass
Eragrostis	*cilianensis*	*leersioides*	Leers	Stink Grass	Love Grass
Eragrostis	*cilianensis*	*major*	Greater	Stink Grass	Love Grass
Eragrostis	*cilianensis*	*megastachya*	Large Spike	Stink Grass	Love Grass
Eragrostis	*cilianensis*	*monodii*	Monod	Stink Grass	Love Grass
Eragrostis	*cilianensis*	*nana*	Dwarf	Stink Grass	Love Grass
Eragrostis	*cilianensis*	*nuttallii*	Nuttall	Stink Grass	Love Grass
Eragrostis	*cilianensis*	*oblonga*	Oblong	Stink Grass	Love Grass
Eragrostis	*cilianensis*	*obtusa*	Obtuse	Stink Grass	Love Grass
Eragrostis	*cilianensis*	*pappii*	Papp	Stink Grass	Love Grass
Eragrostis	*cilianensis*	*pennsylvanica*	Pennsylvania	Stink Grass	Love Grass
Eragrostis	*cilianensis*	*philadelphica*	Philadelphia	Stink Grass	Love Grass
Eragrostis	*cilianensis*	*polyadenia*	Many Gland	Stink Grass	Love Grass
Eragrostis	*cilianensis*	*polymorpha*	Polymorphic	Stink Grass	Love Grass
Eragrostis	*cilianensis*	*polysperma*	Many Seed	Stink Grass	Love Grass
Eragrostis	*cilianensis*	*purpurascens*	Purple	Stink Grass	Love Grass
Eragrostis	*cilianensis*	*roxburghiana*	Roxburgh	Stink Grass	Love Grass
Eragrostis	*cilianensis*	*schweinfurthiana*	Schweinfurth	Stink Grass	Love Grass
Eragrostis	*cilianensis*	*starosselskyi*	Starosselsky	Stink Grass	Love Grass
Eragrostis	*cilianensis*	*subbiloba*	Lesser Two Lobed	Stink Grass	Love Grass
Eragrostis	*cilianensis*	*thyrsiflora*	Thyrse Flower	Stink Grass	Love Grass
Eragrostis	*cilianensis*	*tortuosa*	Twisted	Stink Grass	Love Grass
Eragrostis	*cilianensis*	*triticea*	Tritice	Stink Grass	Love Grass
Eragrostis	*cilianensis*	*violacea*	Violet	Stink Grass	Love Grass
Eragrostis	*cilianensis*	*virletii*	Virlet	Stink Grass	Love Grass
Eragrostis	*ciliaris*	*ciliaris*	Original	Gopher Tail	Love Grass
Eragrostis	*ciliaris*	*arabica*	Arabian	Gopher Tail	Love Grass
Eragrostis	*ciliaris*	*boryana*	Bory	Gopher Tail	Love Grass
Eragrostis	*ciliaris*	*brachystachya*	Short Spike	Gopher Tail	Love Grass
Eragrostis	*ciliaris*	*clarkei*	Short Stack	Gopher Tail	Love Grass

Eragrostis	*ciliaris*	*compta*	Compton	Gopher Tail	Love Grass
Eragrostis	*ciliaris*	*contractus*	Contracted	Gopher Tail	Love Grass
Eragrostis	*ciliaris*	*elegans*	Elegant	Gopher Tail	Love Grass
Eragrostis	*ciliaris*	*elliptica*	Elliptic Leaf	Gopher Tail	Love Grass
Eragrostis	*ciliaris*	*latifolia*	Broad Leaf	Gopher Tail	Love Grass
Eragrostis	*ciliaris*	*laxa*	Limp	Gopher Tail	Love Grass
Eragrostis	*ciliaris*	*lobata*	Lobed	Gopher Tail	Love Grass
Eragrostis	*ciliaris*	*villosa*	Villose	Gopher Tail	Love Grass
Eragrostis	*cumingii*	*cumingii*	Original	Cuming	Love Grass
Eragrostis	*cumingii*	*bleeseri*	Bleeser	Cuming	Love Grass
Eragrostis	*cumingii*	*brownei*	Browne	Cuming	Love Grass
Eragrostis	*cumingii*	*bulbillifera*	Bulbil	Cuming	Love Grass
Eragrostis	*cumingii*	*distans*	Remote	Cuming	Love Grass
Eragrostis	*cumingii*	*kisarensis*	Kisar	Cuming	Love Grass
Eragrostis	*cumingii*	*paniculata*	Panicle	Cuming	Love Grass
Eragrostis	*cumingii*	*ramosa*	Branching	Cuming	Love Grass
Eragrostis	*cumingii*	*reflexa*	Reflexed	Cuming	Love Grass
Eragrostis	*cumingii*	*rindjaniensis*	Rindjani	Cuming	Love Grass
Eragrostis	*cumingii*	*simplex*	Simple	Cuming	Love Grass
Eragrostis	*cumingii*	*vivipara*	Viviparous	Cuming	Love Grass
Eragrostis	*curtipedicellata*	*curtipedicellata*	Original	Short Pedicel	Love Grass
Eragrostis	*curtipedicellata*	*viscosa*	Gummy	Short Pedicel	Love Grass
Eragrostis	*curvula*	*curvula*	Original	Weeping	Love Grass
Eragrostis	*curvula*	*ampla*	Ample	Weeping	Love Grass
Eragrostis	*curvula*	*atrata*	Blackened	Weeping	Love Grass
Eragrostis	*curvula*	*atrovirens*	Dark Green	Weeping	Love Grass
Eragrostis	*curvula*	*capensis*	Africa Cape	Weeping	Love Grass
Eragrostis	*curvula*	*capillifolia*	Capillary Leaf	Weeping	Love Grass
Eragrostis	*curvula*	*chloromelas*	Boer	Weeping	Love Grass
Eragrostis	*curvula*	*conferta*	Crowded	Weeping	Love Grass
Eragrostis	*curvula*	*decolorans*	Discolored	Weeping	Love Grass
Eragrostis	*curvula*	*filiformis*	Thread Leaf	Weeping	Love Grass
Eragrostis	*curvula*	*huillensis*	Huilla	Weeping	Love Grass
Eragrostis	*curvula*	*jeffreysii*	Jeffreys	Weeping	Love Grass
Eragrostis	*curvula*	*poa*	Meadow	Weeping	Love Grass
Eragrostis	*curvula*	*procerior*	Tall	Weeping	Love Grass
Eragrostis	*curvula*	*pubiculmis*	Downy Culm	Weeping	Love Grass
Eragrostis	*curvula*	*robusta*	Robust	Weeping	Love Grass
Eragrostis	*curvula*	*subulata*	Lesser Awn	Weeping	Love Grass
Eragrostis	*curvula*	*thunbergiana*	Thunberg	Weeping	Love Grass
Eragrostis	*curvula*	*valida*	Valid	Weeping	Love Grass
Eragrostis	*cylindriflora*	*cylindriflora*	Original	Cylinder Flower	Love Grass
Eragrostis	*cylindriflora*	*adenocoleos*	Gland	Cylinder Flower	Love Grass
Eragrostis	*cylindriflora*	*agrostoidea*	Field	Cylinder Flower	Love Grass
Eragrostis	*cylindriflora*	*annulata*	Ringed	Cylinder Flower	Love Grass
Eragrostis	*cylindriflora*	*atherstonei*	Atherstone	Cylinder Flower	Love Grass
Eragrostis	*cylindriflora*	*geniculata*	Bent	Cylinder Flower	Love Grass
Eragrostis	*cylindriflora*	*gymnorrhachis*	Nude Rachis	Cylinder Flower	Love Grass
Eragrostis	*cylindriflora*	*henrardii*	Henrard	Cylinder Flower	Love Grass
Eragrostis	*cylindriflora*	*horizontalis*	Horizontal	Cylinder Flower	Love Grass
Eragrostis	*cylindriflora*	*leptocalymma*	Thin Calyx	Cylinder Flower	Love Grass
Eragrostis	*cylindriflora*	*multipilosa*	Pilose	Cylinder Flower	Love Grass
Eragrostis	*cylindriflora*	*quadriflora*	Four Flower	Cylinder Flower	Love Grass
Eragrostis	*cylindriflora*	*rigidior*	Stiff	Cylinder Flower	Love Grass
Eragrostis	*cylindriflora*	*seineri*	Seiner	Cylinder Flower	Love Grass
Eragrostis	*cylindriflora*	*speciosa*	Showy	Cylinder Flower	Love Grass
Eragrostis	*cylindriflora*	*trichophora*	Hairy	Cylinder Flower	Love Grass
Eragrostis	*cylindriflora*	*viscosa*	Sticky	Cylinder Flower	Love Grass
Eragrostis	*elliottii*	*elliottii*	Original	Field	Love Grass
Eragrostis	*elliottii*	*nitida*	Shiny	Field	Love Grass
Eragrostis	*elliottii*	*macropoda*	Large Pod	Field	Love Grass
Eragrostis	*frankii*	*frankii*	Original	Sandbar	Love Grass
Eragrostis	*frankii*	*brevipes*	Short	Sandbar	Love Grass
Eragrostis	*frankii*	*erythrogona*	Red Seed	Sandbar	Love Grass
Eragrostis	*gangetica*	*gangetica*	Original	Slim Flower	Love Grass
Eragrostis	*gangetica*	*cambessediana*	Cambessedes	Slim Flower	Love Grass
Eragrostis	*gangetica*	*dakarensis*	Dakar	Slim Flower	Love Grass
Eragrostis	*gangetica*	*dumasiana*	Dumas	Slim Flower	Love Grass
Eragrostis	*gangetica*	*elegantula*	Elegant	Slim Flower	Love Grass
Eragrostis	*gangetica*	*flamignii*	Flamign	Slim Flower	Love Grass
Eragrostis	*gangetica*	*floribunda*	Abundant Flower	Slim Flower	Love Grass
Eragrostis	*gangetica*	*kalinaensis*	Kalina	Slim Flower	Love Grass
Eragrostis	*gangetica*	*ovina*	Sheep	Slim Flower	Love Grass
Eragrostis	*gangetica*	*pallida*	Pale	Slim Flower	Love Grass
Eragrostis	*gangetica*	*pumila*	Low	Slim Flower	Love Grass
Eragrostis	*gangetica*	*stenoclada*	Thin Scale	Slim Flower	Love Grass
Eragrostis	*gangetica*	*stenophylla*	Thin Leaf	Slim Flower	Love Grass
Eragrostis	*gangetica*	*vinicolor*	Wine Color	Slim Flower	Love Grass
Eragrostis	*gangetica*	*willdenowii*	Willdenow	Slim Flower	Love Grass

Eragrostis	*grandis*	*grandis*	Original	Large Hawaiian	Love Grass
Eragrostis	*grandis*	*oligantha*	Few Flower	Large Hawaiian	Love Grass
Eragrostis	*grandis*	*polyantha*	Many Flower	Large Hawaiian	Love Grass
Eragrostis	*hirsuta*	*hirsuta*	Original	Big Top	Love Grass
Eragrostis	*hirsuta*	*laevivaginata*	Smooth Sheath	Big Top	Love Grass
Eragrostis	*hirsuta*	*sporoboloides*	Seed Bowl	Big Top	Love Grass
Eragrostis	*hypnoides*	*hypnoides*	Original	Teal Creeping	Love Grass
Eragrostis	*hypnoides*	*caespitosa*	Clump	Teal Creeping	Love Grass
Eragrostis	*hypnoides*	*capitata*	Capped	Teal Creeping	Love Grass
Eragrostis	*hypnoides*	*carinata*	Keel	Teal Creeping	Love Grass
Eragrostis	*hypnoides*	*contracta*	Contracted	Teal Creeping	Love Grass
Eragrostis	*hypnoides*	*dioica*	Dioecious	Teal Creeping	Love Grass
Eragrostis	*hypnoides*	*glabra*	Smooth	Teal Creeping	Love Grass
Eragrostis	*hypnoides*	*gouinii*	Gouin	Teal Creeping	Love Grass
Eragrostis	*hypnoides*	*laxior*	Limp	Teal Creeping	Love Grass
Eragrostis	*hypnoides*	*mnioides*	Lead	Teal Creeping	Love Grass
Eragrostis	*hypnoides*	*pygmaea*	Pygmy	Teal Creeping	Love Grass
Eragrostis	*japonica*	*japonica*	Original	Old World Pond	Love Grass
Eragrostis	*japonica*	*aturensis*	Atura	Old World Pond	Love Grass
Eragrostis	*japonica*	*aurea*	Golden	Old World Pond	Love Grass
Eragrostis	*japonica*	*biflora*	Two Flower	Old World Pond	Love Grass
Eragrostis	*japonica*	*brasiliana*	Brazil	Old World Pond	Love Grass
Eragrostis	*japonica*	*brasiliensis*	Brazilian	Old World Pond	Love Grass
Eragrostis	*japonica*	*caudata*	Stem	Old World Pond	Love Grass
Eragrostis	*japonica*	*conferta*	Crowded	Old World Pond	Love Grass
Eragrostis	*japonica*	*confertiflora*	Crowded Flower	Old World Pond	Love Grass
Eragrostis	*japonica*	*depauperata*	Imperfect	Old World Pond	Love Grass
Eragrostis	*japonica*	*diandra*	Two Stamen	Old World Pond	Love Grass
Eragrostis	*japonica*	*diarrhena*	Two Pistel	Old World Pond	Love Grass
Eragrostis	*japonica*	*diplachnoides*	Two Plachnon	Old World Pond	Love Grass
Eragrostis	*japonica*	*elegans*	Elegant	Old World Pond	Love Grass
Eragrostis	*japonica*	*elongata*	Elongated	Old World Pond	Love Grass
Eragrostis	*japonica*	*glomerata*	Cluster	Old World Pond	Love Grass
Eragrostis	*japonica*	*hapalantha*	Two Anther	Old World Pond	Love Grass
Eragrostis	*japonica*	*hochstetteri*	Hochstetter	Old World Pond	Love Grass
Eragrostis	*japonica*	*interrupta*	Interrupted	Old World Pond	Love Grass
Eragrostis	*japonica*	*jubata*	Long Awn	Old World Pond	Love Grass
Eragrostis	*japonica*	*koenigii*	Koenig	Old World Pond	Love Grass
Eragrostis	*japonica*	*kossinskyi*	Kossinsky	Old World Pond	Love Grass
Eragrostis	*japonica*	*laxiflora*	Limp Flower	Old World Pond	Love Grass
Eragrostis	*japonica*	*leprieurii*	Leprieur	Old World Pond	Love Grass
Eragrostis	*japonica*	*leptanthum*	Thin Flower	Old World Pond	Love Grass
Eragrostis	*japonica*	*maritima*	Seaside	Old World Pond	Love Grass
Eragrostis	*japonica*	*micrantha*	Small Flower	Old World Pond	Love Grass
Eragrostis	*japonica*	*milleflora*	Thousand Flower	Old World Pond	Love Grass
Eragrostis	*japonica*	*minutiflora*	Minute Flower	Old World Pond	Love Grass
Eragrostis	*japonica*	*mossulensis*	Mossul	Old World Pond	Love Grass
Eragrostis	*japonica*	*namaquensis*	Namaqua	Old World Pond	Love Grass
Eragrostis	*japonica*	*pallida*	Pale	Old World Pond	Love Grass
Eragrostis	*japonica*	*parviflora*	Few Flower	Old World Pond	Love Grass
Eragrostis	*japonica*	*pilosa*	Pilose	Old World Pond	Love Grass
Eragrostis	*japonica*	*poiformis*	Meadow	Old World Pond	Love Grass
Eragrostis	*japonica*	*robusta*	Robust	Old World Pond	Love Grass
Eragrostis	*japonica*	*sporoboloides*	Sporobol	Old World Pond	Love Grass
Eragrostis	*japonica*	*stricta*	Erect	Old World Pond	Love Grass
Eragrostis	*japonica*	*tenuissima*	Tenuous	Old World Pond	Love Grass
Eragrostis	*japonica*	*verticillata*	Whorled	Old World Pond	Love Grass
Eragrostis	*japonica*	*volgensis*	Volga	Old World Pond	Love Grass
Eragrostis	*japonica*	*walteri*	Walter	Old World Pond	Love Grass
Eragrostis	*leptostachya*	*leptostachya*	Original	Australian	Love Grass
Eragrostis	*leptostachya*	*depauperata*	Pauper	Australian	Love Grass
Eragrostis	*leptostachya*	*hosakai*	Hosaka	Australian	Love Grass
Eragrostis	*leptostachya*	*molybdea*	Molybdenum	Australian	Love Grass
Eragrostis	*leptostachya*	*rara*	Molybdenum	Australian	Love Grass
Eragrostis	*mexicana*	*mexicana*	Original	Mexican	Love Grass
Eragrostis	*mexicana*	*alba*	White	Mexican	Love Grass
Eragrostis	*mexicana*	*densiuscula*	Dense	Mexican	Love Grass
Eragrostis	*mexicana*	*glandulosa*	Gland	Mexican	Love Grass
Eragrostis	*mexicana*	*leptantha*	Thin Flower	Mexican	Love Grass
Eragrostis	*mexicana*	*limbata*	Bordered	Mexican	Love Grass
Eragrostis	*mexicana*	*major*	Greater	Mexican	Love Grass
Eragrostis	*mexicana*	*neomexicana*	New Mexico	Mexican	Love Grass
Eragrostis	*mexicana*	*orcuttiana*	Orcutt Green	Mexican	Love Grass
Eragrostis	*mexicana*	*pauciflora*	Pauper Flower	Mexican	Love Grass
Eragrostis	*mexicana*	*pilosior*	Pilose	Mexican	Love Grass
Eragrostis	*mexicana*	*pygmaea*	Pygmy	Mexican	Love Grass
Eragrostis	*minor*	*minor*	Original	Little	Love Grass
Eragrostis	*minor*	*arenosa*	Sand	Little	Love Grass
Eragrostis	*minor*	*beguinotii*	Beguinot	Little	Love Grass

Eragrostis	*minor*	*borysthenica*	Borysthenia	Little	Love Grass
Eragrostis	*minor*	*brizoides*	Quaking	Little	Love Grass
Eragrostis	*minor*	*canariensis*	Canary Island	Little	Love Grass
Eragrostis	*minor*	*laxiflora*	Limp Flower	Little	Love Grass
Eragrostis	*minor*	*microstachya*	Small Stack	Little	Love Grass
Eragrostis	*minor*	*nana*	Dwarf	Little	Love Grass
Eragrostis	*minor*	*pappiana*	Papp	Little	Love Grass
Eragrostis	*minor*	*poiformis*	Meadow	Little	Love Grass
Eragrostis	*minor*	*pooides*	Meadow Grass	Little	Love Grass
Eragrostis	*minor*	*spiciformis*	Spike	Little	Love Grass
Eragrostis	*minor*	*suaveolens*	Candy	Little	Love Grass
Eragrostis	*minor*	*umbrosa*	Shade Loving	Little	Love Grass
Eragrostis	*minor*	*willdenoviana*	Willdenow	Little	Love Grass
Eragrostis	*palmeri*	*palmeri*	Original	Rio Grande	Love Grass
Eragrostis	*palmeri*	*caudata*	Stem	Rio Grande	Love Grass
Eragrostis	*paupera*	*paupera*	Original	Oahu	Love Grass
Eragrostis	*paupera*	*caumii*	Caum	Oahu	Love Grass
Eragrostis	*paupera*	*whitneyi*	Whitney	Oahu	Love Grass
Eragrostis	*pectinacea*	*pectinacea*	Original	Tufted	Love Grass
Eragrostis	*pectinacea*	*arida*	Desert Miser	Tufted	Love Grass
Eragrostis	*pectinacea*	*brizoides*	Quake Grass	Tufted	Love Grass
Eragrostis	*pectinacea*	*caroliniana*	Carolina	Tufted	Love Grass
Eragrostis	*pectinacea*	*cognata*	Cognate	Tufted	Love Grass
Eragrostis	*pectinacea*	*delicatula*	Delicate	Tufted	Love Grass
Eragrostis	*pectinacea*	*diffusa*	Diffuse	Tufted	Love Grass
Eragrostis	*pectinacea*	*miserrima*	Miser	Tufted	Love Grass
Eragrostis	*pectinacea*	*nuttalliana*	Nuttall	Tufted	Love Grass
Eragrostis	*pectinacea*	*pennsylvanica*	Pennsylvania	Tufted	Love Grass
Eragrostis	*pectinacea*	*purshii*	Pursh	Tufted	Love Grass
Eragrostis	*pectinacea*	*tephrosanthos*	Ash	Tufted	Love Grass
Eragrostis	*pectinacea*	*unionis*	Union	Tufted	Love Grass
Eragrostis	*pilosa*	*pilosa*	Original	Downy	Love Grass
Eragrostis	*pilosa*	*afghanica*	Afghan	Downy	Love Grass
Eragrostis	*pilosa*	*amurensis*	Amur River Valley	Downy	Love Grass
Eragrostis	*pilosa*	*bagdadensis*	Bagdad	Downy	Love Grass
Eragrostis	*pilosa*	*baguirmiensis*	Baguirmia	Downy	Love Grass
Eragrostis	*pilosa*	*bicolor*	Two Color	Downy	Love Grass
Eragrostis	*pilosa*	*bithynica*	Bithynia	Downy	Love Grass
Eragrostis	*pilosa*	*bohemica*	Bohemian	Downy	Love Grass
Eragrostis	*pilosa*	*chilensis*	Chile	Downy	Love Grass
Eragrostis	*pilosa*	*collocarpa*	Column Pod	Downy	Love Grass
Eragrostis	*pilosa*	*delicatior*	Delicate	Downy	Love Grass
Eragrostis	*pilosa*	*filiformis*	Thread Leaf	Downy	Love Grass
Eragrostis	*pilosa*	*glabra*	Smooth	Downy	Love Grass
Eragrostis	*pilosa*	*gracilis*	Slender	Downy	Love Grass
Eragrostis	*pilosa*	*gracillima*	Graceful	Downy	Love Grass
Eragrostis	*pilosa*	*imberbis*	Beardless	Downy	Love Grass
Eragrostis	*pilosa*	*indica*	India	Downy	Love Grass
Eragrostis	*pilosa*	*jeholensis*	Jehol	Downy	Love Grass
Eragrostis	*pilosa*	*laxior*	Limp	Downy	Love Grass
Eragrostis	*pilosa*	*linkii*	Link	Downy	Love Grass
Eragrostis	*pilosa*	*major*	Greater	Downy	Love Grass
Eragrostis	*pilosa*	*mexicana*	Mexican	Downy	Love Grass
Eragrostis	*pilosa*	*multispicula*	Many Spicule	Downy	Love Grass
Eragrostis	*pilosa*	*neglecta*	Neglected	Downy	Love Grass
Eragrostis	*pilosa*	*petersii*	Peters	Downy	Love Grass
Eragrostis	*pilosa*	*plumosa*	Plume	Downy	Love Grass
Eragrostis	*pilosa*	*poiretii*	Poiret	Downy	Love Grass
Eragrostis	*pilosa*	*punctata*	Dotted	Downy	Love Grass
Eragrostis	*pilosa*	*senegalensis*	Senegal	Downy	Love Grass
Eragrostis	*pilosa*	*subspontanea*	Spontaneous	Downy	Love Grass
Eragrostis	*pilosa*	*tenella*	Tenacious	Downy	Love Grass
Eragrostis	*pilosa*	*tenuiflora*	Tenuous Flower	Downy	Love Grass
Eragrostis	*pilosa*	*tenuis*	Tenuous	Downy	Love Grass
Eragrostis	*pilosa*	*verticillata*	Whorled	Downy	Love Grass
Eragrostis	*prolifera*	*prolifera*	Original	Dominican	Love Grass
Eragrostis	*prolifera*	*diversiflora*	Diverse Flower	Dominican	Love Grass
Eragrostis	*prolifera*	*domingensis*	Domingo	Dominican	Love Grass
Eragrostis	*prolifera*	*fascicularis*	Fascicle	Dominican	Love Grass
Eragrostis	*prolifera*	*gigantea*	Giant	Dominican	Love Grass
Eragrostis	*prolifera*	*haenkei*	Haenke	Dominican	Love Grass
Eragrostis	*prolifera*	*maxima*	Great	Dominican	Love Grass
Eragrostis	*prolifera*	*salzmannii*	Salzmann	Dominican	Love Grass
Eragrostis	*refracta*	*refracta*	Original	Coastal Plain	Love Grass
Eragrostis	*refracta*	*acuta*	Sharp	Coastal Plain	Love Grass
Eragrostis	*refracta*	*campestris*	Field	Coastal Plain	Love Grass
Eragrostis	*refracta*	*longiradiata*	Long Ray	Coastal Plain	Love Grass
Eragrostis	*refracta*	*refracta*	Refracted	Coastal Plain	Love Grass
Eragrostis	*refracta*	*virginica*	Virginia	Coastal Plain	Love Grass

Eragrostis	reptans	reptans	Original	Creeping	Love Grass
Eragrostis	reptans	breviflora	Short Flower	Creeping	Love Grass
Eragrostis	reptans	corymbifera	Corymb	Creeping	Love Grass
Eragrostis	reptans	capitata	Capped	Creeping	Love Grass
Eragrostis	reptans	dioeca	Dioecious	Creeping	Love Grass
Eragrostis	reptans	fasciculata	Fascicle	Creeping	Love Grass
Eragrostis	reptans	weigeltiana	Weigelt	Creeping	Love Grass
Eragrostis	scaligera	scaligera	Original	Scaley	Love Grass
Eragrostis	scaligera	laurentiana	Laurent	Scaley	Love Grass
Eragrostis	secundiflora	secundiflora	Original	Red	Love Grass
Eragrostis	secundiflora	beyrichii	Beyrich	Red	Love Grass
Eragrostis	secundiflora	capitata	Capped	Red	Love Grass
Eragrostis	secundiflora	compacta	Compact	Red	Love Grass
Eragrostis	secundiflora	interrupta	Interrupted	Red	Love Grass
Eragrostis	secundiflora	oxylepis	Sharp Thin	Red	Love Grass
Eragrostis	secundiflora	sejuncta	Sejunct	Red	Love Grass
Eragrostis	secundiflora	squarrosa	Scaly	Red	Love Grass
Eragrostis	secundiflora	subfasciculata	Bundle	Red	Love Grass
Eragrostis	secundiflora	veracrucis	Vera Cruz	Red	Love Grass
Eragrostis	secundiflora	yucatana	Yucatan	Red	Love Grass
Eragrostis	sessilispica	sessilispica	Original	Tumble	Love Grass
Eragrostis	sessilispica	rigida	Stiff	Tumble	Love Grass
Eragrostis	spectabilis	spectabilis	Original	Purple	Love Grass
Eragrostis	spectabilis	amabilis	Amiable	Purple	Love Grass
Eragrostis	spectabilis	geyeri	Geyer	Purple	Love Grass
Eragrostis	spectabilis	sparsihirsuta	Petticoat Climbing	Purple	Love Grass
Eragrostis	spectabilis	velutina	Velvet	Purple	Love Grass
Eragrostis	spectabilis	villosa	Villose	Purple	Love Grass
Eragrostis	tef	tef	Original	Teff	Love Grass
Eragrostis	tef	a.l.-whitei	A. L. White	Teff	Love Grass
Eragrostis	tef	alba	White	Teff	Love Grass
Eragrostis	tef	abyssinica	Ethiopian	Teff	Love Grass
Eragrostis	tef	cerealis	Cereal	Teff	Love Grass
Eragrostis	tef	flaccida	Flaccid	Teff	Love Grass
Eragrostis	tef	ivoryi	Ivory	Teff	Love Grass
Eragrostis	tef	radicans	Ray	Teff	Love Grass
Eragrostis	tef	rubra-brunnea	Reddish Brown	Teff	Love Grass
Eragrostis	tef	spiciformis	Spike	Teff	Love Grass
Eragrostis	tef	viridis	Green	Teff	Love Grass
Eragrostis	trichocolea	trichocolea	Original	Hairy Sheath	Love Grass
Eragrostis	trichocolea	floridana	Florida	Hairy Sheath	Love Grass
Eragrostis	trichodes	trichodes	Original	Sand	Love Grass
Eragrostis	trichodes	capillacea	Capillary	Sand	Love Grass
Eragrostis	trichodes	grandiflora	Grand Flower	Sand	Love Grass
Eragrostis	trichodes	pilifera	Pill	Sand	Love Grass
Eragrostis	trichodes	tenuis	Tenuous	Sand	Love Grass
Eragrostis	trichodes	texensis	Texas	Sand	Love Grass
Eragrostis	unioloides	unioloides	Original	Chinese	Love Grass
Eragrostis	unioloides	contracta	Contracted	Chinese	Love Grass
Eragrostis	unioloides	effusa	Effusive	Chinese	Love Grass
Eragrostis	unioloides	euchroa	Tinged	Chinese	Love Grass
Eragrostis	unioloides	formosana	Taiwan	Chinese	Love Grass
Eragrostis	unioloides	indica	India	Chinese	Love Grass
Eragrostis	unioloides	linearis	Strap Leaf	Chinese	Love Grass
Eragrostis	unioloides	ongiemensis	Ongiema	Chinese	Love Grass
Eragrostis	unioloides	prostrata	Prostrate	Chinese	Love Grass
Eragrostis	unioloides	rubens	Red	Chinese	Love Grass
Eragrostis	unioloides	scabriuscula	Scabre	Chinese	Love Grass
Eragrostis	variabilis	variabilis	Original	Kawelu	Love Grass
Eragrostis	variabilis	ciliata	Ciliated	Kawelu	Love Grass
Eragrostis	variabilis	equitans	Equator	Kawelu	Love Grass
Eragrostis	variabilis	hawaiiensis	Hawaiian	Kawelu	Love Grass
Eragrostis	variabilis	hobdyi	Hobdy	Kawelu	Love Grass
Eragrostis	variabilis	niihauensis	Niihau	Kawelu	Love Grass
Eragrostis	variabilis	phleoides	Reed	Kawelu	Love Grass
Eragrostis	variabilis	thyrsoidea	Thyrse	Kawelu	Love Grass
Eragrostis	variabilis	wahowensis	Wahow	Kawelu	Love Grass
Eragrostis	viscosa	viscosa	Original	Sticky	Love Grass
Eragrostis	viscosa	glutinosa	Glutin	Sticky	Love Grass
Eragrostis	viscosa	hirsutissima	Hairy	Sticky	Love Grass
Eragrostis	viscosa	mangalorica	Mangalor	Sticky	Love Grass
Eragrostis	viscosa	pilosissima	Downy	Sticky	Love Grass
Eragrostis	viscosa	retinorrhoea	Retin	Sticky	Love Grass
Eragrostis	viscosa	strigosa	Bristle	Sticky	Love Grass
Eragrostis	viscosa	terebinthacea	Turpentine	Sticky	Love Grass
Eragrostis	viscosa	transvaalensis	Transvaal	Sticky	Love Grass

Millet Foxtail Bristle Grass (*Setaria italica*)

Genus	species	infraspecies	cultivar	specimen	Genome

Setaria	*faberi*	.	.	Chinese Millet	Foxtail Bristle Grass
Setaria	*italica*	.	.	Millet	Foxtail Bristle Grass
Setaria	*italica*	*italica*	Original	Millet	Foxtail Bristle Grass
Setaria	*italica*	*aegyptiaca*	Egyptian	Millet	Foxtail Bristle Grass
Setaria	*italica*	*asiatica*	Asian	Millet	Foxtail Bristle Grass
Setaria	*italica*	*californica*	California	Millet	Foxtail Bristle Grass
Setaria	*italica*	*caudatus*	Stem	Millet	Foxtail Bristle Grass
Setaria	*italica*	*chinense*	Chinese	Millet	Foxtail Bristle Grass
Setaria	*italica*	*colchica*	Crocus	Millet	Foxtail Bristle Grass
Setaria	*italica*	*compacta*	Compact	Millet	Foxtail Bristle Grass
Setaria	*italica*	*elongata*	Elongated	Millet	Foxtail Bristle Grass
Setaria	*italica*	*erythrosperma*	Earth Red Seed	Millet	Foxtail Bristle Grass
Setaria	*italica*	*flavida*	Yellow	Millet	Foxtail Bristle Grass
Setaria	*italica*	*germanica*	German	Millet	Foxtail Bristle Grass
Setaria	*italica*	*globularis*	Globular	Millet	Foxtail Bristle Grass
Setaria	*italica*	*glomerata*	Clustered	Millet	Foxtail Bristle Grass
Setaria	*italica*	*inerme*	Spineless	Millet	Foxtail Bristle Grass
Setaria	*italica*	*intermedia*	Intermediate	Millet	Foxtail Bristle Grass
Setaria	*italica*	*itieri*	Itier	Millet	Foxtail Bristle Grass
Setaria	*italica*	*japonica*	Japanese	Millet	Foxtail Bristle Grass
Setaria	*italica*	*longisetum*	Long Seta	Millet	Foxtail Bristle Grass
Setaria	*italica*	*macrochaeta*	Large Bristle	Millet	Foxtail Bristle Grass
Setaria	*italica*	*major*	Greater	Millet	Foxtail Bristle Grass
Setaria	*italica*	*maritima*	Seaside	Millet	Foxtail Bristle Grass
Setaria	*italica*	*melfrugum*	Melfrug	Millet	Foxtail Bristle Grass
Setaria	*italica*	*melinis*	Honey	Millet	Foxtail Bristle Grass
Setaria	*italica*	*metzgeri*	Metzger	Millet	Foxtail Bristle Grass
Setaria	*italica*	*miliacea*	Italian	Millet	Foxtail Bristle Grass
Setaria	*italica*	*moharica*	Mohar	Millet	Foxtail Bristle Grass
Setaria	*italica*	*multiseta*	Multi Seta	Millet	Foxtail Bristle Grass
Setaria	*italica*	*nigra*	Black	Millet	Foxtail Bristle Grass
Setaria	*italica*	*nigrofructa*	Black Pod	Millet	Foxtail Bristle Grass
Setaria	*italica*	*pachystachya*	Thick Stack	Millet	Foxtail Bristle Grass
Setaria	*italica*	*panis*	Pan	Millet	Foxtail Bristle Grass
Setaria	*italica*	*persica*	Persian	Millet	Foxtail Bristle Grass
Setaria	*italica*	*pumila*	Low	Millet	Foxtail Bristle Grass
Setaria	*italica*	*rubrofructa*	Red Pod	Millet	Foxtail Bristle Grass
Setaria	*italica*	*serotina*	Late Flowering	Millet	Foxtail Bristle Grass
Setaria	*italica*	*sibirica*	Siberian	Millet	Foxtail Bristle Grass
Setaria	*italica*	*stramineofructa*	Straw	Millet	Foxtail Bristle Grass
Setaria	*italica*	*violacea*	Violet	Millet	Foxtail Bristle Grass
Setaria	*viridis*	*viridis*	Original	Green Millet	Foxtail Bristle Grass
Setaria	*viridis*	*breviseta*	Short Bristle	Green Millet	Foxtail Bristle Grass
Setaria	*viridis*	*major*	Greater	Green Millet	Foxtail Bristle Grass
Setaria	*viridis*	*robusta-alba*	Robust White	Green Millet	Foxtail Bristle Grass
Setaria	*viridis*	*robusta-purpurea*	Robust Purple	Green Millet	Foxtail Bristle Grass
Setaria	*viridis*	*weinmannii*	Weinmann	Green Millet	Foxtail Bristle Grass

Sorgo Millet (*Sorghum spp.*)

Genus	species	infraspecies	cultivar	specimen	Genome
Sorghum	*arundinaceum*	*arundinaceum*	Original	Shatter Cane	Sorgo Grass
Sorghum	*arundinaceum*	*abyssinicum*	Abyssinian	Shatter Cane	Sorgo Grass
Sorghum	*arundinaceum*	*aethiopicum*	Ethiopian	Shatter Cane	Sorgo Grass
Sorghum	*arundinaceum*	*astoloniferus*	Stolon	Shatter Cane	Sorgo Grass
Sorghum	*arundinaceum*	*brevicarinatum*	Short Keel	Shatter Cane	Sorgo Grass
Sorghum	*arundinaceum*	*brevifolium*	Short Leaf	Shatter Cane	Sorgo Grass
Sorghum	*arundinaceum*	*castaneum*	Chestnut	Shatter Cane	Sorgo Grass
Sorghum	*arundinaceum*	*congoensis*	Congo	Shatter Cane	Sorgo Grass
Sorghum	*arundinaceum*	*effusum*	Effusive	Shatter Cane	Sorgo Grass
Sorghum	*arundinaceum*	*infrequens*	Infrequent	Shatter Cane	Sorgo Grass
Sorghum	*arundinaceum*	*kinshasanensis*	Kinshasan	Shatter Cane	Sorgo Grass
Sorghum	*arundinaceum*	*lanceolatum*	Lance Leaf	Shatter Cane	Sorgo Grass
Sorghum	*arundinaceum*	*macrochaetum*	Large Bristle	Shatter Cane	Sorgo Grass
Sorghum	*arundinaceum*	*ornatum*	Ornate	Shatter Cane	Sorgo Grass
Sorghum	*arundinaceum*	*panicoides*	Panic Grass	Shatter Cane	Sorgo Grass
Sorghum	*arundinaceum*	*pugionifolium*	Pugion	Shatter Cane	Sorgo Grass
Sorghum	*arundinaceum*	*somaliense*	Somalia	Shatter Cane	Sorgo Grass
Sorghum	*arundinaceum*	*stapfii*	Stapf	Shatter Cane	Sorgo Grass
Sorghum	*arundinaceum*	*swahilorum*	Swahili	Shatter Cane	Sorgo Grass
Sorghum	*arundinaceum*	*usambarense*	Usambar	Shatter Cane	Sorgo Grass
Sorghum	*arundinaceum*	*verticilliflorum*	Whorled Flower	Shatter Cane	Sorgo Grass
Sorghum	*arundinaceum*	*virgatum*	Tunis Grass	Shatter Cane	Sorgo Grass
Sorghum	*arundinaceum*	*vogelianum*	Vogel	Shatter Cane	Sorgo Grass
Sorghum	*bicolor*	*caudatum*	Stalk	Millet	Sorgo Grass
Sorghum	*bicolor-caffrorum*	*caffrorum*	Original	Kaffir Millet	Sorgo Grass
Sorghum	*bicolor-caffrorum*	*alba*	White	Kaffir Millet	Sorgo Grass
Sorghum	*bicolor-caffrorum*	*nigra-caffrorum*	Black	Kaffir Millet	Sorgo Grass
Sorghum	*bicolor-caffrorum*	*brunnea*	Brown	Kaffir Millet	Sorgo Grass
Sorghum	*bicolor-durra*	*durra*	Original	Durra	Sorgo Grass

Sorghum	*bicolor-durra*	*nigra-miloi*	Black Milo 7.0'	Durra	Sorgo Grass
Sorghum	*bicolor-durra*	*gooseneckii*	Gooseneck ?.?'	Durra	Sorgo Grass
Sorghum	*bicolor-miliaceum*	*miliaceum*	Original	Great Millet	Sorgo Grass
Sorghum	*bicolor-miliaceum*	*ambra-alba-sperma*	White Seeded Amber 5.0'	Great Millet	Sorgo Grass
Sorghum	*bicolor-miliaceum*	*blackii*	Black	Great Millet	Sorgo Grass
Sorghum	*bicolor-miliaceum*	*dwarf-grainii*	Dwarf Grain 3.0'	Great Millet	Sorgo Grass
Sorghum	*bicolor-miliaceum*	*early-hegari*	Early Hegar 3.7'	Great Millet	Sorgo Grass
Sorghum	*bicolor-miliaceum*	*golden-shallui*	Golden Shallu	Great Millet	Sorgo Grass
Sorghum	*bicolor-miliaceum*	*egyptian-wheatii*	Egyptian Wheat	Great Millet	Sorgo Grass
Sorghum	*bicolor-miliaceum*	*hadleyi-tetrapoidea*	HadleyTetrapoid	Great Millet	Sorgo Grass
Sorghum	*bicolor-miliaceum*	*jawari*	Jawar ?.?'	Great Millet	Sorgo Grass
Sorghum	*bicolor-miliaceum*	*lesothoensis*	Lesotho 10.'	Great Millet	Sorgo Grass
Sorghum	*bicolor-miliaceum*	*milo-rubra*	Milo Red	Great Millet	Sorgo Grass
Sorghum	*bicolor-miliaceum*	*perennial-M61*	#M61 Perennial 3.5'	Great Millet	Sorgo Grass
Sorghum	*bicolor-miliaceum*	*poppingii*	Popping ?.?'	Great Millet	Sorgo Grass
Sorghum	*bicolor-miliaceum*	*poppingii-tarhahumarai*	Tarahumara Popping	Great Millet	Sorgo Grass
Sorghum	*bicolor-miliaceum*	*poppingii-tepehuana*	Tepehuan Popping	Great Millet	Sorgo Grass
Sorghum	*bicolor-miliaceum*	*rox-orangei*	Rox Orange 7.5'	Great Millet	Sorgo Grass
Sorghum	*bicolor-miliaceum*	*texacoa*	Texaco 4.5'	Great Millet	Sorgo Grass
Sorghum	*bicolor-miliaceum*	*vilgorei*	Vilgore Crookneck	Great Millet	Sorgo Grass
Sorghum	*bicolor-miliaceum*	*white-mountain-apachei*	White Mountain Apache	Great Millet	Sorgo Grass
Sorghum	*bicolor-saccharatum*	*saccharatum*	Original	Molasses	Sorgo Grass
Sorghum	*bicolor-saccharatum*	*africana-alba*	White African 8.0'	Molasses	Sorgo Grass
Sorghum	*bicolor-saccharatum*	*ambra-amesii*	Ames Amber	Molasses	Sorgo Grass
Sorghum	*bicolor-saccharatum*	*ambra-aurantia*	Orange Amber	Molasses	Sorgo Grass
Sorghum	*bicolor-saccharatum*	*ambra-molassesi*	Amber 9.0'	Molasses	Sorgo Grass
Sorghum	*bicolor-saccharatum*	*ambra-molassesi-texas-blackii*	Texas Black Amber	Molasses	Sorgo Grass
Sorghum	*bicolor-saccharatum*	*apache-redii*	Apache Red ?.?'	Molasses	Sorgo Grass
Sorghum	*bicolor-saccharatum*	*brandisi*	Brandis 9.0'	Molasses	Sorgo Grass
Sorghum	*bicolor-saccharatum*	*cana-ganchadoi*	Cana Ganchado 6.0'	Molasses	Sorgo Grass
Sorghum	*bicolor-saccharatum*	*dalei-john-cofferi*	John Coffer Dale 7.0'	Molasses	Sorgo Grass
Sorghum	*bicolor-saccharatum*	*gilai*	Gila	Molasses	Sorgo Grass
Sorghum	*bicolor-saccharatum*	*ho-kii*	Ho K ?.?'	Molasses	Sorgo Grass
Sorghum	*bicolor-saccharatum*	*honey-dripii*	Honey Drip 9.0'	Molasses	Sorgo Grass
Sorghum	*bicolor-saccharatum*	*mennonitei*	Mennonite 8.0'	Molasses	Sorgo Grass
Sorghum	*bicolor-saccharatum*	*missouriensis-rubra*	Missouri Red	Molasses	Sorgo Grass
Sorghum	*bicolor-saccharatum*	*northern-sugar-canei*	Northern Sugar Cane 9.0'	Molasses	Sorgo Grass
Sorghum	*bicolor-saccharatum*	*onavas-redii*	Onavas Red ?.?'	Molasses	Sorgo Grass
Sorghum	*bicolor-saccharatum*	*rioensis*	Rio 7.5'	Molasses	Sorgo Grass
Sorghum	*bicolor-saccharatum*	*salts-redii*	Salts Red ?.?'	Molasses	Sorgo Grass
Sorghum	*bicolor-saccharatum*	*san-felipe-puebloi*	San Felipe Pueblo 5.0'	Molasses	Sorgo Grass
Sorghum	*bicolor-saccharatum*	*sand-mountainii*	Sand Mountain	Molasses	Sorgo Grass
Sorghum	*bicolor-saccharatum*	*santa-fe-redii*	Santa Fe Red	Molasses	Sorgo Grass
Sorghum	*bicolor-saccharatum*	*silver-ribbonii*	Silver Ribbon	Molasses	Sorgo Grass
Sorghum	*bicolor-saccharatum*	*sweet-blackii*	Sweet Black	Molasses	Sorgo Grass
Sorghum	*bicolor-saccharatum*	*tenneessee-tall-girlsi*	Tenneessee Tall Girls 8.5'	Molasses	Sorgo Grass
Sorghum	*bicolor-saccharatum*	*texas-long-sweetii*	Texas Long Sweet 8.0'	Molasses	Sorgo Grass
Sorghum	*bicolor-saccharatum*	*tohono-o'odhamii*	Tohono O'odham	Molasses	Sorgo Grass
Sorghum	*bicolor-saccharatum*	*warihio-indiana*	Warihio Indian	Molasses	Sorgo Grass
Sorghum	*bicolor-technicum*	*technicum*	Original ?.?'	Broomcorn	Sorgo Grass
Sorghum	*bicolor-technicum*	*blackii*	Black 6.0'	Broomcorn	Sorgo Grass
Sorghum	*bicolor-technicum*	*blackii-seededii*	Black Seeded	Broomcorn	Sorgo Grass
Sorghum	*bicolor-technicum*	*blackii-creekii*	Black Creek	Broomcorn	Sorgo Grass
Sorghum	*bicolor-technicum*	*coloredii*	Colored	Broomcorn	Sorgo Grass
Sorghum	*bicolor-technicum*	*hadleyi-deeri*	Deer Hadley 9.0'	Broomcorn	Sorgo Grass
Sorghum	*bicolor-technicum*	*hadleyi-T114*	#114 Hadley	Broomcorn	Sorgo Grass
Sorghum	*bicolor-technicum*	*hadleyi-T115*	#115 Hadley 8.0'	Broomcorn	Sorgo Grass
Sorghum	*bicolor-technicum*	*hudsonii-mixta*	Hudson Mixed	Broomcorn	Sorgo Grass
Sorghum	*bicolor-technicum*	*hungarian-black-seededii*	Hungarian Black Seeded 10.'	Broomcorn	Sorgo Grass
Sorghum	*bicolor-technicum*	*hungarian-redii*	Hungarian Red	Broomcorn	Sorgo Grass
Sorghum	*bicolor-technicum*	*iowa-redii*	Iowa Red 8.0'	Broomcorn	Sorgo Grass
Sorghum	*bicolor-technicum*	*iowa-sweetii*	Iowa Sweet 9.0'	Broomcorn	Sorgo Grass
Sorghum	*bicolor-technicum*	*mayoi*	Mayo ?.?'	Broomcorn	Sorgo Grass
Sorghum	*bicolor-technicum*	*redii*	Red ?.?'	Broomcorn	Sorgo Grass
Sorghum	*bicolor-technicum*	*red-deeri*	Red Deer ?.?'	Broomcorn	Sorgo Grass
Sorghum	*bicolor-technicum*	*red-seededii*	Red Seeded	Broomcorn	Sorgo Grass
Sorghum	*bicolor-technicum*	*seed-saversii-mixta*	Seed Savers Mixed	Broomcorn	Sorgo Grass
Sorghum	*bicolor-technicum*	*shaker-standardii*	Shaker Standard	Broomcorn	Sorgo Grass
Sorghum	*bicolor-technicum*	*yellowii*	Yellow ?.?'	Broomcorn	Sorgo Grass

Family 722.022.4 *Oryzaceae* Rice

Genus	Species	Specimen	Genome	C:Culinary, M:Medicinal, E:Economic
Oryza	*barthii*	Barth	Rice	C:SD(grain):**VG**
Oryza	*glaberrima*	African	Rice	C:SD(grain):**VG**
Oryza	*latifolia*	Broad Leaf	Rice	C:SD(grain):**VG**
Oryza	*longistaminata*	Long Stamen	Rice	C:SD(grain):**VG**
Oryza	*punctata*	Dotted Red	Rice	C:SD(grain):**VG**

Oryza	*rufipogon*	Brown Beard Red	Rice	C:SD(grain):**VG**
Oryza	*sativa*	**Medium Grain**	**Rice**	C:SD(grain):**VG**
Oryza	*sativa-brevisperma*	**Short Grain**	**Rice**	C:SD(grain):**VG**
Oryza	*sativa-longasperma*	**Long Grain**	**Rice**	C:SD(grain):**VG**
Zizania	*aquatica*	**Annual**	**Wild Rice**	C:SD(grain):**VG**
Zizania	*latifolia*	Manchurian	Wild Rice	C:SD(grain):**VG**
Zizania	*miliacea*	Millet	Wild Rice	C:SD(grain):**VG**
Zizania	*palustris*	Marsh	Wild Rice	C:SD(grain):**VG**
Zizania	*aquatica*	Annual	Wild Rice	C:SD(grain):**VG**
Zizania	*texana*	Texas	Wild Rice	C:SD(grain):**VG**

Varieties

Rice (*Oryza sativa*)

Genus	species	Infraspecies	cultivar	Specimen	Genome
Oryza	*sativa*	*fatua*	Fatu	Medium Grain	Rice
Oryza	*sativa*	*indica*	India	Medium Grain	Rice
Oryza	*sativa*	*japonica*	Japanese	Medium Grain	Rice
Oryza	*sativa*	*japonica-cigalon*	Cigalon Japanese	Medium Grain	Rice
Oryza	*sativa*	*rubribarba*	Red Barb	Medium Grain	Rice
Oryza	*sativa*	*savannae*	Savannah	Medium Grain	Rice

Wild Rice (*Zizania aquatica*)

Genus	species	Infraspecies	cultivar	Specimen	Genome
Zizania	*aquatica*	*aquatica*	Original	Annual	Wild Rice
Zizania	*aquatica*	*brevis*	Short	Annual	Wild Rice
Zizania	*aquatica*	*interior*	Inland	Annual	Wild Rice

Family 722.022.5 *Secaleaceae* Rye

Genus	Species	Specimen	Genome	C:Culinary, M:Medicinal, E:Economic
Secale	*africanum*	African	Rye	C:SD(grain):**VG**
Secale	*anatolicum*	Anatolian	Rye	C:SD(grain):**VG**
Secale	*caput-medusae*	Medusa Head	Rye	C:SD(grain):**VG**
Secale	*cereale*	**Cereal**	**Rye**	C:SD(grain):**VG**
Secale	*ciliatiglume*	Hairy Glume	Rye	C:SD(grain):**VG**
Secale	*derzhavinii*	Derzhavin	Rye	C:SD(grain):**VG**
Secale	*hordeaceum*	Barley	Rye	C:SD(grain):**VG**
Secale	*montanum*	Caucasus Mountain	Rye	C:SD(grain):**VG**
Secale	*segetale*	Cornfield	Rye	C:SD(grain):**VG**
Secale	*sylvestre*	Forest	Rye	C:SD(grain):**VG**
Secale	*vavilovii*	Vavilov	Rye	C:SD(grain):**VG**
Secale	*villosum*	Mosquito Grass	Rye	C:SD(grain):**VG**

Varieties

Rye (*Secale cereale*)

Genus	species	Infraspecies	cultivar	Specimen	Genome
Secale	*anatolicum*	*anatolicum*	Original	Anatolian	Rye
Secale	*anatolicum*	*daralagesii*	Daralages	Anatolian	Rye
Secale	*anatolicum*	*perennans*	Perennial	Anatolian	Rye
Secale	*anatolicum*	*rhodopaeum*	Rhodope Mountain	Anatolian	Rye
Secale	*cereale*	*cereale*	Original	Cereal	Rye
Secale	*cereale*	*aestivum*	Summer	Cereal	Rye
Secale	*cereale*	*aidinum*	Aidin	Cereal	Rye
Secale	*cereale*	*ancestrale*	Ancestral	Cereal	Rye
Secale	*cereale*	*arenosum*	Desert	Cereal	Rye
Secale	*cereale*	*arundinaceum*	Reed	Cereal	Rye
Secale	*cereale*	*brevispicatum*	Short Spike	Cereal	Rye
Secale	*cereale*	*compositum*	Composite	Cereal	Rye
Secale	*cereale*	*creticum*	Crete	Cereal	Rye
Secale	*cereale*	*hybernum*	Winter	Cereal	Rye
Secale	*cereale*	*indo-europaeum*	Indoeuropean	Cereal	Rye
Secale	*cereale*	*karaburun*	Karaburun	Cereal	Rye
Secale	*cereale*	*montaniforme*	Mountain	Cereal	Rye
Secale	*cereale*	*ramosum*	Branching	Cereal	Rye
Secale	*cereale*	*rigidum*	Stiff	Cereal	Rye
Secale	*cereale*	*spontaneum*	Spontaneous	Cereal	Rye
Secale	*cereale*	*strictum*	Upright	Cereal	Rye
Secale	*cereale*	*tetraploidum*	Tetraploid	Cereal	Rye
Secale	*cereale*	*triflorum*	Triple Flower	Cereal	Rye
Secale	*cereale*	*tsitsinii*	Tsitsin	Cereal	Rye
Secale	*cereale*	*turkestanicum*	Turkesta	Cereal	Rye
Secale	*cereale*	*vernum*	Spring	Cereal	Rye

Family 722.022.6 *Tripsacaceae* Gamagrass

Wild Maize

Gammagrass (*Tripsacum spp.*) is a Wild Maize and is the ancestor of cultivated Maize below.

Cultivated Maize

Teosinte (*Euchlaena mexicana*) is found in the Mountains of Mexico and Guatemala, where it often hybridizes when it is found close to culivated Corn. Teosinte is the ancestor of modern corn.

Flint/Flour Maize (*Zea huehuetenangensis*) was first cultivated in Mexico in 800 AD. It was found by Christopher Columbus growing in fields on the island of Haiti and the Dominican Republic (also known as Hispanolia Island) on October 16, 1492. It is a primitive corn that is still cultivated today.

Reed Grain Grass (*Coix spp.*) is a Maize genus that has six species. All of the species are native to tropical Asia. The species have been grown for grain in China since 200 AD. Female flowers are enclosed in a hard tear shaped husk used to make rosary beads.

Corn

The cultivation of corn in Central and South America helped feed the Aztecs, Maya, and Inca. American Indians planted corn in the United States. There are several types of cultivated corn based on the shape of the kernel.

Wild Corn (*Zea mays*) pollen was found in fossils in Mexico City, Mexico and dated by archeologists to 60,000 BC. Kernels were found in modern times in Mexico Tecahuan caves dating to 3500 BC. Kernels were found in dry coastal zones in Peru dating to 1000 BC. Kernels were found in the Andes Mountains of Peru dating to 500 BC.

Dent Corn (*Zea mays-denta*) kernels have a indentation on the top, and hard shell top and soft pulp. **Dent Corn** is blue (*Zea mays-denta-blauea*), brown (*Zea mays-denta-brunnea*), green (*Zea mays-denta-gronea*), multicolor (*Zea mays-denta-multicolor*), orange (Zea mays-dental-aurantia), purple (*Zea mays-denta-purpurea*), red (*Zea mays-denta-rubra*), white (*Zea mays-denta-alba*) or yellow (*Zea mays-denta-helva*). When the color is not specified in the source description, the variety is placed under the species Cornmeal Dent Corn (*Zea mays-denta-farina*).

Pop Corn (*Zea mays-evereta*) bursts open when heated. Pop Corn is black (*Zea mays-popa -nigra*), blue (*Zea mays-popa-blauea*), brown (*Zea mays-popa-brunnea*), multicolor (*Zea mays-popa-multicolor*), purple (*Zea mays-popa-purpurea*), red (*Zea mays-popa-rubra*), white (*Zea mays-popa-alba*), or yellow (*Zea mays-popa-helva*).

Flour Corn (Zea *mays-farina*) has soft shell and pulp kernels that are easy to grind. It is blue (*Zea mays-farina-blauea*), multicolor (*Zea mays-farina-multicolor*), purple (*Zea mays-farina-purpurea*), red (*Zea mays-farina-rubra*), white (*Zea mays-farina-alba*), or yellow (*Zea mays-farina-helva*).

Flint Corn (Zea *mays-flinta*) has very hard kernels. Flint corn is blue (*Zea mays-flinta-blauea*), multicolor (*Zea mays-flinta-multicolor*), red (*Zea mays-flinta-rubra*), white (*Zea mays-flinta-alba*), or yellow (*Zea mays-flinta-helva*).

Pod Corn (*Zea mays-parviglumis*) is no longer cultivated but still survives in the wild.

Sweet Corn (*Zea mays-suavis*) kernels lack the ability to convert sugar into starch. Modern Sweet Corn has been bred to be "sugary" or "super-sweet". Today, eating Sweet Corn raw is almost like eating 'Candy Corn". Sweet Corn is black (*Zea mays-suavis -nigra*), blue (*Zea mays-suavis-blauea*), brown (*Zea mays-suavis-brunnea*), multicolor (*Zea mays-suavis-multicolor*), purple (*Zea mays-suavis-purpurea*), red (*Zea mays-suavis-rubra*), sugar (*Zea mays-suavis-sucra*), white (*Zea mays-suavis-alba*), or yellow (*Zea mays-suavis-helva*).

Genus	Species	Specimen	Genome	C:Culinary, M:Medicinal, E:Economic
Tripsacum	*dactyloides*	Finger	Gama Grass	C:SD(grain):**VG**
Tripsacum	*fasciculatum*	Guatemalan	Gama Grass	
Tripsacum	*floridanum*	Florida	Gama Grass	
Tripsacum	*lanceolatum*	Mexican	Gama Grass	
Tripsacum	*latifolium*	Broad Leaf	Gama Grass	
Coix	*angulatis*	Angled	Reed Grain Grass	
Coix	*lacryma-jobi*	Job's Tears	Reed Grain Grass	
Euchlaena	*mexicana*	Mexican	Teosinte	
Zea	*huehuetenangensis*	Flint/Flour Maize	Zeatin Grass	
Zea	*mays-denta-alba*	White Dent Corn	Zeatin Grass	
Zea	*mays-denta-blauea*	Blue Dent Corn	Zeatin Grass	
Zea	*mays-denta-brunnea*	Brown Dent Corn	Zeatin Grass	
Zea	*mays-denta-farina*	Cornmeal Dent Corn	Zeatin Grass	
Zea	*mays-denta-groena*	Green Dent Corn	Zeatin Grass	
Zea	*mays-denta-helva*	Yellow Dent Corn	Zeatin Grass	
Zea	*mays-denta-mixta*	Multicolor Dent Corn	Zeatin Grass	
Zea	*mays-denta-aurantia*	Orange Dent Corn	Zeatin Grass	
Zea	*mays-denta-purpurea*	Purple Dent Corn	Zeatin Grass	
Zea	*mays-denta-rubra*	Red Dent Corn	Zeatin Grass	
Zea	*mays-evereta-alba*	White Popcorn	Zeatin Grass	
Zea	*mays-evereta-blauea*	Blue Popcorn	Zeatin Grass	
Zea	*mays-evereta-brunnea*	Brown Popcorn	Zeatin Grass	
Zea	*mays-evereta-helva*	Yellow Popcorn	Zeatin Grass	
Zea	*mays-evereta-mixta*	Multicolor Popcorn	Zeatin Grass	
Zea	*mays-evereta-purpurea*	Purple Popcorn	Zeatin Grass	
Zea	*mays-evereta-rubra*	Red Popcorn	Zeatin Grass	
Zea	*mays-farina-alba*	White Flour Corn	Zeatin Grass	
Zea	*mays-farina-blauea*	Blue Flour Corn	Zeatin Grass	
Zea	*mays-farina-brunnea*	Brown Flour Corn	Zeatin Grass	
Zea	*mays-farina-mixta*	Multicolor Flour Corn	Zeatin Grass	
Zea	*mays-farina-purpurea*	Purple Flour Corn	Zeatin Grass	
Zea	*mays-farina-rubra*	Red Flour Corn	Zeatin Grass	
Zea	*mays-flinta-alba*	White Flint Corn	Zeatin Grass	
Zea	*mays-flinta-aurantia*	Orange Flint Corn	Zeatin Grass	
Zea	*mays-flinta-blauea*	Blue Flint Corn	Zeatin Grass	
Zea	*mays-flinta-helva*	Yellow Flint Corn	Zeatin Grass	
Zea	*mays-flinta-mixta*	Multicolor Flint Corn	Zeatin Grass	
Zea	*mays-flinta-purpurea*	Purple Flint Corn	Zeatin Grass	
Zea	*mays-flinta-rubra*	Red Flint Corn	Zeatin Grass	
Zea	*mays-suavis -alba*	White Sweet Corn	Zeatin Grass	
Zea	*mays-suavis-blauea*	Blue Sweet Corn	Zeatin Grass	
Zea	*mays-suavis-helva*	Yellow Sweet Corn	Zeatin Grass	
Zea	*mays-suavis-mixta*	Multicolor Sweet Corn	Zeatin Grass	
Zea	*mays-suavis-purpurea*	Purple Sweet Corn	Zeatin Grass	
Zea	*mays-suavis-rubra*	Red Sweet Corn	Zeatin Grass	
Zea	*mays-suavis-sucra*	Sugar Sweet Corn	Zeatin Grass	

Varieties

Teosinte (*Euchlaena mexicana*)

Genus	species	Infraspecies	cultivar	specimen	Genome
Euchlaena	*mexicana*	*baltimorei*	Baltimore	Mexican	Teosinte
Euchlaena	*mexicana*	*diploperennis*	Diploid Perennial	Mexican	Teosinte
Euchlaena	*mexicana*	*luxurians*	Luxurious	Mexican	Teosinte
Euchlaena	*mexicana*	*north-tepehuanii*	North Tepehua	Mexican	Teosinte
Euchlaena	*mexicana*	*perennis*	Perennial	Mexican	Teosinte

Maize, Flint/Flour (*Zea huehetenangensis*)

Genus	species	Infraspecies	cultivar	specimen	Genome
Zea	*huehuetenangensis*	*babyi*	Baby	Flint/Flour Maize	Zeatin Grass
Zea	*huehuetenangensis*	*eastern-tarahumara-tajawei*	Eastern Tarahumara Tajawe	Flint/Flour Maize	Zeatin Grass
Zea	*huehuetenangensis*	*flor-del-rioensis*	Flor Del Rio	Flint/Flour Maize	Zeatin Grass
Zea	*huehuetenangensis*	*gila-pimaensis*	Gila Pima	Flint/Flour Maize	Zeatin Grass
Zea	*huehuetenangensis*	*hispanic-pueblo-redii*	Hispanic Pueblo Red	Flint/Flour Maize	Zeatin Grass
Zea	*huehuetenangensis*	*japonica-striped-maizei*	Japonica Striped	Flint/Flour Maize	Zeatin Grass
Zea	*huehuetenangensis*	*jacarilla-apache-conchoi*	Jacarilla Apache Concho	Flint/Flour Maize	Zeatin Grass
Zea	*huehuetenangensis*	*las-trumpas-conchoi*	Las Tumpas Concho	Flint/Flour Maize	Zeatin Grass
Zea	*huehuetenangensis*	*parch-magentai*	Magenta Parch	Flint/Flour Maize	Zeatin Grass
Zea	*huehuetenangensis*	*pod-cornii*	IncaCeremonialPod	Flint/Flour Maize	Zeatin Grass
Zea	*huehuetenangensis*	*rainbow-incai*	Rainbow Inca	Flint/Flour Maize	Zeatin Grass
Zea	*huehuetenangensis*	*santo-domingo-posolei*	Santo Domingo Posole	Flint/Flour Maize	Zeatin Grass

Genus	species	Infraspecies	cultivar	specimen	Genome
Zea	*huehuetenangensis*	*tarahumara-maiz-chomoi*	Tarahumara Maiz Chomo	Flint/Flour Maize	Zeatin Grass

Corn, Dent (*Zea mays-denta*)

Genus	species	Infraspecies	cultivar	specimen	Genome
Zea	*mays-denta-alba*	*adams-earlyi*	Adams Early	White Dent Corn	Zeatin Grass
Zea	*mays-denta-alba*	*adams-earlyi-improva*	Improved Adams Early	White Dent Corn	Zeatin Grass
Zea	*mays-denta-alba*	*alabama-white-dentii*	Alabama	White Dent Corn	Zeatin Grass
Zea	*mays-denta-alba*	*boone-county-whitei*	Boone County	White Dent Corn	Zeatin Grass
Zea	*mays-denta-alba*	*carolina-gourdseedii*	Carolina Gourdseed	White Dent Corn	Zeatin Grass
Zea	*mays-denta-alba*	*cherokee-white-eaglei*	Cherokee White Eagle	White Dent Corn	Zeatin Grass
Zea	*mays-denta-alba*	*dia-de-san-juanii*	Dia De San Juan	White Dent Corn	Zeatin Grass
Zea	*mays-denta-alba*	*ensilage-eurikai*	Eurika Ensilage	White Dent Corn	Zeatin Grass
Zea	*mays-denta-alba*	*ensilage-goliath-siloi*	Goliath Silo Ensilage	White Dent Corn	Zeatin Grass
Zea	*mays-denta-alba*	*greenfieldii-120*	120 Day Greenfield	White Dent Corn	Zeatin Grass
Zea	*mays-denta-alba*	*hickory-king-whitei*	Hickory King	White Dent Corn	Zeatin Grass
Zea	*mays-denta-alba*	*hopii*	Hopi	White Dent Corn	Zeatin Grass
Zea	*mays-denta-alba*	*mexican-june-whitei*	Mexican June	White Dent Corn	Zeatin Grass
Zea	*mays-denta-alba*	*pride-of-saline-whitei*	Pride Of Saline	White Dent Corn	Zeatin Grass
Zea	*mays-denta-alba*	*silver-kingii*	Silver King	White Dent Corn	Zeatin Grass
Zea	*mays-denta-alba*	*silverminei*	Slivermine	White Dent Corn	Zeatin Grass
Zea	*mays-denta-alba*	*texas-gourdseedii*	Texas Gourdseed	White Dent Corn	Zeatin Grass
Zea	*mays-denta-alba*	*trucker's-favorite-whitei*	Trucker's Favorite	White Dent Corn	Zeatin Grass
Zea	*mays-denta-alba*	*tait's-white-dentii*	Tait's	White Dent Corn	Zeatin Grass
Zea	*mays-denta-alba*	*virginia-gourdseedii*	Virginia Gourdseed	White Dent Corn	Zeatin Grass
Zea	*mays-denta-alba*	*white-surecropperi*	White Surecropper	White Dent Corn	Zeatin Grass
Zea	*mays-denta-alba*	*yeome-vatchii*	Yoeme Vatch Hominy	White Dent Corn	Zeatin Grass
Zea	*mays-denta-aurantia*	*ernest-strubbe's-orangei*	Ernest Strubbe's	Orange Dent Corn	Zeatin Grass
Zea	*mays-denta-blauea*	*blue-claredgei*	Blue Claredge	Blue Dent Corn	Zeatin Grass
Zea	*mays-denta-blauea*	*blue-dentii*	Hopi	Blue Dent Corn	Zeatin Grass
Zea	*mays-denta-blauea*	*ernest-strubbe's-bluei*	Ernest Strubbe's	Blue Dent Corn	Zeatin Grass
Zea	*mays-denta-blauea*	*mccormack's-blue-giantii*	McCormack"s Blue Giant	Blue Dent Corn	Zeatin Grass
Zea	*mays-denta-brunnea*	*ernst-strubbe's-brownii*	Ernest Strubbe's	Brown Dent Corn	Zeatin Grass
Zea	*mays-denta-farina*	*duncan-dentii*	Duncan	Cornmeal Dent Corn	Zeatin Grass
Zea	*mays-denta-farina*	*john-haulkii*	John Haulk	Cornmeal Dent Corn	Zeatin Grass
Zea	*mays-denta-farina*	*norfolk-marketii*	Norfolk Market	Cornmeal Dent Corn	Zeatin Grass
Zea	*mays-denta-farina*	*pencil-cob-cornii*	Pencil Cob	Cornmeal Dent Corn	Zeatin Grass
Zea	*mays-denta-farina*	*tennessee-red-cobii*	Tennessee Red Cob	Cornmeal Dent Corn	Zeatin Grass
Zea	*mays-denta-farina*	*tarahumara-pepitilloi*	Tarahumara Pepitillo	Cornmeal Dent Corn	Zeatin Grass
Zea	*mays-denta-groena*	*oxacaensis*	Oxacan	Green Dent Corn	Zeatin Grass
Zea	*mays-denta-helva*	*abbe-hillsi*	Abbe Hills	Yellow Dent Corn	Zeatin Grass
Zea	*mays-denta-helva*	*buck-lantz-dentii*	Buck Lantz	Yellow Dent Corn	Zeatin Grass
Zea	*mays-denta-helva*	*greenfieldii-113*	113 Day Greenfield	Yellow Dent Corn	Zeatin Grass
Zea	*mays-denta-helva*	*greenfieldii-114*	114 Day Greenfield	Yellow Dent Corn	Zeatin Grass
Zea	*mays-denta-helva*	*henry-moore-yellowi*	Henry Moore	Yellow Dent Corn	Zeatin Grass
Zea	*mays-denta-helva*	*hickory-king-yellowi*	Hickory King	Yellow Dent Corn	Zeatin Grass
Zea	*mays-denta-helva*	*krug's-yellowi*	Krug's	Yellow Dent Corn	Zeatin Grass
Zea	*mays-denta-helva*	*lancaster-surecropii*	Lancaster Surecrop	Yellow Dent Corn	Zeatin Grass
Zea	*mays-denta-helva*	*manzano-yellowi*	Manzano	Yellow Dent Corn	Zeatin Grass
Zea	*mays-denta-helva*	*minnesota-mellum-13*	Minnesota Mellum	Yellow Dent Corn	Zeatin Grass
Zea	*mays-denta-helva*	*mortgage-lifteri*	Mortgage Lifter	Yellow Dent Corn	Zeatin Grass
Zea	*mays-denta-helva*	*mosby'si*	Mosby's	Yellow Dent Corn	Zeatin Grass
Zea	*mays-denta-helva*	*murdockii*	Murdock	Yellow Dent Corn	Zeatin Grass
Zea	*mays-denta-helva*	*nokomis-gold-field-cornii*	Nokomis Gold	Yellow Dent Corn	Zeatin Grass
Zea	*mays-denta-helva*	*nothstinei*	Nothstine	Yellow Dent Corn	Zeatin Grass
Zea	*mays-denta-helva*	*reid's-yellow-dentii*	Reid's	Yellow Dent Corn	Zeatin Grass
Zea	*mays-denta-helva*	*thompson-prolifici*	Thompson Prolific	Yellow Dent Corn	Zeatin Grass
Zea	*mays-denta-helva*	*tohono-o'odham-juneii*	Tohono O'odham June	Yellow Dent Corn	Zeatin Grass
Zea	*mays-denta-helva*	*trucker's-favorite-yellowi*	Trucker's Favorite	Yellow Dent Corn	Zeatin Grass
Zea	*mays-denta-helva*	*yellow-dentii*	Field	Yellow Dent Corn	Zeatin Grass
Zea	*mays-denta-helva*	*yellow-surecropperi*	Yellow Sure Cropper	Yellow Dent Corn	Zeatin Grass
Zea	*mays-denta-helva*	*yumanii*	Yuman	Yellow Dent Corn	Zeatin Grass
Zea	*mays-denta-mixta*	*all-purposei*	All Purpose	Multicolor Dent Corn	Zeatin Grass
Zea	*mays-denta-mixta*	*arkansas-red-&-whitei*	Arkansas Red & White	Multicolor Dent Corn	Zeatin Grass
Zea	*mays-denta-mixta*	*big-chiefii*	Big Chief	Multicolor Dent Corn	Zeatin Grass
Zea	*mays-denta-mixta*	*cherokee-blue-&-whitei*	Cherokee Blue & White	Multicolor Dent Corn	Zeatin Grass
Zea	*mays-denta-mixta*	*chicasaw-dentii*	Chickasaw	Multicolor Dent Corn	Zeatin Grass
Zea	*mays-denta-mixta*	*earth-tonesi*	Earth Tones	Multicolor Dent Corn	Zeatin Grass
Zea	*mays-denta-mixta*	*green-&-goldii*	Green & Gold	Multicolor Dent Corn	Zeatin Grass
Zea	*mays-denta-mixta*	*king-whitei*	King White & Blue	Multicolor Dent Corn	Zeatin Grass
Zea	*mays-denta-mixta*	*mayo-batchii*	Mayo River Batchi Yellow & Pink	Multicolor Dent Corn	Zeatin Grass
Zea	*mays-denta-mixta*	*tuxpenoensis*	Mayo River Valley Tuxpeno	Multicolor Dent Corn	Zeatin Grass
Zea	*mays-denta-mixta*	*mesquakiei*	Mesquakie	Multicolor Dent Corn	Zeatin Grass
Zea	*mays-denta-mixta*	*mexican-junei*	Mexican June Blue & White	Multicolor Dent Corn	Zeatin Grass
Zea	*mays-denta-mixta*	*ohio-calicoi*	Ohio Calico Red & White	Multicolor Dent Corn	Zeatin Grass
Zea	*mays-denta-mixta*	*old-hickory-king-compositei*	Old Hickory King Composite	Multicolor Dent Corn	Zeatin Grass
Zea	*mays-denta-mixta*	*white-eaglei*	White Eagle Blue & White	Multicolor Dent Corn	Zeatin Grass

Zea	*mays-denta-purpurea*	*ernest-strubbe's-purplei*	Ernest Strubbe's	Purple Dent Corn	Zeatin Grass
Zea	*mays-denta-rubra*	*aztec-redii*	Aztec	Red Dent Corn	Zeatin Grass
Zea	*mays-denta-rubra*	*beasley's-red-dentii*	Beasley's	Red Dent Corn	Zeatin Grass
Zea	*mays-denta-rubra*	*bloody-butcheri*	Bloody Butcher	Red Dent Corn	Zeatin Grass
Zea	*mays-denta-rubra*	*bloody-maryi*	Bloody Mary	Red Dent Corn	Zeatin Grass
Zea	*mays-denta-rubra*	*ernest-strubbe's-pinkii*	Ernest Strubbe's Pink	Red Dent Corn	Zeatin Grass
Zea	*mays-denta-rubra*	*northwestern-dentii*	Northwestern	Red Dent Corn	Zeatin Grass

Corn, Popcorn (*Zea mays-evereta*)

Genus	species	Infraspecies	cultivar	specimen	Genome
Zea	*mays-evereta-alba*	*baby-ricei*	Baby Rice	White Popcorn	Zeatin Grass
Zea	*mays-evereta-alba*	*baby-ricei-hullessi*	Hulless Baby Rice	White Popcorn	Zeatin Grass
Zea	*mays-evereta-alba*	*bearpawii*	Bear Paw	White Popcorn	Zeatin Grass
Zea	*mays-evereta-alba*	*japanese-hullessi*	Japanese Hulless	White Popcorn	Zeatin Grass
Zea	*mays-evereta-alba*	*onavenoi*	Onaveno	White Popcorn	Zeatin Grass
Zea	*mays-evereta-alba*	*pennsylvania-dutch-butter-flavoredii*	Pennsylvania Dutch Butter Flavored	White Popcorn	Zeatin Grass
Zea	*mays-evereta-alba*	*pop-whitei*	Pop	White Popcorn	Zeatin Grass
Zea	*mays-evereta-alba*	*reventadori*	Reventador	White Popcorn	Zeatin Grass
Zea	*mays-evereta-alba*	*tom-thumb-hullessi*	Hulless Tom Thumb	White Popcorn	Zeatin Grass
Zea	*mays-evereta-alba*	*white-ricei*	White Rice	White Popcorn	Zeatin Grass
Zea	*mays-evereta-blauea*	*black-azteci*	Aztec Black	Blue Popcorn	Zeatin Grass
Zea	*mays-evereta-blauea*	*blue-boyi*	Blue Boy	Blue Popcorn	Zeatin Grass
Zea	*mays-evereta-blauea*	*blue-popcornii*	Blue	Blue Popcorn	Zeatin Grass
Zea	*mays-evereta-blauea*	*cutie-bluesi*	Cutie Blues	Blue Popcorn	Zeatin Grass
Zea	*mays-evereta-blauea*	*little-boy-bluei*	Little Boy	Blue Popcorn	Zeatin Grass
Zea	*mays-evereta-blauea*	*miniature-bluei*	Miniature	Blue Popcorn	Zeatin Grass
Zea	*mays-evereta-brunnea*	*chapalotei*	Chapalote	Brown Popcorn	Zeatin Grass
Zea	*mays-evereta-brunnea*	*chocolate-cherryi*	Chocolate Cherry	Brown Popcorn	Zeatin Grass
Zea	*mays-evereta-brunnea*	*chocolate-popii*	Chocolate Pop	Brown Popcorn	Zeatin Grass
Zea	*mays-evereta-brunnea*	*cutie-brownii*	Cutie	Brown Popcorn	Zeatin Grass
Zea	*mays-evereta-brunnea*	*mahoganyi*	Mahogany	Brown Popcorn	Zeatin Grass
Zea	*mays-evereta-brunnea*	*pretty-popsi-shades-of-brownii*	Shades Of Brown Pretty Pops	Brown Popcorn	Zeatin Grass
Zea	*mays-evereta-helva*	*chapalote-mayo-yellowii*	Mayo Chapalote	Yellow Popcorn	Zeatin Grass
Zea	*mays-evereta-helva*	*chinese-babyi*	Chinese Baby	Yellow Popcorn	Zeatin Grass
Zea	*mays-evereta-helva*	*chires-baby-cornii*	Chires Baby	Yellow Popcorn	Zeatin Grass
Zea	*mays-evereta-helva*	*cranston-hullessi*	Cranston Hulless	Yellow Popcorn	Zeatin Grass
Zea	*mays-evereta-helva*	*dynamitei*	Dynamite Orange	Yellow Popcorn	Zeatin Grass
Zea	*mays-evereta-helva*	*goldenii*	Golden	Yellow Popcorn	Zeatin Grass
Zea	*mays-evereta-helva*	*goldenii-queenii*	Golden Queen	Yellow Popcorn	Zeatin Grass
Zea	*mays-evereta-helva*	*hullessi*	Hulless	Yellow Popcorn	Zeatin Grass
Zea	*mays-evereta-helva*	*japanese-yellow-hullessi*	Japanese Hulless	Yellow Popcorn	Zeatin Grass
Zea	*mays-evereta-helva*	*mountain-view-farmii*	Mountain View Farm	Yellow Popcorn	Zeatin Grass
Zea	*mays-evereta-helva*	*south-american-giantii-89*	89 Day Giant South American	Yellow Popcorn	Zeatin Grass
Zea	*mays-evereta-helva*	*south-american-giantii-112*	112 Day Giant South American	Yellow Popcorn	Zeatin Grass
Zea	*mays-evereta-helva*	*south-american-dynamitei-102*	102 Day Dynamite S. American	Yellow Popcorn	Zeatin Grass
Zea	*mays-evereta-helva*	*south-american-dynamitei-109*	109 Day Dynamite S. American	Yellow Popcorn	Zeatin Grass
Zea	*mays-evereta-helva*	*tom-thumb-yellowii*	Tom Thumb	Yellow Popcorn	Zeatin Grass
Zea	*mays-evereta-helva*	*yellow-popcornii*	Iowa Hulless	Yellow Popcorn	Zeatin Grass
Zea	*mays-evereta-mixta*	*black-menonnitei*	Menonnite Black	Multicolor Popcorn	Zeatin Grass
Zea	*mays-evereta-mixta*	*calicoi*	Calico	Multicolor Popcorn	Zeatin Grass
Zea	*mays-evereta-mixta*	*carouselii*	Carousel	Multicolor Popcorn	Zeatin Grass
Zea	*mays-evereta-mixta*	*cherokee-long-eari*	Long Ear Cherokee	Multicolor Popcorn	Zeatin Grass
Zea	*mays-evereta-mixta*	*cherokee-trail-of-tearsi*	Trail Of Tears Cherokee	Multicolor Popcorn	Zeatin Grass
Zea	*mays-evereta-mixta*	*cochiti-chapalotei*	Cochiti Chapalote	Multicolor Popcorn	Zeatin Grass
Zea	*mays-evereta-mixta*	*cochiti-puebloi*	Cochiti Pueblo	Multicolor Popcorn	Zeatin Grass
Zea	*mays-evereta-mixta*	*cutie-popsi*	Cutie Pops	Multicolor Popcorn	Zeatin Grass
Zea	*mays-evereta-mixta*	*indian-berriesi*	Indian Berries	Multicolor Popcorn	Zeatin Grass
Zea	*mays-evereta-mixta*	*indian-popping-cornii*	Indian Popping	Multicolor Popcorn	Zeatin Grass
Zea	*mays-evereta-mixta*	*kaudersi*	Kauders	Multicolor Popcorn	Zeatin Grass
Zea	*mays-evereta-mixta*	*ken-asmusi*	Ken Asmus	Multicolor Popcorn	Zeatin Grass
Zea	*mays-evereta-mixta*	*lady-fingeri*	Lady Finger	Multicolor Popcorn	Zeatin Grass
Zea	*mays-evereta-mixta*	*laseri*	Laser	Multicolor Popcorn	Zeatin Grass
Zea	*mays-evereta-mixta*	*miniaturei*	Miniature	Multicolor Popcorn	Zeatin Grass
Zea	*mays-evereta-mixta*	*ornamental-barnesi*	Barnes Ornamental	Multicolor Popcorn	Zeatin Grass
Zea	*mays-evereta-mixta*	*papoosei*	Papoose	Multicolor Popcorn	Zeatin Grass
Zea	*mays-evereta-mixta*	*pretty-popsi*	Pretty Pops	Multicolor Popcorn	Zeatin Grass
Zea	*mays-evereta-mixta*	*red-sunrisei*	Red Sunrise	Multicolor Popcorn	Zeatin Grass
Zea	*mays-evereta-mixta*	*smoke-signalsi*	Smoke Signals	Multicolor Popcorn	Zeatin Grass
Zea	*mays-evereta-mixta*	*spectrum*	Spectrum	Multicolor Popcorn	Zeatin Grass
Zea	*mays-evereta-mixta*	*thanksgiving-bouquetii*	Thanksgiving Bouquet	Multicolor Popcorn	Zeatin Grass
Zea	*mays-evereta-mixta*	*tom-thumb-ornamentalis*	Ornamental Tom Thumb	Multicolor Popcorn	Zeatin Grass
Zea	*mays-evereta-mixta*	*wilda's-pridei*	Wilda's Pride	Multicolor Popcorn	Zeatin Grass
Zea	*mays-evereta-purpurea*	*amethystii*	Amethyst	Purple Popcorn	Zeatin Grass
Zea	*mays-evereta-purpurea*	*little-bow-peepii*	Little Bow Peep Pink	Purple Popcorn	Zeatin Grass
Zea	*mays-evereta-purpurea*	*purple-amishii*	Amish	Purple Popcorn	Zeatin Grass
Zea	*mays-evereta-purpurea*	*wisconsin-blackii*	Wisconsin Black	Purple Popcorn	Zeatin Grass
Zea	*mays-evereta-rubra*	*charleston-blackii*	Charleston Black	Red Popcorn	Zeatin Grass
Zea	*mays-evereta-rubra*	*cutie-pinkii*	Cutie Pink	Red Popcorn	Zeatin Grass

Zea	*mays-evereta-rubra*	*miniature-pinkii*	Minature Pink	Red Popcorn	Zeatin Grass
Zea	*mays-evereta-rubra*	*neon-pinkii*	Neon Pink	Red Popcorn	Zeatin Grass
Zea	*mays-evereta-rubra*	*new-york-red-robinii*	New York Red Robin	Red Popcorn	Zeatin Grass
Zea	*mays-evereta-rubra*	*pink-beautyi*	Pink Beauty	Red Popcorn	Zeatin Grass
Zea	*mays-evereta-rubra*	*pink-earlyi*	Pink Early	Red Popcorn	Zeatin Grass
Zea	*mays-evereta-rubra*	*red-beautyi*	Red Beauty	Red Popcorn	Zeatin Grass
Zea	*mays-evereta-rubra*	*strawberryi*	Strawberry	Red Popcorn	Zeatin Grass

Corn, Flour (*Zea mays-farina*)

Genus	species	Infraspecies	cultivar	specimen	Genome
Zea	*mays-farina-alba*	*acoma-whitei*	Acoma	White Flour Corn	Zeatin Grass
Zea	*mays-farina-alba*	*aztec-giant-whitei*	Aztec Giant	White Flour Corn	Zeatin Grass
Zea	*mays-farina-alba*	*blanco-para-pozolei*	Blanco Para Pozole	White Flour Corn	Zeatin Grass
Zea	*mays-farina-alba*	*cherokee-whitei*	Cherokee	White Flour Corn	Zeatin Grass
Zea	*mays-farina-alba*	*cheyennei-pencilii*	Cheyenne Pencil	White Flour Corn	Zeatin Grass
Zea	*mays-farina-alba*	*cochiti-pueblo-whitei*	Cochiti Pueblo	White Flour Corn	Zeatin Grass
Zea	*mays-farina-alba*	*cancha-amarilloi*	Amarillo Cancha	White Flour Corn	Zeatin Grass
Zea	*mays-farina-alba*	*cancha-chulpii*	Chulpi Cancha	White Flour Corn	Zeatin Grass
Zea	*mays-farina-alba*	*cancha-cuzcoi*	Cuzco Cancha	White Flour Corn	Zeatin Grass
Zea	*mays-farina-alba*	*hopi-hominii*	Hopi Hominy	White Flour Corn	Zeatin Grass
Zea	*mays-farina-alba*	*hopi-whitei*	Hopi	White Flour Corn	Zeatin Grass
Zea	*mays-farina-alba*	*incan-white-giantii*	Inca Giant	White Flour Corn	Zeatin Grass
Zea	*mays-farina-alba*	*iroquois-whitei*	Iroquois	White Flour Corn	Zeatin Grass
Zea	*mays-farina-alba*	*isleta-whitei*	Isleta Pueblo	White Flour Corn	Zeatin Grass
Zea	*mays-farina-alba*	*jemez-whitei*	Jemez Pueblo	White Flour Corn	Zeatin Grass
Zea	*mays-farina-alba*	*maiz-blandoi*	Maiz Blando	White Flour Corn	Zeatin Grass
Zea	*mays-farina-alba*	*maiz-blandoi-mayoi*	Mayo Maize Blando	White Flour Corn	Zeatin Grass
Zea	*mays-farina-alba*	*mayo-tosabatchii*	Mayo Tosabatchi	White Flour Corn	Zeatin Grass
Zea	*mays-farina-alba*	*mojavei*	Mohave	White Flour Corn	Zeatin Grass
Zea	*mays-farina-alba*	*navajo-whitei*	Navajo	White Flour Corn	Zeatin Grass
Zea	*mays-farina-alba*	*posolei*	Posole Hominy	White Flour Corn	Zeatin Grass
Zea	*mays-farina-alba*	*rio-lucio-conchoi*	Rio Lucio Concho	White Flour Corn	Zeatin Grass
Zea	*mays-farina-alba*	*sahuaritensis*	Sahuarita	White Flour Corn	Zeatin Grass
Zea	*mays-farina-alba*	*san-felipe-pueblo-whitei*	San Felipe Pueblo	White Flour Corn	Zeatin Grass
Zea	*mays-farina-alba*	*san-luis-conchoi*	San Luis Concho	White Flour Corn	Zeatin Grass
Zea	*mays-farina-alba*	*santo-domingo-whitei*	Santo Domingo Pueblo	White Flour Corn	Zeatin Grass
Zea	*mays-farina-alba*	*tarahumara-blando-de-sonorensis*	Tarahumara Blando De Sonora	White Flour Corn	Zeatin Grass
Zea	*mays-farina-alba*	*tohono-odham-60-dayi*	60 Day Tohono Oodham	White Flour Corn	Zeatin Grass
Zea	*mays-farina-alba*	*truchas-lumbrosoi*	Truchas Lumbroso	White Flour Corn	Zeatin Grass
Zea	*mays-farina-alba*	*tuscarora-whitei*	Tuscarora	White Flour Corn	Zeatin Grass
Zea	*mays-farina-alba*	*vadito-conchoi*	Concho Flint Vadito Pueblo	White Flour Corn	Zeatin Grass
Zea	*mays-farina-alba*	*vadito-whitei*	Vadito Pueblo	White Flour Corn	Zeatin Grass
Zea	*mays-farina-blauea*	*acoma-bluei*	Acoma	Blue Flour Corn	Zeatin Grass
Zea	*mays-farina-blauea*	*acomita-bluei*	Acomita	Blue Flour Corn	Zeatin Grass
Zea	*mays-farina-blauea*	*big-mountain-bluei*	Big Mountain	Blue Flour Corn	Zeatin Grass
Zea	*mays-farina-blauea*	*blue-squaw-cornii*	Blue Squaw	Blue Flour Corn	Zeatin Grass
Zea	*mays-farina-blauea*	*cochiti-pueblo-bluei*	Cochiti Pueblo	Blue Flour Corn	Zeatin Grass
Zea	*mays-farina-blauea*	*colorado-bluei*	Colorado	Blue Flour Corn	Zeatin Grass
Zea	*mays-farina-blauea*	*encinal-bluei*	Encinal	Blue Flour Corn	Zeatin Grass
Zea	*mays-farina-blauea*	*escondida-bluei*	Escondida	Blue Flour Corn	Zeatin Grass
Zea	*mays-farina-blauea*	*hopi-bluei*	Hopi	Blue Flour Corn	Zeatin Grass
Zea	*mays-farina-blauea*	*isleta-bluei*	Isleta Pueblo	Blue Flour Corn	Zeatin Grass
Zea	*mays-farina-blauea*	*jemez-bluei*	Jemez Pueblo	Blue Flour Corn	Zeatin Grass
Zea	*mays-farina-blauea*	*navajo-bluei*	Navaho	Blue Flour Corn	Zeatin Grass
Zea	*mays-farina-blauea*	*pawnee-bluei*	Pawnee	Blue Flour Corn	Zeatin Grass
Zea	*mays-farina-blauea*	*ponca-blue-flouri*	Ponca	Blue Flour Corn	Zeatin Grass
Zea	*mays-farina-blauea*	*rio-lucio-bluei*	Rio Lucio	Blue Flour Corn	Zeatin Grass
Zea	*mays-farina-blauea*	*san-felipe-pueblo-bluei*	San Felipe Pueblo	Blue Flour Corn	Zeatin Grass
Zea	*mays-farina-blauea*	*santa-ana-pueblo-bluei*	Santa Ana Pueblo	Blue Flour Corn	Zeatin Grass
Zea	*mays-farina-blauea*	*santo-domingo-bluei*	Santo Domingo Pueblo	Blue Flour Corn	Zeatin Grass
Zea	*mays-farina-blauea*	*taos-pueblo-blue-cornii*	Taos Pueblo Black	Blue Flour Corn	Zeatin Grass
Zea	*mays-farina-blauea*	*vadito-bluei*	Vadito Pueblo	Blue Flour Corn	Zeatin Grass
Zea	*mays-farina-blauea*	*velarde-bluei*	Velarde Pueblo	Blue Flour Corn	Zeatin Grass
Zea	*mays-farina-blauea*	*yoeme-bluei*	Yoeme	Blue Flour Corn	Zeatin Grass
Zea	*mays-farina-brunnea*	*osage-brown-flouri*	Osage	Brown Flour Corn	Zeatin Grass
Zea	*mays-farina-helva*	*gaspei*	Gaspe Golden	Yellow Flour Corn	Zeatin Grass
Zea	*mays-farina-helva*	*hopi-yellowii*	Hopi	Yellow Flour Corn	Zeatin Grass
Zea	*mays-farina-helva*	*navajo-goldii*	Navajo Gold	Yellow Flour Corn	Zeatin Grass
Zea	*mays-farina-helva*	*navajo-yellowii*	Navajo	Yellow Flour Corn	Zeatin Grass
Zea	*mays-farina-helva*	*texas-shoepegii*	Texas Shoepeg	Yellow Flour Corn	Zeatin Grass
Zea	*mays-farina-mixta*	*anasazii*	Anasazi	Multicolor Flour Corn	Zeatin Grass
Zea	*mays-farina-mixta*	*assiniboinei*	Assiniboine	Multicolor Flour Corn	Zeatin Grass
Zea	*mays-farina-mixta*	*cherokee-villagei*	Cherokee Village	Multicolor Flour Corn	Zeatin Grass
Zea	*mays-farina-mixta*	*guadalajara-all-purposei*	Guadalajara All Purpose	Multicolor Flour Corn	Zeatin Grass
Zea	*mays-farina-mixta*	*hopi-chin-markii*	Hopi Chin Mark	Multicolor Flour Corn	Zeatin Grass
Zea	*mays-farina-mixta*	*hopi-red,white-&-bluei*	Hopi Red, White & Blue	Multicolor Flour Corn	Zeatin Grass
Zea	*mays-farina-mixta*	*hopi-speckledii*	Hopi Speckled	Multicolor Flour Corn	Zeatin Grass
Zea	*mays-farina-mixta*	*iroqouis-calicoi*	Iroqouis Calico	Multicolor Flour Corn	Zeatin Grass

Zea	*mays-farina-mixta*	*mandan-bridei*	Mandan Bride	Multicolor Flour Corn	Zeatin Grass
Zea	*mays-farina-mixta*	*navajo-robin's-eggii*	Navajo Robin's Egg	Multicolor Flour Corn	Zeatin Grass
Zea	*mays-farina-mixta*	*painted-mountainii*	Painted Mountain	Multicolor Flour Corn	Zeatin Grass
Zea	*mays-farina-mixta*	*pink-&-whitei*	Pink & White	Multicolor Flour Corn	Zeatin Grass
Zea	*mays-farina-mixta*	*santo-domingo-yellow-&-bluei*	Yellow & Blue Santo Domingo Pueblo	Multicolor Flour Corn	Zeatin Grass
Zea	*mays-farina-mixta*	*tarahumara-harinoso-de-ochoi*	Harinoso De Ocho Tarahumara	Multicolor Flour Corn	Zeatin Grass
Zea	*mays-farina-mixta*	*tarahumara-maiz-gordo-rasarii*	Maiz Gordo Rosari Tarahumara	Multicolor Flour Corn	Zeatin Grass
Zea	*mays-farina-mixta*	*tarahumara-maiz-rojoi*	Maiz Rojo Tarahumara	Multicolor Flour Corn	Zeatin Grass
Zea	*mays-farina-mixta*	*tehama-dawnii*	Tehama Dawn	Multicolor Flour Corn	Zeatin Grass
Zea	*mays-farina-mixta*	*velarde-azul-y-blancoi*	Azul Y Blanco Velarde Pueblo	Multicolor Flour Corn	Zeatin Grass
Zea	*mays-farina-purpurea*	*hopi-greasy-hairii*	Hopi Greasy Hair	Purple Flour Corn	Zeatin Grass
Zea	*mays-farina-purpurea*	*hopi-purplei*	Hopi	Purple Flour Corn	Zeatin Grass
Zea	*mays-farina-purpurea*	*mandan-lavenderi*	Mandan Lavender	Purple Flour Corn	Zeatin Grass
Zea	*mays-farina-purpurea*	*morado-maizii*	Maiz Morado Red	Purple Flour Corn	Zeatin Grass
Zea	*mays-farina-purpurea*	*morado-peruviana*	Peruvian Morado	Purple Flour Corn	Zeatin Grass
Zea	*mays-farina-purpurea*	*quapawii*	Quapaw Magenta	Purple Flour Corn	Zeatin Grass
Zea	*mays-farina-rubra*	*apachei*	Apache	Red Flour Corn	Zeatin Grass
Zea	*mays-farina-rubra*	*cheyenne-redii*	Cheyenne	Red Flour Corn	Zeatin Grass
Zea	*mays-farina-rubra*	*hopi-kokoma*	Hopi Kokoma	Red Flour Corn	Zeatin Grass
Zea	*mays-farina-rubra*	*hopi-orange-redii*	Hopi Orange	Red Flour Corn	Zeatin Grass
Zea	*mays-farina-rubra*	*hopi-pinkii*	Hopi Pink	Red Flour Corn	Zeatin Grass
Zea	*mays-farina-rubra*	*hopi-redii*	Hopi	Red Flour Corn	Zeatin Grass
Zea	*mays-farina-rubra*	*mandan-redii*	Mandan Black	Red Flour Corn	Zeatin Grass
Zea	*mays-farina-rubra*	*morado-mexicana*	Mexican Morado Pink	Red Flour Corn	Zeatin Grass
Zea	*mays-farina-rubra*	*parch-supai-redii*	Parch Supai	Red Flour Corn	Zeatin Grass
Zea	*mays-farina-rubra*	*red-midgetii*	Red Midget	Red Flour Corn	Zeatin Grass
Zea	*mays-farina-rubra*	*siebel's-redii*	Siebel's	Red Flour Corn	Zeatin Grass

Corn, Flint (*Zea mays-flinta*)

Genus	species	Infraspecies	cultivar	specimen	Genome
Zea	*mays-flinta-alba*	*mayo-corrientei*	Mayo Corriente	White Flint Corn	Zeatin Grass
Zea	*mays-flinta-alba*	*mount-pima-cristalino-de-chihuahuaensis*	Mount Pima Cristalino De Chihuahua	White Flint Corn	Zeatin Grass
Zea	*mays-flinta-alba*	*nambe-whitei*	Nambe Pueblo	White Flint Corn	Zeatin Grass
Zea	*mays-flinta-alba*	*saskatoon-whitei*	Saskatoon	White Flint Corn	Zeatin Grass
Zea	*mays-flinta-aurantia*	*longafellow-flintii*	Longfellow	Orange Flint Corn	Zeatin Grass
Zea	*mays-flinta-blauea*	*farina(bluei)*	Flour	Blue Flint Corn	Zeatin Grass
Zea	*mays-flinta-blauea*	*casados-bluei*	Casados	Blue Flint Corn	Zeatin Grass
Zea	*mays-flinta-blauea*	*hernandez-bluei*	Hernandez	Blue Flint Corn	Zeatin Grass
Zea	*mays-flinta-blauea*	*lenni-lanape-delawarensis*	Delaware Lenni Lanape	Blue Flint Corn	Zeatin Grass
Zea	*mays-flinta-helva*	*apache-yellow-flintii*	Apache	Yellow Flint Corn	Zeatin Grass
Zea	*mays-flinta-helva*	*byron-yellow-flintii*	Byron	Yellow Flint Corn	Zeatin Grass
Zea	*mays-flinta-helva*	*fort-kent-goldenii*	Fort Kent Golden	Yellow Flint Corn	Zeatin Grass
Zea	*mays-flinta-helva*	*guarijio-maiz-amarilloi*	Guarijio Maiz Amarillo	Yellow Flint Corn	Zeatin Grass
Zea	*mays-flinta-helva*	*tarahumara-chiquitai*	Chiquita Tarahumara	Yellow Flint Corn	Zeatin Grass
Zea	*mays-flinta-helva*	*tarahumara-golden-cristalinoi*	Golden Crystalino Tarahumara	Yellow Flint Corn	Zeatin Grass
Zea	*mays-flinta-helva*	*tarahumara-serranoi*	Serrano Tarahumara	Yellow Flint Corn	Zeatin Grass
Zea	*mays-flinta-helva*	*vermont-yellow-flintii*	Vermont	Yellow Flint Corn	Zeatin Grass
Zea	*mays-flinta-mixta*	*alabama-coschattai*	Alabama Coschatta	Multicolor Flint Corn	Zeatin Grass
Zea	*mays-flinta-mixta*	*bear-island-chippewai*	Bear Island Chippewa	Multicolor Flint Corn	Zeatin Grass
Zea	*mays-flinta-mixta*	*cheyenne-agency-stripedii*	Cheyenne Agency Striped	Multicolor Flint Corn	Zeatin Grass
Zea	*mays-flinta-mixta*	*chinook-farina*	Chinook Flour	Multicolor Flint Corn	Zeatin Grass
Zea	*mays-flinta-mixta*	*dwarf-multicoloredii*	Dwarf	Multicolor Flint Corn	Zeatin Grass
Zea	*mays-flinta-mixta*	*early-indian-farina*	Early Indian Flour	Multicolor Flint Corn	Zeatin Grass
Zea	*mays-flinta-mixta*	*inca-rainbowi*	Inca Rainbow	Multicolor Flint Corn	Zeatin Grass
Zea	*mays-flinta-mixta*	*indian-fingersi*	Indian Fingers	Multicolor Flint Corn	Zeatin Grass
Zea	*mays-flinta-mixta*	*indian-flintii*	Indian	Multicolor Flint Corn	Zeatin Grass
Zea	*mays-flinta-mixta*	*indian-ornamental-rainbowi*	Rainbow Indian Ornamental	Multicolor Flint Corn	Zeatin Grass
Zea	*mays-flinta-mixta*	*indian-ornamental-thanksgivingii*	Thanksgiving Indian Ornamental	Multicolor Flint Corn	Zeatin Grass
Zea	*mays-flinta-mixta*	*indian-jung's-littlei*	Jung's Little Indian	Multicolor Flint Corn	Zeatin Grass
Zea	*mays-flinta-mixta*	*indian-jung's-littlei-improvedii*	Improved Jung's Little Indian	Multicolor Flint Corn	Zeatin Grass
Zea	*mays-flinta-mixta*	*indian-littlei*	Little Indian	Multicolor Flint Corn	Zeatin Grass
Zea	*mays-flinta-mixta*	*indian-miniaturei*	Minature Indian	Multicolor Flint Corn	Zeatin Grass
Zea	*mays-flinta-mixta*	*indian-purple-huskii*	Purple Husk Indian	Multicolor Flint Corn	Zeatin Grass
Zea	*mays-flinta-mixta*	*indian-seneca-cornii*	Seneca Indian	Multicolor Flint Corn	Zeatin Grass
Zea	*mays-flinta-mixta*	*indian-seneca-minii*	Mini Seneca Indian	Multicolor Flint Corn	Zeatin Grass
Zea	*mays-flinta-mixta*	*indian-squaw-cornii*	Squaw Indian	Multicolor Flint Corn	Zeatin Grass
Zea	*mays-flinta-mixta*	*kaladia-ornamental-flintii*	Kaladia Ornamental	Multicolor Flint Corn	Zeatin Grass
Zea	*mays-flinta-mixta*	*lenorei*	Lenore	Multicolor Flint Corn	Zeatin Grass
Zea	*mays-flinta-mixta*	*little-jewelsi*	Little Jewels	Multicolor Flint Corn	Zeatin Grass
Zea	*mays-flinta-mixta*	*osage-redii*	Osage Red	Multicolor Flint Corn	Zeatin Grass
Zea	*mays-flinta-mixta*	*pipestonei*	Pipestone	Multicolor Flint Corn	Zeatin Grass
Zea	*mays-flinta-mixta*	*rainbovii*	Rainbow	Multicolor Flint Corn	Zeatin Grass
Zea	*mays-flinta-mixta*	*roy's-calaisi*	Roy''s Calais	Multicolor Flint Corn	Zeatin Grass
Zea	*mays-flinta-mixta*	*santo-domingo-mixii*	Santo Domingo Mix	Multicolor Flint Corn	Zeatin Grass
Zea	*mays-flinta-mixta*	*seedway-elitei*	Seedway Elite	Multicolor Flint Corn	Zeatin Grass
Zea	*mays-flinta-mixta*	*seneca-blue-bear-dancei*	Seneca Blue Bear Dance	Multicolor Flint Corn	Zeatin Grass
Zea	*mays-flinta-mixta*	*seneca-red-stalkeri*	Seneca Red Stalker	Multicolor Flint Corn	Zeatin Grass
Zea	*mays-flinta-mixta*	*tarahumara-apachitoi*	Apachito Tarahumara	Multicolor Flint Corn	Zeatin Grass

Zea	*mays-flinta-mixta*	*tarahumara-eastern-pepitilloi*	Pepitillo Eastern Tarahumara	Multicolor Flint Corn	Zeatin Grass
Zea	*mays-flinta-mixta*	*tarahumara-maiz-azulii*	Maiz Azul Tarahumara	Multicolor Flint Corn	Zeatin Grass
Zea	*mays-flinta-mixta*	*tarahumara-maiz-pintoi*	Maiz Pinto Tarahumara	Multicolor Flint Corn	Zeatin Grass
Zea	*mays-flinta-mixta*	*tarahumara-serapei*	Serape Tarahumara	Multicolor Flint Corn	Zeatin Grass
Zea	*mays-flinta-mixta*	*tepehuan-maiz-coloradoensis*	Maize Colorado Tepehuan	Multicolor Flint Corn	Zeatin Grass
Zea	*mays-flinta-mixta*	*wampum*	Wampum	Multicolor Flint Corn	Zeatin Grass
Zea	*mays-flinta-purpurea*	*mandan-blackii*	Mandan Black	Purple Flint Corn	Zeatin Grass
Zea	*mays-flinta-rubra*	*king-philip-improvedii*	Improved King Philip	Red Flint Corn	Zeatin Grass
Zea	*mays-flinta-rubra*	*mandan-clay-redii*	Mandan Clay	Red Flint Corn	Zeatin Grass
Zea	*mays-flinta-rubra*	*tarahumara-maiz-color-de-rosai*	Tarahumara Maiz Color De Rosa	Red Flint Corn	Zeatin Grass

Corn, Sweet (*Zea mays-suavis*)

Genus	species	Infraspecies	cultivar	specimen	Genome
Zea	*mays-suavis-alba*	*art-verrell'si*	Art Verrell's	White Sweet Corn	Zeatin Grass
Zea	*mays-suavis-alba*	*aunt-mary'si*	Aunt Mary's	White Sweet Corn	Zeatin Grass
Zea	*mays-suavis-alba*	*country-gentlemanii*	Country Gentleman	White Sweet Corn	Zeatin Grass
Zea	*mays-suavis-alba*	*earliest-whitei*	Earliest	White Sweet Corn	Zeatin Grass
Zea	*mays-suavis-alba*	*early-albertaensis*	Early Alberta	White Sweet Corn	Zeatin Grass
Zea	*mays-suavis-alba*	*early-pearlii*	Early Pearl	White Sweet Corn	Zeatin Grass
Zea	*mays-suavis-alba*	*evergreen-earlyi*	Evergreen Early	White Sweet Corn	Zeatin Grass
Zea	*mays-suavis-alba*	*honey-creamii*	Honey Cream	White Sweet Corn	Zeatin Grass
Zea	*mays-suavis-alba*	*hooker's-sweet-indianii*	Hooker Indian	White Sweet Corn	Zeatin Grass
Zea	*mays-suavis-alba*	*hopi-bantamii-80*	80 Day Hopi Bantam	White Sweet Corn	Zeatin Grass
Zea	*mays-suavis-alba*	*hopi-bantamii-90*	90 Day Hopi Bantam	White Sweet Corn	Zeatin Grass
Zea	*mays-suavis-alba*	*iroquoi-whitei*	Iroquois	White Sweet Corn	Zeatin Grass
Zea	*mays-suavis-alba*	*pease-crosbyi*	Pease Crosby	White Sweet Corn	Zeatin Grass
Zea	*mays-suavis-alba*	*seneca-snow-princei*	Seneca Snow Prince	White Sweet Corn	Zeatin Grass
Zea	*mays-suavis-alba*	*shaffer-8-rowii*	Shaffer 8 Row	White Sweet Corn	Zeatin Grass
Zea	*mays-suavis-alba*	*six-shooteri*	Six Shooter	White Sweet Corn	Zeatin Grass
Zea	*mays-suavis-alba*	*stowell's-evergrenii*	Stowell's Evergreen	White Sweet Corn	Zeatin Grass
Zea	*mays-suavis-alba*	*true-platinum*	True Platinum	White Sweet Corn	Zeatin Grass
Zea	*mays-suavis-alba*	*whipple's-whitei*	Whipple's	White Sweet Corn	Zeatin Grass
Zea	*mays-suavis-blauea*	*black-azteci*	Aztec Black	Blue Sweet Corn	Zeatin Grass
Zea	*mays-suavis-blauea*	*black-mexicana*	Mexican Black	Blue Sweet Corn	Zeatin Grass
Zea	*mays-suavis-blauea*	*bellingham-bluei*	Bellingham	Blue Sweet Corn	Zeatin Grass
Zea	*mays-suavis-blauea*	*blue-jadei*	Blue Jade	Blue Sweet Corn	Zeatin Grass
Zea	*mays-suavis-blauea*	*hopi-massiquai*	Hopi Massiqa	Blue Sweet Corn	Zeatin Grass
Zea	*mays-suavis-blauea*	*lindsey-meyer-bluei*	Lindsey Meyer	Blue Sweet Corn	Zeatin Grass
Zea	*mays-suavis-blauea*	*midnight-snackii*	Midnight Snack	Blue Sweet Corn	Zeatin Grass
Zea	*mays-suavis-brunnea*	*nuettai*	Nuetta	Brown Sweet Corn	Zeatin Grass
Zea	*mays-suavis-helva*	*ashworthii*	Ashworth	Yellow Sweet Corn	Zeatin Grass
Zea	*mays-suavis-helva*	*buhlii*	Buhl	Yellow Sweet Corn	Zeatin Grass
Zea	*mays-suavis-helva*	*clem-bennetii*	Clem Bennet	Yellow Sweet Corn	Zeatin Grass
Zea	*mays-suavis-helva*	*dorinnyi*	Dorinny	Yellow Sweet Corn	Zeatin Grass
Zea	*mays-suavis-helva*	*early-large-adamii*	Early Large Adam	Yellow Sweet Corn	Zeatin Grass
Zea	*mays-suavis-helva*	*fisher's-earliestii*	Fisher's Earliest	Yellow Sweet Corn	Zeatin Grass
Zea	*mays-suavis-helva*	*golden-bantamii*	8 Row Golden Bantam	Yellow Sweet Corn	Zeatin Grass
Zea	*mays-suavis-helva*	*golden-bantamii-improvedii*	12 Row Golden Bantam	Yellow Sweet Corn	Zeatin Grass
Zea	*mays-suavis-helva*	*golden-bantamii-earlyi*	Early 8 Row Golden Bantam	Yellow Sweet Corn	Zeatin Grass
Zea	*mays-suavis-helva*	*golden-bantamii-earlyi-dwarfii*	Dwarf Early 8 Row Golden Bantam	Yellow Sweet Corn	Zeatin Grass
Zea	*mays-suavis-helva*	*golden-bantamii-extra-earlyi*	Extra Early 8 Row Golden Bantam	Yellow Sweet Corn	Zeatin Grass
Zea	*mays-suavis-helva*	*golden-bantamii-extra-earlyi-montanaensis*	Montana Extra Early 8 Row Golden Bantam	Yellow Sweet Corn	Zeatin Grass
Zea	*mays-suavis-helva*	*golden-midgetii*	Golden Midget	Yellow Sweet Corn	Zeatin Grass
Zea	*mays-suavis-helva*	*guarijio-sweetii*	Guarijio Orange	Yellow Sweet Corn	Zeatin Grass
Zea	*mays-suavis-helva*	*hartline's-8-rowii*	Hartline's 8 Row	Yellow Sweet Corn	Zeatin Grass
Zea	*mays-suavis-helva*	*howling-mobii*	Howling Mob	Yellow Sweet Corn	Zeatin Grass
Zea	*mays-suavis-helva*	*jubileei*	Jubilee	Yellow Sweet Corn	Zeatin Grass
Zea	*mays-suavis-helva*	*kievitii*	Kievit	Yellow Sweet Corn	Zeatin Grass
Zea	*mays-suavis-helva*	*kodiacensis*	Kodiac	Yellow Sweet Corn	Zeatin Grass
Zea	*mays-suavis-helva*	*little-giantii*	Little Giant	Yellow Sweet Corn	Zeatin Grass
Zea	*mays-suavis-helva*	*luther-hillii*	Luther Hill	Yellow Sweet Corn	Zeatin Grass
Zea	*mays-suavis-helva*	*malcombsi*	Malcombs	Yellow Sweet Corn	Zeatin Grass
Zea	*mays-suavis-helva*	*martian-jewelsi*	Martian Jewels	Yellow Sweet Corn	Zeatin Grass
Zea	*mays-suavis-helva*	*mayo-cacabatchii*	Mayo Cacabatchi	Yellow Sweet Corn	Zeatin Grass
Zea	*mays-suavis-helva*	*mt.-pima-yellowii*	Mt. Pima	Yellow Sweet Corn	Zeatin Grass
Zea	*mays-suavis-helva*	*orchard-babyi*	Orchard Baby	Yellow Sweet Corn	Zeatin Grass
Zea	*mays-suavis-helva*	*puget-goldii*	Puget Gold	Yellow Sweet Corn	Zeatin Grass
Zea	*mays-suavis-helva*	*queen-annei*	Queen Anne	Yellow Sweet Corn	Zeatin Grass
Zea	*mays-suavis-helva*	*rangoldii*	Rangold	Yellow Sweet Corn	Zeatin Grass
Zea	*mays-suavis-helva*	*simmonetii*	Simmonet	Yellow Sweet Corn	Zeatin Grass
Zea	*mays-suavis-helva*	*sunseekeri*	Sunseeker	Yellow Sweet Corn	Zeatin Grass
Zea	*mays-suavis-helva*	*sweet-nothingsi*	Sweet Nothings	Yellow Sweet Corn	Zeatin Grass
Zea	*mays-suavis-helva*	*texas-honey-junei*	Texas Honey June	Yellow Sweet Corn	Zeatin Grass
Zea	*mays-suavis-helva*	*true-goldii*	True Gold	Yellow Sweet Corn	Zeatin Grass
Zea	*mays-suavis-helva*	*whipple's-yellowii*	Whipple's	Yellow Sweet Corn	Zeatin Grass
Zea	*mays-suavis-helva*	*yukon-chiefii*	Yukon Chief	Yellow Sweet Corn	Zeatin Grass
Zea	*mays-suavis-helva*	*yukon-supremei*	Yukon Supreme	Yellow Sweet Corn	Zeatin Grass
Zea	*mays-suavis-helva*	*yuman-yellowii*	Yuman	Yellow Sweet Corn	Zeatin Grass

Zea	*mays-suavis-mixta*	*multicolor*	Original	Multicolor Sweet Corn	Zeatin Grass
Zea	*mays-suavis-mixta*	*anasazi-sweetii*	Anasazi	Multicolor Sweet Corn	Zeatin Grass
Zea	*mays-suavis-mixta*	*cocopahii*	Cocopah	Multicolor Sweet Corn	Zeatin Grass
Zea	*mays-suavis-mixta*	*double-playi*	Double Play	Multicolor Sweet Corn	Zeatin Grass
Zea	*mays-suavis-mixta*	*double-standardii*	Double Standard	Multicolor Sweet Corn	Zeatin Grass
Zea	*mays-suavis-mixta*	*honey-rainbowii*	Honey Rainbow	Multicolor Sweet Corn	Zeatin Grass
Zea	*mays-suavis-mixta*	*maricopai*	Maricopa	Multicolor Sweet Corn	Zeatin Grass
Zea	*mays-suavis-mixta*	*pow-wowii*	Pow Wow	Multicolor Sweet Corn	Zeatin Grass
Zea	*mays-suavis-mixta*	*triple-playi*	Triple Play	Multicolor Sweet Corn	Zeatin Grass
Zea	*mays-suavis-purpurea*	*double-red-sweetii*	Double Red	Purple Sweet Corn	Zeatin Grass
Zea	*mays-suavis-purpurea*	*guatemalan-purplei*	Guatemalan	Purple Sweet Corn	Zeatin Grass
Zea	*mays-suavis-purpurea*	*jerde's-redii*	Jerde's Red	Purple Sweet Corn	Zeatin Grass
Zea	*mays-suavis-purpurea*	*mini-purplei*	Mini	Purple Sweet Corn	Zeatin Grass
Zea	*mays-suavis-rubra*	*guarijio-redii*	Guarijio	Red Sweet Corn	Zeatin Grass
Zea	*mays-suavis-rubra*	*mandan-red-sweetii*	Mandan	Red Sweet Corn	Zeatin Grass

Corn, Sugar (*Zea mays-suavis-sucra*)

Genus	species	Infraspecies	cultivar	specimen	Genome
Zea	*mays-suavis-sucra*	*butter-&-sugari*	Butter & Sugar	Sugar Sweet Corn	Zeatin Grass
Zea	*mays-suavis-sucra*	*candy-mountainii*	Candy Mountain	Sugar Sweet Corn	Zeatin Grass
Zea	*mays-suavis-sucra*	*hawaiiana*	97 Day Hawaiian Supersweet	Sugar Sweet Corn	Zeatin Grass
Zea	*mays-suavis-sucra*	*hawaiiana-supersweetii*	80 Day Hawaiian Supersweet	Sugar Sweet Corn	Zeatin Grass
Zea	*mays-suavis-sucra*	*hawaiiana-supersweetii-9*	#9 Hawaiian Supersweet	Sugar Sweet Corn	Zeatin Grass
Zea	*mays-suavis-sucra*	*honey-&-creamii*	Honey & Cream	Sugar Sweet Corn	Zeatin Grass
Zea	*mays-suavis-sucra*	*how-sweet-it-isi*	How Sweet It Is	Sugar Sweet Corn	Zeatin Grass
Zea	*mays-suavis-sucra*	*kalakoai*	Kalakoa	Sugar Sweet Corn	Zeatin Grass
Zea	*mays-suavis-sucra*	*lorriei*	Lorrie	Sugar Sweet Corn	Zeatin Grass
Zea	*mays-suavis-sucra*	*miraclei*	Miracle	Sugar Sweet Corn	Zeatin Grass
Zea	*mays-suavis-sucra*	*oraclei*	Oracle	Sugar Sweet Corn	Zeatin Grass
Zea	*mays-suavis-sucra*	*peaches-&-creamii*	Peaches & Cream	Sugar Sweet Corn	Zeatin Grass
Zea	*mays-suavis-sucra*	*snowbellei*	Snowbelle	Sugar Sweet Corn	Zeatin Grass
Zea	*mays-suavis-sucra*	*sugar-bunsi*	Sugar Buns	Sugar Sweet Corn	Zeatin Grass
Zea	*mays-suavis-sucra*	*sugar-chapii*	Sugar Chap	Sugar Sweet Corn	Zeatin Grass
Zea	*mays-suavis-sucra*	*sugar-daddyi*	Sugar Daddy	Sugar Sweet Corn	Zeatin Grass
Zea	*mays-suavis-sucra*	*sweet-baby-bluei*	Sweet Baby Blue	Sugar Sweet Corn	Zeatin Grass
Zea	*mays-suavis-sucra*	*topnotchii*	Topnotch	Sugar Sweet Corn	Zeatin Grass
Zea	*mays-suavis-sucra*	*white-sugari*	White	Sugar Sweet Corn	Zeatin Grass

Family 722.022.7 *Triticaceae* Wheat

Genus	Species	Specimen	Genome	C:Culinary, M:Medicinal, E:Economic
Triticum	*aestivum*	**Summer**	**Wheat**	C:SD(grain):**VG**
Triticum	*aethiopicum*	Ethiopian	Wheat	C:SD(grain):**VG**
Triticum	*araraticum*	Ararat Mountain	Wheat	C:SD(grain):**VG**
Triticum	*boeoticum*	Greek Einkorn	Wheat	C:SD(grain):**VG**
Triticum	*carthlicum*	Persian	Wheat	C:SD(grain):**VG**
Triticum	*compactum*	Clubed	Wheat	C:SD(grain):**VG**
Triticum	*dicoccon*	Emmer	Wheat	C:SD(grain):**VG**
Triticum	*durum*	Durum	Wheat	C:SD(grain):**VG**
Triticum	*ispahanicum*	Esfahan	Wheat	C:SD(grain):**VG**
Triticum	*karamyschevii*	Karamyschev	Wheat	C:SD(grain):**VG**
Triticum	*militinae*	Milite	Wheat	C:SD(grain):**VG**
Triticum	*monococcum*	Einkorn	Wheat	C:SD(grain):**VG**
Triticum	*polonicum*	Polish	Wheat	C:SD(grain):**VG**
Triticum	*spelta*	Spelt	Wheat	C:SD(grain):**VG**
Triticum	*timopheevii*	Timopheev	Wheat	C:SD(grain):**VG**
Triticum	*turanicum*	Ural-Altai Mountain	Wheat	C:SD(grain):**VG**
Triticum	*turgidum*	Poulard	Wheat	C:SD(grain):**VG**
Triticum	*urartu*	Urartu	Wheat	C:SD(grain):**VG**
Triticum	*vavilovii*	Vavilov	Wheat	C:SD(grain):**VG**
Triticum	*zhukovskyi*	Zhukovsky	Wheat	C:SD(grain):**VG**

Varieties

Summer Wheat (*Triticum aestivum*)

Genus	species	Infraspecies	cultivar	specimen	Genome
Triticum	*aestivum*	*aestivum*	Original	Summer	Wheat
Triticum	*aestivum*	*album*	White	Summer	Wheat
Triticum	*aestivum*	*amylosum*	Amylose	Summer	Wheat
Triticum	*aestivum*	*antiquorum*	Antiquity	Summer	Wheat
Triticum	*aestivum*	*aristatum*	Awn	Summer	Wheat
Triticum	*aestivum*	*arundinaceum*	Reed	Summer	Wheat
Triticum	*aestivum*	*asiaticum*	Asian	Summer	Wheat
Triticum	*aestivum*	*australianum*	Australian	Summer	Wheat
Triticum	*aestivum*	*brezhnevii*	Brezhnev	Summer	Wheat

Triticum	*aestivum*	*bucharicum*	Bokhara	Summer	Wheat
Triticum	*aestivum*	*caeruleum*	Blue	Summer	Wheat
Triticum	*aestivum*	*caesium*	Light Blue	Summer	Wheat
Triticum	*aestivum*	*cereale*	Cereal	Summer	Wheat
Triticum	*aestivum*	*clavatum*	Club	Summer	Wheat
Triticum	*aestivum*	*compositum*	Branched Spike	Summer	Wheat
Triticum	*aestivum*	*dorofeevii*	Dorofeev	Summer	Wheat
Triticum	*aestivum*	*duriusculum*	Hard	Summer	Wheat
Triticum	*aestivum*	*erinaceum*	Hedgehog	Summer	Wheat
Triticum	*aestivum*	*erythrospermum*	Red Seed	Summer	Wheat
Triticum	*aestivum*	*ferrugineum*	Ferruginous	Summer	Wheat
Triticum	*aestivum*	*hadropyrum*	Adriatic	Summer	Wheat
Triticum	*aestivum*	*hieminflatum*	Heim Swollen	Summer	Wheat
Triticum	*aestivum*	*horstianum*	Horst	Summer	Wheat
Triticum	*aestivum*	*hybernum*	Late	Summer	Wheat
Triticum	*aestivum*	*imberbe*	Beardless	Summer	Wheat
Triticum	*aestivum*	*inflatum*	Swollen	Summer	Wheat
Triticum	*aestivum*	*inflatum-album-capitatum*	Whitecap Swollen	Summer	Wheat
Triticum	*aestivum*	*inflatum-lutescens*	Yellow Swollen	Summer	Wheat
Triticum	*aestivum*	*Inflatum-meridionale*	Noonday Swollen	Summer	Wheat
Triticum	*aestivum*	*inflatum-quasimeridionale*	Afternoon Swollen	Summer	Wheat
Triticum	*aestivum*	*inflatum-rufum*	Rufous	Summer	Wheat
Triticum	*aestivum*	*irano-asiaticum*	Middle Eastern	Summer	Wheat
Triticum	*aestivum*	*ischkaschimicum*	Ischkaschim	Summer	Wheat
Triticum	*aestivum*	*japschorvi*	Japschorv	Summer	Wheat
Triticum	*aestivum*	*koeleri*	Koeler	Summer	Wheat
Triticum	*aestivum*	*labile*	Unstable	Summer	Wheat
Triticum	*aestivum*	*linnaeanum*	Linnaeus	Summer	Wheat
Triticum	*aestivum*	*lorenze*	Lorenzo	Summer	Wheat
Triticum	*aestivum*	*lutescens*	Lute Yellow	Summer	Wheat
Triticum	*aestivum*	*martius*	Marty	Summer	Wheat
Triticum	*aestivum*	*milturum*	Milture	Summer	Wheat
Triticum	*aestivum*	*mupuru*	Mupuru	Summer	Wheat
Triticum	*aestivum*	*pilosum*	Pilose Downy	Summer	Wheat
Triticum	*aestivum*	*poltawense*	Poltawa	Summer	Wheat
Triticum	*aestivum*	*pubescens*	Pubic Downy	Summer	Wheat
Triticum	*aestivum*	*pulverulentum*	Powdery	Summer	Wheat
Triticum	*aestivum*	*quadratum*	Four Rank	Summer	Wheat
Triticum	*aestivum*	*quasiheraticum*	Quasi Heratic	Summer	Wheat
Triticum	*aestivum*	*ramifera*	Branching	Summer	Wheat
Triticum	*aestivum*	*ramosum-alborubrum*	White & Red Branching	Summer	Wheat
Triticum	*aestivum*	*ramosum-coeruleum*	Blue Branching	Summer	Wheat
Triticum	*aestivum*	*ramosum-milturum*	Milture Branching	Summer	Wheat
Triticum	*aestivum*	*ramosum-muticum*	Blunt	Summer	Wheat
Triticum	*aestivum*	*rossicum*	Ross	Summer	Wheat
Triticum	*aestivum*	*ruchczianum*	Ruchcze	Summer	Wheat
Triticum	*aestivum*	*sativum*	Pastry	Summer	Wheat
Triticum	*aestivum*	*segetale*	Cornfield	Summer	Wheat
Triticum	*aestivum*	*sibiricum*	Siberian	Summer	Wheat
Triticum	*aestivum*	*siliginum*	Siligin	Summer	Wheat
Triticum	*aestivum*	*sisianicum*	Sisian	Summer	Wheat
Triticum	*aestivum*	*subfalseum-erythroleucon*	False Red Seed	Summer	Wheat
Triticum	*aestivum*	*subfalseum-graecum*	False Greek	Summer	Wheat
Triticum	*aestivum*	*subtadjicorum*	Tadjikistan	Summer	Wheat
Triticum	*aestivum*	*sunpanii*	Sunpan	Summer	Wheat
Triticum	*aestivum*	*tibeticum*	Tibet	Summer	Wheat
Triticum	*aestivum*	*transcaucasicum*	Caucasia	Summer	Wheat
Triticum	*aestivum*	*tustella*	Tust	Summer	Wheat
Triticum	*aestivum*	*uralicum*	Ural Mountain	Summer	Wheat
Triticum	*aestivum*	*vavilovii*	Vavilov	Summer	Wheat
Triticum	*aestivum*	*vavilovii-milturum*	Milture Valvilov	Summer	Wheat
Triticum	*aestivum*	*velutinum*	Velvet	Summer	Wheat
Triticum	*aestivum*	*vigorovii*	Vigorov	Summer	Wheat
Triticum	*aestivum*	*yunnanense*	Yunnan	Summer	Wheat

Wheat, Ethiopian (*Triticum aethiopicum*)

Genus	species	Infraspecies	cultivar	specimen	Genome
Triticum	*aethiopicum*	*aethiopicum*	Original	Ethiopian	Wheat
Triticum	*aethiopicum*	*arraseita*	Arraseit	Ethiopian	Wheat
Triticum	*aethiopicum*	*compressum*	Compressed	Ethiopian	Wheat
Triticum	*aethiopicum*	*copticum*	Coptic	Ethiopian	Wheat
Triticum	*aethiopicum*	*diversifolium*	Diverse Leaf	Ethiopian	Wheat
Triticum	*aethiopicum*	*recognitum*	Recognized	Ethiopian	Wheat
Triticum	*aethiopicum*	*schimperi*	Schimper	Ethiopian	Wheat
Triticum	*aethiopicum*	*turgidum*	Turgid	Ethiopian	Wheat
Triticum	*aethiopicum*	*venulosum*	Veined	Ethiopian	Wheat

Wheat, Greek Einkorn (*Triticum boeoticum*)

Genus	species	Infraspecies	cultivar	specimen	Genome

Triticum	*boeoticum*	.	.	Greek Einkorn	Wheat
Triticum	*boeoticum*	*boeoticum*	Original	Greek Einkorn	Wheat
Triticum	*boeoticum*	*aegilopoides*	Beard Grass	Greek Einkorn	Wheat
Triticum	*boeoticum*	*crithodium*	Barley	Greek Einkorn	Wheat
Triticum	*boeoticum*	*jerevanii*	Jerevan	Greek Einkorn	Wheat
Triticum	*boeoticum*	*subaznaburticum*	Aznaburt	Greek Einkorn	Wheat
Triticum	*boeoticum*	*subbuluchevskajae*	Buluchevskaya	Greek Einkorn	Wheat
Triticum	*boeoticum*	*sublarionovii*	Larionov	Greek Einkorn	Wheat
Triticum	*boeoticum*	*thaoudar*	Thaoudar	Greek Einkorn	Wheat
Triticum	*boeoticum*	*viridium-albinigrescens*	White & Black Green	Greek Einkorn	Wheat
Triticum	*boeoticum*	*viridium-haussknechtii*	Haussknecht Green	Greek Einkorn	Wheat
Triticum	*boeoticum*	*viridium-nigrireuteri*	Black Reuter Green	Greek Einkorn	Wheat

Wheat, Carthage (*Triticum carthlicum*)

Genus	species	Infraspecies	cultivar	specimen	Genome
Triticum	*carthlicum*	*carthlicum*	Original	Carthage	Wheat
Triticum	*carthlicum*	*ibericum*	Iberian	Carthage	Wheat
Triticum	*carthlicum*	*narpurual*	Narpurual	Carthage	Wheat
Triticum	*carthlicum*	*paradoxum*	Paradox	Carthage	Wheat
Triticum	*carthlicum*	*persicum*	Persian	Carthage	Wheat

Wheat, Clubed (*Triticum compactum*)

Genus	species	Infraspecies	cultivar	specimen	Genome
Triticum	*compactum*	*compactum*	Original	Clubed	Wheat
Triticum	*compactum*	*afghanicum*	Afghan	Clubed	Wheat
Triticum	*compactum*	*bar-darai*	Bar Dara	Clubed	Wheat
Triticum	*compactum*	*bulbosum*	Bulb	Clubed	Wheat
Triticum	*compactum*	*burnaschewii*	Burnaschew	Clubed	Wheat
Triticum	*compactum*	*creticum*	Crete	Clubed	Wheat
Triticum	*compactum*	*griseovavilovianum*	Gray Valvilov	Clubed	Wheat
Triticum	*compactum*	*irano-asiaticum*	Middle Eastern	Clubed	Wheat
Triticum	*compactum*	*nigmatullinii*	Nigmatullin	Clubed	Wheat
Triticum	*compactum*	*sarezicum*	Sarez	Clubed	Wheat
Triticum	*compactum*	*sieberianum*	Sieber	Clubed	Wheat
Triticum	*compactum*	*sparsum*	Sparse	Clubed	Wheat
Triticum	*compactum*	*vysozkyi*	Vysozky	Clubed	Wheat

Wheat, Emmer (*Triticum dicoccon*)

Genus	species	Infraspecies	cultivar	specimen	Genome
Triticum	*dicoccon*	*dicoccon*	Original	Emmer	Wheat
Triticum	*dicoccon*	*ajar*	Ajar	Emmer	Wheat
Triticum	*dicoccon*	*amyleum*	Amylose	Emmer	Wheat
Triticum	*dicoccon*	*armeniacum*	Armenian	Emmer	Wheat
Triticum	*dicoccon*	*arpurunial*	Arpurun	Emmer	Wheat
Triticum	*dicoccon*	*arras*	Arraseit	Emmer	Wheat
Triticum	*dicoccon*	*asiaticum*	Asian	Emmer	Wheat
Triticum	*dicoccon*	*atratum*	Dark	Emmer	Wheat
Triticum	*dicoccon*	*azerbaidzhanicum*	Azerbaizhan	Emmer	Wheat
Triticum	*dicoccon*	*camosum-solencurum*	Solencure Camos	Emmer	Wheat
Triticum	*dicoccon*	*caucasicum*	Caucasia	Emmer	Wheat
Triticum	*dicoccon*	*chwamlicum*	Chwamlia	Emmer	Wheat
Triticum	*dicoccon*	*cienfuegos*	Cienfuego	Emmer	Wheat
Triticum	*dicoccon*	*dicoccoides*	Palestine	Emmer	Wheat
Triticum	*dicoccon*	*dimococcum*	Two Husk	Emmer	Wheat
Triticum	*dicoccon*	*erythromelan*	Dark Red	Emmer	Wheat
Triticum	*dicoccon*	*erythrurum*	Red Tailed	Emmer	Wheat
Triticum	*dicoccon*	*europaeum*	European	Emmer	Wheat
Triticum	*dicoccon*	*eurorum*	Ror	Emmer	Wheat
Triticum	*dicoccon*	*falcatum-coerulescens*	Blue Sickle	Emmer	Wheat
Triticum	*dicoccon*	*falcatum-erythromelan*	Red Sickle	Emmer	Wheat
Triticum	*dicoccon*	*falcatum-melaleucum*	Bottlebrush	Emmer	Wheat
Triticum	*dicoccon*	*falcatum-murciense*	Murca	Emmer	Wheat
Triticum	*dicoccon*	*falcatum-niloticum*	Nile River Valley	Emmer	Wheat
Triticum	*dicoccon*	*farrum*	Farr	Emmer	Wheat
Triticum	*dicoccon*	*fumidum-album*	White Smokey	Emmer	Wheat
Triticum	*dicoccon*	*fumidum-rubrum*	Red Smokey	Emmer	Wheat
Triticum	*dicoccon*	*gaertnerianum*	Gaertner	Emmer	Wheat
Triticum	*dicoccon*	*georgicum*	Georgia	Emmer	Wheat
Triticum	*dicoccon*	*immaturatum*	Immature	Emmer	Wheat
Triticum	*dicoccon*	*ispahanicum*	Esfahan	Emmer	Wheat
Triticum	*dicoccon*	*italicum*	Italian	Emmer	Wheat
Triticum	*dicoccon*	*karamyschevii*	Karamyschev	Emmer	Wheat
Triticum	*dicoccon*	*levimurciense*	Levi Murca	Emmer	Wheat
Triticum	*dicoccon*	*maturatum*	Mature	Emmer	Wheat
Triticum	*dicoccon*	*nigrum*	Black	Emmer	Wheat
Triticum	*dicoccon*	*nigrum-album*	White & Black	Emmer	Wheat
Triticum	*dicoccon*	*nigrum-rubrum*	Red & Black	Emmer	Wheat
Triticum	*dicoccon*	*palaeocolchicum*	Crocus	Emmer	Wheat
Triticum	*dicoccon*	*ramosum-provinciale*	Provincial Branching	Emmer	Wheat

Triticum	*dicoccon*	*ramosum-reichenbachii*	Reichenbach Branching	Emmer	Wheat
Triticum	*dicoccon*	*rufescens*	Rufous	Emmer	Wheat
Triticum	*dicoccon*	*spicatum-album*	White Spike	Emmer	Wheat
Triticum	*dicoccon*	*spicatum-rubrum*	Red Spike	Emmer	Wheat
Triticum	*dicoccon*	*spontaneum*	Spontaneous	Emmer	Wheat
Triticum	*dicoccon*	*tomentosum*	Woolly	Emmer	Wheat
Triticum	*dicoccon*	*tricoccum*	Three Husk	Emmer	Wheat
Triticum	*dicoccon*	*turgidoides*	Turgid	Emmer	Wheat
Triticum	*dicoccon*	*uncinatum*	Hooked	Emmer	Wheat
Triticum	*dicoccon*	*volgense*	Volga	Emmer	Wheat

Wheat, Durum (*Triticum durum*)

Genus	species	Infraspecies	cultivar	specimen	Genome
Triticum	*durum*	*durum*	Original	Durum	Wheat
Triticum	*durum*	*accessorium*	Accessory	Durum	Wheat
Triticum	*durum*	*alatum*	Winged Seed	Durum	Wheat
Triticum	*durum*	*algeriense*	Algerian	Durum	Wheat
Triticum	*durum*	*allendii*	Allend	Durum	Wheat
Triticum	*durum*	*atromarginatum*	Dark Margin	Durum	Wheat
Triticum	*durum*	*bauhinii*	Bauhin	Durum	Wheat
Triticum	*durum*	*bicolor*	Bicolor	Durum	Wheat
Triticum	*durum*	*brachystachyum*	Short Stack	Durum	Wheat
Triticum	*durum*	*breviaristatum*	Short Awn	Durum	Wheat
Triticum	*durum*	*candissimum*	Pure White	Durum	Wheat
Triticum	*durum*	*caucasicum*	Caucasia	Durum	Wheat
Triticum	*durum*	*cevallos*	Cevallo	Durum	Wheat
Triticum	*durum*	*cochleare*	Ear	Durum	Wheat
Triticum	*durum*	*corvalanii*	Corvalan	Durum	Wheat
Triticum	*durum*	*densiusculum*	Dense	Durum	Wheat
Triticum	*durum*	*dubium*	Dubious	Durum	Wheat
Triticum	*durum*	*elongatum*	Elongated	Durum	Wheat
Triticum	*durum*	*falcatoapulicum*	Sickle	Durum	Wheat
Triticum	*durum*	*fastuosum*	Proud	Durum	Wheat
Triticum	*durum*	*heterochromum*	Variable Color	Durum	Wheat
Triticum	*durum*	*hordeiforme*	Barley	Durum	Wheat
Triticum	*durum*	*laxiusculum*	Limp	Durum	Wheat
Triticum	*durum*	*longisemineum*	Long Seed	Durum	Wheat
Triticum	*durum*	*marginatum*	Margined	Durum	Wheat
Triticum	*durum*	*maurorum*	Maury	Durum	Wheat
Triticum	*durum*	*medium*	Medium	Durum	Wheat
Triticum	*durum*	*melanopogon*	Dark Beard	Durum	Wheat
Triticum	*durum*	*molle*	Soft	Durum	Wheat
Triticum	*durum*	*orientale*	Oriental	Durum	Wheat
Triticum	*durum*	*platystachyum*	Flat Stack	Durum	Wheat
Triticum	*durum*	*pruinosum*	Glistening	Durum	Wheat
Triticum	*durum*	*pseudocopticum*	Lesser Coptic	Durum	Wheat
Triticum	*durum*	*pseudoerythromelan*	Lesser Dark Red	Durum	Wheat
Triticum	*durum*	*pyramidale*	Pyramid	Durum	Wheat
Triticum	*durum*	*ramosoleucurum*	Branching	Durum	Wheat
Triticum	*durum*	*rarum*	Rare	Durum	Wheat
Triticum	*durum*	*siculum*	Sicilian	Durum	Wheat
Triticum	*durum*	*sphaeroaffine*	Affine Round	Durum	Wheat
Triticum	*durum*	*sphaeroafricanum*	African Round	Durum	Wheat
Triticum	*durum*	*sphaeroleucurum*	Light Round	Durum	Wheat
Triticum	*durum*	*sphaeromelanopus*	Dark Round	Durum	Wheat
Triticum	*durum*	*tanaiticum*	Tanait	Durum	Wheat
Triticum	*durum*	*tiflisiense*	Tiflisia	Durum	Wheat
Triticum	*durum*	*tomentosum*	Woolly	Durum	Wheat
Triticum	*durum*	*transcaucasicum*	Caucasian	Durum	Wheat
Triticum	*durum*	*trevisium*	Trevis	Durum	Wheat
Triticum	*durum*	*venulosum*	Veined	Durum	Wheat
Triticum	*durum*	*villosum*	Villose	Durum	Wheat

Wheat, Sword Flour (*Triticum macha*)

Genus	species	Infraspecies	cultivar	specimen	Genome
Triticum	*macha*	*macha*	Original	Sword Flour	Wheat
Triticum	*macha*	*imereticum*	Immer	Sword Flour	Wheat
Triticum	*macha*	*tubalicum*	Tubal	Sword Flour	Wheat

Wheat, Einkorn (*Triticum monococcum*)

Genus	species	Infraspecies	cultivar	specimen	Genome
Triticum	*monococcum*	*monococcum*	Original	Einkorn	Wheat
Triticum	*monococcum*	*aegilopoides*	Shield	Einkorn	Wheat
Triticum	*monococcum*	*albanicum*	Albanian	Einkorn	Wheat
Triticum	*monococcum*	*gaterslebense*	Gatersleben	Einkorn	Wheat
Triticum	*monococcum*	*hornemannii*	Hornemann	Einkorn	Wheat
Triticum	*monococcum*	*lasiorrhachis*	Woolly Rachis	Einkorn	Wheat
Triticum	*monococcum*	*mansfeldii*	Mansfeld	Einkorn	Wheat
Triticum	*monococcum*	*pubescens*	Downy	Einkorn	Wheat

Triticum	*monococcum*	*sinskajae*	Sinskaja	Einkorn	Wheat
Triticum	*monococcum*	*spontaneum*	Spontaneous	Einkorn	Wheat
Triticum	*monococcum*	*tenax*	Tenacious	Einkorn	Wheat
Triticum	*monococcum*	*viridivulgare*	Green	Einkorn	Wheat

Wheat, Polish (*Triticum polonicum*)

Genus	species	Infraspecies	cultivar	specimen	Genome
Triticum	*polonicum*	*polonicum*	Original	Polish	Wheat
Triticum	*polonicum*	*abyssinicum*	Ethiopian	Polish	Wheat
Triticum	*polonicum*	*brevissimum*	Short	Polish	Wheat
Triticum	*polonicum*	*glaucum*	Glaucous	Polish	Wheat
Triticum	*polonicum*	*laevissimum*	Smooth	Polish	Wheat
Triticum	*polonicum*	*maximum*	Greater	Polish	Wheat
Triticum	*polonicum*	*mexicana*	Mexican	Polish	Wheat
Triticum	*polonicum*	*petropavlovskyi*	Petropavlovsky	Polish	Wheat
Triticum	*polonicum*	*petropseudobarbarossa*	Barbary	Polish	Wheat
Triticum	*polonicum*	*tibeticum*	Tibet	Polish	Wheat

Wheat, Spelt (*Triticum spelta*)

Genus	species	Infraspecies	cultivar	specimen	Genome
Triticum	*spelta*	*spelta*	Original	Spelt	Wheat
Triticum	*spelta*	*arias*	Arias	Spelt	Wheat
Triticum	*spelta*	*elymoides*	Rye	Spelt	Wheat
Triticum	*spelta*	*forskalei*	Forskale	Spelt	Wheat
Triticum	*spelta*	*griseoturanorecens*	Gray Turin	Spelt	Wheat
Triticum	*spelta*	*ispharosubbaktiaricum*	Ispharo Baktiar	Spelt	Wheat
Triticum	*spelta*	*kuckuckianum*	Kuckuck	Spelt	Wheat
Triticum	*spelta*	*palmovae*	Palmov	Spelt	Wheat
Triticum	*spelta*	*rufescens*	Rufous	Spelt	Wheat
Triticum	*spelta*	*saharae*	Sahara	Spelt	Wheat
Triticum	*spelta*	*subbuldoji*	Buldoj	Spelt	Wheat
Triticum	*spelta*	*tashausicum*	Tashaus	Spelt	Wheat
Triticum	*spelta*	*turanoalborescens*	White Turin	Spelt	Wheat
Triticum	*spelta*	*turanoalefeldii*	Alefeld Turin	Spelt	Wheat
Triticum	*spelta*	*zea*	Zeatin	Spelt	Wheat

Wheat, Roundseed (*Triticum sphaerococcum*)

Genus	species	Infraspecies	cultivar	specimen	Genome
Triticum	*sphaerococcum*	*sphaerococcum*	Original	Roundseed Flour	Wheat
Triticum	*sphaerococcum*	*ischkaschimicum*	Ischkaschim	Roundseed Flour	Wheat
Triticum	*sphaerococcum*	*kostoffii*	Kostoff	Roundseed Flour	Wheat
Triticum	*sphaerococcum*	*popovii*	Popov	Roundseed Flour	Wheat

Wheat, Timopheev (*Triticum timopheevii*)

Genus	species	Infraspecies	cultivar	specimen	Genome
Triticum	*timopheevii*	*timopheevii*	Original	Timopheev	Wheat
Triticum	*timopheevii*	*araraticum*	Ararat Mountain	Timopheev	Wheat
Triticum	*timopheevii*	*armeniacum*	Armenian	Timopheev	Wheat
Triticum	*timopheevii*	*chaldicum*	Chaldian	Timopheev	Wheat
Triticum	*timopheevii*	*coccum*	Roundseed	Timopheev	Wheat
Triticum	*timopheevii*	*kurdistanicum*	Kurdistan	Timopheev	Wheat
Triticum	*timopheevii*	*miguschovae*	Miguschov	Timopheev	Wheat
Triticum	*timopheevii*	*militinae*	Milite	Timopheev	Wheat
Triticum	*timopheevii*	*montanum*	Mountain	Timopheev	Wheat
Triticum	*timopheevii*	*nabelicum*	Nabel	Timopheev	Wheat
Triticum	*timopheevii*	*nikolai*	Nikola	Timopheev	Wheat

Wheat, Ural-Altai Mountain (*Triticum turanicum*)

Genus	species	Infraspecies	cultivar	specimen	Genome
Triticum	*turanicum*	*turanicum*	Original	Ural-Altai Mountain	Wheat
Triticum	*turanicum*	*quasinotabile*	Notable	Ural-Altai Mountain	Wheat
Triticum	*turanicum*	*orientale*	Oriental	Ural-Altai Mountain	Wheat

Wheat, Poulard (*Triticum turgidum*)

Genus	species	Infraspecies	cultivar	specimen	Genome
Triticum	*turgidum*	*turgidum*	Original	Poulard	Wheat
Triticum	*turgidum*	*compositum*	Composite	Poulard	Wheat
Triticum	*turgidum*	*hosteanum*	Host	Poulard	Wheat
Triticum	*turgidum*	*jakubzineri*	Jakubziner	Poulard	Wheat
Triticum	*turgidum*	*pyramidale*	Pyramid	Poulard	Wheat
Triticum	*turgidum*	*sardinicum*	Sardinia	Poulard	Wheat
Triticum	*turgidum*	*villosum*	Villose	Poulard	Wheat

Wheat, Urartu (*Triticum urartu*)

Genus	species	Infraspecies	cultivar	specimen	Genome
Triticum	*urartu*	*urartu*	Original	Urartu	Wheat
Triticum	*urartu*	*binarturutriru*	Binarturutriru	Urartu	Wheat
Triticum	*urartu*	*libanicum*	Lebanon	Urartu	Wheat

Triticum	*urartu*	*michaelii*	Michael	Urartu	Wheat
Triticum	*urartu*	*pubescentinigrum*	Black Downy	Urartu	Wheat
Triticum	*urartu*	*pubescentiurartu*	Downy	Urartu	Wheat
Triticum	*urartu*	*turcicum*	Turkish	Urartu	Wheat

Order 722.024 *Poales* Meadow Grass

Latin: *Poa* grass

FAMILIES

722.024.1 *Poaceae* Meadow Grass
722.024.2 *Agrostidaceae* Bent Grass
722.024.3 *Andropogonaceae* Beard Grass
722.024.4 *Chloridaceae* Windmill Grass
722.024.5 *Festucaceae* Fescue Grass
722.024.6 *Panicaceae* Panic Grass
722.024.7 *Philaridaceae* Canary Grass
722.024.8 *Zoysiaceae* Zoysia Grass

Family 725.024.7 *Phalaridaceae* Canary Grass

Genus	Species	Specimen	Genome	C:Culinary, M:Medicinal, E:Economic
Hierochloe	*odorata*	Vanilla	Sweet Grass	**C**:LF,OL:BF,CF
Vetiveria	*zizanioides*	Cuscus	Vetiver Grass	**C**:RT,OL:BF,CF

Varieties

Vanilla Sweet Grass (*Hierochole odorata*)

Genus	species	Infraspecies	cultivar	specimen	Genome
Hierochloe	*odorata*	*odorata*	Original	Vanilla	Sweet Grass
Hierochloe	*odorata*	*arctica*	Nootka	Vanilla	Sweet Grass
Hierochloe	*odorata*	*borealis*	Boreal	Vanilla	Sweet Grass
Hierochloe	*odorata*	*dahurica*	Dahurian	Vanilla	Sweet Grass
Hierochloe	*odorata*	*nashii*	Nash	Vanilla	Sweet Grass

Cuscus (*Vetiveria zizanioides*)

Genus	species	Infraspecies	cultivar	specimen	Genome
Vetiveria	*zizanioides*	.	.	Cuscus	Vetiver Grass
Vetiveria	*zizanioides*	*zizanioides*	India	Cuscus	Vetiver Grass
Vetiveria	*zizanioides*	*arundinacea*	Jamaica	Cuscus	Vetiver Grass
Vetiveria	*zizanioides*	*muricata*	Trinidad	Cuscus	Vetiver Grass
Vetiveria	*zizanioides*	*odorata*	Java	Cuscus	Vetiver Grass
Vetiveria	*zizanioides*	*odoratissima*	Sri Lanka	Cuscus	Vetiver Grass

Class 722.03 *Pandanopsida* Wood Lily

Latin: *pandana* screw pine

ORDERS

722.031 *Pandanales* Screw Pine
722.032 *Arecales* Palm
722.033 *Arundoales* Reed

Order 722.032 *Arecales* Palm

Latin: *pandana* screw pine

FAMILIES

722.032.1 *Arecaceae* Feather Palm
722.032.2 *Arecavannaceae* Fan Palm
722.032.3 *Pararecaceae* Frond Palm Reed

Family 722.032.1 *Arecaceae* Feather Palm

Genus	Species	Specimen	Genome	C:Culinary, M:Medicinal
Areca	*catechu*	Betel Nut	Feather Palm	**C**:SD:**NT**
Jubaea	*chilensis*	Chilean	Honey Feather Palm	**C**:SA:FF
Parajubaea	*cocoides*	Coco	Sweet Feather Palm	**C**:SA:FF
Parajubaea	*sunkha*	Sunkh	Sweet Feather Palm	**C**:SA:FF
Parajubaea	*torallyi*	Torally	Sweet Feather Palm	**C**:SA:FF

Varieties

Betel Nut (*Areca catechu*)

Genus	species	Infraspecies	cultivar	specimen	Genome
Areca	*catechu*	*catechu*	Original	Betel Nut	Feather Palm
Areca	*catechu*	*faufel*	Faufel	Betel Nut	Feather Palm
Areca	*catechu*	*himalayana*	Himalaya	Betel Nut	Feather Palm
Areca	*catechu*	*hortensis*	Garden	Betel Nut	Feather Palm
Areca	*catechu*	*nigra*	Black	Betel Nut	Feather Palm

Palm Sap, Honey (*Jubaea chilensis*)

Genus	species	Infraspecies	cultivar	specimen	Genome
Jubaea	*chilensis*	.	.	Chilean	Honey Feather Palm
Jubaea	*chilensis*	*chilensis*	Original	Chilean	Honey Feather Palm
Jubaea	*chilensis*	*spectabilis*	Spectacular	Chilean	Honey Feather Palm

Palm Sap, Sweet (*Parajubaea spp.*)

Genus	species	Infraspecies	cultivar	specimen	Genome
Parajubaea	*cocoides*	.	.	Coco	Sweet Feather Palm
Parajubaea	*sunkha*	.	.	Sunkh	Sweet Feather Palm
Parajubaea	*torallyi*	.	.	Torally	Sweet Feather Palm
Parajubaea	*torallyi*	*torallyi*	Original	Torally	Sweet Feather Palm
Parajubaea	*torallyi*	*microcarpa*	Small Fruit	Torally	Sweet Feather Palm

Order 722.033 *Arundoales* Reed

Family 722.033.1 *Arundoaceae* Reed

Genus	Species	Specimen	Genome	C:Culinary, M:Medicinal
Saccharum	*officinarum*	Edible	Sugar Cane	C:SL:**VG**

Varieties

Sugar Cane (*Saccharum officinarum*)

Genus	species	Infraspecies	cultivar	specimen	Genome
Saccharum	*officinarum*	*officinarum*	New Guinea	Edible	Sugar Cane
Saccharum	*officinarum*	*atrorubens*	Dark Red	Edible	Sugar Cane
Saccharum	*officinarum*	*brevipedicellatum*	Short Pedicel	Edible	Sugar Cane
Saccharum	*officinarum*	*fragile*	Fragile	Edible	Sugar Cane
Saccharum	*officinarum*	*genuinum*	Genuine	Edible	Sugar Cane
Saccharum	*officinarum*	*giganteum*	Giant	Edible	Sugar Cane
Saccharum	*officinarum*	*glabrum*	Smooth	Edible	Sugar Cane
Saccharum	*officinarum*	*jamaicense*	Jamaican	Edible	Sugar Cane
Saccharum	*officinarum*	*litteratum*	Litterate	Edible	Sugar Cane
Saccharum	*officinarum*	*litteratum-breve*	Short Litterate	Edible	Sugar Cane
Saccharum	*officinarum*	*luridum*	Lurid	Edible	Sugar Cane
Saccharum	*officinarum*	*luteum-durum*	Yellow Hard	Edible	Sugar Cane
Saccharum	*officinarum*	*luzonicum*	Luzon	Edible	Sugar Cane
Saccharum	*officinarum*	*obscurum*	Obscure	Edible	Sugar Cane
Saccharum	*officinarum*	*oceanicum*	Ocean Island	Edible	Sugar Cane
Saccharum	*officinarum*	*otaheitense*	Otaheite	Edible	Sugar Cane
Saccharum	*officinarum*	*purpureum*	Purple	Edible	Sugar Cane
Saccharum	*officinarum*	*rubicundum*	Ruby Red	Edible	Sugar Cane
Saccharum	*officinarum*	*rubrum-altum*	Tall Red	Edible	Sugar Cane
Saccharum	*officinarum*	*rubrum-humile*	Low Red	Edible	Sugar Cane
Saccharum	*officinarum*	*tahitense*	Tahiti	Edible	Sugar Cane
Saccharum	*officinarum*	*violaceum*	Violet	Edible	Sugar Cane

Superclass *Orchidopsida* Orchid

Class 722.09 *Orchidaustralasiopsida* South Asian Orchid

Order 722.096 *Epiarethusaustralasiales* South Asian Dragon's Mouth Orchid

Family 722.096.3 *Bletiaustralasiaticeae* South Asian Urn Orchid

Genus	Species	Specimen	Genome	C:Culinary, M:Medicinal, E:Economic
Bletilla	*striata*	Hyacinth	Little Urn Orchid	C:SD:**NT**

Varieties

Hyacinth Nut (*Bletilla striata*)

Genus	species	Infraspecies	cultivar	specimen	Genome

Bletilla	*striata*	*striata*	Orginal	Hyacinth	Little Urn Orchid
Bletilla	*striata*	*albomarginata*	White Margined	Hyacinth	Little Urn Orchid
Bletilla	*striata*	*elegantula*	Elegant	Hyacinth	Little Urn Orchid
Bletilla	*striata*	*gebina*	Gebin	Hyacinth	Little Urn Orchid

Class 722.11 *Orchidatropicameriopsida* Tropical America Orchid

Order 722.115 *Vanillatropicameriales* Vanilla

Family 722.115.B *Vanillatropicameriaceae* Vanilla

Genus	Species	Specimen	Genome	C:Culinary, M:Medicinal
Vanilla-tropicameria	*planifolia*	Spice	Tropical American Vanilla	C:SD(Bean):**SP**
Vanilla-tropicameria	*pompona*	Pompon	Tropical American Vanilla	C:SD(Bean):**SP**

Varieties

Vanilla Bean (*Vanilla-tropicameria planifolia*)

Genus	species	Infraspecies	cultivar	specimen	Genome
Vanilla-tropicameria	*planifolia*	*planifolia*	Original	Spice	American Vanilla Orchid
Vanilla-tropicameria	*planifolia*	*aromatica*	Aromatic	Spice	American Vanilla Orchid
Vanilla-tropicameria	*planifolia*	*duckei*	Ducke	Spice	American Vanilla Orchid
Vanilla-tropicameria	*planifolia*	*fragrans*	Fragrant	Spice	American Vanilla Orchid
Vanilla-tropicameria	*planifolia*	*gigantea*	Giant	Spice	American Vanilla Orchid
Vanilla-tropicameria	*planifolia*	*hirsuta*	Hairy	Spice	American Vanilla Orchid
Vanilla-tropicameria	*planifolia*	*sativa*	Sweet	Spice	American Vanilla Orchid
Vanilla-tropicameria	*planifolia*	*sylvestris*	Forest	Spice	American Vanilla Orchid
Vanilla-tropicameria	*planifolia*	*variegata*	Variegated	Spice	American Vanilla Orchid
Vanilla-tropicameria	*planifolia*	*tiarei*	Tiare	Spice	American Vanilla Orchid
Vanilla-tropicameria	*planifolia*	*viridiflora*	Green Flower	Spice	American Vanilla Orchid

PHYLUM 724 *Dicotovegetophyta* Cultivated Dicot

Latin: *Di* diad, dual, two, Latin: *cota* cotyledon, leaf embryo, *phyta* plant

724.01 Class 01 *Asteropsida*	Aster
724.02 Class 02 *Caryophyllopsida*	Cary
724.03 Class 03 *Cornopsida*	Cornel
724.04 Class 04 *Gentiasolopsida*	Gentian & Sunplant
724.05 Class 05 *Lamiopsida*	Mint
724.06 Class 06 *Legumopsida*	Legume
724.07 Class 07 *Magnoliopsida*	Magnolia
724.08 Class 08 *Malvopsida*	Mallow
724.09 Class 09 *Myrtopsida*	Myrtle
724.10 Class 10 *Nymphaeopsida*	Water Lily
724.11 Class 11 *Ranunculopsida*	Buttercup
724.12 Class 12 *Rosopsida*	Rose
724.13 Class 13 *Rutopsida*	Rue
724.14 Class 14 *Santalopsida*	Sandalwood
724.15 Class 15 *Theopsida*	Tea
724.16 Class 16 *Violopsida*	Violet

Class 724.01 *Asteropsida*	**Aster**
Order 1 *Asterales*	Daisy
Family 1 *Asteraceae*	Daisy
Family 2 *Botaceae*	Button Daisy
Family 3 *Conadasiaceae*	Cone Daisy
Family 4 *Dasiellaceae*	Daisyette
Family 5 *Dendranthemaceae*	Mum
Family 6 *Helianthaceae*	Sunflower
Family 7 *Tagetaceae*	Marigold
Order 2 *Campanulales*	Bellflower
Family 1 *Campanulaceae*	Bellflower
Family 2 *Goodeniaceae*	Goodenough Bellflower
Family 3 *Lobeliaceae*	Lobel Bellflower
Family 4 *Pentaphragmataceae*	Cinquefoil
Family 5 *Sphenocleaceae*	Chicken Spike
Family 6 *Stylidiaceae*	Trigger Plant
Order 3 *Compositales*	Aster
Family 1 *Compositaceae*	Aster
Family 2 *Carduaceae*	Thistle
Family 3 *Floretasteraceae*	Floret Aster
Family 4 *Hymenopappaceae*	Pappus Aster
Family 5 *Lactucaceae*	Lettuce
Family 6 *Radiasteraceae*	Ray Aster
Family 7 *Senecioaceae*	Groundsel
Family 8 *Solidagoaceae*	Goldenrod

Class 724.02 *Caryophyllopsida*	**Succulent**
Order 1 *Caryophyllales*	Succulent
Family 1 *Caryophyllaceae*	Succulent
Family 2 *Aizoaceae*	Ice Plant
Family 3 *Amaranthaceae*	Amaranth
Family 4 *Basellaceae*	Basellvine
Family 5 *Chenopodiaceae*	Goosefoot
Family 6 *Molluginaceae*	Carpetweed
Family 7 *Nyctaginaceae*	Night Lumbar
Family 8 *Phytolaccaceae*	Pokeberry
Family 9 *Portulacaceae*	Purslane
Order 2 *Cactales*	Cactus
Family 1 *Cactaceae-agregata*	Clump Cactus
Family 2 *Cactaceae-barrila*	Barrel Cactus
Family 3 *Cactaceae-columna*	Column Cactus
Family 4 *Cactaceae-epiphyitea*	Epiphyte Cactus
Family 5 *Cactaceae-globa*	Globe Cactus
Family 6 *Cactaceae-pendens*	Pendent Cactus
Family 7 *Cactaceae-platya*	Flatstem Cactus
Family 8 *Cactaceae-ramosa*	Branching Cactus
Family 9 *Cactaceae-rosetta*	Rosette Cactus
Order 3 *Polygonales*	Knotweed
Family 1 *Polygonaceae*	Knotweed
Family 2 *Podostemaceae*	Riverweed

Class 724.03 *Cornopsida*	**Cornel**
Order 1 *Cornales*	Cornel
Family 1 *Cornaceae*	Cornel
Family 2 *Alangiaceae*	Alang
Family 3 *Diapensiaceae*	Diapense
Family 4 *Eucommiaceae*	Gutta Percha
Family 5 *Garryaceae*	Silk Tassel
Family 6 *Morinaceae*	False Thistle
Family 7 *Nyssaceae*	Tupelo
Order 2 *Apiales*	Marshwort
Family 1 *Apiaceae*	Marshwort
Family 2 *Araliaceae*	Aral
Family 3 *Rubiaceae*	Rubyroot
Order 3 *Dipsacales*	Teasel
Family 1 *Dipsacaceae*	Teasel
Family 2 *Adoxaceae*	Musk Root
Family 3 *Caprifoliaceae*	Honeysuckle
Family 4 *Valerianaceae*	Valerian
Order 4 *Ericales*	Heath
Family 1 *Ericaceae*	Heath
Family 2 *Clethraceae*	Sweet Pepper Heath
Family 3 *Cyrillaceae*	Hilltop Heath
Family 4 *Empetraceae*	Empeter Heath
Family 5 *Epacridaceae*	Pukiawe Heath
Family 6 *Monotropaceae*	Pipes
Family 7 *Pyrolaceae*	Wintergreen Heath
Family 8 *Sapotaceae*	Sapote
Order 5 *Nepenthales*	Pitcher Plant
Family 1 *Nepenthaceae*	Pitcher Plant
Family 2 *Droseraceae*	Flytrap Pitcher Plant
Family 3 *Sarraceniaceae*	Insect Pitcher Plant

Class 724.04 *Gentiasolopsida*	**Gentian & Sunplant**
Order 1 *Gentianales*	Gentian
Family 1 *Gentianaceae*	Gentian
Family 2 *Apocynaceae*	Dogbane
Family 3 *Asclepiadaceae*	Milkstem
Family 4 *Loganiaceae*	Loganwood
Order 2 *Solanales*	Sunplant
Family 1 *Solanaceae*	Sunplant
Family 2 *Convolvulaceae*	Bindwood
Family 3 *Cuscutaceae*	Dodder Vine
Family 4 *Fouquieriaceae*	Desert Candle
Family 5 *Hydrophyllaceae*	Waterleaf
Family 6 *Menyanthaceae*	Buckbean
Family 7 *Polemoniaceae*	Phlox

Class 724.05 *Lamiopsida*	**Mint**
Order 1 *Lamiales*	Mint
Family 1 *Lamiaceae*	Mint
Family 2 *Boraginaceae*	Borage
Family 3 *Lennoaceae*	Desert Christmaswood
Family 4 *Plantaginaceae*	Plantain
Family 5 *Rhizophoraceae*	Mangrove Mint
Family 6 *Verbenaceae*	Vervain
Order 2 *Callitrichales*	Starwort
Family 1 *Callitrichaceae*	Starwort
Family 2 *Hippuridaceae*	Mare's Tail
Order 3 *Scrophulariales*	Foxglove
Family 1 *Scrophulariaceae*	Foxglove
Family 2 *Acanthaceae*	Prickleleaf Foxglove
Family 3 *Bignoniaceae*	Trumpetwood
Family 4 *Buddlejaceae*	Buddlewood
Family 5 *Gesneriaceae*	Gesner Foxglove
Family 6 *Lentimyoporaceae*	Mastic & Myopore
Family 7 *Oleaceae*	Olive
Family 8 *Orobanchaceae*	Orobanche
Family 9 *Pedaliaceae*	Sesame

Class 724.06 *Legumopsida*	**Legume**
Order 1 *Legumales*	Legume
Family 1 *Legumaceae*	Legume
Family 2 *Caesalpiniaceae*	Caesalpino
Family 3 *Cytisaceae*	Broom
Family 4 *Ebonaceae*	Ebony
Family 5 *Mimosaceae*	Sensitive Plant
Family 6 *Piptadeniaceae*	Droopgland
Order 2 *Fabales*	Bean
Family 1 *Fabaceae*	Bean
Family 2 *Fabafloraceae*	Beanflower
Family 3 *Dolichaceae*	Longpod Bean
Order 3 *Pisales*	Pea
Family 5 *Pisaceae*	Pea
Family 6 *Pisafloraceae*	Peaflower
Family 7 *Trifoliaceae*	Clover
Family 8 *Viciaceae*	Vetch
Family 9 *Wisteriaceae*	Wisteria

Class 724.07 *Magnoliopsida*	**Magnolia**
Order 1 *Magnoliales*	Magnolia
Family 1 *Magnoliaceae*	Magnolia
Family 2 *Annonaceae*	Custard Apple
Family 3 *Aristolochiaceae*	Loch
Family 4 *Canellaceae*	Wild Cinnamon
Family 5 *Myristicaceae*	Nutmeg
Family 6 *Rafflesiaceae*	Stinkflower
Family 7 *Sonneratiaceae*	Sonner
Family 8 *Winteraceae*	Pepper Bark
Order 2 *Illiciales*	Star Anise
Family 1 *Illiciaceae*	Star Anise
Family 2 *Calyceraceae*	Calyce
Family 3 *Schisandraceae*	Star Vine
Order 3 *Laurales*	Laurel
Family 1 *Lauraceae*	Laurel
Family 2 *Calycanthaceae*	Tropicnut
Family 3 *Hernandiaceae*	Hernando
Family 4 *Monimiaceae*	Monimy
Family 5 *Perseaceae*	Avocado
Family 6 *Sassafrasaceae*	Sassafras

Class 724.08 *Malvopsida*	**Mallow**
Order 1 *Malvales*	Mallow
Family 1 *Malvaceae*	Mallow
Family 2 *Bombacaceae*	Silk Cotton
Family 3 *Elaeocarpaceae*	Oilberry
Family 4 *Leitneriaceae*	Corkwood
Family 5 *Plumbaginaceae*	Leadwort
Family 6 *Sterculiaceae*	Boatnut
Family 7 *Tiliaceae*	Linden
Order 2 *Euphorbiales*	Spurge
Family 1 *Euphorbiaceae*	Spurge
Family 2 *Phyllanthaceae*	Leaf Flower Spurge
Order 3 *Rhamnales*	Buckthorn
Family 1 *Rhamnaceae*	Buckthorn
Family 2 *Elaeagnaceae*	Agnes Olive
Family 3 *Vitaceae*	Grape

Class 724.08 *Malvopsida*	**Mallow**
Order 4 *Urticales*	Nettle
Family 1 *Urticaceae*	Nettle
Family 2 *Cannabaceae*	Hemp
Family 3 *Cecropiaceae*	Pumpwood
Family 4 *Moraceae*	Mulberry
Family 5 *Ulmaceae*	Elm
Class 724.09 *Myrtopsida*	**Myrtle**
Order 1 *Myrtales*	Myrtle
Family 1 *Myrtaceae*	Myrtle
Family 2 *Combretaceae*	Mangrove Myrtle
Family 3 *Lythraceae*	Loosestrife
Family 4 *Melastomataceae*	Melastome
Family 5 *Onagraceae*	Onager
Family 6 *Punicaceae*	Pomegranate
Family 7 *Thymelaeaceae*	Mezereon
Family 8 *Trapaceae*	Water Chestnut
Order 2 *Haloragales*	Seaberry
Family 1 *Haloragaceae*	Seaberry
Family 2 *Gunneraceae*	Gunner Seaberry
Order 3 *Proteales*	Protean
Family 1 *Proteaceae*	Protean
Class 724.10 *Nymphaeopsida*	**Water Lily**
Order 1 *Nymphaeales*	Water Lily
Family 1 *Nymphaeaceae*	Water Lily
Family 2 *Cabombaceae*	Shield Water Lily
Family 3 *Ceratophyllaceae*	Horn Water Lily
Family 4 *Hydatellaceae*	Tell Water Lily
Family 5 *Nelumbonaceae*	Water Lotus
Order 2 *Piperales*	Peppercorn
Family 1 *Piperaceae*	Peppercorn
Family 2 *Chloranthaceae*	Sac Peppercorn
Family 3 *Saururaceae*	Lizard's Tail
Class 724.11 *Ranunculopsida*	**Buttercup**
Order 1 *Ranunculales*	Buttercup
Family 1 *Ranunculaceae*	Buttercup
Family 2 *Berberidaceae*	Barberry
Family 3 *Lardizabalaceae*	Sausagefruit
Family 4 *Menispermaceae*	Moonseed
Family 5 *Sabiaceae*	Sab
Order 2 *Papaverales*	Poppy
Family 1 *Papaveraceae*	Poppy
Family 2 *Fumariaceae*	Fumitory
Class 724.12 *Rosopsida*	**Rose**
Order 1 *Rosaeurosales*	Roset
Family 1 *Rosaeurosaceae*	Roset
Family 2 *Rosafloretaceae*	Floret Roset
Family 3 *Rosastellataceae*	Starflower Roset
Order 2 *Rosagardeniales*	Garden Rose
Family 1 *Rosaheritaceae*	Heritage Rose
Family 2 *Rosaeldaceae*	Old Garden Rose
Family 3 *Rosaneoaceae*	Modern Rose
Order 3 *Rosahortales*	Horticultural Rose
Family 1 *Bruniaceae*	Burnet
Family 2 *Chrysobalanaceae*	Golden Balan
Family 3 *Crassulaceae*	Stonecrop
Family 4 *Crossosomataceae*	Rockflower
Family 5 *Cunoniaceae*	Butterknife
Family 6 *Eucryphiaceae*	Leatherwood
Family 7 *Hydrangeaceae*	Hydrange
Family 8 *Pittosporaceae*	Cheesewood
Family 9 *Saxifragaceae*	Saxifrage
Order 4 *Primulales*	Primrose
Family 1 *Primulaceae*	Primrose
Family 2 *Buxaceae*	Boxwood
Family 3 *Diospyraceae*	Persimmon
Family 4 *Hamamelidaceae*	Witch Hazel
Family 5 *Myrsinaceae*	Colicwood
Family 6 *Styracaceae*	Storax
Family 7 *Symplocaceae*	Sweetleaf
Family 8 *Theophrastaceae*	Theophrastes
Order 5 *Rosafructales*	Rosefruit
Family 1 *Rosahipaceae*	Rosehip
Family 2 *Rosabaccaceae*	Roseberry
Family 3 *Grossulariaceae*	Gooseberry
Family 4 *Myricaceae*	Bayberry
Family 5 *Rubaceae*	Brambleberry
Family 6 *Rosadrupceae*	Rosedrupe
Family 7 *Rosapomaceae*	Rosepome
Order 6 *Rosanuxales*	Rosenut
Family 1 *Rosanuxaceae*	Rosenut
Family 2 *Betulaceae*	Birchnut
Family 3 *Casuarinaceae*	Oaknut
Family 4 *Fagaceae*	Beechnut
Family 5 *Juglandaceae*	Walnut

Class 724.13 *Rutopsida*	**Rue**
Order 1 *Rutales*	Rue
Family 1 *Rutaceae*	Rue
Family 2 *Cneoraceae*	Olive Spurge
Family 3 *Melianthaceae*	Honeyflower
Family 4 *Tremandraceae*	Black Eyed Susan
Order 2 *Geraniales*	Geranium
Family 1 *Geraniaceae*	Geranium
Family 2 *Balsaminaceae*	Touch Me Not
Family 3 *Limnanthaceae*	Meadowfoam
Family 4 *Oxalidaceae*	Wood Sorrel
Family 5 *Tropaeolaceae*	Trophy Cress
Order 3 *Linales*	Flax
Family 1 *Linaceae*	Flax
Family 2 *Erythroxylaceae*	Redwood
Order 4 *Malpighiales*	Acerola Cherry
Family 1 *Malpighiaceae*	Acerola Cherry
Family 2 *Centroplacaceae*	Plate Pandawood
Family 3 *Irvingiaceae*	Irving Cherry
Family 4 *Lacistemataceae*	Lacestem
Family 4 *Pandaceae*	Pandawood
Family 5 *Picrodendraceae*	Bitterwood
Family 6 *Putranjivaceae*	Putranjiv
Order 5 *Polygalales*	Milkwort
Family 1 *Polygalaceae*	Milkwort
Family 2 *Krameriaceae*	Ratany
Order 6 *Sapindales*	Soapberry
Family 1 *Sapindaceae*	Soapberry
Family 2 *Aceraceae*	Maple
Family 3 *Anacardiaceae*	Cashew
Family 4 *Burseraceae*	Torchwood
Family 5 *Hippocastanaceae*	Buckeye Chestnut
Family 6 *Meliaceae*	Mohagany
Family 7 *Simaroubaceae*	Sima Rue
Family 8 *Staphyleaceae*	Bladdernut
Family 9 *Zygophyllaceae*	Bean Caper
Class 724.14 *Santalopsida*	**Sandalwood**
Order 1 *Santalales*	Sandalwood
Family 1 *Santalaceae*	Sandalwood
Family 2 *Balanophoraceae*	Crownose
Family 3 *Eremolepidaceae*	Catkin Mistletoe
Family 4 *Loranthaceae*	Loreflower
Family 5 *Olacaceae*	Tallowood
Family 6 *Viscaceae*	Mistletoe
Order 2 *Celastrales*	Spindlewood
Family 1 *Celastraceae*	Spindlewood
Family 2 *Aquifoliaceae*	Holly
Family 3 *Corynocarpaceae*	Kakara Nut
Family 4 *Hippocrateaceae*	Medicine Vine
Family 5 *Icacinaceae*	Icac Vine
Family 6 *Stackhousiaceae*	Stackhouse Vine
Class 724.15 *Theopsida*	**Tea**
Order 1 *Theales*	Tea
Family 1 *Theaceae*	Tea
Family 2 *Actinidiaceae*	Kiwi
Family 3 *Caryocaraceae*	Souary Nut
Family 4 *Clusiaceae*	Attorney
Family 5 *Dipterocarpaceae*	Camphor
Family 6 *Elatinaceae*	Waterwort
Family 7 *Marcgraviaceae*	Shingle Plant
Family 8 *Ochnaceae*	Tea Pear
Order 2 *Dilleniales*	Dillen
Family 1 *Dilleniaceae*	Dillen
Family 2 *Paeoniaceae*	Peony
Order 3 *Lecythidales*	Lecy
Family 1 *Lecythidaceae*	Lecy
Family 2 *Napoleonaceae*	Napoleon
Family 3 *Scytopetalaceae*	Scyte Petal
Class 724.16 *Violopsida*	**Violet**
Order 1 *Violales*	Violet
Family 1 *Violaceae*	Violet
Family 2 *Begoniaceae*	Begonia
Family 3 *Bixaceae*	Annato
Family 4 *Caricaceae*	Papaya
Family 5 *Cistaceae*	Rock Rose
Family 6 *Cucurbitaceae*	Melon
Family 7 *Flacourtiaceae*	Flacourt
Family 8 *Loasaceae*	Stingwood
Family 9 *Turneraceae*	Alder Violet
Order 2 *Capparales*	Caper
Family 1 *Capparaceae*	Caper
Family 2 *Brassicaceae*	Mustard
Family 3 *Moringaceae*	Moringo
Family 4 *Resedaceae*	Mignonette
Order 3 *Salicales*	Willow
Family 1 *Salicaceae*	Willow
Family 2 *Datisaceae*	Akelbir
Family 3 *Frankeniaceae*	Sea Heath
Family 4 *Passifloraceae*	Passion Flower
Family 5 *Stachyuraceae*	Urn Flower
Family 6 *Tamaricaceae*	Tamarisk

Class 724.01 *Asteropsida* Aster

Order 724.011 *Asterales* Daisy

Family 724.011.1 *Asteraceae* Daisy

))

Costmary (*Tanacetum majus*)

Genus	Species	Specimen Genome	C:Culinary, M:Medicinal, E:Economic
Bellis	*perennis*	Perennial English Daisy	C:LF:FF
Chamaemelum	*nobile*	Roman Chamomile	C:LF:TE
Tanacetum	*majus*	Costmary Tansy	C:LF:FF,TE

Varieties

English Daisy (*Bellis perennis*)

Genus	species	infraspecies	cultivar	specimen Genome
Bellis	*perennis*	*perennis*	Original	Perennial English Daisy
Bellis	*perennis*	*alpina*	Alpine	Perennial English Daisy
Bellis	*perennis*	*armena*	Armenian	Perennial English Daisy
Bellis	*perennis*	*croatica*	Croatian	Perennial English Daisy
Bellis	*perennis*	*integrifolia*	Entire Leaf	Perennial English Daisy
Bellis	*perennis*	*margaritifolia*	Margarite Leaf	Perennial English Daisy
Bellis	*perennis*	*minor*	Lesser	Perennial English Daisy
Bellis	*perennis*	*pumila*	Low	Perennial English Daisy
Bellis	*perennis*	*robusta*	Robust	Perennial English Daisy
Bellis	*perennis*	*scaposa*	Scape	Perennial English Daisy
Bellis	*perennis*	*validula*	Valid	Perennial English Daisy

Family 724.011.2 *Botaceae* Button Daisy

))

Dandelion Greens (*Taraxacum officinale*)

Genus	Species	Specimen Genome	C:Culinary, M:Medicinal
Helichrysum	*Italicum*	Curry Plant Strawflower Button Daisy	C:SM:FF
Helichrysum	*serphyllifolium*	Hottentot Tea Strawflower Button Daisy	C:LF:TE
Taraxacum	*officinale*	Blowball Dandelion	C:LF:**VG**

Family 3 *Conadasiaceae* Cone Daisy
Family 4 *Dasiellaceae* Daisyette

Family 724.011.5 *Dendrathemaceae* Mum

Genus	Species	Specimen Genome	C:Culinary, M:Medicinal
Chrysanthemum	*coronarium*	Chopsuey Greens Gold Mum	C:SR:**VG**
Chrysanthemum	*coronarium-shungiku*	Shungiku Chopsuey Greens Gold Mum	C:SR:**VG**

Varieties

Chopsuey Greens (*Chrysanthemum coronarium*)

Genus	species	infraspecies	cultivar	specimen	Genome
Chrysanthemum	*coronarium*	*coronarium*	Original	Chopsuey Greens	Gold Mum
Chrysanthemum	*coronarium*	*asteraceae*	Fine Leaf	Chopsuey Greens	Gold Mum
Chrysanthemum	*coronarium*	*primrose-gemii*	Primrose Gem	Chopsuey Greens	Gold Mum
Chrysanthemum	*coronarium*	*spatiosum*	Crown Daisy	Chopsuey Greens	Gold Mum
Chrysanthemum	*coronarium-shungiku*	*shungiku*	Original	Shungiku Chopsuey Greens	Gold Mum
Chrysanthemum	*coronarium-shungiku*	*grandifolia*	Large Leaf	Shungiku Chopsuey Greens	Gold Mum
Chrysanthemum	*coronarium-shungiku*	*maiko-garlandii*	Maiko Garland	Shungiku Chopsuey Greens	Gold Mum
Chrysanthemum	*coronarium-shungiku*	*serrulata*	Small Serrated	Shungiku Chopsuey Greens	Gold Mum

Family 724.011.6 *Helianthaceae* Sunflower

Genus	Species	Specimen Genome	C:Culinary, M:Medicinal, E:Economic
Helianthus	*annuus*	Annual Sunflower	C:SD:**VG**
Helianthus	*annuus-nana*	Dwarf Annual Sunflower	C:SD:**VG**
Helianthus	*argophyllus*	Silver Leaf Sunflower	C:SD:**VG**
Helianthus	*argophyllus x annuus*	Hybrid Silver Leaf Sunflower	C:SD:**VG**
Helianthus	*debilis*	Cucumber Leaf Sunflower	C:SD:**VG**
Helianthus	*multiflorus*	Many Head Sunflower	C:SD:**VG**

Genus	Species	Specimen Genome	C:Culinary, M:Medicinal, E:Economic
Helianthus	*ornamentalis*	Ornamental Sunflower	inedible
Helianthus	*ornamentalis-florepleno*	Double Ornamental Sunflower	inedible
Helianthus	*ornamentalis-nana*	Dwarf Ornamental Sunflower	inedible
Helianthus	*ornamentalis-nana-ramosus*	Branching Dwarf Ornamental Sunflower	inedible
Helianthus	*ornamentalis-ramosus*	Branching Ornamental Sunflower	inedible
Helianthus	*tuberosus*	Jerusalem Artichoke Sunflower	C:SD:**VG**

Varieties

Sunflower (*Helianthus annuus*)

Genus	species	infraspecies	cultivar	specimen	Genome
Helianthus	*annuus*	*annuus*	Original	Annual	Sunflower
Helianthus	*annuus*	*aridus*	Desert	Annual	Sunflower
Helianthus	*annuus*	*arikarai*	Arikara	Annual	Sunflower
Helianthus	*annuus*	*black-russianii*	Black Russian	Annual	Sunflower
Helianthus	*annuus*	*black-seeded*	Black Seeded	Annual	Sunflower
Helianthus	*annuus*	*black-stripei*	Black Stripe	Annual	Sunflower
Helianthus	*annuus*	*california-valley-yokutii*	California Valley Yokut	Annual	Sunflower
Helianthus	*annuus*	*giganteus*	15" Giant	Annual	Sunflower
Helianthus	*annuus*	*giganteus-uniflora*	Single 12" Giant	Annual	Sunflower
Helianthus	*annuus*	*giganteus-gray-stripei*	Gray Stripe 12" Giant	Annual	Sunflower
Helianthus	*annuus*	*giganteus-russianii*	Russian 12" Giant	Annual	Sunflower
Helianthus	*annuus*	*giganteus-russianii-rostovii*	Rostov Russian 12" Giant	Annual	Sunflower
Helianthus	*annuus*	*giganteus-russianii-sundakii*	Sundak Russian 12" Giant	Annual	Sunflower
Helianthus	*annuus*	*giganteus-russianii-whitei*	White Seeded Russian 12" Giant	Annual	Sunflower
Helianthus	*annuus*	*hidsatai-1*	Hidsata #1	Annual	Sunflower
Helianthus	*annuus*	*israelii*	Israeli	Annual	Sunflower
Helianthus	*annuus*	*large-seeded-tallii*	Large Seeded Tall	Annual	Sunflower
Helianthus	*annuus*	*lenticularis*	Lenticular	Annual	Sunflower
Helianthus	*annuus*	*macrocarpus*	Large Seed	Annual	Sunflower
Helianthus	*annuus*	*mammothii-diane's-strainii*	Diane's 20" Mammoth	Annual	Sunflower
Helianthus	*annuus*	*mammothii-gray-stripei*	Gray Stripe 20" Mammoth	Annual	Sunflower
Helianthus	*annuus*	*mammothii-titanii*	Titan 20" Mammoth	Annual	Sunflower
Helianthus	*annuus*	*miriam-ediblei*	Miriam	Annual	Sunflower
Helianthus	*annuus*	*oilseedii*	Oilseed	Annual	Sunflower
Helianthus	*annuus*	*purple-seededii*	Purple Seeded	Annual	Sunflower
Helianthus	*annuus*	*ruby-eclipsei*	Ruby Eclipse	Annual	Sunflower
Helianthus	*annuus*	*senecai*	Seneca	Annual	Sunflower
Helianthus	*annuus*	*skyscraperi*	Skyscraper	Annual	Sunflower
Helianthus	*annuus*	*sunrich-orangei*	Sunrich Orange	Annual	Sunflower
Helianthus	*annuus*	*sunzilla*	Sunzilla	Annual	Sunflower
Helianthus	*annuus*	*tarahumara-whitei*	Tarahumara White	Annual	Sunflower
Helianthus	*annuus*	*texanus*	Texas	Annual	Sunflower
Helianthus	*annuus-nana*	*nana*	Original	Dwarf Annual	Sunflower
Helianthus	*annuus-nana*	*big-smilei*	Big Smile	Dwarf Annual	Sunflower
Helianthus	*annuus-nana*	*sunspotii*	Sunspot	Dwarf Annual	Sunflower
Helianthus	*annuus-nana*	*wildii*	Wild	Dwarf Annual	Sunflower
Helianthus	*argophyllus*	*argophyllus*	Original	Silver Leaf	Sunflower
Helianthus	*argophyllus*	*supreme-mixii*	Supreme Mix	Silver Leaf	Sunflower
Helianthus	*argophyllus x annuus*	*endurancei*	Endurance	Hybrid Silver Leaf	Sunflower
Helianthus	*debilis*	*debilis*	Original	Cucumber Leaf	Sunflower
Helianthus	*debilis*	*cucumerifolius*	Cucumber Leaf	Cucumber Leaf	Sunflower
Helianthus	*debilis*	*cucumerifolius-italian-whitei*	Italian White Cucumber Leaf	Cucumber Leaf	Sunflower
Helianthus	*debilis*	*cucumerifolius-vanilla-icei*	Vanilla Ice Cucumber Leaf	Cucumber Leaf	Sunflower
Helianthus	*debilis*	*silvestris*	Forest	Cucumber Leaf	Sunflower
Helianthus	*debilis*	*tardiflorus*	Late Flowering	Cucumber Leaf	Sunflower
Helianthus	*debilis*	*vestitus*	Covered	Cucumber Leaf	Sunflower
Helianthus	*debilis*	*vestitus-hirtus*	Hairy Covered	Cucumber Leaf	Sunflower
Helianthus	*multiflorus*	*multiflorus*	Original	Many Head	Sunflower
Helianthus	*multiflorus*	*autumn-beautyi*	Autumn Beauty	Many Head	Sunflower
Helianthus	*multiflorus*	*bellezza-d'autunoi*	Bellezza D'Autuno	Many Head	Sunflower
Helianthus	*multiflorus*	*capenoch-stari*	Capenoch Star	Many Head	Sunflower
Helianthus	*multiflorus*	*indian-blanketii*	Indian Blanket	Many Head	Sunflower
Helianthus	*multiflorus*	*irish-eyesi*	Irish Eyes	Many Head	Sunflower
Helianthus	*multiflorus*	*earth-walkeri*	Earth Walker	Many Head	Sunflower
Helianthus	*multiflorus*	*florepleno*	Double Flower	Many Head	Sunflower
Helianthus	*multiflorus*	*florepleno-soleil-d'oroi*	Soleil D'Or Double Flower	Many Head	Sunflower
Helianthus	*multiflorus*	*gloriosus*	Glory	Many Head	Sunflower
Helianthus	*multiflorus*	*loddon-goldii*	Loddon Gold	Many Head	Sunflower
Helianthus	*multiflorus*	*moonwalkeri*	Moonwalker	Many Head	Sunflower
Helianthus	*multiflorus*	*peace-and-freedomii*	Peace And Freedom	Many Head	Sunflower
Helianthus	*multiflorus*	*red-menacei*	Red Menace	Many Head	Sunflower
Helianthus	*multiflorus*	*red-sunii*	Red Sun	Many Head	Sunflower

Genus	species	infraspecies	cultivar	specimen	Genome
Helianthus	*multiflorus*	*sara'si*	Sara's	Many Head	Sunflower
Helianthus	*multiflorus*	*selma-sunsi*	Selma Suns	Many Head	Sunflower
Helianthus	*multiflorus*	*single-yellowi*	Single Yellow	Many Head	Sunflower
Helianthus	*multiflorus*	*triumphe-de-gandii*	Triumphe De Gand	Many Head	Sunflower

Artichoke, Jerusalem (*Helianthus tuberosus*)

Genus	species	infraspecies	cultivar	specimen	Genome
Helianthus	*tuberosus*	*tuberosus*	Original	Jerusalem Artichoke	Sunflower
Helianthus	*tuberosus*	*brazilianii*	Brazilian	Jerusalem Artichoke	Sunflower
Helianthus	*tuberosus*	*challengeri*	Challenger	Jerusalem Artichoke	Sunflower
Helianthus	*tuberosus*	*clearwateri*	Clearwater	Jerusalem Artichoke	Sunflower
Helianthus	*tuberosus*	*french-mammoth-whitei*	French Mammoth White	Jerusalem Artichoke	Sunflower
Helianthus	*tuberosus*	*fuseaui*	Fuseau	Jerusalem Artichoke	Sunflower
Helianthus	*tuberosus*	*fuseaui-redii*	Red Fuseau	Jerusalem Artichoke	Sunflower
Helianthus	*tuberosus*	*golden-nuggetii*	Golden Nugget	Jerusalem Artichoke	Sunflower
Helianthus	*tuberosus*	*jack's-coppercladii*	Jack's Copperclad	Jerusalem Artichoke	Sunflower
Helianthus	*tuberosus*	*long-redii*	Long Red	Jerusalem Artichoke	Sunflower
Helianthus	*tuberosus*	*magenta-purplei*	Magenta Purple	Jerusalem Artichoke	Sunflower
Helianthus	*tuberosus*	*mark's-chokesi*	Mark's Chokes	Jerusalem Artichoke	Sunflower
Helianthus	*tuberosus*	*nakhodkai*	Nakhodka	Jerusalem Artichoke	Sunflower
Helianthus	*tuberosus*	*purple-giantii*	Purple Giant	Jerusalem Artichoke	Sunflower
Helianthus	*tuberosus*	*roter-topinamburi*	Roter Topinambur	Jerusalem Artichoke	Sunflower
Helianthus	*tuberosus*	*skorospelkai*	Skorospelka	Jerusalem Artichoke	Sunflower
Helianthus	*tuberosus*	*smooth-garnetii*	Smooth Garnet	Jerusalem Artichoke	Sunflower
Helianthus	*tuberosus*	*stampedei*	Stampede	Jerusalem Artichoke	Sunflower
Helianthus	*tuberosus*	*subcanescens*	Hoary	Jerusalem Artichoke	Sunflower
Helianthus	*tuberosus*	*tomentosus*	Woolly	Jerusalem Artichoke	Sunflower
Helianthus	*tuberosus*	*vadimii*	Vadim	Jerusalem Artichoke	Sunflower
Helianthus	*tuberosus*	*volgoi-2*	Volgo 2	Jerusalem Artichoke	Sunflower

Family 7 *Tagetaceae* Marigold

Order 724.012 *Campanulales* Bellflower

Family 724.012.1 *Campanulaceae* Bellflower

Genus	Species	Specimen Genome	C:Culinary, M:Medicinal, E:Economic
Campanula	*rapunculus*	Rampion Bellflower	C:LF:**SN**,C:RT:**VG**
Adenophora	*liliifolia*	Lily Leaf Ladybells	C:RT:**VG**

Rampion (*Campanula rampunculus*)

Genus	species	infraspecies	cultivar	specimen	Genome
Campanula	*rapunculus*	*rapunculus*	Original	Rampion	Bellflower
Campanula	*rapunculus*	*bracteosa*	Bracted	Rampion	Bellflower
Campanula	*rapunculus*	*calycina*	Calyx	Rampion	Bellflower
Campanula	*rapunculus*	*castellana*	Castellan	Rampion	Bellflower
Campanula	*rapunculus*	*coarctata*	Arctic	Rampion	Bellflower
Campanula	*rapunculus*	*cymosospicata*	Cyme Spike	Rampion	Bellflower
Campanula	*rapunculus*	*decurrens*	Leaf Stem	Rampion	Bellflower
Campanula	*rapunculus*	*elatior*	Tall	Rampion	Bellflower
Campanula	*rapunculus*	*esculenta*	Edible	Rampion	Bellflower
Campanula	*rapunculus*	*fastigiata*	Dense Branched	Rampion	Bellflower
Campanula	*rapunculus*	*glandulosa*	Gland	Rampion	Bellflower
Campanula	*rapunculus*	*hirsuta*	Hairy	Rampion	Bellflower
Campanula	*rapunculus*	*hirsutissima*	Very Hairy	Rampion	Bellflower
Campanula	*rapunculus*	*hirta*	Somewhat Hairy	Rampion	Bellflower
Campanula	*rapunculus*	*hyrcania*	Hyrcan	Rampion	Bellflower
Campanula	*rapunculus*	*lambertiana*	Lambert	Rampion	Bellflower
Campanula	*rapunculus*	*losae*	Losa	Rampion	Bellflower
Campanula	*rapunculus*	*micrantha*	Small Flower	Rampion	Bellflower
Campanula	*rapunculus*	*paniculiformis*	Panicle	Rampion	Bellflower
Campanula	*rapunculus*	*racemosopaniculata*	Raceme Panicle	Rampion	Bellflower
Campanula	*rapunculus*	*reclinata*	Bent Backward	Rampion	Bellflower
Campanula	*rapunculus*	*spiciformis*	Spike	Rampion	Bellflower
Campanula	*rapunculus*	*strigulosa*	Bristle	Rampion	Bellflower
Campanula	*rapunculus*	*verruculosa*	Warty	Rampion	Bellflower
Campanula	*rapunculus*	*verus*	Valid	Rampion	Bellflower
Campanula	*rapunculus*	*virgata*	Twiggy	Rampion	Bellflower

Lily Leaf Ladybells (*Adenophora liliifolia*)

Genus	species	infraspecies	cultivar	specimen	Genome

Genus	species	infraspecies	cultivar	specimen	Genome
Adenophora	*liliifolia*	*liliifolia*	Original	Lily Leaf	Ladybells
Adenophora	*liliifolia*	*alba*	White	Lily Leaf	Ladybells
Adenophora	*liliifolia*	*alpini*	Alpin	Lily Leaf	Ladybells
Adenophora	*liliifolia*	*angustata*	Narrow	Lily Leaf	Ladybells
Adenophora	*liliifolia*	*asperula*	Rough	Lily Leaf	Ladybells
Adenophora	*liliifolia*	*botryantha*	Bottom Flower	Lily Leaf	Ladybells
Adenophora	*liliifolia*	*cordata*	Heart Leaf	Lily Leaf	Ladybells
Adenophora	*liliifolia*	*fischeri*	Fischer	Lily Leaf	Ladybells
Adenophora	*liliifolia*	*genuina*	Genuine	Lily Leaf	Ladybells
Adenophora	*liliifolia*	*hirta*	Hairy	Lily Leaf	Ladybells
Adenophora	*liliifolia*	*hungarica*	Hungarian	Lily Leaf	Ladybells
Adenophora	*liliifolia*	*infundibuliformis*	Fan	Lily Leaf	Ladybells
Adenophora	*liliifolia*	*integerrima*	Entire Leaf	Lily Leaf	Ladybells
Adenophora	*liliifolia*	*integrisepala*	Entire Sepal	Lily Leaf	Ladybells
Adenophora	*liliifolia*	*intermedia*	Intermediate	Lily Leaf	Ladybells
Adenophora	*liliifolia*	*latifolia*	Wide	Lily Leaf	Ladybells
Adenophora	*liliifolia*	*liliflora*	Lily Flower	Lily Leaf	Ladybells
Adenophora	*liliifolia*	*mikoi*	Miko	Lily Leaf	Ladybells
Adenophora	*liliifolia*	*periplocifolia*	Silk Vine	Lily Leaf	Ladybells
Adenophora	*liliifolia*	*perpallens*	Pale	Lily Leaf	Ladybells
Adenophora	*liliifolia*	*pilosa*	Downy	Lily Leaf	Ladybells
Adenophora	*liliifolia*	*pocsii*	Pocs	Lily Leaf	Ladybells
Adenophora	*liliifolia*	*polyadenia*	Many Gland	Lily Leaf	Ladybells
Adenophora	*liliifolia*	*polymorpha*	Polymorphic	Lily Leaf	Ladybells
Adenophora	*liliifolia*	*raddeana*	Radde	Lily Leaf	Ladybells
Adenophora	*liliifolia*	*rhomboidea*	Diamond Leaf	Lily Leaf	Ladybells
Adenophora	*liliifolia*	*setulosa*	Bristle	Lily Leaf	Ladybells
Adenophora	*liliifolia*	*spreta*	Spret	Lily Leaf	Ladybells
Adenophora	*liliifolia*	*stylosa*	Style	Lily Leaf	Ladybells
Adenophora	*liliifolia*	*suaveolens*	Sweet	Lily Leaf	Ladybells
Adenophora	*liliifolia*	*subuniflora*	Single Flower	Lily Leaf	Ladybells
Adenophora	*liliifolia*	*umbrosa*	Shade Loving	Lily Leaf	Ladybells
Adenophora	*liliifolia*	*villosula*	Villose	Lily Leaf	Ladybells

Family 2 *Goodeniaceae* Olive Bellflower
Family 3 *Lobeliaceae* Lobel Bellflower
Family 4 *Pentaphragmataceae* Cinquefoil
Family 5 *Sphenocleaceae* Chicken Spike
Family 6 *Stylidiaceae* Trigger Plant
GENUS 1 *Oroxylum* Sword Fruit1

Order 724.013 *Compositales* Aster

Family 1 *Compositaceae* Aster

Family 724.013.2 *Carduaceae* Thistle

Genus	Species	Specimen	Genome	C:Culinary, M:Medicinal, E:Economic
Arctium	*lappa*	Greater	Burdock	C:SL,RT:**VG**
Centaurea	*benedicta*	Blessed Thistle	Knapweed	C:LF,RT:**VG(artichoke)**
Cyanus	*segetum*	Blue	Cornflower	C:FL:**VG**
Cynara	*cardunculus*	Cardoon	Vegetable Thistle	C:SL:**VG**
Cynara	*scolymus*	Globe Artichoke	Vegetable Thistle	C:LF(head):**VG**
Cynara	*syriaca*	Syrian Artichoke	Vegetable Thistle	C:LF(head):**VG**
Scolymus	*hispanicus*	Spanish Golden	Thistle	C:SL:**VG**
Silybum	*marianum*	Blessed Milk	Thistle	C:FL(head):**VG**

Varieties

Burdock (*Arctium lappa*)

Genus	species	infraspecies	cultivar	specimen	Genome
Arctium	*lappa*	*lappa*	Original	Hare Burr	Burdock
Arctium	*lappa*	*adhaerens*	Adhereing	Hare Burr	Burdock
Arctium	*lappa*	*bardana*	Bard	Hare Burr	Burdock
Arctium	*lappa*	*chaorum*	Chaor	Hare Burr	Burdock
Arctium	*lappa*	*gigantea*	Giant	Hare Burr	Burdock
Arctium	*lappa*	*glabra*	Smooth	Hare Burr	Burdock
Arctium	*lappa*	*grandiflorum*	Grand Flower	Hare Burr	Burdock
Arctium	*lappa*	*majus*	Greater	Hare Burr	Burdock
Arctium	*lappa*	*ruderale*	Ruderale	Hare Burr	Burdock
Arctium	*lappa*	*sakigakei*	Sakigake	Hare Burr	Burdock
Arctium	*lappa*	*takinogawa-longii*	Takinogawa Long	Hare Burr	Burdock

Genus	species	infraspecies	cultivar	specimen	Genome
Arctium	*lappa*	*wantinabe-earlyi*	Wantanabe Early	Hare Burr	Burdock
Arctium	*lappa*	*yanagawai*	Yanagawa	Hare Burr	Burdock

Blessed Thistle Knapweed (*Centaurea benedicta*)

Genus	species	infraspecies	cultivar	specimen	Genome
Centaurea	*benedicta*	*benedicta*	Original	Blessed Thistle	Knapweed
Centaurea	*benedicta*	*bulgarica*	Bulgarian	Blessed Thistle	Knapweed
Centaurea	*benedicta*	*centriflora*	Central Flower	Blessed Thistle	Knapweed
Centaurea	*benedicta*	*horridum*	Horrible	Blessed Thistle	Knapweed
Centaurea	*benedicta*	*kotschyi*	Kotschy	Blessed Thistle	Knapweed
Centaurea	*benedicta*	*lanuginosa*	Woolly	Blessed Thistle	Knapweed

Blue Cornflower (*Cyanus segetum*)

Genus	species	infraspecies	cultivar	specimen	Genome
Cyanus	*segetum*	*segetum*	Original	Blue	Cornflower
Cyanus	*segetum*	*hortorus*	Garden	Blue	Cornflower
Cyanus	*segetum*	*lanata*	Woolly	Blue	Cornflower

Cardoon (*Cynara carduncului*)

Genus	species	infraspecies	cultivar	specimen	Genome
Cynara	*carduncului*	*carduncului*	Original	Cardoon	Vegetable Thistle
Cynara	*carduncului*	*bianco-avorioi*	Bianco Avorio	Cardoon	Vegetable Thistle
Cynara	*carduncului*	*cardon-d'algiersi*	Cardon D'Algiers	Cardoon	Vegetable Thistle
Cynara	*carduncului*	*gigantei*	Gigante	Cardoon	Vegetable Thistle
Cynara	*carduncului*	*gobbo-di-nizza*	Hunchback Of Nice	Cardoon	Vegetable Thistle
Cynara	*carduncului*	*ivory-coastii*	Ivory Coast	Cardoon	Vegetable Thistle
Cynara	*carduncului*	*larga-laevigata*	Large Smooth	Cardoon	Vegetable Thistle
Cynara	*carduncului*	*plein-blanc-inormei*	Plein Blanc Inorme	Cardoon	Vegetable Thistle
Cynara	*carduncului*	*tenuacardia*	Tenderheart	Cardoon	Vegetable Thistle

Artichoke, Globe (*Cynara scolymus*)

Genus	species	infraspecies	cultivar	specimen	Genome
Cynara	*scolymus*	*scolymus*	Original	Globe Artichoke	Vegetable Thistle
Cynara	*scolymus*	*early-green-provencei*	Early Green Provence	Globe Artichoke	Vegetable Thistle
Cynara	*scolymus*	*emeraldii*	Emerald	Globe Artichoke	Vegetable Thistle
Cynara	*scolymus*	*green-globe-silveri*	Globe Silver Green	Globe Artichoke	Vegetable Thistle
Cynara	*scolymus*	*green-globe-californica*	Cal Green Globe	Globe Artichoke	Vegetable Thistle
Cynara	*scolymus*	*green-globe-earlyi*	Early Green Globe	Globe Artichoke	Vegetable Thistle
Cynara	*scolymus*	*green-globe-frenchii*	French Green Globe	Globe Artichoke	Vegetable Thistle
Cynara	*scolymus*	*green-globe-improva*	Improved Green Globe	Globe Artichoke	Vegetable Thistle
Cynara	*scolymus*	*green-globe-silveri*	Silver Green Globe	Globe Artichoke	Vegetable Thistle
Cynara	*scolymus*	*imperial-stari*	Imperial Star	Globe Artichoke	Vegetable Thistle
Cynara	*scolymus*	*kiss-of-burgundii*	Kiss Of Burgundy	Globe Artichoke	Vegetable Thistle
Cynara	*scolymus*	*northern-stari*	Northern Star	Globe Artichoke	Vegetable Thistle
Cynara	*scolymus*	*purple-roscoffii*	Purple Roscoff	Globe Artichoke	Vegetable Thistle
Cynara	*scolymus*	*purple-siciliana*	Purple Sicilian	Globe Artichoke	Vegetable Thistle
Cynara	*scolymus*	*romanescoi*	Romanesco	Globe Artichoke	Vegetable Thistle
Cynara	*scolymus*	*viola*	Violet	Globe Artichoke	Vegetable Thistle
Cynara	*scolymus*	*violettoi*	Violetto	Globe Artichoke	Vegetable Thistle
Cynara	*scolymus*	*violettoi-di-jesii*	Violetto Di Romagna	Globe Artichoke	Vegetable Thistle
Cynara	*scolymus*	*violettoi-di-romagnai*	Violetto Di Jesi	Globe Artichoke	Vegetable Thistle

Spanish Golden Thistle (*Scolymus hispanicus*)

Genus	species	infraspecies	cultivar	specimen	Genome
Scolymus	*hispanicus*	*hispanicus*	Original	Spanish	Golden Thistle
Scolymus	*hispanicus*	*grandiflorus*	Large Flower	Spanish	Golden Thistle
Scolymus	*hispanicus*	*tagamina*	Tagarinine	Spanish	Golden Thistle

Blessed Milk Thistle (*Silybum marianum*)

Genus	Species	infraspecies	cultivar	specimen	Genome
Silybum	*marianum*	*marianum*	Original	Blessed	Milk Thistle
Silybum	*marianum*	*dalmatica*	Dalmation	Blessed	Milk Thistle
Silybum	*marianum*	*lactifolium*	Milky Leaf	Blessed	Milk Thistle
Silybum	*marianum*	*maculatum*	Spotted	Blessed	Milk Thistle
Silybum	*marianum*	*pygmaeum*	Pygmy	Blessed	Milk Thistle
Silybum	*marianum*	*versicolor*	Versicolor	Blessed	Milk Thistle

Family 3 *Floretasteraceae* Floret Aster
Family 4 *Solidagoaceae* Goldenrod
Family 5 *Geoselaceae* Groundsel

Family 724.013.6 *Lactucaceae* Lettuce

Genus	Species	Specimen	Genome	C:Culinary, M:Medicinal
Lactuca	*sativa-angustana*	Celtuce	Lettuce	C:LF:**VG**
Lactuca	*sativa-capitata-larga*	12.1"-16" Large Head	Lettuce	C:LF:**VG**
Lactuca	*sativa-capitata-larga-media*	9.1"-12" Medium Large Head	Lettuce	C:LF:**VG**
Lactuca	*sativa-capitata-media*	6.1"-9" Medium Head	Lettuce	C:LF:**VG**
Lactuca	*sativa-capitata-parva*	2"-6" Small Head	Lettuce	C:LF:**VG**
Lactuca	*sativa-capitata-butrya-larga*	Large Butterhead	Lettuce	C:LF:**VG**
Lactuca	*sativa-capitata-butrya-larga-media*	Medium Large Butterhead	Lettuce	C:LF:**VG**
Lactuca	*sativa-capitata-butrya-media*	Medium Butterhead	Lettuce	C:LF:**VG**
Lactuca	*sativa-capitata-butrya-parva*	Small Butterhead	Lettuce	C:LF:**VG**
Lactuca	*sativa-capitata-crispa-larga*	Large Crisphead	Lettuce	C:LF:**VG**
Lactuca	*sativa-capitata-crispa-larga-media*	Medium Large Crisphead	Lettuce	C:LF:**VG**
Lactuca	*sativa-capitata-crispa-media*	Medium Crisphead	Lettuce	C:LF:**VG**
Lactuca	*sativa-capitata-crispa-parva*	Small Crisphead	Lettuce	C:LF:**VG**
Lactuca	*sativa-capitata-icebergii-larga*	Large Iceberg	Lettuce	C:LF:**VG**
Lactuca	*sativa-capitata-icebergii-larga-media*	Medium Large Iceberg	Lettuce	C:LF:**VG**
Lactuca	*sativa-capitata-icebergii-media*	Medium Iceberg	Lettuce	C:LF:**VG**
Lactuca	*sativa-capitata-icebergii-parva*	Small Iceberg	Lettuce	C:LF:**VG**
Lactuca	*sativa-capitata-rubra-larga*	Large Redhead	Lettuce	C:LF:**VG**
Lactuca	*sativa-capitata-rubra-media*	Medium Redhead	Lettuce	C:LF:**VG**
Lactuca	*sativa-capitata-rubra-parva*	Small Redhead	Lettuce	C:LF:**VG**
Lactuca	*sativa-capitata-verticillata-larga*	Large Loosehead	Lettuce	C:LF:**VG**
Lactuca	*sativa-capitata-verticillata-media*	Medium Loosehead	Lettuce	C:LF:**VG**
Lactuca	*sativa-capitata-verticillata-parva*	Small Loosehead	Lettuce	C:LF:**VG**
Lactuca	*sativa-foliosa-larga*	12.1"-16" Large Looseleaf	Lettuce	C:LF:**VG**
Lactuca	*sativa-foliosa-larga-media*	9.1"-12" Medium Large Looseleaf	Lettuce	C:LF:**VG**
Lactuca	*sativa-foliosa-media*	6.1"-9" Medium Looseleaf	Lettuce	C:LF:**VG**
Lactuca	*sativa-foliosa-parva*	2"-6" Small Looseleaf	Lettuce	C:LF:**VG**
Lactuca	*sativa-foliosa-corugata*	Ruffled Looseleaf	Lettuce	C:LF:**VG**
Lactuca	*sativa-foliosa-corugata-rubra*	Red Ruffled Looseleaf	Lettuce	C:LF:**VG**
Lactuca	*sativa-foliosa-crispa*	Curly Looseleaf	Lettuce	C:LF:**VG**
Lactuca	*sativa-foliosa-crispa-rubra*	Red Curly Looseleaf	Lettuce	C:LF:**VG**
Lactuca	*sativa-foliosa-cuttingii*	Cutting Looseleaf	Lettuce	C:LF:**VG**
Lactuca	*sativa-foliosa-fimbria*	Frilly Looseleaf	Lettuce	C:LF:**VG**
Lactuca	*sativa-foliosa-fimbria-rubra*	Red Frilly Looseleaf	Lettuce	C:LF:**VG**
Lactuca	*sativa-foliosa-helva-groena*	Yellow Green Looseleaf	Lettuce	C:LF:**VG**
Lactuca	*sativa-foliosa-marginata-brunia*	Bronze Margined Looseleaf	Lettuce	C:LF:**VG**
Lactuca	*sativa-foliosa-marginata-rubra*	Red Margined Looseleaf	Lettuce	C:LF:**VG**
Lactuca	*sativa-foliosa-pustula*	Blistered Looseleaf	Lettuce	C:LF:**VG**
Lactuca	*sativa-foliosa-querca*	Oakleaf Looseleaf	Lettuce	C:LF:**VG**
Lactuca	*sativa-foliosa-querca-fimbria*	Frilly Oakleaf Looseleaf	Lettuce	C:LF:**VG**
Lactuca	*sativa-foliosa-rubra*	Red Looseleaf	Lettuce	C:LF:**VG**
Lactuca	*sativa-foliosa-rugosa*	Wrinkled Looseleaf	Lettuce	C:LF:**VG**
Lactuca	*sativa-foliosa-rugosa-rubra*	Red Wrinkled Looseleaf	Lettuce	C:LF:**VG**
Lactuca	*sativa-foliosa-savoia*	Savoy Looseleaf	Lettuce	C:LF:**VG**
Lactuca	*sativa-foliosa-savoia-rubra*	Red Savoy Looseleaf	Lettuce	C:LF:**VG**
Lactuca	*sativa-longifolia-larga*	12.1"-16" Large Romaine	Lettuce	C:LF:**VG**
Lactuca	*sativa-longifolia-larga-media*	9.1"-12" Medium Large Romaine	Lettuce	C:LF:**VG**
Lactuca	*sativa-longifolia-media*	6.1"-9" Medium Romaine	Lettuce	C:LF:**VG**
Lactuca	*sativa-longifolia-parva*	2"-6" Small Romaine	Lettuce	C:LF:**VG**
Lactuca	*sativa-longifolia-rubra*	Red Romaine	Lettuce	C:LF:**VG**
Lactuca	*sativa-longifolia-cosa-larga*	12.1"-16" Large Cos	Lettuce	C:LF:**VG**
Lactuca	*sativa-longifolia-cosa-larga-media*	9.1"-12" Medium Large Cos	Lettuce	C:LF:**VG**
Lactuca	*sativa-longifolia-cosa-media*	6.1"-9" Medium Cos	Lettuce	C:LF:**VG**
Lactuca	*sativa-longifolia-cosa-parva*	2"-6" Small Cos	Lettuce	C:LF:**VG**
Lactuca	*sativa-longifolia-cosa-rubra*	Red Cos	Lettuce	C:LF:**VG**
Cichorium	*endivia*	Endive	Chicory	C:LF:**VG**
Cichorium	*endivia-escarolea*	Escarole	Chicory	C:LF:**VG**
Cichorium	*endivia-ruta*	Root Endive	Chicory	C:RT:**VG**
Cichorium	*intybus*	Succory	Chicory	C:LF:**VG**
Cichorium	*intybus-ruta*	Root Succory	Chicory	C:RT:**VG**
Cichorium	*intybus-stipes*	Stem Succory	Chicory	C:SM:**VG**
Scorzonera	*hispanica*	Black Salsify	Scorzon	C:RT:**VG**
Tragopogon	*porrifolius*	Oyster Salsify	Salsify	C:RT:**VG**

Varieties

Celtuce (*Lactuca sativa-angustana*)

Genus	species	infraspecies	cultivar	specimen	Genome
Lactuca	*sativa-angustana*	*angustana*	Original	Celtuce	Lettuce
Lactuca	*sativa-angustana*	*asparagina*	Asparagus	Celtuce	Lettuce

Genus	species	infraspecies	cultivar	specimen	Genome
Lactuca	*sativa-angustana*	*sinocaulis*	Chinese Stem	Celtuce	Lettuce
Lactuca	*sativa-angustana*	*woo-chui*	Woo Chu	Celtuce	Lettuce
Lactuca	*sativa-angustana*	*woo-sunii*	Woo Sun	Celtuce	Lettuce

Lettuce, Head (*Lactuca sativa-capitata*)

Genus	species	infraspecies	cultivar	specimen	Genome
Lactuca	*sativa-capitata-larga*	*audranii*	Audran	12.1"-16" Large Head	Lettuce
Lactuca	*sativa-capitata-larga*	*bazookai*	Bazooka	12.1"-16" Large Head	Lettuce
Lactuca	*sativa-capitata-larga*	*bella-greenii*	Bella Green	12.1"-16" Large Head	Lettuce
Lactuca	*sativa-capitata-larga*	*bountyi*	Bounty	12.1"-16" Large Head	Lettuce
Lactuca	*sativa-capitata-larga*	*climaxii*	Climax	12.1"-16" Large Head	Lettuce
Lactuca	*sativa-capitata-larga*	*coolgreenii*	Coolgreen	12.1"-16" Large Head	Lettuce
Lactuca	*sativa-capitata-larga*	*desertgreenii*	Desertgreen	12.1"-16" Large Head	Lettuce
Lactuca	*sativa-capitata-larga*	*duchessei*	Duchesse	12.1"-16" Large Head	Lettuce
Lactuca	*sativa-capitata-larga*	*hansonii*	Hanson	12.1"-16" Large Head	Lettuce
Lactuca	*sativa-capitata-larga*	*harmonyi*	Harmony	12.1"-16" Large Head	Lettuce
Lactuca	*sativa-capitata-larga*	*imperial-winteri*	Imperial Winter	12.1"-16" Large Head	Lettuce
Lactuca	*sativa-capitata-larga*	*rhapsodyi*	Rhapsody	12.1"-16" Large Head	Lettuce
Lactuca	*sativa-capitata-larga*	*rigolettoi*	Rigoletto	12.1"-16" Large Head	Lettuce
Lactuca	*sativa-capitata-larga*	*soissonsi*	Soissons	12.1"-16" Large Head	Lettuce
Lactuca	*sativa-capitata-larga*	*vanguardii*	Vanguard	12.1"-16" Large Head	Lettuce
Lactuca	*sativa-capitata-larga*	*vanguardii-75*	#75 Vanguard	12.1"-16" Large Head	Lettuce
Lactuca	*sativa-capitata-larga*	*vanguardii-fallgreenii*	Fallgreen Vanguard	12.1"-16" Large Head	Lettuce
Lactuca	*sativa-capitata-larga*	*vanguardii-gilabenii*	Gilaben Vanguard	12.1"-16" Large Head	Lettuce
Lactuca	*sativa-capitata-larga*	*vanguardii-vanmaxii*	Vanmax Vanguard	12.1"-16" Large Head	Lettuce
Lactuca	*sativa-capitata-larga*	*val-d'orgei*	Val D'Orge Overwinter	12.1"-16" Large Head	Lettuce
Lactuca	*sativa-capitata-larga*	*valdori*	Valdor Overwinter	12.1"-16" Large Head	Lettuce
Lactuca	*sativa-capitata-larga*	*vanityi*	Valdor Overwinter	12.1"-16" Large Head	Lettuce
Lactuca	*sativa-capitata-larga*	*vista*	Vista	12.1"-16" Large Head	Lettuce
Lactuca	*sativa-capitata-larga*	*winter-marvelii*	Winter Marvel	12.1"-16" Large Head	Lettuce
Lactuca	*sativa-capitata-larga-media*	*attraction*	Attraction	9.1"-12" Medium Large Head	Lettuce
Lactuca	*sativa-capitata-larga-media*	*bon-jardinier*	Bon Jardinier	9.1"-12" Medium Large Head	Lettuce
Lactuca	*sativa-capitata-larga-media*	*salinas-bayviewi*	Bayview Salinas	9.1"-12" Medium Large Head	Lettuce
Lactuca	*sativa-capitata-larga-media*	*salinas-bixii*	Bix Salinas	9.1"-12" Medium Large Head	Lettuce
Lactuca	*sativa-capitata-larga-media*	*salinas-paradisei*	Paradise Salinas	9.1"-12" Medium Large Head	Lettuce
Lactuca	*sativa-capitata-larga-media*	*salinas-saladinii*	Saladin Salinas	9.1"-12" Medium Large Head	Lettuce
Lactuca	*sativa-capitata-larga-media*	*salinas-SLE-4105*	SLE #4105 Salinas	9.1"-12" Medium Large Head	Lettuce
Lactuca	*sativa-capitata-larga-media*	*salinas-supremei-88*	Supreme Salinas	9.1"-12" Medium Large Head	Lettuce
Lactuca	*sativa-capitata-larga-media*	*salinas-targetii*	Target Salinas	9.1"-12" Medium Large Head	Lettuce
Lactuca	*sativa-capitata-larga-media*	*snowbirdii*	Snowbird	9.1"-12" Medium Large Head	Lettuce
Lactuca	*sativa-capitata-media*	*ace-of-heartsi*	Ace Of Hearts	6.1"-9" Medium Head	Lettuce
Lactuca	*sativa-capitata-media*	*action*	Action	6.1"-9" Medium Head	Lettuce
Lactuca	*sativa-capitata-media*	*all-the-year-roundii*	All The Year Round	6.1"-9" Medium Head	Lettuce
Lactuca	*sativa-capitata-media*	*amazonii*	Amazon	6.1"-9" Medium Head	Lettuce
Lactuca	*sativa-capitata-media*	*arizonica-rz*	Arizona RZ	6.1"-9" Medium Head	Lettuce
Lactuca	*sativa-capitata-media*	*attractivei*	Attractive	6.1"-9" Medium Head	Lettuce
Lactuca	*sativa-capitata-media*	*avon-defiancei*	Avon Defiance	6.1"-9" Medium Head	Lettuce
Lactuca	*sativa-capitata-media*	*beatricei*	Beatrice	6.1"-9" Medium Head	Lettuce
Lactuca	*sativa-capitata-media*	*cavolo-di-napolii*	Cavolo Di Napoli	6.1"-9" Medium Head	Lettuce
Lactuca	*sativa-capitata-media*	*ceremonyi*	Ceremony	6.1"-9" Medium Head	Lettuce
Lactuca	*sativa-capitata-media*	*ceremonyi-1*	#1 Ceremony	6.1"-9" Medium Head	Lettuce
Lactuca	*sativa-capitata-media*	*ceremonyi-2*	#2 Ceremony	6.1"-9" Medium Head	Lettuce
Lactuca	*sativa-capitata-media*	*del-rayi*	Del Ray	6.1"-9" Medium Head	Lettuce
Lactuca	*sativa-capitata-media*	*densityi*	Density	6.1"-9" Medium Head	Lettuce
Lactuca	*sativa-capitata-media*	*du-bon-jardinierei*	Du Bon Jardiniere	6.1"-9" Medium Head	Lettuce
Lactuca	*sativa-capitata-media*	*emperori*	Emperor	6.1"-9" Medium Head	Lettuce
Lactuca	*sativa-capitata-media*	*empirei*	Empire	6.1"-9" Medium Head	Lettuce
Lactuca	*sativa-capitata-media*	*empirei-desert-queenii*	Desert Queen Empire	6.1"-9" Medium Head	Lettuce
Lactuca	*sativa-capitata-media*	*empirei-maverickii*	Maverick Empire	6.1"-9" Medium Head	Lettuce
Lactuca	*sativa-capitata-media*	*flashy-trout-back*	Flashy Trout Back	6.1"-9" Medium Head	Lettuce
Lactuca	*sativa-capitata-media*	*floresta*	Floresta	6.1"-9" Medium Head	Lettuce
Lactuca	*sativa-capitata-media*	*green-headii*	Greenhead	6.1"-9" Medium Head	Lettuce
Lactuca	*sativa-capitata-media*	*greendayi*	Greenday	6.1"-9" Medium Head	Lettuce
Lactuca	*sativa-capitata-media*	*hjerter-esi*	Hjerter Es	6.1"-9" Medium Head	Lettuce
Lactuca	*sativa-capitata-media*	*impactii*	Impact	6.1"-9" Medium Head	Lettuce
Lactuca	*sativa-capitata-media*	*kinemontpas*	Kinimontpas	6.1"-9" Medium Head	Lettuce
Lactuca	*sativa-capitata-media*	*krolowa-majowychii*	Krolowa Majowych	6.1"-9" Medium Head	Lettuce
Lactuca	*sativa-capitata-media*	*lakelandii*	Lakeland	6.1"-9" Medium Head	Lettuce
Lactuca	*sativa-capitata-media*	*lattughino*	Lattughino	6.1"-9" Medium Head	Lettuce
Lactuca	*sativa-capitata-media*	*legacyi*	Legacy	6.1"-9" Medium Head	Lettuce
Lactuca	*sativa-capitata-media*	*liannei*	Lianne	6.1"-9" Medium Head	Lettuce
Lactuca	*sativa-capitata-media*	*magnum*	Magnum	6.1"-9" Medium Head	Lettuce
Lactuca	*sativa-capitata-media*	*marvel-of-4-seasonii*	Marvel Of 4 Seasons	6.1"-9" Medium Head	Lettuce
Lactuca	*sativa-capitata-media*	*melnicky-majii*	Melnicky Maj	6.1"-9" Medium Head	Lettuce

Genus	species	infraspecies	cultivar	specimen	Genome
Lactuca	*sativa-capitata-media*	*meritii*	Merit	6.1"-9" Medium Head	Lettuce
Lactuca	*sativa-capitata-media*	*miura*	Miura	6.1"-9" Medium Head	Lettuce
Lactuca	*sativa-capitata-media*	*morgana*	Morgana	6.1"-9" Medium Head	Lettuce
Lactuca	*sativa-capitata-media*	*morgesi*	Morges	6.1"-9" Medium Head	Lettuce
Lactuca	*sativa-capitata-media*	*muriettai*	Murietta	6.1"-9" Medium Head	Lettuce
Lactuca	*sativa-capitata-media*	*nevadai*	Murietta	6.1"-9" Medium Head	Lettuce
Lactuca	*sativa-capitata-media*	*pennlake-earlyi*	Early Pennlake	6.1"-9" Medium Head	Lettuce
Lactuca	*sativa-capitata-media*	*pennlake-queen-crownii*	Queen Crown Pennlake	6.1"-9" Medium Head	Lettuce
Lactuca	*sativa-capitata-media*	*reine-de-mai*	Regina De Mai	6.1"-9" Medium Head	Lettuce
Lactuca	*sativa-capitata-media*	*sherwoodii*	Sherwood	6.1"-9" Medium Head	Lettuce
Lactuca	*sativa-capitata-media*	*summer-furyi*	Summer Fury	6.1"-9" Medium Head	Lettuce
Lactuca	*sativa-capitata-media*	*summer-seasonii*	Summer Season	6.1"-9" Medium Head	Lettuce
Lactuca	*sativa-capitata-media*	*sun-devilii*	Sun Devil	6.1"-9" Medium Head	Lettuce
Lactuca	*sativa-capitata-media*	*tania*	Tania	6.1"-9" Medium Head	Lettuce
Lactuca	*sativa-capitata-media*	*val-d'orgei*	Val D'Orge Overwinter	6.1"-9" Medium Head	Lettuce
Lactuca	*sativa-capitata-media*	*victoryi*	Victory	6.1"-9" Medium Head	Lettuce
Lactuca	*sativa-capitata-media*	*wayaheadii*	Wayahead	6.1"-9" Medium Head	Lettuce
Lactuca	*sativa-capitata-media*	*winter-harvestii*	Winter Harvest	6.1"-9" Medium Head	Lettuce
Lactuca	*sativa-capitata-media*	*winter-pickii*	Winter Pick	6.1"-9" Medium Head	Lettuce
Lactuca	*sativa-capitata-media*	*winter-pridei*	Winter Pride	6.1"-9" Medium Head	Lettuce
Lactuca	*sativa-capitata-media*	*wintergreenii*	Wintergreen	6.1"-9" Medium Head	Lettuce
Lactuca	*sativa-capitata-media*	*winterhavenii*	Winterhaven	6.1"-9" Medium Head	Lettuce
Lactuca	*sativa-capitata-parva*	*arctic-kingii*	Arctic King	2"-6" Small Head	Lettuce
Lactuca	*sativa-capitata-parva*	*diamond-gemii*	Diamond Gem	2"-6" Small Head	Lettuce
Lactuca	*sativa-capitata-parva*	*el-doradoi*	El Dorado	2"-6" Small Head	Lettuce
Lactuca	*sativa-capitata-parva*	*geminii*	Gemini	2"-6" Small Head	Lettuce
Lactuca	*sativa-capitata-parva*	*little-lomai*	Little Loma	2"-6" Small Head	Lettuce
Lactuca	*sativa-capitata-parva*	*mignonette-bronzei*	Mignonette Bronze	2"-6" Small Head	Lettuce
Lactuca	*sativa-capitata-parva*	*mignonette-greenii*	Mignonette Green	2"-6" Small Head	Lettuce
Lactuca	*sativa-capitata-parva*	*minettoi*	Minetto	2"-6" Small Head	Lettuce
Lactuca	*sativa-capitata-parva*	*okayama-saladii*	Okayama Salad	2"-6" Small Head	Lettuce
Lactuca	*sativa-capitata-parva*	*santa-fei*	Santa Fe	2"-6" Small Head	Lettuce
Lactuca	*sativa-capitata-parva*	*south-bayi*	South Bay	2"-6" Small Head	Lettuce
Lactuca	*sativa-capitata-parva*	*tennis-ballii*	South Bay	2"-6" Small Head	Lettuce
Lactuca	*sativa-capitata-parva*	*thai-saladii*	Thai Salad	2"-6" Small Head	Lettuce
Lactuca	*sativa-capitata-parva*	*tunska-edrai*	Tunska Edra	2"-6" Small Head	Lettuce
Lactuca	*sativa-capitata-butrya-larga*	*bajai*	Baja	Large Butterhead	Lettuce
Lactuca	*sativa-capitata-butrya-larga*	*ben-shemenii*	Ben Shemen	Large Butterhead	Lettuce
Lactuca	*sativa-capitata-butrya-larga*	*bibbii-greenii*	Green Bibb	Large Butterhead	Lettuce
Lactuca	*sativa-capitata-butrya-larga*	*bibbii-merveille-des-quatre-saisonsi*	Merveille Des Quatre Saisons Bibb	Large Butterhead	Lettuce
Lactuca	*sativa-capitata-butrya-larga*	*boston-bigii*	Big Boston	Large Butterhead	Lettuce
Lactuca	*sativa-capitata-butrya-larga*	*boston-capitanii*	Capitan Boston	Large Butterhead	Lettuce
Lactuca	*sativa-capitata-butrya-larga*	*boston-encantoi*	Encanto Boston	Large Butterhead	Lettuce
Lactuca	*sativa-capitata-butrya-larga*	*boston-margarita*	Margarita Boston	Large Butterhead	Lettuce
Lactuca	*sativa-capitata-butrya-larga*	*boston-optima*	Optima Boston	Large Butterhead	Lettuce
Lactuca	*sativa-capitata-butrya-larga*	*boston-whitei*	White Boston	Large Butterhead	Lettuce
Lactuca	*sativa-capitata-butrya-larga*	*butter-kingii*	Butter King	Large Butterhead	Lettuce
Lactuca	*sativa-capitata-butrya-larga*	*cobham-greenii*	Cobham Green	Large Butterhead	Lettuce
Lactuca	*sativa-capitata-butrya-larga*	*divina*	Divina	Large Butterhead	Lettuce
Lactuca	*sativa-capitata-butrya-larga*	*dollyi*	Dolly	Large Butterhead	Lettuce
Lactuca	*sativa-capitata-butrya-larga*	*fatimai*	Fatima	Large Butterhead	Lettuce
Lactuca	*sativa-capitata-butrya-larga*	*kwiekii*	Kwiek	Large Butterhead	Lettuce
Lactuca	*sativa-capitata-butrya-larga*	*mantilia*	Mantilia	Large Butterhead	Lettuce
Lactuca	*sativa-capitata-butrya-larga*	*mantilia-ermosa*	Ermosa Mantilia	Large Butterhead	Lettuce
Lactuca	*sativa-capitata-butrya-larga*	*marvelii*	Marvel	Large Butterhead	Lettuce
Lactuca	*sativa-capitata-butrya-larga*	*maule's-philadelphia-butteri*	Maule's Philadephia Butter	Large Butterhead	Lettuce
Lactuca	*sativa-capitata-butrya-larga*	*wendelii*	Wendel	Large Butterhead	Lettuce
Lactuca	*sativa-capitata-butrya-larga-media*	*bibbii-floribibbii*	Floribibb Bibb	Medium Large Butterhead	Lettuce
Lactuca	*sativa-capitata-butrya-larga-media*	*bibbii-kagran-summeri*	Kagran Summer Bibb	Medium Large Butterhead	Lettuce
Lactuca	*sativa-capitata-butrya-larga-media*	*may-kingii*	May King	Medium Large Butterhead	Lettuce
Lactuca	*sativa-capitata-butrya-larga-media*	*may-queenii*	May Queen	Medium Large Butterhead	Lettuce
Lactuca	*sativa-capitata-butrya-media*	*apolloi*	Apollo	Medium Butterhead	Lettuce
Lactuca	*sativa-capitata-butrya-media*	*batavia-sierrai*	Sierra Batavian	Medium Butterhead	Lettuce
Lactuca	*sativa-capitata-butrya-media*	*batavia-sierrai-slogunii*	Slogun Sierra Batavian	Medium Butterhead	Lettuce
Lactuca	*sativa-capitata-butrya-media*	*bibbii-kagraner-sommeri*	Kagraner Sommer Bibb	Medium Butterhead	Lettuce
Lactuca	*sativa-capitata-butrya-media*	*bibbii-kagraner-sommeri-torenia*	Torenia Kagraner Sommer Bibb	Medium Butterhead	Lettuce
Lactuca	*sativa-capitata-butrya-media*	*bibbii-piratii*	Pirat Bibb	Medium Butterhead	Lettuce
Lactuca	*sativa-capitata-butrya-media*	*bibbii-piratii-redii*	Pirat Red Edged Bibb	Medium Butterhead	Lettuce
Lactuca	*sativa-capitata-butrya-media*	*bibbii-speckles*	Speckles Bibb	Medium Butterhead	Lettuce
Lactuca	*sativa-capitata-butrya-media*	*bibbii-susan's-redii*	Susan's Red Bibb	Medium Butterhead	Lettuce

Genus	species	infraspecies	cultivar	specimen	Genome
Lactuca	sativa-capitata-butrya-media	blushed-butter-oakii	Blushed Butter Oak	Medium Butterhead	Lettuce
Lactuca	sativa-capitata-butrya-media	boston-attrazionnei	Attrazionne Blond Boston	Medium Butterhead	Lettuce
Lactuca	sativa-capitata-butrya-media	boston-burgundyi	Burgundy Boston	Medium Butterhead	Lettuce
Lactuca	sativa-capitata-butrya-media	boston-escortii	Escort Boston	Medium Butterhead	Lettuce
Lactuca	sativa-capitata-butrya-media	boston-dark-greenii	Dark Green Boston	Medium Butterhead	Lettuce
Lactuca	sativa-capitata-butrya-media	boston-nancyi	Nancy Boston	Medium Butterhead	Lettuce
Lactuca	sativa-capitata-butrya-media	boston-ostinata	Ostinata Boston	Medium Butterhead	Lettuce
Lactuca	sativa-capitata-butrya-media	boston-redii	Red Boston	Medium Butterhead	Lettuce
Lactuca	sativa-capitata-butrya-media	boston-redii-jacquelinei	Jacqueline Red Boston	Medium Butterhead	Lettuce
Lactuca	sativa-capitata-butrya-media	boston-redii-ridinghoodii	Red Ridinghood Red Boston	Medium Butterhead	Lettuce
Lactuca	sativa-capitata-butrya-media	boston-rosei	Rose Boston	Medium Butterhead	Lettuce
Lactuca	sativa-capitata-butrya-media	boston-verianii	Verian Boston	Medium Butterhead	Lettuce
Lactuca	sativa-capitata-butrya-media	brune-d'hiveri	Brune D'Hiver	Medium Butterhead	Lettuce
Lactuca	sativa-capitata-butrya-media	butter-beautyi	Butter Beauty	Medium Butterhead	Lettuce
Lactuca	sativa-capitata-butrya-media	butterhead-greenii	Butterhead Green	Medium Butterhead	Lettuce
Lactuca	sativa-capitata-butrya-media	butterhead-redii	Butterhead Red	Medium Butterhead	Lettuce
Lactuca	sativa-capitata-butrya-media	cassandrai	Cassandra	Medium Butterhead	Lettuce
Lactuca	sativa-capitata-butrya-media	craquerelle-du-midii	Craquerelle Du Midi	Medium Butterhead	Lettuce
Lactuca	sativa-capitata-butrya-media	delta	Delta	Medium Butterhead	Lettuce
Lactuca	sativa-capitata-butrya-media	esmeralda	Esmeralda	Medium Butterhead	Lettuce
Lactuca	sativa-capitata-butrya-media	habana	Habana	Medium Butterhead	Lettuce
Lactuca	sativa-capitata-butrya-media	hildei	Hilde	Medium Butterhead	Lettuce
Lactuca	sativa-capitata-butrya-media	hubbard's-marketii	Hubbard's Market	Medium Butterhead	Lettuce
Lactuca	sativa-capitata-butrya-media	julietii	Juliet	Medium Butterhead	Lettuce
Lactuca	sativa-capitata-butrya-media	key-limei	Key Lime	Medium Butterhead	Lettuce
Lactuca	sativa-capitata-butrya-media	kral-majii	Kral Maj	Medium Butterhead	Lettuce
Lactuca	sativa-capitata-butrya-media	lednickyi	Lednicky	Medium Butterhead	Lettuce
Lactuca	sativa-capitata-butrya-media	magnetii	Magnet	Medium Butterhead	Lettuce
Lactuca	sativa-capitata-butrya-media	musettei	Musette	Medium Butterhead	Lettuce
Lactuca	sativa-capitata-butrya-media	north-polei	North Pole	Medium Butterhead	Lettuce
Lactuca	sativa-capitata-butrya-media	pradoi	Prado	Medium Butterhead	Lettuce
Lactuca	sativa-capitata-butrya-media	red-butteri	Red Butter	Medium Butterhead	Lettuce
Lactuca	sativa-capitata-butrya-media	regina-di-maggioi	Regina De Maggio	Medium Butterhead	Lettuce
Lactuca	sativa-capitata-butrya-media	saint-anne's-slow-boltingii	Saint Anne's Slow Bolting	Medium Butterhead	Lettuce
Lactuca	sativa-capitata-butrya-media	sandrina-winteri	Sandrina Winter	Medium Butterhead	Lettuce
Lactuca	sativa-capitata-butrya-media	selma-wisa	Selma Wisa	Medium Butterhead	Lettuce
Lactuca	sativa-capitata-butrya-media	summer-greenii	Summer Green	Medium Butterhead	Lettuce
Lactuca	sativa-capitata-butrya-media	timpurie-de-maii	Timpurie De Mai	Medium Butterhead	Lettuce
Lactuca	sativa-capitata-butrya-parva	akcelii	Akcel	Small Butterhead	Lettuce
Lactuca	sativa-capitata-butrya-parva	bibbii	Bibb	Small Butterhead	Lettuce
Lactuca	sativa-capitata-butrya-parva	bibbii-burpee'si	Burpee's Bibb	Small Butterhead	Lettuce
Lactuca	sativa-capitata-butrya-parva	bibbii-buttercrunchii	Buttercrunch Bibb	Small Butterhead	Lettuce
Lactuca	sativa-capitata-butrya-parva	bibbii-forcingii	Forcing Bibb	Small Butterhead	Lettuce
Lactuca	sativa-capitata-butrya-parva	bibbii-mescheri	Mescher Bibb	Small Butterhead	Lettuce
Lactuca	sativa-capitata-butrya-parva	bibbii-limestonei	Limestone Bibb	Small Butterhead	Lettuce
Lactuca	sativa-capitata-butrya-parva	bibbii-saladii	Salad Bibb	Small Butterhead	Lettuce
Lactuca	sativa-capitata-butrya-parva	bibbii-slow-boltingii	Slow Bolting Bibb	Small Butterhead	Lettuce
Lactuca	sativa-capitata-butrya-parva	bibbii-summeri	Summer Bibb	Small Butterhead	Lettuce
Lactuca	sativa-capitata-butrya-parva	bostonii-little-gemii	Little Gem Boston	Small Butterhead	Lettuce
Lactuca	sativa-capitata-butrya-parva	bostonii-sucrinei	Sucrine Boston	Small Butterhead	Lettuce
Lactuca	sativa-capitata-butrya-parva	grosse-blonde-paresseusei	Grosse Blonde Paresseuse	Small Butterhead	Lettuce
Lactuca	sativa-capitata-butrya-parva	tom-thumbii	Tom Thumb	Small Butterhead	Lettuce
Lactuca	sativa-capitata-crispa-larga	batavia-bord-rouge	Batavia Bord Rouge	Large Crisphead	Lettuce
Lactuca	sativa-capitata-crispa-larga	batavia-laurai	Batavia Laura	Large Crisphead	Lettuce
Lactuca	sativa-capitata-crispa-larga	crispinoi	Crispino	Large Crisphead	Lettuce
Lactuca	sativa-capitata-crispa-larga	kingii	King	Large Crisphead	Lettuce
Lactuca	sativa-capitata-crispa-larga-media	batavia-canasta	Canasta Batavian	Medium Large Crisphead	Lettuce
Lactuca	sativa-capitata-crispa-larga-media	batavia-canasta-victoria	Victoria Canasta Batavian	Medium Large Crisphead	Lettuce
Lactuca	sativa-capitata-crispa-media	balladei	Ballade	Medium Crisphead	Lettuce
Lactuca	sativa-capitata-crispa-media	batavia-atollii	Atoll Batavian	Medium Crisphead	Lettuce
Lactuca	sativa-capitata-crispa-media	batavia-blondei	Blonde Batavian	Medium Crisphead	Lettuce
Lactuca	sativa-capitata-crispa-media	batavia-blondei-6	#6 Blonde Batavian	Medium Crisphead	Lettuce
Lactuca	sativa-capitata-crispa-media	batavia-carnivalii	Carnival Batavian	Medium Crisphead	Lettuce
Lactuca	sativa-capitata-crispa-media	batavia-cheena	Cheena Batavian	Medium Crisphead	Lettuce
Lactuca	sativa-capitata-crispa-media	batavia-grazeri	Grazer Batavian	Medium Crisphead	Lettuce
Lactuca	sativa-capitata-crispa-media	batavia-long-standingii	Long Standing Batavian	Medium Crisphead	Lettuce
Lactuca	sativa-capitata-crispa-media	batavia-michelle	Michelle Batavian	Medium Crisphead	Lettuce
Lactuca	sativa-capitata-crispa-media	batavia-red-grenoblei	Red Grenoble Batavian	Medium Crisphead	Lettuce
Lactuca	sativa-capitata-crispa-media	batavia-royaltyi	Red Royalty Batavian	Medium Crisphead	Lettuce
Lactuca	sativa-capitata-crispa-media	batavia-spartai	Red Sparta Batavian	Medium Crisphead	Lettuce
Lactuca	sativa-capitata-crispa-media	batavia-tahoei	Red Tahoe Batavian	Medium Crisphead	Lettuce
Lactuca	sativa-capitata-crispa-media	blonde-de-parisi	Blonde De Paris	Medium Crisphead	Lettuce

Genus	species	infraspecies	cultivar	specimen	Genome
Lactuca	*sativa-capitata-crispa-media*	*chou-de-naples-bresi*	Chou De Naples Bresi	Medium Crisphead	Lettuce
Lactuca	*sativa-capitata-crispa-media*	*crispy-frillsi*	Crispy Frills	Medium Crisphead	Lettuce
Lactuca	*sativa-capitata-crispa-media*	*diamond-headii*	Diamond Head	Medium Crisphead	Lettuce
Lactuca	*sativa-capitata-crispa-media*	*formidana*	Formidana	Medium Crisphead	Lettuce
Lactuca	*sativa-capitata-crispa-media*	*la-brilliantei*	La Brilliante	Medium Crisphead	Lettuce
Lactuca	*sativa-capitata-crispa-media*	*lentissima-a-montarei-3*	Lentissima A Montare	Medium Crisphead	Lettuce
Lactuca	*sativa-capitata-crispa-media*	*open-heartii*	Open Heart	Medium Crisphead	Lettuce
Lactuca	*sativa-capitata-crispa-media*	*raleighii*	Raleigh	Medium Crisphead	Lettuce
Lactuca	*sativa-capitata-crispa-media*	*raleighii-proi-917*	#917 Pro Raleigh	Medium Crisphead	Lettuce
Lactuca	*sativa-capitata-crispa-media*	*superi-59*	Super #59	Medium Crisphead	Lettuce
Lactuca	*sativa-capitata-crispa-media*	*webb's-wonderfulii*	Webb's Wonderful	Medium Crisphead	Lettuce
Lactuca	*sativa-capitata-crispa-parva*	*wakefield-crunchii*	Wakefield Crunch	Small Crisphead	Lettuce
Lactuca	*sativa-capitata-crispa-parva*	*warpathii*	Warpath	Small Crisphead	Lettuce
Lactuca	*sativa-capitata-icebergii-larga*	*big-benii*	Big Ben	Large Iceberg	Lettuce
Lactuca	*sativa-capitata-icebergii-larga*	*BOS-9032*	Boston #9032	Large Iceberg	Lettuce
Lactuca	*sativa-capitata-icebergii-larga*	*BOS-9091*	Boston #9091	Large Iceberg	Lettuce
Lactuca	*sativa-capitata-icebergii-larga*	*desert-stormii*	Desert Storm	Large Iceberg	Lettuce
Lactuca	*sativa-capitata-icebergii-larga*	*goliathii*	Goliath	Large Iceberg	Lettuce
Lactuca	*sativa-capitata-icebergii-larga*	*great-lakesi-118*	#118 Great Lakes	Large Iceberg	Lettuce
Lactuca	*sativa-capitata-icebergii-larga*	*great-lakesi-118-A*	#118 A Great Lakes	Large Iceberg	Lettuce
Lactuca	*sativa-capitata-icebergii-larga*	*great-lakesi-200-R*	#200 R Great Lakes	Large Iceberg	Lettuce
Lactuca	*sativa-capitata-icebergii-larga*	*great-lakesi-366*	#366 Great Lakes	Large Iceberg	Lettuce
Lactuca	*sativa-capitata-icebergii-larga*	*great-lakesi-407*	#407 Great Lakes	Large Iceberg	Lettuce
Lactuca	*sativa-capitata-icebergii-larga*	*great-lakesi-407-P*	#407 P Great Lakes	Large Iceberg	Lettuce
Lactuca	*sativa-capitata-icebergii-larga*	*great-lakesi-623-8*	#623 8 Great Lakes	Large Iceberg	Lettuce
Lactuca	*sativa-capitata-icebergii-larga*	*great-lakesi-659*	#659 Great Lakes	Large Iceberg	Lettuce
Lactuca	*sativa-capitata-icebergii-larga*	*great-lakesi-659-G*	#659 G Great Lakes	Large Iceberg	Lettuce
Lactuca	*sativa-capitata-icebergii-larga*	*great-lakesi-659-Mesa*	#659 Mesa Great Lakes	Large Iceberg	Lettuce
Lactuca	*sativa-capitata-icebergii-larga*	*great-lakes-calmari*	Calmar Great Lakes	Large Iceberg	Lettuce
Lactuca	*sativa-capitata-icebergii-larga*	*great-lakesi-earlyi*	Early Great Lakes	Large Iceberg	Lettuce
Lactuca	*sativa-capitata-icebergii-larga*	*great-lakesi-earlyi-king-crownii*	King Crown Early Great Lakes	Large Iceberg	Lettuce
Lactuca	*sativa-capitata-icebergii-larga*	*great-lakes-montemari*	Calmar Great Lakes	Large Iceberg	Lettuce
Lactuca	*sativa-capitata-icebergii-larga*	*great-lakesi-premieri*	Premier Great Lakes	Large Iceberg	Lettuce
Lactuca	*sativa-capitata-icebergii-larga*	*new-yorkii*	New York	Large Iceberg	Lettuce
Lactuca	*sativa-capitata-icebergii-larga*	*new-yorkii-12*	#12 New York	Large Iceberg	Lettuce
Lactuca	*sativa-capitata-icebergii-larga*	*redii*	Red	Large Iceberg	Lettuce
Lactuca	*sativa-capitata-icebergii-larga*	*valleygreenii*	Valleygreen	Large Iceberg	Lettuce
Lactuca	*sativa-capitata-icebergii-larga*	*yumai*	Yuma	Large Iceberg	Lettuce
Lactuca	*sativa-capitata-icebergii-larga-media*	*giant-crystalii*	Giant Crystal	Medium Large Iceberg	Lettuce
Lactuca	*sativa-capitata-icebergii-larga-media*	*seagreenii*	Seagreen	Medium Large Iceberg	Lettuce
Lactuca	*sativa-capitata-icebergii-media*	*alpha-dmri*	Alpha Downy Mildew Resistant	Medium Iceberg	Lettuce
Lactuca	*sativa-capitata-icebergii-media*	*burgundy-icei*	Burgundy Ice	Medium Iceberg	Lettuce
Lactuca	*sativa-capitata-icebergii-media*	*burpeei-A*	Burpee A	Medium Iceberg	Lettuce
Lactuca	*sativa-capitata-icebergii-media*	*cerisei*	Cerise	Medium Iceberg	Lettuce
Lactuca	*sativa-capitata-icebergii-media*	*coach-supremei*	Coach Supreme	Medium Iceberg	Lettuce
Lactuca	*sativa-capitata-icebergii-media*	*continuityi*	Continuity	Medium Iceberg	Lettuce
Lactuca	*sativa-capitata-icebergii-media*	*courtii*	Court	Medium Iceberg	Lettuce
Lactuca	*sativa-capitata-icebergii-media*	*crisp-as-icei*	Crisp As Ice	Medium Iceberg	Lettuce
Lactuca	*sativa-capitata-icebergii-media*	*ice-queenii*	Ice Queen	Medium Iceberg	Lettuce
Lactuca	*sativa-capitata-icebergii-media*	*ithacai*	Ithaca	Medium Iceberg	Lettuce
Lactuca	*sativa-capitata-icebergii-media*	*ithacai-989*	#989 Ithaca	Medium Iceberg	Lettuce
Lactuca	*sativa-capitata-icebergii-media*	*ithacai-skylinei*	Skyline Ithaca	Medium Iceberg	Lettuce
Lactuca	*sativa-capitata-icebergii-media*	*ithacai-summertime*	Summertime Ithaca	Medium Iceberg	Lettuce
Lactuca	*sativa-capitata-icebergii-media*	*montelloi*	Montello	Medium Iceberg	Lettuce
Lactuca	*sativa-capitata-icebergii-media*	*miluna*	Miluna	Medium Iceberg	Lettuce
Lactuca	*sativa-capitata-icebergii-media*	*PYB-0136*	#0136 Pybas	Medium Iceberg	Lettuce
Lactuca	*sativa-capitata-icebergii-media*	*PYB-0250*	#0250 Pybas	Medium Iceberg	Lettuce
Lactuca	*sativa-capitata-icebergii-media*	*PYB-0251*	#0251 Pybas	Medium Iceberg	Lettuce
Lactuca	*sativa-capitata-icebergii-media*	*PYB-2144*	#2144 Pybas	Medium Iceberg	Lettuce
Lactuca	*sativa-capitata-icebergii-media*	*rangeri*	Ranger	Medium Iceberg	Lettuce
Lactuca	*sativa-capitata-icebergii-media*	*regina-dei-ghiaccii*	Regina Dei Ghiacci	Medium Iceberg	Lettuce
Lactuca	*sativa-capitata-icebergii-media*	*saladin-triumphii*	Triumph Saladin	Medium Iceberg	Lettuce
Lactuca	*sativa-capitata-icebergii-media*	*sharpshooteri*	Sharpshooter	Medium Iceberg	Lettuce
Lactuca	*sativa-capitata-icebergii-media*	*SSC-1501*	#1501 Saunders Seed Co.	Medium Iceberg	Lettuce
Lactuca	*sativa-capitata-icebergii-media*	*tiberi*	Tiber	Medium Iceberg	Lettuce
Lactuca	*sativa-capitata-icebergii-media*	*wintersetii*	Winterset	Medium Iceberg	Lettuce
Lactuca	*sativa-capitata-icebergii-media*	*yucaipai*	Yucaipa	Medium Iceberg	Lettuce
Lactuca	*sativa-capitata-icebergii-parva*	*anuenuei*	Anuenue	Small Iceberg	Lettuce
Lactuca	*sativa-capitata-icebergii-parva*	*blushii*	Blush	Small Iceberg	Lettuce
Lactuca	*sativa-capitata-icebergii-parva*	*ice-cubei*	Ice Cube	Small Iceberg	Lettuce
Lactuca	*sativa-capitata-icebergii-parva*	*mini-greenii*	Mini Green	Small Iceberg	Lettuce
Lactuca	*sativa-capitata-icebergii-parva*	*reine-des-glacesi*	Reine Des Glaces	Small Iceberg	Lettuce
Lactuca	*sativa-capitata-icebergii-parva*	*rosa*	Rosy	Small Iceberg	Lettuce

Genus	species	infraspecies	cultivar	specimen	Genome
Lactuca	*sativa-capitata-rubra-larga*	*batavia-rogeri*	Roger Batavian	Large Redhead	Lettuce
Lactuca	*sativa-capitata-rubra-larga*	*rossimoi*	Rossimo	Large Redhead	Lettuce
Lactuca	*sativa-capitata-rubra-media*	*batavia-asconai*	Ascona Batavian	Medium Redhead	Lettuce
Lactuca	*sativa-capitata-rubra-media*	*cardinalei*	Cardinale	Medium Redhead	Lettuce
Lactuca	*sativa-capitata-rubra-media*	*carmonai*	Carmona Butterhead	Medium Redhead	Lettuce
Lactuca	*sativa-capitata-rubra-media*	*cinnamon-redii*	Cinnamon Butterhead	Medium Redhead	Lettuce
Lactuca	*sativa-capitata-rubra-media*	*bibbii-rodan-bronzei*	Rodan Bronze Bibb	Medium Redhead	Lettuce
Lactuca	*sativa-capitata-rubra-media*	*batavian-pabloi*	Pablo Batavian Iceberg	Medium Redhead	Lettuce
Lactuca	*sativa-capitata-rubra-media*	*boston-redii-edoxii*	Edox Red Boston	Medium Redhead	Lettuce
Lactuca	*sativa-capitata-rubra-media*	*hyper-red-rumple-wavedii*	Hyper Red Rumple Waved	Medium Redhead	Lettuce
Lactuca	*sativa-capitata-rubra-media*	*lovina*	Lovina	Medium Redhead	Lettuce
Lactuca	*sativa-capitata-rubra-media*	*magenta*	Magenta	Medium Redhead	Lettuce
Lactuca	*sativa-capitata-rubra-media*	*mikola*	Mikola Butterhead	Medium Redhead	Lettuce
Lactuca	*sativa-capitata-rubra-media*	*red-butterworthii*	Red Butterworth	Medium Redhead	Lettuce
Lactuca	*sativa-capitata-rubra-media*	*red-voguei*	Red Vogue Butterhead	Medium Redhead	Lettuce
Lactuca	*sativa-capitata-rubra-media*	*rouge-grenoblaisei*	Rouge Grenoblaise Butterhead	Medium Redhead	Lettuce
Lactuca	*sativa-capitata-rubra-media*	*sangria*	Sangria Butterhead	Medium Redhead	Lettuce
Lactuca	*sativa-capitata-rubra-media*	*sweet-redii*	Sweet Red	Medium Redhead	Lettuce
Lactuca	*sativa-capitata-rubra-parva*	*bacardii*	Bacardi Butterhead	Small Redhead	Lettuce
Lactuca	*sativa-capitata-rubra-parva*	*red-montpelieri*	Red Montpelier Butterhead	Small Redhead	Lettuce
Lactuca	*sativa-capitata-rubra-parva*	*rougette-du-midii*	Rougette Du Midi Butterhead	Small Redhead	Lettuce
Lactuca	*sativa-capitata-verticillata-larga*	*grandpa-admire'si*	Grandpa Admire's	Large Loosehead	Lettuce
Lactuca	*sativa-capitata-verticillata-larga*	*landis-winteri*	Landis Winter	Large Loosehead	Lettuce
Lactuca	*sativa-capitata-verticillata-media*	*aquarius*	Aquarius	Medium Loosehead	Lettuce
Lactuca	*sativa-capitata-verticillata-media*	*brauner-trotzkopfii*	Brauner Trotzkopf	Medium Loosehead	Lettuce
Lactuca	*sativa-capitata-verticillata-media*	*czechoslovakiana*	Czechoslovakian	Medium Loosehead	Lettuce
Lactuca	*sativa-capitata-verticillata-media*	*drunken-woman-fringed-headedii*	Drunken Woman Fringed	Medium Loosehead	Lettuce
Lactuca	*sativa-capitata-verticillata-media*	*musca*	Musca	Medium Loosehead	Lettuce
Lactuca	*sativa-capitata-verticillata-media*	*manoai*	Manoa	Medium Loosehead	Lettuce
Lactuca	*sativa-capitata-verticillata-media*	*ovationii*	Ovation	Medium Loosehead	Lettuce
Lactuca	*sativa-capitata-verticillata-media*	*speckledii*	Speckled	Medium Loosehead	Lettuce
Lactuca	*sativa-capitata-verticillata-media*	*vicky-bingii*	Vicky Bing	Medium Loosehead	Lettuce
Lactuca	*sativa-capitata-verticillata-parva*	*parella-rossai*	Parella Red	Small Loosehead	Lettuce
Lactuca	*sativa-capitata-verticillata-parva*	*parella-verdei*	Parella Green	Small Loosehead	Lettuce
Lactuca	*sativa-capitata-verticillata-parva*	*zumnii*	Zumni	Small Loosehead	Lettuce

Lettuce, Looseleaf (*Lactuca sativa-foliosa*)

Genus	species	infraspecies	cultivar	specimen	Genome
Lactuca	*sativa-foliosa-larga*	*chicken-or-rabbitii*	Chicken Or Rabbit	12.1"-16" Large Looseleaf	Lettuce
Lactuca	*sativa-foliosa-larga*	*four-seasonsi*	Four Seasons Red Tipped	12.1"-16" Large Looseleaf	Lettuce
Lactuca	*sativa-foliosa-larga*	*green-valleyi*	Green Valley	12.1"-16" Large Looseleaf	Lettuce
Lactuca	*sativa-foliosa-larga*	*italienischeri*	Italienisher	12.1"-16" Large Looseleaf	Lettuce
Lactuca	*sativa-foliosa-larga*	*les-orielles-du-diablei*	Les Orielles Du Diable	12.1"-16" Large Looseleaf	Lettuce
Lactuca	*sativa-foliosa-larga-media*	*bolognai*	Bologna	9.1"-12" Medium Large Looseleaf	Lettuce
Lactuca	*sativa-foliosa-larga-media*	*genecorp-greenii*	Green Genecorp	9.1"-12" Medium Large Looseleaf	Lettuce
Lactuca	*sativa-foliosa-larga-media*	*genecorp-shining-stari*	Shining Star Genecorp	9.1"-12" Medium Large Looseleaf	Lettuce
Lactuca	*sativa-foliosa-larga-media*	*glossy-greenii*	Glossy Green	9.1"-12" Medium Large Looseleaf	Lettuce
Lactuca	*sativa-foliosa-larga-media*	*winterwunderi*	Winterwunder	9.1"-12" Medium Large Looseleaf	Lettuce
Lactuca	*sativa-foliosa-media*	*asian-greenii*	Asian Green	6.1"-9" Medium Looseleaf	Lettuce
Lactuca	*sativa-foliosa-media*	*centennialii*	Centennial	6.1"-9" Medium Looseleaf	Lettuce
Lactuca	*sativa-foliosa-media*	*claret-symphonyi*	Claret Symphone	6.1"-9" Medium Looseleaf	Lettuce
Lactuca	*sativa-foliosa-media*	*cressonette-maroainei*	Cressonette Marocaine	6.1"-9" Medium Looseleaf	Lettuce
Lactuca	*sativa-foliosa-media*	*ferrari*	Ferrari	6.1"-9" Medium Looseleaf	Lettuce
Lactuca	*sativa-foliosa-media*	*funlyi*	Funly	6.1"-9" Medium Looseleaf	Lettuce
Lactuca	*sativa-foliosa-media*	*germanii-pale-greenii*	German Pale Green	6.1"-9" Medium Looseleaf	Lettuce
Lactuca	*sativa-foliosa-media*	*grappai*	Grappa	6.1"-9" Medium Looseleaf	Lettuce
Lactuca	*sativa-foliosa-media*	*green-acei*	Green Ace	6.1"-9" Medium Looseleaf	Lettuce
Lactuca	*sativa-foliosa-media*	*jeannei*	Jeanne	6.1"-9" Medium Looseleaf	Lettuce
Lactuca	*sativa-foliosa-media*	*jebousici*	Jebousic	6.1"-9" Medium Looseleaf	Lettuce
Lactuca	*sativa-foliosa-media*	*lasting-greenii*	Lasting Green	6.1"-9" Medium Looseleaf	Lettuce
Lactuca	*sativa-foliosa-media*	*lau's-pointedii*	Lasting Green	6.1"-9" Medium Looseleaf	Lettuce
Lactuca	*sativa-foliosa-media*	*leopardii*	Leopard Pale Green	6.1"-9" Medium Looseleaf	Lettuce
Lactuca	*sativa-foliosa-media*	*mariahii*	Mariah	6.1"-9" Medium Looseleaf	Lettuce
Lactuca	*sativa-foliosa-media*	*merveille-de-maii*	Merveille De Mai	6.1"-9" Medium Looseleaf	Lettuce
Lactuca	*sativa-foliosa-media*	*misticanzai*	Misticanza	6.1"-9" Medium Looseleaf	Lettuce
Lactuca	*sativa-foliosa-media*	*pink-tipii*	Pink Tip	6.1"-9" Medium Looseleaf	Lettuce
Lactuca	*sativa-foliosa-media*	*quedlinburger-dickkopfii*	Quedlinburger Dickkopf	6.1"-9" Medium Looseleaf	Lettuce
Lactuca	*sativa-foliosa-media*	*rebosai*	Rebosa	6.1"-9" Medium Looseleaf	Lettuce
Lactuca	*sativa-foliosa-media*	*red-foxii*	Red Tipped Fox	6.1"-9" Medium Looseleaf	Lettuce
Lactuca	*sativa-foliosa-media*	*rolinai*	Rolina	6.1"-9" Medium Looseleaf	Lettuce
Lactuca	*sativa-foliosa-media*	*spring-feastii*	Spring Feast	6.1"-9" Medium Looseleaf	Lettuce
Lactuca	*sativa-foliosa-media*	*strela-greenii*	Strela Gren	6.1"-9" Medium Looseleaf	Lettuce

Genus	species	infraspecies	cultivar	specimen	Genome
Lactuca	sativa-foliosa-media	summer-acei	Summer Ace	6.1"-9" Medium Looseleaf	Lettuce
Lactuca	sativa-foliosa-media	sunsetii	Sunset	6.1"-9" Medium Looseleaf	Lettuce
Lactuca	sativa-foliosa-media	tangentii	Tangent	6.1"-9" Medium Looseleaf	Lettuce
Lactuca	sativa-foliosa-media	tangoi	Tango	6.1"-9" Medium Looseleaf	Lettuce
Lactuca	sativa-foliosa-parva	marimba-SPX-0071	Marimba	2"-6" Small Looseleaf	Lettuce
Lactuca	sativa-foliosa-parva	paramixii	Paramix	2"-6" Small Looseleaf	Lettuce
Lactuca	sativa-foliosa-parva	salade-de-russiei	Salade De Russie	2"-6" Small Looseleaf	Lettuce
Lactuca	sativa-foliosa-corugata	cracoviensis	Crakov	Ruffled Looseleaf	Lettuce
Lactuca	sativa-foliosa-corugata	grandpa'si	Grandpa's	Ruffled Looseleaf	Lettuce
Lactuca	sativa-foliosa-corugata	krizetii	Krizet	Ruffled Looseleaf	Lettuce
Lactuca	sativa-foliosa-corugata	riccolina-da-taglioi	Riccolina Da Taglio Pale Green	Ruffled Looseleaf	Lettuce
Lactuca	sativa-foliosa-corugata	salad-bowlii	Salad Bowl Pale Green	Ruffled Looseleaf	Lettuce
Lactuca	sativa-foliosa-corugata	salad-bowlii-compactii	Compact Salad Bowl Pale Green	Ruffled Looseleaf	Lettuce
Lactuca	sativa-foliosa-corugata	trocaderoi	Trocadero Pale Green	Ruffled Looseleaf	Lettuce
Lactuca	sativa-foliosa-corugata-rubra	galactic-redii	Galactic Red	Red Ruffled Looseleaf	Lettuce
Lactuca	sativa-foliosa-corugata-rubra	new-red-firei	New Red Fire	Red Ruffled Looseleaf	Lettuce
Lactuca	sativa-foliosa-corugata-rubra	red-sailsi	Red Sails	Red Ruffled Looseleaf	Lettuce
Lactuca	sativa-foliosa-corugata-rubra	royal-redii	Royal Red	Red Ruffled Looseleaf	Lettuce
Lactuca	sativa-foliosa-corugata-rubra	ruby-rufflesi	Ruby Ruffles	Red Ruffled Looseleaf	Lettuce
Lactuca	sativa-foliosa-corugata-rubra	salad-bowlii-mariachi-SXP-0066	Mariachi Salad Bowl	Red Ruffled Looseleaf	Lettuce
Lactuca	sativa-foliosa-corugata-rubra	salad-bowlii-redii	Red Salad Bowl	Red Ruffled Looseleaf	Lettuce
Lactuca	sativa-foliosa-corugata-rubra	salad-bowlii-ruby-redii	Ruby Red Salad Bowl	Red Ruffled Looseleaf	Lettuce
Lactuca	sativa-foliosa-corugata-rubra	tomahawkii	Tomahawk Pink	Red Ruffled Looseleaf	Lettuce
Lactuca	sativa-foliosa-crispa	lollo-biondoi	Cireo Lollo Biondo Pale Green	Curly Looseleaf	Lettuce
Lactuca	sativa-foliosa-crispa	lollo-biondoi-cireo	Cireo Lollo Biondo Pale Green	Curly Looseleaf	Lettuce
Lactuca	sativa-foliosa-crispa	lollo-biondoi-bolzanoi	Bolzano Lollo Biondo Pale Green	Curly Looseleaf	Lettuce
Lactuca	sativa-foliosa-crispa	monetii	Monet	Curly Looseleaf	Lettuce
Lactuca	sativa-foliosa-crispa-rubra	lollo-rossa	Lollo Rossa	Red Curly Looseleaf	Lettuce
Lactuca	sativa-foliosa-crispa-rubra	lollo-rossa-astinai	Astina Lollo Rossa	Red Curly Looseleaf	Lettuce
Lactuca	sativa-foliosa-crispa-rubra	lollo-rossa-bellisimo	Bellisimo Lollo Rossa	Red Curly Looseleaf	Lettuce
Lactuca	sativa-foliosa-crispa-rubra	lollo-rossa-impulsi	Impuls Lollo Rossa	Red Curly Looseleaf	Lettuce
Lactuca	sativa-foliosa-crispa-rubra	lollo-rossa-maderai	Madera Lollo Rossa	Red Curly Looseleaf	Lettuce
Lactuca	sativa-foliosa-crispa-rubra	lollo-rossa-mercuryi	Mercury Lollo Rossa	Red Curly Looseleaf	Lettuce
Lactuca	sativa-foliosa-crispa-rubra	lollo-rossa-navidadii	Navidad Lollo Rossa	Red Curly Looseleaf	Lettuce
Lactuca	sativa-foliosa-crispa-rubra	lollo-rossa-revolutionii	Revolution Lollo Rossa	Red Curly Looseleaf	Lettuce
Lactuca	sativa-foliosa-crispa-rubra	lollo-rossa-sanguinei	Sanguine Lollo Rossa	Red Curly Looseleaf	Lettuce
Lactuca	sativa-foliosa-crispa-rubra	lollo-rossa-selwayi	Selway Lollo Rossa	Red Curly Looseleaf	Lettuce
Lactuca	sativa-foliosa-crispa-rubra	lollo-rossa-sesamii	Sesam Lollo Rossa	Red Curly Looseleaf	Lettuce
Lactuca	sativa-foliosa-crispa-rubra	lollo-rossa-valeria	Valeria Lollo Rossa	Red Curly Looseleaf	Lettuce
Lactuca	sativa-foliosa-cuttingii	american-gatheringii	American Gathering	Cutting Looseleaf	Lettuce
Lactuca	sativa-foliosa-cuttingii	australian-gatheringii	Australian Gathering	Cutting Looseleaf	Lettuce
Lactuca	sativa-foliosa-cuttingii	bionda-liscia	Bionda Liscia	Cutting Looseleaf	Lettuce
Lactuca	sativa-foliosa-cuttingii	biondo-a-folgie-liscei	Biondo A Foglie Lisce	Cutting Looseleaf	Lettuce
Lactuca	sativa-foliosa-cuttingii	drunkardii	Drunkard	Cutting Looseleaf	Lettuce
Lactuca	sativa-foliosa-cuttingii	feuille-de-chene-blondei	Feuille de Chene Blonde Pale Green	Cutting Looseleaf	Lettuce
Lactuca	sativa-foliosa-cuttingii	hungarian-winteri	Hungarian Winter	Cutting Looseleaf	Lettuce
Lactuca	sativa-foliosa-cuttingii	maravilla-der-verano-vonnyi	Maravilla der Verano Vonny	Cutting Looseleaf	Lettuce
Lactuca	sativa-foliosa-cuttingii	merlotii	Merlot	Cutting Looseleaf	Lettuce
Lactuca	sativa-foliosa-cuttingii	rossa-d'ameriquei	Rossa D'amerique	Cutting Looseleaf	Lettuce
Lactuca	sativa-foliosa-fimbria	biondo-ricciolinii	Biondo Ricciolini	Frilly Looseleaf	Lettuce
Lactuca	sativa-foliosa-fimbria	envyi	Envy	Frilly Looseleaf	Lettuce
Lactuca	sativa-foliosa-fimbria	frisbyi	Frisby	Frilly Looseleaf	Lettuce
Lactuca	sativa-foliosa-fimbria	gold-rushii	Gold Rush	Frilly Looseleaf	Lettuce
Lactuca	sativa-foliosa-fimbria	prizeheadii	Pale Green Prizehead	Frilly Looseleaf	Lettuce
Lactuca	sativa-foliosa-fimbria	prizeheadii-earlyi	Early Pale Green Prizehead	Frilly Looseleaf	Lettuce
Lactuca	sativa-foliosa-fimbria	red-tangoi-crissyi	Crissy Red Tango	Frilly Looseleaf	Lettuce
Lactuca	sativa-foliosa-fimbria	grand-rapidsi	Grand Rapids	Frilly Looseleaf	Lettuce
Lactuca	sativa-foliosa-fimbria	grand-rapidsi-forcingii	Forcing Grand Rapids	Frilly Looseleaf	Lettuce
Lactuca	sativa-foliosa-fimbria	grand-rapidsi-garnetii	Garnet Grand Rapids	Frilly Looseleaf	Lettuce
Lactuca	sativa-foliosa-fimbria	grand-rapidsi-green-wavei	Green Wave Grand Rapids	Frilly Looseleaf	Lettuce
Lactuca	sativa-foliosa-fimbria	grand-rapidsi-ritsai	Ritsa Grand Rapids	Frilly Looseleaf	Lettuce
Lactuca	sativa-foliosa-fimbria	grand-rapidsi-royal-greenii	Royal Green Grand Rapids	Frilly Looseleaf	Lettuce
Lactuca	sativa-foliosa-fimbria	grand-rapidsi-seafreshii	Seafresh Grand Rapids	Frilly Looseleaf	Lettuce
Lactuca	sativa-foliosa-fimbria	grand-rapidsi-sloboltii	Slobolt Grand Rapids	Frilly Looseleaf	Lettuce
Lactuca	sativa-foliosa-fimbria	grand-rapidsi-tipburnii	Tipburn Grand Rapids	Frilly Looseleaf	Lettuce
Lactuca	sativa-foliosa-fimbria	grand-rapidsi-waldmann's-dark-greenii	Dark Green Waldmann's Grand Rapids	Frilly Looseleaf	Lettuce
Lactuca	sativa-foliosa-fimbria	grand-rapidsi-waldman	Fanfare Waldmann's	Frilly Looseleaf	Lettuce

Genus	species	infraspecies	cultivar	specimen	Genome
		n's-fanfare-FM-1530	Grand Rapids		
Lactuca	sativa-foliosa-fimbria	grand-rapidsi-waldmann's-greenii	Green Waldmann's Grand Rapids	Frilly Looseleaf	Lettuce
Lactuca	sativa-foliosa-fimbria	grand-rapidsi-waldmann's-marinii	Marin Waldmann's Grand Rapids	Frilly Looseleaf	Lettuce
Lactuca	sativa-foliosa-fimbria	grand-rapidsi-waldmann's-two-stari	Two Star Waldmann's Grand Rapids	Frilly Looseleaf	Lettuce
Lactuca	sativa-foliosa-fimbria	xena	Xena	Frilly Looseleaf	Lettuce
Lactuca	sativa-foliosa-fimbria	xena-GX-514	#514 Green Xena	Frilly Looseleaf	Lettuce
Lactuca	sativa-foliosa-fimbria-rubra	deep-redii	Deep Red	Red Frilly Looseleaf	Lettuce
Lactuca	sativa-foliosa-fimbria-rubra	grand-rapids-red-rapidsi	Red Rapids Grand Rapids	Red Frilly Looseleaf	Lettuce
Lactuca	sativa-foliosa-fimbria-rubra	prizeheadii-flamei	Flame Prizehead	Red Frilly Looseleaf	Lettuce
Lactuca	sativa-foliosa-fimbria-rubra	prizeheadii-prizeleafii	Prizeleaf Maroon Prizehead	Red Frilly Looseleaf	Lettuce
Lactuca	sativa-foliosa-fimbria-rubra	prizeheadii-red-wavei	Red Wave Prizehead	Red Frilly Looseleaf	Lettuce
Lactuca	sativa-foliosa-fimbria-rubra	prizeheadii-redprizei-FM-1559	Redprize Prizehead	Red Frilly Looseleaf	Lettuce
Lactuca	sativa-foliosa-fimbria-rubra	prizeheadii-super-prizei	Super Prize Prizehead	Red Frilly Looseleaf	Lettuce
Lactuca	sativa-foliosa-fimbria-rubra	rubinii	Rubin	Red Frilly Looseleaf	Lettuce
Lactuca	sativa-foliosa-fimbria-rubra	rustica	Rustica	Red Frilly Looseleaf	Lettuce
Lactuca	sativa-foliosa-fimbria-rubra	vulcanii	Vulcan	Red Frilly Looseleaf	Lettuce
Lactuca	sativa-foliosa-helva-groena	australian-yellowi	Australian Wrinkled	Yellow Green Looseleaf	Lettuce
Lactuca	sativa-foliosa-helva-groena	dunselii	Dunsel	Yellow Green Looseleaf	Lettuce
Lactuca	sativa-foliosa-helva-groena	early-curled-simpsonii	Early Curled Simpson	Yellow Green Looseleaf	Lettuce
Lactuca	sativa-foliosa-helva-groena	thai-greenii	Thai Green	Yellow Green Looseleaf	Lettuce
Lactuca	sativa-foliosa-marginata-brunia	bronze-arrowi	Bronze Arrow	Bronze Margined Looseleaf	Lettuce
Lactuca	sativa-foliosa-marginata-brunia	bronze-arrowi-variantii	Variant Bronze Arrow	Bronze Margined Looseleaf	Lettuce
Lactuca	sativa-foliosa-marginata-rubra	aragon-redii	Aragon	Red Margined Looseleaf	Lettuce
Lactuca	sativa-foliosa-marginata-rubra	asian-redii	Asian	Red Margined Looseleaf	Lettuce
Lactuca	sativa-foliosa-marginata-rubra	big-redii	Big	Red Margined Looseleaf	Lettuce
Lactuca	sativa-foliosa-marginata-rubra	bronc-rodonii	Bronc Rodon	Red Margined Looseleaf	Lettuce
Lactuca	sativa-foliosa-pustula	greenii-505	#505 Green	Blistered Looseleaf	Lettuce
Lactuca	sativa-foliosa-pustula	blackjackii	Blackjack Red	Blistered Looseleaf	Lettuce
Lactuca	sativa-foliosa-pustula	tiarai	Tiara	Blistered Looseleaf	Lettuce
Lactuca	sativa-foliosa-pustula	tigris	Tigris	Blistered Looseleaf	Lettuce
Lactuca	sativa-foliosa-querca	belowai	Belowa	Oakleaf Looseleaf	Lettuce
Lactuca	sativa-foliosa-querca	bernicei	Bernice	Oakleaf Looseleaf	Lettuce
Lactuca	sativa-foliosa-querca	blushed-icy-oakii	Blushed Pink Icy Green	Oakleaf Looseleaf	Lettuce
Lactuca	sativa-foliosa-querca	brunia	Brunia Red Tipped Butterhead	Oakleaf Looseleaf	Lettuce
Lactuca	sativa-foliosa-querca	emerald-oakii	Emerald Oak Butterhead	Oakleaf Looseleaf	Lettuce
Lactuca	sativa-foliosa-querca	green-oak-leafii	Green	Oakleaf Looseleaf	Lettuce
Lactuca	sativa-foliosa-querca	green-oak-leafii-babyi	Baby Green	Oakleaf Looseleaf	Lettuce
Lactuca	sativa-foliosa-querca	lingue-de-canarinoi	Lingue De Canarino	Oakleaf Looseleaf	Lettuce
Lactuca	sativa-foliosa-querca	pom-pomii	Pom Pom	Oakleaf Looseleaf	Lettuce
Lactuca	sativa-foliosa-querca	purple-oak-leafii	Purple	Oakleaf Looseleaf	Lettuce
Lactuca	sativa-foliosa-querca	royal-oak-leafii	Royal	Oakleaf Looseleaf	Lettuce
Lactuca	sativa-foliosa-querca	robles-royalei	Robles Royal	Oakleaf Looseleaf	Lettuce
Lactuca	sativa-foliosa-querca	smilei	Smile	Oakleaf Looseleaf	Lettuce
Lactuca	sativa-foliosa-querca	thai-oakleafii-88	Thai #88	Oakleaf Looseleaf	Lettuce
Lactuca	sativa-foliosa-querca-fimbria	antaresi	Antares	Frilly Oakleaf Looseleaf	Lettuce
Lactuca	sativa-foliosa-querca-fimbria	austrian-greenleafii	Austrian Greenleaf	Frilly Oakleaf Looseleaf	Lettuce
Lactuca	sativa-foliosa-querca-fimbria	mascarai	Mascara	Frilly Oakleaf Looseleaf	Lettuce
Lactuca	sativa-foliosa-querca-rubra	arubai	Aruba	Red Oakleaf Looseleaf	Lettuce
Lactuca	sativa-foliosa-querca-rubra	bolchoii	Bolchoi	Red Oakleaf Looseleaf	Lettuce
Lactuca	sativa-foliosa-querca-rubra	bronze-guardii	Bronze Guard	Red Oakleaf Looseleaf	Lettuce
Lactuca	sativa-foliosa-querca-rubra	cocardei	Cocarde	Red Oakleaf Looseleaf	Lettuce
Lactuca	sativa-foliosa-querca-rubra	danoi	Dano	Red Oakleaf Looseleaf	Lettuce
Lactuca	sativa-foliosa-querca-rubra	fire-mountainii	Fire Mountain	Red Oakleaf Looseleaf	Lettuce
Lactuca	sativa-foliosa-querca-rubra	maseratii	Maserati	Red Oakleaf Looseleaf	Lettuce
Lactuca	sativa-foliosa-querca-rubra	mighty-red-oakii	Mighty Red Oak	Red Oakleaf Looseleaf	Lettuce
Lactuca	sativa-foliosa-querca-rubra	oaky-red-splashii	Oaky Red Splash	Red Oakleaf Looseleaf	Lettuce
Lactuca	sativa-foliosa-querca-rubra	red-oak-leafii	Deep Burgundy	Red Oakleaf Looseleaf	Lettuce
Lactuca	sativa-foliosa-querca-rubra	red-ruffled-oaksi	Red Ruffled Oaks	Red Oakleaf Looseleaf	Lettuce
Lactuca	sativa-foliosa-querca-rubra	redder-ruffledii	Redder Ruffled	Red Oakleaf Looseleaf	Lettuce
Lactuca	sativa-foliosa-querca-rubra	raisai	Raisa	Red Oakleaf Looseleaf	Lettuce
Lactuca	sativa-foliosa-querca-rubra	samanthai	Samantha	Red Oakleaf Looseleaf	Lettuce
Lactuca	sativa-foliosa-querca-rubra	sunfirei	Sunfire	Red Oakleaf Looseleaf	Lettuce
Lactuca	sativa-foliosa-rubra	fierei	Fiere Blood	Red Looseleaf	Lettuce
Lactuca	sativa-foliosa-rubra	GML-7	#7 Genetically Modified Leaf	Red Looseleaf	Lettuce
Lactuca	sativa-foliosa-rubra	ibisi	Ibis	Red Looseleaf	Lettuce
Lactuca	sativa-foliosa-rubra	pentaredii	Pentared	Red Looseleaf	Lettuce
Lactuca	sativa-foliosa-rubra	PS-64289	Pentared Siegers #64289	Red Looseleaf	Lettuce
Lactuca	sativa-foliosa-rubra	PS-21192	Pentared Siegers #21192	Red Looseleaf	Lettuce

Genus	species	infraspecies	cultivar	specimen	Genome
Lactuca	sativa-foliosa-rubra	red-bibbii	Red Bibb	Red Looseleaf	Lettuce
Lactuca	sativa-foliosa-rubra	red-coralii	Red Coral	Red Looseleaf	Lettuce
Lactuca	sativa-foliosa-rubra	red-curlii	Red Curl	Red Looseleaf	Lettuce
Lactuca	sativa-foliosa-rubra	red-ragei	Red Rage	Red Looseleaf	Lettuce
Lactuca	sativa-foliosa-rubra	red-velvetii	Red Velvet	Red Looseleaf	Lettuce
Lactuca	sativa-foliosa-rubra	redinai	Redina	Red Looseleaf	Lettuce
Lactuca	sativa-foliosa-rubra	rosa-PSR	Rosa Iceberg PSR	Red Looseleaf	Lettuce
Lactuca	sativa-foliosa-rubra	sanguine-ameliorei	Sanguine Ameliore	Red Looseleaf	Lettuce
Lactuca	sativa-foliosa-rubra	sunshinei	Sunshine	Red Looseleaf	Lettuce
Lactuca	sativa-foliosa-rubra	sweet-valentinei	Sweet Valentine	Red Looseleaf	Lettuce
Lactuca	sativa-foliosa-rubra	triple-redii	Triple Red	Red Looseleaf	Lettuce
Lactuca	sativa-foliosa-rubra	western-redii	Western Red	Red Looseleaf	Lettuce
Lactuca	sativa-foliosa-rugosa	black-seeded-simpsonii	Black Seeded Simpson Pale Green	Wrinkled Looseleaf	Lettuce
Lactuca	sativa-foliosa-rugosa	black-seeded-simpsonii-elitei	Elite Black Seeded Simpson Pale Green	Wrinkled Looseleaf	Lettuce
Lactuca	sativa-foliosa-rugosa-rubra	biscia-rossa	Biscia Rossa	Red Wrinkled Looseleaf	Lettuce
Lactuca	sativa-foliosa-rugosa-rubra	salad-trimii	Salad Trim	Red Wrinkled Looseleaf	Lettuce
Lactuca	sativa-foliosa-savoia	alpinei	Alpine	Savoy Looseleaf	Lettuce
Lactuca	sativa-foliosa-savoia	deer-tonguei	Deer Tongue	Savoy Looseleaf	Lettuce
Lactuca	sativa-foliosa-savoia	deer-tonguei-amishii	Amish Deer Tongue	Savoy Looseleaf	Lettuce
Lactuca	sativa-foliosa-savoia	deer-tonguei-chadwick's-rodanii	Chadwick's Rodan Deer Tongue	Savoy Looseleaf	Lettuce
Lactuca	sativa-foliosa-savoia	goosei	Goose	Savoy Looseleaf	Lettuce
Lactuca	sativa-foliosa-savoia	green-icei	Green Ice	Savoy Looseleaf	Lettuce
Lactuca	sativa-foliosa-savoia	green-visionii	Green Vision	Savoy Looseleaf	Lettuce
Lactuca	sativa-foliosa-savoia	mambai	Mamba	Savoy Looseleaf	Lettuce
Lactuca	sativa-foliosa-savoia	rossa-di-trentoi	Rossa Di Trento	Savoy Looseleaf	Lettuce
Lactuca	sativa-foliosa-savoia-rubra	deer-tonguei-redii	Red Deer Tongue	Red Savoy Looseleaf	Lettuce
Lactuca	sativa-foliosa-savoia-rubra	red-embersi	Red Embers	Red Savoy Looseleaf	Lettuce
Lactuca	sativa-foliosa-savoia-rubra	red-firei	Red Fire	Red Savoy Looseleaf	Lettuce
Lactuca	sativa-foliosa-savoia-rubra	rubyi	Ruby	Red Savoy Looseleaf	Lettuce

Lettuce, Romaine (*Lactuca sativa-longifolia*)

Lactuca	sativa-longifolia-larga	balloonii	Balloon	12.1"-16" Large Romaine	Lettuce
Lactuca	sativa-longifolia-larga	bionda-lentisimai	Bionda Lentissima Pale Green	12.1"-16" Large Romaine	Lettuce
Lactuca	sativa-longifolia-larga	bionda-ortolanii	Dark Green Bionda Ortolani	12.1"-16" Large Romaine	Lettuce
Lactuca	sativa-longifolia-larga	bionda-ortolanii-deglii	Degli Pale Green Bionda Ortolani	12.1"-16" Large Romaine	Lettuce
Lactuca	sativa-longifolia-larga	bionda-ortolanii-deglii-romana	Romana Degli Pale Green Bionda Ortolani	12.1"-16" Large Romaine	Lettuce
Lactuca	sativa-longifolia-larga	caprii	Capri	12.1"-16" Large Romaine	Lettuce
Lactuca	sativa-longifolia-larga	giant-caesari	Giant Caesar	12.1"-16" Large Romaine	Lettuce
Lactuca	sativa-longifolia-larga	lente-a-monteri	Lente A Monter	12.1"-16" Large Romaine	Lettuce
Lactuca	sativa-longifolia-larga	ortolanii-degli-verdei	Verde Degli Ortolani	12.1"-16" Large Romaine	Lettuce
Lactuca	sativa-longifolia-larga	remingtonii	Remington	12.1"-16" Large Romaine	Lettuce
Lactuca	sativa-longifolia-larga	yedi-kule-cinsi	Yedi Kule Cinsi	12.1"-16" Large Romaine	Lettuce
Lactuca	sativa-longifolia-larga-media	barcarollei	Barcarolle	9.1"-12" Medium Large Romaine	Lettuce
Lactuca	sativa-longifolia-larga-media	big-greenii-SXP-9810	Big Green	9.1"-12" Medium Large Romaine	Lettuce
Lactuca	sativa-longifolia-larga-media	brown-winteri	Brown Winter	9.1"-12" Medium Large Romaine	Lettuce
Lactuca	sativa-longifolia-larga-media	chiefii	Chief	9.1"-12" Medium Large Romaine	Lettuce
Lactuca	sativa-longifolia-larga-media	cosmoi	Cosmo	9.1"-12" Medium Large Romaine	Lettuce
Lactuca	sativa-longifolia-larga-media	crisp-mintii	Crisp Mint	9.1"-12" Medium Large Romaine	Lettuce
Lactuca	sativa-longifolia-larga-media	green-forestii	Green Forest	9.1"-12" Medium Large Romaine	Lettuce
Lactuca	sativa-longifolia-larga-media	guzmainei	Guzmaine	9.1"-12" Medium Large Romaine	Lettuce
Lactuca	sativa-longifolia-larga-media	guzmainei-tallii	Tall Guzmaine	9.1"-12" Medium Large Romaine	Lettuce
Lactuca	sativa-longifolia-larga-media	romainei-whitei	Romaine White	9.1"-12" Medium Large Romaine	Lettuce
Lactuca	sativa-longifolia-larga-media	romainei-blondei	Romaine Blonde	9.1"-12" Medium Large Romaine	Lettuce
Lactuca	sativa-longifolia-larga-media	romainei-dark-greenii	Romaine Dark Green	9.1"-12" Medium Large Romaine	Lettuce
Lactuca	sativa-longifolia-larga-media	romainei-greenii	Romaine Green	9.1"-12" Medium Large Romaine	Lettuce
Lactuca	sativa-longifolia-larga-media	roman-queenii	Roman Queen Pale Green	9.1"-12" Medium Large Romaine	Lettuce
Lactuca	sativa-longifolia-larga-media	tritonii	Triton	9.1"-12" Medium Large Romaine	Lettuce
Lactuca	sativa-longifolia-larga-media	verte-mari	Verte Mar	9.1"-12" Medium Large Romaine	Lettuce
Lactuca	sativa-longifolia-larga-media	winter-densityi	Winter Density	9.1"-12" Medium Large Romaine	Lettuce
Lactuca	sativa-longifolia-media	androsi-XP-999	Andros	6.1"-9" Medium Romaine	Lettuce
Lactuca	sativa-longifolia-media	angustus	Angustus	6.1"-9" Medium Romaine	Lettuce
Lactuca	sativa-longifolia-media	baladyi	Balady	6.1"-9" Medium Romaine	Lettuce
Lactuca	sativa-longifolia-media	maraicherei-blondei	Blonde Maraichere Pale Green	6.1"-9" Medium Romaine	Lettuce
Lactuca	sativa-longifolia-media	maraicherei-vertei	Verte Maraichere Green	6.1"-9" Medium Romaine	Lettuce
Lactuca	sativa-longifolia-media	bunyard's-matchlessi	Bunyard's Matchless	6.1"-9" Medium Romaine	Lettuce
Lactuca	sativa-longifolia-media	claremontii	Claremont	6.1"-9" Medium Romaine	Lettuce
Lactuca	sativa-longifolia-media	clementei	Clemente	6.1"-9" Medium Romaine	Lettuce
Lactuca	sativa-longifolia-media	corsairi	Corsair	6.1"-9" Medium Romaine	Lettuce
Lactuca	sativa-longifolia-media	de-morges-braunii	De Morges Braun	6.1"-9" Medium Romaine	Lettuce
Lactuca	sativa-longifolia-media	delle-lunei-7	#7 Delle Lune	6.1"-9" Medium Romaine	Lettuce

Lactuca	*sativa-longifolia-media*	*evergreenii-SXP-0137*	Evergreen	6.1"-9" Medium Romaine	Lettuce
Lactuca	*sativa-longifolia-media*	*forellenschussi*	Scarlet Blotched Forellenschuss	6.1"-9" Medium Romaine	Lettuce
Lactuca	*sativa-longifolia-media*	*forellenschussi-buntei*	Bunte Forellenschuss	6.1"-9" Medium Romaine	Lettuce
Lactuca	*sativa-longifolia-media*	*forellenschussi-frecklesi*	Red Blotched Freckles Forellenschuss	6.1"-9" Medium Romaine	Lettuce
Lactuca	*sativa-longifolia-media*	*jerichoi*	Jericho	6.1"-9" Medium Romaine	Lettuce
Lactuca	*sativa-longifolia-media*	*kostadinovi*	Kostadinov	6.1"-9" Medium Romaine	Lettuce
Lactuca	*sativa-longifolia-media*	*larissai*	Larissa	6.1"-9" Medium Romaine	Lettuce
Lactuca	*sativa-longifolia-media*	*olgai*	Olga	6.1"-9" Medium Romaine	Lettuce
Lactuca	*sativa-longifolia-media*	*redcurlii*	Redcurl	6.1"-9" Medium Romaine	Lettuce
Lactuca	*sativa-longifolia-media*	*romanyi*	Romany	6.1"-9" Medium Romaine	Lettuce
Lactuca	*sativa-longifolia-media*	*spotted-aleppoi*	Red Spotted Aleppo	6.1"-9" Medium Romaine	Lettuce
Lactuca	*sativa-longifolia-media*	*sudia*	Sudia	6.1"-9" Medium Romaine	Lettuce
Lactuca	*sativa-longifolia-media*	*winchesteri*	Winchester	6.1"-9" Medium Romaine	Lettuce
Lactuca	*sativa-longifolia-media*	*winter-wonderlandii*	Winter Wonderland	6.1"-9" Medium Romaine	Lettuce
Lactuca	*sativa-longifolia-parva*	*bambii*	Bambi	2"-6" Small Romaine	Lettuce
Lactuca	*sativa-longifolia-parva*	*craquante-d'avignonii*	Craquante D'Avignon	2"-6" Small Romaine	Lettuce
Lactuca	*sativa-longifolia-parva*	*little-caesari*	Little Caesar	2"-6" Small Romaine	Lettuce
Lactuca	*sativa-longifolia-parva*	*st-blaisei*	St. Blaise	2"-6" Small Romaine	Lettuce
Lactuca	*sativa-longifolia-rubra*	*apachei*	Apache	Red Romaine Loosehead	Lettuce
Lactuca	*sativa-longifolia-rubra*	*cherry-crispii*	Cherry Crisp	Red Romaine Loosehead	Lettuce
Lactuca	*sativa-longifolia-rubra*	*eruptionii*	Eruption	Red Romaine Loosehead	Lettuce
Lactuca	*sativa-longifolia-rubra*	*eva's-bugundy-winter-lettucei*	Eva's Burgundy Winter	Red Romaine Loosehead	Lettuce
Lactuca	*sativa-longifolia-rubra*	*feveri*	Fever	Red Romaine Loosehead	Lettuce
Lactuca	*sativa-longifolia-rubra*	*grenadieri-SXP-0059*	Grenadier	Red Romaine Loosehead	Lettuce
Lactuca	*sativa-longifolia-rubra*	*majestic-redii*	Majestic Red	Red Romaine Loosehead	Lettuce
Lactuca	*sativa-longifolia-rubra*	*petite-rougei*	Petite Rouge	Red Romaine Loosehead	Lettuce
Lactuca	*sativa-longifolia-rubra*	*red-leprechaunii*	Red Leprechaun	Red Romaine Loosehead	Lettuce
Lactuca	*sativa-longifolia-rubra*	*red-romana*	Red Romana	Red Romaine Loosehead	Lettuce
Lactuca	*sativa-longifolia-rubra*	*red-rufflesi*	Red Ruffles	Red Romaine Loosehead	Lettuce
Lactuca	*sativa-longifolia-rubra*	*romaine-redii*	Romaine Red	Red Romaine Loosehead	Lettuce
Lactuca	*sativa-longifolia-rubra*	*rosalitai*	Rosalita	Red Romaine Loosehead	Lettuce
Lactuca	*sativa-longifolia-rubra*	*rouge-de-hiveri*	Rouge De Hiver	Red Romaine Loosehead	Lettuce
Lactuca	*sativa-longifolia-rubra*	*rouge-de-hiveri-cimmaronii*	Cimmaron Rouge De Hiver	Red Romaine Loosehead	Lettuce
Lactuca	*sativa-longifolia-rubra*	*rouge-de-hiveri-little-leprechaunii*	Little Leprechaun Rouge De Hiver	Red Romaine Loosehead	Lettuce
Lactuca	*sativa-longifolia-rubra*	*ruben's-redii*	Ruben's	Red Romaine Loosehead	Lettuce
Lactuca	*sativa-longifolia-rubra*	*shanghaii-20*	Shanghai #20	Red Romaine Loosehead	Lettuce
Lactuca	*sativa-longifolia-rubra*	*sevillei*	Seville	Red Romaine Loosehead	Lettuce

Lettuce, Cos (*Lactuca sativa-longifolia-cosa*)

Genus	species	infraspecies	cultivar	specimen	Genome
Lactuca	*sativa-longifolia-cosa-larga*	*kalurai*	Kalura	12.1"-16" Large Cos	Lettuce
Lactuca	*sativa-longifolia-cosa-larga*	*major-cosi*	Major	12.1"-16" Large Cos	Lettuce
Lactuca	*sativa-longifolia-cosa-larga*	*paris-green-cosi*	Paris Green	12.1"-16" Large Cos	Lettuce
Lactuca	*sativa-longifolia-cosa-larga*	*paris-white-cosi*	Paris White	12.1"-16" Large Cos	Lettuce
Lactuca	*sativa-longifolia-cosa-larga*	*queenslandica*	Queensland	12.1"-16" Large Cos	Lettuce
Lactuca	*sativa-longifolia-cosa-larga*	*sweetiei*	Sweetie	12.1"-16" Large Cos	Lettuce
Lactuca	*sativa-longifolia-cosa-larga-media*	*bath-cosi*	Bath	9.1"-12" Medium Large Cos	Lettuce
Lactuca	*sativa-longifolia-cosa-larga-media*	*berettai*	Beretta	9.1"-12" Medium Large Cos	Lettuce
Lactuca	*sativa-longifolia-cosa-larga-media*	*berettai-weatherbyi*	Weatherby Beretta	9.1"-12" Medium Large Cos	Lettuce
Lactuca	*sativa-longifolia-cosa-larga-media*	*big-countryi*	Big Country	9.1"-12" Medium Large Cos	Lettuce
Lactuca	*sativa-longifolia-cosa-larga-media*	*blushed-butter-cosi*	Blushed Butter	9.1"-12" Medium Large Cos	Lettuce
Lactuca	*sativa-longifolia-cosa-larga-media*	*dark-green-cosi*	Blushed Butter	9.1"-12" Medium Large Cos	Lettuce
Lactuca	*sativa-longifolia-cosa-larga-media*	*ideal-cosi*	Ideal	9.1"-12" Medium Large Cos	Lettuce
Lactuca	*sativa-longifolia-cosa-larga-media*	*lion's-tonguei*	Lion's Tongue	9.1"-12" Medium Large Cos	Lettuce
Lactuca	*sativa-longifolia-cosa-larga-media*	*lobjoit's-green-cosi*	Lobjoit's Green	9.1"-12" Medium Large Cos	Lettuce
Lactuca	*sativa-longifolia-cosa-larga-media*	*parris-islandii-cosi*	Parris Island	9.1"-12" Medium Large Cos	Lettuce
Lactuca	*sativa-longifolia-cosa-larga-media*	*parris-islandii-318*	#318 Parris Island	9.1"-12" Medium Large Cos	Lettuce
Lactuca	*sativa-longifolia-cosa-larga-media*	*parris-islandii-454*	#454 Parris Island	9.1"-12" Medium Large Cos	Lettuce
Lactuca	*sativa-longifolia-cosa-larga-media*	*parris-islandii-green-towersi*	Green Towers Parris Island	9.1"-12" Medium Large Cos	Lettuce
Lactuca	*sativa-longifolia-cosa-larga-media*	*parris-islandii-medallionii*	Medallion Parris Island	9.1"-12" Medium Large Cos	Lettuce
Lactuca	*sativa-longifolia-cosa-larga-media*	*parris-islandii-remus*	Remus Parris Island	9.1"-12" Medium Large Cos	Lettuce
Lactuca	*sativa-longifolia-cosa-larga-media*	*parris-islandii-romulus*	Romulus Parris Island	9.1"-12" Medium Large Cos	Lettuce
Lactuca	*sativa-longifolia-cosa-larga-media*	*parris-islandii-valmainei*	Valmaine Parris Island	9.1"-12" Medium Large Cos	Lettuce
Lactuca	*sativa-longifolia-cosa-larga-media*	*SSC-1128*	Sweet Selection Cos #1128	9.1"-12" Medium Large Cos	Lettuce
Lactuca	*sativa-longifolia-cosa-media*	*BOS-6160-3*	Tall Red Tinged #6160-3 Boston	6.1"-9" Medium Cos	Lettuce
Lactuca	*sativa-longifolia-cosa-media*	*BOS-6161-2*	Red Tinged #6161-2 Boston	6.1"-9" Medium Cos	Lettuce
Lactuca	*sativa-longifolia-cosa-media*	*BOS-9003*	Prostrate #9003 Boston	6.1"-9" Medium Cos	Lettuce
Lactuca	*sativa-longifolia-cosa-media*	*BOS-9021-fresh-heartii*	Fresh Heart #9021 Boston	6.1"-9" Medium Cos	Lettuce
Lactuca	*sativa-longifolia-cosa-media*	*BOS-9052-conceptii*	Concept #9052 Boston	6.1"-9" Medium Cos	Lettuce
Lactuca	*sativa-longifolia-cosa-media*	*BOS-9079*	Tall #9079 Boston	6.1"-9" Medium Cos	Lettuce
Lactuca	*sativa-longifolia-cosa-media*	*BOS-9114*	Blistered #9114 Boston	6.1"-9" Medium Cos	Lettuce
Lactuca	*sativa-longifolia-cosa-media*	*broad-swordii*	Broad Sword	6.1"-9" Medium Cos	Lettuce

Genus	species	infraspecies	cultivar	specimen	Genome
Lactuca	*sativa-longifolia-cosa-media*	*brown-goldingii*	Brown Golding	6.1"-9" Medium Cos	Lettuce
Lactuca	*sativa-longifolia-cosa-media*	*devil's-tonguei*	Devil's Tongue Red	6.1"-9" Medium Cos	Lettuce
Lactuca	*sativa-longifolia-cosa-media*	*darkland-cosi*	Devil's Tongue Red	6.1"-9" Medium Cos	Lettuce
Lactuca	*sativa-longifolia-cosa-media*	*floricosi*	Floricos	6.1"-9" Medium Cos	Lettuce
Lactuca	*sativa-longifolia-cosa-media*	*floricosi-83*	#83 Floricos	6.1"-9" Medium Cos	Lettuce
Lactuca	*sativa-longifolia-cosa-media*	*platoi*	Plato	6.1"-9" Medium Cos	Lettuce
Lactuca	*sativa-longifolia-cosa-media*	*platoi-2*	Plato II	6.1"-9" Medium Cos	Lettuce
Lactuca	*sativa-longifolia-cosa-media*	*red-splash-cosi*	Plato II	6.1"-9" Medium Cos	Lettuce
Lactuca	*sativa-longifolia-cosa-media*	*romancei*	Romance	6.1"-9" Medium Cos	Lettuce
Lactuca	*sativa-longifolia-cosa-media*	*vaux-self-folding*	Vaux Self Folding	6.1"-9" Medium Cos	Lettuce
Lactuca	*sativa-longifolia-cosa-parva*	*baby-greenii*	Baby Green	2"-6" Small Cos Looseleaf	Lettuce
Lactuca	*sativa-longifolia-cosa-parva*	*little-gemii-cosi*	Little Gem	2"-6" Small Cos Looseleaf	Lettuce
Lactuca	*sativa-longifolia-cosa-parva*	*little-gemii-bubblesi*	Bubbles Little Gem	2"-6" Small Cos Looseleaf	Lettuce
Lactuca	*sativa-longifolia-cosa-rubra*	*red-cosi*	Heirloom	Red Cos Looseleaf	Lettuce
Lactuca	*sativa-longifolia-cosa-rubra*	*red-eye-cosi*	Red Eye	Red Cos Looseleaf	Lettuce
Lactuca	*sativa-longifolia-cosa-rubra*	*rustyi*	Rusty	Red Cos Looseleaf	Lettuce
Lactuca	*sativa-longifolia-cosa-rubra*	*wavy-red-cosi*	Wavy	Red Cos Looseleaf	Lettuce

Endive (*Cichorium endivia*)

Genus	species	infraspecies	cultivar	specimen	Genome
Cichorium	*endivia*	*endivia*	Original	Endive	Chicory
Cichorium	*endivia*	*bianca-riccia*	Riccia Bianca	Endive	Chicory
Cichorium	*endivia*	*bianca-saladii*	Salad Bianca	Endive	Chicory
Cichorium	*endivia*	*bossa*	Bossa	Endive	Chicory
Cichorium	*endivia*	*bubikopfii-3*	#3 Bubikopf	Endive	Chicory
Cichorium	*endivia*	*coralis*	Coral	Endive	Chicory
Cichorium	*endivia*	*elodiei*	Elodie	Endive	Chicory
Cichorium	*endivia*	*erosi*	Eros	Endive	Chicory
Cichorium	*endivia*	*frisee-bellesquei*	Bellesque Frisee	Endive	Chicory
Cichorium	*endivia*	*frisee-fine-curledii*	Fine Curled Frisee	Endive	Chicory
Cichorium	*endivia*	*frisee-fine-curledii-louviersi*	Louviers Fine Curled Frisee	Endive	Chicory
Cichorium	*endivia*	*frisee-frisanii*	Frisan Frisee	Endive	Chicory
Cichorium	*endivia*	*frisee-galia*	Galia Frisee	Endive	Chicory
Cichorium	*endivia*	*frisee-leafy-endivei*	Leafy Frisee	Endive	Chicory
Cichorium	*endivia*	*full-heartii-salad-kingii*	Salad King Full Heart	Endive	Chicory
Cichorium	*endivia*	*full-heartii-salad-kingii-improvedii*	Improved Salad King Full Heart	Endive	Chicory
Cichorium	*endivia*	*gloire-de-l'expositionii*	Gloire De L'Exposition	Endive	Chicory
Cichorium	*endivia*	*grosse-boulceei*	Grosse Bouclee	Endive	Chicory
Cichorium	*endivia*	*large-green-curled-white-ribii*	Large Green Curled White Rib	Endive	Chicory
Cichorium	*endivia*	*lorcai*	Lorca	Endive	Chicory
Cichorium	*endivia*	*markantii*	Markant	Endive	Chicory
Cichorium	*endivia*	*moss-curledii*	Moss Curled	Endive	Chicory
Cichorium	*endivia*	*natalyi*	Nataly	Endive	Chicory
Cichorium	*endivia*	*neosi*	Neos	Endive	Chicory
Cichorium	*endivia*	*pancalieri-gradoi*	Grado Pancalier	Endive	Chicory
Cichorium	*endivia*	*pancalieri-green-curledii*	Green Curled Pancalier	Endive	Chicory
Cichorium	*endivia*	*pax-1003*	PAX #1003	Endive	Chicory
Cichorium	*endivia*	*perfectii*	Perfect	Endive	Chicory
Cichorium	*endivia*	*red-endivei*	Red	Endive	Chicory
Cichorium	*endivia*	*riccia-coure-d'oroi*	Coure D'Oro Riccia	Endive	Chicory
Cichorium	*endivia*	*riccia-pancalieri*	Pancalier Riccia	Endive	Chicory
Cichorium	*endivia*	*riccia-romanesco-da-taglioi*	Romanesco Da Taglio Riccia	Endive	Chicory
Cichorium	*endivia*	*riccia-verde*	Verde Riccia	Endive	Chicory
Cichorium	*endivia*	*salancai*	Salanca	Endive	Chicory
Cichorium	*endivia*	*samanthai*	Samantha	Endive	Chicory
Cichorium	*endivia*	*taglioi*	Taglio	Endive	Chicory
Cichorium	*endivia*	*tres-fine-maraicherei*	Tres Fine Marachiere	Endive	Chicory
Cichorium	*endivia*	*tres-fine-maraicherei-coquettei*	Coquette Tres Fine Marachiere	Endive	Chicory
Cichorium	*endivia*	*tres-fine-maraicherei-rhodos*	Rhodos Tres Fine Marachiere	Endive	Chicory
Cichorium	*endivia*	*tres-fine-maraicherei-tosca*	Tosca Tres Fine Marachiere	Endive	Chicory
Cichorium	*endivia*	*traviata*	Traviata	Endive	Chicory
Cichorium	*endivia*	*wallonei*	Wallone	Endive	Chicory
Cichorium	*endivia*	*wallonei-tasosi*	Tasos Wallone	Endive	Chicory
Cichorium	*endivia*	*white-curledii*	White Curled	Endive	Chicory
Cichorium	*endivia*	*wivolii*	Wivol	Endive	Chicory

Escarole (*Cichorium endivia-escarolea*)

Genus	species	infraspecies	cultivar	specimen	Genome
Cichorium	*endivia-escarolea*	*escarolea*	Original	Escarole	Chicory
Cichorium	*endivia-escarolea*	*cornet-d'anjoui*	Cornet D'Anjou	Escarole	Chicory
Cichorium	*endivia-escarolea*	*cornet-d'bordeauxii*	Cornet De Bordeaux	Escarole	Chicory
Cichorium	*endivia-escarolea*	*full-heartii-batavia*	Batavian Full Heart	Escarole	Chicory
Cichorium	*endivia-escarolea*	*full-heartii-batavia-broad-leavedii*	Broad Leaved Batavian Full Heart	Escarole	Chicory
Cichorium	*endivia-escarolea*	*full-heartii-batavia-broad-leavedii-5*	#5 Broad Leaved Batavian Full Heart	Escarole	Chicory

Genus	species	infraspecies	cultivar	specimen	Genome
Cichorium	*endivia-escarolea*	*full-heartii-a-cuore-pienoi-bionda*	Bionda A Coure Pieno Full Heart	Escarole	Chicory
Cichorium	*endivia-escarolea*	*full-heartii-a-cuore-pienoi-verde*	Verde A Coure Pieno Full Heart	Escarole	Chicory
Cichorium	*endivia-escarolea*	*full-heartii-floridana*	Florida Full Heart	Escarole	Chicory
Cichorium	*endivia-escarolea*	*full-heartii-floridana-deep-heartii*	Deep Heart Florida Full Heart	Escarole	Chicory
Cichorium	*endivia-escarolea*	*full-heartii-presidentii*	President Full Heart	Escarole	Chicory
Cichorium	*endivia-escarolea*	*full-heartii-ruffec-green-curled-rubblei*	Rubble Green Curled Ruffec Full Heart	Escarole	Chicory
Cichorium	*endivia-escarolea*	*full-heartii-ruffec-green-curled-ribbedii*	Ribbed Green Curled Ruffec Full Heart	Escarole	Chicory
Cichorium	*endivia-escarolea*	*headingii*	Heading	Escarole	Chicory
Cichorium	*endivia-escarolea*	*sincoi*	Sinco	Escarole	Chicory

Chicory, Succory (*Cichorium intybus*)

Genus	species	infraspecies	cultivar	specimen	Genome
Cichorium	*intybus*	*intybus*	Original	Succory	Chicory
Cichorium	*intybus*	*belgian-di-bruxelles*	Di Bruxelles Belgian	Succory	Chicory
Cichorium	*intybus*	*catalogna-a-foglia-frastagliata*	A Foglia Frastagliata Catalogna	Succory	Chicory
Cichorium	*intybus*	*catalogna-a-foglia-liscia*	A Foglia Lisca Catalogna	Succory	Chicory
Cichorium	*intybus*	*catalogna-italian-dandelion-redii*	Red Italian Dandelion Catalogna	Succory	Chicory
Cichorium	*intybus*	*catalogna-italian-dandelion-red-dandyi*	Dandy Red Italian Dandelion Catalogna	Succory	Chicory
Cichorium	*intybus*	*catalogna-italian-dandelion-red-ribbedii*	Ribbed Red Italian Dandelion Catalogna	Succory	Chicory
Cichorium	*intybus*	*catalogna-italian-dandelion-san-pasqualei*	San Pasquale Italian Dandelion Catalogna	Succory	Chicory
Cichorium	*intybus*	*catalogna-italian-dandelion-san-pasqualei-baxter's-specialis*	Baxter's Special San Pasquale Italian Dandelion Catalogna	Succory	Chicory
Cichorium	*intybus*	*catalogna-della-radichettai*	Della Radichette Catalogna	Succory	Chicory
Cichorium	*intybus*	*catalogna-pugliesei*	Pugliese Catalogna	Succory	Chicory
Cichorium	*intybus*	*catalogna-specialis*	Special Catalogna	Succory	Chicory
Cichorium	*intybus*	*chioggia-carmenii*	Carmen Chioggia	Succory	Chicory
Cichorium	*intybus*	*chioggia-marina*	Marina Chioggia	Succory	Chicory
Cichorium	*intybus*	*chioggia-marina-sottoi*	Sotto Marina Chioggia	Succory	Chicory
Cichorium	*intybus*	*chioggia-palla-rossa-di-fuocoi*	Di Fuoco Palla Rossa Chioggia	Succory	Chicory
Cichorium	*intybus*	*chioggia-precosei*	Precose Chioggia	Succory	Chicory
Cichorium	*intybus*	*chioggia-rubico*	Rubico Chioggia	Succory	Chicory
Cichorium	*intybus*	*chioggia-variegata*	Variegated Chioggia	Succory	Chicory
Cichorium	*intybus*	*chioggia-variegata-dicastlefranco*	Castle France Variegated Chioggia	Succory	Chicory
Cichorium	*intybus*	*franchi-bianca-a-bergamoi*	Franchi Bianca A Bergamo	Succory	Chicory
Cichorium	*intybus*	*friseei*	Frisee	Succory	Chicory
Cichorium	*intybus*	*friseei-cour-d'oroi*	Cour D'Oro Frisee	Succory	Chicory
Cichorium	*intybus*	*gromolo-biondai*	Bionda Grumolo	Succory	Chicory
Cichorium	*intybus*	*gromolo-biondo-goldenii*	Biondo Golden Grumolo	Succory	Chicory
Cichorium	*intybus*	*gromolo-cerioloi*	Ceriole Gromolo	Succory	Chicory
Cichorium	*intybus*	*gromolo-verdei*	Verde Grumolo	Succory	Chicory
Cichorium	*intybus*	*italian-barba-di-cappuccino*	Barba Di Capuccino Italian	Succory	Chicory
Cichorium	*intybus*	*italian-bionda-foglie-larghei*	Bionda A Foglie Larghe Italian	Succory	Chicory
Cichorium	*intybus*	*italian-large-leaf(folha-larga)*	Large Leaf Italian	Succory	Chicory
Cichorium	*intybus*	*italian-mantovana*	Mantovana Italian	Succory	Chicory
Cichorium	*intybus*	*italian-montmagnyi*	Montmagny Italian	Succory	Chicory
Cichorium	*intybus*	*italian-orchidea-rossai*	Orchidea Rossa Italian	Succory	Chicory
Cichorium	*intybus*	*italian-red-augustoi*	Augusto Red Italian	Succory	Chicory
Cichorium	*intybus*	*italian-selvatica-di-campoi*	Selvatica Di Campo Italian	Succory	Chicory
Cichorium	*intybus*	*italian-trieste-sweetii*	Triste Sweet Italian	Succory	Chicory
Cichorium	*intybus*	*italian-triestina-da-taglio*	Triestina Da Taglio Italian	Succory	Chicory
Cichorium	*intybus*	*italian-variegata-di-lusia*	Variegata Di Lusia Italian	Succory	Chicory
Cichorium	*intybus*	*milano-bianca-di-milanoi*	Bianca Di Milano	Succory	Chicory
Cichorium	*intybus*	*milano-milanesei*	Milanese Milano	Succory	Chicory
Cichorium	*intybus*	*palla-rossa*	Palla Rossa	Succory	Chicory
Cichorium	*intybus*	*palla-rossa-precocei*	Precose Palla Rossa	Succory	Chicory
Cichorium	*intybus*	*palla-rossa-specialis*	Special Palla Rossa	Succory	Chicory
Cichorium	*intybus*	*radicchio-alouttea*	Alouette Radicchio	Succory	Chicory
Cichorium	*intybus*	*radicchio-beaconii*	Beacon Radicchio	Succory	Chicory
Cichorium	*intybus*	*radicchio-castelfrancoi*	Castelfranco Radicchio	Succory	Chicory
Cichorium	*intybus*	*radicchio-castelfrancoi-librai*	Libra Castelfranco Radicchio	Succory	Chicory
Cichorium	*intybus*	*radicchio-castelfrancoi-variegata*	Variegata Castelfranco Radicchio	Succory	Chicory
Cichorium	*intybus*	*radicchio-fieroi*	Fiero Radicchio	Succory	Chicory
Cichorium	*intybus*	*radicchio-firebirdii*	Firebird Radicchio	Succory	Chicory
Cichorium	*intybus*	*radicchio-fossa-verona-tardivai*	Fossa Verona Tardiva Radicchio	Succory	Chicory
Cichorium	*intybus*	*radicchio-indigoi*	Indigo Radicchio	Succory	Chicory
Cichorium	*intybus*	*radicchio-infernoi*	Inferno Radicchio	Succory	Chicory
Cichorium	*intybus*	*radicchio-livrettei*	Livrette Radicchio	Succory	Chicory
Cichorium	*intybus*	*radicchio-melrosei*	Melrose Radicchio	Succory	Chicory
Cichorium	*intybus*	*radicchio-milanensis*	Milan Radicchio	Succory	Chicory
Cichorium	*intybus*	*radicchio-palla-rossa-3*	#3 Palla Rossa Radicchio	Succory	Chicory
Cichorium	*intybus*	*radicchio-palla-rossa-agenai*	Agena Palla Rossa Radicchio	Succory	Chicory
Cichorium	*intybus*	*radicchio-prima-rossa*	Prima Rossa Radicchio	Succory	Chicory
Cichorium	*intybus*	*radicchio-red-devilii*	Red Devil Radicchio	Succory	Chicory
Cichorium	*intybus*	*radicchio-romainei*	Romaine Radicchio	Succory	Chicory
Cichorium	*intybus*	*radicchio-treviso-rossa-dii*	Rossa Di Treviso Raddichio	Succory	Chicory

Genus	species	infraspecies	cultivar	specimen	Genome
Cichorium	*intybus*	*radicchio-treviso-rpssa-dii-earlyi*	Early Rossa Di Treviso Raddichio	Succory	Chicory
Cichorium	*intybus*	*radicchio-treviso-rossa-dii-palla-tardiva*	Tardiva Palla Rossa Di Treviso Radicchio	Succory	Chicory
Cichorium	*intybus*	*radicchio-treviso-rossa-sveltai*	Svelta Rossa Di Treviso Raddichio	Succory	Chicory
Cichorium	*intybus*	*radicchio-verona-rossa-dii*	Rossa Di Vernoa Radicchio	Succory	Chicory
Cichorium	*Intybus*	*radicchio-verona-rossa-dii-arca*	Arca Rossa Di Vernoa Radicchio	Succory	Chicory
Cichorium	*intybus*	*radicchio-verona-rossa-dii-cesarei*	Cesare Rossa Di Vernoa Radicchio	Succory	Chicory
Cichorium	*intybus*	*radicchio-verona-rossa-dii-foglia-larga*	Foglia Larga Rossa Di Vernoa Radicchio	Succory	Chicory
Cichorium	*intybus*	*radicchio-verona-rossa-dii-guilio*	Guilio Rossa Di Vernoa Radicchio	Succory	Chicory
Cichorium	*intybus*	*radicchio-verona-rossa-dii-palla*	Palla Rossa Di Vernoa Radicchio	Succory	Chicory
Cichorium	*intybus*	*radicchio-verona-rossa-dii-rouge*	Rouge Rossa Di Vernoa Radicchio	Succory	Chicory
Cichorium	*intybus*	*radicchio-rubello*	Rubello Radicchio	Succory	Chicory
Cichorium	*intybus*	*radicchio-rubicon*	Rubicon Radicchio	Succory	Chicory
Cichorium	*intybus*	*radicchio-vesuvioi*	Vesuvio Radicchio	Succory	Chicory
Cichorium	*intybus*	*radicchio-wildfirei*	Wildfire Radichhio	Succory	Chicory
Cichorium	*intybus*	*sugar-loaf-crystal-hatii*	Crystal Hat Sugar Loaf	Succory	Chicory
Cichorium	*intybus*	*sugar-loaf-francoi*	French Sugar Loaf	Succory	Chicory
Cichorium	*intybus*	*sugar-loaf-greenlofii*	Greenlof Sugar Loaf	Succory	Chicory
Cichorium	*intybus*	*sugar-loaf-pan-di-zuccheroi*	Pan Di Zucherro Sugar Loaf	Succory	Chicory
Cichorium	*intybus*	*sugar-loaf-pan-di-zuccheroi-borca*	Borca Pan Di Zucherro Sugar Loaf	Succory	Chicory
Cichorium	*intybus*	*sugar-loaf-snowflakei*	Snowflake Sugar Loaf	Succory	Chicory
Cichorium	*intybus*	*sugar-loaf-spadona-da-taglioi*	Da Taglio Spadona Sugar Loaf	Succory	Chicory
Cichorium	*intybus*	*sugar-loaf-spadona-dog's-tonguei*	Dog's Tongue Spadona Sugar Loaf	Succory	Chicory
Cichorium	*Intybus*	*whiteleaf-brusselsensis*	Brussels Witloof	Succory	Chicory
Cichorium	*intybus*	*whiteleaf-improvedii*	Improved Witloof	Succory	Chicory
Cichorium	*intybus*	*whiteleaf-mechelsei*	Mechelse Witloof	Succory	Chicory
Cichorium	*intybus*	*whiteleaf-R'C'*	Red Counterpart Witloof	Succory	Chicory
Cichorium	*intybus-ruta*	*intybus-ruta*	Original	Root Succory	Chicory
Cichorium	*intybus-ruta*	*coffee-rooted-chichoryi*	Coffee	Root Succory	Chicory
Cichorium	*intybus-ruta*	*coffee-large-rootedii*	Large Rooted Coffee	Root Succory	Chicory
Cichorium	*intybus-ruta*	*coffee-magdeburgii*	Magdeburg Coffee	Root Succory	Chicory
Cichorium	*intybus-ruta*	*soncinoi*	Soncino	Root Succory	Chicory
Cichorium	*intybus-stipes*	*brindisina*	Brindisina	Stem Succory	Chicory
Cichorium	*intybus-stipes*	*catalogna-asparagus-chicoryi*	Asparagus Catalogna	Stem Succory	Chicory
Cichorium	*intybus-stipes*	*catalogna-asparagus-dentarella*	Dentarella Asparagus Catalogna	Stem Succory	Chicory
Cichorium	*intybus-stipes*	*catalogna-asparagus-italico-rossoi*	Italico Rosso Asparagus Catalogna	Stem Succory	Chicory
Cichorium	*intybus-stipes*	*catalogna-asparagus-radichettai*	Radichetta Asparagus Catalogna	Stem Succory	Chicory
Cichorium	*intybus-stipes*	*galatina*	Galatina	Stem Succory	Chicory
Cichorium	*intybus-stipes*	*galatina-abruzzese-o-di-galatina*	Abruzzese O Di Galatina	Stem Succory	Chicory
Cichorium	*intybus-stipes*	*puntarella*	Puntarella	Stem Succory	Chicory

Scorzon, Black Salsify (*Scorzonera hispanica*)

Genus	species	infraspecies	cultivar	specimen	Genome
Scorzonera	*hispanica*	*hispanica*	Original ?.?"	Black Salsify	Scorzon
Scorzonera	*hispanica*	*belstar-superi*	Belstar Super 10."	Black Salsify	Scorzon
Scorzonera	*hispanica*	*duplex*	Duplex 9.0"	Black Salsify	Scorzon
Scorzonera	*hispanica*	*geante-noire-de-russiei*	Geante Noir De Russie	Black Salsify	Scorzon
Scorzonera	*hispanica*	*gigantea-russica*	Giant Russian ?.?"	Black Salsify	Scorzon
Scorzonera	*hispanica*	*hoffman's-lange-pfahl*	Hoffman's Lange Pfahl	Black Salsify	Scorzon
Scorzonera	*hispanica*	*lange-janii*	Lange Jan ?.?"	Black Salsify	Scorzon
Scorzonera	*hispanica*	*long-blackii*	Long ?.?"	Black Salsify	Scorzon

Salsify, Oyster (*Tragopogon porrifolius*)

Genus	species	infraspecies	cultivar	specimen	Genome
Tragopogon	*porrifolius*	*porrifolius*	Original		Oyster Salsify
Tragopogon	*porrifolius*	*hawwerwurtzelii*	Hawwerwurtzel 7.0"		Oyster Salsify
Tragopogon	*porrifolius*	*mammoth-whitei*	White Mammoth ?.?"		Oyster Salsify
Tragopogon	*porrifolius*	*mammoth-sandwich-islandii*	Sandwich Island Mammoth 8.5"		Oyster Salsify

Family 7 *Pappaceae* Groundsel
Family 8 *Radiasteraceae* Ray Aster

Class 724.02 *Caryophyllopsida* Succulent

Order 1 *Caryophyllales* Succulent
Family 1 *Caryophyllaceae* Succulent

Family 724.021.2 *Aizoaceae* Ice Plant

Carnation (*Dianthus caryophyllus*)

Genus	Species	Specimen Genome	C:Culinary, M:Medicinal, E:Economic
Dianthus	*caryophyllus*	Carnation Pink	C:FL:AF,BF,**VG**
Tetragonia	*tetragonioides*	New Zealand Spinach Ice Plant	C:LF:**VG**

Varieties

Spinach Ice Plant, New Zealand (*Tetragonia tetragonoides*)

Genus	species	infraspecies	cultivar	specimen	Genome
Tetragonia	*tetragonioides*	*tetragonioides*	Original	New Zealand	Spinach Ice Plant
Tetragonia	*tetragonioides*	*expansa*	Warragal Greens	New Zealand	Spinach Ice Plant

Family 724.021.3 *Amaranthaceae* Amaranth

Genus	Species	Specimen Genome	C:Culinary, M:Medicinal
Amaranthus	*blitum*	Purple Amaranth	C:LF:**VG**
Amaranthus	*caudatus*	Tassel Flower Amaranth	C:LF:**VG**
Amaranthus	*chihuahuensis*	Chihuahua Greens Amaranth	C:LF:**VG**
Amaranthus	*cristata*	Cockscomb Amaranth	C:LF:**VG**
Amaranthus	*cruentus*	Red Amaranth	C:LF:**VG**
Amaranthus	*cruentus x powellii*	Pigweed Red Amaranth	C:LF:**VG**
Amaranthus	*edulis*	Edible Amaranth	C:LF:**VG**
Amaranthus	*gangeticus*	Ganges Spinach Amaranth	C:LF:**VG**
Amaranthus	*hypochondriacus*	Cereal Grain Amaranth	C:SD:**VG**
Amaranthus	*hypochondriacus x hybridus*	Pigweed Cereal Grain Amaranth	C:SD:**VG**
Amaranthus	*mangostana*	Tiger Leaf Amaranth	C:LF:**VG**
Amaranthus	*retroflexus*	Redroot Pigweed Amaranth	C:LF:**VG**
Amaranthus	*tricolor*	Joseph's Coat Spinach Amaranth	C:LF:**VG**
Amaranthus	*viridis*	Green Amaranth	C:LF:**VG**
Achyranthes	*aspera*	Devils' Horsewhip Chaff Flower	C:LF:**VG**

Varieties

Amaranth (*Amaranthus spp.*)

Genus	species	infraspecies	cultivar	specimen Genome
Amaranthus	*blitum*	*blitum*	Original	Purple Amaranth
Amaranthus	*blitum*	*ascendens*	Ascending	Purple Amaranth
Amaranthus	*blitum*	*inchoii*	Inchoi	Purple Amaranth
Amaranthus	*blitum*	*lividus*	Livid	Purple Amaranth
Amaranthus	*blitum*	*polygonoides*	Polygon	Purple Amaranth
Amaranthus	*caudatus*	*caudatus*	Original	Tassel Flower Amaranth
Amaranthus	*caudatus*	*love-lies-bleeding-greenii*	Green Love Lies Bleeding	Tassel Flower Amaranth
Amaranthus	*caudatus*	*love-lies-bleeding-redii*	Red Love Lies Bleeding	Tassel Flower Amaranth
Amaranthus	*caudatus*	*pendens*	Pink Love Lies Bleeding	Tassel Flower Amaranth
Amaranthus	*caudatus*	*pinangensis*	Pinang	Tassel Flower Amaranth
Amaranthus	*caudatus*	*putehii*	Puteh	Tassel Flower Amaranth
Amaranthus	*caudatus*	*viridis-pollex*	Green Thumb	Tassel Flower Amaranth
Amaranthus	*caudatus*	*viridis-urophyllus*	Green Tails	Tassel Flower Amaranth
Amaranthus	*cristata*	*cristata*	Original	Cockscomb Amaranth
Amaranthus	*cristata*	*mano-de-obispo*	Mano De Obispo	Cockscomb Amaranth
Amaranthus	*cruentus*	*cruentus*	Original	Red Amaranth
Amaranthus	*cruentus*	*alba-sperma*	White Seeded	Red Amaranth
Amaranthus	*cruentus*	*alegria*	Alegria	Red Amaranth
Amaranthus	*cruentus*	*golden-giantii*	Golden Giant	Red Amaranth
Amaranthus	*cruentus*	*himalayan-handsi*	Himalayan Hands	Red Amaranth
Amaranthus	*cruentus*	*kinnauri-dhankari*	Kinnauri Dhankar	Red Amaranth
Amaranthus	*cruentus*	*manalii*	Manali	Red Amaranth
Amaranthus	*cruentus*	*mayo-indian-grainii*	Mayo Indian Grain	Red Amaranth
Amaranthus	*cruentus*	*mexican-grainii*	Mexican Grain	Red Amaranth
Amaranthus	*cruentus*	*morai*	Mora	Red Amaranth
Amaranthus	*cruentus*	*multicolori*	Multicolor	Red Amaranth
Amaranthus	*cruentus*	*nepalensis*	Nepalese	Red Amaranth
Amaranthus	*cruentus*	*opopeoensis*	Opopeo	Red Amaranth

Genus	species	infraspecies	cultivar	specimen	Genome
Amaranthus	*cruentus*	*orange-giantii*	Orange Giant	Red	Amaranth
Amaranthus	*cruentus*	*oscar-blancoi*	Oscar Blanco	Red	Amaranth
Amaranthus	*cruentus*	*paiutei*	Paiute	Red	Amaranth
Amaranthus	*cruentus*	*paniculatus*	Prince's Feather	Red	Amaranth
Amaranthus	*cruentus*	*poppingii*	Popping	Red	Amaranth
Amaranthus	*cruentus*	*rama's-fingeri*	Rama's Finger	Red	Amaranth
Amaranthus	*cruentus*	*red-amaranth-R158*	Refined #158	Red	Amaranth
Amaranthus	*cruentus*	*southern-tepehuan-bautei*	Southern Tepehuan Baute	Red	Amaranth
Amaranthus	*cruentus*	*vietnamese-redii*	Vietnamese	Red	Amaranth
Amaranthus	*cruentus x powellii*	*powellii*	Original	Pigweed Red	Amaranth
Amaranthus	*cruentus x powellii*	*hopi-red-dyei*	Hopi Red Dye	Pigweed Red	Amaranth
Amaranthus	*edulis*	*edulis*	Original	Edible	Amaranth
Amaranthus	*edulis*	*black-leafii*	Black Leaf	Edible	Amaranth
Amaranthus	*edulis*	*boliviensis-153*	Boliva #153	Edible	Amaranth
Amaranthus	*edulis*	*calilui*	Calilu	Edible	Amaranth
Amaranthus	*edulis*	*goldenii*	Golden	Edible	Amaranth
Amaranthus	*edulis*	*groena-folia*	Green Leaf	Edible	Amaranth
Amaranthus	*edulis*	*groena-summa*	Green Top	Edible	Amaranth
Amaranthus	*edulis*	*kalilui*	Kalilu	Edible	Amaranth
Amaranthus	*edulis*	*MT-3*	MT-3	Edible	Amaranth
Amaranthus	*edulis*	*prima-nepalensis*	Prima Nepal	Edible	Amaranth
Amaranthus	*edulis*	*red-garnetii*	Red Garnet	Edible	Amaranth
Amaranthus	*edulis*	*red-leafii*	Red Leaf	Edible	Amaranth
Amaranthus	*edulis*	*red-leafii-grainii*	Red Leaf Grain	Edible	Amaranth
Amaranthus	*edulis*	*rodalei-K432*	Rodale K432	Edible	Amaranth
Amaranthus	*edulis*	*rose-beautyi*	Rose Beauty	Edible	Amaranth
Amaranthus	*edulis*	*sunrisei*	Sunrise	Edible	Amaranth
Amaranthus	*edulis*	*sunset-orangei*	Sunset Orange	Edible	Amaranth
Amaranthus	*edulis*	*tarahumara-okitei*	Tarahumara Okite	Edible	Amaranth
Amaranthus	*edulis*	*tender-leafii*	Tender Leaf	Edible	Amaranth
Amaranthus	*edulis*	*tiger-eyei*	Tiger Eye	Edible	Amaranth
Amaranthus	*edulis*	*white-leafii*	White Leaf	Edible	Amaranth
Amaranthus	*gangeticus*	*gangeticus*	Original	Ganges Spinach	Amaranth
Amaranthus	*gangeticus*	*elephant's-headii*	Elephants' Head	Ganges Spinach	Amaranth
Amaranthus	*gangeticus*	*fote-tei*	Fote Te	Ganges Spinach	Amaranth
Amaranthus	*gangeticus*	*graeca*	Greek	Ganges Spinach	Amaranth
Amaranthus	*gangeticus*	*hartman's-giantii*	Hartman's Giant	Ganges Spinach	Amaranth
Amaranthus	*gangeticus*	*joseph's-coatii*	Joseph's Coat	Ganges Spinach	Amaranth
Amaranthus	*gangeticus*	*kahulu-leafii*	Kahulu	Ganges Spinach	Amaranth
Amaranthus	*gangeticus*	*lotus-purplei*	Kahulu	Ganges Spinach	Amaranth
Amaranthus	*gangeticus*	*molten-firei*	Molten Fire	Ganges Spinach	Amaranth
Amaranthus	*gangeticus*	*red-stripe-leafii*	Red Stripe Leaf	Ganges Spinach	Amaranth
Amaranthus	*gangeticus*	*ruby-redii*	Ruby Red	Ganges Spinach	Amaranth
Amaranthus	*gangeticus*	*vletai*	Vleta	Ganges Spinach	Amaranth
Amaranthus	*gangeticus*	*white-round-leafii*	White Round Leaf	Ganges Spinach	Amaranth
Amaranthus	*hypochondriacus*	*hypochondriacus*	Original	Cereal Grain	Amaranth
Amaranthus	*hypochondriacus*	*alba-sperma*	White Seeded	Cereal Grain	Amaranth
Amaranthus	*hypochondriacus*	*all-redii*	All Red	Cereal Grain	Amaranth
Amaranthus	*hypochondriacus*	*burgundia*	Burgundy	Cereal Grain	Amaranth
Amaranthus	*hypochondriacus*	*dreadicus-san-martinii*	San Martin	Cereal Grain	Amaranth
Amaranthus	*hypochondriacus*	*dreadicus-mercadoi*	Mercado	Cereal Grain	Amaranth
Amaranthus	*hypochondriacus*	*golden-grainii*	Golden Grain	Cereal Grain	Amaranth
Amaranthus	*hypochondriacus*	*intense-purplei*	Intense Purple	Cereal Grain	Amaranth
Amaranthus	*hypochondriacus*	*leucocarpus*	Prince Of Wales Feather	Cereal Grain	Amaranth
Amaranthus	*hypochondriacus*	*mannai*	Manna	Cereal Grain	Amaranth
Amaranthus	*hypochondriacus*	*neomexicana*	New Mexico	Cereal Grain	Amaranth
Amaranthus	*hypochondriacus*	*plainsmanii*	Plainsman	Cereal Grain	Amaranth
Amaranthus	*hypochondriacus*	*polandica*	Polish	Cereal Grain	Amaranth
Amaranthus	*hypochondriacus*	*pygmaea-fax*	Pygmy Torch	Cereal Grain	Amaranth
Amaranthus	*hypochondriacus*	*rio-san-lorenzoensis*	Rio San Lorenzo	Cereal Grain	Amaranth
Amaranthus	*hypochondriacus*	*scarlet-torchii*	Scarlet Torch	Cereal Grain	Amaranth
Amaranthus	*hypochondriacus*	*viridis-pollex*	Green Thumb	Cereal Grain	Amaranth
Amaranthus	*hypochondriacus*	*warihioi*	Warihio	Cereal Grain	Amaranth
Amaranthus	*hypochondriacus x hybridus*	*hybridus*	Original	Pigweed Cereal Grain	Amaranth
Amaranthus	*hypochondriacus x hybridus*	*guarijioi*	Guarijio Indian	Pigweed Cereal Grain	Amaranth
Amaranthus	*tricolor*	*tricolor*	Original	Joseph's Coat Spinach	Amaranth
Amaranthus	*tricolor*	*bayamii*	Bayam	Joseph's Coat Spinach	Amaranth
Amaranthus	*tricolor*	*chinensis*	Chinese	Joseph's Coat Spinach	Amaranth
Amaranthus	*tricolor*	*chinensis-hijaui*	Hijau Chinese	Joseph's Coat Spinach	Amaranth
Amaranthus	*tricolor*	*chinensis-hinn-choi*	Hinn Choi Chinese	Joseph's Coat Spinach	Amaranth
Amaranthus	*tricolor*	*early-spendori*	Early Splendor	Joseph's Coat Spinach	Amaranth
Amaranthus	*tricolor*	*illumination*	Illumination	Joseph's Coat Spinach	Amaranth
Amaranthus	*tricolor*	*indica*	India	Joseph's Coat Spinach	Amaranth
Amaranthus	*tricolor*	*merahii*	Merah Magenta	Joseph's Coat Spinach	Amaranth

Genus	species	infraspecies	cultivar	specimen	Genome
Amaranthus	*tricolor*	*mirahii*	Mirah Green	Joseph's Coat Spinach	Amaranth
Amaranthus	*tricolor*	*perfecta*	Perfect	Joseph's Coat Spinach	Amaranth
Amaranthus	*tricolor*	*salicifolius*	Willow Leaf	Joseph's Coat Spinach	Amaranth
Amaranthus	*tricolor*	*salicifolius-flaming-fountainsi*	Flaming Fountains Willow Leaf	Joseph's Coat Spinach	Amaranth
Amaranthus	*tricolor*	*tampalai*	Tampala	Joseph's Coat Spinach	Amaranth
Amaranthus	*viridis*	*viridis*	Original	Green	Amaranth
Amaranthus	*viridis*	*gracilis*	Slender	Green	Amaranth

Devil's Horsewhip (*Achyranthes aspera*)

Genus	species	infraspecies	cultivar	specimen	Genome
Achyranthes	*aspera*	*aspera*	Original	Devils' Horsewhip	Chaff Flower
Achyranthes	*aspera*	*argentea*	Silver	Devils' Horsewhip	Chaff Flower
Achyranthes	*aspera*	*obtusifolia*	Blunt Leaf	Devils' Horsewhip	Chaff Flower

Family 724.021.4 *Basellaceae* Basellvine

Genus	Species	Specimen	Genome	C:Culinary, M:Medicinal, E:Economic
Basella	*alba*	Malabar Spinach	Basellvine	C:LF:**VG**
Basella	*rubra*	Sri Lanka Spinach	Basellvine	C:LF:**VG**
Ullucus	*tuberosus*	Andes Mountain	Ulluco Potato	C:TU:**VG**

Varieties

Malabar Spinach (*Basella alba*)

Genus	species	infraspecies	cultivar	specimen	Genome
Basella	*alba*	*Alba*	Original	Malabar Spinach	Basellvine
Basella	*alba*	*land-kelpii*	Land Kelp	Malabar Spinach	Basellvine
Basella	*alba*	*poi-sagii*	Poi Sag	Malabar Spinach	Basellvine

Sri Lanka Spinach (*Basella rubra*)

Genus	species	infraspecies	cultivar	specimen	Genome
Basella	*rubra*	.		Sri Lanka Spinach	Basellvine

Andes Mountain Ulluco Potato (*Ullucus tuberosus*)

Genus	species	infraspecies	cultivar	specimen	Genome
Ullucus	*tuberosus*	.		Andes Mountain	Ulluco Potato

Family 724.021.5 *Chenopodiaceae* Goosefoot

Genus	Species	Specimen	Genome	C:Culinary
Chenopodium	*album*	Lamb's Quarters	Goosefoot	C:LF:**VG**
Chenopodium	*berlandieri*	Pitseed	Goosefoot	C:LF:**VG**
Chenopodium	*bonus-henricus*	Good King Henry	Goosefoot	C:LF:**VG**
Chenopodium	*capitatum*	Strawberry Spinach	Goosefoot	C:LF:**VG**
Chenopodium	*murale*	Nettle Leaf	Goosefoot	C:LF:**VG**
Chenopodium	*nuttalliae*	Huauzontle Celery	Goosefoot	C:SL:**VG**
Chenopodium	*pallidicaule*	Canihua Grain	Goosefoot	C:SD:**VG**
Chenopodium	*quinoa*	Quinoa Grain	Goosefoot	C:SD:**VG**
Atriplex	*hortensis*	Mountain Spinach	Orache	C:LF:**VG**
Atriplex	*patula*	Spear	Orache	C:LF:**VG**
Atriplex	*prostrata*	Hastate	Orache	C:LF:**VG**
Beta	*vulgaris-cicla-ameria-alba*	White Midrib American Chard	Beet	C:LF:**VG**
Beta	*vulgaris-cicla-ameria-aurantia*	Orange Midrib American Chard	Beet	C:LF:**VG**
Beta	*vulgaris-cicla-ameria-groena*	Green Midrib American Chard	Beet	C:LF:**VG**
Beta	*vulgaris-cicla-ameria-helva*	Yellow Midrib American Chard	Beet	C:LF:**VG**
Beta	*vulgaris-cicla-ameria-multicolori*	Multicolor Midrib American Chard	Beet	C:LF:**VG**
Beta	*vulgaris-cicla-ameria-pinca*	Pink Midrib American Chard	Beet	C:LF:**VG**
Beta	*vulgaris-cicla-ameria-purpurea-rubra*	Purple Red Midrib American Chard	Beet	C:LF:**VG**
Beta	*vulgaris-cicla-ameria-rubra*	Red Midrib American Chard	Beet	C:LF:**VG**
Beta	*vulgaris-cicla-asiatica*	Asian Chard	Beet	C:LF:**VG**
Beta	*vulgaris-cicla-australis*	Australian Chard	Beet	C:LF:**VG**
Beta	*vulgaris-cicla-europa-alba*	White Midrib European Chard	Beet	C:LF:**VG**
Beta	*vulgaris-cicla-europa-groena*	Green Midrib European Chard	Beet	C:LF:**VG**
Beta	*vulgaris-cicla-europa-helva*	Yellow Midrib European Chard	Beet	C:LF:**VG**
Beta	*vulgaris-cicla-europa-rubra*	Red Midrib European Chard	Beet	C:LF:**VG**
Beta	*vulgaris-cicla-svedica-alba*	White Midrib Swiss Chard	Beet	C:LF:**VG**
Beta	*vulgaris-cicla-svedica-groena*	Green Midrib Swiss Chard	Beet	C:LF:**VG**
Beta	*vulgaris-cicla-svedica-helva*	Yellow Midrib Swiss Chard	Beet	C:LF:**VG**
Beta	*vulgaris-cicla-svedica-rubra*	Red Midrib Swiss Chard	Beet	C:LF:**VG**
Beta	*vulgaris-conditiva-alba*	White Globe	Beet	C:BR:**VG**

Genus	Species	Specimen	Genome	C:Culinary
Beta	*vulgaris-conditiva-helva*	Yellow Globe	Beet	C:BR:**VG**
Beta	*vulgaris-conditiva-purpurea*	Purple Globe	Beet	C:BR:**VG**
Beta	*vulgaris-conditiva-purpurea-rubra*	Purple Red Globe	Beet	C:BR:**VG**
Beta	*vulgaris-conditiva-rubra*	Red Globe	Beet	C:BR:**VG**
Beta	*vulgaris-conditiva-rubra-nana*	Baby Red Globe	Beet	C:BR:**VG**
Beta	*vulgaris-crassa-alba*	White Long Sugar	Beet	C:BR:**VG**
Beta	*vulgaris-crassa-helva*	Yellow Long Sugar	Beet	C:BR:**VG**
Beta	*vulgaris-crassa-rubra*	Red Long Sugar	Beet	C:BR:**VG**
Blitum	*foliosum*	Spinach Strawberry	Blite	C:LF:**VG**
Salicornia	*maritima*	Glasswort	Saltwort	C:LF:**VG**
Salicornia	*rubra*	Pacific Glasswort	Saltwort	C:LF:**VG**
Spinacia	*oleracea-aculeasperma*	Smooth Prickly Seed	Spinach	C:LF:**VG**
Spinacia	*oleracea-aculeasperma-savoia*	Savoy Prickly Seed	Spinach	C:LF:**VG**
Spinacia	*oleracea-glabra*	Smooth	Spinach	C:LF:**VG**
Spinacia	*oleracea-savoia*	Savoy	Spinach	C:LF:**VG**
Spinacia	*oleracea-semisavoia*	Semi Savoy	Spinach	C:LF:**VG**

Varieties

Lamb's Quarters (*Chenopodium album*)

Genus	Species	infraspecies	cultivar	specimen	Genome
Chenopodium	*album*	*album*	Original	Lamb's Quarters	Goosefoot
Chenopodium	*album*	*amaranticolor*	Amaranth	Lamb's Quarters	Goosefoot
Chenopodium	*album*	*californicum-gabrielinoi*	Gabrielino California	Lamb's Quarters	Goosefoot
Chenopodium	*album*	*californicum-yokutsi*	Yokuts California	Lamb's Quarters	Goosefoot
Chenopodium	*album*	*candicans*	Virgin White	Lamb's Quarters	Goosefoot
Chenopodium	*album*	*centrorubrum*	Red Center	Lamb's Quarters	Goosefoot
Chenopodium	*album*	*edulis*	Spinach	Lamb's Quarters	Goosefoot
Chenopodium	*album*	*edulis-morton's-selectii*	Morton Select Spinach	Lamb's Quarters	Goosefoot
Chenopodium	*album*	*gabrielino*	Gabriel Spinach	Lamb's Quarters	Goosefoot
Chenopodium	*album*	*giganteum*	Giant Spinach	Lamb's Quarters	Goosefoot
Chenopodium	*album*	*giganteum-magentaspreenii*	MagentaSpreen Giant Spinach	Lamb's Quarters	Goosefoot
Chenopodium	*album*	*lanceolatum*	Lance Leaf	Lamb's Quarters	Goosefoot
Chenopodium	*album*	*microphyllum*	Tiny Leaf	Lamb's Quarters	Goosefoot
Chenopodium	*album*	*missouriense*	Missouri	Lamb's Quarters	Goosefoot
Chenopodium	*album*	*missouriense-paganum*	Pagan Missouri	Lamb's Quarters	Goosefoot
Chenopodium	*album*	*mrs.-hawkin's-spinachii*	Mrs. Hawkin's Spinach	Lamb's Quarters	Goosefoot
Chenopodium	*album*	*polymorphum*	Polymorphic	Lamb's Quarters	Goosefoot
Chenopodium	*album*	*stevensii*	Stevens	Lamb's Quarters	Goosefoot
Chenopodium	*album*	*stevensii-dacoticum*	Stevens	Lamb's Quarters	Goosefoot
Chenopodium	*album*	*stevensii-fallax*	Stevens	Lamb's Quarters	Goosefoot
Chenopodium	*album*	*striatum*	Striped	Lamb's Quarters	Goosefoot
Chenopodium	*album*	*striatum-glaucophyllum*	Glaucous Striped	Lamb's Quarters	Goosefoot
Chenopodium	*album*	*suecicum*	Suez	Lamb's Quarters	Goosefoot
Chenopodium	*album*	*yokutsii*	Yokuts Spinach	Lamb's Quarters	Goosefoot

Pitseed Goosefoot (*Chenopodium berlandieri*)

Genus	Species	infraspecies	cultivar	specimen	Genome
Chenopodium	*berlandieri*	*berlandieri*	Original	Pitseed	Goosefoot
Chenopodium	*berlandieri*	*aztec-red-spinachia*	Aztec Red Spinach	Pitseed	Goosefoot
Chenopodium	*berlandieri*	*boscianum*	Bosc	Pitseed	Goosefoot
Chenopodium	*berlandieri*	*bushianum*	Village	Pitseed	Goosefoot
Chenopodium	*berlandieri*	*bushianum-acutidentatum*	Sharp Toothed Village	Pitseed	Goosefoot
Chenopodium	*berlandieri*	*chualii*	Chual Spinach	Pitseed	Goosefoot
Chenopodium	*berlandieri*	*huazontlei*	Huazontle	Pitseed	Goosefoot
Chenopodium	*berlandieri*	*macrocalycium*	Large Calyx	Pitseed	Goosefoot
Chenopodium	*berlandieri*	*nuttalliae*	Chia Roja	Pitseed	Goosefoot
Chenopodium	*berlandieri*	*quelitei*	Quelite Spinach	Pitseed	Goosefoot
Chenopodium	*berlandieri*	*sinuatum*	Sinuous	Pitseed	Goosefoot
Chenopodium	*berlandieri*	*sinuatum-pseudopetiolare*	Petiole Sinuous	Pitseed	Goosefoot
Chenopodium	*berlandieri*	*zschackii*	Zschack	Pitseed	Goosefoot
Chenopodium	*berlandieri*	*zschackii-acerifolium*	Maple Leaf Zschack	Pitseed	Goosefoot

Genus	Species	infraspecies	cultivar	specimen	Genome
Chenopodium	*berlandieri*	*zschackii-farinosum*	Mealy Zschack	Pitseed	Goosefoot
Chenopodium	*berlandieri*	*zschackii-platyphyllum*	Flat Leaf Zschack	Pitseed	Goosefoot

Strawberry Spinach (*Chenopodium capitatum*)

Genus	Species	infraspecies	cultivar	specimen	Genome
Chenopodium	*capitatum*	*capitatum*	Original	Strawberry Spinach	Goosefoot
Chenopodium	*capitatum*	*beetberryi*	Beetberry	Strawberry Spinach	Goosefoot

Nettle Leaf Goosefoot (*Chenopodium murale*)

Genus	Species	Infraspecies	cultivar	specimen	Genome
Chenopodium	*murale*	*Murale*	Original	Nettle Leaf	Goosefoot
Chenopodium	*murale*	*indian-greensi*	Indian Greens	Nettle Leaf	Goosefoot
Chenopodium	*murale*	*o'odham-onk-iwaki*	O'odham Onk Iwaki	Nettle Leaf	Goosefoot

Quinoa Grain (*Chenopodium quinoa*)

Genus	Species	Infraspecies	cultivar	specimen	Genome
Chenopodium	*quinoa*	*Quinoa*	Original	Quinoa Grain	Goosefoot
Chenopodium	*quinoa*	*407-aureum*	#407 Golden	Quinoa Grain	Goosefoot
Chenopodium	*quinoa*	*407-davei*	# 407 Brown Dave Cusack	Quinoa Grain	Goosefoot
Chenopodium	*quinoa*	*407-nigrum*	#407 Black	Quinoa Grain	Goosefoot
Chenopodium	*quinoa*	*apelowai*	Red & Green Apelawa	Quinoa Grain	Goosefoot
Chenopodium	*quinoa*	*cahuila*	Light Green Cahuil	Quinoa Grain	Goosefoot
Chenopodium	*quinoa*	*cherry-vanillai*	White Cherry Vanilla	Quinoa Grain	Goosefoot
Chenopodium	*quinoa*	*chesanensis*	Yellow White Chesan	Quinoa Grain	Goosefoot
Chenopodium	*quinoa*	*davei*	Orange Dave Cusack	Quinoa Grain	Goosefoot
Chenopodium	*quinoa*	*equadorensis*	Pale Equador	Quinoa Grain	Goosefoot
Chenopodium	*quinoa*	*faroi*	White Faro	Quinoa Grain	Goosefoot
Chenopodium	*quinoa*	*faroi-rubrum*	Red Top White Faro	Quinoa Grain	Goosefoot
Chenopodium	*quinoa*	*gossi*	Multicolor Goss	Quinoa Grain	Goosefoot
Chenopodium	*quinoa*	*islugaensis-helvum*	Yellow Isluga	Quinoa Grain	Goosefoot
Chenopodium	*quinoa*	*kasalaensis*	Kaslala	Quinoa Grain	Goosefoot
Chenopodium	*quinoa*	*kcoitoensis*	Magenta & Gray Kcoito	Quinoa Grain	Goosefoot
Chenopodium	*quinoa*	*lake-titicacaensis*	Multicolor Lake Titicaca	Quinoa Grain	Goosefoot
Chenopodium	*quinoa*	*kiloi*	Kilo	Quinoa Grain	Goosefoot
Chenopodium	*quinoa*	*melangensis*	Melang	Quinoa Grain	Goosefoot
Chenopodium	*quinoa*	*multi-huedii*	Multi-hued	Quinoa Grain	Goosefoot
Chenopodium	*quinoa*	*orange-headii*	Orange Head	Quinoa Grain	Goosefoot
Chenopodium	*quinoa*	*pisonensis*	Yellow & White Pison	Quinoa Grain	Goosefoot
Chenopodium	*quinoa*	*pettyi*	Petty Rice	Quinoa Grain	Goosefoot
Chenopodium	*quinoa*	*rainbowi*	Rainbow	Quinoa Grain	Goosefoot
Chenopodium	*quinoa*	*redheadii*	Yellow White Seed Red Head	Quinoa Grain	Goosefoot
Chenopodium	*quinoa*	*temucoensis*	Golden & White Temuco	Quinoa Grain	Goosefoot
Chenopodium	*quinoa*	*wild-blackii*	Wild Black	Quinoa Grain	Goosefoot
Chenopodium	*quinoa*	*wild-blackii-boliviensis*	Bolivian Wild Black	Quinoa Grain	Goosefoot

Orache (*Atriplex hortensis*)

Genus	Species	infraspecies	cultivar	specimen	Genome
Atriplex	*hortensis*	*hortensis*	Original	Orach	Saltbush
Atriplex	*hortensis*	*atrosanguinea*	Dark Red	Orach	Saltbush
Atriplex	*hortensis*	*aurora-mixta*	Aurora Mix	Orach	Saltbush
Atriplex	*hortensis*	*cuptorosea*	Copper Red	Orach	Saltbush
Atriplex	*hortensis*	*fire-redii*	Fire Red	Orach	Saltbush
Atriplex	*hortensis*	*goldenii*	Golden	Orach	Saltbush
Atriplex	*hortensis*	*green-velvetii*	Green Velvet	Orach	Saltbush
Atriplex	*hortensis*	*helva*	Yellow	Orach	Saltbush
Atriplex	*hortensis*	*magenta-magici*	Magenta Magic	Orach	Saltbush
Atriplex	*hortensis*	*maroonii*	Maroon	Orach	Saltbush
Atriplex	*hortensis*	*montana*	Mountain Spinach	Orach	Saltbush
Atriplex	*hortensis*	*oraclei*	Oracle	Orach	Saltbush
Atriplex	*hortensis*	*plumea*	Plume	Orach	Saltbush
Atriplex	*hortensis*	*purpurea*	Purple	Orach	Saltbush
Atriplex	*hortensis*	*purpurea-savoia*	Savoyed Purple	Orach	Saltbush
Atriplex	*hortensis*	*purpurea-tripla*	Triple Purple	Orach	Saltbush
Atriplex	*hortensis*	*rosea*	Dark Red	Orach	Saltbush
Atriplex	*hortensis*	*rubra*	Blood Red	Orach	Saltbush

Chard (*Beta vulgaris-cicla*)

Genus	Species	infraspecies	cultivar	specimen	Genome
Beta	*vulgaris-cicla-ameria-alba*	*acelga*	Acelga	White Midrib American Chard	Beet

Genus	Species	infraspecies	cultivar	specimen	Genome
Beta	vulgaris-cicla-ameria-alba	broadstem-greenii	Green Leaf Broadstem	White Midrib American Chard	Beet
Beta	vulgaris-cicla-ameria-alba	chadwick's-choicei	Chadwick's Choice	White Midrib American Chard	Beet
Beta	vulgaris-cicla-ameria-alba	chismahoi	Chismaho	White Midrib American Chard	Beet
Beta	vulgaris-cicla-ameria-alba	dark-green-white-ribbedii	Dark Green White Ribbed	White Midrib American Chard	Beet
Beta	vulgaris-cicla-ameria-alba	fordhookii-giantii	FordhookGiant	White Midrib American Chard	Beet
Beta	vulgaris-cicla-ameria-alba	large-smooth-white-ribbedii	Large Smooth	White Midrib American Chard	Beet
Beta	vulgaris-cicla-ameria-alba	large-white-broad-ribbedii	Large Broad Ribbed	White Midrib American Chard	Beet
Beta	vulgaris-cicla-ameria-alba	large-white-ribbed-dark-greenii	Large Ribbed Dark Green Leaf	White Midrib American Chard	Beet
Beta	vulgaris-cicla-ameria-alba	orea	Orea	White Midrib American Chard	Beet
Beta	vulgaris-cicla-ameria-alba	silver-giantii	Silver Giant	White Midrib American Chard	Beet
Beta	vulgaris-cicla-ameria-alba	snow-queenii	Snow Queen	White Midrib American Chard	Beet
Beta	vulgaris-cicla-ameria-alba	white-cloudii	White Cloud	White Midrib American Chard	Beet
Beta	vulgaris-cicla-ameria-aurantia	cicla-ameria-aurantia	Original	Orange Midrib American Chard	Beet
Beta	vulgaris-cicla-ameria-aurantia	orange-fantasia	Orange Fantasia	Orange Midrib American Chard	Beet
Beta	vulgaris-cicla-ameria-aurantia	oriolei	Oriole	Orange Midrib American Chard	Beet
Beta	vulgaris-cicla-ameria-aurantia	swiss-goldii	Swiss Gold	Orange Midrib American Chard	Beet
Beta	vulgaris-cicla-ameria-groena	cicla-ameria	Original	Green Midrib American Chard	Beet
Beta	vulgaris-cicla-ameria-groena	parrota	Parrot	Green Midrib American Chard	Beet
Beta	vulgaris-cicla-ameria-groena	seafoamii	Seafoam	Green Midrib American Chard	Beet
Beta	vulgaris-cicla-ameria-groena	stari-1801	Star #1801	Green Midrib American Chard	Beet
Beta	vulgaris-cicla-ameria-groena	witerbi-mangoldii	Witerbi Mangold	Green Midrib American Chard	Beet
Beta	vulgaris-cicla-ameria-helva	cicla-ameria-helva	Original	Yellow Midrib American Chard	Beet
Beta	vulgaris-cicla-ameria-helva	bright-yellowi	Bright Yellow	Yellow Midrib American Chard	Beet
Beta	vulgaris-cicla-ameria-helva	canaryi	Canary	Yellow Midrib American Chard	Beet
Beta	vulgaris-cicla-ameria-helva	goldenii-sunrisei	Golden Sunrise	Yellow Midrib American Chard	Beet
Beta	vulgaris-cicla-ameria-helva	yellowi	Yellow Stalk	Yellow Midrib American Chard	Beet
Beta	vulgaris-cicla-ameria-multicolori	cicla-ameria-multicolori	Original	Multicolor Midrib American Chard	Beet
Beta	vulgaris-cicla-ameria-multicolori	breeders-rainbowi-1995	Breeders Rainbow 1995	Multicolor Midrib American Chard	Beet
Beta	vulgaris-cicla-ameria-multicolori	breeders-rainbowi-2000	Breeders Rainbow 2000	Multicolor Midrib American Chard	Beet
Beta	vulgaris-cicla-ameria-multicolori	bright-lightsi	Bright Lights	Multicolor Midrib American Chard	Beet
Beta	vulgaris-cicla-ameria-multicolori	neon-lightsi	Neon Lights	Multicolor Midrib American Chard	Beet
Beta	vulgaris-cicla-ameria-pinca	cicla-ameria-pinca	Original	Pink Midrib American Chard	Beet
Beta	vulgaris-cicla-ameria-pinca	flamingo	Flamingo	Pink Midrib American Chard	Beet
Beta	vulgaris-cicla-ameria-pinca	pink-lipstickii	Pink Lipstick	Pink Midrib American Chard	Beet
Beta	vulgaris-cicla-ameria-purpurea-rubra	cicla-ameria-purpurea-rubra	Original	Purple Red Midrib American Chard	Beet
Beta	vulgaris-cicla-ameria-purpurea-rubra	burgundy-giantii	Burgundy Giant	Purple Red Midrib American Chard	Beet
Beta	vulgaris-cicla-ameria-purpurea-rubra	magenta-mixii	Magenta Mix	Purple Red Midrib American Chard	Beet
Beta	vulgaris-cicla-ameria-purpurea-rubra	magenta-sunsetii	Magenta Sunset	Purple Red Midrib American Chard	Beet
Beta	vulgaris-cicla-ameria-purpurea-rubra	rhubarbii	Rhubarb	Purple Red Midrib American Chard	Beet
Beta	vulgaris-cicla-ameria-rubra	cicla-ameria-rubra	Original	Red Midrib American Chard	Beet
Beta	vulgaris-cicla-ameria-rubra	cardinalis	Cardinal	Red Midrib American Chard	Beet
Beta	vulgaris-cicla-ameria-rubra	charlottei	Charlotte	Red Midrib American Chard	Beet
Beta	vulgaris-cicla-ameria-rubra	red-ribii	Red Rib	Red Midrib American Chard	Beet
Beta	vulgaris-cicla-ameria-rubra	red-stalkii	Red Stalk	Red Midrib American Chard	Beet
Beta	vulgaris-cicla-ameria-rubra	ruby-redii	Ruby Red	Red Midrib American Chard	Beet
Beta	vulgaris-cicla-asiatica	japonica	Japanese Green Stalk	Asian Chard	Beet
Beta	vulgaris-cicla-australis	candy-stripei	Candy Stripe	Australian Chard	Beet
Beta	vulgaris-cicla-australis	rainbowi	Five Color Rainbow	Australian Chard	Beet
Beta	vulgaris-cicla-europa-alba	argentata	Argentata	White Midrib European Chard	Beet
Beta	vulgaris-cicla-europa-alba	argentata-costa	Costa Argentata	White Midrib European Chard	Beet
Beta	vulgaris-cicla-europa-alba	doratii	Dorat	White Midrib European Chard	Beet
Beta	vulgaris-cicla-europa-alba	frenchii	French	White Midrib European Chard	Beet
Beta	vulgaris-cicla-europa-alba	french-greenii	French Green Leaf	White Midrib European Chard	Beet
Beta	vulgaris-cicla-europa-alba	italian-silver-ribii	Italian Silver Rib	White Midrib European Chard	Beet
Beta	vulgaris-cicla-europa-alba	lucullus	Lucullus	White Midrib European Chard	Beet
Beta	vulgaris-cicla-europa-alba	lucullus-giantii	Giant Lucullus	White Midrib European Chard	Beet
Beta	vulgaris-cicla-europa-alba	lucullus-silverado	Silverado Lucullus	White Midrib European Chard	Beet
Beta	vulgaris-cicla-europa-alba	lucullus-white-kingii	White King Lucullus	White Midrib European Chard	Beet
Beta	vulgaris-cicla-europa-alba	monstruosoi	Monstruoso	White Midrib European Chard	Beet
Beta	vulgaris-cicla-europa-alba	monstruosoi-virgoi	Virgo Monstruoso	White Midrib European Chard	Beet
Beta	vulgaris-cicla-europa-alba	paris-greenii	Paris Green Leaf	White Midrib European Chard	Beet
Beta	vulgaris-cicla-europa-alba	rex-wide-ribbedii	Rex Wide Ribbed	White Midrib European Chard	Beet
Beta	vulgaris-cicla-europa-alba	verde-da-costai	Verde Da Costa	White Midrib European Chard	Beet
Beta	vulgaris-cicla-europa-alba	verte-a-carde-blanchei	Verte A Carde Blanche	White Midrib European Chard	Beet
Beta	vulgaris-cicla-europa-groena	bietola-da-coste-bionda-de-lionei	Bietola Da Coste Bionda De Lione	Green Midrib European Chard	Beet
Beta	vulgaris-cicla-europa-groena	conradii	Conrad	Green Midrib European Chard	Beet
Beta	vulgaris-cicla-europa-groena	erbette-leafii	Erbette Leaf	Green Midrib European Chard	Beet
Beta	vulgaris-cicla-europa-groena	greenii	Green Stalk	Green Midrib European Chard	Beet
Beta	vulgaris-cicla-europa-groena	palak-durgai	Palak Durga	Green Midrib European Chard	Beet
Beta	vulgaris-cicla-europa-groena	pavichii	Pavich	Green Midrib European Chard	Beet
Beta	vulgaris-cicla-europa-groena	perpetualis	Perpetual Spinach Beet	Green Midrib European Chard	Beet
Beta	vulgaris-cicla-europa-groena	san-francisco-wildii	San Francisco Wild	Green Midrib European Chard	Beet

Genus	Species	infraspecies	cultivar	specimen	Genome
Beta	*vulgaris-cicla-europa-groena*	*verde-dataglio*	Verde Da Taglio	Green Midrib European Chard	Beet
Beta	*vulgaris-cicla-europa-helva*	*goldenii*	Golden	Yellow Midrib European Chard	Beet
Beta	*vulgaris-cicla-europa-rubra*	*feurio*	Feurio	Red Midrib European Chard	Beet
Beta	*vulgaris-cicla-europa-rubra*	*rhubarb-supremei*	Rhubarb Supreme	Red Midrib European Chard	Beet

Swiss Chard (*Beta vulgaris-cicla-svedica*)

Genus	Species	infraspecies	cultivar	specimen	Genome
Beta	*vulgaris-cicla-svedica-alba*	*glater-silberi*	Glatter Silber	White Midrib Swiss Chard	Beet
Beta	*vulgaris-cicla-svedica-alba*	*parosi*	Paros	White Midrib Swiss Chard	Beet
Beta	*vulgaris-cicla-svedica-groena*	*svedica*	Thick	Green Midrib Swiss Chard	Beet
Beta	*vulgaris-cicla-svedica-groena*	*baresensis*	Barese	Green Midrib Swiss Chard	Beet
Beta	*vulgaris-cicla-svedica-groena*	*genevensis*	Geneva	Green Midrib Swiss Chard	Beet
Beta	*vulgaris-cicla-svedica-helva*	*gold-glebei*	Gold Glebe	Yellow Midrib Swiss Chard	Beet
Beta	*vulgaris-cicla-svedica-rubra*	*vulcanii*	Vulcan	Red Midrib Swiss Chard	Beet

Beet, Globe (*Beta vulgaris-conditiva*)

Genus	Species	infraspecies	cultivar	specimen	Genome
Beta	*vulgaris-conditiva-alba*	*albina-vereduna*	Albina Vereduna	White Globe	Beet
Beta	*vulgaris-conditiva-alba*	*albino*	Albino	White Globe	Beet
Beta	*vulgaris-conditiva-alba*	*blankoma-JBT-1002*	Johnny Blankoma Table #1002	White Globe	Beet
Beta	*vulgaris-conditiva-alba*	*snow-whitei*	Snow White	White Globe	Beet
Beta	*vulgaris-conditiva-alba*	*snowfleshii*	Snowflesh	White Globe	Beet
Beta	*vulgaris-conditiva-helva*	*golden-beetii*	Golden	Yellow Globe	Beet
Beta	*vulgaris-conditiva-helva*	*golden-burpee'si*	Burpee's Golden	Yellow Globe	Beet
Beta	*vulgaris-conditiva-helva*	*golden-detroitii*	Detroit Golden	Yellow Globe	Beet
Beta	*vulgaris-conditiva-helva*	*golden-tankardii*	Tankard Golden	Yellow Globe	Beet
Beta	*vulgaris-conditiva-helva*	*golden-wonderi*	Wonder Golden	Yellow Globe	Beet
Beta	*vulgaris-conditiva-purpurea*	*bordo*	Bordo	Purple Globe	Beet
Beta	*vulgaris-conditiva-purpurea*	*paonazza-d'egittoi*	Paonazza d'Egitto	Purple Globe	Beet
Beta	*vulgaris-conditiva-purpurea*	*round-purplei*	Round Purple	Purple Globe	Beet
Beta	*vulgaris-conditiva-purpurea-rubra*	*crimson-globei*	Crimson Globe	Purple Red Globe	Beet
Beta	*vulgaris-conditiva-purpurea-rubra*	*crimson-kingii*	Crimson King	Purple Red Globe	Beet
Beta	*vulgaris-conditiva-purpurea-rubra*	*drumii*	Drum	Purple Red Globe	Beet
Beta	*vulgaris-conditiva-purpurea-rubra*	*early-wonderi-tall-topii*	Tall Top Early Wonder	Purple Red Globe	Beet
Beta	*vulgaris-conditiva-purpurea-rubra*	*edmund's-bloodii*	Edmund's Blood	Purple Red Globe	Beet
Beta	*vulgaris-conditiva-purpurea-rubra*	*flat-egyptianii*	Flat Egyptian	Purple Red Globe	Beet
Beta	*vulgaris-conditiva-rubra*	*always-tenderi*	Always Tender	Red Globe	Beet
Beta	*vulgaris-conditiva-rubra*	*big-topii*	Big Top	Red Globe	Beet
Beta	*vulgaris-conditiva-rubra*	*boltardyi*	Bolt Hardy	Red Globe	Beet
Beta	*vulgaris-conditiva-rubra*	*bull's-bloodii*	Bull's Blood	Red Globe	Beet
Beta	*vulgaris-conditiva-rubra*	*carillon*	Carillon	Red Globe	Beet
Beta	*vulgaris-conditiva-rubra*	*chioggia*	Chioggia	Red Globe	Beet
Beta	*vulgaris-conditiva-rubra*	*crosby's-egyptianii*	Crosby's Egyptian	Red Globe	Beet
Beta	*vulgaris-conditiva-rubra*	*crosby's-green-topii*	Crosby's Green Top	Red Globe	Beet
Beta	*vulgaris-conditiva-rubra*	*deacon-danii*	Deacon Dan	Red Globe	Beet
Beta	*vulgaris-conditiva-rubra*	*detroit -crimson-globei*	Crimson Detroit	Red Globe	Beet
Beta	*vulgaris-conditiva-rubra*	*detroit-dark-redii*	Dark Red Detroit	Red Globe	Beet
Beta	*vulgaris-conditiva-rubra*	*detroit-dark-red-243*	#243 Dark Red Detroit	Red Globe	Beet
Beta	*vulgaris-conditiva-rubra*	*detroit-dark-red-black-knightii*	Black Knight Dark Red Detroit	Red Globe	Beet
Beta	*vulgaris-conditiva-rubra*	*detroit-dark-red-excaliberi*	Excaliber Dark Red Detroit	Red Globe	Beet
Beta	*vulgaris-conditiva-rubra*	*detroit-dark-red-garnetii*	Garnet Dark Red Detroit	Red Globe	Beet
Beta	*vulgaris-conditiva-rubra*	*detroit-dark-red-improvedii*	Improved Dark Red Detroit	Red Globe	Beet
Beta	*vulgaris-conditiva-rubra*	*detroit-dark-red-iridei*	Iride Dark Red Detroit	Red Globe	Beet
Beta	*vulgaris-conditiva-rubra*	*detroit-dark-red-medium-topii*	Medium Top Dark Red Detroit	Red Globe	Beet
Beta	*vulgaris-conditiva-rubra*	*detroit-dark-red-perfectedii*	Perfected Dark Red Detroit	Red Globe	Beet
Beta	*vulgaris-conditiva-rubra*	*detroit-dark-red-rubidus-006*	Rubidus #006 Dark Red Detroit	Red Globe	Beet
Beta	*vulgaris-conditiva-rubra*	*detroit-dark-red-sangria*	Sangria Dark Red Detroit	Red Globe	Beet
Beta	*vulgaris-conditiva-rubra*	*detroit-dark-red-short-topii*	Short Top Dark Red Detroit	Red Globe	Beet
Beta	*vulgaris-conditiva-rubra*	*detroit-dark-red-supremei*	Supreme Dark Red Detroit	Red Globe	Beet
Beta	*vulgaris-conditiva-rubra*	*detroit-dark-red-sweetheartii*	Sweetheart Dark Red Detroit	Red Globe	Beet
Beta	*vulgaris-conditiva-rubra*	*early-wonderi*	Early Wonder	Red Globe	Beet
Beta	*vulgaris-conditiva-rubra*	*early-wonderi-dark-strainii*	Dark Red Early Wonder	Red Globe	Beet
Beta	*vulgaris-conditiva-rubra*	*early-wonderi-green-topii*	Green Top Early Wonder	Red Globe	Beet
Beta	*vulgaris-conditiva-rubra*	*early-wonderi-smooth-leafii*	Smooth Leaf Early Wonder	Red Globe	Beet
Beta	*vulgaris-conditiva-rubra*	*early-wonderi-stays-greenii*	Stays Green Early Wonder	Red Globe	Beet
Beta	*vulgaris-conditiva-rubra*	*feuer-kugelii*	Feuer Kugel	Red Globe	Beet
Beta	*vulgaris-conditiva-rubra*	*first-cropii*	First Crop	Red Globe	Beet
Beta	*vulgaris-conditiva-rubra*	*green-top-bunching-letherman'si*	Letherman's Green Top Bunching	Red Globe	Beet
Beta	*vulgaris-conditiva-rubra*	*indianii*	Indian	Red Globe	Beet
Beta	*vulgaris-conditiva-rubra*	*moneta*	Monet	Red Globe	Beet
Beta	*vulgaris-conditiva-rubra*	*mongrama*	Monogram	Red Globe	Beet
Beta	*vulgaris-conditiva-rubra*	*monopolyi*	Monopoly	Red Globe	Beet
Beta	*vulgaris-conditiva-rubra*	*prontoi*	Pronto	Red Globe	Beet
Beta	*vulgaris-conditiva-rubra*	*punakerai*	Punakera	Red Globe	Beet

Genus	Species	infraspecies	cultivar	specimen	Genome
Beta	*vulgaris-conditiva-rubra*	*red-ball-burpee's*	Burpee's Red Ball	Red Globe	Beet
Beta	*vulgaris-conditiva-rubra*	*red-successi*	Red Success	Red Globe	Beet
Beta	*vulgaris-conditiva-rubra*	*red-titanii*	Red Titan	Red Globe	Beet
Beta	*vulgaris-conditiva-rubra*	*red-velvetii-FM-269*	Red Velvet	Red Globe	Beet
Beta	*vulgaris-conditiva-rubra*	*rossa-quarantina-d'egittoi*	Red Quarantina d'Egitto	Red Globe	Beet
Beta	*vulgaris-conditiva-rubra*	*rote-kugelii*	Rote Kugel	Red Globe	Beet
Beta	*vulgaris-conditiva-rubra*	*ruby-queenii*	Ruby Queen	Red Globe	Beet
Beta	*vulgaris-conditiva-rubra*	*ruby-queenii-improvedii*	Ruby Queen Improved	Red Globe	Beet
Beta	*vulgaris-conditiva-rubra*	*special-earlyi-stokesi*	Stokes Special Early	Red Globe	Beet
Beta	*vulgaris-conditiva-rubra*	*sweet-supra*	SweetSuper	Red Globe	Beet
Beta	*vulgaris-conditiva-rubra*	*tardelii*	Tardel	Red Globe	Beet
Beta	*vulgaris-conditiva-rubra-nana*	*baby-ballii*	Baby Ball	Baby Red Globe	Beet
Beta	*vulgaris-conditiva-rubra-nana*	*dwergina*	Dwergin	Baby Red Globe	Beet
Beta	*vulgaris-conditiva-rubra-nana*	*gladiatori*	Gladiator	Baby Red Globe	Beet
Beta	*vulgaris-conditiva-rubra-nana*	*little-ballii*	Little Ball	Baby Red Globe	Beet
Beta	*vulgaris-conditiva-rubra-nana*	*little-mini-ballii*	Little Mini Ball	Baby Red Globe	Beet
Beta	*vulgaris-conditiva-rubra-nana*	*mona-lisai(mona)*	Mona Lisa	Baby Red Globe	Beet
Beta	*vulgaris-conditiva-rubra-nana*	*spinel-baby-beetii*	Spinel	Baby Red Globe	Beet

Beet, Sugar (*Beta vulgaris-crassa*)

Genus	Species	infraspecies	cultivar	specimen	Genome
Beta	*vulgaris-crassa-alba*	*giant-half-sugari*	Giant Half Sugar	White Long Sugar	Beet
Beta	*vulgaris-crassa-alba*	*klein-wanzleben-sugar-beetii*	Klein Wanzleben	White Long Sugar	Beet
Beta	*vulgaris-crassa-helva*	*giant-western-sugar-beetii*	Giant Western	Yellow Long Sugar	Beet
Beta	*vulgaris-crassa-helva*	*cylindrical-yellowi*	Yellow Cylinderical	Yellow Long Sugar	Beet
Beta	*vulgaris-crassa-helva*	*intermediate-yellowi*	Yellow Intermediate	Yellow Long Sugar	Beet
Beta	*vulgaris-crassa-helva*	*mangel-yellowi*	Yellow Mangel	Yellow Long Sugar	Beet
Beta	*vulgaris-crassa-rubra*	*cheltenhamen-green-topii*	Cheltenham Green Top	Red Long Sugar	Beet
Beta	*vulgaris-crassa-rubra*	*cylinder-deepii*	Cylinder Deep	Red Long Sugar	Beet
Beta	*vulgaris-crassa-rubra*	*cylinder-long-redii*	Cylinder	Red Long Sugar	Beet
Beta	*vulgaris-crassa-rubra*	*cyndori*	Cyndor	Red Long Sugar	Beet
Beta	*vulgaris-crassa-rubra*	*early-blood-turnipii*	Early Blood Turnip	Red Long Sugar	Beet
Beta	*vulgaris-crassa-rubra*	*early-blood-turnipii-dewing'si*	Dewing's Early Blood Turnip	Red Long Sugar	Beet
Beta	*vulgaris-crassa-rubra*	*formanova-cook's-delightii*	Cooks' Delight Formanova	Red Long Sugar	Beet
Beta	*vulgaris-crassa-rubra*	*formanova-tendersweetii*	Tendersweet Formanova	Red Long Sugar	Beet
Beta	*vulgaris-crassa-rubra*	*foronai*	Forona	Red Long Sugar	Beet
Beta	*vulgaris-crassa-rubra*	*giant-red-sugar-beetii*	Giant	Red Long Sugar	Beet
Beta	*vulgaris-crassa-rubra*	*lutzii-germanii*	German Lutz	Red Long Sugar	Beet
Beta	*vulgaris-crassa-rubra*	*lutzii-greenleaf*	Green Leaf Lutz	Red Long Sugar	Beet
Beta	*vulgaris-crassa-rubra*	*lutzii-winter-keeperi*	Winter Keeper Lutz	Red Long Sugar	Beet
Beta	*vulgaris-crassa-rubra*	*macgregor's-favoritei*	MacGregor's Favorite	Red Long Sugar	Beet
Beta	*vulgaris-crassa-rubra*	*mangel-colossal-long-redii*	Colossal Mangel	Red Long Sugar	Beet
Beta	*vulgaris-crassa-rubra*	*mangel-mammoth-long-redii*	Mammoth Mangel	Red Long Sugar	Beet
Beta	*vulgaris-crassa-rubra*	*mangel-wurtzelii*	Wurtzel Mangel	Red Long Sugar	Beet
Beta	*vulgaris-crassa-rubra*	*new-centuryi*	New Century	Red Long Sugar	Beet
Beta	*vulgaris-crassa-rubra*	*red-crapaudinii*	Red Crapaudine	Red Long Sugar	Beet
Beta	*vulgaris-crassa-rubra*	*rocketii*	Rocket	Red Long Sugar	Beet
Beta	*vulgaris-crassa-rubra*	*sugar-beetii*	Original	Red Long Sugar	Beet
Blitum	*foliosum*	*foliosum*	Original	Strawberry Spinach	Strawberry Blite
Blitum	*foliosum*	*parvicapitatum*	Small Head	Strawberry Spinach	Strawberry Blite
Blitum	*foliosum*	*virgatum*	Twiggy	Strawberry Spinach	Strawberry Blite
Salicornia	*maritima*	*maritima*	Turtleweed	Glasswort	Saltwort
Salicornia	*maritima*	*europaea*	Marsh Samphire	Glasswort	Saltwort
Salicornia	*maritima*	*herbacea*	Sea Asparagus	Glasswort	Saltwort
Salicornia	*maritima*	*prostrata*	Prostrate	Glasswort	Saltwort
Salicornia	*maritima*	*ramosissima*	Branching	Glasswort	Saltwort
Salicornia	*rubra*	*rubra*	Original	Pacific Glasswort	Saltwort
Salicornia	*rubra*	*prona*	Samphire	Pacific Glasswort	Saltwort
Salicornia	*rubra*	*saximontana*	Rocky Mountain	Pacific Glasswort	Saltwort

Spinach (*Spinachia oleracea*)

Genus	Species	infraspecies	cultivar	specimen	Genome
Spinacia	*oleracea-aculeasperma*	*latifolia*	Broad Leaf	Smooth Prickly Seed	Spinach
Spinacia	*oleracea-aculeasperma*	*erste-erniei*	Erste Ernie	Smooth Prickly Seed	Spinach
Spinacia	*oleracea-aculeasperma*	*munsterlanderi*	Munsterlander	Smooth Prickly Seed	Spinach
Spinacia	*oleracea-aculeasperma*	*munsterlanderi-rotunda*	Round Munsterlander	Smooth Prickly Seed	Spinach
Spinacia	*oleracea-aculeasperma*	*nobelii*	Nobel	Smooth Prickly Seed	Spinach
Spinacia	*oleracea-aculeasperma*	*nobelii-gigantea*	Giant Nobel	Smooth Prickly Seed	Spinach
Spinacia	*oleracea-aculeasperma*	*nordica-4*	#4 Nordic	Smooth	Spinach
Spinacia	*oleracea-aculeasperma*	*king-of-denmarkii-kosanii*	Konsan King Of Denmark	Smooth Prickly Seed	Spinach
Spinacia	*oleracea-aculeasperma*	*vikingii*	Viking	Smooth Prickly Seed	Spinach
Spinacia	*oleracea-aculeasperma*	*vikingii-dominans*	Dominant Viking	Smooth Prickly Seed	Spinach
Spinacia	*oleracea-aculeasperma*	*vikingii-faresi*	Fares Viking	Smooth Prickly Seed	Spinach
Spinacia	*oleracea-aculeasperma*	*vikingii-lorelei*	Lorelei Viking	Smooth Prickly Seed	Spinach

Genus	Species	infraspecies	cultivar	specimen	Genome
Spinacia	*oleracea-aculeasperma*	*vikingii-mediana*	Mediana Viking	Smooth Prickly Seed	Spinach
Spinacia	*oleracea-aculeasperma*	*vikingii-mediana-rotunda*	Round Mediana Viking	Smooth Prickly Seed	Spinach
Spinacia	*oleracea-aculeasperma-savoia*	*king-of-denmarkii*	King Of Denmark	Savoy Prickly Seed	Spinach
Spinacia	*oleracea-glabra*	*bergonia*	Bergonia	Smooth	Spinach
Spinacia	*oleracea-glabra*	*estivatoi*	Estivato	Smooth	Spinach
Spinacia	*oleracea-glabra*	*F-380*	Form (F) 380	Smooth	Spinach
Spinacia	*oleracea-glabra*	*firstcropii*	First Crop	Smooth	Spinach
Spinacia	*oleracea-glabra*	*germanica*	German	Smooth	Spinach
Spinacia	*oleracea-glabra*	*hivernai*	Hiverna	Smooth	Spinach
Spinacia	*oleracea-glabra*	*latifolia-saumar*	Summer Broadleaf	Smooth	Spinach
Spinacia	*oleracea-glabra*	*longstanding-gamma*	Gamma Longstanding	Smooth	Spinach
Spinacia	*oleracea-glabra*	*longstanding-roundii*	Longstanding Round	Smooth	Spinach
Spinacia	*oleracea-glabra*	*matadori*	Matador	Smooth	Spinach
Spinacia	*oleracea-glabra*	*matadori-matarnoi*	Matarno Matador	Smooth	Spinach
Spinacia	*oleracea-glabra*	*matadori-nobioi*	Nobio Matador	Smooth	Spinach
Spinacia	*oleracea-glabra*	*monnopa*	Monnopa	Smooth	Spinach
Spinacia	*oleracea-glabra*	*monnopa-low-acidii*	Low Acid Monnopa	Smooth	Spinach
Spinacia	*oleracea-glabra*	*monstrueux-de-viroflayi-1866*	1866 Monstrous Viroflay	Smooth	Spinach
Spinacia	*oleracea-glabra*	*monstrueux-de-viroflayi-1885*	1885 Monstrous Viroflay	Smooth	Spinach
Spinacia	*oleracea-glabra*	*monstrueux-de-viroflayi-resistoflayi*	Resistoflay Monstrous Viroflay	Smooth	Spinach
Spinacia	*oleracea-glabra*	*oleracea*	Original	Smooth	Spinach
Spinacia	*oleracea-glabra*	*spacei*	Space	Smooth	Spinach
Spinacia	*oleracea-glabra*	*spring-gigantea*	Spring Giant	Smooth	Spinach
Spinacia	*oleracea-glabra*	*steadfastii*	Steadfast	Smooth	Spinach
Spinacia	*oleracea-glabra*	*summer-gigantea-L*	Summer Giant Late (L)	Smooth	Spinach
Spinacia	*oleracea-glabra*	*supra-veranoi*	Super Verano	Smooth	Spinach
Spinacia	*oleracea-glabra*	*triathlonii*	Triathlon	Smooth	Spinach
Spinacia	*oleracea-glabra*	*wintergreenii*	Wintergreen	Smooth	Spinach
Spinacia	*oleracea-glabra*	*winterriesen-wiri*	Winterriesen Wiri	Smooth	Spinach
Spinacia	*oleracea-savoia*	*bloomesdalei*	Bloomsdale	Savoy	Spinach
Spinacia	*oleracea-savoia*	*bloomesdalei-americana*	American Bloomsdale	Savoy	Spinach
Spinacia	*oleracea-savoia*	*bloomesdalei-dark-greenii*	Dark Green Bloomsdale	Savoy	Spinach
Spinacia	*oleracea-savoia*	*bloomesdalei-extra*	Extra Bloomsdale	Savoy	Spinach
Spinacia	*oleracea-savoia*	*bloomesdalei-merlo-neroi*	Merlo Nero Bloomsdale	Savoy	Spinach
Spinacia	*oleracea-savoia*	*bloomesdalei-resistans*	Resistant Bloomsdale	Savoy	Spinach
Spinacia	*oleracea-savoia*	*bloomesdalei-vintera*	Winter Bloomsdale	Savoy	Spinach
Spinacia	*oleracea-savoia*	*butterflayi*	Butterflay	Savoy	Spinach
Spinacia	*oleracea-savoia*	*gigante-d'invernoi*	Gigante D'Inverno	Savoy	Spinach
Spinacia	*oleracea-savoia*	*norfolkensis*	Norfolk	Savoy	Spinach
Spinacia	*oleracea-savoia*	*old-dominionensis*	Old Dominion	Savoy	Spinach
Spinacia	*oleracea-savoia*	*picnica*	Picnic	Savoy	Spinach
Spinacia	*oleracea-semisavoia*	*balloonii*	Balloon	Semi Savoy	Spinach
Spinacia	*oleracea-semisavoia*	*earlyi-7*	#7 Early	Semi Savoy	Spinach
Spinacia	*oleracea-semisavoia*	*fall-greenii-AR-82-5*	Fall Green	Semi Savoy	Spinach
Spinacia	*oleracea-semisavoia*	*giant-winteri*	Giant Winter	Semi Savoy	Spinach
Spinacia	*oleracea-semisavoia*	*sigmaleafii*	Sigmaleaf	Semi Savoy	Spinach
Spinacia	*oleracea-semisavoia*	*supergreenii*	Super Green	Semi Savoy	Spinach

Family 6 *Molluginaceae* Carpetweed
Family 7 *Nyctaginaceae* Night Lumbar

Family 724.021.8 *Phytolaccaceae* Milk Plant

Genus	Species	Specimen Genome	C:Culinary, M:Medicinal, E:Economic
Phytolacca	*americana*	American Pokeberry	C:LF:**VG**

Varieties

Poke Salad (*Phytolacca americana decandra*)

Genus	Species	infraspecies	cultivar	specimen	Genome
Phytolacca	*americana*	*americana*	Original	American	Pokeberry
Phytolacca	*americana*	*decandra*	Poke Salad	American	Pokeberry
Phytolacca	*americana*	*rigida*	Stiff	American	Pokeberry

Family 724.021.9 *Portulacaceae* Purslane

Genus	Species	Specimen Genome	C:Culinary, M:Medicinal, E:Economic
Portulaca	*oleracea*	Hogweed Purslane	C:LF:**VG**
Montia	*perfoliata*	Edible Miner's Lettuce	

Varieties

Hogweed Purslane (*Portulaca oleracea*)

Genus	Species	infraspecies	cultivar	specimen	Genome
Portulaca	*oleracea*	*oleracea*	Original	Hogweed	Purslane
Portulaca	*oleracea*	*aurea*	Golden	Hogweed	Purslane
Portulaca	*oleracea*	*aurea-alta*	Tall Golden	Hogweed	Purslane
Portulaca	*oleracea*	*aurea-grandifolia*	Large Leaf Golden	Hogweed	Purslane
Portulaca	*oleracea*	*erect-large-leafii*	Erect Large Leaf	Hogweed	Purslane
Portulaca	*oleracea*	*goldbergeri*	Goldberger	Hogweed	Purslane
Portulaca	*oleracea*	*granulatostellulata*	Grainy Star	Hogweed	Purslane
Portulaca	*oleracea*	*green-gardenii*	Green	Hogweed	Purslane
Portulaca	*oleracea*	*green-jadei*	Jade Green	Hogweed	Purslane
Portulaca	*oleracea*	*impolita*	Impolite	Hogweed	Purslane
Portulaca	*oleracea*	*matweedii*	Matweed	Hogweed	Purslane
Portulaca	*oleracea*	*neglecta*	Neglected	Hogweed	Purslane
Portulaca	*oleracea*	*nicaraguensis*	Nicaragua	Hogweed	Purslane
Portulaca	*oleracea*	*nitida*	Shiny	Hogweed	Purslane
Portulaca	*oleracea*	*papillatostellulata*	Downy Star	Hogweed	Purslane
Portulaca	*oleracea*	*retusa*	Notched Tip	Hogweed	Purslane
Portulaca	*oleracea*	*stellata*	Star	Hogweed	Purslane

Miner's Lettuce (*Montia perfoliata*)

Genus	Species	infraspecies	cultivar	specimen	Genome
Montia	*perfoliata*	*perfoliata*	Original	Leafy	Miner's Lettuce
Montia	*perfoliata*	*angustifolia*	Narrow Leaf	Leafy	Miner's Lettuce
Montia	*perfoliata*	*emerald-greenii*	Emerald Green	Leafy	Miner's Lettuce
Montia	*perfoliata*	*intermontana*	Intermountain	Leafy	Miner's Lettuce
Montia	*perfoliata*	*mexicana*	Mexican	Leafy	Miner's Lettuce
Montia	*perfoliata*	*miner's-pickii*	Miner's Pick	Leafy	Miner's Lettuce

Order 2 *Cactales* Cactus
- Family 01 *Cactaceae-barrila* Barrel Cactus
- Family 02 *Cactaceae-nana <1'* Dwarf Column Cactus
- Family 03 *Cactaceae-media 1'-2'* Medium Column Cactus
- Family 04 *Cactaceae-larga 2'-6'* Large Column Cactus
- Family 05 *Cactaceae-alta 6'-16'* Tall Column Cactus
- Family 06 *Cactaceae-supralta 16'-64'* Tower Column Cactus
- Family 07 *Cactaceae-globa* Globe Cactus
- Family 08 *Cactaceae-pendens* Pendent Cactus
- Family 09 *Cactaceae-platya* Flat Cactus
- Family 10 *Cactaceae-rosetta* Rosette Cactus
- Family 11 *Cactaceae-vigna* Vine Cactus
- Family 12 *Cactaceae-arbusta* Clump Cactus < 2'
- Family 13 *Cactaceae-bruscia* Bush Cactus 2'-6'
- Family 14 *Cactaceae-scrubia* Shrub Cactus 6'-16'
- Family 15 *Cactaceae-drysia* Tree Cactus > 16'

Order 724.023 *Polygonales* Jointstem

Family 724.023.1 *Polygonaceae* Jointstem

Buckwheat (*Fagopyrum esculentum*)

Genus	Species	Specimen Genome	C:Culinary, M:Medicinal, E:Economic
Fagopyrum	*esculentum*	Edible Buckwheat	C:SD:**VG**
Fagopyrum	*diobotrys*	Himalayan Buckwheat	C:SD:**VG**
Fagopyrum	*tataricum*	Tatarian Buckwheat	C:SD:**VG**
Rheum	*rhabarbarum*	Mongolian Rhubarb	C:SL:**VG**
Rumex	*acetosella*	Sheep Sorrel Dock	C:LF:**VG**

Varieties

Rhubarb (*Rheum rhabarbarum*)

Genus	Species	infraspecies	cultivar	specimen	Genome
Rheum	*rhabarbarum*	*rhabarbarum*	Original	Mongolian	Rhubarb
Rheum	*rhabarbarum*	*appleton's-forcingii*	Appleton's Forcing	Mongolian	Rhubarb
Rheum	*rhabarbarum*	*canada-redii*	Canada Red	Mongolian	Rhubarb
Rheum	*rhabarbarum*	*canada-redii-chipman'si*	Chipman's Canada Red	Mongolian	Rhubarb

Genus	Species	infraspecies	cultivar	specimen	Genome
Rheum	*rhabarbarum*	*cawood-castlei*	Cawood Castle	Mongolian	Rhubarb
Rheum	*rhabarbarum*	*cherry-redii*	Cherry Red	Mongolian	Rhubarb
Rheum	*rhabarbarum*	*comunei*	Comune	Mongolian	Rhubarb
Rheum	*rhabarbarum*	*crimson-redii*	Crimson Red	Mongolian	Rhubarb
Rheum	*rhabarbarum*	*early-victoria*	Early Victoria	Mongolian	Rhubarb
Rheum	*rhabarbarum*	*glaskin's-perpetual*	Glaskin's Perpetual	Mongolian	Rhubarb
Rheum	*rhabarbarum*	*green-victoria*	Green Victoria	Mongolian	Rhubarb
Rheum	*rhabarbarum*	*hawkes-champagnei*	Hawkes Champagne	Mongolian	Rhubarb
Rheum	*rhabarbarum*	*macdonaldii*	MacDonald	Mongolian	Rhubarb
Rheum	*rhabarbarum*	*pinetree-select-redii*	Pinetree Select Red	Mongolian	Rhubarb
Rheum	*rhabarbarum*	*queen-victoria*	Queen Victoria	Mongolian	Rhubarb
Rheum	*rhabarbarum*	*red-rhubarbii*	Red	Mongolian	Rhubarb
Rheum	*rhabarbarum*	*rubyi*	Ruby	Mongolian	Rhubarb
Rheum	*rhabarbarum*	*starkrimsonii*	Starkrimson	Mongolian	Rhubarb
Rheum	*rhabarbarum*	*stockbridge-cropperi*	Stockbridge Cropper	Mongolian	Rhubarb
Rheum	*rhabarbarum*	*the-suttonii*	The Sutton	Mongolian	Rhubarb
Rheum	*rhabarbarum*	*tildenii*	Tilden	Mongolian	Rhubarb
Rheum	*rhabarbarum*	*valentinei*	Valentine	Mongolian	Rhubarb
Rheum	*rhabarbarum*	*victoria*	Victoria	Mongolian	Rhubarb

Sheep Sorrel Dock (*Rumex acetosella*)

Genus	Species	infraspecies	cultivar	specimen	Genome
Rumex	*acetosella*	*acetosella*	Original	Sheep Sorrel	Dock
Rumex	*acetosella*	*angiocarpus*	Angled Fruit	Sheep Sorrel	Dock
Rumex	*acetosella*	*pyrenaeus*	Pyrenees	Sheep Sorrel	Dock
Rumex	*acetosella*	*tenuifolius*	Weak Leaf	Sheep Sorrel	Dock

Family 2 *Podostemaceae* Riverweed

Class 724.03 *Cornopsida* Cornel

Order 1 *Cornales* Cornel
Family 1 *Cornaceae* Cornel
Family 2 *Alangiaceae* Alang
Family 3 *Diapensiaceae* Diapense
Family 4 *Eucommiaceae* Gutta Percha
Family 5 *Garryaceae* Silk Tassel
Family 6 *Morinaceae* False Thistle
Family 7 *Nyssaceae* Tupelo

Order 724.032 *Apiales* Marshwort

Family 724.032.1 *Apiaceae* Marshwort

Caraway Root (*Carum carvi*)

Gotu Kola (*Centella asiatica*)

Sea Fennel (*Crithmum maritimum*)

Cow Parsnip (*Heracleum sphondylium*)

Lovage (*Levisticum officinale*)

Lovage, Alpine (*Ligusticum scoticum*)

Genus	Species	Specimen	Genome	C:Culinary, M:Medicinal
Apium	*graveolens-dulce-alba*	White Garden	Celery	C:SL:**VG**
Apium	*graveolens-dulce-groena*	Green Garden	Celery	C:SL:**VG**
Apium	*graveolens-dulce-helva*	Yellow Garden	Celery	C:SL:**VG**
Apium	*graveolens-dulce-rubra*	Red Garden	Celery	C:SL:**VG**
Apium	*graveolens-rapaceum*	Celeriac	Celery	C:CR:**VG**
Apium	*graveolens-secalinum*	Leaf Garden	Celery	C:SL:**VG**
Aegopodium	*podagraria*	Bishopwort	Goutwort	C:LF:**VG**
Angelica	*archangelica*	Archangel	Angelic	C:LF,SL:**VG**
Bunium	*bulbocastanum*	Great	Pignut	C:SD:**NT**
Carum	*carvi*	Persian	Caraway	C:SD:**SP**,C:RT:**VG**,
Centella	*asiatica*	Gotu Kola	Hundredleaf	C:LF:**VG**
Coriandrum	*sativum*	Parsley	Coriander	C:LF:**VG**, C:SD:**SP**
Crithmum	*maritimum*	Sea Fennel	Samphire	C:LF:**VG**
Cryptotaenia	*africana*	African	Honewort	C:LF:**VG**
Cryptotaenia	*canadensis*	Canadian Chervil	Honewort	C:LF:**VG**
Cryptotaenia	*japonica*	Japanese Parsley	Honewort	C:LF:**VG**
Daucus	*carota-sativus-cylindra*	Cylindrical	Carrot	C:RT:**VG**
Daucus	*carota-sativus-nana*	Baby	Carrot	C:RT:**VG**
Daucus	*carota-sativus-rotunda*	Round	Carrot	C:RT:**VG**
Daucus	*carota-sativus-tapera-acuta*	Pointed Tapered	Carrot	C:RT:**VG**
Daucus	*carota-sativus-autumna-kingii*	Autumn King Blunt Tapered	Carrot	C:RT:**VG**
Daucus	*carota-sativus-chantenayi*	Chantenay Blunt Tapered	Carrot	C:RT:**VG**
Daucus	*carota-sativus-danversi*	Danvers Blunt Tapered	Carrot	C:RT:**VG**
Daucus	*carota-sativus-imperator*	Imperator Blunt Tapered	Carrot	C:RT:**VG**
Daucus	*carota-sativus-nantes*	Nantes Blunt Tapered	Carrot	C:RT:**VG**
Daucus	*pusillus*	Puny Wild	Carrot	C:RT:**VG**
Foeniculum	*azoricum*	Florence	Fennel	C:CR,SL:**VG**
Foeniculum	*vulgare*	Sweet	Fennel	C:LF,RT:**VG**, C:SD:**SP**
Foeniculum	*vulgare-piperitum*	Cartucci	Fennel	C:SH:**VG**
Heracleum	*sphondylium*	Keck	Cow Parsnip	C:SL,CF,C:SH:**VG**
Levisticum	*officinale*	Love Parsley	Lovage	C:LF:FF,C:SL:CF,C:RT:**VG**
Ligusticum	*scoticum*	Scots Seaside Alpine	Lovage	C:LF,SL:**VG**,
Pastinaca	*pratensis*	Meadow	Parsnip	C:RT:**VG**
Pastinaca	*sativa*	European	Parsnip	C:RT:**VG**
Petroselinum	*crispum*	Curly Leaf	Parsley	C:LF:**VG**
Petroselinum	*hortense*	Turnip Root	Parsley	C:LF:**VG**
Petroselinum	*neapolitanum*	Italian Flat Leaf	Parsley	C:LF:**VG**
Petroselinum	*vulgare*	Flat Leaf	Parsley	C:LF:**VG**
Sium	*carsonii*	Carson	Water Parsnip	C:RT:**VG**
Sium	*latifolium*	Broad Leaf	Water Parsnip	C:RT:**VG**
Sium	*sisarum*	Skirret	Water Parsnip	C:RT:**VG**
Sium	*suave*	Sweet	Water Parsnip	C:RT:**VG**

Varieties

Celery (*Apium graveolens-dulce*)

Genus	species	infraspecies	cultivar	specimen	Genome
Apium	*graveolens-dulce-alba*	*solid-whitei*	Solid White	White Garden	Celery
Apium	*graveolens-dulce-alba*	*white-queenii*	White Queen	White Garden	Celery
Apium	*graveolens-dulce-groena*	*afina*	Afina	Green Garden	Celery
Apium	*graveolens-dulce-groena*	*amsterdam-finei*	Amsterdam Fine	Green Garden	Celery
Apium	*graveolens-dulce-groena*	*celebration*	Celebration	Green Garden	Celery
Apium	*graveolens-dulce-groena*	*chinese-shicunii*	Shicun Chinese	Green Garden	Celery
Apium	*graveolens-dulce-groena*	*chinese-green-giantii*	Green Giant Chinese	Green Garden	Celery
Apium	*graveolens-dulce-groena*	*chinese-kan-tsaii*	Kan Tsai Chinese	Green Garden	Celery
Apium	*graveolens-dulce-groena*	*chinese-kin-tsaii*	Kin Tsai Chinese	Green Garden	Celery
Apium	*graveolens-dulce-groena*	*coquistadori*	Conquistador	Green Garden	Celery
Apium	*graveolens-dulce-groena*	*early-dellii*	Early Dell	Green Garden	Celery
Apium	*graveolens-dulce-groena*	*floridensis-683*	Marvelous #683 Florida	Green Garden	Celery
Apium	*graveolens-dulce-groena*	*floridensis-683-K*	Fusarium Resistant #683 Florida	Green Garden	Celery
Apium	*graveolens-dulce-groena*	*floridensis-683-utah-typei*	Utah Type #683 Florida	Green Garden	Celery
Apium	*graveolens-dulce-groena*	*fordhook-emperori*	Emperor Fordhook	Green Garden	Celery
Apium	*graveolens-dulce-groena*	*fordhook-giantii*	Giant Fordhook	Green Garden	Celery
Apium	*graveolens-dulce-groena*	*fordhook-burpee'si*	Burpee's Fordhook	Green Garden	Celery
Apium	*graveolens-dulce-groena*	*french-dinantii*	French Dinant	Green Garden	Celery
Apium	*graveolens-dulce-groena*	*galaxii*	Galaxy	Green Garden	Celery
Apium	*graveolens-dulce-groena*	*green-bayi*	Green Bay	Green Garden	Celery
Apium	*graveolens-dulce-groena*	*hauseri*	Hauser	Green Garden	Celery
Apium	*graveolens-dulce-groena*	*hopkins-fenlanderi*	Hopkins Fenlander	Green Garden	Celery
Apium	*graveolens-dulce-groena*	*june-bellei*	June Belle	Green Garden	Celery
Apium	*graveolens-dulce-groena*	*pascal-giantii*	Giant Pascal	Green Garden	Celery
Apium	*graveolens-dulce-groena*	*pascal-giantii-selectii*	Select Giant Pascal	Green Garden	Celery
Apium	*graveolens-dulce-groena*	*pascal-giantii-tendracrispii*	Tendercrisp Giant Pascal	Green Garden	Celery
Apium	*graveolens-dulce-groena*	*pascal-greenii*	Green Pascal	Green Garden	Celery
Apium	*graveolens-dulce-groena*	*pascal-summeri*	Summer Pascal	Green Garden	Celery
Apium	*graveolens-dulce-groena*	*martha-warde'si*	Martha Warde's	Green Garden	Celery
Apium	*graveolens-dulce-groena*	*matadori*	Matador	Green Garden	Celery
Apium	*graveolens-dulce-groena*	*matadori-M9*	#9 Matador	Green Garden	Celery
Apium	*graveolens-dulce-groena*	*par-celii*	Par Cel	Green Garden	Celery
Apium	*graveolens-dulce-groena*	*petoi-285*	98 Day Peto Seed	Green Garden	Celery
Apium	*graveolens-dulce-groena*	*petoi-seedii-PS-285*	100 to 120 Day Peto Seed	Green Garden	Celery
Apium	*graveolens-dulce-groena*	*picadori-A865*	Picador	Green Garden	Celery
Apium	*graveolens-dulce-groena*	*picadori-sabrosoi*	Sabroso Picador	Green Garden	Celery
Apium	*graveolens-dulce-groena*	*promisei*	Promise	Green Garden	Celery
Apium	*graveolens-dulce-groena*	*rispensi-PYC-9112*	PYC #9112 Rispens	Green Garden	Celery
Apium	*graveolens-dulce-groena*	*rispensi-XP-266*	XP #266 Rispens	Green Garden	Celery
Apium	*graveolens-dulce-groena*	*summertime-FM-1646*	Summertime#FM1646	Green Garden	Celery
Apium	*graveolens-dulce-groena*	*tehama-sunrisei*	TehamaSunrise	Green Garden	Celery
Apium	*graveolens-dulce-groena*	*utah-5270-giantii*	Giant #5270 Utah	Green Garden	Celery
Apium	*graveolens-dulce-groena*	*utah-5270-giantii-improvedii*	Improved Giant #5270 Utah	Green Garden	Celery
Apium	*graveolens-dulce-groena*	*utah-5270-stokesi*	Stokes #5270 Utah	Green Garden	Celery
Apium	*graveolens-dulce-groena*	*utah-5270-triumphii*	Triumph #5270 Utah	Green Garden	Celery
Apium	*graveolens-dulce-groena*	*utah-5270-heartii*	Heart #5270 Utah	Green Garden	Celery
Apium	*graveolens-dulce-groena*	*utah-5270-heartii-keeperi*	Keeper Heart #5270 Utah	Green Garden	Celery
Apium	*graveolens-dulce-groena*	*utah-5270-resistantii*	Resistant #5270 Utah	Green Garden	Celery
Apium	*graveolens-dulce-groena*	*utah-5270-resistantii-improvedii*	Resistant Improved #5270 Utah	Green Garden	Celery
Apium	*graveolens-dulce-groena*	*utah-5275-tallii*	Tall #5275 Utah	Green Garden	Celery
Apium	*graveolens-dulce-groena*	*utah-EA-special-strainii*	EA Special Strain Utah	Green Garden	Celery
Apium	*graveolens-dulce-groena*	*utah-salt-lakei-15*	Salt Lake Utah	Green Garden	Celery
Apium	*graveolens-dulce-groena*	*utah-starletii*	Starlet Utah	Green Garden	Celery
Apium	*graveolens-dulce-groena*	*utah-summitii*	Summit Utah	Green Garden	Celery
Apium	*graveolens-dulce-groena*	*utah-tallii(utahensis)*	Tall Utah	Green Garden	Celery
Apium	*graveolens-dulce-groena*	*utah-tallii-green*	Green Tall Utah	Green Garden	Celery
Apium	*graveolens-dulce-groena*	*utah-venturai*	Ventura Utah	Green Garden	Celery
Apium	*graveolens-dulce-groena*	*wild-irishii*	WildIrish	Green Garden	Celery
Apium	*graveolens-dulce-helva*	*golden-chinensis*	Chinese Golden	Yellow Garden	Celery
Apium	*graveolens-dulce-helva*	*golden-medium*	Medium Chinese Golden	Yellow Garden	Celery
Apium	*graveolens-dulce-helva*	*golden-pascalii*	Pascal Golden	Yellow Garden	Celery
Apium	*graveolens-dulce-helva*	*golden-plumei*	Plume Golden	Yellow Garden	Celery
Apium	*graveolens-dulce-helva*	*golden-self-blanchingii-doratoi*	Dorato Self Blanching Golden	Yellow Garden	Celery
Apium	*graveolens-dulce-helva*	*golden-self-blanching-dwarfii*	Dwarf Self Blanching Golden	Yellow Garden	Celery
Apium	*graveolens-dulce-helva*	*golden-self-blanching-ivory-toweri*	Ivory Tower Self Blanching Golden	Yellow Garden	Celery
Apium	*graveolens-dulce-helva*	*golden-self-blanching-tallii*	Tall Self Blanching Golden	Yellow Garden	Celery
Apium	*graveolens-dulce-helva*	*golden-self-blanching-yellowi*	Perron's Self Blanching Golden	Yellow Garden	Celery
Apium	*graveolens-dulce-helva*	*golden-self-blanching-parisensis*	Paris Self Blanching Golden	Yellow Garden	Celery
Apium	*graveolens-dulce-rubra*	*pink-hardyi(pinkii)*	Hardy Pink	Red Garden	Celery
Apium	*graveolens-dulce-rubra*	*pink-plumei*	Plume Pink	Red Garden	Celery

Genus	species	infraspecies	cultivar	specimen	Genome
Apium	*graveolens-dulce-rubra*	*red-brydon-prizei*	Brydon Prize	Red Garden	Celery
Apium	*graveolens-dulce-rubra*	*red-giantii*	Giant	Red Garden	Celery
Apium	*graveolens-dulce-rubra*	*red-giantii-reselectionii*	Reselection Giant	Red Garden	Celery
Apium	*graveolens-dulce-rubra*	*red-heirloomii(redii)*	Heirloom	Red Garden	Celery
Apium	*graveolens-dulce-rubra*	*red-stalkii*	Red Stalk	Red Garden	Celery
Apium	*graveolens-dulce-rubra*	*red-stalkii-giantii*	Red Stalk Giant	Red Garden	Celery
Apium	*graveolens-dulce-rubra*	*red-venturei*	Red Venture	Red Garden	Celery

Celeriac (*Apium graveolens-rapaceum*)

Genus	species	infraspecies	cultivar	specimen	Genome
Apium	*graveolens-rapaceum*	*alba-alabasteri*	Alabaster White	Celeriac	Celery
Apium	*graveolens-rapaceum*	*alba-winter-soupii*	Winter Soup White	Celeriac	Celery
Apium	*graveolens-rapaceum*	*balderi*	Balder	Celeriac	Celery
Apium	*graveolens-rapaceum*	*balderi-blancoi*	Blanco Balder	Celeriac	Celery
Apium	*graveolens-rapaceum*	*baliensis*	Bali	Celeriac	Celery
Apium	*graveolens-rapaceum*	*brillians*	Brilliant	Celeriac	Celery
Apium	*graveolens-rapaceum*	*diamantii*	Diamont	Celeriac	Celery
Apium	*graveolens-rapaceum*	*dolvii*	Dolvi	Celeriac	Celery
Apium	*graveolens-rapaceum*	*iramii*	Iram	Celeriac	Celery
Apium	*graveolens-rapaceum*	*josei*	Jose	Celeriac	Celery
Apium	*graveolens-rapaceum*	*juvelii*	Juvel	Celeriac	Celery
Apium	*graveolens-rapaceum*	*knob-rootii(celeriaci)*	Knob Root	Celeriac	Celery
Apium	*graveolens-rapaceum*	*marble-ballii*	Marble Ball	Celeriac	Celery
Apium	*graveolens-rapaceum*	*mentori*	Mentor	Celeriac	Celery
Apium	*graveolens-rapaceum*	*monostopalyii*	Monostopalyi	Celeriac	Celery
Apium	*graveolens-rapaceum*	*monstorpolgii*	Monstorpolgi	Celeriac	Celery
Apium	*graveolens-rapaceum*	*praguei*	Prague	Celeriac	Celery
Apium	*graveolens-rapaceum*	*praguei-giantii*	Giant Prague	Celeriac	Celery
Apium	*graveolens-rapaceum*	*praguei-large-smoothii*	Large Smooth Prague	Celeriac	Celery
Apium	*graveolens-rapaceum*	*praguei-modelii*	Model Prague	Celeriac	Celery
Apium	*graveolens-rapaceum*	*praguei-monarchii*	Monarch Prague	Celeriac	Celery
Apium	*graveolens-rapaceum*	*presidentii*	President	Celeriac	Celery
Apium	*graveolens-rapaceum*	*sedano-r.-bianco-venetoi*	Sedano R. Bianco Veneto	Celeriac	Celery
Apium	*graveolens-rapaceum*	*snow-whitei*	Snow White	Celeriac	Celery
Apium	*graveolens-rapaceum*	*snow-whitei-snevidei*	Snehvide Snow White	Celeriac	Celery
Apium	*graveolens-rapaceum*	*tellus*	Tell	Celeriac	Celery
Apium	*graveolens-secalinum*	*celery-leafii*	Fragrant	Leaf Garden	Celery
Apium	*graveolens-secalinum*	*celeri-a-couperi*	Couper	Leaf Garden	Celery
Apium	*graveolens-secalinum*	*cutting-leaf-celeryi*	Cutting	Leaf Garden	Celery
Apium	*graveolens-secalinum*	*E-Z-leafii*	E-Z	Leaf Garden	Celery
Apium	*graveolens-secalinum*	*zwolsche-krulii*	ZwolscheKrul	Leaf Garden	Celery

Coriander (*Coriandrum sativum*)

Genus	species	infraspecies	cultivar	specimen	Genome
Coriandrum	*sativum*	*chinensis*	Chinese	Parsley	Coriander
Coriandrum	*sativum*	*cilantrum*	Cilantro	Parsley	Coriander
Coriandrum	*sativum*	*cilantrum-cal-long-standingii*	Cal Long Standing Cilantro	Parsley	Coriander
Coriandrum	*sativum*	*cilantrum-yen-sai*	Yen Sai Cilantro	Parsley	Coriander
Coriandrum	*sativum*	*leisurei*	Leisure	Parsley	Coriander
Coriandrum	*sativum*	*marocana*	Moroccan	Parsley	Coriander
Coriandrum	*sativum*	*ostindica*	East Indies	Parsley	Coriander
Coriandrum	*sativum*	*santoi*	Santo	Parsley	Coriander
Coriandrum	*sativum*	*slo-boltii*	Slo-bolt	Parsley	Coriander
Coriandrum	*sativum*	*standbyi*	Standby	Parsley	Coriander
Coriandrum	*sativum*	*tian-chinii*	Tian Chin	Parsley	Coriander
Coriandrum	*sativum*	*vietnamensis*	Vietnamese	Parsley	Coriander

Honewort (*Cryptotaenia spp.*)

Genus	species	infraspecies	cultivar	specimen	Genome
Cryptotaenia	*canadensis*	*alba*	White	Canadian Chervil	Honewort
Cryptotaenia	*japonica*	*atropurpurea*	Dark Purple Red	Japanese Parsley	Honewort
Cryptotaenia	*japonica*	*mitsuba*	Mitsuba	Japanese Parsley	Honewort

Carrot (*Daucus carota-sativus*)

Genus	species	infraspecies	cultivar	specimen	Genome
Daucus	*carota-sativus-cylindra*	*amsterdam-abk*	Bunch Keep Amsterdam 6.0"	Cylindrical	Carrot
Daucus	*carota-sativus-cylindra*	*amsterdam-corelessi*	Coreless Amsterdam ?.?"	Cylindrical	Carrot
Daucus	*carota-sativus-cylindra*	*amsterdam-forcingii-bak*	Bak Forcing Amsterdam 6.5"	Cylindrical	Carrot
Daucus	*carota-sativus-cylindra*	*amsterdam-minicori*	Minicor Amsterdam 6.5"	Cylindrical	Carrot
Daucus	*carota-sativus-cylindra*	*autumn-king-flakkee-grosa*	Grosa Flakkee Autumn King 10."	Cylindrical	Carrot
Daucus	*carota-sativus-cylindra*	*baby-longii*	Baby Long 6.5"	Cylindrical	Carrot
Daucus	*carota-sativus-cylindra*	*berlicum-berjoi*	Berjo Berlicum 8.0"	Cylindrical	Carrot
Daucus	*carota-sativus-cylindra*	*berlicum-berlicummeri*	Berlicummer Berlicum 10."	Cylindrical	Carrot

Genus	species	infraspecies	cultivar	specimen	Genome
Daucus	*carota-sativus-cylindra*	*berlicum-cardinalis*	Cardinal Berlicum ?.?"	Cylindrical	Carrot
Daucus	*carota-sativus-cylindra*	*berlicum-camberleyi*	Camberley Berlicum 8.0"	Cylindrical	Carrot
Daucus	*carota-sativus-cylindra*	*chantenay-gold-king-royalis*	Gold King Royal Chantenay 6.5"	Cylindrical	Carrot
Daucus	*carota-sativus-cylindra*	*chantenay-imperialis*	Imperial Chantenay 6.0"	Cylindrical	Carrot
Daucus	*carota-sativus-cylindra*	*maincrop-nandrina*	Nandrin Maincrop ?.?"	Cylindrical	Carrot
Daucus	*carota-sativus-cylindra*	*maincrop-oranza*	Oranza Maincrop 8.5"	Cylindrical	Carrot
Daucus	*carota-sativus-cylindra*	*manchester-tablei*	ManchesterTable 10."	Cylindrical	Carrot
Daucus	*carota-sativus-cylindra*	*nantes-brilliancei*	Brilliance Nantes ?.?"	Cylindrical	Carrot
Daucus	*carota-sativus-cylindra*	*nantes-corelessi*	Coreless Nantes 7.0"	Cylindrical	Carrot
Daucus	*carota-sativus-cylindra*	*nantes-corelessi-improvedii*	Improved Coreless Nantes 6.5"	Cylindrical	Carrot
Daucus	*carota-sativus-cylindra*	*nantes-corelessi-scarletii*	Scarlet Coreless Nantes 7.0"	Cylindrical	Carrot
Daucus	*carota-sativus-cylindra*	*nantes-corelessi-touchonii*	Touchon Coreless Nantes 7.2"	Cylindrical	Carrot
Daucus	*carota-sativus-cylindra*	*nantes-corelessi-touchon-amtouii*	Amtou Touchon Coreless Nantes 6.5"	Cylindrical	Carrot
Daucus	*carota-sativus-cylindra*	*nantes-corelessi-touchonii-deluxei*	Touchon Deluxe Coreless Nantes 7.0"	Cylindrical	Carrot
Daucus	*carota-sativus-cylindra*	*nantes-dukei*	Duke Nantes 7.5"	Cylindrical	Carrot
Daucus	*carota-sativus-cylindra*	*nantes-dukei-1003*	#1003 Duke Nantes 7.5"	Cylindrical	Carrot
Daucus	*carota-sativus-cylindra*	*nantes-fancyi*	Fancy Nantes 6.5"	Cylindrical	Carrot
Daucus	*carota-sativus-cylindra*	*nantes-fortoi*	Forto Nantes 6.5"	Cylindrical	Carrot
Daucus	*carota-sativus-cylindra*	*nantes-improvedii*	Improved Nantes 6.0"	Cylindrical	Carrot
Daucus	*carota-sativus-cylindra*	*nantes-landai*	Landa Nantes 7.5"	Cylindrical	Carrot
Daucus	*carota-sativus-cylindra*	*nantes-longii*	Long Cylindrical Nantes 6.5"	Cylindrical	Carrot
Daucus	*carota-sativus-cylindra*	*nantes-strong-topii*	Strong Top Nantes 6.5"	Cylindrical	Carrot
Daucus	*carota-sativus-cylindra*	*nantes-tiptopii-earlyi*	Early Tip Top Nantes 7.0"	Cylindrical	Carrot
Daucus	*carota-sativus-cylindra*	*nantes-tiptopii-nantaisei*	Nantaise Tip Top Nantes 7.2"	Cylindrical	Carrot
Daucus	*carota-sativus-cylindra*	*fancyi*	Fancy 7.5"	Cylindrical	Carrot
Daucus	*carota-sativus-cylindra*	*presto*	Presto	Cylindrical	Carrot
Daucus	*carota-sativus-cylindra*	*scarlet-keeperi*	Scarlet Keeper 8.5"	Cylindrical	Carrot
Daucus	*carota-sativus-cylindra*	*sweetheartii*	Sweetheart	Cylindrical	Carrot
Daucus	*carota-sativus-nana*	*amsterdam-amini*	Mini Amsterdam 3.0"	Baby	Carrot
Daucus	*carota-sativus-nana*	*amsterdam-amstelii*	Amstel Amsterdam 4.0"	Baby	Carrot
Daucus	*carota-sativus-nana*	*amsterdam-babettei*	Babette Amsterdam 5.0"	Baby	Carrot
Daucus	*carota-sativus-nana*	*amsterdam-baby-spikei*	Baby Spike Amsterdam 4.0"	Baby	Carrot
Daucus	*carota-sativus-nana*	*amsterdam-forcingii*	Forcing Amsterdam 5.0"	Baby	Carrot
Daucus	*carota-sativus-nana*	*amsterdam-forcing-indus*	Indu Forcing Amsterdam 5.0"	Baby	Carrot
Daucus	*carota-sativus-nana*	*amsterdam-lady-fingeri*	Lady Finger Amsterdam 3.5"	Baby	Carrot
Daucus	*carota-sativus-nana*	*baby-fingeri*	Baby Finger 4.0"	Baby	Carrot
Daucus	*carota-sativus-nana*	*bambina*	Bambina ?.?"	Baby	Carrot
Daucus	*carota-sativus-nana*	*early-marketii*	Early Market 4.0"	Baby	Carrot
Daucus	*carota-sativus-nana*	*mignonii*	Mignon 3.0"	Baby	Carrot
Daucus	*carota-sativus-nana*	*mini-sweetii*	Mini Sweet 4.0"	Baby	Carrot
Daucus	*carota-sativus-nana*	*nantes-little-fingeri*	Little Finger Nantes 3.2"	Baby	Carrot
Daucus	*carota-sativus-nana*	*nantes-little-fingeri-improva*	Improved Little Finger Nantes 3.5"	Baby	Carrot
Daucus	*carota-sativus-nana*	*scarlet-horn-earlyi-60*	60 Day Early Scarlet Horn 4.0"	Baby	Carrot
Daucus	*carota-sativus-nana*	*scarlet-horn-earlyi-68*	68 Day Early Scarlet Horn 4.0"	Baby	Carrot
Daucus	*carota-sativus-nana*	*sucramii*	Sucram 3.5"	Baby	Carrot
Daucus	*carota-sativus-nana*	*sukoi*	Suko 2.5"	Baby	Carrot
Daucus	*carota-sativus-nana*	*sweet-n-shortii*	Sweet 'n Short 3.5"	Thick	Carrot
Daucus	*carota-sativus-rotunda*	*bolinai*	Bolina 2.0"	Round	Carrot
Daucus	*carota-sativus-rotunda*	*early-french-framei*	Early French Frame 1.0"	Round	Carrot
Daucus	*carota-sativus-rotunda*	*kundulus*	Kundulus ?.?"	Round	Carrot
Daucus	*carota-sativus-rotunda*	*orbitii*	Orbit 1.0"	Round	Carrot
Daucus	*carota-sativus-rotunda*	*oxheartii*	Oxheart 6.0"	Round	Carrot
Daucus	*carota-sativus-rotunda*	*parabela*	Parabel 1.2"	Round	Carrot
Daucus	*carota-sativus-rotunda*	*paris-market-redii*	Red Paris Market ?.?"	Round	Carrot
Daucus	*carota-sativus-rotunda*	*paris-market-parisian-ballii*	Parisian Ball Paris Market 1.0"	Round	Carrot
Daucus	*carota-sativus-rotunda*	*paris-market-pariseri-4*	#4 Pariser Paris Market ?.?"	Round	Carrot
Daucus	*carota-sativus-rotunda*	*paris-market-parisiennei*	Parisienne Paris Market 1.5"	Round	Carrot
Daucus	*carota-sativus-rotunda*	*paris-market-parmexii*	Parmex Paris Market 1.2"	Round	Carrot
Daucus	*carota-sativus-rotunda*	*paris-market-rondoi*	Rondo Paris Market ?.?"	Round	Carrot
Daucus	*carota-sativus-rotunda*	*romeoi*	Romeo ?.?"	Round	Carrot
Daucus	*carota-sativus-rotunda*	*thumbelina*	Thumbelina 1.0"	Round	Carrot
Daucus	*carota-sativus-rotunda*	*tonda-di-parigii*	Tonda Di Parigi 1.5"	Round	Carrot
Daucus	*carota-sativus-tapera-acuta*	*24-karatii*	24 Karat ?.?"	Pointed Tapered	Carrot
Daucus	*carota-sativus-tapera-acuta*	*berlicum-fina*	Fina Berlicum 9.5"	Pointed Tapered	Carrot
Daucus	*carota-sativus-tapera-acuta*	*berlicum-noboi*	Nobo Berlicum 8.5"	Pointed Tapered	Carrot
Daucus	*carota-sativus-tapera-acuta*	*berlicum-robila*	Robila Berlicum 7.0"	Pointed Tapered	Carrot
Daucus	*carota-sativus-tapera-acuta*	*berlicum-rona*	Rona Berlicum 8.5"	Pointed Tapered	Carrot
Daucus	*carota-sativus-tapera-acuta*	*beta-3*	Beta-III	Pointed Tapered	Carrot
Daucus	*carota-sativus-tapera-acuta*	*brazilia*	Brazilia	Pointed Tapered	Carrot
Daucus	*carota-sativus-tapera-acuta*	*carentanii*	Carenton	Pointed Tapered	Carrot
Daucus	*carota-sativus-tapera-acuta*	*condor*	Condor	Pointed Tapered	Carrot
Daucus	*carota-sativus-tapera-acuta*	*imperator-centennialis*	Centennial Imperator ?.?"	Pointed Tapered	Carrot
Daucus	*carota-sativus-tapera-acuta*	*imperator-tendersweet-pointedii*	Pointed Tendersweet Imperator 8.5"	Pointed Tapered	Carrot
Daucus	*carota-sativus-tapera-acuta*	*japanese-longii-imperialis*	Imperial Japanese Long 10."	Pointed Tapered	Carrot

Genus	species	infraspecies	cultivar	specimen	Genome
Daucus	*carota-sativus-tapera-acuta*	*japanese-longii*	Japanese Long 24."	Pointed Tapered	Carrot
Daucus	*carota-sativus-tapera-acuta*	*juwarotii*	Juwarot 8.0"	Pointed Tapered	Carrot
Daucus	*carota-sativus-tapera-acuta*	*long-orangei*	Long Orange 12."	Pointed Tapered	Carrot
Daucus	*carota-sativus-tapera-acuta*	*long-orangei-improvedii*	Improved Long Orange 11."	Pointed Tapered	Carrot
Daucus	*carota-sativus-tapera-acuta*	*lubiana*	Lubiana ?.?"	Pointed Tapered	Carrot
Daucus	*carota-sativus-tapera-acuta*	*lunga-di-san-valerioi*	Lunga Di San Velario ?.?"	Pointed Tapered	Carrot
Daucus	*carota-sativus-tapera-acuta*	*maincrop-jamesi-scarletii*	James Scarlet Maincrop ?.?"	Pointed Tapered	Carrot
Daucus	*carota-sativus-tapera-acuta*	*maincrop-oranza-falconii-2*	Falcon II Oranza Maincrop 7.0"	Pointed Tapered	Carrot
Daucus	*carota-sativus-tapera-acuta*	*maincrop-red-intermediate-newi*	Red Intermediate New Maincrop 11."	Pointed Tapered	Carrot
Daucus	*carota-sativus-tapera-acuta*	*red-atomica*	Atomic Red ?.?"	Pointed Tapered	Carrot
Daucus	*carota-sativus-tapera-acuta*	*red-kintoki-earlyi*	Early Kintoki Red 8.0"	Pointed Tapered	Carrot
Daucus	*carota-sativus-tapera-acuta*	*red-kintoki-kyotoi*	Kyoto Kintoki Red 8.0"	Pointed Tapered	Carrot
Daucus	*carota-sativus-tapera-acuta*	*nairobiensis*	Nairobi ?.?"	Pointed Tapered	Carrot
Daucus	*carota-sativus-tapera-acuta*	*napoli*	Napoli ?.?"	Pointed Tapered	Carrot
Daucus	*carota-sativus-tapera-acuta*	*nelsonii*	Nelson ?.?"	Thick	Carrot
Daucus	*carota-sativus-tapera-acuta*	*paranoi*	Parano ?.?"	Pointed Tapered	Carrot
Daucus	*carota-sativus-tapera-acuta*	*primoi*	Primo ?.?"	Pointed Tapered	Carrot
Daucus	*carota-sativus-tapera-acuta*	*PSR-462*	Pacific Strongtop Resistant 11."	Pointed Tapered	Carrot
Daucus	*carota-sativus-tapera-acuta*	*purple-afghanensis*	Afghan Purple 9.0"	Pointed Tapered	Carrot
Daucus	*carota-sativus-tapera-acuta*	*purple-cosmici*	Cosmic Purple ?.?"	Pointed Tapered	Carrot
Daucus	*carota-sativus-tapera-acuta*	*purple-dragonii*	Dragon Purple ?.?"	Pointed Tapered	Carrot
Daucus	*carota-sativus-tapera-acuta*	*red-kurnai*	Kurna Red 6.5"	Pointed Tapered	Carrot
Daucus	*carota-sativus-tapera-acuta*	*red-elephantii*	Elephant Red ?.?"	Pointed Tapered	Carrot
Daucus	*carota-sativus-tapera-acuta*	*red-muscadei*	Muscade Red ?.?"	Pointed Tapered	Carrot
Daucus	*carota-sativus-tapera-acuta*	*red-redcai*	Redca Red ?.?"	Pointed Tapered	Carrot
Daucus	*carota-sativus-tapera-acuta*	*rothildii*	Rothild 9.0"	Pointed Tapered	Carrot
Daucus	*carota-sativus-tapera-acuta*	*rothildii-rodlikai*	Rodelika Rothild 7.0"	Pointed Tapered	Carrot
Daucus	*carota-sativus-tapera-acuta*	*saint-valeryi*	Saint Valery 11."	Pointed Tapered	Carrot
Daucus	*carota-sativus-tapera-acuta*	*saint-valeryi-improvedii*	Improved Saint Valery 9.5"	Pointed Tapered	Carrot
Daucus	*carota-sativus-tapera-acuta*	*saint-valeryi-james-scarletii*	James Scarlet Saint Valery 11."	Pointed Tapered	Carrot
Daucus	*carota-sativus-tapera-acuta*	*scarlet-wonderi*	Scarlet Wonder 13."	Pointed Tapered	Carrot
Daucus	*carota-sativus-tapera-acuta*	*tokita's-scarletii*	Tokita's Scarlet 7.0"	Pointed Tapered	Carrot
Daucus	*carota-sativus-tapera-acuta*	*topweightii*	Top Weight 11."	Pointed Tapered	Carrot
Daucus	*carota-sativus-tapera-acuta*	*white-belgica*	Belgium White 9.0"	Pointed Tapered	Carrot
Daucus	*carota-sativus-tapera-acuta*	*white-belgica-blanchei*	Blanche Belgium White 9.0"	Pointed Tapered	Carrot
Daucus	*carota-sativus-tapera-acuta*	*white-kuttigeri*	Kuttiger White 7.0"	Pointed Tapered	Carrot
Daucus	*carota-sativus-tapera-acuta*	*white-lunari*	Lunar White ?.?"	Pointed Tapered	Carrot
Daucus	*carota-sativus-tapera-acuta*	*white-frenchii*	French White	Pointed Tapered	Carrot
Daucus	*carota-sativus-tapera-acuta*	*white-parlseyi*	Parsley White	Pointed Tapered	Carrot
Daucus	*carota-sativus-tapera-acuta*	*yellow-amarilloi*	Amarillo Yellow ?.?"	Pointed Tapered	Carrot
Daucus	*carota-sativus-tapera-acuta*	*yellow-juane-obtusei*	Jaune Obtuse Yellow ?.?"	Pointed Tapered	Carrot
Daucus	*carota-sativus-tapera-acuta*	*yellow-long-red-surreyi*	Long Red Surrey Yellow 24."	Pointed Tapered	Carrot
Daucus	*carota-sativus-tapera-acuta*	*yellow-solari*	Solar Yellow ?.?"	Pointed Tapered	Carrot
Daucus	*carota-sativus-tapera-acuta*	*yellow-uberlandia*	Uberland Yellow ?.?"	Pointed Tapered	Carrot
Daucus	*carota-sativus-autumna-kingii*	*redii-#2*	#2 Red 11."	Autumn King Blunt Tapered	Carrot
Daucus	*carota-sativus-autumna-kingii*	*campestra*	Field ??."	Autumn King Blunt Tapered	Carrot
Daucus	*carota-sativus-autumna-kingii*	*flakkeei*	Flakkee 10."	Autumn King Blunt Tapered	Carrot
Daucus	*carota-sativus-autumna-kingii*	*flakkee-flakkoi*	Flakko Flakkee 9.0"	Autumn King Blunt Tapered	Carrot
Daucus	*carota-sativus-autumna-kingii*	*flakkee-long-red-giantii*	Long Red Giant Flakkee 24."	Autumn King Blunt Tapered	Carrot
Daucus	*carota-sativus-autumna-kingii*	*flakkee-regolii*	Regol Flakkee 9.5"	Autumn King Blunt Tapered	Carrot
Daucus	*carota-sativus-autumna-kingii*	*flakkee-trofeoi*	Trofeo Flakkee 5.5"	Autumn King Blunt Tapered	Carrot
Daucus	*carota-sativus-autumna-kingii*	*giganta*	Giganta 8.0"	Autumn King Blunt Tapered	Carrot
Daucus	*carota-sativus-autumna-kingii*	*gold-kingii*	Gold King 6.0"	Autumn King Blunt Tapered	Carrot
Daucus	*carota-sativus-autumna-kingii*	*improvedii*	Improved 8.0"	Autumn King Blunt Tapered	Carrot
Daucus	*carota-sativus-chantenayi*	*demilonguei*	Demi-Longue 5.5"	Chantenay Blunt Tapered	Carrot
Daucus	*carota-sativus-chantenayi*	*red-core-cluseedii*	Cluseed Red Core ?.?"	Chantenay Blunt Tapered	Carrot
Daucus	*carota-sativus-chantenayi*	*red-core-cometii*	Comet Red Core ?.?"	Chantenay Blunt Tapered	Carrot
Daucus	*carota-sativus-chantenayi*	*red-core-earlyi*	Early Red Core 5.5"	Chantenay Blunt Tapered	Carrot
Daucus	*carota-sativus-chantenayi*	*red-core-supremei*	Supreme Red Core ?.?"	Chantenay Blunt Tapered	Carrot
Daucus	*carota-sativus-chantenayi*	*kinkoi-4"*	Kinko 4.0"	Chantenay Blunt Tapered	Carrot
Daucus	*carota-sativus-chantenayi*	*kinkoi-6"*	Kinko 6.0"	Chantenay Blunt Tapered	Carrot
Daucus	*carota-sativus-chantenayi*	*kuroda*	Kuroda 6.0"	Chantenay Blunt Tapered	Carrot
Daucus	*carota-sativus-chantenayi*	*kuroda-kono-shin-gosunii*	Kono Shin Goson Kuroda ?.?"	Chantenay Blunt Tapered	Carrot
Daucus	*carota-sativus-chantenayi*	*kuroda-nova*	Nova Kuroda 7.5"	Chantenay Blunt Tapered	Carrot
Daucus	*carota-sativus-chantenayi*	*kuroda-shinnii-5.0"*	Shinn Kuroda 5.0"	Chantenay Blunt Tapered	Carrot
Daucus	*carota-sativus-chantenayi*	*kuroda-shinnii-8.5"*	Shinn Kuroda 8.5"	Chantenay Blunt Tapered	Carrot
Daucus	*carota-sativus-chantenayi*	*longii*	Long 7.0"	Chantenay Blunt Tapered	Carrot
Daucus	*carota-sativus-chantenayi*	*redii*	Red ?.?"	Chantenay Blunt Tapered	Carrot
Daucus	*carota-sativus-chantenayi*	*redcoi*	Redco 6.5"	Chantenay Blunt Tapered	Carrot
Daucus	*carota-sativus-chantenayi*	*red-coredii*	Red Cored 6.0"	Chantenay Blunt Tapered	Carrot
Daucus	*carota-sativus-chantenayi*	*rola*	Rola Royal 7.5"	Chantenay Blunt Tapered	Carrot
Daucus	*carota-sativus-chantenayi*	*royalis*	Royal 6.0"	Chantenay Blunt Tapered	Carrot
Daucus	*carota-sativus-danversi*	*danroi*	Danro 7.0"	Danvers Blunt Tapered	Carrot
Daucus	*carota-sativus-danversi*	*danroi-fakkel-mixii*	Fakkel Mix Danro 6.5"	Danvers Blunt Tapered	Carrot

Genus	species	infraspecies	cultivar	specimen	Genome
Daucus	*carota-sativus-danversi*	*half-longii*	Half Long 7.0"	Danvers Blunt Tapered	Carrot
Daucus	*carota-sativus-danversi*	*red-corei-126*	#126 Red Core 8.0"	Danvers Blunt Tapered	Carrot
Daucus	*carota-sativus-danversi*	*red-corei-126-improvedii*	#126 Improved Red Core 7.0"	Danvers Blunt Tapered	Carrot
Daucus	*carota-sativus-danversi*	*red-corei-half-longii*	Half Long Red Core 7.0"	Danvers Blunt Tapered	Carrot
Daucus	*carota-sativus-imperator*	*#58*	#58 10."	Imperator Blunt Tapered	Carrot
Daucus	*carota-sativus-imperator*	*#58-improva*	Improved #58 8.0"	Imperator Blunt Tapered	Carrot
Daucus	*carota-sativus-imperator*	*gold-pakii*	Gold Pak 9.0"	Imperator Blunt Tapered	Carrot
Daucus	*carota-sativus-imperator*	*gold-pakii-28*	#28 Gold Pak 9.0"	Imperator Blunt Tapered	Carrot
Daucus	*carota-sativus-imperator*	*j.n.-cellobunch*	J. N. Cellobunch 8.0"	Imperator Blunt Tapered	Carrot
Daucus	*carota-sativus-imperator*	*j.o.caropakii*	J. O. Caropak ?.?"	Imperator Blunt Tapered	Carrot
Daucus	*carota-sativus-imperator*	*tendersweetii*	Tendersweet 8.5"	Imperator Blunt Tapered	Carrot
Daucus	*carota-sativus-imperator*	*lange-rote-stumpfe-zinoi-2*	Lange Rote Stumpfe Zino #2 Red 10."	Imperator Blunt Tapered	Carrot
Daucus	*carota-sativus-imperator*	*lavalii*	Laval 9.0"	Imperator Blunt Tapered	Carrot
Daucus	*carota-sativus-imperator*	*longa-bunchingii*	Long Bunching 10."	Imperator Blunt Tapered	Carrot
Daucus	*carota-sativus-imperator*	*nevadaensis*	Nevada ?.?"	Imperator Blunt Tapered	Carrot
Daucus	*carota-sativus-imperator*	*nutri-redii*	Nutri-Red 8.5"	Imperator Blunt Tapered	Carrot
Daucus	*carota-sativus-imperator*	*phalzeri*	Phalzer Yellow ?.?"	Imperator Blunt Tapered	Carrot
Daucus	*carota-sativus-imperator*	*top-pakii*	Top Pak 8.0"	Imperator Blunt Tapered	Carrot
Daucus	*carota-sativus-nantes*	*armstrongii*	Armstrong 7.5"	Nantes Blunt Tapered	Carrot
Daucus	*carota-sativus-nantes*	*corelessi-earlyi*	Early Coreless 6.5"	Nantes Blunt Tapered	Carrot
Daucus	*carota-sativus-nantes*	*corelessi-topascorei*	Top Score Coreless 6.0"	Nantes Blunt Tapered	Carrot
Daucus	*carota-sativus-nantes*	*di-chioggia*	DiChioggi ?.?"	Nantes Blunt Tapered	Carrot
Daucus	*carota-sativus-nantes*	*douceuri*	Douceur 6.5"	Nantes Blunt Tapered	Carrot
Daucus	*carota-sativus-nantes*	*empirei*	Empire 7.5"	Nantes Blunt Tapered	Carrot
Daucus	*carota-sativus-nantes*	*expressi*	Express 6.5"	Nantes Blunt Tapered	Carrot
Daucus	*carota-sativus-nantes*	*half-longii*	Half Long 6.5"	Nantes Blunt Tapered	Carrot
Daucus	*carota-sativus-nantes*	*merida*	Merida 7.5"	Nantes Blunt Tapered	Carrot
Daucus	*carota-sativus-nantes*	*mexicana-sunblestii*	Sunblest Mexican ?.?"	Nantes Blunt Tapered	Carrot
Daucus	*carota-sativus-nantes*	*mexicana-strongii*	Strong Mexican ?.?"	Nantes Blunt Tapered	Carrot
Daucus	*carota-sativus-nantes*	*nandori*	Nandor ?.?"	Nantes Blunt Tapered	Carrot
Daucus	*carota-sativus-nantes*	*nantiasei*	Nantaise 7.5"	Nantes Blunt Tapered	Carrot
Daucus	*carota-sativus-nantes*	*nantiasei-gemii*	Gem Nantaise ?.?"	Nantes Blunt Tapered	Carrot
Daucus	*carota-sativus-nantes*	*nantiasei-gemii-improvedii*	Improved Gem Nantaise ?.?"	Nantes Blunt Tapered	Carrot
Daucus	*carota-sativus-nantes*	*nantaisei-tim-tomii*	Tim Tom Nantaise 5.5"	Nantes Blunt Tapered	Carrot
Daucus	*carota-sativus-nantes*	*nantesa-superiori*	Superior Nantesa 5.5"	Nantes Blunt Tapered	Carrot
Daucus	*carota-sativus-nantes*	*phil-barber'si*	Phil Barber's White ?.?"	Nantes Blunt Tapered	Carrot
Daucus	*carota-sativus-nantes*	*rumbai*	Rumba 6.5"	Nantes Blunt Tapered	Carrot
Daucus	*carota-sativus-nantes*	*specialis-616*	Special #616 7.0"	Nantes Blunt Tapered	Carrot
Daucus	*carota-sativus-nantes*	*titanii*	Titan ?.?"	Nantes Blunt Tapered	Carrot
Daucus	*carota-sativus-nantes*	*touching-earth-2*	Touching Earth #2 7.0"	Nantes Blunt Tapered	Carrot

Florence Fennel (*Foeniculum azoricum*)

Genus	species	infraspecies	cultivar	specimen	Genome
Foeniculum	*azoricum*	*autumn-giantii*	Perfection	Florence	Fennel
Foeniculum	*azoricum*	*dulce*	Sweet	Florence	Fennel
Foeniculum	*azoricum*	*earlyi*	Early	Florence	Fennel
Foeniculum	*azoricum*	*finoi*	Fino	Florence	Fennel
Foeniculum	*azoricum*	*finoi-zefoi*	Zefa Fino	Florence	Fennel
Foeniculum	*azoricum*	*heraldii*	Herald	Florence	Fennel
Foeniculum	*azoricum*	*mammothii*	Mammoth	Florence	Fennel
Foeniculum	*azoricum*	*montebiancoi*	Montebianco	Florence	Fennel
Foeniculum	*azoricum*	*montovanoi*	Montovano	Florence	Fennel
Foeniculum	*azoricum*	*napoletanoi*	Napoletano	Florence	Fennel
Foeniculum	*azoricum*	*parma-sel-pradoi*	Parma Sel Prado	Florence	Fennel
Foeniculum	*azoricum*	*perfection*	Perfection	Florence	Fennel
Foeniculum	*azoricum*	*romanescoi*	Romanesco	Florence	Fennel
Foeniculum	*azoricum*	*romyi*	Romy	Florence	Fennel
Foeniculum	*azoricum*	*wadenromen-grossoi*	Grosso Wadenromen	Florence	Fennel
Foeniculum	*azoricum*	*wadenromen-superi*	Super Wadenromen	Florence	Fennel
Foeniculum	*azoricum*	*white-mountainii*	White Mountain	Florence	Fennel

Fennel (*Foeniculum vulgare*)

Genus	species	infraspecies	cultivar	specimen	Genome
Foeniculum	*vulgare*	*anisatum*	Anise Seed	Sweet	Fennel
Foeniculum	*vulgare*	*piperatum*	Pepper Seed	Sweet	Fennel
Foeniculum	*vulgare*	*purpureum*	Bronze Leaf	Sweet	Fennel

Parsnip (*Pastinaca sativa*)

Pastinaca	*sativa*	*alba-gemma*	White Gem ?.?"	European	Parsnip
Pastinaca	*sativa*	*alba-coniga*	White King ?.?"	European	Parsnip
Pastinaca	*sativa*	*allo-americana*	All American Hollow Crown 11."	European	Parsnip
Pastinaca	*sativa*	*andoverensis*	Andover 11."	European	Parsnip
Pastinaca	*sativa*	*arrovii*	Arrow ?.?"	European	Parsnip

Pastinaca	*sativa*	*avonensis-resistiva*	Avon Resistor 3.0"	European	Parsnip
Pastinaca	*sativa*	*codhamii-improva-malva*	Codham Improved Marrow 8.0"	European	Parsnip
Pastinaca	*sativa*	*dlouhyi-bilyi*	Dlouhy Bily ?.?"	European	Parsnip
Pastinaca	*sativa*	*guernseii*	Guernsey ?.?"	European	Parsnip
Pastinaca	*sativa*	*guernseii-cava-corona*	Hollow Crown Guernsey 12."	European	Parsnip
Pastinaca	*sativa*	*guernseii-cava-corona-improva*	Improved Hollow Crown Guernsey 11."	European	Parsnip
Pastinaca	*sativa*	*guernseii-cava-corona-gaze'si*	Gaze's Hollow Crown Guernsey 10."	European	Parsnip
Pastinaca	*sativa*	*harrisi-modelii*	Harris Model 11."	European	Parsnip
Pastinaca	*sativa*	*harrisi-modelii-lanceri*	Lancer Harris Model 11."	European	Parsnip
Pastinaca	*sativa*	*harrisi-modelii-praecox*	Early Harris Model 11."	European	Parsnip
Pastinaca	*sativa*	*kralii-russica*	Kral Russian ?.?"	European	Parsnip
Pastinaca	*sativa*	*modelii-cava-corona*	Model Hollow Crown ?.?"	European	Parsnip
Pastinaca	*sativa*	*oatneyi*	Oatney Long ?.?"	European	Parsnip
Pastinaca	*sativa*	*offenhamii*	Offenham ?.?"	European	Parsnip
Pastinaca	*sativa*	*short-thickii*	Short Thick ?.?"	European	Parsnip
Pastinaca	*sativa*	*studens*	The Student 15."	European	Parsnip
Pastinaca	*sativa*	*survivori*	Survivor ?.?"	European	Parsnip
Pastinaca	*sativa*	*tender-&-truei*	Tender & True ?.?"	European	Parsnip
Pastinaca	*sativa*	*turga*	Turga	European	Parsnip

Parsley (*Petroselinum crispum*)

Genus	species	infraspecies	cultivar	specimen	Genome
Petroselinum	*crispum*	*crispum*	Original	Curly Leaf	Parsley
Petroselinum	*crispum*	*afroi*	Afro	Curly Leaf	Parsley
Petroselinum	*crispum*	*aphroditea*	Aphrodite	Curly Leaf	Parsley
Petroselinum	*crispum*	*banquetii*	Banquet	Curly Leaf	Parsley
Petroselinum	*crispum*	*bravouri*	Bravour	Curly Leaf	Parsley
Petroselinum	*crispum*	*clivii*	Clivy	Curly Leaf	Parsley
Petroselinum	*crispum*	*curlina*	Curlina	Curly Leaf	Parsley
Petroselinum	*crispum*	*darca-groena-crispa*	Dark Green Curled	Curly Leaf	Parsley
Petroselinum	*crispum*	*darci*	Darky	Curly Leaf	Parsley
Petroselinum	*crispum*	*decora*	Decora	Curly Leaf	Parsley
Petroselinum	*crispum*	*delicata-originalis*	Delicate Original	Curly Leaf	Parsley
Petroselinum	*crispum*	*duplicata-crispa*	Double Curled	Curly Leaf	Parsley
Petroselinum	*crispum*	*envia*	Envy	Curly Leaf	Parsley
Petroselinum	*crispum*	*esquirei*	Esquire	Curly Leaf	Parsley
Petroselinum	*crispum*	*frisca*	Frisca	Curly Leaf	Parsley
Petroselinum	*crispum*	*groena-crispa*	Green Curled	Curly Leaf	Parsley
Petroselinum	*crispum*	*groena-riparia*	Green River	Curly Leaf	Parsley
Petroselinum	*crispum*	*groena-stellata*	Green Star	Curly Leaf	Parsley
Petroselinum	*crispum*	*krausa*	Kraus Market	Curly Leaf	Parsley
Petroselinum	*crispum*	*musca-addita-crispa-nana*	Extra Dwarf Moss Curled	Curly Leaf	Parsley
Petroselinum	*crispum*	*musca-addita-tripla-crispa*	Extra Triple Moss Curled	Curly Leaf	Parsley
Petroselinum	*crispum*	*musca-champion-crispa*	Champion Moss Curled	Curly Leaf	Parsley
Petroselinum	*crispum*	*musca-favorita*	Favorite Moss Curled	Curly Leaf	Parsley
Petroselinum	*crispum*	*musca-petra*	Petra Moss Curled	Curly Leaf	Parsley
Petroselinum	*crispum*	*musca-rina*	Rina Moss Curled	Curly Leaf	Parsley
Petroselinum	*crispum*	*musca-rosetta*	Rosette Moss Curled	Curly Leaf	Parsley
Petroselinum	*crispum*	*optima*	Optima	Curly Leaf	Parsley
Petroselinum	*crispum*	*pagoda*	Pagoda	Curly Leaf	Parsley
Petroselinum	*crispum*	*picnici*	Picnic	Curly Leaf	Parsley
Petroselinum	*crispum*	*sempervirens*	Evergreen	Curly Leaf	Parsley
Petroselinum	*crispum*	*sherwoodii*	Sherwood	Curly Leaf	Parsley
Petroselinum	*crispum*	*suavis-crispa*	Sweet	Curly Leaf	Parsley
Petroselinum	*crispum*	*sylvatica-groena*	Forest Green Moss	Curly Leaf	Parsley
Petroselinum	*crispum*	*sylvatica-groena-crispa*	Curled Forest Green Moss	Curly Leaf	Parsley
Petroselinum	*crispum*	*sylvatica-groena-serta*	Garland Forest Green Moss	Curly Leaf	Parsley
Petroselinum	*crispum*	*sylvatica-groena-paramonta*	Paramount Forest Green Moss	Curly Leaf	Parsley
Petroselinum	*crispum*	*sylvatica-groena-paramonta-parus*	Parus Paramount Forest Green Moss	Curly Leaf	Parsley
Petroselinum	*crispum*	*sylvatica-groena-perfecta*	Perfection Forest Green Moss	Curly Leaf	Parsley
Petroselinum	*crispum*	*titania*	Titan	Curly Leaf	Parsley
Petroselinum	*crispum*	*tripla-crispa*	Triple Curled	Curly Leaf	Parsley
Petroselinum	*crispum*	*unicrispa*	Unicurl	Curly Leaf	Parsley
Petroselinum	*crispum*	*vaughanii-xxx*	Vaughan's XXX	Curly Leaf	Parsley
Petroselinum	*hortense*	*fakiri*	Fakir	Turnip Root	Parsley
Petroselinum	*hortense*	*hamburgensis*	Hamburg	Turnip Root	Parsley
Petroselinum	*hortense*	*praecox-sucra*	Early Sugar	Turnip Root	Parsley
Petroselinum	*neapolitanum*	*neapolitanum*	Original	Italian Flat Leaf	Parsley
Petroselinum	*neapolitanum*	*catalognoi*	Catalogna	Italian Flat Leaf	Parsley
Petroselinum	*neapolitanum*	*catalognoi-gigantea*	Giant Catalogna	Italian Flat Leaf	Parsley
Petroselinum	*neapolitanum*	*crispum*	Crisp Plain Leaf	Italian Flat Leaf	Parsley
Petroselinum	*neapolitanum*	*darca-groena-italica*	Dark Green	Italian Flat Leaf	Parsley
Petroselinum	*neapolitanum*	*darca-groena-survivorum*	Survivor Dark Green	Italian Flat Leaf	Parsley
Petroselinum	*neapolitanum*	*dura-italica*	Hardy	Italian Flat Leaf	Parsley
Petroselinum	*neapolitanum*	*prezzemoloi-communis-2*	#2 Prezzemolo	Italian Flat Leaf	Parsley

Genus	species	infraspecies	cultivar	specimen	Genome
Petroselinum	*neapolitanum*	*prezzemoloi-gigantea-italica*	Giant Prezzemolo	Italian Flat Leaf	Parsley
Petroselinum	*neapolitanum*	*gigantea-napoliensis*	Giant Naples	Italian Flat Leaf	Parsley
Petroselinum	*neapolitanum*	*italica-suavis*	Italian Sweet	Italian Flat Leaf	Parsley
Petroselinum	*neapolitanum*	*planifolium*	Plain	Italian Flat Leaf	Parsley
Petroselinum	*vulgare*	*bartowichii-longa*	Bartowich Long	Flat Leaf	Parsley
Petroselinum	*vulgare*	*budapestensis-macella*	Budapest Market	Flat Leaf	Parsley
Petroselinum	*vulgare*	*communis-aromatica-bolognesensis*	Aromatic Bolognese	Flat Leaf	Parsley
Petroselinum	*vulgare*	*hosszui*	Hosszu	Flat Leaf	Parsley
Petroselinum	*vulgare*	*thailandica*	Thai	Flat Leaf	Parsley
Petroselinum	*vulgare*	*turcensis*	Turkish	Flat Leaf	Parsley
Sium	*suave*	*cicutifolium*	Hemlock	Sweet	Water Parsnip
Sium	*suave*	*floridanum*	Florida	Sweet	Water Parsnip

Family 724.032.2 *Araliaceae* Aral

Genus	Species	Specimen Genome	C:Culinary, M:Medicinal, E:Economic
Aralia	*cordata*	Udo Aral	C:SL:**VG**
Panax	*quinquefolius*	American Ginseng	C:RT,RZ:**VG**

Varieties

Udo (*Aralia cordata*)

Genus	Species	infraspecies	cultivar	specimen	Genome
Aralia	*cordata*	*edulis*	Garden	Udo	Aral
Aralia	*cordata*	*nutans*	Nodding	Udo	Aral
Aralia	*cordata*	*sachalinensis*	Sachalin	Udo	Aral
Aralia	*cordata*	*schmidtii*	Russian	Udo	Aral
Aralia	*cordata*	*taiwaniana*	Taiwan	Udo	Aral

Family 724.032.3 *Rubiaceae* Rubyroot

Bean, Coffee (*Coffea*)

Genus	Species	Specimen Genome	C:Culinary, M:Medicinal, E:Economic
Coffea	*arabica*	Arabian Coffee Bean	C:SD:**VG**
Coffea	*canephora*	Congo Coffee Bean	C:SD:**VG**
Coffea	*catura*	Katura Coffee Bean	C:SD:**VG**
Coffea	*konaensis*	Kona Coffee Bean	C:SD:**VG**
Coffea	*racemosa*	Raceme Coffee Bean	C:SD:**VG**
Galium	*aparine*	Sticky Willie Bedstraw	C:LF,SM:**VG**
Galium	*odoratum*	Sweet Woodruff Bedstraw	C:FL,LF:AF,BF, C:LF,SM:**VG**

Varieties

Bean, Coffee (*Coffea arabica*)

Genus	Species	infraspecies	cultivar	specimen	Genome
Coffea	arabica	arabica	Original	Arabian	Coffee Bean
Coffea	arabica	nana	Dwarf	Arabian	Coffee Bean

Order 724.033 *Dipsacales* Teasel

Family 1 *Dipsacaceae* Teasel
Family 2 *Adoxaceae* Musk Root
Family 3 *Caprifoliaceae* Honeysuckle

Family 724.033.4 *Valerianaceae* Valerian

Genus	Species	Specimen Genome	C:Culinary, M:Medicinal, E:Economic
Valerianella	*eriocarpa*	Italian Cornsalad	C:LF:**VG**
Valerianella	*locusta*	Lewiston Cornsalad	C:LF:**VG**

Varieties

Cornsalad, Italian (*Valerianella eriocarpa*)

Genus	Species	infraspecies	cultivar	specimen	Genome
Valerianella	*eriocarpa*	*eriocarpa*	Original	Italian	Cornsalad
Valerianella	*eriocarpa*	*piedmontii*	Piedmont	Italian	Cornsalad

Cornsalad, Lewiston (*Valerianella locusta*)

Valerianella	*locusta*	*locusta*	Original	Lewiston Cornsalad
Valerianella	*locusta*	*bavalii*	Baval	Lewiston Cornsalad
Valerianella	*locusta*	*blonde-shell-leafii*	Blonde Shell Leaf	Lewiston Cornsalad
Valerianella	*locusta*	*broad-leafii*	Broad Leaf	Lewiston Cornsalad
Valerianella	*locusta*	*broad-leaf-dutchii*	Dutch Broad Leaf	Lewiston Cornsalad
Valerianella	*locusta*	*cavalloi*	Cavallo	Lewiston Cornsalad
Valerianella	*locusta*	*coquille-de-louviersi*	Coquille De Louviers	Lewiston Cornsalad
Valerianella	*locusta*	*elanii*	Elan	Lewiston Cornsalad
Valerianella	*locusta*	*fetticus*	Large Round Leaf Lamb's Lettuce	Lewiston Cornsalad
Valerianella	*locusta*	*gaylai*	Gayla	Lewiston Cornsalad
Valerianella	*locusta*	*grosse-grainii*	Grosse Grain	Lewiston Cornsalad
Valerianella	*locusta*	*large-dutchii*	Large Dutch	Lewiston Cornsalad
Valerianella	*locusta*	*large-leavedii*	Large Leaved	Lewiston Cornsalad
Valerianella	*locusta*	*medallionii*	Medallion	Lewiston Cornsalad
Valerianella	*locusta*	*ronde-maraicherei*	Ronde Maraichere	Lewiston Cornsalad
Valerianella	*locusta*	*oleracea*	Lamb's Lettuce	Lewiston Cornsalad
Valerianella	*locusta*	*olitoria*	Lewiston	Lewiston Cornsalad
Valerianella	*locusta*	*valgrosii*	Valgros	Lewiston Cornsalad
Valerianella	*locusta*	*verte-d'cambrai*	Verte D'Cambrai	Lewiston Cornsalad
Valerianella	*locusta*	*verte-d'etampesi*	Verte D'Etampes	Lewiston Cornsalad
Valerianella	*locusta*	*verte-d'orlandai*	Verte D'Cambrai	Lewiston Cornsalad
Valerianella	*locusta*	*vitii*	Vit	Lewiston Cornsalad

Order 4 *Ericales* Heath
Family 1 *Ericaceae* Heath
Family 2 *Clethraceae* Sweet Pepper Heath
Family 3 *Cyrillaceae* Hilltop Heath
Family 4 *Empetraceae* Empeter Heath
Family 5 *Epacridaceae* Pukiawe Heath
Family 6 *Monotropaceae* Pipes
Family 7 *Pyrolaceae* Wintergreen Heath
Family 8 *Sapotaceae* Sapote

Order 5 *Nepenthales* Pitcher Plant
Family 1 *Nepenthaceae* Pitcher Plant
Family 2 *Droseraceae* Flytrap Pitcher Plant
Family 3 *Sarraceniaceae* Insect Pitcher Plant

Class 724.04 *Gentiasolopsida* Gentian & Sunplant

Order 724.041 *Gentianales* Gentian

Family 1 *Gentianaceae* Gentian
Family 2 *Apocynaceae* Dogbane

Family 724.041.3 *Asclepiadaceae* Milkstem

Genus	Species	Specimen Genome	C:Culinary, M:Medicinal, E:Economic
Asclepias	*tuberosa*	Pleurisy Root Milkstem	C:RT:**VG**

Varieties

Pleurisy Root (*Asclepias tuberosa*)

Genus	Species	Infraspecies	cultivar	specimen	Genome
Asclepias	*tuberosa*	*tuberosa*	Original	Pleurisy Root	Milkstem
Asclepias	*tuberosa*	*indian-paintbrushii*	Indian Paintbrush	Pleurisy Root	Milkstem
Asclepias	*tuberosa*	*interior*	Inland	Pleurisy Root	Milkstem
Asclepias	*tuberosa*	*rolfsii*	Rolfs	Pleurisy Root	Milkstem
Asclepias	*tuberosa*	*terminalis*	Terminal	Pleurisy Root	Milkstem

Family 4 *Loganiaceae* Loganwood

Order 724.042 *Solanales* Sunplant

Family 724.042.1 *Solanaceae* Sunplant

GENERA *Solanum, Acnistus, Atropa, Bouchetia, Browallia, Brugmansia, Brunfelsia, Brunsfelsia, Calibrachoa, Capsicum, Cestrum, Chamaesaracha, Cyphomandra, Datura, Duboisia, Fabiana, Goetzea, Guincula, Hunzikeria, Hyoscyamus, Jaborosa, Jaltomata, Leucophysalis, Lycianthes, Lycium, Lycopersicon, Mandragora, Margaranthus, Nectouxia, Nicandra, Nicotiana, Nierembergia, Nothocestrum, Oryctes, Petunia, Physalis, Quincula, Salpichroa, Salpiglossis, Saracha, Schizanthus, Scopolia, Solandra, Streptosolen, Withania*

Genus	Species	Specimen Genome	C:Culinary, M:Medicinal
Solanum	*aethiopicum*	African Eggplant Sunplant	C:FR:**VG**
Solanum	*aviculare*	Australian Cherry Tomato Sunplant	C:FR:**VG**
Solanum	*brasiliensis*	Brazilian Eggplant Sunplant	C:FR:**VG**
Solanum	*burbankii*	Sunberry Sunplant	C:FR:**VG**
Solanum	*gilo*	Gilo Eggplant Sunplant	C:FR:**VG**
Solanum	*integrifolium*	Hmong Eggplant Sunplant	C:FR:**VG**
Solanum	*lycopersicum-esculentum*	Tree Tomato Sunplant	C:FR:**VG**
Solanum	*lycopersicum-cerasiforme*	American Cherry Tomato Sunplant	C:FR:**VG**
Solanum	*macrocarpon*	Gbogname Eggplant Sunplant	C:FR:**VG**
Solanum	*maglia*	Blue Potato Sunplant	C:TU:**VG**
Solanum	*melanocerasum*	Huckleberry Sunplant	C:FR:**VG**
Solanum	*melongena-alba*	White Eggplant Sunplant	C:FR:**VG**
Solanum	*melongena-aurantia*	Orange Eggplant Sunplant	C:FR:**VG**
Solanum	*melongena-groena*	Green Eggplant Sunplant	C:FR:**VG**
Solanum	*melongena-helva*	Yellow Eggplant Sunplant	C:FR:**VG**
Solanum	*melongena-hybrida-F1*	F1 Hybrid Eggplant Sunplant	C:FR:**VG**
Solanum	*melongena-marmorata*	Marbled Eggplant Sunplant	C:FR:**VG**
Solanum	*melongena-nigra*	Black Eggplant Sunplant	C:FR:**VG**
Solanum	*melongena-nigra-purpurea*	Purple Black Eggplant Sunplant	C:FR:**VG**
Solanum	*melongena-pinca*	Pink Eggplant Sunplant	C:FR:**VG**
Solanum	*melongena-purpurea*	Purple Eggplant Sunplant	C:FR:**VG**
Solanum	*melongena-purpurea-darca*	Dark Purple Eggplant Sunplant	C:FR:**VG**
Solanum	*melongena-purpurea-lavendula*	Lavender Purple Eggplant Sunplant	C:FR:**VG**
Solanum	*melongena-purpurea-rubra*	Red Purple Eggplant Sunplant	C:FR:**VG**
Solanum	*melongena-purpurea-viola*	Violet Purple Sunplant	C:FR:**VG**
Solanum	*melongena-rotunda*	Round Eggplant Sunplant	C:FR:**VG**
Solanum	*melongena-rubra*	Red Eggplant Sunplant	C:FR:**VG**
Solanum	*nigrum*	Black Huckleberry Sunplant	C:FR:**VG**
Solanum	*quitoense*	Naranjilla Sunplant	C:FR:**VG**
Solanum	*sessiliflorum*	Cocona Sunplant	C:FR:**VG**
Solanum	*sisymbriifolium*	Lychee Tomato Sunplant	C:FR:**VG**
Solanum	*tuberosum-album*	Brown Skin White Potato Sunplant	C:TU:**VG**
Solanum	*tuberosum-album-albadermis*	White Skin White Potato Sunplant	C:TU:**VG**
Solanum	*tuberosum-album-blaueadermis*	Blue Skin White Potato Sunplant	C:TU:**VG**
Solanum	*tuberosum-album-helvadermis*	Yellow Skin White Potato Sunplant	C:TU:**VG**
Solanum	*tuberosum-album-pincadermis*	Pink Skin White Potato Sunplant	C:TU:**VG**

Genus	Species	Specimen Genome	C:Culinary, M:Medicinal
Solanum	tuberosum-album-purpureadermis	Purple Skin White Potato Sunplant	C:TU:**VG**
Solanum	tuberosum-album-rubradermis	Red Skin White Potato Sunplant	C:TU:**VG**
Solanum	tuberosum-album-russetii	Russet White Potato Sunplant	C:TU:**VG**
Solanum	tuberosum-blaueum	Blue Potato Sunplant	C:TU:**VG**
Solanum	tuberosum-helvum	Yellow Potato Sunplant	C:TU:**VG**
Solanum	tuberosum-purpureum	Purple Potato Sunplant	C:TU:**VG**
Solanum	tuberosum-rubrum	Red Potato Sunplant	C:TU:**VG**
Capsicum	annuum-aviculare	.1" to 1"Hot Bird Pepper	(Dierbach) D'Arcy & Eshbaugh
Capsicum	annuum-baccatum	Hot Finger Pepper	GSI 5th Ed., page 535 #15
Capsicum	annuum-baccatum-pendulum	Aji Hot Finger Pepper	GSI 6th Ed., page 329 #24
Capsicum	annuum-cerasiforme	>1" Hot Cherry Pepper	Flora, 2004, page 326, #05
Capsicum	annuum-chinense	Hot Torpedo Pepper	Flora, 2004, page 327, #02
Capsicum	annuum-chinense-habanerum	Hot Habanero Pepper	Flora, 2004, page 327, #05
Capsicum	annuum-conoides	Hot Cone Pepper	Flora, 2004, page 326, #08
Capsicum	annuum-frutescens	Hot Pencil Pepper	(L.) Kuntze
Capsicum	annuum-frutescens-chilium	Hot Chili Pepper	RHBOV, 1993, page 164
Capsicum	annuum-frutescens-chilium-aviculare	Bird Hot Chili Pepper	(L.) Kuntze
Capsicum	annuum-frutescens-tabascum	Hot Tabasco Pepper	Flora, 2004, page 327, #07
Capsicum	annuum-grossum	Hot Bell Pepper	GSI, 6th Ed., page 310, #08
Capsicum	annuum-longum-arienum	Hot Banana Pepper	GSI, 6th Ed., page 308, #21
Capsicum	annuum-longum-cayenneum	Hot Cayenne Pepper	Flora, 2004, page 326, #25
Capsicum	annuum-longum-jalapenum	Hot Jalapeno Pepper	Flora, 2004, page 326, #10
Capsicum	annuum-longum-papricum	Hot Paprika Pepper	GSI 6th Ed., page 312, #10
Capsicum	baccatum	Finger Pepper	L.
Capsicum	baccatum-pendulum	Aji Finger Pepper	(Willd.) Eshb.
Capsicum	cerasiforme-mitis	Mild Cherry Pepper	C:FR:**VG**
Capsicum	cerasiforme-sucrum	Sweet Cherry Pepper	C:FR:**VG**
Capsicum	chinense	Torpedo Pepper	Jacq.
Capsicum	exile	Hot Bullet Pepper	GSI 6th Ed., page 330, #13
Capsicum	eximum	Hot Special Pepper	GSI 5th Ed., page 542, #05
Capsicum	fasciculatum-mitis	Mild Cone Pepper	Botanica, 2004, page 189, #07
Capsicum	fasciculatum-sucrum	Sweet Cone Pepper	Botanica, 2004, page 189, #07
Capsicum	frutescens-mitis	Mild Pencil Pepper	Flora, page 327, #06
Capsicum	frutescens-sucrum	Sweet Pencil Pepper	Flora, page 327, #06
Capsicum	glabriusculum	Bird Pepper	(Dunal) Heiser & Pickersgill
Capsicum	grossum-mitis	Mild Bell Pepper	Flora, 2004, page 326, #15
Capsicum	grossum-sucrum	Sweet Bell Pepper	Flora, 2004, page 326, #15
Capsicum	grossum-supra-sucrum	Very Sweet Bell Pepper	Flora, 2004, page 326, #15
Capsicum	longum-mitis	Mild Banana Pepper	C:FR:**VG**
Capsicum	longum-papricum-mitis	Mild Paprika Pepper	C:FR:**VG**
Capsicum	longum-papricum-sucrum	Sweet Paprika Pepper	C:FR:**VG**
Capsicum	longum-pimentum-mitis	Mild Pimento Pepper	Flora, 2004, page 327, #01
Capsicum	longum-pimentum-sucrum	Sweet Pimento Pepper	GSI 6th Ed., page 325, #13
Capsicum	longum-sucrum	Sweet Banana Pepper	Flora, 2004, page 326, #26
Capsicum	pubescens	Hot Rocoto Pepper	Ruiz & Pavon
Cyphomandra	betacea	Tamarillo Tree Tomato	C:FR:**VG**
Lycopersicon	bovistigum-album	White Beefsteak Tomato	C:FR:**VG**
Lycopersicon	bovistigum-aurantium	Orange Beefsteak Tomato	C:FR:**VG**
Lycopersicon	bovistigum-aurantium-helvum	Yellow Orange Beefsteak Tomato	C:FR:**VG**
Lycopersicon	bovistigum-groenum	Green Beefsteak Tomato	C:FR:**VG**
Lycopersicon	bovistigum-helvum	Yellow Beefsteak Tomato	C:FR:**VG**
Lycopersicon	bovistigum-helvum-aurantium	Orange Yellow Beefsteak Tomato	C:FR:**VG**
Lycopersicon	bovistigum-pincum	Pink Beefsteak Tomato	C:FR:**VG**
Lycopersicon	bovistigum-pincum-purpureum	Purplish Pink Beefsteak Tomato	C:FR:**VG**
Lycopersicon	bovistigum-rubrum-giganteum	Giant 3lb 5.0" Red Beefsteak Tomato	C:FR:**VG**
Lycopersicon	bovistigum-rubrum-largum	Large 1.5-2lb 4.0" Red Beefsteak Tomato	C:FR:**VG**
Lycopersicon	bovistigum-rubrum-medium	Medium 1/2lb 2.5" Red Beefsteak Tomato	C:FR:**VG**
Lycopersicon	bovistigum-rubrum-parvum	Small 4oz-6oz 2" Red Beefsteak Tomato	C:FR:**VG**
Lycopersicon	bovistigum-rubrum-aurantium	Orange Red Beefsteak Tomato	C:FR:**VG**
Lycopersicon	bovistigum-striatum	Striped Beefsteak Tomato	C:FR:**VG**
Lycopersicon	cerasiforme-album	White Cherry Tomato	C:FR:**VG**
Lycopersicon	cerasiforme-aurantium	Orange Cherry Tomato	C:FR:**VG**
Lycopersicon	cerasiforme-groenum	Green Cherry Tomato	C:FR:**VG**
Lycopersicon	cerasiforme-helvum	Yellow Cherry Tomato	C:FR:**VG**
Lycopersicon	cerasiforme-helvum-aurantium	Orange Yellow Cherry Tomato	C:FR:**VG**
Lycopersicon	cerasiforme-nigrum	Black Cherry Tomato	C:FR:**VG**
Lycopersicon	cerasiforme-pincum	Pink Cherry Tomato	C:FR:**VG**
Lycopersicon	cerasiforme-rubrum-gigantium	Giant 3.5oz 1.75" Red Cherry Tomato	C:FR:**VG**
Lycopersicon	cerasiforme-rubrum-largum	Large 3.0oz 1.5" Red Cherry Tomato	C:FR:**VG**
Lycopersicon	cerasiforme-rubrum-medium-largum	Large Medium 2oz 1.0" Red Cherry Tomato	C:FR:**VG**
Lycopersicon	cerasiforme-rubrum-medium	Medium 1.5oz 3/4" Red Cherry Tomato	C:FR:**VG**
Lycopersicon	cerasiforme-rubrum-medium-parvum	Small Medium 3/4oz-1oz 1/2" Red Cherry Tomato	C:FR:**VG**
Lycopersicon	cerasiforme-rubrum-parvum	Small 1/2oz 1/4" Red Cherry Tomato	C:FR:**VG**

Genus	Species	Specimen	Genome	C:Culinary, M:Medicinal
Lycopersicon	*cerasiforme-rubrum-minutum*	Tiny 1/4oz 1/8" Red Cherry	Tomato	C:FR:**VG**
Lycopersicon	*cerasiforme-rubrum-aurantium*	Orange Red Cherry	Tomato	C:FR:**VG**
Lycopersicon	*cerasiforme-striatum*	Striped Cherry	Tomato	C:FR:**VG**
Lycopersicon	*cheesmanii*	Cheesman	Tomato	C:FR:**VG**
Lycopersicon	*esculentum-album*	White	Tomato	C:FR:**VG**
Lycopersicon	*esculentum-aurantium*	Orange	Tomato	C:FR:**VG**
Lycopersicon	*esculentum-aurantium-helvum*	Yellow Orange	Tomato	C:FR:**VG**
Lycopersicon	*esculentum-aurantium-rubrum*	Red Orange	Tomato	C:FR:**VG**
Lycopersicon	*esculentum-brunneum*	Brown	Tomato	C:FR:**VG**
Lycopersicon	*esculentum-groenum*	Green	Tomato	C:FR:**VG**
Lycopersicon	*esculentum-helvum*	Yellow	Tomato	C:FR:**VG**
Lycopersicon	*esculentum-helvum-aurantium*	Orange Yellow	Tomato	C:FR:**VG**
Lycopersicon	*esculentum-nigrum*	Black	Tomato	C:FR:**VG**
Lycopersicon	*esculentum-pincum*	Pink	Tomato	C:FR:**VG**
Lycopersicon	*esculentum-pincum-purpureum*	Purplish Pink	Tomato	C:FR:**VG**
Lycopersicon	*esculentum-rubrum-cavitum*	Hollow Red	Tomato	C:FR:**VG**
Lycopersicon	*esculentum-rubrum-giganteum*	Giant 3lb 5.0" Red	Tomato	C:FR:**VG**
Lycopersicon	*esculentum-rubrum-largum*	Large 1.5lb-2lb 4.0" Red	Tomato	C:FR:**VG**
Lycopersicon	*esculentum-rubrum-largum-medium*	Medium Large 1lb 3.5" Red	Tomato	C:FR:**VG**
Lycopersicon	*esculentum-rubrum-medium-largum*	Large Medium 8oz-12oz 3" Red	Tomato	C:FR:**VG**
Lycopersicon	*esculentum-rubrum-medium*	Medium 6oz-8oz 2.5" Red	Tomato	C:FR:**VG**
Lycopersicon	*esculentum-rubrum-parvum*	mall 4oz-6oz 2.0" Red	Tomato	C:FR:**VG**
Lycopersicon	*esculentum-rubrum-aurantium-largum*	Large 2lb 4.0" Orange Red	Tomato	C:FR:**VG**
Lycopersicon	*esculentum-rubrum-aurantium-largum-medium*	Medium Large 1lb 3.5" Orange Red	Tomato	C:FR:**VG**
Lycopersicon	*esculentum-rubrum-aurantium-medium-largum*	Large Medium 3/4lb 3.0" Orange Red	Tomato	C:FR:**VG**
Lycopersicon	*esculentum-rubrum-aurantium-medium*	Medium 1/2lb 2.5" Orange Red	Tomato	C:FR:**VG**
Lycopersicon	*esculentum-rubrum-aurantium-parvum*	Small 1/4lb 2.0" Orange Red	Tomato	C:FR:**VG**
Lycopersicon	*esculentum-striatum*	Striped	Tomato	C:FR:**VG**
Lycopersicon	*longum-aurantium*	Orange Long	Tomato	C:FR:**VG**
Lycopersicon	*longum-groenum*	Green Long	Tomato	C:FR:**VG**
Lycopersicon	*longum-helvum*	Yellow Long	Tomato	C:FR:**VG**
Lycopersicon	*longum-pincum*	Pink Long	Tomato	C:FR:**VG**
Lycopersicon	*longum-rubrum*	Red Long	Tomato	C:FR:**VG**
Lycopersicon	*longum-striatum*	Striped Long	Tomato	C:FR:**VG**
Lycopersicon	*oxacardium-aurantium*	Orange Oxheart	Tomato	C:FR:**VG**
Lycopersicon	*oxacardium-helvum*	Yellow Oxheart	Tomato	C:FR:**VG**
Lycopersicon	*oxacardium-pincum*	Pink Oxheart	Tomato	C:FR:**VG**
Lycopersicon	*oxacardium-rubrum-largum*	Large 1.5lb-2lb 4.0" Red Oxheart	Tomato	C:FR:**VG**
Lycopersicon	*oxacardium-rubrum-largum-medium*	Medium Large 1lb 3.5" Red Oxheart	Tomato	C:FR:**VG**
Lycopersicon	*oxacardium-rubrum-medium-largum*	Large Medium 3/4lb 3.0" Red Oxheart	Tomato	C:FR:**VG**
Lycopersicon	*oxacardium-rubrum-medium*	Medium 1/2lb 2.5" Red Oxheart	Tomato	C:FR:**VG**
Lycopersicon	*oxacardium-rubrum-parvum*	Small 4oz-6oz 2.0" Red Oxheart	Tomato	C:FR:**VG**
Lycopersicon	*oxacardium-rubrum-aurantium*	Orange Red Oxheart	Tomato	C:FR:**VG**
Lycopersicon	*pimpinellifolium-helvum*	Yellow Currant	Tomato	C:FR:**VG**
Lycopersicon	*pimpinellifolium-rubrum*	Red Currant	Tomato	C:FR:**VG**
Lycopersicon	*platycarpum-album*	White Flat	Tomato	C:FR:**VG**
Lycopersicon	*platycarpum-pincum*	Pink Flat	Tomato	C:FR:**VG**
Lycopersicon	*platycarpum-rubrum-largum*	Large 1.5lb 4" x 3" Red Flat	Tomato	C:FR:**VG**
Lycopersicon	*platycarpum-rubrum-largum-medium*	Medium Large 1lb 3" x 2.5" Red Flat	Tomato	C:FR:**VG**
Lycopersicon	*platycarpum-rubrum-medium*	Medium 3/4lb 3" x 2" Red Flat	Tomato	C:FR:**VG**
Lycopersicon	*platycarpum-rubrum-medium-parvum*	Small Medium 1/2lb 2.5" x 2" Red Flat	Tomato	C:FR:**VG**
Lycopersicon	*platycarpum-rubrum-parvum*	Small 4oz-6oz 2" x 1.5" Red Flat	Tomato	C:FR:**VG**
Lycopersicon	*platycarpum-rubrum-auranteum*	Orange Red Flat	Tomato	C:FR:**VG**
Lycopersicon	*pyriforme-album*	White Plum	Tomato	C:FR:**VG**
Lycopersicon	*pyriforme-aurantium*	Orange Plum	Tomato	C:FR:**VG**
Lycopersicon	*pyriforme-aurantium-helvum*	Yellow Orange Plum	Tomato	C:FR:**VG**
Lycopersicon	*pyriforme-helvum*	Yellow Plum	Tomato	C:FR:**VG**
Lycopersicon	*pyriforme-nigrum*	Black Plum	Tomato	C:FR:**VG**
Lycopersicon	*pyriforme-pincum*	Pink Plum	Tomato	C:FR:**VG**
Lycopersicon	*pyriforme-pincum-purpureum*	Purplish Pink Plum	Tomato	C:FR:**VG**
Lycopersicon	*pyriforme-roma-helvum*	Yellow Roma Plum	Tomato	C:FR:**VG**
Lycopersicon	*pyriforme-roma-auranteum*	Orange Roma Plum	Tomato	C:FR:**VG**
Lycopersicon	*pyriforme-roma-rubrum*	Red Roma Plum	Tomato	C:FR:**VG**
Lycopersicon	*pyriforme-rubrum-largum*	Large 1lb-2lb 4"-6" Red Plum	Tomato	C:FR:**VG**
Lycopersicon	*pyriforme-rubrum-largum-medium*	Medium Large 1/2lb-3/4lb 3"-4" Red Plum	Tomato	C:FR:**VG**
Lycopersicon	*pyriforme-rubrum-medium*	Medium 6oz 2.5" Red Plum	Tomato	C:FR:**VG**
Lycopersicon	*pyriforme-rubrum-parvum*	Small 4oz 2.0" Red Plum	Tomato	C:FR:**VG**
Lycopersicon	*pyriforme-rubrum-cerasiforme*	Cherry 2oz-3oz 1"-1.5" Red Plum	Tomato	C:FR:**VG**
Lycopersicon	*pyriforme-rubrum-aurantium*	Orange Red Plum	Tomato	C:FR:**VG**
Lycopersicon	*pyriforme-striatum*	Striped Plum	Tomato	C:FR:**VG**
Physalis	*edenensis*	Eden	Ground Cherry	C:FR:**VG**
Physalis	*ixocarpa-groena*	Green Tomatillo	Ground Cherry	C:FR:**VG**
Physalis	*ixocarpa-helva*	Yellow Tomatillo	Ground Cherry	C:FR:**VG**
Physalis	*ixocarpa-purpurea*	Purple Tomatillo	Ground Cherry	C:FR:**VG**

Genus	Species	Specimen Genome	C:Culinary, M:Medicinal
Physalis	*ixocarpa-rubra*	Red Tomatillo Ground Cherry	C:FR:**VG**
Physalis	*pensylvanica*	Pennsylvania Ground Cherry	C:FR:**VG**
Physalis	*peruviana*	Peruvian Ground Cherry	C:FR:**VG**
Physalis	*philadelphica*	Philadelphia Tomatillo Ground Cherry	C:FR:**VG**
Physalis	*polandica*	Polish Tomatillo Ground Cherry	C:FR:**VG**
Physalis	*pruinosa-groena*	Green Husk Tomatillo Ground Cherry	C:FR:**VG**
Physalis	*pruinosa-helva*	Yellow Husk Tomatillo Ground Cherry	C:FR:**VG**
Physalis	*pruinosa-purpurea*	Purple Husk Tomatillo Ground Cherry	C:FR:**VG**
Physalis	*pruinosa-rubra*	Red Husk Tomatillo Ground Cherry	C:FR:**VG**
Physalis	*pubescens-groena*	Green Husk Tomato Ground Cherry	C:FR:**VG**
Physalis	*pubescens-helva*	Yellow Husk Tomato Ground Cherry	C:FR:**VG**
Physalis	*pubescens-purpurea*	Purple Husk Tomato Ground Cherry	C:FR:**VG**
Physalis	*pubescens-rubra*	Red Husk Tomato Ground Cherry	C:FR:**VG**

Varieties

Genus	Species	Infraspecies	cultivar	specimen	Genome
Solanum	*aethiopicum*	*aethiopicum*	.	African Eggplant	Sunplant
Solanum	*aethiopicum*	*african-eggii*	African Egg	African Eggplant	Sunplant
Solanum	*aethiopicum*	*ambonii*	Amboni	African Eggplant	Sunplant
Solanum	*aethiopicum*	*aubergine-du-maliensis*	Aubergine du Mali	African Eggplant	Sunplant
Solanum	*aethiopicum*	*burkina-faso-redii*	Burkina Faso Red Orange	African Eggplant	Sunplant
Solanum	*aethiopicum*	*goyo-kumbai*	Goyo Kumba Red	African Eggplant	Sunplant
Solanum	*aethiopicum*	*jaxatu-soxnai*	Jaxatu Soxna	African Eggplant	Sunplant
Solanum	*aviculare*	*aviculare*	Original	Bird Cherry Tomato	Sunplant
Solanum	*aviculare*	*laciniatum*	Large	Bird Cherry Tomato	Sunplant
Solanum	*aviculare*	*uporo-cannibalii*	Cannibal Poro Poro	Bird Cherry Tomato	Sunplant
Solanum	*aviculare*	*uporo-kangarooi*	Kangaroo Apple Poro Poro	Bird Cherry Tomato	Sunplant
Solanum	*gilo*	*gilo*	Original	Gilo Eggplant	Sunplant
Solanum	*gilo*	*black-stemii*	Black Stem Red	Gilo Eggplant	Sunplant
Solanum	*integrifolium*	*integrifolium*	.	Hmong Eggplant	Sunplant
Solanum	*integrifolium*	*africana*	African	Hmong Eggplant	Sunplant
Solanum	*integrifolium*	*ruffled-redii*	Ruffled Red Orange	Hmong Eggplant	Sunplant
Solanum	*integrifolium*	*turkish-italian-orangei*	Turkish Italian Orange	Hmong Eggplant	Sunplant
Solanum	*integrifolium*	*turkish-italian-orangei-redii*	Turkish Italian Orange Red	Hmong Eggplant	Sunplant
Solanum	*lycopersicum-esculentum*	*giant-treei*	Giant Tree 1.5lb-2lb 4.0"	Tree Tomato	Sunplant
Solanum	*lycopersicum-esculentum*	*punta-banda*	Punta Banda 6oz-8oz 2.5"	Tree Tomato	Sunplant
Solanum	*lycopersicum-esculentum*	*zlutakyticei*	Zlutakytice Yellow ?.?"	Tree Tomato	Sunplant
Solanum	*lycopersicum-cerasiforme*	*chiapas-wildii*	Wild Chiapas	American Cherry Tomato	Sunplant
Solanum	*lycopersicum-cerasiforme*	*ciudad-victoria*	Ciudad Victoria	American Cherry Tomato	Sunplant
Solanum	*lycopersicum-cerasiforme*	*columbianum*	Columbian	American Cherry Tomato	Sunplant
Solanum	*macrocarpon*	*macrocarpon*	Original	Gbogname Eggplant	Sunplant
Solanum	*macrocarpon*	*eggplant-collardsi*	Collards	Gbogname Eggplant	Sunplant
Solanum	*macrocarpon*	*gbomai*	Gboma	Gbogname Eggplant	Sunplant
Solanum	*melanocerasum*	*melanocerasum*	Original	Huckleberry	Sunplant
Solanum	*melanocerasum*	*fructa*	Fruity	Huckleberry	Sunplant
Solanum	*melanocerasum*	*intrusum*	Intrusive	Huckleberry	Sunplant
Solanum	*melanocerasum*	*mrs.-b'si*	Mrs. B's	Huckleberry	Sunplant
Solanum	*melanocerasum*	*ogomoi*	Ogomo	Huckleberry	Sunplant

Eggplant (*Solanum melongena*)

Genus	Species	Infraspecies	cultivar	specimen	Genome
Solanum	*melongena-alba*	*alba*	Original	White Eggplant	Sunplant
Solanum	*melongena-alba*	*egg-treei*	Egg Tree	White Eggplant	Sunplant
Solanum	*melongena-alba*	*antiguaensis*	Antigua 8.0"	White Eggplant	Sunplant
Solanum	*melongena-alba*	*aubergine-mohicana*	Aubergine Mohican	White Eggplant	Sunplant
Solanum	*melongena-alba*	*bianca-ovalei*	Easter Egg 3.0"	White Eggplant	Sunplant
Solanum	*melongena-alba*	*bridei*	Bride	White Eggplant	Sunplant
Solanum	*melongena-alba*	*casperi*	Casper 8.0"	White Eggplant	Sunplant
Solanum	*melongena-alba*	*chinese-whitei*	Chinese 7.0"	White Eggplant	Sunplant
Solanum	*melongena-alba*	*dourgai*	Dourga 8.0"	White Eggplant	Sunplant
Solanum	*melongena-alba*	*lao-green-stripei*	Lao Green Stripe 2.0"	White Eggplant	Sunplant
Solanum	*melongena-alba*	*listada-de-gandia*	Listada De Gandia Purple Stripe 6.0"	White Eggplant	Sunplant
Solanum	*melongena-alba*	*little-spookyi*	Little Spooky 7.0"	White Eggplant	Sunplant
Solanum	*melongena-alba*	*long-lavender-whitei*	Long Lavender Stripe 8.0"	White Eggplant	Sunplant
Solanum	*melongena-alba*	*manjri-gotai*	Manjri Gota Purple Stripe 5.0"	White Eggplant	Sunplant
Solanum	*melongena-alba*	*oriental-whitei*	Oriental	White Eggplant	Sunplant
Solanum	*melongena-alba*	*paris-whitei*	Paris 2.0"	White Eggplant	Sunplant
Solanum	*melongena-alba*	*snowyi*	Snowy 7.0"	White Eggplant	Sunplant
Solanum	*melongena-alba*	*thai-whitei*	Thai 6.5"	White Eggplant	Sunplant
Solanum	*melongena-alba*	*thomas-jeffersonii*	Thomas Jefferson	White Eggplant	Sunplant
Solanum	*melongena-alba*	*white-beautyi*	White Beauty	White Eggplant	Sunplant
Solanum	*melongena-alba*	*white-eggii*	Japanese White Egg 2.5"	White Eggplant	Sunplant

Genus	Species	Infraspecies	cultivar	specimen	Genome
Solanum	*melongena-alba*	*white-swordii*	Lavender Streaked White Sword	White Eggplant	Sunplant
Solanum	*melongena-alba*	*white-swordii-longa*	Long White Sword 9.0"	White Eggplant	Sunplant
Solanum	*melongena-aurantia*	*aurantia*	Original	Orange Eggplant	Sunplant
Solanum	*melongena-aurantia*	*cookstownensis*	Cookstown	Orange Eggplant	Sunplant
Solanum	*melongena-aurantia*	*red-china-eggplantii*	Red China	Orange Eggplant	Sunplant
Solanum	*melongena-aurantia*	*comprido-verde-claroi*	Comprido Verde Claro 3.5"	Orange Eggplant	Sunplant
Solanum	*melongena-aurantia*	*turkish-orangei*	Turkish 3.0"	Orange Eggplant	Sunplant
Solanum	*melongena-groena*	*applegreenii*	Applegreen	Green Eggplant	Sunplant
Solanum	*melongena-groena*	*green-beautyi*	Green Beauty 11."	Green Eggplant	Sunplant
Solanum	*melongena-groena*	*green-eggsi*	Green Eggs 3.0"	Green Eggplant	Sunplant
Solanum	*melongena-groena*	*green-giant-earlyi*	Green Giant Early	Green Eggplant	Sunplant
Solanum	*melongena-groena*	*green-grapei*	Green Grape	Green Eggplant	Sunplant
Solanum	*melongena-groena*	*jade-sweetii*	Jade Sweet	Green Eggplant	Sunplant
Solanum	*melongena-groena*	*louisiana-green-ovalis*	Louisiana Green Oval	Green Eggplant	Sunplant
Solanum	*melongena-groena*	*louisiana-long-greenii*	Louisiana Long 8.0"	Green Eggplant	Sunplant
Solanum	*melongena-groena*	*thai-green-longii*	Thai Long 10"	Green Eggplant	Sunplant
Solanum	*melongena-helva*	*insanum*	Prickly Bitter	Yellow Eggplant	Sunplant
Solanum	*melongena-helva*	*thai-yellow-eggii*	Thai Egg	Yellow Eggplant	Sunplant
Solanum	*melongena-hybrida-F1*	*black-enorma*	Black Enorma Black	F1 Hybrid Eggplant	Sunplant
Solanum	*melongena-hybrida-F1*	*black-princei*	Black Prince Purple Black	F1 Hybrid Eggplant	Sunplant
Solanum	*melongena-hybrida-F1*	*elondoi*	Elondo Purple Black	F1 Hybrid Eggplant	Sunplant
Solanum	*melongena-hybrida-F1*	*slice-rightii*	Slice Rite Black	F1 Hybrid Eggplant	Sunplant
Solanum	*melongena-marmorata*	*baby-marbledii*	Baby 3.5"	Marbled Eggplant	Sunplant
Solanum	*melongena-marmorata*	*rotonda-bianca-sfumata-di-rosa*	Rotonda Bianca Sfumata Di Rosa	Marbled Eggplant	Sunplant
Solanum	*melongena-marmorata*	*udamalpetii*	Udamalpet 3.0"	Marbled Eggplant	Sunplant
Solanum	*melongena-nigra*	*nigra*	Original 5.0"	Black Eggplant	Sunplant
Solanum	*melongena-nigra*	*black-egg-earlyi-4*	Early Black Egg 4.0"	Black Eggplant	Sunplant
Solanum	*melongena-nigra*	*black-egg-earlyi-6*	Early Black Egg 6.0"	Black Eggplant	Sunplant
Solanum	*melongena-nigra*	*highbush-selectii*	Highbush Select	Black Eggplant	Sunplant
Solanum	*melongena-nigra*	*new-york-improvedii*	New York Improved	Black Eggplant	Sunplant
Solanum	*melongena-nigra-purpurea*	*black-beautyi*	Black Beauty 6.5"	Purple Black Eggplant	Sunplant
Solanum	*melongena-nigra-purpurea*	*black-beautyi-imperialis*	Imperial Black Beauty 8.0"	Purple Black Eggplant	Sunplant
Solanum	*melongena-nigra-purpurea*	*cook's-strainii*	Cook's Strain	Purple Black Eggplant	Sunplant
Solanum	*melongena-nigra-purpurea*	*de-barbentanensis*	De Barbentane	Purple Black Eggplant	Sunplant
Solanum	*melongena-nigra-purpurea*	*florida-marketii*	Florida Market 9.5"	Purple Black Eggplant	Sunplant
Solanum	*melongena-nigra-purpurea*	*hibush-specialis*	HiBush Special 8.0"	Purple Black Eggplant	Sunplant
Solanum	*melongena-nigra-purpurea*	*kurume-longii*	Kurume Long 11."	Purple Black Eggplant	Sunplant
Solanum	*melongena-nigra-purpurea*	*little-fingersi*	Little Fingers 5.5"	Purple Black Eggplant	Sunplant
Solanum	*melongena-nigra-purpurea*	*millionairei*	Millionaire 11."	Purple Black Eggplant	Sunplant
Solanum	*melongena-nigra-purpurea*	*osaka-honnogai*	Osaka Honnoga 7.5"	Purple Black Eggplant	Sunplant
Solanum	*melongena-nigra-purpurea*	*pompano-marketii*	Pompano Market	Purple Black Eggplant	Sunplant
Solanum	*melongena-nigra-purpurea*	*purple-dragonii*	Purple Dragon 9.0"	Purple Black Eggplant	Sunplant
Solanum	*melongena-nigra-purpurea*	*redondai*	Redonda	Purple Black Eggplant	Sunplant
Solanum	*melongena-nigra-purpurea*	*tycoon-orientalis*	Tycoon Oriental 7.5"	Purple Black Eggplant	Sunplant
Solanum	*melongena-nigra-purpurea*	*viserba*	Viserba	Purple Black Eggplant	Sunplant
Solanum	*melongena-pinca*	*italian-pink-bicolori*	Italian Pink Bicolor	Pink Eggplant	Sunplant
Solanum	*melongena-pinca*	*pandora-striped-rosei*	Pandora Striped Rose	Pink Eggplant	Sunplant
Solanum	*melongena-pinca*	*rosa-biancai*	Rosa Bianca 5.0"	Pink Eggplant	Sunplant
Solanum	*melongena-pinca*	*rositai*	Rosita 7.0"	Pink Eggplant	Sunplant
Solanum	*melongena-purpurea*	*chice-de-ciervai*	Chice De Cierva	Purple Eggplant	Sunplant
Solanum	*melongena-purpurea*	*florida-high-bushii*	Florida High Bush	Purple Eggplant	Sunplant
Solanum	*melongena-purpurea*	*gitana*	Gitana	Purple Eggplant	Sunplant
Solanum	*melongena-purpurea*	*japanese-early-purplei*	Japanese Early	Purple Eggplant	Sunplant
Solanum	*melongena-purpurea*	*japanese-purple-picklingii*	Japanese Pickling 7.0"	Purple Eggplant	Sunplant
Solanum	*melongena-purpurea*	*malaysian-dark-red-longii*	Malaysian Dark Red Long 6.0"	Purple Eggplant	Sunplant
Solanum	*melongena-purpurea*	*ping-tung-longa*	Ping Tung Long 14."	Purple Eggplant	Sunplant
Solanum	*melongena-purpurea*	*ping-tung-longa-improva*	Improved Ping Tung Long 9.0"	Purple Eggplant	Sunplant
Solanum	*melongena-purpurea*	*slim-jimii*	Slim Jim 5.0"	Purple Eggplant	Sunplant
Solanum	*melongena-purpurea*	*taiwan-longa*	Taiwan Long	Purple Eggplant	Sunplant
Solanum	*melongena-purpurea*	*violetta-longuei*	Violetta Longue	Purple Eggplant	Sunplant
Solanum	*melongena-purpurea*	*violetta-siciliana*	Violetta Siciliana	Purple Eggplant	Sunplant
Solanum	*melongena-purpurea*	*violetta-tondai*	Violetta Tonda	Purple Eggplant	Sunplant
Solanum	*melongena-purpurea-darca*	*almazii*	Almaz	Dark Purple Eggplant	Sunplant
Solanum	*melongena-purpurea-darca*	*billeaui*	Billeau 3.0"	Dark Purple Eggplant	Sunplant
Solanum	*melongena-purpurea-darca*	*blue-devilii*	Blue Devil 9.0"	Dark Purple Eggplant	Sunplant
Solanum	*melongena-purpurea-darca*	*chinese-longa*	Chinese Long 12."	Dark Purple Eggplant	Sunplant
Solanum	*melongena-purpurea-darca*	*hito-kuchi-nasui*	Hito-Kuchi Nasu 2.0"	Dark Purple Eggplant	Sunplant
Solanum	*melongena-purpurea-darca*	*long-purple-italiana*	Italian Long 9.0"	Dark Purple Eggplant	Sunplant
Solanum	*melongena-purpurea-darca*	*long-purple-italiana-vitorria*	Vitorria Italian Long 9.0"	Dark Purple Eggplant	Sunplant
Solanum	*melongena-purpurea-darca*	*morden-midgetii*	Morden Midget	Dark Purple Eggplant	Sunplant
Solanum	*melongena-purpurea-darca*	*purple-long-slenderi*	Purple Long Slender 8.0"	Dark Purple Eggplant	Sunplant

Genus	Species	Infraspecies	cultivar	specimen	Genome
Solanum	*melongena-purpurea-darca*	*rima*	Rima 8.0"	Dark Purple Eggplant	Sunplant
Solanum	*melongena-purpurea-darca*	*thai-long-purplei*	Thai Long	Dark Purple Eggplant	Sunplant
Solanum	*melongena-purpurea-darca*	*thai-oblong-purplei*	Thai Oblong	Dark Purple Eggplant	Sunplant
Solanum	*melongena-purpurea-darca*	*ukrainian-beautyi*	Urkrainian Beauty	Dark Purple Eggplant	Sunplant
Solanum	*melongena-purpurea-darca*	*waimanalo-longa*	WaiManalo Long15."	Dark Purple Eggplant	Sunplant
Solanum	*melongena-purpurea-darca*	*zaharai*	ZaHara 6.0"	Dark Purple Eggplant	Sunplant
Solanum	*melongena-purpurea-lavendula*	*fengyuan-purplei*	Fengyuan 16"	Lavender Purple Eggplant	Sunplant
Solanum	*melongena-purpurea-lavendula*	*florentine-silkii*	Florentine Silk	Lavender Purple Eggplant	Sunplant
Solanum	*melongena-purpurea-lavendula*	*lao-lavenderi*	Lao 3.0"	Lavender Purple Eggplant	Sunplant
Solanum	*melongena-purpurea-lavendula*	*long-purple-japonica*	Japanese Long 9.0"	Lavender Purple Eggplant	Sunplant
Solanum	*melongena-purpurea-lavendula*	*taiwanagai*	Taiwanaga 9.0"	Lavender Purple Eggplant	Sunplant
Solanum	*melongena-purpurea-lavendula*	*violette-di-firenzei*	Violette Di Firenze	Lavender Purple Eggplant	Sunplant
Solanum	*melongena-purpurea-rubra*	*dark-long-redii*	Dark Long	Red Purple Eggplant	Sunplant
Solanum	*melongena-purpurea-viola*	*diamondii*	Diamond 8.0"	Violet Purple Eggplant	Sunplant
Solanum	*melongena-purpurea-viola*	*long-purple-earlyi*	Early Italian Long 10."	Violet Purple Eggplant	Sunplant
Solanum	*melongena-rotunda*	*foo-chow-roundii*	Foo Chow	Round Eggplant	Sunplant
Solanum	*melongena-rotunda*	*lebanese-bunchingii*	Lebanese Bunching	Round Eggplant	Sunplant
Solanum	*melongena-rotunda*	*oriental-princessi*	Oriental Princess	Round Eggplant	Sunplant
Solanum	*melongena-rotunda*	*pink-bicolori*	Pink Bicolor	Round Eggplant	Sunplant
Solanum	*melongena-rotunda*	*prospera*	Prospera Violet	Round Eggplant	Sunplant
Solanum	*melongena-rotunda*	*purple-ballii*	Purple Ball	Round Eggplant	Sunplant
Solanum	*melongena-rotunda*	*pusa-purple-clusteri*	Pusa Purple Cluster 2.0"	Round Eggplant	Sunplant
Solanum	*melongena-rotunda*	*round-mauvei*	Round Mauve 2.5"	Round Eggplant	Sunplant
Solanum	*melongena-rotunda*	*thai-green-peai*	Thai Green Pea1/4"	Round Eggplant	Sunplant
Solanum	*melongena-rotunda*	*thai-round-brinjalii*	Thai Brinjal	Round Eggplant	Sunplant
Solanum	*melongena-rotunda*	*thai-round-light-greenii*	Thai Light Green 2.5"	Round Eggplant	Sunplant
Solanum	*melongena-rotunda*	*thai-round-purplei*	Thai Purple	Round Eggplant	Sunplant
Solanum	*melongena-rubra*	*red-eggii*	Orange Red Egg	Red Eggplant	Sunplant
Solanum	*melongena-rubra*	*red-rufflesi*	Orange Red Ruffles 2.0"	Red Eggplant	Sunplant
Solanum	*melongena-rubra*	*ruffled-redii*	Ruffled Orange Red 2.5"	Red Eggplant	Sunplant
Solanum	*melongena-rubra*	*sweet-redii*	Sweet	Red Eggplant	Sunplant
Solanum	*nigrum*	*nigrum*	Original	Black Huckleberry	Sunplant
Solanum	*nigrum*	*chichiquilitei*	Chichiquilite	Black Huckleberry	Sunplant
Solanum	*sessiliflorum*	*sessiliflorum*	Original	Cocona	Sunplant
Solanum	*sessiliflorum*	*topiro*	Topiro	Cocona	Sunplant

Potato (*Solanum tuberosum*)

Solanum	*tuberosum-album*	*accentii*	Accent	Brown Skin White Potato	Sunplant
Solanum	*tuberosum-album*	*ailsa*	Ailsa	Brown Skin White Potato	Sunplant
Solanum	*tuberosum-album*	*alaschensis-frostlessi*	Alaska Frostless	Brown Skin White Potato	Sunplant
Solanum	*tuberosum-album*	*anoka*	Anoka	Brown Skin White Potato	Sunplant
Solanum	*tuberosum-album*	*atlanticum*	Atlantic	Brown Skin White Potato	Sunplant
Solanum	*tuberosum-album*	*ausonia*	AusonYellowPotato	Brown Skin White Potato	Sunplant
Solanum	*tuberosum-album*	*austriana-crawlersi*	Australia Crawlers	Brown Skin White Potato	Sunplant
Solanum	*tuberosum-album*	*bananai-album*	Banana	Brown Skin White Potato	Sunplant
Solanum	*tuberosum-album*	*beauty-of-hebronii*	Beauty Of Hebron	Brown Skin White Potato	Sunplant
Solanum	*tuberosum-album*	*beltsvillensis*	Beltsville	Brown Skin White Potato	Sunplant
Solanum	*tuberosum-album*	*candy-stripei*	Candy Stripe	Brown Skin White Potato	Sunplant
Solanum	*tuberosum-album*	*carltonii*	Carlton	Brown Skin White Potato	Sunplant
Solanum	*tuberosum-album*	*castilensis*	Castile	Brown Skin White Potato	Sunplant
Solanum	*tuberosum-album*	*chaleuri*	Chaleur	Brown Skin White Potato	Sunplant
Solanum	*tuberosum-album*	*chippewai*	Chippewa	Brown Skin White Potato	Sunplant
Solanum	*tuberosum-album*	*conestogai*	Conestoga	Brown Skin White Potato	Sunplant
Solanum	*tuberosum-album*	*dazoci*	Dazoc	Brown Skin White Potato	Sunplant
Solanum	*tuberosum-album*	*denalii*	Denali	Brown Skin White Potato	Sunplant
Solanum	*tuberosum-album*	*dr.-macintoshii*	Dr. MacIntosh	Brown Skin White Potato	Sunplant
Solanum	*tuberosum-album*	*duke-of-yorkii*	Duke Of York	Brown Skin White Potato	Sunplant
Solanum	*tuberosum-album*	*early-ohioensis*	Early Ohio	Brown Skin White Potato	Sunplant
Solanum	*tuberosum-album*	*elbai*	Elba	Brown Skin White Potato	Sunplant
Solanum	*tuberosum-album*	*eramosa*	Eramosa	Brown Skin White Potato	Sunplant
Solanum	*tuberosum-album*	*great-northernii*	Great Northern	Brown Skin White Potato	Sunplant
Solanum	*tuberosum-album*	*home-guardii*	HomeGuard	Brown Skin White Potato	Sunplant
Solanum	*tuberosum-album*	*indian-pitii*	Indian Pit	Brown Skin White Potato	Sunplant
Solanum	*tuberosum-album*	*itasca*	Itasca	Brown Skin White Potato	Sunplant
Solanum	*tuberosum-album*	*jemsegii*	Jemseg	Brown Skin White Potato	Sunplant
Solanum	*tuberosum-album*	*kanona*	Kanona	Brown Skin White Potato	Sunplant
Solanum	*tuberosum-album*	*keswickii*	Keswick	Brown Skin White Potato	Sunplant
Solanum	*tuberosum-album*	*la-chipperi*	La Chipper	Brown Skin White Potato	Sunplant
Solanum	*tuberosum-album*	*magic-mollyi*	Magic Molly	Brown Skin White Potato	Sunplant
Solanum	*tuberosum-album*	*marfona*	Marfona	Brown Skin White Potato	Sunplant
Solanum	*tuberosum-album*	*mein's-earlyi*	Mein's Early	Brown Skin White Potato	Sunplant
Solanum	*tuberosum-album*	*mona-lisa*	Mona Lisa	Brown Skin White Potato	Sunplant
Solanum	*tuberosum-album*	*nadinei*	Nadine	Brown Skin White Potato	Sunplant
Solanum	*tuberosum-album*	*nooksackii*	Nootsack	Brown Skin White Potato	Sunplant

Solanum	tuberosum-album	onawayi	Onaway	Brown Skin White Potato Sunplant
Solanum	tuberosum-album	penta	Penta	Brown Skin White Potato Sunplant
Solanum	tuberosum-album	pentlandii-crownii	Crown Pentland	Brown Skin White Potato Sunplant
Solanum	tuberosum-album	pentlandii-dellii	Dell Pentland	Brown Skin White Potato Sunplant
Solanum	tuberosum-album	pentlandii-javelinii	Javelin Pentland	Brown Skin White Potato Sunplant
Solanum	tuberosum-album	pentlandii-kirstyi	Kirsty Pentland	Brown Skin White Potato Sunplant
Solanum	tuberosum-album	pentlandii-squirei	Squire Pentland	Brown Skin White Potato Sunplant
Solanum	tuberosum-album	purple-eyed-seedlingii	Purple Eyed Seedling	Brown Skin White Potato Sunplant
Solanum	tuberosum-album	sefton-wonderi	Sefton Wonder	Brown Skin White Potato Sunplant
Solanum	tuberosum-album	sequoia	Sequoia	Brown Skin White Potato Sunplant
Solanum	tuberosum-album	sherinei	Sherine	Brown Skin White Potato Sunplant
Solanum	tuberosum-album	tobiquei	Tobique	Brown Skin White Potato Sunplant
Solanum	tuberosum-album	tolaasi	Tolaas	Brown Skin White Potato Sunplant
Solanum	tuberosum-album	up-to-datei	Up To Date	Brown Skin White Potato Sunplant
Solanum	tuberosum-album-albadermis	arran-pilotii	Arran Pilot	White Skin White Potato Sunplant
Solanum	tuberosum-album-albadermis	cal-whitei	Cal White	White Skin White Potato Sunplant
Solanum	tuberosum-album-albadermis	cascadei	Cascade	White Skin White Potato Sunplant
Solanum	tuberosum-album-albadermis	cherokeensis	Cherokee	White Skin White Potato Sunplant
Solanum	tuberosum-album-albadermis	chipeta	Chipeta	White Skin White Potato Sunplant
Solanum	tuberosum-album-albadermis	epicurei	Epicure	White Skin White Potato Sunplant
Solanum	tuberosum-album-albadermis	cobbleri-album	White Cobbler	White Skin White Potato Sunplant
Solanum	tuberosum-album-albadermis	cobbleri-irelandica	Irish Cobbler	White Skin White Potato Sunplant
Solanum	tuberosum-album-albadermis	katahdinii	Katahdin	White Skin White Potato Sunplant
Solanum	tuberosum-album-albadermis	kennebecensis	Kennebec	White Skin White Potato Sunplant
Solanum	tuberosum-album-albadermis	maris-peeri	Maris Peer	White Skin White Potato Sunplant
Solanum	tuberosum-album-albadermis	maris-piperi	Maris Piper	White Skin White Potato Sunplant
Solanum	tuberosum-album-albadermis	norchipii	Norchip	White Skin White Potato Sunplant
Solanum	tuberosum-album-albadermis	prince-hairyi-NYL-235-4	Prince Hairy	White Skin White Potato Sunplant
Solanum	tuberosum-album-albadermis	sebagoi	Sebago	White Skin White Potato Sunplant
Solanum	tuberosum-album-albadermis	shepodyi	Shepody	White Skin White Potato Sunplant
Solanum	tuberosum-album-albadermis	snowdenii	Snowden	White Skin White Potato Sunplant
Solanum	tuberosum-album-albadermis	superiori	Superior	White Skin White Potato Sunplant
Solanum	tuberosum-album-albadermis	tundra	Tundra	White Skin White Potato Sunplant
Solanum	tuberosum-album-albadermis	warbai	Warba	White Skin White Potato Sunplant
Solanum	tuberosum-album-albadermis	white-rosei	White Rose	White Skin White Potato Sunplant
Solanum	tuberosum-album-blaueadermis	blue-kerr's-pinkii	Blue Kerr's Pink	Blue Skin White Potato Sunplant
Solanum	tuberosum-album-blaueadermis	blue-maci	Blue Mac	Blue Skin White Potato Sunplant
Solanum	tuberosum-album-blaueadermis	blue-victori	Blue Victor	Blue Skin White Potato Sunplant
Solanum	tuberosum-album-blaueadermis	caribei	Caribe	Blue Skin White Potato Sunplant
Solanum	tuberosum-album-blaueadermis	caribe-sportii	Caribe Sport	Blue Skin White Potato Sunplant
Solanum	tuberosum-album-helvadermis	alleghanyi	Alleghany	Yellow Skin White Potato Sunplant
Solanum	tuberosum-album-helvadermis	blue-eyed-rusica	Blue Eyed Russian	Yellow Skin White Potato Sunplant
Solanum	tuberosum-album-helvadermis	blue-eyed-rusica-catriona	Catriona Blue Eyed Russian	Yellow Skin White Potato Sunplant
Solanum	tuberosum-album-helvadermis	island-sunshinei	Island Sunshine	Yellow Skin White Potato Sunplant
Solanum	tuberosum-album-helvadermis	la-rota	La Rota	Yellow Skin White Potato Sunplant
Solanum	tuberosum-album-helvadermis	warbai-pink-eyei	Warba Pink Eye	Yellow Skin White Potato Sunplant
Solanum	tuberosum-album-pincadermis	antari	Antar	Pink Skin White Potato Sunplant
Solanum	tuberosum-album-pincadermis	diana	Diana	Pink Skin White Potato Sunplant
Solanum	tuberosum-album-pincadermis	early-rosei	Early Rose	Pink Skin White Potato Sunplant
Solanum	tuberosum-album-pincadermis	fanfarei	Fanfare	Pink Skin White Potato Sunplant
Solanum	tuberosum-album-pincadermis	kerr's-pinkii	Kerr's Pink	Pink Skin White Potato Sunplant
Solanum	tuberosum-album-pincadermis	kondori	Kondor	Pink Skin White Potato Sunplant
Solanum	tuberosum-album-pincadermis	morvenii	Morven	Pink Skin White Potato Sunplant
Solanum	tuberosum-album-pincadermis	pentlandii-beautyi	Pentland Beauty	Pink Skin White Potato Sunplant
Solanum	tuberosum-album-pincadermis	pink-duke-of-yorkii	Pink Duke Of York	Pink Skin White Potato Sunplant
Solanum	tuberosum-album-pincadermis	pink-pearlii	Pink Pearl	Pink Skin White Potato Sunplant
Solanum	tuberosum-album-pincadermis	stroma	Stroma	Pink Skin White Potato Sunplant
Solanum	tuberosum-album-purpureadermis	angus-beautyi	Angus Beauty	Purple Skin White Potato Sunplant
Solanum	tuberosum-album-purpureadermis	divernonii	Di Vernon	Purple Skin White Potato Sunplant
Solanum	tuberosum-album-purpureadermis	german-blackii	German Black	Purple Skin White Potato Sunplant
Solanum	tuberosum-album-purpureadermis	longlaci	Longlac	Purple Skin White Potato Sunplant
Solanum	tuberosum-album-purpureadermis	mauve-queenii	Mauve Queen	Purple Skin White Potato Sunplant
Solanum	tuberosum-album-purpureadermis	old-spanishii	Old Spanish	Purple Skin White Potato Sunplant
Solanum	tuberosum-album-purpureadermis	purple-chiefii	Purple Chief	Purple Skin White Potato Sunplant
Solanum	tuberosum-album-purpureadermis	purple-vikingii	Purple Viking	Purple Skin White Potato Sunplant
Solanum	tuberosum-album-purpureadermis	seneca-hornii	Seneca Horn	Purple Skin White Potato Sunplant
Solanum	tuberosum-album-rubradermis	A79543-4R	#79543-4R Alpha	Red Skin White Potato Sunplant
Solanum	tuberosum-album-rubradermis	alhambra	Alhambra	Red Skin White Potato Sunplant
Solanum	tuberosum-album-rubradermis	bisonii	Bison	Red Skin White Potato Sunplant
Solanum	tuberosum-album-rubradermis	bliss-triumphii	Bliss Triumph	Red Skin White Potato Sunplant
Solanum	tuberosum-album-rubradermis	buffaloi	Buffalo	Red Skin White Potato Sunplant
Solanum	tuberosum-album-rubradermis	cal-redii	Cal Red	Red Skin White Potato Sunplant
Solanum	tuberosum-album-rubradermis	cavalier-redii	Cavalier	Red Skin White Potato Sunplant
Solanum	tuberosum-album-rubradermis	cherry-redii	Cherry Red	Red Skin White Potato Sunplant

Solanum	*tuberosum-album-rubradermis*	*chieftanii*	Chieftan	Red Skin White Potato Sunplant
Solanum	*tuberosum-album-rubradermis*	*durango-redii*	Durango	Red Skin White Potato Sunplant
Solanum	*tuberosum-album-rubradermis*	*gladstonei*	Gladstone	Red Skin White Potato Sunplant
Solanum	*tuberosum-album-rubradermis*	*ida-rosei*	Ida Rose	Red Skin White Potato Sunplant
Solanum	*tuberosum-album-rubradermis*	*king-edwardii*	King Edward	Red Skin White Potato Sunplant
Solanum	*tuberosum-album-rubradermis*	*la-rougei*	La Rouge	Red Skin White Potato Sunplant
Solanum	*tuberosum-album-rubradermis*	*nordonna*	NorDonna	Red Skin White Potato Sunplant
Solanum	*tuberosum-album-rubradermis*	*norlandii*	Norland	Red Skin White Potato Sunplant
Solanum	*tuberosum-album-rubradermis*	*norlandii-darca-rubrum*	Dark Red Norland	Red Skin White Potato Sunplant
Solanum	*tuberosum-album-rubradermis*	*pimpernelii*	Pimpernel	Red Skin White Potato Sunplant
Solanum	*tuberosum-album-rubradermis*	*pontiacensis*	Pontiac	Red Skin White Potato Sunplant
Solanum	*tuberosum-album-rubradermis*	*poorlanderi*	Poorlander	Red Skin White Potato Sunplant
Solanum	*tuberosum-album-rubradermis*	*red-cloudii*	Red Cloud	Red Skin White Potato Sunplant
Solanum	*tuberosum-album-rubradermis*	*red-dalei*	Red Dale	Red Skin White Potato Sunplant
Solanum	*tuberosum-album-rubradermis*	*red-la-soda*	Red La Soda	Red Skin White Potato Sunplant
Solanum	*tuberosum-album-rubradermis*	*red-sunii*	Red Sun	Red Skin White Potato Sunplant
Solanum	*tuberosum-album-rubradermis*	*reda*	Reda	Red Skin White Potato Sunplant
Solanum	*tuberosum-album-rubradermis*	*redsenii*	Redsen	Red Skin White Potato Sunplant
Solanum	*tuberosum-album-rubradermis*	*rideaui*	Rideau	Red Skin White Potato Sunplant
Solanum	*tuberosum-album-rubradermis*	*romanoi*	Romano	Red Skin White Potato Sunplant
Solanum	*tuberosum-album-rubradermis*	*ruby-redii*	Ruby Red	Red Skin White Potato Sunplant
Solanum	*tuberosum-album-rubradermis*	*salad-redii*	Salad Red	Red Skin White Potato Sunplant
Solanum	*tuberosum-album-rubradermis*	*sangrei*	Sangre	Red Skin White Potato Sunplant
Solanum	*tuberosum-album-rubradermis*	*vikingii*	Viking	Red Skin White Potato Sunplant
Solanum	*tuberosum-album-rubradermis*	*vikingii-redii*	Red Viking	Red Skin White Potato Sunplant
Solanum	*tuberosum-album-rubradermis*	*warbai-redii*	Warba Red	Red Skin White Potato Sunplant
Solanum	*tuberosum-album-russetii*	*acadia-russetii*	Acadia	Russet White Potato Sunplant
Solanum	*tuberosum-album-russetii*	*bake-kingii*	Bake King	Russet White Potato Sunplant
Solanum	*tuberosum-album-russetii*	*belarusica*	Belrus	Russet White Potato Sunplant
Solanum	*tuberosum-album-russetii*	*buttei*	Butte	Russet White Potato Sunplant
Solanum	*tuberosum-album-russetii*	*buttei-frontieri*	Frontier Butte	Russet White Potato Sunplant
Solanum	*tuberosum-album-russetii*	*centennial-russetii*	Centennial	Russet White Potato Sunplant
Solanum	*tuberosum-album-russetii*	*century-russetii*	Century	Russet White Potato Sunplant
Solanum	*tuberosum-album-russetii*	*coastal-russetii*	Coastal	Russet White Potato Sunplant
Solanum	*tuberosum-album-russetii*	*frontier-russetii*	Frontier	Russet White Potato Sunplant
Solanum	*tuberosum-album-russetii*	*gem-russetii*	Gem	Russet White Potato Sunplant
Solanum	*tuberosum-album-russetii*	*golden-wonderi*	Golden Wonder	Russet White Potato Sunplant
Solanum	*tuberosum-album-russetii*	*green-mountainii*	Green Mountain	Russet White Potato Sunplant
Solanum	*tuberosum-album-russetii*	*highlitei*	Highlite	Russet White Potato Sunplant
Solanum	*tuberosum-album-russetii*	*keystonei*	Keystone	Russet White Potato Sunplant
Solanum	*tuberosum-album-russetii*	*krantzii*	Krantz	Russet White Potato Sunplant
Solanum	*tuberosum-album-russetii*	*lemhi-russetii*	Lemhi	Russet White Potato Sunplant
Solanum	*tuberosum-album-russetii*	*norkotahii*	Norkotah	Russet White Potato Sunplant
Solanum	*tuberosum-album-russetii*	*ranger-russetii*	Ranger	Russet White Potato Sunplant
Solanum	*tuberosum-album-russetii*	*russet-burbankii*	Russet Burbank	Russet White Potato Sunplant
Solanum	*tuberosum-album-russetii*	*russet-burbankii-netted-gemii*	Netted Gem Russet Burbank	Russet White Potato Sunplant
Solanum	*tuberosum-album-russetii*	*russet-nuggetii*	Russet Nugget	Russet White Potato Sunplant
Solanum	*tuberosum-album-russetii*	*sierra*	Sierra	Russet White Potato Sunplant
Solanum	*tuberosum-album-russetii*	*silverton-russetii*	Silverton	Russet White Potato Sunplant
Solanum	*tuberosum-blaueum*	*all-bluei*	All Blue	Blue Potato Sunplant
Solanum	*tuberosum-blaueum*	*lavenderi*	Lavender	Blue Potato Sunplant
Solanum	*tuberosum-blaueum*	*rusica-blauea*	Russian Blue	Blue Potato Sunplant
Solanum	*tuberosum-blaueum*	*salad-bluei*	Salad Blue	Blue Potato Sunplant
Solanum	*tuberosum-blaueum*	*rantonnetii*	Rantonnet	Blue Potato Sunplant
Solanum	*tuberosum-blaueum*	*royal-robei*	Royal Robe	Blue Potato Sunplant
Solanum	*tuberosum-helvum*	*adorai*	Adora	Yellow Potato Sunplant
Solanum	*tuberosum-helvum*	*alby's-goldii*	Alby's Gold	Yellow Potato Sunplant
Solanum	*tuberosum-helvum*	*augsburgii-goldii*	Augsburg Gold	Yellow Potato Sunplant
Solanum	*tuberosum-helvum*	*austriana-crescens*	Austrian Crescent	Yellow Potato Sunplant
Solanum	*tuberosum-helvum*	*bananai*	Russian Banana	Yellow Potato Sunplant
Solanum	*tuberosum-helvum*	*bananai-russica*	Baltic Russian Banana	Yellow Potato Sunplant
Solanum	*tuberosum-helvum*	*bintjei*	Bintje	Yellow Potato Sunplant
Solanum	*tuberosum-helvum*	*blazei*	Blaze	Yellow Potato Sunplant
Solanum	*tuberosum-helvum*	*brigus*	Brigus	Yellow Potato Sunplant
Solanum	*tuberosum-helvum*	*butterfingeri*	Butterfinger	Yellow Potato Sunplant
Solanum	*tuberosum-helvum*	*candy-canei*	Candy Cane	Yellow Potato Sunplant
Solanum	*tuberosum-helvum*	*carlotta*	Carlotta	Yellow Potato Sunplant
Solanum	*tuberosum-helvum*	*carola*	Carola	Yellow Potato Sunplant
Solanum	*tuberosum-helvum*	*charlottei*	Charlotte	Yellow Potato Sunplant
Solanum	*tuberosum-helvum*	*costella*	Costella	Yellow Potato Sunplant
Solanum	*tuberosum-helvum*	*daisy-goldii*	Daisy Gold	Yellow Potato Sunplant
Solanum	*tuberosum-helvum*	*desireei*	Desiree	Yellow Potato Sunplant
Solanum	*tuberosum-helvum*	*dunluce*	Dunluce	Yellow Potato Sunplant
Solanum	*tuberosum-helvum*	*early-bluei*	Early Blue	Yellow Potato Sunplant
Solanum	*tuberosum-helvum*	*edgecote-purplei*	Edgecote Purple	Yellow Potato Sunplant

Solanum	*tuberosum-helvum*	*estima*	EstimaLight	Yellow Potato Sunplant
Solanum	*tuberosum-helvum*	*flava*	Flava	Yellow Potato Sunplant
Solanum	*tuberosum-helvum*	*french-fingerlingii*	French Fingerling	Yellow Potato Sunplant
Solanum	*tuberosum-helvum*	*germanica-butterballii*	German Butterball	Yellow Potato Sunplant
Solanum	*tuberosum-helvum*	*germanica-fingeri*	German Finger	Yellow Potato Sunplant
Solanum	*tuberosum-helvum*	*germanica-fingeri-ladiga*	German Lady Finger	Yellow Potato Sunplant
Solanum	*tuberosum-helvum*	*germanica-fingerlingii*	German Fingerling	Yellow Potato Sunplant
Solanum	*tuberosum-helvum*	*germanica-helvum*	Yellow German	Yellow Potato Sunplant
Solanum	*tuberosum-helvum*	*gold-nuggetii*	Gold Nugget	Yellow Potato Sunplant
Solanum	*tuberosum-helvum*	*gold-raei*	Gold Rae	Yellow Potato Sunplant
Solanum	*tuberosum-helvum*	*goldrushii*	Goldrush	Yellow Potato Sunplant
Solanum	*tuberosum-helvum*	*inca-goldii*	Inca Gold	Yellow Potato Sunplant
Solanum	*tuberosum-helvum*	*kingstonii*	Kingston	Yellow Potato Sunplant
Solanum	*tuberosum-helvum*	*kipfelii*	Kipfel	Yellow Potato Sunplant
Solanum	*tuberosum-helvum*	*la-rattei*	La Ratte	Yellow Potato Sunplant
Solanum	*tuberosum-helvum*	*la-rattei-princessi*	Princess La Ratte	Yellow Potato Sunplant
Solanum	*tuberosum-helvum*	*lola*	Lola	Yellow Potato Sunplant
Solanum	*tuberosum-helvum*	*mainensis-bananai*	Maine Banana	Yellow Potato Sunplant
Solanum	*tuberosum-helvum*	*manna*	Manna	Yellow Potato Sunplant
Solanum	*tuberosum-helvum*	*morning-goldii*	Morning Gold	Yellow Potato Sunplant
Solanum	*tuberosum-helvum*	*nicola*	Nicola	Yellow Potato Sunplant
Solanum	*tuberosum-helvum*	*norgoldii-M*	Modified (M) Norgold	Yellow Potato Sunplant
Solanum	*tuberosum-helvum*	*norgoldii-russetii*	Russet Norgold	Yellow Potato Sunplant
Solanum	*tuberosum-helvum*	*nosebagii*	Nosebag Fingerling	Yellow Potato Sunplant
Solanum	*tuberosum-helvum*	*OAC-royal-goldii*	Royal Gold Ontario Agriculture Canada	Yellow Potato Sunplant
Solanum	*tuberosum-helvum*	*OAC-ruby-goldii*	Ruby Gold Ontario Agriculture Canada	Yellow Potato Sunplant
Solanum	*tuberosum-helvum*	*ozettei*	Ozette	Yellow Potato Sunplant
Solanum	*tuberosum-helvum*	*ozettei-anna-cheeka'si*	Anna Cheeka's Ozette	Yellow Potato Sunplant
Solanum	*tuberosum-helvum*	*ozettei-kasaanii*	Kasaan Ozette	Yellow Potato Sunplant
Solanum	*tuberosum-helvum*	*peanutii*	Peanut	Yellow Potato Sunplant
Solanum	*tuberosum-helvum*	*ptarmiganii*	Ptarmigan	Yellow Potato Sunplant
Solanum	*tuberosum-helvum*	*ptarmiganii-a.c.*	A. C. Ptarmigan	Yellow Potato Sunplant
Solanum	*tuberosum-helvum*	*red-duke-of-yorkii*	Red Duke Of York	Yellow Potato Sunplant
Solanum	*tuberosum-helvum*	*red-goldii*	Red Gold	Yellow Potato Sunplant
Solanum	*tuberosum-helvum*	*rose-finn-applei*	Rose Finn Apple	Yellow Potato Sunplant
Solanum	*tuberosum-helvum*	*rose-goldii*	Rose Gold	Yellow Potato Sunplant
Solanum	*tuberosum-helvum*	*rote-earstlingii*	Rote Erstling	Yellow Potato Sunplant
Solanum	*tuberosum-helvum*	*rubinia*	Rubinia	Yellow Potato Sunplant
Solanum	*tuberosum-helvum*	*ruby-crescentii*	Ruby Cresent	Yellow Potato Sunplant
Solanum	*tuberosum-helvum*	*saginaw-goldii*	Saginaw Gold	Yellow Potato Sunplant
Solanum	*tuberosum-helvum*	*santei*	Sante	Yellow Potato Sunplant
Solanum	*tuberosum-helvum*	*sieglindei*	Sieglinde	Yellow Potato Sunplant
Solanum	*tuberosum-helvum*	*pink-fir-applei*	Pink Fir Apple	Yellow Potato Sunplant
Solanum	*tuberosum-helvum*	*urgenta*	Urgenta	Yellow Potato Sunplant
Solanum	*tuberosum-helvum*	*wilja*	Wilja	Yellow Potato Sunplant
Solanum	*tuberosum-helvum*	*yellow-finlandica*	Yellow Finn	Yellow Potato Sunplant
Solanum	*tuberosum-helvum*	*yukon-goldii*	Yukon Gold	Yellow Potato Sunplant
Solanum	*tuberosum-purpureum*	*blue-cloudii*	Blue Cloud	Purple Potato Sunplant
Solanum	*tuberosum-purpureum*	*cow-hornii*	Cow Horn	Purple Potato Sunplant
Solanum	*tuberosum-purpureum*	*purple-markeri*	Purple Marker	Purple Potato Sunplant
Solanum	*tuberosum-purpureum*	*purple-peruviana*	Purple Peruvian	Purple Potato Sunplant
Solanum	*tuberosum-rubrum*	*alaschensis-sweetheartii*	Alaska Sweetheart	Red Potato Sunplant
Solanum	*tuberosum-rubrum*	*all-pinkii*	All Pink	Red Potato Sunplant
Solanum	*tuberosum-rubrum*	*all-redii*	All Red	Red Potato Sunplant
Solanum	*tuberosum-rubrum*	*blossomii*	Blossom	Red Potato Sunplant
Solanum	*tuberosum-rubrum*	*blushingii*	Blushing	Red Potato Sunplant
Solanum	*tuberosum-rubrum*	*cinderella*	Cinderella	Red Potato Sunplant
Solanum	*tuberosum-rubrum*	*cinnabari*	Cinnabar	Red Potato Sunplant
Solanum	*tuberosum-rubrum*	*deep-redii*	Deep Red	Red Potato Sunplant
Solanum	*tuberosum-rubrum*	*erikii*	Erik The Red	Red Potato Sunplant
Solanum	*tuberosum-rubrum*	*garnet-chilei*	Garnet Chile	Red Potato Sunplant
Solanum	*tuberosum-rubrum*	*huckleberryi*	Huckleberry	Red Potato Sunplant
Solanum	*tuberosum-rubrum*	*maroon-bellsi*	Maroon Bells	Red Potato Sunplant
Solanum	*tuberosum-rubrum*	*mountain-rosei*	Mountain Rose	Red Potato Sunplant
Solanum	*tuberosum-rubrum*	*NDC-4069-4*	North Dakota-Colorado	Red Potato Sunplant
Solanum	*tuberosum-rubrum*	*pink-champaignei*	Pink Champaigne	Red Potato Sunplant
Solanum	*tuberosum-rubrum*	*red-thumbii*	Red Thumb	Red Potato Sunplant
Solanum	*tuberosum-rubrum*	*rojoi*	Rojo	Red Potato Sunplant
Solanum	*tuberosum-rubrum*	*rosiei*	Rosie	Red Potato Sunplant
Solanum	*tuberosum-rubrum*	*sunsetii*	Sunset	Red Potato Sunplant

Pepper, Hot Bird (*Capsicum annuum-aviculare*)

Genus	Species	Infraspecies	cultivar	specimen	Genome

Genus	Species	Infraspecies	cultivar	specimen	Genome
Capsicum	*annuum-aviculare*	*afra-surprisa*	African Surprise ?.?"	Hot Bird	Pepper
Capsicum	*annuum-aviculare*	*ammazzoi*	Ammazzo .75"	Hot Bird	Pepper
Capsicum	*annuum-aviculare*	*auranteum-blossomii*	Orange Blossom 0.5"	Hot Bird	Pepper
Capsicum	*annuum-aviculare*	*aves-oculum*	Birds' Eye .25"	Hot Bird	Pepper
Capsicum	*annuum-aviculare*	*bellingrathensis*	Bellingrath Gardens 1.0"	Hot Bird	Pepper
Capsicum	*annuum-aviculare*	*cabeza-de-lagartoi*	Lizard Head 1.0"	Hot Bird	Pepper
Capsicum	*annuum-aviculare*	*cambuci*	Cambuci ?.?"	Hot Bird	Pepper
Capsicum	*annuum-aviculare*	*chinensis-ornamentalis*	Chinese Ornamental 0.5"	Hot Bird	Pepper
Capsicum	*annuum-aviculare*	*cobaneroi-cobana*	Coban Cobanero 0.9"	Hot Bird	Pepper
Capsicum	*annuum-aviculare*	*cobaneroi-maya-lovei*	Mayan Love Cobanero 0.5"	Hot Bird	Pepper
Capsicum	*annuum-aviculare*	*de-rata-coloi*	Turd Of The Rat ?.?"	Hot Bird	Pepper
Capsicum	*annuum-aviculare*	*ecuadorensis*	Ecuador 1.0"	Hot Bird	Pepper
Capsicum	*annuum-aviculare*	*embersi*	Embers ?.?"	Hot Bird	Pepper
Capsicum	*annuum-aviculare*	*firecrackeri*	Firecracker 1.0"	Hot Bird	Pepper
Capsicum	*annuum-aviculare*	*fireworksi*	Fireworks 1.0"	Hot Bird	Pepper
Capsicum	*annuum-aviculare*	*fluorescens-purpureum*	Fluorescent Purple 0.9"	Hot Bird	Pepper
Capsicum	*annuum-aviculare*	*gardenia-nugeteum*	Garden Nugget 0.3"	Hot Bird	Pepper
Capsicum	*annuum-aviculare*	*grandpa's-home-pepperi*	Grandpa's Home ?.?"	Hot Bird	Pepper
Capsicum	*annuum-aviculare*	*guam-booniesi*	Guam Boonies 1.0"	Hot Bird	Pepper
Capsicum	*annuum-aviculare*	*gumdropii*	Gumdrop 0.5"	Hot Bird	Pepper
Capsicum	*annuum-aviculare*	*inca-red-dropii*	Inca Red Drop ?.?"	Hot Bird	Pepper
Capsicum	*annuum-aviculare*	*kalia*	Kalia 1.0"	Hot Bird	Pepper
Capsicum	*annuum-aviculare*	*koreana*	White Korean 0.5"	Hot Bird	Pepper
Capsicum	*annuum-aviculare*	*little-elfii*	Little Elf ?.?"	Hot Bird	Pepper
Capsicum	*annuum-aviculare*	*malibuensis-purpureum*	Malibu Purple 0.7"	Hot Bird	Pepper
Capsicum	*annuum-aviculare*	*magenta*	Magenta 0.6"	Hot Bird	Pepper
Capsicum	*annuum-aviculare*	*marblesii*	Marbles 0.4"	Hot Bird	Pepper
Capsicum	*annuum-aviculare*	*masqueradei*	Masquerade ?.?"	Hot Bird	Pepper
Capsicum	*annuum-aviculare*	*mcmahon's-texas-bird-pepperi*	McMahon's Texas ?.?"	Hot Bird	Pepper
Capsicum	*annuum-aviculare*	*mexicana-drysium*	Mexican Tree 0.7"	Hot Bird	Pepper
Capsicum	*annuum-aviculare*	*nigra-clustra*	Black Cluster 0.5"	Hot Bird	Pepper
Capsicum	*annuum-aviculare*	*nigra-princeps*	Black Prince 0.5"	Hot Bird	Pepper
Capsicum	*annuum-aviculare*	*nosegayi*	Nosegay 0.4"	Hot Bird	Pepper
Capsicum	*annuum-aviculare*	*ordonoi*	Ordono 1.0"	Hot Bird	Pepper
Capsicum	*annuum-aviculare*	*petinei*	Petine 0.7"	Hot Bird	Pepper
Capsicum	*annuum-aviculare*	*piment-bec-d'oiseau-noir*	Blackbird Beak 0.5"	Hot Bird	Pepper
Capsicum	*annuum-aviculare*	*prairie-firei*	Prairie Fire ?.?"	Hot Bird	Pepper
Capsicum	*annuum-aviculare*	*pulchellum-in-purpureum*	Pretty In Purple 0.7"	Hot Bird	Pepper
Capsicum	*annuum-aviculare*	*purpureum-pepperi*	Purple 1.0"	Hot Bird	Pepper
Capsicum	*annuum-aviculare*	*purpureum-princeps*	Purple Prince 1.0"	Hot Bird	Pepper
Capsicum	*annuum-aviculare*	*rojo-moradoi*	Rojo Morado ?.?"	Hot Bird	Pepper
Capsicum	*annuum-aviculare*	*rooster-combii*	Rooster Comb 0.5"	Hot Bird	Pepper
Capsicum	*annuum-aviculare*	*st.-helena-acornii*	St. Helena Acorn ?.?"	Hot Bird	Pepper
Capsicum	*annuum-aviculare*	*tri-fettii*	Tri Fetti ?.?"	Hot Bird	Pepper
Capsicum	*annuum-aviculare*	*venezuelensis*	Venezuelan ?.?"	Hot Bird	Pepper
Capsicum	*annuum-aviculare*	*vestro-virginica-pisum*	West Virigina Pea 1.0"	Hot Bird	Pepper
Capsicum	*annuum-aviculare*	*white-hotii*	White Hot 1.0"	Hot Bird	Pepper
Capsicum	*annuum-aviculare*	*wild-dynamoi*	Wild Dynamo 0.2"	Hot Bird	Pepper
Capsicum	*annuum-aviculare*	*wirri-wirrii*	Wirri Wirri 0.5"	Hot Bird	Pepper
Capsicum	*annuum-aviculare*	*xigolei*	Xigole ?.?"	Hot Bird	Pepper

Pepper, Hot Finger (*Capsicum annuum-baccatum*)

Genus	Species	Infraspecies	cultivar	specimen	Genome
Capsicum	*annuum-baccatum*	*baccatum*	Original 1.0"	Hot Finger	Pepper
Capsicum	*annuum-baccatum*	*amarilleari*	Amarillear ?.?"	Hot Finger	Pepper
Capsicum	*annuum-baccatum*	*barbadosensis-bishop's-crownii*	Bishop's Crown Barbados ?.?"	Hot Finger	Pepper
Capsicum	*annuum-baccatum*	*barbadosensis-joker's-hatii*	Joker's Hat Barbados ?.?"	Hot Finger	Pepper
Capsicum	*annuum-baccatum*	*barbadosensis-willing'si*	Willing's Barbados ?.?"	Hot Finger	Pepper
Capsicum	*annuum-baccatum*	*hollandica-hot-finger-helvum*	Yellow Holland 6.5"	Hot Finger	Pepper
Capsicum	*annuum-baccatum*	*hollandica-hot-finger-rubrum*	Red Holland 6.5"	Hot Finger	Pepper
Capsicum	*annuum-baccatum*	*limon-dropii*	Lemon Drop 2.0"	Hot Finger	Pepper
Capsicum	*annuum-baccatum*	*PI-260574*	Heart ?.?"	Hot Finger	Pepper
Capsicum	*annuum-baccatum*	*pisum*	Pea 0.2"	Hot Finger	Pepper
Capsicum	*annuum-baccatum*	*praetermissum*	Neglected 1.5"	Hot Finger	Pepper
Capsicum	*annuum-baccatum*	*stella-pisces*	Brazilian Starfish ?.?"	Hot Finger	Pepper
Capsicum	*annuum-baccatum-pendulum*	*pendulum*	Original 3.0"	Aji Hot Finger	Pepper
Capsicum	*annuum-baccatum-pendulum*	*amarilloi*	Amarillo 2.5"	Aji Hot Finger	Pepper
Capsicum	*annuum-baccatum-pendulum*	*andesensis*	Andean ?.?"	Aji Hot Finger	Pepper
Capsicum	*annuum-baccatum-pendulum*	*benitoensis*	Benito ?.?"	Aji Hot Finger	Pepper
Capsicum	*annuum-baccatum-pendulum*	*citoi*	Cito ?.?"	Aji Hot Finger	Pepper
Capsicum	*annuum-baccatum-pendulum*	*habaneroi*	Habanero 2.5"	Aji Hot Finger	Pepper
Capsicum	*annuum-baccatum-pendulum*	*omnicolori*	Omnicolor 2.5"	Aji Hot Finger	Pepper

Genus	Species	Infraspecies	cultivar	specimen	Genome
Capsicum	*annuum-baccatum-pendulum*	*peruviensis*	Peruvian 2.5"	Aji Hot Finger	Pepper
Capsicum	*annuum-baccatum-pendulum*	*peruviensis-brunneum*	Brown Peruvian 4.0"	Aji Hot Finger	Pepper
Capsicum	*annuum-baccatum-pendulum*	*peruviensis-purpureum*	Purple Peruvian 1.0"	Aji Hot Finger	Pepper
Capsicum	*annuum-baccatum-pendulum*	*peruviensis-helvum*	Yellow Peruvian 2.5"	Aji Hot Finger	Pepper
Capsicum	*annuum-baccatum-pendulum*	*peruviensis-helvum-kellu-uchui*	Kellu Uchu Yellow Peruvian 1.5"	Aji Hot Finger	Pepper
Capsicum	*annuum-baccatum-pendulum*	*peruviensis-rubrum*	Red Peruvian 4.0"	Aji Hot Finger	Pepper
Capsicum	*annuum-baccatum-pendulum*	*pucu-uchui*	Puca Uchu ?.?"	Aji Hot Finger	Pepper
Capsicum	*annuum-baccatum-pendulum*	*umbai*	Umba ?.?"	Aji Hot Finger	Pepper
Capsicum	*annuum-baccatum-pendulum*	*verdei*	Verde 4.5"	Aji Hot Finger	Pepper
Capsicum	*annuum-baccatum-pendulum*	*yuquitania*	Yucatan ?.?"	Aji Hot Finger	Pepper

Pepper, Hot Cherry (*Capsicum anuum-cerasiforme*)

Genus	Species	Infraspecies	cultivar	specimen	Genome
Capsicum	*annuum-cerasiforme*	*album-fuerum*	White Fire ?.?"	Hot Cherry	Pepper
Capsicum	*annuum-cerasiforme*	*biaco-di-satana*	Satan's Kiss ?.?"	Hot Cherry	Pepper
Capsicum	*annuum-cerasiforme*	*cherokeei*	Cherokee ?.?"	Hot Cherry	Pepper
Capsicum	*annuum-cerasiforme*	*holiday-cheeri*	Holiday Cheer ?.?"	Hot Cherry	Pepper
Capsicum	*annuum-cerasiforme*	*hungaricum-cerasiforme*	Hungarian ?.?"	Hot Cherry	Pepper
Capsicum	*annuum-cerasiforme*	*joe's-rotundum*	Joe's Round .75"	Hot Cherry	Pepper
Capsicum	*annuum-cerasiforme*	*nigra-prunus*	Black Plum 1.5"	Hot Cherry	Pepper
Capsicum	*annuum-cerasiforme*	*pimenta-di-cheiroi*	Pimenta Di Ciero ?.?"	Hot Cherry	Pepper
Capsicum	*annuum-cerasiforme*	*rotunda-chillyi*	Round Chilly 1.5"	Hot Cherry	Pepper
Capsicum	*annuum-cerasiforme*	*rubrum*	Red 1.3"	Hot Cherry	Pepper
Capsicum	*annuum-cerasiforme*	*rubrum-largum*	Large Red 1.5"	Hot Cherry	Pepper
Capsicum	*annuum-cerasiforme*	*rubrum-parvum*	Small Red 1.2"	Hot Cherry	Pepper

Pepper, Hot Torpedo (*Capsicum annuum-chinense*)

Genus	Species	Infraspecies	cultivar	specimen	Genome
Capsicum	*annuum-chinense*	*belleum-nepalensis*	Nepalese Belle 3.5"	Hot Torpedo	Pepper
Capsicum	*annuum-chinense*	*chinensis-yunanensis*	Chinese Yunan 3.0"	Hot Torpedo	Pepper
Capsicum	*annuum-chinense*	*congo-brunnea*	Brown Congo ?.?"	Hot Torpedo	Pepper
Capsicum	*annuum-chinense*	*cleo'si-dracona*	Cleo's Dragon ?.?"	Hot Torpedo	Pepper
Capsicum	*annuum-chinense*	*cumari*	Cumari 0.2"	Hot Torpedo	Pepper
Capsicum	*annuum-chinense*	*desi-teekhii*	Desi Teekhi ?.?"	Hot Torpedo	Pepper
Capsicum	*annuum-chinense*	*fatalii*	Fatali 2.5"	Hot Torpedo	Pepper
Capsicum	*annuum-chinense*	*gambiensis*	Gambia Squash ?.?"	Hot Torpedo	Pepper
Capsicum	*annuum-chinense*	*hot-paper-lanternii*	Hot Paper Lantern 3.5"	Hot Torpedo	Pepper
Capsicum	*annuum-chinense*	*jamaicensis*	Jamaican ?.?"	Hot Torpedo	Pepper
Capsicum	*annuum-chinense*	*jamaicensis-aureum*	Gold Jamaican ?.?"	Hot Torpedo	Pepper
Capsicum	*annuum-chinense*	*jamaicensis-scotica-bindum*	Scotch Bonnet Jamaican ?.?"	Hot Torpedo	Pepper
Capsicum	*annuum-chinense*	*limoi*	Limo ?.?"	Hot Torpedo	Pepper
Capsicum	*annuum-chinense*	*mushroomii*	Mushroom 2.0"	Hot Torpedo	Pepper
Capsicum	*annuum-chinense*	*mushroomii-rubrum-rocotilloi*	Rocotillo Redcap Mushroom	Hot Torpedo	Pepper
Capsicum	*annuum-chinense*	*purpureum-tigrina*	Purple Tiger ?.?"	Hot Torpedo	Pepper
Capsicum	*annuum-chinense*	*rubrum-torpedoi*	Red 4.0"	Hot Torpedo	Pepper
Capsicum	*annuum-chinense*	*royalis-nigrum*	Royal Black ?.?"	Hot Torpedo	Pepper
Capsicum	*annuum-chinense*	*scotica-bonnetii-auranteum*	Orange Scotch Bonnet ?.?"	Hot Torpedo	Pepper
Capsicum	*annuum-chinense*	*scotica-bonnetii-chocolateum*	Chocolate Scotch Bonnet 1.5"	Hot Torpedo	Pepper
Capsicum	*annuum-chinense*	*scotica-bonnetii-helvum*	Yellow Scotch Bonnet ?.?"	Hot Torpedo	Pepper
Capsicum	*annuum-chinense*	*scotica-bonnetii-helvum-PA-353*	#353 Yellow Scotch Bonnet ?.?"	Hot Torpedo	Pepper
Capsicum	*annuum-chinense*	*scotica-bonnetii-praecox*	Early Scotch Bonnet 0.8"	Hot Torpedo	Pepper
Capsicum	*annuum-chinense*	*scotica-bonnetii-rubrum*	Red Scotch Bonnet ?.?"	Hot Torpedo	Pepper
Capsicum	*annuum-chinense*	*scotica-bonnetii-rubrum-PA-398*	#398 Red Scotch Bonnet ?.?"	Hot Torpedo	Pepper
Capsicum	*annuum-chinense*	*tears-of-firei*	Tears Of Fire 1.5"	Hot Torpedo	Pepper
Capsicum	*annuum-chinense*	*tenneessensis-teardropsi*	Tennessee Teardrops 2.0"	Hot Torpedo	Pepper
Capsicum	*annuum-chinense*	*trinidadensis-congoi*	Congo Trinidad 2.0"	Hot Torpedo	Pepper
Capsicum	*annuum-chinense*	*varingata*	Varingata ?.?"	Hot Torpedo	Pepper

Pepper, Hot Habenero (*Capsicum annuum-chinense-habanerum*)

Genus	Species	Infraspecies	cultivar	specimen	Genome
Capsicum	*annuum-chinense-habanerum*	*habanerum*	Original 1.2"	Hot Habanero	Pepper
Capsicum	*annuum-chinense-habanerum*	*album*	White 1.5"	Hot Habanero	Pepper
Capsicum	*annuum-chinense-habanerum*	*album-bullum*	Bullet White 1.0"	Hot Habanero	Pepper
Capsicum	*annuum-chinense-habanerum*	*auranteum*	Orange ?.?"	Hot Habanero	Pepper
Capsicum	*annuum-chinense-habanerum*	*aureum*	Golden 3.0"	Hot Habanero	Pepper
Capsicum	*annuum-chinense-habanerum*	*brunneum*	Brown ?.?"	Hot Habanero	Pepper
Capsicum	*annuum-chinense-habanerum*	*caribeum-rubrum*	Caribbean Red 2.0"	Hot Habanero	Pepper
Capsicum	*annuum-chinense-habanerum*	*chocolate-congoi*	Chocolate Congo 2.0"	Hot Habanero	Pepper
Capsicum	*annuum-chinense-habanerum*	*datilii*	Datil 2.0"	Hot Habanero	Pepper
Capsicum	*annuum-chinense-habanerum*	*mustardii*	Mustard ?.?"	Hot Habanero	Pepper
Capsicum	*annuum-chinense-habanerum*	*peachii*	Peach ?.?"	Hot Habanero	Pepper
Capsicum	*annuum-chinense-habanerum*	*rubrum-lucifer's-dreamii*	Lucifer's Dream Red ?.?"	Hot Habanero	Pepper
Capsicum	*annuum-chinense-habanerum*	*rubrum-savinai*	Savina Red ?.?"	Hot Habanero	Pepper

Genus	Species	Infraspecies	cultivar	specimen	Genome
Capsicum	annuum-chinense-habanerum	tazmaniensis	Tazmanian ?.?"	Hot Habanero	Pepper

Pepper, Hot Cone (*Capsicum annuum-conoides*)

Genus	Species	Infraspecies	cultivar	specimen	Genome
Capsicum	annuum-conoides	autopickii	Autopick 3.5"	Hot Cone	Pepper
Capsicum	annuum-conoides	baladai	Balada 4.5"	Hot Cone	Pepper
Capsicum	annuum-conoides	beaveri-damii	Beaver Dam ?.?"	Hot Cone	Pepper
Capsicum	annuum-conoides	bulgariana-carrotii	Bulgarian Carrot 2.5"	Hot Cone	Pepper
Capsicum	annuum-conoides	candelalichtum	Candlelight 1.0"	Hot Cone	Pepper
Capsicum	annuum-conoides	chilacatei	Chilacate 5.5"	Hot Cone	Pepper
Capsicum	annuum-conoides	chinensis-multicolorum	Chinese Multicolor 1.0"	Hot Cone	Pepper
Capsicum	annuum-conoides	coccineum-caldorum	Crimson 6.2"	Hot Cone	Pepper
Capsicum	annuum-conoides	cochiti-puebloensis	Cochiti Pueblo 3.5"	Hot Cone	Pepper
Capsicum	annuum-conoides	criolla-des-tres-cantosi	Tres Cantos Criolla ?.?"	Hot Cone	Pepper
Capsicum	annuum-conoides	criolla-sellai	Sella Criolla 3.0"	Hot Cone	Pepper
Capsicum	annuum-conoides	cyklonii	Cyklon 4.5"	Hot Cone	Pepper
Capsicum	annuum-conoides	diablo-grandei	Diablo Grande 7.0"	Hot Cone	Pepper
Capsicum	annuum-conoides	draconum-cheilum	Dragon's Claw 10."	Hot Cone	Pepper
Capsicum	annuum-conoides	dunsoi	Dunso ?.?"	Hot Cone	Pepper
Capsicum	annuum-conoides	eastern-rocketii	Eastern Rocket 4.5"	Hot Cone	Pepper
Capsicum	annuum-conoides	el-pasoensis	El Paso 7.5"	Hot Cone	Pepper
Capsicum	annuum-conoides	embersi-fijiensis	Fiji Embers ?.?"	Hot Cone	Pepper
Capsicum	annuum-conoides	fipsi	Fips 1.0"	Hot Cone	Pepper
Capsicum	annuum-conoides	floralis-gemmatum	Floral Gem 2.2"	Hot Cone	Pepper
Capsicum	annuum-conoides	georgiensis-flameum	Georgia Flame 7.0"	Hot Cone	Pepper
Capsicum	annuum-conoides	guilinii	Guilin 2.5"	Hot Cone	Pepper
Capsicum	annuum-conoides	hot-portugalensis	Hot Portugal 6.5"	Hot Cone	Pepper
Capsicum	annuum-conoides	hungarica-nigrum	Black Hungarian 3.0"	Hot Cone	Pepper
Capsicum	annuum-conoides	jerusalemensis	Jerusalem 1.5"	Hot Cone	Pepper
Capsicum	annuum-conoides	longa-groena-buddhai	Long Green Buddha ?.?"	Hot Cone	Pepper
Capsicum	annuum-conoides	largoi-purpureum	Largo Purple ?.?"	Hot Cone	Pepper
Capsicum	annuum-conoides	martin's-carrotii	Martin's Carrot 2.0"	Hot Cone	Pepper
Capsicum	annuum-conoides	mirasolii	Mirasol 3.5"	Hot Cone	Pepper
Capsicum	annuum-conoides	mirasolii-cascabella-1	Pickling Cascabel Mirasol 1.5"	Hot Cone	Pepper
Capsicum	annuum-conoides	mirasolii-cascabella-2	Sauce Cascabel Mirasol 1.5"	Hot Cone	Pepper
Capsicum	annuum-conoides	mirasolii-neomexicana-improva	NuMex Improved Mirasol 4.7"	Hot Cone	Pepper
Capsicum	annuum-conoides	moraviana	Moravian 6.0"	Hot Cone	Pepper
Capsicum	annuum-conoides	orozcoi-aureum-auranticum	Golden Orange Orozco ?.?"	Hot Cone	Pepper
Capsicum	annuum-conoides	orozcoi-purpureum	Purple Orozco ?.?"	Hot Cone	Pepper
Capsicum	annuum-conoides	ot-chi-theinii	Ot ChiThien ?.?"	Hot Cone	Pepper
Capsicum	annuum-conoides	patagonia	Patagonia ?.?"	Hot Cone	Pepper
Capsicum	annuum-conoides	pepperoncini	Pepperoncini 3.0"	Hot Cone	Pepper
Capsicum	annuum-conoides	pepperoncini-aurea-graeaca	Golden Greek Pepperoncini ?.?"	Hot Cone	Pepper
Capsicum	annuum-conoides	pizzai-pepperi	Pizza Pepper 3.5"	Hot Cone	Pepper
Capsicum	annuum-conoides	puchellum-purpureum	Pretty Purple ?.?"	Hot Cone	Pepper
Capsicum	annuum-conoides	rainbow-birdii	Rainbow Bird 1.5"	Hot Cone	Pepper
Capsicum	annuum-conoides	ram's-hornii	Ram's Horn 7.0"	Hot Cone	Pepper
Capsicum	annuum-conoides	relleno	Relleno 6.5"	Hot Cone	Pepper
Capsicum	annuum-conoides	serranoi	Serrano 2.0"	Hot Cone	Pepper
Capsicum	annuum-conoides	serranoi-amandoi	Amando Serrano 3.5"	Hot Cone	Pepper
Capsicum	annuum-conoides	serranoi-hildagoi-tamai	Tam Hidalgo Serrano ?.?"	Hot Cone	Pepper
Capsicum	annuum-conoides	serrano-hildagoi-tam-pequinoi	Pequino Tam Hidalgo Serrano ?.?"	Hot Cone	Pepper
Capsicum	annuum-conoides	serrano-hildagoi-tam-pequinoi-74	#74 Pequino Tam Hidalgo Serrano ?.?"	Hot Cone	Pepper
Capsicum	annuum-conoides	serranoi-huastecoi	Huasteco Serrano 2.5"	Hot Cone	Pepper
Capsicum	annuum-conoides	serranoi-purpureum	Purple Serrano 2.2"	Hot Cone	Pepper
Capsicum	annuum-conoides	tenneessensis-spiceum	Tennessee Spice ?.?"	Hot Cone	Pepper
Capsicum	annuum-conoides	treasuresi-rubrum	Treasures Red 8.0"	Hot Cone	Pepper
Capsicum	annuum-conoides	wenk'si-helvum-caldorum	Wenk's Yellow Hots 3.0"	Hot Cone	Pepper

Pepper, Hot Pencil (*Capsicum annuum-frutescens*)

Genus	Species	Infraspecies	cultivar	specimen	Genome
Capsicum	annuum-frutescens	assamensis	Assam 4.5"	Hot Pencil	Pepper
Capsicum	annuum-frutescens	azri	Azr 2.5"	Hot Pencil	Pepper
Capsicum	annuum-frutescens	boliviana-rainbowii	Bolivian Rainbow ?.?"	Hot Pencil	Pepper
Capsicum	annuum-frutescens	bradley's-bahamiana	Bradley's Bahamian 1.2"	Hot Pencil	Pepper
Capsicum	annuum-frutescens	cabai-apii	Api Cabai 2.5"	Hot Pencil	Pepper
Capsicum	annuum-frutescens	cabai-burongii	Burong Cabai ?.?"	Hot Pencil	Pepper
Capsicum	annuum-frutescens	cobrai	Cobra 3.5"	Hot Pencil	Pepper
Capsicum	annuum-frutescens	dagger-podii	Dagger Pod 2.5"	Hot Pencil	Pepper
Capsicum	annuum-frutescens	de-rata	Rat 4.5"	Hot Pencil	Pepper
Capsicum	annuum-frutescens	fiesta	Fiesta 2.0"	Hot Pencil	Pepper
Capsicum	annuum-frutescens	indica-PC1	#PC1 India 2.5"	Hot Pencil	Pepper

Genus	Species	Infraspecies	cultivar	specimen Genome
Capsicum	*annuum-frutescens*	*jwalai*	Jwala 4.0"	Hot Pencil Pepper
Capsicum	*annuum-frutescens*	*jwalai-pusai*	Pusa Jwala 3.0"	Hot Pencil Pepper
Capsicum	*annuum-frutescens*	*luteum*	Lute 2.5"	Hot Pencil Pepper
Capsicum	*annuum-frutescens*	*pico-de-galloi*	Rooster's Beak 3.0"	Hot Pencil Pepper
Capsicum	*annuum-frutescens*	*pico-de-pajaroi*	Bird's Beak 5.7"	Hot Pencil Pepper
Capsicum	*annuum-frutescens*	*pungens-rubrum*	Pungent Red ?.?"	Hot Pencil Pepper
Capsicum	*annuum-frutescens*	*punjabensis-parvum-caldorum*	Punjab Small 2.0"	Hot Pencil Pepper
Capsicum	*annuum-frutescens*	*purirai*	Purira 2.0"	Hot Pencil Pepper
Capsicum	*annuum-frutescens*	*rat-tailii*	Rat Tail ?.?"	Hot Pencil Pepper
Capsicum	*annuum-frutescens*	*toothpickii-arizonensis*	Arizona Toothpick 2.5"	Hot Pencil Pepper
Capsicum	*annuum-frutescens*	*toothpickii-vietnamensis*	Vietnamese Toothpick ?.?"	Hot Pencil Pepper
Capsicum	*annuum-frutescens*	*truptii*	Trupti 4.5"	Hot Pencil Pepper
Capsicum	*annuum-frutescens*	*yatsafusai*	Yatsafusa 3.0"	Hot Pencil Pepper

Pepper, Hot Chili (*Capsicum annuum-frutescens-chilium*)

Genus	Species	Infraspecies	cultivar	specimen Genome
Capsicum	*annuum-frutescens-chilium*	*abeytasi*	Abeytas 4.5"	Hot Chili Pepper
Capsicum	*annuum-frutescens-chilium*	*agua-blanca*	Agua Blanca 10."	Hot Chili Pepper
Capsicum	*annuum-frutescens-chilium*	*alcaldei*	Alcalde 4.0"	Hot Chili Pepper
Capsicum	*annuum-frutescens-chilium*	*anaheimensis*	Anaheim 7.0"	Hot Chili Pepper
Capsicum	*annuum-frutescens-chilium*	*anaheimensis-collegei-64L*	#64L College Anaheim 7.0"	Hot Chili Pepper
Capsicum	*annuum-frutescens-chilium*	*anaheimensis-neomexiana-hatchii*	New Mexico Hatch Anaheim 5.0"	Hot Chili Pepper
Capsicum	*annuum-frutescens-chilium*	*anaheimensis-M*	Medium Hot Anaheim 7.0"	Hot Chili Pepper
Capsicum	*annuum-frutescens-chilium*	*anaheimensis-ortega*	Ortega Anaheim 6.5"	Hot Chili Pepper
Capsicum	*annuum-frutescens-chilium*	*anaheimensis-puebloi*	Pueblo Anaheim 6.0"	Hot Chili Pepper
Capsicum	*annuum-frutescens-chilium*	*anaheimensis-sandia*	Sandia Anaheim 6.5"	Hot Chili Pepper
Capsicum	*annuum-frutescens-chilium*	*anaheimensis-sonora*	Sonora Anaheim 8.0"	Hot Chili Pepper
Capsicum	*annuum-frutescens-chilium*	*anaheimensis-TMR*	Tobacco Mosaic Resistant Anaheim 7.5"	Hot Chili Pepper
Capsicum	*annuum-frutescens-chilium*	*anaheimensis-TMR23*	#23 Tobacco Mosaic Resistant Anaheim 7.5"	Hot Chili Pepper
Capsicum	*annuum-frutescens-chilium*	*poblano*	Poblano Ancho ?.?"	Hot Chili Pepper
Capsicum	*annuum-frutescens-chilium*	*poblano-anchoi*	Ancho 4.0"	Hot Chili Pepper
Capsicum	*annuum-frutescens-chilium*	*poblano-anchoi-101*	#101 Ancho 4.0"	Hot Chili Pepper
Capsicum	*annuum-frutescens-chilium*	*poblano-anchoi-costeno*	Costeno Ancho 6.0"	Hot Chili Pepper
Capsicum	*annuum-frutescens-chilium*	*poblano-anchoi-gigantea*	Giant Ancho 4.0"	Hot Chili Pepper
Capsicum	*annuum-frutescens-chilium*	*poblano-anchoi-san-luisensis*	San Luis Ancho 5.5"	Hot Chili Pepper
Capsicum	*annuum-frutescens-chilium*	*apachei*	Apache ?.?"	Hot Chili Pepper
Capsicum	*annuum-frutescens-chilium*	*bahamensis-chilium*	Bahamian 1.5"	Hot Chili Pepper
Capsicum	*annuum-frutescens-chilium*	*barkeri*	Barker 6.0"	Hot Chili Pepper
Capsicum	*annuum-frutescens-chilium*	*burning-bushii*	Burning Bush 2.0"	Hot Chili Pepper
Capsicum	*annuum-frutescens-chilium*	*caribensis*	Caribe 2.5"	Hot Chili Pepper
Capsicum	*annuum-frutescens-chilium*	*caribensis-santa-fe-grandenesis*	Santa Fe Grand Caribe 3.5"	Hot Chili Pepper
Capsicum	*annuum-frutescens-chilium*	*caribensis-hot-gold-spikei*	Hot Gold Spike Caribe ?.?"	Hot Chili Pepper
Capsicum	*annuum-frutescens-chilium*	*chimayoensis*	Chimayo ?.?"	Hot Chili Pepper
Capsicum	*annuum-frutescens-chilium*	*coloradensis-caldorum*	Hot Colorado ?.?"	Hot Chili Pepper
Capsicum	*annuum-frutescens-chilium*	*coloradensis-mitis*	Mild Colorado ?.?"	Hot Chili Pepper
Capsicum	*annuum-frutescens-chilium*	*coloradensis-poblanoi*	Poblano Colorado 3.7"	Hot Chili Pepper
Capsicum	*annuum-frutescens-chilium*	*cow-hornii*	Cow Horn 8.0"	Hot Chili Pepper
Capsicum	*annuum-frutescens-chilium*	*craig'si-chilium*	Craig's ?.?"	Hot Chili Pepper
Capsicum	*annuum-frutescens-chilium*	*de-arbolum*	De Arbol 2.0"	Hot Chili Pepper
Capsicum	*annuum-frutescens-chilium*	*de-arbolum-de-baja-california-surensis*	Baha California Sur De Arbol 4.0"	Hot Chili Pepper
Capsicum	*annuum-frutescens-chilium*	*de-arbolum-purpureum*	Purple De Arbol 2.5"	Hot Chili Pepper
Capsicum	*annuum-frutescens-chilium*	*espanola*	Espanola 6.5"	Hot Chili Pepper
Capsicum	*annuum-frutescens-chilium*	*espanola-improva*	Improved Espanola 5.5"	Hot Chili Pepper
Capsicum	*annuum-frutescens-chilium*	*fons-flameum*	Flame Fountain ?.?"	Hot Chili Pepper
Capsicum	*annuum-frutescens-chilium*	*fons-rubrum*	Red Fountain ?.?"	Hot Chili Pepper
Capsicum	*annuum-frutescens-chilium*	*fresnoensis-chilium*	Fresno 2.5"	Hot Chili Pepper
Capsicum	*annuum-frutescens-chilium*	*fresnoensis-chilium-grandeum*	Grand Frenso 2.5"	Hot Chili Pepper
Capsicum	*annuum-frutescens-chilium*	*grandeum*	Grande 3.0"	Hot Chili Pepper
Capsicum	*annuum-frutescens-chilium*	*grandeum-longum-crassum-rubrum*	Long Thick Red Grande 3.5"	Hot Chili Pepper
Capsicum	*annuum-frutescens-chilium*	*guajilloi*	Guajillo 4.0"	Hot Chili Pepper
Capsicum	*annuum-frutescens-chilium*	*hahongii-kochoi*	Hahong Kocho 2.5"	Hot Chili Pepper
Capsicum	*annuum-frutescens-chilium*	*hahongii-kochoi-rooster-spuri*	Rooster Spur Hahong Kocho 1.2"	Hot Chili Pepper
Capsicum	*annuum-frutescens-chilium*	*hahongii-takanotsumei*	Takanotsume Hahong Kocho 1.5"	Hot Chili Pepper
Capsicum	*annuum-frutescens-chilium*	*hanoiensis-helva*	Yellow Hanoi ?.?"	Hot Chili Pepper
Capsicum	*annuum-frutescens-chilium*	*hanoiensis-rubra*	Red Hanoi ?.?"	Hot Chili Pepper
Capsicum	*annuum-frutescens-chilium*	*inayaguei*	Inayague 1.5"	Hot Chili Pepper

Genus	Species	Infraspecies	cultivar	specimen	Genome
Capsicum	*annuum-frutescens-chilium*	*indica-chilium*	India 2.5"	Hot Chili	Pepper
Capsicum	*annuum-frutescens-chilium*	*isletensis*	Isleta ?.?"	Hot Chili	Pepper
Capsicum	*annuum-frutescens-chilium*	*jaralesi*	Jarales ?.?"	Hot Chili	Pepper
Capsicum	*annuum-frutescens-chilium*	*jellybeanii*	Jellybean ?.?"	Hot Chili	Pepper
Capsicum	*annuum-frutescens-chilium*	*jemezensis*	Jemez 3.0"	Hot Chili	Pepper
Capsicum	*annuum-frutescens-chilium*	*lombakensis*	Lombak 5.5"	Hot Chili	Pepper
Capsicum	*annuum-frutescens-chilium*	*longum-caldorum-chilium*	Long 7.5"	Hot Chili	Pepper
Capsicum	*annuum-frutescens-chilium*	*mexicana-chilium-improva*	Improved Mexican 4.7"	Hot Chili	Pepper
Capsicum	*annuum-frutescens-chilium*	*mexicana-rubrum-caldorum*	Red Hot Mexican 2.5"	Hot Chili	Pepper
Capsicum	*annuum-frutescens-chilium*	*negro-devallei*	Negro De Valle 6.0"	Hot Chili	Pepper
Capsicum	*annuum-frutescens-chilium*	*neomexicana-6*	#6 NuMex 6.0"	Hot Chili	Pepper
Capsicum	*annuum-frutescens-chilium*	*neomexicana-6-longum*	Long #6 NuMex 7.0"	Hot Chili	Pepper
Capsicum	*annuum-frutescens-chilium*	*neomexicana-6-mitis*	Mild #6 NuMex 6.0"	Hot Chili	Pepper
Capsicum	*annuum-frutescens-chilium*	*neomexicana-6-4*	#6-4 NuMex 5.0"	Hot Chili	Pepper
Capsicum	*annuum-frutescens-chilium*	*neomexicana-6-4L*	#6-4L NuMex 6.0"	Hot Chili	Pepper
Capsicum	*annuum-frutescens-chilium*	*neomexicana-big-jimii*	Big Jim NuMex 7.7"	Hot Chili	Pepper
Capsicum	*annuum-frutescens-chilium*	*neomexicana-centennialis*	Centennial NuMex ?.?"	Hot Chili	Pepper
Capsicum	*annuum-frutescens-chilium*	*neomexicana-eclipsei*	Eclipse NuMex 5.0"	Hot Chili	Pepper
Capsicum	*annuum-frutescens-chilium*	*neomexicana-extra-caldorum*	Extra Hot NuMex 6.7"	Hot Chili	Pepper
Capsicum	*annuum-frutescens-chilium*	*neomexicana-joe-parkeri*	Joe Parker NuMex 6.5"	Hot Chili	Pepper
Capsicum	*annuum-frutescens-chilium*	*neomexicana-joe-parkeri-e.*	E. Joe Parker NuMex 6.5"	Hot Chili	Pepper
Capsicum	*annuum-frutescens-chilium*	*neomexicana-r.-nakyama*	R. Naky NuMex ?.?"	Hot Chili	Pepper
Capsicum	*annuum-frutescens-chilium*	*neomexicana-suavis-anaheimensis*	Sweet Anaheim NuMex 7.0"	Hot Chili	Pepper
Capsicum	*annuum-frutescens-chilium*	*neomexicana-sunburstii*	Sunburst NuMex 2.5"	Hot Chili	Pepper
Capsicum	*annuum-frutescens-chilium*	*neomexicana-sunflarei*	Sunflare NuMex 2.5"	Hot Chili	Pepper
Capsicum	*annuum-frutescens-chilium*	*neomexicana-sunglovii*	Sunglow NuMex 2.5"	Hot Chili	Pepper
Capsicum	*annuum-frutescens-chilium*	*neomexicana-sunrisei*	Sunrise NuMex 5.0"	Hot Chili	Pepper
Capsicum	*annuum-frutescens-chilium*	*neomexicana-sunsetii*	Sunset NuMex 5.0"	Hot Chili	Pepper
Capsicum	*annuum-frutescens-chilium*	*parvum-chilium*	Small 5.5"	Hot Chili	Pepper
Capsicum	*annuum-frutescens-chilium*	*peteri-pepperi*	Red Peter Pepper 3.7"	Hot Chili	Pepper
Capsicum	*annuum-frutescens-chilium*	*peteri-pepperi-aurantium*	Orange Peter Pepper 4.0"	Hot Chili	Pepper
Capsicum	*annuum-frutescens-chilium*	*peteri-pepperi-helva*	Yellow Peter Pepper 2.5"	Hot Chili	Pepper
Capsicum	*annuum-frutescens-chilium*	*peteri-pepperi-pinguita-de-mono*	Monkey Dick Peter Pepper 1.0"	Hot Chili	Pepper
Capsicum	*annuum-frutescens-chilium*	*pili-pilii*	Pili Pili 3.0"	Hot Chili	Pepper
Capsicum	*annuum-frutescens-chilium*	*riotii*	Riot ?.?"	Hot Chili	Pepper
Capsicum	*annuum-frutescens-chilium*	*rubrum*	Red 2.1"	Hot Chili	Pepper
Capsicum	*annuum-frutescens-chilium*	*santakai*	Santaka 2.7"	Hot Chili	Pepper
Capsicum	*annuum-frutescens-chilium*	*santo-domingoensis*	Santo Domingo 3.0"	Hot Chili	Pepper
Capsicum	*annuum-frutescens-chilium*	*sue-senger'si-chilium*	Sue Senger's ?.?"	Hot Chili	Pepper
Capsicum	*annuum-frutescens-chilium*	*slim-jimii*	Slim Jim 12."	Hot Chili	Pepper
Capsicum	*annuum-frutescens-chilium*	*supra-chilium*	Super Chile 1.5"	Hot Chili	Pepper
Capsicum	*annuum-frutescens-chilium*	*supra-chilium-matchboxii*	Matchbox Super Chile 1.7"	Hot Chili	Pepper
Capsicum	*annuum-frutescens-chilium*	*texensis-chilium*	Texas 3.0"	Hot Chili	Pepper
Capsicum	*annuum-frutescens-chilium*	*valleroi*	Vallero 5.0"	Hot Chili	Pepper
Capsicum	*annuum-frutescens-chilium-aviculare*	*chiltepinii*	Chiltepin ?.?"	Bird Hot Chili	Pepper
Capsicum	*annuum-frutescens-chilium-aviculare*	*chiltepinii-hermosillo-selectii*	Hermosa Chiltepin 0.7"	Bird Hot Chili	Pepper
Capsicum	*annuum-frutescens-chilium-aviculare*	*chiltepinii-mayoensis*	Mayo Chiltepin ?.?"	Bird Hot Chili	Pepper
Capsicum	*annuum-frutescens-chilium-aviculare*	*chiltepinii-purpureum*	Purple Chiltepin ?.?"	Bird Hot Chili	Pepper
Capsicum	*annuum-frutescens-chilium-aviculare*	*chiltepinii-puya*	Puya Chiltepin ?.?"	Bird Hot Chili	Pepper
Capsicum	*annuum-frutescens-chilium-aviculare*	*chiltepinii-sinaloensis*	Sinaloa Chiltepin 0.4"	Bird Hot Chili	Pepper
Capsicum	*annuum-frutescens-chilium-aviculare*	*chiltepinii-tarahumara*	Tarahumara Chiltepin ?.?"	Bird Hot Chili	Pepper
Capsicum	*annuum-frutescens-chilium-aviculare*	*chiltepinii-tepinii*	Tepin Chiltepin ?.?"	Bird Hot Chili	Pepper
Capsicum	*annuum-frutescens-chilium-aviculare*	*chiltepinii-texensis*	Texas Chiltepin ?.?"	Bird Hot Chili	Pepper
Capsicum	*annuum-frutescens-chilium-aviculare*	*pequinii*	Pequin 3.0"	Bird Hot Chili	Pepper
Capsicum	*annuum-frutescens-chilium-aviculare*	*pequinii-afrum*	African Pequin ?.?"	Bird Hot Chili	Pepper
Capsicum	*annuum-frutescens-chilium-aviculare*	*pequinii-largum-nigrum*	Large Black Pequin 0.5"	Bird Hot Chili	Pepper
Capsicum	*annuum-frutescens-chilium-aviculare*	*pequinii-neomexicana-twilightii*	Twilight NuMex Pequin	Bird Hot Chili	Pepper
Capsicum	*annuum-frutescens-chilium-aviculare*	*pequinii-baileyi*	Bailey Pequin 1.0"	Bird Hot Chili	Pepper
Capsicum	*annuum-frutescens-chilium-aviculare*	*pequinii-congo-de-nicaragua*	Congo De Nicaragua Pequin ?.?"	Bird Hot Chili	Pepper
Capsicum	*annuum-frutescens-chilium-aviculare*	*pequinii-pequeno*	Pequeno Pequin 1.5"	Bird Hot Chili	Pepper
Capsicum	*annuum-frutescens-chilium-aviculare*	*pequinii-russica*	Russian Pequin 0.5"	Bird Hot Chili	Pepper
Capsicum	*annuum-frutescens-chilium-aviculare*	*zimbabwensis*	Zimbabwe ?.?"	Bird Hot Chili	Pepper

Pepper, Hot Tabasco (*Capsicum annuum-frutescens-tabascum*)

Genus	Species	Infraspecies	cultivar	specimen	Genome
Capsicum	*annuum-frutescens-tabascum*	*tabascum*	Original 1.5"	Hot Tabasco	Pepper
Capsicum	*annuum-frutescens-tabascum*	*braziliana-malagueta*	Malagueta Brazilian 1.5"	Hot Tabasco	Pepper
Capsicum	*annuum-frutescens-tabascum*	*brevum-helvum*	Short Yellow ?.?"	Hot Tabasco	Pepper
Capsicum	*annuum-frutescens-tabascum*	*costaricensis*	Costa Rica ?.?"	Hot Tabasco	Pepper
Capsicum	*annuum-frutescens-tabascum*	*dr.-greenleaf'si*	Dr. Greenleaf's ?.?"	Hot Tabasco	Pepper
Capsicum	*annuum-frutescens-tabascum*	*hawaiiana*	Hawaiian ?.?"	Hot Tabasco	Pepper
Capsicum	*annuum-frutescens-tabascum*	*hawaiiana-suavis*	Sweet Hawaiian ?.?"	Hot Tabasco	Pepper

Genus	Species	Infraspecies	cultivar	specimen	Genome
Capsicum	annuum-frutescens-tabascum	ludoviciana-arledgei-caldorum	Arledge Louisiana 4.0"	Hot Tabasco	Pepper
Capsicum	annuum-frutescens-tabascum	thailandica-caldorum	Thai .75"	Hot Tabasco	Pepper
Capsicum	annuum-frutescens-tabascum	thailandica-caldorum-aurantium	Orange Thai 2.5"	Hot Tabasco	Pepper
Capsicum	annuum-frutescens-tabascum	thailandica-caldorum-birdii	Bird Thai 1.2"	Hot Tabasco	Pepper
Capsicum	annuum-frutescens-tabascum	thailandica-caldorum-helvum-chilium	Yellow Thai ?.?"	Hot Tabasco	Pepper
Capsicum	annuum-frutescens-tabascum	thailandica-caldorum-largum	Large Thai ?.?"	Hot Tabasco	Pepper
Capsicum	annuum-frutescens-tabascum	thailandica-caldorum-sirichi	Sirichi Thai 3.5"	Hot Tabasco	Pepper
Capsicum	annuum-frutescens-tabascum	thailandica-caldorum-solum	Sun Thai 1.2"	Hot Tabasco	Pepper

Pepper, Hot Bell (*Capsicum annuum-grossum*)

Genus	Species	Infraspecies	cultivar	specimen	Genome
Capsicum	annuum-grossum	balloonii	Balloon 3.0"	Hot Bell	Pepper
Capsicum	annuum-grossum	chilaulcei-negroi	Chilaucle Negro Muffin 2.5"	Hot Bell	Pepper
Capsicum	annuum-grossum	hinklehatzii	Hinklehatz Heart 1.5"	Hot Bell	Pepper
Capsicum	annuum-grossum	hinklehatzii-helvum	Yellow Hinklehatz Heart 1.5"	Hot Bell	Pepper
Capsicum	annuum-grossum	hot-applei	Hot Apple 1.5"	Hot Bell	Pepper
Capsicum	annuum-grossum	karloi	Karlo Top 3.5"	Hot Bell	Pepper
Capsicum	annuum-grossum	little-nubiana	Little Nubian Bell 1.5"	Hot Bell	Pepper
Capsicum	annuum-grossum	mini-bella-caldorum	Mini Bell ?.?"	Hot Bell	Pepper
Capsicum	annuum-grossum	roumaniensis-caldorum	Romanian Blocky 4.0"	Hot Bell	Pepper
Capsicum	annuum-grossum	roumaniensis-semi-caldorum	Semi Hot Romanian Blocky 4.0"	Hot Bell	Pepper
Capsicum	annuum-grossum	roumaniensis-semi-caldorum-szentesi	Szentesi Semi Hot Romanian Blocky 4.5"	Hot Bell	Pepper
Capsicum	annuum-grossum	roumaniensis-semi-caldorum-szentesi-paradicsom-alaku-zoldii	Paradise Szentesi Semi Hot Romanian Blocky 4.5"	Hot Bell	Pepper
Capsicum	annuum-grossum	tomatoi-pepperi	Tomato ?.?"	Hot Bell	Pepper

Pepper, Hot Banana (*Capsicum annuum-longum-arienum*)

Genus	Species	Infraspecies	cultivar	specimen	Genome
Capsicum	annuum-longum-arienum	abyssinica-brunnea	Ethiopian Brown 4.0"	Hot Banana	Pepper
Capsicum	annuum-longum-arienum	achari	Achar 3.5"	Hot Banana	Pepper
Capsicum	annuum-longum-arienum	aci-sivri	Aci Sivri 8.5"	Hot Banana	Pepper
Capsicum	annuum-longum-arienum	afgani-brevum	Afghan Short ?.?"	Hot Banana	Pepper
Capsicum	annuum-longum-arienum	afgani-longum	Afghan Long 3.0"	Hot Banana	Pepper
Capsicum	annuum-longum-arienum	agni-sikhai-2	Agni Sikha II ?.?"	Hot Banana	Pepper
Capsicum	annuum-longum-arienum	apaseoi	Apaseo 9.0"	Hot Banana	Pepper
Capsicum	annuum-longum-arienum	aurora	Aurora 1.5"	Hot Banana	Pepper
Capsicum	annuum-longum-arienum	barneyi	Barney 5.5"	Hot Banana	Pepper
Capsicum	annuum-longum-arienum	beijingensis	Beijing 4.0"	Hot Banana	Pepper
Capsicum	annuum-longum-arienum	berberei	Berbere 6.0"	Hot Banana	Pepper
Capsicum	annuum-longum-arienum	bouquetii	Bouquet ?.?"	Hot Banana	Pepper
Capsicum	annuum-longum-arienum	buenos-carbidei	Buenos Carbide ?.?"	Hot Banana	Pepper
Capsicum	annuum-longum-arienum	calistana	Calistan ?.?"	Hot Banana	Pepper
Capsicum	annuum-longum-arienum	cantina-helvum	Cantina Yellow 2.7"	Hot Banana	Pepper
Capsicum	annuum-longum-arienum	caloro-helvum-vaxum	Yellow Wax Caloro 3.5"	Hot Banana	Pepper
Capsicum	annuum-longum-arienum	casodosensis-nativum	Casados Native 4.0"	Hot Banana	Pepper
Capsicum	annuum-longum-arienum	chacoense	Chaco 4.0"	Hot Banana	Pepper
Capsicum	annuum-longum-arienum	chahuahuensis	Chahua ?.?"	Hot Banana	Pepper
Capsicum	annuum-longum-arienum	chamborateum	Chamborate 3.5"	Hot Banana	Pepper
Capsicum	annuum-longum-arienum	chayusa-rusoi	Cayusa Ruso ?.?"	Hot Banana	Pepper
Capsicum	annuum-longum-arienum	chi-chienii	Chi Chien 2.5"	Hot Banana	Pepper
Capsicum	annuum-longum-arienum	chilcostlei	Chilcostle 5.5"	Hot Banana	Pepper
Capsicum	annuum-longum-arienum	country-girlii	Country Girl ?.?"	Hot Banana	Pepper
Capsicum	annuum-longum-arienum	curryi	Curry 6.0"	Hot Banana	Pepper
Capsicum	annuum-longum-arienum	de-comida	De Comida 5.0"	Hot Banana	Pepper
Capsicum	annuum-longum-arienum	de-oaxacensis	De Oaxaca ?.?"	Hot Banana	Pepper
Capsicum	annuum-longum-arienum	demrei	Demre ?.?"	Hot Banana	Pepper
Capsicum	annuum-longum-arienum	d'espelletei	D'Espelette ?.?"	Hot Banana	Pepper
Capsicum	annuum-longum-arienum	dulce-mediterraneum-helvum	Yellow Dulce Mediterrano 10."	Hot Banana	Pepper
Capsicum	annuum-longum-arienum	hollandica	Dutch ?.?"	Hot Banana	Pepper
Capsicum	annuum-longum-arienum	hungarica-helvum-vaxum-caldorum	Yellow Wax Hot Hungarian 6.5"	Hot Banana	Pepper
Capsicum	annuum-longum-arienum	hungarica-oranger-ungerischer-pfefferonii	Orange Pfefferoni Hungarian 3.0"	Hot Banana	Pepper
Capsicum	annuum-longum-arienum	hungarica-rainbow-wax-blockii	Rainbow Wax Block Hungarian ?.?"	Hot Banana	Pepper
Capsicum	annuum-longum-arienum	hungarica-rainbow-wax-blockii-brevum	Short Rainbow Wax Block Hungarian ?.?"	Hot Banana	Pepper
Capsicum	annuum-longum-arienum	hungarica-semi-hotii	Semi Hot Hungarian 3.5"	Hot Banana	Pepper
Capsicum	annuum-longum-arienum	escondida	Escondida ?.?"	Hot Banana	Pepper
Capsicum	annuum-longum-arienum	expressum-chinensis	Express China ?.?"	Hot Banana	Pepper
Capsicum	annuum-longum-arienum	macskasarga-felis-helvum	Macskasarga Cat Yellow 2.5"	Hot Banana	Pepper
Capsicum	annuum-longum-arienum	filius-blauea	Filius Blue ?.?"	Hot Banana	Pepper
Capsicum	annuum-longum-arienum	fishi-pepperi	Fish 2.5"	Hot Banana	Pepper
Capsicum	annuum-longum-arienum	fuerum	Fire 2.0"	Hot Banana	Pepper

Genus	Species	Infraspecies	cultivar	specimen	Genome
Capsicum	*annuum-longum-arienum*	*goat-hornii*	Goat Horn 5.0"	Hot Banana	Pepper
Capsicum	*annuum-longum-arienum*	*haitiana*	Haitian 3.0"	Hot Banana	Pepper
Capsicum	*annuum-longum-arienum*	*holiday-timei*	Holiday Time 7.0"	Hot Banana	Pepper
Capsicum	*annuum-longum-arienum*	*hot-lemoni*	Hot Lemon 3.5"	Hot Banana	Pepper
Capsicum	*annuum-longum-arienum*	*inca-hotii*	Inca ?.?"	Hot Banana	Pepper
Capsicum	*annuum-longum-arienum*	*inca-glovi*	Inca Glow ?.?"	Hot Banana	Pepper
Capsicum	*annuum-longum-arienum*	*italicum-album-cerum*	Italian White Wax 2.5"	Hot Banana	Pepper
Capsicum	*annuum-longum-arienum*	*jigsawii*	Jigsaw ?.?"	Hot Banana	Pepper
Capsicum	*annuum-longum-arienum*	*jose-luis-majorcensis*	Jose Luis Majorca 4.0"	Hot Banana	Pepper
Capsicum	*annuum-longum-arienum*	*koreana-darca-groenum*	Dark Green Korean 3.5"	Hot Banana	Pepper
Capsicum	*annuum-longum-arienum*	*kori-sitakamei*	Kori Sitakame ?.?"	Hot Banana	Pepper
Capsicum	*annuum-longum-arienum*	*laotiana*	Laotian ?.?"	Hot Banana	Pepper
Capsicum	*annuum-longum-arienum*	*laungii*	Laungi ?.?"	Hot Banana	Pepper
Capsicum	*annuum-longum-arienum*	*marseillais-petitii*	Petite Marseilles 3.5"	Hot Banana	Pepper
Capsicum	*annuum-longum-arienum*	*mirchii-kahari*	Kahari Mirch 2.0"	Hot Banana	Pepper
Capsicum	*annuum-longum-arienum*	*mirchii-silmai*	Silma Mirch ?.?"	Hot Banana	Pepper
Capsicum	*annuum-longum-arienum*	*navajoensis*	Navaho 7.0"	Hot Banana	Pepper
Capsicum	*annuum-longum-arienum*	*nepalensis-aurantium*	Nepali Orange 1.5"	Hot Banana	Pepper
Capsicum	*annuum-longum-arienum*	*nigra-dallasensis*	Black Dallas 5.0"	Hot Banana	Pepper
Capsicum	*annuum-longum-arienum*	*orientalis-caldorum*	Oriental 3.0"	Hot Banana	Pepper
Capsicum	*annuum-longum-arienum*	*pasilla*	Pasilla 11."	Hot Banana	Pepper
Capsicum	*annuum-longum-arienum*	*pasilla-chilaca*	Chilaca Pasilla ?.?"	Hot Banana	Pepper
Capsicum	*annuum-longum-arienum*	*pasilla-mexicana-negroi*	Mexican Negro Pasilla ?.?"	Hot Banana	Pepper
Capsicum	*annuum-longum-arienum*	*pasilla-isleno-mulato*	Mulato Isleno Pasilla 6.0"	Hot Banana	Pepper
Capsicum	*annuum-longum-arienum*	*poinsettia*	Poinsettia 2.5"	Hot Banana	Pepper
Capsicum	*annuum-longum-arienum*	*portugalensis-calderum*	Portugese ?.?"	Hot Banana	Pepper
Capsicum	*annuum-longum-arienum*	*pricky-nui*	Pricky Nu ?.?"	Hot Banana	Pepper
Capsicum	*annuum-longum-arienum*	*pullai*	Pulla 6.0"	Hot Banana	Pepper
Capsicum	*annuum-longum-arienum*	*pungens-prideum*	Pungent Pride ?.?"	Hot Banana	Pepper
Capsicum	*annuum-longum-arienum*	*purpureum-deliteum*	Purple Delight ?.?"	Hot Banana	Pepper
Capsicum	*annuum-longum-arienum*	*purpureum-gloia-in-darca*	Purple Glow In The Dark ?.?"	Hot Banana	Pepper
Capsicum	*annuum-longum-arienum*	*quatro-milpasensis*	Quatro Milpas 7.5"	Hot Banana	Pepper
Capsicum	*annuum-longum-arienum*	*rio-grandensis-caldorum*	Rio Grande 6.0"	Hot Banana	Pepper
Capsicum	*annuum-longum-arienum*	*rouge-de-la-bressei*	Rouge De La Bresse 2.5"	Hot Banana	Pepper
Capsicum	*annuum-longum-arienum*	*san-felipensis*	San Felipe 5.0"	Hot Banana	Pepper
Capsicum	*annuum-longum-arienum*	*san-juan-tsilensis*	San Juan Tsile ?.?"	Hot Banana	Pepper
Capsicum	*annuum-longum-arienum*	*sapporoi*	Sapporo 4.5"	Hot Banana	Pepper
Capsicum	*annuum-longum-arienum*	*sebesensis-aurantium*	Orange Sebes ?.?"	Hot Banana	Pepper
Capsicum	*annuum-longum-arienum*	*sinahuisa*	Sinahuisa 1.7"	Hot Banana	Pepper
Capsicum	*annuum-longum-arienum*	*stripei*	Stripe 2.0"	Hot Banana	Pepper
Capsicum	*annuum-longum-arienum*	*sylvan'si-selectii*	Sylvan's Select ?.?"	Hot Banana	Pepper
Capsicum	*annuum-longum-arienum*	*velardensis*	Valarde 3.7"	Hot Banana	Pepper
Capsicum	*annuum-longum-arienum*	*waialuai*	Waialua ?.?"	Hot Banana	Pepper
Capsicum	*annuum-longum-arienum*	*white-waxii*	White Wax ?.?"	Hot Banana	Pepper
Capsicum	*annuum-longum-arienum*	*yunanensis*	Yunan 3.0"	Hot Banana	Pepper
Capsicum	*annuum-longum-arienum*	*yung-koi*	Yung Ko ?.?"	Hot Banana	Pepper
Capsicum	*annuum-longum-arienum*	*zia-puebloi*	Zia Pueblo 3.0"	Hot Banana	Pepper

Pepper, Hot Cayenne (*Capsicum annuum-longum-cayenneum*)

Genus	Species	Infraspecies	cultivar	specimen	Genome
Capsicum	*annuum-longum-cayenneum*	*cayennensis*	Original 5.5"	Hot Cayenne	Pepper
Capsicum	*annuum-longum-cayenneum*	*annulum-fuerum*	Ring Of Fire 5.5"	Hot Cayenne	Pepper
Capsicum	*annuum-longum-cayenneum*	*aurea*	Golden 6.0"	Hot Cayenne	Pepper
Capsicum	*annuum-longum-cayenneum*	*buist'si-helva*	Buist 4.0"	Hot Cayenne	Pepper
Capsicum	*annuum-longum-cayenneum*	*carolinensis*	Carolina ?.?"	Hot Cayenne	Pepper
Capsicum	*annuum-longum-cayenneum*	*cayvietii*	Cay Viet 1.5"	Hot Cayenne	Pepper
Capsicum	*annuum-longum-cayenneum*	*charlestonensis-caldorum*	Charleston 5.0"	Hot Cayenne	Pepper
Capsicum	*annuum-longum-cayenneum*	*da-nang-market-chileum*	Da Nang Market Chili ?.?"	Hot Cayenne	Pepper
Capsicum	*annuum-longum-cayenneum*	*georgiensis-alba*	Georgia White ?.?"	Hot Cayenne	Pepper
Capsicum	*annuum-longum-cayenneum*	*iberiana-longum*	Iberian Long 7.0"	Hot Cayenne	Pepper
Capsicum	*annuum-longum-cayenneum*	*islandicum-delectarum*	Island Delight ?.?"	Hot Cayenne	Pepper
Capsicum	*annuum-longum-cayenneum*	*kung-paoi*	Kung Pao 4.5"	Hot Cayenne	Pepper
Capsicum	*annuum-longum-cayenneum*	*largum-rubrum-crassum*	Large Red Thick 6.0"	Hot Cayenne	Pepper
Capsicum	*annuum-longum-cayenneum*	*longum-joe'si*	Joe's Long 9.0"	Hot Cayenne	Pepper
Capsicum	*annuum-longum-cayenneum*	*longum-rubrum*	Red Long 5.0"	Hot Cayenne	Pepper
Capsicum	*annuum-longum-cayenneum*	*longum-rubrum-crassum*	Thick Red Long 6.5"	Hot Cayenne	Pepper
Capsicum	*annuum-longum-cayenneum*	*longum-rubrum-gracilum*	Slim Red Long 5.5"	Hot Cayenne	Pepper
Capsicum	*annuum-longum-cayenneum*	*longum-rubrum-stenum*	Thin Red Long 5.0"	Hot Cayenne	Pepper
Capsicum	*annuum-longum-cayenneum*	*maule'si-rubrum-caldorum*	Maule's Red Hot 10."	Hot Cayenne	Pepper
Capsicum	*annuum-longum-cayenneum*	*negroi*	Negro ?.?"	Hot Cayenne	Pepper
Capsicum	*annuum-longum-cayenneum*	*piccante-di-cajenne*	French 4.0"	Hot Cayenne	Pepper
Capsicum	*annuum-longum-cayenneum*	*purpureum*	Purple 6.0"	Hot Cayenne	Pepper
Capsicum	*annuum-longum-cayenneum*	*sudum-calorum*	Southern Heat ?.?"	Hot Cayenne	Pepper
Capsicum	*annuum-longum-cayenneum*	*thomasi-jeffersonii*	Thomas Jefferson ?.?"	Hot Cayenne	Pepper

Genus	Species	Infraspecies	cultivar	specimen	Genome
Capsicum	*annuum-longum-cayenneum*	*turcensis*	Turkish 9.0"	Hot Cayenne	Pepper
Capsicum	*annuum-longum-cayenneum*	*ultra-gracilimum*	Ultra Slim ?.?"	Hot Cayenne	Pepper

Pepper, Hot Japalapeno (*Capsicum-annuum-longum-japapenum*)

Genus	Species	Infraspecies	cultivar	specimen	Genome
Capsicum	*annuum-longum-jalapenum*	*jalapenum*	Original 3.2"	Hot Jalapeno	Pepper
Capsicum	*annuum-longum-jalapenum*	*aureum*	Golden 2.0"	Hot Jalapeno	Pepper
Capsicum	*annuum-longum-jalapenum*	*czechoslovakiana-nigra*	Czechoslovakian Black 2.0"	Hot Jalapeno	Pepper
Capsicum	*annuum-longum-jalapenum*	*gigantium*	Giant 5.0"	Hot Jalapeno	Pepper
Capsicum	*annuum-longum-jalapenum*	*grandpa's-favoritei*	Grandpa's Favorite ?.?"	Hot Jalapeno	Pepper
Capsicum	*annuum-longum-jalapenum*	*guachinangoi*	Guachinango 2.0"	Hot Jalapeno	Pepper
Capsicum	*annuum-longum-jalapenum*	*jaloroi*	Yellow Jaloro ?.?"	Hot Jalapeno	Pepper
Capsicum	*annuum-longum-jalapenum*	*M*	Market (M) 3.5"	Hot Jalapeno	Pepper
Capsicum	*annuum-longum-jalapenum*	*olei-FM-297*	Ole 3.5"	Hot Jalapeno	Pepper
Capsicum	*annuum-longum-jalapenum*	*praecox*	Early 3.0"	Hot Jalapeno	Pepper
Capsicum	*annuum-longum-jalapenum*	*purpureum*	Purple ?.?"	Hot Jalapeno	Pepper
Capsicum	*annuum-longum-jalapenum*	*tama-mitis-1*	#1 Mild Tam ?.?"	Hot Jalapeno	Pepper
Capsicum	*annuum-longum-jalapenum*	*tama-mitis-2*	#2 Mild Tam ?.?"	Hot Jalapeno	Pepper
Capsicum	*annuum-longum-jalapenum*	*tama-caldorum-veracruzensis*	Vera Cruz Hot Tam ?.?"	Hot Jalapeno	Pepper
Capsicum	*annuum-longum-jalapenum*	*travelleri-strainii*	Traveler Strain 3.0"	Hot Jalapeno	Pepper
Capsicum	*annuum-longum-jalapenum*	*vinni'si-caldorum-pepperi*	Vinni's 2.0"	Hot Jalapeno	Pepper

Pepper, Hot Paprika (*Capsicum annuum-longum-papricum*)

Genus	Species	Infraspecies	cultivar	specimen	Genome
Capsicum	*annuum-longum-papricum*	*papricum*	Original 6.0"	Hot Paprika	Pepper
Capsicum	*annuum-longum-papricum*	*alma-paprikai*	Alma Paprika 2.0"	Hot Paprika	Pepper
Capsicum	*annuum-longum-papricum*	*csereszynei*	Csereszyne 1.0"	Hot Paprika	Pepper
Capsicum	*annuum-longum-papricum*	*macskapiros-felis-rubrum*	Macskapiros Cat Red 2.5"	Hot Paprika	Pepper
Capsicum	*annuum-longum-papricum*	*mesai*	Mesa 7.0"	Hot Paprika	Pepper

Pepper, Mild Finger (*Capsicum baccatum*)

Genus	Species	Infraspecies	cultivar	specimen	Genome
Capsicum	*baccatum*	*beni-highlandica*	Beni Highlands ?.?"	Mild Finger	Pepper
Capsicum	*baccatum*	*birgit's-locotoi*	Birgit's Locoto 2.5"	Mild Finger	Pepper
Capsicum	*baccatum*	*christmasii*	Christmas ?.?"	Mild Finger	Pepper
Capsicum	*baccatum*	*deco-de-moca*	Girl's 3.0"	Mild Finger	Pepper
Capsicum	*baccatum*	*grifii-9221*	Grif ?.?"	Mild Finger	Pepper
Capsicum	*baccatum*	*kovinchuii*	Kovinchu 5.0"	Mild Finger	Pepper
Capsicum	*baccatum*	*valentinea*	Valentine ?.?"	Mild Finger	Pepper
Capsicum	*baccatum-pendulum*	*aves*	Bird 0.2"	Aji Mild Finger	Pepper
Capsicum	*baccatum-pendulum*	*brunneum*	Brown 6.0"	Aji Mild Finger	Pepper
Capsicum	*baccatum-pendulum*	*coloradensis*	Colorado 4.0"	Aji Mild Finger	Pepper
Capsicum	*baccatum-pendulum*	*cristalis*	Aji Cristal 3.5"	Aji Mild Finger	Pepper
Capsicum	*baccatum-pendulum*	*dulcei*	Dulce 2.0"	Aji Mild Finger	Pepper
Capsicum	*baccatum-pendulum*	*pancai*	Panca 5.0"	Aji Mild Finger	Pepper

Pepper, Mild & Sweet Cherry (*Capsicum cerasiforme*)

Capsicum	*cerasiforme-mitis*	*cherrytimei*	Cherrytime 1.7"	Mild Cherry	Pepper
Capsicum	*cerasiforme-mitis*	*chocolateum-babyi*	Chocolate Baby ?.?"	Mild Cherry	Pepper
Capsicum	*cerasiforme-mitis*	*cuneoensis*	Cuneo 3.0"	Mild Cherry	Pepper
Capsicum	*cerasiforme-mitis*	*quintishoi*	Quintisho ?.?"	Mild Cherry	Pepper
Capsicum	*cerasiforme-mitis*	*rotunda*	Rotunda 3.5"	Mild Cherry	Pepper
Capsicum	*cerasiforme-mitis*	*spicy-puertoricensis-no-burnii*	Spicy Puerto Rico No-Burn ?.?"	Mild Cherry	Pepper
Capsicum	*cerasiforme-sucrum*	.	Original 1.5"	Sweet Cherry	Pepper
Capsicum	*cerasiforme-sucrum*	*applei*	Apple 3.5"	Sweet Cherry	Pepper
Capsicum	*cerasiforme-sucrum*	*cerasiforme-supra-sucrum*	Super Sweet 1.7"	Sweet Cherry	Pepper
Capsicum	*cerasiforme-sucrum*	*greygoi-babyi*	Greygo Baby 2.0"	Sweet Cherry	Pepper
Capsicum	*cerasiforme-sucrum*	*kari-babyi-cheesei*	Klari Baby Cheese 2.0"	Sweet Cherry	Pepper
Capsicum	*cerasiforme-sucrum*	*sucrum-picklei*	Sweet Pickle 2.0"	Sweet Cherry	Pepper
Capsicum	*cerasiforme-sucrum*	*topepo-rosso*	Topepo Rosso 2.0"	Sweet Cherry	Pepper
Capsicum	*cerasiforme-sucrum*	*topitoi-cheesei*	Topito Cheese ?.?"	Sweet Cherry	Pepper

Pepper, Mild Torpedo (*Capsicum chinense*)

Capsicum	*chinense*	*bodei*	Bode 0.5"	Mild Torpedo	Pepper
Capsicum	*chinense*	*cachuchai*	Cachucha 1.5"	Mild Torpedo	Pepper
Capsicum	*chinense*	*limon*	Lemon 4.5"	Mild Torpedo	Pepper
Capsicum	*chinense*	*limon-chilensis*	Chile Lemon 1.5"	Mild Torpedo	Pepper
Capsicum	*chinense*	*mushroomii-helvum*	Yellow Mushroom 2.2"	Mild Torpedo	Pepper
Capsicum	*chinense*	*mushroomii-rubrum*	Red Mushroom 2.0"	Mild Torpedo	Pepper
Capsicum	*chinense*	*trinidadensis*	Trinidad 1.5"	Mild Torpedo	Pepper
Capsicum	*eximum*	*ulupica*	Ulupica	Hot Eximious	Pepper
Capsicum	*exile*	*cobincho*	Cobincho	Hot Bullet	Pepper

Pepper, Mild & Sweet Cone (*Capsicum fasciculatum*)

Capsicum	*fasciculatum-mitis*	*andeana*	Andean 9.5"	Mild Cone	Pepper
Capsicum	*fasciculatum-mitis*	*antohi-romaniana*	Antohi Romanian 4.0"	Mild Cone	Pepper
Capsicum	*fasciculatum-mitis*	*bulgarica*	Bulgarian 4.0"	Mild Cone	Pepper
Capsicum	*fasciculatum-mitis*	*corno-di-toro-gialloi*	Giallo Bullhorn 8.0"	Mild Cone	Pepper
Capsicum	*fasciculatum-mitis*	*corno-di-toro-rossoi*	Rosso Bullhorn 9.0"	Mild Cone	Pepper
Capsicum	*fasciculatum-mitis*	*corno-di-toro-rubrum-&-helvum*	Red & Yellow Bullhorn 9.0"	Mild Cone	Pepper
Capsicum	*fasciculatum-mitis*	*italicum-groenum-pyrum*	Italian Green Frying 6.0"	Mild Cone	Pepper
Capsicum	*fasciculatum-mitis*	*kapyai*	Kapya 7.0"	Mild Cone	Pepper
Capsicum	*fasciculatum-mitis*	*little-dickensi*	Little Dickens ?.?"	Mild Cone	Pepper
Capsicum	*fasciculatum-mitis*	*longum-rubrum-mitis*	Long Red ?.?"	Mild Cone	Pepper
Capsicum	*fasciculatum-mitis*	*rubrum-belgicum*	Red Belgian 5.0"	Mild Cone	Pepper
Capsicum	*fasciculatum-sucrum*	*alba-regium*	White King 5.0"	Sweet Cone	Pepper
Capsicum	*fasciculatum-sucrum*	*bujana*	Bujan ?.?"	Sweet Cone	Pepper
Capsicum	*fasciculatum-sucrum*	*bulgarica-5*	#5 Bulgarian ?.?"	Sweet Cone	Pepper
Capsicum	*fasciculatum-sucrum*	*chervena-chujskii*	Chervena Chujski 6.0"	Sweet Cone	Pepper
Capsicum	*fasciculatum-sucrum*	*frigitelloi*	Frigitello 3.5"	Sweet Cone	Pepper
Capsicum	*fasciculatum-sucrum*	*gaylei*	Gayle Chocolate 7.0"	Sweet Cone	Pepper
Capsicum	*fasciculatum-sucrum*	*giftii-moldovensis*	Gift From Moldova 4.0"	Sweet Cone	Pepper
Capsicum	*fasciculatum-sucrum*	*golden-treasurei*	Golden Treasure 8.5"	Sweet Cone	Pepper
Capsicum	*fasciculatum-sucrum*	*healthyi*	Healthy 4.0"	Sweet Cone	Pepper
Capsicum	*fasciculatum-sucrum*	*italia*	Italia 8.0"	Sweet Cone	Pepper
Capsicum	*fasciculatum-sucrum*	*italiana-pyrum*	Italian Frying ?.?"	Sweet Cone	Pepper
Capsicum	*fasciculatum-sucrum*	*italiana-sucrum*	Italian 6.0"	Sweet Cone	Pepper
Capsicum	*fasciculatum-sucrum*	*italiana-sucrum-lemme'si*	Lemme's Italian 7.0"	Sweet Cone	Pepper
Capsicum	*fasciculatum-sucrum*	*lipstickii*	Lipstick 5.0"	Sweet Cone	Pepper
Capsicum	*fasciculatum-sucrum*	*odessa-marketii*	Odessa Market 5.0"	Sweet Cone	Pepper
Capsicum	*fasciculatum-sucrum*	*roumaniana-rainbowii*	Rainbow Roumanian 4.5"	Sweet Cone	Pepper
Capsicum	*fasciculatum-sucrum*	*roumaniana-sucrum*	Roumanian 4.2"	Sweet Cone	Pepper
Capsicum	*fasciculatum-sucrum*	*supra-shepherdii*	Super Shepherd 7.5"	Sweet Cone	Pepper
Capsicum	*fasciculatum-sucrum*	*sucrum-melrosei*	Sweet Melrose 4.0"	Sweet Cone	Pepper
Capsicum	*fasciculatum-sucrum*	*sucrum-melrosei-improvum*	Improved Sweet Melrose 3.5"	Sweet Cone	Pepper
Capsicum	*fasciculatum-sucrum*	*swallowii*	Swallow 4.0"	Sweet Cone	Pepper
Capsicum	*fasciculatum-sucrum*	*taorminai*	Taormina 8.0"	Sweet Cone	Pepper
Capsicum	*fasciculatum-sucrum*	*tequila-sunrisei*	Tequila Sunrise 4.5"	Sweet Cone	Pepper
Capsicum	*fasciculatum-sucrum*	*tollie'si-sucrum-italicum*	Tollie's Sweet Italian 4.5"	Sweet Cone	Pepper
Capsicum	*fasciculatum-sucrum*	*helvum-cerum-jugoi*	Yellow Wax Jugo ?.?"	Sweet Cone	Pepper

Pepper, Mild & Sweet Pencil (*Capsicum frutescens*)

Genus	Species	Infraspecies	cultivar	specimen	Genome
Capsicum	*frutescens-mitis*	*bangalorensis*	Bangalore Torpedo 5.0"	Mild Pencil	Pepper
Capsicum	*frutescens-mitis*	*calcuttensis-longum*	Calcutta Long ?.?"	Mild Pencil	Pepper
Capsicum	*frutescens-mitis*	*chandigarhii*	Chandigarh 3.0"	Mild Pencil	Pepper
Capsicum	*frutescens-mitis*	*grovei*	Grove ?.?"	Mild Pencil	Pepper
Capsicum	*frutescens-mitis*	*japonesi*	Japones 1.5"	Mild Pencil	Pepper
Capsicum	*frutescens-mitis*	*madrasi*	Madras 3.0"	Mild Pencil	Pepper
Capsicum	*frutescens-mitis*	*merahii*	Merah ?.?"	Mild Pencil	Pepper
Capsicum	*frutescens-mitis*	*raja's-pridei*	Raja's Pride 3.5"	Mild Pencil	Pepper
Capsicum	*frutescens-mitis*	*suryamukhi-clusteri*	Suryamukhi Cluster 3.5"	Mild Pencil	Pepper
Capsicum	*frutescens-mitis*	*yesil-tatlii*	Yesil Tatli 6.0"	Mild Pencil	Pepper
Capsicum	*frutescens-sucrum*	*demrei*	Demre 6.5"	Sweet Pencil	Pepper
Capsicum	*frutescens-sucrum*	*centinelii*	Centinel 5.0"	Sweet Pencil	Pepper
Capsicum	*frutescens-sucrum*	*corbacii*	Corbaci 9.0"	Sweet Pencil	Pepper
Capsicum	*frutescens-sucrum*	*jimmy-nardello'si*	Jimmy Nardello's 9.0"	Sweet Pencil	Pepper
Capsicum	*frutescens-sucrum*	*shisitoui*	Shishitou 3.0"	Sweet Pencil	Pepper

Pepper, Mild Bird (*Capsicum glabrisculum*)

Genus	Species	Infraspecies	cultivar	specimen	Genome
Capsicum	*glabriusculum*	*centennialis*	Centennial ?.?"	Mild Bird	Pepper
Capsicum	*glabriusculum*	*venezuelana-purpureum*	Venezuelan Purple .25"	Mild Bird	Pepper

Pepper, Mild & Sweet Bell (*Capsicum grossum*)

Genus	Species	Infraspecies	cultivar	specimen	Genome
Capsicum	*grossum-mitis*	*amethystinum*	Amethyst ?.?"	Mild Bell	Pepper
Capsicum	*grossum-mitis*	*belleum-babyi*	Baby Belle 2.5"	Mild Bell	Pepper
Capsicum	*grossum-mitis*	*belleum-georgiensis*	Georgia Belle ?.?"	Mild Bell	Pepper
Capsicum	*grossum-mitis*	*belleum-jung'si-helvum*	Jung's Yellow Belle ?.?"	Mild Bell	Pepper
Capsicum	*grossum-mitis*	*belleum-miniatum-chocolatum*	Miniature Chocolate Belle 2.0"	Mild Bell	Pepper
Capsicum	*grossum-mitis*	*belleum-prima*	Prima Belle ?.?"	Mild Bell	Pepper
Capsicum	*grossum-mitis*	*big-dipperi*	Big Dipper 4.5"	Mild Bell	Pepper
Capsicum	*grossum-mitis*	*californica-wonderi*	Calwonder 4.2"	Mild Bell	Pepper
Capsicum	*grossum-mitis*	*californica-wonderi-300*	#300 Calwonder 4.7"	Mild Bell	Pepper
Capsicum	*grossum-mitis*	*californica-wonderi-purpureum*	Purple Calwonder ?.?"	Mild Bell	Pepper
Capsicum	*grossum-mitis*	*californica-wonderi-earlimontana*	Mountain Early Calwonder ?.?"	Mild Bell	Pepper

Genus	Species	Infraspecies	cultivar	specimen	Genome
Capsicum	*grossum-mitis*	*californica-wonderi-helvum-belleum*	Yellow Belle Calwonder 3.5"		Mild Bell Pepper
Capsicum	*grossum-mitis*	*californica-wonderi-keystonei-resistans*	Resistant Keystone Calwonder ?.?"		Mild Bell Pepper
Capsicum	*grossum-mitis*	*californica-wonderi-keystonei-resistans-giganteum*	Giant Resistant Keystone Calwonder 4.5"		Mild Bell Pepper
Capsicum	*grossum-mitis*	*californica-wonderi-keystonei-resistans-giganteum-3*	#3 Giant Resistant Keystone Calwonder 4.5"		Mild Bell Pepper
Capsicum	*grossum-mitis*	*californica-wonderi-keystonei-resistans-giganteum-4*	#4 Giant Resistant Keystone Calwonder 4.0"		Mild Bell Pepper
Capsicum	*grossum-mitis*	*californica-wonderi-praecox-nordum*	Northern Early Calwonder 4.2"		Mild Bell Pepper
Capsicum	*grossum-mitis*	*californica-wonderi-praecox-sudum*	Southern Early Calwonder 4.2"		Mild Bell Pepper
Capsicum	*grossum-mitis*	*californica-wonderi-selectum*	Select Calwonder 4.0"		Mild Bell Pepper
Capsicum	*grossum-mitis*	*californica-wonderi-tarogii*	Tarog Calwonder ?.?"		Mild Bell Pepper
Capsicum	*grossum-mitis*	*californica-wonderi-TMR*	Tobacco Mosaic Resistant Calwonder 4.5"		Mild Bell Pepper
Capsicum	*grossum-mitis*	*capistranoi*	Capistrano 4.0"		Mild Bell Pepper
Capsicum	*grossum-mitis*	*carmagnolia-rossi*	Cargmagnolia Ross ?.?"		Mild Bell Pepper
Capsicum	*grossum-mitis*	*criolla*	Criolla ?.?"		Mild Bell Pepper
Capsicum	*grossum-mitis*	*diamondii*	Diamond ?.?"		Mild Bell Pepper
Capsicum	*grossum-mitis*	*earlyi-rubrum*	Earlired ?.?"		Mild Bell Pepper
Capsicum	*grossum-mitis*	*earlyi-niagra-giganteum*	Early Niagra Giant 4.5"		Mild Bell Pepper
Capsicum	*grossum-mitis*	*golden-bellsi*	Golden Bells ?.?"		Mild Bell Pepper
Capsicum	*grossum-mitis*	*golden-queenii*	Golden Queen ?.?"		Mild Bell Pepper
Capsicum	*grossum-mitis*	*granny-smithii*	Granny Smith 3.5"		Mild Bell Pepper
Capsicum	*grossum-mitis*	*gustoi*	Gusto 5.0"		Mild Bell Pepper
Capsicum	*grossum-mitis*	*herculesi-helvum-aureum*	Hercules Yellow Golden ?.?"		Mild Bell Pepper
Capsicum	*grossum-mitis*	*kaalai*	Kaala 3.0"		Mild Bell Pepper
Capsicum	*grossum-mitis*	*kandilii*	Kandil 6.0"		Mild Bell Pepper
Capsicum	*grossum-mitis*	*mercedii-TMR*	Tobacco Mosaic Resistant Merced 3.5"		Mild Bell Pepper
Capsicum	*grossum-mitis*	*mercuryi*	Mercury 4.5"		Mild Bell Pepper
Capsicum	*grossum-mitis*	*midnight-dreamii*	Midnight Dream ?.?"		Mild Bell Pepper
Capsicum	*grossum-mitis*	*olena-rubrum*	Olena Red ?.?"		Mild Bell Pepper
Capsicum	*grossum-mitis*	*aurantium-solum*	Orange Sun 4.5"		Mild Bell Pepper
Capsicum	*grossum-mitis*	*ozarkana-giganteum*	Ozark Giant ?.?"		Mild Bell Pepper
Capsicum	*grossum-mitis*	*permagroenum*	Permagreen ?.?"		Mild Bell Pepper
Capsicum	*grossum-mitis*	*pimlicoensis*	Pimlico ?.?"		Mild Bell Pepper
Capsicum	*grossum-mitis*	*pipii*	Pip 4.5"		Mild Bell Pepper
Capsicum	*grossum-mitis*	*rubrum-giganteum*	Giant Red ?.?"		Mild Bell Pepper
Capsicum	*grossum-mitis*	*resistens-gigeanteum-4*	#4 Resistant Giant 4.5"		Mild Bell Pepper
Capsicum	*grossum-mitis*	*staddon'si-selectii*	Staddon's Select 4.0"		Mild Bell Pepper
Capsicum	*grossum-mitis*	*sunrisei*	Sunrise 4.5"		Mild Bell Pepper
Capsicum	*grossum-mitis*	*taurus*	Taurus 6.0"		Mild Bell Pepper
Capsicum	*grossum-mitis*	*ultra-stuffii*	Ultra Stuff 6.0"		Mild Bell Pepper
Capsicum	*grossum-sucrum*	*arianei*	Ariane ?.?"		Sweet Bell Pepper
Capsicum	*grossum-sucrum*	*aureum-summitii*	Golden Summit ?.?"		Sweet Bell Pepper
Capsicum	*grossum-sucrum*	*belleum-aurantium*	Orange Belle 4.0"		Sweet Bell Pepper
Capsicum	*grossum-sucrum*	*belleum-butryum*	Butter Belle 2.7"		Sweet Bell Pepper
Capsicum	*grossum-sucrum*	*belleum-charlestonensis*	Charleston Belle 3.5"		Sweet Bell Pepper
Capsicum	*grossum-sucrum*	*belleum-doehillii-aureum*	Doe Hill Golden Belle 2.2"		Sweet Bell Pepper
Capsicum	*grossum-sucrum*	*belleum-manda-helvum*	Manda Yellow Belle ?.?"		Sweet Bell Pepper
Capsicum	*grossum-sucrum*	*belleum-miniatum-rubrum*	Miniature Red Belle ?.?"		Sweet Bell Pepper
Capsicum	*grossum-sucrum*	*belleum-miniatum-helvum*	Miniature Yellow Belle 2.0"		Sweet Bell Pepper
Capsicum	*grossum-sucrum*	*belleum-yankeei*	Yankee Belle ?.?"		Sweet Bell Pepper
Capsicum	*grossum-sucrum*	*bullnosei*	Bullnose 4.0"		Sweet Bell Pepper
Capsicum	*grossum-sucrum*	*bullnosei-albinoi*	Albino Bullnose ?.?"		Sweet Bell Pepper
Capsicum	*grossum-sucrum*	*bullnosei-emerald-giganteum*	Emerald Giant Bullnose 4.5"		Sweet Bell Pepper
Capsicum	*grossum-sucrum*	*bullnosei-emerald-gigantea-resistantii*	Resistant Emerald Giant Bullnose 4.5"		Sweet Bell Pepper
Capsicum	*grossum-sucrum*	*bullnosei-regis-nordum*	King Of The North Bullnose 5.5"		Sweet Bell Pepper
Capsicum	*grossum-sucrum*	*burana*	Buran 4.0"		Sweet Bell Pepper
Capsicum	*grossum-sucrum*	*californica-wonderi-aureum*	Golden Calwonder ?.?"		Sweet Bell Pepper
Capsicum	*grossum-sucrum*	*californica-wonderi-roma-helvum-belleum*	Roma Yellow Belle Calwonder ?.?"		Sweet Bell Pepper
Capsicum	*grossum-sucrum*	*californica-wonderi-yoloi*	Yolo Calwonder 4.2"		Sweet Bell Pepper
Capsicum	*grossum-sucrum*	*californica-wonderi-yoloi-A*	Type A Yolo Calwonder 4.0"		Sweet Bell Pepper
Capsicum	*grossum-sucrum*	*californica-wonderi-yoloi-B*	Type B Yolo Calwonder 4.0"		Sweet Bell Pepper
Capsicum	*grossum-sucrum*	*californica-wonderi-yoloi-B-carolinensis*	Carolina Type B Yolo Calwonder ?.?"		Sweet Bell Pepper
Capsicum	*grossum-sucrum*	*californica-wonderi-yoloi-B-improvum*	Improved Type B Yolo Calwonder 4.0"		Sweet Bell Pepper
Capsicum	*grossum-sucrum*	*californica-wonderi-yoloi-B-improvum-TMR*	Tobacco Mosiac Resistant Improved Type B Yolo Calwonder 4.0"		Sweet Bell Pepper
Capsicum	*grossum-sucrum*	*californica-wonderi-yoloi-L*	Type L Yolo Calwonder 4.2"		Sweet Bell Pepper

Genus	Species	Infraspecies	cultivar	specimen	Genome
Capsicum	*grossum-sucrum*	*chinensis-giganteum*	Chinese Giant 6.0"	Sweet Bell	Pepper
Capsicum	*grossum-sucrum*	*chocolateum*	Professor Meader's Chocolate 4.0"	Sweet Bell	Pepper
Capsicum	*grossum-sucrum*	*chocolateum-pulchellum*	Chocolate Beauty 4.0"	Sweet Bell	Pepper
Capsicum	*grossum-sucrum*	*corona*	Corona ?.?"	Sweet Bell	Pepper
Capsicum	*grossum-sucrum*	*earliesti-rubrum-sucrum*	Earliest Red 4.0"	Sweet Bell	Pepper
Capsicum	*grossum-sucrum*	*gardenia-solum*	Garden Sunshine 8.0"	Sweet Bell	Pepper
Capsicum	*grossum-sucrum*	*giganteum-sucrum-groenum*	Giant Sweet Green ?.?"	Sweet Bell	Pepper
Capsicum	*grossum-sucrum*	*helvum-babyi*	Baby Yellow 1.2"	Sweet Bell	Pepper
Capsicum	*grossum-sucrum*	*goldiei*	Goldie 2.2"	Sweet Bell	Pepper
Capsicum	*grossum-sucrum*	*jupiteri*	Jupiter 4.5"	Sweet Bell	Pepper
Capsicum	*grossum-sucrum*	*jupiteri-elitea*	Elite Jupiter ?.?"	Sweet Bell	Pepper
Capsicum	*grossum-sucrum*	*jupiteri-sterlingii*	Sterling Jupiter ?.?"	Sweet Bell	Pepper
Capsicum	*grossum-sucrum*	*lamayoi-album*	Lamayo White 5.0"	Sweet Bell	Pepper
Capsicum	*grossum-sucrum*	*le-rougei-royalis*	Le Rouge Royal 8.0"	Sweet Bell	Pepper
Capsicum	*grossum-sucrum*	*midwayi*	Midway 4.5"	Sweet Bell	Pepper
Capsicum	*grossum-sucrum*	*montana-wonderi*	Montana Wonder ?.?"	Sweet Bell	Pepper
Capsicum	*grossum-sucrum*	*neapolitana*	Neapolitan ?.?"	Sweet Bell	Pepper
Capsicum	*grossum-sucrum*	*padronei*	Padrone 3.0"	Sweet Bell	Pepper
Capsicum	*grossum-sucrum*	*purpureum-pulchellum*	Purple Beauty ?.?"	Sweet Bell	Pepper
Capsicum	*grossum-sucrum*	*quadrati-d'asti-giallo-rossaeum*	Red Quadrati D'Asti Giallo ?.?"	Sweet Bell	Pepper
Capsicum	*grossum-sucrum*	*quadrati-d'asti-giallo-helvum*	Yellow Quadrati D'Asti Giallo ?.?"	Sweet Bell	Pepper
Capsicum	*grossum-sucrum*	*rubium-giganteum*	Giant Ruby 5.0"	Sweet Bell	Pepper
Capsicum	*grossum-sucrum*	*rubium-giganteum-world-beateri*	World Beater Giant Ruby 5.0"	Sweet Bell	Pepper
Capsicum	*grossum-sucrum*	*rubium-regium*	King Ruby 5.0"	Sweet Bell	Pepper
Capsicum	*grossum-sucrum*	*rubrum-babyi*	Baby Red 1.2"	Sweet Bell	Pepper
Capsicum	*grossum-sucrum*	*sucrum-chocolateum*	Sweet Chocolate ?.?"	Sweet Bell	Pepper
Capsicum	*grossum-sucrum*	*sucrum-rubrum*	Sweet Red ?.?"	Sweet Bell	Pepper
Capsicum	*grossum-sucrum*	*sunbrightii*	Sunbright 7.0"	Sweet Bell	Pepper
Capsicum	*grossum-sucrum*	*ta-tongii*	Ta Tong ?.?"	Sweet Bell	Pepper
Capsicum	*grossum-sucrum*	*turcensis-11-B-14*	Turkish 5.0"	Sweet Bell	Pepper
Capsicum	*grossum-sucrum*	*wisconsinensis-lacum*	Wisconsin Lakes ?.?"	Sweet Bell	Pepper
Capsicum	*grossum-supra-sucrum*	*big-redii*	Big Red ?.?"	Very Sweet Bell	Pepper
Capsicum	*grossum-supra-sucrum*	*californica-wonderi-aureum-67*	67 Day Golden Calwonder 5.0"	Very Sweet Bell	Pepper
Capsicum	*grossum-supra-sucrum*	*californica-wonderi-aureum-75*	75 Day Golden Calwonder 5.0"	Very Sweet Bell	Pepper
Capsicum	*grossum-supra-sucrum*	*frankii*	Frank's ?.?"	Very Sweet Bell	Pepper
Capsicum	*grossum-supra-sucrum*	*gourmetii*	Gourmet ?.?"	Very Sweet Bell	Pepper
Capsicum	*grossum-supra-sucrum*	*hungarica-helvum-stuffingii*	Hungarian Yellow Stuffing ?.?"	Very Sweet Bell	Pepper
Capsicum	*grossum-supra-sucrum*	*morgoldii-sucrum*	Morgold 3.5"	Very Sweet Bell	Pepper
Capsicum	*grossum-supra-sucrum*	*napoleana-sucrum*	Napolean 8.0"	Very Sweet Bell	Pepper
Capsicum	*grossum-supra-sucrum*	*super-stuffii*	Super Stuff 6.0"	Very Sweet Bell	Pepper

Pepper, Mild & Sweet Banana (*Capsicum longum*)

Genus	Species	Infraspecies	cultivar	specimen	Genome
Capsicum	*longum-mitis*	*album-cheesei-giganteum*	Giant White Cheese 2.2"	Mild Banana	Pepper
Capsicum	*longum-mitis*	*blanco-del-paisi*	Blanco Del Pais 4.5"	Mild Banana	Pepper
Capsicum	*longum-mitis*	*calusa-indian-moundii*	Calusa Indian Mound ?.?"	Mild Banana	Pepper
Capsicum	*longum-mitis*	*carouselii*	Carousel 4.0"	Mild Banana	Pepper
Capsicum	*longum-mitis*	*cubana*	Cuban 6.0"	Mild Banana	Pepper
Capsicum	*longum-mitis*	*dong-xuan-marketii*	Dong Xuan Market ?.?"	Mild Banana	Pepper
Capsicum	*longum-mitis*	*fushimi-longa-groenum*	Fushimi Long Green 6.0"	Mild Banana	Pepper
Capsicum	*longum-mitis*	*iberiana*	Iberian Long 7.0"	Mild Banana	Pepper
Capsicum	*longum-mitis*	*italicum-giganteum*	Italian Giant 8.0"	Mild Banana	Pepper
Capsicum	*longum-mitis*	*kasmirensis-giganteum-longum*	Kasmir Giant Long ?.?"	Mild Banana	Pepper
Capsicum	*longum-mitis*	*kisseri*	Kisser ?.?"	Mild Banana	Pepper
Capsicum	*longum-mitis*	*lombardoi*	Lombardo ?.?"	Mild Banana	Pepper
Capsicum	*longum-mitis*	*shishitoi*	Shishito 7.0"	Mild Banana	Pepper
Capsicum	*longum-mitis*	*sigaretta-per-sotto-acetii*	Sigaretta Per Soto Aceti ?.?"	Mild Banana	Pepper
Capsicum	*longum-mitis*	*topazii-helvum*	Topaz Yellow ?.?"	Mild Banana	Pepper
Capsicum	*longum-mitis*	*tuste-blancoi*	Tuste Blanco ?.?"	Mild Banana	Pepper
Capsicum	*longum-mitis*	*weaver'si-mennonitei-stuffingii*	Weaver's Mennonite Stuffing ?.?"	Mild Banana	Pepper
Capsicum	*longum-sucrum*	*sucrum*	Original 6.0"	Sweet Banana	Pepper
Capsicum	*longum-sucrum*	*aconcagua*	Aconcagua 9.0"	Sweet Banana	Pepper
Capsicum	*longum-sucrum*	*bulgarica-sucrum*	Bulgarian Sweet 6.5"	Sweet Banana	Pepper
Capsicum	*longum-sucrum*	*carlistonii-sucrum*	Carliston ?.?"	Sweet Banana	Pepper
Capsicum	*longum-sucrum*	*chilei-negroi*	Chile Negro 6.0"	Sweet Banana	Pepper
Capsicum	*longum-sucrum*	*crimsonii-sucrum*	Crimson 6.5"	Sweet Banana	Pepper
Capsicum	*longum-sucrum*	*cubanelleum*	Cubanelle 5.7"	Sweet Banana	Pepper
Capsicum	*longum-sucrum*	*daintyi-sucrum*	Dainty 2.0"	Sweet Banana	Pepper
Capsicum	*longum-sucrum*	*datilii-sucrum*	Datil 1.5"	Sweet Banana	Pepper
Capsicum	*longum-sucrum*	*douce-de-espagnensis*	Spanish ?.?"	Sweet Banana	Pepper
Capsicum	*longum-sucrum*	*dulce-medeterraneoi-rubrum*	Dulce Medeterraneo Red 10."	Sweet Banana	Pepper
Capsicum	*longum-sucrum*	*elephant-trunkii*	Elephant Trunk 9.0"	Sweet Banana	Pepper
Capsicum	*longum-sucrum*	*friarielloi*	Friariello ?.?"	Sweet Banana	Pepper
Capsicum	*longum-sucrum*	*georgescui*	Georgescu ?.?"	Sweet Banana	Pepper

Genus	Species	Infraspecies	cultivar	specimen	Genome
Capsicum	*longum-sucrum*	*georgescui-chocolatei*	Chocolate Georgescu ?.?"	Sweet Banana	Pepper
Capsicum	*longum-sucrum*	*giganteum-helvum*	Giant Yellow 7.0"	Sweet Banana	Pepper
Capsicum	*longum-sucrum*	*goccia-d'oroi*	Goccia D'Oro ?.?"	Sweet Banana	Pepper
Capsicum	*longum-sucrum*	*marconi-aureum*	Golden Marconi 8.5"	Sweet Banana	Pepper
Capsicum	*longum-sucrum*	*marconi-purpureum*	Purple Marconi 5.0"	Sweet Banana	Pepper
Capsicum	*longum-sucrum*	*marconi-rubrum*	Red Marconi 11."	Sweet Banana	Pepper
Capsicum	*longum-sucrum*	*molei-sucrum*	Mole 4.5"	Sweet Banana	Pepper
Capsicum	*longum-sucrum*	*praecox*	Early 6.0"	Sweet Banana	Pepper
Capsicum	*longum-sucrum*	*sucrum-cayennensis*	Sweet Cayenne 11."	Sweet Banana	Pepper
Capsicum	*longum-sucrum*	*sucrum-hispanica*	Sweet Spanish ?.?"	Sweet Banana	Pepper
Capsicum	*longum-sucrum*	*sucrum-rugosum-elda-manii*	Sweet Wrinkled Old Man 2.5"	Sweet Banana	Pepper

Pepper, Mild & Sweet Paprika (*Capsicum longum-papricum*)

Genus	Species	Infraspecies	cultivar	specimen	Genome
Capsicum	*longum-papricum-mitis*	*andrea*	Andrea 3.5"	Mild Paprika	Pepper
Capsicum	*longum-papricum-mitis*	*kishinevii*	Kishinev ?.?"	Mild Paprika	Pepper
Capsicum	*longum-papricum-mitis*	*hispanica*	Spain 6.0"	Mild Paprika	Pepper
Capsicum	*longum-papricum-mitis*	*papri-mitis*	Papri 6.0"	Mild Paprika	Pepper
Capsicum	*longum-papricum-mitis*	*papri-mitis-2*	#2 Papri 6.0"	Mild Paprika	Pepper
Capsicum	*longum-papricum-mitis*	*frutkai*	Frutka 3.0"	Mild Paprika	Pepper
Capsicum	*longum-papricum-mitis*	*papricum-PCR*	Prolific Cone Red 4.0"	Mild Paprika	Pepper
Capsicum	*longum-papricum-mitis*	*szegedensis*	Szeged 9.0"	Mild Paprika	Pepper
Capsicum	*longum-papricum-mitis*	*tennessensis-cheesei*	Tennessee Cheese 3.0"	Mild Paprika	Pepper
Capsicum	*longum-papricum-sucrum*	*boldogii-hungarica-spicei*	Boldog Hungarian Spice 5.5"	Sweet Paprika	Pepper
Capsicum	*longum-papricum-sucrum*	*conquistadorii*	Conquistador 4.2"	Sweet Paprika	Pepper
Capsicum	*longum-papricum-sucrum*	*feher-ozonii*	Feher Ozon 4.5"	Sweet Paprika	Pepper
Capsicum	*longum-papricum-sucrum*	*hungarica*	Hungarian 5.0"	Sweet Paprika	Pepper
Capsicum	*longum-papricum-sucrum*	*kalosca-sucrum-spiceum*	Kalosca Sweet Spice 4.0"	Sweet Paprika	Pepper

Pepper, Mild & Sweet Pimento (*Capsicum longum-pimentum*)

Genus	Species	Infraspecies	cultivar	specimen	Genome
Capsicum	*longum-pimentum-mitis*	*californica-roseum*	Calrose ?.?"	Mild Pimento	Pepper
Capsicum	*longum-pimentum-mitis*	*californica-wonderi-morrowii*	Morrow Calwonder ?.?"	Mild Pimento	Pepper
Capsicum	*longum-pimentum-mitis*	*helvum-cheesei*	Yellow Cheese 3.7"	Mild Pimento	Pepper
Capsicum	*longum-pimentum-mitis*	*largum*	Large (L) 4.5"	Mild Pimento	Pepper
Capsicum	*longum-pimentum-mitis*	*piement-des-landes*	Piement Des Landes 9.0"	Mild Pimento	Pepper
Capsicum	*longum-pimentum-mitis*	*rubrum-cheesei*	Red Cheese 4.0"	Mild Pimento	Pepper
Capsicum	*longum-pimentum-mitis*	*rubrum-ruffledii*	Red Ruffled 3.0"	Mild Pimento	Pepper
Capsicum	*longum-pimentum-mitis*	*rotundum-tomatoi*	Round Tomato ?.?"	Mild Pimento	Pepper
Capsicum	*longum-pimentum-mitis*	*selectii*	Select 3.5"	Mild Pimento	Pepper
Capsicum	*longum-pimentum-mitis*	*tangerinei*	Tangerine 3.0"	Mild Pimento	Pepper
Capsicum	*longum-pimentum-sucrum*	*amishii*	Amish 4.0"	Sweet Pimento	Pepper
Capsicum	*longum-pimentum-sucrum*	*canadensis-cheesei*	Canada Cheese 2.0"	Sweet Pimento	Pepper
Capsicum	*longum-pimentum-sucrum*	*chocoi*	Choco 3.5"	Sweet Pimento	Pepper
Capsicum	*longum-pimentum-sucrum*	*figaroi*	Figaro 3.5"	Sweet Pimento	Pepper
Capsicum	*longum-pimentum-sucrum*	*gamboi*	Gambo 4.0"	Sweet Pimento	Pepper
Capsicum	*longum-pimentum-sucrum*	*grandeum*	Grande ?.?"	Sweet Pimento	Pepper
Capsicum	*longum-pimentum-sucrum*	*hungarica-rotunda*	Round Of Hungary ?.?"	Sweet Pimento	Pepper
Capsicum	*longum-pimentum-sucrum*	*perfectionii*	Perfection 3.0"	Sweet Pimento	Pepper
Capsicum	*longum-pimentum-sucrum*	*pimientoi-ashei-countyi*	Ash County Pimiento ?.?"	Sweet Pimento	Pepper
Capsicum	*longum-pimentum-sucrum*	*sheepnosei*	Sheepnose 4.0"	Sweet Pimento	Pepper
Capsicum	*longum-pimentum-sucrum*	*sunneybrookii*	Sunneybrook 3.0"	Sweet Pimento	Pepper
Capsicum	*longum-pimentum-sucrum*	*supra-rubrum*	Super Red 5.7"	Sweet Pimento	Pepper

Pepper, Hot Roccoto (*Capsicum pubescens*)

Genus	Species	Infraspecies	cultivar	specimen	Genome
Capsicum	*pubescens*	*pubescens*	Original 2.0"	Hot Rocoto	Pepper
Capsicum	*pubescens*	*goat's-weedii*	Goat's Weed 1.5"	Hot Rocoto	Pepper
Capsicum	*pubescens*	*manzano-amarilloi*	Amarillo Manzano 5.0"	Hot Rocoto	Pepper
Capsicum	*pubescens*	*manzano-rojoi*	Red Apple Rojo Manzano ?.?"	Hot Rocoto	Pepper
Capsicum	*pubescens*	*rubrum*	Red 1.0"	Hot Rocoto	Pepper
Cyphomandra	*betacea*	*betacea*	Original	Tamarillo	Tree Tomato
Cyphomandra	*betacea*	*crassifolia*	Crass Leaf	Tamarillo	Tree Tomato
Cyphomandra	*betacea*	*dwarf-championii*	Dwarf Champion Magenta 5oz	Tamarillo	Tree Tomato
Cyphomandra	*betacea*	*helva*	Yellow	Tamarillo	Tree Tomato
Cyphomandra	*betacea*	*ruby-redii*	Ruby Red	Tamarillo	Tree Tomato

Tomato, Beefsteak (*Lycopersicon bovistigum*)

Genus	Species	Infraspecies	cultivar	specimen	Genome
Lycopersicon	*bovistigum-album*	*giant-white-beefsteakii*	Giant White 10oz	White Beefsteak	Tomato
Lycopersicon	*bovistigum-album*	*great-white-beefsteakii*	Great White 15oz	White Beefsteak	Tomato
Lycopersicon	*bovistigum-album*	*halfmoon-chinai*	Halfmoon China 3" 9oz	White Beefsteak	Tomato
Lycopersicon	*bovistigum-aurantium*	*amanaensis*	Amana 5.0"	Orange Beefsteak	Tomato

Genus	Species	Infraspecies	cultivar	specimen	Genome
Lycopersicon	*bovistigum-aurantium*	*big-orangei*	Big 1lb	Orange Beefsteak	Tomato
Lycopersicon	*bovistigum-aurantium*	*elfiei*	Elfie 10oz	Orange Beefsteak	Tomato
Lycopersicon	*bovistigum-aurantium*	*kellogg's-breakfastii*	Kellogg's Breakfast 1.5lb	Orange Beefsteak	Tomato
Lycopersicon	*bovistigum-aurantium*	*mary-reynoldsi*	Mary Reynolds ?.?"	Orange Beefsteak	Tomato
Lycopersicon	*bovistigum-aurantium*	*mennonite-orangei*	Mennonite ?.?"	Orange Beefsteak	Tomato
Lycopersicon	*bovistigum-aurantium*	*orange-kingii*	Orange King 3/4lb	Orange Beefsteak	Tomato
Lycopersicon	*bovistigum-aurantium*	*orange-kingii-94*	#94 Orange King 3/4lb	Orange Beefsteak	Tomato
Lycopersicon	*bovistigum-aurantium*	*orange-queenii*	Orange Queen 1/2lb	Orange Beefsteak	Tomato
Lycopersicon	*bovistigum-aurantium-helvum*	*brimmer-yellowi*	Brimmer ?.?"	Yellow Orange Beefsteak	Tomato
Lycopersicon	*bovistigum-aurantium-helvum*	*faribo-goldheartii*	Faribo Goldheart 3.0"	Yellow Orange Beefsteak	Tomato
Lycopersicon	*bovistigum-aurantium-helvum*	*kentucky-beefsteakii*	Kentucky 2lb	Yellow Orange Beefsteak	Tomato
Lycopersicon	*bovistigum-aurantium-helvum*	*hawaiian-pineapplei*	Hawaiian Pineapple 2lb	Yellow Orange Beefsteak	Tomato
Lycopersicon	*bovistigum-aurantium-helvum*	*spark's-yellowi*	Spark's 3.0"	Yellow Orange Beefsteak	Tomato
Lycopersicon	*bovistigum-aurantium-helvum*	*tangerinei*	Tangerine 9oz	Yellow Orange Beefsteak	Tomato
Lycopersicon	*bovistigum-aurantium-helvum*	*tomatofest-beetsteakii*	Tomatofest 4.0"	Yellow Orange Beefsteak	Tomato
Lycopersicon	*bovistigum-groenum*	*aunt-ruby's-german-greenii*	Aunt"s Ruby German 1lb	Green Beefsteak	Tomato
Lycopersicon	*bovistigum-groenum*	*green-pineapplei*	Green Pineapple ?lb	Green Beefsteak	Tomato
Lycopersicon	*bovistigum-groenum*	*green-velvetii*	Green Velvet 4oz	Green Beefsteak	Tomato
Lycopersicon	*bovistigum-helvum*	*candy's-old-yellowii*	Candy's Old ?.?"	Yellow Beefsteak	Tomato
Lycopersicon	*bovistigum-helvum*	*chuck's-yellowi*	Chuck's ?.?"	Yellow Beefsteak	Tomato
Lycopersicon	*bovistigum-helvum*	*fablonelystynji*	Fablonelystynj ?.?"	Yellow Beefsteak	Tomato
Lycopersicon	*bovistigum-helvum*	*isis*	Isis 4oz	Yellow Beefsteak	Tomato
Lycopersicon	*bovistigum-helvum*	*lenny-&-gracies-kentucky-heirloomii*	Lenny & Gracies Kentucky Heirloom 4"	Yellow Beefsteak	Tomato
Lycopersicon	*bovistigum-helvum*	*limmonyi*	Limmony 4.5"	Yellow Beefsteak	Tomato
Lycopersicon	*bovistigum-helvum*	*lincoln-goldii*	Lincoln Gold ?.?"	Yellow Beefsteak	Tomato
Lycopersicon	*bovistigum-helvum*	*mexican-yellowi*	Mexican 14oz	Yellow Beefsteak	Tomato
Lycopersicon	*bovistigum-helvum*	*mortgage-lifter-bicolor-strainii*	Mortgate Lifter Bicolor Strain 1lb	Yellow Beefsteak	Tomato
Lycopersicon	*bovistigum-helvum*	*persimmonii*	Persimmon 2lb	Yellow Beefsteak	Tomato
Lycopersicon	*bovistigum-helvum*	*whitebeefsteakii*	White	Yellow Beefsteak	Tomato
Lycopersicon	*bovistigum-helvum-aurantium*	*armeniana*	Armenian 1lb	Orange Yellow Beefsteak	Tomato
Lycopersicon	*bovistigum-helvum-aurantium*	*big-golden-deliciousi*	Big Golden Delicious 1lb	Orange Yellow Beefsteak	Tomato
Lycopersicon	*bovistigum-helvum-aurantium*	*caro-richii*	Caro Rich 6oz	Orange Yellow Beefsteak	Tomato
Lycopersicon	*bovistigum-helvum-aurantium*	*elbensis*	Elbe River Valley 1lb	Orange Yellow Beefsteak	Tomato
Lycopersicon	*bovistigum-pincum*	*african-beefsteakii*	African ?.?"	Pink Beefsteak	Tomato
Lycopersicon	*bovistigum-pincum*	*andersonii*	Anderson ?.?"	Pink Beefsteak	Tomato
Lycopersicon	*bovistigum-pincum*	*argentinensis*	Argentina 3.5"	Pink Beefsteak	Tomato
Lycopersicon	*bovistigum-pincum*	*bear-clawi*	Bear Claw ?.?"	Pink Beefsteak	Tomato
Lycopersicon	*bovistigum-pincum*	*blue-riveri*	Blue River ?.?"	Pink Beefsteak	Tomato
Lycopersicon	*bovistigum-pincum*	*butterworth's-potato-leafii*	Butterworths Potato Leaf 1lb	Pink Beefsteak	Tomato
Lycopersicon	*bovistigum-pincum*	*chris-ukrianiana*	Chris Ukrainian Flat ?.?"	Pink Beefsteak	Tomato
Lycopersicon	*bovistigum-pincum*	*cleota-pinkii*	Cleota ?.?"	Pink Beefsteak	Tomato
Lycopersicon	*bovistigum-pincum*	*dr.-buresh-pink-italiana*	Dr. Buresh Pink Italian 1.3lb	Pink Beefsteak	Tomato
Lycopersicon	*bovistigum-pincum*	*german-headii*	German Head 1.5lb	Pink Beefsteak	Tomato
Lycopersicon	*bovistigum-pincum*	*hungariana*	Hungarian 1lb	Pink Beefsteak	Tomato
Lycopersicon	*bovistigum-pincum*	*italian-sweetii*	Italian Sweet ?.?"	Pink Beefsteak	Tomato
Lycopersicon	*bovistigum-pincum*	*julia-childii*	Julia Child 4.0"	Pink Beefsteak	Tomato
Lycopersicon	*bovistigum-pincum*	*mammoth-sandwichii*	Mammoth Sandwich 1.5lb	Pink Beefsteak	Tomato
Lycopersicon	*bovistigum-pincum*	*marianna's-peacei*	Marianna's Peace 1.5lb	Pink Beefsteak	Tomato
Lycopersicon	*bovistigum-pincum*	*marizol-bratkai*	Marizol Bratka 5.0"	Pink Beefsteak	Tomato
Lycopersicon	*bovistigum-pincum*	*missouri-pink-love-applei*	Missouri Love Apple 1lb	Pink Beefsteak	Tomato
Lycopersicon	*bovistigum-pincum*	*mr.-underwoods-pink-german-giantii*	Mr. Underwood's German Giant 3lb	Pink Beefsteak	Tomato
Lycopersicon	*bovistigum-pincum*	*nina's-heirloomii*	Nina's Hierloom ?.?"	Pink Beefsteak	Tomato
Lycopersicon	*bovistigum-pincum*	*richardsonii*	Richardson ?.?"	Pink Beefsteak	Tomato
Lycopersicon	*bovistigum-pincum*	*sandul-moldovanii*	Sandul Moldovan 1lb	Pink Beefsteak	Tomato
Lycopersicon	*bovistigum-pincum*	*shirleyi-S*	Shirley S 2.5lb	Pink Beefsteak	Tomato
Lycopersicon	*bovistigum-pincum*	*west-virginia-strawi*	West Virginia Straw 1lb	Pink Beefsteak	Tomato
Lycopersicon	*bovistigum-pincum*	*yugoslaviana-crnkovici*	Crnkovic Yugoslavian 1lb	Pink Beefsteak	Tomato
Lycopersicon	*bovistigum-pincum-purpureum*	*aunty-ginny's-purplei*	Aunty Ginny's Purple 14oz	Purplish Pink Beefsteak	Tomato
Lycopersicon	*bovistigum-pincum-purpureum*	*beefsteak-watermelonii*	Watermelon 2lb	Purplish Pink Beefsteak	Tomato
Lycopersicon	*bovistigum-pincum-purpureum*	*noire-des-cosebeufii*	Noire Des Cosbuef ?.?"	Purplish Pink Beefsteak	Tomato
Lycopersicon	*bovistigum-pincum-purpureum*	*ponderosa*	Ponderosa Flat 1lb	Purplish Pink Beefsteak	Tomato
Lycopersicon	*bovistigum-pincum-purpureum*	*prudei*	Prude ?.?"	Purplish Pink Beefsteak	Tomato
Lycopersicon	*bovistigum-rubrum-giganteum*	.	.	Giant 3lb 5.0" Red Beefsteak	Tomato
Lycopersicon	*bovistigum-rubrum-giganteum*	*australiensis*	Australia	Giant 3lb 5.0" Red Beefsteak	Tomato
Lycopersicon	*bovistigum-rubrum-giganteum*	*burpee's-deliciousi*	Burpee's Delicious	Giant 3lb 5.0" Red Beefsteak	Tomato
Lycopersicon	*bovistigum-rubrum-giganteum*	*giantii-11*	#11	Giant 3lb 5.0" Red Beefsteak	Tomato
Lycopersicon	*bovistigum-rubrum-giganteum*	*garvey's-beefsteakii*	Garvey's	Giant 3lb 5.0" Red Beefsteak	Tomato
Lycopersicon	*bovistigum-rubrum-giganteum*	*magnum-beefsteakii*	Magnum	Giant 3lb 5.0" Red Beefsteak	Tomato
Lycopersicon	*bovistigum-rubrum-largum*	.	.	Large 1.5-2lb 4.0" Red Beefsteak	Tomato
Lycopersicon	*bovistigum-rubrum-largum*	*aussiei*	Aussie	Large 1.5-2lb 4.0" Red Beefsteak	Tomato
Lycopersicon	*bovistigum-rubrum-largum*	*bragii*	Brag	Large 1.5-2lb 4.0" Red Beefsteak	Tomato
Lycopersicon	*bovistigum-rubrum-largum*	*coustraleei-redii*	Coustralee	Large 1.5-2lb 4.0" Red Beefsteak	Tomato

Genus	Species	Infraspecies	cultivar	specimen	Genome
Lycopersicon	bovistigum-rubrum-largum	*crimson-cushionii*	Cushion Crimson	Large 1.5-2lb 4.0" Red Beefsteak	Tomato
Lycopersicon	bovistigum-rubrum-largum	*marazzi's-giantii*	Marazzi's Giant	Large 1.5-2lb 4.0" Red Beefsteak	Tomato
Lycopersicon	bovistigum-rubrum-largum	*mark-twainii*	Mark Twain	Large 1.5-2lb 4.0" Red Beefsteak	Tomato
Lycopersicon	bovistigum-rubrum-largum	*maryland-large-redii*	Maryland	Large 1.5-2lb 4.0" Red Beefsteak	Tomato
Lycopersicon	bovistigum-rubrum-largum	*mexican-beefsteakii*	Mexican	Large 1.5-2lb 4.0" Red Beefsteak	Tomato
Lycopersicon	bovistigum-rubrum-largum	*old-timeyi*	OldTimey	Large 1.5-2lb 4.0" Red Beefsteak	Tomato
Lycopersicon	bovistigum-rubrum-largum-medium	*bovistigum-rubrum*	Original	Medium Large 1lb 3.5" Red Beefsteak	Tomato
Lycopersicon	bovistigum-rubrum-largum-medium	*burgess-giant-beefsteakii*	Burgess Giant	Medium Large 1lb 3.5" Red Beefsteak	Tomato
Lycopersicon	bovistigum-rubrum-largum-medium	*chuck's-redii*	Chuck's	Medium Large 1lb 3.5" Red Beefsteak	Tomato
Lycopersicon	bovistigum-rubrum-largum-medium	*costoluto-catanesei*	Catanese Costoluto	Medium Large 1lb 3.5" Red Beefsteak	Tomato
Lycopersicon	bovistigum-rubrum-largum-medium	*flavour-steakii*	Flavour Steak	Medium Large 1lb 3.5" Red Beefsteak	Tomato
Lycopersicon	bovistigum-rubrum-largum-medium	*gala*	Gala	Medium Large 1lb 3.5" Red Beefsteak	Tomato
Lycopersicon	bovistigum-rubrum-largum-medium	*germaid-redii*	Germaid	Medium Large 1lb 3.5" Red Beefsteak	Tomato
Lycopersicon	bovistigum-rubrum-largum-medium	*good-old-fashioned-redii*	Good Old Fashioned	Medium Large 1lb 3.5" Red Beefsteak	Tomato
Lycopersicon	bovistigum-rubrum-largum-medium	*great-dividei*	Great Divide	Medium Large 1lb 3.5" Red Beefsteak	Tomato
Lycopersicon	bovistigum-rubrum-largum-medium	*martian-giantii*	Martian Giant	Medium Large 1lb 3.5" Red Beefsteak	Tomato
Lycopersicon	bovistigum-rubrum-largum-medium	*maslov's-giantii*	Maslov's Giant	Medium Large 1lb 3.5" Red Beefsteak	Tomato
Lycopersicon	bovistigum-rubrum-largum-medium	*mt.-athosensis*	Mt. Athos	Medium Large 1lb 3.5" Red Beefsteak	Tomato
Lycopersicon	bovistigum-rubrum-largum-medium	*old-virginiensis*	Old Virigina	Medium Large 1lb 3.5" Red Beefsteak	Tomato
Lycopersicon	bovistigum-rubrum-largum-medium	*page-germanica*	Page German	Medium Large 1lb 3.5" Red Beefsteak	Tomato
Lycopersicon	bovistigum-rubrum-largum-medium	*ponderosa-beefsteakii-improvedii*	Improved Ponderosa	Medium Large 1lb 3.5" Red Beefsteak	Tomato
Lycopersicon	bovistigum-rubrum-largum-medium	*super-beefsteakii-VFN*	Super	Medium Large 1lb 3.5" Red Beefsteak	Tomato
Lycopersicon	bovistigum-rubrum-largum-medium	*vogliottii*	Vogliotti	Medium Large 1lb 3.5" Red Beefsteak	Tomato
Lycopersicon	bovistigum-rubrum-largum-medium	*willow-pond-big-red-beefsteakii*	Willow Pond Big Red	Medium Large 1lb 3.5" Red Beefsteak	Tomato
Lycopersicon	bovistigum-rubrum-medium	*altajskij-urozajnijii*	Altajskij Urozajnij	Medium 1/2lb 2.5" Red Beefsteak	Tomato
Lycopersicon	bovistigum-rubrum-medium	*angelo's-redii*	Angelino's	Medium 1/2lb 2.5" Red Beefsteak	Tomato
Lycopersicon	bovistigum-rubrum-medium	*buckbee's-beefsteakii*	Buckbee's	Medium 1/2lb 2.5" Red Beefsteak	Tomato
Lycopersicon	bovistigum-rubrum-medium	*bush-beefsteakii*	Bush	Medium 1/2lb 2.5" Red Beefsteak	Tomato
Lycopersicon	bovistigum-rubrum-medium	*costoluto-fiorentinoi*	Fiorentino Costoluto	Medium 1/2lb 2.5" Red Beefsteak	Tomato
Lycopersicon	bovistigum-rubrum-medium	*costoluto-genovesei*	Genovese Costoluto	Medium 1/2lb 2.5" Red Beefsteak	Tomato
Lycopersicon	bovistigum-rubrum-medium	*giuseppe's-big-boyi*	Giuseppi's Big Boy	Medium 1/2lb 2.5" Red Beefsteak	Tomato
Lycopersicon	bovistigum-rubrum-medium	*old-ferry-morse-beefsteakii*	Old Ferry Morse	Medium 1/2lb 2.5" Red Beefsteak	Tomato
Lycopersicon	bovistigum-rubrum-medium	*potato-leaf-sugar-beefsteakii*	Potato Leaf Sugar	Medium 1/2lb 2.5" Red Beefsteak	Tomato
Lycopersicon	bovistigum-rubrum-medium	*sicilian-sauceri*	Sicilian Saucer	Medium 1/2lb 2.5" Red Beefsteak	Tomato
Lycopersicon	bovistigum-rubrum-medium	*sugar-beefsteakii*	Sugar	Medium 1/2lb 2.5" Red Beefsteak	Tomato
Lycopersicon	bovistigum-rubrum-medium	*urban-beefsteakii*	Urban	Medium 1/2lb 2.5" Red Beefsteak	Tomato
Lycopersicon	bovistigum-rubrum-medium	*yates-beefsteakii*	Yates	Medium 1/2lb 2.5" Red Beefsteak	Tomato
Lycopersicon	bovistigum-rubrum-parvum	*CN-415*	#415 Canada	Small 4oz-6oz 2.0" Red Beefsteak	Tomato
Lycopersicon	bovistigum-rubrum-parvum	*moira*	Moira	Small 4oz-6oz 2.0" Red Beefsteak	Tomato
Lycopersicon	bovistigum-rubrum-parvum	*new-yorkeri*	New Yorker	Small 4oz-6oz 2.0" Red Beefsteak	Tomato
Lycopersicon	bovistigum-rubrum-parvum	*new-yorkeri-transplantii*	Transplant New Yorker	Small 4oz-6oz 2.0" Red Beefsteak	Tomato
Lycopersicon	bovistigum-rubrum-aurantium	*giant-beefsteakii*	Giant 2lb	Orange Red Beefsteak	Tomato
Lycopersicon	bovistigum-rubrum-aurantium	*bielo-russiana*	Bielo Russian ?.?"	Orange Red Beefsteak	Tomato
Lycopersicon	bovistigum-rubrum-aurantium	*copia*	Copia 3.5"	Orange Red Beefsteak	Tomato
Lycopersicon	bovistigum-rubrum-aurantium	*debbiei*	Debbie 12oz	Orange Red Beefsteak	Tomato
Lycopersicon	bovistigum-rubrum-aurantium	*ponderosa-scarletii*	Ponderosa Scarlet 1.5lb	Orange Red Beefsteak	Tomato
Lycopersicon	bovistigum-rubrum-aurantium	*semenarma-ljubjana*	Semenarma Ljubja	Orange Red Beefsteak	Tomato
Lycopersicon	bovistigum-striatum	*flamei*	Flame 2lb	Striped Beefsteak	Tomato
Lycopersicon	bovistigum-striatum	*flamei-oldii*	Old Flame 2lb	Striped Beefsteak	Tomato
Lycopersicon	bovistigum-striatum	*hillbillyi*	Hillbilly 1.5lb	Striped Beefsteak	Tomato
Lycopersicon	bovistigum-striatum	*hillbillyi-potato-leafii*	Potato Leaf Hillbilly 1lb	Striped Beefsteak	Tomato
Lycopersicon	bovistigum-striatum	*hazel-maei*	Hazel Mae 1lb	Striped Beefsteak	Tomato
Lycopersicon	bovistigum-striatum	*pineapplei*	Pineapple 2lb	Striped Beefsteak	Tomato
Lycopersicon	bovistigum-striatum	*red-streakii*	Red Streak 12oz	Striped Beefsteak	Tomato

Tomato, Cherry (*Lycopersicon cerasiforme*)

Genus	Species	Infraspecies	cultivar	specimen	Genome
Lycopersicon	cerasiforme-album	*album*	Original 1oz	White Cherry	Tomato
Lycopersicon	cerasiforme-album	*bianca*	Bianca ?oz	White Cherry	Tomato
Lycopersicon	cerasiforme-album	*ghost-cherryi*	Ghost 2oz	White Cherry	Tomato
Lycopersicon	cerasiforme-album	*mirabelle-blanchei*	Mirabelle Blanche ?oz	White Cherry	Tomato
Lycopersicon	cerasiforme-album	*petite-pomme-blanchei*	Petite Pomme Blanche ?oz	White Cherry	Tomato
Lycopersicon	cerasiforme-album	*snow-white-cherryi*	Snow White 0.5"	White Cherry	Tomato
Lycopersicon	cerasiforme-album	*white-rabbitii*	White Rabbit 1/4"	White Cherry	Tomato
Lycopersicon	cerasiforme-aurantium	*aurantium*	Original	Orange Cherry	Tomato
Lycopersicon	cerasiforme-aurantium	*amishii*	Amish ?.?"	Orange Cherry	Tomato
Lycopersicon	cerasiforme-aurantium	*fruity-orangei*	Fruity	Orange Cherry	Tomato
Lycopersicon	cerasiforme-aurantium	*miniature-orangei*	Minature 1.3"	Orange Cherry	Tomato
Lycopersicon	cerasiforme-aurantium	*orange-cherry-potato-leafii*	Potato Leaf ?.?"	Orange Cherry	Tomato
Lycopersicon	cerasiforme-aurantium	*prize-of-the-trialsi*	Prize Of The Trials 2oz	Orange Cherry	Tomato
Lycopersicon	cerasiforme-aurantium	*sungold-selectii*	Sungold Select 0.8"	Orange Cherry	Tomato
Lycopersicon	cerasiforme-aurantium	*sweet-orangei-2*	#2 Sweet Orange ?.?"	Orange Cherry	Tomato
Lycopersicon	cerasiforme-groenum	*aunt-ruby's-german-cherryi*	Aunt Ruby's German ?oz	Green Cherry	Tomato
Lycopersicon	cerasiforme-groenum	*green-grapei*	Green Grape 1"	Green Cherry	Tomato

Genus	Species	Infraspecies	cultivar	specimen	Genome
Lycopersicon	*cerasiforme-groenum*	*lime-green-saladii*	Lime Green Salad 2.5oz	Green Cherry	Tomato
Lycopersicon	*cerasiforme-groenum*	*neverwillii*	Neverwill 1.5"	Green Cherry	Tomato
Lycopersicon	*cerasiforme-groenum*	*tomalilloi*	Tomalillo ?oz	Green Cherry	Tomato
Lycopersicon	*cerasiforme-helvum*	*helvum*	Original 0.8"	Yellow Cherry	Tomato
Lycopersicon	*cerasiforme-helvum*	*aunt-ruby's-yellowi*	Aunt Ruby's 0.9"	Yellow Cherry	Tomato
Lycopersicon	*cerasiforme-helvum*	*blondkopfchenii*	Blondkopfchen 0.8"	Yellow Cherry	Tomato
Lycopersicon	*cerasiforme-helvum*	*chelloi*	Chello 1.0"	Yellow Cherry	Tomato
Lycopersicon	*cerasiforme-helvum*	*dwarf-gold-treasurei*	Dwarf Gold Treasure 3oz	Yellow Cherry	Tomato
Lycopersicon	*cerasiforme-helvum*	*esther'si*	Esther's ?.?"	Yellow Cherry	Tomato
Lycopersicon	*cerasiforme-helvum*	*galina'si*	Galina 1oz	Yellow Cherry	Tomato
Lycopersicon	*cerasiforme-helvum*	*galina-dr.-carolyn-malei*	Dr. Carolyn Male Galina 1.25"	Yellow Cherry	Tomato
Lycopersicon	*cerasiforme-helvum*	*galina-grandei*	Grande Galina 1.5"	Yellow Cherry	Tomato
Lycopersicon	*cerasiforme-helvum*	*golden-grapei*	Golden Grape ?.?"	Yellow Cherry	Tomato
Lycopersicon	*cerasiforme-helvum*	*green-cagei*	Green Cage 2.5oz	Yellow Cherry	Tomato
Lycopersicon	*cerasiforme-helvum*	*green-gagei*	Green Gage 1.5"	Yellow Cherry	Tomato
Lycopersicon	*cerasiforme-helvum*	*hartman's-yellow-gooseberryi*	Hartman's Yellow Gooseberry 1.0"	Yellow Cherry	Tomato
Lycopersicon	*cerasiforme-helvum*	*isis-candy-cherryi*	Isis Candy 2.0"	Yellow Cherry	Tomato
Lycopersicon	*cerasiforme-helvum*	*livingston-gold-ballii*	Livingston's Gold Ball 1.3"	Yellow Cherry	Tomato
Lycopersicon	*cerasiforme-helvum*	*little-yellow-pot-tomatoi*	Little Yellow Pot ?.?"	Yellow Cherry	Tomato
Lycopersicon	*cerasiforme-helvum*	*lollipopii*	Lollipop ?.?"	Yellow Cherry	Tomato
Lycopersicon	*cerasiforme-helvum*	*mirabellii*	Mirabell 0.5"	Yellow Cherry	Tomato
Lycopersicon	*cerasiforme-helvum*	*olivette-jaunei*	Olivette Juane ?.?"	Yellow Cherry	Tomato
Lycopersicon	*cerasiforme-helvum*	*peach-yellowi*	Peach 1oz	Yellow Cherry	Tomato
Lycopersicon	*cerasiforme-helvum*	*porter's-pridei*	Porter's Pride ?.?"	Yellow Cherry	Tomato
Lycopersicon	*cerasiforme-helvum*	*sarah-goldstar-cherryi*	Sarah Goldstar ?oz	Yellow Cherry	Tomato
Lycopersicon	*cerasiforme-helvum*	*sun-bellei*	Sun Belle 0.4"	Yellow Cherry	Tomato
Lycopersicon	*cerasiforme-helvum*	*sunshinei*	Sunshine ?.?"	Yellow Cherry	Tomato
Lycopersicon	*cerasiforme-helvum*	*sweet-goldii*	Sweet Gold 1.0"	Yellow Cherry	Tomato
Lycopersicon	*cerasiforme-helvum*	*tennessee-yellowi*	Tennessee 3/4"	Yellow Cherry	Tomato
Lycopersicon	*cerasiforme-helvum*	*una's-yellowi*	Una's ?.?"	Yellow Cherry	Tomato
Lycopersicon	*cerasiforme-helvum*	*warren;s-yellowi*	Warren's ?.?"	Yellow Cherry	Tomato
Lycopersicon	*cerasiforme-helvum*	*yellow-canaryi*	Yellow Canary 1.0"	Yellow Cherry	Tomato
Lycopersicon	*cerasiforme-helvum*	*yellow-marblei*	Yellow Marble 0.3"	Yellow Cherry	Tomato
Lycopersicon	*cerasiforme-helvum-aurantium*	*cerise-orangei*	Cerise ?.?"	Orange Yellow Cherry	Tomato
Lycopersicon	*cerasiforme-helvum-aurantium*	*cherry-goldii*	Cherry Gold ?.?"	Orange Yellow Cherry	Tomato
Lycopersicon	*cerasiforme-helvum-aurantium*	*floragold-basketii*	Floragold Basket ?.?"	Orange Yellow Cherry	Tomato
Lycopersicon	*cerasiforme-helvum-aurantium*	*gold-nuggetii*	Gold Nugget 1.2"	Orange Yellow Cherry	Tomato
Lycopersicon	*cerasiforme-helvum-aurantium*	*golden-pearlii*	Golden Pearl 0.4"	Orange Yellow Cherry	Tomato
Lycopersicon	*cerasiforme-helvum-aurantium*	*grape-tressi*	Grape Tress 0.4"	Orange Yellow Cherry	Tomato
Lycopersicon	*cerasiforme-helvum-aurantium*	*reinhard's-tomatei*	Reinhard's Tomate 1.0"	Orange Yellow Cherry	Tomato
Lycopersicon	*cerasiforme-helvum-aurantium*	*yellow-krimii*	Yellow Krim ?.?"	Orange Yellow Cherry	Tomato
Lycopersicon	*cerasiforme-nigrum*	*nigrum*	Original ?oz	Black Cherry	Tomato
Lycopersicon	*cerasiforme-nigrum*	*aunt-ruby's-german-blackii*	Aunt Ruby's German ?oz	Black Cherry	Tomato
Lycopersicon	*cerasiforme-nigrum*	*black-de-barioi*	Black De Bario 3oz	Black Cherry	Tomato
Lycopersicon	*cerasiforme-pincum*	*pincum*	Original 3/4"	Pink Cherry	Tomato
Lycopersicon	*cerasiforme-pincum*	*amish-saladii*	Amish Salad 2oz	Pink Cherry	Tomato
Lycopersicon	*cerasiforme-pincum*	*brandywine-cherryi*	Cherry Amish Brandywine Creek 1.5oz	Pink Cherry	Tomato
Lycopersicon	*cerasiforme-pincum*	*cherry-pearly-pinkii*	Pearly 1.0"	Pink Cherry	Tomato
Lycopersicon	*cerasiforme-pincum*	*cherry-porter's-darkii*	Porter's Dark ?.?"	Pink Cherry	Tomato
Lycopersicon	*cerasiforme-pincum*	*cherry-sara-pinkii*	Sara ?.?"	Pink Cherry	Tomato
Lycopersicon	*cerasiforme-pincum*	*cherry-valley-girlii*	Valley Girl ?.?"	Pink Cherry	Tomato
Lycopersicon	*cerasiforme-pincum*	*ditmarsheri*	Ditmarsher ?.?"	Pink Cherry	Tomato
Lycopersicon	*cerasiforme-pincum*	*fuzzyi*	Fuzzy 3oz	Pink Cherry	Tomato
Lycopersicon	*cerasiforme-pincum*	*grape-pinkii*	Grape 0.5"	Pink Cherry	Tomato
Lycopersicon	*cerasiforme-pincum*	*pear-pinkii*	Pear	Pink Cherry	Tomato
Lycopersicon	*cerasiforme-pincum*	*pink-icei*	Pink Ice ?.?"	Pink Cherry	Tomato
Lycopersicon	*cerasiforme-pincum*	*pink-nibletsi*	Pink Niblets ?.?"	Pink Cherry	Tomato
Lycopersicon	*cerasiforme-pincum*	*pinkpearii*	Pear ?,?"	Pink Cherry	Tomato
Lycopersicon	*cerasiforme-pincum*	*rose-quartzii*	Rose Quartz ?.?"	Pink Cherry	Tomato
Lycopersicon	*cerasiforme-rubrum-gigantium*	*florida-basketii*	Florida Basket	Giant 3.5oz 1.75" Red Cherry	Tomato
Lycopersicon	*cerasiforme-rubrum-gigantium*	*oregonensis-11*	#11 Oregon	Giant 3.5oz 1.75" Red Cherry	Tomato
Lycopersicon	*cerasiforme-rubrum-gigantium*	*oregon-springii*	Oregon Spring	Giant 3.5oz 1.75" Red Cherry	Tomato
Lycopersicon	*cerasiforme-rubrum-gigantium*	*otraolnyi*	Otraolny	Giant 3.5oz 1.75" Red Cherry	Tomato
Lycopersicon	*cerasiforme-rubrum-gigantium*	*peasantii*	Peasant	Giant 3.5oz 1.75" Red Cherry	Tomato
Lycopersicon	*cerasiforme-rubrum-gigantium*	*plum-redii*	Plum	Giant 3.5oz 1.75" Red Cherry	Tomato
Lycopersicon	*cerasiforme-rubrum-gigantium*	*polar-babyi*	Polar Baby	Giant 3.5oz 1.75" Red Cherry	Tomato
Lycopersicon	*cerasiforme-rubrum-gigantium*	*polar-beautyi*	Polar Beauty	Giant 3.5oz 1.75" Red Cherry	Tomato
Lycopersicon	*cerasiforme-rubrum-gigantium*	*polar-gemii*	Polar Gem	Giant 3.5oz 1.75" Red Cherry	Tomato
Lycopersicon	*cerasiforme-rubrum-gigantium*	*polar-stari*	Polar Star	Giant 3.5oz 1.75" Red Cherry	Tomato
Lycopersicon	*cerasiforme-rubrum-gigantium*	*polishii-eggii*	Egg Polish	Giant 3.5oz 1.75" Red Cherry	Tomato
Lycopersicon	*cerasiforme-rubrum-gigantium*	*precocibeci*	Precocibec	Giant 3.5oz 1.75" Red Cherry	Tomato
Lycopersicon	*cerasiforme-rubrum-gigantium*	*PSR-952A*	#952A Peters Seed & Research	Giant 3.5oz 1.75" Red Cherry	Tomato
Lycopersicon	*cerasiforme-rubrum-gigantium*	*PSR-952B*	#952B Peters Seed & Research	Giant 3.5oz 1.75" Red Cherry	Tomato

Genus	Species	Infraspecies	cultivar	specimen	Genome
Lycopersicon	cerasiforme-rubrum-gigantium	*rocketii*	Rocket	Giant 3.5oz 1.75" Red Cherry	Tomato
Lycopersicon	cerasiforme-rubrum-gigantium	*RP-23-2*	#23-2 Row Pac	Giant 3.5oz 1.75" Red Cherry	Tomato
Lycopersicon	cerasiforme-rubrum-gigantium	*siberiana-earlyi*	Early Siberian	Giant 3.5oz 1.75" Red Cherry	Tomato
Lycopersicon	cerasiforme-rubrum-gigantium	*stokes-alaskensis*	Stokes Alaska	Giant 3.5oz 1.75" Red Cherry	Tomato
Lycopersicon	cerasiforme-rubrum-gigantium	*sugar-lumpii*	Sugar Lump	Giant 3.5oz 1.75" Red Cherry	Tomato
Lycopersicon	cerasiforme-rubrum-gigantium	*tanana*	Tanana	Giant 3.5oz 1.75" Red Cherry	Tomato
Lycopersicon	cerasiforme-rubrum-largum	*all-meatii*	All Meat	Large 3oz 1.5" Red Cherry	Tomato
Lycopersicon	cerasiforme-rubrum-largum	*amy's-sugar-gemii*	Amy's Sugar Gem	Large 3oz 1.5" Red Cherry	Tomato
Lycopersicon	cerasiforme-rubrum-largum	*basket-pakii*	Basket Pak	Large 3oz 1.5" Red Cherry	Tomato
Lycopersicon	cerasiforme-rubrum-largum	*bisonii*	Bison Oscar Will	Large 3oz 1.5" Red Cherry	Tomato
Lycopersicon	cerasiforme-rubrum-largum	*bulgariana-7*	#7 Bulgarian	Large 3oz 1.5" Red Cherry	Tomato
Lycopersicon	cerasiforme-rubrum-largum	*bulgariana-dixii*	Dix Bulgarian	Large 3oz 1.5" Red Cherry	Tomato
Lycopersicon	cerasiforme-rubrum-largum	*cal-jii*	California Jointless	Large 3oz 1.5" Red Cherry	Tomato
Lycopersicon	cerasiforme-rubrum-largum	*canary-saladii*	Canary Salad	Large 3oz 1.5" Red Cherry	Tomato
Lycopersicon	cerasiforme-rubrum-largum	*cherry-earlyi*	Early	Large 3oz 1.5" Red Cherry	Tomato
Lycopersicon	cerasiforme-rubrum-largum	*costaricensis*	Costa Rica	Large 3oz 1.5" Red Cherry	Tomato
Lycopersicon	cerasiforme-rubrum-largum	*dropletii*	Droplet	Large 3oz 1.5" Red Cherry	Tomato
Lycopersicon	cerasiforme-rubrum-largum	*grandpa's-golf-ballsi*	Grandpa's Golf Balls	Large 3oz 1.5" Red Cherry	Tomato
Lycopersicon	cerasiforme-rubrum-largum	*italian-winteri*	Italian Winter	Large 3oz 1.5" Red Cherry	Tomato
Lycopersicon	cerasiforme-rubrum-largum	*lima*	Lima	Large 3oz 1.5" Red Cherry	Tomato
Lycopersicon	cerasiforme-rubrum-largum	*nigeriensis*	Nigeria	Large 3oz 1.5" Red Cherry	Tomato
Lycopersicon	cerasiforme-rubrum-largum	*packhousei*	Packhouse	Large 3oz 1.5" Red Cherry	Tomato
Lycopersicon	cerasiforme-rubrum-largum	*petitbeci*	Petitbec	Large 3oz 1.5" Red Cherry	Tomato
Lycopersicon	cerasiforme-rubrum-largum	*puebla*	Puebla	Large 3oz 1.5" Red Cherry	Tomato
Lycopersicon	cerasiforme-rubrum-largum	*quedlindberger-frue-liebei*	Early Love Of Qued Lindberger	Large 3oz 1.5" Red Cherry	Tomato
Lycopersicon	cerasiforme-rubrum-largum	*red-rockii*	Red Rock	Large 3oz 1.5" Red Cherry	Tomato
Lycopersicon	cerasiforme-rubrum-largum	*startii-Z*	Start Zubrowski (Z)	Large 3oz 1.5" Red Cherry	Tomato
Lycopersicon	cerasiforme-rubrum-largum	*sweet-chelseyi*	Sweet Chelsey	Large 3oz 1.5" Red Cherry	Tomato
Lycopersicon	cerasiforme-rubrum-largum	*sweet-n-brightii*	Sweet N Bright	Large 3oz 1.5" Red Cherry	Tomato
Lycopersicon	cerasiforme-rubrum-largum	*swiftii*	Swift	Large 3oz 1.5" Red Cherry	Tomato
Lycopersicon	cerasiforme-rubrum-largum	*tip-topii*	Tip Top	Large 3oz 1.5" Red Cherry	Tomato
Lycopersicon	cerasiforme-rubrum-largum	*urbikanyi*	Urbikany	Large 3oz 1.5" Red Cherry	Tomato
Lycopersicon	cerasiforme-rubrum-largum	*VF134-1-2*	#134-1-2 Verticilium Fusarium Resistant	Large 3oz 1.5" Red Cherry	Tomato
Lycopersicon	cerasiforme-rubrum-largum	*VF-6203*	#6203 Verticilium Fusarium Resistant	Large 3oz 1.5" Red Cherry	Tomato
Lycopersicon	cerasiforme-rubrum-largum	*wilfordii*	Wilford	Large 3oz 1.5" Red Cherry	Tomato
Lycopersicon	cerasiforme-rubrum-largum	*window-boxii*	Window Box	Large 3oz 1.5" Red Cherry	Tomato
Lycopersicon	cerasiforme-rubrum-largum-medium	*florida-petitei*	Florida Petite	Medium Large 2.5oz 1.25" Red Cherry	Tomato
Lycopersicon	cerasiforme-rubrum-largum-medium	*gardener's-delightii*	Gardener's Delight	Medium Large 2.5oz 1.25" Red Cherry	Tomato
Lycopersicon	cerasiforme-rubrum-largum-medium	*hungarian-italianii*	Hungarian Italian	Medium Large 2.5oz 1.25" Red Cherry	Tomato
Lycopersicon	cerasiforme-rubrum-largum-medium	*lucky-leprechaunii*	Lucky Leprechaun	Medium Large 2.5oz 1.25" Red Cherry	Tomato
Lycopersicon	cerasiforme-rubrum-largum-medium	*red-calabashii*	Red Calabash	Medium Large 2.5oz 1.25" Red Cherry	Tomato
Lycopersicon	cerasiforme-rubrum-largum-medium	*riesentraubei*	Giant Bunch Of Grapes	Medium Large 2.5oz 1.25" Red Cherry	Tomato
Lycopersicon	cerasiforme-rubrum-largum-medium	*sandpointii*	Sandpoint	Medium Large 2.5oz 1.25" Red Cherry	Tomato
Lycopersicon	cerasiforme-rubrum-largum-medium	*sauceyi*	Saucey	Medium Large 2.5oz 1.25" Red Cherry	Tomato
Lycopersicon	cerasiforme-rubrum-largum-medium	*skorospelkai*	Skorospelka	Medium Large 2.5oz 1.25" Red Cherry	Tomato
Lycopersicon	cerasiforme-rubrum-largum-medium	*stupendus*	Stupendous	Medium Large 2.5oz 1.25" Red Cherry	Tomato
Lycopersicon	cerasiforme-rubrum-largum-medium	*subarctica-maxi*	Maxi SubArctic	Medium Large 2.5oz 1.25" Red Cherry	Tomato
Lycopersicon	cerasiforme-rubrum-largum-medium	*sweetiei*	Sweetie	Medium Large 2.5oz 1.25" Red Cherry	Tomato
Lycopersicon	cerasiforme-rubrum-medium-largum	.	.	Large Medium 2oz 1.0" Red Cherry	Tomato
Lycopersicon	cerasiforme-rubrum-medium-largum	*largum-medium*	Original	Large Medium 2oz 1.0" Red Cherry	Tomato
Lycopersicon	cerasiforme-rubrum-medium-largum	*alaskensis*	Alaska	Large Medium 2oz 1.0" Red Cherry	Tomato
Lycopersicon	cerasiforme-rubrum-medium-largum	*angora-supersweetii*	Angora Supersweet	Large Medium 2oz 1.0" Red Cherry	Tomato
Lycopersicon	cerasiforme-rubrum-medium-largum	*blockedii*	Blocked	Large Medium 2oz 1.0" Red Cherry	Tomato
Lycopersicon	cerasiforme-rubrum-medium-largum	*cherrioi*	Cherrio	Large Medium 2oz 1.0" Red Cherry	Tomato
Lycopersicon	cerasiforme-rubrum-medium-largum	*cherry-camp-joyi*	Camp Joy	Large Medium 2oz 1.0" Red Cherry	Tomato
Lycopersicon	cerasiforme-rubrum-medium-largum	*cherry-fence-rowi*	Fence Row	Large Medium 2oz 1.0" Red Cherry	Tomato
Lycopersicon	cerasiforme-rubrum-medium-largum	*cherry-koralikii*	Koralik	Large Medium 2oz 1.0" Red Cherry	Tomato
Lycopersicon	cerasiforme-rubrum-medium-largum	*cherry-loomis-potato-leafii*	Loomis Potato Leaf	Large Medium 2oz 1.0" Red Cherry	Tomato
Lycopersicon	cerasiforme-rubrum-medium-largum	*cherry-mildred-cowger's-belgianii*	Mildred Cowger's Belgian	Large Medium 2oz 1.0" Red Cherry	Tomato
Lycopersicon	cerasiforme-rubrum-medium-largum	*chibikkoi*	Chibikko	Large Medium 2oz 1.0" Red Cherry	Tomato
Lycopersicon	cerasiforme-rubrum-medium-largum	*germanica-roperi*	German Roper	Large Medium 2oz 1.0" Red Cherry	Tomato
Lycopersicon	cerasiforme-rubrum-medium-largum	*grandpa's-minnesotensis*	Grandpa's Minnesota	Large Medium 2oz 1.0" Red Cherry	Tomato
Lycopersicon	cerasiforme-rubrum-medium-largum	*imur-prior-beta*	Imur Prior Beta	Large Medium 2oz 1.0" Red Cherry	Tomato
Lycopersicon	cerasiforme-rubrum-medium-largum	*jungle-saladii*	Jungle Salad	Large Medium 2oz 1.0" Red Cherry	Tomato
Lycopersicon	cerasiforme-rubrum-medium-largum	*king-humbertii*	King Humert	Large Medium 2oz 1.0" Red Cherry	Tomato
Lycopersicon	cerasiforme-rubrum-medium-largum	*kotlasensis*	Kotlas	Large Medium 2oz 1.0" Red Cherry	Tomato
Lycopersicon	cerasiforme-rubrum-medium-largum	*morena*	Morena	Large Medium 2oz 1.0" Red Cherry	Tomato
Lycopersicon	cerasiforme-rubrum-medium-largum	*peach-redii*	Peach	Large Medium 2oz 1.0" Red Cherry	Tomato
Lycopersicon	cerasiforme-rubrum-medium-largum	*russiana-33*	#33 Russian	Large Medium 2oz 1.0" Red Cherry	Tomato
Lycopersicon	cerasiforme-rubrum-medium-largum	*sallisaw-caféi*	Sallisaw Café	Large Medium 2oz 1.0" Red Cherry	Tomato
Lycopersicon	cerasiforme-rubrum-medium-largum	*sprintii*	Sprint	Large Medium 2oz 1.0" Red Cherry	Tomato
Lycopersicon	cerasiforme-rubrum-medium-largum	*sweetii-100*	Sweet 100	Large Medium 2oz 1.0" Red Cherry	Tomato

Genus	Species	Infraspecies	cultivar	specimen	Genome
Lycopersicon	cerasiforme-rubrum-medium-largum	tiny-timii	Tiny Tim	Large Medium 2oz 1.0" Red Cherry	Tomato
Lycopersicon	cerasiforme-rubrum-medium-largum	UC-82-B	#82 Beta University Of California	Large Medium 2oz 1.0" Red Cherry	Tomato
Lycopersicon	cerasiforme-rubrum-medium-largum	UC-82-L	#82 Late University Of California	Large Medium 2oz 1.0" Red Cherry	Tomato
Lycopersicon	cerasiforme-rubrum-medium-largum	whippersnapperi	Whippersnapper	Large Medium 2oz 1.0" Red Cherry	Tomato
Lycopersicon	cerasiforme-rubrum-medium	anna-aasai	Anna Aasa	Medium 1.5oz 3/4" Red Cherry	Tomato
Lycopersicon	cerasiforme-rubrum-medium	auld-sodii	Auld Sod	Medium 1.5oz 3/4" Red Cherry	Tomato
Lycopersicon	cerasiforme-rubrum-medium	besseri	Besser	Medium 1.5oz 3/4" Red Cherry	Tomato
Lycopersicon	cerasiforme-rubrum-medium	bicolor	Bicolor	Medium 1.5oz 3/4" Red Cherry	Tomato
Lycopersicon	cerasiforme-rubrum-medium	cherriette-of-firei	Cherriette Of Fire	Medium 1.5oz 3/4" Red Cherry	Tomato
Lycopersicon	cerasiforme-rubrum-medium	cherry-baxter's-early-bushii	Baxter's Early Bush	Medium 1.5oz 3/4" Red Cherry	Tomato
Lycopersicon	cerasiforme-rubrum-medium	cherry-borgo-cellanoi	Borgo Cellano	Medium 1.5oz 3/4" Red Cherry	Tomato
Lycopersicon	cerasiforme-rubrum-medium	cherry-chadwick'si	Chadwick's	Medium 1.5oz 3/4" Red Cherry	Tomato
Lycopersicon	cerasiforme-rubrum-medium	cherry-delightii	Cherry Delight	Medium 1.5oz 3/4" Red Cherry	Tomato
Lycopersicon	cerasiforme-rubrum-medium	cherry-delightii-E	Excellent Cherry Delight	Medium 1.5oz 3/4" Red Cherry	Tomato
Lycopersicon	cerasiforme-rubrum-medium	cherry-delightii-R	Resistant Cherry Delight	Medium 1.5oz 3/4" Red Cherry	Tomato
Lycopersicon	cerasiforme-rubrum-medium	cherry-MS-5	MS-5	Medium 1.5oz 3/4" Red Cherry	Tomato
Lycopersicon	cerasiforme-rubrum-medium	cherry-oregonensis	Oregon	Medium 1.5oz 3/4" Red Cherry	Tomato
Lycopersicon	cerasiforme-rubrum-medium	cherry-sugari	Sugar	Medium 1.5oz 3/4" Red Cherry	Tomato
Lycopersicon	cerasiforme-rubrum-medium	cherry-wicklinensis	Wickline	Medium 1.5oz 3/4" Red Cherry	Tomato
Lycopersicon	cerasiforme-rubrum-medium	crazyi	Crazy	Medium 1.5oz 3/4" Red Cherry	Tomato
Lycopersicon	cerasiforme-rubrum-medium	fire-cheersi	Fire Cheers	Medium 1.5oz 3/4" Red Cherry	Tomato
Lycopersicon	cerasiforme-rubrum-medium	galenii	Galen	Medium 1.5oz 3/4" Red Cherry	Tomato
Lycopersicon	cerasiforme-rubrum-medium	gem-statei	Gem State	Medium 1.5oz 3/4" Red Cherry	Tomato
Lycopersicon	cerasiforme-rubrum-medium	grapei	Grape	Medium 1.5oz 3/4" Red Cherry	Tomato
Lycopersicon	cerasiforme-rubrum-medium	grapei-mobyi	Moby Grape	Medium 1.5oz 3/4" Red Cherry	Tomato
Lycopersicon	cerasiforme-rubrum-medium	immunei	Immune Irion	Medium 1.5oz 3/4" Red Cherry	Tomato
Lycopersicon	cerasiforme-rubrum-medium	leptophyllum	Thin Leaf	Medium 1.5oz 3/4" Red Cherry	Tomato
Lycopersicon	cerasiforme-rubrum-medium	marizol-majici	Marziol Majic	Medium 1.5oz 3/4" Red Cherry	Tomato
Lycopersicon	cerasiforme-rubrum-medium	mexico-midgetii	Mexico Midget	Medium 1.5oz 3/4" Red Cherry	Tomato
Lycopersicon	cerasiforme-rubrum-medium	pineapple-fogii	Pineapple Fog	Medium 1.5oz 3/4" Red Cherry	Tomato
Lycopersicon	cerasiforme-rubrum-medium	rideau-sweetii	Rideau Sweet	Medium 1.5oz 3/4" Red Cherry	Tomato
Lycopersicon	cerasiforme-rubrum-medium	santa	Santa	Medium 1.5oz 3/4" Red Cherry	Tomato
Lycopersicon	cerasiforme-rubrum-medium	sausalito-cocktailii	Sausalito Cocktail	Medium 1.5oz 3/4" Red Cherry	Tomato
Lycopersicon	cerasiforme-rubrum-medium	short-n-sweetii	Short N Sweet	Medium 1.5oz 3/4" Red Cherry	Tomato
Lycopersicon	cerasiforme-rubrum-medium	siberiana	Siberian	Medium 1.5oz 3/4" Red Cherry	Tomato
Lycopersicon	cerasiforme-rubrum-medium	slava	Glory	Medium 1.5oz 3/4" Red Cherry	Tomato
Lycopersicon	cerasiforme-rubrum-medium	small-wonderi	Small Wonder	Medium 1.5oz 3/4" Red Cherry	Tomato
Lycopersicon	cerasiforme-rubrum-medium	subarctica-25	#25 SubArctic	Medium 1.5oz 3/4" Red Cherry	Tomato
Lycopersicon	cerasiforme-rubrum-medium	subarctica-plentyi	Plenty SubArctic	Medium 1.5oz 3/4" Red Cherry	Tomato
Lycopersicon	cerasiforme-rubrum-medium	sweetie-supremei	Sweetie Supreme	Medium 1.5oz 3/4" Red Cherry	Tomato
Lycopersicon	cerasiforme-rubrum-medium	sweetii-100	Sweet #100	Medium 1.5oz 3/4" Red Cherry	Tomato
Lycopersicon	cerasiforme-rubrum-medium	sweetii-100-OP	Open-Pollinated Sweet #100	Medium 1.5oz 3/4" Red Cherry	Tomato
Lycopersicon	cerasiforme-rubrum-medium	sweetii-billionii	Sweet Billion	Medium 1.5oz 3/4" Red Cherry	Tomato
Lycopersicon	cerasiforme-rubrum-medium	sweetii-millionii	Sweet Million	Medium 1.5oz 3/4" Red Cherry	Tomato
Lycopersicon	cerasiforme-rubrum-medium	sweetii-olivei	Sweet Olive	Medium 1.5oz 3/4" Red Cherry	Tomato
Lycopersicon	cerasiforme-rubrum-medium	tonadose-des-coronesi	Tonadose Des Corones	Medium 1.5oz 3/4" Red Cherry	Tomato
Lycopersicon	cerasiforme-rubrum-medium	totemii	Totem	Medium 1.5oz 3/4" Red Cherry	Tomato
Lycopersicon	cerasiforme-rubrum-medium	TSAC	Tender Sweet American Cherry	Medium 1.5oz 3/4" Red Cherry	Tomato
Lycopersicon	cerasiforme-rubrum-medium	vallee-national-oaxaca	Vallee National Oaxaca	Medium 1.5oz 3/4" Red Cherry	Tomato
Lycopersicon	cerasiforme-rubrum-medium-parvum	medium-parvum	Original	Small Medium 3/4oz-1oz 1/2" Red Cherry	Tomato
Lycopersicon	cerasiforme-rubrum-medium-parvum	abc-potato-leafii	ABC Potato Leaf	Small Medium 3/4oz-1oz 1/2" Red Cherry	Tomato
Lycopersicon	cerasiforme-rubrum-medium-parvum	cherry-foxii	Fox	Small Medium 3/4oz-1oz 1/2" Red Cherry	Tomato
Lycopersicon	cerasiforme-rubrum-medium-parvum	cherry-swanderi	Swander	Small Medium 3/4oz-1oz 1/2" Red Cherry	Tomato
Lycopersicon	cerasiforme-rubrum-medium-parvum	cherry-washingtonensis	Washington	Small Medium 3/4oz-1oz 1/2" Red Cherry	Tomato
Lycopersicon	cerasiforme-rubrum-medium-parvum	elfinii	Elfin	Small Medium 3/4oz-1oz 1/2" Red Cherry	Tomato
Lycopersicon	cerasiforme-rubrum-medium-parvum	little-julia	Little Julia	Small Medium 3/4oz-1oz 1/2" Red Cherry	Tomato
Lycopersicon	cerasiforme-rubrum-medium-parvum	pear-shaped-redii	Pear Shaped	Small Medium 3/4oz-1oz 1/2" Red Cherry	Tomato
Lycopersicon	cerasiforme-rubrum-medium-parvum	PSR-100-2975	#100 Peters Seed & Research	Small Medium 3/4oz-1oz 1/2" Red Cherry	Tomato
Lycopersicon	cerasiforme-rubrum-medium-parvum	PSR-958	#958 Peters Seed & Research	Small Medium 3/4oz-1oz 1/2" Red Cherry	Tomato
Lycopersicon	cerasiforme-rubrum-medium-parvum	red-alertii	Red Alert	Small Medium 3/4oz-1oz 1/2" Red Cherry	Tomato
Lycopersicon	cerasiforme-rubrum-medium-parvum	shoshonei	Shoshone	Small Medium 3/4oz-1oz 1/2" Red Cherry	Tomato
Lycopersicon	cerasiforme-rubrum-medium-parvum	subartica-earlyi	Early SubArctic	Small Medium 3/4oz-1oz 1/2" Red Cherry	Tomato
Lycopersicon	cerasiforme-rubrum-parvum	pop-inii	Pop In	Small 1/2oz 1/4" Red Cherry	Tomato
Lycopersicon	cerasiforme-rubrum-parvum	spoonii	Spoon	Small 1/2oz 1/4" Red Cherry	Tomato
Lycopersicon	cerasiforme-rubrum-parvum	subarctica-cherryi	Cherry SubArctic	Small 1/2oz 1/4" Red Cherry	Tomato
Lycopersicon	cerasiforme-rubrum-minutum	adorationii	Adoration	Tiny 1/4oz 1/8" Red Cherry	Tomato
Lycopersicon	cerasiforme-rubrum-minutum	hawaiiana	Hawaiian	Tiny 1/4oz 1/8" Red Cherry	Tomato
Lycopersicon	cerasiforme-rubrum-minutum	IM-83-1-tallii	IM 83-1 Tall	Tiny 1/4oz 1/8" Red Cherry	Tomato
Lycopersicon	cerasiforme-rubrum-minutum	micro-tomii	Micro Tom	Tiny 1/4oz 1/8" Red Cherry	Tomato
Lycopersicon	cerasiforme-rubrum-minutum	red-emperori	Red Emperor	Tiny 1/4oz 1/8" Red Cherry	Tomato
Lycopersicon	cerasiforme-rubrum-aurantium	aurigai	Auriga 2oz	Orange Red Cherry	Tomato
Lycopersicon	cerasiforme-rubrum-aurantium	break-o-dayi	Break Of Day 6oz	Orange Red Cherry	Tomato
Lycopersicon	cerasiforme-rubrum-aurantium	caro-redii	Caro 6oz	Orange Red Cherry	Tomato
Lycopersicon	cerasiforme-rubrum-aurantium	cherry-saladii	Cherry Salad 1.0"	Orange Red Cherry	Tomato

Genus	Species	Infraspecies	cultivar	specimen	Genome
Lycopersicon	*cerasiforme-rubrum-aurantium*	*cherry-stan's-bestii*	Stan's Best 1.5"	Orange Red Cherry	Tomato
Lycopersicon	*cerasiforme-rubrum-aurantium*	*dourne-de-hiveri*	Dourne D'hiver 1.75"	Orange Red Cherry	Tomato
Lycopersicon	*cerasiforme-rubrum-aurantium*	*gladness-of-gardenii*	Gladness Of Garden ?oz	Orange Red Cherry	Tomato
Lycopersicon	*cerasiforme-rubrum-aurantium*	*lee-williams-redii*	Lee Williams 1.5"	Orange Red Cherry	Tomato
Lycopersicon	*cerasiforme-rubrum-aurantium*	*red-robinii*	Red Robin 1.25"	Orange Red Cherry	Tomato
Lycopersicon	*cerasiforme-striatum*	*arctic-lightsi*	Arctic Lights 3oz	Striped Cherry	Tomato
Lycopersicon	*cerasiforme-striatum*	*black-zebrai*	Black Zebra 1.5"	Striped Cherry	Tomato
Lycopersicon	*cerasiforme-striatum*	*centerpiecei*	Centerpiece 1.75oz	Striped Cherry	Tomato
Lycopersicon	*cerasiforme-striatum*	*heavy-stripesi*	Heavy Stripes 2.5oz	Striped Cherry	Tomato
Lycopersicon	*cerasiforme-striatum*	*mr.-stripeyi*	Mr. Stripey 1.5"	Striped Cherry	Tomato
Lycopersicon	*cerasiforme-striatum*	*tigerette-cherryi*	Tigerette ?"	Striped Cherry	Tomato
Lycopersicon	*cerasiforme-striatum*	*tiny-tigeri*	Tiny Tiger 1.0"	Striped Cherry	Tomato
Lycopersicon	*cheesmanii*	*cheesmanii*	Original	Cheesman	Tomato
Lycopersicon	*cheesmanii*	*minor*	Lesser	Cheesman	Tomato

Tomato, White (*Lycopersicon esculentum-album*)

Genus	Species	Infraspecies	cultivar	specimen	Genome
Lycopersicon	*esculentum-album*	*album*	Original ?oz	White	Tomato
Lycopersicon	*esculentum-album*	*beaute-blanche-du-canadai*	Beaute Blanche Du Canada 5oz	White	Tomato
Lycopersicon	*esculentum-album*	*duggin-whitei*	Duggin White 10oz	White	Tomato
Lycopersicon	*esculentum-album*	*etoile-blanche-d'anversi*	Etoile Blanche D'Anvers ?oz	White	Tomato
Lycopersicon	*esculentum-album*	*little-whitei*	Little White ?oz	White	Tomato
Lycopersicon	*esculentum-album*	*oliver's-german-whitei*	Oliver's German White ?oz	White	Tomato
Lycopersicon	*esculentum-album*	*pink-whitei*	Pink White 1.5lb	White	Tomato
Lycopersicon	*esculentum-album*	*potato-leaf-whitei*	Potato Leaf White ?oz	White	Tomato
Lycopersicon	*esculentum-album*	*snow-white-supra*	Super Snow White ?oz	White	Tomato
Lycopersicon	*esculentum-album*	*snowballii*	Snowball 3.25"	White	Tomato
Lycopersicon	*esculentum-album*	*snowballii-white-beautyi*	White Beauty Snowball 8oz	White	Tomato
Lycopersicon	*esculentum-album*	*sun-&-snowii*	Sun & Snow 4oz-8oz	White	Tomato
Lycopersicon	*esculentum-album*	*suttonii*	Sutton 8oz	White	Tomato
Lycopersicon	*esculentum-album*	*transparentii*	Transparent ?oz Flat	White	Tomato
Lycopersicon	*esculentum-album*	*white-bushii*	White Bush 2"	White	Tomato
Lycopersicon	*esculentum-album*	*white-heirloomii*	White Heirloom 4oz-12oz	White	Tomato
Lycopersicon	*esculentum-album*	*white-potato-leafii*	White Potato Leaf 4.5"	White	Tomato
Lycopersicon	*esculentum-album*	*white-princessi*	White Princess 1.5lb	White	Tomato
Lycopersicon	*esculentum-album*	*white-wonderi*	White Wonder 4oz	White	Tomato

Tomato, Orange (*Lycopersicon esculentum-aurantium*)

Genus	Species	Infraspecies	cultivar	specimen	Genome
Lycopersicon	*esculentum-aurantium*	*bursztynii*	Bursztyn Flat 2.5"	Orange	Tomato
Lycopersicon	*esculentum-aurantium*	*carobetta*	Carobetta ?.?"	Orange	Tomato
Lycopersicon	*esculentum-aurantium*	*earl-of-edgecombei*	Earl Of Edgecombe 3.0"	Orange	Tomato
Lycopersicon	*esculentum-aurantium*	*flammei*	Flamme 1.5"	Orange	Tomato
Lycopersicon	*esculentum-aurantium*	*german-largei*	German Large ?.?"	Orange	Tomato
Lycopersicon	*esculentum-aurantium*	*gold-dustii*	Gold Dust ?.?"	Orange	Tomato
Lycopersicon	*esculentum-aurantium*	*gramma-climen-hagenii*	Gramma Climen Hagen 3.0"	Orange	Tomato
Lycopersicon	*esculentum-aurantium*	*jaffai*	Jaffa ?.?"	Orange	Tomato
Lycopersicon	*esculentum-aurantium*	*jubilee-treasurei*	Treasure Jubilee 9oz	Orange	Tomato
Lycopersicon	*esculentum-aurantium*	*jumbo-jim-orangei*	Jumbo Jim 15oz	Orange	Tomato
Lycopersicon	*esculentum-aurantium*	*moon-glowi*	Moon Glow 7oz	Orange	Tomato
Lycopersicon	*esculentum-aurantium*	*nebraska-weddingii*	Nebraska Wedding 3.5"	Orange	Tomato
Lycopersicon	*esculentum-aurantium*	*orangei*	Vavilov Flat 4.0"	Orange	Tomato
Lycopersicon	*esculentum-aurantium*	*orange-crimea*	Orange Crimea 8oz	Orange	Tomato
Lycopersicon	*esculentum-aurantium*	*orange-glowi*	Orange Glow ?.?"	Orange	Tomato
Lycopersicon	*esculentum-aurantium*	*patio-orangei*	Patio Flat ?.?"	Orange	Tomato
Lycopersicon	*esculentum-aurantium*	*peg's-round-orangei*	Peg's Round 2.5"	Orange	Tomato
Lycopersicon	*esculentum-aurantium*	*PSR-941S*	#941S Peters Seed & Research) 1/2lb	Orange	Tomato
Lycopersicon	*esculentum-aurantium*	*PSR-941SS*	#941SS Peters Seed & Research 1/2lb	Orange	Tomato
Lycopersicon	*esculentum-aurantium*	*rosalie's-sweet-horizonii*	Rosalie's Sweet Horizon ?.?"	Orange	Tomato
Lycopersicon	*esculentum-aurantium*	*russian-orangei-117*	Russian #117	Orange	Tomato
Lycopersicon	*esculentum-aurantium*	*russian-persimmonii*	Russian Persimmon 3.0"	Orange	Tomato
Lycopersicon	*esculentum-aurantium*	*sibirishe-orangei*	Sibirishe ?.?"	Orange	Tomato
Lycopersicon	*esculentum-aurantium*	*small-lapii*	Small Lap Flat ?.?"	Orange	Tomato
Lycopersicon	*esculentum-aurantium*	*sundropii*	Sundrop 1.7"	Orange	Tomato
Lycopersicon	*esculentum-aurantium*	*tangella*	Tangella 2.0"	Orange	Tomato
Lycopersicon	*esculentum-aurantium*	*valencia*	Valencia 10oz	Orange	Tomato
Lycopersicon	*esculentum-aurantium*	*woodle-orangei*	Woodle ?.?"	Orange	Tomato

Tomato, Yellow Orange (*Lycopersicon esculentum-aurantium-helvum*)

Genus	Species	Infraspecies	cultivar	specimen	Genome
Lycopersicon	*esculentum-aurantium-helvum*	*alleghany-sunsetii*	Alleghany Sunset	Yellow Orange	Tomato
Lycopersicon	*esculentum-aurantium-helvum*	*arkansas-marvelii*	Arkansas Marvel 1lb	Yellow Orange	Tomato

Genus	Species	Infraspecies	cultivar	specimen	Genome
Lycopersicon	*esculentum-aurantium-helvum*	*big-yellowii*	Big Yellow Flat 9oz	Yellow Orange	Tomato
Lycopersicon	*esculentum-aurantium-helvum*	*djena-lee's-golden-girlii*	Djena Lee's Golden Girl ?.?"	Yellow Orange	Tomato
Lycopersicon	*esculentum-aurantium-helvum*	*garden-peach-fuzzyi*	Garden Peach Fuzzy ?.?"	Yellow Orange	Tomato
Lycopersicon	*esculentum-aurantium-helvum*	*golden-queenii*	Golden Queen 10oz	Yellow Orange	Tomato
Lycopersicon	*esculentum-aurantium-helvum*	*golden-queenii-sunrayi*	Sun Ray Golden Queen ??oz	Yellow Orange	Tomato
Lycopersicon	*esculentum-aurantium-helvum*	*golden-russian-ballii*	Golden Russian Ball ?.?"	Yellow Orange	Tomato
Lycopersicon	*esculentum-aurantium-helvum*	*ida-goldii*	Idaho (Ida) Gold 2.5oz	Yellow Orange	Tomato
Lycopersicon	*esculentum-aurantium-helvum*	*jubileei*	Jubilee Flat 3.0"	Yellow Orange	Tomato
Lycopersicon	*esculentum-aurantium-helvum*	*mountain-goldii*	Mountain Gold 3.5"	Yellow Orange	Tomato
Lycopersicon	*esculentum-aurantium-helvum*	*northern-lightsi*	Northern Lights 8oz-12oz	Yellow Orange	Tomato
Lycopersicon	*esculentum-aurantium-helvum*	*olga's-yellow-round-chickenii*	Olga's Yellow Round Chicken 6oz	Yellow Orange	Tomato
Lycopersicon	*esculentum-aurantium-helvum*	*pink-lemonii*	Pink Lemon 8oz-10oz	Yellow Orange	Tomato
Lycopersicon	*esculentum-aurantium-helvum*	*red-yellow-capii*	Red Yellow Cap 3.5"	Yellow Orange	Tomato
Lycopersicon	*esculentum-aurantium-helvum*	*stor-gulii*	Stor Gul 12oz	Yellow Orange	Tomato
Lycopersicon	*esculentum-aurantium-helvum*	*sunrayi*	Sunray Flat 3.0"	Yellow Orange	Tomato
Lycopersicon	*esculentum-aurantium-rubrum*	*burracker's-favoritei*	Burracker's Favorite 1.5lb	Red Orange	Tomato
Lycopersicon	*esculentum-aurantium-rubrum*	*la-carotina*	La Carotina 2" 7oz	Red Orange	Tomato
Lycopersicon	*esculentum-brunneum*	*black-sea-manii*	Black Sea Man 7oz	Brown	Tomato
Lycopersicon	*esculentum-brunneum*	*backstari*	Blackstar 16oz	Brown	Tomato
Lycopersicon	*esculentum-brunneum*	*bordo*	Bordo 8oz	Brown	Tomato
Lycopersicon	*esculentum-brunneum*	*brown-fleshii*	Brown Flesh ?oz	Brown	Tomato
Lycopersicon	*esculentum-brunneum*	*cherokee-chocolatei*	Cherokee Chocolate ?oz	Brown	Tomato
Lycopersicon	*esculentum-brunneum*	*giant-fioletii*	Giant Fiolet 10oz Flat	Brown	Tomato
Lycopersicon	*esculentum-brunneum*	*indianii*	Indian 8oz	Brown	Tomato
Lycopersicon	*esculentum-brunneum*	*indische-fleischii*	Indische Fleisch 6oz-16oz	Brown	Tomato
Lycopersicon	*esculentum-brunneum*	*mr.-brownii*	Mr. Brown ?oz	Brown	Tomato
Lycopersicon	*esculentum-brunneum*	*nyagousi*	Nyagous 6oz Flat	Brown	Tomato
Lycopersicon	*esculentum-brunneum*	*paul-robesonii*	Paul Robeson ?oz	Brown	Tomato
Lycopersicon	*esculentum-brunneum*	*poll-robsonii-angolana*	Poll Robson Angolan 6oz	Brown	Tomato

Tomato, Green (*Lycopersicon esculentum-groenum*)

Genus	Species	Infraspecies	cultivar	specimen	Genome
Lycopersicon	*esculentum-groenum*	*groenum*	Original	Green	Tomato
Lycopersicon	*esculentum-groenum*	*dorothy's-greenii*	Dorothy's 1lb	Green	Tomato
Lycopersicon	*esculentum-groenum*	*evergreenii*	Evergreen 14oz	Green	Tomato
Lycopersicon	*esculentum-groenum*	*evergreenii-improva*	Improved Evergreen 14oz	Green	Tomato
Lycopersicon	*esculentum-groenum*	*forever-greenii*	Forever Green 7oz	Green	Tomato
Lycopersicon	*esculentum-groenum*	*garden-limei*	Garden Lime 12oz Flat	Green	Tomato
Lycopersicon	*esculentum-groenum*	*gold-n-greenii*	Gold N Green ?oz	Green	Tomato
Lycopersicon	*esculentum-groenum*	*grandma-oliver's-greenii*	Grandma Oliver's Green 10oz	Green	Tomato
Lycopersicon	*esculentum-groenum*	*granny-smithii*	Granny Smith 8oz	Green	Tomato
Lycopersicon	*esculentum-groenum*	*green-&-yellowi*	Green & Yellow Hollow	Green	Tomato
Lycopersicon	*esculentum-groenum*	*green-applei*	Green Apple ?oz	Green	Tomato
Lycopersicon	*esculentum-groenum*	*green-bell-pepperi*	Green Bell Pepper ?oz Hollow	Green	Tomato
Lycopersicon	*esculentum-groenum*	*green-thumbii*	Green Thumb 4oz-6oz	Green	Tomato
Lycopersicon	*esculentum-groenum*	*greenwichii*	Greenwich ?oz	Green	Tomato
Lycopersicon	*esculentum-groenum*	*liberty-bellii*	Liberty Bell 4.5oz	Green	Tomato
Lycopersicon	*esculentum-groenum*	*moldovan-greenii*	Moldovan Green 10oz	Green	Tomato
Lycopersicon	*esculentum-groenum*	*mystery-keeperi*	Mystery Keeper ?oz	Green	Tomato
Lycopersicon	*esculentum-groenum*	*thompson-seedless-grapei*	Thompson Seedless Grape ?oz	Green	Tomato

Tomato, Yellow (*Lycopersicon esculentum-helvum*)

Genus	Species	Infraspecies	cultivar	specimen	Genome
Lycopersicon	*esculentum-helvum*	*ananasi*	Ananas 3.5"	Yellow	Tomato
Lycopersicon	*esculentum-helvum*	*aranyalmai*	Aranyalma 3.5oz	Yellow	Tomato
Lycopersicon	*esculentum-helvum*	*azoychkai*	Azoychka Flat 3.2"	Yellow	Tomato
Lycopersicon	*esculentum-helvum*	*balconi-yellowi*	Balconi ?.?"	Yellow	Tomato
Lycopersicon	*esculentum-helvum*	*basinga*	Basinga 5.0"	Yellow	Tomato
Lycopersicon	*esculentum-helvum*	*brandywinei-yellowi*	Yellow Amish Brandywine Creek 2lb	Yellow	Tomato
Lycopersicon	*esculentum-helvum*	*ceses-russianii*	Ceses Russian 2.5"	Yellow	Tomato
Lycopersicon	*esculentum-helvum*	*dad's-sunsetii*	Dad's Sunset 5oz	Yellow	Tomato
Lycopersicon	*esculentum-helvum*	*dakota-goldii*	Dakota 3.0"	Yellow	Tomato
Lycopersicon	*esculentum-helvum*	*dicoff's-yellowi*	Dicoff's 1lb	Yellow	Tomato
Lycopersicon	*esculentum-helvum*	*donii*	Doni Seedless 1/2lb	Yellow	Tomato
Lycopersicon	*esculentum-helvum*	*dwarf-yellow-championii*	Dwarf Yellow Champion ?lb	Yellow	Tomato
Lycopersicon	*esculentum-helvum*	*fireworks-yellowi*	Fireworks ?.?"	Yellow	Tomato
Lycopersicon	*esculentum-helvum*	*germanii-oldii*	Old German 1lb	Yellow	Tomato
Lycopersicon	*esculentum-helvum*	*gold-medal-quisemberryi*	Quinsenberry Gold Medal 1.5lb	Yellow	Tomato
Lycopersicon	*esculentum-helvum*	*gold-medal-yellowi*	Gold Medal 3.0"	Yellow	Tomato
Lycopersicon	*esculentum-helvum*	*gold-rushii*	Gold Rush ?.?"	Yellow	Tomato
Lycopersicon	*esculentum-helvum*	*golden-boyi*	Golden Boy	Yellow	Tomato
Lycopersicon	*esculentum-helvum*	*golden-delightii*	Golden Delight 1/2lb	Yellow	Tomato
Lycopersicon	*esculentum-helvum*	*golden-dwarf-championii*	Golden Dwarf Champion ?.?"	Yellow	Tomato

Genus	Species	Infraspecies	cultivar	specimen	Genome
Lycopersicon	*esculentum-helvum*	*golden-monarchii*	Golden Monarch 3.0" Flat	Yellow	Tomato
Lycopersicon	*esculentum-helvum*	*golden-queenii*	Golden Queen ?lb	Yellow	Tomato
Lycopersicon	*esculentum-helvum*	*golden-queenii-livingston'si*	Livingston's Golden Queen ?lb	Yellow	Tomato
Lycopersicon	*esculentum-helvum*	*golden-sunrisei*	Golden Sunrise ?.?"	Yellow	Tomato
Lycopersicon	*esculentum-helvum*	*golden-treasurei*	Golden Treasure 2.5"	Yellow	Tomato
Lycopersicon	*esculentum-helvum*	*golden-treasurei-UR*	Uniform Ripening Golden Treasure 2.5"	Yellow	Tomato
Lycopersicon	*esculentum-helvum*	*goldene-konigenii*	Golden King 2.0"	Yellow	Tomato
Lycopersicon	*esculentum-helvum*	*goldiei*	Goldie ?.?" Flat	Yellow	Tomato
Lycopersicon	*esculentum-helvum*	*goldiei-boyi*	Boy Goldie ?.?" Flat	Yellow	Tomato
Lycopersicon	*esculentum-helvum*	*hahnstown-yellowi*	Hahnstown ?.?"	Yellow	Tomato
Lycopersicon	*esculentum-helvum*	*hessii*	Hess 14oz	Yellow	Tomato
Lycopersicon	*esculentum-helvum*	*hessii-esther-yellowi*	Esther Hess 1.5"	Yellow	Tomato
Lycopersicon	*esculentum-helvum*	*holy-land-yellow-strainii*	Holy Land ?lb	Yellow	Tomato
Lycopersicon	*esculentum-helvum*	*hughsi*	Hughs 2.5lb	Yellow	Tomato
Lycopersicon	*esculentum-helvum*	*jaune-negibii*	Juane Negib Flat 3.0"	Yellow	Tomato
Lycopersicon	*esculentum-helvum*	*kimberton-hillsi*	Kimberton Hills ?.?"	Yellow	Tomato
Lycopersicon	*esculentum-helvum*	*lillian's-yellow-heirloomii*	Lillian's Yellow Heirloom 16oz	Yellow	Tomato
Lycopersicon	*esculentum-helvum*	*manyelii*	Many Moons Flat 10oz	Yellow	Tomato
Lycopersicon	*esculentum-helvum*	*marizol-goldii*	Marizol Gold 1.5lb	Yellow	Tomato
Lycopersicon	*esculentum-helvum*	*mary-robinson's-german-bicolor*	Mary Robinson's German Bicolor 1lb	Yellow	Tomato
Lycopersicon	*esculentum-helvum*	*native-sunii*	Native Sun 8oz	Yellow	Tomato
Lycopersicon	*esculentum-helvum*	*new-sunii*	New Sun 12oz	Yellow	Tomato
Lycopersicon	*esculentum-helvum*	*northern-sunii*	Northern Sun ?.?"	Yellow	Tomato
Lycopersicon	*esculentum-helvum*	*old-wyandottei*	Old Wyandotte Flat 9oz	Yellow	Tomato
Lycopersicon	*esculentum-helvum*	*paskal-yellowi*	Paskal ?.?"	Yellow	Tomato
Lycopersicon	*esculentum-helvum*	*ponderosa-sunrisei*	Ponderosa Sunrise ?oz	Yellow	Tomato
Lycopersicon	*esculentum-helvum*	*red-bellyi*	Red Belly 2lb	Yellow	Tomato
Lycopersicon	*esculentum-helvum*	*ruby-rakes-yellowi*	RubyRakes 2.2"	Yellow	Tomato
Lycopersicon	*esculentum-helvum*	*ruffled-yellowi*	Ruffled Yellow Hollow ?.?"	Yellow	Tomato
Lycopersicon	*esculentum-helvum*	*russian-yellowi*	Russian Flat 3.0"	Yellow	Tomato
Lycopersicon	*esculentum-helvum*	*scotland-yellowi*	Scotland 3.5oz	Yellow	Tomato
Lycopersicon	*esculentum-helvum*	*sun-freei*	Sun Free 6oz	Yellow	Tomato
Lycopersicon	*esculentum-helvum*	*taxii*	Taxi 2.5"	Yellow	Tomato
Lycopersicon	*esculentum-helvum*	*tommy-toe-yellowi*	Tommy Toe ?.?"	Yellow	Tomato
Lycopersicon	*esculentum-helvum*	*wapsipinicon-peachii*	Wapsipinicon Peach 2.0"	Yellow	Tomato
Lycopersicon	*esculentum-helvum*	*wendyi*	Wendy 2.0"	Yellow	Tomato
Lycopersicon	*esculentum-helvum*	*winter-goldii*	Winter Gold 4oz-8oz	Yellow	Tomato
Lycopersicon	*esculentum-helvum*	*wonder-lightii*	Wonder Light 2.5"	Yellow	Tomato
Lycopersicon	*esculentum-helvum*	*yellow-deliciousi*	Yellow Delicious ?.?"	Yellow	Tomato
Lycopersicon	*esculentum-helvum*	*yellow-heartstock*	Yellow Heartstock ?oz	Yellow	Tomato
Lycopersicon	*esculentum-helvum*	*yellow-peachii*	Yellow Peach 1.5"	Yellow	Tomato
Lycopersicon	*esculentum-helvum*	*yellow-perfectionii*	Yellow Perfection ?.?"	Yellow	Tomato
Lycopersicon	*esculentum-helvum*	*yellow-perfectionii-unwinsi*	Unwins Yellow Perfection 2oz	Yellow	Tomato
Lycopersicon	*esculentum-helvum*	*yellow-pygmyi*	Yellow Pygmy ?.?"	Yellow	Tomato
Lycopersicon	*esculentum-helvum*	*yellow-rufflesi*	Yellow Ruffles 3.0"	Yellow	Tomato
Lycopersicon	*esculentum-helvum*	*yellow-stufferi*	Yellow Stuffer 3.0"	Yellow	Tomato

Tomato, Orange Yellow (*Lycopersicon esculentum-helvum-aurantium*)

Genus	Species	Infraspecies	cultivar	specimen	Genome
Lycopersicon	*esculentum-helvum-aurantium*	*amber-coloredii*	Amber Colored 2.0"	Orange Yellow	Tomato
Lycopersicon	*esculentum-helvum-aurantium*	*amish-large-yellowi*	Large Yellow Amish 4.0"	Orange Yellow	Tomato
Lycopersicon	*esculentum-helvum-aurantium*	*aunt-gertie's-goldii*	Aunt Gertie's Gold 1lb	Orange Yellow	Tomato
Lycopersicon	*esculentum-helvum-aurantium*	*brandywinei-platfoot-yellowi*	Platfoot Amish Brandywine Creek 3.5"	Orange Yellow	Tomato
Lycopersicon	*esculentum-helvum-aurantium*	*burgess-colossal-goldenii*	Burgess Colossal Flat 1lb	Orange Yellow	Tomato
Lycopersicon	*esculentum-helvum-aurantium*	*czech's-excellent-yellowi*	Czech's Excellent 3.0"	Orange Yellow	Tomato
Lycopersicon	*esculentum-helvum-aurantium*	*de-barao-goldii*	De Barao Gold 2oz	Orange Yellow	Tomato
Lycopersicon	*esculentum-helvum-aurantium*	*dixie-golden-giantii*	Dixie Golden Giant 2.2lb	Orange Yellow	Tomato
Lycopersicon	*esculentum-helvum-aurantium*	*dr.-wyche's-yellowi*	Dr. Wyche's Flat 1lb	Orange Yellow	Tomato
Lycopersicon	*esculentum-helvum-aurantium*	*german-golden-ribbedii*	German Golden Ribbed ?.?"	Orange Yellow	Tomato
Lycopersicon	*esculentum-helvum-aurantium*	*evergreenii*	Evergreen	Orange Yellow	Tomato
Lycopersicon	*esculentum-helvum-aurantium*	*giant-belgium-yellowi*	Giant Belgium Yellow ?.?"	Orange Yellow	Tomato
Lycopersicon	*esculentum-helvum-aurantium*	*golden-sunburstii*	Golden Sunburst 3.0"	Orange Yellow	Tomato
Lycopersicon	*esculentum-helvum-aurantium*	*hazel-goldii*	Hazel Gold 4.0"	Orange Yellow	Tomato
Lycopersicon	*esculentum-helvum-aurantium*	*hourmai*	Hourma 6oz	Orange Yellow	Tomato
Lycopersicon	*esculentum-helvum-aurantium*	*indian-moonii*	Indian Moon ?.?"	Orange Yellow	Tomato
Lycopersicon	*esculentum-helvum-aurantium*	*jersey-goldenii*	Golden Jersey ?.?"	Orange Yellow	Tomato
Lycopersicon	*esculentum-helvum-aurantium*	*goldenponderosa*	Railroad	Orange Yellow	Tomato
Lycopersicon	*esculentum-helvum-aurantium*	*lemon-boyi*	Lemon Boy 3.5"	Orange Yellow	Tomato
Lycopersicon	*esculentum-helvum-aurantium*	*mammoth-german-goldii*	Mammoth German Gold ?.?"	Orange Yellow	Tomato
Lycopersicon	*esculentum-helvum-aurantium*	*mordenii*	Morden ?.?"	Orange Yellow	Tomato
Lycopersicon	*esculentum-helvum-aurantium*	*mortgage-lifter-yellowi*	Mortgage Lifter Flat 4.0"	Orange Yellow	Tomato

Genus	Species	Infraspecies	cultivar	specimen	Genome
Lycopersicon	*esculentum-helvum-aurantium*	*paragon-yellowi*	Paragon 5oz		Orange Yellow Tomato
Lycopersicon	*esculentum-helvum-aurantium*	*polonaise-juanei*	Pononaise Juane 7oz		Orange Yellow Tomato
Lycopersicon	*esculentum-helvum-aurantium*	*ponderosa-goldenii*	Ponderosa Golden 1lb		Orange Yellow Tomato
Lycopersicon	*esculentum-helvum-aurantium*	*potato-leaf-yellowi*	Potato Leaf Flat 2lb		Orange Yellow Tomato
Lycopersicon	*esculentum-helvum-aurantium*	*sweet-gold-MR*	Sweet Gold Mildew Resistant ?.?"		Orange Yellow Tomato
Lycopersicon	*esculentum-helvum-aurantium*	*tiger-pawi*	Tiger Paw Flat ?.?"		Orange Yellow Tomato
Lycopersicon	*esculentum-helvum-aurantium*	*west-viriginia-yellowi*	West Virgina 1lb		Orange Yellow Tomato
Lycopersicon	*esculentum-helvum-aurantium*	*yellow-belgiumii*	Belgium Goldie 14oz		Orange Yellow Tomato
Lycopersicon	*esculentum-helvum-aurantium*	*yellow-ping-pongii*	Yellow Ping Pong 2.0"		Orange Yellow Tomato
Lycopersicon	*esculentum-helvum-aurantium*	*yellow-stonei*	Yellow Stone 12oz		Orange Yellow Tomato

Tomato, Black (*Lycopersicon esculentum-nigrum*)

Genus	Species	Infraspecies	cultivar	specimen	Genome
Lycopersicon	*esculentum-nigrum*	*nigrum*	Original 4oz-16oz Round		Black Tomato
Lycopersicon	*esculentum-nigrum*	*amazon-chocolatei*	Amazon Chocolate 1lb Round		Black Tomato
Lycopersicon	*esculentum-nigrum*	*black-aisbergii*	Black Aisberg 6oz Beefsteak		Black Tomato
Lycopersicon	*esculentum-nigrum*	*black-brandywinei*	Black Amish Brandywine Creek 14oz Plum		Black Tomato
Lycopersicon	*esculentum-nigrum*	*black-cubana*	Black Cuban 6oz Flat		Black Tomato
Lycopersicon	*esculentum-nigrum*	*black-earlyi*	Black Early 9oz Flat		Black Tomato
Lycopersicon	*esculentum-nigrum*	*black-ethiopiana*	Black Ethiopian 5oz Plum		Black Tomato
Lycopersicon	*esculentum-nigrum*	*black-from-tulaensis*	Black From Tula 10oz Flat		Black Tomato
Lycopersicon	*esculentum-nigrum*	*black-germanica*	Black German ?oz		Black Tomato
Lycopersicon	*esculentum-nigrum*	*black-krimii*	Black Krim 10oz Flat		Black Tomato
Lycopersicon	*esculentum-nigrum*	*black-marizolii*	Black Marizol 4"		Black Tomato
Lycopersicon	*esculentum-nigrum*	*black-peari*	Black Pear 8oz Plum		Black Tomato
Lycopersicon	*esculentum-nigrum*	*black-plumii*	Black 2" Plum		Black Tomato
Lycopersicon	*esculentum-nigrum*	*black-princei*	Black Prince 5oz Round		Black Tomato
Lycopersicon	*esculentum-nigrum*	*black-russiana*	Black Russian ?oz Round		Black Tomato
Lycopersicon	*esculentum-nigrum*	*black-sarai*	Black Sara ?oz		Black Tomato
Lycopersicon	*esculentum-nigrum*	*black-sloveniana*	Black Slovenian ?oz		Black Tomato
Lycopersicon	*esculentum-nigrum*	*bluei*	Blue 8oz Round		Black Tomato
Lycopersicon	*esculentum-nigrum*	*carbonii*	Carbon 10oz Flat		Black Tomato
Lycopersicon	*esculentum-nigrum*	*noir-de-crimee*	Noir De Crimee 4oz-8oz		Black Tomato
Lycopersicon	*esculentum-nigrum*	*noire-charbonneusei*	Noire Charbonneuse 8oz-12oz		Black Tomato
Lycopersicon	*esculentum-nigrum*	*siniyi*	Siniy 8oz-12oz		Black Tomato
Lycopersicon	*esculentum-nigrum*	*southern-nightii*	Southern Night 3.5"		Black Tomato

Tomato, Pink (*Lycopersicon esculentum-pincum*)

Genus	Species	Infraspecies	cultivar	specimen	Genome
Lycopersicon	*esculentum-pincum*	*acmei*	Acme 14oz		Pink Tomato
Lycopersicon	*esculentum-pincum*	*baliensis*	Bali 2.5"		Pink Tomato
Lycopersicon	*esculentum-pincum*	*belgian-beautyi*	Belgian Beauty 1.5lb		Pink Tomato
Lycopersicon	*esculentum-pincum*	*black-mountain-pinkii*	Black Mountain 2.5lb		Pink Tomato
Lycopersicon	*esculentum-pincum*	*brandywinei*	Amish Brandywine Creek 3/4lb		Pink Tomato
Lycopersicon	*esculentum-pincum*	*brandywinei-pawer'si*	Pawer's Amish Brandywine Creek ?.?lb		Pink Tomato
Lycopersicon	*esculentum-pincum*	*brandywinei-sudduth'si*	Sudduth's Amish Brandywine Creek 1.5lb		Pink Tomato
Lycopersicon	*esculentum-pincum*	*briannai*	Brianna 4.5"		Pink Tomato
Lycopersicon	*esculentum-pincum*	*brilliant-pinkii*	Brilliant 4.0"		Pink Tomato
Lycopersicon	*esculentum-pincum*	*burcham's-new-generationii*	Burcham's New Generation ?.?"		Pink Tomato
Lycopersicon	*esculentum-pincum*	*bushy-chabaroyskyi*	Bushy Chabaroysky ?.?"		Pink Tomato
Lycopersicon	*esculentum-pincum*	*canabec-rosei*	Canabec Rose 1/2lb		Pink Tomato
Lycopersicon	*esculentum-pincum*	*caspian-pinkii*	Caspian 14oz		Pink Tomato
Lycopersicon	*esculentum-pincum*	*centuryi*	Century ?.?"		Pink Tomato
Lycopersicon	*esculentum-pincum*	*clear-pink-earlyi*	Clear Pink Early 3.0"		Pink Tomato
Lycopersicon	*esculentum-pincum*	*churchii*	Church 2lb		Pink Tomato
Lycopersicon	*esculentum-pincum*	*crimeana*	Crimean 4oz		Pink Tomato
Lycopersicon	*esculentum-pincum*	*cooper's-specialis*	Cooper's Special 3oz		Pink Tomato
Lycopersicon	*esculentum-pincum*	*county-agentii*	County Agent 2lb		Pink Tomato
Lycopersicon	*esculentum-pincum*	*crimson-jackii*	Crimson Jack 12oz		Pink Tomato
Lycopersicon	*esculentum-pincum*	*czech's-excellent-pinkii*	Czech's Excellent ?.?"		Pink Tomato
Lycopersicon	*esculentum-pincum*	*dr.-lylei*	Dr. Lyle 1.8lb		Pink Tomato
Lycopersicon	*esculentum-pincum*	*dufresnei*	Dufresne ?.?"		Pink Tomato
Lycopersicon	*esculentum-pincum*	*dufresnei-2*	#2 Dufresne 3.5"		Pink Tomato
Lycopersicon	*esculentum-pincum*	*early-detroitensis*	Early Detroit ?.?"		Pink Tomato
Lycopersicon	*esculentum-pincum*	*early-wonderi*	Early Wonder 6oz		Pink Tomato
Lycopersicon	*esculentum-pincum*	*enrolladoi*	Enrollado Ribbed ?.?"		Pink Tomato
Lycopersicon	*esculentum-pincum*	*eschelman's-pridei*	Eschelman's Pride ?.?"		Pink Tomato
Lycopersicon	*esculentum-pincum*	*festive-pinkii*	Festive 7oz		Pink Tomato
Lycopersicon	*esculentum-pincum*	*florida-pinkii*	Florida 3lb		Pink Tomato
Lycopersicon	*esculentum-pincum*	*german-johnsonii*	German Johnson 1lb		Pink Tomato
Lycopersicon	*esculentum-pincum*	*german-pinkii*	German 14oz		Pink Tomato

Genus	Species	Infraspecies	cultivar	specimen	Genome
Lycopersicon	*esculentum-pincum*	*giant-belgium*	Giant Belgium 2lb		Pink Tomato
Lycopersicon	*esculentum-pincum*	*gigantesquei*	Gigantesque Orange Red Skin 5.0"		Pink Tomato
Lycopersicon	*esculentum-pincum*	*gillogly-pinkii*	Gillogly Lobed 4.5"		Pink Tomato
Lycopersicon	*esculentum-pincum*	*glazier's-giantii*	Glazier's Giant 3.5"		Pink Tomato
Lycopersicon	*esculentum-pincum*	*glesenen*	Glesener ?.?"		Pink Tomato
Lycopersicon	*esculentum-pincum*	*gogoshai*	Gogosha 1.5lb		Pink Tomato
Lycopersicon	*esculentum-pincum*	*grace-lahman's-pinkii*	Grace Lahman's ?.?"		Pink Tomato
Lycopersicon	*esculentum-pincum*	*grapefruit-pinkii*	Pink Grapefruit 4oz-6oz		Pink Tomato
Lycopersicon	*esculentum-pincum*	*grosse-coteleei*	Gross Cotelee 3.0"		Pink Tomato
Lycopersicon	*esculentum-pincum*	*hendersonii*	Henderson 2lb		Pink Tomato
Lycopersicon	*esculentum-pincum*	*honeyi*	Honey ?.?"		Pink Tomato
Lycopersicon	*esculentum-pincum*	*ispolinii*	Ispolin 1.5lb		Pink Tomato
Lycopersicon	*esculentum-pincum*	*jan's-pinkii*	Jan's ?.?"		Pink Tomato
Lycopersicon	*esculentum-pincum*	*jefferson-giantii*	Jefferson Giant 1.5lb		Pink Tomato
Lycopersicon	*esculentum-pincum*	*jumboi*	Jumbo ?.?"		Pink Tomato
Lycopersicon	*esculentum-pincum*	*june-pinkii*	June ?.?"		Pink Tomato
Lycopersicon	*esculentum-pincum*	*kelly's-pinkii*	Kelly's ?.?"		Pink Tomato
Lycopersicon	*esculentum-pincum*	*livingston's-beautyi*	Beauty Livingston's 4.5oz		Pink Tomato
Lycopersicon	*esculentum-pincum*	*livingston's-globei*	Globe Livingston's 10oz		Pink Tomato
Lycopersicon	*esculentum-pincum*	*livingston's-magnus*	Magnus Livingston's 7oz		Pink Tomato
Lycopersicon	*esculentum-pincum*	*livingston's-maincropii*	Maincrop Livingston's ?.?"		Pink Tomato
Lycopersicon	*esculentum-pincum*	*louisiana-pinkii*	Louisiana 3.5"		Pink Tomato
Lycopersicon	*esculentum-pincum*	*mac-pinkii*	MacDonald 5oz		Pink Tomato
Lycopersicon	*esculentum-pincum*	*maritime-pinkii*	Maritime 6oz		Pink Tomato
Lycopersicon	*esculentum-pincum*	*marizol-pinkii*	Marizol 9oz		Pink Tomato
Lycopersicon	*esculentum-pincum*	*marlowe-charlestonii*	Marlowe Charleston 3.0"		Pink Tomato
Lycopersicon	*esculentum-pincum*	*mary-annii*	Mary Ann 1lb		Pink Tomato
Lycopersicon	*esculentum-pincum*	*mexicana*	Mexican Potato Leaf 5.0"		Pink Tomato
Lycopersicon	*esculentum-pincum*	*mexicana-ribbedii*	Mexican Ribbed ?.?"		Pink Tomato
Lycopersicon	*esculentum-pincum*	*mexicoensis*	Mexico 1lb		Pink Tomato
Lycopersicon	*esculentum-pincum*	*micado-violettor*	Micado Violettor 5oz		Pink Tomato
Lycopersicon	*esculentum-pincum*	*milgrene-rosei*	Milgrene Rose ?.?"		Pink Tomato
Lycopersicon	*esculentum-pincum*	*mission-dykei*	Mission Dyke ?.?"		Pink Tomato
Lycopersicon	*esculentum-pincum*	*mortgage-lifteri-estler'si*	Estler's Mortgage Lifter 1lb		Pink Tomato
Lycopersicon	*esculentum-pincum*	*mortgage-lifteri-mcgarity's*	McGarity's Mortgage Lifter ?lb		Pink Tomato
Lycopersicon	*esculentum-pincum*	*mortgage-lifteri-pinkii*	Pink Mortgage Lifter 1.5lb		Pink Tomato
Lycopersicon	*esculentum-pincum*	*mortgage-lifteri-radiator-charlie'si*	Radiator Charlie's Mortgage Lifter 3lb		Pink Tomato
Lycopersicon	*esculentum-pincum*	*mortgage-lifteri-reiger's*	Reiger's Mortgage Lifter ?lb		Pink Tomato
Lycopersicon	*esculentum-pincum*	*mother-russica*	Mother Russia 11oz		Pink Tomato
Lycopersicon	*esculentum-pincum*	*mrs.-bensonii*	Mrs. Benson 1lb		Pink Tomato
Lycopersicon	*esculentum-pincum*	*nectar-rosei*	Nectar Rose 5oz		Pink Tomato
Lycopersicon	*esculentum-pincum*	*oaxacana*	Oaxacan ?.?"		Pink Tomato
Lycopersicon	*esculentum-pincum*	*olena-ukraniana*	Olena Ukrainian 1.5lb		Pink Tomato
Lycopersicon	*esculentum-pincum*	*opal-essencei*	Opal Essence ?.?"		Pink Tomato
Lycopersicon	*esculentum-pincum*	*ororikoi*	Ororiko 2.7"		Pink Tomato
Lycopersicon	*esculentum-pincum*	*ozark-pinkii-VF*	Ozark Very Fine 8oz		Pink Tomato
Lycopersicon	*esculentum-pincum*	*peach-pinkii*	Peach ?.?"		Pink Tomato
Lycopersicon	*esculentum-pincum*	*ping-pongii*	Ping Pong 2.0"		Pink Tomato
Lycopersicon	*esculentum-pincum*	*ping-pongii-pinkii*	Pink Ping Pong 2.0"		Pink Tomato
Lycopersicon	*esculentum-pincum*	*pink-accordianii*	Pink Accordian ?.?"		Pink Tomato
Lycopersicon	*esculentum-pincum*	*pink-monserratensis*	Pink Monserrat 3.0"		Pink Tomato
Lycopersicon	*esculentum-pincum*	*pink-petticoatii*	Pink Petticoat 4oz		Pink Tomato
Lycopersicon	*esculentum-pincum*	*pomadora-pinkii*	Pommadora ?.?"		Pink Tomato
Lycopersicon	*esculentum-pincum*	*pomme-d'amorei*	Pomee D'Amore 2.5"		Pink Tomato
Lycopersicon	*esculentum-pincum*	*raspberry-coloredii*	Raspberry Colored 2.0"		Pink Tomato
Lycopersicon	*esculentum-pincum*	*rosalie's-big-rosyi*	Rosalie's Big Rosy 3lb		Pink Tomato
Lycopersicon	*esculentum-pincum*	*rose-of bernei*	Rose Of Berne 1/2lb		Pink Tomato
Lycopersicon	*esculentum-pincum*	*rozai*	Roza ?.?"		Pink Tomato
Lycopersicon	*esculentum-pincum*	*ruby-rakesi*	Ruby Rakes 3/4lb		Pink Tomato
Lycopersicon	*esculentum-pincum*	*ruffled-pinkii*	Ruffled ?.?"		Pink Tomato
Lycopersicon	*esculentum-pincum*	*russian-rosei*	Russian Rose 1/2lb		Pink Tomato
Lycopersicon	*esculentum-pincum*	*sabrei*	Sabre 2lb		Pink Tomato
Lycopersicon	*esculentum-pincum*	*sasha's-pinkii*	Sasha's ?.?"		Pink Tomato
Lycopersicon	*esculentum-pincum*	*segleri*	Segler 3oz		Pink Tomato
Lycopersicon	*esculentum-pincum*	*shirveyi*	Shirvey 4.0"		Pink Tomato
Lycopersicon	*esculentum-pincum*	*siberian-pinkii*	Siberian ?.?"		Pink Tomato
Lycopersicon	*esculentum-pincum*	*stump-of-the-worldii*	Stump Of The World 4.5"		Pink Tomato
Lycopersicon	*esculentum-pincum*	*tennessee-peach-fuzzii*	Tennessee Peach Fuzz 2.0"		Pink Tomato
Lycopersicon	*esculentum-pincum*	*the-1884-tomatoi*	The 1884 Tomato ?.?"		Pink Tomato
Lycopersicon	*esculentum-pincum*	*tiffen-mennonitei*	Tiffen Mennonite 4.0"		Pink Tomato
Lycopersicon	*esculentum-pincum*	*togo-treflei*	Togo Trefle 2.0"		Pink Tomato
Lycopersicon	*esculentum-pincum*	*trucker's-favoritei*	Trucker's Favorite 2.0"		Pink Tomato
Lycopersicon	*esculentum-pincum*	*urbanitei*	Urbanite 5.0"		Pink Tomato

Genus	Species	Infraspecies	cultivar	specimen	Genome
Lycopersicon	*esculentum-pincum*	*vermillionii*	Vermillion 1/2lb	Pink	Tomato
Lycopersicon	*esculentum-pincum*	*violaceum-krypni-rozo*	Violaceum Krypni Rozo 2lb	Pink	Tomato
Lycopersicon	*esculentum-pincum*	*zapotec-ribbedii*	Zapotec Ribbed 4.0"	Pink	Tomato
Lycopersicon	*esculentum-pincum*	*zaryanka*	Zaryanka 5oz	Pink	Tomato
Lycopersicon	*esculentum-pincum-purpureum*	*boondocksi*	Boondocks 1lb	Purplish Pink	Tomato
Lycopersicon	*esculentum-pincum-purpureum*	*brandywinei-purplei*	Purple Amish Brandywine Creek 5.0"	Purplish Pink	Tomato
Lycopersicon	*esculentum-pincum-purpureum*	*brimmer-pinkii*	Brimmer Pink 2.5lb	Purplish Pink	Tomato
Lycopersicon	*esculentum-pincum-purpureum*	*cherokee-purplei*	Cherokee Purple Flat 11oz	Purplish Pink	Tomato
Lycopersicon	*esculentum-pincum-purpureum*	*cherokee-purplei-potato-leafii*	Potato Leaf Cherokee Purple Flat 11oz	Purplish Pink	Tomato
Lycopersicon	*esculentum-pincum-purpureum*	*dutchmanii*	Dutchman Flat 3lb	Purplish Pink	Tomato
Lycopersicon	*esculentum-pincum-purpureum*	*eva-purple-ballii*	Eva Purple Ball 5oz	Purplish Pink	Tomato
Lycopersicon	*esculentum-pincum-purpureum*	*gregori's-altai*	Gregori's Altai Mountain 4.0"	Purplish Pink	Tomato
Lycopersicon	*esculentum-pincum-purpureum*	*gulf-state-marketii*	Gulf State Market ?.?"	Purplish Pink	Tomato
Lycopersicon	*esculentum-pincum-purpureum*	*italiana-purplei*	Italian Purple 5.0"	Purplish Pink	Tomato
Lycopersicon	*esculentum-pincum-purpureum*	*marizol-purplei*	Marizol Purple Flat 3/4lb	Purplish Pink	Tomato
Lycopersicon	*esculentum-pincum-purpureum*	*pruden's-purplei*	Pruden's Purple 1lb	Purplish Pink	Tomato
Lycopersicon	*esculentum-pincum-purpureum*	*purplei*	Purple ?.?"	Purplish Pink	Tomato
Lycopersicon	*esculentum-pincum-purpureum*	*purplei-calabashii*	Purple Calabash 6oz	Purplish Pink	Tomato
Lycopersicon	*esculentum-pincum-purpureum*	*purple-perfectii*	Purple Perfect 1/2lb	Purplish Pink	Tomato
Lycopersicon	*esculentum-pincum-purpureum*	*purplei-pricei*	Purple Price ?.?"	Purplish Pink	Tomato
Lycopersicon	*esculentum-pincum-purpureum*	*purplei-princei*	Purple Prince 10oz	Purplish Pink	Tomato
Lycopersicon	*esculentum-pincum-purpureum*	*purplei-russiana*	Purple Russian 6oz	Purplish Pink	Tomato
Lycopersicon	*esculentum-pincum-purpureum*	*purplei-smudgei*	Purple Smudge ?.?"	Purplish Pink	Tomato
Lycopersicon	*esculentum-pincum-purpureum*	*royal-hillbillyi*	Royal Hillbilly 14oz	Purplish Pink	Tomato
Lycopersicon	*esculentum-pincum-purpureum*	*schwarze-sarahii*	Schwarze Sarah ?.?"	Purplish Pink	Tomato
Lycopersicon	*esculentum-pincum-purpureum*	*tibetii*	Tibet 6oz	Purplish Pink	Tomato

Tomato, Red (*Lycopersicon esculentum-rubrum*)

Genus	Species	Infraspecies	cultivar	specimen	Genome
Lycopersicon	*esculentum-rubrum-cavitum*	*ruffled-redii*	Ruffled Red 3.0"	Hollow Red	Tomato
Lycopersicon	*esculentum-rubrum-cavitum*	*stuffingii*	Stuffing 2.75" x 3.0"	Hollow Red	Tomato
Lycopersicon	*esculentum-rubrum-cavitum*	*stuffingii-burgessi*	Burgess Stuffing 2.5" x 3.0"	Hollow Red	Tomato
Lycopersicon	*esculentum-rubrum-giganteum*	.	.	Giant 3lb 5.0" Red	Tomato
Lycopersicon	*esculentum-rubrum-giganteum*	*goliathii*	Goliath	Giant 3lb 5.0" Red	Tomato
Lycopersicon	*esculentum-rubrum-giganteum*	*ingegnoli-giantii*	Ingengnoli	Giant 3lb 5.0" Red	Tomato
Lycopersicon	*esculentum-rubrum-giganteum*	*mongii*	Mong	Giant 3lb 5.0" Red	Tomato
Lycopersicon	*esculentum-rubrum-giganteum*	*orenburg-giantii*	Orenburg	Giant 3lb 5.0" Red	Tomato
Lycopersicon	*esculentum-rubrum-giganteum*	*rio-grandei*	Rio Grande	Giant 3lb 5.0" Red	Tomato
Lycopersicon	*esculentum-rubrum-giganteum*	*rio-grandei-cal-supremei*	California Supreme Rio Grande	Giant 3lb 5.0" Red	Tomato
Lycopersicon	*esculentum-rubrum-giganteum*	*sunset's-red-horizionii*	Sunset's Red Horizion	Giant 3lb 5.0" Red	Tomato
Lycopersicon	*esculentum-rubrum-largum*	*abraham-lincolnii*	Abraham Lincoln	Large 1.5lb-2lb 4.0" Red	Tomato
Lycopersicon	*esculentum-rubrum-largum*	*believe-it-or-notii*	Believe It Or Not	Large 1.5lb-2lb 4.0" Red	Tomato
Lycopersicon	*esculentum-rubrum-largum*	*big-redii*	Big Red	Large 1.5lb-2lb 4.0" Red	Tomato
Lycopersicon	*esculentum-rubrum-largum*	*d'australiei*	D'Australie	Large 1.5lb-2lb 4.0" Red	Tomato
Lycopersicon	*esculentum-rubrum-largum*	*earlibellii*	Earlibell	Large 1.5lb-2lb 4.0" Red	Tomato
Lycopersicon	*esculentum-rubrum-largum*	*germanica-redii*	Amish German	Large 1.5lb-2lb 4.0" Red	Tomato
Lycopersicon	*esculentum-rubrum-largum*	*giant-colossalis*	Giant Colossal	Large 1.5lb-2lb 4.0" Red	Tomato
Lycopersicon	*esculentum-rubrum-largum*	*grandpa's-cocks-plumei*	Grandpa's Cocks Plume	Large 1.5lb-2lb 4.0" Red	Tomato
Lycopersicon	*esculentum-rubrum-largum*	*imperialis*	Imperial	Large 1.5lb-2lb 4.0" Red	Tomato
Lycopersicon	*esculentum-rubrum-largum*	*liberty-treei*	Liberty Tree	Large 1.5lb-2lb 4.0" Red	Tomato
Lycopersicon	*esculentum-rubrum-largum*	*mega-tomii*	Mega Tom	Large 1.5lb-2lb 4.0" Red	Tomato
Lycopersicon	*esculentum-rubrum-largum*	*mike's-bridei*	Mike's Bride	Large 1.5lb-2lb 4.0" Red	Tomato
Lycopersicon	*esculentum-rubrum-largum*	*monfavetii*	Monfavet	Large 1.5lb-2lb 4.0" Red	Tomato
Lycopersicon	*esculentum-rubrum-largum*	*monsteri*	Monster	Large 1.5lb-2lb 4.0" Red	Tomato
Lycopersicon	*esculentum-rubrum-largum*	*northern-ontario-roadsidei*	Northern Ontario Roadside	Large 1.5lb-2lb 4.0" Red	Tomato
Lycopersicon	*esculentum-rubrum-largum*	*olga's-biggestii*	Olga's Biggest	Large 1.5lb-2lb 4.0" Red	Tomato
Lycopersicon	*esculentum-rubrum-largum*	*stakebreakeri*	Stakebreaker	Large 1.5lb-2lb 4.0" Red	Tomato
Lycopersicon	*esculentum-rubrum-largum*	*tapsi*	Taps	Large 1.5lb-2lb 4.0" Red	Tomato
Lycopersicon	*esculentum-rubrum-largum-medium*	*aker's-west-virginiensis*	Aker's West Virginia	Medium Large 1lb 3.5" Red	Tomato
Lycopersicon	*esculentum-rubrum-largum-medium*	*amateur's-dreamii*	Amateur's Dream	Medium Large 1lb 3.5" Red	Tomato
Lycopersicon	*esculentum-rubrum-largum-medium*	*ardwynai*	Ardwyna	Medium Large 1lb 3.5" Red	Tomato
Lycopersicon	*esculentum-rubrum-largum-medium*	*big's-red-siberiana*	Big's Red Siberian	Medium Large 1lb 3.5" Red	Tomato
Lycopersicon	*esculentum-rubrum-largum-medium*	*blue-ridge-mountainii*	Blue Ridge Mountain	Medium Large 1lb 3.5" Red	Tomato
Lycopersicon	*esculentum-rubrum-largum-medium*	*box-car-williei*	Box Car Willie	Medium Large 1lb 3.5" Red	Tomato
Lycopersicon	*esculentum-rubrum-largum-medium*	*brandywinei-glick's-strainii*	Glick's Amish Brandywine Creek	Medium Large 1lb 3.5" Red	Tomato
Lycopersicon	*esculentum-rubrum-largum-medium*	*bushii-506(no.-506)*	#506 Bush	Medium Large 1lb 3.5" Red	Tomato
Lycopersicon	*esculentum-rubrum-largum-medium*	*chapmanii*	Chapman	Medium Large 1lb 3.5" Red	Tomato
Lycopersicon	*esculentum-rubrum-largum-medium*	*champii*	Champ	Medium Large 1lb 3.5" Red	Tomato
Lycopersicon	*esculentum-rubrum-largum-medium*	*comstock's-sauce-&-slicei*	Comstock's Sauce & Slice	Medium Large 1lb 3.5" Red	Tomato
Lycopersicon	*esculentum-rubrum-largum-medium*	*creolei*	Creole	Medium Large 1lb 3.5" Red	Tomato
Lycopersicon	*esculentum-rubrum-largum-medium*	*czech-selectii*	Czech Select	Medium Large 1lb 3.5" Red	Tomato

Genus	Species	Infraspecies	cultivar	specimen	Genome
Lycopersicon	*esculentum-rubrum-largum-medium*	*dieneri*	Diener	Medium Large 1lb 3.5" Red	Tomato
Lycopersicon	*esculentum-rubrum-largum-medium*	*early-gleei*	Early Glee	Medium Large 1lb 3.5" Red	Tomato
Lycopersicon	*esculentum-rubrum-largum-medium*	*epochii*	Epoch	Medium Large 1lb 3.5" Red	Tomato
Lycopersicon	*esculentum-rubrum-largum-medium*	*fuzzy-bombii*	Fuzzy Bomb	Medium Large 1lb 3.5" Red	Tomato
Lycopersicon	*esculentum-rubrum-largum-medium*	*gantii*	Ganti	Medium Large 1lb 3.5" Red	Tomato
Lycopersicon	*esculentum-rubrum-largum-medium*	*hamsoni-DX-52-12*	DX 52-12 Hamson	Medium Large 1lb 3.5" Red	Tomato
Lycopersicon	*esculentum-rubrum-largum-medium*	*imposteri*	Imposter	Medium Large 1lb 3.5" Red	Tomato
Lycopersicon	*esculentum-rubrum-largum-medium*	*killei-7*	#7 Kille	Medium Large 1lb 3.5" Red	Tomato
Lycopersicon	*esculentum-rubrum-largum-medium*	*large-redii-brown'si*	Brown's	Medium Large 1lb 3.5" Red	Tomato
Lycopersicon	*esculentum-rubrum-largum-medium*	*manapalii*	Manapal	Medium Large 1lb 3.5" Red	Tomato
Lycopersicon	*esculentum-rubrum-largum-medium*	*marionii*	Marion	Medium Large 1lb 3.5" Red	Tomato
Lycopersicon	*esculentum-rubrum-largum-medium*	*memorial-polishii*	Memorial Polish	Medium Large 1lb 3.5" Red	Tomato
Lycopersicon	*esculentum-rubrum-largum-medium*	*merveille-des-marchesi*	Merveille Des Marches	Medium Large 1lb 3.5" Red	Tomato
Lycopersicon	*esculentum-rubrum-largum-medium*	*montreal-tastyi*	Montreal Tasty	Medium Large 1lb 3.5" Red	Tomato
Lycopersicon	*esculentum-rubrum-largum-medium*	*mormon-world's-earliestii*	Mormon World's Earliest	Medium Large 1lb 3.5" Red	Tomato
Lycopersicon	*esculentum-rubrum-largum-medium*	*mortgage-lifter-australiana*	Australian Mortgage Lifter	Medium Large 1lb 3.5" Red	Tomato
Lycopersicon	*esculentum-rubrum-largum-medium*	*mountain-boyi*	Mountain Boy	Medium Large 1lb 3.5" Red	Tomato
Lycopersicon	*esculentum-rubrum-largum-medium*	*old-brooksi*	Old Brooks	Medium Large 1lb 3.5" Red	Tomato
Lycopersicon	*esculentum-rubrum-largum-medium*	*pelicanii*	Pelican	Medium Large 1lb 3.5" Red	Tomato
Lycopersicon	*esculentum-rubrum-largum-medium*	*peronii*	Peron	Medium Large 1lb 3.5" Red	Tomato
Lycopersicon	*esculentum-rubrum-largum-medium*	*pritchard's-scarlet-topperi*	Pritchard's Scarlet Topper	Medium Large 1lb 3.5" Red	Tomato
Lycopersicon	*esculentum-rubrum-largum-medium*	*PSR-956*	#956 Peters Seed & Research	Medium Large 1lb 3.5" Red	Tomato
Lycopersicon	*esculentum-rubrum-largum-medium*	*rockyi*	Rocky	Medium Large 1lb 3.5" Red	Tomato
Lycopersicon	*esculentum-rubrum-largum-medium*	*ruby-gold-sportii*	Ruby Gold Sport	Medium Large 1lb 3.5" Red	Tomato
Lycopersicon	*esculentum-rubrum-largum-medium*	*sainte-luciei*	Sainte Lucie	Medium Large 1lb 3.5" Red	Tomato
Lycopersicon	*esculentum-rubrum-largum-medium*	*seattle-best-of-allii*	Seattle Best Of All	Medium Large 1lb 3.5" Red	Tomato
Lycopersicon	*esculentum-rubrum-largum-medium*	*sebastopolii*	Sebastopol	Medium Large 1lb 3.5" Red	Tomato
Lycopersicon	*esculentum-rubrum-largum-medium*	*sementii*	Sementi	Medium Large 1lb 3.5" Red	Tomato
Lycopersicon	*esculentum-rubrum-largum-medium*	*sheriffii*	Sheriff	Medium Large 1lb 3.5" Red	Tomato
Lycopersicon	*esculentum-rubrum-largum-medium*	*scheyennei*	Sheyenne	Medium Large 1lb 3.5" Red	Tomato
Lycopersicon	*esculentum-rubrum-largum-medium*	*shriveri*	Shriver	Medium Large 1lb 3.5" Red	Tomato
Lycopersicon	*esculentum-rubrum-largum-medium*	*toronado*	Toronado	Medium Large 1lb 3.5" Red	Tomato
Lycopersicon	*esculentum-rubrum-largum-medium*	*uncle-ike's-big-redii*	Uncle Ike's Big Red	Medium Large 1lb 3.5" Red	Tomato
Lycopersicon	*esculentum-rubrum-largum-medium*	*urbana*	Urbana	Medium Large 1lb 3.5" Red	Tomato
Lycopersicon	*esculentum-rubrum-largum-medium*	*verticilium-wilt-resistant-moscovensis*	Verticilium Wilt Resistant Moscow	Medium Large 1lb 3.5" Red	Tomato
Lycopersicon	*esculentum-rubrum-largum-medium*	*valiantii*	Valiant	Medium Large 1lb 3.5" Red	Tomato
Lycopersicon	*esculentum-rubrum-largum-medium*	*wanda's-potato-topii*	Wanda's Potato Top	Medium Large 1lb 3.5" Red	Tomato
Lycopersicon	*esculentum-rubrum-largum-medium*	*zapotec-ribbedii*	Zapotec Ribbed	Medium Large 1lb 3.5" Red	Tomato
Lycopersicon	*esculentum-rubrum-medium-largum*	*Benewahii*	Benewah	Large Medium 8oz-12oz 3.0" Red	Tomato
Lycopersicon	*esculentum-rubrum-medium-largum*	*big-setii*	Big Set	Large Medium 8oz-12oz 3.0" Red	Tomato
Lycopersicon	*esculentum-rubrum-medium-largum*	*brandywinei-redii*	Red Amish Brandywine Creek	Large Medium 8oz-12oz 3.0" Red	Tomato
Lycopersicon	*esculentum-rubrum-medium-largum*	*brandywinei-rosea*	Rose Amish Brandywine Creek	Large Medium 8oz-12oz 3.0" Red	Tomato
Lycopersicon	*esculentum-rubrum-medium-largum*	*cardinalis*	Cardinal	Large Medium 8oz-12oz 3.0" Red	Tomato
Lycopersicon	*esculentum-rubrum-medium-largum*	*coralii*	Coral	Large Medium 8oz-12oz 3.0" Red	Tomato
Lycopersicon	*esculentum-rubrum-medium-largum*	*early-chathamii*	Early Chatham	Large Medium 8oz-12oz 3.0" Red	Tomato
Lycopersicon	*esculentum-rubrum-medium-largum*	*forest-firei*	Forest Fire	Large Medium 8oz-12oz 3.0" Red	Tomato
Lycopersicon	*esculentum-rubrum-medium-largum*	*garden-statei*	Garden State	Large Medium 8oz-12oz 3.0" Red	Tomato
Lycopersicon	*esculentum-rubrum-medium-largum*	*german-giant-redii*	German Giant	Large Medium 8oz-12oz 3.0" Red	Tomato
Lycopersicon	*esculentum-rubrum-medium-largum*	*glacieri*	Glacier	Large Medium 8oz-12oz 3.0" Red	Tomato
Lycopersicon	*esculentum-rubrum-medium-largum*	*graeca-domata*	Greek Domata	Large Medium 8oz-12oz 3.0" Red	Tomato
Lycopersicon	*esculentum-rubrum-medium-largum*	*grosse-lissei*	Grosse Lisse	Large Medium 8oz-12oz 3.0" Red	Tomato
Lycopersicon	*esculentum-rubrum-medium-largum*	*grosse-rondei*	Grosse Ronde	Large Medium 8oz-12oz 3.0" Red	Tomato
Lycopersicon	*esculentum-rubrum-medium-largum*	*hilltopii*	Hilltop	Large Medium 8oz-12oz 3.0" Red	Tomato
Lycopersicon	*esculentum-rubrum-medium-largum*	*hollandica*	Holland	Large Medium 8oz-12oz 3.0" Red	Tomato
Lycopersicon	*esculentum-rubrum-medium-largum*	*idai*	Ida	Large Medium 8oz-12oz 3.0" Red	Tomato
Lycopersicon	*esculentum-rubrum-medium-largum*	*kennington's-big-redii*	Kennington's Big Red	Large Medium 8oz-12oz 3.0" Red	Tomato
Lycopersicon	*esculentum-rubrum-medium-largum*	*lillian's-red-kansas-pastei*	Lillian's Red Kansas Paste	Large Medium 8oz-12oz 3.0" Red	Tomato
Lycopersicon	*esculentum-rubrum-medium-largum*	*marizol-redii*	Marizol	Large Medium 8oz-12oz 3.0" Red	Tomato
Lycopersicon	*esculentum-rubrum-medium-largum*	*market-miraclei*	Market Miracle	Large Medium 8oz-12oz 3.0" Red	Tomato
Lycopersicon	*esculentum-rubrum-medium-largum*	*mortgage-lifter-redii*	Red Mortgage Lifter	Large Medium 8oz-12oz 3.0" Red	Tomato
Lycopersicon	*esculentum-rubrum-medium-largum*	*moscovensis*	Moscow	Large Medium 8oz-12oz 3.0" Red	Tomato
Lycopersicon	*esculentum-rubrum-medium-largum*	*mule-teamii*	Mule Team	Large Medium 8oz-12oz 3.0" Red	Tomato
Lycopersicon	*esculentum-rubrum-medium-largum*	*nepalensis*	Nepal	Large Medium 8oz-12oz 3.0" Red	Tomato
Lycopersicon	*esculentum-rubrum-medium-largum*	*northern-exposurei*	Northern Exposure	Large Medium 8oz-12oz 3.0" Red	Tomato
Lycopersicon	*esculentum-rubrum-medium-largum*	*optimus*	Optimus	Large Medium 8oz-12oz 3.0" Red	Tomato
Lycopersicon	*esculentum-rubrum-medium-largum*	*popovichii*	Popovich	Large Medium 8oz-12oz 3.0" Red	Tomato
Lycopersicon	*esculentum-rubrum-medium-largum*	*princei*	Prince	Large Medium 8oz-12oz 3.0" Red	Tomato
Lycopersicon	*esculentum-rubrum-medium-largum*	*productiva*	Productiva	Large Medium 8oz-12oz 3.0" Red	Tomato
Lycopersicon	*esculentum-rubrum-medium-largum*	*red-rosei*	Red Rose	Large Medium 8oz-12oz 3.0" Red	Tomato
Lycopersicon	*esculentum-rubrum-medium-largum*	*rosalie's-large-pastei*	Rosalie's Large Paste	Large Medium 8oz-12oz 3.0" Red	Tomato
Lycopersicon	*esculentum-rubrum-medium-largum*	*sandia-gemii*	Sandia Gem	Large Medium 8oz-12oz 3.0" Red	Tomato

Genus	Species	Infraspecies	cultivar	specimen	Genome
Lycopersicon	*esculentum-rubrum-medium-largum*	*sauvignaci*	Sauvignac	Large Medium 8oz-12oz 3.0" Red	Tomato
Lycopersicon	*esculentum-rubrum-medium-largum*	*siletzii*	Siletz	Large Medium 8oz-12oz 3.0" Red	Tomato
Lycopersicon	*esculentum-rubrum-medium-largum*	*spring-kingii*	Spring King	Large Medium 8oz-12oz 3.0" Red	Tomato
Lycopersicon	*esculentum-rubrum-medium-largum*	*starlingii*	Starling	Large Medium 8oz-12oz 3.0" Red	Tomato
Lycopersicon	*esculentum-rubrum-medium-largum*	*table-talkii-burpee'si*	Burpee's Table Talk	Large Medium 8oz-12oz 3.0" Red	Tomato
Lycopersicon	*esculentum-rubrum-medium-largum*	*taosi*	Taos	Large Medium 8oz-12oz 3.0" Red	Tomato
Lycopersicon	*esculentum-rubrum-medium-largum*	*teresa*	Teresa	Large Medium 8oz-12oz 3.0" Red	Tomato
Lycopersicon	*esculentum-rubrum-medium-largum*	*tartar-from-monolstanii*	Tartar From Mongolstan	Large Medium 8oz-12oz 3.0" Red	Tomato
Lycopersicon	*esculentum-rubrum-medium-largum*	*tuscaninii*	Tuscanini	Large Medium 8oz-12oz 3.0" Red	Tomato
Lycopersicon	*esculentum-rubrum-medium-largum*	*valley-girlii*	Valley Girl	Large Medium 8oz-12oz 3.0" Red	Tomato
Lycopersicon	*esculentum-rubrum-medium-largum*	*willamettei*	Willamette	Large Medium 8oz-12oz 3.0" Red	Tomato
Lycopersicon	*esculentum-rubrum-medium-largum*	*wisconsinensis-55*	#55 Wisconsin	Large Medium 8oz-12oz 3.0" Red	Tomato
Lycopersicon	*esculentum-rubrum-medium-largum*	*winterkeeperi*	WinterKeeper	Large Medium 8oz-12oz 3.0" Red	Tomato
Lycopersicon	*esculentum-rubrum-medium*	*adelia*	Adelia	Medium 6oz-8oz 2.5" Red	Tomato
Lycopersicon	*esculentum-rubrum-medium*	*alicantei*	Alicante	Medium 6oz-8oz 2.5" Red	Tomato
Lycopersicon	*esculentum-rubrum-medium*	*alisa-craigii*	Alisa Craig	Medium 6oz-8oz 2.5" Red	Tomato
Lycopersicon	*esculentum-rubrum-medium*	*alpha*	Alpha Square	Medium 6oz-8oz 2.5" Red	Tomato
Lycopersicon	*esculentum-rubrum-medium*	*american-beautyi*	American Beauty	Medium 6oz-8oz 2.5" Red	Tomato
Lycopersicon	*esculentum-rubrum-medium*	*anahui*	Anahui	Medium 6oz-8oz 2.5" Red	Tomato
Lycopersicon	*esculentum-rubrum-medium*	*aneta*	Aneta	Medium 6oz-8oz 2.5" Red	Tomato
Lycopersicon	*esculentum-rubrum-medium*	*angelina's-italianii*	Angelina's Italian Square	Medium 6oz-8oz 2.5" Red	Tomato
Lycopersicon	*esculentum-rubrum-medium*	*angorensis*	Angora	Medium 6oz-8oz 2.5" Red	Tomato
Lycopersicon	*esculentum-rubrum-medium*	*balconi-redii*	Balconi	Medium 6oz-8oz 2.5" Red	Tomato
Lycopersicon	*esculentum-rubrum-medium*	*berkshire-polishii*	Berkshire Polish	Medium 6oz-8oz 2.5" Red	Tomato
Lycopersicon	*esculentum-rubrum-medium*	*bruinima-produktii*	Bruinima Produckt	Medium 6oz-8oz 2.5" Red	Tomato
Lycopersicon	*esculentum-rubrum-medium*	*budenkovai*	Budenkova	Medium 6oz-8oz 2.5" Red	Tomato
Lycopersicon	*esculentum-rubrum-medium*	*burgundy-traveleri*	Burgundy Traveler	Medium 6oz-8oz 2.5" Red	Tomato
Lycopersicon	*esculentum-rubrum-medium*	*burrell's-specialis*	Burrell's Special	Medium 6oz-8oz 2.5" Red	Tomato
Lycopersicon	*esculentum-rubrum-medium*	*bush-whopperi*	Bush Whopper	Medium 6oz-8oz 2.5" Red	Tomato
Lycopersicon	*esculentum-rubrum-medium*	*carmelloi*	Carmello	Medium 6oz-8oz 2.5" Red	Tomato
Lycopersicon	*esculentum-rubrum-medium*	*castlerockii*	Castlerock	Medium 6oz-8oz 2.5" Red	Tomato
Lycopersicon	*esculentum-rubrum-medium*	*cheersii*	Cheers	Medium 6oz-8oz 2.5" Red	Tomato
Lycopersicon	*esculentum-rubrum-medium*	*cheetam's-potato-leafii*	Cheetham's Potato Leaf	Medium 6oz-8oz 2.5" Red	Tomato
Lycopersicon	*esculentum-rubrum-medium*	*chianti-rosei*	Chianti Rose	Medium 6oz-8oz 2.5" Red	Tomato
Lycopersicon	*esculentum-rubrum-medium*	*clustermatoi*	Clustermato	Medium 6oz-8oz 2.5" Red	Tomato
Lycopersicon	*esculentum-rubrum-medium*	*cobourgii*	Cobourg	Medium 6oz-8oz 2.5" Red	Tomato
Lycopersicon	*esculentum-rubrum-medium*	*colorado-redii*	Colorado	Medium 6oz-8oz 2.5" Red	Tomato
Lycopersicon	*esculentum-rubrum-medium*	*cougar-redii*	Cougar	Medium 6oz-8oz 2.5" Red	Tomato
Lycopersicon	*esculentum-rubrum-medium*	*coursen-roy's-stuffingii*	Coursen Roy's Stuffing	Medium 6oz-8oz 2.5" Red	Tomato
Lycopersicon	*esculentum-rubrum-medium*	*crackproof-burgessi*	Crackproof Burgess	Medium 6oz-8oz 2.5" Red	Tomato
Lycopersicon	*esculentum-rubrum-medium*	*crimson-pastai*	Crimson Pasta	Medium 6oz-8oz 2.5" Red	Tomato
Lycopersicon	*esculentum-rubrum-medium*	*crimson-sprinteri*	Crimson Sprinter	Medium 6oz-8oz 2.5" Red	Tomato
Lycopersicon	*esculentum-rubrum-medium*	*cup-of-moldovensis*	Cup Of Moldova	Medium 6oz-8oz 2.5" Red	Tomato
Lycopersicon	*esculentum-rubrum-medium*	*double-richii*	Double Rich	Medium 6oz-8oz 2.5" Red	Tomato
Lycopersicon	*esculentum-rubrum-medium*	*doucert's-quebec-earlyi*	Doucet's Quebec Early	Medium 6oz-8oz 2.5" Red	Tomato
Lycopersicon	*esculentum-rubrum-medium*	*dunkin's-delightii*	Dunkin's Delight	Medium 6oz-8oz 2.5" Red	Tomato
Lycopersicon	*esculentum-rubrum-medium*	*e.h.-housei*	E. H. House	Medium 6oz-8oz 2.5" Red	Tomato
Lycopersicon	*esculentum-rubrum-medium*	*early-large-redii*	Early Large Red	Medium 6oz-8oz 2.5" Red	Tomato
Lycopersicon	*esculentum-rubrum-medium*	*early-red-chiefii*	Early Red Chief	Medium 6oz-8oz 2.5" Red	Tomato
Lycopersicon	*esculentum-rubrum-medium*	*early-summer-sunrisei*	Early Summer Sunrise	Medium 6oz-8oz 2.5" Red	Tomato
Lycopersicon	*esculentum-rubrum-medium*	*elimii*	Elim	Medium 6oz-8oz 2.5" Red	Tomato
Lycopersicon	*esculentum-rubrum-medium*	*enormousi*	Enormous	Medium 6oz-8oz 2.5" Red	Tomato
Lycopersicon	*esculentum-rubrum-medium*	*enterprizei*	Enterprize	Medium 6oz-8oz 2.5" Red	Tomato
Lycopersicon	*esculentum-rubrum-medium*	*erosii*	Eros	Medium 6oz-8oz 2.5" Red	Tomato
Lycopersicon	*esculentum-rubrum-medium*	*ethel-watkins-bestii*	Ethel Watkins Best	Medium 6oz-8oz 2.5" Red	Tomato
Lycopersicon	*esculentum-rubrum-medium*	*eurekaensis*	Eureka	Medium 6oz-8oz 2.5" Red	Tomato
Lycopersicon	*esculentum-rubrum-medium*	*eurofreshii*	Eurofresh	Medium 6oz-8oz 2.5" Red	Tomato
Lycopersicon	*esculentum-rubrum-medium*	*eyei*	Eye	Medium 6oz-8oz 2.5" Red	Tomato
Lycopersicon	*esculentum-rubrum-medium*	*f.h.-crowi*	F. H. Crow	Medium 6oz-8oz 2.5" Red	Tomato
Lycopersicon	*esculentum-rubrum-medium*	*fai*	Fa	Medium 6oz-8oz 2.5" Red	Tomato
Lycopersicon	*esculentum-rubrum-medium*	*fire-drillii*	Fire Drill	Medium 6oz-8oz 2.5" Red	Tomato
Lycopersicon	*esculentum-rubrum-medium*	*fireworksi*	Fireworks	Medium 6oz-8oz 2.5" Red	Tomato
Lycopersicon	*esculentum-rubrum-medium*	*fireworksi-2*	#2 Fireworks	Medium 6oz-8oz 2.5" Red	Tomato
Lycopersicon	*esculentum-rubrum-medium*	*florida-MH-1*	#1 Florida Mechanical Harvest	Medium 6oz-8oz 2.5" Red	Tomato
Lycopersicon	*esculentum-rubrum-medium*	*graeca-asemina*	Greek Asemina	Medium 6oz-8oz 2.5" Red	Tomato
Lycopersicon	*esculentum-rubrum-medium*	*grandma-mary's-pastei*	Grandma Mary's Paste	Medium 6oz-8oz 2.5" Red	Tomato
Lycopersicon	*esculentum-rubrum-medium*	*harbingeri*	Harbinger	Medium 6oz-8oz 2.5" Red	Tomato
Lycopersicon	*esculentum-rubrum-medium*	*harrisonii-potentatei*	Potentate Harrison	Medium 6oz-8oz 2.5" Red	Tomato
Lycopersicon	*esculentum-rubrum-medium*	*harvestveei*	Harvestvee	Medium 6oz-8oz 2.5" Red	Tomato
Lycopersicon	*esculentum-rubrum-medium*	*healanii*	Healani	Medium 6oz-8oz 2.5" Red	Tomato
Lycopersicon	*esculentum-rubrum-medium*	*heinzii-1765*	#1765 Heinz	Medium 6oz-8oz 2.5" Red	Tomato
Lycopersicon	*esculentum-rubrum-medium*	*heinzii-2274*	#2274 Heinz	Medium 6oz-8oz 2.5" Red	Tomato
Lycopersicon	*esculentum-rubrum-medium*	*heinzii-2653*	#2653 Heinz	Medium 6oz-8oz 2.5" Red	Tomato
Lycopersicon	*esculentum-rubrum-medium*	*heinzii-VF*	Verticillium Fusarium	Medium 6oz-8oz 2.5" Red	Tomato

Genus	Species	Infraspecies	cultivar	specimen	Genome
			Resistant Heinz		
Lycopersicon	*esculentum-rubrum-medium*	*high-carotenei*	High Carotene	Medium 6oz-8oz 2.5" Red	Tomato
Lycopersicon	*esculentum-rubrum-medium*	*hires-rootstockii*	Hires Rootstock	Medium 6oz-8oz 2.5" Red	Tomato
Lycopersicon	*esculentum-rubrum-medium*	*homesweetii-F2*	Homesweet F2	Medium 6oz-8oz 2.5" Red	Tomato
Lycopersicon	*esculentum-rubrum-medium*	*hungarianii*	Hungarian	Medium 6oz-8oz 2.5" Red	Tomato
Lycopersicon	*esculentum-rubrum-medium*	*huntii-100*	#100 Hunt	Medium 6oz-8oz 2.5" Red	Tomato
Lycopersicon	*esculentum-rubrum-medium*	*illini-stari*	Illini Star	Medium 6oz-8oz 2.5" Red	Tomato
Lycopersicon	*esculentum-rubrum-medium*	*italiana-stallionii*	Italian Stallion	Medium 6oz-8oz 2.5" Red	Tomato
Lycopersicon	*esculentum-rubrum-medium*	*jagai*	Jaga	Medium 6oz-8oz 2.5" Red	Tomato
Lycopersicon	*esculentum-rubrum-medium*	*jawori*	Jawor	Medium 6oz-8oz 2.5" Red	Tomato
Lycopersicon	*esculentum-rubrum-medium*	*jerseiensis*	Jersey	Medium 6oz-8oz 2.5" Red	Tomato
Lycopersicon	*esculentum-rubrum-medium*	*john-baeri*	John Baer	Medium 6oz-8oz 2.5" Red	Tomato
Lycopersicon	*esculentum-rubrum-medium*	*john-lossaso's-low-acid-rubyi*	John Lassaso's Low Acid Ruby	Medium 6oz-8oz 2.5" Red	Tomato
Lycopersicon	*esculentum-rubrum-medium*	*katrina*	Katrina	Medium 6oz-8oz 2.5" Red	Tomato
Lycopersicon	*esculentum-rubrum-medium*	*kecskemetii-509*	#509 Kecskemeti	Medium 6oz-8oz 2.5" Red	Tomato
Lycopersicon	*esculentum-rubrum-medium*	*kewaloi*	Kewalo	Medium 6oz-8oz 2.5" Red	Tomato
Lycopersicon	*esculentum-rubrum-medium*	*kootenai*	Kootenai	Medium 6oz-8oz 2.5" Red	Tomato
Lycopersicon	*esculentum-rubrum-medium*	*krakowsie-wczesnei*	Krakowsie Wczesne	Medium 6oz-8oz 2.5" Red	Tomato
Lycopersicon	*esculentum-rubrum-medium*	*kujawskii*	Kujawsi	Medium 6oz-8oz 2.5" Red	Tomato
Lycopersicon	*esculentum-rubrum-medium*	*landisi*	Landis	Medium 6oz-8oz 2.5" Red	Tomato
Lycopersicon	*esculentum-rubrum-medium*	*landrethii*	Landreth	Medium 6oz-8oz 2.5" Red	Tomato
Lycopersicon	*esculentum-rubrum-medium*	*landry's-russiana*	Landry's Russian	Medium 6oz-8oz 2.5" Red	Tomato
Lycopersicon	*esculentum-rubrum-medium*	*larissa*	Larissa	Medium 6oz-8oz 2.5" Red	Tomato
Lycopersicon	*esculentum-rubrum-medium*	*latahii*	Latah	Medium 6oz-8oz 2.5" Red	Tomato
Lycopersicon	*esculentum-rubrum-medium*	*liberatori*	Liberator	Medium 6oz-8oz 2.5" Red	Tomato
Lycopersicon	*esculentum-rubrum-medium*	*lickurichii*	Lichurich	Medium 6oz-8oz 2.5" Red	Tomato
Lycopersicon	*esculentum-rubrum-medium*	*lida-ukrainiana*	Lida Ukrainian	Medium 6oz-8oz 2.5" Red	Tomato
Lycopersicon	*esculentum-rubrum-medium*	*liliana*	Lilian	Medium 6oz-8oz 2.5" Red	Tomato
Lycopersicon	*esculentum-rubrum-medium*	*linei-125*	#125 Line	Medium 6oz-8oz 2.5" Red	Tomato
Lycopersicon	*esculentum-rubrum-medium*	*lisa-kingii*	Lisa King	Medium 6oz-8oz 2.5" Red	Tomato
Lycopersicon	*esculentum-rubrum-medium*	*livingston's-favoritei*	Favorite Livingston's	Medium 6oz-8oz 2.5" Red	Tomato
Lycopersicon	*esculentum-rubrum-medium*	*livingston's-lutescentii*	Lutescent Livingston's	Medium 6oz-8oz 2.5" Red	Tomato
Lycopersicon	*esculentum-rubrum-medium*	*livingston's-marvelousi*	Marvelous Livingston's	Medium 6oz-8oz 2.5" Red	Tomato
Lycopersicon	*esculentum-rubrum-medium*	*livingston's-perfectionii*	Perfection Livingston's	Medium 6oz-8oz 2.5" Red	Tomato
Lycopersicon	*esculentum-rubrum-medium*	*long-keeperi-burpee'si*	Burpee's Long Keeper	Medium 6oz-8oz 2.5" Red	Tomato
Lycopersicon	*esculentum-rubrum-medium*	*lukullus*	Lukullus	Medium 6oz-8oz 2.5" Red	Tomato
Lycopersicon	*esculentum-rubrum-medium*	*lunch-bucketii*	Lunch Bucket	Medium 6oz-8oz 2.5" Red	Tomato
Lycopersicon	*esculentum-rubrum-medium*	*lynnwoodensis*	Lynnwood	Medium 6oz-8oz 2.5" Red	Tomato
Lycopersicon	*esculentum-rubrum-medium*	*malaysiana*	Malaysian	Medium 6oz-8oz 2.5" Red	Tomato
Lycopersicon	*esculentum-rubrum-medium*	*manaluciei*	Manalucie	Medium 6oz-8oz 2.5" Red	Tomato
Lycopersicon	*esculentum-rubrum-medium*	*marsi*	Mars	Medium 6oz-8oz 2.5" Red	Tomato
Lycopersicon	*esculentum-rubrum-medium*	*maule's-earliest-of-all-fieldii-43*	#43 Earliest of All Field Maule's	Medium 6oz-8oz 2.5" Red	Tomato
Lycopersicon	*esculentum-rubrum-medium*	*maule's-successi*	Success Maule's	Medium 6oz-8oz 2.5" Red	Tomato
Lycopersicon	*esculentum-rubrum-medium*	*medfordii*	Medford	Medium 6oz-8oz 2.5" Red	Tomato
Lycopersicon	*esculentum-rubrum-medium*	*memory-to-vavilovi*	Memory To Vavilov	Medium 6oz-8oz 2.5" Red	Tomato
Lycopersicon	*esculentum-rubrum-medium*	*mikado-ecarlatei*	Ecarlate Mikado	Medium 6oz-8oz 2.5" Red	Tomato
Lycopersicon	*esculentum-rubrum-medium*	*mishca*	Mishca	Medium 6oz-8oz 2.5" Red	Tomato
Lycopersicon	*esculentum-rubrum-medium*	*mountain-princessii*	Mountain Princess	Medium 6oz-8oz 2.5" Red	Tomato
Lycopersicon	*esculentum-rubrum-medium*	*mustangii*	Mustang	Medium 6oz-8oz 2.5" Red	Tomato
Lycopersicon	*esculentum-rubrum-medium*	*new-hampshire-surecropii*	New Hampshire Surecrop	Medium 6oz-8oz 2.5" Red	Tomato
Lycopersicon	*esculentum-rubrum-medium*	*new-jersey-championshipii*	New Jersey Campionship	Medium 6oz-8oz 2.5" Red	Tomato
Lycopersicon	*esculentum-rubrum-medium*	*nick'si*	Nick's	Medium 6oz-8oz 2.5" Red	Tomato
Lycopersicon	*esculentum-rubrum-medium*	*northern-lightii*	Northern Light	Medium 6oz-8oz 2.5" Red	Tomato
Lycopersicon	*esculentum-rubrum-medium*	*novinka-kubanii*	Novinka Kubani	Medium 6oz-8oz 2.5" Red	Tomato
Lycopersicon	*esculentum-rubrum-medium*	*novogogosharyi*	Novogogoshary	Medium 6oz-8oz 2.5" Red	Tomato
Lycopersicon	*esculentum-rubrum-medium*	*ohioensis-7870*	#7870 Ohio	Medium 6oz-8oz 2.5" Red	Tomato
Lycopersicon	*esculentum-rubrum-medium*	*old-'49eri*	Old '49er	Medium 6oz-8oz 2.5" Red	Tomato
Lycopersicon	*esculentum-rubrum-medium*	*old-colussus*	Old Colussus 2lb	Medium 6oz-8oz 2.5" Red	Tomato
Lycopersicon	*esculentum-rubrum-medium*	*old's-oldii*	Old's Old	Medium 6oz-8oz 2.5" Red	Tomato
Lycopersicon	*esculentum-rubrum-medium*	*outdoor-girlii*	Outdoor Girl	Medium 6oz-8oz 2.5" Red	Tomato
Lycopersicon	*esculentum-rubrum-medium*	*ozark-championii*	Ozark Champion	Medium 6oz-8oz 2.5" Red	Tomato
Lycopersicon	*esculentum-rubrum-medium*	*paragon-livingstonii*	Paragon Livingston	Medium 6oz-8oz 2.5" Red	Tomato
Lycopersicon	*esculentum-rubrum-medium*	*pepper-tomatoi*	Bell Pepper	Medium 6oz-8oz 2.5" Red	Tomato
Lycopersicon	*esculentum-rubrum-medium*	*perseyi*	Persey	Medium 6oz-8oz 2.5" Red	Tomato
Lycopersicon	*esculentum-rubrum-medium*	*phil's-fantastici*	Phil's Fantastic	Medium 6oz-8oz 2.5" Red	Tomato
Lycopersicon	*esculentum-rubrum-medium*	*photonii*	Photon	Medium 6oz-8oz 2.5" Red	Tomato
Lycopersicon	*esculentum-rubrum-medium*	*piedmontii*	Piedmont	Medium 6oz-8oz 2.5" Red	Tomato
Lycopersicon	*esculentum-rubrum-medium*	*pioneeri-2*	#2 Pioneer	Medium 6oz-8oz 2.5" Red	Tomato
Lycopersicon	*esculentum-rubrum-medium*	*poloi*	Polo	Medium 6oz-8oz 2.5" Red	Tomato
Lycopersicon	*esculentum-rubrum-medium*	*pozdnyakov'si*	Pozdynakov's	Medium 6oz-8oz 2.5" Red	Tomato
Lycopersicon	*esculentum-rubrum-medium*	*prairie-firei*	Prairie Fire	Medium 6oz-8oz 2.5" Red	Tomato
Lycopersicon	*esculentum-rubrum-medium*	*prairie-pridei*	Prairie Pride	Medium 6oz-8oz 2.5" Red	Tomato
Lycopersicon	*esculentum-rubrum-medium*	*preludei*	Prelude	Medium 6oz-8oz 2.5" Red	Tomato

Genus	Species	Infraspecies	cultivar	specimen	Genome
Lycopersicon	*esculentum-rubrum-medium*	*presidentii*	President	Medium 6oz-8oz 2.5" Red	Tomato
Lycopersicon	*esculentum-rubrum-medium*	*PSR-955*	#955 Peters Seed & Research	Medium 6oz-8oz 2.5" Red	Tomato
Lycopersicon	*esculentum-rubrum-medium*	*puckii*	Puck	Medium 6oz-8oz 2.5" Red	Tomato
Lycopersicon	*esculentum-rubrum-medium*	*pusa-rubyi*	Pusa Ruby	Medium 6oz-8oz 2.5" Red	Tomato
Lycopersicon	*esculentum-rubrum-medium*	*quebecensis-1121*	#1121 Quebec	Medium 6oz-8oz 2.5" Red	Tomato
Lycopersicon	*esculentum-rubrum-medium*	*quebecensis-314*	#314 Quebec	Medium 6oz-8oz 2.5" Red	Tomato
Lycopersicon	*esculentum-rubrum-medium*	*quigley's-italian-pastei*	Quigley's Italian Paste	Medium 6oz-8oz 2.5" Red	Tomato
Lycopersicon	*esculentum-rubrum-medium*	*red-chiefii*	Red Chief	Medium 6oz-8oz 2.5" Red	Tomato
Lycopersicon	*esculentum-rubrum-medium*	*red-dawnii*	Red Dawn	Medium 6oz-8oz 2.5" Red	Tomato
Lycopersicon	*esculentum-rubrum-medium*	*red-house-free-standingii*	Red House Free Standing	Medium 6oz-8oz 2.5" Red	Tomato
Lycopersicon	*esculentum-rubrum-medium*	*red-octoberi*	Red October	Medium 6oz-8oz 2.5" Red	Tomato
Lycopersicon	*esculentum-rubrum-medium*	*red-stari*	Red Star	Medium 6oz-8oz 2.5" Red	Tomato
Lycopersicon	*esculentum-rubrum-medium*	*red-supremei*	Red Supreme	Medium 6oz-8oz 2.5" Red	Tomato
Lycopersicon	*esculentum-rubrum-medium*	*rentita*	Rentita	Medium 6oz-8oz 2.5" Red	Tomato
Lycopersicon	*esculentum-rubrum-medium*	*rief's-redii*	Rief's	Medium 6oz-8oz 2.5" Red	Tomato
Lycopersicon	*esculentum-rubrum-medium*	*rio-fuegoi*	Rio Fuego	Medium 6oz-8oz 2.5" Red	Tomato
Lycopersicon	*esculentum-rubrum-medium*	*robert-wolfe'si*	Robert Wolfe's	Medium 6oz-8oz 2.5" Red	Tomato
Lycopersicon	*esculentum-rubrum-medium*	*rockinghamii*	Rockingham	Medium 6oz-8oz 2.5" Red	Tomato
Lycopersicon	*esculentum-rubrum-medium*	*rosabeci*	Rosabec	Medium 6oz-8oz 2.5" Red	Tomato
Lycopersicon	*esculentum-rubrum-medium*	*rousichii*	Rousich	Medium 6oz-8oz 2.5" Red	Tomato
Lycopersicon	*esculentum-rubrum-medium*	*row-paci*	Row Pack	Medium 6oz-8oz 2.5" Red	Tomato
Lycopersicon	*esculentum-rubrum-medium*	*ruby-treasurei*	RubyTreasure	Medium 6oz-8oz 2.5" Red	Tomato
Lycopersicon	*esculentum-rubrum-medium*	*russiana-redii*	Russian Red	Medium 6oz-8oz 2.5" Red	Tomato
Lycopersicon	*esculentum-rubrum-medium*	*rutgersi*	Rutgers	Medium 6oz-8oz 2.5" Red	Tomato
Lycopersicon	*esculentum-rubrum-medium*	*rutgersi-california-supremei*	California Supreme Rutgers	Medium 6oz-8oz 2.5" Red	Tomato
Lycopersicon	*esculentum-rubrum-medium*	*rutgersi-improved-VF*	Verticillatum Fusarium Resistant Improved Rutgers	Medium 6oz-8oz 2.5" Red	Tomato
Lycopersicon	*esculentum-rubrum-medium*	*rutgersi-PS*	Productive Strain Rutgers	Medium 6oz-8oz 2.5" Red	Tomato
Lycopersicon	*esculentum-rubrum-medium*	*rutgersi-selectii*	Select Rutgers	Medium 6oz-8oz 2.5" Red	Tomato
Lycopersicon	*esculentum-rubrum-medium*	*ruth's-perfectii*	Ruth's Perfect	Medium 6oz-8oz 2.5" Red	Tomato
Lycopersicon	*esculentum-rubrum-medium*	*saltspring-sunrisei*	Saltspring Sunrise	Medium 6oz-8oz 2.5" Red	Tomato
Lycopersicon	*esculentum-rubrum-medium*	*santa-clara-canneri*	Santa Clara Canner	Medium 6oz-8oz 2.5" Red	Tomato
Lycopersicon	*esculentum-rubrum-medium*	*sierra-sweetii*	Sierra Sweet	Medium 6oz-8oz 2.5" Red	Tomato
Lycopersicon	*esculentum-rubrum-medium*	*sigma-bushii*	Sigma Bush	Medium 6oz-8oz 2.5" Red	Tomato
Lycopersicon	*esculentum-rubrum-medium*	*smokey-mountainii*	Smokey Mountain	Medium 6oz-8oz 2.5" Red	Tomato
Lycopersicon	*esculentum-rubrum-medium*	*sophie's-choicei*	Sophie's Choice	Medium 6oz-8oz 2.5" Red	Tomato
Lycopersicon	*esculentum-rubrum-medium*	*spanish-pastei*	Spanish Paste	Medium 6oz-8oz 2.5" Red	Tomato
Lycopersicon	*esculentum-rubrum-medium*	*spanish-sunii*	Spanish Sun	Medium 6oz-8oz 2.5" Red	Tomato
Lycopersicon	*esculentum-rubrum-medium*	*stakelessii*	Stakeless	Medium 6oz-8oz 2.5" Red	Tomato
Lycopersicon	*esculentum-rubrum-medium*	*stuffingii-canadiana*	Canadian Stuffing	Medium 6oz-8oz 2.5" Red	Tomato
Lycopersicon	*esculentum-rubrum-medium*	*suncoastii-VFF*	Suncoast Very Fine Florida	Medium 6oz-8oz 2.5" Red	Tomato
Lycopersicon	*esculentum-rubrum-medium*	*sunseedsi-5715*	#5715 Sunseeds	Medium 6oz-8oz 2.5" Red	Tomato
Lycopersicon	*esculentum-rubrum-medium*	*sunseedsi-6002*	#6002 Sunseeds	Medium 6oz-8oz 2.5" Red	Tomato
Lycopersicon	*esculentum-rubrum-medium*	*sybirski-wczesnyi*	Sybirski Wczesny	Medium 6oz-8oz 2.5" Red	Tomato
Lycopersicon	*esculentum-rubrum-medium*	*tadessei*	Tadesse	Medium 6oz-8oz 2.5" Red	Tomato
Lycopersicon	*esculentum-rubrum-medium*	*teepeei*	Teepee	Medium 6oz-8oz 2.5" Red	Tomato
Lycopersicon	*esculentum-rubrum-medium*	*TH318*	#318 Tomato Hybrid	Medium 6oz-8oz 2.5" Red	Tomato
Lycopersicon	*esculentum-rubrum-medium*	*three-sistersi-salada*	Salad Three Sisters	Medium 6oz-8oz 2.5" Red	Tomato
Lycopersicon	*esculentum-rubrum-medium*	*tibet-appelii*	Tibet Appel	Medium 6oz-8oz 2.5" Red	Tomato
Lycopersicon	*esculentum-rubrum-medium*	*unwins-amateuri*	Unwins Amateur	Medium 6oz-8oz 2.5" Red	Tomato
Lycopersicon	*esculentum-rubrum-medium*	*varibestii*	Varibest	Medium 6oz-8oz 2.5" Red	Tomato
Lycopersicon	*esculentum-rubrum-medium*	*vatanii*	Vatan	Medium 6oz-8oz 2.5" Red	Tomato
Lycopersicon	*esculentum-rubrum-medium*	*vendori*	Vendor	Medium 6oz-8oz 2.5" Red	Tomato
Lycopersicon	*esculentum-rubrum-medium*	*viktorina*	Viktorina	Medium 6oz-8oz 2.5" Red	Tomato
Lycopersicon	*esculentum-rubrum-medium*	*visitation-valleyi*	Visitation Valley	Medium 6oz-8oz 2.5" Red	Tomato
Lycopersicon	*esculentum-rubrum-medium*	*vividii*	Vivid	Medium 6oz-8oz 2.5" Red	Tomato
Lycopersicon	*esculentum-rubrum-medium*	*volgogradskiyei-323*	#323 Volgogradskyie	Medium 6oz-8oz 2.5" Red	Tomato
Lycopersicon	*esculentum-rubrum-medium*	*walteri*	Walter	Medium 6oz-8oz 2.5" Red	Tomato
Lycopersicon	*esculentum-rubrum-medium*	*walteri-improvedii*	Improved Walter	Medium 6oz-8oz 2.5" Red	Tomato
Lycopersicon	*esculentum-rubrum-medium*	*walteri-villemairei*	Villemaire Walter	Medium 6oz-8oz 2.5" Red	Tomato
Lycopersicon	*esculentum-rubrum-medium*	*world-recordii*	World Record	Medium 6oz-8oz 2.5" Red	Tomato
Lycopersicon	*esculentum-rubrum-medium*	*yukonensis*	Yukon	Medium 6oz-8oz 2.5" Red	Tomato
Lycopersicon	*esculentum-rubrum-parvum*	*abraham-lincolnii-originalis*	Original Abraham Lincoln	Small 4oz-6oz 2.0" Red	Tomato
Lycopersicon	*esculentum-rubrum-parvum*	*abraham-lincolnii-regularis*	Regular Abraham Lincoln	Small 4oz-6oz 2.0" Red	Tomato
Lycopersicon	*esculentum-rubrum-parvum*	*AH-3*	#3 American Homestead	Small 4oz-6oz 2.0" Red	Tomato
Lycopersicon	*esculentum-rubrum-parvum*	*AH-LO*	Low Acid American Homestead	Small 4oz-6oz 2.0" Red	Tomato
Lycopersicon	*esculentum-rubrum-parvum*	*alberta-peachii*	Alberta Peach	Small 4oz-6oz 2.0" Red	Tomato
Lycopersicon	*esculentum-rubrum-parvum*	*alice-rooseveltii*	Alice Roosevelt	Small 4oz-6oz 2.0" Red	Tomato
Lycopersicon	*esculentum-rubrum-parvum*	*alfresco*	Alfresco	Small 4oz-6oz 2.0" Red	Tomato
Lycopersicon	*esculentum-rubrum-parvum*	*allerbestii*	Allerbest	Small 4oz-6oz 2.0" Red	Tomato
Lycopersicon	*esculentum-rubrum-parvum*	*amish-pastei*	Amish Paste	Small 4oz-6oz 2.0" Red	Tomato
Lycopersicon	*esculentum-rubrum-parvum*	*amish-redii*	Amish Red	Small 4oz-6oz 2.0" Red	Tomato
Lycopersicon	*esculentum-rubrum-parvum*	*arctic-blumii*	Arctic Blum	Small 4oz-6oz 2.0" Red	Tomato
Lycopersicon	*esculentum-rubrum-parvum*	*aurora*	Aurora	Small 4oz-6oz 2.0" Red	Tomato

Genus	Species	Infraspecies	cultivar	specimen	Genome
Lycopersicon	*esculentum-rubrum-parvum*	*AYWC*	AYWC	Small 4oz-6oz 2.0" Red	Tomato
Lycopersicon	*esculentum-rubrum-parvum*	*azteca-B*	Azteca B	Small 4oz-6oz 2.0" Red	Tomato
Lycopersicon	*esculentum-rubrum-parvum*	*basket-veei*	Basket Vee	Small 4oz-6oz 2.0" Red	Tomato
Lycopersicon	*esculentum-rubrum-parvum*	*BC-2*	Bulgarian Crop #2	Small 4oz-6oz 2.0" Red	Tomato
Lycopersicon	*esculentum-rubrum-parvum*	*beautii*	Beaut	Small 4oz-6oz 2.0" Red	Tomato
Lycopersicon	*esculentum-rubrum-parvum*	*beginneri*	Beginner	Small 4oz-6oz 2.0" Red	Tomato
Lycopersicon	*esculentum-rubrum-parvum*	*big-brotheri*	Big Brother	Small 4oz-6oz 2.0" Red	Tomato
Lycopersicon	*esculentum-rubrum-parvum*	*bitchyeh-gertzeh*	Bitchyeh Gertzeh	Small 4oz-6oz 2.0" Red	Tomato
Lycopersicon	*esculentum-rubrum-parvum*	*blizzardii*	Blizzard	Small 4oz-6oz 2.0" Red	Tomato
Lycopersicon	*esculentum-rubrum-parvum*	*bloody-butcheri*	Bloody Butcher	Small 4oz-6oz 2.0" Red	Tomato
Lycopersicon	*esculentum-rubrum-parvum*	*buckbee's-new-fifty-dayi*	Buckbee's New Fifty Day	Small 4oz-6oz 2.0" Red	Tomato
Lycopersicon	*esculentum-rubrum-parvum*	*bulgariana-triumphii*	Triumph Bulgarian	Small 4oz-6oz 2.0" Red	Tomato
Lycopersicon	*esculentum-rubrum-parvum*	*burbankii*	Burbank	Small 4oz-6oz 2.0" Red	Tomato
Lycopersicon	*esculentum-rubrum-parvum*	*cabotii*	Cabot	Small 4oz-6oz 2.0" Red	Tomato
Lycopersicon	*esculentum-rubrum-parvum*	*cal-culteratori*	Cal Culterator	Small 4oz-6oz 2.0" Red	Tomato
Lycopersicon	*esculentum-rubrum-parvum*	*canabec-rosei*	Canada Quebec Rose	Small 4oz-6oz 2.0" Red	Tomato
Lycopersicon	*esculentum-rubrum-parvum*	*canabec-rougei*	Canada Quebec Rouge	Small 4oz-6oz 2.0" Red	Tomato
Lycopersicon	*esculentum-rubrum-parvum*	*canabec-superi*	Canada Quebec Super	Small 4oz-6oz 2.0" Red	Tomato
Lycopersicon	*esculentum-rubrum-parvum*	*cannonballii*	Cannonball	Small 4oz-6oz 2.0" Red	Tomato
Lycopersicon	*esculentum-rubrum-parvum*	*chalks-early-jewelii*	Chalks Early Jewel	Small 4oz-6oz 2.0" Red	Tomato
Lycopersicon	*esculentum-rubrum-parvum*	*charlie-chaplinii*	Charlie Chaplin	Small 4oz-6oz 2.0" Red	Tomato
Lycopersicon	*esculentum-rubrum-parvum*	*charlie's-red-stakeri*	Charlie's Red Staker	Small 4oz-6oz 2.0" Red	Tomato
Lycopersicon	*esculentum-rubrum-parvum*	*chipollinoi*	Chipollino	Small 4oz-6oz 2.0" Red	Tomato
Lycopersicon	*esculentum-rubrum-parvum*	*cold-setii*	Cold Set	Small 4oz-6oz 2.0" Red	Tomato
Lycopersicon	*esculentum-rubrum-parvum*	*columbiensis*	Columbia	Small 4oz-6oz 2.0" Red	Tomato
Lycopersicon	*esculentum-rubrum-parvum*	*coop-redii*	Coop	Small 4oz-6oz 2.0" Red	Tomato
Lycopersicon	*esculentum-rubrum-parvum*	*counteri*	Counter	Small 4oz-6oz 2.0" Red	Tomato
Lycopersicon	*esculentum-rubrum-parvum*	*czech's-bushii*	Czech's Bush	Small 4oz-6oz 2.0" Red	Tomato
Lycopersicon	*esculentum-rubrum-parvum*	*danai*	Dana	Small 4oz-6oz 2.0" Red	Tomato
Lycopersicon	*esculentum-rubrum-parvum*	*dansk-exportii*	Dansk Export	Small 4oz-6oz 2.0" Red	Tomato
Lycopersicon	*esculentum-rubrum-parvum*	*davington's-epicurei*	Davington's Epicure	Small 4oz-6oz 2.0" Red	Tomato
Lycopersicon	*esculentum-rubrum-parvum*	*de-barao-redii*	De Barao	Small 4oz-6oz 2.0" Red	Tomato
Lycopersicon	*esculentum-rubrum-parvum*	*delmari*	Delaware Maryland	Small 4oz-6oz 2.0" Red	Tomato
Lycopersicon	*esculentum-rubrum-parvum*	*deutscher-fleissi*	Deutscher Fleiss	Small 4oz-6oz 2.0" Red	Tomato
Lycopersicon	*esculentum-rubrum-parvum*	*dwarf-earlyi-506*	#506 Dwarf Early	Small 4oz-6oz 2.0" Red	Tomato
Lycopersicon	*esculentum-rubrum-parvum*	*earlibrightii*	Earlibright	Small 4oz-6oz 2.0" Red	Tomato
Lycopersicon	*esculentum-rubrum-parvum*	*early-fruitii*	Early Fruit	Small 4oz-6oz 2.0" Red	Tomato
Lycopersicon	*esculentum-rubrum-parvum*	*early-stonei*	Early Stone	Small 4oz-6oz 2.0" Red	Tomato
Lycopersicon	*esculentum-rubrum-parvum*	*early-swedishii*	Early Swedish	Small 4oz-6oz 2.0" Red	Tomato
Lycopersicon	*esculentum-rubrum-parvum*	*edelrotii*	Edelrot	Small 4oz-6oz 2.0" Red	Tomato
Lycopersicon	*esculentum-rubrum-parvum*	*eurocrossii*	Eurocross	Small 4oz-6oz 2.0" Red	Tomato
Lycopersicon	*esculentum-rubrum-parvum*	*fireballii*	Fireball	Small 4oz-6oz 2.0" Red	Tomato
Lycopersicon	*esculentum-rubrum-parvum*	*first-pickii*	First Pick	Small 4oz-6oz 2.0" Red	Tomato
Lycopersicon	*esculentum-rubrum-parvum*	*floridadei*	Dade County Florida	Small 4oz-6oz 2.0" Red	Tomato
Lycopersicon	*esculentum-rubrum-parvum*	*floridelii*	Del Florida	Small 4oz-6oz 2.0" Red	Tomato
Lycopersicon	*esculentum-rubrum-parvum*	*gandia-lyc-77*	1977 Gandia	Small 4oz-6oz 2.0" Red	Tomato
Lycopersicon	*esculentum-rubrum-parvum*	*garden-peachii*	Fuzzy Garden Peach	Small 4oz-6oz 2.0" Red	Tomato
Lycopersicon	*esculentum-rubrum-parvum*	*garden-peachii-redii*	Smooth Garden Peach	Small 4oz-6oz 2.0" Red	Tomato
Lycopersicon	*esculentum-rubrum-parvum*	*geminii*	Gemini	Small 4oz-6oz 2.0" Red	Tomato
Lycopersicon	*esculentum-rubrum-parvum*	*germanica-bushyi*	German Bushy	Small 4oz-6oz 2.0" Red	Tomato
Lycopersicon	*esculentum-rubrum-parvum*	*gloriana-burpeei*	Burpee Gloriana	Small 4oz-6oz 2.0" Red	Tomato
Lycopersicon	*esculentum-rubrum-parvum*	*greater-baltimorei*	Greater Baltimore	Small 4oz-6oz 2.0" Red	Tomato
Lycopersicon	*esculentum-rubrum-parvum*	*heinzii-1370*	#1370 Heinz	Small 4oz-6oz 2.0" Red	Tomato
Lycopersicon	*esculentum-rubrum-parvum*	*hellfruchtii*	Hellfrucht	Small 4oz-6oz 2.0" Red	Tomato
Lycopersicon	*esculentum-rubrum-parvum*	*hubiei*	Hubie	Small 4oz-6oz 2.0" Red	Tomato
Lycopersicon	*esculentum-rubrum-parvum*	*irelandica*	Irish	Small 4oz-6oz 2.0" Red	Tomato
Lycopersicon	*esculentum-rubrum-parvum*	*italabeci*	Italabec	Small 4oz-6oz 2.0" Red	Tomato
Lycopersicon	*esculentum-rubrum-parvum*	*italiana-pastei*	Italian Paste	Small 4oz-6oz 2.0" Red	Tomato
Lycopersicon	*esculentum-rubrum-parvum*	*joe's-favoritei*	Joe's Favorite	Small 4oz-6oz 2.0" Red	Tomato
Lycopersicon	*esculentum-rubrum-parvum*	*kecskemetii-886*	#886 Kecskemeti	Small 4oz-6oz 2.0" Red	Tomato
Lycopersicon	*esculentum-rubrum-parvum*	*lafayettei*	Lafayette	Small 4oz-6oz 2.0" Red	Tomato
Lycopersicon	*esculentum-rubrum-parvum*	*manitobensis*	Manitoba	Small 4oz-6oz 2.0" Red	Tomato
Lycopersicon	*esculentum-rubrum-parvum*	*manitobensis-improvedii*	Improved Manitoba	Small 4oz-6oz 2.0" Red	Tomato
Lycopersicon	*esculentum-rubrum-parvum*	*maskabeci*	Maskabec	Small 4oz-6oz 2.0" Red	Tomato
Lycopersicon	*esculentum-rubrum-parvum*	*matina*	Matina	Small 4oz-6oz 2.0" Red	Tomato
Lycopersicon	*esculentum-rubrum-parvum*	*memoirsi*	Memoirs	Small 4oz-6oz 2.0" Red	Tomato
Lycopersicon	*esculentum-rubrum-parvum*	*miracle-of-worldii*	Miracle Of World	Small 4oz-6oz 2.0" Red	Tomato
Lycopersicon	*esculentum-rubrum-parvum*	*moneycrossii*	Moneycross	Small 4oz-6oz 2.0" Red	Tomato
Lycopersicon	*esculentum-rubrum-parvum*	*moneymakeri*	Moneymaker	Small 4oz-6oz 2.0" Red	Tomato
Lycopersicon	*esculentum-rubrum-parvum*	*moskvichii*	Moskvich	Small 4oz-6oz 2.0" Red	Tomato
Lycopersicon	*esculentum-rubrum-parvum*	*mountain-girlii-VF*	Mountain Girl Verticillium Fusarium Resistant	Small 4oz-6oz 2.0" Red	Tomato
Lycopersicon	*esculentum-rubrum-parvum*	*myona*	My Own (Myona)	Small 4oz-6oz 2.0" Red	Tomato
Lycopersicon	*esculentum-rubrum-parvum*	*neptunei*	Neptune	Small 4oz-6oz 2.0" Red	Tomato

Genus	Species	Infraspecies	cultivar	specimen	Genome
Lycopersicon	*esculentum-rubrum-parvum*	*northern-delightii*	Northern Delight	Small 4oz-6oz 2.0" Red	Tomato
Lycopersicon	*esculentum-rubrum-parvum*	*nota*	Nota	Small 4oz-6oz 2.0" Red	Tomato
Lycopersicon	*esculentum-rubrum-parvum*	*odessa*	Odessa	Small 4oz-6oz 2.0" Red	Tomato
Lycopersicon	*esculentum-rubrum-parvum*	*odessa-2*	#2 Odessa	Small 4oz-6oz 2.0" Red	Tomato
Lycopersicon	*esculentum-rubrum-parvum*	*olkhon-islandica*	Olkhon Island	Small 4oz-6oz 2.0" Red	Tomato
Lycopersicon	*esculentum-rubrum-parvum*	*onyxii*	Onyx	Small 4oz-6oz 2.0" Red	Tomato
Lycopersicon	*esculentum-rubrum-parvum*	*patio-kingii*	Patio King	Small 4oz-6oz 2.0" Red	Tomato
Lycopersicon	*esculentum-rubrum-parvum*	*payettei*	Payette	Small 4oz-6oz 2.0" Red	Tomato
Lycopersicon	*esculentum-rubrum-parvum*	*peremogai*	Peremoga	Small 4oz-6oz 2.0" Red	Tomato
Lycopersicon	*esculentum-rubrum-parvum*	*petoi-76*	#76 Peeling Tomato	Small 4oz-6oz 2.0" Red	Tomato
Lycopersicon	*esculentum-rubrum-parvum*	*petoi-86*	#86 Peeling Tomato	Small 4oz-6oz 2.0" Red	Tomato
Lycopersicon	*esculentum-rubrum-parvum*	*petoi-mechii*	Mechanical Harvest Peeling Tomato	Small 4oz-6oz 2.0" Red	Tomato
Lycopersicon	*esculentum-rubrum-parvum*	*petoi-mechii-2*	#2 Mechanical Harvest Peeling Tomato	Small 4oz-6oz 2.0" Red	Tomato
Lycopersicon	*esculentum-rubrum-parvum*	*plainsmanii*	Plainsman	Small 4oz-6oz 2.0" Red	Tomato
Lycopersicon	*esculentum-rubrum-parvum*	*plumpton-kingii*	Plumpton King	Small 4oz-6oz 2.0" Red	Tomato
Lycopersicon	*esculentum-rubrum-parvum*	*porteri*	Porter	Small 4oz-6oz 2.0" Red	Tomato
Lycopersicon	*esculentum-rubrum-parvum*	*porteri-improvedii*	Improved Porter	Small 4oz-6oz 2.0" Red	Tomato
Lycopersicon	*esculentum-rubrum-parvum*	*PSR-037*	#037 Peters Seed & Research	Small 4oz-6oz 2.0" Red	Tomato
Lycopersicon	*esculentum-rubrum-parvum*	*PSR-286*	#286 Peters Seed & Research	Small 4oz-6oz 2.0" Red	Tomato
Lycopersicon	*esculentum-rubrum-parvum*	*PSR-546*	#546 Peters Seed & Research	Small 4oz-6oz 2.0" Red	Tomato
Lycopersicon	*esculentum-rubrum-parvum*	*PSR-933V*	Vendor #933 Peters Seed & Research	Small 4oz-6oz 2.0" Red	Tomato
Lycopersicon	*esculentum-rubrum-parvum*	*PSR-951*	#951 Peters Seed & Research	Small 4oz-6oz 2.0" Red	Tomato
Lycopersicon	*esculentum-rubrum-parvum*	*PSR-953*	#953 Peters Seed & Research	Small 4oz-6oz 2.0" Red	Tomato
Lycopersicon	*esculentum-rubrum-parvum*	*PSR-954*	#954 Peters Seed & Research	Small 4oz-6oz 2.0" Red	Tomato
Lycopersicon	*esculentum-rubrum-parvum*	*PSR-957*	#957 Peters Seed & Research	Small 4oz-6oz 2.0" Red	Tomato
Lycopersicon	*esculentum-rubrum-parvum*	*PSR-959*	#958 Peters Seed & Research	Small 4oz-6oz 2.0" Red	Tomato
Lycopersicon	*esculentum-rubrum-parvum*	*PSR-1071*	#1071 Peters Seed & Research	Small 4oz-6oz 2.0" Red	Tomato
Lycopersicon	*esculentum-rubrum-parvum*	*quintei*	Quinte	Small 4oz-6oz 2.0" Red	Tomato
Lycopersicon	*esculentum-rubrum-parvum*	*red-cupii*	Red Cup	Small 4oz-6oz 2.0" Red	Tomato
Lycopersicon	*esculentum-rubrum-parvum*	*red-kingii*	Red King	Small 4oz-6oz 2.0" Red	Tomato
Lycopersicon	*esculentum-rubrum-parvum*	*rodadei*	Round Dade County	Small 4oz-6oz 2.0" Red	Tomato
Lycopersicon	*esculentum-rubrum-parvum*	*ronaclavei*	Ronaclave	Small 4oz-6oz 2.0" Red	Tomato
Lycopersicon	*esculentum-rubrum-parvum*	*san-francisco-fogii*	San Francisco Fog	Small 4oz-6oz 2.0" Red	Tomato
Lycopersicon	*esculentum-rubrum-parvum*	*santiamii*	Santiam	Small 4oz-6oz 2.0" Red	Tomato
Lycopersicon	*esculentum-rubrum-parvum*	*santiamii-PL*	Peters Late Santiam	Small 4oz-6oz 2.0" Red	Tomato
Lycopersicon	*esculentum-rubrum-parvum*	*scotia*	Scotia	Small 4oz-6oz 2.0" Red	Tomato
Lycopersicon	*esculentum-rubrum-parvum*	*sequoia-alpinei*	Sequoia Alpine	Small 4oz-6oz 2.0" Red	Tomato
Lycopersicon	*esculentum-rubrum-parvum*	*sergey's-ukrainian-redii*	Sergey's Ukrainian Red	Small 4oz-6oz 2.0" Red	Tomato
Lycopersicon	*esculentum-rubrum-parvum*	*shirleyi*	Shirley	Small 4oz-6oz 2.0" Red	Tomato
Lycopersicon	*esculentum-rubrum-parvum*	*siberica*	Siberia	Small 4oz-6oz 2.0" Red	Tomato
Lycopersicon	*esculentum-rubrum-parvum*	*siouxii*	Sioux	Small 4oz-6oz 2.0" Red	Tomato
Lycopersicon	*esculentum-rubrum-parvum*	*siouxii-supra*	Super Sioux	Small 4oz-6oz 2.0" Red	Tomato
Lycopersicon	*esculentum-rubrum-parvum*	*st.-pierrei*	St. Pierre	Small 4oz-6oz 2.0" Red	Tomato
Lycopersicon	*esculentum-rubrum-parvum*	*stupicei*	Stupice	Small 4oz-6oz 2.0" Red	Tomato
Lycopersicon	*esculentum-rubrum-parvum*	*subarctic-pastei*	Subarctic Paste	Small 4oz-6oz 2.0" Red	Tomato
Lycopersicon	*esculentum-rubrum-parvum*	*subarctic-plentyi*	Subarctic Plenty	Small 4oz-6oz 2.0" Red	Tomato
Lycopersicon	*esculentum-rubrum-parvum*	*sunrisei-V*	#5 Sunrise	Small 4oz-6oz 2.0" Red	Tomato
Lycopersicon	*esculentum-rubrum-parvum*	*superfantastica*	SuperFantastic	Small 4oz-6oz 2.0" Red	Tomato
Lycopersicon	*esculentum-rubrum-parvum*	*swiss-alpinea*	Swiss Alpine	Small 4oz-6oz 2.0" Red	Tomato
Lycopersicon	*esculentum-rubrum-parvum*	*tappy's-heritagei*	Tappy's Heritage	Small 4oz-6oz 2.0" Red	Tomato
Lycopersicon	*esculentum-rubrum-parvum*	*thessaloniki*	Thessaloniki	Small 4oz-6oz 2.0" Red	Tomato
Lycopersicon	*esculentum-rubrum-parvum*	*togorifica*	Togorific	Small 4oz-6oz 2.0" Red	Tomato
Lycopersicon	*esculentum-rubrum-parvum*	*tommy-toei*	TommyToe	Small 4oz-6oz 2.0" Red	Tomato
Lycopersicon	*esculentum-rubrum-parvum*	*trauffaut-precocei*	Traffaut Precoce	Small 4oz-6oz 2.0" Red	Tomato
Lycopersicon	*esculentum-rubrum-parvum*	*trimsonii*	Trimson	Small 4oz-6oz 2.0" Red	Tomato
Lycopersicon	*esculentum-rubrum-parvum*	*trophyi*	Trophy	Small 4oz-6oz 2.0" Red	Tomato
Lycopersicon	*esculentum-rubrum-parvum*	*vendori-VFT*	Verticilium Fusarium Tobacco Mosaic Resistant	Small 4oz-6oz 2.0" Red	Tomato
Lycopersicon	*esculentum-rubrum-parvum*	*VF-10*	#10 Verticilium Fusarium Resistant	Small 4oz-6oz 2.0" Red	Tomato
Lycopersicon	*esculentum-rubrum-parvum*	*VF145B-7879*	#145B-7879 Verticilium Fusarium Resistant	Small 4oz-6oz 2.0" Red	Tomato
Lycopersicon	*esculentum-rubrum-parvum*	*yuliana*	Yuliana	Small 4oz-6oz 2.0" Red	Tomato
Lycopersicon	*esculentum-rubrum-parvum*	*zieglers-fleischii*	Zieglers Fleisch	Small 4oz-6oz 2.0" Red	Tomato
Lycopersicon	*esculentum-rubrum-parvum*	*yellow-out-red-inii*	YellowSkin	Small 4oz-6oz 2.0" Red	Tomato

Tomato, Orange Red (*Lycopersicon esculentum-rubrum-aurantium*)

Genus	Species	Infraspecies	cultivar	specimen	Genome
Lycopersicon	*esculentum-rubrum-aurantium-largum*	*coustraleei-orange-redii*	Coulstralee	Large 2lb 4.0" Orange Red	Tomato
Lycopersicon	*esculentum-rubrum-aurantium-largum-medium*	*red-raspii*	Red Rasp	Medium Large 1lb 3.5" Orange Red	Tomato
Lycopersicon	*esculentum-rubrum-aurantium-largum-medium*	*schelleberg's-favoritei*	Schellenberg's Favorite	Medium Large 1lb 3.5" Orange Red	Tomato
Lycopersicon	*esculentum-rubrum-aurantium-medium-largum*	*indian-riveri*	Indian River	Large Medium ¾lb 3.0" Orange Red	Tomato

Genus	Species	Infraspecies	cultivar	specimen	Genome
Lycopersicon	*esculentum-rubrum-aurantium-medium-largum*	*new-round-pastei*	New Round Paste	Large Medium ¾lb 3.0" Orange Red	Tomato
Lycopersicon	*esculentum-rubrum-aurantium-medium*	*big-tigeri*	Big Tiger	Medium ½lb 2.5" Orange Red	Tomato
Lycopersicon	*esculentum-rubrum-aurantium-medium*	*bonny-bestii*	Bonny Best	Medium ½lb 2.5" Orange Red	Tomato
Lycopersicon	*esculentum-rubrum-aurantium-medium*	*bonny-bestii-pollockii*	Pollock Bonny Best	Medium ½lb 2.5" Orange Red	Tomato
Lycopersicon	*esculentum-rubrum-aurantium-medium*	*bountyi*	Bounty	Medium ½lb 2.5" Orange Red	Tomato
Lycopersicon	*esculentum-rubrum-aurantium-medium*	*chesapeakei*	Chesapeake	Medium ½lb 2.5" Orange Red	Tomato
Lycopersicon	*esculentum-rubrum-aurantium-medium*	*dad's-mugii*	Dad's Mug	Medium ½lb 2.5" Orange Red	Tomato
Lycopersicon	*esculentum-rubrum-aurantium-medium*	*glasnostii*	Glasnost	Medium ½lb 2.5" Orange Red	Tomato
Lycopersicon	*esculentum-rubrum-aurantium-medium*	*harzfeueri*	Harzfeuer	Medium ½lb 2.5" Orange Red	Tomato
Lycopersicon	*esculentum-rubrum-aurantium-medium*	*hastings-everbearing-scarlet-globei*	Hastings Everbearing Scarlet Globe ?oz	Medium ½lb 2.5" Orange Red	Tomato
Lycopersicon	*esculentum-rubrum-aurantium-medium*	*italiana*	Italian Sweet	Medium ½lb 2.5" Orange Red	Tomato
Lycopersicon	*esculentum-rubrum-aurantium-medium*	*karzetek-tallinskii*	Karzetek Tallinski	Medium ½lb 2.5" Orange Red	Tomato
Lycopersicon	*esculentum-rubrum-aurantium-medium*	*perestroikai*	Perestroika	Medium ½lb 2.5" Orange Red	Tomato
Lycopersicon	*esculentum-rubrum-aurantium-medium*	*red-lightningii*	Red Lightning	Medium ½lb 2.5" Orange Red	Tomato
Lycopersicon	*esculentum-rubrum-aurantium-medium*	*stonei*	Stone	Medium ½lb 2.5" Orange Red	Tomato
Lycopersicon	*esculentum-rubrum-aurantium-medium*	*stonei-dwarfii*	Dwarf Stone	Medium ½lb 2.5" Orange Red	Tomato
Lycopersicon	*esculentum-rubrum-aurantium-medium*	*sunsetii*	Sunset 5oz-7oz	Medium ½lb 2.5" Orange Red	Tomato
Lycopersicon	*esculentum-rubrum-aurantium-medium*	*sztambowyi-karlikowyi*	Sztambowyi Karlikowy	Medium ½lb 2.5" Orange Red	Tomato
Lycopersicon	*esculentum-rubrum-aurantium-medium*	*wayaheadii-jung's-improvedii*	Jung's Improved Wayahead	Medium ½lb 2.5" Orange Red	Tomato
Lycopersicon	*esculentum-rubrum-aurantium-medium*	*wayaheadii-super-selectii*	Super Select Wayahead	Medium ½lb 2.5" Orange Red	Tomato
Lycopersicon	*esculentum-rubrum-aurantium-parvum*	*donai*	Dona	Small ¼lb 2.0" Orange Red	Tomato

Tomato, Striped (*Lycopersicon esculentum-striatum*)

Genus	Species	Infraspecies	cultivar	specimen	Genome
Lycopersicon	*esculentum-striatum*	*aviuri*	Aviuri 6oz	Striped	Tomato
Lycopersicon	*esculentum-striatum*	*big-rainbowii*	Big Rainbow 1lb	Striped	Tomato
Lycopersicon	*esculentum-striatum*	*big-white-pink-stripesi*	Big White Pink Stripes 1lb	Striped	Tomato
Lycopersicon	*esculentum-striatum*	*deweese-streakedii*	Deweese Streaked ?oz	Striped	Tomato
Lycopersicon	*esculentum-striatum*	*early-yellow-stripei*	Early Yellow Stripe 2"	Striped	Tomato
Lycopersicon	*esculentum-striatum*	*elberta-girlii*	Elberta Girl 2.5"	Striped	Tomato
Lycopersicon	*esculentum-striatum*	*esikos-botermoi*	Esikos Botermo 2.5"	Striped	Tomato
Lycopersicon	*esculentum-striatum*	*eva's-amish-stripei*	Eva's Amish Stripe 2lb	Striped	Tomato
Lycopersicon	*esculentum-striatum*	*georgia-streakii*	Georgia Steak 1.5lb	Striped	Tomato
Lycopersicon	*esculentum-striatum*	*german-goldii*	German Gold 1.5lb	Striped	Tomato
Lycopersicon	*esculentum-striatum*	*german-goldii-mammothii*	Mammoth German Gold 2lb	Striped	Tomato
Lycopersicon	*esculentum-striatum*	*german-striped-stufferi*	German Striped Stuffer ?lb	Striped	Tomato
Lycopersicon	*esculentum-striatum*	*green-zebrai*	Green Zebra 2"	Striped	Tomato
Lycopersicon	*esculentum-striatum*	*marvel-stripedii*	Marvel Striped 2lb	Striped	Tomato
Lycopersicon	*esculentum-striatum*	*old-yellow-candystripei*	Old Yellow Candystripe ?lb	Striped	Tomato
Lycopersicon	*esculentum-striatum*	*olympic-flamei*	Olympic Flame 12oz-24oz	Striped	Tomato
Lycopersicon	*esculentum-striatum*	*peach-blow-suttonii*	Peach Blow Sutton ?oz	Striped	Tomato
Lycopersicon	*esculentum-striatum*	*peppermintii*	Peppermint 8oz-12oz	Striped	Tomato
Lycopersicon	*esculentum-striatum*	*pixie-stripedii*	Pixie Striped 2"	Striped	Tomato
Lycopersicon	*esculentum-striatum*	*red-georgiensis*	Red Georgia 2.5"	Striped	Tomato
Lycopersicon	*esculentum-striatum*	*red-gold-stripei*	Red Gold Stripe 8oz	Striped	Tomato
Lycopersicon	*esculentum-striatum*	*red-zebrai*	Red Zebra ?oz	Striped	Tomato
Lycopersicon	*esculentum-striatum*	*regina's-yellowi*	Regina's Yellow 8oz	Striped	Tomato
Lycopersicon	*esculentum-striatum*	*rose-beautyi*	Rose Beauty ?oz	Striped	Tomato
Lycopersicon	*esculentum-striatum*	*ruby-goldii*	Ruby Gold ?oz	Striped	Tomato
Lycopersicon	*esculentum-striatum*	*schimmeig-cregii*	Creg Schimmeig 4oz-6oz	Striped	Tomato
Lycopersicon	*esculentum-striatum*	*schimmeig-striped-hollowii*	Striped Hollow Schimmeig 3oz-5oz	Striped	Tomato
Lycopersicon	*esculentum-striatum*	*striped-cavernii*	Striped Cavern 8oz	Striped	Tomato
Lycopersicon	*esculentum-striatum*	*striped-germanii*	Striped German 2lb	Striped	Tomato
Lycopersicon	*esculentum-striatum*	*striped-roman-yellowi*	Striped Roman Yellow ?oz	Striped	Tomato
Lycopersicon	*esculentum-striatum*	*striped-stuffer*	Striped Stuffer 6oz	Striped	Tomato
Lycopersicon	*esculentum-striatum*	*tigerella*	Tigerella 1.75"	Striped	Tomato
Lycopersicon	*esculentum-striatum*	*tigeri*	Tiger 2"	Striped	Tomato
Lycopersicon	*esculentum-striatum*	*tigeri-fuzzii*	Tiger Fuzz ?oz	Striped	Tomato
Lycopersicon	*esculentum-striatum*	*tigeri-likei*	Tiger Like 2"	Striped	Tomato
Lycopersicon	*esculentum-striatum*	*tigeri-stripei*	Tiger Stripe ?oz	Striped	Tomato
Lycopersicon	*esculentum-striatum*	*tigeri-tomii*	Tiger Tom 2.5"	Striped	Tomato
Lycopersicon	*esculentum-striatum*	*tigerlyi*	Tigerly ?oz	Striped	Tomato
Lycopersicon	*esculentum-striatum*	*tomangoi*	Tomango Hollow ?oz	Striped	Tomato
Lycopersicon	*esculentum-striatum*	*turkish-striped-monasteryi*	Turkish Striped Monastery 2"	Striped	Tomato
Lycopersicon	*esculentum-striatum*	*wagon-wheelii*	Wagon Wheel 1b	Striped	Tomato
Lycopersicon	*esculentum-striatum*	*ziganii*	Zigan ?oz	Striped	Tomato

Tomato, Long (*Lycopersicon esculentum-longum*)

Genus	Species	Infraspecies	cultivar	specimen	Genome
Lycopersicon	*longum-aurantium*	*bananai*	Banana 2" x 4"	Orange Long	Tomato
Lycopersicon	*longum-aurantium*	*bananai-orangei*	Orange Banana 2" x 4"	Orange Long	Tomato
Lycopersicon	*longum-aurantium*	*pirkstine-orangei*	Perkstine ?" x ?"	Orange Long	Tomato
Lycopersicon	*longum-groenum*	*green-sausagei*	Green Sausage ?" x ?"	Green Long	Tomato

Genus	Species	Infraspecies	cultivar	specimen	Genome
Lycopersicon	*longum-helvum*	*banana-legsi*	Legs Banana 2" x 4"	Yellow Long	Tomato
Lycopersicon	*longum-helvum*	*brown's-yellow-giantii*	Brown's Yellow Giant 3" x 5"	Yellow Long	Tomato
Lycopersicon	*longum-helvum*	*pirkstine-yellowi*	Perkstine ?" x ?"	Yellow Long	Tomato
Lycopersicon	*longum-helvum*	*roman-candlei*	Roman Candle 2" x 4"	Yellow Long	Tomato
Lycopersicon	*longum-helvum*	*russian-limei*	Russian Lime 2" x 4"	Yellow Long	Tomato
Lycopersicon	*longum-helvum*	*yellow-bellii*	Yellow Bell 1.5" x 4"	Yellow Long	Tomato
Lycopersicon	*longum-pincum*	*large-pinkii*	Large 2.5" x 4"	Pink Long	Tomato
Lycopersicon	*longum-pincum*	*khaborovskyi*	Khaborovsky 6oz	Pink Long	Tomato
Lycopersicon	*longum-pincum*	*khaborovskyi-308*	#308 Khaborovsky 6oz	Pink Long	Tomato
Lycopersicon	*longum-pincum*	*pink-stufferi*	Pink Stuffer ?" x ?"	Pink Long	Tomato
Lycopersicon	*longum-rubrum*	*andesensis*	Andes 2" x 5"	Red Long	Tomato
Lycopersicon	*longum-rubrum*	*aunt-lucy's-italian-pastei*	Aunt Lucy's Italian Paste ?" x ?"	Red Long	Tomato
Lycopersicon	*longum-rubrum*	*aunt-mary's-pastei*	Aunt Mary's Paste 1.5" x 3.0"	Red Long	Tomato
Lycopersicon	*longum-rubrum*	*bananai-1-x-6*	Banana 1" x 6"	Red Long	Tomato
Lycopersicon	*longum-rubrum*	*beach-boyi*	Beach Boy Orange Red 2" x 6"	Red Long	Tomato
Lycopersicon	*longum-rubrum*	*bearoi*	Bearo ?" x ?"	Red Long	Tomato
Lycopersicon	*longum-rubrum*	*berwick-germanii*	Berwick German 2" x 4"	Red Long	Tomato
Lycopersicon	*longum-rubrum*	*big-italian-pastei*	Big Italian Paste 2" x 4"	Red Long	Tomato
Lycopersicon	*longum-rubrum*	*big-italian-plumii*	Big Italian Plum 2" x 4"	Red Long	Tomato
Lycopersicon	*longum-rubrum*	*big-ray's-argentina-pastei*	Big Ray's Argentina Paste 2" x 6"	Red Long	Tomato
Lycopersicon	*longum-rubrum*	*bull-saci*	Bull Sac 3" x 6"	Red Long	Tomato
Lycopersicon	*longum-rubrum*	*carrotii*	Carrot 1" x 4"	Red Long	Tomato
Lycopersicon	*longum-rubrum*	*centennial-rocketii*	Centennial Rocket ?" x ?"	Red Long	Tomato
Lycopersicon	*longum-rubrum*	*chinensis-rubrum*	Chinese 2" x 7"	Red Long	Tomato
Lycopersicon	*longum-rubrum*	*corne-de-bouci*	Corne De Bouc ?" x 6"	Red Long	Tomato
Lycopersicon	*longum-rubrum*	*dad's-barber-pastei*	Dad's Barber Paste ?" x 4"	Red Long	Tomato
Lycopersicon	*longum-rubrum*	*dix-doights-de-naplesi*	Dix Doights De Naples ?" x ?"	Red Long	Tomato
Lycopersicon	*longum-rubrum*	*eleanori*	Eleanor 2" x 4"	Red Long	Tomato
Lycopersicon	*longum-rubrum*	*ermakii*	Ermak ?" x ?"	Red Long	Tomato
Lycopersicon	*longum-rubrum*	*ernie's-pointedii*	Ernie's Pointed ?" x ?"	Red Long	Tomato
Lycopersicon	*longum-rubrum*	*federelei*	Federle ?" x 6"	Red Long	Tomato
Lycopersicon	*longum-rubrum*	*gilbertie-pastei*	Gilbertie Paste ?" x 7"	Red Long	Tomato
Lycopersicon	*longum-rubrum*	*grandma-maryi*	Grandma Mary ?" x 3"	Red Long	Tomato
Lycopersicon	*longum-rubrum*	*harry's-italian-plumii*	Harry's Italian Plum ?" x 7"	Red Long	Tomato
Lycopersicon	*longum-rubrum*	*horvath-plumii*	Horvath Plum ?" x 4"	Red Long	Tomato
Lycopersicon	*longum-rubrum*	*howard-germanii*	Howard German ?" x ?"	Red Long	Tomato
Lycopersicon	*longum-rubrum*	*italiana-bananai*	Italian Banana ?" x 6"	Red Long	Tomato
Lycopersicon	*longum-rubrum*	*italiana-pepper-shapedii*	Italian Pepper ?" x 5"	Red Long	Tomato
Lycopersicon	*longum-rubrum*	*jersey-devilii*	Jersey Devil ?" x 5"	Red Long	Tomato
Lycopersicon	*longum-rubrum*	*jersey-giantii*	Jersey Giant ?" x 5"	Red Long	Tomato
Lycopersicon	*longum-rubrum*	*long-tomii*	Long Tom 2" x 5"	Red Long	Tomato
Lycopersicon	*longum-rubrum*	*marianne'si*	Marianne's 1.75" x 6"	Red Long	Tomato
Lycopersicon	*longum-rubrum*	*market-kingii*	Market King 1" x 2.5"	Red Long	Tomato
Lycopersicon	*longum-rubrum*	*opalka*	Olpaka ?" x 5"	Red Long	Tomato
Lycopersicon	*longum-rubrum*	*pete's-italian-plumii*	Pete's Italian Plum 2" x 6"	Red Long	Tomato
Lycopersicon	*longum-rubrum*	*red-rocketii*	Red Rocket ?" x ?"	Red Long	Tomato
Lycopersicon	*longum-rubrum*	*red-sausagei*	Red Sausage ?" x 6"	Red Long	Tomato
Lycopersicon	*longum-rubrum*	*roprecoi*	Ropreco 2" x 4"	Red Long	Tomato
Lycopersicon	*longum-rubrum*	*san-marzanoi*	San Marzano 1.5" x 4"	Red Long	Tomato
Lycopersicon	*longum-rubrum*	*san-marzanoi-la-padrino*	La Padrino San Marzano 1.5" x 4"	Red Long	Tomato
Lycopersicon	*longum-rubrum*	*san-marzanoi-redorta*	Redorta San Marzano 1.5" x 4"	Red Long	Tomato
Lycopersicon	*longum-rubrum*	*san-marzanoi-supra*	Super San Marzano 1" x 5"	Red Long	Tomato
Lycopersicon	*longum-rubrum*	*seache's-italian*	Seache's Italian 2" x 6"	Red Long	Tomato
Lycopersicon	*longum-rubrum*	*spitzii*	Spitz 2" x 4.5"	Red Long	Tomato
Lycopersicon	*longum-rubrum*	*stusei*	Stuse 1.75" x 3"	Red Long	Tomato
Lycopersicon	*longum-rubrum*	*supra-italiana-pastei*	Super Italian Paste 6" x ?"	Red Long	Tomato
Lycopersicon	*longum-rubrum*	*vilmsi*	Vilms 2.5" x ?"	Red Long	Tomato
Lycopersicon	*longum-rubrum*	*vita-29*	#29 Vita 2" x 3"	Red Long	Tomato
Lycopersicon	*longum-rubrum*	*you-goi*	You Go 4" x 5"	Red Long	Tomato
Lycopersicon	*longum-striatum*	*speckled-romanii*	Speckled Roman 3" x 5"	Striped Long	Tomato

Tomato, Oxheart (*Lycopersicon oxacardium*)

Genus	Species	Infraspecies	cultivar	specimen	Genome
Lycopersicon	*oxacardium-aurantium*	*aurantium*	Original 11oz	Orange Oxheart	Tomato
Lycopersicon	*oxacardium-aurantium*	*german-orange-strawberryi*	German Orange Strawberry ?.?"	Orange Oxheart	Tomato
Lycopersicon	*oxacardium-aurantium*	*orange-heartii*	Orange Heart 6oz	Orange Oxheart	Tomato
Lycopersicon	*oxacardium-aurantium*	*orange-russian-117*	Orange #117 Russian 1.5lb	Orange Oxheart	Tomato
Lycopersicon	*oxacardium-aurantium*	*verna-orangei*	Verna 3lb	Orange Oxheart	Tomato
Lycopersicon	*oxacardium-helvum*	*helvum*	Original	Yellow Oxheart	Tomato
Lycopersicon	*oxacardium-helvum*	*anna-hermannii*	Anna Hermann ?.?"	Yellow Oxheart	Tomato
Lycopersicon	*oxacardium-helvum*	*farnsworthii*	Farnsworth ?.?"	Yellow Oxheart	Tomato
Lycopersicon	*oxacardium-helvum*	*herman's-yellowi*	Herman's ?.?"	Yellow Oxheart	Tomato
Lycopersicon	*oxacardium-pincum*	*oxacardium-pincum*	Original 7oz	Pink Oxheart	Tomato
Lycopersicon	*oxacardium-pincum*	*amana-pinkii*	Amana 1.5lb	Pink Oxheart	Tomato

Genus	Species	Infraspecies	cultivar	specimen	Genome
Lycopersicon	*oxacardium-pincum*	*anna-russianii*	Anna Russian 1lb	Pink Oxheart	Tomato
Lycopersicon	*oxacardium-pincum*	*bull's-heartii*	Bull's Heart 4.0"	Pink Oxheart	Tomato
Lycopersicon	*oxacardium-pincum*	*butter-&-bull-heartii*	Butter & Bull Heart 3/4lb	Pink Oxheart	Tomato
Lycopersicon	*oxacardium-pincum*	*german-heartii*	German Heart Flat ?.?"	Pink Oxheart	Tomato
Lycopersicon	*oxacardium-pincum*	*happy-jackii*	Happy Jack 14oz	Pink Oxheart	Tomato
Lycopersicon	*oxacardium-pincum*	*heart-of-the-bullii*	Heart Of The Bull ?.?"	Pink Oxheart	Tomato
Lycopersicon	*oxacardium-pincum*	*hungarian-heartii*	Hungarian Heart 1lb	Pink Oxheart	Tomato
Lycopersicon	*oxacardium-pincum*	*italiana-mike'si*	Mike's Italian ?.?"	Pink Oxheart	Tomato
Lycopersicon	*oxacardium-pincum*	*italiana-newtown-plumii*	Newtown Plum Italian ?.?"	Pink Oxheart	Tomato
Lycopersicon	*oxacardium-pincum*	*italiana-wagner's*	Wagner's Italian ?.?"	Pink Oxheart	Tomato
Lycopersicon	*oxacardium-pincum*	*japanese-oxheartii*	Japanese 14oz	Pink Oxheart	Tomato
Lycopersicon	*oxacardium-pincum*	*nicky-crainii*	Nicky Crain 1lb	Pink Oxheart	Tomato
Lycopersicon	*oxacardium-pincum*	*oxheart-giantii*	Giant 2lb	Pink Oxheart	Tomato
Lycopersicon	*oxacardium-pincum*	*oxheart-light-pinkii*	Light Pink 12oz	Pink Oxheart	Tomato
Lycopersicon	*oxacardium-pincum*	*palestiniana*	Palestinian 1lb	Pink Oxheart	Tomato
Lycopersicon	*oxacardium-pincum*	*pink-russiana-117*	#117 Pink Russian 12oz	Pink Oxheart	Tomato
Lycopersicon	*oxacardium-pincum*	*polish-heartii*	Heart Polish ?.?"	Pink Oxheart	Tomato
Lycopersicon	*oxacardium-pincum*	*red-butter-heartii*	Red Butter Heart 5.0"	Pink Oxheart	Tomato
Lycopersicon	*oxacardium-pincum*	*russian-heartii*	Russian Heart 14oz	Pink Oxheart	Tomato
Lycopersicon	*oxacardium-pincum*	*sochulakii*	Sochulak ?.?"	Pink Oxheart	Tomato
Lycopersicon	*oxacardium-pincum*	*ukranian-heartii*	Ukrainian Heart 1.5lb	Pink Oxheart	Tomato
Lycopersicon	*oxacardium-pincum*	*winstead-oxheartii*	Winsted ?.?"	Pink Oxheart	Tomato
Lycopersicon	*oxacardium-pincum*	*wolford's-wonderi*	Wolford's Wonder 2lb	Pink Oxheart	Tomato
Lycopersicon	*oxacardium-pincum*	*yugoslaviana-yashai*	Yasha Yugoslavian 1.2lb	Pink Oxheart	Tomato
Lycopersicon	*oxacardium-rubrum-largum*	*cuore-de-toroi*	Cuore De Toro	Large 1.5lb-2lb 4.0" Red Oxheart	Tomato
Lycopersicon	*oxacardium-rubrum-largum*	*dinner-platei*	Dinner Plate	Large 1.5lb-2lb 4.0" Red Oxheart	Tomato
Lycopersicon	*oxacardium-rubrum-largum*	*jewishii*	Jewish	Large 1.5lb-2lb 4.0" Red Oxheart	Tomato
Lycopersicon	*oxacardium-rubrum-largum*	*oregon-pridei*	Oregon Pride	Large 1.5lb-2lb 4.0" Red Oxheart	Tomato
Lycopersicon	*oxacardium-rubrum-largum*	*oregon-starii*	Oregon Star	Large 1.5lb-2lb 4.0" Red Oxheart	Tomato
Lycopersicon	*oxacardium-rubrum-largum*	*red-butter-heartii*	Red Butter Heart	Large 1.5lb-2lb 4.0" Red Oxheart	Tomato
Lycopersicon	*oxacardium-rubrum-largum*	*russiana-117*	#177 Russian	Large 1.5lb-2lb 4.0" Red Oxheart	Tomato
Lycopersicon	*oxacardium-rubrum-largum*	*shilling-giantii*	Schilling Giant	Large 1.5lb-2lb 4.0" Red Oxheart	Tomato
Lycopersicon	*oxacardium-rubrum-largum*	*sylvan-guamei*	Sylvan Guame	Large 1.5lb-2lb 4.0" Red Oxheart	Tomato
Lycopersicon	*oxacardium-rubrum-largum-medium*	*bisignanoi*	Bisignano	Medium Large 1lb 3.5" Red Oxheart	Tomato
Lycopersicon	*oxacardium-rubrum-largum-medium*	*bisignanoi-2*	#2 Bisignano	Medium Large 1lb 3.5" Red Oxheart	Tomato
Lycopersicon	*oxacardium-rubrum-largum-medium*	*enormous-plumii*	Enormous Plum	Medium Large 1lb 3.5" Red Oxheart	Tomato
Lycopersicon	*oxacardium-rubrum-largum-medium*	*giant-syrianii*	Giant Syrian	Medium Large 1lb 3.5" Red Oxheart	Tomato
Lycopersicon	*oxacardium-rubrum-largum-medium*	*painted-ukrainiana*	Painted Ukrainian	Medium Large 1lb 3.5" Red Oxheart	Tomato
Lycopersicon	*oxacardium-rubrum-largum-medium*	*reif-italian-heartii*	Reif Italian Heart	Medium Large 1lb 3.5" Red Oxheart	Tomato
Lycopersicon	*oxacardium-rubrum-medium-largum*	*grightmire's-pridei*	Grightmire's Pride	Large Medium 3/4lb 3.0" Red Oxheart	Tomato
Lycopersicon	*oxacardium-rubrum-medium-largum*	*ponder-heartii-redii*	Ponder Heart	Large Medium 3/4lb 3.0" Red Oxheart	Tomato
Lycopersicon	*oxacardium-rubrum-medium*	*amish-oxheartii*	Amish	Medium 1/2lb 2.5" Red Oxheart	Tomato
Lycopersicon	*oxacardium-rubrum-medium*	*berkshire-oxheartii*	Berkshire	Medium 1/2lb 2.5" Red Oxheart	Tomato
Lycopersicon	*oxacardium-rubrum-medium*	*bobbiei*	Bobbie	Medium 1/2lb 2.5" Red Oxheart	Tomato
Lycopersicon	*oxacardium-rubrum-medium*	*bull's-heartii-selectii*	Bull's Heart Select	Medium 1/2lb 2.5" Red Oxheart	Tomato
Lycopersicon	*oxacardium-rubrum-medium*	*cuor-di-buei*	Cuor Di Bue	Medium 1/2lb 2.5" Red Oxheart	Tomato
Lycopersicon	*oxacardium-rubrum-medium*	*grandfather-barnini's-oxheartii*	Grandfather Barnini's	Medium 1/2lb 2.5" Red Oxheart	Tomato
Lycopersicon	*oxacardium-rubrum-medium*	*pomadorai*	Pomadora	Medium 1/2lb 2.5" Red Oxheart	Tomato
Lycopersicon	*oxacardium-rubrum-medium*	*teton-de-venus*	Teton De Venus	Medium 1/2lb 2.5" Red Oxheart	Tomato
Lycopersicon	*oxacardium-rubrum-medium*	*tom-patti's-italian-pastei*	Tom Patti's Italian Paste	Medium 1/2lb 2.5" Red Oxheart	Tomato
Lycopersicon	*oxacardium-rubrum-parvum*	*brandywine-heart-shapedii*	Heart Shaped Amish Brandywine Creek	Small 4oz-6oz 2.0" Red Oxheart	Tomato
Lycopersicon	*oxacardium-rubrum-parvum*	*forme-de-coeuri*	Forme De Coeur	Small 4oz-6oz 2.0" Red Oxheart	Tomato
Lycopersicon	*oxacardium-rubrum-parvum*	*redskinii*	Redskin	Small 4oz-6oz 2.0" Red Oxheart	Tomato
Lycopersicon	*oxacardium-rubrum-aurantium*	.	.	Orange Red Oxheart	Tomato
Lycopersicon	*oxacardium-rubrum-aurantium*	*hog-heartii*	Hog Heart 7"	Orange Red Oxheart	Tomato
Lycopersicon	*oxacardium-rubrum-aurantium*	*justinei*	Justine ?"	Orange Red Oxheart	Tomato

Tomato, Currant (*Lycopersicon pimpinellifolium*)

Genus	Species	Infraspecies	cultivar	specimen	Genome
Lycopersicon	*pimpinellifolium-helvum*	*helvum*	Original	Yellow Currant	Tomato
Lycopersicon	*pimpinellifolium-helvum*	*broad-ripple-yellow-currantii*	Broad Ripple Yellow 0.5"	Yellow Currant	Tomato
Lycopersicon	*pimpinellifolium-helvum*	*currant-lemon-dropii*	Lemon Drop 0.4"	Yellow Currant	Tomato
Lycopersicon	*pimpinellifolium-helvum*	*currant-yellow-cherryi*	Yellow Cherry 0.5"	Yellow Currant	Tomato
Lycopersicon	*pimpinellifolium-rubrum*	*rubrum*	Wild Cherry 0.5"	Red Currant	Tomato
Lycopersicon	*pimpinellifolium-rubrum*	*cherry-wildii*	Wild Cherry 0.5"	Red Currant	Tomato
Lycopersicon	*pimpinellifolium-rubrum*	*cherry-wildii-matt'si*	Matt's Wild Cherry 0.5"	Red Currant	Tomato
Lycopersicon	*pimpinellifolium-rubrum*	*currant-drops-fruitii*	Drops Fruit ?.?"	Red Currant	Tomato
Lycopersicon	*pimpinellifolium-rubrum*	*currant-hawaiiana*	Hawaiian	Red Currant	Tomato
Lycopersicon	*pimpinellifolium-rubrum*	*currant-holds-fruitii*	Holds Fruit (HF) 0.5"	Red Currant	Tomato
Lycopersicon	*pimpinellifolium-rubrum*	*peacevinei*	Peacevine	Red Currant	Tomato
Lycopersicon	*pimpinellifolium-rubrum*	*russian-currantii*	Russian	Red Currant	Tomato
Lycopersicon	*pimpinellifolium-rubrum*	*texas-wildii*	Texas Wild .25"-.5"	Red Currant	Tomato

Tomato, Flat (*Lycopersicon platycarpum*)

Lycopersicon platycarpum-album	*big-white*	Big White 10oz	White Flat Tomato
Lycopersicon platycarpum-album	*shahii*	Shah 8oz-12oz	White Flat Tomato
Lycopersicon platycarpum-album	*white-queenii*	White Queen 3" 8oz-12oz	White Flat Tomato
Lycopersicon platycarpum-pincum	*platycarpum-pincum*	Original 4.0"	Pink Flat Tomato
Lycopersicon platycarpum-pincum	*afghanistanensis*	Afghanistan 3.5"	Pink Flat Tomato
Lycopersicon platycarpum-pincum	*arkansas-travelleri*	85 Day Arkansas Traveler ?oz	Pink Flat Tomato
Lycopersicon platycarpum-pincum	*arkansas-travelleri-76*	76 Day Arkansas Traveler 7oz	Pink Flat Tomato
Lycopersicon platycarpum-pincum	*bradleyi*	Bradley 6oz	Pink Flat Tomato
Lycopersicon platycarpum-pincum	*brandywinei-joyce'si*	Joyce's Amish Brandywine Creek ?.?lb	Pink Flat Tomato
Lycopersicon platycarpum-pincum	*bulgarian-large-pinkii*	Large Pink Bulgarian 3/4lb	Pink Flat Tomato
Lycopersicon platycarpum-pincum	*dr.-nealii*	Dr. Neal 4.0"	Pink Flat Tomato
Lycopersicon platycarpum-pincum	*gerigii*	Gerig ?.?"	Pink Flat Tomato
Lycopersicon platycarpum-pincum	*german-giantii*	German Giant 5.0"	Pink Flat Tomato
Lycopersicon platycarpum-pincum	*german-queenii*	German Queen 1lb	Pink Flat Tomato
Lycopersicon platycarpum-pincum	*giant-roseyi*	Giant Rosy 14oz	Pink Flat Tomato
Lycopersicon platycarpum-pincum	*giuseppe's-big-boyi*	Giuseppe's Big Boy 3/4lb	Pink Flat Tomato
Lycopersicon platycarpum-pincum	*holmes-mexicana*	Holmes Mexican 1lb	Pink Flat Tomato
Lycopersicon platycarpum-pincum	*italiana-genuinei*	Genuine Italian 5.2"	Pink Flat Tomato
Lycopersicon platycarpum-pincum	*kornesevsijei*	Kornesevsije 1lb	Pink Flat Tomato
Lycopersicon platycarpum-pincum	*mr.-hawkinsi*	Mr. Hawkins 5.0"	Pink Flat Tomato
Lycopersicon platycarpum-pincum	*nectarinei*	Nectarine 5oz	Pink Flat Tomato
Lycopersicon platycarpum-pincum	*new-zealand-pastei*	New Zealand Paste ?.?"	Pink Flat Tomato
Lycopersicon platycarpum-pincum	*omar's-lebanensensis*	Omar's Lebanese 1.2lb	Pink Flat Tomato
Lycopersicon platycarpum-pincum	*ozark-pinkii*	Ozark 7oz	Pink Flat Tomato
Lycopersicon platycarpum-pincum	*peruensis*	Peru 2.5"	Pink Flat Tomato
Lycopersicon platycarpum-pincum	*pink-climberi*	Pink Climber 14oz	Pink Flat Tomato
Lycopersicon platycarpum-pincum	*pink-potato-topii*	Piink Potato Top 5.2"	Pink Flat Tomato
Lycopersicon platycarpum-pincum	*pink-russiana*	Pink Russian 14oz	Pink Flat Tomato
Lycopersicon platycarpum-pincum	*pink-sweetii*	Pink Sweet 1lb	Pink Flat Tomato
Lycopersicon platycarpum-pincum	*polishii-C*	C Polish ?.?"	Pink Flat Tomato
Lycopersicon platycarpum-pincum	*polishii-non-acidii*	Non-Acid Polish ?.?"	Pink Flat Tomato
Lycopersicon platycarpum-pincum	*ponder-heartii-pinkii*	Ponder Heart ?.?"	Pink Flat Tomato
Lycopersicon platycarpum-pincum	*potato-leaf-pinkii*	Potato Leaf 5.0"	Pink Flat Tomato
Lycopersicon platycarpum-pincum	*redfield-beautyi*	Redfield Beauty ?.?"	Pink Flat Tomato
Lycopersicon platycarpum-pincum	*soldacki*	Soldacki 1lb	Pink Flat Tomato
Lycopersicon platycarpum-pincum	*tappy's-finestii*	Tappy's Finest ?.?"	Pink Flat Tomato
Lycopersicon platycarpum-pincum	*wins-allii*	Wins All 2lb	Pink Flat Tomato
Lycopersicon platycarpum-rubrum-largum	*cosmonaut-volkov-redii*	Cosmonaut Volkov	Large 1.5lb 4" x 3" Red Flat Tomato
Lycopersicon platycarpum-rubrum-largum	*early-a-paci*	Early A Pack	Large 1.5lb 4" x 3" Red Flat Tomato
Lycopersicon platycarpum-rubrum-largum	*elizabeth-crook'si*	Elizabeth Crook's	Large 1.5lb 4" x 3" Red Flat Tomato
Lycopersicon platycarpum-rubrum-largum	*druzbai*	Druzba	Large 1.5lb 4" x 3" Red Flat Tomato
Lycopersicon platycarpum-rubrum-largum	*early-anniei*	Early Annie	Large 1.5lb 4" x 3" Red Flat Tomato
Lycopersicon platycarpum-rubrum-largum	*kanner-hoellii*	Kanner Hoell	Large 1.5lb 4" x 3" Red Flat Tomato
Lycopersicon platycarpum-rubrum-largum	*early-rougei*	Early Rouge	Large 1.5lb 4" x 3" Red Flat Tomato
Lycopersicon platycarpum-rubrum-largum	*langada*	Langada	Large 1.5lb 4" x 3" Red Flat Tomato
Lycopersicon platycarpum-rubrum-largum	*redheartii*	Redheart	Large 1.5lb 4" x 3" Red Flat Tomato
Lycopersicon platycarpum-rubrum-largum	*trip-L-cropii-burgessi*	Burgess Triple L Crop	Large 1.5lb 4" x 3" Red Flat Tomato
Lycopersicon platycarpum-rubrum-largum	*trip-L-cropii-climbingii*	Climbing Triple L Crop	Large 1.5lb 4" x 3" Red Flat Tomato
Lycopersicon platycarpum-rubrum-largum	*zogolai*	Zogola	Large 1.5lb 4" x 3" Red Flat Tomato
Lycopersicon platycarpum-rubrum-largum-medium	*andrew-rahart's-jumbo-redii*	Andrew Rahart's Jumbo	Medium Large 1lb 3" x 2.5" Red Flat Tomato
Lycopersicon platycarpum-rubrum-largum-medium	*big-a-sandyi*	Big A Sandy	Medium Large 1lb 3" x 2.5" Red Flat Tomato
Lycopersicon platycarpum-rubrum-largum-medium	*big-boyi*	Big Boy	Medium Large 1lb 3" x 2.5" Red Flat Tomato
Lycopersicon platycarpum-rubrum-largum-medium	*disarsa*	Disarsa	Medium Large 1lb 3" x 2.5" Red Flat Tomato
Lycopersicon platycarpum-rubrum-largum-medium	*fanii-3*	#3 Fan	Medium Large 1lb 3" x 2.5" Red Flat Tomato
Lycopersicon platycarpum-rubrum-largum-medium	*hanky-redii*	Hanky Red	Medium Large 1lb 3" x 2.5" Red Flat Tomato
Lycopersicon platycarpum-rubrum-largum-medium	*indica*	India	Medium Large 1lb 3" x 2.5" Red Flat Tomato
Lycopersicon platycarpum-rubrum-largum-medium	*italiana-giantii*	Italian Giant	Medium Large 1lb 3" x 2.5" Red Flat Tomato
Lycopersicon platycarpum-rubrum-largum-medium	*large-redii*	Large Red	Medium Large 1lb 3" x 2.5" Red Flat Tomato
Lycopersicon platycarpum-rubrum-largum-medium	*brandywinei-OTV*	Off The Vine Amish Brandywine Creek	Medium Large 1lb 3" x 2.5" Red Flat Tomato
Lycopersicon platycarpum-rubrum-largum-medium	*matchlessi*	Matchless	Medium Large 1lb 3" x 2.5" Red Flat Tomato
Lycopersicon platycarpum-rubrum-largum-medium	*matchlessi-austinii*	Austin Matchless	Medium Large 1lb 3" x 2.5" Red Flat Tomato
Lycopersicon platycarpum-rubrum-largum-medium	*matchlessi-burpee'si*	Burpee's Matchless	Medium Large 1lb 3" x 2.5" Red Flat Tomato
Lycopersicon platycarpum-rubrum-largum-medium	*mexican-saladii*	Mexican Salad	Medium Large 1lb 3" x 2.5" Red Flat Tomato
Lycopersicon platycarpum-rubrum-largum-medium	*opal's-homesteadii*	Opal's Homestead	Medium Large 1lb 3" x 2.5" Red Flat Tomato
Lycopersicon platycarpum-rubrum-largum-medium	*pakmori-B*	Type B Packmore	Medium Large 1lb 3" x 2.5" Red Flat Tomato
Lycopersicon platycarpum-rubrum-largum-medium	*readheartii*	Readheart	Medium Large 1lb 3" x 2.5" Red Flat Tomato
Lycopersicon platycarpum-rubrum-largum-medium	*september-dawnii*	September Dawn	Medium Large 1lb 3" x 2.5" Red Flat Tomato
Lycopersicon platycarpum-rubrum-largum-medium	*shirleyi-amish-redii*	Shirley Amish	Medium Large 1lb 3" x 2.5" Red Flat Tomato
Lycopersicon platycarpum-rubrum-largum-medium	*silvery-fir-treei*	Silvery Fir Tree	Medium Large 1lb 3" x 2.5" Red Flat Tomato
Lycopersicon platycarpum-rubrum-medium	*atkinsonii*	Atkinson	Medium 3/4lb 3" x 2" Red Flat Tomato
Lycopersicon platycarpum-rubrum-medium	*culstragei*	Culstrage	Medium 3/4lb 3" x 2" Red Flat Tomato
Lycopersicon platycarpum-rubrum-medium	*currieri*	Currier	Medium 3/4lb 3" x 2" Red Flat Tomato

Lycopersicon	platycarpum-rubrum-medium	daydreamii	Daydream	Medium 3/4lb 3" x 2" Red Flat	Tomato
Lycopersicon	platycarpum-rubrum-medium	dombitoi	Dumbito	Medium 3/4lb 3" x 2" Red Flat	Tomato
Lycopersicon	platycarpum-rubrum-medium	early-pakii	Early Pak	Medium 3/4lb 3" x 2" Red Flat	Tomato
Lycopersicon	platycarpum-rubrum-medium	early-pakii-7	#7 Early Pak	Medium 3/4lb 3" x 2" Red Flat	Tomato
Lycopersicon	platycarpum-rubrum-medium	ecuadorensis	Ecuador	Medium 3/4lb 3" x 2" Red Flat	Tomato
Lycopersicon	platycarpum-rubrum-medium	germanica-bushii	German Bush	Medium 3/4lb 3" x 2" Red Flat	Tomato
Lycopersicon	platycarpum-rubrum-medium	giant-pepperviewi	Giant Pepperview	Medium 3/4lb 3" x 2" Red Flat	Tomato
Lycopersicon	platycarpum-rubrum-medium	glamouri	Glamour	Medium 3/4lb 3" x 2" Red Flat	Tomato
Lycopersicon	platycarpum-rubrum-medium	grosso-rosso-nostralii	Grosso Rosso Nostral	Medium 3/4lb 3" x 2" Red Flat	Tomato
Lycopersicon	platycarpum-rubrum-medium	ham-green-favoritei	Ham Green Favorite	Medium 3/4lb 3" x 2" Red Flat	Tomato
Lycopersicon	platycarpum-rubrum-medium	hollandica-rubrum	Holland	Medium 3/4lb 3" x 2" Red Flat	Tomato
Lycopersicon	platycarpum-rubrum-medium	korai	Kora	Medium 3/4lb 3" x 2" Red Flat	Tomato
Lycopersicon	platycarpum-rubrum-medium	large-redii	Large Red	Medium 3/4lb 3" x 2" Red Flat	Tomato
Lycopersicon	platycarpum-rubrum-medium	legendii	Legend	Medium 3/4lb 3" x 2" Red Flat	Tomato
Lycopersicon	platycarpum-rubrum-medium	manzana	Manzana	Medium 3/4lb 3" x 2" Red Flat	Tomato
Lycopersicon	platycarpum-rubrum-medium	marmandei-superi	Super Marmande	Medium 3/4lb 3" x 2" Red Flat	Tomato
Lycopersicon	platycarpum-rubrum-medium	monte-verdei	Monte Verde	Medium 3/4lb 3" x 2" Red Flat	Tomato
Lycopersicon	platycarpum-rubrum-medium	muchamielii	Muchamiel	Medium 3/4lb 3" x 2" Red Flat	Tomato
Lycopersicon	platycarpum-rubrum-medium	pearsonii	Pearson	Medium 3/4lb 3" x 2" Red Flat	Tomato
Lycopersicon	platycarpum-rubrum-medium	pearsonii-improvedii	Improved Pearson	Medium 3/4lb 3" x 2" Red Flat	Tomato
Lycopersicon	platycarpum-rubrum-medium	quebecensis-059	#059 Quebec	Medium 3/4lb 3" x 2" Red Flat	Tomato
Lycopersicon	platycarpum-rubrum-medium	red-beautyi	Red Beauty	Medium 3/4lb 3" x 2" Red Flat	Tomato
Lycopersicon	platycarpum-rubrum-medium	three-sistersi-flatii	Flat Three Sisters	Medium 3/4lb 3" x 2" Red Flat	Tomato
Lycopersicon	platycarpum-rubrum-medium	yucutanii	Yucutan	Medium 3/4lb 3" x 2" Red Flat	Tomato
Lycopersicon	platycarpum-rubrum-medium-parvum	.	.	Small Medium 1/2lb 2.5" x 2" Red Flat	Tomato
Lycopersicon	platycarpum-rubrum-medium-parvum	acei	Campbell Soup Ace	Small Medium 1/2lb 2.5" x 2" Red Flat	Tomato
Lycopersicon	platycarpum-rubrum-medium-parvum	acei-55-VF	#55-VF Ace	Small Medium 1/2lb 2.5" x 2" Red Flat	Tomato
Lycopersicon	platycarpum-rubrum-medium-parvum	acei-55-VF-calii	Cal #55-VF Ace	Small Medium 1/2lb 2.5" x 2" Red Flat	Tomato
Lycopersicon	platycarpum-rubrum-medium-parvum	acei-55-VF-improved	Improved #55-VF Ace	Small Medium 1/2lb 2.5" x 2" Red Flat	Tomato
Lycopersicon	platycarpum-rubrum-medium-parvum	alpinei	Alpine	Small Medium 1/2lb 2.5" x 2" Red Flat	Tomato
Lycopersicon	platycarpum-rubrum-medium-parvum	bonny-bestii-improvedii	Improved Bonny Best	Small Medium 1/2lb 2.5" x 2" Red Flat	Tomato
Lycopersicon	platycarpum-rubrum-medium-parvum	brookpactii	Brookpact	Small Medium 1/2lb 2.5" x 2" Red Flat	Tomato
Lycopersicon	platycarpum-rubrum-medium-parvum	brookpactii-porter's-earlyi	Porter's Early Brookpact	Small Medium 1/2lb 2.5" x 2" Red Flat	Tomato
Lycopersicon	platycarpum-rubrum-medium-parvum	cal-lakei-4	#4 Cal Lake	Small Medium 1/2lb 2.5" x 2" Red Flat	Tomato
Lycopersicon	platycarpum-rubrum-medium-parvum	calypsoi	Calypso	Small Medium 1/2lb 2.5" x 2" Red Flat	Tomato
Lycopersicon	platycarpum-rubrum-medium-parvum	campbellii	Campbell	Small Medium 1/2lb 2.5" x 2" Red Flat	Tomato
Lycopersicon	platycarpum-rubrum-medium-parvum	campbellii-33	#0033 Campbell	Small Medium 1/2lb 2.5" x 2" Red Flat	Tomato
Lycopersicon	platycarpum-rubrum-medium-parvum	campbellii-146	#0146 Campbell	Small Medium 1/2lb 2.5" x 2" Red Flat	Tomato
Lycopersicon	platycarpum-rubrum-medium-parvum	campbellii-1327	#1327 Campbell	Small Medium 1/2lb 2.5" x 2" Red Flat	Tomato
Lycopersicon	platycarpum-rubrum-medium-parvum	caribei	Caribe	Small Medium 1/2lb 2.5" x 2" Red Flat	Tomato
Lycopersicon	platycarpum-rubrum-medium-parvum	cherokeei	Cherokee	Small Medium 1/2lb 2.5" x 2" Red Flat	Tomato
Lycopersicon	platycarpum-rubrum-medium-parvum	CPC-2	CPC-2	Small Medium 1/2lb 2.5" x 2" Red Flat	Tomato
Lycopersicon	platycarpum-rubrum-medium-parvum	frosty-f.-housei	Frosty F. House	Small Medium 1/2lb 2.5" x 2" Red Flat	Tomato
Lycopersicon	platycarpum-rubrum-medium-parvum	heinzii-1350	#1350 Heinz	Small Medium 1/2lb 2.5" x 2" Red Flat	Tomato
Lycopersicon	platycarpum-rubrum-medium-parvum	homesteadii	Homestead	Small Medium 1/2lb 2.5" x 2" Red Flat	Tomato
Lycopersicon	platycarpum-rubrum-medium-parvum	homesteadii-24	#24 Homestead	Small Medium 1/2lb 2.5" x 2" Red Flat	Tomato
Lycopersicon	platycarpum-rubrum-medium-parvum	homesteadii-500	#500 Homestead	Small Medium 1/2lb 2.5" x 2" Red Flat	Tomato
Lycopersicon	platycarpum-rubrum-medium-parvum	homesteadii-improved-VF	Verticillium Fusarium Resistant Homestead	Small Medium 1/2lb 2.5" x 2" Red Flat	Tomato
Lycopersicon	platycarpum-rubrum-medium-parvum	landino-di-panachia	Landino Di Panachia	Small Medium 1/2lb 2.5" x 2" Red Flat	Tomato
Lycopersicon	platycarpum-rubrum-medium-parvum	lebanonensis	Lebanon	Small Medium 1/2lb 2.5" x 2" Red Flat	Tomato
Lycopersicon	platycarpum-rubrum-medium-parvum	marglobei	Maryland Globe	Small Medium 1/2lb 2.5" x 2" Red Flat	Tomato
Lycopersicon	platycarpum-rubrum-medium-parvum	marglobei-improvedii	Improved Maryland Globe	Small Medium 1/2lb 2.5" x 2" Red Flat	Tomato
Lycopersicon	platycarpum-rubrum-medium-parvum	marglobei-PS	Productive Strain Maryland Globe	Small Medium 1/2lb 2.5" x 2" Red Flat	Tomato
Lycopersicon	platycarpum-rubrum-medium-parvum	marglobei-selectii	Select Maryland Globe	Small Medium 1/2lb 2.5" x 2" Red Flat	Tomato
Lycopersicon	platycarpum-rubrum-medium-parvum	marglobei-supremei	Supreme Maryland Globe	Small Medium 1/2lb 2.5" x 2" Red Flat	Tomato
Lycopersicon	platycarpum-rubrum-medium-parvum	pakmori-VF	Verticillium Fusarium Resistant Packmore	Small Medium 1/2lb 2.5" x 2" Red Flat	Tomato
Lycopersicon	platycarpum-rubrum-medium-parvum	red-cloudii	Red Cloud	Small Medium 1/2lb 2.5" x 2" Red Flat	Tomato
Lycopersicon	platycarpum-rubrum-medium-parvum	sasha's-altai	Sahsa's Altai	Small Medium 1/2lb 2.5" x 2" Red Flat	Tomato
Lycopersicon	platycarpum-rubrum-medium-parvum	sasha's-pridei	Sasha's Pride	Small Medium 1/2lb 2.5" x 2" Red Flat	Tomato
Lycopersicon	platycarpum-rubrum-medium-parvum	tropica	Tropic	Small Medium 1/2lb 2.5" x 2" Red Flat	Tomato
Lycopersicon	platycarpum-rubrum-medium-parvum	zarnitsai	Zarnitsa	Small Medium 1/2lb 2.5" x 2" Red Flat	Tomato
Lycopersicon	platycarpum-rubrum-parvum	.	.	Small 4oz-6oz 2" x 1.5" Red Flat	Tomato
Lycopersicon	platycarpum-rubrum-parvum	belyi-naivii	Belyi Naliv	Small 4oz-6oz 2" x 1.5" Red Flat	Tomato
Lycopersicon	platycarpum-rubrum-parvum	brazilensis	Brazil Flat	Small 4oz-6oz 2" x 1.5" Red Flat	Tomato
Lycopersicon	platycarpum-rubrum-parvum	ceylonensis	Ceylon	Small 4oz-6oz 2" x 1.5" Red Flat	Tomato
Lycopersicon	platycarpum-rubrum-parvum	chinensis	Chinese	Small 4oz-6oz 2" x 1.5" Red Flat	Tomato
Lycopersicon	platycarpum-rubrum-parvum	earliana	Earliana	Small 4oz-6oz 2" x 1.5" Red Flat	Tomato
Lycopersicon	platycarpum-rubrum-parvum	earliana-sunneybrookii	Sunnybrook Earliana	Small 4oz-6oz 2" x 1.5" Red Flat	Tomato
Lycopersicon	platycarpum-rubrum-parvum	earliana-spark's-improvedii	Spark's Improved Earliana	Small 4oz-6oz 2" x 1.5" Red Flat	Tomato
Lycopersicon	platycarpum-rubrum-parvum	earlirougei	Earlirouge	Small 4oz-6oz 2" x 1.5" Red Flat	Tomato
Lycopersicon	platycarpum-rubrum-parvum	hayslipii	Hayslip	Small 4oz-6oz 2" x 1.5" Red Flat	Tomato
Lycopersicon	platycarpum-rubrum-parvum	heinzii-1439	#1439 Heinz	Small 4oz-6oz 2" x 1.5" Red Flat	Tomato
Lycopersicon	platycarpum-rubrum-parvum	lerica	Lerica	Small 4oz-6oz 2" x 1.5" Red Flat	Tomato

Lycopersicon	*platycarpum-rubrum-parvum*	*marmandei*	Marmande	Small 4oz-6oz 2" x 1.5" Red Flat	Tomato
Lycopersicon	*platycarpum-rubrum-parvum*	*marmandei-gourmetii-VF*	Verticillum Fusarium Resistant Gourmet Marmande	Small 4oz-6oz 2" x 1.5" Red Flat	Tomato
Lycopersicon	*platycarpum-rubrum-parvum*	*maria*	Maria	Small 4oz-6oz 2" x 1.5" Red Flat	Tomato
Lycopersicon	*platycarpum-rubrum-parvum*	*market-at-huachinango-puebla-mexicoi*	Huachinango Pueblo Market	Small 4oz-6oz 2" x 1.5" Red Flat	Tomato
Lycopersicon	*platycarpum-rubrum-parvum*	*olomovici*	Olomovic	Small 4oz-6oz 2" x 1.5" Red Flat	Tomato
Lycopersicon	*platycarpum-rubrum-parvum*	*picardyi*	Picardy	Small 4oz-6oz 2" x 1.5" Red Flat	Tomato
Lycopersicon	*platycarpum-rubrum-parvum*	*saladmasteri*	Saladmaster	Small 4oz-6oz 2" x 1.5" Red Flat	Tomato
Lycopersicon	*platycarpum-rubrum-parvum*	*santorini-pastei*	Santorini Paste	Small 4oz-6oz 2" x 1.5" Red Flat	Tomato
Lycopersicon	*platycarpum-rubrum-parvum*	*starfirei*	Starfire	Small 4oz-6oz 2" x 1.5" Red Flat	Tomato
Lycopersicon	*platycarpum-rubrum-parvum*	*sunrisei-I*	#1 Sunrise	Small 4oz-6oz 2" x 1.5" Red Flat	Tomato
Lycopersicon	*platycarpum-rubrum-parvum*	*sunrisei-VII*	#7 Sunrise	Small 4oz-6oz 2" x 1.5" Red Flat	Tomato
Lycopersicon	*platycarpum-rubrum-parvum*	*tlacolula-ribbedii*	Tlacolula Ribbed	Small 4oz-6oz 2" x 1.5" Red Flat	Tomato
Lycopersicon	*platycarpum-rubrum-parvum*	*victoria*	Victoria	Small 4oz-6oz 2" x 1.5" Red Flat	Tomato
Lycopersicon	*platycarpum-rubrum-auranteum*	*mikado-scarletii*	Mikado Scarlet	Orange Red Flat	Tomato
Lycopersicon	*platycarpum-rubrum-auranteum*	*victori*	Victor	Orange Red Flat	Tomato
Lycopersicon	*platycarpum-rubrum-auranteum*	*victoryi*	Victory 6oz-8oz	Orange Red Flat	Tomato

Tomato, Plum (*Lycopersicon pyriforme*)

Genus	Species	Infraspecies	cultivar	specimen	Genome
Lycopersicon	*pyriforme-album*	*ivory-eggii*	Ivory Egg ?oz	White Plum	Tomato
Lycopersicon	*pyriforme-aurantium*	*glory-of-moldovai*	Glory Of Moldova 4oz	Orange Plum	Tomato
Lycopersicon	*pyriforme-aurantium*	*kaki-coingii*	Kaki Coing 3.0"	Orange Plum	Tomato
Lycopersicon	*pyriforme-aurantium*	*mandarinii*	Mandarin 2.5oz	Orange Plum	Tomato
Lycopersicon	*pyriforme-aurantium*	*orange-strawberryi*	Orange Strawberry 3.0"	Orange Plum	Tomato
Lycopersicon	*pyriforme-aurantium*	*qiyanai-huangii*	Qiyanai Huang ?.?"	Orange Plum	Tomato
Lycopersicon	*pyriforme-aurantium-helvum*	*mandarinii-crossi*	Mandarin Cross 8oz	Yellow Orange Plum	Tomato
Lycopersicon	*pyriforme-aurantium-helvum*	*roughwood-golden-plumii*	Roughwood Golden ?.?"	Yellow Orange Plum	Tomato
Lycopersicon	*pyriforme-aurantium-helvum*	*sun-cheersi-MR*	Sun Cheers Mildew Resistant ?.?"	Yellow Orange Plum	Tomato
Lycopersicon	*pyriforme-helvum*	*helvum*	Original	Yellow Plum	Tomato
Lycopersicon	*pyriforme-helvum*	*amish-goldii*	Amish Gold 1.75"	Yellow Plum	Tomato
Lycopersicon	*pyriforme-helvum*	*beam's-yellow-peari*	Beam's 1.5"	Yellow Plum	Tomato
Lycopersicon	*pyriforme-helvum*	*coeur-de-pigeon-juanei*	Coeur De Pigeon Juane ?.?"	Yellow Plum	Tomato
Lycopersicon	*pyriforme-helvum*	*d'anjou-peari*	D'anjou ?.?"	Yellow Plum	Tomato
Lycopersicon	*pyriforme-helvum*	*fargoi*	Fargo 2.5"	Yellow Plum	Tomato
Lycopersicon	*pyriforme-helvum*	*golden-eggii*	Golden Egg 1.5"	Yellow Plum	Tomato
Lycopersicon	*pyriforme-helvum*	*ildii*	Ilde ?.?"	Yellow Plum	Tomato
Lycopersicon	*pyriforme-helvum*	*ilse's-yellow-latviana*	Ilse's Yellow Latvian 3.0"	Yellow Plum	Tomato
Lycopersicon	*pyriforme-helvum*	*italiana-quadratic-yellowi*	Italian Quadratic 4.0"	Yellow Plum	Tomato
Lycopersicon	*pyriforme-helvum*	*new-moonii*	New Moon 2.5"	Yellow Plum	Tomato
Lycopersicon	*pyriforme-helvum*	*old-ivory-eggii*	Old Ivory Egg 2.0"	Yellow Plum	Tomato
Lycopersicon	*pyriforme-helvum*	*plum-lemonii*	Lemon 3.0"	Yellow Plum	Tomato
Lycopersicon	*pyriforme-helvum*	*powers-heirloomii*	Powers Heirloom 3.5"	Yellow Plum	Tomato
Lycopersicon	*pyriforme-helvum*	*red-marbled-yellow-bellii*	Red Marbled Yellow Bell ?oz	Yellow Plum	Tomato
Lycopersicon	*pyriforme-helvum*	*sugar-peari*	Sugar Pear 2.5"	Yellow Plum	Tomato
Lycopersicon	*pyriforme-helvum*	*yellow-peari*	Yellow Pear 1.8"	Yellow Plum	Tomato
Lycopersicon	*pyriforme-nigrum*	*Japanese-trifele-blackii*	Japanese Trifele	Black Plum	Tomato
Lycopersicon	*pyriforme-pincum*	*airy-leafii*	Airy Leaf 4.0"	Pink Plum	Tomato
Lycopersicon	*pyriforme-pincum*	*aker's-plum-pinkii*	Aker's ?.?"	Pink Plum	Tomato
Lycopersicon	*pyriforme-pincum*	*balladii*	Ballad 1lb	Pink Plum	Tomato
Lycopersicon	*pyriforme-pincum*	*brandywine-sportii*	Sport Amish Brandywine Creek 4oz	Pink Plum	Tomato
Lycopersicon	*pyriforme-pincum*	*god-lovei*	God Love 2.0"	Pink Plum	Tomato
Lycopersicon	*pyriforme-pincum*	*hungarian-pink-kalman'si*	Kalman's Hungarian 4.0"	Pink Plum	Tomato
Lycopersicon	*pyriforme-pincum*	*jahamato-pink-pastei*	Jahamato Pink Paste 6oz	Pink Plum	Tomato
Lycopersicon	*pyriforme-pincum*	*kingumbertoi*	KingUmberto	Pink Plum	Tomato
Lycopersicon	*pyriforme-pincum*	*lacketai*	Laketa ?.?"	Pink Plum	Tomato
Lycopersicon	*pyriforme-pincum*	*madagascarensis*	Madagascar ?.?"	Pink Plum	Tomato
Lycopersicon	*pyriforme-pincum*	*new-zealand-peari*	New Zealand Pear 9oz	Pink Plum	Tomato
Lycopersicon	*pyriforme-pincum*	*oli-rose-de-st.-dominiquei*	Oli Rose De St. Dominique ?.?"	Pink Plum	Tomato
Lycopersicon	*pyriforme-pincum*	*polishii-eckertii*	Eckert Polish ?.?"	Pink Plum	Tomato
Lycopersicon	*pyriforme-pincum*	*sheboyganii*	Sheboygan ?.?"	Pink Plum	Tomato
Lycopersicon	*pyriforme-pincum*	*speckled-peachii*	Speckled Peach 2.0"	Pink Plum	Tomato
Lycopersicon	*pyriforme-pincum*	*thai-pinkii*	Thai 2oz	Pink Plum	Tomato
Lycopersicon	*pyriforme-pincum*	*wheatley's-frost-resistantii*	Wheatley's Frost Resistant 1.5"	Pink Plum	Tomato
Lycopersicon	*pyriforme-pincum-purpureum*	*hungarian-ovalii*	Hungarian Oval 5.0"	Purplish Pink Plum	Tomato
Lycopersicon	*pyriforme-pincum-purpureum*	*jeff-davisi*	Jeff Davis 14oz	Purplish Pink Plum	Tomato
Lycopersicon	*pyriforme-pincum-purpureum*	*pear-ukrainiana*	Ukraine Pear 3.5"	Purplish Pink Plum	Tomato
Lycopersicon	*pyriforme-roma-helvum*	*golden-roma*	Golden 3.0"	Yellow Roma Plum	Tomato
Lycopersicon	*pyriforme-roma-helvum*	*goldminei*	Goldmine ?.?"	Yellow Roma Plum	Tomato
Lycopersicon	*pyriforme-roma-helvum*	*italiana-goldii*	Italian Gold ?.?"	Yellow Roma Plum	Tomato
Lycopersicon	*pyriforme-roma-auranteum*	*orange-roma*	Original ?.?"	Orange Roma Plum	Tomato
Lycopersicon	*pyriforme-roma-auranteum*	*sweet-orangei*	Sweet Orange ?.?"	Orange Roma Plum	Tomato
Lycopersicon	*pyriforme-roma-rubrum*	*cherry-roma*	Cherry 1.0"	Red Roma Plum	Tomato

Genus	Species	Infraspecies	cultivar	specimen	Genome
Lycopersicon	pyriforme-roma-rubrum	balleri	Baller 3.0"	Red Roma Plum	Tomato
Lycopersicon	pyriforme-roma-rubrum	blue-beechii	Blue Beech 7oz	Red Roma Plum	Tomato
Lycopersicon	pyriforme-roma-rubrum	de-barraoi-2	#2 De Barrao 7oz 3.0"	Red Roma Plum	Tomato
Lycopersicon	pyriforme-roma-rubrum	dominick's-pastei	Dominick's Paste ?oz	Red Roma Plum	Tomato
Lycopersicon	pyriforme-roma-rubrum	italiana-roma	Italian 3.0"	Red Roma Plum	Tomato
Lycopersicon	pyriforme-roma-rubrum	la-roma	La Roma ?oz	Red Roma Plum	Tomato
Lycopersicon	pyriforme-roma-rubrum	long-roma	Long 5oz	Red Roma Plum	Tomato
Lycopersicon	pyriforme-roma-rubrum	mama-mia'si	Mama Mia's ?oz	Red Roma Plum	Tomato
Lycopersicon	pyriforme-roma-rubrum	martino's-roma	Martino's 2.5oz	Red Roma Plum	Tomato
Lycopersicon	pyriforme-roma-rubrum	mt.-roma	Mt. Roma 2.5oz	Red Roma Plum	Tomato
Lycopersicon	pyriforme-roma-rubrum	nova	Nova 2.0oz	Red Roma Plum	Tomato
Lycopersicon	pyriforme-roma-rubrum	pantano-romanescoi	Panto Romanesco 8oz-12oz	Red Roma Plum	Tomato
Lycopersicon	pyriforme-roma-rubrum	puree-roma	Puree ?oz	Red Roma Plum	Tomato
Lycopersicon	pyriforme-roma-rubrum	romeoi	Romeo 2lb	Red Roma Plum	Tomato
Lycopersicon	pyriforme-roma-rubrum	romeoi-&-juliettei	Romeo & Juliette ?oz	Red Roma Plum	Tomato
Lycopersicon	pyriforme-roma-rubrum	rossolii	Rossol ?oz	Red Roma Plum	Tomato
Lycopersicon	pyriforme-roma-rubrum	rosollii-VF-N	Verticillium Fusarium Nematode Resistant Rossol ?oz	Red Roma Plum	Tomato
Lycopersicon	pyriforme-roma-rubrum	royal-chicoi	Royal Chico 3oz	Red Roma Plum	Tomato
Lycopersicon	pyriforme-roma-rubrum	russian-big-roma	Russian Big 4.0"	Red Roma Plum	Tomato
Lycopersicon	pyriforme-roma-rubrum	supra-roma	SuperRomaPaste	Red Roma Plum	Tomato
Lycopersicon	pyriforme-roma-rubrum	three-sistersi-roma	Roma Three Sisters	Red Roma Plum	Tomato
Lycopersicon	pyriforme-roma-rubrum	tolli-roma	Tolli 3oz	Red Roma Plum	Tomato
Lycopersicon	pyriforme-roma-rubrum	uplinei	Upline	Red Roma Plum	Tomato
Lycopersicon	pyriforme-roma-rubrum	veeroma	VeeRoma	Red Roma Plum	Tomato
Lycopersicon	pyriforme-roma-rubrum	VF-roma	Verticillium Fusarium Resistant 2.5oz 3.0"	Red Roma Plum	Tomato
Lycopersicon	pyriforme-roma-rubrum	VF-N-roma	Verticillium Fusarium Nematode Resistant 2.5oz 3.0"	Red Roma Plum	Tomato
Lycopersicon	pyriforme-roma-rubrum	VF-super-roma	Super Verticillium Fusarium Resistant 2.5oz 3.0"	Red Roma Plum	Tomato
Lycopersicon	pyriforme-roma-rubrum	warm-fuzziesi	Warm Fuzzies	Red Roma Plum	Tomato
Lycopersicon	pyriforme-rubrum-largum	.	.	Large 1lb-2lb 4"-6" Red Plum	Tomato
Lycopersicon	pyriforme-rubrum-largum	gallo-plumii	Gallo	Large 1lb-2lb 4"-6" Red Plum	Tomato
Lycopersicon	pyriforme-rubrum-largum	pittman-valley-plumii	Pittman Valley	Large 1lb-2lb 4"-6" Red Plum	Tomato
Lycopersicon	pyriforme-rubrum-largum	polishii	Brick Red Polish	Large 1lb-2lb 4"-6" Red Plum	Tomato
Lycopersicon	pyriforme-rubrum-largum	polishii-giantii	Giant Polish	Large 1lb-2lb 4"-6" Red Plum	Tomato
Lycopersicon	pyriforme-rubrum-largum	polishii-pastei	Paste Polish	Large 1lb-2lb 4"-6" Red Plum	Tomato
Lycopersicon	pyriforme-rubrum-largum	yugoslaviana	Yugoslavian	Large 1lb-2lb 4"-6" Red Plum	Tomato
Lycopersicon	pyriforme-rubrum-largum-medium	ed's-fat-plumii	Ed's Fat	Medium Large 1/2lb-3/4lb 3"-4" Red Plum	Tomato
Lycopersicon	pyriforme-rubrum-largum-medium	egyptiana	Egyptian	Medium Large 1/2lb-3/4lb 3"-4" Red Plum	Tomato
Lycopersicon	pyriforme-rubrum-largum-medium	giant-italian-plumii	Giant Italian	Medium Large 1/2lb-3/4lb 3"-4" Red Plum	Tomato
Lycopersicon	pyriforme-rubrum-largum-medium	giant-italian-potato-leafii	Potato Leaf Giant Italian	Medium Large 1/2lb-3/4lb 3"-4" Red Plum	Tomato
Lycopersicon	pyriforme-rubrum-largum-medium	gilbert-italian-plumii	Gilbert Italian	Medium Large 1/2lb-3/4lb 3"-4" Red Plum	Tomato
Lycopersicon	pyriforme-rubrum-largum-medium	graham's-good-keeperi	Graham's Good Keeper	Medium Large 1/2lb-3/4lb 3"-4" Red Plum	Tomato
Lycopersicon	pyriforme-rubrum-largum-medium	joe's-plumii	Joe's Plum	Medium Large 1/2lb-3/4lb 3"-4" Red Plum	Tomato
Lycopersicon	pyriforme-rubrum-largum-medium	mexican-pastei	Mexican Paste	Medium Large 1/2lb-3/4lb 3"-4" Red Plum	Tomato
Lycopersicon	pyriforme-rubrum-largum-medium	mikarda(micarda)	Mikarda	Medium Large 1/2lb-3/4lb 3"-4" Red Plum	Tomato
Lycopersicon	pyriforme-rubrum-largum-medium	mikarda-sweetii	Sweet Mikarda	Medium Large 1/2lb-3/4lb 3"-4" Red Plum	Tomato
Lycopersicon	pyriforme-rubrum-largum-medium	msia	Msia	Medium Large 1/2lb-3/4lb 3"-4" Red Plum	Tomato
Lycopersicon	pyriforme-rubrum-largum-medium	polishii-linguisai	Linguisa Polish	Medium Large 1/2lb-3/4lb 3"-4" Red Plum	Tomato
Lycopersicon	pyriforme-rubrum-largum-medium	sojourner-south-americana	Sojourner South American	Medium Large 1/2lb-3/4lb 3"-4" Red Plum	Tomato
Lycopersicon	pyriforme-rubrum-largum-medium	tyborski-giant-plumii	Tyborski Giant	Medium Large 1/2lb-3/4lb 3"-4" Red Plum	Tomato
Lycopersicon	pyriforme-rubrum-medium	aker's-plum-redii	Aker's	Medium 6oz 2.5" Red Plum	Tomato
Lycopersicon	pyriforme-rubrum-medium	bradley-redii	Bradley	Medium 6oz 2.5" Red Plum	Tomato
Lycopersicon	pyriforme-rubrum-medium	britain's-breakfastii	Britain's Breakfast	Medium 6oz 2.5" Red Plum	Tomato
Lycopersicon	pyriforme-rubrum-medium	burkina-fasoensis	Burkina Faso	Medium 6oz 2.5" Red Plum	Tomato
Lycopersicon	pyriforme-rubrum-medium	chicoi-3	#3 Chico Grande	Medium 6oz 2.5" Red Plum	Tomato
Lycopersicon	pyriforme-rubrum-medium	czech-plumii	Czech Plum	Medium 6oz 2.5" Red Plum	Tomato
Lycopersicon	pyriforme-rubrum-medium	denalii	Denali	Medium 6oz 2.5" Red Plum	Tomato
Lycopersicon	pyriforme-rubrum-medium	earliest-pastei	Earliest Paste	Medium 6oz 2.5" Red Plum	Tomato
Lycopersicon	pyriforme-rubrum-medium	emie's-plumpii	Ernie's Plump	Medium 6oz 2.5" Red Plum	Tomato
Lycopersicon	pyriforme-rubrum-medium	friscoi	Frisco	Medium 6oz 2.5" Red Plum	Tomato
Lycopersicon	pyriforme-rubrum-medium	german-red-strawberryi	German Red Strawberry	Medium 6oz 2.5" Red Plum	Tomato
Lycopersicon	pyriforme-rubrum-medium	giant-pastei	Giant Paste	Medium 6oz 2.5" Red Plum	Tomato
Lycopersicon	pyriforme-rubrum-medium	giles-mullis-plumii	Giles Mullis	Medium 6oz 2.5" Red Plum	Tomato
Lycopersicon	pyriforme-rubrum-medium	grappoli-d'invernoi	Grappoli D'Inverno	Medium 6oz 2.5" Red Plum	Tomato
Lycopersicon	pyriforme-rubrum-medium	grozneyi-91	#91 Grozney	Medium 6oz 2.5" Red Plum	Tomato
Lycopersicon	pyriforme-rubrum-medium	grushovka	Grushovka	Medium 6oz 2.5" Red Plum	Tomato
Lycopersicon	pyriforme-rubrum-medium	heidii	Heidi	Medium 6oz 2.5" Red Plum	Tomato
Lycopersicon	pyriforme-rubrum-medium	himmelssturmeri	Himmelssturmer	Medium 6oz 2.5" Red Plum	Tomato
Lycopersicon	pyriforme-rubrum-medium	hssaio-hungshih	Hssaio Hungshih	Medium 6oz 2.5" Red Plum	Tomato
Lycopersicon	pyriforme-rubrum-medium	italiana-peari	Italian Pear	Medium 6oz 2.5" Red Plum	Tomato
Lycopersicon	pyriforme-rubrum-medium	jitomate-bulitoi	Jitomate Bulito	Medium 6oz 2.5" Red Plum	Tomato
Lycopersicon	pyriforme-rubrum-medium	jubileum	Jubileum	Medium 6oz 2.5" Red Plum	Tomato

Genus	Species	Infraspecies	cultivar	specimen	Genome
Lycopersicon	*pyriforme-rubrum-medium*	*lebyajinskyi*	Lebyaijinsky	Medium 6oz 2.5" Red Plum	Tomato
Lycopersicon	*pyriforme-rubrum-medium*	*maceroi-2*	#2 Macero	Medium 6oz 2.5" Red Plum	Tomato
Lycopersicon	*pyriforme-rubrum-medium*	*maria-augustina*	Maria Augustina	Medium 6oz 2.5" Red Plum	Tomato
Lycopersicon	*pyriforme-rubrum-medium*	*MH-6203-VF*	Mechanical Harvest #6203 Verticillium Fusarium Resistant	Medium 6oz 2.5" Red Plum	Tomato
Lycopersicon	*pyriforme-rubrum-medium*	*minelli-plumii*	Minelli	Medium 6oz 2.5" Red Plum	Tomato
Lycopersicon	*pyriforme-rubrum-medium*	*northhampton-italian-plumii*	Northampton Italian	Medium 6oz 2.5" Red Plum	Tomato
Lycopersicon	*pyriforme-rubrum-medium*	*plum-seedlessi*	Seedless	Medium 6oz 2.5" Red Plum	Tomato
Lycopersicon	*pyriforme-rubrum-medium*	*russian-giantii*	Russian Giant	Medium 6oz 2.5" Red Plum	Tomato
Lycopersicon	*pyriforme-rubrum-medium*	*santa-cruz-angela*	Angel Santa Cruz	Medium 6oz 2.5" Red Plum	Tomato
Lycopersicon	*pyriforme-rubrum-medium*	*santa-cruz-kada*	Kada Santa Cruz	Medium 6oz 2.5" Red Plum	Tomato
Lycopersicon	*pyriforme-rubrum-medium*	*sunrisei-III*	#3 Sunrise	Medium 6oz 2.5" Red Plum	Tomato
Lycopersicon	*pyriforme-rubrum-medium*	*sunseedsi-1642*	#1642 Sunseeds	Medium 6oz 2.5" Red Plum	Tomato
Lycopersicon	*pyriforme-rubrum-medium*	*sutton's-ornamentalis*	Sutton's Ornamental	Medium 6oz 2.5" Red Plum	Tomato
Lycopersicon	*pyriforme-rubrum-medium*	*teardropii*	Teardrop	Medium 6oz 2.5" Red Plum	Tomato
Lycopersicon	*pyriforme-rubrum-medium*	*tondino-di-manduria*	Tondino Di Manduria	Medium 6oz 2.5" Red Plum	Tomato
Lycopersicon	*pyriforme-rubrum-medium*	*ventura*	Ventura	Medium 6oz 2.5" Red Plum	Tomato
Lycopersicon	*pyriforme-rubrum-medium*	*viola-italiana*	Viola Italian	Medium 6oz 2.5" Red Plum	Tomato
Lycopersicon	*pyriforme-rubrum-parvum*	.	.	Small 4oz 2.0" Red Plum	Tomato
Lycopersicon	*pyriforme-rubrum-parvum*	*alaskan-fancyi*	Alaska Fancy	Small 4oz 2.0" Red Plum	Tomato
Lycopersicon	*pyriforme-rubrum-parvum*	*austin's-red-peari*	Austin's Red Pear	Small 4oz 2.0" Red Plum	Tomato
Lycopersicon	*pyriforme-rubrum-parvum*	*azurei*	Azure	Small 4oz 2.0" Red Plum	Tomato
Lycopersicon	*pyriforme-rubrum-parvum*	*bacconei*	Baccone	Small 4oz 2.0" Red Plum	Tomato
Lycopersicon	*pyriforme-rubrum-parvum*	*baylor-pastei*	Baylor Paste	Small 4oz 2.0" Red Plum	Tomato
Lycopersicon	*pyriforme-rubrum-parvum*	*bellstari*	Bellstar	Small 4oz 2.0" Red Plum	Tomato
Lycopersicon	*pyriforme-rubrum-parvum*	*bellstari-cover-upii*	Cover Up Bellstar	Small 4oz 2.0" Red Plum	Tomato
Lycopersicon	*pyriforme-rubrum-parvum*	*bijskij-zeltyii*	Bijskij Zeltyi	Small 4oz 2.0" Red Plum	Tomato
Lycopersicon	*pyriforme-rubrum-parvum*	*chicoi*	Chico Grande	Small 4oz 2.0" Red Plum	Tomato
Lycopersicon	*pyriforme-rubrum-parvum*	*ding-wall-scottyi*	Ding Wall Scotty	Small 4oz 2.0" Red Plum	Tomato
Lycopersicon	*pyriforme-rubrum-parvum*	*eggii*	Egg	Small 4oz 2.0" Red Plum	Tomato
Lycopersicon	*pyriforme-rubrum-parvum*	*ethiopia-roi-humbortii*	Ethiopia Roi Humbart	Small 4oz 2.0" Red Plum	Tomato
Lycopersicon	*pyriforme-rubrum-parvum*	*IPK-T-0120*	Italian Paste King Type 0120	Small 4oz 2.0" Red Plum	Tomato
Lycopersicon	*pyriforme-rubrum-parvum*	*IPK-T-0912*	Italian Paste King Type 0912	Small 4oz 2.0" Red Plum	Tomato
Lycopersicon	*pyriforme-rubrum-parvum*	*IPK-T-0955*	Italian Paste King Type 0955	Small 4oz 2.0" Red Plum	Tomato
Lycopersicon	*pyriforme-rubrum-parvum*	*IPK-T-0983*	Italian Paste King Type 0983	Small 4oz 2.0" Red Plum	Tomato
Lycopersicon	*pyriforme-rubrum-parvum*	*IPK-T-0985*	Italian Paste King Type 0985	Small 4oz 2.0" Red Plum	Tomato
Lycopersicon	*pyriforme-rubrum-parvum*	*IPK-T-1118*	Italian Paste King Type 1118	Small 4oz 2.0" Red Plum	Tomato
Lycopersicon	*pyriforme-rubrum-parvum*	*plumitoi*	Plumito	Small 4oz 2.0" Red Plum	Tomato
Lycopersicon	*pyriforme-rubrum-parvum*	*prometeoi*	Prometeo	Small 4oz 2.0" Red Plum	Tomato
Lycopersicon	*pyriforme-rubrum-parvum*	*reigartii*	Reigart	Small 4oz 2.0" Red Plum	Tomato
Lycopersicon	*pyriforme-rubrum-parvum*	*square-pastei-80*	Square Paste	Small 4oz 2.0" Red Plum	Tomato
Lycopersicon	*pyriforme-rubrum-parvum*	*wuhibii*	Wuhib	Small 4oz 2.0" Red Plum	Tomato
Lycopersicon	*pyriforme-rubrum-cerasiforme*	*almetia*	Almetia	Cherry 2oz-3oz 1"-1.5" Red Plum	Tomato
Lycopersicon	*pyriforme-rubrum-cerasiforme*	*canestrinoi-1*	#1 Canestrino	Cherry 2oz-3oz 1"-1.5" Red Plum	Tomato
Lycopersicon	*pyriforme-rubrum-cerasiforme*	*canestrinoi-2*	#2 Canestrino	Cherry 2oz-3oz 1"-1.5" Red Plum	Tomato
Lycopersicon	*pyriforme-rubrum-cerasiforme*	*fond-red-mini-plumii*	Fond Red Mini	Cherry 2oz-3oz 1"-1.5" Red Plum	Tomato
Lycopersicon	*pyriforme-rubrum-cerasiforme*	*napoliensis*	Napoli	Cherry 2oz-3oz 1"-1.5" Red Plum	Tomato
Lycopersicon	*pyriforme-rubrum-cerasiforme*	*oroma*	Oregon Roma	Cherry 2oz-3oz 1"-1.5" Red Plum	Tomato
Lycopersicon	*pyriforme-rubrum-cerasiforme*	*peeloi*	Peelo	Cherry 2oz-3oz 1"-1.5" Red Plum	Tomato
Lycopersicon	*pyriforme-rubrum-cerasiforme*	*principe-borghesei*	Principe Borghese	Cherry 2oz-3oz 1"-1.5" Red Plum	Tomato
Lycopersicon	*pyriforme-rubrum-cerasiforme*	*red-figii*	Red Fig	Cherry 2oz-3oz 1"-1.5" Red Plum	Tomato
Lycopersicon	*pyriforme-rubrum-cerasiforme*	*romancei*	Romance	Cherry 2oz-3oz 1"-1.5" Red Plum	Tomato
Lycopersicon	*pyriforme-rubrum-aurantium*	.	.	Orange Red Plum	Tomato
Lycopersicon	*pyriforme-rubrum-aurantium*	*extra-eros-zlatolaska*	Extra Eros Zlatolaska 3oz	Orange Red Plum	Tomato
Lycopersicon	*pyriforme-rubrum-aurantium*	*gary's-late-canneri*	Gary's Late Canner ?.?"	Orange Red Plum	Tomato
Lycopersicon	*pyriforme-rubrum-aurantium*	*high-countryi*	High Country 3.5oz	Orange Red Plum	Tomato
Lycopersicon	*pyriforme-rubrum-aurantium*	*mankin-plumii*	Mankin ?.?"	Orange Red Plum	Tomato
Lycopersicon	*pyriforme-rubrum-aurantium*	*pear-redii*	Italian Pear	Orange Red Plum	Tomato
Lycopersicon	*pyriforme-rubrum-aurantium*	*tony's-italiana*	Tony's Italian 4oz-6oz	Orange Red Plum	Tomato
Lycopersicon	*pyriforme-striatum*	*potato-red-skinsi*	Potato Red Skins ?oz	Striped Plum	Tomato
Lycopersicon	*pyriforme-striatum*	*tonneletii*	Tonnelet ?oz	Striped Plum	Tomato

Ground Cherry (*Physalis spp.*)

Genus	Species	Infraspecies	cultivar	specimen	Genome
Physalis	*ixocarpa-groena*	*cisinerosa*	Cisineros 2.0"	Green Tomatillo	Ground Cherry
Physalis	*ixocarpa-groena*	*grande-rio-verdei*	Grande Rio Verde 3oz.	Green Tomatillo	Ground Cherry
Physalis	*ixocarpa-groena*	*large-greenii*	Large 2.5"	Green Tomatillo	Ground Cherry
Physalis	*ixocarpa-helva*	*giant-yellowi*	Giant 1.2"	Yellow Tomatillo	Ground Cherry
Physalis	*ixocarpa-purpurea*	*merida-marketii*	Merida Market ?.?"	Purple Tomatillo	Ground Cherry
Physalis	*ixocarpa-purpurea*	*miltomato-locoi*	Loco Miltomato 1/4"	Purple Tomatillo	Ground Cherry
Physalis	*ixocarpa-purpurea*	*miltomato-vallistoi*	Vallisto Miltomato 1/4"	Purple Tomatillo	Ground Cherry
Physalis	*ixocarpa-rubra*	*indian-strainii*	Indian Strain ?.?"	Red Tomatillo	Ground Cherry

Genus	Species	Infraspecies	cultivar	specimen	Genome
Physalis	*ixocarpa-rubra*	*indian-summeri*	Indian Summer ?.?"	Red Tomatillo	Ground Cherry
Physalis	*ixocarpa-rubra*	*mexican-strainii*	Mexican Strain 2.0"	Red Tomatillo	Ground Cherry
Physalis	*ixocarpa-rubra*	*pineapplei*	Pineapple ?.?"	Red Tomatillo	Ground Cherry
Physalis	*pensylvanica*	*pensylvanica*	Original	Pennsylvania	Ground Cherry
Physalis	*pensylvanica*	*hanoverensis*	Hanover	Pennsylvania	Ground Cherry
Physalis	*pensylvanica*	*sinuatodentata*	Sinuous Tooth	Pennsylvania	Ground Cherry
Physalis	*peruviana*	*peruviana*	Original	Peruvian	Ground Cherry
Physalis	*peruviana*	*edulis*	Cape Gooseberry	Peruvian	Ground Cherry
Physalis	*peruviana*	*gigantea*	Golden Pohaberry	Peruvian	Ground Cherry
Physalis	*peruviana*	*pohai*	Golden Pohaberry	Peruvian	Ground Cherry
Physalis	*philadelphica*	*philadelphica*	Original	Philadelphia Tomatillo	Ground Cherry
Physalis	*philadelphica*	*golden-nuggetii*	Golden Nugget	Philadelphia Tomatillo	Ground Cherry
Physalis	*philadelphica*	*immaculata*	Immaculate	Philadelphia Tomatillo	Ground Cherry
Physalis	*philadelphica*	*purple-de-milpai*	Purple De Milpa	Philadelphia Tomatillo	Ground Cherry
Physalis	*philadelphica*	*tepehuana*	Tepehuan	Philadelphia Tomatillo	Ground Cherry
Physalis	*philadelphica*	*zunii*	Zuni	Philadelphia Tomatillo	Ground Cherry
Physalis	*polandica*	*aunt-molly'si*	Aunt Molly's	Polish Tomatillo	Ground Cherry
Physalis	*pruinosa-groena*	*puebla-verdei*	Puebla Verde 1.25"	Green Husk Tomatillo	Ground Cherry
Physalis	*pruinosa-groena*	*tomate-verdei*	Tomate Verde 2.5"	Green Husk Tomatillo	Ground Cherry
Physalis	*pruinosa-groena*	*tomatillo-green-huskii*	Tomatillo ?.?"	Green Husk Tomatillo	Ground Cherry
Physalis	*pruinosa-helva*	*goldiei*	Goldie	Yellow Husk Tomatillo	Ground Cherry
Physalis	*pruinosa-purpurea*	*purpurea*	Original 1.25"	Purple Husk Tomatillo	Ground Cherry
Physalis	*pruinosa-purpurea*	*purpurea-demilpa*	Purple DeMilpa	Purple Husk Tomatillo	Ground Cherry
Physalis	*pruinosa-rubra*	*cossack-pineapplei*	Cossack Pineapple	Red Husk Tomatillo	Ground Cherry
Physalis	*pruinosa-rubra*	*huberschmidtii*	Huberschmidt	Red Husk Tomatillo	Ground Cherry
Physalis	*pubescens-helva*	*yellow-huskii*	Florida	Yellow Husk Tomato	Ground Cherry
Physalis	*pubescens-helva*	*goldenstrawberryi*	Golden Strawberry	Yellow Husk Tomato	Ground Cherry
Physalis	*pubescens-helva*	*goldenyellowstrawberryi*	Golden Yellow Strawberry	Yellow Husk Tomato	Ground Cherry
Physalis	*pubescens-helva*	*sweetamberi*	Asian Sweet Amber	Yellow Husk Tomato	Ground Cherry
Physalis	*pubescens-helva*	*sweet-amberi*	Florida Sweet Amber	Yellow Husk Tomato	Ground Cherry
Physalis	*pubescens-helva*	*yellowstrawberryi*	Yellow Strawberry	Yellow Husk Tomato	Ground Cherry
Physalis	*pubescens-rubra*	*barbadensis*	Barbados	Red Husk Tomato	Ground Cherry
Physalis	*pubescens-rubra*	*floridana*	Florida	Red Husk Tomato	Ground Cherry
Physalis	*pubescens-rubra*	*integrifolia*	Entire Leaf	Red Husk Tomato	Ground Cherry
Physalis	*pubescens-rubra*	*fragaria*	Strawberry	Red Husk Tomato	Ground Cherry
Physalis	*pubescens-rubra*	*vintera-cerasus*	Winter Cherry	Red Husk Tomato	Ground Cherry

Family 724.042.2 *Convolvulaceae* Bindwood

Genus	Species	Specimen	Genome	C:Culinary, M:Medicinal, E:Economic
Ipomoea	*aquatica*	Water Spinach	Morning Glory	C:LF:**VG**
Ipomoea	*batatas-alba*	White Sweet Potato	Morning Glory	C:TU:**VG**
Ipomoea	*batatas-aurantia*	Orange Sweet Potato	Morning Glory	C:TU:**VG**
Ipomoea	*batatas-aurantia-helva*	Orange Yellow Sweet Potato	Morning Glory	C:TU:**VG**
Ipomoea	*batatas-brunnea*	Brown Sweet Potato	Morning Glory	C:TU:**VG**
Ipomoea	*batatas-helva*	Yellow Sweet Potato	Morning Glory	C:TU:**VG**
Ipomoea	*batatas-purpurea*	Purple Sweet Potato	Morning Glory	C:TU:**VG**
Ipomoea	*cairica*	Mile A Minute Tuber	Morning Glory	C:TU:**VG**
Ipomoea	*conzattii*	Jicama Montes	Morning Glory	C:TU:**VG**

Varieties

Water Spinach (*Ipomoea aquatica*)

Genus	Species	infraspecies	cultivar	specimen	Genome
Ipomoea	*aquatica*	*aquatica*	Original	Water Spinach	Morning Glory
Ipomoea	*aquatica*	*enstai*	Enstai	Water Spinach	Morning Glory
Ipomoea	*aquatica*	*reptans*	Creeping	Water Spinach	Morning Glory
Ipomoea	*aquatica*	*ung-choi*	Ung Choy	Water Spinach	Morning Glory

Sweet Potato (*Ipomoea batatas*)

Ipomoea	*batatas-alba*	.	.	White Sweet Potato	Morning Glory
Ipomoea	*batatas-alba*	*amish-pinkii*	Amish Pink Skin	White Sweet Potato	Morning Glory
Ipomoea	*batatas-alba*	*batasii*	Batas White Skin	White Sweet Potato	Morning Glory
Ipomoea	*batatas-alba*	*bermuda-pinca*	Bermuda Pink Splotch White Skin	White Sweet Potato	Morning Glory
Ipomoea	*batatas-alba*	*bermuda-whitei*	Bermuda White Skin	White Sweet Potato	Morning Glory
Ipomoea	*batatas-alba*	*braziliana*	Brazilian White Skin	White Sweet Potato	Morning Glory
Ipomoea	*batatas-alba*	*brinkley-whitei*	Brinkley White Skin	White Sweet Potato	Morning Glory
Ipomoea	*batatas-alba*	*crystal-whitei*	Crystal Cream Skin	White Sweet Potato	Morning Glory
Ipomoea	*batatas-alba*	*cubana*	Cuban Pink Purple Skin	White Sweet Potato	Morning Glory
Ipomoea	*batatas-alba*	*frazieri*	Frazier White Skin	White Sweet Potato	Morning Glory
Ipomoea	*batatas-alba*	*hayman-whitei*	Hayman White Skin	White Sweet Potato	Morning Glory

Ipomoea	*batatas-alba*	*ivis-white-creamii*	Ivis White Cream Skin	White Sweet Potato Morning Glory
Ipomoea	*batatas-alba*	*jersey-yellowi*	Jersey Yellow Cream Skin	White Sweet Potato Morning Glory
Ipomoea	*batatas-alba*	*korean-purplei*	Korean Purple Skin	White Sweet Potato Morning Glory
Ipomoea	*batatas-alba*	*kumarai*	Kumara Purple Skin	White Sweet Potato Morning Glory
Ipomoea	*batatas-alba*	*o'henryi*	O'Henry White Skin	White Sweet Potato Morning Glory
Ipomoea	*batatas-alba*	*okinawaensis*	Okinawa White Skin	White Sweet Potato Morning Glory
Ipomoea	*batatas-alba*	*old-kentuckyi*	Old Kentucky White Skin	White Sweet Potato Morning Glory
Ipomoea	*batatas-alba*	*poplar-rootii*	Poplar Root White Skin	White Sweet Potato Morning Glory
Ipomoea	*batatas-alba*	*purple-yamii*	Yam Purple Skin	White Sweet Potato Morning Glory
Ipomoea	*batatas-alba*	*southern-queenii*	Southern Queen White Skin	White Sweet Potato Morning Glory
Ipomoea	*batatas-alba*	*tainungii-65*	#65 Tainung Rose Skin	White Sweet Potato Morning Glory
Ipomoea	*batatas-alba*	*violetta*	Violetta Purple Skin	White Sweet Potato Morning Glory
Ipomoea	*batatas-alba*	*white-bunchii*	White Bunch White Skin	White Sweet Potato Morning Glory
Ipomoea	*batatas-alba*	*white-crystalis*	White Crystal White Skin	White Sweet Potato Morning Glory
Ipomoea	*batatas-alba*	*white-delightii*	White Delight Purple Skin	White Sweet Potato Morning Glory
Ipomoea	*batatas-alba*	*white-queenii*	White Queen White Skin	White Sweet Potato Morning Glory
Ipomoea	*batatas-alba*	*white-stari*	White Star White Skin	White Sweet Potato Morning Glory
Ipomoea	*batatas-alba*	*white-triumphii*	White Triumph White Skin	White Sweet Potato Morning Glory
Ipomoea	*batatas-alba*	*white-yamii*	White Yam White Skin	White Sweet Potato Morning Glory
Ipomoea	*batatas-aurantia*	*amish-redii*	Amish Red Pink Skin	Orange Sweet Potato Morning Glory
Ipomoea	*batatas-aurantia*	*B-18*	Simcoe Breed #18 Orange Skin	Orange Sweet Potato Morning Glory
Ipomoea	*batatas-aurantia*	*beauregardii*	Beauregard Orange Skin	Orange Sweet Potato Morning Glory
Ipomoea	*batatas-aurantia*	*carolina-bunchii*	Carolina Bunch Copper Skin	Orange Sweet Potato Morning Glory
Ipomoea	*batatas-aurantia*	*carolina-rosei*	Carolina Rose Skin	Orange Sweet Potato Morning Glory
Ipomoea	*batatas-aurantia*	*carteri*	Carter	Orange Sweet Potato Morning Glory
Ipomoea	*batatas-aurantia*	*carveri*	Carver Rose Skin	Orange Sweet Potato Morning Glory
Ipomoea	*batatas-aurantia*	*centennialis*	Centennial Copper Skin	Orange Sweet Potato Morning Glory
Ipomoea	*batatas-aurantia*	*centennialis-redii*	Centennial Red Skin	Orange Sweet Potato Morning Glory
Ipomoea	*batatas-aurantia*	*copper-jewelii*	Copper Jewel	Orange Sweet Potato Morning Glory
Ipomoea	*batatas-aurantia*	*cordneri*	Cordner	Orange Sweet Potato Morning Glory
Ipomoea	*batatas-aurantia*	*darbyi*	Darby Red Skin	Orange Sweet Potato Morning Glory
Ipomoea	*batatas-aurantia*	*excelii*	Excel Copper Skin	Orange Sweet Potato Morning Glory
Ipomoea	*batatas-aurantia*	*fastigiata*	Branching	Orange Sweet Potato Morning Glory
Ipomoea	*batatas-aurantia*	*garnet-redii*	Garnet Red Skin	Orange Sweet Potato Morning Glory
Ipomoea	*batatas-aurantia*	*georgia-jetii*	Georgia Jet Red Skin	Orange Sweet Potato Morning Glory
Ipomoea	*batatas-aurantia*	*georgia-jetii-semibushii*	Semibush Georgia Jet Red Skin	Orange Sweet Potato Morning Glory
Ipomoea	*batatas-aurantia*	*georgia-redii*	Georgia Red Skin	Orange Sweet Potato Morning Glory
Ipomoea	*batatas-aurantia*	*georgia-yamii*	Georgia Yam Orange Skin	Orange Sweet Potato Morning Glory
Ipomoea	*batatas-aurantia*	*ginseng-redii*	Ginseng Red Skin	Orange Sweet Potato Morning Glory
Ipomoea	*batatas-aurantia*	*golden-slipperi*	Golden Slipper Orange Pink Skin	Orange Sweet Potato Morning Glory
Ipomoea	*batatas-aurantia*	*goldmari*	Gold Skin Market (Goldmar)	Orange Sweet Potato Morning Glory
Ipomoea	*batatas-aurantia*	*gold-nuggetii*	Gold Nugget Red Skin	Orange Sweet Potato Morning Glory
Ipomoea	*batatas-aurantia*	*gold-rushii*	Gold Rush Copper Skin	Orange Sweet Potato Morning Glory
Ipomoea	*batatas-aurantia*	*goldstari*	Goldstar Pink Red Skin	Orange Sweet Potato Morning Glory
Ipomoea	*batatas-aurantia*	*hernandezii*	Hernandez Copper Skin	Orange Sweet Potato Morning Glory
Ipomoea	*batatas-aurantia*	*jamaican-redii*	Jamaican Red	Orange Sweet Potato Morning Glory
Ipomoea	*batatas-aurantia*	*japanesei*	Japanese Pink Red Skin	Orange Sweet Potato Morning Glory
Ipomoea	*batatas-aurantia*	*jewelii*	Jewel Copper Skin	Orange Sweet Potato Morning Glory
Ipomoea	*batatas-aurantia*	*jumboi*	Jumbo Semibush	Orange Sweet Potato Morning Glory
Ipomoea	*batatas-aurantia*	*laceleafii*	Laceleaf Pink Skin	Orange Sweet Potato Morning Glory
Ipomoea	*batatas-aurantia*	*mahonii*	Mahon Rose Skin	Orange Sweet Potato Morning Glory
Ipomoea	*batatas-aurantia*	*martinsi*	Martins Pink Skin	Orange Sweet Potato Morning Glory
Ipomoea	*batatas-aurantia*	*marylandica-810*	#810 Maryland Red Skin	Orange Sweet Potato Morning Glory
Ipomoea	*batatas-aurantia*	*memphis-pridei*	Memphis Pride	Orange Sweet Potato Morning Glory
Ipomoea	*batatas-aurantia*	*oakleafii*	Oakleaf Pink Skin	Orange Sweet Potato Morning Glory
Ipomoea	*batatas-aurantia*	*oklahoma-redii*	Oklahoma Red Skin	Orange Sweet Potato Morning Glory
Ipomoea	*batatas-aurantia*	*old-orangei*	Old Orange Skin	Orange Sweet Potato Morning Glory
Ipomoea	*batatas-aurantia*	*painteri*	Painter Orange Skin	Orange Sweet Potato Morning Glory
Ipomoea	*batatas-aurantia*	*paramutai*	Paramuai	Orange Sweet Potato Morning Glory
Ipomoea	*batatas-aurantia*	*puerto-rico-bushii*	Puerto Rico Bush Copper Skin	Orange Sweet Potato Morning Glory
Ipomoea	*batatas-aurantia*	*red-ivy-leafii*	Red Ivy Leaf Red Skin	Orange Sweet Potato Morning Glory
Ipomoea	*batatas-aurantia*	*red-jewelii*	Red Jewel Red Skin	Orange Sweet Potato Morning Glory
Ipomoea	*batatas-aurantia*	*redmari*	Red Skin Market (Redmar)	Orange Sweet Potato Morning Glory
Ipomoea	*batatas-aurantia*	*red-wine-velvetii*	Red Wine Velvet Violet Red Skin	Orange Sweet Potato Morning Glory
Ipomoea	*batatas-aurantia*	*regalis*	Regal Purple Red Skin	Orange Sweet Potato Morning Glory
Ipomoea	*batatas-aurantia*	*resistoi*	Resisto	Orange Sweet Potato Morning Glory
Ipomoea	*batatas-aurantia*	*rose-centennialis*	Rose Centennial Rose Skin	Orange Sweet Potato Morning Glory
Ipomoea	*batatas-aurantia*	*southern-delightii*	Southern Delight Rose Skin	Orange Sweet Potato Morning Glory
Ipomoea	*batatas-aurantia*	*spanish-whitei*	Spanish White Skin	Orange Sweet Potato Morning Glory
Ipomoea	*batatas-aurantia*	*superiori*	Superior	Orange Sweet Potato Morning Glory
Ipomoea	*batatas-aurantia*	*tiliacea*	Linden Leaf	Orange Sweet Potato Morning Glory
Ipomoea	*batatas-aurantia*	*triloba*	Triple Lobe	Orange Sweet Potato Morning Glory
Ipomoea	*batatas-aurantia*	*tuskegeensis*	Tuskegee	Orange Sweet Potato Morning Glory
Ipomoea	*batatas-aurantia*	*vardamanii*	Vardaman Gold Yellow Skin	Orange Sweet Potato Morning Glory
Ipomoea	*batatas-aurantia-helva*	*shoregoldii*	Shoregold Orange Skin	Orange Yellow Sweet Potato Morning Glory

Ipomoea	*batatas-aurantia-helva*	*tennessee-topmarkii*	Tennessee Top Mark	Orange Yellow Sweet Potato	Morning Glory
Ipomoea	*batatas-aurantia-helva*	*tokatoka-gold*	Tokatoka Gold Brown Skin	Orange Yellow Sweet Potato	Morning Glory
Ipomoea	*batatas-brunnea*	*red-yamsii*	Red Yams Red Skin	Brown Sweet Potato	Morning Glory
Ipomoea	*batatas-helva*	*all-goldii*	All Gold Yellow Skin	Yellow Sweet Potato	Morning Glory
Ipomoea	*batatas-helva*	*betty'si*	Betty's Red Skin	Yellow Sweet Potato	Morning Glory
Ipomoea	*batatas-helva*	*edna-evansi*	Edna Evans Light Pink Skin	Yellow Sweet Potato	Morning Glory
Ipomoea	*batatas-helva*	*nancy-hallii*	Nancy Hall Yellow Skin	Yellow Sweet Potato	Morning Glory
Ipomoea	*batatas-helva*	*puerto-rico-runningii*	Puerto Rico Running Copper Pink Skin	Yellow Sweet Potato	Morning Glory
Ipomoea	*batatas-helva*	*sumori*	Sumor	Yellow Sweet Potato	Morning Glory
Ipomoea	*batatas-helva*	*sunneysidei*	Sunneyside Cream Yellow Skin	Yellow Sweet Potato	Morning Glory
Ipomoea	*batatas-helva*	*toka-goldii*	Toka Gold Yellow Skin	Yellow Sweet Potato	Morning Glory
Ipomoea	*batatas-purpurea*	*purpurea*	Original	Purple Sweet Potato	Morning Glory

Family 3 *Cuscutaceae* Dodder Vine
Family 4 *Fouquieriaceae* Desert Candle
Family 5 *Hydrophyllaceae* Waterleaf
Family 6 *Menyanthaceae* Buckbean
Family 7 *Polemoniaceae* Phlox

Class 724.05 *Lamiopsida* Mint

Order 724.051 *Lamiales* Mint

Family 724.051.1 *Lamiaceae* Mint

Genus	Species	Specimen Genome	C:Culinary, M:Medicinal, E:Economic
Perilla	*frutescens*	Shiso Beefsteak Mint	C:LF:**VG**
Salvia	*columbariae*	Chia Sage	C:SR:**VG**
Salvia	*tiliifolia*	Linden Leaf Chia Sage	C:SD:**VG**
Stachys	*affinis*	Chinese Artichoke Hedge Nettle	C:TU:**VG**

Varieties

Shizo (*Perilla frutescens*)

Genus	Species	infraspecies	cultivar	specimen	Genome
Perilla	*frutescens*	*groena*	Green	Shiso	Beefsteak Mint
Perilla	*frutescens*	*groena-cumin-scentedii*	Cumin Scented Green	Shiso	Beefsteak Mint
Perilla	*frutescens*	*hojisoi*	Hojiso	Shiso	Beefsteak Mint
Perilla	*frutescens*	*ocymoides*	Basil Leaf	Shiso	Beefsteak Mint
Perilla	*frutescens*	*purpurea*	Purple	Shiso	Beefsteak Mint
Perilla	*frutescens*	*rubra*	Red	Shiso	Beefsteak Mint

Chia (*Salvia columbaria, Salvia tiliifolia*)

Genus	Species	infraspecies	cultivar	specimen	Genome
Salvia	*tiliifolia*	*tiliifolia*	Original	Linden Leaf Chia	Sage
Salvia	*tiliifolia*	*tarahumarai*	Tarahumara	Linden Leaf Chia	Sage

Chinese Artichoke (*Stachys affinis*)

Genus	Species	infraspecies	cultivar	specimen	Genome
Stachys	*affinis*	*affinis*	Original	Chinese Artichoke	Hedge Nettle
Stachys	*affinis*	*charogii*	Charogi	Chinese Artichoke	Hedge Nettle
Stachys	*affinis*	*crosnes*	Crosnes	Chinese Artichoke	Hedge Nettle
Stachys	*affinis*	*knotrootii*	Knotroot	Chinese Artichoke	Hedge Nettle
Stachys	*affinis*	*sieboldii*	Siebold	Chinese Artichoke	Hedge Nettle
Stachys	*affinis*	*spiral-rootsi*	Spiral Roots	Chinese Artichoke	Hedge Nettle
Stachys	*affinis*	*tuberifera*	Tuber	Chinese Artichoke	Hedge Nettle

Family 724.051.2 *Boraginaceae* Borage

Borage (*Borago officinalis*)

Viper's Bugloss (*Echium vulgare*)

Comfrey (*Symphytum australiensis*)

Genus	Species	Specimen Genome	C:Culinary, M:Medicinal, E:Economic
Borago	*officinalis*	Mediterranean Borage	C:LF:**VG**
Echium	*lycopsis*	Purple Viper's Bugloss	C:FL:BF,CF,C:LF:**VG**
Echium	*vulgare*	Blue Devil Viper's Bugloss	C:FL:BF,CF,C:LF:**VG**
Symphytum	*australiensis*	Australian Comfrey	C:LF:**VG**

Family 3 *Lennoaceae* Desert Christmaswood

Family 724.051.4 *Plantaginaceae* Plantain

Genus	Species	Specimen Genome	C:Culinary, M:Medicinal, E:Economic
Plantago	*coronopus*	Buck's Horn Plantain	C:FL-BD:**VG**

Varieties

Buck's Horn Plantain (*Plantago coronopus*)

Genus	Species	infraspecies	cultivar	specimen	Genome
Plantago	*coronopus*	*coronopus*	Original	Buck's Horn	Plantain
Plantago	*coronopus*	*commutata*	Cut Leaf	Buck's Horn	Plantain
Plantago	*coronopus*	*minutina*	Minutina	Buck's Horn	Plantain

Family 5 *Rhizophoraceae* Mangrove Mint

Family 6 *Verbenaceae* Vervain
Order 2 *Callitrichales* Starwort
Family 1 *Callitrichaceae* Starwort
Family 2 *Hippuridaceae* Mare's Tail

Order 724.053 *Scrophulariales* Foxglove
Family 724.053.1 *Scrophulariaceae* Foxglove

Empress Olive (*Paulownia spp.*)

Genus	Species	Specimen Genome	C:Culinary, M:Medicinal, E:Economic
Paulownia	*fargesii*	Sichuan Empress Olive	C:FR:**VG**
Paulownia	*fortunei*	White Flower Empress Olive	C:FR:**VG**
Paulownia	*kawakamii*	Kawakami Empress Olive	C:FR:**VG**
Paulownia	*taiwaniana*	Taiwan Empress Olive	C:FR:**VG**
Paulownia	*tomentosa*	Princess Tree Empress Olive	C:FR:**VG**

Varieties

Genus	Species	infraspecies	cultivar	specimen	Genome
Paulownia	*tomentosa*	*tomentosa*	Original	Empress Tree	Paulowna Olive
Paulownia	*tomentosa*	*lilacina*	Lilac	Empress Tree	Paulowna Olive
Paulownia	*tomentosa*	*sapphire-dragonii*	Sapphire Dragon	Empress Tree	Paulowna Olive

Family 2 *Acanthaceae* Prickleleaf Foxglove
Family 3 *Bignoniaceae* Trumpetwood
Family 4 *Buddlejaceae* Buddlewood
Family 5 *Gesneriaceae* Gesner Foxglove
Family 6 *Lentimyoporaceae* Mastic & Myopore

Family 724.053.7 *Oleaceae* Olive

Genus	Species	Specimen Genome	C:Culinary, M:Medicinal, E:Economic
Olea	*africana*	African Black Olive	C:FR:**VG**
Olea	*capensis*	Ironwood Black Olive	C:FR:**VG**
Olea	*chrysophylla*	Golden Leaf Olive	C:FR:**VG**
Olea	*europaea*	European Olive	C:FR:**VG**
Olea	*hansineensis*	Hansine Olive	C:FR:**VG**
Olea	*paniculata*	Australian Black Olive	C:FR:**VG**
Osmanthus	*fragrans*	Sweet Devilwood Olive	C:FR:**VG**

Varieties

Olive (*Olea europaea*)

Genus	Species	infraspecies	cultivar	specimen Genome
Olea	*europaea*	*europaea*	.	European Olive
Olea	*europaea*	*aglandaui*	Aglandau	European Olive
Olea	*europaea*	*amphissis*	Amphissa	European Olive
Olea	*europaea*	*arabana*	Araban	European Olive
Olea	*europaea*	*arbequina*	Arbequin	European Olive
Olea	*europaea*	*arbequina-I-18*	Arbequin #I-18	European Olive
Olea	*europaea*	*arbosana*	Arbosan	European Olive
Olea	*europaea*	*arbosana-I-43*	Arbosan #I-43	European Olive
Olea	*europaea*	*ascolana*	Ascolano	European Olive
Olea	*europaea*	*ascolana-tenera*	Ascolano Light Green	European Olive
Olea	*europaea*	*barounii*	Barouni	European Olive
Olea	*europaea*	*belgentieroisea*	Belgentieroise	European Olive
Olea	*europaea*	*bella-di-cerignolai*	Bella Di Cerignola	European Olive
Olea	*europaea*	*biancolilla*	Biancolillo	European Olive
Olea	*europaea*	*bosana*	Bosan	European Olive
Olea	*europaea*	*bouteillana*	Bouteillan	European Olive
Olea	*europaea*	*bouteillana*	Bouteillan	European Olive
Olea	*europaea*	*cailletieri*	Cailletier	European Olive
Olea	*europaea*	*caninoi*	Canino	European Olive
Olea	*europaea*	*carboncella*	Carbogna	European Olive
Olea	*europaea*	*casaliva*	Casalivo	European Olive
Olea	*europaea*	*coratina*	Corato	European Olive
Olea	*europaea*	*empeltrea*	Empeltre	European Olive
Olea	*europaea*	*farga*	Farga	European Olive
Olea	*europaea*	*frangiventoi*	Frangivento	European Olive
Olea	*europaea*	*frantolioi*	Frantolio	European Olive
Olea	*europaea*	*gailletieri*	Gailletier	European Olive

Genus	Species	infraspecies	cultivar	specimen	Genome
Olea	*europaea*	*grapoloi*	Grapolo	European	Olive
Olea	*europaea*	*grignanii*	Grignan	European	Olive
Olea	*europaea*	*grossannea*	Grossanne	European	Olive
Olea	*europaea*	*hojiblanca*	Hojiblanca	European	Olive
Olea	*europaea*	*kalamata*	Kalamata	European	Olive
Olea	*europaea*	*koroneikii*	Koroneiki	European	Olive
Olea	*europaea*	*koroneikii-I-38*	Koroneiki #I-38	European	Olive
Olea	*europaea*	*leccinoi*	Leccino	European	Olive
Olea	*europaea*	*lechin-di-savilla*	Black Sevielle	European	Olive
Olea	*europaea*	*little-olliei*	Little Ollie	European	Olive
Olea	*europaea*	*luccai*	Lucca	European	Olive
Olea	*europaea*	*lucquesi*	Lucques	European	Olive
Olea	*europaea*	*manzanilla*	Manzanillo	European	Olive
Olea	*europaea*	*maurinoi*	Maurino	European	Olive
Olea	*europaea*	*megaronii*	Megaron	European	Olive
Olea	*europaea*	*missionii*	Mission	European	Olive
Olea	*europaea*	*moraioloi*	Moraiolo	European	Olive
Olea	*europaea*	*negrettea*	Negrette	European	Olive
Olea	*europaea*	*nocellara-del-belicei*	Nocellara Del Belice	European	Olive
Olea	*europaea*	*oleaster*	Thin	European	Olive
Olea	*europaea*	*pendolinoi*	Pendolino	European	Olive
Olea	*europaea*	*picholinea*	Picholine	European	Olive
Olea	*europaea*	*picualis*	Picual	European	Olive
Olea	*europaea*	*poumalii*	Poumal	European	Olive
Olea	*europaea*	*queen-manzanilla*	Queen Manzanilla	European	Olive
Olea	*europaea*	*sabinea*	Sabine	European	Olive
Olea	*europaea*	*salonenquei*	Salonenque	European	Olive
Olea	*europaea*	*santa-caterina*	Santa Caterina	European	Olive
Olea	*europaea*	*saracenensis*	Saracena	European	Olive
Olea	*europaea*	*sevillanoi*	Sevillano	European	Olive
Olea	*europaea*	*taggiasca*	Taggiasca	European	Olive
Olea	*europaea*	*tanchea*	Tanche	European	Olive
Olea	*europaea*	*verdalea*	Verdale	European	Olive

Sweet Devilwood Olive (*Osmanthus fragrans*)

Genus	Species	infraspecies	cultivar	specimen	Genome
Osmanthus	*fragrans*	*fragrans*	Original	Sweet	Devilwood Olive
Osmanthus	*fragrans*	*aurantiacus*	Orange	Sweet	Devilwood Olive

Family 8 *Orobanchaceae* Orobanche

Family 724.053.9 *Pedaliaceae* Sesame

Genus	Species	Specimen	Genome	C:Culinary, M:Medicinal, E:Economic
Sesamum	*alatum*	Red Seed	Sesame	C:SD:**VG**
Sesamum	*indicum*	White Seed	Sesame	C:SD:**VG**
Sesamum	*goma*	Yellow Seed	Sesame	C:SD:**VG**
Sesamum	*orientale*	Black Seed	Sesame	C:SD:**VG**
Sesamum	*radiatum*	Brown Seed	Sesame	C:SD:**VG**
Proboscidea	*althaeifolia*	Straight Tube	Unicorn Plant	C:SD:**VG**
Proboscidea	*louisianica*	Ram's Horn	Unicorn Plant	C:SD:**VG**
Proboscidea	*parviflora*	Devil's Claw	Unicorn Plant	C:SD:**VG**
Proboscidea	*parviflora-hohokamiana*	Hohokam	Unicorn Plant	C:SD:**VG**
Proboscidea	*sabulosa*	Sand Dune	Unicorn Plant	C:SD:**VG**
Proboscidea	*spicata*	New Mexico	Unicorn Plant	C:SD:**VG**

Varieties

Sesame Seed (*Sesamum spp.*)

Genus	Species	infraspecies	cultivar	specimen	Genome
Sesamum	*indicum*	*indicum*	Original	White Seed	Sesame
Sesamum	*indicum*	*bennei*	Benne	White Seed	Sesame
Sesamum	*indicum*	*ginglii*	Gingli	White Seed	Sesame
Sesamum	*indicum*	*linei-31*	Line #31	White Seed	Sesame
Sesamum	*goma*	*goma*	Original	Yellow Seed	Sesame
Sesamum	*goma*	*kingoma*	Kin Goma	Yellow Seed	Sesame
Sesamum	*goma*	*kurogoma*	Kuro Goma	Yellow Seed	Sesame
Sesamum	*goma*	*shorogoma*	Shiro Goma	Yellow Seed	Sesame
Sesamum	*goma*	*tohum*	Toh Goma Golden	Yellow Seed	Sesame
Sesamum	*orientale*	*orientale*	Original	Black Seed	Sesame
Sesamum	*orientale*	*thailandica*	Thai	Black Seed	Sesame
Sesamum	*radiatum*	*radiatum*	Original	Brown Seed	Sesame

Genus	Species	infraspecies	cultivar	specimen	Genome
Sesamum	*radiatum*	*afghanica*	Afghani Tan	Brown Seed	Sesame
Sesamum	*radiatum*	*mixta*	Landrace Mix Tan	Brown Seed	Sesame
Sesamum	*radiatum*	*anatolia-brunnea*	Turkish	Brown Seed	Sesame
Sesamum	*radiatum*	*anatolia-tana*	Turkish Tan	Brown Seed	Sesame
Proboscidea	*althaeifolia*	*althaeifolia*	Original	Straight Tube	Unicorn Plant
Proboscidea	*althaeifolia*	*arenaria*	Desert	Straight Tube	Unicorn Plant
Proboscidea	*louisianica*	*louisianica*	Original	Ram's Horn	Unicorn Plant
Proboscidea	*louisianica*	*fragrans*	Chihuahua	Ram's Horn	Unicorn Plant
Proboscidea	*louisianica*	*jussieui*	Jussieu	Ram's Horn	Unicorn Plant
Proboscidea	*parviflora*	*parviflora*	Original	Devil's Claw	Unicorn Plant
Proboscidea	*parviflora*	*crassibracteata*	Thick Bract	Devil's Claw	Unicorn Plant
Proboscidea	*parviflora*	*eagle-creekii*	Eagle Creek	Devil's Claw	Unicorn Plant
Proboscidea	*parviflora*	*gracillima*	Slender	Devil's Claw	Unicorn Plant
Proboscidea	*parviflora*	*multiclavii*	Multiclawed	Devil's Claw	Unicorn Plant
Proboscidea	*parviflora*	*o'odhamii*	Eagle Creek	Devil's Claw	Unicorn Plant
Proboscidea	*parviflora*	*sinaloensis*	Sinaloan	Devil's Claw	Unicorn Plant
Proboscidea	*parviflora*	*sinaloensis-pima-bajoi*	Pima Bajo Sinaloan	Devil's Claw	Unicorn Plant
Proboscidea	*parviflora-hohokamiana*	*domesticata*	Domesticated	Hohokam	Unicorn Plant
Proboscidea	*parviflora-hohokamiana*	*hopii*	Hopi	Hohokam	Unicorn Plant
Proboscidea	*parviflora-hohokamiana*	*morelosensis*	Morelos	Hohokam	Unicorn Plant
Proboscidea	*parviflora-hohokamiana*	*paiutei*	Paiute	Hohokam	Unicorn Plant
Proboscidea	*parviflora-hohokamiana*	*po'onompii*	Po'onomp	Hohokam	Unicorn Plant
Proboscidea	*parviflora-hohokamiana*	*san-carlos-apachei*	San Carlos Apache	Hohokam	Unicorn Plant

Class 724.06 *Legumopsida* Legume

Order 1 *Legumales* Legume

Family 1 *Legumaceae* Legume

GENERA *Leguma, Acosmium, Adenopodia, Adesmia, Aenictophyton(Synonym for Leguma), Affonsea, Afgekia, Aganope, Alantsilodendron, Aldina, Alexa, Alistilus, Almaleea, Amblygonocarpus(Synonym for Tetrapleura), Amburana, Ammodendron, Amphimas, Amphithalea, Androcalymma(Synonym for Leguma), Angylocalyx, Antheroporum, Anthonotha, Aotus, Aphanocalyx, Aphyllodium, Apoplanesia, Apuleia(Synonym for Leguma), Apurimacia, Arapatiella, Archidendron, Archidendropsis, Arcoa(Synonym for Leguma), Argyrolobium, Arthrocarpum, Arthroclianthus(Synonym for Leguma), Astracantha, Ateleia, Aubrevillea, Augouardia(Synonym for Bikinia), Austrosteenisia, Baikiaea, Batesia(Synonym for Leguma), Baudouinia, Behaimia(Synonym for Leguma), Bergeronia(Synonym for Leguma), Berlinia, Bikinia, Bocoa, Bolusia, Bossiaea, Bowdichia(Synonym for Diplotropis), Bowringia(Synonym for Leucomphalos), Brachycylix(Synonym for Heterostemon), Brachystegia, Brandzeia(Synonym for Leguma), Brenierea(Synonym for Leguma), Brodriguesia(Synonym for Leguma), Bryaspis, Buchenroedera, Burkea(Synonym for Leguma), Burkilliodendron(Synonym for Leguma), Bussea, Cadia, Calicotome, Calispepla(Synonym for Leguma), Callerya, Callistachys(Synonym for Oxylobium), Calophaca, Calpocalyx, Camoensia, Campsiandra, Camptosema, Carissoa(Synonym for Leguma), Cascaronia(Synonym for Leguma), Cenostigma, Centrolobium, Chadsia, Chaetocalyx, Chesneya, Chidlowia(Synonym for Leguma), Clathrotropis, Cleobulia, Coelidium, Collaea, Cordyla, Craibia, Cranocarpus, Craspedolobium(Synonym for Leguma), Cratylia, Cruddasia(Synonym for Ophrestia), Cryptosepalum, Cyclolobium, Cyclopia, Cylicodiscus(Synonym for Newtonia), Dahlstedtia(Synonym for Camptosema), Dalhousiea, Deguelia, Dendrolobium, Detarium, Dewevrea, Dicerma(Synonym for Aphyllodium), Dichilus, Dicorynia, Dicraeopetalum, Dicymbe, Didelotia, Dimorphandra, Dinizia(Synonym for Leguma), Diphyllarium(Synonym for Leguma), Diplotropis, Diptychandra(Synonym for Leguma), Discolobium, Distemonanthus(Synonym for Leguma), Droogmansia, Dumasia, Duparquetia(Synonym for Leguma), Dussia, Ecuadendron(Synonym for Leguma), Elephantorrhiza, Eligmocarpus(Synonym for Leguma), Elizabetha, Eminia, Endertia(Synonym for Leguma), Englerodendron(Synonym for Anthonotha), Eperua, Erichsenia(Synonym for Leguma), Euchilopsis(Synonym for Leguma), Euchresta, Eurypetalum, Eutaxia, Eversmannia, Exostyles, Fiebrigiella(Synonym for Leguma), Fillaeopsis(Synonym for Leguma), Fissicalyx(Synonym for Monopteryx), Fordia, Gagnebina, Gastrolobium, Geissaspis, Gigasiphon, Gilbertiodendron, Gilletiodendron, Goniorrhachis, Gossweilerodendron, Griffonia, Gueldenstaedtia, Hardwickia(Synonym for Kingiodendron), Harleyodendron, Harpalyce, Hegnera(Synonym for Leguma), Hesperothamnus, Heterostemon, Holocalyx(Synonym for Leguma), Hovea, Humboldtia, Humularia, Hydrochorea, Hylodendron(Synonym for Leguma), Hymenolobium, Hymenostegia, Hypocalyptus, Icuria(Synonym for Leguma), Indigastrum, Isoberlinia, Isotropis, Jacksonia, Jacqueshuberia, Jansonia(Synonym for Leguma), Julbernardia, Kalappia(Synonym for Leguma), Kingiodendron, Klugiodendron(Synonym for Leguma), Koompassia, Kotschya, Kunstleria, Labichea, Lamprolobium, Latrobea, Lebeckia, Lecointea, Lemurodendron, Lennea, Leonardoxa, Leptodesmia, Leptosema, Lessertia, Leucochloron, Leucomphalos, Librevillea(Synonym for Brachystegia), Liparia, Loesenera, Lophocarpinia(Synonym for Cenostigma), Lotononis, Luetzelburgia, Luzonia(Synonym for Leguma), Macrolobium, Macropsychanthus, Maniltoa, Margaritolobium(Synonym for Leguma), Marmaroxylon, Mecopus(Synonym for Uraria), Melanoxylon, Melliniella(Synonym for Leguma), Melolobium, Mendoravia(Synonym for Leguma), Michelsonia(Synonym for Tetraberlinia), Microberlinia(Synonym for Berlinia), Microcharis, Mildbraediodendron(Synonym for Leguma), Mirbelia, Moldenhawera, Monopteryx, Mora, Muelleranthus, Myrocarpus, Nemcia, Neoapaloxylon, Neochevalierodendron(Synonym for Hymenostegia), Neoharmsia(Synonym for Cadia), Neorautanenia, Nephrodesmus, Nesphostylis, Newtonia, Nogra, Oddoniodendron, Ophrestia, Ormocarpopsis, Ormocarpum, Orphanodendron(Synonym for Leguma), Otholobium, Otoptera, Ottleya, Oxylobium, Oxystigma, Pachyelasma(Synonym for Leguma), Painteria, Paloue, Paloveopsis(Synonym for Leguma), Panurea(Synonym for Leguma), Paracalyx, Paramacrolobium(Synonym for Macrolobium), Pearsonia, Pellegriniodendron(Synonym for Macrolobium), Peltiera, Peltogyne, Periandra, Petaladenium, Petalostylis, Phylacium, Phyllocarpus, Phyllota, Phylloxylon, Placolobium, Plagiocarpus(Synonym for Leguma), Plagiosiphon, Plathymenia, Platycelyphium(Synonym for Leguma), Platycyamus, Platymiscium, Platypodium, Platysepalum, Podolobium(Synonym for Oxylobium), Poecilanthe, Poeppigia(Synonym for Leguma), Poiretia, Polhillia, Polystemonanthus(Synonym for Leguma), Pomaria, Priestleya, Prioria(Synonym for Kingiodendron), Pseudarthria, Pseudeminia, Pseudoeriosema, Pseudomacrolobium(Synonym for Berlinia), Pterodon(Synonym for Acosmium), Pterogyne(Synonym for Leguma), Pterolobium, Ptycholobium, Ptychosema, Pultenaea, Pyranthus, Racosperma, Rafnia, Ramirezella, Ramorinoa, Recordoxylon, Rhodopis, Riedeliella, Robynsiophyton, Sakoanala, Salweenia(Synonym for Leguma), Sartoria(Synonym for Leguma), Schefflerodendron, Schleinitzia, Sclerolobium, Sellocharis(Synonym for Leguma), Sesbania, Shuteria, Sindora, Smithia, Soemmeringia(Synonym for Leguma), Spathionema(Synonym for Leguma), Spatholobus, Sphaerolobium, Stachyothyrsus, Stemonocoleus(Synonym for Leguma), Stenodrepanum(Synonym for Leguma), Stonesiella(Synonym for Leguma), Storckiella, Streblorrhiza(Synonym for Leguma), Stuhlmannia(Synonym for Leguma), Sweetia(Synonym for Acosmium), Sylvichadsia, Sympetalandra, Syrmatium, Tachigali, Tadehagi, Talbotiella, Taverniera, Terua(Synonym for Leguma), Tessmannia, Tetraberlinia, Tetrapleura, Tetrapterocarpon, Thylacanthus(Synonym for Leguma), Tibetia, Uleanthus(Synonym for Leguma), Umtiza(Synonym for Leguma), Uraria, Urariopsis(Synonym for Uraria), Uribea(Synonym for Leguma), Urodon(Synonym for Pultenaea), Vatairea, Vataireopsis, Vaughania, Viguieranthus, Viminaria(Synonym for Leguma), Vouacapoua, Wajira(Synonym for Leguma), Weberbauerella, Wiborgia, Willardia, Xanthocercis(Synonym for Cadia), Xeroderris(Synonym for Aganope), Xipotheca, Xylia, Zenia(Synonym for Leguma), Zenkerella, Zollernia, Zygia*

Family 2 *Caesalpiniaceae* Caesalpino

GENERA *Caesalpinia, Acrocarpus, Adipera(Synonym), Amherstia, Balizia(Synonym for Pithecellobium), Balsamocarpon(Synonym for Caesalpinia), Barklya, Biancaea(Synonym), Brownea, Browneopsis, Cassia,* ***Ceratonia*(Carob)**, ***Cercidium(Palo Verde Horse Bean),*** *Cercis, Chamaefistula(Synonym for Senna), Chamaesenna(Synonym for Senna), Chloroleucon, Cojoba, Colvillea, Conzattia, Delonix, Dialium, Ditremexa(Synonym for Senna), Emelista(Synonym for Senna), Erythrophleum, Eysenhardtia, Gaumerocassia(Synonym), Gleditsia, Guilandina(Synonym for Caesalpinia),* ***Haematoxylum*(Bloodwood Herb)**, *Halimodendron, Herpetica(Synonym for Senna), Hoffmannseggia, Ichthyomethia(Synonym for* ***Piscidia****), Isandrina(Synonym for Senna), Leucostegane(Synonym for Saraca), Libidibia(Synonym for Caesalpinia), Lysidice, Lysiloma, Lysiphyllum, Mezoneuron(Synonym for Caesalpinia), Moullava(Synonym for Caesalpinia), Obolinga(Synonym for Cojoba), Pararchidendron(Synonym for Pithecellobium), Parkinsonia, Peiranisia(Synonym for Senna), Peltophorum,* ***Piscidia(Stinkwood Herb)****, Pithecellobium, Poinciana(Synonym for Delonix), Saraca, Schizolobium,* ***Schotia*(African Walnut)**, *Sciacassia(Synonym for Senna),* ***Senna(Senna Herb)***, ***Tamarindus indica* (India Date Tamarind)**, *Ticanto, Uittienia(Synonym for Dialium), Wallaceodendron(Synonym for Pithecellobium), Zuccagnia(Synonym for Hoffmannseggia)*

Family 3 *Cytisaceae* Broom

GENERA ***Cytisus(Broom)***, *Adenocarpus, Anarthrophyllum, Argyrocytisus,* ***Aspalathus(Peaflower Broom)***, *Borbonia(Synonym Aspalathus), Calpurnia, Carmichaelia, Chamaecystis, Chordospartium, Corallospartium, Cytisophyllum, Cytisopsis, Echinospartum, Eremosparton, Erinacea, Errazurizia,* ***Genista(Pea Broom)***, *Genistidium, Gonocytisus, Hesperolaburnum, Lembotropis, Laburnocytisus,* ***Laburnum(Golden Chain Broom)***, *Notospartium, Parryella, Petteria(Synonym for Cytisus), Podocytisus, Retama, Sarothamnus(Synonym for Cytisus), Smirnowia(Synonym for Eremosparton), Spartidium, Spartium, Teline(Synonym for Genista)*

Family 4 *Ebenaceae* Ebony

GENERA *Ebenus, Adenolobus(Synonym for Bauhinia), Bauhinia(FalseOrchid), Bobgunnia(Synonym for Swartzia), Brya* (Coccuswood), *Candolleodendron(Synonym for Swartzia), Cyathostegia(Synonym for Swartzia), Dalbergia*(Dalberg Ebony), *Dalbergiella, Ebenopsis, Etaballia(Synonym for Dalbergia), Grazielodendron(Synonym for Dalbergia), Swartzia*(Wamara) *Piliostigma(Synonym for Bauhinia), Spirotropis(Synonym for Swartzia)*

Family 5 *Mimosaceae* Sensitive Plant

GENERA *Mimosa, **Acacia(Wattle),** Acaciella(Synonym for Acacia), **Albizia(Koko),** Calliandra, Calliandropsis, Faidherbia, Falcataria(Synonym for Albizia), Hesperalbizia, Leptoglottis(Synonym for Mimosa), Macrosamanea, Mimosopsis(Synonym for Mimosa), Mimozyganthus, Morongia(Synonym for Mimosa), Neltuma(Synonym for Prosopis), Neptunia, Paraserianthes, Poponax(Synonym for Acacia), Prosopidastrum, Prosopis, Pseudoprosopis, Pseudosamanea, Samanea, Schrankia, Senegalia(Synonym for Acacia), Serianthes, Vachellia(Synonym for Acacia), Xerocladia(Synonym for Prosopidastrum),*

Family 6 *Piptadeniaceae* Droopgland

GENERA *Piptadenia, Ammopiptanthus, Cedrelinga(Synonym for Piptadenia), Indopiptadenia, Microlobius(Synonym for Piptadenia), Parapiptadenia, Piptadeniastrum, Piptadeniopsis, Piptanthus, Pseudopiptadenia*

Order 724.062 *Fabales* Bean

Family 724.062.1 *Fabaceae* Bean

GENERA ***Faba (Broad Bean), Apios(Potato Bean),*** *Arcyphyllum(Synonym), Cacara(Synonym),* ***Canavalia(Jack Bean),*** *Cathormion (Synonym for Inga), Cochlianthus (Synonym for Mucuna),* ***Cordeauxia (Cordeaux Bean), Cyamopsis(Guar Bean), Dipogon(Okie Bean),*** *Dipteryx(Tonka Bean), Dunbaria (Dunbar Bean), Eleiotis (Synonym for Glycine monophyllos), Fagelia(Synonym for Glycine),* ***Glycine(Soy Bean), Gymnocladus(Coffee Bean), Inga(Ice Cream Bean), Lablab(Lablab Bean), Lens(Lentil Bean), Leucaena* (Jumbie Bean),** *Mastersia(Synonym for Mucuna),* ***Mucuna(Velvet Bean),*** *Mysanthus(Synonym for Phaseolus), Neonotonia(Synonym for Glycine), Nomismia(Synonym for Rhynchosia),* ***Pachyrhizus(Yam Bean), Phaseolus acutifolius-latifolius(Tepary Bean),Phaseolus coccineus(Runner Bean), Phaseolus lunatus(Lima Bean), Phaseolus vulgaris (Bean),*** *Physostigma(Calabar Bean),* ***Psophocarpus (Exploding Bean), Pueraria(Kudzu Potato) Rhynchosia(Snout Bean),*** *Soja(Synonym Glycine max soja), Teyleria(Synonym for Pueraria),* ***Tylosema(Buck Bean),*** *Vatovaea(Synonym for Lablab),* ***Vigna(Vine Bean)***

Genus	Species	Specimen	Genome	C:Culinary
Faba	*brevipoda*	Short Pod (1-4 Bean)	Broad Bean	C:PD:**VG**
Faba	*mediapoda*	Medium Pod (4-6 Bean)	Broad Bean	C:PD:**VG**
Faba	*longapoda*	Long Pod (6-12 Bean)	Broad Bean	C:PD:**VG**
Faba	*sativa*	Sweet Pod	Broad Bean	C:PD:**VG**
Apios	*americana*	Groundnut	Potato Bean	C:TU:**VG**
Canavalia	*ensiformis*	American Sword Bean	Jack Bean	C:PD:**VG**
Canavalia	*gladiata*	Asian Sword Bean	Jack Bean	C:PD:**VG**
Cordeauxia	*edulis*	Yeheb	Cordeaux Bean	C:PD:**VG**
Cyamopsis	*tetragonoloba*	Java	Guar Bean	C:PD:**VG**
Dipogon	*lignosus*	Mile-A-Minute	Okie Bean	C:PD:**VG**
Glycine	*max-alba*	White Soybean	Glycine Bean	C:PD:**VG**
Glycine	*max-brunnea*	Brown Soybean	Glycine Bean	C:PD:**VG**
Glycine	*max-groena*	Green Soybean	Glycine Bean	C:PD:**VG**
Glycine	*max-helva*	Yellow Soybean	Glycine Bean	C:PD:**VG**
Glycine	*max-nigra*	Black Soybean	Glycine Bean	C:PD:**VG**
Glycine	*max-soja*	Wild Soybean	Glycine Bean	C:PD:**VG**
Inga	*edulis*	Edible	Ice Cream Bean	C:PD:**VG**
Inga	*oerstediana*	Oersted Edible	Ice Cream Bean	C:PD:**VG**
Lablab	*purpureus*	Hyacinth Bean	Lablab Bean	C:PD:**VG**
Lens	*culinaris*	Culinary	Lentil Bean	C:SD:**VG**
Leucaena	*esculenta*	Edible	Jumbie Bean	C:PD:**VG**
Mucuna	*pruriens*	Edible	Velvet Bean	C:PD:**VG**
Pachyrhizus	*erosus*	Jicama	Yam Bean	C:PD:**VG**
Phaseolus	*acutifolius-latifolius-alba*	White	Tepary Bean	C:SD:**VG**
Phaseolus	*acutifolius-latifolius-blauea*	Blue	Tepary Bean	C:SD:**VG**
Phaseolus	*acutifolius-latifolius-brunnea*	Brown	Tepary Bean	C:SD:**VG**
Phaseolus	*acutifolius-latifolius-nigra*	Black	Tepary Bean	C:SD:**VG**
Phaseolus	*acutifolius-latifolius-tana*	Tan	Tepary Bean	C:SD:**VG**
Phaseolus	*coccineus-alba*	White	Runner Bean	C:PD:**VG**
Phaseolus	*coccineus-brunnea*	Brown	Runner Bean	C:PD:**VG**
Phaseolus	*coccineus-flora*	Scarlet Flower	Runner Bean	C:PD:**VG**
Phaseolus	*coccineus-nigra*	Black	Runner Bean	C:PD:**VG**
Phaseolus	*coccineus-pinca*	Pink	Runner Bean	C:PD:**VG**
Phaseolus	*coccineus-purpurea*	Purple	Runner Bean	C:PD:**VG**
Phaseolus	*coccineus-tana*	Tan	Runner Bean	C:PD:**VG**
Phaseolus	*lunatus-bruscia-alba*	White Seed Bush	Lima Bean	C:SD:**VG**
Phaseolus	*lunatus-bruscia-aurantea*	Orange Seed Bush	Lima Bean	C:SD:**VG**
Phaseolus	*lunatus-bruscia-grisea*	Gray Seed Bush	Lima Bean	C:SD:**VG**
Phaseolus	*lunatus-bruscia-groena*	Green Seed Bush	Lima Bean	C:SD:**VG**
Phaseolus	*lunatus-bruscia-nana-alba*	White Baby Bush	Lima Bean	C:SD:**VG**
Phaseolus	*lunatus-bruscia-nana-groena*	Green Baby Bush	Lima Bean	C:SD:**VG**
Phaseolus	*lunatus-bruscia-rubra*	Red Seed Bush	Lima Bean	C:SD:**VG**
Phaseolus	*lunatus-bruscia-tana*	Tan Seed Bush	Lima Bean	C:SD:**VG**
Phaseolus	*lunatus-vigna-alba*	White Seed Vine	Lima Bean	C:SD:**VG**
Phaseolus	*lunatus-vigna-aurantea*	White Seed Vine	Lima Bean	C:SD:**VG**
Phaseolus	*lunatus-vigna-groena*	Green Seed Vine	Lima Bean	C:SD:**VG**
Phaseolus	*lunatus-vigna-helva*	Yellow Seed Vine	Lima Bean	C:SD:**VG**
Phaseolus	*lunatus-vigna-mixta*	Mixed Seed Vine	Lima Bean	C:SD:**VG**
Phaseolus	*lunatus-vigna-nana-alba*	White Baby Vine	Lima Bean	C:SD:**VG**
Phaseolus	*lunatus-vigna-nana-groena*	Green Baby Vine	Lima Bean	C:SD:**VG**

Genus	Species	Specimen	Genome	C:Culinary
Phaseolus	*lunatus-vigna-nigra*	Black Seed Vine Lima	Bean	C:SD:**VG**
Phaseolus	*lunatus-vigna-rubra*	Red Seed Vine Lima	Bean	C:SD:**VG**
Phaseolus	*lunatus-vigna-tana*	Tan Seed Vine Lima	Bean	C:SD:**VG**
Phaseolus	*vulgaris-bruscia-alba*	White Dry Bush	Bean	C:SD:**VG**
Phaseolus	*vulgaris-bruscia-baca*	Baking Dry Bush	Bean	C:SD:**VG**
Phaseolus	*vulgaris-bruscia-brunnea*	Brown Dry Bush	Bean	C:SD:**VG**
Phaseolus	*vulgaris-bruscia-chili*	Chili Dry Bush	Bean	C:SD:**VG**
Phaseolus	*vulgaris-bruscia-grisea*	Gray Dry Bush	Bean	C:SD:**VG**
Phaseolus	*vulgaris-bruscia-groena*	Green Dry Bush	Bean	C:SD:**VG**
Phaseolus	*vulgaris-bruscia-helva*	Yellow Dry Bush	Bean	C:SD:**VG**
Phaseolus	*vulgaris-bruscia-hortus*	Shell Dry Bush	Bean	C:SD:**VG**
Phaseolus	*vulgaris-bruscia-navalis*	Navy Dry Bush	Bean	C:SD:**VG**
Phaseolus	*vulgaris-bruscia-nigra*	Black Dry Bush	Bean	C:SD:**VG**
Phaseolus	*vulgaris-bruscia-pinca*	Pink Dry Bush	Bean	C:SD:**VG**
Phaseolus	*vulgaris-bruscia-pintoi*	Pinto Dry Bush	Bean	C:SD:**VG**
Phaseolus	*vulgaris-bruscia-purpurea*	Purple Dry Bush	Bean	C:SD:**VG**
Phaseolus	*vulgaris-bruscia-rena*	Kidney Dry Bush	Bean	C:SD:**VG**
Phaseolus	*vulgaris-bruscia-rubra*	Red Dry Bush	Bean	C:SD:**VG**
Phaseolus	*vulgaris-bruscia-tana*	Tan Dry Bush	Bean	C:SD:**VG**
Phaseolus	*vulgaris-bruscia-snapa-alba*	White Seed Snap Bush	Bean	C:PD:**VG**
Phaseolus	*vulgaris-bruscia-snapa-astrenga*	Stringless Round Snap Bush	Bean	C:PD:**VG**
Phaseolus	*vulgaris-bruscia-snapa-brunnea*	Brown Seed Snap Bush	Bean	C:PD:**VG**
Phaseolus	*vulgaris-bruscia-snapa-filetii*	Filet Snap Bush	Bean	C:PD:**VG**
Phaseolus	*vulgaris-bruscia-snapa-grisea*	Gray Seed Snap Bush	Bean	C:PD:**VG**
Phaseolus	*vulgaris-bruscia-snapa-groena*	Green Round Snap Bush	Bean	C:PD:**VG**
Phaseolus	*vulgaris-bruscia-snapa-helva*	Yellow Pod Snap Bush	Bean	C:PD:**VG**
Phaseolus	*vulgaris-bruscia-snapa-hortus*	Shell Snap Bush	Bean	C:PD:**VG**
Phaseolus	*vulgaris-bruscia-snapa-nigra*	Black Seed Snap Bush	Bean	C:PD:**VG**
Phaseolus	*vulgaris-bruscia-snapa-platya*	Flat Snap Bush	Bean	C:PD:**VG**
Phaseolus	*vulgaris-bruscia-snapa-platya-astrenga*	Stringless Flat Snap Bush	Bean	C:PD:**VG**
Phaseolus	*vulgaris-bruscia-snapa-purpurea*	Purple Pod Snap Bush	Bean	C:PD:**VG**
Phaseolus	*vulgaris-bruscia-snapa-stena-alba*	White Thin Round Snap Bush	Bean	C:PD:**VG**
Phaseolus	*vulgaris-bruscia-snapa-stena-groena*	Green Thin Round Snap Bush	Bean	C:PD:**VG**
Phaseolus	*vulgaris-bruscia-vaxa-alba*	White Seed Wax Bush	Bean	C:PD:**VG**
Phaseolus	*vulgaris-bruscia-vaxa-groena*	Green Pod Wax Bush	Bean	C:PD:**VG**
Phaseolus	*vulgaris-bruscia-vaxa-helva*	Yellow Pod Wax Bush	Bean	C:PD:**VG**
Phaseolus	*vulgaris-bruscia-vaxa-helva-albasperma*	White Seed Yellow Pod Wax Bush	Bean	C:PD:**VG**
Phaseolus	*vulgaris-bruscia-vaxa-helva-astrenga*	Stringless Yellow Pod Wax Bush	Bean	C:PD:**VG**
Phaseolus	*vulgaris-bruscia-vaxa-helva-astrenga-alba*	White Seed Stringless Yellow Pod Wax Bush	Bean	C:PD:**VG**
Phaseolus	*vulgaris-bruscia-vaxa-helva-platya*	Flat Yellow Pod Wax Bush	Bean	C:PD:**VG**
Phaseolus	*vulgaris-bruscia-vaxa-helva-platya-astrenga*	Stringless Flat Yellow Pod Wax Bush	Bean	C:PD:**VG**
Phaseolus	*vulgaris-bruscia-vaxa-helva-stena*	Thin Yellow Pod Wax Bush	Bean	C:PD:**VG**
Phaseolus	*vulgaris-vigna-alba*	White Seed Dry Vine	Bean	C:PD:**VG**
Phaseolus	*vulgaris-vigna-grisea*	Gray Seed Dry Vine	Bean	C:PD:**VG**
Phaseolus	*vulgaris-vigna-hortus*	Shell Dry Vine	Bean	C:PD:**VG**
Phaseolus	*vulgaris-vigna-mixta*	Mixed Dry Vine	Bean	C:PD:**VG**
Phaseolus	*vulgaris-vigna-nigra*	Black Seed Dry Vine	Bean	C:PD:**VG**
Phaseolus	*vulgaris-vigna-purpurea*	Purple Pod Dry Vine	Bean	C:PD:**VG**
Phaseolus	*vulgaris-vigna-tana*	Tan Seed Dry Vine	Bean	C:PD:**VG**
Phaseolus	*vulgaris-vigna-tarahumara*	Tarahumara Dry Vine	Bean	C:PD:**VG**
Phaseolus	*vulgaris-vigna-snapa-alba*	White Seed Snap Vine	Bean	C:PD:**VG**
Phaseolus	*vulgaris-vigna-snapa-astrenga*	Stringless Round Snap Vine	Bean	C:PD:**VG**
Phaseolus	*vulgaris-vigna-snapa-astrenga-alba*	White Seed Stringless Round Snap Vine	Bean	C:PD:**VG**
Phaseolus	*vulgaris-vigna-snapa-groena*	Green Round Snap Vine	Bean	C:PD:**VG**
Phaseolus	*vulgaris-vigna-snapa-helva*	Yellow Pod Snap Vine	Bean	C:PD:**VG**
Phaseolus	*vulgaris-vigna-snapa-hortus*	Shell Snap Vine	Bean	C:PD:**VG**
Phaseolus	*vulgaris-vigna-snapa-mixta*	Mixed Snap Vine	Bean	C:PD:**VG**
Phaseolus	*vulgaris-vigna-snapa-nigra*	Black Seed Snap Vine	Bean	C:PD:**VG**
Phaseolus	*vulgaris-vigna-snapa-platya*	Flat Snap Vine	Bean	C:PD:**VG**
Phaseolus	*vulgaris-vigna-snapa-platya-astrenga*	Stringless Flat Snap Vine	Bean	C:PD:**VG**
Phaseolus	*vulgaris-vigna-snapa-purpurea*	Purple Pod Snap Vine	Bean	C:PD:**VG**
Phaseolus	*vulgaris-vigna-snapa-rubra*	Red Seed Snap Vine	Bean	C:PD:**VG**
Phaseolus	*vulgaris-vigna-snapa-stena*	Thin Round Snap Vine	Bean	C:PD:**VG**
Phaseolus	*vulgaris-vigna-snapa-stena-astrenga*	Stringless Thin Round Snap Vine	Bean	C:PD:**VG**
Phaseolus	*vulgaris-vigna-snapa-vaxa-helva*	Yellow Wax Round Snap Vine	Bean	C:PD:**VG**
Phaseolus	*vulgaris-vigna-snapa-vaxa-helva-alba*	White Seed Yellow Wax Snap Vine	Bean	C:PD:**VG**
Phaseolus	*vulgaris-vigna-snapa-vaxa-helva-astrenga*	Stringless Yellow Wax Snap Vine	Bean	C:PD:**VG**
Phaseolus	*vulgaris-vigna-snapa-vaxa-helva-platya-astrenga*	Stringless Flat Yellow Wax Snap Vine	Bean	C:PD:**VG**
Psophocarpus	*tetragonolobus*	Winged Bean	Exploding Bean	C:PD:**VG**
Pueraria	*edulis*	Edible	Kudzu Potato Bean	C:PD:**VG**

Genus	Species	Specimen Genome	C:Culinary
Rhynchosia	*edulis*	Chihuahuan Snout Bean	C:PD:**VG**
Tylosema	*esculentum*	Edible Buck Bean	C:PD:**VG**
Vigna	*aconitifolia*	Moth Vine Bean	C:PD:**VG**
Vigna	*angularis*	Adzuki Vine Bean	C:PD:**VG**
Vigna	*mungo*	Gram Mung Vine Bean	C:PD:**VG**
Vigna	*radiata*	Mung Vine Bean	C:PD:**VG**
Vigna	*umbellata*	Rice Bean Vine Bean	C:PD:**VG**
Vigna	*unguiculata-alba*	White Seed Cowpea Vine Bean	C:PD:**VG**
Vigna	*unguiculata-brunnea*	Brown Seed Cowpea Vine Bean	C:PD:**VG**
Vigna	*unguiculata-creamii*	Cream Seed Cowpea Vine Bean	C:PD:**VG**
Vigna	*unguiculata-crowderi*	Crowder Cowpea Vine Bean	C:PD:**VG**
Vigna	*unguiculata-dekindtiana*	Black Eyed Cowpea Vine Bean	C:PD:**VG**
Vigna	*unguiculata-grisea*	Gray Seed Cowpea Vine Bean	C:PD:**VG**
Vigna	*unguiculata-groena*	Green Pod Cowpea Vine Bean	C:PD:**VG**
Vigna	*unguiculata-marmorata*	Mottled Seed Cowpea Vine Bean	C:PD:**VG**
Vigna	*unguiculata-multicolor*	Multicolor Seed Cowpea Vine Bean	C:PD:**VG**
Vigna	*unguiculata-nigra*	Black Seed Cowpea Vine Bean	C:PD:**VG**
Vigna	*unguiculata-rubra*	Red Seed Cowpea Vine Bean	C:PD:**VG**
Vigna	*unguiculata-sesquipedalis*	Yard Long Cowpea Vine Bean	C:PD:**VG**
Vigna	*unguiculata-sesquipedalis-asparagus*	Asparagus Bean Cowpea Vine Bean	C:PD:**VG**
Vigna	*unguiculata-tana*	Tan Seed Cowpea Vine Bean	C:PD:**VG**

Varieties

Bean, Fava (*Faba spp.*)

Genus	Species	infraspecies	cultivar	specimen	Genome
Faba	*brevipoda*	*aztec-yellowii*	Aztec Yellow	Short Pod (1-4 Bean)	Broad Bean
Faba	*brevipoda*	*bellii*	Bell Peanut Tic	Short Pod (1-4 Bean)	Broad Bean
Faba	*brevipoda*	*brunettei*	Brunette 4.4"	Short Pod (1-4 Bean)	Broad Bean
Faba	*brevipoda*	*cairo-marketii*	Cairo Market	Short Pod (1-4 Bean)	Broad Bean
Faba	*brevipoda*	*expressii*	Express 7.2"	Short Pod (1-4 Bean)	Broad Bean
Faba	*brevipoda*	*iant's-yellowi*	Iant's Yellow	Short Pod (1-4 Bean)	Broad Bean
Faba	*brevipoda*	*jumboi*	Jumbo	Short Pod (1-4 Bean)	Broad Bean
Faba	*brevipoda*	*la-mesillensis*	La Mesilla	Short Pod (1-4 Bean)	Broad Bean
Faba	*brevipoda*	*nintoku-giantii*	Nintoku Giant	Short Pod (1-4 Bean)	Broad Bean
Faba	*brevipoda*	*ojitoi*	Ojito	Short Pod (1-4 Bean)	Broad Bean
Faba	*brevipoda*	*threefold-whitei*	Threefold White	Short Pod (1-4 Bean)	Broad Bean
Faba	*brevipoda*	*vaditoi*	Vadito	Short Pod (1-4 Bean)	Broad Bean
Faba	*brevipoda*	*vroma*	Vroma 3.5"	Short Pod (1-4 Bean)	Broad Bean
Faba	*brevipoda*	*walter-krivda'si*	Walter Krivda's	Short Pod (1-4 Bean)	Broad Bean
Faba	*brevipoda*	*windsor-giantii*	Windsor Giant	Short Pod (1-4 Bean)	Broad Bean
Faba	*mediapoda*	*redii-MB-5-164*	#164 Medium Bean Red	Medium Pod (4-6 Bean)	Broad Bean
Faba	*mediapoda*	*aprovecho-selectii*	Aprovecho Select	Medium Pod (4-6 Bean)	Broad Bean
Faba	*mediapoda*	*aquadulcei-claudia-8*	Claudia Aquadulce 8.0"	Medium Pod (4-6 Bean)	Broad Bean
Faba	*mediapoda*	*banneri*	Banner	Medium Pod (4-6 Bean)	Broad Bean
Faba	*mediapoda*	*blackii*	Black	Medium Pod (4-6 Bean)	Broad Bean
Faba	*mediapoda*	*bonny-ladii*	Bonny Lad 5.5"	Medium Pod (4-6 Bean)	Broad Bean
Faba	*mediapoda*	*cambridge-scarletii*	Cambridge Scarlet	Medium Pod (4-6 Bean)	Broad Bean
Faba	*mediapoda*	*chak'rusgai*	Chak'rusga	Medium Pod (4-6 Bean)	Broad Bean
Faba	*mediapoda*	*copperi*	Copper	Medium Pod (4-6 Bean)	Broad Bean
Faba	*mediapoda*	*crimson-floweri*	Crimson Flower	Medium Pod (4-6 Bean)	Broad Bean
Faba	*mediapoda*	*equina*	Equina	Medium Pod (4-6 Bean)	Broad Bean
Faba	*mediapoda*	*espagnola-valleyi*	Espagnola Valley	Medium Pod (4-6 Bean)	Broad Bean
Faba	*mediapoda*	*feligreenii*	Feligreen	Medium Pod (4-6 Bean)	Broad Bean
Faba	*mediapoda*	*fonsecai*	Fonseca	Medium Pod (4-6 Bean)	Broad Bean
Faba	*mediapoda*	*foul-muddammai*	Foul Muddamma	Medium Pod (4-6 Bean)	Broad Bean
Faba	*mediapoda*	*fredericksii*	Fredericks	Medium Pod (4-6 Bean)	Broad Bean
Faba	*mediapoda*	*frog-island-nationii*	Frog Island Nation	Medium Pod (4-6 Bean)	Broad Bean
Faba	*mediapoda*	*frostproofii*	Frostproof Lima	Medium Pod (4-6 Bean)	Broad Bean
Faba	*mediapoda*	*guatemalan-purplei*	Guatemalan Purple	Medium Pod (4-6 Bean)	Broad Bean
Faba	*mediapoda*	*habas-chiquitasi*	Chiquitas Habas	Medium Pod (4-6 Bean)	Broad Bean
Faba	*mediapoda*	*habas-del-nortei*	Del Norte Habas	Medium Pod (4-6 Bean)	Broad Bean
Faba	*mediapoda*	*habas-jergonasi*	Jergonas Habas	Medium Pod (4-6 Bean)	Broad Bean
Faba	*mediapoda*	*habas-tarahumarai*	Tarahuarma Habas	Medium Pod (4-6 Bean)	Broad Bean
Faba	*mediapoda*	*havai*	Hava	Medium Pod (4-6 Bean)	Broad Bean
Faba	*mediapoda*	*italica*	Italian	Medium Pod (4-6 Bean)	Broad Bean
Faba	*mediapoda*	*itei*	Ite 6.0"	Medium Pod (4-6 Bean)	Broad Bean
Faba	*mediapoda*	*jadei*	Jade	Medium Pod (4-6 Bean)	Broad Bean
Faba	*mediapoda*	*masterpiecei*	Masterpiece 8.0"	Medium Pod (4-6 Bean)	Broad Bean
Faba	*mediapoda*	*medesii*	Medes	Medium Pod (4-6 Bean)	Broad Bean
Faba	*mediapoda*	*mr.-barton'si*	Mr. Barton's	Medium Pod (4-6 Bean)	Broad Bean
Faba	*mediapoda*	*primoi*	Primo 6.5"	Medium Pod (4-6 Bean)	Broad Bean

Genus	Species	infraspecies	cultivar	specimen	Genome
Faba	*mediapoda*	*red-epicurei*	Red Epicure	Medium Pod (4-6 Bean)	Broad Bean
Faba	*mediapoda*	*reina-morai*	Reina Mora	Medium Pod (4-6 Bean)	Broad Bean
Faba	*mediapoda*	*san-luisensis*	San Luis	Medium Pod (4-6 Bean)	Broad Bean
Faba	*mediapoda*	*sanuki-kotsubai*	Sanuki Kotsuba	Medium Pod (4-6 Bean)	Broad Bean
Faba	*mediapoda*	*suprifinii*	Suprifin 8.0"	Medium Pod (4-6 Bean)	Broad Bean
Faba	*mediapoda*	*sweet-loranei*	Sweet Lorane	Medium Pod (4-6 Bean)	Broad Bean
Faba	*mediapoda*	*swissi*	Swiss	Medium Pod (4-6 Bean)	Broad Bean
Faba	*mediapoda*	*suttonii*	The Sutton 5.4"	Medium Pod (4-6 Bean)	Broad Bean
Faba	*mediapoda*	*totoi*	Toto 8.0"	Medium Pod (4-6 Bean)	Broad Bean
Faba	*mediapoda*	*windsor-broadii*	Windsor 7.0"	Medium Pod (4-6 Bean)	Broad Bean
Faba	*mediapoda*	*windsor-broadii-improved-long-podii*	Improved Windsor 7.0"	Medium Pod (4-6 Bean)	Broad Bean
Faba	*mediapoda*	*windsor-english-long-podii*	English Windsor 7.0"	Medium Pod (4-6 Bean)	Broad Bean
Faba	*mediapoda*	*windsor-exhibition-long-podii*	Exhibition Windsor 7.0"	Medium Pod (4-6 Bean)	Broad Bean
Faba	*mediapoda*	*windsor-faba*	Fava Windsor 7.0"	Medium Pod (4-6 Bean)	Broad Bean
Faba	*mediapoda*	*windsor-long-podii*	Red Brown Lima Windsor 7.0"	Medium Pod (4-6 Bean)	Broad Bean
Faba	*mediapoda*	*windsor-long-podii-early-improvedii*	Early Improved Windsor 7.0"	Medium Pod (4-6 Bean)	Broad Bean
Faba	*mediapoda*	*windsor-witkiem-improvedii*	Improved Witkiem Windsor	Medium Pod (4-6 Bean)	Broad Bean
Faba	*longapoda*	*aguadulcei*	Aguadulce 14."	Long Pod (6-12 Bean)	Broad Bean
Faba	*longapoda*	*aquadulcei-claudia-15*	Claudia Aquadulce 15."	Long Pod (6-12 Bean)	Broad Bean
Faba	*longapoda*	*cona-amorei*	Con Amore 8.0"	Long Pod (6-12 Bean)	Broad Bean
Faba	*longapoda*	*histalii*	Histal	Long Pod (6-12 Bean)	Broad Bean
Faba	*longapoda*	*hylonii*	Hylon	Long Pod (6-12 Bean)	Broad Bean
Faba	*longapoda*	*imperial-green-longpodii*	Imperial Green 18"	Long Pod (6-12 Bean)	Broad Bean
Faba	*longapoda*	*polaris*	Polar	Long Pod (6-12 Bean)	Broad Bean
Faba	*longapoda*	*verdyi*	VerdyGreen	Long Pod (6-12 Bean)	Broad Bean
Faba	*longapoda*	*windsor-bunyard's-exhibitionii*	Bunyard's Exhibition Windsor 13."	Long Pod (6-12 Bean)	Broad Bean
Faba	*longapoda*	*windsor-conquerori*	Conqueror Windsor	Long Pod (6-12 Bean)	Broad Bean
Faba	*longapoda*	*windsor-jubileei*	Jubilee Windsor	Long Pod (6-12 Bean)	Broad Bean
Faba	*longapoda*	*windsor-jubileei-hysori*	Hysor Jubilee Windsor	Long Pod (6-12 Bean)	Broad Bean
Faba	*longapoda*	*windsor-reloni*	Relon Windsor 20."	Long Pod (6-12 Bean)	Broad Bean
Faba	*longapoda*	*windsor-witkeim-majorii*	Witkeim Major Windsor 10."	Long Pod (6-12 Bean)	Broad Bean
Faba	*sativa*	*sativa*	Original	Sweet Pod	Broad Bean
Faba	*sativa*	*amphicarpa*	Two Bean	Sweet Pod	Broad Bean
Faba	*sativa*	*cordata*	Heart Leaf	Sweet Pod	Broad Bean
Faba	*sativa*	*devia*	Devious	Sweet Pod	Broad Bean
Faba	*sativa*	*incisa*	Incised	Sweet Pod	Broad Bean
Faba	*sativa*	*linearis*	Strap Leaf	Sweet Pod	Broad Bean
Faba	*sativa*	*macrocarpa*	Large Bean	Sweet Pod	Broad Bean
Faba	*sativa*	*nigra*	Black Bean	Sweet Pod	Broad Bean
Faba	*sativa*	*nigra-angustifolia*	Narrow Leaf Black Bean	Sweet Pod	Broad Bean
Faba	*sativa*	*nigra-consobrina*	Cousin Black Bean	Sweet Pod	Broad Bean
Faba	*sativa*	*nigra-cuneata*	Wedge Leaf Black Bean	Sweet Pod	Broad Bean
Faba	*sativa*	*nigra-heterophylla*	Variable Leaf Black Bean	Sweet Pod	Broad Bean
Faba	*sativa*	*nigra-minor*	Lesser Black Bean	Sweet Pod	Broad Bean
Faba	*sativa*	*notata*	Notable	Sweet Pod	Broad Bean
Faba	*sativa*	*platysperma*	Flat Bean	Sweet Pod	Broad Bean
Faba	*sativa*	*platysperma-leganyana*	Legany Flat Bean	Sweet Pod	Broad Bean
Faba	*sativa*	*platysperma-lentisperma*	Lentil Flat Bean	Sweet Pod	Broad Bean
Faba	*sativa*	*terana*	Winter Tare	Sweet Pod	Broad Bean
Apios	*americana*	*americana*	Original	Groundnut	Potato Bean
Apios	*americana*	*tuberosa*	Tuber	Groundnut	Potato Bean
Apios	*americana*	*turrigera*	Turret	Groundnut	Potato Bean
Canavalia	*ensiformis*	*ensiformis*	Original	American Sword Bean	Jack Bean
Canavalia	*ensiformis*	*chickasaw-lima-beanii*	Chickasaw Lima Bean	American Sword Bean	Jack Bean
Canavalia	*gladiata*	*gladiata*	Original	Asian Sword Bean	Jack Bean
Canavalia	*gladiata*	*mame-akanata*	Akanata Mame	Asian Sword Bean	Jack Bean
Canavalia	*gladiata*	*mame-shironata*	Shironata Mame	Asian Sword Bean	Jack Bean
Cyamopsis	*tetragonoloba*	*tetragonoloba*	Original	Java	Guar Bean
Cyamopsis	*tetragonoloba*	*psoraloides*	Cluster	Java	Guar Bean

Bean, Soybean (*Glycine max*)

Genus	Species	infraspecies	cultivar	specimen	Genome
Glycine	*max-alba*	*velutina*	Velvet ?.?'	White Soybean	Glycine Bean
Glycine	*max-alba*	*white-lionii*	White Lion 2.0'	White Soybean	Glycine Bean
Glycine	*max-brunnea*	*agatei*	Agate 1.0'	Brown Soybean	Glycine Bean
Glycine	*max-brunnea*	*black-eyebrowi*	Black Eyebrow 3.0'	Brown Soybean	Glycine Bean
Glycine	*max-brunnea*	*brun-hatif-rouestii-B-006*	#006 Brun Hatif Rouest ?.?'	Brown Soybean	Glycine Bean
Glycine	*max-brunnea*	*canattoa*	Canatto ?.?'	Brown Soybean	Glycine Bean
Glycine	*max-brunnea*	*crestii*	Crest ?.?'	Brown Soybean	Glycine Bean
Glycine	*max-brunnea*	*daizui*	Daizu ?.?'	Brown Soybean	Glycine Bean
Glycine	*max-brunnea*	*daizui-kura-kakei*	Kura Kake Daizu ?.?'	Brown Soybean	Glycine Bean
Glycine	*max-brunnea*	*fiskebyi*	Fiskeby ?.?'	Brown Soybean	Glycine Bean

Genus	Species	infraspecies	cultivar	specimen	Genome
Glycine	*max-brunnea*	*jacquesi-brownii*	Jacques 2.7'	Brown Soybean	Glycine Bean
Glycine	*max-brunnea*	*kahalai*	Kahala ?.?'	Brown Soybean	Glycine Bean
Glycine	*max-brunnea*	*kegonii*	Kegon ?.?'	Brown Soybean	Glycine Bean
Glycine	*max-brunnea*	*manitoba-brownii*	Manitoba ?.?'	Brown Soybean	Glycine Bean
Glycine	*max-brunnea*	*misono-greenii*	Misono Green Pod 2.0'	Brown Soybean	Glycine Bean
Glycine	*max-brunnea*	*morlanvia*	Morlanvia ?.?'	Brown Soybean	Glycine Bean
Glycine	*max-brunnea*	*normanii*	Norman ?.?'	Brown Soybean	Glycine Bean
Glycine	*max-brunnea*	*ozziei*	Ozzie 3.0'	Brown Soybean	Glycine Bean
Glycine	*max-brunnea*	*rouestii-N-03-13-A1-2*	Rouest ?.?'	Brown Soybean	Glycine Bean
Glycine	*max-brunnea*	*sapporo-midori*	Sapporo Midori ?.?'	Brown Soybean	Glycine Bean
Glycine	*max-brunnea*	*si-lu-beanii*	Si Lou Kidney ?.?	Brown Soybean	Glycine Bean
Glycine	*max-brunnea*	*taiwamei*	Taiwame ?.?'	Brown Soybean	Glycine Bean
Glycine	*max-brunnea*	*ugra-sajai*	Ugra Saja ?.?'	Brown Soybean	Glycine Bean
Glycine	*max-brunnea*	*yusuzumi*	Yusuzumi ?.?'	Brown Soybean	Glycine Bean
Glycine	*max-groena*	*amish-greenii*	Amish Green ?.?'	Green Soybean	Glycine Bean
Glycine	*max-groena*	*beer-friendii*	Beer Friend 2.7'	Green Soybean	Glycine Bean
Glycine	*max-groena*	*blue-ribbonii*	Blue Ribbon ?.?'	Green Soybean	Glycine Bean
Glycine	*max-groena*	*butterbeansi*	Butterbeans 2.0'	Green Soybean	Glycine Bean
Glycine	*max-groena*	*edamamei*	Edamame ?.?'	Green Soybean	Glycine Bean
Glycine	*max-groena*	*edamamei-saint-itai*	Saint Ita Edamame ?.?'	Green Soybean	Glycine Bean
Glycine	*max-groena*	*edamamei-sayamusumei*	Sayamusmume Edamame 2.0'	Green Soybean	Glycine Bean
Glycine	*max-groena*	*edamamei-yedamamei*	Yedamame Edamame ?.?'	Green Soybean	Glycine Bean
Glycine	*max-groena*	*envyi*	Envy 2.0'	Green Soybean	Glycine Bean
Glycine	*max-groena*	*frostbeateri*	Frostbeater ?.?'	Green Soybean	Glycine Bean
Glycine	*max-groena*	*green-legendii*	Green Legend ?.?'	Green Soybean	Glycine Bean
Glycine	*max-groena*	*lancoi*	Lanco ?.?'	Green Soybean	Glycine Bean
Glycine	*max-groena*	*prizei*	Prize 2.5'	Green Soybean	Glycine Bean
Glycine	*max-groena*	*PI-290156*	Prize Improved ?.?'	Green Soybean	Glycine Bean
Glycine	*max-groena*	*PI-458541*	China Prize Improved ?.?'	Green Soybean	Glycine Bean
Glycine	*max-groena*	*PI-522192*	Russia Prize Improved ?.?'	Green Soybean	Glycine Bean
Glycine	*max-groena*	*shironomai*	Shironomai ?.?'	Green Soybean	Glycine Bean
Glycine	*max-groena*	*verde*	Verde 3.0'	Green Soybean	Glycine Bean
Glycine	*max-helva*	*helva*	Original 2.0'	Yellow Soybean	Glycine Bean
Glycine	*max-helva*	*altona*	Altona ?.?'	Yellow Soybean	Glycine Bean
Glycine	*max-helva*	*chicoi*	Chico ?.?'	Yellow Soybean	Glycine Bean
Glycine	*max-helva*	*evan's-mutationii*	Evan's Mutation 3.0'	Yellow Soybean	Glycine Bean
Glycine	*max-helva*	*grandforksensis*	Grand Forks ?.?'	Yellow Soybean	Glycine Bean
Glycine	*max-helva*	*hackucho-earlyi*	Hakucho Early 1.0'	Yellow Soybean	Glycine Bean
Glycine	*max-helva*	*hedgei*	Hedge ?.?'	Yellow Soybean	Glycine Bean
Glycine	*max-helva*	*jewelii*	Jewel 2.5'	Yellow Soybean	Glycine Bean
Glycine	*max-helva*	*maple-arrowii*	Maple Arrow 2.5'	Yellow Soybean	Glycine Bean
Glycine	*max-helva*	*mccallii*	McCall 2.2'	Yellow Soybean	Glycine Bean
Glycine	*max-helva*	*nattoi*	Natto ?.?'	Yellow Soybean	Glycine Bean
Glycine	*max-helva*	*syncomii*	Synergy Commercial ?.?'	Yellow Soybean	Glycine Bean
Glycine	*max-helva*	*vintonii-81*	#81 Vinton 3.0'	Yellow Soybean	Glycine Bean
Glycine	*max-nigra*	*beijingensis-blackii*	Beijing ?.?'	Black Soybean	Glycine Bean
Glycine	*max-nigra*	*black-jetii*	Black Jet 2.0'	Black Soybean	Glycine Bean
Glycine	*max-nigra*	*black-pearlii*	Black Pearl 2.7'	Black Soybean	Glycine Bean
Glycine	*max-nigra*	*dieckman-blackii*	Dieckman ?.?'	Black Soybean	Glycine Bean
Glycine	*max-nigra*	*hokkoaido-blackii*	Hokkaido ?.?'	Black Soybean	Glycine Bean
Glycine	*max-nigra*	*jacquesi-blackii*	Jacques ?.?'	Black Soybean	Glycine Bean
Glycine	*max-nigra*	*koreana*	Korean ?.?'	Black Soybean	Glycine Bean
Glycine	*max-nigra*	*kuromanei*	Kuromane ?.?'	Black Soybean	Glycine Bean
Glycine	*max-nigra*	*kuromanei-panthera*	Panther Kuromane ?.?'	Black Soybean	Glycine Bean
Glycine	*max-nigra*	*lammer's-blackii*	Lammer's 2.0'	Black Soybean	Glycine Bean
Glycine	*max-nigra*	*laredoensis*	Laredo ?.?'	Black Soybean	Glycine Bean
Glycine	*max-nigra*	*late-giant-blackii*	Late Giant ?.?'	Black Soybean	Glycine Bean
Glycine	*max-nigra*	*wisconsin-blackii*	Wisconsin ?.?'	Black Soybean	Glycine Bean
Glycine	*max-soja*	*soja*	Original	Wild Soybean	Glycine Bean
Glycine	*max-soja*	*brevifolia*	Short Leaf ?.?'	Wild Soybean	Glycine Bean
Glycine	*max-soja*	*formosana*	Taiwan ?.?'	Wild Soybean	Glycine Bean
Glycine	*max-soja*	*javanica*	Java ?.?'	Wild Soybean	Glycine Bean
Glycine	*max-soja*	*ovata*	Oval ?.?'	Wild Soybean	Glycine Bean
Glycine	*max-soja*	*shirofumi*	Shirofumi Green 3.0'	Wild Soybean	Glycine Bean
Glycine	*max-soja*	*tohyai*	Tohya Green ?.?'	Wild Soybean	Glycine Bean
Glycine	*max-soja*	*ussuriensis*	Ussuri River Valley ?.?'	Wild Soybean	Glycine Bean
Inga	*edulis*	*edulis*	Original	Edible	Ice Cream Bean
Inga	*edulis*	*benthamiana*	Bentham	Edible	Ice Cream Bean
Inga	*edulis*	*monkey-tailii*	Monkey Tail	Edible	Ice Cream Bean
Inga	*edulis*	*parviflora*	Small Flower	Edible	Ice Cream Bean
Inga	*edulis*	*scabriuscula*	Scabre	Edible	Ice Cream Bean
Inga	*edulis*	*ynga*	Ynga	Edible	Ice Cream Bean
Inga	*oerstediana*	*oerstediana*	Original	Oersted Edible	Ice Cream Bean
Inga	*oerstediana*	*chartana*	Chart	Oersted Edible	Ice Cream Bean

Genus	Species	infraspecies	cultivar	specimen	Genome
Inga	*oerstediana*	*chiriquensis*	Chirique	Oersted Edible	Ice Cream Bean
Inga	*oerstediana*	*culagana*	Culagan	Oersted Edible	Ice Cream Bean
Inga	*oerstediana*	*minutula*	Tiny	Oersted Edible	Ice Cream Bean
Inga	*oerstediana*	*pamplonae*	Pamplon	Oersted Edible	Ice Cream Bean
Lablab	*purpureus*	*purpureus*	Original	Hyacinth Bean	Lablab Bean
Lablab	*purpureus*	*albiflora*	White Flowered	Hyacinth Bean	Lablab Bean
Lablab	*purpureus*	*bengalensis*	Bengal	Hyacinth Bean	Lablab Bean
Lablab	*purpureus*	*crenatifructus*	Scalloped	Hyacinth Bean	Lablab Bean
Lablab	*purpureus*	*cultratus*	Cultured	Hyacinth Bean	Lablab Bean
Lablab	*purpureus*	*fujimame-akahana*	Aka Hana Fujimame	Hyacinth Bean	Lablab Bean
Lablab	*purpureus*	*fujimame-murasakiirohana*	Murasakiiro Hana Fujimame	Hyacinth Bean	Lablab Bean
Lablab	*purpureus*	*fujimame-shirohana*	Shiro Hana Fujimame	Hyacinth Bean	Lablab Bean
Lablab	*purpureus*	*niger*	Black Bean	Hyacinth Bean	Lablab Bean
Lablab	*purpureus*	*uncinatus*	Hooked	Hyacinth Bean	Lablab Bean

Bean, Lentil (*Lens culinaris*)

Genus	Species	infraspecies	cultivar	specimen	Genome
Lens	*culinaris*	*culinaris*	Original	Culinary	Lentil Bean
Lens	*culinaris*	*abyssinica*	Ethiopian	Culinary	Lentil Bean
Lens	*culinaris*	*belugai*	Beluga	Culinary	Lentil Bean
Lens	*culinaris*	*breweri*	Brewer	Culinary	Lentil Bean
Lens	*culinaris*	*chilensis*	Chilean	Culinary	Lentil Bean
Lens	*culinaris*	*coccineus*	Crimson	Culinary	Lentil Bean
Lens	*culinaris*	*coccineus-petita*	Petite Crimson	Culinary	Lentil Bean
Lens	*culinaris*	*du-puyi*	Du Puy	Culinary	Lentil Bean
Lens	*culinaris*	*du-puyi-vertei*	Verte Du Puy	Culinary	Lentil Bean
Lens	*culinaris*	*equinus*	Horse	Culinary	Lentil Bean
Lens	*culinaris*	*franco-groena*	French Green	Culinary	Lentil Bean
Lens	*culinaris*	*hispanica-pardina*	Spanish Pardina	Culinary	Lentil Bean
Lens	*culinaris*	*masoori*	Masoor	Culinary	Lentil Bean
Lens	*culinaris*	*microsperma*	Tiny Seed	Culinary	Lentil Bean
Lens	*culinaris*	*odemensis*	Odem	Culinary	Lentil Bean
Lens	*culinaris*	*orientalis*	Oriental	Culinary	Lentil Bean
Lens	*culinaris*	*rubra-princeps*	Red Chief	Culinary	Lentil Bean
Lens	*culinaris*	*rubra-sperma*	Red Seeded	Culinary	Lentil Bean
Lens	*culinaris*	*rubra-syriaca*	Red Syrian	Culinary	Lentil Bean
Lens	*culinaris*	*tarahumara-pinksi*	Tarahumara Pinks	Culinary	Lentil Bean
Lens	*culinaris*	*tomentosus*	Woolly	Culinary	Lentil Bean
Leucaena	*esculenta*	*esculenta*	Original	Edible	Jumbie Bean
Leucaena	*esculenta*	*confusa*	Confused	Edible	Jumbie Bean
Leucaena	*esculenta*	*doylei*	Doyle	Edible	Jumbie Bean
Mucuna	*pruriens*	*pruriens*	Original	Edible	Velvet Bean
Mucuna	*pruriens*	*floridana-maculatus*	Florida Speckled	Edible	Velvet Bean
Mucuna	*pruriens*	*hirsuta*	Hairy	Edible	Velvet Bean
Mucuna	*pruriens*	*novo-guineensis*	New Guinea	Edible	Velvet Bean
Mucuna	*pruriens*	*thekkadiensis*	Thekkadian	Edible	Velvet Bean
Mucuna	*pruriens*	*utilis*	Useful	Edible	Velvet Bean
Pachyrhizus	*erosus*	*erosus*	Original	Jicama	Yam Bean
Pachyrhizus	*erosus*	*angulatus*	Angled	Jicama	Yam Bean
Pachyrhizus	*erosus*	*articulatus*	Articulated	Jicama	Yam Bean
Pachyrhizus	*erosus*	*bulbosus*	Bulbous	Jicama	Yam Bean
Pachyrhizus	*erosus*	*palmatilobus*	Palm Leaf	Jicama	Yam Bean
Pachyrhizus	*erosus*	*strigosus*	Bristle	Jicama	Yam Bean

Bean, Tepary (*Phaseolus acutifolius-latifolius*)

Genus	Species	infraspecies	cultivar	specimen	Genome
Phaseolus	*acutifolius-latifolius-alba*	*big-fieldsensis-alba*	Big Fields	White Tepary	Bean
Phaseolus	*acutifolius-latifolius-alba*	*cocopah-alba*	Cocopah	White Tepary	Bean
Phaseolus	*acutifolius-latifolius-alba*	*durangoensis*	Durango	White Tepary	Bean
Phaseolus	*acutifolius-latifolius-alba*	*guajiro-alba*	Guajiro	White Tepary	Bean
Phaseolus	*acutifolius-latifolius-alba*	*hopi-alba*	Hopi	White Tepary	Bean
Phaseolus	*acutifolius-latifolius-alba*	*kickapo-alba*	Kickapoo	White Tepary	Bean
Phaseolus	*acutifolius-latifolius-alba*	*mayo-alba*	Mayo	White Tepary	Bean
Phaseolus	*acutifolius-latifolius-alba*	*namaquipe-alba*	Namaquipe	White Tepary	Bean
Phaseolus	*acutifolius-latifolius-alba*	*o'odhamii-alba*	O'odham	White Tepary	Bean
Phaseolus	*acutifolius-latifolius-alba*	*santa-rosa-alba*	Santa Rosa	White Tepary	Bean
Phaseolus	*acutifolius-latifolius-alba*	*san-felipe-puelbo-alba*	San Felipe Pueblo	White Tepary	Bean
Phaseolus	*acutifolius-latifolius-alba*	*san-ignacio-alba*	San Ignacio	White Tepary	Bean
Phaseolus	*acutifolius-latifolius-alba*	*sonorensis-alba*	Sonoran	White Tepary	Bean
Phaseolus	*acutifolius-latifolius-alba*	*tohono-o'odham-alba*	Tohono O'odham	White Tepary	Bean
Phaseolus	*acutifolius-latifolius-alba*	*warihio-alba*	Warihio	White Tepary	Bean
Phaseolus	*acutifolius-latifolius-blauea*	*speca-blauea*	Speckled	Blue Tepary	Bean
Phaseolus	*acutifolius-latifolius-blauea*	*speca-blauea-grisea*	Blue Gray Speckled	Blue Tepary	Bean
Phaseolus	*acutifolius-latifolius-brunnea*	*bigfieldsensis-brunnea*	Big Fields	Brown Tepary	Bean

Genus	Species	infraspecies	cultivar	specimen	Genome
Phaseolus	*acutifolius-latifolius-brunnea*	*cocopah-brunnea*	Cocopah	Brown Tepary	Bean
Phaseolus	*acutifolius-latifolius-brunnea*	*cumpasensis*	Cumpas	Brown Tepary	Bean
Phaseolus	*acutifolius-latifolius-brunnea*	*light-brown-seededii*	Light Brown Seeded	Brown Tepary	Bean
Phaseolus	*acutifolius-latifolius-brunnea*	*little-tuscon-brownii*	Little Tuscon	Brown Tepary	Bean
Phaseolus	*acutifolius-latifolius-brunnea*	*o'odhamii-brunnea*	O'odham	Brown Tepary	Bean
Phaseolus	*acutifolius-latifolius-brunnea*	*paiute-mixta*	Paiute Mixed	Brown Tepary	Bean
Phaseolus	*acutifolius-latifolius-brunnea*	*papagoa-brunnea*	Papago	Brown Tepary	Bean
Phaseolus	*acutifolius-latifolius-brunnea*	*sacaton-brunnea*	Sacaton	Brown Tepary	Bean
Phaseolus	*acutifolius-latifolius-brunnea*	*san-pedroensis*	San Pedro	Brown Tepary	Bean
Phaseolus	*acutifolius-latifolius-brunnea*	*sonorensis*	Sonoran	Brown Tepary	Bean
Phaseolus	*acutifolius-latifolius-brunnea*	*speca*	Speckled	Brown Tepary	Bean
Phaseolus	*acutifolius-latifolius-brunnea*	*speca-brunnea*	Speckled	Brown Tepary	Bean
Phaseolus	*acutifolius-latifolius-brunnea*	*tohono-o'odham-brunnea*	Tohono O'odham	Brown Tepary	Bean
Phaseolus	*acutifolius-latifolius-brunnea*	*wild-willow-leafii*	Wild Willow Leaf	Brown Tepary	Bean
Phaseolus	*acutifolius-latifolius-nigra*	*milta-nigra*	Mitla	Black Tepary	Bean
Phaseolus	*acutifolius-latifolius-nigra*	*pinacatensis-nigra*	Pinacate	Black Tepary	Bean
Phaseolus	*acutifolius-latifolius-nigra*	*san-pablo-balleza*	San Pablo Balleza	Black Tepary	Bean
Phaseolus	*acutifolius-latifolius-nigra*	*tucsonensis-nigra*	Tucson	Black Tepary	Bean
Phaseolus	*acutifolius-latifolius-tana*	*speca-colonia-morelos*	Colonia Morelos Speckled	Tan Tepary	Bean
Phaseolus	*acutifolius-latifolius-tana*	*paiutensis-helva*	Paiute Yellow	Tan Tepary	Bean
Phaseolus	*acutifolius-latifolius-tana*	*pima-beige-&-brownii*	Pima Beige & Brown	Tan Tepary	Bean
Phaseolus	*acutifolius-latifolius-tana*	*pinacatensis-tana*	Pinacate	Tan Tepary	Bean
Phaseolus	*acutifolius-latifolius-tana*	*sonorensis-goldii*	Sonoran Gold	Tan Tepary	Bean
Phaseolus	*acutifolius-latifolius-tana*	*virus-free-yellowi*	Virus Free Yellow	Tan Tepary	Bean
Phaseolus	*acutifolius-latifolius-tana*	*yaquiensis*	Yaqui Beige	Tan Tepary	Bean
Phaseolus	*acutifolius-latifolius-tana*	*yoeme-brownii*	Yoeme Brown	Tan Tepary	Bean

Bean, Runner (*Phaseolus coccineus*)

Genus	Species	infraspecies	cultivar	specimen	Genome
Phaseolus	*coccineus-alba*	*aleppo-arabii*	Aleppo Arab	White Runner	Bean
Phaseolus	*coccineus-alba*	*aztec-dwarf-alba*	Aztec Dwarf	White Runner	Bean
Phaseolus	*coccineus-alba*	*bettyi*	Betty Potato Lima	White Runner	Bean
Phaseolus	*coccineus-alba*	*czari*	Czar	White Runner	Bean
Phaseolus	*coccineus-alba*	*donnai*	Donna Potato Lima	White Runner	Bean
Phaseolus	*coccineus-alba*	*emergoi*	Emergo	White Runner	Bean
Phaseolus	*coccineus-alba*	*jack-&-the-beanstalkii*	Jack & The Beanstalk	White Runner	Bean
Phaseolus	*coccineus-alba*	*jerusalemensis*	Jerusalem	White Runner	Bean
Phaseolus	*coccineus-alba*	*tarahumara-bordalii*	Tarahumara Bordal	White Runner	Bean
Phaseolus	*coccineus-alba*	*white-dutchii*	White Dutch	White Runner	Bean
Phaseolus	*coccineus-alba*	*white-ladyi*	White Lady	White Runner	Bean
Phaseolus	*coccineus-brunnea*	*apricot-runneri*	Apricot Flower	Brown Runner	Bean
Phaseolus	*coccineus-brunnea*	*butleri*	Garden Runner	Brown Runner	Bean
Phaseolus	*coccineus-brunnea*	*desireei*	Desiree	Brown Runner	Bean
Phaseolus	*coccineus-brunnea*	*emperori*	Emperor	Brown Runner	Bean
Phaseolus	*coccineus-brunnea*	*erecta*	Erect Snijproker	Brown Runner	Bean
Phaseolus	*coccineus-brunnea*	*grammy-tilleyi*	Grammy Tilley Lima	Brown Runner	Bean
Phaseolus	*coccineus-brunnea*	*gulliveri*	Gulliver Stringless	Brown Runner	Bean
Phaseolus	*coccineus-brunnea*	*hestia*	Hest Stringless	Brown Runner	Bean
Phaseolus	*coccineus-brunnea*	*kelvedon-marvelii*	Kelvedon Marvel	Brown Runner	Bean
Phaseolus	*coccineus-brunnea*	*lady-diana*	Lady Di	Brown Runner	Bean
Phaseolus	*coccineus-brunnea*	*royal-standardii*	Roya lStandard	Brown Runner	Bean
Phaseolus	*coccineus-brunnea*	*sadie's-horse-beanii*	Sadie Horse Bean	Brown Runner	Bean
Phaseolus	*coccineus-brunnea*	*salt-&-pepperi*	Salt & Pepper	Brown Runner	Bean
Phaseolus	*coccineus-brunnea*	*streamlinei*	Streamline	Brown Runner	Bean
Phaseolus	*coccineus-brunnea*	*sun-brightii*	Sun Bright	Brown Runner	Bean
Phaseolus	*coccineus-brunnea*	*sunsetii*	Sunset	Brown Runner	Bean
Phaseolus	*coccineus-brunnea*	*zembylasii*	Zembylas	Brown Runner	Bean
Phaseolus	*coccineus-brunnea*	*tocamres-chocolatei*	Tocamres Chocolate	Brown Runner	Bean
Phaseolus	*coccineus-flora*	*butonii*	Buton	Scarlet Flower Runner	Bean
Phaseolus	*coccineus-flora*	*coccineus*	Original	Scarlet Flower Runner	Bean
Phaseolus	*coccineus-flora*	*droitwich-championii*	Drotwich Champion	Scarlet Flower Runner	Bean
Phaseolus	*coccineus-flora*	*dwarf-beesi*	Dwarf Bees	Scarlet Flower Runner	Bean
Phaseolus	*coccineus-flora*	*enorma*	Enorma	Scarlet Flower Runner	Bean
Phaseolus	*coccineus-flora*	*formosus*	Handsome	Scarlet Flower Runner	Bean
Phaseolus	*coccineus-flora*	*formosus-sylvestris*	Forest Handsome	Scarlet Flower Runner	Bean
Phaseolus	*coccineus-flora*	*griseus*	Gray Seed	Scarlet Flower Runner	Bean
Phaseolus	*coccineus-flora*	*hammond-scarlet-runner-bushii*	Hammond	Scarlet Flower Runner	Bean
Phaseolus	*coccineus-flora*	*leiosepalus*	Lion Sepal	Scarlet Flower Runner	Bean
Phaseolus	*coccineus-flora*	*multiflorus*	.	Scarlet Flower Runner	Bean
Phaseolus	*coccineus-flora*	*obvallatus*	.	Scarlet Flower Runner	Bean
Phaseolus	*coccineus-flora*	*polyanthus*	Many Flower	Scarlet Flower Runner	Bean
Phaseolus	*coccineus-flora*	*polyanthus-flavescens*	Yellow Many Flower	Scarlet Flower Runner	Bean
Phaseolus	*coccineus-flora*	*polyanthus-harmsianus*	Harms Many Flower	Scarlet Flower Runner	Bean
Phaseolus	*coccineus-flora*	*red-knightii*	Red Knight	Scarlet Flower Runner	Bean

Genus	Species	infraspecies	cultivar	specimen	Genome
Phaseolus	*coccineus-flora*	*red-rumii*	Red Rum	Scarlet Flower Runner	Bean
Phaseolus	*coccineus-flora*	*striatus*	.	Scarlet Flower Runner	Bean
Phaseolus	*coccineus-flora*	*strigillosus*	.	Scarlet Flower Runner	Bean
Phaseolus	*coccineus-flora*	*superbus*	.	Scarlet Flower Runner	Bean
Phaseolus	*coccineus-nigra*	*best-of-allii*	Best Of All	Black Runner	Bean
Phaseolus	*coccineus-nigra*	*blackcoatii*	Blackcoat	Black Runner	Bean
Phaseolus	*coccineus-nigra*	*black-knightii*	Black Knight	Black Runner	Bean
Phaseolus	*coccineus-pinca*	*ivanhoei*	Ivanhoe Lilac	Pink Runner	Bean
Phaseolus	*coccineus-pinca*	*painted-ladyi*	Painted Lady Brown	Pink Runner	Bean
Phaseolus	*coccineus-purpurea*	*achievement-meritii*	Achievement Merit	Purple Runner	Bean
Phaseolus	*coccineus-purpurea*	*goliathii*	Goliath	Purple Runner	Bean
Phaseolus	*coccineus-purpurea*	*granny's-scarlet-runneri*	Granny's	Purple Runner	Bean
Phaseolus	*coccineus-purpurea*	*pickwickii*	Pickwick	Purple Runner	Bean
Phaseolus	*coccineus-purpurea*	*prizewinneri*	Prizewinner Violet	Purple Runner	Bean
Phaseolus	*coccineus-purpurea*	*scarlet-emperori*	Scarlet Emperor	Purple Runner	Bean
Phaseolus	*coccineus-purpurea*	*seneca-bear-pawii*	Seneca Bear Paw	Purple Runner	Bean
Phaseolus	*coccineus-purpurea*	*tarahumarai*	Tarahumara	Purple Runner	Bean
Phaseolus	*coccineus-purpurea*	*tarahumara-tekomari*	Tekomari Tarahumara	Purple Runner	Bean
Phaseolus	*coccineus-tana*	*barnetii*	Barnet	Tan Runner	Bean

Bean, Lima Bush (*Phaseolus lunatus-bruscia*)

Genus	Species	infraspecies	cultivar	specimen	Genome
Phaseolus	*lunatus-bruscia-alba*	*burpee's-bush-limai*	Burpee's 4.5" Pod	White Seed Bush Lima	Bean
Phaseolus	*lunatus-bruscia-alba*	*fordhookii-0242*	Potato Fordhook 3.7" Pod	White Seed Bush Lima	Bean
Phaseolus	*lunatus-bruscia-alba*	*fordhookii-0242-excelii*	Excel Fordhook 3.7" Pod	White Seed Bush Lima	Bean
Phaseolus	*lunatus-bruscia-alba*	*kingii*	King ?.?" Pod	White Seed Bush Lima	Bean
Phaseolus	*lunatus-bruscia-alba*	*simmons-red-streakii*	Red Streaked White Simmons	White Seed Bush Lima	Bean
Phaseolus	*lunatus-bruscia-aurantea*	*hopi-orangei*	Orange Seed Hopi	Orange Seed Bush Lima	Bean
Phaseolus	*lunatus-bruscia-grisea*	*hopi-lima-grayi*	Gray Seed Hopi	Gray Seed Bush Lima	Bean
Phaseolus	*lunatus-bruscia-groena*	*bipunctatus*	Dotted	Green Seed Bush Lima	Bean
Phaseolus	*lunatus-bruscia-groena*	*burpee's-improvedii*	Burpee's Improved 5.0" Pod	Green Seed Bush Lima	Bean
Phaseolus	*lunatus-bruscia-groena*	*buttergreenii*	Buttergreen 6.0" Pod	Green Seed Bush Lima	Bean
Phaseolus	*lunatus-bruscia-groena*	*early-bush-limai*	Early ?.?" Pod	Green Seed Bush Lima	Bean
Phaseolus	*lunatus-bruscia-groena*	*early-giant-bush-lamai*	Early Giant 5.5" Pod	Green Seed Bush Lima	Bean
Phaseolus	*lunatus-bruscia-groena*	*fordhookii*	Potato Fordhook 4.2" Pod	Green Seed Bush Lima	Bean
Phaseolus	*lunatus-bruscia-groena*	*fordhookii-0242-1072*	Fordhook 3.7" Pod	Green Seed Bush Lima	Bean
Phaseolus	*lunatus-bruscia-groena*	*fordhookii-0901*	Fordhook 4.0" Pod	Green Seed Bush Lima	Bean
Phaseolus	*lunatus-bruscia-groena*	*giant-improved-bush-limai*	Giant Improved 5.5" Pod	Green Seed Bush Lima	Bean
Phaseolus	*lunatus-bruscia-groena*	*ilocanus*	Ilocan	Green Seed Bush Lima	Bean
Phaseolus	*lunatus-bruscia-groena*	*improved-bush-limai*	Improved 5.0" Pod	Green Seed Bush Lima	Bean
Phaseolus	*lunatus-bruscia-groena*	*inamoenus*	Unpleasant	Green Seed Bush Lima	Bean
Phaseolus	*lunatus-bruscia-groena*	*leei*	Lee ?.?" Pod	Green Seed Bush Lima	Bean
Phaseolus	*lunatus-bruscia-groena*	*limensis*	Lima	Green Seed Bush Lima	Bean
Phaseolus	*lunatus-bruscia-groena*	*packersi-FM-263*	Packers FM Light	Green Seed Bush Lima	Bean
Phaseolus	*lunatus-bruscia-groena*	*portoricensis*	Puerto Rico	Green Seed Bush Lima	Bean
Phaseolus	*lunatus-bruscia-groena*	*puberulus*	Downy	Green Seed Bush Lima	Bean
Phaseolus	*lunatus-bruscia-groena*	*rosei*	Rose's	Green Seed Bush Lima	Bean
Phaseolus	*lunatus-bruscia-groena*	*saccharatus*	Sugar	Green Seed Bush Lima	Bean
Phaseolus	*lunatus-bruscia-groena*	*silvester*	Silvester	Green Seed Bush Lima	Bean
Phaseolus	*lunatus-bruscia-groena*	*superba-giant-poddedii*	Superb Giant	Green Seed Bush Lima	Bean
Phaseolus	*lunatus-bruscia-groena*	*tunkinensis*	Tunkin	Green Seed Bush Lima	Bean
Phaseolus	*lunatus-bruscia-groena*	*UC-92*	UC #82	Green Seed Bush Lima	Bean
Phaseolus	*lunatus-bruscia-groena*	*vexillatus*	Whorled	Green Seed Bush Lima	Bean
Phaseolus	*lunatus-bruscia-groena*	*viridis*	Verdant	Green Seed Bush Lima	Bean
Phaseolus	*lunatus-bruscia-groena*	*xuaresii*	Xuares	Green Seed Bush Lima	Bean
Phaseolus	*lunatus-bruscia-nana*	*butterpeai*	Butterpea ?.?" Pod	Baby Bush Lima	Bean
Phaseolus	*lunatus-bruscia-nana*	*butterpeai-dixie-speckledii*	Speckled Brown Seed Dixie Butterpea 3.2" Pod	Baby Bush Lima	Bean
Phaseolus	*lunatus-bruscia-nana*	*hendersonii-improvedii*	White Seed Improved Henderson 3.0" Pod	Baby Bush Lima	Bean
Phaseolus	*lunatus-bruscia-nana*	*hendersonii-redii*	Red Seed Henderson	Baby Bush Lima	Bean
Phaseolus	*lunatus-bruscia-nana*	*verna-shirkii*	Maroon Verna Shirk	Baby Bush Lima	Bean
Phaseolus	*lunatus-bruscia-nana*	*wood's-prolifici*	Wood's Prolific	Baby Bush Lima	Bean
Phaseolus	*lunatus-bruscia-nana-alba*	*baby-bushii*	Plump Bean 2.5" Pod	White Seed Baby Bush Lima	Bean
Phaseolus	*lunatus-bruscia-nana-alba*	*butterpeai-dixie-alba*	Dixie Butterpea 3.5" Pod	White Seed Baby Bush Lima	Bean
Phaseolus	*lunatus-bruscia-nana-alba*	*eastlandii*	Eastland 3.5" Pod	White Seed Baby Bush Lima	Bean
Phaseolus	*lunatus-bruscia-nana-groena*	*baby-limai-184-85*	#184-85 3.5" Pod	Green Seed Baby Bush Lima	Bean
Phaseolus	*lunatus-bruscia-nana-groena*	*baby-limai-878*	#878 4.0" Pod	Green Seed Baby Bush Lima	Bean
Phaseolus	*lunatus-bruscia-nana-groena*	*bridgetonii*	Bridgeton 3.0" Pod	Green Seed Baby Bush Lima	Bean
Phaseolus	*lunatus-bruscia-nana-groena*	*fordhookii-babyi*	Fordhook 2.8" Pod	Green Seed Baby Bush Lima	Bean
Phaseolus	*lunatus-bruscia-nana-groena*	*genevensis*	Geneva ?.?" Pod	Green Seed Baby Bush Lima	Bean
Phaseolus	*lunatus-bruscia-nana-groena*	*hendersonii-bush-limai*	Henderson 3.1" Pod	Green Seed Baby Bush Lima	Bean
Phaseolus	*lunatus-bruscia-nana-groena*	*hendersonii-cangreenii*	Cangreen Henderson	Green Seed Baby Bush Lima	Bean
Phaseolus	*lunatus-bruscia-nana-groena*	*hendersonii-thorogreenii*	Thorogreen Henderson 3.0" Pod	Green Seed Baby Bush Lima	Bean

Genus	Species	infraspecies	cultivar	specimen	Genome
Phaseolus	*lunatus-bruscia-nana-groena*	*hendersonii-thorogreenii-earlyi*	Early Thorogreen Henderson 2.7" Pod	Green Seed Baby Bush Lima	Bean
Phaseolus	*lunatus-bruscia-nana-groena*	*nemagreenii*	Nemagreen 3.0" Pod	Green Seed Baby Bush Lima	Bean
Phaseolus	*lunatus-bruscia-nana-groena*	*packeri-DM*	Dark Packer 3.0" Pod	Green Seed Baby Bush Lima	Bean
Phaseolus	*lunatus-bruscia-rubra*	*hopi-lima-redii*	Red Seed Hopi	Red Seed Bush Lima	Bean
Phaseolus	*lunatus-bruscia-tana*	*hopi-tanii*	Black & Tan Seed Hopi	Tan Seed Bush Lima	Bean
Phaseolus	*lunatus-bruscia-tana*	*hopi-yellowi*	Tan Yellow Seed Hopi	Tan Seed Bush Lima	Bean
Phaseolus	*lunatus-bruscia-tana*	*jackson-wonderi*	Speckled Buff Jackson Wonder	Tan Seed Bush Lima	Bean
Phaseolus	*lunatus-bruscia-tana*	*pima-beigei*	Beige Seed Pima	Tan Seed Bush Lima	Bean

Bean, Lima Vine (*Phaseolus lunatus-vigna*)

Genus	Species	infraspecies	cultivar	specimen	Genome
Phaseolus	*lunatus-vigna-alba*	*big-frostyi*	Big Frosty ?.?" Pod	White Seed Vine Lima	Bean
Phaseolus	*lunatus-vigna-alba*	*calicoi*	Calico 3.5" Pod	White Seed Vine Lima	Bean
Phaseolus	*lunatus-vigna-alba*	*christmasi*	Christmas 5.0" Pod	White Seed Vine Lima	Bean
Phaseolus	*lunatus-vigna-alba*	*civil-war-alba*	Civil War ?.?" Pod	White Seed Vine Lima	Bean
Phaseolus	*lunatus-vigna-alba*	*civil-war-maroon-whitei*	Civil War Maroon-White	White Seed Vine Lima	Bean
Phaseolus	*lunatus-vigna-alba*	*incan-giant-whitei*	Inca Giant ?.?" Pod	White Seed Vine Lima	Bean
Phaseolus	*lunatus-vigna-alba*	*king-of-the-gardenii*	King Of The Garden 5.0" Pod	White Seed Vine Lima	Bean
Phaseolus	*lunatus-vigna-alba*	*purple-eyei*	Purple Eye ?.?" Pod	White Seed Vine Lima	Bean
Phaseolus	*lunatus-vigna-alba*	*shanty-boatii*	Shanty Boat ?.?" Pod	White Seed Vine Lima	Bean
Phaseolus	*lunatus-vigna-alba*	*snow-on-the-mountainii*	Snow On The Mountain	White Seed Vine Lima	Bean
Phaseolus	*lunatus-vigna-alba*	*white-christmasi*	White Christmas	White Seed Vine Lima	Bean
Phaseolus	*lunatus-vigna-alba*	*willow-leaf-alba*	Willow Leaf ?.?" Pod	White Seed Vine Lima	Bean
Phaseolus	*lunatus-vigna-aurantea*	*pima-orangei*	Pima	Orange Seed Vine Lima	Bean
Phaseolus	*lunatus-vigna-groena*	*burpee's-bestii*	Potato Burpee's Best 4.4" Pod	Green Seed Vine Lima	Bean
Phaseolus	*lunatus-vigna-groena*	*cape-may-giantii*	Cape May Giant	Green Seed Vine Lima	Bean
Phaseolus	*lunatus-vigna-groena*	*dr.-martinii*	Dr. Martin ?.?" Pod	Green Seed Vine Lima	Bean
Phaseolus	*lunatus-vigna-groena*	*illinois-giantii*	Illinois Giant ?.?" Pod	Green Seed Vine Lima	Bean
Phaseolus	*lunatus-vigna-groena*	*north-polei*	North Pole ?.?" Pod	Green Seed Vine Lima	Bean
Phaseolus	*lunatus-vigna-groena*	*prizetakeri*	Prizetaker 6.5" Pod	Green Seed Vine Lima	Bean
Phaseolus	*lunatus-vigna-groena*	*sieva*	Sieva 3.5" Pod	Green Seed Vine Lima	Bean
Phaseolus	*lunatus-vigna-groena*	*winfieldii*	Winfield ?.?" Pod	Green Seed Vine Lima	Bean
Phaseolus	*lunatus-vigna-helva*	*buttermanii*	Butterman ?.?" Pod	Yellow Seed Vine Lima	Bean
Phaseolus	*lunatus-vigna-helva*	*florida-butteri*	Florida Butter 3.2" Pod	Yellow Seed Vine Lima	Bean
Phaseolus	*lunatus-vigna-helva*	*smithii-butterbeanii*	Speckled Cream Seed Smith Butterbean	Yellow Seed Vine Lima	Bean
Phaseolus	*lunatus-vigna-mixta*	*ganymedei*	Red Black & White Seed Ganymede	Mixed Seed Vine Lima	Bean
Phaseolus	*lunatus-vigna-mixta*	*lima-o-del-papai*	O Del Papa	Mixed Seed Vine Lima	Bean
Phaseolus	*lunatus-vigna-mixta*	*lynch-butterbeanii*	Lynch Butterbean	Mixed Seed Vine Lima	Bean
Phaseolus	*lunatus-vigna-mixta*	*willow-leaf-coloredii*	Willow Leaf ?.?" Pod	Mixed Seed Vine Lima	Bean
Phaseolus	*lunatus-vigna-nana*	*alma's-limasi*	Purple Black Alma's Limas	Baby Vine Lima	Bean
Phaseolus	*lunatus-vigna-nana*	*black-stari*	Black Star ?.?" Pod	Baby Vine Lima	Bean
Phaseolus	*lunatus-vigna-nana*	*carolinensis*	Carolina	Baby Vine Lima	Bean
Phaseolus	*lunatus-vigna-nigra*	*blackii*	Purple Black	Black Seed Vine Lima	Bean
Phaseolus	*lunatus-vigna-nigra*	*loudermilkii*	Black & White Loudermilk	Black Seed Vine Lima	Bean
Phaseolus	*lunatus-vigna-nigra*	*civil-war-maroon-black*	Civil War Maroon-Black	Black Seed Vine Lima	Bean
Phaseolus	*lunatus-vigna-rubra*	*indian-red-polei*	Indian Red Pole	Red Seed Vine Lima	Bean
Phaseolus	*lunatus-vigna-rubra*	*johnny's-redii*	Johnny's	Red Seed Vine Lima	Bean
Phaseolus	*lunatus-vigna-rubra*	*pennsylvania-german-redii*	Pennsylvania German	Red Seed Vine Lima	Bean
Phaseolus	*lunatus-vigna-rubra*	*red-calicoi*	Red Calico ?.?" Pod	Red Seed Vine Lima	Bean
Phaseolus	*lunatus-vigna-rubra*	*red-speckled-polei*	Red Speckled Pole	Red Seed Vine Lima	Bean
Phaseolus	*lunatus-vigna-rubra*	*red-sabai*	Red Saba ?.?" Pod	Red Seed Vine Lima	Bean
Phaseolus	*lunatus-vigna-rubra*	*wochester-indian-red-polei*	Worchester Indian Red Pole	Red Seed Vine Lima	Bean
Phaseolus	*lunatus-vigna-tana*	*cliff-dwelleri*	Cliff Dweller	Tan Seed Vine Lima	Bean
Phaseolus	*lunatus-vigna-tana*	*florida-speckledii*	Florida Speckled 3.4" Pod	Tan Seed Vine Lima	Bean

Bean, Dry Bush (*Phaseolus vulgaris-bruscia*)

Phaseolus	*vulgaris-bruscia-alba*	*algarroboi*	Algarrobo	White Dry Bush	Bean
Phaseolus	*vulgaris-bruscia-alba*	*alluviasi*	Alluvias	White Dry Bush	Bean
Phaseolus	*vulgaris-bruscia-alba*	*aunt-emma'si*	Aunt Emma's	White Dry Bush	Bean
Phaseolus	*vulgaris-bruscia-alba*	*aurorai*	Aurora	White Dry Bush	Bean
Phaseolus	*vulgaris-bruscia-alba*	*aztec-michiganensis*	Michigan Aztec	White Dry Bush	Bean
Phaseolus	*vulgaris-bruscia-alba*	*aztec-anasazii*	Anasazi Half Runner Aztec	White Dry Bush	Bean
Phaseolus	*vulgaris-bruscia-alba*	*beautifulii*	Beautiful Maroon Splashed	White Dry Bush	Bean
Phaseolus	*vulgaris-bruscia-alba*	*bloweri*	Blower	White Dry Bush	Bean
Phaseolus	*vulgaris-bruscia-alba*	*boy-soldieri*	Boy Soldier	White Dry Bush	Bean
Phaseolus	*vulgaris-bruscia-alba*	*bumblebeei*	Bumblebee	White Dry Bush	Bean
Phaseolus	*vulgaris-bruscia-alba*	*canadian-wild-goosei*	Canadian Wild Goose	White Dry Bush	Bean
Phaseolus	*vulgaris-bruscia-alba*	*dog-beanii*	Dog Purple Spotted	White Dry Bush	Bean
Phaseolus	*vulgaris-bruscia-alba*	*draboi*	Drabo	White Dry Bush	Bean
Phaseolus	*vulgaris-bruscia-alba*	*duane-baptiste's-potato-beanii*	Duane Baptiste's Potato	White Dry Bush	Bean
Phaseolus	*vulgaris-bruscia-alba*	*kahl-alba*	Kahl	White Dry Bush	Bean

Phaseolus	*vulgaris-bruscia-alba*	*larry-locke'si*	Larry Locke's	White Dry Bush Bean
Phaseolus	*vulgaris-bruscia-alba*	*maccaronei*	Maccarone Dappled	White Dry Bush Bean
Phaseolus	*vulgaris-bruscia-alba*	*maine-sunsetii*	Maine Sunset Ivory	White Dry Bush Bean
Phaseolus	*vulgaris-bruscia-alba*	*marble-whitei*	Marble	White Dry Bush Bean
Phaseolus	*vulgaris-bruscia-alba*	*paintii*	Paint Yellow Eyed	White Dry Bush Bean
Phaseolus	*vulgaris-bruscia-alba*	*spauldingii*	Spaulding	White Dry Bush Bean
Phaseolus	*vulgaris-bruscia-alba*	*stevenson-blue-eyei*	Stevenson Blue Eye	White Dry Bush Bean
Phaseolus	*vulgaris-bruscia-alba*	*yellow-eye-dotii*	Dot Yellow Eye	White Dry Bush Bean
Phaseolus	*vulgaris-bruscia-alba*	*yellow-eye-molasses-facei*	Molasses Face Yellow Eye	White Dry Bush Bean
Phaseolus	*vulgaris-bruscia-alba*	*yellow-eye-kenearlyi*	Kenearly Yellow Eye	White Dry Bush Bean
Phaseolus	*vulgaris-bruscia-alba*	*yellow-eye-mainensis*	Maine Yellow Eye	White Dry Bush Bean
Phaseolus	*vulgaris-bruscia-alba*	*yellow-eye-mainensis-improvedii*	Maine Improved Yellow Eye	White Dry Bush Bean
Phaseolus	*vulgaris-bruscia-alba*	*yellow-eye-steubenii*	Steuben Yellow Eye	White Dry Bush Bean
Phaseolus	*vulgaris-bruscia-baca*	*brown-beautyi*	Brown Beauty Brown	Baking Dry Bush Bean
Phaseolus	*vulgaris-bruscia-baca*	*calypsoi*	Calypso Black & White	Baking Dry Bush Bean
Phaseolus	*vulgaris-bruscia-baca*	*chickasawi*	Chickasaw Lime Green	Baking Dry Bush Bean
Phaseolus	*vulgaris-bruscia-baca*	*dutch-brownii*	Pennsylvania Dutch Brown	Baking Dry Bush Bean
Phaseolus	*vulgaris-bruscia-baca*	*gauchoi*	Gaucho Tan	Baking Dry Bush Bean
Phaseolus	*vulgaris-bruscia-baca*	*moneyi*	Money Red Speckled	Baking Dry Bush Bean
Phaseolus	*vulgaris-bruscia-baca*	*pawneei*	Pawnee Brown	Baking Dry Bush Bean
Phaseolus	*vulgaris-bruscia-baca*	*rucklei(walchersei)*	Ruckle Walcherse White	Baking Dry Bush Bean
Phaseolus	*vulgaris-bruscia-baca*	*soldieri(johnsonii)*	Johnson Soldier White	Baking Dry Bush Bean
Phaseolus	*vulgaris-bruscia-baca*	*swedish-brownii*	Swedish Brown	Baking Dry Bush Bean
Phaseolus	*vulgaris-bruscia-baca*	*ukrainianii*	Ukranian Marrowfat	Baking Dry Bush Bean
Phaseolus	*vulgaris-bruscia-baca*	*white-marrowfatii*	White Marrowfat	Baking Dry Bush Bean
Phaseolus	*vulgaris-bruscia-brunnea*	*african-primera*	African Primier Rusty	Brown Dry Bush Bean
Phaseolus	*vulgaris-bruscia-brunnea*	*arikara-yellowi*	Arikara Yellow	Brown Dry Bush Bean
Phaseolus	*vulgaris-bruscia-brunnea*	*beka-brunnea*	Beka	Brown Dry Bush Bean
Phaseolus	*vulgaris-bruscia-brunnea*	*fradinhoi*	Fradinho	Brown Dry Bush Bean
Phaseolus	*vulgaris-bruscia-brunnea*	*norwegian-brownii*	Norwegian	Brown Dry Bush Bean
Phaseolus	*vulgaris-bruscia-brunnea*	*scandanavica*	Scandanavian	Brown Dry Bush Bean
Phaseolus	*vulgaris-bruscia-brunnea*	*zertii*	Zert	Brown Dry Bush Bean
Phaseolus	*vulgaris-bruscia-chili*	*dark-red-kidneyi*	Dark Red Kidney	Chili Dry Bush Bean
Phaseolus	*vulgaris-bruscia-chili*	*red-kidneyi*	Mahogany Red Kidney	Chili Dry Bush Bean
Phaseolus	*vulgaris-bruscia-chili*	*pinca*	Pink	Chili Dry Bush Bean
Phaseolus	*vulgaris-bruscia-chili*	*scarlet-beautyi*	Scarlet Beauty	Chili Dry Bush Bean
Phaseolus	*vulgaris-bruscia-chili*	*small-redii*	Small Red	Chili Dry Bush Bean
Phaseolus	*vulgaris-bruscia-grisea*	*italian-zebrai*	Italian Zebra	Gray Dry Bush Bean
Phaseolus	*vulgaris-bruscia-grisea*	*kutzner's-russian-soupii*	Kutzner's Russian Soup	Gray Dry Bush Bean
Phaseolus	*vulgaris-bruscia-grisea*	*wild-pigeon-beanii*	Wild Pigeon Iroquios	Gray Dry Bush Bean
Phaseolus	*vulgaris-bruscia-groena*	*aborigineus*	Aboriginal	Green Dry Bush Bean
Phaseolus	*vulgaris-bruscia-groena*	*bakerii*	Baker	Green Dry Bush Bean
Phaseolus	*vulgaris-bruscia-groena*	*candyi*	Candy	Green Dry Bush Bean
Phaseolus	*vulgaris-bruscia-groena*	*cayugai*	Cayuga	Green Dry Bush Bean
Phaseolus	*vulgaris-bruscia-groena*	*compessus*	Compressed	Green Dry Bush Bean
Phaseolus	*vulgaris-bruscia-groena*	*cuetzalanii*	Cuetzalan	Green Dry Bush Bean
Phaseolus	*vulgaris-bruscia-groena*	*familyi*	Family	Green Dry Bush Bean
Phaseolus	*vulgaris-bruscia-groena*	*flora-de-rioi*	Flora De Rio	Green Dry Bush Bean
Phaseolus	*vulgaris-bruscia-groena*	*geril'si*	Geril's	Green Dry Bush Bean
Phaseolus	*vulgaris-bruscia-groena*	*gotlands-speckledii*	Gotlands Speckled	Green Dry Bush Bean
Phaseolus	*vulgaris-bruscia-groena*	*grandma-turner's*	Grandma Turner's	Green Dry Bush Bean
Phaseolus	*vulgaris-bruscia-groena*	*hignell's-italianii*	Hignell's Italian	Green Dry Bush Bean
Phaseolus	*vulgaris-bruscia-groena*	*humason's-bestii*	Humason's Best	Green Dry Bush Bean
Phaseolus	*vulgaris-bruscia-groena*	*hutterite-soupii*	Hutterite Soup Light	Green Dry Bush Bean
Phaseolus	*vulgaris-bruscia-groena*	*ireland-creek-annie'si*	Ireland Creek Annie's	Green Dry Bush Bean
Phaseolus	*vulgaris-bruscia-groena*	*japanese-twineri*	Japanese Twiner	Green Dry Bush Bean
Phaseolus	*vulgaris-bruscia-groena*	*jellyi*	Jelly	Green Dry Bush Bean
Phaseolus	*vulgaris-bruscia-groena*	*kahl-yellow-greenii*	Kahl Yellow Green	Green Dry Bush Bean
Phaseolus	*vulgaris-bruscia-groena*	*kivai*	Kiva	Green Dry Bush Bean
Phaseolus	*vulgaris-bruscia-groena*	*lightningii*	Lighting	Green Dry Bush Bean
Phaseolus	*vulgaris-bruscia-groena*	*mennonitei*	Mennonite Olive	Green Dry Bush Bean
Phaseolus	*vulgaris-bruscia-groena*	*missouri-bill'si*	Missouri Bill's	Green Dry Bush Bean
Phaseolus	*vulgaris-bruscia-groena*	*mrs.-kramer'si*	Mrs. Kramer's	Green Dry Bush Bean
Phaseolus	*vulgaris-bruscia-groena*	*murdockii*	Murdock	Green Dry Bush Bean
Phaseolus	*vulgaris-bruscia-groena*	*nanus*	Dwarf	Green Dry Bush Bean
Phaseolus	*vulgaris-bruscia-groena*	*native-whispering*	Native Whispering	Green Dry Bush Bean
Phaseolus	*vulgaris-bruscia-groena*	*navajoi*	Navaho	Green Dry Bush Bean
Phaseolus	*vulgaris-bruscia-groena*	*old-rumanianii*	Old Rumanian	Green Dry Bush Bean
Phaseolus	*vulgaris-bruscia-groena*	*peruvian-goosei*	Peruvian Goose	Green Dry Bush Bean
Phaseolus	*vulgaris-bruscia-groena*	*pretzelii*	Pretzel	Green Dry Bush Bean
Phaseolus	*vulgaris-bruscia-groena*	*rhodenizeri*	Rhodenizer	Green Dry Bush Bean
Phaseolus	*vulgaris-bruscia-groena*	*ricei*	Rice	Green Dry Bush Bean
Phaseolus	*vulgaris-bruscia-groena*	*russian-soupii*	Russian Soup	Green Dry Bush Bean
Phaseolus	*vulgaris-bruscia-groena*	*senecai*	Seneca	Green Dry Bush Bean
Phaseolus	*vulgaris-bruscia-groena*	*senecai-tonawandai*	Seneca Tonawanda	Green Dry Bush Bean

Phaseolus	*vulgaris-bruscia-groena*	*stingeri*	Stinger	Green Dry Bush Bean
Phaseolus	*vulgaris-bruscia-groena*	*stockbridge-indianii*	Stockbridge Indian	Green Dry Bush Bean
Phaseolus	*vulgaris-bruscia-groena*	*tuteloi*	Tutelo	Green Dry Bush Bean
Phaseolus	*vulgaris-bruscia-groena*	*world-warii*	World War	Green Dry Bush Bean
Phaseolus	*vulgaris-bruscia-helva*	*adventista*	Adventist	Yellow Dry Bush Bean
Phaseolus	*vulgaris-bruscia-helva*	*azoresensis*	Azores	Yellow Dry Bush Bean
Phaseolus	*vulgaris-bruscia-helva*	*baha-azufradosi*	Baha Azufrados	Yellow Dry Bush Bean
Phaseolus	*vulgaris-bruscia-helva*	*canaryi*	Canary	Yellow Dry Bush Bean
Phaseolus	*vulgaris-bruscia-helva*	*chinese-yellowi*	Chinese Yellow	Yellow Dry Bush Bean
Phaseolus	*vulgaris-bruscia-helva*	*four-corners-goldii*	Four Corners Gold	Yellow Dry Bush Bean
Phaseolus	*vulgaris-bruscia-helva*	*granny-hagler-half-runneri*	Granny Hagler	Yellow Dry Bush Bean
Phaseolus	*vulgaris-bruscia-helva*	*indian-women-squaw-helva*	Indian Woman Squaw	Yellow Dry Bush Bean
Phaseolus	*vulgaris-bruscia-helva*	*marfaxii*	Marfax Gold	Yellow Dry Bush Bean
Phaseolus	*vulgaris-bruscia-helva*	*pepa-de-zapalloi*	Pepa De Zapallo Gold	Yellow Dry Bush Bean
Phaseolus	*vulgaris-bruscia-helva*	*scott's-choicei*	Scott Choice Golden	Yellow Dry Bush Bean
Phaseolus	*vulgaris-bruscia-helva*	*summerfeltii*	Summerfelt Golden	Yellow Dry Bush Bean
Phaseolus	*vulgaris-bruscia-hortus*	*coco-praguei*	Coco Prague Kidney Red	Shell Dry Bush Bean
Phaseolus	*vulgaris-bruscia-hortus*	*coco-rubicoi*	Coco Rubico Kidney Buff	Shell Dry Bush Bean
Phaseolus	*vulgaris-bruscia-hortus*	*cranberryi*	Cranberry Rusty Brown	Shell Dry Bush Bean
Phaseolus	*vulgaris-bruscia-hortus*	*etnai*	Etna Buff Red	Shell Dry Bush Bean
Phaseolus	*vulgaris-bruscia-hortus*	*european-soldieri*	European Soldier	Shell Dry Bush Bean
Phaseolus	*vulgaris-bruscia-hortus*	*flageoletii*	Gourmet White Flageolet	Shell Dry Bush Bean
Phaseolus	*vulgaris-bruscia-hortus*	*flageoletii-flambeaui*	Flambeau Green Flageolet	Shell Dry Bush Bean
Phaseolus	*vulgaris-bruscia-hortus*	*flageoletii-green-&-whitei*	Green & White Flageolet	Shell Dry Bush Bean
Phaseolus	*vulgaris-bruscia-hortus*	*flageoletii-green-chevrieri*	Chevrier Green Flageolet	Shell Dry Bush Bean
Phaseolus	*vulgaris-bruscia-hortus*	*flageoletii-yellow-seedii*	Yellow & Buff Flageolet	Shell Dry Bush Bean
Phaseolus	*vulgaris-bruscia-hortus*	*horticultural-dwarfii*	Dwarf Pink Buff	Shell Dry Bush Bean
Phaseolus	*vulgaris-bruscia-hortus*	*horticultural-frenchii*	French Purple	Shell Dry Bush Bean
Phaseolus	*vulgaris-bruscia-hortus*	*horticultural-robin's-eggii*	Robin's Egg Violet	Shell Dry Bush Bean
Phaseolus	*vulgaris-bruscia-hortus*	*horticultural-scarlet-beauty-elitei*	Scarlet Beauty Elite Pink	Shell Dry Bush Bean
Phaseolus	*vulgaris-bruscia-hortus*	*horticultural-taylori*	Taylor Dwarf Red Splashed Buff	Shell Dry Bush Bean
Phaseolus	*vulgaris-bruscia-hortus*	*horticultural-vulcanoi*	Volcano Buff Red	Shell Dry Bush Bean
Phaseolus	*vulgaris-bruscia-hortus*	*horticultural-whitei*	White	Shell Dry Bush Bean
Phaseolus	*vulgaris-bruscia-hortus*	*iroquois-cranberryi*	Iroquois Cranberry	Shell Dry Bush Bean
Phaseolus	*vulgaris-bruscia-hortus*	*uncle-willie'si*	Uncle Willie's Cranberry Crimson Streaked Cream	Shell Dry Bush Bean
Phaseolus	*vulgaris-bruscia-hortus*	*john's-old-beanii*	John's Old Bean Purple Mottled Tan	Shell Dry Bush Bean
Phaseolus	*vulgaris-bruscia-hortus*	*krolii*	Krol	Shell Dry Bush Bean
Phaseolus	*vulgaris-bruscia-hortus*	*limelightii*	Limelight White	Shell Dry Bush Bean
Phaseolus	*vulgaris-bruscia-hortus*	*low's-championii*	Low's Champion Cranberry	Shell Dry Bush Bean
Phaseolus	*vulgaris-bruscia-hortus*	*platai*	Plata Brown	Shell Dry Bush Bean
Phaseolus	*vulgaris-bruscia-hortus*	*proto-granadai*	Poroto Granada Cranberry	Shell Dry Bush Bean
Phaseolus	*vulgaris-bruscia-hortus*	*red-peanut-beanii*	Red Pod Peanut	Shell Dry Bush Bean
Phaseolus	*vulgaris-bruscia-hortus*	*serenei*	Serene Violet Gray	Shell Dry Bush Bean
Phaseolus	*vulgaris-bruscia-hortus*	*speckled-baysi*	Speckled Bays Cream	Shell Dry Bush Bean
Phaseolus	*vulgaris-bruscia-hortus*	*tierra-del-fuego*	Tierra Del Fuego Cream	Shell Dry Bush Bean
Phaseolus	*vulgaris-bruscia-hortus*	*tongue-of-firei*	Tongue Of Fire Ivory Tan	Shell Dry Bush Bean
Phaseolus	*vulgaris-bruscia-hortus*	*vermont-cranberryi*	Vermont Cranberry	Shell Dry Bush Bean
Phaseolus	*vulgaris-bruscia-mixta*	*appaloosa-neo-mexicana-marona*	New Mexico Appaloosa Maroon & White	Mixed Dry Bush Bean
Phaseolus	*vulgaris-bruscia-mixta*	*appaloosa-neo-mexicana-nigra*	New Mexico Appaloosa Black & White	Mixed Dry Bush Bean
Phaseolus	*vulgaris-bruscia-mixta*	*appaloosa-neo-mexicana-rubra*	New Mexico Appaloosa Red & White	Mixed Dry Bush Bean
Phaseolus	*vulgaris-bruscia-mixta*	*appaloosa-vermontensis-gelda*	Vermont Appaloosa Gold & White	Mixed Dry Bush Bean
Phaseolus	*vulgaris-bruscia-mixta*	*appaloosa-vermontensis-marona*	Vermont Appaloosa Maroon & White	Mixed Dry Bush Bean
Phaseolus	*vulgaris-bruscia-mixta*	*arkansas-bruscia*	Arkansas Bush Tan & Brown	Mixed Dry Bush Bean
Phaseolus	*vulgaris-bruscia-mixta*	*bert-goodwin'si*	Bert Goodwin's Tan & Brown	Mixed Dry Bush Bean
Phaseolus	*vulgaris-bruscia-mixta*	*boston-beautyi*	Boston Beauty Maroon & White	Mixed Dry Bush Bean
Phaseolus	*vulgaris-bruscia-mixta*	*kilham-goosei*	Kilham Goose Purple On White	Mixed Dry Bush Bean
Phaseolus	*vulgaris-bruscia-mixta*	*mimbres-cavei*	Mimbres Cave Maroon Brown Mottled White	Mixed Dry Bush Bean
Phaseolus	*vulgaris-bruscia-mixta*	*mojavensis*	Mohave Maroon & White	Mixed Dry Bush Bean
Phaseolus	*vulgaris-bruscia-mixta*	*odawa-indianii*	Odawa Indian Tan Marked White	Mixed Dry Bush Bean
Phaseolus	*vulgaris-bruscia-mixta*	*odawa-soupii*	Odawa Soup Pinto & White	Mixed Dry Bush Bean
Phaseolus	*vulgaris-bruscia-mixta*	*orcai*	Orca Black & White	Mixed Dry Bush Bean
Phaseolus	*vulgaris-bruscia-mixta*	*painted-ponyi*	Painted Pony Brown Over White	Mixed Dry Bush Bean
Phaseolus	*vulgaris-bruscia-mixta*	*peregionii*	Peregion Tan Streaked Brown	Mixed Dry Bush Bean
Phaseolus	*vulgaris-bruscia-mixta*	*piros-feheri*	Piros Feher Red Marked White	Mixed Dry Bush Bean
Phaseolus	*vulgaris-bruscia-mixta*	*rajman-himalayanii*	Rajman Himalayan Red Mottled Cream	Mixed Dry Bush Bean
Phaseolus	*vulgaris-bruscia-mixta*	*rockwellii*	Rockwell Maroon Marked White	Mixed Dry Bush Bean
Phaseolus	*vulgaris-bruscia-mixta*	*rowleyi*	Rowley Black Spot White	Mixed Dry Bush Bean

Phaseolus	vulgaris-bruscia-mixta	sacrament-beanii	Sacrament Maroon Marked White	Mixed Dry Bush Bean
Phaseolus	vulgaris-bruscia-mixta	tender-frostii	Tender Frost Brown & White	Mixed Dry Bush Bean
Phaseolus	vulgaris-bruscia-mixta	toni's-redii	Toni's Half Rose Red & Half White	Mixed Dry Bush Bean
Phaseolus	vulgaris-bruscia-navalis	navalis	Original	White Navy Dry Bush Bean
Phaseolus	vulgaris-bruscia-navalis	centralia	Centralia	White Navy Dry Bush Bean
Phaseolus	vulgaris-bruscia-navalis	clem-&-sarah's-big-beanii	Clem & Sarah's Big	White Navy Dry Bush Bean
Phaseolus	vulgaris-bruscia-navalis	fleetwood-navyi	Fleetwood	White Navy Dry Bush Bean
Phaseolus	vulgaris-bruscia-navalis	french-navyi	French	White Navy Dry Bush Bean
Phaseolus	vulgaris-bruscia-navalis	great-northern-gerald's-whitei	Gerald's Great Northern	White Navy Dry Bush Bean
Phaseolus	vulgaris-bruscia-navalis	great-northern-montana-whitei	Montana Great Northern	White Navy Dry Bush Bean
Phaseolus	vulgaris-bruscia-navalis	lucas-navyi	Lucas	White Navy Dry Bush Bean
Phaseolus	vulgaris-bruscia-navalis	michilite-navyi	Michilite	White Navy Dry Bush Bean
Phaseolus	vulgaris-bruscia-navalis	pea-beanii	Pea Bean	White Navy Dry Bush Bean
Phaseolus	vulgaris-bruscia-navalis	sanalic-navyi	Sanalic	White Navy Dry Bush Bean
Phaseolus	vulgaris-bruscia-navalis	seafarer-navyi	Seafarer	White Navy Dry Bush Bean
Phaseolus	vulgaris-bruscia-navalis	white-hailstones-navyi	White Hailstones	White Navy Dry Bush Bean
Phaseolus	vulgaris-bruscia-nigra	blackii-cocoa	Black Coco	Black Dry Bush Bean
Phaseolus	vulgaris-bruscia-nigra	blackii-jasonii	Black Jason	Black Dry Bush Bean
Phaseolus	vulgaris-bruscia-nigra	blackii-mexicana	Black Mexican	Black Dry Bush Bean
Phaseolus	vulgaris-bruscia-nigra	blackii-turtlei	Black Turtle Soup	Black Dry Bush Bean
Phaseolus	vulgaris-bruscia-nigra	blackii-turtlei-midnightii	Midnight Black Turtle Soup	Black Dry Bush Bean
Phaseolus	vulgaris-bruscia-nigra	blacked-twineri	Blacked Twiner	Black Dry Bush Bean
Phaseolus	vulgaris-bruscia-nigra	cubanosi	Cubanos	Black Dry Bush Bean
Phaseolus	vulgaris-bruscia-nigra	delgado-black-beanii	Delgado Shawl Dye Small	Black Dry Bush Bean
Phaseolus	vulgaris-bruscia-nigra	mantoba-nigra	Manitoba	Black Dry Bush Bean
Phaseolus	vulgaris-bruscia-nigra	mitla-nigra	Mitla	Black Dry Bush Bean
Phaseolus	vulgaris-bruscia-nigra	monos-negrosi	Black Monkeys	Black Dry Bush Bean
Phaseolus	vulgaris-bruscia-nigra	negro-taxumalii	Negro Taxumal	Black Dry Bush Bean
Phaseolus	vulgaris-bruscia-nigra	pi-194582-guatemala	PI #194582 Guatemala	Black Dry Bush Bean
Phaseolus	vulgaris-bruscia-nigra	pueblo-nigra	Pueblo	Black Dry Bush Bean
Phaseolus	vulgaris-bruscia-nigra	ravenii	Raven	Black Dry Bush Bean
Phaseolus	vulgaris-bruscia-nigra	sarah-ross-black-beanii	Sarah Ross	Black Dry Bush Bean
Phaseolus	vulgaris-bruscia-nigra	shortcut-queryi	Shortcut Query	Black Dry Bush Bean
Phaseolus	vulgaris-bruscia-nigra	spanish-nigra	Spanish Small	Black Dry Bush Bean
Phaseolus	vulgaris-bruscia-pinca	bjlovar-beautyi	Bjlovar Beauty	Pink Dry Bush Bean
Phaseolus	vulgaris-bruscia-pinca	o'odham-pinca	O'odham Desert	Pink Dry Bush Bean
Phaseolus	vulgaris-bruscia-pinca	santa-maria-pinquitoi	Santa Maria Pinquito	Pink Dry Bush Bean
Phaseolus	vulgaris-bruscia-pinca	texas-pinca	Texas	Pink Dry Bush Bean
Phaseolus	vulgaris-bruscia-pinca	tigrei	Tigre Maroon Streaked	Pink Dry Bush Bean
Phaseolus	vulgaris-bruscia-pintoi	pintoi	Original	Pinto Dry Bush Bean
Phaseolus	vulgaris-bruscia-pintoi	agassiz-pintoi	Agassiz Buff	Pinto Dry Bush Bean
Phaseolus	vulgaris-bruscia-pintoi	agate-pintoi	Agate Buff	Pinto Dry Bush Bean
Phaseolus	vulgaris-bruscia-pintoi	blue-pintoi	Blue Black	Pinto Dry Bush Bean
Phaseolus	vulgaris-bruscia-pintoi	buckskinii	Buckskin	Pinto Dry Bush Bean
Phaseolus	vulgaris-bruscia-pintoi	burlescynai	Burlescyna Purple Brown	Pinto Dry Bush Bean
Phaseolus	vulgaris-bruscia-pintoi	juan-gonzalesi	Brown Marked Buff Juan Gonzales	Pinto Dry Bush Bean
Phaseolus	vulgaris-bruscia-pintoi	mayo-colima-pintoi	Mayo Colima Cream, Brown & Gray	Pinto Dry Bush Bean
Phaseolus	vulgaris-bruscia-pintoi	mexican-pintoi	Mexican	Pinto Dry Bush Bean
Phaseolus	vulgaris-bruscia-pintoi	minnesota-cranberryi	Minnesota Cranberry	Pinto Dry Bush Bean
Phaseolus	vulgaris-bruscia-pintoi	nodak-pintoi	Nodak	Pinto Dry Bush Bean
Phaseolus	vulgaris-bruscia-pintoi	othelloi	Othello	Pinto Dry Bush Bean
Phaseolus	vulgaris-bruscia-pintoi	pinacatei	Brown Speckled Beige Pinacate	Pinto Dry Bush Bean
Phaseolus	vulgaris-bruscia-pintoi	pintoi-3	Field III Shell	Pinto Dry Bush Bean
Phaseolus	vulgaris-bruscia-pintoi	pintoi-improvedii	Pinto Improved Stringless	Pinto Dry Bush Bean
Phaseolus	vulgaris-bruscia-pintoi	pintoi-UI-111	Univ. Of Idaho #111 Buff	Pinto Dry Bush Bean
Phaseolus	vulgaris-bruscia-pintoi	red-eyei	Red Eye Small White	Pinto Dry Bush Bean
Phaseolus	vulgaris-bruscia-pintoi	red-jasonii	Red Jason Red Patch White	Pinto Dry Bush Bean
Phaseolus	vulgaris-bruscia-pintoi	rio-bavispensis	Brown Speckled Beige Rio Bavispe	Pinto Dry Bush Bean
Phaseolus	vulgaris-bruscia-pintoi	sonoran-pintoi	Sonoran Beige	Pinto Dry Bush Bean
Phaseolus	vulgaris-bruscia-pintoi	spanish-tolosana-pintoi	Spanish Tolosana White Splotch Maroon	Pinto Dry Bush Bean
Phaseolus	vulgaris-bruscia-purpurea	goose-cranberryi	Goose Cranberry Red	Purple Dry Bush Bean
Phaseolus	vulgaris-bruscia-purpurea	flor-de-mayo-purpurea	Flor De Mayo	Purple Dry Bush Bean
Phaseolus	vulgaris-bruscia-purpurea	mansell-magici	Mansell Magic Dark	Purple Dry Bush Bean
Phaseolus	vulgaris-bruscia-rena	bean-soup-specialis	Bean Soup Special Red	Kidney Dry Bush Bean
Phaseolus	vulgaris-bruscia-rena	boston-favoritei	Boston Favorite Pink	Kidney Dry Bush Bean
Phaseolus	vulgaris-bruscia-rena	carminea	Carmine Red	Kidney Dry Bush Bean
Phaseolus	vulgaris-bruscia-rena	golden-heirloomii	Golden Heirloom Tan	Kidney Dry Bush Bean
Phaseolus	vulgaris-bruscia-rena	jacob's-cattlei-anasazii(anasazii)	Anasazi Maroon & White Jacob's Cattle	Kidney Dry Bush Bean
Phaseolus	vulgaris-bruscia-rena	jacob's-cattlei-dalmationii	Dalmation Red & Orange Mottled White Jacob's Cattle	Kidney Dry Bush Bean
Phaseolus	vulgaris-bruscia-rena	jacob's-cattlei-gaslessi	Gasless Maroon & White Jacob's Cattle	Kidney Dry Bush Bean
Phaseolus	vulgaris-bruscia-rena	jacob's-cattlei-	Golondrinas Gold &	Kidney Dry Bush Bean

		golandrinasi	White Jacob's Cattle	
Phaseolus	vulgaris-bruscia-rena	jacob's-cattlei-goldii(jacob's-goldii)	Gold Yellow Eyed White Jacob's Cattle	Kidney Dry Bush Bean
Phaseolus	vulgaris-bruscia-rena	jacob's-cattlei-mt.-pima-pintadosi	Mt.Pima Pintados Plum & White Jacob's Cattle	Kidney Dry Bush Bean
Phaseolus	vulgaris-bruscia-rena	jacob's-cattlei-palominoi(palominoi)	Palomino Yellow Brown & White Jacob's Cattle	Kidney Dry Bush Bean
Phaseolus	vulgaris-bruscia-rena	jacob's-cattlei-troutii(jacob's-cattlei)	Trout Maroon & White Jacob's Cattle	Kidney Dry Bush Bean
Phaseolus	vulgaris-bruscia-rena	jacob's-cattlei-trout-black-&-whitei	Trout Black & White Jacob's Cattle	Kidney Dry Bush Bean
Phaseolus	vulgaris-bruscia-rena	jacob's-cattlei-zuni-shalakoi	Jacob's Cattle Zuni Shalako Gold & White	Kidney Dry Bush Bean
Phaseolus	vulgaris-bruscia-rena	immigrantii	Immigrant Golden Yellow	Kidney Dry Bush Bean
Phaseolus	vulgaris-bruscia-rena	aztec-red-kidneyi	Aztec Vibrant Red	Kidney Dry Bush Bean
Phaseolus	vulgaris-bruscia-rena	brick-red-kidneyi	Brick Red	Kidney Dry Bush Bean
Phaseolus	vulgaris-bruscia-rena	californica-red-kidneyi	California Red	Kidney Dry Bush Bean
Phaseolus	vulgaris-bruscia-rena	charlevoix-dark-red-kidneyi	Charlevoix Dark Red	Kidney Dry Bush Bean
Phaseolus	vulgaris-bruscia-rena	large-red-kidneyi	Large Red	Kidney Dry Bush Bean
Phaseolus	vulgaris-bruscia-rena	large-royal-kidneyi	Large Royal	Kidney Dry Bush Bean
Phaseolus	vulgaris-bruscia-rena	light-red-kidneyi	Pink Mahogany Light Red	Kidney Dry Bush Bean
Phaseolus	vulgaris-bruscia-rena	mull-kidneyi	Mull Deep Maroon	Kidney Dry Bush Bean
Phaseolus	vulgaris-bruscia-rena	pink-kidneyi	Pink	Kidney Dry Bush Bean
Phaseolus	vulgaris-bruscia-rena	purple-kidneyi	Maroon Purple	Kidney Dry Bush Bean
Phaseolus	vulgaris-bruscia-rena	ralph-dutcher-white-kidneyi	Ralph Dutcher White	Kidney Dry Bush Bean
Phaseolus	vulgaris-bruscia-rena	troomly's-dark-red-kidneyi	Troomly's Dark Red	Kidney Dry Bush Bean
Phaseolus	vulgaris-bruscia-rena	white-kidneyi	White	Kidney Dry Bush Bean
Phaseolus	vulgaris-bruscia-rena	montcalmii	Montcalm Dark Red	Kidney Dry Bush Bean
Phaseolus	vulgaris-bruscia-rena	suppenbohne-wybelsumii	Suppenbohne Wybelsum Pink Beige	Kidney Dry Bush Bean
Phaseolus	vulgaris-bruscia-rubra	red-king-of-the-earlyi	King Of The Early Mottled	Red Dry Bush Bean
Phaseolus	vulgaris-bruscia-rubra	red-kintokii	Kintoki	Red Dry Bush Bean
Phaseolus	vulgaris-bruscia-rubra	red-montezuma-redii	Montezuma	Red Dry Bush Bean
Phaseolus	vulgaris-bruscia-rubra	red-niccaraguan-bushii	Nicaraguan Bush	Red Dry Bush Bean
Phaseolus	vulgaris-bruscia-rubra	red-rojo-de-sedai	Rojo De Seda Red Silk	Red Dry Bush Bean
Phaseolus	vulgaris-bruscia-rubra	red-sangre-de-toroi	Sangre De Toro Bull's Blood	Red Dry Bush Bean
Phaseolus	vulgaris-bruscia-rubra	red-six-nationsi	Six Nations Iroquois White Speckled	Red Dry Bush Bean
Phaseolus	vulgaris-bruscia-rubra	red-smith-river-super-specklei	Smith River Super Speckle	Red Dry Bush Bean
Phaseolus	vulgaris-bruscia-rubra	red-taos-rubra	Taos	Red Dry Bush Bean
Phaseolus	vulgaris-bruscia-rubra	red-tzutuhil-redii	Tzutuhil Burgundy	Red Dry Bush Bean
Phaseolus	vulgaris-bruscia-tana	andrew-kentii	Andrew Kent Buff	Tan Dry Bush Bean
Phaseolus	vulgaris-bruscia-tana	auntie-brownii	Auntie Brown Buff	Tan Dry Bush Bean
Phaseolus	vulgaris-bruscia-tana	bolitas-coloradoensis	Colorado Bolitas	Tan Dry Bush Bean
Phaseolus	vulgaris-bruscia-tana	bolitas-garciai	Garcia Bolitas	Tan Dry Bush Bean
Phaseolus	vulgaris-bruscia-tana	bolitas-neomexicana	New Mexico Bolitas	Tan Dry Bush Bean
Phaseolus	vulgaris-bruscia-tana	bolitas-ojitoi	Ojito Bolitas	Tan Dry Bush Bean
Phaseolus	vulgaris-bruscia-tana	bolitas-vaditoensis	Tan Beige Vadito Bolitas	Tan Dry Bush Bean
Phaseolus	vulgaris-bruscia-tana	bolitas-tani (boletai)	Pinto Bolitas Light Brown	Tan Dry Bush Bean
Phaseolus	vulgaris-bruscia-tana	del-nortei	Del Norte Brown Yellow	Tan Dry Bush Bean
Phaseolus	vulgaris-bruscia-tana	fisher-beanii	Fisher Iroquois Light	Tan Dry Bush Bean
Phaseolus	vulgaris-bruscia-tana	esther's-swedishii	Esther's Swedish	Tan Dry Bush Bean
Phaseolus	vulgaris-bruscia-tana	granny-beanii	Granny Cream	Tan Dry Bush Bean
Phaseolus	vulgaris-bruscia-tana	mary-ison's-little-brown-bunchii	Mary Ison's Little Brown Bunch	Tan Dry Bush Bean
Phaseolus	vulgaris-bruscia-tana	mayocobai	Mayocoba Beige	Tan Dry Bush Bean
Phaseolus	vulgaris-bruscia-tana	mayocobai-bolasi	Bolas Mayocoba Beige	Tan Dry Bush Bean
Phaseolus	vulgaris-bruscia-tana	morenoi	Moreno	Tan Dry Bush Bean
Phaseolus	vulgaris-bruscia-tana	mrociumerei	Mrociumere	Tan Dry Bush Bean
Phaseolus	vulgaris-bruscia-tana	nez-percei	Nez Perce Yellow Brown	Tan Dry Bush Bean
Phaseolus	vulgaris-bruscia-tana	wild-goosei	Wild Goose Tan Mottled Dark	Tan Dry Bush Bean

Bean, Snap Bush (*Phaseolus vulgaris-bruscia-snapa*)

Genus	Species	infraspecies	cultivar	specimen Genome
Phaseolus	vulgaris-bruscia-snapa-alba	.	.	White Seed Snap Bush Bean
Phaseolus	vulgaris-bruscia-snapa-alba	astroi	Astro 6.1"	White Seed Snap Bush Bean
Phaseolus	vulgaris-bruscia-snapa-alba	blue-lake-bruscia	Blue Lake #5065 6.5"	White Seed Snap Bush Bean
Phaseolus	vulgaris-bruscia-snapa-alba	blue-lake-bruscia-0047	Bliss Blue Lake #0047 6.0"	White Seed Snap Bush Bean
Phaseolus	vulgaris-bruscia-snapa-alba	blue-lake-bruscia-0091G	Oregon Blue Lake #0091G	White Seed Snap Bush Bean
Phaseolus	vulgaris-bruscia-snapa-alba	blue-lake-bruscia-0109	Blue Lake #0109 5.0"	White Seed Snap Bush Bean
Phaseolus	vulgaris-bruscia-snapa-alba	blue-lake-bruscia-0274BBL	Big Blue Lake #0274 6.0"	White Seed Snap Bush Bean
Phaseolus	vulgaris-bruscia-snapa-alba	blue-lake-bruscia-1993	Envy Blue Lake #1993 ?.?"	White Seed Snap Bush Bean
Phaseolus	vulgaris-bruscia-snapa-alba	blue-lake-bruscia-1994	Select Blue Lake #1994 ?.?"	White Seed Snap Bush Bean
Phaseolus	vulgaris-bruscia-snapa-alba	blue-lake-bruscia-1999	Minuette Blue Lake #1999 ?.?"	White Seed Snap Bush Bean
Phaseolus	vulgaris-bruscia-snapa-alba	blue-lake-bruscia-2117	Improved Blue Lake #2117 6.5"	White Seed Snap Bush Bean
Phaseolus	vulgaris-bruscia-snapa-alba	blue-lake-bruscia-4156	Earliserve Blue Lake #4156 4.0"	White Seed Snap Bush Bean
Phaseolus	vulgaris-bruscia-snapa-alba	blue-lake-bruscia-4848	Rapid Green Blue Lake #4848 ?.?"	White Seed Snap Bush Bean
Phaseolus	vulgaris-bruscia-snapa-alba	blue-lake-bruscia-4855	Venture Blue Lake #4855 6.5"	White Seed Snap Bush Bean

Genus	Species	infraspecies	cultivar	specimen	Genome
Phaseolus	*vulgaris-bruscia-snapa-alba*	*blue-lake-bruscia-5153*	Tenderlake Blue Lake #5153 6.0"	White Seed Snap Bush	Bean
Phaseolus	*vulgaris-bruscia-snapa-alba*	*blue-lake-bruscia-5454*	True Blue Blue Lake #5454 5.5"	White Seed Snap Bush	Bean
Phaseolus	*vulgaris-bruscia-snapa-alba*	*blue-lake-bruscia-5555E*	Endurance Blue Lake #5555-E 6.5"	White Seed Snap Bush	Bean
Phaseolus	*vulgaris-bruscia-snapa-alba*	*blue-lake-bruscia-5555M*	Mirada Blue Lake #5555-M 6.2"	White Seed Snap Bush	Bean
Phaseolus	*vulgaris-bruscia-snapa-alba*	*blue-lake-bruscia-5555O*	Oregon Trail Blue Lake #5555-O ?.?"	White Seed Snap Bush	Bean
Phaseolus	*vulgaris-bruscia-snapa-alba*	*blue-lake-bruscia-5556*	Tenderette Blue Lake #5556 6.0"	White Seed Snap Bush	Bean
Phaseolus	*vulgaris-bruscia-snapa-alba*	*blue-lake-bruscia-5560*	Blue Lagoon Blue Lake #5560 5.5"	White Seed Snap Bush	Bean
Phaseolus	*vulgaris-bruscia-snapa-alba*	*blue-lake-bruscia-5575*	E-Z Pick Blue Lake #5575 6.2"	White Seed Snap Bush	Bean
Phaseolus	*vulgaris-bruscia-snapa-alba*	*blue-lake-bruscia-5656*	Festina Blue Lake #5656 6.5"	White Seed Snap Bush	Bean
Phaseolus	*vulgaris-bruscia-snapa-alba*	*blue-lake-bruscia-5858*	Largo Blue Lake #5858 6.5"	White Seed Snap Bush	Bean
Phaseolus	*vulgaris-bruscia-snapa-alba*	*broncoi*	Bronco 5.5"	White Seed Snap Bush	Bean
Phaseolus	*vulgaris-bruscia-snapa-alba*	*capricornii*	Capricorn 6.0"	White Seed Snap Bush	Bean
Phaseolus	*vulgaris-bruscia-snapa-alba*	*chow-chow-eva's-beansi*	Eva's Chow Chow ?.?"	White Seed Snap Bush	Bean
Phaseolus	*vulgaris-bruscia-snapa-alba*	*chow-chow-grandma-stober's*	Grandma Stober's Chow Chow	White Seed Snap Bush	Bean
Phaseolus	*vulgaris-bruscia-snapa-alba*	*cloudburstii*	Cloudburst 6.0"	White Seed Snap Bush	Bean
Phaseolus	*vulgaris-bruscia-snapa-alba*	*daytonai*	Daytona 6.5"	White Seed Snap Bush	Bean
Phaseolus	*vulgaris-bruscia-snapa-alba*	*distinctionii*	Distinction 6.0"	White Seed Snap Bush	Bean
Phaseolus	*vulgaris-bruscia-snapa-alba*	*dona-bobolinkii*	Dona Bobolink Burgundy	White Seed Snap Bush	Bean
Phaseolus	*vulgaris-bruscia-snapa-alba*	*duskyi*	Dusky 6.0"	White Seed Snap Bush	Bean
Phaseolus	*vulgaris-bruscia-snapa-alba*	*early-gallatinii*	Early Gallatin 5.7"	White Seed Snap Bush	Bean
Phaseolus	*vulgaris-bruscia-snapa-alba*	*fandangoi*	Fandango 5.5"	White Seed Snap Bush	Bean
Phaseolus	*vulgaris-bruscia-snapa-alba*	*firstmarkii*	Firstmark 6.0"	White Seed Snap Bush	Bean
Phaseolus	*vulgaris-bruscia-snapa-alba*	*florencei*	Florence 6.5"	White Seed Snap Bush	Bean
Phaseolus	*vulgaris-bruscia-snapa-alba*	*gator-greenii*	Gator Green 6.5"	White Seed Snap Bush	Bean
Phaseolus	*vulgaris-bruscia-snapa-alba*	*gator-greenii-15*	Thin Oval Gator Green 6.5"	White Seed Snap Bush	Bean
Phaseolus	*vulgaris-bruscia-snapa-alba*	*gator-greenii-shorei*	Eastern Shore Gator Green 5.5"	White Seed Snap Bush	Bean
Phaseolus	*vulgaris-bruscia-snapa-alba*	*gator-greenii-hialeahii*	Hialeah Gator Green 6.0"	White Seed Snap Bush	Bean
Phaseolus	*vulgaris-bruscia-snapa-alba*	*green-islei*	Green Isle 7.0"	White Seed Snap Bush	Bean
Phaseolus	*vulgaris-bruscia-snapa-alba*	*grenoblei*	Grenoble 5.7"	White Seed Snap Bush	Bean
Phaseolus	*vulgaris-bruscia-snapa-alba*	*half-runneri-mississippii*	Mississippi White Seed Half Runner 4.3"	White Seed Snap Bush	Bean
Phaseolus	*vulgaris-bruscia-snapa-alba*	*half-runneri-mountaineeri*	Mountaineer White Seed Half Runner 4.5"	White Seed Snap Bush	Bean
Phaseolus	*vulgaris-bruscia-snapa-alba*	*half-runneri-volunteeri*	Volunteer White Seed Half Runner 5.0"	White Seed Snap Bush	Bean
Phaseolus	*vulgaris-bruscia-snapa-alba*	*haricot-nain-daisyi*	Daisy Haricot Nain 6.0"	White Seed Snap Bush	Bean
Phaseolus	*vulgaris-bruscia-snapa-alba*	*hull-white-&-greenil*	White & Green Hull ?.?"	White Seed Snap Bush	Bean
Phaseolus	*vulgaris-bruscia-snapa-alba*	*hull-white-bunch-beanii*	Bunch White Hull ?.?"	White Seed Snap Bush	Bean
Phaseolus	*vulgaris-bruscia-snapa-alba*	*hurricanei*	Hurricane 5.5"	White Seed Snap Bush	Bean
Phaseolus	*vulgaris-bruscia-snapa-alba*	*javelinii*	Javelin 6.0"	White Seed Snap Bush	Bean
Phaseolus	*vulgaris-bruscia-snapa-alba*	*landmarkii*	Landmark 7.0"	White Seed Snap Bush	Bean
Phaseolus	*vulgaris-bruscia-snapa-alba*	*landstari*	Landstar 6.7"	White Seed Snap Bush	Bean
Phaseolus	*vulgaris-bruscia-snapa-alba*	*lizzie-milleri*	Lizzie Miller Tan Pattern	White Seed Snap Bush	Bean
Phaseolus	*vulgaris-bruscia-snapa-alba*	*mammyi*	Mammy 4.0"	White Seed Snap Bush	Bean
Phaseolus	*vulgaris-bruscia-snapa-alba*	*matadori*	Matador 6.2"	White Seed Snap Bush	Bean
Phaseolus	*vulgaris-bruscia-snapa-alba*	*mercuryi*	Mercury ?.?"	White Seed Snap Bush	Bean
Phaseolus	*vulgaris-bruscia-snapa-alba*	*mrs.-neidigh's-six-week-beanii*	Mrs. Neidigh's Six Week 4.5"	White Seed Snap Bush	Bean
Phaseolus	*vulgaris-bruscia-snapa-alba*	*opus-XPB-223*	Opus 5.5"	White Seed Snap Bush	Bean
Phaseolus	*vulgaris-bruscia-snapa-alba*	*ozark-bunchii*	Ozark Bunch Cream Seed	White Seed Snap Bush	Bean
Phaseolus	*vulgaris-bruscia-snapa-alba*	*processori*	Processor ?.?"	White Seed Snap Bush	Bean
Phaseolus	*vulgaris-bruscia-snapa-alba*	*prosperityi*	Prosperity ?.?"	White Seed Snap Bush	Bean
Phaseolus	*vulgaris-bruscia-snapa-alba*	*provider-white-seedii*	Provider 6.0"	White Seed Snap Bush	Bean
Phaseolus	*vulgaris-bruscia-snapa-alba*	*rapidsii*	Rapids 5.7"	White Seed Snap Bush	Bean
Phaseolus	*vulgaris-bruscia-snapa-alba*	*rhapsodyi*	Rhapsody 5.5"	White Seed Snap Bush	Bean
Phaseolus	*vulgaris-bruscia-snapa-alba*	*RS-1384*	RS #1384 5.2"	White Seed Snap Bush	Bean
Phaseolus	*vulgaris-bruscia-snapa-alba*	*RS-5402*	RS #5402 5.8"	White Seed Snap Bush	Bean
Phaseolus	*vulgaris-bruscia-snapa-alba*	*sevillensis*	Seville 6.0"	White Seed Snap Bush	Bean
Phaseolus	*vulgaris-bruscia-snapa-alba*	*taverai*	Tavera 4.5"	White Seed Snap Bush	Bean
Phaseolus	*vulgaris-bruscia-snapa-alba*	*tenderpick-burpee'si*	Burpee's Tenderpick 5.5"	White Seed Snap Bush	Bean
Phaseolus	*vulgaris-bruscia-snapa-alba*	*thibodeau-du-comte-beaucei*	Thibodeau Du Compte Beauce Maroon Spotted	White Seed Snap Bush	Bean
Phaseolus	*vulgaris-bruscia-snapa-alba*	*zanei*	Zane ?.?"	White Seed Snap Bush	Bean
Phaseolus	*vulgaris-bruscia-snapa-alba*	*zodiaci*	Zodiac ?.?"	White Seed Snap Bush	Bean
Phaseolus	*vulgaris-bruscia-snapa-astrenga*	*baffinii*	Baffin 4.0"	Stringless Round Snap Bush	Bean
Phaseolus	*vulgaris-bruscia-snapa-astrenga*	*kentucky-wonder-125*	#125 Kentucky Wonder 6.5"	Stringless Round Snap Bush	Bean
Phaseolus	*vulgaris-bruscia-snapa-astrenga*	*kentucky-wonder-commodorei*	Commodore Kentucky Wonder 8.0"	Stringless Round Snap Bush	Bean
Phaseolus	*vulgaris-bruscia-snapa-astrenga*	*kentucky-wonder-commodorei-improvedii*	Improved Commodore Kentucky Wonder 7.0"	Stringless Round Snap Bush	Bean
Phaseolus	*vulgaris-bruscia-snapa-astrenga*	*kentucky-wonder-dreameri*	Dreamer Kentucky Wonder 6.5"	Stringless Round Snap Bush	Bean
Phaseolus	*vulgaris-bruscia-snapa-astrenga*	*kentucky-wonder-greencropii*	Greencrop Kentucky Wonder 7.5"	Stringless Round Snap Bush	Bean
Phaseolus	*vulgaris-bruscia-snapa-astrenga*	*kentucky-wonder-magnum-XPB-211*	Magnum Kentucky Wonder 7.0"	Stringless Round Snap Bush	Bean

Genus	Species	infraspecies	cultivar	specimen	Genome
Phaseolus	vulgaris-bruscia-snapa-astrenga	contenderi	Contender 7.0"	Stringless Round Snap Bush	Bean
Phaseolus	vulgaris-bruscia-snapa-astrenga	cordobaensis	Cordoba 6.5"	Stringless Round Snap Bush	Bean
Phaseolus	vulgaris-bruscia-snapa-astrenga	dutch-princessi	Dutch Princess 4.0"	Stringless Round Snap Bush	Bean
Phaseolus	vulgaris-bruscia-snapa-astrenga	dutch-princessi-montanoi	Montano Dutch Princess 4.0"	Stringless Round Snap Bush	Bean
Phaseolus	vulgaris-bruscia-snapa-astrenga	dutch-princessi-recordii	Record Dutch Princess 4.0"	Stringless Round Snap Bush	Bean
Phaseolus	vulgaris-bruscia-snapa-astrenga	early-bush-italianii	Early Italian 5.5"	Stringless Round Snap Bush	Bean
Phaseolus	vulgaris-bruscia-snapa-astrenga	empressi	Empress White Seed 5.8"	Stringless Round Snap Bush	Bean
Phaseolus	vulgaris-bruscia-snapa-astrenga	fowleri	Fowler Thin 5.0"	Stringless Round Snap Bush	Bean
Phaseolus	vulgaris-bruscia-snapa-astrenga	giant-stringlessi-12	Giant 10" to 12"	Stringless Round Snap Bush	Bean
Phaseolus	vulgaris-bruscia-snapa-astrenga	giant-stringlessi-burpee's-green-podii	Burpee's Giant Green Pod 5.5"	Stringless Round Snap Bush	Bean
Phaseolus	vulgaris-bruscia-snapa-astrenga	giant-stringlessi-green-podii	Giant Green Pod 6.5"	Stringless Round Snap Bush	Bean
Phaseolus	vulgaris-bruscia-snapa-astrenga	gitanai(pros-gitanai)	Pros Gitana 4.0"	Stringless Round Snap Bush	Bean
Phaseolus	vulgaris-bruscia-snapa-astrenga	green-beanii	Green Bean 8.0"	Stringless Round Snap Bush	Bean
Phaseolus	vulgaris-bruscia-snapa-astrenga	half-runneri-old-dutchii	Old Dutch Stringless Half Runner 5.0"	Stringless Round Snap Bush	Bean
Phaseolus	vulgaris-bruscia-snapa-astrenga	half-runneri-parkeri	Parker Stringless Half Runner 5.0"	Stringless Round Snap Bush	Bean
Phaseolus	vulgaris-bruscia-snapa-astrenga	half-runneri-pink-eyei	Pink Eye Half Runner 3.2"	Stringless Round Snap Bush	Bean
Phaseolus	vulgaris-bruscia-snapa-astrenga	half-runneri-pinkii	Pink Peanut Half Runner 4.5"	Stringless Round Snap Bush	Bean
Phaseolus	vulgaris-bruscia-snapa-astrenga	half-runneri-spartanii	Spartan Tan Striped Seed Half Runner 4.8"	Stringless Round Snap Bush	Bean
Phaseolus	vulgaris-bruscia-snapa-astrenga	half-runneri-statei	State White Seed Stringless Flat Half Runner 4.0"	Stringless Round Snap Bush	Bean
Phaseolus	vulgaris-bruscia-snapa-astrenga	harvesteri	Harvester 5.5"	Stringless Round Snap Bush	Bean
Phaseolus	vulgaris-bruscia-snapa-astrenga	hawksburyensis	Hawkesbury Wonder	Stringless Round Snap Bush	Bean
Phaseolus	vulgaris-bruscia-snapa-astrenga	loch-nessensis	Loch Ness 6.0"	Stringless Round Snap Bush	Bean
Phaseolus	vulgaris-bruscia-snapa-astrenga	montana-greenii	Montana Green Buff Seed 6.5"	Stringless Round Snap Bush	Bean
Phaseolus	vulgaris-bruscia-snapa-astrenga	nomadii	Nomad 5.2"	Stringless Round Snap Bush	Bean
Phaseolus	vulgaris-bruscia-snapa-astrenga	operai	Opera 4.5"	Stringless Round Snap Bush	Bean
Phaseolus	vulgaris-bruscia-snapa-astrenga	preludei	Prelude ?.?"	Stringless Round Snap Bush	Bean
Phaseolus	vulgaris-bruscia-snapa-astrenga	royal-burgundyi	Royal Burgundy Pod 5.5"	Stringless Round Snap Bush	Bean
Phaseolus	vulgaris-bruscia-snapa-astrenga	royalty-purple-podii	Royalty Purple Pod 5.2"	Stringless Round Snap Bush	Bean
Phaseolus	vulgaris-bruscia-snapa-astrenga	spartan-arrowi	Spartan Arrow Tan Seed 6.0"	Stringless Round Snap Bush	Bean
Phaseolus	vulgaris-bruscia-snapa-astrenga	spritei	Sprite 5.2"	Stringless Round Snap Bush	Bean
Phaseolus	vulgaris-bruscia-snapa-astrenga	stringless-green-podii-4545	Giant Green Pod #4545 6.5"	Stringless Round Snap Bush	Bean
Phaseolus	vulgaris-bruscia-snapa-astrenga	stringless-green-podii-5054	Green Pod #5054 5.5"	Stringless Round Snap Bush	Bean
Phaseolus	vulgaris-bruscia-snapa-astrenga	stringless-green-podii-5054B	Burpee's Green Pod #5054-B 6.5"	Stringless Round Snap Bush	Bean
Phaseolus	vulgaris-bruscia-snapa-astrenga	stringless-green-podii-5253	Improved Green Pod #5253 6.0"	Stringless Round Snap Bush	Bean
Phaseolus	vulgaris-bruscia-snapa-astrenga	stringless-green-podii-5255	Landreth's Green Pod #5255 5.5"	Stringless Round Snap Bush	Bean
Phaseolus	vulgaris-bruscia-snapa-astrenga	sunrayi	Sunray Brown Seed 6.2"	Stringless Round Snap Bush	Bean
Phaseolus	vulgaris-bruscia-snapa-astrenga	tendergreenii	Tendergreen Purple Seed 6.2"	Stringless Round Snap Bush	Bean
Phaseolus	vulgaris-bruscia-snapa-astrenga	tendergreenii-improvedii	Improved Tendergreen Purple Seed 6.2"	Stringless Round Snap Bush	Bean
Phaseolus	vulgaris-bruscia-snapa-astrenga	tenderpodii	Tenderpod 5.0"	Stringless Round Snap Bush	Bean
Phaseolus	vulgaris-bruscia-snapa-astrenga	topcropii	Topcrop Brown Seed 6.5"	Stringless Round Snap Bush	Bean
Phaseolus	vulgaris-bruscia-snapa-astrenga	topcropii-joyi	Joy Topcrop Brown Seed 5.5"	Stringless Round Snap Bush	Bean
Phaseolus	vulgaris-bruscia-snapa-astrenga	yer-fasulyasi	Yer Fasulasi Brown Seed	Stringless Round Snap Bush	Bean
Phaseolus	vulgaris-bruscia-snapa-brunnea	broncoi-mustangii-XPB-202	Mustang Bronco Brown Seed 5.5"	Brown Seed Snap Bush	Bean
Phaseolus	vulgaris-bruscia-snapa-brunnea	couch'si	Couch's Brown Seed	Brown Seed Snap Bush	Bean
Phaseolus	vulgaris-bruscia-snapa-brunnea	gill's-relici	Gill's Relic Brown Seed 6.0"	Brown Seed Snap Bush	Bean
Phaseolus	vulgaris-bruscia-snapa-brunnea	jo-beanii	Jo Tan & Brown Seed 3.0"	Brown Seed Snap Bush	Bean
Phaseolus	vulgaris-bruscia-snapa-brunnea	regal-saladii	Regal Salad Bronze Pod	Brown Seed Snap Bush	Bean
Phaseolus	vulgaris-bruscia-snapa-brunnea	rushmorei	Rushmore Brown Seed 5.5"	Brown Seed Snap Bush	Bean
Phaseolus	vulgaris-bruscia-snapa-brunnea	saxai	Saxa Tan Seed 5.5"	Brown Seed Snap Bush	Bean
Phaseolus	vulgaris-bruscia-snapa-brunnea	seminolei	Seminole Tan Seed	Brown Seed Snap Bush	Bean
Phaseolus	vulgaris-bruscia-snapa-brunnea	stormii	Storm Brown Seed 5.5"	Brown Seed Snap Bush	Bean
Phaseolus	vulgaris-bruscia-snapa-brunnea	tanya's-pink-podii	Tanya's Pink Pod Tan Seed	Brown Seed Snap Bush	Bean
Phaseolus	vulgaris-bruscia-snapa-brunnea	temai-XPB-201	Tema Brown Seed 5.0"	Brown Seed Snap Bush	Bean
Phaseolus	vulgaris-bruscia-snapa-brunnea	vanderpool-six-weekii	Vanderpool Six Week Brown Seed	Brown Seed Snap Bush	Bean
Phaseolus	vulgaris-bruscia-snapa-filetii	aramisi	Aramis Stringless 5.5"	Filet Snap Bush	Bean
Phaseolus	vulgaris-bruscia-snapa-filetii	astrelii	Astrel Stringless 4.0"	Filet Snap Bush	Bean
Phaseolus	vulgaris-bruscia-snapa-filetii	bahaloresii	Bahalores Stringless 4.5"	Filet Snap Bush	Bean
Phaseolus	vulgaris-bruscia-snapa-filetii	canadian-filetii	Canadian ?.?"	Filet Snap Bush	Bean
Phaseolus	vulgaris-bruscia-snapa-filetii	carusoi	Caruso 8.0"	Filet Snap Bush	Bean
Phaseolus	vulgaris-bruscia-snapa-filetii	decibelii	Decibel Stringless 5.0"	Filet Snap Bush	Bean
Phaseolus	vulgaris-bruscia-snapa-filetii	fin-des-bangolsi	Fin Des Bagnols Shoestring	Filet Snap Bush	Bean
Phaseolus	vulgaris-bruscia-snapa-filetii	frenchii	French 6.0"	Filet Snap Bush	Bean
Phaseolus	vulgaris-bruscia-snapa-filetii	haricot-vert-dandyi	Dandy Haricot Vert Stringless 4.0"	Filet Snap Bush	Bean
Phaseolus	vulgaris-bruscia-snapa-filetii	haricot-vert-cupidonii	Dandy Haricot Vert Stringless 7.0"	Filet Snap Bush	Bean
Phaseolus	vulgaris-bruscia-snapa-filetii	haricot-vert-delinelii	Delinel Haricot Vert Stringless 6.7"	Filet Snap Bush	Bean
Phaseolus	vulgaris-bruscia-snapa-filetii	haricot-vert-maxibelii	Maxibel Haricot Vert Stringless 7.5"	Filet Snap Bush	Bean
Phaseolus	vulgaris-bruscia-snapa-filetii	haricot-vert-morganei	Morgane Haricot Vert Stringless 7.2"	Filet Snap Bush	Bean
Phaseolus	vulgaris-bruscia-snapa-filetii	haricot-vert-nickelii	Nickel Haricot Vert Stringless 4.0"	Filet Snap Bush	Bean

Genus	Species	infraspecies	cultivar	specimen	Genome
Phaseolus	vulgaris-bruscia-snapa-filetii	haricot-vert-regalfinii	Regalfin Haricot Vert Stringless ?.?"	String Snap Bush	Bean
Phaseolus	vulgaris-bruscia-snapa-filetii	haricot-vert-rolandei	Rolande Haricot Vert Stringless 6.0"	Filet Snap Bush	Bean
Phaseolus	vulgaris-bruscia-snapa-filetii	haricot-vert-vernandonii	Vernandon Haricot Vert Stringless 8.0"	Filet Snap Bush	Bean
Phaseolus	vulgaris-bruscia-snapa-filetii	la-francei	La France ?.?"	Filet Snap Bush	Bean
Phaseolus	vulgaris-bruscia-snapa-filetii	medinahii	Medinah 5.0"	Filet Snap Bush	Bean
Phaseolus	vulgaris-bruscia-snapa-filetii	morinakai	Morinaka ?.?"	Filet Snap Bush	Bean
Phaseolus	vulgaris-bruscia-snapa-filetii	montpellieri	Montpellier 7.0"	Filet Snap Bush	Bean
Phaseolus	vulgaris-bruscia-snapa-filetii	nerinai	Nerina Mangetout 4.5"	Filet Snap Bush	Bean
Phaseolus	vulgaris-bruscia-snapa-filetii	safari	Safari 5.0"	Filet Snap Bush	Bean
Phaseolus	vulgaris-bruscia-snapa-filetii	straight-n-narrowi	Straight 'N Narrow 5.0"	Filet Snap Bush	Bean
Phaseolus	vulgaris-bruscia-snapa-filetii	totemii	Totem 9.0"	Filet Snap Bush	Bean
Phaseolus	vulgaris-bruscia-snapa-filetii	triomphe-de-farcyi	Triomphe De Farcy 6.5"	Filet Snap Bush	Bean
Phaseolus	vulgaris-bruscia-snapa-filetii	white-settler-filetii	Covelo White Settler	Filet Snap Bush	Bean
Phaseolus	vulgaris-bruscia-snapa-grisea	petit-gris-dieul-fin-precocei	Diuel Fin Precoce Pitit Gris	Gray Seed Snap Bush	Bean
Phaseolus	vulgaris-bruscia-snapa-grisea	petit-gris-hystylei	Hystyle Pitit Gris ?.?"	Gray Seed Snap Bush	Bean
Phaseolus	vulgaris-bruscia-snapa-groena	allurei	Allure ?.?"	Green Round Snap Bush	Bean
Phaseolus	vulgaris-bruscia-snapa-groena	ambrai-HMX-0104	Ambra 6.0"	Green Round Snap Bush	Bean
Phaseolus	vulgaris-bruscia-snapa-groena	amigoi	Amigo ?.?"	Green Round Snap Bush	Bean
Phaseolus	vulgaris-bruscia-snapa-groena	aunt-alleyi	Aunt Alley	Green Round Snap Bush	Bean
Phaseolus	vulgaris-bruscia-snapa-groena	baby-bopii	Baby Bop 4.0"	Green Round Snap Bush	Bean
Phaseolus	vulgaris-bruscia-snapa-groena	baccicia	Baccicia 6.5"	Green Round Snap Bush	Bean
Phaseolus	vulgaris-bruscia-snapa-groena	benchmarkii	Benchmark 6.0"	Green Round Snap Bush	Bean
Phaseolus	vulgaris-bruscia-snapa-groena	bentonii	Benton Tender Crop 6.0"	Green Round Snap Bush	Bean
Phaseolus	vulgaris-bruscia-snapa-groena	blue-bloomii	Blue Bloom 3.0"	Green Round Snap Bush	Bean
Phaseolus	vulgaris-bruscia-snapa-groena	blue-jayi	Blue Jay ?.?"	Green Round Snap Bush	Bean
Phaseolus	vulgaris-bruscia-snapa-groena	blue-wonderi	Blue Wonder 5.5"	Green Round Snap Bush	Bean
Phaseolus	vulgaris-bruscia-snapa-groena	bountyi	Bounty 5.5"	Green Round Snap Bush	Bean
Phaseolus	vulgaris-bruscia-snapa-groena	brokeri	Broker ?.?"	Green Round Snap Bush	Bean
Phaseolus	vulgaris-bruscia-snapa-groena	BX-207	BX #207 5.5"	Green Round Snap Bush	Bean
Phaseolus	vulgaris-bruscia-snapa-groena	canberrai	Canberra 4.0"	Green Round Snap Bush	Bean
Phaseolus	vulgaris-bruscia-snapa-groena	capitolei	Capitole Baby 4.0"	Green Round Snap Bush	Bean
Phaseolus	vulgaris-bruscia-snapa-groena	carr-virginica	Carr Virginia ?.?"	Green Round Snap Bush	Bean
Phaseolus	vulgaris-bruscia-snapa-groena	castanoi	Castano 5.5"	Green Round Snap Bush	Bean
Phaseolus	vulgaris-bruscia-snapa-groena	castelii	Castel Gondolfo ?.?"	Green Round Snap Bush	Bean
Phaseolus	vulgaris-bruscia-snapa-groena	concesai	Concesa 5.7"	Green Round Snap Bush	Bean
Phaseolus	vulgaris-bruscia-snapa-groena	corbett-refugeei	Corbett Refugee ?.?"	Green Round Snap Bush	Bean
Phaseolus	vulgaris-bruscia-snapa-groena	cyrus	Cyrus ?.?"	Green Round Snap Bush	Bean
Phaseolus	vulgaris-bruscia-snapa-groena	dartii	Dart ?.?"	Green Round Snap Bush	Bean
Phaseolus	vulgaris-bruscia-snapa-groena	derbyi-FM-175	Derby 6.5"	Green Round Snap Bush	Bean
Phaseolus	vulgaris-bruscia-snapa-groena	english-long-greenii	English Long Green	Green Round Snap Bush	Bean
Phaseolus	vulgaris-bruscia-snapa-groena	espadai	Espada 6.0"	Green Round Snap Bush	Bean
Phaseolus	vulgaris-bruscia-snapa-groena	flevoroi	Flevoro ?.?"	Green Round Snap Bush	Bean
Phaseolus	vulgaris-bruscia-snapa-groena	fordhook-standardii	Fordhook Standard	Green Round Snap Bush	Bean
Phaseolus	vulgaris-bruscia-snapa-groena	frazier's-choicei	Frazier's Choice 5.5"	Green Round Snap Bush	Bean
Phaseolus	vulgaris-bruscia-snapa-groena	full-measurei	Full Measure ?.?"	Green Round Snap Bush	Bean
Phaseolus	vulgaris-bruscia-snapa-groena	granny-easley'si	Granny Easley's ?.?"	Green Round Snap Bush	Bean
Phaseolus	vulgaris-bruscia-snapa-groena	greensleevesi	Greensleeves ?.?"	Green Round Snap Bush	Bean
Phaseolus	vulgaris-bruscia-snapa-groena	heavyweight-2	Heavyweight II 8.0"	Green Round Snap Bush	Bean
Phaseolus	vulgaris-bruscia-snapa-groena	HMX-5991	HMX #5991 ?.?"	Green Round Snap Bush	Bean
Phaseolus	vulgaris-bruscia-snapa-groena	homestylei	Homestyle ?.?"	Green Round Snap Bush	Bean
Phaseolus	vulgaris-bruscia-snapa-groena	idaho-refugeei	Idaho Refugee ?.?"	Green Round Snap Bush	Bean
Phaseolus	vulgaris-bruscia-snapa-groena	jadei-GB45-131	Jade 6.0"	Green Round Snap Bush	Bean
Phaseolus	vulgaris-bruscia-snapa-groena	leonii	Leon 5.5"	Green Round Snap Bush	Bean
Phaseolus	vulgaris-bruscia-snapa-groena	lodii	Lodi ?.?"	Green Round Snap Bush	Bean
Phaseolus	vulgaris-bruscia-snapa-groena	macedonia-snapii	Macedonia ?.?"	Green Round Snap Bush	Bean
Phaseolus	vulgaris-bruscia-snapa-groena	magpiei	Magpie 7.0"	Green Round Snap Bush	Bean
Phaseolus	vulgaris-bruscia-snapa-groena	marsiellesensis	Marseilles 4.5"	Green Round Snap Bush	Bean
Phaseolus	vulgaris-bruscia-snapa-groena	modus	Modus 6.0"	Green Round Snap Bush	Bean
Phaseolus	vulgaris-bruscia-snapa-groena	mon-petit-cherii	Mon Petit Cheri 4.0"	Green Round Snap Bush	Bean
Phaseolus	vulgaris-bruscia-snapa-groena	normai	Norma ?.?"	Green Round Snap Bush	Bean
Phaseolus	vulgaris-bruscia-snapa-groena	normandiei	Normandie 4.5"	Green Round Snap Bush	Bean
Phaseolus	vulgaris-bruscia-snapa-groena	parsonii	Parson ?.?"	Green Round Snap Bush	Bean
Phaseolus	vulgaris-bruscia-snapa-groena	parisian-stringii	Parisian String ?.?"	Green Round Snap Bush	Bean
Phaseolus	vulgaris-bruscia-snapa-groena	radarii	Radar French String	Green Round Snap Bush	Bean
Phaseolus	vulgaris-bruscia-snapa-groena	remus	Remus 9.0"	Green Round Snap Bush	Bean
Phaseolus	vulgaris-bruscia-snapa-groena	royalnelii	Royalnel ?.?"	Green Round Snap Bush	Bean
Phaseolus	vulgaris-bruscia-snapa-groena	sablei	Sable 5.0"	Green Round Snap Bush	Bean
Phaseolus	vulgaris-bruscia-snapa-groena	saratogai	Saratoga 5.5"	Green Round Snap Bush	Bean
Phaseolus	vulgaris-bruscia-snapa-groena	savannahensis	Savannah 5.7"	Green Round Snap Bush	Bean
Phaseolus	vulgaris-bruscia-snapa-groena	SB-4223	SB #4223 6.0"	Green Round Snap Bush	Bean
Phaseolus	vulgaris-bruscia-snapa-groena	SB-4243	SB #4243 5.2"	Green Round Snap Bush	Bean
Phaseolus	vulgaris-bruscia-snapa-groena	sentryi	Sentry 5.5"	Green Round Snap Bush	Bean
Phaseolus	vulgaris-bruscia-snapa-groena	shamrockii	Shamrock 6.0"	Green Round Snap Bush	Bean
Phaseolus	vulgaris-bruscia-snapa-groena	sonata-FMX-602	Sonata ?.?"	Green Round Snap Bush	Bean

Genus	Species	infraspecies	cultivar	specimen	Genome
Phaseolus	*vulgaris-bruscia-snapa-groena*	*speculatori*	Speculator ?.?"	Green Round Snap Bush	Bean
Phaseolus	*vulgaris-bruscia-snapa-groena*	*st.-andreas-mit-fadenii*	Saint Andreas Mit Faden	Green Round Snap Bush	Bean
Phaseolus	*vulgaris-bruscia-snapa-groena*	*stoltzfus-stringii*	Stoltzfus String ?.?"	Green Round Snap Bush	Bean
Phaseolus	*vulgaris-bruscia-snapa-groena*	*tasmanii*	Tasman ?.?"	Green Round Snap Bush	Bean
Phaseolus	*vulgaris-bruscia-snapa-groena*	*teepee-micheletii*	Michelet Teepee ?.?"	Green Round Snap Bush	Bean
Phaseolus	*vulgaris-bruscia-snapa-groena*	*teseoi*	Teseo 5.5"	Green Round Snap Bush	Bean
Phaseolus	*vulgaris-bruscia-snapa-groena*	*thousand-to-onei*	Thousand To One 5.0"	Green Round Snap Bush	Bean
Phaseolus	*vulgaris-bruscia-snapa-groena*	*topper-burpee'si*	Burpee's Topper 6.0"	Green Round Snap Bush	Bean
Phaseolus	*vulgaris-bruscia-snapa-groena*	*venitoi*	Venito ?.?"	Green Round Snap Bush	Bean
Phaseolus	*vulgaris-bruscia-snapa-groena*	*wrangleri*	Wrangler ?.?"	Green Round Snap Bush	Bean
Phaseolus	*vulgaris-bruscia-snapa-groena*	*xerai*	Xera 5.7"	Green Round Snap Bush	Bean
Phaseolus	*vulgaris-bruscia-snapa-groena*	*yugoslavian-blushii*	Yugoslavian Blush Pink Seed	Green Round Snap Bush	Bean
Phaseolus	*vulgaris-bruscia-snapa-helva*	*comradesi*	Comrades ?.?"	Yellow Pod Snap Bush	Bean
Phaseolus	*vulgaris-bruscia-snapa-helva*	*constanzai*	Constanza White Seed	Yellow Pod Snap Bush	Bean
Phaseolus	*vulgaris-bruscia-snapa-helva*	*goodwin's-bush-butterbeanii*	Goodwin's Butterbean	Yellow Pod Snap Bush	Bean
Phaseolus	*vulgaris-bruscia-snapa-helva*	*soleilii*	Soleil 4.5"	Yellow Pod Snap Bush	Bean
Phaseolus	*vulgaris-bruscia-snapa-helva*	*sulphuri(china-yellowi)*	China Brimestone Sulfer	Yellow Pod Snap Bush	Bean
Phaseolus	*vulgaris-bruscia-snapa-hortus*	.	.	Shell Snap Bush	Bean
Phaseolus	*vulgaris-bruscia-snapa-hortus*	*aunt-lara'si*	Aunt Lara's ?.?"	Shell Snap Bush	Bean
Phaseolus	*vulgaris-bruscia-snapa-hortus*	*chevrier-vertii*	Chevrier Vert ?.?"	Shell Snap Bush	Bean
Phaseolus	*vulgaris-bruscia-snapa-hortus*	*coco-nain-blanc-precocei*	Coco Nain Blanc Precoce	Shell Snap Bush	Bean
Phaseolus	*vulgaris-bruscia-snapa-hortus*	*davis-hortus*	Davis White Seed	Shell Snap Bush	Bean
Phaseolus	*vulgaris-bruscia-snapa-hortus*	*flageolet-flaroi*	Flaro Flageolet ?.?"	Shell Snap Bush	Bean
Phaseolus	*vulgaris-bruscia-snapa-hortus*	*flageolet-frenchii*	French Flageolet ?.?"	Shell Snap Bush	Bean
Phaseolus	*vulgaris-bruscia-snapa-hortus*	*flageolet-rougei*	Rouge Flageolet ?.?"	Shell Snap Bush	Bean
Phaseolus	*vulgaris-bruscia-snapa-hortus*	*flageolet-vernelii*	Vernel Flageolet ?.?"	Shell Snap Bush	Bean
Phaseolus	*vulgaris-bruscia-snapa-hortus*	*hortoi-americana*	Tierra Del Fuego Horto 6.0"	Shell Snap Bush	Bean
Phaseolus	*vulgaris-bruscia-snapa-hortus*	*hortoi-italica*	Tongue Of Fire Horto 6.0"	Shell Snap Bush	Bean
Phaseolus	*vulgaris-bruscia-snapa-hortus*	*potato-patchii*	Potato Patch White Seed 3.0"	Shell Snap Bush	Bean
Phaseolus	*vulgaris-bruscia-snapa-nigra*	*mrs.-fearn'si*	Mrs. Fearn's ?.?"	Black Seed Snap Bush	Bean
Phaseolus	*vulgaris-bruscia-snapa-platya*	*canadian-wonderi*	Canadian Wonder 7.5"	Flat Snap Bush	Bean
Phaseolus	*vulgaris-bruscia-snapa-platya*	*charlie-murphyi*	Charlie Murphy ?.?"	Flat Snap Bush	Bean
Phaseolus	*vulgaris-bruscia-snapa-platya*	*kebarikai*	Kebarika 6.0"	Flat Snap Bush	Bean
Phaseolus	*vulgaris-bruscia-snapa-platya*	*masterpiecei*	Masterpiece 9.0"	Flat Snap Bush	Bean
Phaseolus	*vulgaris-bruscia-snapa-platya*	*oregon-giant-bushii*	Oregon Giant 5.5"	Flat Snap Bush	Bean
Phaseolus	*vulgaris-bruscia-snapa-platya*	*the-princei*	The Prince ?.?"	Flat Snap Bush	Bean
Phaseolus	*vulgaris-bruscia-snapa-platya-astrenga*	*admiresa*	Admires ?.?"	Stringless Flat Snap Bush	Bean
Phaseolus	*vulgaris-bruscia-snapa-platya-astrenga*	*atlantaensis*	Atlanta ?.?"	Stringless Flat Snap Bush	Bean
Phaseolus	*vulgaris-bruscia-snapa-platya-astrenga*	*atlasi*	Atlas 6.0"	Stringless Flat Snap Bush	Bean
Phaseolus	*vulgaris-bruscia-snapa-platya-astrenga*	*bountifulii*	Bountiful Tan Seed 6.5"	Stringless Flat Snap Bush	Bean
Phaseolus	*vulgaris-bruscia-snapa-platya-astrenga*	*marconi-nano-blackii*	Black Marconi Nano	Stringless Flat Snap Bush	Bean
Phaseolus	*vulgaris-bruscia-snapa-platya-astrenga*	*marconi-nano-greenii*	Green Marconi Nano	Stringless Flat Snap Bush	Bean
Phaseolus	*vulgaris-bruscia-snapa-platya-astrenga*	*marconi-nano-whiteii*	White Marconi Nano	Stringless Flat Snap Bush	Bean
Phaseolus	*vulgaris-bruscia-snapa-platya-astrenga*	*romano-bushii*	Romano 5.5"	Stringless Flat Snap Bush	Bean
Phaseolus	*vulgaris-bruscia-snapa-platya-astrenga*	*romano-014*	#014 Romano 6.0"	Stringless Flat Snap Bush	Bean
Phaseolus	*vulgaris-bruscia-snapa-platya-astrenga*	*romano-026*	#026 Romano 6.5"	Stringless Flat Snap Bush	Bean
Phaseolus	*vulgaris-bruscia-snapa-platya-astrenga*	*romano-942*	#942 Romano 6.2"	Stringless Flat Snap Bush	Bean
Phaseolus	*vulgaris-bruscia-snapa-platya-astrenga*	*romano-baromai*	Baroma Bush Romano 6.0"	Stringless Flat Snap Bush	Bean
Phaseolus	*vulgaris-bruscia-snapa-platya-astrenga*	*romano-gaiai*	Gaia Brown Seed Romano 5.5"	Stringless Flat Snap Bush	Bean
Phaseolus	*vulgaris-bruscia-snapa-platya-astrenga*	*romano-ginai*	Gina Romano 5.5"	Stringless Flat Snap Bush	Bean
Phaseolus	*vulgaris-bruscia-snapa-platya-astrenga*	*romano-goldii*	Gold Romano 4.7"	Stringless Flat Snap Bush	Bean
Phaseolus	*vulgaris-bruscia-snapa-platya-astrenga*	*romano-green-ruleri*	Green Ruler Romano	Stringless Flat Snap Bush	Bean
Phaseolus	*vulgaris-bruscia-snapa-platya-astrenga*	*romano-improvedii*	Improved Romano 8.0"	Stringless Flat Snap Bush	Bean
Phaseolus	*vulgaris-bruscia-snapa-platya-astrenga*	*romano-jumboi*	Jumbo Romano 8.0"	Stringless Flat Snap Bush	Bean
Phaseolus	*vulgaris-bruscia-snapa-platya-astrenga*	*romano-milanoi*	Milano Romano 6.0"	Stringless Flat Snap Bush	Bean
Phaseolus	*vulgaris-bruscia-snapa-platya-astrenga*	*romano-primoi-FM-168*	Primo Romano ?.?"	Stringless Flat Snap Bush	Bean
Phaseolus	*vulgaris-bruscia-snapa-platya-astrenga*	*romano-purpiatii*	Purpiat Purple Pod Romano 5.0"	Stringless Flat Snap Bush	Bean
Phaseolus	*vulgaris-bruscia-snapa-platya-astrenga*	*romano-romai*	Roma Romano 5.0"	Stringless Flat Snap Bush	Bean
Phaseolus	*vulgaris-bruscia-snapa-platya-astrenga*	*romano-romai-2*	Roma II Romano 4.7"	Stringless Flat Snap Bush	Bean
Phaseolus	*vulgaris-bruscia-snapa-platya-astrenga*	*romano-romanettei*	Romanette Romano 5.5"	Stringless Flat Snap Bush	Bean
Phaseolus	*vulgaris-bruscia-snapa-platya-astrenga*	*romano-mrs.-marottii*	Mrs. Marotti's Romano	Stringless Flat Snap Bush	Bean
Phaseolus	*vulgaris-bruscia-snapa-platya-astrenga*	*romano-sequoiai*	Sequoia Romano 5.0"	Stringless Flat Snap Bush	Bean
Phaseolus	*vulgaris-bruscia-snapa-platya-astrenga*	*romano-spezia*	Spezia Romano ?.?"	Stringless Flat Snap Bush	Bean
Phaseolus	*vulgaris-bruscia-snapa-platya-astrenga*	*settler-2-green-podii*	Settler II Green Pod 9.0"	Stringless Flat Snap Bush	Bean
Phaseolus	*vulgaris-bruscia-snapa-platya-astrenga*	*settler-4-purple-podii*	Settler IV Purple Pod 5.0"	Stringless Flat Snap Bush	Bean
Phaseolus	*vulgaris-bruscia-snapa-platya-astrenga*	*tennessee-green-podii*	Tennessee Green Pod 7.0"	Stringless Flat Snap Bush	Bean
Phaseolus	*vulgaris-bruscia-snapa-purpurea*	*prestige-saladii*	Prestige Salad	Purple Pod Snap Bush	Bean
Phaseolus	*vulgaris-bruscia-snapa-purpurea*	*provider-purple-seedii*	Provider Purple Seed 7.0"	Purple Pod Snap Bush	Bean
Phaseolus	*vulgaris-bruscia-snapa-purpurea*	*purple-podii*	Purple Pod 5.0"	Purple Pod Snap Bush	Bean
Phaseolus	*vulgaris-bruscia-snapa-purpurea*	*purple-queenii*	Purple Queen 7.5"	Purple Pod Snap Bush	Bean
Phaseolus	*vulgaris-bruscia-snapa-purpurea*	*purple-teepeei*	Purple Teepee 5.5"	Purple Pod Snap Bush	Bean
Phaseolus	*vulgaris-bruscia-snapa-purpurea*	*royal-dukei*	Royal Duke	Purple Pod Snap Bush	Bean
Phaseolus	*vulgaris-bruscia-snapa-purpurea*	*teepee-purplei*	Purple Teepee 5.5"	Purple Pod Snap Bush	Bean
Phaseolus	*vulgaris-bruscia-snapa-purpurea*	*tweed-wonderi*	Tweed Wonder Purple Seed	Purple Pod Snap Bush	Bean

Genus	Species	infraspecies	cultivar	specimen	Genome
Phaseolus	*vulgaris-bruscia-snapa-stena*	*aiguille-vertii*	Aiguille Vert ?.?"	Thin Round Snap Bush	Bean
Phaseolus	*vulgaris-bruscia-snapa-stena*	*alicantei*	Alicante 5.7"	Thin Round Snap Bush	Bean
Phaseolus	*vulgaris-bruscia-snapa-stena*	*boby-biancoi*	Boby Bianco ?.?"	Thin Round Snap Bush	Bean
Phaseolus	*vulgaris-bruscia-snapa-stena*	*brioi*	Brio 5.0"	Thin Round Snap Bush	Bean
Phaseolus	*vulgaris-bruscia-snapa-stena*	*buckskinii-girlii*	Buckskin Girl	Thin Round Snap Bush	Bean
Phaseolus	*vulgaris-bruscia-snapa-stena*	*eaglei*	Eagle 6.0"	Thin Round Snap Bush	Bean
Phaseolus	*vulgaris-bruscia-snapa-stena*	*finaudii*	Finaud 5.0"	Thin Round Snap Bush	Bean
Phaseolus	*vulgaris-bruscia-snapa-stena*	*frenchyi*	Frenchy Gourmet 3.7"	Thin Round Snap Bush	Bean
Phaseolus	*vulgaris-bruscia-snapa-stena*	*labradorensis*	Labrador 5.5"	Thin Round Snap Bush	Bean
Phaseolus	*vulgaris-bruscia-snapa-stena*	*masaii*	Masai 4.5"	Thin Round Snap Bush	Bean
Phaseolus	*vulgaris-bruscia-snapa-stena*	*rapieri*	Rapier 5.0"	Thin Round Snap Bush	Bean
Phaseolus	*vulgaris-bruscia-snapa-stena*	*stallionii*	Stallion Brown Seed 5.5"	Thin Round Snap Bush	Bean
Phaseolus	*vulgaris-bruscia-snapa-stena*	*strikei*	Strike Stringless 5.5"	Thin Round Snap Bush	Bean
Phaseolus	*vulgaris-bruscia-snapa-stena*	*strikei-carloi*	Carlo Strike Stringless 5.0"	Thin Round Snap Bush	Bean
Phaseolus	*vulgaris-bruscia-snapa-stena*	*tendercropii*	Tendercrop Purple Seed 6.0"	Thin Round Snap Bush	Bean
Phaseolus	*vulgaris-bruscia-snapa-stena*	*valentine-stringless-blackii*	Black Seed Valentine Stringless 6.2"	Thin Round Snap Bush	Bean
Phaseolus	*vulgaris-bruscia-snapa-stena*	*valentine-stringless-redii*	Red Seed Valentine Stringless 5.0"	Thin Round Snap Bush	Bean
Phaseolus	*vulgaris-bruscia-snapa-stena*	*vilbelii*	Vilbel 7.0"	Thin Round Snap Bush	Bean
Phaseolus	*vulgaris-bruscia-snapa-stena*	*wadei*	Wade Brown Seed Stringless 6.2"	Thin Round Snap Bush	Bean
Phaseolus	*vulgaris-bruscia-snapa-stena-alba*	*narbonnei*	Narbonne 5.0"	White Seed Thin Snap Bush	Bean
Phaseolus	*vulgaris-bruscia-snapa-stena-alba*	*podasquada*	Podsquad 6.0"	White Seed Thin Snap Bush	Bean
Phaseolus	*vulgaris-bruscia-snapa-stena-alba*	*shadei*	Shade 5.5"	White Seed Thin Snap Bush	Bean
Phaseolus	*vulgaris-bruscia-snapa-stena-alba*	*slendrettei*	Slenderette Stringless 5.5"	White Seed Thin Snap Bush	Bean

Bean, Wax Bush (*Phaseolus vulgaris-bruscia-vaxa*)

Phaseolus	*vulgaris-bruscia-vaxa-alba*	*eurekensis*	Eureka Creamy Pod 6.0"	White Seed Wax Bush	Bean
Phaseolus	*vulgaris-bruscia-vaxa-alba*	*icei*	Ice Green Pod ?.?"	White Seed Wax Bush	Bean
Phaseolus	*vulgaris-bruscia-vaxa-alba*	*unidori*	Unidor ?.?"	White Seed Wax Bush	Bean
Phaseolus	*vulgaris-bruscia-vaxa-groena*	.	.	Green Pod Wax Bush	Bean
Phaseolus	*vulgaris-bruscia-vaxa-groena*	*fortin's-familyi*	Fontin's Family ?.?"	Green Pod Wax Bush	Bean
Phaseolus	*vulgaris-bruscia-vaxa-groena*	*klondykei*	Klondyke 6.0"	Green Pod Wax Bush	Bean
Phaseolus	*vulgaris-bruscia-vaxa-groena*	*la-victoirei*	La Victoire ?.?"	Green Pod Wax Bush	Bean
Phaseolus	*vulgaris-bruscia-vaxa-groena*	*lois-monihan's-candy-waxii*	Lois Monihan's Candy	Green Pod Wax Bush	Bean
Phaseolus	*vulgaris-bruscia-vaxa-helva*	.	.	Yellow Pod Wax Bush	Bean
Phaseolus	*vulgaris-bruscia-vaxa-helva*	*beste-von-allenii*	Beste Von Allen White Seed	Yellow Pod Wax Bush	Bean
Phaseolus	*vulgaris-bruscia-vaxa-helva*	*buerre-de-rocquencourtii*	Buerre De Rocquencourt Black Seed 5.5"	Yellow Pod Wax Bush	Bean
Phaseolus	*vulgaris-bruscia-vaxa-helva*	*brittle-waxii*	Brittle Wax White Seed 6.5"	Yellow Pod Wax Bush	Bean
Phaseolus	*vulgaris-bruscia-vaxa-helva*	*bush-baby-bean-mini-yellowi*	Bush Baby 4.0"	Yellow Pod Wax Bush	Bean
Phaseolus	*vulgaris-bruscia-vaxa-helva*	*buttercrispii*	Buttercrisp 5.5"	Yellow Pod Wax Bush	Bean
Phaseolus	*vulgaris-bruscia-vaxa-helva*	*butterwax-easternii*	New England Butterwax 6.5"	Yellow Pod Wax Bush	Bean
Phaseolus	*vulgaris-bruscia-vaxa-helva*	*butterwax-goldenii*	Golden Butterwax 6.2"	Yellow Pod Wax Bush	Bean
Phaseolus	*vulgaris-bruscia-vaxa-helva*	*century-goldii*	Century Gold White Seed 6.0"	Yellow Pod Wax Bush	Bean
Phaseolus	*vulgaris-bruscia-vaxa-helva*	*dandy-goldii*	Dandy Gold 4.0"	Yellow Pod Wax Bush	Bean
Phaseolus	*vulgaris-bruscia-vaxa-helva*	*ez-goldii*	EZ Gold 5.5"	Yellow Pod Wax Bush	Bean
Phaseolus	*vulgaris-bruscia-vaxa-helva*	*goldcrop-waxii*	Goldcrop White Seed 6.0"	Yellow Pod Wax Bush	Bean
Phaseolus	*vulgaris-bruscia-vaxa-helva*	*golden-rodii*	Golden Rod White Seed 5.5"	Yellow Pod Wax Bush	Bean
Phaseolus	*vulgaris-bruscia-vaxa-helva*	*goldiei*	Goldie 5.5"	Yellow Pod Wax Bush	Bean
Phaseolus	*vulgaris-bruscia-vaxa-helva*	*golditoi*	Goldito 4.7"	Yellow Pod Wax Bush	Bean
Phaseolus	*vulgaris-bruscia-vaxa-helva*	*gold-rushii*	Gold Rush 5.7"	Yellow Pod Wax Bush	Bean
Phaseolus	*vulgaris-bruscia-vaxa-helva*	*gold-rushii-gold-minei*	Gold Mine Gold Rush 5.0"	Yellow Pod Wax Bush	Bean
Phaseolus	*vulgaris-bruscia-vaxa-helva*	*gold-rushii-indy-goldii*	Indy Gold Gold Rush 6.0"	Yellow Pod Wax Bush	Bean
Phaseolus	*vulgaris-bruscia-vaxa-helva*	*hildorai*	Hildora 7.0"	Yellow Pod Wax Bush	Bean
Phaseolus	*vulgaris-bruscia-vaxa-helva*	*kishwaukee-yellowi*	Kishwaukee ?.?"	Yellow Pod Wax Bush	Bean
Phaseolus	*vulgaris-bruscia-vaxa-helva*	*majori*	Major Black Seed 5.7"	Yellow Pod Wax Bush	Bean
Phaseolus	*vulgaris-bruscia-vaxa-helva*	*mont-d'ori*	Golden Butter Mont D'Or 5.7"	Yellow Pod Wax Bush	Bean
Phaseolus	*vulgaris-bruscia-vaxa-helva*	*rapieri*	Rapier 5.0"	Yellow Pod Wax Bush	Bean
Phaseolus	*vulgaris-bruscia-vaxa-helva*	*sunrisei*	Sunrise 6.2"	Yellow Pod Wax Bush	Bean
Phaseolus	*vulgaris-bruscia-vaxa-helva*	*sunraei*	Sunrae 5.7"	Yellow Pod Wax Bush	Bean
Phaseolus	*vulgaris-bruscia-vaxa-helva-albasperma*	*nuggetii*	Nugget 5.5"	White Seed Yellow Pod Wax Bush	Bean
Phaseolus	*vulgaris-bruscia-vaxa-helva-albasperma*	*nuggetii-vald'ori*	Vald'or Nugget 5.5"	White Seed Yellow Pod Wax Bush	Bean
Phaseolus	*vulgaris-bruscia-vaxa-helva-albasperma*	*slenderwaxii*	Slenderwax 5.0"	White Seed Yellow Pod Wax Bush	Bean
Phaseolus	*vulgaris-bruscia-vaxa-helva-albasperma*	*sunburstii*	Sunburst 5.5"	White Seed Yellow Pod Wax Bush	Bean
Phaseolus	*vulgaris-bruscia-vaxa-helva-astrenga*	.	.	Stringless Yellow Pod Wax Bush	Bean
Phaseolus	*vulgaris-bruscia-vaxa-helva-astrenga*	*berggoldii*	Berggold 5.0"	Stringless Yellow Pod Wax Bush	Bean
Phaseolus	*vulgaris-bruscia-vaxa-helva-astrenga*	*dorabelii*	Dorabel Filet 4.7"	Stringless Yellow Pod Wax Bush	Bean
Phaseolus	*vulgaris-bruscia-vaxa-helva-astrenga*	*golden-sinclairii*	Golden Sinclair ?.?"	Stringless Yellow Pod Wax Bush	Bean
Phaseolus	*vulgaris-bruscia-vaxa-helva-astrenga*	*orlinelii*	Orlinel 4.7"	Stringless Yellow Pod Wax Bush	Bean
Phaseolus	*vulgaris-bruscia-vaxa-helva-astrenga*	*pencil-pod-black-waxii*	Pencil Pod Black Seed 6.0"	Stringless Yellow Pod Wax Bush	Bean
Phaseolus	*vulgaris-bruscia-vaxa-helva-astrenga*	*rocquencourt-buerre-dei*	Buerre De Rocquencourt 5.5"	Stringless Yellow Pod Wax Bush	Bean
Phaseolus	*vulgaris-bruscia-vaxa-helva-astrenga*	*rocquencourt-golden-rockyi*	Golden Rocquencourt 6.0"	Stringless Yellow Pod Wax Bush	Bean

Phaseolus	*vulgaris-bruscia-vaxa-helva-astrenga*	*rocquencourt-roc-d'ori*	Roc D'Or Roquencourt 6.2"	Stringless Yellow Pod Wax Bush	Bean
Phaseolus	*vulgaris-bruscia-vaxa-helva-astrenga*	*roma-goldii*	Roma Gold ?.?"	Stringless Yellow Pod Wax Bush	Bean
Phaseolus	*vulgaris-bruscia-vaxa-helva-astrenga*	*sungoldii*	Sungold 5.7"	Stringless Yellow Pod Wax Bush	Bean
Phaseolus	*vulgaris-bruscia-vaxa-helva-astrenga-alba*	*brittle-waxii*	Brittle Wax 6.5"	White Seed Stringless Yellow Pod Wax Bush	Bean
Phaseolus	*vulgaris-bruscia-vaxa-helva-astrenga-alba*	*brittle-waxii-kinghorn-waxii*	Kinghorn Brittle Wax 5.7"	White Seed Stringless Yellow Pod Wax Bush	Bean
Phaseolus	*vulgaris-bruscia-vaxa-helva-astrenga-alba*	*brittle-waxii-round-pod-kidney-waxii*	Round Pod Kidney Brittle Wax 6.0"	White Seed Stringless Yellow Pod Wax Bush	Bean
Phaseolus	*vulgaris-bruscia-vaxa-helva-astrenga-alba*	*earliwaxii*	Earliwax White Seed 5.5"	White Seed Stringless Yellow Pod Wax Bush	Bean
Phaseolus	*vulgaris-bruscia-vaxa-helva-astrenga-alba*	*galagoldii*	Galagold White Seed 5.5"	White Seed Stringless Yellow Pod Wax Bush	Bean
Phaseolus	*vulgaris-bruscia-vaxa-helva-astrenga-alba*	*puregoldii*	Puregold Slim 5.5"	White Seed Stringless Yellow Pod Wax Bush	Bean
Phaseolus	*vulgaris-bruscia-vaxa-helva-platya*	*cherokee-waxii*	Cherokee Stringless 6.0"	Flat Yellow Pod Wax Bush	Bean
Phaseolus	*vulgaris-bruscia-vaxa-helva-platya*	*cherokee-waxii-resistans*	Reisistant Cherokee Stringless 6.0"	Flat Yellow Pod Wax Bush	Bean
Phaseolus	*vulgaris-bruscia-vaxa-helva-platya*	*dracona-langeriei*	Dragon's Tongue Langerie	Flat Yellow Pod Wax Bush	Bean
Phaseolus	*vulgaris-bruscia-vaxa-helva-platya*	*romano-waxii*	Romano 6.0"	Flat Yellow Pod Wax Bush	Bean
Phaseolus	*vulgaris-bruscia-vaxa-helva-platya*	*romano-waxii-264*	#264 Romano 6.5"	Flat Yellow Pod Wax Bush	Bean
Phaseolus	*vulgaris-bruscia-vaxa-helva-platya-astrenga*	*golden-waxii*	White Seed Golden 5.0"	Stringless Flat Yellow Pod Wax Bush	Bean
Phaseolus	*vulgaris-bruscia-vaxa-helva-platya-astrenga*	*golden-waxii-improvedii*	Improved White Seed Golden 6.0"	Stringless Flat Yellow Pod Wax Bush	Bean
Phaseolus	*vulgaris-bruscia-vaxa-helva-platya-astrenga*	*golden-waxii-rustproofii*	Rustproof White Seed Golden 5.7"	Stringless Flat Yellow Pod Wax Bush	Bean
Phaseolus	*vulgaris-bruscia-vaxa-helva-platya-astrenga*	*golden-waxii-topnotchii*	Topnotch White Seed Golden 5.7"	Stringless Flat Yellow Pod Wax Bush	Bean
Phaseolus	*vulgaris-bruscia-vaxa-helva-platya-astrenga*	*settler-3-yellow-podii*	White Seed Settler III 5.5"	Stringless Flat Yellow Pod Wax Bush	Bean
Phaseolus	*vulgaris-bruscia-vaxa-helva-platya-astrenga*	*surecrop-stringless-waxii*	Surecrop 6.2"	Stringless Flat Yellow Pod Wax Bush	Bean
Phaseolus	*vulgaris-bruscia-vaxa-helva-stena*	*goldkistii*	Goldkist 5.7"	Thin Yellow Pod Wax Bush	Bean
Phaseolus	*vulgaris-bruscia-vaxa-helva-stena*	*keygoldii*	Keygold 6.0"	Thin Yellow Pod Wax Bush	Bean
Phaseolus	*vulgaris-bruscia-vaxa-helva-stena*	*pisarecka-zlutoluskei*	Pisarecka Zlutoluske	Thin Yellow Pod Wax Bush	Bean

Bean, Dry Vine (*Phaseolus vulgaris-vigna*)

Genus	Species	infraspecies	cultivar	specimen	Genome
Phaseolus	*vulgaris-vigna-alba*	*1500-year-old-cave-beanii*	Maroon Speckled Cave Anasazi	White Seed Dry Vine	Bean
Phaseolus	*vulgaris-vigna-alba*	*amish-knuttlei*	Purple Dotted Cream Pink Original Amish Knuttle	White Seed Dry Vine	Bean
Phaseolus	*vulgaris-vigna-alba*	*brita's-foot-longii*	Brita's Foot Long	White Seed Dry Vine	Bean
Phaseolus	*vulgaris-vigna-alba*	*cannellini-largei*	Large Canellone	White Seed Dry Vine	Bean
Phaseolus	*vulgaris-vigna-alba*	*cannellini-minestronei*	Minestrone Canellone	White Seed Dry Vine	Bean
Phaseolus	*vulgaris-vigna-alba*	*chapman's-climbing-bumblebeei*	Chapman's Climbing Bumblebee	White Seed Dry Vine	Bean
Phaseolus	*vulgaris-vigna-alba*	*fagi-utanii*	Violet Streaked Cream Fagy Utan	White Seed Dry Vine	Bean
Phaseolus	*vulgaris-vigna-alba*	*frijol-mt.-pima-blancoi*	Blanco White Mt. Pima Frijol	White Seed Dry Vine	Bean
Phaseolus	*vulgaris-vigna-alba*	*hidatsa-shield-figurei*	Shield Figure On White Hidatsa	White Seed Dry Vine	Bean
Phaseolus	*vulgaris-vigna-alba*	*hidatsa-whitei*	White Hidatsa	White Seed Dry Vine	Bean
Phaseolus	*vulgaris-vigna-alba*	*hopi-whitei*	White Hopi	White Seed Dry Vine	Bean
Phaseolus	*vulgaris-vigna-alba*	*tender-frostii*	Tender Frost	White Seed Dry Vine	Bean
Phaseolus	*vulgaris-vigna-grisea*	*amish-knuttlei-corn-hillii*	Purple Gray Corn Hill Amish Knuttle	Gray Seed Dry Vine	Bean
Phaseolus	*vulgaris-vigna-grisea*	*amish-knuttlei-mayfloweri*	Soft Gray Mayflower Amish Knuttle	Gray Seed Dry Vine	Bean
Phaseolus	*vulgaris-vigna-grisea*	*dovei*	Blue Gray Dove	Gray Seed Dry Vine	Bean
Phaseolus	*vulgaris-vigna-hortus*	*bird's-egg-horticulturalii*	Buff Bird's Egg	Shell Dry Vine	Bean
Phaseolus	*vulgaris-vigna-hortus*	*bird's-egg-lina-cisco's*	Lina Cisco's Tan Bird's Egg	Shell Dry Vine	Bean
Phaseolus	*vulgaris-vigna-hortus*	*horticultural-brocktonii*	Red Striped Brockton	Shell Dry Vine	Bean
Phaseolus	*vulgaris-vigna-hortus*	*horticultural-cranberryi*	Red Speckled Buff Cranberry	Shell Dry Vine	Bean
Phaseolus	*vulgaris-vigna-hortus*	*horticultural-cranberryi-gramma-waltersii*	Gramma Walters Cranberry	Shell Dry Vine	Bean
Phaseolus	*vulgaris-vigna-hortus*	*horticultural-cranberryi-truei*	Purple True Cranberry	Shell Dry Vine	Bean
Phaseolus	*vulgaris-vigna-hortus*	*horticultural-cranberry-vermont-polei*	Vermont Pole Cranberry	Shell Dry Vine	Bean
Phaseolus	*vulgaris-vigna-hortus*	*horticultural-kingii*	Stringless King	Shell Dry Vine	Bean
Phaseolus	*vulgaris-vigna-hortus*	*horticultural-worcesteri*	Red Speckled Buff Worcester	Shell Dry Vine	Bean
Phaseolus	*vulgaris-vigna-hortus*	*neabel'si*	Neabel's	Shell Dry Vine	Bean
Phaseolus	*vulgaris-vigna-hortus*	*octoberi*	Tan Streaked Purple October	Shell Dry Vine	Bean
Phaseolus	*vulgaris-vigna-hortus*	*wren's-eggi*	Buff Wren's Egg	Shell Dry Vine	Bean
Phaseolus	*vulgaris-vigna-hortus*	*wren's-eggi-improvedii*	Improved Wren's Egg	Shell Dry Vine	Bean
Phaseolus	*vulgaris-vigna-mixta*	*bald-hornetii*	Purple & White Bald Hornet	Mixed Dry Vine	Bean
Phaseolus	*vulgaris-vigna-mixta*	*flaggii*	Purple Streaked Buff Flagg	Mixed Dry Vine	Bean
Phaseolus	*vulgaris-vigna-mixta*	*frijol-blancoi*	Blanco White & Tan Frijol	Mixed Dry Vine	Bean
Phaseolus	*vulgaris-vigna-mixta*	*frijol-en-seco-pintoi*	En Seco Pinto Frijol	Mixed Dry Vine	Bean
Phaseolus	*vulgaris-vigna-mixta*	*dolloffii*	Gold Swirled Pink Dolloff	Mixed Dry Vine	Bean
Phaseolus	*vulgaris-vigna-mixta*	*davis-blackii*	Black, White & Brown Davis	Mixed Dry Vine	Bean
Phaseolus	*vulgaris-vigna-mixta*	*fred-taylor[si*	Pink Streaked White Fred Taylor's	Mixed Dry Vine	Bean
Phaseolus	*vulgaris-vigna-mixta*	*golden-limai*	Gold On Pink Golden Lima	Mixed Dry Vine	Bean
Phaseolus	*vulgaris-vigna-mixta*	*good-mother-stallardii*	Maroon, Brown, Tan, & White Good Mother Stallard	Mixed Dry Vine	Bean
Phaseolus	*vulgaris-vigna-mixta*	*hopi-pintoi*	Brown & Beige Pinto Hopi	Mixed Dry Vine	Bean
Phaseolus	*vulgaris-vigna-mixta*	*hopi-pintoi-blackii*	Black Mottled Beige Pinto Hopi	Mixed Dry Vine	Bean
Phaseolus	*vulgaris-vigna-mixta*	*nicaraguensis*	Black, Red & Tan Nicaraguan	Mixed Dry Vine	Bean
Phaseolus	*vulgaris-vigna-mixta*	*papa-de-rolai*	Half White & Half Beige Papa De Rola	Mixed Dry Vine	Bean
Phaseolus	*vulgaris-vigna-mixta*	*toni's-redii*	Half White & Half Rose Toni's Red	Mixed Dry Vine	Bean
Phaseolus	*vulgaris-vigna-mixta*	*yin-yangii*	Half White & Half Red Yin Yang	Mixed Dry Vine	Bean

Genus	Species	infraspecies	cultivar	specimen	Genome
Phaseolus	*vulgaris-vigna-nigra*	*frijol-dolores-de-hildago-negroi*	Dolores De Hildago Negro Black Frijol	Black Seed Dry Vine	Bean
Phaseolus	*vulgaris-vigna-nigra*	*guatemalan-highlandsi*	Guatemalan Highlands	Black Seed Dry Vine	Bean
Phaseolus	*vulgaris-vigna-nigra*	*guadalupe-y-calvo-negroi*	Guadalupe Y Calvo Negro	Black Seed Dry Vine	Bean
Phaseolus	*vulgaris-vigna-nigra*	*hopi-blackii*	Black Hopi	Black Seed Dry Vine	Bean
Phaseolus	*vulgaris-vigna-nigra*	*oceana's-summeri*	Charcoal Ocean's Summer	Black Seed Dry Vine	Bean
Phaseolus	*vulgaris-vigna-nigra*	*or-du-rhinii*	Or Du Rhin	Black Seed Dry Vine	Bean
Phaseolus	*vulgaris-vigna-purpurea*	*hopi-string-purplei*	Purple String Hopi	Purple Pod Dry Vine	Bean
Phaseolus	*vulgaris-vigna-rubra*	*blanche-mcgheei*	Maroon Blanche McGhee	Red Seed Dry Vine	Bean
Phaseolus	*vulgaris-vigna-rubra*	*frijol-coloradoi*	Colorado Red Kidney Frijol	Red Seed Dry Vine	Bean
Phaseolus	*vulgaris-vigna-rubra*	*frijol-rojoi(red-chilii)*	Rojo Red Chili Frijol	Red Seed Dry Vine	Bean
Phaseolus	*vulgaris-vigna-rubra*	*hidatsa-redii*	Red Hidatsa	Red Seed Dry Vine	Bean
Phaseolus	*vulgaris-vigna-rubra*	*hopi-pinkii*	Pink Hopi	Red Seed Dry Vine	Bean
Phaseolus	*vulgaris-vigna-tana*	*frijol-de-cerocahuii*	Cerocahui Tan Frijol	Tan Seed Dry Vine	Bean
Phaseolus	*vulgaris-vigna-tana*	*frijol-de-sinaloaensis*	Sinaloa Tan Frijol	Tan Seed Dry Vine	Bean
Phaseolus	*vulgaris-vigna-tana*	*frijol-dolores-de-hildago-canaroi*	Dolores De Hildago Canario Tan Frijol	Tan Seed Dry Vine	Bean
Phaseolus	*vulgaris-vigna-tana*	*frijol-manzanoi*	Manzano Frijol	Tan Seed Dry Vine	Bean
Phaseolus	*vulgaris-vigna-tana*	*frijol-mt.-pima-vapai-bavii*	Mount Pima Vapai Bavi Tan Mt. Pima Frijol	Tan Seed Dry Vine	Bean
Phaseolus	*vulgaris-vigna-tana*	*goose-beanii*	Greenish Tan Goose	Tan Seed Dry Vinc	Bean
Phaseolus	*vulgaris-vigna-tana*	*goose-beanii-mostoller-wildii*	Mostoller Wild Goose	Tan Seed Dry Vine	Bean
Phaseolus	*vulgaris-vigna-tana*	*goose-beanii-snowcapii*	Half White Snowcap Goose	Tan Seed Dry Vine	Bean
Phaseolus	*vulgaris-vigna-tana*	*hopi-beigei*	Beige Hopi	Tan Seed Dry Vine	Bean
Phaseolus	*vulgaris-vigna-tana*	*hopi-goldii*	Gold Brown Hopi	Tan Seed Dry Vine	Bean
Phaseolus	*vulgaris-vigna-tana*	*hopi-redii*	Red Beige Hopi	Tan Seed Dry Vine	Bean
Phaseolus	*vulgaris-vigna-tana*	*hopi-string-beanii*	Beige String Hopi	Tan Seed Dry Vine	Bean
Phaseolus	*vulgaris-vigna-tana*	*hopi-yellow*	Seka Mori Yellow Hopi	Tan Seed Dry Vine	Bean
Phaseolus	*vulgaris-vigna-tana*	*hopi-yellow-lightii*	Light Yellow Ocher Hopi	Tan Seed Dry Vine	Bean
Phaseolus	*vulgaris-vigna-tana*	*indian-moundii*	Indian Mound	Tan Seed Dry Vine	Bean
Phaseolus	*vulgaris-vigna-tana*	*mimbresii*	Mimbres	Tan Seed Dry Vine	Bean
Phaseolus	*vulgaris-vigna-tana*	*montezumai*	Montezuma	Tan Seed Dry Vine	Bean
Phaseolus	*vulgaris-vigna-tana*	*ojo-de-cabra-chihuahuai*	Chihuahuan Ojo De Cabra	Tan Seed Dry Vine	Bean
Phaseolus	*vulgaris-vigna-tana*	*ojo-de-cabra-mt.-pimai*	Mt. Pima Ojo De Cabra	Tan Seed Dry Vine	Bean
Phaseolus	*vulgaris-vigna-tana*	*ojo-de-cabra-san-luis-potosi*	San Luis Potosi Ojo De Cabra	Tan Seed Dry Vine	Bean
Phaseolus	*vulgaris-vigna-tana*	*ojo-de-cabra-tarahumarai*	Tarahumara Ojo De Cabra	Tan Seed Dry Vine	Bean
Phaseolus	*vulgaris-vigna-tana*	*ojo-de-cabra-yeomei*	Yeome Ojo De Cabra	Tan Seed Dry Vine	Bean
Phaseolus	*vulgaris-vigna-tana*	*scotchii*	Butterscotch Scotch	Tan Seed Dry Vine	Bean
Phaseolus	*vulgaris-vigna-tana*	*succotashii*	Succotash	Tan Seed Dry Vine	Bean
Phaseolus	*vulgaris-vigna-tana*	*superba*	Red Splotched Buff Superb	Tan Seed Dry Vine	Bean
Phaseolus	*vulgaris-vigna-tana*	*zapoteci*	Zapotec	Tan Seed Dry Vine	Bean
Phaseolus	*vulgaris-vigna-tarahumara*	*tarahumara-azufradoi*	Yellow & Beige Azufrado	Tarahumara Dry Vine	Bean
Phaseolus	*vulgaris-vigna-tarahumara*	*tarahumara-bakaminai*	Dark Red Bakamina	Tarahumara Dry Vine	Bean
Phaseolus	*vulgaris-vigna-tarahumara*	*tarahumara-black-&-bluei*	Black & Blue	Tarahumara Dry Vine	Bean
Phaseolus	*vulgaris-vigna-tarahumara*	*tarahumara-caféi*	Coffee Brown Café	Tarahumara Dry Vine	Bean
Phaseolus	*vulgaris-vigna-tarahumara*	*tarahumara-canarioi*	Yellow & Beige Canario	Tarahumara Dry Vine	Bean
Phaseolus	*vulgaris-vigna-tarahumara*	*tarahumara-capiramei*	Maroon & Beige Capirame	Tarahumara Dry Vine	Bean
Phaseolus	*vulgaris-vigna-tarahumara*	*tarahumara-carpinterosi*	Black & White Carpinteros	Tarahumara Dry Vine	Bean
Phaseolus	*vulgaris-vigna-tarahumara*	*tarahumara-chokamei*	Black Chokame	Tarahumara Dry Vine	Bean
Phaseolus	*vulgaris-vigna-tarahumara*	*tarahumara-choliwamei*	Gray Mottled Purple Choiwame	Tarahumara Dry Vine	Bean
Phaseolus	*vulgaris-vigna-tarahumara*	*tarahumara-ejotero-negroi*	Black Ejotero Negro	Tarahumara Dry Vine	Bean
Phaseolus	*vulgaris-vigna-tarahumara*	*tarahumara-flor-de-mayoi*	White Rayed Purple Flor De Mayo	Tarahumara Dry Vine	Bean
Phaseolus	*vulgaris-vigna-tarahumara*	*tarahumara-frijol-amarilloi*	Dark Yellow Amarillo Frijol	Tarahumara Dry Vine	Bean
Phaseolus	*vulgaris-vigna-tarahumara*	*tarahumara-frijol-amarilloi-del-nortei*	Gold Amarillo Del Norte Frijol	Tarahumara Dry Vine	Bean
Phaseolus	*vulgaris-vigna-tarahumara*	*tarahumara-frijol-amarilloi-oroi*	Golden Brown Oro Amarillo Frijol	Tarahumara Dry Vine	Bean
Phaseolus	*vulgaris-vigna-tarahumara*	*tarahumara-frijol-negroi*	Black Negro Frijol	Tarahumara Dry Vine	Bean
Phaseolus	*vulgaris-vigna-tarahumara*	*tarahumara-kakamirai*	Red Kakamira	Tarahumara Dry Vine	Bean
Phaseolus	*vulgaris-vigna-tarahumara*	*tarahumara-mt.-pima-burro-&-caballitoi*	Burro & Caballito Brown On White Mt. Pima	Tarahumara Dry Vine	Bean
Phaseolus	*vulgaris-vigna-tarahumara*	*tarahumara-nortenoi*	Beige Norteno	Tarahumara Dry Vine	Bean
Phaseolus	*vulgaris-vigna-tarahumara*	*tarahumara-pintoi-blackii*	Black On Gray & Beige Pinto	Tarahumara Dry Vine	Bean
Phaseolus	*vulgaris-vigna-tarahumara*	*tarahumara-pintoi-mantequillai*	Brown & Beige Mantequilla Pinto	Tarahumara Dry Vine	Bean
Phaseolus	*vulgaris-vigna-tarahumara*	*tarahumara-purplei*	Purple	Tarahumara Dry Vine	Bean
Phaseolus	*vulgaris-vigna-tarahumara*	*tarahumara-purplei-ojosi*	Purple Ojos	Tarahumara Dry Vine	Bean
Phaseolus	*vulgaris-vigna-tarahumara*	*tarahumara-purplei-stari*	White & Purple Star	Tarahumara Dry Vine	Bean
Phaseolus	*vulgaris-vigna-tarahumara*	*tarahumara-redsi*	Purple Red	Tarahumara Dry Vine	Bean
Phaseolus	*vulgaris-vigna-tarahumara*	*tarahumara-sitakamei*	Beige & Purple Sitakame	Tarahumara Dry Vine	Bean
Phaseolus	*vulgaris-vigna-tarahumara*	*tarahumara-stari*	Black & White Star	Tarahumara Dry Vine	Bean
Phaseolus	*vulgaris-vigna-tarahumara*	*tarahumara-vayoi*	Beige Vayo	Tarahumara Dry Vine	Bean
Phaseolus	*vulgaris-vigna-tarahumara*	*tarahumara-vayoi-blancoi*	Pale Beige Vayo	Tarahumara Dry Vine	Bean
Phaseolus	*vulgaris-vigna-tarahumara*	*tarahumara-vayoi-de-sonora-suri*	Pink & Tan De Sonora Sur Vayo	Tarahumara Dry Vine	Bean
Phaseolus	*vulgaris-vigna-tarahumara*	*tarahumara-vayoi-o'odhamii*	Gold & Tan O'odham Vayo	Tarahumara Dry Vine	Bean

Genus	Species	infraspecies	cultivar	specimen	Genome
Phaseolus	*vulgaris-vigna-tarahumara*	*tarahumara-vayoi-o'odhamii-amarilloi*	Gold Amarillo Tohono O'odham Vayo	Tarahumara Dry Vine	Bean
Phaseolus	*vulgaris-vigna-tarahumara*	*tarahumara-vayoi-vayitoi*	Gold Ringed Beige Vayito Vayo	Tarahumara Dry Vine	Bean
Phaseolus	*vulgaris-vigna-tarahumara*	*tarahumara-vayoi-yeomei*	Beige Yeome Vayo	Tarahumara Dry Vine	Bean

Bean, Snap Vine (*Phaseolus vigna-snapa*)

Phaseolus	*vulgaris-vigna-snapa-alba*	.	.	White Seed Snap Vine	Bean
Phaseolus	*vulgaris-vigna-snapa-alba*	*aunt-jean'si*	Maroon Eyed Aunt Jean's	White Seed Snap Vine	Bean
Phaseolus	*vulgaris-vigna-snapa-alba*	*blue-tipii*	Blue Tip ?.?"	White Seed Snap Vine	Bean
Phaseolus	*vulgaris-vigna-snapa-alba*	*cornfield-whitei*	Cornfield ?.?"	White Seed Snap Vine	Bean
Phaseolus	*vulgaris-vigna-snapa-alba*	*early-riseri*	Early Riser 9.0"	White Seed Snap Vine	Bean
Phaseolus	*vulgaris-vigna-snapa-alba*	*malibuensis*	Malibu 7.5"	White Seed Snap Vine	Bean
Phaseolus	*vulgaris-vigna-snapa-alba*	*new-mexico-cavei*	Black Blotched New Mexico Cave	White Seed Snap Vine	Bean
Phaseolus	*vulgaris-vigna-snapa-alba*	*urai*	Ura 6.5"	White Seed Snap Vine	Bean
Phaseolus	*vulgaris-vigna-snapa-alba*	*white-frostii*	White Frost 7.0"	White Seed Snap Vine	Bean
Phaseolus	*vulgaris-vigna-snapa-astrenga*	*blue-lakei-5563*	Black Seeded Blue Lake #5563 6.8"	Stringless Round Snap Vine	Bean
Phaseolus	*vulgaris-vigna-snapa-astrenga*	*blue-lakei-6969*	Blue Ribbon Blue Lake #6969 ?.?"	Stringless Round Snap Vine	Bean
Phaseolus	*vulgaris-vigna-snapa-astrenga*	*brejoi*	Tan Brejo 6.5"	Stringless Round Snap Vine	Bean
Phaseolus	*vulgaris-vigna-snapa-astrenga*	*cascade-giantii*	Cascade Giant 9.0"	Stringless Round Snap Vine	Bean
Phaseolus	*vulgaris-vigna-snapa-astrenga*	*climbing-frenchii*	Climbing French 5.5"	Stringless Round Snap Vine	Bean
Phaseolus	*vulgaris-vigna-snapa-astrenga*	*cobrai*	Cobra 7.5"	Stringless Round Snap Vine	Bean
Phaseolus	*vulgaris-vigna-snapa-astrenga*	*earlipolii*	Earlipol ?.?"	Stringless Round Snap Vine	Bean
Phaseolus	*vulgaris-vigna-snapa-astrenga*	*fasoldii*	Fasold ?.?"	Stringless Round Snap Vine	Bean
Phaseolus	*vulgaris-vigna-snapa-astrenga*	*fawcett'si*	Fawcett's ?.?"	Stringless Round Snap Vine	Bean
Phaseolus	*vulgaris-vigna-snapa-astrenga*	*fortexii*	Fortex Filet 11."	Stringless Round Snap Vine	Bean
Phaseolus	*vulgaris-vigna-snapa-astrenga*	*fortexii-JPB-4396*	JPB #4396 Fortex Filet 11."	Stringless Round Snap Vine	Bean
Phaseolus	*vulgaris-vigna-snapa-astrenga*	*ideal-marketii*	Ideal Market 5.5"	Stringless Round Snap Vine	Bean
Phaseolus	*vulgaris-vigna-snapa-astrenga*	*mantrai*	Mantra 9.0"	Stringless Round Snap Vine	Bean
Phaseolus	*vulgaris-vigna-snapa-astrenga*	*patomacensis*	Potomac Riverbank 6.5"	Stringless Round Snap Vine	Bean
Phaseolus	*vulgaris-vigna-snapa-astrenga*	*purple-peacockii*	Purple Pod Peacock	Stringless Round Snap Vine	Bean
Phaseolus	*vulgaris-vigna-snapa-astrenga*	*selma-zebrai*	Selma Zebra ?.?"	Stringless Round Snap Vine	Bean
Phaseolus	*vulgaris-vigna-snapa-astrenga*	*striped-creasebackii*	Nancy Davis Striped Creaseback 4.7"	Stringless Round Snap Vine	Bean
Phaseolus	*vulgaris-vigna-snapa-astrenga*	*trionfo-violettoi*	Trionfo Violetto 7.0"	Stringless Round Snap Vine	Bean
Phaseolus	*vulgaris-vigna-snapa-astrenga*	*turkey-crawii*	Turkey Craw Cutshort 3.7"	Stringless Round Snap Vine	Bean
Phaseolus	*vulgaris-vigna-snapa-astrenga*	*violet-podded-stringlessi*	Violet Podded ?.?"	Stringless Round Snap Vine	Bean
Phaseolus	*vulgaris-vigna-snapa-astrenga*	*westlandii*	Westland 6.0"	Stringless Round Snap Vine	Bean
Phaseolus	*vulgaris-vigna-snapa-astrenga-alba*	*blue-lakei-5565*	Fillbasket Mosaic Resistant 1 Blue Lake #5565 6.5"	White Seed Stringless Round Snap Vine	Bean
Phaseolus	*vulgaris-vigna-snapa-astrenga-alba*	*blue-lakei-5962*	Stringless 7' Flat Blue Lake #5962 6.0"	White Seed Stringless Round Snap Vine	Bean
Phaseolus	*vulgaris-vigna-snapa-astrenga-alba*	*blue-lakei-6060*	Stringless 7' Round Blue Lake #6060 6.5"	White Seed Stringless Round Snap Vine	Bean
Phaseolus	*vulgaris-vigna-snapa-astrenga-alba*	*blue-lakei-6060-OR*	Oregon Shell Blue Lake #6060-OR 6.5"	White Seed Stringless Round Snap Vine	Bean
Phaseolus	*vulgaris-vigna-snapa-astrenga-alba*	*blue-lakei-6066*	White Seeded Blue Lake #6066 6.5"	White Seed Stringless Round Snap Vine	Bean
Phaseolus	*vulgaris-vigna-snapa-astrenga-alba*	*blue-lakei-6070*	Fillbasket Mosaic Resistant 1 Shell Blue Lake #6070 6.0"	White Seed Stringless Round Snap Vine	Bean
Phaseolus	*vulgaris-vigna-snapa-astrenga-alba*	*blue-lakei-6275*	Pole Blue Lake #6275 6.0"	White Seed Stringless Round Snap Vine	Bean
Phaseolus	*vulgaris-vigna-snapa-astrenga-alba*	*blue-lakei-6666*	Fillbasket Mosaic Resistant 1 Can Blue Lake #6666 5.7"	White Seed Stringless Round Snap Vine	Bean
Phaseolus	*vulgaris-vigna-snapa-astrenga-alba*	*bobis-a-grano-biancoi*	Bobis A Grano Bianco	White Seed Stringless Round Snap Vine	Bean
Phaseolus	*vulgaris-vigna-snapa-astrenga-alba*	*lazy-wifei(hofferi)*	Hoffer Lazy Housewife	White Seed Stringless Round Snap Vine	Bean
Phaseolus	*vulgaris-vigna-snapa-groena*	*big-mamai*	Big Mama ?.?"	Green Round Snap Vine	Bean
Phaseolus	*vulgaris-vigna-snapa-groena*	*berner-lanfrauenii*	Berner Landfrauen 7.0"	Green Round Snap Vine	Bean
Phaseolus	*vulgaris-vigna-snapa-groena*	*blauhildei*	Blauhilde 10.0"	Green Round Snap Vine	Bean
Phaseolus	*vulgaris-vigna-snapa-groena*	*blue-cocoi*	Blue Coco 5.5"	Green Round Snap Vine	Bean
Phaseolus	*vulgaris-vigna-snapa-groena*	*cherokee-trail-of-tearsii*	CherokeeTrail Of Tears 6.0"	Green Round Snap Vine	Bean
Phaseolus	*vulgaris-vigna-snapa-groena*	*cornfield-greasyi*	Greasy Cornfield	Green Round Snap Vine	Bean
Phaseolus	*vulgaris-vigna-snapa-groena*	*cornfield-cherokeei*	Cherokee Cornfield	Green Round Snap Vine	Bean
Phaseolus	*vulgaris-vigna-snapa-groena*	*cornfield-juanita-smith-beansi*	Juanita Smith Cornfield	Green Round Snap Vine	Bean
Phaseolus	*vulgaris-vigna-snapa-groena*	*cornfield-kahnawake-mohawkii*	Kahnawake Mohawk Cornfield 5.0"	Green Round Snap Vine	Bean
Phaseolus	*vulgaris-vigna-snapa-groena*	*cornfield-stripedii*	Striped Cornfield	Green Round Snap Vine	Bean
Phaseolus	*vulgaris-vigna-snapa-groena*	*cornfield-susie'si*	Susie's Cornfield	Green Round Snap Vine	Bean
Phaseolus	*vulgaris-vigna-snapa-groena*	*cutshort-genuinei*	Genuine Cutshort 3.5"	Green Round Snap Vine	Bean
Phaseolus	*vulgaris-vigna-snapa-groena*	*cutshort-greasyi*	Original Greasy Cutshort 3.5"	Green Round Snap Vine	Bean
Phaseolus	*vulgaris-vigna-snapa-groena*	*cutshort-greasy-tennessensis*	Tennessee Greasy Cutshort 3.0"	Green Round Snap Vine	Bean
Phaseolus	*vulgaris-vigna-snapa-groena*	*cutshort-ohioensis*	Ohio Cutshort 3.5"	Green Round Snap Vine	Bean
Phaseolus	*vulgaris-vigna-snapa-groena*	*cutshort-old-timei*	Old Time Cutshort 3.5"	Green Round Snap Vine	Bean
Phaseolus	*vulgaris-vigna-snapa-groena*	*garden-of-edenii*	Garden Of Eden ?.?"	Green Round Snap Vine	Bean

Phaseolus	vulgaris-vigna-snapa-groena	gilaensis	Gila River 6.0" Green Round Snap Vine Bean
Phaseolus	vulgaris-vigna-snapa-groena	glastada	Glastada 6.0" Green Round Snap Vine Bean
Phaseolus	vulgaris-vigna-snapa-groena	grandma-nespeca's	Grandma Nespeca's Green Round Snap Vine Bean
Phaseolus	vulgaris-vigna-snapa-groena	grannyi	Granny ?.?" Green Round Snap Vine Bean
Phaseolus	vulgaris-vigna-snapa-groena	granny-richmondii	Granny Richmond Green Round Snap Vine Bean
Phaseolus	vulgaris-vigna-snapa-groena	H-732A	H #732-A Upright Green Round Snap Vine Bean
Phaseolus	vulgaris-vigna-snapa-groena	hawaiian-wonderi	Hawaiian Wonder Green Round Snap Vine Bean
Phaseolus	vulgaris-vigna-snapa-groena	honeyi	Honey Seed ?.?" Green Round Snap Vine Bean
Phaseolus	vulgaris-vigna-snapa-groena	kapralii	Kapral ?.?" Green Round Snap Vine Bean
Phaseolus	vulgaris-vigna-snapa-groena	landfrauenii	Landfrauen 5.5" Green Round Snap Vine Bean
Phaseolus	vulgaris-vigna-snapa-groena	leather-britchesii	Leather Britches ?.?" Green Round Snap Vine Bean
Phaseolus	vulgaris-vigna-snapa-groena	manoa-wonderi	Manoa Wonder ?.?" Green Round Snap Vine Bean
Phaseolus	vulgaris-vigna-snapa-groena	mazlenk-visoki-zeleni	Mazlenk Visoki Zeleni 8.0" Green Round Snap Vine Bean
Phaseolus	vulgaris-vigna-snapa-groena	mccormick's-wonderi	McCormick's Wonder Green Round Snap Vine Bean
Phaseolus	vulgaris-vigna-snapa-groena	old-timeyi	Old Timey 4.5" Green Round Snap Vine Bean
Phaseolus	vulgaris-vigna-snapa-groena	oregon-giant	Paul Bunyan Oregon Giant 8.0" Green Round Snap Vine Bean
Phaseolus	vulgaris-vigna-snapa-groena	paterge-headii	Paterge Head ?.?" Green Round Snap Vine Bean
Phaseolus	vulgaris-vigna-snapa-groena	poamohoi	Poamoho ?.?" Green Round Snap Vine Bean
Phaseolus	vulgaris-vigna-snapa-groena	priscilla-polishii	Priscilla Polish 7.0" Green Round Snap Vine Bean
Phaseolus	vulgaris-vigna-snapa-groena	ram's-hornii	Ram's Horn ?.?" Green Round Snap Vine Bean
Phaseolus	vulgaris-vigna-snapa-groena	rattlesnakei	Rattlesnake 7.5" Green Round Snap Vine Bean
Phaseolus	vulgaris-vigna-snapa-groena	robsplashii	Robsplash ?.?" Green Round Snap Vine Bean
Phaseolus	vulgaris-vigna-snapa-groena	ruth-biblei	Ruth Bible 3.5" Green Round Snap Vine Bean
Phaseolus	vulgaris-vigna-snapa-groena	socks-beanii	Socks Bean ?.?" Green Round Snap Vine Bean
Phaseolus	vulgaris-vigna-snapa-groena	sultan's-emerald-moonii	Sultan's Emerald Moon Green Round Snap Vine Bean
Phaseolus	vulgaris-vigna-snapa-groena	sunwardii	Sunward 5.7" Green Round Snap Vine Bean
Phaseolus	vulgaris-vigna-snapa-groena	tsing-toui	Tsing Tou ?.?" Green Round Snap Vine Bean
Phaseolus	vulgaris-vigna-snapa-helva	.	. Yellow Pod Snap Vine Bean
Phaseolus	vulgaris-vigna-snapa-helva	fagiolo-meraviglia-venezia	Fagiolo Meraviglia Venezia Yellow Pod Snap Vine Bean
Phaseolus	vulgaris-vigna-snapa-helva	old-time-golden-stickii	Old Time Golden Stick Yellow Pod Snap Vine Bean
Phaseolus	vulgaris-vigna-snapa-hortus	berta-talaskai	Berta Talaska ?.?" Shell Snap Vine Bean
Phaseolus	vulgaris-vigna-snapa-hortus	chinese-beigei	Chinese Beige ?.?" Shell Snap Vine Bean
Phaseolus	vulgaris-vigna-snapa-hortus	country-womanii	Country Woman ?.?" Shell Snap Vine Bean
Phaseolus	vulgaris-vigna-snapa-hortus	heirloom-polei	Heirloom Pole 7.0" Shell Snap Vine Bean
Phaseolus	vulgaris-vigna-snapa-hortus	horticultural-neohampshirensis	New Hamphsire ?.?" Shell Snap Vine Bean
Phaseolus	vulgaris-vigna-snapa-hortus	jembo-polishii	Jembo Polish ?.?" Shell Snap Vine Bean
Phaseolus	vulgaris-vigna-snapa-hortus	missouri-wonderi	Missouri Wonder 7.0" Shell Snap Vine Bean
Phaseolus	vulgaris-vigna-snapa-hortus	ozark-heirloomii	Ozark Heirloom ?.?" Shell Snap Vine Bean
Phaseolus	vulgaris-vigna-snapa-hortus	stregonatai	Stregonata ?.?" Shell Snap Vine Bean
Phaseolus	vulgaris-vigna-snapa-hortus	wren's-eggii	Speckled Cranberry Wren's Egg Flat Snap Vine Bean
Phaseolus	vulgaris-vigna-snapa-hortus	yeome-purple-stringii	Yeome Purple String Shell Snap Vine Bean
Phaseolus	vulgaris-vigna-snapa-mixta	aunt-jean'si-beanii	Half Red & Half White Seed Aunt Jean's Mixed Snap Vine Bean
Phaseolus	vulgaris-vigna-snapa-mixta	borlotti-polei	Red Streaked Tan Pole Borlotto Mixed Snap Vine Bean
Phaseolus	vulgaris-vigna-snapa-mixta	borlotti-bushii	Rose Red & Cream Bush Borlotto Mixed Snap Vine Bean
Phaseolus	vulgaris-vigna-snapa-mixta	borlotti-flamma-rapicantei	Flamma Rampicante Borlotto Mixed Snap Vine Bean
Phaseolus	vulgaris-vigna-snapa-mixta	borlotti-gipsyi	Gipsy Borlotto ?.?" Mixed Snap Vine Bean
Phaseolus	vulgaris-vigna-snapa-mixta	borlotti-lamonii	Rose Mottled White Lamon Borlotto Mixed Snap Vine Bean
Phaseolus	vulgaris-vigna-snapa-mixta	borlotti-lingua-di-fuoco-nanoi	Red Speckled Cream Lingua Di Fuoco Nana Borlotto Mixed Snap Vine Bean
Phaseolus	vulgaris-vigna-snapa-mixta	borlotti-sanguignoi	Red & White Sanguigno Borlotto Mixed Snap Vine Bean
Phaseolus	vulgaris-vigna-snapa-mixta	borlotti-vigevanoi	Vegevano Borlotto Mixed Snap Vine Bean
Phaseolus	vulgaris-vigna-snapa-mixta	cutshort-greasy-brown-speckledii	Brown Speckled Greasy Cutshort 3.5" Mixed Snap Vine Bean
Phaseolus	vulgaris-vigna-snapa-mixta	white-&-green-hullii	White & Green Hull Mixed Snap Vine Bean
Phaseolus	vulgaris-vigna-snapa-nigra	cornfield-blackii	Black Seed Cornfield Black Seed Snap Vine Bean
Phaseolus	vulgaris-vigna-snapa-nigra	georgia-blackii	Georgia Fat ?.?" Black Seed Snap Vine Bean
Phaseolus	vulgaris-vigna-snapa-nigra	grandma-black'si	Grandma Black's 9.0" Black Seed Snap Vine Bean
Phaseolus	vulgaris-vigna-snapa-nigra	griggs-blackii	Griggs ?.?" Black Seed Snap Vine Bean
Phaseolus	vulgaris-vigna-snapa-nigra	marsingill-blackii	Marsingill ?.?" Black Seed Snap Vine Bean
Phaseolus	vulgaris-vigna-snapa-platya	aunt-ada-italian-polei	Aunt Ada ?.?" Flat Snap Vine Bean
Phaseolus	vulgaris-vigna-snapa-platya	celina's-wide-greenii	Celina's Wide Green Flat Snap Vine Bean
Phaseolus	vulgaris-vigna-snapa-platya	dadensis	Green Pod Dade 7.5" Flat Snap Vine Bean
Phaseolus	vulgaris-vigna-snapa-platya	dow-purple-podii	Dow Purple Pod 7.5" Flat Snap Vine Bean
Phaseolus	vulgaris-vigna-snapa-platya	grandma-nellie'si	Grandma Nellie's Flat Snap Vine Bean
Phaseolus	vulgaris-vigna-snapa-platya	grape-beanii	Grape Bean 5.0" Flat Snap Vine Bean
Phaseolus	vulgaris-vigna-snapa-platya	hickman'si	Hickman's 7.0" Flat Snap Vine Bean
Phaseolus	vulgaris-vigna-snapa-platya	hildai	Hilda 12." Flat Snap Vine Bean
Phaseolus	vulgaris-vigna-snapa-platya	italian-polei	Italian Pole 6.0" Flat Snap Vine Bean
Phaseolus	vulgaris-vigna-snapa-platya	kingston-goldii	Kingston Gold ?.?" Flat Snap Vine Bean
Phaseolus	vulgaris-vigna-snapa-platya	kwintus	Qwintus 12." Flat Snap Vine Bean
Phaseolus	vulgaris-vigna-snapa-platya	kwintus-northeasteri	Northeaster Qwintus 8.0" Flat Snap Vine Bean
Phaseolus	vulgaris-vigna-snapa-platya	meraldai	Meralda 7.5" Flat Snap Vine Bean
Phaseolus	vulgaris-vigna-snapa-platya	valena-italianii	Valena Italian ?.?" Flat Snap Vine Bean
Phaseolus	vulgaris-vigna-snapa-platya	white-greasyi-mullins-strainii	Mullins White Seed Greasy 5.5" Flat Snap Vine Bean

Phaseolus	*vulgaris-vigna-snapa-platya-astrenga*	*case-knifei*	White Seed Case Knife 8.0"	Stringless Flat Snap Vine	Bean
Phaseolus	*vulgaris-vigna-snapa-platya-astrenga*	*heldai*	Helda 9.0"	Stringless Flat Snap Vine	Bean
Phaseolus	*vulgaris-vigna-snapa-platya-astrenga*	*hunter-stringlessi*	Hunter 8.0"	Stringless Flat Snap Vine	Bean
Phaseolus	*vulgaris-vigna-snapa-platya-astrenga*	*jeminezii*	Jeminez 7.5"	Stringless Flat Snap Vine	Bean
Phaseolus	*vulgaris-vigna-snapa-platya-astrenga*	*jeminezii-idaho-wonderi*	Idaho Wonder Jeminez 10."	Stringless Flat Snap Vine	Bean
Phaseolus	*vulgaris-vigna-snapa-platya-astrenga*	*kentucky-wonderi*	Kentucky Wonder 8.5"	Stringless Flat Snap Vine	Bean
Phaseolus	*vulgaris-vigna-snapa-platya-astrenga*	*kentucky-wonderi-191*	#191 Kentucky Wonder 6.2"	Stringless Flat Snap Vine	Bean
Phaseolus	*vulgaris-vigna-snapa-platya-astrenga*	*kentucky-wonderi-black-seededii*	Black Seed Kentucky Wonder 7.5"	Stringless Flat Snap Vine	Bean
Phaseolus	*vulgaris-vigna-snapa-platya-astrenga*	*kentucky-wonderi-blue-lakei*	Blue Lake Kentucky Wonder 6.5"	Stringless Flat Snap Vine	Bean
Phaseolus	*vulgaris-vigna-snapa-platya-astrenga*	*kentucky-wonderi-brown-seededii*	Brown Seed Kentucky Wonder 8.5"	Stringless Flat Snap Vine	Bean
Phaseolus	*vulgaris-vigna-snapa-platya-astrenga*	*kentucky-wonderi-kentucky-kingii*	Kentucky King Kentucky Wonder 7.0"	Stringless Flat Snap Vine	Bean
Phaseolus	*vulgaris-vigna-snapa-platya-astrenga*	*kentucky-wonderi-morse's-polei*	Morse's Pole Kentucky Wonder 8.5"	Stringless Flat Snap Vine	Bean
Phaseolus	*vulgaris-vigna-snapa-platya-astrenga*	*kentucky-wonderi-mr.-watersi*	Mr. Waters White Seed Kentucky Wonder	Stringless Flat Snap Vine	Bean
Phaseolus	*vulgaris-vigna-snapa-platya-astrenga*	*kentucky-wonderi-old-homesteadii*	Old Homestead Kentucky Wonder 9.5"	Stringless Flat Snap Vine	Bean
Phaseolus	*vulgaris-vigna-snapa-platya-astrenga*	*kentucky-wonderi-rust-resistantii*	Rust Res. Kentucky Wonder 6.5"	Stringless Flat Snap Vine	Bean
Phaseolus	*vulgaris-vigna-snapa-platya-astrenga*	*kentucky-wonderi-white-seededii*	White Seeded Kentucky Wonder 7.5"	Stringless Flat Snap Vine	Bean
Phaseolus	*vulgaris-vigna-snapa-platya-astrenga*	*mccaslanii*	McCaslan 7.5"	Stringless Flat Snap Vine	Bean
Phaseolus	*vulgaris-vigna-snapa-platya-astrenga*	*mccaslanii-42*	#42 McCaslan 7.7"	Stringless Flat Snap Vine	Bean
Phaseolus	*vulgaris-vigna-snapa-platya-astrenga*	*rampicantei-metro-o-stringai*	Metro O Stringa Rampicante	Stringless Flat Snap Vine	Bean
Phaseolus	*vulgaris-vigna-snapa-platya-astrenga*	*rampicantei-supermarconi-blackii*	Black Seed Super Marconi Rampicante 10."	Stringless Flat Snap Vine	Bean
Phaseolus	*vulgaris-vigna-snapa-platya-astrenga*	*rampicantei-supermarconi-whitei*	WhiteSeed Super Marconi Rampicante 10."	Stringless Flat Snap Vine	Bean
Phaseolus	*vulgaris-vigna-snapa-platya-astrenga*	*rampicantei-violettoi*	Violetto Rampicante	Stringless Flat Snap Vine	Bean
Phaseolus	*vulgaris-vigna-snapa-platya-astrenga*	*romano-green-anellinoi*	Green Anellino Romano	Stringless Flat Snap Vine	Bean
Phaseolus	*vulgaris-vigna-snapa-platya-astrenga*	*romano-green-anellinoi-hillaryi*	Brown Seed Hillary Green Anellino Romano	Stringless Flat Snap Vine	Bean
Phaseolus	*vulgaris-vigna-snapa-platya-astrenga*	*romano-garafal-oroi*	Garafal Oro Romano 12."	Stringless Flat Snap Vine	Bean
Phaseolus	*vulgaris-vigna-snapa-platya-astrenga*	*romano-golden-of-bacaui*	Golden Of Bacau Romano 8.0"	Stringless Flat Snap Vine	Bean
Phaseolus	*vulgaris-vigna-snapa-platya-astrenga*	*romano-musici*	Musica Romano 9.5"	Stringless Flat Snap Vine	Bean
Phaseolus	*vulgaris-vigna-snapa-platya-astrenga*	*romano-polei*	Green Pod Pole Romano 6.0"	Stringless Flat Snap Vine	Bean
Phaseolus	*vulgaris-vigna-snapa-platya-astrenga*	*selka-improvedii*	Selka Improved 11."	Stringless Flat Snap Vine	Bean
Phaseolus	*vulgaris-vigna-snapa-purpurea*	*anderson's-purple-kingii*	Anderson's Purple King	Purple Pod Snap Vine	Bean
Phaseolus	*vulgaris-vigna-snapa-purpurea*	*hillbilly-purplei*	Hillbilly?.?"	Purple Pod Snap Vine	Bean
Phaseolus	*vulgaris-vigna-snapa-purpurea*	*louisiana-purple-podii*	Louisiana 7.5"	Purple Pod Snap Vine	Bean
Phaseolus	*vulgaris-vigna-snapa-purpurea*	*purple-prizei*	Purple Prize ?.?"	Purple Pod Snap Vine	Bean
Phaseolus	*vulgaris-vigna-snapa-purpurea*	*robison-purple-podii*	Robison ?.?"	Purple Pod Snap Vine	Bean
Phaseolus	*vulgaris-vigna-snapa-rubra*	*alice-whitisi*	Rose Seed Alice Whitis	Red Seed Snap Vine	Bean
Phaseolus	*vulgaris-vigna-snapa-rubra*	*ohio-polei*	Maroon Seed Ohio Pole	Red Seed Snap Vine	Bean
Phaseolus	*vulgaris-vigna-snapa-stena*	*domatsu-snap-vinei*	Domatsu 7.5"	Thin Round Snap Vine	Bean
Phaseolus	*vulgaris-vigna-snapa-stena*	*emeritei*	Emerite 8.0"	Thin Round Snap Vine	Bean
Phaseolus	*vulgaris-vigna-snapa-stena*	*peter-bluei*	Peter Blue Purple	Thin Round Snap Vine	Bean
Phaseolus	*vulgaris-vigna-snapa-stena*	*santa-annai*	Santa Anna ?.?"	Thin Round Snap Vine	Bean
Phaseolus	*vulgaris-vigna-snapa-stena-astrenga*	*neckarkonigenii*	Neckarkonigen ?.?"	Stringless Thin Round Snap Vine	Bean

Bean, Wax Vine (*Phaseolus vulgaris-vigna-vaxa*)

Genus	Species	infraspecies	cultivar	specimen	Genome
Phaseolus	*vulgaris-vigna-vaxa-helva*	*appaloosa-goosei*	Pale Gray Seed Appaloosa Goose	Yellow Pod Wax Snap Vine	Bean
Phaseolus	*vulgaris-vigna-vaxa-helva*	*grandma-nellie's-yellow-mushroomii*	Grandma Nellie's Mushroom 5.0"	Yellow Pod Wax Snap Vine	Bean
Phaseolus	*vulgaris-vigna-vaxa-helva*	*ramdori*	Ramdor 7.0"	Yellow Pod Wax Snap Vine	Bean
Phaseolus	*vulgaris-vigna-vaxa-helva*	*white-half-runneri*	White Half Runner	Yellow Pod Wax Snap Vine	Bean
Phaseolus	*vulgaris-vigna-vaxa-helva-alba*	*b.b.-waxii*	B. B. ?.?"	White Seed Yellow Wax Snap Vine	Bean
Phaseolus	*vulgaris-vigna-vaxa-helva-alba*	*goldmariei*	Goldmarie 10."	White Seed Yellow Wax Snap Vine	Bean
Phaseolus	*vulgaris-vigna-vaxa-helva-astrenga*	*goldstrawi*	Goldstraw 7.0"	Stringless Yellow Wax Snap Vine	Bean
Phaseolus	*vulgaris-vigna-vaxa-helva-astrenga*	*neckargoldii*	Neckargold ?.?"	Stringless Yellow Wax Snap Vine	Bean
Phaseolus	*vulgaris-vigna-vaxa-helva-platya-astrenga*	*goldfieldii*	Goldfield 10."	Stringless Flat Yellow Wax Snap Vine	Bean
Phaseolus	*vulgaris-vigna-vaxa-helva-platya-astrenga*	*kentucky-wonderi-yellowi*	Yellow Kentucky Wonder 9.0"	Stringless Flat Yellow Wax Snap Vine	Bean
Phaseolus	*vulgaris-vigna-vaxa-helva-platya-astrenga*	*kentucky-wonderi-waxii*	Wax Kentucky Wonder 7.5"	Stringless Flat Yellow Wax Snap Vine	Bean
Phaseolus	*vulgaris-vigna-vaxa-helva-platya-astrenga*	*meraviglia-di-venezia*	Marvel Of Venice 8.5"	Stringless Flat Yellow Wax Snap Vine	Bean
Phaseolus	*vulgaris-vigna-vaxa-helva-platya-astrenga*	*romano-burro-d'ingegnolii*	Burro D'Ingegnoli Romano 5.0"	Stringless Flat Yellow Wax Snap Vine	Bean
Phaseolus	*vulgaris-vigna-vaxa-helva-platya-astrenga*	*romano-yellow-anillinoi*	Yellow Anellino Romano	Stringless Flat Yellow Wax Snap Vine	Bean
Psophocarpus	*tetragonolobus*	*tetragonolobus*	Original	Winged Bean	Exploding Bean

Genus	Species	infraspecies	cultivar	specimen	Genome
Psophocarpus	*tetragonolobus*	*asparagus*	Asparagus	Winged Bean	Exploding Bean
Psophocarpus	*tetragonolobus*	*day-neutralis*	Day neutral	Winged Bean	Exploding Bean
Psophocarpus	*tetragonolobus*	*princessi*	Princess	Winged Bean	Exploding Bean
Pueraria	*edulis*	*edulis*	Original	Edible	Kudzu Potato Bean
Pueraria	*edulis*	*bicalcarata*	Limestone	Edible	Kudzu Potato Bean
Pueraria	*edulis*	*quadristipellata*	Four Stipple	Edible	Kudzu Potato Bean
Tylosema	*esculentum*	*esculentum*	Original	Edible	Buck Bean
Tylosema	*esculentum*	*camel's-footii*	Camel's Foot	Edible	Buck Bean
Tylosema	*esculentum*	*gemsbuckii*	Gemsbuck	Edible	Buck Bean

Bean, Adzuki Vine (*Vigna angularis*)

Genus	Species	infraspecies	cultivar	specimen	Genome
Vigna	*angularis*	*angularis*	Trailing	Adzuki	Vine Bean
Vigna	*angularis*	*adzuki-buffii*	Buff	Adzuki	Vine Bean
Vigna	*angularis*	*adzuki-expressi*	Express	Adzuki	Vine Bean
Vigna	*angularis*	*adzuki-gray-mottledii*	Gray Mottled	Adzuki	Vine Bean
Vigna	*angularis*	*adzuki-mame-late-tambai*	Late Tamba Mamae	Adzuki	Vine Bean
Vigna	*angularis*	*adzuki-red/whitei*	Red/White	Adzuki	Vine Bean
Vigna	*angularis*	*nipponensis*	Japanese Red	Adzuki	Vine Bean
Vigna	*mungo*	*mungo*	Black	Gram Mung	Vine Bean
Vigna	*mungo*	*groena*	Green	Gram Mung	Vine Bean
Vigna	*mungo*	*urdii*	Urd Bean	Gram Mung	Vine Bean

Bean, Mung Vine (*Vigna radiata*)

Vigna	*radiata*	*radiata*	Original	Mung	Vine Bean
Vigna	*radiata*	*berkenii*	Berken	Mung	Vine Bean
Vigna	*radiata*	*grandiflora*	Grand Flower	Mung	Vine Bean
Vigna	*radiata*	*nigra*	Black	Mung	Vine Bean
Vigna	*radiata*	*nigra-kali-gramii*	Kali Gram Black	Mung	Vine Bean
Vigna	*radiata*	*setulosa*	Bristle	Mung	Vine Bean
Vigna	*radiata*	*setulosa-opistricha*	Hairy Bristle	Mung	Vine Bean
Vigna	*radiata*	*sublobata*	Lesser Lobed	Mung	Vine Bean
Vigna	*radiata*	*sublobata-brachycarpa*	Short Fruit Lesser Lobed	Mung	Vine Bean
Vigna	*radiata*	*sublobata-perrieriana*	Perrier Lesser Lobed	Mung	Vine Bean
Vigna	*umbellata*	*umbellata*	Original	Rice Bean	Vine Bean
Vigna	*umbellata*	*alba*	White	Rice Bean	Vine Bean
Vigna	*umbellata*	*calcarata*	Limestone	Rice Bean	Vine Bean
Vigna	*umbellata*	*gracilis*	Slender	Rice Bean	Vine Bean
Vigna	*umbellata*	*papuana*	Papua New Guinea	Rice Bean	Vine Bean
Vigna	*umbellata*	*rubra*	Red	Rice Bean	Vine Bean

Bean, Cowpea Vine (*Vigna unguiculata*)

Vigna	*unguiculata-alba*	.	.	White Seed Cowpea	Vine Bean
Vigna	*unguiculata-alba*	*browneye-six-weekii*	Six Week Browneye	White Seed Cowpea	Vine Bean
Vigna	*unguiculata-alba*	*early-ladyi*	Early Lady	White Seed Cowpea	Vine Bean
Vigna	*unguiculata-alba*	*guarijio-muni-cafei*	Guarijio Muni Cafe	White Seed Cowpea	Vine Bean
Vigna	*unguiculata-alba*	*ladyi*	Lady	White Seed Cowpea	Vine Bean
Vigna	*unguiculata-alba*	*pima-bejoensis*	Pima Bejo	White Seed Cowpea	Vine Bean
Vigna	*unguiculata-alba*	*pinkeye-purple-hull-whitei*	Purple Hull Pinkeye	White Seed Cowpea	Vine Bean
Vigna	*unguiculata-alba*	*pinkeye-purple-hull-californica*	California Purple Hull Pinkeye	White Seed Cowpea	Vine Bean
Vigna	*unguiculata-alba*	*purple-hull-49-texasensis*	Texas 49 Day Purple Hull	White Seed Cowpea	Vine Bean
Vigna	*unguiculata-alba*	*rice-peai*	Rice Pea	White Seed Cowpea	Vine Bean
Vigna	*unguiculata-alba*	*sunapeei*	Sunapee	White Seed Cowpea	Vine Bean
Vigna	*unguiculata-brunnea*	*chori-redii*	Chori Red	Brown Seed Cowpea	Vine Bean
Vigna	*unguiculata-brunnea*	*chosi-brown-seededii*	Chosi	Brown Seed Cowpea	Vine Bean
Vigna	*unguiculata-brunnea*	*dimpled-brownii*	Dimpled	Brown Seed Cowpea	Vine Bean
Vigna	*unguiculata-brunnea*	*dixie-leei*	Dixie Lee	Brown Seed Cowpea	Vine Bean
Vigna	*unguiculata-brunnea*	*guarijio-frijol-gamuzai*	Guarijio Frijol Gamuza	Brown Seed Cowpea	Vine Bean
Vigna	*unguiculata-brunnea*	*missisippi-shipperi*	Mississippi Shipper	Brown Seed Cowpea	Vine Bean
Vigna	*unguiculata-brunnea*	*purple-hull-speckledii*	Speckled Purple Hull	Brown Seed Cowpea	Vine Bean
Vigna	*unguiculata-brunnea*	*stick-upii*	Stick Up	Brown Seed Cowpea	Vine Bean
Vigna	*unguiculata-brunnea*	*two-crop-brownii*	Two Crop	Brown Seed Cowpea	Vine Bean
Vigna	*unguiculata-creamii*	*bettergreenii*	Bettergreen	Cream Seed Cowpea	Vine Bean
Vigna	*unguiculata-creamii*	*bettersnapii*	Bettersnap	Cream Seed Cowpea	Vine Bean
Vigna	*unguiculata-creamii*	*bigboy-brown-eyedii*	Brown Eyed Big Boy	Cream Seed Cowpea	Vine Bean
Vigna	*unguiculata-creamii*	*bigboy-purple-hullii*	Purple Hull Big Boy	Cream Seed Cowpea	Vine Bean
Vigna	*unguiculata-creamii*	*championii*	Champion	Cream Seed Cowpea	Vine Bean
Vigna	*unguiculata-creamii*	*charleston-nemagreenii*	Charleston Nemagreen	Cream Seed Cowpea	Vine Bean
Vigna	*unguiculata-creamii*	*chori-creamii*	Chori	Cream Seed Cowpea	Vine Bean
Vigna	*unguiculata-creamii*	*creamii-8*	#8 Kidney	Cream Seed Cowpea	Vine Bean
Vigna	*unguiculata-creamii*	*creamii-12*	#12 Round	Cream Seed Cowpea	Vine Bean
Vigna	*unguiculata-creamii*	*creamii-40*	#40 Kidney	Cream Seed Cowpea	Vine Bean
Vigna	*unguiculata-creamii*	*creamii-combinei*	Combine	Cream Seed Cowpea	Vine Bean

Vigna	unguiculata-creamii	elitei	Elite	Cream Seed Cowpea Vine Bean
Vigna	unguiculata-creamii	italian-creamii	Italian	Cream Seed Cowpea Vine Bean
Vigna	unguiculata-creamii	lady-creamii	Lady	Cream Seed Cowpea Vine Bean
Vigna	unguiculata-creamii	mississippi-creamii	Mississippi	Cream Seed Cowpea Vine Bean
Vigna	unguiculata-creamii	pinkeye-creamii	Pinkeye	Cream Seed Cowpea Vine Bean
Vigna	unguiculata-creamii	pinkeye-mississippiensis	Mississippi Pinkeye	Cream Seed Cowpea Vine Bean
Vigna	unguiculata-creamii	pinkeye-purple-hullii-BVR	Bean Mosaic Virus Resistant Purple Hull Pinkeye	Cream Seed Cowpea Vine Bean
Vigna	unguiculata-creamii	pinkeye-purple-hullii-ludiviciana	Louisiana Purple Hull Pinkeye	Cream Seed Cowpea Vine Bean
Vigna	unguiculata-creamii	pinkeye-purple-hullii-texasensis	Texas Purple Hull Pinkeye	Cream Seed Cowpea Vine Bean
Vigna	unguiculata-creamii	pinkeye-quickpickii	Quickpick Pinkeye	Cream Seed Cowpea Vine Bean
Vigna	unguiculata-creamii	running-conchii	Running Conch	Cream Seed Cowpea Vine Bean
Vigna	unguiculata-creamii	running-conchii-mississippi-whitei	Mississippi White Running Conch	Cream Seed Cowpea Vine Bean
Vigna	unguiculata-creamii	sa-dandyi	Sa Dandy Conch	Cream Seed Cowpea Vine Bean
Vigna	unguiculata-creamii	tender-creamii	Tender	Cream Seed Cowpea Vine Bean
Vigna	unguiculata-creamii	texas-big-boyi	Big Boy Texas	Cream Seed Cowpea Vine Bean
Vigna	unguiculata-creamii	texas-creamii-08	#08 Texas	Cream Seed Cowpea Vine Bean
Vigna	unguiculata-creamii	texas-creamii-40	#40 Texas	Cream Seed Cowpea Vine Bean
Vigna	unguiculata-creamii	whippoorwillii-whitei	Whippoorwill	Cream Seed Cowpea Vine Bean
Vigna	unguiculata-creamii	white-acrei	White Acre	Cream Seed Cowpea Vine Bean
Vigna	unguiculata-creamii	yori-muni-mt.-pimaensis	Mt. Pima Yori Muni	Cream Seed Cowpea Vine Bean
Vigna	unguiculata-creamii	yori-muni-sonorensis	Sonoran Yori Muni	Cream Seed Cowpea Vine Bean
Vigna	unguiculata-crowderi	arkansasensis	Arkansas Light Brown	Crowder Cowpea Vine Bean
Vigna	unguiculata-crowderi	blackii	Black	Crowder Cowpea Vine Bean
Vigna	unguiculata-crowderi	blue-goosei	Blue Goose	Crowder Cowpea Vine Bean
Vigna	unguiculata-crowderi	brown-crowderi	Brown	Crowder Cowpea Vine Bean
Vigna	unguiculata-crowderi	brown-crowderi-CT-dimpledii	CT Dimpled Brown	Crowder Cowpea Vine Bean
Vigna	unguiculata-crowderi	brown-crowderi-dimpledii	Dimpled Brown	Crowder Cowpea Vine Bean
Vigna	unguiculata-crowderi	brown-sugar-crowderi	Brown Sugar	Crowder Cowpea Vine Bean
Vigna	unguiculata-crowderi	calico-crowderi	Calico	Crowder Cowpea Vine Bean
Vigna	unguiculata-crowderi	colossus	Colossus Light Brown	Crowder Cowpea Vine Bean
Vigna	unguiculata-crowderi	creamii-crowderi	Cream	Crowder Cowpea Vine Bean
Vigna	unguiculata-crowderi	grandma's-black-crowderi	Grandma's Black	Crowder Cowpea Vine Bean
Vigna	unguiculata-crowderi	knuckle-purplehullii	Knuckle Purple Hull Brown Sugar	Crowder Cowpea Vine Bean
Vigna	unguiculata-crowderi	knucklehullii-VNR	Virus & Nematode Resistant Knuckle Purple Hull Brown Sugar	Crowder Cowpea Vine Bean
Vigna	unguiculata-crowderi	mississippi-purple-hull-crowderi	Mississippi Purple Hull Tan	Crowder Cowpea Vine Bean
Vigna	unguiculata-crowderi	mississippi-purple-hull-silveri	Silver Cross Mississippi Purple Hull Tan	Crowder Cowpea Vine Bean
Vigna	unguiculata-crowderi	mississippi-silveri	Mississippi Silver Hull Green & Cream	Crowder Cowpea Vine Bean
Vigna	unguiculata-crowderi	mississippi-silveri-brownii	Mississippi Silver Hull Brown	Crowder Cowpea Vine Bean
Vigna	unguiculata-crowderi	purple-hull-browneyei	Browneye Purple Hull White	Crowder Cowpea Vine Bean
Vigna	unguiculata-crowderi	silver-tipii	SilverTip Gray Mottled	Crowder Cowpea Vine Bean
Vigna	unguiculata-crowderi	tennessee-white-crowderi	Tennessee White	Crowder Cowpea Vine Bean
Vigna	unguiculata-crowderi	turkey-craw-crowderi	Turkey Craw Tan & Red Brown	Crowder Cowpea Vine Bean
Vigna	unguiculata-crowderi	zipper-creamii	70 to 75 Day Zipper Cream	Crowder Cowpea Vine Bean
Vigna	unguiculata-crowderi	zipper-creamii-crowderi	75 Day Zipper Cream	Crowder Cowpea Vine Bean
Vigna	unguiculata-dekindtiana	blackeye-peai	Crowder Green	Black Eyed Cowpea Vine Bean
Vigna	unguiculata-dekindtiana	blackeye-bettergroi	Bettergro	Black Eyed Cowpea Vine Bean
Vigna	unguiculata-dekindtiana	blackeye-botswanensis	Botswana	Black Eyed Cowpea Vine Bean
Vigna	unguiculata-dekindtiana	blackeye-californica	California Cream	Black Eyed Cowpea Vine Bean
Vigna	unguiculata-dekindtiana	blackeye-californica-5	#5 California Cream	Black Eyed Cowpea Vine Bean
Vigna	unguiculata-dekindtiana	blackeye-californica-46	#46 California Cream	Black Eyed Cowpea Vine Bean
Vigna	unguiculata-dekindtiana	blackeye-royalis	Royal	Black Eyed Cowpea Vine Bean
Vigna	unguiculata-dekindtiana	georgia-longii	Georgia Long	Black Eyed Cowpea Vine Bean
Vigna	unguiculata-dekindtiana	green-dixiei	Green Dixie	Black Eyed Cowpea Vine Bean
Vigna	unguiculata-dekindtiana	huillensis	Huill Wild	Black Eyed Cowpea Vine Bean
Vigna	unguiculata-dekindtiana	queen-annei	Queen Anne	Black Eyed Cowpea Vine Bean
Vigna	unguiculata-grisea	charcoal-grayi	Charcoal	Gray Seed Cowpea Vine Bean
Vigna	unguiculata-grisea	gray-adzukii	Adzuki	Gray Seed Cowpea Vine Bean
Vigna	unguiculata-grisea	gray-speckled-palapyei	Gray Speckled Palapye	Gray Seed Cowpea Vine Bean
Vigna	unguiculata-grisea	paw's-old-gray-peai	Paw's Old	Gray Seed Cowpea Vine Bean
Vigna	unguiculata-groena	aduensis	Adu	Green Pod Cowpea Vine Bean
Vigna	unguiculata-groena	baoulensis	Baoul	Green Pod Cowpea Vine Bean
Vigna	unguiculata-groena	bohemiana	Bohemian	Green Pod Cowpea Vine Bean
Vigna	unguiculata-groena	brachycalyx	Short Calyx	Green Pod Cowpea Vine Bean
Vigna	unguiculata-groena	burundiensis	Burundi	Green Pod Cowpea Vine Bean
Vigna	unguiculata-groena	californica	California	Green Pod Cowpea Vine Bean
Vigna	unguiculata-groena	catjang	Catjang	Green Pod Cowpea Vine Bean
Vigna	unguiculata-groena	clemson-purplei	Clemson Purple Pod	Green Pod Cowpea Vine Bean
Vigna	unguiculata-groena	cylindrica	Horse Gram	Green Pod Cowpea Vine Bean
Vigna	unguiculata-groena	double-green-delightii	Double Green Delight	Green Pod Cowpea Vine Bean
Vigna	unguiculata-groena	green-pixiei	Green Pixie	Green Pod Cowpea Vine Bean
Vigna	unguiculata-groena	hercules	Hercules	Green Pod Cowpea Vine Bean

Vigna	unguiculata-groena	kentucky-heirloomii	Kentucky Heirloom	Green Pod Cowpea Vine Bean
Vigna	unguiculata-groena	letouzeyi	Letouzey	Green Pod Cowpea Vine Bean
Vigna	unguiculata-groena	magnolia	Magnolia	Green Pod Cowpea Vine Bean
Vigna	unguiculata-groena	missouri-heirloomii	Missouri Heirloom	Green Pod Cowpea Vine Bean
Vigna	unguiculata-groena	nigeriensis	Nigerian	Green Pod Cowpea Vine Bean
Vigna	unguiculata-groena	pawekiae	Pawek	Green Pod Cowpea Vine Bean
Vigna	unguiculata-groena	pubescens	Downy	Green Pod Cowpea Vine Bean
Vigna	unguiculata-groena	purple-hull-viningii	Purple Hull Vining	Green Pod Cowpea Vine Bean
Vigna	unguiculata-groena	purple-hull-viningii-calhounii	Calhoun Purple Hull Vining	Green Pod Cowpea Vine Bean
Vigna	unguiculata-groena	saidyi	Saidy	Green Pod Cowpea Vine Bean
Vigna	unguiculata-groena	scabra	Scabre	Green Pod Cowpea Vine Bean
Vigna	unguiculata-groena	sinensis	Chinese	Green Pod Cowpea Vine Bean
Vigna	unguiculata-groena	spontanea	Spontaneous	Green Pod Cowpea Vine Bean
Vigna	unguiculata-groena	stenophylla	Thin Leaf	Green Pod Cowpea Vine Bean
Vigna	unguiculata-groena	stenophylla-hispida	Hairy Thin Leaf	Green Pod Cowpea Vine Bean
Vigna	unguiculata-groena	stenophylla-protracta	Protracted Thin Leaf	Green Pod Cowpea Vine Bean
Vigna	unguiculata-groena	stenophylla-rhomboidea	Rhomboid Thin Leaf	Green Pod Cowpea Vine Bean
Vigna	unguiculata-groena	tenuis	Slender	Green Pod Cowpea Vine Bean
Vigna	unguiculata-groena	tenuis-coerulea	Blue Slender	Green Pod Cowpea Vine Bean
Vigna	unguiculata-groena	tenuis-malosana	Malos Slender	Green Pod Cowpea Vine Bean
Vigna	unguiculata-groena	unguiculata	Original	Green Pod Cowpea Vine Bean
Vigna	unguiculata-groena	worthmorei	Worthmore	Green Pod Cowpea Vine Bean
Vigna	unguiculata-marmorata	africana	African	Mottled Seed Cowpea Vine Bean
Vigna	unguiculata-marmorata	holsteinii	Holstein	Mottled Seed Cowpea Vine Bean
Vigna	unguiculata-marmorata	mayo-colimai-pintoi	Mayo Colima Pinto	Mottled Seed Cowpea Vine Bean
Vigna	unguiculata-marmorata	mayo-colimai-speckledii	Mayo Colima Speckled	Mottled Seed Cowpea Vine Bean
Vigna	unguiculata-marmorata	minnesotensis	Minnesota	Mottled Seed Cowpea Vine Bean
Vigna	unguiculata-marmorata	minnesotensis-13	#13 Minnesota	Mottled Seed Cowpea Vine Bean
Vigna	unguiculata-marmorata	minnesotensis-157	#157 Minnesota	Mottled Seed Cowpea Vine Bean
Vigna	unguiculata-marmorata	ozark-razorbackii	Ozark Razorback	Mottled Seed Cowpea Vine Bean
Vigna	unguiculata-marmorata	papago-cowpeai	Papago	Mottled Seed Cowpea Vine Bean
Vigna	unguiculata-marmorata	papago-surprisei	Papago Suprise	Mottled Seed Cowpea Vine Bean
Vigna	unguiculata-marmorata	PI-189374	#189374 Productive Insect Resistant (PI)	Mottled Seed Cowpea Vine Bean
Vigna	unguiculata-marmorata	PI-225901	#225901 Productive Insect Resistant (PI)	Mottled Seed Cowpea Vine Bean
Vigna	unguiculata-marmorata	PI-257463	#257463 Productive Insect Resistant (PI)	Mottled Seed Cowpea Vine Bean
Vigna	unguiculata-marmorata	PI-339619	#339619 Productive Insect Resistant (PI)	Mottled Seed Cowpea Vine Bean
Vigna	unguiculata-marmorata	pigott-family-heirloomii	Pigott Family Heirloom	Mottled Seed Cowpea Vine Bean
Vigna	unguiculata-marmorata	tetapeche-gray-mottledii	Tetapeche Gray	Mottled Seed Cowpea Vine Bean
Vigna	unguiculata-marmorata	tohonoo-odham-cowpeai	Tohono Oodham	Mottled Seed Cowpea Vine Bean
Vigna	unguiculata-marmorata	whippoorwillii	Whippoorwill	Mottled Seed Cowpea Vine Bean
Vigna	unguiculata-marmorata	whippoorwillii-purple-hullii	Purple Hull Whippoorwill	Mottled Seed Cowpea Vine Bean
Vigna	unguiculata-marmorata	yori-cahuii	Yori Cahui	Mottled Seed Cowpea Vine Bean
Vigna	unguiculata-multicolor	.	.	Multicolor Seed Cowpea Vine Bean
Vigna	unguiculata-multicolor	alstonii	Alston Purple Hull	Multicolor Seed Cowpea Vine Bean
Vigna	unguiculata-multicolor	burmai	Burma	Multicolor Seed Cowpea Vine Bean
Vigna	unguiculata-multicolor	iron-&-clay-mixedii	Iron & Clay Mixed	Multicolor Seed Cowpea Vine Bean
Vigna	unguiculata-multicolor	mayo-colimai	Mayo Colima	Multicolor Seed Cowpea Vine Bean
Vigna	unguiculata-multicolor	nyaritensis	Nyarit	Multicolor Seed Cowpea Vine Bean
Vigna	unguiculata-multicolor	rattlesnakei	Rattlesnake	Multicolor Seed Cowpea Vine Bean
Vigna	unguiculata-multicolor	red-&-blackii	Red & Black	Multicolor Seed Cowpea Vine Bean
Vigna	unguiculata-nigra	bisbee-blackii	Brisbee	Black Seed Cowpea Vine Bean
Vigna	unguiculata-nigra	jet-blackii	Jet	Black Seed Cowpea Vine Bean
Vigna	unguiculata-nigra	rouge-et-noiri	Rouge Et Noir	Black Seed Cowpea Vine Bean
Vigna	unguiculata-nigra	thai-blackii	Thai	Black Seed Cowpea Vine Bean
Vigna	unguiculata-nigra	turtle-peasi	Turtle Peas	Black Seed Cowpea Vine Bean
Vigna	unguiculata-rubra	brisbeei-redii	Brisbee	Red Seed Cowpea Vine Bean
Vigna	unguiculata-rubra	cardinalis	Cardinal	Red Seed Cowpea Vine Bean
Vigna	unguiculata-rubra	chinese-red-beanii	Chinese	Red Seed Cowpea Vine Bean
Vigna	unguiculata-rubra	corrientesi	Corrientes	Red Seed Cowpea Vine Bean
Vigna	unguiculata-rubra	crudupii	Crudup	Red Seed Cowpea Vine Bean
Vigna	unguiculata-rubra	haricot-rougei	Haricot Rouge	Red Seed Cowpea Vine Bean
Vigna	unguiculata-rubra	red-ripperi	Red Ripper	Red Seed Cowpea Vine Bean
Vigna	unguiculata-sesquipedalis	sesquipedalis	Original	Yard Long Cowpea Vine Bean
Vigna	unguiculata-sesquipedalis	yard-long-akasanjakui	Akasanjaku Red Seed	Yard Long Cowpea Vine Bean
Vigna	unguiculata-sesquipedalis	yard-long-doliquei	Dolique	Yard Long Cowpea Vine Bean
Vigna	unguiculata-sesquipedalis	yard-long-extra-long-blackii	Extra Long Black Seed	Yard Long Cowpea Vine Bean
Vigna	unguiculata-sesquipedalis	yard-long-kurojurokui	Kurojuroku Black Seed	Yard Long Cowpea Vine Bean
Vigna	unguiculata-sesquipedalis	yard-long-lianai	Liana	Yard Long Cowpea Vine Bean
Vigna	unguiculata-sesquipedalis	yard-long-long-white-snakei	Long White Seed Snake	Yard Long Cowpea Vine Bean
Vigna	unguiculata-sesquipedalis	yard-long-orient-wonderi	Orient Wonder Brown Seed	Yard Long Cowpea Vine Bean
Vigna	unguiculata-sesquipedalis	yard-long-purple-podii	Purple Pod	Yard Long Cowpea Vine Bean

Vigna	*unguiculata-sesquipedalis*	*yard-long-sabah-snakei*	Sabah Snake	Yard Long Cowpea Vine Bean
Vigna	*unguiculata-sesquipedalis*	*yard-long-white-seededii*	White Seeded	Yard Long Cowpea Vine Bean
Vigna	*unguiculata-sesquipedalis-asparagus*	*asparagus-bean-black-seededii*	Black Seeded	Asparagus Bean Cowpea Vine Bean
Vigna	*unguiculata-sesquipedalis-asparagus*	*asparagus-bean-black-seededii-stripei*	Stripe Black Seeded	Asparagus Bean Cowpea Vine Bean
Vigna	*unguiculata-sesquipedalis-asparagus*	*asparagus-bean-dow-gaukii*	Dow Gauk	Asparagus Bean Cowpea Vine Bean
Vigna	*unguiculata-sesquipedalis-asparagus*	*asparagus-bean-kaohsiungensis*	KaohSiung	Asparagus Bean Cowpea Vine Bean
Vigna	*unguiculata-sesquipedalis-asparagus*	*asparagus-bean-purple-podii*	Purple Pod	Asparagus Bean Cowpea Vine Bean
Vigna	*unguiculata-sesquipedalis-asparagus*	*asparagus-bean-red-seededii*	Red Seeded	Asparagus Bean Cowpea Vine Bean
Vigna	*unguiculata-sesquipedalis-asparagus*	*asparagus-bean-two-tonei*	Two Tone	Asparagus Bean Cowpea Vine Bean
Vigna	*unguiculata-sesquipedalis-asparagus*	*asparagus-bean-white-seedii*	White Seed	Asparagus Bean Cowpea Vine Bean
Vigna	*unguiculata-tana*	*chosi-white-seededii*	Chosi	Tan Seed Cowpea Vine Bean
Vigna	*unguiculata-tana*	*clayi*	Clay	Tan Seed Cowpea Vine Bean
Vigna	*unguiculata-tana*	*ejoteroi*	Ejotero	Tan Seed Cowpea Vine Bean
Vigna	*unguiculata-tana*	*october-peai*	October	Tan Seed Cowpea Vine Bean
Vigna	*unguiculata-tana*	*penny-rilei*	Penny Rile	Tan Seed Cowpea Vine Bean
Vigna	*unguiculata-tana*	*washdayi*	Washday	Tan Seed Cowpea Vine Bean

Family 2 *Fabafloraceae* Beanflower

GENERA *Faba(Synonym), Abarema, Adenanthera, Afzelia, Airyantha(Synonym for Baphia), Amorpha, Anadenanthera, Andira, Baphia, Baphiastrum, Baphiopsis, Butea, Corynella(Synonym), Coumarouna(Synonym), Cynometra, Disynstemon(Synonym for Lonchocarpus), Endosamara(Synonym for Robinia), Entada, Entadopsis(Synonym), Ervum(Synonym), Erythrina, Geoffroea, Gliricidia, Guibourtia, Haplormosia, Hebestigma(Synonym for Gliricidia), Herpyza(Synonym for Teramnus), Hybosema(Synonym for Gliricidia), Intsia(Synonym for Afzelia), Kanaloa, Kerstingiella(Synonym), Lebruniodendron(Synonym for Cynometra), Leucopterum(Synonym), Lonchocarpus, Macroptilium, Meizotropis(Synonym for Butea), Micropteryx(Synonym), Millettia, Myrospermum, Neocollettia(Synonym for Teramnus), Neorudolphia(Synonym for Butea), Niopa(Synonym), Notodon(Synonym for Poitea), Olneya, Ostryocarpus(Synonym for Millettia), Parkia, Pentaclethra, Philenoptera, Pitcheria(Synonym), Poitea, Psorothamnus, Robinia, Sabinea(Synonym), Scorodophloeus(Synonym for Cynometra), Sphenostylis, Stahlia, Stizolobium(Synonym), Strophostyles, Stryphnodendron, Teramnus, Tipuana, Vexibia(Synonym), Voandzeia(Synonym)*

Family 724.062.3 *Dolichaceae* Longpod Bean

GENERA *Dolichos, Adenodolichos, Austrodolichos, Decorsea, Dolicholus(Synonym), Dolichopsis, Dysolobium, Lackeya(Synonym for Dolichos), Macrotyloma, Oryxis(Synonym for Dolichos), Pseudovigna(Synonym for Dolichos), Sinodolichos*

Genus	Species	Specimen Genome	C:Culinary
Dolichos	*fagiolino-dolico-venetoi*	Fagiolino Dolico Veneto Longpod Bean	C:PD:**VG**

Order 724.063 *Pisales* Pea

Family 624.063.1 *Pisaceae* Pea

GENERA ***Pisum(Pea), Abrus(Jequerty Pea), Amphicarpaea(Hog Peanut), Arachis(Peanut),*** *Atylosia(Synonym for Cajanus),* ***Cajanus(Pigeon Pea),*** *Cantharospermum(Synonym for Cajanus),* ***Caragana(Caragan Pea), Castanospermum(Chestnut Pea), Cicer(Chick Pea),*** *Colophospermum(Synonym for Copaifera),* ***Copaifera(Copaiban Balsam Copal Gum), Cullen(Babchi Scurf Pea), Daniellia(Copal Gum), Daviesia (Hops Bitter Pea), Dillwynia(Parrot Pea),*** *Falcata(Synonym for Amphicarpaea),* ***Flemingia(Wild Hops), Galactia(Milk Pea), Galega(Goat Rue), Glycyrrhiza(Licorice Root Herb), Indigofera(Indigo Herb), Inocarpus (Dove Pea), Lemuropisum(Tara Pea), Lupinus(Lupine Pea), Medicago (Alfalfa),*** *Moghania(Synonym for Flemingia),* ***Myroxylon(Myrrhwood), Pediomelum(Breadroot Pea), Platylobium(Flatpod Pea),*** *Pseudosindora(Synonym for Copaifera),* ***Psoralea(Scurf Pea), Pterocarpus(Barwood Herb),*** *Rothia(Synonym for Trigonella), Sindoropsis(Synonym for Copaifera),* ***Sophora(Necklace Pod),*** *Styphnolobium(Synonym for Sophora),* ***Tetragonolobus(Winged Pea), Thermopsis(Golden Banner Herb), Trigonella(Fenugreek Seed), Ulex(Gorse Pea),*** *Vavilovia(Synonym for Pisum)*

Genus	Species	Specimen Genome	C:Culinary
Pisum	*sativum-arvense-brevis*	Short Pod (<2.5") Snap Pea	C:PD:**VG**
Pisum	*sativum-arvense-media*	Medium Pod (2.5"-3.5") Snap Pea	C:PD:**VG**
Pisum	*sativum-arvense-macrocarpon*	Long Pod (>3.5") Snap Pea	C:PD:**VG**
Pisum	*sativum-arvense-sucra-brevis*	Short Pod (<2.5") Sugar Pea	C:PD:**VG**
Pisum	*sativum-arvense-sucra-media*	Medium Pod (2.5"-3.5") Sugar Pea	C:PD:**VG**
Pisum	*sativum-arvense-sucra-longa*	Long Pod (>3.5") Sugar Pea	C:PD:**VG**
Pisum	*sativum-elatius-media*	Medium Pod (5-8 pea) Bush Pea	C:PD:**VG**
Pisum	*sativum-elatius-longa*	Long Pod (7-12 pea) Bush Pea	C:PD:**VG**
Pisum	*sativum-elatius-suavis*	Bush Sweet Pea	C:PD:**VG**
Pisum	*sativum-hortense-media*	Medium Pod (5-8 pea) Vine Pea	C:PD:**VG**
Pisum	*sativum-hortense-longa*	Long Pod (7-12 pea) Vine Pea	C:PD:**VG**
Pisum	*sativum-hortense-schiza*	Vine Split Pea	C:SD:**VG**
Pisum	*sativum-hortense-sopius*	Vine Soup Pea	C:SD:**VG**
Pisum	*sativum-hortense-suavis-media*	Medium Pod (5-8 pea) Vine Sweet Pea	C:SD:**VG**
Pisum	*sativum-hortense-suavis-longa*	Long Pod (7-12 pea) Vine Sweet Pea	C:SD:**VG**
Amphicarpaea	*bracteata*	American Hog Peanut	C:SD:**NT**
Arachis	*hypogaea*	Goober Peanut	C:SD:**NT**
Arachis	*pintoi*	Pinto Peanut	C:SD:**NT**
Cajanus	*cajan*	South Asia Pigeon Pea	C:PD:**VG**
Caragana	*arborescens*	Siberian Pea Tree Caragan Pea	C:PD:**VG**
Caragana	*frutex*	Bushy Caragan Pea	C:PD:**VG**
Caragana	*sinica*	Chinese Caragan Pea	C:PD:**VG**
Castanospermum	*australe*	Moreton Bay Chestnut Pea	C:PD:**VG**
Cicer	*arietinum*	Garbanzo Bean Chick Pea	C:SD:**VG**

Genus	Species	Specimen	Genome	C:Culinary
Cicer	*reticulatum*	Netted	Chick Pea	C:SD:**VG**
Inocarpus	*fagifer*	Tahiti	Dove Pea	C:PD:**VG**
Inocarpus	*papuanus*	Papuan	Dove Pea	C:PD:**VG**
Lemuropisum	*edule*	Edible	Tara Pea	C:PD:**VG**
Lupinus	*albus*	White	Lupine Pea	C:PD:**VG**
Lupinus	*libanica*	Lebanese	Lupine Pea	C:PD:**VG**
Psoralea	*esculenta*	Prairie Turnip	Scurf Pea	C:PD:**VG**
Tetragonolobus	*edulis*	Asparagus Pea	Winged Pea	C:PD:**VG**

Varieties

Pea, Snap Pea (*Pisum sativum-arvense*)

Genus	Species	infraspecies	cultivar	specimen	Genome
Pisum	*sativum-arvense-brevis*	*banana-nanoi*	Banana Nano ?.?'	Short Pod (<2.5") Snap	Pea
Pisum	*sativum-arvense-brevis*	*cukrowy-ilowieckii*	Cukrowy Ilowiecki	Short Pod (<2.5") Snap	Pea
Pisum	*sativum-arvense-brevis*	*little-sweetiei*	Little Sweetie ?.?'	Short Pod (<2.5") Snap	Pea
Pisum	*sativum-arvense-brevis*	*nordlandii*	Nordland 4.0'	Short Pod (<2.5") Snap	Pea
Pisum	*sativum-arvense-brevis*	*small-podded-china-pisa*	Small Pod China 2.2'	Short Pod (<2.5") Snap	Pea
Pisum	*sativum-arvense-brevis*	*snowhitei*	Snowhite 3.0'	Short Pod (<2.5") Snap	Pea
Pisum	*sativum-arvense-brevis*	*sweet-snapii*	Sweet Snap 2.9'	Short Pod (<2.5") Snap	Pea
Pisum	*sativum-arvense-media*	*amish-snapii*	Amish 5.0'	Medium Pod (2.5"-3.5") Snap	Pea
Pisum	*sativum-arvense-media*	*bon-apetitei-2*	Bon Apetite II ?.?'	Medium Pod (2.5"-3.5") Snap	Pea
Pisum	*sativum-arvense-media*	*braziliana*	Brazilian ?.?'	Medium Pod (2.5"-3.5") Snap	Pea
Pisum	*sativum-arvense-media*	*corgii*	Corgi 2.0'	Medium Pod (2.5"-3.5") Snap	Pea
Pisum	*sativum-arvense-media*	*dwarf-white-blossomii*	Dwarf White Blossom 3.0'	Medium Pod (2.5"-3.5") Snap	Pea
Pisum	*sativum-arvense-media*	*golden-edible-podii*	Golden Edible Pod 6.0'	Medium Pod (2.5"-3.5") Snap	Pea
Pisum	*sativum-arvense-media*	*golden-sweetii*	Golden Sweet 7.0'	Medium Pod (2.5"-3.5") Snap	Pea
Pisum	*sativum-arvense-media*	*goliathii*	Goliath 4.0'	Medium Pod (2.5"-3.5") Snap	Pea
Pisum	*sativum-arvense-media*	*hungariana*	Hungarian ?.?'	Medium Pod (2.5"-3.5") Snap	Pea
Pisum	*sativum-arvense-media*	*jadei*	Jade 3.0'	Medium Pod (2.5"-3.5") Snap	Pea
Pisum	*sativum-arvense-media*	*purpurea-poda*	Purple Podded 6.0'	Medium Pod (2.5"-3.5") Snap	Pea
Pisum	*sativum-arvense-media*	*rampicante-mangia-tuttoi*	Mangia Tutto Rampicante	Medium Pod (2.5"-3.5") Snap	Pea
Pisum	*sativum-arvense-media*	*sapporo-expressa*	Saporo Express ?.?'	Medium Pod (2.5"-3.5") Snap	Pea
Pisum	*sativum-arvense-media*	*snow-pea-shootsi*	Snow Pea Shoots	Medium Pod (2.5"-3.5") Snap	Pea

Pea, Long Pod Snap Pea (*Pisum sativum-arvense-macrocarpon*)

Genus	Species	infraspecies	cultivar	specimen	Genome
Pisum	*sativum-arvense-macrocarpon*	*megas*	Mega Sweet 2.7'	Long Pod (>3.5") Snap	Pea
Pisum	*sativum-arvense-macrocarpon*	*oregonesis-gigantea*	Oregon Giant 2.7'	Long Pod (>3.5") Snap	Pea
Pisum	*sativum-arvense-macrocarpon*	*sandyi*	Sandy 3.7'	Long Pod (>3.5") Snap	Pea
Pisum	*sativum-arvense-macrocarpon*	*snappyi*	Snappy 5.0'	Long Pod (>3.5") Snap	Pea
Pisum	*sativum-arvense-macrocarpon*	*yokomoensis-gigantea*	Yokomo Giant ?.?'	Long Pod (>3.5") Snap	Pea

Pea, Sugar Pea (*Pisum sativum-arvense-sucra*)

Genus	Species	infraspecies	cultivar	specimen	Genome
Pisum	*sativum-arvense-sucra-brevis*	*dwarf-white-sugari*	Dwarf White 2.2'	Short Pod (<2.5") Sugar	Pea
Pisum	*sativum-arvense-sucra-brevis*	*seedling-sugari*	Seedling 0.5'	Short Pod (<2.5") Sugar	Pea
Pisum	*sativum-arvense-sucra-media*	*carouby-de-mausannei*	Carouby De Maussane 4.5'	Medium Pod (2.5"-3.5") Sugar	Pea
Pisum	*sativum-arvense-sucra-media*	*cascadia*	Cascadia 3.0'	Medium Pod (2.5"-3.5") Sugar	Pea
Pisum	*sativum-arvense-sucra-media*	*dwarf-gray-sugari*	Dwarf Gray 2.2'	Medium Pod (2.5"-3.5") Sugar	Pea
Pisum	*sativum-arvense-sucra-media*	*dwarf-sugari*	Dwarf 2.4'	Medium Pod (2.5"-3.5") Sugar	Pea
Pisum	*sativum-arvense-sucra-media*	*grandma-hershey's-sugari*	Grandma Hershey's	Medium Pod (2.5"-3.5") Sugar	Pea
Pisum	*sativum-arvense-sucra-media*	*honey-podii*	Honey Pod 1.3'	Medium Pod (2.5"-3.5") Sugar	Pea
Pisum	*sativum-arvense-sucra-media*	*melii-sugari*	Sugar Mel 2.2'	Medium Pod (2.5"-3.5") Sugar	Pea
Pisum	*sativum-arvense-sucra-media*	*nivea-asiatica-blizzardii*	Blizzard Asian Snow Pea 2.5'	Medium Pod (2.5"-3.5") Sugar	Pea
Pisum	*sativum-arvense-sucra-media*	*nivea-svedica-bohlen'si*	Bohlen's Swedish Snow Pea 4.0'	Medium Pod (2.5"-3.5") Sugar	Pea
Pisum	*sativum-arvense-sucra-media*	*nivea-japonica-kawanai*	Kawana Japanese Snow Pea 1.2'	Medium Pod (2.5"-3.5") Sugar	Pea
Pisum	*sativum-arvense-sucra-media*	*nivea-chinensis*	Chinese Snow Pea 1.2'	Medium Pod (2.5"-3.5") Sugar	Pea
Pisum	*sativum-arvense-sucra-media*	*nivea-chinensis-alta*	Tall Chinese Snow Pea 6.0'	Medium Pod (2.5"-3.5") Sugar	Pea
Pisum	*sativum-arvense-sucra-media*	*nivea-chinensis-edula-poda*	Edible Pod Chinese Snow Pea	Medium Pod (2.5"-3.5") Sugar	Pea
Pisum	*sativum-arvense-sucra-media*	*nivea-chinensis-ho-lohn-dowii*	Ho Lohn Dow Chinese Snow Pea	Medium Pod (2.5"-3.5") Sugar	Pea
Pisum	*sativum-arvense-sucra-media*	*nivea-chinensis-manoa-sugari*	Manoa Chinese Snow Pea	Medium Pod (2.5"-3.5") Sugar	Pea
Pisum	*sativum-arvense-sucra-media*	*nivea-russica-sugari*	Russian Snow Pea	Medium Pod (2.5"-3.5") Sugar	Pea
Pisum	*sativum-arvense-sucra-media*	*nivea-snowbirdii*	Snowbird Snow Pea 1.4'	Medium Pod (2.5"-3.5") Sugar	Pea
Pisum	*sativum-arvense-sucra-media*	*nivea-snowwindii*	Snow Wind Snow Pea 1.5'	Medium Pod (2.5"-3.5") Sugar	Pea
Pisum	*sativum-arvense-sucra-media*	*norlii*	Norli 1.5'	Medium Pod (2.5"-3.5") Sugar	Pea
Pisum	*sativum-arvense-sucra-media*	*osaya-endoi*	Osaya Endo ?.?'	Medium Pod (2.5"-3.5") Sugar	Pea
Pisum	*sativum-arvense-sucra-media*	*risser-early-sugari*	Risser Early ?.?'	Medium Pod (2.5"-3.5") Sugar	Pea
Pisum	*sativum-arvense-sucra-media*	*sugar-annii*	Sugar Ann 2.2'	Medium Pod (2.5"-3.5") Sugar	Pea
Pisum	*sativum-arvense-sucra-media*	*sugar-bonii*	Sugar Bon 1.7'	Medium Pod (2.5"-3.5") Sugar	Pea
Pisum	*sativum-arvense-sucra-media*	*sugar-daddyi*	Sugar Daddy 2.2'	Medium Pod (2.5"-3.5") Sugar	Pea

Genus	Species	infraspecies	cultivar	specimen	Genome
Pisum	*sativum-arvense-sucra-media*	*sugar-gemii*	Sugar Gem 2.0'	Medium Pod (2.5"-3.5") Sugar	Pea
Pisum	*sativum-arvense-sucra-media*	*sugar-lacei*	Sugar Lace 2.5'	Medium Pod (2.5"-3.5") Sugar	Pea
Pisum	*sativum-arvense-sucra-media*	*sugar-lacei-2*	#2 Sugar Lace 2.0'	Medium Pod (2.5"-3.5") Sugar	Pea
Pisum	*sativum-arvense-sucra-media*	*sugar-lodei*	Sugar Lode 3.0'	Medium Pod (2.5"-3.5") Sugar	Pea
Pisum	*sativum-arvense-sucra-media*	*sugar-lordii*	Sugar Lord ?.?'	Medium Pod (2.5"-3.5") Sugar	Pea
Pisum	*sativum-arvense-sucra-media*	*sugar-popii*	Sugar Pop 1.5'	Medium Pod (2.5"-3.5") Sugar	Pea
Pisum	*sativum-arvense-sucra-media*	*sugar-raei*	Sugar Rae 2.2'	Medium Pod (2.5"-3.5") Sugar	Pea
Pisum	*sativum-arvense-sucra-media*	*sugar-snapii-cascadia*	Cascadia Sugar Snap 5.0'	Medium Pod (2.5"-3.5") Sugar	Pea
Pisum	*sativum-arvense-sucra-media*	*sugar-snapii-nana*	Dwarf Sugar Snap 1.7'	Medium Pod (2.5"-3.5") Sugar	Pea
Pisum	*sativum-arvense-sucra-media*	*sugar-snapii-praecox*	Early Sugar Snap 1.6'	Medium Pod (2.5"-3.5") Sugar	Pea
Pisum	*sativum-arvense-sucra-media*	*sugar-sprintii*	Sugar Sprint 2.1'	Medium Pod (2.5"-3.5") Sugar	Pea
Pisum	*sativum-arvense-sucra-media*	*sugar-stella*	Sugar Star 2.1'	Medium Pod (2.5"-3.5") Sugar	Pea
Pisum	*sativum-arvense-sucra-media*	*supra-sugar-snapii*	Super Sugar Snap 5.0'	Medium Pod (2.5"-3.5") Sugar	Pea
Pisum	*sativum-arvense-sucra-longa*	*hyogo-sugari*	Hyogo 3.6'	Long Pod (>3.5") Sugar	Pea
Pisum	*sativum-arvense-sucra-longa*	*mammoth-melting-sugari*	Mammoth Melting 4.5'	Long Pod (>3.5") Sugar	Pea
Pisum	*sativum-arvense-sucra-longa*	*melii-sugari-supra*	Super Sugar Mel ?.?'	Long Pod (>3.5") Sugar	Pea
Pisum	*sativum-arvense-sucra-longa*	*melting-sugari*	Melting 4.1'	Long Pod (>3.5") Sugar	Pea
Pisum	*sativum-arvense-sucra-longa*	*nivea-asiatica-snowflakei*	Snowflake Asian Snow Pea 2.2'	Long Pod (>3.5") Sugar	Pea
Pisum	*sativum-arvense-sucra-longa*	*nivea-chinensis-gigantea*	Giant Chinese Snow Pea	Long Pod (>3.5") Sugar	Pea
Pisum	*sativum-arvense-sucra-longa*	*nivea-chinensis-taichungii-11*	#11 Taichung Chinese Snow Pea	Long Pod (>3.5") Sugar	Pea
Pisum	*sativum-arvense-sucra-longa*	*nivea-chinensis-taichungii-13*	#13 Taichung Chinese Snow Pea	Long Pod (>3.5") Sugar	Pea
Pisum	*sativum-arvense-sucra-longa*	*nivea-schweizer-riesenii*	Giant Swiss Snow Pea 5.0'	Long Pod (>3.5") Sugar	Pea
Pisum	*sativum-arvense-sucra-longa*	*oregonensis-gigantea-sugar-podii*	Oregon Giant Sugar Pod 2.7'	Long Pod (>3.5") Sugar	Pea
Pisum	*sativum-arvense-sucra-longa*	*oregonensis-sugar-podii*	Oregon Sugar Pod 2.3'	Long Pod (>3.5") Sugar	Pea
Pisum	*sativum-arvense-sucra-longa*	*oregonensis-sugar-podii-2*	#2 Oregon Sugar Pod 2.3'	Long Pod (>3.5") Sugar	Pea
Pisum	*sativum-arvense-sucra-longa*	*sugar-podii-2*	#2 Sugar Pod ?.?'	Long Pod (>3.5") Sugar	Pea
Pisum	*sativum-arvense-sucra-longa*	*sugaree-snapii*	Sugaree Snap 8.0'	Long Pod (>3.5") Sugar	Pea
Pisum	*sativum-arvense-sucra-longa*	*supra-snappyi*	Super Snappy 2.5'	Long Pod (>3.5") Sugar	Pea
Pisum	*sativum-arvense-sucra-longa*	*taiwan-sugari*	Taiwan ?.?'	Long Pod (>3.5") Sugar	Pea

Pea, Bush Pea (*Pisum sativum-elatius*)

Genus	Species	infraspecies	cultivar	specimen	Genome
Pisum	*sativum-elatius-media*	*arisea*	Arise 1.4'	Medium Pod (5-8 pea) Bush	Pea
Pisum	*sativum-elatius-media*	*aristagroena*	Aristagreen 1.7'	Medium Pod (5-8 pea) Bush	Pea
Pisum	*sativum-elatius-media*	*balmoralis*	Balmoral 1.4'	Medium Pod (5-8 pea) Bush	Pea
Pisum	*sativum-elatius-media*	*beaglei*	Beagle 1.7'	Medium Pod (5-8 pea) Bush	Pea
Pisum	*sativum-elatius-media*	*bountyi*	Bounty 1.2'	Medium Pod (5-8 pea) Bush	Pea
Pisum	*sativum-elatius-media*	*corvalettea*	Corvalette 1.8'	Medium Pod (5-8 pea) Bush	Pea
Pisum	*sativum-elatius-media*	*daybreakii*	Daybreak 1.8'	Medium Pod (5-8 pea) Bush	Pea
Pisum	*sativum-elatius-media*	*felthamii-firstii*	Feltham First 1.5'	Medium Pod (5-8 pea) Bush	Pea
Pisum	*sativum-elatius-media*	*little-marvelii*	Little Marvel 1.5'	Medium Pod (5-8 pea) Bush	Pea
Pisum	*sativum-elatius-media*	*little-marvelii-improva*	Improved Little Marvel 1.5'	Medium Pod (5-8 pea) Bush	Pea
Pisum	*sativum-elatius-media*	*little-marvelii-montanaensis*	Montana Little Marvel 1.5'	Medium Pod (5-8 pea) Bush	Pea
Pisum	*sativum-elatius-media*	*meteori-rotunda-sperma*	Round Seeded Meteor 1.2'	Medium Pod (5-8 pea) Bush	Pea
Pisum	*sativum-elatius-media*	*meteori-tom-thumbii*	Tom Thumb Meteor 0.7'	Medium Pod (5-8 pea) Bush	Pea
Pisum	*sativum-elatius-media*	*monicoi*	Monico 1.8'	Medium Pod (5-8 pea) Bush	Pea
Pisum	*sativum-elatius-media*	*petita-provencalis*	Petit Provincal 1.3'	Medium Pod (5-8 pea) Bush	Pea
Pisum	*sativum-elatius-media*	*sparklei*	Sparkle 1.4'	Medium Pod (5-8 pea) Bush	Pea
Pisum	*sativum-elatius-media*	*springii*	Spring 1.7'	Medium Pod (5-8 pea) Bush	Pea
Pisum	*sativum-elatius-media*	*spritei*	Sprite ?.?'	Medium Pod (5-8 pea) Bush	Pea
Pisum	*sativum-elatius-longa*	*boleroi*	Bolero 1.4'	Long Pod (7-12 pea) Bush	Pea
Pisum	*sativum-elatius-longa*	*burpeeana-praecox*	Burpee Early 1.7'	Long Pod (7-12 pea) Bush	Pea
Pisum	*sativum-elatius-longa*	*knightii*	Knight 1.7'	Long Pod (7-12 pea) Bush	Pea
Pisum	*sativum-elatius-longa*	*laxtonii-blue-bantamii*	Blue Bantum Laxton 1.4'	Long Pod (7-12 pea) Bush	Pea
Pisum	*sativum-elatius-longa*	*laxtonii-early-birdii*	Early Bird Laxton 1.7'	Long Pod (7-12 pea) Bush	Pea
Pisum	*sativum-elatius-longa*	*laxtonii-hundredfoldi*	Hundredfold Laxton 1.7'	Long Pod (7-12 pea) Bush	Pea
Pisum	*sativum-elatius-longa*	*laxtonii-laxtoniana*	Laxtonian Laxton 1.7'	Long Pod (9-12 pea) Shell	Pea
Pisum	*sativum-elatius-longa*	*laxtonii-progressa*	Progress Laxton 1.4'	Long Pod (7-12 pea) Bush	Pea
Pisum	*sativum-elatius-longa*	*laxtonii-progressa-9*	#9 Progress Laxton 1.4'	Long Pod (7-12 pea) Bush	Pea
Pisum	*sativum-elatius-longa*	*laxtonii-progressa-9-morse'si*	Morse's #9 Progress Laxton 1.4'	Long Pod (7-12 pea) Bush	Pea
Pisum	*sativum-elatius-longa*	*laxtonii-progressa-improva*	Improved Progress Laxton 1.4'	Long Pod (7-12 pea) Bush	Pea
Pisum	*sativum-elatius-longa*	*laxtonii-progressa-olympia*	Olympia Laxton Progress 1.4'	Long Pod (7-12 pea) Bush	Pea
Pisum	*sativum-elatius-longa*	*novella-lacyi-ladiga*	Lacy Lady Novell 1.5'	Long Pod (7-12 pea) Bush	Pea
Pisum	*sativum-elatius-longa*	*mesai*	Mesa 1.4'	Long Pod (7-12 pea) Bush	Pea
Pisum	*sativum-elatius-longa*	*patriotii*	Patriot 1.9'	Long Pod (7-12 pea) Bush	Pea
Pisum	*sativum-elatius-suavis*	*americana-vondera*	American Wonder 1.2'	Bush Sweet	Pea
Pisum	*sativum-elatius-suavis*	*allsweetii-3*	Allsweet III 1.8'	Bush Sweet	Pea
Pisum	*sativum-elatius-suavis*	*citadela*	Citadel 1.2'	Bush Sweet	Pea
Pisum	*sativum-elatius-suavis*	*citadela-fristoi*	Fristo Citadel 1.2'	Bush Sweet	Pea
Pisum	*sativum-elatius-suavis*	*kelvedonii-vondera*	Kelvedon Wonder 1.5'	Bush Sweet	Pea
Pisum	*sativum-elatius-suavis*	*payloadii*	Payload 1.9'	Bush Sweet	Pea

Genus	Species	infraspecies	cultivar	specimen	Genome
Pisum	*sativum-elatius-suavis*	*pisello-nono-sole-di-sicilia*	Sicilian Sunshine	Bush Sweet	Pea
Pisum	*sativum-elatius-suavis*	*rallyi*	Rally 1.6'	Bush Sweet	Pea
Pisum	*sativum-elatius-suavis*	*twinklei*	Twinkle 1.8'	Bush Sweet	Pea
Pisum	*sativum-elatius-suavis*	*waverexii*	Waverex 1.5'	Bush Sweet	Pea

Pea, Vine Pea (*Pisum sativum-hortense*)

Genus	Species	infraspecies	cultivar	specimen	Genome
Pisum	*sativum-hortense-media*	*ambassadori*	Ambassador 2.5'	Medium Pod (5-8 pea) Vine	Pea
Pisum	*sativum-hortense-media*	*anoka*	Anoka ?.?'	Medium Pod (5-8 pea) Vine	Pea
Pisum	*sativum-hortense-media*	*annonayensis*	Annonay ?.?'	Medium Pod (5-8 pea) Vine	Pea
Pisum	*sativum-hortense-media*	*argenteum*	Silver Leaf 2.0'	Medium Pod (5-8 pea) Vine	Pea
Pisum	*sativum-hortense-media*	*austriensis*	Austrian Winter ?.?'	Medium Pod (5-8 pea) Vine	Pea
Pisum	*sativum-hortense-media*	*banquetii*	Banquet 2.0'	Medium Pod (5-8 pea) Vine	Pea
Pisum	*sativum-hortense-media*	*cabreei*	Cabree ?.?'	Medium Pod (5-8 pea) Vine	Pea
Pisum	*sativum-hortense-media*	*caractacus*	Caratac ?.?'	Medium Pod (5-8 pea) Vine	Pea
Pisum	*sativum-hortense-media*	*chicharoi-criolloi*	Chick Criole 5.0'	Medium Pod (5-8 pea) Vine	Pea
Pisum	*sativum-hortense-media*	*chicharoi-criolloi-zapoteca*	Zapotec Chick Criole 5.0'	Medium Pod (5-8 pea) Vine	Pea
Pisum	*sativum-hortense-media*	*coralis*	Coral 2.0'	Medium Pod (5-8 pea) Vine	Pea
Pisum	*sativum-hortense-media*	*cukor-borsi*	Cukor Borse ?.?'	Medium Pod (5-8 pea) Vine	Pea
Pisum	*sativum-hortense-media*	*earlyi-frostii*	Early Frosty 2.4'	Medium Pod (5-8 pea) Vine	Pea
Pisum	*sativum-hortense-media*	*earlyi-vondera*	Early Wonder ?.?'	Medium Pod (5-8 pea) Vine	Pea
Pisum	*sativum-hortense-media*	*fairbeard'si-nonpareilii*	Fairbeard's Nonpareil 3.0'	Medium Pod (5-8 pea) Vine	Pea
Pisum	*sativum-hortense-media*	*fortunei*	Fortune ?.?'	Medium Pod (5-8 pea) Vine	Pea
Pisum	*sativum-hortense-media*	*freezeri-69*	#69 Freezer 2.2'	Medium Pod (5-8 pea) Vine	Pea
Pisum	*sativum-hortense-media*	*freezeri-bertholdii*	Berthold Freezer ?.?'	Medium Pod (5-8 pea) Vine	Pea
Pisum	*sativum-hortense-media*	*freezeri-earlyi-680*	Early #680 Freezer	Medium Pod (5-8 pea) Vine	Pea
Pisum	*sativum-hortense-media*	*freezeri-homei*	Home Freezer ?.?'	Medium Pod (5-8 pea) Vine	Pea
Pisum	*sativum-hortense-media*	*frizzyi*	Frizzy ?.?'	Medium Pod (5-8 pea) Vine	Pea
Pisum	*sativum-hortense-media*	*gigantea-a-fiorei-violettoi*	Giant Fire Violet ?.?'	Medium Pod (5-8 pea) Vine	Pea
Pisum	*sativum-hortense-media*	*gloirei-dei-quimperi*	Quimper Glory ?.?'	Medium Pod (5-8 pea) Vine	Pea
Pisum	*sativum-hortense-media*	*hatifa-dea-anonayi*	Hatif De Anonay 2.0'	Medium Pod (5-8 pea) Vine	Pea
Pisum	*sativum-hortense-media*	*jaguari*	Jaguar 2.3'	Medium Pod (5-8 pea) Vine	Pea
Pisum	*sativum-hortense-media*	*karisma*	Karisma ?.?'	Medium Pod (5-8 pea) Vine	Pea
Pisum	*sativum-hortense-media*	*kickhamii*	Kickham ?.?'	Medium Pod (5-8 pea) Vine	Pea
Pisum	*sativum-hortense-media*	*kurume-hi-cropii*	Kurume Hi Crop ?.?'	Medium Pod (5-8 pea) Vine	Pea
Pisum	*sativum-hortense-media*	*markana*	Markana 2.5'	Medium Pod (5-8 pea) Vine	Pea
Pisum	*sativum-hortense-media*	*mrs.-van'si*	Mrs. Van's ?.?'	Medium Pod (5-8 pea) Vine	Pea
Pisum	*sativum-hortense-media*	*multistella*	Multistar 2.5'	Medium Pod (5-8 pea) Vine	Pea
Pisum	*sativum-hortense-media*	*o'odhamii-adapta-importa*	O'odham Adapted Import	Medium Pod (5-8 pea) Vine	Pea
Pisum	*sativum-hortense-media*	*onwardii*	Onward 2.0'	Medium Pod (5-8 pea) Vine	Pea
Pisum	*sativum-hortense-media*	*onwardii-earlyi*	Early Onward 2.0'	Medium Pod (5-8 pea) Vine	Pea
Pisum	*sativum-hortense-media*	*paladiosa*	Paladio 2.1'	Medium Pod (5-8 pea) Vine	Pea
Pisum	*sativum-hortense-media*	*parsleyi*	Parsley ?.?'	Medium Pod (5-8 pea) Vine	Pea
Pisum	*sativum-hortense-media*	*peeweei*	Peewee ?.?'	Medium Pod (5-8 pea) Vine	Pea
Pisum	*sativum-hortense-media*	*picoloi-provenzalei*	Picolo Provenzale 3.0'	Medium Pod (5-8 pea) Vine	Pea
Pisum	*sativum-hortense-media*	*piona*	Pioneer 2.0'	Medium Pod (5-8 pea) Vine	Pea
Pisum	*sativum-hortense-media*	*piona-oregonensis*	Oregon Pioneer 2.0'	Medium Pod (5-8 pea) Vine	Pea
Pisum	*sativum-hortense-media*	*plein-de-panieri*	Full Basket ?.?'	Medium Pod (5-8 pea) Vine	Pea
Pisum	*sativum-hortense-media*	*prince-albertii*	Prince Albert ?.?'	Medium Pod (5-8 pea) Vine	Pea
Pisum	*sativum-hortense-media*	*serpettea*	Serpette ?.?'	Medium Pod (5-8 pea) Vine	Pea
Pisum	*sativum-hortense-media*	*shoshonei*	Shoshone ?.?'	Medium Pod (5-8 pea) Vine	Pea
Pisum	*sativum-hortense-media*	*sicklei*	Risser Sickle Pod	Medium Pod (5-8 pea) Vine	Pea
Pisum	*sativum-hortense-media*	*simai*	Sima ?.?'	Medium Pod (5-8 pea) Vine	Pea
Pisum	*sativum-hortense-media*	*saumar-pisa*	Summer Pea 4.5'	Medium Pod (5-8 pea) Vine	Pea
Pisum	*sativum-hortense-media*	*survivori*	Survivor 2.0'	Medium Pod (5-8 pea) Vine	Pea
Pisum	*sativum-hortense-media*	*sutton's-harbingeri*	Sutton's Harbinger 2.5'	Medium Pod (5-8 pea) Vine	Pea
Pisum	*sativum-hortense-media*	*tacomensis*	Tacoma ?.?'	Medium Pod (5-8 pea) Vine	Pea
Pisum	*sativum-hortense-media*	*victory-freezeri*	Victory Freezer 2.5'	Medium Pod (5-8 pea) Vine	Pea
Pisum	*sativum-hortense-media*	*willet-wonderi*	Willet Wonder ?.?'	Medium Pod (5-8 pea) Vine	Pea
Pisum	*sativum-hortense-media*	*world's-recordii*	World's Record 2.5'	Medium Pod (5-8 pea) Vine	Pea
Pisum	*sativum-hortense-longa*	*aldermanii-telefono*	Telephone Alderman 5.5'	Long Pod (7-12 pea) Vine	Pea
Pisum	*sativum-hortense-longa*	*aldermanii-telefono-alta*	Tall Telephone Alderman 5.5'	Long Pod (7-12 pea) Vine	Pea
Pisum	*sativum-hortense-longa*	*aldermanii-telefono-darca-poda*	Dark Pod Telephone Alderman 5.5'	Long Pod (7-12 pea) Vine	Pea
Pisum	*sativum-hortense-longa*	*aldermanii-telefono-gigantea-stridea*	Giant Stride Telephone Alderman 2.6'	Long Pod (7-12 pea) Vine	Pea
Pisum	*sativum-hortense-longa*	*aldermanii-telefono-nana*	Dwarf Telephone Alderman 2.0'	Long Pod (7-12 pea) Vine	Pea
Pisum	*sativum-hortense-longa*	*big-skyi*	Big Sky ?.?'	Long Pod (7-12 pea) Vine	Pea
Pisum	*sativum-hortense-longa*	*canoea*	Canoe 2.3'	Long Pod (7-12 pea) Vine	Pea
Pisum	*sativum-hortense-longa*	*cariboua*	Caribou 3.0'	Long Pod (7-12 pea) Vine	Pea
Pisum	*sativum-hortense-longa*	*champion-anglia*	Champion Of England Wrinkled	Long Pod (7-12 pea) Vine	Pea
Pisum	*sativum-hortense-longa*	*dualis*	Dual 2.5'	Long Pod (7-12 pea) Vine	Pea
Pisum	*sativum-hortense-longa*	*endeavori*	Endeavor 2.4'	Long Pod (7-12 pea) Vine	Pea
Pisum	*sativum-hortense-longa*	*frostyi*	Frosty 2.4'	Long Pod (7-12 pea) Vine	Pea

Genus	Species	infraspecies	cultivar	specimen	Genome
Pisum	*sativum-hortense-longa*	*galena-FP2237*	Galena ?.?'	Long Pod (7-12 pea) Vine	Pea
Pisum	*sativum-hortense-longa*	*groenasagea*	Greensage 2.4'	Long Pod (7-12 pea) Vine	Pea
Pisum	*sativum-hortense-longa*	*groena-shafta*	Green Arrow 2.3'	Long Pod (7-12 pea) Vine	Pea
Pisum	*sativum-hortense-longa*	*groena-shafta-alta*	Tall Green Arrow ?.?'	Long Pod (7-12 pea) Vine	Pea
Pisum	*sativum-hortense-longa*	*groena-shafta-hispanica*	Spanish Green Arrow	Long Pod (7-12 pea) Vine	Pea
Pisum	*sativum-hortense-longa*	*groena-shafta-hispanica-lincolnii*	Lincoln Spanish Green Arrow	Long Pod (7-12 pea) Vine	Pea
Pisum	*sativum-hortense-longa*	*groena-shafta-hispanica-skyscraperi*	Skyscraper Spanish Green Arrow 5.5'	Long Pod (7-12 pea) Vine	Pea
Pisum	*sativum-hortense-longa*	*groena-shafta-hurstii*	Hurst Wrinkled Green Arrow 2.3'	Long Pod (7-12 pea) Vine	Pea
Pisum	*sativum-hortense-longa*	*homesteaderi*	Homesteader 2.1'	Long Pod (7-12 pea) Vine	Pea
Pisum	*sativum-hortense-longa*	*homesteaderi-lincolnii*	Lincoln Homesteader 2.1'	Long Pod (7-12 pea) Vine	Pea
Pisum	*sativum-hortense-longa*	*laxtonii-freezoniana-thomasi-improva*	Improved Freezonian **Thomas Laxton** 2.8'	Long Pod (7-12 pea) Vine	Pea
Pisum	*sativum-hortense-longa*	*laxtonii-freezoniana-thomasi-maritima*	Maritime Freezonian **Thomas Laxton** 2.8'	Long Pod (7-12 pea) Vine	Pea
Pisum	*sativum-hortense-longa*	*laxtonii-freezoniana-thomasi-rugosa*	Wrinkled **Freezonian** Thomas Laxton 2.8'	Long Pod (7-12 pea) Vine	Pea
Pisum	*sativum-hortense-longa*	*laxtonii-superba*	Superb Laxton 2.1'	Long Pod (7-12 pea) Vine	Pea
Pisum	*sativum-hortense-longa*	*manitoba*	Manitoba ?.?'	Long Pod (7-12 pea) Vine	Pea
Pisum	*sativum-hortense-longa*	*neo-centuria*	New Century 2.7'	Long Pod (7-12 pea) Vine	Pea
Pisum	*sativum-hortense-longa*	*novella-2*	#2 Novell 2.3'	Long Pod (7-12 pea) Vine	Pea
Pisum	*sativum-hortense-longa*	*novella-bikinii*	Bikini Novell 2.0'	Long Pod (7-12 pea) Vine	Pea
Pisum	*sativum-hortense-longa*	*oregonensis-traila*	Oregon Trail ?.?'	Long Pod (7-12 pea) Vine	Pea
Pisum	*sativum-hortense-longa*	*perfecta-darca-groena*	Dark Green Perfection 2.4'	Long Pod (7-12 pea) Vine	Pea
Pisum	*sativum-hortense-longa*	*perfecta-hylitea*	Hylite Perfection 2.4'	Long Pod (7-12 pea) Vine	Pea
Pisum	*sativum-hortense-longa*	*perfecta-praecox*	Early Perfection 2.4'	Long Pod (7-12 pea) Vine	Pea
Pisum	*sativum-hortense-longa*	*rondosa*	Rondo 2.4'	Long Pod (7-12 pea) Vine	Pea
Pisum	*sativum-hortense-longa*	*samishii*	Shelling Amish 2.0'	Long Pod (7-12 pea) Vine	Pea
Pisum	*sativum-hortense-longa*	*saturnus*	Saturn 2.5'	Long Pod (7-12 pea) Vine	Pea
Pisum	*sativum-hortense-longa*	*semileaflessii*	Semileafless 2.5'	Long Pod (7-12 pea) Vine	Pea
Pisum	*sativum-hortense-longa*	*somersetensis*	Somerset ?.?'	Long Pod (7-12 pea) Vine	Pea
Pisum	*sativum-hortense-longa*	*stratagemii*	Stratagem 2.0'	Long Pod (7-12 pea) Vine	Pea
Pisum	*sativum-hortense-longa*	*stratagemii-improva*	Improved Stratagem 2.4'	Long Pod (7-12 pea) Vine	Pea
Pisum	*sativum-hortense-longa*	*pilotii*	Pilot 4.0'	Long Pod (7-12 pea) Vine	Pea
Pisum	*sativum-hortense-longa*	*top-podii*	Top Pod 2.0'	Long Pod (7-12 pea) Vine	Pea
Pisum	*sativum-hortense-longa*	*triple-treatii*	Triple Treat 2.5'	Long Pod (7-12 pea) Vine	Pea
Pisum	*sativum-hortense-longa*	*utrillosa*	Utrillo 2.5'	Long Pod (7-12 pea) Vine	Pea
Pisum	*sativum-hortense-longa*	*wandosa*	Wando 2.2'	Long Pod (7-12 pea) Vine	Pea

Pea, Split Pea (*Pisum sativum-hortense-schiza*)

Genus	Species	infraspecies	cultivar	specimen	Genome
Pisum	*sativum-hortense-schiza*	*alascensis-maestro*	Maestro Alaska 2.2'	Vine Split	Pea
Pisum	*sativum-hortense-schiza*	*alascensis-praeissima*	Earliest Of All Alaska 2.5'	Vine Split	Pea
Pisum	*sativum-hortense-schiza*	*alascensis-praecox*	Early Alaska 2.3'	Vine Split	Pea
Pisum	*sativum-hortense-schiza*	*alascensis-praecox-extra*	Extra Early Alaska 2.5'	Vine Split	Pea

Pea, Soup Pea (*Pisum sativum-hortens-sopius*)

Genus	Species	infraspecies	cultivar	specimen	Genome
Pisum	*sativum-hortense-sopius*	*alverjonii-temporalis*	Temporal Alverjon	Vine Soup	Pea
Pisum	*sativum-hortense-sopius*	*alverjonii-truchas*	Truchas Alverjon ?.?'	Vine Soup	Pea
Pisum	*sativum-hortense-sopius*	*amplissimo-victoria*	Amplissimo Victoria	Vine Soup	Pea
Pisum	*sativum-hortense-sopius*	*attari*	Attar ?.?'	Vine Soup	Pea
Pisum	*sativum-hortense-sopius*	*australiensis*	Australian 5.0'	Vine Soup	Pea
Pisum	*sativum-hortense-sopius*	*austriensis-vintera*	Winter Austrian ?.?'	Vine Soup	Pea
Pisum	*sativum-hortense-sopius*	*bill-jump's-soup-pisa*	Bill Jump's ?.?'	Vine Soup	Pea
Pisum	*sativum-hortense-sopius*	*bisilia-marketii*	Bisilia ?.?'	Vine Soup	Pea
Pisum	*sativum-hortense-sopius*	*capucijnersi*	Capuchin 2.5'	Vine Soup	Pea
Pisum	*sativum-hortense-sopius*	*capucijnersi-alta*	Tall Capuchin ?.?'	Vine Soup	Pea
Pisum	*sativum-hortense-sopius*	*capucijnersi-blauea-poda*	Blue Pod Capuchin 4.5'	Vine Soup	Pea
Pisum	*sativum-hortense-sopius*	*capucijnersi-eroica*	Eroica Capuchin 3.0'	Vine Soup	Pea
Pisum	*sativum-hortense-sopius*	*capucijnersi-franco-nana*	Dwarf French Capuchin 2.0'	Vine Soup	Pea
Pisum	*sativum-hortense-sopius*	*capucijnersi-gruno-rosynii*	Gruno Rosyn Capuchin 3.0'	Vine Soup	Pea
Pisum	*sativum-hortense-sopius*	*capucijnersi-halai*	Hala Capuchin ?.?'	Vine Soup	Pea
Pisum	*sativum-hortense-sopius*	*capucijnersi-hollandica*	Holland Wrinkled Capuchin	Vine Soup	Pea
Pisum	*sativum-hortense-sopius*	*capucijnersi-margaret-mckee's-bakingii*	Margaret McKee's Baking Capuchin 4.0'	Vine Soup	Pea
Pisum	*sativum-hortense-sopius*	*capucijnersi-nederlandica*	Dutch Capuchin 2.0'	Vine Soup	Pea
Pisum	*sativum-hortense-sopius*	*capucijnersi-raisinii*	Raisin Capuchin 2.0'	Vine Soup	Pea
Pisum	*sativum-hortense-sopius*	*carlinii*	Carlin ?.?	Vine Soup	Pea
Pisum	*sativum-hortense-sopius*	*cremella*	Creamette 5.0'	Vine Soup	Pea
Pisum	*sativum-hortense-sopius*	*cremella-centauria*	Century Creamette	Vine Soup	Pea
Pisum	*sativum-hortense-sopius*	*desireei-blauea-poda*	Blue Pod Desire ?.?'	Vine Soup	Pea
Pisum	*sativum-hortense-sopius*	*ervilha-torta-flor-roxa*	Red Flowering ?.?'	Vine Soup	Pea
Pisum	*sativum-hortense-sopius*	*hollandica-brunnea*	Holland Brown ?.?'	Vine Soup	Pea

Genus	Species	infraspecies	cultivar	specimen	Genome
Pisum	*sativum-hortense-sopius*	*elda-scotica*	Old Scotch 5.0'	Vine Soup	Pea
Pisum	*sativum-hortense-sopius*	*kazankijii*	Kazankij 4.0'	Vine Soup	Pea
Pisum	*sativum-hortense-sopius*	*kimberleyi*	Kimberley ?.?'	Vine Soup	Pea
Pisum	*sativum-hortense-sopius*	*king-tutii*	King Tut 4.0'	Vine Soup	Pea
Pisum	*sativum-hortense-sopius*	*marrowfatii*	Marrowfat ?.?'	Vine Soup	Pea
Pisum	*sativum-hortense-sopius*	*mellow-yellowi*	Mellow Yellow ?.?'	Vine Soup	Pea
Pisum	*sativum-hortense-sopius*	*mennonitea*	Mennonite ?.?'	Vine Soup	Pea
Pisum	*sativum-hortense-sopius*	*mexicana-sopia*	Mexican 6.0'	Vine Soup	Pea
Pisum	*sativum-hortense-sopius*	*monk-pisa*	Monk ?.?'	Vine Soup	Pea
Pisum	*sativum-hortense-sopius*	*ojitoi*	Ojito ?.?'	Vine Soup	Pea
Pisum	*sativum-hortense-sopius*	*o'odhamii*	O'odham ?.?'	Vine Soup	Pea
Pisum	*sativum-hortense-sopius*	*prussiana-blauea*	Prussian Blue ?.?'	Vine Soup	Pea
Pisum	*sativum-hortense-sopius*	*purpurea-passion*	Purple Passion 5.0'	Vine Soup	Pea
Pisum	*sativum-hortense-sopius*	*rotunda-groena*	Round Green 2.0'	Vine Soup	Pea
Pisum	*sativum-hortense-sopius*	*russica*	Russian ?.?'	Vine Soup	Pea
Pisum	*sativum-hortense-sopius*	*san-luisensis*	San Luis ?.?'	Vine Soup	Pea
Pisum	*sativum-hortense-sopius*	*st.-hubertii*	St. Hubert ?.?'	Vine Soup	Pea
Pisum	*sativum-hortense-sopius*	*subrufrum*	Mahogany Purple 6.0'	Vine Soup	Pea
Pisum	*sativum-hortense-sopius*	*taosi*	Taos ?.?'	Vine Soup	Pea
Pisum	*sativum-hortense-sopius*	*tepehuana*	Tepehuan ?.?'	Vine Soup	Pea
Pisum	*sativum-hortense-sopius*	*trapperi*	Trapper Field ?.?'	Vine Soup	Pea
Pisum	*sativum-hortense-sopius*	*turcensis-USDA-PL-120630*	Turkish 4.5'	Vine Soup	Pea
Pisum	*sativum-hortense-sopius*	*vaditoi*	Vadito ?.?'	Vine Soup	Pea
Pisum	*sativum-hortense-sopius*	*yoquivo-del-surensis*	Yoquivo Del Sur ?.?'	Vine Soup	Pea
Pisum	*sativum-hortense-sopius*	*zeiner's-gelda*	Zeiner's Gold 6.5'	Vine Soup	Pea

Pea, Sweet Pea (*Pisum sativum-hortense-suavis*)

Genus	Species	infraspecies	cultivar	specimen	Genome
Pisum	*sativum-hortense-suavis-media*	*aurea-suavis-vigna*	Golden Sweet ?.?	Medium Pod (5-8 pea) Vine Sweet	Pea
Pisum	*sativum-hortense-suavis-media*	*brittanica-vondera*	British Wonder 2.6'	Medium Pod (5-8 pea) Vine Sweet	Pea
Pisum	*sativum-hortense-suavis-media*	*caseloadii*	Caseload 2.5'	Medium Pod (5-8 pea) Vine Sweet	Pea
Pisum	*sativum-hortense-suavis-media*	*eclipsea*	Eclipse 3.0'	Medium Pod (5-8 pea) Vine Sweet	Pea
Pisum	*sativum-hortense-suavis-media*	*maxigoltii*	Maxigolt 4.0'	Medium Pod (5-8 pea) Vine Sweet	Pea
Pisum	*sativum-hortense-suavis-media*	*pugetii*	Puget ?.?'	Medium Pod (5-8 pea) Vine Sweet	Pea
Pisum	*sativum-hortense-suavis-media*	*quantum*	Quantum ?.?'	Medium Pod (5-8 pea) Vine Sweet	Pea
Pisum	*sativum-hortense-suavis-media*	*recorda*	Record 3.0'	Medium Pod (5-8 pea) Vine Sweet	Pea
Pisum	*sativum-hortense-suavis-media*	*sierra-madre-del-suri*	Sierra Madre Del Sur	Medium Pod (5-8 pea) Vine Sweet	Pea
Pisum	*sativum-hortense-suavis-longa*	*dakotensis*	Dakota 3.0'	Long Pod (7-12 pea) Vine Sweet	Pea
Pisum	*sativum-hortense-suavis-longa*	*gardenia-suavis*	Garden Sweet 3.5'	Long Pod (7-12 pea) Vine Sweet	Pea
Pisum	*sativum-hortense-suavis-longa*	*mayfairi*	Mayfair 3.5'	Long Pod (7-12 pea) Vine Sweet	Pea
Pisum	*sativum-hortense-suavis-longa*	*miragrouena*	Miragreen 4.0'	Long Pod (7-12 pea) Vine Sweet	Pea
Pisum	*sativum-hortense-suavis-longa*	*mr.-bigii*	Mr. Big 3.5'	Long Pod (7-12 pea) Vine Sweet	Pea
Pisum	*sativum-hortense-suavis-longa*	*norda-suavis*	Northern Sweet 2.1'	Long Pod (7-12 pea) Vine Sweet	Pea
Pisum	*sativum-hortense-suavis-longa*	*petita-poisi*	Petit Pois 2.5'	Long Pod (7-12 pea) Vine Sweet	Pea
Pisum	*sativum-hortense-suavis-longa*	*petita-poisi-elfii*	Elf Petit Pois 2.5'	Long Pod (7-12 pea) Vine Sweet	Pea
Pisum	*sativum-hortense-suavis-longa*	*petita-poisi-giroyi*	Giroy Petit Pois 3.5'	Long Pod (7-12 pea) Vine Sweet	Pea

Pea, Hog Peanut (*Amphicarpaea bracteata*)

Genus	Species	infraspecies	cultivar	specimen	Genome
Amphicarpaea	*bracteata*	*bracteata*	Original	American	Hog Peanut
Amphicarpaea	*bracteata*	*chamaecaulis*	Chamise Stem	American	Hog Peanut
Amphicarpaea	*bracteata*	*ciliata*	Ciliated	American	Hog Peanut
Amphicarpaea	*bracteata*	*comosa*	Tufted	American	Hog Peanut
Amphicarpaea	*bracteata*	*cuspidata*	Cuspidate	American	Hog Peanut
Amphicarpaea	*bracteata*	*deltifolia*	Triange Leaf	American	Hog Peanut
Amphicarpaea	*bracteata*	*edgeworthii*	Edgeworth	American	Hog Peanut
Amphicarpaea	*bracteata*	*elliottii*	Elliot	American	Hog Peanut
Amphicarpaea	*bracteata*	*heterophyla*	Variable Leaf	American	Hog Peanut
Amphicarpaea	*bracteata*	*japonica*	Japanese	American	Hog Peanut
Amphicarpaea	*bracteata*	*monoica*	Monoicious	American	Hog Peanut
Amphicarpaea	*bracteata*	*pitcheri*	Pritcher	American	Hog Peanut
Amphicarpaea	*bracteata*	*sarmentosa*	Runner	American	Hog Peanut
Amphicarpaea	*bracteata*	*villosa*	Villose	American	Hog Peanut

Pea, Peanut (*Arachis hypogaea*)

Genus	Species	infraspecies	cultivar	specimen	Genome
Arachis	*hypogaea*	*hypogaea*	Original	Goober	Peanut
Arachis	*hypogaea*	*arizonensis-nativa*	ArizonaNative	Goober	Peanut
Arachis	*hypogaea*	*carolinensis-nigra*	Carolina Black	Goober	Peanut
Arachis	*hypogaea*	*equadorensis*	Equadorian Purple	Goober	Peanut
Arachis	*hypogaea*	*florunneri*	Florunner	Goober	Peanut
Arachis	*hypogaea*	*garoyi*	Garoy	Goober	Peanut
Arachis	*hypogaea*	*georgiensis-groena*	Georgia Green	Goober	Peanut

Genus	Species	infraspecies	cultivar	specimen	Genome
Arachis	*hypogaea*	*hispanica-valencia*	Valencia Spanish	Goober	Peanut
Arachis	*hypogaea*	*hispanica-improva*	Improved Spanish	Goober	Peanut
Arachis	*hypogaea*	*hispanica-praecox*	Early Spanish	Goober	Peanut
Arachis	*hypogaea*	*hispanica-pronto*	Pronto Spanish	Goober	Peanut
Arachis	*hypogaea*	*hispanica-rubra*	Tennessee Red Spanish	Goober	Peanut
Arachis	*hypogaea*	*hispanica-suavis*	Sweet Spanish	Goober	Peanut
Arachis	*hypogaea*	*jamaicensis-rubra*	Jamaican Red	Goober	Peanut
Arachis	*hypogaea*	*jumbo-improva*	Improved Jumbo	Goober	Peanut
Arachis	*hypogaea*	*jumbo-mammothii*	Mammoth Jumbo	Goober	Peanut
Arachis	*hypogaea*	*nambyquarae*	Nambyquara	Goober	Peanut
Arachis	*hypogaea*	*norda-carolinensis-jumbo-NC-7*	#7 Jumbo North Carolina	Goober	Peanut
Arachis	*hypogaea*	*norda-carolinensis-NC-V11*	#V11 North Carolina	Goober	Peanut
Arachis	*hypogaea*	*oleifera*	Oil Bearing	Goober	Peanut
Arachis	*hypogaea*	*pre-civil-warii*	Pre Civil War	Goober	Peanut
Arachis	*hypogaea*	*purpurea-eaglei*	Purple Eagle	Goober	Peanut
Arachis	*hypogaea*	*suffolkensis-nigra*	Suffolk Black	Goober	Peanut
Arachis	*hypogaea*	*texasensis-rubra-&-alba*	Texas Red & White	Goober	Peanut
Arachis	*hypogaea*	*virginica-buncha*	Bunch Virginia	Goober	Peanut
Arachis	*hypogaea*	*virginica-carwilei*	Carwile Virginia	Goober	Peanut
Arachis	*hypogaea*	*virginica-gregoryi*	Gregory Virginia	Goober	Peanut
Arachis	*hypogaea*	*virginica-improva*	Improved Virginia	Goober	Peanut
Arachis	*hypogaea*	*virginica-jumboi*	Jumbo Virginia	Goober	Peanut
Arachis	*pintoi*	*pintoi*	Original	Pinto	Peanut
Arachis	*pintoi*	*golden-gloryi*	Golden Glory	Pinto	Peanut

Pea, Pigeon Pea (*Cajanus cajan*)

Genus	Species	infraspecies	cultivar	specimen	Genome
Cajanus	*cajan*	*cajan*	Original	Tahiti	Pigeon Pea
Cajanus	*cajan*	*bicolor*	Bicolor	Tahiti	Pigeon Pea
Cajanus	*cajan*	*flavus*	Yellow	Tahiti	Pigeon Pea
Cajanus	*cajan*	*indicus*	India	Tahiti	Pigeon Pea
Cajanus	*cajan*	*inodorus*	Odorless	Tahiti	Pigeon Pea
Cajanus	*cajan*	*luteus*	Lute Yellow	Tahiti	Pigeon Pea
Cajanus	*cajan*	*maculatus*	Spotted	Tahiti	Pigeon Pea
Cajanus	*cajan*	*obcordifolius*	Heart Leaf	Tahiti	Pigeon Pea
Cajanus	*cajan*	*striatus*	Striped	Tahiti	Pigeon Pea
Caragana	*frutex*	*frutex*	Original	Bushy	Caragan Pea
Caragana	*frutex*	*latifolia*	Broad Leaf	Bushy	Caragan Pea

Pea, Chick Pea (*Cicer arietinum*)

Genus	Species	infraspecies	cultivar	specimen	Genome
Cicer	*arietinum*	*arietinum*	Original	Garbanzo Bean	Chick Pea
Cicer	*arietinum*	*album*	White	Garbanzo Bean	Chick Pea
Cicer	*arietinum*	*blonda*	Blond	Garbanzo Bean	Chick Pea
Cicer	*arietinum*	*brunnea*	Brown	Garbanzo Bean	Chick Pea
Cicer	*arietinum*	*channa-kalai*	Kabul	Garbanzo Bean	Chick Pea
Cicer	*arietinum*	*channa-rubra*	Red	Garbanzo Bean	Chick Pea
Cicer	*arietinum*	*del-norte-garbanzos*	Del Norte	Garbanzo Bean	Chick Pea
Cicer	*arietinum*	*dolores-de-hildagoi*	Dolores De Hildago	Garbanzo Bean	Chick Pea
Cicer	*arietinum*	*edessanum*	Edessa	Garbanzo Bean	Chick Pea
Cicer	*arietinum*	*gene-pool-mixta*	Gene Pool Mix	Garbanzo Bean	Chick Pea
Cicer	*arietinum*	*green-seeded*	Green Seeded	Garbanzo Bean	Chick Pea
Cicer	*arietinum*	*grossum*	Large	Garbanzo Bean	Chick Pea
Cicer	*arietinum*	*hannanii*	Hannan	Garbanzo Bean	Chick Pea
Cicer	*arietinum*	*kaboulii*	Kabul	Garbanzo Bean	Chick Pea
Cicer	*arietinum*	*mayo-winteri*	Mayo Winter	Garbanzo Bean	Chick Pea
Cicer	*arietinum*	*mylesi*	Myles	Garbanzo Bean	Chick Pea
Cicer	*arietinum*	*nigrum*	Black Seeded	Garbanzo Bean	Chick Pea
Cicer	*arietinum*	*physodes*	Bladder Pod	Garbanzo Bean	Chick Pea
Cicer	*arietinum*	*rotundum*	Round	Garbanzo Bean	Chick Pea
Cicer	*arietinum*	*sarahii*	Sarah	Garbanzo Bean	Chick Pea
Cicer	*arietinum*	*sativum*	Sweet	Garbanzo Bean	Chick Pea
Cicer	*arietinum*	*sintenisii*	Sintenis	Garbanzo Bean	Chick Pea
Cicer	*arietinum*	*tarahumara*	Sarah	Garbanzo Bean	Chick Pea

Lupine Pea (*Lupinus albus*)

Genus	Species	infraspecies	cultivar	specimen	Genome
Lupinus	*albus*	*albus*	Original	White	Lupine Pea
Lupinus	*albus*	*graecus*	Greek	White	Lupine Pea
Lupinus	*albus*	*sativus-kievensis*	Kiev Sweet	White	Lupine Pea

Alfalfa (*Medicago sativa*)

Genus	Species	infraspecies	Cultivar	specimen	Genome

Genus	Species	infraspecies	Cultivar	specimen	Genome
Medicago	*sativa*			Alfalfa	Medick

Winged Pea (*Tetragonolobus edulis*)

Genus	Species	infraspecies	cultivar	specimen	Genome
Tetragonolobus	*edulis*	*edulis*	Original	Asparagus Pea	Winged Pea
Tetragonolobus	*edulis*	*purpureus*	Purple	Asparagus Pea	Winged Pea

Family 2 *Pisafloraceae* Peaflower

GENERA *Acmispon, Acuan(Synonym), Alhagi, Amerimnon(Synonym), Amicia, Anneslea(Synonym), Antopetitia(Synonym for Ornithopus), Aspalthium(Synonym), Barbieria, Belairia(Synonym for Pictetia), Benthamantha(Synonym), Bituminaria, Blanchetiodendron(Synonym for Enterolobium), Brachysema, Bradburya(Synonym), Brongniartia, Cailliea(Synonym for Dichrostachys), Calopogonium, Casparea(Synonym), Caspareopsis(Synonym), Centrosema, Chamaecrista, Chapmannia, Chorizema, Christia, Clianthus, Clitoria, Clitoriopsis, Cologania, Colutea, Coursetia, Cracca(Synonym), Crotalaria, Crudia, Cymbosema(Synonym for Dioclea), Derris, Desmanthus, Dichrostachys, Dioclea, Diphysa, Drepanocarpus(Synonym), Ecastaphyllum(Synonym), Enterolobium, Eriosema, Erophaca(Synonym for Colutea), Glottidium, Gompholobium, Hardenbergia, Havardia, Hoita, Hosackia, Hymenaea, Kebirita(Synonym for Lotus), Kennedia, Kennedya(Synonym), Leptoderris, Lotus, Lourea(Synonym), Maackia, Machaerium, Martiodendron, Martiusia(Synonym), Mundulea, Nissolia, Onobrychis, Ononis, Orbexilum, Oreophysa(Synonym for Colutea), Ormosia, Ornithopus, Orobus(Synonym), Oxyrhynchus, Paramachaerium, Pericopsis, Pickeringia, Pictetia, Podalyria, Pongamia, Pongamiopsis(Synonym for Pongamia), Psoralidium, Rhytidomene(Synonym), Rhynchotropis(Synonym for Crotalaria), Rupertia, Sarcodum, Scorpiurus, Sphaerophysa, Sphinga, Stenolobium(Synonym), Stirtonanthus(Synonym for Podalyria), Strongylodon, Stylosanthes, Sutherlandia, Swainsonia, Taralea, Templetonia, Trachylobium(Synonym), Vandasina, Virgilia, Xiphocarpus(Synonym), Zapoteca, Zornia*

Family 724.062.3 *Trifoliaceae* Clover

GENERA ***Trifolium(Trefoil Herb),*** *Alysicarpus, Anagyris (Synonym for Trifolium),* ***Baptisia(False Indigo Herb),*** *Barnebyella(Synonym for Dorycnium), Campylotropis, Codariocalyx, Dalea, Desmodiastrum, Desmodium, Dorycnium, Dorycnopsis, Eleiotis(Synonym for Desmodium rottleri), Goodia, Hylodesmum, Kuhnistera(Synonym for Dalea), Kummerowia,* ***Lespedeza (Cespedes Vegetable),*** *Lupinaster(Synonym for Trifolium), Marina, Meibomia(Synonym for Desmodium),* ***Melilotus(Melilot Herb),*** *Parochetus, Parosela(Synonym for Dalea), Petalostemon(Synonym for Dalea), Phyllodium, Pseudolotus(Synonym for Dorycnium), Sagotia(Synonym for Desmodium), Trifidacanthus(Synonym for Desmodium)*

Cespedes Clover (*Lespedeza thunbergii*)

Sweet Clover (*Melilotus spp.*)

Genus	Species	Specimen Genome	C:Culinary
Lespedeza	*thunbergii*	Thunberg Cespedes Clover	C:LF:**VG**
Melilotus	*alba*	White Sweet Clover	C:LF:**VG**
Melilotus	*caerulea*	Blue Sweet Clover	C:WP:FF
Melilotus	*officinalis*	Yellow Sweet Clover	C:LF,SH:FF,TE, **C**:FL,SD:FF

Family 4 *Viciaceae* Vetch

GENERA ***Vicia(Vetch Herb),*** *Aeschynomene, Anthyllis, Aragallus(Synonym for Oxytropus),* ***Astragalus(Milk Vetch Leaf Vegetable),*** *Atelophragma(Synonym for Astragalus), Batidophaca(Synonym for Astragalus), Bonaveria(Synonym for Securigera), Cnemidophacos(Synonym for Astragalus),* ***Coronilla(Crown Vetch Herb),*** *Cyclocarpa(Synonym for Aeschynomene), Diholcos(Synonym for Astragalus), Ervum (Synonym for Vicia), Geoprumnon(Synonym for Astragalus), Hammatolobium, Hedysarum, Hippocrepis, Holcophacos(Synonym for Astragalus), Homalobus(Synonym for Astragalus), Hymenocarpos(Synonym for Anthyllis), Kentrophyta(Synonym for Astragalus), Lathyrus, Ohwia(Synonym for Hedysarum), Ophiocarpus(Synonym for Astragalus), Orophaca,* ***Oxytropis(Locoweed Vetch Herb),*** *Pachecoa(Synonym for Hedysarum), Paratephrosia, Peteria, Phaca(Synonym for Astragalus), Pisophaca(Synonym for Astragalus), Podolotus(Synonym for Astragalus), Pycnospora(Synonym for Hedysarum), Requienia(Synonym for Tephrosia), Secula(Synonym for Aeschynomene), Securigera, Sphaerophysa (Synonym for Vicia), Sphinctospermum(Synonym for Tephrosia), Spiesia(Synonym for Oxytropis), Spongiocarpella, Stracheya(Synonym for Astragalus),* ***Tephrosia(Hoary Pea Herb),*** *Tripodion(Synonym for Hammatolobium), Tium(Synonym for Astragalus), Xiphocarpus (Synonym for Tephrosia) Xylophacos(Synonym for Astragalus)*

Family 5 *Wisteriaceae* Wisteria

GENERA *Wisteria, Bolusanthus, Cladrastis, Hardenbergia, Kraunhia(Synonym for Wisteria), Rehsonia*

Class 724.07 *Magnoliopsida* Magnolia

Order 1 *Magnoliales* Magnolia
Family 1 *Magnoliaceae* Magnolia
Family 2 *Annonaceae* Gourd
Family 3 *Aristolochiaceae* Loch
Family 4 *Canellaceae* Wild Cinnamon
Family 6 *Myristicaceae* Nutmeg
Family 7 *Rafflesiaceae* Stinkflower
Family 8 *Sonneratiaceae* Sonner
Family 9 *Winteraceae* Pepper Bark
Order 2 *Illiciales* Star Anise
Family 1 *Illiciaceae* Star Anise
Family 2 *Calyceraceae* Calyce
Family 3 *Schisandraceae* Star Vine

Order 724.073 *Laurales* Laurel

Family 724.073.1 *Lauraceae* Laurel

Avocado (*Persea spp.*)

Genus	Species	Specimen	Genome	C:Culinary
Persea	*americana*	West Indies & Central America	Avocado	C:PU:**VG**
Persea	*borbonia*	Red Bay Southeast USA	Avocado	C:PU:**VG**
Persea	*caribaea*	West Indies	Avocado	C:PU:**VG**
Persea	*humilis*	Silk Bay Southern USA	Avocado	C:PU:**VG**
Persea	*indica*	India	Avocado	C:PU:**VG**
Persea	*krugii*	Canela West Indies	Avocado	C:PU:**VG**
Persea	*lingue*	Tongue Leaf Chilean	Avocado	C:PU:**VG**
Persea	*mexicola*	Mexican	Avocado	C:PU:**VG**
Persea	*mexicola x caribaea*	West Indies & Mexico	Avocado	C:PU:**VG**
Persea	*nubigena*	Guatemalan	Avocado	C:PU:**VG**
Persea	*nubigena x caribaea*	West Indies & Guatemala	Avocado	C:PU:**VG**
Persea	*palustris*	Swamp Bay Southern USA	Avocado	C:PU:**VG**
Persea	*schiedeana*	Coyo Mexican	Avocado	C:PU:**VG**
Persea	*urbaniana*	Aquatillo Puerto Rico	Avocado	C:PU:**VG**
Sassafras	*albidum*	White	Sassafras	C:RT:**VG**

Varieties

Avocado (*Persea spp.*)

Genus	Species	infraspecies	cultivar	specimen	Genome
Persea	*americana*	*americana*	Original	West Indies & Central America	Avocado
Persea	*americana*	*berneckeri*	Bernecker	West Indies & Central America	Avocado
Persea	*americana*	*beta*	Beta	West Indies & Central America	Avocado
Persea	*americana*	*carrizo-nortei*	Carrizo Norte	West Indies & Central America	Avocado
Persea	*americana*	*catalinai*	Catalina	West Indies & Central America	Avocado
Persea	*americana*	*choquettei*	Choquette	West Indies & Central America	Avocado
Persea	*americana*	*donii*	Doni	West Indies & Central America	Avocado
Persea	*americana*	*drymifolia*	Dryad Leaf	West Indies & Central America	Avocado
Persea	*americana*	*drymifolia-baconii*	Bacon Dryad Leaf	West Indies & Central America	Avocado
Persea	*americana*	*drymifolia-fuertei*	Fuerte Dryad Leaf	West Indies & Central America	Avocado
Persea	*americana*	*drymifolia-pinkertonii*	Pinkerton Dryad Leaf	West Indies & Central America	Avocado
Persea	*americana*	*dukei-7*	Duke #7	West Indies & Central America	Avocado
Persea	*americana*	*fujikawai*	Fujikawa	West Indies & Central America	Avocado
Persea	*americana*	*gladysi*	Gladys	West Indies & Central America	Avocado
Persea	*americana*	*gratissima*	Pleasing	West Indies & Central America	Avocado
Persea	*americana*	*green-goldii*	Green Gold	West Indies & Central America	Avocado
Persea	*americana*	*gwenii*	Gwen	West Indies & Central America	Avocado
Persea	*americana*	*hallii*	Hall	West Indies & Central America	Avocado
Persea	*americana*	*hassi*	Hass	West Indies & Central America	Avocado
Persea	*americana*	*hassi-californica*	California Hass	West Indies & Central America	Avocado
Persea	*americana*	*hassi-floridensis*	Florida Hass	West Indies & Central America	Avocado
Persea	*americana*	*holidayi*	Holiday	West Indies & Central America	Avocado
Persea	*americana*	*hulumanui*	Hulumanu	West Indies & Central America	Avocado
Persea	*americana*	*jimii*	Jim	West Indies & Central America	Avocado
Persea	*americana*	*kahaluui*	Kahaluu	West Indies & Central America	Avocado
Persea	*americana*	*keonei*	Keone	West Indies & Central America	Avocado
Persea	*americana*	*leahii*	Leahi	West Indies & Central America	Avocado

Genus	Species	infraspecies	cultivar	specimen	Genome
Persea	*americana*	*lindai*	Linda	West Indies & Central America	Avocado
Persea	*americana*	*little-cadoi*	Little Cado	West Indies & Central America	Avocado
Persea	*americana*	*malama*	Malama	West Indies & Central America	Avocado
Persea	*americana*	*moritai*	Morita	West Indies & Central America	Avocado
Persea	*americana*	*murashigei*	Murashige	West Indies & Central America	Avocado
Persea	*americana*	*nabalii*	Nabal	West Indies & Central America	Avocado
Persea	*americana*	*nishikawai*	Nishikawa	West Indies & Central America	Avocado
Persea	*americana*	*nogamii*	Nogami	West Indies & Central America	Avocado
Persea	*americana*	*pollockii*	Pollock	West Indies & Central America	Avocado
Persea	*americana*	*reedii*	Reed	West Indies & Central America	Avocado
Persea	*americana*	*russellii*	Russell	West Indies & Central America	Avocado
Persea	*americana*	*seedlessii*	Seedless	West Indies & Central America	Avocado
Persea	*americana*	*semilii-34*	Semil #34	West Indies & Central America	Avocado
Persea	*americana*	*sharwilii*	Sharwil	West Indies & Central America	Avocado
Persea	*americana*	*simmondsi*	Simmonds	West Indies & Central America	Avocado
Persea	*americana*	*stewartii*	Stewart	West Indies & Central America	Avocado
Persea	*americana*	*whitsellii*	Whitsell	West Indies & Central America	Avocado
Persea	*americana*	*yamagatai*	Yamagata	West Indies & Central America	Avocado
Persea	*americana*	*yanii*	Yani	West Indies & Central America	Avocado
Persea	*americana*	*zutanoi*	Zutano	West Indies & Central America	Avocado
Persea	*borbonia*	.	.	Red Bay Southeast USA	Avocado
Persea	*borbonia*	*borbonia*	Original	Red Bay Southeast USA	Avocado
Persea	*borbonia*	*littoralis*	Seaside	Red Bay Southeast USA	Avocado
Persea	*mexicola*	*mexicola*	Original	Mexican	Avocado
Persea	*mexicola*	*grandei*	Grande	Mexican	Avocado
Persea	*mexicola*	*ponchoi*	Poncho	Mexican	Avocado
Persea	*mexicola*	*san-miguelii*	San Miguel	Mexican	Avocado
Persea	*mexicola*	*wilmai*	Wilma	Mexican	Avocado
Persea	*mexicola x caribaea*	*brogdonii*	Brogdon	West Indies & Mexico	Avocado
Persea	*nubigena x caribaea*	*dayi*	Day	West Indies & Guatemala	Avocado
Persea	*nubigena x caribaea*	*lulai*	Lula	West Indies & Guatemala	Avocado
Persea	*nubigena x caribaea*	*marcus*	Marcus	West Indies & Guatemala	Avocado
Persea	*nubigena x caribaea*	*monroei*	Monroe	West Indies & Guatemala	Avocado
Persea	*palustris*	*palustris*	Original	Swamp Bay Southern USA	Avocado
Persea	*palustris*	*pubescens*	Downy	Swamp Bay Southern USA	Avocado

Family 2 *Calycanthaceae* Tropicnut
Family 3 *Hernandiaceae* Hernando
Family 4 *Monimiaceae* Monimy
Family 5 *Perseaceae* Avocado
Family 6 *Sassafrasaceae* Sassafras

Class 724.08 *Malvopsida* Mallow

Order 724.081 *Malvales* Mallow

Family 724.081.1 *Malvaceae* Mallow

Genus	Species	Specimen Genome	C:Culinary
Malva	*crispa*	Curly Mallow	C:LF,SH:FF
Malva	*moschata*	Musk Mallow	C:PD:FF
Malva	*neglecta*	Buttonweed Mallow	C:LF,SH:FF
Malva	*parviflora*	Cheeseweed Mallow	C:LF:**VG**
Malva	*sylvestris*	High Mallow	C:LF,SH:**VG**
Malva	*verticillata*	Farmer's Tobacco Mallow	C:LF,SH:FF
Abelmoschus	*esculentus-alba*	White Okra Musk Mallow	C:PD:**VG**
Abelmoschus	*esculentus-groena*	Green Okra Musk Mallow	C:PD:**VG**
Abelmoschus	*esculentus-rubra*	Red Okra Musk Mallow	C:PD:**VG**
Abelmoschus	*esculentus-alba*	White Okra Musk Mallow	C:PD:**VG**
Abelmoschus	*manihot*	Aibika Musk Mallow	C:PD:**VG**
Abelmoschus	*moschatus*	Gumbo Okra Musk Mallow	C:PD:**VG**
Althaea	*officinalis*	White Marsh Mallow	C:SH:**VG**
Hibiscus	*sabdariffa*	Rozelle Rose Mallow	C:LF:**VG**

Varieties

Curly Mallow (*Malva crispa*)

Genus	Species	infraspecies	cultivar	specimen	Genome
Malva	*crispa*	*crispa*	Original	Curly	Mallow
Malva	*crispa*	*verticillata*	Whorled	Curly	Mallow

Musk Mallow (*Malva moschata*)

Genus	Species	infraspecies	cultivar	specimen	Genome
Malva	*moschata*	*moschata*	Original		Musk Mallow
Malva	*moschata*	*alba*	White		Musk Mallow
Malva	*moschata*	*rosea*	Rose Pink		Musk Mallow

Buttonweed Mallow (*Malva neglecta*)

Genus	Species	infraspecies	cultivar	specimen	Genome
Malva	*neglecta*	*neglecta*	Original	Buttonweed	Mallow
Malva	*neglecta*	*pusilla*	Dwarf	Buttonweed	Mallow
Malva	*neglecta*	*rotundifolia*	Round Leaf	Buttonweed	Mallow

Cheeseweed Mallow (*Malva parviflora*)

Genus	Species	infraspecies	cultivar	specimen	Genome
Malva	*parviflora*	*parviflora*	Original	Cheeseweed	Mallow
Malva	*parviflora*	*verticillata*	Small Whorl	Cheeseweed	Mallow

High Mallow (*Malva sylvestris*)

Genus	Species	infraspecies	cultivar	specimen	Genome
Malva	*sylvestris*	*sylvestris*	Original		High Mallow
Malva	*sylvestris*	*alba*	White		High Mallow
Malva	*sylvestris*	*brave-heartii*	Brave Heart		High Mallow
Malva	*sylvestris*	*cottenham-bluei*	Cottenham Blue		High Mallow
Malva	*sylvestris*	*mauritiana*	Mauritania		High Mallow
Malva	*sylvestris*	*mauritiana-bibor-felhoi*	Bibor Felho Mauritania		High Mallow
Malva	*sylvestris*	*primley-bluei*	Primley Blue		High Mallow
Malva	*sylvestris*	*zebrina*	Zebra		High Mallow

Cluster Mallow (*Malva verticillata*)

Genus	Species	infraspecies	cultivar	specimen	Genome
Malva	*verticillata*	.	.	Cluster	Mallow

Okra (*Abelmoschus esculentus*)

Genus	Species	infraspecies	cultivar	specimen	Genome
Abelmoschus	*esculentus-alba*	*alba*	White 9.0"	White Okra	Musk Mallow
Abelmoschus	*esculentus-alba*	*alba-velutina*	White Velvet Creole 6.5"	White Okra	Musk Mallow
Abelmoschus	*esculentus-alba*	*argenta-quena*	Silver Queen 7.0"	White Okra	Musk Mallow
Abelmoschus	*esculentus-alba*	*blondyi*	Blondy Spineless 3.0"	White Okra	Musk Mallow
Abelmoschus	*esculentus-alba*	*brasilensis*	Brazil ?.?"	White Okra	Musk Mallow
Abelmoschus	*esculentus-groena*	*afra*	African	Green Okra	Musk Mallow
Abelmoschus	*esculentus-groena*	*alice-elliotii*	Alice Elliot 6.0"	Green Okra	Musk Mallow

Genus	Species	infraspecies	cultivar	specimen	Genome
Abelmoschus	*esculentus-groena*	*beck'si-gardenvillensis*	Beck's Gardenville Giant 6.0"	Green Okra	Musk Mallow
Abelmoschus	*esculentus-groena*	*bengalii*	Bengali ?.?"	Green Okra	Musk Mallow
Abelmoschus	*esculentus-groena*	*big-unii*	Big Un ?.?"	Green Okra	Musk Mallow
Abelmoschus	*esculentus-groena*	*bornensis*	Borneo ?.?"	Green Okra	Musk Mallow
Abelmoschus	*esculentus-groena*	*burmaensis*	Burmese Spineless ?.?"	Green Okra	Musk Mallow
Abelmoschus	*esculentus-groena*	*cajun-delightii*	Cajun Delight 3.5"	Green Okra	Musk Mallow
Abelmoschus	*esculentus-groena*	*cajun-jewelii*	Cajun Jewel 7.0"	Green Okra	Musk Mallow
Abelmoschus	*esculentus-groena*	*cherokee-longapoda*	Cherokee Long Pod 15."	Green Okra	Musk Mallow
Abelmoschus	*esculentus-groena*	*choppeei*	Choppee ?.?"	Green Okra	Musk Mallow
Abelmoschus	*esculentus-groena*	*clemsonensis-80*	Early Clemson Spineless 8.0"	Green Okra	Musk Mallow
Abelmoschus	*esculentus-groena*	*clemsonensis-inermis*	Clemson Spineless 8.0"	Green Okra	Musk Mallow
Abelmoschus	*esculentus-groena*	*cowhornii*	Cow Horn 11."	Green Okra	Musk Mallow
Abelmoschus	*esculentus-groena*	*ganawia*	Ganawia 4.0"	Green Okra	Musk Mallow
Abelmoschus	*esculentus-groena*	*goldcoastensis*	Gold Coast ?.?"	Green Okra	Musk Mallow
Abelmoschus	*esculentus-groena*	*grandapai*	Grandpa ?.?"	Green Okra	Musk Mallow
Abelmoschus	*esculentus-groena*	*groena-velutina*	Green Velvet Spineless 7.5"	Green Okra	Musk Mallow
Abelmoschus	*esculentus-groena*	*groena-velutina-ludoviciana*	Louisiana Green Velvet Spineless 6.5"	Green Okra	Musk Mallow
Abelmoschus	*esculentus-groena*	*guarijioi-nescafei*	Guarijio Nescafe Coffee ?.?"	Green Okra	Musk Mallow
Abelmoschus	*esculentus-groena*	*nana-longa-poda*	Dwarf Long Pod 7.5"	Green Okra	Musk Mallow
Abelmoschus	*esculentus-groena*	*sempera-tenera*	Ever Tender 6.0"	Green Okra	Musk Mallow
Abelmoschus	*esculentus-groena*	*smaragda*	Velvet Spinless Emerald 8.0"	Green Okra	Musk Mallow
Abelmoschus	*esculentus-groena*	*ilia*	Jade 6.0"	Green Okra	Musk Mallow
Abelmoschus	*esculentus-groena*	*james-hopperi*	James Hopper 4.0"	Green Okra	Musk Mallow
Abelmoschus	*esculentus-groena*	*leei-nana*	Lee Spinless 6.5"	Green Okra	Musk Mallow
Abelmoschus	*esculentus-groena*	*longa-poda*	Long Pod 7.5"	Green Okra	Musk Mallow
Abelmoschus	*esculentus-groena*	*longacornea*	Longhorn 12."	Green Okra	Musk Mallow
Abelmoschus	*esculentus-groena*	*ludoviciana-brevis*	Louisiana Short 6.0"	Green Okra	Musk Mallow
Abelmoschus	*esculentus-groena*	*mammon-inermis-longapoda*	Mammoth Spineless Long Pod 7.0"	Green Okra	Musk Mallow
Abelmoschus	*esculentus-groena*	*penta-groena*	Penta Green ?.?"	Green Okra	Musk Mallow
Abelmoschus	*esculentus-groena*	*perkinsii-aspina*	Perkins Spineless ?.?"	Green Okra	Musk Mallow
Abelmoschus	*esculentus-groena*	*perkinsi-nana-sola*	Perkins Dwarf Sun 9.0"	Green Okra	Musk Mallow
Abelmoschus	*esculentus-groena*	*perkinsi-mammon*	Perkins Mammoth Spineless 7.0"	Green Okra	Musk Mallow
Abelmoschus	*esculentus-groena*	*stella-davidii*	Star Of David 7.0"	Green Okra	Musk Mallow
Abelmoschus	*esculentus-rubra*	*afra-rubra*	African 8.0"	Red Okra	Musk Mallow
Abelmoschus	*esculentus-rubra*	*afra-striata*	African Striped ?.?"	Red Okra	Musk Mallow
Abelmoschus	*esculentus-rubra*	*alabamensis*	Alabama ?.?"	Red Okra	Musk Mallow
Abelmoschus	*esculentus-rubra*	*artista*	Artist Purple 3.0"	Red Okra	Musk Mallow
Abelmoschus	*esculentus-rubra*	*bowlingii*	Bowling 7.0"	Red Okra	Musk Mallow
Abelmoschus	*esculentus-rubra*	*burgundia*	Burgundy 7.0"	Red Okra	Musk Mallow
Abelmoschus	*esculentus-rubra*	*collis-contra-rubra*	Hill Country 5.0"	Red Okra	Musk Mallow
Abelmoschus	*esculentus-rubra*	*ludoviciana-rubra*	Louisiana 6.5"	Red Okra	Musk Mallow
Abelmoschus	*esculentus-rubra*	*my-joanniea*	My Joannie 7.0"	Red Okra	Musk Mallow
Abelmoschus	*esculentus-rubra*	*pincii*	Pinky 7.5"	Red Okra	Musk Mallow
Abelmoschus	*esculentus-rubra*	*rubra-germanica*	Red Germany ?.?"	Red Okra	Musk Mallow
Abelmoschus	*esculentus-rubra*	*rubra-ocra*	Red 6.5"	Red Okra	Musk Mallow
Abelmoschus	*esculentus-rubra*	*rubra-velutina*	Red Velvet 6.5"	Red Okra	Musk Mallow
Abelmoschus	*esculentus-rubra*	*sarajevo-bamija*	Sarajevo Bamija 3.5"	Red Okra	Musk Mallow
Abelmoschus	*esculentus-rubra*	*UGA-rubra*	University Of Georgia 8.0"	Red Okra	Musk Mallow
Abelmoschus	*esculentus-rubra*	*yuma-rubra*	Yuma Desert 4.5"	Red Okra	Musk Mallow
Abelmoschus	*moschatus*	*moschatus*	Original	Gumbo Okra	Musk Mallow
Abelmoschus	*moschatus*	*mischiefii*	Mischief	Gumbo Okra	Musk Mallow
Abelmoschus	*moschatus*	*oriental-pinkii*	Oriental Pink	Gumbo Okra	Musk Mallow
Abelmoschus	*moschatus*	*pacifica*	Pacific	Gumbo Okra	Musk Mallow

Marsh Mallow (*Althaea officinalis*)

Genus	Species	infraspecies	cultivar	specimen	Genome
Hibiscus	*sabdariffa*	*sabdariffa*	Original	Rozelle	Rose Mallow

Rozelle (*Hibiscus sabdariffa*)

Genus	Species	infraspecies	cultivar	specimen	Genome
Hibiscus	*sabdariffa*	*sabdariffa*	Original	Rozelle	Rose Mallow
Hibiscus	*sabdariffa*	*altissima*	Tall	Rozelle	Rose Mallow
Hibiscus	*sabdariffa*	*florida-cranberryi*	Florida Cranberry	Rozelle	Rose Mallow
Hibiscus	*sabdariffa*	*jamaicensis*	Jamaica Sorrel	Rozelle	Rose Mallow
Hibiscus	*sabdariffa*	*thai-redii*	Thai Red	Rozelle	Rose Mallow

Family 2 *Bombacaceae* Silk Cotton
Family 3 *Elaeocarpaceae* Oilberry
Family 2 *Leitneriaceae* Corkwood

Family 3 *Plumbaginaceae* Leadwort
Family 4 *Sterculiaceae* Boatnut

Family 724.081.5 *Tiliaceae* Linden

Genus	Species	Specimen Genome	C:Culinary
Corchorus	*olitorius*	Nalta Jute	C:LF:**VG**

Varieties

Nalta Jute (*Corchorus olitorius*)

Genus	Species	infraspecies	cultivar	specimen	Genome
Corchorus	*olitorius*	*olitorius*	Original	Nalta	Jute
Corchorus	*olitorius*	*melo-khiya*	Melo Khiya	Nalta	Jute

Order 2 *Euphorbiales* Spurge
Family 1 *Euphorbiaceae* Spurge
Family 2 *Phyllanthaceae* Leaf Flower Spurge

Order 724.083 *Rhamnales* Buckthorn

Family 724.083.1 *Rhamnaceae* Buckthorn

Raisin Tree (*Hovenia dulcis*)

Genus	Species	Specimen Genome	C:Culinary
Hovenia	*dulcis*	Japanese Raisin Tree	C:LF:**VG**

Family 724.083.2 *Elaeagnaceae* Agnes Olive

Genus	Species	Specimen Genome	C:Culinary
Elaeagnus	*angustifolia*	Russian Agnes Olive	C:PU:**VG**
Elaeagnus	*commutata*	Silverberry Agnes Olive	C:PU:**VG**
Elaeagnus	*ebbingei*	Ebbinge Agnes Olive	C:PU:**VG**
Elaeagnus	*glabra*	Bald Agnes Olive	C:PU:**VG**
Elaeagnus	*macrophylla*	Broad Leaf Agnes Olive	C:PU:**VG**
Elaeagnus	*multiflora*	Goumi Cherry Agnes Olive	C:PU:**VG**
Elaeagnus	*orientalis*	Oriental Agnes Olive	C:PU:**VG**
Elaeagnus	*pungens*	Thorny Agnes Olive	C:PU:**VG**
Elaeagnus	*reflexa*	Refexed Agnes Olive	C:PU:**VG**
Elaeagnus	*umbellata*	Autumn Agnes Olive	C:PU:**VG**

Varieties

Olive, Agnes (*Elaeagnus spp.*)

Genus	Species	infraspecies	cultivar	specimen	Genome
Elaeagnus	*angustifolia*	*angustifolia*	Original	Russian	Agnes Olive
Elaeagnus	*angustifolia*	*quicksilveri*	Quicksilver	Russian	Agnes Olive
Elaeagnus	*angustifolia*	*red-kingii*	Red King	Russian	Agnes Olive
Elaeagnus	*ebbingei*	*gilt-edgei*	Gilt Edge	Ebbinge	Agnes Olive
Elaeagnus	*ebbingei*	*limelightii*	Limeling	Ebbinge	Agnes Olive
Elaeagnus	*multiflora*	*multiflora*	Original	Goumi Cherry	Agnes Olive
Elaeagnus	*multiflora*	*rubra-gemma*	Red Gem	Goumi Cherry	Agnes Olive
Elaeagnus	*multiflora*	*suavis-coccinea*	Sweet Scarlet	Goumi Cherry	Agnes Olive
Elaeagnus	*pungens*	*pungens*	Original	Thorny	Agnes Olive
Elaeagnus	*pungens*	*frederici*	Frederick	Thorny	Agnes Olive
Elaeagnus	*pungens*	*fruitlandii*	Fruitland	Thorny	Agnes Olive
Elaeagnus	*pungens*	*goldenii*	Golden	Thorny	Agnes Olive
Elaeagnus	*pungens*	*goldrimii*	Goldrim	Thorny	Agnes Olive
Elaeagnus	*pungens*	*maculata*	Marked	Thorny	Agnes Olive
Elaeagnus	*pungens*	*variegata*	Variegated	Thorny	Agnes Olive
Elaeagnus	*umbellata*	*umbellata*	Original	Autumn	Agnes Olive
Elaeagnus	*umbellata*	*amberi*	Amber	Autumn	Agnes Olive
Elaeagnus	*umbellata*	*brillians-rosea*	Brilliant Rose	Autumn	Agnes Olive
Elaeagnus	*umbellata*	*cardinalis*	Cardinal	Autumn	Agnes Olive
Elaeagnus	*umbellata*	*charlie's-goldenii*	Charlie's Golden	Autumn	Agnes Olive
Elaeagnus	*umbellata*	*delightfulii*	Delightful	Autumn	Agnes Olive
Elaeagnus	*umbellata*	*garnetii*	Garnet	Autumn	Agnes Olive
Elaeagnus	*umbellata*	*jazzboi*	Jazzbo	Autumn	Agnes Olive

Genus	Species	infraspecies	cultivar	specimen	Genome
Elaeagnus	*umbellata*	*jewelii*	Jewel	Autumn	Agnes Olive
Elaeagnus	*umbellata*	*parvifolia*	Oleaster	Autumn	Agnes Olive
Elaeagnus	*umbellata*	*red-autumnii*	Red	Autumn	Agnes Olive
Elaeagnus	*umbellata*	*rubyi*	Ruby	Autumn	Agnes Olive
Elaeagnus	*umbellata*	*sweet-n-tartii*	Sweet N Tart	Autumn	Agnes Olive
Elaeagnus	*umbellata*	*titania*	Titan	Autumn	Agnes Olive

Family 3 *Vitaceae* Grape

Order 724.084 *Urticales* Nettle

Family 1 *Urticaceae* Nettle

Family 724.084.2 *Cannabaceae* Hemp

Hops (*Humulus lupulus*)

Genus	Species	Specimen	Genome	C:Culinary
Humulus	*lupulus*		Beer Hops	**C:**FR:AF,**C:**OL:BF,FF,**C:**LF,SH:**VG**

Family 3 *Cecropiaceae* Pumpwood
Family 4 *Moraceae* Mulberry
Family 5 *Ulmaceae* Elm

Class 724.09 *Myrtopsida* Myrtle

Order 1 *Myrtales* Myrtle
Family 1 *Myrtaceae* Myrtle
Family 2 *Combretaceae* Mangrove Myrtle
Family 3 *Lythraceae* Loosestrife
Family 4 *Melastomataceae* Melastome
Family 5 *Onagraceae* Rodwood
Family 6 *Punicaceae* Onager
Family 7 *Thymelaeaceae* Pomegranate
Family 8 *Trapaceae* Mezereon
Water Chestnut
Order 2 *Haloragales* Seaberry
Family 1 *Haloragaceae* Seaberry
Family 2 *Gunneraceae* Gunner Seaberry
Order 3 *Proteales* Protean
Family 1 *Proteaceae* Protean

Class 724.10 *Nymphaeopsida* Water Lily

Order 724.101 *Nymphaeales* Water Lily

Family 1 *Nymphaeaceae* Water Lily
Family 2 *Cabombaceae* Shield Water Lily
Family 3 *Ceratophyllaceae* Horn Water Lily
Family 4 *Hydatellaceae* Tell Water Lily

Family 724.101.5 *Nelumbonaceae* Water Lotus

Genus	Species	Specimen Genome	C:Culinary
Nelumbo	*lutea*	American Water Lotus	C:LF:**VG**
Nelumbo	*nucifera*	Sacred Water Lotus	C:LF:**VG**

Varieties

Water Lotus (*Nelumbo spp.*)

Genus	Species	Infraspecies	cultivar	specimen	Genome
Nelumbo	*lutea*	*lutea*	Original	American	Water Lotus
Nelumbo	*lutea*	*pentapetala*	Five Petal	American	Water Lotus
Nelumbo	*nucifera*	*nucifera*	Original	Sacred	Water Lotus
Nelumbo	*nucifera*	*alba-grandiflora*	White Large Flower	Sacred	Water Lotus
Nelumbo	*nucifera*	*rosea-plena*	Pink Double Flower	Sacred	Water Lotus
Nelumbo	*nucifera*	*speciosa*	Showy	Sacred	Water Lotus

Order 724.102 *Piperales* Peppercorn

Family 1 *Piperaceae* Peppercorn
Family 2 *Chloranthaceae* Sac Peppercorn

Family 724.102.3 *Saururaceae* Lizard's Tail

Houttuyn Lizard's Tail (*Houttuynia*)

Genus	Species	Specimen Genome	C:Culinary
Houttuynia	*cordata-chinensis*	Coriander Leaf Houttuyn Lizard's Tail	C:LF:**VG**
Houttuynia	*cordata-japnonica*	Orange Scented Houttuyn Lizard's Tail	C:LF:**VG**

Class 724.11 *Ranunculopsida* Buttercup

Order 1 *Ranunculales* Buttercup
Family 1 *Ranunculaceae* Buttercup
Family 2 *Berberidaceae* Barberry
Family 3 *Lardizabalaceae* Sausagefruit
Family 4 *Menispermaceae* Moonseed
Family 5 *Sabiaceae* Sab

Order 724.112 *Papaverales* Poppy

Family 724.112.1 *Papaveraceae* Poppy

Genus	Species	Specimen	Genome	C:Culinary
Papaver	*somniferum*		Opium Water Lotus	C:LF:**VG**

Varieties

Poppy, Breadseed (*Papaver somniferum-sperma*)

Genus	Species	Infraspecies	cultivar	specimen	Genome
Papaver	*somniferum*	*somniferum*	Original	Opium	Poppy
Papaver	*somniferum*	*danebrogii*	Danebrog	Opium	Poppy
Papaver	*somniferum*	*dianthus*	Carnation	Opium	Poppy
Papaver	*somniferum*	*czechoslovakian-whitei*	Czechoslovakian White	Opium	Poppy
Papaver	*somniferum*	*hen-&-chickensi*	Hen & Chickens	Opium	Poppy
Papaver	*somniferum*	*mauveaflora*	Mauve Flowered	Opium	Poppy
Papaver	*somniferum*	*paeoniflorum*	Peony Flowered	Opium	Poppy
Papaver	*somniferum*	*setigerum*	Bristle	Opium	Poppy
Papaver	*somniferum*	*white-cloudii*	White Cloud	Opium	Poppy
Papaver	*somniferum-sperma*	*breadseedii*	Original	Breadseed	Poppy
Papaver	*somniferum-sperma*	*breadseedii-hutteritei*	Hutterite	Breadseed	Poppy
Papaver	*somniferum-sperma*	*breadseedii-przemkoi*	Przemko	Breadseed	Poppy
Papaver	*somniferum-sperma*	*breadseedii-whitei*	White	Breadseed	Poppy
Papaver	*somniferum-sperma*	*breadseedii-whitei-persiana*	Persian White	Breadseed	Poppy

Family 2 *Fumariaceae* Fumitory

Class 724.12 *Rosopsida* Rose

Order 724.121 *Rosaeurosales* Roset

Family 1 *Rosaeurosaceae* Roset

Family 724.121.2 *Rosafloretaceae* Floret Roset

Meadowsweet (*Filipendula spp.*)

Genus	Species	Specimen Genome	C:Culinary
Filipendula	*rubra*	Queen of the Prairie Meadowsweet	C:RT:**VG**
Filipendula	*ulmaria*	Queen of the Meadow Meadowsweet	C:LF:AF,BF
Filipendula	*vulgaris*	Dropwort Meadowsweet	C:RT:**VG**

Family 3 *Rosastellataceae* Starflower Roset
Order 2 *Rosagardeniales* Garden Rose
Family 1 *Rosaheritaceae* Heritage Rose
Family 2 *Rosaeldaceae* Old Garden Rose
Family 3 *Rosaneoaceae* Modern Rose
Order 3 *Rosahortales* Horticultural Roset
Family 1 *Bruniaceae* Burnet
Family 2 *Chrysobalanaceae* Golden Balan
Family 3 *Crassulaceae* Stonecrop
Family 4 *Crossosomataceae* Rockflower
Family 5 *Cunoniaceae* Butterknife
Family 6 *Eucryphiaceae* Leatherwood
Family 7 *Hydrangeaceae* Hydrange
Family 8 *Pittosporaceae* Cheesewood
Family 9 *Saxifragaceae* Saxifrage
Order 4 *Primulales* Primrose
Family 1 *Primulaceae* Primrose
Family 2 *Buxaceae* Boxwood
Family 3 *Diospyraceae* Persimmon
Family 4 *Hamamelidaceae* Witch Hazel
Family 5 *Myrsinaceae* Colicwood
Family 6 *Styracaceae* Storax
Family 7 *Symplocaceae* Sweetleaf
Family 8 *Theophrastaceae* Theophrastes
Order 5 *Rosafructales* Rosefruit
Family 1 *Rosahipaceae*

Family 724.125.2 *Rosabaccaceae* Roseberry

Herb Varieties

Hawthorn Berry, English (*Crataegus laevigata*)

Genus	Species	Infraspecies	cultivar	specimen Genome
Crataegus-edulis	*laevigata*	*laevigata*	Original	English Hawthorn Berry
Crataegus-edulis	*laevigata*	*auriculata*	Ear Leaf	English Hawthorn Berry
Crataegus-edulis	*laevigata*	*autumn-gloryi*	Autumn Glory	English Hawthorn Berry
Crataegus-edulis	*laevigata*	*carnoviensis*	Carnovian	English Hawthorn Berry
Crataegus-edulis	*laevigata*	*coriacea*	Leather Leaf	English Hawthorn Berry
Crataegus-edulis	*laevigata*	*crimson-cloudii*	Crimson Cloud	English Hawthorn Berry
Crataegus-edulis	*laevigata*	*cuneato-trifida*	Wedge Leaf	English Hawthorn Berry
Crataegus-edulis	*laevigata*	*ellipticifolia*	Elliptic Leaf	English Hawthorn Berry
Crataegus-edulis	*laevigata*	*eriocarpa*	Woolly Fruit	English Hawthorn Berry
Crataegus-edulis	*laevigata*	*glabrata*	Smooth	English Hawthorn Berry
Crataegus-edulis	*laevigata*	*globosa*	Globe	English Hawthorn Berry
Crataegus-edulis	*laevigata*	*integrifolia*	Entire Leaf	English Hawthorn Berry
Crataegus-edulis	*laevigata*	*mathei*	Mathe	English Hawthorn Berry
Crataegus-edulis	*laevigata*	*microcarpa*	Small Fruit	English Hawthorn Berry
Crataegus-edulis	*laevigata*	*microphylla*	Small Leaf	English Hawthorn Berry
Crataegus-edulis	*laevigata*	*obtusata*	Blunt Leaf	English Hawthorn Berry
Crataegus-edulis	*laevigata*	*obtusiloba*	Blunt Lobe	English Hawthorn Berry
Crataegus-edulis	*laevigata*	*palmstruchii*	Palmstruch	English Hawthorn Berry
Crataegus-edulis	*laevigata*	*paul's-scarletii*	Paul's Scarlet	English Hawthorn Berry
Crataegus-edulis	*laevigata*	*plena*	Double Flower	English Hawthorn Berry
Crataegus-edulis	*laevigata*	*polygyna*	Polygyne	English Hawthorn Berry

Genus	Species	Infraspecies	cultivar	specimen	Genome
Crataegus-edulis	*laevigata*	*rhenana*	Rhen	English	Hawthorn Berry
Crataegus-edulis	*laevigata*	*rosea*	Rose	English	Hawthorn Berry
Crataegus-edulis	*laevigata*	*rosea-florepleno*	Double Rose	English	Hawthorn Berry
Crataegus-edulis	*laevigata*	*subinermis*	Thornless	English	Hawthorn Berry
Crataegus-edulis	*laevigata*	*vernicosa*	Varnished	English	Hawthorn Berry
Crataegus-edulis	*laevigata*	*walokochiana*	Walokoch	English	Hawthorn Berry

Hawthorn Berry, Quickthorn (*Crataegus monogyna*)

Genus	Species	Infraspecies	cultivar	specimen	Genome
Crataegus-edulis	*monogyna*	*monogyna*	Original	Quickthorn	Hawthorn Berry
Crataegus-edulis	*monogyna*	*acutiloba*	Sharp Lobe	Quickthorn	Hawthorn Berry
Crataegus-edulis	*monogyna*	*aegeica*	Aegean	Quickthorn	Hawthorn Berry
Crataegus-edulis	*monogyna*	*aguilaris*	Aguilar	Quickthorn	Hawthorn Berry
Crataegus-edulis	*monogyna*	*albanthera*	White Flower	Quickthorn	Hawthorn Berry
Crataegus-edulis	*monogyna*	*alboplena*	White Double	Quickthorn	Hawthorn Berry
Crataegus-edulis	*monogyna*	*apiifolia*	Parsley Leaf	Quickthorn	Hawthorn Berry
Crataegus-edulis	*monogyna*	*arborescens*	Woody	Quickthorn	Hawthorn Berry
Crataegus-edulis	*monogyna*	*argenteo-variegata*	Silver Variegated	Quickthorn	Hawthorn Berry
Crataegus-edulis	*monogyna*	*aurea*	Golden	Quickthorn	Hawthorn Berry
Crataegus-edulis	*monogyna*	*azarella*	Azarelle	Quickthorn	Hawthorn Berry
Crataegus-edulis	*monogyna*	*bicolor*	Bicolor	Quickthorn	Hawthorn Berry
Crataegus-edulis	*monogyna*	*biflora*	Glastonberry	Quickthorn	Hawthorn Berry
Crataegus-edulis	*monogyna*	*biserrata*	Sawtooth	Quickthorn	Hawthorn Berry
Crataegus-edulis	*monogyna*	*brevispina*	Short Spine	Quickthorn	Hawthorn Berry
Crataegus-edulis	*monogyna*	*bruanti*	Bruant	Quickthorn	Hawthorn Berry
Crataegus-edulis	*monogyna*	*cabulica*	Kabul	Quickthorn	Hawthorn Berry
Crataegus-edulis	*monogyna*	*calycina*	Calyx	Quickthorn	Hawthorn Berry
Crataegus-edulis	*monogyna*	*compacta*	Compact	Quickthorn	Hawthorn Berry
Crataegus-edulis	*monogyna*	*curvisepala*	Curved Sepal	Quickthorn	Hawthorn Berry
Crataegus-edulis	*monogyna*	*decussata*	Crosswise	Quickthorn	Hawthorn Berry
Crataegus-edulis	*monogyna*	*eriocarpa*	Woolly Fruit	Quickthorn	Hawthorn Berry
Crataegus-edulis	*monogyna*	*ferox*	Ferocious	Quickthorn	Hawthorn Berry
Crataegus-edulis	*monogyna*	*filicifolia*	Fern Leaf	Quickthorn	Hawthorn Berry
Crataegus-edulis	*monogyna*	*fissa*	Fissure	Quickthorn	Hawthorn Berry
Crataegus-edulis	*monogyna*	*flexuosa*	Flexible	Quickthorn	Hawthorn Berry
Crataegus-edulis	*monogyna*	*flore-albo-pleno*	White Double Flower	Quickthorn	Hawthorn Berry
Crataegus-edulis	*monogyna*	*flore-roseo-pleno*	Rose Double Flower	Quickthorn	Hawthorn Berry
Crataegus-edulis	*monogyna*	*flore-rubro-pleno*	Red Double Flower	Quickthorn	Hawthorn Berry
Crataegus-edulis	*monogyna*	*foucaudii*	Foucaud	Quickthorn	Hawthorn Berry
Crataegus-edulis	*monogyna*	*glabra*	Smooth	Quickthorn	Hawthorn Berry
Crataegus-edulis	*monogyna*	*globosa*	Globe	Quickthorn	Hawthorn Berry
Crataegus-edulis	*monogyna*	*granatensis*	Granada	Quickthorn	Hawthorn Berry
Crataegus-edulis	*monogyna*	*gumperi*	Gumper	Quickthorn	Hawthorn Berry
Crataegus-edulis	*monogyna*	*heterophylla*	Variable Leaf	Quickthorn	Hawthorn Berry
Crataegus-edulis	*monogyna*	*hirsuta*	Hairy	Quickthorn	Hawthorn Berry
Crataegus-edulis	*monogyna*	*horrida*	Horrible	Quickthorn	Hawthorn Berry
Crataegus-edulis	*monogyna*	*incisifolia*	Incised Leaf	Quickthorn	Hawthorn Berry
Crataegus-edulis	*monogyna*	*inermis*	Thornless	Quickthorn	Hawthorn Berry
Crataegus-edulis	*monogyna*	*inermis-compacta*	Compact Thornless	Quickthorn	Hawthorn Berry
Crataegus-edulis	*monogyna*	*insegnae*	Insegna	Quickthorn	Hawthorn Berry
Crataegus-edulis	*monogyna*	*insularis*	Island	Quickthorn	Hawthorn Berry
Crataegus-edulis	*monogyna*	*intermedia*	Intermediate	Quickthorn	Hawthorn Berry
Crataegus-edulis	*monogyna*	*kermesina-plena*	Kermes	Quickthorn	Hawthorn Berry
Crataegus-edulis	*monogyna*	*kyrtostyla*	Kyrt Style	Quickthorn	Hawthorn Berry
Crataegus-edulis	*monogyna*	*laciniata*	Lace Leaf	Quickthorn	Hawthorn Berry
Crataegus-edulis	*monogyna*	*lagenaria*	Lager	Quickthorn	Hawthorn Berry
Crataegus-edulis	*monogyna*	*lasiocarpa*	Downy Fruit	Quickthorn	Hawthorn Berry
Crataegus-edulis	*monogyna*	*liocarpa*	Lion Fruit	Quickthorn	Hawthorn Berry
Crataegus-edulis	*monogyna*	*maritima*	Seaside	Quickthorn	Hawthorn Berry
Crataegus-edulis	*monogyna*	*media*	Median	Quickthorn	Hawthorn Berry
Crataegus-edulis	*monogyna*	*microcarpa*	Small Fruit	Quickthorn	Hawthorn Berry
Crataegus-edulis	*monogyna*	*microphylla*	Tiny Leaf	Quickthorn	Hawthorn Berry
Crataegus-edulis	*monogyna*	*orthocalyx*	Straight Calyx	Quickthorn	Hawthorn Berry
Crataegus-edulis	*monogyna*	*pectinata*	Pectin	Quickthorn	Hawthorn Berry
Crataegus-edulis	*monogyna*	*pendula*	Weeping	Quickthorn	Hawthorn Berry
Crataegus-edulis	*monogyna*	*pendula-rosea*	Rose Weeping	Quickthorn	Hawthorn Berry
Crataegus-edulis	*monogyna*	*pilifera*	Downy	Quickthorn	Hawthorn Berry
Crataegus-edulis	*monogyna*	*pinnatifida*	Fern Leaf	Quickthorn	Hawthorn Berry
Crataegus-edulis	*monogyna*	*poloniensis*	Poland	Quickthorn	Hawthorn Berry
Crataegus-edulis	*monogyna*	*polonskiana*	Polonsky	Quickthorn	Hawthorn Berry
Crataegus-edulis	*monogyna*	*polyacantha*	Many Thorn	Quickthorn	Hawthorn Berry
Crataegus-edulis	*monogyna*	*praecox*	Early	Quickthorn	Hawthorn Berry
Crataegus-edulis	*monogyna*	*pteridifolia*	Fern Leaf	Quickthorn	Hawthorn Berry
Crataegus-edulis	*monogyna*	*punicea*	Punic	Quickthorn	Hawthorn Berry

Genus	Species	Infraspecies	cultivar	specimen	Genome
Crataegus-edulis	*monogyna*	*punicea-plena*	Punic Double	Quickthorn	Hawthorn Berry
Crataegus-edulis	*monogyna*	*quercifolia*	Oak Leaf	Quickthorn	Hawthorn Berry
Crataegus-edulis	*monogyna*	*ronnigeri*	Ronniger	Quickthorn	Hawthorn Berry
Crataegus-edulis	*monogyna*	*rosea*	Rose	Quickthorn	Hawthorn Berry
Crataegus-edulis	*monogyna*	*rosiformis*		Quickthorn	Hawthorn Berry
Crataegus-edulis	*monogyna*	*rotundicarpa*	Round Fruit	Quickthorn	Hawthorn Berry
Crataegus-edulis	*monogyna*	*rubra-plena*	Red Double	Quickthorn	Hawthorn Berry
Crataegus-edulis	*monogyna*	*ruscinonensis*	Ruscinon	Quickthorn	Hawthorn Berry
Crataegus-edulis	*monogyna*	*segobricensis*	Segobrica	Quickthorn	Hawthorn Berry
Crataegus-edulis	*monogyna*	*semperflorens*	Ever Flowering	Quickthorn	Hawthorn Berry
Crataegus-edulis	*monogyna*	*sericea*	Silky	Quickthorn	Hawthorn Berry
Crataegus-edulis	*monogyna*	*stenosepala*	Thin Sepal	Quickthorn	Hawthorn Berry
Crataegus-edulis	*monogyna*	*stricta*	Strict	Quickthorn	Hawthorn Berry
Crataegus-edulis	*monogyna*	*subintegra*	Lesser Entire Leaf	Quickthorn	Hawthorn Berry
Crataegus-edulis	*monogyna*	*subpinnatifida*	Lesser Pinnate	Quickthorn	Hawthorn Berry
Crataegus-edulis	*monogyna*	*transalpina*	Transalpine	Quickthorn	Hawthorn Berry
Crataegus-edulis	*monogyna*	*variegata*	Variegated	Quickthorn	Hawthorn Berry

Hawthorn Berry, Sharpthorn (*Crataegus oxyacantha*)

Genus	Species	Infraspecies	cultivar	specimen	Genome
Crataegus-edulis	*oxyacantha*	*oxyacantha*	Original	Sharpthorn	Hawthorn Berry
Crataegus-edulis	*oxyacantha*	*albaplena*	White Double	Sharpthorn	Hawthorn Berry
Crataegus-edulis	*oxyacantha*	*aurea*	Golden	Sharpthorn	Hawthorn Berry
Crataegus-edulis	*oxyacantha*	*auriculata*	Ear Leaf	Sharpthorn	Hawthorn Berry
Crataegus-edulis	*oxyacantha*	*berberifolia*	Beard Leaf	Sharpthorn	Hawthorn Berry
Crataegus-edulis	*oxyacantha*	*candida*	Virgin White	Sharpthorn	Hawthorn Berry
Crataegus-edulis	*oxyacantha*	*candidoplena*	Virgin White Double	Sharpthorn	Hawthorn Berry
Crataegus-edulis	*oxyacantha*	*caucasica*	Caucasian	Sharpthorn	Hawthorn Berry
Crataegus-edulis	*oxyacantha*	*cossonii*	Cosson	Sharpthorn	Hawthorn Berry
Crataegus-edulis	*oxyacantha*	*decussata*	Decussate	Sharpthorn	Hawthorn Berry
Crataegus-edulis	*oxyacantha*	*elegans*	Elegant	Sharpthorn	Hawthorn Berry
Crataegus-edulis	*oxyacantha*	*eriocalyx*	Woolly Caylx	Sharpthorn	Hawthorn Berry
Crataegus-edulis	*oxyacantha*	*eriocarpa*	Woolly Fruit	Sharpthorn	Hawthorn Berry
Crataegus-edulis	*oxyacantha*	*fastigiata*	Dense Branched	Sharpthorn	Hawthorn Berry
Crataegus-edulis	*oxyacantha*	*ferox*	Ferocious	Sharpthorn	Hawthorn Berry
Crataegus-edulis	*oxyacantha*	*filicifolia*	Thread Leaf	Sharpthorn	Hawthorn Berry
Crataegus-edulis	*oxyacantha*	*flexuosa*	Flexible	Sharpthorn	Hawthorn Berry
Crataegus-edulis	*oxyacantha*	*flore-punicea*	Punic Flower	Sharpthorn	Hawthorn Berry
Crataegus-edulis	*oxyacantha*	*fructo-luteo*	Yellow Fruit	Sharpthorn	Hawthorn Berry
Crataegus-edulis	*oxyacantha*	*genuina*	Genuine	Sharpthorn	Hawthorn Berry
Crataegus-edulis	*oxyacantha*	*gireoudii*	Gireoud	Sharpthorn	Hawthorn Berry
Crataegus-edulis	*oxyacantha*	*glabra*	Smooth	Sharpthorn	Hawthorn Berry
Crataegus-edulis	*oxyacantha*	*heterophylla*	Variable Leaf	Sharpthorn	Hawthorn Berry
Crataegus-edulis	*oxyacantha*	*hirsuta*	Hairy	Sharpthorn	Hawthorn Berry
Crataegus-edulis	*oxyacantha*	*horrida*	Horrible	Sharpthorn	Hawthorn Berry
Crataegus-edulis	*oxyacantha*	*incisa*	Incised	Sharpthorn	Hawthorn Berry
Crataegus-edulis	*oxyacantha*	*integerrimus*	Whole Leaf	Sharpthorn	Hawthorn Berry
Crataegus-edulis	*oxyacantha*	*integrifolia*	Entire Leaf	Sharpthorn	Hawthorn Berry
Crataegus-edulis	*oxyacantha*	*laciniata*	Lace Leaf	Sharpthorn	Hawthorn Berry
Crataegus-edulis	*oxyacantha*	*lagenariformis*	Lager	Sharpthorn	Hawthorn Berry
Crataegus-edulis	*oxyacantha*	*liocalyx*	Lion Calyx	Sharpthorn	Hawthorn Berry
Crataegus-edulis	*oxyacantha*	*lobata*	Lobed	Sharpthorn	Hawthorn Berry
Crataegus-edulis	*oxyacantha*	*macrocarpa*	Large Fruit	Sharpthorn	Hawthorn Berry
Crataegus-edulis	*oxyacantha*	*masekii*	Masek	Sharpthorn	Hawthorn Berry
Crataegus-edulis	*oxyacantha*	*mauriannensis*	Maurianne	Sharpthorn	Hawthorn Berry
Crataegus-edulis	*oxyacantha*	*mediterranea*	Mediterranean	Sharpthorn	Hawthorn Berry
Crataegus-edulis	*oxyacantha*	*mutabilis*	Changeable	Sharpthorn	Hawthorn Berry
Crataegus-edulis	*oxyacantha*	*ovata*	Oval Leaf	Sharpthorn	Hawthorn Berry
Crataegus-edulis	*oxyacantha*	*paulii*	Paul	Sharpthorn	Hawthorn Berry
Crataegus-edulis	*oxyacantha*	*pectinata*	Pectin	Sharpthorn	Hawthorn Berry
Crataegus-edulis	*oxyacantha*	*pendula*	Drooping	Sharpthorn	Hawthorn Berry
Crataegus-edulis	*oxyacantha*	*petermannii*	Petermann	Sharpthorn	Hawthorn Berry
Crataegus-edulis	*oxyacantha*	*plena*	Double	Sharpthorn	Hawthorn Berry
Crataegus-edulis	*oxyacantha*	*praecox*	Early	Sharpthorn	Hawthorn Berry
Crataegus-edulis	*oxyacantha*	*pseudolaciniatus*	Lesser Lace Leaf	Sharpthorn	Hawthorn Berry
Crataegus-edulis	*oxyacantha*	*punicea*	Punic	Sharpthorn	Hawthorn Berry
Crataegus-edulis	*oxyacantha*	*punicea-plena*	Double Punic	Sharpthorn	Hawthorn Berry
Crataegus-edulis	*oxyacantha*	*quercifolia*	Oak Leaf	Sharpthorn	Hawthorn Berry
Crataegus-edulis	*oxyacantha*	*reginae*	Regina	Sharpthorn	Hawthorn Berry
Crataegus-edulis	*oxyacantha*	*rigida*	Stiff	Sharpthorn	Hawthorn Berry
Crataegus-edulis	*oxyacantha*	*rosea*	Rose	Sharpthorn	Hawthorn Berry
Crataegus-edulis	*oxyacantha*	*rosea-superba*	Superb Rose	Sharpthorn	Hawthorn Berry
Crataegus-edulis	*oxyacantha*	*rubra*	Red	Sharpthorn	Hawthorn Berry
Crataegus-edulis	*oxyacantha*	*rubro-pleno*	Red Double	Sharpthorn	Hawthorn Berry

Genus	Species	Infraspecies	cultivar	specimen	Genome
Crataegus-edulis	*oxyacantha*	*semperflorens*	Everflowering	Sharpthorn	Hawthorn Berry
Crataegus-edulis	*oxyacantha*	*songarica*	Songar	Sharpthorn	Hawthorn Berry
Crataegus-edulis	*oxyacantha*	*spinosissima*	Spiny	Sharpthorn	Hawthorn Berry
Crataegus-edulis	*oxyacantha*	*splendens*	Splendid	Sharpthorn	Hawthorn Berry
Crataegus-edulis	*oxyacantha*	*stricta*	Erect	Sharpthorn	Hawthorn Berry
Crataegus-edulis	*oxyacantha*	*subintegrifolia*	Lesser Entire Leaf	Sharpthorn	Hawthorn Berry
Crataegus-edulis	*oxyacantha*	*subtriloba*	Lesser Three Lobe	Sharpthorn	Hawthorn Berry
Crataegus-edulis	*oxyacantha*	*sutheriana*	Suther	Sharpthorn	Hawthorn Berry
Crataegus-edulis	*oxyacantha*	*tortuosa*	Twisted	Sharpthorn	Hawthorn Berry
Crataegus-edulis	*oxyacantha*	*trifidus*	Three Part	Sharpthorn	Hawthorn Berry
Crataegus-edulis	*oxyacantha*	*triloba*	Three Lobe	Sharpthorn	Hawthorn Berry
Crataegus-edulis	*oxyacantha*	*variegata*	Varigated	Sharpthorn	Hawthorn Berry
Crataegus-edulis	*oxyacantha*	*vernicosa*	Varnished	Sharpthorn	Hawthorn Berry
Crataegus-edulis	*oxyacantha*	*villosa*	Villose	Sharpthorn	Hawthorn Berry
Crataegus-edulis	*oxyacantha*	*wallrothii*	Wallroth	Sharpthorn	Hawthorn Berry
Crataegus-edulis	*oxyacantha*	*xanthocarpa*	Yellowish Fruit	Sharpthorn	Hawthorn Berry

Family 3 *Grossulariaceae* Gooseberry
Family 4 *Myricaceae* Bayberry
Family 5 *Rubaceae* Brambleberry
Family 6 *Rosadrupaceae* Rosedrupe
Family 7 *Rosapomaceae* Rosepome
Order 6 *Rosanuxales* Rosenut
Family 1 *Rosanuxaceae* Rosenut
Family 2 *Betulaceae* Birchnut
Family 3 *Casuarinaceae* Oaknut
Family 4 *Fagaceae* Beechnut
Family 5 *Juglandaceae* Walnut

Class 724.13 *Rutopsida* Rue

Order 1 *Rutales* Rue
Family 1 *Rutaceae* Rue
Family 2 *Cneoraceae* Olive Spurge
Family 3 *Melianthaceae* Honeyflower
Family 4 *Tremandraceae* Black Eyed Susan
Order 2 *Geraniales* Geranium
Family 1 *Geraniaceae* Geranium
Family 2 *Balsaminaceae* Touch Me Not
Family 3 *Limnanthaceae* Meadowfoam
Family 4 *Oxalidaceae* Wood Sorrel
Family 5 *Tropaeolaceae* Trophy Cress
Order 3 *Linales* Flax
Family 1 *Linaceae* Flax
Family 2 *Erythroxylaceae* Redwood
Order 4 *Malpighiales* Acerola Cherry
Family 1 *Malpighiaceae* Acerola Cherry
Family 2 *Centroplacaceae* Plate Pandawood
Family 3 *Irvingiaceae* Irving Cherry
Family 4 *Pandaceae* Pandawood
Family 5 *Picrodendraceae* Bitterwood
Family 6 *Putranjivaceae* Putranjiv
Order 5 *Polygalales* Milkwort
Family 1 *Polygalaceae* Milkwort
Family 2 *Krameriaceae* Ratany

Order 724.136 *Sapindales* Soapberry

Family 1 *Sapindaceae* Soapberry
Family 2 *Aceraceae* Maple

Family 724.136.3 *Anacardiaceae* Cashew

Cashew (*Anacardium occidentale*)

Genus	Species	Specimen Genome	C:Culinary
Anacardium	*occidentale*	Acajou Cashew	C:SD:**NT**
Schinus	*terebinthifolius*	Brazilian Mastic Peppercorn	C:SD:**VG**

Varieties

Genus	Species	Infraspecies	cultivar	specimen	Genome
Schinus	*terebinthifolius*	*terebinthifolius*	Original	Brazilian	Mastic Peppercorn
Schinus	*terebinthifolius*	*raddianus*	Christmas Berry	Brazilian	Mastic Peppercorn

Family 4 *Burseraceae* Torchwood
Family 5 *Hippocastanaceae* Buckeye Chestnut
Family 6 *Meliaceae* Mohagany
Family 7 *Simaroubaceae* Sima Rue
Family 8 *Staphyleaceae* Bladdernut
Family 9 *Zygophyllaceae* Bean Caper

Class 724.14 *Santalopsida* Sandalwood

Order 1 *Santalales* Sandalwood
Family 1 *Santalaceae* Sandalwood
Family 2 *Balanophoraceae* Crownose
Family 3 *Eremolepidaceae* Catkin Mistletoe
Family 4 *Loranthaceae* Loreflower
Family 5 *Olacaceae* Tallowood
Family 6 *Viscaceae* Mistletoe
Order 2 *Celastrales* Spindlewood
Family 1 *Celastraceae* Spindlewood
Family 2 *Aquifoliaceae* Holly
Family 3 *Corynocarpaceae* Kakara Nut
Family 4 *Hippocrateaceae* Medicine Vine
Family 5 *Icacinaceae* Icac Vine
Family 6 *Stackhousiaceae* Stackhouse Vine

Class 724.15 *Theopsida* Tea

Order 1 *Theales* Tea
Family 1 *Theaceae* Tea
Family 2 *Actinidiaceae* Kiwi
Family 3 *Caryocaraceae* Souary Nut
Family 4 *Clusiaceae* Attorney
Family 5 *Dipterocarpaceae* Camphor
Family 6 *Elatinaceae* Waterwort
Family 7 *Marcgraviaceae* Shingle Plant
Family 8 *Ochnaceae* Tea Pear
Order 2 *Dilleniales* Dillen
Family 1 *Dilleniaceae* Dillen
Family 2 *Paeoniaceae* Peony
Order 3 *Lecythidales* Lecy
Family 1 *Lecythidaceae* Lecy
Family 2 *Napoleonaceae* Napoleon
Family 3 *Scytopetalaceae* Scyte Petal

Class 724.16 *Violopsida* Violet

Order 724.161 *Violales* Violet

Family 1 *Violaceae* Violet
Family 2 *Begoniaceae* Begonia
Family 3 *Bixaceae* Annato
Family 4 *Caricaceae* Papaya
Family 5 *Cistaceae* Rock Rose

Family 724.161.6 *Cucurbitaceae* Melon

Genus	Species	Specimen Genome	C:Culinary
Cucurbita	*digitata*	Fingerleaf Melon	C:PD:**VG**
Cucurbita	*ficifolia*	Figleaf Squash Melon	C:PD:**VG**
Cucurbita	*foetidissima*	Buffalo Gourd Melon	C:PD:**VG**
Cucurbita	*maxima-aurantia*	Orange Winter Neckless Squash Melon	C:PD:**VG**
Cucurbita	*maxima-brunnea*	Brown Winter Neckless Squash Melon	C:PD:**VG**
Cucurbita	*maxima-grisea*	Gray Winter Neckless Squash Melon	C:PD:**VG**
Cucurbita	*maxima-groena*	Green Winter Neckless Squash Melon	C:PD:**VG**
Cucurbita	*maxima-groena-blauea*	Blue Green Winter Neckless Squash Melon	C:PD:**VG**
Cucurbita	*maxima-groena-grisea*	Gray Green Winter Neckless Squash Melon	C:PD:**VG**
Cucurbita	*maxima-helva*	Yellow Winter Neckless Squash Melon	C:PD:**VG**
Cucurbita	*maxima-pepo-alba*	White Winter Pumpkin Melon	C:PD:**VG**
Cucurbita	*maxima-pepo-aurantia*	Orange Winter Pumpkin Melon	C:PD:**VG**
Cucurbita	*maxima-pepo-grisea*	Gray Winter Pumpkin Melon	C:PD:**VG**
Cucurbita	*maxima-pepo-groena*	Green Winter Pumpkin Melon	C:PD:**VG**
Cucurbita	*maxima-pepo-groena-blauea*	Blue Green Winter Pumpkin Melon	C:PD:**VG**
Cucurbita	*maxima-pepo-helva*	Yellow Winter Pumpkin Melon	C:PD:**VG**
Cucurbita	*maxima-pepo-multicolor*	Multicolor Winter Pumpkin Melon	C:PD:**VG**
Cucurbita	*maxima-pepo-pinca*	Pink Winter Pumpkin Melon	C:PD:**VG**
Cucurbita	*maxima-pinca*	Pink Winter Neckless Squash Melon	C:PD:**VG**
Cucurbita	*maxima-rubra*	Red Winter Neckless Squash Melon	C:PD:**VG**
Cucurbita	*mixta*	Cushaw Squash Melon	C:PD:**VG**
Cucurbita	*mixta-sperma*	Seed Cushaw Squash Melon	C:SD:**VG**
Cucurbita	*moschata-aurantia*	Orange Flesh Winter Necked Squash Melon	C:PD:**VG**
Cucurbita	*moschata-aurantia-helva*	Yellow Orange Flesh Winter Necked Squash Melon	C:PD:**VG**
Cucurbita	*moschata-helva*	Yellow Flesh Winter Necked Squash Melon	C:PD:**VG**
Cucurbita	*moschata-pepo-aurantia*	Orange Flesh Flat Winter Pumpkin Melon	C:PD:**VG**
Cucurbita	*moschata-pepo-aurantia-helva*	Yellow Orange Flesh Flat Winter Pumpkin Melon	C:PD:**VG**
Cucurbita	*moschata-pepo-helva*	Yellow Flesh Flat Winter Pumpkin Melon	C:PD:**VG**
Cucurbita	*moschata-sperma*	Seed Winter Necked Squash Melon	C:PD:**VG**
Cucurbita	*okeechobeensis*	Okeechobee Melon	C:PD:**VG**
Cucurbita	*palmata*	Coyote Melon	C:PD:**VG**
Cucurbita	*pepo-alba*	White Summer Squash Melon	C:PD:**VG**
Cucurbita	*pepo-aurantia*	Orange Summer Squash Melon	C:PD:**VG**
Cucurbita	*pepo-brunnea*	Brown Summer Squash Melon	C:PD:**VG**
Cucurbita	*pepo-groena*	Green Summer Squash Melon	C:PD:**VG**
Cucurbita	*pepo-helva*	Yellow Summer Squash Melon	C:PD:**VG**
Cucurbita	*pepo-helva-aurantia*	Orange Yellow Summer Squash Melon	C:PD:**VG**
Cucurbita	*pepo-melopepo-alba*	White Vegetable Marrow Summer Squash Melon	C:PD:**VG**
Cucurbita	*pepo-melopepo-aurantia*	Orange Vegetable Marrow Summer Squash Melon	C:PD:**VG**
Cucurbita	*pepo-melopepo-groena*	Green Vegetable Marrow Summer Squash Melon	C:PD:**VG**
Cucurbita	*pepo-melopepo-helva*	Yellow Vegetable Marrow Summer Squash Melon	C:PD:**VG**
Cucurbita	*pepo-melopepo-zucchini-alba*	White Zucchini Summer Squash Melon	C:PD:**VG**
Cucurbita	*pepo-melopepo-zucchini-groena*	Green Zucchini Summer Squash Melon	C:PD:**VG**
Cucurbita	*pepo-melopepo-zucchini-groena-grisea*	Gray Green Zucchini Summer Squash Melon	C:PD:**VG**
Cucurbita	*pepo-melopepo-zucchini-groena-helva*	Yellow Green Zucchini Summer Squash Melon	C:PD:**VG**
Cucurbita	*pepo-melopepo-zucchini-groena-nigra*	Black Green Zucchini Summer Squash Melon	C:PD:**VG**
Cucurbita	*pepo-melopepo-zucchini-helva*	Yellow Zucchini Summer Squash Melon	C:PD:**VG**
Cucurbita	*pepo-melopepo-zucchini-mixta*	Multicolor Zucchini Summer Squash Melon	C:PD:**VG**
Cucurbita	*pepo-mixta*	Multicolor Summer Squash Melon	C:PD:**VG**
Cucurbita	*pepo-pepon-alba*	White Summer Pumpkin Melon	C:PD:**VG**
Cucurbita	*pepo-pepon-aurantia*	Orange Summer Pumpkin Melon	C:PD:**VG**
Cucurbita	*pepo-pepon-helva*	Yellow Summer Pumpkin Melon	C:PD:**VG**
Cucurbita	*pepo-pepon-helva-aurantia*	Orange Yellow Summer Pumpkin Melon	C:PD:**VG**
Cucurbita	*pepo-pepon-groena*	Green Summer Pumpkin Melon	C:PD:**VG**
Cucurbita	*pepo-pepon-groena-blauea*	Blue Green Summer Pumpkin Melon	C:PD:**VG**
Cucurbita	*pepo-pepon-groena-nigra*	Black Green Summer Pumpkin Melon	C:PD:**VG**

Genus	Species	Specimen	Genome	C:Culinary
Cucurbita	*pepo-pepon-mixta*	Multicolor Summer Pumpkin	Melon	C:PD:**VG**
Cucurbita	*pepo-pepon-sperma*	Seed Summer Pumpkin	Melon	C:PD:**VG**
Cucurbita	*soraria*	Sister Pumpkin	Melon	C:PD:**VG**
Benincasa	*hispida*	Squash Benincasa	Melon	C:PD:**VG**
Benincasa	*cerifera*	Wax Gourd Benincasa	Melon	C:PD:**VG**
Citrullus	*egusi*	Egusi Citron	Melon	C:PD:**VG**
Cucumis	*anguria*	Prickly Gherkin	Cucumber	C:PD:**VG**
Cucumis	*metuliferus*	African Horned	Cucumber	C:PD:**VG**
Cucumis	*sativus-condio*	Pickling	Cucumber	C:PD:**VG**
Cucumis	*sativus-crema*	Cream	Cucumber	C:PD:**VG**
Cucumis	*sativus-groena-darca*	Dark Green	Cucumber	C:PD:**VG**
Cucumis	*sativus-groena-leuca*	Light Green	Cucumber	C:PD:**VG**
Cucumis	*sativus-groena-media*	Medium Green	Cucumber	C:PD:**VG**
Cucumis	*sativus-gurca*	Gherkin	Cucumber	C:PD:**VG**
Cucumis	*sativus-helva*	Yellow	Cucumber	C:PD:**VG**
Cucumis	*sativus-hybrida-F1*	F1 Hybrid	Cucumber	C:PD:**VG**
Cucumis	*sativus-rotunda*	Round	Cucumber	C:PD:**VG**
Cucumis	*sativus-sikkimensis*	Sikkim	Cucumber	C:PD:**VG**
Cucumis	*sativus-spina-alba*	White Spine	Cucumber	C:PD:**VG**
Cucumis	*sativus-spina-nigra*	Black Spine	Cucumber	C:PD:**VG**
Lagenaria	*longissima*	Long Bottle	Gourd	C:PD:**VG**
Lagenaria	*siceraria*	Crookneck Bottle	Gourd	C:PD:**VG**
Luffa	*acutangula*	Ridged	Loofah	C:PD:**VG**
Luffa	*aegyptiaca*	Egyptian	Loofah	C:PD:**VG**
Luffa	*operculata*	Onovas	Loofah	C:PD:**VG**
Momordica	*balsamina*	Balsam Apple Bitter	Melon	C:PD:**VG**
Momordica	*charantia*	Balsam Pear Bitter	Melon	C:PD:**VG**
Momordica	*cochinchinensis*	Cochin Chinese Bitter	Melon	C:PD:**VG**
Momordica	*dioica*	Dioecious Bitter	Melon	C:PD:**VG**
Momordica	*indica-longa*	India Long Bitter	Melon	C:PD:**VG**
Sechium	*edule*	Vegetable Pear	Chayote	C:PD:**VG**
Trichosanthes	*anguina*	Hairy Flower Snake	Gourd	C:PD:**VG**
Trichosanthes	*cucumerina*	Cucumber Snake	Gourd	C:PD:**VG**
Trichosanthes	*dioica*	Pointed Snake	Gourd	C:PD:**VG**

Varieties

Genus	Species	Infraspecies	cultivar	specimen	Genome
Cucurbita	*ficifolia*	*ficifolia*	Original	Figleaf Squash	Melon
Cucurbita	*ficifolia*	*chilacayotei*	Chlacayote	Figleaf Squash	Melon
Cucurbita	*ficifolia*	*chilicayotlii*	Chilicayotl	Figleaf Squash	Melon
Cucurbita	*ficifolia*	*malabarensis*	Malabar Gourd	Figleaf Squash	Melon
Cucurbita	*foetidissima*	*foetidissima*	Original	Buffalo Gourd	Melon
Cucurbita	*foetidissima*	*mock-orangei*	Mock Orange	Buffalo Gourd	Melon

Squash, Winter Neckless (*Cucurbita maxima*)

Genus	Species	Infraspecies	cultivar	specimen	Genome
Cucurbita	*maxima-aurantia*	*bananai-aurantia*	Orange Banana ??lb	Orange Winter Neckless Squash	Melon
Cucurbita	*maxima-aurantia*	*buttercup-kindredii*	Kindred Buttercup 4lb	Orange Winter Neckless Squash	Melon
Cucurbita	*maxima-aurantia*	*buttercup-gold-nuggetii*	Gold Nugget Buttercup 2lb	Orange Winter Neckless Squash	Melon
Cucurbita	*maxima-aurantia*	*buttercup-windsor-cupii*	Windsor Cup Buttercup 4lb	Orange Winter Neckless Squash	Melon
Cucurbita	*maxima-aurantia*	*calabaza-del-nortei*	Calabaza Del Norte ?lb	Orange Winter Neckless Squash	Melon
Cucurbita	*maxima-aurantia*	*essexensis-hybrida*	Hybrid Essex 10lb	Orange Winter Neckless Squash	Melon
Cucurbita	*maxima-aurantia*	*essexensis-turbana*	Turban Essex 14lb	Orange Winter Neckless Squash	Melon
Cucurbita	*maxima-aurantia*	*golden-deliciousi*	Golden Delicious 15lb	Orange Winter Neckless Squash	Melon
Cucurbita	*maxima-aurantia*	*greengoldii*	Greengold 6lb	Orange Winter Neckless Squash	Melon
Cucurbita	*maxima-aurantia*	*hokkaido-aurantia*	Hokkaido 4lb	Orange Winter Neckless Squash	Melon
Cucurbita	*maxima-aurantia*	*hokkaido-kurii-rubra*	Red Kuri Hokkaido 8lb	Orange Winter Neckless Squash	Melon
Cucurbita	*maxima-aurantia*	*hokkaido-kurii-uchikii*	Uchiki Kuri Hokkaido 10lb	Orange Winter Neckless Squash	Melon
Cucurbita	*maxima-aurantia*	*hubbardii-gold-montana*	Gold Mountain Hubbard 15lb	Orange Winter Neckless Squash	Melon
Cucurbita	*maxima-aurantia*	*hubbardii-hidatsai*	Hidatsa Hubbard ?lb	Orange Winter Neckless Squash	Melon
Cucurbita	*maxima-aurantia*	*hubbardii-little-gemii*	Little Gem Hubbard 3lb	Orange Winter Neckless Squash	Melon
Cucurbita	*maxima-aurantia*	*hubbardii-minnie's-apachei*	Minnie's Apache Hubbard ?lb	Orange Winter Neckless Squash	Melon
Cucurbita	*maxima-aurantia*	*hubbardii-mountaineeri*	Mountaineer Hubbard 5lb	Orange Winter Neckless Squash	Melon
Cucurbita	*maxima-aurantia*	*hubbardii-navajoi*	Navaho Hubbard ?lb	Orange Winter Neckless Squash	Melon
Cucurbita	*maxima-aurantia*	*mooregoldii*	Mooregold ?lb	Orange Winter Neckless Squash	Melon
Cucurbita	*maxima-aurantia*	*red-goldii*	Red Gold 8lb	Orange Winter Neckless Squash	Melon
Cucurbita	*maxima-aurantia*	*sweetkeeperi*	Sweetkeeper 11lb	Orange Winter Neckless Squash	Melon
Cucurbita	*maxima-aurantia*	*turban-mini-rubra*	Mini Red Turban ?lb	Orange Winter Neckless Squash	Melon
Cucurbita	*maxima-aurantia*	*turban-rubra-warrenii*	Warren Red Turban ?lb	Orange Winter Neckless Squash	Melon
Cucurbita	*maxima-brunnea*	*hubbardii-anna-swartzii*	Anna Swartz Hubbard 7lb	Brown Winter Neckless Squash	Melon
Cucurbita	*maxima-grisea*	*baby-queenii*	Baby Queen 12"	Gray Winter Neckless Squash	Melon

Genus	Species	Infraspecies	cultivar	specimen	Genome
Cucurbita	*maxima-grisea*	*crown-princei*	Crown Prince 20lb	Gray Winter Neckless Squash	Melon
Cucurbita	*maxima-grisea*	*hubbardii-baby-blauea*	Baby Blue Hubbard 5lb	Gray Winter Neckless Squash	Melon
Cucurbita	*maxima-grisea*	*hubbardii-blauea-balletii*	Ballet Blue Hubbard 6lb	Gray Winter Neckless Squash	Melon
Cucurbita	*maxima-grisea*	*hubbardii-blauea-neoanglica*	New England Blue Hubbard 35lb	Gray Winter Neckless Squash	Melon
Cucurbita	*maxima-grisea*	*hubbardii-blauea-specialis*	Special Blue Hubbard ??lb	Gray Winter Neckless Squash	Melon
Cucurbita	*maxima-grisea*	*hubbardii-dulong-qhii*	Dulong Qhi Blue Hubbard 10lb	Gray Winter Neckless Squash	Melon
Cucurbita	*maxima-grisea*	*hubbardii-sibleyi*	Sibley Blue Hubbard 10lb	Gray Winter Neckless Squash	Melon
Cucurbita	*maxima-grisea*	*hubbardii-sucra*	Sugar Blue Hubbard 17lb	Gray Winter Neckless Squash	Melon
Cucurbita	*maxima-grisea*	*hokkaido-kabocha*	Kabocha Hokkaido 4.5lb	Gray Winter Neckless Squash	Melon
Cucurbita	*maxima-grisea*	*neozelandica-blauea*	New Zealand Blue 8lb	Gray Winter Neckless Squash	Melon
Cucurbita	*maxima-groena*	*buttercup-bitterrootii*	Bitterroot Buttercup 3lb	Green Winter Neckless Squash	Melon
Cucurbita	*maxima-groena*	*buttercup-brown-seededii*	Brown Seeded Buttercup ?lb	Green Winter Neckless Squash	Melon
Cucurbita	*maxima-groena*	*buttercup-burgessii*	Burgess Buttercup 4lb	Green Winter Neckless Squash	Melon
Cucurbita	*maxima-groena*	*buttercup-bushii*	Bush Buttercup 4lb	Green Winter Neckless Squash	Melon
Cucurbita	*maxima-groena*	*buttercup-bushii-emeraldii*	Emerald Bush Buttercup 4lb	Green Winter Neckless Squash	Melon
Cucurbita	*maxima-groena*	*buttercup-chestnutii*	Chestnut Buttercup 3lb	Green Winter Neckless Squash	Melon
Cucurbita	*maxima-groena*	*buttercup-discus-bushii*	Discus Bush Buttercup 3lb	Green Winter Neckless Squash	Melon
Cucurbita	*maxima-groena*	*buttercup-qualityi*	Quality Buttercup ?lb	Green Winter Neckless Squash	Melon
Cucurbita	*maxima-groena*	*buttercup-turk's-capii*	Turk's Cap Buttercup 4lb	Green Winter Neckless Squash	Melon
Cucurbita	*maxima-groena*	*courge-olive-vertei*	Courge Olive Verte 7lb	Green Winter Neckless Squash	Melon
Cucurbita	*maxima-groena*	*deliciosa-groena*	Green Delicious 8lb	Green Winter Neckless Squash	Melon
Cucurbita	*maxima-groena*	*hokkaido-kabocha-black-forestii*	Black Forest Kabocha Hokkaido	Green Winter Neckless Squash	Melon
Cucurbita	*maxima-groena*	*hubbardii*	Hubbard ??lb	Green Winter Neckless Squash	Melon
Cucurbita	*maxima-groena*	*hubbardii-chicago-wartedii*	Chicago Warted Hubbard 14lb	Green Winter Neckless Squash	Melon
Cucurbita	*maxima-groena*	*hubbardii-groena*	Green Hubbard 12lb	Green Winter Neckless Squash	Melon
Cucurbita	*maxima-groena*	*hubbardii-groena-babyi*	Baby Green Hubbard ?lb	Green Winter Neckless Squash	Melon
Cucurbita	*maxima-groena*	*hubbardii-groena-improva*	Improved Green Hubbard 11lb	Green Winter Neckless Squash	Melon
Cucurbita	*maxima-groena*	*hubbardii-groena-mini*	Mini Green Hubbard 2.5lb	Green Winter Neckless Squash	Melon
Cucurbita	*maxima-groena*	*hubbardii-groena-smoothii*	Smooth Green Hubbard 10lb	Green Winter Neckless Squash	Melon
Cucurbita	*maxima-groena*	*hubbardii-truei*	True Hubbard 25lb	Green Winter Neckless Squash	Melon
Cucurbita	*maxima-groena*	*hubbardii-rio-lucio-pumpkinii*	Rio Lucio Hubbard ??lb	Green Winter Neckless Squash	Melon
Cucurbita	*maxima-groena*	*hubbardii-wartedii*	Warted Hubbard 20lb	Green Winter Neckless Squash	Melon
Cucurbita	*maxima-groena*	*hubbardii-wartedii-improva*	Improved Warted Hubbard 13lb	Green Winter Neckless Squash	Melon
Cucurbita	*maxima-groena-blauea*	*candy-roasteri*	Candy Roaster 14lb	Blue Green Winter Neckless Squash	Melon
Cucurbita	*maxima-groena-blauea*	*hokkaido-kuri-blauea*	Blue Kuri Hokkaido 4.5lb	Blue Green Winter Neckless Squash	Melon
Cucurbita	*maxima-groena-blauea*	*hubbardii-azurea*	Azure Hubbard 20lb	Blue Green Winter Neckless Squash	Melon
Cucurbita	*maxima-groena-blauea*	*hubbardii-mormonii*	Mormon Hubbard ??lb	Blue Green Winter Neckless Squash	Melon
Cucurbita	*maxima-groena-blauea*	*old-bluei*	Old Blue 8lb	Blue Green Winter Neckless Squash	Melon
Cucurbita	*maxima-groena-grisea*	*argentinensis*	Argentine ?lb	Gray Green Winter Neckless Squash	Melon
Cucurbita	*maxima-groena-grisea*	*argentinensis-primitiva*	Primitive Argentine 8lb	Gray Green Winter Neckless Squash	Melon
Cucurbita	*maxima-groena-grisea*	*bananai*	Banana 25lb	Gray Green Winter Neckless Squash	Melon
Cucurbita	*maxima-groena-grisea*	*bananai-blauea*	Blue Banana 30lb	Gray Green Winter Neckless Squash	Melon
Cucurbita	*maxima-groena-grisea*	*bananai-blauea-guatemalensis*	Guatemalan Blue Banana 35lb	Gray Green Winter Neckless Squash	Melon
Cucurbita	*maxima-groena-grisea*	*berretina-piacentina*	Berretina Piacentina 4lb	Gray Green Winter Neckless Squash	Melon
Cucurbita	*maxima-groena-grisea*	*blauea-blockersi*	Blue Blockers 6lb	Gray Green Winter Neckless Squash	Melon
Cucurbita	*maxima-groena-grisea*	*hokkaido-groena*	Hokkaido ?lb	Gray Green Winter Neckless Squash	Melon
Cucurbita	*maxima-groena-grisea*	*hokkaido-semi-bushii*	Semi Bush Hokkaido 3lb	Gray Green Winter Neckless Squash	Melon
Cucurbita	*maxima-groena-grisea*	*hubbardii-blauea*	Blue Hubbard 16lb	Gray Green Winter Neckless Squash	Melon
Cucurbita	*maxima-groena-grisea*	*hubbardii-silver-bellii*	Silver Bell Hubbard 6lb	Gray Green Winter Neckless Squash	Melon
Cucurbita	*maxima-groena-grisea*	*taosensis*	Taos 10lb	Gray Green Winter Neckless Squash	Melon
Cucurbita	*maxima-groena-grisea*	*zapallo-macrei*	Macre Zapallo 15lb	Gray Green Winter Neckless Squash	Melon
Cucurbita	*maxima-helva*	*ameria-indica*	American Indian ?lb	Yellow Winter Neckless Squash	Melon
Cucurbita	*maxima-helva*	*australian-butteri*	Australian Butter 14lb	Yellow Winter Neckless Squash	Melon
Cucurbita	*maxima-helva*	*chersonskaya*	Chersonskaya ?lb	Yellow Winter Neckless Squash	Melon
Cucurbita	*maxima-helva*	*delicata*	Delicata ?lb	Yellow Winter Neckless Squash	Melon
Cucurbita	*maxima-helva*	*hokkaido-kabocha-buttercup-verdas*	Verdas Buttercup Kabocha ?lb	Yellow Winter Neckless Squash	Melon
Cucurbita	*maxima-helva*	*mexican-bananai*	Mexican Banana ?lb	Yellow Winter Neckless Squash	Melon
Cucurbita	*maxima-helva*	*mexigoldii*	Mexigold ?lb	Yellow Winter Neckless Squash	Melon
Cucurbita	*maxima-helva*	*quality-winter-squashii*	Quality ?lb	Yellow Winter Neckless Squash	Melon
Cucurbita	*maxima-helva*	*zipinki-campana*	Zipinki Campana 2lb	Yellow Winter Neckless Squash	Melon
Cucurbita	*maxima-helva-aurantia*	*hubbardii-goldenii*	Golden Hubbard 10lb	Orange Yellow Winter Neckless Squash	Melon
Cucurbita	*maxima-helva-aurantia*	*hubbardii-goldenii-NK580*	#NK580 Golden Hubbard ??lb	Orange Yellow Winter Neckless Squash	Melon
Cucurbita	*maxima-helva-aurantia*	*hubbardii-goldenii-deliciousi*	Delicious Golden Hubbard 8lb	Orange Yellow Winter Neckless Squash	Melon
Cucurbita	*maxima-mixta*	*rainbowii*	Rainbow 3lb	Multicolor Winter Neckless Squash	Melon
Cucurbita	*maxima-mixta*	*turban-lakota-siouxii*	Lakota Sioux Turban 5lb	Multicolor Winter Neckless Squash	Melon
Cucurbita	*maxima-mixta*	*turban-tinyturkii*	Tiny Turk Turban 2lb	Multicolor Winter Neckless Squash	Melon
Cucurbita	*maxima-mixta*	*turban-turk's-capii*	Turk's Cap Turban 5lb	Multicolor Winter Neckless Squash	Melon
Cucurbita	*maxima-nigra*	*turban-windsor-blackii*	Windsor Black Turban 5lb	Black Winter Neckless Squash	Melon
Cucurbita	*maxima-nigra*	*verruquex-du-portugalensis*	Verraquex Du Portugal ?lb	Black Winter Neckless Squash	Melon
Cucurbita	*maxima-pinca*	*bananai-pinca*	Pink Banana 10lb	Pink Winter Neckless Squash	Melon
Cucurbita	*maxima-pinca*	*bananai-pinca-jumboi*	Jumbo Pink Banana 70lb	Pink Winter Neckless Squash	Melon
Cucurbita	*maxima-pinca*	*candy-roasteri-georgiensis*	Georgia Candy Roaster 60lb	Pink Winter Neckless Squash	Melon
Cucurbita	*maxima-pinca*	*germanica*	German 12lb	Pink Winter Neckless Squash	Melon

Genus	Species	Infraspecies	cultivar	specimen	Genome
Cucurbita	*maxima-pinca*	*hubbardii-arikara*	Arikara Hubbard 15lb	Pink Winter Neckless Squash	Melon
Cucurbita	*maxima-pinca*	*goldkeeperi*	Goldkeeper ?lb	Pink Winter Neckless Squash	Melon
Cucurbita	*maxima-rubra*	*rubra-chestnutii*	Red Chestnut ?lb	Red Winter Neckless Squash	Melon
Cucurbita	*maxima-rubra*	*rubra-warty-thingii*	Red Warty Thing ?lb	Red Winter Neckless Squash	Melon

Pumpkin, Winter (*Curcurbita maxima-pepo*)

Genus	Species	Infraspecies	cultivar	specimen	Genome
Cucurbita	*maxima-pepo-alba*	*alba-hendersonvillensis*	Hendersonville ?lb	White Winter Pumpkin	Melon
Cucurbita	*maxima-pepo-alba*	*flat-alba-pumpkinii*	Flat 22lb	White Winter Pumpkin	Melon
Cucurbita	*maxima-pepo-alba*	*fortna-alba*	Fortna ?lb	White Winter Pumpkin	Melon
Cucurbita	*maxima-pepo-alba*	*hopii-whitei*	White Hopi 8lb	White Winter Pumpkin	Melon
Cucurbita	*maxima-pepo-alba*	*hopii-whitei-pumpkinii*	White Hopi 35lb	White Winter Pumpkin	Melon
Cucurbita	*maxima-pepo-alba*	*lumina*	Lumina13lb	White Winter Pumpkin	Melon
Cucurbita	*maxima-pepo-alba*	*mayo-blusheri*	Mayo Blusher 9lb	White Winter Pumpkin	Melon
Cucurbita	*maxima-pepo-alba*	*sweet-dumplingii*	Sweet Dumpling ?lb	White Winter Pumpkin	Melon
Cucurbita	*maxima-pepo-alba*	*valencianoi*	Valenciano ?lb	White Winter Pumpkin	Melon
Cucurbita	*maxima-pepo-alba*	*white-cheesequakei*	White Cheesquake ?lb	White Winter Pumpkin	Melon
Cucurbita	*maxima-pepo-aurantia*	.	.	Orange Winter Pumpkin	Melon
Cucurbita	*maxima-pepo-aurantia*	*amish-piei*	Amish Pie 70lb	Orange Winter Pumpkin	Melon
Cucurbita	*maxima-pepo-aurantia*	*atlantic-giantii*	Atlantic Giant 800lb	Orange Winter Pumpkin	Melon
Cucurbita	*maxima-pepo-aurantia*	*atlantic-giantii-dill'si*	Dill's Atlantic Giant 600lb	Orange Winter Pumpkin	Melon
Cucurbita	*maxima-pepo-aurantia*	*big-maxii*	Big Max 100lb	Orange Winter Pumpkin	Melon
Cucurbita	*maxima-pepo-aurantia*	*big-moonii*	Big Moon 150lb	Orange Winter Pumpkin	Melon
Cucurbita	*maxima-pepo-aurantia*	*big-tomii*	Big Tom 18lb	Orange Winter Pumpkin	Melon
Cucurbita	*maxima-pepo-aurantia*	*giant-pumpkinii-burgessi*	Burgess Giant 250lb	Orange Winter Pumpkin	Melon
Cucurbita	*maxima-pepo-aurantia*	*harvest-moonii*	Harvest Moon 16lb	Orange Winter Pumpkin	Melon
Cucurbita	*maxima-pepo-aurantia*	*hopii-aurantia*	Orange Hopi 14lb	Orange Winter Pumpkin	Melon
Cucurbita	*maxima-pepo-aurantia*	*jackpotii*	Jackpot 25lb	Orange Winter Pumpkin	Melon
Cucurbita	*maxima-pepo-aurantia*	*mammoth-larga*	Large Mammoth ??lb	Orange Winter Pumpkin	Melon
Cucurbita	*maxima-pepo-aurantia*	*mammoth-goldii*	Gold Mammoth 50lb	Orange Winter Pumpkin	Melon
Cucurbita	*maxima-pepo-aurantia*	*mammota-regis*	King Mammoth 80lb	Orange Winter Pumpkin	Melon
Cucurbita	*maxima-pepo-aurantia*	*mammota-aurantia-goldii*	Orange Gold Mammoth 75lb	Orange Winter Pumpkin	Melon
Cucurbita	*maxima-pepo-aurantia*	*mavis-sweeteneri*	Mavis Sweetener ?lb	Orange Winter Pumpkin	Melon
Cucurbita	*maxima-pepo-aurantia*	*muscat-de-provencei*	Muscat De Provence 10lb	Orange Winter Pumpkin	Melon
Cucurbita	*maxima-pepo-aurantia*	*nanicoke-indiana*	Nanicoke Indian 5lb	Orange Winter Pumpkin	Melon
Cucurbita	*maxima-pepo-aurantia*	*ol-zeb'si*	Ol Zeb's 25lb	Orange Winter Pumpkin	Melon
Cucurbita	*maxima-pepo-aurantia*	*quintale-seme-gialloi*	Quintale Seme Gaillo 5lb	Orange Winter Pumpkin	Melon
Cucurbita	*maxima-pepo-aurantia*	*rellekei*	Relleke ??lb	Orange Winter Pumpkin	Melon
Cucurbita	*maxima-pepo-aurantia*	*rouge-vif-d'etampesi*	Rouge Vif D'etampes 40lb	Orange Winter Pumpkin	Melon
Cucurbita	*maxima-pepo-aurantia*	*trojanii*	Trojan 25lb	Orange Winter Pumpkin	Melon
Cucurbita	*maxima-pepo-aurantia*	*vert-d'espagnei*	Vert D'Espagne 35lb	Orange Winter Pumpkin	Melon
Cucurbita	*maxima-pepo-aurantia*	*weeks-north-carolina-giant-pumpkinii*	Weeks North Calolina Giant 210lb	Orange Winter Pumpkin	Melon
Cucurbita	*maxima-pepo-grisea*	*butterballii*	Butterball ?lb	Gray Winter Pumpkin	Melon
Cucurbita	*maxima-pepo-grisea*	*hopii-pale-grisea*	Pale Gray Hopi 8lb	Gray Winter Pumpkin	Melon
Cucurbita	*maxima-pepo-grisea*	*jarrahdalei*	Jarrahdale 16lb	Gray Winter Pumpkin	Melon
Cucurbita	*maxima-pepo-grisea*	*jarrahdalei-redlands-trailblazeri*	Redlands Trailblazer Jarradale ??lb	Gray Winter Pumpkin	Melon
Cucurbita	*maxima-pepo-grisea*	*marbleheadii*	Marblehead 20lb	Gray Winter Pumpkin	Melon
Cucurbita	*maxima-pepo-grisea*	*marbleheadii-umatilla*	Umatilla Marblehead ??lb	Gray Winter Pumpkin	Melon
Cucurbita	*maxima-pepo-grisea*	*marbleheadii-yakima*	Yakima Marblehead 16lb	Gray Winter Pumpkin	Melon
Cucurbita	*maxima-pepo-grisea*	*marina-di-chioggia*	Marina Di Chioggia 10lb	Gray Winter Pumpkin	Melon
Cucurbita	*maxima-pepo-grisea*	*nepalensis*	Nepalese 18lb	Gray Winter Pumpkin	Melon
Cucurbita	*maxima-pepo-grisea*	*penascoensis*	Penasco 8lb	Gray Winter Pumpkin	Melon
Cucurbita	*maxima-pepo-grisea*	*sweet-meatii*	Sweet Meat 15lb	Gray Winter Pumpkin	Melon
Cucurbita	*maxima-pepo-grisea*	*triamblei*	Triamble 15lb	Gray Winter Pumpkin	Melon
Cucurbita	*maxima-pepo-grisea*	*wangaparoa-crown-pumpkinii*	Whangaparoa Crown ?lb	Gray Winter Pumpkin	Melon
Cucurbita	*maxima-pepo-groena*	*larga-turca-turbana*	Large Turk's Turban ??lb	Green Winter Pumpkin	Melon
Cucurbita	*maxima-pepo-groena-blauea*	*crown-pumpkinii*	Crown ?lb	Blue Green Winter Pumpkin	Melon
Cucurbita	*maxima-pepo-groena-blauea*	*indiana-pumpkinii*	Indian ??lb	Blue Green Winter Pumpkin	Melon
Cucurbita	*maxima-pepo-groena-blauea*	*queenslandica-blauea*	Queensland Blue 15lb	Blue Green Winter Pumpkin	Melon
Cucurbita	*maxima-pepo-groena-blauea*	*sweetmeatii-katy's-super-sweetii*	Katy's Super Sweet Sweetmeat 18lb	Blue Green Winter Pumpkin	Melon
Cucurbita	*maxima-pepo-groena-blauea*	*zapallo-plomoi*	Plomo Zapallo ?lb	Blue Green Winter Pumpkin	Melon
Cucurbita	*maxima-pepo-helva*	*juaune-gros-de-parisi*	Jaune Gros De Paris 100lb	Yellow Winter Pumpkin	Melon
Cucurbita	*maxima-pepo-helva*	*hundred-weightii*	Hundred Weight ???lb	Yellow Winter Pumpkin	Melon
Cucurbita	*maxima-pepo-helva*	*mammoth-regis*	King Of Mammoth 65lb	Yellow Winter Pumpkin	Melon
Cucurbita	*maxima-pepo-mixta*	*iranica*	Iran 25lb	Multicolor Winter Pumpkin	Melon
Cucurbita	*maxima-pepo-mixta*	*kakai*	Kakai 6lb	Multicolor Winter Pumpkin	Melon
Cucurbita	*maxima-pepo-mixta*	*padanai*	Padana ?lb	Multicolor Winter Pumpkin	Melon
Cucurbita	*maxima-pepo-mixta*	*show-kingii*	Show King 200lb	Multicolor Winter Pumpkin	Melon
Cucurbita	*maxima-pepo-pinca*	*galeux-d'eysinesi*	Galeux D'Eysines ?lb	Pink Winter Pumpkin	Melon
Cucurbita	*maxima-pepo-pinca*	*great-pumpkinii*	Great Pumpkin 100lb	Pink Winter Pumpkin	Melon
Cucurbita	*maxima-pepo-pinca*	*lower-salmon-riverii*	Lower Salmon River 13lb	Pink Winter Pumpkin	Melon
Cucurbita	*maxima-pepo-pinca*	*my-best-piei*	My Best Pie ??lb	Pink Winter Pumpkin	Melon

Genus	Species	Infraspecies	cultivar	specimen	Genome
Cucurbita	*maxima-pepo-pinca*	*wickersham-sweet-potatoi*	Wickersham Sweet Potato 40lb	Pink Winter Pumpkin	Melon

Squash, Cushaw (*Curcurbita mixta*)

Genus	Species	Infraspecies	cultivar	specimen	Genome
Cucurbita	*mixta*	*albino-pepita*	Albino Pepita 25lb	Cushaw Squash	Melon
Cucurbita	*mixta*	*calabaza-de-las-aguasi*	Calabaza De Las Aguas ?lb	Cushaw Squash	Melon
Cucurbita	*mixta*	*campechensis*	Campeche 6lb	Cushaw Squash	Melon
Cucurbita	*mixta*	*chochiti-puebloensis*	Cochiti Pueblo 2lb	Cushaw Squash	Melon
Cucurbita	*mixta*	*chompai*	Chompa ?lb	Cushaw Squash	Melon
Cucurbita	*mixta*	*gila*	Gila 11lb	Cushaw Squash	Melon
Cucurbita	*mixta*	*gila-clivicola*	Cliff Dweller Gila 10lb	Cushaw Squash	Melon
Cucurbita	*mixta*	*gold-striata*	Gold Striped 18lb	Cushaw Squash	Melon
Cucurbita	*mixta*	*golden-striata*	Golden Striped 28lb	Cushaw Squash	Melon
Cucurbita	*mixta*	*groena-striata*	Green Striped 14lb	Cushaw Squash	Melon
Cucurbita	*mixta*	*hopi*	Hopi 15lb	Cushaw Squash	Melon
Cucurbita	*mixta*	*hopi-nigra-groena*	Black Green Hopi 50lb	Cushaw Squash	Melon
Cucurbita	*mixta*	*hopi-taosensis*	Taos Hopi 25lb	Cushaw Squash	Melon
Cucurbita	*mixta*	*hopi-teardropii*	Teardrop Hopi 15lb	Cushaw Squash	Melon
Cucurbita	*mixta*	*hopi-vatnga*	Vatnga Hopi 20lb	Cushaw Squash	Melon
Cucurbita	*mixta*	*illinoisensis*	Illinois ??lb	Cushaw Squash	Melon
Cucurbita	*mixta*	*japanese-piei*	Japanese Pie 12lb	Cushaw Squash	Melon
Cucurbita	*mixta*	*magdalena-stripedii*	Magdalena Striped ??lb	Cushaw Squash	Melon
Cucurbita	*mixta*	*mayo-arrotei*	Mayo Arrote 15lb	Cushaw Squash	Melon
Cucurbita	*mixta*	*mexican-xtopii*	Mexican Xtop ??lb	Cushaw Squash	Melon
Cucurbita	*mixta*	*papalotei*	Papalote 7lb	Cushaw Squash	Melon
Cucurbita	*mixta*	*papalotei-ranchii*	Ranch Papalote 10lb	Cushaw Squash	Melon
Cucurbita	*mixta*	*parralii*	Parral 10lb	Cushaw Squash	Melon
Cucurbita	*mixta*	*salem-whitei*	Salem White ??lb	Cushaw Squash	Melon
Cucurbita	*mixta*	*san-bernardoensis*	San Bernardo ??lb	Cushaw Squash	Melon
Cucurbita	*mixta*	*san-juan-pubeloensis*	San Juan Pubelo 15lb	Cushaw Squash	Melon
Cucurbita	*mixta*	*san-pedroensis*	San Pedro ??lb	Cushaw Squash	Melon
Cucurbita	*mixta*	*santo-domingoensis*	Santo Domingo 14lb	Cushaw Squash	Melon
Cucurbita	*mixta*	*tennessee-sweet-potatoi*	Tennessee Sweet Potato 12lb	Cushaw Squash	Melon
Cucurbita	*mixta*	*tricolor*	Tricolor 14lb	Cushaw Squash	Melon
Cucurbita	*mixta*	*tohono-o'odhamii*	Tohona O'odham ??lb	Cushaw Squash	Melon
Cucurbita	*mixta*	*white-johnathanii*	White Johnathan 30lb	Cushaw Squash	Melon
Cucurbita	*mixta*	*vintera-vigna*	Winter Vining 14lb	Cushaw Squash	Melon
Cucurbita	*mixta*	*yeome-yaqui-kama*	Yeome Yaqui Kama ??lb	Cushaw Squash	Melon
Cucurbita	*mixta*	*zapoteci*	Zapotec ??lb	Cushaw Squash	Melon
Cucurbita	*mixta*	*zebra-mysteryi*	Zebra Mystery ??lb	Cushaw Squash	Melon
Cucurbita	*mixta-sperma*	*silver-edgedii*	Silver Edged ??lb	Seed Cushaw Squash	Melon
Cucurbita	*mixta-sperma*	*silver-seed-thick-marginii*	Silver Seed Thick Margin ??lb	Seed Cushaw Squash	Melon
Cucurbita	*mixta-sperma*	*striped-belizensis*	Striped Belize ??lb	Seed Cushaw Squash	Melon
Cucurbita	*mixta-sperma*	*tamala*	Tamala ??lb	Seed Cushaw Squash	Melon
Cucurbita	*mixta-sperma*	*tamala-de-carnei*	De Carne Tamala ??lb	Seed Cushaw Squash	Melon
Cucurbita	*mixta-sperma*	*tamala-de-huesoi*	De Hueso Tamala ??lb	Seed Cushaw Squash	Melon
Cucurbita	*mixta-sperma*	*tequila-blackii*	Tequila Black 10lb	Seed Cushaw Squash	Melon
Cucurbita	*mixta-sperma*	*veracruz-pepita*	Veracruz Pepita ??lb	Seed Cushaw Squash	Melon

Squash, Winter Necked (*Curcurbita moschata*)

Genus	Species	Infraspecies	cultivar	specimen	Genome
Cucurbita	*moschata-aurantia*	*butternutii*	Butternut 4lb	Orange Flesh Winter Necked Squash	Melon
Cucurbita	*moschata-aurantia*	*butternutii-africana*	African Butternut 7lb	Orange Flesh Winter Necked Squash	Melon
Cucurbita	*moschata-aurantia*	*butternutii-butterbushii*	Butterbush Butternut 1.7lb	Orange Flesh Winter Necked Squash	Melon
Cucurbita	*moschata-aurantia*	*butternutii-earlyi*	Early Butternut ?lb	Orange Flesh Winter Necked Squash	Melon
Cucurbita	*moschata-aurantia*	*butternutii-hopii-tanii*	Hopi Tan Butternut 8lb	Orange Flesh Winter Necked Squash	Melon
Cucurbita	*moschata-aurantia*	*butternutii-long-of-naplesensis*	Long Of Naples Butternut 28lb	Orange Flesh Winter Necked Squash	Melon
Cucurbita	*moschata-aurantia*	*butternutii-ponca*	Ponca Butternut 2lb	Orange Flesh Winter Necked Squash	Melon
Cucurbita	*moschata-aurantia*	*butternutii-quaker-pie-squashii*	Quaker Pie Butternut 7lb	Orange Flesh Winter Necked Squash	Melon
Cucurbita	*moschata-aurantia*	*butternutii-trombonei*	Trombone Butternut ?lb	Orange Flesh Winter Necked Squash	Melon
Cucurbita	*moschata-aurantia*	*calabaza-la-primera*	La Primera Calabaza 22lb	Orange Flesh Winter Necked Squash	Melon
Cucurbita	*moschata-aurantia*	*crookneckii-canadensis*	Canada Crookneck ?lb	Orange Flesh Winter Necked Squash	Melon
Cucurbita	*moschata-aurantia*	*cushaw-orangei*	Orange Cushaw 15lb	Orange Flesh Winter Necked Squash	Melon
Cucurbita	*moschata-aurantia*	*cushaw-orangei-striata*	Striped Orange Cushaw 8lb	Orange Flesh Winter Necked Squash	Melon
Cucurbita	*moschata-aurantia*	*huicha*	Huicha ?lb	Orange Flesh Winter Necked Squash	Melon
Cucurbita	*moschata-aurantia*	*mayo-arrotei-rio-fuertei*	Rio Fuerte Mayo Arrote ?lb	Orange Flesh Winter Necked Squash	Melon
Cucurbita	*moschata-aurantia*	*north-falkland-islandica*	North Falkland Island ?lb	Orange Flesh Winter Necked Squash	Melon
Cucurbita	*moschata-aurantia*	*old-fashioned-Tennessee-viningii*	Old Fashioned Tennessee Vining 15lb	Orange Flesh Winter Necked Squash	Melon
Cucurbita	*moschata-aurantia*	*segualca-guarijioi*	Guarijio Segualca ?lb	Orange Flesh Winter Necked Squash	Melon
Cucurbita	*moschata-aurantia*	*segualca-mayoi*	Mayo Segualca ?lb	Orange Flesh Winter Necked Squash	Melon
Cucurbita	*moschata-aurantia*	*segualca-mayoi-rioi*	Rio Mayo Segualca ?lb	Orange Flesh Winter Necked Squash	Melon
Cucurbita	*moschata-aurantia*	*segualca-mayoi-rioi-fuertei*	Fuerte Rio Mayo Segualca ?lb	Orange Flesh Winter Necked Squash	Melon
Cucurbita	*moschata-aurantia*	*segualca-piedras-verdesensis*	Piedras Verdes Segualca ?lb	Orange Flesh Winter Necked Squash	Melon

Genus	Species	Infraspecies	cultivar	specimen	Genome
Cucurbita	*moschata-aurantia*	*segualca-yeomei*	Yeome Segualca ?lb	Orange Flesh Winter Necked Squash	Melon
Cucurbita	*moschata-aurantia*	*yokohamensis*	Yokohama ?lb	Orange Flesh Winter Necked Squash	Melon
Cucurbita	*moschata-aurantia-helva*	*butternutii-walthamii*	Waltham Butternut 5lb	Yellow Orange Flesh Winter Necked Squash	Melon
Cucurbita	*moschata-aurantia-helva*	*butternutii-canadai-mezoides*	Canadai Mezoides Butternut ?lb	Yellow Orange Flesh Winter Necked Squash	Melon
Cucurbita	*moschata-aurantia-helva*	*butternutii-tahitian-melon-squashii*	Tahitian Melon Butternut 24lb	Yellow Orange Flesh Winter Necked Squash	Melon
Cucurbita	*moschata-helva*	*butternutii-babyi*	Baby Butternut ?lb	Yellow Flesh Winter Necked Squash	Melon
Cucurbita	*moschata-helva*	*butternutii-buglei*	Bugle Butternut ?lb	Yellow Flesh Winter Necked Squash	Melon
Cucurbita	*moschata-helva*	*butternutii-improva*	Improved Butternut 2lb	Yellow Flesh Winter Necked Squash	Melon
Cucurbita	*moschata-helva*	*butternutii-bushii*	Bush Butternut 2lb	Yellow Flesh Winter Necked Squash	Melon
Cucurbita	*moschata-helva*	*butternutii-cubana*	Cuban Butternut ?lb	Yellow Flesh Winter Necked Squash	Melon
Cucurbita	*moschata-helva*	*butternutii-nicklow's-delightii*	Nicklow's Delight Butternut ?lb	Yellow Flesh Winter Necked Squash	Melon
Cucurbita	*moschata-helva*	*butternutii-puritanii*	Puritan Butternut ?lb	Yellow Flesh Winter Necked Squash	Melon
Cucurbita	*moschata-helva*	*butternutii-zenithii*	Zenith Butternut 1kg	Yellow Flesh Winter Necked Squash	Melon
Cucurbita	*moschata-helva*	*butternutii-zucca-piena-di-napoli*	Zucca Piena Di Napoli Butternut 4.5lb	Yellow Flesh Winter Necked Squash	Melon
Cucurbita	*moschata-helva*	*calabaza-cubensis*	Cuban Calabaza ?lb	Yellow Flesh Winter Necked Squash	Melon
Cucurbita	*moschata-helva*	*calabaza-segualca*	Segualca Calabaza ?lb	Yellow Flesh Winter Necked Squash	Melon
Cucurbita	*moschata-helva*	*calabaza-united-nations-plazai*	United Nations Plaza Calabaza 5lb	Yellow Flesh Winter Necked Squash	Melon
Cucurbita	*moschata-helva*	*carrizoi*	Carrizo ?lb	Yellow Flesh Winter Necked Squash	Melon
Cucurbita	*moschata-helva*	*choctaw-sweet-potatoi*	Choctaw Sweet Potato ?lb	Yellow Flesh Winter Necked Squash	Melon
Cucurbita	*moschata-helva*	*cushaw-goldenii*	Golden Cushaw 11lb	Yellow Flesh Winter Necked Squash	Melon
Cucurbita	*moschata-helva*	*flat-tan-fieldii*	Flat Tan Field ?lb	Yellow Flesh Winter Necked Squash	Melon
Cucurbita	*moschata-helva*	*german-above-ground-sweet-potatoi*	German Above Ground Sweet Potato ?lb	Yellow Flesh Winter Necked Squash	Melon
Cucurbita	*moschata-helva*	*kikuza-early-whitei*	Kikuza Early White 6lb	Yellow Flesh Winter Necked Squash	Melon
Cucurbita	*moschata-helva*	*middle-rio-conchosi*	Middle Rio Conchos ?lb	Yellow Flesh Winter Necked Squash	Melon
Cucurbita	*moschata-helva*	*paw-pawii*	Paw Paw 2.5lb	Yellow Flesh Winter Necked Squash	Melon
Cucurbita	*moschata-helva*	*pima-bajoi*	Pima Bajo ?lb	Yellow Flesh Winter Necked Squash	Melon
Cucurbita	*moschata-helva*	*sucrine-du-berryi*	Sucrine Du Berry 2lb	Yellow Flesh Winter Necked Squash	Melon
Cucurbita	*moschata-helva*	*tromboncinoi*	Tromboncino ?lb	Yellow Flesh Winter Necked Squash	Melon
Cucurbita	*moschata-helva*	*tromboncinoi-zucchetta-rampicantei*	Rampicante Zucchetta Tromboncino ?lb	Yellow Flesh Winter Necked Squash	Melon
Cucurbita	*moschata-helva*	*xhosai*	Xhosa 11lb	Yellow Flesh Winter Necked Squash	Melon
Cucurbita	*moschata-pepo-aurantia*	*butternutii-creolei*	Creole Butternut 45lb	Orange Flesh Flat Winter Pumpkin	Melon
Cucurbita	*moschata-pepo-aurantia*	*butternutii-neckii*	Neck Butternut ??lb	Orange Flesh Flat Winter Pumpkin	Melon
Cucurbita	*moschata-pepo-aurantia*	*cheesei*	Cheese ??lb	Orange Flesh Flat Winter Pumpkin	Melon
Cucurbita	*moschata-pepo-aurantia*	*cheesei-dickinsonii*	Dickinson Cheese 40lb	Orange Flesh Flat Winter Pumpkin	Melon
Cucurbita	*moschata-pepo-aurantia*	*cheesei-grisea*	Gray Cheese 12lb	Orange Flesh Flat Winter Pumpkin	Melon
Cucurbita	*moschata-pepo-aurantia*	*cheesei-landrethii*	Landreth Cheese ??lb	Orange Flesh Flat Winter Pumpkin	Melon
Cucurbita	*moschata-pepo-aurantia*	*cheesei-long-islandensis*	Long Island Cheese 10lb	Orange Flesh Flat Winter Pumpkin	Melon
Cucurbita	*moschata-pepo-aurantia*	*cheesei-long-islandensis-milk-pumpkinii*	Milk Long Island Cheese 15lb	Orange Flesh Flat Winter Pumpkin	Melon
Cucurbita	*moschata-pepo-aurantia*	*cheesei-magdalena-bigii*	Magdelena Big Cheese ??lb	Orange Flesh Flat Winter Pumpkin	Melon
Cucurbita	*moschata-pepo-aurantia*	*cheesei-milkii*	Milk Cheese 18lb	Orange Flesh Flat Winter Pumpkin	Melon
Cucurbita	*moschata-pepo-aurantia*	*cheesei-mrs.-anderson's-pumpkinii*	Mrs.Amerson's Cheese ??lb	Orange Flesh Flat Winter Pumpkin	Melon
Cucurbita	*moschata-pepo-aurantia*	*cheesei-shumway's-sweet-potatoei*	Shumway's Sweet Potato Cheese 10lb	Orange Flesh Flat Winter Pumpkin	Melon
Cucurbita	*moschata-pepo-aurantia*	*cheesei-siciliana*	Sicilian Cheese ??lb	Orange Flesh Flat Winter Pumpkin	Melon
Cucurbita	*moschata-pepo-aurantia*	*cheesei-tanii*	Tan Cheese 10lb	Orange Flesh Flat Winter Pumpkin	Melon
Cucurbita	*moschata-pepo-aurantia*	*japa*	Jap ?lb	Orange Flesh Flat Winter Pumpkin	Melon
Cucurbita	*moschata-pepo-aurantia*	*papaya-pumpkinii*	Papaya 2lb	Orange Flesh Flat Winter Pumpkin	Melon
Cucurbita	*moschata-pepo-aurantia*	*seminole-acornii*	Acorn Seminole ?lb	Orange Flesh Flat Winter Pumpkin	Melon
Cucurbita	*moschata-pepo-aurantia*	*seminole-pumpkinii*	Seminole ?lb	Orange Flesh Flat Winter Pumpkin	Melon
Cucurbita	*moschata-pepo-aurantia*	*texas-indian-moschata*	Texas Indian 12lb	Orange Flesh Flat Winter Pumpkin	Melon
Cucurbita	*moschata-pepo-aurantia*	*white-rind-sugari*	White Rind Sugar 15lb	Orange Flesh Flat Winter Pumpkin	Melon
Cucurbita	*moschata-pepo-aurantia-helva*	*chirimenii*	Chirimen 8lb	Yellow Orange Flesh Flat Winter Pumpkin	Melon
Cucurbita	*moschata-pepo-aurantia-helva*	*futtsu-early-blackii*	Futtsu Early Black 6lb	Yellow Orange Flesh Flat Winter Pumpkin	Melon
Cucurbita	*moschata-pepo-aurantia-helva*	*cheesei-kentucky-fieldii*	Kentucky Field Cheese 12lb	Yellow Orange Flesh Flat Winter Pumpkin	Melon
Cucurbita	*moschata-pepo-helva*	*thai-large-pumpkinii*	Thai Large 10lb	Yellow Flesh Flat Winter Pumpkin	Melon
Cucurbita	*moschata-pepo-helva*	*thai-small-pumpkinii*	Thai Small ?lb	Yellow Flesh Flat Winter Pumpkin	Melon
Cucurbita	*okeechobeensis*	*okeechobeensis*	Original	Okeechobee	Melon
Cucurbita	*okeechobeensis*	*martinezii*	Martinez	Okeechobee	Melon
Cucurbita	*palmata*	*palmata*	Original	Coyote	Melon
Cucurbita	*palmata*	*californica*	California	Coyote	Melon

Squash, Summer (*Curcurbita pepo*)

Cucurbita	*pepo-alba*	*acorn-nivea-alba*	Snow White Acorn ?.?lb	White Summer Squash	Melon
Cucurbita	*pepo-alba*	*acorn-swan-alba*	Swan White Acorn 2.0lb	White Summer Squash	Melon
Cucurbita	*pepo-alba*	*austriana-bruscia*	Austrian Bush ?.?lb	White Summer Squash	Melon
Cucurbita	*pepo-alba*	*little-booi*	Little Boo ?.?lb	White Summer Squash	Melon
Cucurbita	*pepo-alba*	*mammoth-white-bush-scallopii*	Mammoth White Bush Scallop ?.?lb	White Summer Squash	Melon
Cucurbita	*pepo-alba*	*manteca-large-whitei*	Manteca Large ?.?lb	White Summer Squash	Melon
Cucurbita	*pepo-alba*	*mayeras*	Mayeras 5.0lb	White Summer Squash	Melon
Cucurbita	*pepo-alba*	*nest-eggii*	Nest Egg ?.?lb	White Summer Squash	Melon

Cucurbita	pepo-alba	scallop-early-white-bushii	Early White Bush Scallop 2.5lb	White Summer Squash Melon
Cucurbita	pepo-alba	scallop-earliest-prolific-wood'si	Wood's Earliest Prolific Scallop ?.?lb	White Summer Squash Melon
Cucurbita	pepo-alba	scallop-patty-panii	Patty Pan Scallop ?.?lb	White Summer Squash Melon
Cucurbita	pepo-alba	scallop-sacred-white-rattlei	Sacred White Rattle Scallop ?.?lb	White Summer Squash Melon
Cucurbita	pepo-alba	scallop-white-bushii	White Bush Scallop 2.5lb	White Summer Squash Melon
Cucurbita	pepo-alba	straightneck-early-prolifici-improva	Improved Early Prolific Straightneck ?.?lb	White Summer Squash Melon
Cucurbita	pepo-alba	vegetable-spaghettii	Vegetable Spaghetti 10."	White Summer Squash Melon
Cucurbita	pepo-alba	yugoslavensis	Yugoslavia Finger Fruit Cream	White Summer Squash Melon
Cucurbita	pepo-aurantia	acorn-paydon-heirloomii	Paydon Acorn ?.?lb	Orange Summer Squash Melon
Cucurbita	pepo-aurantia	acorn-table-queenii-royalis	Royal Table Queen Acorn 3.5lb	Orange Summer Squash Melon
Cucurbita	pepo-aurantia	acorn-uconnii	Uconn Acorn ?lb	Orange Summer Squash Melon
Cucurbita	pepo-aurantia	african-wartyi	African Warty ?lb	Orange Summer Squash Melon
Cucurbita	pepo-aurantia	big-red-california-sugari	Big Red California Sugar 3.5lb	Orange Summer Squash Melon
Cucurbita	pepo-aurantia	dinosauria-eggii	Dinosaur Egg ?lb	Orange Summer Squash Melon
Cucurbita	pepo-aurantia	menomineei	Menominee ?lb	Orange Summer Squash Melon
Cucurbita	pepo-aurantia	minnesota-sweetii	Minnesota Sweet 3.0lb	Orange Summer Squash Melon
Cucurbita	pepo-aurantia	morganii	Morgan ?lb	Orange Summer Squash Melon
Cucurbita	pepo-aurantia	omahaensis	Omaha Indian 3.5lb	Orange Summer Squash Melon
Cucurbita	pepo-aurantia	orange-gourdii	Orange Gourd ?.?lb	Orange Summer Squash Melon
Cucurbita	pepo-aurantia	orange-wartedii	Orange Warted ?.?lb	Orange Summer Squash Melon
Cucurbita	pepo-aurantia	ovifera	Oval	Orange Summer Squash Melon
Cucurbita	pepo-aurantia	ozarkana	Ozark	Orange Summer Squash Melon
Cucurbita	pepo-aurantia	rebeccai	Rebecca 4.8"	Orange Summer Squash Melon
Cucurbita	pepo-aurantia	scallop-yellow-custardii	Yellow Custard Scallop ?.?lb	Orange Summer Squash Melon
Cucurbita	pepo-aurantia	small-applei	Small Apple ?lb	Orange Summer Squash Melon
Cucurbita	pepo-aurantia	small-orange-ballii	Small Orange Ball ?.?lb	Orange Summer Squash Melon
Cucurbita	pepo-aurantia	straightneck-saffron-prolifici	Saffron Prolific Straightneck	Orange Summer Squash Melon
Cucurbita	pepo-aurantia	texana	Texas	Orange Summer Squash Melon
Cucurbita	pepo-aurantia	turner's-selectii	Turner's Select 6lb	Orange Summer Squash Melon
Cucurbita	pepo-aurantia	tweet-ee-oo-baker'si	Tweet Ee Oo Baker's	Orange Summer Squash Melon
Cucurbita	pepo-aurantia	vegetable-spaghettii-orangettii	Orangetti Vegetable Spaghetti	Orange Summer Squash Melon
Cucurbita	pepo-aurantia	xochitian-puebloi	Xochitian Pueblo ?.?lb	Orange Summer Squash Melon
Cucurbita	pepo-brunnea	acorn-fordhookii	Fordhook Acorn ?lb	Brown Summer Squash Melon
Cucurbita	pepo-groena	acorn-fisher's	Fisher's Acorn ?.?lb	Green Summer Squash Melon
Cucurbita	pepo-groena	acorn-table-queenii	Table Queen Acorn 1.6lb	Green Summer Squash Melon
Cucurbita	pepo-groena	acorn-table-queenii-bruscia	Bush Table Queen Acorn	Green Summer Squash Melon
Cucurbita	pepo-groena	acorn-table-queenii-mammothii	Mammoth Table Queen Acorn 3.5lb	Green Summer Squash Melon
Cucurbita	pepo-groena	acorn-tuffyi	Tuffy Acorn 2.0lb	Green Summer Squash Melon
Cucurbita	pepo-groena	acorn-zapallito-del-troncoi	Zapallito Del Tronco Acorn 3.0"	Green Summer Squash Melon
Cucurbita	pepo-groena	courgettei-arlesai	Arlesa Courgette ?.?lb	Green Summer Squash Melon
Cucurbita	pepo-groena	kojaci	Kojac ?.?lb	Green Summer Squash Melon
Cucurbita	pepo-groena	lebanensis-light-greenii	Light Green Lebanese 2.0lb	Green Summer Squash Melon
Cucurbita	pepo-groena	odessa	Odessa ?.?lb	Green Summer Squash Melon
Cucurbita	pepo-groena	redondo-di-troncoi	Redondo Di Tronco 1.0lb	Green Summer Squash Melon
Cucurbita	pepo-groena	scallop-benning's-green-tintii	Benning's Green Tint Scallop	Green Summer Squash Melon
Cucurbita	pepo-groena	tahitiensis	Tahitian Small ?.?lb	Green Summer Squash Melon
Cucurbita	pepo-groena	tatumei	Tatume ?.?lb	Green Summer Squash Melon
Cucurbita	pepo-helva	acorn-gill's-golden-pippinii	Gill's Golden Pippin Acorn 1.5lb	Yellow Summer Squash Melon
Cucurbita	pepo-helva	crookneck-dixiei	Dixie Crookneck ?lb	Yellow Summer Squash Melon
Cucurbita	pepo-helva	crookneck-giant-summeri	Giant Crookneck ?lb	Yellow Summer Squash Melon
Cucurbita	pepo-helva	crookneck-goldarchii	Goldarch Crookneck ?lb	Yellow Summer Squash Melon
Cucurbita	pepo-helva	crookneck-yellow-summer-earlyi	Early Crookneck ?lb	Yellow Summer Squash Melon
Cucurbita	pepo-helva	crookneck-yellow-summer-improva	Improved Crookneck ?lb	Yellow Summer Squash Melon
Cucurbita	pepo-helva	lemonii	Lemon ?lb	Yellow Summer Squash Melon
Cucurbita	pepo-helva	straightneck-early-prolifici	Early Prolific Straightneck ?lb	Yellow Summer Squash Melon
Cucurbita	pepo-helva	straightneck-early-yellowi	Early Yellow Straightneck ?lb	Yellow Summer Squash Melon
Cucurbita	pepo-helva-aurantia	acorn-goldenii	Golden Acorn 6.0"	Orange Yellow Summer Squash Melon
Cucurbita	pepo-helva-aurantia	acorn-jerseyensis	Jersey Golden Acorn ?lb	Orange Yellow Summer Squash Melon
Cucurbita	pepo-helva-aurantia	acorn-table-goldii	Table Gold Acorn ?lb	Orange Yellow Summer Squash Melon
Cucurbita	pepo-helva-aurantia	acorn-thelma-sanders-sweet-potatoi	Thelma Sanders Sweet Potato Acorn	Orange Yellow Summer Squash Melon
Cucurbita	pepo-helva-aurantia	golden-scallopini-bushii	Golden Scallopini Bush ?lb	Orange Yellow Summer Squash Melon
Cucurbita	pepo-helva-aurantia	lebanensis	Lebanese ?lb	Orange Yellow Summer Squash Melon
Cucurbita	pepo-helva-aurantia	piena-di-napolii	Piena Di Napoli 28lb	Orange Yellow Summer Squash Melon
Cucurbita	pepo-helva-aurantia	scallop-golden-bushii	Golden Bush Scallop ?lb	Orange Yellow Summer Squash Melon
Cucurbita	pepo-helva-aurantia	scallop-sunburstii	Sunburst Scallop ?lb	Orange Yellow Summer Squash Melon
Cucurbita	pepo-helva-aurantia	scallop-yellow-bushii	Yellow Bush Scallop ?lb	Orange Yellow Summer Squash Melon
Cucurbita	pepo-mixta	acorn-scarchuk's-supremei	Scarchuk Supreme Acorn ?.?lb	Multicolor Summer Squash Melon
Cucurbita	pepo-mixta	autumn-wingsi	Autumn Wings 8.0"	Multicolor Summer Squash Melon
Cucurbita	pepo-mixta	bakery'si	Bakery's 2.5lb	Multicolor Summer Squash Melon
Cucurbita	pepo-mixta	black-sweet-potatoi	Black Sweet Potato	Multicolor Summer Squash Melon
Cucurbita	pepo-mixta	calabaza-mercado-verdei	Calabaza Mercado Verde 2.0lb	Multicolor Summer Squash Melon
Cucurbita	pepo-mixta	crown-of-thorns-shenotii	Shenot Crown Of Thorns	Multicolor Summer Squash Melon

Cucurbita	pepo-mixta	delicata	Delicatessen ?.?lb	Multicolor Summer Squash	Melon
Cucurbita	pepo-mixta	delicata-cornell's-bushii	Cornell's Bush Delicatessen	Multicolor Summer Squash	Melon
Cucurbita	pepo-mixta	delicata-JS	Johnny's Seeds Delicatessen 1.7lb	Multicolor Summer Squash	Melon
Cucurbita	pepo-mixta	delicata-zeppelinii	Zeppelin Delicatessen	Multicolor Summer Squash	Melon
Cucurbita	pepo-mixta	delicata-honeyboatii	Honeyboat Delicatessen	Multicolor Summer Squash	Melon
Cucurbita	pepo-mixta	delicata-sugar-loafii	Sugar Loaf Delicatessen	Multicolor Summer Squash	Melon
Cucurbita	pepo-mixta	delicata-sweet-potatoi	Sweet Potato Delicatessen	Multicolor Summer Squash	Melon
Cucurbita	pepo-mixta	eat-allii	Eat All ?lb	Multicolor Summer Squash	Melon
Cucurbita	pepo-mixta	crown-of-thomsi	Crown Of Thorns ?lb	Multicolor Summer Squash	Melon
Cucurbita	pepo-mixta	dancing-gourdii	Dancing Gourd ?lb	Multicolor Summer Squash	Melon
Cucurbita	pepo-mixta	lakotai	Lakota Sioux 6lb	Multicolor Summer Squash	Melon
Cucurbita	pepo-mixta	mandanii	Mandan ?lb	Multicolor Summer Squash	Melon
Cucurbita	pepo-mixta	mandanii-strainii-1	Strain #1 Mandan ?lb	Multicolor Summer Squash	Melon
Cucurbita	pepo-mixta	mandanii-yellowi	Yellow Mandan ?lb	Multicolor Summer Squash	Melon
Cucurbita	pepo-mixta	miniature-ballii	Miniature Ball ?lb	Multicolor Summer Squash	Melon
Cucurbita	pepo-mixta	patissons-golden-marbrei	Golden Marbre Patissons	Multicolor Summer Squash	Melon
Cucurbita	pepo-mixta	patissons-panache-blanc-et-vertii	Blanc Et Vert Panache Patissons	Multicolor Summer Squash	Melon
Cucurbita	pepo-mixta	patissons-panache-juane-et-vertii	Jaune Et Vert Panache Patissons	Multicolor Summer Squash	Melon
Cucurbita	pepo-mixta	pear-bicolor	Pear Bicolor ?lb	Multicolor Summer Squash	Melon
Cucurbita	pepo-mixta	small-flat-stripei	Small Flat Striped ?lb	Multicolor Summer Squash	Melon
Cucurbita	pepo-mixta	small-spoon-gourdii	Small Spoon Gourd	Multicolor Summer Squash	Melon
Cucurbita	pepo-mixta	sweet-dumplingii	Sweet Dumpling 1/2 lb	Multicolor Summer Squash	Melon
Cucurbita	pepo-mixta	striped-peari	Striped Pear ?lb	Multicolor Summer Squash	Melon
Cucurbita	pepo-mixta	ten-commandmentsi	Ten Commandments	Multicolor Summer Squash	Melon
Cucurbita	pepo-mixta	tepehuana	Tepehuan ?lb	Multicolor Summer Squash	Melon

Squash, Vegetable Marrow (*Curcubita pepo-melopepo*)

Genus	Species	Infraspecies	cultivar	specimen	Genome
Cucurbita	pepo-melopepo-alba	lubnani	Lubnani 1.5lb	White Vegetable Marrow Squash	Melon
Cucurbita	pepo-melopepo-alba	vegeta-marrova-anglica	English 17."	White Vegetable Marrow Squash	Melon
Cucurbita	pepo-melopepo-alba	vegeta-marrova-bruscia	Bush 8.0"	White Vegetable Marrow Squash	Melon
Cucurbita	pepo-melopepo-alba	vegeta-marrova-libanica	Lebanese 7.0"	White Vegetable Marrow Squash	Melon
Cucurbita	pepo-melopepo-alba	vegeta-marrova-long-white-bushii	Long White Bush 12."	White Vegetable Marrow Squash	Melon
Cucurbita	pepo-melopepo-aurantia	bostonensis	Boston 15lb	Orange Vegetable Marrow Squash	Melon
Cucurbita	pepo-melopepo-groena	all-green-bushii	All Green Bush ??."	Green Vegetable Marrow Squash	Melon
Cucurbita	pepo-melopepo-groena	clarita	Clarita ??."	Green Vegetable Marrow Squash	Melon
Cucurbita	pepo-melopepo-groena	minipakii	Minipak 12."	Green Vegetable Marrow Squash	Melon
Cucurbita	pepo-melopepo-groena	supremoi	Supremo ??"	Green Vegetable Marrow Squash	Melon
Cucurbita	pepo-melopepo-groena	tender-&-truei	Tender & True 6.0"	Green Vegetable Marrow Squash	Melon
Cucurbita	pepo-melopepo-groena	tiger-crossi	Tiger Cross 11."	Green Vegetable Marrow Squash	Melon
Cucurbita	pepo-melopepo-groena	trailing-greenii	Trailing ??."	Green Vegetable Marrow Squash	Melon
Cucurbita	pepo-melopepo-groena	vegeta-marrovi-danica	Danish ??."	Green Vegetable Marrow Squash	Melon
Cucurbita	pepo-melopepo-groena	vegeta-marrovi-green-bruscia	Green Bush 15."	Green Vegetable Marrow Squash	Melon
Cucurbita	pepo-melopepo-helva	condensa	Condensed	Yellow Vegetable Marrow Squash	Melon
Cucurbita	pepo-melopepo-helva	long-yellowi	Long ??lb	Yellow Vegetable Marrow Squash	Melon

Squash, Zucchini (*Cucurbita pepo-melopepo-zucchini*)

Genus	Species	Infraspecies	cultivar	specimen	Genome
Cucurbita	pepo-melopepo-zucchini-alba	.	.	White Zucchini Squash	Melon
Cucurbita	pepo-melopepo-zucchini-alba	alba	Original 8.0"	White Zucchini Squash	Melon
Cucurbita	pepo-melopepo-zucchini-alba	alba-aegyptica	Egyptian ?.?"	White Zucchini Squash	Melon
Cucurbita	pepo-melopepo-zucchini-alba	french-white-bushii	French White Bush 10."	White Zucchini Squash	Melon
Cucurbita	pepo-melopepo-zucchini-alba	lungo-blancoi	Lungo Blanco ?.?"	White Zucchini Squash	Melon
Cucurbita	pepo-melopepo-zucchini-alba	odessai	Odessa ?.?lb	White Zucchini Squash	Melon
Cucurbita	pepo-melopepo-zucchini-groena	alberello-di-saranza	Alberello Di Saranza	Green Zucchini Squash	Melon
Cucurbita	pepo-melopepo-zucchini-groena	bolognesensis	Bolognese 4.0"	Green Zucchini Squash	Melon
Cucurbita	pepo-melopepo-zucchini-groena	cereberus-bushii	Cereberus Bush ?.?"	Green Zucchini Squash	Melon
Cucurbita	pepo-melopepo-zucchini-groena	cocozelle	Cocozelle 11."	Green Zucchini Squash	Melon
Cucurbita	pepo-melopepo-zucchini-groena	cocozelle-bruscia-marrowii	Bush Marrow Cocozelle	Green Zucchini Squash	Melon
Cucurbita	pepo-melopepo-zucchini-groena	cocozelle-caserta	Caserta Cocozelle 6.0"	Green Zucchini Squash	Melon
Cucurbita	pepo-melopepo-zucchini-groena	cocozelle-groena-bruscia	Bush Green Cocozelle 6.0"	Green Zucchini Squash	Melon
Cucurbita	pepo-melopepo-zucchini-groena	cocozelle-groena-striata	Striped Green Cocozelle 12."	Green Zucchini Squash	Melon
Cucurbita	pepo-melopepo-zucchini-groena	cocozelle-longa	Long Cocozelle 7.0"	Green Zucchini Squash	Melon
Cucurbita	pepo-melopepo-zucchini-groena	cocozelle-italica	Italian Cocozelle 3.0lb	Green Zucchini Squash	Melon
Cucurbita	pepo-melopepo-zucchini-groena	costata-romanesca	Ribbed Romanesque 6.0"	Green Zucchini Squash	Melon
Cucurbita	pepo-melopepo-zucchini-groena	christoforoi	Cristoforo ?.?"	Green Zucchini Squash	Melon
Cucurbita	pepo-melopepo-zucchini-groena	dark-greenii	Dark Green 8.0"	Green Zucchini Squash	Melon
Cucurbita	pepo-melopepo-zucchini-groena	gray-fleckedii	Gray Flecked 8.0"	Green Zucchini Squash	Melon
Cucurbita	pepo-melopepo-zucchini-groena	green-satinii	GreenSatin ?.?"	Green Zucchini Squash	Melon
Cucurbita	pepo-melopepo-zucchini-groena	nano-verde-di-milanoensis	Nano Verde De Milano	Green Zucchini Squash	Melon
Cucurbita	pepo-melopepo-zucchini-groena	nimba	Nimba ?.?"	Green Zucchini Squash	Melon
Cucurbita	pepo-melopepo-zucchini-groena	rondini	Rondini Round 2.0"	Green Zucchini Squash	Melon

Genus	Species	Infraspecies	cultivar	specimen	Genome
Cucurbita	pepo-melopepo-zucchini-groena	ronde-de-nicei	Ronde De Nice Round 1.0"	Green Zucchini Squash	Melon
Cucurbita	pepo-melopepo-zucchini-groena	sakizii	Sakiz ?.?"	Green Zucchini Squash	Melon
Cucurbita	pepo-melopepo-zucchini-groena	silcilianoi	Siciliano 12."	Green Zucchini Squash	Melon
Cucurbita	pepo-melopepo-zucchini-groena	small-green-algeriana	Small Green Algerian	Green Zucchini Squash	Melon
Cucurbita	pepo-melopepo-zucchini-groena	striato-d'italia	Striato D'Italia 9.0"	Green Zucchini Squash	Melon
Cucurbita	pepo-melopepo-zucchini-groena	tonda-di-paicenzai	Tonda Di Piacenza Round 2.5"	Green Zucchini Squash	Melon
Cucurbita	pepo-melopepo-zucchini-groena	verde-d'italia	Verde D'Italia ?.?"	Green Zucchini Squash	Melon
Cucurbita	pepo-melopepo-zucchini-groena	zuboda	Zuboda ?.?"	Green Zucchini Squash	Melon
Cucurbita	pepo-melopepo-zucchini-groena-grisea	grise-de-algiersi	Grise De Algiers 6.0"	Gray Green Zucchini Squash	Melon
Cucurbita	pepo-melopepo-zucchini-groena-grisea	round-bushii	Round Bush 2.5"	Gray Green Zucchini Squash	Melon
Cucurbita	pepo-melopepo-zucchini-groena-helva	genovesensis	Genovese ?.?"	Yellow Green Zucchini Squash	Melon
Cucurbita	pepo-melopepo-zucchini-groena-nigra	black-beautyi	Black Beauty 7.0"	Black Green Zucchini Squash	Melon
Cucurbita	pepo-melopepo-zucchini-groena-nigra	black-magici	Black Magic ?.?"	Black Green Zucchini Squash	Melon
Cucurbita	pepo-melopepo-zucchini-groena-nigra	black-shellii	Black Shell 7.0"	Black Green Zucchini Squash	Melon
Cucurbita	pepo-melopepo-zucchini-groena-nigra	black-shelledii	Black Shelled ?.?"	Black Green Zucchini Squash	Melon
Cucurbita	pepo-melopepo-zucchini-groena-nigra	burpee's-fordhookii	Burpee's Fordhook 12."	Black Green Zucchini Squash	Melon
Cucurbita	pepo-melopepo-zucchini-groena-nigra	midnightii	Midnight 8.0"	Black Green Zucchini Squash	Melon
Cucurbita	pepo-melopepo-zucchini-groena-nigra	milano-blackii	Milano 8.0"	Black Green Zucchini Squash	Melon
Cucurbita	pepo-melopepo-zucchini-helva	helva	Original 7.0"	Yellow Zucchini Squash	Melon
Cucurbita	pepo-melopepo-zucchini-helva	burpee's-goldenii	Burpee's Golden ?.?"	Yellow Zucchini Squash	Melon
Cucurbita	pepo-melopepo-zucchini-helva	gold-satinii	Gold Satin ?.?"	Yellow Zucchini Squash	Melon
Cucurbita	pepo-melopepo-zucchini-helva	golden-bushii	Golden Bush 10."	Yellow Zucchini Squash	Melon
Cucurbita	pepo-melopepo-zucchini-mixta	vegeta-marrovi-bruscia	Vegetable Marrow Bush ?lb	Multicolor Zucchini Squash	Melon
Cucurbita	pepo-melopepo-zucchini-mixta	table-daintyi	Table Dainty 6"	Multicolor Zucchini Squash	Melon

Pumpkin, Summer (*Cucurbita pepo-pepon*)

Genus	Species	Infraspecies	cultivar	specimen	Genome
Cucurbita	pepo-pepon-alba	baby-booi	Baby Boo 3.oz	White Summer Pumpkin	Melon
Cucurbita	pepo-pepon-alba	casperi	Casper 15.lb	White Summer Pumpkin	Melon
Cucurbita	pepo-pepon-alba	chinensis-miniata-alba	Chinese Miniature	White Summer Pumpkin	Melon
Cucurbita	pepo-pepon-alba	jack-o-lanternii-hopi-whitei	Hopi White Jack O'Lantern 14lb	White Summer Pumpkin	Melon
Cucurbita	pepo-pepon-alba	montana-frostii	Montana Frost ??lb	White Summer Pumpkin	Melon
Cucurbita	pepo-pepon-aurantia	amish-piei	Amish Pie 70lb	Orange Summer Pumpkin	Melon
Cucurbita	pepo-pepon-aurantia	ancestorsi	Ancestors 11lb	Orange Summer Pumpkin	Melon
Cucurbita	pepo-pepon-aurantia	baby-beari	Baby Bear 2.0lb	Orange Summer Pumpkin	Melon
Cucurbita	pepo-pepon-aurantia	baby-pamii	Baby Pam ?.?lb	Orange Summer Pumpkin	Melon
Cucurbita	pepo-pepon-aurantia	chinensis-miniata-aurantia	Chinese Miniature	Orange Summer Pumpkin	Melon
Cucurbita	pepo-pepon-aurantia	cheyenne-bushii	Cheyenne Bush 6.5lb	Orange Summer Pumpkin	Melon
Cucurbita	pepo-pepon-aurantia	cinderella	Cinderella 22.lb	Orange Summer Pumpkin	Melon
Cucurbita	pepo-pepon-aurantia	connecticut-sweet-piei	Connecticut Sweet Pie 7lb	Orange Summer Pumpkin	Melon
Cucurbita	pepo-pepon-aurantia	comfield-pumpkinii	Cornfield 11lb	Orange Summer Pumpkin	Melon
Cucurbita	pepo-pepon-aurantia	cow-pumpkinii	Cow ?lb	Orange Summer Pumpkin	Melon
Cucurbita	pepo-pepon-aurantia	fairytalei	Fairytale 20lb	Orange Summer Pumpkin	Melon
Cucurbita	pepo-pepon-aurantia	fat-boyi	Fat Boy 20lb	Orange Summer Pumpkin	Melon
Cucurbita	pepo-pepon-aurantia	field-connecticutii	Connecticut Field 20lb	Orange Summer Pumpkin	Melon
Cucurbita	pepo-pepon-aurantia	field-connecticutii-howdenii	Howden Connecticut Field 22lb	Orange Summer Pumpkin	Melon
Cucurbita	pepo-pepon-aurantia	field-connecticutii-howdenii-biggiei	Biggie Howden Connecticut Field 45lb	Orange Summer Pumpkin	Melon
Cucurbita	pepo-pepon-aurantia	field-pumpkinii	Field ??lb	Orange Summer Pumpkin	Melon
Cucurbita	pepo-pepon-aurantia	field-pankow'si	Pankow's Field 25lb	Orange Summer Pumpkin	Melon
Cucurbita	pepo-pepon-aurantia	ghost-rideri	Ghost Rider 18lb	Orange Summer Pumpkin	Melon
Cucurbita	pepo-pepon-aurantia	half-moonii	Half Moon 16lb	Orange Summer Pumpkin	Melon
Cucurbita	pepo-pepon-aurantia	idaho-gemii	Idaho Gem 5lb	Orange Summer Pumpkin	Melon
Cucurbita	pepo-pepon-aurantia	jack-o-lanternii	Jack O'Lantern 14lb	Orange Summer Pumpkin	Melon
Cucurbita	pepo-pepon-aurantia	jack-o-lanternii-halloweenii	Halloween Jack O'Lantern 14lb	Orange Summer Pumpkin	Melon
Cucurbita	pepo-pepon-aurantia	jack-o-lanternii-happy-jackii	Happy Jack Jack O'Lantern 18lb	Orange Summer Pumpkin	Melon
Cucurbita	pepo-pepon-aurantia	jack-o-lanternii-ichabodii	Ichabod Jack O'Lantern ??lb	Orange Summer Pumpkin	Melon
Cucurbita	pepo-pepon-aurantia	jack-o-lanternii-jack-be-littlei	Jack Be Little Jack O'Lantern 1/2lb	Orange Summer Pumpkin	Melon
Cucurbita	pepo-pepon-aurantia	jack-o-lanternii-jack-be-quick	Jack Be Quick Jack O'Lantern 1/4lb	Orange Summer Pumpkin	Melon
Cucurbita	pepo-pepon-aurantia	jack-o-lanternii-jack-o-litei	Jack O'Lite Jack O'Lantern 14lb	Orange Summer Pumpkin	Melon
Cucurbita	pepo-pepon-aurantia	jack-o-lanternii-jumpin-jackii	Jumpin Jack Jack O'Lantern 40lb	Orange Summer Pumpkin	Melon
Cucurbita	pepo-pepon-aurantia	jack-o-lanternii-little-lanternii	Little Lantern Jack O'Lantern ?/?lb	Orange Summer Pumpkin	Melon
Cucurbita	pepo-pepon-aurantia	jack-o-lanternii-mini-jackii	Mini Jack Jack O'Lantern ?/?lb	Orange Summer Pumpkin	Melon
Cucurbita	pepo-pepon-aurantia	jack-o-lanternii-montana-jackii	Montana Jack Jack O'Lantern 12lb	Orange Summer Pumpkin	Melon
Cucurbita	pepo-pepon-aurantia	kumikumi	Kumikumi ?lb	Orange Summer Pumpkin	Melon
Cucurbita	pepo-pepon-aurantia	kumikuri	Kumikuri ?lb	Orange Summer Pumpkin	Melon
Cucurbita	pepo-pepon-aurantia	little-octoberi	Little October 3/4lb	Orange Summer Pumpkin	Melon
Cucurbita	pepo-pepon-aurantia	long-pie-pumpkinii	Long Pie 4lb	Orange Summer Pumpkin	Melon
Cucurbita	pepo-pepon-aurantia	munchkinii	Munchkin 1/4lb	Orange Summer Pumpkin	Melon
Cucurbita	pepo-pepon-aurantia	new-engand-piei	New England Pie 7lb	Orange Summer Pumpkin	Melon
Cucurbita	pepo-pepon-aurantia	new-engand-piei-small-sugari	Small Sugar New England Pie 7lb	Orange Summer Pumpkin	Melon
Cucurbita	pepo-pepon-aurantia	new-engand-piei-small-sugari-asgrovii	Asgrow Small Sugar New England Pie 5lb	Orange Summer Pumpkin	Melon
Cucurbita	pepo-pepon-aurantia	new-engand-piei-spookiei	Spookie New England Pie 8lb	Orange Summer Pumpkin	Melon

Genus	Species	Infraspecies	cultivar	specimen	Genome
Cucurbita	*pepo-pepon-aurantia*	*new-engand-piei-storagei*	Storage New England Pie ?lb	Orange Summer Pumpkin	Melon
Cucurbita	*pepo-pepon-aurantia*	*new-engand-piei-sugar-piei*	Sugar Pie New England Pie 7lb	Orange Summer Pumpkin	Melon
Cucurbita	*pepo-pepon-aurantia*	*new-engand-piei-young's-beautyi*	Young's Beauty New England Pie 11lb	Orange Summer Pumpkin	Melon
Cucurbita	*pepo-pepon-aurantia*	*northern-bushii*	Northern Bush 7lb	Orange Summer Pumpkin	Melon
Cucurbita	*pepo-pepon-aurantia*	*old-timey-flat-pumpkinii*	Old Timey Flat ?lb	Orange Summer Pumpkin	Melon
Cucurbita	*pepo-pepon-aurantia*	*potimarronii*	Potimarron ?lb	Orange Summer Pumpkin	Melon
Cucurbita	*pepo-pepon-aurantia*	*rocky-mountain-piei*	Rocky Mountain Pie 6lb	Orange Summer Pumpkin	Melon
Cucurbita	*pepo-pepon-aurantia*	*schooltimei*	Schooltime 9lb	Orange Summer Pumpkin	Melon
Cucurbita	*pepo-pepon-aurantia*	*sweetie-piei*	Sweetie Pie 5oz	Orange Summer Pumpkin	Melon
Cucurbita	*pepo-pepon-aurantia*	*tallmanii*	Tallman 22lb	Orange Summer Pumpkin	Melon
Cucurbita	*pepo-pepon-aurantia*	*thomas-halloweenii*	Thomas Halloween 25lb	Orange Summer Pumpkin	Melon
Cucurbita	*pepo-pepon-aurantia*	*tom-foxii*	Tom Fox ??lb	Orange Summer Pumpkin	Melon
Cucurbita	*pepo-pepon-aurantia*	*turneri-familyi*	Turner Family 12lb	Orange Summer Pumpkin	Melon
Cucurbita	*pepo-pepon-helva*	*hopi-pumpkinii*	Hopi ?lb	Yellow Summer Pumpkin	Melon
Cucurbita	*pepo-pepon-helva*	*juane-gros-de-parisi*	Juane Gros De Paris 20lb	Yellow Summer Pumpkin	Melon
Cucurbita	*pepo-pepon-helva*	*pacheco-pumpkinii*	Pacheco ?lb	Yellow Summer Pumpkin	Melon
Cucurbita	*pepo-pepon-helva*	*peter-panii*	Peter Pan 4.0"	Yellow Summer Pumpkin	Melon
Cucurbita	*pepo-pepon-helva*	*tarahumara-pumpkinii*	Tarahumara ?lb	Yellow Summer Pumpkin	Melon
Cucurbita	*pepo-pepon-helva-aurantia*	.	.	Orange Yellow Summer Pumpkin	Melon
Cucurbita	*pepo-pepon-helva-aurantia*	*buskinii*	Bushkin 9.0lb	Orange Yellow Summer Pumpkin	Melon
Cucurbita	*pepo-pepon-helva-aurantia*	*gold-rushii*	Gold Rush 30lb	Orange Yellow Summer Pumpkin	Melon
Cucurbita	*pepo-pepon-helva-aurantia*	*luxury-pie-pumpkinii*	Luxury Pie 7lb	Orange Yellow Summer Pumpkin	Melon
Cucurbita	*pepo-pepon-helva-aurantia*	*melonnette-de-vendeei*	Melonnette De Vendee 3lb	Orange Yellow Summer Pumpkin	Melon
Cucurbita	*pepo-pepon-helva-aurantia*	*northern-goldii*	Northern Gold 9lb	Orange Yellow Summer Pumpkin	Melon
Cucurbita	*pepo-pepon-helva-aurantia*	*vintera-luxuria*	Winter Luxury 8lb	Orange Yellow Summer Pumpkin	Melon
Cucurbita	*pepo-pepon-helva-aurantia*	*wee-b-littlei*	Wee-B-Little 1lb	Orange Yellow Summer Pumpkin	Melon
Cucurbita	*pepo-pepon-groena*	*acoma*	Acoma 12.lb	Green Summer Pumpkin	Melon
Cucurbita	*pepo-pepon-groena*	*acorn-autumn-princeps*	Autumn Prince Acorn 1.7lb	Green Summer Pumpkin	Melon
Cucurbita	*pepo-pepon-groena*	*acorn-table-queenii*	Table Queen Acorn	Green Summer Pumpkin	Melon
Cucurbita	*pepo-pepon-groena*	*little-gemii*	Little Gem Round ?/?lb	Green Summer Pumpkin	Melon
Cucurbita	*pepo-pepon-groena*	*sweet-mamai*	Sweet Mama ?.?lb	Green Summer Pumpkin	Melon
Cucurbita	*pepo-pepon-groena-blauea*	*acoma-pumpkinii*	Acoma Blue 12lb	Blue Green Summer Pumpkin	Melon
Cucurbita	*pepo-pepon-groena-nigra*	*acorn-table-kingii*	Table King Acorn 2.0lb	Black Green Summer Pumpkin	Melon
Cucurbita	*pepo-pepon-groena-nigra*	*acorn-table-queenii-ebonii*	Ebony Table Queen Acorn 1.7lb	Black Green Summer Pumpkin	Melon
Cucurbita	*pepo-pepon-groena-nigra*	*scallopinii*	Scallopini ?.?"	Black Green Summer Pumpkin	Melon
Cucurbita	*pepo-pepon-mixta*	*mt.-pima-vavulii*	Mt. Pima Vavuli 10lb	Multicolor Summer Pumpkin	Melon
Cucurbita	*pepo-pepon-mixta*	*southern-mineri*	Southern Miner ??lb	Multicolor Summer Pumpkin	Melon
Cucurbita	*pepo-pepon-mixta*	*tarahumara*	La Bufa Tarahumara	Multicolor Summer Pumpkin	Melon
Cucurbita	*pepo-pepon-sperma*	*lady-godivai*	Lady Godiva 10lb	Seed Summer Pumpkin	Melon
Cucurbita	*pepo-pepon-sperma*	*lady-godivai-bushii*	Bush Lady Godiva 15lb	Seed Summer Pumpkin	Melon
Cucurbita	*pepo-pepon-sperma*	*salvadorensis-ayotei*	Salvadorean Ayote ??lb	Seed Summer Pumpkin	Melon
Cucurbita	*pepo-pepon-sperma*	*sonorensis-hullessi*	Sonoran Hulless ??lb	Seed Summer Pumpkin	Melon
Cucurbita	*pepo-pepon-sperma*	*snack-'r-jackii*	Snack 'R Jack 1.5lb	Seed Summer Pumpkin	Melon
Cucurbita	*pepo-pepon-sperma*	*triple-treatii*	Triple Treat 7lb	Seed Summer Pumpkin	Melon
Cucurbita	*pepo-pepon-sperma*	*styrian-hullessi*	Styrian Hulless 15lb	Seed Summer Pumpkin	Melon
Cucurbita	*pepo-pepon-sperma*	*jack-o-lanternii-streaker-jackii*	Streaker Jack Jack O'Lantern	Seed Summer Pumpkin	Melon
Cucurbita	*soraria*	*soraria*	Original	Sister Pumpkin	Melon
Cucurbita	*soraria*	*mexicana*	Mexican 8lb	Sister Pumpkin	Melon

Wax Gourd, Squash (*Benicasa hispida*)

Genus	Species	Infraspecies	cultivar	specimen	Genome
Benincasa	*hispida*	.	.	Squash	Wax Gourd
Benincasa	*hispida*	*hispida*	Original	Squash	Wax Gourd
Benincasa	*hispida*	*ash-pumpkinii*	Ash Pumpkin	Squash	Wax Gourd
Benincasa	*hispida*	*cham-kwai*	Cham Kwa	Squash	Wax Gourd
Benincasa	*hispida*	*japonica-fuzzyi-gourdii*	Japanese Fuzzy	Squash	Wax Gourd
Benincasa	*hispida*	*chinensis-fuzzyi-gourdii*	Chinese Fuzzy	Squash	Wax Gourd
Benincasa	*hispida*	*chinensis-fuzzyi-mao-kwai*	Mao Kwa Chinese Fuzzy 3.0" x 6.0"	Squash	Wax Gourd
Benincasa	*hispida*	*chinensis-perserving-melonii*	Chinese Perserving Melon	Squash	Wax Gourd
Benincasa	*hispida*	*larga-longa*	Long Large 8" x 36"	Squash	Wax Gourd
Benincasa	*hispida*	*larga-rotunda*	Rounded Large 12" x 24"	Squash	Wax Gourd

Wax Gourd (*Benincasa cerifera*)

Genus	Species	Infraspecies	cultivar	specimen	Genome
Benincasa	*cerifera*	*cerifera*	Original	Hairy	Wax Gourd
Benincasa	*cerifera*	*chinese-wintermelonii*	Chinese Wintermelon	Hairy	Wax Gourd
Benincasa	*cerifera*	*tung-kwai(doan-gwai)*	Tung Kwa 9.0" x 9.0"	Hairy	Wax Gourd
Benincasa	*cerifera*	*tung-kwai-oblongii*	Oblong Tung Kwa 18." x 30."	Hairy	Wax Gourd
Benincasa	*cerifera*	*tung-kwai-roundii*	Round Tung Kwa 12." x 12."	Hairy	Wax Gourd
Benincasa	*cerifera*	*wax-gourd-oblongii*	Oblong 12." x 24."	Hairy	Wax Gourd
Benincasa	*cerifera*	*wax-gourd-smallii*	Small	Hairy	Wax Gourd

Gherkin, Prickly (*Cucumis anguria*)

Genus	Species	Infraspecies	cultivar	specimen	Genome
Cucumis	anguria	anguria	West Indies	Prickly Gherkin	Cucumber
Cucumis	anguria	africana	African	Prickly Gherkin	Cucumber
Cucumis	anguria	beit-alpha-CMVR	Cucumber Mosaic Virus Resistant Beit Alpha	Prickly Gherkin	Cucumber
Cucumis	anguria	beit-alpha-ellamii	Ellam Beit Alpha	Prickly Gherkin	Cucumber
Cucumis	anguria	beit-alpha-MR	Mosaic Resistant Beit Alpha	Prickly Gherkin	Cucumber
Cucumis	anguria	longaculeatus	Long Spine	Prickly Gherkin	Cucumber
Cucumis	anguria	jamaicensis	Jamaican	Prickly Gherkin	Cucumber
Cucumis	metuliferus	metuliferus	Original	African Horned	Cucumber
Cucumis	metuliferus	jelly-melonii	Jelly Melon	African Horned	Cucumber
Cucumis	metuliferus	kiwanoi	Kiwano	African Horned	Cucumber

Cucumber (*Cucumis sativus*)

Genus	Species	Infraspecies	cultivar	specimen	Genome
Cucumis	sativus-crema	bianco-lungo-di-parigii	Bianco Lungo Di Parigi White	Cream	Cucumber
Cucumis	sativus-crema	de-bouenilii	De Bouenil White	Cream	Cucumber
Cucumis	sativus-crema	long-whitei	Long White	Cream	Cucumber
Cucumis	sativus-crema	poona-kheerai	Poona Kheera	Cream	Cucumber
Cucumis	sativus-crema	precoce-grosso-bianco-crema	Precoce Grosso Bianco Crema	Cream	Cucumber
Cucumis	sativus-crema	white-plumpii	Plump White	Cream	Cucumber
Cucumis	sativus-crema	white-cucumberi	Small White	Cream	Cucumber
Cucumis	sativus-crema	white-wonderi	White Wonder 8.0"	Cream	Cucumber
Cucumis	sativus-groena-darca	addisi	Addis	Dark Green	Cucumber
Cucumis	sativus-groena-darca	athensi	Athens	Dark Green	Cucumber
Cucumis	sativus-groena-darca	black-diamondii	Black Diamond 8.0"	Dark Green	Cucumber
Cucumis	sativus-groena-darca	burplessi	Burpless 9.0"	Dark Green	Cucumber
Cucumis	sativus-groena-darca	chinese-longii	Chinese Long 28."	Dark Green	Cucumber
Cucumis	sativus-groena-darca	cirrus	Cirrus 7.5"	Dark Green	Cucumber
Cucumis	sativus-groena-darca	early-fortunei	Early Fortune 8.5"	Dark Green	Cucumber
Cucumis	sativus-groena-darca	longfellowii	Longfellow 13."	Dark Green	Cucumber
Cucumis	sativus-groena-darca	marketeri	Marketer 8.5"	Dark Green	Cucumber
Cucumis	sativus-groena-darca	marketeri-longii	Marketer Long 8.5"	Dark Green	Cucumber
Cucumis	sativus-groena-darca	marketmorei	Marketmore 9.0"	Dark Green	Cucumber
Cucumis	sativus-groena-darca	marketmorei-70	#70 Marketmore 8.5"	Dark Green	Cucumber
Cucumis	sativus-groena-darca	marketmorei-76	#76 Marketmore 8.5"	Dark Green	Cucumber
Cucumis	sativus-groena-darca	marketmorei-80	#80 Marketmore 8.5"	Dark Green	Cucumber
Cucumis	sativus-groena-darca	marketmorei-86	#86 Marketmore 8.5"	Dark Green	Cucumber
Cucumis	sativus-groena-darca	marketmorei-97	#97 Marketmore 10."	Dark Green	Cucumber
Cucumis	sativus-groena-darca	masterpiecei	Masterpiece 8.0"	Dark Green	Cucumber
Cucumis	sativus-groena-darca	medalistii	Medalist	Dark Green	Cucumber
Cucumis	sativus-groena-darca	oriental-burplessi	Burpless Oriental	Dark Green	Cucumber
Cucumis	sativus-groena-darca	oriental-tasty-greenii	Tasty Green Oriental	Dark Green	Cucumber
Cucumis	sativus-groena-darca	spacemasteri	Spacemaster	Dark Green	Cucumber
Cucumis	sativus-groena-darca	spacemasteri-80	#80 Spacemaster	Dark Green	Cucumber
Cucumis	sativus-groena-darca	telegraphii	Telegraph 16"	Dark Green	Cucumber
Cucumis	sativus-groena-darca	telegraphii-improvedii	Improved Telegraph 16"	Dark Green	Cucumber
Cucumis	sativus-groena-darca	telegraphii-englishii	English Telegraph 16"	Dark Green	Cucumber
Cucumis	sativus-groena-darca	telegraphii-rollinson'si	Rollinson's Telegraph 16"	Dark Green	Cucumber
Cucumis	sativus-groena-darca	yamato-extra-longii	Yamato Extra Long 20"	Dark Green	Cucumber
Cucumis	sativus-groena-leuca	delikatessi	Delicatess Warty	Light Green	Cucumber
Cucumis	sativus-groena-leuca	kyotoensis	Kyoto 11.5"	Light Green	Cucumber
Cucumis	sativus-groena-leuca	per-sotto-acetii	Per Sotto Aceti	Light Green	Cucumber
Cucumis	sativus-groena-leuca	sanjiaku-kiurii	Sanjiaku Kiuri 18."	Light Green	Cucumber
Cucumis	sativus-groena-leuca	surprisei	Surprise	Light Green	Cucumber
Cucumis	sativus-groena-leuca	totarello-baresei	Tortarello Barese	Light Green	Cucumber
Cucumis	sativus-groena-leuca	uzbekskii	Uzbekski 7.0"	Light Green	Cucumber
Cucumis	sativus-groena-media	amirai-2	Amira II	Medium Green	Cucumber
Cucumis	sativus-groena-media	bedfordshire-prizei	Bedfordshire Prize	Medium Green	Cucumber
Cucumis	sativus-groena-media	burplessi-muncheri	Burpless Muncher	Medium Green	Cucumber
Cucumis	sativus-groena-media	bush-championii	Bush Champion 10."	Medium Green	Cucumber
Cucumis	sativus-groena-media	bush-cropii	Bush Crop 7.0"	Medium Green	Cucumber
Cucumis	sativus-groena-media	bush-cucumberi	Bush 6.0"	Medium Green	Cucumber
Cucumis	sativus-groena-media	bush-whopperi	Bush Whopper 10."	Medium Green	Cucumber
Cucumis	sativus-groena-media	early-quebecensis	Early Quebec	Medium Green	Cucumber
Cucumis	sativus-groena-media	heiwa-green-prolifici	Heiwa Prolific 16."	Medium Green	Cucumber
Cucumis	sativus-groena-media	langeland-giantii	Langeland Giant 12."	Medium Green	Cucumber
Cucumis	sativus-groena-media	mideast-prolifici	Mideast Prolific 7.0"	Medium Green	Cucumber
Cucumis	sativus-groena-media	mirella	Mirella	Medium Green	Cucumber
Cucumis	sativus-groena-media	muncheri	Muncher 9.0"	Medium Green	Cucumber
Cucumis	sativus-groena-media	perfectionii	Perfection 9.0"	Medium Green	Cucumber
Cucumis	sativus-groena-media	qualitasi	Qualitas	Medium Green	Cucumber
Cucumis	sativus-groena-media	reisenschalii	Riesenschal	Medium Green	Cucumber

Genus	Species	Infraspecies	cultivar	specimen	Genome
Cucumis	*sativus-groena-media*	*shintokiwai*	Shintokiwa 12."	Medium Green	Cucumber
Cucumis	*sativus-groena-media*	*straight-eightii*	Straight Eight 8.0"	Medium Green	Cucumber
Cucumis	*sativus-groena-media*	*straight-ninei*	Straight Nine 9.0"	Medium Green	Cucumber
Cucumis	*sativus-groena-media*	*super-bush-slicersi*	Super Bush Slicers	Medium Green	Cucumber
Cucumis	*sativus-groena-media*	*thai-greenii*	Thai Green 7.0"	Medium Green	Cucumber
Cucumis	*sativus-groena-media*	*timun-hijanii*	Timun Hijan 5.0"	Medium Green	Cucumber
Cucumis	*sativus-groena-media*	*vorgebirstaugenii*	Vorgebirstauben	Medium Green	Cucumber
Cucumis	*sativus-groena-media*	*windermoor-wonderi*	Windermoor Wonder 10."	Medium Green	Cucumber
Cucumis	*sativus-groena-media*	*windermoor-wonderi-long-greenii*	Long Green Windermoor Wonder 10."	Medium Green	Cucumber
Cucumis	*sativus-groena-media*	*windermoor-wonderi-long-greenii-improvedii*	Improved Long Green Windermoor Wonder 10."	Medium Green	Cucumber
Cucumis	*sativus-groena-media*	*zeppelinii*	Zeppelin	Medium Green	Cucumber
Cucumis	*sativus-helva*	*chinese-yellowii*	Chinese 12."	Yellow	Cucumber
Cucumis	*sativus-helva*	*gele-trosii*	Gele Tros	Yellow	Cucumber
Cucumis	*sativus-helva*	*humong-redii*	Humong Red	Yellow	Cucumber
Cucumis	*sativus-helva*	*landis-whitei*	Landis White	Yellow	Cucumber
Cucumis	*sativus-helva*	*mordon-earlyi*	Mordon Early	Yellow	Cucumber
Cucumis	*sativus-helva*	*vyaznikoskyi*	Vyaznikosky	Yellow	Cucumber
Cucumis	*sativus-helva*	*yellow-submarinei*	Yellow Submarine 8.0"	Yellow	Cucumber
Cucumis	*sativus-hybrida-F1*	*athenei*	Athene Dark Green	F1 Hybrid	Cucumber
Cucumis	*sativus-hybrida-F1*	*burplessi-tasty-greenii*	Burpless Tasty Green	F1 Hybrid	Cucumber
Cucumis	*sativus-hybrida-F1*	*condai*	Conda Pickling	F1 Hybrid	Cucumber
Cucumis	*sativus-hybrida-F1*	*dianai*	Diana	F1 Hybrid	Cucumber
Cucumis	*sativus-rotunda*	*apple-greenii*	Apple Green	Round	Cucumber
Cucumis	*sativus-rotunda*	*crystal-apple-lemonii*	Lemon Crystal Apple	Round	Cucumber
Cucumis	*sativus-rotunda*	*crystal-apple-whitei*	White Crystal Apple	Round	Cucumber
Cucumis	*sativus-rotunda*	*maduriana*	Mandurian	Round	Cucumber
Cucumis	*sativus-rotunda*	*richmond-green-applei*	Richmond Green Apple 3.0"	Round	Cucumber
Cucumis	*sativus-spina-alba*	*ashleyi*	Ashley Dark Green 7.5"	White Spine	Cucumber
Cucumis	*sativus-spina-alba*	*burplessi-26*	#26 Burpless Dark Green 9.0"	White Spine	Cucumber
Cucumis	*sativus-spina-alba*	*homemade-picklesi*	Homemade Pickles 5.5"	White Spine	Cucumber
Cucumis	*sativus-spina-alba*	*long-green-ridgei*	Long Green Ridge	White Spine	Cucumber
Cucumis	*sativus-spina-alba*	*poinsettii*	Poinsett	White Spine	Cucumber
Cucumis	*sativus-spina-alba*	*poinsettii-76*	Improved Poinsett	White Spine	Cucumber
Cucumis	*sativus-spina-alba*	*sumpteri*	Sumpter	White Spine	Cucumber
Cucumis	*sativus-spina-alba*	*suyo-longii*	Suyo Long 14."	White Spine	Cucumber
Cucumis	*sativus-spina-alba*	*tendergreen-burplessi*	Tendergreen Burpless 10."	White Spine	Cucumber
Cucumis	*sativus-spina-alba*	*tokiwa-jihaii*	Tokiwa Jihai 12."	White Spine	Cucumber
Cucumis	*sativus-spina-alba*	*white-spine-improvedii*	Improved 7.7"	White Spine	Cucumber
Cucumis	*sativus-spina-alba*	*white-spine-earlyi*	Early 9.0"	White Spine	Cucumber
Cucumis	*sativus-spina-nigra*	*boothby's-blondei*	Boothby's Blonde 7.0"	Black Spine	Cucumber
Cucumis	*sativus-spina-nigra*	*early-clusteri*	Early Custer Pickling 5.5"	Black Spine	Cucumber
Cucumis	*sativus-spina-nigra*	*early-ochiaii*	Early Ochiai 7.5"	Black Spine	Cucumber
Cucumis	*sativus-spina-nigra*	*japanese-climbingii*	Japanese Climbing 8.0"	Black Spine	Cucumber
Cucumis	*sativus-spina-nigra*	*national-picklingii*	National Pickling 6.5"	Black Spine	Cucumber
Cucumis	*sativus-spina-nigra*	*national-picklingii-improvedii*	Improved National Pickling 6.5"	Black Spine	Cucumber
Cucumis	*sativus-spina-nigra*	*nejinskyi*	Nejinsky 5.5"	Black Spine	Cucumber
Cucumis	*sativus-spina-nigra*	*northern-picklingii*	Northern Pickling	Black Spine	Cucumber
Cucumis	*sativus-spina-nigra*	*pickler-burpeei*	Pickler Burpee	Black Spine	Cucumber
Cucumis	*sativus-spina-nigra*	*rhinish-picklei*	Rhinish Pickle	Black Spine	Cucumber
Cucumis	*sativus-spina-nigra*	*rhinish-picklei-ridui*	Ridu Rhinish Pickle	Black Spine	Cucumber
Cucumis	*sativus-spina-nigra*	*wisconsinensis-SMR*	Squash Mosaic Resistant Wisconsin 6.0"	Black Spine	Cucumber
Cucumis	*sativus-spina-nigra*	*wisconsinensis-SMR-18*	# 18 Squash Mosaic Resistant Wisconsin 6.0"	Black Spine	Cucumber
Cucumis	*sativus-spina-nigra*	*wisconsinensis-SMR-58*	# 58 Squash Mosaic Resistant Wisconsin 6.0"	Black Spine	Cucumber

Pickle (*Cucumis sativus-condio*)

Genus	Species	Infraspecies	cultivar	specimen	Genome
Cucumis	*sativus-condio*	*A-&-C-picklingii*	Abbot & Cobb 9.0"	Pickling	Cucumber
Cucumis	*sativus-condio*	*aunt-rita's-monestaryi*	Aunt Rita's Monestary	Pickling	Cucumber
Cucumis	*sativus-condio*	*boston-picklingii*	Boston 5.7"	Pickling	Cucumber
Cucumis	*sativus-condio*	*boston-picklingii-improva*	Improved Boston 5.7"	Pickling	Cucumber
Cucumis	*sativus-condio*	*bushyi*	Bushy 5.2"	Pickling	Cucumber
Cucumis	*sativus-condio*	*caroline's-pickleri*	Caroline's Pickler	Pickling	Cucumber
Cucumis	*sativus-condio*	*carosello-mezzo-lungo-baresei*	Carosello Mezzo Lungo Barese	Pickling	Cucumber
Cucumis	*sativus-condio*	*chicago-picklingii*	Chicago 6.5"	Pickling	Cucumber
Cucumis	*sativus-condio*	*clintonii*	Clinton 5.5"	Pickling	Cucumber
Cucumis	*sativus-condio*	*dekahii*	Dekah 6.0"	Pickling	Cucumber
Cucumis	*sativus-condio*	*double-yield-picklingii*	Double Yield 5.5"	Pickling	Cucumber
Cucumis	*sativus-condio*	*early-prolific-short-greenii*	Early Prolific Short Green 7.0"	Pickling	Cucumber

Genus	Species	Infraspecies	cultivar	specimen	Genome
Cucumis	*sativus-condio*	*early-russianii*	Early Russian 5.0"		Pickling Cucumber
Cucumis	*sativus-condio*	*everbearingii*	Everbearing 5.0"		Pickling Cucumber
Cucumis	*sativus-condio*	*everbearingii-SMR-58*	#58 Squash Mosaic Resistant Everbearing 5.0"		Pickling Cucumber
Cucumis	*sativus-condio*	*fancy-pickling-shumwayi*	Fancy Shumway 5.0"		Pickling Cucumber
Cucumis	*sativus-condio*	*japanese-long-picklingii*	Japanese Long 14."		Pickling Cucumber
Cucumis	*sativus-condio*	*japanese-picklingii*	Japanese 7.0"		Pickling Cucumber
Cucumis	*sativus-condio*	*jersey-picklei*	Jersey Pickle 6.0"		Pickling Cucumber
Cucumis	*sativus-condio*	*katsura-giant-picklingii*	Katsura Giant 12."		Pickling Cucumber
Cucumis	*sativus-condio*	*monastici*	Monastic		Pickling Cucumber
Cucumis	*sativus-condio*	*north-carolina-heirloom-picklingii*	North Carolina Heirloom		Pickling Cucumber
Cucumis	*sativus-condio*	*paradei*	Parade 5.0"		Pickling Cucumber
Cucumis	*sativus-condio*	*persica*	Persian		Pickling Cucumber
Cucumis	*sativus-condio*	*piccolo-di-parigii*	Piccolo Di Parigi 7.0"		Pickling Cucumber
Cucumis	*sativus-condio*	*russica*	Russian 6.0"		Pickling Cucumber
Cucumis	*sativus-condio*	*smart-picklei*	Smart Pickle 5.5"		Pickling Cucumber
Cucumis	*sativus-condio*	*snow's-fancy-picklingii*	Snow's Fancy 5.5"		Pickling Cucumber
Cucumis	*sativus-condio*	*spring-of-wateri*	Spring Of Water		Pickling Cucumber
Cucumis	*sativus-condio*	*ukranian-picklingii*	Ukrainian		Pickling Cucumber
Cucumis	*sativus-condio*	*wautomai*	Wautoma		Pickling Cucumber

Gherkin (*Cucumis sativus-gurca*)

Genus	Species	Infraspecies	cultivar	specimen	Genome
Cucumis	*sativus-gurca*	*mexican-sour-gherkinii*	Mexican Sour 1.5"		Gherkin Cucumber
Cucumis	*sativus-gurca*	*bianco-primaticcio*	Bianco Primaticcio 4.5"		Gherkin Cucumber
Cucumis	*sativus-gurca*	*bush-picklei*	Bush Pickle 4.5"		Pickling Cucumber
Cucumis	*sativus-gurca*	*cornichon-de-agrestalis*	Cornichon 2.0"		Gherkin Cucumber
Cucumis	*sativus-gurca*	*cornichon-de-bourbonnei*	Bourbonne Cornichon 2.0"		Gherkin Cucumber
Cucumis	*sativus-gurca*	*cornichon-de-parisi*	Small Paris Cornichon 2.0"		Gherkin Cucumber
Cucumis	*sativus-gurca*	*cornichon-fine-meauxii*	Fine Meaux Cornichon 3.0"		Gherkin Cucumber
Cucumis	*sativus-gurca*	*cornichon-vert-de-massyi*	Vert De Massy Cornichon 2.0"		Gherkin Cucumber
Cucumis	*sativus-gurca*	*cornichon-vert-petit-de-paris*	Vert Petit de Paris 2.0"		Gherkin Cucumber
Cucumis	*sativus-gurca*	*earnest-familyi*	Earnest Family 4.5"		Gherkin Cucumber
Cucumis	*sativus-gurca*	*edmonsonii*	Edmonson 4.0"		Gherkin Cucumber
Cucumis	*sativus-gurca*	*jamaicensis*	Jamaican		Gherkin Cucumber
Cucumis	*sativus-gurca*	*little-leafii*	Little Leaf 4.0"		Gherkin Cucumber
Cucumis	*sativus-gurca*	*mincu-babyi*	Baby Mincu 4.0"		Gherkin Cucumber
Cucumis	*sativus-gurca*	*mincu-midgetii*	Midget Mincu 2.5"		Gherkin Cucumber
Cucumis	*sativus-gurca*	*miniature-whitei*	Miniature White 3.0"		Gherkin Cucumber
Cucumis	*sativus-gurca*	*pickle-bushii*	Picklebush 4.5"		Gherkin Cucumber
Cucumis	*sativus-gurca*	*vert-de-massyi*	Vert De Massy 4.0"		Gherkin Cucumber

Bottle Gourd (*Lagenaria spp.*)

Genus	Species	Infraspecies	cultivar	specimen	Genome
Lagenaria	*longissima*	*calabash-longii*	Calabash 6.0" x 36."	Long	Bottle Gourd
Lagenaria	*longissima*	*cucuzzii-italian-edible-gourdii*	Italian Edible Squash Cucuzzi 5.0" x 48."	Long	Bottle Gourd
Lagenaria	*longissima*	*dipperi-extra-long-handlei*	Extra Long Handle 6.0" x 36."	Long	Bottle Gourd
Lagenaria	*siceraria*	*african-giant-wateri*	African Giant Water 9.0" x 24."	Crookneck	Bottle Gourd
Lagenaria	*siceraria*	*applei*	Apple 7.0" x 7.0"	Crookneck	Bottle Gourd
Lagenaria	*siceraria*	*bananai*	Banana 4.0" x 10."	Crookneck	Bottle Gourd
Lagenaria	*siceraria*	*basketii*	Basket	Crookneck	Bottle Gourd
Lagenaria	*siceraria*	*kettle-africana*	African Kettle 8.0" x 16."	Crookneck	Bottle Gourd
Lagenaria	*siceraria*	*kettle-africana-big-winei*	Big Wine African Kettle 10." x 20."	Crookneck	Bottle Gourd
Lagenaria	*siceraria*	*leucantha*	Crookneck	Crookneck	Bottle Gourd
Lagenaria	*siceraria*	*black-seeded-bottlei*	Black Seeded 3.5" x 7.0"	Crookneck	Bottle Gourd
Lagenaria	*siceraria*	*bottleneckii*	Bottleneck 10." x 20."	Crookneck	Bottle Gourd
Lagenaria	*siceraria*	*bowlii*	Bowl 4.5" x 9.0"	Crookneck	Bottle Gourd
Lagenaria	*siceraria*	*bulei*	French Bule 5.5" x 7.0"	Crookneck	Bottle Gourd
Lagenaria	*siceraria*	*bulei-mayo-giantii*	Giant Mayo Bule 7.0" x 17."	Crookneck	Bottle Gourd
Lagenaria	*siceraria*	*bulei-mayo-wartyi*	Warty Mayo Bule 10." x 12."	Crookneck	Bottle Gourd
Lagenaria	*siceraria*	*bulei-tarahumara-smallii*	Small Tarahumara Bule 5.5" x 7.0"	Crookneck	Bottle Gourd
Lagenaria	*siceraria*	*bushelii*	Bushel 48." x 48."	Crookneck	Bottle Gourd
Lagenaria	*siceraria*	*calabash-med-longii*	Med-Long Calabash 4.0" x 15."	Crookneck	Bottle Gourd
Lagenaria	*siceraria*	*calabash-powderhornii*	Powderhorn Calabash 5.0" x 14."	Crookneck	Bottle Gourd
Lagenaria	*siceraria*	*calabash-roundii*	Round Calabash 12." x 12."	Crookneck	Bottle Gourd
Lagenaria	*siceraria*	*cannon-ballii*	Cannon Ball	Crookneck	Bottle Gourd
Lagenaria	*siceraria*	*canteen-tarahumarai*	Tarahumara Canteen 6.0" x 10."	Crookneck	Bottle Gourd
Lagenaria	*siceraria*	*chinese-bottlei*	Chinese 4.0" x 5.0"	Crookneck	Bottle Gourd
Lagenaria	*siceraria*	*clubii*	Club 10." x 30."	Crookneck	Bottle Gourd
Lagenaria	*siceraria*	*clubii-caveman'si*	Caveman's Club 6.0" x 11."	Crookneck	Bottle Gourd
Lagenaria	*siceraria*	*clubii-caveman'si-dinosaur-gourdii*	Dinosaur Caveman's Club 4.0" x 14."	Crookneck	Bottle Gourd
Lagenaria	*siceraria*	*clubii-caveman'si-marankai*	Maranka Caveman's Club 7.0" x 13."	Crookneck	Bottle Gourd
Lagenaria	*siceraria*	*cork-&-twinei*	Cork & Twine	Crookneck	Bottle Gourd
Lagenaria	*siceraria*	*corsican-flatii*	Flat Corsican	Crookneck	Bottle Gourd

Genus	Species	Infraspecies	cultivar	specimen	Genome
Lagenaria	*siceraria*	*corsican-kettlei*	Kettle Corsican	Crookneck	Bottle Gourd
Lagenaria	*siceraria*	*cucuzzii*	India Squash Cucuzzi 1.0" x 6.0"	Crookneck	Bottle Gourd
Lagenaria	*siceraria*	*cucuzzii-caravazzii*	Caravazzi Squash Cucuzzi 2.0" x 12."	Crookneck	Bottle Gourd
Lagenaria	*siceraria*	*cucuzzii-new-guinea-butter-vinei*	New Guinea Butter Squash Cucuzzi	Crookneck	Bottle Gourd
Lagenaria	*siceraria*	*dipperi*	Dipper 8.0" x 24."	Crookneck	Bottle Gourd
Lagenaria	*siceraria*	*dipperi-apachei*	Apache Dipper 4.0" x 12."	Crookneck	Bottle Gourd
Lagenaria	*siceraria*	*dipperi-mini*	Mini Dipper 1.0" x 4.0"	Crookneck	Bottle Gourd
Lagenaria	*siceraria*	*dipperi-o'odhamii*	O'Odham Dipper 4.0" x 13."	Crookneck	Bottle Gourd
Lagenaria	*siceraria*	*dipperi-payote-ceremonialii*	Peyote Ceremonial Dipper 1.0" x 3.0"	Crookneck	Bottle Gourd
Lagenaria	*siceraria*	*dipperi-short-handlei*	Short Handle Dipper 5.0" x 14."	Crookneck	Bottle Gourd
Lagenaria	*siceraria*	*dipperi-short-handlei-mottledii*	Mottled Short Handle Dipper 5.0" x 17."	Crookneck	Bottle Gourd
Lagenaria	*siceraria*	*dipperi-south-sea-islandii*	South Sea Island Dipper 4.5" x 17."	Crookneck	Bottle Gourd
Lagenaria	*siceraria*	*disc-nantesi*	Disc Nantes	Crookneck	Bottle Gourd
Lagenaria	*siceraria*	*dudhi-bottlei*	Dudhi 10." x 30."	Crookneck	Bottle Gourd
Lagenaria	*siceraria*	*early-long-greenii*	Early Long Green 4.0" x 18."	Crookneck	Bottle Gourd
Lagenaria	*siceraria*	*gakhaai*	Gakhaa 7.0" x 19."	Crookneck	Bottle Gourd
Lagenaria	*siceraria*	*hawaiian-dance-maskii*	Hawaiian Dance Mask 8.0" x 10."	Crookneck	Bottle Gourd
Lagenaria	*siceraria*	*healing-squashii*	Healing Squash 6.0" x 18."	Crookneck	Bottle Gourd
Lagenaria	*siceraria*	*indian-bottlei*	Indian	Crookneck	Bottle Gourd
Lagenaria	*siceraria*	*japanese-long-gourdii*	Japanese Long 6.0" x 24."	Crookneck	Bottle Gourd
Lagenaria	*siceraria*	*japanese-roundii*	Japanese Round 13." x 13."	Crookneck	Bottle Gourd
Lagenaria	*siceraria*	*large-bottlei*	Large 5.0" x 11."	Crookneck	Bottle Gourd
Lagenaria	*siceraria*	*legendary-bottlei*	Legendary 10." x 18."	Crookneck	Bottle Gourd
Lagenaria	*siceraria*	*little-manii*	Little Man 5.0" x 5.0"	Crookneck	Bottle Gourd
Lagenaria	*siceraria*	*lump-in-the-neck-bottle-gourdii*	Lump In The Neck	Crookneck	Bottle Gourd
Lagenaria	*siceraria*	*maritius-wartedii*	Maritius Island Warted 4.0" x 10."	Crookneck	Bottle Gourd
Lagenaria	*siceraria*	*mayo-bilobalii*	Mayo Bilobal	Crookneck	Bottle Gourd
Lagenaria	*siceraria*	*mayo-gooseneckii*	Mayo Gooseneck 10." x 24."	Crookneck	Bottle Gourd
Lagenaria	*siceraria*	*mexican-bottlei*	Mexican 6.0" x 12."	Crookneck	Bottle Gourd
Lagenaria	*siceraria*	*mini-bottlei*	Mini 2.0" x 7.5"	Crookneck	Bottle Gourd
Lagenaria	*siceraria*	*nigerian-sayboi*	Nigerian Saybo	Crookneck	Bottle Gourd
Lagenaria	*siceraria*	*rattle-acomai*	Acoma Rattle 5.0" x 6.0"	Crookneck	Bottle Gourd
Lagenaria	*siceraria*	*rattle-ceremonialis*	Ceremonial Rattle 2.0" x 4.0"	Crookneck	Bottle Gourd
Lagenaria	*siceraria*	*rattle-mayo-deer-dancei*	Mayo Deer Dance Rattle	Crookneck	Bottle Gourd
Lagenaria	*siceraria*	*rattle-hopii*	Hopi Rattle 4.0" x 5.0"	Crookneck	Bottle Gourd
Lagenaria	*siceraria*	*siphonii*	Siphon 5.5" x 8.0"	Crookneck	Bottle Gourd
Lagenaria	*siceraria*	*snakei-newi*	New Snake 3.0" x 18."	Crookneck	Bottle Gourd
Lagenaria	*siceraria*	*snakei-SSE*	Seed Saver's Snake 3.0" x 20."	Crookneck	Bottle Gourd
Lagenaria	*siceraria*	*speckled-swanii*	Speckled Swan 7.5" x 15."	Crookneck	Bottle Gourd
Lagenaria	*siceraria*	*sugar-bowl-peruensis*	Peru Sugar Bowl 6.0" x 6.0"	Crookneck	Bottle Gourd
Lagenaria	*siceraria*	*sugar-bowl-tobacco-boxii*	Tobacco Box Sugar Bowl 3.0" x 6.0"	Crookneck	Bottle Gourd
Lagenaria	*siceraria*	*sugar-trough-balii*	Bali Sugar Trough 8.0" x 19."	Crookneck	Bottle Gourd
Lagenaria	*siceraria*	*sugar-trough-birdhousei*	Birdhouse Sugar Trough 11." x 14."	Crookneck	Bottle Gourd
Lagenaria	*siceraria*	*sugar-trough-martinhousei*	Martinhouse Sugar Trough 11." x 15."	Crookneck	Bottle Gourd
Lagenaria	*siceraria*	*tindai*	Tinda	Crookneck	Bottle Gourd
Lagenaria	*siceraria*	*vollyballii*	Volleyball	Crookneck	Bottle Gourd
Lagenaria	*siceraria*	*water-jugii*	Water Jug 9.0" x 22."	Crookneck	Bottle Gourd
Lagenaria	*siceraria*	*water-ladle-giantii*	Water Ladle 11." x 11."	Crookneck	Bottle Gourd
Lagenaria	*siceraria*	*zaire-roundii*	Zaire Round 14." x 14."	Crookneck	Bottle Gourd
Lagenaria	*siceraria*	*zucca-melon-gourdii*	Sheep's Nose Zucca 12." x 36."	Crookneck	Bottle Gourd
Lagenaria	*siceraria*	*zucca-sicillanoi*	Sicillano Zucca 3.0" x 30."	Crookneck	Bottle Gourd

Loofah (*Luffa spp.*)

Genus	Species	Infraspecies	cultivar	specimen	Genome
Luffa	*acutangula*	*acutangula*	Original	Ridged	Loofah
Luffa	*acutangula*	*sinkwa-towelii*	Sinkwa Towel	Ridged	Loofah
Luffa	*aegyptiaca*	*aegyptiaca*	Original	Egyptian	Loofah
Luffa	*aegyptiaca*	*angulata*	Angled	Egyptian	Loofah
Luffa	*aegyptiaca*	*cee-gwai*	Cee Gwa	Egyptian	Loofah
Luffa	*aegyptiaca*	*cylindrica*	Cylinder	Egyptian	Loofah
Luffa	*aegyptiaca*	*laevigata*	Smooth	Egyptian	Loofah

Bitter Melon (*Momordica spp.*)

Genus	Species	Infraspecies	cultivar	specimen	Genome
Momordica	*charantia*	*charantia*	Original	Balsam Pear	Bitter Melon
Momordica	*charantia*	*batu-improva*	Improved Batu	Balsam Pear	Bitter Melon
Momordica	*charantia*	*fontana*	Fontana	Balsam Pear	Bitter Melon
Momordica	*charantia*	*foo-gwai*	Foo Gwa	Balsam Pear	Bitter Melon
Momordica	*charantia*	*formosa-larga*	Taiwan Large	Balsam Pear	Bitter Melon
Momordica	*charantia*	*hong-congensis*	Hong Kong	Balsam Pear	Bitter Melon
Momordica	*charantia*	*thailandica*	Thailand	Balsam Pear	Bitter Melon
Momordica	*indica-longa*	*indica-longa*	Original	India Long	Bitter Melon
Momordica	*indica-longa*	*alba*	White	India Long	Bitter Melon
Momordica	*indica-longa*	*groena*	Green	India Long	Bitter Melon

Family 7 *Flacourtiaceae* Flacourt
Family 8 *Loasaceae* Stingwood
Family 9 *Turneraceae* Alder Violet

Order 724.162 *Capparales* Caper

Family 724.162.1 *Capparaceae* Caper

Genus	Species	Specimen Genome	C:Culinary
Capparis	*brevispina*	Short Spine Caper	C:FL-BD:**VG**
Capparis	*decidua*	Deciduous Caper	C:FL-BD:**VG**
Capparis	*rupestris*	Rock Dwelling Caper	C:FL-BD:**VG**
Capparis	*spinosa*	Spiny Caper	C:FL-BD:**VG**
Capparis	*zeylanica*	Ceylon Caper	C:FL-BD:**VG**
Cleome	*serrulata*	Rocky Mountain Bee Spiderflower Caper	C:FL-BD:**VG**

Varieties

Caper (*Caparis spinosa*)

Genus	Species	Infraspecies	cultivar	specimen Genome
Capparis	*spinosa*	*spinosa*	Original	Spiny Caper
Capparis	*spinosa*	*inermis*	Spineless	Spiny Caper
Cleome	*serrulata*	*serrulata*	Original	Rocky Mountain Bee Spiderflower Caper
Cleome	*serrulata*	*angusta*	Narrow	Rocky Mountain Bee Spiderflower Caper
Cleome	*serrulata*	*guacoi*	Guaco	Rocky Mountain Bee Spiderflower Caper

Family 724.162.2 *Brassicaceae* Mustard

Genus	Species	Specimen	Genome	C:Culinary
Brassica	*brevispina*	Chinese Kale	Mustard	C:LF:**VG**
Brassica	*bourgeaui*	Bourgeau Cabbage	Mustard	C:LF:**VG**
Brassica	*carinata*	Ethiopian Cabbage	Mustard	C:LF:**VG**
Brassica	*chinensis*	Pak Choy	Mustard	C:LF:**VG**
Brassica	*cretica*	Crete Cauliflower	Mustard	C:SD:**VG**
Brassica	*elongata*	Elongated	Mustard	C:LF:**VG**
Brassica	*fruticulosa*	Mediterranean Cabbage	Mustard	C:LF:**VG**
Brassica	*hilarionis*	Hilary Cabbage	Mustard	C:LF:**VG**
Brassica	*incana*	Hoary Cabbage	Mustard	C:LF:**VG**
Brassica	*insularis*	Island Cabbage	Mustard	C:LF:**VG**
Brassica	*juncea-chinensis*	Chinese Mustard Greens	Mustard	C:LF:**VG**
Brassica	*juncea-crispifolia*	Curly Mustard Greens	Mustard	C:LF:**VG**
Brassica	*juncea-indica*	Brown Seed	Mustard	C:LF:**VG**
Brassica	*juncea-indochinensis*	Indochina Mustard Greens	Mustard	C:LF:**VG**
Brassica	*juncea-integrifolia*	Lettuce Mustard Greens	Mustard	C:LF:**VG**
Brassica	*juncea-involutus*	Wrapped Heart Cabbage	Mustard	C:LF:**VG**
Brassica	*juncea-megarrhiza*	Swollen Root	Mustard	C:SL:**VG**
Brassica	*juncea-multiceps*	In The Snow Greens	Mustard	C:LF:**VG**
Brassica	*juncea-platyfolia*	Smooth Mustard Greens	Mustard	C:LF:**VG**
Brassica	*juncea-sturnata*	Swollen Petiole Mustard Greens	Mustard	C:SL:**VG**
Brassica	*juncea-tumida*	Swollen Stem Mustard Greens	Mustard	C:SL:**VG**
Brassica	*napus*	Nape Rape	Mustard	C:LF,RT:**VG**
Brassica	*napus-napobrassica-alba*	White Rutabaga	Mustard	C:RT:**VG**
Brassica	*napus-napobrassica-aurantia*	Orange Rutabaga	Mustard	C:RT:**VG**
Brassica	*napus-napobrassica-groena*	Green Rutabaga	Mustard	C:RT:**VG**
Brassica	*napus-napobrassica-helva*	Yellow Rutabaga	Mustard	C:RT:**VG**
Brassica	*napus-napobrassica-purpurea*	Purple Rutabaga	Mustard	C:RT:**VG**
Brassica	*napus-pabularia*	Blue Green Siberian Kale	Mustard	C:LF:**VG**
Brassica	*narinosa*	Vitamin Greens	Mustard	C:LF:**VG**
Brassica	*nigra*	Black Seed	Mustard	C:SD:**VG**
Brassica	*oleracea-acephala-blauea*	Blue Green Kale	Mustard	C:LF:**VG**
Brassica	*oleracea-acephala-colea*	Collard Greens	Mustard	C:LF:**VG**
Brassica	*oleracea-acephala-crispa*	Dark Green Curly Kale	Mustard	C:LF:**VG**
Brassica	*oleracea-acephala-groena*	Green Kale	Mustard	C:LF:**VG**
Brassica	*oleracea-acephala-laeva*	Smooth Kale	Mustard	C:LF:**VG**
Brassica	*oleracea-acephala-longata*	Walking Stick Kale	Mustard	C:LF:**VG**
Brassica	*oleracea-acephala-mixta*	Multicolor Kale	Mustard	C:LF:**VG**
Brassica	*oleracea-acephala-purpurea*	Purple Kale	Mustard	C:LF:**VG**
Brassica	*oleracea-acephala-purpurea-rubra*	Purple Red Kale	Mustard	C:LF:**VG**
Brassica	*oleracea-acephala-rubra*	Red Kale	Mustard	C:LF:**VG**
Brassica	*oleracea-acephala-sabellica*	Borecole Blue Green Curly Kale	Mustard	C:LF:**VG**
Brassica	*oleracea-acephala-surcula*	Sprout Kale	Mustard	C:SR:**VG**

Brassica	*oleracea-acephala-tumida*	Swollen Stem Kale	Mustard	C:SL:**VG**
Brassica	*oleracea-alboglabra*	Multihead Broccoli	Mustard	C:SD:**VG**
Brassica	*oleracea-botrytis-brocca-60*	60 Day Summer Broccoli	Mustard	C:SD:**VG**
Brassica	*oleracea-botrytis-brocca-70*	70 Day Summer Broccoli	Mustard	C:SD:**VG**
Brassica	*oleracea-botrytis-brocca-80*	80 Day Summer Broccoli	Mustard	C:SD:**VG**
Brassica	*oleracea-botrytis-brocca-150*	150 Day Winter Broccoli	Mustard	C:SD:**VG**
Brassica	*oleracea-botrytis-brocca-220*	220 Day Overwinter Broccoli	Mustard	C:SD:**VG**
Brassica	*oleracea-botrytis-cauliflora-50*	50 Day Spring Cauliflower	Mustard	C:SD:**VG**
Brassica	*oleracea-botrytis-cauliflora-60*	60 Day Summer Cauliflower	Mustard	C:SD:**VG**
Brassica	*oleracea-botrytis-cauliflora-70*	70 Day Summer Cauliflower	Mustard	C:SD:**VG**
Brassica	*oleracea-botrytis-cauliflora-80*	80 Day Summer Cauliflower	Mustard	C:SD:**VG**
Brassica	*oleracea-botrytis-cauliflora-90*	90 Day Summer Cauliflower	Mustard	C:SD:**VG**
Brassica	*oleracea-botrytis-cauliflora-110*	110 Day Fall Cauliflower	Mustard	C:SD:**VG**
Brassica	*oleracea-botrytis-cauliflora-240*	240 Day Overwinter Cauliflower	Mustard	C:SD:**VG**
Brassica	*oleracea-botrytis-italica-60*	60 Day Italian Broccoli	Mustard	C:SD:**VG**
Brassica	*oleracea-botrytis-italica-70*	70 Day Italian Broccoli	Mustard	C:SD:**VG**
Brassica	*oleracea-botrytis-italica-80*	80 Day Italian Broccoli	Mustard	C:SD:**VG**
Brassica	*oleracea-botrytis-italica-220*	220 Day Italian Broccoli	Mustard	C:SD:**VG**
Brassica	*oleracea-capitata-50*	50 Day Spring Green Cabbage	Mustard	C:LF:**VG**
Brassica	*oleracea-capitata-60*	60 Day Summer Green Cabbage	Mustard	C:LF:**VG**
Brassica	*oleracea-capitata-70*	70 Day Summer Green Cabbage	Mustard	C:LF:**VG**
Brassica	*oleracea-capitata-75*	75 Day Summer Green Cabbage	Mustard	C:LF:**VG**
Brassica	*oleracea-capitata-80*	80 Day Summer Green Cabbage	Mustard	C:LF:**VG**
Brassica	*oleracea-capitata-90*	90 Day Summer Green Cabbage	Mustard	C:LF:**VG**
Brassica	*oleracea-capitata-100*	100 Day Summer Green Cabbage	Mustard	C:LF:**VG**
Brassica	*oleracea-capitata-110*	110 Day Fall Green Cabbage	Mustard	C:LF:**VG**
Brassica	*oleracea-capitata-125*	125 Day Winter Cabbage	Mustard	C:LF:**VG**
Brassica	*oleracea-capitata-180*	180 Day Winter Cabbage	Mustard	C:LF:**VG**
Brassica	*oleracea-capitata-220*	220 Day Overwinter Cabbage	Mustard	C:LF:**VG**
Brassica	*oleracea-capitata-purpurea*	Purple Cabbage	Mustard	C:LF:**VG**
Brassica	*oleracea-capitata-rubra*	Red Cabbage	Mustard	C:LF:**VG**
Brassica	*oleracea-capitata-sabanda*	Savoy Cabbage	Mustard	C:LF:**VG**
Brassica	*oleracea-tronchuda*	Sea Kale Cabbage	Mustard	C:LF:**VG**
Brassica	*oleracea-gemmifera-85*	85 Day Brussels Sprout	Mustard	C:LF:**VG**
Brassica	*oleracea-gemmifera-110*	110 Day Brussels Sprout	Mustard	C:LF:**VG**
Brassica	*oleracea-gemmifera-220*	220 Day Brussels Sprout	Mustard	C:LF:**VG**
Brassica	*oleracea-gongylodes-50*	50 Day Kohlrabi	Mustard	C:CM:**VG**
Brassica	*oleracea-gongylodes-60*	60 Day Kohlrabi	Mustard	C:CM:**VG**
Brassica	*oleracea-gongylodes-65*	65 Day Kohlrabi	Mustard	C:CM:**VG**
Brassica	*oleracea-gongylodes-130*	130 Day Kohlrabi	Mustard	C:CM:**VG**
Brassica	*oleracea-gongylodes-x-napus-napobrassica*	Kohlbaga	Mustard	C:CM:**VG**
Brassica	*oleracea-italica*	Italian Cabbage	Mustard	C:LF:**VG**
Brassica	*parachinensis*	False Pak Choy	Mustard	C:LF:**VG**
Brassica	*perviridis*	Komatsuna Spinach	Mustard	C:LF:**VG**
Brassica	*rapa-campestris*	Field Rape	Mustard	C:LF:**VG**
Brassica	*rapa-chinensis*	Chinese Rape Greens	Mustard	C:LF:**VG**
Brassica	*rapa-japonica*	Japanese Rape Greens	Mustard	C:LF:**VG**
Brassica	*rapa-japonica-laciniata*	Mizuna Greens	Mustard	C:LF:**VG**
Brassica	*rapa-oleifera*	Oil Seed Rape	Mustard	C:SD:**VG**
Brassica	*rapa-parapekinens*	Loose Leaf Chinese Cabbage	Mustard	C:LF:**VG**
Brassica	*rapa-pekinensis*	Head Chinese Cabbage	Mustard	C:LF:**VG**
Brassica	*rapa-rapifera-alba-30*	30 Day White Turnip	Mustard	C:RT:**VG**
Brassica	*rapa-rapifera-alba-35*	35 Day White Turnip	Mustard	C:RT:**VG**
Brassica	*rapa-rapifera-alba-40*	40 Day White Turnip	Mustard	C:RT:**VG**
Brassica	*rapa-rapifera-alba-50*	50 Day White Turnip	Mustard	C:RT:**VG**
Brassica	*rapa-rapifera-alba-55*	55 Day White Turnip	Mustard	C:RT:**VG**
Brassica	*rapa-rapifera-alba-65*	65 Day White Turnip	Mustard	C:RT:**VG**
Brassica	*rapa-rapifera-alba-70*	70 Day White Turnip	Mustard	C:RT:**VG**
Brassica	*rapa-rapifera-alba-80*	80 Day White Turnip	Mustard	C:RT:**VG**
Brassica	*rapa-rapifera-foliosa-30*	30 Day Turnip Greens	Mustard	C:LF:**VG**
Brassica	*rapa-rapifera-foliosa-40*	40 Day Turnip Greens	Mustard	C:LF:**VG**
Brassica	*rapa-rapifera-foliosa-50*	50 Day Turnip Greens	Mustard	C:LF:**VG**
Brassica	*rapa-rapifera-foliosa-60*	60 Day Turnip Greens	Mustard	C:LF:**VG**
Brassica	*rapa-rapifera-helva-40*	40 Day Yellow Turnip	Mustard	C:RT:**VG**
Brassica	*rapa-rapifera-helva-70*	70 Day Yellow Turnip	Mustard	C:RT:**VG**
Brassica	*ruvo-30*	30 Day Broccoli Raab Kale	Mustard	C:LF,SH:**VG**
Brassica	*ruvo-40*	40 Day Broccoli Raab Kale	Mustard	C:LF,SH:**VG**
Brassica	*ruvo-50*	50 Day Broccoli Raab Kale	Mustard	C:LF,SH:**VG**
Brassica	*ruvo-60*	60 Day Broccoli Raab Kale	Mustard	C:LF,SH:**VG**
Brassica	*ruvo-85*	85 Day Broccoli Raab Kale	Mustard	C:LF,SH:**VG**
Brassica	*ruvo-220*	220 Day Broccoli Raab Kale	Mustard	C:LF,SH:**VG**
Brassica	*septiceps*	Seven Leaf	Mustard	C:LF:**VG**
Brassica	*sisymbrioides*	Tumble	Mustard	C:LF:**VG**
Brassica	*tournefortii*	Bird Rape	Mustard	C:LF:**VG**
Brassica	*villosa*	Downy Cabbage	Mustard	C:LF:**VG**

Amoracia	*rusticana*	Red Cole	Horseradish	C:LF,RT:**VG**
Barbarea	*orthoceras*	Land	St. Barbara Cress	C:LF:**VG**
Barbarea	*rupicola*	Curly Moss	St. Barbara Cress	C:LF:**VG**
Barbarea	*verna*	Spring	St. Barbara Cress	C:LF:**VG**
Bunias	*erucago*	Corn Rocket	Warty Cabbage	C:LF,SH:**VG**
Bunias	*orientalis*	Turkish Rocket	Warty Cabbage	C:LF,SH:**VG**
Capsella	*bursa-pastoris*	Shepherd's Purse	Capsell Mustard	C:LF:**VG**
Cardamine	*pratensis*	Meadow	Bitter Cress	C:LF:**VG**
Crambe	*abyssinica*	Ethiopian	Sea Kale	C:LF:**VG**
Crambe	*hispanica*	Spanish	Sea Kale	C:LF:**VG**
Crambe	*maritima*	Maritime	Sea Kale	C:LF:**VG**
Diplotaxis	*erucoides*	Rocket Salad	Wall Rocket Cress	C:LF:**VG**
Diplotaxis	*muralis*	Annual	Wall Rocket Cress	C:LF:**VG**
Diplotaxis	*siifolia*	Si Leaf	Wall Rocket Cress	C:LF:**VG**
Diplotaxis	*tenuifolia*	Perennial	Wall Rocket Cress	C:LF:**VG**
Eruca	*eruca*	Rucola	Rocket Cress	C:LF:**VG**
Eruca	*sativa*	Roquette	Rocket Cress	C:LF:**VG**
Eruca	*vesicaria*	Arugula	Rocket Cress	C:LF:**VG**
Lepidium	*sativum*	Mediterranean	Pepper Cress	C:LF:**VG**
Nasturtium	*nigeriensis*	Nigerian	Water Cress	C:LF:**VG**
Nasturtium	*officinale*	Burning	Water Cress	C:LF:**VG**
Raphanus	*caudatus*	Mongri	Radish	C:RT:**VG**
Raphanus	*raphanistrum*	Wild	Radish	C:RT:**VG**
Raphanus	*sativus-albus-longus*	Long White	Radish	C:RT:**VG**
Raphanus	*sativus-albus-rotundus*	Round White	Radish	C:RT:**VG**
Raphanus	*sativus-bruneus-longus*	Long Brown	Radish	
Raphanus	*sativus-bruneus-rotundus*	Round Brown	Radish	
Raphanus	*sativus-groenus-longus*	Long Green	Radish	
Raphanus	*sativus-groenus-rotundus*	Round Green	Radish	
Raphanus	*sativus-nigrus-longus*	Long Black	Radish	
Raphanus	*sativus-nigrus-rotundus*	Round Black	Radish	
Raphanus	*sativus-purpureus-longus*	Long Purple	Radish	
Raphanus	*sativus-purpureus-rotundus*	Round Purple	Radish	
Raphanus	*sativus-rubrus-longus*	Long Red	Radish	
Raphanus	*sativus-rubrus-rotundus*	Round Red	Radish	
Raphanus	*sativus-spinachius*	Spinach Leaf	Radish	
Raphanus	*sativus-surculus*	Sprouting	Radish	
Sinapis	*alba*	White Seed	Sinape Mustard	C:SD:**SP**
Sinapis	*arvensis*	Charlock	Sinape Mustard	C:SD:**SP**

Varieties

Pak Choy (*Brassica chinensis*)

Genus	Species	Infraspecies	cultivar	specimen	Genome
Brassica	*chinensis*	*chinensis*	Original	Pak Choy	Mustard
Brassica	*chinensis*	*bok-choi*	Bok Choy	Pak Choy	Mustard
Brassica	*chinensis*	*bok-choi-babyi*	Baby Bok Choy	Pak Choy	Mustard
Brassica	*chinensis*	*bok-choi-purple-stalkii*	Purple Stalk Bok Choy	Pak Choy	Mustard
Brassica	*chinensis*	*bok-choi-shanghaii*	Shanghai Bok Choy	Pak Choy	Mustard
Brassica	*chinensis*	*bok-choi-spoonii*	Spoon Bok Choy	Pak Choy	Mustard
Brassica	*chinensis*	*bok-choi-sum-flora*	Flowering Sum Bok Choy	Pak Choy	Mustard
Brassica	*chinensis*	*bok-choi-sum-long-white-petiolei*	Long White Petiole Sum Bok Choy	Pak Choy	Mustard
Brassica	*chinensis*	*bok-choi-sum-short-white-petiolei*	Short White Petiole Sum Bok Choy	Pak Choy	Mustard
Brassica	*chinensis*	*bok-choi-takii-greenii*	Taki Green Baby Bok Choy	Pak Choy	Mustard
Brassica	*chinensis*	*bok-choi-tropicalis*	Tropical Bok Choy	Pak Choy	Mustard
Brassica	*chinensis*	*bok-choi-tsai-shimii*	Tsai Shim Bok Choy	Pak Choy	Mustard
Brassica	*chinensis*	*cantonii*	Canton	Pak Choy	Mustard
Brassica	*chinensis*	*cantonii-nana*	Dwarf Canton	Pak Choy	Mustard
Brassica	*chinensis*	*cantonii-short-white-petiolei*	Short White Petiole Canton	Pak Choy	Mustard
Brassica	*chinensis*	*china-choi*	China Choy	Pak Choy	Mustard
Brassica	*chinensis*	*ching-chiang-short-green-petiolei*	Short Green Petiole Ching Chiang	Pak Choy	Mustard
Brassica	*chinensis*	*dai-gai-choi*	Dai Gai Choy	Pak Choy	Mustard
Brassica	*chinensis*	*entsai*	Entsai	Pak Choy	Mustard
Brassica	*chinensis*	*gai-choi*	Gai Choy	Pak Choy	Mustard
Brassica	*chinensis*	*green-stalkii*	Green Stalk	Pak Choy	Mustard
Brassica	*chinensis*	*hon-tsai-tai*	Hon Tsai Tai	Pak Choy	Mustard
Brassica	*chinensis*	*hon-tsai-tai-flora-purpurea*	Flowering Purple Hon Tsai Tai	Pak Choy	Mustard
Brassica	*chinensis*	*hon-tsai-tai-flora-groena*	Flowering Green Hon Tsai Tai	Pak Choy	Mustard
Brassica	*chinensis*	*hung-chin-cabbagei*	Hung Chin	Pak Choy	Mustard
Brassica	*chinensis*	*hy-sawi-manisi*	Hy Saw Manis	Pak Choy	Mustard
Brassica	*chinensis*	*joi-choi*	Joi Choy	Pak Choy	Mustard
Brassica	*chinensis*	*kaisin-hakusai*	Kaisin Hakusai	Pak Choy	Mustard
Brassica	*chinensis*	*koganei*	Kogane	Pak Choy	Mustard
Brassica	*chinensis*	*kosai-tai*	Kosai Tai	Pak Choy	Mustard
Brassica	*chinensis*	*lei-choi*	Lei Choy	Pak Choy	Mustard
Brassica	*chinensis*	*mei-quingii*	Mei Quing	Pak Choy	Mustard

Brassica	*chinensis*	*ming-choi*	Ming Choy	Pak Choy	Mustard
Brassica	*chinensis*	*narinosa*	Tat Soi	Pak Choy	Mustard
Brassica	*chinensis*	*parachinensis*	Sum	Pak Choy	Mustard
Brassica	*chinensis*	*qing-caii*	Qing Cai	Pak Choy	Mustard
Brassica	*chinensis*	*shia-li-honii*	Shia Li Hon	Pak Choy	Mustard
Brassica	*chinensis*	*tah-tsai*	Tah Tsai	Pak Choy	Mustard
Brassica	*chinensis*	*tah-tsoi*	Tah Tsoi	Pak Choy	Mustard
Brassica	*chinensis*	*tai-sai-giantii*	Giant Tai Sai	Pak Choy	Mustard
Brassica	*chinensis*	*tsoi-simii*	Tsoi Sim	Pak Choy	Mustard
Brassica	*elongata*	*elongata*	Original	Elongated	Mustard
Brassica	*elongata*	*integrifolia*	Entire Leaf	Elongated	Mustard

Mustard Greens (*Brassica juncea*)

Genus	Species	Infraspecies	cultivar	specimen	Genome
Brassica	*juncea-chinensis*	*broad-leavedii*	Broad Leaved	Chinese Mustard Greens	Mustard
Brassica	*juncea-chinensis*	*chinese-nan-fongii*	Nan-Fong	Chinese Mustard Greens	Mustard
Brassica	*juncea-chinensis*	*common-leaved-mustardii*	Common Leaved Mustard	Chinese Mustard Greens	Mustard
Brassica	*juncea-chinensis*	*leaf-heading-mustardii*	Leaf Heading Mustard	Chinese Mustard Greens	Mustard
Brassica	*juncea-chinensis*	*megarrhiza*	Tuber Root	Chinese Mustard Greens	Mustard
Brassica	*juncea-chinensis*	*red-leafii*	Red Leaf	Chinese Mustard Greens	Mustard
Brassica	*juncea-chinensis*	*san-ho-giantii*	San Ho Giant	Chinese Mustard Greens	Mustard
Brassica	*juncea-crispifolia*	*fordhook-fancyi*	Fordhook Fancy	Curly Mustard Greens	Mustard
Brassica	*juncea-crispifolia*	*giant-curledii*	Giant Curled	Curly Mustard Greens	Mustard
Brassica	*juncea-crispifolia*	*giant-southern-curledii*	Giant Southern Curled	Curly Mustard Greens	Mustard
Brassica	*juncea-crispifolia*	*green-wavei*	Green Wave	Curly Mustard Greens	Mustard
Brassica	*juncea-crispifolia*	*magmai*	Magma	Curly Mustard Greens	Mustard
Brassica	*juncea-crispifolia*	*old-fashionii*	Old Fashion	Curly Mustard Greens	Mustard
Brassica	*juncea-crispifolia*	*purple-wavei*	Purple Wave	Curly Mustard Greens	Mustard
Brassica	*juncea-crispifolia*	*sawtoothii*	Sawtooth	Curly Mustard Greens	Mustard
Brassica	*juncea-crispifolia*	*spicy-curlsi*	Spicy Curls	Curly Mustard Greens	Mustard
Brassica	*juncea-indochinensis*	.	.	Indochina Mustard Greens	Mustard
Brassica	*juncea-indochinensis*	*india-mustardii*	India Mustard	Indochina Mustard Greens	Mustard
Brassica	*juncea-indochinensis*	*india-mustardii-redii*	Red India Mustard	Indochina Mustard Greens	Mustard
Brassica	*juncea-indochinensis*	*takana-akai*	Aka Takana	Indochina Mustard Greens	Mustard
Brassica	*juncea-indochinensis*	*takana-aoi*	Ao Takana	Indochina Mustard Greens	Mustard
Brassica	*juncea-indochinensis*	*takana-chinensis*	China Takana	Indochina Mustard Greens	Mustard
Brassica	*juncea-indochinensis*	*takana-kekkyui*	Kekkyu Takana	Indochina Mustard Greens	Mustard
Brassica	*juncea-indochinensis*	*takana-yanagawa*	YanagawaTakana	Indochina Mustard Greens	Mustard
Brassica	*juncea-indochinensis*	*thailandica*	Thailand	Indochina Mustard Greens	Mustard
Brassica	*juncea-integrifolia*	*integrifolia*	Original	Lettuce Mustard Greens	Mustard
Brassica	*juncea-integrifolia*	*chirimenii*	Chirimen	Lettuce Mustard Greens	Mustard
Brassica	*juncea-integrifolia*	*chirimenii-redii*	Red Chirimen	Lettuce Mustard Greens	Mustard
Brassica	*juncea-integrifolia*	*foliosa*	Swatow	Lettuce Mustard Greens	Mustard
Brassica	*juncea-integrifolia*	*osaka-purplei*	OsakaPurple	Lettuce Mustard Greens	Mustard
Brassica	*juncea-integrifolia*	*rugosa*	Giant Red	Lettuce Mustard Greens	Mustard
Brassica	*juncea-involutus*	*involutus*	Original	Wrapped Heart Cabbage	Mustard
Brassica	*juncea-involutus*	*bau-sinii*	Bau Sin	Wrapped Heart Cabbage	Mustard
Brassica	*juncea-multiceps*	*greenii(hsueh-li-hungii)*	Hsueh Li Hung Green	In The Snow Greens	Mustard
Brassica	*juncea-multiceps*	*luxuriens*	Luxurient	In The Snow Greens	Mustard
Brassica	*juncea-multiceps*	*redii*	Red	In The Snow Greens	Mustard
Brassica	*juncea-platyfolia*	*black-winteri*	Black Winter	Smooth Mustard Greens	Mustard
Brassica	*juncea-platyfolia*	*cherokee-bluei*	Cherokee Blue	Smooth Mustard Greens	Mustard
Brassica	*juncea-platyfolia*	*florida-broad-leafii*	Florida Broad Leaf	Smooth Mustard Greens	Mustard
Brassica	*juncea-platyfolia*	*florida-giantii*	Florida Giant	Smooth Mustard Greens	Mustard
Brassica	*juncea-platyfolia*	*mayo-quelitei*	Mayo Quelite	Smooth Mustard Greens	Mustard
Brassica	*juncea-platyfolia*	*mostaza-roja*	Mostaz Roja	Smooth Mustard Greens	Mustard
Brassica	*juncea-platyfolia*	*oak-firei*	Oak Fire	Smooth Mustard Greens	Mustard
Brassica	*juncea-platyfolia*	*south-african-spinach-mustardii*	South African Spinach	Smooth Mustard Greens	Mustard
Brassica	*juncea-platyfolia*	*spinach-mustardii*	Spinach Mustard	Smooth Mustard Greens	Mustard
Brassica	*juncea-platyfolia*	*tarahumara-espinacai*	Tarahumara Espinaca	Smooth Mustard Greens	Mustard
Brassica	*juncea-sturmata*	*sturmata*	Original	Swollen Petiole Mustard Greens	Mustard
Brassica	*juncea-sturmata*	*parcel-&-pocket-mustardii*	Parcel & Pocket	Swollen Petiole Mustard Greens	Mustard
Brassica	*juncea-tumida*	*tumida*	Original	Swollen Stem Mustard Greens	Mustard
Brassica	*juncea-tumida*	*cha-tsoii*	Cha Tsoi	Swollen Stem Mustard Greens	Mustard
Brassica	*juncea-tumida*	*big-stem-mustardii*	Big Stem	Swollen Stem Mustard Greens	Mustard
Brassica	*juncea-tumida*	*bamboo-mustardii*	Bamboo	Swollen Stem Mustard Greens	Mustard
Brassica	*juncea-tumida*	*celery-mustardii*	Celery Mustard	Swollen Stem Mustard Greens	Mustard
Brassica	*juncea-tumida*	*bouquetii*	Bouquet	Swollen Stem Mustard Greens	Mustard
Brassica	*juncea-tumida*	*heavy-purple-stemii*	Heavy Purple Stem	Swollen Stem Mustard Greens	Mustard
Brassica	*juncea-tumida*	*granatii*	Granat Celery	Swollen Stem Mustard Greens	Mustard
Brassica	*juncea-tumida*	*horned-mustardii*	Horned Mustard	Swollen Stem Mustard Greens	Mustard
Brassica	*juncea-tumida*	*japanese-white-celery-mustardii*	Japanese White Celery	Swollen Stem Mustard Greens	Mustard
Brassica	*juncea-tumida*	*sareptana*	Sareptan	Swollen Stem Mustard Greens	Mustard

Mustard, Brown Seed (*Brassica juncea-indica*)

Genus	Species	Infraspecies	cultivar	specimen	Genome
Brassica	*juncea-indica*	*indica*	Original	Brown Seed	Mustard
Brassica	*juncea-indica*	*indischeri*	Indischer	Brown Seed	Mustard

Brassica	*juncea-indica*	*mostaza-indica*	Mostaza India	Brown Seed	Mustard
Brassica	*juncea-indica*	*moutarde-de-chinei*	Moutarde De Chine	Brown Seed	Mustard
Brassica	*juncea-indica*	*senape-indiana*	Senape Indiana	Brown Seed	Mustard

Rutabaga (*Brassica napus-napobrassica*)

Genus	Species	Infraspecies	cultivar	specimen	Genome
Brassica	*napus*	*napus*	Original	Nape Rape	Mustard
Brassica	*napus*	*rapa-salada*	Salad	Nape Rape	Mustard
Brassica	*napus*	*rapa-dwarf-essexii*	Dwarf Essex	Nape Rape	Mustard
Brassica	*napus-napobrassica-alba*	.	.	White Rutabaga	Mustard
Brassica	*napus-napobrassica-alba*	*bristol-whitei*	Bristol	White Rutabaga	Mustard
Brassica	*napus-napobrassica-alba*	*macomberi*	MacomberSweet	White Rutabaga	Mustard
Brassica	*napus-napobrassica-alba*	*merrickii*	Merrick	White Rutabaga	Mustard
Brassica	*napus-napobrassica-alba*	*tona-a-colletto-violai*	Tona A Coletto Viola	White Rutabaga	Mustard
Brassica	*napus-napobrassica-aurantia*	*wilhelmsburgeri-green-topii*	Green Top Wilhelmsburger	Orange Rutabaga	Mustard
Brassica	*napus-napobrassica-groena*	.	.	Green Rutabaga	Mustard
Brassica	*napus-napobrassica-groena*	*aspasi*	Aspas	Green Rutabaga	Mustard
Brassica	*napus-napobrassica-groena*	*doonii*	Doon	Green Rutabaga	Mustard
Brassica	*napus-napobrassica-groena*	*lizzyi*	Lizzy	Green Rutabaga	Mustard
Brassica	*napus-napobrassica-groena*	*maryi*	Mary	Green Rutabaga	Mustard
Brassica	*napus-napobrassica-groena*	*monarchii*	Monarch	Green Rutabaga	Mustard
Brassica	*napus-napobrassica-groena*	*old-jakei*	Old Jake	Green Rutabaga	Mustard
Brassica	*napus-napobrassica-groena*	*purple-topii-gaze's-specialis*	Gaze's Special Purple Top	Green Rutabaga	Mustard
Brassica	*napus-napobrassica-groena*	*purple-topii-swedei*	Swede Purple Top	Green Rutabaga	Mustard
Brassica	*napus-napobrassica-helva*	.	.	Yellow Rutabaga	Mustard
Brassica	*napus-napobrassica-helva*	*altasweetii*	Altasweet	Yellow Rutabaga	Mustard
Brassica	*napus-napobrassica-helva*	*american-purple-topii*	American Purple Top	Yellow Rutabaga	Mustard
Brassica	*napus-napobrassica-helva*	*american-purple-topii-improvedii*	Improved American Purple Top	Yellow Rutabaga	Mustard
Brassica	*napus-napobrassica-helva*	*best-of-allii*	Best Of All	Yellow Rutabaga	Mustard
Brassica	*napus-napobrassica-helva*	*canadian-gemii*	Canadian Gem	Yellow Rutabaga	Mustard
Brassica	*napus-napobrassica-helva*	*fortin's-familyi*	Fortin's Family	Yellow Rutabaga	Mustard
Brassica	*napus-napobrassica-helva*	*golden-necklessi*	Golden Neckless	Yellow Rutabaga	Mustard
Brassica	*napus-napobrassica-helva*	*joanii*	Joan	Yellow Rutabaga	Mustard
Brassica	*napus-napobrassica-helva*	*laurentiana*	Golden Laurentian	Yellow Rutabaga	Mustard
Brassica	*napus-napobrassica-helva*	*laurentiana-certifiedii-1*	Certified #1 Laurentian	Yellow Rutabaga	Mustard
Brassica	*napus-napobrassica-helva*	*laurentiana-necklessi*	Neckless Laurentian	Yellow Rutabaga	Mustard
Brassica	*napus-napobrassica-helva*	*laurentiana-OCS*	Ontario Certified Stock Laurentian	Yellow Rutabaga	Mustard
Brassica	*napus-napobrassica-helva*	*laurentiana-thompsonii*	Thompson Laurentian	Yellow Rutabaga	Mustard
Brassica	*napus-napobrassica-helva*	*pikei*	Pike	Yellow Rutabaga	Mustard
Brassica	*napus-napobrassica-helva*	*purple-top-yellowi*	Purple Top	Yellow Rutabaga	Mustard
Brassica	*napus-napobrassica-helva*	*purple-top-yellowi-improvedii*	Improved Purple Top	Yellow Rutabaga	Mustard
Brassica	*napus-napobrassica-helva*	*virtuei*	Virtue	Yellow Rutabaga	Mustard
Brassica	*napus-napobrassica-helva*	*wilhelmsburgeri-gelbei*	Gelbe Wilhelmsburger	Yellow Rutabaga	Mustard
Brassica	*napus-napobrassica-helva*	*yorkii*	York	Yellow Rutabaga	Mustard
Brassica	*napus-napobrassica-purpurea*	*mariana*	Marian	Purple Rutabaga	Mustard

Kale, Siberian (*napus-pabularia*)

Genus	Species	Infraspecies	cultivar	specimen	Genome
Brassica	*napus-pabularia*	*hanover-earlyi-premier*	Primier Early Hanover	Blue Green Siberian Kale	Mustard
Brassica	*napus-pabularia*	*hanover-earlyi-springii*	Spring Early Hanover	Blue Green Siberian Kale	Mustard
Brassica	*napus-pabularia*	*hanover-long-seasonsi*	Long Seasons Hanover	Blue Green Siberian Kale	Mustard
Brassica	*napus-pabularia*	*hanover-saladii*	Salad Hanover	Blue Green Siberian Kale	Mustard
Brassica	*napus-pabularia*	*siberiana*	Early Curled	Blue Green Siberian Kale	Mustard
Brassica	*napus-pabularia*	*siberiana-bluei*	Waxy	Blue Green Siberian Kale	Mustard
Brassica	*napus-pabularia*	*siberiana-dwarfii*	Dwarf	Blue Green Siberian Kale	Mustard
Brassica	*napus-pabularia*	*siberiana-improvedii*	Improved Early Curled	Blue Green Siberian Kale	Mustard

Mustard, Black Seed (*Brassica nigra*)

Genus	Species	Infraspecies	cultivar	specimen	Genome
Brassica	*nigra*	*nigra*	Original	Black Seed	Mustard
Brassica	*nigra*	*mostaza-negra*	Mostaza Negra	Black Seed	Mustard
Brassica	*nigra*	*moutarde-noirei*	Moutarde Noire	Black Seed	Mustard
Brassica	*nigra*	*senape-nera*	Senape Nera	Black Seed	Mustard
Brassica	*nigra*	*schwartzeri*	Schwartzer	Black Seed	Mustard

Kale (*Brassica oleracea-acephala*)

Genus	Species	Infraspecies	cultivar	specimen	Genome
Brassica	*oleracea-acephala-blauea*	*lacinatoi-niro-de-philoi*	Nero De Philo Lacinato	Blue Green Kale	Mustard
Brassica	*oleracea-acephala-blauea*	*lacinatoi-niro-di-toscana*	Nero Di Toscana Lacinato	Blue Green Kale	Mustard
Brassica	*oleracea-acephala-blauea*	*westlandse-verdura*	Verdura Westlandsei	Blue Green Kale	Mustard
Brassica	*oleracea-acephala-crispa*	*blackii(savoyi)*	Savoy Black	Dark Green Curly Kale	Mustard
Brassica	*oleracea-acephala-crispa*	*bona*	Bona	Dark Green Curly Kale	Mustard
Brassica	*oleracea-acephala-crispa*	*curly-greenii*	Curly Green	Dark Green Curly Kale	Mustard
Brassica	*oleracea-acephala-crispa*	*dwarf-green-curledii*	Dwarf Green Curled	Dark Green Curly Kale	Mustard
Brassica	*oleracea-acephala-crispa*	*dwarf-green-curledii-scotchii*	Scotch Dwarf Green Curled	Dark Green Curly Kale	Mustard
Brassica	*oleracea-acephala-crispa*	*frosty-curly-kalei*	Frosty	Dark Green Curly Kale	Mustard
Brassica	*oleracea-acephala-crispa*	*grun-kohl-lerchenzungenii*	Grun Kohl Lerchenzungen	Dark Green Curly Kale	Mustard
Brassica	*oleracea-acephala-crispa*	*harvesterii*	Harvester	Dark Green Curly Kale	Mustard
Brassica	*oleracea-acephala-crispa*	*judy'si*	Judy"s	Dark Green Curly Kale	Mustard

Brassica	oleracea-acephala-crispa	konservai	Konserva	Dark Green Curly Kale	Mustard
Brassica	oleracea-acephala-crispa	lavoi	Lavo	Dark Green Curly Kale	Mustard
Brassica	oleracea-acephala-crispa	tall-green-curledii	TallGreenCurled	Dark Green Curly Kale	Mustard
Brassica	oleracea-acephala-crispa	westlandse-semi-dwarfii	Westlandse Semi Dwarf	Dark Green Curly Kale	Mustard
Brassica	oleracea-acephala-groena	white-flowered-kalei	White Flowered	Green Kale	Mustard
Brassica	oleracea-acephala-groena	yellow-flowered-kalei	Yellow Flowered	Green Kale	Mustard
Brassica	oleracea-acephala-laeva	black-sea-man-turkishii	Black Sea Man Turkish	Smooth Kale	Mustard
Brassica	oleracea-acephala-laeva	caulet-de-flandrei	Cautlet De Flandre	Smooth Kale	Mustard
Brassica	oleracea-acephala-laeva	smoothii	Redwood City	Smooth Kale	Mustard
Brassica	oleracea-acephala-laeva	smoothii-long-standingii	Long Standing	Smooth Kale	Mustard
Brassica	oleracea-acephala-longata	chou-cavalieri	Chou Cavalier	Walking Stick Kale	Mustard
Brassica	oleracea-acephala-longata	giant-jerseyi	Giant Jersey	Walking Stick Kale	Mustard
Brassica	oleracea-acephala-longata	giant-walking-stickii	Giant	Walking Stick Kale	Mustard
Brassica	oleracea-acephala-longata	palm-tree-cabbagei	Savoy Palm Tree	Walking Stick Kale	Mustard
Brassica	oleracea-acephala-mixta	flowering-kale-miniaturei	Flowering Miniature	Multicolor Kale	Mustard
Brassica	oleracea-acephala-mixta	coral-queenii	Coral Queen	Multicolor Kale	Mustard
Brassica	oleracea-acephala-mixta	lacinatoi-rainbowi	Rainbow Lacinato	Multicolor Kale	Mustard
Brassica	oleracea-acephala-mixta	ornamentalis	Ornamental	Multicolor Kale	Mustard
Brassica	oleracea-acephala-purpurea	delawayi	Delaway	Purple Kale	Mustard
Brassica	oleracea-acephala-purpurea	russica-greenpeacei	Greenpeace Russian	Purple Kale	Mustard
Brassica	oleracea-acephala-purpurea-rubra	crimson-gardenii	Crimson Garden	Purple Red Kale	Mustard
Brassica	oleracea-acephala-purpurea-rubra	ozark-blackii	Ozark Black	Purple Red Kale	Mustard
Brassica	oleracea-acephala-rubra	bora	RedBor	Red Kale	Mustard
Brassica	oleracea-acephala-rubra	russian-frillyi	Frilly Russian	Red Kale	Mustard
Brassica	oleracea-acephala-rubra	russian-ragged-jackii	Ragged Jack Russian	Red Kale	Mustard
Brassica	oleracea-acephala-rubra	russian-redii	Red Russian	Red Kale	Mustard
Brassica	oleracea-acephala-rubra	russian-spring-sweetii	Spring Sweet Russian	Red Kale	Mustard
Brassica	oleracea-acephala-rubra	russian-ursa-redii	Ursa Russian	Red Kale	Mustard
Brassica	oleracea-acephala-rubra	russian-whitei	White Midrib Russian	Red Kale	Mustard
Brassica	oleracea-acephala-rubra	russian-wild-redii	Wild Red Russian	Red Kale	Mustard
Brassica	oleracea-acephala-rubra	willy'si	Willy's	Red Kale	Mustard
Brassica	oleracea-acephala-rubra	winter-redii	Winter	Red Kale	Mustard
Brassica	oleracea-acephala-sabellica	dwarf-blue-curledii	Dwarf Blue Curled	Borecole Blue Green Curly Kale	Mustard
Brassica	oleracea-acephala-sabellica	dwarf-blue-curledii-scotchii	Scotch Dwarf Blue Curled	Borecole Blue Green Curly Kale	Mustard
Brassica	oleracea-acephala-sabellica	dwarf-blue-curledii-vatesi	Vates Dwarf Blue Curled	Borecole Blue Green Curly Kale	Mustard
Brassica	oleracea-acephala-sabellica	dwarf-blue-curledii-vatesii-improvedii	Improved Vates Dwarf Blue Curled	Borecole Blue Green Curly Kale	Mustard
Brassica	oleracea-acephala-sabellica	dwarf-curled	Dwarf Curled	Borecole Blue Green Curly Kale	Mustard
Brassica	oleracea-acephala-sabellica	squirea	Squire	Borecole Blue Green Curly Kale	Mustard
Brassica	oleracea-acephala-sabellica	thousand-headedii	Thousand Headed	Borecole Blue Green Curly Kale	Mustard
Brassica	oleracea-acephala-surcula	asparagus	Asparagus	Sprout Kale	Mustard
Brassica	oleracea-acephala-surcula	pentland-brigii	Pentland Brig	Sprout Kale	Mustard
Brassica	oleracea-acephala-surcula	purple-napinii	Purple Napini	Sprout Kale	Mustard
Brassica	oleracea-acephala-tumida	maris-kestrelii	Maris Kestrel	Swollen Stem Kale	Mustard
Brassica	oleracea-acephala-tumida	marrow-stemii	Marrow Stem	Swollen Stem Kale	Mustard

Collard Greens (*Brassica oleracea-colea*)

Brassica	oleracea-acephala-colea	afghanistanensis	Afghanistan	Collard Greens	Mustard
Brassica	oleracea-acephala-colea	cabbage-collardsi	Cabbage	Collard Greens	Mustard
Brassica	oleracea-acephala-colea	championii	Champion	Collard Greens	Mustard
Brassica	oleracea-acephala-colea	georgia-truei	True Georgia	Collard Greens	Mustard
Brassica	oleracea-acephala-colea	georgia-blue-stemii	Blue Stem Georgia	Collard Greens	Mustard
Brassica	oleracea-acephala-colea	georgia-greenii	Green Georgia	Collard Greens	Mustard
Brassica	oleracea-acephala-colea	giant-purple-flat-polei	GiantPurpleFlatPole	Collard Greens	Mustard
Brassica	oleracea-acephala-colea	green-glazei	Greasy Green Glaze	Collard Greens	Mustard
Brassica	oleracea-acephala-colea	green-glazei-mccormack'si	McCormack's Green Glaze	Collard Greens	Mustard
Brassica	oleracea-acephala-colea	mesic-zero-degreesi	Mesic Zero Degrees	Collard Greens	Mustard
Brassica	oleracea-acephala-colea	morris-headingii	Morris Heading	Collard Greens	Mustard
Brassica	oleracea-acephala-colea	morris-headingii-improvedii	Improved Morris Heading	Collard Greens	Mustard
Brassica	oleracea-acephala-colea	syriana	Syrian	Collard Greens	Mustard
Brassica	oleracea-acephala-colea	variegata	Variegated	Collard Greens	Mustard
Brassica	oleracea-acephala-colea	vatesii	Blue Stem Vates	Collard Greens	Mustard

Broccoli, Chinese (*Brassica oleracea-alboglabra*)

Genus	Species	Infraspecies	cultivar	specimen	Genome
Brassica	oleracea-alboglabra	alboglabra	Original	Chinese Broccoli	Mustard
Brassica	oleracea-alboglabra	chinese-broccolii	Chinese	Chinese Broccoli	Mustard
Brassica	oleracea-alboglabra	gai-choii	Gai Choi	Chinese Broccoli	Mustard
Brassica	oleracea-alboglabra	gai-lanii	Gai Lan	Chinese Broccoli	Mustard
Brassica	oleracea-alboglabra	gai-lohnii	Gai Lohn	Chinese Broccoli	Mustard
Brassica	oleracea-alboglabra	green-lancei	Green Lance	Chinese Broccoli	Mustard
Brassica	oleracea-alboglabra	kaai-laanii	Kaai Laan	Chinese Broccoli	Mustard
Brassica	oleracea-alboglabra	little-guyi	Little Guy	Chinese Broccoli	Mustard
Brassica	oleracea-alboglabra	scopei	Scope	Chinese Broccoli	Mustard
Brassica	oleracea-alboglabra	torazirohii	Torazirhoh	Chinese Broccoli	Mustard

Broccoli (*Brassica oleracea-botrytis-brocca*)

Genus	Species	Infraspecies	cultivar	specimen	Genome

Brassica	*oleracea-botrytis-brocca-60*	*atlantica*	Blue Green Atlantic	60 Day Summer Broccoli	Mustard
Brassica	*oleracea-botrytis-brocca-60*	*boramii-92*	Boram	60 Day Summer Broccoli	Mustard
Brassica	*oleracea-botrytis-brocca-60*	*dandy-earlyi*	Dandy Early	60 Day Summer Broccoli	Mustard
Brassica	*oleracea-botrytis-brocca-60*	*green-goliathii*	Green Goliath	60 Day Summer Broccoli	Mustard
Brassica	*oleracea-botrytis-brocca-60*	*muchkinii*	Blue Geen Munchkin	60 Day Summer Broccoli	Mustard
Brassica	*oleracea-botrytis-brocca-60*	*prontoi*	Pronto	60 Day Summer Broccoli	Mustard
Brassica	*oleracea-botrytis-brocca-60*	*rosalindii*	Purple Rosalind	60 Day Summer Broccoli	Mustard
Brassica	*oleracea-botrytis-brocca-60*	*umpoquai*	Umpquah	60 Day Summer Broccoli	Mustard
Brassica	*oleracea-botrytis-brocca-70*	*albertii*	Dark Green Albert	70 Day Summer Broccoli	Mustard
Brassica	*oleracea-botrytis-brocca-70*	*early-greenii*	Early Green	70 Day Summer Broccoli	Mustard
Brassica	*oleracea-botrytis-brocca-70*	*green-mountainii*	Green Mountain	70 Day Summer Broccoli	Mustard
Brassica	*oleracea-botrytis-brocca-70*	*health-plusi*	Health Plus	70 Day Summer Broccoli	Mustard
Brassica	*oleracea-botrytis-brocca-70*	*nutribudii*	Nutribud	70 Day Summer Broccoli	Mustard
Brassica	*oleracea-botrytis-brocca-80*	*enchantmentii*	Blue Green Enchantment	80 Day Summer Broccoli	Mustard
Brassica	*oleracea-botrytis-brocca-80*	*second-wavei*	Second Wave	80 Day Summer Broccoli	Mustard
Brassica	*oleracea-botrytis-brocca-80*	*supradomei*	Super Dome	80 Day Summer Broccoli	Mustard
Brassica	*oleracea-botrytis-brocca-80*	*thompsonii*	Medium Thompson	80 Day Summer Broccoli	Mustard
Brassica	*oleracea-botrytis-brocca-80*	*thompsonii-92*	Large Thompson	80 Day Summer Broccoli	Mustard
Brassica	*oleracea-botrytis-brocca-80*	*toori-92*	Toor	80 Day Summer Broccoli	Mustard
Brassica	*oleracea-botrytis-brocca-80*	*white-sproutingii*	White Sprouting	80 Day Summer Broccoli	Mustard
Brassica	*oleracea-botrytis-brocca-150*	*CR-1*	Clubroot Resistant #1	150 Day Winter Broccoli	Mustard
Brassica	*oleracea-botrytis-brocca-150*	*purple-sproutingii-rudolphii*	Rudolph Purple Sprouting	150 Day Winter Broccoli	Mustard
Brassica	*oleracea-botrytis-brocca-220*	*purple-sproutingii*	Late Purple Sprouting	220 Day Overwinter Broccoli	Mustard
Brassica	*oleracea-botrytis-brocca-220*	*purple-sproutingii-christmasi*	Christmas Purple Sprouting	220 Day Overwinter Broccoli	Mustard
Brassica	*oleracea-botrytis-brocca-220*	*purple-sproutingii-earlyi*	Early Purple Sprouting	220 Day Overwinter Broccoli	Mustard
Brassica	*oleracea-botrytis-brocca-220*	*purple-sproutingii-redheadii*	Redhead Purple Sprouting	220 Day Overwinter Broccoli	Mustard
Brassica	*oleracea-botrytis-brocca-220*	*purple-sproutingii-red-arrowi*	Red Arrow Purple Sprouting	220 Day Overwinter Broccoli	Mustard
Brassica	*oleracea-botrytis-brocca-220*	*purple-sproutingii-spring-royaltyi*	Spring Royalty Purple Sprouting	220 Day Overwinter Broccoli	Mustard
Brassica	*oleracea-botrytis-brocca-220*	*white-sproutingii-latei*	Late White Sprouting	220 Day Overwinter Broccoli	Mustard

Cauliflower (*Brassica oleracea-botrytis-cauliflora*)

Genus	Species	Infraspecies	cultivar	specimen	Genome
Brassica	*oleracea-botrytis-cauliflora-50*	*alertii*	Alert	50 Day Spring Cauliflower	Mustard
Brassica	*oleracea-botrytis-cauliflora-50*	*masteri*	Master	50 Day Spring Cauliflower	Mustard
Brassica	*oleracea-botrytis-cauliflora-50*	*mechelsei-earlyi*	Mechelse Early	50 Day Spring Cauliflower	Mustard
Brassica	*oleracea-botrytis-cauliflora-50*	*snowballii-new-earlyi*	New Early Snowball	50 Day Spring Cauliflower	Mustard
Brassica	*oleracea-botrytis-cauliflora-50*	*snowballii-danish-perfectionii*	Danish Perfection Snowball	50 Day Spring Cauliflower	Mustard
Brassica	*oleracea-botrytis-cauliflora-50*	*snowballii-earlyi*	Early Snowball	50 Day Spring Cauliflower	Mustard
Brassica	*oleracea-botrytis-cauliflora-50*	*snowballii-T2-earlyi*	Type T2 Early Snowball	50 Day Spring Cauliflower	Mustard
Brassica	*oleracea-botrytis-cauliflora-50*	*zeroi*	Zero	50 Day Spring Cauliflower	Mustard
Brassica	*oleracea-botrytis-cauliflora-60*	*snowballii-superi-burpeeana*	Burpeeana Super Snowball	60 Day Summer Cauliflower	Mustard
Brassica	*oleracea-botrytis-cauliflora-60*	*snowballii*	Snowball	60 Day Summer Cauliflower	Mustard
Brassica	*oleracea-botrytis-cauliflora-60*	*snowballii-A*	Type A Snowball	60 Day Summer Cauliflower	Mustard
Brassica	*oleracea-botrytis-cauliflora-60*	*snowballii-A-earlyi*	Early Type A Snowball	60 Day Summer Cauliflower	Mustard
Brassica	*oleracea-botrytis-cauliflora-60*	*snowballii-A-improvedii*	Improved Type A Snowball	60 Day Summer Cauliflower	Mustard
Brassica	*oleracea-botrytis-cauliflora-60*	*snowballii-extra-earlyi*	SnowballExtraEarly	60 Day Summer Cauliflower	Mustard
Brassica	*oleracea-botrytis-cauliflora-60*	*snowballii-imperialis-10-6*	Imperial Snowball	60 Day Summer Cauliflower	Mustard
Brassica	*oleracea-botrytis-cauliflora-60*	*snowballii-self-blanchingii*	Self Blanching Snowball	60 Day Summer Cauliflower	Mustard
Brassica	*oleracea-botrytis-cauliflora-60*	*snowballii-snowconei*	Snowcone Snowball	60 Day Summer Cauliflower	Mustard
Brassica	*oleracea-botrytis-cauliflora-60*	*snowballii-snowdriftii*	Snowdrift Snowball	60 Day Summer Cauliflower	Mustard
Brassica	*oleracea-botrytis-cauliflora-60*	*snowballii-superi*	Super Snowball	60 Day Summer Cauliflower	Mustard
Brassica	*oleracea-botrytis-cauliflora-60*	*snowballii-T2*	Type T2 Snowball	60 Day Summer Cauliflower	Mustard
Brassica	*oleracea-botrytis-cauliflora-60*	*snowballii-X*	Type X Snowball	60 Day Summer Cauliflower	Mustard
Brassica	*oleracea-botrytis-cauliflora-70*	*alpha-begum*	Begum Alpha	70 Day Summer Cauliflower	Mustard
Brassica	*oleracea-botrytis-cauliflora-70*	*alpha-brio*	Brio Alpha	70 Day Summer Cauliflower	Mustard
Brassica	*oleracea-botrytis-cauliflora-70*	*alpha-fortados*	Fortados Alpha	70 Day Summer Cauliflower	Mustard
Brassica	*oleracea-botrytis-cauliflora-70*	*alpha-harvesteri-S*	Harvester S Alpha	70 Day Summer Cauliflower	Mustard
Brassica	*oleracea-botrytis-cauliflora-70*	*alpha-idolii*	Idol Alpha	70 Day Summer Cauliflower	Mustard
Brassica	*oleracea-botrytis-cauliflora-70*	*alpha-kingii*	King Alpha	70 Day Summer Cauliflower	Mustard
Brassica	*oleracea-botrytis-cauliflora-70*	*andesensis*	Andes	70 Day Summer Cauliflower	Mustard
Brassica	*oleracea-botrytis-cauliflora-70*	*avalanchei*	Avalanche	70 Day Summer Cauliflower	Mustard
Brassica	*oleracea-botrytis-cauliflora-70*	*brocoverde*	Brocoverde Green	70 Day Summer Cauliflower	Mustard
Brassica	*oleracea-botrytis-cauliflora-70*	*cauliflower-of-maceratai*	Macerata Green	70 Day Summer Cauliflower	Mustard
Brassica	*oleracea-botrytis-cauliflora-70*	*danovai*	Danova	70 Day Summer Cauliflower	Mustard
Brassica	*oleracea-botrytis-cauliflora-70*	*dominans*	Dominant	70 Day Summer Cauliflower	Mustard
Brassica	*oleracea-botrytis-cauliflora-70*	*flora-blanca-18*	Flora Blanca #18	70 Day Summer Cauliflower	Mustard
Brassica	*oleracea-botrytis-cauliflora-70*	*iglooi-casablancai*	Casablanca Igloo	70 Day Summer Cauliflower	Mustard
Brassica	*oleracea-botrytis-cauliflora-70*	*iglooi-stove-pipei*	Stove Pipe Igloo	70 Day Summer Cauliflower	Mustard
Brassica	*oleracea-botrytis-cauliflora-70*	*purple-queenii*	Purple Queen	70 Day Summer Cauliflower	Mustard
Brassica	*oleracea-botrytis-cauliflora-70*	*purple-rosalindii*	Purple Rosalind	70 Day Summer Cauliflower	Mustard
Brassica	*oleracea-botrytis-cauliflora-70*	*sicilian-purplei*	Sicilian Purple	70 Day Summer Cauliflower	Mustard
Brassica	*oleracea-botrytis-cauliflora-70*	*sierra-nevadensis*	Sierra Nevada	70 Day Summer Cauliflower	Mustard
Brassica	*oleracea-botrytis-cauliflora-70*	*snowballii-016*	#016 Snowball	70 Day Summer Cauliflower	Mustard
Brassica	*oleracea-botrytis-cauliflora-70*	*snowballii-123*	#123 Snowball	70 Day Summer Cauliflower	Mustard
Brassica	*oleracea-botrytis-cauliflora-70*	*snowballii-crossi*	Cross Snowball	70 Day Summer Cauliflower	Mustard
Brassica	*oleracea-botrytis-cauliflora-70*	*snowballii-D*	Type D Snowball	70 Day Summer Cauliflower	Mustard
Brassica	*oleracea-botrytis-cauliflora-70*	*snowballii-Y*	Type Y Snowball	70 Day Summer Cauliflower	Mustard
Brassica	*oleracea-botrytis-cauliflora-70*	*snowballii-Y-improvedii*	Type Y Improved Snowball	70 Day Summer Cauliflower	Mustard
Brassica	*oleracea-botrytis-cauliflora-70*	*snowballii-floweri*	Snowflower Snowball	70 Day Summer Cauliflower	Mustard

Brassica	oleracea-botrytis-cauliflora-70	snowballii-white-rockii	White Rock Snowball	70 Day Summer Cauliflower	Mustard
Brassica	oleracea-botrytis-cauliflora-70	white-topii	WhiteTop	70 Day Summer Cauliflower	Mustard
Brassica	oleracea-botrytis-cauliflora-80	absolutei	Absolute	80 Day Summer Cauliflower	Mustard
Brassica	oleracea-botrytis-cauliflora-80	alaschensis	Alaska	80 Day Summer Cauliflower	Mustard
Brassica	oleracea-botrytis-cauliflora-80	dania	Danian	80 Day Summer Cauliflower	Mustard
Brassica	oleracea-botrytis-cauliflora-80	erfurta	Erfurt	80 Day Summer Cauliflower	Mustard
Brassica	oleracea-botrytis-cauliflora-80	erfurta-erfu	Erfu Erfurt	80 Day Summer Cauliflower	Mustard
Brassica	oleracea-botrytis-cauliflora-80	erfurta-grodana	Grodan Erfurt	80 Day Summer Cauliflower	Mustard
Brassica	oleracea-botrytis-cauliflora-80	harvesteri	Harvester	80 Day Summer Cauliflower	Mustard
Brassica	oleracea-botrytis-cauliflora-80	headmanii	Multiheaded Headman	80 Day Summer Cauliflower	Mustard
Brassica	oleracea-botrytis-cauliflora-80	palla-di-nevei	Palla Di Neve	80 Day Summer Cauliflower	Mustard
Brassica	oleracea-botrytis-cauliflora-80	puakeai	Puakea	80 Day Summer Cauliflower	Mustard
Brassica	oleracea-botrytis-cauliflora-80	snowballii-pybasi-184	Pybas #184 Snowball	80 Day Summer Cauliflower	Mustard
Brassica	oleracea-botrytis-cauliflora-80	starlai	Starla	80 Day Summer Cauliflower	Mustard
Brassica	oleracea-botrytis-cauliflora-80	violetii(purple-broccolii	Violet Purple Broccoli	80 Day Summer Cauliflower	Mustard
Brassica	oleracea-botrytis-cauliflora-90	alverdai	Alverda Green	90 Day Summer Cauliflower	Mustard
Brassica	oleracea-botrytis-cauliflora-90	all-the-year-roundii	All The Year Round	90 Day Summer Cauliflower	Mustard
Brassica	oleracea-botrytis-cauliflora-90	amazingii	Amazing	90 Day Summer Cauliflower	Mustard
Brassica	oleracea-botrytis-cauliflora-90	corvilia	Corvilia	90 Day Summer Cauliflower	Mustard
Brassica	oleracea-botrytis-cauliflora-90	early-pearlii	Early Pearl	90 Day Summer Cauliflower	Mustard
Brassica	oleracea-botrytis-cauliflora-90	garantii	Mini Garant	90 Day Summer Cauliflower	Mustard
Brassica	oleracea-botrytis-cauliflora-90	ice-capii	Ice Cap	90 Day Summer Cauliflower	Mustard
Brassica	oleracea-botrytis-cauliflora-90	latemanii	Lateman	90 Day Summer Cauliflower	Mustard
Brassica	oleracea-botrytis-cauliflora-90	predominans	Mini Predominant	90 Day Summer Cauliflower	Mustard
Brassica	oleracea-botrytis-cauliflora-90	purple-giantii	Purple Giant	90 Day Summer Cauliflower	Mustard
Brassica	oleracea-botrytis-cauliflora-90	purple-headii	Purple Head	90 Day Summer Cauliflower	Mustard
Brassica	oleracea-botrytis-cauliflora-90	self-blanchei	Self Blanche	90 Day Summer Cauliflower	Mustard
Brassica	oleracea-botrytis-cauliflora-90	romanesco-shannonii	Shannon Romanesco	90 Day Summer Cauliflower	Mustard
Brassica	oleracea-botrytis-cauliflora-90	snowballii-090	#90 Snowball	90 Day Summer Cauliflower	Mustard
Brassica	oleracea-botrytis-cauliflora-90	snowballii-snowmanii	Snowman Snowball	90 Day Summer Cauliflower	Mustard
Brassica	oleracea-botrytis-cauliflora-90	snowballii-snowpackii	Snowpak Snowball	90 Day Summer Cauliflower	Mustard
Brassica	oleracea-botrytis-cauliflora-90	veitch's-self-protectingii	Veitch Self Protecting	90 Day Summer Cauliflower	Mustard
Brassica	oleracea-botrytis-cauliflora-90	violetta-italia	Violetta Italia Purple	90 Day Summer Cauliflower	Mustard
Brassica	oleracea-botrytis-cauliflora-110	autumn-giantii	Autumn Giant	110 Day Fall Cauliflower	Mustard
Brassica	oleracea-botrytis-cauliflora-110	igloory	Igloory	110 Day Fall Cauliflower	Mustard
Brassica	oleracea-botrytis-cauliflora-110	limelightii	Limelight Green	110 Day Fall Cauliflower	Mustard
Brassica	oleracea-botrytis-cauliflora-110	winneri	Winner	110 Day Fall Cauliflower	Mustard
Brassica	oleracea-botrytis-cauliflora-110	violet-queenii	Violet Queen Purple	110 Day Fall Cauliflower	Mustard
Brassica	oleracea-botrytis-cauliflora-240	arcadei	Arcade	240 Day Overwinter Cauliflower	Mustard
Brassica	oleracea-botrytis-cauliflora-240	early-purple-siciliana	Early Purple Sicilian	240 Day Overwinter Cauliflower	Mustard
Brassica	oleracea-botrytis-cauliflora-240	fleurlii	Fleurly	240 Day Overwinter Cauliflower	Mustard
Brassica	oleracea-botrytis-cauliflora-240	incai	Inca	240 Day Overwinter Cauliflower	Mustard
Brassica	oleracea-botrytis-cauliflora-240	purple-capei	Purple Cape	240 Day Overwinter Cauliflower	Mustard
Brassica	oleracea-botrytis-cauliflora-240	purple-capei-oasisi	Oasis Purple Cape	240 Day Overwinter Cauliflower	Mustard
Brassica	oleracea-botrytis-cauliflora-240	snow's-winter-whitei	Snow's Winter White	240 Day Overwinter Cauliflower	Mustard
Brassica	oleracea-botrytis-cauliflora-240	walcheren-alsmeeri	Aalsmeer Walcheren	240 Day Overwinter Cauliflower	Mustard
Brassica	oleracea-botrytis-cauliflora-240	walcheren-armado-3	#3 Armado Walcheren	240 Day Overwinter Cauliflower	Mustard
Brassica	oleracea-botrytis-cauliflora-240	walcheren-armado-aprilii	April Armado Walcheren	240 Day Overwinter Cauliflower	Mustard
Brassica	oleracea-botrytis-cauliflora-240	walcheren-armado-mayi	May Armado Walcheren	240 Day Overwinter Cauliflower	Mustard
Brassica	oleracea-botrytis-cauliflora-240	walcheren-armado-tardoi	Tardo Armado Walcheren	240 Day Overwinter Cauliflower	Mustard
Brassica	oleracea-botrytis-cauliflora-240	walcheren-galleonii	Galleon Walcheren	240 Day Overwinter Cauliflower	Mustard
Brassica	oleracea-botrytis-cauliflora-240	walcheren-markantai	Markanta Walcheren	240 Day Overwinter Cauliflower	Mustard
Brassica	oleracea-botrytis-cauliflora-240	walcheren-maystari	Maystar Walcheren	240 Day Overwinter Cauliflower	Mustard
Brassica	oleracea-botrytis-cauliflora-240	walcheren-winter-pilgrimii	Winter Pilgrim Walcheren	240 Day Overwinter Cauliflower	Mustard

Broccoli, Italian (*Brassica oleracea-botrytis-italica*)

Genus	Species	Infraspecies	cultivar	specimen	Genome
Brassica	oleracea-botrytis-italica-60	calabrese-cavolo-broccolo-ramosa-grandei	Cavolo Ramosa Grande Calabrese	60 Day Italian Broccoli	Mustard
Brassica	oleracea-botrytis-italica-60	calabrese-citationii	Citation Calabrese	60 Day Italian Broccoli	Mustard
Brassica	oleracea-botrytis-italica-60	calabrese-cruiseri	Cruiser Calabrese	60 Day Italian Broccoli	Mustard
Brassica	oleracea-botrytis-italica-60	calabrese-el-centroi	El Centro Calabrese	60 Day Italian Broccoli	Mustard
Brassica	oleracea-botrytis-italica-60	calabrese-greenii-sproutingii	Green Sprouting Calabrese	60 Day Italian Broccoli	Mustard
Brassica	oleracea-botrytis-italica-60	calabrese-italian-green-sproutingii	Italian Green Sprouting Calabrese	60 Day Italian Broccoli	Mustard
Brassica	oleracea-botrytis-italica-60	calabrese-lancelatii	Lancelot Calabrese	60 Day Italian Broccoli	Mustard
Brassica	oleracea-botrytis-italica-60	calabrese-mercedesi	Mercedes Calabrese	60 Day Italian Broccoli	Mustard
Brassica	oleracea-botrytis-italica-60	calabrese-walthamii-29	Waltham #29 Green Sprouting Calabrese	60 Day Italian Broccoli	Mustard
Brassica	oleracea-botrytis-italica-60	de-ciccoi	DeCicco	60 Day Italian Broccoli	Mustard
Brassica	oleracea-botrytis-italica-60	de-ciccoi-earlyi	Early DeCicco	60 Day Italian Broccoli	Mustard
Brassica	oleracea-botrytis-italica-70	spgarielloi	Spigariello Non-heading	70 Day Italian Broccoli	Mustard
Brassica	oleracea-botrytis-italica-80	calabrese-emperori	Emperor Calabrese	80 Day Italian Broccoli	Mustard
Brassica	oleracea-botrytis-italica-80	calabrese-green-cometii	Green Comet Calabrese	80 Day Italian Broccoli	Mustard
Brassica	oleracea-botrytis-italica-80	romanesco-gialastroi	Gialastro Romanesco	80 Day Italian Broccoli	Mustard
Brassica	oleracea-botrytis-italica-80	romanesco-improvedii	Improved Romanesco	80 Day Italian Broccoli	Mustard
Brassica	oleracea-botrytis-italica-80	romanesco-minaretii	Lime Green Minaret Romanesco	80 Day Italian Broccoli	Mustard
Brassica	oleracea-botrytis-italica-220	calabrese-shogunii	Blue Green Shogun	220 Day Italian Broccoli	Mustard

Cabbage, Green (*Brassica oleracea-capitata*)

Genus	Species	Infraspecies	cultivar	specimen	Genome
Brassica	oleracea-capitata-50	ballhead-bergkabisi	Bergkabis Ballhead	50 Day Spring Green Cabbage	Mustard

Brassica	oleracea-capitata-50	conehead-capehornensis	Cape Horn Conehead	50 Day Spring Green Cabbage	Mustard
Brassica	oleracea-capitata-50	conehead-hispii	Hispi Conehead	50 Day Spring Green Cabbage	Mustard
Brassica	oleracea-capitata-60	ballhead-derby-dayi	Derby Day Ballhead	60 Day Summer Green Cabbage	Mustard
Brassica	oleracea-capitata-60	ballhead-early-acrei-earliana	Earliana Golden Acre Ballhead	60 Day Summer Green Cabbage	Mustard
Brassica	oleracea-capitata-60	ballhead-emerald-acrei	Emerald Acre Ballhead	60 Day Summer Green Cabbage	Mustard
Brassica	oleracea-capitata-60	ballhead-first-of-junei	First Of June Ballhead	60 Day Summer Green Cabbage	Mustard
Brassica	oleracea-capitata-60	ballhead-primoi	Primo Ballhead	60 Day Summer Green Cabbage	Mustard
Brassica	oleracea-capitata-60	ballhead-primoi-delphii	Delphi Primo Ballhead	60 Day Summer Green Cabbage	Mustard
Brassica	oleracea-capitata-60	conehead-spitfirei	Spitfire	60 Day Summer Green Cabbage	Mustard
Brassica	oleracea-capitata-60	minihead-marner-allfrohii	Marner Allfroh Minihead	60 Day Summer Green Cabbage	Mustard
Brassica	oleracea-capitata-70	ballhead-copenhagen-marketii	Copenhagen Market Ballhead	70 Day Summer Green Cabbage	Mustard
Brassica	oleracea-capitata-70	ballhead-copenhagen-marketii-improvedii	Improved Copenhagen Market Ballhead	70 Day Summer Green Cabbage	Mustard
Brassica	oleracea-capitata-70	ballhead-delicatessei	Delicatesse Ballhead	70 Day Summer Green Cabbage	Mustard
Brassica	oleracea-capitata-70	ballhead-early-marvelii	Early Marvel Ballhead	70 Day Summer Green Cabbage	Mustard
Brassica	oleracea-capitata-70	ballhead-golden-acrei	Golden Acre Ballhead	70 Day Summer Green Cabbage	Mustard
Brassica	oleracea-capitata-70	ballhead-golden-acrei-earlyi	Early Golden Acre Ballhead	70 Day Summer Green Cabbage	Mustard
Brassica	oleracea-capitata-70	ballhead-golden-acrei-elitei	Elite Golden Acre Ballhead	70 Day Summer Green Cabbage	Mustard
Brassica	oleracea-capitata-70	ballhead-golden-acrei-primaxii	Primax Golden Acre Ballhead	70 Day Summer Green Cabbage	Mustard
Brassica	oleracea-capitata-70	ballhead-golden-acrei-wisconsinensis	Wisconsin Golden Acre Ballhead	70 Day Summer Green Cabbage	Mustard
Brassica	oleracea-capitata-70	ballhead-golden-acrei-yellows-resistantii	Resistant Golden Acre Ballhead	70 Day Summer Green Cabbage	Mustard
Brassica	oleracea-capitata-70	conehead-coeur-de-boeuf-moyen-de-la-hollei	Coeur De Boeuf Moyen Conehead	70 Day Summer Green Cabbage	Mustard
Brassica	oleracea-capitata-70	conehead-cuor-di-buei	Cuor Di Bue Conehead	70 Day Summer Green Cabbage	Mustard
Brassica	oleracea-capitata-70	conehead-cuor-di-buei-grossoi	Grosso Cuor Di Bue Conehead	70 Day Summer Green Cabbage	Mustard
Brassica	oleracea-capitata-70	conehead-greyhoundii	Greyhound Conehead	70 Day Summer Green Cabbage	Mustard
Brassica	oleracea-capitata-70	minihead-cabbage-sproutsi	Cabbage Sprouts Minihead	70 Day Summer Green Cabbage	Mustard
Brassica	oleracea-capitata-75	ballhead-columbiensis	Columbia Ballhead	75 Day Summer Green Cabbage	Mustard
Brassica	oleracea-capitata-75	conehead-charleston-wakefieldii	Charleston Wakefield Conehead	75 Day Summer Green Cabbage	Mustard
Brassica	oleracea-capitata-75	conehead-charleston-wakefieldii-henderson'si	Henderson's Charleston Wakefield Conehead	75 Day Summer Green Cabbage	Mustard
Brassica	oleracea-capitata-75	conehead-charleston-wakefieldii-largei	Large Charleston Wakefield Conehead	75 Day Summer Green Cabbage	Mustard
Brassica	oleracea-capitata-75	conehead-jersey-wakefieldii	Jersey Wakefield Conehead	75 Day Summer Green Cabbage	Mustard
Brassica	oleracea-capitata-75	conehead-jersey-wakefieldii-earlyi	Early Jersey Wakefield Conehead	75 Day Summer Green Cabbage	Mustard
Brassica	oleracea-capitata-75	conehead-jersey-wakefieldii-pewai	Pew Jersey Wakefield Conehead	75 Day Summer Green Cabbage	Mustard
Brassica	oleracea-capitata-75	conehead-mountain-wakefieldii-earlyi	Mountain Wakefield Conehead	75 Day Summer Green Cabbage	Mustard
Brassica	oleracea-capitata-80	ballhead-castelloi	Castello Ballhead	80 Day Summer Green Cabbage	Mustard
Brassica	oleracea-capitata-80	ballhead-dutch-earlyi	Early Dutch Ballhead	80 Day Summer Green Cabbage	Mustard
Brassica	oleracea-capitata-80	ballhead-glory-of-enkhuizenii	Glory Of Enkhouse Ballhead	80 Day Summer Green Cabbage	Mustard
Brassica	oleracea-capitata-80	ballhead-marion-marketii	Marion Market Ballhead	80 Day Summer Green Cabbage	Mustard
Brassica	oleracea-capitata-80	ballhead-wisconsin-hollanderi-8	Wisconsin Hollander Ballhead	80 Day Summer Green Cabbage	Mustard
Brassica	oleracea-capitata-80	conehead-hornspii	Hornspi Conehead	80 Day Summer Green Cabbage	Mustard
Brassica	oleracea-capitata-80	drumhead-jumboi	Jumbo Drumhead	80 Day Summer Green Cabbage	Mustard
Brassica	oleracea-capitata-80	erfurter-zwergii	Erfurter Zwerg	80 Day Summer Green Cabbage	Mustard
Brassica	oleracea-capitata-80	flathead-dutch-earlyi	Early Dutch Flathead	80 Day Summer Green Cabbage	Mustard
Brassica	oleracea-capitata-80	storage-marneri	Storage Marner	80 Day Summer Green Cabbage	Mustard
Brassica	oleracea-capitata-80	sugar-loafii	Sugar Loaf	80 Day Summer Green Cabbage	Mustard
Brassica	oleracea-capitata-80	winningstadtensis	Winningstadt	80 Day Summer Green Cabbage	Mustard
Brassica	oleracea-capitata-90	ballhead-head's-up-II-greenii	Heads Up II Green Ballhead	90 Day Summer Green Cabbage	Mustard
Brassica	oleracea-capitata-90	drumhead-brunswickii	Brunswick Drumhead	90 Day Summer Green Cabbage	Mustard
Brassica	oleracea-capitata-90	drumhead-all-seasonsi	All Seasons Drumhead	90 Day Summer Green Cabbage	Mustard
Brassica	oleracea-capitata-90	drumhead-all-seasonsi-wisconsinensis	Wisconsin All Seasons Drumhead	90 Day Summer Green Cabbage	Mustard
Brassica	oleracea-capitata-90	flathead-dutch-earlyi-stein'si	Stein's Early Dutch Flathead	90 Day Summer Green Cabbage	Mustard
Brassica	oleracea-capitata-90	aureacreacrosa	GoldenAcreCross	90 Day Summer Green Cabbage	Mustard
Brassica	oleracea-capitata-90	grouenaexpresa	GreenExpress	90 Day Summer Green Cabbage	Mustard
Brassica	oleracea-capitata-100	ballhead-houston-evergreenii	Houston Evergreen Ballhead	100 Day Summer Green Cabbage	Mustard
Brassica	oleracea-capitata-100	drumhead-quickstepii	Quickstep Drumhead	100 Day Summer Green Cabbage	Mustard
Brassica	oleracea-capitata-100	flathead-dutch-latei	Late Dutch Flathead	100 Day Summer Green Cabbage	Mustard
Brassica	oleracea-capitata-100	flathead-dutch-latei-premium	Premium Late Dutch Flathead	100 Day Summer Green Cabbage	Mustard
Brassica	oleracea-capitata-110	ballhead-april-greenii	April Green Ballhead	110 Day Fall Green Cabbage	Mustard
Brassica	oleracea-capitata-110	ballhead-danica	Danish Ballhead	110 Day Fall Green Cabbage	Mustard
Brassica	oleracea-capitata-110	ballhead-danica-burpee'si	Burpee's Danish Ballhead	110 Day Fall Green Cabbage	Mustard
Brassica	oleracea-capitata-110	ballhead-danica-penn-statei	Penn State Danish Ballhead	110 Day Fall Green Cabbage	Mustard
Brassica	oleracea-capitata-110	ballhead-danica-shortstemii	Short Stem Danish Ballhead	110 Day Fall Green Cabbage	Mustard
Brassica	oleracea-capitata-110	drumhead-christmasi	Christmas Drumhead	110 Day Fall Green Cabbage	Mustard
Brassica	oleracea-capitata-110	drumhead-northern-giantii	McFayden Northern Giant Drumhead	110 Day Fall Green Cabbage	Mustard
Brassica	oleracea-capitata-110	drumhead-quintal-alsacensis	Quintal Alsace Drumhead	110 Day Fall Green Cabbage	Mustard
Brassica	oleracea-capitata-125	flathead-amager-green-storagei	Storage Amager Green Flathead	125 Day Winter Cabbage	Mustard
Brassica	oleracea-capitata-125	conehead-offenhamii	Offenham Green Conehead	125 Day Winter Cabbage	Mustard
Brassica	oleracea-capitata-125	conehead-offenhamii-hardyi	Hardy Offenham Green Conehead	125 Day Winter Cabbage	Mustard
Brassica	oleracea-capitata-125	conehead-offenheimii-3	#3 Offenheim Green Conehead	125 Day Winter Cabbage	Mustard
Brassica	oleracea-capitata-180	ballhead-january-kingii	January King Ballhead	180 Day Winter Cabbage	Mustard
Brassica	oleracea-capitata-180	ballhead-january-kingii-cometii	Comet January King Ballhead	180 Day Winter Cabbage	Mustard
Brassica	oleracea-capitata-180	ballhead-january-kingii-hardy-late-stockii-3	Hardy Late Stock #3 January King Ballhead	180 Day Winter Cabbage	Mustard
Brassica	oleracea-capitata-180	ballhead-january-kingii-wintertonii	Winterton January King Ballhead	180 Day Winter Cabbage	Mustard
Brassica	oleracea-capitata-220	ballhead-spring-heroi	Spring Hero Ballhead	220 Day Overwinter Cabbage	Mustard
Brassica	oleracea-capitata-220	conehead-aprilii	April Conehead	220 Day Overwinter Cabbage	Mustard
Brassica	oleracea-capitata-220	conehead-aprillii-pixiei	Pixie April Conehead	220 Day Overwinter Cabbage	Mustard
Brassica	oleracea-capitata-220	loosehead-first-early-marketii	First Early Market Loosehead	220 Day Overwinter Cabbage	Mustard
Brassica	oleracea-capitata-220	loosehead-first-early-marketii-218	#218 First Early Market Loosehead	220 Day Overwinter Cabbage	Mustard
Brassica	oleracea-capitata-220	loosehead-durham-earlyi	Durham Early Loosehead	220 Day Overwinter Cabbage	Mustard
Brassica	oleracea-capitata-220	minihead-mayatt's-offenham-compacta	Myatt's Offenham Compact Minihead	220 Day Overwinter Cabbage	Mustard

Cabbage, Purple (*Brassica oleracea-capitata-purpurea*)

Genus	Species	Infraspecies	cultivar	specimen	Genome
Brassica	oleracea-capitata-purpurea	loosehead-late-flat-polei	Late Flat Poll Loosehead	Purple Cabbage	Mustard
Brassica	oleracea-capitata-purpurea	ballhead-kissendrupii	Kissendrup Ballhead	Purple Cabbage	Mustard
Brassica	oleracea-capitata-purpurea	ballhead-meteori-purplei	Purple Meteor Ballhead	Purple Cabbage	Mustard

Cabbage, Red (*Brassica oleracea-capitata-rubra*)

Genus	Species	Infraspecies	cultivar	specimen	Genome
Brassica	oleracea-capitata-rubra	ballhead-acre-redii	Red Acre Ballhead	Red Cabbage	Mustard
Brassica	oleracea-capitata-rubra	ballhead-acre-redii-earlyi	Early Red Acre Ballhead	Red Cabbage	Mustard
Brassica	oleracea-capitata-rubra	ballhead-blackii	Black Ballhead	Red Cabbage	Mustard
Brassica	oleracea-capitata-rubra	ballhead-cabeza-negrai	Cabeza Negra Ballhead	Red Cabbage	Mustard
Brassica	oleracea-capitata-rubra	ballhead-danica-redii	Red Danish Ballhead	Red Cabbage	Mustard
Brassica	oleracea-capitata-rubra	ballhead-landedijkeri	Langedijker Ballhead	Red Cabbage	Mustard
Brassica	oleracea-capitata-rubra	ballhead-landedijkeri-summer-debutii	Debut Summer Langedijker Ballhead	Red Cabbage	Mustard
Brassica	oleracea-capitata-rubra	ballhead-landedijkeri-summer-lassoi	Lasso Summer Langedijker Ballhead	Red Cabbage	Mustard
Brassica	oleracea-capitata-rubra	ballhead-meteori-redii	Red Meteor Ballhead	Red Cabbage	Mustard
Brassica	oleracea-capitata-rubra	ballhead-mohrenkopfii	Mohrenkopf Ballhead	Red Cabbage	Mustard
Brassica	oleracea-capitata-rubra	ballhead-redii-early	Early Red Ballhead	Red Cabbage	Mustard
Brassica	oleracea-capitata-rubra	ballhead-redii-rockii	Red Rock Ballhead	Red Cabbage	Mustard
Brassica	oleracea-capitata-rubra	ballhead-redii-rodanii	Red Rodan Ballhead	Red Cabbage	Mustard
Brassica	oleracea-capitata-rubra	ballhead-rodyndai	Rodynda Ballhead	Red Cabbage	Mustard
Brassica	oleracea-capitata-rubra	ballhead-soroi	Soro Ballhead	Red Cabbage	Mustard
Brassica	oleracea-capitata-rubra	ballhead-vestai	Vesta Ballhead	Red Cabbage	Mustard
Brassica	oleracea-capitata-rubra	drumhead-danica-mammoth-red-rockii	Mammoth Red Rock Danish Drumhead	Red Cabbage	Mustard
Brassica	oleracea-capitata-rubra	drumhead-heads-up-redii-2	Heads Up II Drumhead	Red Cabbage	Mustard
Brassica	oleracea-capitata-rubra	drumhead-redii	Red Drumhead	Red Cabbage	Mustard
Brassica	oleracea-capitata-rubra	loosehead-pozky-koppus	Pozky Koppus Loosehead	Red Cabbage	Mustard
Brassica	oleracea-capitata-rubra	loosehead-red-verona-savoyi	Red Verona Savoy Loosehead	Red Cabbage	Mustard
Brassica	oleracea-capitata-rubra	loosehead-rougettei	Rougette Loosehead	Red Cabbage	Mustard
Brassica	oleracea-capitata-rubra	ovalhead-langedijkeri-winter-keeper-redii	Winter Keeper Langedijker Ovalhead	Red Cabbage	Mustard
Brassica	oleracea-capitata-rubra	ovalhead-red-expressi	Red Express Ovalhead	Red Cabbage	Mustard

Cabbage, Savoy (*Brassica oleracea-capitata-sabanda*)

Genus	Species	Infraspecies	cultivar	specimen	Genome
Brassica	oleracea-capitata-sabanda	ballhead-amager-tall-gramii	Amager Tall Grami Ballhead	Savoy Cabbage	Mustard
Brassica	oleracea-capitata-sabanda	ballhead-amager-tall-wintai	Amager Tall Winta Ballhead	Savoy Cabbage	Mustard
Brassica	oleracea-capitata-sabanda	ballhead-best-of-allii	Best Of All Ballhead	Savoy Cabbage	Mustard
Brassica	oleracea-capitata-sabanda	ballhead-brunswick-unirai	Unira Brunswick Ballhead	Savoy Cabbage	Mustard
Brassica	oleracea-capitata-sabanda	ballhead-des-vertus	Des Vertus Ballhead	Savoy Cabbage	Mustard
Brassica	oleracea-capitata-sabanda	ballhead-ditmarsch-eqa-409	Eqa Ditmarsch Ballhead	Savoy Cabbage	Mustard
Brassica	oleracea-capitata-sabanda	ballhead-ditmarsch-midii-84	Midi Ditmarsch Ballhead	Savoy Cabbage	Mustard
Brassica	oleracea-capitata-sabanda	ballhead-ditmarsch-specialis-212	Special Ditmarsch Ballhead	Savoy Cabbage	Mustard
Brassica	oleracea-capitata-sabanda	ballhead-holland-whitei	Holland White Ballhead	Savoy Cabbage	Mustard
Brassica	oleracea-capitata-sabanda	ballhead-julius	Julius Ballhead	Savoy Cabbage	Mustard
Brassica	oleracea-capitata-sabanda	ballhead-langedijkerdauer-duralii	Dauer Dural Langedijker Ballhead	Savoy Cabbage	Mustard
Brassica	oleracea-capitata-sabanda	ballhead-langedijker-earlyi	Early Langedijker Ballhead	Savoy Cabbage	Mustard
Brassica	oleracea-capitata-sabanda	ballhead-langedijker-WK-savoyi	Winter Keeper Langedijker Ballhead	Savoy Cabbage	Mustard
Brassica	oleracea-capitata-sabanda	ballhead-lisboai	Lisboa Ballhead	Savoy Cabbage	Mustard
Brassica	oleracea-capitata-sabanda	ballhead-midwayi	Midway Ballhead	Savoy Cabbage	Mustard
Brassica	oleracea-capitata-sabanda	ballhead-ormskirk-earlyi	Early Ormskirk Ballhead	Savoy Cabbage	Mustard
Brassica	oleracea-capitata-sabanda	ballhead-ormskirk-latei	Late Ormskirk Ballhead	Savoy Cabbage	Mustard
Brassica	oleracea-capitata-sabanda	ballhead-perfectionii	Perfection Ballhead	Savoy Cabbage	Mustard
Brassica	oleracea-capitata-sabanda	ballhead-rhum-von-enkhuizenii	Ruhm Von Enkhuisen Ballhead	Savoy Cabbage	Mustard
Brassica	oleracea-capitata-sabanda	ballhead-sagai	Saga Ballhead	Savoy Cabbage	Mustard
Brassica	oleracea-capitata-sabanda	ballhead-santanai	Santana	Savoy Cabbage	Mustard
Brassica	oleracea-capitata-sabanda	ballhead-testa-di-ferroi	Testa Di Ferro Ballhead	Savoy Cabbage	Mustard
Brassica	oleracea-capitata-sabanda	ballhead-tundrai	Tundra Ballhead	Savoy Cabbage	Mustard
Brassica	oleracea-capitata-sabanda	ballhead-verza-di-veronai	Verza Di Verona Ballhead	Savoy Cabbage	Mustard
Brassica	oleracea-capitata-sabanda	ballhead-verza-montovanoi	Verza Montovano Ballhead	Savoy Cabbage	Mustard
Brassica	oleracea-capitata-sabanda	ballhead-winter-kingii	Winter King Ballhead	Savoy Cabbage	Mustard
Brassica	oleracea-capitata-sabanda	ballhead-winterfurstii	Winterfurst Ballhead	Savoy Cabbage	Mustard
Brassica	oleracea-capitata-sabanda	ballhead-wivoyi	Wivoy Ballhead	Savoy Cabbage	Mustard
Brassica	oleracea-capitata-sabanda	conehead-erstling-lingai	Erstling Linga Conehead	Savoy Cabbage	Mustard
Brassica	oleracea-capitata-sabanda	drumhead-savoyi	Drumhead	Savoy Cabbage	Mustard
Brassica	oleracea-capitata-sabanda	drumhead-aubervilliersi	Aubervilliers Drumhead	Savoy Cabbage	Mustard
Brassica	oleracea-capitata-sabanda	drumhead-blue-maxii	Blue Max Drumhead	Savoy Cabbage	Mustard
Brassica	oleracea-capitata-sabanda	drumhead-chieftain-savoyi	Chieftan Drumhead	Savoy Cabbage	Mustard
Brassica	oleracea-capitata-sabanda	drumhead-perfection-savoyi	Perfection Drumhead	Savoy Cabbage	Mustard
Brassica	oleracea-capitata-sabanda	drumhead-vertus	Vertus Drumhead	Savoy Cabbage	Mustard
Brassica	oleracea-capitata-sabanda	loosehead-wintergreenii	Wintergreen Loosehead	Savoy Cabbage	Mustard
Brassica	oleracea-capitata-sabanda	loosehead-wintessai	Wintessa Loosehead	Savoy Cabbage	Mustard
Brassica	oleracea-capitata-sabanda	minihead-elams-earlyi	Elams Early Minihead	Savoy Cabbage	Mustard

Kale, Sea Kale Cabbage (*Brassica oleracea-tronchuda*)

Brassica	oleracea-tronchuda	couvei	Couve	Sea Kale Cabbage	Mustard
Brassica	oleracea-tronchuda	gloria-de-portugalensis	Gloria De Portugal	Sea Kale Cabbage	Mustard
Brassica	oleracea-tronchuda	manteiga	Manteiga	Sea Kale Cabbage	Mustard

Brassica	oleracea-tronchuda	pencuda-da-povoa	Da Povoa Pencuda	Sea Kale Cabbage	Mustard
Brassica	oleracea-tronchuda	pencuda-espanhola	Spanish Pencuda	Sea Kale Cabbage	Mustard
Brassica	oleracea-tronchuda	portuguesa	Portuguese	Sea Kale Cabbage	Mustard
Brassica	oleracea-tronchuda	valhascosa	Valhascos	Sea Kale Cabbage	Mustard

Brussels Sprout (*Brassica oleracea-gemmifera*)

Genus	Species	Infraspecies	cultivar	specimen	Genome
Brassica	*oleracea-gemmifera-85*	*bedford-fillbasketii*	Bedford Fillbasket	85 Day Brussels Sprout	Mustard
Brassica	*oleracea-gemmifera-85*	*bedfordshire-darkmari-21*	Darkmar Bedfordshire	85 Day Brussels Sprout	Mustard
Brassica	*oleracea-gemmifera-85*	*cambridgensis-5*	Cambridge #5	85 Day Brussels Sprout	Mustard
Brassica	*oleracea-gemmifera-85*	*catskillensis*	Catskill	85 Day Brussels Sprout	Mustard
Brassica	*oleracea-gemmifera-85*	*catskillensis-improvedii*	CatskillImproved	85 Day Brussels Sprout	Mustard
Brassica	*oleracea-gemmifera-85*	*dark-beautyi*	Dark Beauty	85 Day Brussels Sprout	Mustard
Brassica	*oleracea-gemmifera-85*	*early-half-tallii*	Early Half Tall	85 Day Brussels Sprout	Mustard
Brassica	*oleracea-gemmifera-85*	*evesham-specialis*	Special Evesham	85 Day Brussels Sprout	Mustard
Brassica	*oleracea-gemmifera-85*	*mallardii*	Mallard	85 Day Brussels Sprout	Mustard
Brassica	*oleracea-gemmifera-85*	*montgomerii*	Montgomery	85 Day Brussels Sprout	Mustard
Brassica	*oleracea-gemmifera-85*	*noisettei*	Noisette	85 Day Brussels Sprout	Mustard
Brassica	*oleracea-gemmifera-85*	*oliveri*	Oliver	85 Day Brussels Sprout	Mustard
Brassica	*oleracea-gemmifera-85*	*rubine-redii*	Rubine Red	85 Day Brussels Sprout	Mustard
Brassica	*oleracea-gemmifera-110*	*danica-early-dwarfii*	Early Dwarf Danish	110 Day Brussels Sprout	Mustard
Brassica	*oleracea-gemmifera-110*	*groningeri*	Groninger	110 Day Brussels Sprout	Mustard
Brassica	*oleracea-gemmifera-110*	*long-islandii*	Long Island	110 Day Brussels Sprout	Mustard
Brassica	*oleracea-gemmifera-110*	*long-islandii-improvedii*	Improved Long Ilsand	110 Day Brussels Sprout	Mustard
Brassica	*oleracea-gemmifera-110*	*mezzo-nanoi*	Mezzo Nano	110 Day Brussels Sprout	Mustard
Brassica	*oleracea-gemmifera-110*	*paris-marketii*	Paris Market	110 Day Brussels Sprout	Mustard
Brassica	*oleracea-gemmifera-110*	*seven-hillsi*	Seven Hills	110 Day Brussels Sprout	Mustard
Brassica	*oleracea-gemmifera-220*	*falstaffii*	Falstaff	220 Day Brussels Sprout	Mustard

Kohlrabi (*Brassica oleracea-gongylodes*)

Genus	Species	Infraspecies	cultivar	specimen	Genome
Brassica	*oleracea-gongylodes-50*	*azure-stari*	Azure Star Purple	50 Day Kohlrabi	Mustard
Brassica	*oleracea-gongylodes-50*	*blaroi*	Blaro Purple	50 Day Kohlrabi	Mustard
Brassica	*oleracea-gongylodes-50*	*vienna-early-purplei*	Early Vienna Purple	50 Day Kohlrabi	Mustard
Brassica	*oleracea-gongylodes-50*	*vienna-early-whitei*	Early Vienna White	50 Day Kohlrabi	Mustard
Brassica	*oleracea-gongylodes-50*	*granlibakkenii*	Granlibakken	50 Day Kohlrabi	Mustard
Brassica	*oleracea-gongylodes-50*	*logoi*	Logo White	50 Day Kohlrabi	Mustard
Brassica	*oleracea-gongylodes-50*	*rapidii*	Rapid Purple Skin White	50 Day Kohlrabi	Mustard
Brassica	*oleracea-gongylodes-60*	*delicatesse-bluei*	Delicatesse Blue	60 Day Kohlrabi	Mustard
Brassica	*oleracea-gongylodes-60*	*delicatesse-whiteii*	Delicatesse White	60 Day Kohlrabi	Mustard
Brassica	*oleracea-gongylodes-60*	*delicacy-early-whitei*	Early Delicacy White	60 Day Kohlrabi	Mustard
Brassica	*oleracea-gongylodes-60*	*delicacy-purplei*	Delicacy Purple	60 Day Kohlrabi	Mustard
Brassica	*oleracea-gongylodes-60*	*pekingensis*	Peking White	60 Day Kohlrabi	Mustard
Brassica	*oleracea-gongylodes-60*	*super-schmelzii*	Super Schmelz White	60 Day Kohlrabi	Mustard
Brassica	*oleracea-gongylodes-65*	*blustai*	Blusta Purple Blue	65 Day Kohlrabi	Mustard
Brassica	*oleracea-gongylodes-65*	*green-globei*	Globe Green	65 Day Kohlrabi	Mustard
Brassica	*oleracea-gongylodes-65*	*purple-speckii*	Purple Speck Purple Blue	65 Day Kohlrabi	Mustard
Brassica	*oleracea-gongylodes-65*	*rolanoi*	Rolano	65 Day Kohlrabi	Mustard
Brassica	*oleracea-gongylodes-130*	*gigantei*	Giant	130 Day Kohlrabi	Mustard
Brassica	*perviridis*	*perviridis*	Original	Komatsuna Spinach	Mustard
Brassica	*perviridis*	*big-topii*	Big Top	Komatsuna Spinach	Mustard
Brassica	*perviridis*	*tendergreenii*	Tendergreen	Komatsuna Spinach	Mustard
Brassica	*perviridis*	*torasanii*	Torasan	Komatsuna Spinach	Mustard

Rape Greens, Chinese (*Brassica rapa-chinensis*)

Genus	Species	Infraspecies	cultivar	specimen	Genome
Brassica	*rapa-campestris*	*campestris*	Original	Field Rape	Mustard
Brassica	*rapa-campestris*	*tarahumara-mostaza*	Tarahumara Mostaza	Field Rape	Mustard
Brassica	*rapa-campestris*	*mocosalii*	Mocosali	Field Rape	Mustard
Brassica	*rapa-chinensis*	*chinensis*	Original	Chinese Rape Greens	Mustard
Brassica	*rapa-chinensis*	*chinese-broad-leafii*	Elephant Ear Broad Leaf	Chinese Rape Greens	Mustard
Brassica	*rapa-chinensis*	*chinese-flat-cabbagei*	Flat Cabbage	Chinese Rape Greens	Mustard
Brassica	*rapa-chinensis*	*chinese-smooth-leafii*	Smooth Leaf	Chinese Rape Greens	Mustard
Brassica	*rapa-chinensis*	*grelas-de-santiagoi*	Grelas De Santiago	Chinese Rape Greens	Mustard
Brassica	*rapa-chinensis*	*green-vegetablei*	Green Vegetable	Chinese Rape Greens	Mustard
Brassica	*rapa-chinensis*	*kate's-mustard-lettucei*	Kate's Mustard Lettuce	Chinese Rape Greens	Mustard
Brassica	*rapa-chinensis*	*molokheiyai*	Molokheiya	Chinese Rape Greens	Mustard
Brassica	*rapa-chinensis*	*peacock-tailii*	Peacock Tail	Chinese Rape Greens	Mustard
Brassica	*rapa-chinensis*	*radish-rooted-turnipii*	Radish Rooted Turnip	Chinese Rape Greens	Mustard
Brassica	*rapa-chinensis*	*santungii-rotundaflora*	Round Leaf Santung	Chinese Rape Greens	Mustard
Brassica	*rapa-chinensis*	*seoul-cabbagei*	Seoul Cabbage	Chinese Rape Greens	Mustard
Brassica	*rapa-chinensis*	*small-cabbage-round-leavedii*	Round Leaf Small Cabbage	Chinese Rape Greens	Mustard
Brassica	*rapa-chinensis*	*tyfonii*	Tyfon	Chinese Rape Greens	Mustard
Brassica	*rapa-chinensis*	*tyfonii-holland-greensi*	Holland Tyfon	Chinese Rape Greens	Mustard

Rape Greens, Japanese (*Brassica rapa-japonica*)

Genus	Species	Infraspecies	cultivar	specimen	Genome
Brassica	*rapa-japonica*	*beka-marui*	Beka Maru	Japanese Rape Greens	Mustard
Brassica	*rapa-japonica*	*hiroshima-greenii*	Pickling Hiroshima	Japanese Rape Greens	Mustard

Brassica	*rapa-japonica*	*hiroshimana*	Chard Hiroshima	Japanese Rape Greens	Mustard
Brassica	*rapa-japonica*	*kyonai*	Kyona	Japanese Rape Greens	Mustard
Brassica	*rapa-japonica*	*mibuensis*	Mibuna	Japanese Rape Greens	Mustard
Brassica	*rapa-japonica*	*miikei-giantii*	Miike Giant	Japanese Rape Greens	Mustard
Brassica	*rapa-japonica*	*miikei-long-shootii*	Miike Long Shoot	Japanese Rape Greens	Mustard
Brassica	*rapa-japonica*	*miikei-purplei*	Miike Purple	Japanese Rape Greens	Mustard
Brassica	*rapa-japonica*	*pot-herb-mustardii*	Pot Herb	Japanese Rape Greens	Mustard
Brassica	*rapa-japonica*	*red-giantii*	Red Giant	Japanese Rape Greens	Mustard
Brassica	*rapa-japonica*	*santo-marubai*	Maruba Santo	Japanese Rape Greens	Mustard
Brassica	*rapa-japonica*	*santo-minatoi*	Minato Santo	Japanese Rape Greens	Mustard
Brassica	*rapa-japonica*	*shironai*	Shirona	Japanese Rape Greens	Mustard
Brassica	*rapa-japonica*	*tokyo-beaui*	Tokyo Beau	Japanese Rape Greens	Mustard
Brassica	*rapa-japonica*	*tokyo-bekanai)*	Tokyo Bekana	Japanese Rape Greens	Mustard
Brassica	*rapa-japonica*	*vitaminnai*	Vitaminna	Japanese Rape Greens	Mustard
Brassica	*rapa-japonica*	*yukina-savoyi*	Yukina Savoy	Japanese Rape Greens	Mustard
Brassica	*rapa-japonica-laciniata*	*kyona-mizunai*	Kyona	Mizuna Greens	Mustard
Brassica	*rapa-japonica-laciniata*	*kyoto-mizunai*	Kyoto	Mizuna Greens	Mustard
Brassica	*rapa-japonica-laciniata*	*mizspoona-mixta*	Mix Mizspoona	Mizuna Greens	Mustard
Brassica	*rapa-japonica-laciniata*	*mizspoona-salad-selectii*	Salad Select Mizspoona	Mizuna Greens	Mustard
Brassica	*rapa-japonica-laciniata*	*pink-petiole-mix-mizunai*	Pink Petiole Mix	Mizuna Greens	Mustard
Brassica	*rapa-japonica-laciniata*	*purpurea*	Purple	Mizuna Greens	Mustard
Brassica	*rapa-japonica-laciniata*	*sakatai*	Sakata	Mizuna Greens	Mustard
Brassica	*rapa-oleifera*	*amplexicaulis*	Clasping	Oil Seed Rape	Mustard
Brassica	*rapa-oleifera*	*campestris*	Field	Oil Seed Rape	Mustard
Brassica	*rapa-oleifera*	*dichotoma*	India	Oil Seed Rape	Mustard
Brassica	*rapa-oleifera*	*sarson*	Brown Sarson	Oil Seed Rape	Mustard
Brassica	*rapa-oleifera*	*silvestris*	Colza	Oil Seed Rape	Mustard
Brassica	*rapa-oleifera*	*trilocularis*	Yellow Sarson	Oil Seed Rape	Mustard

Cabbage, Chinese Loose Leaf (*Brassica rapa-parapekinensis*)

Genus	Species	Infraspecies	cultivar	specimen	Genome
Brassica	*rapa- parapekinensis*	*parapekinensis*	Original	Loose Leaf Chinese Cabbage	Mustard
Brassica	*rapa-parapekinensis*	*korean-spring-&-fallii*	Korean Spring & Fall	Loose Leaf Chinese Cabbage	Mustard
Brassica	*rapa-parapekinensis*	*lettucy-typei*	Lettucy Type	Loose Leaf Chinese Cabbage	Mustard
Brassica	*rapa-parapekinensis*	*nagodai*	Nagoda	Loose Leaf Chinese Cabbage	Mustard
Brassica	*rapa-parapekinensis*	*santoi*	Santo	Loose Leaf Chinese Cabbage	Mustard
Brassica	*rapa-parapekinensis*	*santoi-rotundaflora*	Round Leaf Santo	Loose Leaf Chinese Cabbage	Mustard
Brassica	*rapa-parapekinensis*	*shantungii*	Shantung	Loose Leaf Chinese Cabbage	Mustard
Brassica	*rapa-parapekinensis*	*shantungii-4S*	Shantung 4S	Loose Leaf Chinese Cabbage	Mustard

Cabbage, Chinese Head (*Brassica rapa-pekinensis*)

Genus	Species	Infraspecies	cultivar	specimen	Genome
Brassica	*rapa-pekinensis*	*pekinensis*	Original	Head Chinese Cabbage	Mustard
Brassica	*rapa-pekinensis*	*aichi-haksaii*	Aichi Haksai	Head Chinese Cabbage	Mustard
Brassica	*rapa-pekinensis*	*amplexicaulis*	Won Bok	Head Chinese Cabbage	Mustard
Brassica	*rapa-pekinensis*	*celery-chinese-cabbagei*	Celery	Head Chinese Cabbage	Mustard
Brassica	*rapa-pekinensis*	*challengeri*	Challenger	Head Chinese Cabbage	Mustard
Brassica	*rapa-pekinensis*	*che-fooi*	Che Foo	Head Chinese Cabbage	Mustard
Brassica	*rapa-pekinensis*	*chihilii*	Chihili	Head Chinese Cabbage	Mustard
Brassica	*rapa-pekinensis*	*chihilii-longa*	Long Chihili	Head Chinese Cabbage	Mustard
Brassica	*rapa-pekinensis*	*chinese-stem-mustardii*	Chinese Stem	Head Chinese Cabbage	Mustard
Brassica	*rapa-pekinensis*	*choi-prizei*	Prize Choy	Head Chinese Cabbage	Mustard
Brassica	*rapa-pekinensis*	*choi-yu*	Yu Choy	Head Chinese Cabbage	Mustard
Brassica	*rapa-pekinensis*	*choi-yu-50*	50 Day Yu Choy	Head Chinese Cabbage	Mustard
Brassica	*rapa-pekinensis*	*choi-yu-60*	60 Day Yu Choy	Head Chinese Cabbage	Mustard
Brassica	*rapa-pekinensis*	*choi-yu-80*	80 Day Yu Choy	Head Chinese Cabbage	Mustard
Brassica	*rapa-pekinensis*	*eskimoi*	Eskimo	Head Chinese Cabbage	Mustard
Brassica	*rapa-pekinensis*	*houshui*	Houshu	Head Chinese Cabbage	Mustard
Brassica	*rapa-pekinensis*	*jade-pagodai*	Jade Pagoda	Head Chinese Cabbage	Mustard
Brassica	*rapa-pekinensis*	*jade-pagodai-earlyi*	Early Jade Pagoda	Head Chinese Cabbage	Mustard
Brassica	*rapa-pekinensis*	*kasumii*	Kasumi	Head Chinese Cabbage	Mustard
Brassica	*rapa-pekinensis*	*kyotoi-3*	Kyoto #3	Head Chinese Cabbage	Mustard
Brassica	*rapa-pekinensis*	*market-pridei*	Market Pride	Head Chinese Cabbage	Mustard
Brassica	*rapa-pekinensis*	*matsushimai*	Matsushima	Head Chinese Cabbage	Mustard
Brassica	*rapa-pekinensis*	*matsushimai-2*	Matsushima #2	Head Chinese Cabbage	Mustard
Brassica	*rapa-pekinensis*	*michihlii*	Michihli	Head Chinese Cabbage	Mustard
Brassica	*rapa-pekinensis*	*napai*	Napa	Head Chinese Cabbage	Mustard
Brassica	*rapa-pekinensis*	*napai-1*	Napa #1	Head Chinese Cabbage	Mustard
Brassica	*rapa-pekinensis*	*nozaki-babyi*	Nozaki Baby	Head Chinese Cabbage	Mustard
Brassica	*rapa-pekinensis*	*nozaki-earlyi*	Nozaki Baby	Head Chinese Cabbage	Mustard
Brassica	*rapa-pekinensis*	*orient-expressi*	Orient Express	Head Chinese Cabbage	Mustard
Brassica	*rapa-pekinensis*	*pe-tsai*	Pe Tsai	Head Chinese Cabbage	Mustard
Brassica	*rapa-pekinensis*	*wong-bokii*	Wong Bok	Head Chinese Cabbage	Mustard

Turnip (*Brassica rapa-rapifera*)

Genus	Species	Infraspecies	cultivar	specimen	Genome
Brassica	*rapa-rapifera-alba-30*	*all-seasonsi*	All Seasons	30 Day White Turnip	Mustard
Brassica	*rapa-rapifera-alba-30*	*tokyo-marketii*	Tokyo Market	30 Day White Turnip	Mustard
Brassica	*rapa-rapifera-alba-35*	*milan-purple-topii*	Purple Top Milan	35 Day White Turnip	Mustard

Brassica	rapa-rapifera-alba-35	milan-red-topii	Red Top Milan	35 Day White Turnip	Mustard
Brassica	rapa-rapifera-alba-35	milan-rouge-topii	Rouge Top Milan	35 Day White Turnip	Mustard
Brassica	rapa-rapifera-alba-35	milan-white-topii	White Top Milan	35 Day White Turnip	Mustard
Brassica	rapa-rapifera-alba-35	ivoryi	Ivory	35 Day White Turnip	Mustard
Brassica	rapa-rapifera-alba-40	snowballii	Snowball	40 Day White Turnip	Mustard
Brassica	rapa-rapifera-alba-40	snowballii-earlyi	Early Snowball	40 Day White Turnip	Mustard
Brassica	rapa-rapifera-alba-40	hinona-kabu-picklingii	Hinona Kabu Pickling	40 Day White Turnip	Mustard
Brassica	rapa-rapifera-alba-40	yori-springii	Yori Spring	40 Day White Turnip	Mustard
Brassica	rapa-rapifera-alba-50	croissyi	Croissy	50 Day White Turnip	Mustard
Brassica	rapa-rapifera-alba-50	de-berlinensis	De Berlin	50 Day White Turnip	Mustard
Brassica	rapa-rapifera-alba-50	italian-early-white-red-topii	Early Red Top Italian	50 Day White Turnip	Mustard
Brassica	rapa-rapifera-alba-50	ochsenhorneri	Purple Top Oxhorn	50 Day White Turnip	Mustard
Brassica	rapa-rapifera-alba-50	scarlet-ballii	Scarlet Ball	50 Day White Turnip	Mustard
Brassica	rapa-rapifera-alba-50	white-globei	White Globe	50 Day White Turnip	Mustard
Brassica	rapa-rapifera-alba-55	des-vertus-marteaui	Des Vertus Marteau	55 Day White Turnip	Mustard
Brassica	rapa-rapifera-alba-55	des-vertus-marteaui-navetii	Navet Des Vertus Marteau	55 Day White Turnip	Mustard
Brassica	rapa-rapifera-alba-55	ohno-red-roundii	Red Round Ohno	55 Day White Turnip	Mustard
Brassica	rapa-rapifera-alba-55	ohno-scarletii	Scarlet Ohno	55 Day White Turnip	Mustard
Brassica	rapa-rapifera-alba-55	ohno-scarletii-revivalii	Revival Scarlet Ohno	55 Day White Turnip	Mustard
Brassica	rapa-rapifera-alba-55	purple-top-globei	Globe Purple Top	55 Day White Turnip	Mustard
Brassica	rapa-rapifera-alba-55	purple-top-strap-leafii	Strap Leaf Purple Top	55 Day White Turnip	Mustard
Brassica	rapa-rapifera-alba-55	purple-top-white-globei	White Globe Purple Top	55 Day White Turnip	Mustard
Brassica	rapa-rapifera-alba-55	waldoboro-green-neckii	Green Neck Waldoboro	55 Day White Turnip	Mustard
Brassica	rapa-rapifera-alba-55	white-eggii	White Egg	55 Day White Turnip	Mustard
Brassica	rapa-rapifera-alba-65	red-globei	Red Globe	65 Day White Turnip	Mustard
Brassica	rapa-rapifera-alba-65	red-globei-veitch'si	Veitch's Red Globe	65 Day White Turnip	Mustard
Brassica	rapa-rapifera-alba-70	shogoinii-roundii	Shogoin Round	70 Day White Turnip	Mustard
Brassica	rapa-rapifera-alba-80	gilfeatheri	Gilfeather	80 Day White Turnip	Mustard
Brassica	rapa-rapifera-alba-80	vermontensis	Vermont	80 Day White Turnip	Mustard
Brassica	rapa-rapifera-helva-40	petrowskii	Petrowski	40 Day Yellow Turnip	Mustard
Brassica	rapa-rapifera-helva-70	amber-globei	Amber Globe	70 Day Yellow Turnip	Mustard
Brassica	rapa-rapifera-helva-70	golden-ballii	Golden Ball	70 Day Yellow Turnip	Mustard
Brassica	rapa-rapifera-helva-70	golden-ballii-orange-jellyi	Orange Jelly Golden Ball	70 Day Yellow Turnip	Mustard
Brassica	rapa-rapifera-helva-70	yellow-globei	Yellow Globe	70 Day Yellow Turnip	Mustard
Brassica	rapa-rapifera-helva-70	yellow-globei-largei	Large Yellow Globe	70 Day Yellow Turnip	Mustard

Turnip Greens (*Brassica rapa-rapifera-foliosa*)

Genus	Species	Infraspecies	cultivar	specimen	Genome
Brassica	rapa-rapifera-foliosa-30	shogoinii	Shogoin	30 Day Turnip Greens	Mustard
Brassica	rapa-rapifera-foliosa-40	yori-springii	Yori Spring	40 Day Turnip Greens	Mustard
Brassica	rapa-rapifera-foliosa-50	crawfordii	Crawford	50 Day Turnip Greens	Mustard
Brassica	rapa-rapifera-foliosa-50	septiceps	Seven Top	50 Day Turnip Greens	Mustard
Brassica	rapa-rapifera-foliosa-50	turnip-top-greensi	Turnip Top	50 Day Turnip Greens	Mustard
Brassica	rapa-rapifera-foliosa-50	turnip-top-greensi-nozawanai	Nozawana Turnip Top	50 Day Turnip Greens	Mustard
Brassica	rapa-rapifera-foliosa-60	fortress	Fortess	60 Day Turnip Greens	Mustard
Brassica	rapa-rapifera-foliosa-60	quarantina	Quarantina	60 Day Turnip Greens	Mustard
Brassica	rapa-rapifera-foliosa-60	sessantina	Sessantine	60 Day Turnip Greens	Mustard

Kale, Broccoli Raab (*Brassica ruvo*)

Genus	Species	Infraspecies	cultivar	specimen	Genome
Brassica	ruvo-30	di-rapai-novatina	Novatina Di Rapa	30 Day Broccoli Raab Kale	Mustard
Brassica	ruvo-30	rapa-da-foglia-senza-testai	Rapa Da Foglia Senza Testa	30 Day Broccoli Raab Kale	Mustard
Brassica	ruvo-40	sparachetti	Sparachetti	40 Day Broccoli Raab Kale	Mustard
Brassica	ruvo-40	quarantinoi	Quarantino	40 Day Broccoli Raab Kale	Mustard
Brassica	ruvo-40	rappone-saladii	Salad Rappone	40 Day Broccoli Raab Kale	Mustard
Brassica	ruvo-40	sessatina-grossai	Sessatina Grossa	40 Day Broccoli Raab Kale	Mustard
Brassica	ruvo-40	sorrentoi	Sorrento	40 Day Broccoli Raab Kale	Mustard
Brassica	ruvo-50	di-rapai-maceratesei	Novatina Di Rapa	50 Day Broccoli Raab Kale	Mustard
Brassica	ruvo-50	italian-turnipii	Italian Turnip	50 Day Broccoli Raab Kale	Mustard
Brassica	ruvo-50	zambonii	Zamboni	50 Day Broccoli Raab Kale	Mustard
Brassica	ruvo-60	rappone-latei	Late Rappone	60 Day Broccoli Raab Kale	Mustard
Brassica	ruvo-60	rapini-springii	Spring Rapini	60 Day Broccoli Raab Kale	Mustard
Brassica	ruvo-60	rapini-superi	Super Rapini	60 Day Broccoli Raab Kale	Mustard
Brassica	ruvo-85	di-rapai	Di Rapa	85 Day Broccoli Raab Kale	Mustard
Brassica	ruvo-220	sparachettii-fallii	Fall Sparachetti	220 Day Broccoli Raab Kale	Mustard
Brassica	tournefortii	tournefortii	Original	Bird Rape	Mustard
Brassica	tournefortii	sisymbrioides	Turnip	Bird Rape	Mustard

Horseradish (*Amoracia rusticana*)

Genus	Species	Infraspecies	cultivar	specimen	Genome
Amoracia	rusticana	rusticana	Original	Red Cole	Horseradish
Amoracia	rusticana	bohemiana	Bohemian	Red Cole	Horseradish
Amoracia	rusticana	bohemiana-maliner-krenii	Maliner Kren Bohemian	Red Cole	Horseradish
Amoracia	rusticana	lapathifolia	Lap Leaf	Red Cole	Horseradish
Amoracia	rusticana	variegata	Variegated	Red Cole	Horseradish

Cress, St. Barbara (*Barbarea spp.*)

Genus	Species	Infraspecies	cultivar	specimen	Genome
Barbarea	verna	verna	Original	Spring	St. Barbara Cress

Barbarea	*verna*	*uplandia*	Upland	Spring	St. Barbara Cress
Barbarea	*verna*	*uplandia-specialis*	Special Upland	Spring	St. Barbara Cress
Barbarea	*vulgaris*	*vulgaris*	Original	Winter	St. Barbara Cress
Barbarea	*vulgaris*	*variegata*	Variegated	Winter	St. Barbara Cress

Spinach, Shepherds Purse *(Capsella bursa-pastoris)*

Genus	Species	Infraspecies	cultivar	specimen	Genome
Capsella	*bursa-pastoris*	*bursa-pastoris*	Original	Shepherd's Purse	Capsell Mustard
Capsella	*bursa-pastoris*	*auriculata*	Witches Pouches	Shepherd's Purse	Capsell Mustard
Capsella	*bursa-pastoris*	*rubella*	Red	Shepherd's Purse	Capsell Mustard

Kale, Sea Kale (*Crambe maritima*)

Genus	Species	Infraspecies	cultivar	specimen	Genome
Crambe	*maritima*	*maritima*	Original	Maritime	Sea Kale
Crambe	*maritima*	*lily-whitei*	Lily White	Maritime	Sea Kale

Cress, Roquette (*Eruca sativa*)

Genus	Species	Infraspecies	cultivar	specimen	Genome
Eruca	*sativa*	*sativa*	Original	Roquette	Rocket Cress
Eruca	*sativa*	*francoensis*	French	Roquette	Rocket Cress
Eruca	*sativa*	*italica*	Italian	Roquette	Rocket Cress
Eruca	*sativa*	*slow-boltii*	Slow Bolt	Roquette	Rocket Cress

Cress, Arugula (*Eruca vesicaria*)

Genus	Species	Infraspecies	cultivar	specimen	Genome
Eruca	*vesicaria*	*vesicaria*	Original	Arugula	Rocket Cress
Eruca	*vesicaria*	*graeca*	Greek	Arugula	Rocket Cress
Eruca	*vesicaria*	*italiana*	Italian	Arugula	Rocket Cress
Eruca	*vesicaria*	*sputnikii*	Sputnik	Arugula	Rocket Cress
Eruca	*vesicaria*	*sylvatica*	Sylvetta	Arugula	Rocket Cress

Cress, Pepper (*Lepidium sativum*)

Genus	Species	Infraspecies	cultivar	specimen	Genome
Lepidium	*sativum*	*sativum*	Original	Mediterranean	Pepper Cress
Lepidium	*sativum*	*crispum*	Curled	Mediterranean	Pepper Cress
Lepidium	*sativum*	*crispum-gramineum*	Peppergrass Curled	Mediterranean	Pepper Cress
Lepidium	*sativum*	*gerard'si*	Gerard's	Mediterranean	Pepper Cress
Lepidium	*sativum*	*graecensis*	Greek	Mediterranean	Pepper Cress
Lepidium	*sativum*	*latifolum*	Broad Leaf	Mediterranean	Pepper Cress
Lepidium	*sativum*	*latifolum-reforma*	Reform Broad Leaf	Mediterranean	Pepper Cress
Lepidium	*sativum*	*mahantongo*	Shallot	Mediterranean	Pepper Cress
Lepidium	*sativum*	*persica*	Persian	Mediterranean	Pepper Cress
Lepidium	*sativum*	*persica x rugosum x crispum*	Wrinkled Crinkled Crumpled	Mediterranean	Pepper Cress
Lepidium	*sativum*	*rugosum*	Crinkly	Mediterranean	Pepper Cress
Lepidium	*sativum*	*simplex*	Plain	Mediterranean	Pepper Cress
Lepidium	*sativum*	*tricrispum*	Triple Curled	Mediterranean	Pepper Cress
Lepidium	*sativum*	*victoria*	Victoria	Mediterranean	Pepper Cress

Cress, Water (*Nasturtium officinale*)

Genus	Species	Infraspecies	cultivar	specimen	Genome
Nasturtium	*officinale*	*aquaticum*	Original	Burning	Water Cress
Nasturtium	*officinale*	*cressida*	Cressida	Burning	Water Cress
Nasturtium	*officinale*	*lacustre*	Lakeside	Burning	Water Cress
Nasturtium	*officinale*	*latifiolium*	Broad Leaf	Burning	Water Cress
Nasturtium	*officinale*	*persica-gardenia*	Persian	Burning	Water Cress
Nasturtium	*officinale*	*persica-latifolia*	Broad Leaf Persian	Burning	Water Cress
Nasturtium	*officinale*	*rishadii*	Rishad	Burning	Water Cress
Nasturtium	*officinale*	*siifolium*	Si Leaf	Burning	Water Cress
Nasturtium	*officinale*	*turcensis*	Turkish	Burning	Water Cress
Nasturtium	*officinale*	*validum*	Valid	Burning	Water Cress

Radish (*Raphanus spp.*)

Genus	Species	Infraspecies	cultivar	specimen	Genome
Raphanus	*caudatus*	*edible-longpodii*	Edible Longpod 12."	Mongri	Radish
Raphanus	*caudatus*	*madras-sweet-edible-podii*	Sweet Madras 2.5"	Mongri	Radish
Raphanus	*caudatus*	*madras-munchen-bieri*	Munchen Bier Madras 2.5"	Mongri	Radish
Raphanus	*caudatus*	*madras-munchener-bieri*	Munchener Bier Madras 2.5"	Mongri	Radish
Raphanus	*caudatus*	*rat's-tailii*	Rat's Tail 6.0"	Mongri	Radish
Raphanus	*raphanistrum*	*raphanistrum*	Original	Wild	Radish
Raphanus	*raphanistrum*	*maratimus*	Seaside	Wild	Radish
Raphanus	*raphanistrum*	*raphanistroides*	Japanese	Wild	Radish
Raphanus	*raphanistrum*	*rostratus*	Beaked	Wild	Radish
Raphanus	*sativus-albus-longus*	*albus-longus*	Original ?.?"	Long White	Radish
Raphanus	*sativus-albus-longus*	*april-crossi*	April Cross 6.0"	Long White	Radish
Raphanus	*sativus-albus-longus*	*californica-mammoth-whitei*	California Mammoth 10."	Long White	Radish
Raphanus	*sativus-albus-longus*	*candela-di-ghiaccioi*	Candela Di Ghiaccio 9.0"	Long White	Radish
Raphanus	*sativus-albus-longus*	*chinensis*	Chinese 6.0"	Long White	Radish
Raphanus	*sativus-albus-longus*	*chinensis-albus*	China White 7.0"	Long White	Radish

Raphanus	sativus-albus-longus	chinensis-albus-celestialis	Chinese White Celestial 8.0"	Long White	Radish
Raphanus	sativus-albus-longus	daikonii	Daikon 15."	Long White	Radish
Raphanus	sativus-albus-longus	daikonii-all-seasonsi	All Seasons Diakon 12."	Long White	Radish
Raphanus	sativus-albus-longus	daikonii-earlyi-40	40 Day Early Daikon 15."	Long White	Radish
Raphanus	sativus-albus-longus	diakonii-earlyi-mino	Mino Early Daikon 18."	Long White	Radish
Raphanus	sativus-albus-longus	diakonii-sproutingii	Sprouting Daikon ??."	Long White	Radish
Raphanus	sativus-albus-longus	daikonii-tokinashi	Tokinashi Daikon 12."	Long White	Radish
Raphanus	sativus-albus-longus	daikonii-vinterus	Winter Diakon ??."	Long White	Radish
Raphanus	sativus-albus-longus	GV	Green Vegetable (GV) 15."	Long White	Radish
Raphanus	sativus-albus-longus	iciclei	Icicle 5.5"	Long White	Radish
Raphanus	sativus-albus-longus	iciclei-earlyi	Early Icicle 5.0"	Long White	Radish
Raphanus	sativus-albus-longus	iciclei-short-topii	Short Top Icicle 5.5"	Long White	Radish
Raphanus	sativus-albus-longus	iciclei-syla	Syla Icicle ?.?"	Long White	Radish
Raphanus	sativus-albus-longus	iciclei-whitei	White Icicle 5.0"	Long White	Radish
Raphanus	sativus-albus-longus	iciclei-whitei-longii	Long White Icicle 6.0"	Long White	Radish
Raphanus	sativus-albus-longus	iciclei-whitei-short-topii	Short Top White Icicle 4.5"	Long White	Radish
Raphanus	sativus-albus-longus	iciclei-whitei-strassburgii	Strassburg White Icicle ?.?"	White	Radish
Raphanus	sativus-albus-longus	japonica	15 Pound Japanese ??."	Long White	Radish
Raphanus	sativus-albus-longus	indica-albus	India White 30."	Long White	Radish
Raphanus	sativus-albus-longus	lobokii	Lo Bok 5.5"	Long White	Radish
Raphanus	sativus-albus-longus	mammoth-whitei	Mammoth ?.?"	Long White	Radish
Raphanus	sativus-albus-longus	mikadoi	Mikado 24."	Long White	Radish
Raphanus	sativus-albus-longus	minowasi	Minowasi 18."	Long White	Radish
Raphanus	sativus-albus-longus	minowasi-2	#2 Summer Minowasi 18."	Long White	Radish
Raphanus	sativus-albus-longus	miyashigei	Miyashige 14."	Long White	Radish
Raphanus	sativus-albus-longus	miyashigei-white-neck-longii	White Neck Miyashige 15."	Long White	Radish
Raphanus	sativus-albus-longus	nirmea	Nirema 25."	Long White	Radish
Raphanus	sativus-albus-longus	ohkura	Ohkura 16."	Long White	Radish
Raphanus	sativus-albus-longus	risoi	Riso 16."	Long White	Radish
Raphanus	sativus-albus-longus	shunkyoh-semi-longii	Shunkyoh Semilong 5.0"	Long White	Radish
Raphanus	sativus-albus-longus	tai-mei-hwai	Tai Mei Hwa 6.0"	Long White	Radish
Raphanus	sativus-albus-longus	wakayama-whitei	Wkayama 13."	Long White	Radish
Raphanus	sativus-albus-longus	white-winteri	White Winter 7.5"	Long White	Radish
Raphanus	sativus-albus-longus	winter-king-miyakoi	Miyako Winter King 19."	Long White	Radish
Raphanus	sativus-albus-longus	winter-queen-ohkurai	Ohkura Winter Queen 17."	Long White	Radish
Raphanus	sativus-albus-rotundus	albus-rotundus	Original ?.?"	Round White	Radish
Raphanus	sativus-albus-rotundus	beer-gardenii	Beer Garden 3.0"	Round White	Radish
Raphanus	sativus-albus-rotundus	blanc-de-sezannei	Blanc De Sezanne ?.?"	Round White	Radish
Raphanus	sativus-albus-rotundus	burpee-whitei	Burpee 0.8"	Round White	Radish
Raphanus	sativus-albus-rotundus	hailstonei	Hailstone ?.?"	Round White	Radish
Raphanus	sativus-albus-rotundus	japonica-ballii	Japanese Ball 3.5"	Round White	Radish
Raphanus	sativus-albus-rotundus	munchener-bieri	Munchener Bier 3.7"	Round White	Radish
Raphanus	sativus-albus-rotundus	philadelphia-white-boxii	Philadelphia White Box ?.?"	Round White	Radish
Raphanus	sativus-albus-rotundus	sakurajima	Sakurajima 12."	Round White	Radish
Raphanus	sativus-albus-rotundus	sezannei	Sezanne ?.?"	Round White	Radish
Raphanus	sativus-albus-rotundus	shogoinii	Shogoin 5.7"	Round White	Radish
Raphanus	sativus-albus-rotundus	silver-dollari	Silver Dollar 1.5"	Round White	Radish
Raphanus	sativus-albus-rotundus	snow-bellei	Snow Belle ?.?"	Round White	Radish
Raphanus	sativus-albus-rotundus	snow-bellei-white-globei	White Globe Snow Belle ?.?"	Round White	Radish
Raphanus	sativus-albus-rotundus	tsukushi-spring-crossi	Spring Cross Tsukushi 6.0"	Round White	Radish
Raphanus	sativus-albus-rotundus	white-ballii	White Ball ?.?"	Round White	Radish
Raphanus	sativus-albus-rotundus	white-globei-giantii	Giant White Globe ?.?"	Round White	Radish
Raphanus	sativus-albus-rotundus	white-prince-bigii	Big White Prince ?.?"	White	Radish
Raphanus	sativus-bruneus-longus	french-goldenii	French Golden 4.5"	Long Brown	Radish
Raphanus	sativus-bruneus-longus	pfitzer's-maindreieekii	Pfitzers Maindreeek Long 3.5"	Long Brown	Radish
Raphanus	sativus-bruneus-rotundus	heliosi	Helios 1.4"	Round Brown	Radish
Raphanus	sativus-groenus-longus	groenus-meatii	Green Meat 10."	Long Green	Radish
Raphanus	sativus-groenus-longus	groenus-skinii	Green Skin 6.0"	Long Green	Radish
Raphanus	sativus-groenus-longus	korea-groenus	Korea Green 10."	Long Green	Radish
Raphanus	sativus-groenus-longus	misato-greenii-fleshii	Misato Green Flesh 10."	Long Green	Radish
Raphanus	sativus-groenus-longus	shangtung-green-skinii	Shangtung Green Skin 12."	Green	Radish
Raphanus	sativus-groenus-rotundus	chinensis-groenus-goddessi	Chinese Green Goddess ?.?"	Round Green	Radish
Raphanus	sativus-groenus-rotundus	groenus-skinii-rubrus-fleshii	Red Flesh Green Skin 4.2"	Round Green	Radish
Raphanus	sativus-groenus-rotundus	misato-groenus	Misato 5.0"	Round Green	Radish
Raphanus	sativus-groenus-rotundus	uzbekistanensis	Uzbekistan 3.0"	Round Green	Radish
Raphanus	sativus-groenus-rotundus	valentinei	Valentine ?.?"	Round Green	Radish
Raphanus	sativus-nigrus-longus	black-french-longii	French 11."	Long Black	Radish
Raphanus	sativus-nigrus-longus	black-spanish-longii	Spanish 8.5"	Long Black	Radish
Raphanus	sativus-nigrus-rotundus	black-winteri-roundii	Italian ?.?"	Round Black	Radish
Raphanus	sativus-nigrus-rotundus	black-spanish-roundii	Spanish 3.5"	Round Black	Radish
Raphanus	sativus-purpureus-longus	purpureus	Original 3.0"	Long Purple	Radish
Raphanus	sativus-purpureus-longus	violet-de-gournayi	Violet Gournay 9.0"	Long Purple	Radish
Raphanus	sativus-purpureus-longus	greenskin&purplefleshii	GreenSkinPurpleFlesh	Long Purple	Radish
Raphanus	sativus-purpureus-rotundus	purpureus-plumii	Purple Plum 1.5"	Round Purple	Radish
Raphanus	sativus-purpureus-rotundus	purpureus-olive-shapedii	Olive Shaped ?.?"	Round Purple	Radish
Raphanus	sativus-rubrus-longus	rubrus-longus	Original	Long Red	Radish
Raphanus	sativus-rubrus-longus	bartender-red-mammothii	Bartender Red Mammoth 9.0"	Long Red	Radish
Raphanus	sativus-rubrus-longus	china-rosei	China Rose 7.0"	Long Red	Radish
Raphanus	sativus-rubrus-longus	d'avignonii	D'Avignon 3.5"	Long Red	Radish

Raphanus	sativus-rubrus-longus	demi-longue-de-18-joursii	18 Day Half Long ?.?"	Long Red	Radish
Raphanus	sativus-rubrus-longus	demi-longue-eclarate-a-bout-blancai	Flamboyant Half Long 2.0"	Long Red	Radish
Raphanus	sativus-rubrus-longus	demi-longue-flamboyantii-3	#3 Flamboyant Half Long 3.5"	Long Red	Radish
Raphanus	sativus-rubrus-longus	demi-longue-flamboyantii-25	25 Day Flamboyant Half Long	Long Red	Radish
Raphanus	sativus-rubrus-longus	demi-longue-flamboyantii-sabinai	Sabina Flamboyant Half Long	Long Red	Radish
Raphanus	sativus-rubrus-longus	demi-longue-flamboyantii-vidanii	Vidan Flamboyant Half Long	Long Red	Radish
Raphanus	sativus-rubrus-longus	demi-longue-french-breakfastii	French Breakfast Half Long 2.0"	Long Red	Radish
Raphanus	sativus-rubrus-longus	demi-longue-french-breakfastii-25-LD	25 Day French Breakfast Half Long	Long Red	Radish
Raphanus	sativus-rubrus-longus	demi-longue-french-breakfastii-red-flamei	Red Flame French Breakfast Half Long 3.0"	Long Red	Radish
Raphanus	sativus-rubrus-longus	demi-longue-french-dressingii	French Dressing Half Long 2.0"	Long Red	Radish
Raphanus	sativus-rubrus-longus	demi-longue-scarlet-white-tipii-035	#035 White Tip Scarlet Half Long	Long Red	Radish
Raphanus	sativus-rubrus-longus	demi-longue-scarlet-white-tipii-393	#393 White Tip Scarlet Half Long	Long Red	Radish
Raphanus	sativus-rubrus-longus	demi-longue-scarlet-white-tipii-875	# 875 White Tip Scarlet Half Long	Long Red	Radish
Raphanus	sativus-rubrus-longus	demi-longue-scarlet-white-tipii-fotai	Fota White Tip Scarlet Half Long	Long Red	Radish
Raphanus	sativus-rubrus-longus	fire-candlei	Fire Candle 6.0"	Long Red	Radish
Raphanus	sativus-rubrus-longus	flamivilii	Flamivil Half Long ?.?"	Long Red	Radish
Raphanus	sativus-rubrus-longus	fluoi	Fluo ?.?"	Long Red	Radish
Raphanus	sativus-rubrus-longus	long-scarletii-cincinnati-marketii	Cincinnati Market Long Scarlet 6.0"	Long Red	Radish
Raphanus	sativus-rubrus-longus	long-scarletii-short-topii	Short Top Long Scarlet 5.5"	Long Red	Radish
Raphanus	sativus-rubrus-longus	pontvilii	Pontvil 2.5"	Long Red	Radish
Raphanus	sativus-rubrus-longus	salad-rosei	Salad Rose 8.0"	Long Red	Radish
Raphanus	sativus-rubrus-longus	scarlet-longus	Scarlet Long 5.5"	Long Red	Radish
Raphanus	sativus-rubrus-longus	wood's-framei	Wood's Frame ?.?"	Long Red	Radish
Raphanus	sativus-rubrus-rotundus	altabellei	Altabelle ?.?"	Round Red	Radish
Raphanus	sativus-rubrus-rotundus	altaglobei	Altaglobe ?.?"	Round Red	Radish
Raphanus	sativus-rubrus-rotundus	arabica-rubrus	Arabic ?.?"	Round Red	Radish
Raphanus	sativus-rubrus-rotundus	beaconii	Beacon 1.0"	Round Red	Radish
Raphanus	sativus-rubrus-rotundus	brightest-breakfastii	Brightest Breakfast 1.5"	Round Red	Radish
Raphanus	sativus-rubrus-rotundus	cavalieri	Cavalier ?.?"	Round Red	Radish
Raphanus	sativus-rubrus-rotundus	cavalrondoi	Caval Rondo ?.?"	Round Red	Radish
Raphanus	sativus-rubrus-rotundus	cherry-bellei	Cherry Belle 1.2"	Round Red	Radish
Raphanus	sativus-rubrus-rotundus	cherry-bellei-championii	Champion Cherry Belle 1.2"	Round Red	Radish
Raphanus	sativus-rubrus-rotundus	cherry-bellei-extra-short-topii	Extra Short Top Cherry Belle 1.2"	Round Red	Radish
Raphanus	sativus-rubrus-rotundus	cherry-bellei-gala	Gala Cherry Belle 1.2"	Round Red	Radish
Raphanus	sativus-rubrus-rotundus	cherry-bellei-red-beretii	Red Beret Cherry Belle 2.0"	Round Red	Radish
Raphanus	sativus-rubrus-rotundus	cherry-bombii	Cherry Bomb ?.?"	Round Red	Radish
Raphanus	sativus-rubrus-rotundus	cometii	Comet 0.9"	Round Red	Radish
Raphanus	sativus-rubrus-rotundus	copenhagen-market-demai	Copenhagen Market Dema	Round Red	Radish
Raphanus	sativus-rubrus-rotundus	crimson-giantii	Crimson Giant 2.5"	Round Red	Radish
Raphanus	sativus-rubrus-rotundus	danra	Danra ?.?"	Round Red	Radish
Raphanus	sativus-rubrus-rotundus	desert-cherryi	Desert Cherry ?.?"	Round Red	Radish
Raphanus	sativus-rubrus-rotundus	fancy-redii	Fancy Red ?.?"	Round Red	Radish
Raphanus	sativus-rubrus-rotundus	fuegoi	Fuego 2.6"	Round Red	Radish
Raphanus	sativus-rubrus-rotundus	gaudryi-2	Gaundry II ?.?"	Round Red	Radish
Raphanus	sativus-rubrus-rotundus	german-giantii	German Giant 2.0"	Round Red	Radish
Raphanus	sativus-rubrus-rotundus	globalii	Global 1.2"	Round Red	Radish
Raphanus	sativus-rubrus-rotundus	jumboi	Jumbo 3.0"	Round Red	Radish
Raphanus	sativus-rubrus-rotundus	kim-giantii	Kim Giant 2.5"	Round Red	Radish
Raphanus	sativus-rubrus-rotundus	lady-slipperi	Lady Slipper Oblong ?.?"	Round Red	Radish
Raphanus	sativus-rubrus-rotundus	marabellei	Marabelle ?.?"	Round Red	Radish
Raphanus	sativus-rubrus-rotundus	misato-rubrus	Misato 5.0"	Round Red	Radish
Raphanus	sativus-rubrus-rotundus	misato-rose-fleshii	Misato Rose Flesh 4.0"	Round Red	Radish
Raphanus	sativus-rubrus-rotundus	nationalis	National ?.?"	Round Red	Radish
Raphanus	sativus-rubrus-rotundus	non-plus-ultra-406	Non Plus Ultra ?.?"	Round Red	Radish
Raphanus	sativus-rubrus-rotundus	noviredii	Novired ?.?"	Round Red	Radish
Raphanus	sativus-rubrus-rotundus	paratii	Parat ?.?"	Round Red	Radish
Raphanus	sativus-rubrus-rotundus	pink-beautyi	Pink Beauty ?.?"	Round Red	Radish
Raphanus	sativus-rubrus-rotundus	pink-celebrationii	Pink Celerbration 1.1"	Round Red	Radish
Raphanus	sativus-rubrus-rotundus	pokeri	Poker ?.?"	Round Red	Radish
Raphanus	sativus-rubrus-rotundus	rapid-red-sanova	Rapid Red Sanova ?.?"	Round Red	Radish
Raphanus	sativus-rubrus-rotundus	rebelii	Rebel ?.?"	Round Red	Radish
Raphanus	sativus-rubrus-rotundus	red-ballii	Red Ball ?.?"	Round Red	Radish
Raphanus	sativus-rubrus-rotundus	red-boyi	Red Boy ?.?"	Round Red	Radish
Raphanus	sativus-rubrus-rotundus	red-cherryi	Red Cherry ?.?"	Round Red	Radish
Raphanus	sativus-rubrus-rotundus	red-devilii	Red Devil ?.?"	Round Red	Radish
Raphanus	sativus-rubrus-rotundus	red-devilii-B	Short Top Red Devil ?.?"	Round Red	Radish
Raphanus	sativus-rubrus-rotundus	red-fortressi	Red Fortress ?.?"	Round Red	Radish
Raphanus	sativus-rubrus-rotundus	red-kingii	Red King ?.?"	Round Red	Radish
Raphanus	sativus-rubrus-rotundus	red-meatii	Red Meat 3.2"	Round Red	Radish
Raphanus	sativus-rubrus-rotundus	red-princei	Red Prince ?.?"	Round Red	Radish
Raphanus	sativus-rubrus-rotundus	red-princei-improvedii	Improved Red Prince ?.?"	Round Red	Radish
Raphanus	sativus-rubrus-rotundus	red-princei-red-baronii	Red Baron Red Prince ?.?"	Round Red	Radish
Raphanus	sativus-rubrus-rotundus	red-silkii	Red Silk 1.0"	Round Red	Radish
Raphanus	sativus-rubrus-rotundus	red-skoaringii	Red Skoaring ?.?"	Round Red	Radish
Raphanus	sativus-rubrus-rotundus	reggaei	Reggae ?.?"	Round Red	Radish
Raphanus	sativus-rubrus-rotundus	revosa	Revosa ?.?"	Round Red	Radish
Raphanus	sativus-rubrus-rotundus	rovii	Rovi ?.?"	Round Red	Radish
Raphanus	sativus-rubrus-rotundus	rudolfii	Rudolf ?.?"	Round Red	Radish
Raphanus	sativus-rubrus-rotundus	saxa	Saxa ?.?"	Round Red	Radish

Raphanus	*sativus-rubrus-rotundus*	*saxa-galahadii*	Galahad Saxa ?.?"	Round Red	Radish
Raphanus	*sativus-rubrus-rotundus*	*saxa-korto*	Korto Saxa ?.?"	Round Red	Radish
Raphanus	*sativus-rubrus-rotundus*	*scarlet-globei*	Scarlet Globe 1.0"	Round Red	Radish
Raphanus	*sativus-rubrus-rotundus*	*scarlet-globei-earlyi*	Early Scarlet Globe ?.?"	Round Red	Radish
Raphanus	*sativus-rubrus-rotundus*	*scarlet-globei-earlyi-shortii*	Short Top Early Scarlet Globe	Round Red	Radish
Raphanus	*sativus-rubrus-rotundus*	*scarlet-globei-medium-topii*	Medium Top Scarlet Globe	Round Red	Radish
Raphanus	*sativus-rubrus-rotundus*	*scarlet-globei-specialii*	Special Scarlet Globe ?.?"	Round Red	Radish
Raphanus	*sativus-rubrus-rotundus*	*scarlet-globei-white-tippedii*	White Tipped Scarlet Globe 1.5"	Round Red	Radish
Raphanus	*sativus-rubrus-rotundus*	*scarlet-knightii*	Scarlet Knight ?.?"	Round Red	Radish
Raphanus	*sativus-rubrus-rotundus*	*scarlet-turnip-white-tipii*	Scarlet Turnip White Tip ?.?"	Round Red	Radish
Raphanus	*sativus-rubrus-rotundus*	*shinrimei*	Shinrimei Red Flesh ?.?"	Round Red	Radish
Raphanus	*sativus-rubrus-rotundus*	*sora*	Sora ?.?"	Round Red	Radish
Raphanus	*sativus-rubrus-rotundus*	*sparkleri*	Sparkler 1.1"	Round Red	Radish
Raphanus	*sativus-rubrus-rotundus*	*tondo-rosso-a-punta-bianca*	Tondo rosso A Punta Bianca	Round Red	Radish
Raphanus	*sativus-rubrus-rotundus*	*vintagei-HMX-0764*	Vintage ?.?"	Round Red	Radish
Raphanus	*sativus-rubrus-rotundus*	*tricolouri*	Tricolour ?.?"	Round Red	Radish
Raphanus	*sativus-spinachius*	*fourseasonsi*	Four Seasons ?.?"	Spinach Leaf	Radish
Raphanus	*sativus-surculus*	*surculus*	Sprouts	Sprouting	Radish
Raphanus	*sativus-surculus*	*todo-redii*	Todo Red	Sprouting	Radish

Mustard, White Seed (*Sinapis alba*)

Genus	Species	Infraspecies	cultivar	specimen	Genome
Sinapis	*alba*	*alba*	Original	White Seed	Sinape Mustard
Sinapis	*alba*	*hirta*	Hairy	White Seed	Sinape Mustard
Sinapis	*alba*	*hirta-burgondei*	Burgonde Hairy	White Seed	Sinape Mustard
Sinapis	*alba*	*hirta-helva*	Yellow Hairy	White Seed	Sinape Mustard
Sinapis	*alba*	*finei*	Fine	White Seed	Sinape Mustard
Sinapis	*alba*	*helva*	Yellow	White Seed	Sinape Mustard
Sinapis	*alba*	*mostaza-silvestrei*	Moustaza Silvestre	White Seed	Sinape Mustard
Sinapis	*alba*	*moutarde-blanchei*	Moutarde Blanche	White Seed	Sinape Mustard
Sinapis	*alba*	*sinape-bianca*	Sinape Bianca	White Seed	Sinape Mustard
Sinapis	*alba*	*weisseri*	Weisser	White Seed	Sinape Mustard

Family 724.162.3 *Moringaceae* Moringo

Genus	Species	Specimen	Genome	C:Culinary
Moringa	*oleifera*		Drumstick Moringo	C:LF:**VG**

Varieties

Moringo (*Moringa oleifera*)

Genus	Species	Infraspecies	cultivar	specimen	Genome
Moringa	*oleifera*	*oleifera*	Original		Drumstick Moringo
Moringa	*oleifera*	*pterygosperma*	Wing Seed		Drumstick Moringo

Family 4 *Resedaceae* Mignonette
Order 3 *Salicales* Willow
Family 1 Salicaceae Willow
Family 2 Datisaceae Akelbir
Family 3 Frankeniaceae Sea Heath
Family 4 Passifloraceae Passion Flower
Family 5 Stachyuraceae
Family 6 Tamaricaceae Tamarisk

CD Order Form

Mail to: **Lawrence J. Fusillo**
280 Picadilly Sq. Apt. E-2
Athens, GA 30605

Item Number			
N01	FNCS Master CD	includes contents of CD N01 through CD N17	$100.00
N02	Genera & Higher Taxa CD	includes all genus names for microbes, plants, and animals	$10.00
N03	Nonliving Matter Species CD	includes Elements, Minerals, Soils, Sediments & Rocks	$10.00
N04	Virus Species CD	includes all 5000 known microbe, plant & animal Viruses	$10.00
N05	Microbe Species CD	includes all Bacteria, Protophytes (Plankton) & Protozoans	$10.00
N06	Orchid Species CD	includes all known species and varieties of Orchids	$10.00
N07	Spore Plant Species CD	includes all Algae, Fungus, Moss & Ferns	$10.00
N08	Fruit Variety CD	includes all known edible Fruit Varieties	$10.00
N09	Herb & Vegetable Variety CD	includes all known Vegetables & Herbs	$10.00
N10	Flower Variety CD	includes florist quality Flower Plant Varieties	$10.00
N11	Wildflower Species CD	includes flower species that grow wild	$10.00
N12	Inv. Worm Species Checklist CD	includes all Seanettle, Coral, Sponge, Snail & Worm Species	$10.00
N13	Invertebrate Shellfish Species CD	includes Sea Urchins, Crustaceans, Mollusks, Gastropods	$10.00
N14	Invertebrate Beetle Species CD	includes 144,324 of the 315,000 beetle species & varieties	$10.00
N15	Invertebrate Arthropod Species CD	includes Chelicers (mites, ticks, spiders, scorpions) & Insects	$10.00
N16	Vertebrate Fish CD	includes all known vertebrate fish	$10.00
N17	V. Amphibian, Reptile,Bird & Mammal CD	includes all known higher vertebrates	$10.00

Ship To: Name ______________________________
Address ______________________________
____
City ____________________ State______ Zip__________

Quantity	Item Number	CD Title	Price

Georgia Resident 6.6% Sales Tax $ __________
Postage & Handling $3.50
Amount Enclosed $ __________

CHECK OR MONEY ORDER ONLY
Make Check Payable to Lawrence J. Fusillo

About the Author

Lawrence Fusillo was born in Greenbelt Maryland on May 22 1949. While in elementary school, he was introduced to plant life, by his neighbor, Anna Belle Owens, who had a large garden in her back yard. His interest in biology continued in junior high school in a biology club, and in high school when he was a volunteer in a microbiology laboratory at the Veteran's Administration hospital in Washington DC. He received a B. A. in Physics from John's Hopkins University in 1971, and an M. S. in Applied Physics from the University of New Orleans in 1981.

He became interested in the names of colors in 1996, working at the Naval Air Development Center in Warminster Pennsylvania while developing software to display the colors of seafloor sediment. He found that most colors were named after plants that exhibited the color in the form of flowers. He began studying the relationship between plants, animals, minerals, and color names and discovered that the Linnaeus *Genus species* Latin classification system for plants, animals and minerals had not been updated since 1758.

After a 26 year career designing magnetometer, sonar and radar software for the US Navy first as a government employee (1972-1981) and later as a US Navy contractor (1982-1998), he designed air traffic control software as a contractor for the Federal Aviation Administration (1999-2001). He began background research for writing scientific books in May of 2001.

The "Fusillo Nature Classification System Fruit Variety Checklist" with category names in both Latin & English, is designed to replace the Linnaeus Classification System and all subsequent classification schemes. The categories Domain, Kingdom, Subkingdom, Phylum, Class, Order, Family, Genus, species and variety were assigned both in English and in Latin. A companion CD "N07 Fusillo Nature Classification System Fruit Variety Checklist" contains a Microsoft Excel Spreadsheet (.xls) implementation of the Fruit Varieties which can be imported into database programs.

Books by Lawrence J. Fusillo

The Fusillo Nature Classification System (FNCS) containing all living organisms was published in;

	Fusillo Nature Classification System (FNCS) Title		Year
1	"Anna Belle Nature Trail Guide 1st Edition"	ISBN:0-7414-4201-9	Jul 2007
2	"Fusillo Nature Classification System Genus & Higher Taxa 1st Edition"	ISBN:0-7414-4495-3	Jan 2008
3	"FNCS Nonliving Matter Species Checklist 1st Edition"	ISBN:0-7414-4459-3	Mar 2008
4	"FNCS Virus Species Checklist 1st Edition"	ISBN:0-7414-5214-6	Mar 2009
5	"FNCS Microbe Species Checklist 1st Edition"	ISBN:0-7414-5475-0	Aug 2009
6	"FNCS Orchid Species Checklist 1st Edition"	ISBN:0-7414-5668-0	Dec 2009
7	"FNCS Spore Plant Species Checklist 1st Edition"	ISBN:0-7414-5878-0	May 2010
8	"FNCS Fruit Variety Checklist 1st Edition"	ISBN:0-7414-6441-1	Mar 2011
9	"FNCS Herb & Vegetable Variety Checklist 1st Edition"	ISBN:0-7414-9666-6	Aug 2013
10	"FNCS Flower Variety Checklist 1st Edition"	future	
11	"FNCS Wildflower Species Checklist 1st Edition"	future	
12	"FNCS Invertebrate Worm Species Checklist 1st Edition"	future	
13	"FNCS Invertebrate Shellfish Species Checklist 1st Edition"	future	
14	"FNCS Invertebrate Beetle Species Checklist 1st Edition"	future	
15	"FNCS Invertebrate Chelicer & Insect Species Checklist 1st Edition"	future	
16	"FNCS Vertebrate Fish Species Checklist"	future	
17	"FNCS Amphibian, Reptile, Bird, & Mammal Species Checklist 1st Edition"	future	
1	:"Fusillo Knowledge Infrastructure Knowledge Classification System 1st Edition"	ISBN:0-7414-6647-3	Jun 2011
2	"FKI Gardening By Mail 1st Edition"	ISBN:0-7414-6792-5	Sep 2011

Made in United States
Cleveland, OH
06 November 2026

47787696R00297